ASTRONOMY AND ASTROPHYSICS ABSTRACTS

A Publication of the Astronomisches Rechen-Institut Heidelberg

Member of the International Council
for Scientific and Technical Information

Astronomy and Astrophysics Abstracts is Prepared
Under the Auspices of the International Astronomical Union

Volume 44
Literature 1987, Part 2

Edited by
U. Esser H. Hefele I. Heinrich W. Hofmann
D. Krahn V. R. Matas L. D. Schmadel G. Zech

Springer-Verlag Berlin Heidelberg GmbH 1988

Astronomisches Rechen-Institut, Mönchhofstraße 12–14,
D-6900 Heidelberg 1, F.R. Germany
Telex: 461 336 ARIHD D
Director: Prof. Dr. Roland Wielen

Astronomy and Astrophysics Abstracts
Department head: Dr. Lutz D. Schmadel
Editors-in-Chief: Inge Heinrich, Dr. Lutz D. Schmadel

ISBN 978-3-662-12363-8   ISBN 978-3-662-12361-4 (eBook)
DOI 10.1007/978-3-662-12361-4

Library of Congress Catalog Card Number 72-104650.
Media conversion: Daten- und Lichtsatz-Service, Würzburg.
2156/3150-543210

In memoriam
Professor Dr.Dr. h.c. mult. Walter Fricke
1 April 1915–21 March 1988

former director of the Astronomisches Rechen-Institut
and professor emeritus of the University of Heidelberg

He died at Heidelberg, shortly before his 73th birthday, after a long illness. We mourn his death.

Walter Fricke did outstanding work in many fields of astronomy. In this publication, we shall especially appreciate his merits of having established the international bibliography *Astronomy and Astrophysics Abstracts.*

Under Fricke's directorship, the Astronomisches Rechen-Institut first continued to produce, as it did since 1905, the bibliography *Astronomischer Jahresbericht.* It became, however, clear during the years that the structure of this publication was no longer optimal for a comprehensive international bibliography: The use of the German language in the abstracts made them difficult to understand for most users, and for many papers, no abstract was given at all. The latter defect was caused by the rapidly increasing number of published papers and the long-standing tradition of the Jahresbericht, not to use the author's abstract but to ask a number of qualified astronomers to write a short "review" of the paper.

Walter Fricke did not only clearly recognize these weak points of the Jahresbericht, he also had the ability and the courage to change the long tradition of the bibliographic work of the Institute. Mainly together with our former colleague, Miss Frieda Henn, he developed the new concept of *Astronomy and Astrophysics Abstracts:* to use the English language whenever possible, to basically use the author's abstract directly in the bibliography, to avoid a double proofreading by producing a camera-ready manuscript, and to speed up the publication of the bibliography significantly.

In order to help the new publication of the Astronomisches Rechen-Institut to a good start, he himself devoted for some years a large portion of his time to the preparation of published papers for the bibliography, i.e. to classify the papers into subject groups, to find appropriate key words for them, and to shorten the authors' abstracts in a meaningful manner. After the successful introduction of the new bibliography, he always continued to support this bibliographical work whenever necessary and possible, and he followed new developments for it with keen interest and advice.

Beside all his other scientific and organizational merits, we shall remember Walter Fricke as a pioneer of modern astronomical bibliography.

Roland Wielen
Director, Astronomisches Rechen-Institut

Preface

Astronomy and Astrophysics Abstracts aims to present a comprehensive documentation of the literature concerning all aspects of astronomy, astrophysics, and their border fields. It is devoted to the recording, summarizing, and indexing of the relevant publications throughout the world. *Astronomy and Astrophysics Abstracts* is prepared by a special department of the *Astronomisches Rechen-Institut* under the auspices of the *International Astronomical Union*.

Volume 44 records literature published in 1987 and received before February 15, 1988. Some older documents which we received late and which are not surveyed in earlier volumes are included too. We acknowledge with thanks contributions of our colleagues all over the world. We also express our gratitude to all organizations, observatories, and publishers which provide us with complimentary copies of their publications.

Dr. Siegfried Böhme retired from his duties as co-editor of *Astronomy and Astrophysics Abstracts* on December 31, 1987. Since 1950 he participated in the bibliographic work of the institute. He served as a reviewer for the *Astronomischer Jahresbericht* and became one of the editors of *Astronomy and Astrophysics Abstracts* in 1969. After his retirement in 1975 he took care of, particularly, the Russian literature on a voluntary basis for 12 years. It is a pleasure to thank Siegfried Böhme for his valuable contributions.

Starting with Volume 33, all the recording, correction, and data processing work was done by means of computers. The recording was done by our technical staff members Ms. Helga Ballmann, Ms. Christiane Jehn, Ms. Monika Kohl, Ms. Sylvia Matyssek, Ms. Doris Schmitz-Braunstein, Ms. Utta-Barbara Stegemann. Mr. Jochen Heidt and Mr. Kristopher Polzine supported our task by careful proofreading. It is a pleasure to thank them all for their encouragement.

Heidelberg, April 1988   **The Editors**

Contents

Introduction

Astronomical Bibliographies

Astronomy and Astrophysics Abstracts started documentation and abstracting work in 1969 as the direct successor of the Astronomischer Jahresbericht. For information on astronomical literature before this date consultation of one of the following bibliographies is suggested:

(1) J. J. de Lalande, Bibliographie Astronomique, Paris 1803 (this work covers the time from 480 B. C. to the year 1803, VIII + 966 pages).

(2) J. C. Houzeau, A. Lancaster, Bibliographie générale de l'astronomie, Volume I (in two parts), Bruxelles 1887, 1889, Volume II, Bruxelles 1882. The complete title of Volume II is "Bibliographie générale de l'astronomie ou catalogue méthodique des ouvrages, des mémoires et des observations astronomiques, publiés depuis l'origine de l'imprimerie jusqu'en 1880". A new edition of these volumes was prepared by D. W. Dewhirst in 1964.

(3) Bibliography of Astronomy, 1881–1898. The literature of this period was recorded on standard slips by the Observatoire Royal de Belgique. From the material (some 52,000 items) a microfilm version was produced by University Microfilms Limited, Tylers Green, High Wycombe, Buckinghamshire, England, in 1970.

(4) Astronomischer Jahresbericht, 1899 gegründet von Walter Wislicenus, herausgegeben vom Astronomischen Rechen-Institut in Heidelberg (formerly in Berlin), Verlag W. de Gruyter, Berlin. For the period from 1899 to 1968 sixty-eight volumes were published, each of which, in general, covers the literature of one year.

Concept of *Astronomy and Astrophysics Abstracts*

This abstracting service aims to present a comprehensive documentation of the literature in all fields of astronomy and astrophysics and their border fields. It appears in semi-annual volumes. Two of these volumes cover the literature of one calendar year. Every effort will be made to ensure that the average time interval between the receiving date of the original documents and publication of the abstracts will not exceed eight months. This time interval is near to that achieved by monthly abstracting journals, compared to which our system of accumulation of information over six months offers the advantage of greater convenience for the user.

The main characteristics of the concept of *Astronomy and Astrophysics Abstracts* may be summarized as follows:

(1) The subdivision of astronomy and its border fields into subject categories is facilitated by the fact that the astronomical objects appear to be particularly well suited for the formation of categories. It may be assumed that such subdivisions can be maintained for a long period. Experience shows, however, that progress in research might imply minor changes in the classification scheme.

(2) Each paper has been classified into one of 106 numbered subject categories and given a serial number within the category. In this way each item is numbered by six figures: the first three indicate the number of the category, the following three the serial number within the category. Reference to an abstract in Volume 1 is indicated by "01" before the number of the category; for example: 01.074.028, denotes Volume 1, category 074, abstract 028. A paper might be classified into more than one

category. In this case, its abstract is placed only in one category, whereas in the other categories only cross references are given. These are listed at the end of each category.

(3) Authors' abstracts are used whenever possible. Popular articles are not abstracted.

(4) If possible, titles of papers and abstracts are given in English. A special reference is made to titles which we have not taken in the original language.

The whole material was recorded by means of modified ITT 3030 microcomputers. All text recording programs and other data processing software were developed by Multicom GmbH, Gröbenzell, F. R. Germany and by our staff members as well. The index computations were carried out on the IBM 3090-180 computer of the University of Heidelberg.

Classification Systems

The two most common and widely used classification systems in astronomy and astrophysics are given by Class 9 of the revised edition of the International Classification System for Physics, published by the International Council of Scientific Unions Abstracting Board (Second edition 1978. ICSU-AB, 17 Rue Mirabeau, 75017 Paris, France, ISSN 0305-9618), and the *Astronomy and Astrophysics Abstracts* classification. In order to facilitate literature searches, we introduce a concordance relation between these two very different systems. This solution is only a unilateral one. Starting from the third hierarchical level of the PHYS-Classification Scheme 1987, the appropriate *Astronomy and Astrophysics Abstracts* chapter numbers are listed. This cannot imply an identical content of the respective chapters in both systems. In many cases there is only a rather partial concordance, and therefore the *Astronomy and Astrophysics Abstracts* numbers are enclosed in parentheses.

Transliteration Scheme for the Russian Alphabet

The transliteration of the Russian alphabet in use in *Astronomy and Astrophysics Abstracts* is presented here.

А	а	a	П	п	p
Б	б	b	Р	р	r
В	в	v	С	с	s
Г	г	g	Т	т	t
Д	д	d	У	у	u
Е	е	e	Ф	ф	f
Ё	ё	e	Х	х	kh
Ж	ж	zh	Ц	ц	ts
З	з	z	Ч	ч	ch
И	и	i	Ш	ш	sh
Й	й	j	Щ	щ	shch
К	к	k	Ы	ы	y
Л	л	l	Ь	ь	'
М	м	m	Э	э	eh
Н	н	n	Ю	ю	yu
О	о	o	Я	я	ya

This transliteration was recommended by the Abstracting Board of the International Council of Scientific Unions in 1969. It corresponds essentially to the transliteration proposed by the Academy of Sciences, Moscow. In this case the letters can be read and printed by usual data processing machines. If the names of Russian authors in the literature are transliterated in a different scheme, we present the names as they are given in the references cited and in addition in brackets according to our transliteration table.

Sources of Information

The majority of sources of information for this volume is given in category **001 Periodicals** and in category **008 Observatories, Institutes.** It may be noted that the titles of the periodicals are given in the original languages, and that Russian titles have been transliterated applying the transliteration scheme given above. Category 008 records publication series of observatories and astronomical institutes. Titles of the periodicals have been given following the recommendations of the "International List of Periodical Title Word Abbreviations" and its additions (see also **Abbreviations**). In most cases they permit recognition of the full title without recourse to the key in category 001. If other secondary sources have been consulted, we cite these papers and give reference to the respective services.

The total number of papers (some do not give names of authors) recorded in this volume amounts to 9,782.

Author, Subject, and Object Indexes

The subject category and the serial number have been used as a reference in all three indexes. These references are more precise than page references and offer considerable advantages in indexing by means of computers.

The author index of this volume contains 11,869 names.

We consider the subject index as an approximation to an optimal index covering all fields of astronomy and astrophysics. Starting with Volume 18, the subject index was enlarged to a certain extent in order to provide a thesaurus of astronomical and astrophysical terms. At present, the *Astronomy and Astrophysics Abstracts Vocabulary* 17.88, containing 2,343 key words, is in use. This is done not only for the users' convenience, but also with the intention to propose the use of special key words to authors and publishers.

While each volume is scheduled to contain an author index and a subject index, the magnetic tapes containing the index information will be used to produce separate index volumes (authors and subjects) at intervals of five years.

Beginning with Vol. 39 *Astronomy and Astrophysics Abstracts* will additionally contain an object index providing a further key to the documents abstracted. For detailed information concerning selection, standardization, sorting, etc, see introduction to the object index.

The sorting program for the indexes is based on the IBM SORT/MERGE Program. This program sorts blank before hyphen and before letters. Apostrophes are ignored by a special routine.

The users are requested to inform us on spelling errors within the author index in order to assist us in eliminating mistakes in future cumulative indexes.

Astronomy and Astrophysics Abstracts is prepared at the *Astronomisches Rechen-Institut*, Heidelberg under the auspices of the *International Astronomical Union* on a non-profit basis. The editors urge publishers of literature related to astronomy and astrophysics to provide our service in due time with complimentary copies of their material.

Publications should be mailed to:

Astronomy and Astrophysics Abstracts
Astronomisches Rechen-Institut
Moenchhofstrasse 12–14
D 6900 Heidelberg 1
F. R. Germany
Telex: 461 336 ARIHD D

Concordance Relation

between the PHYS-Classification Scheme 1987
and the *Astronomy and Astrophysics Abstracts* Classification Scheme

PHYS Classification Scheme	AAA Classification Scheme
0 General	
0100 Communication, education, history, and philosophy	
0130 Physics literature and publications	002, 003, 012, 046
0140 Education	014
0150 Educational aids	014
0160 Biographical, historical, and personal notes	004–007
0190 Other topics of general interest	015
9 Geophysics, Astronomy, and Astrophysics	
9100 Solid Earth physics	
9110 Geodesy and gravity	044, 045, 081
9125 Geomagnetism and paleomagnetism; geoelectricity	084 (081)
9135 Earth's interior structure and properties	081
9190 Other topics in solid Earth physics	081
9400 Aeronomy	
9410 Physics of the neutral atmosphere	082 (063, 084)
9420 Physics of the ionosphere	083 (062, 084)
9430 Physics of the magnetosphere	084 (062, 106)
9440 Cosmic-ray interactions with the Earth	078, 144 (085, 105)
9500 Fundamental astronomy and astrophysics, instrumentation, techniques, and astronomical observations	
9510 Fundamental astronomy	041–043 (052, 079, 095, 096)
9520 Historical and archaeoastronomy	004

PHYS **Classification Scheme**	AAA **Classification Scheme**
9530 Fundamental aspects of astrophysics	022, 061–063, 066
9540 Artificial Earth satellites	051, 053
9545 Observatories	008, 009
9555 Astronomical and space-research instrumentation	032–035 (031, 036, 053)
9575 Observation and reduction techniques	036 (021)
9580 Catalogues, atlases, etc.	002 (046)

9600 Solar system

9610 General, solar nebula, and cosmogony	107 (091)
9620 Moon	094
9630 Planets and satellites (excluding the Moon)	091–093, 097, 099–101
9640 Cosmic rays	078, 144 (106)
9650 Interplanetary space	074, 098, 102–106
9660 Solar physics	071–080

9700 Stars

9710 Stellar characteristics and properties	064, 065, 111–116, 131, (061, 062, 063)
9720 Normal stars (by class); general or individual	113–116, 121, 126
9730 Variable and peculiar stars (including novae)	112, 116, 117, 122–124
9760 Late stages of stellar evolution (including black holes)	067, 125, 126 (065)
9780 Binary and multiple stars (including extrasolar planetary systems)	117–120

**9800 Stellar systems; galactic and extragalactic
objects and systems; the Universe**

9810 Stellar dynamics	151
9820 Stellar clusters and associations	152–154
9840 Interstellar matter and nebulae	125, 131–134
9850 Galaxies, extragalactic objects and systems	155–160 (066, 161)
9870 Other objects and background radiations of unknown origin or distances	133, 141–144, 161
9880 Cosmology	161 (061, 066)

Abbreviations

Abbreviations used in *Astronomy and Astrophysics Abstracts* are primarily based on the 'International List of Periodical Title Word Abbreviations', prepared for the UNISIST/ICSU-AB Working Group on Bibliographic Descriptions (1970).

A.A.B.	Associazione Astrofili Bolognesi
Aarg.	Aargang
AAS	American Astronomical Society
AAVSO	American Association of Variable Star Observers
Abh.	Abhandlung −
Abstr.	Abstract −
Abt.	Abteilung
Acad.	Academi −, Academy
Accad.	Accademi −
Act.	Active, Activit −
Adm.	Administr −
Adv.	Advanc −
Aehron.	Aehronomi −
Aeron.	Aeronom −
Aeronaut.	Aeronauti −
Aerosp.	Aerospace
Afr.	Africa −
AG	Astronomische Gesellschaft
AIAA	American Institute of Aeronautics and Astronautics
AJB	Astronomischer Jahresbericht
Akad.	Akadem −
Ala.	Alabama
Alm.	Almanac −
Am.	America −, Amerika −, Amerique −
Amat.	Amateur −
Amst.	Amsterdam
An.	Anais, Anale −, Anali −, Anals
Anal.	Analis −, Analit −, Analys −, Analyt −
Angew.	Angewandt −
Ann.	Annaes, Annal −
Annu.	Annu −
Anst.	Anstalt
Anu.	Anual −, Anuar −
Anz.	Anzeiger
Appl.	Applied
Arb.	Arbeit
Arch.	Archiv −
Årg.	Årgang
Argent	Argentin −
Ariz.	Arizona
Ark.	Arkiv −
Arkh.	Arkhiv −
Artif.	Artifici −
ASA	Astronomical Society of Australia
Asoc.	Asocia −
ASP	Astronomical Society of the Pacific
ASSA	Astronomical Society of Southern Africa
Assem.	Assembl −
Assoc.	Associ −
Assoz.	Assozi −
Astrofis.	Astrofisic −
Astrofiz.	Astrofizi −
Astrometr.	Astrometr −
Astron.	Astronom −
Astronaut.	Astronauti −, Astronauty −
Astrophys.	Astrophys −
ASV	Astronomical Society of Victoria
ASWA	Astronomical Society of Western Australia
At.	Atom −

Atmos.	Atmosf −, Atmosph −
Aust.	Australi −
BAA	British Astronomical Association
Barc.	Barcelona
Bayer.	Bayerisch −
Beitr.	Beitrag, Beiträge
Belg.	Belge −, Belgi −
Beob.	Beobacht −
Beogr.	Beograd −
Ber.	Bericht −
Bibl.	Bibliot −
Bibliogr.	Bibliograf −, Bibliograph −
BIH	Bureau International de l'Heure
Bimest.	Bimestr −
Bl.	Blatt, Blätter
Bol.	Boletin
Boll.	Bolletino
Br.	British
Bras.	Brasil −
Brun.	Brunens −
Bruss.	Brussel, s
Brux.	Bruxelles
Bul.	Buleten −, Buletin −, Bulten
Bulg.	Bulgar −
Bull.	Bulletin −, Bullettino
Bur.	Bureau −
Byul.	Byuleten −, Byuletin −
Byull.	Byulleten −
C. R.	Comptes Rendus
Cah.	Cahier −
Calif.	California
Camb.	Cambridge
Can.	Canadi −, Canada
Carol.	Carolina −
Cas.	Casopis
Cat.	Catalog −
Celest.	Celestial
Cent.	Center, Central, Centrale, Centrally, Centre
Cercet.	Cercetari
Cesk.	Ceskoslov −
Chem.	Chemi −
Chim.	Chimi −
Chin.	Chinese
Chron.	Chronic −, Chronik, Chronique
Chronom.	Chronometr −
Cie.	Compagnie
Cienc.	Ciencia −
Cient.	Cientific −
Circ.	Circolar −, Circolo, Circolaire −, Circular −, Circulo
Cirk.	Cirkulaer −
Cl.	Clasa, Classe −
Co.	Companies, Company
Coll.	College
Collect.	Collect −
Colloq.	Colloqui −
Colo.	Colorado

Comet.	Cometary
Commentat.	Commentat –
Commun.	Communica –
Comput.	Computation, Computer –, Computing
Comun.	Comunica –
Conf.	Conferen –
Congr.	Congres –
Conn.	Connecticut
Contract.	Contract –
Contrib.	Contribu –
Copenh.	Copenhagen
Cosmochim.	Cosmochimi –
COSPAR	Committee on Space Research
Crystallogr.	Crystallograph –
CSIRO	Commonwealth Scientific and Industrial Research Organization
Cult.	Cultur –, Cultuur
Curr.	Current
Czech.	Czechoslovak –
D. C.	District of Columbia
DDR	Deutsche Demokratische Republik
Del.	Delaware
Dep.	Departament, Département, Department
Dev.	Development –, Développement –
Dig.	Digest
Dir.	Director –
Diss.	Disserta –
Div.	Divis –
Doc.	Document –
Dok.	Dokument –
Dokl.	Doklad –
Dom.	Dominion
Dtsch.	Deutsch
Ed.	Edit –
Edinb.	Edinburgh
Ehksp.	Ehksperiment –
Eidg.	Eidgenössisch –
Eksp.	Eksperiment –
Electron.	Electroni –
Eng.	Engineer –
Environ.	Environment –
Equip.	Equipement, Equipment
Ergeb.	Ergebnis –
ESA	European Space Agency
ESO	European Southern Observatory
Espec.	Especial –
ESRO	European Space Research Organization
Eur.	Europ –
Eval.	Evaluation –
Exp.	Experiment –
Extraterr.	Extraterrestr –
F. R. Germany	Federal Republic of Germany
Fac.	Facolt –, Faculd –, Facult –
Fak.	Fakult
Fasc.	Fascicul –
Fenn.	Fenni –
Finn.	Finni –
Fis.	Fisic –, Fisik –
Fiz.	Fizic –, Fizik –, Fizyk –
Fla.	Florida
Fluid.	Fluidi –
Fond.	Fondation –, Fondazione
Fortschr.	Fortschritt –
Fotogr.	Fotograf –
Found.	Foundation –
Fr.	Français –
Freq.	Frequen –
Fundam.	Fundamenta –
Fys.	Fysik –, Fysisch, Fysisk –
Fyz.	Fyzik –

G.	Giornale
Ga.	Georgia
Gaz.	Gazeta, Gazette
Gazz.	Gazzetta
Gen.	General
Geochem.	Geochem –
Geochim.	Geochim –
Geod.	Geodaes –, Geodaet –, Geodes –, Geodet –, Geodez –
Geofis.	Geofis –
Geofiz.	Geofiz –
Geofys.	Geofys –
Geogr.	Geograf –, Geograph –
Geokhim.	Geokhim –
Geol.	Geolog –, Geolosk –
Geomagn.	Geomagneti –
Geophys.	Geophys –
Ges.	Gesellschaft
Gesch.	Geschichte
Gl.	Glavno –
Glas.	Glasnik
Gos.	Gosudarst –
Gov.	Government –
Grenzgeb.	Grenzgebiet –
GSFC	Goddard Space Flight Center
H.M.	Her Majesty's, His Majesty's
Hamb.	Hamburg
Handb.	Handbook, Handbuch
Heidelb.	Heidelberg –
Helv.	Helveti –
Her.	Herald –
Hist.	History
Hochsch.	Hochschule
Hoegsk.	Hoegskol –
HR-diagram	Hertzsprung-Russell diagram
Hung.	Hungar –
Hydrogr.	Hydrograf –, Hydrograph –
IAF	International Astronautical Federation
IAU	International Astronomical Union
IBM	International Business Machines Corporation
ICSTI	International Council for Scientific and Technical Information
ICSU	International Council of Scientific Unions
ICSU-AB	International Council of Scientific Unions – Abstracting Board
IEEE	Institute of Electrical and Electronics Engineers
Ill.	Illinois
Inc.	Incorporated
Ind.	Industr –
Inf.	Informat –, Informaz –, Informe –
Ing.	Ingenieur
INIS	International Nuclear Information System
INSPEC	International Information Services for the Physics and Engineering Communities
Inst.	Institut –, Instytut –
Instrum.	Instrument –
Int.	International, Internazional –
Intell.	Intelligenc –
Inter.	Intérieur –, Interior
Interplanet.	Interplanetary
Intez.	Intezet –
Invest.	Investiga –
Ionos.	Ionosfer –, Ionospher –
Ir.	Irish
Iskusstv.	Iskusstvenn –
Isr.	Israel –
Issled.	Issledovan –
Ist.	Istitut
Ital.	Itali –
Izd.	Izdatel –
Izv.	Izvesti –

J.	Joernaal –, Jornal –, Journal –	**Muench.**	Muenchen
Jaarb.	Jaarboek –	**Mus.**	Museum
Jahr.	Jahrbuch, Jahrbücher		
Jahresber.	Jahresbericht –	**N. C.**	North Carolina
Jahresschr.	Jahresschrift	**N. D.**	North Dakota
Jahrg.	Jahrgang	**N. H.**	New Hampshire
JPL	Jet Propulsion Laboratory	**N. J.**	New Jersey
Jpn.	Japan –	**N. M.**	New Mexico
		N. Y.	New York
K.	Königlich –, Koninkljik –, Kunglig –	**N. Z.**	New Zealand
Kans.	Kansas	**Nablyud.**	Nablyudeni –
Kartogr.	Kartograf –	**Nac.**	Nacion –
Kernforsch.	Kernforschung	**Nachr.**	Nachricht –
Kernphys.	Kernphysik –	**NASA**	National Aeronautics and Space
Khem.	Khemyi –		Administration
Khim.	Khimi –	**Nat.**	Natur –
Kim.	Kimija –, Kimya	**Natl.**	National –
Kl.	Klass –	**Naturforsch.**	Naturforsch –
Kolloq.	Kolloquium –	**Naturwiss.**	Naturwissenschaft –
Komet.	Kometnyj	**Natuurkd.**	Natuurkunde
Komm.	Kommission –	**Nauchn.**	Nauchny –
Konf.	Konfer –	**Nauk.**	Nauka, Naukite, Naukov –, Naukow –
Kongr.	Kongress	**Naut.**	Nautic –
Kosm.	Kosmich –	**Nav.**	Naval –
Kosmog.	Kosmogon –	**Navig.**	Navigat –
Kozp.	Kozponti	**Naz.**	Nazion –
KPNO	Kitt Peak National Observatory	**Nebr.**	Nebraska
Ky.	Kentucky	**Ned.**	Nederland –
		Nev.	Nevada
La.	Louisiana	**Newsl.**	Newsletter –
Lab.	Laborato –	**Not.**	Notationes, Notic –, Notise, Notizi –
LEST	Large European Solar Telescope	**Nouv.**	Nouveau –, Nouvell –
Lett.	Letter –, Lettra, Lettre	**Nov.**	Novoe
Libr.	Librair –, Librar –	**Nucl.**	Nucléaire –, Nuclear –, Nucl –
		Nukl.	Nukle –
		Numer.	Numeri –
Madr.	Madrid		
Mag.	Magasin, Magazin –		
Magn.	Magneti –, Magnitn –	**O-va**	Obshchestva
Mar.	Marin –	**O-vo**	Obshchestvo
Mass.	Massachusetts	**Obs.**	Observ –
Mat.	Matemaat –, Matemat –	**Österr.**	Österreich –
Mater.	Material –	**Off.**	Offic –
Math.	Mathemat –	**Okla.**	Oklahoma
Md.	Maryland	**Opt.**	Optic –, Optik –, Optique
Meas.	Measur –	**Oreg.**	Oregon
Mec.	Mecani –	**Oss.**	Osserva –
Mech.	Mechani –		
Medd.	Meddelande –, Meddelelse		
Meded.	Mededeeling, Mededeling –	**Pa.**	Pennsylvania
Mekh.	Mekhani –	**Pac.**	Pacific
Mem.	Memento –, Memoir –, Memori –,	**Paleontol.**	Paleontolog –
	Memory –, Memuary	**Pap.**	Paper –, Papier
Memo.	Memorand –	**Part.**	Particle
Mens.	Mensile, Mensual –, Mensuel –	**Pekin.**	Pekinens –
Messtech.	Messtechni –	**Perem.**	Peremenn –
Meteorol.	Meteorolog –	**Period.**	Periodi –
Mex.	Mexic –	**Petrol.**	Petrolog –
Mich.	Michigan	**Philos.**	Philosoph –
Micromec.	Micromecaniq –	**Photogr.**	Photograf –, Photograph –
Miner.	Mineral, Minerale –, Minerali –	**Photogramm.**	Photogrammetr –
Mineral.	Mineralog –	**Photom.**	Photometr –
Minn.	Minnesota	**Phys.**	Physic –, Physik –, Physique –, Physisch –
Miss.	Mississippi	**Pict.**	Picture –
MIT	Massachusetts Institute of Technology	**Planet.**	Planetary
Mitt.	Mitteilung –	**Pol.**	Polish, Polon –
Mo.	Missouri	**Pr.**	Prac –
Mod.	Modern –	**Prelim.**	Prelimin –
Mol.	Molecul –, Molekul –	**Prepr.**	Preprint
Mon.	Monat, Monatlich –, Month –	**Prib.**	Pribor –
Monogr.	Monograph –	**Prikl.**	Prikladn –
Mont.	Montana	**Prilozh.**	Prilozhen –
MPI	Max-Planck-Institut	**Prir.**	Prirodn –
Mt.	Mount	**Prirodoved.**	Prirodoved –

Probl.	Problem—		**Sitzungsber.**	Sitzungsbericht—
Proc.	Proceedings		**Skr.**	Skrift—
Prod.	Prodott—, Produc—, Produkt,		**Smithson.**	Smithsonian
Prog.	Progres—		**Soc.**	Sociedad—, Societ—
Propag.	Propagation		**Sol.**	Solar
Prospect.	Prospecting		**Soln.**	Solnechn—
Prov.	Provinc—, Provints—, Provinz—		**Sonderdr.**	Sonderdruck—
Pubbl.	Pubblicazion—		**Soobshch.**	Soobshchen—
Publ.	Publicac—, Publicas—, Publicat—,		**South.**	Southern
	Publikas—, Publikat—		**Spacecr.**	Spacecraft
			Spat.	Spatial—
			Spec.	Special—
Q.	Quarterly		**Spectrosc.**	Spectroscop—
Quant.	Quantit—		**Spectrosk.**	Spectroskop—
			Spets.	Spetsial—
			Spez.	Spezial—, Speziell—
R.	Royal		**SSR**	Sovetskaya Sotsialisticheskaya
R. I.	Rhode Island			Respublika
Radiat.	Radiati—		**SSSR**	Soyuz Sovetskikh Sotsialisticheskikh
Radioact.	Radioactiv—, Radioaktiv—			Respublik
Radioisot.	Radioisotop—		**St.**	Saint—, Sankt—, Sant—
Rap.	Raport—		**—St.**	—Straße, Street
Rapp.	Rapport—		**Stand.**	Standard—, Standart—
RAS	Royal Astronomical Society		**Sternw.**	Sternwarte—
Rec.	Record—		**Stiint.**	Stiintific—
Rech.	Recherche—		**Stn.**	Station, Stazione
Ref.	Referat—, Reference—, Referieren		**Stud.**	Studia, Studie—, Studii
Relat.	Related, Relation—		**Supl.**	Suplement—, Supliment—
Relativ.	Relativit—		**Suppl.**	Supplement—
Rend.	Rendicont—		**Surv.**	Survey—
Rep.	Report—		**Syd.**	Sydney
Repr.	Reprint—		**Symp.**	Sympos—, Sympoz—
Repub.	Republi—		**Syst.**	System—
Res.	Research—		**Sz.**	Szemle
Result.	Resultad—, Resultat—			
Rev.	Review—, Revisio, Revista, Revue—		**Teach.**	Teacher—, Teaching
Rezul't.	Rezul'tat—		**Tec.**	Tecni—
Ric.	Ricerca, Ricerche		**Tech.**	Techni—
Riv.	Rivist—		**Technol.**	Technolog—
Roum.	Roumain—		**Tecnol.**	Tecnolog—
Rundsch.	Rundschau		**Teh.**	Tehnic—, Tehnika, Tehnisk—
			Tehnol.	Tehnolog—, Tehnolosk—
			Tek.	Tekni—
S. Afr.	South Africa		**Tekh.**	Tekhni—
S. C.	South Carolina		**Tekhnol.**	Tekhnolog—
S. D.	South Dakota		**Teknol.**	Teknolog—
SAF	Société Astronomique de France		**Telesc.**	Telescop—
SAI	Società Astronomica Italiana		**Telev.**	Television—
Samml.	Sammlung—		**Tenn.**	Tennessee
SAO	Smithsonian Astrophysical Observatory		**Teor.**	Teoret—, Teori—
SAS	Société Astronomique de Suisse		**Terr.**	Terrestr—
Satell.	Satellite		**Test.**	Testing
Sb.	Sbornik—		**Tex.**	Texas
Scand.	Scandinavi—		**TH**	Technische Hochschule
Sch.	Schul—		**Theor.**	Theoret—, Theori—
Schr.	Schrift—		**Tidschr.**	Tidschrift—
Schriftenr.	Schriftenreihe		**Tidskr.**	Tidskrift—
Schweiz.	Schweizer—		**Tidsskr.**	Tidsskrift—
Sci.	Scienc—, Scient—, Scienz—		**Top.**	Topic—
Scr.	Scripta, Scritt—		**Torun.**	Torunensis
Secc.	Seccion—		**Tr.**	Trudy
Sect.	Secti—		**Trans.**	Transactions, Transazione
Sekc.	Sekci—, Sekcj—		**Tsentr.**	Tsentral—
Sekt.	Sektion—, Sektor—		**Tsirk.**	Tsirkulyar—
Sekts.	Sektsi—		**TU**	Technical University
Sel.	Seleccion—, Select—, Selek—, Selezione			
Selsk.	Selskab—, Selskap—			
Semin.	Séminair—, Seminar—			
Sep.	Separat—		**Uch.**	Uchen—
Ser.	Seria—, Serie—, Seriya		**Uchebn.**	Uchebn—
Serv.	Servic—, Serviz—		**UK**	United Kingdom
Sess.	Sessi—		**Umsch.**	Umschau
Signal.	Signalétique—		**UN**	United Nations
Simp.	Simpoz—		**Univ.**	Universidad—, Universit—, Univerzitet—
Sin.	Sinica		**Ups.**	Upsaliens—

US	United States	**Vyp.**	Vypusk—
USA	United States of America	**Vyssh.**	Vyssh—
USSR	Union of Soviet Socialist Republics	**Vyzk.**	Vyzkum—
Utr.	Utrecht		
		W. Va.	West Virginia
Va.	Virginia	**Wash.**	Washington
Var.	Various	**West.**	Western
Ver.	Verein—, Verenig—	**Wet.**	Wetenschap—, Wetenskap—
Veränderl.	Veränderlich—	**Wis.**	Wisconsin
Verh.	Verhandl—	**Wiss.**	Wissenschaft—
Vermess.	Vermessung—	**Wyo.**	Wyoming
Vermessungswes.	Vermessungswesen		
Veröff.	Veröffentlich—	**Yad.**	Yadern—
Vesn.	Vesnik		
Vestn.	Vestnik		
Vetensk.	Vetenskap—	**Z.**	Zeitschrift—
Vgl.	Vergleich—	**ZA**	Zero Age
Vidensk.	Videnskab—, Videnskap	**ZAED**	Zentralstelle für Atomkernenergie-
Vierteljahresschr.	Vierteljahresschrift—		Dokumentation
Vierteljahrsschr.	Vierteljahrsschrift	**Zap.**	Zapisk—, Zapyisk—
VLB	Very Long Baseline	**Zaved.**	Zaveden—
Volcanol.	Volcanolog—	**Zent.**	Zentral
Vopr.	Vopros—	**Zentralbl.**	Zentralblatt
Vortr.	Vorträge	**Zesz.**	Zeszyt
Vses.	Vsesoyuzn—	**Zh.**	Zhurnal—
Vt.	Vermont	**Zirk.**	Zirkular

Periodicals, Proceedings, Books, Activities

001 Periodicals

A.A.O. Newsl.
A.A.O. Newsletter. Anglo–Australian Observatory, PO Box 296, Epping, N.S.W. 2121, Australia.

Abastumanskaya Astrofiz. Obs., Byull.
Abastumanskaya Astrofizicheskaya Observatoriya, Gora Kanobili. Byulleten'. Akademiya Nauk Gruzinskoj SSR. Izdatel'stvo Metsniereba, Tbilisi, USSR. ISSN 0375–6644.

Abh. Akad. Wiss. DDR
Abhandlungen der Akademie der Wissenschaften der DDR. Abteilung Mathematik, Naturwissenschaften, Technik. Akademie–Verlag, Berlin, German Democratic Republic.

Abh. Hamb. Sternw.
Abhandlungen aus der Hamburger Sternwarte. Hamburger Sternwarte, Universität Hamburg, Gojenbergsweg 2, D–2050 Hamburg 80, F.R. Germany. ISSN 0374–1583.

Abstr. Submitted Pap.
Abstracts of Submitted Papers. Institute of Astronomy, The Observatories, Madingley Road, Cambridge CB3 0HA, England.

Acad. R. Belg., Bull. Cl. Sci.
Académie Royale de Belgique, Bulletin de la Classe des Sciences (Koninklijke Academie van België, Mededelingen van de Klasse der Wetenschappen). 5e Série, Palais des Académies, Bruxelles, Belgium.

Acad. R. Belg., Mém. Cl. Sci.
Académie Royale de Belgique, Mémoires de la Classe des Sciences. Collection in 8°, 2e Série, Palais des Académies, Bruxelles, Belgium.

Acad. Sci. Est. SSR, Div. Phys. Math. Tech. Sci., Prepr.
Academy of Sciences of the Estonian SSR, Division of Physical, Mathematical and Technical Sciences, Preprints. Institute of Astrophysics and Atmospheric Physics, Academy of Sciences of the Estonian SSR, 202444 Tartu, Tôravere, Estonian SSR, USSR.

Acta Acust.
Acta Acustica. Science Press, Beijing, People's Republic of China. Subscription address: Guozi Shudian, PO Box 399, Beijing, People's Republic of China.

Acta Astron.
Acta Astronomica. An International Quarterly Journal. Polska Akademia Nauk, Komitet Astronomii, Państwowe Wydawnictwo Naukowe, Warszawa–Kraków, Poland. Subscription address: Ars Polona, 00–068 Warszawa, Krakowskie Przedmieście 7, Poland. ISSN 0001–5237.

Acta Astron. Sin.
Acta Astronomica Sinica. Purple Mountain Observatory, Academia Sinica. Nanjing, People's Republic of China. English translation in Chin. Astron. Astrophys. ISSN 0001–5245.

Acta Astronaut.
Acta Astronautica. Journal of the International Academy of Astronautics. Pergamon Press, Oxford – New York – Toronto – Paris – Frankfurt – Sydney. ISSN 0094–5765.

Acta Astrophys. Sin.
Acta Astrophysica Sinica. Beijing Astronomical Observatory, Academia Sinica, Beijing, People's Republic of China. Subscription address: Guozi Shudian, PO Box 399, Beijing, People's Republic of China. English translation in Chin. Astron. Astrophys.

Acta Cosmologica
Acta Cosmologica. Zeszyty Naukowe Uniwersytetu Jagiellońskiego. Państwowe Wydawnictwo Naukowe, Warszawa – Kraków, Poland. ISSN 0137–2386.

Acta Fac. Rerum Nat. Univ. Comenianae, Astron. Geophys.
Acta Facultatis Rerum Naturalium Universitatis Comenianae, Astronomia et Geophysica. Published for the Komenský University by Slovenské pedagogické nakladatel'stvo, 89112 Bratislava, Czechoslovakia.

Acta Geod. Geophys.
Acta Geodaetica et Geophysica. Institute of Geodesy and Geophysics, Academia Sinica, Beijing, People's Republic of China. Subscription address: Guozi Shudian, PO Box 399, Beijing, People's Republic of China.

Acta Geod. Geophys. Montan.
Acta Geodaetica, Geophysica et Montanistica. Akademiai Kiado, H–1054 Budapest, Alkotmany utca 21, Hungary. ISSN 0374–1842.

Acta Geophys. Pol.
Acta Geophysica Polonica. Subscription address: Ars Polona, 00–068 Warszawa, Krakowskie Przedmieście 7, Poland. ISSN 0001–5725.

Acta Geophys. Sin.
Acta Geophysica Sinica. Department of Geophysical Research, Academia Sinica, Beijing, People's Republic of China. Subscription address: Guozi Shudian, PO Box 399, Beijing, People's Republic of China. ISSN 0001–5733.

Acta Math. Sin.
Acta Mathematica Sinica. Academia Sinica, Beijing, People's Republic of China. Subscription address: Guozi Shudian, PO Box 399, Beijing, People's Republic of China.

Acta Mech. Sin.
Acta Mechanica Sinica. Academia Sinica, Beijing, People's Republic of China. Subscription address: Guozi Shudian, PO Box 399, Beijing, People's Republic of China. ISSN 0254–3060.

Acta Meteorol. Sin.
Acta Meteorologica Sinica. Subscription address: Guozi Shu-dian, PO Box 399, Beijing, People's Republic of China. ISSN 0577–6619.

Acta Phys. Austriaca
Acta Physica Austriaca. Springer–Verlag, Wien, Austria. ISSN 0001–6713.

Acta Phys. Hung.
Acta Physica Hungarica. Akademiai Kiado, Publishing House of the Hungarian Academy of Sciences, H–1054 Budapest, Alkotmany U. 21, Hungary. ISSN 0231–4428.

Acta Phys. Pol., Ser. A
Acta Physica Polonica, Ser. A. (General Physics, Solid State Physics, Atomic and Molecular Spectroscopy, Applied Phys-ics). Polska Akademia Nauk, Warszawa, Poland. Subscription address: Ars Polona, 00–068 Warszawa, Krakowskie Przedmieście 7, Poland. ISSN 0587–4246.

Acta Phys. Pol., Ser. B
Acta Physica Polonica, Ser. B (Elementary Particle Physics, Nuclear Physics, Theory of Relativity, Field Theory). Polska Akademia Nauk, Warszawa, Poland. Subscription address: Ars Polona, 00–068 Warszawa, Krakowskie Przedmieście 7, Poland. ISSN 0587–4254.

Acta Phys. Sin.
Acta Physica Sinica. Institute of Physics, Academia Sinica, Beijing, People's Republic of China. ISSN 0372–736X.

Acta Phys. Slovaca
Acta Physica Slovaca. VEDA Publishing House of the Slovak Academy of Sciences, 89530 Bratislava, Klemensova 19, Czechoslovakia. ISSN 0323–0465.

Acta Polytech. Scand., Appl. Phys. Ser.
Acta Polytechnica Scandinavia, Applied Physics Series. Finn-ish Academy of Technical Sciences, Kansakoulukatu 10A, SF–00100 Helsinki, Finland. ISSN 0355–2721.

Acta Sci. Nat. Univ. Pekin.
Acta Scientiarum Naturalium Universitatis Pekinensis. Beijing, People's Republic of China. ISSN 0479–8023.

Acta Sci. Nat. Univ. Sunyatseni
Acta Scientiarum Naturalium Universitatis Sunyatseni (Zhongshandaxue Xuebao). Canton Post Office, Canton, People's Republic of China.

Acta Tech. Acad. Sci. Hung.
Acta Technica Academiae Scientiarum Hungaricae. Akade-miai Kiado, Budapest 1363, PO Box 24, Hungary. ISSN 0001–7035.

Acta Tech. ČSAV
Acta Technica Československá akademie věd. Academia, Publishing House of the Czechoslovak Academy of Sciences, Vodičkova 40, 112 29 Praha 1, Czechoslovakia. Subscription address: John Benjamins N.V., Periodical Trade, Warmoesstraat 54, Amsterdam, The Netherlands. ISSN 0001–7043.

Acta Univ. Carol. Math. Phys.
Acta Universitatis Carolinae. Mathematica et Physica. Fakulta matematicko–fyzikální, Karlova universita, Praha, Czechoslovakia. ISSN 0001–7140.

A.D.I.O.N. Bull.
A.D.I.O.N. Bulletin. Association pour le Développement International de l'Observatoire de Nice. Observatoire de Nice, B.P. 252, F–06007 Nice Cedex, France.

Adv. Astronaut. Sci.
Advances in the Astronautical Sciences. American Astronauti-cal Society, Publications Office, Univelt, Inc., PO Box 28130, San Diego, Calif. 92128, USA. ISSN 0065–3438.

Adv. Phys.
Advances in Physics. Taylor & Francis Ltd., London, En-gland. ISSN 0001–8732.

Adv. Space Res.
Advances in Space Research. The official journal of the Committee on Space Research (COSPAR). Pergamon Press, Oxford – New York – Toronto – Sydney – Paris – Frankfurt. ISSN 0273–1177.

Aéronaut. Astronaut.
L'Aéronautique et l'Astronautique. Editions Air et Cosmos, 6 Rue Anatole de la Forge, F–75017 Paris, France. ISSN 0001–9275.

AIAA J.
AIAA Journal. American Institute of Aeronautics and Astro-nautics, 1290 Avenue of the Americas, New York, N.Y. 10019, USA. ISSN 0001–1452.

AIP Conf. Proc.
AIP Conference Proceedings. American Institute of Physics, 335 East 45th Street, New York, N.Y. 10017, USA. ISSN 0094–243X.

Akad. Wiss. DDR, Jahrb.
Akademie der Wissenschaften der DDR, Jahrbuch. Akade-mie–Verlag, Leipziger Straße 3–4, DDR–108 Berlin, German Democratic Republic. ISSN 0304–2154.

Algoritm. Nebesnoj Mekh.
Algoritmy Nebesnoj Mekhaniki (Materialy Matematichesko-go Obespecheniya EhVM). Institut Teoreticheskoj Astronomii Akademii Nauk SSSR. 191187 Leningrad, D–187, nab. Kutuzova, dom–10. USSR.

Alta Freq.
Alta Frequenza. Ufficio Centrale AEI–CEI, Viale Monza 259, I–20126 Milano, Italy. ISSN 0002–6557.

Am. Assoc. Variable Star Obs. Bull.
The American Association of Variable Star Observers Bulletin. The American Association of Variable Star Observers, 187 Concord Avenue, Cambridge, Mass. 02138, USA.

Am. Assoc. Variable Star Obs. Circ.
American Association of Variable Star Observers Circular. Subscription address: AAVSO, C. E. Scovil, Stamford Obser-vatory, 39 Scofieldtown Rd., Stamford, CT 06903, USA.

Am. Assoc. Variable Star Obs. Rep.
The American Association of Variable Star Observers Report. The American Association of Variable Star Observers, 187 Concord Avenue, Cambridge, Mass. 02138, USA.

Am. Assoc. Variable Star Obs. Sol. Bull.
The American Association of Variable Star Observers Solar Bulletin. AAVSO – Solar Division, 25 Birch Street, Cam-bridge, Mass. 02138–1205, USA.

Am. J. Phys.
American Journal of Physics. Published for the American Association of Physics Teachers by the American Institute of Physics, 335 East 45th Street, New York, N.Y. 10017, USA. ISSN 0002–9505.

Am. Mineral.
American Mineralogist. Mineralogical Society of America, 2000 Florida Avenue, N.W., Washington, D.C. 20009, USA. ISSN 0003–004X.

Am. Sci.
American Scientist. Society of Sigma XI, 345 Whitney Avenue, New Haven, Conn. 06510, USA. ISSN 0003–0996.

An. Acad. Bras. Cienc.
Anais da Academia Brasileira de Ciencias. Caixa Postal 229, ZC–00 Rio de Janeiro gb, Brazil. ISSN 0001–3765.

An. Fis., Ser. A
Anales de Fisica, Serie A (Fenomenos e Interacciones). Real Sociedad Espanola de Fisica, Facultades de Ciencias, Ciudad Universitaria, Madrid 3, Spain. ISSN 0211–6243.

An. Fis., Ser. B
Anales de Fisica, Serie B (Aplicaciones, Metodes e Instrumentos). Real Sociedad Espanola de Fisica, Facultades de Ciencias, Ciudad Universitaria, Madrid 3, Spain. ISSN 0211–6251.

An. Obs. Astron. Univ. Coimbra
Anais do Observatório Astronómico da Universidade de Coimbra. Observatório Astronómico da Universidade de Coimbra, Coimbra, Portugal. ISSN 0870–2853.

Anal. Chem.
Analytical Chemistry. American Chemical Society, 1155 16th Street N.W., Washington, D.C. 20036, USA. ISSN 0003–2700.

Anglo–Aust. Obs., Prepr.
Anglo–Australian Observatory, Preprints (AAO PP). Anglo–Australian Observatory, PO Box 296, Epping, N.S.W. 2121, Australia.

Anglo–Aust. Telesc., Annu. Rep.
Anglo–Australian Telescope, Annual Report. Anglo–Australian Observatory, PO Box 296, Epping, N.S.W. 2121, Australia.

Ann. Acad. Sci. Fenn., Ser. A VI
Annales Academiae Scientiarum Fennicae, Series A VI (Physica). Academia Scientiarum Fennica, Snellmaninkato 9–11, 00 170 Helsinki 17, Finland. ISSN 0066–2003.

Ann. Fond. Louis de Broglie
Annales de la Fondation Louis de Broglie. Fondation Louis de Broglie, 1 rue Montgolfier, F–75003 Paris, France.

Ann. Geophys., Ser. A
Annales Geophysicae, Series A (Upper Atmosphere and Space Sciences). European Geophysical Society. Gauthier–Villars. Subscription address: C.D.R. Centrale des Revues, 11 rue Gossin, F–92543 Montrouge Cedex, France. Supersedes in part Ann. Geophys.

Ann. Inst. Henri Poincaré Phys. Théor.
Annales de l'Institut Henri Poincaré Physique Théorique. Gauthier–Villars, C.D.R., Centrale des Revues, 11, rue Gossin, F–92543 Montrouge Cedex, France. ISSN 0246–0211.

Ann. Isr. Phys. Soc.
Annals of the Israel Physical Society. c/o Department of Physics, Bar–Ilan University, Ramat–Gan, Israel. Adam Hilger Ltd., Bristol, England. ISSN 0309–8710.

Ann. Nucl. Energy
Annals of Nuclear Energy. Pergamon Press, Oxford – New York – Toronto – Paris – Frankfurt – Sydney. ISSN 0306–4549.

Ann. N.Y. Acad. Sci.
Annals of the New York Academy of Sciences. The New York Academy of Sciences, 2 East 63rd Street, New York, N.Y., USA. ISSN 0077–8923.

Ann. Obs. Astron. Alger
Annales de l'Observatoire Astronomique d'Alger. Observatoire Astronomique de l'Université d'Alger, Algiers, Algeria.

Ann. Phys. (Leipzig)
Annalen der Physik (Leipzig). Johann Ambrosius Barth, Salomonstr. 18B, Leipzig 701, German Democratic Republic. ISSN 0003–3804.

Ann. Phys. (N.Y.)
Annals of Physics (New York). Academic Press Inc., New York – London. ISSN 0003–4916.

Ann. Phys. (Paris)
Annales de Physique (Paris). Masson et Cie S.A., 120 Boulevard Saint–Germain, F–75280 Paris Cedex 06, France. ISSN 0003–4169.

Ann. Sci.
Annals of Science. Taylor & Francis Ltd., London, England. ISSN 0003–3790.

Ann. Sci. Univ. Besançon, Phys.
Annales Scientifiques de l'Université de Besançon, Physique. Institut des Sciences Naturelles, Place Leclerc, F–25000 Besançon, France. ISSN 0365–6543.

Ann. Shanghai Obs., Acad. Sin.
Annals of Shanghai Observatory, Academia Sinica. Shanghai Scientific and Technical Publishers, Shanghai, Rei Jing Er Street 450, People's Republic of China.

Ann. Soc. Sci. Brux., Sér. I
Annales de la Société Scientifique de Bruxelles, Série I (Sciences Mathématiques, Astronomiques et Physiques). Rue de Bruxelles 61, B–5000 Namur, Belgium. ISSN 0037–959X.

Ann. Tokyo Astron. Obs., Second Ser.
Annals of the Tokyo Astronomical Observatory, Second Series. Tokyo Astronomical Observatory, University of Tokyo, Mitaka, Tokyo 181, Japan. ISSN 0082–4704.

Ann. Univ.–Sternw. Wien
Annalen der Universitäts–Sternwarte Wien. Institut für Astronomie der Universität Wien, Türkenschanzstr. 17, A–1180 Wien, Austria. Published by Ferd. Dümmlers Verlag, Bonn, F.R. Germany. ISSN 0342–4030.

Annu. Rep. Astron. Inst. Greece
Annual Reports of the Astronomical Institutes of Greece. Published by the Greek National Committee for Astronomy, Athens, Greece.

Annu. Rep. (B.I.H.)
Annual Report (B.I.H.). Bureau International de l'Heure, 61, avenue de l'Observatoire, F–75014 Paris, France.

Annu. Rep. Dir., Mt. Wilson Las Campanas Obs.
Annual Report of the Director, The Mt. Wilson and Las Campanas Observatories. Mt. Wilson and Las Campanas Observatories of the Carnegie Institution of Washington, 813 Santa Barbara Street, Pasadena, Calif. 91101–1292, USA.

Annu. Rep. (ERP)
Annual Report (ERP). Xu–Jia–Hui Section, Shanghai Observatory, Academia Sinica. Subscription address: Shanghai Scientific and Technical Publishers, Shanghai, Rei Jing Er Street 450, People's Republic of China.

Annu. Rep. Geophys. Obs.
Annual Report of Geophysical Observations. The International Latitude Observatory of Mizusawa, Mizusawa–Shi, Iwate–Ken, Japan. ISSN 0579–5958.

Annu. Rep. Int. Polar Motion Serv.
Annual Report of the International Polar Motion Service. Central Bureau of the International Polar Motion Service, International Latitude Observatory of Mizusawa, Mizusawa–Shi, Iwate–Ken, Japan. ISSN 0074–7432.

Annu. Rep. Meteorol. Obs. Int. Latitude Obs. Mizusawa
Annual Report of the Meteorological Observations made at the International Latitude Observatory of Mizusawa. The International Latitude Observatory of Mizusawa, Mizusawa–Shi, Iwate–Ken, Japan. ISSN 0303–8378.

Annu. Rev. Astron. Astrophys.
Annual Review of Astronomy and Astrophysics. Annual Reviews Inc., 4139 El Camino Way, Palo Alto, Calif. 94306, USA. ISSN 0066–4146.

Annu. Rev. Earth Planet. Sci.
Annual Review of Earth and Planetary Sciences. Annual Reviews Inc., 4139 El Camino Way, Palo Alto, Calif. 94306, USA. ISSN 0084–6597.

Annu. Rev. Fluid Mech.
Annual Review of Fluid Mechanics. Annual Reviews Inc., 4139 El Camino Way, Palo Alto, Calif. 94306, USA. ISSN 0066–4189.

Annu. Rev. Nucl. Part. Sci.
Annual Review of Nuclear and Particle Science. Annual Reviews Inc., 4139 El Camino Way, Palo Alto, Calif. 94306, USA. ISSN 0163–8998.

Annu. Univ. Sofia
Annuaire de l'Université de Sofia. Faculté de Physique. Bibliothèque de l'Université, Sofia, Bulgaria. ISSN 0584–0279.

Antenna
L'Antenna. Via Monte Generoso 6/a, I–20155 Milano, Italy. ISSN 0003–5386.

Anz. Österr. Akad. Wiss., Math.–Naturwiss. Kl.
Anzeiger der Österreichischen Akademie der Wissenschaften, Mathematisch–Naturwissenschaftliche Klasse. Springer–Verlag, Wien, Austria. ISSN 0065–535X.

Appl. Opt.
Applied Optics. A monthly publication of the Optical Society of America. American Institute of Physics, 335 East 45th Street, New York, N.Y. 10017, USA. ISSN 0003–6935.

Appl. Phys., B
Applied Physics, B (Photophysics and Laser Chemistry). Springer–Verlag, Berlin – Heidelberg – New York – Tokyo. ISSN 0721–7269.

Appl. Phys. Lett.
Applied Physics Letters. American Institute of Physics, 335 East 45th Street, New York, N.Y. 10017, USA. ISSN 0003–6951.

Appl. Spectrosc.
Applied Spectroscopy. Society for Applied Spectroscopy, 428 East Preston Street, Baltimore, Md. 21202, USA. ISSN 0003–7028.

Appl. Spectrosc. Rev.
Applied Spectroscopy Reviews. Marcel Dekker Inc., 270 Madison Avenue, New York, N.Y. 10016, USA. ISSN 0570–4928.

Arch. Elektrotech.
Archiv für Elektrotechnik. Springer–Verlag, Berlin – Heidelberg – New York – Tokyo. ISSN 0003–9039.

Arch. Hist. Exact Sci.
Archive for History of Exact Sciences. Springer–Verlag, Berlin – Heidelberg – New York – Tokyo. ISSN 0003–9519.

Arch. Int. Hist. Sci.
Archives Internationales d'Histoire des Sciences. Istituto dell'Enciclopedia Italiana, fondata da Giovanni Treccani, Roma, Italy. ISSN 0003–9810.

Arch. Sci.
Archives des Sciences. Société de Physique et d'Histoire Naturelle de Genève. Subscription address: Librairie Payot, 6 Rue Grenus, CH–1211 Geneva 11, Switzerland. ISSN 0003–9705.

Archaeoastronomy (U.K.)
Archaeoastronomy. Supplement to Journal for the History of Astronomy. Science History Publications Ltd., Halfpenny Furze, Mill Lane, Chalfont St Giles, Bucks. England, HP8 4NR. ISSN 0142–7253.

Archaeoastronomy (U.S.A.)
Archaeoastronomy. The Bulletin of the Center for Archaeoastronomy, Space Sciences Building, University of Maryland, College Park, Md. 20742, USA. ISSN 0190–9940.

Archaeometry
Archaeometry. Research Laboratory for Archaeology and the History of Art, Oxford University, 6 Keble Road, Oxford OX1 3QJ, England. ISSN 0003–813X.

Archenhold–Sternw. Berlin–Treptow, Sonderdr.
Archenhold–Sternwarte Berlin–Treptow, Sonderdruck. Archenhold–Sternwarte, DDR–1193 Berlin, Alt Treptow 1, German Democratic Republic.

Archenhold–Sternw. Berlin–Treptow, Vortr. Schr.
Archenhold–Sternwarte Berlin–Treptow, Vorträge und Schriften. Archenhold–Sternwarte, DDR–1193 Berlin, Alt Treptow 1, German Democratic Republic. ISSN 0570–6262.

Arecibo Obs./NAIC, Newsl.
Arecibo Observatory/NAIC, Newsletter. National Astronomy and Ionosphere Center, PO Box 995, Arecibo, P.R. 00613, Puerto Rico.

Ark. Fys. Semin. Trondheim
Arkiv for det Fysiske Seminar i Trondheim. c/o Institutt for Teoretisk Fysikk, Universitetet i Trondheim, NTH, N–7034 Trondheim, Norway. ISSN 0365–2459.

Ark. Mat.
Arkiv för Matematik. Institut Mittag–Leffler, Auravägen 17, S–182 62 Djursholm, Sweden. ISSN 0004–2080.

Armagh Obs., Prepr. Ser.
Armagh Observatory, Preprint Series. Armagh Observatory, College Hill, Armagh BT61 9DG, Northern Ireland.

Armagh Obs., Publ.
Armagh Observatory, Publications. Armagh Observatory, College Hill, Armagh BT61 9DG, Northern Ireland.

Artif. Satell.
Artificial Satellites. Polish Academy of Sciences, Space Research Committee. Państwowe Wydawnictwo Naukowe, Warszawa–Lódź, Poland. Subscription address: Ars Polona, 00–068 Warszawa, Krakowskie Przedmieście 7, Poland. ISSN 0571–205X.

Astrofiz. Issled. Izv. Spets. Astrofiz. Obs.
Astrofizicheskie Issledovaniya. Izvestiya Spetsial'noj Astrofizicheskoj Observatorii. Akademiya Nauk SSSR. Izdatel'stvo Nauka, Leningradskoe Otdelenie, Leningrad. 199164 Leningrad, V–164, Mendeleevskaya l., 1, USSR. English translation in Bull. Spec. Astrophys. Obs. – North Caucasus. ISSN 0324–1459.

Astrofizika
Astrofizika. Izdatel'stvo Akademii Nauk Armyanskoj SSR, Erevan, USSR. English translation in Astrophysics. ISSN 0571–7132.

Astrofys. Inst., Vrije Univ. Bruss., Overdruk
Astrofysisch Instituut, Vrije Universiteit Brussel, Overdruk. Astrofysisch Instituut, Vrije Universiteit Brussel, Pleinlaan 2, B–1050 Brussel, Belgium.

Astron. Astrophys.
Astronomy and Astrophysics. A European Journal. Springer–Verlag, Berlin – Heidelberg – New York – Tokyo. ISSN 0004–6361.

Astron. Astrophys., Suppl. Ser.
Astronomy and Astrophysics, Supplement Series. A European Journal. Les Editions de Physique, Z.I. de Courtaboeuf, B.P. 112, F–91944 Les Ulis Cedex, France. ISSN 0365–0138.

Astron. Bull., Carter Obs.
Astronomical Bulletin, Carter Observatory. Carter Observatory, PO Box 2909, Wellington 1, New Zealand. ISSN 0373–7268.

Astron. Circ.
Astronomical Circular. Edited by the Chinese Astronomical Society. Compiled by the Editors of Acta Astronomica Sinica, Purple Mountain Observatory, Academia Sinica, Nanjing, People's Republic of China.

Astron. Contrib. Univ. Manchester, Ser. II: Jodrell Bank Repr.
Astronomical Contributions from the University of Manchester, Series II: Jodrell Bank Reprints. Nuffield Radio Astronomy Laboratories, Jodrell Bank, Macclesfield, Cheshire SK11 9DL, England.

Astron. Contrib. Univ. Manchester, Ser. III
Astronomical Contributions from the University of Manchester, Series III. Department of Astronomy, University of Manchester, Oxford Road, Manchester M13 9PL, England.

Astron. Data Cent. Bull.
AstronomicaL Data Center Bulletin. National Space Science Data Center/World Data Center A for Rockets and Satellites. National Aeronautics and Space Administration, Goddard Space Flight Center, Greenbelt, Md. 20771, USA.

Astron. Express
Astronomy Express. Cambridge University Press, London – New York – New Rochelle – Melbourne – Sydney. ISSN 0265–5365.

Astron. Her.
Astronomical Herald. Astronomical Society of Japan, Tokyo Astronomical Observatory, Oosawa Mitaka, Tokyo, Japan. ISSN 0374–2466.

Astron. Inst. "Anton Pannekoek", Univ. Amst., Repr.
Astronomical Institute "Anton Pannekoek", University of Amsterdam, Reprint.

Astron. Inst. Univ. Brno, Contrib.
Astronomical Institute of the University Brno, Contributions. Astronomical Institute, University Brno, Kotlářská 2, Brno, Czechoslovakia.

Astron. Inst. Univ. Brno, Publ.
Astronomical Institute of the University Brno, Publications. Astronomical Institute, University Brno, Kotlářská 2, Brno, Czechoslovakia.

Astron. J.
The Astronomical Journal. Published for the American Astronomical Society by the American Institute of Physics, 335 East 45th Street, New York, N.Y. 10017, USA. ISSN 0004–6256.

Astron. Mitt. Wien
Astronomische Mitteilungen Wien. Institut für Astronomie der Universität Wien, Türkenschanzstraße 17, A–1180 Wien, Austria.

Astron. Nachr.
Astronomische Nachrichten. Akademie–Verlag, Leipziger Straße 3–4, DDR–108 Berlin, German Democratic Republic. ISSN 0004–6337.

Astron. Now
Astronomy Now. Intra Press, 16 Garway Road, London W2 4NH, England.

Astron. Obs. Trieste, Publ.
Astronomical Observatory Trieste, Publications. Osservatorio Astronomico di Trieste, Via G.B. Tiepolo 11, I–34131 Trieste, Italy.

Astron. Pap.
Astronomical Papers prepared for the use of the American Ephemeris and Nautical Almanac. Published by the Nautical Almanac Office, U.S. Naval Observatory by direction of the Secretary of the Navy and under the authority of Congress. U.S. Government Printing Office, Washington, D.C., USA.

Astron. Q.
The Astronomy Quarterly. Pachart Publishing House, 1130 San Lucas Circle, Tucson, Ariz. 85704, USA. ISSN 0364–9229.

Astron. Raumfahrt
Astronomie und Raumfahrt. Kulturbund der DDR, Zentrale Kommission Astronomie und Raumfahrt. Redaktionssitz: 9630 Crimmitschau, Pionier– und Jugendsternwarte "Johannes Kepler", Strasse der Jugend 8. Available from Zeitungsvertriebsamt, Abt. Export, 1004 Berlin, Strasse der Pariser Kommune 3–4, German Democratic Republic. ISSN 0587–565X.

Astron. Rechen–Inst. Heidelb., Mitt., Ser. A
Astronomisches Rechen–Institut Heidelberg, Mitteilungen, Serie A. Astronomisches Rechen–Institut, Mönchhofstraße 12–14, D–6900 Heidelberg, F.R. Germany.

Astron. Rechen–Inst. Heidelb., Mitt., Ser. B
Astronomisches Rechen–Institut Heidelberg, Mitteilungen, Serie B. Astronomisches Rechen–Institut, Mönchhofstraße 12–14, D–6900 Heidelberg, F.R. Germany.

Astron. Rechen–Inst. Heidelb., Prepr. Ser.
Astronomisches Rechen–Institut Heidelberg, Preprint Series. Astronomisches Rechen–Institut, Mönchhofstraße 12–14, D–6900 Heidelberg, F.R. Germany.

Astron. Rep.
The Astronomical Reports. Polskie Towarzystwo Miłośników Astronomii, Polish Amateur Astronomical Society, ul. Ludwika Solskiego 30/8, PL–31–027 Kraków, Poland.

Astron. Sch.
Astronomie in der Schule. Verlag Volk und Wissen, DDR–1086 Berlin, Krausenstraße 50, Postfach 1213, German Democratic Republic. ISSN 0004–6310.

Astron. Tidsskr.
Astronomisk Tidsskrift. Astronomisk Selskab, Kobenhavn; Norsk Astronomisk Selskap, Oslo; Svenska Astronomiska Sällskapet, Stockholm. Subscription address: Svenska Astronomiska Sällskapet, Stockholms Observatorium, S–13300 Saltsjöbaden, Sweden. ISSN 0004–6345.

Astron. Tsirk.
Astronomicheskij Tsirkulyar. Izdavaemyj Byuro Astronomicheskikh Soobshchenij Akademii Nauk SSSR. ISSN 0365–7248.

Astron. Vestn.
Astronomicheskij Vestnik. Izdatel'stvo Nauka, 918, Moskva, K–9, USSR. English translation in Sol. Syst. Res. ISSN 0320–930X.

Astron. Zh.
Astronomicheskij Zhurnal. Akademiya Nauk SSSR. Izdatel'stvo Nauka, Moskva, 103717, GSP, Moskva, K–62, Podsosenskij per., 21, USSR. English translation in Soviet Astron. ISSN 0004–6299.

Astronomia
Astronomia. Periodico trimestrale dell'Unione Astrofili Italiani. Subscription address: L. Baldinelli, C.P. 1630, I–40100 Bologna A.D., Italy. ISSN 0392–2308.

Astronomie
L'Astronomie et Bulletin de la Société Astronomique de France. Société Astronomique de France, 3, Rue Beethoven, F–75016 Paris, France. ISSN 0004–6302.

Astronomy
Astronomy. Astro Media Corp., 625 E. St. Paul Avenue, PO Box 92788 Milwaukee, Wis. 53202, USA. ISSN 0091–6358.

Astrophys. J.
The Astrophysical Journal. Published by The University of Chicago Press for the American Astronomical Society. The University of Chicago Press, 5801 S. Ellis Avenue, Chicago, Ill. 60637, USA. ISSN 0004–637X.

Astrophys. J., Lett. Ed.
The Astrophysical Journal, Letters to the Editor. Published by The University of Chicago Press for the American Astronomical Society. The University of Chicago Press, 5801 S. Ellis Avenue, Chicago, Ill. 60637, USA. ISSN 0571–7248.

Astrophys. J., Suppl. Ser.
The Astrophysical Journal, Supplement Series. Published by The University of Chicago Press for the American Astronomical Society. The University of Chicago Press, 5801 S. Ellis Avenue, Chicago, Ill. 60637, USA. ISSN 0067–0049.

Astrophys. Lett. Commun.
Astrophysical Letters and Communications. Gordon and Breach Science Publishers, New York – London – Paris – Montreux – Tokyo – Melbourne. ISSN 0888–6512.

Astrophys. Prepr. Ser.
Astrophysics Preprint Series. Astronomy and Astrophysics Group, Faculty of Physics, Pontificia Universidad Católica de Chile, Casilla 6014, Santiago de Chile, Chile.

Astrophys. Relativ., Prepr. Ser.
Astrophysics and Relativity, Preprint Series. Department of Applied Mathematics and Astronomy, University College, PO Box 78, Cardiff CF1 1XL, England.

Astrophys. Space Phys. Rev.
Astrophysics and Space Physics Reviews. Soviet Scientific Reviews, Section E. Harwood Academic Publishers GmbH, PO Box 786, Cooper Station, New York, N.Y. 10276, USA. ISSN 0143–0432.

Astrophys. Space Sci.
Astrophysics and Space Science. An International Journal of Cosmic Physics. D. Reidel Publishing Company, Dordrecht – Boston. ISSN 0004–640X.

Astrophysics
Astrophysics. A cover–to–cover translation of Astrofizika of the Academy of Sciences of the Armenian SSR. Consultants Bureau, 227 West 17th Street, New York, N.Y. 10011, USA. ISSN 0004–6396.

Atmos. Environ.
Atmospheric Environment. Pergamon Press, Oxford – New York – Toronto – Paris – Frankfurt – Sydney. ISSN 0004–6981.

Atomkernenerg. Kerntech.
Atomkernenergie und Kerntechnik. Verlag Karl–Thiemig AG, Postfach 900740, D–8000 München 90, F.R. Germany. ISSN 0171–5747.

Atti Accad. Naz. Lincei, Rend. Adunanze Solenni
Atti della Accademia Nazionale dei Lincei. Rendiconti delle Adunanze Solenni. Accademia Nazionale dei Lincei, Roma, Italy.

Atti Accad. Naz. Lincei, Ser. Ottava, Rend.
Atti della Accademia Nazionale dei Lincei. Serie Ottava, Rendiconti. Classe di Scienze fisiche, matematiche e naturali. Accademia Nazionale dei Lincei, Roma, Italy. ISSN 0392–7881.

Atti Accad. Sci. Torino I
Atti della Accademia delle Scienze di Torino I. Classe di Scienze fisiche, matematiche e naturali. Via Accademia delle Scienze 6, Via Maria Vittoria 3, Torino (208), Italy. ISSN 0001–4419.

Atti Fond. Giorgio Ronchi
Atti della Fondazione Giorgio Ronchi. Largo Enrico Fermi 1, I–50125 Arcetri–Firenze, Italy. ISSN 0015–606X.

Aust. Comput. J.
Australian Computer Journal. Australian Trade Publications, 28 Chippen Street, Chippendale, N.S.W. 2008, Australia. ISSN 0004–8917.

Aust. J. Astron.
Australian Journal of Astronomy. Astral Press, PO Box 107, Wembley, W.A. 6014, Australia. ISSN 0814–5628.

Aust. J. Geod. Photogramm. Surv.
Australian Journal of Geodesy, Photogrammetry, and Surveying. School of Surveying, University of New South Wales, PO Box 1, Kensington, NSW 2033, Australia. ISSN 0313–9220.

Aust. J. Phys.
Australian Journal of Physics. Commonwealth Scientific and Industrial Research Organization (CSIRO), 314 Albert Street, East Melbourne, Victoria 3002, Australia. ISSN 0004–9506.

Aust. Phys.
Australian Physicist. Australian Institute of Physics, Science Centre, 35–43 Clarence Street, Sydney, NSW 2000, Australia. ISSN 0004–9972.

Austrian Pap. Asteroids
Austrian Papers on Asteroids. Institut für Astronomie, Universitätsplatz 5, A–8010 Graz, Austria.

Automatica
Automatica. Pergamon Press, Oxford – New York – Frankfurt – Paris – Sydney – Toronto – Tokyo. ISSN 0005–1098.

BAV Mitt.
BAV Mitteilungen. Berliner Arbeitsgemeinschaft für Veränderliche Sterne e.V., Sternwarte, Munsterdamm 90, D–1000 Berlin 41, F.R. Germany.

BAV Rundbrief
BAV Rundbrief. Mitteilungsblatt der Berliner Arbeitsgemeinschaft für Veränderliche Sterne. BAV Berliner Arbeitsgemeinschaft für Veränderliche Sterne e.V., Sternwarte, Munsterdamm 90, D–1000 Berlin 41, F.R. Germany. ISSN 0405–5497.

BBSAG Bull.
Bedeckungsveränderlichen Beobachter der Schweizerischen Astronomischen Gesellschaft Bulletin. Available from K. Locher, Rebrain 39, 8624 Grüt, Switzerland.

Be Star Newsl.
Be Star Newsletter. Space Telescope – European Coordinating Facility, European Southern Observatory, Karl–Schwarzschild–Straße 2, D–8046 Garching bei München, F.R. Germany. ISSN 0296–3140.

Beitr. Plasma Phys.
Beiträge aus der Plasma Physik. Akademie–Verlag, Leipziger Straße 3–4, DDR–108 Berlin, German Democratic Republic. ISSN 0005–8025.

B.I.H. Circ.
Bureau International de l'Heure (B.I.H.) Circulars A, D, E. 61, avenue de l'Observatoire, F–75014 Paris, France.

Biul. Obs. Astron. Uniw. M. Kopernika Toruniu
Biuletyn Obserwatorium Astronomicznego Uniwersytetu M. Kopernika w Toruniu. Institute of Astronomy, Nicolaus Copernicus University, Chopina 12/18, PL–87–100 Toruń, Poland.

Bol. Acad. Cienc. Fis. Mat. Nat.
Boletin de la Academia de Ciencias Fisicas Matematicas y Naturales. Academia de Ciencias Fisicas Matematicas y Naturales, Apartado de Correo 1421, Caracas 1010–A, Venezuela. ISSN 0366–1652.

Bol. Asoc. Argent. Astron.
Boletin de la Asociación Argentina de Astronomía, La Plata, Argentina. ISSN 0571–3285.

Bol. Astron.
Boletin Astronômico. Observatório do Capricórnio, Prefeitura Municipal de Campinas–SP, Brazil.

Bol. Astron. Obs. Madr.
Boletín Astronómico del Observatorio de Madrid. Instituto Geografico Nacional, General Ibáñez de Ibero, 3. Madrid 3, Spain. ISSN 0373–7101.

Bol. Inst. Obs. Mar.
Boletin Instituto y Observatorio de Marina. Instituto y Observatorio de Marina, San Fernando (Cádiz), Spain.

Bol. Inst. Tonantzintla
Boletin del Instituto de Tonantzintla. Instituto Nacional de Astrofisica, Optica y Electronica, Apartados Postales Nos. 216 y 51, Puebla, Pue., Mexico. ISSN 0303–7584.

Bol. Obs. Astron. Quito
Boletin del Observatorio Astronomico de Quito. Escuela Politécnica Nacional, Observatorio Astronomico de Quito, Quito, Ecuador.

Bol. Obs. Ebro
Boletín del Observatorio del Ebro. Consejo Superior de Investigaciones Científicas, Observatorio del Ebro, Roquetas (Tarragona), Spain. ISSN 0211–5166.

Boll. Geofis. Teor. Appl.
Bollettino di Geofisica Teorica ed Applicada. Osservatorio Geofisico Sperimentale, I–34123 Trieste, Italy. ISSN 0006–6729.

Boundary–Layer Meteorol.
Boundary–Layer Meteorology. D. Reidel Publishing Company, Dordrecht – Boston – London. ISSN 0006–8314.

Boyden Obs., Occas. Publ.
Boyden Observatory, Occasional Publication. Boyden Observatory, Institute and Department of Astronomy, University of the Orange Free State, PO Box 339, Bloemfontein 9300, South Africa.

Boyden Obs., Repr.
Boyden Observatory, Reprint. Boyden Observatory, Astronomy Department, University of the Orange Free State, PO Box 339, Bloemfontein 9300, South Africa.

Br. Astron. Assoc. Circ.
British Astronomical Association Circular. British Astronomical Association, Burlington House, Piccadilly, London W1V 9AG, England. ISSN 0264–4185.

Br. J. Philos. Sci.
British Journal for the Philosophy of Science. Cambridge University Press, Cambridge – London – New York – New Rochelle – Melbourne – Sydney. ISSN 0007–0882.

Br. J. Photogr.
British Journal of Photography. Henry Greenwood & Co. Ltd., 28 Great James Street, London WC1N 3HL, England. ISSN 0007–1196.

Bulg. J. Phys.
Bulgarian Journal of Physics. Bulgarian Academy of Sciences, Faculty of Physics, 5 Anton Ivanov Blvd., 1126 Sofia, Bulgaria. ISSN 0323–9217.

Bull. Am. Astron. Soc.
Bulletin of the American Astronomical Society. Published by the American Institute of Physics, 335 East 45th Street, New York, N.Y. 10017, USA. ISSN 0002–7537.

Bull. Assoc. Fr. Obs. Etoiles Variables
Bulletin de l'Association Française des Observateurs d'Etoiles Variables. Revue trimestrielle. A.F.O.E.V. Observatoire de Lyon, F–69230 Saint Genis Laval, France. ISSN 0153–9949.

Bull. Astron.
Bulletin Astronomique. Observatoire Royal de Belgique (Astronomisch Bulletin. Koninklijke Sterrenwacht van België). Observatoire Royal de Belgique, 3, avenue Circulaire, Uccle, B–1180 Bruxelles, Belgium.

Bull. Astron. Inst. Czech.
Bulletin of the Astronomical Institutes of Czechoslovakia. Academia, Publishing House of the Czechoslovak Academy of Sciences, Vodičkova 40, 11229 Praha 1, Czechoslovakia. ISSN 0004–6248.

Bull. Astron. Soc. India
Bulletin of the Astronomical Society of India. Astronomical Society of India, Osmania University, Hyderabad 500 007, India. ISSN 0304–9523.

Bull. Aust. Math. Soc.
Bulletin of the Australian Mathematical Society. University of Queensland Press, St. Lucia, Queensland 4067, Australia. ISSN 0004–9727.

Bull. Cl. Sci., Acad. R. Belg.
Bulletin de la Classe des Sciences, Académie Royale de Belgique. Académie Royale des Sciences des Lettres et de Beaux–Arts de Belgique, Brussels, Belgium. ISSN 0001–4141.

Bull. Crimean Astrophys. Obs.
Bulletin of the Crimean Astrophysical Observatory. A cover-to–cover translation of Izv. Krymskoj Astrofiz. Obs. Allerton Press, Inc. 150 5th Avenue, New York, N.Y. 10011, USA.

Bull. Etoiles Tardives Spectre Particulier
Bulletin sur les Etoiles Tardives à Spectre Particulier (Newsletter of Chemically Peculiar Late–type Stars). Observatoire de Strasbourg, 11, rue de l'Université, F–67000 Strasbourg, France. ISSN 0296–3132.

Bull. Géod.
Bulletin Géodésique. The Journal of the International Association of Geodesy. Bureau Central de l'Association Internationale de Géodésie, 39 rue Gay–Lussac, F–75005 Paris, France. ISSN 0007–4632.

Bull. Geogr. Surv. Inst.
Bulletin of the Geographical Survey Institute. Geographical Survey Institute, Ministry of Construction, Kitasato–1, Yatabe–Machi, Tsukuba–Gun, Ibaraki–ken, Japan. ISSN 0373–7160.

Bull. Geophys.
Bulletin of Geophysics. National Central University, Chung–Li, Taiwan, Republic of China. ISSN 0253–4800.

Bull. Inf. Cent. Données Stellaires
Bulletin d'Information du Centre de Données Stellaires. Observatoire de Strasbourg, 11, rue de l'Université, F–67000 Strasbourg, France. ISSN 0242–6536.

Bull. Obs. Astron. Belgr.
Bulletin de l'Observatoire Astronomique de Belgrade. Observatoire Astronomique de Belgrade, Beograd, Volgina 7, Yugoslavia. ISSN 0373–3734.

Bull. Res. Inst. Sci. Meas., Tôhoku Univ.
Bulletin of the Research Institute for Scientific Measurements, Tôhoku University, Sendai, Japan. ISSN 0040–8689.

Bull. Soc. R. Sci. Liège
Bulletin de la Société Royale des Sciences de Liège. L'Université, 15 Avenue des Tilleurs, Liège, Belgium. ISSN 0037–9565.

Bull. Spec. Astrophys. Obs. – North Caucasus
Bulletin of the Special Astrophysical Observatory – North Caucasus. A cover–to–cover translation of Astrofiz. Issled. Izv. Spets. Astrofiz. Obs. Allerton Press, Inc., 150 5th Avenue, New York, N.Y. 10011, USA.

Bull., Time Serv. Mizusawa Obs.
Bulletins, Time Service of the Mizusawa Observatory. The International Latitude Observatory of Mizusawa, Mizusawa–Shi, Iwate–Ken, Japan. ISSN 0580–6585.

Bull. Tokyo Gakugei Univ., Ser. IV
Bulletin of Tokyo Gakugei University, Series IV (Mathematics and Natural Sciences) 4–1–1 Nukui–kita–machi, Koganei, Tokyo, Japan. ISSN 0371–6813.

Byull. Inst. Astrofiz.
Byulleten' Instituta Astrofiziki. Akademiya Nauk Tadzhikskoj SSR. Izdatel'stvo Donish, Dushanbe, USSR. ISSN 0568–6865.

Byull. Inst. Teor. Astron.
Byulleten' Instituta Teoreticheskoj Astronomii. Akademiya Nauk SSSR. Leningradskoe Otdelenie, Izdatel'stvo Nauka, Leningrad V–164, Mendeleevskaya l., 1, USSR. ISSN 0002–3302.

C. R. Acad. Sci., Sér. Gén., Vie Sci.
Comptes Rendus de l'Académie des Sciences, Série Générale, La Vie des Sciences. Académie des Sciences, Paris. Subscription address: Gauthier–Villars, C.D.R., Centrale des Revues, 11, rue Gossin, F–92543 Montrouge Cedex, France.

C. R. Acad. Sci., Sér. I
Comptes Rendus de l'Académie des Sciences, Série I: Mathématique. Académie des Sciences, Paris. Subscription address: Gauthier–Villars, C.D.R., Centrale des Revues, 11, rue Gossin, F–92543 Montrouge Cedex, France. ISSN 0249–6291.

C. R. Acad. Sci., Sér. II
Comptes Rendus de l'Académie des Sciences, Série II: Mécanique, Physique, Chimie, Sciences de l'Univers, Sciences de la Terre. Académie des Sciences, Paris. Subscription address: Gauthier–Villars, C.D.R., Centrale des Revues, 11, rue Gossin, F–92543 Montrouge Cedex, France. ISSN 0249–6305.

Can. J. Earth Sci.
Canadian Journal of Earth Sciences. National Research Council of Canada, Ottawa KIA OR6, Canada. ISSN 0008–4077.

Can. J. Phys.
Canadian Journal of Physics. National Research Council of Canada, Ottawa K1A 0R6, Canada. ISSN 0008–4204.

Carter Obs., Repr. Ser.
Carter Observatory, Reprint Series. Carter Observatory, PO Box 2909, Wellington 1, New Zealand.

Cartes Synoptiques
Cartes Synoptiques de la Chromosphère Solaire et Catalogues des Filaments et des Centres d'Activité. Observatoire de Paris, Section d'Astrophysique, F–92190 Meudon, France.

Celest. Mech.
Celestial Mechanics. An International Journal of Space Dynamics. D. Reidel Publishing Company, Dordrecht – Boston. ISSN 0008–8714.

Cent. Astron. Sci. Spat., Obs. Sol.
Centre de l'Astronomie et des Sciences Spatiales, Observations Solaires. Centre de l'Astronomie et des Sciences Spatiales, Academiei Republicii Socialiste România, Bucuresti, Rumania.

Cent. Astrophys., Prepr. Ser.
Center for Astrophysics, Preprint Series. Center for Astrophysics, 60 Garden St., Cambridge, Mass. 02138, USA.

Centaurus
Centaurus. International Magazine of the History of Mathematics, Science, and Technology. Munksgaard Ltd., Copenhagen, Denmark. ISSN 0008–8994.

Česk. Čas. Fyz., Sekce A
Československý časopis pro fyziku, Sekce A. Academia Publishing House of the Czechoslovak Adademy of Sciences, Vodičkova 40, 11229 Praha 1, Czechoslovakia. ISSN 0009–0700.

Chem. Phys.
Chemical Physics. North–Holland Publishing Company, PO Box 211, 1000 AE Amsterdam, The Netherlands. ISSN 0301–0104.

Chem. Phys. Lett.
Chemical Physics Letters. North–Holland Publishing Company, PO Box 211, 1000 AE Amsterdam, The Netherlands. ISSN 0009–2614.

Chin. Astron. Astrophys.
Chinese Astronomy and Astrophysics. A selected translation from the current issues of Acta Astronomica Sinica, Acta Astrophysica Sinica and Chinese Journal of Space Science. Pergamon Press, Oxford – New York – Toronto – Paris – Frankfurt – Sydney. ISSN 0275–1062.

Chin. J. Phys.
Chinese Journal of Physics. Physical Society of the Republic of China, Physics Department, National Taiwan University, Taipei, Taiwan, Republic of China. ISSN 0577–9073.

Chin. J. Space Sci.
Chinese Journal of Space Science. Subscription address: China International Book Trading Corporation, Guozi Shudian, PO Box 2820, Beijing, People's Republic of China.

Chin. Phys.
Chinese Physics. Selected translations from current issues of major Chinese physics and astronomy journals. American Institute of Physics, 335 East 45th Street, New York, N.Y. 10017, USA. ISSN 0273–429X.

Chin. Phys. Lett.
Chinese Physics Letters. Science Press, 137 Chaoyangmennei Street, Beijing, People's Republic of China.

Ciel
Le Ciel. Bulletin de la Société Astronomique de Liège. Société Astronomique de Liège, B–4410 Vottem, Belgium.

Ciel Espace
Ciel et Espace. Association Française d'Astronomie, 115 rue de Charenton, F–75012 Paris, France.

Ciel Terre
Ciel et Terre. Bulletin de la Société Royale Belge d'Astronomie, de Météorologie et de Physique du Globe. Société Royale Belge d'Astronomie, de Météorologie et de Physique du Globe, 3, avenue Circulaire, B–1180 Bruxelles, Belgium. ISSN 0009–6709.

Circ. Czech. Obs., Time Latitude
Circular of the Czechoslovak Observatories, Time and Latitude. Czechoslovak Academy of Sciences, Astronomical Institute, Prague, Czechoslovakia.

Circ. Inf.
Circulaire d'Information. Union Astronomique Internationale. Commission des Etoiles Doubles. Observatoire de Paris, F–92190 Meudon, France.

Circ. Stn. Astron. Int. Latitudine, Carloforte–Cagliari
Circolari della Stazione Astronomica Internazionale di Latitudine, Carloforte–Cagliari. Series A, B, C. Stazione Astronomica Internazionale di Latitudine, Carloforte–Cagliari, Italy.

Circ. Time Latitude Serv.
Circular Time and Latitude Service. Polish Academy of Sciences, Astronomical Latitude Observatory, Borowiec, Poland.

Classical Quantum Gravity
Classical and Quantum Gravity. Institute of Physics, 47 Belgrave Square, London SW1X 8QX, England.

Clim. Change
Climatic Change. D. Reidel Publishing Company, Dordrecht – Boston – London. ISSN 0165–0009.

CODATA Bull.
CODATA Bulletin. Committee on Data for Science and Technology of the International Council of Scientific Unions. Pergamon Press, Oxford – New York – Toronto – Sydney – Paris – Frankfurt. ISSN 0366–757X.

Commentat. Phys.–Math.
Commentationes Physico–Mathematicae. Societas Scientiarum Fennica, Snellmaninkatu 9–11, SF–00170 Helsinki, Finland. ISSN 0069–6609.

Comments Astrophys.
Comments on Astrophysics. A Journal of Critical Discussion of the Current Literature. Comments on Modern Physics: Part C. Gordon and Breach Science Publishers, New York – London. ISSN 0146–2970.

Comments At. Mol. Phys.
Comments on Atomic and Molecular Physics. Gordon and Breach Science Publishers, New York – London. ISSN 0010–2687.

Comments Nucl. Part. Phys.
Comments on Nuclear and Particle Physics. Gordon and Breach Science Publishers, New York – London. ISSN 0010–2709.

Comments Plasma Phys. Controlled Fusion
Comments on Plasma Physics and Controlled Fusion. Gordon and Breach Science Publishers, New York – London. ISSN 0374–2806.

Commun. Astron. Dep., Univ. Ankara
Communications from the Astronomical Department, University of Ankara. Fen Fakültesi, Ankara, Turkey.

Commun. Fac. Sci. Univ. Ankara, Sér. A3
Communications de la Faculté des Sciences de l'Université d'Ankara, Série A3 (Astronomie). Ankara Üniversitesi, Fen Fakültesi, Besevler–Ankara, Turkey.

Commun. Konkoly Obs.
Communications from the Konkoly Observatory of the Hungarian Academy of Sciences, Budapest, Hungary. ISSN 0324–2234.

Commun. Math. Phys.
Communications in Mathematical Physics. Springer–Verlag, Berlin – Heidelberg – New York – Tokyo. ISSN 0010–3616.

Commun. Theor. Phys.
Communications in Theoretical Physics. Huazhong University of Science and Technology Press, Wuhan, People's Republic of China.

Commun. Univ. Lond. Obs.
Communications from the University of London Observatory. University of London Observatory, Mill Hill Park, London NW7 2QS, England. ISSN 0458–2128.

Commun. Univ. Obs., St. Andrews
Communications from the University Observatory, St. Andrews. University Observatory, Buchanan Gardens, St. Andrews, Fife KY16 9LZ, Scotland.

Comput. Graphics
Computer Graphics. Association for Computing Machinery, 1133 Avenue of the Americas, New York, N.Y. 10036, USA. ISSN 0097–8930.

Comput. Methods Appl. Mech. Eng.
Computer Methods in Applied Mechanics and Engineering. North–Holland Publishing Company, PO Box 211, 1000 AE Amsterdam, The Netherlands. ISSN 0045–7825.

Comput. Phys. Commun.
Computer Physics Communications. North–Holland Publishing Company, Amsterdam, The Netherlands. ISSN 0010–4655.

Comput. Syst.
Computer Systems. Techpress Publishing Company Ltd., Walton House, 93 High Street, Bromley BR1 1JW, England. ISSN 0264–4193.

Computer
Computer. The Institute of Electrical and Electronic Engineers Computer Society, 5855 Naples Plaza, Suite 301, Long Beach, Calif. 90803, USA. ISSN 0018–9162.

Comun. Obs. Astron. Univ. Coimbra
Comunicações do Observatório Astronomico da Universidade de Coimbra, Portugal.

Contemp. Phys.
Contemporary Physics. Taylor and Francis Ltd., 10–14 Macklin Street, London, WC2B 5NF, England. ISSN 0010–7514.

Contrib. Astron. Obs. Skalnaté Pleso
Contributions of the Astronomical Observatory Skalnaté Pleso. VEDA, vydavatel'stvo Slovenskej akadémie vied, Bratislava, Czechoslovakia.

Contrib. Atmos. Phys.
Contributions to Atmospheric Physics. (Beiträge zur Physik der Atmosphäre). Friedrich Vieweg & Sohn Verlagsgesellschaft mbH, Postfach 5829, D–6200 Wiesbaden, F.R. Germany. ISSN 0005–8173.

Contrib. Bosscha Obs.
Contributions from the Bosscha Observatory. Bosscha Observatory, Bandung Institute of Technology, Department of Science, Lembang, Java, Indonesia.

Contrib. Dep. Astron., Univ. Kyoto
Contributions from the Department of Astronomy, University of Kyoto. Department of Astronomy, Faculty of Science, University of Kyoto, Sakyo–ku, Kyoto 608, Japan. ISSN 0388–0230.

Contrib. Dep. Astron., Univ. Tokyo
Contributions from the Department of Astronomy, University of Tokyo. Department of Astronomy, University of Tokyo, Bunko–ku, Tokyo 113, Japan. ISSN 0563–8038.

Contrib. Dep. Geod. Astron., Univ. Thessaloniki
Contributions from the Department of Geodetic Astronomy, University of Thessaloniki. Department of Geodetic Astronomy, University of Thessaloniki, Thessaloniki, Greece.

Contrib. Dom. Astrophys. Obs.
Contributions from the Dominion Astrophysical Observatory. Dominion Astrophysical Observatory, National Research Council of Canada, 5071 West Saanich Road, Victoria, B.C. V8X 4M6, Canada.

Contrib. Dunsink Obs.
Contributions from the Dunsink Observatory. Dunsink Observatory, Castleknock, County Dublin, Republic of Ireland.

Contrib. Geophys. Inst. Slovak Acad. Sci.
Contributions of the Geophysical Institute of the Slovak Academy of Sciences. Slovak Academy of Sciences, Dubravská 9, 84228 Bratislava, Czechoslovakia. ISSN 0586–4607.

Contrib. Inst. Argent. Radioastron.
Contribuciones del Instituto Argentino de Radioastronomia. Instituto Argentino de Radioastronomia, Casilla de Correo No. 5, RA–1894 Villa Elisa, Prov. de Buenos Aires, Argentina.

Contrib. Inst. Astron. Res.
Contributions of the Institute for Astronomical Research. Institute for Astronomical Research, Inc., PO Box 15854, Baton Rouge, La. 70895, USA.

Contrib. Kwasan Hida Obs., Univ. Kyoto
Contributions from the Kwasan and Hida Observatories, University of Kyoto. Kwasan Observatory, University of Kyoto, Yamashina, Kyoto, 607, Japan. Hida Observatory, University of Kyoto, Kamitakara, Gifu–ken, 506–13, Japan. ISSN 0388–2349.

Contrib. La. State Univ. Obs.
Contributions of the Louisiana State University Observatory. Louisiana State University Observatory, Baton Rouge, La. 70803, USA.

Contrib. Lick Obs.
Contributions from the Lick Observatory. Lick Observatory, Board of Studies in Astronomy and Astrophysics, University of California at Santa Cruz, Santa Cruz, Calif. 95064, USA.

Contrib. Nicholas Copernicus Obs. Planetarium Brno
Contributions of the Nicholas Copernicus Observatory and Planetarium in Brno. Nicholas Copernicus Observatory and Planetarium Brno, Czechoslovakia.

Contrib. Nizamiah Japal–Rangapur Obs.
Contributions from Nizamiah and Japal–Rangapur Observatories. Centre of Advanced Study in Astronomy, Osmania University, Hyderabad–500 007, India.

Contrib. Oss. Astron. Torino
Contributi dell'Osservatorio Astronomico di Torino (Pino Torinese). Osservatorio Astronomico di Torino, I–10025 Pino Torinese (Torino), Italy.

Contrib. Van Vleck Obs.
Contributions from the Van Vleck Observatory. Van Vleck Observatory, Middletown, Conn., USA.

Contrib. Wrocław Astron. Obs.
Contributions from the Wrocław Astronomical Observatory. Wrocław University Observatory, Wrocław, Poland.

Copenh. Univ. Obs., Repr.
Copenhagen University Observatory, Reprints. University Observatory, Øster Voldgade 3, DK–1350 Copenhagen K, Denmark.

Cosmic Res.
Cosmic Research. A cover–to–cover translation of Kosm. Issled. Consultants Bureau, 227 West 17th Street, New York, N.Y. 10011, USA. ISSN 0010–9525.

Cracow Obs. Repr.
Cracow Observatory Reprints. Cracow Observatory, Jagellonian University, Cracow, Poland.

CSELT Rapp. Tec.
CSELT Rapporti Tecnici. Centro Studi e Laboratori Telecomunicazioni, Via Guglielmo Reiss Romoli 274, Torino, Italy. ISSN 0390–1815.

Curr. Sci.
Current Science. Current Science Association, Raman Research Institute, Bangalore 560006, India. ISSN 0011–3891.

Curr. Top. Chin. Sci., Sect. E: Astron.
Current Topics in Chinese Science, Section E: Astronomy. An annual selection of papers published in Scientia Sinica and Kexue Tongbao. Gordon and Breach Science Publishers, Inc., London – New York – Paris. ISSN 0732–4421.

Czech. J. Phys., Sect. B
Czechoslovak Journal of Physics, Section B. Academia Publishing House of the Czechoslovak Academy of Sciences, Vodičkova 40, 11229 Praha 1, Czechoslovakia. ISSN 0011–4626.

Data Rep. Hydrogr. Obs., Ser. Astron. Geod.
Data Report of Hydrographic Observations, Series of Astronomy and Geodesy. Hydrographic Department of Japan, Tsukiji–5, Chuo–ku, Tokyo 104, Japan.

Debrecen Heliophys. Obs. Hung. Acad. Sci., Repr.
Debrecen Heliophysical Observatory of the Hungarian Academy of Sciences, Reprint. Heliophysical Observatory of the Hungarian Academy of Sciences, 4010 Debrecen, Hungary.

Dějiny Věd Tech.
Dějiny věd a techniky. Academia Praha, Praha, Czechoslovakia. ISSN 0300–4414.

Dep. Astron. McDonald Obs. Univ. Tex., Repr.
Department of Astronomy and McDonald Observatory of the University of Texas, Reprints. Astronomy Department, R.L.M. 15.220, University of Texas, Austin, Tex. 78712, USA.

Dep. Astrophys., Univ. Oxford, Publ.
Department of Astrophysics, University of Oxford, Publication. Department of Astrophysics, University of Oxford, South Parks Road, Oxford OX1 3RQ, England.

Diss. Abstr. Int., Sect. B
Dissertation Abstracts International – Section B (The Sciences and Engineering). University Microfilms International, Ann Arbor, Michigan 48106, USA. ISSN 0419–4217.

Dokl. Akad. Nauk BSSR
Doklady Akademii Nauk BSSR. Akademiya Nauk, Minsk, Belorussian SSR. ISSN 0002–354X.

Dokl. Akad. Nauk SSSR. Ser. Mat. Fiz.
Doklady Akademii Nauk SSSR. Seriya Matematika, Fizika. Izdatel'stvo Nauka, Moskva. Editorial address: 117874 GSP–7 Moskva, B–485, Profsoyuznaya Ul., 90 Kom. 533, USSR. ISSN 0002–3264.

Dokl. Bolg. Akad. Nauk
Doklady Bolgarskoj Akademii Nauk. Sofiya, Bulgaria. ISSN 0366–8681.

Dtsch. Geod. Komm. Bayer. Akad. Wiss., Reihe A
Deutsche Geodätische Kommission bei der Bayerischen Akademie der Wissenschaften, Reihe A: Theoretische Geodäsie. Deutsche Geodätische Kommission, Marstallplatz 8, D–8000 München 22, F.R. Germany. ISSN 0065–5309.

Dtsch. Geod. Komm. Bayer. Akad. Wiss., Reihe B
Deutsche Geodätische Kommission bei der Bayerischen Akademie der Wissenschaften, Reihe B: Angewandte Geodäsie. Deutsche Geodätische Kommission, Marstallplatz 8, D–8000 München 22, F.R. Germany. ISSN 0065–5317.

Dtsch. Geod. Komm. Bayer. Akad. Wiss., Reihe C
Deutsche Geodätische Kommission bei der Bayerischen Akademie der Wissenschaften, Reihe C: Dissertationen. Deutsche Geodätische Kommission, Marstallplatz 8, D–8000 München 22, F.R. Germany. ISSN 0065–5325.

Dtsch. Geod. Komm. Bayer. Akad. Wiss., Reihe D
Deutsche Geodätische Kommission bei der Bayerischen Akademie der Wissenschaften, Reihe D: Tafelwerke. Deutsche Geodätische Kommission, Marstallplatz 8, D–8000 München 22, F.R. Germany.

Dtsch. Geod. Komm. Bayer. Akad. Wiss., Reihe E
Deutsche Geodätische Kommission bei der Bayerischen Akademie der Wissenschaften, Reihe E: Geschichte und Entwicklung der Geodäsie. Deutsche Geodätische Kommission, Marstallplatz 8, D–8000 München 22, F.R. Germany. ISSN 0065–5341.

Dtsch. Hydrogr. Inst. Hamb., Zeit–Breitendienst
Deutsches Hydrographisches Institut Hamburg, Zeit– und Breitendienst. Deutsches Hydrographisches Institut, D–2000 Hamburg, F.R. Germany.

Dudley Obs. Rep.
Dudley Observatory Report. The Dudley Observatory, 69 Union Avenue, Schenectady, N.Y. 12308, USA.

Dunsink Obs. Repr.
Dunsink Observatory Reprints. Dunsink Observatory, Castleknock, County Dublin, Republic of Ireland.

Earth, Moon, Planets
Earth, Moon, and Planets. An International Journal of Comparative Planetology. D. Reidel Publishing Company, Dordrecht – Boston. ISSN 0167–9295.

Earth–Oriented Appl. Space Technol.
Earth–Oriented Applications of Space Technology. Pergamon Press, Oxford – New York – Frankfurt. ISSN 0277–4488.

Earth Planet. Sci. Lett.
Earth and Planetary Science Letters. Elsevier Scientific Publishing Company, PO Box 211, 1000 AE Amsterdam, The Netherlands. ISSN 0012–821X.

Earth–Sci. Rev.
Earth–Science Reviews. Elsevier Scientific Publishing Company, PO Box 330, 1000 AH Amsterdam, The Netherlands. ISSN 0012–8252.

Edinb. Astron. Prepr.
Edinburgh Astronomy Preprint. Royal Observatory, Blackford Hill, Edinburgh EW9 3HJ, Scotland.

Educ. Earth Sci.
Education of Earth Science. Japan Society of Earth Science Education, c/o Tokyo Gakugei University, Koganei–shi, Tokyo 184, Japan. ISSN 0009–3831.

Electron. Lett.
Electronics Letters. Institution of Electrical Engineers, Savoy Place, London, WC2R 0BL, England. ISSN 0013–5194.

Electronics
Electronics. McGraw–Hill Publishing Company, 1221 Avenue of the Americas, New York, N.Y. 10020, USA. ISSN 0013–5070.

Elektronik
Elektronik. Franzis–Verlag GmbH, D–8000 München 37, Postfach 3701 20, Karlstraße 37, F.R. Germany. ISSN 0013–5658.

Elettron. Telecomun.
Elettronica e Telecomunicazioni. Via Arsenale 41, I–10121 Torino, Italy. ISSN 0013–6123.

Endeavour New Ser.
Endeavour New Series. Pergamon Press, Oxford – New York – Toronto – Paris – Frankfurt – Sydney. ISSN 0013–7162.

EOS Trans. Am. Geophys. Union
EOS Transactions of the American Geophysical Union. American Geophysical Union, 2000 Florida Avenue N.W., Washington, D.C. 20009, USA. ISSN 0096–3941.

ESA Brochure
ESA Brochure (ESA BR). European Space Agency, Scientific and Technical Publications Branch, ESTEC, Postbus 299, 2200 AG Noordwijk, The Netherlands. ISSN 0250–1589.

ESA Bull.
ESA Bulletin. European Space Agency, Scientific and Technical Publications Branch, ESTEC, Postbus 299, 2200 AG Noordwijk, The Netherlands. ISSN 0376–4265.

ESA IUE Newsl.
ESA IUE Newsletters. The ESA IUE Observatory. Apartado 540 65, Madrid, Spain. Subscription address: European Space Agency, 8–10 rue Mario–Nikis, F–75738 Paris Cedex 15, France.

ESA J.
ESA Journal. European Space Agency, Scientific and Technical Publications Branch, ESTEC, Postbus 299, 2200 AG Noordwijk, The Netherlands. ISSN 0379–2285.

ESA SCI
European Space Agency SCI. European Space Agency, Scientific and Technical Publications Branch, ESTEC, Postbus 299, 2200 AG Noordwijk, The Netherlands.

ESA Sci. Tech. Rep.
ESA Scientific and Technical Reports. European Space Agency, Scientific and Technical Publications Branch, ESTEC, Postbus 299, NL–2200 AG Noordwijk, The Netherlands.

ESA Spec. Publ.
ESA Special Publication (ESA SP). European Space Agency, Scientific and Technical Publications Branch, ESTEC, Postbus 299, 2200 AG Noordwijk, The Netherlands. ISSN 0379–6566.

ESO Annu. Rep.
ESO Annual Report. European Southern Observatory, Karl–Schwarzschild–Straße 2, D–8046 Garching bei München, F.R. Germany. ISSN 0531–4496.

ESO Conf. Workshop Proc.
ESO Conference and Workshop Proceedings. European Southern Observatory, Karl–Schwarzschild–Straße 2, D–8046 Garching bei München, F.R. Germany.

ESO Sci. Prepr.
ESO Scientific Preprint. European Southern Observatory, Karl–Schwarzschild–Straße 2, D–8046 Garching bei München, F.R. Germany.

ESO Sci. Rep.
ESO Scientific Report. European Southern Observatory, Karl–Schwarzschild–Straße 2, D–8046 Garching bei München, F.R. Germany.

ESO Tech. Rep.
ESO Technical Report. European Southern Observatory, Karl–Schwarzschild–Straße 2, D–8046 Garching bei München, F.R. Germany.

Eur. J. Phys.
European Journal of Physics. Institute of Physics, 47 Belgrave Square, London SW1X 8QX, England. ISSN 0143–0807.

Europhys. Lett.
Europhysics Letters, European Physical Society, PO Box 69, CH–1213 Petit–Lancy 2, Switzerland. ISSN 0302–072X. Formed by merger of Lett. Nuovo Cimento and J. Phys. Lett.

Europhys. News
Europhysics News. European Physical Society, PO Box 69, CH–1213 Petit–Lancy 2, Switzerland. ISSN 0531–7479.

EXOSAT Express
EXOSAT Express. ESA, EXOSAT Observatory, ESOC, Robert–Bosch–Str. 5, D–6100 Darmstadt, F.R. Germany.

Feingerätetechnik
Feingerätetechnik. VEB Verlag Technik, Oranienburger Straße 13/14, DDR–1020 Berlin, German Democratic Republic. ISSN 0014–9683.

Fis. Tecnol.
Fisica e Tecnologia. Societa Italiana di Fisica, Via Loderingo Degli Andalo 2, I–40124 Bologna, Italy. ISSN 0391–9757.

Fiz. Sz.
Fizikai Szemle. Kiadja a Lapkiado Vallalat, Budapest VII, Lenin korut 9–11, Hungary. ISSN 0015–3257.

Fizika
Fizika. Mladost Export–Import, 41000 Zagreb, Ilica 30, Yugoslavia. ISSN 0015–3206.

Folia Fac. Sci. Nat. Univ. Purkynianae Brun., Phys.
Folia Facultatis Scientiarum Naturalium Universitatis Purkynianae Brunensis, Physica. University J. E. Purkyne, 61137 Brno–Kotlarska 2, Czechoslovakia. ISSN 0323–0287.

Fortschr. Phys.
Fortschritte der Physik. Akademie–Verlag, DDR–108 Berlin, Leipziger Straße 3–4, German Democratic Republic. ISSN 0015–8208.

Found. Phys.
Foundations of Physics. Plenum Publishing Corporation, 227 West 17th Street, New York, N.Y. 10011, USA. ISSN 0015–9018.

Fra Fys. Verden
Fra Fysikkens Verden. Fysisk Institutt, Universitetet i Trondheim, Norges Laererhogskole, N–7000 Trondheim, Norway. ISSN 0015–9247.

Fujitsu Sci. Tech. J.
Fujitsu Scientific and Technical Journal. Fujitsu Ltd., 1015 Kamikodanaka, Nakahara–ku, Kawasaki 211, Kanagawa, Japan. ISSN 0016–2523.

Fundam. Cosmic Phys.
Fundamentals of Cosmic Physics. Gordon and Breach Science Publishers, New York – London – Paris. ISSN 0094–5846.

Funkschau
Funkschau. Franzis–Verlag GmbH, Postfach 370120, Karlstr. 37, D–8000 München 2, F.R. Germany. ISSN 0016–2841.

Fys. Tidsskr.
Fysisk Tidsskrift. Subscription address: Jul. Gjellerups Boghandel, Solvgade 87, DK–1307 Kobenhavn, Denmark. ISSN 0016–3392.

G. A.A.B.
Giornale dell'A.A.B. Associazione Astrofili Bolognesi, Casella Postale 1630, I–40100 Bologna A.D., Italy. ISSN 0392–3932.

G. Astron.
Giornale di Astronomia. Società Astronomica Italiana, Largo E. Fermi, 5, I–50125 Firenze, Italy. ISSN 0390–1106.

Gemini
Gemini. Newsletter of the Royal Greenwich Observatory. Royal Greenwich Observatory, Herstmonceux Castle, Hailsham, East Sussex BN27 1RP, England.

Gen. Relativ. Gravitation
General Relativity and Gravitation. Published under the auspices of the International Committee on General Relativity and Gravitation GRG. Plenum Publishing Corporation, 233 Spring Street, New York, N.Y. 10013, USA. ISSN 0001–7701.

Geochim. Cosmochim. Acta
Geochimica et Cosmochimica Acta. Journal of the Geochemical Society and the Meteoritical Society. Pergamon Press, New York – Oxford – Toronto – Paris – Frankfurt – Sydney. ISSN 0016–7037.

Geod.–Geophys. Arb. Schweiz
Geodätisch–Geophysikalische Arbeiten in der Schweiz. Schweizerische Geodätische Kommission, Institut für Geodäsie und Photogrammetrie, ETH–Hönggerberg, CH–8093 Zürich, Switzerland.

Geod. Geophys. Veröff., Reihe II
Geodätische und Geophysikalische Veröffentlichungen, Reihe II (Solar–terrestrische Beziehungen und Physik der Atmosphäre). Nationalkomitee für Geodäsie und Geophysik bei der Akademie der Wissenschaften der Deutschen Demokratischen Republik, DDR–1500 Potsdam, Telegrafenberg, German Democratic Republic.

Geod. Geophys. Veröff., Reihe III
Geodätische und Geophysikalische Veröffentlichungen, Reihe III (Physik der festen Erde). Nationalkomitee für Geodäsie und Geophysik bei der Akademie der Wissenschaften der Deutschen Demokratischen Republik, DDR–1500 Potsdam, Telegrafenberg, German Democratic Republic. ISSN 0435–6187.

Geod. Kartogr.
Geodezja i Kartografia. Komitet Geodezji Polskiej Akademii Nauk. Publisher: Państwowe Wydawnictwo Naukowe, Warszawa, Poland. ISSN 0016–7134.

Geomagn. Aehron.
Geomagnetizm i Aehronomiya. Akademiya Nauk SSSR. Izdatel'stvo Nauka, Moskva. English translation in Geomagn. Aeron. ISSN 0016–7940.

Geomagn. Aeron.
Geomagnetism and Aeronomy. A cover–to–cover translation of Geomagn. Aehron. American Geophysical Union, 2000 Florida Avenue, N.W., Washington, D. C. 20009, USA. ISSN 0016–7932.

Geophys. Astrophys. Fluid Dyn.
Geophysical and Astrophysical Fluid Dynamics. Gordon and Breach Science Publishers, New York – London – Paris. ISSN 0309–1929.

Geophys. J. R. Astron. Soc.
Geophysical Journal of the Royal Astronomical Society. Published for the Royal Astronomical Society by Blackwell Scientific Publications, Oxford – London – Edinburgh – Boston – Melbourne. ISSN 0016–8009.

Geophys. Norv.
Geophysica Norvegica. Universitetsforlaget, PO Box 2959, Toyen, Oslo 6, Norway. ISSN 0332–5903.

Geophys. Prospect.
Geophysical Prospecting. European Association of Exploration Geophysicists, PO Box 162, 2501 AN The Hague, The Netherlands. ISSN 0016–8025.

Geophys. Res. Lett.
Geophysical Research Letters. American Geophysical Union, 2000 Florida Avenue, N.W., Washington, D.C. 20009, USA. ISSN 0094–8276.

Geophysics
Geophysics. Society of Exploration Geophysicists, PO Box 3098, Tulsa, Okla. 74101, USA. ISSN 0016–8033.

GEOS Circ.
GEOS Circular. Series: Cep (Cepheids), EB (eclipsing binaries), RR (RR Lyrae type variables), SA (small–amplitude variables), SR (red variables). GEOS (Groupe d'Etude et d'Observation Stellaire), 12, rue Bezout, F–75014 Paris, France.

Gerlands Beitr. Geophys.
Gerlands Beiträge zur Geophysik. Akademische Verlagsgesellschaft Geest & Portig KG, DDR–7010 Leipzig, Sternwartenstraße 8, German Democratic Republic. ISSN 0016–8696.

Glas. Mat., Ser. III
Glasnik Matematički, Serija III. Društvo matematičara i fizičara SRH, Marulićev trg 19, YU–41001 Zagreb, P.P. 187, Yugoslavia. ISSN 0017–095X.

Greenwich Time Rep.
Greenwich Time Report. Royal Greenwich Observatory, Time and Latitude Service, Herstmonceux Castle, Hailsham, East Sussex BN27 1RP, England. ISSN 0264–4177.

Hadronic J.
Hadronic Journal. Hadronic Press Inc., Nonantum, Mass. 02195, USA. ISSN 0162–5519.

Heavens
The Heavens. The Oriental Astronomical Association, Ōtsushi, Shiga–ken, Japan. In Japanese.

Helv. Phys. Acta
Helvetica Physica Acta. Schweizerische Physikalische Gesellschaft. Birkhäuser Verlag, Elisabethenstraße 19, CH–4000 Basel 10, Switzerland. ISSN 0018–0238.

H.M. Naut. Alm. Off., Libr. Repr.
H.M. Nautical Almanac Office, Library Reprint. H.M. Nautical Almanac Office, Royal Greenwich Observatory, Herstmonceux Castle, Hailsham, East Sussex BN27 1RP, England.

Hvar Obs. Bull.
Hvar Observatory Bulletin. Faculty of Geodesy, University of Zagreb, Kačićeva 26, YU–41000 Zagreb, Yugoslavia. ISSN 0351–2651.

Hyperfine Interactions
Hyperfine Interactions. J. C. Baltzer AG, Scientific Publishing Co., Wettsteinplatz 10, CH–4058, Basel, Switzerland. ISSN 0304–3843.

I.A.P.P.P. Commun.
International Amateur–Professional Photoelectric Photometry Communication. Dyer Observatory, Vanderbilt University, Nashville, Tenn. 37235, USA. ISSN 0886–6961.

IAU Circ.
International Astronomical Union, Circular. Central Bureau for Astronomical Telegrams, Smithsonian Astrophysical Observatory, 60 Garden Street, Cambridge, Mass. 02138, USA. ISSN 0081–0304.

IAU Inf. Bull.
International Astronomical Union Information Bulletin. IAU–UAI Secrétariat, 61, Avenue de l'Observatoire, F–75014 Paris, France. Published by D. Reidel Publishing Company, Dordrecht – Boston – London.

Icarus
Icarus. International Journal of Solar System Studies. Academic Press Inc., San Diego – Orlando – New York – Austin – London – Montreal – Sydney – Tokyo – Toronto. ISSN 0019–1035.

IEE Proc. F
IEE Proceedings F (Communications, Radar and Signal Processing). Institution of Electrical Engineers, PO Box 8, Southgate House, Stevenage, Herts. SG1 1HQ, England. ISSN 0143–7070.

IEEE J. Solid–State Circuits
IEEE Journal of Solid–State Circuits. The Institute of Electrical and Electronics Engineers, 345 East 47th Street, New York, N.Y. 10017, USA. ISSN 0018–9200.

IEEE Spectrum
IEEE Spectrum. The Institute of Electrical and Electronics Engineers, 345 East 47th Street, New York, N.Y. 10017, USA. ISSN 0018–9235.

IEEE Trans. Aerosp. Electron. Syst.
IEEE Transactions on Aerospace and Electronic Systems. The Institute of Electrical and Electronics Engineers, 345 East 47th Street, New York, N.Y. 10017, USA. ISSN 0018–9251.

IEEE Trans. Antennas Propag.
IEEE Transactions on Antennas and Propagation. The Institute of Electrical and Electronics Engineers, 345 East 47th Street, New York, N.Y. 10017, USA. ISSN 0018–926X.

IEEE Trans. Commun.
IEEE Transactions on Communications. The Institute of Electrical and Electronics Engineers, 345 East 47th Street, New York, N.Y. 10017, USA. ISSN 0090–6778.

IEEE Trans. Geosci. Remote Sensing
IEEE Transactions on Geoscience and Remote Sensing. The Institute of Electrical and Electronics Engineers, 345 East 47th Street, New York, N.Y. 10017, USA. ISSN 0196–2892.

IEEE Trans. Ind. Electron.
IEEE Transactions on Industrial Electronics. The Institute of Electrical and Electronics Engineers, 345 East 47th Street, New York, N.Y. 10017, USA. ISSN 0278–0046.

IEEE Trans. Instrum. Meas.
IEEE Transactions on Instrumentation and Measurement. The Institute of Electrical and Electronics Engineers, 345 East 47th Street, New York, N.Y. 10017, USA. ISSN 0018–9456.

IEEE Trans. Microwave Theory Tech.
IEEE Transactions on Microwave Theory and Techniques. The Institute of Electrical and Electronics Engineers, 345 East 47th Street, New York, N.Y. 10017, USA. ISSN 0018–9480.

IEEE Trans. Nucl. Sci.
IEEE Transactions on Nuclear Science. The Institute of Electrical and Electronics Engineers, 345 East 47th Street, New York, N.Y. 10017, USA. ISSN 0018–9499.

IEEE Trans. Plasma Sci.
IEEE Transactions on Plasma Science. The Institute of Electrical and Electronics Engineers, 345 East 47th Street, New York, N.Y. 10017, USA. ISSN 0093–3813.

IHW Newsl.
The International Halley Watch Newsletter. Jet Propulsion Laboratory, California Institute of Technology, Pasadena, Calif. 91109, USA.

IMA J. Appl. Math.
IMA Journal of Applied Mathematics. Academic Press Inc., London – New York. ISSN 0272–4960.

Indian East. Eng.
Indian and Eastern Engineer. "Piramal Mansion", 235 Dr. D. Naoroji Road, Bombay 400 001, India. ISSN 0019–4352.

Indian Inst. Astrophys. Prepr.
Indian Institute of Astrophysics, Preprints. Indian Institute of Astrophysics, Sarjapur Road, Bangalore – 560 034, India.

Indian J. Cryog.
Indian Journal of Cryogenics. Indian Cryogenics Council, Jadavpur University, Calcutta 700 032, India. ISSN 0379–0479.

Indian J. Hist. Sci.
Indian Journal of History of Science. Indian National Science Academy, Bahadur Shah Zafar Marg, New Delhi 110 002, India.

Indian J. Phys., Part B
Indian Journal of Physics, Part B. Indian Association for the Cultivation of Science, 2 & 3 Raja Subodh Chandra Mallik Road, Calcutta 700 032, India. ISSN 0374–3330.

Indian J. Pure Appl. Math.
Indian Journal of Pure and Applied Mathematics. Indian National Science Academy, Bahadur Shah Zafar Marg, New Delhi 110 002, India. ISSN 0019–5588.

Indian J. Pure Appl. Phys.
Indian Journal of Pure and Applied Physics. Council of Scientific & Industrial Research. Publications & Information Directorate, Hillside Road, New Delhi 110 012, India. ISSN 0019–5596.

Indian J. Radio Space Phys.
Indian Journal of Radio and Space Physics. Council of Scientific & Industrial Research. Publications & Information Directorate, Hillside Road, New Delhi 110 012, India. ISSN 0367–8393.

Indian J. Theor. Phys.
Indian Journal of Theoretical Physics. Institute of Theoretical Physics, Bognan Kutir, 4–1 Mohan Bagan Lane, Calcutta 700 004, India. ISSN 0019–5693.

Inf. Bull. Variable Stars
Information Bulletin on Variable Stars. Commission 27 of the IAU. Konkoly Observatory, Budapest, Hungary. ISSN 0374–0676.

Infrared Phys.
Infrared Physics. An International Research Journal. Pergamon Press, Oxford – New York – Toronto – Sydney – Paris – Frankfurt. ISSN 0020–0891.

Inst. Astron. Astrophys. Tech. Univ. Berlin, Mitt.
Institut für Astronomie und Astrophysik der Technischen Universität Berlin, Mitteilungen. Institut für Astronomie und Astrophysik der Technischen Universität Berlin, Hardenbergstraße 36, D–1000 Berlin 12, F.R. Germany.

Inst. Astron. Fis. Espacio, Tirada Aparte
Instituto de Astronomía y Fisica del Espacio, Tirada Aparte. Instituto de Astronomía y Fisica del Espacio, Casilla de Correo 67, Sucursal 28, 1428 Buenos Aires, Argentina.

Inst. Astron. Geod., Univ. Madr., Publ.
Instituto de Astronomia y Geodesia, Universidad Complutense, Facultad de Ciencias Matemáticas, Madrid, Publicación. Universidad Complutense, Madrid, Spain. ISSN 0213–6198.

Inst. Astron., Univ. Camb., Annu. Rep.
Institute of Astronomy, University of Cambridge, Annual Report. The Observatories, Madingley Road, Cambridge CB3 0HA, England.

Inst. Astrophys. Paris, Pré–Publ.
Institut d'Astrophysique de Paris, Pré–Publication. Institut d'Astrophysique, 98 bis, Boulevard Arago, F–75014 Paris, France.

Inst. Obs. Mar., Bol. Astron.
Instituto y Observatorio de Marina, Boletin Astronomico. Instituto y Observatorio de Marina, San Fernando (Cadiz), Spain.

Inst. Obs. Mar., Bol. Ser. B
Instituto y Observatorio de Marina, Boletin Serie B. (Pequeños Planetas). Instituto y Observatorio de Marina, San Fernando (Cadiz), Spain. ISSN 0558–3993.

Inst. Obs. Mar., Bol. Ser. C
Instituto y Observatorio de Marina, Boletin Serie C. (Rotacion de la Tierra). Instituto y Observatorio de Marina, San Fernando (Cadiz), Spain. ISSN 0210–6485.

Inst. Obs. Mar., Bol. Ser. D
Instituto y Observatorio de Marina, Boletin Serie D. (Ocultaciones). Instituto y Observatorio de Marina, San Fernando (Cadiz), Spain.

Inst. Obs. Mar., Mem. Act.
Instituto y Observatorio de Marina, Memoria de las Actividades. Instituto y Observatorio de Marina, San Fernando (Cadiz), Spain.

Inst. Teor. Astrofys., Blindern–Oslo, Småtrykk
Institutt for Teoretisk Astrofysikk, Blindern–Oslo, Småtrykk. Institute of Theoretical Astrophysics, University of Oslo, PO Box 1029, Blindern, N–0315 Oslo 3, Norway.

Inst. Theor. Astrophys., Blindern–Oslo, Rep.
Institute of Theoretical Astrophysics, Blindern–Oslo, Reports. Institute of Theoretical Astrophysics, University of Oslo, PO Box 1029, Blindern, N–0315 Oslo 3, Norway. ISSN 0078–6780.

Inst. Theor. Astrophys., Blindern–Oslo, Repr.
Institute of Theoretical Astrophysics, Blindern–Oslo, Reprints. Institute of Theoretical Astrophysics, University of Oslo, PO Box 1029, Blindern, N–0315 Oslo 3, Norway.

Inst. Theor. Astrophys., Univ. Oslo, Publ. Ser.
Institute of Theoretical Astrophysics, University of Oslo, Publication Series. Institute of Theoretical Astrophysics, University of Oslo, PO Box 1029, Blindern, N–0315 Oslo 3, Norway. ISSN 0800–6652.

Inst. Theor. Phys. Sternw. Univ. Kiel, Repr.
Institut für Theoretische Physik und Sternwarte der Universität Kiel, Reprints. Institut für Theoretische Physik und Sternwarte der Universität Kiel, D–2300 Kiel, F.R. Germany.

Int. Comet Q.
The International Comet Quarterly. Department of Physics and Astronomy, Appalachian State University, Boone, N.C. 28608, USA. ISSN 0736–6922.

Int. J. Gen. Syst.
International Journal of General Systems. Gordon and Breach Science Publishers Inc., New York – London – Paris. ISSN 0308–1079.

Int. J. Heat Mass Transfer
International Journal of Heat and Mass Transfer. Pergamon Press, Oxford – New York – Toronto – Sydney – Paris – Frankfurt. ISSN 0017–9310.

Int. J. Infrared Millimeter Waves
International Journal of Infrared and Millimeter Waves. Plenum Publishing Corporation, 227 West 17th Street, New York, N.Y. 10011, USA. ISSN 0195–9271.

Int. J. Mass Spectrom. Ion Processes
International Journal of Mass Spectrometry and Ion Processes. Elsevier Scientific Publishing Company, PO Box 330, 1000 AH Amsterdam, The Netherlands. ISSN 0020–7381.

Int. J. Mod. Phys. A
International Journal of Modern Physics A. World Scientific, Singapore. ISSN 0217–751X.

Int. J. Non–Linear Mech.
International Journal of Non–Linear Mechanics. Pergamon Press, Oxford – New York – Toronto – Paris – Frankfurt – Sydney. ISSN 0020–7462.

Int. J. Theor. Phys.
International Journal of Theoretical Physics. Plenum Publishing Corporation, 233 Spring Street, New York, N.Y. 10013, USA. ISSN 0020–7748.

Interdisciplinary Sci. Rev.
Interdisciplinary Science Reviews. Heyden & Son Ltd., Spectrum House, Hillview Gardens, London NW4 2JQ, England. ISSN 0308–0188.

Ir. Astron. J.
The Irish Astronomical Journal. A Half–Yearly Publication under the Auspices of the Observatories of Armagh and Dunsink. Armagh Observatory, Armagh BT61 9DG, Northern Ireland. ISSN 0021–1052.

IRIS Bull. A
IRIS Bulletin A (Earth Orientation). Subcommission International Radio Interferometric Surveying (IRIS) Steering Committee. Available from National Geodetic Survey, National Oceanic and Atmospheric Administration, N/CG114, Rockville, Md. 20852, USA.

ISIS
ISIS. An International Review devoted to the History of Science and its Cultural Influences. Department of History and Sociology of Science, University of Pennsylvania, Philadelphia, Pa. 19104, USA. ISSN 0021–1753.

Issled. Solntsa Krasnykh Zvezd
Issledovaniya Solntsa i Krasnykh Zvezd. (Investigations of the Sun and Red Stars). Akademiya Nauk Latvijskoj SSR, Radioastrofizicheskaya Observatoriya. Izdatel'stvo Zinatne, Riga, USSR. ISSN 0135–1303.

Izv. Akad. Nauk Arm. SSR, Ser. Fiz.
Izvestiya Akademii Nauk Armyanskoj SSR, Seriya Fizika. Akademiya Nauk Armyanskoj SSR, Erevan. 375019 Erevan, ul. Barekamutyan, 24, USSR. ISSN 0002–3035.

Izv. Astron. Ehngel'gardt. Obs.
Izvestiya Astronomicheskoj Ehngel'gardtovskoj Observatorii. Izdatel'stvo Kazanskogo Universiteta, Kazan, ul. Lenina, d. 4/5, USSR.

Izv. Glav. Astron. Obs. Pulkovo
Izvestiya Glavnoj Astronomicheskoj Observatorii v Pulkove. Akademiya Nauk SSSR. Leningradskoe Otdelenie, Izdatel'stvo Nauka, Leningrad, V–164, 199164 Leningrad, Mendeleevskaya l., 1, USSR. ISSN 0367–7966.

Izv. Krymskoj Astrofiz. Obs.
Izvestiya Ordena Trudovogo Krasnogo Znameni Krymskoj Astrofizicheskoj Observatorii. Akademiya Nauk SSSR. Izdatel'stvo 'Nauka', Moskva, USSR. English translation in Bull. Crimean Astrophys. Obs. ISSN 0367–8466.

Izv. Vyssh. Uchebn. Zaved., Radiofiz.
Izvestiya Vysshikh Uchebnikh Zavedenij, Radiofizika. Gor'kovskij Universitet. Gor'kij, ul. Lyadova 25, USSR. English translation in Radiophys. Quantum Electron. ISSN 0021–3462.

J. Am. Assoc. Variable Star Obs.
The Journal of the American Association of Variable Star Observers. The American Association of Variable Star Observers, 187 Concord Avenue, Cambridge, Mass. 02138, USA.

J. Am. Chem. Soc.
Journal of the American Chemical Society. American Chemical Society, 1155 16th Street N.W., Washington, D.C. 20036, USA. ISSN 0002–7863.

J. Appl. Meteorol.
Journal of Applied Meteorology. American Meteorological Society, 45 Beacon Street, Boston, Mass. 02108, USA. ISSN 0021–8952.

J. Appl. Phys.
Journal of Applied Physics. American Institute of Physics, 335 East 45th Street, New York, N.Y. 10017, USA. ISSN 0021–8979.

J. Astron. Fr.
Le Journal des Astronomes Français. Société Française des Spécialistes d'Astronomie. Edité à l'Observatoire du Pic–du–Midi et de Toulouse, 14, avenue Ed. Belin, F–31400 Toulouse, France. ISSN 0181–9429.

J. Astron. Soc. Egypt
Journal of the Astronomical Society of Egypt. Astronomical Society of Egypt, Astronomy Department, Faculty of Sciences, Cairo University, Cairo, Egypt.

J. Astron. Soc. West. Aust.
Journal of the Astronomical Society of Western Australia. Astronomical Society of Western Australia, PO Box 421, Subiaco 6008, W.A.

J. Astronaut. Sci.
Journal of the Astronautical Sciences. American Astronautical Society, 6060 Duke Street, Alexandria, Va. 22304, USA. ISSN 0021–9142.

J. Astrophys. Astron.
Journal of Astrophysics and Astronomy. Indian Academy of Sciences, PO Box 8005, Bangalore 560 080, India. ISSN 0250–6335.

J. Atmos. Sci.
Journal of the Atmospheric Sciences. American Meteorological Society, 45 Beacon Street, Boston, Mass. 02108, USA. ISSN 0022–4928.

J. Atmos. Terr. Phys.
Journal of Atmospheric and Terrestrial Physics. Pergamon Press, Oxford – New York – Frankfurt. ISSN 0021–9169.

J. Aust. Math. Soc., Ser. B
Journal of the Australian Mathematical Society, Series B. (Applied Mathematics). Department of Mathematics, University of Queensland, St. Lucia, QLD 4067, Australia. ISSN 0334–2700.

J. Br. Astron. Assoc.
Journal of the British Astronomical Association. The British Astronomical Association, Burlington House, Piccadilly, London, W1V 9AG, England. ISSN 0007–0297.

J. Br. Interplanet. Soc.
Journal of the British Interplanetary Society. The British Interplanetary Society, 27–29 South Lambeth Road, London, SW8 1SZ, England. ISSN 0007–084X.

J. Chem. Educ.
Journal of Chemical Education. American Chemical Society, 119 West 24th Street, New York, N.Y. 10011, USA. ISSN 0021–9584.

J. Chem. Phys.
Journal of Chemical Physics. American Institute of Physics, 335 East 45th Street, New York, N.Y. 10017, USA. ISSN 0021–9606.

J. Comput. Phys.
Journal of Computational Physics. Academic Press Inc., 111 5th Avenue, New York, N.Y. 10003, USA. ISSN 0021–9991.

J. Differ. Equations
Journal of Differential Equations. Academic Press Inc., New York – London. ISSN 0022–0396.

J. Electron Spectrosc. Relat. Phenom.
Journal of Electron Spectroscopy and Related Phenomena. Elsevier Scientific Publishing Company, PO Box 330, 1000 AH Amsterdam, The Netherlands. ISSN 0368–2048.

J. Electrostat.
Journal of Electrostatics. Elsevier Scientific Publishing Company, PO Box 330, 1000 AH Amsterdam, The Netherlands. ISSN 0304–3886.

J. Environ. Sci.
Journal of Environmental Sciences. Institute of Environmental Sciences, 940 East Northwest Highway, Mt. Prospect, Ill. 60056, USA. ISSN 0022–0906.

J. Fluid Mech.
Journal of Fluid Mechanics. Cambridge University Press, Bentley House, 200 Euston Road, London, NW1 2DB, England. ISSN 0022–1120.

J. Geodyn.
Journal of Geodynamics. Geophysical Press, Brouwersgracht 236, 1013 HE Amsterdam, The Netherlands. ISSN 0264–3707.

J. Geol.
Journal of Geology. The University of Chicago Press, 5801 S. Ellis Avenue, Chicago, Ill. 60637, USA. ISSN 0022–1376.

J. Geomagn. Geoelectr.
Journal of Geomagnetism and Geoelectricity. University of Tokyo Press, c/o Business Centre for Academic Societies, 4–16 Yayoi 2–chome, Bunkyo–ku, Tokyo 113, Japan. ISSN 0022–1392.

J. Geophys.
Journal of Geophysics. Springer–Verlag, Berlin – Heidelberg – New York – Tokyo. ISSN 0340–062X.

J. Geophys. Res.
Journal of Geophysical Research. Section A: Space Physics, Section B: Solid Earth and Planets, Section C: Ocean, Section D: Atmosphere. American Geophysical Union, 2000 Florida Avenue, N.W., Washington, D.C. 20009, USA. ISSN 0148–0227.

J. Guid. Control Dyn.
Journal of Guidance, Control, and Dynamics. American Institute of Aeronautics and Astronautics, 1290 Avenue of the Americas, New York, N.Y. 10104, USA. ISSN 0731–5090.

J. Hist. Astron.
Journal for the History of Astronomy. Published by Science History Publications Ltd., Halfpenny Furze, Mill Lane, Chalfont St Giles, Bucks. HP8 4NR, England. ISSN 0021–8286.

J. Inst. Electron. Commun. Eng. Jpn.
Journal of the Institute of Electronics and Communications Engineers of Japan. Institute of Electronics and Communications Engineers of Japan, Denshi Tsushin Gakkai, Kikai–Shinko–Kaikan, 5–8, Shibakoen 3 Chome Minato–ku, Tokyo 105, Japan. ISSN 0373–6121.

J. Inst. Electron. Telecommun. Eng.
Journal of the Institution of Electronics and Telecommunication Engineers. Institution of Electronics and Telecommunication Engineers, 2 Institutional Area, Lodi Road, New Delhi 110 003, India. ISSN 0377–2063.

J. Korean Astron. Soc.
The Journal of the Korean Astronomical Society. The Korean Astronomical Society, Seoul, Korea.

J. Magn. Magn. Mater.
Journal of Magnetism and Magnetic Materials. North-Holland Publishing Company, PO Box 211, 1000 AE Amsterdam, The Netherlands. ISSN 0304–8853.

J. Math. Phys.
Journal of Mathematical Physics. American Institute of Physics, 335 East 45th Street, New York, N.Y. 10017, USA. ISSN 0022–2488.

J. Math. Phys. Sci.
Journal of Mathematical and Physical Sciences. Indian Institute of Tech., Madras, India. ISSN 0047–2557.

J. Mec. Theor. Appl.
Journal de Mécanique Théorique et Appliquée. Centrale des Revues, 11 rue Gossin, F–92543 Montrouge Cedex, France. ISSN 0750–7240.

J. Mech. Eng. Lab.
Journal of Mechanical Engineering Laboratory. Namiki Sakura–mura, Niihari–gun, Ibaraki, Japan. ISSN 0388–4252.

J. Microcomput. Appl.
Journal of Microcomputer Applications. Academic Press Inc. Ltd., 24 – 28 Oval Road, London NW1 7DX, England. ISSN 0143–3792.

J. Mol. Spectrosc.
Journal of Molecular Spectroscopy. Academic Press Inc., 111 5th Avenue, New York, N.Y. 10003, USA. ISSN 0022–2852.

J. Nanjing Univ.
Journal of Nanjing University. (Natural Sciences Edition). Nanjing University, Nanjing, People's Republic of China.

J. Navig.
The Journal of Navigation. The Royal Institute of Navigation. Cambridge University Press, Cambridge – London – New York – New Rochelle – Melbourne – Sydney. ISSN 0020–3009.

J. Non–Cryst. Solids
Journal of Non–Crystalline Solids. North–Holland Publishing Company, PO Box 211, 1000 AE Amsterdam, The Netherlands. ISSN 0022–3093.

J. Opt. (Calcutta)
Journal of Optics. Optical Society of India, Department of Applied Physics, University of Calcutta, 92 Acharya Prafulla Chandra Road, Calcutta 700 009, India. ISSN 0304–5811.

J. Opt. (Paris)
Journal of Optics. Masson Editeur, 120 Boulevard Saint–Germain, F–75280 Paris Cedex 06, France. ISSN 0150–536X.

J. Opt. Soc. Am. A
Journal of the Optical Society of America A. Optics and Image Science. Published for the Optical Society of America by the American Institute of Physics, 335 East 45th Street, New York, N.Y. 10017, USA. ISSN 0740–3232.

J. Opt. Soc. Am. B
Journal of the Optical Society of America B. Optical Physics. Published for the Optical Society of America by the American Institute of Physics, 335 East 45th Street, New York, N.Y. 10017, USA. ISSN 0740–3224.

J. Phys.
Journal de Physique. Les Editions de Physique, Z. I. de Courtabœuf, B.P. 112, F–91944 Les Ulis Cedex, France. ISSN 0302–0738.

J. Phys. A
Journal of Physics A. (Mathematical and General Physics). Institute of Physics, 47 Belgrave Square, London SW1X 8QX, England. ISSN 0305–4470.

J. Phys. B
Journal of Physics B. (Atomic and Molecular Physics). Institute of Physics, 47 Belgrave Square, London SW1X 8QX, England. ISSN 0022–3700.

J. Phys. Chem.
Journal of Physical Chemistry. American Chemical Society, 1155 16th Street N.W., Washington, D.C. 20036, USA. ISSN 0022–3654.

J. Phys. Chem. Ref. Data
Journal of Physical and Chemical Reference Data. American Chemical Society, 1155 16th Street, N. W., Washington, D. C. 20036, USA. ISSN 0047–2689.

J. Phys. Colloq.
Journal de Physique Colloque. Les Editions de Physique, Z.I. de Courtabœuf, B.P. 112, F–91944 Les Ulis Cedex, France. ISSN 0449–1947.

J. Phys. D
Journal of Physics D. (Applied Physics). Institute of Physics, 47 Belgrave Square, London SW1X 8QX, England. ISSN 0022–3727.

J. Phys. E
Journal of Physics E. (Scientific Instruments). Institute of Physics, 47 Belgrave Square, London SW1X 8QX, England. ISSN 0022–3735.

J. Phys. Earth
Journal of Physics of the Earth. University of Tokyo Press. Subscription address: Japan Publications Trading Company Ltd., C.P.O. 722, Tokyo, Japan. ISSN 0022–3743.

J. Phys. F
Journal of Physics F. (Metal Physics). Institute of Physics, 47 Belgrave Square, London SW1X 8QX, England. ISSN 0305–4608.

J. Phys. G
Journal of Physics G. (Nuclear Physics). Institute of Physics, 47 Belgrave Square, London SW1X 8QX, England. ISSN 0305–4616.

J. Phys. Soc. Jpn.
Journal of the Physical Society of Japan. Room 211, Kikai Shinko Building, 3–5–8 Shiba Koen, Minato–ku, Tokyo 105, Japan. ISSN 0031–9015.

J. Plasma Phys.
Journal of Plasma Physics. Cambridge University Press, Cambridge – London – New York – New Rochelle – Melbourne – Sydney. ISSN 0022–3778.

J. Proc. R. Soc. N.S.W.
Journal and Proceedings of the Royal Society of New South Wales. Science Centre, 35 Clarence Street, Sydney, N.S.W. 2000, Australia. ISSN 0035–9173.

J. Quant. Spectrosc. Radiat. Transfer
Journal of Quantitative Spectroscopy and Radiative Transfer. Pergamon Press, Oxford – New York – Toronto – Paris – Frankfurt – Sydney. ISSN 0022–4073.

J. R. Astron. Soc. Can.
The Journal of the Royal Astronomical Society of Canada. The Royal Astronomical Society of Canada, 124 Merton Street, Toronto, Ont. M4S 2Z2, Canada. ISSN 0035–872X.

J. Radio Res. Lab.
Journal of the Radio Research Laboratories. Radio Research Laboratories, Ministry of Posts & Telecommunications, Nukui–Kitamachi, Konganei–shi, Tokyo 184, Japan. ISSN 0033–8001.

J. Res. Natl. Bur. Stand.
Journal of Research of the National Bureau of Standards. Subscription address: US Government Printing Office, Washington, D.C. 20402, USA. ISSN 0091–0635.

J. Sci. Ind. Res.
Journal of Scientific and Industrial Research. Council of Scientific & Industrial Research, Publications & Information Directorate, Hillside Road, New Delhi 110 012, India. ISSN 0022–4456.

J. Sci. Res. Banaras Hindu Univ.
Journal of Scientific Research of the Banaras Hindu University. Department of Physics, Faculty of Science, PO Banaras Hindu University, Varnasi 221 005, India. ISSN 0447–9483.

J. Soc. Instrum. Control Eng.
Journal of the Society of Instrument and Control Engineers. Society of Instrument and Control Engineers, Hongo, 1–35–28–303, Bunkyo–ku, Tokyo 113, Japan. ISSN 0453–4662.

J. Spacecr. Rockets
Journal of Spacecraft and Rockets. American Institute of Aeronautics and Astronautics, 1290 Avenue of the Americas, New York, N.Y. 10104, USA. ISSN 0022–4650.

J. Spectrosc. Soc. Jpn.
Journal of the Spectroscopical Society of Japan. Spectroscopical Society of Japan, Room 301, Clean Building, 1–13, Kanda–Awaji–cho, Chiyoda–ku, Tokyo 101, Japan. ISSN 0038–7002.

J. Stat. Phys.
Journal of Statistical Physics. Plenum Publishing Corporation, 227 West 17th Street, New York, N.Y. 10011, USA. ISSN 0022–4715.

J. Toyo Univ., Gen. Educ., Nat. Sci.
Journal of the Toyo University, General Education, Natural Science. Toyo University, 28, Hakusan 5–chôme, Bunkyo–ku, Tokyo, Japan.

Jenaer Rundsch.
Jenaer Rundschau. Jenoptik Jena GmbH. VEB Verlag Technik, Oranienburger Str. 13/14, DDR–1020 Berlin, German Democratic Republic. ISSN 0368–203X.

JETP Lett.
JETP Letters. A cover–to–cover translation of Pis'ma v Zhurnal Ehksperimental'noj i Teoreticheskoj Fiziki. American Institute of Physics, 335 East 45th Street, New York, N.Y. 10017, USA. ISSN 0021–3640.

Johns Hopkins APL Tech. Dig.
Johns Hopkins APL Technical Digest. The Johns Hopkins University Applied Physics Laboratory. Johns Hopkins Road, Laurel, Md. 20707, USA. ISSN 0270–5214.

Jpn. J. Appl. Phys., Part 1
Japanese Journal of Applied Physics, Part 1 (Regular Papers and Short Notes). Publication Office, Daini Toyokaiji Building, 24–8, Shinbashi 4–chome, Minato–ku, Tokyo 105, Japan. ISSN 0021–2922.

Jpn. J. Appl. Phys., Part 2
Japanese Journal of Applied Physics, Part 2 (Letters). Publication Office, Daini Toyokaiji Building, 24–8, Shinbashi 4–chome, Minato–ku, Tokyo 105, Japan. ISSN 0021–4922.

Kandilli Obs., Heliophys. Serv. Publ., Second Ser.
Kandilli Observatory, Heliophysics Service Publications, Second Series. Kandilli Observatory, Istanbul, Turkey.

Kapteyn Astron. Inst., Annu. Rep.
Kapteyn Astronomical Institute, Annual Report. Department of Astronomy Rijksuniversiteit Groningen, Groningen, The Netherlands.

KDD Tech. J.
KDD Technical Journal. International Communications Research Institute, Tokyo, Japan. ISSN 0452-3431.

Kexue Tongbao
Kexue Tongbao (Science Bulletin). Science Press, No. 137, Chaoyangmennei Street, Beijing, People's Republic of China. ISSN 0250-7862.

Kinematika Fiz. Nebesn. Tel
Kinematika i Fizika Nebesnykh Tel. Akademiya Nauk Ukrainskoj SSR. Otdelenie Fiziki i Astronomii. Glavnaya Astronomicheskaya Observatoriya Akademii Nauk USSR. Izdatel'stvo Naukova Dumka, 252127 Kiev 127, Goloseevo, USSR. English translation in Kinematics Phys. Celest. Bodies. ISSN 0233-7665.

Kodaikanal Obs. Bull.
Kodaikanal Observatory Bulletins. Indian Institute of Astrophysics, Bangalore 560 034, India.

Kodaikanal Obs. Repr.
Kodaikanal Observatory Reprints. Indian Institute of Astrophysics, Bangalore 560 034, India.

Komet. Tsirk.
Kometnyj Tsirkulyar. Gruppa po Issledovaniyu Komet Sektsii 'Solnechnaya Sistema' Astronomicheskogo Soveta AN SSSR. Kievskij Universitet im. T.G. Shevchenko. Glavnaya Astronomicheskaya Observatoriya AN USSR.

Komety Meteory
Komety i Meteory. Akademiya Nauk Tadzhikskoj SSR. Astronomicheskij Sovet Akademii Nauk SSSR. Izdatel'stvo 'Donish', 734029 Dushanbe, ul. Ajni, 121, korp. 2, USSR. ISSN 0568-6199.

Kosm. Issled.
Kosmicheskie Issledovaniya. Akademiya Nauk SSSR. Izdatel'stvo 'Nauka', Moskva. Editorial address: 103717 GSP, Moskva, K–62, Podsosenskij per., 21. USSR. English translation in Cosmic Res., Consultants Bureau, New York, N.Y., USA. ISSN 0023-4206.

Kosm. Luchi
Kosmicheskie Luchi. Rezul'taty Issledovanij po Mezhdunarodnym Geofizicheskim Proektam. Mezhduvedomstvennyj Geofizicheskij Komitet pri Prezidiume Akademii Nauk SSSR. Izdatel'stvo Sovetskoe radio, Moskva, Glavpochtamt, a/ya No. 693. USSR.

Kozmos
Kozmos. Populárno–vedecký astronomický časopis Slovenského ústredia amatérskej astronómie v Hurbanove (Slovak popular astronomical journal), Obzor, Bratislava, Czechoslovakia.

KPM
Kometen Planetoiden Meteore. Jost Jahn, Rosenweg 2, D–2410 Mölln, Lübeck, F.R. Germany. Subscription address: Michael Möller, Steiluferallee 7, D–2408 Timmendorfer Strand, F.R. Germany. ISSN 0930-102X.

Laser Part. Beams
Laser and Particle Beams. Cambridge University Press, Cambridge – London – New York – New Rochelle – Melbourne – Sydney. ISSN 0263-0346.

Latitude Circ.
Latitude Circular. Astronomic–Geodetical Observatory at Józefosław. Warsaw Technical University, Warsaw, Poland.

LEST Found., Annu. Rep.
Large Earth–based Solar Telescope Foundation, Annual Report. Institute of Theoretical Astrophysics, University of Oslo, PO Box 1029, Blindern, N–0315 Oslo 3, Norway. ISSN 0800-7799.

LEST Found., Tech. Rep.
Large Earth–based Solar Telescope Foundation, Technical Report. Institute of Theoretical Astrophysics, University of Oslo, PO Box 1029, Blindern, N–0315 Oslo 3, Norway. ISSN 0800-7780.

Lett. Math. Phys.
Letters in Mathematical Physics. D. Reidel Publishing Company, Dordrecht – Boston. ISSN 0377-9017.

Lick Obs. Bull.
Lick Observatory Bulletin. Lick Observatory, Board of Studies in Astronomy and Astrophysics, University of California at Santa Cruz, Santa Cruz, Calif. 95064, USA.

Lohrmann–Obs., Tech. Univ. Dresden, Zirk.
Lohrmann–Observatorium, Technische Universität Dresden, Zirkular. Lohrmann–Observatorium, Technische Universität Dresden, DDR–8027 Dresden, Mommsenstraße 13, German Democratic Republic.

Lowell Obs. Bull.
Lowell Observatory Bulletin. Lowell Observatory, Flagstaff, Ariz., USA.

Madà
Madà (Science). The Weizmann Science Press of Israel, Jerusalem, Israel. ISSN 0368-833X.

Magy. Geofiz.
Magyar Geofizika. Lapkiado Vallalat, H–1073 Budapest, Lenin korut 9–11, Hungary. ISSN 0025-0120.

Manuscr. Geod.
Manuscripta Geodaetica. Springer–Verlag, Berlin – Heidelberg – New York – London – Paris – Tokyo. ISSN 0340-8825.

Math. Modelling
Mathematical Modelling. Pergamon Press, New York – Oxford – Toronto – Paris – Frankfurt – Sydney. ISSN 0270-0255.

Math. Proc. Camb. Philos. Soc.
Mathematical Proceedings of the Cambridge Philosophical Society. Cambridge University Press, Cambridge – London – New York – New Rochelle – Melbourne – Sydney. ISSN 0305-0041.

Meccanica
Meccanica. Pitagora Editrice, via Zamboni 57, I–40126 Bologna, Italy. ISSN 0025-6455.

Meded. K. Acad. Wet., Lett. Schone Kunsten Belg., Kl. Wet.
Mededelingen van de Koninklijke Academie voor Wetenschappen, Letteren en Schone Kunsten van Belgie, Klasse der Wetenschappen. Koninklijke Academie voor Wetenschappen, Letteren en Schone Kunsten van Belgie. Paleis der Academien, Hertogsstraat 1, B–1000 Bruxelles, Belgium.

Mem. Astron. Soc. India
Memoirs of the Astronomical Society of India. Astronomical Society of India, Department of Astronomy, Osmania University, Hyderabad 500 007, India.

Mem. Fac. Eng., Kyoto Univ.
Memoirs of the Faculty of Engineering, Kyoto University. Faculty of Engineering, Kyoto University, Kyoto, Japan. ISSN 0023–6063.

Mem. Fac. Eng., Osaka City Univ.
Memoirs of the Faculty of Engineering, Osaka City University. Faculty of Engineering, Osaka City University, 459 Sugimoto–cho, Sumi Yoshi–kum, Osaka, Japan. ISSN 0078–6659.

Mem. Fac. Sci., Kyoto Univ., Ser. Phys., Astrophys., Geophys., Chem.
Memoirs of the Faculty of Science, Kyoto University, Series of Physics, Astrophysics, Geophysics, and Chemistry. Faculty of Science, University of Kyoto, Kyoto 608, Japan. ISSN 0368–9689.

Mem. Jpn. Astron. Study Assoc.
Memoirs of the Japan Astronomical Study Association. National Science Museum, Ueno Park, Taito–ku, Tokyo, Japan.

Mem. Soc. Astron. Ital.
Memorie della Società Astronomica Italiana. Società Astronomica Italiana, Largo Fermi, 5, I–50125 Firenze, Italy. ISSN 0037–8720.

Mercury
Mercury. The Journal of the Astronomical Society of the Pacific. The Astronomical Society of the Pacific, 1290 24th Avenue, San Francisco, Calif. 94122, USA. ISSN 0047–6773.

Meres Autom.
Meres es Automatika. Lapkiado Vallalat, Budapest VII, Lenin Korut 9 – 11, Hungary. ISSN 0025–9993.

Messenger
The Messenger – El Mensajero. European Southern Observatory, Karl–Schwarzschild–Straße 2, D–8046 Garching bei München, F.R. Germany. ISSN 0722–6691.

Meteoritics
Meteoritics. The Journal of the Meteoritical Society. Center for Meteorite Studies, Arizona State University, Tempe, Ariz. 85287, USA. ISSN 0026–1114.

Meteoritika
Meteoritika. Akademiya Nauk SSSR. Komitet po Meteoritam. Izdatel'stvo 'Nauka', Moskva. 117864 GSP–7, Moskva V–485, Profsoyuznaya ul., d. 90, USSR. ISSN 0369–2507.

Meteorol. Mag.
Meteorological Magazine. General Meteorological Office, London Road, Bracknell, Berks. RG12 2SZ, England. ISSN 0026–1149.

Metrologia
Métrologia. Springer–Verlag, Berlin – Heidelberg – New York. ISSN 0026–1394.

Micro Syst.
Micro Systèmes. Société Parisienne d'Edition, 2 à 12 rue de Bellevue, F–75940 Paris Cedex 19, France. ISSN 0183–5084.

Mikrocomput. Z.
Mikrocomputer Zeitschrift. Franzis–Verlag GmbH, Karlstr. 37, D–8000 München 2, F.R. Germany. ISSN 0720–4442.

Mikrowellen Mag.
Mikrowellen Magazin. Sprechsaal–Verlag, PO Box 401, D–8630 Coburg, F.R. Germany. ISSN 0722–8244.

Milano Prepr. Ser. Astrophys.
Milano Preprint Series in Astrophysics. Osservatorio Astronomico di Brera, Istituto di Fisica Cosmica del C.N.R., Dipartimento di Fisica dell' Università, Milano, Italy.

Minor Planet Bull.
The Minor Planet Bulletin. Bulletin of the Minor Planets Section of the Association of Lunar and Planetary Observers. Editorial Office: R. P. Binzel, Department of Astronomy, University of Texas, Austin, Tex. 78712, USA.

Minor Planet Circ.
The Minor Planet Circulars/Minor Planets and Comets. Minor Planet Center, Smithsonian Astrophysical Observatory, 60, Garden Street, Cambridge, Mass. 02138, USA. ISSN 0736–6884.

Mitsubishi Denki Giho
Mitsubishi Denki Giho. Mitsubishi Electric Corporation, Mitsubishi Denki Building, Marunouchi, Tokyo 100, Japan. ISSN 0369–2302.

Mitt. Archenhold–Sternw. Berlin–Treptow
Mitteilungen der Archenhold–Sternwarte Berlin–Treptow. Archenhold–Sternwarte, DDR–1193 Berlin, Alt Treptow 1, German Democratic Republic.

Mitt. Astron. Ges.
Mitteilungen der Astronomischen Gesellschaft, Hamburg. Subscription address: Astronomisches Institut der Universität Bochum, Postfach 102148, D–4630 Bochum, F.R. Germany. ISSN 0172–5483.

Mitt. Astrophys. Obs. Potsdam
Mitteilungen des Astrophysikalischen Wissenschaften der DDR, Zentralinstitut für Astrophysik. Astrophysikalisches Observatorium Potsdam, DDR–1500 Potsdam, Telegrafenberg, German Democratic Republic.

Mitt. Geod. Inst. Rheinischen Friedrich–Wilhelms–Univ. Bonn
Mitteilungen aus den Geodätischen Instituten der Rheinischen Friedrich–Wilhelms–Universität Bonn. Geodätische Institute der Rheinischen Friedrich–Wilhelms–Universität Bonn, Nußallee 17, D–5300 Bonn 1, F.R. Germany. ISSN 0723–4325.

Mitt. Inst. Astron. Phys. Geod. Tech. Hochsch. Münch.
Mitteilungen aus dem Institut für Astronomische und Physikalische Geodäsie der Technischen Hochschule München. Institut für Astronomische und Physikalische Geodäsie der Technischen Hochschule München, D–8000 München, F.R. Germany.

Mitt. Karl–Schwarzschild–Obs. Tautenburg
Mitteilungen des Karl–Schwarzschild–Observatoriums Tautenburg. Akademie der Wissenschaften der DDR, Zentralinstitut für Astrophysik. Karl–Schwarzschild–Observatorium Tautenburg, DDR–6901 Tautenburg, German Democratic Republic.

Mitt. Lohrmann–Obs. Tech. Univ. Dresden
Mitteilungen des Lohrmann–Observatoriums der Technischen Universität Dresden. Lohrmann–Observatorium, Technische Universität Dresden, DDR–8027 Dresden, Mommsenstraße 13, German Democratic Republic.

Mitt. Satell.–Beobachtungsstn. Zimmerwald
Mitteilungen der Satelliten-Beobachtungsstation Zimmerwald. Druckerei der Universität Bern, Switzerland.

Mitt. Sonnenobs. Kanzelhöhe
Mitteilungen des Sonnenobservatoriums Kanzelhöhe. Sonnenobservatorium Kanzelhöhe, A–9520 Sattendorf, Austria.

Mitt. Sternw. Münch.
Mitteilungen der Sternwarte München. Institut für Astronomie und Astrophysik der Universität München, Universitäts–Sternwarte, Scheinerstr. 1, D–8000 München 80, F.R. Germany.

Mitt. Sternw. Sonneberg
Mitteilungen der Sternwarte zu Sonneberg. Akademie der Wissenschaften der DDR, Zentralinstitut für Astrophysik. Sternwarte Sonneberg, DDR–6400 Sonneberg, German Democratic Republic.

Mitt. Univ.–Sternw. Innsb.
Mitteilungen der Universitäts–Sternwarte Innsbruck. Institut für Astronomie der Universität Innsbruck, Universitätsstraße 4, A–6020 Innsbruck, Austria.

Mitt. Univ.–Sternw. Jena
Mitteilungen der Universitäts–Sternwarte zu Jena. Universitätssternwarte Jena, Schillergäßchen 2, DDR–6900 Jena, German Democratic Republic.

Mitt. Universitätssternw. Graz
Mitteilungen der Universitätssternwarte Graz. Institut für Astronomie, Universitätsplatz 5, A–8010 Graz, Austria.

Mitt. Veränderliche Sterne
Mitteilungen über Veränderliche Sterne. Akademie der Wissenschaften der DDR, Zentralinstitut für Astrophysik, Sternwarte Sonneberg, DDR–6400 Sonneberg, German Democratic Republic.

Mitt. Zentralinst. Phys. Erde
Mitteilungen des Zentralinstituts für Physik der Erde. Akademie der Wissenschaften der DDR, Zentralinstitut für Physik der Erde, DDR–1500 Potsdam, Telegrafenberg A17, German Democratic Republic.

Mod. Geol.
Modern Geology. Gordon and Breach Science Publishers, New York – London. ISSN 0026–7775.

Mon. Not. R. Astron. Soc.
Monthly Notices of the Royal Astronomical Society. Published for the Royal Astronomical Society by Blackwell Scientific Publications, Oxford – London – Edinburgh – Boston – Melbourne. ISSN 0035–8711.

Mon. Notes Astron. Soc. S. Afr.
Monthly Notes of the Astronomical Society of Southern Africa. South African Astronomical Observatory, PO Box 9, Observatory, 7935 Cape, South Africa. ISSN 0024–8266.

Mon. Notes Int. Polar Motion Serv.
Monthly Notes of the International Polar Motion Service. Central Bureau of the International Polar Motion Service, International Latitude Observatory of Mizusawa, Mizusawashi, Iwate–ken, Japan. ISSN 0020–8337.

Mons Astrophys. Pap.
Mons Astrophysical Papers. (Communications du Département d'Astrophysique de la Faculté des Sciences de Mons. Departement d'Astrophysique, Université de Mons, B–7000 Mons, Belgium.

MPA Rep.
Max–Planck–Institut für Physik und Astrophysik, Institut für Astrophysik, Reports. Max–Planck–Institut für Physik und Astrophysik, Institut für Astrophysik, Karl–Schwarzschild–Straße 1, D–8046 Garching bei München, F.R. Germany.

MPE Contrib.
Max–Planck–Institut für Physik und Astrophysik, Institut für Extraterrestrische Physik, Contributions. Max–Planck–Institut für Extraterrestrische Physik, Giessenbachstraße, D–8046 Garching bei München, F.R. Germany.

MPE Intern. Rep.
Max–Planck–Institut für Physik und Astrophysik, Institut für Extraterrestrische Physik, Internal Report. Max–Planck–Institut für Extraterrestrische Physik, Giessenbachstraße, D–8046 Garching bei München, F.R. Germany. ISSN 0178–0719.

MPE Prepr.
Max–Planck–Institut für Physik und Astrophysik, Institut für Extraterrestrische Physik, Preprints. Max–Planck–Institut für Extraterrestrische Physik, Giessenbachstraße, D–8046 Garching bei München, F.R. Germany. ISSN 0340–8922.

MPE Rep.
Max–Planck–Institut für Physik und Astrophysik, Institut für Extraterrestrische Physik, Reports. Max–Planck–Institut für Extraterrestrische Physik, Giessenbachstraße, D–8046 Garching bei München, F.R. Germany. ISSN 0173–699X.

MPE Repr.
Max–Planck–Institut für Physik und Astrophysik, Institut für Extraterrestrische Physik, Reprints. Max–Planck–Institut für Extraterrestrische Physik, Giessenbachstraße, D–8046 Garching bei München, F.R. Germany.

MPG Spiegel
MPG Spiegel. Max–Planck–Gesellschaft zur Förderung der Wissenschaften, Residenzstraße 1a, D–8000 München 2, F.R. Germany. ISSN 0341–7727.

MSN Microwave Syst. News
MSN Microwave Systems News. EW Communications Inc., 1170 East Meadow Drive, Palo Alto, Calif. 94303, USA. ISSN 0164–3371.

Mt. Stromlo Siding Spring Obs., Repr.
Mount Stromlo and Siding Spring Observatories, Reprints. Mount Stromlo and Siding Spring Observatories, Research School of Physical Sciences, The Australian National University, Private Bag, Woden PO, ACT 2606, Australia.

Nablyud. Iskusstv. Nebesn. Tel
Nablyudeniya Iskusstvennykh Nebesnykh Tel. Published by Astronomicheskij Sovet Akademii Nauk SSSR, Moskva. Moskva Zh–17, ul. Pyatnitskaya, 48, Astronomicheskij Sovet AN SSSR.

Nachr. Akad. Wiss. Göttingen. II
Nachrichten der Akademie der Wissenschaften in Göttingen. II. Mathematisch–Physikalische Klasse. Vandenhoeck und Ruprecht, Göttingen, F.R. Germany. ISSN 0065–5295.

Nachr. Karten–Vermessungswes., Reihe I
Nachrichten aus dem Karten– und Vermessungswesen, Reihe I: Orginalbeiträge. Verlag des Instituts für Angewandte Geodäsie, Frankfurt a.M., F.R. Germany. ISSN 0469–4236.

Nachr. Olbers–Ges. Bremen
Nachrichten der Olbers–Gesellschaft Bremen. Dr. Walter Stein, Werderstr. 73, Bremen, F.R. Germany.

Nanjing Univ. Obs., Publ.
Nanjing University Observatory, Publications. Department of Astronomy and Astrophysics, Nanjing University, Nanjing, People's Republic of China.

NASA Conf. Publ.
NASA Conference Publication. National Aeronautics and Space Administration, Scientific and Technical Information Branch, Washington, D.C. 20546, USA. For sale by the National Technical Information Service, Springfield, Va. 22161, USA.

NASA Contract. Rep.
NASA Contractor Report. National Aeronautics and Space Administration, Scientific and Technical Information Branch, Washington, D.C. 20546, USA. For sale by the National Technical Information Service, Springfield, Va. 22161, USA. ISSN 0565-7059.

NASA IUE Newsl.
NASA IUE Newsletter. National Aeronautics and Space Administration, Goddard Space Flight Center, Greenbelt, Maryland 20771, USA. ISSN 0738-2677.

NASA Ref. Publ.
NASA Reference Publication. National Aeronautics and Space Administration, Scientific and Technical Information Branch, Washington, D.C. 20546, USA. For sale by the National Technical Information Service, Springfield, Va. 22161, USA.

NASA Spec. Publ.
NASA Special Publication. National Aeronautics and Space Administration, Scientific and Technical Information Branch, Washington, D.C. 20546, USA. For sale by the National Technical Information Service, Springfield, Va. 22161, USA. ISSN 0091-0805.

NASA Tech. Brief
NASA Technical Brief. NASA Technology Utilization Program, Technology Transfer Division, PO Box 8757, Baltimore/Washington International Airport, Md. 21240, USA. ISSN 0096-7494.

NASA Tech. Memo.
NASA Technical Memorandum. National Aeronautics and Space Administration, Scientific and Technical Information Branch, Washington, D.C. 20546, USA. For sale by the National Technical Information Service, Springfield, Va. 22161, USA. ISSN 0499-9320.

NASA Tech. Note
NASA Technical Note. National Aeronautics and Space Administration, Scientific and Technical Information Branch, Washington, D.C. 20546, USA. For sale by the National Technical Information Service, Springfield, Va. 22161, USA. ISSN 0499-9339.

NASA Tech. Pap.
NASA Technical Paper. National Aeronautics and Space Administration, Scientific and Technical Information Branch, Washington, D.C. 20546, USA. For sale by the National Technical Information Service, Springfield, Va. 22161, USA.

Natl. Acad. Sci. Lett.
National Academy Science Letters. National Academy of Sciences, 5-Lajpat Rai Road, Allahabad 211 002, India. ISSN 0250-541X.

Natl. Astron. Ionos. Cent., Astron. Prepr.
National Astronomy and Ionosphere Center, Astronomy Preprints. National Astronomy and Ionosphere Center, Space Sciences Building, Cornell University, Ithaca, N.Y. 14853, USA.

Natl. Astron. Ionos. Cent., Astron. Publ.
National Astronomy and Ionosphere Center, Astronomy Publications. National Astronomy and Ionosphere Center, Space Sciences Building, Cornell University, Ithaca, N.Y. 14853, USA.

Natl. Electron. Rev.
National Electronics Reviews. National Electronics Council, Abell House, John Islip Street, London SW1P 4LN, England. ISSN 0305-2257.

Natl. Geogr.
National Geographic. National Geographic Society, 17th and M Sts. N.W., Washington, D.C. 20036, USA. ISSN 0027-9358.

Natl. Radio Astron. Obs., Repr., Ser. A
National Radio Astronomy Observatory, Reprints, Series A. National Radio Astronomy Observatory, PO Box 2, Green Bank, W.Va. 24944, USA.

Natl. Radio Astron. Obs., Repr., Ser. B
National Radio Astronomy Observatory, Reprints, Series B. National Radio Astronomy Observatory, PO Box 2, Green Bank, W.Va. 24944, USA.

Nature
Nature. Macmillan Journals Ltd., London, England. Subscription address: Nature, Circulation Dept., Brunel Road, Basingstoke, Hants RG21 2XS, England. ISSN 0028-0836.

Naturens Verden
Naturens Verden. Naturens Verdens Fond. Rhodos Internationalt Forlag for Litteratur, Kunst og Videnskab, Strandgade 36, DK-1401 København, Denmark. ISSN 0028-0895.

Naturwissenschaften
Die Naturwissenschaften. Springer-Verlag, Berlin - Heidelberg - New York - Tokyo. ISSN 0028-1042.

Nauchn. Inf.
Nauchnye Informatsii. Astronomicheskij Sovet Akademii Nauk SSSR, Moskva, USSR. ISSN 0130-9773.

Naučna Misao
Naučna Misao (Scientific Idea). Society for Promotion and Propagation of Science, 41000 Zagreb, Pavleka Miškine 37, Yugoslavia. ISSN 0467-0468.

Navigation (Paris)
Navigation. Institut Française de Navigation, 3, avenue Octave-Greard, F-75340 Paris Cedex 07, France. ISSN 0028-1530.

Navigation (Wash.)
Navigation. Journal of the Institute of Navigation. Institute of Navigation, Suite 832. 815 15th Street, N.W., Washington, D.C. 20005, USA. ISSN 0028-1522.

Ned. Tijdschr. Natuurkd. A
Nederlands Tijdschrift voor Natuurkunde A. Nederlandse Natuurkundige Vereniging, Princetonplein 5, NL-3508 TA Utrecht, The Netherlands. ISSN 0378-6374.

NEOS Earth Orientation Bull.
National Earth Orientation Service Earth Orientation Bulletin. Incorporating U.S. Naval Observatory, Time Service Publications, Series 7. U.S. Naval Observatory, Time Service Division (62C), Washington, D.C. 20390, USA.

New Phys., Korean Phys. Soc.
New Physics, Korean Physical Society. Korea Institute for Industrial Economics & Technology, PO Box 205, Chungryang, Seoul, Republic of Korea. ISSN 0374–4914.

New Sci.
New Scientist. New Science Publications, Commonwealth House, 1–19 New Oxford Street, London WC1A 1NG, England. ISSN 0028–6664.

News Lett. Astron. Soc. N.Y.
News Letter of the Astronomical Society of New York. Astronomial Society of New York, 1125 Oxford Place, Schenectady, N.Y. 12308, USA.

Nizamiah Rangapur Obs. Dep. Astron. Osmania Univ., Repr.
Nizamiah and Rangapur Observatories and Department of Astronomy of Osmania University, Reprint. Centre of Advanced Study in Astronomy, Osmania University, Hyderabad 500 007, India.

Nova Acta Leopoldina
Nova Acta Leopoldina. Abhandlungen der Deutschen Akademie der Naturforscher Leopoldina. Deutsche Akademie der Naturforscher Leopoldina, DDR–4010 Halle (Saale), German Democratic Republic. Johann Ambrosius Barth, DDR–7010 Leipzig. ISSN 0369–5034.

Nova Acta Regiae Soc. Sci. Ups., Ser. V: A
Nova Acta Regiae Societatis Scientiarum Upsaliensis, Ser. V: A (Astronomy and Mathematical Sciences). Royal Society of Sciences of Uppsala. Subscripton address: Almqvist & Wiksell International, Stockholm, Sweden. ISSN 0346–6253.

NRAO Workshop
NRAO Workshop. National Radio Astronomy Observatory, PO Box 2, Green Bank, WV 24944–0002, USA.

NSSDC Newsl.
National Space Science Data Center Newsletter. National Space Science Data Center, NASA/Goddard Space Flight Center, Greenbelt, Md. 20771, USA.

NSSDC/WDC–A–R&S, Publ.
NSSDC/WDC–A–R&S, Publication. National Space Science Data Center/World Data Center A for Rockets and Satellites, National Aeronautics and Space Administration, Goddard Space Flight Center, Greenbelt, Md. 20771, USA.

Nucl. Instrum. Methods Phys. Res., Sect. A
Nuclear Instruments and Methods in Physics Research, Section A. Accelerators, Spectrometers, Detectors and Associated Equipment. North–Holland Publishing Company, PO Box 211, 1000 AE Amsterdam, The Netherlands. ISSN 0167–5087.

Nucl. Instrum. Methods Phys. Res., Sect. B
Nuclear Instruments and Methods in Physics Research, Section B. Beam Interactions with Materials and Atoms. North–Holland Publishing Company, PO Box 211, 1000 AE Amsterdam, The Netherlands. ISSN 0168–583X.

Nucl. Phys. A
Nuclear Physics A. North–Holland Publishing Company, PO Box 211, 1000 AE Amsterdam, The Netherlands. ISSN 0029–5582.

Nucl. Phys. B, Part. Phys.
Nuclear Physics B, Particle Physics. North–Holland Publishing Company, PO Box 211, 1000 AE Amsterdam, The Netherlands. ISSN 0550–3213.

Nucl. Sci. Appl., Sect. A
Nuclear Science Applications, Section A. Harwood Academic Publishers GmbH, Poststraße 22, CH–7000 Chur, Switzerland. ISSN 0191–1686.

Nucl. Tracks Radiat. Meas.
Nuclear Tracks and Radiation Measurements. Pergamon Press, Oxford – New York – Toronto – Paris – Frankfurt – Sydney. ISSN 0191–278X.

Numer. Math.
Numerische Mathematik. Springer–Verlag, Berlin – Heidelberg – New York – Tokyo. ISSN 0029–599X.

Nuovo Cimento A
Il Nuovo Cimento della Società Italiana di Fisica A. (Nuclei, particles and fields). Società Italiana di Fisica, Editrice Compositori, viale XII Giugno, 1, I–40124 Bologna, Italy. ISSN 0369–3546.

Nuovo Cimento B
Il Nuovo Cimento della Società Italiana di Fisica B. (General physics, relativity, astronomy and mathematical physics and methods). Società Italiana di Fisica, Editrice Compositori, viale XII Giugno, 1, I–40124 Bologna, Italy. ISSN 0369–3554.

Nuovo Cimento C
Il Nuovo Cimento della Società Italiana di Fisica C. (Geophysics and space physics). Società Italiana di Fisica, Editrice Compositori, viale XII Giugno, 1, I–40124 Bologna, Italy. ISSN 0390–5551.

Nuovo Cimento D
Il Nuovo Cimento della Società Italiana di Fisica D. (Condensed matter, atomic, molecular and chemical physics, biophysics). Società Italiana di Fisica, Editrice Compositori, viale XII Giugno, 1, I–40124 Bologna, Italy. ISSN 0392–6737.

N.Z. J. Sci.
New Zealand Journal of Science. New Zealand Department of Scientific and Industrial Research. Subscription address: Science Information Division, DSIR, PO Box 9741, Wellington, New Zealand. ISSN 0028–8365.

OAA Comput. Sect. Circ.
OAA Computing Section Circular. S. Nakano, PO Box No. 32, Sumoto Post Office, Hyogo-Ken, 656–91, Japan.

Obs. Astron. Antares, Contrib. Cient.
Observatório Astronômico Antares, Contribuição Cientifica. Universidade Estadual de Feira de Santana, Feira de Santana, Brazil.

Obs. Astron. Córdoba, Tirada Aparte
Observatorio Astronomico Córdoba, Tirada Aparte. Observatorio Astronomico, Laprida 854, 5000 Córdoba, Argentina.

Obs. Astron. La Plata, Sep. Astron.
Observatorio Astronómico La Plata, Separata Astronomica. Observatorio Astronómico, Universidad Nacional de La Plata, La Plata, Argentina.

Obs. Astron. Nac., Cerro Calan, Pre–publ.
Observatorio Astronomico Nacional, Cerro Calan, Pre–publicación. Departamento de Astronomía, Facultad de Ciencias Fisicas y Matematicas, Universidad de Chile, Casilla 36–D, Santiago de Chile, Chile.

Obs. Astron. Univ. Nac. La Plata, Ser. Astron.
Observatorio Astronómico de la Universidad Nacional de La Plata, Serie Astronómica. Observatorio Astronómico, Universidad Nacional de La Plata, La Plata, Argentina. ISSN 0325–3163.

Obs. Astron. Univ. Nac. La Plata, Ser. Espec.
Observatorio Astronómico de la Universidad Nacional de La Plata, Serie Especial. Observatorio Astronómico, Universidad Nacional de La Plata, La Plata, Argentina. ISSN 0325–3015.

Obs. Astrophys. Lab., Univ. Helsinki, Rep.
Observatory and Astrophysics Laboratory, University of Helsinki, Report. Observatory and Astrophysics Laboratory, University of Helsinki, Tähtitorninmäki, SF–00130 Helsinki 13, Finland. ISSN 0355–9289.

Obs. Haute Provence, Pré–Publ.
Observatoire de Haute Provence, Pré–Publication. Observatoire de Haute Provence, F–04870 St. Michel l'Observatoire, France.

Obs. Lyon Prepr.
Observatoire de Lyon Preprints. Observatoire de Lyon, F–69230 St. Genis Laval, France.

Obs. Lyon Repr.
Observatoire de Lyon Reprint. Observatoire de Lyon, F–69230 St. Genis Laval, France.

Obs. R. Belg., Commun., Sér. A
Observatoire Royal de Belgique, Communications, Série A. (Koninklijke Sterrenwacht van België, Mededelingen, Reeks A). Observatoire Royale de Belgique, 3, avenue Circulaire, Uccle, B–1180 Bruxelles, Belgium.

Obs. R. Belg., Commun., Sér. B
Observatoire Royal de Belgique, Communications, Série B. (Koninklijke Sterrenwacht van België, Mededelingen, Reeks B). Observatoire Royale de Belgique, 3, avenue Circulaire, Uccle, B–1180 Bruxelles, Belgium.

Obs. Satell.
Observations of Satellites. Finnish Meteorological Institute. Subscription address: Government Printing Centre, Marketing Department, PO Box 516, SF–00101 Helsinki 10, Finland. ISSN 0355–2004.

Obs. Trav.
Observations et Travaux. Société Astronomique de France, 3, rue Beethoven, F–75016 Paris, France.

Obs. Variable Stars, Rep.
Observations of Variable Stars, Report. Nederlandse Vereniging voor Weer–en Sterrenkunde. Kapteyn Astronomical Institute, Postbus 800, 9700 AV Groningen, The Netherlands.

Observatory
The Observatory. A Review of Astronomy. Royal Greenwich Observatory, Herstmonceux Castle, Hailsham Sussex, BN27 1RP England, ISSN 0029–7704.

Occas. Rep. R. Obs., Edinb.
Occasional Reports of the Royal Observatory, Edinburgh. Royal Observatory, Blackford Hill, Edinburgh EH9 3HJ, Scotland. ISSN 0309–099X.

Occultation Newsl.
Occultation Newsletter. International Occultation Timing Association (IOTA), PO Box 596, Tinley Park, Ill. 60477, USA.

Österr. Z. Vermessungswes. Photogramm.
Österreichische Zeitschrift für Vermessungswesen und Photogrammetrie. Österreichischer Verein für Vermessungswesen und Photogrammetrie, Friedrich–Schmidt–Platz 3, A–1082 Wien, Austria.

Onsala Space Obs., Prepr.
Onsala Space Observatory, Preprint. Chalmers University of Technology, Onsala Space Observatory, S–43900 Onsala, Sweden.

Opt. Acta
Optica Acta. Taylor and Francis Ltd., 4 John Street, London WC1N 2ET, England. ISSN 0030–3909.

Opt. Commun.
Optics Communications. North–Holland Publishing Company, PO Box 211, 1000 AE Amsterdam, The Netherlands. ISSN 0030–4018.

Opt. Eng.
Optical Engineering. Society of Photo–Optical Instrumentation Engineers (SPIE), PO Box 10, 405 Fieldston Road, Bellingham, Wash. 98225, USA. ISSN 0091–3286.

Opt. Lett.
Optics Letters. Published for the Optical Society of America by the American Institute of Physics, 335 East 45th Street, New York, N.Y. 10017, USA. ISSN 0146–9592.

Opt. Pura Apl.
Optica Pura y Aplicada. Sociedad Española de Optica, Serrano 121, Madrid 6, Spain. ISSN 0030–3917.

Origins Life
Origins of Life and Evolution of the Biosphere. The Journal of the International Society for the Study of the Origin of Life. D. Reidel Publishing Company, Dordrecht – Boston – Lancaster – Tokyo. ISSN 0302–1688.

Orion
Orion. Zeitschrift der Schweizerischen Astronomischen Gesellschaft (SAG). Zentralsekretariat, Hirtenhofstraße 9, CH–6005 Luzern, Switzerland. ISSN 0030–557X.

Orione
Orione. Rivista Trimestrale di Astronomia. Rivista Orione, via Roma 6, I–10025 Pino Torinese, Italy.

Oss. Astrofis. Catania, Pubbl.
Osservatorio Astrofisico di Catania, Pubblicazione. Osservatorio Astrofisico, città universitaria, I–95125 Catania, Italy.

Oss. Astron. Milano–Merate, Contrib.
Osservatorio Astronomico Milano–Merate, Contributo. Osservatorio Astronomico, Via Emilio Bianchi 46, I–22055 Merate, Como, Italy.

Oss. Mem. Oss. Astrofis. Arcetri
Osservazioni e Memorie dell'Osservatorio Astrofisico di Arcetri. Università Degli Studi di Firenze, Firenze, Italy.

Oss. Privato, Specola "Ariel", Pubbl.
Osservatorio Privato, Specola "Ariel", Pubblicazione. Osservatorio Privato, Specola "Ariel", Via S.Antonio 7, I–3100 Treviso, Italy.

Oyo Buturi
Oyo Buturi. Japan Society of Applied Physics, Room No. 209–2, Kikai–Shinko Building, 3 Shiba–Koen Minato–ku, Tokyo, Japan. ISSN 0369–8009.

Part. Accel.
Particle Accelerators. Gordon and Breach Science Publishers Inc., New York – London – Paris. ISSN 0031–2460.

Pascal Explore
Pascal Explore. E48 (Environnement cosmique terrestre, astronomie et géologie extraterrestre). Centre National de la Recherche Scientifique, Centre de Documentation Scientifique et Technique, 26, rue Boyer, F–75971 Paris Cedex 20, France. ISSN 0761–2109.

Patrika
Patrika. Newsletter of the Indian Academy of Sciences. Indian Academy of Sciences, Bangalore 560 080, India.

Perem. Zvezdy
Peremennye Zvezdy (Variable Stars). Sbornik statej, izdavaemyj Astronomicheskim Sovetom Akademii Nauk SSSR. Editorial address: Sternberg State Astronomical Institute of the Moscow University, Universitetskij prospekt, 13, 117234 Moscow, USSR. ISSN 0373–7683.

Perem. Zvezdy, Prilozh.
Peremennye Zvezdy, Prilozhenie (Variable Stars, Supplement). Sbornik statej, izdavaemyj Astronomicheskim Sovetom Akademii Nauk SSSR. Astronomicheskij Sovet Akademii Nauk SSSR, Moskva, USSR.

Perth Obs. Commun.
Perth Observatory Communications. Perth Observatory, Bickley, Australia.

Philos. Trans. R. Soc. London, Ser. A
Philosophical Transactions of the Royal Society of London, Series A: Mathematical and Physical Sciences. The Royal Society, 6 Carlton House Terrace, London SW1Y 5AG, England. ISSN 0080–4614.

Photogr. Sci. Eng.
Photographic Science and Engineering. Society of Photographic Scientists and Engineers, Suite 204, 1330 Massachusetts Avenue N.W., Washington, D.C. 20005, USA. ISSN 0031–8760.

Photogramm. Eng. Remote Sensing
Photogrammetric Engineering and Remote Sensing. American Society of Photogrammetry, 105 North Virginia Avenue, Falls Church, Va. 22046, USA. ISSN 0099–1112.

Photonics Spectra
Photonics Spectra. The Magazine of Optical/Electro–Optical/Laser Technology. Published by the Optical Publishing Co., Inc., PO Box 1146, Pittsfield, Mass. 01201, USA. ISSN 0191–0647.

Photorin
Photorin, Mitteilungen der Lichtenberg–Gesellschaft e.V.

Phys. Abstr.
Physics Abstracts. Science Abstracts Series A. An INSPEC Publication published by the Institution of Electrical Engineers in association with the Institute of Electrical and Electronics Engineers Inc. Subscription address: INSPEC Marketing Department, IEE, Station House, Nightingale Road, Hitchin, Herts. SG5 1RJ, England. ISSN 0036–8091.

Phys. Bl.
Physikalische Blätter. Physik–Verlag GmbH, Postfach 1260/1280, D–6940 Weinheim, F.R. Germany. ISSN 0031–9279.

Phys. Briefs
Physics Briefs. Physikalische Berichte. Edited by Deutsche Physikalische Gesellschaft und Fachinformationszentrum Energie, Physik, Mathematik in cooperation with American Institute of Physics. Physik–Verlag GmbH, Postfach 1260/1280, D–6940 Weinheim, F.R. Germany. ISSN 0170–7434.

Phys. Bull.
Physics Bulletin. Institute of Physics, 47 Belgrave Square, London SW1X 8QX, England. ISSN 0031–9112.

Phys. Chem. Miner.
Physics and Chemistry of Minerals. Springer–Verlag, Berlin – Heidelberg – New York – Tokyo. ISSN 0342–1791.

Phys. Earth Planet. Inter.
Physics of the Earth and Planetary Interiors. Elsevier Scientific Publishing Company, PO Box 211, 1000 AE Amsterdam, The Netherlands. ISSN 0031–9201.

Phys. Educ.
Physics Education. Institute of Physics. 47 Belgrave Square, London SW1X 8QX, England. ISSN 0031–9120.

Phys. Energ. Fortis Phys. Nucl.
Physica Energiae Fortis et Physica Nuclearis. Science Press, Beijing. Subscription address: Guozi Shudian, PO Box 399, Beijing, People's Republic of China. ISSN 0254–3052.

Phys. Fluids
Physics of Fluids. American Institute of Physics, 335 East 45th Street, New York, N.Y. 10017, USA. ISSN 0031–9171.

Phys. Lett. A
Physics Letters A. (General, Atomic, and Solid State Physics). North–Holland Publishing Company, PO Box 211, 1000 AE Amsterdam, The Netherlands. ISSN 0031–9163.

Phys. Lett. B
Physics Letters B. (Nuclear, Elementary Particle, and High–Energy Physics). North–Holland Publishing Company, PO Box 211, 1000 AE Amsterdam, The Netherlands. ISSN 0031–9163.

Phys. Rep.
Physics Reports. A review section of Physics Letters. North–Holland Publishing Company, PO Box 211, NL–1000 AE Amsterdam, The Netherlands. ISSN 0370–1573.

Phys. Rev. A
Physical Review A. (General Physics). Published for the American Physical Society by the American Institute of Physics, 335 East 45th Street, New York, N.Y. 10017, USA. ISSN 0556–2791.

Phys. Rev. B
Physical Review B. (Condensed Matter). Published for the American Physical Society by the American Institute of Physics, 335 East 45th Street, New York, N.Y. 10017, USA. ISSN 0163–1829.

Phys. Rev. C
Physical Review C. (Nuclear Physics). Published for the American Phyiscal Society by the American Institute of Physics, 335 East 45th Street, New York, N.Y. 10017, USA. ISSN 0556–2813.

Phys. Rev. D
Physical Review D. (Particles and Fields). Published for the American Physical Society by the American Institute of Physics, 335 East 45th Street, New York, N.Y. 10017, USA. ISSN 0556–2821.

Phys. Rev. Lett.
Physical Review Letters. Published for the American Physical Society by the American Institute of Physics, 335 East 45th Street, New York, N.Y. 10017, USA. ISSN 0031–9007.

Phys. Scr.
Physica Scripta. An International Journal for Experimental and Theoretical Physics. The Royal Swedish Academy of Sciences, Publications Department, Box 50005, S–104 05 Stockholm, Sweden. ISSN 0031–8949.

Phys. Teach.
Physics Teacher. American Institute of Physics, 335 East 45th Street, New York, N.Y. 10017, USA. ISSN 0031–921X.

Phys. Technol.
Physics in Technology. Institute of Physics, 47 Belgrave Square, London SW1X 8QX, England. ISSN 0305–4624.

Phys. Today
Physics Today. American Institute of Physics, 335 East 45th Street, New York, N.Y. 10017, USA. ISSN 0031–9228.

Phys. Unserer Zeit
Physik in unserer Zeit. Verlag Chemie GmbH, Postfach 1260/ 1280, D–6940 Weinheim, F.R. Germany. ISSN 0031–9252.

Physica A
Physica A. North–Holland Publishing Company, PO Box 211, 1000 AE Amsterdam, The Netherlands. ISSN 0378–4371.

Physica B, C
Physica B and C (Low Temperature and Solid State Physics; Atomic, Molecular and Plasma Physics; Optics). North–Holland Publishing Company, PO Box 211, 1000 AE Amsterdam, The Netherlands. ISSN 0378–4363.

Physica D
Physica D. North–Holland Publishing Company, PO Box 211, 1000 AE Amsterdam, The Netherlands. ISSN 0167–2789.

Pis'ma Astron. Zh.
Pis'ma v Astronomicheskij Zhurnal. Akademiya Nauk SSSR. Izdatel'stvo 'Nauka', Moskva. 103717, GSP, Moskva, K–62, Podsosenskij per., 21. USSR. English translation in Soviet Astron. Lett. ISSN 0320–0108.

Pis'ma Zh. Ehksp. Teor. Fiz.
Pis'ma v Zhurnal Ehksperimental'noj i Teoreticheskoj Fiziki. Akademiya Nauk SSSR. Izdatel'stvo Nauka, Moskva, Leninskij Prospekt, 14, USSR. English translation in JETP Lett. ISSN 0370–274X.

Planet. Space Sci.
Planetary and Space Science. Pergamon Press, Oxford – New York – Beijing – Frankfurt – São Paulo – Sydney – Tokyo – Toronto. ISSN 0032–0633.

Plasma Phys. Controlled Fusion
Plasma Physics and Controlled Fusion. Pergamon Press, Oxford – New York – Toronto – Paris – Frankfurt – Sydney. ISSN 0032–1028.

Pokroky Mat., Fyz. Astron.
Pokroky matematiky, fyziky a astronomie. Academia, Praha, Czechoslovakia. ISSN 0032–2423.

Port. Phys.
Portugaliae Physica. Sociedade Portuguesa de Fisica, Avenida de Republica, 37–40, P–1000 Lisboa, Portugal. ISSN 0048–4903.

Postepy Astron.
Postepy Astronomii. Czasopismo Poświecone Upowszechnianiu Wiedzy Astronomicznej. Polskie Towarzystwo Astronomiczne, Warszawa. Printed in Poland by Pánstwowe Wydawnictwo Naukowe, Warszawa–Lódź, Poland. ISSN 0032–5414.

Postepy Fiz.
Postepy Fiziyki. Polskie Towarzystwo Fizykczne, 00–681 Warszawa, ul. Hoza 69, Poland. ISSN 0032–5430.

Pramāna
Pramāna. Indian Academy of Sciences, Bangalore 560 006, India. ISSN 0304–4289.

Prepr. Astron. Inst. Univ. Basel
Preprint of the Astronomical Institute of the University of Basel. Astronomisches Institut der Universität Basel, Venusstraße 7, CH–4102 Binningen, Switzerland.

Prepr. Steward Obs.
Preprints of the Steward Observatory. Steward Observatory, The University of Arizona, Tucson, Ariz. 85721, USA.

Priroda
Priroda. Izdatel'stvo 'Nauka', Moskva. Editorial address: 117049, Moskva, GSP–1, Maronovskij per., 26, USSR. ISSN 0032–874X.

Probl. Kosm. Fiz.
Problemy Kosmicheskoj Fiziki. Respublikanskij Mezhvedomstvennyj Nauchnyj Sbornik. Izdatel'stvo pri Kievskom Gosudarstvennom Universitete Izdatel'skogo Obedineniya 'Vishcha Shkola'. Kiev, USSR. ISSN 0555–2796.

Proc. Astron. Soc. Aust.
Proceedings of the Astronomical Society of Australia. Astronomical Society of Australia, Sydney. Subscription address: Division of Radiophysics, PO Box 76, Epping, N.S.W. 2121, Australia. ISSN 0066–9997.

Proc. Cosmic–Ray Res. Lab. Nagoya Univ.
Proceedings of the Cosmic–Ray Research Laboratory of Nagoya University. Cosmic Ray Research Laboratory and Department of Physics, Nagoya University, Chikusa–ku, Nagoya, Japan. ISSN 0250–5444.

Proc. IEEE
Proceedings of the IEEE. The Institute of Electrical and Electronics Engineers, 345 East 47th Street, New York, N.Y. 10017, USA. ISSN 0018–9219.

Proc. Indian Acad. Sci., Earth Planet. Sci.
Proceedings of the Indian Academy of Sciences, Earth and Planetary Sciences. Indian Academy of Sciences, Bangalore 560 080, India. ISSN 0370–0089.

Proc. Indian Acad. Sci., Eng. Sci.
Proceedings of the Indian Academy of Sciences, Engineering Sciences. Indian Academy of Sciences, Bangalore 560 080, India. ISSN 0253–4096.

Proc. Indian Natl. Sci. Acad., Part A
Proceedings of the Indian National Science Academy, Part A. (Physical Sciences). The Indian National Science Academy, Bahadur Shah Zafar Marg, New Delhi 110 002, India. ISSN 0370–0046.

Proc. Int. Latitude Obs. Mizusawa
Proceedings of the International Latitude Observatory of Mizusawa. The International Latitude Observatory of Mizusawa, Mizusawa–Shi, Iwate–Ken, Japan. ISSN 0536–3403.

Proc. Jpn. Acad., Ser. A
Proceedings of the Japan Academy, Series A. (Mathematical Sciences). Japan Academy, Ueno Park, Tokyo 110, Japan. ISSN 0386–2194.

Proc. Jpn. Acad., Ser. B
Proceedings of the Japan Academy, Series B. (Physical and Biological Sciences). Japan Academy, Ueno Park, Tokyo 110, Japan. ISSN 0386–2208.

Proc. K. Ned. Akad. Wet., Ser. B
Proceedings of the Koninklijke Nederlandse Akademie van Wetenschappen, Series B. (Physical Sciences). North–Holland Publishing Company, PO Box 211, 1000 AE Amsterdam, The Netherlands. ISSN 0023–3366.

Proc. Natl. Acad. Sci. India, Sect. A
Proceedings of the National Academy of Sciences of India, Section A (Physical Sciences). National Academy of Sciences of India, 5 Lajpatrai Road, Allahabad–2, India. ISSN 0369–8203.

Proc. Natl. Acad. Sci. U.S.A.
Proceedings of the National Academy of Sciences of the United States of America. National Academy of Sciences, 2101 Constitution Avenue, Washington, D.C. 20418, USA. ISSN 0027–8424.

Proc. R. Soc. Edinb., Sect. A
Proceedings of the Royal Society of Edinburgh, Section A (Mathematical and Physical Sciences). Royal Society of Edinburgh, 22 George Street, Edinburgh EH2 2PQ, Scotland. ISSN 0308–2105.

Proc. R. Soc. London, Ser. A
Proceedings of the Royal Society of London, Series A. (Mathematical and Physical Sciences). The Royal Society, 6 Carlton House Terrace, London, SW1Y 5AG, England. ISSN 0080–4630.

Proc. SPIE Int. Soc. Opt. Eng.
Proceedings of the SPIE – The International Society for Optical Engineering. The International Society for Optical Engineering, PO Box 10, Bellingham, Wash. 98227, USA. ISSN 0277–786X.

Prog. Part. Nucl. Phys.
Progress in Particle and Nuclear Physics. Pergamon Press, New York – Toronto – Paris – Frankfurt – Sydney. ISSN 0146–6410.

Prog. Theor. Phys.
Progress of Theoretical Physics. Research Institute for Fundamental Physics and the Physical Society of Japan. Subscription address: Yukawa Hall, Kyoto University, 606 Kyoto, Japan. ISSN 0033–068X.

Prog. Theor. Phys., Suppl.
Progress of Theoretical Physics, Supplement. Research Institute for Fundamental Physics and the Physical Society of Japan. Subscription address: Yukawa Hall, Kyoto University, 606 Kyoto, Japan. ISSN 0375–9687.

PTB Mitt.
PTB Mitteilungen. Forschen + Prüfen. Amts– und Mitteilungsblatt der Physikalisch–Technischen Bundesanstalt Braunschweig – Berlin. Friedr. Vieweg & Sohn Verlagsgesellschaft mbH, Faulbrunnenstraße 13, D–6200 Wiesbaden, F.R. Germany. ISSN 0030–834X.

Pubbl. Stn. Astron. Int. Latitudine, Carloforte–Cagliari, Nuov. Ser.
Pubblicazioni della Stazione Astronomica Internazionale di Latitudine, Carloforte–Cagliari, Nuova Serie. Stazione Astronomica Internazionale di Latitudine, Carloforte–Cagliari, Italy.

Pubbl. Varie Fuori Ser. Oss. Astron. Torino (Pino Torinese)
Pubblicazioni Varie Fuori Serie dell'Osservatorio Astronomico di Torino (Pino Torinese). Osservatorio Astronomico di Torino, I–10025 Pino Torinese (Torino), Italy.

Publ. Astron. Dep. Eötvös Univ.
Publications of the Astronomy Department of the Eötvös University. Department of Astronomy, Eötvös Lorand University, Budapest, Hungary.

Publ. Astron. Inst. Czech. Acad. Sci.
Publications of the Astronomical Institute of the Czechoslovak Academy of Sciences. Astronomical Institute, Czechoslovak Academy of Sciences, Budečská 6, 120 23 Praha 2, Czechoslovakia.

Publ. Astron. Obs. Sarajevo
Publications of the Astronomical Observatory of Sarajevo. Astronomical Observatory of Sarajevo, Sarajevo, M. Tita 44, Yugoslavia. ISSN 0351–4587.

Publ. Astron. Opservatorije Beogr.
Publikacija Astronomske Opservatorije u Beogradu (Publications de l'Observatoire Astronomique de Beograd). Astronomska Opservatorija u Beogradu, YU–11050 Beograd, Volgina 7, Yugoslavia. ISSN 0373–3742.

Publ. Astron. Soc. Jpn.
Publications of the Astronomical Society of Japan. Astronomical Society of Japan, Tokyo Astronomical Observatory. Mitaka–shi, Tokyo 181, Japan. ISSN 0004–6264.

Publ. Astron. Soc. Pac.
Publications of the Astronomical Society of the Pacific. The Astronomical Society of the Pacific, 1290 24th Avenue, San Francisco, Calif. 94122, USA. ISSN 0004–6280.

Publ. Astron. Soc. "Rudjer Bošković"
Publication of the Astronomical Society "Rudjer Bošković". Astronomical Society "Rudjer Bošković", Beograd, Yugoslavia. ISSN 0506–4295.

Publ. Astrophys. Obs. Potsdam
Publikationen des Astrophysikalischen Observatoriums zu Potsdam. Akademie der Wissenschaften der DDR, Zentralinstitut für Astrophysik. Astrophysikalisches Observatorium Potsdam, DDR–1500 Potsdam, Telegrafenberg, German Democratic Republic.

Publ. Beijing Astron. Obs.
Publications of the Beijing Astronomical Observatory. Beijing Astronomical Observatory, Academia Sinica, Beijing, People's Republic of China.

Publ. Bosscha Obs.
Publications of the Bosscha Observatory. Bandung Institute of Technology, Department of Science, Lembang, Indonesia.

Publ. Debrecen Heliophys. Obs.
Publications of Debrecen Heliophysical Observatory. Hungarian Academy of Sciences, H–4010 Debrecen, Hungary. ISSN 0209–7567.

Publ. Dep. Astron., Univ. Beogr.
Publications of the Department of Astronomy, University of Beograd. Department of Astronomy, University of Beograd, Beograd, Yugoslavia. ISSN 0350–3283.

Publ. Dep. Astron., Univ. Cape Town
Publications of the Department of Astronomy of the University of Cape Town. Astronomy Department, University of Cape Town, Rondebosch, 7700 Cape, South Africa.

Publ. Dep. Astron., Univ. Chile
Publicaciones Departamento de Astronomía, Universidad de Chile. Observatorio Astronomico Nacional, Cerro Calan, Santiago de Chile, Chile.

Publ. Dep. Geod. Astron., Univ. Thessaloniki
Publications of the Department of Geodetic Astronomy, University of Thessaloniki. Department of Geodetic Astronomy, University of Thessaloniki, Thessaloniki, Greece.

Publ. Dom. Astrophys. Obs.
Publications of the Dominion Astrophysical Observatory. Dominion Astrophysical Observatory, National Research Center of Canada, 5071 West Saanich Road, Victoria, B.C. V8X 4M6, Canada. ISSN 0078–6950.

Publ. Ege Univ. Obs.
Publications of Ege University Observatory. Ege University Observatory, P.K. 21, Bornova–Izmir, Turkey.

Publ. Espec. Obs. Astron. Quito
Publicacion Especial del Observatorio Astronomico de Quito. Observatorio Astronomico de Quito, Escuela Politécnica Nacional, Quito, Ecuador.

Publ. Finn. Geod. Inst.
Publications of the Finnish Geodetic Institute. Finnish Geodetic Institute, Helsinki, Finland. ISSN 0085–6932.

Publ. Inst. Geophys., Pol. Acad. Sci., Ser. F (Planet Geod.)
Publications of the Institute of Geophysics, Polish Academy of Sciences, Series F (Planetary Geodesy). Państwowe Wydawnictwo Naukowe, Warszawa–Łódź, Poland. ISSN 0318–0222.

Publ. Int. Latitude Obs. Mizusawa
Publications of the International Latitude Observatory of Mizusawa. The International Latitude Observatory of Mizusawa, Mizusawa–shi, Iwate–ken, Japan. ISSN 0386–0779.

Publ. Istanbul Univ. Obs.
Publications of the Istanbul University Observatory. Istanbul University Observatory, Istanbul, Turkey.

Publ. Lick Obs.
Publications of the Lick Observatory. Lick Observatory, Board of Studies in Astronomy and Astrophysics, University of California at Santa Cruz, Santa Cruz, Calif. 95064, USA.

Publ. Obs. Astron. Nac., Univ. Colombia
Publicaciones del Observatorio Astronomico Nacional. Observatorio Astronomico Nacional, Universidad Nacional de Colombia, Apartado Aéreo 2584, Bogotá, Colombia.

Publ. Obs. Astron. "Prof. Manuel de Barros", Fac. Cienc. Porto
Publicações do Observatório Astronómico "Prof. Manuel de Barros", da Faculdade de Ciências do Porto. Observatório Astronómico "Prof. Manuel de Barros", Universidade do Porto, Monte da Virgem – Vila Nova de Gaia, Portugal.

Publ. Obs. Astron. Univ. Cluj
Publicatiile Observatorului Astronomic al Universitatii din Cluj. Astronomical Observatory, University Babes–Bolyai, Cluj–Napoca, Romania.

Publ. Obs. Bordeaux, Nouv. Sér.
Publications de l'Observatoire de Bordeaux, Nouvelle Série. Observatoire de l'Université de Bordeaux I, F–33270 Floirac, France.

Publ. Obs. Ebro
Publicaciones del Observatorio del Ebro (Miscelanea). Consejo Superior de Investigaciones Cientificas, Observatorio del Ebro, Roquetas (Tarragona), Spain.

Publ. Obs. Genève, Sér. A
Publications de l'Observatoire de Genève, Série A. Observatoire de Genève, Chemin des Maillettes 51, CH–1290 Sauverny, Switzerland.

Publ. Obs. Genève, Sér. B
Publications de l'Observatoire de Genève, Série B. Observatoire de Genève, Chemin des Maillettes 51, CH–1290 Sauverny, Switzerland.

Publ. Obs. Genève, Sér. C: Pré–Publ.
Publications de l'Observatoire de Genève, Série C: Pré–Publications. Observatoire de Genève, Chemin des Maillettes 51, CH–1290 Sauverny, Switzerland.

Publ. Purple Mt. Obs.
Publications of the Purple Mountain Observatory. Purple Mountain Observatory, Academia Sinica, Nanking, People's Republic of China.

Publ. R. Obs., Edinb.
Publications of the Royal Observatory, Edinburgh. Science Research Council. Royal Observatory, Blackford Hill, Edinburgh EH9 3HJ, Scotland. ISSN 0305–2001.

Publ. Shaanxi Astron. Obs.
Publications of the Shaanxi Astronomical Observatory. Shaanxi Astronomical Observatory, Academia Sinica, Lintong, Xian, People's Republic of China.

Publ. Spéc. Cent. Données Stellaires
Publication Spéciale du Centre de Données Stellaires. Observatoire de Strasbourg, 11, rue de l'Université, F–67000 Strasbourg, France.

Publ. U.S. Nav. Obs., Second Ser.
Publications of the United States Naval Observatory, Second Series. U.S. Government Printing Office, Washington, D.C. 20402, USA.

Publ. Variable Star Sect., R. Astron. Soc. N.Z.
Publications of Variable Star Section, Royal Astronomical Society of New Zealand. Astronomical Research Ltd., PO Box 3093, Greerton, Tauranaga, New Zealand.

Publ. Warner Swasey Obs.
Publications of the Warner and Swasey Observatory. Warner and Swasey Observatory, Case Western Reserve University, Department of Astronomy, Smith Bldg., Cleveland, Ohio 44106, USA. ISSN 0160–2500.

Publ. Yunnan Obs.
Publications of Yunnan Observatory. Yunnan Observatory, Academia Sinica, PO Box 110, Kunming, Yunnan Province, People's Republic of China.

Pure Appl. Geophys.
Pure and Applied Geophysics. Birkhäuser Boston Inc., 380 Green Street, Cambridge, Mass. 02139, USA. ISSN 0033–4553.

Q. Appl. Math.
Quarterly of Applied Mathematics. American Mathematical Society, PO Box 1571, Providence, R.I. 02901, USA. ISSN 0033–569X.

Q. Bull. Sol. Act.
Quarterly Bulletin on Solar Activity. International Astronomical Union. Published by the Tokyo Astronomical Observatory, University of Tokyo, Mitaka, Tokyo 181, Japan.

Q. J. R. Astron. Soc.
The Quarterly Journal of the Royal Astronomical Society. Published for the Royal Astronomical Society by Blackwell Scientific Publications, Oxford – London – Edinburgh – Boston – Melbourne. ISSN 0035–8738.

Q. J. R. Meteorol. Soc.
Quarterly Journal of the Royal Meteorological Society. Cromwell House, High Street, Bracknell, Berks., England. ISSN 0035–9009.

R. Greenwich Obs., Annu. Rep.
Royal Greenwich Observatory, Annual Report. Royal Greenwich Observatory, Herstmonceux Castle, Hailsham, East Sussex BN27 1RP, England. ISSN 0308–3322.

R. Greenwich Obs., Bull.
Royal Greenwich Observatory, Bulletins. Royal Greenwich Observatory, Herstmonceux Castle, Hailsham, East Sussex BN27 1RP, England. ISSN 0308–5074.

R. Greenwich Obs., Prepr.
Royal Greenwich Observatory, Preprints. Royal Greenwich Observatory, Herstmonceux Castle, Hailsham, East Sussex BN27 1RP, England.

R. Obs. Ann.
Royal Observatory Annals. Royal Greenwich Observatory, Herstmonceux Castle, Hailsham, East Sussex BN27 1RP, England. ISSN 0080–4371.

R. Obs. Edinb., Res. Facilities
Royal Observatory Edinburgh, Research and Facilities. Royal Observatory, Blackford Hill, Edinburgh EH9 3HJ, Scotland.

Radiant
Radiant. Journal of the Dutch Meteor Society. Federation of European Meteor Astronomers (FEMA), Morssingel 35a, 2312 AZ Leiden, The Netherlands.

Radiat. Eff.
Radiation Effects. Gordon and Breach Science Publishers Inc., New York – London – Paris. ISSN 0033–7579.

Radio Sci.
Radio Science. American Geophysical Union, 2000 Florida Avenue, N.W., Washington, D.C. 20009, USA. ISSN 0048–6604.

Radiochem. Radioanal. Lett.
Radiochemical and Radioanalytical Letters. Elsevier Sequoia S.A., PO Box 851, CH–1001 Lausanne, Switzerland. ISSN 0079–9483.

Rapp. Act. Obs. Paris
Rapport d'Activité de l'Observatoire de Paris. Observatoire de Paris, F–92190 Meudon, France.

Rech. Aerosp.
Recherche Aerospatiale. Office National d'Etudes et de Recherches Aerospatiales, 29 Avenue de la Division Leclerc, F–92320 Chatillon, France. ISSN 0034–1223.

Recherche
Recherche. Société d'Editions Scientifiques, 57, rue de Seine, F–75006 Paris, France. ISSN 0029–5671.

Ref. Zh., 51. Astron.
Referativnyj Zhurnal, 51. Astronomiya. Gosudarstvennyj Komitet SSSR po Nauke i Tekhnike. Vsesoyuznyj Institut Nauchnoj i Tekhnicheskoj Informatsii. Akademiya Nauk SSSR. Moskva. 125219, Moskva, A–219, Baltijskaya ul., 14, USSR. ISSN 0486–2236.

Ref. Zh., 52. Geod. Aehrosemka
Referativnyj Zhurnal, 52. Geodeziya i Aehrosemka. Gosudarstvennyj Komitet SSSR po Nauke i Tekhnike. Vsesoyuznyj Institut Nauchnoj i Tekhnicheskoj Informatsii. Akademiya Nauk SSSR. Moskva. 125219, Moskva, A–219, Baltijskaya ul., 14, USSR. ISSN 0375–9717.

Ref. Zh., 62. Issled. Kosm. Prostranstva
Referativnyj Zhurnal, 62. Issledovanie Kosmicheskogo Prostranstva. Gosudarstvennyj Komitet SSSR po Nauke i Tekhnike. Vsesoyuznyj Institut Nauchnoj i Tekhnicheskoj Informatsii. Akademiya Nauk SSSR. Moskva. 125219, Moskva, A–219, Baltijskaya ul., 14, USSR. ISSN 0034–2408.

Rep. Finn. Geod. Inst.
Reports of the Finnish Geodetic Institute. Finnish Geodetic Institute, Helsinki, Finland. ISSN 0355–1962.

Rep. Inst. Phys. Chem. Res.
Reports of the Institute of Physical and Chemical Research. Institute of Physical and Chemical Research, Rikagaku Kenkyushu, Wako–shi, Saitama 351, Japan. ISSN 0020–3084.

Rep. Obs. Lund
Reports from the Observatory of Lund. Institutionen för Astronomi, Lunds Universitet, Box 1107, 22104 Lund, Sweden. ISSN 0349–4217.

Rep. Prog. Phys.
Reports on Progress in Physics. Published by the Institute of Physics, 47 Belgrave Square, London SW1X 8QX, England. ISSN 0034–4885.

Rep. Ser., Dep. Phys. Sci., Univ. Turku
Report Series, Department of Physical Sciences, University of Turku. Department of Physical Sciences, University of Turku, SF–20500 Turku 50, Finland. ISSN 0356–9896.

Repr. Theor. Astrophys. Group, Dep. Math., UMIST
Reprints from the Theoretical Astrophysics Group, Department of Mathematics, UMIST. Department of Mathematics, UMIST, PO Box 88, Manchester M6O–1QD, England.

Res. Bull. Meisei Univ., Phys. Sci. Eng.
Research Bulletin of Meisei University, Physical Sciences and Engineering. Meisei University, Hino–Shi, Tokyo, Japan. ISSN 0388–130X.

Res. Cent. Astron. Appl. Math., Acad. Athens, Contrib. Ser. I
Research Center for Astronomy and Applied Mathematics, Academy of Athens, Contributions Series I (Astronomy). Research Center for Astronomy and Applied Mathematics, Academy of Athens, Athens, Greece.

Res. Lab. Electron. Onsala Space Obs., Res. Rep.
Research Laboratory of Electronics and Onsala Space Observatory, Research Report. Chalmers University of Technology, Research Laboratory of Electronics, Fack, S–40220 Göteborg 5, Sweden.

Rev. Astron.
Revista Astronomica. Organo de la Asociación Argentina Amigos de la Astronomia. Asociación Argentina Amigos de la Astronomia, C. C. 369–C. Central, 1000–Buenos Aires, Argentina. ISSN 0044–9253.

Rev. Bras. Fis.
Revista Brasiliera de Fisica. Sociedade Brasileira de Fisica, Caixa Postal 20553, Sao Paulo SP, Brazil. ISSN 0374–4922.

Rev. Fac. Sci. Univ. Istanbul, Ser. C
Review of Faculty of Science University of Instanbul, Serie C. (Üniversitesi fen Fakültesi Mecmuasi, Seri C). Fen Fakültesi Basimevi, Istanbul, Turkey. ISSN 0444–7298.

Rev. Geophys.
Reviews of Geophysics. American Geophysical Union, 2000 Florida Avenue, N.W., Washington, D.C. 20009, USA. ISSN 8755–1209.

Rev. Hist. Sci.
Revue d'Histoire des Sciences. Presses Universitaires de France, Paris, France. ISSN 0151–4105.

Rev. Mex. Astron. Astrofis.
Revista Mexicana de Astronomia y Astrofisica. Instituto de Astronomia, Universidad Nacional Autónoma de México, Apartado Postal 70–264, Mexico 20, D.F., Mexico. ISSN 0185–1101.

Rev. Mex. Fis.
Revista Mexicana de Fisica. Sociedad Mexicana de Fisica, Apartado Postal No. 20–364, Mexico 20, D.F., Mexico. ISSN 0035–001X.

Rev. Mod. Phys.
Reviews of Modern Physics. Published for the American Physical Society by the American Institute of Physics, 335 East 45th Street, New York, N.Y. 10017, USA. ISSN 0034–6861.

Rev. Phys. Appl.
Revue de Physique Appliquée. Editions de Physique, Z.I. de Courtabœuf, B.P. 112, F–91944 Les Ulis Cedex, France. ISSN 0035–1687.

Rev. Quest. Sci.
Revue des Questions Scientifiques. Société Scientifique de Bruxelles, 61, rue de Bruxelles, B–5000 Namur, Belgium. ISSN 0035–2160.

Rev. Radio Res. Lab.
Review of the Radio Research Laboratories. Ministry of Posts & Telecommunications, Nukui–Kitamachi, Konganei-shi, Tokyo 184, Japan. ISSN 0033–801X.

Rev. Roum. Phys.
Revue Roumaine de Physique. Academie Republicii Populare Romine, B.P. 134–135, Bucuresti, Rumania. ISSN 0035–4090.

Rev. Roum. Sci. Tech., Sér. Mec. Appl.
Revue Roumaine des Sciences Techniques, Série de Mécanique Appliquée. Academia R.S.R., Str. Constantin Mille 15, Bucuresti, Rumania. ISSN 0035–4074.

Rev. Sci. Instrum.
Review of Scientific Instruments. American Institute of Physics, 335 East 45th Street, New York, N.Y. 10017, USA. ISSN 0034–6748.

Rezul't. Nablyud. Iskusstv. Sputnikov Zemli
Rezul'taty Nablyudenij Iskusstvennykh Sputnikov Zemli. Published by Astronomicheskij Sovet Akademii Nauk SSSR, Ryazanskij Gosudarstvennyj Pedagogicheskij Institut, Ryazan'. 390000 GSP, Ryazan, ul. Svobody, 46, USSR. ISSN 0131–8586.

Ric. Astron.
Ricerche Astronomiche. Specola Vaticana, I–00120 Città del Vaticano.

Rijksuniv. Gent, Sterrenkundig Obs., Meded.
Rijksuniversiteit Gent, Sterrenkundig Observatorium, Mededeling. Astronomical Observatory, State University of Ghent, Krijgslaan 271, B–9000 Ghent, Belgium.

Říše hvězd
Říše hvězd. Populárně vědecký astronomický časopis (Czech popular astronomical journal), Panorama, Praha, Czechoslovakia. ISSN 0035–5550.

Riv. Nuovo Cimento
La Rivista del Nuovo Cimento della Società Italiana di Fisica. Società Italiana di Fisica, Editrice Compositori, viale XII Giugno 1, I–40124 Bologna, Italy. ISSN 0035–5917.

Rutherford Appleton Lab., Rep.
Rutherford Appleton Laboratory, Report. Rutherford Appleton Laboratory, Chilton, Didcot, Oxfordshire OX11 0QX, England.

SAAO Newsl.
SAAO Newsletter. South African Astronomical Observatory, PO Box 9, Observatory 7935, Cape, South Africa.

Sac Peak Update
Sac Peak Update. Sacramento Peak Observatory, Sunspot, N.M. 88349, USA.

S.Afr. Astron. Obs., Annu. Rep.
South African Astronomical Observatory, Annual Report. South African Astronomical Observatory, PO Box 9, Observatory, 7935 Cape, South Africa. ISSN 0250–0671.

S.Afr. Astron. Obs., Circ.
South African Astronomical Observatory, Circulars. South African Astronomical Observatory, PO Box 9, Observatory, 7935 Cape, South Africa.

S.Afr. J. Phys.
South African Journal of Physics. Bureau for Scientific Publications, PO Box 1758, Pretoria 0001, South Africa. ISSN 0379–4377.

Schweiz. Tech. Z.
Schweizerische Technische Zeitschrift. Schweizerischer Technischer Verband, Weinbergstraße 41, CH–8023 Zürich, Switzerland. ISSN 0040–151X.

Sci. Am.
Scientific American. Scientific American, Inc., 415 Madison Avenue, New York, N.Y. 10017, USA. ISSN 0036–8733.

Sci. Atmos. Sin.
Scientia Atmospherica Sinica. Science Press, Peking. Subscription address: Guozi Shudian, PO Box 399, Beijing, People's Republic of China.

Sci. Dimension
Science Dimension. National Research Council of Canada, Ottawa K1A OR6, Canada. ISSN 0036–830X.

Sci. Pap. Inst. Phys. Chem. Res.
Scientific Papers of the Institute of Physical and Chemical Research. Rikagaku Kenkyusho, Wako–shi, Saitama 351, Japan. ISSN 0020–3092.

Sci. Prog.
Science Progress. Blackwell Scientific Publications, Oxford – London – Edinburgh – Boston – Melbourne. ISSN 0036–8504.

Sci. Rep. Tôhoku Univ., Eighth Ser.
The Science Reports of the Tôhoku University, Eighth Series (Physics and Astronomy). Faculty of Science, Tôhoku University, Sendai 980, Japan. ISSN 0388–5607.

Sci. Sin., Ser. A
Scientia Sinica, Series A (Mathematical, Physical, Astronomical and Technical Sciences). Academia Sinica, Beijing. Science Press, No. 137, Chaoyangmennei Street, Beijing, People's Republic of China. Subscription address: Scientific and Technical Books Service Ltd., PO Box 197, London WC2N 4DE, England. ISSN 0253–5831.

Science
Science. American Association for the Advancement of Science, 1515 Massachusetts Avenue, N.W., Washington, D.C. 20005, USA. ISSN 0036–8075.

Scr. Fac. Sci. Nat. Univ. Purkynianae Brun., Phys.
Scripta Facultatis Scientiarum Naturalium Universitatis Purkynianae Brunensis, Physica. University J.E. Purkyne, 61137 Brno–Kotlarská 2, Czechoslovakia. ISSN 0231–6129.

Sendai Astron. Rap.
Sendai Astronomiaj Raportoj. Astronomical Institute, Tôhoku University, Sendai 980, Japan.

Shaanxi Astron. Obs., Repr.
Shaanxi Astronomical Observatory, Reprints. Shaanxi Astronomical Observatory, Academia Sinica, Lintong, Xian, People's Republic of China.

Shanghai Obs., Acad. Sin., Prepr.
Shanghai Observatory, Academia Sinica, Preprint. Shanghai Observatory, Academia Sinica, 80 Nandan Road, Shanghai, People's Republic of China.

Sitzungsber. Akad. Wiss. DDR, Math. Naturwiss. Tech.
Sitzungsberichte der Akademie der Wissenschaften der DDR, Mathematik – Naturwissenschaften – Technik. Akademie-Verlag, DDR–1086 Berlin, Leipziger Straße 3–4, German Democratic Republic. ISSN 0138–3956.

Sitzungsber. Heidelb. Akad. Wiss.
Sitzungsberichte der Heidelberger Akademie der Wissenschaften. Mathematisch–Naturwissenschaftliche Klasse. Springer-Verlag, Berlin – Heidelberg – New York – Tokyo. ISSN 0371–0165.

Sitzungsber., Österr. Akad. Wiss., Math.–Naturwiss. Kl. Abt. II
Sitzungsberichte, Österreichische Akademie der Wissenschaften, Mathematisch–Naturwissenschaftliche Klasse, Abteilung II (Mathematik, Astronomie, Physik, Meteorologie und Technik). Springer–Verlag, Wien, Austria. ISSN 0029–8816.

Sky Telesc.
Sky and Telescope. Sky Publishing Corporation, 49 Bay State Rd., Cambridge, Mass. 02238–1290, USA. ISSN 0037–6604.

Smithson. Astrophys. Obs., Spec. Rep.
Smithsonian Astrophysical Observatory, Special Report. Available from the Publications Division, Distribution Section, Smithsonian Astrophysical Observatory, Cambridge, Mass. 02138, USA.

Sol. Energy
Solar Energy. Pergamon Press, Oxford – New York – Toronto – Paris – Frankfurt – Sydney. ISSN 0038–092X.

Sol. Maps Act.
Solar Maps and Activity. Manila Observatory, Solar Division, Manila, Philippines.

Sol. Phys.
Solar Physics. A Journal for Solar Research and the Study of Solar Terrestrial Physics. D. Reidel Publishing Company, Dordrecht – Boston – Lancaster – Tokyo. ISSN 0038–0938.

Sol. Radio Data
Solar Radio Data. OSRA, Observatorium für solare Radioastronomie, Zentralinstitut für Astrophysik, Akademie der Wissenschaften der DDR, DDR–1501 Tremsdorf/Potsdam, German Democratic Republic.

Sol. Syst. Res.
Solar System Research. Translation of Astron. Vestn. Consultants Bureau, 227 West 17th Street, New York, N.Y. 10011, USA. ISSN 0038–0946.

Sol. Terr. Environ. Res. Jpn.
Solar Terrestrial Environmental Research in Japan. Institute of Space and Aeronautical Science, University of Tokyo, 4–6–1, Komaba, Meguro–ku, Tokyo 153, Japan. ISSN 0386–5444.

Solid State Phys.
Solid State Physics. Agne Gijutsu Center, Kitamura Building, 5–1–25, Minamiayama, Minatoku, Tokyo, Japan. ISSN 0454–4544.

Soln. Dannye, Byull.
Solnechnye Dannye, Byulleten' (Solar Data). Akademiya Nauk SSSR. Astronomicheskij Sovet i Glavnaya Astronomicheskaya Observatoriya. Izdatel'stvo 'Nauka', Leningradskoe Otdelenie, Leningrad. 199164 Leningrad, V–164, Mendeleevskaya l. 1, USSR. ISSN 0552–5829.

Sonne
Sonne. Mitteilungsblatt der Amateursonnenbeobachter. Wilhelm–Foerster–Sternwarte, Munsterdamm 90, D–1000 Berlin 41, F.R. Germany. ISSN 0721–0094.

Soobshch. Akad. Nauk Gruz. SSR
Soobshcheniya Akademii Nauk Gruzinskoj SSR. Akademiya Nauk Gruzinskoj SSR. ISSN 0002–3167.

Soobshch. Byurakan. Obs.
Soobshcheniya Byurakanskoj Observatorii. Akademiya Nauk Armyanskoj SSR, Erevan. 375019 Erevan, Barekamutyan, 24, USSR. ISSN 0370–8691.

Soobshch. Gos. Astron. Inst. Shternberg
Soobshcheniya Gosudarstvennogo Astronomicheskogo Instituta im. P.K. Shternberga. Moskovskij Gosudarstvennyj Universitet im. M.V. Lomonosova. Izdatel'stvo Moskovskogo Universiteta, Moskva. Moskva, K–9, ul. Gertsena, 5/7, USSR. ISSN 0038–1489.

Soobshch. Spets. Astrofiz. Obs.
Soobshcheniya Spetsial'noj Astrofizicheskoj Observatorii. Akademiya Nauk SSSR. Izdanie Spetsial'noj Astrofizicheskoj Observatorii, USSR.

South. Stars
Southern Stars. Journal of the Royal Astronomical Society of New Zealand (Inc.). Royal Astronomical Society of New Zealand, PO Box 3181, Wellington, New Zealand. ISSN 0049–1640.

Sov. Astron.
Soviet Astronomy. A translation of Astron. Zh. American Institute of Physics, 335 East 45th Street, New York, N.Y. 10017, USA. ISSN 0038–5301.

Sov. Astron. Lett.
Soviet Astronomy Letters. A translation of Pis'ma Astron. Zh. American Institute of Physics, 335 East 45th Street, New York, N.Y. 10017, USA. ISSN 0360–0327.

Space Sci. Rev.
Space Science Reviews. D. Reidel Publishing Company, Dordrecht – Boston – Lancaster – Tokyo. ISSN 0038–6308.

Space Telesc. Sci. Inst., Instrum. Sci. Rep.
Space Telescope Science Institute, Instrument Science Report. Space Telescope Science Institute, 3700 San Martin Drive, Homewood Campus, Baltimore, Md. 21218, USA.

Space Telesc. Sci. Inst., Newsl.
Space Telescope Science Institute, Newsletter. Space Telescope Science Institute, 3700 San Martin Drive, Homewood Campus, Baltimore, Md. 21218, USA.

Space Telesc. Sci. Inst., Prepr. Ser.
Space Telescope Science Institute, Preprint Series. Space Telescope Science Institute, 3700 San Martin Drive, Homewood Campus, Baltimore, Md. 21218, USA.

Spaceflight
Spaceflight. A Publication of The British Interplanetary Society. The British Interplanetary Society, 27/29 South Lambeth Road, London SW8 1SZ, England. ISSN 0038–6340.

Spectrosc. Lett.
Spectroscopy Letters. Marcel Dekker Inc., 270 Madison Avenue, New York, N.Y. 10016, USA. ISSN 0038–7010.

Speculations Sci. Technol.
Speculations in Science and Technology. Elsevier Sequoia, S.A., PO Box 851, CH–1001 Lausanne 1, Switzerland. ISSN 0155–7785.

ST–ECF Newsl.
Space Telescope–European Coordinating Facility Newsletter. Space Telescope–European Facility, European Southern Observatory, Karl–Schwarzschild–Straße 2, D–8046 Garching bei München, F.R. Germany.

Stand. Star Newsl.
Standard Star Newsletter. Working Group on Standard Stars (IAU Commissions 29, 30, 45). Dipartimento di Fisica, Università degli Studi di Milano, Milano, Italy.

Sterne
Die Sterne. Zeitschrift für alle Gebiete der Himmelskunde. Johann Ambrosius Barth, DDR–7010 Leipzig, Salomonstraße 18b, German Democratic Republic. ISSN 0039–1255.

Sterne Weltraum
Sterne und Weltraum. Astronomische Monatsschrift. Verlag Sterne und Weltraum Dr. Vehrenberg GmbH, Portiastraße 10, D–8000 München 90, F.R. Germany, ISSN 0039–1263.

Sternenbote
Der Sternenbote. Österreichische Astronomische Monatsschrift. Astronomisches Büro, Hasenwartgasse 32, A–1238 Wien, Austria. ISSN 0039–1271.

Stockholms Obs., Prepr.
Stockholms Observatorium, Preprint. Stockholms Observatorium, S–13300 Saltsjöbaden, Sweden.

Stockholms Obs., Rep.
Stockholms Observatorium, Report. Stockholms Observatorium, S–13300 Saltsjöbaden, Sweden.

Strolling Astron.
The Strolling Astronomer. The Journal of The Association of Lunar and Planetary Observers. A.L.P.O., J. E. Westfall, PO Box 16131, San Francisco, Calif. 94116, USA. ISSN 0039–2502.

Stud. Cercet. Fiz.
Studii si Cercetari de Fizica. Academia Republicii Populare Romine, PO Box 134–5, Calca Victoriei 126, Bucuresti, Rumania, ISSN 0039–3940.

Stud. Geophys. Geod.
Studia Geophysica et Geodaetica. Geophysical Institute of the Czechoslovak Academy of Sciences. Published by Academia, Prague, Czechoslovakia.

Stud. Hist. Philos. Sci.
Studies in History and Philosophy of Science. Pergamon Press, Oxford – New York – Toronto – Paris – Frankfurt – Sydney. ISSN 0039–3681.

Stud. Soc. Sci. Torun., Sect. F
Studia Societatis Scientiarum Torunensis, Sectio F (Astronomia). Państwowe Wydawnictwo Naukowe, Warszawa – Poznań – Toruń, Poland. ISSN 0082–5573.

Surv. Geophys.
Surveys in Geophysics. An International Review Journal of Geophysics and Planetary Sciences. D. Reidel Publishing Company, Dordrecht – Boston – Lancaster – Tokyo. ISSN 0169–3298.

Surv. High Energy Phys.
Surveys in High Energy Physics. Harwood Academic Publishers GmbH, Poststraße 22, CH–7000 Chur, Switzerland. ISSN 0142–2413.

Swarthmore Astron. Repr.
Swarthmore Astronomical Reprints. Sproul Observatory, Swarthmore College, Swarthmore, Pa. 19081, USA.

Syd. Obs. Pap.
Sydney Observatory Papers. School of Physics, University of Sydney, Sydney, N.S.W. 2006, Australia.

Syst. Int.
Systems International. IPC Electrical Electronic Press Ltd., Quadrant House, The Quadrant, Sutton, Surrey SM2 5AS, England. ISSN 0309–1171.

Tartu Astrofüüs. Obs. Publ.
W. Struve nimelise Tartu Astrofüüsika Observatooriumi, Publikatsioonid. Eesti NSV Teaduste Akadeemia, Tartu.

Tartu Astrofüüs. Obs. Teated
Tartu Astrofüüsika Observatoorium Teated. Eesti NSV Teaduste Akadeemia W. Struve nim. Tartu Astrofüüsika Observatoorium. Valgus, Tallinn.

Tech. Mess. – TM
Technisches Messen – TM. Archiv für Technisches Messen. R. Oldenbourg Verlag GmbH, Rosenheimer Str. 145, D–8000 München 80, F.R. Germany. ISSN 0171–8096.

Tech. Mitt. Krupp Forschungsber.
Technische Mitteilungen Krupp Forschungsberichte. Fachbücherei Krupp, Postfach 917, D–4300 Essen 1, F.R. Germany. ISSN 0494–9382.

Tectonophysics
Tectonophysics. International Journal of Geotectonics and the Geology and Physics of the Interior of the Earth. Elsevier Publishing Company, PO Box 211, 1000 AE Amsterdam, The Netherlands. ISSN 0040–1951.

Telecommun. J.
Telecommunication Journal. (English Edition). International Telecommunications Union. Place des Nations, CH–1221 Genève 20, Switzerland. ISSN 0497–137X.

Tellus, Ser. B
Tellus, Series B. (Chemical and Physical Meteorology). Swedish Geophysical Society, Arrhenius Laboratoriet, S–10691 Stockholm, Sweden. ISSN 0280–6509.

Tex. J. Sci.
Texas Journal of Science. The Talley Press, San Angelo, Tex., USA. ISSN 0040–4403.

THEOCHEM
THEOCHEM. Elsevier Scientific Publishing Company, PO Box 330, 1000 AH Amsterdam, The Netherlands. ISSN 0166–1280.

Theor. Pap.
Theoretic Papers. The Blindern Theoretic Research Team, P.B. 33, Blindern, Oslo 3, Norway. ISSN 0801–3128.

Time Freq. Serv., Bull.
Time and Frequency Services, Bulletin. Shaanxi Astronomical Observatory, Academia Sinica, Lintong, Xian, People's Republic of China.

Time Serv. Bull.
Time Service Bulletin. Osservatorio Astronomico di Torino, I–10025 Pino Torinese (Torino), Italy.

Tôhoku Geophys. J.
Tôhoku Geophysical Journal. Science Reports of the Tôhoku University, Fifth Series. Faculty of Science, Tôhoku University, Sendai 980, Japan. ISSN 0040–8794.

Tokyo Astron. Bull., Second Ser.
Tokyo Astronomical Bulletin, Second Series. Tokyo Astronomical Observatory, University of Tokyo, Mitaka, Tokyo 181, Japan. ISSN 0082–4690.

Tokyo Astron. Obs., Kiso Inf. Bull.
Tokyo Astronomical Observatory, Kiso Information Bulletin. Tokyo Astronomical Observatory, University of Tokyo, Mitaka, Tokyo 181, Japan. ISSN 0286–1380.

Tokyo Astron. Obs. Rep.
Tokyo Astronomical Observatory Report. Tokyo Astronomical Observatory, University of Tokyo, Mitaka, Tokyo 181, Japan. ISSN 0374–4639.

Tokyo Astron. Obs. Repr.
Tokyo Astronomical Observatory Reprints. Tokyo Astronomical Observatory, University of Tokyo, Mitaka, Tokyo 181, Japan. ISSN 0082–4712.

Tokyo Astron. Obs., Time Latitude Bull.
Tokyo Astronomical Observatory, Time and Latitude Bulletins. Tokyo Astronomical Observatory, University of Tokyo, Mitaka, Tokyo 181, Japan. ISSN 0388–3701.

Top. Astrophys. Astron. Space Sci.
Topics in Astrophysics, Astronomy, and Space Sciences. Centre for Astronomy and Space Sciences, Central Institute of Physics. CIP Press, Cutitul de Argint 5, 75 212 Bucharest, Romania.

Tr. Astrofiz. Inst. Alma–Ata
Trudy Astrofizicheskogo Instituta. Akademiya Nauk Kazakhskoj SSR. Izdatel'stvo Nauka Kazakhskoj SSR, Alma–Ata. 480021, Alma–Ata, ul. Shevchenko, 28, USSR.

Tr. Astron. Obs., Leningrad
Trudy Astronomicheskoj Observatorii. Uchenye Zapiski Gosudarstvennogo Universiteta im. A.A. Zhdanova, Seriya matematicheskikh nauk = Izdatel'stvo Leningradskogo Gosudarstvennogo Universiteta A.A. Zhdanova, Leningrad. 199164, Leningrad V–164, Universitetskij nab., 7/9, USSR. ISSN 0136–8109; ISSN 0136–8141.

Tr. Glav. Astron. Obs. Pulkovo, Ser. 2
Trudy Glavnoj Astronomicheskoj Observatorii v Pulkove, Seriya 2. Akademiya Nauk SSSR. Leningradskoe Otdelenie, Nauka, Leningrad. 199164, Leningrad V–164, Mendeleevskaya l., 1, USSR.

Tr. Gos. Astron. Inst. Shternberg
Trudy Gosudarstvennogo Astronomicheskogo Instituta im. P.K. Shternberga. Izdatel'stvo Moskovskogo Universiteta, Moskva, USSR. ISSN 0371–6791.

Tr. Inst. Teor. Astron., Leningrad
Trudy Instituta Teoreticheskoj Astronomii, Akademiya Nauk SSSR. Izdatel'stvo Nauka, Leningrad. 199164, Leningrad V–164, Mendeleevskaya l., 1, USSR. ISSN 0568–6016.

Tr. Kazan. Gorod. Astron. Obs.
Trudy Kazanskoj Gorodskoj Astronomicheskoj Observatorii. Izdatel'stvo Kazanskogo Universiteta, Kazan. Kazan, ul. Lenina, 2, USSR. ISSN 0371–8247.

Tr. Tashkent. Astron. Obs.
Trudy Ordena Trudovogo Krasnogo Znameni Astronomicheskogo Instituta Akademii Nauk Uzbekskoj SSR. Izdatel'stvo FAN Uzbekskoj SSR, Tashkent. Tashkent, ul. Gololya, 70, USSR.

Trans. Am. Nucl. Soc.
Transactions of the American Nuclear Society. American Nuclear Society, 555 North Kensington Avenue, La Grange Park, Ill. 60525, USA. ISSN 0003–018X.

Trans. Astron. Obs. Yale Univ.
Transactions of the Astronomical Observatory of Yale University. Yale University Observatory, 260 Whitney Avenue, New Haven, Conn. 06511, USA.

Trans. Bose Res. Inst.
Transactions of the Bose Research Institute. Bose Research Institute, 93/1 Acharya Prafulla Chandra Road, Calcutta 9, India. ISSN 0006–7903.

Trans. IAU
Transactions of the International Astronomical Union. Published on behalf of the IAU by D. Reidel Publishing Company, Dordrecht – Boston – London.

Trans. Inst. Electron. Commun. Eng. Jpn., Part C
Transactions of the Institute of Electronic and Communication Engineers of Japan, Part C. Denshi Tsushin Gakkai, Kikai–Shinko–Kaikan, 5–8 Shibakoen 3 Chome, Minato–ku, Tokyo 105, Japan. ISSN 0373–6113.

Transp. Theory Stat. Phys.
Transport Theory and Statistical Physics. Marcel Dekker Inc., 270 Madison Avenue, New York, N.Y. 10016, USA. ISSN 0041–1450.

Tsirk. Astron. Inst. Tashkent
Tsirkulyar Astronomicheskogo Instituta, Tashkent. Akademiya Nauk Uzbekskoj SSR. Izdatel'stvo FAN Uzbekskoj SSR, Tashkent. Tashkent, ul. Gololya, 70, USSR. ISSN 0373–7675.

Tsirk., Astron. Obs. L'vov
Tsirkulyar, Astronomicheskaya Observatoriya L'vov. L'vovskij Ordena Lenina Gosudarstvennyj Universitet im. Ivana Franko. Izdatel'stvo L'vovskogo Universiteta, L'vov. 290005, L'vov–5, ul. Lomonosova, 8, USSR. ISSN 0374–0722.

Tsirk. Shemakh. Astrofiz. Obs.
Tsirkulyar Shemakhinskoj Astrofizicheskoj Observatorii. Akademiya Nauk Azerbajdzhanskoj SSR. Izdatel'stvo Ehlm, Baku. 370143 Baku–143, prospekt Narimanova, 31. ISSN 0135–0420.

Turku Univ. Obs., Informeto
Turku University Observatory, Informeto. Turku University Observatory, Tuorla, SF–2150 Piikkiö, Finland.

Turku Univ. Obs., Informo
Turku University Observatory, Informo. Turku University Observatory, Tuorla, SF–21500 Piikkiö, Finland.

Umschau
Die Umschau. Das Wissenschaftsmagazin. Postfach 110262, Stuttgarter Straße 18–24, D–6000 Frankfurt am Main 11, F.R. Germany. ISSN 0722–8562.

Univ. Barc., Dep. Fis. Tierra Cosmos, Publ.
Universidad de Barcelona, Departamento de Fisica de la Tierra y del Cosmos, Publicacion. Departamento de Fisica de la Tierra y del Cosmos, Universidad de Barcelona, Spain.

Univ. Hannover, Astron. Stn., Veröff.
Universität Hannover, Astronomische Station, Veröffentlichungen. Universität Hannover, Astronomische Station, Hannover, F.R. Germany.

Univ. Obs., St. Andrews, Repr.
University Observatory, St. Andrews, Reprint. University Observatory, Buchanan Gardens, St. Andrews, Fife KY16 9LZ, Scotland.

Univ. Tex., Monogr. Astron.
The University of Texas, Monographs in Astronomy. Department of Astronomy, University of Texas, Austin, Tex. 78712, USA.

Univ. Tex., Publ. Astron.
The University of Texas, Publications in Astronomy. Department of Astronomy, University of Texas, Austin, Tex. 78712, USA. ISSN 0276–1106.

Upps. Astron. Obs. Ann.
Uppsala Astronomiska Observatoriums Annaler. Astronomiska Observatoriet, Box 515, S–75120 Uppsala, Sweden.

Upps. Astron. Obs. Rep.
Uppsala Astronomical Observatory, Report. Astronomiska Observatoriet, Box 515, S–75120 Uppsala, Sweden.

U.S. Nav. Obs., Circ.
U.S. Naval Observatory, Circular. U.S. Naval Observatory, Washington, D.C. 20390, USA.

U.S. Nav. Obs., Time Serv. Publ., Ser. 01
U.S. Naval Observatory, Time Service Publications, Series 1 (Worldwide Primary Time and Frequency VLF and HF Transmissions). U.S. Naval Observatory, Time Service Division (62C), Washington, D.C. 20390, USA.

U.S. Nav. Obs., Time Serv. Publ., Ser. 04
U.S. Naval Observatory, Time Service Publications, Series 4 (Daily Time Differences and Relative Phase Values). U.S. Naval Observatory, Time Service Division (62C), Washington, D.C. 20390, USA.

U.S. Nav. Obs., Time Serv. Publ., Ser. 06
U.S. Naval Observatory, Time Service Publications, Series 6 (A.1–UT1 Data). U.S. Naval Observatory, Time Service Division (62C), Washington, D.C. 20390, USA.

U.S. Nav. Obs., Time Serv. Publ., Ser. 10
U.S. Naval Observatory, Time Service Publications, Series 10 (Astronomical Programs). U.S. Naval Observatory, Time Service Division (62C), Washington, D.C. 20390, USA.

U.S. Nav. Obs., Time Serv. Publ., Ser. 11
U.S. Naval Observatory, Time Service Publications, Series 11 (Time Service Report). U.S. Naval Observatory, Time Service Division (62C), Washington, D.C. 20390, USA.

U.S. Nav. Obs., Time Serv. Publ., Ser. 14
U.S. Naval Observatory, Time Service Publications, Series 14 (Time Service Announcement). U.S. Naval Observatory, Time Service Division (62C), Washington, D.C. 20390, USA.

Utr. Sterrekundige Overdrukken
Utrechtse Sterrekundige Overdrukken. Sonnenborgh Observatory, Zonnenburg 2, NL–3512 Utrecht, The Netherlands.

Uttar Pradesh State Obs., Repr.
Uttar Pradesh State Observatory, Reprint. Uttar Pradesh State Observatory, Manora Peak, Naini Tal 263 139, India.

V–Gram
V–Gram. Jet Propulsion Laboratory, California Institute of Technology, 4800 Oak Grove Drive, Pasadena, CA 91109, USA.

Variable Star Bull.
Variable Star Bulletin. Variable Star Observers League in Japan. National Science Museum, Ueno Park, Taito–ku, Tokyo 110, Japan.

Vasiona
Vasiona. Revue d'Astronomie. Bulletin de la Société Astronomique 'R. Bošković'. Vasiona, Narodna opservatorija, Kalemegdan, Gornji Grad, Beograd, Yugoslavia. ISSN 0506–4295.

Vatican Obs. Publ.
Vatican Observatory Publications. Specola Vaticana, I–00120 Città del Vaticano.

Vatican Obs. Publ., Spec. Ser., Studi Galileiani
Vatican Observatory Publications, Special Series, Studi Galileiani. Specola Vaticana, I–00120 Città del Vaticano.

Veröff. Archenhold–Sternw. Berlin–Treptow
Veröffentlichungen der Archenhold-Sternwarte Berlin-Treptow. Archenhold-Sternwarte, DDR–1193 Berlin, Alt-Treptow 1, German Democratic Republic.

Veröff. Astron. Inst. Bonn
Veröffentlichungen der Astronomischen Institute Bonn. Ferd. Dümmler Verlag, Kaiserstraße 31–37, D–5300 Bonn 1, F.R. Germany. ISSN 0340–9821.

Veröff. Astron. Rechen–Inst. Heidelb.
Veröffentlichungen des Astronomischen Rechen–Instituts Heidelberg. Verlag G. Braun, Karl–Friedrich–Straße 14–18, D–7500 Karlsruhe 1, F.R. Germany.

Veröff. Bayer. Komm. Int. Erdmessung Bayer. Akad. Wiss., Astron.–Geod. Arb.
Veröffentlichungen der Bayerischen Kommission für die Internationale Erdmessung der Bayerischen Akademie der Wissenschaften, Astronomisch–Geodätische Arbeiten. Bayerische Akademie der Wissenschaften, Marstallplatz 8, D–8000 München 22, F.R. Germany. ISSN 0340–7691.

Veröff. Forschungsinst. Dtsch. Mus. Gesch. Naturwiss. Tech., Reihe A
Veröffentlichungen des Forschungsinstituts des Deutschen Museums für die Geschichte der Naturwissenschaften und der Technik, Reihe A. Forschungsinstitut des Deutschen Museums für die Geschichte der Naturwissenschaften und der Technik, D–8000 München, F.R. Germany. ISSN 0418–9949.

Veröff. Remeis–Sternw. Bamberg
Veröffentlichungen der Remeis-Sternwarte Bamberg. Astronomisches Institut der Universität Erlangen–Nürnberg. Remeis–Sternwarte, Sternwartstraße 7, D–8600 Bamberg, F.R. Germany.

Veröff. Sternw. Pulsnitz
Veröffentlichungen der Sternwarte Pulsnitz. Sternwarte Pulsnitz, DDR–8514 Pulsnitz, German Democratic Republic.

Veröff. Sternw. Sonneberg
Veröffentlichungen der Sternwarte in Sonneberg. Akademie der Wissenschaften der DDR, Zentralinstitut für Astrophysik, Sternwarte Sonneberg, DDR–6400 Sonneberg, German Democratic Republic. ISSN 0373–6199.

Veröff. Wilhelm–Foerster–Sternw.
Veröffentlichungen der Wilhelm–Foerster–Sternwarte. Wilhelm–Foerster–Sternwarte, Munsterdamm 90, D–1000 Berlin 41, F.R. Germany.

Veröff. Zentralinst. Phys. Erde
Veröffentlichungen des Zentralinstituts für Physik der Erde. Akademie der Wissenschaften der DDR, Zentralinstitut für Physik der Erde, DDR–1500 Potsdam, Telegrafenberg A17, German Democratic Republic. ISSN 0514–8790.

Vesmír
Vesmír. Přírodovědecký časopis Československé a Slovenské akademie věd (Natural scientific journal), Academia, Praha, Czechoslovakia. ISSN 0042–4544.

Vestn. Kiev. Univ. Astron.
Vestnik Kievskogo Universiteta. Astronomiya. Izdatel'stvo pri Kievskom Gosudarstvennom Universitete Izdatel'skogo Obedineniya Vishcha Shkola, Kiev. 252001, Kiev–1, Kreshchatik, 4, USSR. ISSN 0203–7319; ISSN 0321–3927.

Vestn. Leningr. Univ., Mat. Mekh. Astron.
Vestnik Leningradskogo Universiteta Matematika, Mekhanika, Astronomiya. Leningradskij Universitet, Universitetskaya Nab. 7/9 Leningrad B–164, USSR. ISSN 0024–0850.

Vestn. Mosk. Univ., Ser. 3. Fiz. Astron.
Vestnik Moskovskogo Universiteta, Seriya 3. Fizika, Astronomiya. Izdatel'stvo Moskovskogo Universiteta, 103009, Moskva, Ul.Gertsena 5/7, USSR. English translation in Mosc. Univ. Phys. Bull. ISSN 0579–9392.

Villanova Univ. Obs. Contrib.
Villanova University Observatory Contributions. Department of Astronomy, Villanova University, Villanova, Pa. 19085, USA.

Vilniaus Astron. Obs. Biul.
Vilniaus Astronomijos Observatorijos Biuletenis. Vilniaus Valstybinis V. Kapsuko Universitetas, LTSR Mokslu Akademijos Fizikos Institutas, Vilnius. 232031, Vilnius 31, Čiurlionio 29, Lithuania, USSR. ISSN 0136–3697.

Vistas Astron.
Vistas in Astronomy. An International Review Journal. Pergamon Press, Oxford – New York – Frankfurt. ISSN 0083–6656.

Warner Swasey Obs., Repr.
Warner and Swasey Observatory, Reprints. Warner and Swasey Observatory, Case Western Reserve University, Department of Astronomy, Smith Bldg., Cleveland, Ohio 44106, USA.

Weather
Weather. James Glaisher House, Grenville Place, Bracknell, Berks, RG12 1BX, England. ISSN 0043–1656.

WGN
WGN. Werkgroepnieuws – meteoren. The international circular for meteor observers. Tweemaandelijks tijdschrift. Vereniging voor Sterrenkunde, P. Roggemans, Dellingstraat 25, B–2800 Mechelen, Belgium.

Wiss. Z. Friedrich–Schiller–Univ. Jena, Math.–Naturwiss. Reihe
Wissenschaftliche Zeitschrift der Friedrich–Schiller–Universität Jena, Mathematisch–Naturwissenschaftliche Reihe. Friedrich–Schiller–Universität Jena. Subscription address: Zeitungsvertriebsamt, Abt. Export, DDR–1004 Berlin, Straße der Pariser Kommune 3/4, German Democratic Republic. ISSN 0043–6836.

Wiss. Z. Humboldt–Univ. Berlin, Math.–Naturwiss. Reihe
Wissenschaftliche Zeitschrift der Humboldt–Universität zu Berlin, Mathematisch–Naturwissenschaftliche Reihe. Humboldt–Universität Berlin. Subscription address: Zeitschriftenvertriebsamt, Abt. Export, DDR–1004 Berlin, Straße der Pariser Kommune 3/4, German Democratic Republic. ISSN 0522–9863.

Working Group Stars with Extended Atmospheres, Prepr.
Working Group Stars with Extended Atmospheres, Preprints. Working Group Stars with Extended Atmospheres, Beneluxlaan 21, NL–3527 HS Utrecht, The Netherlands.

Wrocław Astron. Obs., Repr.
Wrocław Astronomical Observatory, Reprint. Wrocław University Observatory, Wrocław, Poland.

Wuli
Wuli. Science Press, Beijing. Subscription address: Guozi Shudian, PO Box 399, Beijing, People's Republic of China. ISSN 0379–4148.

Yad. Fiz.
Yadernaya Fizika. Izdatel'stvo Nauka, 103717 GSP Moskva K–62. Podsosenskij Per. 24, SSSR. English translation in Sov. J. Nucl. Phys. ISSN 0044–0027.

Yamamoto Circ.
Yamamoto Circular. Yamamoto Observatory, 289, Kamitanakami–kiryutyo, OTU, Sigaken, 520–21 Japan.

Z. Angew. Math. Mech.
Zeitschrift für angewandte Mathematik und Mechanik. Akademie–Verlag GmbH, DDR–108 Berlin, Leipziger Str. 3–4, German Democratic Republic. ISSN 0044–2267.

Z. Angew. Math. Phys.
Zeitschrift für Angewandte Mathematik und Physik. Verlag Birkhäuser, Postfach 4000, Basel 24, Switzerland. ISSN 0044–2275.

Z. Naturforsch., A
Zeitschrift für Naturforschung, Section A. A Journal of Physical Sciences. Verlag der Zeitschrift für Naturforschung, PO Box 2645, D–7400 Tübingen, F.R. Germany. ISSN 0340–4811.

Z. Phys., A
Zeitschrift für Physik, A. (Atoms und Nuclei). Springer–Verlag, Berlin – Heidelberg – New York – Tokyo. ISSN 0340–2193.

Z. Phys., B
Zeitschrift für Physik, B (Condensed Matter). Springer–Verlag, Berlin – Heidelberg – New York – Tokyo. ISSN 0722–3277.

Z. Phys., C
Zeitschrift für Physik, C (Particles and Fields). Springer–Verlag, Berlin – Heidelberg – New York – Tokyo. ISSN 0170–9739.

Z. Vermessungswes.
Zeitschrift für Vermessungswesen. Verlag Konrad Wittner KG, Postfach 147, D–7000 Stuttgart 1, F.R. Germany. ISSN 0513–9155.

Zeiss Inf.
Zeiss Information. Carl Zeiss, Oberkochen. F.R. Germany. ISSN 0044-2054.

Zeit–Breitenbestimmungen
Zeit- und Breitenbestimmungen, Empfangszeiten von Zeitsignalen, Präzisionszeitvergleiche. Akademie der Wissenschaften der DDR, Zentralinstitut für Physik der Erde, DDR–1500 Potsdam, Telegrafenberg A 17, German Democratic Republic.

Zemlya Vselennaya
Zemlya i Vselennaya. Astronomiya, Geofizika. Issledovaniya Kosmicheskogo Prostranstva. Nauchno–Populyarnyj Zhurnal Akademii Nauk SSSR. Izdatel'stvo Nauka, Moskva, USSR. ISSN 0044-3948.

Zenit
Zenit. Populair–wetenschappelijk maandblad over sterrenkunde, weerkunde, ruimtevaart, ruimte–onderzoek en aanverwante wetenschappen en technieken. Stichting De Koepel, Nachtegaalstraat 82 bis, Utrecht, The Netherlands. ISSN 0165-0211.

Zentralbl. Math. Grenzgeb. – Math. Abstr.
Zentralblatt für Mathematik und ihre Grenzgebiete – Mathematics Abstracts. Heidelberger Akademie der Wissenschaften und Fachinformationszentrum Energie Physik Mathematik GmbH, Karlsruhe. Springer–Verlag, Berlin – Heidelberg – New York – Tokyo. ISSN 0044-4235.

Zentralinst. Astrophys., Sternw. Babelsberg, Mitt.–Neue Folge
Zentralinstitut für Astrophysik, Sternwarte Babelsberg, Mitteilungen–Neue Folge. Akademie der Wissenschaften der DDR. Sternwarte Babelsberg, DDR–1502 Potsdam–Babelsberg, Rosa–Luxemburg–Straße 17a, German Democratic Republic.

Zh. Ehksp. Teor. Fiz.
Zhurnal Ehksperimental'noj i Teoreticheskoj Fiziki. Akademiya Nauk SSSR, Leninskij Prosp. 14, Moskva, USSR. English translation in Sov. Phys.–JETP. ISSN 0044-4510.

Zvaigžnota Debess
Zvaigžnota Debess. Latvijas PSR Zinātnu Akadémijas Radioastrofizikas Observatorijas Populārzinatnisks Rakstu Krāgums. Izdevnieciba Zinātne, Riga. 226004 Riga, Vienibas Gatve 11, USSR. ISSN 0135-129X.

Journals Abstracted Completely

A selected number of journals listed in category 001 (periodicals) are central to the subject scope of *Astronomy and Astrophysics Abstracts*. Depending on their relevance, almost all papers of the journals listed below are abstracted in our service.

Acta Astron.
Acta Astron. Sin.
Acta Astrophys. Sin.
Acta Cosmologica
Algoritm. Nebesnoj Mekh.
Am. Assoc. Variable Star Obs. Bull.
Annu. Rev. Astron. Astrophys.
Archaeoastronomy (U.K.)
Archaeoastronomy (U.S.A.)
Astrofizika
Astron. Astrophys.
Astron. Astrophys., Suppl. Ser.
Astron. Circ.
Astron. Data Cent. Bull.
Astron. J.
Astron. Nachr.
Astron. Q.
Astron. Tidsskr.
Astron. Tsirk.
Astron. Vestn.
Astron. Zh.
Astronomia
Astronomie
Astrophys. J.
Astrophys. J., Lett. Ed.
Astrophys. J., Suppl. Ser.
Astrophys. Lett. Commun.
Astrophys. Space Sci.
Aust. J. Astron.

BAV Rundbrief
BBSAG Bull.
Bol. Asoc. Argent. Astron.
Bol. Astron.
Br. Astron. Assoc. Circ.
Bull. Am. Astron. Soc.
Bull. Assoc. Fr. Obs. Etoiles Variables
Bull. Astron. Inst. Czech.
Bull. Astron. Soc. India
Bull. Inf. Cent. Données Stellaires

Celest. Mech.
Chin. Astron. Astrophys.
Ciel
Ciel Terre
Circ. Inf.
Comments Astrophys.
Curr. Top. Chin. Sci., Sect. E: Astron.

Earth, Moon, Planets

Fundam. Cosmic Phys.

G. Astron.
Gen. Relativ. Gravitation
GEOS Circ.

I.A.P.P.P. Commun.
IAU Circ.
Icarus
IHW Newsl.
Inf. Bull. Variable Stars
Int. Comet Q.
Ir. Astron. J.
Issled. Solntsa Krasnykh Zvezd

J. Am. Assoc. Variable Star Obs.
J. Astrophys. Astron.
J. Br. Astron. Assoc.
J. Hist. Astron.
J. Korean Astron. Soc.
J. R. Astron. Soc. Can.

Kinematika Fiz. Nebesn. Tel
Komet. Tsirk.
Komety Meteory

Mem. Astron. Soc. India
Mem. Soc. Astron. Ital.
Mercury
Messenger
Meteoritics
Meteoritika
Minor Planet Bull.
Minor Planet Circ.
Mitt. Astron. Ges.
Mon. Not. R. Astron. Soc.
Mon. Notes Astron. Soc. S.Afr.

Nablyud. Iskusstv. Nebesn. Tel
Nauchn. Inf.
News Lett. Astron. Soc. N.Y.

Observatory
Occultation Newsl.
Orion
Orione

Perem. Zvezdy
Perem. Zvezdy, Prilozh.
Pis'ma Astron. Zh.
Postępy Astron.
Probl. Kosm. Fiz.
Proc. Astron. Soc. Aust.
Publ. Astron. Soc. Jpn.
Publ. Astron. Soc. Pac.
Publ. Variable Star Sect., R. Astron. Soc. N.Z.

Q. J. R. Astron. Soc.

Rev. Astron.
Rev. Mex. Astron. Astrofis.

Sky Telesc.
Sol. Phys.
Soln. Dannye Byull.
South. Stars
Space Sci. Rev.
Sterne
Sterne Weltraum
Strolling Astron.

Trans. IAU

Vasiona
Vestn. Kiev. Univ. Astron.
Vistas Astron.

Yamamoto Circ.

Zenit
Zvaigžnota Debess

Publications of Observatories and Astronomical Institutes

Reports, communications, publications, numbered series of reprints and preprints of observatories and astronomical institutes (listed in category 001) which are scanned completely in our service are listed below.

A.A.O. Newsl.
Abastumanskaya Astrofiz. Obs., Byull.
Abh. Hamb. Sternw.
Abstr. Submitted Pap.
Acad. Sci. Est. SSR, Div. Phys. Math. Tech. Sci., Prepr.
A.D.I.O.N. Bull.
An. Obs. Astron. Univ. Coimbra
Anglo-Aust. Obs., Prepr.
Anglo-Aust. Telesc., Annu. Rep.
Ann. Obs. Astron. Alger
Ann. Shanghai Obs., Acad. Sin.
Ann. Tokyo Astron. Obs., Second Ser.
Ann. Univ.-Sternw. Wien
Annu. Rep. Astron. Inst. Greece
Annu. Rep. (B.I.H.)
Annu. Rep. Dir., Mt. Wilson Las Campanas Obs.
Annu. Rep. Geophys. Obs.
Annu. Rep. Int. Polar Motion Serv.
Annu. Rep. Meteorol. Obs. Int. Latitude Obs. Mizusawa
Archenhold-Sternw. Berlin-Treptow, Sonderdr.
Archenhold-Sternw. Berlin-Treptow, Vortr. Schr.
Arecibo Obs./NAIC, Newsl.
Armagh Obs., Prepr. Ser.
Armagh Obs., Publ.
Armagh Obs., Repr.
Astrofiz. Issled. Izv. Spets. Astrofiz. Obs.
Astrofys. Inst., Vrije Univ. Bruss., Overdruk
Astron. Bull., Carter Obs.
Astron. Contrib. Univ. Manchester, Ser. II: Jodrell Bank Repr.
Astron. Contrib. Univ. Manchester, Ser. III
Astron. Inst., "Anton Pannekoeck", Univ. Amst., Repr.
Astron. Inst. Univ. Brno, Contrib.
Astron. Inst. Univ. Brno, Publ.
Astron. Mitt. Wien
Astron. Obs. Trieste, Publ.
Astron. Pap.
Astron. Rechen-Inst. Heidelb., Mitt., Ser. A
Astron. Rechen-Inst. Heidelb., Mitt., Ser. B
Astron. Rechen-Inst. Heidelb., Prepr. Ser.
Astron. Zeit-Breitenbestimmungen
Astrophys. Prepr. Ser.
Astrophys. Relativ., Prepr. Ser.
Austrian Pap. Asteroids

BAV Mitt.
B.I.H. Circ.
Biul. Obs. Astron. Uniw. M. Kopernika Toruniu
Bol. Astron. Obs. Madr.
Bol. Inst. Obs. Mar.
Bol. Inst. Tonantzintla
Bol. Obs. Astron. Quito
Bol. Obs. Ebro
Boyden Obs., Occas. Publ.
Boyden Obs., Repr.
Bull. Astron.
Bull. Etoiles Tardives Spectre Particulier
Bull. Inf. Cent. Données Stellaires
Bull. Obs. Astron. Belgr.
Bull., Time Serv. Mizusawa Obs.
Byull. Inst. Astrofiz.
Byull. Inst. Teor. Astron.

Carter Obs., Repr. Ser.
Cartes Synoptiques
Cent. Astron. Sci. Spat., Obs. Sol.
Cent. Astrophys., Prepr. Ser.
Circ. Czech. Obs., Time Latitude
Circ. Stn. Astron. Int. Latitudine, Carloforte-Cagliari

Circ. Time Latitude Serv.
Commun. Astron. Dep., Univ. Ankara
Commun. Konkoly Obs.
Commun. Univ. Lond. Obs.
Commun. Univ. Obs., St. Andrews
Comun. Obs. Astron. Univ. Coimbra
Contrib. Astron. Obs. Skalnaté Pleso
Contrib. Bosscha Obs.
Contrib. Dep. Astron., Univ. Kyoto
Contrib. Dep. Astron., Univ. Tokyo
Contrib. Dep. Geod. Astron., Univ. Thessaloniki
Contrib. Dom. Astrophys. Obs.
Contrib. Dunsink Obs.
Contrib. Inst. Argent. Radioastron.
Contrib. Inst. Astron. Res.
Contrib. Kwasan Hida Obs., Univ. Kyoto
Contrib. La. State Univ. Obs.
Contrib. Lick Obs.
Contrib. Nicholas Copernicus Obs. Planetarium Brno
Contrib. Nizamiah Japal-Rangapur Obs.
Contrib. Oss. Astron. Torino
Contrib. Van Vleck Obs.
Contrib. Wrocław Astron. Obs.
Copenh. Univ. Obs., Repr.
Cracow Obs. Repr.

Data Rep. Hydrogr. Obs., Ser. Astron. Geod.
Debrecen Heliophys. Obs., Hung. Acad. Sci., Repr.
Dep. Astron. McDonald Obs. Univ. Tex., Repr.
Dep. Astrophys., Univ. Oxford, Publ.
Dudley Obs. Rep.
Dunsink Obs. Repr.

ESO Annu. Rep.
ESO Conf. Workshop Proc.
ESO Sci. Prepr.
ESO Sci. Rep.
ESO Tech. Rep.

Gemini
Greenwich Time Rep.

H.M. Naut. Alm. Off., Libr. Repr.
Hvar Obs. Bull.

Inst. Astron. Astrophys. Tech. Univ. Berlin, Mitt.
Inst. Astron. Fis. Espacio, Tirada Aparte
Inst. Astron. Geod., Univ. Madr., Publ.
Inst. Astron., Univ. Camb., Annu. Rep.
Inst. Astrophys. Paris, Pré-Publ.
Inst. Obs. Mar., Bol. Astron.
Inst. Obs. Mar., Bol. Ser. B
Inst. Obs. Mar., Bol. Ser. C
Inst. Obs. Mar., Bol. Ser. D
Inst. Obs. Mar., Mem. Act.
Inst. Teor. Astrofys., Blindern-Oslo, Småtrykk
Inst. Theor. Astrophys., Blindern-Oslo, Rep.
Inst. Theor. Astrophys., Blindern-Oslo, Repr.
Inst. Theor. Astrophys., Univ. Oslo, Publ. Ser.
Inst. Theor. Phys. Sternw. Univ. Kiel, Repr.
Izv. Astron. Ehngel'gardt. Obs.
Izv. Glav. Astron. Obs. Pulkovo
Izv. Krymskoj Astrofiz. Obs.

Kandilli Obs., Heliophys. Serv. Publ., Second Ser.
Kapteyn Astron. Inst., Annu. Rep.

Kodaikanal Obs. Bull.
Kodaikanal Obs. Repr.

Latitude Circ.
Lick Obs. Bull.
Lohrmann-Obs., Tech. Univ. Dresden, Zirk.
Lowell Obs. Bull.

Messenger
Milano Prepr. Ser. Astrophys.
Mitt. Archenhold-Sternw. Berlin-Treptow
Mitt. Astrophys. Obs. Potsdam
Mitt. Geod. Inst. Rheinischen Friedrich-Wilhelms-Univ. Bonn
Mitt. Karl-Schwarzschild Obs. Tautenburg
Mitt. Lohrmann-Obs. Tech. Univ. Dresden
Mitt. Satell.-Beobachtungsstn. Zimmerwald
Mitt. Sonnenobs. Kanzelhöhe
Mitt. Sternw. Münch.
Mitt. Sternw. Sonneberg
Mitt. Universitätssternw. Graz
Mitt. Univ.-Sternw. Innsb.
Mitt. Univ.-Sternw. Jena
Mitt. Veränderliche Sterne
Mitt. Zentralinst. Phys. Erde
Mon. Notes Int. Polar Motion Serv.
Mons Astrophys. Pap.
MPE Contrib.
MPE Intern. Rep.
MPE Prepr.
MPE Rep.
MPE Repr.
Mt. Stromlo Siding Spring Obs., Repr.

Nanjing Univ. Obs., Publ.
NASA IUE Newsl.
Natl. Astron. Ionos. Cent., Astron. Prepr.
Natl. Astron. Ionos. Cent., Astron. Publ.
Natl. Radio Astron. Obs., Repr., Ser. A
Natl. Radio Astron. Obs., Repr., Ser. B
NEOS Earth Orientation Bull.
Nizamiah Rangapur Obs. Dep. Astron. Osmania Univ., Repr.
NSSDC Newsl.

Obs. Astron. Antares, Contrib. Cient.
Obs. Astron. Córdoba, Tirada Aparte
Obs. Astron. La Plata, Sep. Astron.
Obs. Astron. Univ. Nac. La Plata, Ser. Astron.
Obs. Astron. Univ. Nac. La Plata, Ser. Espec.
Obs. Astrophys. Lab., Univ. Helsinki, Rep.
Obs. Haute Provence, Pré-Publ.
Obs. Lyon Prepr.
Obs. Lyon Repr.
Obs. R. Belg., Commun., Sér. A
Obs. R. Belg., Commun., Sér. B
Obs. Variable Stars, Rep.
Occas. Rep. R. Obs., Edinb.
Onsala Space Obs., Prepr.
Oss. Astrofis. Catania, Publ.
Oss. Astron. Milano-Merate, Contrib.

Perth Obs. Commun.
Prepr. Astron. Inst. Univ. Basel
Prepr. Steward Obs.
Proc. Int. Latitude Obs. Mizusawa
Pubbl. Stn. Astron. Int. Latitudine Carloforte-Cagliari, Nuov. Ser.
Pubbl. Varie Fuori Ser. Oss. Astron. Torino (Pino Torinese)
Publ. Astron. Dep. Eötvös Univ.
Publ. Astron. Inst. Czech. Acad. Sci.
Publ. Astron. Obs. Sarajevo
Publ. Astron. Opservatorije Beogr.
Publ. Astrophys. Obs. Potsdam
Publ. Beijing Astron. Obs.
Publ. Bosscha Obs.
Publ. Debrecen Heliophys. Obs.

Publ. Dep. Astron. Univ. Beogr.
Publ. Dep. Astron. Univ. Cape Town
Publ. Dep. Astron., Univ. Chile
Publ. Dep. Geod. Astron., Univ. Thessaloniki
Publ. Dom. Astrophys. Obs.
Publ. Ege Univ. Obs.
Publ. Inst. Geophys., Pol. Acad. Sci., Ser. F (Planet. Geod.)
Publ. Int. Latitude Obs. Mizusawa
Publ. Istanbul Univ. Obs.
Publ. Lick Obs.
Publ. Obs. Astron. Nac., Univ. Colombia
Publ. Obs. Astron. "Prof. Manuel de Barros", Fac. Cienc. Porto
Publ. Obs. Astron. Univ. Cluj
Publ. Obs. Bordeaux, Nouv. Ser.
Publ. Obs. Ebro
Publ. Obs. Genève, Sér. A
Publ. Obs. Genève, Sér. B
Publ. Obs. Genève, Sér. C: Pré-Publ.
Publ. Purple Mt. Obs.
Publ. R. Obs., Edinb.
Publ. Shaanxi Astron. Obs.
Publ. Spéc. Cent. Données Stellaires
Publ. U.S. Nav. Obs., Second Ser.
Publ. Warner Swasey Obs.
Publ. Yunnan Obs.

Q. Bull. Sol. Act.

R. Greenwich Obs., Annu. Rep.
R. Greenwich Obs., Bull.
R. Obs. Ann.
R. Obs. Edinb., Res. Facilities
Rapp. Act. Obs. Paris
Rep. Obs. Lund
Rep. Ser., Dep. Phys. Sci., Univ. Turku
Repr. Theor. Astrophys. Group, Dep. Math., UMIST
Res. Cent. Astron. Appl. Math., Acad. Athens, Contrib. Ser. I
Res. Lab. Electron. Onsala Space Obs., Res. Rep.
Ric. Astron.
Rijksuniv. Gent, Sterrenkundig Obs., Meded.
Rutherford Appleton Lab., Rep.

SAAO Newsl.
Sac Peak Update
S.Afr. Astron. Obs., Annu. Rep.
S.Afr. Astron. Obs., Circ.
Sci. Rep. Tôhoku Univ., Eighth Ser.
Sendai Astron. Rap.
Shaanxi Astron. Obs., Repr.
Shanghai Obs., Acad. Sin., Prepr.
Smithson. Astrophys. Obs., Spec. Rep.
Sol. Maps Act.
Sol. Radio Data
Soobshch. Byurakan. Obs.
Soobshch. Gos. Astron. Inst. Shternberg
Soobshch. Spets. Astrofiz. Obs.
Space Telesc. Sci. Inst., Instrum. Sci. Rep.
Space Telesc. Sci. Inst., Newsl.
Space Telesc. Sci. Inst., Prepr. Ser.
ST-ECF Newsl.
Stockholms Obs., Prepr.
Stockholms Obs., Rep.
Stud. Soc. Sci. Torun., Sect. F
Swarthmore Astron. Repr.
Syd. Obs. Pap.

Tartu Astrofüüs. Obs. Publ.
Tartu Astrofüüs. Obs. Teated
Theor. Pap.
Time Freq. Serv., Bull.
Time Serv. Annu. Rep.
Time Serv. Bull.
Tokyo Astron. Bull., Second Ser.
Tokyo Astron. Obs., Kiso Inf. Bull.
Tokyo Astron. Obs. Rep.

Tokyo Astron. Obs. Repr.
Tokyo Astron. Obs., Time Latitude Bull.
Tr. Astrofiz. Inst. Alma-Ata
Tr. Astron. Obs., Leningrad
Tr. Glav. Astron. Obs. Pulkovo, Ser. 2
Tr. Gos. Astron. Inst. Shternberg
Tr. Inst. Teor. Astron., Leningrad
Tr. Kazan. Gorod. Astron. Obs.
Tr. Tashkent. Astron. Obs.
Trans. Astron. Obs. Yale Univ.
Tsirk. Astron. Inst. Tashkent
Tsirk. Astron. Obs. L'vov
Tsirk. Shemakh. Astrofiz. Obs.
Turku Univ. Obs., Informeto
Turku Univ. Obs., Informo

UKIRT Rep.
Univ. Barc., Dep. Fis. Tierra Cosmos, Publ.
Univ. Hannover, Astron. Stn., Veröff.
Univ. Obs., St. Andrews, Repr.
Univ. Tex., Monogr. Astron.
Univ. Tex., Publ. Astron.
Upps. Astron. Obs. Ann.
Upps. Astron. Obs., Rep.
U.S. Nav. Obs., Circ.
U.S. Nav. Obs., Time Serv. Publ., Ser. 1

U.S. Nav. Obs., Time Serv. Publ., Ser. 4
U.S. Nav. Obs., Time Serv. Publ., Ser. 6
U.S. Nav. Obs., Time Serv. Publ., Ser. 10
U.S. Nav. Obs., Time Serv. Publ., Ser. 11
U.S. Nav. Obs., Time Serv. Publ., Ser. 14
Utr. Sterrekundige Overdrukken
Uttar Pradesh State Obs., Repr.

Vatican Obs. Publ.
Vatican Obs. Publ., Spec. Ser., Studi Galileiani
Veröff. Archenhold-Sternw. Berlin-Treptow
Veröff. Astron. Inst. Bonn
Veröff. Astron. Rechen-Inst. Heidelb.
Veröff. Remeis-Sternw. Bamberg
Veröff. Sternw. Pulsnitz
Veröff. Sternw. Sonneberg
Veröff. Wilhelm-Foerster-Sternw.
Veröff. Zentralinst. Phys. Erde
Villanova Univ. Obs. Contrib.
Vilniaus Astron. Obs. Biul.

Warner Swasey Obs., Repr.
Working Group Stars with Extended Atmospheres, Prepr.
Wrocław Astron. Obs., Repr.

Zentralinst. Astrophys., Sternw. Babelsberg, Mitt.-Neue Folge

002 Bibliographical Publications, Documentation, Catalogues, Data Bases

002.001 Atlas of reflectance spectra of terrestrial, lunar and meteoritic powders and frosts from 92 to 1800 nm.
J. Wagner, B. Hapke, E. Wells.
NASA Tech. Memo., NASA TM–89810, p. 205 (1987).
Abstract. – See Abstr. 003.001.

002.002 MicroVAX–based data management and reduction system for the regional planetary image facilities.
R. Arvidson, E. Guinness, S. Slavney, B. Weiss.
NASA Tech. Memo., NASA TM–89810, p. 543 – 544 (1987). – See Abstr. 003.001.

002.003 Planetary nomenclature.
M. E. Strobell, H. Masursky.
NASA Tech. Memo., NASA TM–89810, p. 551 – 553 (1987). – See Abstr. 003.001.

002.004 Catalogue of cataclysmic binaries, low–mass X–ray binaries and related objects (fourth edition).
H. Ritter.
Astron. Astrophys., Suppl. Ser., Vol. 70, No. 3, p. 335 – 367 (1987).
The catalogue lists coordinates, magnitudes, orbital parameters, stellar parameters of the components and other characteristic properties of 116 cataclysmic binaries, 30 low–mass X–ray binaries and 23 related objects with known or suspected orbital periods together with a comprehensive selection of the relevant literature that appeared after 1983. In addition the catalogue contains a list of references to published finding charts for 164 of the 169 objects. A cross–reference list of objects designations concludes the catalogue.

002.005 An atlas and catalogue of northern dwarf novae.
A. Bruch, F.–J. Fischer, U. Wilmsen.
Astron. Astrophys., Suppl. Ser., Vol. 70, No. 3, p. 481 – 516 (1987).
The authors present an atlas of 90 northern dwarf novae ($\delta > 20°$). It is meant to supplement the Atlas of Southern and Equatorial Dwarf Novae of Vogt and Bateson (1982). Together, these publications represent a comprehensive collection of dwarf nova finding charts for the whole sky. In addition to the atlas coordinates are given for most of the dwarf novae with accuracies significantly higher than published hitherto.

002.006 One hundred and fifty–three diatomic molecules, molecular ions, and radicals of astrophysical interest.
M. Singh, J. P. Chaturvedi.
Astrophys. Space Sci., Vol. 135, No. 1, p. 1 – 74 (1987).
A critical analysis of the 583 available references in literature has been made to select 153 diatomic molecules, molecular ions, and radicals of astrophysical significance. The compilation contains various information for each molecule, such as the dissociation energy, spectral region, transition levels, astrophysical objects where the respective molecules have been detected (say, comet, meteorite, Sun, planet, star, interstellar matter, Galaxy, etc.); computed theoretical parameters (i.e. FCFs, transition probabilities, $\bar{r}$–centroids, PE curves), and available laboratory data with respective references. Thirty–one new diatomic molecules/molecular ions/radicals of astrophysical significance have also been listed.

002.007 Litterae Uraniae: de constellationibus.
A. Heck.
Ciel Terre, Vol. 103, No. 4, p. 113 – 114 (1987).

002.008 Zodiacal Catalog J2000 printed in Japan.
D. W. Dunham.
Occultation Newsl., Vol. 4, No. 5, p. 97 – 99 (1987).

002.009 Hvordan får planeter og måner navn?
K. Aksnes.
Astron. Tidsskr., Årg. 20, Nr. 3, p. 116 – 132 (1987).

002.010 Interstellar travel and communication bibliography – 1986 update.
Z. Paptrony, J. Lehmann, J. Prytz.
J. Br. Interplanet. Soc., Vol. 40, No. 8, p. 353 – 364 (1987).

002.011 Catalogues of coordinates of reference faint, bright and double stars south of –47° and corrections to FK4 positions as observed by the Pulkovo astronomers with the meridian circle at the Cerro–Calan Observatory (Chile) in 1963 – 1968.
G. D. Baturina, V. S. Bedin, K. G. Gnevysheva, M. S. Zverev, A. A. Naumova, A. I. Plyugina, D. D. Polozhentsev, T. A. Polozhentseva, E. A. Stepanova.
Tr. Glav. Astron. Obs. Pulkovo, Ser. 2, Tom 86, p. 4 – 158 (1986). In Russian.
The work of the Pulkovo expedition in the southern hemisphere is reported: the observational programs, instruments, the technique of the observations and organization for their reduction. Besides the series of the reference and other stars under study, special series of fundamental stars along the entire length of the meridian, including lower culmination were regularly observed.

002.012 A catalogue of declinations of 3554 stars in the FK4 system for the observing epoch and equinox 1950.0.
B. K. Bagil'dinskij, G. S. Kosin, L. I. Medvedeva, V. A. Fomin, M. S. Chubej.
Tr. Glav. Astron. Obs. Pulkovo, Ser. 2, Tom 86, p. 159 – 200 (1986). In Russian.
The catalogue of declinations of stars (northern hemisphere) from programs for zenith telescopes and PZT results as a fulfillment of recommendations of the 14th Astrometric Conference of the USSR and IAU Assembly X. The observations were made with the Struve–Ertel vertical circle. Each star was observed at two clamps. The latter fact (exclusion of the zenith point) considerably simplified the reduction of observations and derivation of stars declinations. 574 FK4 stars in the declination zone from $+7°$ to $+84°$ were taken as reference stars. Errors of one observation of reference stars and latitude stars proved to be practically the same ($\pm 0\overset{''}{.}32$ and $\pm 0\overset{''}{.}34$ for a reference star and a latitude star, respectively). The standard deviation of the catalogue declination is $\pm 0\overset{''}{.}21$, the mean epoch 1972.2, systematic FK4 $\Delta\delta_\delta$ and $\Delta\delta_\alpha$ differences are given. A catalogue of individual corrections to FK4 declinations is also presented.

002.013 Second Byurakan Spectral Sky Survey. V. Results for the region centered on $\alpha = 15^h30^m$, $\delta = +59°00'$.
B. E. Markarian, D. A. Stepanyan, L. K. Erastova.
Astrophysics, Vol. 25, No. 2, p. 551 – 562 (1987). English translation of 42.002.075.

002.014 A technical and scientific bulletin board system.
R. G. Deyoe.
I.A.P.P.P. Commun., No. 29, p. 14 17 (1987).
A computer bulletin board system has been established for the exchange of technical and scientific information. The system is located in San Bernardino, California and is available 24 hours a day.

002.015 The EXOSAT results database.
EXOSAT Express, No. 18, p. 4 – 19 (1987).
The EXOSAT database will contain the principal results obtained from a standard analysis performed on every observation. The objective is to provide a computer accessible overview of each EXOSAT observation. This will include a summary of

the results from each instrument, as well as the option of obtaining the data in a reduced form. A data base management system will allow manipulation of the results summary files (e.g. cross–correlating parameters). The reprocessing of all EXOSAT data using second generation software is now well underway and the full database system is expected to be available by the end of 1988.

002.016 The European Space Physics Analysis Network.
T. R. Sanderson, M. Albrecht, J. Franks, C. Harvey,
N. v. d. Heijden, J. L. Green, G. Veldman.
EXOSAT Express, No. 18, p. 20 – 23 (1987).

002.017 Bibliography on Be stars.
A. M. Hubert, J. Jugaku, P. Koubsky, M. Ruusalepp,
A. Slettebak.
Be Star Newsl., No. 16, p. 31 – 33 (1987).
Bibliography of papers on Be stars published in 1986/87.

002.018 Nieuwe steratlas van Wil Tirion. Uranometria 2000.0.
G. Schilling.
Zenit, 14. Jaarg., Nr. 10, p. 336 – 339 (1987).

002.019 A study of the publishing activity of astronomers since 1969.
E. Davoust, L. D. Schmadel.
Publ. Astron. Soc. Pac., Vol. 99, No. 617, p. 700 – 710 (1987).
This is an analysis of the scientific production of astronomers worldwide over the past 17 years. The inflation in astronomical literature is due to the increasing number of astronomers. They are becoming more productive, but there are also more authors per paper. The total production per astronomer has been decreasing, but the trend may be changing. The frequency distribution of productivity falls more steeply than Lotka's law, hence the more prolific authors do not contribute much to the total production. The proportion of more productive astronomers is increasing with time: astronomers' productivity seems to grow with age. Sociological rather than scientific or technical causes are probably responsible for most of the identified trends.

002.020 A catalog of stellar angular diameters measured by lunar occultation.
N. M. White, B. H. Feierman.
Astron. J., Vol. 94, No. 3, p. 751 – 770 (1987).
A catalog of 348 published measurements (124 stars) of stellar angular diameters by the lunar–occultation method is presented. The overall precision and accuracy are discussed.

002.021 On the general catalogue of fundamental faint stars (FKSZ).
M. S. Zverev, A. N. Kur'yanova, A. D. Polozhentsev,
D. D. Polozhentsev, Ya. S. Yatskiv.
Kinematika Fiz. Nebesn. Tel, Tom 3, No. 4, p. 3 – 6 (1987). In Russian. English translation in Kinematics Phys. Celest. Bodies.
Five stages of the compilation of the FKSZ catalogue are described. The general FKSZ catalogue is formed on the basis of individual catalogues of Fundamental Faint Stars (FSZ): 17 catalogues in right ascension and 15 catalogues in declination, as well as the observations of these stars according to the AGK3R and SRS programmes. The final positions and proper motions of 931 faint stars are in the system of the FK4 for the equinox and the epoch of 1950.0. The mean values of the root mean square errors of these star positions are $\pm 0''050$ in right ascension and $\pm 0''074$ in declination. The mean values of the root mean square errors of proper motions are ± 0.26 and $\pm 0''32$ per century in α and δ, respectively (for the declination zone from -20 to $+90°$).

002.022 On the general catalogue of stellar proper motions with respect to galaxies in the areas of the Galaxy main meridional section.
N. V. Kharchenko.
Kinematika Fiz. Nebesn. Tel, Tom 3, No. 4, p. 7 – 10 (1987). In Russian. English translation in Kinematics Phys. Celest. Bodies.
In the framework of the programme of complex study of the main meridional section of the Galaxy (MEGA) the general catalogue of proper motions of 14111 stars with respect to 206 galaxies and their equatorial coordinates is compiled. The data from the published catalogue KSZ, AGK3 and SAO are used. The proper motions are brought to the general catalogue system with allowance for the magnitude equation error. The root mean square errors are $\pm 0''009$/yr for μ_x; $\pm 0''08$/yr for μ_y; $\pm 0''3$ for α, δ; $\pm 0^m3$ for m_{pg}.

002.023 Supernovae.
A. Fraknoi.
Mercury, Vol. 16, No. 4, p. 122 – 123 (1987).

002.024 On the structure of the subsystem of an Automatic Bank of Astronomical Data (ABAD).
I. G. Zhurkin, O. V. Kochetkov.
Sostoyanie i perspektivy razvitiya geod. i kartogr. Mater. Vses. nauchn.–tekh. konf. Moskva, 5 – 7 sent., 1984. Moskva, p. 148 – 151 (1986). In Russian. Abstr. in Ref. Zh., 52. Geod. Aehrosemka, 8.52.99 (1987).

002.025 American Astronomical Society 1988 Membership Directory.
AAS Executive Office, 2000 Florida Ave, N.W., Washington, DC 20009, USA. 129 pp. Price US$ 8.00 (1987).

002.026 Surveying the northern sky.
J. Schombert.
Sky Telesc., Vol. 74, No. 2, p. 128 – 131 (1987).

002.027 An index to variable star finder charts in the Journal of the AAVSO 1972 – 1985.
M. Morel.
J. Am. Assoc. Variable Star Obs., Vol. 16, No. 1, p. 23 – 28 (1987).
An index is presented of variable star identification charts which have been published in the Journal of the American Association of Variable Star Observers, Volumes 1 – 14 (1972 – 1985). Additional information is provided about the identification of some variables.

002.028 Neue Namen von Kleinplaneten.
J. Jahn.
KPM, Jahrg. 2, Nr. 6, p. 9 – 11 (1987).

002.029 Valinhos 2.2 micron survey of the southern galactic plane. II. Near–IR photometry, IRAS identifications and nature of the sources.
N. Epchtein, T. Le Bertre, J. R. D. Lépine,
P. Marques dos Santos, O. T. Matsuura, E. Picazzio.
Astron. Astrophys., Suppl. Ser., Vol. 71, No. 1, p. 39 – 55 (1987). With a correction in Vol. 71, No. 2, p. 411 (1987).
This paper reports on broad band near–IR photometric observations of 630 objects found in the Valinhos 2.2 μm survey of the southern galactic plane. Ninety percent of them are identified with IRAS Point Sources. Most of them are typical of late–type stars surrounded by circumstellar dust shells. A classification of the objects based on three colour indices is defined. Several interesting peculiar sources are briefly discussed and proposed for further multi–frequency observations.

002.030 _UBV_ photoelectric catalogue (1986). II Analysis of the data.
J.–C. Mermilliod.
Astron. Astrophys., Suppl. Ser., Vol. 71, No. 1, p. 119 – 124 (1987).
The _UBV_ photoelectric data of the stars presenting several entries in the 1986 edition of the _UBV_ catalogue have been systematically intercompared and this paper presents a discussion of the stars for which discrepancies larger than 0.2 mag were found. Thirty–six probably variable stars have been detected, among which 18 are Be stars. Sixty further stars present differences in the _V_ magnitude larger than 0.2 mag. Sixteen stars already appear in the NSV catalogue. Although many problems are probably due to poor observations, new (eclipsing) variable

stars may be found in this sample. Complete disagreement is found between the values published in two independent sources in 34 cases. A first analysis of the quality of the *UBV* data shows that 65 percent of the differences in the *V* magnitude and in *U–B* colour, for respectively 11500 and 7200 stars with two sources of data, are smaller than 0.04. The scatter on the *B–V* index appears to be smaller, since the same percentage reaches 79%.

002.031   UBV photoelectric photometry catalogue (1986). I. The original data (magnetic tape).
J.–C. Mermilliod.
Astron. Astrophys., Suppl. Ser., Vol. 71, No. 3, p. 413 – 420 (1987).

A new and extensive catalogue of UBV photoelectric photometry has been compiled which collects all data published from 1953 to the end of 1985 in the Johnson and Morgan system. In addition, data obtained in closely related systems using UBV filters have also been collected and are available in a separate file. All the data have been merged, yielding a final catalogue of 136,719 entries concerning 87,267 stars and components. The data have been taken from 1413 (UBV) and 127 (other) references found in 73 different periodicals. Great care has been taken to detect and correct all kinds of errors. For the analysis of the data see Abstr. 002.030 (Paper II).

002.032   A catalog of precise reference star positions for the astrometry network of the international comet P/Halley campaign.
C. de Vegt, N. Zacharias.
Astron. Astrophys., Suppl. Ser., Vol. 71, No. 3, p. 525 – 530 (1987).

A catalog of 506 reference stars in the FK4 system, covering the P/Halley pre–perihelion cometary path approximately between November 1985 and January 1986, has been derived from plates taken with the Hamburg 23–cm astrograph in fall 1985. Stars have been selected from AGK3 within a strip of about 1.5 degrees width along the cometary path.

002.033   Atlas of comet Halley 1910 II.
B. Donn, J. Jahe, J. C. Brandt.
NASA Spec. Publ., NASA SP–488, 12 + 600 pp. (1986). For sale by the Superintendent of Documents, U.S. Governement Printing Office, Washington, DC 20402, USA.

Contents: 1. Visual observations of comet Halley 1835 III. 2. Visual observations of comet Halley 1910 II. 3. Photographs of comet Halley 1910 II. 4. Comparison of visual and photographic observations of comet Halley 1910 II. 5. Digitally processed photographs. 6. Spectra of comet Halley 1910 II.

002.034   On systematic errors of proper motions of stars contained in the catalogue of the USSR Time Service.
V. V. Vityazev.
Vestn. Leningr. Univ., Mat. Mekh. Astron., No. 2, p. 78 – 85 (1987). In Russian. Abstr. in Ref. Zh., 51. Astron., 9.51.115 (1987).

002.035   The Green Bank Third (GB3) Survey of extragalactic radio sources at 1400 MHz.
S. Ryś, J. Machalski.
Acta Astron., Vol. 37, No. 2, p. 163 – 178 (1987).

The NRAO 91–m telescope was used to make a 1400 MHz sky survey covering an area of 0.0988 sr at declinations $70° \leqslant \delta_{1950} \leqslant 76°8$ with 10.1×10.5 arcmin resolution. This survey ends the series of smaller than 1–sr surveys made at 1400 MHz with that telescope. A catalogue of 502 radiosources is presented, statistically complete to 112 mJy, which is about five times the rms noise and extragalactic confusion. The observations and data reduction are briefly summarized; the position and flux density errors are discussed.

002.036   Regarding a proposal for a speckle databank.
E. K. Hege.
Interferometric imaging in astronomy, p. 255 – 256 (1987). – See Abstr. 012.035.

002.037   The archives of Armagh Observatory.
J. Butler, M. Hoskin.
J. Hist. Astron., Vol. 18, Part 4, p. 295 – 307 (1987).

002.038   The Data Inventory of Space–Based Celestial Observations (DISCO), Version 2.0.
L. E. Brotzman, A. C. Raugh, J. M. Mead.
Bull. Am. Astron. Soc., Vol. 19, No. 2, p. 690 (1987). Abstract. – See Abstr. 010.061.

002.039   The Bright Star Catalogue, 5th revised edition.
W. H. Warren Jr., D. Hoffleit.
Bull. Am. Astron. Soc., Vol. 19, No. 2, p. 733 – 734 (1987). Abstract. – See Abstr. 010.061.

002.040   A distributed data system for astronomy: an update.
E. J. Schreier.
Bull. Am. Astron. Soc., Vol. 19, No. 2, p. 738 (1987). Abstract. – See Abstr. 010.061.

002.041   Astronomical archives: dealing with used data.
R. Albrecht, F. Ochsenbein, A. Richmond, G. Russo.
Bull. Am. Astron. Soc., Vol. 19, No. 2, p. 744 (1987). Abstract. – See Abstr. 010.061.

002.042   The Astronomical Data Center (ADC) network catalog request service.
L. E. Brotzman, J. M. Mead, W. H. Warren Jr.
Bull. Am. Astron. Soc., Vol. 19, No. 2, p. 744 (1987). Abstract. – See Abstr. 010.061.

002.043   The guide star catalog. I. Overview, history, and prospective.
B. M. Lasker, H. Jenkner, J. L. Russell.
Space Telesc. Sci. Inst., Prepr. Ser., No. 187, p. 1 – 5 (1987). To appear in IAU Colloq. No. 133.

002.044   The guide star catalog. II. Astrometric and photometric algorithms and precision.
J. L. Russell, B. M. Lasker, H. Jenkner.
Space Telesc. Sci. Inst., Prepr. Ser., No. 187, p. 6 – 8 (1987). To appear in IAU Colloq. No. 133.

002.045   The guide star catalog. III. Structure and publication, status and plans.
H. Jenkner, J. L. Russell, B. M. Lasker.
Space Telesc. Sci. Inst., Prepr. Ser., No. 187, p. 9 – 13 (1987). To appear in IAU Colloq. No. 133.

002.046   *The Bright Star Catalogue*, 5th revised edition.
D. Hoffleit, W. H. Warren Jr.
Astron. Data Cent. Bull., Vol. 1, No. 4, p. 285 – 294 (1987).

Since the publication of the fourth edition of the catalog (1982) and the release of its machine version, errors reported by colleagues and discovered by us have been collected along with new data from a number of papers published in the astronomical literature. These changes have now been incorporated into the machine version to produce a fifth edition of the computerized catalog.

002.047   IRAS point source associations with other astronomical catalogs.
J. Chillemi.
Astron. Data Cent. Bull., Vol. 1, No. 4, p. 295 – 298 (1987).

002.048   Suggestions for machine–readable catalogs: a second look.
R. S. Hill.
Astron. Data Cent. Bull., Vol. 1, No. 4, p. 302 – 305 (1987).

Nowadays, the standards for ease in computing are set by the better commercial products for personal computers. All users, including astronomers, have rising expectations of product correctness and of clear documentation. The makers and providers of computer–readable versions of astronomical cata-logs, whether they like it or not, will find themselves being judged by the same standards. Careful procedures for handling catalogs, therefore, are more necessary than ever.

002.049   Overlay plots for the ESO/SERC Southern Sky Atlas.
L. E. Brotzman.
Astron. Data Cent. Bull., Vol. 1, No. 4, p. 306 – 307 (1987).

002.050   New version of the Bidelman–Parsons Spectroscopic and Bibliographical Catalog.
S. B. Parsons, R. S. Hill.
Astron. Data Cent. Bull., Vol. 1, No. 4, p. 308 – 311 (1987).

002.051   Data Inventory of Space–based Celestial Observations, Version 1.0.
L. E. Brotzman, R. S. Hill, J. M. Mead.
Astron. Data Cent. Bull., Vol. 1, No. 4, p. 312 – 317 (1987).
The Data Inventory of Space–based Celestial Observations (DISCO) is a directory to data contained in sixteen catalogs dealing exclusively with observations from space. DISCO tells whether an object has been observed by a space instrument. It also describes the instruments and indicates the type and quality of the data, as stated in each catalog. The catalogs covered to date contain X–ray, far–ultraviolet, and ultraviolet data.

002.052   Computerization of the *Cape Photographic Durchmu-sterung*.
B. N. Rappaport, W. H. Warren Jr.
Astron. Data Cent. Bull., Vol. 1, No. 4, p. 318 – 319 (1987).
The methodology and error analysis used during the transfer of the CPD data to machine–readable form are discussed. Both computer and manual checking were used in the error analysis. The final estimated error percentage of the transferred data is 0.020 percent, while a projected estimate of the number of errors still present yields 545, or 0.014 percent.

002.053   A revised machine version of the *General Catalogue of 33342 stars for the epoch 1950 (GC)*.
N. G. Roman, W. H. Warren Jr.
Astron. Data Cent. Bull., Vol. 1, No. 4, p. 320 – 323 (1987).
A revised and corrected version of the machine–readable GC (Boss 1937) has been prepared. Cross identifications of the GC stars to the HD and DM catalogs have been replaced by data from the new *SAO–HD–GC–DM Cross Index* (Roman, Warren and Schofield 1983), including component identifications for multiple SAO entries having identical DM numbers in the SAO Catalog, supplemental *Bonner Durchmusterung* stars (lower case letter designations) and codes for multiple HD stars. Additional individual corrections have been incorporated based upon errors found during analyses of other catalogs.

002.054   Bibliographical Star Index retrieval system, Version 4.0.
R. S. Hill, L. E. Brotzman, J. M. Mead.
Astron. Data Cent. Bull., Vol. 1, No. 4, p. 324 – 325 (1987).
The Bibliographical Star Index is an extensive list of astronom-ical literature references indexed according to the identifiers of stars referred to in either the title or the body of each paper.

002.055   The machine–readable version of the ESO/Uppsala Survey of the ESO(B) atlas.
L. E. Brotzman, R. S. Hill.
Astron. Data Cent. Bull., Vol. 1, No. 4, p. 326 – 329 (1987).
The authors describe the survey itself and how they modified the machine–readable version of the catalog.

002.056   A new machine–readable version of *The Henry Draper Catalogue*.
N. G. Roman, W. H. Warren Jr.
Astron. Data Cent. Bull., Vol. 1, No. 4, p. 330 – 335 (1987).
An improved and uniformly formatted version of the ma-chine–readable catalog has been prepared for archiving and distribution. In addition to the correction of all known errors in the previous machine edition, basic improvements include

expansion of the Durchmusterung numbers to full representa-tion, distinction of multiple–star components and BD supple-mental stars, and revision of the codes used for variable and missing magnitudes.

002.057   Availability of the Interim Supplement to the General Catalogue of Radial Velocities.
R. S. Hill, L. E. Brotzman.
Astron. Data Cent. Bull., Vol. 1, No. 4, p. 336 – 337 (1987).
The Interim Supplement to the General Catalogue of Radial Velocities (ISGCRV; Evans 1979) contains radial velocities for stars with right ascensions from 0 through 20 hours. The ISGCRV is intended as a continuation of the General Catalogue of Radial Velocities (GCRV; Wilson 1953).

002.058   New and revised catalogs available from the Astronomi-cal Data Center.
W. H. Warren Jr., R. S. Hill, L. E. Brotzman.
Astron. Data Cent. Bull., Vol. 1, No. 4, p. 338 – 366 (1987).
Machine–readable astronomical catalogs and data sets re-ceived and/or revised since the publication of the last issue of this Bulletin are described briefly. Although the authors have tried to indicate the data content of new catalogs, the published editions should be consulted for detailed information.

002.059   Astronomical Data Center status report indexes.
W. H. Warren Jr., Y. W. Kang, A. C. Raugh.
Astron. Data Cent. Bull., Vol. 1, No. 4, p. 376 – 394 (1987).

002.060   Status report on machine–readable astronomical cata-logs: Astronomical Data Center, NASA–Goddard Space Flight Center.
W. H. Warren Jr., J. M. Mead, L. E. Brotzman, R. S. Hill.
Astron. Data Cent. Bull., Vol. 1, No. 4, p. 395 – 440 (1987).

002.061   Current status of the Second Cape Photographic Cata-logue.
C. de Vegt, N. Zacharias, M. J. Penston, C. A. Murray.
R. Greenwich Obs., Prepr., No. 58, p. 16 – 20 (1987). To appear in IAU Symp. No. 133.

002.062   An atlas of the thorium–argon spectrum for the ESO Echelle Spectrograph in the $\lambda\lambda 3400 - 9000$ Å region.
S. D'Odorico, M. Ghigo, D. Ponz.
ESO Sci. Rep., No. 6, 170 pp. (1987).

002.063   Astronomy and Astrophysics Abstracts. Vol. 43. Litera-ture 1987, Part 1.
S. Böhme, U. Esser, H. Hefele, I. Heinrich, W. Hofmann, D. Krahn, V. R. Matas, L. D. Schmadel, G. Zech (Editors).
Published for Astronomisches Rechen–Institut by Springer–Verlag, Berlin. 10 + 1291 pp. Price DM 228.00, US$ 138.00 [Subscription Price DM 182.40, US$ 110.40] (1987). ISBN 3–540–18640–9, F.R. Germany, ISBN 0–387–18640–9 (USA).

002.064   The Third Catalogue of Nearby Stars with special emphasis on wide binaries.
W. Gliese, H. Jahreiß.
Astron. Rechen–Inst. Heidelb., Prepr. Ser., No. 12, 15 pp. (1987). To appear in IAU Symp. No. 97.

002.065   The 105 cm Schmidt plates (Nos. 4901 – 5350).
Tokyo Astron. Obs., Kiso Inf. Bull., Vol. 2, No. 4, p. 69 – 78 (1987).

002.066   Tokyo PMC catalog 85: catalog of positions of 1007 stars observed in 1985 with Tokyo Photoelectric Meridi-an Circle.
M. Yoshizawa, S. Suzuki, R. Fukaya.
Ann. Tokyo Astron. Obs., Second Ser., Vol. 21, No. 4, p. 399 – 421 (1987).
The catalog of positions of 1007 stars (792 FK4 and FK4S stars, 57 OB stars, 49 NPZT stars, and 109 SAO stars) is presented. They were observed during the period from December

1984 to September 1985 with the Tokyo Photoelectric Meridian Circle (Tokyo PMC). The positions in the catalog are referred to the equinox and equator of J2000, and are based on the FK4 system. The comparison of the positions of the FK4 stars in the present catalog with those of the FK4 catalog shows the significant differences $\Delta\alpha_\delta$ and $\Delta\delta_\delta$ in some declination zones. Some of those differences are commonly found in other recent catalogs. Thus they may be considered to be the real systematic errors in the FK4 system.

002.067   Triennial report on globular star cluster research (1984 – 1986).
R. E. White.
Prepr. Steward Obs., No. 765, 26 pp. (1987). Prepared for IAU Commission 37.

002.068   Asteroid reference catalogue.
C. I. Lagerkvist.
Upps. Astron. Obs. Rep., No. 44, 35 + 144 pp. (1987).

002.069   The Pulkovo sky survey in the galactic coordinate system.
N. V. Bystrova, L. I. Yagudin.
Astrofiz. Issled. Izv. Spets. Astrofiz. Obs., Tom 24, p. 108 – 131 (1987). In Russian. English translation in Bull. Spec. Astrophys. Obs. – North Caucasus.
The maps of the Pulkovo sky survey in the galactic coordinate system are presented. They are made for the two components of interstellar neutral hydrogen radio emission: the maps of the antenna (BPR) temperature distribution for 10 radial velocities are given within $b = \pm 70°$. The map of the antenna temperature mean values within the whole radial velocity interval is presented also.

002.070   The Pulkovo sky survey in Magellanic and ecliptic coordinate systems.
N. V. Bystrova.
Astrofiz. Issled. Izv. Spets. Astrofiz. Obs., Tom 24, p. 132 – 177 (1987). In Russian. English translation in Bull. Spec. Astrophys. Obs. – North Caucasus.
The maps of the Pulkovo sky survey in Magellanic and ecliptic coordinate systems are presented. They are made separately for two components of the interstellar neutral hydrogen radio emission.

002.071   Galactic and Magellanic polar areas in the Pulkovo H I survey.
N. A. Bystrova, L. I. Yagudin.
Astrofiz. Issled. Izv. Spets. Astrofiz. Obs., Tom 25, p. 176 – 193 (1987). In Russian. English translation in Bull. Spec. Astrophys. Obs. – North Caucasus, Vol. 25.
The contour maps of the interstellar neutral hydrogen distribution are given for the polar areas till the latitudes $\pm 50°$ in the galactic and Magellanic coordinate systems. The maps in the velocity interval 4.2 km/s are given for 10 velocities between –21.8 and + 25.6 km/s and for the mean values for the two components of the H I radio emission.

002.072   The formation of an astrometric data bank in the framework of the complex programme of studying the main meridional section of the Galaxy (MEGA).
N. V. Kharchenko.
Tartu Astrofüüs. Obs. Teated, Nr. 84, p. 4 – 7 (1987).
The procedures of preparing a data bank consisting of absolute proper motion data for 15000 stars in the Galactic main meridional section are outlined.

002.073   On a special compiled catalogue of absolute proper motions of stars.
S. P. Rybka.
Red. Zh. Kinematika Fiz. Nebesn. Tel, Kiev, 21 pp. (1987). In Russian. Abstr. in Ref. Zh., 51. Astron. 11.51.67 (1987).

002.074   Neutron capture cross sections for s–process studies.
Z. Y. Bao, F. Kappeler.
At. Data Nucl. Data Tables, Vol. 36, No. 3, p. 411 – 451 (1987). Abstr. in Phys. Abstr., Vol. 90, No. 1311, Entry 95114 (1987).

002.075   Beta–decay rates of highly ionized heavy atoms in stellar interiors.
K. Takahashi, K. Yokoi.
At. Data Nucl. Data Tables, Vol. 36, No. 3, p. 375 – 409 (1987). Abstr. in Phys. Abstr., Vol. 90, No. 1311, Entry 101767 (1987).

002.076   The rare and antiquarian book collection of the Armagh Observatory.
J. McFarland.
Ir. Astron. J., Vol. 18, No. 2, p. 102 – 115 (1987).

002.077   Errata: 'A reference catalogue and atlas of galactic novae', [Space Sci. Rev., Vol 45, Nos.1 + 2, p. 1 – 212 (1987)].
H. W. Duerbeck.
Space Sci. Rev., Vol. 45, Nos. 3 + 4, p. 405 (1987). See Abstr. 43.002.037.

002.078   A list of reference stars in 252 areas with extragalactic radio sources.
I. I. Kumkova, P. F. Lazorenko, A. S. Kharin.
Spets. Astrofiz. Obs. Akad. Nauk SSSR, 28 pp. (1987). In Russian. Abstr. in Ref. Zh., 51. Astron., 12.51.60 (1987).

002.079   A subject index to astronomy articles in Scientific American magazine January 1979 – September 1987.
A. Fraknoi.
Mercury, Vol. 16, No. 5, p. 155 – 158 (1987).

002.080   Atlas of time–resolved spectrophotometry of cataclysmic variables.
R. K. Honeycutt, R. H. Kaitchuck, E. M. Schlegel.
Astrophys. J., Suppl. Ser., Vol. 65, No. 3, p. 451 – 458 (1987). With plates 2 – 24.
Whole–orbit, time–resolved spectrophotometry of 18 cataclysmic variables is reduced and displayed in a homogeneous fashion. Each set of digital data is binned into a phase–dependent gray–scale image which resembles a single–trailed photographic spectrogram. This technique permits convenient comparison among the different systems of phase–dependent spectral behavior. The approximate wavelength interval 4250 – 4950 Å was observed at 2.5 Å resolution, an interval containing the accretion disk emission lines of Hβ, Hγ, He I λ4471, and He II λ4686. Most of the systems are of high orbital inclination, i.e., eclipsing or nearly eclipsing.

002.081   An atlas of optical spectrophotometry of Wolf–Rayet carbon and oxygen stars.
A. V. Torres, P. Massey.
Astrophys. J., Suppl. Ser., Vol. 65, No. 3, p. 459 – 483 (1987).
This atlas contains a homogeneous set of optical spectrophotometric observations (3300 – 7300 Å) at moderate resolution (~ 10 Å) of almost all WC and WO stars in the Galaxy, the LMC, and the SMC. The data are presented in the form of spectral tracings (in magnitude units) arranged by subtype, with no correction for interstellar reddening. A montage of prototype stars of each spectral class is also shown. Comprehensive line identifications are given for the optical lines of WC and WO spectra, with major contributions tabulated and unidentified lines noted. Fluxes of individual stars can be obtained from the Astronomical Data Center at NASA/Goddard Space Flight Center.

002.082   Catálogo SRS de San Fernando, zona –10° a –30°, equinoccio 1950.0.
Real Instituto y Observatorio de La Armada, San Fernando, 84 pp. (1987).
The San Fernando SRS (Southern Reference Star) Catalogue has been formed with the results of the visual observations of

3678 stars of the SRS Programme with declinations $-10° > \delta > -30°$ carried out with the Grubb Parsons Meridian Circle of the Instituto y Observatorio de Marina from 1963 through 1973. The Catalogue lists the positions for the mean epoch of observation referred to the equinox 1950.0. The reference system is that of the FK4.

002.083   A few thoughts on astronomical databases.
   H. Smith Jr.
Bull. Inf. Cent. Données Stellaires, No. 33, p. 63 – 67 (1987).
   The interconnections among certain kinds of astronomical data suggest the creation of unified databases with built–in, statistically rigorous ways of combining related data and capable of growth to incorporate new data.

002.084   A proposed universal reference code.
   A. Martinez.
Bull. Inf. Cent. Données Stellaires, No. 33, p. 69 – 75 (1987).
   A system is proposed whereby any resolved celestial object can be given a unique hexadecimal code. Extended objects can be labelled by truncating the code, and superimposed objects by extending it. The coded objects can be named easily by referring to a table of proper names.

002.085   Enquiry on data centers in astronomy.
   C. Jaschek.
Bull. Inf. Cent. Données Stellaires, No. 33, p. 81 – 107 (1987).

002.086   Corrections and additions to CDS Catalogue 4017 (cross–identificatins for stars in open clusters).
   R. F. Griffin.
Bull. Inf. Cent. Données Stellaires, No. 33, p. 117 – 118 (1987).

002.087   Compilation of the fifth volume of the Morphological Catalogue of Galaxies on magnetic tape.
   N. G. Kogoshvili.
Bull. Inf. Cent. Données Stellaires, No. 33, p. 121 – 122 (1987).

002.088   Catalogue of photographic, photovisual and photored magnitudes for 3124 stars in the Selected Area 40.
   L. N. Kolesnik.
Bull. Inf. Cent. Données Stellaires, No. 33, p. 123 – 125 (1987).

002.089   The catalogue of Trapezium–Type Multiple Systems in machine–readable form.
   G. N. Salukvadze.
Bull. Inf. Cent. Données Stellaires, No. 33, p. 127 – 130 (1987).
   The present catalogue of Trapezium–Type Multiple Systems was compiled on the basis of the Index–Catalogue of Visual Binary Stars.

002.090   A catalogue of seven–colour photometry of 752 stars in the Vilnius system for a region of two square degrees in the direction of the globular cluster M56.
   P. Smriglio, R. P. Boyle, V. Straizys, R. Janulis, K. Nandy, H. MacGillivray, L. Rossi, C. Francini.
Bull. Inf. Cent. Données Stellaires, No. 33, p. 135 – 136 (1987).

002.091   Machine–readable version of the Tonantzintla catalogue of the Pleiades flare stars.
   M. K. Tsvetkov, K. Ya. Stavrev, K. P. Tsevtkova.
Bull. Inf. Cent. Données Stellaires, No. 33, p. 137 – 139 (1987).
   A machine–readable version of the Tonantzintla catalogue of flare stars in the Pleiades region containing data for 1531 flares of 519 stars is described.

002.092   The magnetic tape version of the LDS catalogue.
   J. L. Halbwachs.
Bull. Inf. Cent. Données Stellaires, No. 33, p. 141 – 143 (1987).
   The lists of the double stars with common proper motions discovered by Luyten are now available on magnetic tape.

002.093   Activities in documentation and astronomical data.
   G. A. Wilkins.
Bull. Inf. Cent. Données Stellaires, No. 33, p. 157 – 162 (1987).
= IAU Commission 5, Newsl. No. 2.

002.094   Star catalogs and files available at the Stellar Data Center.
Bull. Inf. Cent. Données Stellaires, No. 33, p. 163 – 168 (1987).
   This list is a supplement to the one published in Bull. Inf. Cent. Données Stellaires, No. 32, p. 113 (1987). – See Abstr. 43.002.082.

002.095   Catalogs to become available from Centre de Données Stellaires.
Bull. Inf. Cent. Données Stellaires, No. 33, p. 169 (1987).

002.096   Catalogs currently available on microfiche from Centre de Données Stellaires.
Bull. Inf. Cent. Données Stellaires, No. 33, p. 171 – 175 (1987).

002.097   On a version of star catalogues for photographic astrometry.
   N. N. Matveev.
Astron. Tsirk., No. 1475, p. 7 – 8 (1987). In Russian.

002.098   List of dissertations of the Moscow State University in 1986.
   L. N. Bondarenko.
Astron. Tsirk., No. 1479, p. 7 – 8 (1987). In Russian.

002.099   Data base for complex use of astronomical catalogues on computers of ES series.
   G. I. Kondratenko, N. D. Kostyuk, O. Yu. Malkov,
   V. I. Myakutin, A. Eh. Piskunov.
Nauchn. Inf., Vyp. 59, p. 3 – 12 (1986). In Russian.
   The authors shortly describe the aim, structure and resources of the astronomical data base, created at the Soviet Center of Astronomical Data using machine–readable catalogues received from the Strasbourg Stellar Data Center.

002.100   List of catalogues available at the Soviet Center of Astronomical Data (first supplement).
Nauchn. Inf., Vyp. 59, p. 31 – 60 (1986). In Russian.

002.101   Catalogue of determinations of metallicity, velocity components and orbital elements of F5–K5 dwarfs in the neighbourhood of 80 parsecs from the Sun.
   V. A. Marsakov, Yu. G. Shevelev.
Nauchn. Inf., Vyp. 59, p. 61 – 63 (1986). In Russian.
   Metallicities, distances, components of spatial velocities and elements of osculating orbits are calculated for 1065 F5–K5 dwarfs from the galactic disk with UBV data, proper motions and radial velocities.

002.102   Catalogue of determinations of [Fe/H], velocity components and orbital elements of F stars.
   Yu. G. Shevelev.
Nauchn. Inf., Vyp. 59, p. 64 – 66 (1986). In Russian.
   [Fe/H] values, distances, components of spatial velocities and elements of osculating orbits have been calculated for 1304 non-evolved F stars with homogeneous uvbyβ data and proper motions.

002.103   Catalogue of determinations of [Fe/H], distances, velocity components and orbital elements of F stars of the southern Galactic pole.
   Yu. G. Shevelev, V. A. Marsakov.
Nauchn. Inf., Vyp. 59, p. 67 – 69 (1986). In Russian.
   Normal photometric indices, absolute magnitudes, [Fe/H] values, distances, components of spatial velocities in the galactic coordinate system, eccentricities of osculating orbits as well as apo– and perigalactic distances have been calculated for 556 non-evolved F stars with homogeneous uvbyβ data and proper motions near the southern Galactic pole.

002.104   A catalog of bright $uvby\beta$ standard stars.
C. L. Perry, E. H. Olsen, D. L. Crawford.
Publ. Astron. Soc. Pac., Vol. 99, No. 621, p. 1184 – 1200 (1987).
= Contrib. Louisiana State Univ. Obs., No. 208.
An all–sky catalog of bright standard stars is presented for the $uvby\beta$ photometric systems.

002.105   The distribution of astronomical research among the United States over the past half century.
H. A. Thronson Jr.
Publ. Astron. Soc. Pac., Vol. 99, No. 621, p. 1209 – 1213 (1987).
The geographical distribution of astronomical research papers produced in the United States over about the past 70 years was determined to test the hypothesis that this science has become widely distributed over the past decades and to provide grist for speculation about the development of modern American astronomy.

002.106   A 1.49 GHz atlas of spiral galaxies with $B_T \leqslant +12$ and $\delta \geqslant -45°$.
J. J. Condon.
Astrophys. J., Suppl. Ser., Vol. 65, No. 4, p. 485 – 541 (1987).
The VLA has been used in its most compact D– and C/D– configurations to make low–resolution ($\theta \approx 0\overset{''}{.}9$ FWHM) 1.49 GHz maps of the spiral galaxies north of $\delta = -45°$ and brighter than $B_T = +12$, the completeness limit of the Revised Shapley–Ames Catalog. At least 94% of the galaxies were detected with $S \geqslant 1$ mJy. An atlas of contour maps, a table of total flux densities plus other radio source parameters, and references to published radio maps are given.

002.107   A 1.49 GHz supplementary atlas of spiral galaxies with H–magnitudes.
J. J. Condon, Q. F. Yin, D. Burstein.
Astrophys. J., Suppl. Ser., Vol. 65, No. 4, p. 543 – 553 (1987).
The VLA has been used in its most compact D– and C/D– configurations to make low–resolution ($\theta \approx 0\overset{''}{.}9$) 1.49 GHz maps of spiral galaxies with measured infrared magnitudes $H_{-0.5}$ and H I velocity widths from the 1982 list of Aaronson and coworkers. Many appear in the 1.49 GHz atlas of all spiral galaxies brighter than $B_T = +12$ and north of $\delta = -45°$; this supplementary atlas presents maps and radio source parameters of the 55 fainter galaxies observed.

002.108   A catalog of spectroscopically identified white dwarfs.
G. P. McCook, E. M. Sion.
Astrophys. J., Suppl. Ser., Vol. 65, No. 4, p. 603 – 671 (1987).
A catalog of 1279 spectroscopically identified white dwarfs is presented, complete to 1987 January. For each degenerate star, the catalog lists a coordinate designation, in order of increasing right ascension; the full coordinates for 1950.0; the spectral type on the new white dwarf classification system; a symbol denoting binary membership; most known names; proper motion and position angle; broad–band, narrow–band, and multichannel colors; a best available absolute magnitude; trigonometric parallax; and radial velocity. A Notes section and a coded Reference Key are presented, as well as a table cross–referencing all names to catalog coordinate designation.

002.109   Katalog orbital'nykh ehlementov, mass i svetimostej tesnykh dvojnykh zvezd. (*Catalogue of orbital elements, masses and luminosities of close binaries*).
M. A. Svechnikov.
Izdatel'stvo Irkutskogo Universiteta, Irkutsk, USSR. 226 pp. Price 1 Rbl. 40 Kop. (1986).
Contents: Part I: Description of the catalogue.
Part II: Catalogue of relative and absolute elements of 246 eclipsing binaries.
Part III: Remarks on the catalogue and bibliographical references.

002.110   The solar system: an introductory bibliography.
A. Fraknoi.
Mercury, Vol. 16, No. 6, p. 186 – 190 (1987).

002.111   Catalog of Infrared Observations. Part I: Data. Part II: Appendixes.
D. Y. Gezari, M. Schmitz, J. M. Mead.
NASA Ref. Publ., NASA RP–1196, 22 + 603 + 332 pp. (1987).
The Catalog of Infrared Observations (CIO) is a compilation of infrared astronomical data obtained from an extensive literature search of astronomical journals and major astronomical catalogs and surveys. The literature searches are complete for years 1965 through 1986. The Catalog is published in two parts, with the observational data (roughly 200,000 observations of 20,000 individual sources) listed in Part I, and supporting appendices in Part II. The expanded Second Edition contains a new feature: complete IRAS 4–band data for all CIO sources detected, listed with the main Catalog observations, as well as in complete detail in the Appendix. The appendices include an atlas of infrared source positions, two bibliographies of infrared literature upon which the search was based, and an atlas of infrared spectral ranges, and IRAS data for the CIO sources. The complete CIO database is available in printed microfiche and magnetic tape formats.

002.112   Celestial images: astronomical charts from 1500 to 1900.
Boston University Art Gallery, Boston, MA, USA. 64 pp. Price US$ 10.00 (1985).
Review in J. Hist. Astron., Vol. 18, Part 4, p. 287 – 288, 1987 (*W. B. Ashworth Jr.*).

002.113   Catalogus codicum manuscriptorum medii aevi latinorum qui in Bibliotheca Jagellonica Cracovie asservantur, Vol. 3.
M. Kowalczyk, A. Kozlowska, M. Markowski, S. Włodek, G. Zathey, M. Zwiercan.
Polish Academy of Sciences and the Jagellonian Library, Wrocław, Poland. 18 + 508 pp. (1984).
Review in J. Hist. Astron., Vol. 18, Part 3, p. 229, 1987 (*O. Gingerich*).

002.114   1726 – 1799 catalogo della corrispondenza degli astronomi di Brera, Vol. 1.
A. Mandrino, G. Tagliaferri, P. Tucci.
Università degli Studi di Milano, Milan, Italy. 427 pp. (1986).
Review in J. Hist. Astron., Vol. 18, Part 3, p. 229 – 230, 1987 (*O. Gingerich*).

002.115   Manuscripts of the Dibner Collection in the Dibner Library of the History of Science and Technology of the Smithsonian Institution Libraries.
Smithsonian Institution Libraries, Research Guide No. 5. Smithsonian Institution Libraries, Washington, DC, USA. 9 + 145 pp. Price US$ 35.00 (1985).
Review in J. Hist. Astron., Vol. 18, Part 3, p. 229, 1987 (*O. Gingerich*).

002.116   Handbook for Astronomical Societies 1987.
B. Jones (Editor).
Federation of Astronomical Societies, Attention Mr Ken Marcus, 5 Cedars Gardens, Brighton, East Sussex BN1 6YD, England. 92 pp. Price £ 2.50 (1987).
From Sky Telesc., Vol. 74, No. 6, p. 612 (1987).

002.117   Norton's star atlas and reference handbook.
17th edition. Longman Group Ltd., Harlow, UK. 8 + 153 pp. Price £ 12.95 (1986). ISBN 0–582–98898–5.
Review in J. Br. Astron. Assoc., Vol. 97, No. 5, p. 296, 1987 (*N. Bone*).

002.118   Zur Altersstruktur der Autoren bedeutender Forschungsergebnisse in der Astronomie des 20. Jahrhunderts (1900 – 1975).
D. B. Herrmann.
NTM–Schriftenr. Gesch. Naturwiss., Technik, Med. (*Leipzig*), Vol. 24, No. 2, p. 77 – 80 (1987).

Canon der verduisteringen.
See Abstr. 004.026.

A tape version of Ptolemy's star catalogue.
See Abstr. 004.115.

ROSAT, data centers, and X–ray data analysis, a new era?
See Abstr. 013.051.

Applications of AI technology in astronomical research II: What can be done, what should be done?
See Abstr. 013.055.

The Soviet Center of Astronomical Data.
See Abstr. 013.080.

Sampling criteria in multicollection searching.
See Abstr. 021.018.

The Greenwich and Oxford astrographic telescopes 1958 – 1987.
See Abstr. 032.027.

Identification of a constellation from a position.
See Abstr. 036.044.

Compare – a catalog access/comparison algorithm.
See Abstr. 036.150.

Set of programmes for fulfilling requests automatically by the Center of Astronomical Data.
See Abstr. 036.228.

Plotting of stellar fields.
See Abstr. 036.229.

A system for morphological analysis and classification of astronomical objects.
See Abstr. 036.238.

Rigorous compilation of the Northern International Reference Stars.
See Abstr. 041.004.

Precise proper motions and positions of stars from a combination of fundamental catalogues with the HIPPARCOS catalogue.
See Abstr. 041.009.

Análisis del campo de distorsión del Catálogo Astrográfico.
See Abstr. 041.014.

Provisional values of internal group corrections for the VL3 astrolabe catalogue.
See Abstr. 041.015.

Representation of systematic differences astrolabe–catalog.
See Abstr. 041.016.

Systematic errors in visual and photographic latitude star catalogues.
See Abstr. 041.023.

Construction of fundamental and reference systems.
See Abstr. 043.003.

Fundamental catalogues.
See Abstr. 043.004.

Construction of the system of positions and proper motions in the FK5.
See Abstr. 043.005.

Hipparcos. The Input Catalogue.
See Abstr. 051.043.

A catalogue of sunspot observations from 165 BC to AD 1684.
See Abstr. 072.001.

A catalogue of non–telescopic sunspot observations from 165 BC to AD 1684.
See Abstr. 072.063.

Catalogue of LDE (*Long Duration Event*) flares (January 1969 – March 1986).
See Abstr. 073.071.

Catalogue of orbital elements of 3516 minor planets numbered by November 16, 1987, on magnetic tape.
See Abstr. 098.085.

UBV photometry of stars whose positions are accurately known. V.
See Abstr. 113.025.

Ultraviolet observations of cataclysmic variables: the IUE archive.
See Abstr. 117.142.

Candidates for spectroscopic binaries found in the Mount Wilson Halo–Mapping Program.
See Abstr. 120.009.

A database of RR Lyr star radial velocities.
See Abstr. 122.060.

Study of Fernie's photometric catalogue of supergiants.
See Abstr. 131.276.

An atlas of IUE spectra of planetary nebulae and related objects.
See Abstr. 134.028.

A catalogue of concentric aperture UBVRI photoelectric photometry of globular clusters.
See Abstr. 154.049.

Catalogue of the coordinates of 3600 stars in the globular cluster M15 for the period 1896 – 1910.
See Abstr. 154.086.

A southern atlas of galactic hydrogen (The region $0° \leqslant l \leqslant 12°$, $-3° \geqslant b \geqslant -17°$).
See Abstr. 155.021.

A catalog of LMC star clusters outside the Hodge–Wright atlas.
See Abstr. 156.030.

Standard photometric diameters of galaxies. III. Reduction of the diameters in the ESO–B and SGC catalogues to the standard diameter system at the 25 mag arcsec^{-2} brightness level.
See Abstr. 157.119.

Quasar and stellar objects in the Byurakan Surveys.
See Abstr. 159.040.

List of clusters of galaxies with published redshifts.
See Abstr. 160.111.

Catalogue of counterparts of Abell and Zwicky clusters of galaxies on magnetic tape.
See Abstr. 160.112.

003 Books

003.001   Reports of planetary geology and geophysics program – 1986.
With a foreword by J. Boyce.
NASA Tech. Memo., NASA TM–89810. 20 + 558 pp. Price US$ 42.90 (North America), US$ 85.90 (Foreign) (1987).
The individual contributions within the subject scope of Astronomy and Astrophysics Abstracts are included in their corresponding subject categories – see abstracts 002.001 – 002.003, 021.001, 021.002, 022.001 – 022.030, 035.001, 035.002, 036.007 – 036.009, 063.002, 081.001 – 081.003, 084.001, 084.002, 091.001 – 091.016, 092.001, 092.002, 093.001 – 093.008, 094.001 – 094.017, 097.001 – 097.057, 098.001 – 098.006, 099.001 – 099.014, 100.001 – 100.007, 101.001 – 101.015, 102.001 – 102.003, 103.401, 105.001 – 105.004, 106.001, 107.001 – 107.010, 131.018.

003.002   Astrophysics and Space Physics Reviews. Volume 5 (1986).
R. A. Syunyaev (Editor).
Soviet Scientific Reviews, Section E. Astrophys. Space Phys. Rev., Vol. 5, 12 + 263 pp. Price US$ 170.00 (1987). ISBN 3–7186–0214–8.
This volume contains papers originally published in Russian in ”Itogi Nauki i Tekhniki, Seriya Astronomiya, Tom 31” (See Abstr. 42.003.054).
The individual contributions are included in their corresponding subject categories – see abstracts 033.013, 035.021, 065.029, 114.024, 161.061.

003.003   Annual Review of Astronomy and Astrophysics, Volume 25, 1987.
G. Burbidge, D. Layzer, J. G. Phillips (Editors).
Annu. Rev. Astron. Astrophys., Vol. 25, 10 + 684 pp. Price US$ 44.00 (USA), US$ 47.00 (elsewhere) (1987). ISBN 0–8243–0925–1.
The individual contributions are included in their corresponding subject categories – see abstracts 004.024, 065.033, 075.009, 080.017, 102.019, 114.029, 116.021, 122.033, 131.066, 131.067, 133.007, 151.034, 151.035, 155.023 – 155.025, 161.074.

003.004   Itogi Nauki i Tekhniki. Seriya Astronomiya. Tom 32. Astrophysics and space physics.
R. A. Syunyaev (Editor).
Vzesoyuznyj Institut Nauchnoj i Tekhnicheskoj Informatsii, Moskva. 332 pp. Price 3 Rbl. 40 Kop. (1987). In Russian. ISSN 0202–0742.
The individual contributions are included in their corresponding subject categories – see abstracts 062.032, 080.019, 093.017, 125.031, 143.017, 161.080, 161.081.

003.005   Zvezdnye atmosfery (*Stellar atmospheres*).
Odesskij Universitet, Odessa. 180 pp. (1987). In Russian.
Review in Ref. Zh., 51. Astron., 9.51.45 (1987).
See abstracts 064.021 – 064.027, 114.033 – 114.036, 114.038, 117.103 – 117.106, 119.017, 122.037 – 122.039, 155.031.

003.006   1988 Yearbook of Astronomy.
P. Moore (Editor).
Sidgwick & Jackson Ltd., London, UK. 200 pp. Price £ 9.95 cloth, £ 6.95 paper (1987). ISBN 0–283–99474–6 cloth, ISBN 0–283–99475–4 paper.
Review in Astron. Now, Vol. 1, No. 5, p. 37, 1987 (*H. Hatfield*).
Contents: Part 1: Monthly charts and astronomical phenomena.
Part 2: Article section; for the individual contributions see abstracts 032.013, 079.002, 097.077, 125.222, 161.101, 161.102.
Part 3: Miscellaneous.

003.007   The solar wind and the Earth.
S.–I. Akasofu, Y. Kamide (Editors).
Geophysics and Astrophysics Monographs, Vol. 30. Terra Scientific Publishing Co., Tokyo and D. Reidel Publishing Co., Dordrecht, The Netherlands, 8 + 318 pp. Price Dfl. 220.00, US$ 99.00, £ 74.00 (1987). ISBN 90–277–2471–7 (cloth), ISBN 90–277–2472–5 (paper).
The individual contributions are included in their corresponding subject categories – see abstracts 051.034, 053.003, 072.043, 074.037, 082.028, 082.029, 083.007, 084.017 – 084.022, 144.021.

003.008   Itogi Nauki i Tekhniki. Seriya Astronomiya. Tom 30. New methods for construction of coordinate systems.
V. V. Podobed (Editor).
Vsesoyuznyj Institut Nauchnoj i Tekhnicheskoj Informatsii, Moskva, 164 pp. Price 1 Rbl. 65 Kop. (1987). In Russian. ISSN 0202–0742.
The individual contributions are included in their corresponding subject categories – see abstracts 032.014, 043.003.

003.009   Itogi Nauki i Tekhniki. Seriya Issledovanie Zemli iz Kosmosa. Tom 1. Physical foundations, methods and means of investigations of the earth from space.
Ya. L. Ziman (Editor).
Vsesoyuznyj Institut Nauchnoj i Tekhnicheskoj Informatsii, Moskva, 196 pp. Price 1 Rbl. 50 Kop. (1987). In Russian. ISSN 0234–9671.
Contents: Physical foundations of remote sensing (*B. S. Zhukov*), p. 6 – 78. Methodical problems of remote sensing (*V. V. Egorov, B. S. Zhukov*), p. 79 – 130. Means for obtaining aerocosmic information on the earth (*V. V. Egorov, V. A. Kottsov*), p. 131 – 179. Problems, programs and space systems for investigation of the earth (*V. V. Egorov, Z. K. Fedotova*), p. 180 – 194.

003.010   Spectroscopy of astrophysical plasmas.
A. Dalgarno, D. Layzer (Editors).
Cambridge University Press, Cambridge, UK. 14 + 357 pp. Price US$ 59.50 cloth, US$ 19.95 paper (1987). ISBN 0–521–26315–8 cloth, ISBN 0–521–26927–X paper.
The individual contributions are included in their corresponding subject categories – see abstracts 022.082, 061.035, 062.060, 073.051, 074.044, 112.114, 131.123, 131.124, 132.014, 132.015, 155.076, 159.039.

003.011   Geomagnetism. Vol. 2.
J. A. Jacobs (Editor).
Academic Press, London, UK. 20 + 579 pp. Price £ 62.00, US$ 112.00 (1987). ISBN 0–12–378672–X.
The individual contributions are included in their corresponding subject categories – see abstracts 081.014 – 081.016, 091.029, 094.030, 105.031.

003.012   Three hundred years of gravitation.
S. W. Hawking, W. Israel (Editors).
Cambridge University Press, Cambridge, UK. 14 + 684 pp. Price £ 45.00 (1987). ISBN 0–521–34312–7.
Reviews in Nature, Vol. 329, No. 6142, p. 771 – 772, 1987 (*J. D. Barrow*); Phys. Bl., 44 Jahrg., Heft 2, p. 51, 1988 (*J. Ehlers*).
The individual contributions are included in their corresponding subject categories – see abstracts 004.040 – 004.042, 061.036 – 061.038, 066.061 – 066.065, 067.079, 161.132 – 161.135.

003.013 **Halley's comet. Based on the Symposium on the exploration of Halley's comet, held in Heidelberg, F.R. Germany, 27 – 31 October 1986.**
With an editorial by F. Praderie and M. Grewing.
Astron. Astrophys., Vol. 187, No. 1/2, 20+936 pp. (1987).
For the proceedings see 43.012.015.
The individual contributions are included in their corresponding subject categories – see abstracts 004.056, 022.105, 022.106, 035.067 – 035.069, 051.042, 074.056, 102.046 – 102.051, 103.029 – 103.036, 103.102, 103.127, 103.128, 103.465 – 103.595, 104.070 – 104.076.

003.014 **Annual Review of Fluid Mechanics, Volume 19, 1987.**
J. L. Lumley, M. Van Dyke, H. L. Reed (Editors).
Annu. Rev. Fluid Mech., Vol. 19, 626 pp. (1987). ISBN 0–8243–0719–4.
Review in Phys. Abstr., Vol. 90, No. 1311, Entry 95112 (1987). See abstract 062.089.

003.015 **Problemy fiziki kosmicheskikh luchej. (*Problems of cosmic ray physics*).**
Compiled by E. V. Gorchakov.
Nauka, Moskva. 288 pp. (1987).
Review in Ref. Zh., 62. Issled. Kosm. Prostranstva, 11.62.95 (1987).
See abstracts 061.103, 078.013, 084.075, 144.051, 144.052.

003.016 **Numerical radiative transfer.**
W. Kalkofen (Editor).
Cambridge University Press, Cambridge, U.K., 8+373 pp. Price £ 27.50, US$ 49.50 (1987). ISBN 0–521–34100–0.
The individual contributions are included in their corresponding subject categories – see abstracts 063.084 – 063.097.

003.017 **Itogi Nauki i Tekhniki. Seriya Astronomiya. Tom 29. Clusters of galaxies.**
V. G. Gorbatskij, A. G. Kritsuk; edited by
I. S. Shcherbina–Samojlova, D. I. Nagirner.
Vsesoyuznyj Institut Nauchnoj i Tekhnicheskoj Informatsii, Moskva, 112 pp. Price 1 Rbl. 20 Kop. (1987). In Russian. ISSN 0202–0742.
The individual contributions are included in their corresponding subject categories – see abstracts 160.113, 160.114.

003.018 **Formes et couleurs dans l'univers. Nébuleuses – amas d'étoiles – galaxies.**
A. Acker, avec la collaboration de E. Alt, E. Brodkorb, K. Rihm, J. Rusche, R. Mosser, J.–M. Roques, A. Rihn.
Masson S.A., Paris, France. 215 pp. Price FF 220.00 (1987). ISBN 2–225–80798–1.
Review in J. Br. Interplanet. Soc., Vol. 40, No. 12, p. 565 (1987).
Contents: 1. Techniques de la photographie astronomique (*E. Alt, E. Brodkorb, K. Rihm, J. Rusche*): Les instruments. Les techniques photographiques. L'astrophotographie en couleur. 2. Des formes et des couleurs (*A. Acker*): Les formes. Les couleurs.
3. Les nébuleuses, nuages de gaz entre les étoiles (*A. Acker*): Un cadre grandiose. Un milieu interstellaire complexe et agité.
4. Quelques astres choisis (*A. Acker*): Les nébuleuses diffuses. Les restes d'étoiles. Les amas d'étoiles. Les galaxies.

003.019 **Problema poiska zhizni vo vselennoj (*Searching for life in the universe*).**
V. A. Ambartsumyan, N. S. Kardashev, V. S. Troitskij (Editors).
Nauka, Moskva, USSR. 256 pp. Price 1 Rbl. 70 Kop. (1986).
Review in Priroda, No. 12, p. 117 (1987).

003.020 **Quasars, redshifts and controversies.**
H. Arp.
Interstellar Media, 2153 Russell St., Berkeley, CA 94705, USA. 198 pp. Price US$ 19.95 (1987). ISBN 0–941325–00–8.
Review in Sky Telesc., Vol. 75, No. 1, p. 38 – 43, 1988 (*G. Burbidge*).

003.021 **Der große JRO Atlas der Astronomie.**
J. Audouze, G. Israel (Editors).
JRO Kartographische Verlagsanstalt, München, F.R. Germany. 432 pp. Price DM 198.00 (1987). ISBN 3–87504–977–2.
Review in Sterne Weltraum, 26. Jahrg., Nr. 12, p. 725, 1987 (*T. Neckel*).
This book is a German language version of "*Le Grand Atlas de l'Astronomie*" and "*The Cambridge Atlas of Astronomy*" published in 1985 (39.003.018).

003.022 **Maya city planning and the calendar.**
A. Aveni, H. Hartung.
Transactions of the American Philosophical Society, Vol. 76, Part 7. American Philosophical Society, 104 S. Fifth St., Philadelphia, PA 19106, USA. 87 pp. Price US$ 15.00 (1986). ISBN 0–87169–767–X.
From Sky Telesc., Vol. 74, No. 6, p. 613 (1987).

003.023 **Das Lob der Sternkunst. Astronomie in der deutschen Aufklärung.**
R. Baasner.
Abhandlungen der Akademie der Wissenschaften in Göttingen, Mathematisch–Physikalische Klasse, Dritte Folge, Nr. 40. Vandenhoeck & Ruprecht, Göttingen, F.R. Germany. 240 pp. + 16 plates. Price DM 84.00 (1987). ISBN 3–525–82117–4.
Contents: 1. Einleitung. 2. Die Astronomie im Rahmen der Aufklärungs–Physik. 3. Das Lob der Sternkunst. 4. Ein Blick auf die Sternwarten. 5. Allgemeine Darstellungen der Sternkunde. 6. Schleppende Rezeption: Das kopernikanische Weltbild. 7. Himmelsphysik: Die Debatte um die causa gravitatis. 8. Theorie der Himmelskörper. 9. Die Erde als Gegenstand der Astronomie. 10. Die Sonne. 11. Der Mond. 12. Die Planeten. 13. Die Kometen. 14. Die Fixsterne. 15. Die Entstehung der Welt. 16. Beiträge der Astrotheologie. 17. Der Kampf gegen die Astrologen.

003.024 **Meteory i ikh nablyudenie (*Meteors and their observations*).**
P. B. Babadzhanov.
Seriya "Biblioteka lyubitelya astronomii". Nauka, Moskva, USSR. Price 35 Kop. (1987).
Review in Priroda, No. 11, p. 123 (1987).

003.025 **The divided circle. A history of instruments for astronomy, navigation and surveying.**
J. A. Bennett.
Phaidon–Christie's Ltd., Oxford, UK. 224 pp. Price £ 45.00, US$ 75.00 (1987). ISBN 0–7148–8038–8.
Review in Nature, Vol. 332, No. 6159, p. 28, 1988 (*C. Stott*).
Contents: 1. Foundations in astronomy. 2. The beginnings of oceanic navigation. 3. The impact of geometry on surveying. 4. The new science of the seventeenth century. 5. The early specialist trade. 6. The heroic age. 7. The growth of observatories. 8. The longitude found. 9. Geodetic and commonplace surveying. 10. The industrial age. 11. Astronomical circles. 12. Reform of navigational practice. 13. New standards in surveying practice. 14. A practical postscript.

003.026 **Isaac Newton: De la gravitation ou les fondements de la mécanique classique.**
M.–F. Biarnais.
Editions Les Belles Lettres, Paris, France. 191 pp. Price FF 180.00 (1985).
Review in J. Hist. Astron., Vol. 18, Part 4, p. 291 – 292, 1987 (*C. Wilson*).

003.027 **Issac Newton: Principia mathematica.**
M.–F. Biarnais.
Collection "Epistémé". Editions Christian Bourgois, Paris, France. 376 pp.
Review in Astronomie, Vol. 101, p. 449, 1987 (*S. Débarbat*).

003.028 **Galactic dynamics.**
J. Binney, S. Tremaine.
Princeton Series in Astrophysics. Princeton University Press, Princeton, NJ, USA. 16+733 pp. Price US$ 75.00 cloth, US$ 25.00 paper (1987). ISBN 0–691–08444–0 cloth, ISBN 0–691–08445–9 paper.
Contents: 1. Introduction: An overview of the observations.
2. Potential theory: Spherical systems. Potential–density pairs for flattened systems. Ellipsoidal systems. Multipole expansion. Potential energy tensors. Potentials of disks. The potential of our Galaxy. Numerical methods.
3. The orbits of stars: Orbits in static spherical potentials. Orbits in axisymmetric potentials. Orbits in planar non–axisymmetric potentials. Orbits in three–dimensional triaxial potentials. The phase–space structure of orbits. Slowly varying potentials.
4. Equilibria of collisionless systems: The collisionless Boltzmann equation. The Jeans equations. The virial equations. The Jeans theorems and spherical systems. Axisymmetric systems. Triaxial systems. The choice of equilibrium.
5. Stability of collisionless systems: The Jeans instability. The stability of spherical systems. The stability of uniformly rotating systems. Summary.
6. Disk dynamics and spiral structure: Introduction. Wave mechanics of differentially rotating disks. Global stability of differentially rotating disks. Theories of spiral structure. Bars. Warps.
7. Collisions and encounters of stellar systems: Dynamical friction. High–speed encounters. Tidal radii. Mergers. Encounters in stellar disks.
8. Kinetic theory: Exact results. The gravothermal catastrophe. The Fokker–Planck approximation. The evolution of spherical stellar systems. Summary.
9. Stellar evolution in galaxies: Luminosity and color evolution of galaxies. Chemical evolution of disk galaxies. Early evolution of spheroidal components.
10. Dark matter: Dark matter in individual galaxies. Dark matter in systems of galaxies. Dark matter in cosmology. The composition of the dark matter. Summary.

003.029 **Vénus dévoilée. Voyage autour d'une planète.**
J. Blamont.
Editions Odile Jacob, 15 rue Soufflot, 75005 Paris, France. 370 pp. Price FF 130.00 (1987). ISBN 2–02–009643–9.
Reviews in Ciel Terre, Vol. 103, No. 4, p. 123, 1987 (*C. Muller*); Orion, 45. Jahrg., Nr. 223, p. 233, 1987 (*N. Cramer*).

003.030 **Götter in Planeten und Monden.**
J. Blunck.
Verlag Harri Deutsch, Frankfurt/Main, F.R. Germany. 215 pp. Price DM 44.00 (1987). ISBN 3–8171–1003–0.
After an historical account of the nomenclature of the bodies of the solar system this book presents an illustrated description of the planets and their satellites and recounts the mythological legends associated with their names.

003.031 **Sternbilder – *Sternsagen*. Mythen und Legenden um Sternbilder.**
A. Bonov.
Weltbild–Bücherdienst, Augsburg, F.R. Germany; under licence from Urania–Verlag, Leipzig, GDR. 288 pp. Price DM 16.80 (1987). ISBN 3–926187–03–04.
This book is a German translation (translator L. Korniljew) of the Bulgarian original "*Mitove i legendi za s'zvezdiyata*"; Sofia 1978.
It describes in detail how stars and star constellations were named and presents accounts of the mythological legends associated with these names.

003.032 **Mit den Wellen des Lichts. Ursprünge und Entwicklung der Optik im süddeutschen Raum.**
A. Brachner.
Günter Olzog Verlag, München, F.R. Germany. 176 pp. Price DM 48.00 (1987).
Review in Sternenbote, 30. Jahrg., Nr. 10, p. 209 – 210 (1987).

003.033 **Meteory, meteority, meteoroidy (*Meteors, meteorites, meteoroids*).**
V. A. Bronshtehn, edited by P. B. Babadzhanov.
Nauka, Moskva, USSR. 176 pp. Price 65 Kop. (1987).
Reviews in Priroda, No. 7, p. 123 (1987); Ref. Zh., 51. Astron., 7.51.40 (1987).

003.034 **History of the Earth's atmosphere.**
M. I. Budyko, A. B. Ronov, A. L. Yanshin.
Springer–Verlag, Berlin, F.R. Germany. 8+139 pp. Price DM 88.00 (1987). ISBN 3–540–17235–1 (F.R. Germany), ISBN 0–387–17235–1 (USA).
This book is an English translation of the Russian original "*Istoriya atmosferi*", published in 1985 by Gidrometeoizdat, Moscow.
Contents: 1. Introduction: The modern atmosphere. Cycles of atmospheric gases. Studies of the evolution of the atmosphere.
2. Methods for determining changes in the composition of the atmosphere: Sedimentary layer of the Earth's crust. Carbon in the sedimentary layer. The dependence of amounts of CO_2 and O_2 in the atmosphere on carbon mass in sediments. 3. The evolution of the chemical composition of the atmosphere: Carbon dioxide. Oxygen. Past and future of the atmosphere. Conclusion.

003.035 **Gauss. Eine biographische Studie.**
W. K. Bühler.
Springer–Verlag, Berlin, F.R. Germany. 8+191 pp. Price DM 56.00 (1987). ISBN 3–540–16883–4 (F.R. Germany), ISBN 0–387–16883–4 (USA).
This book is a German version of the biography first published in English in 1982 (29.003.030).

003.036 **X–ray astronomy. Selected reprints.**
C. R. Canizares (Editor).
American Association of Physics Teachers, 5110 Roanoke Place, Suite 101, College Park, MD 20740, USA. 6+151 pp. Price US$ 10.00 (1986). ISBN 0–917853–21–0.
This book is a collection of 19 research papers in X–ray astronomy. These papers are organized in the chapters: 1. General references. 2. Stellar coronae. 3. Galactic X–ray sources. 4. Extragalactic X–ray sources.

003.037 **Theory of planetary atmospheres. An introduction to their physics and chemistry.**
J. W. Chamberlain, in collaboration with D. M. Hunten.
International Geophysics Series, Vol. 36. Academic Press Inc., Orlando, FL, USA. 14+481 pp. Price £ 41.50 (1987). ISBN 0–12–167251–4.
Second revised and extended edition of the book first published in 1978 (22.003.032).
Contents: 1. Vertical structure of an atmosphere. 2. Hydrodynamics of atmospheres. 3. Chemistry and dynamics of Earth's stratosphere. 4. Planetary astronomy. 5. Ionospheres. 6. Airglows, auroras, and aeronomy. 7. Stability of planetary atmospheres.

003.038 **Stars and planets.**
M. Chown.
Macdonald Educational, London, UK. 31 pp. Price £ 4.50 (1987). ISBN 0–356–11877–0.
Review in Observatory, Vol. 107, No. 1081, p. 283 – 284, 1987 (*T. King*).

003.039 **The particle explosion.**
F. Close, M. Marten, C. Sutton.
Oxford University Press, Oxford, UK. 239 pp. Price £ 15.00, US$ 35.00 (1987). ISBN 0–19–851965–6.
Review in Sky Telesc., Vol. 74, No. 6, p. 609 – 610, 1987 (*P. Davies*).

003.040   **The birth of a new physics.**
I. B. Cohen.
Penguin Books Ltd., Harmondsworth, UK and W. W. Norton &
Co. Inc., New York, NY, USA. 14+258 pp. Price £ 4.95,
US$ 5.95 (1987). ISBN 0–14–022694–X (UK).
Review in Sci. Am., Vol. 257, No. 4, p. 170 – 172, 1987
(*P. Morrison*).
Second, revised and updated edition of a monograph first
published in 1961.
    Contents: 1. The physics of a moving Earth. 2. The old physics.
3. The Earth and the universe. 4. Exploring the depths of the
universe. 5. Towards an inertial physics. 6. Kepler's celestial
music. 7. The grand design – a new physics.

003.041   **Chto takoe pul'sary?**
        G. Dautcourt, translated by V. V. Kalenichenko, edit-
ed by Yu. I. Izotov.
Naukova Dumka, Kiev, USSR. 168 pp. Price 35 Kop. (1986).
Review in Priroda, No. 8, p. 123 (1987).
This book is a Russian translation of the German original *"Was
sind Pulsare?"* published in 1981 (32.003.033).

003.042   **Superforce. Recherches pour une théorie unifiée de
        l'univers.**
P. Davies.
Editions Payot, Paris, France. 322 pp. Price FF 140.00 (1987).
Review in Ciel Terre, Vol. 103, No. 4, p. 124, 1987
(*J. Vandermeulen*).
This book is a French translation of the English original
"Superforce. The search for a Grand Unified Theory of nature"
published in 1984 (39.003.079).

003.043   **Guide de l'astronomie d'amateur.**
P. de la Cotardière.
Editions Hachette, Paris, France. 235 pp. Price FF 55.00 (1987).
Reviews in Astronomie, Vol. 101, p. 604, 1987 (*M. Gros*); Ciel
Terre, Vol. 103, No. 4, p. 123, 1987 (*J. Manfroid*).

003.044   **Larousse astronomy.**
P. de la Cotardière (Editor).
Hamlyn Publishing Group Ltd., London, UK. 326 pp. Price
£ 14.94 (1987). ISBN 0–600–50108–6.
Review in Astron. Now, Vol. 1, No. 5, p. 39, 1987 (*C. Ronan*).

003.045   **Exploring the night sky. The Equinox astronomy guide
        for beginners.**
T. Dickinson.
Camden House/Firefly Books Ltd., London, UK. Price US$ 9.95
(1987).
Review in Sci. Am., Vol. 257, No. 6, p. 104, 1987 (*Philip
Morrison, Phyllis Morrison*).

003.046   **Analisi delle onde elettromagnetiche in astrofisica.**
M. E. Dilaghi Pestellini (Editor).
Quaderni della Società Astronomica Italiana. Available from
Osservatorio di Arcetri, Largo Enrico Fermi 5, I–50125 Firenze,
Italy. 328 pp. Price L 10000.00.
Review in G. Astron., Vol. 13, N. 1 – 2, p. 48, 1987
(*G. De Filippo*).

003.047   **Die Beziehungen zwischen Johannes Kepler und dem
        Leipziger Mathematikprofessor Philipp Müller.**
D. Döring.
Sitzungsberichte der Sächsischen Akademie der Wissenschaften
zu Leipzig, Philologisch–historische Klasse, Band 126, Heft 6.
Akademie–Verlag, Berlin, GDR. 54 pp. Price DM 7.00 (1986).
Review in Sterne, 63. Band, Heft 5, p. 311 – 312, 1987
(*V. Bialas*).

003.048   **Building and using an astronomical observatory.**
P. Doherty.
Patrick Stephens Ltd., Wellingborough, UK. 232 pp. Price
£ 6.99, US$ 12.95 (1986). ISBN 0–85059–808–7.
Reviews in J. Br. Astron. Assoc., Vol. 97, No. 5, p. 297, 1987
(*A. Young*); Observatory, Vol. 107, No. 1079, p. 173 (1987); Sky
Telesc., Vol. 74, No. 2, p. 155 (1987).

003.049   **Galilée.**
        S. Drake, translated by J.–P. Scheidecker.
Actes Sud – Hubert Nyssen Editeur, Arles, France. 144 pp. Price
FF 89.00 (1987).
Review in Ciel Terre, Vol. 103, No. 6, p. 163, 1987 (*A. Heck*).
This book is a French translation of the English original
"Galileo" published in 1980 (28.003.040).

003.050   **Guide–SAF de l'astronome amateur.**
M. Dumont.
Editions Société Astronomique de France, Paris, France. 186 pp.
Price FF 102.00 (1986).
Review in Ciel Terre, Vol. 103, No. 4, p. 123, 1987
(*A. Koeckelenbergh*).

003.051   **Astronomie pratique et informatique.**
C. Dumoulin, J.–P. Parisot.
Editions Masson, Paris, France. 12+401 pp. Price FF 145.00
(1987). ISBN 2–225–81142–3.
From J. Br. Astron. Assoc., Vol. 97, No. 6, p. 321 (1987).

003.052   **Astronomie für Einsteiger.**
S. Dunlop.
Kosmos, Gesellschaft für Naturfreunde. Franckh'sche Verlags-
handlung, Stuttgart, F.R. Germany. 191 pp. Price DM 24.00
(1987).
Review in Astron. Tidsskr., Årg. 20, Nr. 3, p. 143 (1987).

003.053   **Vvedenie v prikladnuyu radiatsionnuyu nebesnuyu me-
        khaniku (Introduction to radiation effects in celestial
        mechanics).**
N. D. Dzhumanaliev, M. I. Kiselev.
Akademiya Nauk Kirgizskoj SSR, Institut Fiziki, Izdatel'stvo
Ilim, Frunze. 202 pp. Price 1 Rbl. 90 Kop. (1986).
Review in Ref. Zh., 62. Issled. Kosm. Prostranstva, 8.62.124
(1987).

003.054   **The expanding universe.**
        A. S. Eddington, with a foreword to the reissue by
W. McCrea.
Cambridge Science Classics. Cambridge University Press,
Cambridge, UK. 24+128 pp. Price £ 7.95, US$ 12.95 (1987).
ISBN 0–521–34976–1.
Reprint of a monograph first published in 1933.
    This classic book investigates the experimental determination
of one of the fundamental constants of astrophysics and its
significance for astronomy. The equations of general relativity
include a constant lambda in their solution. If lambda is non–
zero and positive, this represents the phenomenon of *cosmic
repulsion*. In this book Eddington discussed the implications of
this repulsion for models of the universe.
    Contents: 1. The recession of the galaxies. 2. Spherical space.
3. Features of the expanding universe. 4. The universe and the
atom.

003.055   **Gravitation: Raum–Zeit–Struktur und Wechselwirkung.**
        Mit einer Einführung von J. Ehlers und G. Börner.
Spektrum der Wissenschaft: Verständliche Forschung. Spektrum
der Wissenschaft Verlagsgesellschaft mbH & Co., Heidelberg,
F.R. Germany. 191 pp. Price DM 42.00 (1987). ISBN 3–922508–
42–1.
This book is a selection of 15 papers published in the journal
"Spektrum der Wissenschaft". The English versions of these
papers were originally published in "Scientific American".

003.056   **Zadachi i metody obrabotki kosmicheskoj informatsii** (*Problems and methods of data reduction in space science*).
P. E. Ehl'yasberg (Editor).
Nauka, Moskva, USSR. 188 pp. (1987).
Review in Ref. Zh., 62. Issled. Kosm. Prostranstva, 8.62.121 (1987).

003.057   **Optical telescopes.**
H. Everett.
Bradford Astronomical Society, 17 Haveloc Street, Bradford, UK. 84 pp. Price £ 2.80 (1986).
From J. Br. Astron. Assoc., Vol. 97, No. 5, p. 299 (1987).

003.058   **L'evoluzione del sistema solare.**
G. Favero.
Editione Curcio. Price L 35000.00.
Review in G. Astron., Vol. 13, N. 1 – 2, p. 48, 1987 (*G. Romano*).

003.059   **The nature of time.**
R. Flood, M. Lockwood (Editors).
Basil Blackwell Ltd., Oxford, UK. 187 pp. Price US$ 19.95 (1986). ISBN 0–631–14807–8.
From Phys. Today, Vol. 40, No. 10, p. 133 (1987).

003.060   **The Greek cosmologists. Vol. 1: The formation of the atomic theory and its earliest critics.**
D. Furley.
Cambridge University Press, Cambridge, UK. 220 pp. Price £ 25.00, US$ 34.50 (1987). ISBN 0–521–33328–8.
From Phys. Today, Vol. 40, No. 10, p. 133 (1987).

003.061   **Cosmology, physics, and philosophy. Including a new theory of aesthetics.**
B. Gal–Or.
2nd edition. Springer–Verlag, New York, NY, USA. 36 + 522 pp. Price US$ 39.95, DM 89.00 (1987). ISBN 0–387–96526–2 (USA), ISBN 3–540–96526–2 (F.R. Germany).
Second, revised and extended edition of a monograph first published in 1981 (31.003.047).

003.062   **Teoreticheskaya fizika i astrofizika. Dopolnitel'nye glavy** (*Theoretical physics and astrophysics*).
V. L. Ginzburg.
Third edition. Nauka, Moskva, USSR. 488 pp. (1987).
Review in Ref. Zh., 51. Astron., 9.51.43 (1987).
A revised and extended new edition of the monograph published in 1981 (30.003.065).

003.063   **Geheimnisvolle Sonne.**
R. G. Giovanelli, translated by R. Beck.
VCH Verlagsgesellschaft, Weinheim, F.R. Germany. 136 pp. Price DM 68.00 (1987). ISBN 3–527–26501–5.
Review in Sterne Weltraum, 26. Jahrg., Nr. 9, p. 521, 1987 (*M. Beetz*).
This book is a German translation of the English original "*Secrets of the sun*" published in 1984 (38.003.013).

003.064   **Echelles de temps atomique.**
M. Granveaud.
Collection des monographies du Bureau National de Métrologie. Editions Chiron, Paris, France. 300 pp. Price FF 180.00 (1987).
Review in Astronomie, Vol. 101, p. 603, 1987 (*S. Débarbat*).

003.065   **Planetary landscapes.**
R. Greeley.
Allen & Unwin Ltd., London, UK. 275 pp. Price £ 14.95 (1987). ISBN 0–04–551081–4.
Review in Astron. Now, Vol. 1, No. 4, p. 44, 1987 (*G. Fielder*).
Paperback version of the 1985 hardbound edition (41.003.049).

003.066   **The omega point. The search for the missing mass and the ultimate fate of the universe.**
J. Gribbin.
William Heinemann Ltd., London, UK. 10 + 245 pp. Price £ 12.95 (1987). ISBN 0–434–30591–X.
Review in Nature, Vol. 330, No. 6145, p. 294 – 295, 1987 (*R. H. Sanders*).
Contents: 1. The arrow of time. 2. The universe in a nutshell. 3. Time and the universe. 4. Elementary evidence. 5. Dynamic factors. 6. Close to critical. 7. The solar connection. 8. A choice of futures. Appendix: SUSY, GUTs and string.

003.067   **Lines to the mountain gods: Nazca and the mysteries of Peru.**
E. Hadingham.
Random House Inc., New York, NY, USA. 307 pp. Price US$ 22.50 (1987). ISBN 0–394–54235–5.
Reviews in Archaeoastronomy (U.K.), No. 11, p. 570 – 571, 1987 (*W. E. Shawcross*); Sky Telesc., Vol. 74, No. 5, p. 493 (1987).

003.068   **Découvrir le ciel. Le guide de l'astronomie facile.**
J.–L. Halbwachs.
Editions Bueb et Reumaux, Strasbourg, France. 145 pp. Price FF 72.00 (1987).
Review in Ciel Terre, Vol. 103, No. 6, p. 162, 1987 (*J. Manfroid*).

003.069   **Astrologie – Tochter der Astronomie?**
J. Hamel.
"Akzent" Taschenbuch Nr. 85. Urania–Verlag, Leipzig, GDR. 128 pp. DM 6.50 (1987). ISBN 3–332–00128–0.

003.070   **Darkness at night. A riddle of the universe.**
E. Harrison.
Harvard University Press, Cambridge, MA, USA. 10 + 293 pp. Price US$ 25.00, £ 19.95 (1987). ISBN 0–674–19270–2.
Review in Nature, Vol. 330, No. 6145, p. 288, 1987 (*O. Gingerich*).
Contents: 1. Why is the sky dark at night?
Part I. The riddle begins: 2. Three rival systems. 3. Celestial light. 4. The starry message.
Part II. The riddle develops: 5. The Cartesian system. 6. Newton's needles and Halley's shells. 7. A forest of stars. 8. The misty forest. 9. Worlds on worlds. 10. Revelations of chaos.
Part III. The riddle continues: 11. The fractal universe. 12. The visible universe. 13. The golden walls of Edgar Allen Poe. 14. Lord Kelvin sees the light. 15. Ether voids, curved space, and a midnight Sun. 16. The expanding universe. 17. The cosmic redshift. 18. Energy in the universe. Epilogue.

003.071   **Astronomy: the cosmic journey.**
W. K. Hartmann.
Fourth edition. Wadsworth Publishing Co., Belmont, CA, USA. 552 pp. Price £ 30.95 (1987). ISBN 0–534–07938–5.
Review in Astron. Now, Vol. 2, No. 3, p. 52 – 53, 1988 (*T. J. C. A. Moseley*).
Fourth, revised edition of a textbook first published in 1978. The third edition appeared in 1985 (40.003.121).

003.072   **The Galileo connection.**
C. E. Hummel.
Intervarsity Press, Downer's Grove, IL, USA. 293 pp. Price US$ 8.95 (1986).
From J. Hist. Astron., Vol. 18, Part 3, p. 230 (1987).

003.073   **Principles of plasma diagnostics.**
I. H. Hutchinson.
Cambridge University Press, Cambridge, UK. 16 + 364 pp. Price £ 50.00, US$ 65.00 (1987). ISBN 0–521–32622–2.
Contents: 1. Plasma diagnostics. 2. Magnetic diagnostics. 3. Plasma particle flux. 4. Refractive–index measurements. 5. Electromagnetic emission by free electrons. 6. Electromagnetic

radiation from bound electrons. 7. Scattering of electromagnetic radiation. 8. Ion processes.

003.074 Life in the universe.
F. Jackson, P. Moore.
Routledge & Kegan Paul Ltd., London, UK. 10 + 162 pp. Price £ 9.95 (1987). ISBN 0–7102–0948–7.
Review in Astron. Now, Vol. 2, No. 1, p. 46 – 47, 1988 (N. Bone).
Contents: 1. The universe around us. 2. The nature and origin of living organisms. 3. Myths or men? 4. The Moon. 5. Mars. 6. Venus. 7. Other worlds of the solar system. 8. Comets and meteorites. 9. Planets of other stars. 10. Alien life. 11. Overview.

003.075 The M–type stars.
H. R. Johnson, F. R. Querci, with the collaboration of S. Jordan, R. N. Thomas, L. Goldberg, J.–C. Pecker.
Monograph Series on Nonthermal Phenomena in Stellar Atmospheres. Centre National de la Recherche Scientifique, Paris, France and National Aeronautics and Space Administration, Washington, DC, USA. NASA Spec. Publ., NASA SP–492, 82 + 502 pp. (1986).
Contents: Perspective (R. N. Thomas). 1. Basic properties and photometric variability (F. Querci). 2. Spectroscopy and non-thermal processes (M. Querci). 3. Circumstellar radio molecular lines (Nguyen–Quang–Rieu). 4. Circumstellar shells, the formation of grains, and radiation transfer (J. Lefèvre). 5. Mass loss (L. Goldberg). 6. Circumstellar chemistry (A. E. Glassgold, P. J. Huggins). 7. Thermal atmospheric models (H. R. Johnson). 8. Quasi–thermal models (R. de la Reza). 9. The atmospheres of M dwarfs: observations (M. Rodonò). 10. M dwarfs: theoretical work (D. J. Mullan).

003.076 Laboratory experiments for astronomy.
P. E. Johnson, R. Canterna.
Saunders Golden Sunburst Series. Saunders College Publishing, Philadelphia, PA, USA. 16 + 233 pp. Price US$ 17.00 (1987). ISBN 0–03–009677–4.
Contents: 1. Quantitative techniques. 2. Properties of angles. 3. Celestial positions and time. 4. Modeling the Earth–Sun system. 5. Phases of the Moon. 6. Planetary motion. 7. Observing. 8. Comparative planetology. 9. Measurement and experimental error. 10. The inverse–square law. 11. Blackbody radiation and spectra. 12. The hydrogen atom. 13. The Hertzsprung–Russell diagram. 14. Distances to the stars. 15. The Andromeda galaxy. 16. Cosmology. 17. Galaxies with active nuclei.

003.077 Beyond Einstein. The cosmic quest for the theory of the universe.
M. Kaku, J. Trainer.
Bantam Books Inc., New York, NY, USA. 225 pp. Price US$ 9.95 (1987). ISBN 0–553–34349–1.
From Phys. Today, Vol. 40, No. 8, p. 77 (1987).

003.078 Fundamental astronomy.
H. Karttunen, P. Kröger, H. Oja, M. Poutanen, K. J. Donner (Editors).
Springer–Verlag, Berlin, F.R. Germany. 14 + 478 pp. Price DM 128.00, US$ 45.00 (1987). ISBN 3–540–17264–5 (F.R. Germany), ISBN 0–387–17264–5 (USA).
Review in Nature, Vol. 332, No. 6160, p. 187, 1988 (R. Connon Smith).
This book is an English translation of the Finnish original "Tähtitieteen perusteet" published in 1984. It is intended as a comprehensive introductory text for first– and second–year university students and for serious amateurs in the field.
Contents: 1. Introduction. 2. Spherical astronomy. 3. Observations and instruments. 4. Photometric concepts and magnitudes. 5. Radiation mechanisms. 6. Temperatures. 7. Celestial mechanics. 8. The solar system. 9. Stellar spectra. 10. Binary stars and stellar masses. 11. Stellar structure. 12. Stellar evolution. 13. The Sun. 14. Variable stars. 15. Compact stars. 16. The interstellar medium. 17. Star clusters and associations. 18. The Milky Way. 19. Galaxies. 20. Cosmology.

003.079 Meteoritnye kratery na Zemle (Meteorite craters on the Earth).
L. P. Khryanina.
Nedra, Moskva, USSR. 112 pp. (1987).
Review in Ref. Zh., 51. Astron., 8.51.10 (1987).

003.080 Islamic mathematical astronomy.
D. A. King.
Variorum Reprints, London, UK. 350 pp. Price £ 34.00 (1986).
From J. Hist. Astron., Vol. 18, Part 3, p. 230 (1987).
Collection of research papers originally published in various journals during the period 1973 to 1983.

003.081 Satellite orbits in an atmosphere. Theory and applications.
D. King–Hele.
Blackie and Son Ltd., Glasgow, UK. 11 + 291 pp. Price £ 49.00 (1987). ISBN 0–216–92252–6.
Reviews in J. Br. Interplanet. Soc., Vol. 40, No. 12, p. 565 (1987); Mon. Notes Astron. Soc. S. Afr., Vol. 46, Nos. 11 – 12, p. 174 – 175, 1987 (G. Roberts).

003.082 Begegnung mit Halley.
G. Klaus.
Published by G. Klaus, Waldeggstr. 10, CH–2540 Grenchen, Switzerland. 88 pp. (1987).
Review in Ciel Terre, Vol. 103, No. 4, p. 125, 1987 (A. Heck).

003.083 Exercises in astronomy.
J. Kleczek (Editor).
Revised and extended edition of "Practical work in elementary astronomy" by M. G. J. Minnaert. D. Reidel Publishing Co., Dordrecht, The Netherlands. 24 + 339 pp. Price Dfl. 140.00, US$ 64.00, £ 49.50 cloth; Dfl. 59.00, US$ 19.50, £ 18.00 paper (1987). ISBN 90–277–2409–1 cloth, ISBN 90–277–2423–7 paper.
This book is an updated and considerably extended version of Minnaert's work published in 1969 (01.003.051).
Many new exercises referring to new observational techniques and methods have been incorporated by the editor in collaboration with the contributing authors D. A. Allen, Z. Ceplecha, S. Ferraz Mello, K. J. Gordon, L. Houziaux, C. Jaschek, Z. Kopal, J. Manfroid, J. Palouš, J. Podolský, G. R. Quast, J. Surdej, A. B. Underhill, J. M. Vreux, D. G. Wentzel.
The exercises are organized in the following sections:
A. The planetary system: 1. Space and time, instruments. 2. The motions of celestial bodies. 3. Planets and satellites.
B. The stars: 1. The Sun. 2. Stars and nebulae. 3. Stellar systems.

003.084 Otkrytie vselennoj (Exploring the universe).
I. A. Klimishin.
Nauka, Moskva, USSR. 320 pp. (1987).
Review in Ref. Zh., 51. Astron., 11.51.16 (1987).

003.085 Fizika svecheniya atmosfer planet i komet (Radiation processes in planetary and cometary atmospheres).
V. A. Krasnopol'skij.
Nauka, Moskva, USSR. 304 pp. (1987).
Reviews in Priroda, No. 7, p. 122 – 123 (1987); Ref. Zh., 51. Astron., 7.51.39 (1987).
Contents: 1. Atomic spectra. 2. Molecular spectra. 3. Elementary processes and excitation mechanisms of radiation of atmospheres. 4. Some problems of aeronomy. 5. Observational methods. 6. Radiation of the earth's atmosphere. 7. Mars. 8. Radiation of Venus. 9. Radiation of giant planets. 10. Comets.

003.086 **Dimensies in de natuur.**
W. Kruit, G. Schilling.
Aramith Uitgevers, Amsterdam, The Netherlands. 184 pp. Price
Dfl. 32.50 (1987). ISBN 90–6834–026–3.
Review in Zenit, 14. Jaarg., Nr. 11, p. 378 (1987).

003.087 **Pavel Karlovich Shternberg 1865–1920.**
P. G. Kulikovskij, edited by U. Sagitov.
Second, revised edition. Nauka, Moskva, USSR. 125 pp. Price
45 Kop. (1987).
Reviews in Priroda, No. 11, p. 97 (1987); Ref. Zh., 51. Astron.,
11.51.15 (1987).

003.088 **Exploring the southern sky. A pictorial atlas from the
European Southern Observatory (ESO).**
S. Laustsen, C. Madsen, R. M. West.
Springer–Verlag, Berlin, F.R. Germany. 6+274 pp. + 1 fold–
out plate. Price US$ 39.00, DM 128.00 (1987). ISBN 3–540–
17735–3 (F.R. Germany), ISBN 0–387–17735–3 (USA).
Reviews in Nature, Vol. 330, No. 6149, p. 618, 1987
(*D. W. Hughes*); Sky Telesc., Vol. 74, No. 6, p. 608–609, 1987,
(*C. Raymo*).
Contents: 1. The universe and its galaxies: A look into Fornax.
An ordered sequence of galaxies. The Local Group. The Sculptor
Group. Multiple galaxies. Clusters of galaxies. Peculiar galaxies.
2. The Milky Way galaxy: Panorama of the Milky Way. The
Milky Way from Orion to Puppis. The Milky Way from Vela to
Carina. The Milky Way from Crux to Norma. The Milky Way
from Scorpius to Scutum. Milky Way objects at high galactic
latitude.
3. Minor bodies in the solar system: Meteoroids and minor
planets. Comets.
4. The southern sky and ESO: A European organization for
astronomy. The La Silla Observatory. The headquarters in
Garching. The next generation of telescopes.

003.089 **Entdeckungen am Südhimmel. Ein Bildatlas der Euro-
päischen Südsternwarte (ESO).**
S. Laustsen, C. Madsen, R. M. West, translated by H.–M. Hahn.
Springer–Verlag, Berlin, F.R. Germany and Birkhäuser Verlag,
Basel, Switzerland. 6+274 pp. + 1 fold–out plate. Price
DM 128.00. ISBN 3–7643–1896–1.
Review in Sterne Weltraum, 26. Jahrg., Nr. 12, p. 724, 1987
(*J. Staude*).
This book is a German–language version of *"Exploring the
southern sky"* (44.003.088).

003.090 **A travers la Voie Lactée.**
J. Lequeux.
"Sciences et découvertes". Editions Le Rocher, Paris, France.
126 pp. (1987).
Review in Astronomie, Vol. 101, p. 502, 1987 (*J. Mouette*).

003.091 **Osnovy relyativisticheskoj teorii gravitatsii (*Founda-
tions of a relativistic theory of gravitation*).**
A. A. Logunov, M. A. Mestvirishvili.
Second edition. Izdatel'stvo Moskovskogo Gosudarstvennogo
Universiteta, Moskva, USSR. 308 pp. (1986).
Review in Ref. Zh., 51. Astron., 9.51.44 (1987).
A revised and extended edition of a monograph first published
under the title *"Teoriya gravitatsii"* in 1985 (41.003.129).

003.092 **Die Planeten des Sonnensystems.**
M. J. Marow (*M. Ya. Marov*).
Verlag MIR, Moskau und BSB B. G. Teubner Verlagsgesell-
schaft, Leizig, GDR. 376 pp. Price DM 19.80 (1987). ISBN 3–
322–00316–7.
This book is a German translation of the Russian original
"Planety solnechnoj sistemy" published in 1981 (30.003.096).
Contents: 1. Einige allgemeine Angaben über das Sonnensy-
stem. 2. Die grundlegenden mechanischen Charakteristika der
Planeten und die Besonderheiten ihrer Bewegung. 3. Die Plane-
tenoberflächen. 4. Der innere Aufbau und die thermische Entwicklung. 5. Die Planetenatmosphären.

003.093 **Space. The next twenty–five years.**
T. R. McDonough.
Wiley Interscience Editions. John Wiley & Sons Inc., New York,
NY, USA. 16+237 pp. Price US$ 17.95, £ 14.25 (1987). ISBN 0–
471–85671–1.
Review in Astron. Now, Vol. 1, No. 5, p. 39, 1987 (*K. Gatland*).
Contents: Introduction: Space, the final frontier. 1. From fire
arrows to Neil Armstrong: The history of space exploration.
2. Orbital truck: The space shuttle. 3. A nice place to visit: Earth.
4. Star wars: The strategic defense initiative. 5. Captain Kirk,
here we come! Humans vs. robots in space. 6. A room with a
view: Space stations. 7. The man and woman in the Moon: A
lunar base. 8. Who stole the canals? Mars. 9. Dinosaur killers?
Asteroids and comets. 10. Near neighbors: The inner solar
system. 11. Giant worlds: The outer planets. 12. Glimpsing
infinity: The universe. 13. How to build a starship: Interstellar
travel. 14. In search of E. T.: The search for extraterrestrial
intelligence. 15. Footsteps into the universe: The future.

003.094 **Physics and astronomy.**
D. McGillivray.
Macmillan Publishers Ltd., London, UK. 186 pp. Price £ 6.95
(1987). ISBN 0–333–42861–7.
Review in Astron. Now, Vol. 1, No. 4, p. 44–45, 1987
(*H. Miles*).

003.095 **Astronomy. A self–teaching guide.**
D. L. Moché.
Third edition. John Wiley & Sons Ltd.. Chichester, UK.
11+291 pp. Price £ 10.65 (1987). ISBN 0–471–85297–X.
Reviews in Astron. Tidsskr., Årg. 20, Nr. 4, p. 192 (1987); Ciel,
Vol. 49, p. 362–363, 1987 (*J. Manfroid*).

003.096 **They dance in the sky. Native American star myths.**
J. G. Monroe, R. A. Williamson.
Houghton Mifflin Co., Boston, MA, USA. 130 pp. Price
US$ 12.95 (1987). ISBN 0–395–39970–X.
Review in Sky Telesc., Vol. 74, No. 4, p. 376 (1987).

003.097 **Astronomers' stars.**
P. Moore.
Routledge & Kegan Paul Ltd., London, UK. 10+164 pp. Price
£ 9.95 (1987). ISBN 0–7102–1287–9. Review in Astron. Now,
Vol. 2, No. 3, p. 53, 1988 (*R. Pickard*).
Contents: 1. Introducing the stars. 2. 61 Cygni: the flying star.
3. Mizar: the horse and his rider. 4. Betelgeux: the lives of the
stars. 5. Sirius: the dog–star and the pup. 6. Vega: other solar
systems? 7. Algol: the demon star. 8. Epsilon Aurigæ: the
mysterious 'kid'. 9. Mira: the 'wonderful star'. 10. Delta Cephei:
standard candle in space. 11. CN Tauri: birth of the Crab.
12. Eta Carinæ: erratic star. 13. SS433: the cosmic lawn–
sprinkler. 14. Alcyone: brightest of the sisters. 15. Becklin's star:
the star we can never see. 16. S Andromedae: the new star in the
spiral. 17. 3C 273: the star that is not a star.

003.098 **The astronomy encyclopaedia.**
P. Moore (Editor).
Mitchell Beazley International Ltd., London, UK. 464 pp. Price
£ 14.95 (1987). ISBN 0–85533–604–8.
Review in Astron. Now, Vol. 1, No. 5, p. 38, 1987 (*R. Prout*).

003.099 **Observatório Astronômico – um século de história
(1827–1927).**
H. Morize.
Coleção Documentos de História da Ciência, 1. Museu de
Astronomia e Ciências Afins, Rio de Janeiro, Brazil. 179 pp.
(1987).

003.100 **The amateur astronomer's handbook.**
J. Muirden.
Third edition. Harper & Row Publishers, New York, NY, USA.
8+472 pp. Price US$ 10.95 (1987). ISBN 0–06–091426–2 paper.
Paperback version of the 1983 hardbound edition (34.003.113).

003.101 Das Universum. Faszination der modernen Astronomie.
I. Nicolson, P. Moore, mit einer Einführung von
J. Meadows.
Mosaik Verlag, München, F.R. Germany. 256 pp. Price
DM 68.00 (1987). ISBN 3–570–07135–9.
This book is a German translation of the English original "The
universe" published in 1985 (40.003.155).
Contents: 1. Das Sonnensystem. 2. Die Sonne. 3. Vermessung
des Universums. 4. Die Natur der Sterne. 5. Die Natur der
Galaxien. 6. Ursprung und Evolution.

003.102 Universet.
I. Nicolson, P. Moore.
Politikens Forlag A/S, København, Denmark. 256 pp. Price
D.kr. 197.00 (1986).
Review in Astron. Tidsskr., Årg. 20, Nr. 3, p. 141, 1987
(*T. Johnsen*).
This book is a Danish translation of the English original "The
universe" published in 1985 (40.003.155).

**003.103 The light of nature. Essays in the history and philosophy
of science presented to A. C. Crombie.**
J. D. North, J. J. Roche (Editors).
International Archives of the History of Ideas, Vol. 110. Martin-
us Nijhoff Publishers, Dordrecht, The Netherlands. 8 + 471 pp.
Price US$ 82.50 (1985).
From J. Hist. Astron., Vol. 18, Part 3, p. 231 (1987).

003.104 Dr. Luis Ugueto – ingeniero, astronomo y profesor.
A. E. Olivares.
Academia de Ciencias Físicas, Matemáticas y Naturales, Cara-
cas, Venezuela. 309 pp. (1986). ISBN 980–265–456–6.

003.105 Early astronomy from Babylonia to Copernicus.
W. M. O'Neil.
Sydney University Press, Sydney, Australia. 14 + 214 pp.
US$ 27.50 (1986). ISBN 0–424–00117–9.
Review in Sky Telesc., Vol. 75, No. 3, p. 277 – 278, 1988
(*J. Evans*).

003.106 Geohistory. Global evolution of the Earth.
M. Ozima, translated from the Japanese by
J. Wakabayashi.
Springer-Verlag, Berlin, F.R. Germany. 8 + 165 pp. Price
DM 48.00 (1987). ISBN 3–540–16595–9 (F.R. Germany), ISBN
0–387–16595–9 (USA).
Contents: 1. Geohistory as a discipline. 2. The Earth as a
planet in the solar system: Pre–solar history. Condensation
theory – from nebular gas to crystal particles. Moon, meteorites,
and other planets – the key to an understanding of early
geohistory. 3. Evolution of the Earth: The driving force behind
the Earth's evolution. Composition of the Earth – the meteorite
analogy. The layered structure of the Earth. Formation of the
layered structure. The time of core formation, based on
Pb isotopic ratio data. Mantle differentiation. The age of the
mantle. Origin and evolution of the atmosphere and oceans. The
primordial mantle and the differentiated mantle. 4. Changes in
the Earth's crust: Rock magnetism and paleomagnetism. Ocean
floor spreading, continental drift, and plate tectonics. Exchange
of material between the mantle and the Earth's crust. Geochronology. 5. Man and geohistory: Bolide impacts: mass extinction
of life? The fate of radioactive waste – the Oklo phenomenon.
Epilog.

003.107 Den kosmiska koden.
H. R. Pagels.
ICA Bokförlag, Västerås, Sweden. 331 pp. Price Sv.kr. 227.00
(1986).
Review in Astron. Tidsskr., Årg. 20, Nr. 4, p. 192 (1987).
This book is a Swedish translation of the English original "The
cosmic code" published in 1982 (32.003.093).

003.108 Search for a supertheory. From atoms to superstrings.
B. Parker.
Plenum Press, New York, USA. 10 + 292 pp. Price US$ 21.95
North America, US$ 26.30 elsewhere (1987). ISBN 0–306–
42702–8.
Contents: 1. Introduction. 2. Probing the atom. 3. Particle
accelerators. 4. Organizing the particle zoo. 5. Overcoming infinity. 6. Building a universe. 7. Gauging the universe. 8. Adding
color. 9. Adding charm. 10. Search for the W. 11. Unifying.
12. Looking deeper. 13. Supergravity. 14. Adding more dimensions. 15. Superstrings: tying it all together. 16. Cosmic strings
and inflation. 17. Epilogue.

003.109 Astronomy: from the earth to the universe.
J. M. Pasachoff.
Third edition. Saunders College Publishing, New York, NY,
USA. 647 pp. Price US$ 23.25 (1987). ISBN 0–03–008114–9.
Review in Sky Telesc., Vol. 74, No. 5, p. 487 – 493, 1987
(*H. P. Coyle*).

**003.110 Spinors and space–time. Vol. 1: Two–spinor calculus and
relativistic fields.**
R. Penrose, W. Rindler.
Cambridge University Press, Cambridge, UK. 458 pp. Price
£ 17.50, US$ 29.95 (1987).
Reviews in Astrophys. Space Sci., Vol. 139, No. 1, p. 195, 1987
(*L. Dvořák*); Observatory, Vol. 107, No. 1081, p. 275 – 277, 1987
(*J. Peacock*).
Paperback version of the 1985 hardbound edition (39.003.151).

003.111 Theodor Brorsen, astronom.
H. R. Petersen.
H. C. Lorenzens Forlag, Nordborg, Denmark. 90 pp. Price
D.kr. 68.50 (1986).
Review in Astron. Tidsskr., Årg. 20, Nr. 3, p. 142 – 143, 1987
(*P. Darnell*).

003.112 Les méthodes nouvelles de la mécanique céleste.
H. Poincaré, preface by J. Kovalevsky.
Les Grands Classiques de Gauthier–Villars. Available from
Librairie Scientifique et Technique – Albert Blanchard, 9, rue de
Médicis, F–75006 Paris, France.
Tome I: Solutions périodiques. – Non–existence des intégrales
uniformes. Solutions asymptotiques. 12 + 385 pp.
Tome II: Méthodes de MM. Newcomb, Gyldén, Lindstedt et
Bohlin. 12 + 479 pp.
Tome III: Invariants intégraux. – Solutions périodiques du deu-
xième genre. Solutions doublement asymptotiques. 6 + 414 pp.
Price (3 vols.) FF 544.00 (1987). ISBN 2–85367–093–7.
Facsimile reprint of the classical work first published in the years
1892 (Tome I), 1893 (Tome II), and 1899 (Tome III).

**003.113 Planetarnye tumannosti. Izuchenie pozdnikh stadij
zvezdnoj ehvolyutsii.**
S. R. Pottasch.
MIR, Moskva, USSR. 351 pp. (1987).
Review in Ref. Zh., 51. Astron., 8.51.11 (1987).
This book is a Russian translation of the English original
"Planetary nebulae. A study of late stages of stellar evolution"
published in 1984 (37.003.003).

**003.114 The recursive universe. Cosmic complexity and the limits
of scientific knowledge.**
W. Poundstone.
Oxford University Press, Oxford, UK. 250 pp. Price £ 5.95
(1987).
Review in Observatory, Vol. 107, No. 1081, p. 281 – 282, 1987
(*C. Benn*).

003.115 The evolution of relativity.
C. Ray.
Adam Hilger, IOP Publishing Ltd., Bristol, UK. 12 + 211 pp.
Price £ 29.50 (1987). ISBN 0–85274–423–4.
Contents: Introduction. 1. Ernst Mach and the search for
simplicity. 2. The foundations of the general theory of relativity.

3. Mainstream classical relativity. 4. Classical and quantum relativity. 5. Relativity – dead or alive?

003.116   Galileo: Heretic.
P. Redondi, tranlated by R. Rosenthal.
Princeton University Press, Princeton, NJ, USA. 356 pp. Price US$ 29.95 (1987).
Review in Nature, Vol. 330, No. 6149, p. 617 – 618, 1987 (*R. Naylor*).

003.117   Ehtyudy o vselennoj.
T. Regge, translated by J. B. Pontekorvo, edited by B. M. Pontekorvo.
MIR, Moskva, USSR. 189 pp. (1985).
Review in Priroda, No. 7, p. 121 – 123, 1987 (*I. Yu. Kobzarev*).
This book is a Russian translation of the Italian original *"Cronache dell' universo"* published in 1981 (31.003.107).

003.118   Extraterrestrials. Science and alien intelligence.
E. Regis Jr. (Editor).
Cambridge University Press, Cambridge, UK. 10 + 278 pp. Price £ 7.95, US$ 12.95 (1987). ISBN 0–521–34852–8.
From J. Br. Astron. Assoc., Vol. 97, No. 6, p. 320 (1987).
Paperback version of the 1985 hardbound edition (40.003.008).

003.119   Longman illustrated dictionary of astronomy and astronautics.
I. Ridpath.
Longman Group Ltd., Harlow, UK. 224 pp. Price £ 4.50 (1987). ISBN 0–582–89381–X.
Review in J. Br. Astron. Assoc., Vol. 97, No. 6, p. 320, 1987 (*J. Mitton*).

003.120   The monthly sky guide.
I. Ridpath, W. Tirion.
Cambridge University Press, Cambridge, UK. 63 pp. Price £ 4.95, US$ 7.95 (1987). ISBN 0–521–33921–9.
Review in Astron. Now, Vol. 2, No. 4, p. 50 – 51, 1988 (*N. Bone*).
This book contains maps of the sky for each month of the year, realistically depicting the stars visible to the naked eye, and uniquely including a 'fist' symbol to show the scale. With these charts one can identify the stars and constellations visible on any night of the year in the Northern hemisphere between 30 and 60 degrees North. In addition, specific constellations of interest are treated in detail each month, with a map and descriptions of selected objects, stars, clusters, nebulae and galaxies, for viewing with binoculars or a small telescope. There are notes on meteor showers, and the positions of the planets are given for a five–year period.

003.121   Kosmos Astronomiegeschichte. Astronomen, Instrumente, Entdeckungen.
G. D. Roth.
Kosmos–Gesellschaft der Naturfreunde. Franckh'sche Verlagshandlung, Stuttgart, F.R. Germany. 190 pp. Price DM 48.00 (1987). ISBN 3–440–05800–X.

003.122   Geometriya, dinamika, vselennaya (*Geometry, dynamics and the universe*).
I. L. Rozental'.
Nauka, Moskva, USSR. 145 pp. (1987).
Review in Ref. Zh., 51. Astron., 9.51.46 (1987).

003.123   Principles of statistical radiophysics 1. Elements of random process theory.
S. M. Rytov, Yu. A. Kravtsov, V. I. Tatarskii.
Springer–Verlag, Berlin, F.R. Germany. 10 + 253 pp. Price DM 124.00 (1987). ISBN 3–540–12562–0 (F.R. Germany), ISBN 0–387–12562–0 (USA).
This book is an English translation of the Russian original *"Vvedenie v statisticheskuyu radiofiziku I: Sluchainuie protsessui"*, 2nd edition, published by Nauka, Moscow 1976.
Contents: 1. General introduction. 2. The Bernoulli problem. 3. Random pulses. 4. Random functions. 5. Markov processes. 6. Stochastic differential equations.

003.124   Joseph von Fraunhofer. Forscher – Erfinder – Unternehmer.
H.–P. Sang.
Verlag Dr. Peter Glas oHG, Tegernseer Landstraße 161, D–8000 München 90, F.R. Germany. 163 pp. Price DM 27.80 (1987). ISBN 3–89004–38.
Review in Phys. Bl., 44. Jahrg., Heft 3, p. 90, 1988 (*G. Hellbardt*).
This book is a biography of the German optician and inventor Joseph von Fraunhofer (1787 – 1826). Among a broad range of scientific and technical activities it describes also Fraunhofer's work in astronomical optics, culminating in the design and manufacturing of the famous "Dorpat Refraktor".

003.125   Physics of the Galaxy and interstellar matter.
H. Scheffler, H. Elsässer, translated by A. H. Armstrong.
Astronomy and Astrophysics Library. Springer–Verlag, Berlin, F.R. Germany. 12 + 492 pp. Price DM 118.00, US$ 69.50 (1987). ISBN 3–540–17314–5 (F.R. Germany), ISBN 0–387–17314–5 (USA).
This book is an English translation of the German original *"Bau und Physik der Galaxis"* published in 1982 (32.003.108).
Contents: 1. Introductory survey.
2. Positions, motions and distances of the stars – concepts and methods.
3. Structure and kinematics of the stellar system: Apparent distribution of the stars. The local galactic star field. Large scale distribution of the stars. Large scale motion of the stars. General summary, stellar populations.
4. Interstellar phenomena: The generally distributed medium. Interstellar clouds.
5. Physics of the interstellar matter: Radiation in the interstellar gas. State of the interstellar gas. The interstellar dust grains. Distribution and motion of the interstellar matter.
6. Dynamics of the Galaxy: Stellar dynamics. Gravitational theory of the spiral structure. Dynamics of the interstellar gas.

003.126   Prostranstvenno–vremennaya struktura vselennoj.
E. Schrödinger.
Nauka, Moskva, USSR. 224 pp. (1986).
Review in Ref. Zh., 51. Astron., 8.51.12 (1987)
This book is a Russian translation of the monograph *"Space–time structure"* first published in 1950 and reprinted in 1985 (42.003.066).

003.127   Meteority Ukrainy (*Meteorites of the Ukraine*).
V. P. Semenenko, Eh. V. Sobotovich, B. V. Tertychnaya.
Naukova Dumka, Kiev, USSR. 220 pp. (1987).
Review in Ref. Zh., 51. Astron., 12.51.29 (1987).

003.128   Spacelab. Forschung im Weltraum.
D. Shapland, M. Rycroft.
VCH Verlagsgesellschaft, Weinheim, F.R. Germany. 194 pp. Price DM 68.00 (1986). ISBN 3–527–26500–7.
Review in Sterne Weltraum, 26. Jahrg., Nr. 7 – 8, p. 450, 1987 (*A. M. Quetsch*).
This book is a German translation of the English original *"Spacelab. Research in earth orbit"* published in 1984 (38.003.049).

003.129   A career in astronomy.
H. Shipman.
Available from H. Shipman, Education Office, American Astronomical Society, Sharp Laboratory, University of Delaware, Newark, DE 19711, USA. 20 pp. Price US$ 0.25.
Review in Strolling Astron., Vol. 32, Nos. 3 – 4, p. 90 (1987).

003.130   Dynamical evolution of globular clusters.
L. Spitzer Jr.
Princeton Series in Astrophysics. Princeton University Press. Princeton, NJ, USA. 12 + 180 pp. Price US$ 35.00 cloth, US$ 14.50 paper (1987). ISBN 0 691–08309–6 cloth, ISBN 0–691–08460–2 paper.
Contents: 1. Overview: Observations. Stationary equilibria. Perturbations and their effects.

2. Velocity changes produced by stellar encounters: Diffusion coefficients. Fokker–Planck equation. Adiabatic invariants.
3. Evolution of idealized models: Evaporation from an isolated uniform cluster. Evaporation from a tidally limited cluster. Collapse of an isothermal sphere. Mass stratification instability.
4. Dynamical evolution of the standard model.
5. Effects of external fields: Tidal galactic field. Time–dependent fields.
6. Encounters with binary stars: Encounters of a single star with a binary. Encounters between binaries. Formation of binary stars.
7. Late phases of cluster evolution: Binaries as an energy source in clusters. Termination of collapse. Post–collapse evolution.

003.131   Kepler's physical astronomy.
B. Stephenson.
Studies in the History of Mathematics and Physical Sciences, Vol. 13. Springer–Verlag, New York, NY, USA. 8 + 217 pp. Price DM 118.00 (1987). ISBN 0–387–96541–6 (USA), ISBN 3–540–96541–6 (F.R. Germany).
Contents: 1. Introduction. 2. *Mysterium Cosmographicum.* 3. *Astronomia nova.* 4. Epitome of Copernican astronomy. 5. Kepler and the development of modern science.

003.132   Konstruirovanie i tekhnologiya izgotovleniya kosmiches-kikh priborov (*Design and technology of space instrumentation*).
S. R. Tabaldyev (Editor).
Nauka, Moskva, USSR. 173 pp. (1987).
Review in Ref. Zh., 62. Issled. Kosm. Prostranstva, 10.62.131 (1987).

003.133   Orbits for amateurs with a microcomputer. Volume II.
D. Tattersfield.
Stanley Thornes (Publishers) Ltd., Leckhampton, Cheltenham GL53 0DN, UK. 12 + 159 pp. Price £ 16.95 (1987). ISBN 0–85950–664–9.
Review in Mon. Notes Astron. Soc. S. Afr., Vol. 46, Nos. 7 – 8, p. 111, 1987 (*P. Mack*).
Contents: 1. Construction of an ephemeris for a comet. 2. Determining the elements of a comet orbit from optical sightings. 3. Perturbations of a comet orbit. 4. The orbit of a meteor stream. 5. Artificial Earth–satellite orbits.

003.134   The adjustment and testing of telescope objectives.
H. D. Taylor.
Adam Hilger, Bristol, UK. 95 + 7 pp. Price £ 7.50 (1986). ISBN 0–85274–765–X.
Review in Orion, 45. Jahrg., Nr. 222, p. 192, 1987 (*A. Tarnutzer*).
Reprint of the fifth edition 1983 (37.003.156).

003.135   Astronomical memoirs...
J. Tebutt.
Hawkesbury Shire Council, Windsor, UK. 22 + 154 pp. Price US$ 20.00 cloth, US$ 10.00 paper (1986). ISBN 0–949694–09–6 cloth, ISBN 0–949694–08–8 paper.
Review in Aust. J. Astron., Vol. 2, No. 2, p. 79 – 80, 1987 (*W. Orchiston*).

003.136   Introduction to space dynamics.
W. T. Thomson.
Dover Publications Inc., Mineola, NY, USA. 16 + 317 pp. Price US$ 8.00 (1986). ISBN 0–486–65113–4.
This Dover edition is an unabridged, corrected republication of the work first published by John Wiley & Sons Inc., New York, in 1961.
Contents: 1. Introduction. 2. Kinematics. 3. Transformation of coordinates. 4. Particle dynamics (satellite orbits). 5. Gyrodynamics. 6. Dynamics of gyroscopic instruments. 7. Space vehicle motion. 8. Performance and optimization. 9. Generalized theories of mechanics.

003.137   Relativity, thermodynamics and cosmology.
R. C. Tolman.
Dover, Mineola, NY, USA. 501 pp. Price US$ 11.95 (1987). ISBN 0–486–65383–8.
From Phys. Today, Vol. 40, No. 11, p. 104 (1987).
Dover reprint of a 1934 monograph.

003.138   Meditations at 10,000 feet: a scientist in the mountains.
J. Trefil.
Charles Scribner's Sons, New York, NY, USA. 236 pp. Price US$ 16.95 (1986). ISBN 0–684–18627–6.
From Phys. Today, Vol. 40, No. 12, p. 95 (1987).

003.139   Meditations at sunset: a scientist looks at the sky.
J. Trefil.
Charles Scribner's Sons, New York, NY, USA. 208 pp. Price US$ 16.95 (1987). ISBN 0–684–18787–6.
From Phys. Today, Vol. 40, No. 12, p. 95 (1987).

003.140   Korotkoperiodnye tsikly solnechnoj aktivnosti (*Short-period cycles of solar activity*).
A. V. Vinitskij.
Severnyj–Vostochnyj Kompleks NII DVNTs SSSR, Magadan, USSR. 48 pp. (1987).
Review in Ref. Zh., 51. Astron., 11.51.18 (1987).

003.141   Webb Society deep–sky observer's handbook. Vol. 6: Anonymous galaxies.
M. J. Thomson, R. J. Morales, edited by K. G. Jones, with a foreword by W. S. Houston.
Enslow Publishers, Inc., Hillside, NJ, USA; Aldershot, Hants, UK. 18 + 137 pp. Price US$ 12.95 (1987). ISBN 0–89490–133–8.

003.142   Webb Society deep–sky observer's handbook. Vol. 7: The southern sky.
S. J. Hynes, edited by K. G. Jones, with a foreword by G. de Vaucouleurs.
Enslow Publishers, Inc., Hillside, NJ, USA; Aldershot, Hants, UK. 18 + 198 pp. Price US$ 16.95 (1987). ISBN 0–89490–134–6.
Review in Observatory, Vol. 108, No. 1082, p. 20 – 21, 1988 (*D. Jones*).

003.143   Living the sky: the cosmos of the North American Indian.
R. A. Williamson.
University of Oklahoma Press, 1005 Asp Ave., Norman, OK 73019, USA. 366 pp. Price US$ 14.95 (1987). ISBN 0–8061–2034–7.
From Sky Telesc., Vol. 74, No. 6, p. 613 (1987).
Paperback reprint of the 1984 hardbound edition (39.003.187).

003.144   Trajectoires spatiales.
O. Zarrouati.
Cepadues–Editions, 111 rue Nicholas–Vauquelin, F–31100 Toulouse, France. 522 pp. Price FF 199.00 (1987). ISBN 2–85428–166–7.
From Phys. Today, Vol. 40, No. 8, p. 74 (1987).

003.145   Interior structure of the Earth and planets.
V. N. Zharkov, translated by W. B. Hubbard and R. A. Masteler.
Harwood Academic Publishers, Chur, Switzerland. 18 + 436 pp. Price US$ 143.00 (1986). ISBN 3–7186–0067–3.
This book is a revised and extended English translation of the Russian original "*Vnutrennee stroenie Zemli i planet*" published in 1983 (37.003.170).
Contents: Part I: Structure of the solid Earth – 1. Seismology. 2. Gravimetry. 3. Free oscillations of the Earth. 4. Magnetism and electrical conductivity of the Earth. 5. Geothermics, temperature distribution, heat flow from the Earth's interior. 6. Study of geophysical materials at high pressures. 7. A model of the Earth's interior structure. 8. Plate tectonics.
Part II: Structure of the Moon and planets – 9. Structure of the terrestrial planets. 10. Interior structure of the giant planets.

11. Interior structure of the Moon. 12. Structure of the Galilean satellites of Jupiter, Titan, and the icy satellites of Saturn.

003.146   Aratea.
    Facsimile edition. Facsimile–Verlag, Luzern, Switzerland. 200 + 160 pp. Price DM 4950.00 (1987).
Review in Sterne Weltraum, 26. Jahrg., Nr. 12, p. 728, 1987 (G. D. Roth).
High–quality facsimile edition of a famous astronomical manuscript from the Carolingian epoch. The original has been produced in the second quarter of the 9th century and is in possession of the University Library at Leiden, The Netherlands.

003.147   Modulyatsiya kosmicheskikh luchej v solnechnoj sisteme (*Cosmic ray modulation in the solar system*).
Collected papers of the Institute of Cosmophysical Research and Aeronomy, Yakutsk, USSR. 96 pp. (1986).
Review in Ref. Zh., 51. Astron., 7.51.42 (1987).

003.148   Vsemirnoe tyagotenie i teorii prostranstva i vremeni (*Gravitation and the theory of space and time*).
Universitet Druzhby Narodov, Moskva, USSR. 164 pp. (1987).
Review in Ref. Zh., 51. Astron., 12.51.31 (1987).
A collection of papers dedicated to the 300th anniversary of Isaac Newton's "*Principia mathematica*" in 1687.

004 History of Astronomy

004.001   Sonnenfinsternisse auf prähistorischen Kultplätzen durch Felsritzungen dokumentiert.
W. Brunner–Bosshard.
Orion, 45. Jahrg., Nr. 221, p. 132 – 135 (1987).

004.002   On the origin of the Ptolemaic star catalogue: Part I.
    J. Evans.
J. Hist. Astron., Vol. 18, Part 3, p. 155 – 171 (1987).

004.003   Corrigendum: "On the origin of Horrocks's lunar theory" [J. Hist. Astron., Vol. 18, Part 2, p. 77 – 94 (1987)].
C. Wilson.
J. Hist. Astron., Vol. 18, Part 3, p. 172 (1987). See Abstr. 43.004.057.

004.004   Observing with the armillary astrolabe.
    J. Włodarczyk.
J. Hist. Astron., Vol. 18, Part 3, p. 173 – 195 (1987).
    The author investigates the merits and limitations of the armillary astrolabe, which served for direct observations of the ecliptic longitudes and latitudes of the heavenly bodies from Antiquity to the end of the sixteenth century. Observations made with a modern replica of the instrument are compared with historical astrolabic observations as reported by Ptolemy in the Almagest and with measurements made in 1503–4 by Bernard Walther in Nuremberg.

004.005   Nineteenth–century Italian contributions to galactic theory.
M. Turchetta, G. Gavazzi.
J. Hist. Astron., Vol. 18, Part 3, p. 196 – 208 (1987).

004.006   Flamsteed's missing stars.
    M. Wagman.
J. Hist. Astron., Vol. 18, Part 3, p. 209 – 223 (1987).

004.007   De Aratea: uniek handschrift uit de negende eeuw.
    G. Schilling.
Zenit, 14. Jahrg., Nr. 9, p. 276 – 280 (1987).

004.008   Newtons rationella mekanik 300 år. I. Gravitationsmekaniken.
U. Uhlhorn.
Astron. Tidsskr., Årg. 20, Nr. 3, p. 97 – 115 (1987).

004.009   D'Alembert versus Euler on the precession of the equinoxes and the mechanics of rigid bodies.
C. Wilson.
Arch. Hist. Exact Sci., Vol. 37, No. 3, p. 233 – 273 (1987).

004.010   Astronomy and "*Les Philosophes*".
    J. N. Brown.
J. Br. Astron. Assoc., Vol. 97, No. 5, p. 267 – 269 (1987).
    This paper aims to investigate some of the main ideas of the late seventeenth century French writers on astronomy, Pierre Bayle and Bernard le Bouvier de Fontenelle. Bayle's *Thoughts on the Comet*, in which he is highly critical of anything superstitious or not founded on reason and observation, landed him in hot water with contemporary religious authorities. Fontenelle's famous work *Conversations on the Plurality of Worlds* was, from the outset, a best seller on astronomy. In it, he reviews the Copernican teachings of his century and the telescopic observations of Galileo and J.–D. Cassini.

004.011   Astronomy in Poland during the Second World War. Memories of a participating astronomer.
K. Walter.
J. Br. Astron. Assoc., Vol. 97, No. 5, p. 270 – 273 (1987).
    In 1941 the author was invited by the German authorities to direct the work of the observatories in the Polish cities of Warsaw, Cracow and Lwow. Despite the enormous difficulties of working in war-torn central Europe, the astronomers at these observatories continued with what research they could, particularly the observation of variable stars and the computation of solar system orbits.

004.012   A medieval reference to the Andromeda Nebula.
    P. Kunitzsch.
Messenger, No. 49, p. 42 – 43 (1987).

004.013   Zur Geschichte der populären Astronomie in Berlin.
J. Hamel.
Astron. Sch., 24. Jahrg., Heft 4, p. 80 – 82 (1987).

004.014   La mecánica cuántica y la realidad del universo.
    P. Lustgarten.
Bol. Acad. Cienc. Fis. Mat. Nat., Tomo 45, Nos. 139 – 140, p. 53 – 66 (1985).

004.015   Il y a 300 ans, la gravitation ...
    S. Débarbat.
Astronomie, Vol. 101, p. 421 – 425 (1987).

004.016   Histoire de la Grande Ourse ou les métamorphoses d'une constellation.
A. Le Bœuffle.
Astronomie, Vol. 101, p. 431 – 436 (1987).

004.017   Herscheliana.
    B. Warner.
Vistas Astron., Vol. 30, Part 1, p. 85 – 96 (1987).

004.018 The mysterious nebulae, 1610 – 1924.
O. Gingerich.
J. R. Astron. Soc. Can., Vol. 81, No. 4, p. 113 – 127 (1987).
The history of our knowledge of nebulae is traced from the earliest visual observations to the recognition that spiral nebulae were indeed "island universes".

004.019 Carl Wirtz und die Flucht der Spiralnebel.
W. Priester, R. Schaaf.
Sterne Weltraum, 26. Jahrg., Nr. 7 – 8, p. 376 – 377 (1987).

004.020 Der Refraktor der Sternwarte Pulkowa. Eine traurige Geschichte.
G. Hartl.
Sterne Weltraum, 26. Jahrg., Nr. 7 – 8, p. 397 – 404 (1987).

004.021 Zwischen Himmel und Erde. Ein sächsischer Komet.
G. Rüdiger.
Sterne Weltraum, 26. Jahrg., Nr. 9, p. 470 – 473 (1987).
Report on life and work of Johann Georg Palitzsch.

004.022 The Shang dynasty's supernova.
Z.-R. Wang.
Recherche, Vol. 18, No. 193, p. 1416 – 1417 (1987). In French.

004.023 Die Ahlhorner Heide – ein frühes Großobservatorium?
P. Schmitz.
Sterne Weltraum, 26. Jahrg., Nr. 10, p. 576 – 579 (1987).

004.024 Clustering of astronomers.
W. H. McCrea.
Annu. Rev. Astron. Astrophys., Vol. 25, p. 1 – 22 (1987). – See Abstr. 003.003.
The author describes and discusses some clusters of astronomers that he has been able to observe in his own experience. He examines the clustering phenomenon as it affects astronomers at levels from undergraduate students to established astronomers of note. The examples are selected for their aptness in illustrating features of the phenomenon, not because of the reputations of the persons involved.
Contents: Cambridge: late 1920s, 1952 – 53. Göttingen 1928 – 29. Edinburgh 1930 – 32. Imperial College 1932 – 36. Queen's University, Belfast 1936 – 44. Admiralty 1943 – 45. Royal Holloway College 1944 – 66. Berkeley 1956, 1967. Warner and Swasey Observatory: Case Institute of Technology 1964. Sussex from 1965. Other places.

004.025 Schroeters Mars–Forschungen.
D. Gerdes.
Nachr. Olbers–Ges. Bremen, Nr. 141, p. 4 – 5 (1987).

004.026 Canon der verduisteringen.
W. Carton.
Zenit, 14. Jaarg., Nr. 11, p. 348 – 355 (1987).

004.027 Nachbau antiker astronomischer Instrumente.
G. Aulenbacher.
Sterne Weltraum, 26. Jahrg., Nr. 11, p. 606 – 611 (1987).

004.028 300 years mathematical sciences and celestial mechanics.
V. I. Arnol'd.
Priroda, No. 8, p. 5 – 15 (1987). In Russian.

004.029 The physical conception of time from Newton to today.
A. D. Chernin.
Priroda, No. 8, p. 27 – 37 (1987). In Russian.

004.030 Newton's *Principia*: a retrospective.
G. E. Christianson.
Sky Telesc., Vol. 74, No. 1, p. 18 – 20 (1987).

004.031 Charting the southern sky.
J. Lankford.
Sky Telesc., Vol. 74, No. 3, p. 243 – 246 (1987).

004.032 A mysterious woodcut.
D. W. Hughes.
Sky Telesc., Vol. 74, No. 3, p. 252 (1987).

004.033 Perfecting the modern reflector.
W. Tobin.
Sky Telesc., Vol. 74, No. 4, p. 358 – 359 (1987).

004.034 The earliest infrared light curves.
D. Hoffleit.
J. Am. Assoc. Variable Star Obs., Vol. 16, No. 1, p. 29 – 33 (1987).
In 1932 John S. Hall, a graduate student at Yale, built a caesium–oxide photoelectric photometer giving an effective wavelength of about 8000 Å. With this he observed the Cepheid ζ Gem. Later Dr. A. L. Bennett used the same equipment to observe many variables.

004.035 The discovery of Sirius B: a case of strategy or serendipity?
B. L. Welther.
J. Am. Assoc. Variable Star Obs., Vol. 16, No. 1, p. 34 (1987).

004.036 Chronometrische Dienste Berliner Sternwarten (1787 – 1913).
K.-H. Tiemann.
Astron. Raumfahrt, 25. Jahrg., Heft 5, p. 151 – 152 (1987).

004.037 Amphitrite and the astronomy of Regent's Park.
A. Wilson.
Astron. Now, Vol. 1, No. 5, p. 43 – 45 (1987).

004.038 The Roman fireball of 76 BC.
R. B. Stothers.
Observatory, Vol. 107, No. 1080, p. 211 – 213 (1987).

004.039 The Chinese "candle star" of 76 BC.
Y.-L. Huang.
Observatory, Vol. 107, No. 1080, p. 213 – 217 (1987).
The unique Chinese "candle star" reported in 76 BC was previously suggested to be a comet or a nova. Some even interpret it as the most brilliant of the known historical novae. However, an extensive study of ancient literature shows that a fireball with its trajectory parallel to the line of sight may be the most plausible explanation for the event.

004.040 Newton's *Principia*.
S. W. Hawking.
Three hundred years of gravitation, p. 1 – 4 (1987). – See Abstr. 003.012.

004.041 Newtonianism and today's physics.
S. Weinberg.
Three hundred years of gravitation, p. 5 – 16 (1987). – See Abstr. 003.012.

004.042 Dark stars: the evolution of an idea.
W. Israel.
Three hundred years of gravitation, p. 199 – 276 (1987). – See Abstr. 003.012.
Contents: 1. Introduction. 2. Early speculations (1784 – 1921). 3. White dwarfs: the first compact massive objects (1910 – 1926). 4. The Chandrasekhar limit (1929 – 1935). 5. Eddington's intervention (1935). 6. Neutron stars and gravitational collapse (1934 – 1959). 7. The Schwarzschild "singularity" (1916 – 1966). 8. Quasars and relativistic astrophysics (1951 – 1972). 9. Non–spherical collapse: from frozen star to black hole (1964 – 1971). 10. Towards the quantum era: the thermodynamics of black holes (1970 – 1974).

004.043 Foucault's invention of the silvered–glass reflecting telescope and the history of his 80–cm reflector at the Observatoire de Marseille.
W. Tobin.
Vistas Astron., Vol. 30, Part 2, p. 153 – 184 (1987).
Léon Foucault's serendipitous development of the silvered–glass reflecting telescope is described. Foucault's largest mirror has an 80 cm diameter and was installed under the clear skies of Marseille in 1864. The story of this telescope's hundred years of service is sketched out; famous observers with the instrument were Stéphan, Fabry and Jonckheere.

004.044 Die "Principia": ein Buch verändert die Welt. Newton und sein Einfluß auf die Entwicklung in Deutschland.
K. von Meyenn.
Phys. Bl., 43. Jahrg., Heft 12, p. 441 – 445 (1987).
Der Autor beschäftigt sich mit der Aufnahme und den Wirkungen der newtonschen Physik in der Gelehrtenwelt Deutschlands.

004.045 On the origin of the Ptolemaic star catalogue: Part 2.
J. Evans.
J. Hist. Astron., Vol. 18, Part 4, p. 233 – 278 (1987).

004.046 Astronomy in Renaissance Hungary.
M. Vargha, E. Both.
J. Hist. Astron., Vol. 18, Part 4, p. 279 – 283 (1987).

004.047 Easter Island's "sun stones": a re–evaluation.
G. Lee, W. Liller.
Archaeoastronomy (U.K.), No. 11, p. S1 – S11 (1987).

004.048 On the astronomical origin of the offset street grid at Teotihuacan.
C. W. Peterson, B. C. Chiu.
Archaeoastronomy (U.K.), No. 11, p. S13 – S18 (1987).

004.049 Heliacal rise phenomena.
B. E. Schaefer.
Archaeoastronomy (U.K.), No. 11, p. S19 – S33 (1987).

004.050 The Borana calendar: some observations.
C. L. N. Ruggles.
Archaeoastronomy (U.K.), No. 11, p. S35 – S53 (1987).

004.051 Astronomical orientations of five megalithic tombs at Madau, near Fonni in Sardinia.
E. Proverbio, G. Romano, A. Aveni.
Archaeoastronomy (U.K.), No. 11, p. S55 – S65 (1987).

004.052 Dating "Almagest" from proper motions of stars.
Yu. N. Efremov, E. D. Pavlovskaya.
Dokl. Akad. Nauk SSSR. Ser. Mat. Fiz., Tom 294, No. 2, p. 310 – 313 (1987). In Russian. Abstr. in Ref. Zh., 51. Astron., 10.51.7 (1987).

004.053 Another reports for observation of Venus transit by Avicenna and its effects on ancient astronomy.
S. M. H. Hadavi.
Bull. Am. Astron. Soc., Vol. 19, No. 2, p. 690 – 691 (1987). Abstract. – See Abstr. 010.061.

004.054 Planetary position records in ancient China.
S.–x. Wu, C.–y. Liu.
Publ. Shaanxi Astron. Obs., Vol. 9, No. 1, p. 1 – 4 (1986). In Chinese.

004.055 Sobre a história e desenvolvimento da astronomia em Portugal.
J. Pereira Osório.
Hist. Desenvolvimento Cienc. Portugal, Vol. 1, p. 111 – 142 (1986). = Publ. Obs. Astron. Fac. Cienc. Porto, No. 38.

004.056 New information on comet P/Halley as depicted by Giotto di Bondone and other Western artists.
R. J. M. Olson, J. M. Pasachoff.
Astron. Astrophys., Vol. 187, No. 1/2, p. 1 – 11 (1987). – See Abstr. 003.013.
Artists' depictions of comets provide the only visual evidence of historical comets, most notably of Halley's comet. In this paper the authors discuss the visual evidence of comet P/Halley at several passages through that of 1301 and compare it with descriptions and modern images. Since it was first recognized that Giotto di Bondone painted a comet in place of the Star of Bethlehem and suggested that this was a portrait of the 1301 apparition of comet Halley, a great deal of new information has come to light. The authors present a synopsis of the textual, visual, and astronomical evidence to support the theory that when Giotto painted his comet in the Scrovegni Chapel he was reflecting his viewing of comet Halley in 1301.

004.057 The problem of the Copernican revolution and of the propagation of Copernican ideas.
B. Rajtsmen.
Istor.–Astron. Issled., Moskva, No. 19, p. 295 – 310 (1987). In Russian. Abstr. in Ref. Zh., 51. Astron. 11.51.4 (1987).

004.058 Reflections on Copernicus' star catalogue.
E. P. Fedorov.
Istor.–Astron. Issled., Moskva, No. 19, p. 87 – 102 (1987). In Russian. Abstr. in Ref. Zh., 51. Astron. 11.51.5 (1987).

004.059 La astronomía en México hacia su etapa actual.
P. Pismis.
Rev. Mex. Astron. Astrofis., Vol. 14, No. 1, p. 35 – 42 (1987). – See Abstr. 012.042.
An account is given of the development of astronomical endeavors in Mexico starting with the founding of the National Astronomical Observatory in the late eighties of the last century. An important contribution of the early period, along with activity in various branches of classical astronomy, has been the completion of the Astrographic Catalogue of the Tacubaya Zone. The development, throughout the past few decades, leading to the present state of astronomy in Mexico is presented, based largely on personal experience of the author.

004.060 On the astronomy in the mesoamerican calendar (preliminary remarks).
L. Maupomé.
Rev. Mex. Astron. Astrofis., Vol. 14, No. 1, p. 43 – 44 (1987). – See Abstr. 012.042.
The mesoamerican calendar gathers astronomical commensurabilities by means of several artificial cycles, based on the sacred calendar of 260 days. The periods which are built from it, are expressions which cypher, to the highest accuracy, the motions of the solar system. Interrelationships between mesoamerican numbers found in inscriptions, codices, and the calendar, and astronomical periods and dates, are discussed. It is observed that several of these numbers are members of Pythagorean triples, and that they may express relation with binomial expansion.

004.061 Trajectoires et impasses de la solution de Schwarzschild.
J. Eisenstaedt.
Arch. Hist. Exact Sci., Vol. 37, No. 4, p. 275 – 357 (1987).

004.062 Discovering M31's spiral shape.
G. de Vaucouleurs.
Sky Telesc., Vol. 74, No. 6, p. 595 – 598 (1987).

004.063 Accretion research: how it started.
M. A. Abramowicz, C. Marsi.
Observatory, Vol. 107, No. 1081, p. 245 – 247 (1987).
The theoretical work of Bondi, Hoyle, Lyttleton and others in Cambridge, and Shakura and Sunyaev in Moscow stimulated the development of research on accretion on to compact objects as much as discoveries of quasars, pulsars and X–ray sources.

004.064 **Le calendrier des Gaulois.**
P.–M. Duval.
C. R. Acad. Sci., Sér. Gén., Vie Sci., Tome 4, No. 4, p. 333 – 348 (1987).

004.065 **L'atomisme, face cachée de la condamnation de Galilée?**
P. Costabel.
C. R. Acad. Sci., Sér. Gén., Vie Sci., Tome 4, No. 4, p. 349 – 365 (1987).

004.066 **Contribution of the Lands Department to the development of astronomy in New South Wales during the nineteenth century.**
W. Orchiston.
Aust. J. Astron., Vol. 2, No. 2, p. 65 – 74 (1987).
During the last three decades of the nineteenth century, astronomy was part of the official charter of the Lands Department. Initially, good relations existed between the Department and Sydney Observatory, the Colony's official astronomical centre, but this situation changed dramatically during the 1880s as the Observatory turned increasingly to meteorology. In this environment, the Lands Department was able to become a significant force in New South Wales astronomy.

004.067 **Of whirls and molten gold. An introduction to Fontenelle's "Entretiens" (1686).**
R. M. West.
Messenger, No. 50, p. 40 – 43 (1987).

004.068 **Four Korean "guest stars" observed in AD 1592.**
F. R. Stephenson, K. K. C. Yau.
Q. J. R. Astron. Soc., Vol. 28, No. 4, p. 431 – 444 (1987).
During the year 1592, four "guest stars" were sighted by the Korean court astronomers. The authors present here an evaluation of the suitability of these stars as potential novae or supernovae. The investigation shows that three of the stars are strong nova candidates. Two of these objects appeared in the constellation of Cassiopeia and one in Cetus.

004.069 **On the trail of meteor trains.**
M. Beech.
Q. J. R. Astron. Soc., Vol. 28, No. 4, p. 445 – 455 (1987).
The study of meteoric phenomena is an ancient one. The long history of meteoric observations has led to a vast, almost folkloric, array of ideas and theories describing these transient night–time flashes. On rare occasions some meteors, particularly very bright ones, leave a brightly coloured train, which can sometimes on very rare occasions endure for up to an hour. These are truely "strange" phenomena among the meteors, and it is to these that the author turns. This article is, hence concerned with the historic observation and debate of these "strange", colourful and enduring apparitions.

004.070 **L'Observatoire de la Babote à Montpellier.**
J. M. Faidit.
Obs. Trav., No. 12, p. 45 – 59 (1987).

004.071 **Orientamenti astronomici di alcuni monumenti nel Veneto.**
G. Romano.
Astronomia, N. 3 – 4, p. 12 – 18 (1987).

004.072 **Archäo–astronomische Betrachtungen zur Fundstätte Glozel in Frankreich.**
H.–R. Hitz, H. Schilt, W. Knaus, H. Jäger.
Orion, 45. Jahrg., Nr. 223, p. 228 – 231 (1987). In German and French.

004.073 **The William Herschel 14–foot telescope.**
B. Warner.
Mon. Notes Astron. Soc. S. Afr., Vol. 46, Nos. 11 – 12, p. 158 – 163 (1987).

004.074 **On a singular Chinese portable sundial.**
E. Proverbio, G. Bertuccioli.
Istituto e Museo di Storia della Scienza Firenze, Nuncius, Annali di Storia della Scienza, Anno 1, Fasc. 1, p. 47 – 58 (1986). = Pubbl. Stn. Astron. Int. Latitudine, Carloforte–Cagliari, Nuov. Ser., N. 128.
The production of portable sundials in China presumably goes back to the XVII–XVIII centuries following the diffusion of sundials of European production. A singular portable Chinese sundial conserved in the Siamese Museum of Cagliari is described. It presents quite special characteristics and is presumed to have been built and used for divinatory purposes on the basis of ancient Chinese astrological and geomantic doctrines.

004.075 **Notes on the history of the Tashkent Astronomical and Physical Observatory.**
N. F. Bulaevskij.
Istor.–Astron. Issled., Moskva, No. 19, p. 325 – 340 (1987). In Russian. Abstr. in Ref. Zh., 51. Astron., 12.51.9 (1987).

004.076 **Astronomy in Kazakhstan during the Second World War.**
Z. V. Karyagina.
Istor.–Astron. Issled., Moskva, No. 19, p. 203 – 218 (1987). In Russian. Abstr. in Ref. Zh., 51. Astron., 12.51.10 (1987).

004.077 **50 years Astronomical Council of the Academy of Sciences of the USSR.**
B. M. Shustov.
Istor.–Astron. Issled., Moskva, No. 19, p. 185 – 202 (1987). In Russian. Abstr. in Ref. Zh., 51. Astron., 12.51.11 (1987).

004.078 **Kirik Novgorodets in N. V. Stepanov's correspondence with A. A. Shakhmatov.**
A. M. Pashkov, R. A. Simonov.
Istor.–Astron. Issled., Moskva, No. 19, p. 311 – 324 (1987). In Russian. Abstr. in Ref. Zh., 51. Astron., 12.51.16 (1987).

004.079 **La astronomía griega. II (Fase helenística).**
M. Ruffo.
Rev. Astron., Tomo 58, No. 240, p. 6 – 13 (1987).

004.080 **Main characteristics and achievements of ancient Indian astronomy in historical perspective.**
K. S. Shukla.
History of Oriental astronomy, p. 9 – 22 (1987). – See Abstr. 012.060 (IAU Colloq. No. 91).
The ancient Indian astronomers did not possess the telescope. They made their observations with the naked eye using suitable devices for measuring angles. Their astronomy therefore remained confined to the study of the Sun, Moon and the planets.

004.081 **The asterisms.**
A. K. Chakravarty.
History of Oriental astronomy, p. 23 – 28 (1987). – See Abstr. 012.060 (IAU Colloq. No. 91).
A system of 27 asterisms plays an important role in Indian astronomy and calendrical science. The present convention is that the ecliptic is divided into 27 equal parts each $13°20'$ long commencing from one initial point.

004.082 **Vedic literature vis–a–vis mathematical astronomy.**
R. Sarkar.
History of Oriental astronomy, p. 29 – 32 (1987). – See Abstr. 012.060 (IAU Colloq. No. 91).
The present paper restricts to the analysis of some passages from the Vedic literature from the view point of mathematical astronomy.

004.083 **The characteristics of ancient Chinas's astronomy.**
Z. Xi.
History of Oriental astronomy, p. 33 – 40 (1987). – See Abstr. 012.060 (IAU Colloq. No. 91).

004.084   **Ancient Chinese auroral records: interpretation problems and methods.**
M. Teboul.
History of Oriental astronomy, p. 41 – 50 (1987). – See Abstr. 012.060 (IAU Colloq. No. 91).
In recent years a number of auroral catalogues have been compiled from Oriental records, but the auroral identifications given vary greatly from one catalogue to another. This is due to the fact that, up to now, no reliable criteria have been found which could help us discriminate between auroral and non-auroral ancient records. In this paper two new criteria are propounded and their usefulness illustrated by the solution of a few currently much debated questions.

004.085   **Work of art of the Han Dynasty unearthed in China and observations of solar phenomena.**
Z.-t. Xu.
History of Oriental astronomy, p. 51 – 56 (1987). – See Abstr. 012.060 (IAU Colloq. No. 91).

004.086   **Research on scale and precision of the Water Clock in ancient China.**
H. Quan.
History of Oriental astronomy, p. 57 – 61 (1987). – See Abstr. 012.060 (IAU Colloq. No. 91).

004.087   **Greek astronomers and their neighbours.**
O. Pedersen.
History of Oriental astronomy, p. 63 – 73 (1987). – See Abstr. 012.060 (IAU Colloq. No. 91).

004.088   **Revision of *yugas* and *yuga*–constants in Indian astronomy.**
K. V. Sarma.
History of Oriental astronomy, p. 77 – 83 (1987). – See Abstr. 012.060 (IAU Colloq. No. 91).

004.089   **System of astronomical constants in Hindu astronomy.**
A. Bandyopadhyay, A. K. Bhatnagar.
History of Oriental astronomy, p. 85 – 89 (1987). – See Abstr. 012.060 (IAU Colloq. No. 91).
Astronomical constants such as the length of the solar year, sidereal and synodic periods of revolutions of the Moon and five brighter planets have been computed using the system of astronomy in ancient and mediaeval India and a comparison made with their modern values.

004.090   **The meridians of reference of Indian astronomical canons.**
R. Mercier.
History of Oriental astronomy, p. 97 – 107 (1987). – See Abstr. 012.060 (IAU Colloq. No. 91).
The canons of Sanskrit astronomy depend on mean motions which are normally postulated to refer to the central meridian of Ujjain. The present work is a statistical analysis of these mean motions designed to discover the optimum position of the meridian, by comparison with modern mean motions. This follows earlier work done by Billard in determining the optimum year.

004.091   **Periodic nature of cometary motions as known to Indian astronomers before eleventh century A.D.**
S. D. Sharma.
History of Oriental astronomy, p. 109 – 112 (1987). – See Abstr. 012.060 (IAU Colloq. No. 91).
Apparitions of comets were thought to be a bad omen in earlier times in almost all the old civilizations. This led to correlating these apparitions with some particular events which took place simultaneously. Although the information was collected and recorded merely for astrological purposes, yet these records are in no way less important from astronomical points of view.

004.092   **Planetary theories in Sanskrit astronomical texts.**
S. N. Sen.
History of Oriental astronomy, p. 113 – 124 (1987). – See Abstr. 012.060 (IAU Colloq. No. 91).

004.093   **An examination of al–Bīrūnī's knowledge of Indian astronomy.**
M. S. Khan.
History of Oriental astronomy, p. 139 – 145 (1987). – See Abstr. 012.060 (IAU Colloq. No. 91).

004.094   **An unknown Arabic source for star names.**
P. Kunitzsch.
History of Oriental astronomy, p. 155 – 163 (1987). – See Abstr. 012.060 (IAU Colloq. No. 91).

004.095   **Al–Suphi's star–atlases and middle Europe.**
K. A. F. Fischer.
History of Oriental astronomy, p. 165 – 167 (1987). – See Abstr. 012.060 (IAU Colloq. No. 91).

004.096   **On the solar model and the precession of the equinoxes in the Alphonsine zīj and its Arabic sources.**
J. Samsó.
History of Oriental astronomy, p. 175 – 183 (1987). – See Abstr. 012.060 (IAU Colloq. No. 91).

004.097   **Nāsir ad–Dīn on determination of the declination function.**
J. Hamadani–Zadeh.
History of Oriental astronomy, p. 185 – 189 (1987). – See Abstr. 012.060 (IAU Colloq. No. 91).

004.098   **A note on some Sanskrit manuscripts on astronomical instruments.**
Y. Ohashi.
History of Oriental astronomy, p. 191 – 195 (1987). – See Abstr. 012.060 (IAU Colloq. No. 91).

004.099   **Mādhava's rule for finding angle between the ecliptic and the horizon and Āryabhata's knowledge of it.**
R. C. Gupta.
History of Oriental astronomy, p. 197 – 202 (1987). – See Abstr. 012.060 (IAU Colloq. No. 91).

004.100   **Bibliographical notes on Islamic astronomy, the results of a study of the exact sciences among the Jews of Yemen.**
Y. T. Langermann.
History of Oriental astronomy, p. 203 – 206 (1987). – See Abstr. 012.060 (IAU Colloq. No. 91).

004.101   **La théorie astronomique selon Jabir ibn Aflah.**
H. Hugonnard–Roche.
History of Oriental astronomy, p. 207 – 208 (1987). In English. – See Abstr. 012.060 (IAU Colloq. No. 91).

004.102   **Trigonometry in two sixteenth century works: the De Revolutionibus Orbium Coelestium and the Sidra al-Muntahā.**
S. Tekeli.
History of Oriental astronomy, p. 209 – 214 (1987). – See Abstr. 012.060 (IAU Colloq. No. 91).

004.103   **Two treatises on astronomical instruments by ᶜAbd al–Munᶜ im al–ᶜĀmilī & Qāsim ᶜAlī al–Qāyinī.**
S. M. R. Ansari, S. A. Khan Ghori.
History of Oriental astronomy, p. 215 – 225 (1987). – See Abstr. 012.060 (IAU Colloq. No. 91).

004.104   **A Moorish astrolabe from Granada.**
M. G. Firneis.
History of Oriental astronomy, p. 227 – 232 (1987). – See Abstr. 012.060 (IAU Colloq. No. 91).

004.105 **Astronomical efforts of Sawai Jai Singh – a review.**
V. N. Sharma.
History of Oriental astronomy, p. 233 – 240 (1987). – See Abstr.
012.060 (IAU Colloq. No. 91).

004.106 **Sky calendars of the Indo–Malay archipelago.**
G. Ammarell.
History of Oriental astronomy, p. 241 – 247 (1987). – See Abstr.
012.060 (IAU Colloq. No. 91).

004.107 **Astronomical aspects of "Pranotomongso" of the 19th century Central Java.**
N. Daldjoeni, B. Hidayat.
History of Oriental astronomy, p. 249 – 252 (1987). – See Abstr.
012.060 (IAU Colloq. No. 91).

004.108 **Uses of ancient data in modern astronomy.**
J. A. Eddy.
History of Oriental astronomy, p. 253 – 260 (1987). – See Abstr.
012.060 (IAU Colloq. No. 91).
Ancient astronomical data are a limited resource that can fill unique needs in modern research. Records from the Orient are of particular value because they were taken more or less continuously. The author reviews three problems in modern astronomy that lean heavily on ancient records from the Middle East and Orient: the study of solar variability; studies of the variable rotation of the earth; and studies of the occurrence and physics of novae, supernovae and comets.

004.109 **Old sunspot records.**
V. R. Venugopal.
History of Oriental astronomy, p. 261 – 264 (1987). – See Abstr.
012.060 (IAU Colloq. No. 91).
The works of Eddy (1976) and Clark and Stephenson (1978) on the ancient sunspot records are reviewed and a plea is made for the search for ancient records of astronomical events and phenomena in India.

004.110 **Guest stars: historical supernovae and remnants.**
T. Velusamy.
History of Oriental astronomy, p. 265 – 270 (1987). – See Abstr.
012.060 (IAU Colloq. No. 91).

004.111 **Les satellites galiléens de Jupiter de 1610 à 1985: observations et mouvements.**
B. Morando.
Ann. Phys. (Paris), Vol. 12, Colloq. No. 1, p. 3 – 6 (1987). – See Abstr. 012.061.
As soon as the Galilean Satellites were discovered in 1610, they have been extensively observed in order to establish tables of their motions. Visual eclipse observations were replaced by photographic observations in the XIXth century but mutual events give the best accuracy. The most important theories were Laplace's and Sampson's. New approaches will soon lead to better ephemerides.

004.112 **Explications antiques de la Voie Lactée.**
A. Le Bœuffle.
Astronomie, Vol. 101, p. 575 – 580 (1987).

004.113 **Tales from the first international Halley watch (1755 – 59): 2. Thomas Stevenson of Barbados and his two–comet theory.**
C. B. Waff, S. Skinner.
IHW Newsl., No. 10, p. 3 – 9 (1987).

004.114 **100 Jahre "Carte du Ciel"–Projekt.**
W. R. Dick.
Astron. Raumfahrt, 25. Jahrg., Heft 6, p. 162 – 166 (1987).

004.115 **A tape version of Ptolemy's star catalogue.**
C. Jaschek.
Bull. Inf. Cent. Données Stellaires, No. 33, p. 145 (1987).

004.116 **Newton and the apple.**
F. E. L. Priestley.
J. R. Astron. Soc. Can., Vol. 81, No. 6, p. 185 – 194 (1987).

004.117 **L'archeoastronomia: scopi, ricerche, risultati.**
A. F. Aveni.
G. Astron., Vol. 13, N. 1 – 2, p. 15 – 18 (1987).

004.118 **Astronomia e divinazione in Mesopotamia: l'età neo–assira.**
G. B. Lanfranchi.
G. Astron., Vol. 13, N. 1 – 2, p. 19 – 21 (1987).

004.119 **Classification of comets recorded in ancient China (300 BC – 1800 AD) and statistics of type–I tails.**
L.–s. Yan, Z.–w. Hu, H.–j. Quan.
ESA Spec. Publ., ESA SP–278, p. 649 – 651 (1987). – See Abstr. 012.075.
A total of 373 comets were recorded in ancient China from the 3rd century BC to the 18th century AD. According to their morphology, colour and length of tail, the authors have found 92 comets that belong to type–I, i.e. about 25%.

004.120 **The temporary Belgrade Astronomical and Meteorological Observatory.**
M. Jéličić.
Vasiona, Année 35, No. 3 – 4, p. 78 – 88 (1987). In Serbo–Croatian.
After his return from further studies in Paris in 1887, the astronomer and meteorologist Milan Nedeljković founded the temporary Astronomical and Meteorological Observatory. Four years later, on May 1, 1891, a building dedicated to this observatory received a permanent building specially designed for this purpose, and the initial house was rented out. The observatory functioned as the head office of the Serbian Network of Meterological Stations. Included are facts regarding the observatory building as well as its owner.

004.121 **How the instruments for the Observatory at the High School and University of Belgrade were obtained.**
M. Dokić.
Vasiona, Année 35, No. 3 – 4, p. 91 – 95 (1987). In Serbo–Croatian.

004.122 **Short chronology of Soviet astronomy.**
P. G. Kulikovskij.
Zemlya Vselennaya, No. 6, p. 6 – 11 (1987). In Russian.

004.123 **Kepler's laws of planetary motion, before and after Newton's Principia: an essay on the transformation of scientific problems.**
B. S. Baigrie.
Stud. Hist. Philos. Sci., Vol. 18, No. 2, p. 177 – 208 (1987). Abstr. in Phys. Abstr., Vol. 90, No. 1317, Entry 134424 (1987).

004.124 **The periodisation of the development of astronomy based on the idea of invisible information.**
J. Marek.
Atti Fond. Giorgio Ronchi, Vol. 41, No. 6, p. 749 – 752 (1986). Abstr. in Phys. Abstr., Vol. 90, No. 1318, Entry 140405 (1987).

004.125 **The first optical identifications of radio sources.**
W. T. Sullivan III.
Publ. Astron. Soc. Pac., Vol. 99, No. 621, p. 1152 (1987). Abstract. – See Abstr. 010.281.

004.126 **Orientation of the Slavic burial–grounds in South Moravia.**
R. Rajchl.
Říše hvězd, Vol. 68, No. 9, p. 170 – 174 (1987). In Czech.

004.127 **Kant's cosmogony re–evaluated.**
S. Palmquist.
Stud. Hist. Philos. Sci., Vol. 18, No. 3, p. 255 – 269 (1987).

004.128   Hour tables and Thule in Pliny's *Natural History*.
C. H. Roseman.
Centaurus, Vol. 30, No. 2, p. 93 – 105 (1987).

004.129   La *Tribiblos astronomique* de Theodore Méliténiote (*Vat.gr. 792*).
R. Leurquin.
Janus, Vol. 72, No. 4, p. 257 – 282 (1985).
   Résumer le contenu d'un ouvrage aussi vaste que la *Tribiblos astronomique* serait une entreprise bien hasardeuse. Aussi a–t–il semblé plus opportun à l'auteur de reproduire les sommaires établis par Méliténiote au début de chacun des trois livres, en les commentant, quand ils n'étaient pas suffisamment explicites; il a également épinglé quelques passages moins techniques mais plus significatifs de la personnalité de l'auteur.

004.130   Celestial orbs in the Latin Middle Ages.
E. Grant.
ISIS, Vol. 78, p. 153 – 173 (1987).
   The author attempts to determine whether, during the late Middle Ages, the celestial orbs were conceived of as hard and rigid or fluid and soft.

004.131   On Kepler's awareness of the problem of experimental error.
G. Hon.
Ann. Sci., Vol. 44, p. 545 – 591 (1987).
   This paper is an account of Kepler's awareness of the problem of experimental error. As a study of the *Astronomia nova* shows, Kepler exploited his awareness of the occurrences of experimental errors to guide him to the right conclusion. Errors were thus employed, so to speak, perhaps for the first time, to bring about a major physical discovery: Kepler's laws of planetary motion.

004.132   The astronomer Abu 'l–Husayn al–Sūfī and his book on the constellations.
P. Kunitzsch.
Z. Gesch. Arabisch–Islamischen Wiss., Band 3, p. 56 – 81 (1986).

Maya city planning and the calendar.
See Abstr. 003.022.

Das Lob der Sternkunst. Astronomie in der deutschen Aufklärung.
See Abstr. 003.023.

The divided circle. A history of instruments for astronomy, navigation and surveying.
See Abstr. 003.025.

Isaac Newton: De la gravitation ou les fondements de la mécanique classique.
See Abstr. 003.026.

Issac Newton: Principia mathematica.
See Abstr. 003.027.

Götter in Planeten und Monden.
See Abstr. 003.030.

Sternbilder – *Sternsagen*. Mythen und Legenden um Sternbilder.
See Abstr. 003.031.

The birth of a new physics.
See Abstr. 003.040.

Die Beziehungen zwischen Johannes Kepler und dem Leipziger Mathematikprofessor Philipp Müller.
See Abstr. 003.047.

Galilée.
See Abstr. 003.049.

Lines to the mountain gods: Nazca and the mysteries of Peru.
See Abstr. 003.067.

Darkness at night. A riddle of the universe.
See Abstr. 003.070.

The Galileo connection.
See Abstr. 003.072.

Pavel Karlovich Shternberg 1865 – 1920.
See Abstr. 003.087.

They dance in the sky. Native American star myths.
See Abstr. 003.096.

Observatório Astronômico – um século de história (1827 – 1927).
See Abstr. 003.099.

The light of nature. Essays in the history and philosophy of science presented to A. C. Crombie.
See Abstr. 003.103.

Early astronomy from Babylonia to Copernicus.
See Abstr. 003.105.

Kosmos Astronomiegeschichte. Astronomen, Instrumente, Entdeckungen.
See Abstr. 003.121.

Joseph von Fraunhofer. Forscher – Erfinder – Unternehmer.
See Abstr. 003.124.

Kepler's physical astronomy.
See Abstr. 003.131.

Living the sky: the cosmos of the North American Indian.
See Abstr. 003.143.

Vsemirnoe tyagotenie i teorii prostranstva i vremeni (*Gravitation and the theory of space and time*).
See Abstr. 003.148.

The remarkable extragalactic research of Erik Holmberg: a glimpse from Santa Cruz.
See Abstr. 005.005.

David Gill and celestial photography.
See Abstr. 005.006.

Dr. William Bone, and the role of the amateur observatory in Australian astronomy.
See Abstr. 005.009.

D. F. J. Arago and the development of instrumental astrophysics. On the occasion of the 200th anniversary of the scientist's birthday.
See Abstr. 005.010.

Delisle's scientific dynasty.
See Abstr. 005.011.

Theodor Ritter von Oppolzer und die OKIE (*Österreichische Kommission für die Internationale Erdmessung*).
See Abstr. 005.012.

Marcel Nicolet and cometary research in the late thirties.
See Abstr. 005.014.

The Pulkovo Observatory and the Belgrade Observatory.
See Abstr. 009.027.

Schrödinger: centenary celebration of a polymath. Papers presented at a conference held at Imperial College, London, UK, April 1987.
See Abstr. 012.082.

Giant atoms in the cosmos.
See Abstr. 013.064.

The date and time of the vernal equinox: a graphical representation of the Gregorian calendar.
See Abstr. 014.064.

Indian calendars.
See Abstr. 046.030.

The *Hsiu–Yao Ching* and its Sanskrit sources.
See Abstr. 046.031.

The position of the *Futian* calendar on the history of east–west intercourse of astronomy.
See Abstr. 046.032.

Newton, quantum theory and reality.
See Abstr. 061.036.

A catalogue of non–telescopic sunspot observations from 165 BC to AD 1684.
See Abstr. 072.063.

The shadow chasers.
See Abstr. 079.002.

Lunar domes – first observed by Johann Hieronymus Schroeter.
See Abstr. 094.022.

The history of Halley's comet.
See Abstr. 103.425.

Associations between ancient comets and meteor showers.
See Abstr. 104.076.

Die Meteoritensammlung E. F. F. Chladnis.
See Abstr. 105.053.

Ancient Guest Stars as harbingers of neutron star formation.
See Abstr. 125.011.

General historical introduction (*concerning elliptical galaxies*).
See Abstr. 157.147.

005 Biography

005.001   Wie fand ich zur Astronomie?
J. Hoppe.
Astron. Sch., 24. Jahrg., Heft 4, p. 83 – 84 (1987).

005.002   On life and activities of S. P. Korolev (1906 – 1966), on the occasion of the 80th anniversary of his birthday.
A. Yu. Ishlinskij.
16 Gagarin. nauchn. chteniya po kosmonavt. i aviatsii, 1986. Moskva, p. 7 – 15 (1987). In Russian. Abstr. in Ref. Zh., 62. Issled. Kosm. Prostranstva, 8.62.1 (1987).

005.003   The legacy of E. E. Barnard.
G. S. Mumford.
Sky Telesc., Vol. 74, No. 1, p. 30 – 34 (1987).

005.004   Joseph Fraunhofer (1787 – 1826).
I. Howard–Duff.
J. Br. Astron. Assoc., Vol. 97, No. 6, p. 339 – 347 (1987).
   This paper commemorates the bicentenary of Fraunhofer's birth in 1787 and summarises his life, work and achievements in the context of early nineteenth century optics and telescope making.

005.005   The remarkable extragalactic research of Erik Holmberg: a glimpse from Santa Cruz.
H. J. Rood.
Publ. Astron. Soc. Pac., Vol. 99, No. 619, p. 921 – 951 (1987).
   The major extragalactic research contributions of Erik Holmberg are related to the present astronomical situation. A complete bibliography of Holmberg's scientific publications is contained. This bibliography has been used to select a list of galaxy publications which could provide a basis for a university course.

005.006   David Gill and celestial photography.
C. A. Murray.
R. Greenwich Obs., Prepr., No. 58, p. 10 – 15 (1987). To appear in IAU Symp. No. 133.

005.007   On the occasion of the 200th anniversary of M. V. Lomonosov's birthday.
V. I. Vernadskij.
Istor.–Astron. Issled., Moskva, No. 19, p. 18 – 21 (1987). In Russian. Abstr. in Ref. Zh., 51. Astron. 11.51.7 (1987).

005.008   Laureates of the F. A. Bredikhin prize of the USSR Academy of Sciences.
A. G. Masevich, A. K. Terent'eva.
Istor.–Astron. Issled., Moskva, No. 19, p. 219 – 255 (1987). In Russian. Abstr. in Ref. Zh., 51. Astron. 11.51.8 (1987).

005.009   Dr. William Bone, and the role of the amateur observatory in Australian astronomy.
W. Orchiston.
South. Stars, Vol. 32, No. 4, p. 111 – 128 (1987).
   Dr. William Bone's contribution to Australian astronomy is examined. During the second half of the nineteenth century he was one of a number of accomplished amateur astronomers who actively promoted astronomy at the public level by operating his Castlemaine Observatory in Victoria as a de facto city observatory.

005.010   D. F. J. Arago and the development of instrumental astrophysics. On the occasion of the 200th anniversary of the scientist's birthday.
A. I. Eremeeva.
Istor.–Astron. Issled., Moskva, No. 19, p. 256 – 272 (1987). In Russian. Abstr. in Ref. Zh., 51. Astron. 12.51.12 (1987).

005.011   Delisle's scientific dynasty.
M. G. Novlyanskaya, G. E. Pavlova.
Istor.–Astron. Issled., Moskva, No. 19, p. 273 – 294 (1987). In Russian. Abstr. in Ref. Zh., 51. Astron. 12.51.15 (1987).

005.012   Theodor Ritter von Oppolzer und die OKIE (*Österreichische Kommission für die Internationale Erdmessung*).
K. Rinner.
Vorträge beim Theodor Ritter von Oppolzer–Gedächtnissymposium, p. 8 – 17 (1987). – See Abstr. 012.073.

005.013 **Leben und Wirken des Theodor Ritter von Oppolzer (1841 – 1886).**
M. G. Firneis.
Vorträge beim Theodor Ritter von Oppolzer–Gedächtnissymposium, p. 18 – 36 (1987). – See Abstr. 012.073.

005.014 **Marcel Nicolet and cometary research in the late thirties.**
R. Lüst.
ESA Spec. Publ., ESA SP–278, p. 3 – 4 (1987). – See Abstr. 012.075.

005.015 **Marcel Nicolet: aeronomist.**
D. R. Bates.
ESA Spec. Publ., ESA SP–278, p. 5 – 6 (1987). – See Abstr. 012.075.

005.016 **Marcel Nicolet: mentor to many.**
P. Mange.
ESA Spec. Publ., ESA SP–278, p. 7 – 11 (1987). – See Abstr. 012.075.

005.017 **Bagrat Konstantinovich Ioannisiani (1911 – 1985), on the occasion of the 75th anniversary of his birthday.**
L. V. Mirzoyan, E. K. Kharadze.
Zemlya Vselennaya, No. 4, p. 35 – 40 (1987). In Russian.

005.018 **V. Erhart – master of astronomical optics.**
L. Hutta.
Kozmos, Vol. 18, No. 5, p. 162 – 164 (1987). In Slovak.

005.019 **Daniel Kirkwood.**
F. K. Edmondson.
Proceedings of the Indiana Academy of Science, Vol. 95, p. 363 – 365 (1986).
   This paper has four major parts: 1. Indiana University prior to Kirkwood's arrival, 2. chronology of Kirkwood's life, 3. Daniel Kirkwood at Indiana University and his work as a scientist, and 4. Daniel Kirkwood and the Indiana Academy of Science.

Gauss. Eine biographische Studie.
See Abstr. 003.035.

Die Beziehungen zwischen Johannes Kepler und dem Leipziger Mathematikprofessor Philipp Müller.
See Abstr. 003.047.

Galilée.
See Abstr. 003.049.

Pavel Karlovich Shternberg 1865 – 1920.
See Abstr. 003.087.

Dr. Luis Ugueto – ingeniero, astronomo y profesor.
See Abstr. 003.104.

Theodor Brorsen, astronom.
See Abstr. 003.111.

Galileo: Heretic.
See Abstr. 003.116.

Joseph von Fraunhofer. Forscher – Erfinder – Unternehmer.
See Abstr. 003.124.

Astronomical memoirs...
See Abstr. 003.135.

L'atomisme, face cachée de la condamnation de Galilée?
See Abstr. 004.065.

Astronomical efforts of Sawai Jai Singh – a review.
See Abstr. 004.105.

Tales from the first international Halley watch (1755 – 59): 2. Thomas Stevenson of Barbados and his two–comet theory.
See Abstr. 004.113.

Newton and the apple.
See Abstr. 004.116.

On Kepler's awareness of the problem of experimental error.
See Abstr. 004.131.

The astronomer Abu 'l–Husayn al–Sūfī and his book on the constellations.
See Abstr. 004.132.

Schrödinger: centenary celebration of a polymath. Papers presented at a conference held at Imperial College, London, UK, April 1987.
See Abstr. 012.082.

006 Personal Notes

006.001 **Dr. Paul Ahnert zum 90. Geburtstag.**
S. Marx.
Astron. Raumfahrt, 25. Jahrg., Heft 5, p. 146 (1987).

006.002 **Dr. Paul Ahnert zum 90. Geburtstag.**
W. Wenzel.
Sterne, 63. Band, Heft 5, p. 251 – 253 (1987). = Mitt. Sternw. Sonneberg, Nr. 70.

006.003 **Joseph Ashbrook received the Klumpke–Roberts Award 1987 of the Astronomical Society of the Pacific.**
J. E. Hesser, A. Fraknoi.
Mercury, Vol. 16, No. 5, p. 147 (1987).

006.004 **S. Jocelyn Bell Burnell received the Beatrice M. Tinsley Prize of the American Astronomical Society.**
Phys. Today, Vol. 40, No. 8, Part 1, p. 67 (1987).

006.005 **Jean Delhaye received the Prix Janssen 1987.**
Astronomie, Vol. 101, p. 643 (1987).

006.006 **H. Melvin Dyck received the Muhlmann Prize 1986 of the Astronomical Society of the Pacific.**
Phys. Today, Vol. 40, No. 7, p. 85 (1987).

006.007 **Charles Federer, Jr. received the Klumpke–Roberts Award 1987 of the Astronomical Society of the Pacific.**
J. E. Hesser, A. Fraknoi.
Mercury, Vol. 16, No. 5, p. 147 (1987).

006.008 **Prof. K. Ferrari d'Occhieppo feiert seinen 80. Geburtstag.**
H. Haupt.
Sternenbote, 30. Jahrg., Nr. 12, p. 254 – 255 (1987).

006.009   **Clinton B. Ford received the Amateur Achievement Award 1987 of the Astronomical Society of the Pacific.**
J. E. Hesser, A. Fraknoi.
Mercury, Vol. 16, No. 5, p. 148 (1987).

006.010   **Herbert Friedman received the 1987 Physics Prize of the Wolf Foundation.**
Phys. Today, Vol. 40, No. 8, Part 1, p. 81 (1987).

006.011   **Herbert Friedman received the 1987 Wolf Prize for Physics.**
Sol. Phys., Vol. 110, No. 1, p. IV (1987).

006.012   **Riccardo Giacconi received the 1987 Physics Prize of the Wolf Foundation.**
Phys. Today, Vol. 40, No. 8, Part 1, p. 82 (1987).

006.013   **John M. Hill received the Trumpler Award 1986 of the Astronomical Society of the Pacific.**
Phys. Today, Vol. 40, No. 7, p. 85 (1987).

006.014   **Address of the President (Professor R. D. Davies) on the presentation of the Hannah Jackson (née Gwilt) Medal to David Malin on Friday 1987 May 8.**
R. D. Davies.
Q. J. R. Astron. Soc., Vol. 28, No. 4, p. 429 (1987).

006.015   **An irreverent history of space science. (Remarks on the occasion of the 60th birthday of Frank B. McDonald).**
J. E. Naugle.
NASA Conf. Publ., NASA CP–2464, p. 413 – 423 (1987). – See Abstr. 012.004.

006.016   **Address of the President (Professor R. D. Davies) on the presentation of the Gold Medal of the Royal Astronomical Society to Professor Martin Rees on Friday 1987 March 13.**
Q. J. R. Astron. Soc., Vol. 28, No. 3, p. 185 – 186 (1987).

006.017   **Leif Robinson received the Klumpke–Roberts Award 1987 of the Astronomical Society of the Pacific.**
J. E. Hesser, A. Fraknoi.
Mercury, Vol. 16, No. 5, p. 147 (1987).

006.018   **Bruno Benedetto Rossi received the 1987 Physics Prize of the Wolf Foundation.**
Phys. Today, Vol. 40, No. 8, Part 1, p. 81 – 82 (1987).

006.019   **Edwin E. Salpeter received the Catherine Wolfe Bruce Medal 1987 of the Astronomical Society of the Pacific.**
J. E. Hesser, A. Fraknoi.
Mercury, Vol. 16, No. 5, p. 144 – 145 (1987).

006.020   **Edwin E. Salpeter received the Catherine Bruce Medal 1987 of the Astronomical Society of the Pacific.**
Phys. Today, Vol. 40, No. 7, p. 83 (1987).

006.021   **Stephan Schneider received the Trumpler Award 1987 of the Astronomical Society of the Pacific.**
J. E. Hesser, A. Fraknoi.
Mercury, Vol. 16, No. 5, p. 145 – 146 (1987).

006.022   **Stephan E. Schneider received the Robert J. Trumpler Award 1987 of the Astronomical Society of the Pacific.**
Phys. Today, Vol. 40, No. 7, p. 83 (1987).

006.023   **Alan N. Stockton received the Muhlmann Prize 1987 of the Astronomical Society of the Pacific.**
J. E. Hesser, A. Fraknoi.
Mercury, Vol. 16, Nol 5, p. 146 – 147 (1987).

006.024   **Alan N. Stockton received the Muhlmann Prize 1987 of the Astronomical Society of the Pacific.**
Phys. Today, Vol. 40, No. 7, p. 83 – 84 (1987).

006.025   **Fred Lawrence Whipple received the Catherine Bruce Medal 1986 of the Astronomical Society of the Pacific.**
Phys. Today, Vol. 40, No. 7, p. 84 – 85 (1987).

006.026   **Albert Edward Whitford, in honor of his 80th year.**
S. M. Faber.
Nearly normal galaxies. From the Planck time to the present, p. V – VII (1987). – See Abstr. 012.031.

006.027   **Clifford M. Will received the 1987 AIP Science–Writing–Award in Physics and Astronomy.**
R. Hart.
Phys. Today, Vol. 40, No. 12, p. 77 – 78 (1987).

006.028   **Benjamin M. Zuckerman received the Muhlmann Prize 1986 of the Astronomical Society of the Pacific.**
Phys. Today, Vol. 40, No. 7, p. 85 (1987).

007 Obituaries

007.001   **Marc Aaronson (1950 – 1987).**
W. Priester.
Sterne Weltraum, 26. Jahrg., Nr. 9, p. 465 (1987).

007.002   **Bart Jan Bok, 28 April 1906 – 5 August 1983.**
C. J. Lada.
Q. J. R. Astron. Soc., Vol. 28, No. 4, p. 539 – 542 (1987).

007.003   **Johannes Classen, 30. Oktober 1908 – 4. August 1987.**
J. Helfricht.
Astron. Raumfahrt, 25. Jahrg., Heft 6, p. 187 – 188 (1987).

007.004   **Giuseppe Colombo, uomo e scienziato, 1920 – 20 February 1984.**
G. Puppi.
Atti Accad. Naz. Lincei, Ser. Ottava, Rend., Vol. 78, Fasc. 3, p. 119 – 126 (1985).

007.005   **Antoinette de Vaucouleurs (1921 – 1987).**
P. Baize.
Astronomie, Vol. 101, p. 566 (1987).

007.006   **Antoinette de Vaucouleurs, 14 November 1921 – 29 August 1987.**
L. J. Robinson.
Sky Telesc., Vol. 74, No. 6, p. 598 (1987).

007.007   **Paul Adrien Maurice Dirac, 8 August 1902 – 20 October 1984.**
G. C. Wick.
Atti Accad. Naz. Lincei, Ser. Ottava, Rend., Vol. 78, Fasc. 4, p. 181 – 193 (1985).

007.008   **Fran Dominko, 26 July 1903 – 22 February 1987.**
P. Razinger.
Vasiona, Année 35, No. 2, p. 38 – 39 (1987). In Croatian.

007.009   James Harland Duthie died 13 October 1986.
B. Thompson.
South. Stars, Vol. 32, No. 3, p. 102 – 103 (1987).

007.010   Johannes Hoppe, 30.4.1907 – 20.4.1987.
S. Marx.
Astron. Raumfahrt, 25. Jahrg., Heft 5, p. 155 (1987).

007.011   Carolyn J. Hurless, 1934 – 1987: AAVSO's enthusiastic ambassador.
J. A. Mattei.
J. Am. Assoc. Variable Star Obs., Vol. 16, No. 1, p. 35 – 36 (1987).

007.012   Demetrios Kotsakis (1909 – 9 May 1986).
G. Contopoulos.
Q. J. R. Astron. Soc., Vol. 28, No. 3, p. 397 – 398 (1987).

007.013   Konstantin Alekseevich Kulikov, 21 October 1902 – 26 July 1987.
Astron. Zh., Tom 64, Vyp. 6, p. 1338 – 1339 (1987). In Russian. English translation in Sov. Astron., Vol. 31, No. 6.

007.014   Harold William Newton, 13 October 1893 – 12 December 1985.
P. Wayman.
Q. J. R. Astron. Soc., Vol. 28, No. 4, p. 542 – 544 (1987).

007.015   Frederick J. O'Connor, 1907 – 1987.
P. A. Wayman.
Ir. Astron. J., Vol. 18, No. 2, p. 132 (1987).

007.016   Kirill Fedorovich Ogorodnikov, 30 June 1900 – 29 June 1985.
Tr. Astron. Obs., Leningrad, Tom 41, p. 194 – 196 (1987). Uch. Zap. LGU, No. 420, Ser. Mat. Nauk, Vyp. 63. In Russian.

007.017   Michael William Ovenden, 21 May 1926 – 15 March 1987.
A. E. Roy.
J. Br. Astron. Assoc., Vol. 98, No. 1, p. 53 (1987).

007.018   Michael William Ovenden, 21 May 1926 – 15 March 1987.
A. H. Batten.
J. R. Astron. Soc. Can., Vol. 81, No. 4, p. 109 – 112 (1987).

007.019   Professor Michael William Ovenden died 15 March 1987.
Vistas Astron., Vol. 30, Part 1, p. III (1987).

007.020   Stefan Leon Piotrowski (11 April 1910 – 17 January 1985).
Z. Kopal.
Q. J. R. Astron. Soc., Vol. 28, No. 3, p. 399 – 400 (1987).

007.021   A. B. Severnyj, 11 May 1913 – 4 April 1987.
Astron. Zh., Tom 64, Vyp. 4, p. 891 – 892 (1987). In Russian. English translation in Sov. Astron., Vol. 31, No. 4.

007.022   A. B. Severnyj, 11 May 1913 – 4 April 1987.
Pis'ma Astron. Zh., Tom 13, No. 7, p. 638 – 640 (1987). In Russian. English translation in Sov. Astron. Lett., Vol. 13.

007.023   In memoriam: Andrej Borisovich Severnyj (11 May 1913 – 4 April 1987.
Zemlya Vselennaya, No. 4, p. 33 – 34 (1987). In Russian.

007.024   Roman U. Sexl, 19 October 1939 – 10 July 1986.
P. C. Aichelburg, H. Kühnetl, A. P. French.
Phys. Today, Vol. 40, No. 7, p. 88 (1987).

007.025   Iosif Samuilovich Shklovskij: in memoriam (1 July 1916 – 3 March 1985).
Sov. Astron., Vol. 30, No. 5, p. 495 – 497 (1986). English translation of 42.007.031.

007.026   Bengt Strömgren, 21. 1. 1908 – 4. 7. 1987.
M. Rudkjøbing.
Astron. Tidsskr., Årg. 20, Nr. 4, p. 175 – 177 (1987).

007.027   Bengt Strömgren (1908 – 1987).
L. Woltjer.
Messenger, No. 49, p. 1, 43 – 44 (1987).

007.028   Bengt Strömgren (1908 – 1987).
W. Priester.
Sterne Weltraum, 26. Jahrg., Nr. 12, p. 677 (1987).

007.029   Dorde Teleki, 20 August 1928 – 23 February 1987.
S. Sadžakov.
Vasiona, Année 35, No. 2, p. 33 – 35 (1987). In Croatian.

007.030   Albert G. Velghe, 22 September 1916 – 27 October 1986.
A. H. Jarrett.
Mon. Notes Astron. Soc. S. Afr., Vol. 46, Nos. 7 – 8, p. 101 – 102 (1987).

007.031   Reginald Lawson Waterfield, 12 April 1900 – 10 June 1986.
M. J. Hendrie.
Q. J. R. Astron. Soc., Vol. 28, No. 4, p. 544 – 546 (1987).

007.032   Professor Sir Richard Woolley, OBE, ScD, FRS, 24 April 1906 – 24 December 1986.
D. Lynden–Bell.
Q. J. R. Astron. Soc., Vol. 28, No. 4, p. 546 – 551 (1987).

007.033   Sir Richard van der Riet Woolley died 24 December 1986.
D. J. Stickland, E. Budding.
South. Stars, Vol. 32, No. 3, p. 100 – 101 (1987).

008 Publications of Observatories, Institutes

Reports, communications, and publications of observatories and astronomical institutes are recorded in this section; included are numbered series of reprints. Whenever possible, the numbers of the abstracts referring to the publications are given. The places of observatories and institutes are listed in alphabetical order. If only the formal name of an observatory or institute is known, its place can be found in the following index list.

Aerospace Corporation	**El Segundo**, Calif.
Algonquin Radio Observatory	**Lake Traverse**, Canada
Allegheny Observatory	**Pittsburgh**, Pa.
Anglo-Australian Observatory	**Epping**, Australia
Antares Observatory	**Feira de Santana**, Brazil
Applied Research Corporation	**Landover**, Md.
Archenhold Observatory	**Berlin** (East)
Argentine Institute of Radio Astronomy	**Villa Elisa**, Argentina
Argentine Radio Astronomy Institute	**Pereyra**, Argentina
Arizona State University	**Tempe**, Ariz.
Astronomisches Rechen-Institut	**Heidelberg**, F.R. Germany
Babelsberg Observatory	**Potsdam**, German Democratic Republic
Bartol Research Foundation	**Newark**, Del.
Battelle Memorial Institute	**Richland**, Wash.
Bell Telephone Laboratories	**Holmdel**, N.J.
Bergedorf Observatory	**Hamburg**, F.R. Germany
Blindern Institute of Theoretical Astrophysics	**Oslo**, Norway
Bosscha Observatory	**Lembang**, Indonesia
Bouzareah Observatory	**Algiers**, Algeria
Boyden Observatory	**Bloemfontein**, South Africa
Bureau International de l'Heure	**Paris**, France
C. E. Kenneth Mees Observatory	**Rochester**, N.Y.
California Institute of Technology	**Pasadena**, Calif.
Cantonal Observatory	**Neuchâtel**, Switzerland
Carlsberg Automatic Meridian Circle	**La Palma**, Spain
Carter Observatory	**Wellington**, New Zealand
Center for Astrophysics	**Cambridge**, Mass.
Center for High Angular Resolution Astronomy, Georgia State University	**Atlanta**, Ga.
Centre de Données Stellaires	**Strasbourg**, France
Centro Astronomico Hispano-Aleman	**Calar Alto**, Spain
Centro de Investigación de Astronomía (CIDA)	**Mérida**, Venezuela
Cerro Calan National Astronomical Observatory	**Santiago**, Chile
Cerro Tololo Interamerican Observatory	**La Serena**, Chile
Chamberlin Observatory	**Denver**, Colo.
Charles University Astronomical Institute	**Prague**, Czechoslovakia
Climenhaga Observatory	**Victoria**, Canada
Cointe Observatory	**Liège**, Belgium
Computer Sciences Corporation	**Silver Spring**, Md.
Cornell University	**Ithaca**, N.Y.
Crawford Hill Observatory	**Holmdel**, N.J.
David Dunlap Observatory	**Richmond Hill**, Canada
Dearborn Observatory	**Evanston**, Ill.
Department of Astronomy and Space Sciences, Punjabi University	**Patiala**, India
Department of Mathematics, University of Poona	**Poona**, India
Deutsches Hydrographisches Institut	**Hamburg**, F.R. Germany
Dominion Astrophysical Observatory	**Ottawa**, Canada
Dominion Astrophysical Observatory	**Victoria**, Canada
Dominion Radio Astrophysical Observatory	**Penticton**, Canada
Dudley Observatory	**Schenectady**, N.Y.
Dunsink Observatory	**Dublin**, Ireland
Dyer Observatory	**Nashville**, Tenn.
Ebro Observatory	**Roquetas**, Spain
Effelsberg Radio Observatory	**Bonn**, F.R. Germany
Ege University Observatory	**Izmir**, Turkey
Electronics Research Laboratory	**Göteborg**, Sweden
Engelhardt Observatory	**Kazan**, USSR
Erwin W. Fick Observatory	**Ames**, Iowa
European Southern Observatory	**Garching**, F.R. Germany
European Southern Observatory	**La Silla**, Chile

Fabra Observatory	**Barcelona,** Spain
Felix Aguilar Observatory	**San Juan,** Argentina
Fernbank Science Center	**Atlanta,** Ga.
Figl Observatory	**Vienna,** Austria
Five College Astronomy Department	**Amherst,** Mass.
Floirac Observatory	**Bordeaux,** France
Florida State University Radio Observatory	**Tallahassee,** Fla.
Flower and Cook Observatory	**Malvern,** Pa.
Fort Skala Station	**Cracow,** Poland
Franko State University Astronomical Observatory	**Lvov,** USSR
Georgia State University	**Atlanta,** Ga.
Goddard Space Flight Center	**Greenbelt,** Md.
Goethe Link Observatory	**Bloomington,** Indiana
Hale Observatories	**Pasadena,** Calif.
Hartebeesthoek Radio Astronomy Observatory	**Johannesburg,** South Africa
Haute Provence Observatory	**Saint Michel l'Observatoire,** France
Haystack Observatory	**Westford,** Mass.
Heinrich-Hertz-Institut	**Berlin** (East)
Heliophysical Observatory	**Debrecen,** Hungary
Herzberg Institute of Astrophysics	**Ottawa,** Canada
High Altitude Observatory	**Boulder,** Colo.
H.M. Nautical Almanac Office	**Greenwich,** England
Hoher List Observatory	**Bonn,** F.R. Germany
Hopkins Observatory	**Williamstown,** Mass.
Horn d'Arturo Observatory	**Bologna,** Italy
Hvar Observatory	**Zagreb,** Yugoslavia
IBM Thomas J. Watson Research Center	**Yorktown Heights,** N.Y.
Indian Institute of Astrophysics	**Bangalore,** India
Infrared Telescope Facility	**Honolulu,** Hawaii
Institut Astrofiziki	**Dushanbe,** USSR
Instituto Nacíonal de Astrofisica, Optica y Electronica	**Puebla,** Pue., México
International Latitude Observatory	**Carloforte-Cagliari,** Italy
International Latitude Observatory	**Mizusawa,** Japan
IUE Observatory	**Villafranca,** Spain
James Mims Observatory	**Baton Rouge,** La.
Jet Propulsion Laboratory	**Pasadena,** Calif.
Jodrell Bank Radio Observatory	**Manchester,** England
Judson B. Coit Observatory	**Boston,** Mass.
Kandilli Observatory	**Istanbul,** Turkey
Kanzelhöhe Solar Observatory	**Graz,** Austria
Kapteyn Astronomical Laboratory	**Groningen,** The Netherlands
Karl Schwarzschild Observatory	**Tautenburg,** German Democratic Republic
Kiepenheuer Institut	**Freiburg,** F.R. Germany
Kiso Observatory	**Tokyo,** Japan
Kitt Peak National Observatory	**Tucson,** Ariz.
Kodaikanal Observatory	**Bangalore,** India
Königstuhl Observatory	**Heidelberg,** F.R. Germany
Konkoly Observatory	**Budapest,** Hungary
Korean National Astronomical Observatory	**Seoul,** Korea
Kwasan and Hida Observatories	**Kyoto,** Japan
Laboratory for High Energy Astrophysics	**Greenbelt,** Md.
Las Campanas Observatory	**Pasadena,** Calif.
Latitude Station of the Polish Academy of Sciences	**Borowiec,** Poland
Lawrence Livermore National Laboratory	**Livermore,** Calif.
Leander McCormick Observatory	**Charlottesville,** Va.
Lick Observatory	**Santa Cruz,** Calif.
Lindheimer Astronomical Research Center	**Evanston,** Ill.
Llano del Hato Observatory	**Merida,** Venezuela
Lockheed Palo Alto Research Laboratory	**Palo Alto,** Calif.
Lockheed Solar Observatory	**Saugus,** Calif.
Lohrmann Observatory	**Dresden,** German Democratic Republic
Louisiana State University Observatory	**Baton Rouge,** La.
Lowell Observatory	**Flagstaff,** Ariz.
Lunar and Planetary Laboratory	**Tucson,** Ariz.
Main Astronomical Observatory of the USSR Academy of Sciences	**Pulkovo,** USSR
Max-Planck-Institut für Astronomie	**Calar Alto,** Spain

Max-Planck-Institut für Astronomie	**Heidelberg,** F.R. Germany
Max-Planck-Institut für Kernphysik	**Heidelberg,** F.R. Germany
Max-Planck-Institut für Physik und Astrophysik	**Garching,** F.R. Germany
Max-Planck-Institut für Radioastronomie	**Bonn,** F. R. Germany
McDonald Observatory	**Fort Davis,** Tex.
McDonnell Center for the Space Sciences	**St. Louis,** Mo.
Meudon Observatory	**Paris,** France
Michigan State University Observatory	**East Lansing,** Mich.
Millstone Hill Radar Observatory	**Westford,** Mass.
Minor Planet Center	**Cambridge,** Mass.
Molonglo Radio Observatory	**Sydney,** Australia
Monterey Institute for Research in Astronomy	**Carmel Valley,** Calif.
Mount Hamilton	**Santa Cruz,** Calif.
Mount John University Observatory	**Lake Tekapo,** New Zealand
Mount Palomar Observatory	**Pasadena,** Calif.
Mount Stromlo Observatory	**Canberra,** Australia
Mount Wilson Observatory	**Pasadena,** Calif.
Mullard Radio Astronomy Observatory	**Cambridge,** England
Mullard Space Science Laboratory	**London,** England
Multiple Mirror Telescope Observatory	**Tucson,** Ariz.
N. Copernicus Astronomical Center	**Warsaw,** Poland
N. Copernicus University Observatory	**Torun,** Poland
Narrabri Observatory	**Sydney,** Australia
NASA Headquarters	**Washington,** D.C.
National Astronomy and Ionosphere Center	**Ithaca,** N.Y.
National Bureau of Standards	**Washington,** D.C.
National Radio Astronomy Observatory	**Green Bank,** W.Va.
National Radio Astronomy Observatory	**Socorro,** N.M.
National Solar Observatory	**Sunspot,** N.M.
Naval Observatory	**San Fernando** (Cadiz), Spain
New Mexico State University Observatory	**Las Cruces,** N.M.
Nicholas Copernicus Observatory	**Brno,** Czechoslovakia
Nizamiah and Japal-Rangapur Observatories	**Hyderabad,** India
Nuffield Radio Astronomy Laboratories	**Manchester,** England
Oak Ridge Observatory	**Cambridge,** Mass.
Observatorio del Ebro	**Roquetas,** Spain
Ole Roemer Observatory	**Aarhus,** Denmark
Onsala Space Observatory	**Göteborg,** Sweden
Owens Valley Radio Observatory	**Big Pine,** Calif.
Pacific Northwest Laboratory	**Richland,** Wash.
Pennsylvania State University	**University Park,** Pa.
Perkins Observatory	**Delaware,** Ohio
Perth Observatory	**Bickley,** Australia
Pic du Midi Observatory	**Toulouse,** France
Pino Torinese Observatory	**Turin,** Italy
Purple Mountain Observatory	**Nanking,** China
Raman Research Institute	**Bangalore,** India
Rattlesnake Mountain Observatory	**Richland,** Wash.
Remeis Observatory	**Bamberg,** F.R. Germany
Rensselaer Observatory	**Troy,** N.Y.
Ritter Astrophysical Research Center	**Toledo,** Ohio
Rosemary Hill Observatory	**Bronson,** Fla.
Rothney Astrophysical Observatory	**Calgary,** Canada
Royal Observatory	**Edinburgh,** Scotland
Royal Observatory	**Greenwich,** England
Royal Observatory of Belgium	**Uccle,** Belgium
Rutgers University	**Piscataway,** N.J.
Rutherford Appleton Laboratory	**Chilton,** England
Sagamore Hill Radio Observatory	**Hanscom,** Mass.
Saltsjöbaden Observatory	**Stockholm,** Sweden
San Vittore Observatory	**Bologna,** Italy
Satelliten-Beobachtungsstation	**Zimmerwald,** Switzerland
Shaanxi Astronomical Observatory	**Lintong,** China
Smithsonian Astrophysical Observatory	**Cambridge,** Mass.
Sommers-Bausch Observatory	**Boulder,** Colo.
Sonnenborgh Observatory	**Utrecht,** The Netherlands
Sonoma State University	**Rohnert Park,** Calif.
South African Astronomical Observatory	**Cape Town,** South Africa
Space Telescope Science Institute	**Baltimore,** Md.
Special Astrophysical Observatory	**Zelenchukskaya,** USSR

Sproul Observatory	**Swarthmore,** Pa.
Stanford Center for Radar Astronomy	**Menlo Park,** Calif.
Stellar Data Center	**Strasbourg,** France
Sternberg State Astronomical Institute	**Moscow,** USSR
Steward Observatory	**Tucson,** Ariz.
Stockert Radio Observatory	**Bonn,** F.R. Germany
Struve Astrophysical Observatory	**Tartu,** USSR
Tata Institute of Fundamental Research	**Bombay,** India
Tôhoku University Observatory	**Sendai,** Japan
United Kingdom Infrared Telescope (UKIRT)	**Hilo,** Hawaii
University of Sussex	**Brighton,** England
U.S. Naval Observatory	**Washington,** D.C.
UT Radio Astronomy Observatory	**Marfa,** Tex.
Uttar Pradesh State Observatory	**Naini Tal,** India
Van Vleck Observatory	**Middletown,** Conn.
Vatican Observatory	**Castel Gandolfo,** Vatican City
Warner and Swasey Observatory	**Cleveland,** Ohio
Washburn Observatory	**Madison,** Wis.
Wesleyan Radio Observatory	**Delaware,** Ohio
Western Ontario University Observatory	**London,** Canada
Whipple Observatory (MMT)	**Tucson,** Ariz.
Wilhelm Foerster Observatory	**Berlin** (West)
Yale University Astronomical Observatory	**New Haven,** Conn.
Yerkes Observatory	**Williams Bay,** Wis.
Yunnang Obs.	**Kunming,** China
Zentralinstitut für Physik der Erde	**Potsdam,** German Democratic Republic

008.018   Alma–Ata

Trudy Astrofizicheskogo Instituta.
Tom 47 (44.153.038, 44.042.043 – 44.042.046, 44.117.221,
44.042.047, 44.062.085, 44.062.086, 44.155.121, 44.151.121,
44.151.122, 44.161.239, 44.161.240, 44.066.106, 44.066.107).
Tom 48 (44.132.032, 44.134.038, 44.132.033, 44.131.220,
44.131.221, 44.064.062, 44.112.149, 44.116.056, 44.157.282,
44.157.283, 44.034.099, 44.036.164, 44.031.050, 44.031.051,
44.034.100, 44.031.052, 44.102.053).

008.057   Baltimore, Md.

Space Telescope Science Institute, Preprint Series.
Nos. 175 (44.160.086), 177 (44.117.196), 178 (44.158.228), 179
(44.103.459), 180 (44.124.201), 181 (44.063.067), 182
(44.134.035, 44.114.075), 183 (44.114.076, 44.154.045,
44.154.046), 184 (44.151.111), 185 (44.122.075), 186
(44.153.036), 187 (44.002.043 – 44.002.045), 188 (44.159.070),
189 (44.161.202), 190 (44.158.229), 191 (44.036.146), 192
(44.158.230), 193 (44.125.285), 194 (44.117.197), 195
(44.159.071), 196 (44.117.198), 197 (44.124.009), 198
(44.126.075), 199 (44.117.200), 200 (44.131.198), 201
(44.035.058), 202 (44.064.055), 203 (44.121.038), 204
(44.157.264, 44.160.087), 205 (44.160.088), 206 (44.155.109),
207 (44.161.203), 208 (44.161.204), 209 (44.156.025), 210
(44.125.286), 211 (44.067.092), 212 (44.159.072), 213
(44.126.076), 214 (44.080.047), 215 (44.158.231), 220
(44.158.232), 221 (44.158.233), 222 (44.062.076, 44.131.199).

Space Telescope Science Institute, Newsletter.
Vol. 4, Nos. 3 – 4 (1987).

008.063   Bangalore

**Indian Institute of Astrophysics. Report for the period 1985
April 1 – 1986 March 31.**
Bull. Astron. Soc. India, Vol. 15, Nos. 2 – 3, p. 121 – 162
(1987).

**Raman Research Institute. Report for the period 1985 April 1 –
1986 March 31.**
Bull. Astron. Soc. India, Vol. 15, Nos. 2 – 3, p. 163 – 165
(1987).

008.084   Berlin (East)

**Heinrich–Hertz–Institut für Atmosphärenforschung und
Geomagnetismus. Report for 1986.**
Akad. Wiss. DDR, Jahrb. 1986, p. 102 – 104, wiss. Veröff.
p. 154 – 157 (1987).

Institut für Kosmosforschung. Report for 1986.
Akad. Wiss. DDR, Jahrb. 1986, p. 109, wiss. Veröff.
p. 165 – 167 (1987).

008.087   Berlin (West)

**Institut für Astronomie und Astrophysik der Technischen
Universität Berlin, Mitteilungen.**
No. 8 (43.008.087).

008.096   Bloemfontein

Boyden Observatory, Occasional Publication.
No. 4 (44.125.284).

Boyden Observatory, Reprint.
Nos. 64 (43.032.036), 65 (44.125.283).

008.147   Budapest

Communications from the Konkoly Observatory.
Nos. 87 (44.122.071), 88 (44.122.072), 89 (44.122.073), 90
(44.122.074).

Publications of the Astronomy Department of the Eötvös
University.
No. 8 (44.012.081).

008.162   Cambridge, England

Institute of Astronomy, University of Cambridge, Annual Report
1986 – 1987.
D. Lynden–Bell.
Inst. Astron., Univ. Camb., Annu. Rep., 45 pp. (1987).

008.165   Cambridge, Mass.

The Minor Planet Circulars/Minor Planets and Comets.
Nos. 11887 – 12624 (1987).

IAU Circulars.
Nos. 4413 – 4521 (1987).

008.171   Cape Town

South African Astronomical Observatory, Circulars.
No. 11 (44.120.010, 44.120.011, 44.113.040 – 44.113.042,
44.154.048, 44.113.043, 44.113.044, 44.122.077 – 44.122.079).

SAAO Newsletter.
No. 10 (1987).

008.180   Carloforte–Cagliari

Circolari della Stazione Astronomica Internazionale di
Latitudine, Carloforte–Cagliari.
N. 32 (44.044.035), 33 (44.082.066), 34 (44.082.067), 35
(44.082.068), 36 (44.044.036), 37 (44.044.037).

Pubblicazioni della Stazione Astronomica Internazionale di
Latitudine, Carloforte–Cagliari, Nuova Serie.
N. 120, 125 (41.013.053), 126 (42.044.006), 127 (41.044.097),
128 (44.004.074), 129 (44.082.065), 136 (43.004.065), 139
(44.041.002), 140 (44.013.069).

Stazione Astronomica Internazionale di Latitudine, Carloforte–
Cagliari, Note Tecniche.
N. 7 (44.036.184).

008.186   Castel Gandolfo

Annual Report 1987.
G. V. Coyne.
Vatican Obs. Publ., Vol. 2, No. 11, p. 139 – 152 (1987).

Vatican Observatory Publications.
Vol. 2, No. 11 (44.008.186).

008.201   Chilton

Rutherford Appleton Laboratory, Report.
RAL 87–032 (44.036.147), 87–043 (44.012.010), 87–044
(44.084.047), 87–045 (44.062.077), 87–058 (44.036.148), 87–068
(44.154.047).

008.237   Crimea

Chronicle.
Izv. Krymskoj Astrofiz. Obs., Tom 76, p. 204 – 206 (1987). In
Russian. English translation in Bull. Crimean Astrophys. Obs.,
Vol. 76.

Izvestiya Ordena Trudovogo Krasnogo Znameni Krymskoj
Astrofizicheskoj Observatorii.
Tom 76 (44.122.054, 44.122.055, 44.114.065, 44.114.066,
44.119.028, 44.113.032, 44.113.033, 44.116.039, 44.117.168,
44.113.034, 44.117.169, 44.158.179, 44.158.180, 44.159.050,
44.126.068, 44.071.019 – 44.071.022, 44.075.021 – 44.075.024,
44.080.040, 44.074.050, 44.034.054, 44.031.032, 44.008.237),
Tom 77 (44.073.060, 44.073.061, 44.072.055, 44.072.056,
44.073.062, 44.080.042 – 44.080.045, 44.077.043, 44.080.046,
44.117.189, 44.112.139, 44.158.219 – 44.158.222, 44.036.139,
44.031.033, 44.031.034, 44.034.061, 44.032.024).

Bulletin of the Crimean Astrophysical Observatory.
Vol. 74 (44.114.007, 44.122.015, 44.114.008 – 44.114.010,
44.116.007, 44.116.008, 44.022.050, 44.117.035, 44.072.008,
44.073.008, 44.073.009, 44.074.020, 44.080.007, 44.084.011,
44.071.005, 44.034.011, 44.063.011, 44.031.003),
Vol. 75 (44.073.023, 44.073.024, 44.080.015, 44.066.056,
44.080.016, 44.073.025 – 44.073.027, 44.077.015 – 44.077.017,
44.091.022, 44.122.028, 44.117.067, 44.117.068, 44.063.024,
44.122.029, 44.064.014, 44.117.069, 44.121.012, 44.158.078,
44.032.010, 44.031.022).

008.240   Debrecen

Publications of Debrecen Heliophysical Observatory.
Vol. 6, No. 1 (44.072.057 – 44.072.059).

008.249   Dresden

Lohrmann–Observatorium, Technische Universität Dresden,
Zirkular.
Nr. 119 – 120 (44.044.016).

008.252   Dublin

Dunsink Observatory in 1986. Report for the year 1986.
P. A. Wayman.
Ir. Astron. J., Vol. 18, No. 2, p. 116 – 121 (1987).

008.267   Edinburgh

Royal Observatory Edinburgh, Research and Facilities 1987.
M. S. Longair.
R. Obs. Edinb., Res. Facilities, 4 + 117 pp. (1987).

Edinburgh Astronomy Preprint.
Nos. 10/87 (44.032.026), 11/87 (44.161.205), 12/87 (44.157.265),
13/87 (44.157.266), 14/87 (44.158.235), 15/87 (44.158.236), 16/
87 (44.158.237), 17/87 (44.156.026), 19/87 (44.112.142), 20/87
(44.158.238), 21/87 (44.158.239), 22/87 (44.134.036), 23/87
(44.131.200), 24/87 (44.158.240), 25/87 (44.151.112).

008.273   **Epping**

Anglo–Australian Telescope 1986 – 87. Report of the Anglo–Australian Telescope Board, 1 July 1986 to 30 June 1987.
Australian Government Publishing Service, Canberra,
Australia, 8 + 64 pp. (1987).

Anglo–Australian Observatory, Preprints.
Nos. 216 (44.160.089), 217 (44.036.149), 218 (44.125.288), 219 (44.159.073), 220 (44.160.090), 221 (44.034.071), 222 (44.117.199).

AAO Newsletter.
Nos. 42, 43 (1987).

008.285   **Farnborough**

Royal Aircraft Establishment, Farnborough, Geophysical Studies in Space Department. Report for the year ending 1987 March 31.
Q.J.R. Astron. Soc., Vol. 28, No. 4, p. 497 – 501 (1987).

008.306   **Garching**

ESO's Directors General: retrospect and prospect.
L. Woltjer, H. van der Laan.
Messenger, No. 50, p. 1 – 3 (1987).

ESO Scientific Report.
No. 6 (44.002.062).

ESO Scientific Preprint.
Nos. 514 (44.158.243), 515 (44.157.268), 516 (44.125.066), 517 (44.151.114), 518 (44.125.821), 519 (44.161.207), 520 (44.036.151), 521 (44.132.028), 522 (44.103.460), 523 (44.125.067), 524 (44.159.074), 525 (44.125.068), 526 (44.154.050), 527 (44.122.076), 528 (44.154.051), 529 (44.122.080, 44.122.081), 530 (44.154.052), 531 (44.158.244), 532 (44.158.245), 533 (44.158.246), 534 (44.159.075), 535 (44.161.208), 536 (44.158.247, 44.158.248), 537 (44.161.209, 44.160.091, 44.161.210), 538 (44.131.204), 539 (44.161.211), 540 (44.132.029), 541 (44.125.291), 542 (44.154.053), 543 (44.154.054), 544 (44.159.076), 545 (44.159.077), 546 (44.114.077), 547 (44.158.249), 548 (44.117.204), 549 (44.131.205), 550 (44.157.269), 551 (44.112.143), 552 (44.158.250), 553 (44.154.055), 554 (44.125.292), 555 (44.133.015), 556 (44.132.030), 557 (44.125.293).

The Messenger – El Mensajero.
Nos. 49, 50 (1987).

ST–ECF Newsletter.
No. 8 (1987).

MPA Reports.
Nos. 304 (44.067.094), 305 (44.131.203), 306 (44.067.095), 307 (44.067.096), 308 (44.157.267), 309 (44.064.056), 310 (44.080.048), 311 (44.161.206), 312 (44.125.289), 313 (44.062.078), 314 (44.151.113), 315 (44.067.097), 316 (44.062.079), 317 (44.062.080), 318 (44.066.096).

MPE Reports.
Nos. 203 (44.067.093), 204 (44.117.203).

MPE Contributions.
Nos. 2486 (43.035.001), 2488, 2494, 2501 (42.157.204), 2503 (43.158.040), 2505 (43.117.036), 2511, 2512, 2514 (43.103.464), 2515 (43.074.013), 2516 (43.074.014), 2517 (43.074.015), 2518 (43.067.044), 2522 (43.117.176), 2523, 2525 (43.155.029), 2530, 2537, 2539 (43.084.010), 2541 (43.062.062), 2542 (43.131.297), 2545, 2546, 2547 (43.117.277).

008.315   **Glasgow**

Department of Astronomy, University of Glasgow. Report for the period 1983 October to 1986 July.
A. E. Roy.
Q.J.R. Astron. Soc., Vol. 28, No. 4, p. 502 – 509 (1987).

008.318   **Goeteborg**

Onsala Space Observatory, Preprint.
No. 87:59 (44.131.201).

008.327   **Graz**

Mitteilungen der Universitätssternwarte Graz.
Nr. 116 (40.098.020), 117 (41.098.002), 118 (42.098.036), 119 (41.098.020), 120 (41.098.029), 121 (41.098.030), 122 (42.117.067), 123 (42.080.003), 124 (42.098.035), 125 (42.098.032), 126 (43.098.016).

Mitteilungen des Sonnenobservatoriums Kanzelhöhe.
Nr. 40 (41.002.059), 41 (41.080.039), 42 (41.072.002), 43 (44.073.063), 44 (43.072.004), 45 (42.080.048), 46 (41.073.138).

008.330   **Green Bank, W.Va.**

NRAO Workshops.
Nos. 15 (44.012.076), 16 (44.012.015), 17 (44.012.033), 18 (44.012.009).

008.333   **Greenbelt, Md.**

Astronomical Data Center (ADC). Annual Report for 1983.
W. H. Warren Jr., J. M. Mead.
Astron. Data Cent. Bull., Vol. 1, No. 4, p. 367 – 370 (1987).

Astronomical Data Center (ADC). Annual Report for 1984.
W. H. Warren Jr., J. M. Mead.
Astron. Data Cent. Bull., Vol. 1, No. 4, p. 371 – 374 (1987).

Astronomical Data Center Bulletin.
Vol. 1, No. 4 (1987).

008.336   **Greenwich**

Royal Greenwich Observatory, Preprints.
Nos. 55 (44.117.201), 56 (44.113.045), 57 (44.141.016), 58 (44.032.027, 44.041.008, 44.005.006, 44.002.061, 44.013.058), 59 (43.158.066), 62 (44.158.241), 63 (44.131.202), 64 (44.061.063), 65 (44.013.059), 66 (44.158.242), 67 (44.117.202).

Gemini.
No. 17 (1987).

008.354   **Heidelberg**

Astronomisches Rechen–Institut Heidelberg, Mitteilungen, Serie A.
Nr. 186 – 188, 189 (42.111.008), 190 (43.151.122), 191 (43.041.028).

Astronomisches Rechen–Institut Heidelberg, Mitteilungen Serie B.
Nr. 151 (44.002.019), 152 (44.103.436), 153 (44.041.019).

Astronomisches Rechen–Institut Heidelberg, Preprint Series.
Nos. 7 (44.043.004), 8 (44.043.005), 9 (44.041.009), 10
(44.036.152), 11 (44.041.010), 12 (44.002.064), 13 (44.155.110),
14 (44.151.115).

Apparent Places of Fundamental Stars 1988.
(44.046.014).

Astronomy and Astrophysics Abstracts.
Vol. 43 (44.002.063).

**Max–Planck–Institut für Astronomie, Heidelberg–Königstuhl,
Preprints.**
(44.157.270, 44.131.206, 44.133.016, 44.117.205, 44.121.039).

008.414 Kunming, Yunnan

Publications of Yunnan Observatory.
No. 1/1987 (44.067.101, 44.161.234, 44.036.153, 44.034.080 –
44.034.082, 44.033.022, 44.034.083, 44.036.154 – 44.036.156,
44.082.058, 44.103.462, 44.080.050, 44.071.025, 44.034.084,
44.032.028).
No. 2/1987 (44.034.085, 44.032.086, 44.036.157, 44.034.087,
44.034.088, 44.085.013, 44.078.008, 44.072.061, 44.073.069,
44.080.051, 44.118.022, 44.158.254, 44.161.235, 44.032.029,
44.131.214, 44.103.463, 44.034.089, 44.034.090, 44.082.059,
44.013.060).

008.417 Kyoto

**Contributions from the Kwasan and Hida Observatories,
University of Kyoto.**
Nos. 261 (40.073.044), 262 (40.099.117), 263 (40.072.005), 264
(41.097.025), 265 (41.073.005), 266 (41.072.094), 267
(41.074.113), 268 (43.036.262), 269 (42.071.015), 270
(42.073.040), 271 (42.074.028), 272 (43.073.038), 273
(42.073.110), 274 (44.073.064), 275 (43.097.008), 276
(43.073.186), 277 (43.071.020).

**Contributions from the Department of Astronomy, University of
Kyoto.**
Nos. 207 (41.062.129), 208 (41.121.061), 209 (42.131.013), 210
(41.120.004), 211 (42.112.010), 212 (42.158.010), 213
(42.157.057), 214 (41.119.056), 215 (42.125.058), 216
(42.151.107), 217 (42.160.071), 218 (42.120.019), 219
(43.157.015), 220 (43.119.012), 221 (43.021.045), 222
(43.125.070), 223 (43.062.096), 224 (44.151.009), 225
(44.067.044).

008.435 Lake Tekapo

**Astronomy at the University of Canterbury, Physics Department
and the Mount John University Observatory. Annual Report
1986.**
B. G. Wybourne.
South. Stars, Vol. 32, No. 4, p. 130 – 133 (1987).

008.456 Leningrad

Chronicle.
Byull. Inst. Teor. Astron., Tom 15, No. 9, p. 544 – 548 (1986).
In Russian.

Chronicle.
Tr. Astron. Obs. Leningrad, Tom 41, p. 197 – 201 (1987)
= Uch. Zap. LGU, No. 420, Ser. Mat. Nauk, Vyp. 3. In
Russian.

Byulleten' Instituta Teoreticheskoj Astronomii.
Tom 15, No. 9 (44.052.014, 44.046.020, 44.052.015, 44.094.037,
44.045.006, 44.042.036, 44.100.021, 44.066.095, 44.008.456).

Trudy Astronomicheskoj Observatorii.
Tom 41 (44.077.056, 44.118.031, 44.158.333, 44.036.214,
44.122.106, 44.080.068, 44.155.155, 44.032.050, 44.042.061,
44.007.016, 44.008.456).

008.465 Lintong

Publications of the Shaanxi Astronomical Observatory.
Vol. 9, No. 1 (44.004.054, 44.085.012, 44.044.023, 44.044.024,
44.034.075 – 44.034.079).

Time and Frequency Services, Bulletin.
Nos. 94 – 99 (44.044.017).

008.474 London, England

**Mullard Space Science Laboratory, University College, London.
Report for the period 1985 November 1 to 1986 October 31.**
J. L. Culhane.
Q.J.R. Astron. Soc., Vol. 28, No. 3, p. 355 – 365 (1987).

008.483 Lund

Reports from the Observatory of Lund.
No. 20 (44.153.037).

008.495 Madrid

Instituto de Astronomia y Geodesia, Publicación.
Nos. 143 (44.044.018), 144 (44.044.019), 151, 152 (44.044.020).

008.501 Manchester

**Nuffield Radio Astronomy Laboratories, University of
Manchester. Report on the researches in radio astronomy at
Jodrell Bank for the year ending December 31st 1986.**
F. G. Smith.
Q.J.R. Astron. Soc., Vol. 28, No. 3, p. 366 – 385 (1987).

**Astronomical Contributions from the University of Manchester,
Series II: Jodrell Bank Reprints.**
Nos. 786 (41.159.173), 787 (42.159.112), 788 (42.158.104), 789
(42.033.033), 790 (41.159.046), 791 (42.008.501), 792
(42.131.125), 793 (42.158.007), 794 (42.159.113), 795
(41.124.013), 796 (43.158.005), 797 (42.117.191), 798
(43.158.006), 799 (43.157.018), 800 (43.033.001), 801
(42.159.149), 802 (43.160.011), 803 (43.159.010), 804
(42.013.074), 805 (43.126.009), 806 (43.158.019), 807
(43.112.017), 808 (41.158.301), 809 (44.033.020), 810
(43.112.031), 811 (43.013.065), 812 (43.158.115), 813
(43.159.039), 814 (43.134.017), 815 (43.126.032), 816
(43.131.031), 817 (44.008.501), 818 (43.159.030), 819
(43.161.122), 820 (44.117.127), 821 (44.158.087), 822
(44.158.092), 823 (44.112.053), 824 (44.159.066), 825
(44.126.057), 826 (44.126.059), 827, 828 (44.131.247), 829
(44.117.253), 830, 831.

008.504 Manila

Solar Maps and Activity.
(44.072.060).

008.519   **Milano**

Milano Preprint Series in Astrophysics.
Nos. 23 (44.160.092), 24 (44.160.093).

008.525   **Mizusawa**

Publications of the International Latitude Observatory of Mizusawa.
Vol. 19, Nos. 1 – 2 (44.045.007).

Monthly Notes of the International Polar Motion Service.
Nos. 5 – 11 (1987) (44.044.022).

Bulletins, Time Service of the Mizusawa Observatory.
Vol. 30 (44.044.021).

008.549   **Nanking**

Publications of Purple Mountain Observatory.
Vol. 6, Nos. 2 – 3 (1987).

008.576   **Oslo**

Theoretic Papers.
Vol. 7, Nr. 7 – 10.

008.585   **Oxford**

Department of Astrophysics, University of Oxford. Report for the period 1985 August 1 – 1986 July 31.
D. E. Blackwell.
Q.J.R. Astron. Soc., Vol. 28, No. 3, p. 386 – 395 (1987).

008.591   **Paris**

Institut d'Astrophysique de Paris, Pré–Publication.
Nos. 192 (44.157.276), 193 (44.112.146), 194 (44.159.080), 195 (44.158.255), 196 (44.159.081), 197 (44.116.054), 198 (44.061.070), 199 (44.159.082), 200 (44.158.256), 201 (44.156.029), 202 (44.063.070), 203 (44.034.092), 204 (44.032.030), 205 (44.117.208), 206 (44.117.209), 207 (44.117.210), 207a (44.125.302), 208 (44.159.083), 209 (44.158.257), 210 (44.151.119), 211 (44.159.084).

Bureau International de l'Heure (B.I.H.) Circulars.
D248 – D253 (44.044.026),
E14 (44.044.027).

008.627   **Porto**

Publicações do Observatório Astronómico "Prof. Manuel de Barros", da Faculdade de Ciências do Porto.
Nos. 37 (44.155.117), 38 (44.004.055).

008.633   **Potsdam**

Zentralinstitut für Astrophysik. Report for 1986.
Akad. Wiss. DDR, Jahrb. 1986, p. 100 – 102, wiss. Veröff.
p. 152 – 154 (1987).

Zeit– und Breitenbestimmungen, Zeitsysteme, Präzisionszeitvergleiche.
Jahrg. 1985, Nos. 1 – 4 (44.044.029).

Solar Radio Data.
Part II, III, IV 1987 (44.077.002).

Zentralinstitut für Physik der Erde. Report for 1986.
Akad. Wiss. DDR, Jahrb. 1986, p. 104 – 105, wiss. Veröff., p. 157 – 159 (1987).

Veröffentlichungen des Zentralinstituts für Physik der Erde.
Nr. 84.

008.636   **Prague**

Publications of the Astronomical Institute of the Czechoslovak Academy of Sciences.
No. 64 (44.035.064, 44.035.065).

Circular of the Czechoslovak Observatories, Time and Latitude.
July 1986 – March 1987 (44.044.028).

008.684   **Saint Michel l'Observatoire**

Observatoire de Haute Provence, Pré–Publication.
Nos. 22 (44.134.037), 23 (44.111.015), 24 (44.122.086), 25 (44.125.581), 26 (44.125.300), 27 (44.142.035), 28 (44.122.087), 29 (44.125.301).

008.690   **San Fernando**

Catáloga SRS de San Fernando, zona −10° a −30°, equinoccio 1950.0.
See Abstr. 002.082.

008.699   **Santa Cruz, Calif.**

Contributions from the Lick Observatory.
Nos. 434 (38.154.055), 438 (41.080.089), 439 (42.131.066), 440 (42.107.016), 441 (43.064.010), 442 (44.034.203), 443 (44.042.010).

Lick Observatory Bulletin.
Nos. 1027 (41.159.140), 1030 (41.158.262), 1032 (41.067.111), 1033 (42.080.009), 1034 (42.107.010), 1037 (42.153.012), 1038 (42.107.019), 1040 (42.121.054), 1041 (42.041.015), 1042 (42.065.040), 1043 (42.158.077), 1044 (42.157.004), 1045 (42.107.027), 1046 (42.121.086), 1047 (43.151.019), 1048 (42.121.050), 1049 (43.158.044), 1050 (43.133.010), 1051 (42.121.078), 1052 (43.114.017), 1053 (43.157.047), 1054 (43.134.010), 1055 (43.121.027), 1056 (44.122.003), 1057 (43.034.136), 1058 (43.156.020), 1059 (44.158.036), 1060 (43.121.088), 1061 (43.034.182), 1062 (44.157.198), 1063 (43.114.103), 1064 (43.114.104), 1065 (43.155.151), 1066 (44.036.185), 1067 (44.125.215), 1068 (44.111.004), 1069 (44.111.005).

008.702   **Santiago**

Astrophysics Preprint Series.
Nos. 17 (44.160.094), 18 (44.112.148).

008.717   **Sendai**

Sendai Astronomiaj Raportoj.
Nos. 308 (43.157.128), 310 (43.064.015), 311 (43.158.291), 312 (43.082.073), 314 (43.117.351), 315 (43.117.352), 316 (43.158.379), 317 (44.065.031), 318 (43.131.160), 319 (44.158.052).

008.732   **Skalnaté Pleso**

Contributions of the Astronomical Observatory Skalnaté Pleso.
Vol. 16 (44.113.046, 44.118.023, 44.074.055, 44.034.098,
44.114.080, 44.114.081, 44.117.217, 44.073.071, 44.119.041,
44.119.042).

008.735   **Socorro, N.M.**

**Astrophysics Research Center, New Mexico Institute of Mining
and Technology, Socorro, New Mexico 87801. Report for the
period 1985 September – 1986 October.**
S. N. Shore.
Bull. Am. Astron. Soc., Vol. 19, No. 2, p. 771 – 773 (1987).

008.738   **Sonneberg**

Veröffentlichungen der Sternwarte in Sonneberg.
Band 10, Heft 2 (44.123.004).

Mitteilungen der Sternwarte zu Sonneberg.
Nos. 70 (44.006.002), 71 (44.153.044), 72 (44.121.047), 73
(44.117.237), 74 (44.119.052), 75 (44.117.238), 76 (44.122.098).

Mitteilungen über Veränderliche Sterne.
Band 11, Heft 1 (44.123.010, 44.122.099, 44.123.011,
44.122.100, 44.122.101, 44.117.254, 44.123.012, 44.122.102).
Band 11, Heft 2 (44.117.213, 44.119.039, 44.119.040,
44.123.005, 44.124.142, 44.122.084, 44.117.214).
Band 11, Heft 3 (44.122.085, 44.117.215, 44.117.216,
44.123.006, 44.123.007).

008.741   **St. Andrews**

Communications from the University Observatory, St. Andrews.
Nos. 43 (38.122.082), 44 (38.119.047), 45 (38.117.206), 46
(39.113.009), 47 (40.122.037), 48 (42.122.044), 49 (42.065.107),
50 (42.117.192), 51 (42.065.108), 52 (43.113.006), 53
(43.117.061), 54 (42.122.169), 55 (43.119.004), 56 (44.113.016).
This series terminates with No. 56.

University Observatory, St. Andrews, Reprint.
Nos. 101 (38.117.024), 102 (37.117.245), 103 (37.122.201), 104
(38.122.025), 105 (38.114.102), 106 (39.117.061), 107
(40.122.117), 108 (41.117.011), 109 (40.111.004), 110
(41.065.088), 111 (42.154.011), 112 (42.118.012), 113
(42.119.042), 114 (42.117.194), 115 (42.117.195), 116
(42.117.196), 117 (43.155.003), 118 (41.157.250), 119
(44.117.043), 120 (43.111.010), 121 (44.122.049), 122
(43.119.099), 123 (43.117.362), 124 (44.117.044), 125
(44.117.042), 126 (43.034.216).
This series terminates with No. 126.

008.753   **Stockholm**

Stockholms Observatorium, Preprint.
1987:2 (44.131.217).

008.759   **Strasbourg**

Bulletin d'Information du Centre de Données Stellaires.
No. 33 (1987).

Bulletin sur les Etoiles Tardives à Spectre Particulier.
No. 5 (1987).

008.777   **Tartu**

Tartu Astrofüüsika Observatoorium Teated.
Nr. 84 (44.002.072), 44.111.016, 44.036.163, 44.041.012).
Nr. 85 (44.063.071).

008.792   **Tokyo**

Annals of the Tokyo Astronomical Observatory, Second Series.
Vol. 21, No. 4 (44.031.049, 44.158.258, 44.117.211, 44.002.066,
44.155.118, 44.157.277).

Tokyo Astronomical Bulletin, Second Series.
No. 280 (44.103.464).

Tokyo Astronomical Observatory Report.
Vol. 21, No. 1 (44.033.023, 44.031.047, 44.031.048, 44.036.161,
44.035.066, 44.033.024, 44.073.070, 44.072.064).

Tokyo Astronomical Observatory Reprints.
Nos. 807 (42.073.009), 808 (41.073.147), 813 (42.072.053), 814
(42.131.356), 815 (42.036.184), 816 (43.112.051), 828
(43.112.005), 836 (43.134.033), 837 (44.157.060), 838
(44.158.043), 839 (44.103.111), 840 (44.031.002), 841
(43.036.202), 842 (44.157.091), 843 (44.157.092), 844
(44.155.019), 845 (44.155.020), 846 (44.074.026), 847
(44.131.021), 848 (44.158.309), 849 (44.155.148), 850
(44.131.248), 851 (44.064.070), 856 (44.157.197), 857
(44.035.055), 858 (44.151.033).

Tokyo Astronomical Observatory, Kiso Information Bulletin.
Vol. 2, No. 4 (44.002.065, 44.034.093).

Tokyo Astronomical Observatory, Time and Latitude Bulletins.
Vol. 61, Nos. 1 – 3 (44.044.030).

Quarterly Bulletin on Solar Activity.
Vol. 24, Part V (44.077.045).

008.810   **Trieste**

Astronomical Observatory Trieste, Publications.
Nos. 1087 – 1090, 1091 (43.114.023), 1092 – 1095, 1096
(43.112.023), 1097 (43.112.033), 1098 (43.114.008), 1099
(43.036.097), 1100 – 1102, 1103 (43.073.005), 1104 (43.103.211),
1105 (43.125.224), 1106 (43.063.011), 1107 (43.117.161), 1108,
1109 (44.131.032), 1110 (44.112.017), 1111 (43.131.346), 1112
(43.131.348), 1113 (43.114.113), 1114 – 1117, 1118 (44.117.141),
1119 (44.114.016), 1120 (44.114.017), 1121 (44.077.046), 1122
(44.073.072), 1123, 1124, 1125 (44.073.029), 1126 (44.073.030),
1127 (44.077.022), 1128 (44.112.121), 1129 (43.142.015), 1130
(43.159.032), 1131 (43.159.062), 1132 – 1134, 1135 (44.114.103),
1136 (44.114.104), 1137 (43.117.327), 1138, 1139 (43.125.213),
1140.

Department of Astronomy, Trieste University, Publications.
Nos. 99 (42.051.102), 100 (43.112.023), 101 (43.158.093), 102
(43.103.211), 103 – 105, 106 (43.114.008), 107 (44.114.017), 108
(43.117.161), 109 (43.119.104), 110 (43.112.222), 111
(43.113.061), 112 (44.112.017), 113, 114, 115 (44.073.030), 116
(43.157.198), 117 (43.160.073), 118, 119, 120 (44.114.095).

008.816   **Tucson, Ariz.**

Preprints of the Steward Observatory.
Nos. 739 (44.158.259), 740 (44.157.278), 741 (44.158.260), 742
(44.034.094), 743 (44.034.095), 744 (44.158.261), 745
(44.114.079), 746 (44.157.279), 747 (44.155.119), 748
(44.034.096), 749 (44.125.298), 750 (44.161.236), 751
(44.131.216), 752 (44.101.039), 753 (44.061.071), 754
(44.061.072), 755 (44.051.041), 756 (44.066.105), 757

(44.156.030), 758 (44.159.085), 759 (44.082.060), 760
(44.034.097), 761 (44.159.086), 762 (44.157.280), 763
(44.157.281), 764 (44.158.262), 765 (44.002.067), 766
(44.125.299), 767 (44.151.120), 768 (44.126.077), 769
(44.067.102), 770 (44.112.147), 771 (44.117.212), 772
(44.161.237).

008.822   Turku

Turku University Observatory, Informo.
Nos. 124 (44.131.218), 125 (44.158.263), 126 (44.158.264), 127
(44.161.238), 128 (44.158.265), 129 (44.155.120).

Report Series, Department of Physical Sciences, University of Turku.
FTL–R130 (44.131.218), FTL–R132 (44.158.263), FTL–R133
(44.158.264), FTL–R135 (44.161.238), FTL–R136 (44.158.265),
FTL–R137 (44.155.120).

008.834   Uppsala

Uppsala Astronomical Observatory, Report.
No. 44 (44.002.068).

008.840   Utrecht

Working Group Stars with Extended Atmospheres, Preprint.
Nos. 103 (44.113.047), 104 (44.064.060, 44.114.082,
44.064.061).

008.846   Victoria

Dominion Astrophysical Observatory, Victoria, British Columbia. Report for the year 1986 April 1 – 1987 March 31.
J. E. Hesser.
Q.J.R. Astron. Soc., Vol. 28, No. 4, p. 510 – 532 (1987).

Publications of the Dominion Astrophysical Observatory.
Vol. 16, No. 16 (44.125.069).

008.864   Warsaw

Latitude Circular.
Nos. 104 – 106 (44.044.034).

008.867   Washington, D.C.

U.S. Naval Observatory, Time Service Publications, Series 4.
Nos. 1065 – 1091 (44.044.031).

NEOS, Earth Orientation Bulletin.
Vol. 2, Nos. 27 – 53 (44.044.032).

U.S. Naval Observatory, Time Service Announcement, Series 14.
No. 45 (44.044.033).

008.870   Wellington

Carter Observatory. Report for the year ending 31 March 1987.
R. J. Dodd.
South. Stars, Vol. 32, No. 4, p. 134 – 142 (1987).

008.894   Zagreb

Hvar Observatory Bulletin.
Vol. 10, No. 1 (44.117.218, 44.077.046, 44.073.072, 44.022.074).

008.897   Zelenchukskaya

Chronicle.
Astrofiz. Issled. Izv. Spets. Astrofiz. Obs., Tom 24,
p. 219 – 221 (1987). In Russian. English translation in Bull.
Spec. Astrophys. Obs. – North Caucasus.

Chronicle.
Bull. Spec. Astrophys. Obs. – North Caucasus, Vol. 22,
p. 119 – 120 (1987). English translation of 41.008.876.

Astrofizicheskie Issledovaniya. Izvestiya Spetsial'noj Astrofizicheskoj Observatorii.
Tom 24 (44.158.266, 44.158.267, 44.111.043, 44.114.083,
44.022.110, 44.131.219, 44.141.017, 44.002.069, 44.002.070,
44.158.268, 44.033.025 – 44.033.028, 44.008.897),
Tom 25 (44.114.084, 44.114.085, 44.116.055, 44.117.219,
44.120.012, 44.114.086, 44.114.087, 44.158.269, 44.125.070,
44.072.065, 44.033.029 – 44.033.031, 44.002.071).

Soobshcheniya Spetsial'noj Astrofizicheskoj Observatorii.
Vyp. 52 (44.117.220, 44.160.095, 44.041.011, 44.036.162,
44.033.032),
Vyp. 53 (44.012.008).

Bulletin of the Special Astrophysical Observatory – North Caucasus.
Vol. 22 (44.114.051, 44.114.052, 44.122.048, 44.116.031,
44.114.053, 44.158.152, 44.114.054, 44.034.039, 44.036.070,
44.036.071, 44.034.040 – 44.034.043, 44.032.016, 44.033.017,
44.033.018, 44.008.897).

009 Notes on Observatories, Planetaria, Exhibitions

009.001 **A visit to the Pic du Midi Observatory.**
J. Dragesco, R. McKim.
J. Br. Astron. Assoc., Vol. 97, No. 5, p. 280 – 287 (1987).

009.002 **25 jaar ESO.**
G. Schilling.
Zenit, 14. Jaarg., Nr. 10, p. 316 – 325 (1987).

009.003 **History of the Sternberg State Astronomical Institute.**
E. P. Aksenov.
Istor. Astron. Obs. Mosk. Univ. i GAISh, Moskva, p. 4 – 61 (1986). In Russian. Abstr. in Ref. Zh., 51. Astron., 7.51.15 (1987).

009.004 **The Astronomical Observatory of the Moscow University during the pre–revolutionary period.**
P. G. Kulikovskij.
Istor. Astron. Obs. Mosk. Univ. i GAISh, Moskva, p. 62 – 73 (1986). In Russian. Abstr. in Ref. Zh., 51. Astron., 7.51.16 (1987).

009.005 **Astrometrical investigations at the Sternberg State Astronomical Institute.**
V. V. Podobed, N. S. Blinov, A. P. Gulyaev, V. V. Nesterov.
Istor. Astron. Obs. Mosk. Univ. i GAISh, Moskva, p. 74 – 85 (1986). In Russian. Abstr. in Ref. Zh., 51. Astron., 7.51.17 (1987).

009.006 **Celestial mechanics at the Sternberg State Astronomical Institute.**
A. A. Orlov.
Istor. Astron. Obs. Mosk. Univ. i GAISh, Moskva, p. 86 – 102 (1986). In Russian. Abstr. in Ref. Zh., 51. Astron., 7.51.18 (1987).

009.007 **Investigations in the field of stellar astronomy.**
Yu. N. Efremov.
Istor. Astron. Obs. Mosk. Univ. i GAISh, Moskva, p. 103 – 124 (1986). In Russian. Abstr. in Ref. Zh., 51. Astron., 7.51.19 (1987).

009.008 **Astrophysical investigations at the Sternberg State Astronomical Institute.**
N. G. Bochkarev, Eh. A. Dibaj, Yu. P. Pskovskij, N. I. Shakura.
Istor. Astron. Obs. Mosk. Univ. i GAISh, Moskva, p. 125 – 156 (1986). In Russian. Abstr. in Ref. Zh., 51. Astron., 7.51.20 (1987).

009.009 **History of gravimetry at the Sternberg State Astronomical Observatory.**
M. U. Sagitov.
Istor. Astron. Obs. Mosk. Univ. i GAISh, Moskva, p. 157 – 182 (1986). In Russian. Abstr. in Ref. Zh., 51. Astron., 7.51.21 (1987).

009.010 **Big and bright: a brief history of the McDonald Observatory.**
D. S. Evans, J. D. Mulholland.
Mercury, Vol. 16, No. 4, p. 98 – 107 (1987).

009.011 **Kölner 3–m–Radioteleskop in den Walliser Alpen: Gornergrat.**
G. Winnewisser.
Orion, 45. Jahrg., Nr. 222, p. 174, 179 – 180 (1987).

009.012 **Raumflugplanetarium für Madrid eingeweiht.**
M. Oemler, I. Stark.
Jenaer Rundsch., 32. Jahrg., Heft 2, p. 97 (1987).

009.013 **Die europäische Südsternwarte (ESO). Zur Ausstellung im Planetarium der Stadt Wien ab 17. Dezember 1987.**
A. Schnell.
Sternenbote, 30. Jahrg., Nr. 12, p. 234 – 247 (1987).

009.014 **Watchcroft Observatory. The most westerly dome in England?**
T. Tempest.
J. Br. Astron. Assoc., Vol. 97, No. 6, p. 330 – 333 (1987).
This paper is an account of the construction of an amateur–built observatory. It details all stages through to the completion of the attractive and functional dome.

009.015 **Großplanetarium COSMORAMA für Berlin.**
M. Schumacher, S. Pieler.
Jenaer Rundsch., 32. Jahrg., Heft 3, p. 146 – 147 (1987).

009.016 **Enhancing connections and promoting works of scientific and technological developments.**
M.–h. Jiang, Y.–b. Wang, Y. Liang.
Publ. Purple Mt. Obs., Vol. 6, No. 2, p. 200 – 205 (1987). In Chinese.
This paper gives a brief summary of scientific works and technological developments carried out in the Purple Mountain Observatory in the past two years.

009.017 **Das Planetarium in Jena. Eine Bilddokumentation anläßlich seines 60jährigen Jubiläums.**
L. Meier.
Sterne Weltraum, 26. Jahrg., Nr. 12, p. 678 – 682 (1987).

009.018 **25 Jahre European Southern Observatory.**
M. Rosa, R. M. West.
Sterne Weltraum, 26. Jahrg., Nr. 12, p. 690 – 695 (1987).

009.019 **No need for chemists.**
G. Eiby.
South. Stars, Vol. 32, No. 4, p. 105 – 110 (1987).
A review is given of the difficult plight of Carter Observatory, seen in its historical context, and leading up to the present precarious situation. A plea is made for a united presentation of a case for support to Government for the astronomical cause represented by Carter Observatory.

009.020 **Recent activities at RT (*Radiotelescope*)–Wettzell.**
R. Kilger.
Mitt. Geod. Inst. Rheinischen Friedrich–Wilhelms–Univ. Bonn, Nr. 71, p. 3 – 5 (1987). – See Abstr. 012.066.

009.021 **The Medicina VLBI station.**
F. Mantovani.
Mitt. Geod. Inst. Rheinischen Friedrich–Wilhelms–Univ. Bonn, Nr. 71, p. 7 – 9 (1987). – See Abstr. 012.066.

009.022 **Station report: Hartebeesthoek Radio Astronomy Observatory.**
A. Nothnagel.
Mitt. Geod. Inst. Rheinischen Friedrich–Wilhelms–Univ. Bonn, Nr. 71, p. 11 – 12 (1987). – See Abstr. 012.066.

009.023 **Station report for the radio telescope in Effelsberg.**
W. Alef.
Mitt. Geod. Inst. Rheinischen Friedrich–Wilhelms–Univ. Bonn, Nr. 71, p. 27 (1987). – See Abstr. 012.066.

009.024 **Status report of the Vega/Venus balloons VLBI experiment.**
G. Petit.
Mitt. Geod. Inst. Rheinischen Friedrich–Wilhelms–Univ. Bonn, Nr. 71, p. 31 (1987). – See Abstr. 012.066.

009.025 On the centenary of the Astronomical and Meteorological Observatory.
M. Mitrović, D. Milićević.
Vasiona, Année 35, No. 3 – 4, p. 57 – 63 (1987). In Serbo–Croatian.
The first part contains the description of activities organized on the centenary of the foundation of the Astronomical and Meteorological Observatory. The second part of the article contains a short review of present tasks of the Astronomical Observatory in Belgrade and of the Republic Hydro–Meteorological Institute of the SR Serbia which developed from the initially common observatory.

009.026 Centenary of the Astronomical Observatory in Belgrade.
V. Protić–Benišek.
Vasiona, Année 35, No. 3 – 4, p. 64 – 70 (1987). In Serbo–Croatian.
The history of the Astronomical Observatory in Belgrade is presented.

009.027 The Pulkovo Observatory and the Belgrade Observatory.
S. A. Tolchel'nikova–Muri, S. Sadžakov.
Vasiona, Année 35, No. 3 – 4, p. 100 – 106 (1987). In Serbo–Croatian.
A review of the development of the Pulkovo Observatory and the Belgrade Observatory is given from their foundations to the present time. The collaboration of these observatories and their joint work on international projects within IAU is described.

009.028 The Bernardino Rivadavia meteorite collection.
M. Rocca.
Meteoros (GB), Vol. 17, No. 4, p. 69 – 72 (1987). Abstr. in Phys. Abstr., Vol. 90, No. 1318, Entry 145903 (1987).

The archives of Armagh Observatory.
See Abstr. 002.037.

Observatório Astronômico – um século de história (1827 – 1927).
See Abstr. 003.099.

Notes on the history of the Tashkent Astronomical and Physical Observatory.
See Abstr. 004.075.

The temporary Belgrade Astronomical and Meteorological Observatory.
See Abstr. 004.120.

How the instruments for the Observatory at the High School and University of Belgrade were obtained.
See Abstr. 004.121.

Short chronology of Soviet astronomy.
See Abstr. 004.122.

New developments at Table Mountain.
See Abstr. 013.025.

Astronomie auf neuen Wegen – Forschungsergebnisse des Max–Planck–Instituts für Astronomie.
See Abstr. 013.042.

First fully automatic telescope on La Silla.
See Abstr. 032.041.

Present state and future of geodetic correlation in Bonn.
See Abstr. 033.041.

010 Societies, Associations, Organizations

American Association of Variable Star Observers (AAVSO)

010.021 The American Association of Variable Star Observers Bulletin.
No. 51 (1987).

010.022 The American Association of Variable Star Observers Solar Bulletin.
Vol. 43, Nos. 6 – 12 (1987).

010.023 American Association of Variable Star Observers Circular.
Nos. 201 – 206 (1987).

010.024 Meetings and activities of the Society, committee reports.
J. Am. Assoc. Variable Star Obs., Vol. 16, No. 1, p. 55 – 64 (1987).

010.025 The Journal of the American Association of Variable Star Observers.
Vol. 16, No. 1 (1987).

American Astronomical Society (AAS)

010.061 Joint meeting of the American Astronomical Society and the Canadian Astronomical Society, 14 – 18 June 1987, Vancouver, Canada. Abstracts of presented papers.
Bull. Am. Astron. Soc., Vol. 19, No. 2, p. 655 – 768 (1987).

010.062 Bulletin of the American Astronomical Society.
Vol. 19, No. 2 (1987).

Association Française des Observateurs d'Etoiles Variables

010.101 Activité de l'A.F.O.E.V. en 1986.
E. Schweitzer.
Bull. Assoc. Fr. Obs. Etoiles Variables, No. 41, p. 1 – 7 (1987).
Contents: Activité générale. Variables du type Mira et semi–régulières.

010.102 Activité de l'A.F.O.E.V. en 1986 (2ème partie).
E. Schweitzer.
Bull. Assoc. Fr. Obs. Etoiles Variables, No. 42, p. 1 – 11 (1987).
Contents: Variables du type UG. SS Cygni. Variables du type Z Cam. Variables du type RCB. Variables du type Z And. Novae et novae récurrentes. Étoile particulière: AM Herculis.

010.103 La vie de l'Association.
E. Schweitzer, J. Minois.
Bull. Assoc. Fr. Obs. Etoiles Variables, No. 41, p. 17, No. 42, p. 21 – 22 (1987).

010.104 Bulletin de l'Association Française des Observateurs d'Etoiles Variables.
Nos. 41, 42 (1987).

Association of Lunar and Planetary Observers (A.L.P.O.)

010.121 The A.L.P.O.'s fortieth birthday.
J. E. Westfall.
Strolling Astron., Vol. 32, Nos. 3 – 4, p. 49 – 50 (1987).

010.122   **The Strolling Astronomer. The Journal of the Association of Lunar and Planetary Observers.**
Vol. 32, Nos. 3 – 4, 5 – 6 (1987).

010.123   **The Minor Planet Bulletin.**
Vol. 14, Nos. 3, 4 (1987).

Astronomical Society of India

010.201   **Bulletin of the Astronomical Society of India.**
Vol. 15, Nos. 2 – 3 (1987).

Astronomical Society of Japan

010.221   **Publications of the Astronomical Society of Japan.**
Vol. 39, Nos. 3 – 6 (1987).

010.222   **The Astronomical Herald.**
Vol. 80, Nos. 7 – 12 (1987). In Japanese.

Astronomical Society of Southern Africa (ASSA)

010.261   **Annual General Meeting 1987.**
H. E. Krumm.
Mon. Notes Astron. Soc. S. Afr., Vol. 46, Nos. 9 – 10, p. 121 – 123 (1987).

010.262   **Report of Council for the year ending 1987 June 30.**
H. E. Krumm.
Mon. Notes Astron. Soc. S. Afr., Vol. 46, Nos. 9 – 10, p. 124 – 127 (1987).

010.263   **Centre reports: Bloemfontein Centre, Harare Centre, Natal Centre, Natal Midlands Centre, Pretoria Centre, Transvaal Centre.**
H. I. Terblanche, R. Fleet, S. Calvert, R. M. Jarmain, W. F. Wargau, B. D. Fraser.
Mon. Notes Astron. Soc. S. Afr., Vol. 46, Nos. 11 – 12, p. 166 – 174 (1987).

010.264   **Section reports.**
Mon. Notes Astron. Soc. S. Afr., Vol. 46, Nos. 9 – 10, p. 128 – 141 (1987).

010.265   **Monthly Notes of the Astronomical Society of Southern Africa.**
Vol. 46, Nos. 7 – 12 (1987).

Astronomical Society of the Pacific (ASP)

010.281   **Abstracts of papers presented at the summer scientific meeting of the Astronomical Society of the Pacific at Pomona College, Claremont, California, 14 – 16 July 1987.**
Publ. Astron. Soc. Pac., Vol. 99, No. 621, p. 1144 – 1152 (1987).

010.282   **Publications of the Astronomical Society of the Pacific.**
Vol. 99, Nos. 617 – 621 (1987).

010.283   **Mercury. The Journal of the Astronomical Society of the Pacific.**
Vol. 16, Nos. 4 – 6 (1987).

Astronomical Society of Western Australia (ASWA)

010.301   **Journal of the Astronomical Society of Western Australia.**
Vol. 38, Nos. 1 – 5 (1987).

British Astronomical Association (BAA)

010.381   **Report of the Council for the session 1986 August 1 to 1987 July 31.**
J. Br. Astron. Assoc., Vol. 97, No. 6, p. 353 – 365 (1987).

010.382   **British Astronomical Association Circular.**
Nos. 669 – 672, 674 (1987).

010.383   **Meetings and activities of the Association.**
J. Br. Astron. Assoc., Vol. 97, No. 5, p. 303 – 306, 366 – 368, No. 6, p. 51 – 53 (1987).

010.384   **Journal of the British Astronomical Association.**
Vol. 97, Nos. 5, 6, Vol. 98, No. 1 (1987).

010.385   **The British Astronomical Association Newsletter.**
Nos. 25 – 27 (1987).

British Interplanetary Society (BIS)

010.401   **JBIS. Journal of the British Interplanetary Society.**
Vol. 40, Nos. 7 – 12 (1987).

010.402   **Spaceflight. A publication of the British Interplanetary Society.**
Vol. 29, Nos. 7 – 12 (1987).

Canadian Astronomical Society

010.421   **Joint meeting of the American Astronomical Society and the Canadian Astronomical Society, 14 – 18 June 1987, Vancouver, Canada.**
Bull. Am. Astron. Soc., Vol. 19, No. 2, p. 655 – 768 (1987).

European Space Agency (ESA)

010.481   **ESA Bulletin.**
Nos. 51, 52 (1987).

010.482   **ESA Journal.**
Vol. 11, Nos. 2, 3 (1987).

010.483   **ESA IUE Newsletter.**
No. 28 (1987).

010.484   **EXOSAT Express.**
No. 18 (1987).

010.485   **ESA Special Publication.**
ESA SP–273 (012.044), ESA SP–278 (012.075), ESA SP–1093 (012.072).

International Amateur–Professional Photoelectric Photometry

010.501   **I.A.P.P.P. Communication.**
Nos. 29, 30 (1987).

International Astronomical Union (IAU)

010.541   **Commission 27 of the IAU, Information Bulletin on Variable Stars.**
Nos. 3036 – 3126 (1987).

010.542   **IAU Circulars.**
Nos. 4413 – 4521 (1987).

010.543   **Circulaire d'Information.**
No. 103 (1987).

010.544   **The Minor Planet Circulars/Minor Planets and Comets.**
Nos. 11887 – 12624 (1987).

010.545   **IAU Symposia.**
Nos. 125 (012.001), 127 (012.032).

010.546   **IAU Colloquia.**
Nos. 91 (012.060), 92 (012.030).

010.547   **Quarterly Bulletin on Solar Activity.**
Vol. 24, Part V.

International Occultation Timing Association (IOTA)

010.561   **Occultation Newsletter.**
Vol. 4, No. 5 (1987).

Korean Astronomical Society

010.621   **The Journal of the Korean Astronomical Society.**
Vol. 20, No. 2 (1987).

LEST Foundation

010.641   **LEST Foundation, Technical Report.**
Nos. 25 – 27 (1987).

Oriental Astronomical Association

010.721   **The Heavens.**
Vol. 68, Nos. 7 – 12 (1987). In Japanese.

Royal Astronomical Society (RAS)

010.741   **The origin of the solar system. Summary of the discussion meeting held 1987 March 13.**
I. P. Williams.
Observatory, Vol. 107, No. 1080, p. 184 – 185 (1987).

010.742   **Astronomy in Britain since the Second World War. II. Summary of the RAS specialist discussion, held 1987 May 9 in the Scientific Societies' Lecture Theatre, Savile Row.**
Observatory, Vol. 107, No. 1081, p. 239 – 244 (1987).

010.743   **Meetings of the Society.**
Observatory, Vol. 107, No. 1079, p. 137 – 146, No. 1080, p. 177 – 184, No. 1081, p. 231 – 238 (1987).

010.744   **Meetings and activities of the Society.**
Q. J. R. Astron. Soc., Vol. 28, No. 3, p. 401 – 426 (1987).

010.745   **Monthly Notices of the Royal Astronomical Society.**
Vol. 227, Nos. 1 – 4, Vol. 228, Nos. 1 – 4, Vol. 229, Nos. 1 – 4 (1987).

010.746   **Geophysical Journal of the Royal Astronomical Society.**
Vol. 90, Nos. 1 – 3, Vol. 91, Nos. 1 – 3 (1987).

010.747   **The Quarterly Journal of the Royal Astronomical Society.**
Vol. 28, Nos. 3, 4 (1987).

Royal Astronomical Society of Canada

010.761   **The Royal Astronomical Society of Canada.**
L. Enright.
J. R. Astron. Soc. Can., Vol. 81, No. 5, p. 165 – 174 (1987).

010.762   **The Journal of the Royal Astronomical Society of Canada.**
Vol. 81, Nos. 4 – 6 (1987).

010.763   **National Newsletter. Supplement to the Journal of the Royal Astronomical Society of Canada.**
Vol. 81, Nos. 4 – 6 (1987).

Royal Astronomical Society of New Zealand

010.781   **Southern Stars. Journal of the Royal Astronomical Society of New Zealand.**
Vol. 32, Nos. 3, 4 (1987).

Schweizerische Astronomische Gesellschaft (SAG)

010.801   **BBSAG Bulletin.**
Nos. 84 – 85 (1987).

010.802   **Mitteilungen.**
Orion, 45. Jahrg., Nr. 221, p. 137 – 140, Nr. 222, p. 175 – 178, Nr. 223, p. 215 – 218 (1987).

010.803   **Orion. Zeitschrift der Schweizerischen Astronomischen Gesellschaft. Revue de la Société Astronomique de Suisse.**
45. Jahrg., Nr. 221 – 223 (1987).

Società Astronomica Italiana (S.A.It.)

010.821   **Giornale di Astronomia.**
Vol. 13, N. 1 – 2 (1987).

Société Astronomique de France

010.841   **Souvenirs d'un centenaire.**
J. Minois.
Astronomie, Vol. 101, p. 618 – 622 (1987).

010.842   **Centenaire de la Société Astronomique de France. Conférence du 23 juin 1987.**
B. Clouet.
Astronomie, Vol. 101, p. 636 – 642 (1987).

010.843   **Séances, commissions, activités de la Société.**
Astronomie, Vol. 101, p. 547 – 552 (1987).

010.844   **L'Astronomie et Bulletin de la Société Astronomique de France.**
Vol. 101, juillet – décembre (1987).

010.845   **Observations et Travaux.**
Nos. 11, 12 (1987).

Société Astronomique de Liège

010.861   **Le Ciel.**
Vol. 49, septembre – décembre, p. 249 – 364 (1987).

Société Royale Belge d'Astronomie

010.901   **Ciel et Terre. Bulletin de la Société Royale Belge d'Astronomie, de Météorologie et de Physique du Globe.**
Vol. 103, Nos. 4 – 6 (1987).

Vereinigung der Sternfreunde e.V. (VdS)

010.941   **Sonne. Mitteilungsblatt der Amateursonnenbeobachter.**
Jahrg. 11, Nr. 42 – 44 (1987).

010.942   **Nachrichten der Vereinigung der Sternfreunde e.V.**
Sterne Weltraum, 26. Jahrg., Nr. 7 – 8, p. 444 – 445, Nr. 9, p. 516 – 519, Nr. 10, p. 582 – 586, Nr. 11, p. 655 – 659, Nr. 12, p. 720 – 723 (1987).

010.943   **BAV Rundbrief.**
36. Jahrg., Nr. 3, 4 (1987).

010.944   **BAV Mitteilungen.**
Nr. 44 – 47 (1987).

011   Reports on Colloquia, Congresses, Meetings, Symposia, Expeditions

011.001   **Solar Cycle Workshop. A review.**
P. R. Wilson.
Sol. Phys., Vol. 110, No. 1, p. 1 – 9 (1987). – See Abstr. 012.005.

011.002   **The 1987 American Workshop on Cometary Astronomy.**
C. S. Morris.
Int. Comet Q., Vol. 9, No. 2, p. 64 (1987).

011.003   **Die Supernova 1987A – das astronomische Jahrhundertereignis. Neues vom ESO–Workshop in Garching, BRD.**
H. M. Maitzen.
Sternenbote, 30. Jahrg., Nr. 9, p. 178 – 184 (1987).

011.004   **Les potins d'Uranie: La place des amateurs.**
Al Nath (*A. Heck*).
Ciel, Vol. 49, p. 254 – 259 (1987).
Du 20 au 24 juin 1986 s'est tenu à Paris le colloque No. 98 de l'Union Astronomique Internationale intitulé "La contribution des astronomes amateurs à l'astronomie".

011.005   **Supernova 1987A. A summary of the ESO Workshop held from 6 – 8 July 1987.**
S. van den Bergh.
Messenger, No. 49, p. 32 – 33 (1987).

011.006   **Le colloque No. 98 de l'UAI à Paris.**
P. Bacchus.
Astronomie, Vol. 101, p. 459 – 460 (1987).

011.007   **The 1987 Big Bear I.A.P.P.P. Symposium.**
R. A. Jones, R. G. Deyoe, R. Wasson.
I.A.P.P.P. Commun., No. 29, p. 1 – 4 (1987).
Report on the sixth annual I.A.P.P.P. Big Bear Symposium, held 1987 May 20 – 22 at Big Bear Lake, California.

011.008   **The 1987 APT Service summer workshop.**
D. L. Crawford, W. Pearce, R. J. Dukes Jr., D. R. Genet, R. M. Genet, R. M. Genet, D. S. Hayes, J. Hayes, K. E. Kissell, J. Lee, W. Keel, D. P. Smith.
I.A.P.P.P. Commun., No. 29, p. 23 – 28 (1987).

011.009   **Detectors for Lyman and SOHO.**
M. Penston.
ESA IUE Newsl., No. 28, p. 47 – 50 (1987).
Report on the workshop "Open–window detectors for ultraviolet astronomy", held in Herstmonceux, 17 – 19 November 1986.

011.010   **Comet notes: X. The Fourth American Workshop on Cometary Astronomy.**
D. H. Levy.
Strolling Astron., Vol. 32, Nos. 3 – 4, p. 79 – 80 (1987).

011.011   **All–Union conference on "Wolf–Rayet–type stars and related objects", held in the Estonian SSR (near the Tõravere Observatory), 14 – 17 October 1986.**
T. A. Nugis.
Astron. Zh., Tom 64, Vyp. 4, p. 888 – 890 (1987). In Russian. English translation in Sov. Astron., Vol. 31, No. 4.

011.012   **Conference of the working group "Galaxy" of the section "Physics and evolution of galaxies and the Metagalaxy" of the Astronomical Council of the USSR Academy of Sciences, held at the Abastumani Astrophysical Observatory, 17 – 20 September 1986.**
I. G. Kolesnik, S. A. Silich.
Kinematika Fiz. Nebesn. Tel, Tom 3, No. 4, p. 94 – 96 (1987). In Russian. English translation in Kinematics Phys. Celest. Bodies.

011.013   **IAU Colloquium 98: the contribution of amateur astronomers to astronomy.**
M. D. Overbeek, J. A. S. Campos.
Mon. Notes Astron. Soc. S. Afr., Vol. 46, Nos. 9 – 10, p. 117 – 120 (1987).

011.014   **Second Tartu Cosmology Seminar: missing mass, large–scale structure, microwave background, held at Tõravere, 10 – 14 June, 1985.**
L. A. Kofman, A. A. Starobinskij, S. F. Shandarin, J. E. Einasto.
Sov. Astron., Vol. 30, No. 6, p. 729 – 731 (1987). English translation of 42.011.046.

011.015   **Horizons of present–day meteoritics.**
Yu. A. Shukolyukov.
Priroda, No. 9, p. 28 – 30 (1987). In Russian.

011.016   **Amateurs triumph in Paris.**
S. J. O'Meara.
Sky Telesc., Vol. 74, No. 5, p. 480 – 483 (1987).

011.017   Infrared excess stirs cosmologists.
D. Lindley.
Nature, Vol. 328, No. 6128, p. 291 (1987).
This note reports on some new observational results on the extragalactic infrared background, presented at the Third International IRAS Conference held at London, UK, 6 – 10 July 1987.

011.018   Probing the early universe.
R. F. Carswell.
Nature, Vol. 328, No. 6129, p. 376 – 378 (1987).
This note reports on the meeting "QSO absorption lines: probing the universe", held at the Space Telescope Science Institute, Baltimore, MD, USA, 10 – 21 May 1987.

011.019   Models of quasars reappraised.
M. Elvis.
Nature, Vol. 328, No. 6133, p. 762 – 763 (1987).
This note reports on a conference on "Emission lines in quasars and active galaxies", held at the Rutherford Appleton Laboratory, Chilton, Didcot, UK, 27 – 29 May 1987.

011.020   Conference on geochronology, cosmochronology and isotopic geology, held in Cambridge, England, 30 June – 4 July 1986.
Yu. A. Shukolyukov.
Vestn. Akad. Nauk SSSR, No. 2, p. 118 – 122 (1987). In Russian. Abstr. in Ref. Zh., 51. Astron., 9.51.16 (1987).

011.021   Scientific session of the Department of General Physics and Astronomy and the Department of Nuclear Physics of the USSR Academy of Sciences, 24 – 25 September 1986.
Usp. Fiz. Nauk, Tom 151, No. 4, p. 715 – 724 (1987). In Russian. Abstr. in Ref. Zh., 51. Astron., 9.51.25 (1987).

011.022   Frontiers and the conquest of space. The philosopher's touch stone.
M. de Groot.
Ir. Astron. J., Vol. 18, No. 2, p. 65 – 83 (1987).
Report on an international colloquium, held in Paris from 13th to 16th January, 1987.

011.023   Reconnection MIST meeting, 1986 May 9 "Reconnection in solar system plasmas".
H. Rishbeth.
Q. J. R. Astron. Soc., Vol. 28, No. 3, p. 339 – 343 (1987).

011.024   Scientific session of the Department of General Physics and Astronomy and of the Department of Nuclear Physics of the USSR Academy of Sciences (26 – 27 November 1986).
Usp. Fiz. Nauk, Tom 152, Vyp. 2, p. 333 – 344 (1987). In Russian. Abstr. in Ref. Zh., 51. Astron., 10.51.38 (1987).

011.025   Scientific session of the Department of General Physics and Astronomy and of the Department of Nuclear Physics of the USSR Academy of Sciences (24 – 25 December 1986).
Usp. Fiz. Nauk, Tom 152, Vyp. 2, p. 345 – 346 (1987). In Russian. Abstr. in Ref. Zh., 51. Astron., 10.51.39 (1987).

011.026   The XXth All–Union meteoritic conference held in Tallinn, 10 – 12 February 1987.
Yu. A. Shukolyukov, A. V. Ivanov.
Astron. Vestn., Tom 21, No. 4, p. 344 – 347 (1987). In Russian. English translation in Sol. Syst. Res.

011.027   A joint meeting of the Canadian & American Astronomical Societies held in Vancouver, June 14 – 19, 1987.
C. Aikman.
J. R. Astron. Soc. Can., Vol. 81, No. 5, p. 161 – 163 (1987).

011.028   The contribution of amateur astronomers to astronomy.
M. de Groot.
Ir. Astron. J., Vol. 18, No. 2, p. 124 – 126 (1987).
Report on the IAU Colloquium No. 98, held in Paris from 20th to 24th June, 1987.

011.029   Autumn MIST Meeting, 1986.
P. A. Hadjiry, M. J. Laird.
Q. J. R. Astron. Soc., Vol. 28, No. 4, p. 472 – 476 (1987).

011.030   Les potins d'Uranie: La place des amateurs.
A. Nath (*A. Heck*).
Orion, 45. Jahrg., Nr. 223, p. 226 – 227 (1987). Reprint of 011.004.

011.031   Meteor section meeting, 1986 June 14, held at the Rupert Beckett Lecture Theatre, Leeds University, Woodhouse Lane, Leeds 2.
M. D. Taylor.
J. Br. Astron. Assoc., Vol. 98, No. 1, p. 51 – 53 (1987).

011.032   The interactions between amateur and professional astronomers. A report of the joint meeting of the British Astronomical Association and the Royal Astronomical Society held in London, on 1987 March 14.
N. Henbest.
J. Br. Astron. Assoc., Vol. 98, No. 1, p. 55 – 64 (1987).

011.033   Supernova 1987A: The parent and its environs.
P. Murdin.
Nature, Vol. 329, No. 6134, p. 12 – 13 (1987).
This paper reports on the ESO Workshop on supernova 1987A, Garching, F.R. Germany, 6 – 8 July 1987. This conference focused on the developing interaction of SN 1987A with the surrounding material as well as the nature of the progenitor star.

011.034   Japanese Comet Conference, 1988 March 20 – 21.
Yamamoto Circ., No. 2096 (1987). In Japanese.

011.035   Conference on "Physics and dynamics of meteors", held in Katsiveli (Crimea) on 14 – 17 October 1986.
V. A. Bronshtehn, A. K. Terent'eva.
Zemlya Vselennaya, No. 4, p. 41 – 43 (1987). In Russian.

011.036   Second Orlov conference held in Poltava, 3rd October 1986 on the 60th anniversary of the Poltava Gravimetric Observatory which was established by the famous astronomer and geophysicist A. Ya. Orlov.
I. A. Dychko.
Zemlya Vselennaya, No. 5, p. 58 – 60 (1987). In Russian.

012 Proceedings of Colloquia, Congresses, Meetings, Symposia

012.001 **The origin and evolution of neutron stars. Proceedings of the 125th IAU Symposium, held in Nanjing, China, 26 – 30 May 1986.**
D. J. Helfand, J.–H. Huang (Editors), with introductory remarks by T. Lu.
D. Reidel Publishing Co., Dordrecht, The Netherlands. 27 + 572 pp. Price Dfl. 215.00, US$ 99.00, £ 74.00 cloth (1987). ISBN 90–277–2537–3 cloth; ISBN 90–277–2538–1 paper.
The individual contributions are included in their corresponding subject categories – see abstracts 034.001, 051.001, 063.001, 067.001 – 067.036, 117.001 – 117.008, 125.001 – 125.011, 126.001 – 126.035, 142.001 – 142.009, 143.001 – 143.008, 144.001, 154.001, 155.001, 157.001.

012.002 **Star formation in galaxies. Proceedings of a conference, held at the California Institute of Technology, Pasadena, California, USA, 16 – 19 June 1986.**
C. J. Lonsdale Persson (Editor).
NASA Conf. Publ., NASA CP–2466, 20 + 788 pp. (1987).
The individual contributions are included in their corresponding subject categories – see abstracts 051.002, 131.001 – 131.017, 133.001 – 133.003, 155.002 – 155.005, 156.001 – 156.004, 157.002 – 157.044, 158.001 – 158.034, 160.001 – 160.003.

012.003 **The SHIRSOG Workshop. Proceedings of a workshop on prospects for a new synoptic high resolution spectroscopic observing facility, held at the National Solar Observatory, National Optical Astronomy Observatories, Tucson, Arizona, USA, 3 September 1986.**
M. S. Giampapa (Editor), with welcoming remarks by J. T. Jefferies and keynote address by J. L. Linsky.
National Optical Astronomy Observatories, Tucson, Arizona, USA. 6 + 149 pp. (1986).
The individual contributions are included in their corresponding subject categories – see abstracts 013.001 – 013.006, 032.001, 032.002, 034.002 – 034.006, 036.001 – 036.006, 071.001, 112.001, 119.001, 121.001, 122.001, 122.002.

012.004 **Essays in space science. Proceedings of a symposium held in honour of Frank B. McDonald at NASA Goddard Space Flight Center, Greenbelt, MD, USA, 23 April 1985.**
R. Ramaty, T. L. Cline, J. F. Ormes (Editors), with concluding remarks by P. Freier.
NASA Conf. Publ., NASA CP–2464, 12 + 423 pp. Price US$ 36.95 USA, US$ 73.90 Foreign (1987).
The individual contributions are included in their corresponding subject categories – see abstracts 006.015, 013.007, 035.003, 073.002, 074.003, 074.004, 084.003, 101.016, 106.003, 142.011, 142.012, 143.009 – 143.011, 144.002 – 144.005.

012.005 **Solar Cycle Workshop. Proceedings of the first meeting held at Big Bear Solar Observatory, Pasadena, CA, USA, 17 – 20 August, 1986.**
P. R. Wilson (Editor).
Sol. Phys., Vol. 110, No. 1, p. 1 – 128 (1987).
The individual contributions are included in their corresponding subject categories – see abstracts 006.000, 011.001, 072.002, 072.003, 075.002 – 075.006, 080.002 – 080.004.

012.006 **Gravitational collapse and relativity. Proceedings of Yamada Conference XIV, held at Kyoto, Japan, 7 – 11 April 1986.**
H. Sato, T. Nakamura (Editors), with a summary by J. Ehlers.
World Scientific Publishing Co., Singapore. 14 + 512 pp. Price US$ 77.00, £ 61.65 (1986). ISBN 9971–50–207–0.
The individual contributions are included in their corresponding subject categories – see abstracts 065.012, 066.003 – 066.031, 067.048 – 067.053, 117.021, 151.013, 161.018 – 161.023.

012.007 **The Jack S. Josey Centennial Professorship Symposium in honor of David S. Evans, held at the University of Texas at Austin, 18 – 19 September 1986.**
T. G. Barnes III (Editor).
Vistas Astron., Vol. 30, Part 1 (special issue), p. 1 – 96 (1987).
The individual contributions are included in their corresponding subject categories – see abstracts 004.017, 013.014, 096.007, 111.003, 116.005, 116.006, 118.005.

012.008 **Proceedings of the Seminar on "The large–scale structure of the Universe", held at the Special Astrophysical Observatory of the USSR (Nizhnij Arkhyz), 15 – 21 September 1986.**
S. N. Fabrika (Editor), M. G. Mingaliev, A. I. Shapovalova.
Soobshch. Spets. Astrofiz. Obs., Vyp. 53, 104 pp. (1987).
The individual contributions are included in their corresponding subject categories – see abstracts 036.024, 036.025, 141.004, 157.077, 161.033 – 161.047.

012.009 **Radio astronomy from space. Proceedings of a workshop held at the National Radio Astronomy Observatory, Green Bank, W.Va., USA, 30 September – 2 October 1986.**
K. W. Weiler (Editor), with a foreword by K. W. Weiler, B. K. Dennison, J. O. Burns.
NRAO Workshop No. 18. National Radio Astronomy Observatory, P.O. Box 2, Green Bank, W.Va. 24944–0002, USA. 4 + 336 pp. Price US$ 13.00 (1987).
The individual contributions are included in their corresponding subject categories – see abstracts 013.015 – 013.022, 022.051, 033.004 – 033.011, 034.012, 035.007 – 035.014, 036.019, 051.010 – 051.013, 074.021, 091.020, 131.036 – 131.038, 141.001, 161.029.

012.010 **Rutherford Appleton Laboratory workshop on advanced technology reflectors for space instrumentation, held at the Cosener's House, Abingdon, UK, 16 – 18 June 1986.**
M. Grande (Editor).
Rutherford Appleton Lab., Rep., RAL 87–043, 12 + 272 pp. (1987).
The individual contributions are included in their corresponding subject categories – see abstracts 022.052 – 022.054, 031.004 – 031.021, 035.015 – 035.020.

012.011 **Selected topics on data analysis in astronomy. General lectures given at the II workshop on data analysis in astronomy, held at Erice, Sicily, Italy, 20 – 30 April 1986.**
V. Di Gesù, L. Scarsi, P. Crane (Editors).
World Scientific Publishing Co., Singapore. 7 + 170 pp. Price US$ 39.75, £ 31.80 (1987). ISBN 9971–50–262–3.
The individual contributions are included in their corresponding subject categories – see abstracts 013.028, 021.008, 036.035 – 036.040.

012.012 **Instabilities in luminous early type stars. Proceedings of a workshop in honour of Professor Cees de Jager on the occasion of his 65th birthday, held at Lunteren, The Netherlands, 21 – 24 April 1986.**
H. J. G. L. M. Lamers, C. W. H. de Loore (Editors).
Astrophysics and Space Science Library, Vol. 136. D. Reidel Publishing Co., Dordrecht, The Netherlands. 16 + 290 pp. Price Dfl. 145.00, US$ 69.00, £ 49.00 (1987). ISBN 90–277–2522–5.
The individual contributions are included in their corresponding subject categories – see abstracts 063.022, 063.023, 064.009 – 064.013, 065.021 – 065.024, 112.038 – 112.048, 115.004, 115.005, 116.014, 117.062, 122.024 – 122.027, 157.079.

012.013 **ESA Horizon 2000. Le programme à long terme de l'Agence Spatiale Européenne. 3ème Forum SFSA, held at Université Paris–Sud, Orsay, 17 – 18 Mars 1987.**
J. Astron. Fr., Suppl. au No. 30, 20 pp. (1987).

012.014   **Unification of fundamental interactions. Proceedings of Nobel Symposium 67, held at Marstrand, Sweden, 2 – 7 June 1986.**
L. Brink, R. Marnelius, J. S. Nilsson, P. Salomonson, B.–S. Skagerstam (Editors).
Physica Scripta, Topical Issue, Vol. T15, 209 pp.   Price SEK 430.00, US$ 75.00 (1987). ISBN 91–87308–02–9.
The individual contributions within the subject scope of Astronomy and Astrophysics Abstracts are included in their corresponding categories – see abstracts 061.002 – 061.004, 066.034, 161.053, 161.054.

012.015   **Radio continuum processes in clusters of galaxies. Proceedings of a workshop held at the National Radio Astronomy Observatory, Green Bank, West Virginia, USA, 4 – 8 August 1986.**
C. P. O'Dea, J. M. Uson (Editors), with a workshop summary by R. D. Ekers.
NRAO Workshop, No. 16, 8 + 349 pp. (1986).
The individual contributions are included in their corresponding subject categories – see abstracts 157.089, 158.079 – 158.085, 160.024 – 160.052, 161.063.

012.016   **Particle acceleration and trapping in solar flares. Selected contributions to the workshop held at Aubigny–sur–Nère (Bourges), France, 23 – 26 June 1986.**
G. Trottet, M. Pick (Editors).
Sol. Phys., Vol. 111, No. 1, 8 + 234 pp. (1987). Available also as hardbound edition, Price Dfl. 130.00, US$ 64.00, £ 40.25. ISBN 90–277–2609–4.
The individual contributions are included in their corresponding subject categories – see abstract 062.030, 062.031, 073.028 – 073.030, 074.027, 074.028, 076.003, 076.004, 077.018 – 077. 029.

012.017   **20th All–Union meteoritic conference, held in Tallinn, 10 – 12 February 1987.**
Part 1: Meteoritic craters and impactites.
Part 2: Meteoritic matter and conditions of its formation.
Part 3: Astronomical and physical aspects of meteoritics.
Institut Geokhimii i Analiticheskoj Khimii. Moskva. 68 pp. + 150 pp. + 47 pp. (1987). In Russian.
From Ref. Zh., 51. Astron., 7.51.41, 8.51.8, 8.51.9 (1987).
See abstracts 082.041, 094.024, 098.042, 098.043, 103.421, 104.034 – 104.038, 104.042, 104.044 – 104.048, 105.013 – 105.019, 105.032 – 105.034.

012.018   **Problems of solar flares. Publications of the Third Annual Seminar organized by the Astronomical Council of the USSR Academy of Sciences, 12 – 14 November 1984.**
V. V. Fomichev (Editor).
Institut zemnogo magnetizma, ionosfery i rasprostraneniya radiovoln, Moskva. 215 pp. (1986). In Russian.
Review in Ref. Zh., 51. Astron., 7.51.33 (1987).
See abstracts 072.032, 072.033, 073.031 – 073.040, 075.010, 075.011, 076.005, 078.004 – 078.006, 080.018.

012.019   **Variability of galactic and extragalactic X–ray sources. Proceedings of an international symposium held at Villa Olmo, Como, Italy, 20 – 22 October 1986.**
A. Treves (Editor).
Associazione per l'Avanzamento dell'Astronomia, Publication Division, Via Misa, 10, 40137 Bologna, Italy. 14 + 279 pp. Price L 72000.00 (1987).
The individual contributions are included in their corresponding subject categories – see abstracts 035.023, 036.057, 051.019, 063.031, 063.032, 067.071, 117.092 – 117.099, 124.004, 158.098 – 158.111.

012.020   **The Advanced X–Ray Astrophysics Facility. Invited papers presented at the 167th meeting of the American Astronomical Society, High Energy Astrophysics Division, held at Houston, Texas, USA, 5 – 9 January 1986.**
Astrophys. Lett. Commun., Vol. 26, Nos. 1 – 2, 4 + 151 pp. (1987).
The individual contributions are included in their corresponding subject categories – see abstracts 035.024 – 035.030, 112.057, 126.053, 142.018, 157.103, 160.057.

012.021   **Astrophysical jets and their engines. Proceedings of a NATO Advanced Study Institute, held at Erice, Sicily, Italy, 17 – 25 September 1986.**
W. Kundt (Editor).
NATO Advanced Science Institutes, Ser. C: Mathematical and Physical Sciences, Vol. 208. D. Reidel Publishing Co., Dordrecht, The Netherlands, 12 + 256 pp. Price Dfl. 125.00, US$ 59.50, £ 44.00 (1987). ISBN 90–277–2548–9.
The individual contributions are included in their corresponding subject categories – see abstracts 062.038 – 062.047, 067.073, 117.102, 121.015, 155.029, 158.112 – 158.117, 159.029, 161.089.

012.022   **Environments of planetary bodies and Shuttle. Proceedings of Workshops I and XII of the COSPAR Twenty–sixth Plenary Meeting held in Toulouse, France, 30th June – 11th July 1986.**
J. Oró, T. Owen, F. Raulin, G. G. Fazio (Editors).
Adv. Space Res., Vol. 7, No. 5, 6 + 235 pp. (1987).
The individual contributions are inclued in their corresponding subject categories – see abstracts 022.071, 022.072, 036.059, 051.020 – 051.031, 091.023 – 091.025, 097.074, 097.075, 100.013 – 100.016, 102.024 – 102.026, 103.422, 107.016.

012.023   **Scientific ballooning –V. Proceedings of Symposium 10 of the COSPAR Twenty–sixth Plenary Meeting held in Toulouse, France, 30th June – 11th July 1986.**
W. Riedler, K. Torkar (Editors).
Adv. Space Res., Vol. 7, No. 7, 6 + 133 pp. (1987).
The individual contributions within the subject scope of Astronomy and Astrophysics Abstracts are included in their corresponding categories – see abstracts 013.034, 035.032, 035.033, 051.032, 082.026.
   This new volume of "Scientific ballooning" documents recent developments in the fields of scientific balloon technology, systems and applications.

012.024   **The early universe and its evolution. Proceedings of the International School of Nuclear Physics, held at Erice, Sicily, Italy, 2 – 14 April 1986.**
A. Faessler (Editor).
Prog. Part. Nucl. Phys., Vol. 17. 7 + 465 pp. Price £ 94.50, US$ 140.00 (1986). ISBN 0–08–035185–9.
The individual contributions are included in their corresponding subject categories – see abstracts 022.075, 061.012 – 061.017, 065.038, 065.039, 107.017, 125.035 – 125.037, 157.105, 161.094 – 161.098, 161.233.

012.025   **Neutrinos and the present–day universe. Proceedings of a topical meeting during the national plenary meeting of the French Physical Society, held at Nice, France, 9 – 13 September 1985.**
T. Montmerle, M. Spiro (Editors).
Centre d'Etudes Nucléaires de Saclay, F–91191 Gif–sur–Yvette, France. 9 + 194 pp. (1986). ISBN 2–7272–0103–6.
The individual contributions are included in their corresponding subject categories – see abstracts 013.038, 034.032, 034.033, 061.018 – 061.022, 080.021, 080.022.

012.026   **Diffuse matter in the solar system: comet Halley and other studies. Proceedings of a Royal Society Discussion Meeting, held at London, UK, 21 – 22 May 1986.**
G. Turner, C. T. Pillinger (Editors), with introductory remarks by W. McCrea and a general discussion by D. McNally.
Philos. Trans. R. Soc. London, Ser. A, Vol. 323, No. 1572, p. 247 – 447 (1987). Also available as hardbound separate print from: The Royal Society Publications Sales, 6 Carlton House Terrace, London, SW1Y 5AG, UK. Price £ 39.50 (UK), £ 42.00 (overseas). ISBN 0–85403–325–4.
The individual contributions are included in their corresponding subject categories – see abstracts 061.025, 098.027, 102.030, 102.031, 103.123, 103.425 – 103.428, 105.020, 105.021, 106.016, 106.017, 131.086.

012.027   **Physical processes in interstellar clouds. Proceedings of a NATO Advanced Study Institute, held at Irsee, F.R. Germany, 18 – 28 August 1986.**
G. E. Morfill, M. Scholer (Editors).
NATO Advanced Science Institutes, Ser. C: Mathematical and Physical Sciences, Vol. 210. D. Reidel Publ. Co., Dordrecht, The Netherlands. 10 + 554 pp. Price Dfl. 230.00, US$ 92.00, £ 64.00 (1987). ISBN 90–277–2563–2.
The individual contributions are included in their corresponding subject categories – see abstracts 063.037, 064.029, 131.087 – 131.109, 155.035 – 155.037.

012.028   **Cosmology and particle physics. Proceedings of a theoretical workshop held at Lawrence Berkeley Laboratory, Berkeley, CA, USA, 28 July – 15 August 1986.**
I. Hinchliffe (Editor).
World Scientific Publishing Co., Singapore. 10 + 205 pp. Price US$ 59.60, £ 47.70 (1987). ISBN 9971–50–213–5.
The individual contributions are included in their corresponding subject categories – see abstracts 061.026 – 061.031, 161.107 – 161.117.

012.029   **The galactic center. Proceedings of the symposium honoring C. H. Townes, held at Berkeley, CA, USA, 25 October 1986.**
D. C. Backer (Editor)
AIP Conf. Proc., No. 155. 10 + 204 pp. Price US$ 49.25 (1987). ISBN 0–88318–355–2.
The individual contributions are included in their corresponding subject categories – see abstracts 041.003, 111.009, 155.042 – 155.075, 157.126, 158.128.

012.030   **Physics of Be stars. Proceedings of the 92nd Colloquium of the International Astronomical Union, held at Boulder, CO, USA, 18 – 22 August 1986.**
A. Slettebak, T. P. Snow (Editors).
Cambridge University Press, Cambridge, U.K. 18 + 557 pp. Price US$ 79.50 (1987). ISBN 0–521–33078–5.
The individual contributions are included in their corresponding subject categories – see abstracts 013.043, 064.035 – 064.042, 065.053, 065.054, 112.067 – 112.113, 113.021, 114.045 – 114.047, 115.010, 116.027 – 116.030, 117.128 – 117.139, 153.016 – 153.018.

012.031   **Nearly normal galaxies. From the Planck time to the present. The Eighth Santa Cruz Summer Workshop in Astronomy and Astrophysics, held at the Lick Observatory, Santa Cruz, CA ,USA, 21 July – 1 August 1986.**
S. M. Faber (Editor), with a summary by J. E. Gunn.
Santa Cruz Summer Workshops in Astronomy and Astrophysics. Springer–Verlag, New York, NY, USA. 26 + 464 pp. Price DM 72.00. ISBN 0–387–96521–1 (USA), ISBN 3–540–96521–1 (F.R. Germany).
The individual contributions are included in their corresponding subject categories – see abstracts 006.026, 131.125, 151.047 – 151.050, 154.019, 155.077 – 155.080, 157.129 – 157.146, 160.062 – 160.064, 161.137 – 161.154.

012.032   **Structure and dynamics of elliptical galaxies. Proceedings of the 127th Symposium of the International Astronomical Union, held in Princeton, NJ, USA, 27 – 31 May 1986.**
T. de Zeeuw (Editor), with a summary by S. D. Tremaine.
D. Reidel Publishing Co., Dordrecht, The Netherlands. 16 + 579 pp. (1987). ISBN 90–277–2585–3 cloth, ISBN 90–277–2586–1 paper.
The individual contributions are included in their corresponding subject categories – see abstracts 151.051 – 151.091, 154.020, 155.081, 157.147 – 157.193, 158.133 – 158.144, 160.065 – 160.067, 161.155 – 161.157.

012.033   **Cometary radio astronomy. Proceedings of an NRAO Workshop held at the National Radio Astronomy Observatory, Green Bank, WV, USA, 24 – 26 September 1986.**
W. M. Irvine, F. P. Schloerb, L. E. Tacconi–Garman (Editors).
NRAO Workshop No. 17. National Radio Astronomy Observatory, Green Bank, WV, USA. 10 + 162 pp. (1987).
The individual contributions are included in their corresponding subject categories – see abstracts 102.034, 103.013 – 103.018, 103.438 – 103.454.

012.034   **Chastitsy i kosmologiya. (*Particles and cosmology*.) Publications of the Third All–Union School, April 1985.**
Institut Yadernykh Issledovanij Akademii Nauk SSSR, 148 pp. (1987).
From Ref. Zh., 51. Astron., 10.51.34 (1987). In Russian.
See abstracts 065.060, 080.036, 080.037, 143.024, 161.158 – 161.160, 161.189, 161.192.

012.035   **Interferometric imaging in astronomy. Proceedings of the joint workshop on high resolution imaging from the ground using interferometric techniques, held at Oracle, AZ, USA, 12 – 15 January 1987.**
J. W. Goad (Editor).
European Southern Observatory, Garching, F.R. Germany, and National Optical Astronomy Observatories, Tucson, AZ, USA. 20 + 262 pp. (1987).
The individual contributions are included in their corresponding subject categories – see abstracts 002.036, 032.018 – 032.021, 034.045, 034.046, 036.073 – 036.122, 099.037, 121.026, 158.158.

012.036   **Proceedings of the meeting held by the Astronomical Science Group of Ireland at Trinity College, Dublin on 6th April, 1987.**
Ir. Astron. J., Vol. 18, No. 2, p. 84 – 101 (1987).
The individual contributions are included in their corresponding subject categories – see abstracts 034.118, 106.031, 112.164, 117.239, 153.045, 161.313.

012.037   **The Nandy Symposium on the interstellar medium and hot stars, held in Edinburgh on 29 – 30 September 1986.**
With an introduction by R. D. Wolstencroft.
Q. J. R. Astron. Soc., Vol. 28, No. 3, p. 207 – 337 (1987).
The individual contributions are included in their corresponding subject categories – see abstracts 112.116 – 112.119, 114.057, 122.049, 131.134 – 131.144, 134.023, 155.086, 156.014, 157.204.

012.038   **General relativity and gravitational physics. Proceedings of the 7th Italian Conference on General Relativity and Gravitational Physics, held at Rapallo (Genoa), Italy, 3 – 6 September 1986.**
U. Bruzzo, R. Cianci, E. Massa (Editors).
World Scientific Publishing Co., Singapore. 13 + 512 pp. (1987). ISBN 9971–50–257–7, cloth, ISBN 9971–50–285–5 paper.
The individual contributions are included in their corresponding subject categories – see abstracts 034.050 – 034.052, 051.038, 061.040 – 061.060, 062.065, 062.066, 066.080 – 066.091, 067.085, 161.170 – 161.177.

012.039   **Infrared technology XII. Proceedings of a conference held at San Diego, CA, USA, 19 – 20 August 1986.**
Proc. SPIE Int. Soc. Opt. Eng., Vol. 685 (1986).
Review in Phys. Abstr., Vol. 90, No. 1309, Entry 80164 (1987).
See abstracts 035.060, 035.061, 133.017.

012.040   **Infrared detectors, sensors, and focal plane arrays. Proceedings of a conference held at San Diego, CA, USA, 21 – 22 August 1986.**
Proc. SPIE Int. Soc. Opt. Eng., Vol. 686 (1986).
Review in Phys. Abstr., Vol. 90, No. 1309, Entry 80165 (1987).
See abstracts 034.072, 035.063.

012.041   **The evolution of the small bodies of the solar system. Proceedings of the International School of Physics "Enrico Fermi", Course 98, held at Varenna, Italy, 5 – 10 August 1985.**
M. Fulchignoni, Ľ. Kresák (Editors).
North–Holland Physics Publishing, Amsterdam, The Netherlands. 14 + 307 pp. Price US$ 90.25, £ 55.45, Dfl. 157.25 (1987). ISBN 0–444–87035–0.
The individual contributions are included in their corresponding subject categories – see abstracts 021.011, 091.037, 098.051 – 098.058, 099.046, 102.041 – 102.045, 104.069, 106.028, 107.026.

012.042   **Memorias de la Quinta Reunión Regional Latinoamericana de Astronomía, held in Mérida, Mexico, 6 – 10 October 1986.**
S. Torres–Peimbert, I. Cruz–González (Editors).
Rev. Mex. Astron. Astrofis., Vol. 14, Nos. 1 – 2, 788 pp. (1987).
The individual contributions are included in their corresponding subject categories – see abstracts 004.059, 004.060, 013.062, 022.112, 031.053 – 031.057, 032.031, 032.032, 033.035, 034.102 – 034.110, 036.167, 041.014 – 041.018, 042.050 – 042.053, 051.043, 063.078, 063.079, 064.064, 065.081, 065.082, 066.108, 066.109, 074.058, 074.059, 078.009, 091.040, 100.026, 101.043, 102.055, 103.596 – 103.605, 111.017 – 111.019, 112.150 – 112.156, 113.048, 113.049, 114.089 – 114.097, 117.224 – 117.228, 119.044, 119.045, 120.013, 120.014, 121.042 – 121.045, 122.088 – 122.093, 125.071, 125.072, 131.222 – 131.231, 132.034 – 132.043, 133.019 – 133.021, 134.039 – 134.046, 141.018, 144.042, 151.123, 151.124, 153.039 – 153.043, 154.057, 155.123 – 155.130, 156.031 – 156.036, 157.285 – 157.296, 158.271 – 158.284, 159.088, 160.096 – 160.098, 161.246 – 161.250.

012.043   **The early universe. Proceedings of a NATO Advanced Study Institute, held at Victoria, B.C., Canada, 17 – 30 August 1986.**
W. G. Unruh, G. W. Semenoff (Editors).
NATO Advanced Science Institutes, Ser. C: Mathematical and Physical Sciences, Vol. 219. D. Reidel Publishing Co., Dordrecht, The Netherlands. 12 + 421 pp. Price Dfl. 185.00, US$ 89.00, £ 57.25 (1988). ISBN 90–277–2619–1.
The individual contributions are included in their corresponding subject categories – see abstracts 061.073, 080.053, 159.089, 161.251 – 161.260.

012.044   **ESA workshop on optical interferometry in space. Proceedings of an ESA workshop held at Granada, Spain, 16 – 18 June 1987.**
N. Longdon, V. David (Editors), with a summary and recommendations of the workshop by S. Volonté.
ESA Spec. Publ., ESA SP–273, 4 + 239 pp. Price Dfl. 37.00, US$ 12.00, £ 9.00 (1987).
The individual contributions are included in their corresponding subject categories – see abstracts 031.058, 032.033 – 032.037, 034.111, 035.070 – 035.081, 036.168 – 036.176, 042.054, 051.044, 074.060, 074.061, 126.078, 158.285, 158.286.

012.045   **Proceedings of the 8th International Conference on Vacuum Ultraviolet Radiation Physics, held at Lund, Sweden, 4 – 8 August 1986.**
P.–O. Nilsson, J. Nordgren (Editors).
Physica Scripta, Topical Issue, Vol. T17, 245 pp. Price SEK 500.00, US$ 80.00 (1987). ISBN 91–87308–08–8.
The individual contribution within the subject scope of Astronomy and Astrophysics Abstracts is included in its corresponding category – see abstract 112.157.

012.046   **Cosmology and particle physics. Proceedings of the XVII. GIFT International Seminar on Theoretical Physics, held at Peñiscola, Castellón, Spain, 2 – 7 June 1986.**
E. Alvarez, R. Domínguez Tenreiro, J. M. Ibáñez Cabanell, M. Quirós (Editors).
World Scientific Publishing Co., Singapore. 16 + 283 pp. Price US$ 69.55 (1987). ISBN 9971–50–259–3.
The individual contributions are included in their corresponding subject categories – see abstracts 061.074, 161.261 – 161.266.

012.047   **Resonance ionization spectroscopy 1986. Proceedings of the Third International Symposium on Resonance Ionization Spectroscopy, held at Swansea, Wales, 7 – 12 September 1986.**
IOP Publishing, Bristol, UK. 12 + 369 pp. (1987). ISBN 0–85498–175–6.
Review in Phys. Abstr., Vol. 90, No. 1310, Entry 88744 (1987).
See abstracts 022.114, 080.058, 080.059.

012.048   **Grazing incidence optics. Proceedings of a conference held at Orlando, FL, USA, 3 – 4 April 1986.**
Proc. SPIE Int. Soc. Opt. Eng., Vol. 640 (1986).
Review in Phys. Abstr., Vol. 90, No. 1311, Entry 95085 (1987).
See abstract 031.059 – 031.064, 035.083 – 035.085, 035.087.

012.049   **Symposium on computational physics. Proceedings of a conference held at Amsterdam, The Netherlands, 1 – 2 October 1986.**
Comput. Phys. Commun., Vol. 44, No. 3 (1987).
Review in Phys. Abstr., Vol. 90, No. 1312, Entry 101996 (1987).
See abstracts 062.092, 125.073.

012.050   **Probing the standard model. Proceedings of a summer institute on particle physics, held at Stanford, CA, USA, 28 July – 8 August 1986.**
Stanford Linear Accel. Center, Stanford, CA, USA, SLAC R–312. 725 pp. (1987).
Review in Phys. Abstr., Vol. 90, No. 1312, Entry 102016 (1987).
See abstracts 061.089, 144.047.

012.051   **Stellar evolution and dynamics in the outer halo of the Galaxy. Proceedings of an ESO workshop held at Garching, F.R. Germany, 7 – 9 April 1987.**
M. Azzopardi, F. Matteucci (Editors), with a summary by J. Lequeux.
ESO Conf. Workshop Proc., No. 27, 10 + 712 pp. Price DM 50.00 (1987). ISBN 3–923524–27–7.
The individual contributions are included in their corresponding subject categories – see abstracts 065.086, 065.087, 113.050, 114.098 – 114.106, 122.094, 122.095, 126.080, 134.047 – 134.051, 154.058 – 154.074, 155.132 – 155.144, 156.037 – 156.053, 157.297 – 157.302, 160.099.

012.052   **Seminar on "Astrophysical problems of planetary cosmogony", held by the Astronomical Council of the USSR of Sciences, April 1, 1987.**
Astron. Vestn., Tom 21, No. 4, p. 298 – 334 (1987). In Russian. English translation in Sol. Syst. Res.
The individual contributions are included in their corresponding subject categories – see abstracts 065.088 – 065.090, 102.057, 107.030 – 107.033, 112.159, 131.233 – 131.235.

012.053   **Superluminal radio sources. Proceedings of a workshop in honor of Professor Marshall H. Cohen, held at Big Bear Solar Observatory, Calif., USA, 28 – 30 October 1986.**
J. A. Zensus, T. J. Pearson (Editors), with a foreword by J. L. Greenstein.
Cambridge University Press, Cambridge, UK. 15 + 361 pp. Price £ 30.00, US$ 49.50 cloth, £ 15.00, US$ 24.95 paper (1987). ISBN 0–521–34560–X cloth, ISBN 0–521–35962–7 paper.
The individual contributions are included in their corresponding subject categories – see abstracts 036.183, 067.122, 158.292 – 158.308, 159.094 – 159.113.

012.054   **International Reference Ionosphere – status 1986/87. Proceedings of Workshop XI of the COSPAR Twenty-sixth Plenary Meeting held in Toulouse, France, 30th June – 11th July 1986.**
K. Rawer, P. A. Bradley (Editors).
Adv. Space Res., Vol. 7, No. 6, 7 + 129 pp. (1987).
The individual contributions within the subject scope of Astronomy and Astrophysics Abstracts are included in their corresponding categories – see abstracts 083.012 – 083.019, 084.054.

012.055   **Magnetosphere, ionosphere and thermosphere. Proceedings of the Topical Meeting of the COSPAR Interdisciplinary Scientific Commission C (Meeting C1) of the COSPAR Twenty–sixth Plenary Meeting held in Toulouse, France, 30th June – 11th July 1986.**
E. R. Schmerling (Editor).
Adv. Space Res., Vol. 7, No. 8, 6 + 89 pp. (1987).
The individual contributions within the subject scope of Astronomy and Astrophysics Abstracts are included in their corresponding categories – see abstracts 013.070, 082.070, 084.055 – 084.057.

012.056   **Middle atmosphere trace constituents. Proceedings of Workshop IX of the COSPAR Twenty–sixth Plenary Meeting held in Toulouse, France, 30th June – 11th July 1986.**
G. M. Keating (Editor).
Adv. Space Res., Vol. 7, No. 9, 6 + 141 pp. (1987).
The individual contributions within the subject scope of Astronomy and Astrophysics Abstracts are included in their corresponding categories – see abstracts 082.071 – 082.080.

012.057   **The earth's middle and upper atmosphere. Proceedings of the Topical Meeting of the COSPAR Interdisciplinary Scientific Commission C (Meeting C2) and of Workshops XIII, XV and XVI of the COSPAR Twenty–sixth Plenary Meeting held in Toulouse, France, 30th June – 11th July 1986.**
K. U. Grossmann, K. S. W. Champion, M. Roemer, W. L. Oliver, T. A. Blix (Editors).
Adv. Space Res., Vol. 7, No. 10, 8 + 364 pp.(1987).
The individual contributions within the subject scope of Astronomy and Astrophysics Abstracts are included in their corresponding categories – see abstracts 082.081 – 082.089, 084.058.

012.058   **Materials of the Soviet–American conference on the problem of detection of solar neutrinos, held on 1 – 2 October 1985 in Moscow.**
Ivz. Akad. Nauk SSSR. Ser. Fiz., Tom 51, No. 6, p. 1225 – 1242 (1987).
From Ref. Zh., 51. Astron., 12.51.20 (1987). – See Abstr. 080.065.

012.059   **Quarks – 86. Publications of the All–Union seminar, held in Tbilisi, 15 – 17 April 1986.**
Moskva. 506 pp. (1987). In Russian and English.
Review in Ref. Zh., 51. Astron., 10.51.25 (1987). – See abstracts 161.323, 161.324.

012.060   **History of Oriental astronomy. Proceedings of the 91st Colloquium of the International Astronomical Union, held at New Delhi, India, 13 – 16 November 1985.**
G. Swarup, A. K. Bag, K. S. Shukla (Editors), with an introductory lecture by E. S. Kennedy and concluding remarks by O. Gingerich.
Cambridge University Press, Cambridge, UK. 17 + 289 pp. Price £ 27.50, US$ 54.50 (1987). ISBN 0–521–34659–2.
The individual contributions are included in their corresponding subject categories – see abstracts 004.080 – 004.110, 046.030 – 046.034.

012.061   **Journées PHEMU85 sur l'observation des phénomènes mutuels des satellites de Jupiter en 1985.**
J.–E. Arlot (Editor).
Ann. Phys. (Paris), Vol. 12, Colloq. No. 1, 241 pp. (1987). ISBN 2–86883–049–8.
The individual contributions are included in their corresponding subject categories – see abstracts 004.111, 032.044, 034.125 – 034.127, 036.188 – 036.195, 098.068 – 098.069, 099.054 – 099.067, 101.048.

012.062   **Infrared, adaptive, and synthetic aperture optical systems. Proceedings of two conferences held at Arlington, VA, USA, 8 April 1985 and Orlando, FL, USA, 1 – 2 April 1986.**
Proc. SPIE Int. Soc. Opt. Eng., Vol. 643 (1986).
Review in Phys. Abstr., Vol. 90, No. 1313, Entry 108025 (1987).
See abstracts 031.066 – 031.068, 032.045, 032.046, 035.098, 036.198.

012.063   **Optical system design, analysis, and production for advanced technology systems. Proceedings of a conference, held at Innsbruck, Austria, 15 – 17 April 1986.**
Proc. SPIE Int. Soc. Opt. Eng., Vol. 655 (1986).
Review in Phys. Abstr., Vol. 90, No. 1313, Entry 108026 (1987).
See abstracts 031.069, 035.095, 035.099, 051.054.

012.064   **NASA conference on double layers in plasmas, held at Huntsville, AL, USA, March 1986.**
Laser Part. Beams, Vol. 5, Part 2 (1987).
Review in Phys. Abstr., Vol. 90, No. 1314, Entry 115677 (1987).
See abstracts 062.108 – 062.120, 084.068 – 084.072.
Further constrictions are given in NASA CP–2469 see 43.012.108.

012.065   **Classical and quantum aspects of gravitation. Proceedings of the symposium on general relativity and its applications to astrophysics and cosmology, held at Burdwan, India, 26 – 28 December 1985.**
Burdwan University, Burdwan, India. 85 pp. (1986).
Review in Phys. Abstr., Vol. 90, No. 1314, Entry 115694 (1987).
See abstracts 066.146, 066.148, 067.140, 161.351, 161.352.

012.066   **Proceedings of the 5th working meeting on European VLBI for geodesy and astrometry, held at Wettzell, F.R. Germany, 7 – 8 November 1986.**
J. Campbell, H. Schuh (Editors).
Mitt. Geod. Inst. Rheinischen Friedrich–Wilhelms–Univ. Bonn, Nr. 71, 138 pp. (1987).
The individual contributions are included in their corresponding subject categories – see abstracts 009.020 – 009.024, 013.072 – 013.075, 033.040, 033.041, 036.199, 036.200, 041.021, 041.022, 044.038, 045.011 – 045.013, 159.123.

012.067   **Remote sensing from space. Proceedings of Symposium 3, Workshop V and of the COSPAR Interdisciplinary Scientific Commission A (Meeting A2) of the COSPAR Twenty–sixth Plenary Meeting held in Toulouse, France, 30th June – 11th July 1986.**
S. N. Goward, S. G. Ungar, J. Nithack, A, F. Hasler (Editors).
Adv. Space Res., Vol. 7, No. 11, 8 + 391 pp. (1987).
Contents:Terrestrial patterns and processes: perspectives from space (Symp. 3). Quantitative radar remote sensing of land and

oceanic surface features (Workshop V). The use of satellite observations for weather prediction (Mtg A2).

012.068   Studies of the middle atmosphere. A discussion arranged by the British National Committees on Space Research, Solar–Terrestrial Physics, and Geodesy and Geophysics. The discussion was held at 4 and 5 December 1986.
J. A. Pyle, L. Thomas, R. Wilson (Editors).
Philos. Trans. R. Soc. London, Ser. A, Vol. 323, No. 1575, p. 519 – 710 (1987).
The individual contributions within the subject scope of Astronomy and Astrophysics Abstracts are included in their corresponding categories – see abstracts 082.094 – 082.096.

012.069   Astrophysical radiation hydrodynamics. Based on the proceedings of a NATO Advanced Research Workshop, held at Garching, F.R. Germany, 2 – 13 August 1982.
K.–H. A. Winkler, M. L. Norman (Editors).
NATO Advanced Science Institutes, Ser. C: Mathematical and Physical Sciences, Vol. 188. D. Reidel Publishing Co., Dordrecht, The Netherlands. 8 + 590 pp. Price Dfl. 235.00, US$ 98.00, £ 69.95 (1986). ISBN 90–277–2335–4.
The individual contributions are included in their corresponding subject categories – see abstracts 061.112, 062.125 – 062.132, 063.101 – 063.104, 066.152 – 066.154.

012.070   Infrared astronomy with arrays. Proceedings of the workshop on ground–based astronomical observations with infrared array detectors, held at Hilo, HI, USA, 24 – 26 March 1987.
C. G. Wynn–Williams, E. E. Becklin, L. H. Good (Editors), with a summary of highlights by D. N. B. Hall and closing remarks by G. W. van Citters.
Univ. of Hawaii, Institute of Astronomy, Honolulu, HI, USA. 12 + 551 pp. (1987).
The individual contributions are included in their corresponding subject categories –see abstracts 034.128 – 034.176, 035.100 – 035.104, 036.202 – 036.207, 117.262, 121.052, 131.271 – 131.275, 132.046, 134.057, 134.058, 155.153, 155.154, 157.320 – 157.324, 158.330 – 158.332.

012.071   Site testing. Meeting on the LEST site survey held at Roque de los Muchachos, La Palma, 18 – 19 February 1987.
O. Engvold, Ø. Hauge (Editors).
LEST Found., Tech. Rep., No. 26, 96 pp. (1987)
The individual contributions are included in their corresponding subject categories – see abstracts 013.076, 031.070, 032.047, 034.177, 034.178, 036.208, 082.097 – 082.103.

012.072   COPE. Co–orbiting platform elements. Proceedings of an ESA workshop held at ESTEC, 7 – 8 April 1987.
T.–D. Guyenne (Editor), with introductory remarks by P. Goldsmith.
ESA Spec. Publ., ESA SP–1093, 71 pp. Price DM 33.00, US$ 12.00 (1987).
The individual contributions within the subject scope of Astronomy and Astrophysics Abstracts are included in their corresponding subject categories – see abstracts 013.077, 051.055.

012.073   Vorträge beim Theodor Ritter von Oppolzer–Gedächtnissymposium am 3.12.1986.
K. Bretterbauer, J. Zeger (Editors), with a preface by K. Rinner and a welcome address by F. Rotter.
Geod. Arb. Österr. Int. Erdmessung, Neue Folge, Band 5, 122 pp. (1987).
The individual contributions within the subject scope of Astronomy and Astrophysics Abstracts are included in their corresponding categories – see abstracts 005.012, 005.013, 013.079, 044.039, 079.003, 079.004.

012.074   The standard model. The supernova 1987A. Proceedings of the XXIInd Rencontre de Moriond, Leptonic Session, Vol. 1, held at Les Arcs, Savoie, France, March 8 – 15, 1987.
J. Trân Thanh Vân (Editor), with a summary by L.–L. Chau.
Editions Frontières, 91190 Gif sur Yvette, France. 11 + 796 pp. (1987). ISBN 2–86332–047–5.
The individual contributions within the subject scope of Astronomy and Astrophysics Abstracts are included in their corresponding categories – see abstracts 061.113, 080.067, 125.329 – 125.336, 161.357.

012.075   Symposium on the diversity and similarity of comets. Proceedings of an international symposium, held in Brussels, Belgium, 6 – 9 April 1987.
E. J. Rolfe, B. Battrick (Editors), with a foreword by M. Ackerman, M. Scherer and R. Reinhard.
ESA Spec. Publ., ESA SP–278, 13 + 763 pp. Price Dfl. 100.00 (1987).
The individual contributions are included in their corresponding subject categories – see abstracts 004.119, 005.014 – 005.016, 013.082, 022.128 – 022.132, 036.216, 051.056 – 051.060, 052.026, 062.137, 102.063 – 102.081, 103.039 – 103.046, 103.129 – 103.130, 103.175, 103.623 – 103.690, 103.931, 104.096 – 104.097, 106.035.

012.076   The use of supercomputers in observational astronomy. Proceedings of a workshop held in Minneapolis, Minn., USA, 4 – 6 November 1985.
T. J. Cornwell (Editor).
NRAO Workshop, No. 15. 6 + 148 pp. Price US$ 13.00 (1988).
The individual contributions are included in their corresponding subject categories – see abstracts 013.083, 021.020 – 021.024, 033.046 – 033.048, 036.217 – 036.220, 126.093, 155.159, 161.372.

012.077   Advances in turbulence. Proceedings of the First European Turbulence Conference, held at Lyon, France, 1 – 4 July 1986.
Springer–Verlag, Berlin (West). 16 + 585 pp. (1987). ISBN 3–540–17586–5.
Review in Phys. Abstr., Vol. 91, No. 1319, Entry 24 (1988).
See abstracts 062.141, 080.070.

012.078   Surface characterization and testing. Proceedings of a conference held at San Diego, CA, USA, 21 – 22 August 1986.
Proc SPIE Int. Soc. Opt. Eng., Vol. 680 (1987).
Review in Phys. Abstr., Vol. 91, No. 1320, Entry 4548 (1988).
See abstracts 031.074, 031.075.

012.079   Proceedings of the international topical meeting on image detection and quality, held at Paris, France, 16 – 18 July 1986.
SPIE, Bellingham, WA, USA, 7 + 394 pp. (1986). ISBN 2–900195–09–8.
Review in Phys. Abstr., Vol. 91, No. 1320, Entry 4559 (1988).
See abstracts 031.077, 034.204 – 034.208, 035.119, 035.120, 036.241 – 036.245, 051.068.

012.080   Proceedings of the 1986 summer study on the physics of the superconducting supercollider, held at Snowmass, CO, USA, 23 June – 11 July 1986.
Argonne Nat. Lab., Argonne, IL, USA. 14 + 753 pp. (1987).
Review in Phys. Abstr., Vol. 91, No. 1320, Entry 4560 (1988).
See abstracts 034.200, 061.139 – 061.141, 061.143, 080.071, 080.072, 117.276, 144.078, 157.329, 161.441, 161.442.

012.081 **Star clusters and associations. Proceedings of a sympo-sium, held in Sofia, Bulgaria, 28 – 30 May 1985.**
B. A. Balázs, G. Szécsényi–Nagy (Editors).
Publ. Astron. Dep. Eötvös Univ., No. 8, 149 pp. (1986).
The individual contributions are included in their corresponding subject categories – see abstracts 131.288, 131.289, 151.150, 152.003, 152.004, 153.053 – 153.059, 154.086, 157.348, 157.349.

012.082 **Schrödinger: centenary celebration of a polymath. Papers presented at a conference held at Imperial College, London, UK, April 1987.**
C. W. Kilmister (Editor).
Cambridge University Press, Cambridge, UK. 253 pp. Price £ 30.00, US$ 54.50 (1987). ISBN 0–521–34017–9.
Reviews in Observatory, Vol. 107, No. 1081, p. 284 (1987); Sky Telesc., Vol. 74, No. 5, p. 493 (1987).

012.083 **String theory, quantum cosmology and quantum gravity, integrable and conformal invariant theories. Proceedings of a colloquium held at Paris and Meudon, France, September 1986.**
H. J. de Vega, N. Sánchez (Editors).
World Scientific Publishing Co., Singapore. 511 pp. Price US$ 64.00 cloth, US$ 34.00 paper (1987). ISBN 9971–50–286–0 cloth, ISBN 9971–50–299–2 paper.
From Phys. Today, Vol. 40, No. 11, p. 104 (1987).

012.084 **Superstrings, unified theories and cosmology. Proceedings of a workshop held at Trieste, Italy, June 1986.**
G. Furlan, R. Jengo, J. C. Pati, D. W. Sciama, E. Sezgin, Q. Shafi (Editors).
ICTP Series in Theoretical Physics, Vol. 3. World Scientific Publishing Co., Singapore. 454 pp. Price US$ 68.00 cloth, US$ 28.00 paper (1987). ISBN 9971–50–271–2 cloth, ISBN 9971–50–280–1 paper.
From Phys. Today, Vol. 40, No. 11, p. 104 (1987).

012.085 **Martian geomorphology and its relation to subsurface volatiles. Proceedings of a MECA special session at the XVII. Lunar and Planetary Science Conference, held at Houston, TX, USA, 17 March 1986 (Abstracts of papers).**
S. M. Clifford, L. A. Rossbacher, J. R. Zimbelman (Editors).
LPI–TR–87–2. Lunar and Planetary Institute, Houston, TX, USA. 51 pp. (1987).
Review in Phys. Abstr., Vol. 90, No. 1317, Entry 134369 (1987).

012.086 **Volny v atmosfere Solntsa (*Oscillations and waves on the Sun*). Third seminar of the working group "Waves in the atmosphere of the Sun", held at Novosibirsk, USSR, 30 March – 1 April 1987.**
Novosibirsk, USSR. 30 pp. (1987).
Review in Ref. Zh., 51. Astron., 10.51.33 (1987).

013 Reports on Astronomy in Various Countries and Particular Fields

013.001 **Asteroseismology and SYNOP.**
J. Harvey.
The SHIRSOG Workshop, p. 5 (1986). – See Abstr. 012.003.

013.002 **Scheduling and other issues for a solar–stellar synoptic observatory.**
J. S. Gallagher III.
The SHIRSOG Workshop, p. 94 – 101 (1986). – See Abstr. 012.003.

013.003 **The High Altitude Observatory–Lowell Observatory Solar–Stellar Spectrophotometry Project.**
R. L. Gilliland.
The SHIRSOG Workshop, p. 104 – 116 (1986). – See Abstr. 012.003.
The author proposes to study chromospheric activity on time scales of rotation and activity cycles in the Sun and solar–like stars using a high–dispersion, dedicated fiber–fed spectrograph facility currently being completed. Using CCD detectors one will record the line intensities reflecting chromospheric activity. To study magnetic activity, near nightly observations of about 40 stars over a 120 day time base for several seasons will be obtained. It is hoped to extend the detection of rotation periods using rotational modulation of chromospheric emission to later spectral types and other stellar population groups than are currently available. To study stellar cycles observations of about 200 stars once per month over a long time base will be obtained. It is planned to define cycle properties (amplitude, shape, morphology) as a function of stellar mass, age, rotation rate, chemical composition, etc.

013.004 **The McMath Solar–Stellar Nighttime Program.**
M. A. Smith.
The SHIRSOG Workshop, p. 124 – 132 (1986). – See Abstr. 012.003.
On October 1, 1984 the NSO officially commenced its operation of a spectroscopic facility at the McMath solar telescope on Kitt Peak. This instrument is dedicated to the synoptic observations of "solar–stellar" phenomena in stars and appears to be the only one accessible to the general astronomical community. The author describes the spectrograph itself, prospects for upgrades, NSO's operational and scheduling policies, and the kinds of programs currently serviced.

013.005 **The solar–stellar connection at low spectral resolution.**
R. R. Radick.
The SHIRSOG Workshop, p. 133 – 135 (1986). – See Abstr. 012.003.
Observations at low spectral resolution (i.e., less than a few thousand) have played an important role in the development of the subdiscipline called the solar–stellar connection. The author indicates briefly some examples of recent results, primarily from a photometric program at Lowell Observatory, but also from the H–K spectrophotometric program at Mount Wilson, in order to illustrate what is currently being done in this area, and then he outlines a concept for a new approach aimed at studying the rotation and activity of very large samples of stars.

013.006 **A proposal to use and manage The Mt. Wilson 100–inch Hooker Telescope for the study of stellar interiors.**
R. K. Ulrich.
The SHIRSOG Workshop, p. 138 – 139 (1986). – See Abstr. 012.003.

013.007 **Infrared astronomy.**
M. G. Hauser.
NASA Conf. Publ., NASA CP–2464, p. 395 – 401 (1987). – See Abstr. 012.004.

013.008 **Astrophysics in the Orient.**
J. B. Hearnshaw.
South. Stars, Vol. 32, No. 3, p. 73 – 86 (1987).
A personal account is given of a study and lecture tour to East and South–East Asia. Highlights of some current operations in astronomy encountered en route are recalled.

013.009 Experiences on my recent study tour of astronomical facilities in the USA.
W. H. Allen.
South. Stars, Vol. 32, No. 3, p. 87 – 95 (1987).
A personal account of a recent study tour in the South West of the United States is given. Impressions of noteworthy astronomical facilities are recalled.

013.010 Thirty years of the Mark IA radio telescope at Jodrell Bank.
T. Jones.
J. Br. Astron. Assoc., Vol. 97, No. 5, p. 261 – 262 (1987).
Even after 30 years, Jodrell Bank's Mark IA radio telescope, now called the Lovell Telescope, is still the second largest fully steerable disk in the world. New instruments have been added to the observatory, the latest being part of the MERLIN network, but the Mark IA remains the top attraction for visitors.

013.011 A stay at Bonn University for VLBI research.
T. Yoshino.
Astron. Her., Vol. 80, No. 8, p. 223 – 227 (1987). In Japanese.

013.012 Simple theory of stellar structure. Research with Professor Hayashi.
H. Sato.
Astron. Her., Vol. 80, No. 9, p. 256 – 260 (1987). In Japanese.

013.013 Five nights on a bare mountain – an outsider's look at La Silla.
G. Schilling.
Messenger, No. 49, p. 39 – 41 (1987).

013.014 The future of high angular resolution astronomy: seeing the unseen.
H. A. McAlister.
Vistas Astron., Vol. 30, Part 1, p. 27 – 38 (1987). – See Abstr. 012.007.
Technological advances during the last quarter century are leading to the development of new techniques for high angular resolution astronomy and are allowing the implementation of methods orginally suggested in the last century. This paper describes the achievements and limitations of speckle interferometry and imaging with large, single aperture telescopes and explores the scientific potential implied by milli–arcsecond resolution that will result from the ground–based optical array projects now under development. Interferometry from space–based instruments will eventually permit unprecedented resolution when baselines of many kilometers are employed at optical wavelengths.

013.015 Current problems in astrophysics needing space–based radio astronomy.
C. A. Norman.
Radio astronomy from space, p. 15 – 19 (1987). – See Abstr. 012.009.
There are a number of topics of current interest in astrophysics that could benefit considerably from space–based radio astronomy and the associated VLBI networks. This talk shall briefly focus on specific examples from the four major areas of cosmology, active galactic nuclei and starbursts, the interstellar and intergalactic medium, and molecular clouds and associated star formation.

013.016 The need for space–based radio astronomy.
K. J. Johnston, K. W. Weiler.
Radio astronomy from space, p. 21 – 27 (1987). – See Abstr. 012.009.
There is a vital need for new opportunities for space research in radio astronomy. Astrophysical processes require study in all wavelength ranges and for radio astronomy this means developing and deploying spacecraft to investigate those parts of the radio spectrum denied to earth based observers by absorption, reflection, and refraction in the earth's atmosphere and ionosphere. Also, even at frequencies observable from the ground,

radio astronomers must reach beyond the baseline limits imposed by the diameter of the earth.

013.017 Radio astronomy on the moon.
J. O. Burns, J. Asbell.
Radio astronomy from space, p. 29 – 39 (1987). – See Abstr. 012.009.
The authors review the advantages and opportunities for radio astronomy on the moon during the early to mid 21st century. In particular, they argue that the lack of atmosphere, the extremely low seismic activity, the low radio–frequency background, and the natural cryogenic environment make the moon (particularly the far side and the poles) a nearly ideal locale for submillimeter/far–IR to very low frequency (< 10 MHz) radio astronomy.

013.018 A coordinated approach to ground–, airplane–, balloon–, and space–based millimeter and sub–millimeter astronomy.
R. M. Hjellming.
Radio astronomy from space, p. 43 – 50 (1987). – See Abstr. 012.009.
The author discusses balanced or coordinated approachs to millimeter and sub–millimeter astronomy, in which ground–, airplane–, balloon–, and space–based facilities are constructed and operated to support and complement each other. Six levels of instrumental accessibility are defined with primary and secondary roles for each level.

013.019 Status report on RADIOASTRON.
R. T. Schilizzi.
Radio astronomy from space, p. 153 – 155 (1987). – See Abstr. 012.009.
A brief report on the mission concept for the USSR RADIOASTRON Space VLBI project is given.

013.020 Collaborative VLBI experiments with RADIOASTRON.
D. L. Jones, R. A. Preston, J. F. Jordan, R. P. Linfield.
Radio astronomy from space, p. 157 – 163 (1987). – See Abstr. 012.009.
The Soviet Union is planning to launch a 10 meter diameter radio telescope into Earth orbit for use in VLBI observations. This mission (RADIOASTRON) will be the first opportunity for astronomically important VLBI experiments with baselines much longer than can be obtained between telescopes on the Earth. This paper describes the potential scientific advantages of combining data from the orbiting telescope with data from some of the very sensitive radio telescopes in western Europe, Australia, Japan, and the United States.

013.021 QUASAT – European status report.
R. T. Schilizzi.
Radio astronomy from space, p. 164 – 169 (1987). – See Abstr. 012.009.
The current status of the QUASAT Space VLBI mission is described. A brief review of the mission concept and the main scientific objectives is also given.

013.022 A review of decametric radio astronomy: instruments and science.
W. C. Erickson, H. V. Cane.
Radio astronomy from space, p. 255 – 263 (1987). – See Abstr. 012.009.
The authors review galactic and extragalactic radio astronomy at wavelengths greater than 10 meters and discuss the instruments designed for work in this wavelength range. The results of these studies have implications concerning programs that might be undertaken by a Low Frequency Space Array.

013.023 Status of the Perseus optical flasher.
G. J. Corso, F. A. Ringwald, R. W. Harris.
Astron. Astrophys., Vol. 183, No. 1, p. L9 (1987).
The position of the short duration optical flashes reported by Katz et al. (1986) was monitored with a variety of instruments

and detectors for 76 hours between Oct. 1986 and Mar. 1987. No optical flashes were detected.

013.024 Opacity Project: astrophysical and fusion applications.
A. K. Pradhan.
Phys. Scr., Vol. 35, No. 6, p. 840 – 845 (1987).

An overview is presented of a project to calculate large quantities of accurate atomic data for radiative processes of importance in the precise determination of opacities in stellar atmospheres, and for astrophysical and laboratory applications in general. Work is in progress on the oscillator strengths, photoionization cross sections, damping constants, etc., for all atoms and ions in hydrogen through neon isoelectronic sequences going up to iron.

013.025 New developments at Table Mountain.
R. J. Chambers.
I.A.P.P.P. Commun., No. 29, p. 5 – 13 (1987).

013.026 Astrophysik auf dem Gornergrat. Teil 2: Erste Ergebnisse.
G. Winnewisser.
Sterne Weltraum, 26. Jahrg., Nr. 7–8, p. 382–386 (1987).

013.027 Astronomie, eine interdisziplinäre Wissenschaft.
W. H. Kegel.
Sterne Weltraum, 26. Jahrg., Nr. 9, p. 466 – 469 (1987).

013.028 Information in astrophysics.
V. Castellani.
Selected topics on data analysis in astronomy, p. 1 – 11 (1987). – See Abstr. 012.011.

The author gives a general overview on the meaning and procedures of astrophysics.

013.029 La cosmologie en France.
M. Lachieze–Rey.
J. Astron. Fr., No. 30, p. 8 – 10 (1987).

013.030 Formation à et par la recherche en astronomie: mythes et réalités.
J. Heyvaerts.
J. Astron. Fr., No. 30, p. 10 – 15 (1987).

013.031 The Automatic Photoelectric Telescope Service.
R. M. Genet, L. J. Boyd, K. E. Kissell,
D. L. Crawford, D. S. Hall, D. S. Hayes, S. L. Baliunas.
Publ. Astron. Soc. Pac., Vol. 99, No. 617, p. 660 – 667 (1987).

Automatic observatories have the potential of gathering sizable amounts of high–quality astronomical data at low cost. The Automatic Photoelectric Telescope Service (APT Service) has realized this potential and is routinely making photometric observations of a large number of variable stars. However, without observers to provide on–site monitoring, it was necessary to incorporate special quality checks into the operation of the APT Service at its multiple automatic telescope installation on Mount Hopkins.

013.032 History of cosmic ray investigations in the USSR.
I. V. Dorman.
Kosm. luchi, Moskva, No. 24, p. 5 – 31 (1987). In Russian. Abstr. in Ref. Zh., 51. Astron., 7.51.14 (1987).

013.033 Achievements and problems of the Department of General Physics and Astronomy of the USSR Academy of Sciences.
A. M. Prokhorov.
Vestn. Akad. Nauk SSSR, No. 11, p. 56 – 61 (1986). In Russian. Abstr. in Ref. Zh., 51. Astron., 7.51.35 (1987).

013.034 Development of balloon technology in China.
Y. Gu, K. Yuan.
Adv. Space Res., Vol. 7, No. 7, p. 71 – 76 (1987). – See Abstr. 012.023.

In this paper the authors present the recent progress in China on balloon materials, balloon design and manufacture procedures, launch and recovery techniques, telemetry and telecommand facilities etc.

013.035 Publicising astronomy.
C. R. G. Turk.
Mon. Notes Astron. Soc. S. Afr., Vol. 46, Nos. 9 – 10, p. 141 – 143 (1987).

013.036 Proposal for a large telescope in South Africa.
I. S. Glass.
Mon. Notes Astron. Soc. S. Afr., Vol. 46, Nos. 9 – 10, p. 147 – 152 (1987).

013.037 L'astronomie dans le monde.
J. Manfroid.
Ciel, Vol. 49, p. 300 – 302 (1987).

013.038 The DUMAND project.
P. K. F. Grieder.
Neutrinos and the present–day universe, p. 171 – 194 (1986). – See Abstr. 012.025.

013.039 The Soviet astronomy in 1977 – 1987. To the 70th anniversary of the Great October.
Astron. Zh., Tom 64, Vyp. 5, p. 897 – 899 (1987). In Russian. English translation in Sov. Astron., Vol. 31, No. 5.

013.040 Satellite geodesy investigations in Poland in 1983 – 1986.
B. Kołaczek.
Artif. Satell., Vol. 22, No. 2, p. 5 – 14 (1987).

013.041 Les potins d'Uranie: l'ICSU.
Al Nath (*A. Heck*).
Ciel, Vol. 49, p. 319 – 321 (1987).

013.042 Astronomie auf neuen Wegen – Forschungsergebnisse des Max–Planck–Instituts für Astronomie.
H. Elsässer.
Sterne, 63. Band, Heft 3, p. 131 – 144 (1987).

013.043 Panel discussion of future research on Be stars.
G. J. Peters, A. G. Hearn, H. F. Henrichs, T. Kogure, C. T. Bolton, M. A. Smith, M. J. Plavec.
Physics of Be stars, p. 535 – 552 (1987). – See Abstr. 012.030 (IAU Colloq. No. 92).

013.044 What do we learn from space? Space science in Japan.
M. Oda.
Phys. Today, Vol. 40, No. 12, p. 26 – 33 (1987).

A philosophy of modest but well–defined projects and a strategy of frequent launches have made Japan a major participant in space research – particularly in X–ray astronomy.

013.045 Observational neutrino astrophysics.
M.–T. Koshiba.
Phys. Today, Vol. 40, No. 12, p. 38 – 42 (1987).

Pioneering measurements of the solar neutrino flux and detailed observations of the neutrino burst from SN 1987a and of solar boron–8 neutrinos have signaled the birth of observational neutrino astrophysics.

013.046 Optical astrometry in the UK.
C. A. Murray.
Q. J. R. Astron. Soc., Vol. 28, No. 3, p. 347 – 353 (1987).

013.047   **New directions for space astronomy.**
R. A. Brown, R. Giacconi.
Science, Vol. 238, No. 4827, p. 617 – 619 (1987).

013.048   **Status report on the SYNOP project to monitor stars with high–resolution spectroscopy.**
J. L. Linsky, M. S. Giampapa.
Bull. Am. Astron. Soc., Vol. 19, No. 2, p. 701 (1987). Abstract. – See Abstr. 010.061.

013.049   **IRAF status report.**
F. Valdes.
Bull. Am. Astron. Soc., Vol. 19, No. 2, p. 738 (1987). Abstract. – See Abstr. 010.061.

013.050   **Status report: science data analysis software for the Hubble Space Telescope.**
R. J. Hanisch.
Bull. Am. Astron. Soc., Vol. 19, No. 2, p. 738 (1987). Abstract. – See Abstr. 010.061.

013.051   **ROSAT, data centers, and X–ray data analysis, a new era?**
R. L. Pisarski, R. D. Price, D. M. Worrall, S. S. Murray.
Bull. Am. Astron. Soc., Vol. 19, No. 2, p. 739 (1987). Abstract. – See Abstr. 010.061.

013.052   **Status report on the VAX–based IUE Regional Data Analysis Facility at GSFC.**
C. A. Grady, R. W. Thompson, A. Michalitsianos.
Bull. Am. Astron. Soc., Vol. 19, No. 2, p. 739 (1987). Abstract. – See Abstr. 010.061.

013.053   **Computer graphics and images display devices for astronomy.**
H. P. Murphy.
Bull. Am. Astron. Soc., Vol. 19, No. 2, p. 742 – 743 (1987). Abstract. – See Abstr. 010.061.

013.054   **Applications of AI technology in astronomical research I: An expert data analysis assistant.**
R. Albrecht, M. Johnston.
Bull. Am. Astron. Soc., Vol. 19, No. 2, p. 743 (1987). Abstract. – See Abstr. 010.061.

013.055   **Applications of AI technology in astronomical research II: What can be done, what should be done?**
R. Albrecht, R. Rampazzo, H.–M. Adorf, R. Fosbury, M. Johnston.
Bull. Am. Astron. Soc., Vol. 19, No. 2, p. 743 (1987). Abstract. – See Abstr. 010.061.

013.056   **Telescience at the U.C. Berkeley Space Sciences Laboratory.**
S. Chakrabarti, C. A. Dobson, J. G. Jernigan.
Bull. Am. Astron. Soc., Vol. 19, No. 2, p. 744 (1987). Abstract. – See Abstr. 010.061.

013.057   **Toward a network for astronomy – status report.**
P. M. B. Shames.
Bull. Am. Astron. Soc., Vol. 19, No. 2, p. 744 (1987). Abstract. – See Abstr. 010.061.

013.058   **Astrographic catalogues of British observatories.**
N. P. J. O'Hora.
R. Greenwich Obs., Prepr., No. 58, p. 21 – 24 (1987). To appear in IAU Symp. No. 133.

013.059   **Isaac Newton Telescope observations 1984 – 86.**
C. R. Benn, R. Martin.
R. Greenwich Obs., Prepr., No. 65, 20 pp. (1987). To appear in Q. J. R. Astron. Soc.

013.060   **Progress in infrared astronomy.**
P.–s. Chen.
Publ. Yunnan Obs., No. 2, p. 108 – 120 (1987). In Chinese.
The historical development and the importance of the infrared astronomy are described in this paper. The up–date observational techniques of the infrared astronomy, especially the IRAS mission are introduced in detail. The development of the infrared astronomy in China is also reviewed.

013.061   **Main problems of present–day cosmophysics.**
G. F. Krymskij.
Metodol. probl. razvitiya nauk. v regione, Novosibirsk, p. 175 – 186 (1987). In Russian. Abstr. in Ref. Zh., 51. Astron. 11.51.2 (1987).

013.062   **Mesa redonda "Perspectivas de la astronomia en Latinoamerica ante los problemas economicos de la region".**
J. Sahade, S. M. Viegas–Aldrovandi, J. Maza, M. T. Ruiz, M. Peimbert, G. Bruzual, A. Poveda.
Rev. Mex. Astron. Astrofis., Vol. 14, No. 1, p. 20 – 32 (1987). – See Abstr. 012.042.

013.063   **India's space program.**
S. M. Monin.
Indiya, 1985 – 1986. Ezhegodnik, p. 68 – 87 (1987). In Russian. Abstr. in Ref. Zh., 62. Issled. Kosm. Prostranstva, 11.62.86 (1987).

013.064   **Giant atoms in the cosmos.**
R. L. Sorochenko, A. E. Salomonovich.
Priroda, No. 11, p. 82 – 94 (1987). In Russian.
Concerns the discovery of the atoms in the cosmos and the possibilities for radio astronomical investigations.

013.065   **The Australia Telescope project.**
J. B. Whiteoak.
Aust. J. Astron., Vol. 2, No. 2, p. 54 – 55 (1987).

013.066   **Astronomy (and some other things) in China.**
A. H. Batten.
J. R. Astron. Soc. Can., Vol. 81, No. 5, p. 147 – 156 (1987).
The writer describes his recent visit to China and gives some account of the major observatories in that country.

013.067   **UK allocations of *EISCAT* observing time.**
H. Rishbeth, D. M. Willis.
Q. J. R. Astron. Soc., Vol. 28, No. 4, p. 477 – 480 (1987).

013.068   **Isaac Newton Telescope observations 1984–6.**
C. R. Benn, R. Martin.
Q. J. R. Astron. Soc., Vol. 28, No. 4, p. 481 – 496 (1987).

013.069   **The new satellite laser ranging system at Cagliari Observatory.**
A. Banni, V. Capoccia.
Pubbl. Stn. Astron. Int. Latitudine, Carloforte–Cagliari, Nuov. Ser., N. 140, 12 pp. (1987).
The new station is expected to become operative in the first months of 1987.

013.070   **Space research – an international endeavor.**
E. R. Schmerling.
Adv. Space Res., Vol. 7, No. 8, p. 1 – 2 (1987). – See Abstr. 012.055.

013.071   **Project MERIT.**
B. Kołaczek.
Postepy Astron., Tom 34, Zesz. 4, p. 219 – 248 (1986). In Polish.
Purposes, organization and achievements of the Project MERIT (Monitoring Earth Rotation and Intercompare the Techniques of Observations and Analysis) are presented in the article.

013.072   **French involvement in geodetic VLBI.**
C. Boucher.
Mitt. Geod. Inst. Rheinischen Friedrich–Wilhelms–Univ. Bonn,
Nr. 71, p. 29 – 30 (1987). – See Abstr. 012.066.

013.073   **Ongoing geodetic/astrometric VLBI activities.**
F. J. J. Brouwer.
Mitt. Geod. Inst. Rheinischen Friedrich–Wilhelms–Univ. Bonn,
Nr. 71, p. 33 (1987). – See Abstr. 012.066.

013.074   **NASA MDSCC upgrade: status report and plans concerning geodetic VLBI in Europe.**
A. Rius.
Mitt. Geod. Inst. Rheinischen Friedrich–Wilhelms–Univ. Bonn,
Nr. 71, p. 35 – 40 (1987). – See Abstr. 012.066.
During the next months the NASA Madrid Deep Space Communications Complex (MDSCC) will be equipped with instrumentation compatible with the standard geodetic VLBI Data Acquisition Terminals providing new resources for geodynamical research in Europe. In this report the author summarizes this observing capabilities as well as plans in this field.

013.075   **European VLBI for geodesy and astrometry: goals and objectives.**
J. Campbell.
Mitt. Geod. Inst. Rheinischen Friedrich–Wilhelms–Univ. Bonn,
Nr. 71, p. 41 – 52 (1987). – See Abstr. 012.066.
The present document is intended to serve as a reference work collecting the full scope of goals that support the use of VLBI in geodesy, geophysics and astrometry. At the same time it should be useful to open ways for and facilitate the coordination of the activities of the European VLBI groups.

013.076   **Specification of LEST site survey program 1987 – 88.**
P. N. Brandt, A. Erasmus, J. Fuentes Gandia,
U. Kusoffsky.
LEST Found., Tech. Rep., No. 26, p. 89 – 96 (1987). – See Abstr.
012.071.
In June 1986 a proposal for the LEST site testing program in the Canary Islands and Hawaii was presented to the Scientific and Technical Advisory Committee of LEST. The site investigation team and other interested persons met in La Palma on February 18 and 19, 1987 to define specific objectives and procedures of the LEST site survey in the Canary Islands and Hawaii. The program for the first phase has been finalized. The program for the second phase was discussed fairly extensively, but no specific plans are presented in this report.

013.077   **"Solid Earth" panel report.**
P. Paquet, S. Hieber.
ESA Spec. Publ., ESA SP–1093, p. 57 – 60 (1987). – See Abstr.
012.072.

013.078   **Work on peculiar red giants at South African Astronomical Observatory (SAAO).**
Bull. Etoiles Tardives Spectre Particulier, No. 5, p. 5 – 6 (1987).

013.079   **Der Beitrag der Wiener Universitätssternwarte zum Projekt MERIT.**
E. Göbel.
Vorträge beim Theodor Ritter von Oppolzer–Gedächtnissymposium, p. 53 – 59 (1987). – See Abstr. 012.073.

013.080   **The Soviet Center of Astronomical Data.**
O. B. Dluzhnevskaya.
Astron. Tsirk., No. 1476, p. 3 – 6 (1987). In Russian.
The Center of Astronomical Data of the Astronomical Council of the USSR Academy of Sciences has been operating for 5 years. 380 catalogues on magnetic tapes and 45 catalogues on microfiches are available in the Center. During this period more than 40 catalogues have been compiled in the USSR and 25 of them have been sent to the Centre de Données Stellaires (CDS) at Strasbourg. As a branch of the CDS Soviet Astronomical Data

Center fulfills requests for copying catalogues and prepares various samples by means of the Data Base of the Center.

013.081   **Neutrons, the sun and Canada.**
P. Jedicke.
J. R. Astron. Soc. Can., Vol. 81, No. 6, p. 221 – 225 (1987).
The problem of detecting solar neutrinos is described and the discrepancy between their observed numbers and the theoretical expectations is pointed out. Plans to set up a heavy–water neutrino detector in an old Canada nickel mine are described.

013.082   **International scientific cooperation: past and future.**
J. G. Roederer.
ESA Spec. Publ., ESA SP–278, p. 13 – 15 (1987). – See Abstr.
012.075.
This article addresses some non–scientific, yet no less significant, aspects of international cooperation in science, focuses on the social responsibility of the scientists engaged in cooperative research, and relates this to Marcel Nicolet's role in and contributions to international programs.

013.083   **N–body work on supercomputers.**
R. H. Miller.
NRAO Workshop, No. 15, p. 29 – 31 (1988). – See Abstr.
012.076.

013.084   **Program support of ES–1033 computer of the Astronomical Council of the USSR Academy of Sciences.**
S. V. Vereshchagin, V. I. Ponomareva, L. P. Ustinova.
Nauchn. Inf., Vyp. 59, p. 82 – 88 (1986). In Russian.
A report on the activities of the Programme Support Team during 1979 – 1985 is presented. Opportunities provided to users by the Computing Center of the Astronomical Council of the USSR Academy of Sciences are described.

013.085   **Evading the zone of avoidance.**
V. Trimble.
Nature, Vol. 330, No. 6145, p. 212 (1987).
Nearly one–fifth of the extragalactic sky lurks out of sight, obscured by dust and crowded by images of stars in the disk of our own Milky Way. Astronomers have long wished to penetrate this zone of avoidance with infrared and radio observations. Now both the infrared–driven and blind 21–cm techniques are known to work, deliberate searches can be mounted for continuations and bridges from known clusters and filaments into the zone of avoidance.

013.086   **Soviet astronomy in the last decade.**
A. A. Boyarchuk, L. I. Antipova.
Kozmos, Vol. 18, No. 6, p. 181 – 186 (1987). In Slovak.

013.087   **Siberian style of solar exploration.**
M. Rybanský.
Kozmos, Vol. 18, No. 6, p. 206 – 207 (1987). In Czech.

013.088   **Thirty years of the space era, astronautics in the year 1986.**
M. Grün, P. Koubský.
Říše hvězd, Vol. 68, No. 10, p. 192 – 196 (1987). In Czech.

013.089   **Possibility of the Czechoslovak participation in the International Inter–Disciplinary Programme: geosphere – biosphere and their global variations.**
M. Kopecký.
Říše hvězd, Vol. 68, No. 11, p. 214 – 216 (1987). In Czech.

013.090   **Highlights of astronomy 1986.**
J. Grygar.
Říše hvězd, Vol. 68, Nos. 7 – 10, p. 122 – 125, 147 – 150, 163 – 166, 187 – 191 (1987). In Czech.

Short chronology of Soviet astronomy.
See Abstr. 004.122.

The remarkable extragalactic research of Erik Holmberg: a glimpse from Santa Cruz.
See Abstr. 005.005.

Celestial mechanics at the Sternberg State Astronomical Institute.
See Abstr. 009.006.

Investigations in the field of stellar astronomy.
See Abstr. 009.007.

Astrophysical investigations at the Sternberg State Astronomical Institute.
See Abstr. 009.008.

History of gravimetry at the Sternberg State Astronomical Observatory.
See Abstr. 009.009.

Soft X–ray optics at King's College London.
See Abstr. 031.011.

Some developments in optical telescope instrumentation and data reduction systems in Japan.
See Abstr. 032.031.

A 32–m radio telescope for VLBI research in Poland.
See Abstr. 033.038.

Instrumentación astronómica en México.
See Abstr. 034.102.

GONG: to see inside our sun.
See Abstr. 080.028.

Bilan de la campagne "PHÉMU 85" (*phénomènes mutuels*): la participation des amateurs.
See Abstr. 099.015.

The rebirth of stellar radial–velocity research.
See Abstr. 111.003.

Spectral radio astronomical observations in the 2 – 4 mm wavelength range.
See Abstr. 131.022.

General historical introduction (*concerning elliptical galaxies*).
See Abstr. 157.147.

014 Teaching in Astronomy

014.001 Draaibare sterrenkaart in de computer.
A. Nagel.
Zenit, Jaarg. 14, Nr. 7/8, p. 253 (1987).

014.002 Sterrenkunde op de huiscomputer: Afstanden aan de hemelbol.
Zenit, 14. Jahrg., Nr. 9, p. 293 (1987).

014.003 Projet: spectres stellaires.
Obs. Trav., No. 11, p. 17 – 22 (1987).

014.004 Teaching of astronomy in Fukushima University.
T.Ōki.
Astron. Her., Vol. 80, No. 8, p. 230 – 234 (1987). In Japanese.

014.005 Zur Stoffeinheit "Überblick über das Sonnensystem".
H. Bernhard.
Astron. Sch., 24. Jahrg., Heft 4, p. 84 – 86 (1987).

014.006 Zur Stoffeinheit "Planeten".
K. Ullerich.
Astron. Sch., 24. Jahrg., Heft 4, p. 86 – 90 (1987).

014.007 Teaching space.
T. W. Becker.
Spaceflight, Vol. 29, Suppl. No. 1, p. 8 – 9 (1987).
  Space technology teaching needs to be introduced at all levels of the educational process. The author already has wide experience in meeting this need in the USA, particularly in the St. Louis area, and is author of the teacher training manual, 'Space Education in the Classroom'.

014.008 How big is the sky?
R. Maddison.
Astron. Now, Vol. 1, No. 3, p. 27 – 30 (1987).

014.009 Einführung in die Sonnenbeobachtung. I.
H. Hilbrecht, W. Paech.
Sonne, Jahrg. 11, Nr. 43, p. 63 – 71 (1987).

014.010 Drei Minuten Instant–Astronomie. Teil 1: Der Sonnendurchmesser als Schlüsselwert in der Schulastronomie.
W. Schlosser.
Sterne Weltraum, 26. Jahrg., Nr. 7 – 8, p. 406 – 409 (1987).

014.011 Computereinsatz in der Amateurradioastronomie.
J. Sänger, B. Gährken.
Sterne Weltraum, 26. Jahrg., Nr. 7 – 8, p. 422 – 424 (1987).

014.012 Drei Minuten Instant–Astronomie. Teil 2: Der Sonnendurchmesser als Schlüsselwert in der Schulastronomie.
W. Schlosser.
Sterne Weltraum, 26. Jahrg., Nr. 9, p. 474, 476 – 477 (1987).

014.013 Einiges über die Verwendung der drehbaren Sternkarte.
H. Zeuner.
Sterne Weltraum, 26. Jahrg., Nr. 10, p. 552 – 555 (1987).

014.014 Status reports on astronomical education at Hakodate College, Hokkaido University of Education.
T. Okuda.
Astron. Her., Vol. 80, No. 10, p. 292 – 296 (1987). In Japanese.

014.015 Choosing your first telescope.
B. Crabb.
Astron. Now, Vol. 1, No. 4, p. 50 – 55 (1987).

014.016 Observational astrology.
B. Mayer.
Mercury, Vol. 16, No. 4, p. 111 – 118 (1987).

014.017 Zur Anwendung des Taschenrechners im Astronomieunterricht.
P. Klein.
Astron. Sch., 24. Jahrg., Heft 5, p. 105 – 108 (1987).

014.018 Backyard astronomy – Observing from the city.
A. MacRobert.
Sky Telesc., Vol. 74, No. 1, p. 35 – 37 (1987).

014.019 **Astronomical computing: the orbit of a binary star.**
M. P. Greaney.
Sky Telesc., Vol. 74, No. 1, p. 71 – 72 (1987).

014.020 **Astronomical computing: Do orbits change in 100 million years?**
R. W. Sinnott.
Sky Telesc., Vol. 74, No. 2, p. 182 – 183 (1987).

014.021 **Astronomical computing: Stars and spikes.**
R. W. Sinnott.
Sky Telesc., Vol. 74, No. 3, p. 294 – 296 (1987).

014.022 **Backyard astronomy: the art of planetary observing – I.**
D. C. Parker, T. A. Dobbins.
Sky Telesc., Vol. 74, No. 4, p. 370 – 372 (1987).

014.023 **Astronomical computing: Fun with stereographic projections.**
R. A. Mulford.
Sky Telesc., Vol. 74, No. 4, p. 407 – 408 (1987).

014.024 **Astronomical computing: The tide at Tarawa.**
D. W. Olson.
Sky Telesc., Vol. 74, No. 5, p. 526 – 529 (1987).

014.025 **Data für den rechnenden Sternfreund. Kalender.**
K.–H. Bücke.
Astron. Raumfahrt, 25. Jahrg., Heft 5, p. 149 – 150 (1987).

014.026 **Spectroscopy: the analysis of starlight.**
R. Maddison.
Astron. Now, Vol. 1, No. 5, p. 54 – 59 (1987).

014.027 **Sterrenkunde op de huiscomputer: Werken met een sferometer.**
J. Loonen.
Zenit, 14. Jaarg., Nr. 12, p. 405 (1987).

014.028 **A clock–driven planisphere.**
A. A. Mills.
J. Br. Astron. Assoc., Vol. 97, No. 6, p. 348 – 349 (1987).
An illuminated planisphere suitable for mounting on a wall is described. Driven at an accurate sidereal rate by an adapted quartz clock movement, it shows the stars currently above the observer's horizon, and their relationship to the Sun.

014.029 **Adding astronomy to the Ontario school science curriculum.**
J. R. Percy.
Bull. Am. Astron. Soc., Vol. 19, No. 2, p. 715 (1987). Abstract. – See Abstr. 010.061.

014.030 **Science teaching through its astronomical roots: project STAR.**
K. Brecher, P. Sadler, I. I. Shapiro.
Bull. Am. Astron. Soc., Vol. 19, No. 2, p. 715 (1987). Abstract. – See Abstr. 010.061.

014.031 **Planetary design in teaching astronomy.**
S. Simon.
Bull. Am. Astron. Soc., Vol. 19, No. 2, p. 716 (1987). Abstract. – See Abstr. 010.061.

014.032 **Calculation of hydrogen and helium Strömgren zones for blackbodies and model stars.**
H. J. Augensen, M. R. McKee.
Bull. Am. Astron. Soc., Vol. 19, No. 2, p. 716 (1987). Abstract. – See Abstr. 010.061.

014.033 **The Runge–Lenz vector and Einstein perihelion precession.**
T. Garavaglia.
Am. J. Phys., Vol. 55, No. 2, p. 164 – 165 (1987). Abstr. in Phys. Abstr., Vol. 90, No. 1309, Entry 80232 (1987).

014.034 **Tidal disruption of a solid body.**
J. Conwell, D. Wilkins.
Am. J. Phys., Vol. 55, No. 2, p. 165 – 166 (1987). Abstr. in Phys. Abstr., Vol. 90, No. 1309, Entry 80233 (1987).

014.035 **Stellar sky as seen from the vicinity of a black hole.**
J. Schastok, M. Soffel, H. Ruder, M. Schneider.
Am. J. Phys., Vol. 55, No. 4, p. 336 – 341 (1987). Abstr. in Phys. Abstr., Vol. 90, No. 1309, Entry 80253 (1987).

014.036 **The use of amorphous metallic ribbon as a torsion spring for the measurement of the gravitational constant.**
R. A. Dunlap.
Am. J. Phys., Vol. 55, No. 4, p. 380 (1987). Abstr. in Phys. Abstr., Vol. 90, No. 1309, Entry 80268 (1987).

014.037 **When did the universe begin?**
J. Rosen.
Am. J. Phys., Vol. 55, No. 6, p. 498 – 499 (1987). Abstr. in Phys. Abstr., Vol. 90, No. 1311, Entry 95131 (1987).

014.038 **Ground trajectories of geosynchronous satellites and the analemma.**
J. S. Chalmers.
Am. J. Phys., Vol. 55, No. 6, p. 548 – 552 (1987). Abstr. in Phys. Abstr., Vol. 90, No. 1311, Entry 95142 (1987).

014.039 **Does the Earth rotate?**
M. M. Payne.
Phys. Teach., Vol. 25, No. 2, p. 86 – 87 (1987). Abstr. in Phys. Abstr., Vol. 90, No. 1311, Entry 95156 (1987).

014.040 **Cosmic strings. Gravitation without local curvature.**
T. M. Helliwell, D. A. Konkowski.
Am. J. Phys., Vol. 55, No. 5, p. 401 – 407 (1987). Abstr. in Phys. Abstr., Vol. 90, No. 1312, Entry 102044 (1987).

014.041 **The black hole as a gravitational ”lens”.**
H. C. Ohanian.
Am. J. Phys., Vol. 55, No. 5, p. 428 – 432 (1987). Abstr. in Phys. Abstr., Vol. 90, No. 1312, Entry 102049 (1987).

014.042 **A close look at a canonical semiclassical derivation of the gravitational redshift.**
J. Goldstein.
Am. J. Phys., Vol. 55, No. 5, p. 476 – 477 (1987). Abstr. in Phys. Abstr., Vol. 90, No. 1312, Entry 102057 (1987).

014.043 **Backyard astronomy: The art of planetary observing. II.**
D. C. Parker, T. A. Dobbins.
Sky Telesc., Vol. 74, No. 6, p. 603 – 607 (1987).

014.044 **Astronomical computing: Sundials on walls.**
W. S. Maddux.
Sky Telesc., Vol. 74, No. 6, p. 646 – 648 (1987).

014.045 **Ein Periodensuchprogramm für Veränderliche Sterne.**
W. Quester, W. Quehl.
Sterne Weltraum, 26. Jahrg., Nr. 12, p. 704 – 705 (1987).

014.046 **La divulgazione e la didattica dell'astronomia.**
P. Ranfagni.
Astronomia, N. 3 – 4, p. 27 – 32 (1987).

014.047 **Zur Gestaltung des Unterrichts im Stoffgebiet ”Sterne, Sternsysteme, Metagalaxis”.**
K. Lindner.
Astron. Sch., 24. Jahrg., Heft 6, p. 127 – 135 (1987).

014.048 **Zur Anwendung des Taschenrechners im Astronomieunterricht. II.**
P. Klein.
Astron. Sch., 24. Jahrg., Heft 6, p. 140 – 143 (1987).

014.049 **Secretos de la observacion visual.**
A. MacRobert.
Rev. Astron., Tomo 58, No. 240, p. 2 – 5 (1987).

014.050 **Fotometría fotográfica.**
M. Lopez Alvarez.
Rev. Astron., Tomo 58, No. 240, p. 14 – 17 (1987).

014.051 **Microcomputación y astronomía: Notas sobre el programa de cálculo de hora sidérea.**
C. Rusquellas.
Rev. Astron., Tomo 58, No. 240, p. 18 – 19 (1987).

014.052 **Microcomputación y astronomía: Salida, paso y puesta.**
C. Rusquellas.
Rev. Astron., Tomo 58, No. 240, p. 19 – 21 (1987).

014.053 **An expectation value formulation of the perturbed Kepler problem.**
L. J. Curtis, R. R. Haar, M. Kummer.
Am. J. Phys., Vol. 55, No. 7, p. 627 – 631 (1987). Abstr. in Phys. Abstr., Vol. 90, No. 1313, Entry 108083 (1987).

014.054 **Einführung in die Sonnenbeobachtung. II.**
S. Hammerschmidt, H. Hilbrecht.
Sonne, Jahrg. 11, Nr. 44, p. 96 – 101 (1987).

014.055 **Data für den rechnenden Sternfreund: Zeit.**
K.–H. Bücke.
Astron. Raumfahrt, 25. Jahrg., Heft 6, p. 177 – 178 (1987).

014.056 **Behind the event horizon.**
R. H. Good.
Eur. J. Phys., Vol. 8, No. 3, p. 174 – 177 (1987). Abstr. in Phys. Abstr., Vol. 90, No. 1315, Entry 120734 (1987).

014.057 **Astronomy on an Apple Macintosh.**
J. E. Mosley.
J. Comput. Math. Sci. Teach., (USA), Vol. 6, No. 3, p. 55 – 57 (1987). Abstr. in Phys. Abstr., Vol. 90, No. 1315, Entry 120746 (1987).

014.058 **Problemi educativi e proposte didattiche nella scuola: il ruolo della SAIt.**
G. Astron., Vol. 13, N. 1 – 2, p. 22 – 24 (1987).

014.059 **Real perspectives for astronomy at school.**
E. P. Levitan.
Zemlya Vselennaya, No. 6, p. 60 – 65 (1987). In Russian.

014.060 **A student experiment for accurate measurements of the Newtonian gravitational constant.**
J.–C. Dousse, C. Rheme.
Am. J. Phys., Vol. 55, No. 8, p. 706 – 711 (1987). Abstr. in Phys. Abstr., Vol. 90, No. 1316, Entry 128212 (1987).

014.061 **A new derivation of the areal velocity.**
T. Yoshida.
Am. J. Phys., Vol. 55, No. 8, p. 752 – 753 (1987). Abstr. in Phys. Abstr., Vol. 90, No. 1316, Entry 128221 (1987).

014.062 **Order–of–magnitude "theory" of stellar structure.**
G. Greenstein.
Am. J. Phys., Vol. 55, No. 9, p. 804 – 810 (1987). Abstr. in Phys. Abstr., Vol. 90, No. 1317, Entry 134393 (1987).

014.063 **Automatic recording for the Cavendish balance.**
C. W. Fischer, J. L. Hunt, P. Sawatzky.
Am. J. Phys., Vol. 55, No. 9, p. 855 – 856 (1987). Abstr. in Phys. Abstr., Vol. 90, No. 1317, Entry 134403 (1987).

014.064 **The date and time of the vernal equinox: a graphical representation of the Gregorian calendar.**
R. L. Reese, G. Y. Chang.
Am. J. Phys., Vol. 55, No. 9, p. 848 – 849 (1987). Abstr. in Phys. Abstr., Vol. 90, No. 1317, Entry 134414 (1987).

014.065 **The Rutherford cross section and the perihelion shift of Mercury with the Runge–Lenz vector.**
C. Farina, M. Machado.
Am. J. Phys., Vol. 55, No. 10, p. 921 – 923 (1987). Abstr. in Phys. Abstr., Vol. 91, No. 1320, Entry 4601 (1988).

Laboratory experiments for astronomy.
See Abstr. 003.076.

Exercises in astronomy.
See Abstr. 003.083.

A career in astronomy.
See Abstr. 003.129.

015 Miscellanea (Philosophical Aspects, Extraterrestrial Life, etc.)

015.001 **Examination of the embargo hypothesis as an explanation for the Great Silence.**
J. W. Deardorff.
J. Br. Interplanet. Soc., Vol. 40, No. 8, p. 373 – 379 (1987).
 The embargo or quarantine hypothesis for explaining the 'Great Silence' is reviewed and found to be more plausible than the view that, at most, we might expect to receive radio messages from some distant star. The latter hypothesis is shown to be compatible with extraterrestrial technologies only a few hundred years in advance of our own, whereas the embargo hypothesis more reasonably infers that they should be tens of thousands of years in advance and in control of any contact with humanity. Reasons why the embargo hypothesis has received insufficient attention are presented

015.002 **Worldships – at the end of time.**
G. Matloff.
Spaceflight, Vol. 29, Suppl. No. 1, p. 40 – 42 (1987).
 Scientists and philosophers have been struggling to produce an answer to the famous Fermi paradox. The author presents a new approach: the Galaxy is not yet old enough to prompt civilisations to spread among the stars.

015.003 **Astronomische Feldposthefte.**
C. Kaspari.
Sterne Weltraum, 26. Jahrg., Nr. 10, p. 570 – 571 (1987).

015.004 **Les potins d'Uranie: Juste ciel!**
Al Nath (A. Heck).
Ciel, Vol. 49, p. 291 – 292 (1987).

015.005   **A record breaking problem.**
R. E. Hill.
J. Am. Assoc. Variable Star Obs., Vol. 16, No. 1, p. 45 – 47 (1987).
Tests were made to examine the fading of a number of inks with exposure to sunlight. Tests were also made to examine the stability of electro–static copies (xeroxes). The results have important implications for amateur astronomers and all others who maintain written records.

015.006   **A quick dark–adapt technique.**
V. Stryker.
J. Am. Assoc. Variable Star Obs., Vol. 16, No. 1, p. 48 (1987).
A simple and quick technique for adapting one's eyes to the dark is described.

015.007   **Geological cyclicity and its sources.**
V. B. Smirnov.
Ufim. Neft. Inst., Ufa, 8 pp. (1987). In Russian. Abstr. in Ref. Zh., 51. Astron., 10.51.1 (1987).

015.008   **The riddle of Phanerozoic catastrophes.**
V. B. Smirnov.
Ufim. Neft. Inst., Ufa, 8 pp. (1987). In Russian. Abstr. in Ref. Zh., 51. Astron., 10.51.2 (1987).

015.009   **The media and supernova Shelton 1987A: science writers needed!**
R. F. Garrison.
Bull. Am. Astron. Soc., Vol. 19, No. 2, p. 716 (1987). Abstract. – See Abstr. 010.061.

015.010   **The Colorado Scale Model Solar System.**
J. O. Bennett, T. R. Ayres, K. Center, R. Bass, M. Carter.
Bull. Am. Astron. Soc., Vol. 19, No. 2, p. 750 (1987). Abstract. – See Abstr. 010.061.

015.011   **Odd regularities in planetary astronomy (On the evolution of understanding natural laws).**
Yu. V. Chajkovskij.
Istor.–Astron. Issled., Moskva, No. 19, p. 69 – 86 (1987). In Russian. Abstr. in Ref. Zh., 51. Astron. 11.51.1 (1987).

015.012   **Shakespeare's astronomy.**
D. S. Evans.
Mon. Notes Astron. Soc. S. Afr., Vol. 46, Nos. 11 – 12, p. 164 – 166 (1987).

015.013   **Newtonianism, reductionism and the art of congressional testimony.**
S. Weinberg.
Nature, Vol. 330, No. 6147, p. 433 – 437 (1987).
This paper presents the author's talk given at the Tercentenary Celebration of Newton's "Principia" at the University of Cambridge.

Problema poiska zhizni vo vselennoj (*Searching for life in the universe***).**
See Abstr. 003.019.

Life in the universe.
See Abstr. 003.074.

Extraterrestrials. Science and alien intelligence.
See Abstr. 003.118.

Europa, tidally heated oceans, and habitable zones around giant planets.
See Abstr. 091.025.

Comets and life.
See Abstr. 102.026.

The anthropic Universe.
See Abstr. 161.385.

Applied Mathematics, Physics

021 Mathematical Papers Related to Astronomy and Astrophysics, Computing

021.001 Application of numerical methods to planetary radio-wave scattering.
R. A. Simpson, G. L. Tyler.
NASA Tech. Memo., NASA TM–89810, p. 253 (1987). – See Abstr. 003.001.
Continued advances in computer technology make numerical solution of certain scattering problems feasible. The authors' work includes investigation of existing techniques to determine those which might be applicable to planetary surface studies with the goal of improving the interpretation of radar data from Venus, Mars, the moon, and icy satellites.

021.002 Computer simulations of 10–km–diameter asteroid impacts into oceanic and continental sites – preliminary results on atmospheric passage, cratering, and ejecta dynamics.
D. J. Roddy, S. H. Schuster, M. Rosenblatt, L. B. Grant,
P. J. Hassig, K. N. Kreyenhagen.
NASA Tech. Memo., NASA TM–89810, p. 377 – 379 (1987). – See Abstr. 003.001.

021.003 A unified presentation of the Voigt functions.
H. M. Srivastava, E. A. Miller.
Astrophys. Space Sci., Vol. 135, No. 1, p. 111 – 118 (1987).
This paper aims at presenting a unified study of the Voigt functions $K(x, y)$ and $L(x, y)$ which play a rather important role in several diverse fields of physics such as astrophysical spectroscopy and the theory of neutron reactions. Explicit expressions for these functions are given in terms of relatively more familiar special functions of one and two variables; indeed, each of these representations will naturally lead to various other needed properties of the Voigt functions.

021.004 Astronomy and personal computers.
J. B. Dunham.
Occultation Newsl., Vol. 4, No. 5, p. 109 – 111 (1987).

021.005 Die Berechnung einiger planetarer Phänomene.
J. Meeus.
Astron. Raumfahrt, 25. Jahrg., Heft 4, p. 107 – 111 (1987).

021.006 Generalized numerical error analysis with applications to geochronology and thermodynamics.
J. C. Roddick.
Geochim. Cosmochim. Acta, Vol. 51, No. 8, p. 2129 – 2135 (1987).
A general numerical error propagation procedure is developed to calculate the error in a quantity derived from measurements which are subject to errors. It is an alternative to computer intensive techniques such as Monte Carlo and can be applied to quite complex analytical problems where correlation among the measurement errors and among the final errors in results are present.

021.007 Sharp PC 1360 + PEGASUS. Ein Taschencomputer als astronomische Beobachtungshilfe.
O. Montenbruck.
Sterne Weltraum, 26. Jahrg., Nr. 7 – 8, p. 432 – 433 (1987).

021.008 Evolution of architectures for data processing.
V. Cantoni.
Selected topics on data analysis in astronomy, p. 87 – 109 (1987). – See Abstr. 012.011.
In order to satisfy the continuous growth and evolution of computational demands, many attempts to develop circuits and processor components that operate faster have been followed and of course continue to be persued. Major improvements depend on computer architecture and processing techniques. Advanced computer architectures are centered around the concept of parallel processing. The most significant paradigms of new computer architectures are described.

021.009 A second order, three–dimensional hydrodynamic computer code.
H. A. Williams, J. E. Tohline.
Bull. Am. Astron. Soc., Vol. 19, No. 2, p. 744 (1987). Abstract. – See Abstr. 010.061.

021.010 The first integrals and stability of ordinary differential equations.
T.–x. Weng.
Publ. Purple Mt. Obs., Vol. 6, No. 2, p. 139 – 150 (1987). In Chinese.

021.011 Perturbation computations and numerical modelling experiments.
A. Carusi, E. Perozzi, G. B. Valsecchi.
The evolution of the small bodies of the solar system, p. 191 – 201 (1987). – See Abstr. 012.041.
Contents: 1. The integration of orbits (the n–body problem. Cowell's and Encke's equations, numerical integration methods in general, DVDQ, RADAU, other integrators). 2. The reliability of cometary orbital integrations. 3. Dynamical modelling experiments (chaotic orbits in the solar system, modelling).

021.012 Remarks on the Test–Run Output by Jain and Thompson.
S. Chandra.
Astrophys. Space Sci., Vol. 138, No. 1, p. 221 – 226 (1987).
The Test–Run Output of erratum notice published by Jain and Thompson (1985) for their paper (Jain and Thompson, 1983) appears to be inconsistent. Hence, their computer programme needs a further modification.

021.013 The analytical operations system GRATOS. Part 1: conceptions and possibilities.
S. V. Tarasevich, A. V. Krivov, V. B. Titov.
Algoritm. Nebesnoj Mekh., No. 53, 43 pp. (1987). In Russian.
A subroutine package for analytical manipulations with tensor expressions in the 4– or 3–dimensional Riemannian space is described. There are two purposes of this paper: to describe the general possibilities of GRATOS as the specialised system designed to solve problems of general relativity and to serve as the manual for practical application of GRATOS.

021.014 **The analytical operations system GRATOS. Part 2: BESM–6 and ES implementations.**
S. V. Tarasevich, A. V. Krivov, V. B. Titov.
Algoritm. Nebesnoj Mekh., No. 54, 35 pp. (1987). In Russian.

021.015 **FORTRAN–Programme für die Berechnung von zeitvariablen Meereshöhen und Bahnfehlern aus Satelliten–Altimeterdaten.**
P. Arnold.
Mitt. Geod. Inst. Rheinischen Friedrich–Wilhelms–Univ. Bonn, Nr. 73, 141 pp. (1987).

021.016 **Effiziente Rechenverfahren für umfangreiche geodätische Parameterschätzungen.**
U. Grepel.
Mitt. Geod. Inst. Rheinischen Friedrich–Wilhelms–Univ. Bonn, Nr. 75, 128 pp. (1987). Diss. Hohen Landwirtschaftlichen Fakultät der Rheinischen Friedrich–Wilhelms–Universität Bonn.

021.017 **Die Analyse astronomischer Zeitreihen und ihre Anwendung durch Amateure.**
R. Schult.
Sterne, 63. Band, Heft 6, p. 344 – 350 (1987).

021.018 **Sampling criteria in multicollection searching.**
A. Gilio, R. Scozzafava, P. G. Marchetti.
ESA J., Vol. 11, No. 3, p. 343 – 352 (1987).
In the first stage of the document retrieval process, no information concerning relevance of a particular document is available. On the other hand, computer implementation requires that the analysis be made only for a sample of retrieved documents. This paper addresses the significance and suitability of two different sampling criteria for a multicollection online search facility. The inevitability of resorting to a logarithmic criterion in order to achieve a "spread of representativeness" from the multicollection is demonstrated.

021.019 **A five–level program for ions of astrophysical interest.**
M. M. De Robertis, R. J. Dufour, R. W. Hunt.
J. R. Astron. Soc. Can., Vol. 81, No. 6, p. 195 – 220 (1987).
An interactive FORTRAN program is presented which determines the physical conditions in a nebula given an appropriate emission–line intensity ratio, or the line emissivities given a density and temperature. A method for estimating ionic abundances based on the Hβ emissivity is also included. The relevant underlying physics is reviewed and a detailed description of the operation of the program given. Uncertainties in the interpretation of the results are discussed briefly.

021.020 **What are we missing?**
V. Icke.
NRAO Workshop, No. 15, p. 19 – 28 (1988). – See Abstr. 012.076.
There are at present an embarrassingly large number of astrophysical problems for which (1) observations are so good that we know there's something worth working on, and we have a feeling for the features that require explanation; and (2) we have the physics on hand to do the explaining, but (3) computational limitations prevent this. As a practical example, the author discusses the construction of a model of high energy clouds in nonthermal radio sources, and argues that the merits of such models should be assessed by studying them with what he has called a "numerical astrophysical observatory".

021.021 **Alternatives to class VI computing.**
H. L. Johnson.
NRAO Workshop, No. 15, p. 33 – 52 (1988). – See Abstr. 012.076.
This paper deals with the alternatives to supercomputing, identifying the state–of–the–art, and discussing the needs of this technology.

021.022 **Hypercube multicomputers and their use in observational and theoretical astrophysics.**
D. L. Meier.
NRAO Workshop, No. 15, p. 53 – 62 (1988). – See Abstr. 012.076.
The architecture, evolution, and use in astrophysics of hypercube multicomputers is described. Because of their potential for great speed at reduced cost, and because they can be utilized with high efficiency, multicomputers promise to be an important class of machines for astrophysics as parallel processing becomes increasingly important in the near future.

021.023 **Progress in using a Cray X–MP for radio astronomy.**
R. T. Duquet.
NRAO Workshop, No. 15, p. 69 – 71 (1988). – See Abstr. 012.076.

021.024 **Certification and benchmarking of AIPS (*Astronomical Image Processing System*) on the Convex C–1 and Alliant FX/8.**
K. C. Hilldrup, D. C. Wells, W. D. Cotton.
NRAO Workshop, No. 15, p. 73 – 97 (1988). – See Abstr. 012.076.

021.025 **Transformation of geocentric to geodetic coordinates without approximations.**
K. M. Borkowski.
Astrophys. Space Sci., Vol. 139, No. 1, p. 1 – 4 (1987).
An exact and relatively simple analytical transform of the rectangular coordinates to the geodetic coordinates is presented. It does not involve any approximation and the accuracy of practical calculations depends exclusively on the round–off errors. The algorithm is based on one solution to the quartic equation in $\tan(45° - \psi/2)$, where ψ is the parametric (or eccentric) latitude.

021.026 **A numerical scheme for one–dimensional mechanical problems.**
P. J. Pascual, L. Vazquez.
Hadronic J., Vol. 9, No. 6, p. 307 – 310 (1987). Abstr. in Phys. Abstr., Vol. 90, No. 1317, Entry 134466 (1987).

021.027 **Programme for the calculation of apparent places of the Moon.**
S. Svoboda.
Říše hvězd, Vol. 68, Nos. 7 – 9, p. 135 – 136, 158 – 159, 182 – 183 (1987). In Czech.

021.028 **RESPECT – software package for reduction of astronomical spectra.**
T. P. Prabhu, G. C. Anupama, S. Giridhar.
Bull. Astron. Soc. India, Vol. 15, Nos. 2 – 3, p. 98 – 115 (1987).
The interactive software package RESPECT developed for reducing astronomical spectroscopic data at the Vainu Bappu Observatory, Kavalur, is described.

FORWARD: extension for multiple stations.
See Abstr. 036.209.

A non–recursive method for identifying erroneous observations.
See Abstr. 036.250.

Exponential instability of collision orbit in the anisotropic Kepler problem.
See Abstr. 042.005.

An application of the scaled Runge–Kutta algorithms to some problems of celestial mechanics.
See Abstr. 042.029.

Bahnbestimmung mit dem Heimcomputer.
See Abstr. 042.033.

A comparison of numerical integration methods in the equatorial magnetic–binary problem.
See Abstr. 042.076.

Several problems on the numerical integration of celestial orbits.
See Abstr. 042.082.

Tables for the trigonometric series representations of the orbital inclination function.
See Abstr. 052.031.

Finite element methods for time dependent problems.
See Abstr. 062.129.

Description and philosophy of spectral methods.
See Abstr. 062.130.

Numerical modeling of subgrid–scale flow in turbulence, rotation, and convection.
See Abstr. 062.131.

Particle methods.
See Abstr. 062.132.

The characteristic initial value problem in general relativity.
See Abstr. 066.154.

Saros et périodicité des éclipses.
See Abstr. 095.001.

Performance characteristics of tree codes.
See Abstr. 151.030.

Observing models of the universe on a supercomputer.
See Abstr. 161.372.

022 Physical Papers Related to Astronomy and Astrophysics

022.001 Experimental studies on the impact properties of water ice.
F. G. Bridges, D. N. C. Lin, A. P. Hatzes.
NASA Tech. Memo., NASA TM–89810, p. 60 – 62 (1987). – See Abstr. 003.001.

022.002 High pressure cosmochemistry of major planetary interiors: laboratory studies of the water–rich region of the system ammonia–water.
M. Nicol, M. Johnson, S. Boone, H. Cynn.
NASA Tech. Memo., NASA TM–89810, p. 133 – 135 (1987). – See Abstr. 003.001.

022.003 Experiments pertaining to the formation and equilibration of planetary cores. I. High–pressure metallization of FeO and implications for the Earth's core.
E. Knittle, R. Jeanloz.
NASA Tech. Memo., NASA TM–89810, p. 136 (1987). Abstract. – See Abstr. 003.001.

022.004 Experiments pertaining to the formation and equilibration of planetary cores. II. The melting curve of iron to over 100 GPa and of iron sulfide to over 60 GPa.
Q. Williams, R. Jeanloz.
NASA Tech. Memo., NASA TM–89810, p. 137 (1987). Abstract. – See Abstr. 003.001.

022.005 Melting of troilite at high pressure in a diamond cell by laser heating.
W. A. Bassett, M. S. Weathers.
NASA Tech. Memo., NASA TM–89810, p. 138 – 140 (1987). – See Abstr. 003.001.

022.006 Reflectance spectra of mafic silicates and phyllosilicates from 0.6 to 4.6 μm.
T. L. Roush, R. B. Singer, T. B. McCord.
NASA Tech. Memo., NASA TM–89810, p. 187 – 189 (1987). – See Abstr. 003.001.

022.007 Spectral effects of dehydration on phyllosilicates.
E. A. Bruckenthal, R. B. Singer.
NASA Tech. Memo., NASA TM–89810, p. 190 – 192 (1987). – See Abstr. 003.001.
Six phyllosilicates have been progressively dehydrated under controlled conditions in an effort to study the spectral effects of their dehydration. Justification for the study may be found in both terrestrial and planetary soil science applications.

022.008 Studies of the scattering/absorption properties of minerals.
R. N. Clark.
NASA Tech. Memo., NASA TM–89810, p. 193 – 195 (1987). – See Abstr. 003.001.
Reflectance spectra of planetary surfaces are most affected by the weight fraction and grain sizes of the minerals in the surface. The reflectance can range from 1.0 to about 0.01 by changing the grain size or weight fraction, a factor of 100. Viewing geometry changes the reflectance by about 25% or less.

022.009 Thermal–infrared spectral observations of geological materials in emission.
P. R. Christensen, S. J. Luth.
NASA Tech. Memo., NASA TM–89810, p. 199 – 201 (1987). – See Abstr. 003.001.

022.010 Deconvolution of spectra for intimate mixtures.
J. F. Mustard, C. M. Pieters, S. F. Pratt.
NASA Tech. Memo., NASA TM–89810, p. 206 – 207 (1987). – See Abstr. 003.001.

022.011 Measurements of the dielectric constants for planetary volatiles.
V. G. Anicich, W. T. Huntress Jr.
NASA Tech. Memo., NASA TM–89810, p. 251 – 252 (1987). – See Abstr. 003.001.

022.012 Wind ripples in low density atmospheres.
J. S. Miller, J. R. Marshall, R. Greeley.
NASA Tech. Memo., NASA TM–89810, p. 268 – 270 (1987). – See Abstr. 003.001.

022.013 Mass transport by aeolian saltation on Earth, Mars and Venus: the effects of full saltation cloud development and choking.
S. H. Williams, R. Greeley.
NASA Tech. Memo., NASA TM–89810, p. 274 – 275 (1987). – See Abstr. 003.001.
The purpose of this study is to be able to predict the characteristics of particle motion and the quantity transported by the wind under a variety of planetary environmental conditions.

022.014   **Development of wind tunnel techniques for the solution of problems in planetary aeolian processes.**
R. Sullivan, J. Lee, R. Greeley.
NASA Tech. Memo., NASA TM–89810, p. 276 – 278 (1987). –
See Abstr. 003.001.
The objective of this study is to evaluate wind tunnel experiments in predicting full–scale field results. Such a direct comparison between wind tunnel scale models and full–scale field results can identify working guidelines for a broad range of boundary layer geological modelling applications on Earth, but is especially relevant and critical for the planetary context.

022.015   **Aeolian abrasion on Venus: preliminary results from the Venus simulator.**
J. R. Marshall, R. Greeley, D. W. Tucker, J. B. Pollack.
NASA Tech. Memo., NASA TM–89810, p. 279 – 281 (1987). –
See Abstr. 003.001.

022.016   **Reynolds number effects on surface shear stress patterns around isolated hemispheres.**
J. A. Lee, R. Greeley.
NASA Tech. Memo., NASA TM–89810, p. 282 – 284 (1987). –
See Abstr. 003.001.
Obstacles projecting into the wind stream alter the shear stress on the surface around them, thus altering the erosion, transportation, and deposition of aeolian sediment. This study is concerned with the effect of Reynolds number on the pattern of shear stress on the surface around an isolated hemisphere.

022.017   **Determination of surface shear stress with the naphthalene sublimation technique.**
J. A. Lee, R. Greeley.
NASA Tech. Memo., NASA TM–89810, p. 285 – 287 (1987). –
See Abstr. 003.001.
Aeolian entrainment and transport are functions of surface shear stress and particle characteristics. Measuring surface shear stress is difficult, however, where logarithmic wind profiles are not found, such as regions around large roughness elements. Presented here is an outline of a method whereby shear stress can be mapped on the surface around an object.

022.018   **Groundwater sapping channels: summary of effects of experiments with varied stratigraphy.**
R. C. Kochel, D. W. Simmons.
NASA Tech. Memo., NASA TM–89810, p. 291 – 293 (1987). –
See Abstr. 003.001.
Experiments in the authors' recirculating flume sapping box have modelled valley formation by groundwater sapping processes in a number of settings. They have examined the effects of the following parameters on sapping channel morphology: (1) surface slope; (2) stratigraphic variations in permeability cohesion and dip; and (3) structure–joints and dikes.

022.019   **Non–equilibrium freezing of water–ice in sandy basaltic regoliths and implications for fluidized debris flows on Mars.**
J. L. Gooding.
NASA Tech. Memo., NASA TM–89810, p. 305 – 306 (1987). –
See Abstr. 003.001.
Many geomorphic features on Mars have been attributed to Earth–analogous, cold–climate processes involving movement of water– or ice–lubricated debris. New experiments have been performed with sand–sized samples of natural "basaltic" regoliths in order to further elucidate how water/regolith interactions depend upon grain size and mineralogy.

022.020   **Impacts of hemispherical granular targets: implications for global impacts.**
P. H. Schultz, D. E. Gault, D. Crawford.
NASA Tech. Memo., NASA TM–89810, p. 380 – 381 (1987). –
See Abstr. 003.001.

022.021   **Impact vaporization: late time phenomena from experiments.**
P. H. Schultz, D. E. Gault.
NASA Tech. Memo., NASA TM–89810, p. 382 – 383 (1987). –
See Abstr. 003.001.

022.022   **Oblique impact: projectile ricochet, concomitant ejecta, and momentum transfer.**
D. E. Gault, P. H. Schultz.
NASA Tech. Memo., NASA TM–89810, p. 384 – 385 (1987). –
See Abstr. 003.001.

022.023   **Momentum transfer from oblique impacts.**
P. H. Schultz, D. E. Gault.
NASA Tech. Memo., NASA TM–89810, p. 386 – 387 (1987). –
See Abstr. 003.001.

022.024   **Experimental studies of collision and fragmentation phenomena.**
W. K. Hartmann, D. R. Davis, S. J. Weidenschilling.
NASA Tech. Memo., NASA TM–89810, p. 388 – 390 (1987). –
See Abstr. 003.001.

022.025   **Centrifuge impact cratering experiments: scaling laws for non–porous targets.**
R. M. Schmidt.
NASA Tech. Memo., NASA TM–89810, p. 391 (1987).
Abstract. – See Abstr. 003.001.

022.026   **Experimental evidence for non–proportional growth of large craters.**
P. H. Schultz, D. E. Gault.
NASA Tech. Memo., NASA TM–89810, p. 394 – 395 (1987). –
See Abstr. 003.001.

022.027   **Centrifuge impact cratering experiments: scaling laws for non–porous targets.**
R. M. Schmidt.
NASA Tech. Memo., NASA TM–89810, p. 396 – 398 (1987). –
See Abstr. 003.001.

022.028   **The effect of overburden thickness on tension fracture patterns above an uplifting dome.**
H. Pranger II.
NASA Tech. Memo., NASA TM–89810, p. 497 – 498 (1987). –
See Abstr. 003.001.

022.029   **Creep of ice: further studies.**
H. C. Heard, W. B. Durham, S. H. Kirby.
NASA Tech. Memo., NASA TM–89810, p. 511 – 513 (1987). –
See Abstr. 003.001.

022.030   **Observations of industrial sulfur flows and implications for Io.**
S. W. Lee, D. A. Crown, N. Lancaster, R. Greeley.
NASA Tech. Memo., NASA TM–89810, p. 514 – 516 (1987). –
See Abstr. 003.001.
Recent observations of industrial sulfur flows, which are much larger than are possible to produce experimentally, may provide important information concerning natural sulfur flows on both Earth and Io.

022.031   **Excitation and dissociation of molecular hydrogen in shock waves at interstellar densities.**
J. E. Dove, A. C. M. Rusk, P. H. Cribb, P. G. Martin.
Astrophys. J., Vol. 318, No. 1, p. 379 – 391 (1987).
The rate constant k_D for the thermal dissociation of molecular para–hydrogen by collisions with helium at kinetic temperatures of $200 – 10,000K$ is calculated for the whole range of densities appropriate to interstellar molecular clouds. Collisional transitions among the rotation–vibration levels, collision induced dissociation out of all levels, and infrared quadrupole emission are included in a solution of the master equation for the dissociation reaction. The population distribution among the

rotation–vibration levels can show large deviations from a Boltzmann distribution; under certain circumstances there are even population inversions.

022.032   Collisional excitation of interstellar sulfur dioxide.
A. Palma.
Astrophys. J., Suppl. Ser., Vol. 64, No. 3, p. 565 – 579 (1987).

The authors have computed state–to–state rotational excitation rates for the asymmetric top molecule SO_2 in collisions with low–energy He atoms. The intermolecular forces were obtained from an electron gas model, and collision dynamics were treated with the infinite–order sudden approximation. The errors introduced by these approximations are discussed.

022.033   Stark broadening trends along homologous sequences.
M. S. Dimitrijević, A. A. Mihajlov, M. M. Popović.
Astron. Astrophys., Suppl. Ser., Vol. 70, No. 1, p. 57 – 61 (1987).

Using a semiclassical formalism, a detailed analysis of Stark broadening parameters as a function of the impact electron angular momentum quantum number and the temperature is carried out for resonance lines of the alkalis. On the basis of the results obtained, the differences between the contribution of elastic, inelastic and strong collisions to the line widths for analogous transitions in homologous atoms are discussed.

022.034   Tabulated extinction efficiencies for various types of submicron amorphous carbon grains in the wavelength range 1000 Å – 300 μm.
E. Bussoletti, L. Colangeli, A. Borghesi, V. Orofino.
Astron. Astrophys., Suppl. Ser., Vol. 70, No. 2, p. 257 – 268 (1987).

The authors present, in a tabular form, the extinction efficiencies for various types of submicron amorphous carbon particles. Homogeneous samples have been morphologically characterized and spectroscopically analysed in the extended wavelength range: 1000 Å – 300 μm. Previous measurements performed in narrower ranges have been repeated and validated. New data are presented from 1000 Å to 3000 Å and from 35 μm to 300 μm in addition. A comparative analysis of the data with experimental results reported in the literature allows to estimate the properties of amorphous carbon grains able to account for portions of the interstellar extinction curve previously attributed to graphite.

022.035   Numerical fits to the electron impact transition rate coefficients for atomic hydrogen as a function of electron temperature.
C. Giovanardi, A. Natta, F. Palla.
Astron. Astrophys., Suppl. Ser., Vol. 70, No. 2, p. 269 – 280 (1987).

Using the most recent theoretical and experimental estimates of the cross–sections, the authors present numerical approximations for the value of the atomic hydrogen bound–bound effective collision strength due to electron impact as a function of electron temperature. The effective collision strength Γ is computed for transitions between the first 15 levels of principal quantum number n; for the first 4 principal levels Γ is given separately for sublevels of different orbital quantum number l. The numerical fits can be used for the evaluation of collisional transition rates in the range of electron temperatures between 5000K and 500,000K.

022.036   Laboratory studies on cometary molecules.
M. Singh, J. P. Chaturvedi.
Earth, Moon, Planets, Vol. 38, No. 3, p. 253 – 261 (1987).

The authors show results for some new bands of C_2, CN, $N_2{}^+$, CO^+, NH, OH, and CH hitherto unidentified, but expected to be present in the spectrum of comets by the analysis of Franck–Condon factors. Vibrational transition probabilities, Franck–Condon factors have been evaluated by an approximate analytical method for the A–X system of C_2, A–X, and B–X systems of CN, B–X system of $N_2{}^+$, A–X, and B–A systems of CO^+, A–X system of NH and A–X system of OH.

022.037   High resolution absorption cross–sections and band oscillator strengths of the Schumann–Runge bands of oxygen at 79K.
K. Yoshino, D. E. Freeman, J. R. Esmond, W. H. Parkinson.
Planet. Space Sci., Vol. 35, No. 8, p. 1067 – 1075 (1987).

Cross–sections of O_2 at 79K have been obtained from photoabsorption measurements at various pressures throughout the wavelength region 179.3 – 198.0 nm with a 6.65 m photoelectric scanning spectrometer equipped with a 2400 lines mm^{-1} grating and having an instrumental width (FWHM) of 0.0013 nm. The measured absorption cross–sections of the Schumann–Runge bands (12, 0) through (2, 0) are independent of the instrumental width. The measured cross–sections are presented. Band oscillator strengths of these bands have been determined by direct numerical integration of the measured cross–sections.

022.038   Fe II level populations in a hollow cathode discharge.
R. S. Hudson, L. L. Skrumeda, W. Whaling.
J. Quant. Spectrosc. Radiat. Transfer, Vol. 38, No. 1, p. 1 – 4 (1987).

Relative populations of 232 excited levels of the Fe^+ ion have been measured in a hollow cathode discharge in neon and in argon. The population distribution shows that thermal charge exchange between noble gas ions and ground level Fe atoms is responsible for most of the excited Fe^+ population. The cross section for thermal charge exchange is enhanced when the excited Fe^+ final level has the same $3d^6(^5D)$ core configuration as the initial Fe atom ground level.

022.039   Effects of the close approach of potential curves in photoabsorption by diatomic molecules – I. Theory and computational procedures.
L. Torop, D. G. McCoy, A. J. Blake, J. Wang, T. Scholz.
J. Quant. Spectrosc. Radiat. Transfer, Vol. 38, No. 1, p. 9 – 18 (1987).

A procedure is given for the calculation of continuum photoabsorption cross sections of diatomic molecules in a spectral region corresponding to the avoided crossing of molecular potentials. The technique involves the simultaneous solution of coupled Schrödinger equations and the numerical procedures adopted for their solution are outlined. Results of calculations using simple analytic potentials similar to those of the $B^3\Sigma_u{}^-$ and $E^3\Sigma_u{}^-$ states of oxygen are presented. It was found that non–adiabatic effects are significant as much as 1 eV below the crossing energy.

022.040   Effects of the close approach of potential curves in photoabsorption by diatomic molecules – II. Temperature dependence of the O_2 cross section in the region 130 – 160 nm.
J. Wang, D. G. McCoy, A. J. Blake, L. Torop.
J. Quant. Spectrosc. Radiat. Transfer, Vol. 38, No. 1, p. 19 – 27 (1987).

The photoabsorption cross section of oxygen has been measured at temperatures of 295 and 575K over the wavelength range 130 – 160 nm. The temperature coefficient shows strong structure in the part of the Schumann–Runge continuum below 136 nm. The observed data have been fitted with theoretical calculations that include a $^3\Pi_u$ continuum and take account of the coupling between the valence and Rydberg $^3\Sigma_u{}^-$ states. Potential curves and transition moments for these states obtained from the fitting procedure are given. The diabatic electronic coupling constant for the $^3\Sigma_u{}^-$ states is found to be 0.485 eV.

022.041   Absorption properties of a high temperature nitrogen plasma.
R. D. Taylor, A. W. Ali.
J. Quant. Spectrosc. Radiat. Transfer, Vol. 38, No. 1, p. 29 – 36 (1987).

A detailed description of the absorption cross sections in a nitrogen plasma is given for electron temperatures of 1.0 – 3.0 eV and a plasma density of 0.78 times normal atmospheric density.

Results are presented for partial and total absorption coefficients, where the components are the bound–bound, bound–free and free–free absorptions. Rosseland and Planck mean opacities are also computed.

022.042   Stark broadening of neutral potassium lines.
M. S. Dimitrijević, S. Sahal–Bréchot.
J. Quant. Spectrosc. Radiat. Transfer, Vol. 38, No. 1, p. 37 – 45 (1987).

Using a semiclassical approach, the authors have calculated electron–proton and Ar II impact line widths and shifts of 50 neutral potassium lines. The comprehensive set of results obtained is used for investigation of Stark–broadening–parameter regularities within the spectral series.

022.043   Relativistic radiative transition rates and energies for the Li–like isoelectronic sequence: electric and magnetic dipole and quadrupole radiation.
A. D. Steiger.
J. Quant. Spectrosc. Radiat. Transfer, Vol. 38, No. 2, p. 81 – 87 (1987).

For Li–like ions in the isoelectronic sequence between the nuclear charges $Z = 26$ and 94, energies, oscillator strengths and probabilities of electric and magnetic dipole and quadrupole transitions between the first 22 atomic levels were calculated by means of the multiconfigurational Dirac–Fock model. The effects of nonlocal Breit interaction, electron self–energy and vacuum polarization were included in calculating energy levels and radiative rates. The trends of transition energies and rates along the isoelectronic sequence, in particular, the enhancement of the forbidden transitions with increasing atomic number Z, are discussed.

022.044   Analytic vibration–rotational matrix elements for high Δv infrared transitions of diatomic molecules.
J. P. Bouanich.
J. Quant. Spectrosc. Radiat. Transfer, Vol. 38, No. 2, p. 89 – 112 (1987).

Theoretical expressions for the vibration–rotational matrix elements corresponding to i.r. transitions $vJ \rightarrow v'J'$ with $\Delta v = v' - v = 4, 5, 6, 7$ are given in terms of polynomials in m and m^2, including contributions from sixth–order Dunham potential–energy parameters and from dipole–momentum expansion coefficients M_0 to M_7. The formalism has been applied to vibrational transitions of HCl.

022.045   The pressure dependence of the Herzberg photoabsorption continuum of oxygen.
A. J. Blake, D. G. McCoy.
J. Quant. Spectrosc. Radiat. Transfer, Vol. 38, No. 2, p. 113 – 120 (1987).

The origin of pressure dependence in the dipole forbidden Herzberg continuum of oxygen is discussed in terms of the formation of oxygen dimers and the collision of free molecules. The small temperature dependence of the pressure coefficient indicates that the collisions of free molecules have the dominant influence. Consideration of the selection rules applying to the Herzberg systems and the strengths of the associated band systems lead to the conclusion that the pressure dependence results from enforced dipole transitions in the Herzberg III $(A'^3\Delta_u - X^3\Sigma_g^-)$ system.

022.046   A model for ionization balance and L–shell spectroscopy of non–LTE plasmas.
Y. T. Lee.
J. Quant. Spectrosc. Radiat. Transfer, Vol. 38, No. 2, p. 131 – 145 (1987).

The author has developed a model for calculating ionic charge–state distribution and level populations in steady–state non–local thermal equilibrium (non–LTE) plasmas. He uses this model to normalize the relative populations for a single ionization stage so that he can compute spectral line intensities as functions of electron temperature and density. These relative populations are determined by the balance of three processes:

collisional excitation, de–excitation, and radiative spontaneous emission. As an application of the model, the author has computed spectral–line intensities for all the $\Delta n = 0$ transitions of nitrogen–, oxygen– and fluorine–like argon ions, and he has used these results to analyze recent experiments.

022.047   Oscillator strength measurements in the vacuum–ultraviolet. II. The strong 1260, 1277, 1329, 1463, 1561 and 1657 Å multiplets of neutral carbon.
C. Goldbach, G. Nollez.
Astron. Astrophys., Vol. 181, No. 1, p. 203 – 209 (1987).

Oscillator strengths of 23 lines (or strongly blended lines) of neutral carbon, belonging to the strong vacuum–ultraviolet multiplets at 1261, 1277, 1329, 1463, 1561 and 1657 Å, are measured in emisson with a wall–stabilized arc. Optical depth profiles are derived, while boundary layer absorption is carefully minimized. The uncertainty on the oscillator strength values is between ± 10 and $\pm 15\%$. With the exception of the multiplet at 1329 Å, the agreement with the theoretical values of Nussbaumer and Storey (1984) is within 5%.

022.048   A theoretical study of the $H_3^+ + CO$ protonation process. I. The formation of HCO^+.
D. Talbi, F. Pauzat.
Astron. Astrophys., Vol. 181, No. 2, p. 394 – 397 (1987).

The potential hypersurface for protonation of CO by H_3^+ has been calculated in order to determine whether an activation barrier is to be found on the formation path of HCO^+ by direct proton transfer. The calculations confirm the hypothesis of no barrier at all. The potential hypersurface for protonation of CO by H^+ has also been plotted at an equivalent level of calculation, so that to test the validity of such a protonation model. In the region of activation barriers where bumps are expected to appear, the two curves are found strictly parallel, which justifies such a type of modelization.

022.049   Refined diatomic partition functions. I. Calculational methods and H_2 and CO results.
A. W. Irwin.
Astron. Astrophys., Vol. 182, No. 2, p. 348 – 358 (1987).

A new partition function approximation is derived that corrects the Tatum approximation for a number of effects including first order centrifugal distortion. This new approximation reduces the H_2 partition function errors from 6% to 2% and the CO partition function errors from 1% to 0.004% at 4000K. The most accurate partition functions of H_2 and CO are calculated using direct energy summation methods. From 1000 to 9000K the maximum systematic errors of these calculations are estimated to be 0.01% in the partition function, 0.1% in the related contribution to the internal energy, and 1% in the related contribution to the specific heat at constant volume. These standards of accuracy are maintained in polynomial fits to the H_2 and CO partition functions.

022.050   Oscillator strengths for the lines of ionized iron.
A. A. Boyarchuk, I. S. Savanov.
Bull. Crimean Astrophys. Obs., Vol. 74, p. 48 – 64 (1987). English translation of 41.022.101.

022.051   The old electron handbook.
D. E. Harris.
Radio astronomy from space, p. 265 – 275 (1987). – See Abstr. 012.009.

022.052   Sputtering of wide bandgap semiconductors.
R. P. Howson.
Rutherford Appleton Lab., Rep., RAL 87–043, p. 193 – 196 (1987). With 9 figures. – See Abstr. 012.010.

022.053   Production of thin films by beam deposition.
J. Franks.
Rutherford Appleton Lab., Rep., RAL 87–043, p. 197 – 200 (1987). – See Abstr. 012.010.
The structures of thin films produced by beam sputtering, plasma sputtering and evaporation are compared.

022.054   Use of broad beam ion sources for ion assisted deposition and sputter deposition.
T. W. Jolly.
Rutherford Appleton Lab., Rep., RAL 87–043, p. 201 (1987). Abstract. – See Abstr. 012.010.

022.055   The dipole moment of C_3H_2.
R. D. Brown, P. D. Godfrey, R. P. A. Bettens.
Mon. Not. R. Astron. Soc., Vol. 227, No. 2, p. 19P – 20P (1987).
The electric dipole moment of the interstellar molecule cyclopropenylidene C_3H_2 has been measured to be 3.32(5) Debye from the Stark effect in the microwave spectrum of the $1_{10} \leftarrow 1_{01}$ and $2_{20} \leftarrow 2_{11}$ transitions.

022.056   Identification of intercombination transitions in Fe XIV and Fe XIII in the spectra of foil–excited ions and solar flares.
E. Träbert, R. Hutton, I. Martinson.
Mon. Not. R. Astron. Soc., Vol. 227, No. 3, p. 27P – 31P (1987).
The authors have recorded prompt and time–delayed beam–foil spectra of iron in the VUV region, using 24 MeV ions of Fe. The time–delayed spectra, observed 17 ns after excitation by the foil, showed intercombination lines in Al–like Fe XIV and Si–like Fe XIII which have not been classified earlier. The measured wavelengths coincide with those of some unidentified transitions, observed in high–resolution spectra of solar flares. The solar data are used to obtain information on the excitation energies of the quartet and quintet term systems in Fe XIV and Fe XIII, respectively.

022.057   A comparison of relativistic and quasirelativistic line strengths.
A. K. Mohanty, D. H. Sampson.
Phys. Scr., Vol. 35, No. 5, p. 645 – 649 (1987).
A comparison is made between radiative line strengths, or more precisely the square of matrix elements of the electron position vector, calculated fully relativistically and with two quasirelativistic approaches. The purpose of the work is to find the optimum quasirelativistic approach to use for rapid calculation of line strengths and collision strengths for highly charged ions and to obtain some indication of the accuracy of the approach.

022.058   C–type transitions in methyl formate.
G. M. Plummer, E. Herbst, F. C. De Lucia.
Astrophys. J., Vol. 318, No. 2, p. 873 – 875 (1987).
In a previous laboratory analysis of the millimeter–wave spectrum of the asymmetric top methyl formate in its ground ($v_t = 0$) torsional E (degenerate) substate, the authors predicted the transition frequencies and intensities of a large number of rotational transitions below 300 GHz. They have recently found that there are some additional moderately strong spectral lines – called c–type transitions – which would be forbidden in methyl formate, since it does not possess a dipole moment along the c–principal axis, except for the perturbative influence of internal rotation (torsional motion). The authors have calculated the frequencies and intensities of these c–type transitions, as well as other "forbidden" transitions labeled as x–type. The stronger c–type transitions below 300 GHz have been included in a list of spectral frequencies presented in this paper.

022.059   Collision–induced rototranslational absorption spectra of CH_4–CH_4 pairs at temperatures from 50 to 300K.
A. Borysow, L. Frommhold.
Astrophys. J., Vol. 318, No. 2, p. 940 – 943 (1987).
In a previous paper, rigorous quantum computations of the rototranslational absorption spectra of CH_4–CH_4 pairs have been communicated which closely reproduce existing laboratory measurements at temperatures from 124 to 300K, and over a range of frequencies from 0 to 750 cm^{-1}. The authors describe how the spectra can be reproduced from simple, analytical functions that closely model the quantum profiles and, thus, the laboratory measurements. The functions apply for temperatures from 50 to 300K.

022.060   The laboratory millimeter– and submillimeter–wave spectrum of ^{13}C methanol.
T. Anderson, E. Herbst, F. C. De Lucia.
Astrophys. J., Suppl. Ser., Vol. 64, No. 4, p. 703 – 714 (1987).
Three hundred previously unmeasured rotational transitions of the internal rotor $^{13}CH_3OH$ (^{13}C methanol) involving the rotational quantum number $J \leqslant 9$ have been studied in the millimeter– and submillimeter–wave regions of the spectrum. These data, consisting mainly but not exclusively of transitions in the lowest ($v_t = 0$) torsional state, have been combined with previous data at lower frequencies into a global data set which has been analyzed via an internal axis method. The resulting rotational constants have been utilized to predict the frequencies of numerous additional lines of interest to astronomers.

022.061   Excitation of [O I] and [C I] fine structure transitions by He and H_2: a neglected selection rule.
T. S. Monteiro, D. R. Flower.
Mon. Not. R. Astron. Soc., Vol. 228, No. 1, p. 101 – 107 (1987).
There exists a selection rule forbidding (to first order) the 0–1 fine structure transition in collisions involving $O(^3P)$ or $C(^3P)$ and He or H_2. Quantal close coupling calculations for $O(^3P) – He(^1S)$ show this selection rule to be quantitatively significant, reducing the rate coefficient for the 0–1 transition by an order of magnitude relative to the 0–2 and 1–2 transitions. The rate coefficients for the latter two transitions are found to be similar in magnitude to those computed by Launay & Roueff for atomic hydrogen impact.

022.062   The method for identification of technetium ion spectra excited in a coaxial plasma accelerator.
N. A. Afanas'eva, L. A. Korostyleva, Yu. P. Dontsov.
Kinematika Fiz. Nebesn. Tel, Tom 3, No. 4, p. 52 – 56 (1987). In Russian. English translation in Kinematics Phys. Celest. Bodies.
A new design of plasma accelerator discharge chamber is used for the investigation of radioactive elements spectra. In the region $\lambda\lambda 210 – 290$ nm 880 molybdenum lines and 440 technetium lines are recorded. On the basis of the Doppler shift values the spectral lines are divided into groups corresponding to various charge species. It is established that molybdenum and technetium ions possessing the same charges have similar velocities in the plasma jet. For the complex spectrum analysis it is possible to take into account the different spectral line intensity distribution of various ions over the plasma volume.

022.063   The S_{21} lines of the $A^2\Sigma^+(v' = 1) \leftarrow X^2\Pi$ $(v'' = 0)$ transitions of OH and OD.
S.–R. Lin, S.–T. Lee, Y.–P. Lee.
J. Quant. Spectrosc. Radiat. Transfer, Vol. 38, No. 3, p. 163 – 166 (1987).
The S_{21} lines of the OH radical for the $A^2\Sigma^+$ ($v' = 1$)$\leftarrow X^2\Pi$ ($v'' = 0$) transition have been observed in the 278 – 280 nm region from the laser–excitation spectrum. The observed peak wavenumbers are in excellent agreement with those predicted for the S_{21} bands of the A ($v' = 1$)$\leftarrow X$ ($v'' = 0$) transition from the known spectroscopic parameters. Observation of the corresponding S_{21} system for OD in the 286.0 – 286.9 nm region further confirms this assignment. The relative absorption cross–sections of about one–tenth those of the P_1 lines are in good agreement with theoretical considerations.

022.064   Diode laser measurements of CO line widths at planetary atmospheric temperatures.
P. Varanasi, S. Chudamani, S. Kapur.
J. Quant. Spectrosc. Radiat. Transfer, Vol. 38, No. 3, p. 167 – 171 (1987).
Hydrogen–broadened half–widths and nitrogen–broadened half–widths of eight lines between P(1) and P(15) in the CO

fundamental have been measured at several temperatures between 94 and 298K using a tunable diode laser spectrometer and the sweep integration technique.

022.065   **Intensities and H_2–broadened half–widths of germane lines around 4.7 μm at temperatures relevant to Jupiter's atmosphere.**
P. Varanasi, S. Chudamani.
J. Quant. Spectrosc. Radiat. Transfer, Vol. 38, No. 3, p. 173 – 177 (1987).
Absolute intensities and H_2–broadened half–widths of R(0) and R(1) of $^{72}GeH_4$, have been measured at 119.5, 150.3, 199.0, 250.0, and 296.0K using a tunable diode laser and the sweep integration technique. The combined intensity of the ν_3 bands of all of the five isotopic species of GeH_4 as well as that of the Q–branches alone have been measured independently at 294K using the Wilson–Wells–Penner–Weber technique.

022.066   **Measurements on 4.7 μm CH_3D lines broadened by H_2 and N_2 at temperatures relevant to planetary atmospheres.**
S. Chudamani, P. Varanasi.
J. Quant. Spectrosc. Radiat. Transfer, Vol. 38, No. 3, p. 179 – 181 (1987).
Absolute intensities, H_2–broadened half–widths and N_2–broadened half–widths of P(5,2), P(5,3), P(7,2), P(7,3) and P(8,3) in the ν_2–fundamental band of $^{12}CH_3D$ have been measured at several temperatures between 94 and 300K using a tunable diode laser and the sweep integration technique.

022.067   **Self– and foreign–gas broadening of ethane lines determined from diode laser measurements at 12 μm.**
W. E. Blass, G. W. Halsey, D. E. Jennings.
J. Quant. Spectrosc. Radiat. Transfer, Vol. 38, No. 3, p. 183 – 184 (1987).
Self– and foreign–gas broadening of ethane lines have been measured in the ν_9 band at 12 μm. A coefficient of 0.125 cm^{-1}atm^{-1} was determined for self broadening. Foreign–gas broadening coefficients determined are (in cm^{-1}atm^{-1}) 0.090 for N_2, 0.069 for He, 0.068 for Ar, 0.108 for H_2, and 0.096 for CH_4. Results are given for a sample temperature of 296K.

022.068   **Hydrogen and nitrogen broadening of the lines of C_2H_2 at 14 μm.**
W. E. Blass, V. W. L. Chin.
J. Quant. Spectrosc. Radiat. Transfer, Vol. 38, No. 3, p. 185 – 188 (1987).
Collision broadening of 14 μm acetylene transitions was studied using a diode laser spectrometer. Fifteen $P(J)$ and $Q(J)$ transitions in the ν_5 fundamental band, broadened by hydrogen and nitrogen, were used in the study. The results for the nitrogen–broadened lines are 0.0853 cm^{-1}atm^{-1} for the collision broadening coefficient a at $J = 0$, $b = -1.63 \times 10^{-3}$, and $c = 3.92 \times 10^{-5}$cm^{-1}atm^{-1}. The results for the hydrogen broadened lines in cm^{-1}atm^{-1} are 0.0812, -1.16×10^{-3}, and 2.93×10^{-5}, respectively.

022.069   **Measurement and analysis of the far infrared absorption spectrum of the gaseous mixture H_2–CH_4.**
G. Birnbaum, A. Borysow, H. G. Sutter.
J. Quant. Spectrosc. Radiat. Transfer, Vol. 38, No. 3, p. 189 – 199 (1987).
The collision–induced absorption of H_2–CH_4 mixtures was measured from ~20 to 900 cm^{-1} at 195 and 297K. By subtracting the absorption due to H_2–H_2 and CH_4–CH_4 collisions from that of the mixture, the absorption due to H_2–CH_4 collisions was obtained. This spectrum was analyzed using the BC model line shape to provide a way of estimating the far–i.r. spectrum of H_2–CH_4 for various concentrations of H_2 and CH_4. Theoretical spectral moments were computed with different potential functions and compared with experimental values.

022.070   **Remote sensing of atomic oxygen: some observational difficulties in the use of the forbidden O I λ1173 Å and O I λ1641 Å transitions.**
P. W. Erdman, E. C. Zipf.
J. Geophys. Res., Vol. 92, No. A9, p. 10,140 – 10,144 (1987).
The authors have made a detailed, high–resolution study of the far ultraviolet emission features in the regions surrounding the atomic oxygen transitions at λ1172.6 Å and λ1641.3 Å. These spectra, which were excited by electron impact on O_2 and N_2, are presented in an attempt to display some potential sources of interference in aeronomical measurements of these O I lines. Both atomic and molecular emissions are found, and the spectral resolution necessary to make unambiguous measurements is discussed.

022.071   **Trapping of gases by water ice and implications for icy bodies.**
A. Bar–Nun, D. Prialnik, D. Laufer, E. Kochavi.
Adv. Space Res., Vol. 7, No. 5, p. 45 – 47 (1987). – See Abstr. 012.022.
The trapping of various gases by water ice at low temperatures (20 – 80K) and their release from the ice upon warming, was studied experimentally. The results of these experiments, together with a computation of the thermal evolution of a cometary nucleus, can explain the gas and dust jets which were observed to emanate from the nucleus of P/Halley. The experimental results are important also to the gas content of Titan.

022.072   **Laboratory experiments in the study of the chemistry of the outer planets.**
T. W. Scattergood.
Adv. Space Res., Vol. 7, No. 5, p. 99 – 108 (1987). – See Abstr. 012.022.
Alone, and in conjunction with modeling, laboratory experiments will continue to be used to further our understanding of the outer solar system, and some experiments that need to be done are listed.

022.073   **Fluorescent and collisional excitation in diatomic molecules.**
R. Gredel.
Diss. Naturwiss.–Math. Gesamtfak., Ruprecht–Karls–Univ., Heidelberg, F.R. Germany, 9+173 pp. (1987).

022.074   **Quasistatic absorption coefficient of two–component gases.**
R. Logožar.
Hvar Obs. Bull., Vol. 10, No. 1, p. 23 – 34 (1986).
The asymptotic calculations of adiabatic potential curves and quasistatic absorption coefficients for quasimolecular systems consisting of two neutral atoms are outlined. The application in an analysis of spectral lines originating in low density photospheres (atmospheres) of cool stars (T_{eff} below 3000K) and other relative cool astrophysical objects is suggested.

022.075   **Laboratory approaches of nuclear reactions involved in primordial and stellar nucleosynthesis.**
C. Rolfs.
Prog. Part. Nucl. Phys., Vol. 17, p. 365 – 391 (1986). – See Abstr. 012.024.
The grand concept of elemental nucleosynthesis will not be truly established until we attain a deeper and more precise understanding of the many nuclear processes operating in various astrophysical environments. The desired cross sections of the nuclear processes are among the smallest measured in the nuclear laboratory, often requiring new technologies and long data collection times with painstaking attention to background. The laboratory approaches in studying charged–particle–induced nuclear reactions are the subject of this paper.

022.076   **Collisional excitation of interstellar cyclopropenylidene.**
S. Green, D. J. DeFrees, A. D. McLean.
Astrophys. J., Suppl. Ser., Vol. 65, No. 1, p. 175 – 191 (1987).

Theoretical rotational excitation rates were computed for C_3H_2 in collisions with He atoms at temperatures from 30 to 120K. The intermolecular forces were obtained from accurate self–consistent field and perturbation theory calculations, and collision dynamics were treated within the infinite–order sudden approximation. The accuracy of the latter was examined by comparing with the more exact coupled states approximation.

022.077   **Collisional effects in He I lines and helium abundances in planetary nebulae.**
R. E. S. Clegg.
Mon. Not. R. Astron. Soc., Vol. 229, No. 1, p. 31P – 39P (1987).

Attention is drawn to new, 19–state quantal calculations for collisional excitation by electron impact in neutral helium. Recommended empirical formulae are given for the collisional contribution to He I recombination lines such as $\lambda\lambda$4471, 5876 Å in gaseous nebulae. Collisional ionization of metastable (2^3S) He I is significant for high temperature nebulae. Collisional transfers provide significant cooling in nebulae with low heavy-element abundances. Revised mean He/H ratios for three large samples of planetary nebulae (PN) are given. It is shown that the hypothesis, that He abundances in PNs should not depend on the nebular temperature and density, is better satisfied when corrections for collisional effects are made.

022.078   **Determination of $O_2(a^1\Delta g)$ and $O_2(b^1\Sigma^+g)$ yields in the reaction $O + ClO \rightarrow Cl + O_2$: implications for photochemistry in the atmosphere of Venus.**
M.–T. Leu, Y. L. Yung.
Geophys. Res. Lett., Vol. 14, No. 9, p. 949 – 952 (1987).

A discharge flow apparatus with chemiluminescence detector has been used to study the reaction $O + ClO \rightarrow Cl + O_2^*$, where $O_2^* = O_2(a^1\Delta g)$ or $O_2(b^1\Sigma^+g)$. The observed $O_2(a^1\Delta g)$ airglow of Venus cannot be explained in the context of standard photochemistry using the authors' experimental results and those reported in recent literature. The possibility of an alternative source of O atoms derived from SO_2 photolysis in the mesosphere of Venus is suggested.

022.079   **Optical evolution of laboratory–produced organics: applications to Phoebe, Iapetus, outer belt asteroids and cometary nuclei.**
G. Andronico, G. A. Baratta, F. Spinella, G. Strazzulla.
Astron. Astrophys., Vol. 184, No. 1/2, p. 333 – 336 (1987).

Optical and NIR spectra ($0.3 - 2.5 \mu m$) of organic materials, synthesized in the laboratory by ion beam bombardment, are presented. The spectral response of the organics changes as the ion fluence increases. They become darker and darker with increasing fluence. The authors suggest that bombardment by solar ions may produce both organic materials similar to those on D–type asteroids (and on the Iapetus leading hemisphere) and carbonaceous materials similar to those on C–type asteroids (and on Phoebe). The relevance of these experimental results in understanding the darkness of (some?) cometary nuclei is also outlined.

022.080   **On the applicability of the Sobolev approximation to the calculation of profiles and integral intensities of emission lines in optically thick slowly expanding media.**
D. V. Bisikalo.
Nauchn. Inf., Vyp. 63, p. 152 – 160 (1987). In Russian.

The applicability of the Sobolev approximation to the calculation of the profiles and integral intensities of emission lines, arising in optically thick media expanding with velocities, exceeding only by few times the thermal ones, are discussed. Expressions for the relative errors of these quantities introduced by the method are obtained. As an example the errors for a typical rotational line of H_2O, which is formed in a cometary coma, are estimated.

022.081   **Optical efficiencies of lightning in planetary atmospheres.**
W. J. Borucki, C. P. McKay.
Nature, Vol. 328, No. 6130, p. 509 – 510 (1987).

Spacecraft observations show that the presence of lightning activity is not confined to the terrestrial atmosphere, but is also found in the atmospheres of Venus, Jupiter and Saturn. The authors report the first simulations of lightning in planetary atmospheres by laser–induced plasmas. These simulations show that the fraction of the energy in lightning discharge channels that is radiated in the visible spectrum is similar for Earth, Venus and Titan, but quite different for Jupiter. One implication of these results is that the amount of trace gases produced by lightning in the jovian atmosphere must be larger than previously estimated.

022.082   **Laboratory astrophysics: atomic spectroscopy.**
W. H. Parkinson.
Spectroscopy of astrophysical plasmas, p. 302 – 353 (1987). – See Abstr. 003.010.

Contents: 1. Introduction. 2. Basic formulae and definitions: Einstein probability coefficients. The oscillator strength or f–value. Electron impact excitation cross sections. Bound–free transition parameters. 3. Measurement of bound–bound radiative transition parameters. 4. Measurement of bound–free transition parameters.

022.083   **Electron excitation of fine–structure transitions within the ground $3s^23p^3$ configuration in Fe XII.**
S. S. Tayal, R. J. W. Henry, A. K. Pradhan.
Astrophys. J., Vol. 319, No. 2, p. 951 – 956 (1987).

LS–coupled K matrices calculated in an R–matrix approach are recoupled to obtain collision strengths for fine–structure transitions within the ground $3s^23p^3$ configuration in an intermediate coupling scheme. The target states are represented by extensive configuration–interaction wave functions. The relativistic effects in the target are allowed in the Breit–Pauli formulation. The effective collision strengths are obtained by integrating the collision strengths over a Maxwellian distribution of electron energies. These are listed over a wide temperature range (400,000 – 3,000,000K).

022.084   **Radiative lifetimes of the $4p^4P^0_{5/2,3/2}$ levels in Ar II from beam–foil experiments.**
F. J. Coetzer, T. C. Kotze, P. van der Westhuizen.
J. Quant. Spectrosc. Radiat. Transfer, Vol. 38, No. 4, p. 253 – 260 (1987).

The beam–foil excitation technique and the ANDC (arbitrary normalized decay curve) methods were used to obtain radiative lifetimes of the $4p^4P^0_{5/2}$ and $4p^4P^0_{3/2}$ energy levels in Ar II. The lifetimes are compared with experimental and theoretical values obtained by other investigators. Discrepancies between the results obtained with different techniques are discussed.

022.085   **Quantum defect method applied to oscillator strengths.**
R. E. H. Clark, A. L. Merts.
J. Quant. Spectrosc. Radiat. Transfer, Vol. 38, No. 4, p. 287 – 293 (1987).

The authors use ideas from quantum defect theory to obtain approximate energy levels and oscillator strengths. The results are easily obtained, require little computer capability, and typically agree with detailed calculations to better than 5%. The method can be applied to cases where hydrogenic and scaling techniques are inappropriate.

022.086   **Determination of transition probability for the 655–nm Tl line.**
D. Karabourniotis, S. Couris, J. J. Damelincourt.
J. Quant. Spectrosc. Radiat. Transfer, Vol. 38, No. 4, p. 303 – 310 (1987).

Studies of high–pressure Hg–Tl I a.c. (50 Hz) arc plasmas have been used to verify the validity of Boltzmann statistics at the moment of maximum electron density (5 ms) by applying LTE criteria. For a known plasma temperature, the transition probability of the optically–thin 655–nm line of Tl was derived from emission measurements by using the self–reversed 535–nm line of Tl as reference $[A_{655} = (3.74 \pm 0.37) \times 10^6 s^{-1}]$.

022.087 Dielectronic recombination rates for ions of the magnesium sequence – II.
M. P. Dube, K. J. LaGattuta.
J. Quant. Spectrosc. Radiat. Transfer, Vol. 38, No. 4, p. 311 – 318 (1987).
The dielectronic recombination (DR) rate coefficient α^{DR} is explicitly calculated for Ar, Fe and Mo target ions of the Mg isoelectronic sequence (12 electrons). The $2p$ transitions are dominant at high temperatures and are considered in detail with full LS coupling. This work extends the authors' previous study in which both the $3s$, $\Delta n = 0$ and $3s$, $\Delta n \neq 0$ transitions were considered. Scaling of α^{DR} with free–electron temperature is also discussed.

022.088 Laboratory study of the rotational spectrum of vibrationally excited C_2H.
D. R. Woodward, J. C. Pearson, C. A. Gottlieb, M. Guélin, P. Thaddeus.
Astron. Astrophys., Vol. 186, No. 1/2, p. L14 – L16 (1987).
The $N = 1 \rightarrow 2$, $2 \rightarrow 3$ and $3 \rightarrow 4$ millimeter–wave rotational transitions in the $v_2 = 1$ level of $X^2\Sigma^+$ C_2H have been detected in a laboratory discharge through helium and acetylene. Accurate values for the parameters A, B, D, γ, p, p_D, q, q_D, a, b_F, c and d were determined from a least–squares fit of the measured frequencies to an effective Hamiltonian; the hyperfine parameters are the first reported for a vibrationally excited carbon chain radical. An upper limit of $T_R \leqslant 0.3K$ in IRC + 10216 has been established with the IRAM 30 m telescope for two transitions in the $N = 1 \rightarrow 2$ manifold. The accurately measured laboratory frequencies will be used in astronomical searches for vibrationally excited C_2H.

022.089 High–explosive simulation of supernovae.
W. K. Brown.
Publ. Astron. Soc. Pac., Vol. 99, No. 618, p. 858 – 861 (1987).
Comparison of photographs of explosive experiments to the Cas A supernova remnant reveals a striking similarity.

022.090 Erratum: "Collision–induced rototranslational absorption spectra of N_2–N_2 pairs for temperatures from 50 to 300K" [Astrophys. J., Vol. 311, No. 2, p. 1043 – 1057 (1986)].
A. Borysow, L. Frommhold.
Astrophys. J., Vol. 320, No. 1, p. 437 (1987). See Abstr. 42.022.194.

022.091 Infrared spectrum of quenched carbonaceous composite (QCC). II. A new identification of the 7.7 and 8.6 micron unidentified infrared emission bands.
A. Sakata, S. Wada, T. Onaka, A. T. Tokunaga.
Astrophys. J., Lett. Ed., Vol. 320, No. 1, p. L63 – L67 (1987).
An infrared spectrum of "oxidized" quenched carbonaceous composite (QCC) is presented. In addition to the features seen in "unoxidized" QCC, new features appear at 7.7 and 8.6 μm and the strengths of the features at 6.2, 7.3, and 11.4 μm increase. The infrared features of "oxidized" QCC are in good agreement with nine of 11 members of the unidentified infrared (UIR) emission bands. The absorption bands of QCC are all broad without fine structure, clearly different from sharp bands of molecules. The present results indicate that oxygen can play an important role in the structures of the UIR emitting material as well as carbon and hydrogen.

022.092 Laboratory and astronomical identification of sulfur–containing carbon–chain molecules, CCS and C_3S.
S. Yamamoto, S. Saito, K. Kawaguchi, H. Suzuki, M. Ohishi, N. Kaifu.
Bull. Am. Astron. Soc., Vol. 19, No. 2, p. 690 (1987). Abstract. – See Abstr. 010.061.

022.093 Heating by PAH molecules or small grains.
S. Lepp, A. Delgarno.
Bull. Am. Astron. Soc., Vol. 19, No. 2, p. 724 (1987). Abstract. – See Abstr. 010.061.

022.094 High resolution vacuum–ultraviolet spectra and photoabsorption cross sections of carbon monoxide.
P. L. Smith, G. Stark, K. Yoshino, W. H. Parkinson, K. Ito.
Bull. Am. Astron. Soc., Vol. 19, No. 2, p. 725 (1987). Abstract. – See Abstr. 010.061.

022.095 Some astronomical applications of the sequential fragmentation theory.
W. K. Brown.
Bull. Am. Astron. Soc., Vol. 19, No. 2, p. 751 (1987). Abstract. – See Abstr. 010.061.

022.096 Accurate wavelengths in O II.
K. B. S. Eriksson.
J. Opt. Soc. Am. B, Vol. 4, No. 9, p. 1369 – 1371 (1987).
The wavelengths of 71 strong O II lines between 4676 and 525 Å have been measured. Improved energy levels have been derived. Smoothed wavelengths have been recalculated from these. The uncertainty is about 1 mÅ.

022.097 A laboratory measurement of the Doppler broadened N II $\lambda5005.15$ Å emission line–width produced by electron impact excitation of N_2.
P. W. Erdman, E. C. Zipf.
Planet. Space Sci., Vol. 35, No. 11, p. 1471 – 1474 (1987).

022.098 Rotational partition functions for spherical–top molecules.
R. S. McDowell.
J. Quant. Spectrosc. Radiat. Transfer, Vol. 38, No. 5, p. 337 – 346 (1987).
An accurate closed–form expression is obtained for the rotational partition function of spherical–top molecules at moderate and low temperatures.

022.099 Tunable diode laser measurements of absolute line-strengths in the HNO_3 v_2 band near 5.8 μm.
R. D. May, C. R. Webster, L. T. Molina.
J. Quant. Spectrosc. Radiat. Transfer, Vol. 38, No. 5, p. 381 – 388 (1987).

022.100 Electrostatic charge on a dust size distribution in a plasma.
H. L. F. Houpis, E. C. Whipple Jr.
J. Geophys. Res., Vol. 92, No. A11, p. 12,057 – 12,068 (1987).
The capacitance of a grain immersed in a steady state plasma containing a size distribution of dust particles is calculated. Assuming the equilibrium potential has been obtained by a simple balance of electron and ion collection currents, the grain charge is also obtained.

022.101 Comments on absolute absorption coefficients of atmospheric water vapor at CO_2 laser wavelengths.
Infrared Phys., Vol. 27, No. 5, p. 345 – 347 (1987).

022.102 Extinction coefficients of carbon grains in far–infrared.
J. Wirsich.
Infrared Phys., Vol. 27, No. 6, p. 399 – 405 (1987).
With a simple model it is shown that the wavelength-dependence of the extinction of crystalline and amorphous carbon grains in the FIR region is caused by the vibrational density of states $g_c(\omega)$ and $g_d(\omega)$, respectively. Both are approximated by the well known expressions of the model of the linear chain and the susceptibilities in the FIR are calculated by the equation of Kubo. The result is that the extinction in FIR of the crystalline carbon grains has a λ^{-2}–dependence and the amorphous modification has a λ^{-1}–dependence. The results are compared with laboratory measurements, astrophysical observations and other theoretical models.

022.103 Statistical phase space theory of ion–polar molecule systems: application to the reaction $H_2O \cdot H_3O^+ \to H_2O + H_3O^+$.
L. M. Bass, M. T. Bowers.
J. Chem. Phys., Vol. 86, No. 5, p. 2611 – 2616 (1987). Abstr. in Phys. Abstr., Vol. 90, No. 1309, Entry 86593 (1987).

022.104 Interstellar problems and matrix solutions.
L. J. Allamandola.
J. Mol. Struct., Vol. 157, No. 1 – 3, p. 255 – 273 (1987). Abstr. in Phys. Abstr., Vol. 90, No. 1309, Entry 88102 (1987).

022.105 Electronic spectroscopy and relaxation of some molecular cations of cometary interest.
S. Leach.
Astron. Astrophys., Vol. 187, No. 1/2, p. 195 – 200 (1987). – See Abstr. 003.013, 43.022.044.
Recent laboratory studies of the electronic spectra and nonradiative relaxation properties of a number of singly–charged molecular cations are discussed . Their interest for cometary physics is emphasized. Spectroscopic and relevant cometary information is given on CS^+, SH^+, CO_2^+, and H_2S^+. Coincidence techniques have been used for studying radiationless transitions in CO_2^+, NH_3^+ and other polyatomic ions. The results have been amplified by the use of laser excitation methods in the case of CO_2^+. Some discussion is given on recent cometary observations of CO_2^+ emission. The fluorescence quantum yields are measured for several molecular ions which could be formed from parent molecules likely to exist as frozen gases in the nucleus. The possibility of spectral observation of these cations in comets is considered.

022.106 Charging of dust particles in comets and in interplanetary space.
P. Notni, H. Tiersch.
Astron. Astrophys., Vol. 187, No. 1/2, p. 796 – 800 (1987). – See Abstr. 003.013, 43.022.045.
The authors calculate the electric potential which is to be expected on dust particles in a plasma, covering a wide range of plasma temperature, density and relative velocity between plasma and dust particle, two plasma compositions (H_2O^+ and H^+) and some selected values of photoemission and secondary electron emission coefficients. An error concerning the high temperature region in the graphs of the authors' paper (Notni and Tiersch, 1986) is corrected. Systematic trends in the results are discussed, and a crude estimate is given of the influence of a charge on the motion of dust in comets.

022.107 Sulfur in vacuum: sublimation effects on frozen melts, and applications to Io's surface and torus.
D. B. Nash.
Icarus, Vol. 72, No. 1, p. 1 – 34 (1987).
The author has found from laboratory experiments that vacuum sublimation has a profound effect on the molecular composition, microtexture, bulk density (porosity), and the UV/visible spectral reflectance of the surface of solid sulfur samples, both when the sulfur is in the form of frozen or quenched melts and as laboratory–grade sulfur powder. These sublimation effects produce a unique surface material, the understanding of which may have important implications for deciphering the many enigmatic optical and textural properties of the surface of Jupiter's satellite Io. This planetary body is thought to have a surface greatly enriched in volcanically produced elemental sulfur and sulfur compounds and to have a surface atmospheric pressure with an upper limit of $\sim 10^{-7}$ atm, comparable to a good laboratory vacuum, and surface hotspots at temperatures of about 300K covering about 0.3% of its global surface.

022.108 Laboratory measurements of the microwave opacity of gaseous ammonia (NH_3) under simulated conditions for the Jovian atmosphere.
P. G. Steffes, J. M. Jenkins.
Icarus, Vol. 72, No. 1, p. 35 – 47 (1987).
Gaseous ammonia (NH_3) has long been recognized as a primary source .of microwave opacity in the atmosphere of Jupiter. In order to more accurately infer the abundance and distribution of ammonia from radio emission measurements in the 1– to 20–cm wavelength range and radio occultation measurements at 3.6 and 13 cm, the authors have made measurements of the microwave opacity from gaseous ammonia under simulated conditions for the Jovian atmosphere. Measurements of ammonia absorptivity were made at five frequencies from 1.62 to 21.7 GHz (wavelengths from 18.5 to 1.38 cm), at temperatures from 178 to 300K, and at pressures from 1 to 6 atm, in a 90% hydrogen/10% helium atmosphere.

022.109 Electric discharge synthesis of HCN in simulated Jovian atmospheres.
R. Stribling, S. L. Miller.
Icarus, Vol. 72, No. 1, p. 48 – 52 (1987).
HCN energy yields (moles J^{-1}) were measured using corona discharge for gas mixtures containing H_2, CH_4, NH_3, with H_2/CH_4 ratios from 4.4 to 1585. The yields are approximately proportional to the mole fraction of methane in the gas mixture. Assuming that the 3/1 ratio of corona discharge to lightning energy on the Earth applies to Jupiter, HCN column densities from corona discharge could account for approximately 10% of the observed HCN. These estimates are very dependent on the values used for the energy available as lightning on Jupiter and the eddy diffusion coefficients in the region of synthesis.

022.110 Hydrogen molecule ion in the magnetic field of a neutron star. Probabilities of vibration–rotation transitions.
V. K. Khersonskij.
Astrofiz. Issled. Izv. Spets. Astrofiz. Obs., Tom 24, p. 68 – 84 (1987). In Russian. English translation in Bull. Spec. Astrophys. Obs. – North Caucasus.
Probabilities of vibration–rotation transitions in the ground electron state of the molecule ion H_2^+ in a strong magnetic field typical for the surface of a neutron star ($B = 10^{12} - 10^{13}$G) are calculated. The author discusses the selection rules for projection of the angular momentum on the magnetic field direction and considers polarization characteristics and angular distribution of radiation. The obtained data may be used to study the formation conditions of H_2^+ lines in the region of a strong magnetic field near the neutron star's surface.

022.111 Self– and N_2–broadened spectra of water vapor between 7.5 and 14.5 μm.
P. Varanasi, S. Chudamani.
J. Quant. Spectrosc. Radiat. Transfer, Vol. 38, No. 6, p. 407 – 412 (1987).
Self– and N_2–broadened absorption spectra of water vapor have been measured in the infrared between 7.5 and 14.5 μm at 294, 311, 333 and 339K. The so–called continuum aspect of the observed absorption coefficients is interpreted in terms of existing theoretical models.

022.112 Quasi–statistical solution of early homogeneous nucleation phases.
S. C. F. Rossi, P. Benevides–Soares.
Rev. Mex. Astron. Astrofis., Vol. 14, No. 2, p. 624 (1987). Abstract. – See Abstr. 012.042.

022.113 Time arrow in quantum theory.
E. Joos.
J. Non–Equilibrium Thermodyn., Vol. 12, No. 1, p. 27 – 43 (1987). Abstr. in Phys. Abstr., Vol. 90, No. 1310, Entry 89025 (1987).

022.114 Application of RIMS to the study of noble gases in meteorites.
G. Turner.
Resonance ionization spectroscopy 1986, p. 51 – 58 (1987). Abstr. in Phys. Abstr., Vol. 90, No. 1310, Entry 93451 (1987). – See Abstr. 012.047.

022.115 Radiation chemistry under unconventional conditions: dosimetry and aqueous radiolysis relevant to comet nuclei and early Earth structure.
I. G. Draganic, Z. D. Draganic.
Radiat. Phys. Chem., Vol. 29, No. 3, p. 227 – 230 (1987). Abstr. in Phys. Abstr., Vol. 90, No. 1310, Entry 94656 (1987).

022.116 Photoabsorption and photodissociation cross sections of CO between 88.5 and 115 nm.
C. Letzelter, M. Eidelsberg, F. Rostas, J. Breton, B. Thieblemont.
Chem. Phys., Vol. 114, No. 2, p. 273 – 288 (1987). Abstr. in Phys. Abstr., Vol. 90, No. 1312, Entry 103592 (1987).

022.117 Mass–independent isotopic fractionation in nonadiabatic molecular collisions.
J. J. Valentini.
J. Chem. Phys., Vol. 86, No. 12, p. 6757 – 6765 (1987). Abstr. in Phys. Abstr., Vol. 90, No. 1312, Entry 103597 (1987).

022.118 Inelastic electron scattering on strongly–excited atoms.
N. I. Rovenskaya.
Kinematika Fiz. Nebesn. Tel, Tom 3, No. 6, p. 68 – 76 (1987). In Russian. English translation in Kinematics Phys. Celest. Bodies.
Semi–classical transition probabilities, the cross–sections and rates of the inelastic scattering are determined by analytic methods for some asymptotic states of Rydberg's atoms ($n \gg 1$, $m/l \ll 1$, $m/l \sim 1$, $l/n \ll 1$, $l/n \sim 1$). Calculated effective collision widths of lines allow the detectable broadening of decameter recombination lines from the cold cosmic plasma components to be interpreted.

022.119 Effective collision strengths for fine–structure forbidden transitions in the $3p^3$ configuration of Ar IV.
C. J. Zeippen, K. Butler, J. Le Bourlot.
Astron. Astrophys., Vol. 188, No. 1, p. 251 – 257 (1987).
Effective collision strengths for all the fine–structure forbidden transitions within the $3p^3$ ground configuration of Ar IV have been calculated in a 7–state close–coupling approximation including configuration interaction in the target representation. A comparison with data available in the literature shows that the refinements used in the present work result in important changes, which should considerably affect astrophysical diagnostics, for example in nebulae. This is illustrated with the case of the electron density–dependent line intensity ratio $I(\lambda 4741.5)/I(\lambda 4712.7)$.

022.120 Electron impact polarization of resonance lines from lithium–like ions.
S. Chandra, A. W. Joshi.
Astrophys. Space Sci., Vol. 138, No. 2, p. 333 – 339 (1987).
The Oppenheimer–Penney theory, as developed by Percival and Seaton (1958), is applied to calculate the polarization of resonance lines from Li–like ions. Two laws for the pitch–angle distribution of electrons around the magnetic field are accounted. The degrees of polarization are averaged over the energy of non–thermal electrons generated during the initial phase of solar flares. It is found that for the full space pitch–angle distribution, as adopted by Chandra and Joshi (1984), the degrees of polarization are nearly independent of the atomic number of ion. Whereas for the forward–cone distribution used by Haug (1981), they depend on the choice of the free parameter E_0. The polarization of the resonance lines from Li–like ions is two times larger than that of the Lα radiations from H–like ions. Hence, under favourable conditions, it may be detected during solar flares.

022.121 Theoretical emission line ratios for Si XIII compared to solar observations.
S. M. McCann, F. P. Keenan.
Sol. Phys., Vol. 112, No. 1, p. 83 – 88 (1987).

022.122 The 1986 adjustment of the fundamental physical constants.
E. R. Cohen, B. N. Taylor.
Rev. Mod. Phys., Vol. 59, No. 4, p. 1121 – 1148 (1987).
The first "Recommended Consistent Values of the Fundamental Physical Constants" prepared by CODATA appeared in 1973 and was subsequently adopted by most international and national bodies in their own recommendations. This paper presents a revised version of these recommendations which takes into account the significant advances in metrology that have occurred since the 1973 analysis. It represents a 5–year effort involving experts from the major metrological laboratories of the world.

022.123 System of energy levels and wavelengths of 26757 lines of neutral iron Fe I (1622 – 99948 Å). Part 1: Introduction, energy levels, Grotrian diagram and description.
A. G. Gasanalizade.
Shemakhin. Astrofiz. Obs. Akad. Nauk AzSSR, Baku, 53 pp. (1987). In Russian. Abstr. in Ref. Zh., 51. Astron., 12.51.101 (1987).

022.124 System of energy levels and wavelengths of 26757 lines of neutral iron Fe I (1622 – 99948 Å). Part 2: Wavelengths from 1622 to 9486 Å.
A. G. Gasanalizade.
Shemakhin. Astrofiz. Obs. Akad. Nauk AzSSR, Baku, 245 pp. (1987). In Russian. Abstr. in Ref. Zh., 51. Astron., 12.51.102 (1987).

022.125 System of energy levels and wavelengths of 25757 lines of neutral iron Fe I (1622 – 99948 Å). Part 3: Wavelengths from 9486 to 99948 Å.
A. G. Gasanalizade.
Shemakhin. Astrofiz. Obs. Akad. Nauk AzSSR, Baku, 244 pp. (1987). In Russian. Abstr. in Ref. Zh., 51. Astron., 12.51.103 (1987).

022.126 Amorphous ice. A microporous solid: astrophysical implications.
E. Mayer, R. Pletzer.
J. Phys. Colloq., Vol. 48, No. C–1, p. 581 – 586 (1987). Abstr. in Phys. Abstr., Vol. 90, No. 1314, Entry 117819 (1987).

022.127 Erratum: "Ion–dipolar molecule rate coefficients" [Astrophys. J., Vol. 314, No. 2, p. 817 – 821 (1987)].
W. L. Morgan, D. R. Bates.
Astrophys. J., Vol. 321, No. 2, p. 1049 (1987). See Abstr. 43.022.101.

022.128 Simulation of cometary nuclei.
E. Grün, H. Kochan, K. Roessler, D. Stöffler.
ESA Spec. Publ., ESA SP–278, p. 501 – 508 (1987). – See Abstr. 012.075.
Simulation experiments of cometary nuclei are planned in the satellite test facility (Space Simulator) of the DFVLR. The program is directed to test cometary models on the base of the recent observations, to assist future cometary missions and to provide knowlege of sample handling at low temperatures. This paper reports the historical background, the activity status and the first results of the simulation experiments.

022.129 Different trapping mechanisms of gases by water ice and their relevance for comet nuclei.
B. Schmitt, J. Klinger.
ESA Spec. Publ., ESA SP–278, p. 613 – 619 (1987). – See Abstr. 012.075.
The large variability of the CO productions compared to the fairly constant CN/H_2O, C_2/CN and C_3/CN ratios may have its origin in the evolution processes of the nucleus. In order to test this possibility the authors try to experimentally study all the physical processes that may occur inside a comet nucleus.

022.130   Laboratory amorphous carbon: a possible analogue of cometary dust.
A. Blanco, A. Borghesi, S. Fonti, V. Orofino, E. Bussoletti, L. Colangeli.
ESA Spec. Publ., ESA SP–278, p. 677 – 679 (1987). – See Abstr. 012.075.

Spacecraft and ground based observations of Halley's comet in the near IR have revealed an emission band around 3.4 μm. Laboratory samples of hydrogenated amorphous carbon grains (HAC) (diameter $\cong$ 80 Å) have shown a similar feature at 3.4 μm. In this paper the authors discuss the possibility that the band can be attributed to small HAC particles.

022.131   Physical–mechanical properties of matrixes on the comet nuclei surface models.
Kh. I. Ibadinov, S. A. Aliev, A. A. Rahmonov (*A. A. Rakhmonov*).
ESA Spec. Publ., ESA SP–278, p. 713 – 716 (1987). – See Abstr. 012.075.

On purpose of investigation of physical–mechanical properties of comet ice nucleus surfaces matrix the authors carried out laboratory experiments with comet nuclei models under the conditions similar to natural ones. The present paper describes some results.

022.132   Sublimation characteristics of H_2O comet nucleus with CO_2 impurities.
Kh. I. Ibadinov, S. A. Aliev.
ESA Spec. Publ., ESA SP–278, p. 717 – 719 (1987). – See Abstr. 012.075.

By the method of laboratory modelling the sublimation of H_2O ice nucleus model containing up to 4.4% CO_2 was studied. It was found that the destruction of the model surface takes place in the form of microexplosions with ejection of ice particles up to 1–2 mm in size. It is concluded that under comet nuclear surface layer conditions the CO_2 concentration in H_2O ices could not be more than some per cents.

022.133   Spectral properties of plagioclase and pyroxene mixtures and the interpretation of lunar soil spectra.
D. A. Crown, C. M. Pieters.
Icarus, Vol. 72, No. 3, p. 492 – 506 (1987).

Several laboratory experiments concerning the spectral properties of terrestrial minerals and mineral mixtures which bear on the interpretation of remote lunar spectra are summarized here to place the present work in the context of recent studies. The spectral properties of lunar soils, the most widespread surface material on the Moon, are controlled by many variables. Examined here are the effects of particle size and mineral proportions, two of the most important of the many parameters which characterize particulate mixtures.

022.134   Undercooled water in basaltic regoliths and implications for fluidized debris flows on Mars.
J. L. Gooding.
Icarus, Vol. 72, No. 3, p. 519 – 527 (1987).

This paper summarizes an exploratory investigation of the freezing and melting of ice mixed with sand–sized soils from a natural basaltic regolith. Because of the previously proposed analogy between the cold–desert soils of Mauna Kea, Hawaii, and those on Mars (Ugolini 1976, Japp and Gooding 1980), soils representing the volcanic and glacial pedogenic processes on Mauna Kea were chosen for study. Results suggest that a Martian regolith composed of mechanically comminuted mafic or ultramafic rocks should permit short–term departures of water from its equilibrium behavior. Water–lubricated mass movements should be favored by the fact that most igneous minerals are relatively poor nucleators of ice so that liquid water can be substantially undercooled before freezing.

022.135   Rate constant for the reaction of atomic oxygen with phosphine at 298K.
L. J. Stief, W. A. Payne, D. F. Nava.
J. Chem. Phys., Vol. 87, No. 4, p. 2112 – 2115 (1987). Abstr. in Phys. Abstr., Vol. 90, No. 1316, Entry 132954 (1987).

022.136   Grain formation through the nucleation process in astrophysical environments. II. Nucleation and grain growth accompanied by chemical reaction.
T. Kozasa, H. Hasegawa.
Prog. Theor. Phys., Vol. 77, No. 6, p. 1402 – 1410 (1987). Abstr. in Phys. Abstr., Vol. 90, No. 1316, Entry 134129 (1987).

022.137   Cyclic and linear isomers of $C_3H_2^+$ and $C_3H_3^+$. The $C_3H^+ + H_2$ reaction.
D. Smith, N. G. Adams.
Int. J. Mass Spectrom. Ion Processes, Vol. 76, No. 3, p. 307 – 317 (1987). Abstr. in Phys. Abstr., Vol. 90, No. 1317, Entry 138972 (1987).

022.138   Clustering reactions of H_2CN^+ ions with HCN.
B. K. Chatterjee, R. Johnsen.
J. Chem. Phys., Vol. 87, No. 4, p. 2399 – 2400 (1987). Abstr. in Phys. Abstr., Vol. 90, No. 1317, Entry 138976 (1987).

022.139   Rotational and vibration–rotational intensities of CS isotopes.
A. L. Pineiro, R. H. Tipping, C. Chackerian Jr.
J. Mol. Spectrosc., Vol.125, No. 1, p. 91 – 98 (1987). Abstr. in Phys. Abstr., Vol. 91, No. 1320, Entry 5898 (1988).

022.140   Emission bands of AlH ($X^1\Sigma^+$): the 2–0 sequence.
J. L. Deutsch, W. S. Neil, D. A. Ramsay.
J. Mol. Spectrosc., Vol.125, No. 1, p. 115 – 121 (1987). Abstr. in Phys. Abstr., Vol. 91, No. 1320, Entry 5945 (1988).

022.141   Thermally activated release of stored chemical energy in cryogenic media.
J. M. Carpenter.
Nature, Vol. 330, No. 6146, p. 358 – 360 (1987).

The author has observed very rapid spontaneous heating of the cold solid methane moderator when irradiated by fast neutrons at low temperatures in the Intense Pulsed Neutron Source. An explanation for these observations is that energy stored in the form of reactive species (H, CH_3,...) accumulates in solid methane irradiated at low temperatures, and is released by thermally activated diffusion and subsequent reaction. Similar effects may occur in other cryogenic moderator materials, and may be responsible for the jets observed from cometary nuclei.

022.142   Vapour pressure of amorphous H_2O ice and its astrophysical implications.
A. Kouchi.
Nature, Vol. 330, No. 6148, p. 550 – 552 (1987).

Because an understanding of the physical properties of vapour–deposited amorphous H_2O, I_{as}, is important for the discussion of the evolution of the Solar System, they have been extensively studied. The author presents the results of experiments demonstrating that the vapour pressure of I_{as} is one or two orders of magnitude larger than that of crystalline H_2O and depends greatly on the condensation temperature and the rate of condensation. He also discusses the evaporation of I_{as} in space.

022.143   Static compression of H_2O–ice to 128 GPa (1.28 Mbar).
R. J. Hemley, A. P. Jephcoat, H. K. Mao, C. S. Zha, L. W. Finger, D. E. Cox.
Nature, Vol. 330, No. 6150, p. 737 – 740 (1987).

The high pressure behaviour of H_2O is of fundamental importance in both condensed matter and planetary physics. The authors have compressed ice in a diamond–anvil cell to 128 GPa and measured the molar volume as a function of pressure by synchrotron X–ray diffraction techniques. The measured equation of state indicates that ice is less compressible at very high pressures than is suggested by recent experiments in the

30 – 50 GPa range, but more compressible than statistical electron and recent pair–potential models predict.

022.144 **Laboratory spectroscopy of astrophysically interesting molecules.**
S. P. Davis.
Publ. Astron. Soc. Pac., Vol. 99, No. 621, p. 1105 – 1114 (1987).
= Invited paper presented at the Symposium on cool stars and the structure of galaxies, at the 99th Annual Scientific Meeting of the Astronomical Society of the Pacific, at Pomona College, Claremont, California, July 1987.
A systematic program on the laboratory analyses of selected molecular spectra, suggested by Charlotte M. Sitterly in 1956 and started in 1958 by Francis A. Jenkins and John G. Phillips, continues to the present day. The program has included tabulations of molecular spectra, analyses, calculations of molecular parameters, measurements of radiative lifetimes, and determinations of transition strengths. Work has been completed or is in progress on the spectra of ArH^+, CN, C_2, CS, CaCl, CaH, FeH, HgH, HgD, InI, OH, SH, Si_2, SiC_2, TiCl, TiO, TiO^+, ZrCl, ZrO, and ZrS.

022.145 **Laboratory analysis of the red and orange bands of CaCl.**
J. E. Littleton, S. P. Davis.
Publ. Astron. Soc. Pac., Vol. 99, No. 621, p. 1148 (1987).
Abstract. – See Abstr. 010.281.

022.146 **Absolute oscillator strengths for 108 lines of Si I between 163 and 410 nanometers.**
P. L. Smith, M. C. E. Huber, G. P. Tozzi, H. E. Griesinger, B. L. Cardon, G. G. Lombardi.
Astrophys. J., Vol. 322, No. 1, p. 573 – 583 (1987).
Measurements of neutral silicon oscillator strengths (f–values) made by both absorption and emission techniques have been combined, using the technique of Cardon, Smith, and Whaling, to produce 108 f–values for Si I lines between 163 and 410 nm. The absolute scale was established by referring to beam–foil–lifetime measurements. The measurements have uncertainties of about 0.07 dex ($\pm 16\%$) at the 1 σ level of confidence.

022.147 **The near–infrared spectrum of the FeH molecule.**
J. G. Phillips, S. P. Davis, B. Lindgren, W. J. Balfour.
Astrophys. J., Suppl. Ser., Vol. 65, No. 4, p. 721 – 778 (1987).
A rotational analysis has been carried out on seven bands of the $^4\Delta$–$^4\Delta$ system of the FeH molecule in the 7786–13400 Å region. Molecular constants, term values, spin splittings, and Λ–type doublings have been derived. Calculated wavelengths and wavenumbers are presented for two additional bands with heads at 10639 and 14927 Å. Finding lists of lines in the observed bands are provided to aid in astronomical applications.

022.148 **Ionization equilibrium of atomic hydrogen in a strong magnetic field.**
V. K. Khersonskij.
Sov. Astron., Vol. 31, No. 2, p. 225 – 227 (1987). English translation of 43.022.057.

022.149 **Spectral parameters of the interstellar molecule H_3O^+.**
V. K. Khersonskij, D. A. Varshalovich.
Astrofizika, Tom 27, Vyp. 2, p. 325 – 334 (1987). In Russian. English translation in Astrophysics, Vol. 27, No. 2.
The line strengths and the probabilities of inversion–rotational transitions of the hydroxonium ion H_3O^+ are calculated. Ex021cped optical thicknesses of typical interstellar clouds are estimated for H_3O^+ lines of radio astronomical interest ($\lambda = 0.8 - 1.0$ mm).

022.150 **Vibration–rotational intensities for the $X\,^1\Sigma^+$ state of AlH and AlD.**
R. H. Tipping, A. Lopez Piñeiro, C. Chackerian Jr.
Astrophys. J., Vol. 323, No. 2, p. 810 – 813 (1987).
Vibration–rotational line strengths (Einstein A–coefficients) for the ground electronic state ($X\,^1\Sigma^+$) of AlH and AlD have been calculated for the following sets of vibrational and rotational quantum numbers: $0 \leqslant \Delta v \leqslant 5$ for $0 \leqslant v' \leqslant 15$, and $J' - J = \pm 1$ for $0 \leqslant J' \leqslant 50$. Representative numerical results are presented, and the possibility of observing these transitions in stellar spectra is briefly discussed.

022.151 **Laboratory detection of HC_3NH^+ by infrared difference frequency laser spectroscopy.**
S. K. Lee, T. Amano.
Astrophys. J., Lett. Ed., Vol. 323, No. 2, p. L145 – L148 (1987).
The v_1 fundamental band (N–H stretch) of protonated cyanoacetylene HC_3NH^+ has been observed in absorption for the first time with a difference frequency laser as a radiation source around 3.1 μm. The ion was generated in a modulated hollow cathode discharge through a gas mixture of H_2 and HC_3N. The molecular constants in the ground state as well as in the excited state have been determined. The predicted rotational transition frequencies will enable searches to be made for this species in interstellar space.

022.152 **Laboratory microwave spectroscopy of the vibrational satellites for the v_7 and $2v_7$ states of C_4H and their astronomical identification.**
S. Yamamoto, S. Saito, M. Guélin, J. Cernicharo, H. Suzuki, M. Ohishi.
Astrophys. J., Lett. Ed., Vol. 323, No. 2, p. L149 – L153 (1987).
The rotational transitions of the C_4H radical ($X^2\Sigma^+$) in the lowest bending state, v_7, and its overtone state [$2v_7, l = 0(\Sigma)$ and $2(\Delta)$] were observed by laboratory microwave spectroscopy and were analyzed to determine the molecular constants. Based on the observed and calculated frequencies, 22 previously unidentified lines in IRC +10216 are assigned to transitions in the v_7 state. Furthermore, the series of line doublets recently detected by Guélin et al. toward IRC +10216 is definitely assigned to transitions in the $2v_7(\Sigma)$ state.

One hundred and fifty–three diatomic molecules, molecular ions, and radicals of astrophysical interest.
See Abstr. 002.006.

An atlas of the thorium–argon spectrum for the ESO Echelle Spectrograph in the $\lambda\lambda 3400 - 9000$ Å region.
See Abstr. 002.062.

Opacity Project: astrophysical and fusion applications.
See Abstr. 013.024.

Effects of low–frequency instability on the Hall conductivity in plasma: applications to astrophysics and space physics.
See Abstr. 062.022.

Si IV line ratios in laboratory plasmas: a comparison of experimental data and theoretical computations.
See Abstr. 062.057.

Optimization of detector arrays for measuring the plasma distribution function in space.
See Abstr. 062.058.

Compositional information for the moon: some characteristics of current near–IR spectra (telescopic and laboratory).
See Abstr. 094.007.

Populations of the rotational levels of molecules in clouds with large redshifts.
See Abstr. 131.024.

How abundant are complex interstellar molecules?
See Abstr. 131.048.

Infrared emission from interstellar PAHs.
See Abstr. 131.102.

Measurements of the fragmentation cross sections of relativistic heavy nuclei and their application to cosmic ray propagation.
See Abstr. 144.035.

Dark matter in the Universe ... and in the laboratory.
See Abstr. 161.357.

Astronomical Instruments and Techniques

031 Astronomical Optics

031.001 Principes de polissage et de contrôle des miroirs d'amateurs. II.
G. Philippon.
Obs. Trav., No. 11, p. 23 – 48 (1987).

031.002 F/1 camera for a spectrograph.
K. Nariai, Y. Yamashita.
Publ. Astron. Soc. Jpn., Vol. 39, No. 3, p. 505 – 516 (1987).
A two–mirror system with a corrector is studied. Aspheric coefficients of surfaces below the sixth order are determined on the basis of the aberration theory. Ray traces are used for the determination of the coefficients of the eighth or higher order. Numerical data are given for the systems with $f = 100$ mm and $F/1$.

031.003 Ritchey–Chrétien system with a pre–focal achromatic meniscus for a wide spectral range.
G. M. Popov.
Bull. Crimean Astrophys. Obs., Vol. 74, p. 150 – 159 (1987). English translation of 41.031.057.

031.004 Mirror fabrication and metrology for ROSAT.
K. Beckstette.
Rutherford Appleton Lab., Rep., RAL 87–043, p. 118 – 126 (1987). – See Abstr. 012.010.

031.005 Multilayer mirrors for X rays and the extreme UV.
J. H. Underwood.
Rutherford Appleton Lab., Rep., RAL 87–043, p. 127 (1987). Abstract. – See Abstr. 012.010.
This paper has already appeared in Optics News, Vol. 12, No. 3.

031.006 Progress with LSM optics for solar observations within the French space program.
J. P. Delaboudiniere.
Rutherford Appleton Lab., Rep., RAL 87–043, p. 133 – 137 (1987). – See Abstr. 012.010.
Development of Layered Synthetic Microstructures for XUV optics has begun a few years ago in several French laboratories. The author's laboratory has been concerned with potential applications of this new technique to observations of the solar corona in the EUV wavelength range with normal incidence optical systems. This activity was recently bolstered by perspectives opened by the projected ESA solar mission SOHO.

031.007 Platinum – carbon multilayer X–ray reflectors.
B. J. Kent, B. L. Evans.
Rutherford Appleton Lab., Rep., RAL 87–043, p. 138 – 142 (1987). With 10 figures. – See Abstr. 012.010.
The authors have shown that the modulated ion beam sputtering system using an inert gas is capable of the precision control necessary to produce commensurate multilayer structures. Precision control over the deposition parameters is essential in order to avoid thickness errors which will ultimately cause loss of stack periodicity and consequent loss of performance.

031.008 Monochromatization by multilayered optics on a cylindrical reflector and on an ellipsoidal focusing ring.
G. F. Marshall.
Rutherford Appleton Lab., Rep., RAL 87–043, p. 143 – 153 (1987). = Opt. Eng., Vol. 25, No. 8, p. 922 – 932 (1986). – See Abstr. 012.010.
The isoangle and isobragg are new concepts used in mapping loci of incident and reflected rays associated with a dispersive reflector. An isoangle is a line joining the points of incidence at which the angle of incidence of rays from a point source is the same. An isobragg is the line joining the points of intersection at any surface by rays of the same wavelength that have been reflected according to Bragg's law. Their patterns provide an understanding of why and where we use dispersive multilayered optics in X–ray monochromators and spectrometers with Rowland circle geometry. They have led to the design of an X–ray focusing ring for electron spectroscopy for chemical analysis instrumentation and the possibility of a design for a scanning X–ray microscope.

031.009 The transmission electron microscopy of multilayers with specific reference to X–ray mirrors.
K. B. Alexander, C. S. Baxter, J. E. Evetts, R. E. Somekh, W. M. Stobbs.
Rutherford Appleton Lab., Rep., RAL 87–043, p. 154 – 160 (1987). – See Abstr. 012.010.

031.010 UHV DC magnetron sputtering of multilayers.
R. E. Somekh, K. A. Alexander, C. S. Baxter, W. M. Stobbs, J. E. Evetts.
Rutherford Appleton Lab., Rep., RAL 87–043, p. 161 – 166 (1987). With 4 figures. – See Abstr. 012.010.

031.011 Soft X–ray optics at King's College London.
A. G. Michette.
Rutherford Appleton Lab., Rep., RAL 87–043, p. 167 – 171 (1987). With 5 figures. – See Abstr. 012.010.

031.012 Multilayers – their characterisation and use with synchrotron radiation.
M. P. Bruijn, G. van der Laan, J. B. Goedkoop, A. A. MacDowell.
Rutherford Appleton Lab., Rep., RAL 87–043, p. 172 – 192 (1987). With 9 figures. – See Abstr. 012.010.
The authors report results of characterization of a double "crystal" monochromator for synchrotron radiation in the region of 200 – 900 eV in which two Ni/C multilayer coatings (2d = 62 Å) are used. The energy resolution is measured to be about 1/75 by reflection measurements with a KAP crystal and by recording absorption edges and "white lines" of 3d–transition metals. The potential of multilayer coatings is further illustrated by measurements of reflection profiles of other ReW/C coatings. The authors also report results on an approach in which a double "crystal" monochromator is equipped with an organic crystal (RbAP) preceeded by a single multilayer with a layer thickness matched to the 2d–spacing of the organic crystal. The use of multilayers in other concepts to monochromate synchrotron radiation is also discussed.

031.013 The precision machining of grazing incidence X–ray mirror substrates.
W. J. Wills–Moren.
Rutherford Appleton Lab., Rep., RAL 87–043, p. 202 – 213 (1987). – See Abstr. 012.010.
A brief description is given of the successful development of a large computer controlled machine for diamond turning aluminium alloy X–ray mirror substrates (1.5 m dia., 600 mm axial length). Examples are based on the ROSAT and SXT mirrors for which the axial and radial location/mounting surfaces are produced with precise reference to the reflecting surfaces.

031.014 The production of super–smooth surfaces by dip–coating using acrylic lacquers.
L. Jalota, R. Willingale.
Rutherford Appleton Lab., Rep., RAL 87–043, p. 214 – 222 (1987). With 7 figures. – See Abstr. 012.010.
Research at Leicester University into the use of acrylic lacquers to produce super–smooth surfaces for X–ray optics is reported. The coating technique, substrates chosen, and use of metrology are discussed. Comparison with calibrated test flats show very smooth finishes with r.m.s. surface heights of 10 Å or better over lengths of a few hundred microns only.

031.015 Replicated X–ray mirrors for a multi–mirror array.
D. K. Bedford, W. R. Purcell, G. M. Simnett, M. P. Ulmer, Y. Matsui.
Rutherford Appleton Lab., Rep., RAL 87–043, p. 223 – 230 (1987). With 9 figures. – See Abstr. 012.010.
Electroformed replica Wolter I mirrors have been developed for use in X–ray astronomical telescopes. Nested mirror shells, two at present but with a provision for six, have been evaluated with white light and with monochromatic X–rays in the 1 – 8 keV band. An angular resolution of <20 arcsec FWHM and reflectivities of 50 – 88% (1 – 8 keV) have been achieved with development mirrors. Preliminary vibration testing has proved the mechanical integrity of the mirror shells when suitably mounted.

031.016 Computer–controlled polishing machine for supersmooth aspheric surfaces.
G. Doughty, J. Smith.
Rutherford Appleton Lab., Rep., RAL 87–043, p. 231 – 240 (1987). – See Abstr. 012.010.
A new polishing machine was developed. Colloidal silica was used to polish lithium niobate surfaces with aspheric departure of 77 μm from the closest sphere of 10 mm radius of curvature. Their profiles were maintained within 0.5 μm of diamond turning. Surface finish was better than 1 nm Ra, and sub–surface damage was low enough to permit the formation of optical waveguides (by titanium–diffusion) whose main losses were due to curvature rather than scatter.

031.017 A review of lightweight mirror technology for large space observatories.
M. A. Cutter.
Rutherford Appleton Lab., Rep., RAL 87–043, p. 241 – 244 (1987). – See Abstr. 012.010.
This paper briefly summarises the manufacturing techniques and concepts for generating low density mirror blanks, discusses the applicability of various substrate materials and summarises the findings and conclusions of the study.

031.018 The mirror material test programme at the Daresbury SRS.
A. A. MacDowell, J. B. West, T. Koide.
Rutherford Appleton Lab., Rep., RAL 87–043, p. 245 – 253 (1987). With 3 figures. – See Abstr. 012.010.
A mirror test programme, in collaboration with the Photon Factory, Japan, has been under way on the Synchrotron Radiation Source at Daresbury Laboratory for two years. Many samples, ranging from glasses, through ceramics to metals such as molybdenum, have been exposed to radiation densities up to 70 watts/sq.cm., and long term changes in their optical properties

measured. Considerable effort has been devoted towards developing a cooled mount satisfactory for use in an ultra high vacuum environment. Recommendations are made for the most suitable materials for future use. Finally, a programme for more detailed measurements on the effect of high density heating and radiation on the optical performance of mirrors is described.

031.019 The metrology of X–ray optical components: mapping the limits of measuring instruments.
M. Stedman.
Rutherford Appleton Lab., Rep., RAL 87–043, p. 254 – 258 (1987). With 5 figures. – See Abstr. 012.010.

031.020 Mirror surface microripple: its generation, measurement and avoidance.
K. Lindsey.
Rutherford Appleton Lab., Rep., RAL 87–043, p. 259 – 264 (1987). – See Abstr. 012.010.

031.021 Measurement of surface roughness of mirrored surfaces, using a triple–axis perfect–crystal X–ray diffractometer.
F. E. Christensen, A. Hornstrup, H. W. Schnopper.
Rutherford Appleton Lab., Rep., RAL 87–043, p. 265 – 272 (1987). With 15 figures. – See Abstr. 012.010.
X–ray scattering measurements and integrated reflectivity measurements from mirrored surfaces are presented.

031.022 Diffraction images for aberration–free objectives with annular apertures.
G. M. Popov, N. V. Steshenko, M. B. Popova.
Bull. Crimean Astrophys. Obs., Vol. 75, p. 174 – 184 (1987). English translation of 42.031.062.

031.023 Wide–field conversions for reflecting telescopes.
R. V. Willstrop.
Mon. Not. R. Astron. Soc., Vol. 229, No. 1, p. 143 – 155 (1987).
Auxiliary optics have been designed which would give the Anglo–Australian Telescope a field 3° in diameter, suitable for multiple–object spectroscopy with optical fibres. A new secondary mirror and an achromatic aspheric corrector are required. These could be built relatively quickly and cheaply. An alternative design using two new mirrors (Paul–Baker system) would give smaller images, but only over a field of 2°, and there would be severe practical difficulties. For the William Herschel Telescope the first design must be modified to include two corrector plates, but both would need to be deeply aspheric, and the field is limited to 2°.

031.024 The first images from optical aperture synthesis.
C. A. Haniff, C. D. Mackay, D. J. Titterington, D. Sivia, J. E. Baldwin, P. J. Warner.
Nature, Vol. 328, No. 6132, p. 694 – 696 (1987).
The attainment of diffraction–limited images with large ground–based optical telescopes is an important objective for many astronomical programmes. The limited resolution set by atmospheric fluctuations in refractive index can be overcome by applying aperture synthesis and phase–closure techniques to short–exposure images taken through non–redundant aperture masks. The first optical images obtained with this technique are presented here. The diffraction limit is achieved with good dynamic range even when the number of photons in each short exposure is small ($\ll 100$).

031.025 Surface finish measurements of diamond–turned electroless–nickel–plated mirrors.
J. S. Taylor, C. K. Syn, T. T. Saito, R. R. Donaldson.
Proc. SPIE Int. Soc. Opt. Eng., Vol. 571, p. 10 – 21 (1986). – See Abstr. 43.012.094.
Surface roughness data are presented for a matrix of diamond–turned electroless–nickel samples having a combination of six phosphorous contents and four heat treatments. The results are discussed in terms of the material composition and heat treatment, plus other factors having an observed influence on the surface roughness.

031.026 Aspheric figure generation using feedback from an infrared phase–shifting interferometer.
H. P. Stahl, D. Ketelsen.
Proc. SPIE Int. Soc. Opt. Eng., Vol. 571, p. 22 – 29 (1986). – See Abstr. 43.012.094.
This paper discusses the usefulness of the infrared phase–shifting interferometric system for providing figure correcting feedback to the optician during the generation of the off–axis parabolic segments and how it is affected by the surface roughness produced by each generator tool.

031.027 Proposed method of producing large optical mirrors: single point diamond crushing followed by polishing with a small area tool.
G. Wright, J. B. Bryan.
Proc. SPIE Int. Soc. Opt. Eng., Vol. 571, p. 30 – 34 (1986). – See Abstr. 43.012.094.
Improved production of large optical mirrors may result from combining single point diamond crushing of the glass with polishing using a small area tool to smooth the surface and remove the damaged layer. Diamond crushing allows an accurate surface contour to be generated and the small area polishing tool allows the surface roughness to be removed without destroying the contour.

031.028 Parameters for large optics generation.
R. E. Parks.
Proc. SPIE Int. Soc. Opt. Eng., Vol. 571, p. 35 – 38 (1986). – See Abstr. 43.012.094.
The ability to generate aspheric components to micrometer tolerances while leaving a surface relatively free of subsurface damage is quite recent. This paper describes the optimum parameters for generating high–quality aspherics.

031.029 Experiences in the precision machining of grazing incidence X–ray mirror substrates.
P. A. McKeown, R. F. J. Read, W. J. Wills–Moren.
Proc. SPIE Int. Soc. Opt. Eng., Vol. 571, p. 42 – 50 (1986). – See Abstr. 43.012.094.
A brief description is given of the successful development of a large computer controlled machine for diamond turning aluminium alloy X–ray mirror substrates. The in situ metrology facility for axial profile, roundness and diameter measurement is described, together with the technique for automatic programming of the final tool path to compensate for errors of workpiece deflection, tool setting and edge radius, etc. Examples are based on the ROSAT and SXT mirrors for which the axial and radial location/mounting surfaces are produced with precise reference to the reflecting surfaces.

031.030 Lightweight large mirror blanks of Zerodur.
A. J. Marker III, H. Fuhrmann, H. Tietze, W. Froehlich.
Proc. SPIE Int. Soc. Opt. Eng., Vol. 571, p. 51 – 59 (1986). – See Abstr. 43.012.094.
The goal of this article is to describe the manufacture of lightweighted, 4 m diameter mirror blanks made of Zerodur in a fast and economical manner. Basic experiments have been performed involving production techniques and processes which may be usable to manufacture lightweighted mirrors in a mass–production format.

031.031 Design aspects of future very large telescopes.
A. Y. S. Cheng.
Diss. Abstr. Int., Sect. B, Vol. 48, No. 1, p. 160–B (1987). Thesis, The University of Arizona, 246 pp. (1987). Order No. DA8709888.

031.032 Camera lenses for fast–speed spectrographs.
G. M. Popov, M. B. Popova.
Izv. Krymskoj Astrofiz. Obs., Tom 76, p. 200 – 203 (1987). In Russian. English translation in Bull. Crimean Astrophys. Obs., Vol. 76.
Two samples of simple fast–speed lenses are designed. Field of view is flat, lenses are expected to cover a 12°field. Lenses consist of four rear lenses and two rear lenses are cemented to reduce light loss and scattering. There is no vignetting in these lenses. All surfaces are spherical that is why it is easy to make lenses. These lenses may be used in fast–speed spectrographs, for changing focal length, to observe with color glasses and for meteor observations with prism installed in front of lens.

031.033 Modulation–transfer function of an absolute ring astronomical telescope.
V. K. Prokof'ev.
Izv. Krymskoj Astrofiz. Obs., Tom 77, p. 166 – 171 (1987). In Russian. English translation in Bull. Crimean Astrophys. Obs., Vol. 77.
The structure of the modulation transfer function is analyzed for ring nonfast astronomical telescopes at large values of central vignetting $\varepsilon \geqslant 0.5$. The diffraction luminosity distribution has been adopted as intitial, which was thoroughly investigated in a recent paper. It is shown that a high modulation transfer function in a low frequency band corresponds to a central group maximum of diffraction distribution.

031.034 Fast systems composed of four spherical mirrors with external focus.
G. M. Popov, M. B. Popova.
Izv. Krymskoj Astrofiz. Obs., Tom 77, p. 172 – 178 (1987). In Russian. English translation in Bull. Crimean Astrophys. Obs., Vol. 77.
The systems composed of four spherical mirrors, the first and the fourth being concave, the second and the third–convex, are discussed. A method of computation for such systems has been proposed, which reduces the calculation to that of a precise equation. As an example the authors present the computation of particular systems with relative aperture up to 1:0.85 at the field–of–view angle 1.5 – 4°. Relative to the flux the field–of–view is concave. Such systems can be implemented either for ground–based, or for space observations within a wide spectral range.

031.035 Degeneracy in the Fraunhofer diffraction of truncated Gaussian beams.
Y. Li.
J. Opt. Soc. Am. A, Vol. 4, No. 7, p. 1237 – 1242 (1987).
The variations of the Fraunhofer diffraction pattern formed by a Gaussian beam passing through a circular aperture are described. It is shown that a discontinuous process of several steps occurs as the beam truncation by the aperture decreases. This sequence repeats continuously and finally leads to the degeneration of the Airy pattern into a single spot, i.e., the Gaussian pattern corresponding to the Fraunhofer diffraction of a nontruncated Gaussian beam.

031.036 Analysis of diffraction reduction by use of a Lyot stop.
B. R. Johnson.
J. Opt. Soc. Am. A, Vol. 4, No. 8, p. 1376 – 1384 (1987).
The efficiency of a Lyot stop in reducing diffraction effects in a simple optical system is analyzed. The result is a simple formula for calculating the diffraction–reduction efficiency of the Lyot stop. A comparison of this work with other published work is also presented.

031.037 Third–order design of refractive Offner compensators.
J. R. Moya, J. E. A. Landgrave.
Appl. Opt., Vol. 26, No. 13, p. 2667 – 2672 (1987).
From first–order optics and the condition for zero third–order spherical aberration, a simple set of equations can be derived to produce a good preliminary design of a refractive Offner compensator. An advantage of these formulas is that they provide a design that satisfies specified values for the diameter of

the compensating lens and the focal ratio of the cone of observation. An example is given to show how the formulas are used.

031.038   Dummy lens for the computer optimization of autostigmatic null correctors.
J. E. A. Landgrave, J. R. Moya.
Appl. Opt., Vol. 26, No. 13, p. 2673 – 2675 (1987).
The computer optimization of an autostigmatic null corrector can be simplified by using a plano–convex lens as a dummy element. The lens is placed in contact with the aspheric surface and rays traced from star space. The exact shape of the convex surface is found parametrically for an arbitrary aspheric surface or as an accurate approximation for a conicoid with a focal ratio > 2.

031.039   Bessel annular apodizers: imaging characteristics.
J. Ojeda–Castañeda, L. R. Berriel–Valdos, E. Montes.
Appl. Opt., Vol. 26, No. 14, p. 2770 – 2772 (1987).
Apodizers with relatively high transmittance over an annular region of the exit pupil can reduce the sensitivity to defocusing and to spherical aberration. The authors analyze the imaging properties (pupil functions, point spread functions, optical transfer functions, and Strehl ratios) of the Bessel type of annular apodizers. They also show some computer–simulated images, obtained with and without this kind of annular apodizer.

031.040   Heterodyne Fizeau interferometer for testing flat surfaces.
T. H. Barnes.
Appl. Opt., Vol. 26, No. 14, p. 2804 – 2809 (1987).
A heterodyne Fizeau interferometer, which uses a rotating radial grating to achieve the required optical frequency shift, is described. An analysis of the effects of grating ruling errors shows that they may be nearly eliminated by averaging the interferometer phase readings over integral numbers of grating revolutions. Experimental tests indicate that the interferometer is capable of measuring with a reproducibility of $\lambda/200$ (λ = 632.8 nm), limited by temperature effects.

031.041   Numerical design method for aberration–reduced concave grating spectrometers.
W. R. McKinney, C. Palmer.
Appl. Opt., Vol. 26, No. 15, p. 3108 – 3118 (1987).
A general method is described for the design of single concave grating optical systems. Proper parameter and variable sets are defined, and a straightforward technique leads to a final set of optimized variable values which minimize a merit function based on spectroscopic performance.

031.042   Adaptive optics for optimization of image resolution.
J. P. Gaffard, C. Boyer.
Appl. Opt., Vol. 26, No. 18, p. 3772 – 3777 (1987).
Adaptive optics can be used to improve the quality of optical imaging instruments otherwise limited in resolution by atmospheric turbulence. The authors have calculated the mean value of the optical transfer function (OTF) of an instrument corrected by adaptive optics using the OTF as a criterion for resolution optimization and describing perturbations through their phase structure function. Assuming that the optics diameter and the perturbation Fried's parameter are known, it then becomes possible to optimize the system by a proper choice of the number of actuators and of the influence diameter, a parameter which characterizes the size of the required corrections.

031.043   Diffracted electromagnetic fields in the neighborhood of the focus of a paraboloidal mirror having a central obscuration.
R. Barakat.
Appl. Opt., Vol. 26, No. 18, p. 3790 – 3795 (1987).
The structure of the diffracted electromagnetic field generated by a paraboloidal mirror having a possible central obscuration without wavefront aberrations, but allowing for defocus, is studied. Contour plots are shown; as expected they are not rotationally symmetric. Finally, the author introduces and numerically evaluates the encircled time average electric energy density.

031.044   Considerations on balloon–borne far infrared telescopes.
L. Piccirillo, A. Moleti, S. Masi.
Infrared Phys., Vol. 27, No. 4, p. 215 – 225 (1987).
The design of a telescope devoted to infrared and millimetric differential measurements is described. The optics are well corrected for aberrations, either in the on–axis or in the off–axis configurations, for tilt angles ~ 1° of the secondary mirror. The diffraction pattern at large angles is calculated as a function of the surface quality and the in flight background at balloon altitude is estimated.

031.045   Meniscus–Gregorian telescope.
S.–j. Yang.
Publ. Purple Mt. Obs., Vol. 6, No. 2, p. 151 – 155 (1987). In Chinese.

031.046   Aluminizing a 60 cm mirror by using a DM–700A type vacuum coating machine.
Y.–b. Guo.
Publ. Purple Mt. Obs., Vol. 6, No. 2, p. 196 – 199 (1987). In Chinese.

031.047   Preliminary study on decentring errors and optical alignment of reflecting telescopes (Correction of decentring coma and residual astigmatism).
Y. Yamashita, M. Nakagiri.
Tokyo Astron. Obs. Rep., Vol. 21, No. 1, p. 14 – 32 (1987). In Japanese.

031.048   An experimental casting of mirror blanks with honeycomb structure.
N. Ohshima, S. Isobe, H. Kamo.
Tokyo Astron. Obs. Rep., Vol. 21, No. 1, p. 33 – 49 (1987). In Japanese.

031.049   Correctors for the primary focus of 7.5 m telescope.
K. Nariai, Y. Yamashita, M. Nakagiri.
Ann. Tokyo Astron. Obs., Second Ser., Vol. 21, No. 4, p. 357 – 362 (1987).
Three lens correctors with 10 mm plane–parallel plate for 7.5 m main mirror are optimized for the blue– and the red wavelength regions. Aspheric coefficients A_6 and A_8 of the main mirror are also used in the optimization of the blue corrector. The Ritchey–Chretien secondary mirror that corresponds to this main mirror is also presented.

031.050   New types of interferometers for control of astronomical and space optics.
E. G. Popov.
Tr. Astrofiz. Inst. Alma–Ata, Tom 48, p. 117 – 122 (1987). In Russian.

031.051   Taking into account flexures in the calculation of optical systems.
E. G. Popov, B. A. Bulibekov.
Tr. Astrofiz. Inst. Alma–Ata, Tom 48, p. 123 – 128 (1987). In Russian.

031.052   Bowen's system and its modification.
G. M. Popov, B. A. Bulibekov.
Tr. Astrofiz. Inst. Alma–Ata, Tom 48, p. 136 – 148 (1987). In Russian.

031.053   The Féry spectrograph updated.
E. E. Mendoza V., C. X. Mendoza G.
Rev. Mex. Astron. Astrofis., Vol. 14, No. 2, p. 763 – 765 (1987). – See Abstr. 012.042.
The Féry spectrograph uses a single quartz optical unit that combines the characteristics of prism, mirror, and lenses. Hence,

it presents its greatest advantage in the ultraviolet, where loss of light is minimized by the small number of air–quartz surfaces and the rather small thickness of optical material used. A few attractive astronomical applications are described, as well as how to overcome its vertical astigmatism.

031.054 Sistemas correctores de campo para el telescopio Casse-grain IAC80.
M. J. Galán, F. J. Cobos.
Rev. Mex. Astron. Astrofis., Vol. 14, No. 2, p. 767 (1987). Abstract. – See Abstr. 012.042.

031.055 Análisis de alteraciones en la imágen debidas a descoli-mación de un telescopio.
F. J. Cobos, M. J. Galán.
Rev. Mex. Astron. Astrofis., Vol. 14, No. 2, p. 768 (1987). Abstract. – See Abstr. 012.042.

031.056 Determinación de caracteristicas ópticas del telescopio OAN150.
M. J. Galán, F. J. Cobos.
Rev. Mex. Astron. Astrofis., Vol. 14, No. 2, p. 769 (1987). Abstract. – See Abstr. 012.042.

031.057 Sistemas correctores de campo para el telescopio Rit-chey–Chrétien UNAM212.
F. J. Cobos, M. J. Galán.
Rev. Mex. Astron. Astrofis., Vol. 14, No. 2, p. 770 (1987). Abstract. – See Abstr. 012.042.

031.058 ESA technological research activities on lightweight mirrors.
A. Connolly.
ESA Spec. Publ., ESA SP–273, p. 103 – 110 (1987). – See Abstr. 012.044.
Hardware research and development activities in ESA's Technological Research Programme relating to lightweight mirrors are described. These include programmes aimed at demonstrating the technology of imaging quality "semi-passive" normal incidence mirrors of up to 4 metres in diameter, and developing a replication technique for large scale production of grazing incidence mirrors. Also included is a recent survey of manufacturing capabilities in passive lightweight mirrors.

031.059 Grazing–incidence optics for synchrotron–radiation in-sertion–device beams.
V. Rehn.
Proc. SPIE Int. Soc. Opt. Eng., Vol. 640, p. 106 – 115 (1986). Abstr. in Phys. Abstr., Vol. 90, No. 1311, Entry 95706 (1987). – See Abstr. 012.048.

031.060 Automated figure formation for a Kirkpatrick–Baez X–ray mirror.
D. Fabricant, M. Conroy, L. Cohen, P. Gorenstein.
Proc. SPIE Int. Soc. Opt. Eng., Vol. 640, p. 164 – 169 (1986). Abstr. in Phys. Abstr., Vol. 90, No. 1311, Entry 95709 (1987). – See Abstr. 012.048.

031.061 Transverse ray aberrations of Wolter type 1 telescopes.
T. T. Saha.
Proc. SPIE Int. Soc. Opt. Eng., Vol. 640, p. 10 – 19 (1986). Abstr. in Phys. Abstr., Vol. 90, No. 1311, Entry 96945 (1987). – See Abstr. 012.048.

031.062 Design and development of conical X–ray imaging mirrors.
R. Petre, P. J. Serlemitsos.
Proc. SPIE Int. Soc. Opt. Eng., Vol. 640, p. 98 – 104 (1986). Abstr. in Phys. Abstr., Vol. 90, No. 1311, Entry 97116 (1987). – See Abstr. 012.048.

031.063 Simulation of free–abrasive grinding of grazing–inci-dence mirrors with vertical–honing and flexible blades.
F. D. Powell.
Proc. SPIE Int. Soc. Opt. Eng., Vol. 640, p. 85 – 90 (1986). Abstr. in Phys. Abstr., Vol. 90, No. 1311, Entry 97180 (1987). – See Abstr. 012.048.

031.064 Extreme Ultraviolet Explorer: mirror optical tests and results.
C. H. Gillespie, D. F. Edwards, M. A. Nichols, D. S. Finley.
Proc. SPIE Int. Soc. Opt. Eng., Vol. 640, p. 182 – 187 (1986). Abstr. in Phys. Abstr., Vol. 90, No. 1311, Entry 97183 (1987). – See Abstr. 012.048.

031.065 Active optics. I. A system for optimizing the optical quality and reducing the costs of large telescopes.
R. N. Wilson, F. Franza, L. Noethe.
J. Mod. Opt., Vol. 34, No. 4, p. 485 – 509 (1987). Abstr. in Phys. Abstr., Vol. 90, No. 1312, Entry 107720 (1987).

031.066 A new approach for phasing optical arrays.
K. P. Bechis, G. O. Sauermann, G. Megaloudis.
Proc. SPIE Int. Soc. Opt. Eng., Vol. 643, p. 25 – 34 (1986). Abstr. in Phys. Abstr., Vol. 90, No. 1313, Entry 110505 (1987). – See Abstr. 012.062.

031.067 A parametric study of various synthetic aperture tele-scope configurations for coherent imaging applications.
J. E. Harvey, A. B. Wissinger, A. N. Bunner.
Proc. SPIE Int. Soc. Opt. Eng., Vol. 643, p. 194 – 205 (1986). Abstr. in Phys. Abstr., Vol. 90, No. 1313, Entry 110511 (1987). – See Abstr. 012.062.

031.068 Survey of material for an infrared–opaque coating.
S. M. Smith, R. V. Howitt.
Proc. SPIE Int. Soc. Opt. Eng., Vol. 643, p. 53 – 62 (1986). Abstr. in Phys. Abstr., Vol. 90, No. 1313, Entry 110532 (1987). – See Abstr. 012.062.

031.069 Low mass mirrors in large optics.
W. W. Ernst.
Proc. SPIE Int. Soc. Opt. Eng., Vol. 655, p. 405 – 408 (1986). Abstr. in Phys. Abstr., Vol. 90, No. 1313, Entry 115244 (1987). – See Abstr. 012.063.

031.070 A scheme for measuring differential image motion (Proposal for LEST site testing).
P. N. Brandt.
LEST Found., Tech. Rep., No. 26, p. 67 – 69 (1987). – See Abstr. 012.071.
The principle of an extremely simple optical scheme is described which in combination with two linear diode arrays and a suitably programmed microprocessor allows the measurement of sub–arcsec solar image motion in conditions of severe telescope vibrations. Its merits and limitations are discussed briefly.

031.071 On an analysis of the optics of large telescopes.
Eh. A. Vitrichenko, L. A. Pushnoj.
Astron. Tsirk., No. 1476, p. 1 – 3 (1987). In Russian.
An analysis of the large telescope contour maps suggests that the main type of the optical surface errors is a nonsymmetrical one.

031.072 Testing of hyperbolic mirrors in the Ritchey autocolli-mated scheme.
N. N. Fashchevskij.
Astron. Tsirk., No. 1477, p. 4 – 5 (1987). In Russian.

031.073   A laser reflecting system for geophysical satellites.
S. K. Tatevyan, D. T. Matveev, I. K. Makhalov,
N. Georgiev, A. Hadjiiski (*A. Khadzhijkij*), G. Krystev.
Nauchn. Inf., Vyp. 58, p. 31 – 38 (1986). In Russian.
    The laser reflecting system was constructed in the Central
Laboratory of Geodesy of the Bulgarian Academy of Sciences.

031.074   Testing the primary mirror of the W. M. Keck Observatory.
E. Stryjewski, R. J. Zielinski, J. T. Smith.
Proc. SPIE Int. Soc. Opt. Eng., Vol. 680, p. 54 – 58 (1987). Abstr.
in Phys. Abstr., Vol. 91, No. 1320, Entry 6465 (1988). – See
Abstr. 012.078.

**031.075   Intensity distribution in out–of–focus images of a
rotationally symmetric optical system.**
S. N. Wong, R. E. Parks, L.–z. Shao.
Proc. SPIE Int. Soc. Opt. Eng., Vol. 680, p. 62 – 70 (1987). Abstr.
in Phys. Abstr., Vol. 91, No. 1320, Entry 6466 (1988). – See
Abstr. 012.078.

031.076   In–process metrology for X–ray optics.
A. F. Slomba, L. A. Montagnino.
Proc. SPIE Int. Soc. Opt. Eng., Vol. 676, p. 80 – 89 (1987). Abstr.
in Phys. Abstr., Vol. 91, No. 1320, Entry 6597 (1988).

**031.077   Adaptive optics for the European Very Large
Telescope (VLT).**
F. Merkle.
Proceedings of the international topical meeting on image
detection and quality, p. 251 – 254 (1986). Abstr. in Phys. Abstr.,
Vol. 91, No. 1320, Entry 10408 (1988). – See Abstr. 012.079.

031.078   Liquid mirror telescopes: present and future.
E. F. Borra.
Publ. Astron. Soc. Pac., Vol. 99, No. 621, p. 1229 – 1240 (1987).
    The work done with liquid mirrors at Laval University is
briefly reviewed. The concept appears sound. It is argued that
almost any type of astronomical observations could be carried
out with liquid mirror telescopes. As an example, it is suggested
how spectroscopy could be implemented on small and large

LMTs. Some technical considerations relevant to liquid mirror
telescopes are briefly discussed. It is argued that, because the
advent of very cheap zenith LMTs should decrease the cost per
photon by two orders of magnitude, it could bring about a major
increase to the global collecting area of optical telescopes.

031.079   Grinding of astronomical mirrors (V).
V. Přibyl.
Kozmos, Vol. 18, No. 5, p. 166 – 168 (1987). In Slovak.

The adjustment and testing of telescope objectives.
See Abstr. 003.134.

Adjustment of the prime focus optics of the GAT.
See Abstr. 032.016.

An update on the 15–m National New Technology telescope.
See Abstr. 032.046.

Design criteria, considerations, objective and tradeoffs.
See Abstr. 034.004.

ROSAT mirror system features of design, assembly and qualification.
See Abstr. 035.018.

XUV reflectivity measurements on the ROSAT WFC mirrors.
See Abstr. 035.019.

**Development of pure aluminium mirror surfaces for the Lyman far–
ultraviolet astronomical satellite mission.**
See Abstr. 035.020.

**Parametric study of stable optical references for large flexible
structures in space.**
See Abstr. 035.038.

On–orbit active alignment of the SOT observatory.
See Abstr. 035.039.

Geometrical aberration of a generalized Wolter type I telescope.
See Abstr. 035.055.

032 Astronomical Instruments

032.001   The spectroscopic survey telescope.
L. W. Ramsey, D. W. Weedman.
The SHIRSOG Workshop, p. 117 – 123 (1986). – See Abstr.
012.003.
    The authors describe the motivation for and general design of a
large special purpose spectroscopic survey telescope (SST). The
telescope is based on an array of 73 spherically figured segments
about a meter in diameter with a radius of curvature of 26 meters
making up a primary mirror with an effective aperture exceeding
7 meters. All tracking motions are in a lightweight focal surface
system that can track objects for up to 40 minutes using a two
element reflecting corrector for spherical aberration. The telescope has a fixed tilt to the vertical which allows access to a
48 degree strip of declination through an azimuth angle rotation.
Instrumentation is coupled to the focus by means of fiber optics.
Multiple object tracking over the 50 deg^2 field is possible.

032.002   The Multiple Object Spectroscopic Telescope.
C. A. Pilachowski.
The SHIRSOG Workshop, p. 136 – 137 (1986). – See Abstr.
012.003.

032.003   Das Fernrohr des Sternfreundes: Linsenfernrohre.
H. Scholze.
Astron. Raumfahrt, 25. Jahrg., Heft 4, p. 112 – 116 (1987).

032.004   De Vixen super polarismontering.
L. Aerts.
Zenit, Jaarg. 14, Nr. 7/8, p. 250 – 252 (1987).

032.005   Upgrading of the ESO 1.52–m telescope.
D. Alloin.
Messenger, No. 49, p. 18 – 19 (1987).

**032.006   Optical design for very large telescopes: the TEMOS
concept.**
A. Baranne, G. Lemaitre.
C. R. Acad. Sci., Sér. II, Tome 305, No. 6, p. 445 – 450 (1987). In
French.
    The authors have discovered a curious optomechanical proper-
ty of deformations which is particularly easy to apply in the case
of a "fond de vase". This makes it possible to increase the
collecting area of a very large telescope to a size corresponding to

a classical telescope of more than 8 m diameter without any innovation, at least as far as the optics are concerned.

032.007 **The 0.75–meter automatic telescope system.**
R. M. Genet, L. J. Boyd, D. S. Hayes,
D. L. Crawford, D. R. Genet.
I.A.P.P.P. Commun., No. 29, p. 42 – 48 (1987).

032.008 **Elektrische Fernrohrantriebe.**
K. Güssow.
Sterne Weltraum, 26. Jahrg., Nr. 7 – 8, p. 440 – 443 (1987).

032.009 **Le guidage automatique du télescope de deux mètres du Pic–du–Midi.**
F. Beigbeder.
J. Astron. Fr., No. 30, p. 17 – 18 (1987).

032.010 **Photoguidance accuracy for the tower solar telescope: photodiode array measurements.**
L. V. Didkovskij.
Bull. Crimean Astrophys. Obs., Vol. 75, p. 168 – 173 (1987). English translation of 42.036.065.

032.011 **A new mounting system for very large telescopes – with particular application to the classical Schmidt.**
V. C. Reddish, E. W. Simmonds.
Mon. Not. R. Astron. Soc., Vol. 228, No. 2, p. 537 – 543 (1987).
The high cost of conventional telescope structures arises from three factors: the practice of building vertically with consequent accumulation of load–bearing problems downwards; the use of cantilevers to carry loads indirectly to earth; and variations in the loads due to changing orientation with respect to gravity. The mounting system proposed in this paper overcomes these problems by building horizontally, and by taking loads directly to earth through servo controlled jacks on carriages running on tracks. Apart from the optics the system uses low cost engineering and has considerable advantages in simplicity of construction and operation. To provide a costed example it is applied to a classical Schmidt telescope of aperture 5 m (primary mirror 7.5 m) which could cost £ 9 m.

032.012 **System of a big multipurpose solar telescope. Requirements and the ways of their realization.**
V. N. Karpinskij, N. N. Mikhel'son, M. A. Sosnina.
Kinematika Fiz. Nebesn. Tel, Tom 3, No. 4, p. 80 – 84 (1987). In Russian. English translation in Kinematics Phys. Celest. Bodies.
A perspective big photospheric–chromospheric solar telescope must be of 0.5 – 0."25 resolution (possibly up to 0."1), have a tracking power not worse than 1″ and a small effect of instrumental polarization ($10^{-3} – 10^{-5}$). The 2–m telescope proposed is automatically pointing to the Sun, operates with an open (without vacuum or helium pumping) short (less than 10 m) path of rays of large aperture on the three–mirror Cassegrain–Gregory scheme. Polarization analyser is set in the Cassegrain focus. The image stability is reached by using the coelostat system at the exit after the elliptical mirror.

032.013 **The William Herschel Telescope.**
P. Murdin, A. Boksenberg.
1988 Yearbook of Astronomy, p. 125 – 134 (1987). – See Abstr. 003.006.

032.014 **Modern astrometric instruments.**
V. G. Shamaev.
Itogi Nauki i Tekhniki. Seriya Astronomiya. New methods for construction of coordinate systems, Tom 30, p. 3 – 78 (1987). In Russian. – See Abstr. 003.008.
The author describes the construction of a new instrumental basis for fundamental astrometry, automatic meridian instruments, photoelectric meridian circles, horizontal meridian instruments, and projects for new original constructions of astrometric instruments.

032.015 **Das Fernrohr des Sternfreundes. Spiegelfernrohre I.**
H. Scholze.
Astron. Raumfahrt, 25. Jahrg., Heft 5, p. 147 – 149 (1987).

032.016 **Adjustment of the prime focus optics of the GAT.**
V. Ya. Vajnberg, N. A. Vikul'ev, L. I. Snezhko.
Bull. Spec. Astrophys. Obs. – North Caucasus, Vol. 22, p. 96 – 102 (1987). English translation of 41.032.024.

032.017 **Comparison of the value of revolution [of a micrometer screw] from observations with three methods.**
V. V. Lapaeva, I. A. Urasina, N. N. Chudinov.
Kazan. Univ. Kazan', 5 pp. (1987). In Russian. Abstr. in Ref. Zh., 52. Geod. Aehrosemka, 9.52.68 (1987).

032.018 **Status of the Mark III interferometer.**
M. Shao, M. M. Colavita.
Interferometric imaging in astronomy, p. 115 – 119 (1987). – See Abstr. 012.035.

032.019 **A fiber–linked ground–based array.**
P. Connes, S. Shaklan, F. Roddier.
Interferometric imaging in astronomy, p. 165 – 168 (1987). – See Abstr. 012.035.
The authors give a preliminary discussion of what a multi–telescope fiber–linked ground–based array might look like, and an account of some encouraging fiber tests performed at NOAO.

032.020 **Interferometry with the European Very Large Telescope.**
P. Léna, F. Merkle.
Interferometric imaging in astronomy, p. 169 – 170 (1987). – See Abstr. 012.035.
The European Very Large Telescope is an array of four 8 meters diameter telescopes, to be installed from 1993 on a yet unselected site in Chile. The four telescopes can be either operated individually, or incoherently co–added, or coherently cophased and combined. The latter mode, called the interferometric mode, is briefly reported.

032.021 **Interferometric capabilities of the NNTT.**
J. M. Beckers.
Interferometric imaging in astronomy, p. 171 (1987). Abstract. – See Abstr. 012.035.

032.022 **ESO's Very Large Telescope: en andere telescopen van de toekomst.**
G. Schilling.
Zenit, 14. Jaarg., Nr. 12, p. 388 – 397 (1987).

032.023 **A 25–inch automated telescope for photoelectric photometry.**
J. B. Rafert.
Bull. Am. Astron. Soc., Vol. 19, No. 2, p. 746 (1987). Abstract. – See Abstr. 010.061.

032.024 **Transistor gate circuits for gamma–telescope electric drive.**
V. G. Shitov.
Izv. Krymskoj Astrofiz. Obs., Tom 77, p. 190 – 197 (1987). In Russian. English translation in Bull. Crimean Astrophys. Obs., Vol. 77.
The peculiarities of transistors operation in gate mode are discussed. Power losses in gate circuit elements are analyzed. Samples of driver circuits synthesis with overload protection high–power transistors are described. A bridge network for gamma telescope step motor control is proposed.

032.025 **Large chopping secondary mirror for the 15–m submillimeter James Clerk Maxwell telescope.**
H. van de Stadt, J. Verkerk.
Appl. Opt., Vol. 26, No. 16, p. 3446 – 3454 (1987).
A 75–cm diam chopping secondary mirror has been developed for the 15–m diam James Clerk Maxwell telescope. The large

focal ratio ($D/F = 1{:}0.36$) requires a highly convex secondary mirror, which was manufactured out of a solid slab of high–grade aluminum. The mirror surface approaches a predescribed hyperboloid with a rms precision better than 9 μm, which is adequate for use in the submillimeter wavelength region. The design aspects, the performance of a two–axis chopping mechanism, and the construction of the mirror are described.

032.026 **A new mounting system for very large telescopes – with particular application to the classical Schmidt.**
V. C. Reddish, E. W. Simmonds.
Edinb. Astron. Prepr., No. 10/87, 12 pp. (1987). To appear in Mon. Not. R. Astron. Soc.

032.027 **The Greenwich and Oxford astrographic telescopes 1958 – 1987.**
D. H. P. Jones.
R. Greenwich Obs., Prepr., No. 58, p. 1 – 4 (1987). To appear in IAU Symp. No. 133.

032.028 **Technique and calculation for the improvement of the transmission system in the four–axis telescope.**
J.–m. Yu.
Publ. Yunnan Obs., No. 1, p. 123 – 134 (1987). In Chinese.

032.029 **Statistics of changes of the Cassegrain focus with temperature.**
X.–d. Liu.
Publ. Yunnan Obs., No. 2, p. 69 – 70 (1987). In Chinese.
According to the observation data obtained during 1980 – 1985, the author concluded an empirical relationship between the temperature and the changes of the Cassegrain focus of the 1 meter telescope.

032.030 **Improvement measurements of scattered light level behind occulting systems.**
S. Koutchmy, M. Belmahdi.
Inst. Astrophys. Paris, Pré–Publ., No. 204, 15 pp. (1987). To appear in J. Opt. (Paris).

032.031 **Some developments in optical telescope instrumentation and data reduction systems in Japan.**
S. Isobe.
Rev. Mex. Astron. Astrofis., Vol. 14, No. 2, p. 722 – 735 (1987). – See Abstr. 012.042.
The author reviews Japanese efforts on the following topics, although there are many other topics under studying: (1) design work on the 7.5–m telescope, (2) experiments of honeycomb mirror production, (3) development of accurate and high sensitivity detectors, (4) study of an analysis system for speckle interferometric observations, and (5) development of a two–dimensional image processing system.

032.032 **Calidad de imágen del telescopio UNAM212.**
F. J. Cobos, C. Tejada de Vargas.
Rev. Mex. Astron. Astrofis., Vol. 14, No. 2, p. 766 (1987). Abstract. – See Abstr. 012.042.

032.033 **Ground–based interferometry with the European Very Large Telescope (VLT).**
F. Merkle.
ESA Spec. Publ., ESA SP–273, p. 127 – 138 (1987). – See Abstr. 012.044.
In the linear array concept of ESO's Very Large Telescope (VLT) project four independent telescopes of the 8–meter class are coupled together. This offers the unique possibility to operate the telescopes as a long baseline interferometer in the IR and visible with approximately 104 m baseline. Two smaller movable telescopes are added to the linear array for spatial frequency filling reasons. Adaptive optics will be applied for a real–time partial or full phase compensation of atmospheric distortions.

032.034 **CERGA high angular resolution optical network.**
G. Schumacher, P. Bourlon, A. Boischot, P. Cruzalèbes, M. Dugué, L. Koechlin, Y. Rabbia, F. Vakili.
ESA Spec. Publ., ESA SP–273, p. 139 – 143 (1987). – See Abstr. 012.044.
The new CERGA project of high angular resolution imaging prototype is described. It consists of a 4 phased telescope interferometer working at visible wavelengths. The apertures of the telescopes are limited to 26 cm, and they will move on two perpendicular baselines in a cross N/S and E/W configuration. The optical path compensation is done by delay lines, and the 4 beams are recombined on a fixed central table. The 6 fringe patterns are recorded by a short exposure, photon–counting camera in order to freeze the atmospheric turbulence.

032.035 **Practical limits to ground based interferometry and its impact on space interferometry.**
M. Shao.
ESA Spec. Publ., ESA SP–273, p. 145 – 149 (1987). – See Abstr. 012.044.
The Mark III interferometer is a phase coherent stellar interferometer designed to make astrometric and stellar diameter measurements in a fully automatic mode. A future absolute interferometer with 1 meter elements can make photon starved measurements of fringe amplitude and closure phase on objects as faint as 15 mag, competitive with many first generation space interferometers, with 30 – 50 cm apertures.

032.036 **Optical 16–telescope interferometer at Erlangen.**
F. Fleischmann, F. Grieger, G. Weigelt.
ESA Spec. Publ., ESA SP–273, p. 155 – 156 (1987). – See Abstr. 012.044.
An optical interferometer which consists of sixteen 11 cm–telescopes on a MMT mount. The 16–telescope array is a linear array with 7.5 m baseline. The advantage of the large number of telescopes is the fact that many baselines are measured simultaneously. Therefore, 1–dimensional images can be obtained in the snapshot mode and 2–dimensional images can be obtained by using the earth rotation. The authors plan to interfere all beams in one central station to obtain speckle interferograms. They plan to use speckle masking for image reconstruction.

032.037 **Performances of an actively stabilized stellar interferometer (ASSI): faint magnitudes, low fringe contrast measurements and operationality.**
L. Damé, M. Faucherre.
ESA Spec. Publ., ESA SP–273, p. 205 – 215 (1987). – See Abstr. 012.044.
The authors describe the performances of a new approach to long baseline Michelson interferometry using a real time active stabilization of the central fringe position ("fringe tracking"). In this approach the monitoring function (star pointing and optical path delay information) is clearly dissociated (and flux optimized) from the scientific analysis function, in which time integration and stability are favored, for evident reasons of precision on the fringe contrast measurement.

032.038 **Hawaii: a picture window on the Universe.**
I. Anderson, M. Redfern.
New Sci., Vol. 114, No. 1558, p. 46 – 50 (1987). Abstr. in Phys. Abstr., Vol. 90, No. 1310, Entry 94659 (1987).

032.039 **An expedition solar telescope.**
Eh. V. Kononovich, Eh. Ya. Kononov, Yu. A. Kupryakov, O. B. Smirnova.
Astron. Vestn., Tom 21, No. 4, p. 340 – 343 (1987). In Russian. English translation in Sol. Syst. Res.
A description is given of an expedition solar telescope for the observation of the Sun in continuum (region about 550 μm) and in the calcium Ca II K line.

032.040   **The VLT – genesis of a project.**
D. Enard.
Messenger, No. 50, p. 30 – 32 (1987).

032.041   **First fully automatic telescope on La Silla.**
R. Florentin Nielsen, P. Nørregaard, E. H. Olsen.
Messenger, No. 50, p. 45 – 46 (1987).

032.042   **An interferometric mode for the VLT.**
P. Lena.
Messenger, No. 50, p. 53 – 55 (1987).

032.043   **The William Herschel Telescope.**
J. V. Wall.
J. Br. Astron. Assoc., Vol. 98, No. 1, p. 8 – 13 (1987).
The first observations have been made with the 4.2–metre William Herschel Telescope at the international Observatorio del Roque de los Muchachos. Eventually, the telescope will be equipped with a range of versatile instruments enabling it to observe planets, stars and the most distant galaxies known, just as Herschel did.

032.044   **La lunette équatoriale de 38 cm de l'Observatoire de Paris.**
J.–D. Wahiche.
Ann. Phys. (Paris), Vol. 12, Colloq. No. 1, p. 21 – 25 (1987). – See Abstr. 012.061.
The equatorial 38 cm refracting telescope of the Paris Observatory, set up on the East Tower in 1856, was put back into service again after sixteen years without use, for the PHEMU85 campaign. A description of the instrumentation itself, and the 12 m diameter dome is given, followed by a survey of the history of this instrument and of its utilization over the last century.

032.045   **Optical synthesis telescopes.**
G. W. Swenson Jr., C. S. Gardner, R. H. T. Bates.
Proc. SPIE Int. Soc. Opt. Eng., Vol. 643, p. 129 – 140 (1986).
Abstr. in Phys. Abstr., Vol. 90, No. 1313, Entry 115240 (1987). – See Abstr. 012.062.

032.046   **An update on the 15–m National New Technology telescope.**
L. D. Barr.
Proc. SPIE Int. Soc. Opt. Eng., Vol. 643, p. 234 – 243 (1986).
Abstr. in Phys. Abstr., Vol. 90, No. 1313, Entry 115242 (1987). – See Abstr. 012.062.

032.047   **Site testing telescope configurations.**
R. B. Dunn.
LEST Found., Tech. Rep., No. 26, p. 79 – 85 (1987). – See Abstr. 012.071.
For an optical site survey for a solar telescope, one might like to observe or measure: (1) Granulation at the center of the disk, (2) R_0, or some other characteristic of the distribution of turbulence cells across the image of the aperture, and (3) Size of isoplanatic patch. The author discusses some telescope options for these three measurements.

032.048   **An optical telescope for site testing.**
D. Bonaccini.
LEST Found., Tech. Rep., No. 27, 50 pp. (1987).
The LEST site testing telescope is designed in order to provide simultaneously an image of the solar granulation at very short exposure times, "knife edges" tests to monitor the phase changes of the wavefront in the pupil, and the correlation of the limb fluctuations in the solar image. The site–testing telescope is designed to match the various requirements of the different tests.

032.049   **Das Fernrohr des Sternfreundes. Spiegelfernrohre II.**
H. Scholze.
Astron. Raumfahrt, 25. Jahrg., Heft 6, p. 178 – 181 (1987).

032.050   **On thermal instrumental errors in astrometric observations.**
M. P. Mishchenko.
Tr. Astron. Obs., Leningrad, Tom 41, p. 162 – 175 (1987). Uch. Zap. LGU, No. 420, Ser. Mat. Nauk, Vyp. 63. In Russian.
The influence of the external medium on the astronomical time determinations with a photoelectric transit instrument is studied. It is shown that the protection of the instrument from external thermal effects reduces the systematic variation of the instrumental system caused by the effect of the thermal flexure.

032.051   **Consideration of the instrument's temperature in processing astrolabe observations.**
N. M. Gaftonyuk, A. L. Shcherbanovskij.
Nauchn. Inf., Vyp. 58, p. 45 – 50 (1986). In Russian.
A systematic linear shift of the group adjustment to the instrumental zenith distance has been displayed while processing the results of the equal–altitude method observations obtained with the Danjon astrolabe OPL–23. The authors propose to introduce a correction for the instrument's temperature when reducing the observational results.

032.052   **The theory of vertical mounting.**
Yu. Kh. Zhagar, M. B. Dimitrova.
Nauchn. Inf., Vyp. 58, p. 107 – 113 (1986). In Russian.
This paper shows the advantages of the vertical mounting associated with the operating electronic computer. Its mathematical model taking into account instrumental errors is described.

032.053   **The first 10 years of operation of the BTA.**
L. I. Snezhko.
Zemlya Vselennaya, No. 6, p. 12 – 17 (1987). In Russian.

032.054   **The raised yoke mounting of the 1.56 m telescope.**
S.–s. Gong.
Acta Astron. Sin., Vol. 28, No. 2, p. 204 – 210 (1987). In Chinese.
Designing principles and deflection measurements for a raised yoke mount of a 1.56 m astrometrical telescope are briefly described. The pointing accuracy amounts to only several arcsec when the telescope is moved from 0^h to 6^h. The lowest eigenfrequency of the telescope is 6.5 Hz. The use of welded structures obviously can produce a higher damping.

032.055   **Pointing error correction for the 3.8 m United Kingdom Infrared Telescope.**
J.–q. Cheng.
Acta Astron. Sin., Vol. 28, No. 3, p. 308 – 314 (1987). In Chinese.
Mathematical principles of pointing error corrections of telescopes are developed. The method is applied to the 3.8 m UKIRT.

Itogi Nauki i Tekhniki. Seriya Astronomiya. Tom 30. New methods for construction of coordinate systems.
See Abstr. 003.008.

The adjustment and testing of telescope objectives.
See Abstr. 003.134.

Herscheliana.
See Abstr. 004.017.

Foucault's invention of the silvered–glass reflecting telescope and the history of his 80–cm reflector at the Observatoire de Marseille.
See Abstr. 004.043.

Scheduling and other issues for a solar–stellar synoptic observatory.
See Abstr. 013.002.

The Automatic Photoelectric Telescope Service.
See Abstr. 013.031.

Proposal for a large telescope in South Africa.
See Abstr. 013.036.

Wide–field conversions for reflecting telescopes.
See Abstr. 031.023.

Sistemas correctores de campo para el telescopio Cassegrain IAC80.
See Abstr. 031.054.

Determinación de caracteristicas ópticas del telescopio OAN150.
See Abstr. 031.056.

Sistemas correctores de campo para el telescopio Ritchey–Chrétien UNAM212.
See Abstr. 031.057.

Active optics. I. A system for optimizing the optical quality and reducing the costs of large telescopes.
See Abstr. 031.065.

A parametric study of various synthetic aperture telescope configurations for coherent imaging applications.
See Abstr. 031.067.

On an analysis of the optics of large telescopes.
See Abstr. 031.071.

Adaptive optics for the European Very Large Telescope (VLT).
See Abstr. 031.077.

Liquid mirror telescopes: present and future.
See Abstr. 031.078.

The Lick Observatory TV autoguider.
See Abstr. 034.069.

Guidelines for allocation of observing time and scheduling.
See Abstr. 036.005.

The automatic photoelectric telescope service – an update.
See Abstr. 036.131.

Two–color method for optical astrometry: theory and preliminary measurements with the Mark III stellar interferometer.
See Abstr. 036.141.

La campagne PHEMU85 à la lunette équatoriale de 38 cm de l'Observatoire de Paris.
See Abstr. 036.188.

Mapping the sky with the Carlsberg Automatic Meridian Circle.
See Abstr. 041.008.

Atmospheric phase measurements with the Mark III stellar interferometer.
See Abstr. 082.049.

The photometric UBVR system of the 50/70 cm Schmidt telescope of the National Astronomical Observatory Rozhen.
See Abstr. 113.057.

Structure of the extended emission in the infrared celestial background.
See Abstr. 133.017.

033 Radio Telescopes and Equipment

033.001 Interferometers, aberration and the Sagnac effect.
M. J. Kesteven.
Aust. J. Phys., Vol. 40, No. 3, p. 425 – 439 (1987).
In this note the author shows that the usual expression for diurnal aberration is not valid for some types of radio interferometers. The difference (usual minus correct) is a constant of the instrument, and is absorbed in the calibration procedure. The analysis of multibaseline VLBI may also be affected.

033.002 New radiotelescopes open era of submillimeter astronomy.
W. Sweet.
Phys. Today, Vol. 40, No. 8, Part 1, p. 65 – 67 (1987).

033.003 Project Radioastron: an earth–space interferometer.
V. V. Andreyanov, N. S. Kardashev, M. V. Popov, V. A. Rudakov, R. Z. Sagdeev, V. I. Slysh, G. S. Tsarevskij.
Sov. Astron., Vol. 30, No. 5, p. 504 – 508 (1986). English translation of 42.033.008.

033.004 The current concept for the proposed NRAO millimeter array.
R. M. Hjellming.
Radio astronomy from space, p. 107 – 110 (1987). – See Abstr. 012.009.
The National Radio Astronomy Observatory is developing a proposal for a large, ground–based, millimeter wavelength imaging facility.

033.005 The SAO submillimeter–wavelength telescope array.
J. M. Moran, M. S. Elvis, G. G. Fazio, P. T. P. Ho, P. C. Myers, M. J. Reid, S. P. Willner.
Radio astronomy from space, p. 111 – 116 (1987). – See Abstr. 012.009.
The Smithsonian Astrophysical Observatory is investigating the design and construction of a submillimeter–wavelength interferometer. The preliminary design is for an array of six 6–m diameter telescopes operating in the wavelength range from 1.3 to 0.35 mm with an angular resolution of one arcsecond or finer.

033.006 VLBA status report.
J. D. Romney.
Radio astronomy from space, p. 185 – 189 (1987). – See Abstr. 012.009.
The salient characteristics and status of construction of the VLBA project are reviewed. Aspects particularly relevant to VLBI observations with an orbiting element such as the proposed Quasat mission are emphasized.

033.007 Self–correction of telescope surface errors using a correlating focal plane array.
T. J. Cornwell, P. J. Napier.
Radio astronomy from space, p. 215 – 219 (1987). – See Abstr. 012.009.
The effects on the performance of a large radio telescope of aberrations such as reflector surface errors, defocussing, coma and pointing errors can be removed if the telescope is equipped with an array feed in its focal plane. If the cross correlations between all possible pairs of array elements are measured, then aberration–free images of radio sources can be obtained.

033.008 Beyond the diameter–wavelength–ratio of reflector antennas – a film lens antenna.
Y. Chikada.
Radio astronomy from space, p. 221 – 224 (1987). – See Abstr. 012.009.

A possibility is investigated to realize a super–large aperture telescope in short wavelength region. As is widely known, it is very hard to build a reflector antenna of diameter–wavelength ratio better than 10^4. On the other hand, it will be easy with a thin film lens antenna because surface error perpendicular to the film lens does not affect its performances and also tilt of the lens does not affect its pointing accuracy.

033.009 Cambridge observations at 38 – 151 MHz and their implications for space astronomy.
R. Saunders.
Radio astronomy from space, p. 293 – 300 (1987). – See Abstr. 012.009.

The Cambridge low–frequency telescopes (38 – 151 MHz) are described, particularly from the viewpoint of calibrating out the ionosphere. Experience with the telescopes places strong constraints on what it is possible to do from the ground. There are also important implications for space–based low–frequency telescopes concerning knowledge of primary beam, aperture coverage and ISS.

033.010 A low–frequency synthesis array in Earth orbit.
D. L. Jones, R. A. Preston, T. B. H. Kuiper.
Radio astronomy from space, p. 301 – 304 (1987). – See Abstr. 012.009.

The possibility of observing the universe at radio frequencies below the ionospheric plasma frequency is exciting, as it opens up a new spectral window many octives wide. To obtain high angular resolution at long wavelengths, an interferometer array in space is required. This paper presents a summary of the current JPL design for such an array.

033.011 The Low Frequency Space Array.
B. Dennison, K. W. Weiler, K. J. Johnston,
R. S. Simon, J. H. Spencer, L. M. Hammarstrom,
P. G. Wilhelm, W. C. Erickson, M. L. Kaiser, M. D. Desch,
J. Fainberg, L. W. Brown, R. G. Stone.
Radio astronomy from space, p. 305 – 313 (1987). – See Abstr. 012.009.

The Low Frequency Space Array (LFSA) is a conceptual mission to survey the entire sky and to image individual sources at frequencies between 1.5 and 26 MHz, a frequency range over which the Earth's ionosphere transmits poorly or not at all. A number of major scientific goals can be pursued with such a mission.

033.012 A solar 230 – 300 MHz radio acousto–optic spectrograph.
J.–s. Wang, Z.–g. Xia, J.–y. Chen, S.–y. Jiang, M.–l. Min,
B.–h. Xu, G.–c. Huang, G.–x. Ren.
Sci. Sin., Ser. A, Vol. 30, No. 8, p. 853 – 859 (1987).

For the purpose of observing dynamic spectra of the solar metre–wave radiation, a digital solar metre–wave radio acousto–optic spectrograph has been established. Its design performance is suitable for observations of the solar burst of type I.

033.013 Experiment Cold: the first deep sky survey with the RATAN–600 radio telescope.
Yu. N. Parijskij, D. V. Korol'kov.
Astrophys. Space Phys. Rev., Vol. 5, p. 39 – 179 (1987). – See Abstr. 003.002.

Contents: 1. Introduction. 2. The RATAN–600 radio telescope. 3. Experimental technique. 4. Limiting factors and characteristics of the RATAN. 5. Astrophysical problems studied in Experiment Cold. 6. Data reduction methods used in experiment cold. 7. Results of the discrete source observations. 8. The Galaxy. 9. Search for anisotropy in the 3–degree background. 10. What next?
For the Russian original see 42.033.050.

033.014 VLBI correlator can perform better.
K. M. Borkowski.
Acta Astron., Vol. 37, No. 1, p. 89 – 98 (1987).

Possibilities of improving the performance of digital VLBI correlators by introducing an effective 5–level fringe rotator in place of the usual 3–level one are studied.

033.015 On the influence of ocean tides on radiointerferometric observations.
V. M. Gorban'.
Kinematika Fiz. Nebesn. Tel, Tom 3, No. 5, p. 92 – 94 (1987). In Russian. English translation in Kinematics Phys. Celest. Bodies.

An algorithm is considered to obtain corrections for the ocean tides influence on the geometric signal delay and interference frequency in astrometric VLBI observations. Some results of correction computations for the continental and intercontinental bases are presented.

033.016 A 6 × 320–MHz 1024–channel FFT cross–spectrum analyzer for radio astronomy.
Y. Chikada, M. Ishiguro, H. Hirabayashi, M. Morimoto,
K.–I. Morita, T. Kanzawa, H. Iwashita, K. Nakazima,
S.–I. Ishikawa, T. Takahashi, K. Handa, T. Kasuga,
S. Okumura, T. Miyazawa, T. Nakazuru, K. Miura,
S. Nagasawa.
Proc. IEEE, Vol. 75, No. 9, p. 1203 – 1210 (1987).

A wide–band FFT spectrum analyzer, which is called FX, has been in operation since 1983 at the Nobeyama Radio Observatory for spectroscopy of radio waves from interstellar molecules. The FX incorporates about 4500 newly developed CMOS LSI chips. They are designed using CAD (computer–aided design) and have 3900 or 2000 gates/chip, operate at a clock rate of 10 MHz.

033.017 Determination of the parameters of the focusing ring system of the RATAN–600 radiotelescope for observations in the zenith.
E. K. Majorova, A. A. Stotskij.
Bull. Spec. Astrophys. Obs. – North Caucasus, Vol. 22, p. 103 – 110 (1987). English translation of 41.033.014.

033.018 Geodetic adjustment of the RATAN–600 radiotelescope.
Yu. K. Zverev.
Bull. Spec. Astrophys. Obs. – North Caucasus, Vol. 22, p. 111 – 118 (1987). English translation of 41.033.015.

033.019 Very long baseline interferometry.
J. Campbell, R. Kilger, H. Seeger.
Dtsch. Geod. Komm. Bayer. Akad. Wiss., Reihe B, Heft Nr. 284, p. 119 – 132 (1987). In German and English.

033.020 Resurfacing the Jodrell Bank Mk II radio telescope.
R. E. Spencer, R. J. Melling, J. S. Haggis,
I. Morrison, R. J. Davis, X. Q. Cui.
Fifth International Conference on Antennas and Propagation ICAP 87, held at York, UK, 30 March – 2 April 1987, Part 1. IEE Conf. Publ., No. 274. IEE, London, UK, p. 137 – 140 (1987). = Astron. Contrib. Univ. Manchester, Ser. II: Jodrell Bank Repr., No. 809.

The project to improve the short wavelength performance of an old radio telescope has proved to be very successful, achieving a final r.m.s. profile error of 0.6 mm. This success has stemmed from the invention of an inexpensive technique of panel construction and measurement, combined with the use of radio astronomical holographic techniques to measure the telescope under actual operational conditions.

033.021 Design considerations for a multibeam receiver for millimeter wave astronomy.
A. van Adenne.
Int. J. Infrared Millimeter Waves, Vol. 8, No. 2, p. 107 – 117 (1987). Abstr. in Phys. Abstr., Vol. 90, No. 1309, Entry 88110 (1987).

033.022   **Designing principles and block diagram of a metrewave solar radio spectral receiver.**
J.-s. Wang.
Publ. Yunnan Obs., No. 1, p. 57 – 59 (1987). In Chinese.

The fundamental designing principles and block diagram of a wideband superheterodyne receiver for the 230 – 300 MHz solar radio acousto–optic spectrograph are reported.

033.023   **A pointing calibration system developed for the Nobeyama millimeter array.**
T. Kanzawa, K.–I. Morita, M. Ishiguro.
Tokyo Astron. Obs. Rep., Vol. 21, No. 1, p. 1 – 13 (1987). In Japanese.

033.024   **Control system of the Nobeyama millimeter–wave array.**
K.–I. Morita, T. Kanazawa, T. Takahashi, M. Ishiguro.
Tokyo Astron. Obs. Rep., Vol. 21, No. 1, p. 82 – 100 (1987). In Japanese.

033.025   **Architecture, hardware and software for the net of the measuring–computing complexes at the RATAN–600 radio telescope.**
B. L. Erukhimov, V. N. Chernenkov.
Astrofiz. Issled. Izv. Spets. Astrofiz. Obs., Tom 24, p. 183 – 190 (1987). In Russian. English translation in Bull. Spec. Astrophys. Obs. – North Caucasus.

Organization principles, technical and program realization of the radial net of local measure–computing complexes (LMCC) are considered. LMCC is the base element of the future system for collective use of the radio telescope.

033.026   **Control equipment for two spectrometers at the RATAN–600 radio telescope.**
V. G. Mogileva, A. S. Morozov, V. A. Prozorov, N. F. Ryzhkov.
Astrofiz. Issled. Izv. Spets. Astrofiz. Obs., Tom 24, p. 191 – 204 (1987). In Russian. English translation in Bull. Spec. Astrophys. Obs. – North Caucasus.

033.027   **The control system for moving the RATAN–600 radio telescope feed based on a microcomputer.**
G. S. Golubchin, S. F. Gol'dshmidt, Eh. S. Muchnik, E. K. Nizhel'skaya.
Astrofiz. Issled. Izv. Spets. Astrofiz. Obs., Tom 24, p. 205 – 209 (1987). In Russian. English translation in Bull. Spec. Astrophys. Obs. – North Caucasus.

The hardware and the program facilities of the automatic control system of the RATAN–600 radio telescope feed, based on a computer are described. The characteristics of this system answer the demands of carrying out the methodical and observational experiments at the radio telescope.

033.028   **False signals in a radiometer with a noise pilot–signal.**
N. F. Ryzhkov, V. A. Prozorov.
Astrofiz. Issled. Izv. Spets. Astrofiz. Obs., Tom 24, p. 210 – 218 (1987). In Russian. English translation in Bull. Spec. Astrophys. Obs. – North Caucasus.

The formulas for false signals caused by an instability of the parameters of blocks of a radiometer with a noise pilot–signal and gain modulation are obtained. It is shown that the radiometer with a noise pilot–signal is intermediate between total–power radiometer and Dicke radiometer with regard to instability of the parameters of their blocks.

033.029   **On the possibility of operation of the RATAN–600 radio telescope at millimeter wavelengths.**
E. K. Majorova.
Astrofiz. Issled. Izv. Spets. Astrofiz. Obs., Tom 25, p. 135 – 142 (1987). In Russian. English translation in Bull. Spec. Astrophys. Obs. – North Caucasus, Vol. 25.

The influence of a systematic phase error, caused by deviation of cylindric reflecting elements from the perfectly conical surface on the antenna in millimetre wavelengths is investigated. The expected values of effective areas in the wavelength range of 2 – 8 mm are estimated.

033.030   **Autocollimation adjustment and investigation of the stability of the RATAN–600 radio telescope.**
A. A. Stotskij, G. N. Kalikhevich, T. N. Osina, G. A. Pinchuk.
Astrofiz. Issled. Izv. Spets. Astrofiz. Obs., Tom 25, p. 143 – 167 (1987). In Russian. English translation in Bull. Spec. Astrophys. Obs. – North Caucasus, Vol. 25.

The autocollimation method used for adjusting the radio telescope (relative positioning of the main mirror screens) and for some other measurements is described. The autocollimation method of adjustment is being used for providing observations at the radio telescope for more than 10 years. Stability of the radio telescope focusing system and reliability of counting–adjusting mechanisms of screens are estimated on the basis of measurement results obtained during this time.

033.031   **Operative presentation of astrophysical results of observations of the Sun at the RATAN–600 radio telescope.**
V. A. Shatilov.
Astrofiz. Issled. Izv. Spets. Astrofiz. Obs., Tom 25, p. 168 – 175 (1987). In Russian. English translation in Bull. Spec. Astrophys. Obs. – North Caucasus, Vol. 25.

A complex of programs for operative processing of multiwave solar observations with the RATAN–600 radio telescope has been developed. The data are processed at the computer complex accomodated in the feed just after the observations. In addition the problems arising at the experiment procedure are solved to get its high accuracy.

033.032   **Autocollimation adjustment of the RATAN–600 radio telescope in the regime with total circular aperture.**
A. D. Dibizhev, E. K. Majorova, G. A. Pinchuk, V. I. Sinyanskij, A. A. Stotskij, V. B. Khajkin.
Soobshch. Spets. Astrofiz. Obs., Vyp. 52, p. 77 – 83 (1987). In Russian.

A procedure of radiotechnical adjustment of the RATAN–600 radio telescope main reflector in the observational regime of the near–zenith region is described. The main characteristics of the focusing system at the wavelength of 8.2 cm based on the results of autocollimation measurements and radio astronomical observations are presented.

033.033   **The antenna system "Radioastron".**
V. I. Altunin, V. V. Andreyanov, E. I. Bitushan, O. A. Kuz'min, A. I. Perlov, B. A. Prigoda, Yu. I. Churilov.
Inst. Kosm. Issled. Akad. Nauk SSSR, Prepr., No. 1237, p. 3 – 33 (1987). In Russian. Abstr. in Ref. Zh., 51. Astron. 11.51.942 (1987).

033.034   **Construction of the antenna of a radio telescope.**
E. I. Bitushan, S. N. Sayapin, A. A. Drakov, V. Kh. Gataullin, I. V. Shishov, V. D. Loginov, V. I. Altunin, A. G. Trubnikov.
Inst. Kosm. Issled. Akad. Nauk SSSR, Prepr., No. 1243, 20 pp. (1987). In Russian. Abstr. in Ref. Zh., 51. Astron. 11.51.943 (1987).

033.035   **Reactivación del radio interferómetro solar de Tonantzintla.**
E. Mendoza T.
Rev. Mex. Astron. Astrofis., Vol. 14, No. 2, p. 759 – 762 (1987). – See Abstr. 012.042.

The author discusses the behavior of some components of the interferometer, that makes it possible to investigate several characteristics of the Sun's signal, like global flux, polarization, active regions flux, etc.

033.036   Low–frequency radioastronomical observations during the Spacelab 2 plasma depletion experiment.
G. R. A. Ellis, A. Klekociuk, A. C. Woods, G. Reber,
G. T. Goldstone, G. Burns, P. Dyson, E. Essex, M. Mendillo.
Aust. Phys., Vol. 24, No. 3, p. 56 – 58 (1987). Abstr. in Phys. Abstr., Vol. 90, No. 1310, Entry 94667 (1987).

033.037   Measurement of the reflector accuracy of the 30 m millimeter radio telescope.
J. W. M. Baars, W. Harth, A. Greve, D. Morris.
NTG–Fachber., Vol. 99, p. 305 – 309 (1987). In German, English. Abstr. in Phys. Abstr., Vol. 90, No. 1311, Entry 101686 (1987).

033.038   A 32–m radio telescope for VLBI research in Poland.
K. M. Borkowski.
Postepy Astron., Tom 34, Zesz. 3, p. 201 – 214 (1986). In Polish.
The article describes basic characteristics of a proposed 32–meter fully–steerable telescope for the Toruń Radio Astronomy Observatory. The design meets all requirements of the VLBI technique. It is an altitude–azimuth instrument with the Cassegrain optics for work at frequencies above about 1 GHz. For use at longer wavelengths the subreflector is removed to allow for the prime focus work. The overall reflector surface accuracy and structural rigidity are expected to be good for observations at frequencies of up to 43 GHz.

033.039   Heterodyne mixing experiments with NbN–based SIS junctions.
C. Letrou, D. Crete, J.–C. Pernot, A. Rabhi, P. Encrenaz.
Int. J. Infrared Millimeter Waves, Vol. 8, No. 4, p. 333 – 353 (1987). Abstr. in Phys. Abstr., Vol. 90, No. 1314, Entry 120277 (1987).

033.040   Status of the SFB 78 Water Vapor Radiometer.
G. Reichert.
Mitt. Geod. Inst. Rheinischen Friedrich–Wilhelms–Univ. Bonn, Nr. 71, p. 13 – 18 (1987). – See Abstr. 012.066.

033.041   Present state and future of geodetic correlation in Bonn.
A. Müskens.
Mitt. Geod. Inst. Rheinischen Friedrich–Wilhelms–Univ. Bonn, Nr. 71, p. 19 – 25 (1987). – See Abstr. 012.066.

033.042   On the significance of Doppler antennae calibration results.
F. J. Lohmar, H. Seeger.
Mitt. Geod. Inst. Rheinischen Friedrich–Wilhelms–Univ. Bonn, Nr. 72, p. 49 – 58 (1987). Paper presented at the Third United Nations Regional Cartographic Conference for the Americas, held 19 February – 1 March 1985.
The goal of this report is to resume the results of several Doppler antennae calibration campaigns and to investigate the reproducibility of the results. Finally the coordinates of a recent campaign are used to demonstrate the improving effect of applying especially height offsets.

033.043   The VLBI antenna of the Medicina radioastronomy station – the Mark II digital signal processing system.
A. Gallerani.
Elettrotecnica (*Italy*), Vol. 74, No. 2, p. 149 – 156 (1987). In Italian. Abstr. in Phys. Abstr., Vol. 90, No. 1315, Entry 127720 (1987).

033.044   A low noise broadband 125 – 175 GHz SIS receiver for radioastronomy observations.
J. Ibruegger, M. Carter, R. Blundell.
Int. J. Infrared Millimeter Waves, Vol. 8, No. 6, p. 595 – 607 (1987). Abstr. in Phys. Abstr., Vol. 90, No. 1315, Entry 127722 (1987).

033.045   A low noise 230 GHz SIS receiver.
B. N. Ellison, R. E. Miller.
Int. J. Infrared Millimeter Waves, Vol. 8, No. 6, p. 609 – 625 (1987). Abstr. in Phys. Abstr., Vol. 90, No. 1315, Entry 127723 (1987).

033.046   The NRAO supercomputing initiative.
R. D. Ekers, P. vanden Bout.
NRAO Workshop, No. 15, p. 63 – 67 (1988). – See Abstr. 012.076.
Because of the relatively long wavelength of radio waves a conventional monolithic radio telescope with the angular resolution of an optical telescope would require an aperture many kilometers in diameter. It would be prohibitively expensive and probably impossible to construct. The development of the digital computer made it possible to achieve this resolution with "aperture synthesis" radio telescopes. These telescope use arrays of relatively small antennas to measure the coherence of the wave front over a large area; an image is then formed in a digital computer, which now becomes an essential component of the telescope.

033.047   The large computing problems for the proposed NRAO Millimeter Array.
R. M. Hjellming.
NRAO Workshop, No. 15, p. 99 – 102 (1988). – See Abstr. 012.076.
While it can be confidently stated that a large fraction of the scientific problems that one would do with a millimeter array can be done with VAX–class computers, it can also be confidently stated that there will be a group of problems requiring supercomputer–class facilities. Particularly if the goal is to allow astronomer–friendly data reduction, that is, spectral line aperture synthesis or image processing which takes up a small enough fraction of an astronomer's time so that it is possible to be primarily a researcher, teacher, etc.

033.048   Plans for the Berkeley–Illinois Array.
R. M. Crutcher.
NRAO Workshop, No. 15, p. 103 – 106 (1988). – See Abstr. 012.076.
The proposed Berkeley–Illinois Array is a joint project between the Radio Astronomy Laboratory of the University of California, Berkeley and the University of Illinois. The proposal is to expand the present 3–antenna millimeter–wave interferometer at Hat Creek, California to a 6–antenna array and to establish a supercomputer–based data reduction and image processing facility at the University of Illinois.

033.049   Low–noise SIS heterodyne receiver at 230 GHz.
S. R. Davies, L. T. Little, C. T. Cunningham.
Electron. Lett., Vol. 23, No. 18, p. 946 – 948 (1987). Abstr. in Phys. Abstr., Vol. 90, No. 1316, Entry 134132 (1987).

033.050   The telescope that never sleeps (*VLA*).
L. A. Shore.
Astronomy, Vol. 15, No. 8, p. 14 – 22 (1987). Abstr. in Phys. Abstr., Vol. 90, No. 1317, Entry 139920 (1987).

Thirty years of the Mark IA radio telescope at Jodrell Bank.
See Abstr. 013.010.

Status report on RADIOASTRON.
See Abstr. 013.019.

Collaborative VLBI experiments with RADIOASTRON.
See Abstr. 013.020.

QUASAT – European status report.
See Abstr. 013.021.

The Australia Telescope project.
See Abstr. 013.065.

Evading the zone of avoidance.
See Abstr. 013.085.

Using TDRSS as an orbiting VLBI observatory.
See Abstr. 035.009.

Link calibrations for the TDRSS orbiting VLBI experiment.
See Abstr. 035.010.

TDRSS OVLBI experiment: data correlation and results.
See Abstr. 035.011.

Astro–array: a space–based, coherent radio interferometer array.
See Abstr. 035.012.

Analysis of misalignment and thermal distortion effects in the FIRST antenna.
See Abstr. 035.062.

Fiber–linked telescope arrays on the ground and in space.
See Abstr. 035.074.

Quasat program: the ESA reflector.
See Abstr. 035.086.

Imaging in radio astronomy.
See Abstr. 036.035.

Near zenith tracking limits for altitude–azimuth telescopes.
See Abstr. 036.052.

On the resolving power of the method of interplanetary scintillations.
See Abstr. 036.186.

Kalman filtering VLA phase data with a supercomputer.
See Abstr. 036.220.

Image analysis of the observation of the DA240 region with a meter–wave synthesis radio telescope.
See Abstr. 036.249.

Geodetic VLBI with large antennas.
See Abstr. 045.015.

Differential Doppler tracking of interplanetary spacecraft.
See Abstr. 045.023.

A method of reduction of one–dimensional scans of the Sun.
See Abstr. 077.007.

Ionospheric refraction correction in radio astronomy.
See Abstr. 083.021.

Application of an acousto–optical TeO_2 detector in radio astronomy.
See Abstr. 131.214.

Microwave background measurements from space at 15 to 90 GHz.
See Abstr. 161.029.

034 Auxiliary Instrumentation, Photographic Materials, Clocks

034.001 **An experiment for observing VHE gamma ray sources.**
Y. L. Jang, C. X. He, C. X. Xu, Y. K. Yuan,
Y. G. Li, D. B. Chen, R. T. Zheng, A. X. Huo, M. H. Ye,
S. Z. Tang, L. J. Zhang, S. X. Meng, H. L. Shi, S. Y. Jiang,
Q. B. Li, J. Yang, H. Q. Zhang.
The origin and evolution of neutron stars, p. 553 (1987).
Abstract. – See Abstr. 012.001 (IAU Symp. No. 125).

034.002 **The Penn State fiber coupled CCD/echelle spectrograph.**
L. W. Ramsey.
The SHIRSOG Workshop, p. 80 – 87 (1986). – See Abstr. 012.003.
The author describes a versatile echelle spectrograph which is coupled to their 1.6 meter telescope via an optical fiber. Observations obtained since winter 1985 have shown the instrument to have excellent flat–field and radial velocity performance properties with quite respectable throughput.

034.003 **Fiber optic uses with the spectrograph.**
S. C. Barden.
The SHIRSOG Workshop, p. 88 (1986). – See Abstr. 012.003.

034.004 **Design criteria, considerations, objective and tradeoffs.**
R. B. Dunn.
The SHIRSOG Workshop, p. 89 – 90 (1986). – See Abstr. 012.003.

034.005 **CCD detectors for synoptic spectroscopic monitoring.**
R. F. Green.
The SHIRSOG Workshop, p. 91 – 92 (1986). – See Abstr. 012.003.

034.006 **Report of the instrumentation sub–panel.**
R. Dunn, L. Ramsey, M. Smith.
The SHIRSOG Workshop, p. 93 – 94 (1986). – See Abstr. 012.003.

034.007 **Additive Dreifarben–Photographie mit dem Film Kodak 2415.**
E. Meyer.
Sternenbote, 30. Jahrg., Nr. 8, p. 158 – 165 (1987).

034.008 **Determination de l'échelle dans un micromètre à fils.**
O. S. Angelo.
Obs. Trav., No. 11, p. 49 – 52 (1987).
L'auteur analyse le procédé employé pour déterminer l'échelle du micromètre appliqué au réfracteur, diamètre 150 mm, longueur focale 3000 mm, de l'Observatoire privé E. Dembowski de Polpenazze (Brescia)–Italie. Il pense que cette expérience peut être utile à tous ceux qui voudront se mesurer avec l'observation des étoiles doubles visuelles.

034.009 **Unit–power "finders".**
A. S. White.
J. Br. Astron. Assoc., Vol. 97, No. 5, p. 291 – 292 (1987).
Finding an object in the sky can be the most difficult part of observational astronomy for the beginner. A single aid that enables the observer to align his telescope on the area of interest is described. Its advantages are simplicity and low cost.

034.010 **A new device for performing high–speed polarimetric measurements.**
K. Metz, D. Kunze, M. Roth, D. Hofstadt.
Messenger, No. 49, p. 24 – 25 (1987).

034.011 Preliminary data obtained with the integrating spectrometer mounted on the solar tower telescope BST–2 of the Crimean Astrophysical Observatory.
L. V. Didkovskij, I. E. Kozhevatov, N. N. Stepanyan.
Bull. Crimean Astrophys. Obs., Vol. 74, p. 133 – 146 (1987).
English translation of 41.034.097.

034.012 A hydrogen maser clock for space: clocks in future possible and improbable applications.
R. F. C. Vessot.
Radio astronomy from space, p. 195 – 213 (1987). – See Abstr. 012.009.

Development has continued on a space–borne atomic hydrogen maser clock for long–term continuous operation in space. The basic design of the successful space–borne clock used in the 1976 space probe to test Einstein's general theory of relativity has been evolved to include many new ideas. The design of a new clock is described and data of equivalent earth–based clocks are discussed.

034.013 Erratum: "The image photon counting system: performance in detail, and quest for high accuracy" [Mon. Not. R. Astron. Soc., Vol. 226, No. 2, p. 341 – 360 (1987)].
C. R. Jenkins.
Mon. Not. R. Astron. Soc., Vol. 227, No. 2, p. 543 (1987). See Abstr. 43.034.105.

034.014 The Faint Object Spectrograph for the 2.5–m Isaac Newton Telescope.
J. M. Breare, G. C. Cox, R. S. Ellis, G. P. Martin, I. R. Parry, A. Purvis, N. R. Waltham, J. Webster, R. A. E. Fosbury, D. W. Gellatly, P. R. Jorden, C. M. Lowne, W. F. Lupton, J. R. Powell, D. J. Thorne, I. G. van Breda, S. P. Worswick, C. G. Wynne.
Mon. Not. R. Astron. Soc., Vol. 227, No. 4, p. 909 – 919 (1987).

The Faint Object Spectrograph (FOS) is a highly efficient fixed format spectrograph for the 2.5–m Isaac Newton Telescope. It uses a CCD detector for low resolution (15 – 20 Å fwhm) spectrophotometry over the spectral range 4000 – 10500 Å. An outline of the mechanical and electronic arrangement of FOS is given, together with a discussion of its modes of operation and facilities. Data reduction software which has been written specifically for FOS is described. Commissioning tests on the spectrograph have been carried out during 1984 and 1985 and the performance of the instrument is assessed.

034.015 The multichannel astrometric photometer and atmospheric limitations in the measurement of relative positions.
G. D. Gatewood.
Astron. J., Vol. 94, No. 1, p. 213 – 224 (1987).

The operational multichannel astrometric photometer (MAP) now in use in the Allegheny Observatory astrometric program is the culmination of a decade of design and development effort. A detailed description of the system and its related software is followed by analysis of data acquired in four stellar regions. The study indicates an accuracy (in the sense of conformity to the best model), per night, for stars of the eighth magnitude or brighter, of 0.003 arcsec or better. These data points each have approximately twice the precision of the annual normal points obtained in the author's photographic program. The probable performance at more favorable sites is discussed briefly.

034.016 A simple 7226A/micro–computer photometer interface.
J. Gordon.
I.A.P.P.P. Commun., No. 29, p. 18 – 20 (1987).

034.017 Multichannel birefringent filter. (I). Principle and video spectrograph.
G.–x. Ai, Y.–f. Hu.
Sci. Sin., Ser. A, Vol. 30, No. 8, p. 868 – 876 (1987).

When a compound polarizing beam splitter replaces the traditional polaroid in an element of birefringent filter, the output of the element includes two beams, which are orthogonal in both polarizing and propagating orientations, but supplemental to each other in the transmitting spectra. When the compound polarizing beam splitters are used in each element of the filter, the multichannel birefringent filter can be constructed. When the thickest element is placed in the fore part of the filter followed by thinner elements one after another, a new spectrograph, the video spectrograph, can be obtained.

034.018 ME calibration update.
A. Smith, A. N. Parmar.
EXOSAT Express, No. 18, p. 32 – 34 (1987).

034.019 La Palma's "multi–purpose fotometer" and Be stars.
J. Tinbergen.
Be Star Newsl., No. 16, p. 22 – 25 (1987).

034.020 Eine fehlerfreie Reise–Kameranachführung.
K.–P. Schröder, W. Rössner.
Sterne Weltraum, 26. Jahrg., Nr. 7 – 8, p. 437 – 439 (1987).

034.021 Een TV–systeem voor meteoren.
K. Jobse, M. de Lignie.
Radiant, Jaarg. 9, Nr. 3, p. 38 – 41 (1987).

034.022 Photographie Planetarischer Nebel.
G. Reus.
Sterne Weltraum, 26. Jahrg., Nr. 10, p. 580 – 581 (1987).

034.023 Feasibility of cosmic neutrino detectors based on coherent interaction with superconductors.
P. F. Smith, J. D. Lewin.
Astrophys. J., Vol. 318, No. 2, p. 738 – 743 (1987).

The authors discuss in detail the feasibility of detecting a dominant neutrino component of the Galactic halo by coherent interaction with electrons in a superconducting target. Because the neutrino force is equivalent to an electric field $< 10^{-36} V$ cm^{-1}, the problem of transferring sufficient energy to a SQUID measuring coil leads to stringent requirements for the target coil. These requirements are analyzed. It is concluded that, although conceptually possible, the potential feasibility of superconducting detectors for relic low–energy neutrinos has not yet been demonstrated, and no realistic proposals for the construction of such detectors would be possible in the immediate future.

034.024 An off–axis guider for astrophotography.
D. Ratledge.
Astron. Now, Vol. 1, No. 4, p. 15 – 17 (1987).

With the advent of high–speed colour films, spectacular photographs can now be obtained of deep sky objects too faint to see, using nothing more than the amateur's equatorially–mounted Newtonian reflecting telescope. However, accurate guiding during the exposure is essential. One method of achieving this is to build an off–axis guider.

034.025 Improvements in spectroscopic continuum noise with fiber–optics illumination of a reticon array.
M. A. Smith, J. E. Graves, D. B. Jaksha, C. L. Plymate, L. W. Ramsey.
Publ. Astron. Soc. Pac., Vol. 99, No. 617, p. 654 – 659 (1987).

It is shown that under certain observing conditions the pixel–to–pixel spectroscopic noise of a Reticon detector can be reduced somewhat by the uniform illumination of the pixels by a fiber–optic image scrambler. These gains can nearly offset the transmission losses in the fiber imaging system.

034.026 Improving the blue and UV response of silicon CCD detectors.
R. W. Leach, M. P. Lesser.
Publ. Astron. Soc. Pac., Vol. 99, No. 617, p. 668 – 671 (1987).

A UV flooding technique derived from original work at JPL which boosts the response of two TI CCDs to nearly reflection limited values in the blue and UV is discussed along with some theory concerning silicon surface charging. This technique produces a photometrically stable detector that is well suited for use with ground–based optical telescopes.

034.027 A panoramic photon–counting detector system.
D. Durand, E. Hardy, J. Couture.
Publ. Astron. Soc. Pac., Vol. 99, No. 617, p. 686 – 694 (1987).
The authors describe the characteristics and implementation of the Laval Image Photon–Counting System (LIPS), a flexible, compact, and transportable 2–D detector system which is currently in use in the spectroscopic mode on the 1.6–m telescope at the Mount Mégantic Observatory. They stress the modularity and flexibility of the system and the richness of its software environment which allows real–time display and manipulation of the data. The present limitations in terms of quantum efficiency and resolution are discussed and an example of applications of LIPS to long–slit spectroscopy of elliptical galaxies to faint surface–brightness limits is presented.

034.028 One–channel spectrometers for earth–bound astronomy.
S. K. Panteleev.
Red. Zh. Kinematika Fiz. Nebesn. Tel, Kiev, 18 pp. (1987). In Russian. Abstr. in Ref. Zh., 51. Astron., 7.51.842 (1987).

034.029 Speckle–interferometry on the AZT–8 telescope at the Astronomical Observatory of the Kharkov University.
V. D. Bakhtin, V. G. Vakulik, A. P. Zheleznyak, V. V. Konichek, I. E. Sinel'nikov, S. A. Stepanov.
Kinematika Fiz. Nebesn. Tel, Tom 3, No. 4, p. 90 – 93 (1987). In Russian. English translation in Kinematics Phys. Celest. Bodies.
The speckle camera based on the image intentifier YM–92 is described. It is applied for observations of binary stars. The field of view is 30″, limiting stellar magnitude is 6^m. The possibility of a high–precision speckle interferometric measurements of binary stars with separations up to 30″ is demonstrated.

034.030 Ein Infrarot–Gitterspektrograph für die Calar Alto Teleskope.
A. Krabbe.
Diss. Naturwiss.–Math. Gesamtfak., Ruprecht–Karls–Univ., Heidelberg, F.R. Germany, 6 + 100 pp. (1987).

034.031 "Pegasus" – ein neues Modul für astronomische Berechnungen.
H. Spaude.
Sterne Weltraum, 26. Jahrg., Nr. 11, p. 637 (1987).

034.032 Astrophysical and terrestrial neutrinos in supernova detectors.
P. O. Lagage.
Neutrinos and the present–day universe, p. 127 – 143 (1986). – See Abstr. 012.025.
Supernova (SN) explosions are the place of very fundamental phenomena, whose privileged messengers are neutrinos. But such events are very rare, and SN detection has to be combined with other purposes. The recent developments of SN detectors have been associated with developments of underground particle physics (proton decay, monopoles ...). But here, the author will restrict himself to discuss the possibilities for a supernova detector to be sensitive to other sources of neutrinos, astrophysical or terrestrial.

034.033 Preliminary results on atmospheric neutrinos and Cygnus X–3 in the Fréjus detector.
P. Bareyre, R. Barloutaud, L. Behr, C. Berger, R. Bland, G. Chardin, H. Daum, B. Degrange, S. Demski, G. Deuzet, L. di Ciaccio, B. Dudelzak, D. Edmunds, J. Ernwein, P. Eschtruth, G. Gerbier, R. Hinners, A. Hofmann, M. A. Jabiol, S. Jullian, W. Kohrs, W. Kolton, B. Kuznick, D. Lalanne, F. Laplanche, C. Longuemare, R. Mayer, H. Meyer, L. Mosca, L. Moscoso, U. Nguyen–Khac, D. Ortmann, B. Pietryscz, C. Paulot, J. Peters, F. Raupach, P. Roy, G. Schmitz, M. Schubnell, P. Serri, G. Szklarz, J. Thierjung, S. Tisserant, J. Tutas, B. Voigtlander.
Neutrinos and the present–day universe, p. 146 – 169 (1986). Submitted to Nature. – See Abstr. 012.025.
The Fréjus nucleon decay detector is a 900 ton fine grain calorimeter located in the Fréjus road tunnel in the Alps at an average depth of 4800 mwe. The status of the experiment about neutrino interactions and cosmic ray events is presented, with a special emphasis on Cygnus X–3.

034.034 The spectro–interferometer of the Arcetri Solar Tower.
F. Cavallini, G. Ceppatelli, A. Righini, M. Meco, S. Paloschi, F. Tantulli.
Astron. Astrophys., Vol. 184, No. 1/2, p. 386 – 392 (1987).
The authors describe the spectro–interferometer installed at the Arcetri Observatory Solar Tower. This instrument basically consists of a Fabry–Perot interferometer mounted in tandem with a medium sized grating spectrograph, acting as order sorter. This mounting allows the measurement of solar absorption lines in the range 5500 – 6500 Å with high wavelength stability (0.08 mÅ rms in 12 h) and high spectral resolution (900,000 at 6328 Å). An image guider allows the pointing of an assigned solar region with an accuracy better than 2″ rms. This instrument, suitable for investigating line shifts and asymmetries, has been extensively used for studying such solar problems as meridional mass motions, line asymmetry "5–min" oscillations, and convective effects in solar active regions.

034.035 Instrumental profile of the HSFA–type spectrograph in the near infrared.
M. Sobotka, P. Kotrč.
Bull. Astron. Inst. Czech., Vol. 38, No. 5, p. 272 – 275 (1987).
A method of determining the instrumental profile of the horizontal solar spectrograph (HSFA) using an infrared sensitized photographic emulsion is described. Instrumental profiles, which include the effects of the non–zero slit width, diffraction and photographic emulsion along with possible distortions of the theoretical instrumental profile were determined by numerical convolution and deconvolution.

034.036 Sky on a chip: the fabulous CCD.
J. Janesick, M. Blouke.
Sky Telesc., Vol. 74, No. 3, p. 238 – 242 (1987).

034.037 The wandering stars of Allegheny.
R. W. Sinnott.
Sky Telesc., Vol. 74, No. 4, p. 360 – 363 (1987).

034.038 Measurement of the instrumental response function of the Mount Stromlo coudé échelle spectrograph.
I. A. Crawford, P. C. T. Rees, F. Diego.
Observatory, Vol. 107, No. 1079, p. 147 – 153 (1987).
The authors have calculated the theoretical instrumental response function of the Mount Stromlo spectrograph operating with the 130–inch camera. Measurements made with a narrow–line lamp and models based upon them have indicated that this theoretical profile is a fair approximation to the actual one.

034.039 Two–processor scanner for the GAT. I. New possibilities and description of operation.
S. V. Drabek, I. M. Kopylov, N. N. Somov, T. A. Somova.
Bull. Spec. Astrophys. Obs. – North Caucasus, Vol. 22, p. 54 – 61 (1987). English translation of 41.034.065.

034.040 Efficiency of M–300 and 2415 high–resolution photographic emulsions sensitized in hydrogen.
N. A. Tikhonov, M. F. Shabanov.
Bull. Spec. Astrophys. Obs. – North Caucasus, Vol. 22, p. 76 – 80 (1987). English translation of 41.034.066.

034.041 The most efficient astronomical photographic emulsions after sensitization in hydrogen.
M. F. Shabanov.
Bull. Spec. Astrophys. Obs. – North Caucasus, Vol. 22, p. 81 – 88 (1987). English translation of 41.034.067.

034.042 Spectrophotometry without dispersing elements.
B. I. Gil'man, V. K. Dubrovich, S. L. Liberman.
Bull. Spec. Astrophys. Obs. – North Caucasus, Vol. 22, p. 89 – 92 (1987). English translation of 41.034.068.

034.043 Contact transfer of the image from the screen of an image converter to a photographic emulsion.
L. V. Gyavgyanen, V. S. Rylov, T. A. Skosyrskaya.
Bull. Spec. Astrophys. Obs. – North Caucasus, Vol. 22, p. 93 – 95 (1987). English translation of 41.034.069.

034.044 Three–stage photoelectric unit for measuring the moments of meridian passages.
B. Bal'zhinova, D. Ojdov.
Nauchn. apparatura, Tom 1, No. 4, p. 89 – 94 (1986). In Russian. Abstr. in Ref. Zh., 51. Astron., 9.51.1107 (1987).

034.045 The multi–spectral fringe detector for the Mk III astrometric interferometer.
M. M. Colavita.
Interferometric imaging in astronomy, p. 125 – 127 (1987). – See Abstr. 012.035.

034.046 Groundbased high resolution imaging laboratory (GHRIL).
J. E. Noordam.
Interferometric imaging in astronomy, p. 223 – 224 (1987). Abstract. – See Abstr. 012.035.

034.047 Making a sidereal clock.
J. Watson.
J. Br. Astron. Assoc., Vol. 97, No. 6, p. 327 – 329 (1987).

034.048 Deep–sky photography.
R. Arbour.
J. Br. Astron. Assoc., Vol. 97, No. 6, p. 334 – 338 (1987).
Despite the poor climatic conditions in the UK, photography is now a technique used by most deep–sky observers. There is a plethora of instrumentation, materials and techniques available and this paper attempts to aid the amateur with his or her choice so that costly mistakes can be avoided.

034.049 A high–speed CCD photometer.
R. J. Stover, S. L. Allen.
Publ. Astron. Soc. Pac., Vol. 99, No. 618, p. 877 – 886 (1987). = Lick Obs. Bull., No. 1070.
A new high–speed photometer which employs a CCD as the detector is described. The CCD images are reduced and analyzed in real time and only the photometric measurements are recorded and displayed. The photometer can operate continuously with a 12–msec dead time between integrations and can do accurate differential photometry on up to six stars simultaneously. The signal–to–noise ratio of the photometric measurements is maximized by the use of a profile–weighting function. Because the sky brightness is accurately measured in each integration, accurate differential photometry through thin clouds and haze is possible.

034.050 Search for neutrino bursts from gravitational stellar collapses.
P. Galeotti.
General relativity and gravitational physics, p. 439 – 448 (1987). – See Abstr. 012.038.
Contents: 1. Neutrinos from collapsing stars. 2. The Mont Blanc Liquid Scintillation Detector (LSD). 3. The Gran Sasso Large Volume Detector (LVD).

034.051 The seismic noise challenge to the detection of low frequency gravitational waves.
A. Giazotto.
General relativity and gravitational physics, p. 449 – 463 (1987). – See Abstr. 012.038.
The seismic noise is producing a very large low frequency background in the test masses of an interferometric antenna for gravitational wave detection. Results are presented of an experiment performed in Pisa on the active seismic noise reduction. An experiment on the passive seismic noise reduction is also presented.

034.052 Operation of the cryogenic detector of the Rome group at CERN.
G. V. Pallottino.
General relativity and gravitational physics, p. 465 – 477 (1987). – See Abstr. 012.038.
In November 1985 the cryogenic gravitational wave detector of the Rome group at CERN has started to operate. The antenna is an aluminum cylinder with mass of 2270 kg and length of 3 m, cooled to 4.2K. The transducer is a resonant capacitor. The amplifier is a dc–SQUID. The detector obtained an effective noise temperature of 12 mK, corresponding to a metric tensor impulse variation with Fourier transform $H = 8.8 \times 10^{-22} \text{Hz}^{-1}$.

034.053 Resonant–mass detectors of gravitational radiation.
P. F. Michelson, J. C. Price, R. C. Taber.
Science, Vol. 237, No. 4811, p. 150 – 157 (1987).
A network of second–generation low–temperature gravitational radiation detectors is nearing completion. These detectors, sensitive to mechanical strains of order 10^{-18}, are possible because of a variety of technical innovations that have been made in cryogenics, low–noise superconducting instrumentation, and vibration isolation techniques. Another five orders of magnitude improvement in energy sensitivity of resonant–mass detectors is possible before the linear amplifier quantum limit is encountered.

034.054 New image dissector spectrometer of the Crimean Astrophysical Observatory.
A. B. Bukach, I. V. Il'in, A. E. Tarasov, A. G. Shcherbakov.
Izv. Krymskoj Astrofiz. Obs., Tom 76, p. 192 – 200 (1987). In Russian. English translation in Bull. Crimean Astrophys. Obs., Vol. 76.
An image dissector spectrometer (IDS) based on one–stage fiber–optic image tube attached to dissector unit is designed for stellar spectra registration in coude focus of the 2.6–m telescope of the Crimean Observatory. It permits to attain 4000 – 8000 Å spectral region with dispersion of 0.42 Å/pix and to place 256×64 pixels with dimension 0.1×0.4 mm on dissector unit. The spectral resolution of IDS equals to 2 pixels. The IDS is controlled by microcomputer MERA–60 with RT–60 operational system. Some preliminary results of standard and variable star observations are presented.

034.055 System design and high–accuracy astrometry.
C. E. KenKnight.
Bull. Am. Astron. Soc., Vol. 19, No. 2, p. 690 (1987). Abstract. – See Abstr. 010.061.

034.056 A magneto–optical filter for solar oscillation measurements.
S. Tomczyk, A. Cacciani, E. J. Rhodes Jr.
Bull. Am. Astron. Soc., Vol. 19, No. 2, p. 701 (1987). Abstract. – See Abstr. 010.061.

034.057 Differential non–linearity in analog–to–digital converters: photometric consequences for astronomical microdensitometry.
H. M. Heckathorn.
Bull. Am. Astron. Soc., Vol. 19, No. 2, p. 746 – 747 (1987). Abstract. – See Abstr. 010.061.

034.058 Initial stellar diameter measurements with the Mark III interferometer.
D. J. Hutter, M. Shao, M. M. Colavita.
Bull. Am. Astron. Soc., Vol. 19, No. 2, p. 748 (1987). Abstract. – See Abstr. 010.061.

034.059 The DDO photon counting spectrometer: first results.
S. W. Mochnacki, S. Chew, K. Kamper, W. Kunowski, F. Hawker, D. Blyth, A. Ridder.
Bull. Am. Astron. Soc., Vol. 19, No. 2, p. 748 (1987). Abstract. – See Abstr. 010.061.

034.060 **X–ray reflectivity measurements of high line density gratings.**
M. P. Kowalski, J. F. Meekins, R. G. Cruddace.
Bull. Am. Astron. Soc., Vol. 19, No. 2, p. 749 (1987). Abstract. –
See Abstr. 010.061.

034.061 **Geometrical characteristics of light detectors and their influence on registered angular dimensions of Čerenkov flashes on gamma telescopes.**
Yu. L. Zyskin.
Izv. Krymskoj Astrofiz. Obs., Tom 77, p. 179 – 189 (1987). In Russian. English translation in Bull. Crimean Astrophys. Obs., Vol. 77.

In the majority of modern ground–based gamma–telescopes operating at very high energy bands while chosing γ–events from the cosmic rays nuclear component background, one needs to analyse the dimension, shape and direction of Čerenkov pulse image in the focal plane of a narrow angle detector, the image tube being used as a light detector. The range of registered pulses depends on the angular aperture, number and size of image tubes and the implementation of light detectors. The influence of these factors on the searched parameters and the confidence level of γ–showers is analyzed and the optimal characteristics of a light detector for energies $\sim 10^{12}$eV are found. Particularly, one of the best designs comprises 37 image tubes with light guides and aperture angle $\sim 3°$.

034.062 **Electrooptic Fabry–Perot filter: development for the study of solar oscillations.**
C. H. Burton, A. J. Leistner, D. M. Rust.
Appl. Opt., Vol. 26, No. 13, p. 2637 – 2642 (1987).

Observations of nonradial oscillations require Doppler velocity measurement at many points over the photosphere with a velocity resolution better than 1 m/s. An attractive form of imaging spectrophotometer for such a task utilizes a thin, solid, electrically tunable Fabry–Perot interference filter or etalon made of an electrooptic material such as lithium niobate ($LiNbO_3$). The problems to be overcome in producing such an etalon for an imaging spectrophotometer are discussed and practical solution demonstrated on the basis of measurements made on prototype devices.

034.063 **Quantum efficiencies of imaging detectors with alkali halide photocathodes. 1: microchannel plates with separate and integral CsI photocathodes.**
G. R. Carruthers.
Appl. Opt., Vol. 26, No. 14, p. 2925 – 2930 (1987).

The author has measured and compared the quantum efficiencies of microchannel plate (MCP) detectors in the far–UV (below 2000–Å) wavelength range using CsI photocathodes (a) deposited on the front surfaces of microchannel plates and (b) deposited on solid substrates as opaque photocathodes with the resulting photoelectrons input to microchannel plates. Typical efficiencies are $\sim 15\%$ at 1216 Å for a CsI–coated MCP compared with 65% for an opaque CsI photocathode MCP detector. Special processing has yielded an efficiency as high as 20% for a CsI–coated MCP. At present there still remains a factor of at least 3 quantum efficiency advantage in the separate opaque CsI photocathode configuration.

034.064 **Blaze angle measurements of 31.6–g/mm and 79.01–g/mm R2 echelle gratings from Bausch & Lomb.**
F. Diego.
Appl. Opt., Vol. 26, No. 22, p. 4714 – 4716 (1987).

034.065 **Acoustooptic filters for astronomical photometry: design and fabrication.**
B. Bates, D. R. Halliwell, S. McNoble, Y. Li, M. Catney.
Appl. Opt., Vol. 26, No. 22, p. 4783 – 4787 (1987).

Using calculations based on the acoustooptic properties of the crystal material a procedure is outlined for the design of filters having a particular bandwidth specification and reduced sensitivity to the incidence angle of the input beam. Relevant aspects of the fabrication techniques are given. The optical performance of

a prototype filter having a 0.5 – 2–nm bandwidth over the 400 – 700–nm tuning range based on the design methods outlined is satisfactory for the chosen application, but it is not optimum.

034.066 **Photometric characteristics of the Vega 1 and Vega 2 CCD cameras for the observation of comet Halley.**
A. Abergel, J. L. Bertaux, G. A. Avanessov (*G. A. Avanesov*),
V. I. Tarnopolsky (*V. I. Tarnopol'skij*), B. S. Zhukov.
Appl. Opt., Vol. 26, No. 20, p. 4457 – 4468 (1987).

The first pictures of the nucleus of comet Halley were returned from the CCD TV system placed onboard the two Soviet spacecraft Vega 1 and 2. After a brief description of the experiment, the on–ground calibration tests are discussed. Many images were registered and processed to obtain standard correcting images and absolute calibration. Photometric performance could also be checked during flight with observations of Jupiter; in–flight and on–ground performances are compared.

034.067 **A 10–μm infrared camera.**
J. F. Arens, J. G. Jernigan, M. C. Peck,
C. A. Dobson, E. Kilk, J. Lacy, S. Gaalema.
Appl. Opt., Vol. 26, No. 18, p. 3846 – 3851 (1987).

An IR camera has been built for astronomical observations. The camera has been used primarily for high angular resolution imaging at mid–IR wavelengths. In the observations the system has been used as an imager with interference coated and Fabry–Perot filters. These measurements have demonstrated a sensitivity consistent with photon shot noise, showing that the system is limited by the radiation from the telescope and atmosphere. Measurements of read noise, crosstalk, and hysteresis have been made.

034.068 **Progress toward a multiobject radial–velocity spectrometer.**
T. E. Lutz, T. Ingerson, W. G. Weller.
Astron. J., Vol. 94, No. 6, p. 1686 – 1693 (1987).

The authors show that it is possible to obtain stellar radial velocities of high accuracy using a fiber–fed echelle spectrograph with no cross disperser. Although the resulting spectrum contains many overlapping orders, the radial–velocity information is still present. Velocities may be obtained by cross correlation with reference data obtained with the same instrument. The accuracy is a function of the signal–to–noise ratio, and ranges from ±4 km/s to ±1 km/s. Velocities are obtained with exposure times just sufficient to obtain the barest trace of a spectrum using the same spectrograph and CCD detector with a cross disperser. Thus it appears possible to obtain precision similar to that obtained with an order separator, but with considerably better sensitivity, due to a multiplexing gain.

034.069 **The Lick Observatory TV autoguider.**
R. Kibrick, L. Robinson.
Publ. Astron. Soc. Pac., Vol. 99, No. 619, p. 1014 – 1021 (1987). = Lick Obs. Bull., No. 1076.

A microprocessor–based telescope autoguider is described that uses the signal from an existing field acquisition TV camera to generate tracking–error information for an optical telescope.

034.070 **Erratum: "The silicon linear–array camera of the Vienna Observatory" [Publ. Astron. Soc. Pac., Vol. 99, No. 614, p. 304 – 306 (1987)].**
W. W. Weiss, A. Schalk, V. Ogris.
Publ. Astron. Soc. Pac., Vol. 99, No. 620, p. 1102 (1987). See Abstr. 43.034.138

034.071 **An ultra–high resolution spectrograph in a hurry.**
P. Gillingham.
Anglo–Aust. Obs., Prepr., No. 221, 16 pp. (1987).

034.072   The University of California at San Diego near–infrared charge–coupled device (CCD).
S. D. Friedman, B. Jones, R. C. Puetter.
Proc. SPIE Int. Soc. Opt. Eng., Vol. 686, p. 96 – 100 (1986). Abstr. in Phys. Abstr., Vol. 90, No. 1309, Entry 80880 (1987). – See Abstr. 012.040.

034.073   A sidereal drive corrector for small telescopes.
N. M. Ashok, T. Chandrasekhar, K. S. B. Manian.
J. Phys. E, Vol. 20, No. 4, p. 469 – 470 (1987). Abstr. in Phys. Abstr., Vol. 90, No. 1309, Entry 88116 (1987).

034.074   Wide bandwidth signal processor for removing dispersion distortion from pulsar radio signals.
T. H. Hankins, J. M. Rajkowski.
Rev. Sci. Instrum., Vol. 58, No. 4, p. 674 – 680 (1987). Abstr. in Phys. Abstr., Vol. 90, No. 1309, Entry 88124 (1987).

034.075   Use of the U.S. TRANSIT navigation satellite FTM.
A.–x. Ma.
Publ. Shaanxi Astron. Obs., Vol. 9, No. 1, p. 25 – 31 (1986). In Chinese.

034.076   BPM time signal in Wu Chang.
H.–q. Chen, L.–x. Ma.
Publ. Shaanxi Astron. Obs., Vol. 9, No. 1, p. 32 – 37 (1986). In Chinese.

034.077   Flying clock measure longwave timing accuracy in Guangzhou Man–Made Satellite Observing Station.
J.–f. Peng.
Publ. Shaanxi Astron. Obs., Vol. 9, No. 1, p. 38 – 43 (1986). In Chinese.

034.078   Technique of ”scanning and display” with time mark used for the frequency calibration and timing.
R.–m. Fan.
Publ. Shaanxi Astron. Obs., Vol. 9, No. 1, p. 48 – 50 (1986). In Chinese.

034.079   Introduction to the CSAO–1 standard time signal receiver.
H.–y. Wang.
Publ. Shaanxi Astron. Obs., Vol. 9, No. 1, p. 51 – 52 (1986). In Chinese.

034.080   Determination of stellar positions and ERP with a stellar interferometer. I. Synopsis of all kinds of interferometers.
C. Li.
Publ. Yunnan Obs., No. 1, p. 23 – 32 (1987). In Chinese.
    The development of stellar interferometry and the principles of various stellar interferometers, as well as the theories which the interferometers depend on, are discussed.

034.081   Determination of stellar positions and ERP with a stellar interferometer. II. Geometric principles.
C. Li, W. Mao.
Publ. Yunnan Obs., No. 1, p. 33 – 49 (1987). In Chinese.
    From basic principles and derivation of formulae, the authors analyze the use of the stellar interferometer to determine stellar positions and earth rotation parameters (ERP).

034.082   Determination of stellar positions and ERP with a stellar interferometer. III. Achieving of the observed quantity τ.
C. Li.
Publ. Yunnan Obs., No. 1, p. 50 – 56 (1987). In Chinese.
    Two examples are briefly given and depending on their outcome, the possibility of achieving the observed quantity – time delay τ – is analyzed.

034.083   Improvement of the performance of a digital acousto–optic spectrograph.
Z.–g. Xia, S.–y. Jiang, J.–y. Chen.
Publ. Yunnan Obs., No. 1, p. 60 – 64 (1987). In Chinese.

034.084   Report on experimental tests of a near infrared detector system.
Y. Zhang, G.–z. Xie, R.–w. Lu, N.–p. Xue, J. Sun, G.–l. Zen.
Publ. Yunnan Obs., No. 1, p. 114 – 122 (1987). In Chinese.

034.085   Grating form of the micrometer of a low–latitude meridian circle. II.
L.–k. Wang, Y. Fan, W. Mao.
Publ. Yunnan Obs., No. 2, p. 1 – 6 (1987). In Chinese.

034.086   Determination of the irregularity of the pivot of the horizontal axis of a meridian circle by means of an axis collimator.
Y. Fan, L.–k. Wang, W. Mao.
Publ. Yunnan Obs., No. 2, p. 7 – 12 (1987). In Chinese.

034.087   The analysis of short wave time receiving data.
X.–j. Chen.
Publ. Yunnan Obs., No. 2, p. 17 – 20 (1987). In Chinese.

034.088   The stability of the time delay of short wave transmissions and its adopted values.
X.–j. Chen.
Publ. Yunnan Obs., No. 2, p. 21 – 24 (1987). In Chinese.

034.089   The control circuit of a telescope for the observation of Halley's comet.
Z.–z. Liu.
Publ. Yunnan Obs., No. 2, p. 83 – 86 (1987). In Chinese.
    The control circuit of the four–axis telescope, which is used to observe Halley's comet in Yunnan Province, is introduced in this paper. A CMOS integrated circuit with high anti–interference and lower power consumption is used in the control circuit and a step motor is utilized as the driver. The rotational speed is adjustable and stable, thereby making the mechanical rotation system much simpler.

034.090   The prediction of the four–axis tracking of Halley's comet.
J.–s. Wang.
Publ. Yunnan Obs., No. 2, p. 87 – 89 (1987). In Chinese.

034.091   Hypersensitization of Kodak spectroscopic plates of types IIa–O and IIIa–J by hydrogen soaking.
C.–s. Zhang, H. Ke.
Publ. Purple Mt. Obs., Vol. 6, No. 2, p. 156 – 157 (1987). In Chinese.

034.092   Signal–to–noise ratio and astronomical Fourier transform spectroscopy.
J. P. Maillard.
Inst. Astrophys. Paris, Pré–Publ., No. 203, 8 pp. (1987). To appear in IAU Symp. No. 132.

034.093   A new table of constants of the step wedges in log intensity.
Tokyo Astron. Obs., Kiso Inf. Bull., Vol. 2, No. 4, p. 90 – 91 (1987).

034.094   Germanium diodes as high performance near infrared detectors.
G. H. Rieke, R. J. Elston, M. J. Lebofsky, C. E. Walker.
Prepr. Steward Obs., No. 742, 5 pp. (1987). To appear in Proc. Hilo Conference on Infrared Arrays.

034.095   Rockwell HgCdTe arrays as imagers.
M. J. Rieke, G. H. Rieke, E. F. Montgomery.
Prepr. Steward Obs., No. 743, 8 pp. (1987). To appear in Proc. Hilo Conference on Infrared Arrays.

034.096 A CCD–based imaging polarimeter for astronomy.
D. P. Clemens, R. W. Leach.
Prepr. Steward Obs., No. 748, 12 pp. (1987). To appear in Opt. Eng.

034.097 The MX spectrometer: operational results.
J. M. Hill, M. P. Lesser.
Prepr. Steward Obs., No. 760, 17 pp. (1987). To appear in "Instrumentation for ground–based optical astronomy: present and future".

034.098 Automated photometer at the Skalnaté Pleso Observatory.
Ľ. Klocok, J. Žižňovský, J. Zverko.
Contrib. Astron. Obs. Skalnaté Pleso, Vol. 16, p. 43 – 46 (1987).
An automated interactive photometric measuring device, fed by an 0.6–m reflector and built about a programmable desk calculator, is described.

034.099 The photometric system of the electropolarimeter of the 1–m telescope.
A. V. Didenko, F. K. Rspaev.
Tr. Astrofiz. Inst. Alma–Ata, Tom 48, p. 109 – 111 (1987). In Russian.
Concerns the 1–m telescope at the Assy–Turg. Obs. (Kazakh SSR).

034.100 Four–channel electrophotometer on the base of the UM–92 photomultiplier.
I. G. Babkin.
Tr. Astrofiz. Inst. Alma–Ata, Tom 48, p. 129 – 135 (1987). In Russian.

034.101 Superwide–angle astronomical camera.
I. N. Nikanorova.
Glav. Astron. Obs. Akad. Nauk SSSR. Leningrad, 7 pp. (1987). In Russian. Abstr. in Ref. Zh., 51. Astron. 11.51.918 (1987).

034.102 Instrumentación astronómica en México.
E. Ruiz Schneider.
Rev. Mex. Astron. Astrofis., Vol. 14, No. 2, p. 717 – 721 (1987). – See Abstr. 012.042.
The progress of astronomy is mainly based on quality and quantity of observations made with astronomical instruments. This work describes the development and design of some of them mainly used to solve the control and handling problems of telescopes, the acquisition and transfer of massive data and the development of new devices used to detect with high efficiency and resolution the information of bidimensional optical incidency.

034.103 Guiador excéntrico para el telescopio de 2.1 mts. en Sn. Pedro Martir, B.C.
A. Angeles, E. Carrasco, S. Cuevas, R. Enríquez, L. Gutiérrez, R. Langarica, E. Ruiz, E. Sacristán.
Rev. Mex. Astron. Astrofis., Vol. 14, No. 2, p. 736 – 738 (1987). – See Abstr. 012.042.
The operational aspects of the optic, mechanic and electronic systems of the offset guider for the UNAM212 telescope are presented.

034.104 Integración del guiador excéntrico al telescopio UNAM212 y su repercusión en las imágenes obtenidas con diversos instrumentos.
F. J. Cobos, R. Langarica.
Rev. Mex. Astron. Astrofis., Vol. 14, No. 2, p. 739 – 743 (1987). – See Abstr. 012.042.

034.105 El sistema CCD de Tonantzintla. Pruebas y planes futuros.
O. Cardona, E. Chavira, I. Furenlid, B. Iriarte.
Rev. Mex. Astron. Astrofis., Vol. 14, No. 2, p. 744 – 748 (1987). – See Abstr. 012.042.
The authors present results of the laboratory tests of the CCD camera system recently acquired by INAOE, also the theoretical and observational performance of the instrument with the 1m–telescope of UNAM. The system has a TI 4849 CCD with 390 × 584 pixels.

034.106 Sistema de adquisición de datos para un fotómetro doble basado en una microcomputadora.
F. Angeles.
Rev. Mex. Astron. Astrofis., Vol. 14, No. 2, p. 749 – 750 (1987). – See Abstr. 012.042.
An IBM PC compatible microcomputer based acquisition system for the double photometer of Tonantzintla, Puebla, is proposed.

034.107 Un interferómetro de rotación de frente de onda con modulación.
S. Cuevas.
Rev. Mex. Astron. Astrofis., Vol. 14, No. 2, p. 753 – 755 (1987). – See Abstr. 012.042.
A modulating rotation shearing interferometer is proposed. The optical path is modulated by an oscillating glass parallel faces plate. The image processing with this interferometer is simplified.

034.108 Diseño de un interferómetro de rotación de frente de onda para obtención de imágenes de alta resolución espacial.
S. Tinoco, S. Cuevas.
Rev. Mex. Astron. Astrofis., Vol. 14, No. 2, p. 756 – 758 (1987). – See Abstr. 012.042.
The mechanical and optical design of an instrument for high resolution imaging is described. It consists of a Roddier rotation shearing interferometer working on the telescope pupil plane and one additional path allowing the acquisition of the image plane speckles.

034.109 Deposición de un fotocátodo S11 para uso astronómico.
G. F. Bisiacchi, C. Firmani.
Rev. Mex. Astron. Astrofis., Vol. 14, No. 2, p. 771 – 775 (1987). – See Abstr. 012.042.
Experiments have been done to produce an antimony–cesium photocathod (S11). The procedure generally used to produce this kind of thin film has been modified. The quantum efficiency obtained as function of the wavelength is good compared with that obtained by other authors.

034.110 Resolution and sensibility increase in CR–39 plastic cosmic ray detector.
A. Laville, J. Pérez–Peraza, D. López, M. Balcazar–García, A. López.
Rev. Mex. Astron. Astrofis., Vol. 14, No. 2, p. 776 – 782 (1987). – See Abstr. 012.042.

034.111 A ground–based interferometric system for the near–infrared.
J.–M. Mariotti, S. Ridgway, P. Léna, Y. Rabbia, F. Sibille.
ESA Spec. Publ., ESA SP–273, p. 175 (1987). Abstract. – See Abstr. 012.044.

034.112 Research of Q values of some materials for gravitational wave antennae at low temperatures.
S.–h. Chen, M.–x. Tang.
Acta Sci. Nat. Univ. Sunyatseni, No. 4, p. 89 – 93 (1986). In Chinese. Abstr. in Phys. Abstr., Vol. 90, No. 1310, Entry 89101 (1987).

034.113   **A piezoceramic position detector.**
R. Weilguny, G. Fasching.
Feinwerktech. Messtech., Vol. 95, No. 1, p. 35 – 37 (1987). In German. Abstr. in Phys. Abstr., Vol. 90, No. 1311, Entry 101683 (1987).

034.114   **Ultra–high–energy gamma–ray astronomy using atmospheric Cerenkov detectors at large zenith angles.**
P. Sommers, J. W. Elbert.
J. Phys. G, Vol. 13, No. 4, p. 553 – 566 (1987). Abstr. in Phys. Abstr., Vol. 90, No. 1312, Entry 107724 (1987).

034.115   **Zur Theorie der analemmatischen Sonnenuhr.**
H. Lippold.
Sterne, 63. Band, Heft 4, p. 225 – 229 (1987).

034.116   **A new CCD camera for the Echelec spectrograph.**
A. Gilliotte, P. Magain.
Messenger, No. 50, p. 46 – 47 (1987).

034.117   **First results from remote control observations with CAT/CES.**
P. François, E. Brocato.
Messenger, No. 50, p. 47 – 48 (1987).

034.118   **Calibration of a charge–coupled device.**
J. Whyte, B. McBreen, L. Metcalfe.
Ir. Astron. J., Vol. 18, No. 2, p. 89 – 91 (1987). – See Abstr. 012.036.
A CCD camera system has been designed and used in a programme to study blue compact galaxies. A calibration system is described here that has been designed and constructed so that all the parameters of the CCD can be determined in the laboratory.

034.119   **A simple device for occultation observations.**
P. Lee, C. Watson, H. Williams, R. Miller, R. James, W. Coskrey.
I.A.P.P.P. Commun., No. 30, p. 5 – 14 (1987).

034.120   **Interfacing the Starlight–1 photometer to the Osbourne Executive computer.**
E. Moore, G. Denn.
I.A.P.P.P. Commun., No. 30, p. 20 – 24 (1987).

034.121   **Les intensificateurs d'image: utilisation en astronomie amateur.**
T. Midavaine.
Obs. Trav., No. 12, p. 5 – 22 (1987).

034.122   **Mesure de la méridienne de Le Monnier à St. Sulpice**
A. Gotteland.
Obs. Trav., No. 12, p. 23 – 36 (1987).

034.123   **Astrophotographie mit Farbnegativfilmen.**
J. Alean.
Orion, 45. Jahrg., Nr. 223, p. 221 – 224 (1987). In German and French.

034.124   **An adiabatic demagnetization cooled bolometer system for millimeter continuum astronomy.**
L. Lesyna.
Diss. Abstr. Int., Sect. B, Vol. 48, No. 6, p. 1711–B (1987). Thesis, Stanford University, 130 pp. (1987). Order No. DA8720411.

034.125   **Participation du CERGA et de l'Observatoire de Nice à la campagne PHEMU85.**
M. Froeschlé, C. Meyer, G. Helmer, D. Pinte, B. Chauvineau.
Ann. Phys. (Paris), Vol. 12, Colloq. No. 1, p. 39 – 50 (1987). – See Abstr. 012.061.
Nice Observatory and CERGA have participated to the PHEMU85 campaign of observations of the mutual phenomena.

In this paper the authors describe the photometric instrumentations used for these observations and they give some light–curves obtained.

034.126   **Observations PHEMU85 à l'Observatoire de Bordeaux.**
G. Dourneau.
Ann. Phys. (Paris), Vol. 12, Colloq. No. 1, p. 51 – 58 (1987). – See Abstr. 012.061.
Mutual phenomena have been observed at the Bordeaux Observatory. Here, the author gives a description of the photometric instrumentation, and of the observations made.

034.127   **Un photomètre adapté aux phénomènes mutuels.**
F. Sèvre.
Ann. Phys. (Paris), Vol. 12, Colloq. No. 1, p. 175 – 194 (1987). – See Abstr. 012.061.
The author describes several types of photoelectric photometers, and some problems related to their utilization. He also gives indications to build a photometer well–adapted to the specific problem of the mutual phenomena of the Galilean satellites.

034.128   **Infrared arrays for ground–based astronomy.**
F. C. Gillett.
Infrared astronomy with arrays, p. 3 – 12 (1987). – See Abstr. 012.070.
Ground–based infrared astronomy is currently going through a revolution as a result of the introduction of integrating array detectors. The background limited imaging and moderate resolution spectroscopic performance of a large ground–based telescope in the 1 to 20 μm range is evaluated. The corresponding array requirements are derived.

034.129   **SWIR HgCdTe focal plane arrays for astronomy.**
J. P. Rode, J. D. Blackwell, M. A. Blessinger, K. Vural.
Infrared astronomy with arrays, p. 13 – 20 (1987). – See Abstr. 012.070.
Short–wavelength infrared focal plane arrays (FPAs) for astronomical uses are under development at Rockwell Science Center. These FPAs use photovoltaic HgCdTe in a hybrid structure which utilizes a Si multiplexer. The current generation of arrays, which are in use in several telescopes, is represented by a 64 × 64 array on 52 μm centers. This paper discusses both current and future FPAs.

034.130   **Short wavelength HgCdTe photovoltaic detector arrays for low background astronomy applications.**
N. Hartle, J. Stobie, A. Hairston, P. Zimmermann, A. Sood, R. Capps, D. Depoy, D. Hall.
Infrared astronomy with arrays, p. 21 – 28 (1987). – See Abstr. 012.070.
The performance of short wavelength (1 – 3 μm) photovoltaic HgCdTe detector/focal plane arrays are presented. BLIP performance for low background space astronomy applications has been demonstrated with HgCdTe mosaic detector arrays which have RoA products of 10^{12} ohm cm^2 at 77K and quantum efficiencies greater than 70% at 175K. Multiplexer dark currents of 20 e/s were measured and multiplexer read noise of 30 rms electrons at 77K are obtainable.

034.131   **Operation and calibration of self–integrating multiplexed arrays.**
A. W. Hoffman.
Infrared astronomy with arrays, p. 29 – 36 (1987). – See Abstr. 012.070.
Multiplexed infrared detector arrays that use a direct, self–integrating input SFD (Source Follower per Detector) circuit have become available to the astronomy community. Because of its simplicity and flexibility, this circuit is used in several array designs, including the 58–by–62 element arrays offered commercially by Santa Barbara Research Center. The SFD circuit has been successfully coupled to both photovoltaic and photoconductive detectors. This paper discusses the theory of operation of the SFD input and demonstrates that a calibrated two–

dimensional image can be attained with a dark slide and a single flat–field source.

034.132 **Testing of a 58 × 62 InSb engineering array.**
Z. Ninkov, W. J. Forrest, J. L. Pipher.
Infrared astronomy with arrays, p. 37 – 40 (1987). – See Abstr. 012.070.
The authors report on results of their detector evaluation program on a high doped InSb 58 × 62 hybridized detector array obtained from Santa Barbara Research Center for engineering checkout.

034.133 **Fabrication of low–capacitance InSb detector.**
H. Murakami, K. Noguchi, T. Matsumoto, M. Noda, H. Fujisada, K. Hara, T. Sasase.
Infrared astronomy with arrays, p. 41 – 44 (1987). – See Abstr. 012.070.

034.134 **Performance measurements on low noise infrared detector arrays for astronomical applications.**
M. Hewitt, D. Randall.
Infrared astronomy with arrays, p. 45 – 51 (1987). – See Abstr. 012.070.
The description and performance of an infrared charge–coupled device imager with low noise (250 e⁻) and large dynamic range (5×10^4) are presented.

034.135 **FET switch multiplexers for large area infrared focal planes.**
G. C. Bailey.
Infrared astronomy with arrays, p. 52 – 59 (1987). – See Abstr. 012.070.

034.136 **Evaluation of a selfscanned linear InSb detector array.**
G. Finger, M. Meyer, A. F. M. Moorwood.
Infrared astronomy with arrays, p. 60 – 68 (1987). – See Abstr. 012.070.
A noise model for detectors operated in the capacitive discharge mode is presented. It is used to analyse the noise performance of the ESO nested timing readout technique applied to a linear 32 element InSb array which is multiplexed by a silicon switched–FET shift register.

034.137 **Germanium diodes as high performance near infrared detectors.**
G. H. Rieke, R. J. Elston, M. J. Lebofsky, C. E. Walker.
Infrared astronomy with arrays, p. 69 – 74 (1987). – See Abstr. 012.070.
Germanium photodiodes manufactured by Ford Aerospace have dark currents below 20 electrons/sec and high quantum efficiency. Coupled with integrating JFET readouts, these detectors can perform background–limited spectroscopy between 1 and 1.6 μm.

034.138 **Solid state line scanners employing PtSi Schottky–Barrier detectors.**
G. P. Weckler, L. R. Hudson, H.–F. Tseng, W. L. Wang.
Infrared astronomy with arrays, p. 75 – 82 (1987). – See Abstr. 012.070.
Self–scanned linear arrays with sensitivities in the mid–wave infrared are described. These devices are similar to existing visible line scanners except the p–n junction detector is replaced by a PtSi Schottky–Barrier detector. Two architectures are described, one that employs a digital MOS scan–generator readout, and one that employs an analog CCD readout.

034.139 **Performance of a Litton 64 × 64 channel CCD multiplexer.**
D. Enders, D. Pommerrening, M. E. Gramer.
Infrared astronomy with arrays, p. 83 – 86 (1987). – See Abstr. 012.070.
Design features and data are presented on a 64 × 64 surface channel CCD multiplexer developed by Litton Electron Devices Division, Tempe, Arizona. These data include: noise vs. signal

level and temperature, conversion gain vs. temperature and transfer efficiency vs. temperature.

034.140 **Si–Ga array detectors optimized for the 10 μm atmospheric window.**
P. O. Lagage, L. Audaire, C. Cesarsky, C. Lucas, J. L. Monin, F. Sibille.
Infrared astronomy with arrays, p. 87 – 90 (1987). – See Abstr. 012.070.
Si–Ga DRO detector arrays, specifically designed for 10 μm ground–based observations, are under development at the LETI/LIR. These detectors will be integrated, at the end of this year, in an IR camera currently built for INSU by the astrophysical laboratories of Lyon, Saclay and Grenoble.

034.141 **Development of close packed low temperature bolometer arrays.**
F. J. Low, T. Nishimura, A. W. Davidson, M. Alwardi.
Infrared astronomy with arrays, p. 91 – 96 (1987). – See Abstr. 012.070.
Using the high yield and nearly "perfect" performance of Si bolometers with ion implanted contacts and metalized glass supports, a 1 × 8 element close packed array was built for super resolution of the Kuiper Airborne Observatory.

034.142 **Solid state photomultiplier: a high performance detector for astronomy.**
R. Bharat, M. D. Petroff, M. G. Stapelbroek.
Infrared astronomy with arrays, p. 97 – 102 (1987). – See Abstr. 012.070.
A solid state photomultiplier (SSPM) developed by Rockwell has been used to detect single photons at wavelengths between 0.45 and 20 microns on a continuous basis. Data are presented on the performance of a silicon SSPM under various operating conditions.

034.143 **Detector arrays for high resolution spectroscopy from 5 – 28 microns.**
G. Wiedemann, D. E. Jennings, S. H. Moseley, G. Lamb.
Infrared astronomy with arrays, p. 103 – 107 (1987). – See Abstr. 012.070.
A linear Si:As BIB detector array (Rockwell International) is being implemented in a postdispersion detection system for ground based Fourier transform spectrometers. The array version can be used as a multichannel narrow band filter for extended spectral coverage or for imaging with a narrow bandpass. A Si:As solid state photomultiplier array (Rockwell) is evaluated for use in high resolution infrared spectrometers. Test results and applications are discussed.

034.144 **Development of optimized low–background photoconductors.**
E. T. Young, F. J. Low, G. H. Rieke, E. E. Haller, J. Beeman.
Infrared astronomy with arrays, p. 121 – 127 (1987). – See Abstr. 012.070.
The authors report on the photoconductor array development effort in support of future space–borne infrared missions. Measurements of detector properties are presented for Si:In, Si:Ga, Si:B, Ge:Be, and Ge:Ga. Together, these materials span the wavelength range from 2 μm to 120 μm. The detector arrays described are interfaced to the latest generation of J–Fet integrators. The authors present measurements of the performance of these amplifiers that demonstrate an equivalent read noise of less than 20 e for a 100 s integration time.

034.145 **Test results with Rockwell SiBIBIB hybrid arrays.**
T. Herter, C. Fuller, G. E. Gull, J. R. Houck.
Infrared astronomy with arrays, p. 128 – 135 (1987). – See Abstr. 012.070.
Initial test results for a Rockwell International hybrid array under low background conditions are presented. This hybrid consists of a 10 × 50 Si:As Back–Illuminated Blocked–Impurity–Band (SiBIBIB) detector pumb–bonded to a low–noise switched MOSFET multiplexer. Measurements of linearity, read noise,

dark current, responsivity, and quantum efficiency have been performed for different clocking frequencies, integration times, and sampling modes.

034.146   Recent data on Si:As impurity band conduction multiplexed arrays.
J. D. Merriam, R. K. Bentley.
Infrared astronomy with arrays, p. 136 – 139 (1987). – See Abstr. 012.070.
Data is presented from recent measurements of two Si:As impurity band conduction multiplexed arrays. The devices were produced by different manufacturers. The devices were measured at a temperature of 12K, and at a background of 1.5×10^{12} photons/sec cm^2. Both devices show high responsivity and excellent uniformity.

034.147   58 × 62 Si:Sb DRO infrared detector array performance.
M. E. McKelvey, J. H. Goebel, C. R. McCreight, N. N. Moss.
Infrared astronomy with arrays, p. 140 – 143 (1987). – See Abstr. 012.070.
Preliminary test results from the evaluation of two 58 × 62 element Si:Sb infrared detector arrays produced by Hughes Aircraft Co./Santa Barbara Research Center are presented. The arrays are hybridized to silicon direct readout (DRO) multiplexers which allow random–access and non–destructive readout.

034.148   Initial test results on a 58 × 62 pixel Si:Ga DRO array detector for SIRTF/IRAC band II (5 – 15 μm).
G. Lamb, P. Shu, J. Mather, A. Ewin, J. Bowser.
Infrared astronomy with arrays, p. 144 – 155 (1987). – See Abstr. 012.070.

034.149   Infrared hybrid array test facility.
K. Shivanandan, P. Schwartz, S. Odenwald, H. Thronson, R. Puetter.
Infrared astronomy with arrays, p. 156 – 159 (1987). – See Abstr. 012.070.

034.150   High throughput bit–slice front–end processor for the new Goddard 10 μm 58 × 62 DRO array camera.
G. Chin, D. Gezari.
Infrared astronomy with arrays, p. 160 – 168 (1987). – See Abstr. 012.070.

034.151   Ground based imaging with a 32 × 32 InSb array.
W. J. Forrest.
Infrared astronomy with arrays, p. 171 – 179 (1987). – See Abstr. 012.070.
Some of the experiences in applying an infrared array detector to astronomical broadband and spectral imaging in the 1 to 5 micron region are discussed with the aim of helping other such efforts.

034.152   Results with the UKIRT infrared camera.
I. S. McLean.
Infrared astronomy with arrays, p. 180 – 192 (1987). – See Abstr. 012.070.
A wide range of observational results from the commissioning of the first infrared camera on UKIRT are presented. The camera, called IRCAM 1, employs the 62 × 58 InSb DRO array from SBRC in an otherwise general purpose system which is briefly described. Several imaging modes are possible including staring, chopping and a high–speed snapshot mode. Infrared imaging polarimetry is also possible. Results to be presented include the first true high resolution images at IR wavelengths of the entire Orion nebula. The experiences with using this array on the telescope since October 1986 are discussed and future plans outlined.

034.153   Simulation and modelling of infrared camera systems.
M. J. McCaughrean, I. S. McLean.
Infrared astronomy with arrays, p. 193 – 196 (1987). – See Abstr. 012.070.
The authors describe a software modelling and simulation package (SIRCAM) implemented to aid the design and optimisation of new infrared instrumentation based around solid state 2–d detector arrays.

034.154   The NOAO infrared imagers: description and performance.
A. M. Fowler, F. C. Gillett, B. Gregory, R. R. Joyce, R. G. Probst, R. Smith.
Infrared astronomy with arrays, p. 197 – 203 (1987). – See Abstr. 012.070.
NOAO has constructed infrared imaging systems for use at Kitt Peak and Cerro Tololo which utilize SBRC 58 × 62 InSb arrays. The authors briefly describe the two systems and give examples of telescope performance.

034.155   Infrared imaging with JPL's linear array camera.
M. S. Hanner, P. N. Kupferman, G. Bailey, J. C. Zarnecki.
Infrared astronomy with arrays, p. 205 – 211 (1987). – See Abstr. 012.070.
JPL's 128 element InSb line array was used at the 3.9–m Anglo Australian Telescope in April 1986. Images of comet Halley and the Galactic Center are presented. Distinct changes in the morphology of Halley's inner dust coma are apparent between images taken a few hours apart and from night to night.

034.156   Rockwell HgCdTe arrays as imagers.
M. J. Rieke, G. H. Rieke, E. F. Montgomery.
Infrared astronomy with arrays, p. 213 – 221 (1987). – See Abstr. 012.070.
During the last two years, a hybrid focal plane array manufactured by Rockwell International has been characterized in the laboratory and used extensively at the telescope. Different observing strategies such as beam–switching and time–delay–and–integrate have been used and their strengths and weaknesses are described.

034.157   Ground–based applications for the JPL/SISEX SWIR array.
R. W. Capps, K. W. Hodapp, D. N. B. Hall, E. E. Becklin, D. A. Simons, G. C. Bailey, V. G. Wright.
Infrared astronomy with arrays, p. 222 – 231 (1987). – See Abstr. 012.070.
The SISEX focal plane arrays are 64 × 64, 2.5 micron cut–off, HgCdTe hybrid devices which were developed by Rockwell International for JPL's Shuttle Imaging Spectrometer Experiment (SISEX). One such device was installed in the UH IR Camera System and used successfully for high resolution imaging at the f/35 Cass focus of the UH 2.2 meter and NASA IRTF telescopes. This system was also used for extensive testing of the SISEX device under low light level conditions. Several techniques for optimizing the performance were investigated and linear photometric performance over at least 10 stellar magnitudes was demonstrated.

034.158   A 32 × 32 infrared 2 – 5 μm HgCdTe/CCD camera. First results.
J.–L. Monin, I. Vauglin, F. Sibille, S. Guilloteau, P. Merlin.
Infrared astronomy with arrays, p. 232 – 236 (1987). – See Abstr. 012.070.
The authors describe an infrared camera base on an hybrid device designed for high background applications. Its main characteristics are given and the observing procedure is described. The whole system is optimized for efficient use in the L' and M bands where the background flux is high. Recent observational results are presented.

034.159   "CIRCUS": a 32 × 32 CID InSb common–users infrared camera for large telescopes.
D. Tiphene, D. Rouan, F. Lacombe, M. Combes.
Infrared astronomy with arrays, p. 237 – 240 (1987). – See Abstr. 012.070.

A 32 × 32 infrared camera based on a CID InSb array has been designed and built to be a common–users instrument for work on large telescopes. The main technical features of the short wavelength channel, which is now in qualification phase, are presented.

034.160   Astronomical observations at 10 micron wavelength with the NASA/GSFC array camera system.
W. F. Hoffmann, G. G. Fazio, R. Tresch–Fienberg,
L. K. Deutsch, D. Y. Gezari, G. M. Lamb, P. Shu,
C. R. McCreight.
Infrared astronomy with arrays, p. 241 – 255 (1987). – See Abstr. 012.070.

An infrared array camera has been developed at the NASA/ Goddard Space Flight Center for astronomical observations in the 8 – 13 μm wavelength region using an Aerojet Electrosystems Co. 16 × 16 pixel Si:Bi accumulation mode charge injection device. The camera has been used successfully at the University of Arizona Steward Observatory 2.3–meter telescope and the NASA Infrared Telescope Facility 3–meter telescope to produce high resolution maps at several wavelengths, as well as color temperature and optical depths maps of the Orion Nebula BN/ KL region, the planetary nebulae NGC 7027, the Egg Nebula (CRL 2688), M8E, the central 1.5 pc of the Galactic center, and the central kiloparsec of the Seyfert galaxy NGC 1068.

034.161   Berkeley–Hughes infrared camera.
J. F. Arens, J. G. Jernigan, R. Ball, M. C. Peck,
S. Gaalema, J. Lacy.
Infrared astronomy with arrays, p. 256 – 262 (1987). – See Abstr. 012.070.

The Berkeley–Hughes infrared camera has been rebuilt for reduced size and complexity and has been used to make astronomical observations. Seeing motion can be removed when viewing bright objects, and superresolved images with 0.25″ FWHM features have been produced. The stability of the observing system, including the atmosphere, is being investigated, but a preliminary result indicates that astronomical observations can be performed successfully without the use of a rapidly chopping secondary mirror. Additionally, the read noise has been reduced to a level of 50 electrons per read.

034.162   A ^{3}He cooled bolometer array for the IRAM 30 m telescope.
E. Kreysa, G. Haslam, E. E. Haller.
Infrared astronomy with arrays, p. 263 – 267 (1987). – See Abstr. 012.070.

Arrays of bolometers for submm–mm continuum mapping are being developed at MPIfR. The bolometers are composite bolometers with neutron–transmutation–doped germanium thermometers for operation at 0.3K. A dedicated data acquisition system is being developed on the basis of a IBM–AT2. The authors report the status of this project.

034.163   High spatial resolution imaging with linear arrays.
E. E. Bloemhof, W. C. Danchi, R. A. McLaren.
Infrared astronomy with arrays, p. 268 – 271 (1987). – See Abstr. 012.070.

The authors have used scanned linear arrays at large ground– based telescopes to image with high spatial resolution in the thermal infrared. They describe the experimental technique, discuss some of the practical complications, and illustrate the method by presenting a sample image profile of a bright source with an extended core, the carbon star IRC + 10216.

034.164   Use of a 32–element Reticon array for 1 – 5 micrometer spectroscopy.
A. T. Tokunaga, R. G. Smith, E. Irwin.
Infrared astronomy with arrays, p. 367 – 378 (1987). – See Abstr. 012.070.

A 32–element InSb Reticon array has been installed and used in a Cooled–Grating Array Spectrometer at the NASA Infrared Telescope Facility, Mauna Kea, Hawaii. The sensitivity, in terms of limiting magnitude obtainable at the telescope, is slightly better than that of the best feedback amplifier circuits for single detector InSb systems. It is imperative, however, to carefully design the read–out electronics to achieve optimum performance, and the techniques utilized are described.

034.165   IRSPEC: design, performance and first scientific results.
A. F. M. Moorwood.
Infrared astronomy with arrays, p. 379 – 387 (1987). – See Abstr. 012.070.

IRSPEC is a cooled grating spectrometer currently equipped with a monolithic, linear, array of InSb diodes coupled to a Reticon multiplexer. The array is sensitive between 1 μm and 5 μm, has a pixel size of 200 μm corresponding to a 6 arcsec slit on the ESO 3.6 m telescope and yields resolving powers between 1000 and 2500 depending on wavelength and grating order. The overall performance is illustrated by a variety of astronomical spectra and a brief report on recent scientific results from observations of supernova remnants and galaxy nuclei.

034.166   Near infrared spectroscopy with the array spectrometer of the Anglo–Australian telescope.
A. R. Hyland, P. J. McGregor.
Infrared astronomy with arrays, p. 388 – 392 (1987). – See Abstr. 012.070.

The infrared cooled grating spectrometer of the AAT (FIGS) uses an InSb detector array with 16 discrete elements. The system has been optimised for use in low background conditions by concentrating on the low noise performance of the detectors and preamplifier electronics. The authors present spectra covering the 2.0 – 2.4 μm region, of a variety of interesting objects with K magnitudes to 12, to demonstrate the capabilities of this instrument. They also show new spectra of Nova Cen 1986, obtained approximately 15 days following the initial formation of a dust shell around the nova.

034.167   A new IR–grating spectrograph for the Calar Alto Observatory.
A. Krabbe, R. Lenzen, S. Weil.
Infrared astronomy with arrays, p. 393 – 397 (1987). – See Abstr. 012.070.

An infrared grating spectrometer for the 1 to 5 μm wavelength range has been developed. The spectrometer is completely cooled down to 50K. A one–dimensional array of 32 elements is scheduled for focal instrumentation. The authors developed a 32 channel parallel readout electronics which has been tested. The spectrometer and the focal instrumentation with readout electronics are described.

034.168   Compact grating spectrometer with an InSb array detector.
H. Suto, K. Mizutani, T. Maihara, T. Nakajima.
Infrared astronomy with arrays, p. 398 – 401 (1987). – See Abstr. 012.070.

A compact cooled grating spectrometer was built and has currently been put into use on several telescopes for near– infrared spectrophotometric observations. The spectrometer with the resolving power between 150 and 300 has been installed in an HD–3 dewar, utilizing an InSb array detector with eight elements. The detectivity when attached to a 2–m class telescope is characterized by 3 σ detection limits of about 1×10^{-20} W cm^{-2} in the K–band and 6×10^{-20} W cm^{-2} in the L–band by the typical integration time of 300 sec.

034.169 A mid–infrared cryogenic echelle spectrometer.
J. H. Lacy, J. F. Arens, M. C. Peck, S. D. Gaalema.
Infrared astronomy with arrays, p. 402 – 410 (1987). – See Abstr. 012.070.

An echelle spectrometer has been constructed for astronomical use at mid–infrared wavelengths. It contains a liquid helium cooled echelle grating and a 10 × 64 element Si:Ga detector array, providing sampling of ten points spaced by $0.5'' – 3''$ along a slit and 64 spectral elements with 3000 – 30000 spectral resolving power. The instrument has been used on the McDonald Observatory 2.7 m telescope and proved to be photon noise limited with a sensitivity of each pixel equal to that of the best comparable single pixel spectrometer.

034.170 Recent astronomical results obtained with the AFGL ten micron Array Spectrometer.
P. D. LeVan, P. C. Tandy.
Infrared astronomy with arrays, p. 411 – 421 (1987). – See Abstr. 012.070.

The authors describe the Array Spectrometer electronics and optics, the testing of the array in the laboratory, and the results of using the spectrometer on the University of Wyoming 92" telescope. Results include spectral and spatial scans for IRC + 10216 and evaluation of instrumental performance.

034.171 An advanced infrared array spectrometer for large telescopes.
C. M. Mountain, T. J. Lee, D. J. Robertson, R. Wade.
Infrared astronomy with arrays, p. 434 – 437 (1987). – See Abstr. 012.070.

The authors are currently building common user cooled grating spectrometers for UKIRT with the principle goal of fully utilizing all current and foreseeable 2–D arrays in the $1 \mu m - 5 \mu m$ region.

034.172 A rotation shearing interferometer for the near infrared.
J.–M. Mariotti, A. Zadrozny, I. Vauglin, J.–L. Monin.
Infrared astronomy with arrays, p. 468 – 472 (1987). – See Abstr. 012.070.

The authors designed and built a rotation shearing interferometer for the 2 – 5 micron range. The shear angle can be set to 0° for calibration or 180° for full spatial resolution. The interferometer has two outputs and can be operated in flat field mode or with fringes. The detector is a 32 × 32 CCD array. Initial tests in the laboratory and at the 152 cm telescope of Haute Provence Observatory are described and show promising results.

034.173 Matching infrared array instruments to future large telescopes.
R. D. Gehrz.
Infrared astronomy with arrays, p. 499 – 507 (1987). – See Abstr. 012.070.

The author considers the design and performance requirements for versatile imaging systems for very large aperture telescopes. The performance specifications are driven by scientific requirements suggested by recent studies addressing the applications of these telescopes. A specific effort to design an imaging system for the National Optical Astronomy Observatory four–barrel MMT–style National New Technology Telescope is discussed to illustrate the physical constraints that are encountered in practical implementation.

034.174 Infrared array instruments for the Keck Telescope.
B. Jones.
Infrared astronomy with arrays, p. 508 – 516 (1987). – See Abstr. 012.070.

The infrared instruments planned for the Keck Telescope are discussed. These are (1) a near infrared spectrometer using a 2 × 512 array of indium antimonide detectors, (2) a mid infrared camera using a 20 × 64 array of arsenic doped silicon detectors, (3) a mid infrared spectrometer also using an array of 20 × 64 arsenic doped silicon detectors, and (4) a shortwavelength camera. Details of the optical, mechanical and electronic

interface between the telescope and instruments are discussed. Instrument designs and expected performance levels are presented.

034.175 Desired characteristics of 2–D arrays for SWIR imagers and spectrographs.
R. Wade, T. J. Lee, I. S. McLean, C. M. Mountain, M. J. McCaughrean, I. Baker.
Infrared astronomy with arrays, p. 517 – 521 (1987). – See Abstr. 012.070.

High performance arrays are needed for the short wavelength infrared (SWIR) range, 1 – 2.5 microns, where requirements are rather special because of the low background in astronomy applications. The background as a function of wavelength for UKIRT on Mauna Kea is discussed and properties of existing arrays are used to derive desired characteristics for arrays for typical observations.

034.176 A program of instrumentation for large telescopes using 2–D infrared arrays.
T. J. Lee, C. M. Mountain, R. Wade.
Infrared astronomy with arrays, p. 522 – 526 (1987). – See Abstr. 012.070.

A program is in progress for the design and build of imagers and spectrographs suitable for the 3.8 m UKIRT and the 4.2 m William–Herschel Telescope covering the wavelength range $1 \mu m - 14 \mu m$. All of these instruments will require detector arrays. The program, its aims and requirements are described.

034.177 Instrumentation of the initial LEST site testing campaign.
A. Erasmus, C. M. Barreto Cabrera, U. Kusoffsky.
LEST Found., Tech. Rep., No. 26, p. 57 – 61 (1987). – See Abstr. 012.071.

034.178 Intercalibration of two solar limb motion meters.
P. N. Brandt.
LEST Found., Tech. Rep., No. 26, p. 63 – 65 (1987). – See Abstr. 012.071.

Two solar limb motion meters are planned to be used for comparative measurements of the seeing at the Swedish solar telescope (La Palma) and at the Gregory telescope (Tenerife). Since the two devices had been built by different workshops, it seemed advisable to calibrate them electronically and to check them for equal performance in a beam splitting arrangement. In November 1986 a first attempt in this direction was undertaken at the Swedish turret telescope in La Palma. The results are presented briefly.

034.179 Stellar differential photoelectric photometry with a simultaneous two–telescope system: detection of short–term variations in M supergiants.
F. Querci, M. Querci, C. Gregory, B. Fontaine.
Bull. Etoiles Tardives Spectre Particulier, No. 5, p. 7 – 8 (1987).

034.180 Low–dispersion TV spectrometer.
A. N. Abramenko, L. M. Sharipova, A. V. Bagrov, M. A. Smirnov.
Astron. Tsirk., No. 1464, p. 3 – 5 (1986). In Russian.

034.181 Comparison of a coherence interferometer and a photoelectric seeing monitor.
A. Eh. Gur'yanov, M. S. Pekur, B. N. Irkaev, A. A. Semenikin, A. A. Tokovinin, P. V. Shcheglov.
Astron. Tsirk., No. 1467, p. 1 – 3 (1986). In Russian.

Comparison of an image folding interferometer and a PSM device was made in 1984 and 85 near the 6–m telescope and at Mt. Sanglock. No systematic errors were found in both methods.

034.182 **Use of a Fabry–Perot etalon for observations of absorption spectra.**
L. I. Shestakova.
Astron. Tsirk., No. 1467, p. 4 – 6 (1986). In Russian.
The basic requirements on the concordance Fabry–Perot etalon and interference filter for observations of the absorption lines in the optical scheme with the wide and the narrow beams are considered.

034.183 **On an investigation of some characteristics and optimization of operating conditions of PMT–136 used in astronomical practice.**
O. V. Benashvili, Yu. D. Mateshvili, A. G. Avsadzhanishvili.
Astron. Tsirk., No. 1475, p. 5 – 7 (1987). In Russian.

034.184 **Development of a new model of the Vilnius electronographic camera.**
J. Jukonis, R. Drazdys, G. Vilkajtis, V. Vansevicius.
Astron. Tsirk., No. 1481, p. 5 – 6 (1987). In Russian.
The results of tests of a new model of the Vilnius electronographic camera, VEK–3, are given. In 1985 and 1986 VEK–3 was tested on the 60–cm and 6–m telescopes. The main technical characteristics of the VEK–3 are presented.

034.185 **A new method to identify muons in an extensive air shower array.**
J. Linsley, S. Mikocki, J. Poirier.
J. Phys. G, Vol. 13, No. 8, p. L163 – L168 (1987). Abstr. in Phys. Abstr., Vol. 90, No. 1315, Entry 127733 (1987).

034.186 **A high–resolution Fabry–Pérot spectrometer for emission line studies in planetary nebulae and other extended astronomical objects.**
D. P. K. Banerjee, B. G. Anandarao, J. N. Desai, N. S. Jog, P. K. Kikani, R. K. Mahadkar, K. S. B. Manian, F. M. Pathan, N. C. Shah, M. Thomas.
Astrophys. Space Sci., Vol. 139, No. 2, p. 327 – 335 (1987).
The authors describe here a scanning piezo–electric Fabry–Pérot spectrometer (operating in the photon–counting mode and whose plate spacing and parallelism are maintained by a servo–controlled system to ensure high accuracy) for the study of emission lines from extended astronomical objects in the spectral range 4500 – 7000 Å. Details of the optical set–up and the Data Acquisition System are described. Its performance at the Cassegrain focus of the 1 m telescope at Kavalur is discussed. Some line profiles on planetary nebulae studied with the above spectrometer are also presented.

034.187 **A device for the automatic operation of the shutter of the VAU camera when taking photographs of the geostationary satellites.**
A. N. Biryukov.
Nauchn. Inf., Vyp. 58, p. 81 – 86 (1986). In Russian.
The device operates by quartz clock signals. The process of observations and a further reduction require less time and effort.

034.188 **A new model of circumzenithal VUGTK 100/1000.**
V. D. Krajchev.
Nauchn. Inf., Vyp. 58, p. 131 – 140 (1986). In Russian.
A description of a new circumzenithal is presented. It is a unique astronomical device for a simultaneous determination of both astronomical coordinates by the equal altitudes method. The external mean quadratic error in the location of the instrument determined within one night is 0".15.

034.189 **Programme of observations on the circumzenithal for the Geodynamic Observatory "Plana".**
V. D. Krajchev.
Nauchn. Inf., Vyp. 58, p. 141 – 151 (1986). In Russian.
A programme for observations by the method of equal heights on the circumzenithal "VUGTK 100/1000" for the Geodynamic Observatory Plana of the Bulgarian Academy of Sciences is compiled. The requirements which were taken into account in the process of the programme compilation are described.

034.190 **Modern radiation detectors.**
A. S. Kutyrev.
Zemlya Vselennaya, No. 5, p. 43 – 49 (1987). In Russian.

034.191 **Cosmic–ray monopole search at IBM–BNL using superconducting induction detectors.**
S. Bermon.
IEEE Trans. Magn., Vol. MAG–23, No. 2, p. 441 – 449 (1987). Abstr. in Phys. Abstr., Vol. 90, No. 1316, Entry 134140 (1987).

034.192 **Design and performance of a 0.18 m² inductive detector for cosmic magnetic monopoles.**
J. C. Schouten, A. D. Caplin, C. N. Guy, M. Hardiman, M. Koratzinos, W. S. Steer.
J. Phys. E, Vol. 20, No. 7, p. 850 – 861 (1987). Abstr. in Phys. Abstr., Vol. 90, No. 1316, Entry 134142 (1987).

034.193 **Report on the Stanford octagonal magnetic monopole detector.**
M. E. Huber, B. Cabrera, M. A. Taber, R. D. Gardner.
IEEE Trans. Magn., Vol. MAG–23, No. 2, p. 1134 – 1137 (1987). Abstr. in Phys. Abstr., Vol. 90, No. 1317, Entry 135736 (1987).

034.194 **Cosmic ray tests of 7.6 m drift–tube counters and the readout electronics system of the VENUS muon detector.**
Y. Asano, S. Mori, M. Moriya, M. Shioden, Y. Yamagishi, M. Yoshida, Y. Ikegami, M. Kichise, I. Nakano, Y. Shimomura.
Nucl. Instrum. Methods Phys. Res., Sect. A, Vol. A259, No. 3, p. 430 – 437 (1987). Abstr. in Phys. Abstr., Vol. 90, No. 1317, Entry 139932 (1987).

034.195 **Monte Carlo simulations of effects due to delta rays in aluminum drift–tube counters.**
Y. Asano, S. Mori, M. Moriya, M. Shioden, Y. Yamagishi, Y. Ikegami, I. Nakano.
Nucl. Instrum. Methods Phys. Res., Sect. A, Vol. A259, No. 3, p. 438 – 446 (1987). Abstr. in Phys. Abstr., Vol. 90, No. 1317, Entry 139933 (1987).

034.196 **A heavy water detector to resolve the solar neutrino problem.**
G. Aardsma, R. C. Allen, J. D. Anglin, M. Bercovitch, A. L. Carter, H. H. Chen, W. F. Davidson, P. J. Doe, E. D. Earle, H. C. Evans, G. T. Ewan, E. D. Hallman, C. K. Hargrove, P. Jagam, D. Kessler, H. W. Lee, J. R. Leslie, J. D. MacArthur, H.–B. Mak, A. B. MacDonald, W. McLatchie, B. C. Robertson, J. J. Simpson, D. Sinclair, P. Skensved, R. S. Storey.
Phys. Lett. B, Vol. 194, No. 2, p. 321 – 325 (1987). Abstr. in Phys. Abstr., Vol. 90, No. 1318, Entry 141400 (1987).

034.197 **Development of low cost liquid scintillator counters for cosmic ray experiments.**
S. Bultena, D. Hanna, K. Murthy.
Nucl. Instrum. Methods Phys. Res., Sect. A, Vol. A260, No. 1, p. 247 – 253 (1987). Abstr. in Phys. Abstr., Vol. 90, No. 1318, Entry 145833 (1987).

034.198 **Charge–coupled–device X–ray detector performance model.**
M. W. Bautz, G. E. Berman, J. P. Doty, G. R. Ricker.
Opt. Eng., Vol. 26, No. 8, p. 757 – 765 (1987). Abstr. in Phys. Abstr., Vol. 91, No. 1320, Entry 5073 (1988).

034.199 **X–ray measurements of charge diffusion effect in EEV Ltd. charge–coupled devices.**
D. H. Lumb, E. G. Chowanietz, A. Wells.
Opt. Eng., Vol. 26, No. 8, p. 773 – 778 (1987). Abstr. in Phys. Abstr., Vol. 91, No. 1320, Entry 5074 (1988).

034.200 A combined cosmic ray muon spectrometer and high energy air shower array.
M. L. Cherry, D. S. Ayres, F. Halzen.
Physics of the superconducting supercollider, p. 655 – 659 (1987).
Abstr. in Phys. Abstr., Vol. 91, No. 1320, Entry 10381 (1988). – See Abstr. 012.080.

034.201 Evaluation of the RCA 640 × 1024 charge–coupled–device imager for astronomical use.
P. Waddell, C. Christian.
Opt. Eng., Vol. 26, No. 8, p. 734 – 741 (1987). Abstr. in Phys. Abstr., Vol. 91, No. 1320, Entry 10405 (1988).

034.202 Four–shooter: a large format charge–coupled–device camera for the Hale telescope.
J. E. Gunn, M. Carr, G. E. Danielson, E. O. Lorenz,
R. Lucinio, V. E. Nenow, J. D. Smith, J. A. Westphal,
P. D. Schneider, B. A. Zimmerman.
Opt. Eng., Vol. 26, No. 8, p. 779 – 787 (1987). Abstr. in Phys. Abstr., Vol. 91, No. 1320, Entry 10415 (1988).

034.203 Lick Observatory charge–coupled–device data acquisition system.
L. B. Robinson, R. J. Stover, J. Osborne, J. S. Miller,
S. S. Vogt, S. L. Allen.
Opt. Eng., Vol. 26, No. 8, p. 795 – 805 (1987). = Contrib. Lick Obs., No. 442.
The Lick Observatory CCD data acquisition system is described, with some observational results to illustrate the system capability.

034.204 40 mm electrostatic electronic camera for spectrography with the Canada–France–Hawaii Cassegrain telescope.
M. Duchesne, C. Joubert, J. M. Le Flohic, B. Servan,
G. Wlerick, A. Sellier, A. Baranne.
Proceedings of the international topical meeting on image detection and quality, p. 77 – 79 (1986). Abstr. in Phys. Abstr., Vol. 91, No. 1320, Entry 10420 (1988). – See Abstr. 012.079.

034.205 Astronomical detectors for .001 to 300000 Å, an overview.
W. C. Livingston.
Proceedings of the international topical meeting on image detection and quality, p. 221 – 231 (1986). Abstr. in Phys. Abstr., Vol. 91, No. 1320, Entry 10421 (1988). – See Abstr. 012.079.

034.206 Astronomical application of the large–field electronic camera with the Canada–France–Hawaii telescope.
G. Wlerick, B. Servan, L. Renard, D. Horville, J. Fromage,
G. Lelievre, A. Bijaoui.
Proceedings of the international topical meeting on image detection and quality, p. 237 – 240 (1986). Abstr. in Phys. Abstr., Vol. 91, No. 1320, Entry 10422 (1988). – See Abstr. 012.079.

034.207 A 40 mm photon counting camera.
A. Blazit.
Proceedings of the international topical meeting on image detection and quality, p. 259 – 263 (1986). Abstr. in Phys. Abstr., Vol. 91, No. 1320, Entry 10423 (1988). – See Abstr. 012.079.

034.208 A high sensitivity InSb CID camera for infrared astronomy.
F. Lacombe, D. Tiphene, D. Rouan.
Proceedings of the international topical meeting on image detection and quality, p. 271 – 274 (1986). Abstr. in Phys. Abstr., Vol. 91, No. 1320, Entry 10424 (1988). – See Abstr. 012.079.

034.209 A high–precision multichannel photometer for planetary detection.
L. E. Allen, W. J. Borucki, S. W. Taylor, S. J. Kleinman.
Publ. Astron. Soc. Pac., Vol. 99, No. 621, p. 1145 (1987). Abstract. – See Abstr. 010.281.

034.210 Shall we rotate a sundial face?
R. H. Garstang.
Publ. Astron. Soc. Pac., Vol. 99, No. 621, p. 1147 (1987). Abstract. – See Abstr. 010.281.

034.211 The Lick Observatory Hamilton Echelle Spectrometer.
S. S. Vogt.
Publ. Astron. Soc. Pac., Vol. 99, No. 621, p. 1214 – 1228 (1987). = Lick Obs. Bull. No. 1085.
The Hamilton Echelle Spectrometer, installed at the coudé focus of the Shane 3–m telescope, is a high–dispersion spectrograph optimized for use with today's largest available CCDs, and for the even larger CCDs expected in the future. This paper describes the Hamilton spectrograph and discusses the philosophy behind its design. Several examples of the performance of the system are given.

034.212 A new photoelectric photometer for the 60 cm telescope of Beijing Observatory.
C. Shi, B. Du, L. Gao, Z. Jiang, X. Wang.
Acta Astrophys. Sin., Vol. 7, No. 3, p. 230 – 237 (1987). In Chinese.
This paper describes a new photoelectric photometer controlled by a microcomputer. The filters, stars and sky background can be changed by the computer according to the present control on the keyboard. The photometer is used for real–time data acquisition, processing and display during observations. It can carry out two kinds of work: one is photon–counting and the other is DC amplifying–V/F converting. The authors give one typical light curve obtained with this photometer which has been routinely used since 16 Nov. 1983.

034.213 On the principles of the white–light solar magnetometer.
S. Ye, J. Jin.
Acta Astrophys. Sin., Vol. 7, No. 3, p. 238 – 242 (1987). In Chinese.
On the basis of the profiles of Stokes parameters calculated from the numerical solution of the system of equations of transfer and in view of the magneto–optical effect, the authors investigate the causes of the asymmetry of the V–profiles and the net circular polarization. This allows them to demonstrate the principles of the white–light solar magnetometer.

034.214 The multichannel birefringent filter. III. The multichannel head an the multichannel solar telescope.
G. Ai, Y. Hu.
Acta Astrophys. Sin., Vol. 7, No. 4, p. 305 – 311 (1987). In Chinese.

034.215 Some new ideas concerning birefringent filters.
D.–q. Su.
Acta Astron. Sin., Vol. 28, No. 3, p. 296 – 302 (1987). In Chinese.

Echelles de temps atomique.
See Abstr. 003.064.

On a singular Chinese portable sundial.
See Abstr. 004.074.

Research on scale and precision of the Water Clock in ancient China.
See Abstr. 004.086.

Scheduling and other issues for a solar–stellar synoptic observatory.
See Abstr. 013.002.

The McMath Solar–Stellar Nighttime Program.
See Abstr. 013.004.

The DUMAND project.
See Abstr. 013.038.

Status report on the SYNOP project to monitor stars with high–resolution spectroscopy.
See Abstr. 013.048.

Platinum – carbon multilayer X–ray reflectors.
See Abstr. 031.007.

Camera lenses for fast–speed spectrographs.
See Abstr. 031.032.

The spectroscopic survey telescope.
See Abstr. 032.001.

The Multiple Object Spectroscopic Telescope.
See Abstr. 032.002.

Operation, performance, and reliability testing of charge–coupled devices for star trackers.
See Abstr. 035.118.

Two–dimensional spectrophotometry of planetary nebulae by CCD imaging.
See Abstr. 036.043.

A calibration method for measurements of the longitudinal magnetic field and differential radial velocity.
See Abstr. 036.051.

Astrometrie mit einem Mikrodensitometer PDS 1010 A.
See Abstr. 036.061.

Rotating waveplates as polarization modulators for Stokes polarimetry of the sun: evaluation of seeing–induced crosstalk errors.
See Abstr. 036.143.

Development and observations of the stellar speckle camera.
See Abstr. 036.161.

A Macintosh based data system for array spectrometers.
See Abstr. 036.202.

Deep infrared surveys.
See Abstr. 036.205.

Enquiry on rapid automatic measuring machines.
See Abstr. 036.213.

Gravitational wave astronomy – potential and possible realisation.
See Abstr. 066.066.

Coincidence probabilities for a network of interferometric detectors of gravitational waves.
See Abstr. 066.068.

Analysis of 18 months of data of the GEOGRAV experiment.
See Abstr. 066.131.

Limits on sensitivity of large silicon bolometers for solar neutrino detection.
See Abstr. 080.039.

First results with a transmission echelle grating on the ESO Faint Object Spectrograph: observations of the SN 1986a in NGC 3367 and of the nucleus of the galaxy.
See Abstr. 125.821.

035 Space Instrumentation

035.001 A compilation system for Venus radar mission (Magellan).
S. S. C. Wu, F. J. Schafer, A.–E. Howington.
NASA Tech. Memo., NASA TM–89810, p. 233 – 236 (1987). – See Abstr. 003.001.

035.002 The Mars Observer Camera.
M. C. Malin, G. E. Danielson, A. P. Ingersoll, H. Masursky, J. Veverka, T. Soulanille, M. Ravine.
NASA Tech. Memo., NASA TM–89810, p. 548 – 550 (1987). – See Abstr. 003.001.

035.003 X–ray astronomical spectroscopy.
S. S. Holt.
NASA Conf. Publ., NASA CP–2464, p. 379 – 394 (1987). – See Abstr. 012.004.
This paper describes the development of a new X–ray "microcalorimeter" for the AXAF study payload. This technology combines a resolving power of order 10^3 with quantum efficiency of order unity and is likely to revolutionize the study of cosmic X–ray spectra.

035.004 The on–board software of the Vega TV system.
G. A. Avanesov, K. G. Sukhanov, Yu. K. Zaikon (Yu. K. Zajko), S. I. Zatsepin, E. Dénes, I. Manno, G. Pintér, L. Várhalmi, M. Zsenei.
ESA J., Vol. 11, No. 2, p. 239 – 244 (1987).
In 1984, two spacecraft were launched in the framework of the Vega project to make scientific observations of the planet Venus and of Halley's Comet. Onboard the probes, a revolving mount ensured that the TV system and some of the scientific instruments could always look at the comet. This article describes the onboard software that guided and controlled this TV system.

035.005 Gamma 1: a telescope for 50 – 5000 MeV astronomy.
B. Agrinier, V. V. Akimov, M. Avignon, V. M. Balebanov, A. R. Bazer–Bachi, A. S. Belousov, I. D. Blokhintsev, A. Bouere, F. Cardon, V. Yu. Chesnokov, E. I. Chujkin, F. Cotin, J. Cretolle, M. B. Dobriyan, C. Doulade, G. Ducros, J. Durand, D. Fournier, M. I. Fradkin, V. I. Fuks, A. M. Gal'per, I. A. Gerasimov, V. A. Grigor'ev, M. Gros, M. V. Guzenko, C. Hugot, J. P. Joli, L. F. Kalinkin, P. Keirle, V. G. Kirillov–Ugryumov, S. V. Koldashev, V. D. Kozlov, L. V. Kurnosova, J. M. Lavigne, A. Lecomte, N. G. Leikov, J. P. Leray, P. Mandrou, P. Masse, A. A. Moiseev, B. Mougin, J. Mouli, Yu. I. Nagornykh, V. E. Nesterov, M. Nobileau, E. Orsal, Yu. V. Ozerov, B. Parlier, J. A. Paul, M. Poivillier, V. P. Poluehktov, A. V. Popov, O. F. Prilutskij, V. L. Prokhin, A. Raviart, V. G. Rodin, V. A. Rud'ko, M. F. Runtso, M. A. Rusakovich, A. V. Serov, G. Serra, F. Soroka, S. R. Tabaldyev, N. P. Topchiev, G. Vedrenne, S. A. Voronov, Yu. T. Yurkin, V. M. Zemskov, V. G. Zverev.
Sov. Astron., Vol. 30, No. 5, p. 508 – 514 (1986). English translation of 42.035.017.

035.006 A cooled submillimeter telescope.
G. B. Sholomitskij, V. M. Balebanov, V. D. Gromov, M. Z. Khokhlov, I. A. Maslov, Yu. V. Nikol'skij, A. S. Petukhov, O. F. Prilutskij, V. G. Rodin, V. A. Shaposhnikov, V. A. Soglasnova.
Sov. Astron., Vol. 30, No. 5, p. 514 – 518 (1986). English translation of 42.035.018.

035.007 **The Cosmic Background Explorer (COBE) satellite.**
C. L. Bennett.
Radio astronomy from space, p. 51 – 59 (1987). – See Abstr. 012.009.

035.008 **The Large Deployable Reflector (LDR): plans and progress.**
P. N. Swanson.
Radio astronomy from space, p. 67 – 80 (1987). – See Abstr. 012.009.
The history of LDR from its beginnings in 1977, to the present, is discussed. The science objectives of LDR are summarized along with a comparison of the estimated performance of LDR against other astronomical observatories such as HST, SIRTF, the KAO and IRTF. The system requirements, derived from the scientific objectives, are summarized as well as the key technologies that must be developed.

035.009 **Using TDRSS as an orbiting VLBI observatory.**
J. S. Ulvestad.
Radio astronomy from space, p. 119 – 123 (1987). – See Abstr. 012.009.
Successful Very Long Baseline Interferometry (VLBI) observations at 2.3 GHz have been made using an earth–orbiting spacecraft as one of the telescopes. These observations employed the first deployed satellite (TDRSE) of the Tracking and Data Relay Satellite System (TDRSS). Fringes were found for 3 radio sources on baselines between TDRSE and telescopes in Australia and Japan. This paper describes the purpose of the experiment, the characteristics of the spacecraft that are related to the VLBI observations, and the requirements and procedures of the experiment. A brief summary of the results is also included.

035.010 **Link calibrations for the TDRSS orbiting VLBI experiment.**
C. D. Edwards.
Radio astronomy from space, p. 125 – 129 (1987). – See Abstr. 012.009.
The first successful interferometric observations of extragalactic radio sources using an orbiting antenna as one of the observing stations were achieved in July and August, 1986, using the Tracking and Data Relay Satellite System. The technical obstacles to maintaining phase coherence between the orbiting antenna and the ground stations are reviewed, with an emphasis on the effects of spacecraft motion. An analysis of the interferometric delay and phase reveals the signature of errors in the spacecraft ephemeris. Various calibration schemes are discussed, including the use of a ground beacon at White Sands to calibrate the communications link between White Sands and the TDRSE satellite.

035.011 **TDRSS OVLBI experiment: data correlation and results.**
R. P. Linfield.
Radio astronomy from space, p. 131 – 135 (1987). – See Abstr. 012.009.
The use of an orbiting antenna in a VLBI experiment complicates the data correlation, due to the spacecraft motion and the presence of the satellite–to–ground link. For this experiment, using a TDRSS satellite, a multi–step method was used to calculate delay and phase for correlation and fringe-fitting. Data from three sources: 1510–089, NRAO 530 (1730–130), and 1741–038 were successfully correlated.

035.012 **Astro–array: a space–based, coherent radio interferometer array.**
K. W. Weiler, J. H. Spencer, K. J. Johnston.
Radio astronomy from space, p. 175 – 184 (1987). – See Abstr. 012.009.
With the proven techniques of radio astronomy and the rapidly advancing technology of space science, it is clear that the expansion of radio astronomy arrays beyond the limits of the Earth's surface is called for. Therefore, a preliminary discussion of the design and capability of a purely space based array is presented.

035.013 **Cooled–HEMT microwave front–ends for space astronomy.**
S. Weinreb.
Radio astronomy from space, p. 191 – 194 (1987). – See Abstr. 012.009.

035.014 **The first radio astronomy from space: RAE.**
M. L. Kaiser.
Radio astronomy from space, p. 227 – 238 (1987). – See Abstr. 012.009.
The first dedicated radio astronomy mission in space was the Radio Astronomy Explorer (RAE) mission nearly two decades ago. The general characteristics of the spacecraft and instrumentation are described and the scientific accomplishments and failures are briefly discussed.

035.015 **Low orbit effects on optical coatings and materials as noted on early shuttle flights.**
T. R. Gull, H. Herzig, J. F. Osantowski, A. R. Toft.
Rutherford Appleton Lab., Rep., RAL 87–043, p. 1 – 11 (1987). – See Abstr. 012.010.
Properties of some material samples flown on early flights were found to change significantly. The most notable changes were the loss of material from Kapton insulation, total loss of osmium and carbon–deposited films, and modification of silver–deposited film to be a transparent silver oxide coating.

035.016 **A study of the mirror contamination on Ariel VI.**
A. M. Cruise, B. J. Kellett.
Rutherford Appleton Lab., Rep., RAL 87–043, p. 12 – 14 (1987). – See Abstr. 012.010.

035.017 **High reflectance coatings for space science applications in the vacuum ultraviolet.**
J. F. Osantowski, H. Herzig, R. A. M. Keski–Kuha, A. R. Toft.
Rutherford Appleton Lab., Rep., RAL 87–043, p. 15 – 94 (1987). – See Abstr. 012.010.

035.018 **ROSAT mirror system features of design, assembly and qualification.**
D. Reinhardt.
Rutherford Appleton Lab., Rep., RAL 87–043, p. 95 – 103 (1987). – See Abstr. 012.010.
The paper describes the 4nested Wolter I mirror system and the assembly method used. Qualification tests were successfully performed on a verification model and the structure thermal model based on it. All mirrors for the flight model have already been completed.

035.019 **XUV reflectivity measurements on the ROSAT WFC mirrors.**
S. R. Milward, M. Sims, R. Willingale.
Rutherford Appleton Lab., Rep., RAL 87–043, p. 104 – 109 (1987). With 8 figures. – See Abstr. 012.010.
The ROSAT Wide Field Camera is an imaging instrument operating in the energy band 12 – 250 eV, due for launch in the late 1980s as part of the German ROSAT X–ray astronomy satellite. A cut–away diagram of the WFC instrument and a schematic of the principal optical components, i.e. the mirrors, filters, and the CsI sensitised Microchannel Plate detectors are shown. The mirrors operate at grazing incidence and are of the Wolter–Schwarzschild type I configuration. Analysis of scattering measurements indicate a surface roughness of 0.8 nm rms over spatial wavelengths less than 100 μm. Two attempts are illustrated to predict the reflectivity of gold surfaces, both models use measured absorption data and anomalous dispersion theory to determine optical constants from which the reflectivity is predicted using the Fresnel equations.

035.020 Development of pure aluminium mirror surfaces for the Lyman far–ultraviolet astronomical satellite mission.
W. M. Burton, M. Grande, M. C. W. Sandford.
Rutherford Appleton Lab., Rep., RAL 87–043, p. 110 – 117 (1987). With 4 figures. – See Abstr. 012.010.

The proposed Lyman astronomical satellite mission requires reflecting surfaces optimised for the wavelength region 90 – 120 nm. A possible method for producing mirror surfaces with high reflectance at normal incidence in this spectral range is based on the use of pure unoxidised aluminium mirror coatings produced by evaporation in orbit within the spectrometer. This paper reports on progress in the development of technology needed to produce and maintain pure aluminium surfaces in a spacecraft environment.

035.021 The ultraviolet telescope on the Astron satellite.
A. A. Boyarchuk.
Astrophys. Space Phys. Rev., Vol. 5, p. 225 – 239 (1987). – See Abstr. 003.002.

On 23 March 1983 in the USSR, the Astron astrophysical satellite, with the largest ultraviolet telescope in the world (main mirror diameter 80 cm) and a set of X–ray instruments on board, was placed in a high-apogee orbit. The design of the ultraviolet telescope and the results of some of the observations carried out with it are described.
For the Russian original see 42.035.145.

035.022 Mass analyzer for investigation of the Venus atmosphere.
V. F. Samodurov, V. V. Petrov, L. R. Diament.
Metod. apparatura analit. veshchestva dlya kosm. issled., Ryazan', p. 12 – 18 (1986). In Russian. Abstr. in Ref. Zh., 62. Issled. Kosm. Prostranstva, 7.62.175 (1987).

035.023 All Sky Supernova and Transient Explorer (ASTRE).
P. Gorenstein.
Variability of galactic and extragalactic X–ray sources, p. 253 – 263 (1987). – See Abstr. 012.019.

A new type of wide field camera is described that provides a moderately large effective area (~ 100 cm²) and good angular resolution on every object in the night sky. This permits two types of studies; the detection and positioning of non–repetitive transients from arbitrary directions and measurements of longer term temporal behavior of X–ray sources on time scales from a day to a year. The instrument images in two dimensions, one by focussing, the other by a pseudo random distribution of slits.

035.024 The Advanced X–Ray Astrophysics Facility: an overview.
M. C. Weisskopf.
Astrophys. Lett. Commun., Vol. 26, Nos. 1 – 2, p. 1 – 6 (1987). – See Abstr. 012.020.

NASA'S Advanced X–Ray Astrophysics Facility (AXAF) is planned for launch in the 1990's. This paper presents an overview of the capability of the AXAF.

035.025 The AXAF CCD Imaging Spectrometer Experiment (ACIS).
J. A. Nousek, G. P. Garmire, G. R. Ricker, S. A. Collins, G. R. Reigler.
Astrophys. Lett. Commun., Vol. 26, Nos. 1 – 2, p. 35 – 41 (1987). – See Abstr. 012.020.

The ACIS experiment consists of an array of charge coupled device (CCD) chips placed at the focus of the AXAF mirror assembly. The instrument offers a powerful combination of the best qualities of the entire generation of Einstein Observatory detectors. The authors describe the technical capabilities of using CCD's to detect X–rays, and then consider a subset of the exciting astrophysical questions which can be attacked by ACIS.

035.026 X–ray spectroscopy of AGN with the AXAF "microcalorimeter".
S. S. Holt.
Astrophys. Lett. Commun., Vol. 26, Nos. 1 – 2, p. 61 – 71 (1987). – See Abstr. 012.020.

A novel technique for X–ray spectroscopy has been configured as part of the definition payload of the AXAF Observatory. It is basically a calorimeter which, operating at 0.1K, senses the total conversion of single photoelectrically absorbed X–rays via the differential temperature rise of the absorber. The technique promises to achieve < 10 eV FWHM with near–unit efficiency simultaneously over the entire AXAF bandpass. This combination of high resolution and high efficiency allows for the possibility of investigating thermal, fluorescent and absorption X–ray line features in many types of X–ray source, including a large sample of active galactic nuclei.

035.027 Low energy X–ray transmission grating spectrometer for AXAF.
A. C. Brinkman, J. J. van Rooijen, J. A. M. Bleeker, J. H. Dijkstra, J. Heise, P. A. J. de Korte, R. Mewe, F. Paerels.
Astrophys. Lett. Commun., Vol. 26, Nos. 1 – 2, p. 73 – 86 (1987). – See Abstr. 012.020.

The proposed grating spectrometer for the Advanced X–ray Astrophysics Facility (AXAF) covers the wavelength region between 2 and 140 Å. The wavelength resolution $\Delta\lambda = 0.05$ Å. The effective sensitive area as a function of wavelength is discussed. To illustrate the expected performance of the spectrometer some simulated spectra of a few interesting astrophysical objects are presented.

035.028 The MIT spectroscopy investigation of AXAF and the study of supernova remnants.
C. R. Canizares, H. V. D. Bradt, G. W. Clark, A. C. Fabian, P. C. Joss, A. M. Levine, W. H. G. Lewin, T. H. Markert, W. Mayer, G. R. Ricker, M. L. Schattenburg, H. I. Smith, B. E. Woodgate.
Astrophys. Lett. Commun., Vol. 26, Nos. 1 – 2, p. 87 – 98 (1987). – See Abstr. 012.020.

The investigation involves two complementary dispersive instruments, a Bragg Crystal Spectrometer and a high energy transmission grating. Together, these give high spectral resolution and good sensitivity for both point and extended sources in the energy range of 0.4 – 9 keV. The authors describe briefly each of the instruments and illustrate its capabilites for performing plasma diagnostics of supernova remnants.

035.029 The AXAF High Resolution Camera (HRC) and its use for observations of distant clusters of galaxies.
S. S. Murray, J. H. Chappell, M. S. Elvis, W. R. Forman, J. E. Grindlay, F. R. Harnden Jr., C. F. Jones, T. Maccacaro, H. D. Tananbaum, G. S. Vaiana, K. A. Pounds, G. W. Fraser, J. P. Henry.
Astrophys. Lett. Commun., Vol. 26, Nos. 1 – 2, p. 113 – 125 (1987). With plates IV – V.– See Abstr. 012.020.

The authors briefly describe the High Resolution Camera being developed for AXAF, comparing it with its predecessor the Einstein HRI and providing preliminary performance data. The overall sensitivity of the HRC on AXAF is about 50 times that of the Einstein HRI. An example of the power of this instrument is discussed in the context of studies of distant clusters of galaxies. The HRC can detect individual galaxies in a cluster of redshift ~ 1, if they are similar to the early–type galaxies in nearby clusters that have been observed by the Einstein instruments. The HRC will be used to measure the luminosity function for clusters of galaxies as a function of their distance up to a redshift of 1.

035.030 Expected AXAF mirror characteristics and their implications for measurements of the Hubble constant using the Sunyaev Zel'dovich effect.
L. P. van Speybroeck.
Astrophys. Lett. Commun., Vol. 26, Nos. 1 – 2, p. 127 – 146 (1987). – See Abstr. 012.020.

This paper contains a discussion of mirror technology relevant to the AXAF program, including the test results obtained with the developmental Technology Mirror Assembly and an estimate of the accuracies which can be expected in determining the Hubble constant by combining measurements of the Sunyaev Zel'dovich effect with X–ray measurements of the properties of distant clusters of galaxies.

035.031   **The gas scintillation proportional counter in the Spacelab environment: in–flight performance and post–flight calibration.**
P. Lamb, G. Manzo, S. Re, G. Boella, G. Villa, R. Andresen, M. R. Sims, G. F. Clark.
Astrophys. Space Sci., Vol. 136, No. 2, p. 369 – 378 (1987).

The authors describe the in–orbit performance of the gas scintillation proportional counter which formed part of the Spacelab–1 payload. Discontinuities in the instrument gain are observed (similar to those of the xenon–filled GSPC's on the EXOSAT and TENMA satellites). A post–flight recalibration of the instrument was performed using synchrotron radiation, which found the discontinuities to be coincident with the xenon L edges.

035.032   **A high sensitivity phoswich scintillator X–ray telescope for hard X–ray (20 – 120 keV) astronomy from balloon platform.**
S. V. Damle, A. T. Kothare, P. K. Kunte, J. P. Malkar, S. Naranan, B. V. Sreekantan, D. Venkatesan.
Adv. Space Res., Vol. 7, No. 7, p. 115 – 119 (1987). – See Abstr. 012.023.

A large area (400 cm^2) low background X–ray telescope consisting of four collimated Na I/Cs I scintillator phoswich detectors (each 100 cm^2) was built and successfully flown several times during 1980 – 84. The instrument characteristics, relevant details on the pointing system, detector system, associated electronics and telemetry, and in–flight performance are presented.

035.033   **Fabrication and flight performance of a large area balloon borne hard X–ray telescope.**
A. R. Rao, P. C. Agrawal, R. K. Manchanda, M. R. Shah.
Adv. Space Res., Vol. 7, No. 7, p. 129 – 131 (1987). – See Abstr. 012.023.

The authors describe the fabrication and flight performance of a balloon–borne large area hard X–ray (20 – 100 keV) telescope for spectral studies of discrete cosmic X–ray sources. The telescope consists of two multi–wire xenon filled proportional counters of effective area 1200 cm^2 each, mounted on an orientable platform. It can be pre–programmed to track any celestial source with a pointing accuracy of 0.5 degrees. For one hour of observation the telescope has a 5 σ detection sensitivity of 10^{-5}ph cm^{-2}s^{-1}. The laboratory test results and the performance in a series of balloon flights conducted in 1984 – 86 period are discussed and the preliminary results obtained for some X–ray sources are presented.

035.034   **X–ray astronomy satellite "Ginga".**
K. Makishima.
Astron. Her., Vol. 80, No. 11, p. 316 – 321 (1987). In Japanese.

035.035   **A gamma–ray pinhole camera.**
W. J. Wild.
Sky Telesc., Vol. 74, No. 2, p. 126 – 127 (1987).

035.036   **UV–spectrometer for investigating the Venus atmosphere.**
S. A. Ignatenko, A. V. Izherovskij, A. V. Rabinkov, A. P. Ehkonomov, R. M. Yakhin.
Konstruir. i tekhnol. izgotovleniya kosm. priborov. Moskva, p. 123 – 128 (1987). In Russian. Abstr. in Ref. Zh., 62. Issled. Kosm. Prostranstva, 9.62.149 (1987).

035.037   **UV–photometer for investigating the Venus atmosphere.**
Yu. M. Golovin, A. P. Ehkonomov, A. M. Sasov, V. M. Bedrosov.
Konstruir. i tekhnol. izgotovleniya kosm. priborov. Moskva, p. 128 – 132 (1987). In Russian. Abstr. in Ref. Zh., 62. Issled. Kosm. Prostranstva, 9.62.150 (1987).

035.038   **Parametric study of stable optical references for large flexible structures in space.**
M. J. Clayton, A. L. Wertheimer.
Proc. SPIE Int. Soc. Opt. Eng., Vol. 571, p. 149 – 157 (1986). – See Abstr. 43.012.094.

The authors discuss several concepts to keep track of small relative angular motions between a reference point and a remote platform when the two are located at separate points on a flexible structure. The goal is to accurately transfer knowledge of the orientation of the remote platform relative to the reference point. First–order equations and order–of–magnitude calculations are presented for image centroiding and interference fringe projection approaches for Space Station applications.

035.039   **On–orbit active alignment of the SOT observatory.**
M. Yellin.
Proc. SPIE Int. Soc. Opt. Eng., Vol. 571, p. 196 – 202 (1986). – See Abstr. 43.012.094.

The Solar Optical Telescope (SOT) is a Shuttle–borne telescope that will point deep ultraviolet science instruments at the sun with sub–arc second stability. The telescope will be actively aligned in orbit while viewing the sun. This paper describes the unique active compensation techniques used on SOT to satisfy the on–orbit resolution requirements.

035.040   **X–ray photographs of a solar active region with a multilayer telescope at normal incidence.**
J. H. Underwood, M. E. Bruner, B. M. Haisch, W. A. Brown, L. W. Acton.
Science, Vol. 238, No. 4823, p. 61 – 64 (1987).

A photograph was obtained with a multilayer X–ray telescope. A 4–centimeter tungsten–carbon multilayer mirror was flown as part of an experimental solar rocket payload, and successful images were taken of the sun at normal incidence at a wavelength of 44 angstroms. Coronal Si–XII emission from an active region was recorded on film; as expected, the structure is very similar to that observed at O–VIII wavelengths by the Solar Maximum Mission flat crystal spectrometer at the same time. The small, simple optical system used in this experiment appears to have achieved a resolution of 5 to 10 arc seconds.

035.041   **SOFIA: stratospheric observatory for infrared astronomy.**
L. J. Caroff, E. F. Erickson, G. W. Thorley.
Bull. Am. Astron. Soc., Vol. 19, No. 2, p. 686 – 687 (1987). Abstract. – See Abstr. 010.061.

035.042   **Rocket spectrometer to measure the submillimeter CBR spectrum.**
M. Halpern, E. Wishnow, H. P. Gush.
Bull. Am. Astron. Soc., Vol. 19, No. 2, p. 688 (1987). Abstract. – See Abstr. 010.061.

035.043   **The circumstellar imaging telescope – direct detection of extra–solar planets.**
C. Ftaclas, E. T. Siebert, R. J. Terrile.
Bull. Am. Astron. Soc., Vol. 19, No. 2, p. 689 (1987). Abstract. – See Abstr. 010.061.

035.044   **Extragalactic imaging with the Infrared Array Camera, IRCAM, on UKIRT.**
M. G. Smith, I. S. McLean, C. M. Telesco, M. J. Ward, N. Devereux.
Bull. Am. Astron. Soc., Vol. 19, No. 2, p. 689 (1987). Abstract. – See Abstr. 010.061.

035.045 **Thin film filters to be used on the Extreme Ultraviolet Explorer satellite.**
J. V. Vallerga, O. H. W. Siegmund, P. Jelinsky, R. F. Malina.
Bull. Am. Astron. Soc., Vol. 19, No. 2, p. 690 (1987). Abstract. – See Abstr. 010.061.

035.046 **Wide Field/Planetary Camera–II for the Hubble Space Telescope.**
J. T. Trauger, D. Crisp, S. R. Federman, R. E. Griffiths, J. G. Hoessel.
Bull. Am. Astron. Soc., Vol. 19, No. 2, p. 746 (1987). Abstract. – See Abstr. 010.061.

035.047 **Cosmic ray events observed in Space Telescope Wide Field Camera exposures.**
S. P. Ewald, R. Griffiths, J. W. MacKenty.
Bull. Am. Astron. Soc., Vol. 19, No. 2, p. 746 (1987). Abstract. – See Abstr. 010.061.

035.048 **A detector for the Ultraviolet Imaging Telescope.**
P. C. Chen.
Bull. Am. Astron. Soc., Vol. 19, No. 2, p. 746 (1987). Abstract. – See Abstr. 010.061.

035.049 **Enhancement of data from the High Resolution Spectrograph using a block iterative image restoration algorithm.**
D. Ebbets.
Bull. Am. Astron. Soc., Vol. 19, No. 2, p. 747 (1987). Abstract. – See Abstr. 010.061.

035.050 **Current state of the HST Faint Object Spectrograph.**
R. J. Harms.
Bull. Am. Astron. Soc., Vol. 19, No. 2, p. 757 (1987). Abstract. – See Abstr. 010.061.

035.051 **The Goddard High Resolution Spectrograph (GHRS) for the Hubble Space Telescope (HST): status June 1987.**
J. Brandt, S. Heap, K. Carpenter, D. Ebbets, D. Lindler.
Bull. Am. Astron. Soc., Vol. 19, No. 2, p. 757 – 758 (1987). Abstract. – See Abstr. 010.061.

035.052 **Advanced scientific instruments for the Hubble Space Telescope.**
J. T. Clarke.
Bull. Am. Astron. Soc., Vol. 19, No. 2, p. 758 (1987). Abstract. – See Abstr. 010.061.

035.053 **Bragg imaging of extended cosmic X–ray sources: the objective crystal spectrometer.**
H. W. Schnopper, B. P. Byrnak.
Appl. Opt., Vol. 26, No. 14, p. 2871 – 2876 (1987).
A large flat objective crystal couples nicely to a modest resolution but high throughput X–ray concentrating telescope to produce high dispersion images of diffuse X–ray sources in each of the several lines present in the spectrum. Unprecedented spectral resolution is provided for point sources and, provided that they do not vary on rapid time scales, a spectrum can be scanned over a wide energy range with unparalleled sensitivity.

035.054 **X–ray optics: a technique for high resolution imaging.**
W. Cash.
Appl. Opt., Vol. 26, No. 14, p. 2915 – 2920 (1987).
Both scattering and figure errors in grazing incidence optics are larger in the plane of incidence than out–of–plane by a factor equal to $1/\sin\theta$, where θ is the graze angle. When the full annular aperture of a grazing incidence telescope is stopped down, the point spread function becomes highly elliptical with a width as much as $\sin\theta$ times narrower than the full image. In practice this means that improvements in resolution of up to 100 times can be achieved, and effective resolution can approach the diffraction limit. Laboratory data demonstrating the effect are presented.

035.055 **Geometrical aberration of a generalized Wolter type I telescope.**
K. Nariai.
Appl. Opt., Vol. 26, No. 20, p. 4428 – 4432 (1987).
Using two hyperboloids, one can control spherical aberration and coma of an imaging X–ray telescope. The coma–free condition is satisfied only at a particular radius of the first mirror. The best performance is obtained when we allow the coma–free condition at the outer edge and cancel about half of the defocused image for the largest height of an object by introducing spherical aberration.

035.056 **Electroformed grazing incidence X–ray mirrors for a mirror array telescope.**
M. P. Ulmer, Y. Matsui, D. K. Bedford, G. M. Simnett, P. Z. Takacs.
Appl. Opt., Vol. 26, No. 18, p. 3852 – 3857 (1987).
Grazing incidence Wolter type I mirrors for higher–energy X rays have been replicated from two superpolished mandrels by electroforming. The authors present the design of the mandrels, mirror mounting scheme, and results of the X–ray test. The microroughnesses of the mirrors measured using an optical profilometer were compared with the X–ray test results.

035.057 **A liquid–helium–cooled far–infrared grating spectrometer for a balloon–borne infrared telescope.**
H. Takami, T. Maihara, K. Mizutani, N. Hiromoto, H. Shibai.
Publ. Astron. Soc. Pac., Vol. No. 619, p. 1022 – 1026 (1987).
A liquid–helium–cooled far–infrared grating spectrometer has been developed for a 50–cm balloon–borne infrared telescope. The spectral coverage is from 50 μm to 110 μm, with the spectral resolution of 0.5 μm to 0.35 μm. The diaphragm aperture is 2 arc min in diameter when attached to the telescope. This spectrometer was used in two balloon observations made in March 1985 and in August 1986 in Australia.

035.058 **The ultraviolet calibration of the Hubble Space Telescope: II. A correction for the change in sensitivity of the SWP camera on IUE.**
R. C. Bohlin, C. J. Grillmair.
Space Telesc. Sci. Inst., Prepr. Ser., No. 201, 27 pp. (1987). To appear in Astrophys. J., Suppl. Ser.

035.059 **Ultraviolet astrophysical experiment on board a stratospheric balloon. II. Detector and signal–to–noise ratio.**
M. Rego, J. Zamorano, G. Rodriguez Caderot, M. Cornide.
An. Fis., Ser. B, Vol. 82, No. 3, p. 343 – 349 (1986). In Spanish. Abstr. in Phys. Abstr., Vol. 90, No. 1309, Entry 88108 (1987).

035.060 **An imaging spectrometer for the investigation of Mars.**
J. E. Duval.
Proc. SPIE Int. Soc. Opt. Eng., Vol. 685, p. 6 – 15 (1986). Abstr. in Phys. Abstr., Vol. 90, No. 1309, Entry 88117 (1987). – See Abstr. 012.039.

035.061 **Test of IR arrays on the Kuiper Airborne Observatory.**
R. W. Russell, G. S. Rossano, D. K. Lynch, G. T. Colon–Bonet, J. A. Hackwell, T. C. Morse, R. H. Macklin, D. Murray, D. A. Retig, C. J. Rice, D. A. Roux, R. M. Young.
Proc. SPIE Int. Soc. Opt. Eng., Vol. 685, p. 88 – 98 (1986). Abstr. in Phys. Abstr., Vol. 90, No. 1309, Entry 88118 (1987). – See Abstr. 012.039.

035.062 **Analysis of misalignment and thermal distortion effects in the FIRST antenna.**
A. Garcia Pino, M. Calvo, C. E. Montesano.
Fifth International Conference on Antennas and Propagation ICAP 87, held at York, UK, 30 March – 2 April 1987, Vol. 1. IEE Conf. Publ., No. 274, IEE, London, UK, p. 198 – 201 (1987). Abstr. in Phys. Abstr., Vol. 90, No. 1309, Entry 88121 (1987).

035.063   Detector arrays for low–background space infrared astronomy.
C. R. McCreight, M. E. McKelvey, J. H. Goebel,
G. M. Anderson, J. H. Lee.
Proc. SPIE Int. Soc. Opt. Eng., Vol. 686, p. 66 – 75 (1986). Abstr. in Phys. Abstr., Vol. 90, No. 1309, Entry 88122 (1987). – See Abstr. 012.040.

035.064   The paraboloid–paraboloid microscopic optical X–ray system: first experience. 1. Production and tests.
R. Hudec, B. Valníček, R. Peřestý, I. Šolc, L. Lochman,
L. Svátek, V. Landa, V. Jelínek, I. Zhitnik, V. Krutov,
W. Burkert, H. Bräuninger, P. Predehl.
Publ. Astron. Inst. Czech. Acad. Sci., No. 64, p. 1 – 15 (1987).

The microscopic optical X–ray system of the paraboloid–paraboloid type is based on a galvanoplastic replica. The study describes the production procedure and the first tests in optical and X–ray light. Briefly discussed are the results obtained as well as the possibility of using mirrors of this type for mapping X–ray plasma during nuclear fusion.

035.065   X–ray optical properties of galvanoplastic grazing incidence mirrors: full aperture tests. (I).
R. Hudec, B. Valníček, H. Bräuninger, W. Burkert, P. Predehl.
Publ. Astron. Inst. Czech. Acad. Sci., No. 64, p. 16 – 49 (1987).

Two different replica technologies were developed and used for manufacturing X–ray grazing incidence mirrors. Five of the mirrors were tested in the X–ray long test facility at MPI Garching in 1984. The results are presented and briefly discussed.

035.066   Metallic mirror telescope aboard the sounding rocket 520–8CN.
M. Nakagiri, T. Onaka, W. Tanaka, A. Yamaguchi, T. Kono.
Tokyo Astron. Obs. Rep., Vol. 21, No. 1, p. 67 – 81 (1987). In Japanese.

035.067   The dependence of mass resolution and sensitivity of the PUMA instrument on the energy spread of ions produced by hypervelocity impacts.
R. Z. Sagdeev, J. Kissel, E. N. Evlanov, M. N. Fomenkova,
N. A. Inogamov, V. N. Khromov, G. G. Managadze,
O. F. Prilutski (O. F. Prilutskij), V. D. Shapiro, I. Y. Shutyaev
(I. Yu. Shutyaev), B. V. Zubkov.
Astron. Astrophys., Vol. 187, No. 1/2, p. 179 – 182 (1987). – See Abstr. 003.013, 43.035.018.

Measurements of the element composition of the dust particles in comet Halley were made by the PUMA–1 and PUMA–2 instruments onboard the Vega spacecraft. The time–of–flight analysis of ions produced by the hypervelocity particle impact on the target was applied with a wide ($W = 100$ eV) and a narrow ($W = 20$ eV) energy window of the instruments. The analysis of measured spectra shows that the mass resolution for "small" particles ($Q < 10^{-13}$C) was high enough in both modes to separate individual mass lines. For large particles the resolution did not change for the narrow window, it decreased, however, for the wide one (to about 50). A relative trasnmission Q_w/Q_n dependence on the mass number for the wide and narrow window has been found.

035.068   Calibration of the DIDSY–IPM dust detector and application to other impact ionisation detectors on board the P/Halley probes.
J. R. Göller, E. Grün, D. Maas.
Astron. Astrophys., Vol. 187, No. 1/2, p. 693 – 698 (1987). – See Abstr. 003.013.

Results from calibration tests with the DIDSY–IPM dust detector, flown on the Giotto mission, are reported and applied to the impact ionisation dust detectors SP–1 and SP–2, flown on the two Vega Halley probes.

035.069   An attempt to evaluate the structure of cometary dust particles.
V. N. Smirnov, O. L. Vaisberg (O. L. Vajsberg), S. Anisimov.
Astron. Astrophys., Vol. 187, No. 1/2, p. 774 – 778 (1987). – See Abstr. 003.013, 43.035.014.

A comparison of the counting rates of two sensors of the SP–1 plasma impact detector, one open and one shielded by a thin foil is used to roughly estimate the mean density of dust particles. Some inconsistencies between the current understanding of the penetration of film by particles and the observed behavior are found. A mass density of $\lesssim 1$ g cm^{-3} does not contradict to measurements in the low mass range ($\lesssim 10^{-14}$g). A lower density value is suggested for heavier particles (m $\gtrsim 10^{-12}$g), implying a fluffy structure of these particles.

035.070   SAMSI – a spacecraft array for Michelson spatial interferometry.
R. V. Stachnik, M. Faucherre.
ESA Spec. Publ., ESA SP–273, p. 61 (1987). Abstract. – See Abstr. 012.044.

035.071   COSMIC (Coherent Optical System of Modular Imaging Collectors).
W. A. Traub.
ESA Spec. Publ., ESA SP–273, p. 63 – 65 (1987). – See Abstr. 012.044.

The design goals of COSMIC which most distinguish it from other current telescope array concepts are these: a high degree of structural stiffness, a wide field of view, and a broad wavelength band. These properties directly enhance the scientific productivity and technical reliability of the array. If a sufficiently stiff structure can be designed, COSMIC can operate as a phased array; if the structure relaxes before it can be rephased on a reference star, it degrades to a coherent array.

035.072   Binary Star Explorer.
W. A. Traub.
ESA Spec. Publ., ESA SP–273, p. 67 – 68 (1987). – See Abstr. 012.044.

The Binary Star Explorer (BSE) is a proposed precursor space interferometer which could make fundamental astrophysical measurements with a relatively simple instrument. The BSE will determine distances and masses of a large number of binary systems in our Galaxy and in the LMC. These binaries include many Cepheid systems, so that the fundamental distance scale can be calibrated accurately in terms of the directly measured quantities of angular separation and radial velocity.

035.073   ISIS: Imaging Speckle Interferometer in Space.
G. Weigelt.
ESA Spec. Publ., ESA SP–273, p. 69 – 72 (1987). – See Abstr. 012.044.

The author proposes the construction of a large multiple–mirror interferometer in space. For example, at $\lambda \sim 100$ nm and with a baseline of 20 m, a resolution of 0″.001 can be obtained. The following three interferometer types are very attractive: (a) linear 14–m array (launched by the Space Shuttle), (b) 2–dimensional, deployable 20 m array, and (c) array of 6 to 20 free–flying telescopes with baselines up to 40 km and resolution of 10^{-6}″ at $\lambda \sim 200$ nm. The limiting magnitude of optical long–baseline interferometry in space is $\sim 24^m$ or even fainter.

035.074   Fiber–linked telescope arrays on the ground and in space.
P. Connes, F. Roddier, S. Shaklan, E. Ribak.
ESA Spec. Publ., ESA SP–273, p. 73 – 83 (1987). – See Abstr. 012.044.

The use of single–mode optical fibers in telescope arrays is further developed. Two proposals are described: a ground–based array of small optical telescopes supported on a radio dish, and a similar space array. The control system is almost the same in both cases, hence the ground–based array can be considered as a test–bench for the space device.

035.075 **Proposed studies of a 30 meter imaging interferometer concept.**
R. T. Stebbins, P. L. Bender, J. E. Faller.
ESA Spec. Publ., ESA SP–273, p. 85 – 91 (1987). – See Abstr. 012.044.

An attractive concept for an imaging interferometer in space is based on the use of roughly 15 m sections of graphite–epoxy truss structure to form the basic mechanical support system. One simple design is a Y–shaped array of 3 coplanar arms, each 15 m long and 1.5 m in diameter, together with a perpendicular mast of similar length. Roughly 15 observing telescopes of 0.5 m diameter would be used, with laser interferometers controlling the optical pathlengths.

035.076 **Space Station based interferometry.**
H. Olthof.
ESA Spec. Publ., ESA SP–273, p. 93 – 102 (1987). – See Abstr. 012.044.

035.077 **NASA plans for interferometry in space.**
R. Stachnik.
ESA Spec. Publ., ESA SP–273, p. 111 (1987). Abstract. – See Abstr. 012.044.

035.078 **NASA/JPL study on optical imaging interferometry in space.**
S. P. Synnott, R. E. Freeland, E. Ribak, E. F. Tubbs.
ESA Spec. Publ., ESA SP–273, p. 113 – 115 (1987). – See Abstr. 012.044.

035.079 **Aperture synthesis in space: technical problems.**
D. Morancais, P. Roussel.
ESA Spec. Publ., ESA SP–273, p. 177 – 185 (1987). – See Abstr. 012.044.

The authors describe a "schematic" space interferometer by its different tasks or modules. Then, they derive the requirements on these different modules in order to achieve ideal performances. Starting point will be an angular resolution of $10^{-4} - 10^{-6}$arcsec, a magnitude above 10 and operating in the visible ($\lambda = 0.5$ microns).

035.080 **Solar interferometry with a 4–aperture non–redundant and stabilized network.**
L. Damé, C. Aime, M. Faucherre, J. Heyvaerts.
ESA Spec. Publ., ESA SP–273, p. 189 – 195 (1987). – See Abstr. 012.044.

The design of a solar interferometer is intrinsically complex since many requirements, often found separately, and difficult by themselves, are brought together: UV spectral range, limb observations, resolved structures (low contrast) and time resolution. The stabilized interferometry technique, applied to a non–redundant array of 4 telescopes, provides an elegant solution to those complex problems.

035.081 **A test–bed for space interferometry: SPI.**
L. Damé, M. Faucherre, R. V. Stachnik, W. A. Traub.
ESA Spec. Publ., ESA SP–273, p. 197 – 204 (1987). – See Abstr. 012.044.

SPI (Space Platform Interferometer) is a 20 meter, two–mirror Michelson interferometer which can reach magnitude 14 at UV and visible wavelengths. SPI is attached to a platform serviced from the Space Station.

035.082 **Use of a tunable quasimonoenergetic gamma–ray beam for the calibration of the EGRET gamma–ray telescope in the range 20 – 24000 MeV.**
J. R. Mattox, R. Hofstadter, E. B. Hughes, Y. C. Lin, P. L. Nolan, A. H. Walker.
Nucl. Instrum. Methods Phys. Res., Sect. B, Vol. B24–B25, Part 2, p. 888 – 892 (1987). Abstr. in Phys. Abstr., Vol. 90, No. 1310, Entry 94660 (1987).

035.083 **Hybrid X–ray telescope systems.**
D. L. Shealy, R. B. Hoover.
Proc. SPIE Int. Soc. Opt. Eng., Vol. 640, p. 28 – 44 (1986). Abstr. in Phys. Abstr., Vol. 90, No. 1311, Entry 95702 (1987). – See Abstr. 012.048.

035.084 **Technology Mirror Assembly mirror quality requirements and achievements.**
P. Glenn, A. Slomba, R. Babish.
Proc. SPIE Int. Soc. Opt. Eng., Vol. 640, p. 45 – 58 (1986). Abstr. in Phys. Abstr., Vol. 90, No. 1311, Entry 95703 (1987). – See Abstr. 012.048.

035.085 **Assembly and alignment of the Technology Mirror Assembly.**
N. A. De Filippis, P. Glenn, R. Cahil.
Proc. SPIE Int. Soc. Opt. Eng., Vol. 640, p. 155 – 163 (1986). Abstr. in Phys. Abstr., Vol. 90, No. 1311, Entry 95708 (1987). – See Abstr. 012.048.

035.086 **Quasat program: the ESA reflector.**
G. G. Reibaldi, M. C. Bernasconi.
Acta Astronaut., Vol. 15, No. 3, p. 181 – 187 (1987). Abstr. in Phys. Abstr., Vol. 90, No. 1311, Entry 101680 (1987).

035.087 **OSAC analysis of the far ultraviolet spectroscopic explorer (FUSE) telescope.**
T. T. Saha, D. A. Thomas, J. F. Osantowski.
Proc. SPIE Int. Soc. Opt. Eng., Vol. 640, p. 79 – 84 (1986). Abstr. in Phys. Abstr., Vol. 90, No. 1311, Entry 101687 (1987). – See Abstr. 012.048.

035.088 **The high–flying Kvant module.**
J. K. Beatty.
Sky Telesc., Vol. 74, No. 6, p. 599 – 601 (1987).

035.089 **All–sky monitors for X–ray astronomy.**
S. S. Holt, W. Priedhorsky.
Space Sci. Rev., Vol. 45, Nos. 3+4, p. 269 – 289 (1987).

The authors discuss the rationale for a semi–permanent all–sky X–ray monitor, and investigate a variety of options for its implementation. They conclude that the Space Station offers an excellent opportunity for hosting such a monitor, and that a set of pinhole cameras can be configured to provide an effective and economical monitor system. A baseline of six independent pinhole modules, each of which requires approximately one cubic foot, 30 pounds, 2 watts, and 100 bits per second, can provide full sky coverage with scientifically interesting sensitivities. The baseline system can locate bright sources to a few arc min, and can simultaneously measure each of the several hundred sources in the sky brighter than a few thousandths the intensity of the Crab nebula every day for decades.

035.090 **Coded aperture imaging in X– and gamma–ray astronomy.**
E. Caroli, J. B. Stephen, G. di Cocco, L. Natalucci, A. Spizzichino.
Space Sci. Rev., Vol. 45, Nos. 3+4, p. 349 – 403 (1987).

Coded aperture imaging in high energy astronomy represents an important technical advance in instrumentation over the full energy range from X– to γ–rays and is playing a unique role in those spectral ranges where other techniques become ineffective or impracticable due to limitations connected to the physics of interactions of photons with matter. The theory underlying this method of indirect imaging is of strong relevance both in design optimization of new instruments and in the data analysis process. The coded aperture imaging method is herein reviewed with emphasis on topics of mainly practical interest along with a description of already developed and forthcoming implementations.

035.091   Calibration of the EGRET gamma ray telescope with a back–scattered laser beam.
J. R. Mattox.
Diss. Abstr. Int., Sect. B, Vol. 48, No. 6, p. 1711–B (1987). Thesis, Stanford University, 152 pp. (1987). Order No. DA8720416.

035.092   Preliminary observations from the auroral and iono-spheric remote sensing imager.
C. I. Meng, R. E. Huffman.
Johns Hopkins APL Tech. Dig., Vol. 8, No. 3, p. 303 – 307 (1987).
The Auroral and Ionospheric Remote Sensing Experiment on board the Polar BEAR satellite uses a four–color imager covering selected wavelengths in the visible/near–UV and vacuum ultraviolet (110.0– to 180.0–nm ranges). This device is capable of imaging global auroral display and the distribution of atmospheric emissions in both dark and sunlit regions. In addition to providing images, it can also operate in either spectrometric or photometric modes. This article describes the techniques employed and gives preliminary observations from the experiment.

035.093   Auroral images from space: imagery, spectroscopy, and photometry.
F. W. Schenkel, B. S. Ogorzalek.
Johns Hopkins APL Tech. Dig., Vol. 8, No. 3, p. 308 – 317 (1987).
The Polar BEAR Mission required a multimodal instrument (comprising imagery, spectroscopy, and photometry) known as the Auroral Ionospheric Remote Sensor. The sensor produces auroral images in both dark and sunlit hemispheres and enables the remote sensing of ionospheric airglows to aid in the detection of ionospheric electron–density profiles and atmospheric background emissions.

035.094   The Polar BEAR magnetic field experiment.
P. F. Bythrow, T. A. Potemra, L. J. Zanetti,
F. F. Mobley, L. Scheer, W. E. Radford.
Johns Hopkins APL Tech. Dig., Vol. 8, No. 3, p. 318 – 323 (1987).
A primary route for the transfer of solar wind energy to the earth's ionosphere is via large–scale currents that flow along geomagnetic field lines. The currents, called "Birkeland" or "field aligned", are associated with complex plasma processes that produce ionospheric scintillations, joule heating, and auroral emissions. The Polar BEAR magnetic field experiment, in conjunction with auroral imaging and radio beacon experiments, provides a way to evaluate the role of Birkeland currents in the generation of auroral phenomena.

035.095   Design manufacture and test of the HIPPARCOS beam combiner.
R. Geyl.
Proc. SPIE Int. Soc. Opt. Eng., Vol. 655, p. 396 – 401 (1986). Abstr. in Phys. Abstr., Vol. 90, No. 1313, Entry 110537 (1987). – See Abstr. 012.063.

035.096   Hungarian results in the exploration of Halley's comet.
I. Apathy, P. Bereczki, G. Endroczy, J. Ero Jr.,
I. Gombos, A. Gschwind, L. Lohonyai, G. Kozma, I. Naday,
I. Renyi, A. Somogyi, F. Szabó, L. Szabó, S. Szalai, K. Szegö,
A. Varga, P. Zalan.
Meres Autom., Vol. 35, No. 3, p. 73 – 84 (1987). In Hungarian. Abstr. in Phys. Abstr., Vol. 90, No. 1313, Entry 115198 (1987).

035.097   The performance of a multistep proportional counter for use in X–ray astronomy.
B. D. Ramsey, M. C. Weisskopf.
IEEE Trans. Nucl. Sci., Vol. NS–34, No. 3, p. 672 – 675 (1987). Abstr. in Phys. Abstr., Vol. 90, No. 1313, Entry 115234 (1987).

035.098   Optical arrays for future astronomical telescopes in space.
A. N. Brunner.
Proc. SPIE Int. Soc. Opt. Eng., Vol. 643, p. 180 – 188 (1986). Abstr. in Phys. Abstr., Vol. 90, No. 1313, Entry 115241 (1987). – See Abstr. 012.062.

035.099   Optical design of the HIPPARCOS telescope.
J. J. Arnoux, D. Dubet, M. Fruit.
Proc. SPIE Int. Soc. Opt. Eng., Vol. 655, p. 380 – 387 (1986). Abstr. in Phys. Abstr., Vol. 90, No. 1313, Entry 115243 (1987). – See Abstr. 012.063.

035.100   Infrared detectors for space applications.
J. R. Houck.
Infrared astronomy with arrays, p. 108 – 115 (1987). – See Abstr. 012.070.
NASA–sponsored detector development efforts are investigating infrared detectors for space applications over the range from 1 to 1,000 microns. The conditions under which space–based detectors must work and a summary of the various efforts to develop suitable detectors are discussed. The performance of a spectrograph which is under consideration for inclusion in SIRTF is presented.

035.101   Review of arrays developed for the Infrared Space Observatory (ISO).
F. Sibille, C. Cesarsky, D. Rouan.
Infrared astronomy with arrays, p. 116 – 120 (1987). – See Abstr. 012.070.
This paper gives first a short description of the ISO satellite and of its four instruments. Two 32×32 array detectors are currently developed for the $3 - 17\,\mu m$ camera ISOCAM: an InSb CID at SAT, and a Si:Ga DRO at LETI/LTR. Both their achieved and expected performances are presented.

035.102   The Short–Wavelength Spectrometer for ISO.
T. de Graauw, D. A. Beintema, W. Luinge,
G. Ploeger, P. R. Wesselius, K. Wildeman, J. Wijnbergen,
S. W. Drapatz, R. Genzel, G. Haerendel, L. Haser,
R. Katterloher, F. Melzner, J. Stöcker, T. M. Kamperman,
K. A. van der Hucht, W. C. A. van Dijkhuizen.
Infrared astronomy with arrays, p. 438 – 442 (1987). – See Abstr. 012.070.
The Short–Wavelength Spectrometer for ISO comprises two grating spectrometers with a resolving power of 1000 in the range 2.3 to 45 μm. Fabry–Perot etalons can boost the resolution to 30,000 in the range 15 to 35 μm. The baseline detector configuration has Si:In, Si:Ga, Si:P and Ge:Be detectors from the Battelle–Institut.

035.103   Grism spectroscopy with infrared array detectors.
R. I. Thompson.
Infrared astronomy with arrays, p. 443 – 447 (1987). – See Abstr. 012.070.
This paper outlines the properties and uses of grisms for infrared spectroscopy. With area detectors, grisms produce spectra of all objects in the field of view. In a properly designed system the spectra appear centered on the focal position of the object. Examples from the NICMOS instrument for HST indicate possible designs and constraints.

035.104   Future space and sub–orbital programs that will complement ground–based astronomy with infrared arrays.
M. W. Werner, E. F. Erickson.
Infrared astronomy with arrays, p. 527 – 535 (1987). – See Abstr. 012.070.
The status and prospects for future space and sub–orbital programs for infrared astronomy are reviewed. The programs to be discussed include ESA's Infrared Space Observatory (ISO) and NASA's Space Infrared Telescope Facility (SIRTF), which are cryogenically cooled observatory–class facilities for space infrared astronomy; the second generation instruments for infrared imaging and spectroscopy under definition study for the

Hubble Space Telescope; and new initiatives for stratospheric observations – including large balloon–borne telescopes and the large airborne telescope, the Stratospheric Observatory for Infrared Astronomy (SOFIA).

035.105   **System implications of aperture–shade design for the SIRTF observatory.**
J. H. Lee, W. F. Brooks, S. Maa.
J. Spacecr. Rockets, Vol. 24, No. 2, p. 162 – 168 (1987). Abstr. in Phys. Abstr., Vol. 90, No. 1315, Entry 127724 (1987).

035.106   **The Hubble Space Telescope.**
A. A. Tokovinin.
Zemlya Vselennaya, No. 4, p. 49 – 55 (1987). In Russian.

035.107   **A hard X–ray telescope and its observation of the Crab pulsar.**
C.-j. Dai, M. Wu, Y.-q. Ma, Z.-g. Lu, G.-h. Li, C.-m. Zhang, Z.-z. Fan, C.-x. Xu, Y.-d. Gu, X.-y. Zhang, T.-p. Li.
Chin. Astron. Astrophys., Vol. 11, No. 3, p. 179 – 185 (1987). English translation of 43.035.037.

035.108   **ASTROMAG: A superconducting particle astrophysics magnet facility for the Space Station.**
G. F. Smoot, R. L. Golden, M. H. Israel, R. Kephart, R. Niemann, R. A. Mewalt, J. F. Ormes, P. Spillantini, M. E. Widenbeck.
IEEE Trans. Magn., Vol. MAG–23, No. 2, p. 1240 – 1243 (1987). Abstr. in Phys. Abstr., Vol. 90, No. 1316, Entry 134141 (1987).

035.109   **The Giotto magnetometer experiment.**
F. M. Neubauer, M. H. Acuña, L. F. Burlaga, B. Franke, B. Gramkow, F. Mariani, G. Musmann, N. F. Ness, H. U. Schmidt, R. Terenzi, E. Ungstrup, M. Wallis.
J. Phys. E, Vol. 20, No. 6, Part 2, p. 714 – 720 (1987). Abstr. in Phys. Abstr., Vol. 90, No. 1317, Entry 139883 (1987).

035.110   **The Giotto electron plasma experiment.**
H. Rème, F. Cotin, A. Cros, J. L. Medale, J. A. Sauvaud, C. d'Uston, K. A. Anderson, C. W. Carlson, D. W. Curtis, R. P. Lin, A. Korth, A. K. Richter, A. Loidl, D. A. Mendis.
J. Phys. E, Vol. 20, No. 6, Part 2, p. 721 – 731 (1987). Abstr. in Phys. Abstr., Vol. 90, No. 1317, Entry 139884 (1987).

035.111   **The Giotto dust impact detection system.**
J. A. M. McDonnell.
J. Phys. E, Vol. 20, No. 6, Part 2, p. 741 – 758 (1987). Abstr. in Phys. Abstr., Vol. 90, No. 1317, Entry 139885 (1987).

035.112   **The ion mass spectrometer on Giotto.**
H. Balsiger, K. Altwegg, J. Benson, F. Bühler, J. Fischer, J. Geiss, B. E. Goldstein, R. Goldstein, P. Hemmerich, G. Kulzer, A. J. Lazarus, A. Meier, M. Neugebauer, U. Rettenmund, H. Rosenbauer, K. Sager, T. Sanders, R. Schwenn, E. G. Shelley, D. Simpson, D. T. Young.
J. Phys. E, Vol. 20, No. 6, Part 2, p. 759 – 767 (1987). Abstr. in Phys. Abstr., Vol. 90, No. 1317, Entry 139886 (1987).

035.113   **The heavy ion analyser PICCA for the comet Halley fly–by with Giotto.**
A. Korth, A. K. Richter, A. Loidl, W. Güttler, K. A. Anderson, C. W. Carlson, D. W. Curtis, R. P. Lin, H. Rème, F. Cotin, A. Cros, J. L. Medale, J. A. Sauvaud, C. d'Uston, D. A. Mendis.
J. Phys. E, Vol. 20, No. 6, Part 2, p. 787 – 792 (1987). Abstr. in Phys. Abstr., Vol. 90, No. 1317, Entry 139887 (1987).

035.114   **The Giotto three–dimensional positive ion analyser.**
A. D. Johnstone, A. J. Coates, B. Wilken, W. Studemann, W. Weiss, R. Cerulli Irelli, V. Formisano, H. Borg, S. Olsen, J. D. Winningham, D. A. Bryant, S. J. Kellock.
J. Phys. E, Vol. 20, No. 6, Part 2, p. 795 – 805 (1987). Abstr. in Phys. Abstr., Vol. 90, No. 1317, Entry 139888 (1987).

035.115   **The Halley multicolour camera.**
H. U. Keller, W. K. H. Schmidt, K. Wilhelm, C. Becker, W. Curdt, W. Engelhardt, H. Hartwig, J. R. Kramm, H. J. Meyer, R. Schmidt, F. Gliem, E. Krahn, H. P. Schmidt, G. Schwarz, J. J. Turner, P. Bouyries, S. Cazes, F. Angrilli, G. Bianchini, G. Fanti, P. Brunello, A. Delamere, H. Reitsema, C. Jamar, A. Cucchiaro.
J. Phys. E, Vol. 20, No. 6, Part 2, p. 807 – 820 (1987). Abstr. in Phys. Abstr., Vol. 90, No. 1317, Entry 139889 (1987).

035.116   **The lightweight energetic particle detector EPONA and its performance on Giotto.**
S. McKenna–Lawlor, E. Kirsch, A. Thompson, D. O'Sullivan, K.-P. Wenzel.
J. Phys. E, Vol. 20, No. 6, Part 2, p. 732 – 740 (1987). Abstr. in Phys. Abstr., Vol. 90, No. 1317, Entry 139923 (1987).

035.117   **The Giotto implanted ion spectrometer (IIS): physics and techniques of detection.**
B. Wilken, W. Weiss, W. Studemann, N. Hasebe.
J. Phys. E, Vol. 20, No. 6, Part 2, p. 778 – 785 (1987). Abstr. in Phys. Abstr., Vol. 90, No. 1317, Entry 139924 (1987).

035.118   **Operation, performance, and reliability testing of charge–coupled devices for star trackers.**
G. R. Hopkinson, R. A. Cockshott, D. J. Purll, M. D. Skipper, B. Taylor.
Opt. Eng., Vol. 26, No. 8, p. 725 – 733 (1987). Abstr. in Phys. Abstr., Vol. 91, No. 1320, Entry 10404 (1988).

035.119   **Image corrections of the detector of the gamma–ray telescope SIGMA.**
P. Laudet, J. P. Roques.
Proceedings of the international topical meeting on image detection and quality, p. 233 – 236 (1986). Abstr. in Phys. Abstr., Vol. 91, No. 1320, Entry 10407 (1988). – See Abstr. 012.079.

035.120   **The Endeavour Get–Away–Special experiment.**
E. Roberts, T. Stapinski, M. Dopita.
Proceedings of the international topical meeting on image detection and quality, p. 281 – 284 (1986). Abstr. in Phys. Abstr., Vol. 91, No. 1320, Entry 10410 (1988). – See Abstr. 012.079.

Konstruirovanie i tekhnologiya izgotovleniya kosmicheskikh priborov (*Design and technology of space instrumentation*).
See Abstr. 003.132.

The Advanced X–Ray Astrophysics Facility. Invited papers presented at the 167th meeting of the American Astronomical Society, High Energy Astrophysics Division, held at Houston, Texas, USA, 5 – 9 January 1986.
See Abstr. 012.020.

Scientific ballooning –V. Proceedings of Symposium 10 of the COSPAR Twenty–sixth Plenary Meeting held in Toulouse, France, 30th June – 11th July 1986.
See Abstr. 012.023.

The future of high angular resolution astronomy: seeing the unseen.
See Abstr. 013.014.

Development of balloon technology in China.
See Abstr. 013.034.

Mirror fabrication and metrology for ROSAT.
See Abstr. 031.004.

Progress with LSM optics for solar observations within the French space program.
See Abstr. 031.006.

Experiences in the precision machining of grazing incidence X–ray mirror substrates.
See Abstr. 031.029.

Considerations on balloon–borne far infrared telescopes.
See Abstr. 031.044.

ESA technological research activities on lightweight mirrors.
See Abstr. 031.058.

Grazing–incidence optics for synchrotron–radiation insertion–device beams.
See Abstr. 031.059.

Automated figure formation for a Kirkpatrick–Baez X–ray mirror.
See Abstr. 031.060.

Transverse ray aberrations of Wolter type 1 telescopes.
See Abstr. 031.061.

Design and development of conical X–ray imaging mirrors.
See Abstr. 031.062.

Simulation of free–abrasive grinding of grazing–incidence mirrors with vertical–honing and flexible blades.
See Abstr. 031.063.

Extreme Ultraviolet Explorer: mirror optical tests and results.
See Abstr. 031.064.

Survey of material for an infrared–opaque coating.
See Abstr. 031.068.

Low mass mirrors in large optics.
See Abstr. 031.069.

In–process metrology for X–ray optics.
See Abstr. 031.076.

Low–frequency radioastronomical observations during the Spacelab 2 plasma depletion experiment.
See Abstr. 033.036.

A hydrogen maser clock for space: clocks in future possible and improbable applications.
See Abstr. 034.012.

Photometric characteristics of the Vega 1 and Vega 2 CCD cameras for the observation of comet Halley.
See Abstr. 034.066.

Modern radiation detectors.
See Abstr. 034.190.

Imaging strategies for a space–borne interferometer.
See Abstr. 036.169.

Astronomical image reconstruction via space slit aperture telescope.
See Abstr. 036.198.

Relative motion near the triangular libration points in the Earth–Moon system.
See Abstr. 042.054.

Observations of neutron stars planned by the High Speed Photometer team using Space Telescope.
See Abstr. 051.001.

A long–duration balloon system for middle–atmosphere measurements.
See Abstr. 051.032.

The status of the Cosmic Background Explorer mission.
See Abstr. 051.040.

Verification of the HIPPARCOS payload short term stability performances.
See Abstr. 051.054.

The Giotto mission to comet Halley.
See Abstr. 051.067.

Mass spectrometer for investigation of the chemical composition of the Venus atmosphere.
See Abstr. 093.015.

Radar images of Venus (Results of Venera–15 and Venera–16 space probes).
See Abstr. 093.017.

An investigation of stellar coronae with AXAF.
See Abstr. 112.057.

Gamma–ray observations of the Crab region using a coded–aperture telescope.
See Abstr. 143.040.

036 Methods of Observation and Reduction, Data Processing

036.001 Techniques for observing stellar oscillations.
    J. W. Harvey.
The SHIRSOG Workshop, p. 6 – 20 (1986). – See Abstr. 012.003.
    Although stellar oscillations have been observed for more than two centuries, the demands of asteroseismology require new observations of substantially higher precision. Two major techniques are reviewed: Doppler spectroscopy and photometry. Fundamental limitations are described using the sun as a representative stellar target. The current state of the art is limited by lack of light in the case of Doppler methods and by atmospheric noise in the case of photometry. Prospects for improvements in both of these techniques are good and it is possible to expect someday to be able to detect solar–like oscillations of stars as faint as 10th magnitude.

036.002 Observational requirements for a synoptic spectroscopic study of nonradial pulsations in OB stars.
M. A. Smith.
The SHIRSOG Workshop, p. 21 – 33 (1986). – See Abstr. 012.003.

036.003 **Doppler imaging.**
J. E. Neff.
The SHIRSOG Workshop, p. 34 – 43 (1986). – See Abstr. 012.003.

During the past few years, several powerful techniques to indirectly probe the spatial (two–dimensional) structure of stellar atmospheres have been developed. The author briefly reviews these techniques and describes some recent results. By overlaying "maps" of an atmosphere at many different levels (obtained by using a wide range of spectral diagnostics), one can study the three–dimensional structure of the atmosphere. With observations obtained over a long timescale, one can also study the evolution of the atmospheric structure.

036.004 **The detection of magnetic fields on late–type stars: progress, problems, and future needs.**
S. H. Saar.
The SHIRSOG Workshop, p. 44 – 52 (1986). – See Abstr. 012.003.

036.005 **Guidelines for allocation of observing time and scheduling.**
J. L. Linsky.
The SHIRSOG Workshop, p. 102 (1986). – See Abstr. 012.003.

036.006 **Further guidelines for allocation of observing time and scheduling.**
R. L. Gilliland.
The SHIRSOG Workshop, p. 103 (1986). – See Abstr. 012.003.

036.007 **Mid–infrared spectroscopic investigation.**
J. W. Salisbury, L. Walter, N. Vergo.
NASA Tech. Memo., NASA TM–89810, p. 196 – 198 (1987). – See Abstr. 003.001.

036.008 **Cartography of irregularly shaped satellites.**
R. M. Batson, K. Edwards.
NASA Tech. Memo., NASA TM–89810, p. 525 – 526 (1987). – See Abstr. 003.001.

Irregularly shaped satellites, such as Phobos and Amalthea, do not lend themselves to mapping by conventional methods because mathematical projections of their surfaces fail to convey an accurate visual impression of landforms and because large and irregular scale changes make their features difficult to measure on maps. A digital mapping technique has therefore been developed by which maps are compiled from digital topographic and spacecraft image files.

036.009 **Enhanced Landsat images of Antarctica and planetary exploration.**
B. K. Lucchitta, J. A. Bowell, K. Edwards, E. M. Eliason, H. M. Ferguson.
NASA Tech. Memo., NASA TM–89810, p. 554 (1987). Abstract. – See Abstr. 003.001.

036.010 **Let's report occultation timings on diskettes.**
D. W. Dunham.
Occultation Newsl., Vol. 4, No. 5, p. 92 – 97 (1987).

036.011 **The ground segment's vital role in the Hipparcos scientific mission.**
J. Van der Ha.
ESA Bull., No. 51, p. 27 – 33 (1987).

ESA's astrometry satellite Hipparcos is presently at an advanced stage of integration and testing. Design of the dedicated ground–segment functions required to support this mission has been completed and the various support elements are now being implemented. The paper provides a global description of the important role of the ground segment in supporting the scientific mission operations and in guaranteeing the quality of the data products to be delivered to the scientific community.

036.012 **Meteorstrombeobachtung: Amateurprogramm.**
E. Filimon.
Sternenbote, 30. Jahrg., Nr. 7, p. 140 – 147 (1987).

036.013 **Is de centrale ster in de Ringnevel te zien?**
H. Feijth.
Zenit, Jaarg. 14, Nr. 7/8, p. 252 (1987).

036.014 **Things to see and do in the dark.**
D. Malin.
J. Br. Astron. Assoc., Vol. 97, No. 5, p. 288 – 290 (1987).

We do not usually see much colour in the night sky but colour is there, especially in the brighter stars and it can be photographed with very simple techniques and equipment. This article covers in more detail the technique outline by David Malin in his talk to the BAA Exhibition Meeting in London, 1987 May 16.

036.015 **In search of other planetary systems.**
D. C. Black.
Spaceflight, Vol. 29, Suppl. No. 1, p. 36 – 37 (1987).

036.016 **Image restoration using the point spread function of the Halley Multicolour Camera.**
J. R. Kramm, W. Möhring, H. U. Keller.
Geophys. Res. Lett., Vol. 14, No. 7, p. 677 – 680 (1987).

The optical resolution of the Halley Multicolour Camera is limited by its finite aperture, by minute mechanical degradations within the instrument, and by the high speed time delay and integration illumination method adapted to the spin of the spacecraft. The sum of all implies a visible loss of image sharpness. The smearing effect has been calibrated using images of a star which can be considered as a point source. The point spread function has been determined and applied to the received images using Fourier methods. The effectiveness of the scheme is demonstrated on images of the earth and comet Halley.

036.017 **The calibration problem. III. First–order solution for mean absolute magnitude and dispersion.**
H. Smith Jr.
Astron. Astrophys., Vol. 181, No. 2, p. 391 – 393 (1987).

An approximate first–order method is derived for estimating the mean and intrinsic dispersion of absolute magnitudes for a homogeneous stellar group using trigonometric parallaxes.

036.018 **The objective function implicit in the CLEAN algorithm.**
K. A. Marsh, J. M. Richardson.
Astron. Astrophys., Vol. 182, No. 1, p. 174 – 178 (1987).

In the problem of deconvolving the synthesized beam from a "dirty" image representing the Fourier transform of an incompletely sampled spatial frequency function, it is well known that an infinite number of solutions is possible. The authors have addressed the question as to which particular solution is obtained by CLEAN. They find that when CLEAN is applied to intensity images, it may be regarded as an approximate method for minimizing an objective function represented by Σx_j, where x_j is the intensity in the jth pixel, subject to $x_j \geqslant 0$ and subject to consistency with the data. The approximation is particularly good in the case of point–like images. The CLEAN algorithm thus contains a built–in bias towards the solution corresponding to minimum total flux.

036.019 **Imaging and data processing with the Low Frequency Space Array.**
R. S. Simon, J. H. Spencer, B. K. Dennison, K. W. Weiler, K. J. Johnston, L. M. Hammarstrom, P. G. Wilhelm, W. C. Erickson, M. L. Kaiser, M. D. Desch, J. Fainberg, L. W. Brown, R. G. Stone.
Radio astronomy from space, p. 315 – 319 (1987). – See Abstr. 012.009.

The Low Frequency Space Array (LFSA) is being designed to image the entire sky at extremely low radio frequencies with arcminute to sub–arcminute resolution. To accomplish this goal, data from LFSA will be continuously integrated for many months and then be used with aperture synthesis techniques to produce images. After transforming the data to produce an initial image, it is possible to remove low–level sidelobe responses remaining in the image and thereby produce a high dynamic-range image.

036.020   Image selection and binning for improved atmospheric calibration of infrared speckle data.
J. C. Christou, D. W. McCarthy Jr., M. L. Cobb.
Astron. J., Vol. 94, No. 2, p. 516 – 522 (1987).

Using one–dimensional scans of unresolved sources, observed at a wavelength of 2.2 μm, the authors have investigated three image–sharpness criteria and have compared them to instantaneous estimates of the Fried parameter r_0 as determined by fitting the low–frequency domain of the power spectra. One of these criteria is strongly correlated with these seeing measurements, allowing to expediently sort the specklegrams according to image quality. In addition, the authors have investigated the "spectral ratio", a seeing–dependent, object–independent parameter used by von der Lühe (1984). They confirm its seeing dependence and also show that it can be used to calibrate visibilities when a resolved object and an unresolved calibrator are observed under different seeing conditions. Finally, the authors show that computing visibilities from seeing–matched bins requires ≈ 0.25 cm bin widths in r_0.

036.021   Statistical problems about the use of the ordinary least–squares method in astrometry. Application to the Paris–Astrolabe data.
M. L. Bougeard.
Astron. Astrophys., Vol. 183, No. 1, p. 156 – 166 (1987).

The Ordinary Least–Squares (OLS) method is commonly used in many fields and for a long time in optical astrometry. The author discusses the implicit assumptions under which the OLS fit is valid, with an application to the Paris–Astrolabe data. The statistical analysis developed here may be extended, through slight modifications, to many other astronomical data.

036.022   A direct surface smoothing procedure for Fourier image reconstruction in radiophysics.
I. Koch, R. S. Anderssen.
Astron. Astrophys., Vol. 183, No. 1, p. 170 – 176 (1987).

A direct surface smoothing procedure is proposed which efficiently estimates the sky image f from noisy and incomplete visibility observations z in the Fourier domain of f. The formulation of the reconstruction problem involves two steps: The construction of the pseudo–data d, a suitable image domain representation of the visibilities z, and the choice of an appropriate stabilizer S. The approximation $\bar{f}$ to f is then found directly as the unique maximiser of $\{\lambda S(g) - \Sigma|d_i - g_i|^2\}$. Efficient computational methods for direct surface smoothing are described. The observation times associated with radio astronomy measurements are used in the construction of d.

036.023   A faint object processing software: description and testing.
L. Infante.
Astron. Astrophys., Vol. 183, No. 1, p. 177 – 184 (1987).

A description and the performance of techniques employed to analyze faint data recorded on deep 4 m size telescopes' photographic plates or CCD images are reported. By simulating data (i.e. stars and galaxies on noisy frames), it was possible to test the performance of these algorithms. The tests were designed so that the simulations resemble as much as possible deep digitized data from a 4 m type telescope prime focus plate. Nevertheless, the following results are applicable to data from CCD frames or other size telescopes, provided an appropriate scaling is done. (1) The detection was found to be 95% complete up to $J = 23.7$ mag for galaxies at a mean redshift $z = 0.5$. (2) The r_{-2} star/galaxy classifier proved to be reliable up to $J \approx 22$ and (3) the photometry as done with the "first moment" method yields uncertainties of the order of ± 0.17 mag at $J = 23.0$ mag.

036.024   Computer model of the experiment "COLD". Evaluation of the microwave background inhomogeneity.
V. R. Amirkhanyan.
Soobshch. Spets. Astrofiz. Obs., Vyp. 53, p. 96 – 98 (1987). – See Abstr. 012.008.

The computer model of the experiment "COLD" which was carried out with the RATAN–600 radio telescope is constructed. It turned out that the published results ($\Delta T/T < 10^{-5}$ and $S_{min} = 1$ mJy) are not attainable in principle in this experiment which has limits $\Delta T/T \geq (3 – 5) \times 10^{-4}$, and $S_{min} = 10$ mJy.

036.025   On computer modeling of the experiment "COLD".
Yu. N. Parijskij, A. B. Berlin, V. V. Vitkovskij.
Soobshch. Spets. Astrofiz. Obs., Vyp. 53, p. 99 – 103 (1987). – See Abstr. 012.008.

It is shown that the computer model of the initial row of data in the experiment "COLD" used in the Amirkhanyan paper (036.024) is incorrect. The computed beam differs greatly from the real one at large distances from the axis of the main direction of the RATAN–600 beam axis. The method of data reduction differs also from that used in Parijskij et al. (1984). Additional experimental data published recently and confirming the main results of the authors are presented.

036.026   Die Bestimmung der ZHR bei Meteorbeobachtungen. Teil 1: Die Grenzgrößenkorrektur.
B. Koch.
KPM, Jahrg. 2, No. 5, p. 38 – 41 (1987).

036.027   Photography: see what you will get.
C. Steyaert.
WGN, Vol. 15, Nr. 4, p. 112 – 114 (1987).

A method is described to determine in advance which part of the sky will be covered by a camera during a photographical meteor watch. This is in fact the opposite of the astrometric problem.

036.028   Timing binary star eclipses.
A. D. Mallama.
I.A.P.P.P. Commun., No. 29, p. 33 – 41 (1987).

The purpose of this paper is to explain how to obtain a photoelectric record of a binary star eclipse light curve that will yield a reliable timing of minimum brightness. The emphasis is on procedures to use at the telescope and ways to control errors of measurement.

036.029   Data compression for very sparse integer arrays.
J. R. Sternberg.
EXOSAT Express, No. 18, p. 25 – 26 (1987).

036.030   Position determination and the Fine Sun Sensor.
J. Osborne.
EXOSAT Express, No. 18, p. 35 – 40 (1987).

036.031   Refinements to the LEIT calibrations.
H. van der Woerd, J. Osborne.
EXOSAT Express, No. 18, p. 41 – 44 (1987).

036.032   The LEIT off–axis point spread function and the long term detection efficiency.
P. Giommi, L. Angelini.
EXOSAT Express, No. 18, p. 45 – 49 (1987).

036.033   Nomogramm zum Bestimmen der Vergrößerung von Abbildern flächenhaft erscheinender Himmelskörper.
K. Schiefer, U. Schiefer.
Sterne Weltraum, 26. Jahrg., Nr. 7 – 8, p. 425 – 430 (1987).

036.034   Visuele radiantbepaling.
A. Scholten.
Radiant, Jaarg. 8, Nr. 6, p. 116 – 117, 119 (1986).

036.035   **Imaging in radio astronomy.**
L. Feretti, M. Vigotti.
Selected topics on data analysis in astronomy, p. 13 – 38 (1987). –
See Abstr. 012.011.
The authors present a review of some of the methods and
techniques used in the production and analysis of images
obtained with radio telescopes.

036.036   **An introduction to X–ray astronomy data analysis.**
M. Morini.
Selected topics on data analysis in astronomy, p. 39 – 55 (1987). –
See Abstr. 012.011.
This paper is intended to provide a very brief introduction to
X–ray astronomy, with some emphasis on the data analysis
techniques.

036.037   **Computational methods.**
P. Grosbøl.
Selected topics on data analysis in astronomy, p. 57 – 85 (1987). –
See Abstr. 012.011.
Astronomical image processing applies a large variety of
numerical methods to extract scientific results from observed
data. The standard computational techniques used in this process
are discussed with special emphasis on the problems and
advantanges associated to them when applied to astronomical
data. The methods are presented in the order in which they are
applied in a typical reduction sequence.

036.038   **Hardware for graphics and image display.**
I. De Lotto, M. Savini.
Selected topics on data analysis in astronomy, p. 111 – 122
(1987). – See Abstr. 012.011.
The hardware components and architectures for graphics and
image display are shortly reviewed. A display system is seen as
made up of three main parts: the display itself, the image memory
and the image processors. For each of these a short description of
the performances and features is presented.

036.039   **Data analysis systems.**
D. C. Wells.
Selected topics on data analysis in astronomy, p. 123 – 151
(1987). – See Abstr. 012.011.
The author reports on the design, management and operation
of data analysis systems, as well as the stategy of how one
organizes the projects and why one even does these projects.

036.040   **Crystal gazing v. computer system technology projections.**
D. C. Wells.
Selected topics on data analysis in astronomy, p. 153 – 170
(1987). – See Abstr. 012.011.

036.041   **Use of a minimum rate of change formalism to quantify
variability of extragalactic X–ray sources.**
D. A. Schwartz.
Astrophys. J., Vol. 318, No. 2, p. 568 – 576 (1987).
The author suggests a method to obtain rigorous, quantitative
constraints on the rate of variability. Conceptually the data are
fitted to a function of time and a set of free parameters. The
method of Lagrange multipliers is used to solve for that
parameter set which minimizes the value of the derivative at time
t_m. The result can be related physically to the minimum efficiency
with which rest mass must be converted to radiation energy.
Published data for OX 169 are used to show how this method
relaxes previously stated constraints, and analysis of the variability time scale of the BL Lac object H0323+022 is presented.

036.042   **DUMAND: a detector of both neutrinos and gamma
rays?**
V. S. Berezinskij, G. Cini–Castagnoli, V. A. Kudryavtsev,
O. G. Ryazhskaya, G. T. Zatsepin.
Sov. Astron. Lett., Vol. 12, No. 5, p. 296 – 300 (1987). English
translation of 42.036.015.

036.043   **Two–dimensional spectrophotometry of planetary nebulae by CCD imaging.**
G. H. Jacoby, R. J. Quigley, J. L. Africano.
Publ. Astron. Soc. Pac., Vol. 99, No. 617, p. 672 – 685 (1987).
With a correction in Vol. 99, No. 619, p. 1027 – 1029 (1987).
The authors have derived the spatial distribution of the
electron temperature and density and the ionic abundances of
O^+, O^{++}, N^+, and S^+ from CCD images of the planetary
nebulae NGC 40 and NGC 6826 taken in the important emission
lines of [O II], [O III], Hβ, [N II], and [S II]. Advantages of the
imaging technique include complete spatial coverage and excellent spatial resolution, good light throughput, reduced effects of
spatial averaging over filamented and knotted regions, elimination of the effects of atmospheric dispersion, and absolute
spectrophotometry for objects having angular sizes of several
arcmin. The disadvantages include additional effort to analyze
the data, limited spectral coverage, and added difficulties in
deblending nearby lines, removing the nebular continuum
contribution, and correcting for nonphotometric conditions.

036.044   **Identification of a constellation from a position.**
N. G. Roman.
Publ. Astron. Soc. Pac., Vol. 99, No. 617, p. 695 – 699 (1987).
A table permits rapid determination of the constellation in
which an object is located from its 1875.0 position.

036.045   **The significance of stellar magnetic field measurements
obtained with the photographic technique: the spurious
magnetic field of the supergiant Canopus.**
M. J. Stift.
Mon. Not. R. Astron. Soc., Vol. 228, No. 1, p. 109 – 118 (1987).
The author proposes a number of straightforward statistical
tests aimed at establishing the significance of stellar magnetic
field measurements obtained with Babcock's photographic technique. The power of these methods is illustrated with three
Ap stars for which the presence of a magnetic field can be
established at an exceedingly high significance level. Application
to the supergiant Canopus on the contrary gives no support to
the claims for the detection of a kilogauss surface field. Further
investigations involving classical scaling and an error analysis of
the MSHIFT correlation method favour the view that instrumental instabilities are at the origin of the apparent magnetic
variations in Canopus.

036.046   **Determining approximate values for the angular ele-
ments of the outer orientation of star images.**
I. A. Stolbov.
Izv. Vuzov. Geod. Aehrofotosemka, No. 6, p. 77 – 79 (1986). In
Russian. Abstr. in Ref. Zh., 52. Geod. Aehrosemka, 7.52.93
(1987).

036.047   **On the reduction of observations made with photograph-
ic zenith tubes to the meridian.**
Z. M. Malkin.
Probl. Opredeleniya Parametrov Vrashcheniya Zemli, Vladivos-
tok, p. 70 – 77 (1986). In Russian. Abstr. in Ref. Zh., 52. Geod.
Aehrosemka, 7.52.94 (1987).

036.048   **Methods of remote sensing with automatic interplanetary stations in mapping planets.**
Yu. S. Tyuflin.
Sostoyanie i perspektivy razvitiya Geod., Kartogr. Mater. Vses.
nauchn.– tekh. konf., Moskva 5 – 7 sent., 1984. Moskva,
p. 82 – 89 (1986). In Russian. Abstr. in Ref. Zh., 52. Geod.
Aehrosemka, 7.52.258 (1987).

036.049   **Processing of Venus radar images with the analyzer
"Madzhiskan–2".**
A. Ya. Danil'chenko, M. S. Markov, M. V. Ostrovskij,
Yu. S. Tyuflin, D. Ya. Choporov.
Geod. Kartogr., No. 2, p. 51 – 55 (1987). In Russian. Abstr. in
Ref. Zh., 52. Geod. Aehrosemka, 7.52.259 (1987).

036.050   **Complex of programs for processing telemetric data of the "VEGA" experiment in real time.**
O. V. Balakina.
Inst. Kosm. Issled. Akad. Nauk SSSR, Prepr., No. 1183, 18 pp. (1986). In Russian. Abstr. in Ref. Zh., 62. Issled. Kosm. Prostranstva, 7.62.236 (1987).

036.051   **A calibration method for measurements of the longitudinal magnetic field and differential radial velocity.**
M. L. Demidov, N. I. Kobanov, V. M. Grigor'ev.
Kinematika Fiz. Nebesn. Tel, Tom 3, No. 4, p. 85 – 89 (1987). In Russian. English translation in Kinematics Phys. Celest. Bodies.
A new calibration method of magnetographs and instruments designed for measuring the differential radial velocity is suggested. The method provides an improved calibration accuracy, can be carried out under virtually the same conditions as routine observations, and offers some better operational characteristics.

036.052   **Near zenith tracking limits for altitude–azimuth telescopes.**
K. M. Borkowski.
Acta Astron., Vol. 37, No. 1, p. 79 – 88 (1987).
The diurnal rotation of the sky, when viewed in the horizon system of coordinates, exhibits a singularity at the zenith where the angular azimuthal speed and acceleration are infinite. This prevents all telescopes of the altitude–azimuth mounting, no matter how good is their performance, to be able to track celestial objects in a small region of the sky about the zenith. A simple algorithm to determine the shape and size of this blind spot is presented. An example is given for the projected 32–m radio telescope to be located near Toruń.

036.053   **Ultraviolet detection of very low–surface–brightness objects.**
R. W. O'Connell.
Astron. J., Vol. 94, No. 4, p. 876 – 882 (1987).
The night–sky surface brightness at excellent ground–based sites is compared to the sky background in space. In directions typical of extragalactic pointings, the background in the space ultraviolet reaches μ_λ(2000 Å) = 26 mag arcsec^{-2}, which is a factor of 40 darker than at any wavelength on the ground. This represents an important new "window" for the study of extragalactic systems with low surface brightnesses. The author finds that in certain favorable circumstances UV photometry may permit the detection of regions with equivalent V band surface brightnesses as low as 35 mag arcsec^{-2}. He considers applications of UV surface photometry to the study of circumgalactic regions, dwarf galaxies, low–surface–brightness spirals, and the detection of primeval galaxies, and briefly discusses the usefulness of existing space instrumentation for such problems.

036.054   **A survey of proper–motion stars. II. Extracting metallicities from high–resolution, low–S/N spectra.**
B. W. Carney, J. B. Laird, D. W. Latham, R. L. Kurucz.
Astron. J., Vol. 94, No. 4, p. 1066 – 1076 (1987).
The authors discuss the calculations of a grid of high–resolution synthetic spectra covering $\lambda\lambda$5150 – 5250, T_{eff} = 4750, 5000, ..., 6500K, [M/H] = +0.5, 0, ..., –3.0, and gravities of $\log g$ = 4.5 or 4.0. They describe a method using χ^2 fits with these spectra as templates in the determination of mean line strengths from observed spectra. The authors estimate metallicities for 48 stars whose abundances have been previously determined from high–resolution, high–S/N spectra and conventional fine–analysis techniques. None of the observed spectra are high S/N, yet they result in metallicities with internal scatter of typically 0.1 dex. Comparison with published fine analyses shows excellent agreement (–0.03 ± 0.02 dex) for metal–poor stars, and a small systematic error at higher abundances, which is easily corrected. Metallicities with σ < 0.2 dex are still obtained even when S/N is as low as 3.

036.055   **Fourier removal of stripe artifacts in *IRAS* images.**
D. Van Buren.
Astron. J., Vol. 94, No. 4, p. 1092 – 1094 (1987).
By working in the Fourier plane, approximate removal of stripe artifacts in *IRAS* images can be effected. The image of interest is smoothed and subtracted from the original, giving the high–spatial–frequency part. This "filtered" image is then clipped to remove point sources and then Fourier transformed. Subtracting the Fourier components contributing to the stripes in this image from the Fourier transform of the original and transforming back to the image plane yields substantial removal of the stripes.

036.056   **The Newton–Gauss regularized method: application to point–spread–function determination in CCD frames.**
O. Bendinelli, G. Parmeggiani, A. Piccioni, F. Zavatti.
Astron. J., Vol. 94, No. 4, p. 1095 – 1100 (1987).
Modification of the Newton–Gauss linearization method in the Tikhonov regularization sense is described. Its ability to give reliable estimates of a large number of parameters is shown by application to the PSF determination from CCD frames. Extension of the Van Altena and Auer star–image model using a weighted sum of two Gaussians, and explicitly taking its integration on the pixel into account, enables the authors to determine the PSF up to about 10 mag below the central value with an error fit in the range 0.01 – 0.03 mag arcsec^{-2}.

036.057   **Search for flux variability in a sample of X–ray sources dominated by Poisson statistics.**
S. Mereghetti, B. Garilli, T. Maccacaro.
Variability of galactic and extragalactic X–ray sources, p. 225 – 231 (1987). – See Abstr. 012.019.
Many objects have been repeatedly observed with the IPC on board the Einstein Observatory. The authors have analyzed the repeated IPC exposures to search for flux variability in a large number of sources for which 2 or more observations exist. For this purpose they have developed a test which is based on Poisson statistics and allows the use of upper limits on the source flux. They briefly describe this method and present some results of its application to a sample of about 300 sources.

036.058   **Techniques for the analysis of data from coded–mask X–ray telescopes.**
G. K. Skinner, T. J. Ponman, A. P. Hammersley, C. J. Eyles.
Astrophys. Space Sci., Vol. 136, No. 2, p. 337 – 349 (1987).
Several techniques useful in the analysis of data from coded–mask telescopes are presented. Methods of handling changes in the instrument pointing direction are reviewed and ways of using FFT techniques to do the deconvolution considered. Emphasis is on techniques for optimally–coded systems, but it is shown that the range of systems included in this class can be extended through the new concept of "partial cycle averaging".

036.059   **The radar–glory theory for icy moons with implications for radar mapping.**
V. R. Eshleman.
Adv. Space Res., Vol. 7, No. 5, p. 133 – 136 (1987). – See Abstr. 012.022.
The anomalous radar echoing properties of three ice–clad moons of Jupiter appear to be due to glory–like backscattering from buried craters. The enormous glare from these sources would impair geologic studies based on standard methods of radar mapping. It is not known whether similar or different problems will arise in the radar study of other icy surfaces in the outer solar system, or of the unseen surface of Titan. In any event, the results from the moons of Jupiter illustrate the role of exploratory measurements and the importance of possible bistatic radar–mapping techniques based on the use of separated transmitters and receivers.

036.060   Reduction of the high–dispersion spectrograms of late–type stars with computers.
L. A. Yakovina.
Red. Zh. Kinematika Fiz. Nebesn. Tel. Kiev, 22 pp. (1987). In Russian. Abstr. in Ref. Zh., 51. Astron., 8.51.944 (1987).

036.061   Astrometrie mit einem Mikrodensitometer PDS 1010 A.
F. Klefenz.
Diplomarbeit, Max–Planck–Institut für Astronomie, Heidelberg, F.R. Germany, 109 pp. (1987).

036.062   GNOMPLOT – ein Hilfsprogramm für Meteorbeobachter.
D. Heinlein.
Sterne Weltraum, 26. Jahrg., Nr. 11, p. 652 – 654 (1987).

036.063   On the choice of a coordinate system for the reduction of overlapping plates.
A. I. Yatsenko.
Kinematika Fiz. Nebesn. Tel, Tom 3, No. 5, p. 25 – 29 (1987). In Russian. English translation in Kinematics Phys. Celest. Bodies.

Using a mathematical model it is shown that for the reduction of a great number of overlapping plates the equiinterval azimuthal polar projection is better to apply to the projection of the celestial sphere on the auxiliary plane. The coordinates of stars on the formal plate (with outset in the North Pole of celestial equator) are determined from equatorial coordinates by expressions: $U = (\pi/2-\delta)\sin \alpha$; $V = -(\pi/2-\delta)\cos \alpha$. These coordinates are connected with tangential polar coordinates U', V' by expressions $U = U'(\pi/2-\delta)\mathrm{tg} \delta$, $V = V'(\pi/2-\delta)\mathrm{tg} \delta$.

036.064   The regression models in photographic astronomy.
S. G. Valeev, M. G. Shamarin, I. A. Dautov, I. E. Tselishchev.
Kinematika Fiz. Nebesn. Tel, Tom 3, No. 5, p. 30 – 35 (1987). In Russian. English translation in Kinematics Phys. Celest. Bodies.

For an astrographic set of observations the structure of reduction models obtained by using the normal scheme of regression analysis is studied. The correctness of the application of "variable" structure model is justified and the model consisting of the constant and variable parts changing the structure from plate to plate is suggested.

036.065   Comment on the "Signal loss due to imperfect fringe rotation in VLBI correlators revisited".
P. A. Friedman (*P. A. Fridman*).
Astrophys. Space Sci., Vol. 137, No. 1, p. 189 (1987). See Abstr. 42.036.136.

036.066   Data reduction and spectrophotometric performances of PUMA 1: an on–line multiaperture spectroscopic system used at the CFHT.
G. Soucail, Y. Mellier, B. Fort, J. P. Picat, M. Cailloux.
Astron. Astrophys., Vol. 184, No. 1/2, p. 361 – 372 (1987).

This paper presents the on–line multiaperture spectroscopic system called PUMA 1, which was first used at the 3.60 m Canada–France–Hawaii Telescope. Following a short description of the observing procedure on the telescope, the software package developed for the reduction of multi–spectroscopic data is given in details. On the basis of data coming from a run in September 1985, the spectrophotometric performances obtained for extended objects like galaxies in high redshift clusters ($z = 0.4$) are discussed (from signal–to–noise ratio of the spectra to accuracy on radial velocity measurements). Basic instrumental and reduction references are given and a possible evolution is described.

036.067   Computerized monthly reports.
E. A. Halbach.
J. Am. Assoc. Variable Star Obs., Vol. 16, No. 1, p. 37 – 41 (1987).

Computerized monthly variable star reports usually start with the manual recording of data at the telescope, followed by the transcription of observations into a computer data bank. A computer program was written to allow direct data entry at the telescope, thus avoiding the later transcription of data.

036.068   Plotting variables with a home computer.
G. P. Dyck.
J. Am. Assoc. Variable Star Obs., Vol. 16, No. 1, p. 42 – 44 (1987).

In this paper the author describes how AAVSO observations can be processed and plotted with an inexpansive home computer. Sample screen tracings of six cataclysmic variables demonstrate that data from a single observer can yield viable light curves.

036.069   Magnification test program.
R. J. Bouma.
KPM, Jahrg. 2, Nr. 6, p. 6 – 8 (1987).

036.070   Two–processor scanner for the GAT. II. SIPRAN – a specialized programming language.
N. N. Somov.
Bull. Spec. Astrophys. Obs. – North Caucasus, Vol. 22, p. 62 – 64 (1987). English translation of 41.036.095.

036.071   Two–processor scanner for the GAT. III. Automated fast processing of stellar spectra.
I. M. Kopylov, N. N. Somov, T. A. Somova.
Bull. Spec. Astrophys. Obs. – North Caucasus, Vol. 22, p. 65 – 75 (1987). English translation of 41.036.096.

036.072   Interferometric imaging: a numerical simulation.
J. B. Zirker.
Sol. Phys., Vol. 111, No. 2, p. 235 – 242 (1987).

Non–redundant arrays offer the possibility of reconstructing optical solar images to the diffraction limit of resolution, with minimal data processing. A particular algorithm, due to W. T. Rhodes, for the analysis of the fringes produced by a set of non–redundant arrays, has ben evaluated numerically. The algorithm can give satisfactory results with $S/N \sim 100$, but can also break down at moments of unfavorable seeing.

036.073   Single aperture interferometry: general introduction.
F. Roddier.
Interferometric imaging in astronomy, p. 1 – 8 (1987). – See Abstr. 012.035.

The paper deals with interferometric techniques used to reconstructed turbulence degraded images through a single telescope. The author reviews the physical basis of these techniques and describes briefly the present state of the art without trying to be exhaustive.

036.074   Calibration problems in solar speckle interferometry.
O. von der Lühe.
Interferometric imaging in astronomy, p. 9 – 12 (1987). – See Abstr. 012.035.

036.075   Infrared speckle calibration methods.
C. Perrier.
Interferometric imaging in astronomy, p. 13 – 16 (1987). – See Abstr. 012.035.

Infrared speckle interferometry has produced a number of results in the form of visibility amplitudes. These data are subject to various causes of degradation affecting more or less seriously the quality of the extracted information. The corrections of such degradations require different calibrations which have been progressively made necessary by the improvement of performances with time. This paper is mainly issued from the practice of the observation with the ESO infrared specklegraph installed at the 3.6 m telescope at La Silla.

036.076   Image selection and binning for improved atmospheric calibration of infrared speckle data.
J. C. Christou, D. W. McCarthy Jr., M. L. Cobb.
Interferometric imaging in astronomy, p. 17 – 20 (1987). – See Abstr. 012.035.

036.077   The University of Maryland program on multi–aperture amplitude interferometry.
D. G. Currie.
Interferometric imaging in astronomy, p. 21 – 24 (1987). – See Abstr. 012.035.

036.078   Processing of interferograms.
C. Roddier, F. Roddier.
Interferometric imaging in astronomy, p. 25 – 28 (1987). – See Abstr. 012.035.
Since 1976, the authors have developed algorithms to map both the amplitude and the phase of the fringes in pupil–plane interferograms. These algorithms are essentially based on fast Fourier transforms. The authors describe here the latest improvements.

036.079   The modified Knox–Thompson algorithm.
J. C. Fontanella, A. Sève.
Interferometric imaging in astronomy, p. 29 – 35 (1987). – See Abstr. 012.035.
The purpose of this method is to recover the object Fourier transform from a set of speckle images. The authors briefly expose what modifications can be done to the well–known Knox–Thompson algorithm to get diffraction–limited images of extended objects.

036.080   Application of the Knox–Thompson method to solar observations.
O. von der Lühe.
Interferometric imaging in astronomy, p. 37 – 40 (1987). – See Abstr. 012.035.

036.081   Application of the Knox–Thompson method to IR observations.
R. R. Howell.
Interferometric imaging in astronomy, p. 41 – 42 (1987). – See Abstr. 012.035.

036.082   Speckle masking and speckle spectroscopy.
G. Weigelt, K.–H. Hofmann.
Interferometric imaging in astronomy, p. 43 – 45 (1987). – See Abstr. 012.035.

036.083   Application of triple correlation to one–dimensional infrared speckle data.
J. D. Freeman, J. C. Christou, F. Roddier,
D. W. McCarthy Jr., M. L. Cobb.
Interferometric imaging in astronomy, p. 47 – 50 (1987). – See Abstr. 012.035.
Triple correlation analysis, also known as bispectrum or speckle masking, is a recently introduced technique to recover object phases from atmospherically degraded short–exposure images (specklegrams). In this paper the authors discuss how they have applied the bispectrum analysis to 1–D infrared speckle data, taking into account additive detector noise, to obtain object phases and to compare these phases to those obtained using Knox–Thompson analysis. The authors also compare three methods of obtaining the object phases from the bispectrum phases, one of which is a rigorous statistical analysis using a minimum variance unbiased estimator.

036.084   The weighted shift–and–add method.
J. C. Christou, J. D. Freeman, E. K. Hege.
Interferometric imaging in astronomy, p. 51 – 54 (1987). – See Abstr. 012.035.
The authors describe a variation on the shift–and–add techniques of image reconstruction from astronomical speckle interferometric data which yields diffraction–limited images which both scale with wavelength and are self–calibrating for seeing. They also discuss its applicability to both unresolved sources and extended objects.

036.085   Filtered, weighted shift–and–add: theory and practice.
E. Ribak.
Interferometric imaging in astronomy, p. 55 – 58 (1987). – See Abstr. 012.035.

036.086   Phase–gradient estimation for speckle image reconstruction.
G. J. M. Aitken.
Interferometric imaging in astronomy, p. 59 – 61 (1987). – See Abstr. 012.035.

036.087   Probability imaging of double and multiple stars.
C. Aime.
Interferometric imaging in astronomy, p. 63 – 66 (1987). – See Abstr. 012.035.
It is shown that the second probability density function of a stellar speckle pattern observed at the focus of a large telescope can be used for imaging double and multiple stars.

036.088   Image reconstruction from Fourier modulus samples.
J. R. Fienup.
Interferometric imaging in astronomy, p. 67 – 70 (1987). – See Abstr. 012.035.

036.089   Automatic deconvolution and phase retrieval.
R. H. T. Bates, R. G. Lane.
Interferometric imaging in astronomy, p. 71 – 73 (1987). – See Abstr. 012.035.

036.090   Comparison between image plane phase reconstruction methods in optical interferometry.
A. Chelli.
Interferometric imaging in astronomy, p. 75 – 77 (1987). – See Abstr. 012.035.
The author has carried out a quantitative and a qualitative analysis intercomparing image plane phase reconstruction methods in optical interferometry.

036.091   Phase closure with rotational shear interferometers.
F. Roddier, C. Roddier.
Interferometric imaging in astronomy, p. 79 – 82 (1987). – See Abstr. 012.035.
The authors show that phase closure relations can be obtained from rotational shear interferograms and they discuss the necessary conditions for the object phases to be recovered from such relations.

036.092   Triple shearing interferometry and shearing spectroscopy.
K.–H. Hofmann, G. Weigelt.
Interferometric imaging in astronomy, p. 83 – 84 (1987). – See Abstr. 012.035.

036.093   Phase retrieval from a polychromatic speckle analysis.
C. Aime.
Interferometric imaging in astronomy, p. 85 – 88 (1987). – See Abstr. 012.035.
A technique based on one–dimensional space–wavelength analysis of speckles is proposed for the recovery of turbulence–degraded images.

036.094   Submilliarcsecond imaging of rotating stars using differential speckle interferometry.
R. G. Petrov.
Interferometric imaging in astronomy, p. 89 – 92 (1987). – See Abstr. 012.035.
Differential speckle interferometry (DSI) is a high resolution technique based on the cross analysis of different short exposure astronomical images affected by the same atmospheric turbulence. This technique can be applied to a wide class of astronomical candidates but the author focuses here on its application to rotating stars, following a brief discussion of the various ways of implementing a DSI experiment and their resulting signal–to–noise ratios.

036.095   **A method for multispectral infrared interferometry.**
S. T. Ridgway, J.–M. Mariotti.
Interferometric imaging in astronomy, p. 93 – 96 (1987). – See Abstr. 012.035.

The fringes produced in the classical Michelson pupil plane interferometer may be processed to produce both spatial and spectral information. Observational techniques developed for Fourier transform spectroscopy may be readily adapted to this observing mode. The authors have tested this observing technique at two telescopes with several instrumental configurations, and find that it holds excellent promise for operation at all spectral resolutions from very low to very high, and may be ideally suited for multiple telescope interferometry. They present here a description of the method, and describe initial results.

036.096   **Multiple aperture interferometry: towards the Optical Very Large Array.**
A. Labeyrie, I. Bosc, D. Mourard.
Interferometric imaging in astronomy, p. 97 – 104 (1987). – See Abstr. 012.035.

036.097   **Phased array imaging with the Multiple Mirror Telescope.**
E. K. Hege, D. W. McCarthy Jr., J. C. Hebden, J. C. Christou.
Interferometric imaging in astronomy, p. 105 – 108 (1987). – See Abstr. 012.035.

036.098   **The CERGA small interferometer.**
L. Koechlin, F. Vakili, Y. Rabbia,
G. P. di Benedetto, G. C. Conti, C. Thom, P. Granes,
P. Nisenson, C. Papaliolios, M. Lacasse, P. Cruzalebes,
G. Schumacher.
Interferometric imaging in astronomy, p. 109 – 113 (1987). – See Abstr. 012.035.

Contents: 1. Stellar diameters at $\lambda = 0.6\,\mu$m (1985 – 86). 2. Stellar diameters at $\lambda = 2.2\,\mu$m and 1.6 μm. 3. γ Cassiopeiae envelope in the Hα emission line. 4. Observations of α Cyg with the PAPA detector at $\lambda = 0.6\,\mu$m.

036.099   **The Sydney University stellar interferometry programme: a progress report.**
J. Davis.
Interferometric imaging in astronomy, p. 121 – 124 (1987). – See Abstr. 012.035.

036.100   **Magnitude limit of the group delay fringe tracking method for long baseline interferometry.**
P. Nisenson, W. Traub.
Interferometric imaging in astronomy, p. 129 – 133 (1987). – See Abstr. 012.035.

036.101   **Signal–to–noise ratios and beam combination.**
F. Roddier.
Interferometric imaging in astronomy, p. 135 – 138 (1987). – See Abstr. 012.035.

036.102   **Imaging by optical aperture synthesis.**
J. E. Baldwin.
Interferometric imaging in astronomy, p. 139 – 141 (1987). – See Abstr. 012.035.

Optical aperture synthesis is the analogue of the well established radio case in which high resolution images are reconstructed from a number of samples, often well separated, of the spatial coherence function. A few assertions are made about it.

036.103   **A comparison of the determination of closure phase in optical interferometry with fully filled apertures and non–redundant aperture masks.**
A. C. S. Readhead.
Interferometric imaging in astronomy, p. 143 – 152 (1987). – See Abstr. 012.035.

036.104   **TOAST, a terrestrial optical aperture synthesis technique.**
A. H. Greenaway, D. P. Cheese, J. D. Bregman,
J. E. Noordam.
Interferometric imaging in astronomy, p. 153 – 156 (1987). – See Abstr. 012.035.

036.105   **Image reconstruction from long–baseline interferograms by speckle masking.**
K.–H. Hofmann, T. Reinheimer, G. Weigelt.
Interferometric imaging in astronomy, p. 157 – 159 (1987). – See Abstr. 012.035.

036.106   **Bi–spectrum imaging in radio interferometry.**
T. J. Cornwell.
Interferometric imaging in astronomy, p. 161 – 163 (1987). – See Abstr. 012.035.

036.107   **Towards multiple aperture interferometry.**
S. T. Ridgway.
Interferometric imaging in astronomy, p. 173 – 175 (1987). – See Abstr. 012.035.

036.108   **The practice of deconvolution.**
T. J. Cornwell.
Interferometric imaging in astronomy, p. 177 – 182 (1987). – See Abstr. 012.035.

036.109   **Phase retrieval with the Maximum Entropy Method.**
R. Narayan.
Interferometric imaging in astronomy, p. 183 – 186 (1987). – See Abstr. 012.035.

The Maximum Entropy Method (MEM) is known to deconvolve the effects of point spread functions quite well. The author considers a more difficult application of the MEM, viz. when Fourier amplitudes of the image are given but not the phases. The topology of the entropy surface in the space of unknown phases is displayed for a simple model. Multiple extrema are found to be the rule, and the DC level of the image (i.e. zero Fourier coefficient) is seen to play a crucial role. The implications of the results for potential numerical algorithms are explored. Preliminary results with simulations on two–dimensional images are described.

036.110   **Stability conditions and regularization procedures.**
A. Lannes.
Interferometric imaging in astronomy, p. 187 – 190 (1987). – See Abstr. 012.035.

The aim of this communication is to indicate how to choose the regularization principle by taking account of the specificity of the reconstruction problems encountered in aperture synthesis.

036.111   **On the three M's of image enhancement: MAP (*maximum a posteriori restoration*), ME (*maximum entropy restoration*) and MW (*median window restoration*).**
B. R. Frieden.
Interferometric imaging in astronomy, p. 191 – 196 (1987). – See Abstr. 012.035.

036.112   **Fourier inversion and deconvolution methods.**
M. L. Cobb, D. W. McCarthy Jr.
Interferometric imaging in astronomy, p. 197 – 200 (1987). – See Abstr. 012.035.

The goal of producing images and not just visibility amplitudes, adds another level of complications to a high angular resolution experiment. The authors take a brief look at some the subtle problems of producing final images from high angular resolution data.

036.113 **Atmospheric phase measurements with the Mk II and Mk III interferometers at Mt. Wilson.**
M. M. Colavita, M. Shao.
Interferometric imaging in astronomy, p. 205 – 208 (1987). – See Abstr. 012.035.

The fundamental limitations to ground–based interferometric astrometry are the phase fluctuations introduced by the turbulent atmosphere. With an active, phase–coherent instrument, i.e., the Mk II or Mk III astrometric interferometer, measurements of the atmospheric phase process are readily obtained.

036.114 **Deconvolution from wavefront sensing.**
J. C. Fontanella, J. Primot.
Interferometric imaging in astronomy, p. 209 – 214 (1987). – See Abstr. 012.035.

Atmospheric turbulence dramatically affects the image quality of large telescopes observing from the ground and generally limits their effective angular resolution to about one arc–second at optical wavelengths. To overcome this limitation two classes of methods have been proposed: speckle imaging in conjunction with post–processing algorithms (shift–and–add, Knox–Thompson, triple correlation, ...) and adaptive optics. An intermediate solution is proposed here which could be called "adaptive optics without adaptive optics".

036.115 **Phase relations in a rotational shear interferogram.**
E. Ribak.
Interferometric imaging in astronomy, p. 215 – 218 (1987). – See Abstr. 012.035.

Phase closure is a radio astronomical technique that enables recovering Fourier transform phases that would otherwise be corrupted by atmospheric and instrumental errors. The application of aperture synthesis and phase closure to the visible region has been suggested and demonstrated. Proposed here is a simple and efficient way of achieving phase relations by means of rotational shear interferometry.

036.116 **Computationally–cost–effective speckle imaging.**
R. H. T. Bates, B. L. K. Davey.
Interferometric imaging in astronomy, p. 219 – 222 (1987). – See Abstr. 012.035.

036.117 **Study of sizes, brightnesses and dynamics of solar facular points.**
O. von der Lühe.
Interferometric imaging in astronomy, p. 225 – 228 (1987). – See Abstr. 012.035.

This paper presents first results of an ongoing project to study the structure and the dynamics of small faculae.

036.118 **Speckle imaging at CfA.**
P. Nisenson, M. Karovska.
Interferometric imaging in astronomy, p. 229 – 230 (1987). Abstract. – See Abstr. 012.035.

036.119 **Infrared speckle interferometry on Calar Alto.**
C. Leinert, M. Haas.
Interferometric imaging in astronomy, p. 233 – 236 (1987). – See Abstr. 012.035.

036.120 **Imaging of low mass binary companions and circumstellar disks.**
D. W. McCarthy Jr.
Interferometric imaging in astronomy, p. 237 – 240 (1987). – See Abstr. 012.035.

036.121 **Two dimensional high angular resolution infrared imaging of circumstellar shells.**
M. L. Cobb, D. W. McCarthy Jr., J. F. Arens.
Interferometric imaging in astronomy, p. 241 – 244 (1987). – See Abstr. 012.035.

Preliminary results of a high angular resolution experiment using the ten micron camera developed at the University of California, Berkeley are presented. Three OH/IR stars have been imaged by means of a deconvolution algorithm within the 9.7 μm silicate feature and at 8.5 μm in the nearby continuum. All sources appear partially resolved at both wavelengths. The sources OH 26.5+0.6 and NML Cyg show signs of bi–polar outflow with elongated images having position angles different by 90° at the two wavelengths. The source IRC +10420 is elongated in the N–S position but its images are similar at the two wavelengths suggesting a different shell chemistry or geometry is presented for this source.

036.122 **Infrared interferometric studies of circumstellar dust shells.**
S. T. Ridgway.
Interferometric imaging in astronomy, p. 245 – 246 (1987). – See Abstr. 012.035.

036.123 **The identification of vignetted sources in coded aperture imaging.**
J. B. Stephen, E. Caroli, G. Di Cocco, P. P. Maggioli, L. Natalucci, A. Spizzichino.
Astron. Astrophys., Vol. 185, No. 1/2, p. 343 – 348 (1987).

The next generation of gamma–ray telescopes will utilise the technique of coded apertures in order to provide high resolution images of the celestial sphere. This method, however, suffers from a disadvantage wherein vignetted sources are reconstructed in false positions with incorrect intensities. Most techniques of identifying these sources involve the use of time consuming iterative computer algorithms. One recent suggestion, however, allows fast decoding which is also accurate when there are a limited number of sources in the field of view and a high background level, as is the case for contemporary low energy gamma–ray instruments. This method has been examined in detail by means of Monte–Carlo simulation in order to assess its accuracy and sensitivity.

036.124 **On the inversion of the Baade–Wesselink technique.**
N. R. Simon.
Publ. Astron. Soc. Pac., Vol. 99, No. 618, p. 868 – 876 (1987).

A general framework is given for the inversion of a "two–point" formulation of the Baade–Wesselink technique. The inversion method is used to evaluate the surface brightness/$(V–R)$ relation in two classical Cepheids, V482 Sco and R TrA. A proposal is made for employing the B–W inversion to construct a parameterization for brightness and temperature. This leads to the determination of radii and, ultimately, of temperatures and luminosities as well.

036.125 **The detection probability for emission–line objects in slitless spectrum surveys.**
R. G. Gratton, P. S. Osmer.
Publ. Astron. Soc. Pac., Vol. 99, No. 618, p. 899 – 903 (1987).

The authors present a simple model for the detectability of emission lines in slitless spectrum surveys. The model introduces the concept of detection probability for emission lines superposed on a continuum. They illustrate the detection probabilities for different equivalent widths and continuum magnitudes in the original CTIO Curtis Schmidt and 4–m surveys; they indicate that strong–lined quasars can be detected at magnitudes significantly fainter than what would normally be the limit of completeness. The model provides a useful framework for future work based on surveys which utilize digital detectors or digitized scans of photographic plates.

036.126 **The MIDAS image processing system.**
K. Banse, C. Ounnas, D. Ponz, P. Grosbøl, R. Warmels.
Bull. Am. Astron. Soc., Vol. 19, No. 2, p. 738 (1987). Abstract. – See Abstr. 010.061.

036.127 **New VAX–based IUE spectral image processing system for NASA/GSFC.**
J. Nichols–Bohlin, D. F. Stone, L. Smith.
Bull. Am. Astron. Soc., Vol. 19, No. 2, p. 739 (1987). Abstract. – See Abstr. 010.061.

036.128 **The use of AIPS for analysis of IRAS skyflux images.**
R. A. White.
Bull. Am. Astron. Soc., Vol. 19, No. 2, p. 743 (1987). Abstract. –
See Abstr. 010.061.

036.129 **STMODEL – a MIDAS application package to simulate HST data.**
M. Rosa, T. Courvoisier, F. Murtagh, S. di Serego Alighieri,
D. Baade, K. Banse, D. Ponz, R. Fosbury.
Bull. Am. Astron. Soc., Vol. 19, No. 2, p. 743 (1987). Abstract. –
See Abstr. 010.061.

036.130 **A general purpose nonlinear least squares subroutine.**
D. H. Gudehus.
Bull. Am. Astron. Soc., Vol. 19, No. 2, p. 744 (1987). Abstract. –
See Abstr. 010.061.

036.131 **The automatic photoelectric telescope service – an update.**
K. E. Kissell, R. M. Genet, L. J. Boyd, S. L. Baliunas,
D. S. Hall.
Bull. Am. Astron. Soc., Vol. 19, No. 2, p. 747 (1987). Abstract. –
See Abstr. 010.061.

036.132 **The effect of deconvolution on detectability of faint galaxies.**
D. J. Lindler.
Bull. Am. Astron. Soc., Vol. 19, No. 2, p. 747 (1987). Abstract. –
See Abstr. 010.061.

036.133 **A numerical method for hard X–ray spectra deconvolution.**
V. F. Polcaro.
Bull. Am. Astron. Soc., Vol. 19, No. 2, p. 747 (1987). Abstract. –
See Abstr. 010.061.

036.134 **10 μm imaging of circumstellar dust.**
E. E. Bloemhof, W. C. Danchi.
Bull. Am. Astron. Soc., Vol. 19, No. 2, p. 748 (1987). Abstract. –
See Abstr. 010.061.

036.135 **Speckle imaging at CfA.**
P. Nisenson, M. Karovska, R. Noyes, C. Papaliolios,
R. Stachnik, S. Strom, S. Edwards.
Bull. Am. Astron. Soc., Vol. 19, No. 2, p. 748 (1987). Abstract. –
See Abstr. 010.061.

036.136 **Acousto–optic Fourier transform processors for astronomical applications.**
J. P. Norris, K. S. Wood, H. W. Smathers.
Bull. Am. Astron. Soc., Vol. 19, No. 2, p. 748 (1987). Abstract. –
See Abstr. 010.061.

036.137 **Application of bispectrum analysis to 1–D infrared speckle data.**
J. C. Christou, J. D. Freeman, F. Roddier, D. W. McCarthy,
M. L. Cobb.
Bull. Am. Astron. Soc., Vol. 19, No. 2, p. 749 (1987). Abstract. –
See Abstr. 010.061.

036.138 **Automatic MK–style classification of objective–prism spectra.**
J. LaSala.
Bull. Am. Astron. Soc., Vol. 19, No. 2, p. 749 (1987). Abstract. –
See Abstr. 010.061.

036.139 **A periodogram of observational data series formed by segments separated by arbitrary time intervals.**
B. A. Burnasheva, O. P. Gollandskij.
Izv. Krymskoj Astrofiz. Obs., Tom 77, p. 157 – 165 (1987). In
Russian. English translation in Bull. Crimean Astrophys. Obs.,
Vol. 77.
It is shown that the sum of modified periodograms is a
distributed statistics with $n \lesssim 2k$ degrees of freedom, where k is

the number of summands. The phase shift between the summands does not affect the sum. Therefore the time interval
between the segments of observational data whose periodograms
are calculated may be arbitrary under the condition that the
observed process is stationary. But if the phase shift occurs inside
the segment, then the conditions of detecting the signal will be
deteriorated.

036.140 **Redundant versus nonredundant beam recombination in an aperture synthesis with coherent optical arrays.**
F. Roddier.
J. Opt. Soc. Am. A, Vol. 4, No. 8, p. 1396 – 1401 (1987).
Signal–to–noise ratios for the amplitude of the object Fourier
components are compared assuming either redundant or nonredundant beam recombination. A general condition is given for
the object brightness below which redundant beam recombination is superior. A similar condition is found when the variance of
the closure phases is considered.

036.141 **Two–color method for optical astrometry: theory and preliminary measurements with the Mark III stellar interferometer.**
M. M. Colavita, M. Shao, D. H. Staelin.
Appl. Opt., Vol. 26, No. 19, p. 4113 – 4122 (1987).
The two–color method for interferometric astrometry provides
a means of reducing the error in a stellar position measurement
attributable to atmospheric turbulence. The primary limitation
of the method is shown to be turbulent water vapor fluctuations.
Secondary atmospheric effects caused by diffraction from small
refractive–index inhomogeneities and differential refraction for
the observation of stars away from zenith are shown to introduce
errors that behave as white noise and which should usually not be
significant. Other potential error sources due to photon noise,
systematic instrumental effects, and imperfect data reduction are
also considered. Some preliminary two–color measurements with
the Mark III interstellar interferometer at Mt. Wilson are presented which demonstrate a factor of ~5 reduction in the
amplitude of the atmospheric fluctuations in a stellar position
measurement.

036.142 **Phase–gradient reconstruction from photon–limited stellar speckle images.**
G. J. M. Aitken, R. Johnson.
Appl. Opt., Vol. 26, No. 19, p. 4246 – 4249 (1987).
The phase–gradient process for reconstructing stellar speckle
images is implemented in a photon–address mode for application
to images containing small numbers of photons. Comparisons
are made with an address–mode Knox–Thompson process which
show that the phase–gradient approach has computational
advantages and better SNR performance. Reconstruction from
simulated data and real data for the binary source Beta Delphini
are presented.

036.143 **Rotating waveplates as polarization modulators for Stokes polarimetry of the sun: evaluation of seeing–induced crosstalk errors.**
B. W. Lites.
Appl. Opt., Vol. 26, No. 18, p. 3838 – 3845 (1987).
A formalism for estimating the crosstalk error among Stokes
I, Q, U, V introduced by seeing–induced image motion is presented. This formalism is applied to several modulation schemes for
polarization involving rotating waveplates, and it is evaluated
using an observed power spectrum of image motion. It is shown
that rotating waveplates offer an acceptable alternative for
measurements of absorption line polarization of features observed on the solar disk, provided the detection can be carried out
at video frame rates or faster.

036.144 **Fast decoding algorithm for uniformly redundant arrays.**
J. P. Roques.
Appl. Opt., Vol. 26, No. 18, p. 3862 – 3865 (1987).
Coded mask aperture imaging techniques are now used in
various fields, in particular in X–ray and gamma–ray astronomy.
With such systems it is possible to construct apertures which have

autocorrelation functions with perfectly flat sidelobes. For quadratic residue arrays (uniformly redundant arrays), the author has developed fast deconvolution algorithms based on conventional decoding methods but taking advantage of the mathematical properties of the decoding matrices. Balanced decoding, delta decoding, and finally sampled decoding are discussed.

036.145 Performance of maximum entropy image restoration method.
R. Lieu, R. B. Hicks, C. J. Bland.
Mon. Not. R. Astron. Soc., Vol. 229, No. 3, p. 49p – 53p (1987).

The performance of a maximum entropy image restoration scheme due to Gull, Daniell, Skilling and Burch, is considered in quantitative terms as follows. By means of a random number generator, it is possible to obtain data samples of a given signal contaminated with noise of known characteristics. Each of these data samples is then processed by the restoration scheme. For simple signal patterns, it is found that the majority of samples bear less resemblance (in terms of χ^2) to the original signal after being "restored".

036.146 Fourier removal of stripe artifacts in IRAS images.
D. Van Buren.
Space Telesc. Sci. Inst., Prepr. Ser., No. 191, 4 pp. (1987). To appear in Astron. J.

036.147 Single–Gaussian curve of growth abundance determinations from UV interstellar absorption line data.
A. W. Harris.
Rutherford Appleton Lab., Rep., RAL–87–032, 18 pp. (1987). To appear in Astrophys. J.

036.148 METRIC: an algorithm for maximum entropy reconstruction using iterative multiple constraints.
M. W. Johnson.
Rutherford Appleton Lab., Rep., RAL–87–058, 9 pp. (1987).

The METRIC algorithm is described which determines the maximum entropy image consistent with a series of linear or non-linear constraints. It owes its origins to the ART algorithm and may be used in a variety of deconvolution and image reconstruction applications.

036.149 The superimposition of multiple plates.
D. Malin.
Anglo–Aust. Obs., Prepr., No. 217, 6 pp. (1987). Paper presented at the 1987 IAU Working Group on Astronomical Photography.

036.150 Compare – a catalog access/comparison algorithm.
P. Hacking, J. Chillemi.
Astron. Data Cent. Bull., Vol. 1, No. 4, p. 299 – 301 (1987).

036.151 From low–noise observations to high–quality data.
D. Baade.
ESO Sci. Prepr., No. 520, 12 pp. (1987). Paper presented at the 27th Liège Astrophys. Colloq., June 1987.

036.152 A program for a new reduction of plates of the Astrographic Catalogue.
S. Röser, H. Jahreiß.
Astron. Rechen–Inst. Heidelb., Prepr. Ser., No. 10, 4 pp. (1987). To appear in IAU Symp. No. 133.

036.153 The inertial coordinate system based on a CCD (V) experiment for measuring the relative positions of galaxies with respect to reference stars.
X.–j. Guo, S. Xu, G.–j. Wu, R.–w. Lu, W. Mao.
Publ. Yunnan Obs., No. 1, p. 13 – 22 (1987). In Chinese.

The process of the experimental observations is described and the program of the calculation is also given. The precision of the relative position of a galaxy with respect to two reference stars measured by CCD is $\pm 0\rlap{.}''050$, the standard deviation is $\pm 0\rlap{.}''022$ and the accuracy is $\pm 0\rlap{.}''044$. Thereby, it is proved feasible to accurately determine the relative position of a galaxy with respect

to only two reference stars by using the "overlap exposure method" with the CCD.

036.154 Program design for spectral observations of type I solar radio bursts.
J.–y. Chen, Z.–g. Xia.
Publ. Yunnan Obs., No. 1, p. 65 – 70 (1987). In Chinese.

036.155 Data reduction of photographic plates observed by SBG.
J.–s. Wang, C.–h. Sha.
Publ. Yunnan Obs., No. 1, p. 71 – 76 (1987). In Chinese.

036.156 The automatic selection of stars and data processing on the computer PDP–11/23.
J.–s. Wang, W. Wang.
Publ. Yunnan Obs., No. 1, p. 77 – 82 (1987). In Chinese.

036.157 Estimation of the accuracy of measuring the position of a stellar image by means of an array device.
W.–x. Rong.
Publ. Yunnan Obs., No. 2, p. 13 – 16 (1987). In Chinese.

036.158 Optimum choice of parameters of a coded aperture imaging system.
H.–c. Pan, H.–r. Hang.
Publ. Purple Mt. Obs., Vol. 6, No. 2, p. 158 – 165 (1987). In Chinese.

036.159 Imaging properties of several proposed coded masks.
H.–c. Pan, H.–r. Hang.
Publ. Purple Mt. Obs., Vol. 6, No. 2, p. 166 – 173 (1987). In Chinese.

036.160 Communication between IBM–PC and VAX–11.
M. Dong, Y.–m. Zheng, X.–h. Di, Y.–l. Chen.
Publ. Purple Mt. Obs., Vol. 6, No. 2, p. 174 – 179 (1987). In Chinese.

036.161 Development and observations of the stellar speckle camera.
M. Noguchi, S. Isobe, Y. Norimoto, Y. Iizuka, N. Baba.
Tokyo Astron. Obs. Rep., Vol. 21, No. 1, p. 50 – 66 (1987). In Japanese.

036.162 The three–dimensional data–processing techniques in the research survey.
M. G. Larionov.
Soobshch. Spets. Astrofiz. Obs., Vyp. 52, p. 68 – 76 (1987). In Russian.

Optimum filtering of survey data is considered using the one–dimensional Davis method. Fluxes and coordinates of radio sources are estimated together with differential survey completeness.

036.163 Direct reduction of astrometric data in terms of centroids and variances.
H. Eelsalu.
Tartu Astrofüüs. Obs. Teated, Nr. 84, p. 18 – 24 (1987).

The importance of obtaining independent estimates to various kinematic distribution moments is emphasized. Equations for direct estimation of centroids and variances of the proper motion distribution are presented and commented.

036.164 Reduction of electrophotometric observations of artificial celestial bodies at the Astrophysical Institute of the Academy of Sciences of the Kazakh SSR.
A. V. Didenko, L. A. Usol'tseva.
Tr. Astrofiz. Inst. Alma–Ata, Tom 48, p. 112 – 116 (1987). In Russian.

036.165 **Restoration of astronomical images with the maximum entropy method.**
V. V. Makarov.
Red. Zh. Vestn. LGU. Mat., Mekh., Astron., Leningrad, 23 pp. (1987). In Russian. Abstr. in Ref. Zh., 51. Astron. 11.51.953 (1987).

036.166 **Signal storage in observation of an astronomical object through a turbulent atmosphere.**
Yu. V. Kornienko, V. N. Uvarov.
Dokl. Akad. Nauk USSR, A, No. 4, p. 60 – 63 (1987). In Russian. Abstr. in Ref. Zh., 51. Astron. 11.51.954 (1987).

036.167 **Comparison between image plane phase reconstruction methods in optical interferometry.**
A. Chelli.
Rev. Mex. Astron. Astrofis., Vol. 14, No. 2, p. 751 – 752 (1987). – See Abstr. 012.042.
The author has carried out a quantitative and a qualitative analysis intercomparing image plane phase reconstruction methods in optical interferometry.

036.168 **A comparison of interferometry from space and ground.**
A. H. Greenaway.
ESA Spec. Publ., ESA SP–273, p. 5 – 10 (1987). – See Abstr. 012.044.
It is argued that interferometry in space is superior to that from the ground only at visible frequencies ($\lambda < 1~\mu$m). In the visible regime interferometry in space shows a clear advantage over interferometry from the ground only when observations on objects fainter than about $m_v > 10$ are considered. It is further argued that mechanical tolerances on maintained geometry are fairly relaxed provided that they may be monitored to a very high precision.

036.169 **Imaging strategies for a space–borne interferometer.**
F. Roddier.
ESA Spec. Publ., ESA SP–273, p. 23 – 29 (1987). – See Abstr. 012.044.
The author describes two main strategies: (1) a large filled aperture consisting of a mosaic of light–weight mirrors. Internal referencing is used to coalign and approximately cophase the array. Diffraction–limited imaging is achieved by means of interferometric techniques such as roll deconvolution or pupil–plane interferometry. (2) A diluted array of diffraction–limited telescopes. Such an array can be exactly coaligned and cophased using internal and/or external references. In case of either strategy, pointing accuracy and mechanical vibrations will limit the system performances.

036.170 **Imaging at radio wavelengths.**
T. J. Cornwell.
ESA Spec. Publ., ESA SP–273, p. 31 – 36 (1987). – See Abstr. 012.044.
Modern interferometric arrays such as the VLA can produce images rivalling those from conventional optical telescopes both in quality and complexity. The recent advances contributing to this state of affairs are numerous, but the major ones are in electronic instrumentation, computing hardware capabilities, and algorithms.

036.171 **ISIS: image reconstruction methods and signal–to–noise ratio investigations.**
K.–H. Hofmann, G. Weigelt.
ESA Spec. Publ., ESA SP–273, p. 37 – 39 (1987). – See Abstr. 012.044.
In this paper the signal–to–noise ratio of the two image reconstruction methods optical phase–closure imaging and speckle masking (triple correlation) are discussed. The results show that (1) phase–closure imaging yields images with higher signal–to–noise ratio than speckle masking for bright objects and that (2) speckle masking yields images with higher signal–to–noise ratio than phase–closure imaging for faint objects.

036.172 **ISIS: image reconstruction experiments and comparison of various array configurations.**
T. Reinheimer, K.–H. Hofmann, G. Weigelt.
ESA Spec. Publ., ESA SP–273, p. 41 – 43 (1987). – See Abstr. 012.044.
The application of speckle masking (triple correlation processing) to coherent telescope arrays in space can yield images with very high angular resolution. The authors show computer simulations of optical aperture synthesis by speckle masking. They describe simulations of a 2–dimensional ring–shaped array and of a linear 1–dimensional array and discuss the dependence of the signal–to–noise ratio in the reconstructed image on photon noise.

036.173 **On the concept of resolution ellipse in aperture synthesis matched deconvolution with error analysis.**
A. Lannes, S. Roques.
ESA Spec. Publ., ESA SP–273, p. 45 – 48 (1987). – See Abstr. 012.044.
The regularization procedures to be implemented in aperture synthesis lead to a certain compromise between resolution and robustness. The authors show how to control the choice of the corresponding parameters. The regularization and deconvolution procedure under consideration is applied to the high–resolution study of the B component of the double QSO 2345 + 007.

036.174 **Ground–based optical interferometry.**
S. T. Ridgway.
ESA Spec. Publ., ESA SP–273, p. 119 – 125 (1987). – See Abstr. 012.044.
Interferometric methods have progressed rapidly, leading to image reconstruction of numerous extended sources. Phase closure has been demonstrated. Several two–telescope systems have obtained first fringes and first results. Major development projects are underway solely or partly for the support of optical interferometry. Within the next decade substantial new facilities will be operational for optical interferometry.

036.175 **Milliarcsecond ground–based imaging with single telescopes.**
C. A. Haniff.
ESA Spec. Publ., ESA SP–273, p. 171 – 173 (1987). – See Abstr. 012.044.
A number of images reconstructed from data obtained with a four–element Michelson interferometer are presented. These demonstrate the importance of closure phase measurement at optical wavelengths and indicate that true diffraction–limited imaging with large ground–based telescopes is relatively straight-forward.

036.176 **Image sharpening observations of active galactic nuclei.**
A. Campos Aguilar, I. Perez–Fournon, J. M. Carranza.
ESA Spec. Publ., ESA SP–273, p. 221 – 225 (1987). – See Abstr. 012.044.
It is shown that recentering, selecting and adding up sharp short–exposure images obtained with a photon–counting detector working in "time–resolved" mode can yield a gain in angular resolution better than a factor of 2. A cross–correlation selection algorithm, appropriate for faint and extended objects, is developed and compared with other image sharpness criteria. The limiting magnitude of the technique is about $m_v = 15$ under normal conditions (seeing ~ 1 arcsec). Preliminary results for a sample of bright active galaxies are presented.

036.177 **Finding high–redshift quasars using low–resolution spectra.**
P. Hewett, M. Irwin.
Pattern Recognition Lett., Vol. 5, No. 2, p. 113 – 117 (1987). Abstr. in Phys. Abstr., Vol. 90, No. 1310, Entry 94670 (1987).

036.178   **The calibration problem. IV. The Lutz–Kelker correction.**
H. Smith Jr.
Astron. Astrophys., Vol. 188, No. 1, p. 233 – 238 (1987).

The corrections to be applied to absolute magnitudes computed from trigonometric parallaxes are evaluated, both for individual stars and for specified samples of a particular stellar type assuming a gaussian luminosity function. For simplicity it is also assumed that the stars are distributed with uniform density and that there is no observational selection according to proper motion. The range of validity of the Lutz–Kelker corrections is explored, and certain features of those corrections are discussed.

036.179   **A three–dimensional extended Kolmogorov–Smirnov test as a useful tool in astronomy.**
E. Gosset.
Astron. Astrophys., Vol. 188, No. 1, p. 258 – 264 (1987).

The authors derive a three–dimensional version of the well-known Kolmogorov–Smirnov test. Such a test is of great practical interest when one wishes to investigate the spatial distribution of a set of data points, particularly for the case of small size samples. A comparison with assumed three–dimensional density laws is made possible for most of current applications. A table of critical values of the new statistic is given for usual significance levels; empirical formulate to simulate the asymptotic behaviour are given as well. The tree–dimensional extended two–sided Kolmogorov–Smirnov test is shown to be sufficiently distribution–free.

036.180   **Inverse photometric problem for spotted stars.**
D. P. Kjurkchieva.
Astrophys. Space Sci., Vol. 138, No. 1, p. 141 – 145 (1987).

On the basis of the analytical solution of the direct photometric problem for spotted stars a solution of the corresponding inverse problem is obtained. Expressions are given for determination of the location, size, and temperature of the spot and of the inclination of the rotational axis of the spotted star with regard to the line–of–sight.

036.181   **Automated rectification techniques and their application to velocity measurements.**
A. J. Adamson.
Observatory, Vol. 107, No. 1081, p. 252 – 258 (1987).

The application of automated spectrum–rectification techniques to stellar radial–velocity determinations is described in the context of (1) large–scale survey work and (2) large–light–ratio binary stars. Improvements to the methods are described in cases where flux conservation and equivalent–width accuracy are not required, and the dependence of the results upon signal–to–noise is demonstrated.

036.182   **Elektronische Bildauswertung am Beispiel von H II–Emissionsgebieten.**
C. Wöhler, W. Köhler.
Sterne Weltraum, 26. Jahrg., Nr. 12, p. 709 – 711 (1987).

036.183   **Imaging superluminal sources: prospects for the next decade.**
P. N. Wilkinson.
Superluminal radio sources, p. 211 – 216 (1987). – See Abstr. 012.053.

The author is reasonably confident that in ten years time VLBI arrays will be achieving what the VLA is achieving now for extended jets, viz., a dynamic range of 10^5:1 and excellent fidelity over 100 – 200 beams. This must surely lead to a more profound understanding of the superluminal phenomenon.

036.184   **Hardware and software features of the acquisition–control microcomputer of Cagliari SLR (*Satellite Laser Ranging*) system.**
A. Banni, V. Capoccia.
Stn. Astron. Int. Latitudine, Carloforte–Cagliari, Note Tec., N. 7, 9 pp. (1986).

036.185   **Doppler images of rotating stars using maximum entropy image reconstruction.**
S. S. Vogt, G. D. Penrod, A. P. Hatzes.
Astrophys. J., Vol. 321, No. 1, p. 496 – 515 (1987). = Lick Obs. Bull., No. 1066.

The authors present an improved version of the Doppler imaging technique, which now uses the principles of maximum entropy image reconstruction to derive spatially resolved images of rapidly rotating stars. The authors examine the effects that noise, finite spectral resolution, and uncertain stellar parameters are likely to have on real data sets, and demonstrate through a variety of test cases that the technique is efficient and accurate at recovering test images from realistic synthetic spectral data. The authors are currently using the technique to study cool starspots on RS CVn and FK Com stars, and the surface abundance distributions on Ap stars.

036.186   **On the resolving power of the method of interplanetary scintillations.**
V. S. Artyukh, T. D. Shishova.
Red. Zh. Izv. Vuzov. Radiofiz., Gor'kij, 17 pp. (1987). In Russian. Abstr. in Ref. Zh., 51. Astron., 12.51.1157 (1987).

036.187   **Atmospheric noise on the bispectrum in optical speckle interferometry.**
S. N. Karbelkar, R. Nityananda.
J. Astrophys. Astron., Vol. 8, No. 3, p. 271 – 274 (1987).

Based on a simple picture of speckle phenomena in optical interferometry it is shown that the recent signal–to–noise ratio estimate for the so called bispectrum, due to Wirnitzer (1985), does not possess the right limit when photon statistics is unimportant. In this wave–limit, which is true for bright sources, his calculations over–estimate the signal–to–noise ratio for the bispectrum by a factor of the order of the square root of the number of speckles.

036.188   **La campagne PHEMU85 à la lunette équatoriale de 38 cm de l'Observatoire de Paris.**
D. Briot.
Ann. Phys. (Paris), Vol. 12, Colloq. No. 1, p. 27 – 32 (1987). – See Abstr. 012.061.

Mutual phenomena of the satellites of Jupiter were observed with the 38 cm refracting telescope of Paris Observatory. A description of the technical experiment and data acquisition especially elaborated for this campaign, is given. Some examples of observed phenomena are presented.

036.189   **Observations vidéo de phénomènes mutuels faites à Meudon, au Pic du Midi et à Nice.**
W. Thuillot, P. Laques.
Ann. Phys. (Paris), Vol. 12, Colloq. No. 1, p. 33 – 38 (1987). – See Abstr. 012.061.

Some mutual phenomena have been observed with video material at the observatories of Meudon, Pic du Midi and Nice. This type of observation allows to observe in spite of bad observational conditions (Jupiter very low above the horizon, mist, twilight ...). Since these observations are easy to make, 16 observers obtained 19 observations at Pic du Midi Observatory, 23 at Meudon and 4 at Nice.

036.190   **Observations effectuées par le "GEA" (Group d'Estudis Astronomics).**
J. M. Gomez–Forrellad.
Ann. Phys. (Paris), Vol. 12, Colloq. No. 1, p. 77 – 95 (1987). – See Abstr. 012.061.

A group of Spanish amateur observers of variable stars describe their instruments, explain their methods and give the light–curves they obtained for some mutual phenomena.

036.191 **La campagne PHEMU85 a l'ESO.**
C. Gouiffes, P. Bouchet, F. X. Schmider.
Ann. Phys. (Paris), Vol. 12, Colloq. No. 1, p. 97 – 108 (1987). –
See Abstr. 012.061.
46 phenomena were observed at La Silla and 25 "good"
lightcurves were obtained. The 50 cm and 1 m telescopes of ESO
(f/13.6) were used. Absolute photometry was not made since the
air mass was often too large. It was noted that the predicted
values of the magnitude drop do not agree with the observations.
Neither the penumbra, nor the phase defect on the satellites may
explain this difference since these are either positive or negative.

036.192 **Considérations sur les systèmes d'observation utilisés et
la réduction des données.**
V. D'Ambrosio.
Ann. Phys. (Paris), Vol. 12, Colloq. No. 1, p. 129 – 137 (1987). –
See Abstr. 012.061.
Several problems related to the method of observation and
reduction using photoelectric photometers are developed. Sug-
gestions are made for the next campaign of observations.

036.193 **Les informations à extraire de l'observation des phénomènes mutuels et leur utilisation.**
J.–E. Arlot.
Ann. Phys. (Paris), Vol. 12, Colloq. No. 1, p. 167 – 173 (1987). –
See Abstr. 012.061.
Once the observation has been reduced for the observational
effects, it is possible to analyse the light–curves in order to deduce
three main parameters: the date for the minimum of light, the
value of the magnitude drop, and the duration of the phenomenon. This paper gives elements to explain how to deal with these
parameters in order to obtain theoretical corrections of the
motions of the satellites.

036.194 **La numérisation des images vidéo.**
G. Coupinot, L. Bergeal.
Ann. Phys. (Paris), Vol. 12, Colloq. No. 1, p. 201 – 205 (1987). –
See Abstr. 012.061.
This paper gives a description of the digitization system which
is realized at the "Observatoire du Pic du Midi et de Toulouse".
Future developments, and applications are also described.

036.195 **Projet d'observation des occultations par la Pleine Lune.**
J. Bourgeois.
Ann. Phys. (Paris), Vol. 12, Colloq. No. 1, p. 221 – 223 (1987). –
See Abstr. 012.061.
Observations of lunar occultations are not often made during
Full Moon: the aim of this communication is to present a
technique allowing such observations in order to improve the
distribution of the observations along the lunar month.

036.196 **Artificial guide stars for adaptive imaging.**
J. D. H. Pilkington.
Nature, Vol. 330, No. 6144, p. 116 (1987).

036.197 **Reply to "Artificial guide stars for adaptive imaging" by
J. D. H. Pilkington.**
L. Thompson, C. Gardner.
Nature, Vol. 330, No. 6144, p. 116 (1987). See Abstr. 036.196.

036.198 **Astronomical image reconstruction via space slit aper-
ture telescope.**
F. Martin, A. Bijaoui, H. Touma, C. Aime.
Proc. SPIE Int. Soc. Opt. Eng., Vol. 643, p. 189 – 193 (1986).
Abstr. in Phys. Abstr., Vol. 90, No. 1313, Entry 115251 (1987). –
See Abstr. 012.062.

036.199 **Description and (tentative) validation of a VLBI soft
correlator.**
G. Petit.
Mitt. Geod. Inst. Rheinischen Friedrich–Wilhelms–Univ. Bonn,
Nr. 71, p. 77 – 78 (1987). – See Abstr. 012.066.

036.200 **Geodetic VLBI–monitoring of the milliarcsecond struc-
tures of extragalactic radio sources.**
C. J. Schalinski, W. Alef, A. Witzel, J. Campbell, R. Wynands,
H. Schuh, G. Zeppenfeld.
Mitt. Geod. Inst. Rheinischen Friedrich–Wilhelms–Univ. Bonn,
Nr. 71, p. 103 – 105 (1987). – See Abstr. 012.066.

036.201 **Data analysis of Project ERIDOC (European Radio
Interferometry and DOppler Campaign).**
F. J. J. Brouwer, G. J. Husti, W. Beyer, J. Campbell, B. Hasch,
F. J. Lohmar, H. Seeger, G. Lundqvist, B. Rönnäng,
A. van Ardenne, R. T. Schilizzi, B. Anderson, R. Booth,
P. N. Wilkinson, P. Richards, S. Tallquist.
Mitt. Geod. Inst. Rheinischen Friedrich–Wilhelms–Univ. Bonn,
Nr. 72, p. 1 – 37 (1987). Paper presented at the IUGG/AIG
General Assembly, Hamburg, F.R. Germany, August 1983.
In this report the multi–station solutions of both the VLBI and
the Doppler measurements are presented. Taking into account
the ground survey ties, a comparison is made between the
coordinates of the VLBI phase centres and the Doppler phase
centres. From this, transformation parameters between the VLBI
and the Doppler coordinate system are determined.

036.202 **A Macintosh based data system for array spectrometers.**
J. Bregeman, N. Moss.
Infrared astronomy with arrays, p. 448 – 451 (1987). – See Abstr.
012.070.
An interactive data aquisition and reduction system has been
assembled by combining a Macintosh computer with an instrument controller (an Apple II computer) via an RS–232 interface.
The data system provides flexibility for operating different linear
array spectrometers. The standard Macintosh interface is used to
provide ease of operation and to allow transferring the reduced
data to commercial graphics software.

036.203 **Array imaging at high angular resolution.**
P. Léna.
Infrared astronomy with arrays, p. 455 – 463 (1987). – See Abstr.
012.070.
The author analyzes how array detectors, combined with the
current methods of interferometric imaging, can improve large
telescope performances and provide diffraction–limited imaging.

036.204 **High angular resolution infrared imaging at NOAO.**
J. C. Christou, J. M. Beckers, F. Roddier,
S. Ridgway, R. Probst, J. D. Freeman, D. W. McCarthy Jr.,
M. L. Cobb.
Infrared astronomy with arrays, p. 464 – 467 (1987). – See Abstr.
012.070.
The authors discuss how the SBRC 58×62 InSb infrared
imaging array will be used for 2–D infrared speckle interferometry (IRSI) at NOAO. The 2–D IRSI is a logical extension of
the 1–D slit–scanning technique currently employed by a number
of groups.

036.205 **Deep infrared surveys.**
S. J. Lilly, L. L. Cowie.
Infrared astronomy with arrays, p. 473 – 482 (1987). – See Abstr.
012.070.
The two–dimensional near–infrared array detectors that are
now being characterised on large telescopes offer for the first time
an opportunity to carry out surveys of selected regions of the sky
to faint flux levels at infrared wavelengths.

036.206 **Techniques for faint object imaging at 1 micron.**
J. A. Tyson.
Infrared astronomy with arrays, p. 483 – 488 (1987). – See Abstr.
012.070.
The purpose of this brief paper is twofold: (1) to review
techniques developed for processing optical/near IR CCD images
to 0.01% of background, and (2) to review 1 micron counts of
galaxies from a survey of 12 high galactic latitude fields, and
estimate the expected K–band counts.

036.207   Galactic infrared spectroscopy with large multiplexed arrays.
T. R. Geballe.
Infrared astronomy with arrays, p. 489 – 498 (1987). – See Abstr. 012.070.

The impact of large one- and two–dimensional arrays in spectrometers for astronomical use is discussed. Recent examples of infrared spectroscopy of solar system objects, stars, nebulae, and interstellar matter serve as illustrations of the enormous potential of infrared spectrometers containing large detector arrays. The examples also demonstrate the importance of the availability to the observer of a wide range of spectral resolutions.

036.208   The Foucault test for solar telescopes.
T. A. Darvann, R. B. Dunn.
LEST Found., Tech. Rep., No. 26, p. 71 – 78 (1987). – See Abstr. 012.071.

The purpose of this note is to describe the Foucault technique and to review and encourage its application to solar telescopes.

036.209   FORWARD: extension for multiple stations.
C. Steyaert.
WGN, Vol. 15, Nr. 6, p. 193 – 195 (1987).

The calculation of the Observability Function given earlier was implicitly for one transmitter. In this article, a generalization from multiple broadcasting stations is presented.

036.210   La définition de la magnitude des objets étendus.
A. Bijaoui.
Bull. Inf. Cent. Données Stellaires, No. 33, p. 3 – 7 (1987).

Nowadays, an extensive work is performed in two–dimensional photometry, specially using the CCD cameras. After having defined the units, the author briefly reviews the main instrumentation used to measure the magnitude of extended objects. He indicates some classical recipes used to delimitate the objects, to map the sky and to delete the spurious objects.

036.211   Uniformisation de la notion de magnitude.
G. Paturel, L. Bottinelli, P. Fouqué, A. Fruscione, L. Gouguenheim.
Bull. Inf. Cent. Données Stellaires, No. 33, p. 9 – 15 (1987).

In practice, it is often better to use magnitudes instead of fluxes in exotic units. Moreover, for galaxies, it is possible to convert some quantities, like apparent diameters, into a parameter equivalent to a magnitude. The authors give relations to calculate an apparent magnitude of a galaxy from the isophotal diameter and the brightness code.

036.212   Choix d'un système de filtres pour la photométrie des galaxies en cosmologie.
G. Mathez.
Bull. Inf. Cent. Données Stellaires, No. 33, p. 27 – 32 (1987).

The contemporary observational cosmology needs a continuous increase in photometric data on larger and larger samples of fainter and fainter galaxies. The use of CCD cameras is increasing but, as a consequence of their high sensitivity in the red, difficulties arise due to the night sky emission. Narrow or intermediate band multicolor photometry allows the determination of both the redshift and spectral type of galaxies. Concertation is needed to define a suitable system of filters and to measure faint spectrophotometric standards.

036.213   Enquiry on rapid automatic measuring machines.
C. Jaschek.
Bull. Inf. Cent. Données Stellaires, No. 33, p. 109 – 115 (1987).

036.214   Code for digital image processing of extended objects.
V. V. Makarov, V. P. Reshetnikov, V. A. Yakovleva.
Tr. Astron. Obs., Leningrad, Tom 41, p. 112 – 136 (1987). Uch. Zap. LGU, No. 420, Ser. Mat. Nauk, Vyp. 63. In Russian.

The code for the photometric processing of large–scale photographic plates of extended objects is described. The modern techniques of digital image restoration, sky background reduction and standardization are used. Surface brightness distributions of the object in various colour bands, distributions of the colour indices, integral magnitudes and other characteristics may be obtained with this code.

036.215   COLOUR program for data processing of UBVR photometry of variable stars.
L. N. Berdnikov, G. Sh. Rojzman.
Astron. Tsirk., No. 1484, p. 3 – 5 (1987). In Russian.

036.216   Image processing on VEGA pictures.
A. Abergel, J. L. Bertaux, E. Dimarellis.
ESA Spec. Publ., ESA SP–278, p. 689 – 694 (1987). – See Abstr. 012.075.

The raw TV images of comet Halley received from the two spacecraft VEGA 1 and 2 generally do not clearly display the contour of the nucleus nor any structure in the coma. In fact, the quality of these images is affected by several kinds of noise. The periodic components of this noise are first eliminated by two–dimensional Fourier filtering. Then the contrast of the filtered images is locally enhanced to detect with no ambiguity the limb of the sunlit nucleus and some spectacular structures in the coma images obtained far from the nucleus.

036.217   Supercomputers in optical astronomy.
N. A. Sharp.
NRAO Workshop, No. 15, p. 107 – 114 (1988). – See Abstr. 012.076.

There is a clear, perceived and growing need for the processing power of (and beyond) currently existing supercomputer. This power is needed not just for the theoretical interpretation of observational data, but simply to process the data into usable scientific forms, including incorporating such abilities into new instrumentation, and to post–process data so as to make the most efficient use of existing facilities.

036.218   Real–time signal processing requirements for diffraction limited optical imaging.
E. K. Hege, J. C. Christou.
NRAO Workshop, No. 15, p. 115 – 124 (1988). – See Abstr. 012.076.

Diffraction limited optical images can now be produced using large ground–based astronomical telescopes. This is quite new, with convincing validation being achieved only in the past year. The computational requirements necessary to implement this capability for routine observations already challenge the capacity of present superminicomputers. Extension to larger telescopes will require supercomputer support from the beginning.

036.219   Imaging the Sun at 1.4 GHz.
T. S. Bastian, G. A. Dulk.
NRAO Workshop, No. 15, p. 125 – 132 (1988). – See Abstr. 012.076.

The authors briefly describe the maximum entropy–like image reconstruction techniques employed in reconstructing images of the quiet Sun and present a representative example.

036.220   Kalman filtering VLA phase data with a supercomputer.
Y. Zheng, J. P. Basart.
NRAO Workshop, No. 15, p. 141 – 148 (1988). – See Abstr. 012.076.

The development of modeling and filtering method is limited by the computation time and memory size. This problem can be solved with the aid of a supercomputer. Experimental results show that the execution of a filtering program, which was optimized and vectorized, on the BCS CRAY X–MP/24 is 32.8 times faster than that on the VLA DEC–10 computer. This encourages one to develop an adaptive Kalman filtering procedure for the VLA image processing. In order to take advantage of vector processing machines, the program should be vectorized.

036.221 **Image recovering from the modulated CCD imagers.**
D. A. Ralys, V. Dadurkevičius.
Astrophys. Space Sci., Vol. 139, No. 1, p. 155 – 162 (1987).

The CCD imagers have a spatial resolution comparable to the size of a constituent photosensitive cell. The modulation by means of a rotating Ronchi grating helps to exceed this limit. While the grating rotates in front of the CCD imager the counts of each cell are modulated thus producing a time–varying modulation pattern. The small systematic variations of the form of the pattern encode the fragment of the image pertaining to the cell. The images can be recovered from the modulation patterns by means of the least–squares as well as maximum entropy reconstruction technique. Both methods are able to decode the modulation patterns having high signal–to–noise ratio, however, maximum entropy reconstruction seems to be more robust. A gain in spatial resolution may be of the order of magnitude under favourable circumstances.

036.222 **Astrometric use of the second short exposure of Tautenburg pilot program plates.**
R.-D. Scholz.
Astron. Nachr., Vol. 308, No. 6, p. 375 – 378 (1987). In German.

To overcome a possible magnitude equation the Tautenburg Schmidt plates of the Lohrmann programme were taken with a long 20–minute exposure (System A) and a short 20–second exposure (System C) displaced about 0.5 minute of arc in declination from the long exposure. The distance between the two images in dependence on the magnitude of a star and its position on the plate was analysed. The transformation of coordinates from system C to A can be performed by using first order polynomials. No higher order terms were found to be necessary. The coefficients of the polynomials can not be assumed as constant for all plates but must be computed from least squares solutions involving a sufficient number of stars on each plate. For three of four plates investigated significant orthogonal linear terms were detected.

036.223 **Use of the statistic analysis for preliminary processing of laser ranging results of satellites.**
S. K. Tatevyan, O. A. Petrova, V. V. Kirichuk,
O. A. Abrikosov, A. N. Marchenko.
Nauchn. Inf., Vyp. 58, p. 3 – 8 (1986). In Russian.

The least–squares collocation method is applied to estimate the accuracy of laser ranging of satellites. The optimal form of the model covariance function is selected and described by the decreasing cosine equation. The analysis of the covariance function parameters is performed for the cases when the observation density is different, namely for Lageos and ICB–1300 satellites. The filtering and interpolation methods are described by the least–squares collocation.

036.224 **A method of astrometrical reduction of negatives of the VAU–camera.**
A. N. Biryukov.
Nauchn. Inf., Vyp. 58, p. 66 – 80 (1986). In Russian.

The accuracy of the method is estimated and remains quite high when a larger region of the negative is reduced.

036.225 **A method of observations of geostationary satellites on the TV–guide of the camera VAU.**
F. N. Masumi, V. I. Chumachenko, Yu. N. Ehsaulov.
Nauchn. Inf., Vyp. 58, p. 96 – 99 (1986). In Russian.

The accuracy of the position is $\pm 6''$, the brightness of satellites is $\pm 0\overset{m}{.}1$.

036.226 **A method of photographic observations of geostationary satellites with the VAU–camera.**
O. Naimov.
Nauchn. Inf., Vyp. 58, p. 100 – 102 (1986). In Russian.

A method of photographic observations of geostationary satellites is suggested. It simplifies the elaboration of the plates received, increases the time of the work of the rapid shutter, and allows saving photomaterials.

036.227 **Changes in the programme complex of astrometrical reduction of satellite photographs by AFU–75 and VAU cameras.**
L. A. Yurova.
Nauchn. Inf., Vyp. 58, p. 128 – 130 (1986). In Russian.

036.228 **Set of programmes for fulfilling requests automatically by the Center of Astronomical Data.**
S. V. Vereshchagin.
Nauchn. Inf., Vyp. 59, p. 13 – 30 (1986). In Russian.

036.229 **Plotting of stellar fields.**
N. D. Kostyuk.
Nauchn. Inf., Vyp. 59, p. 70 – 75 (1986). In Russian.

Main ideas of the programme for plotting maps of stellar fields using data from the SAO and CSI catalogues are described.

036.230 **Service programmes for expanding possibilities of programming in FORTRAN language in OS ES.**
V. Yu. Lukin, N. E. Piskunov.
Nauchn. Inf., Vyp. 59, p. 89 – 98 (1986). In Russian.

036.231 **General principles of computer image processing.**
V. Yu. Lukin, K. N. Mit'kin, N. E. Piskunov.
Nauchn. Inf., Vyp. 59, p. 99 – 112 (1986). In Russian.

036.232 **A method to search k shortest routes in a graph.**
A. M. Sobolev.
Nauchn. Inf., Vyp. 59, p. 113 – 117 (1986). In Russian.

The text in FORTRAN programme and a test example are given.

036.233 **Monte–Carlo programme to search the routes of higher probability in a graph.**
A. M. Sobolev.
Nauchn. Inf., Vyp. 59, p. 118 – 123 (1986). In Russian.

The text of FORTRAN IV programme and a test example are given. When a detailed study of the routes of population transfer in the multilevel quantum system is needed, the Monte Carlo procedure requires essentially lesser computer time than the sort method.

036.234 **Methods and algorithms for reducing television images of the nucleus of comet Halley. Some results.**
R. Z. Sagdeev, G. A. Avanesov, Ya. L. Ziman, V. A. Shamis,
V. A. Krasikov.
Kosm. Issled., Tom 25, Vyp. 6, p. 820 – 830 (1987). In Russian.
English translation in Cosm. Res.

036.235 **Laser and stellar speckle.**
E. R. Mendez.
Sci. Prog., Vol. 71, No. 283, Part 3, p. 365 – 380 (1987). Abstr. in Phys. Abstr., Vol. 90, No. 1316, Entry 134148 (1987).

036.236 **Algorithms for image reconstruction after nonuniform sampling.**
M. Piacentini, C. Cafforio, F. Rocca.
IEEE Trans. Acoust., Speech Signal Process., Vol. ASSP–35, No. 8, p. 1185 – 1189 (1987). Abstr. in Phys. Abstr., Vol. 90, No. 1317, Entry 139942 (1987).

036.237 **Measurement technique for the Giotto radio science experiment.**
P. Edenhofer, M. K. Bird, H. Buschert, P. B. Esposito,
H. Porsche, H. Volland.
J. Phys. E, Vol. 20, No. 6, Part 2, p. 768 – 776 (1987). Abstr. in Phys. Abstr., Vol. 90, No. 1317, Entry 139943 (1987).

036.238   A system for morphological analysis and classification of astronomical objects.
F. Pasian, M. Pucillo, P. Santin.
Image analysis and processing. Proceedings of the Third International Conference, Rapallo, Italy, 30 September – 2 October 1985. Plenum, New York, USA, p. 197 – 204 (1986). Abstr. in Phys. Abstr., Vol. 90, No. 1317, Entry 139945 (1987).

036.239   An application of L_1 to astronomy.
P. J. Rousseeuw.
Statistical data analysis based on the L_1–norm and related methods. First International Conference, Neuchâtel, Switzerland, 31 August – 4 September 1987. North – Holland, Amsterdam, The Netherlands, p. 437 – 445 (1987). Abstr. in Phys. Abstr., Vol. 90, No. 1317, Entry 139946 (1987).

036.240   Comparison of methods for determining the centers of extensive air showers.
J. Poirier, E. Funk, S. Mikocki, N. Rohrer.
Nucl. Instrum. Methods Phys. Res., Sect. A, Vol. A260, No. 1, p. 280 – 282 (1987). Abstr. in Phys. Abstr., Vol. 90, No. 1318, Entry 145834 (1987).

036.241   Some critical cases of astronomical image detection.
A. Bijaoui, G. Wlerick.
Proceedings of the international topical meeting on image detection and quality, p. 241 – 244 (1986). Abstr. in Phys. Abstr., Vol. 91, No. 1320, Entry 10438 (1988). – See Abstr. 012.079.

036.242   Image quality evaluation from wavefront sensing.
J. C. Fontanella, R. Deron, G. Rousset.
Proceedings of the international topical meeting on image detection and quality, p. 245 – 249 (1986). Abstr. in Phys. Abstr., Vol. 91, No. 1320, Entry 10439 (1988). – See Abstr. 012.079.

036.243   High resolution interferometric imaging using a large optical telescope.
R. H. Frater, J. G. Robertson, J. D. O'Sullivan.
Proceedings of the international topical meeting on image detection and quality, p. 255 – 258 (1986). Abstr. in Phys. Abstr., Vol. 91, No. 1320, Entry 10440 (1988). – See Abstr. 012.079.

036.244   Infrared differential speckle interferometry with Fourier's spectral selectivity.
J. Gay, D. Mekarnia.
Proceedings of the international topical meeting on image detection and quality, p. 265 – 269 (1986). Abstr. in Phys. Abstr., Vol. 91, No. 1320, Entry 10441 (1988). – See Abstr. 012.079.

036.245   Digital stacking of short exposures to improve photometric accuracy and extend dynamic range of panoramic detectors used in astronomical imagery.
H. M. Heckathorn, C. B. Opal.
Proceedings of the international topical meeting on image detection and quality, p. 285 – 288 (1986). Abstr. in Phys. Abstr., Vol. 91, No. 1320, Entry 10442 (1988). – See Abstr. 012.079.

036.246   Effect of source structure on the precision of baseline measurements.
F.-j. Zhang, R.-x. Zhou.
Chin. Astron. Astrophys., Vol. 11, No. 4, p. 348 – 354 (1987). English translation of Acta Astron. Sin., Vol. 28, No. 2, p. 173 – 181 (1987).
The authors used brightness distribution of compact radio sources to calculate the delay and fringe rate produced by source structure, and used MK III VLBI data to calculate the effect of source structure on the precision of baseline measurements. For an 8000-km baseline, the effect can sometimes reach the order of centimetres.

036.247   The effect of C/O ratio on the Blackwell–Shallis method of determining stellar temperatures for cool evolved stars.
G. C. Augason, H. R. Johnson, D. R. Alexander.
Publ. Astron. Soc. Pac., Vol. 99, No. 621, p. 1145 – 1146 (1987). Abstract. – See Abstr. 010.281.

036.248   An improvement of the method of plate constants for the determination of stellar trigonometric parallaxes.
H.-q. Zhao.
Acta Astron. Sin., Vol. 28, No. 2, p. 190 – 196 (1987). In Chinese.
The author introduces so–called self-dependences of reference stars into the plate constant method for the determination of trigonometric parallaxes. By finding and rejecting reference stars with unusually large proper motions or parallaxes he improves the classical procedure. The parallax of BD + 70°68 is given and some related results are discussed.

036.249   Image analysis of the observation of the DA240 region with a meter–wave synthesis radio telescope.
M.-z. Wei, S.-j. Qian.
Acta Astron. Sin., Vol. 28, No. 3, p. 250 – 259 (1987). In Chinese.
The data of the DA240 region observed at 232 MHz with the Miyun radio telescope are analysed in detail. Errors and effects which degrade the quality of the maps are analysed and corrected by image processing methods. Several maps of the radio source have been obtained.

036.250   A non–recursive method for identifying erroneous observations.
P.-z. Jia.
Acta Astron. Sin., Vol. 28, No. 3, p. 288 – 295 (1987). In Chinese.

036.251   A statistical method for the processing of CCD images.
J.-s. Chen, X.-m. Fan, X.-y. Tan.
Acta Astron. Sin., Vol. 28, No. 3, p. 303 – 307 (1987). In Chinese.

MicroVAX–based data management and reduction system for the regional planetary image facilities.
See Abstr. 002.002.

The guide star catalog. II. Astrometric and photometric algorithms and precision.
See Abstr. 002.044.

Suggestions for machine–readable catalogs: a second look.
See Abstr. 002.048.

A few thoughts on astronomical databases.
See Abstr. 002.083.

Scheduling and other issues for a solar–stellar synoptic observatory.
See Abstr. 013.002.

The future of high angular resolution astronomy: seeing the unseen.
See Abstr. 013.014.

IRAF status report.
See Abstr. 013.049.

Status report: science data analysis software for the Hubble Space Telescope.
See Abstr. 013.050.

Status report on the VAX–based IUE Regional Data Analysis Facility at GSFC.
See Abstr. 013.052.

Computer graphics and images display devices for astronomy.
See Abstr. 013.053.

Applications of AI technology in astronomical research I: An expert data analysis assistant.
See Abstr. 013.054.

Applications of AI technology in astronomical research II: What can be done, what should be done?
See Abstr. 013.055.

Einführung in die Sonnenbeobachtung. I.
See Abstr. 014.009.

Einführung in die Sonnenbeobachtung. II.
See Abstr. 014.054.

Evolution of architectures for data processing.
See Abstr. 021.008.

Progress in using a Cray X–MP for radio astronomy.
See Abstr. 021.023.

RESPECT – software package for reduction of astronomical spectra.
See Abstr. 021.028.

The first images from optical aperture synthesis.
See Abstr. 031.024.

Adaptive optics for optimization of image resolution.
See Abstr. 031.042.

Status of the Mark III interferometer.
See Abstr. 032.018.

A fiber–linked ground–based array.
See Abstr. 032.019.

Interferometry with the European Very Large Telescope.
See Abstr. 032.020.

Interferometric capabilities of the NNTT.
See Abstr. 032.021.

Ground–based interferometry with the European Very Large Telescope (VLT).
See Abstr. 032.033.

CERGA high angular resolution optical network.
See Abstr. 032.034.

Practical limits to ground based interferometry and its impact on space interferometry.
See Abstr. 032.035.

Optical 16–telescope interferometer at Erlangen.
See Abstr. 032.036.

Performances of an actively stabilized stellar interferometer (ASSI): faint magnitudes, low fringe contrast measurements and operationality.
See Abstr. 032.037.

Consideration of the instrument's temperature in processing astrolabe observations.
See Abstr. 032.051.

Architecture, hardware and software for the net of the measuring–computing complexes at the RATAN–600 radio telescope.
See Abstr. 033.025.

Control equipment for two spectrometers at the RATAN–600 radio telescope.
See Abstr. 033.026.

The control system for moving the RATAN–600 radio telescope feed based on a microcomputer.
See Abstr. 033.027.

Operative presentation of astrophysical results of observations of the Sun at the RATAN–600 radio telescope.
See Abstr. 033.031.

The NRAO supercomputing initiative.
See Abstr. 033.046.

Plans for the Berkeley–Illinois Array.
See Abstr. 033.048.

Design criteria, considerations, objective and tradeoffs.
See Abstr. 034.004.

A panoramic photon–counting detector system.
See Abstr. 034.027.

Differential non–linearity in analog–to–digital converters: photometric consequences for astronomical microdensitometry.
See Abstr. 034.057.

Electrooptic Fabry–Perot filter: development for the study of solar oscillations.
See Abstr. 034.062.

Interfacing the Starlight–1 photometer to the Osbourne Executive computer.
See Abstr. 034.120.

TDRSS OVLBI experiment: data correlation and results.
See Abstr. 035.011.

ISIS: Imaging Speckle Interferometer in Space.
See Abstr. 035.073.

Aperture synthesis in space: technical problems.
See Abstr. 035.079.

Solar interferometry with a 4–aperture non–redundant and stabilized network.
See Abstr. 035.080.

Rigorous compilation of the Northern International Reference Stars.
See Abstr. 041.004.

Dependence of the results of determining Universal Time with photographic zenith tubes in Kitab on the configuration of the dome.
See Abstr. 044.006.

The mean–square collocation for the laser ranging normal points determiation.
See Abstr. 045.020.

Les programmes du service des observations lasers du satellite artificiel de la Terre.
See Abstr. 045.022.

OASIS (Optical Aperture Synthesis in Space): a mission concept.
See Abstr. 051.044.

Operating of programme dividers in the satellite tracing system.
See Abstr. 053.010.

Gravitational wave astronomy – potential and possible realisation.
See Abstr. 066.066.

Investigation of linear polarization of moustaches.
See Abstr. 073.061.

About the interest of solar interferometric observations.
See Abstr. 074.060.

Diagnostics of solar coronal loops at interferometric angular resolution.
See Abstr. 074.061.

Method for automated initial reduction of results obtained by observations of the Sun with the RATAN–600 radio telescope and foundation of archives.
See Abstr. 077.042.

A means to sharper images.
See Abstr. 082.037.

Experiments on laser guide stars at Mauna Kea Observatory for adaptive imaging in astronomy.
See Abstr. 082.038.

Day–time seeing statistics at Sacramento Peak Observatory.
See Abstr. 082.064.

Comparison of tropospheric models used for VLBI data analysis.
See Abstr. 082.092.

Precision in determination of astronomical refraction from aerological data.
See Abstr. 082.110.

Spherical harmonic modeling of the geomagnetic field using the fast Fourier transform.
See Abstr. 084.036.

Radar images of Venus (Results of Venera–15 and Venera–16 space probes).
See Abstr. 093.017.

A study of the efficiency of some inversion techniques applied to a simple model of the Moon.
See Abstr. 094.027.

Near IR imaging of Io.
See Abstr. 099.037.

Teleskopische Meteorbeobachtung.
See Abstr. 104.003.

Observational results 1986 – 87 from Florida, USA.
See Abstr. 104.093.

Sulla caduta di meteoriti, la loro ricerca e classificazione.
See Abstr. 105.049.

Discussion of trigonometric parallaxes determined along right ascension and declination.
See Abstr. 111.021.

Instrumental effects and the Strömgren photometric system.
See Abstr. 113.024.

Stellar photometry in crowded fields.
See Abstr. 113.038.

Detection of carbon stars in two independent surveys.
See Abstr. 114.106.

Identification of lines in the satellite ultraviolet: the spectrum of Tau Scorpii.
See Abstr. 114.111.

Millisecond X–ray binary pulsars.
See Abstr. 117.097.

Astrophysical results on young stars and active objects.
See Abstr. 121.026.

Search for optical counterparts of gamma–ray burst sources.
See Abstr. 143.038.

Integrated magnitudes and colours of clusters in the LMC with the COSMOS machine.
See Abstr. 156.050.

High spatial resolution IR observations and variability of the nuclear region of NGC 1068: structure and nature of the inner 100 parsec.
See Abstr. 158.158.

Galactic nuclei and quasars at high angular resolution.
See Abstr. 158.286.

The accuracy of cross–correlation estimates of quasar emission–line region sizes.
See Abstr. 159.033.

Positional Astronomy, Celestial Mechanics

041 Astrometry

041.001 Comparisons of positions of extragalactic compact radio sources.
P. Brosche, D. Sinachopoulos.
Mon. Not. R. Astron. Soc., Vol. 227, No. 2, p. 341 – 346 (1987).
The existence of systematic differences is shown between recent precise lists of VLBI positions. They depend simultaneously on both spherical coordinates and reach amplitudes of a few milliarcseconds. The influence on fundamental astrometric quantities can most easily be estimated if the differences are represented by a small rotation. Significant rotational components also reach a few milliarcseconds. A list of precise positions of the optical counterparts of compact radio sources was given by de Vegt & Gehlich (1982), and the differences from the radio positions are again discussed in this paper.

041.002 Right ascension corrections to 120 FK 4–stars by the analysis of Time observations obtained with the Photoelectric Transit Instrument at Torino Observatory.
G. Chiumiento, M. Sarasso, A. Poma.
Astron. Astrophys., Vol. 183, No. 2, p. 403 – 410 (1987).
Right ascension corrections to 120 FK 4–stars have been calculated by the analysis of Time observations obtained with the Photoelectric Transit Instrument at Torino Observatory in the period 1980.3 – 1985.3. The method performed has been based both on internal adjustment of the stars in each group, by a method of successive approximations (Yokoyama, 1968), and on the simultaneous determination of group corrections and local seasonal variations through a least–squares solution according to the model by Feissel (1980). The mean error of the corrections has proved to be better than 0″.004 for most stars. Systematic deviations as a function of right ascension, declination and magnitude have been preliminary derived. The reduction of Time observations with the revised star positions has been accomplished and an improvement in weight of 1.76 times on internal adjustment and 1.30 times on external agreement with respect to the uncorrected data series has been obtained.

041.003 Astrometric position of IRS–7 in the galactic center.
E. E. Becklin, H. Dinerstein, I. Gatley, M. W. Werner, B. Jones.
AIP Conf. Proc., No. 155, p. 162 (1987). – See Abstr. 012.029.

041.004 Rigorous compilation of the Northern International Reference Stars.
C. S. Cole.
Diss. Abstr. Int., Sect. B, Vol. 48, No. 4, p. 1071–B (1987). Thesis, University of Florida, 146 pp. (1986). Order No. DA8715985.

041.005 Right ascensions of major planets and the Sun obtained at the Tashkent meridian circle in 1978 – 1980.
G. G. Khodak.
Astron. Inst. Akad. Nauk UzSSR, Tashkent, 2 pp. (1987). In Russian. Abstr. in Ref. Zh., 51. Astron., 10.51.105 (1987).

041.006 Determination of right ascensions of the Sun, Venus and Mercury in Tashkent in the period 1981 – 1983.
G. G. Khodak.
Astron. Inst. Akad. Nauk UzSSR, Tashkent, 24 pp. (1987). In Russian. Abstr. in Ref. Zh., 51. Astron., 10.51.106 (1987).

041.007 First astrometric results with the Mount Wilson Mark III optical interferometer.
D. Mozurkewich, D. J. Hutter, K. J. Johnston, R. S. Simon, M. M. Colavita, M. Shao, J. L. Hershey, J. A. Hughes, G. H. Kaplan.
Bull. Am. Astron. Soc., Vol. 19, No. 2, p. 749 (1987). Abstract. – See Abstr. 010.061.

041.008 Mapping the sky with the Carlsberg Automatic Meridian Circle.
L. V. Morrison, L. Helmer, L. Quijano.
R. Greenwich Obs., Prepr., No. 58, p. 5 – 9 (1987). To appear in IAU Symp. No. 133.

041.009 Precise proper motions and positions of stars from a combination of fundamental catalogues with the HIPPARCOS catalogue.
R. Wielen.
Astron. Rechen–Inst. Heidelb., Prepr. Ser., No. 9, 6 pp. (1987). To appear in IAU Symp. No. 133.

041.010 Optical positions and proper motions of selected radio stars.
H. M. Schwerdtfeger, R. Hering, H. G. Walter, H. Jahreiß.
Astron. Rechen–Inst. Heidelb., Prepr. Ser., No. 11, 5 pp. (1987). To appear in IAU Symp. No. 133.

041.011 Declination improvements in surveys with the RATAN–600 radio telescope.
M. G. Larionov.
Soobshch. Spets. Astrofiz. Obs., Vyp. 52, p. 62 – 67 (1987). In Russian.
The improvement of declinations in the supplemental survey with the West sector of the RATAN–600 is considered. The position errors and the program of optical identifications are discussed.

041.012 An HIPPARCOS astrometric programme for spectral classification standard stars.
V. Malyuto, G. Jimsheleishvili (*G. N. Dzhimshelejshvili*).
Tartu Astrofüüs. Obs. Teated, Nr. 84, p. 25 – 31 (1987).
A list of F–G–K stars based on various sources has been drawn up for a wide range of T_{eff}, M_V and [Fe/H]. The list has been proposed to the astrometric observatory HIPPARCOS with the intent of enhancing the accuracy and homogeneity of spectral classification systems.

041.013 Right ascensions of the Sun, Mercury and Venus observed with the transit instrument in Nikolaev in 1986.
A. G. Petrov, G. M. Petrov, R. T. Fedorova, P. N. Fedorov.
Glav. Astron. Obs. Akad. Nauk SSSR, Leningrad, 11 pp. (1987). In Russian. Abstr. in Ref. Zh., 51. Astron. 11.51.69 (1987).

041.014 **Análisis del campo de distorsión del Catálogo Astrográfico.**
J. Stock, C. Abad.
Rev. Mex. Astron. Astrofis., Vol. 14, No. 2, p. 448 – 452 (1987). – See Abstr. 012.042.

The method for the reduction of astrometric plates proposed by Stock (1980) was extended to include the determination of field distortion. In a project in collaboration with the Hamburg Observatory the method was applied to two groups of plates of the Astrographic Catalogue, including part of the Paris–, Oxford–, Potsdam–, and Helsingfors zones.

041.015 **Provisional values of internal group corrections for the VL3 astrolabe catalogue.**
L. B. F. Clauzet, M. E. Mattos.
Rev. Mex. Astron. Astrofis., Vol. 14, No. 2, p. 453 (1987). Abstract. – See Abstr. 012.042.

041.016 **Representation of systematic differences astrolabe–catalog.**
P. Benevides–Soares.
Rev. Mex. Astron. Astrofis., Vol. 14, No. 2, p. 454 (1987). Abstract. – See Abstr. 012.042.

041.017 **Optical and radio positions of α Scorpii.**
L. B. F. Clauzet, S. Débarbat, F. Chollet.
Rev. Mex. Astron. Astrofis., Vol. 14, No. 2, p. 455 (1987). Abstract. – See Abstr. 012.042.

041.018 **Determinação astrométrica de posições com filmes do "ESO–B Survey".**
R. Vieira–Martins, C. H. Veiga.
Rev. Mex. Astron. Astrofis., Vol. 14, No. 2, p. 456 – 459 (1987). – See Abstr. 012.042.

The positions of astronomical objects with blue magnitudes ranging from 6 to 20 are obtained using the film copies of the "ESO(B) Survey". The mean uncertainty is about 0.5 arcsec. Special reduction procedures were developed for treating bright objects, particularly the reference stars, as these present saturated images as well as optical deformations caused by the plate support of the Schmidt Camera. These procedures can be used to determine secondary Catalogues which have relative positions with errors smaller than 0.05 arcsec.

041.019 **Weights of star positions in meridian circle catalogues.**
R. Bien.
Astron. Astrophys., Vol. 188, No. 1, p. 225 – 232 (1987).

The author proposes an application of the method of maximum likelihood to the problem of weighting the positions of meridian circle catalogues. For instance, weights play an important role for the individual improvement of the Fourth Fundamental Catalogue. The method is described and compared with the conventional procedure. Experiments with simulated data are analyzed, and applications to modern catalogues are discussed. It turns out that the new method is superior to the existing procedure. In particular, the weights are less uncertain, and the individual errors of the fundamental stars can be taken into account more precisely.

041.020 **Right ascensions of the Sun, Mercury and Venus observed at the transit instrument in Nikolaev in 1985.**
G. M. Petrov, R. T. Fedorova, P. N. Fedorov.
Glav. Astron. Obs. Akad. Nauk SSSR, Leningrad, 9 pp. (1987). In Russian. Abstr. in Ref. Zh., 51. Astron., 12.51.83 (1987).

041.021 **High precision astrometry via VLBI.**
G.–q. Tang.
Mitt. Geod. Inst. Rheinischen Friedrich–Wilhelms–Univ. Bonn, Nr. 71, p. 97 – 102 (1987). – See Abstr. 012.066.

041.022 **Recent radiometric observations of radiostars from the Hipparcos Input Catalogue.**
J. M. Paredes, R. Estalella, A. Rius.
Mitt. Geod. Inst. Rheinischen Friedrich–Wilhelms–Univ. Bonn, Nr. 71, p. 107 – 113 (1987). – See Abstr. 012.066.

041.023 **Systematic errors in visual and photographic latitude star catalogues.**
S. N. Sadžakov.
Astron. Nachr., Vol. 308, No. 6, p. 363 – 374 (1987).

The point discussed is the importance of investigation of the systematic errors of $\Delta\delta_\alpha$ and $\Delta\mu_\alpha$ types and their effects on stellar positions as obtained from visual and photographic observations.

041.024 **Observations of radio stars and radio reference stars with the eight–inch transit circle at Flagstaff.**
S. E. Browne.
Publ. Astron. Soc. Pac., Vol. 99, No. 621, p. 1146 – 1147 (1987). Abstract. – See Abstr. 010.281.

Catalogues of coordinates of reference faint, bright and double stars south of –47° and corrections to FK4 positions as observed by the Pulkovo astronomers with the meridian circle at the Cerro–Calan Observatory (Chile) in 1963 – 1968.
See Abstr. 002.011.

A catalogue of declinations of 3554 stars in the FK4 system for the observing epoch and equinox 1950.0.
See Abstr. 002.012.

On the general catalogue of fundamental faint stars (FKSZ).
See Abstr. 002.021.

A catalog of precise reference star positions for the astrometry network of the international comet P/Halley campaign.
See Abstr. 002.032.

The guide star catalog. II. Astrometric and photometric algorithms and precision.
See Abstr. 002.044.

Computerization of the *Cape Photographic Durchmusterung*.
See Abstr. 002.052.

A revised machine version of the *General Catalogue of 33342 stars for the epoch 1950 (GC)*.
See Abstr. 002.053.

New and revised catalogs available from the Astronomical Data Center.
See Abstr. 002.058.

Current status of the Second Cape Photographic Catalogue.
See Abstr. 002.061.

Tokyo PMC catalog 85: catalog of positions of 1007 stars observed in 1985 with Tokyo Photoelectric Meridian Circle.
See Abstr. 002.066.

The formation of an astrometric data bank in the framework of the complex programme of studying the main meridional section of the Galaxy (MEGA).
See Abstr. 002.072.

A list of reference stars in 252 areas with extragalactic radio sources.
See Abstr. 002.078.

On a version of star catalogues for photographic astrometry.
See Abstr. 002.097.

Itogi Nauki i Tekhniki. Seriya Astronomiya. Tom 30. New methods for construction of coordinate systems.
See Abstr. 003.008.

On the solar model and the precession of the equinoxes in the Alphonsine zij and its Arabic sources.
See Abstr. 004.096.

Astrometrical investigations at the Sternberg State Astronomical Institute.
See Abstr. 009.005.

Optical astrometry in the UK.
See Abstr. 013.046.

Astrographic catalogues of British observatories.
See Abstr. 013.058.

European VLBI for geodesy and astrometry: goals and objectives.
See Abstr. 013.075.

On thermal instrumental errors in astrometric observations.
See Abstr. 032.050.

The multichannel astrometric photometer and atmospheric limitations in the measurement of relative positions.
See Abstr. 034.015.

Three-stage photoelectric unit for measuring the moments of meridian passages.
See Abstr. 034.044.

Determination of stellar positions and ERP with a stellar interferometer. II. Geometric principles.
See Abstr. 034.081.

Programme of observations on the circumzenithal for the Geodynamic Observatory "Plana".
See Abstr. 034.189.

The ground segment's vital role in the Hipparcos scientific mission.
See Abstr. 036.011.

Statistical problems about the use of the ordinary least-squares method in astrometry. Application to the Paris-Astrolabe data.
See Abstr. 036.021.

Astrometrie mit einem Mikrodensitometer PDS 1010 A.
See Abstr. 036.061.

A program for a new reduction of plates of the Astrographic Catalogue.
See Abstr. 036.152.

The inertial coordinate system based on a CCD (V) experiment for measuring the relative positions of galaxies with respect to reference stars.
See Abstr. 036.153.

Astrometric use of the second short exposition of Tautenburg pilot program plates.
See Abstr. 036.222.

Construction of fundamental and reference systems.
See Abstr. 043.003.

Fundamental catalogues.
See Abstr. 043.004.

Construction of the system of positions and proper motions in the FK5.
See Abstr. 043.005.

The connection of the HIPPARCOS reference system to extragalactic objects by photographic astrometry.
See Abstr. 043.007.

Contributions to the conventional inertial system by VLBI and by the Tautenburg Schmidt-telescope.
See Abstr. 043.008.

Hipparcos. The Input Catalogue.
See Abstr. 051.043.

On some conditions of meridian observations of the Sun at the High-Altitude Station near Kislovodsk.
See Abstr. 082.061.

Radar astrometry of near-earth asteroids.
See Abstr. 098.020.

Catalogue of astrometric observations of comet P/Halley at its apparition 1909 – 1911.
See Abstr. 103.436.

Systematic and external errors of trigonometric parallaxes.
See Abstr. 111.001.

Lick Northern Proper Motion program. I. Goals, organization, and methods.
See Abstr. 111.004.

Lick Northern Proper Motion program. II. Solar motion and galactic rotation.
See Abstr. 111.005.

Vitesses radiales pour le programme HIPPARCOS. II.
See Abstr. 111.015.

The Tautenburg field of the programme of studying the main meridional section of the Galaxy.
See Abstr. 111.016.

An astrometric search for a stellar companion to the Sun.
See Abstr. 111.020.

Discussion of trigonometric parallaxes determined along right ascension and declination.
See Abstr. 111.021.

UBVRI photometry of FKSZ stars. I.
See Abstr. 113.006.

New carbon stars in selected regions of the Milky Way.
See Abstr. 114.004.

Programme Hipparcos: courbes de lumière.
See Abstr. 122.006.

Radio positions and optical identifications for a complete sample of southern flat-spectrum radio sources – I. Region 06^h to 18^h.
See Abstr. 141.003.

Catalogue of the coordinates of 3600 stars in the globular cluster M15 for the period 1896 – 1910.
See Abstr. 154.086.

042 Celestial Mechanics, Figures of Celestial Bodies

042.001 Collisions of two solid bodies in Keplerian orbits; new orbits and systematic effects.
S. Clairemidi, F. Depasse.
Earth, Moon, Planets, Vol. 39, No. 1, p. 37 – 50 (1987).

It has been shown in various papers dealing with systems of colliding bodies in a Keplerian field that the dynamical evolution does not depend only on the initial orbital conditions. This is a consequence of the wide range of orbits generated by the collision process. From the study of a few pairs of orbits the authors examine the factors which produce that variety of orbits, and search for systematic effects. The roles of the positions along the orbits, of inelasticity, of size, of mass and of relative inclination are emphasized.

042.002 On the restricted circular three–charged–body problem.
D. D. Dionysiou, D. A. Vaiopoulos.
Astrophys. Space Sci., Vol. 135, No. 2, p. 253 – 260 (1987).

Two–charged bodies M_1 and M_2 revolve round their centre of mass in circular orbits under Newton's inverse–square law and the so similar Coulomb's law. A third–charged–body M, without mass and charge (i.e., such that it is attracted or repulsed by M_1 and M_2, but does not influence their motion), moves in their field. The existence and location of the collinear and equilateral Lagrangian points or solutions is discussed and the interpretation of them is given.

042.003 The linear stability of libration points of the photogravitational restricted three–body problem when the smaller primary is an oblate spheroid.
R. K. Sharma.
Astrophys. Space Sci., Vol. 135, No. 2, p. 271 – 281 (1987).

This paper deals with the stationary solutions of the planar restricted three–body problem when the more massive primary is a source of radiation and the smaller primary is an oblate spheroid with its equatorial plane coincident with the plane of motion. The collinear equilibria have conditional retrograde elliptical periodic orbits around them in the linear sense, while the triangular points have long– or short–periodic retrograde elliptical orbits for the mass parameter $0 \leqslant \mu < \mu_{crit}$, the critical mass parameter, which decreases with the increase in oblateness and radiation force. Through special choice of initial conditions, retrograde elliptical periodic orbits exist for the case $\mu = \mu_{crit}$, whose eccentricity increases with oblateness and decreases with radiation force for non–zero oblateness.

042.004 Chaos in a quartic dynamical model.
N. Caranicolas, C. Vozikis.
Celest. Mech., Vol. 40, No. 1, p. 35 – 49 (1987).

The authors study the orbital characteristics of a time independent, two dimensional quartic dynamical model with two exact periodic orbits that displays always closed zero velocity curves. It is shown that the stability of the periodic orbits depends on the value of the coupling parameter α. Computer calculations suggest that the degree of stochasticity is small for the values of α in the range $1 < \alpha < 3$ while it grows rapidly when $\alpha > 3$. The authors also compute the Lyapunov characteristic exponents for different values of the coupling parameter.

042.005 Exponential instability of collision orbit in the anisotropic Kepler problem.
H. Yoshida.
Celest. Mech., Vol. 40, No. 1, p. 51 – 66 (1987).

The straight–line collision solution in the anisotropic Kepler problem is extended to a periodic solution by means of Sundman's analytic continuation. It is shown that this collision periodic solution is always exponentially unstable.

042.006 On Goudas' surfaces in the magnetic–binary problem.
V. Banfi.
Celest. Mech., Vol. 40, No. 1, p. 67 – 76 (1987).

The containment property of Goudas' surfaces has been studied in this paper. Physical implications of this property are discussed. A general relationship is obtained, which connects the sizes of those surfaces, the magnetic moments of two stars and their mean motion.

042.007 The two–body problem in the (truncated) PPN theory.
M. Soffel, H. Ruder, M. Schneider.
Celest. Mech., Vol. 40, No. 1, p. 77 – 85 (1987).

The solution of the two–body problem in the (truncated) PPN theory is presented. It is given in two different analytical forms (the Wagoner–Will and Brumberg representation) and by the method of osculating elements.

042.008 Long–term numerical integrations and synthetic theories for the motion of the outer planets.
M. Carpino, A. Milani, A. M. Nobili.
Astron. Astrophys., Vol. 181, No. 1, p. 182 – 194 (1987).

The LONGSTOP research project has been set up to investigate the dynamical stability of the Solar System over timescales comparable to its lifetime by means of numerical integrations. This paper contains a synthetic secular perturbation theory for the nonsingular elements h, k (related to the eccentricities) and p, q (related to the inclinations) as deduced from the LONGSTOP 1A integration of the outer planets for 9.3 million years. The results are compared with the existing analytical theories, showing fair agreement for the main spectral lines and a significant discrepancy for the smaller combination lines. The methods devised and tested in this work can be applied to process the output of longer numerical integrations such as the 100 million year LONGSTOP 1B.

042.009 Restrictions on the motion in the general four–body problem.
R. Sergysels, A. Loks.
Astron. Astrophys., Vol. 182, No. 1, p. 163 – 166 (1987).

The authors used the integrals of energy and angular momentum as was done by Marchal and Saari in 1975 for the general three–body problem and applied them to the general four–body problem. This leads to the equation of surfaces in three-dimensional space, limiting regions where motion cannot occur. Restrictions on the angles between the orbital planes and the invariable plane are also found.

042.010 On the confinement of planetary arcs.
D. N. C. Lin, J. C. B. Papaloizou, S. P. Ruden.
Mon. Not. R. Astron. Soc., Vol. 227, No. 1, p. 75 – 95 (1987). = Lick Obs. Contrib. No. 443.

The authors adopt the conventional model for the azimuthal confinement of incomplete planetary rings or arcs to be due to trapping of particles around potential equilibria located at the corotation resonances of hypothetical satellites. Although particle orbits around these equilibria are dynamically stable, they are secularly unstable because the gravitational potential attains local maxima at the equilibria. Energy dissipation due to particle collisions induces a general tendency towards spreading in the radial and azimuthal directions. The authors show that arc confinement is possible if particles near the corotation resonance are shepherded by a discrete Lindblad resonance of another distant satellite. They discuss the possibility that the non–axisymmetric potentials, required for arc confinement, may be generated by modes of planetary non–radial oscillation as well as by satellites.

042.011 **Expansion of the disturbing force–function for the study of high–eccentricity librations.**
S. Ferraz–Mello.
Astron. Astrophys., Vol. 183, No. 2, p. 397 – 402 (1987).
The function of the force acting on a resonant asteroid, due to Jupiter, is averaged and expanded about a high–eccentricity libration center. This expansion is suitable to study the motion of high–eccentricity librators as the Hildas (3:2, $e_{av} = 0.12 - 0.24$) and the Griquas (2:1, $e_{av} = 0.36 - 0.48$) as well as some planetary satellites in similar conditions, as Hyperion (4:3, $e_{av} = 0.1044$). The technique used in this expansion follows that introduced by Woltjer in 1928, to study the motion of Hyperion. The actual application to the orbit of the mentioned celestial bodies gives better results than those obtained when using classical expansions.

042.012 **Analysis of the dynamics of a free Poincaré planet model.**
M. Shen, T. Huang, C. Zhang.
J. Nanjing Univ., Vol. 23, No. 2, p. 271 – 289 (1987).
The rotation of a planet with an ideal liquid core and a rigid mantle is discussed. Three families of equilibrium solutions are obtained and discussed in detail. The motions near the equilibriums are explored. It is shown that the angular momentum axis tends to the figure axis because of the existence of a liquid core, and the frequencies of the free wobble are determined by the size of the core and the dynamical flattening.

042.013 **The two–body problem in general relativity.**
T. Damour, G. Schäfer.
C. R. Acad. Sci., Sér. II, Tome 305, No. 10, p. 839 – 842 (1987). In French.
The authors study the dynamics of a system of two comparable masses, at the second post–Newtonian approximation of general relativity (i.e. at the last level where the system is still conservative). They obtain explicit expressions of the radial period and secular periastron advance in terms of energy and angular momentum. The authors then deduce the expression of the periastron advance in terms of the masses and directly observable quantities. The latter result is of observational significance in view of the high precision now obtained in measuring the binary pulsar PSR 1913 + 16.

042.014 **Integrable cases of the Hamilton–Jacobi equation and the straight–line restricted three–body problem with variable mass.**
A. A. Bekov.
Astron. Zh., Tom 64, Vyp. 4, p. 850 – 859 (1987). In Russian. English translation in Sov. Astron., Vol. 31, No. 4.
New integrable cases for the Hamilton–Jacobi equation have been found. As an application, the straight–line restricted three–body problem with variable mass is considered. The exact solutions of the problem in which the motion of two finite bodies is determined by the classical non–stationary Gylden–Mestschersky problem, have been obtained.

042.015 **The plane unrestricted three–body problem.**
V. N. Tkhaj.
Astron. Zh., Tom 64, Vyp. 4, p. 860 – 864 (1987). In Russian. English translation in Sov. Astron., Vol. 31, No. 4.
The Lagrangian coordinates of the unrestricted three–body problem with forces proportional to r_{ij}^{α} (r_{ij} are mutual distances between the bodies and α is an arbitrary real number not equal to -1) are chosen in such a way to make one of the equations of motion to be a Lagrange–Jacobi one. Cases of the unlimitedly increasing polar moment of inertia are specially considered. The paper contains the geometrical interpretation for the plane problem.

042.016 **Stability of the coplanar libration points in the photo-gravitational three–body problem. II.**
A. T. Tureshbaev.
Sov. Astron. Lett., Vol. 12, No. 5, p. 303 – 304 (1987). English translation of 42.042.013.

042.017 **New families of simple symmetric motions in the equatorial magnetic–binary problem.**
T. J. Kalvouridis.
Astrophys. Space Sci., Vol. 136, No. 1, p. 21 – 41 (1987).
The author presents 14 new families of simple symmetric motions in the equatorial magnetic–binary problem. The study of the families found so far leads to some interesting remarks which characterize the totality of the solutions provided that the hypothesis of Poincaré (1892) holds in this case.

042.018 **On a particular solution of the averaged elliptical Hill problem.**
B. N. Khaimov.
Red. Zh. Izv. Akad. Nauk TadzhSSR. Otd–nie Fiz.–Mat., Khim. Geol. Nauk. Dushanbe, 7 pp. (1987). In Russian. Abstr. in Ref. Zh., 51. Astron., 7.51.73 (1987).

042.019 **Periodic orbits in the neighbourhood of a collinear libration point in the planar elliptical three–body problem.**
V. P. Evteev.
Dokl. Akad. Nauk TadzhSSR, Tom 29, No. 9, p. 520 – 522 (1986). In Russian. Abstr. in Ref. Zh., 51. Astron., 7.51.74 (1987).

042.020 **Coordinate invariance of the Schwarzschild shift of the pericenter.**
A. V. Krivov.
Vestn. Leningr. Univ., Mat. Mekh. Astron., No. 1, p. 102 – 108 (1987). In Russian. Abstr. in Ref. Zh., 51. Astron., 7.51.81 (1987).

042.021 **Symmetry of trajectories relative to the straight line in the restricted three–body problem.**
V. P. Semenko.
Tr. 19 Chtenij, posvyashch. nauchn. naslediya i razvitiyu idej K. Eh. Tsiolkovskogo, Kaluga 13 – 16 sent. 1984. Sekts. mekh. kosm. poleta, Moskva, p. 63 – 67 (1986). In Russian. Abstr. in Ref. Zh., 62. Issled. Kosm. Prostranstva, 7.62.274 (1987).

042.022 **On a geometric method for determining the non–perturbed orbit from observations using the transformation groups.**
V. B. Titov.
Kinematika Fiz. Nebesn. Tel, Tom 3, No. 4, p. 26 – 29 (1987). In Russian. English translation in Kinematics Phys. Celest. Bodies.
The transformation groups of phase trajectories of three–dimensional two–body problem are considered. These groups are used for development of geometric ("angle only" observations) method of orbit determination. The equations for orbit plane determination are given.

042.023 **Rotationally–perturbed orbital elements of close binary systems.**
B. Zafiropoulos, F. Zafiropoulos.
Astrophys. Space Sci., Vol. 136, No. 2, p. 211 – 219 (1987).
The aim of this investigation is to present the secular and periodic perturbations of the six orbital elements of a close binary system due to rotational distortion. The authors consider very small inclinations ι of the orbital plane of the system, whereas the eccentricity of the orbit may assume any value between $0 < e < 1$. The final formulae for the various elements have been expressed by means of the unperturbed true anomaly measured from the ascending node.

042.024 **Theorem on the nonexistence of additional analytical first integrals in the problem of motion of a satellite of a triaxial planet.**
P. F. Sevryukov.
Moskovskij Gosudarstvennyj Universitet, Moskva, 24 pp. (1987). In Russian. Abstr. in Ref. Zh., 51. Astron., 8.51.52 (1987).

042.025   **Hyperresonances of a planetary system.**
V. V. Petrunenko.
Belorus. Politekh. Inst. Minsk, 8 pp. (1987). In Russian. Abstr. in Ref. Zh., 51. Astron., 8.51.56 (1987).

042.026   **Translational and rotational motion of a test rigid body in relativistic celestial mechanics.**
N. N. Vasil'ev, A. V. Voinov.
Kinematika Fiz. Nebesn. Tel, Tom 3, No. 5, p. 18 – 24 (1987). In Russian. English translation in Kinematics Phys. Celest. Bodies.

Equations of translational and rotational motion of a test body are obtained in terms of variables closely related with astronomically observable quantities. These equations are derived by two different methods. The first one is based on the equivalence principle and the equation of the second geodetic deviation. The second method involves the multipole formalism expansions.

042.027   **Evolution of asteroid orbits in the case of second– and third–order commensurabilities.**
I. A. Gerasimov.
Sov. Astron., Vol. 30, No. 6, p. 717 – 720 (1987). English translation of 42.042.085.

042.028   **Stability of Lagrangian points in the restricted, photogravitational three–body problem.**
L. G. Luk'yanov.
Sov. Astron., Vol. 30, No. 6, p. 720 – 724 (1987). English translation of 42.042.086.

042.029   **An application of the scaled Runge–Kutta algorithms to some problems of celestial mechanics.**
G. Papageorgiou, T. Kalvouridis, T. Simos.
Astrophys. Space Sci., Vol. 137, No. 1, p. 129 – 138 (1987).

New Runge–Kutta algorithms are applied for determining the solution of a system of ordinary differential equations at any point within a given integrating step. In this paper the authors propose the application of these new algorithms in order to determine, with the smallest possible cost, the exact point of intersection of a symmetric orbit, with the axis or plane of symmetry, which appear in various problems of celestial mechanics.

042.030   **Perturbations in close binary systems produced by the tides lagging in latitude.**
B. Zafiropoulos.
Astrophys. Space Sci., Vol. 137, No. 1, p. 139 – 149 (1987).

The aim of this investigation is to present the periodic and secular perturbations of the orbital elements of close binary systems due to tidal lag in latitude. The variational equations of the problem of plane motion are set up in terms of the rectangular components R, S, and W of the disturbing accelerations. These equations are highly nonlinear with respect to the orbital elements and the author presents analytic approximations to the effects produced by the perturbing acceleration due to dynamical tides lagging in latitude. The perturbed elements of the orbit have been expressed by means of Hansen coefficients in the compact form of summations.

042.031   **Particular solutions of the problem of three bodies possessing two mutually perpendicular planes of symmetries.**
L. Yu. Khorseva.
Astron. Zh., Tom 64, Vyp. 5, p. 1120 – 1123 (1987). In Russian. English translation in Sov. Astron., Vol. 31, No. 5.

Conditions of existence of rectilinear motions of mass centres of three bodies, which have two mutually perpendicular planes of dynamical and geometric symmetries are obtained using the accurate value of the force function in integral form. Regular rotations of the bodies around the axes normal to the orbital plane accompany these motions.

042.032   **On the solution of the problem of motion of a point mass in the gravitational field of a rigid body.**
A. A. Kochiev.
Astron. Zh., Tom 64, Vyp. 5, p. 1124 – 1128 (1987). In Russian. English translation in Sov. Astron., Vol. 31, No. 5.

A special force function is constructed which, being added to the force function of an arbitrary axisymmetrical body, leads to integrable equations of motion of the point mass. A new integrable case has been found for the equations of motion of the point mass in one of the three mutually perpendicular planes of geometrical and dynamical symmetry of a body with an arbitrary axisymmetrical part and with a specially chosen longitudinal one.

042.033   **Bahnbestimmung mit dem Heimcomputer.**
H.–G. Reimann.
Sterne, 63. Band, Heft 3, p. 145 – 157 (1987).

042.034   **An approximate solution to the energy change of a circular binary in a parabolic three–body encounter.**
T.–Y. Huang, M. J. Valtonen.
Mon. Not. R. Astron. Soc., Vol. 229, No. 2, p. 333 – 344 (1987).

The authors study parabolic encounters between a circular binary and a third body of similar mass. Numerical orbit calculations, about 18000 in total, are compared with predictions from perturbation theory. Even though the perturbation theory is not strictly applicable near the stability boundary, the authors are able to use the functional forms from the theory and to fit the unknown coefficients from the numerical data. In this way analytic functions are constructed, valid near the stability boundary, which give the relative energy change of the binary $\Delta E/E$ and its final eccentricity e as a function of the seven initial parameters of the problem. Applications to the stability of hierarchical triple systems are discussed.

042.035   **Rotational dynamics of irregularly shaped natural satellites.**
J. Wisdom.
Astron. J., Vol. 94, No. 5, p. 1350 – 1360 (1987).

All irregularly shaped natural satellites must tumble chaotically before being captured into synchronous rotation.

042.036   **On the representation of stationary time series by the sum of harmonic terms.**
M. Yu. Klokacheva.
Byull. Inst. Teor. Astron., Tom 15, No. 9, p. 524 – 533 (1986). In Russian.

A method concerned with searching for hidden periodicities and the representation of stationary time series by the sum of harmonic terms is considered. The method is based on spectral analysis. An example is given for the approximation of the Moon's coordinates by the sum of sine terms.

042.037   **A method for constructing conditionally–periodic solutions of resonance problems of celestial mechanics in different systems of elements.**
S. G. Zhuravlev, A. A. Zlenko.
Mosk. avtomobil.–dorozhn. Inst., Moskva, 59 pp. (1987). In Russian. Abstr. in Ref. Zh., 51. Astron., 10.51.90 (1987).

042.038   **Application of an averaging method to a solution of the perturbation problem of R. B. Barrar.**
V. A. Padalka.
Frunz. Politekh. Inst., Frunze, 4 pp. (1987). In Russian. Abstr. in Ref. Zh., 51. Astron., 10.51.91 (1987).

042.039   **On a method for solving the problem of perturbed motion of a distant satellite.**
V. A. Padalka.
Frunz. Politekh. Inst., Frunze, 4 pp. (1987). In Russian. Abstr. in Ref. Zh., 51. Astron., 10.51.92 (1987).

042.040 **On a method for the approximate representation of coordinates of a distant satellite.**
V. A. Padalka.
Frunz. Politekh. Inst., Frunze, 4 pp. (1987). In Russian. Abstr. in Ref. Zh., 51. Astron., 10.51.93 (1987).

042.041 **Stability regions in the elliptic restricted problem of the three bodies.**
J. G. Ries.
Bull. Am. Astron. Soc., Vol. 19, No. 2, p. 751 – 752 (1987). Abstract. – See Abstr. 010.061.

042.042 **Efficient inverse solution of Kepler's equation.**
F. W. Boltz.
J. Astronaut. Sci., Vol. 34, No. 4, p. 431 – 443 (1986). Abstr. in Phys. Abstr., Vol. 90, No. 1309, Entry 88085 (1987).

042.043 **On analogs of Jacobi's integral in the restricted problem of three bodies of variable mass.**
A. A. Bekov.
Tr. Astrofiz. Inst. Alma–Ata, Tom 47, p. 12 – 29 (1987). In Russian.

042.044 **On the generalized problem of two centers with variable center distance.**
A. A. Bekov.
Tr. Astrofiz. Inst. Alma–Ata, Tom 47, p. 30 – 41 (1987). In Russian.

042.045 **Motion of a mass point around a gravitating body of constant form; variable dimensions and mass. II.**
M. D. Minglibaev.
Tr. Astrofiz. Inst. Alma–Ata, Tom 47, p. 42 – 47 (1987). In Russian.

042.046 **Energy variation of a gravitating system with variable composition and mass.**
M. D. Minglibaev.
Tr. Astrofiz. Inst. Alma–Ata, Tom 47, p. 48 – 50 (1987). In Russian.

042.047 **The system of osculating elements in the two–body problem with variable Lapin masses.**
R. K. Mukhametkalieva.
Tr. Astrofiz. Inst. Alma–Ata, Tom 47, p. 60 – 66 (1987). In Russian.

042.048 **The Shirokov effect in equatorial circular orbits in the gravitational field of the oblate Earth.**
V. I. Khlebnikov.
Gravitasiya i teor. otnositel'nosti, Kazan', No. 24, p. 93 – 99 (1987). In Russian. Abstr. in Ref. Zh., 51. Astron. 11.51.133 (1987).

042.049 **Motion of a test particle in the gravitational field of the Sun taking into account the modification of the Newton potential.**
A. D. Beriev.
Vses. Zaoch. Mashinostroit. Inst. Moskva, 5 pp. (1987). In Russian. Abstr. in Ref. Zh., 51. Astron. 11.51.135 (1987).

042.050 **Aspects of the phase–portrait of resonant problems.**
W. Sessin.
Rev. Mex. Astron. Astrofis., Vol. 14, No. 2, p. 631 – 635 (1987). – See Abstr. 012.042.
The paper analyses some aspects of the phase–portrait of resonant systems. The method of Hori and the extended Delaunay method are used in this study.

042.051 **Some instabilities in the 3/1 resonant problem.**
T. Yokoyama, M. Sato, J. Koiller, J. M. Balthazar.
Rev. Mex. Astron. Astrofis., Vol. 14, No. 2, p. 636 – 638 (1987). – See Abstr. 012.042.
By using the Laplace–Lagrange method, a very simple criterion of instability is obtained for the 3/1 planar resonance problem.

042.052 **Sôbre algumas técnicas de perturbação utilizadas no problema ressonante 3/1.**
J. M. Balthazar, J. L. Sagnier, S. Ferraz Mello, J. Koiller, T. Yokoyama.
Rev. Mex. Astron. Astrofis., Vol. 14, No. 2, p. 639 – 640 (1987). – See Abstr. 012.042.
This work concerns with the study of a particular 3/1 resonant problem for which the authors have determined formal solutions according to the model belonging to the domain of the restricted elliptic problem of three bodies.

042.053 **A first order analytical elliptic planar model for 3/1 resonance problem.**
M. Sato, T. Yokoyama.
Rev. Mex. Astron. Astrofis., Vol. 14, No. 2, p. 641 (1987). Abstract. – See Abstr. 012.042.

042.054 **Relative motion near the triangular libration points in the Earth–Moon system.**
C. Simó, R. Martinez, G. Gómez, J. Llibre, W. Flury, J. Rodriguez–Canabal.
ESA Spec. Publ., ESA SP–273, p. 157 – 170 (1987). – See Abstr. 012.044.
As the triangular libration points in the Earth–Moon system are of potential interest for optical interferometry a study has been conducted to analyse their dynamical property adopting realistic models. First, quasi–periodic orbits have been identified, and second, their stability investigated. It turns out that relative motion leads to baseline variation of 1 – 20 cm/s, which may be unacceptably large for optical interferometry.

042.055 **On orbitally stable resonance surfaces in the planetary problem of many bodies.**
V. N. Shinkin.
Tsentral'nyj Nauchno–Issledovatel'skij Institut TEhIpriborostr., Moskva, 48 pp. (1987). In Russian. Abstr. in Ref. Zh., 51. Astron., 12.51.47 (1987).

042.056 **The first integrals and orbit equation for the Kepler problem with drag.**
P. G. L. Leach.
J. Phys. A, Vol. 20, No. 8, p. 1997 – 2002 (1987). Abstr. in Phys. Abstr., Vol. 90, No. 1313, Entry 115217 (1987).

042.057 **Relaxation–chaos phenomena in celestial mechanics. I. On wisdom's model for the 3/1 Kirkwood gap.**
J. Koiller, J. M. Balthazar, T. Yokoyama.
Physica D, Vol. 26D, No. 1 – 3, p. 85 – 122 (1987). Abstr. in Phys. Abstr., Vol. 90, No. 1313, Entry 115218 (1987).

042.058 **The Keplerian metric of space–time.**
M. D. Leko.
Z. Angew. Math. Mech., Vol. 67, No. 5, p. T426 – T427 (1987). Abstr. in Phys. Abstr., Vol. 90, No. 1314, Entry 120258 (1987).

042.059 **On the stability of the coplanar libration points in the restricted photogravitational three–body problem.**
L. G. Luk'yanov.
Astron. Zh., Tom 64, Vyp. 6, p. 1291 – 1299 (1987). In Russian. English translation in Sov. Astron., Vol. 31, No. 6.
It is shown that the libration points L_8 und L_9 are always unstable. For the libration points L_6 and L_7 the stability regions are obtained for any values of the three parameters: of the ratio of the masses of the primaries and of the two reduction coefficients, describing the light pressure of the primaries.

042.060   Periodic rotational motion of a rigid body near the points of libration of the circular restricted three–body problem.
L. Yu. Khorseva.
Astron. Zh., Tom 64, Vyp. 6, p. 1320 – 1323 (1987). In Russian. English translation in Sov. Astron., Vol. 31, No. 6.

Periodic solutions of the problem of rotational motion of a rigid body possessing three mutually perpendicular planes of form and structure symmetries are constructed by the Poincaré small–parameter method. The small parameter is the dynamical oblateness of the body. The main bodies are axially symmetric ones having the plane of symmetry normal to the axis of symmetry.

042.061   On the exact solution of the N–body problem in the domain of large energies.
L. L. Sokolov, K. V. Kholshevnikov.
Tr. Astron. Obs., Leningrad, Tom 41, p. 175 – 193 (1987). Uch. Zap. LGU, No. 420, Ser. Mat. Nauk, Vyp. 63. In Russian.

The N–body problem in the domain of hyperbolic and hyperbolic–elliptic motions is examined under the condition that close approaches between corresponding points and point pairs are absent. The equations of motion in this domain are integrated exactly: the existence of a solution in the infinite time interval is proved; the qualitative behaviour of the phase flow is established; the sequence convergent to the solution uniformly with respect to the time–axis multiplied by an initial data domain is constructed.

042.062   The libration points in the restricted problem of three bodies with variable mass.
A. A. Bekov.
Astron. Tsirk., No. 1468, p. 3 – 5 (1986). In Russian.

New (complanar) libration points in the restricted three–body problem with variable masses, where masses of primary bodies change isotropically according the united Meshcherskij's law, are found.

042.063   The Ideal Resonance Problem. A comparison of two formal solutions. III.
A. H. Jupp.
Celest. Mech., Vol. 40, No. 2, p. 87 – 93 (1987).

This is the last article in a series of the same title. The two formal solutions of the Ideal Resonance Problem, developed respectively by Garfinkel and Jupp, were compared and contrasted in the earlier papers. It was stated there that the principal shortcoming of Jupp's analytical solution was the occurrence of a singularity at the separatrix. The purpose of this contribution is to demonstrate how this singularity may readily be removed. Accordingly, modified solutions are presented for the libration and circulation regions.

042.064   Arbitrary order numerical solutions conserving the Jacobi constant in the motion near the equilibrium points.
A. Marciniak.
Celest. Mech., Vol. 40, No. 2, p. 95 – 110 (1987).

In the paper a modification of the polynomial extrapolation for solving the problem of motion nearby the equilibrium points is presented. It appears that the modification yields a better approximation of the exact solution than the convential polynomial extrapolation and other methods. Moreover, the modification conserves the Jacobi constant of motion. Computer examples for orbits nearby the equilibrium points of the Sun–Jupiter system are given.

042.065   Intrinsic stability of periodic orbits.
M. E. Hough.
Celest. Mech., Vol. 40, No. 2, p. 111 – 153 (1987).

Families of orbits of a conservative, two degree–of–freedom system are represented by an unsteady velocity field with components $u(x,y,t)$ and $v(x,y,t)$. It is shown that a necessary condition for stable periodic orbits is satisfied when the orbit–averaged divergence is zero, which results in bounded normal variations. A sufficient condition for stability is derived from the

requirement that tangential variations do not exhibit secular growth. In a steady, divergence–free field, velocity component functions $u(x,y)$ and $v(x,y)$ may be continued analytically from any initial condition, except when velocity is parallel to ∇U or at equilibria. In an unsteady field, the orbit–averaged divergence is zero when the vorticity function is periodic. When such a field exists, initial conditions for stable periodic orbits may be determined analytically.

042.066   On the stability of the triangular libration points for the photogravitational circular restricted problem of three bodies when both of the attracting bodies are radiating as well.
V. Kumar, R. K. Choudhry.
Celest. Mech., Vol. 40, No. 2, p. 155 – 170 (1987).

The stability of triangular libration points when both the attracting bodies are radiating as well has been investigated under the non–resonance cases. It is found that except for some cases for all values of the radiation reduction factors and for all values of $\mu < 0.0285954...$, the motion will be stable.

042.067   Symmetric motions in the Equatorial Magnetic–Binary problem.
T. J. Kalvouridis, A. G. Mavraganis.
Celest. Mech., Vol. 40, No. 2, p. 177 – 196 (1987).

Classical numerical techniques are applied to find families of symmetric periodic orbits in the Equatorial Magnetic–Binary Problem. Stability for each orbit is also studied by means of variational methods. Finally an example in a concrete system is given to verify the procedure proposed.

042.068   From rotations and inclinations to zero configurational velocity surfaces. II. The best possible configurational velocity surfaces.
D. G. Saari.
Celest. Mech., Vol. 40, Nos. 3 – 4, p. 197 – 223 (1987).

The best possible zero configurational velocity surfaces for the general N–body problem in three space are derived. The basic construction of these surfaces is described in detail for the three body problem and for other flat configurations. The construction for nonflat configurations is outlined.

042.069   Bifurcation of a central configuration.
K. R. Meyer.
Celest. Mech., Vol. 40, Nos. 3 – 4, p. 273 – 282 (1987).

This paper presents a relatively simple example of a bifurcation of a central configuration in the four body problem.

042.070   Second special case of the restricted N–body problem.
G. N. Doubochine (G. N. Duboshin).
Celest. Mech., Vol. 40, No. 3 – 4, p. 283 – 292 (1987). In French.

On envisage ici le problème du mouvement d'une masse passive sous l'action de n masses actives, qui restent toujours sur une droite qui tourne uniformement autour du leurs baricentre.

042.071   Application of the extended Delaunay method to the ideal resonance problem.
W. Sessin.
Celest. Mech., Vol. 40, Nos. 3 – 4, p. 293 – 301 (1987).

An application of the extended Delaunay method is made to the ideal resonance problem. The author shows how the theory of integration proposed in a preceding paper works in a simple problem, and discusses how to proceed in more complicated situations.

042.072   The solution of Kepler's equation, III.
J. M. A. Danby.
Celest. Mech., Vol. 40, Nos. 3 – 4, p. 303 – 312 (1987).

Recently proposed methods of iteration and initial guesses are discussed, including the method of Laguerre–Conway. Tactics for a more refined initial guess for use with universal variables over a small time interval are described.

042.073   **Generalized elliptic anomalies.**
J. M. Ferrándiz, S. Ferrer, M. L. Sein–Echaluce.
Celest. Mech., Vol. 40, Nos. 3 – 4, p. 315 – 328 (1987).
A two–parameter time transformation $dt = r^{3/2} (\alpha_0 + \alpha_1 r)^{-1/2} d\tau$ is proposed, where r is the radial distance. In Keplerian systems, the quadrature implied by the transformation may be carried out by elliptic functions. When $\alpha_0 = 0$, τ is the eccentric anomaly; if $\alpha_1 = 0$, then τ is the intermediate or elliptic anomaly. Considering several values of α_0 and α_1, numerical examples of the relation of the generalized elliptic anomaly τ with the classical and elliptic anomalies are given. Application of this transformation to some perturbed Kepler problems is briefly outlined.

042.074   **A cubic approximation for Kepler's equation.**
S. Mikkola.
Celest. Mech., Vol. 40, Nos. 3 – 4, p. 329 – 334 (1987).
The author derives a new method to obtain an approximate solution for Kepler's equation. By means of an auxiliary variable it is possible to obtain a starting approximation correct to about three figures. A high order iteration formula then corrects the solution to high precision at once. The method can be used for all orbit types, including hyperbolic. To obtain this solution the trigonometric or hyperbolic functions must be evaluated only once.

042.075   **Consecutive collision orbits in the limiting case $\mu = 0$ of the elliptic restricted problem.**
K. C. Howell.
Celest. Mech., Vol. 40, Nos. 3 – 4, p. 393 – 407 (1987).
A timing condition for consecutive collision orbits in the planar, circular three–body problem has been extended to the elliptic restricted problem for $\mu = 0$. The expression developed relates eccentric anomalies at the time of collision. Some families of solutions are presented.

042.076   **A comparison of numerical integration methods in the equatorial magnetic–binary problem.**
T. Kalvouridis, G. Papageorgiou, T. Simos.
Astrophys. Space Sci., Vol. 139, No. 1, p. 21 – 35 (1987).
The authors compare several Runge–Kutta type methods of fourth–order when they are applied in the Equatorial Magnetic–Binary Problem (EMBP) for which the analytical solutions are not known. The results of the comparison are presented with tables and diagrams, and the most efficient method is proposed according to the used criteria which are described.

042.077   **The motion of a satellite in a ring potential.**
B. Zafiropoulos.
Astrophys. Space Sci., Vol. 139, No. 2, p. 353 – 364 (1987).
This investigation presents the orbital elements of a satellite moving in a circular ring potential. The ring is considered to be of infinitesimal thickness and of unit radius. The components of the perturbing accelerations due to the ring potential have been substituted into the Gauss form of Lagrange's planetary equations to yield the first–order approximations. The elements of the orbit have been expressed by means of Hansen coefficients. The results include the effects produced by the 2nd, 4th, 6th, and 8th spherical harmonics.

042.078   **The solution of the N–body problem of celestial mechanics by means of the successive approximation method.**
R. Frischbier.
Z. Angew. Math. Mech., Vol. 67, No. 5, p. T419 – T420 (1987). In German. Abstr. in Phys. Abstr., Vol. 90, No. 1317, Entry 139893 (1987).

042.079   **The cometary cloud in the solar system and the Resibois–Prigogine singular invariants of motion.**
T. Y. Petrosky.
J. Stat. Phys., Vol. 48, No. 5 – 6, p. 1363 – 1372 (1987). Abstr. in Phys. Abstr., Vol. 91, No. 1319, Entry 4288 (1988).

042.080   **Asymptotic expansion of the general solution of the regularized Hill's problem.**
V. A. Kuz'minykh.
Sov. Astron., Vol. 31, No. 2, p. 221 – 223 (1987). English translation of 43.042.027.

042.081   **Libration points in the generalized Hill problem.**
B. N. Khaimov.
Dokl. Akad. Nauk TadzhSSR, Tom 29, No. 11, p. 659 – 661 (1986). In Russian. Abstr. in Ref. Zh., 51. Astron. 11.51.44 (1987).

042.082   **Several problems on the numerical integration of celestial orbits.**
L. Liu, X.–h. Liao.
Acta Astron. Sin., Vol. 28, No. 3, p. 215 – 225 (1987). In Chinese.

Orbits for amateurs with a microcomputer. Volume II.
See Abstr. 003.133.

Planetary theories in Sanskrit astronomical texts.
See Abstr. 004.092.

Celestial mechanics at the Sternberg State Astronomical Institute.
See Abstr. 009.006.

Astronomical computing: Do orbits change in 100 million years?
See Abstr. 014.020.

The Runge–Lenz vector and Einstein perihelion precession.
See Abstr. 014.033.

An expectation value formulation of the perturbed Kepler problem.
See Abstr. 014.053.

A new derivation of the areal velocity.
See Abstr. 014.061.

A numerical scheme for one–dimensional mechanical problems.
See Abstr. 021.026.

Multi–body force function: geodetic aspects of astrodynamics.
See Abstr. 052.013.

Rotational evolution of a symmetric gyrostat with visco–elastic bars around the center of mass in a circular orbit.
See Abstr. 052.027.

Note on Cid's radial intermediary and the method of averaging.
See Abstr. 052.028.

Radial, transverse and normal satellite position perturbations due to the geopotential.
See Abstr. 052.029.

The motion of a satellite in an axi–symmetric gravitational field.
See Abstr. 052.030.

Tables for the trigonometric series representations of the orbital inclination function.
See Abstr. 052.031.

Perturbations in the perigee distance due to atmospheric drag for artificial Earth satellites.
See Abstr. 052.032.

Maclaurin π–ellipsoid in parametrized post–Newtonian formalism.
See Abstr. 066.142.

Urey Prize Lecture: Chaotic dynamics in the solar system.
See Abstr. 091.039.

A numerical suimulation of planetary rings. II. Monte Carlo model.
See Abstr. 091.041.

Colombo's Top.
See Abstr. 091.049.

First order planetary perturbations with elliptic functions.
See Abstr. 091.050.

An investigation of the motions of the node and perihelion of Mercury.
See Abstr. 092.003.

On the stability of rotation of the Moon according to Cassini's laws.
See Abstr. 094.029.

Averaging the elliptic asteroidal problem near a first–order resonance.
See Abstr. 098.021.

A simple analytical model for the secular resonance v_6 in the asteroidal belt.
See Abstr. 098.087.

Corrections to the theory of the orbit of Saturn's satellite Hyperion.
See Abstr. 100.012.

First order perturbation in the Enceladus–Dione system.
See Abstr. 100.026.

A semi–analytical solution for the eccentricities and longitudes of the pericenter of the Uranian satellites.
See Abstr. 101.025.

Comparison of Bretagnon's VSOP 82 theory with observations of Neptune.
See Abstr. 101.034.

GUST 86. An analytical ephemeris of the Uranian satellites.
See Abstr. 101.044.

Baanelementen: een meteoorbaan om de zon.
See Abstr. 104.026.

The probability of falling of Apollo–Amor–type interplanetary bodies on the surface of planets.
See Abstr. 104.078.

Planetesimal rotations induced by collisions.
See Abstr. 107.012.

Perturbed orbital elements of close binary systems due to tidal lag in longitude.
See Abstr. 117.071.

Was dürfen Doppelsterne?
See Abstr. 118.030.

Systems of colliding bodies in a gravitational field: impact velocity, inelasticity and size distributions.
See Abstr. 151.004.

Violent relaxation and mixing in 1–D gravitational systems.
See Abstr. 151.086.

043 Astronomical Constants, Reference Systems

043.001   Tetrad description of astrometric reference systems.
Yu. P. Vyblyj.
Mater. 2 Konf. mol. uchenykh Inst. prikl. probl. mekh. i mat. Akad. Nauk USSR, L'vov, 1 – 3 Okt. 1985. Chast' 1. Inst. prikl. probl. mekh. i mat. Akad. Nauk USSR, L'vov, p. 29 – 33 (1987). In Russian. Abstr. in Ref. Zh., 51. Astron., 7.51.114 (1987).

043.002   Comparison of different terrestrial reference frame realizations based on artificial satellites laser ranging.
V. Ya. Cholij.
Kinematika Fiz. Nebesn. Tel, Tom 3, No. 4, p. 75 – 79 (1987). In Russian. English translation in Kinematics Phys. Celest. Bodies.
  This paper deals with the comparison of different realizations of satellite Terrestrial Reference Frame. It is shown that within the same processing method the coincidence of the results is good. The results may be used to construct the Combined Terrestrial Reference Frame.

043.003   Construction of fundamental and reference systems.
A. P. Gulyaev.
Itogi Nauki i Tekhniki. Seriya Astronomiya. New methods for construction of coordinate systems, Tom 30, p. 79 – 161 (1987). In Russian. – See Abstr. 003.008.
  The new fundamental catalogue FK5 is described; astrometrical foundations for the construction of coordinate systems are given; compiled catalogues are listed; the present–day approach to construction and compilation of future fundamental systems is outlined.

043.004   Fundamental catalogues.
W. Gliese.
Astron. Rechen–Inst. Heidelb., Prepr. Ser., No. 7, 6 pp. (1987). To appear in IAU Symp. No. 133.

043.005   Construction of the system of positions and proper motions in the FK5.
H. Schwan.
Astron. Rechen–Inst. Heidelb., Prepr. Ser., No. 8, 5 pp. (1987). To appear in IAU Symp. No. 133.

043.006   Normalized units.
T. Kiang.
Q. J. R. Astron. Soc., Vol. 28, No. 4, p. 456 – 471 (1987).
  Normalized units are such units in terms of which a certain physical constant or constants have the value 1. The normalization of k constants involving n ($\geqslant k$) physical dimensions leads to an (n,k) system of normalized units consisting of $n–k$ arbitrarily chosen units and k defined units. Nine such systems used by experts in diverse topics are collected in a table.

043.007   The connection of the HIPPARCOS reference system to extragalactic objects by photographic astrometry.
A. I. Yatsenko, S. P. Rybka, R.–D. Scholz.
Astron. Nachr., Vol. 308, No. 5, p. 319 – 322 (1987).
  The expected error of the connection of the HIPPARCOS system with an extragalactic reference system by means of ground–based photographic astrometry was calculated. Two different catalogues of proper motions with respect to extragalactic objects and the HIPPARCOS proper motions were simulated. An accuracy of $0\rlap{.}''14$ per century using only 10 fields

with 119 link stars of the Tautenburg Schmidt telescope (134/200/400) and 0″.07 per century using 93 fields with 336 link stars of the Kiev long focus astrograph (40/550) were obtained.

043.008   Contributions to the conventional inertial system by VLBI and by the Tautenburg Schmidt–telescope.
K.–G. Steinert.
Gerlands Beitr. Geophys., Vol. 96, No. 3 – 4, p. 218 – 221 (1987).
Abstr. in Phys. Abstr., Vol. 90, No. 1317, Entry 139435 (1987).

043.009   Untersuchung zur radioastronomischen Bestimmung des Äquinoktiums und der Ekliptikschiefe aus Pulsarbeobachtungen.
H.–H. Bernstein.
Dtsch. Geod. Komm. Bayer. Akad. Wiss., Reihe C, Heft Nr. 331, 4 + 77 pp. (1987). ISBN 3–7696–9380–9. = Diss. Fachbereich 12 – Vermessungswesen der Technischen Hochschule Darmstadt, F.R. Germany.
Die vorliegende Arbeit untersucht ein Verfahren zur Bestim-

mung des Äquinoktiums sowie der Ekliptikschiefe unter Verwendung von pulse timing Messungen und VLBI zu Pulsaren. Neben Meßmethode und Datenreduktion von pulse timing wird ein Ausgleichsmodell zur Bestimmung der Pulsarkoordinaten und weiterer Pulsarparameter entwickelt. Zur Behandlung der timing noise Prozesse wird die Kollokationsmethode genutzt.

The 1986 adjustment of the fundamental physical constants.
See Abstr. 022.122.

Rotation de la Terre et systèmes de référence.
See Abstr. 044.003.

Optimum positioning of attracting masses in the Heyl–type determination of the Newtonian gravitational constant.
See Abstr. 066.166.

044   Time and Latitude Determination, Earth Rotation, Polar Motion

044.001   Statistical investigations on atmospheric angular momentum functions and on their effects on polar motion.
A. Brzeziński.
Manuscr. Geod., Vol. 12, No. 4, p. 268 – 281 (1987).
The short term atmospheric excitation of polar motion is analysed using the available three series of the atmospheric "effective angular momentum" (EAM) vector. It is shown that the pressure terms χ^P_1 and χ^P_2 at periods less than 100 days can be adequately modeled by a 3rd order autoregressive process, whose striking feature is pseudoharmonic oscillation with a period of about 9 to 13 days and a very short relaxation time of 2 to 4 days. Different estimates of the same components of the equatorial EAM functions are compared in order to compute their signal and noise statistics. Finally, the atmospheric effects on polar motion are discussed using various statistical models for the excitation process.

044.002   A detection of short–period terms in Universal Time of Chinese Joint System.
Y.–r. Ding, N.–y. Xiao.
J. Nanjing Univ., Vol. 23, No. 2, p. 290 – 297 (1987).
The purpose of this paper is to detect the tidal short–period terms in Universal Time (UT) of Chinese Joint System (CJS). A reduction of the relevant data in "Annual Report of Chinese Time Service" and "Annual Report of BIH" gives a consecutive series of UT of CJS at the interval from 1967 to 1978. Then an auto–correlation analysis detects the short–period terms in this series and shows an absence of the expected terms.

044.003   Rotation de la Terre et systèmes de référence.
M. Feissel.
J. Astron. Fr., No. 30, p. 5 – 7 (1987).

044.004   Random variations of the Earth's rotation rate.
N. S. Blinov, V. E. Zharov.
Astron. Zh., Tom 64, Vyp. 4, p. 876 – 880 (1987). In Russian. English translation in Sov. Astron., Vol. 31, No. 4.
The 1984 results of Universal Time determinations by very–long–baseline radio interferometers and by classical methods are analysed. It is shown that in the Earth's rotation rate, there are irregular random fluctuations with amplitudes of several tenths of a millisecond.

044.005   Some statistical characteristics of observational series of the Pulkovo Time Service.
V. L. Gorshkov, O. V. Kotreleva.
Probl. Opredeleniya Parametrov Vrashcheniya Zemli, Vladivostok, p. 60 – 67 (1986). In Russian. Abstr. in Ref. Zh., 52. Geod. Aehrosemka, 7.52.95 (1987).

044.006   Dependence of the results of determining Universal Time with photographic zenith tubes in Kitab on the configuration of the dome.
Z. M. Malkin.
Probl. Opredeleniya Parametrov Vrashcheniya Zemli, Vladivostok, p. 68 – 70 (1986). In Russian. Abstr. in Ref. Zh., 52. Geod. Aehrosemka, 7.52.96 (1987).

044.007   Results of observations made in Paris with the astrolabe. Time and latitude 1986.
F. Chollet, S. Débarbat, J.–C. Hascoët, S. K. Lam, J. Mangombi dei Ilonga, P. Texier.
Astron. Astrophys., Suppl. Ser., Vol. 71, No. 1, p. 109 – 114 (1987). In French.
Results are given for the observations made with the astrolabe during the year 1986. No change has been made in the star catalogue (since 1956) and in the computation (since 1972). The results are in the intermediate system requested by the IAU for the astronomical reduction (IAU 1976/1980). The system is the FK5 system but with positions of the stars still given by the FK4 catalogue. The height of the pillar has been changed (but not its coordinates): it is now possible to observe at 30° (with the usual prism) and at 60° (with a special reflecting prism).

044.008   A matter of time.
E. Marshall.
Science, Vol. 238, No. 4834, p. 1641 – 1643 (1987).
Bowing to celestial time, the world's timekeepers will slow their atomic clocks with a "leap second" at the New Year 1988.

044.009   Global geodynamics.
H. Drewes, E. Geiß.
Dtsch. Geod. Komm. Bayer. Akad. Wiss., Reihe B, Heft Nr. 284, p. 248 – 256 (1987). In German and English.

044.010   Numerical and analytical methods of the general theory of the earth's rotation.
Zh. S. Erzhanov, A. A. Kalybaev, Kh. U. Abdeshov.
EhVM i nauchn.–tekh. progress. Kaz. Univ. Alma–Ata, p. 38 – 42 (1987). In Russian. Abstr. in Ref. Zh., 51. Astron., 10.51.107 (1987).

044.011   Changes in the Earth's rotation and low–degree gravitational field induced by earthquakes.
B. F. Chao, R. S. Gross.
Geophys. J. R. Astron. Soc., Vol. 91, No. 3, p. 569 – 596 (1987).
The change in the Earth's density distribution caused by an earthquake dislocation will change the Earth's rotation and gravitational field. In this paper the authors first develop the analytical formulae based on the normal–mode theory. Equipped with these formulae and using a spherically symmetric earth model (1066B) and the centroid–moment tensor solutions for earthquake sources, they then compute the earthquake–induced changes in the Earth's rotation (polar motion and length of day) and low–degree harmonics of the gravitational field for the period 1977 – 85 (altogether 2146 earthquakes). The authors then conduct simple spectral and statistical analyses on these changes.

044.012   Apparent polar wander of the mean–lithosphere reference frame.
R. G. Gordon, R. A. Livermore.
Geophys. J. R. Astron. Soc., Vol. 91, No. 3, p. 1049 – 1057 (1987).
Apparent polar wander in the mean–lithosphere (= no–net–rotation = no–net–torque uniform drag) reference frame is compared with apparent polar wander in the hotspot reference frame over the past 100 Myr. Palaeomagnetic poles and plate rotations previously used to determine an apparent polar wander path for the hotspot reference frame are here used to determine an apparent polar wander path in the mean–lithosphere reference frame.

044.013   Snow load effect on the earth's rotation and gravitational field, 1979 – 1985.
B. F. Chao, W. P. O'Connor, A. T. C. Chang, D. K. Hall, J. L. Foster.
J. Geophys. Res., Vol. 92, No. B9, p. 9415 – 9422 (1987).
A global, monthly snow depth data set has been generated from the Nimbus 7 satellite observations using passive microwave remote–sensing techniques. In this paper the authors analyze 7 years of data, 1979 – 1985, to compute the snow load effects on the earth's rotation and low–degree zonal gravitational field.

044.014   Short–period variations in the length of day: atmospheric angular momentum and tidal components.
S.–f. Luo, D.–w. Zheng, D. S. Robertson, W. E. Carter.
J. Geophys. Res., Vol. 92, No. B11, p. 11657 – 11661 (1987).
Using very long baseline interferometry (VLBI) observations of extragalactic radio sources, it is now possible to determine the rotational phase angle of the Earth (UT1) with an accuracy of 0.1 ms in an observing period of about 1 hour. The authors analyzed a 3–month series of daily length–of–day (LOD) values produced from VLBI observations collected under project International Radio Interferometric Surveying (IRIS) during April – July 1984. Comparison of the IRIS LOD values with variations in the atmospheric angular momentum (AAM) during the same time period revealed quasi–periodic components in both series at about 9, 33, and 58 days, with closely matching phases.

044.015   Relationships between changes in the length of day and the 40– to 50–day oscillation in the tropics.
R. A. Madden.
J. Geophys. Res., Vol. 92, No. D7, p. 8391 – 8399 (1987).
Attempts are made to link observed variations of the length–of–day with a period near 50 days to the atmospheric 40– to 50–day oscillation in the tropics. Because the earth–atmosphere system maintains constant angular momentum a longer length of day (slower rotation rate of the earth) goes with higher momentum of the atmosphere. It is shown that the longest length of day and greatest atmospheric momentum occurs when tropical convection associated with the 40– to 50–day oscillation is beginning to weaken near or east of the date line. The author presents some evidence suggesting that, over a broad period range bridging 40 – 50 days, a strengthening of the easterlies east of the convection increases wind stress and frictional torques, and serves as an important mechanism for exchange of momentum from earth to atmosphere.

044.016   Breitenbestimmungen. Januar – Juli 1987.
Lohrmann–Obs., Tech. Univ. Dresden, Zirk., Nos. 119 – 120 (1987).

044.017   Time and Frequency Services, Bulletin. 1987 July – December.
Time Freq. Serv., Bull., Nos. 94 – 99 (1987).

044.018   Hamiltonian formulation of the polar motion for an elastic Earth's model.
M. J. Sevilla, P. Romero.
Inst. Astron. Geod., Univ. Madr., Publ., No. 143, 26 pp. (1986).
A canonical formulation of the rotational motion for an elastic Earth's model is presented. From the canonical equations for the precession and nutation motion in an inertial frame, the equations in an Earth fixed frame have been deduced. The linearized equations deduced for polar motion are equivalents to those obtained using the Liouville's equations.

044.019   The Sasao–Okubo–Saito equations by Hamilton theory. First results.
P. Romero, M. J. Sevilla.
Inst. Astron. Geod., Univ. Madr., Publ., No. 144, 22 pp. (1986).
A canonical formulation of the rotational motion for an elastic Earth model has been developed. From the canonical equations for the precessional and nutational motions in an inertial frame, the equations in an Earth fixed frame are deduced. The linearized equations obtained for polar motion and liquid core motion are equivalents to the Sasao–Okubo–Saito equations.

044.020   Tratamiento canonico del problema de Poincaré. Movimiento del polo.
P. Romero, M. J. Sevilla.
Inst. Astron. Geod., Univ. Madr., Publ., No. 152, 15 pp. (1986).
The rotational motion for a rigid Earth model with a homogeneous liquid core has been obtained using Hamilton's equations. From the canonical equations for the precessional und nutational motions in an inertial frame, the corresponding equations in an Earth fixed frame are deduced. The linearized equations obtained for polar motion and liquid core motion are equivalent to the Moritz's equations (1980).

044.021   Bulletins, Time Service of the Mizusawa Observatory in 1985.
Bull., Time Serv. Mizusawa Obs., Vol. 30, 2 + 20 pp. (1987).

044.022   Monthly Notes of the International Polar Motion Service.
Mon. Notes Int. Polar Motion Serv., Nos. 5 – 11, p. 69 – 183 (1987).

044.023   The change of Earth's rotation since 1983 and predictions of UT1–UTC.
Q.–l. Zhang, H.–q. Yang, X.–i. Jiang.
Publ. Shaanxi Astron. Obs., Vol. 9, No. 1, p. 11 – 17 (1986). In Chinese.

044.024   Data management system of the Joint Atomic Time Scale.
L.–z. Tu, X.–p. Pan.
Publ. Shaanxi Astron. Obs., Vol. 9, No. 1, p. 18 – 24 (1986). In Chinese.

044.025   **The determination of the latitude of the datum mark of the millimeter wavelength antenna at Delingha Field Station.**
J.–s. Yao, S.–z. Wang, W. Shi.
Publ. Purple Mt. Obs., Vol. 6, No. 2, p. 127 – 130 (1987). In Chinese.
The authors have determined the latitude of the datum mark of the millimeter wavelength antenna at Delingha Field Station using the Talcott method with the Askania transit instrument.

044.026   **Bureau International de l'Heure (B.I.H.), Circular D.**
B.I.H. Circ., D248 – D253 (1987).
Contents: UTC, TAI, UT and coordinates of the pole, rotation of the Earth; Data for 1987, May – October.

044.027   **UTC time step on the 1st of January 1988.**
B. Guinot.
B.I.H. Circ., E 14 (1987).

044.028   **Time and Latitude. July 1986 – March 1987.**
V. Ptáček, J. Vondrák, R. Weber.
Circ. Czech. Obs., Time Latitude, 21 + 19 + 19 pp. (1987).

044.029   **Zeit– und Breitenbestimmungen, Zeitsysteme, Präzisionszeitvergleiche. Januar – Juni 1985.**
Zeit–Breitenbestimmungen, Jahrg. 1985, Nr. 1 – 4, 20 + 15 + 18 + 15 pp. (1987).

044.030   **Time and Latitude Bulletins. 1987 January – September.**
Tokyo Astron. Obs., Time Latitude Bull., Vol. 61, No. 1 – 3, p. 1 – 62 (1987).

044.031   **Daily time differences and relative phase values. 1987 July – December.**
U.S. Nav. Obs., Time Serv. Publ., Ser. 04, Nos. 1065 – 1091 (1987).

044.032   **NEOS. Earth Orientation. 1987 July – December.**
NEOS Earth Orientation Bull., Vol. 2, Nos. 27 – 53 (1987).

044.033   **Time Service Announcement.**
G. M. R. Winkler.
U.S. Nav. Obs., Time Serv. Publ., Ser. 14, No. 45 (1987).

044.034   **Results of determination of latitude in Józefosław, January – September 1987, by observations of the Horrebow–Talcott pairs.**
L. Pieczyński.
Latitude Circ., No. 104 – 106, 4 + 4 pp. (1987).

044.035   **Time service for the year 1985.**
L. Mureddu.
Circ. Stn. Astron. Int. Latitudine, Carloforte–Cagliari, N. 32 (Ser. A(12)), 35 pp. (1986).

044.036   **Reduction in the MERIT standards of the observations made with the astrolabe Danjon at Cagliari.**
A. Poma, V. Gusai.
Circ. Stn. Astron. Int. Latitudine, Carloforte–Cagliari, N. 36 (Ser. B(12)), 32 pp. (1987).

044.037   **Time service for the year 1986.**
L. Mureddu.
Circ. Stn. Astron. Int. Latitudine, Carloforte–Cagliari, N. 37 (Ser. A(13)), 37 pp. (1987).

044.038   **Analysis of UT1 observations by VLBI for the determination of the Love number k.**
H. Schuh.
Mitt. Geod. Inst. Rheinischen Friedrich–Wilhelms–Univ. Bonn, Nr. 71, p. 61 – 72 (1987). – See Abstr. 012.066.

044.039   **Oppolzer und die Definition des Himmelspols.**
H. Moritz.
Vorträge beim Theodor Ritter von Oppolzer–Gedächtnissymposium, p. 60 – 68 (1987). – See Abstr. 012.073.

044.040   **Electromagnetic core–mantle coupling and the Earth's rotation.**
H. Greiner–Mai.
Gerlands Beitr. Geophys., Vol. 96, No. 3 – 4, p. 230 – 238 (1987). Abstr. in Phys. Abstr., Vol. 90, No. 1317, Entry 139436 (1987).

044.041   **Tidal evolution of the Earth–Sun system.**
M. Burša.
Stud. Geophys. Geod., Vol. 31, No. 4, p. 331 – 334 (1987).
The tidal decrease in the Earth's heliocentric longitude generated by the Sun has been computed. It represents the increase in the length of year $\sim 10^{-7}$s per century. The resonant angular velocity of the Earth's rotation is approximately equal to the present Earth's mean motion, however, for the model used, i.e., considering the Sun as the point–mass.

044.042   **Secular tidal and nontidal variations in the Earth's rotation.**
M. Burša.
Stud. Geophys. Geod., Vol. 31, No. 3, p. 219 – 224 (1987).
On the basis of the angular momentum balance in the Earth–Moon–Sun system and with the use of the observed secular variation in the Moon's mean motion and the variation in the second zonal geopotential harmonic, the tidal and nontidal variations in the angular velocity of the Earth's rotation are computed and different values describing the field of tidal forces estimated.

044.043   **Polar motion caused by the variation of air mass distribution.**
B.–x. Gao.
Acta Astron. Sin., Vol. 28, No. 2, p. 197 – 203 (1987). In Chinese.
The author discusses different effects on polar motion which are produced by excitations of different periods. The Fourier series of the excitation function of the air mass distribution, the annual polar motion, and the forced Chandler wobble produced by the variation of the air mass distribution is analyzed. It is shown that the variation of the air mass distribution not only produces predominant effects to the annual polar motion but also important influence to the Chandler wobble.

044.044   **On the error equations for determining the Earth rotation parameters with a VLBI network.**
Y.–f. Xia.
Acta Astron. Sin., Vol. 28, No. 3, p. 282 – 287 (1987). In Chinese.

044.045   **The IRIS Earth orientation parameters.**
IRIS Bull. A, Nos. 41 – 46 (1987).

Echelles de temps atomique.
See Abstr. 003.064.

The physical conception of time from Newton to today.
See Abstr. 004.029.

Project MERIT.
See Abstr. 013.071.

Der Beitrag der Wiener Universitätssternwarte zum Projekt MERIT.
See Abstr. 013.079.

Does the Earth rotate?
See Abstr. 014.039.

On thermal instrumental errors in astrometric observations.
See Abstr. 032.050.

Determination of stellar positions and ERP with a stellar interferometer. I. Synopsis of all kinds of interferometers.
See Abstr. 034.080.

Determination of stellar positions and ERP with a stellar interferometer. II. Geometric principles.
See Abstr. 034.081.

Determination of stellar positions and ERP with a stellar interferometer. III. Achieving of the observed quantity τ.
See Abstr. 034.082.

Right ascension corrections to 120 FK 4-stars by the analysis of Time observations obtained with the Photoelectric Transit Instrument at Torino Observatory.
See Abstr. 041.002.

Der Canon der Finsternisse und die Rotationsdauer der Erde.
See Abstr. 079.003.

Secular variations in the curvature of equipotential surfaces and directions of verticals due to geodynamic phenomena.
See Abstr. 081.037.

Geomagnetic secular variation, core motions and implications for the earth's wobbles.
See Abstr. 084.043.

Secular deceleration of the Moon and of the Earth's rotation and variation in the zonal geopotential harmonic.
See Abstr. 094.028.

The tidal evolution of the Earth–Moon system.
See Abstr. 094.045.

045 Astronomical Geodesy, Satellite Geodesy, Navigation

045.001 Problem of the elimination of the refractional effects in Doppler positioning.
I. Gougoutoudis.
Geod. Kartogr., Tom 35, Zesz. 3 – 4, p. 91 – 99 (1986). In Polish.
The influence of the tropospheric refraction on the Doppler positioning is discussed. It is found that the differences of coordinates resulting from the use of standard atmospheric parameters instead of real ones could amount to 0.60 m for single point positioning and 0.20 m for multilocation. The necessity of registration of the real meteorologic parameters at the Doppler station is confirmed.

045.002 The vector Borowiec – Wettzell from Doppler observations.
W. Jakś.
Artif. Satell., Vol. 22, No. 2, p. 15 – 27 (1987).
The paper presents the results of a homogeneous elaboration of the results of common Doppler observations carried out at Borowiec and Wettzell stations during different campaigns. The influence of the orbital error model on the determined vector is examined. Obtained results are compared with results of other authors.

045.003 Readjustment of the Polish experimental Doppler network.
J. B. Rogowski.
Artif. Satell., Vol. 22, No. 2, p. 29 – 45 (1987).

045.004 Signal and noise in S.L.R. (*Satellite Laser Ranging*) data.
B. Betti, M. Carpino, F. Migliaccio, F. Sansò.
Bull. Géod., Vol. 61, No. 3, p. 235 – 260 (1987).

045.005 Geodetic astronomy.
K. Kaniuth, G. Soltau.
Dtsch. Geod. Komm. Bayer. Akad. Wiss., Reihe B, Heft Nr. 284, p. 60 – 68 (1987). In German and English.
This report describes the activities performed in the Federal Republic of Germany between spring 1983 and spring 1987 in the field of geodetic astronomy. It is restricted to the classical optical procedures and does not include geodetic radioastronomy, which is covered otherwise. The present report is divided into the sections (1) instruments and methods of observation, (2) observation activities, (3) analysis of observation results, (4) theoretical fundamentals, and concludes with a list of publications.

045.006 I. D. Zhongolovich's ideas of the construction of a terrestrial coordinate system.
V. I. Valyaev, A. A. Malkov, T. B. Sabanina.
Byull. Inst. Teor. Astron., Tom 15, No. 9, p. 519 – 523 (1986). In Russian.
The development of Zhongolovich's ideas of construction of a terrestrial and celestial coordinate system are considered under historical aspects. Models of solid polyhedrons are given provided that the chords of these polyhedrons have been obtained by the VLBI method. The evaluations of the parameters for the construction of a terrestrial coordinate system have been obtained.

045.007 Non–tidal motions between Mizusawa and Washington, D.C. with the aid of the PZT observations.
C. Kakuta, D. D. McCarthy, T. Hara, K. Sato, K. Yokoyama, S. Manabe, S. Sakai, H. Kitago, K. Iwadate, A. K. Babcock, I. W. Lindenblad, L. Hinnov.
Publ. Int. Latitude Obs. Mizusawa, Vol. 19, No. 1 – 2, p. 1 – 13 (1986).
Results of the PZT observations at the ILOM and the USNO based on a common star system (MC3) are studied from non-tidal geophysical points of view. Relative displacements of Mizusawa to Washington, which are derived from residuals of time and latitude are compared with Minster and Jordan's plate motion (1978). Differences of optical observations between Mizusawa and Washington show a secular variation associated with a periodic variation over a 4 year period. The secular variation shows a similar direction as that of Minster and Jordan's plate motion. But the magnitude of the optical observation is found to be larger than that of the plate motion derived by Minster and Jordan.

045.008 The preprocessing of the tracking data of the MX 1502 satellite surveyor.
Y. Ding, Y. Xia.
Publ. Purple Mt. Obs., Vol. 6, No. 2, p. 131 – 138 (1987). In Chinese.

045.009 Microwave geodesy by interferometric radiometry.
A. J. Skalafuris.
Astrophys. Space Sci., Vol. 137, No. 2, p. 317 – 345 (1987).
The aim of this paper is to establish a new method – first introduced by Soviet scientists in 1972 – for mapping the Earth in the microwave region passively using the interference pattern which results by correlating the signals from two radiometers which presumably can be mounted in two or more moving

satellites. In such a procedure, the image is enhanced progressively by synthetic composition similar to that achieved by air–borne synthetic aperture radar.

045.010   Report on the geodetic activities in the years 1983 to 1987.
Swiss Geodetic Commission and Federal Office of Topography. 33 + 11 pp. (1987).
Contents: 1. Positioning: RETrig and the Swiss National Triangulation Net: European Triangulation Net (RETrig) (*B. Bürki, E. Gubler*). Electronic distance measurements (*A. Elmiger, H. Chablais*). Alpine traverse St. Gotthard (*A. Elmiger*). Diagnostic adjustment (*H. Chablais*). UELN and the Swiss National Levelling Net (*E. Gubler*). Rapid Precision Levelling System (RPLS) (*H. Ingensand*). Refraction (*F. Chaperon, R. Köchle*). Inertial techniques for geodesy (*A. Wiget, H.–G. Kahle*). TRANSIT Doppler activities: SWISSDOC (*A. Geiger, H.–G. Kahle, A. Wiget*). 3D test net Turtmann: Objectives and installation of the net (*D. Schneider*). Classical geodetic observations (*D. Schneider*). GPS campaigns (*G. Beutler*). Integrated geodesy (*M. Müller, H.–G. Kahle, A. Geiger*).
2. Advanced space technology: Satellite observation station Zimmerwald (*I. Bauersima, W. Gurtner*). GPS test measurements and tropospheric error modelling: GPS test meaurements with WM101 and TI4100 receivers (*A. Geiger, H.–G. Kahle, M. Cocard*). Atmospheric effects on geodetic space measurements (*A. Geiger*).
3. Determination of the gravity field: Gravity base network and calibration (*W. Fischer*). Geoid determinations: Gravimetric geoid of Switzerland (*A. Geiger, H.–G. Kahle*). Geodetic astronomy (*B. Bürki, B. Wirth, U. Marti*). Determination of the geoid in Europe: ALGEDOP (*A. Wiget, A. Geiger, H.–G. Kahle*). Gravimetric evaluation and interpretation (*H.–G. Kahle*). Gravity surveys for applied geophysics (*E. Klingelé*).
4. General theory and methodology: Geodetic and geometric modelling (*R. Conzett, W. Kuhn, B. Studemann*). Data processing at the Federal Office of Topography (*A. Carosio*).
5. Geodynamics: Gravity anomalies and geodynamics of mountain belts (*H.–G. Kahle, S. Ott*). Physical interpretation of gravity anomalies (*A. Geiger, H.–G. Kahle*). WEGENER–MEDLAS Project: the Swiss trans–Alpine laser experiment Monte Generoso–Jungfraujoch–Zimmerwald (*H.–G. Kahle, B. Bürki, A. Geiger, W. Gurtner, S. Müller*). The Swiss National Research Project NFP 20: exploration of the deep geological structure of Switzerland (*B. Wirth, H.–G. Kahle, E. Gubler, I. Bauersima*). The geological structure of the zone of Ivrea–Verbano (*B. Wirth, B. Bürki, H.–G. Kahle*). Strain energy and recent crustal movements in the Alpine–Mediterranean region analyzed in the Swiss Alps (*A. Geiger, H.–G. Kahle, E. Gubler*). Earth tide measurements (*A. Wiget, R. Edge, E. Klingelé, H.–G. Kahle*).

045.011   Results of the first MARK III VLBI measurements with the Hartebeesthoek Radio Astronomy Observatory.
A. Nothnagel, G. D. Nicolson, H. Schuh, J. Campbell, H. Cloppenburg.
Mitt. Geod. Inst. Rheinischen Friedrich–Wilhelms–Univ. Bonn, Nr. 71, p. 53 – 60 (1987). – See Abstr. 012.066.

045.012   Preliminary results of the post–VEGA Mk II–VLBI experiments (GRIG–2).
G. Petit.
Mitt. Geod. Inst. Rheinischen Friedrich–Wilhelms–Univ. Bonn, Nr. 71, p. 73 – 76 (1987). – See Abstr. 012.066.

045.013   Methods to correct for the wet path delay in geodetic VLBI.
G. Elgered, J. Johansson.
Mitt. Geod. Inst. Rheinischen Friedrich–Wilhelms–Univ. Bonn, Nr. 71, p. 79 – 90 (1987). – See Abstr. 012.066.

045.014   VLBI for geodynamics involving the Hartebeesthoek Radio Astronomy Observatory in South Africa.
A. Nothnagel, G. D. Nicolson, H. Schuh, J. Campbell, A. Rius.
Mitt. Geod. Inst. Rheinischen Friedrich–Wilhelms–Univ. Bonn, Nr. 72, p. 39 – 47 (1987). Paper presented at the 1984 AGU Spring Meeting, Cincinnati, Ohio, USA, 1984 May 14 – 18.
The high geodetic and geophysical interest of baseline determinations between South Africa and adjacent continents has prompted efforts at the Hartebeesthoek Radio Astronomy Observatory (HartRAO) to intensify its geodetic VLBI activities. In order to explore the exciting geometric potentials of the long N–S baselines connecting Europe and South Africa two pilot experiments using the Mk II BWS technique have been carried out at S–band (2.3 GHz) between HartRAO and the NASA DSN stations 61 and 63 near Madrid. The baseline results which show meter accuracy are discussed in relation to the further accuracy improvements needed for crustal motion detection.

045.015   Geodetic VLBI with large antennas.
A. Rius, J. Rodríguez, J. Campbell.
Mitt. Geod. Inst. Rheinischen Friedrich–Wilhelms–Univ. Bonn, Nr. 72, p. 59 – 67 (1987). Paper presented at the XII General Assembly of the European Geophysical Society held in Strasbourg, France, April 9 – 14, 1987.

045.016   Zeitvariable Meereshöhen und Bahnfehler aus Satelliten–Altimetermessungen.
P. Arnold.
Mitt. Geod. Inst. Rheinischen Friedrich–Wilhelms–Univ. Bonn, Nr. 74, 119 pp. (1987). Diss. Hohen Landwirtschaftlichen Fakultät der Rheinischen Friedrich–Wilhelms–Universität Bonn.

045.017   A comparison of the accuracy of the observations of artificial satellites by laser ranging radar "Intercosmos" and the tracing camera AFU–75.
E. A. Yurov.
Nauchn. Inf., Vyp. 58, p. 14 – 22 (1986). In Russian.

045.018   Accuracy investigation of geostationary satellite's position obtained with the VAU camera.
A. M. Basharin, V. G. Reva, Yu. V. Rusin.
Nauchn. Inf., Vyp. 58, p. 23 – 30 (1986). In Russian.
The accuracy of GSS observations obtained with the VAU camera is estimated. The mean square error of an observation is not more than 1″.5.

045.019   The amplitude distribution of impulses during laser ranging of the Lageos satellite with the laser radar "Intercosmos".
L. S. Stirberg (*L. S. Shtirberg*).
Nauchn. Inf., Vyp. 58, p. 39 – 44 (1986). In Russian.
The experimental signal and noise distributions are given. A single photoelectronic regime of the laser radar is discussed.

045.020   The mean–square collocation for the laser ranging normal points determiation.
O. A. Soboleva.
Nauchn. Inf., Vyp. 58, p. 121 – 127 (1986). In Russian.

045.021   Algorithm of a more accurate definition of the distances between laser ranging stations.
S. P. Oraevskaya, B. M. Shustov.
Nauchn. Inf., Vyp. 58, p. 162 – 168 (1986). In Russian.
An algorithm of the distance determination between laser ranging stations is described. Recommendations for a priori weighting of separate arcs based on the results of simulations as well as the observation of the Lageos satellite are given.

045.022   Les programmes du service des observations lasers du satellite artificiel de la Terre.
L. A. Yurova.
Nauchn. Inf., Vyp. 59, p. 76 – 81 (1986). In Russian.

**045.023   Differential Doppler tracking of interplanetary space-
craft.**
L. Iess, P. Bonifazi, B. Bertotti, G. Comoretto.
Nuovo Cimento C, Vol. 10C, Ser. 1, No. 2, p. 235 – 246 (1987).
Abstr. in Phys. Abstr., Vol. 90, No. 1316, Entry 134143 (1987).

**045.024   Geometrical correction of satellite imagery using the
collinearity equation.**
N. Shu.
Bull. Soc. Fr. Photogramm. Teledetect. (*France*), No. 105,
p. 27 – 40 (1987). In French. Abstr. in Phys. Abstr., Vol. 91,
No. 1319, Entry 3914 (1988).

**045.025   The requirement of the orbit accuracy of GPS for precise
positioning and orbit determination within a local area.**
Q.-f. Xu.
Acta Astron. Sin., Vol. 28, No. 3, p. 226 – 236 (1987). In Chinese.
   The Global Positioning System (GPS) provides a new possibility to solve some problems of geodesy and geodynamics. An
empirical formula is applied to estimate the influence of orbit
errors on baseline determinations. An analytical formula is
derived and some simulations are made. These show that the
analytical formula is more precise than the empirical one. The
advantages of GPS satellite distribution for the orbit determination within a local area are discussed.

Satellite geodesy investigations in Poland in 1983 – 1986.
See Abstr. 013.040.

European VLBI for geodesy and astrometry: goals and objectives.
See Abstr. 013.075.

**FORTRAN–Programme für die Berechnung von zeitvariablen
Meereshöhen und Bahnfehlern aus Satelliten–Altimeterdaten.**
See Abstr. 021.015.

**Effiziente Rechenverfahren für umfangreiche geodätische Parame-
terschätzungen.**
See Abstr. 021.016.

**Transformation of geocentric to geodetic coordinates without
approximations.**
See Abstr. 021.025.

A laser reflecting system for geophysical satellites.
See Abstr. 031.073.

**Programme of observations on the circumzenithal for the Geody-
namic Observatory "Plana".**
See Abstr. 034.189.

**Geodetic VLBI–monitoring of the milliarcsecond structures of
extragalactic radio sources.**
See Abstr. 036.200.

**Data analysis of Project ERIDOC (European Radio Interfer-
ometry and Doppler Campaign).**
See Abstr. 036.201.

**Use of the statistic analysis for preliminary processing of laser
ranging results of satellites.**
See Abstr. 036.223.

**A method of observations of geostationary satellites on the TV–
guide of the camera VAU.**
See Abstr. 036.225.

**A method of photographic observations of geostationary satellites
with the VAU–camera.**
See Abstr. 036.226.

Multi–body force function: geodetic aspects of astrodynamics.
See Abstr. 052.013.

046 Ephemerides, Almanacs, Calendars, Chronology

**046.001   Ephémérides astronomiques 1988. Calendriers – soleil –
lune – planètes – astéroides – satellites – comètes –
étoiles.**
Annuaire du Bureau des Longitudes.
Gauthier–Villars, Paris, France. 18 + 280 pp. (1987). ISBN 2–04–
016974–1.

046.002   Ephémérides 1988.
Supplément à l'Astronomie de Janvier 1988.
18 + 280 pp. (1987). ISBN 2–04–016974–1.

046.003   Régularités et irrégularités du calendrier.
F. Mignard.
Astronomie, Vol. 101, p. 411 – 419 (1987).

**046.004   The Handbook of the British Astronomical Association
1988.**
Prepared by the Computing Section of the Association under the
supervision of G. E. Taylor.
Office of the Association, Burlington House, Piccadilly,
London, W1V 9AG, England. 99 pp. Price £ 6.00 (1987).
ISSN 0068–130X.

**046.005   Annuaire de l'Observatoire Royal de Belgique (Jaarboek
van de Koninklijke Sterrenwacht van België), 1988.**
L'Observatoire Royal de Belgique, Avenue Circulaire 3, B–1180
Bruxelles, Belgium. 155ᵉ année, 223 pp. (1987). ISSN 0373–4900.

**046.006   Kalender für Sternfreunde 1988. Kleines astronomisches
Jahrbuch.**
P. Ahnert.
Johann Ambrosius Barth, Leipzig, German Democratic Repub-
lic. 175 + 16 pp. (1987). ISBN 3–335–00059–5, ISSN 0075–4706.

**046.007   Éphémerides astronomiques et calendrier des événements
célestes pour 1988.**
P. Cugnon, A. Koeckelenbergh, G. Evrard.
Ciel Terre, Vol. 103, No. 5, 76 pp. (1987).

046.008   Observer's Handbook 1988.
R. L. Bishop (Editor).
The Royal Astronomical Society of Canada, 136 DuPont Street,
Toronto, Ontario M5R 1V2, Canada. 212 pp. (1987). ISSN
0080–4193.

**046.009   Himmelskalender 1988. Ein kleines astronomisches
Jahrbuch für Österreich.**
H. Mucke.
Astro–Verein, Astronomisches Büro, Hasenwartgasse 32, 1238
Wien, Austria. 32. Jahrg., 132 pp. (1987).

046.010   Efemérides Astronómicas 1988.
Real Instituto y Observatorio de la Armada en San
Fernando (Cádiz), Spain, Vol. 197, 6 + 458 pp. (1987). ISBN 84–
7469–035–8, ISSN 0080–5971.

046.011   **Almanaque Náutico, 1988. Con Suplemento para la Navegación Aerea.**
Instituto y Observatorio de Marina, San Fernando (Cádiz), Spain. 414 + 30 + 6 pp. (1986). ISBN 84–7469–036–6, ISSN 0210–735X.

046.012   **Dados Astronómicos para os Almanaques de 1988 para Portugal.**
Observatório Astronómico de Lisboa, Lisboa, Portugal. 55 pp. Price Esc. 200.00 (1987). ISSN 0870–3434.

046.013   **Sternführer 1988. Ein astronomisches Jahrbuch.**
B. Koch, T. Jurriens, J. Meeus.
Treugesell–Verlag Dr. Vehrenberg KG, Schillerstr. 17, D–4000 Düsseldorf 1, F.R. Germany. 2 + 191 pp. Price DM 29.00 (1987). ISBN 3–87974–088–7.

046.014   **Apparent Places of Fundamental Stars 1988 for 1535 stars from the Fifth Fundamental Catalogue (FK5).**
R. Wielen, T. Lederle, H. Schwan (Editors).
Published for Astronomisches Rechen–Institut, Heidelberg, F.R. Germany by Verlag G. Braun, Karl–Friedrich–Straße 14 – 18, D–7500 Karlsruhe 1, F.R. Germany. 44 + 510 + 30 pp. Price DM 59.00 (1987). ISBN 3–7650–0088–4, ISSN 0174–254X.

046.015   **Corrections FK5 – FK4 to be applied to the published ''Apparent Places of Fundamental Stars (APFS)'' for the years 1984 to 1987.**
R. Wielen, T. Lederle, H. Schwan (Editors).
Apparent Places of Fundamental Stars 1988, p. A1 – A30 (1987). – See Abstr. 046.014. Available also as separate reprint.

046.016   **1988 Nautical Almanac, Pub. No. 681.**
Hydrographic Department, Maritime Safety Agency, Tokyo, Japan. 6 + 467 + 12 pp. (1987). ISSN 0910–0407.

046.017   **1988 Abridged Nautical Almanac. Pub. No. 683.**
Hydrographic Department, Maritime Safety Agency, Tokyo, Japan. 4 + 243 + 8 pp. (1987). ISSN 0910–0415.

046.018   **Der Sternenhimmel 1988. Astronomisches Jahrbuch für Sternfreunde.**
E. Hügli, H. Roth, K. Städeli (Editors).
48. Jahrg. Verlag Sauerländer, Aarau, Switzerland; Otto Salle Verlag, Frankfurt, F.R. Germany. 4 + 210 pp. Price SF 28.00, DM 29.80 (1987). ISBN 3–7941–2950–4 (Sauerländer), ISBN 3–7935–5028–1 (Salle).

046.019   **The Indian Astronomical Ephemeris for the year 1988.**
Prepared by the Positional Astronomy Centre, India Meteorological Department, New Alipore, Calcutta–700053, India.
Available from The Controller of Publications, Civil Lines, Delhi–110054, India; The High Commission of India, India House, Aldwych, London, W.C. 2, England. 6 + 582 pp. Price Rs. 100.00, £ 11.66, $ 36.00 (1987).

046.020   **Computing ephemerides for the Sun, the Moon and the planets based on modern theories.**
N. I. Glebova.
Byull. Inst. Teor. Astron., Tom 15, No. 9, p. 486 – 504 (1986). In Russian.
Methods of computing new ephemerides on the basis of both the American DE200/LE200 and French theories VSOP82/LP2000 are presented. Various algorithms for taking into account the relativistic deflection of light are considered. The results of a comparison of the new ephemerides computed in accordance with the American and French theories are given as well as those of new ephemerides with the theories by Newcomb for the inner planets and numerical theories for the outer planets over the time span 1980 – 1990.

046.021   **Astronomiskais Kalendārs 1988.**
J. Bikše, I. Daube, M. Dirikis, J. Francmanis, V. Freijs, J. Miezis.
Latvijas PSR Zinātnu Akadēmija, Radioastrofizikas Observatorija, Vissavienibas Astronomijas un Geodēzijas Biedribas Latvijas Nodala, Zinātne, Riga. 188 pp. Price 55 Kop. (1987).

046.022   **International Geophysical Calendar 1988.**
J. Atmos. Terr. Phys., Vol. 49, Nos. 11/12, p. 1169 – 1173 (1987).

046.023   **Nautisches Jahrbuch 1988.**
Seehydrographischer Dienst der Deutschen Demokratischen Republik, Rostock, German Democratic Republic. 38. Jahrg., 8721. 45 + 366 pp. (1987). ISSN 0433–681X.

046.024   **Rocznik Astronomiczny Obserwatorium Krakowskiego 1988. International Supplement Nr. 59.**
K. Rudnicki, E. Danielkiewicz–Krośniak, M. Krośniak (Editors).
Państwowe Wydawnictwo Naukowe, Warszawa–Kraków, Poland. 6 + 138 pp. + 4 tables. Price zł 480.00 (1987). ISBN 83–01–08372–7, ISSN 0075–7047. – See abstracts 046.025, 119.037, 119.038, 122.070.

046.025   **Geocentric ephemerides for the year 1988 of the libration points L_4 and L_5 in the Earth–Moon system and in the Sun–Venus system.**
J. Kordylewski, R. Szafraniec.
Rocznik Astronomiczny Obserwatorium Krakowskiego 1988, p. 133 – 137 (1987). – See Abstr. 046.024.

046.026   **Efemérides Astronômicas 1988.**
Ano CIV. CNPq Conselho Nacional de Desenvolvimento Científico e Tecnológico, Observatório Nacional, Rio de Janeiro, Brazil. 18 + 541 pp. (1987).

046.027   **Astronomical Yearbook of the USSR for the year 1989.**
V. K. Abalakin (Editor).
Institut Teoreticheskoj Astronomi Akademii Nauk SSSR. Izdatel'stvo Nauka, Leningradskoe Otdelenie, Leningrad. 692 pp. Price 14 Rbl. 40 Kop. (1987). In Russian.

046.028   **Hvězdářská ročenka 1988.**
V. Vanýsek (Editor).
Ročnik 64. Academia, nakladatelství Československé akademie věd, Praha, Czechoslovakia. 205 pp. Price 28.00 Kčs (1987).

046.029   **The Nautical Almanac for the year 1989.**
Issued by Her Majesty's Nautical Almanac Office, London and Nautical Almanac Office, United States Naval Observatory, Washington.
N. P. 314 – 89. UK Edition: Her Majesty's Stationery Office, London, England. US Edition: Superintendent of Documents, US Government Printing Office, Washington, D.C. 20402, USA. A4 + 318 + 36pp. Price £ 10.95 (1988). ISBN 0–11–772355–X.

046.030   **Indian calendars.**
S. K. Chatterjee.
History of Oriental astronomy, p. 91 – 95 (1987). – See Abstr. 012.060 (IAU Colloq. No. 91).

046.031   **The *Hsiu–Yao Ching* and its Sanskrit sources.**
M. Yano.
History of Oriental astronomy, p. 125 – 134 (1987). – See Abstr. 012.060 (IAU Colloq. No. 91).

046.032   **The position of the *Futian* calendar on the history of east–west intercourse of astronomy.**
S. Nakayama.
History of Oriental astronomy, p. 135 – 138 (1987). – See Abstr. 012.060 (IAU Colloq. No. 91).

046.033   **The ancients' criterion of earliest visibility of the lunar crescent: how good is it?**
M. Ilyas.
History of Oriental astronomy, p. 147 – 152 (1987). – See Abstr. 012.060 (IAU Colloq. No. 91).
Earliest visibility of the lunar crescent is an important calendrical element. It was needed in all early calendars and remains in use on some lunar calendars today. An astronomical criterion of earliest lunar visibility was therefore evolved quite early, using observations, right from the Babylonian era. Recently, an improved and comprehensive global criterion of earliest visibility, developed by the author, has been used to generate an extensive inverted moonset lag data set. These data, as a function of latitude and season, all for the first time provide a useful comparison with the simple ancient criterion.

046.034   **A study of some astronomical data in Muslim calendar.**
M. Chen.
History of Oriental astronomy, p. 169 – 174 (1987). – See Abstr. 012.060 (IAU Colloq. No. 91).

046.035   **System of calculation of the ephemerides of the Moon and the eight major planets by means of the theory of the Bureau des Longitudes (France).**
V. I. Skripnichenko.
Algoritm. Nebesnoj Mekh., No. 52, 19 pp. (1987). In Russian.
A system suitable for the calculation on the BESM–6 computer of the ephemerides of the eight major planets (Mercury – Neptune) and the Moon by means of the expansions in Tchebyschev's polynomials is described. The ephemerides are extended over the period from JD 2334544.5 to JD 2455760.5 [1679 August 31, 2011 July 18].

046.036   **Sur la Sainte Luce et les autres dates du calendrier.**
R. Sagot, D. Savoie.
Astronomie, Vol. 101, p. 651 – 656 (1987).

046.037   **Chinese Astronomical Ephemeris 1988.**
Purple Mountain Observatory, Academia Sinica, Nanking, People's Republic of China. 6 + 539 pp. (1987). In Chinese.

046.038   **Philippine Astronomical Handbook 1987.**
Prepared by the Astronomical Observation Division of the National Geophysical and Astronomical Office.
Philippine Atmospheric, Geophysical and Astronomical Services Administration, Quezon City, Philippines. 12 + 63 pp. (1987). ISSN 0115-1207.

046.039   **Astronomical calendar for the year 1988.**
B. Kovachev (Editor).
Izdatelstvo na Blgarskata Akademiya na Naukite, Sofiya. 35th year of publication. 171 pp. Price 2.85 Lv, (1988). In Bulgarian.

046.040   **Über die mißverstandene Dekanenlehre in Rheinischen Kalendern (*Missunderstanding of the decane–system and medieval calendars of the Rhine–Area.*)**
K. A. F. Fischer.
Privatdruck, Karl Fischer, Rott im Elsass, France, 3 pp. (1987).

046.041   **Tables of sunrise, sunset, twilight, moonrise, & moonset 1988.**
Prepared by the Astronomical Publication Unit, Astronomical Research and Development Section, Atmospheric, Geophysical and Space Science Branch.
Philippine Atmospheric, Geophysical and Astronomical Services Administration, Quezon City, Philippines, 11 + 57 pp. (1987). ISSN 0115-3307.

046.042   **Astronomische Kurzkalender 1900 – 2000. Der erste Himmelskalender für hundert Jahre.**
H. Mucke.
Astronomisches Büro, Hasenwartgasse 32, 1238 Wien, Austria. Price öS 250.00 (1988).
Review in Sternenbote, 30. Jahrg., Nr. 12, p. 259 (1987).

046.043   **Anuarul Astronomic 1988.**
Centrul de Astronomie si Stiinte Spatiale, Bucuresti. Editura Academiei Republicii Socialiste România, R. 79717, Bucuresti, Calea Victoriei 125, Rumania. 357 pp. Price Lei 17.00 (1987).

1988 Yearbook of Astronomy.
See Abstr. 003.006.

The Borana calendar: some observations.
See Abstr. 004.050.

Sky calendars of the Indo–Malay archipelago.
See Abstr. 004.106.

The date and time of the vernal equinox: a graphical representation of the Gregorian calendar.
See Abstr. 014.064.

Die Berechnung einiger planetarer Phänomene.
See Abstr. 021.005.

Jupiter without satellites.
See Abstr. 099.053.

1988 Comet Handbook.
See Abstr. 103.022.

Space Research

051 Extraterrestrial Research Related to Astronomy and Astrophysics

051.001 **Observations of neutron stars planned by the High Speed Photometer team using Space Telescope.**
J. F. Dolan.
The origin and evolution of neutron stars, p. 556 (1987). Abstract. – See Abstr. 012.001 (IAU Symp. No. 125).

051.002 **Using SIRTF to study extragalactic star formation.**
E. L. Wright.
NASA Conf. Publ., NASA CP–2466, p. 629 (1987). Abstract. – See Abstr. 012.002.

051.003 **Inter–agency coordination of solar–terrestrial science projects.**
R. Reinhard.
ESA Bull., No. 51, p. 8 – 21 (1987).

051.004 **Medium–sized astronomy missions under study for ESA's Horizon 2000 programme.**
S. Volonté.
ESA Bull., No. 51, p. 22 – 26 (1987).
An important element of ESA's Space Science: Horizon 2000' Long–Term Programme consists of projects of the medium–class category. The present selection cycle within the Space Science Programme includes five mission concepts that will be in competition for approval in 1988. Three are astronomy missions, covering the areas of gamma–ray and ultra–violet observations ('GRASP' and 'Lyman', respectively) and space radio interferometry ('Quasat').

051.005 **International Space Report.**
Spaceflight, Vol. 29, No. 8, p. 270 – 279 (1987).
1987 August review of space news and events.

051.006 **Space at JPL.**
W. McLaughlin.
Spaceflight, Vol. 29, No. 8, p. 301 – 306 (1987).
Contents: To the moon on the cheap. Scanning tunnelling microsope. Galileo mission.

051.007 **International Space Report.**
Spaceflight, Vol. 29, No. 9, p. 318 – 323 (1987).
1987 September review of space news and events.

051.008 **Space at JPL.**
W. McLaughlin.
Spaceflight, Vol. 29, No. 9, p. 346 – 350 (1987).
Contents: Our galactic home. Atmospheric molecules.

051.009 **The Phobos project.**
V. Balebanov, A. Zakharov, V. Linkin.
Astron. Now, Vol. 1, No. 3, p. 6 – 11 (1987).
Next year will see the start of an international expedition to Mars and its satellites. The Soviet Union is to launch two robot probes, and the larger of the two Martian satellites, Phobos, will be studied from very close range.

051.010 **The Space Station Millimeter Facility.**
K. W. Weiler, B. K. Dennison, R. M. Bevilacqua, J. H. Spencer, K. J. Johnston.
Radio astronomy from space, p. 81 – 96 (1987). – See Abstr. 012.009.
The design and use of a large millimeter telescope to be constructed on the planned Space Station (the Space Station Millimeter Facility, SSMF) are described. Such a facility will have manifold applications in both basic and applied research and will be the premier instrument in the world at high radio frequencies.

051.011 **The Astronomical and Atmospheric Spectroscopy Explorer.**
L. J. Rickard.
Radio astronomy from space, p. 97 – 100 (1987). – See Abstr. 012.009.
The proposed Astronomical and Atmospheric Spectroscopy Explorer attempts to exploit the interrelation of advances in millimeter–wavelength and far–infrared spectroscopic studies of the interstellar medium and the earth's atmosphere.

051.012 **An Ionized Carbon atomic ISM Explorer (*ICE*).**
S. R. Kulkarni, D. M. Watson.
Radio astronomy from space, p. 101 – 106 (1987). – See Abstr. 012.009.
The authors propose an Explorer–class satellite for an all–sky survey in the [C II] 157.7 μm line at low angular resolution and high spectral resolution with complete Galactic velocity coverage.

051.013 **QUASAT.**
J. F. Jordan.
Radio astronomy from space, p. 171 – 173 (1987). – See Abstr. 012.009.

051.014 **Space: looking to year 2001.**
M. Wilhite.
Spaceflight, Vol. 29, Suppl. No. 2, p. 48 – 53 (1987).

051.015 **30 years of space exploration.**
Spaceflight, Vol. 29, Suppl. No. 2, p. 54 – 79 (1987).

051.016 **Weltraummissionen zu Venus und Mars. Teil 1: Missionen zur Venus.**
H.–H. Altfeld.
Sterne Weltraum, 26. Jahrg., Nr. 10, p. 546 – 551 (1987).

051.017 **XMM (mission de spectroscopie X) pierre angulaire du programme Horizon 2000 de l'E.S.A.**
J. Astron. Fr., No. 30, p. 15 – 16 (1987).

051.018 **Potential of remote sensing for the study of global change. COSPAR report to the International Council of Scientific Unions (ICSU).**
S. I. Rasool (Editor).
Adv. Space Res., Vol. 7, No. 1, 6 + 97 pp. (1987).

051.019   The SAX mission.
L. Scarsi.
Variability of galactic and extragalactic X–ray sources, p. 233 – 252 (1987). – See Abstr. 012.019.

The SAX mission, acronym for "Satellite for Astronomy in X–rays", has the aim of carrying out spectroscopic, spectral and time variability studies on celestial X–ray sources, with a broad energy range measurement capability, covering the region from 0.1 keV to 200 keV, continuing and expanding upon previous observations. The mission is based on a Free Flyer, of the one ton class on a low altitude (550 km) circular orbit near equatorial.

051.020   The natural background at Shuttle altitudes.
M. R. Torr, J. K. Owens, J. W. Eun, D. G. Torr, P. G. Richards.
Adv. Space Res., Vol. 7, No. 5, p. 141 – 151 (1987). – See Abstr. 012.022.

Optical measurements made from the Space Shuttle include several sources of emission, each modified according to viewing configuration, Shuttle altitude, solar activity, local time, and latitude. In this paper the authors present a model spectrum for one of these components, the natural airglow background. The spectrum is modeled over a wavelength range extending from the extreme ultraviolet to the near infrared. The effect of different viewing configurations is illustrated, together with day to night variations.

051.021   The Shuttle induced background: gaseous constituents.
G. R. Carignan, E. R. Miller.
Adv. Space Res., Vol. 7, No. 5, p. 153 – 160 (1987). – See Abstr. 012.022.

051.022   The particle environment on–orbit: observations, calculations, and implications.
B. D. Green.
Adv. Space Res., Vol. 7, No. 5, p. 161 – 168 (1987). – See Abstr. 012.022.

The full potential for making remote observations from space free from atmospheric attenuations and distortions may not be realized due to the residual environment surrounding orbital experiments: particulates could overwhelm or severely complicate remote astronomical or atmospheric sounding observations. Small particles are lifted into space by the observatory and its carrier and take considerable time to evolve from surfaces. Single near–field particles have been observed which produce irradiance levels larger than the brightest stars and brighter than the emission from the entire earth limb airglow layer. The existing data bases are reviewed in this paper.

051.023   Space vehicle optical contamination by ram glow.
S. B. Mende, G. R. Swenson, E. J. Llewellyn.
Adv. Space Res., Vol. 7, No. 5, p. 169 – 178 (1987). – See Abstr. 012.022.

051.024   Plasma interactions in the Space Shuttle Orbiter environment.
W. J. Raitt.
Adv. Space Res., Vol. 7, No. 5, p. 179 – 188 (1987). – See Abstr. 012.022.

In this review, the author puts particular emphasis on some of the plasma effects resulting from the interactions between the Space Shuttle Orbiter system and its environment which have been observed by experiments flown on scientific Space Shuttle Orbiter flights to date.

051.025   Comments about observing conditions for UV astronomy aboard Spacelab.
G. Courtes, M. Viton, J. P. Simon, A. Gary, R. Decher.
Adv. Space Res., Vol. 7, No. 5, p. 189 – 193 (1987). – See Abstr. 012.022.

The Spacelab missions provided a good opportunity for a first general UV survey of astrophysical objects. The UV observations carried out on Spacelab 1 yielded some new results, but revealed at the same time observing and operation constraints which must be considered for planning and preparing for future Spacelab experiments. Experience gained with two astronomical experiments, the UV Very Wide Field Camera (VWFC, 1ES022) and FAUST Far Ultraviolet Telescope (1NS–005) are discussed.

051.026   The extreme and far ultraviolet environment at shuttle altitudes.
S. Chakrabarti.
Adv. Space Res., Vol. 7, No. 5, p. 195 – 202 (1987). – See Abstr. 012.022.

The astronomical data obtained by the Far Ultraviolet Space Telescope (FAUST) and the Very Wide Field Camera (VWFC) on board the Spacelab I mission have triggered questions on the natural and induced extreme and far ultraviolet (EUV and FUV) environment of the space shuttle. Moreover, the recent discovery of ~1k Rayleighs N_2 Lyman Birge Hopfield (LBH) nightglow emissions by the U.S. Air Force's S3–4 satellite, and subsequent confirmation by the Imaging Spectrometric Observatory (ISO) experiment on the Spacelab I mission have serious implications for the astronomical observations from the shuttle. Since both ISO and S3–4 experiments were conducted from shuttle altitudes, the implied EUV and FUV environment for astronomical observations can be severe. In order to address the question of the suitability of the shuttle as an astronomical platform, the author has examined data from FAUST and other experiments. He concludes that the FAUST background is most likely due to the observation of tropical UV arcs, a natural airglow phenomenon. Strategies for future shuttle experiments to overcome this and other natural emissions are discussed.

051.027   Optical observations from the Space Shuttle.
J. L. Weinberg.
Adv. Space Res., Vol. 7, No. 5, p. 203 – 205 (1987). – See Abstr. 012.022.

A brief survey is given of UV and visual observations made from the Space Shuttle. These include five astronomy experiments, four UV and one visual, some of which were affected by contamination originating with the Shuttle and its environment.

051.028   Search for ultraviolet Shuttle glow.
P. D. Tennyson, P. D. Feldman, R. C. Henry.
Adv. Space Res., Vol. 7, No. 5, p. 207 – 210 (1987). – See Abstr. 012.022.

The Space Shuttle Columbia flown in January 1986 carried two ultraviolet experiments (UVX) designed to observe very weak diffuse emission from various astronomical sources at wavelengths below 3200 Å with moderate spectral resolution. Such observations are extremely sensitive to the presence of any shuttle induced ultraviolet glow, since the wavelength range, 1200 – 3200 Å, includes strong emission lines or bands of species such as O, NO, and OH which are predicted to radiate strongly by models of the shuttle glow. The UVX spectrometers are sensitive to emission features as faint as 0.1 Rayleighs. Emissions from O_2, O and NO are detected and shown to be consistent with an atmospheric origin.

051.029   Infrared observations of contaminants from Shuttle flight 51–F.
D. G. Koch, G. G. Fazio, W. Hoffmann, G. Melnick, G. Rieke, J. Simpson, F. Witteborn, E. Young.
Adv. Space Res., Vol. 7, No. 5, p. 211 – 221 (1987). – See Abstr. 012.022.

A small helium cooled infrared telescope, IRT, was flown on the Shuttle in July/August 1985. The principle astrophysical objectives were to measure the large scale structure of sources and the background radiation. A cold shutter was incorporated to permit absolute flux measurements. Additionally, the engineering objectives included setting upper limits on the infrared radiation from the local environment. Even though the local background overwhelmed the astrophysical background, astronomical sources were still detectable superimposed on this background radiation. Data are presented covering the spectral range from 2 μm to 120 μm. The spatial, spectral and temporal variations are described. Based on the spectral character and variability in different wavelength bands, the background radiation does not appear to have a single origin. In this paper the results on the Shuttle environment are presented.

051.030 X–ray observations from the Space Shuttle.
G. K. Skinner, C. J. Eyles, A. P. Willmore,
D. Bertram, M. J. Church, P. K. S. Harper, J. R. H. Herring,
J. C. M. Peden, A. M. T. Pollock, T. J. Ponman, M. P. Watt.
Adv. Space Res., Vol. 7, No. 5, p. 223 – 230 (1987). – See Abstr. 012.022.
The flight of the University of Birmingham X–ray Telescope, which took place between 29 July and 6 August 1985 is reviewed. The background of spurious events in the detector, due mainly to energetic particles, was found to be low and well-behaved, except for occasional events which are not easy to distinguish from X–ray bursts. The spectral intensity of unrejected background varies from about $2 \times 10^{-3}/\text{cm}^2\text{sec keV}$ at 8 keV to $5 \times 10^{-4}/\text{cm}^2\text{sec keV}$ at 25 keV.

051.031 Measurements of background gamma radiation on Spacelab 2.
G. J. Fishman, W. S. Paciesas, J. C. Gregory.
Adv. Space Res., Vol. 7, No. 5, p. 231 – 234 (1987). – See Abstr. 012.022.
The Nuclear Radiation Monitor measured background gamma rays and charged particles nearly continuously during the Spacelab 2 mission. Preliminary investigation of the spectra and time histories shows a variety of expected effects resulting from incident charged particles, Earth albedo radiation, and secondary radiation produced within the vehicle itself. While the first two of these are well known from previous experiments, the latter component is unique to the Shuttle/Spacelab system and represents a large fraction of the total. Work is currently underway to characterize the secondary radiation in a manner which will be useful in the design of future Spacelab experiments.

051.032 A long–duration balloon system for middle–atmosphere measurements.
J. H. Smalley, N. E. Carlson.
Adv. Space Res., Vol. 7, No. 7, p. 41 – 51 (1987). – See Abstr. 012.023.
The National Center for Atmospheric Research (NCAR) collaborated with several universities on a superpressure balloon flight program for studies of the electrodynamics of the middle atmosphere (EMA). Measurements were made at 26 km for periods ranging from 15 to 60 days. All flights took place in the southern hemisphere. This paper discusses the unique features of the EMA vehicle as a platform for long–duration measurements at stratospheric altitudes.

051.033 ”VEGA” in action.
V. M. Kovtunenko.
16 Gagarin. nauchn. chteniya po kosmonavt. i aviatsii, 1986. Moskva, p. 42 – 54 (1987). In Russian. Abstr. in Ref. Zh., 62. Issled. Kosm. Prostranstva, 8.62.147 (1987).

051.034 Active plasma experiments.
G. Haerendel.
The solar wind and the Earth, p. 214 – 242 (1987). – See Abstr. 003.007.
Contents: Comets. Barium. Motions. Ion jets. Structure. Space as a laboratory. Stimulation of equatorial spread–F. Critical velocity ionization. Artificial comets.

051.035 International Space Report.
Spaceflight, Vol. 29, No. 11, p. 380 – 395 (1987).
1987 November review of space news and events.

051.036 Space at JPL.
W. McLaughlin.
Spaceflight, Vol. 29, No. 11, p. 388 – 393 (1987).
Contents: Computer aided engineering. Report from India. Imaging Neptune. Mars Rover Sample Return.

051.037 Space at JPL.
W. McLaughlin.
Spaceflight, Vol. 29, No. 12, p. 433 – 437 (1987).
Contents: Antarctic ozone. International Halley Watch. Space artist.

051.038 The Italian satellite for X–ray astronomy (SAX).
G. Spada.
General relativity and gravitational physics, p. 479 – 496 (1987). – See Abstr. 012.038.
The Italian satellite for X–ray astronomy to be launched in late 1990 will be devoted to the study of galactic and extragalactic X–ray sources in the energy interval 0.1 – 200 keV. The unique characteristics of SAX are its ability to image X–ray sources with moderate angular resolution (about 1 arcmin) over the energy range 1 – 10 keV including the Fe lines complex. The instrumentation of SAX consists of four X–ray imaging concentrators (nested double cones, approximation to Wolter I type grazing optics), a high pressure gas scintillation proportional counter, a phoswich detector and two wide field cameras.

051.039 End to end software system for the Extreme Ultraviolet Explorer Mission.
C. A. Dobson, H. L. Marshall, R. F. Malina.
Bull. Am. Astron. Soc., Vol. 19, No. 2, p. 739 (1987). Abstract. – See Abstr. 010.061.

051.040 The status of the Cosmic Background Explorer mission.
J. C. Mather, M. G. Hauser, T. Kelsall, G. F. Smoot.
Bull. Am. Astron. Soc., Vol. 19, No. 2, p. 749 – 750 (1987). Abstract. – See Abstr. 010.061.

051.041 Infrared cosmology from space.
C. J. Hogan.
Prepr. Steward Obs., No. 755, 8 pp. (1987). To appear in Astrophys. Lett. Commun.

051.042 The Sakigake/Suisei encounter with comet P/Halley.
K. Hirao, T. Itoh.
Astron. Astrophys., Vol. 187, No. 1/2, p. 39 – 46 (1987). – See Abstr. 003.013.
Two spacecraft, Sakigake and Suisei, which were launched by the Institute of Space and Astronautical Science of Japan, flied successfully by the comet P/Halley at minimum distances of 7 million and 150,000 km, respectively. The scientific objectives of these spacecraft were to explore the environment of the comet directly taking the opportunity of the re–revisiting of the first class comet, Halley. Plasma and neutral hydrogen environments were selected for scientific exploration. This article gives an overview of recent scientific results of the missions.

051.043 Hipparcos. The Input Catalogue.
A. E. Gómez, J. Torra.
Rev. Mex. Astron. Astrofis., Vol. 14, No. 2, p. 441 – 447 (1987). – See Abstr. 012.042.
The main aspects of the European Space Agency's Hipparcos mission are described. The authors briefly explain the procedure used to prepare the Input Catalogue, and show its present content for different types of stars.

051.044 OASIS (*Optical Aperture Synthesis in Space*): a mission concept.
J. E. Noordam, A. H. Greenaway, J. D. Bregman,
R. S. le Poole.
ESA Spec. Publ., ESA SP–273, p. 51 – 59 (1987). – See Abstr. 012.044.
OASIS is a proposal for a multi–aperture optical interferometer, mounted on a single structure in space. The design is guided by radio aperture synthesis experience. The essential element is the use of a bright nearby reference object to equalise optical pathlengths in the array during long integrations on faint objects. An example for a 20 m array with 9 telescopes has been worked out in detail.

051.045   IUE: nine years of astronomy.
L. A. Shore.
Astronomy, Vol. 15, No. 4, p. 14 – 17 (1987). Abstr. in Phys. Abstr., Vol. 90, No. 1311, Entry 101654 (1987).

051.046   Voyager into the third dimension [*Ulysses mission*].
G. L. Bennett.
Astronomy, Vol. 15, No. 5, p. 14 – 22 (1987). Abstr. in Phys. Abstr., Vol. 90, No. 1311, Entry 101655 (1987).

051.047   Japanese Halley's Comet probes.
T. Hayashi, M. Hashimoto.
J. Inst. Electron. Commun. Eng. Jpn., Vol. 69, No. 9, p. 928 – 933 (1986). In Japanese. Abstr. in Phys. Abstr., Vol. 90, No. 1312, Entry 107690 (1987).

051.048   The Magellan 1989 Type IV mission design.
E. Cutting.
V–Gram, No. 10, p. 4 – 8 (1987). Abstr. in Phys. Abstr., Vol. 90, No. 1312, Entry 107691 (1987).

051.049   Magellan data products.
R. E. Arvidson.
V–Gram, No. 10, p. 9 – 13 (1987). Abstr. in Phys. Abstr., Vol. 90, No. 1312, Entry 107692 (1987).

051.050   The Magellan altimetry experiment.
G. H. Pettengill, P. Ford.
V–Gram, No. 10, p. 13 – 20 (1987). Abstr. in Phys. Abstr., Vol. 90, No. 1312, Entry 107693 (1987).

051.051   Phobos – a multi–purpose mission for investigation of Phobos, Mars, the Sun and cosmic space.
V. M. Kovtunenko, R. S. Kremnev, G. N. Rogovskij, K. G. Sukhanov.
Astron. Vestn., Tom 21, No. 4, p. 278 – 285 (1987). In Russian. English translation in Sol. Syst. Res.
The most important characteristics of the multi–purpose Phobos mission designed for the investigation of the outer space, of Phobos itself and the Sun (the Phobos program), are described. The mission scenario, including the convergence with Phobos, and the design of the space vehicle as well as its airborne systems are presented.

051.052   Landmarks in the cosmic era (on the occasion of the 30th anniversary of launching the first artificial satellite of the Earth).
Ya. S. Yatskiv, R. M. West.
Ocherki istorii estestvoznaniya i tekhniki, Vyp. 33, p. 30 – 42 (1987). In Russian.

051.053   An update on the Active Magnetospheric Particle Tracer Explorers (AMPTE) program.
R. W. McEntire.
Johns Hopkins APL Tech. Dig., Vol. 8, No. 3, p. 340 – 347 (1987).
On August 16, 1984, the three spacecraft of the AMPTE mission were launched from Cape Canaveral into earth orbit. With both active experiments and a new and much more capable generation of instruments and data handling systems, AMPTE represents a significant advance in space plasma physics. The third anniversary of the launch seems a good time to reflect on the AMPTE mission and to review progress to date.

051.054   Verification of the HIPPARCOS payload short term stability performances.
J. P. Camus.
Proc. SPIE Int. Soc. Opt. Eng., Vol. 655, p. 372 – 379 (1986). Abstr. in Phys. Abstr., Vol. 90, No. 1313, Entry 115203 (1987). – See Abstr. 012.063.

051.055   Use of Space Station for space science.
G. P. Haskell.
ESA Spec. Publ., ESA SP–1093, p. 25 – 26 (1987). – See Abstr. 012.072.

051.056   The Comet Rendezvous Asteroid Flyby mission.
M. Neugebauer.
ESA Spec. Publ., ESA SP–278, p. 517 – 522 (1987). – See Abstr. 012.075.
The Comet Rendezvous Asteroid Flyby mission (CRAF) is planned to be launched in February, 1993. After gravity assists from Venus and Earth, the spacecraft will fly by the asteroid 46 Hestia en route to a rendezvous with P/Tempel 2 in November, 1996, when the comet is near aphelion. The scientific results expected to be obtained from this experiment are described.

051.057   The Giotto extended mission.
R. Reinhard.
ESA Spec. Publ., ESA SP–278, p. 523 – 529 (1987). – See Abstr. 012.075.
About one week after the Halley flyby Giotto's orbit was changed so that it can return to the Earth. Using an Earth swingby on 2 July 1990 Giotto could be targeted to encounter a second comet. The short–period comet Grigg–Skjellerup was selected as the most suitable target. The encounter would occur on 10 July 1992, 12 days before its perihelion passage. Before the decision can be made to extend the Giotto mission the spacecraft and all experiments will be carefully checked out.

051.058   Rosetta: a mission to sample the nucleus of a comet.
J. A. Wood.
ESA Spec. Publ., ESA SP–278, p. 531 – 537 (1987). – See Abstr. 012.075.
A joint ESA/NASA committee has considered the opportunities and problems of a spacecraft mission that would land on a comet nucleus, take samples, and return them in a suitably refrigerated state for study in terrestrial laboratories. These studies would greatly advance both planetary science and astrophysics.

051.059   Retrieving samples from comet nuclei.
E. Stuhlinger, H. Bassner, H. Fechtig, E. Igenbergs, H. Kuczera, H. Loeb, D. Schobert.
ESA Spec. Publ., ESA SP–278, p. 539 – 546 (1987). – See Abstr. 012.075.
Material samples from a comet nucleus for analysis in earth–based laboratories should be collected continuously to a depth of about 3 meters below the surface, and at a solar distance of at least 2.5 AU where the comet surface is inactive. On the proposed mission, the spacecraft will be propelled by chemical and also by electric thrusters.

051.060   Proposition for a mission to comet P/Schwassmann–Wachmann 1.
D. Benest, J. Klinger.
ESA Spec. Publ., ESA SP–278, p. 733 – 736 (1987). – See Abstr. 012.075.
Possible spaceprobe transfer orbits from Earth to the 1995 and 2002 passage–to–node (i.e. ecliptic–crossing) positions of P/Schwassmann–Wachmann 1 are presented, with return trajectories.

051.061   Project "Phobos" – new expedition to Mars.
A. V. Zakharov, G. N. Rogovskij.
Zemlya Vselennaya, No. 4, p. 7 – 14 (1987). In Russian.

051.062   Program "INTERKOSMOS" – 20 years.
V. A. Egorov.
Zemlya Vselennaya, No. 4, p. 44 – 48 (1987). In Russian.

051.063   **Scientific results of the Vega mission.**
V. I. Moroz.
Kosm. Issled., Tom 25, Vyp. 5, p. 643 – 648 (1987). In Russian.
English translation in Cosm. Res.

051.064   **Balloons for Mars missions.**
J. Blamont.
Acta Astronaut., Vol. 15, No. 8, p. 523 – 525 (1987). Abstr. in
Phys. Abstr., Vol. 90, No. 1316, Entry 134106 (1987).

051.065   **Communications from the limits of the solar system.**
E. C. Posner, R. L. Hortorr, T. L. Grant.
Elettrotecnica (*Italy*), Vol. 74, No. 4, p. 351 – 359 (1987). In
Italian. Abstr. in Phys. Abstr., Vol. 90, No. 1316, Entry 134107
(1987).

051.066   **Deep space optical communications development program.**
J. R. Lesh.
Proc. SPIE Int. Soc. Opt. Eng., Vol. 756, p. 8 – 11 (1987). Abstr.
in Phys. Abstr., Vol. 90, No. 1316, Entry 134133 (1987).

051.067   **The Giotto mission to comet Halley.**
R. Reinhard.
J. Phys. E, Vol. 20, No. 6, Part 2, p. 700 – 712 (1987). Abstr. in
Phys. Abstr., Vol. 90, No. 1317, Entry 139882 (1987).

051.068   **Analysis of simulated images from the Extreme Ultraviolet Explorer.**
H. L. Marshall, C. A. Dobson, R. F. Malina, S. Bowyer.
Proceedings of the international topical meeting on image
detection and quality, p. 275 – 280 (1986). Abstr. in Phys. Abstr.,
Vol. 91, No. 1320, Entry 10409 (1988). – See Abstr. 012.079.

051.069   **Energija – the first return rocket.**
M. Grün.
Kozmos, Vol. 18, No. 5, p. 145 – 146 (1987). In Slovak.

051.070   **On the necessity of the space mission to comet Encke.**
I. A. Simoniya.
Komet. Tsirk., No. 370 (1987). In Russian.

Space. The next twenty–five years.
See Abstr. 003.093.

Status report on RADIOASTRON.
See Abstr. 013.019.

Collaborative VLBI experiments with RADIOASTRON.
See Abstr. 013.020.

QUASAT – European status report.
See Abstr. 013.021.

What do we learn from space? Space science in Japan.
See Abstr. 013.044.

**Cambridge observations at 38 – 151 MHz and their implications
for space astronomy.**
See Abstr. 033.009.

A low–frequency synthesis array in Earth orbit.
See Abstr. 033.010.

The Low Frequency Space Array.
See Abstr. 033.011.

The Cosmic Background Explorer (COBE) satellite.
See Abstr. 035.007.

The Advanced X–Ray Astrophysics Facility: an overview.
See Abstr. 035.024.

Space Station based interferometry.
See Abstr. 035.076.

The ground segment's vital role in the Hipparcos scientific mission.
See Abstr. 036.011.

A comparison of interferometry from space and ground.
See Abstr. 036.168.

**Precise proper motions and positions of stars from a combination of
fundamental catalogues with the HIPPARCOS catalogue.**
See Abstr. 041.009.

**Detection of low frequency gravitational waves with interplanetary
spacecraft.**
See Abstr. 066.088.

**A new laser ranged satellite, LAGEOS 3, to measure the gravito-
magnetic field.**
See Abstr. 066.089.

The Sun's spots and flares.
See Abstr. 072.043.

Satellite data alters view on Earth–space environment.
See Abstr. 083.002.

Future exploration of Mars.
See Abstr. 097.077.

Vitesses radiales pour le programme HIPPARCOS. II.
See Abstr. 111.015.

**The anisotropy of the microwave background: space experiment
"RELICT".**
See Abstr. 161.044.

052 Astrodynamics, Navigation of Space Vehicles

052.001 Low–thrust navigation to comets or asteroids.
M. Noton, S. V. Salehi.
ESA J., Vol. 11, No. 2, p. 215 – 231 (1987).
The use of electric propulsion at very low thrust levels for rendezvous missions to comets and asteroids requires special techniques, both for the generation of the nominal trajectories and for the navigation in the vicinity of such trajectories. This paper is concerned with orbit determination and guidance by ground–based measurements and command for the heliocentric transfer trajectories. The orbit determination is affected by stochastic effects from the motor, which also gives rise to the need for guidance by means of a feedback law.

052.002 Analysis of the orbit of Intercosmos 10 rocket (1973–82B) in its last 15 days.
A. N. Winterbottom, W. J. Boulton.
Planet. Space Sci., Vol. 35, No. 7, p. 921 – 936 (1987).
Intercosmos 10 rocket (1973–82B) was launched on 30 October 1973 into an orbit of eccentricity 0.08 inclined at 74° to the equator, and decayed on 8 October 1977. Orbits have been determined daily in the last 15 days of its life from NORAD and U.S. Navy observations; and an orbital accuracy of about 55 m radially and 70 m cross–track was obtained. Analysis of the decrease in perigee distance has provided two values of atmospheric density scale height at heights of 190 and 200 km. The variations in orbital inclination due to atmospheric rotation and 16th–order resonance with the geopotential have been analysed, and give a morning value for the atmospheric rotation rate of 0.90 ± 0.06 rev day^{-1} at a height of approximately 200 km and two values for lumped 16th–order harmonic coefficients with accuracies equivalent to 1.5 and 5.0 cm in geoid height.

052.003 An analytic satellite theory using gravity and a dynamic atmosphere.
F. R. Hoots, R. G. France.
Celest. Mech., Vol. 40, No. 1, p. 1 – 18 (1987).
An analytical solution is given for the motion of an artificial Earth satellite under the combined influences of gravity and atmospheric drag. The gravitational effects of the zonal harmonics J_2, J_3, and J_4 are included, and the drag effects of any arbitrary dynamic atmosphere are included. By a dynamic atmosphere, the authors mean any of the modern empirical models which use various observed solar and geophysical parameters as inputs to produce a dynamically varying atmosphere model. The subtleties of using such an atmosphere model with an analytic theory are explored, and real world data is used to determine the optimum implementation. Performance is measured by predictions against real world satellites. As a point of reference, predictions against a special perturbations model are also given.

052.004 The LAGEOS Lense–Thirring precession and the LAGEOS non–gravitational nodal perturbations – I.
I. Ciufolini.
Celest. Mech., Vol. 40, No. 1, p. 19 – 33 (1987).
After a brief description of the experiment to detect the gravitomagnetic field using high altitude laser ranged artificial satellites, the author studies several non–gravitational perturbations that affect the LAGEOS nodal longitude. It is shown that the error in the calculated value of the secular nodal precession or the value of the secular nodal precession itself is, for each perturbation, less than 1% of the gravitomagnetic drag.

052.005 Analytical formulae of the perturbations of the Delaunay orbital elements of a triaxial satellite moving in a central gravitational field.
D. Z. Koenov.
Izv. Akad. Nauk TadzhSSR. Otd–nie Fiz.–Mat., Khim. Geol. Nauk, No. 2, p. 61 – 67 (1986). In Russian. Abstr. in Ref. Zh., 51. Astron., 7.51.88; 62. Issled. Kosm. Prostranstva, 7.62.278 (1987).

052.006 Conditions for constraints on the orbit of a space vehicle in the neighbourhood of the collinear libration center of the restricted elliptical three–body problem.
P. E. Ehl'yasberg, T. A. Timikhova.
Dokl. Akad. Nauk SSSR. Ser. Mat. Fiz., Tom 293, No. 1, p. 55 – 58 (1987). In Russian. Abstr. in Ref. Zh., 62. Issled. Kosm. Prostranstva, 7.62.275 (1987).

052.007 On libration points of a few bodies of the solar system.
Yu. V. Berkin, Yu. M. Gubarev.
Tr. 19 Chtenij, posvyashch. nauchn. naslediya i razvitiyu idej K. Eh. Tsiolkovskogo, Kaluga 13 – 16 sent. 1984. Sekts. mekh. kosm. poleta, Moskva, p. 82 – 84 (1986). In Russian. Abstr. in Ref. Zh., 62. Issled. Kosm. Prostranstva, 7.62.279 (1987).

052.008 On the improvement of the parameters of motion of artificial earth satellites.
G. I. Smievskaya, A. A. Pyarnpuu.
Mat. zadachi priklad. aehron., p. 41 – 59 (1987). In Russian. Abstr. in Ref. Zh., 62. Issled. Kosm. Prostranstva, 7.62.282 (1987).

052.009 Application of spline interpolation in the Laplace method.
M. G. Ber, S. M. Poleshchikov.
Komi Filial Akad. Nauk SSSR. Syktyvkar, 12 pp. (1987). In Russian. Abstr. in Ref. Zh., 51. Astron., 8.51.60 (1987).

052.010 The estimation of the influence of tesseral harmonics perturbations on the orbits of some geodetic satellites.
E. Wnuk, I. Wytrzyszczak.
Artif. Satell., Vol. 22, No. 2, p. 47 – 64 (1987).
The analysis of the influence of particular tesseral harmonics of the earth gravity field as well as the whole field of tesseral harmonics on the motion of a few satellites is discussed.

052.011 The observation of artificial satellites and the determination of the orbit.
H. Hauck, E. Reinhart, P. Wilson.
Dtsch. Geod. Komm. Bayer. Akad. Wiss., Reihe B, Heft Nr. 284, p. 94 – 118 (1987). In German and English.

052.012 Satellite–to–satellite tracking, satellite gravity gradiometry.
K. H. Ilk.
Dtsch. Geod. Komm. Bayer. Akad. Wiss., Reihe B, Heft Nr. 284, p. 133 – 138 (1987). In German and English.

052.013 Multi–body force function: geodetic aspects of astrodynamics.
K. H. Ilk.
Dtsch. Geod. Komm. Bayer. Akad. Wiss., Reihe B, Heft Nr. 284, p. 191 – 198 (1987). In German and English.

052.014 On the choice of the reference system in investigations of the motion of high satellites.
I. S. Gayazov, A. S. Sochilina.
Byull. Inst. Teor. Astron., Tom 15, No. 9, p. 481 – 485 (1986). In Russian.
In the investigation of the motion of high satellites under the influence of the geopotential and the luni–solar attraction it is possible to introduce an analogy of the Laplacian plane. This plane is moving with the plane of the equator of date; the proposed coordinate system is not an equatorial one. The perturbations in the orbital elements of the satellite due to the motion of the adopted system have been studied.

052.015   **On nonlinear random effects in astrodynamics.**
V. G. Degtyarev, N. Yu. Dodonov,
S. Yu. Shishkovskij.
Byull. Inst. Teor. Astron., Tom 15, No. 9, p. 505 – 510 (1986). In Russian.

Results are given for the approximate determination of the distribution of the random vector function describing a nonlinear random astrodynamic effect. Their application for study of the distribution of the semi–major axis and eccentricity of an orbit as functions of random initial coordinates and velocities, distributed according to the normal law, are considered. Inequalities for estimation of the proximity of distribution laws according to the proximity of the generating functions are established. A method for determining the distribution law of the random vector-argument function by the use of its expansion in Fourier series with respect to Chebyshev–Hermite functions is developed.

052.016   **Motion of a mass point in the gravitational field of a spheroidal body.**
V. L. Lisitsyn.
Sistemy upr. let. apparatov. Khar'kov, p. 183 – 189 (1986). In Russian. Abstr. in Ref. Zh., 62. Issled. Kosm. Prostranstva, 10.62.199 (1987).

052.017   **Geopotential orbit variations: applications to error analysis.**
C. A. Wagner.
J. Geophys. Res., Vol. 92, No. B8, p. 8136 – 8146 (1987).

The position and velocity errors of a near–circular orbit due to geopotential uncertainty are developed in Fourier series. Application of the development is made to (1) the prediction of errors for a future altimeter satellite, (2) the capability of earth–based tracking of a close Mars orbiter in resolving that planet's gravity potential, and (3) improvement of the resolution of the low–degree geopotential from the Geopotential Research Mission using tracking data from the Global Positioning System.

052.018   **Earth anisotropic reflection and the orbit of LAGEOS.**
D. P. Rubincam, P. Knocke, V. R. Taylor,
S. Blackwell.
J. Geophys. Res., Vol. 92, No. B11, p. 11662 – 11668 (1987).

The authors find that anisotropic reflection of sunlight by the oceans cannot explain the large fluctuations in the anomalous along–track acceleration of LAGEOS. Even for the idealized case of a cloudless earth this mechanism accounts for only half the magnitude of the observed accelerations. A more realistic earth with a highly symmetric cloud cover reduces the accelerations even further. A guess would place them an order of magnitude below those observed.

052.019   **Strategies for high–precision global positioning system orbit determination.**
S. M. Lichten, J. S. Border.
J. Geophys. Res., Vol. 92, No. B12, p. 12751 – 12762 (1987).

High–precision orbit determination of Global Positioning System (GPS) satellites is a key requirement for GPS–based precise geodetic measurements and precise low–Earth orbiter tracking. The authors explore different strategies for orbit determination with data from 1985 GPS field experiments.

052.020   **Orbits for meridian observation.**
N. Nagarajan, S. Akila.
J. Astronaut. Sci., Vol. 34, No. 4, p. 421 – 430 (1986). Abstr. in Phys. Abstr., Vol. 90, No. 1309, Entry 88075 (1987).

052.021   **Orbit transfer error analysis for multiple, finite perigee burn, ascent trajectories.**
N. J. Adams, R. G. Melton.
J. Astronaut. Sci., Vol. 34, No. 4, p. 355 – 373 (1986). Abstr. in Phys. Abstr., Vol. 90, No. 1309, Entry 88084 (1987).

052.022   **The contraction of satellite orbits under the influence of air drag. VII. Orbits of high eccentricity, with scale height dependent on altitude.**
D. G. King–Hele, D. M. C. Walker.
Proc. R. Soc. London, Ser. A, Vol. 411, No. 1840, p. 1 – 17 (1987). Abstr. in Phys. Abstr., Vol. 90, No. 1310, Entry 94632 (1987).

052.023   **An improved theory for determining changes in satellite orbits caused by meridional winds.**
D. G. King–Hele, D. M. C. Walker.
Proc. R. Soc. London, Ser. A, Vol. 411, No. 1840, p. 19 – 33 (1987). Abstr. in Phys. Abstr., Vol. 90, No. 1310, Entry 94633 (1987).

052.024   **The motion of a geosynchronous satellite. II.**
K. B. Bhatnagar, M. Mehra.
Indian J. Pure Appl. Math., Vol. 18, No. 5, p. 461 – 477 (1987). Abstr. in Phys. Abstr., Vol. 90, No. 1312, Entry 107689 (1987).

052.025   **Autonomous orbit determination.**
R. W. Day, J. S. Curiale.
1987 IEEE Aerospace Applications Conference Digest, Vail, CO, USA, 8 – 13 February 1987. IEEE, New York, USA, 16 pp. (1987). Abstr. in Phys. Abstr., Vol. 90, No. 1313, Entry 115211 (1987).

052.026   **Alternative cometary targets for the Giotto extended mission.**
R. W. Farquhar, D. W. Dunham, S. C. Hsu.
ESA Spec. Publ., ESA SP–278, p. 727 – 731 (1987). – See Abstr. 012.075.

The Giotto spacecraft, which is now operating in a "hibernation" configuration, is scheduled to be reactivated in November 1989. If it is found that the imaging system can still provide useful data, serious consideration will be given to sending Giotto to another comet. The Giotto Science Working Team has tentatively selected the Grigg–Skjellerup encounter on July 10, 1992, for the proposed extended–mission phase.

052.027   **Rotational evolution of a symmetric gyrostat with visco–elastic bars around the center of mass in a circular orbit.**
S. A. A. E.–H. Aly.
Celest. Mech., Vol. 40, Nos. 3 – 4, p. 225 – 231 (1987).

The motion of a gyrostat in a circular orbit in a Newtonian field of force is considered. The gyrostat has four homogeneous viscoelastic bars attached to it. Rotation of the symmetric rotor inside the rigid body is statically and dynamically balanced. Bending deformations of the bars, accompanied by dissipation of energy, are the cause of the evolution of the system's rotational motion. Approximate equations describing this evolution are derived, together with averaged equations in Andoyer variables.

052.028   **Note on Cid's radial intermediary and the method of averaging.**
A. Deprit, S. Ferrer.
Celest. Mech., Vol. 40, Nos. 3 – 4, p. 335 – 343 (1987).

In the main problem of artificial satellite theory, the difference between the Hamiltonian and Cid's radial intermediary is a function of the argument of latitude whose average over the mean anomaly is zero.

052.029   **Radial, transverse and normal satellite position perturbations due to the geopotential.**
G. W. Rosborough, B. D. Tapley.
Celest. Mech., Vol. 40, Nos. 3 – 4, p. 409 – 421 (1987).

Perturbations in the position of a satellite due to the Earth's gravitational effects are presented. The perturbations are given in the radial, transverse (or along–track) and normal (or cross–track) components. The solution is obtained by projecting the Kepler element perturbations obtained by Kaula (1966) into each of the three components.

052.030   **The motion of a satellite in an axi–symmetric gravitational field.**
B. Zafiropoulos.
Astrophys. Space Sci., Vol. 139, No. 1, p. 111 – 128 (1987).

The equations for the variation of the osculating elements of a satellite moving in an axi–symmetric gravitational field are integrated to yield the complete first–order perturbations for the elements of the orbit. The expressions obtained include the effects produced by the second to eighth spherical harmonics. The orbital elements are presented in the most general form of summations by means of Hansen coefficients. This paper gives the respective general expressions for the secular perturbations of the orbital elements. The formulae presented should be useful for the reductions of Earth–satellite observations and geopotential studies based on them.

052.031   **Tables for the trigonometric series representations of the orbital inclination function.**
M. A. Sharaf, M. S. Ella, M. S. Abo–Elazm.
Astrophys. Space Sci., Vol. 139, No. 2, p. 199 – 231 (1987).

In this paper, tables for the trigonometric series representations of the orbital inclination function $F_{lmp}(i)$ in multiples of cosines or sines are presented for $l = 2(1)10$; $m = 0(1)l$; $p = 0(1)l$.

052.032   **Perturbations in the perigee distance due to atmospheric drag for artificial Earth satellites.**
Y. E. Helali.
Bull. Astron. Inst. Czech., Vol. 38, No. 6, p. 329 – 334 (1987).

The mathematical formulation of the problem of the perturbations in the perigee distance due to atmospheric drag and its analytical solution, taking into consideration the TD model (Total Density) which was formulated by Sehnal (1986) is presented. Using the TD model, the theory is valid for altitudes ranging from 200 to 500 km above the Earth's surface and for the solar 10.7 cm flux from 69 to 220 (in 10^{-22}Watt m^{-2}Hz^{-1} bandwidth units). Numerical examples are given to show the perturbing effect of air drag on the perigee distance for artificial Earth satellites.

052.033   **Earth gravity field and high satellite orbits.**
J. Klokočnik, J. Kostelecký.
Bull. Astron. Inst. Czech., Vol. 38, No. 6, p. 334 – 344 (1987).

Matrices of sensitivity for high orbits of satellites LAGEOS 1, 2, POPSAT, GPS, GLONASS and of geostationary satellites are computed. The use of these matrices to determine the harmonic geopotential coefficients from the individual satellites is discussed.

052.034   **A method for approximating the function of variation of AES orbital elements.**
G. S. Kurbasova.
Nauchn. Inf., Vyp. 58, p. 9 – 13 (1986). In Russian.

A method of satellite's orbit elements prediction for ephemeris calculation is described. A use of cubic spline functions is suggested.

052.035   **Description of the relative motion of an artificial satellite in a coorbiting coordinate system.**
M. V. Kuznetsov.
Nauchn. Inf., Vyp. 58, p. 87 – 90 (1986). In Russian.

The paper deals with the relative motion of two artificial satellites and the determination of the gravity field in their neighborhood.

052.036   **Radio interferometry and space navigation.**
L. I. Matveenko.
Zemlya Vselennaya, No. 4, p. 20 – 26 (1987). In Russian.

052.037   **Analytical dispersion analysis of a strategy of orbital manœuvres.**
P. Escudier.
Acta Astronaut., Vol. 15, No. 5, p. 249 – 252 (1987). Abstr. in Phys. Abstr., Vol. 90, No. 1317, Entry 139879 (1987).

052.038   **Singular fuel–optimal space trajectories based on linearization about a point in circular orbit.**
T. E. Carter.
J. Optimiz. Theory Appl. (USA), Vol. 54, No. 3, p. 447 – 470 (1987). Abstr. in Phys. Abstr., Vol. 90, No. 1318, Entry 145806 (1987).

052.039   **Optimal control laws for orbiting tethered platform systems.**
P. M. Bainum, S. E. Woodard, J.–N. Juang.
J. Astronaut. Sci., Vol. 35, No. 2, p. 135 – 153 (1987). Abstr. in Phys. Abstr., Vol. 91, No. 1320, Entry 10384 (1988).

052.040   **Analytic study of the solution families of the extended Godal's time equation for Lambert problem.**
F. T. Sun, N. X. Vinh, T.–J. Chern.
J. Astronaut. Sci., Vol. 35, No. 2, p. 213 – 234 (1987). Abstr. in Phys. Abstr., Vol. 91, No. 1320, Entry 10385 (1988).

052.041   **An analytical theory on the disturbed motion of an artificial Earth satellite.**
M. T. R. Fitzgibbon, R. Vilhena de Moraes.
J. Astronaut. Sci., Vol. 35, No. 2, p. 235 – 244 (1987). Abstr. in Phys. Abstr., Vol. 91, No. 1320, Entry 10386 (1988).

052.042   **Atmospheric tidal perturbations on the orbits of artificial satellites.**
C. Huang, M.–f. He.
Acta Astron. Sin., Vol. 28, No. 3, p. 237 – 241 (1987). In Chinese.

An analytical expression for atmospheric tidal perturbations is derived and its order of magnitude is estimated. The relative order of magnitude of atmospheric tidal perturbations is 5×10^{-11}, or 2.6% of the solid tidal perturbations. Therefore, this effect must be taken into account for precise orbit determinations of artificial satellites.

Satellite orbits in an atmosphere. Theory and applications.
See Abstr. 003.081.

Orbits for amateurs with a microcomputer. Volume II.
See Abstr. 003.133.

Introduction to space dynamics.
See Abstr. 003.136.

Ground trajectories of geosynchronous satellites and the analemma.
See Abstr. 014.038.

Position determination and the Fine Sun Sensor.
See Abstr. 036.030.

Generalized elliptic anomalies.
See Abstr. 042.073.

Values of 16th–order harmonics in the geopotential from analysis of resonant orbits.
See Abstr. 081.005.

The accuracy of a Goddard TOPEX gravity model as seen by independent resonant data.
See Abstr. 081.006.

Models of the thermosphere total density for satellite dynamics.
See Abstr. 082.089.

The structure of ULF waves produced by a tethered satellite system.
See Abstr. 106.020.

053 Artificial Satellites, Space Probes

053.001 Satellite Digest–204.
R. D. Christy.
Spaceflight, Vol. 29, No. 8, p. 278 – 279 (1987).
1987 August listing of satellite and spacecraft launches.

053.002 Satellite Digest–205.
R. D. Christy.
Spaceflight, Vol. 29, No. 9, p. 320 (1987).
1987 September listing of satellite and spacecraft launches.

053.003 Solar power satellites.
R. R. Vondrak.
The solar wind and the Earth, p. 286 – 307 (1987). – See Abstr. 003.007.

053.004 Satellite Digest–206.
R. D. Christy.
Spaceflight, Vol. 29, No. 11, p. 381, 383 (1987).
1987 November listing of satellite and spacecraft launches.

053.005 Satellite Digest–207.
R. D. Christy.
Spaceflight, Vol. 29, No. 12, p. 406 – 407 (1987).
December 1987 listing of satellite and spacecraft launches.

053.006 The Polar BEAR spacecraft.
M. R. Peterson, D. G. Grant.
Johns Hopkins APL Tech. Dig., Vol. 8, No. 3, p. 295 – 302 (1987).
The Polar BEAR spacecraft was developed to measure auroral and ionospheric parameters and their effects on RF wave propagation. It provides coverage of the auroral oval in an area different from that covered from previous spacecraft, and the data gathered will complement the research being carried out by earlier programs. This article provides a technical description of the spacecraft and its initial in–orbit performance.

053.007 Spectral observations of a geostationary satellite in Simeiz.
A. V. Bagrov.
Astron. Tsirk., No. 1478, p. 3 – 5 (1987). In Russian.
The sequence of 33 spectra of geostationary satellite "INTELSAT–5F" showed spectral variations with phase angle. Preliminary results are discussed.

053.008 Observations of geostationary satellites by the TV–method.
V. I. Chumachenko.
Nauchn. Inf., Vyp. 58, p. 91 – 95 (1986). In Russian.
The method and results of complex observations of geostationary satellites on TV–systems are considered. The observations were carried out on the 70–cm reflector AZT–8.

053.009 Shape determination of an artificial satellite by photometric observations.
S. A. Severnyj, M. A. Smirnov, A. V. Bagrov.
Nauchn. Inf., Vyp. 58, p. 103 – 106 (1986). In Russian.
Photometric data allow the determination of the shape and properties of a satellite's surface. Some characteristics of a geostationary satellite are derived.

053.010 Operating of programme dividers in the satellite tracing system.
Yu. Kh. Zhagar, M. K. Abele.
Nauchn. Inf., Vyp. 58, p. 114 – 120 (1986). In Russian.

053.011 Investigation of artificial satellite's shape by photometric BVR observations.
A. V. Bagrov, M. A. Smirnov.
Nauchn. Inf., Vyp. 58, p. 152 – 161 (1986). In Russian.
The interpretation of artificial satellite's photometrical data is reported. The total photometric curve was calculated as superposition of individual reflections from parts of a simple shape. The reconstructed shape of the satellite is presented.

053.012 Preliminary reduction of observations of the IKB–1300 satellite.
O. M. Bulygina, I. A. Yankovskaya.
Nauchn. Inf., Vyp. 58, p. 169 – 170 (1986). In Russian.
Preliminary calculations of orbital elements of the IKB–1300 satellite by photographic and laser observations have been made.

053.013 The automatic stations Vega 1 and Vega 2. Operation of the instruments at the encounter with comet Halley.
V. P. Dolgopolov, V. P. Karyagin, V. M. Kovtunenko,
R. S. Kremnev, O. V. Pankov, K. M. Pichkhadze,
G. N. Rogovskij, R. Z. Sagdeev, K. G. Sukhanov.
Kosm. Issled., Tom 25, Vyp. 5, p. 655 – 658 (1987). In Russian.
English translation in Cosm. Res.

053.014 Visual observations, horizontal and equatorial (1950.0) coordinates of Arkad 3.
Rezul't. Nablyud. Iskusstv. Sputnikov Zemli, No. 110 (250), 48 pp. (1986). In Russian.
Concerning Arkad 3–1981–94–1, February – October, December 1984.

053.015 Visual observations, horizontal coordinates, and equatorial coordintes of artificial satellites.
Rezul't. Nablyud. Iskusstv. Sputnikov Zemli, No. 111 (251), 45 pp. (1986). In Russian.
Concerning Intercosmos 19–1979–20–1, September, October, December 1983; Arkad 3–1981–94–1, September, November, December 1983 (horizontal coordinates). Arkad 3–1981–94–1, December 1983, January – April 1984 (equatorial coordinates).

Mirror fabrication and metrology for ROSAT.
See Abstr. 031.004.

A study of the mirror contamination on Ariel VI.
See Abstr. 035.016.

ROSAT mirror system features of design, assembly and qualification.
See Abstr. 035.018.

XUV reflectivity measurements on the ROSAT WFC mirrors.
See Abstr. 035.019.

Development of pure aluminium mirror surfaces for the Lyman far–ultraviolet astronomical satellite mission.
See Abstr. 035.020.

The Sakigake/Suisei encounter with comet P/Halley.
See Abstr. 051.042.

Theoretical Astrophysics

061 General Aspects (Nucleosynthesis, Elementary Particles, Neutrino Astronomy, etc.)

061.001 La matière de quarks étrange et l'astronomie.
F. Grassi, J.–L. Masnou.
Astronomie, Vol. 101, p. 473 – 485 (1987).

061.002 Strings and superspace.
R. Kallosh.
Phys. Scr., Vol. T15, p. 118 – 125 (1987). – See Abstr. 012.014.
The scale–invariant superspace formulation of a $d = 10$ Einstein–Yang–Mills system is considered which gives a most economical and clear presentation of the full on–shell content of the theory. A geometrical formulation of the heterotic string theory in superspace with $10 + 496$ bosonic and 16 fermionic coordinates is presented. The relation between this superspace and the anomaly–free $d = 10$ Einstein–Yang–Mills supergravity is discussed.

061.003 Higher curvature supergravity and superstrings.
S. Ferrara.
Phys. Scr., Vol. T15, p. 132 – 137 (1987). – See Abstr. 012.014.
Recent attempts to construct higher curvature supergravity theories are reported and some implications concerning the effective low–energy action for superstring theories discussed.

061.004 Gravity from strings.
S. Deser.
Phys. Scr., Vol. T15, p. 138 – 142 (1987). – See Abstr. 012.014.
The author obtains the Einstein action plus quadratic curvature corrections generated by closed bosonic, heterotic and supersymmetric strings by matching the four–graviton amplitude (to first order in the slope parameter and fourth power of momenta) with an effective local gravitational action. The properties of solutions to these corrected theories are examined. First neglecting dilatons, the explicit "Schwarzschild" metrics are found. Both asymptotically flat and de Sitter solutions are present. When dilatons are included, the cosmological vacua are gratifyingly excluded.

061.005 Mass and dual mass: a gravitational analogue of the Dirac quantization rule and its physical implications.
A. Magnon, H.–Y. Guo.
Gen. Relativ. Gravitation, Vol. 19, No. 8, p. 809 – 825 (1987).
It is shown that (asymptotically multi–NUT) gravitational magnetic monopoles, which can be described by an S^3/Z_N principal Hopf–bundle structure at conformal null infinity (Z_N is a cyclic subgroup of order N of Z), provide a gravitational analogue of the Dirac quantization rule, which involves the total magnetic (dual) mass of the space–time – a measurement of the first Chern class of the bundle – and the mass of a test particle located in the rest frame defined at infinity by the Bondi (or dual Bondi) 4–momentum.

061.006 Tantalum–180 production by (p, n) reactions.
V. G. Batij, E. A. Skakun, Yu. N. Rakivnenko, O. A. Rastrepin.
Sov. Astron. Lett., Vol. 12, No. 5, p. 337 – 338 (1987). English translation of 42.061.021.

061.007 Superstrings.
J. H. Schwarz.
Phys. Today, Vol. 40, No. 11, p. 33 – 40 (1987).
Considering the elementary building blocks of nature to be strings rather than point particles allows one to construct consistent quantum theories that unify gravity with the other known forces.

061.008 Cyclotron radiation in astrophysics. Review.
V. V. Zheleznyakov.
Izv. Vuzov. Radiofiz., Tom 30, No. 2, p. 144 – 160 (1987). In Russian. Abstr. in Ref. Zh., 51. Astron., 8.51.97 (1987).

061.009 On a possibility of recording a stellar collapse from the interaction of neutrinos and antineutrinos with carbon–12.
E. I. Chujkin.
Fiz.–Tekh. Inst. Akad. Nauk SSSR, Prepr., No. 1106, p. 3 – 29 (1987). In Russian. Abstr. in Ref. Zh., 51. Astron., 8.51.352 (1987).

061.010 Possible appearance of the anomalous 4ν–interaction.
A. V. Berkov, Yu. P. Nikitin, A. L. Sudarikov, M. Yu. Khlopov.
Inst. Prikl. Mat. Akad. Nauk SSSR, Prepr., No. 59, 15 pp. (1987). In Russian. Abstr. in Ref. Zh., 51. Astron., 8.51.857 (1987).

061.011 Elementaire deeltjes en krachten in de natuur: de hoofdrolspelers in de kosmische schouwburg.
A. Achterberg.
Zenit, 14. Jaarg., Nr. 11, p. 368 – 373 (1987).

061.012 Grand unified theories and the double beta–decay.
A. Faessler.
Prog. Part. Nucl. Phys., Vol. 17, p. 85 – 96 (1986). – See Abstr. 012.024.
Grand unified and the superstring theories suggest that the neutrino is a Majorana particle identical with its antiparticle. The same theories indicate that it has a small finite mass and that it also interacts by a right handed current. These properties can be tested by the neutrinoless double beta–decay. The author shows that due to the virtual intermediate neutrino in the neutrinoless double–beta decay the classification according to allowed, first forbidden and second forbidden transitions does not reflect anymore the strength of the transition probability.

061.013 Nuclear matter under extreme conditions.
H. Müther.
Prog. Part. Nucl. Phys., Vol. 17, p. 97 – 116 (1986). – See Abstr. 012.024.
The paper discusses the present states of the attempts to evaluate the energy of nuclear matter specifically at high densities in a microscopic way. One of the problems involved in such calculations is the solution of the classical nuclear many–body problem. Relativistic effects, subnucleonic degrees of freedom and a many–body theory to account for these degrees of freedom

in the appropriate way must be considered to obtain reliable results.

061.014   Binary stars as sources of iron and of s–process isotopes.
I. Iben Jr.
Prog. Part. Nucl. Phys., Vol. 17, p. 287 – 294 (1986). – See Abstr. 012.024.
The autor reviews our understanding of how far which stars in which stages of development are the major contributors of specific elements such as carbon, iron, and the s–process isotopes.

061.015   The origin of the light nuclides.
M. Arnould.
Prog. Part. Nucl. Phys., Vol. 17, p. 305 – 347 (1986). – See Abstr. 012.024.
The theory of nucleosynthesis aims at interpreting the present composition of the universe and of its various constituting objects, as well as the variations with time of that composition. The light nuclides are especially interesting and powerful tools for investigating a quite incredible variety of very important physical and astrophysical questions, including particle physics, cosmology, cosmic–ray physics, stellar hydrodynamics and evolution, and galaxy chemical evolution.

061.016   The s–process of stellar nucleosynthesis.
G. Schatz.
Prog. Part. Nucl. Phys., Vol. 17, p. 393 – 417 (1986). – See Abstr. 012.024.
The present status of our understanding of the s–process of stellar nucleosynthesis is reviewed from the point of view of nuclear physics. Observed abundances of chemical elements and their isotopes are combined with information on neutron capture cross sections and nuclear structure. Quantitative information on neutron flux, temperature, and density during the synthesis of heavy elements by slow neutron capture is derived. The (inconclusive) status of the ^{176}Lu nucleochronometer for determining the time of s–process nucleosynthesis is discussed.

061.017   Nuclear beta strength, neutrino mass and cosmology.
H. V. Klapdor.
Prog. Part. Nucl. Phys., Vol. 17, p. 419 – 455 (1986). – See Abstr. 012.024.
The improvement in our understanding of nuclear beta decay achieved in recent years and consequences for the extraction of the neutrino mass from double beta decay experiments are described. Then consequences for astrophysics and cosmology are discussed: Synthesis of heavy elements, age of the universe, cosmological constant, solar and galactic neutrinos.

061.018   The electroweak interaction.
R. Turlay.
Neutrinos and the present–day universe, p. 1 – 27 (1986). – See Abstr. 012.025.
The author presents the very successful model of the unification of electromagnetic and weak interactions. He spends a longer time on the history and the successive developments of the unification of these two interactions.

061.019   The Zürich neutrino mass experiment.
M. Fritschi, E. Holzschuh, W. Kündig,
J. W. Petersen, R. E. Pixley, H. Stüssi.
Neutrinos and the present–day universe, p. 29 – 37 (1986). – See Abstr. 012.025.

061.020   Nuclear double beta decay.
P. Hubert, P. Mennrath.
Neutrinos and the present–day universe, p. 39 – 48 (1986). – See Abstr. 012.025.
The processes of double beta decay with and without emission of neutrinos are briefly reviewed. After the definitions of the processes and implications for the neutrino properties, the present status of the experimental results is discussed. The authors conclude with a description of the Bordeaux–Zaragoza–Strasbourg experiment which will run in the Frejus tunnel.

061.021   Neutrino emission by supernova explosions.
R. Schaeffer.
Neutrinos and the present–day universe, p. 85 – 99 (1986). – See Abstr. 012.025.
Neutrinos emitted by supernova explosions are predicted to be responsible for most (99%) of the energy output. Their time structure, a double burst followed by a more quiescent emission, as well as their energy spectrum and expected flux are analysed in the light of recent progress in understanding these events. Current studies and projects aiming at detecting these neutrinos in terrestrial experiments are discussed.

061.022   Neutrino transport in stellar matter.
J. L. Basdevant, J. P. Chieze, P. Mellor.
Neutrinos and the present–day universe, p. 101 – 126 (1986). – See Abstr. 012.025.
The authors reconsider the neutrino transport problem in dense stellar matter which has a variety of applications among which the participation of neutrinos to the dynamics of type II supernova explosions. They describe the position of the problem and make some criticism of previously used approximation methods. They then propose a method which is capable of handling simultaneously the optically thick, optically thin, and intermediate regimes, which is of crucial importance in such problems.

061.023   Elemental technetium and promethium as cosmic–ray clocks.
J. Drach, M. H. Salamon.
Astrophys. J., Vol. 319, No. 1, p. 237 – 246 (1987).
The possibility of using elemental Tc $(Z = 43)$ and Pm $(Z = 61)$ as clocks to measure the mean cosmic–ray confinement time in the Galaxy, τ_e, is considered. For this purpose it is necessary to estimate the unknown β^+ decay half–lives of several Tc and Pm isotopes; these estimates are obtained using β–decay systematics. In the case of Tc it is possible to estimate the half–lives sufficiently well and show that this element can indeed be used as a cosmic–ray clock; in the case of Pm the half–lives are too uncertain to permit any conclusion.

061.024   Radiative alpha–capture rates leading to $A = 7$ nuclei: applications to the solar neutrino problem and big bang nucleosynthesis.
T. Kajino, H. Toki, S. M. Austin.
Astrophys. J., Vol. 319, No. 1, p. 531 – 540 (1987).
Recent resonating group calculations provide a good description of the cross sections for the radiative capture reactions $^3He(\alpha, \gamma)^7Be$ and $^3H(\alpha, \gamma)^7Li$. The authors have reviewed the available experimental data for these reactions and extrapolated them to zero energy by normalizing to the resonating group results in a consistent way. Applying these rates to the production of solar neutrinos, the authors conclude that remaining uncertainties in the nuclear reaction rates are not a likely explanation of the low observed detection rate in the ^{37}Cl solar neutrino detector. It is also found that the reaction rates are sufficiently accurate to predict big bang production of ^7Li within $\pm 35\%$ at all relevant densities.

061.025   Nucleosynthesis contributions to the solar nebula.
M. Arnould.
Philos. Trans. R. Soc. London, Ser. A, Vol. 323, No. 1572, p. 251 – 267 (1987). – See Abstr. 012.026.
The discovery of isotopic anomalies in meteorites suggests that the solar system is made of material from compositionally different and imperfectly mixed reservoirs. One of them, which comprises the bulk solar system material, is considered to be made of the well–homogenized ashes of many nucleosynthesis events. Its composition can be studied through models of the chemical evolution of the Galaxy. The main nucleosynthetic agents responsible for that evolution are very briefly reviewed, as well as the level of reliability of the model predictions.

061.026 **Bounds on galactic cold dark matter particle candidates and solar axions from a Ge–spectrometer.**
G. Gelmini.
Cosmology and particle physics, p. 80 – 93 (1987). – See Abstr. 012.028.

The ultralow background Ge spectrometer developed by the USC/PNL group is used as a detector of cold dark matter candidates from the halo of our Galaxy and of solar axions (and other light bosons), yielding interesting bounds. Some of them are: heavy standard Dirac neutrinos with mass $20 \text{ GeV} \leqslant m \leqslant 1 \text{ TeV}$ are excluded as main components of the halo of our Galaxy; Dine–Fischler–Srednicki axion models with $F/2x_e' \lesssim 0.5 \times 10^7 \text{GeV}$ are excluded.

061.027 **Does the photino decay?**
L. J. Hall.
Cosmology and particle physics, p. 106 – 115 (1987). – See Abstr. 012.028.

The stability of the lightest superpartner is a crucial aspect of many experimental searches for supersymmetry and of supersymmetric dark matter candidates. It is shown that R parity may occur in operators of dimension four or less as an accidental consequence of an exact Z_N symmetry. In this case the lightest superpartner can decay via higher dimension operators.

061.028 **Superstring candidates for dark matter.**
K. A. Olive.
Cosmology and particle physics, p. 133 – 147 (1987). – See Abstr. 012.028.

The author discusses some of the possible candidates which naturally arise in low energy superstring–inspired models. In addition, the author reviews the possibility of detecting cold dark matter using under–ground proton–decay detectors searching for high energy ($E > 1$ GeV) neutrinos from the annihilation of dark matter inside the Sun.

061.029 **On the present mass density of relic photinos.**
K. A. Olive, M. Srednicki, J. Silk.
Cosmology and particle physics, p. 148 – 152 (1987). – See Abstr. 012.028.

The authors calculate the minimum mass density of relic photinos assuming various choices for slepton masses.

061.030 **The effects of resonant neutrino oscillations on the solar neutrino experiments.**
S. J. Parke.
Cosmology and particle physics, p. 153 – 165 (1987). – See Abstr. 012.028.

Analytic results are derived for the electron neutrino survival probability after passage through a resonant oscillation region. This survival probability together with a sophisticated model of the production distribution of the solar neutrino sources and the solar electron number density are used to study the effects of resonant neutrino oscillation in the solar interior on the current and proposed solar electron neutrino experiments.

061.031 **Cosmological analysis of R_p–breaking.**
P. Salati.
Cosmology and particle physics, p. 176 – 186 (1987). – See Abstr. 012.028.

The breaking of R–parity allows the lightest supersymmetric particle (LSP) to decay. The author studies the cosmological bounds on its mass and lifetime. These bounds can be translated into a lower limit on the neutrino mass.

061.032 **The light–element synthesis in a two–temperature astrophysical plasma.**
F. A. Agaronyan, R. A. Syunyaev.
Astrofizika, Tom 27, Vyp. 1, p. 131 – 145 (1987). In Russian. English translation in Astrophysics, Vol. 27, No. 1.

The efficiency of light–element production in a two–temperature astrophysical plasma is investigated. It has been shown that the light element nucleosynthesis breaks down, depending on the mode of radiative cooling of electrons in different stages of

evolution. The cases with optically thin and optically thick plasma are considered.

061.033 **Radio–dating the Galaxy.**
G. Gilmore.
Nature, Vol. 328, No. 6126, p. 111 (1987).

This note comments on a paper by H. R. Butcher discussing the possible use of the nuclide ^{232}Th as chronometer for stellar and galactic evolution (see abstr. 061.034).

061.034 **Thorium in G–dwarf stars as a chronometer for the Galaxy.**
H. R. Butcher.
Nature, Vol. 328, No. 6126, p. 127 – 131 (1987).

Observation of the radioactive nuclide ^{232}Th (half life 14 Gyr) in stars of various ages offers the possibility of directly relating the timescale for nucleosynthesis to that for stellar and galactic evolution. An initial set of such observations reveals no detectable evolution of the thorium abundance with respect to a stable element. This result is seen as evidence for a short timescale for galactic evolution and also for a problem with the stellar age scale.

061.035 **Chemical abundances.**
L. H. Aller.
Spectroscopy of astrophysical plasmas, p. 89 – 124 (1987). – See Abstr. 003.010.

Contents: 1. Introduction. 2. Importance of composition studies. 3. The Sun and the primordial solar system composition (PSSC). 4. Stellar abundance determinations, curve of growth. 5. Breakdown of LTE and hydrostatic equilibria in stellar atmospheres. 6. Abundances in gaseous nebulae. 7. Summaries of some further perspectives on problems of elemental abundances.

061.036 **Newton, quantum theory and reality.**
R. Penrose.
Three hundred years of gravitation, p. 17 – 49 (1987). – See Abstr. 003.012.

Contents: 1. Newton's corpuscular–undulatory theory and reality. 2. Stated reasons for rejection of wave theory. 3. Newton and relativity. 4. Newton's route to an undulatory–corpuscular picture. 5. Quantum mechanics. 6. Physical reality. 7. Reality of the state vector. 8. Quantum non–locality. 9. Quantum mechanics and macroscopic physics. 10. Linearity and time–evolution. 11. Quantum gravity and time–asymmetry. 12. Time asymmetry of state–vector reduction. 13. Reduction and the (longitudinal) graviton count. 14. Non–locality in quantum geometry.

061.037 **Superstring unification.**
J. H. Schwarz.
Three hundred years of gravitation, p. 652 – 675 (1987). – See Abstr. 003.012.

Contents: 1. Introduction. 2. Classification of string theories. 3. Feynman diagrams. 4. String field theory. 5. Anomalies. 6. Compactification. 7. Remaining problems and conclusions.

061.038 **Covariant description of canonical formalism in geometrical theories.**
C. Crnkovic, E. Witten.
Three hundred years of gravitation, p. 676 – 684 (1987). – See Abstr. 003.012.

The essence of the canonical formalism for quantization of field theories can be developed in a way that manifestly preserves all relevant symmetries, including Poincaré invariance. The purpose of the present paper is to carry this out in the case of non–abelian gauge theories and general relativity.

061.039 **A unified approach to the gauging of space–time and internal symmetries.**
E. A. Lord.
Gen. Relativ. Gravitation, Vol. 19, No. 10, p. 983 – 1002 (1987).

The properties of the manifold of a Lie group G, fibered by the cosets of a subgroup H, are exploited to obtain a geometrical description of gauge theories in space–time G/H. Gauge potentials and matter fields are pullbacks of equivariant fields on G.

061.040   Intrinsic ADM formalism for Lagrangian field theories in fibered manifolds.
M. Ferraris, M. Francaviglia.
General relativity and gravitational physics, p. 55 – 66 (1987). – See Abstr. 012.038.
An intrinsic approach to the ADM formalism for field theories is presented. The approach is based on the formal theory of calculus of variations in fibered manifolds and on the existence of global Poincaré–Cartan forms associated to an arbitrary Lagrangian density.

061.041   A new macroscopic force?
R. Barbieri.
General relativity and gravitational physics, p. 113 – 125 (1987). – See Abstr. 012.038.
The author describes an attempt to update the suggestion, made by J. Scherk in 1977, that a new interaction with a range accessible to detection by laboratory or geophysical experiments might be the manifestation of an underlying supergravity theory. The paper stresses the possible connection of a new force of finite range with the fate of symmetries like baryon or lepton number conservation.

061.042   Field theory and strings.
L. Bonora, P. Cotta–Ramusino, M. Rinaldi.
General relativity and gravitational physics, p. 127 – 141 (1987). – See Abstr. 012.038.
This paper discusses the possibility of a field theoretic truncation of a superstring theory in 10–D, and, subsequently, of a compactification to 4–D.

061.043   Integration on supermanifolds.
U. Bruzzo.
General relativity and gravitational physics, p. 143 – 155 (1987). – See Abstr. 012.038.
The author reviews the current status of the theory of integration on supermanifolds. Applications to the formulation of field theories on supermanifolds are discussed.

061.044   Algebraic geometry and string theory.
R. Catenacci.
General relativity and gravitational physics, p. 157 – 162 (1987). – See Abstr. 012.038.
This paper discusses the holomorphy properties of the p–loop partition function for the bosonic Polyakov string in 26 dimensions.

061.045   Supermanifolds, field theories and the Cartan Kähler integration theorem.
R. Cianci.
General relativity and gravitational physics, p. 163 – 178 (1987). – See Abstr. 012.038.
After presenting some applications of the Cartan Kähler integration theorem in the framework of classical field theories, the author describes an extension of this theorem to infinitely generated supermanifolds by means of a generalization of the Cauchy Kowalewsky theorem.

061.046   Curvature and torsion as induced quantum effects.
G. Denardo, E. Spallucci.
General relativity and gravitational physics, p. 179 – 188 (1987). – See Abstr. 012.038.

061.047   Recent developments in supergravity theories.
L. Girardello.
General relativity and gravitational physics, p. 189 – 211 (1987). – See Abstr. 012.038.
This paper reviews theories of extended supergravity and discusses the spontaneous breaking of local supersymmetry in flat space–time. Then it analyzes the structure of $N = 1$ theories compatible with superstring compactification.

061.048   Local gauge invariant Lagrangians.
L. Mangiarotti.
General relativity and gravitational physics, p. 227 – 239 (1987). – See Abstr. 012.038.
In the framework of jet spaces, the author presents new geometrical techniques and new results on the Lagrangian formulation of classical gauge theories. The main tool is a canonical jet shift covariant differential.

061.049   Global aspects of string theory.
C. Reina.
General relativity and gravitational physics, p. 245 – 255 (1987). – See Abstr. 012.038.
The author gives an introductory account of some mathematical aspects of the construction of path integrals for strings.

061.050   Topology and quantization.
F. Salmistraro.
General relativity and gravitational physics, p. 257 – 268 (1987). – See Abstr. 012.038.
A recent approach to the problem of charge quantization, based on cohomology theory, is reviewed and discussed.

061.051   Heat kernel expansion in a Riemann–Cartan space–time.
R. Brunelli, G. Cognola, S. Zerbini.
General relativity and gravitational physics, p. 269 – 273 (1987). – See Abstr. 012.038.

061.052   Modular transformations of spin structures on surfaces with boundary.
L. Dabrowski, R. Percacci.
General relativity and gravitational physics, p. 275 – 278 (1987). – See Abstr. 012.038.

061.053   A uniqueness result for quantum field theory on space–times with bifurcate Killing horizons.
B. S. Kay, R. M. Wald.
General relativity and gravitational physics, p. 279 – 286 (1987). – See Abstr. 012.038.
The authors describe a rigorous mathematical theorem, which throws some new light on some old issues in quantum–field–theory–in–curved–spacetime connected with the Hawking effect.

061.054   Spinor and gauge connections over oriented spheres.
G. Landi.
General relativity and gravitational physics, p. 287 – 291 (1987). – See Abstr. 012.038.

061.055   Effective action for the Schwinger model in curved spacetime.
F. Legovini, E. Spallucci.
General relativity and gravitational physics, p. 293 – 297 (1987). – See Abstr. 012.038.

061.056   Multitemporal Newtonian mechanics and its quantization.
L. Lusanna.
General relativity and gravitational physics, p. 299 – 302 (1987). – See Abstr. 012.038.

061.057   Renormalization group techniques in lattice gravity.
M. Martellini, A. Marzuoli.
General relativity and gravitational physics, p. 303 – 307 (1987). – See Abstr. 012.038.

061.058   Covariant canonical formalism for polynomial supergravity in any dimension.
J. E. Nelson.
General relativity and gravitational physics, p. 309 – 314 (1987). – See Abstr. 012.038.

061.059   Clifford algebras and magnetic monopoles.
E. Recami.
General relativity and gravitational physics, p. 315 – 321 (1987).
– See Abstr. 012.038.

061.060   Continuous product representations in quantum gravity.
C. Rovelli.
General relativity and gravitational physics, p. 323 – 327 (1987).
– See Abstr. 012.038.

061.061   Radiative transitions in ^{27}Al and their relevance to the ^{27}Al$(p,\gamma)^{27}$Si reaction under stellar conditions.
C. Rangacharyulu, G. Kuechler, A. Richter, E. Spamer.
Astrophys. J., Vol. 320, No. 1, p. 405 – 408 (1987).
The ^{27}Al(e,e') reaction measured at low momentum transfers had been used to determine the ground–state radiative transition strengths from levels in the excitation region between about 6 and 8 MeV. The transition strengths can be directly correlated with the corresponding transitions in the mirror nucleus ^{27}Si, where they are of relevance for the depletion of ^{26}Al by radiative proton capture into ^{27}Si. Reaction rates under stellar conditions are deduced from the present data.

061.062   An experimental search for low energy astrophysical neutrino sources.
D. W. Joutras.
Diss. Abstr. Int., Sect. B, Vol. 48, No. 3, p. 793–B (1987). Thesis, The University of Wisconsin–Madison, 172 pp. (1987). Order No. DA8711032.

061.063   The origin and distribution of the elements.
B. E. J. Pagel.
R. Greenwich Obs., Prepr., No. 64, 19 pp. (1987). Paper presented at the American Chemical Society symposium on "Origin and distribution of the elements", New Orleans, USA, September 1987.

061.064   Analytic treatments of matter–enhanced solar neutrino oscillations.
W. C. Haxton.
Phys. Rev. D, Vol. 35, No. 8, p. 2352 – 2364 (1987). Abstr. in Phys. Abstr., Vol. 90, No. 1309, Entry 81078 (1987).

061.065   On the cosmological constant in the heterotic string theory.
E. Gava, R. Iengo.
Phys. Lett. B, Vol. 187, No. 1 – 2, p. 22 – 24 (1987). Abstr. in Phys. Abstr., Vol. 90, No. 1309, Entry 81149 (1987).

061.066   Nuclear structure of ^{49}Ca above 5 MeV excitation from n + ^{48}Ca and astrophysics for 30 keV neutrons.
R. F. Carlton, J. A. Harvey, R. L. Macklin, C. H. Johnson, B. Castel.
Nucl. Phys. A, Vol. A465, No. 2, p. 274 – 290 (1987). Abstr. in Phys. Abstr., Vol. 90, No. 1309, Entry 81382 (1987).

061.067   Cross sections for production of the 15.10 MeV and other atrophysically significant gamma–ray lines through excitation and spallation of ^{12}C and ^{16}O with protons.
F. L. Lang, C. W. Werntz, C. J. Crannell, J. I. Trombka, C. C. Chang.
Phys. Rev. C, Vol. 35, No. 4, p. 1214 – 1227 (1987). Abstr. in Phys. Abstr., Vol. 90, No. 1309, Entry 81388 (1987).

061.068   Laboratory limits on solar axions from an ultralow–background germanium spectrometer.
F. T. Avignone III, R. L. Brodzinski, S. Dimopoulos, G. D. Starkman, A. K. Drukier, D. N. Spergel, G. Gelmini, B. W. Lynn.
Phys. Rev. D, Vol. 35, No. 9, p. 2752 – 2757 (1987). Abstr. in Phys. Abstr., Vol. 90, No. 1309, Entry 88051 (1987).

061.069   Further remarks on particle–like solutions in spinor–connection theory.
J. T. Lynch.
Phys. Rev. D, Vol. 35, No. 8, p. 2372 – 2377 (1987). Abstr. in Phys. Abstr., Vol. 90, No. 1309, Entry 88720 (1987).

061.070   Implications of bimodal star formation on the chemical evolution of the Galaxy: the evolution of deuterium.
E. Vangioni–Flam, J. Audouze.
Inst. Astrophys. Paris, Pré–Publ., No. 198, 20 pp. (1987). To appear in Astron. Astrophys.

061.071   Cosmological QCD, neutron diffusion, and the production of primordial heavy elements.
J. H. Applegate, C. J. Hogan, R. J. Scherrer.
Prepr. Steward Obs., No. 753, 22 pp. (1987). Submitted to Astrophys. J.

061.072   Neutron tori and the origin of r–process elements.
C. J. Hogan, J. H. Applegate.
Prepr. Steward Obs., No. 754, 6 pp. (1987). Submitted to Nature.

061.073   Detecting cold dark matter candidates.
A. K. Drukier.
The early universe, p. 361 – 391 (1988). – See Abstr. 012.043.
The growing synergy between astrophysics, particle physics, and low background experiments strengthens the possibility of detecting astrophysical non–baryonic matter. The author reports two tests designed to study the practicality of a superheated superconducting colloid detector using a SQUID readout system. Furthermore, he shows that in case of particles with spin interactions, one should consider detectors based on compounds of boron, lithium and fluorine.

061.074   The axion couplings.
P. Sikivie.
Cosmology and particle physics, p. 143 – 169 (1987). – See Abstr. 012.046.
A pedagogical derivation of the couplings of the axion to quarks, leptons, photons, gluons and nucleons is given. The effective Lagrangian method is used throughout. The results are applicable to all axion models.

061.075   Semiclassical gravitational effects near cosmic strings.
W. A. Hiscock.
Phys. Lett. B, Vol. 188, No. 3, p. 317 – 320 (1987). Abstr. in Phys. Abstr., Vol. 90, No. 1310, Entry 89093 (1987).

061.076   Baryon asymmetry of the universe in standard electro–weak theory.
M. E. Shaposhnikov.
Nucl. Phys. B, Part. Phys., Vol. B287, No. 4, p. 757 – 775 (1987). Abstr. in Phys. Abstr., Vol. 90, No. 1310, Entry 89411 (1987).

061.077   Observational limits on the time evolution of extra spatial dimensions.
J. D. Barrow.
Phys. Rev. D, Vol. 35, No. 6, p. 1805 – 1810 (1987). Abstr. in Phys. Abstr., Vol. 90, No. 1310, Entry 89418 (1987).

061.078   Compactification of the twisted heterotic string.
V. P. Nair, A. Shapere, A. Strominger, F. Wilczek.
Nucl. Phys. B, Part. Phys., Vol. B287, No. 3, p. 402 – 418 (1987). Abstr. in Phys. Abstr., Vol. 90, No. 1310, Entry 89467 (1987).

061.079 **Limits on the neutrino lifetime.**
J. M. LoSecco, R. M. Bionta, G. Blewitt,
C. B. Bratton, D. Casper, R. Claus, B. Cortez, S. Errede,
G. Foster, W. Gajewski, K. S. Ganezer, M. Goldhaber,
T. J. Haines, T. W. Jones, D. Kielczewska, W. R. Kropp,
J. G. Learned, E. Lehmann, H. S. Park, F. Reines, J. Schultz,
S. Seidel, E. Shumard, D. Sinclair, H. W. Sobel, J. L. Stone,
L. Sulak, R. Svoboda, J. C. Van der Velde, C. Wuest.
Phys. Rev. D, Vol. 35, No. 7, p. 2073 – 2076 (1987). Abstr. in
Phys. Abstr., Vol. 90, No. 1310, Entry 94622 (1987).

061.080 **"Just so" neutrino oscillations.**
S. L. Glashow, L. M. Krauss.
Phys. Lett. B, Vol. 190, No. 1 – 2, p. 199 – 207 (1987). Abstr. in
Phys. Abstr., Vol. 90, No. 1311, Entry 95807 (1987).

061.081 **E_6 based mechanism for the generation of fermion electric dipole moments: an application to the solar neutrino puzzle.**
J. A. Grifols, J. Sola.
Phys. Lett. B, Vol. 189, No. 1 – 2, p. 63 – 67 (1987). Abstr. in
Phys. Abstr., Vol. 90, No. 1311, Entry 95907 (1987).

061.082 **Superconducting cosmic strings with massive fermions.**
C. T. Hill, L. M. Widrow.
Phys. Lett. B, Vol. 189, No. 1 – 2, p. 17 – 22 (1987). Abstr. in
Phys. Abstr., Vol. 90, No. 1311, Entry 101975 (1987).

061.083 **Decay of gravitinos and photo–destruction of light elements.**
M. Kawasaki, K. Sato.
Phys. Lett. B, Vol. 189, No. 1 – 2, p. 23 – 28 (1987). Abstr. in
Phys. Abstr., Vol. 90, No. 1311, Entry 101976 (1987).

061.084 **Nongravitational decay of cosmic strings.**
M. Srednicki, S. Theisen.
Phys. Lett. B, Vol. 189, No. 4, p. 397 – 400 (1987). Abstr. in Phys.
Abstr., Vol. 90, No. 1311, Entry 101977 (1987).

061.085 **Light pseudoscalars, particle physics and cosmology.**
J. E. Kim.
Phys. Rep., Vol. 150, No. 1 – 2, p. 1 – 177 (1987). Abstr. in Phys.
Abstr., Vol. 90, No. 1311, Entry 101978 (1987).

061.086 **SU(2, 2/1, 1) supergravity and N = 2 supersymmetry with arbitrary cosmological constant.**
B. de Wit, A. Zwartkruis.
Classical Quantum Gravity, Vol. 4, No. 3, p. L59 – L66 (1987).
Abstr. in Phys. Abstr., Vol. 90, No. 1312, Entry 102279 (1987).

061.087 **Casimir effect in supergravity theories and the quantum birth of the Universe with non–trivial topology.**
Yu. P. Goncharov, A. A. Bytsenko.
Classical Quantum Gravity, Vol. 4, No. 3, p. 555 – 571 (1987).
Abstr. in Phys. Abstr., Vol. 90, No. 1312, Entry 102280 (1987).

061.088 **Neutrino oscillations and the solar–neutrino problem.**
A. Dar, A. Mann, Y. Melina, D. Zajfman.
Phys. Rev. D, Vol. 35, No. 12, p. 3607 – 3620 (1987). Abstr. in
Phys. Abstr., Vol. 90, No. 1312, Entry 102661 (1987).

061.089 **Resonant neutrino oscillations.**
S. J. Parke.
Probing the standard model, p. 349 – 360 (1987). Abstr. in Phys.
Abstr., Vol. 90, No. 1312, Entry 102668 (1987). – See Abstr.
012.050.

061.090 **How reliable are neutrino mass limits derived from SN1987A?**
E. W. Kolb, A. J. Stebbins, M. S. Turner.
Phys. Rev. D, Vol. 35, No. 12, p. 3598 – 3606 (1987). Abstr. in
Phys. Abstr., Vol. 90, No. 1312, Entry 107714 (1987).

061.091 **Stellar reaction rate of $^{14}N(p,\gamma)^{15}O$ and hydrogen burning in massive stars.**
U. Schröder, H. W. Becker, G. Bogaert, J. Gorres, C. Rolfs,
H. P. Trautvetter, R. E. Azuma, C. Campbell, J. D. King,
J. Vise.
Nucl. Phys. A, Vol. A467, No. 2, p. 240 – 260 (1987). Abstr. in
Phys. Abstr., Vol. 90, No. 1312, Entry 107812 (1987).

061.092 **Astrophysical aspects of the $^{25}Mg(p,\gamma)^{26}Al$ reaction.**
P. M. Endt, C. Rolfs.
Nucl. Phys. A, Vol. A467, No. 2, p. 261 – 272 (1987). Abstr. in
Phys. Abstr., Vol. 90, No. 1312, Entry 107813 (1987).

061.093 **Astrophysical S(E) factor of $^3He(^3He, 2p)^4He$ at solar energies.**
A. Krauss, H. W. Becker, H. P. Trautvetter, C. Rolfs.
Nucl. Phys. A, Vol. A467, No. 2, p. 273 – 290 (1987). Abstr. in
Phys. Abstr., Vol. 90, No. 1312, Entry 107814 (1987).

061.094 **Current status of nuclear astrophysics.**
C. Rolfs, H. P. Trautvetter, W. S. Rodney.
Rep. Prog. Phys., Vol. 50, No. 3, p. 233 – 325 (1987). Abstr. in
Phys. Abstr., Vol. 90, No. 1312, Entry 107815 (1987).

061.095 **Gravitomagnetic monopoles and the quantisation of frequency.**
P. W. Forder.
Classical Quantum Gravity, Vol. 4, No. 3, p. 703 – 710 (1987).
Abstr. in Phys. Abstr., Vol. 90, No. 1312, Entry 107994 (1987).

061.096 **Elemental abundances for stellar s–process studies.**
R. A. Malaney.
Astrophys. Space Sci., Vol. 137, No. 2, p. 251 – 255 (1987).
Nuclear reaction network calculations of heavy element enhancements are tabulated for different values of the mean neutron exposure τ_0. These tables are useful for comparison with observed s–process abundance patterns which appear to span a wide range in τ_0.

061.097 **Particle physics for everybody.**
P. Davies.
Sky Telesc., Vol. 74, No. 6, p. 582 – 587, 589 (1987).

061.098 **Prospects for fifth force fade.**
J. Maddox.
Nature, Vol. 329, No. 6137, p. 283 (1987).
Conflicting observations of the gravitational–like interactions which have engendered talk about the fifth force might have been reconcile if it were not for another set of data.

061.099 **Th/Nd abundance ratio in the surfaces of G–dwarfs.**
D. D. Clayton.
Nature, Vol. 329, No. 6138, p. 397 – 398 (1987). See Abstr.
061.034.

061.100 **Reply to "Th/Nd abundance ratio in the surfaces of G–dwarfs" by D. D. Clayton.**
H. R. Butcher.
Nature, Vol. 329, No. 6138, p. 398 (1987). See Abstr. 061.099.

061.101 **Weakly interacting massive particle distribution in and evaporation from the Sun.**
A. Gould.
Astrophys. J., Vol. 321, No. 1, p. 560 – 570 (1987).
This paper analyzes the evaporation of weakly interacting massive particles (WIMPs) from the Sun, both analytically and numerically. First, an analytic formula is derived for evaporation from a truncated Maxwell–Boltzmann distribution of WIMPs due to interaction with a truly thermal gas of nuclei. Then, the actual (nonthermal) distribution of Dirac neutrino WIMPs in the Sun is calculated numerically for WIMPs of mass 1 – 7 GeV. It is found that the evaporation mass for a solar lifetime is 3.7 GeV, and for an "annihilation lifetime," 2.9 GeV.

061.102   Resonant enhancements in weakly interacting massive particle capture by the Earth.
A. Gould.
Astrophys. J., Vol. 321, No. 1, p. 571 – 585 (1987).

The exact formulae for the capture of weakly interacting massive particles (WIMPs) by a massive body are derived. Capture by the Earth is found to be significantly enhanced whenever the WIMP mass is roughly equal to the nuclear mass of an element present in the Earth in large quantities. For Dirac neutrino WIMPs of mass 10 – 90 GeV, the capture rate is 10 – 300 times that previously believed. Capture rates for the Sun are also recalculated. The Earth alone or the Earth in combination with the Sun is found to give a much stronger annihilation signal from Dirac neutrino WIMPs than the Sun alone over a very large mass range.

061.103   High–energy neutrino astrophysics.
V. S. Berezinskij, G. T. Zatsepin.
Problems of cosmic ray physics, p. 99 – 121 (1987). In Russian. Abstr. in Ref. Zh., 51. Astron., 12.51.100 (1987). – See Abstr. 003.015.

061.104   Resonance oscillations and limitations on the parameters of neutrinos by a possible observation of a v–flare from the gravitational collapse of a star.
S. P. Mikheev, A. Yu. Smirnov.
Pis'ma Zh. Ehksp. Teor. Fiz., Tom 46, No. 1, p. 11 – 13 (1987). In Russian. Abstr. in Ref. Zh., 51. Astron., 12.51.559 (1987).

061.105   Strings in four dimensions.
J. Ellis.
Nature, Vol. 329, No. 6139, p. 488 – 489 (1987).

061.106   Permutation symmetry and the fermion mass matrix.
S. Pakvasa.
Indian J. Phys., Part B, Vol. 61B, No. 3, p. 302 – 308 (1987). Abstr. in Phys. Abstr., Vol. 90, No. 1313, Entry 108823 (1987).

061.107   Neutrino oscillations and the Landau–Zener formula.
C. W. Kim, W. K. Sze, S. Nussinov.
Phys. Rev. D, Vol. 35, No. 12, p. 4014 – 4017 (1987). Abstr. in Phys. Abstr., Vol. 90, No. 1313, Entry 108829 (1987).

061.108   Stellar krypton cross sections at $kT = 25$ and 52 keV.
F. Käppeler, A. A. Naqvi, M. Al–Ohali.
Phys. Rev. C, Vol. 35, No. 3, p. 936 – 941 (1987). Abstr. in Phys. Abstr., Vol. 90, No. 1313, Entry 109166 (1987).

061.109   Astrophysical S–factor of $^3H(\alpha, \gamma)^7Li$.
U. Schröder, A. Redder, C. Rolfs, R. E. Azuma, L. Buchmann, C. Campbell, J. D. King, T. R. Donoghue.
Phys. Lett. B, Vol. 192, No. 1 – 2, p. 55 – 58 (1987). Abstr. in Phys. Abstr., Vol. 90, No. 1313, Entry 115229 (1987).

061.110   Dynamical symmetry breaking in a de Sitter–invariant vacuum.
S. Kawati, A. Kokado.
Phys. Rev. D, Vol. 35, No. 10, p. 3092 – 3099 (1987). Abstr. in Phys. Abstr., Vol. 90, No. 1314, Entry 116095 (1987).

061.111   Opportunities for particle physics: solar neutrinos and superstrings.
S. Weinberg.
Int. J. Mod. Phys. A, Vol. 2, No. 2, p. 301 – 317 (1987). Abstr. in Phys. Abstr., Vol. 90, No. 1314, Entry 116102 (1987).

061.112   Neutrino transport in relativity.
J. R. Wilson.
Astrophysical radiation hydrodynamics, p. 477 – 490 (1986). – See Abstr. 012.069.

Two methods of solving the neutrino transport equation in general relativity are presented. The first method is for the spherical collapse of stars to black holes. It uses a Lagrangian hydrodynamic formulation as a framework. The method is of mediocre accuracy, but it can handle all cases efficiently. The second method of solving the transport equation is for use in cosmology. In this case all velocities are of the same size and an explicit monotonic second order in space differencing is used. This latter method includes the effects of neutrino rest mass.

061.113   Supersymmetric dark matter.
J. S. Hagelin.
The standard model. The supernova 1987A, p. 679 – 688 (1987). – See Abstr. 012.074.

The author shows that substantial dark matter abundances are a generic feature of supersymmetric models. He computes the flux of high energy (0.1 – 10 GeV) neutrinos produced by the annihilation of supersymmetric dark matter trapped in the sun, and concludes that many supersymmetric dark matter candidates, inlcuding photinos and Higgsinos, lead to potentially observable neutrino fluxes.

061.114   Multiloop modular invariance and the cosmological constant.
F. Gliozzi.
Phys. Lett. B, Vol. 194, No. 1, p. 30 – 34 (1987). Abstr. in Phys. Abstr., Vol. 90, No. 1315, Entry 120904 (1987).

061.115   Analysis of no–scale supergravity models leading to inflationary scenarios.
J. M. Molera, M. Quiros.
Phys. Rev. D, Vol. 36, No. 2, p. 375 – 385 (1987). Abstr. in Phys. Abstr., Vol. 90, No. 1315, Entry 120918 (1987).

061.116   On the non adiabatic neutrino oscillations in matter.
S. T. Petcov.
Phys. Lett. B, Vol. 191, No. 3, p. 299 – 303 (1987). Abstr. in Phys. Abstr., Vol. 90, No. 1315, Entry 121402 (1987).

061.117   Sphalerons, small fluctuations, and baryon–number violation in electroweak theory.
P. Arnold, L. McLerran.
Phys. Rev. D, Vol. 36, No. 2, p. 581 – 595 (1987). Abstr. in Phys. Abstr., Vol. 90, No. 1315, Entry 121409 (1987).

061.118   Mikheyev–Smirnov–Wolfenstein enhancement of oscillations as a possible solution to the solar neutrino problem.
S. P. Rosen, J. M. Gelb.
Proceedings of the International Nuclear Physics Conference, Harrogate, UK, 25 – 30 August 1986. IOP, Bristol, UK, Vol. 1, p. 512 (1987). Abstr. in Phys. Abstr., Vol. 90, No. 1315, Entry 121414 (1987).

061.119   Low–energy cross sections for $^{11}B(p, 3\alpha)$.
H. W. Becker, C. Rolfs, H. P. Trautvetter.
Z. Phys., A, Vol. 327, No. 3, p. 341 – 355 (1987). Abstr. in Phys. Abstr., Vol. 90, No. 1315, Entry 121848 (1987).

061.120   Low–energy $^2H(d, \gamma)^4He$ reaction and the D–state admixture in the ^4He ground state.
H. J. Assenbaum, K. Langanke.
Phys. Rev. C, Vol. 36, No. 1, p. 17 – 20 (1987). Abstr. in Phys. Abstr., Vol. 90, No. 1315, Entry 121859 (1987).

061.121   Spontaneous e^+e^- pair creation in heavy ion collisions.
D. Carrier, L. M. Krauss.
Phys. Lett. B, Vol. 194, No. 1, p. 141 – 146 (1987). Abstr. in Phys. Abstr., Vol. 90, No. 1315, Entry 121895 (1987).

061.122   A search for strongly interacting dark matter.
J. Rich, R. Rocchia, M. Spiro.
Phys. Lett. B, Vol. 194, No. 1, p. 173 – 176 (1987). Abstr. in Phys. Abstr., Vol. 90, No. 1315, Entry 127663 (1987).

061.123 Paraphotons and axions: similarities in stellar emission and detection.
S. Hoffmann.
Phys. Lett. B, Vol. 193, No. 1, p. 117 – 122 (1987). Abstr. in Phys. Abstr., Vol. 90, No. 1315, Entry 127717 (1987).

061.124 On the possibility of avoiding singularities by dilaton emission.
I. Antoniadis, G. F. R. Ellis, J. Ellis, C. Kounnas, D. V. Nanopoulos.
Phys. Lett. B, Vol. 191, No. 4, p. 393 – 398 (1987). Abstr. in Phys. Abstr., Vol. 90, No. 1316, Entry 128419 (1987).

061.125 Neutrinos.
D. H. Perkins.
Electron. Power (GB), Vol. 33, No. 7, p. 433 – 436 (1987). Abstr. in Phys. Abstr., Vol. 90, No. 1316, Entry 128877 (1987).

061.126 Decay of ^{80}Zn: implications for shell structure and r–process nucleosynthesis.
J. A. Winger, J. C. Hill, F. K. Wohn, R. Moreh, R. L. Gill, R. F. Casten, D. D. Warner, A. Piotrowski, H. Mach.
Phys. Rev. C, Vol. 36, No. 2, p. 758 – 764 (1987). Abstr. in Phys. Abstr., Vol. 90, No. 1316, Entry 129002 (1987).

061.127 On the vanishing of the cosmological constant in four–dimensional superstring models.
D. Arnaudon, C. P. Bachas, V. Rivasseau, P. Vegreville.
Phys. Lett. B, Vol. 195, No. 2, p. 167 – 176 (1987). Abstr. in Phys. Abstr., Vol. 90, No. 1317, Entry 134610 (1987).

061.128 Adjusting the cosmological constant dynamically: cosmons and a new force weaker than gravity.
R. D. Peccei, J. Sola, C. Wetterich.
Phys. Lett. B, Vol. 195, No. 2, p. 183 – 190 (1987). Abstr. in Phys. Abstr., Vol. 90, No. 1317, Entry 134611 (1987).

061.129 Relaxing the cosmological bound on axions.
G. Lazarides, C. Panagiotakopoulos, Q. Shafi.
Phys. Lett. B, Vol. 192, No. 3 – 4, p. 323 – 326 (1987). Abstr. in Phys. Abstr., Vol. 90, No. 1317, Entry 135074 (1987).

061.130 Hadron cross sections at ultrahigh energies and unitarity bounds on diffraction.
T. K. Gaisser, U. P. Sukhatme, G. B. Yodh.
Phys. Rev. D, Vol. 36, No. 5, p. 1350 – 1357 (1987). Abstr. in Phys. Abstr., Vol. 90, No. 1317, Entry 135171 (1987).

061.131 Neutrino mass and mixing constrained from the LMC supernova burst.
J. Arafune, M. Fukugita, T. Yanagida, M. Yoshimura.
Phys. Lett. B, Vol. 194, No. 4, p. 477 – 481 (1987). Abstr. in Phys. Abstr., Vol. 90, No. 1317, Entry 139912 (1987).

061.132 Resonant neutrino oscillations and the neutrino signature of supernovae.
T. P. Walker, D. N. Schramm.
Phys. Lett. B, Vol. 195, No. 3, p. 331 – 336 (1987). Abstr. in Phys. Abstr., Vol. 90, No. 1317, Entry 140136 (1987).

061.133 A simple solution to the solar neutrino and missing mass problems.
S. Raby, G. B. West.
Nucl. Phys. B, Part. Phys., Vol. B292, No. 4, p. 793 – 812 (1987). Abstr. in Phys. Abstr., Vol. 90, No. 1317, Entry 140337 (1987).

061.134 Bounds on neutrino masses from neutrino decay rates, cosmology and the see–saw mechanism.
H. Harari, Y. Nir.
Nucl. Phys. B, Part. Phys., Vol. B292, No. 2, p. 251 – 297 (1987). Abstr. in Phys. Abstr., Vol. 90, No. 1318, Entry 140848 (1987).

061.135 Effects of matter oscillations on supernova neutrino flux.
L. Wolfenstein.
Phys. Lett. B, Vol. 194, No. 2, p. 197 – 199 (1987). Abstr. in Phys. Abstr., Vol. 90, No. 1318, Entry 140852 (1987).

061.136 Thermodynamics of non–relativistic free quark gas.
S. P. Singh, M. Hasan.
Indian J. Theor. Phys., Vol. 32, No. 3, p. 209 – 218 (1984). Abstr. in Phys. Abstr., Vol. 90, No. 1318, Entry 141006 (1987).

061.137 Effects of electron screening on low–energy fusion cross sections.
H. J. Assenbaum, K. Langanke, C. Rolfs.
Z. Phys., A, Vol. 327, No. 4, p. 461 – 468 (1987). Abstr. in Phys. Abstr., Vol. 90, No. 1318, Entry 141122 (1987).

061.138 No future for the fourth generation?
M. Drees, K. Enqvist, D. V. Nanopoulos.
Nucl. Phys. B, Part. Phys., Vol. B294, No. 1, p. 1 – 29 (1987). Abstr. in Phys. Abstr., Vol. 91, No. 1319, Entry 196 (1988).

061.139 Testing superstrings with cosmology and the SSC.
D. V. Nanopoulos, K. A. Olive.
Physics of the superconducting supercollider, p. 284 – 288 (1987). Abstr. in Phys. Abstr., Vol. 91, No. 1319, Entry 441 (1988). – See Abstr. 012.080.

061.140 A review of matter oscillations and solar neutrinos.
S. P. Rosen.
Physics of the superconducting supercollider, p. 681 – 685 (1987). Abstr. in Phys. Abstr., Vol. 91, No. 1319, Entry 442 (1988). – See Abstr. 012.080.

061.141 A possible experimental method for measurement of the neutrino flux from the reaction e + ^{7}Be → ^{7}Li + ν_e in the Sun.
A. K. Mann.
Physics of the superconducting supercollider, p. 701 – 703 (1987). Abstr. in Phys. Abstr., Vol. 91, No. 1319, Entry 560 (1988). – See Abstr. 012.080.

061.142 Exact analytical solution of the two–neutrino evolution equation in matter with exponentially varying density.
S. Toshev.
Phys. Lett. B, Vol. 196, No. 2, p. 170 – 174 (1987). Abstr. in Phys. Abstr., Vol. 91, No. 1319, Entry 4253 (1988).

061.143 Extrasolar neutrino astronomy.
V. J. Stenger.
Physics of the superconducting supercollider, p. 660 – 664 (1987). Abstr. in Phys. Abstr., Vol. 91, No. 1319, Entry 4255 (1988). – See Abstr. 012.080.

061.144 Tabulation of astrophysical constraints on axions and Nambu–Goldstone bosons.
H.–Y. Cheng.
Phys. Rev. D, Vol. 36, No. 6, p. 1649 – 1656 (1987). Abstr. in Phys. Abstr., Vol. 91, No. 1319, Entry 4278 (1988).

061.145 A vortex–line model for infinite straight cosmic strings.
B. Linet.
Phys. Lett. A, Vol. 124, No. 4 – 5, p. 240 – 242 (1987). Abstr. in Phys. Abstr., Vol. 91, No. 1320, Entry 4778 (1988).

061.146 The decay of the new isotope ^{180}Yb and the search for the r–process path to ^{180m}Ta.
E. Runte, W.–D. Schmidt–Ott, W. Eschner, I. Rosner, R. Kirchner, O. Klepper, K. Rykaczewski.
Z. Phys., A, Vol. 328, No. 1, p. 119 – 122 (1987). Abstr. in Phys. Abstr., Vol. 91, No. 1320, Entry 5455 (1988).

061.147   Limits on cold dark matter candidates from an ultralow background germanium spectrometer.
S. P. Ahlen, F. T. Avignone III, R. L. Brodzinski,
A. K. Drukier, G. Gelmini, D. N. Spergel.
Phys. Lett. B, Vol. 195, No. 4, p. 603 – 608 (1987). Abstr. in Phys. Abstr., Vol. 91, No. 1320, Entry 10395 (1988).

061.148   Axion mediated forces in the early universe.
H. E. Haber, M. Sher.
Phys. Lett. B, Vol. 196, No. 1, p. 33 – 38 (1987). Abstr. in Phys. Abstr., Vol. 91, No. 1320, Entry 10396 (1988).

061.149   Galactic chronology of thorium/neodymium.
W. A. Fowler.
Nature, Vol. 330, No. 6150, p. 703 – 704 (1987). See Abstr. 061.034.

061.150   Reply to "Galactic chronology of thorium/neodymium" by W. A. Fowler.
H. Butcher.
Nature, Vol. 330, No. 6150, p. 704 (1987).

061.151   Galactic chronology of thorium/neodymium.
P. D. Noerdlinger.
Nature, Vol. 330, No. 6150, p. 704 (1987).

061.152   Reply to "Galactic chronology of thorium/neodymium" by P. D. Noerdlinger.
H. Butcher.
Nature, Vol. 330, No. 6150, p. 704 (1987).

061.153   Superstring thermodynamics and its application to cosmology.
N. Matsuo.
Z. Phys., C, Vol. 36, No. 2, p. 289 – 304 (1987). Abstr. in Phys. Abstr., Vol. 91, No. 1320, Entry 10692 (1988).

061.154   Resonant neutrino oscillations and stellar collapse.
G. M. Fuller, R. W. Mayle, J. R. Wilson,
D. N. Schramm.
Astrophys. J., Vol. 322, No. 2, p. 795 – 803 (1987).
The Mikheyev–Smirnov–Wolfenstein mechanism for resonant amplification of neutrino oscillations is shown to occur in collapsing presupernova stellar cores if there exist massive unstable neutrinos which mix with the electron neutrino. The relevant massive neutrino mass range is 200 eV to 25 keV, and the required vacuum mixing angle is $\theta \geqslant 10^{-6}$rad. It is shown that adiabatic conversion of electron neutrinos into such massive neutrinos would occur during stellar collapse with resultant readjustment of lepton numbers and small entropy generation. These changes have implications for the supernova explosion mechanism.

061.155   The initial neutrino events from supernovae as evidence for matter versus antimatter.
A. V. Barnes, T. J. Weiler, S. Pakvasa.
Astrophys. J., Lett. Ed., Vol. 323, No. 1, p. L31 – L33 (1987).
It is shown that information in the initial neutrino burst from a supernova can (1) reveal whether the source is made of matter or antimatter and (2) yield very stringent limits on the mass of the electron neutrino. Characteristics of the first Kamiokande II event suggest that their initial interacting particle was a ν_e rather than a $\bar{\nu}_e$; if this neutrino is identified with the neutronization burst of SN 1987A, it constitutes evidence that the collapsing star was made of matter rather than antimatter.

061.156   On the asymptotic mass spectrum of interacting particles.
G. V. Pechernikova.
Astron. Tsirk., No. 1468, p. 5 – 8 (1986). In Russian.
The asymptotic solution for the coagulation equation with the kernel $A \propto (m+m'^{\alpha})(m^{\beta}+m'^{\beta})$ is obtained in the form $n(m) \propto m^{-q(\alpha, \beta)}$.

061.157   Nuclear chronometers from the r–process and the age of the Galaxy.
J. J. Cowan, F.-K. Thielemann, J. W. Truran.
Astrophys. J., Vol. 323, No. 2, p. 543 – 552 (1987).
The cosmochronologically important nuclear production ratios ^{232}Th/^{238}U, ^{235}U/^{238}U, and ^{244}Pu/^{238}U are calculated in the context of a specific model for r–process nucleosynthesis which predicts an overall abundance pattern in good agreement with the distribution of abundances of the r–process nuclei in solar system matter. The calculated production ratios, when interpreted within a simple exponential model of Galactic evolution allowing values of initial disk enrichment ranging from approximately 10% to 30%, lead to Galactic ages ranging from 12.4 to 14.7×10^9yr. More complex Galactic evolution models with infall and without instantaneous initial metal enrichment could increase the age of the Galaxy by as much as approximately $2 - 3 \times 10^9$yr.

061.158   Urto materia antimateria per capire l'Universo.
G. Salvini.
Atti Accad. Naz. Lincei, Rend. Adunanze Solenni, Vol. 8, Fasc. 8, p. 403 – 419 (1984). With plates 1 – 5.

Neutron capture cross sections for s–process studies.
See Abstr. 002.074.

Beta–decay rates of highly ionized heavy atoms in stellar interiors.
See Abstr. 002.075.

The particle explosion.
See Abstr. 003.039.

The early universe and its evolution. Proceedings of the International School of Nuclear Physics, held at Erice, Sicily, Italy, 2 – 14 April 1986.
See Abstr. 012.024.

String theory, quantum cosmology and quantum gravity, integrable and conformal invariant theories. Proceedings of a colloquium held at Paris and Meudon, France, September 1986.
See Abstr. 012.083.

Superstrings, unified theories and cosmology. Proceedings of a workshop held at Trieste, Italy, June 1986.
See Abstr. 012.084.

Observational neutrino astrophysics.
See Abstr. 013.045.

Laboratory approaches of nuclear reactions involved in primordial and stellar nucleosynthesis.
See Abstr. 022.075.

Feasibility of cosmic neutrino detectors based on coherent interaction with superconductors.
See Abstr. 034.023.

Search for neutrino bursts from gravitational stellar collapses.
See Abstr. 034.050.

Report on the Stanford octagonal magnetic monopole detector.
See Abstr. 034.193.

Magnetic monopole plasma oscillations and the survival of galactic magnetic fields.
See Abstr. 062.103.

Neutrinos from gravitational collapse.
See Abstr. 065.003.

The effects of neutrino transport on the collapse of iron stellar cores.
See Abstr. 065.025.

On the origin of the solar system s–process abundances.
See Abstr. 065.077.

Analytical results connecting stellar structure parameters and neutrino fluxes.
See Abstr. 065.083.

The s–process nucleosynthesis of barium stars.
See Abstr. 065.100.

A note on the linear stellar model.
See Abstr. 065.109.

Positron annihilation gamma rays from novae.
See Abstr. 065.119.

Can mini–superspace quantization be justified?
See Abstr. 066.029.

Strings from gravity.
See Abstr. 066.034.

Ghost neutrino fields in flat space–time.
See Abstr. 066.040.

In pursuit of the fifth force.
See Abstr. 066.060.

Gravitational interactions of cosmic strings.
See Abstr. 066.065.

Thermal relaxation time of a mixture of relativistic electrons and neutrinos.
See Abstr. 066.108.

The gravity of cosmic loops.
See Abstr. 066.110.

A Lagrangian formulation of high–spin fields coupled to gravity.
See Abstr. 066.115.

The neutron stars with magnetic charges.
See Abstr. 067.032.

The incompressibility of hot, neutron–rich nuclear matter.
See Abstr. 067.054.

Particle production in de Sitter spacetime.
See Abstr. 067.100.

Neutron tori and the origin of r–process elements.
See Abstr. 067.167.

Early neutron stars and quark matter.
See Abstr. 067.168.

The sun and its neutrinos.
See Abstr. 080.029.

Can the solar neutrino problem be the first detected signature of dark matter from the halo of our galaxy? (Candidates for solar cosmions).
See Abstr. 080.053.

Dark matter induced neutrinos from the Sun: theory versus experiment.
See Abstr. 080.054.

Limits on neutrino oscillation parameters from the chlorine solar–neutrino experiment.
See Abstr. 080.056.

^{16}O excesses in Murchison and Murray hibonites: a case against a late supernova injection origin of isotopic anomalies in O, Mg, Ca, and Ti.
See Abstr. 105.059.

Neutron–exposure variations in MS and S stars, and the implications for s–process nucleosynthesis.
See Abstr. 114.078.

Production and interaction of high energy neutrinos in close X–ray binaries.
See Abstr. 117.008.

Time profile of the neutrino burst from SN 1987A in the Large Magellanic Cloud.
See Abstr. 125.205.

Neutrinos from SN 1987A.
See Abstr. 125.217.

Neutrinos from supernova 1987a.
See Abstr. 125.226.

Regularly pulsed neutrinos from supernova SN 1987A?
See Abstr. 125.237.

The neutrino burst from Supernova 1987A: a search for periodicities.
See Abstr. 125.246.

On the event observed in the Mont Blanc Underground Neutrino Observatory during the occurrence of supernova 1987a.
See Abstr. 125.303.

Comments on the two events observed in neutrino detectors during the supernova 1987A outburst.
See Abstr. 125.304.

SN 1987A supernova: a black–hole precursor?
See Abstr. 125.312.

Particle acceleration and production of energetic photons in SN 1987A.
See Abstr. 125.314.

Neutrino spectroscopy of supernova 1987A.
See Abstr. 125.320.

The neutrino emission of SN 1987A.
See Abstr. 125.323.

A limit on the neutrino–neutrino scattering cross section from the supernova.
See Abstr. 125.325.

Neutrino observations from supernova 1987a.
See Abstr. 125.330.

Observation of a neutrino burst from the supernova SN 1987A.
See Abstr. 125.331.

Observation of a neutrino burst from supernova SN 1987A.
See Abstr. 125.332.

On possible detection of a neutrino burst on February 23, 1987 at the Baksan underground scintillation telescope.
See Abstr. 125.333.

MSW effect and (anti)neutrinos from SN 1987A.
See Abstr. 125.335.

Neutrino mass determinations from the supernova SN 1987A bursts.
See Abstr. 125.336.

Neutrino observations from supernova 1987A.
See Abstr. 125.343.

Information on neutrino masses obtained from supernova 1987A data.
See Abstr. 125.344.

Neutrino mass and supernova 1987A.
See Abstr. 125.345.

Neutrinos from supernova 1987A.
See Abstr. 125.346.

Total energy of the neutrino burst from the supernova 1987A and the mass of the neutron star just born.
See Abstr. 125.347.

Severe restrictions of neutrino masses and mixing angles from SN 1987A.
See Abstr. 125.348.

Making the most of SN 1987A.
See Abstr. 125.355.

Measurements of the ^{3}He abundance in the interstellar medium.
See Abstr. 132.049.

Nucleosynthesis and astrophysical gamma ray spectroscopy.
See Abstr. 143.010.

High energy cosmic rays from young neutron stars.
See Abstr. 144.001.

Extensive air showers and the nature of charge to neutral ratio of hadrons at high energies: some comments.
See Abstr. 144.012.

Cosmic ray kinetics in a strong magnetic field with large–scale inhomogeneities.
See Abstr. 144.013.

Compound nuclei in primary cosmic rays as sources of γ–quanta of superhigh energy.
See Abstr. 144.015.

Search for superheavy grand unified magnetic monopoles in cosmic rays.
See Abstr. 144.040.

Galactic center positronium fraction: observations and simulations.
See Abstr. 155.082.

Chemical evolution of elliptical galaxies.
See Abstr. 157.208.

Photons in curved space–time.
See Abstr. 161.012.

Particle physics and inflationary cosmology.
See Abstr. 161.015.

Unstable neutrinos can do it!
See Abstr. 161.056.

Cosmic particles.
See Abstr. 161.094.

Cosmological aspects of superstring models.
See Abstr. 161.107.

Cosmic strings, galaxy formation and peculiar velocities.
See Abstr. 161.109.

Galaxy formation with baryonic infall: implications for galaxy dynamics, decaying dark matter and dark matter detection.
See Abstr. 161.110.

Cosmic string searches.
See Abstr. 161.113.

A slow rollover phase transition in the Schrödinger picture.
See Abstr. 161.115.

Cold dark matter candidates.
See Abstr. 161.117.

Fundamental physics and dark matter.
See Abstr. 161.145.

The quark–hadron phase transition and primordial nucleosynthesis.
See Abstr. 161.148.

Supersymmetrical particles, cosmology and astrophysics.
See Abstr. 161.158.

Neutrinos and cosmology.
See Abstr. 161.159.

Op zoek naar de eenheid van het heelal.
See Abstr. 161.163.

Cosmology with twice compactified internal space and higher–order gravitational Lagrangian.
See Abstr. 161.172.

The quark–hadron phase transition and primordial nucleosynthesis.
See Abstr. 161.212.

The monopole problem and the primordial black hole problem in the inflationary universe.
See Abstr. 161.216.

Late baryogenesis in superstring models.
See Abstr. 161.221.

Entropy production in tepid inflation.
See Abstr. 161.223.

The constancy of physics.
See Abstr. 161.233.

Cosmology and particle physics.
See Abstr. 161.252.

Supersymmetry and the early universe.
See Abstr. 161.253.

Cosmic strings.
See Abstr. 161.264.

String theory and quantum cosmology.
See Abstr. 161.265.

Cosmology and extra dimensions.
See Abstr. 161.266.

Cosmologies based on Lyra's geometry.
See Abstr. 161.269.

Abelian anisotropic cosmological models in 11–dimensional super-gravity.
See Abstr. 161.290.

Leptonic and hadronic mass scales – a cosmic connection?
See Abstr. 161.293.

Neutrino families: the early Universe meets elementary particle/acceleration physics.
See Abstr. 161.294.

Primordial nucleosynthesis and $\Omega_B \sim 1$ cosmologies with interacting radiation and matter.
See Abstr. 161.312.

Nuclear processes in the early universe. I. The thermal history of matter. II. The cosmic quark–hadron phase transition.
See Abstr. 161.327.

Light neutrinos as cold dark matter.
See Abstr. 161.338.

Nucleosynthesis versus the mirror universe.
See Abstr. 161.341.

Gravitational particle creation and inflation.
See Abstr. 161.344.

Dark matter in the Universe ... and in the laboratory.
See Abstr. 161.357.

Formation of the first systems in the wakes of moving cosmic strings.
See Abstr. 161.364.

Amplification of correlation functions by gravity in the cosmic string model.
See Abstr. 161.365.

Decay of massive particles in Robertson–Walker universes with statically bounded expansion laws.
See Abstr. 161.381.

Quantum cosmology of superstrings.
See Abstr. 161.387.

Superconducting cosmic strings – energy loss by plasma dissipation.
See Abstr. 161.401.

On the evolution of global strings in the early universe.
See Abstr. 161.402.

Decaying axion and the cosmic UV background.
See Abstr. 161.403.

A model for baryogenesis in superstring unification.
See Abstr. 161.414.

On the decay of cosmic string loops.
See Abstr. 161.419.

Cosmic strings and an improved upper bound on the energy density during inflation.
See Abstr. 161.420.

Baryogenesis in chaotic inflationary cosmology.
See Abstr. 161.421.

Baryogenesis at the MeV era.
See Abstr. 161.422.

Fragmentation of cosmic string loops.
See Abstr. 161.423.

Numerical simulation of cosmic–string evolution in flat spacetime.
See Abstr. 161.424.

Cosmic balls of trapped neutrinos.
See Abstr. 161.426.

Ultra–high–energy cosmic rays from superconducting cosmic strings.
See Abstr. 161.427.

The consistency probems of large scale structure.
See Abstr. 161.441.

Galaxy distribution in a cold dark matter universe.
See Abstr. 161.443.

Can nucleons close the Universe?
See Abstr. 161.446.

062 Hydrodynamics, Magnetohydrodynamics, Plasma

062.001 **Spatial stability of relativistic jets: application to 3C 345.**
P. E. Hardee.
Astrophys. J., Vol. 318, No. 1, p. 78 – 92 (1987).
The relativistic dispersion relation describing growth and the propagation of the harmonic components of a perturbation to the surface of a cylindrical relativistic jet are analyzed numerically for relativistic flows. The results are used to extend previous expressions for resonant frequencies, wavelengths, and the maximum growth rate at resonance valid for nonrelativistic flows to relativistic flows. The cases of an isothermal and an adiabatic expanding jet are presented. These two cases provide limits on the possible helical trajectories. This model is then applied to observations of the motion of components along the inner part of the jet in 3C 345.

062.002 **Rate of change of vorticity covariance in turbulence.**
N. Kishore, A. Sinha.
Astrophys. Space Sci., Vol. 135, No. 1, p. 191 – 193 (1987).
The present paper deals with the turbulent flow of an incompressible viscous fluid which is isotropic and spatially homogeneous. The expression for the rate of change of vorticity covariance is derived. The derived result shows that the defining scalars $\alpha(r, t)$ and $\beta(r, t)$ of the rate of change of vorticity covariance solely depend upon the defining scalar Q of the two–point velocity correlation.

062.003 **Current gradient–driven linear and nonlinear electromagnetic waves in a magnetized electron–positron plasma.**
R. Bharuthram, P. K. Shukla, M. Y. Yu, V. N. Pavlenko.
Astrophys. Space Sci., Vol. 135, No. 2, p. 211 – 218 (1987).

A set of coupled nonlinear differential equations which govern the dynamics of finite amplitude electromagnetic waves in the presence of an external current gradient in a magnetized electron–positron plasma has been derived. It is shown that the current gradient can make shear Alfvén–like waves unstable. A quasi–stationary solution of the mode–coupling equations is the well–localized dipole vortex. Application of the results to plasma transport in the pulsar magnetosphere is briefly discussed.

062.004   Unsteady MHD free–convection flows with time–dependent heating in a rotating medium.
P. K. Kythe, P. Puri.
Astrophys. Space Sci., Vol. 135, No. 2, p. 219 – 228 (1987).

Unsteady free–convection flows near an infinite vertical plate in a rotating medium in the presence of a constant transverse magnetic field are investigated under an arbitrary time–dependent heating of the plate. By using the Laplace transform technique, the Green's function of the problem is determined and exact solutions are obtained for special cases of the time–dependent heating effect. The thermal influence on skin friction at the plate and the displacement thickness of the boundary layer are determined, and the structure of the thermal wave trains is discussed.

062.005   Rayleigh–Taylor instability of rotating fluid in the presence of a vertical magnetic field.
P. D. Ariel.
Astrophys. Space Sci., Vol. 135, No. 2, p. 365 – 375 (1987).

The Rayleigh–Taylor instability of two rotating superposed fluids in the presence of a vertical magnetic field has been investigated. It is shown that n^2 is purely real, where n is the growth rate of a perturbation. In the basis of this fact it is shown that a unique dispersion relation exists if the lighter fluid lies beneath the heavier one. However, if the heavier fluid lies beneath the lighter fluid, then no unique dispersion relation exists. The effect of rotation is to slow the rate at which potentially unstable stratification departs from the equilibrium position.

062.006   Magnetogasdynamic plane shock motion.
R. C. Srivastava, K. G. Roesner, D. Leutloff.
Astrophys. Space Sci., Vol. 135, No. 2, p. 399 – 407 (1987).

In the present work the reflection of a plane shock wave is studied in order to achieve a high pressure and temperature state by a reflected shock wave. The authors consider the plane geometry and solve the one–dimensional, time–dependent system of hyperbolic equations by Rusanov's method.

062.007   Numerical modelling of three–dimensional plasma cloud evolution in crossed E × B fields.
S. P. Voskoboynikov, V. A. Rozhansky (*V. A. Rozhanskij*), L. D. Tsendin.
Planet. Space Sci., Vol. 35, No. 7, p. 835 – 844 (1987).

A numerical solution of the complete three–dimensional equation system for plasma inhomogeneity evolution in crossed **E × B** fields is presented. The approximation of homogeneous infinite ambient plasma is considered. The model values of transport coefficients were used. The physical pattern of processes is discussed. The simple analytical formulae for plasma evolution at various stages are derived. These formulae are in good agreement with numerical results.

062.008   The effect of electron screening on thermonuclear reaction rates.
D. G. Yakovlev, D. A. Shalybkov.
Pis'ma Astron. Zh., Tom 13, No. 8, p. 730 – 736 (1987). In Russian. English translation in Sov. Astron. Lett., Vol. 13.

The enhancement factor of thermonuclear reaction rates due to the effects of plasma screening is calculated for a degenerate weakly non–ideal electron gas and multi–component strongly coupled ion liquid. The screening effects without and with the account for the polarizability of electron gas are considered. The results may be applied to degenerate cores of white dwarfs and envelopes of neutron stars.

062.009   Compact radio sources as a plasma turbulent reactor. I. Formation of Maxwell–like spectra of relativistic electrons at the acceleration on resonant Langmuir waves.
A. M. Atoyan, A. Nagapetyan.
Astrofizika, Tom 26, Vyp. 3, p. 527 – 541 (1987). In Russian. English translation in Astrophysics, Vol. 26, No. 3, p. 318 – 327 (1987).

For extragalactic compact radio sources such as QSOs and AGNs a self–consistent plasma turbulent reactor model is proposed wherein the acceleration of relativistic electrons responsible for the electromagnetic radiation of these objects on resonant Langmuir waves in thermal plasma occurs. It is shown that this stochastic acceleration process is dominant even if an extremely small fraction of the energy density of resonant turbulence $W_l^{(res)}$ is present in the total energy density of the turbulence $W_l \sim W_H \lesssim nk_BT$. Self–consistent consideration of the acceleration and radiative losses of the electrons leads to the Maxwell–like spectra of relativistic electrons with the characteristic Lorentz–factor $\gamma_0 \sim 10^3$.

062.010   The stability of differentially rotating self–gravitating gas clouds. II. Polytropic configurations.
F. Schmitz, R. Ebert.
Astron. Astrophys., Vol. 181, No. 1, p. 41 – 49 (1987).

The stability of differentially rotating gas clouds with polytropic equations of state is investigated. The equilibrium states are self–consistent. The structure of the infinitely extended configurations is very simple. Perpendicular to the rotation axis, the density distribution and the differential rotational velocity are described by power laws. In the limit of vanishing rotation the authors obtain the familiar polytropic gas spheres. Only marginal axisymmetric perturbations are considered. The structure of the equilibrium states and the form of the perturbations are both calculated with the approximation method presented in paper I.

062.011   Relative emission line strengths for Fe VII in astrophysical plasmas.
F. P. Keenan, P. H. Norrington.
Astron. Astrophys., Vol. 181, No. 2, p. 370 – 372 (1987).

Recent R–matrix calculations of electron impact excitation rates in Fe VII are used to derive relative emission line strengths for a range of electron temperatures ($8000K \leqslant T_e \leqslant 120000K$) and densities ($10^4 cm^{-3} \leqslant N_e \leqslant 10^{10} cm^{-3}$) applicable to gaseous nebulae. The results are found to be significantly different from those of Nussbaumer and Storey, especially at low temperatures.

062.012   Unresolved dielectronic satellite lines of Ly α Ca XX resonance lines in high temperature plasmas.
S. Volonté, J. Lion, P. Faucher, J. Dubau.
Astron. Astrophys., Vol. 182, No. 1, p. 167 – 173 (1987).

New calculations have been carried out for the high dielectronic satellite lines of the type $1snl–2l'nl$ for $n \geqslant 5$ in calcium. These calculations based on hydrogenic wavefunctions offer the advantage of analyticity in the derivation of the energies and transition probabilities whereas the autoionization rates are obtained from Quantum Defect Theory. The effect of the $3l'nl$ resonant states on the direct excitation of the Ca XX Ly α doublet has also been estimated. The set of new data is used together with previous results for $2 \leqslant n \leqslant 5$ (Blanchet et al. 1985) to improve the diagnostic capabilities provided by the satellite to resonance line intensities in high temperature plasmas.

062.013   Autowave dissipation of forceless magnetic fields.
A. A. Solov'ev.
Astrophysics, Vol. 25, No. 2, p. 587 – 595 (1987). English translation of 42.062.123.

062.014   Correlation properties of fine–structured magnetic fields.
N. I. Kliorin, A. A. Ruzmajkin, D. D. Sokolov.
Soln. Dannye, Byull., 1987, No. 2, p. 59 – 66 (1987). In Russian.

The correlation structure of a small–scale magnetic field in a random velocity field is studied in the framework of the dynamo theory. Three characteristic scales are noted: l, $lR_m^{-1/2}$ and

$lR_m^{-1/4}$, where l is a characteristic scale of a turbulent cell, R_m is the magnetic Reynolds number. The results are suggested to be compared with the observed solar magnetic field.

062.015 Thermal conduction and self–similar accretion flows.
M. Anderson.
Mon. Not. R. Astron. Soc., Vol. 227, No. 3, p. 623 – 638 (1987).

A series of self–similar rotating accretion flows is constructed with the inclusion of a weak shear viscosity and a simple thermal conduction law. Comparison is made with previously published solutions, and the models analysed for convective stability. The solutions, all of which show either slow infall or outflow along the polar axis, are found to share the same convective instability at moderate to high latitudes that besets other self–similar models.

062.016 Thermal overstability of hydromagnetic surface waves.
P. S. Joarder, M. H. Gokhale, P. Venkatakrishnan.
Sol. Phys., Vol. 110, No. 2, p. 255 – 270 (1987).

The authors investigate the effects of radiative heat losses and thermal conductivity on the hydromagnetic surface waves along a magnetic discontinuity in a plasma of infinite electrical conductivity. They show that the effects of radiative heat losses on such surface waves are appreciable only when values of the plasma pressure on the two sides of the discontinuity are substantially different. Overstability of a surface wave requires that the medium in which it gives larger first–order compression should satisfy the criterion of Field (1965). Possible applications of the study to magnetic discontinuities in solar corona are briefly discussed.

062.017 On magnetohydrodynamic solitons in jets.
B. Roberts.
Astrophys. J., Vol. 318, No. 2, p. 590 – 594 (1987).

Nonlinear solitary wave propagation in a compressible magnetic beam model of an extragalactic radio jet is examined and shown to lead to solitons of the Benjamin–Ono type. A number of similarities between such magnetic beam models of jets and models of solar photospheric flux tubes are pointed out and exploited. A single soliton has the appearance of a symmetric bulge on the jet which propagates faster than the jet's flow.

062.018 Magnetic reorientation and the spontaneous formation of tangential discontinuities in deformed magnetic fields.
E. N. Parker.
Astrophys. J., Vol. 318, No. 2, p. 876 – 887 (1987).

This paper provides an explicit illustration of the formation of tangential discontinuities (current sheets) in a force–free magnetic field whose footpoints have been subjected to bounded continuous displacement and shuffling, so that the lines of force are wound about each other in complex but continuous patterns. The discontinuities appear spontaneously because of the reorientation of the field under the enhanced pressure where two regions of different field topology are pressed together by the general winding and wrapping.

062.019 Propagation of compressive waves through fibril magnetic fields. II. Scattering from a slab of magnetic flux tubes.
T. J. Bogdan.
Astrophys. J., Vol. 318, No. 2, p. 888 – 895 (1987).

The reflection/transmission of an acoustic plane wave from/through a slab of parallel, randomly distributed, magnetic flux tubes is worked out in the long wavelength (thin flux tube) limit. These results are then contrasted with the reflection/transmission problem for (1) a uniformly magnetized homogeneous slab and (2) an infinite half–space of parallel, randomly distributed, flux tubes.

062.020 Propagation of compressive waves through fibril magnetic fields. III. Waves that propagate along the magnetic field.
T. J. Bogdan.
Astrophys. J., Vol. 318, No. 2, p. 896 – 903 (1987).

An isothermal fibril magnetic field composed of a uniform distribution of parallel magnetic flux tubes embedded in an unmagnetized fluid is shown to support compressive modes that propagate along the flux tube axes with phase velocities below the sound speed. These modes are simply the familiar tube waves of a single isolated magnetic flux tube that are modified by the presence of the many individual flux tubes that constitute the fibril magnetic field. This modification is worked out for an ensemble of randomly placed parallel magnetic flux tubes distributed uniformly throughout all space.

062.021 Adiabatic consideration of the motion of charged particles in superposed dipole and uniform magnetic fields.
Z. M. Psillakis, G. A. Katsiaris.
Astrophys. Space Sci., Vol. 136, No. 1, p. 43 – 62 (1987).

The motion of a charged particle is studied within a magnetic field. This field consists of two separate fields; a dipole and a uniform magnetic field, parallel to dipole's magnetic moment. The present study is maintained by means of the adiabatic theory. The authors use a numerical integration of the equations of motion and give comparative results between the adiabatic theory and the numerical integration. The previous results are applied to the case of the Earth's open magnetosphere. Diagrams and tables support this application.

062.022 Effects of low–frequency instability on the Hall conductivity in plasma: applications to astrophysics and space physics.
C. L. Xaplanteris.
Astrophys. Space Sci., Vol. 136, No. 1, p. 171 – 181 (1987).

The author studied the effect of low–frequency drift wave instability on Hall conductivity in plasma. Using an external oscillation the variation on Hall conductivity is observed. The effect is probably to be attributed to electron trapping by the waves potential. Good agreement between experimental and calculated values of azimuthal drift currents near and away from resonance lead the author to believe that the proposed explanation by electron trapping is correct. In addition, the interaction of plasma with the magnetic field is important in a large variety of astrophysical phenomena. A large class of solar and magnetospheric phenomena involve the conversion of stored magnetic energy to thermal and kinetic energy of the plasma with mechanism in which important role have the plasma's conductivity. Accordingly, this experimental work must be considered as a good laboratory simulation to solar plasma devices.

062.023 Rarefied gas flow near an asymmetric plane stagnation point.
A. Raptis, H. S. Takhar.
Astrophys. Space Sci., Vol. 136, No. 1, p. 187 – 190 (1987).

A study of the flow of a rarefied gas near an asymmetric plane stagnation point is considered. The solution of the problem is obtained numerically.

062.024 Physics of modes in a differentially rotating system – analysis of the shearing sheet.
R. Narayan, P. Goldreich, J. Goodman.
Mon. Not. R. Astron. Soc., Vol. 228, No. 1, p. 1 – 41 (1987).

The authors analyse the linear non–vortical modes of the shearing sheet, a model compressible two–dimensional fluid system with constant density, constant shear, and Coriolis force. This model has several features found in differentially rotating systems of interest in astrophysics, such as disc galaxies, accretion tori, planetary rings, protostellar nebulae, and possibly even rotating stars.

062.025   On the stability of compressible differentially rotating cylinders – II.
W. Glatzel.
Mon. Not. R. Astron. Soc., Vol. 228, No. 1, p. 77 – 100 (1987).
    The stability of differentially rotating fluid cylinders is re-examined. One kind of instability occurring is shown to be due to resonant interaction of modes belonging to different boundaries. A detailed description of the mode interaction is given applicable both to surface and sound waves. The author classifies the modes due to their physical origin and provides a general scheme of their order. The second kind of instability is due to energy loss from the system, in particular due to acoustic radiation. A physical interpretation is given for both of the instabilities in terms of the energy of a mode.

062.026   Mode classification and wave propagation in a magnetically structured medium.
S. S. Hasan, Y. Sobouti.
Mon. Not. R. Astron. Soc., Vol. 228, No. 2, p. 427 – 451 (1987).
= Contrib. No. 14, Biruni Obs.
    The authors examine the structure of motions that can occur in a vertical magnetic flux tube with a rectangular cross–section. A polytropic stratification is assumed in the vertical direction. The authors use a gauged version of Helmholtz's theorem, the decompose the perturbations into an irrotational component and a solenoidal component, which they further split into the sum of poloidal and toroidal components. These components are identified with p, g and toroidal modes of a fluid. The normal modes of the tube are determined using a Rayleigh–Ritz variational technique. The authors' technique efficiently isolates all the modes to high orders. They first consider some special cases, in order to highlight some interesting properties of the modes. Then, they choose a parameter range to study the properties of oscillations in intense flux tubes on the Sun.

062.027   Mass loaded astronomical flows – III. The structure of supernova remnants and the local soft X–ray background.
J. E. Dyson, T. W. Hartquist.
Mon. Not. R. Astron. Soc., Vol. 228, No. 2, p. 453 – 461 (1987).
    The authors derive a similarity solution for a supernova remnant whose mass is dominated by material mixed in by hydrodynamic ablation from embedded clumps of gas. They use this solution to calculate the properties of a remnant which can reproduce both the relative and absolute count rates in the M, B, and C soft X–ray bands measured for the local supernova remnant. They also show that it is possible to produce the same soft X–ray counts in a remnant whose mass is dominated by material injected by conductive mixing. The authors conclude that it is not possible to distinguish uniquely between physically very different remnant structures on the basis of soft X–ray counts alone.

062.028   Reaction rates and spectra in relativistic plasmas.
M. Baring.
Mon. Not. R. Astron. Soc., Vol. 228, No. 3, p. 681 – 693 (1987).
    Analytic expressions for total reaction rates, particle production and absorption spectral rates, and rates of energy exchange are derived for arbitrary interactions in relativistic plasmas. The collisions are assumed to involve two incident particles and produce two particles, conserving energy. The expressions are then evaluated for the case of two interacting thermal distribution functions at different temperatures, resulting in single integrals for each rate except for the production spectrum, where a scattering angle integration also appears. Comparison is made with known results where possible.

062.029   The effect of cooling processes on the tail of a relativistic thermal pair distribution.
M. Baring.
Mon. Not. R. Astron. Soc., Vol. 228, No. 3, p. 695 – 712 (1987).
    The maintenance of the population in the Maxwellian tail of a relativistic thermal pair plasma by binary collisions is investigated, and the effect of perturbations on the tail of the distribution due to cooling by bremsstrahlung, inverse Compton scattering of soft photons, synchrotron emission and pair annihilation is considered. A comparison is made between the spectral time–scales for two–body thermalization and cooling at different pair energies.

062.030   Quasi–monochromatic wave–particle interactions in magnetospheric plasmas.
D. Le Quéau, A. Roux.
Sol. Phys., Vol. 111, No. 1, p. 59 – 80 (1987). – See Abstr. 012.016.
    The natural waves observed within magnetospheres are often emitted over a very narrow spectral range. They thus can interact coherently with the charged particles forming the weakly collisional magnetospheric plasmas. The authors review some aspects of these quasi–coherent wave particle interactions (WPI), usually known as trapping processes. A particular emphasis is put on the situation effectively encountered with natural plasmas. Hence, the effects of plasma inhomogeneities on the resonant WPI are thoroughly studied.

062.031   Plasma emission: a review.
D. B. Melrose.
Sol. Phys., Vol. 111, No. 1, p. 89 – 101 (1987). – See Abstr. 012.016.
    The theory of plasma emission is reviewed emphasizing general concepts rather than the details of the analysis. The generation of the Langmuir turbulence, its evolution due to nonlinear process-es, and the plasma emission processes are described. Several outstanding difficulties in the applications to solar radio bursts are discussed, concentrating on those with implications on local density inhomogeneities in the sources.

062.032   Plasma turbulence in space.
V. D. Shapiro, V. I. Shevchenko.
Itogi Nauki i Tekhniki. Seriya Astronomiya. Tom 32. Astrophys-ics and space physics, p. 235 – 319 (1987). In Russian. – See Abstr. 003.004.

062.033   The configuration of slow–mode shocks.
R. Wolfson.
J. Geophys. Res., Vol. 92, No. A9, p. 9875 – 9884 (1987).
    This paper presents a simple method for computing slow shock flows in an infinitely conducting plasma, based on the solution of a free boundary problem for the shock configuration that matches a post shock potential flow to a uniform preshock flow. The method is applied to shocks arising from flow about spherical and paraboloidal obstacles. A parametric study of shock formation as a function of sonic and Alfvén Mach numbers shows standoff distance increasing as either Mach number decreases. Streamlines and shock configurations for several slow shock flows are also presented.

062.034   The guidance of oblique whistler mode waves along magnetospheric field lines.
S. S. Sazhin.
Astrophys. Space Sci., Vol. 136, No. 2, p. 221 – 224 (1987).
    An approximate analytical expression for the wave normal angle $\theta_g \neq 0$, at which the whistler–mode group velocity is directed along magnetic field line in a hot anisotropic plasma, is derived. This expression is simple enough for practical applica-tions in the magnetospheric conditions and is compatible with the results of numerical analysis for wave frequencies below but not close to the half of electron gyrofrequency.

062.035   Group Doppler effect in anisotropic plasmas.
P. L. Leung, P. C. W. Fung.
Astrophys. Space Sci., Vol. 136, No. 2, p. 259 – 279 (1987).
    In anisotropic plasmas, the radiative power emitted and the power observed per unit solid angle should be calculated along the direction of the group velocity v_g. The two power functions referred differ by a product of two factors: one is the group Doppler factor and the other is the "squeezing effect" of the radiative energy due to the dependence v_g on direction. In this

paper, the group Doppler factor is derived using two different methods, and the relevant physical concepts are analyzed in detail. A number of numerical examples pertaining to astrophysical situations are presented to illustrate the significance of the group Doppler effect with respect to the "wave Doppler effect" which is valid in isotropic media.

062.036   Structure of magnetogasdynamic weak shock waves.
M. A. Khidr, S. S. Elghabaty, M. A. A. Mahmoud.
Astrophys. Space Sci., Vol. 136, No. 2, p. 379 – 391 (1987).

The problem of the detailed structure of magnetogasdynamic shock waves is investigated. It is assumed that the flow takes place under normal magnetic field H_0 and the conductivity of the medium is considered infinite. An approximate analytical solution of the nonlinear differential equations describing the phenomena is obtained. The suggested analytical results in this paper are in good agreement with the previous numerical computations for the thickness and the velocity distribution inside the transition region. In addition, the enthalpy distribution inside the shock front is predicted.

062.037   The Chandrasekhar–Kendall functions.
S. Chandra.
Astrophys. Space Sci., Vol. 136, No. 2, p. 409 – 412 (1987).

The author finds that the Chandrasekhar–Kendall functions (A) do not satisfy the identity $(A \cdot \nabla)A = (\nabla \times A) \times A + (1/2)\nabla A^2$ and, therefore, the results, in magnetohydrodynamics and other fields, obtained by two independent authors may differ from one another.

062.038   The sources and their models.
W. Kundt.
Astrophysical jets and their engines, p. 1 – 13 (1987). – See Abstr. 012.021.

The similarities are stressed between the different astrophysical jet sources, or bipolar flows, which apparently make up four classes: active galactic nuclei, young stellar objects, binary neutron stars and binary white dwarfs. This similarity is remarkable in view of the vastly different environments and contrasts with the heterogeneous models which have been advanced for their description.

062.039   The central engine.
W. Kundt.
Astrophysical jets and their engines, p. 13 – 20 (1987). – See Abstr. 012.021.

The literature of the past 10 years has favoured black holes over non–collapsed (super–)massive objects as the central engines of AGN behaviour. But bipolar flows and jets from compact stars are similar and driven by the cores of accretion disks. Some difficulties of the black–hole model are discussed, in particular the problem of the missing ashes. The (super–)massive magnetised core model of an accretion disk is revisited.

062.040   The jets.
W. Kundt.
Astrophysical jets and their engines, p. 21 – 28 (1987). – See Abstr. 012.021.

Jets may simply be the outlets of overpressure bubbles filled with relativistic pair plasma. They ram channels through media of various densities but can be destroyed by neutral intruders, such as photons and hydrogen atoms. The possibility of in–situ acceleration of electrons in their hotspots, the correct form of the relativistic beaming law and the confinement problem are discussed. It is argued that all jets are supersonic, that the flow in the cocoon depends on the ambient density distribution and that many extragalactic jets interact visibly with massive shells in the halos of their host galaxies.

062.041   Numerical simulation of jets.
S. A. E. G. Falle.
Astrophysical jets and their engines, p. 151 – 161 (1987). – See Abstr. 012.021.

The reliability of published numerical simulations of fluid jets is assessed. It is shown that the resolution used in these calculations is much too low for the results to be of any use.

062.042   The gas dynamics of jets.
S. A. E. G. Falle.
Astrophysical jets and their engines, p. 163 – 170 (1987). – See Abstr. 012.021.

This article explores some of the consequences of assuming that both stellar and extragalactic jets behave like a fluid. General similarity arguments suggest that some of the observed morphology of extragalactic jets is due to the form of the ambient density distribution. The series of knots seen in both stellar and extragalactic jets appear to be related to the shock cell structure found in laboratory jets. It is shown that there can be quite a large pressure ratio between the jet and the outside medium even if there is no magnetic confinement.

062.043   Unified beaming models and compact radio sources.
J. A. Peacock.
Astrophysical jets and their engines, p. 185 – 196 (1987). – See Abstr. 012.021.

062.044   Magnetic equilibria of jets.
G. Benford.
Astrophysical jets and their engines, p. 197 – 204 (1987). – See Abstr. 012.021.

062.045   Stability of magnetic jet equilibria.
G. Benford.
Astrophysical jets and their engines, p. 205 – 210 (1987). – See Abstr. 012.021.

062.046   Dynamical effects of large–scale magnetic fields in jets.
A. Achterberg.
Astrophysical jets and their engines, p. 211 – 222 (1987). – See Abstr. 012.021.

The possible dynamical effect of large–scale magnetic fields in astrophysical jets is discussed. They range from collimation and refocussing of jets to interaction between adjacent current–carrying jets.

062.047   Particle acceleration in astrophysical jets.
A. Achterberg.
Astrophysical jets and their engines, p. 223 – 236 (1987). – See Abstr. 012.021.

The different mechanisms for the acceleration of particles in astrophysical jets are discussed. The formation of power–law particle distributions is considered by the processes of Fermi acceleration, gyro–resonant acceleration and shock acceleration.

062.048   Nonequilibrium populations of rotational levels of emitting molecules in a low–density gas.
B. F. Gordiets, A. N. Stepanovich.
Mat. zadachi prikl. aehron., Moskva, 1987, p. 111 – 125 (1987). In Russian. Abstr. in Ref. Zh., 51. Astron., 8.51.86 (1987).

062.049   An expanding plasma cloud irradiated by gamma–quanta: its structure, optical emission and influence on gamma–burst spectra in the "soft region".
G. S. Bisnovatyj–Kogan, A. F. Illarionov.
Inst. Kosm. Issled. Akad. Nauk SSSR, Prepr., No. 1184, 38 pp. (1986). In Russian. Abstr. in Ref. Zh., 51. Astron., 8.51.109 (1987).

062.050   Helicity conservation laws for fluids and plasmas.
J. D. Bekenstein.
Astrophys. J., Vol. 319, No. 1, p. 207 – 214 (1987).

The author studies helicity conservation laws in fluid dynamics and magnetohydrodynamics. Ertel and Katz demonstrated that "potential vorticity" is invariant along streamlines in nonisentropic flow of a perfect fluid. The authors show, in covariant language, that although Moffatt's fluid helicity is not conserved under such circumstances, a related conserved quantity, fluid chirality, can be defined. The author also presents the covariant version of Woltjer's law of conservation of mixed fluid–magnetic helicity in perfect magnetohydrodynamics. A new kind of conserved fluid–magnetic helicity, distinct from Woltjer's, is displayed. Finally, the author considers departures from perfect magnetohydrodynamics by studying in some detail a special class of flows with entrained electromagnetic fields, "potential–field flows".

062.051 Hydrodynamical constraints on cosmic–ray acceleration in relativistic shocks.
G. M. Webb.
Astrophys. J., Vol. 319, No. 1, p. 215 – 236 (1987).

A two–fluid hydrodynamical model governing the transport of cosmic rays in a relativistically moving background plasma is developed. The equations are used to discuss the time asymptotic structure of a relativistic, plane–parallel shock wave significantly modified by first–order Fermi acceleration of cosmic rays. The model allows for an anisotropic cosmic–ray pressure tensor. Astrophysical implications of the results are briefly discussed.

062.052 On point explosion in a nonuniform medium with a symmetry plane.
I. G. Kovalenko.
Kinematika Fiz. Nebesn. Tel, Tom 3, No. 5, p. 78 – 83, 96 (1987). In Russian. English translation in Kinematics Phys. Celest. Bodies.

The problem of propagation of a shock wave from a point explosion in medium with exponential density distribution and with a symmetry plane is considered. Using the Kompaneets method solutions of an 8–shaped expanding shock envelope are obtained. In case of an explosion above the symmetry plane the formation of toroidal structures of increased density is possible, which can be observed in X–rays. It is shown that a part of energy which generates a shock wave in the galactic plane does not exceed 10^{-7} if the wave reaches $r \geqslant 1$ kpc.

062.053 First–order Fermi acceleration in the two–stream limit.
T. J. Bogdan, G. M. Webb.
Mon. Not. R. Astron. Soc., Vol. 229, No. 1, p. 41 – 52 (1987).

A study of the first–order Fermi mechanism for accelerating cosmic rays at relativistic and non–relativistic shocks is carried out by using the two–stream approximation. Exact steady–state analytic solutions illustrating the shock acceleration process in the test–particle limit in which monoenergetic (relativistic) seed particles enter the shock through an upstream free–escape boundary are obtained.

062.054 The evolution of spectral densities in weakly inhomogeneous plasmas.
K. Rönnmark.
J. Geophys. Res., Vol. 92, No. A10, p. 11,053 – 11,058 (1987).

Theoretical studies of waves and instabilities are often performed for an infinite, homogeneous plasma, and the results are given in terms of eigenmodes (plane waves) extending over all space. Real plasmas are, however, always finite and inhomogeneous, and observations of waves in space are normally presented as frequency power spectra which depend on the time and place of observation. The purpose of this study is to reconcile these apparently disparate descriptions of plasma waves.

062.055 Propagating of sound and thermal waves in a reacting fluid.
M. H. Ibáñez S., C. A. Mendoza B.
Astrophys. Space Sci., Vol. 137, No. 1, p. 1 – 15 (1987).

The propagation of linear sound and thermal waves in a reacting fluid, in which the heating and cooling processes can be represented by a heat–loss function $L(\varrho, T, \xi)$ is studied. A complex dispersion relation is found, from which the phase velocity and the scale length for damping (or amplification), of the above two–wave mode are calculated. Wave amplification may occur in reacting locally stable fluids. Results are applied to a hydrogen plasma model assumed to be heated at a non–specified constant rate and cooled by recombination, excitation, and ionization by collisions, and free–free transitions. The phase velocity $\tilde{v}$, the scale–length for damping $\tilde{l}$, and the relevant relaxation times are calculated as functions of the dimensionless frequency $\tilde{\omega}$ for temperatures ranging from those at which the hydrogen plasma is neutral to those at which it becomes completely ionized.

062.056 The geometry of magnetic fieldlines in magnetogeostrophic flows.
S. N. Singh, B. P. Singh, D. D. Tripathi.
Astrophys. Space Sci., Vol. 137, No. 1, p. 23 – 31 (1987).

It is proved that the only circulation preserving magnetogeostrophic flows whose current density is lamellar, and bears a constant on a current density vector have (1) a plane motion of constant current density (on which certain unsteady potential motions may be superposed) and (2) a particular circular helical motion.

062.057 Si IV line ratios in laboratory plasmas: a comparison of experimental data and theoretical computations.
M. Finkenthal, T. L. Yu, S. L. Allen, L. K. Huang, S. Lippmann, H. W. Moos, B. C. Stratton, P. L. Dufton, A. E. Kingston.
Astron. Astrophys., Vol. 184, No. 1/2, p. 337 – 340 (1987).

Time resolved, absolutely calibrated spectra for transitions between the $n = 3$ levels in Si IV have been measured from both TEXT tokamak and TMX–U mirror device. In both plasmas, the electron density (n_e) and temperature (T_e) were measured independently by non–spectroscopic methods, the Si IV ions being located in regions with $n_e \sim 10^{11} - 10^{12} \text{cm}^{-3}$ and $T_e \sim 5 - 20$ eV. The experimental line ratios are in good agreement with theoretical predictions, providing evidence that the discrepancies between previous computations and solar observations were most probably due to the blending of the observed lines. Consequently, these transitions can be used for electron temperature diagnostics in astrophysical or laboratory plasmas.

062.058 Optimization of detector arrays for measuring the plasma distribution function in space.
M. Macháček.
Bull. Astron. Inst. Czech., Vol. 38, No. 5, p. 297 – 303 (1987).

The author shows that the one–particle distribution function of plasma is best described by a set of mean values of low–order Hermite polynomials of particle velocities. He then describes how these Hermite moments are calculated from detector signals, and how the detectors should be optimized to minimize the errors in the Hermite moments. The optimization procedure can also be used with other sets of physical quantities describing the distribution function.

062.059 On the thermodynamics of a relativistic Fermi–gas.
S. I. Blinnikov.
Pis'ma Astron. Zh., Tom 13, No. 9, p. 820 – 823 (1987). In Russian. English translation in Sov. Astron. Lett., Vol. 13.

The equation of state of an equilibrium ideal gas of fermions is presented as a series converging for any degree of degeneracy. The first terms of the series provide high accuracy at relativistic density or temperature and the rest are summed up and presented as a finite integral of a regular function.

062.060 Astrophysical shocks in diffuse gas.
C. F. McKee.
Spectroscopy of astrophysical plasmas, p. 226 – 254 (1987). – See Abstr. 003.010.

Contents: 1. The nature of astrophysical shocks. 2. Shock jump conditions: Derivation. Non–radiative jump conditions (J–shocks). Radiative shocks (J– and C–shocks). Effects of cosmic rays on J–shocks. 3. J–shocks: Structure of the shock front. Radiative J–shocks: maximum velocity. Radiative J–shocks: structure and spectrum. Emission from non–radiative J–shocks.

4. C–shocks: Physical conditions in C–shocks. Application to Orion.

062.061 Strong adiabatic shock waves in arbitrarily inhomogeneous media. Analytic approach.
B. I. Gnatyk.
Astrophysics, Vol. 26, No. 1, p. 66 – 77 (1987). English translation of 43.062.022.

062.062 On negative viscosity in global circulations.
A. S. Monin.
Dokl. Akad. Nauk SSSR. Ser. Mat. Fiz., Tom 293, No. 1, p. 70 – 73 (1987). In Russian. Abstr. in Ref. Zh., 51. Astron., 9.51.157 (1987).

062.063 Calculation of the moments of the charge state distribution in hot and dense plasmas using the Thomas–Fermi models.
H. Szichman, S. Eliezer, D. Salzmann.
J. Quant. Spectrosc. Radiat. Transfer, Vol. 38, No. 4, p. 281 – 286 (1987).
An algorithm is proposed to calculate the first and second moments of the charge distribution, $\bar{Z}$ and $\bar{Z}^2$, within the framework of the Thomas–Fermi (TF) and the Thomas–Fermi–Dirac (TFD) models. Results obtained by applying this algorithm for copper and lead plasmas are in good agreement with those obtained from the Saha model. This scheme is generalized to calculate higher order moments and the charge-state distributions in TF and TFD models.

062.064 The origin and cosmogonic implications of seed magnetic fields.
M. J. Rees.
Q. J. R. Astron. Soc., Vol. 28, No. 3, p. 197 – 206 (1987).
The purpose of this paper is to air several inter–related issues relating to cosmic magnetic fields which seem to have received rather little attention in the literature: (1) Where did the seed fields in galaxies come from, and how strong were they? (2) Is intergalactic space pervaded by a magnetic field, and what are its effects? (3) What difference would it make to, for instance, star formation if there were no magnetic field (or if the field were dynamically negligible), as might be expected early in the history of a galaxy?

062.065 On hydrodynamics of radiatively driven flows.
R. Turolla.
General relativity and gravitational physics, p. 387 – 398 (1987). – See Abstr. 012.038.
The basic properties of relativistic radiation hydrodynamics are briefly reviewed. In particular the issue of radiatively accelerated flows is discussed in connection with astrophysical applications. The existence of regularity constraints for the transonic solution is stressed both in the adiabatic and in the diffusion limit.

062.066 Non–linear stability of a 3–dimensional spatially symmetric solution of the relativistic Poisson Vlasov equation.
C. Marchioro, E. Pagani.
General relativity and gravitational physics, p. 405 – 409 (1987). – See Abstr. 012.038.

062.067 A numerical model for the formation and propagation of supersonic radio jets.
J. J. Mitteldorf.
Diss. Abstr. Int., Sect. B, Vol. 48, No. 3, p. 793–B (1987). Thesis, University of Pennsylvania, 224 pp. (1987). Order No. DA8714094.

062.068 X–ray pulsars: accretion flow deceleration.
G. S. Miller.
Diss. Abstr. Int., Sect. B, Vol. 48, No. 4, p. 1072–B (1987). Thesis, Cornell University, 151 pp. (1987). Order No. DA8708911.

062.069 Kinetics of a hot plasma. Equations of equilibrium and velocity of elementary particles.
I. L. Bejgman.
Fiz. Inst. Akad. Nauk SSSR, Prepr., No. 78, p. 1 – 19 (1987). In Russian. Abstr. in Ref. Zh., 51. Astron., 10.51.133 (1987).

062.070 Existence of a force acting on a cosmic plasma cloud moving in a weak magnetic field.
V. M. Antonov, L. M. Toptunova.
Kramat. Industr. Inst. Kramatorsk, 10 pp. (1987). In Russian. Abstr. in Ref. Zh., 51. Astron., 10.51.136 (1987).

062.071 Hydromagnetic dynamo as source of planetary, solar and galactic magnetism.
Ya. B. Zel'dovich, A. A. Ruzmajkin.
Usp. Fiz. Nauk, Tom 152, No. 2, p. 263 – 284 (1987). In Russian. Abstr. in Ref. Zh., 51. Astron., 10.51.150 (1987).

062.072 PIC simulations of electrodynamic models for astrophysical jets.
G. R. Gisler, R. V. E. Lovelace, M. E. Sulkanen, M. L. Norman.
Bull. Am. Astron. Soc., Vol. 19, No. 2, p. 685 (1987). Abstract. – See Abstr. 010.061.

062.073 Spatial stability analysis of slab and cylindrical jets.
P. E. Hardee, M. L. Norman.
Bull. Am. Astron. Soc., Vol. 19, No. 2, p. 701 (1987). Abstract. – See Abstr. 010.061.

062.074 Radiative reconnection – a fast start for a flare?
G. Van Hoven, T. Tachi, D. D. Schnack.
Bull. Am. Astron. Soc., Vol. 19, No. 2, p. 741 (1987). Abstract. – See Abstr. 010.061.

062.075 Spiral shocks and accretion in discs.
H. C. Spruit, T. Matsuda, M. Inoue, K. Sawada.
Mon. Not. R. Astron. Soc., Vol. 229, No. 3, p. 517 – 527 (1987).
Recent numerical and analytical results on disc–like accretion with shock waves as the only dissipation mechanism are compared. The global properties of the process are similar to those of the viscous (α) disc model, but precise values of the effective α value as a function of the accretion rate can be calculated. At low values of the ratio of specific heats ($\gamma < 1.45$) accretion is possible without radiative losses. Such adiabatic accretion can occur in practice at high accretion rates on to low mass objects and may be important in the formation of planets. Following Donner, and Lynden–Bell, it is pointed out that non–axisymmetric perturbations in the outer parts of a disc increase in amplitude as they propagate in and cause spiral shocks more easily in a disc than perturbations originating in the inner parts. It is suggested for this reason that the cause of spiral structure in normal spiral galaxies lies in moderate non–axisymmetries in their gaseous outer discs.

062.076 The theory of bipolar flows.
C. A. Norman.
Space Telesc. Sci. Inst., Prepr. Ser., No. 222, p. 1 – 8 (1987). To appear in Proc. 10th Europ. Regional Meeting of the IAU "Structure of galaxies and star formation", Prague, August 1987.

062.077 Self–modulation of large amplitude Langmuir waves.
R. Bingham, U. de Angelis, W. Mori, P. K. Shukla.
Rutherford Appleton Lab., Rep., RAL–87–045, 9 pp. (1987).

062.078 Numerical studies of the dynamical stability of differentially rotating tori.
J. Frank, J. A. Robertson.
MPA Rep., No. 313, 40 pp. (1987). Submitted to Mon. Not. R. Astron. Soc.

062.079 **Fermi acceleration at shocks with arbitrary velocity profiles.**
P. Schneider, J. G. Kirk.
MPA Rep., No. 316, 14 pp. (1987). To appear in Astrophys. J., Lett. Ed.

062.080 **Cometary gas and plasma flow with detailed chemistry.**
H. U. Schmidt, R. Wegmann, W. F. Huebner, D. C. Boice.
MPA Rep., No. 317, 42+28 pp. (1987). Submitted to Comput. Phys. Commun.

062.081 **Energetic particle acceleration in spherically symmetric accretion flows and shocks.**
G. M. Webb, T. J. Bogdan.
Astrophys. J., Vol. 320, No. 2, p. 683 – 698 (1987).
   Steady state spherically symmetric solutions of the cosmic–ray transport equation describing the acceleration of energetic particles in galactic accretion flows onto neutron stars, black holes, white dwarfs, and protostars are studied. The particles are accelerated both by first–order Fermi acceleration at the stand–off accretion shock and by adiabatic compression in the upstream flow. The solutions also include the effects of particle energy losses via a loss term proportional to the paricle momentum. Analytic solutions are obtained for the case where particles are injected monenergetically at the shock. The solutions are then used to investigate the particle redistribution, and the flow and energy change patterns of the accelerated particles for a range of physical parameters.

062.082 **Magnetic braking in weakly ionized media.**
A. Königl.
Astrophys. J., Vol. 320, No. 2, p. 726 – 740 (1987).
   This paper presents an analytic solution of the magnetic braking problem for a weakly ionized, rigidly rotating disk whose angular velocity and magnetic field vectors are aligned with the symmetry axis. This solution illustrates the effects of the relative azimuthal drift of neutrals and ions. The effects of radial drift are discussed and some general comments are made concerning the ratio of the characteristic ambipolar diffusion and magnetic braking time scales in high–mass and low–mass disks. The combined action of these two processes is illustrated by means of a numerical calculation. Implications of these results for accreting protostars are considered.

062.083 **Resonance absorption of magnetohydrodynamic surface waves: viscous effects.**
J. V. Hollweg.
Astrophys. J., Vol. 320, No. 2, p. 875 – 883 (1987).
   The author considers the effects of viscosity on the resonance absorption of incompressible MHD surface waves, which occurs when the waves are supported by a thin "transition layer", rather than by a discontinuous surface. Two forms of the viscous stress tensor are considered: (1) classical viscosity; (2) the stress tensor in a highly magnetized plasma. The author calculates the net viscous heating which occurs due to the velocity shears in a thin "energy–containing" layer inside the transition layer, and finds that the net steady state heating is independent of the viscosity coefficient. Implications of the results for the heating problem of the solar corona are discussed.

062.084 **Some aspects of shock–wave research.**
I. I. Glass.
AIAA J., Vol. 25, No. 2, p. 214 – 229 (1987). Abstr. in Phys. Abstr., Vol. 90, No. 1309, Entry 83248 (1987).

062.085 **Gravitational stability of a torus in the presence of a magnetic field.**
I. L. Genkin, T. N. Nuzhnova.
Tr. Astrofiz. Inst. Alma–Ata, Tom 47, p. 67 – 73 (1987). In Russian.

062.086 **Influence of a magnetic field on the gravitational stability of a cylinder.**
I. L. Genkin, T. N. Nuzhnova.
Tr. Astrofiz. Inst. Alma–Ata, Tom 47, p. 74 – 78 (1987). In Russian.

062.087 **Excitation of non–axisymmetrical modes of the mean magnetic field of the Sun.**
A. A. Ruzmajkin, D. D. Sokolov, S. V. Starchenko.
Inst. Zemn. Magn., Ionos. Rasprostr. Radiovoln Akad. Nauk SSSR, Prepr., No. 9/698, p. 3 – 17 (1987). In Russian. Abstr. in Ref. Zh., 51. Astron. 11.51.353 (1987).

062.088 **Multiple charge ions in astrophysics.**
T. G. Heil.
Nucl. Instrum. Methods Phys. Res., Sect. B, Vol. B23, No. 1 – 2, p. 222 – 225 (1987). Abstr. in Phys. Abstr., Vol. 90, No. 1310, Entry 91285 (1987).

062.089 **Isolated eddy models in geophysics.**
G. R. Flierl.
Annu. Rev. Fluid Mech., Vol. 19, p. 493 – 530 (1987). Abstr. in Phys. Abstr., Vol. 90, No. 1311, Entry 101678 (1987). – See Abstr. 003.014.

062.090 **Numerical simulation of a magnetic reconnection process.**
S. Wang, A. Wang.
Kexue Tongbao, Vol. 32, No. 4, p. 228 – 230 (1987). Abstr. in Phys. Abstr., Vol. 90, No. 1311, Entry 101756 (1987).

062.091 **Resonant mode conversion of a shear Alfvén wave.**
B. K. Shivamoggi.
Can. J. Phys., Vol. 65, No. 4, p. 357 – 358 (1987). Abstr. in Phys. Abstr., Vol. 90, No. 1312, Entry 104389 (1987).

062.092 **Axisymmetric accretion flows in astrophysics.**
J. A. Robertson.
Comput. Phys. Commun., Vol. 44, No. 3, p. 279 – 288 (1987). Abstr. in Phys. Abstr., Vol. 90, No. 1312, Entry 107712 (1987). – See Abstr. 012.049.

062.093 **A simplified cascade model for M.H.D. turbulence.**
V. Carbone, P. Veltri.
Astron. Astrophys., Vol. 188, No. 1, p. 239 – 250 (1987).
   The authors built up a simplified cascade model for M.H.D. turbulence, in which nonlinear interactions are limited to neighbouring wave numbers. It is shown that (1) initial conditions with an imbalance between the energies injected in the two possible modes of Alfvén wave propagation, relax toward a state in which the only mode present is that initially dominant, and such a final state is a static one characterized by parallel magnetic and velocity fields; (2) a stationary state for this system is possible also in the presence of a source asymmetrical with respect to the two possible wave propagation modes. The following picture is suggested by the results: a process corresponding to the observed self–organization of M.H.D. turbulence can be obtained as a direct consequence of the hypothesis used to build up the model; on the contrary, the early stages of evolution of an M.H.D. cascade are perhaps best described as the development of a linear parametric instability.

062.094 **Self–organization of electromagnetic waves into vortices in a magnetized electron–positron plasma.**
T. D. Kaladze, P. K. Shukla.
Astrophys. Space Sci., Vol. 137, No. 2, p. 293 – 296 (1987).
   This paper presents a new class of well localized dipolar vortex solutions to the newly derived set of coupled nonlinear equations governing the dynamics of low–frequency electromagnetic waves in a strongly magnetized electron–positron plasma.

062.095 An amended magnetohydrodynamic equation which predicts field–aligned current sheets.
N. A. Salingaros.
Astrophys. Space Sci., Vol. 137, No. 2, p. 385 – 395 (1987).
A derivation of the physical conditions for magnetohydrodynamic equilibrium from first principles establishes a set of equations which differ slightly from the usual ones. The difference is only relevant in the special case when the current is parallel to the magnetic field. What is noteworthy is that these modified equations predict that the simplest magnetohydrodynamic equilibria can only exist in regions of field–aligned current sheets. This prediction is entirely consistent with the observed large–scale structures in the Earth's magnetosphere.

062.096 Two–photon annihilation radiation in strong magnetic field: the case of small longitudinal velocities of electrons and positrons.
A. D. Kaminker, G. G. Pavlov, P. G. Mamradze.
Astrophys. Space Sci., Vol. 138, No. 1, p. 1 – 18 (1987).
Spectra, angular distributions, and polarization of two–photon annihilation radiation in a magnetic field are studied in detail in the case of small longitudinal velocities of annihilating electrons and positrons which occupy the ground Landau level. Magnetic field essentially affects the annihilation if its magnitude B is not very low in comparison with $B_{cr} = 4.4 \times 10^{13}$G, which may take place near the surface of a neutron star. The magnetic field broadens the spectra and leads to their asymmetry. The angular distribution may be highly anisotropic, being fan–like or pencil–like for different photon energies ω. The total annihilation rate is suppressed by the magnetic field. The radiation is linearly polarized; the degree and orientation of the polarization depend on $\mathbf{B}$, ϑ and ω. The polarization may reach several tens percent.

062.097 Electrostatic waves at frequencies close to the harmonics of electron gyrofrequency.
S. S. Sazhin.
Astrophys. Space Sci., Vol. 138, No. 1, p. 99 – 103 (1987).
An analytical solution of the dispersion equation for electrostatic waves propagating in a plasma consisting of cold electrons and the electrons with the loss cone distribution function is obtained for wave frequencies close to the harmonics of electron gyrofrequency. Electrostatic emissions at these frequencies were observed in the magnetospheres of the Earth, Jupiter, and Saturn.

062.098 Self–similar stong shock waves with thermal radiation in an atmosphere of exponentially–decreasing density.
S. N. Ojha.
Astrophys. Space Sci., Vol. 138, No. 2, p. 303 – 313 (1987).
Similarity solutions describing the flow of a perfect gas behind a strong plane shock wave propagating in an exponentially decreasing atmosphere with radiation heat flux are investigated. The Planck's diffusion approximation has been taken into account to observe the effects of radiation heat flux.

062.099 Rotational circulation–preserving magnetogeostrophic flows.
H. P. Singh, D. D. Tripathi, R. B. Mishra.
Astrophys. Space Sci., Vol. 138, No. 2, p. 315 – 321 (1987).
It has been shown that the only steady, inviscid, magnetogeostrophic rotational circulation–preserving motion whose magnetic field line pattern is that of the irrotational motion is a complex–lamellar motion whose magnetic field magnitude bears a constant value on a magnetic field line.

062.100 Relativistic thermal plasmas: time development of electron–positron pair concentration.
M. Kusunose.
Astrophys. J., Vol. 321, No. 1, p. 186 – 198 (1987).
The author investigates properties of an effectively thin relativistic thermal plasma with proton number density 10^{10}cm^{-3} by numerically calculating the time–dependent radiative transfer with pair processes in a spherically symmetric geometry. The parameters chosen are appropriate to plasmas in

active galactic nuclei. Properties of a pair equilibrium plasma and its spatial structure are described for a uniform distribution of protons and uniform heating. The effects of impulsive heating on the plasma state are also investigated.

062.101 Oblique magnetohydrodynamic shock waves in molecular clouds.
M. Wardle, B. T. Draine.
Astrophys. J., Vol. 321, No. 1, p. 321 – 333 (1987).
The structure of oblique, steady, plane–parallel, radiative, reacting, C–type shocks in diffuse molecular clouds is examined. Numerical models are presented for typical parameters: $n(H_2) = 20$ cm^{-3}, $n(H) = 10$ cm^{-3}, $v_s = 10$ km s^{-1}. C–type shock structure is weakly dependent on the angle θ between the upstream magnetic field B_0 and the direction of shock propagation. Models differing only in the component of B_0 parallel to the direction of shock propagation (i.e., having identical transverse components of B_0) can have markedly different hydrodynamic structures.

062.102 Shocked relativistic magnetohydrodynamic flows with application to pulsar winds.
R. T. Emmering, R. A. Chevalier.
Astrophys. J., Vol. 321, No. 1, p. 334 – 348 (1987).
The authors consider the time–dependent behavior of a shocked, spherically symmetric, relativistic fluid with tangential magnetic field in the case where the boundaries of the shocked fluid move at constant velocity so that self–similar solutions exist. The behavior of the fluid in the ultrarelativistic regime is compared to that in the nonrelativistic regime. Analytic expressions are given which describe the flow. The solutions can be applied to the evolution of shocked relativistic pulsar winds which are probably observed as Crab–like supernova remnants.

062.103 Magnetic monopole plasma oscillations and the survival of galactic magnetic fields.
E. N. Parker.
Astrophys. J., Vol. 321, No. 1, p. 349 – 354 (1987).
This paper explores the general nature of magnetic monopole plasma oscillations as a theoretical possibility for the observed galactic magnetic field in the presence of a high abundance of magnetic monopoles. The modification of the hydromagnetic induction equation by the monopole oscillations produces the *half–velocity effect*, in which the magnetic field is transported bodily with a velocity midway between the motion of the conducting fluid and the monopole plasma. Observational studies of the magnetic fields in the Galaxy, and in other galaxies, exclude the half–velocity effect, indicating that the magnetic field is not associated with monopole oscillations.

062.104 Jacobi ellipsoid with magnetic field.
N. P. Bondarenko.
Kiev. Politekh. Inst., Kiev, 15 pp. (1987). In Russian. Abstr. in Ref. Zh., 51. Astron., 12.51.126 (1987).

062.105 Coherent structures of kinetic propagation of electromagnetic waves in a plasma.
A. A. Vodyanitskij, N. S. Erokhin, S. S. Moiseev.
Inst. Kosm. Issled. Akad. Nauk SSSR, Prepr., No. 1240, p. 1 – 52 (1987). In Russian. Abstr. in Ref. Zh., 51. Astron., 12.51.129 (1987).

062.106 MHD instabilities of a cylindrical plasma with a realistic energy equation.
G. Torricelli–Ciamponi, V. Ciampolini, C. Chiuderi.
J. Plasma Phys., Vol. 37, Part 2, p. 175 – 184 (1987). Abstr. in Phys. Abstr., Vol. 90, No. 1313, Entry 111192 (1987).

062.107 Nonlinear waves in two dimensional Keplerian flows.
E. Siregar, J. Leorat, J. P. Poyet, A. Pouquet.
Geophys. Astrophys. Fluid Dyn., Vol. 38, No. 2, p. 69 – 104 (1987). Abstr. in Phys. Abstr., Vol. 90, No. 1314, Entry 117624 (1987).

062.108 Ion phase–space vortices and their relation to small amplitude double–layers.
H. L. Pecseli.
Laser Part. Beams, Vol. 5, Part 2, p. 211 – 217 (1987). Abstr. in Phys. Abstr., Vol. 90, No. 1314, Entry 117718 (1987). – See abstracts 012.064, 43.062.183.

062.109 Current driven weak double layers.
G. Chanteur.
Laser Part. Beams, Vol. 5, Part 2, p. 177 – 189 (1987). Abstr. in Phys. Abstr., Vol. 90, No. 1314, Entry 117726 (1987). – See abstracts 012.064, 43.062.184.

062.110 The formation of a double layer leading to the critical velocity phenomenon.
A. C. Williams.
Laser Part. Beams, Vol. 5, Part 2, p. 197 – 202 (1987). Abstr. in Phys. Abstr., Vol. 90, No. 1314, Entry 117727 (1987). – See abstracts 012.064, 43.062.191.

062.111 Formation mechanisms of laboratory double layers in triple plasma devices.
C. Chan.
Laser Part. Beams, Vol. 5, Part 2, p. 219 – 231 (1987). Abstr. in Phys. Abstr., Vol. 90, No. 1314, Entry 117728 (1987). – See abstracts 012.064, 43.062.177.

062.112 Electric fields and double layers in plasmas.
N. Singh, H. Thiemann, R. W. Schunk.
Laser Part. Beams, Vol. 5, Part 2, p. 233 – 255 (1987). Abstr. in Phys. Abstr., Vol. 90, No. 1314, Entry 117729 (1987). – See abstracts 012.064, 43.062.185.

062.113 Puming potential wells.
N. Hershkowitz, C. Forest, E. Y. Wang, T. Intrator.
Laser Part. Beams, Vol. 5, Part 2, p. 257 – 267 (1987). Abstr. in Phys. Abstr., Vol. 90, No. 1314, Entry 117730 (1987). – See abstracts 012.064, 43.062.179.

062.114 A new hydrodynamic analysis of double layers.
M. P. Goldsworthy, F. Green, H. Hora.
Laser Part. Beams, Vol. 5, Part 2, p. 269 – 286 (1987). Abstr. in Phys. Abstr., Vol. 90, No. 1314, Entry 117731 (1987). – See abstracts 012.064, 43.062.182.

062.115 Linear Vlasov stability in one–dimensional double layers.
J. Teichmann.
Laser Part. Beams, Vol. 5, Part 2, p. 287 – 293 (1987). Abstr. in Phys. Abstr., Vol. 90, No. 1314, Entry 117732 (1987). – See Abstr. 012.064.

062.116 Some aspects of double layer formation in a plasma constrained by a magnetic mirror.
W. Lennartsson.
Laser Part. Beams, Vol. 5, Part 2, p. 315 – 324 (1987). Abstr. in Phys. Abstr., Vol. 90, No. 1314, Entry 117733 (1987). – See abstracts 012.064, 43.062.188.

062.117 Some dynamical properties of very strong double layers in a triple plasma device.
T. Carpenter, S. Torven.
Laser Part. Beams, Vol. 5, Part 2, p. 325 – 337 (1987). Abstr. in Phys. Abstr., Vol. 90, No. 1314, Entry 117734 (1987). – See abstracts 012.064, 43.062.178.

062.118 A laboratory investigation of potential double layers.
P. Leung.
Laser Part. Beams, Vol. 5, Part 2, p. 339 – 349 (1987). Abstr. in Phys. Abstr., Vol. 90, No. 1314, Entry 117735 (1987). – See abstracts 012.064, 43.062.180.

062.119 Particle simulation of auroral double layers.
B. L. Smith, H. Okuda.
Laser Part. Beams, Vol. 5, Part 2, p. 367 – 380 (1987). Abstr. in Phys. Abstr., Vol. 90, No. 1314, Entry 117736 (1987). – See Abstr. 012.064.

062.120 Double layers and plasma–wave resistivity in extragalactic jets: cavity formation and radio–wave emission.
J. E. Borovsky.
Laser Part. Beams, Vol. 5, Part 2, p. 169 – 175 (1987). Abstr. in Phys. Abstr., Vol. 90, No. 1314, Entry 120271 (1987). – See abstracts 012.064, 43.062.190.

062.121 Quasi–isentropic non–linear waves in thermally unstable gas.
K. V. Krasnobaev, V. Yu. Tarev.
Astron. Zh., Tom 64, Vyp. 6, p. 1210 – 1219 (1987). In Russian. English translation in Sov. Astron., Vol. 31, No. 6.
By means of the two–scale expansion method, it is found that the propagation of plane, cylindrical and spherical waves in a thermally unstable medium is accompanied by their breaking and shock wave formation. It is shown how the simple wave solution can be generalized for the case of the shocks. It is established that, in contrast to the case of adiabatic motion, the amplitude growth rate and the distance of the wave breaking can both increase and decrease in an inhomogeneous gas for the waves of the fixed propagation direction (in comparison with their values of the gas with constant initial density). Stationary waves of finite amplitude can exist in the presence of dissipative effects under certain initial conditions. The obtained results confirm the conclusion about the unstable travelling waves as source of inhomogeneities and shock waves in H_2 clouds and in the neighbourhood of compact H II regions.

062.122 Generation of plasma waves by thick–target electron beams, and the expected radiation signature.
R. J. Hamilton, V. Petrosian.
Astrophys. J., Vol. 321, No. 2, p. 721 – 734 (1987).
The production of plasma waves by a nonthermal beam of high–energy electrons injected into a background thermal plasma is investigated. The coupled kinetic equations for the plasma wave and particle distributions are used to place an upper bound on the energy density and spectrum of the plasma waves generated by this process. The situation of an inhomogeneous electron beam is considered. It is shown that the wave–particle interactions have a significant, but not dominant, effect on the overall distribution of the electrons and that it is unlikely that such effects can be discerned in the observed bremsstrahlung or synchrotron radiation of the nonthermal electrons.

062.123 A hybrid jet of beam–type and wind–type models.
J. Fukue.
Publ. Astron. Soc. Jpn., Vol. 39, No. 6, p. 895 – 905 (1987).
A unified jet model of the beam model confined by the spheroidal gaseous envelope and of the wind–type jet model under the gravitational field of the central object is presented. It is analyzed under the steady assumption and under the self–similar treatment. This hybrid flow is responsible for bipolar jets from a protostar embedded in a dense molecular cloud.

062.124 Critical Mach numbers in classical magnetohydrodynamics.
C. F. Kennel.
J. Geophys. Res., Vol. 92, No. A12, p. 13,427 – 13,437 (1987).
The author uses stationary point analysis to compute generalized critical Mach numbers for finite amplitude fast and slow shocks in classical MHD fluids. He pays particular attention to the case where the resistive and thermal conduction dissipation scale lengths are comparable and much larger than the viscous scale lengths. With both resistivity and thermal conduction, the critical Mach number at which viscosity must be invoked is determined by the condition that the downstream flow speed equal the isothermal sound speed. The author also shows that resistivity and thermal conduction can provide convergent

stationary point solutions for nearly all slow shocks, except perhaps switch–off shocks.

062.125   Implicit 2D–radiation hydrodynamics.
W. M. Tscharnuter.
Astrophysical radiation hydrodynamics, p. 181 – 185 (1986). – See Abstr. 012.069.

062.126   2–D Eulerian hydrodynamics with fluid interfaces, self–gravity and rotation.
M. L. Norman, K.–H. A. Winkler.
Astrophysical radiation hydrodynamics, p. 187 – 221 (1986). – See Abstr. 012.069.
The purpose of this paper is to describe in detail the numerical approach the authors have developed over the past five years for solving 2–dimensional gas–dynamical problems in astrophysics involving inviscid·compressible flow, self–gravitation, rotation, and fluid instabilities of the Rayleigh–Taylor and Kelvin–Helmholtz types. The computer code to be described has been applied most recently to modeling jets in radio galaxies and is an outgrowth of a code developed for studying rotating protostellar collapse.

062.127   MUNACOLOR: understanding high–resolution gas dynamical simulations through color graphics.
K.–H. A. Winkler, M. L. Norman.
Astrophysical radiation hydrodynamics, p. 223 – 243 (1986). – See Abstr. 012.069.

062.128   PPM: piecewise–parabolic methods for astrophysical fluid dynamics.
P. R. Woodward.
Astrophysical radiation hydrodynamics, p. 245 – 326 (1986). – See Abstr. 012.069.
A general description of some modern numerical techniques for the simulation of astrophysical fluid flow is presented. The methods are introduced with a thorough discussion of the especially simple case of advection. Attention is focussed on the piecewise–parabolic method (PPM). A description of the SLIC method for treating multifluid problems is also given. The discussion is illustrated by a number of advection and hydrodynamics test problems. Finally, a study of Kelvin–Helmholtz instability of supersonic jets using PPM with SLIC fluid interfaces is presented.

062.129   Finite element methods for time dependent problems.
D. F. Griffiths.
Astrophysical radiation hydrodynamics, p. 327 – 357 (1986). – See Abstr. 012.069.
By the term finite element method (FEM) the author means the Galerkin formulation or its extension, the Petrov–Galerkin FEM though several other variations are well established, for example the Ritz, Mixed and Hybrid FEMs. The aim is to give a description of the basic method along with some of its pertinent properties.

062.130   Description and philosophy of spectral methods.
P. S. Marcus.
Astrophysical radiation hydrodynamics, p. 359 – 386 (1986). – See Abstr. 012.069.
The author describes the use of spectral methods in computational fluid dynamics. Spectral methods are generally more accurate and often faster than finite–differences. The author shows how to choose basis functions that give fast convergence and outlines the differences between Galerkin, tau, modal, collocation, and pseudo–spectral methods.

062.131   Numerical modeling of subgrid–scale flow in turbulence, rotation, and convection.
P. S. Marcus.
Astrophysical radiation hydrodynamics, p. 387 – 414 (1986). – See Abstr. 012.069.
The author shows that it is impossible to simulate numerically all of the length–scales in astrophysical turbulence. He looks at the effects of ignoring the unresolvable, small–scale flow, and shows how numerical simulations that neglect the subgrid–scale motions produce erroneous solutions. Then he discusses the "quick fix" remedy of introducing a numerical or eddy–viscosity. The author sketches how the analytic theories of turbulence attempt to model the large–scale effects of the small–scale motions. He examines four anisotropic, inhomogeneous flows of astrophysical interest for which numerical and eddy–viscosities produce incorrect solutions.

062.132   Particle methods.
J. W. Eastwood.
Astrophysical radiation hydrodynamics, p. 415 – 447 (1986). – See Abstr. 012.069.
There are many methods of transforming the differential equations to discrete forms suitable for numerical computations. In this paper the author shows how particle methods relate to other approaches (finite difference, finite element and spectral methods) and surveys particle methods. He focuses respectively on particle–mesh methods for collisionless systems, particle–particle/ particle–mesh methods for correlated systems and fluid particle methods for fluids and magnetofluids.

062.133   Unsteady flow of a slightly rarefied radiating gas in a rotating channel.
A. R. Bestman.
J. Mec. Theor. Appl., Vol. 6, No. 2, p. 167 – 178 (1987). Abstr. in Phys. Abstr., Vol. 90, No. 1315, Entry 123683 (1987).

062.134   Weak compressive ion acoustic double layers in presence of negative ions on the auroral field lines.
K. S. Goswami, S. Bujarbarua.
J. Phys. Soc. Jpn., Vol. 56, No. 7, p. 2396 – 2400 (1987). Abstr. in Phys. Abstr., Vol. 90, No. 1315, Entry 123872 (1987).

062.135   Qualitative theory of unsteady spherically symmetric self–gravitating gas flows.
D. Summers, A. Whitworth.
Nuovo Cimento B, Vol. 99B, Ser. 11, No. 1, p. 1 – 14 (1987). Abstr. in Phys. Abstr., Vol. 90, No. 1315, Entry 127716 (1987).

062.136   A model of an isolated magnetic flux tube in a stratified atmosphere.
W.–R. Hu.
J. Plasma Phys., Vol. 37, Part 3, p. 323 – 333 (1987). Abstr. in Phys. Abstr., Vol. 90, No. 1315, Entry 127812 (1987).

062.137   Time dependent MHD models for the cometary magnetosphere.
M. Schmidt–Voigt.
ESA Spec. Publ., ESA SP–278, p. 127 – 131 (1987). – See Abstr. 012.075.
The author presents three dimensional MHD models for the comet–solar wind interaction. The results are generally in fairly good agreement with those of other authors. A difference to Fedder et al. (1983) is found in the flattening of the ion tail which is the main reason for discrepancies in the simulation of time dependent effects due to discontinuities in the solar wind magnetic field. In his models the author cannot reproduce streamers of tail rays by this mechanism.

062.138   Effect of low–frequency instability on Hall conductivity in plasma.
C. L. Xaplanteris.
Astrophys. Space Sci., Vol. 139, No. 2, p. 233 – 242 (1987).
The author studies the effect of low–frequency drift wave instability on Hall conductivity in plasma. Using an external oscillation it is possible to affect the drift wave amplitude (mainly around resonance) and the variation on Hall conductivity is observed. The effect is probably attributed to electron trapping by the wave potential. Good agreement between experimental and calculated values of azimuthal drift currents on the resonance and away from resonance lead the author to believe that the proposed explanation by electron trapping is correct.

062.139   Gravitational stability of finitely conducting two–component plasma through porous medium.
R. K. Chhajlani, D. S. Vaghela.
Astrophys. Space Sci., Vol. 139, No. 2, p. 337 – 352 (1987).
The gravitational stability of magnetized self–gravitating two–component plasma of finite conductivity flowing through porous medium is studied. Effects of magnetic field, porosity, viscosity, finite conductivity, and neutral gas friction are considered. On the basis of Hurwitz criterion, the stability of the system is discussed. It is found that Jeans's criterion determines the stability of the system.

062.140   A composite stellar model of geostrophic flows. I.
H. P. Singh, D. D. Tripathi, S. C. Singh, R. B. Mishra.
Astrophys. Space Sci., Vol. 139, No. 2, p. 413 – 419 (1987).
The geometric study and solutions of the electromagnetogeostrophic flows, and the geometry of magnetic and current lines are discussed.

062.141   Geophysical and astrophysical turbulence.
H. K. Moffatt.
Advances in turbulence, p. 228 – 244 (1987). Abstr. in Phys. Abstr., Vol. 90, No. 1316, Entry 130653 (1987). – See Abstr. 012.077.

062.142   Numerical observations of dynamic behavior in two–dimensional compressible convection.
G. P. Ginet, R. N. Sudan.
Phys. Fluids, Vol. 30, No. 6, p. 1667 – 1677 (1987). Abstr. in Phys. Abstr., Vol. 90, No. 1316, Entry 130704 (1987).

062.143   Hydrodynamics in curvilinear coordinates.
B. Kämpfer, B. Lukacs.
Acta Phys. Hung., Vol. 61, No. 3 – 4, p. 317 – 341 (1987). Abstr. in Phys. Abstr., Vol. 90, No. 1317, Entry 134515 (1987).

062.144   The interaction of quasiperpendicular shock waves in a collisionless plasma.
P. J. Cargill, C. C. Goodrich.
Phys. Fluids, Vol. 30, No. 8, p. 2504 – 2517 (1987). Abstr. in Phys. Abstr., Vol. 90, No. 1317, Entry 137083 (1987).

062.145   Particle acceleration at astrophysical shocks.
R. Blandford, D. Eichler.
Phys. Rep., Vol. 154, No. 1, p. 1 – 75 (1987). Abstr. in Phys. Abstr., Vol. 90, No. 1318, Entry 145792 (1987).

062.146   A mechanism for differential rotation based on angular momentum transport by compressible convection.
L. L. Kichatinov.
Geophys. Astrophys. Fluid Dyn., Vol. 38, No. 4, p. 273 – 292 (1987). Abstr. in Phys. Abstr., Vol. 90, No. 1318, Entry 145817 (1987).

062.147   On the computation of steady, self–consistent spherical dynamos.
D. R. Fearn, M. R. E. Proctor.
Geophys. Astrophys. Fluid Dyn., Vol. 38, No. 4, p. 293 – 325 (1987). Abstr. in Phys. Abstr., Vol. 90, No. 1318, Entry 145818 (1987).

062.148   The onset of eruptive processes in plasmas. I.
M. Hesse, M. Kiessling.
Phys. Fluids, Vol. 30, No. 9, p. 2720 – 2728 (1987). Abstr. in Phys. Abstr., Vol. 90, No. 1318, Entry 145819 (1987).

062.149   Interdiffusion in binary ionic mixtures.
D. B. Boercker, E. L. Pollock.
Phys. Rev. A, Vol. 36, No. 4, p. 1779 – 1785 (1987). Abstr. in Phys. Abstr., Vol. 91, No. 1319, Entry 1789 (1988).

062.150   Marfes: radiative condensation in tokamak edge plasma.
J. F. Drake.
Phys. Fluids, Vol. 30, No. 8, p. 2429 – 2433 (1987). Abstr. in Phys. Abstr., Vol. 91, No. 1319, Entry 1851 (1988).

062.151   Solitons in strongly magnetized electron–positron plasmas and pulsar microstructure.
U. A. Mofiz, J. Podder.
Phys. Rev. A, Vol. 36, No. 4, p. 1811 – 1814 (1987). Abstr. in Phys. Abstr., Vol. 91, No. 1319, Entry 4276 (1988).

062.152   The scattering of Alfvén waves by density fluctuations.
H.–S. Li, E. G. Zweibel.
Astrophys. J., Vol. 322, No. 1, p. 248 – 255 (1987).
The authors consider the propagation of an Alfvén wave packet through a medium containing time–dependent random density fluctuations. The Alfvén wave interaction with these density irregularities causes the transfer of wave power to both shear (Alfvénic) and compressive (magnetosonic) disturbances. The latter are dissipated and heat the plasma. The authors compute the Alfvén wave decay rate in the limit of short–wavelength density fluctuations and apply the results to the solar corona and interstellar medium.

062.153   Particle acceleration at shocks: a Monte Carlo method.
J. G. Kirk, P. Schneider.
Astrophys. J., Vol. 322, No. 1, p. 256 – 265 (1987).
A Monte Carlo method is presented for the problem of the acceleration of test particles at relativistic shocks. The particles are assumed to diffuse in pitch angle as a result of scattering off magnetic irregularities frozen into the fluid. Several tests are perfomed using the analytic results available for both relativistic and nonrelativistic shock speeds. The authors investigate the acceleration at relativistic shocks under the influence of radiation losses, and include the effects of a momentum dependence in the diffusion coefficient.

062.154   On the relativistic theory of Alfvén solitons.
D. I. Dzhavakhishvili, O. V. Chediya.
Astrofizika, Tom 27, Vyp. 2, p. 347 – 357 (1987). In Russian. English translation in Astrophysics, Vol. 27, No. 2.
The authors have considered a collisionless magnetoactive electron–ion plasma in which the propagation of an Alfvén wave with large amplitude results in relativistically high velocities of the plasma components, comparable with the light velocity (a possible situation for the plasma of a number of cosmic objects). It is shown that the nonlinear stage of the Alfvén wave propagation can be described by the Schrödinger equation with nonlinearity of the derivative complemented by nonzero boundary conditions at infinity. Simple soliton structures as well as simple–wave–type solutions of this equation are considered. It has been shown that relativistic effects strongly influence the process of formation of nonlinear waves in a plasma.

062.155   A simple model to account for the effects of plasma screening on thermonuclear reaction rate.
D. A. Shalybkov, D. G. Yakovlev.
Astrofizika, Tom 27, Vyp. 2, p. 383 – 393 (1987). In Russian. English translation in Astrophysics, Vol. 27, No. 2.
The Thomas–Fermi model in a high–density regime is used for analytic evaluation of the enhancement factor of the thermonuclear reaction rate due to plasma screening in a weakly nonideal degenerate electron gas and strongly coupled two–component ion liquid with large ion charge. The density and temperature domain is found in which the screening effect produced by the electron gas compressibility is important. It is pointed out that the latter effect may be influenced by strong magnetic fields. The results may be applied to degenerate cores of white dwarfs and envelopes of neutron stars.

062.156 Nonlinear magnetohydrodynamic waves in a steady zonal circulation for a shallow fluid shell on the surface of a rotating sphere.
Y. Q. Lou.
Astrophys. J., Vol. 322, No. 2, p. 862 – 869 (1987).

The author considers two–dimensional nonlinear magneto-hydrodynamic (MHD) waves of large horizontal spatial scales for a thin magnetofluid layer on the surface of a rotating sphere. This simple model may be of interest in astrophysical and geophysical applications. The author generalizes the "shallow fluid" hydrodynamic equations to include the effects of magnetic fields. Two kinds of finite–amplitude MHD waves are obtained; the first is an inertial wave of the Rossby–Haurwitz type, modified by the presence of the background zonal magnetic field, while the second is a magnetic Alfvén–like wave which is modified by the uniform rotation of the background atmosphere.

062.157 On the controversy concerning turbulent bremsstrahlung.
D. B. Melrose, J. Kuijpers.
Astrophys. J., Vol. 323, No. 1, p. 338 – 345 (1987).

It is shown that the derivation of a nonzero growth rate for turbulent bremsstrahlung involves an incomplete summation of the nonlinear responses. When the classically correct expression is used, the growth rate is identically zero. Next the authors show that the radiative correction to the resonant emission of ion sound waves in the presence of nonresonant Langmuir waves is derived explicitly. Finally, they reply to criticisms by Nambu (1986) of an earlier paper and show that the inclusion of a background magnetic field gives a zero growth rate for this proposed kind of turbulent bremsstrahlung.

062.158 Electric current sheet formation in a magnetic field induced by continuous magnetic footpoint displacements.
B. C. Low.
Astrophys. J., Vol. 323, No. 1, p. 358 – 367 (1987).

This paper presents two analytical examples to illustrate the formation of an electric current sheet in a magnetic field without neutral points, as the result of continuous boundary displacement of the magnetic footpoints. The effect can be demonstrated with potential magnetic fields, treated as special cases of the general force–free magnetic field. The results and their implications are discussed.

062.159 The α–effect generated by random Alfvén waves in a stratified atmosphere and particle acceleration.
T. Namikawa, H. Hamabata, H. Wada.
Astrophys. J., Vol. 323, No. 1, p. 414 – 421 (1987).

The mean electromotive and ponderomotive forces generated by random Alfvén waves propagating along the vertical magnetic field in a stratified atmosphere are evaluated using the simple, two–layer model of the solar atmosphere. It is shown that the α–effect can exist due to the stratification of the medium even if the magnetic diffusivity is zero and that the α–effect in the chromosphere is a possible acceleration mechanism of solar cosmic rays.

062.160 Fermi acceleration at shocks with arbitrary velocity profiles.
P. Schneider, J. G. Kirk.
Astrophys. J., Lett. Ed., Vol. 323, No. 1, p. L87 – L90 (1987).

Diffusive particle acceleration in modified shock fronts is considered. The authors outline a method by which one can obtain the resulting particle spectrum for an arbitrary flow profile; this method is based on a reformulation of the transport equation as a nonlinear first–order equation. Besides the shape of the flow profile, the spectral index of particles depends on the Péclet number of the shock. Results are given for some sample profiles.

062.161 Fragmentation of rapidly rotating gas clouds. I. A universal criterion for fragmentation.
I. Hachisu, J. E. Tohline, Y. Eriguchi.
Astrophys. J., Vol. 323, No. 2, p. 592 – 613 (1987).

A pressure–supported, collapsing gas cloud evolves toward an axisymmetric equilibrium state if one exists. Therefore, one can explore the fragmentation problem, in part, by studying the structure and stability of axisymmtric equilibrium states. The authors formulate a pair of criteria for rotating gas clouds that must be satisfied simultaneously if the final collapsed object is going to be unstable toward fragmentation. The first criterion involves the ratio of rotational energy T to gravitational energy W. Centrally condensed objects become dynamically unstable to non–axisymmetric modes if $T/|W| = 0.27$. Self–gravitating toroidal systems, however, become dynamically unstable at $T/|W| = 0.14$. The second criterion is related to the strength of the Coriolis force and states that the condition for fragmentation is coincident with the condition for ring formation.

062.162 A new example of the Papaloizou–Pringle instability.
M. A. Abramowicz, O. M. Blaes, P. Ghosh.
Astrophys. J., Vol. 323, No. 2, p. 629 – 633 (1987).

A class of differentially rotating incompressible annuli is shown to be unstable to modes which strongly resemble the classical Kelvin–Helmholtz instability.

Principles of plasma diagnostics.
See Abstr. 003.073.

Absorption properties of a high temperature nitrogen plasma.
See Abstr. 022.041.

A model for ionization balance and L–shell spectroscopy of non–LTE plasmas.
See Abstr. 022.046.

Electrostatic charge on a dust size distribution in a plasma.
See Abstr. 022.100.

The light–element synthesis in a two–temperature astrophysical plasma.
See Abstr. 061.032.

Some physical processes influencing the polarization of continuum and line radiation.
See Abstr. 063.010.

Dynamical models of radiative shocks – III. Spectra.
See Abstr. 063.017.

Magnetic field and synchrotron radiation in mildly relativistic shocks.
See Abstr. 063.019.

Asymmetry of the H_β central part measured in a T–tube.
See Abstr. 063.028.

Widths and shifts of some plasma–broadened oxygen and carbon multiplets.
See Abstr. 063.029.

Special features of cyclotron radiation in anisotropic plasmas.
See Abstr. 063.038.

Compact radio sources as a plasma turbulent reactor. II. General characteristics of electromagnetic radiation spectra.
See Abstr. 063.041.

An approximate method for calculating Planck and Rosseland mean opacities in hot, dense plasmas.
See Abstr. 063.065.

Radiation hydrodynamics in astrophysics.
See Abstr. 063.078.

Problems in astrophysical radiation hydrodynamics.
See Abstr. 063.101.

The equations of radiation hydrodynamics.
See Abstr. 063.102.

WH80s: numerical radiation hydrodynamics.
See Abstr. 063.103.

Evolution of the axial electron cyclotron maser instability, with applications to solar microwave spikes.
See Abstr. 063.108.

Partial redistribution in high–density, highly ionized plasmas.
See Abstr. 063.117.

The thermodynamics of anisotropic radiating systems.
See Abstr. 063.119.

Physical constraints on variable Eddington factors for the gray problem in radiative transfer.
See Abstr. 063.120.

On pumping the strong water maser sources.
See Abstr. 063.121.

Cherenkov damping of surface waves.
See Abstr. 063.123.

Magnetogasdynamic cylindrical shock wave model in an optically thin atmosphere.
See Abstr. 064.001.

Nonlinear dynamics of instabilities in line–driven stellar winds.
See Abstr. 064.013.

Thermal–convective instability of a composite plasma in a stellar atmosphere with rotation and magnetic field.
See Abstr. 064.068.

Wave energy in white dwarf atmospheres. I. Magneto-hydrodynamic energy spectra for homogeneous DB and layered DA stars.
See Abstr. 064.076.

Cosmic–ray–modified stellar winds. I Solution topologies and singularities.
See Abstr. 064.083.

Inhibition of degeneracy by intense magnetic fields: derivation and astrophysical application.
See Abstr. 065.007.

Statistical mechanics of partially ionized stellar plasmas: the Planck–Larkin partition function, polarization shifts, and simulations of optical spectra.
See Abstr. 065.043.

The dynamical oscillation and propulsion of magnetic fields in the convective zone of a star. II. Thermal shadows.
See Abstr. 065.101.

The dynamical oscillation and propulsion of magnetic fields in the convective zone of a star. III. Accumulation of heat and the onset of the Rayleigh–Taylor instability.
See Abstr. 065.102.

The Pomeau–Manneville intermittent transition to chaos in hydro-dynamic pulsation models.
See Abstr. 065.108.

On the application of a hydrodynamic analogy to the general theory of relativity.
See Abstr. 066.106.

The general relativistic {2+2} formulation of magnetohydrody-namics.
See Abstr. 066.134.

Why ultrarelativistic numerical hydrodynamics is difficult.
See Abstr. 066.152.

Numerical relativistic gravitational collapse with spatial time slices.
See Abstr. 066.153.

Hydromagnetic flows from rapidly rotating compact objects. II. The relativistic axisymmetric jet equilibrium.
See Abstr. 067.076.

Grand Unified Models.
See Abstr. 067.122.

Electromagnetic jets from compact objects.
See Abstr. 067.144.

Microwave radiation from a dense magneto–active plasma.
See Abstr. 073.013.

Stability of current–sheet models of quiescent prominences and equatorial disks.
See Abstr. 073.055.

Current loop coalescence model of solar flares.
See Abstr. 073.090.

Heating of coronal loops by phase–mixed shear Alfvén waves.
See Abstr. 074.078.

Energy buildup in coronal magnetic flux tubes.
See Abstr. 074.079.

The origin of rigidly rotating magnetic field patterns on the Sun.
See Abstr. 075.016.

Alfvén wave dissipation in the solar atmosphere.
See Abstr. 080.023.

Magnetohydrodynamics of the Earth's core.
See Abstr. 081.014.

Origin of the main field: kinematics.
See Abstr. 081.015.

Origin of the main field: dynamics.
See Abstr. 081.016.

On the origin of magnetic fields of the earth and celestial bodies.
See Abstr. 091.034.

The Jupiter–Io connection: an Alfvén engine in space.
See Abstr. 099.040.

Alfvén wave plasma turbulence during solar wind–comet interac-tion.
See Abstr. 102.060.

Nonlinear propagation of Alfvén waves in cometary plasmas.
See Abstr. 102.084.

Diffusive shock acceleration: comparison of a unified shock model to bow shock observations.
See Abstr. 106.002.

The structure of ULF waves produced by a tethered satellite system.
See Abstr. 106.020.

On the gravitational instability in thin gaseous Kepler disks.
See Abstr. 107.038.

Observations of quasi–periodic oscillations in Cyg X–2.
See Abstr. 117.003.

Time–dependent bow shocks and the condensation structure of Herbig–Haro objects.
See Abstr. 121.063.

Non–spherical supernova remnants. IV. Sequential explosions in OB associations.
See Abstr. 125.018.

Cloud fluid models of gas dynamics and star formation in galaxies.
See Abstr. 131.017.

The structure of time–dependent interstellar shocks and grain destruction in the interstellar medium.
See Abstr. 131.051.

Stability of radiative shocks with time dependent cooling.
See Abstr. 131.172.

A simulation of high–energy cosmic–ray propagation.
See Abstr. 144.028.

Antiproton production of propagating cosmic rays under distributed reacceleration.
See Abstr. 144.039.

The evolution of clumpy gas in young elliptical galaxies.
See Abstr. 151.014.

Hot gas evolution in nearly normal elliptical galaxies.
See Abstr. 157.137.

Velocities in radio galaxies and quasars.
See Abstr. 158.114.

Polarization and magnetic field structure.
See Abstr. 158.115.

Kinetic Alfvén waves in extended radio sources. I. Reacceleration.
See Abstr. 158.290.

Kinetic Alfvén waves in extended radio sources. II. Electric currents, collimated jets, and inhomogeneities.
See Abstr. 158.291.

The effect of a quasi–stellar object on its host galaxy: dynamical and physical processes in the interstellar medium around a quasi–stellar object.
See Abstr. 159.141.

The evolution of cooling flows. I. Self–similar cluster flows.
See Abstr. 160.004.

Hydrogen molecules and the radiative cooling of pregalactic shocks.
See Abstr. 161.003.

A hydrodynamical three–component model of the universe during the recombination era.
See Abstr. 161.246.

Hydrogen molecules and the radiative cooling of pregalactic shocks II: low velocity shocks at high redshift.
See Abstr. 161.247.

063 Radiative Transfer, Scattering

063.001 **Inverse Compton model of gamma ray burst spectra.**
W. M. Howard, E. P. Liang.
The origin and evolution of neutron stars, p. 547 (1987). Abstract. – See Abstr. 012.001 (IAU Symp. No. 125).

063.002 **Bidirectional reflectance spectroscopy. 4. The extinction coefficient and the opposition effect.**
B. Hapke.
NASA Tech. Memo., NASA TM–89810, p. 219 (1987). Abstract. – See Abstr. 003.001.
    For the full paper see 42.063.020.

063.003 **Erratum: "Anomalous Zeeman effect: moments and expansion coefficients" [Astron. Astrophys., Suppl. Ser., Vol. 67, No. 3, p. 557 – 568 (1987)].**
G. Mathys, J. O. Stenflo.
Astron. Astrophys., Suppl. Ser., Vol. 70, No. 1, p. 142 (1987). See Abstr. 43.063.012.

063.004 **The influence of relativistic electrons on a photoionized gaseous cloud.**
R. B. Gruenwald, S. M. Viegas–Aldrovandi.
Astron. Astrophys., Suppl. Ser., Vol. 70, No. 2, p. 143 – 156 (1987).
    A grid of models for a dilute photoionized gas cloud reached by a flux of relativistic electrons is presented. Their influence on the physical conditions of the emitting gas and on the emission

line intensities is analysed. The input parameters are the elemental abundances, the hydrogen density, the ionization parameter and the flux of the relativistic electrons. These models can describe the narrow emission line region of the active galactic nuclei showing radio emission.

063.005 **The Neumann solution of the multiple scattering problem in a plane–parallel medium – IIa. Semi–infinite spaces and H–function.**
B. Rutily, J. Bergeat.
J. Quant. Spectrosc. Radiat. Transfer, Vol. 38, No. 1, p. 47 – 60 (1987).
    The aim of this paper is first to expand the H–function in power series in the albedo a and then to derive the general equations involving the nth order terms in this expansion. When multiplied by a^n and summed over n from $n = 1$ to $+\infty$, they yield the classical equations of the existing theory including the H–equation. This approach is thus a new way to tackle the theory of multiple scattering in a half–space.

063.006 **The Neumann solution of the multiple scattering problem in a plane–parallel medium – IIb. The resolvent function in a semi–infinite space.**
B. Rutily, J. Bergeat.
J. Quant. Spectrosc. Radiat. Transfer, Vol. 38, No. 1, p. 61 – 69 (1987).
    The authors solve the problem of multiple scattering of radiation in a plane–parallel, semi–infinite medium by means of

its Neumann solution. The nth iterated function is given for any $n \geqslant 1$, which leads to a rigorous description of the n–times scattered radiation field in a half–space. The resolvent function is then calculated as the sum of its Neumann series.

063.007   The Neumann solution of the multiple scattering problem in a plane–parallel medium – IIc. The specific intensity in a semi–infinite space.
B. Rutily, J. Bergeat.
J. Quant. Spectrosc. Radiat. Transfer, Vol. 38, No. 1, p. 71 – 78 (1987).
Using a new formulation of the resolvent function, the authors solve the Milne integral equation for the mean intensity in a semi–infinite plane–parallel atmosphere. The specific intensity is then derived at any optical depth. By expanding the intensity in power series in the albedo, the authors derive the corresponding intensities of a n–times scattered radiation field propagating in a half–space ($n \geqslant 1$).

063.008   Application of an extended ESFT method to calculation of solar heating rates by water vapor absorption.
S. Asano, A. Uchiyama.
J. Quant. Spectrosc. Radiat. Transfer, Vol. 38, No. 2, p. 147 – 158 (1987).
A new method for the exponential–sum fitting of transmissions (ESFT) is developed. The authors determine a set of equivalent absorption coefficients by a successive correction technique with a set of weights preselected empirically or according to some quadrature weights. The ESFT method is then applied to transmission functions of water vapor bands in the near i.r. region deduced from the AFGL absorption–line parameters. Numerical values of the ESFT parameters are presented for each of the water vapor bands and for the one–band parameterization for total absorption. Solar radiative heating rates computed from the present ESFT parameters are compared with those in the literature.

063.009   Transfer of resonant line photons in spherically accelerating envelopes.
S. Beckwith, A. Natta.
Astron. Astrophys., Vol. 181, No. 1, p. 57 – 70 (1987).
Numerical calculations of the scattering of resonance photons in spherically symmetric, outward accelerating envelopes with turbulence are used to discuss common features of the line shapes and brightness distributions. The velocity and density distributions of scatterers encompass most of the forms proposed for different mass–loss processes from stellar atmospheres.

063.010   Some physical processes influencing the polarization of continuum and line radiation.
K. N. Nagendra, A. Peraiah.
Astron. Astrophys., Vol. 181, No. 1, p. 71 – 76 (1987).
Some physical mechanisms which affect the continuum and line polarization are studied. The physical conditions of the plasma selected for this purpose represent different astrophysical situations of interest, particularly the magnetic stars. The pure absorption polarization transfer equation is solved individually taking these effects into account.

063.011   Numerical algorithm for the solution of the equations of radiative transfer in a magnetic field.
D. N. Rachkovskij.
Bull. Crimean Astrophys. Obs., Vol. 74, p. 147 – 149 (1987). English translation of 41.063.074.

063.012   Saturated comptonization in a superstrong magnetic field.
Yu. Eh. Lyubarskij.
Astrophysics, Vol. 25, No. 2, p. 577 – 586 (1987). English translation of 42.063.095.

063.013   "Reflection function" for an infinite atmosphere in the case of incoherent scattering.
A. G. Nikogosyan.
Astrophysics, Vol. 25, No. 2, p. 596 – 602 (1987). English translation of 42.063.096.

063.014   On the central part of the Hα–line profile in the gas–dynamic model of a flare.
E. A. Bruevich, M. A. Livshits.
Soln. Dannye, Byull., 1987, No. 2, p. 87 – 90 (1987). In Russian.
A study of the central part of the Hα line profile in bright flares based on a gas–dynamical model of a flare is presented. A temperature gradient of about 0.5K/cm for top levels of the condensation in the low temperature flare emission is found. The radiation transfer equation for the Hα line is solved using two approximate methods for different $n(T)$ at top levels of condensation. The emission of the Lα line is also estimated for the same conditions. The value of VT and $n(T)$ proved to be chosen fairly well. Top levels of condensation were shown to play an important role in forming the centre of the Hα line profile.

063.015   Non–thermal pair production in compact X–ray sources: first–order Compton cascades in soft radiation fields.
R. Svensson.
Mon. Not. R. Astron. Soc., Vol. 227, No. 2, p. 403 – 451 (1987).
Non–thermal pair cascades, where injected relativistic electrons (or pairs) cool by inverse Compton scattering on soft photons in the Thomson limit and where the resulting hard photons pair produce on other up–scattered photons, are described by an integral equation. A general integral equation for all relevant optical depths is obtained by using a simple prescription for the radiative transfer effects due to photon–photon absorption and to Compton scattering by the cooled pairs. The integral equation is solved analytically in the limit of large injection compactness L_l/R (L_l is the injected particle power and R the size of the injection region), where every γ–ray photon produces a pair (saturated pair production). Each photon generation is described separately and has a well–determined spectral index at small photon energies, being $\alpha = 1/2$ and $3/4$ for the first and second generation, respectively. Analytical and numerical solutions of the general integral are discussed in detail.

063.016   A link between X–ray variability and absorption in active galactic nuclei.
M. D. Smith.
Mon. Not. R. Astron. Soc., Vol. 227, No. 3, p. 783 – 795 (1987).
The author investigates the requirements for a hot ($10^7 – 10^8$K) Compton–heated wind to be able to cool into dense clouds. He studies in detail the case where the hot wind blows off the disc intermittently (i.e. during stages of high X–ray luminosity) and is allowed to cool when the X–ray source is in a low state.

063.017   Dynamical models of radiative shocks – III. Spectra.
D. E. Innes, J. R. Giddings, S. A. E. G. Falle.
Mon. Not. R. Astron. Soc., Vol. 227, No. 4, p. 1021 – 1053 (1987).
The authors have investigated the spectral characteristics of steady and unsteady radiative shock models in the context of Supernova Remnants (SNRs). By means of detailed, multi–level radiative transfer calculations, they have generated emission and absorption line profiles for selected shock models. Geometrical effects are simulated by forming the plane models into a thin spherical shell, which is then observed along different lines–of–sight. The authors show that the nature of unsteady shocks can best be studied through observations of absorption and emission line profiles along lines–of–sight near to the centre of the SNR. Also, they show why emission maps in the light of optically thick

transitions should not exhibit the bright filaments associated with optically thin lines.

063.018 Multidimensional radiative transfer in stratified atmospheres. V. Energy transport by radiation.
F. Kneer, J. Trujillo–Bueno.
Astron. Astrophys., Vol. 183, No. 1, p. 91 – 97 (1987).

Energy transport by radiation is an important contribution to the energy budget in stellar atmospheric structures. In this paper, radiative relaxation of small–scale structures is investigated. The authors show in a linear analysis: (1) Already at structural lengths of 10 opacity scale heights, horizontal photon exchange is important for the energy budget. (2) In atmospheric layers near continuum optical depth $\tau_c = 1$ and below, the continuum absorption and emission processes dominate the radiative relaxation. (3) Weak spectral lines or lines with $\sigma_l \varepsilon \leqslant 1$ have little influence on the energy exchange. (4) At large heights, transport in few spectral lines with $\sigma_l \varepsilon \gg 1$ can compete with continuum processes.

063.019 Magnetic field and synchrotron radiation in mildly relativistic shocks.
T. J.–L. Courvoisier, M. Camenzind.
Astron. Astrophys., Vol. 183, No. 1, p. 167 – 169 (1987).

The magnetic field behind a mildly relativistic shock is calculated using the electrodynamic properties of the shock. The authors present a heuristic argument showing that the field is close to equipartition with the kinetic energy of the incoming flow. The result is similar to results obtained with detailed simulations of solar system shocks. The electric field present in the shock accelerates some electrons to relativistic energies. The spectral parameters of the synchrotron radiation resulting from the interaction of the relativistic electrons with the magnetic field is related to the shock density and the inflow velocity. It is shown that it is possible to accelerate electrons to gamma factors of about 10^2–10^3. These electrons are the natural candidates for the source of the synchrotron emission which is observed in the mm–infrared spectral domain of active galactic nuclei.

063.020 A unified treatment of polarized light emerging from a homogeneous plane–parallel atmosphere.
J. W. Hovenier.
Astron. Astrophys., Vol. 183, No. 2, p. 363 – 370 (1987).

The reflection and transmission matrices are combined into one 4×4–matrix which is called the exit matrix. Symmetry considerations are employed to derive an integral equation for the exit matrix. This equation contains only three double integrals although it carries the same information as the pair of coupled integral equations for the reflection and transmission matrices in which 16 double integrals are involved. A great deal of the redundancy occurring in the traditional approach is avoided. The integral equation for the exit matrix is decomposed in a set of equations for the components in a Fourier series with respect to azimuth. The azimuth–independent equation is, by way of example, further analyzed for scattering by molecules, which includes Rayleigh scattering. In this case solutions are obtained in terms of matrices depending on only one angular variable and the close analogy between finite and semi–infinite atmospheres is revealed.

063.021 The adding method for multiple scattering calculations of polarized light.
J. F. de Haan, P. B. Bosma, J. W. Hovenier.
Astron. Astrophys., Vol. 183, No. 2, p. 371 – 391 (1987).

The authors consider a vertically inhomogeneous atmosphere bounded below by a reflecting surface. They calculate the intensity and state of polarization of the light emerging at the top and the bottom of the atmosphere when internal light sources can be ignored. They adopt the usual procedure of dividing the atmosphere in a number of homogeneous layers and use a combination of adding and doubling to obtain the desired quantities.

063.022 Sobolev type line profile in case of non radial wind density perturbations.
R. Scuflaire, J.–M. Vreux.
Instabilities in luminous early type stars, p. 217 – 220 (1987). – See Abstr. 012.012.

The authors have investigated the modifications induced on the P Cygni line profiles of an outwards accelerating wind by density fluctuations modulated by non radial pulsations. The results obtained in a first approach of the problem compare favourably with some observed time dependent profiles of ultraviolet lines.

063.023 Radiation pressure in acoustic wave calculations of early–type stars.
B. E. Wolf, P. Ulmschneider.
Instabilities in luminous early type stars, p. 263 – 266 (1987). – See Abstr. 012.012.

A new method to treat radiation pressure in spectral lines has been developed which avoids the Sobolev–approximation. This method has been used for time–dependent acoustic wave calculations in early–type stars.

063.024 Radiation losses from hydrogen gas in a moving medium showing predominantly collisional atomic excitation and ionization.
A. S. Mitskevich.
Bull. Crimean Astrophys. Obs., Vol. 75, p. 130 – 136 (1987). English translation of 42.063.059.

063.025 The formation of iron features in the cooling spectrum of X–ray bursts.
A. J. Foster, R. R. Ross, A. C. Fabian.
Mon. Not. R. Astron. Soc., Vol. 228, No. 1, p. 259 – 268 (1987).

A detailed calculation of the formation and detection of iron spectral features in X–ray bursts is presented. Observation of these features will enable the mass and radius of a neutron star to be estimated separately. Comptonization and free–free absorption and emission processes are taken into account, as well as the effects of iron in its last four states of ionization. The authors estimate the size of detector, with response similar to that of the EXOSAT ME, needed in order to observe the iron features. The large–area detectors on GINGA should meet this requirement. The authors show that the absorption feature reported at 4.1 keV from MXB 1636–536 has an equivalent width much larger than expected from Fe XXV.

063.026 Energy dependent delay measurements of quasi–periodic oscillations in low–mass X–ray binaries.
R. A. M. J. Wijers, W. van Paradijs, W. H. G. Lewin.
Mon. Not. R. Astron. Soc., Vol. 228, No. 2, p. 17P – 21P (1987).

The authors present results of Monte Carlo simulations of the Comptonization of photons in a finite spherical Comptonizing cloud. The energy dependence of the arrival times of the Comptonized photons is very different for different input photon spectra. If the delays observed in the high–frequency QPO of Cyg X–2 and GX 5–1 are the result of Comptonization, multi–channel time–delay measurements may constrain the energy of the input photons and thus provide important information on the accretion process in low–mass X–ray binaries.

063.027 Scattering of light by dielectric spheroids. V.
N. V. Voshchinnikov, V. G. Farafonov.
Vestn. Leningr. Univ., Mat. Mekh. Astron., No. 1, p. 90 – 95 (1987). In Russian. Abstr. in Ref. Zh., 51. Astron., 7.51.120 (1987).

063.028 Asymmetry of the H_β central part measured in a T–tube.
Z. Mijatović, M. Pavlov, S. Djurović.
J. Quant. Spectrosc. Radiat. Transfer, Vol. 38, No. 3, p. 209 – 210 (1987).

The asymmetry parameter δI of the H_β central part has been measured as a function of the plasma electron density. The results are compared with recently published theories and measurements. The plasma source was a small electromagnetic T–shaped shock tube. The electron density varied between 0.7×10^{23} and $3 \times 10^{23} \mathrm{m}^{-3}$, while the temperatures were 20,000K and more.

063.029   **Widths and shifts of some plasma–broadened oxygen and carbon multiplets.**
A. Goly, S. Weniger.
J. Quant. Spectrosc. Radiat. Transfer, Vol. 38, No. 3, p. 225 – 230 (1987).
Profile measurements for some multiplets of neutral oxygen and carbon have been carried out in a wall–stabilized d.c. cascade arc. The electron–impact half–widths and shifts are compared with theoretical values and other experimental data. Reasonable agreement between theoretical and experimental results are found for O I 3947 Å, O I 4368 Å, O I 5436 Å, and C I 5798 Å. For the multiplet O I 5330 Å, the available theoretical values differ considerably and the difference between experimentally determined and calculated shift is too large. Griem's theoretical half–width is close to the experimental value.

063.030   **Relation between multidimensional radiative transfer in cylindrical and rectangular coordinates with anisotropic scattering.**
A. L. Crosbie, L. C. Lee.
J. Quant. Spectrosc. Radiat. Transfer, Vol. 38, No. 3, p. 231 – 241 (1987).

063.031   **Variability from pair atmospheres.**
A. C. Fabian.
Variability of galactic and extragalactic X–ray sources, p. 111 – 119 (1987). – See Abstr. 012.019.
An atmosphere of electron–positron pairs is expected to form around compact sources of X– and gamma radiation. The manner in which they significantly change the emergent spectrum and variability of the source is discussed. Observational estimates of the compactness of sources may be underestimated due to these effects.

063.032   **Relativistic plasmas in active galactic nuclei.**
A. A. Zdziarski, A. P. Lightman.
Variability of galactic and extragalactic X–ray sources, p. 121 – 135 (1987). – See Abstr. 012.019.
The models proposed for the formation of X–ray and γ–ray spectra of active galactic nuclei are discussed and their predictions are compared with observations. Basic relevant radiative processes are briefly reviewed. A comparison is made between thermal and nonthermal classes of models of power law X–ray and γ–ray emission. The authors conclude that the current observations do not exclude either of these classes of models.

063.033   **Red–shifted annihilation lines in thermal synchrotron model of gamma–ray bursts.**
T.–y. Shi, T. Lu.
Astrophys. Space Sci., Vol. 136, No. 2, p. 363 – 368 (1987).
The origins of the gravitationally–red–shifted annihilation lines of gamma–ray bursts in thermal synchrotron–radiation models are discussed in this paper. It is shown that the positrons produced by high–energy photons which are obtained by extrapolation from the continua through strong magnetic fields could not be the main source of the annihilation lines, while that escaping from the hot thin region might account for the emission features.

063.034   **Effect of ion correlations on free–free opacities.**
N. R. Dagdeviren, S. E. Koonin.
Astrophys. J., Vol. 319, No. 1, p. 192 – 194 (1987).
The authors investigate the effect of ion–ion correlations on the free–free photoabsorption process. While the quantum correlation effects are negligible, classical Debye correlations reduce the free–free absorption cross section by $\sim 3\%$ in the center of the Sun. This decreases the solar free–free Rosseland mean opacity by $\sim 1.6\%$.

063.035   **Emission–line ratios for O III in gaseous nebulae and a comparison between theory and observation.**
F. P. Keenan, K. M. Aggarwal.
Astrophys. J., Vol. 319, No. 1, p. 403 – 406 (1987).
Recent R–matrix calculations of electron excitation rates in O III are used to derive electron temperature and density–sensitive emission–line ratios for a range of T_e (7500 – 40,000K) and N_e ($10^2 - 10^7 \mathrm{cm}^{-3}$) applicable to gaseous nebulae. Electron temperatures and densities deduced from these calculations and from observed values of the line ratios for several planetary nebulae and a Seyfert galaxy in general show good internal agreement and also compare favorably with results deduced from the line ratios of other species.

063.036   **The "Lα/Fe II problem" – solved by fluorescence?**
M. V. Penston.
Mon. Not. R. Astron. Soc., Vol. 229, No. 1, p. 1P – 5P (1987).
The models of Fe II line emission from active galactic nuclei calculated by Wills, Netzer & Wills predict too little Fe II emission. This can be accounted for by the Lα fluorescence process of Johansson & Jordan which both strengthens the Fe II emission above that predicted by collisional excitation, the only process included in the models, and somewhat weakens the Lα line.

063.037   **Structural features and scattering properties of dust particles.**
K. O. Thielheim.
Physical processes in interstellar clouds, p. 413 – 421 (1987). – See Abstr. 012.027.
First, the author discusses preliminary studies performed by him on the scattering of light by dielectric needles and disks. Then, he discusses the scattering properties of an ensemble of dielectric homogeneous large particles which are convex on a greater scale and exhibit an irregular though smooth surface. After that, the author discusses particles of the type which on their surface exhibit what is called either macro–roughness or else microroughness structures. Progressing to more complex surface structures he discusses a composite model of roughness. Finally, a new scheme of a scattering theory for an ensemble of particles of irregular shape and arbitrary size based on the method of perturbed boundary conditions is presented.

063.038   **Special features of cyclotron radiation in anisotropic plasmas.**
P. C. W. Fung, P. L. Leung.
Astrophys. Space Sci., Vol. 137, No. 1, p. 33 – 62 (1987).
The authors analyze the special features of cyclotron radiation in four different modes radiated by a mildly relativistic electron current in an anisotropic plasma, taking into consideration that the radiation is transmitted along the group velocity, rather than the wave normal direction. A systematic series of numerical analysis is carried out, to demonstrate the characteristics of the focussing effect and Doppler effect of the radiation, arising from anisotropy. The parameters used pertain to plasmas and radiators of the types encountered in the terrestrial upper atmosphere and the solar corona.

063.039   **A moment method for radiative transfer in an anisotropically–scattering slab medium with space–dependent albedo $\omega(x)$.**
S. J. Wilson, F. S. Wan.
Astrophys. Space Sci., Vol. 137, No. 1, p. 107 – 113 (1987).
The moment method is used to solve the radiative transfer problem in an anisotropic scattering plane medium with arbitrary space–dependent albedo $\omega(x)$. The results are compared with those obtained recently by Cengel and Özisik.

063.040   **Model atmospheres for type I supernovae: curvature effects.**
R. López, E. Simonneau, J. Isern.
Astron. Astrophys., Vol. 184, No. 1/2, p. 249 – 255 (1987).
Type I supernovae are thought to be the result of the thermonuclear explosion of a carbon–oxygen white dwarf in a

close binary system. Previous models have shown that the mean free path of photons is of the same order as the size of the configuration and that the curvature effects can no longer be ignored. In this paper the authors propose a simple and accurate generalization of the Eddington relationship and analyze the role played by the curvature of the layers on the radiative transfer.

063.041 Compact radio sources as a plasma turbulent reactor. II. General characteristics of electromagnetic radiation spectra.
A. M. Atoyan, A. Nagapetyan.
Astrofizika, Tom 27, Vyp. 1, p. 117 – 130 (1987). In Russian. English translation in Astrophysics, Vol. 27, No. 1.

The electromagnetic radiation spectra of a homogeneous source wherein the relativistic electron acceleration on the Langmuir waves leads to the formation of Maxwell–like spectra with characteristic value of the Lorentz–factor $\gamma_0 \sim 10^3$ are considered. It has been shown that due to synchrotron radiation of relativistic electrons, usually observed from CRSs flat radio spectra, gradually steepening at submillimeter wavelengths are naturally formed in the optically thin range of frequences. The electromagnetic radiation at the scattering of the electrons on the turbulence produces significant nonthermal infrared radiation. Inverse Compton scattering of the relativistic electrons on the radio–infrared photons leads to the production of X–rays. The characteristics of the electromagnetic radiation spectra obtained in the model are compared with the observational ones.

063.042 Anisotropic light scattering in an inhomogeneous atmosphere. The radiation field for nearly conservative scattering.
N. N. Fomin, Eh. G. Yanovitskij.
Astron. Zh., Tom 64, Vyp. 5, p. 992 – 1003 (1987). In Russian. English translation in Sov. Astron., Vol. 31, No. 5.

For the case of nearly conservative scattering, the problem of determination of the radiation field in an inhomogeneous semi–infinite atmosphere as well as in an optically thick slab is considered. The equation of transfer in an infinite medium and in Milne's problem is investigated. The asymptotic formulae for the intensity of the radiation field are obtained. The solution of the characteristic equation is presented for the case of isotropic scattering with the single scattering albedo $\lambda(\tau) = \exp(-m\tau)$. This solution is compared with the results of exact numerical calculations.

063.043 Scattering of Lα photons in an infinite expanding medium when there is absorption in the continuum.
N. N. Chugaj.
Astrophysics, Vol. 26, No. 1, p. 53 – 57 (1987). English translation of 43.063.020.

063.044 Formation of X–ray spectra in close binary systems. Reflection effects.
L. G. Titarchuk.
Astrophysics, Vol. 26, No. 1, p. 57 – 66 (1987). English translation of 43.063.095.

063.045 Formation of spectral lines when there is partial frequency redistribution.
D. I. Nagirner.
Astrophysics, Vol. 26, No. 1, p. 90 – 115 (1987). English translation of 43.063.021.

063.046 Quantized synchrotron radiation in strong magnetic fields.
A. K. Harding, R. Preece.
Astrophys. J., Vol. 319, No. 2, p. 939 – 950 (1987).

Radiative transition rates for relativistic electrons in strong magnetic fields are calculated for Landau states up to $n = 500$. Using these rates, one can directly evaluate the accuracy of classical spectra at high magnetic fields and electron energies which have been used to fit gamma–ray burst data. The authors first discuss the behavior of the transition rates as a function of initial and final spin state and final Landau state of the electron.

Monte Carlo spectra for monoenergetic and thermal electron distributions as well as for the case of steady state injection are then calculated from the quantum transition rates and compared to classical and asymptotic quantum spectra.

063.047 The two–flux approximations for radiative transfer in scattering media.
K. Kamiuto.
J. Quant. Spectrosc. Radiat. Transfer, Vol. 38, No. 4, p. 261 – 265 (1987).

Several two–flux approximations for predicting hemispherical transmittance and reflectance of a plane–parallel scattering medium are examined. The relation between the two–flux and P_1 approximations is discussed.

063.048 The discrete spectrum for radiative transfer with polarization.
R. D. M. Garcia, C. E. Siewert.
J. Quant. Spectrosc. Radiat. Transfer, Vol. 38, No. 4, p. 295 – 301 (1987).

Elementary considerations are used to define and analyze the discrete spectrum for a general radiative transfer model that includes polarization effects.

063.049 Erratum: "Matrix formulations for the transfer of solar radiation in a plane–parallel scattering atmosphere" [J. Quant. Spectrosc. Radiat. Transfer, Vol. 35, No. 1, p. 13 – 21 (1986)].
T. Nakajima, M. Tanaka.
J. Quant. Spectrosc. Radiat. Transfer, Vol. 38, No. 4, p. 323 (1987). See Abstr. 41.063.010.

063.050 Probabilistic interpretation of radiative transfer. I. The $\sqrt{\varepsilon}$–law.
I. Hubeny.
Astron. Astrophys., Vol. 185, No. 1/2, p. 332 – 335 (1987).

A simple physical explanation of the long–known result of the standard problem of line formation theory (a semi–infinite, isothermal atmosphere, two–level atoms, complete frequency redistribution), stating that the source function at the surface is given by $(\sqrt{\varepsilon})B$, is presented. The basis of the approach consists in a recognition that the frequency averaged mean intensity of radiation and the source function, at a given depth, are simply proportional to the probability of ultimate thermalization of a photon emitted or absorbed at this depth. The $\sqrt{\varepsilon}$–law follows then simply from an application of the principle of invariance.

063.051 Probabilistic interpretation of radiative transfer. II. Rybicki equation.
I. Hubeny.
Astron. Astrophys., Vol. 185, No. 1/2, p. 336 – 342 (1987).

By extending the approach developed in the previous part of this paper, the author presents a physical explanation of the Rybicki generalization of the $\sqrt{\varepsilon}$–law to all depths of an atmosphere. To this end, various probabilistic concepts, in particular the frequency– and angle–dependent thermalization probabilities, are introduced, and relations between them and the specific intensity of radiation are studied in detail.

063.052 The Poynting–Robertson effect of a rotating star.
E. T. Rusk.
Astrophys. J., Vol. 320, No. 1, p. 315 – 318 (1987).

An expression is derived for the force on an orbiting particle due to radiation from a rotating star. The expression is then compared to an expression for the case involving a nonrotating star derived by Guess (1962). One effect of this additional component is to decrease slowly the inclination of orbiting particles. This effect is investigated as a possible mechanism for aligning circumstellar dust into rings. It is found that the effect is not sufficient to form these rings and, in the solar system, is small compared to planetary perturbations.

063.053 **Mathematical modelling of light scattering by surfaces in the polarimetry of atmosphereless cosmic bodies.**
L. O. Kolokolova.
Inst. Teor. Fiz. Akad. Nauk USSR, Prepr., No. 70 R, p. 2 – 24 (1987). In Russian. Abstr. in Ref. Zh., 51. Astron., 10.51.155 (1987).

063.054 **Radiation from charged particles moving in the magnetic fields of extragalactic jets.**
W. K. Rose.
Bull. Am. Astron. Soc., Vol. 19, No. 2, p. 692 (1987). Abstract. – See Abstr. 010.061.

063.055 **Gamma ray production by comptonization of electron beams in pulsar magnetospheres.**
J. K. Daugherty.
Bull. Am. Astron. Soc., Vol. 19, No. 2, p. 693 (1987). Abstract. – See Abstr. 010.061.

063.056 **X–ray spectra from the boundary–layer region of a (disk) accreting weakly magnetized neutron star.**
R. A. Saenz, W. Kluzniak, R. V. Wagoner.
Bull. Am. Astron. Soc., Vol. 19, No. 2, p. 693 (1987). Abstract. – See Abstr. 010.061.

063.057 **The pion boiler as a model of gamma–ray sources.**
C. D. Dermer.
Bull. Am. Astron. Soc., Vol. 19, No. 2, p. 693 (1987). Abstract. – See Abstr. 010.061.

063.058 **Electron–positron pairs in accretion flows.**
B. Tritz, S. Tsuruta.
Bull. Am. Astron. Soc., Vol. 19, No. 2, p. 694 – 695 (1987). Abstract. – See Abstr. 010.061.

063.059 **Power–law X–ray emission from electron–positron pair winds.**
K. Leighly, B. Tritz, S. Tsuruta.
Bull. Am. Astron. Soc., Vol. 19, No. 2, p. 695 (1987). Abstract. – See Abstr. 010.061.

063.060 **Formation of multicomponent blends in envelopes of supernovae. I. Effects of nonlocal radiative coupling.**
O. S. Bartunov, A. L. Mozgovoj.
Astrophysics, Vol. 26, No. 2, p. 136 – 147 (1987). English translation of 43.063.049.

063.061 **Radiation field in an infinite dust nebula illuminated by a star.**
A. K. Kolesov, V. Yu. Perov.
Astrophysics, Vol. 26, No. 2, p. 147 – 154 (1987). English translation of 43.063.050.

063.062 **Determination of the characteristics of interstellar turbulence from data on the diffusion coefficient of cosmic rays.**
V. N. Fedorenko, V. M. Ostryakov.
Astrophysics, Vol. 26, No. 2, p. 175 – 184 (1987). English translation of 43.063.051.

063.063 **Short characteristic solution of the non–LTE transfer problem by operator perturbation. I. The one–dimensional planar slab.**
G. L. Olson, P. B. Kunasz.
J. Quant. Spectrosc. Radiat. Transfer, Vol. 38, No. 5, p. 325 – 336 (1987).
The authors present the formal solution of the transfer problem in terms of the exponential short characteristic method and derive approximate operators that allow for the iterative solution of the non–LTE two–level atom problem. An eigenvalue analysis for the convergence rate of these operators and several approximate operators proposed by other authors is presented. The family of operators presented for the short characteristic

approach range from local diagonal approximations to tridiagonal and pentadiagonal operators. The extension to multidimensions of the several proposed approximate operators is discussed.

063.064 **A fast method for computing the integrals of the relativistic Compton scattering kernel for radiative transfer.**
D. S. Kershaw.
J. Quant. Spectrosc. Radiat. Transfer, Vol. 38, No. 5, p. 347 – 352 (1987).
For various computer simulation applications, one needs the integrals of the Compton scattering kernel over its parameters. An efficient and accurate method for evaluating these integrals is described and the corresponding software is available upon request.

063.065 **An approximate method for calculating Planck and Rosseland mean opacities in hot, dense plasmas.**
G. D. Tsakiris, K. Eidmann.
J. Quant. Spectrosc. Radiat. Transfer, Vol. 38, No. 5, p. 353 – 368 (1987).
A method is described for computing the Rosseland and Planck mean opacities of an element as a function of temperature and density. The method is based on the average ion representation for calculating the occupation numbers of the various electronic levels. The photoabsorption coefficients are calculated in the hydrogenic approximation. To account for the important contribution of line absorption, a simple phenomenological method has been developed, which takes into consideration effects such as broadening of the upper electronic levels into bands and band overlapping in the heaviest of the elements. Calculations are presented for various elements and compared with the results of more elaborate methods. Power–law scaling relations for the dependence on temperature and density are derived for a number of elements of practical importance.

063.066 **Effects of parameter variations in the radiative transfer equation: application to the lower stratosphere in the far infrared.**
P. Rabache, B. Rebours, D. Leclerc.
Infrared Phys., Vol. 27, No. 6, p. 379 – 388 (1987).
Numerical data for the relative variations of the atmospheric and spectroscopic parameters of the radiative transfer equation are computed from spectral representations. The analysis is applied to the lower stratosphere in the FIR where many high–resolution measurements have been made. The data shows that the concentration of the minor constituents in the medium is a measurable quantity but, to make the determination, the other radiative transfer parameters must be known with a 1% precision, at the time the measurement is being made.

063.067 **On pumping the strong water maser sources.**
N. D. Kylafis, C. Norman.
Space Telesc. Sci. Inst., Prepr. Ser., No. 181, 21 pp. (1987). To appear in Astrophys. J.

063.068 **Analytic and numerical calculations of quantum synchrotron spectra from relativistic electron distributions.**
J. J. Brainerd, V. Petrosian.
Astrophys. J., Vol. 320, No. 2, p. 703 – 713 (1987).
The authors calculate numerically and analytically synchrotron spectra for thermal and power–law electron distributions using the single–particle synchrotron power spectrum derived from quantum electrodynamics. The photon energy at which quantum effects appear is proportional to temperature and independent of field strength for thermal spectra; quantum effects introduce an exponential roll–off away from the classical spectra. For power–law spectra, the photon energy at which quantum effects appear is inversely proportional to the magnetic field strength; quantum effects produce a steeper power law than is found classically.

063.069 Quantum synchrotron spectra from semirelativistic electrons in teragauss magnetic fields.
J. J. Brainerd.
Astrophys. J., Vol. 320, No. 2, p. 714 – 725 (1987).

Synchrotron spectra are calculated from quantum electrodynamic transition rates for thermal and power–law electron distributions. Quantum effects appear in thermal spectra when the photon energy is greater than the electron temperature, and in power–law spectra when the electron energy in units of the electron rest mass times the magnetic field strength in units of the critical field strength is of order unity. The author compares these spectra with spectra calculated from the ultrarelativistic approximation for synchrotron emission.

063.070 Resonance line transfer and transport of excited atoms. II. Self–consistent solutions (2).
O. Atanackovic, J. Borsenberger, J. Oxenius, E. Simonneau.
Inst. Astrophys. Paris, Pré–Publ., No. 202, 61 pp. (1987). To appear in J. Quant. Spectrosc. Radiat. Transfer.

063.071 Calculation of the radiation field in a homogeneous anisotropically scattering plane–parallel atmosphere.
T. Viik.
Tartu Astrofüüs. Obs. Teated, Nr. 85, 46 pp. (1987). In Russian.

The radiation field in a homogeneous anisotropically scattering plane–parallel atmosphere is found by using the Sobolev method of decomposition and approximating the resolvent function by a superposition of exponents. Using the package of subroutines in the appendix the following problems can be solved: (1) The main (or the standard) problem, i.e. determining the radiation field in a planetary atmosphere which is illuminated by parallel rays of the Sun; (2) Determining the radiation field in the deep regions of a semi–infinite atmosphere; (3) Determining the radiation field for the Milne problem, i.e. when there are sources of radiation at infinity.

063.072 Integral relations in the theory of polarized radiative transfer.
V. V. Sobolev.
Dokl. Akad. Nauk SSSR. Ser. Mat. Fiz., Tom 295, No. 1, p. 60 – 63 (1987). In Russian. Abstr. in Ref. Zh., 51. Astron. 11.51.98 (1987).

063.073 Radiation transfer in absorbing, emitting, isotropically scattering, homogeneous cylindrical media.
S. T. Thynell, M. N. Özisik.
J. Quant. Spectrosc. Radiat. Transfer, Vol. 38, No. 6, p. 413 – 426 (1987).

One–dimensional radiative transfer in an absorbing, emitting, gray, isotropically scattering, homogeneous, solid cylinder having a diffusively reflecting boundary surface and externally incident radiation is solved by using the appropriate expansion functions. Expressions for the radiation intensity, incident radiation and net radiative heat flux are presented. The method also has the potential for generalization to problems involving anisotropic scattering.

063.074 Resonance line transfer and transport of excited atoms. III. Self–consistent solutions (2).
O. Atanackovič, J. Borsenberger, J. Oxenius, E. Simonneau.
J. Quant. Spectrosc. Radiat. Transfer, Vol. 38, No. 6, p. 427 – 446 (1987).

. In this last part of their study on non–LTE line transfer with convective transport of excited atoms, the authors present self–consistent solutions of the radiative transfer equation and the kinetic equation of the excited two–level atoms when the excited atoms undergo elastic velocity–changing collisions. They assume pure Doppler broadening of the spectral line and investigate reflecting and destroying boundaries for the excited atoms. Concerning elastic collisions of the excited atoms, the study covers all cases, from a collisionless gas (free particle streaming) discussed in Part II of this series of papers to a collision–dominated gas with the limiting case of complete redistribution.

063.075 Integral form of the time–dependent radiation transfer equation. I. Inhomogeneous slabs.
A. Munier.
J. Quant. Spectrosc. Radiat. Transfer, Vol. 38, No. 6, p. 447 – 455 (1987).

The formal solution of the time–dependent radiation transfer equation in an inhomogeneously emitting, absorbing and non–scattering slab is given in Cartesian coordinates. The optical depth and retardation are introduced together with a separation parameter to separate boundary and initial conditions. The method uses integration of the equation along its characteristics.

063.076 Integral form of the time–dependent radiation transfer equation. II. Inhomogeneous spherical media.
A. Munier.
J. Quant. Spectrosc. Radiat. Transfer, Vol. 38, No. 6, p. 457 – 474 (1987).

The integral form of the radiation transfer equation, with spherical symmetry for a non–isotropic, time–dependent, inhomogeneous, non–scattering medium is given for arbitrary initial and boundary conditions. The optical depth is generalized to an optical retardation and a separation parameter is introduced to separate boundary and initial conditions. The static solution is identical to the asymptotic form of the time–dependent solution.

063.077 Integral form of the time–dependent radiation transfer equation. III. Moving boundaries.
A. Munier.
J. Quant. Spectrosc. Radiat. Transfer, Vol. 38, No. 6, p. 475 – 487 (1987).

The theory of characteristics is used to give the general time–dependent integral form of the transfer equation in a time–dependent, inhomogeneous medium, submitted to arbitrary boundary and initial conditions. The medium emits and absorbs radiation, but scattering is neglected. The boundary condition is applied to a moving surface. The general solution is given analytically in Cartesian and spherical coordinates for a 1–D configuration.

063.078 Radiation hydrodynamics in astrophysics.
M. H. Ibáñez S.
Rev. Mex. Astron. Astrofis., Vol. 14, No. 2, p. 573 – 586 (1987). – See Abstr. 012.042.

Inertial and comoving–frame equations of radiation hydrodynamics are analyzed, starting from their general covariant form. The relative importance of their different terms, in two limit regimes of interest in astrophysics, is discussed: (1) in the free–streaming regime during radiative and fluid–flow time scales; (2) in static and dynamic diffusion regimes. Two kinds of general solutions are discussed: linear solutions (stability and wave propagation), and non–linear ones. Several open questions of increasing interest in astrophysics are also examined.

063.079 A model for the thermal radio continuum emission produced by a shock wave and its application to the Herbig–Haro objects 1 and 2.
S. Curiel, J. Cantó, L. F. Rodríguez.
Rev. Mex. Astron. Astrofis., Vol. 14, No. 2, p. 595 – 602 (1987). – See Abstr. 012.042.

The authors present a model for the thermal radio continuum emission produced by a shock wave and its application to the observations at 20, 6 and 2 cm of the HH 1–2 system.

063.080 The energy–dependent radiation transfer in a homogeneous finite sphere.
S. A. El Wakil, H. Machali, M. T. Attia, E. A. Saad.
Astrophys. Space Sci., Vol. 137, No. 2, p. 239 – 250 (1987).

Radiation transfer problem in the slowing–down region is solved for a spherical geometry. Modelled kernels are used to represent the slowing–down kernel. Calculations are performed for the radiation flux at the boundary of the sphere for two cases: (1) in the presence of internal source, (2) if the radiation is incident on the sphere. Numerical results are obtained by using the Padé approximant technique.

063.081   Vibrational transition probabilities, r̄–centroids and PE–curves for OH and PN.
M. Singh.
Astrophys. Space Sci., Vol. 138, No. 1, p. 79 – 86 (1987).

Vibrational transition probabilities namely Franck–Condon factors and r̄–centroids have been evaluated using an approximate analytical method for the $A–X$ systems of OH and PN. KKRV potential energy curves for $X^2\Pi_i$, $X^1\Sigma^+$, $A^2\Sigma^+$, and $A^1\Pi$ states of OH and PN have been constructed using the latest spectroscopic data. The value of r̄–centroids for the band have been found to increase linearly with the corresponding wavelengths. The author shows results for six new bands and OH and eight new bands of PN in the spectra of astronomical objects.

063.082   Radiation transfer due to a point–source in an inhomogeneous finite sphere.
S. J. Wilson.
Astrophys. Space Sci., Vol. 138, No. 1, p. 191 – 196 (1987).

A simple method for solving the radiation transfer due to a point source in an isotropically–scattering sphere with space–dependent albedo is presented. The results are compared with the only other available results obtained by Thynell and Özisik (1986) for inhomogenous sphere. The present method is applicable for a wider class of albedo functions.

063.083   Decomposition of product of certain functions relevant to the solution of transfer equations by Wiener–Hopf technique.
S. Das Gupta, S. R. Das Gupta.
Astrophys. Space Sci., Vol. 138, No. 2, p. 403 – 418 (1987).

Decomposition of a product of functions obtained by Das Gupta (1978) and relevant to the solution of equations of radiative transfer or of transfer problems in finite media by Wiener–Hopf technique, is reviewed and transformed to quite simple integral forms amenable to easy numerical evaluations. The same forms are then shown to be directly obtainable in one step under a slightly stronger condition consistent with practical cases.

063.084   Survey of operator perturbation methods.
W. Kalkofen.
Numerical radiative transfer, p. 23 – 34 (1987). – See Abstr. 003.016.

This paper surveys operator perturbation methods in radiative transfer. It discusses the salient features of methods employing integral as well as differential equations, and it emphasizes the physical principles on which the approximate operators are based.

063.085   Line formation in expanding atmospheres: multi–level calculations using approximate lambda operators.
W.–R. Hamann.
Numerical radiative transfer, p. 35 – 65 (1987). – See Abstr. 003.016.

The author studies multi–level non–LTE spectral formation in spherically expanding atmospheres, using a perturbation technique. The method employs approximate lambda operators which act only locally. Adequate operators are defined for continua and lines, the latter being constructed from considerations in the comoving frame. The method is tested for a two–level atom and then generalized to the multi–level case. Test calculations are presented for a typical Wolf–Rayet star atmosphere consisting of pure helium, with the ground state of He I, ten bound levels and 45 lines of He II, and He III. Convergence is found to be satisfactory.

063.086   Stellar atmospheres in non–LTE: model construction and line formation calculations using approximate lambda operators.
K. Werner.
Numerical radiative transfer, p. 67 – 99 (1987). – See Abstr. 003.016.

The author describes a method of calculating plane–parallel model atmospheres subject to the constraints of radiative, hydrostatic, and statistical equilibrium. Simple modifications make this method equally suitable for multi–level line formation calculations. It is based on the use of local, approximate lambda operators and a perturbation technique yielding exact solutions for radiative transfer problems. Up to 100 non–LTE levels can be treated. As an example, line formation calculations are presented for a five–level hydrogen atom with continuum in six different model atmospheres. In addition, several pure–hydrogen model atmospheres have been computed, demonstrating the power of the method.

063.087   Acceleration of convergence.
L. Auer.
Numerical radiative transfer, p. 101 – 109 (1987). – See Abstr. 003.016.

The author presents an inexpensive method which can double or even triple the efficiency of any linearly convergent method. Successive iterations of the unaccelerated scheme are used to predict the accurate solution. As an example, the technique is applied to accelerating simple iterative solution techniques for the two–level atom non–coherent scattering problem.

063.088   Line formation in a time–dependent atmosphere.
W. Kalkofen.
Numerical radiative transfer, p. 111 – 134 (1987). – See Abstr. 003.016.

The author describes an operator perturbation method for solving radiative transfer problems in moving media; the transfer equation is solved in the quasi–static approximation but the structure of the medium may be time–dependent. The characteristic feature of the method is that it solves transfer problems by separating the calculation into two parts: that of the error made by a solution in satisfying the constraint of statistical equilibrium, and that of corrections to the solution. The properties of the method are illustrated in two problems of a line in a moving medium, and a variant using integral equations in the correction calculations is discussed.

063.089   Iterative solution of multilevel transfer problems.
E. H. Avrett, R. Loeser.
Numerical radiative transfer, p. 135 – 161 (1987). – See Abstr. 003.016.

The authors show how the "equivalent two–level atom" equations can be used to solve multilevel transfer problems. Iterative convergence can be very slow if the coupled equations are not formulated in a way that directly requires the solution to be self consistent. The authors give an example in which convergence is obtained after a few iterations or requires more than 100 iterations, depending on how the equations are formulated. They also examine the accuracy of using the escape–probability approximation in multilevel problems instead of solving the transfer equations.

063.090   An algorithm for the simultaneous solution of thousands of transfer equations under global constraints.
L. S. Anderson.
Numerical radiative transfer, p. 163 – 190 (1987). – See Abstr. 003.016.

The author derives an algorithm for the solution of many equations of transfer coupled by integral constraints such as statistical, radiative, and hydrostatic equilibrium. The formalism uses complete linearization, variable Eddington factors, and the Feautrier transfer algorithm to reduce non–linear systems to a set of matrix equations for corrections to dependent variables which is tridiagonal in depth. A new feature makes use of the redundancy of spectral information to reduce the number of dependent radiation densities from the thousands required to resolve the spectrum to at most about one hundred. The method is particularly well suited for the study of nonLTE line blanketing in planar and spherical geometries.

063.091   **Operator perturbation for differential equations.**
W. Kalkofen.
Numerical radiative transfer, p. 191 – 211 (1987). – See Abstr. 003.016.

Cannon's suggestion to solve transfer problems by perturbing the exact differential equations about lower–order, approximate equations provided the impetus for the powerful and efficient modern methods in numerical radiative transfer. The author discusses a general method of deriving perturbation equations in differential form from the exact differential equations, using integral equations in an intermediary step, and applies the procedure to an isotropic scattering problem where the temperature is given and to a transfer problem where the temperature is determined from the radiative equilibrium constraint.

063.092   **A gentle introduction to polarized radiative transfer.**
D. E. Rees.
Numerical radiative transfer, p. 213 – 239 (1987). – See Abstr. 003.016.

The author discusses the representation and measurement of polarized light in terms of the Stokes parameters. Possible sources of error because of ambiguities about sign conventions are highlighted by an analysis of the LTE transfer equations for the Stokes parameters of a normal Zeeman triplet. A simple derivation of the analytic solution of these equations is given for a Milne–Eddington model atmosphere. The discussion is complemented by a brief review of developments in non–LTE polarized transfer.

063.093   **Non–LTE polarized radiative transfer in spectral lines.**
D. E. Rees, G. A. Murphy.
Numerical radiative transfer, p. 241 – 264 (1987). – See Abstr. 003.016.

Numerical methods applicable to two kinds of non–LTE spectral line formation with polarization are discussed. First the authors describe two techniques for formal integration of the Stokes vector transfer equation of a Zeeman split line: one is a generalization of the Feautrier (1964) finite–difference method; the other involves the lambda operator associated with the diagonal element of the absorption matrix. Then the authors consider the problem of resonance line polarization in the absence of magnetic fields. An iterative solution method is outlined. This is an adaptation of Scharmer's (1981, 1983) approximate lambda operator method, the polarization being treated as a perturbation.

063.094   **Transfer of polarized radiation, using 4 × 4 matrices.**
E. Landi Degl'Innocenti.
Numerical radiative transfer, p. 265 – 278 (1987). – See Abstr. 003.016.

The main characteristics of the radiative transfer equations for polarized radiation, including the symmetry properties and the physical interpretation of the various terms of the absorption matrix, are briefly reviewed. A formal solution of the transfer equations is presented for the case where both the absorption matrix and the source–function vector are given functions of optical depth. A suitable algorithm is presented to obtain from the formal solution a numerical solution, and a brief comparison is presented with different numerical methods that are often used for the solution of the same set of equations. Finally, an iteration scheme is suggested for attacking a large variety of non–LTE problems for polarized radiation according to a perturbative scheme.

063.095   **Radiative transfer in the presence of strong magnetic fields.**
A. A. van Ballegooijen.
Numerical radiative transfer, p. 279 – 304 (1987). – See Abstr. 003.016.

The author reviews the transfer equation for polarized radiation in a Zeeman–split spectral line. Then he rewrites the transfer equation in terms of complex 2 × 2 matrices, and derives a formal solution analogous to the familiar integral expression for unpolarized light. This solution is used to formulate an integral equation for the line source function. Finally, the author describes a numerical method for the evaluation of the integral operator, and a matrix equation for the source function is obtained.

063.096   **An integral operator technique of radiative transfer in spherical symmetry.**
A. Peraiah.
Numerical radiative transfer, p. 305 – 340 (1987). – See Abstr. 003.016.

The integral operator technique, by which the solution of the radiative transfer equation is obtained, is reviewed. Three applications of this method have been presented: (1) solution of transfer equation with coherent, isotropic scattering (2) solution of line transfer in the comoving frame and (3) solution of transfer equation with aberration and advection terms included.

063.097   **Discrete ordinate matrix methods.**
M. Schmidt, R. Wehrse.
Numerical radiative transfer, p. 341 – 371 (1987). – See Abstr. 003.016.

Methods for the solution of the angle and frequency discretized one–dimensional transfer equation by means of transition matrices are described. It is shown that expressions for the transition and reflection operators as well as the source terms can be derived which do not contain increasing exponentials and which can be evaluated by numerical standard techniques. Several examples demonstrate the applicability of these methods to complex analytical and numerical problems.

063.098   **Effects of acoustic and gravity waves on the curve–of–growth.**
C. Marmolino.
Sol. Phys., Vol. 112, No. 2, p. 211 – 226 (1987).

Unresolved motions, or 'microturbulence', play a very important role in determining the Doppler width of the line–absorption coefficient. The concept of microturbulence was introduced because observed solar and stellar lines are broader than is explicable within the framework of other line–broadening mechanisms. By constructing a curve–of–growth in the presence of a dynamical model for simple acoustic and gravity waves, this paper evaluates the errors introduced by the use of kinematic models for line–broadening (microturbulence) into the determinations of non–thermal velocity fields and element abundances. Distinct differences between gravity and acoustic waves are apparent in the trend of the line asymmetry with the excitation potential as well as in the magnitude of the errors introduced by the temperature and pressure variations associated with the velocity field.

063.099   **Broadening and shift of Fe I lines perturbed by atomic hydrogen.**
M. T. Gomez, C. Marmolino, G. Roberti, G. Severino.
Sol. Phys., Vol. 112, No. 2, p. 227 – 232 (1987).

The broadening and shift parameters for a number of Fe I lines perturbed by atomic hydrogen are computed using the interatomic potential due to Hindmarsh et al (1967, 1970). It is also shown that the rms radius and the effective radius of the radiating atom, which determine the force constants in the interatomic potential, can be simply related to each other, depending on the orbital quantum number of the atomic level.

063.100   **Nearly–exact analytical solutions of the problem of radiation transfer in a plane layer for a model of complete frequency redistribution.**
R. G. Gabrielyan, A. R. Mkrtchyan, Kh. V. Kotandzhyan, M. A. Mnatsakanyan.
Izv. Akad. Nauk Arm. SSR, Ser. Fiz., Tom 22, Vyp. 4, p. 191 – 197 (1987). In Russian.

Approximate, nearly–exact analytical solutions of the most commonly formulated problem of incoherent transfer of radiation in a three–dimensional medium of finite thickness were obtained for the model of complete frequency redistribution.

063.101   **Problems in astrophysical radiation hydrodynamics.**
J. I. Castor.
Astrophysical radiation hydrodynamics, p. 1 – 43 (1986). – See
Abstr. 012.069.

The basic equations of radiation hydrodynamics are discussed
in the regime that the radiation is dynamically as well as
thermally important. Particular attention is paid to the question
of what constitutes an acceptable approximate non–relativistic
system of dynamical equations for matter and radiation in this
regime. Further discussion is devoted to two classes of application of these ideas. The first class consists of problems dominated
by line radiation, which is sensitive to the velocity field through
the Doppler effect. The second class is of problems in which the
advection of radiation by moving matter dominates radiation
diffusion.

063.102   **The equations of radiation hydrodynamics.**
D. Mihalas.
Astrophysical radiation hydrodynamics, p. 45 – 69 (1986). – See
Abstr. 012.069.

The purpose of this paper is to give an overview of the role of
radiation in the transport of energy and momentum in a
combined matter–radiation fluid. The transport equation for a
moving radiating fluid is presented in both a fully Eulerian and a
fully Lagrangean formulation, along with conservation equations describing the dynamics of the fluid. Special attention is
paid to the problem of deriving equations that are mutually
consistent in each frame, and between frames, to O(v/c). A
detailed analysis is made to show that in situations of broad
interest, terms that are formally of O(v/c) actually dominate the
solution.

063.103   **WH80s: numerical radiation hydrodynamics.**
K.–H. A. Winkler, M. L. Norman.
Astrophysical radiation hydrodynamics, p. 71 – 139 (1986). – See
Abstr. 012.069.

An implicit, adaptive–mesh technique for radiation hydrodynamics computations in one spatial dimension is described in
detail.

063.104   **Numerical solution of the equation of radiation transfer.**
H. W. Yorke.
Astrophysical radiation hydrodynamics, p. 141 – 179 (1986). –
See Abstr. 012.069.

The author discusses the numerical solution of the equation of
radiation transfer in an astrophysical context. Rather than try to
review this complex subject, he is principally concerned with
including radiation transfer effects in numerical "model" calculations. The author demonstrates how to obtain the "exact"
numerical solution of the continuum and line transfer equations
in spherical symmetry. He also shows how to calculate the
expected appearance of models for which the source function and
extinction coefficient are known. Examples of such calculations
are presented and discussed.

063.105   **Radiation transfer in inhomogeneous dispersive media.**
E. A. Saad, M. S. Abdel Krim, A. A. El Ghazaly.
Astrophys. Space Sci., Vol. 139, No. 1, p. 75 – 82 (1987).

Equations for the angular density of radiation, reflected and
transmitted intensities associated with radiation scattered by
inhomogeneous dispersive media are obtained. The Padé approximant technique is used to calculate these intensities in inhomogeneous and homogeneous media. The results for the [0/1] Padé
approximant lead to numerical results that compared with the
exact results.

063.106   **Radiative transfer in an infinite cylinder.**
S. A. El Wakil, E. A. Saad.
Astrophys. Space Sci., Vol. 139, No. 2, p. 321 – 326 (1987).

The problem of radiation transfer in a cylinder with diffuse
reflectivity and containing an energy source is connected with the
source–free radiation transfer problem with isotropic boundary
conditions. An equation for the radiation heat flux is obtained
for a polynomial source. In the special case of isotropic

scattering, the radiation heat flux is given in terms of the albedo
of the second problem. An expression is also given for the net
radiation heat flux.

063.107   **Predicted [Ni II] infrared line strengths in the Crab
Nebula and IC 443.**
R. B. C. Henry.
Astrophys. J., Vol. 322, No. 1, p. 399 – 402 (1987).

The author has performed calculations for a 17–level Ni^+ ion
over a range in T_e and N_e representative of conditions found in
gaseous nebulae in order to find additional observable [Ni II]
emission lines which could be used to verify the assumption that
the strong $\lambda 7378$ line seen in numerous objects is primarily
excited by electron–ion collisions. It is found that lines occurring
at 1.19 μm, 6.63 μm and 10.7 μm should range in strength from
roughly 0.1 to several times that of $\lambda 7378$ in an unreddened
spectrum. Predicted strengths of these three lines relative to
$\lambda 7378$ are presented for five filament positions in the Crab
Nebula plus one location in the supernova remnant IC 443.

063.108   **Evolution of the axial electron cyclotron maser instabili-
ty, with applications to solar microwave spikes.**
L. Vlahos, P. Sprangle.
Astrophys. J., Vol. 322, No. 1, p. 463 – 472 (1987).

The nonlinear evolution of cyclotron radiation from streaming
and gyrating electrons in an external magnetic field is analyzed.
The nonlinear dynamics of both the fields and the particles are
treated fully relativistically and self–consistently. The model
includes a background plasma and electrostatic effects. The
analytical and numerical results show that a substantial portion
of the beam particle energy can be converted to electromagnetic
wave energy at frequencies far above the electorn cyclotron
frequency. The authors explore in detail the possibility of using
this model to explain the intense radio emission observed from
the Sun.

063.109   **Stokes profile analysis and vector magnetic fields.
I. Inversion of photospheric lines.**
A. Skumanich, B. W. Lites.
Astrophys. J., Vol. 322, No. 1, p. 473 – 482 (1987).

The authors consider improvements to the Auer, Heasley, and
House method for the analytic inversion of Stokes profiles via
nonlinear least squares. In the application of this method to
actual sunspot observations, the authors have found that its
simplifications often yield erroneous solutions or nonconvergent
behavior. By including damping wings and magneto–optical
birefringence and by decoupling the intensity profile from the
three–vector polarization profile in the analysis, the authors
develop a more robust inversion method that provides a more
reliable and accurate estimate of sunspot vector magnetic fields
without significant loss of economy.

063.110   **The polarization properties of model sunspots: the
broad–band polarization signature of the Schlüter–
Temesvary representation.**
A. Skumanich, B. W. Lites.
Astrophys. J., Vol. 322, No. 1, p. 483 – 493 (1987).

The general properties and diagnostic content of net, i.e.,
integrated over wavelength, Zeeman–induced polarization in
spectral lines formed in sunspots are considered. The authors
consider the net circular polarization (NCP) signature and study
the effects of a magnetic field gradient and anomalous dispersion
(magneto–optical birefringence), as well as thermal models,
model atom, and velocity structure. The net linear polarization
(NLP) is found to be significantly more sensitive to the spot
thermodynamic structure and less sensitive to the magnetic–
velocity field gradients than the NCP.

063.111   **Reexamination of the metal contribution to astrophysi-
cal opacity.**
C. A. Iglesias, F. J. Rogers, B. G. Wilson.
Astrophys. J., Lett. Ed., Vol. 322, No. 1, p. L45 – L48 (1987).

Results from a new opacity code are presented which show that
the Rosseland mean opacity for metals (atomic number > 2) is

significantly larger than the standard opacity for densities and temperatures important to Cepheid models. A large part of the difference is due to an improved treatment of the atomic physics. Spectral line broadening effects are found to be important.

063.112   Radiation pressure on the plasma above degenerate stars with a strong magnetic field.
V. V. Zheleznyakov, A. A. Litvinchuk.
Sov. Astron., Vol. 31, No. 2, p. 159 – 166 (1987). English translation of 43.063.030.

063.113   The statistical description of a radiation field on the basis of the invariance principle. IV. The results of numerical calculations.
G. A. Arutyunyan, A. G. Nikogosyan.
Astrofizika, Tom 27, Vyp. 2, p. 335 – 345 (1987). In Russian. English translation in Astrophysics, Vol. 27, No. 2.
Results of numerical calculations for the mean number of scattering and the mean time of photon travel in a medium, considered in previous papers, Nikogosyan (1984, 1986), are given. The dependence of the mentioned mean quantities on the scattering mechanism, the original characteristics of a photon, as well as the values of the parameters λ and β, have been revealed.

063.114   On the stability of turbulent synchrotron sources with respect to e^+-e^- pair creation.
F. A. Agaronyan, A. M. Atoyan.
Astrofizika, Tom 27, Vyp. 2, p. 371 – 381 (1987). In Russian. English translation in Astrophysics, Vol. 27, No. 2.

063.115   Synchrotron emission from shock waves in active galactic nuclei.
P. L. Biermann, P. A. Strittmatter.
Astrophys. J., Vol. 322, No. 2, p. 643 – 649 (1987).
The origin of the sharp near–infrared cutoff in the continuous energy distribution of many compact nonthermal sources (BL Lacs, OVV's, red quasars, and certain jets) is considered under the assumption that particle acceleration takes place in shocks. Energy losses due to synchrotron emission and photon interactions are taken into account. In these circumstances the upstream disturbance to the flow is dominated by the most energetic protons which are postulated to excite a turbulent wave spectrum of Kolmogorov type in this region. This in turn sets the relative acceleration times for all particles as a function of energy.

063.116   Inverse Compton scattering of ambient radiation by a cold relativistic jet: a source of beamed, polarized continuum in blazars?
M. C. Begelman, M. Sikora.
Astrophys. J., Vol. 322, No. 2, p. 650 – 661 (1987).
The authors present a general formalism for computing the intensity and polarization properties of unpolarized radiation scattered by a flow of cold electrons with relativistic bulk velocities. They consider the case of isotropic incident radiation with a power–law spectrum, and develop the "head–on" approximation to describe the nearly unidirectional character of the incident radiation in the electron rest frame. The authors argue that Comptonization of unbeamed radiation by a relativistic jet in active galactic nuclei can account for the polarization and the overall spectral shape of the IR–optical continuum in blazars, and speculate that the mechanism may also produce soft X–ray and γ–ray excesses in some objects.

063.117   Partial redistribution in high–density, highly ionized plasmas.
B. Talin, R. Stamm, V. P. Kaftandjian, L. Klein.
Astrophys. J., Vol. 322, No. 2, p. 804 – 811 (1987).
The conditions for which partial redistribution functions must be used for radiation transport in high–Z, high–density, laser–produced plasmas are examined. A previously developed two–photon formalism based on the model microfield method is used to calculate redistribution functions including electron and ion Stark broadening with ion dynamic effects. The competition between the relaxation rates and spontaneous emission is shown

to determine the conditions for partial redistribution. The redistribution function for the Lyα transition of Ar XVIII is presented for typical plasma conditions.

063.118   The infrared diagnostic of a dusty plasma with applications to supernova remnants.
E. Dwek.
Astrophys. J., Vol. 322, No. 2, p. 812 – 821 (1987).
IRAS observations of infrared emission from supernova remnants constitute the first observational evidence of shock–heated dust in the interstellar medium. The author examines the possibilities and limitations of using the infrared observations as a diagnostic for the shocked gas, and as a means of determining remnant parameters. The paper presents improved calculations for the cooling rate of a dusty plasma by means of gas–grain collisions, and the equilibrium temperature of the dust for a variety of plasma conditions. These results are then used to define the density–temperature parameter space of the plasma. The analysis is applied to the remnants Cas A and the Cygnus Loop.

063.119   The thermodynamics of anisotropic radiating systems.
A. Fu.
Astrophys. J., Vol. 323, No. 1, p. 211 – 226 (1987).
A formalism is presented which employs the principles of statistical mechanics to approximate the state of a radiation field that is out of equilibrium with the material medium through which it passes. Central to the theory is the statistically constrained angular distribution of the photons and the *statistically independent* energy distribution. The author is addressing the problem of finding a statistical distribution for the *photons*. He argues that the physics underlying the derived specific intensity of the radiation field is valid for gray transport, and he discusses possibilities of adapting the theory to nongray transport. The author shows that the functional form of the distribution function leads to a natural closure of the hierarchy of radiative moment equations at the third angular order. A few examples of currently interesting applications in high–energy astrophysics are briefly discussed. The consequences of this theory for some basic thermodynamic relations, and a similar treatment for neutrino transport, are also discussed.

063.120   Physical constraints on variable Eddington factors for the gray problem in radiative transfer.
A. Fu.
Astrophys. J., Vol. 323, No. 1, p. 227 – 242 (1987).
An analysis of the solutions to the gray radiative transfer problem in planar and spherical geometry reveals important physical features of the anisotropic behavior of the radiation field in light of a theory of radiating systems previously presented by the author. The planar gray problem is solved using the formalism, and numerical solutions to the Kosirev problem are found to lie within the space of solutions given by the theory. The author concludes that the results provide a significant physical basis for certain variable Eddington factors and also provide tools for developing improved techniques of handling radiative transfer flow problems in astrophysics.

063.121   On pumping the strong water maser sources.
N. D. Kylafis, C. Norman.
Astrophys. J., Vol. 323, No. 1, p. 346 – 357 (1987).
The authors have studied the collisional pumping of water masers. They find that the strong water masers can be pumped collisionally in regions where the charged particles are at a significantly higher temperature than that of the neutral particles. Such conditions exist in the precursors of MHD shocks and in massive starburst galaxies and active nuclei. The ionization fraction required for this pumping mechanism to work is roughly equal to the ratio of the typical neutral particle collision rate to that of the charged particles. A physical explanation of why population inversion in H_2O molecules occurs is given under the above conditions and compute the total power emitted. The authors determine the efficiency for masing as a function of ionization fraction and temperature difference between the charged and neutral particles.

063.122   Proton–electron bremsstrahlung polarization: solar flare X–ray and gamma–ray polarization.
D. Heristchi.
Astrophys. J., Vol. 323, No. 1, p. 391 – 398 (1987).

The polarization of proton–electron bremsstrahlung is studied. Appreciable polarization occurs near both the low– and high– energy ends of the photon spectrum. It is also shown that the maximum polarization is obtained in the direction perpendicular to the particle trajectory for low–energy photons. The resulting polarization from a proton spectrum in a thick target is also given. The degree and direction of the polarization of solar flare X–rays are predicted from our model (spiraling protons in the solar magnetic field). These results are discussed in connection with previously measured values.

063.123   Cherenkov damping of surface waves.
A. S. Assis, J. Busnardo–Neto.
Astrophys. J., Vol. 323, No. 1, p. 399 – 405 (1987).

The authors calculate the Cherenkov damping (Landau damping and transit–time magnetic pumping) of waves with an arbitrary wave vector, and in particular that of low–frequency surface waves. It is shown that for surface waves transit–time magnetic pumping is the only collisionless damping mechanism. It is also shown that in some situations surface waves can deposit energy and momentum in the corona via transit–time magnetic pumping as efficiently as kinetic Alfvén waves do via Landau damping.

063.124   Positron annihilation at the Galactic center.
B. J. Carrigan, J. I. Katz.
Astrophys. J., Vol. 323, No. 2, p. 557 – 564 (1987).

The authors calculate the efficiency of production of positrons by γ–γ interaction under conditions which may occur in the Galactic center. Their annihilation line and three–photon annihilation continuum contribute to the emitted radiation. If a geometrical model is assumed, it will be possible to determine the parameters of the γ–ray source region. By comparing the computed spectra to observations a rough estimate is made using extant data.

063.125   Inverse Comptonization by one–dimensional relativistic electrons.
E. Canfield, W. M. Howard, E. P. Liang.
Astrophys. J., Vol. 323, No. 2, p. 565 – 574 (1987).

The authors study gamma–ray spectra produced by the inverse Compton upscattering of soft photons by relativistic electrons with a one–dimensional momentum distribution using the Monte Carlo technique. Such electron distributions may be obtained in strong magnetic fields ($B > 10^{10}$G) when the synchrotron cooling time of transverse energy becomes much shorter than the isotropization time via Coulomb or Compton collisions. Gamma–ray spectra emerging from such Compton upscatterings may be relevant to cosmic gamma–ray bursts if they originate from strongly magnetized neutron stars.

063.126   Effects of electron scattering on the oscillations of an X–ray source.
N. D. Kylafis, G. S. Klimis.
Astrophys. J., Vol. 323, No. 2, p. 678 – 684 (1987).

The authors have determined analytically the time variability observed at infinity due to a variable point source at the center of a spherical cloud of radius R and optical depth to electron scattering τ. Various assumptions for the emission pattern of the source and its time variability are investigated. The results of these calculations reveal the conditions under which quasi– periodic oscillations (QPO) *can* be observed from X–ray sources while periodic oscillations are completely smeared out. Furthermore, one can use the results to study the X–ray oscillations of such sources as Her X–1, Cyg X–3, and the Vela pulsar, which are believed to be embedded in scattering clouds.

063.127   Transmission and reflection operators of radiative transfer equation with aberration and advection terms. II. Line radiation in spherical symmetry.
A. Peraiah.
Bull. Astron. Soc. India, Vol. 15, Nos. 2 – 3, p. 70 – 85 (1987).

A formal solution of the equation of radiative transfer is presented with aberration and advection terms corresponding to the lines included. The operators of reflection and transmission in a radially expanding spherically symmetric shell are derived.

Opacity Project: astrophysical and fusion applications.
See Abstr. 013.024.

Stark broadening trends along homologous sequences.
See Abstr. 022.033.

On the applicability of the Sobolev approximation to the calculation of profiles and integral intensities of emission lines in optically thick slowly expanding media.
See Abstr. 022.080.

Cyclotron radiation in astrophysics. Review.
See Abstr. 061.008.

Unresolved dielectronic satellite lines of Ly α Ca XX resonance lines in high temperature plasmas.
See Abstr. 062.012.

The effect of cooling processes on the tail of a relativistic thermal pair distribution.
See Abstr. 062.029.

Plasma emission: a review.
See Abstr. 062.031.

Group Doppler effect in anisotropic plasmas.
See Abstr. 062.035.

Nonequilibrium populations of rotational levels of emitting molecules in a low–density gas.
See Abstr. 062.048.

An expanding plasma cloud irradiated by gamma–quanta: its structure, optical emission and influence on gamma–burst spectra in the "soft region".
See Abstr. 062.049.

On hydrodynamics of radiatively driven flows.
See Abstr. 062.065.

Relativistic thermal plasmas: time development of electron– positron pair concentration.
See Abstr. 062.100.

Generation of plasma waves by thick–target electron beams, and the expected radiation signature.
See Abstr. 062.122.

Implicit 2D–radiation hydrodynamics.
See Abstr. 062.125.

On the controversy concerning turbulent bremsstrahlung.
See Abstr. 062.157.

Line profiles from moving spherical shells.
See Abstr. 064.008.

The radiation–hydrodynamics of expanding atmospheres.
See Abstr. 064.052.

Polarized radiation from extended magnetic polar caps.
See Abstr. 064.058.

Radiative scattering in ion lines as the main cause of formation and heating of the stellar winds of hot stars.
See Abstr. 064.062.

Polarization line radiative transfer in the atmospheres of magnetic white dwarfs.
See Abstr. 064.073.

Spectral flux from low–density photospheres: low–temperature results.
See Abstr. 064.078.

Effect of Balmer continuum absorption on line emission from Be stars.
See Abstr. 064.079.

OH and CH continuous opacity in solar and stellar atmospheres.
See Abstr. 064.080.

Formation of chromospheric lines in cool dwarf stars.
See Abstr. 064.082.

An inverse Compton scattering model for the spectra of X–ray pulsars.
See Abstr. 067.009.

Inverse Compton scattering in strong magnetic fields: applied to the radiation mechanism of PSR 0531 + 21.
See Abstr. 067.027.

Pair production and Compton scattering in compact sources and comparison to observations of active galactic nuclei.
See Abstr. 067.080.

Electron injection by relativistic protons in active galactic nuclei.
See Abstr. 067.098.

On nonthermal models for active galactic nuclei.
See Abstr. 067.128.

A pulsar emission model: observational tests.
See Abstr. 067.173.

Model of the spectral albedo for atmosphereless cosmic bodies.
See Abstr. 091.026.

Study of some statistical relationships for polarimetric quantities of atmosphereless bodies.
See Abstr. 091.027.

Light scattering in cometary dust comae.
See Abstr. 102.005.

A model for the excitation of water in comets.
See Abstr. 102.016.

Influence of the spectrum's shape on the effect of the "relativistic searchlight" in close binaries.
See Abstr. 117.144.

LTE models of the emission lines of the dwarf nova Z Cha.
See Abstr. 117.201.

On the nonthermal radio emission of double stars with relativistic components.
See Abstr. 117.247.

Radiative transfer in type I supernovae atmospheres.
See Abstr. 125.072.

Expected X–ray radiation spectra of supernova 1987A. Monte–Carlo computations.
See Abstr. 125.240.

Photon–photon pair production and the opacity of SN 1987A to TeV and PeV γ–rays.
See Abstr. 125.313.

Particle acceleration and production of energetic photons in SN 1987A.
See Abstr. 125.314.

The polarization spectrum of supernova 1987A interpreted in terms of shape asymmetry.
See Abstr. 125.315.

X–rays expected from supernova 1987A compared with the source discovered by the Ginga satellite.
See Abstr. 125.352.

Thermal X–ray emission from supernova 1987A.
See Abstr. 125.353.

The effects of inverse Compton scattering on the pulsars' radiation.
See Abstr. 126.017.

Percolation of ionizing photons in an inhomogenous medium.
See Abstr. 131.170.

Are the high–velocity molecular flows really clumpy?
See Abstr. 131.263.

H_2O line emission from shocked gas.
See Abstr. 131.296.

Radio recombination lines from fast shocks in molecular clouds, with application to bipolar flows.
See Abstr. 131.297.

Fluorescent excitation of interstellar H_2.
See Abstr. 131.299.

Dust emission and star formation in compact H II regions.
See Abstr. 132.005.

Relativistic bulk acceleration in blazar jets. Observational indications and model calculations.
See Abstr. 158.110.

Formation of low ionization lines in active galactic nuclei.
See Abstr. 158.123.

A nonthermal model for the XUV, soft X–ray emission of AGN and QSOs.
See Abstr. 158.173.

What heats the hot phase in active nuclei?
See Abstr. 158.348.

Constraints on quasar accretion disks from the optical/ultraviolet/soft X–ray big bump.
See Abstr. 159.116.

New considerations on the broad–line regions of quasars.
See Abstr. 159.140.

The Sunyaev–Zel'dovich effect: measurements and implications.
See Abstr. 160.044.

New results on the Sunyaev–Zel'dovich effect in 0016 + 16, Abell 665 and Abell 2218.
See Abstr. 160.045.

064 Stellar Atmospheres, Stellar Envelopes, Mass Loss, Accretion

064.001   Magnetogasdynamic cylindrical shock wave model in an optically thin atmosphere.
S. N. Ojha.
Astrophys. Space Sci., Vol. 135, No. 1, p. 175 – 185 (1987).

Similarity solutions for one–dimensional unsteady isothermal flow of a perfect gas behind a magnetogasdynamic shock wave including the effects of thermal radiation has been investigated in a uniform thin atmosphere. The flow is caused by an expanding piston and the total energy of the flow is assumed to be constant. Radiation pressure and energy have been neglected in comparison to radiation heat flux and the gas is assumed to be grey and opaque.

064.002   Turbulent heat transport and latitudinal temperature variations in rotating convective envelopes.
L. L. Kichatinov.
Pis'ma Astron. Zh., Tom 13, No. 8, p. 723 – 729 (1987). In Russian. English translation in Sov. Astron. Lett., Vol. 13.

It is shown that allowance for nonlinear effects in the problem of turbulent heat transport in rotating convective envelopes reduces substantially the dependence of temperature on latitude. The evaluated equator–to–pole temperature difference in the solar photosphere is about 1K that does not contradict observational data.

064.003   The wings of the calcium infrared triplet lines in solar–type stars.
G. Smith, J. J. Drake.
Astron. Astrophys., Vol. 181, No. 1, p. 103 – 111 (1987).

Profiles of the infrared triplet lines of ionized calcium (8498, 8542 and 8662 Å) have been calculated in the LTE approximation using model atmospheres representative of solar–type stars. The calculations span the ranges 5000 to 6600K in effective temperature, 4.0 to 4.5 in logarithm of the surface gravity and –0.5 to 0.0 in metallicity (logarithmic abundance of metallic elements relative to the Sun). The depth of absorption in the line wings is found to be particularly insensitive to surface gravity. Provided that the relative abundance of calcium is consistent with the metallicitiy of the model atmosphere, the depth of absorption becomes more sensitive to metallicity with increasing effective temperature. The conclusions have been tested against accurate measurements of infrared triplet line profiles in two bright stars, τ Ceti (G8 V) and η Cas A (G0 V).

064.004   An extension to the wavelength coincidence statistics for spectral line identification.
S. G. Ansari.
Astron. Astrophys., Vol. 181, No. 2, p. 328 – 332 (1987).

The wavelength coincidence statistics method has been extended by including theoretical line strength estimates based on the Kurucz and Peytremann line data and an appropriate model atmosphere. The authors illustrate this new technique with the spectrum of the HgMn star ϕ Her which is based on the average of 11 photographic spectra of reciprocal dispersion of 2.4 Å/mm. The 4 elements for which the WCS method gives the highest significance, viz. Ti II, Mn II, Fe II, and Y II, were singled out for detailed conventional abundance investigation. Only unblended lines which were correctly identified down to the estimated lower identification threshold, were used in the subsequent final analysis. The results are compared to those derived by previous authors.

064.005   Structure and kinematics of stellar wind bubbles.
H. Hanami, S. Sakashita.
Astron. Astrophys., Vol. 181, No. 2, p. 343 – 350 (1987).

The authors investigate the structure and kinematics of stellar wind bubbles for various scales. They classify the stellar wind bubbles into two types by the criterion of whether the radius of the bubble is smaller or larger than the typical scale of the cloud in which the wind producing star is embedded. For the former case, the standard wind bubble model of Weaver et al. (1977) is generalized by taking into account the density gradient of the ambient matter surrounding the star. When the radius of the wind bubble becomes larger than the size of the surrounding cloud, the cloudlets of interstellar matter are engulfed by the hot wind region and then the evaporation of the cloudlets considerably modifies the structure of the wind bubble. This reproduces well the observational data on the kinematic character and the structure of extended ring nebulae.

064.006   Erratum: "Expanding chromospheres of late G and K supergiants" [Mon. Not. R. Astron. Soc., Vol. 222, No. 2, p. 307 – 322 (1986)].
S. V. Mallik.
Mon. Not. R. Astron. Soc., Vol. 227, No. 2, p. 543 (1987). See Abstr. 42.064.033.

064.007   Radiation driven winds of hot luminous stars. III. Detailed statistical equilibrium calculations for hydrogen to zinc.
A. Pauldrach.
Astron. Astrophys., Vol. 183, No. 2, p. 295 – 313 (1987).

Detailed multi–level statistical equilibrium calculations for the dominant ions of H to Zn in hot star stellar winds are presented. The calculations comprise in total 133 ionization stages of 26 elements with altogether 4000 levels and 10,000 radiative bound–bound transitions based on the line list of Abbott (1982). Electron collisions are included and the correct continuous radiation field obtained from the solution of the spherical transfer equation is taken into account for the bound–free transitions. The calculations are applied on two objects, the O4 f–star ζ Pup and the O9.5 V–star τ Sco.

064.008   Line profiles from moving spherical shells.
C. Bertout, C. Magnan.
Astron. Astrophys., Vol. 183, No. 2, p. 319 – 323 (1987).

Spherically symmetric moving shells that surround stars create very different spectral line profiles, depending on the local line width, the velocity gradient, and the emitting layer's geometrical extension and optical thickness. Analytical formulae are given for deriving the main profile characteristics in several limiting cases. An investigation follows of the conditions required for a transition from a quasi–rectangular or parabolic profile to a bi–triangular (saw–toothed) profile with a flux minimum at line center.

064.009   Shock waves in luminous early–type stars.
J. I. Castor.
Instabilities in luminous early type stars, p. 159 – 173 (1987). – See Abstr. 012.012.

Shock waves that occur in stellar atmospheres have their origin in some hydrodynamic instability of the atmosphere itself or of the stellar interior. In luminous early–type stars these two possibilities are represented by shocks due to an unstable radiatively–accelerated wind, and to shocks generated by the non–radial pulsations known to be present in many or most OB stars. This review is concerned with the structure and development of the shocks in these two cases, and especially with the mass loss that may be due specifically to the shocks.

064.010   Radiation driven instabilities.
G. B. Rybicki.
Instabilities in luminous early type stars, p. 175 – 189 (1987). – See Abstr. 012.012.

Various radiation driven instabilities have been shown to operate in the atmospheres and winds of early–type luminous stars. The strongest of these occur in the supersonic parts of the winds, where as many as one hundred e–folds of linear growth can occur during a typical outflow time. The nonlinear growth of such instabilities can possibly account for the observed superionization and X-ray emission in these stars. Developments in the linear theory of these instabilities is reviewed.

064.011   Stellar instabilities in the upper part of the Hertzsprung–Russell diagram.
C. de Jager.
Instabilities in luminous early type stars, p. 191 – 198 (1987). – See Abstr. 012.012.
Summary of the workshop on "Instabilites in luminous early type stars".

064.012   The influence of photospheric turbulence on stellar mass loss.
C. de Jager, H. Nieuwenhuijzen.
Instabilities in luminous early type stars, p. 267 – 268 (1987). – See Abstr. 012.012.
The authors show that the stellar rate of mass loss is positively correlated with the average microturbulent photospheric velocity, and that the energy contained in the microturbulent motions is of the same order of magnitude as the wind energies.

064.013   Nonlinear dynamics of instabilities in line–driven stellar winds.
S. P. Owocki, J. I. Castor, G. B. Rybicki.
Instabilities in luminous early type stars, p. 269 – 272 (1987). – See Abstr. 012.012.
The authors have been developing a numerical radiation-hydrodynamics program in order to study the nonlinear evolution of instabilities in line–driven winds from luminous, early-type stars. Initial tests of the code indicate that the velocity structure of nonlinear pulses in such a wind may be quite different than assumed in previous analyses.

064.014   A model atmosphere and the chemical composition for the Ap star 73 Dra.
L. S. Lyubimkov.
Bull. Crimean Astrophys. Obs., Vol. 75, p. 143 – 150 (1987). English translation of 42.064.049.

064.015   Instability of isentropic geometrically thin disks due to corotation resonance.
S. Kato.
Publ. Astron. Soc. Jpn., Vol. 39, No. 4, p. 645 – 666 (1987).
Dynamical instability of isentropic, geometrically thin disks against nonaxisymmetric perturbations is examined by use of the WKBJ method, mainly when the disks are semiinfinite [i.e., they have a reflecting edge on one (inner or outer) side but have no such boundary on the other side]. The author's attention is restricted to adiabatic perturbations with no node in the vertical direction. The results show that the semiinfinite disks are unstable, if the reflecting edge is on the side (of the corotation radius) where overreflection occurs. If the edge is on the opposite side, the disks are stable. These results imply that waves amplified between the overreflecting corotation resonance and the reflecting edge really lead the system to instability, if the other side of the disk has no boundary.

064.016   Astrophysical jets in SS433.
J. Fukue.
Publ. Astron. Soc. Jpn., Vol. 39, No. 4, p. 679 – 683 (1987).
Hydrodynamical wind–type jets emanated from the funnel formed by an accretion torus account well for the velocity field and temperature distribution of the SS433 jets obtained by recent X–ray and optical observations. The jets attain their terminal velocity before exiting the funnel – the confinement region – and have constant velocity through the X–ray and optical–line emitting regions. Beyond the exit of the funnel the temperature of jets drops quickly due to adiabatic expansion. The sonic point is located deep inside the funnel at about 26 Schwarzschild radii from the central object. Hence, the jets are highly supersonic in the observed line–emitting regions.

064.017   Non–LTE model atmospheres for central stars of planetary nebulae.
R. E. S. Clegg, D. Middlemass.
Mon. Not. R. Astron. Soc., Vol. 228, No. 3, p. 759 – 778 (1987).
A grid of hot non–LTE model atmospheres with H and He opacity sources has been calculated. The emergent fluxes, together with integrals required for Zanstra and Energy–Balance methods, are presented. These are needed for the analyses and modelling of planetary nebulae. The grid spans the following ranges: T_{eff}: $40 – 180 \times 10^3$K, $\log g$: $4.0 – 8.0$, He/H: 10 and 1 per cent by number. The low helium abundance models are included to study the effects of gravitational settling in some central stars. The grid of models has been chosen so that the models cover adequately, in the temperature–gravity plane, Schönberner's evolutionary tracks for central stars. Some results on the non–LTE populations of He$^+$ levels and the formation of the He II continuum are described.

064.018   Stellar winds and magnetic fields in the helium peculiar stars.
S. N. Shore.
Astron. J., Vol. 94, No. 3, p. 731 – 736 (1987).
This note discusses the consequences of the interaction of radiatively driven mass loss with magnetic fields in upper–main-sequence helium–peculiar stars. The field constrains mass loss to occur in narrow cones above the magnetic poles. These jets can explain the observed behavior of UV resonance lines in the helium–peculiar stars, expecially the sn and helium–rich stars. Suppression of mass outflow at the magnetic equator produces a region in which radiatively driven diffusive mass transport can levitate helium in the hottest stars, producing helium enrichment of the atmosphere. In the cooler stars, the surface becomes progressively more helium poor.

064.019   On the possibility of interpretation of Mg II profiles in hybrid stars.
M. Jasiński.
Acta Astron., Vol. 37, No. 1, p. 1 – 15 (1987).
Semiempirical models of expanding chromospheres have been constructed for five hybrid stars by fitting of the computed Mg II profiles to the observational data. The mass loss rates obtained from these models range from 1.34×10^{-9} to $3.14 \times 10^{-7} M_\odot$/year. On the basis of energetic considerations a possible mechanism accelerating winds in the chromospheres of hybrid stars is indicated.

064.020   The curve of growth for stars in the effective temperature range 4000 – 8500K.
M. Villada, L. Rossi.
Astrophys. Space Sci., Vol. 136, No. 2, p. 351 – 361 (1987).
A comparison of different curves of growth derived from available model atmospheres has been done. The authors found that, in the effective temperature range (4000 – 8500K), the differences between models with the same $(T_{eff}, \log g)$ from different authors, do not affect the shape of the curve of growth. Moreover, the shifts needed to superpose the curves, for different elements, differ only slightly when models from different authors are used. Then, in the low–temperature and for the gravities the authors considered, the particular model atmosphere used to draw the curve of growth is not the main constrain of the method.

064.021   Structure of the atmospheres of cool giants.
N. S. Komarov.
Stellar atmospheres, p. 2 – 16 (1987). In Russian. Abstr. in Ref. Zh., 51. Astron., 8.51.385 (1987). – See Abstr. 003.005.

064.022   Thermochemical equilibrium in the atmospheres of cool stars. Molecules.
N. S. Komarov, V. V. Tsymbal.
Stellar atmospheres, p. 63 – 82 (1987). In Russian. Abstr. in Ref. Zh., 51. Astron., 8.51.386 (1987). – See Abstr. 003.005.

064.023   Analysis of the atmospheres of K–giants.
T. V. Mishenina, N. S. Komarov, V. E. Panchuk.
Stellar atmospheres, p. 17 – 22 (1987). In Russian. Abstr. in Ref. Zh., 51. Astron., 8.51.387 (1987). – See Abstr. 003.005.

064.024   Ionization equilibrium and accuracy of determining the chemical abundance of the atmospheres of K–giants.
Ya. V. Pavlenko, A. V. Shavrina.
Stellar atmospheres, p. 23 – 29 (1987). In Russian. Abstr. in Ref. Zh., 51. Astron., 8.51.388 (1987). – See Abstr. 003.005.

064.025   The influence of a shock–heated zone on the physical conditions in stellar atmospheres.
S. M. Andrievskij, G. A. Garbuzov, I. S. Savanov.
Stellar atmospheres, p. 153 – 169 (1987). In Russian. Abstr. in Ref. Zh., 51. Astron., 8.51.389 (1987). – See Abstr. 003.005.

064.026   Location of the temperature minimum in the atmosphere of an M–giant depending on the chromosphere and theoretical profiles of strong Mg I lines.
Ya. V. Pavlenko.
Stellar atmospheres, p. 41 – 47 (1987). In Russian. Abstr. in Ref. Zh., 51. Astron., 8.51.390 (1987). – See Abstr. 003.005.

064.027   Retrothermal effect in the atmospheres of carbon stars.
D. N. Dojkov.
Stellar atmospheres, p. 83 – 87 (1987). In Russian. Abstr. in Ref. Zh., 51. Astron., 8.51.392 (1987). – See Abstr. 003.005.

064.028   Nonlinear selfsimilar problems of the non–stationary disc accretion.
Yu. Eh. Lyubarskij, N. I. Shakura.
Pis'ma Astron. Zh., Tom 13, No. 10, p. 917 – 928 (1987). In Russian. English translation in Sov. Astron. Lett., Vol. 13.
Similarity solutions of a fundamental equation describing non–stationary disc accretion are obtained.

064.029   Dust formation in stellar winds.
H.–P. Gail, E. Sedlmayr.
Physical processes in interstellar clouds, p. 275 – 303 (1987). – See Abstr. 012.027.
Some observational constraints for circumstellar dust condensation and the chemical composition of the gas phase prior to dust condensation are discussed. The principle methods for treating dust condensation and the chemical pathways for dust formation in cases with M–stars and C–stars are outlined.

064.030   Mass–loss of globular cluster red giants. A semi–empirical estimation.
C. Martinez Roger, E. Paez.
Astron. Astrophys., Vol. 184, No. 1/2, p. 155 – 163 (1987).
Limits to the total mass–loss in red giants in globular clusters are calculated on the basis of Sweigart and Gross (1976) Horizontal Branch models by means of Sweigart and Gross (1978) Red Giant Branch stellar models. The upper limit is obtained through the condition that the stars do not leave their nuclei bare. The lower limit results from the requirement that, when reaching the Zero Age Horizontal Branch, they do not exceed the giant branch to the right. Both limits are calculated for several values of the metal and helium contents and age of the galactic globular cluster system. Mass–loss rates are also estimated using the Sweigart and Gross (1978) life times of red giants; the results obtained are in the range $0 < dM/dt < 6 \times 10^{-8} M_\odot \mathrm{yr}^{-1}$. From the present results it ensues that, for fixed chemical composition, older systems are associated with lower total mass–loss. The helium content is also inversely related to the total mass–loss. Finally, the opposite applies for the metallicity of the clusters: higher metal content is positively correlated with higher total mass–loss and higher mass–loss rates.

064.031   Stationary shocks in accretion disks.
H. C. Spruit.
Astron. Astrophys., Vol. 184, No. 1/2, p. 173 – 184 (1987).
Special solutions of the equations of motion and continuity near the midplane of a thin accretion disk are obtained by analytic and numerical means. They have the form of stationary, self–similar flows containing two or more spiral shaped shock waves. They represent shocks initiated by a disturbance at the outer edge of the disk. Self–similar shocks excited at the center of the disk do not seem to exist. No solutions were found containing only one shock. Adiabatically accreting solutions exist in which there is a unique relation between the ratio of specific heats γ and the opening angle of the shock. When radiative losses from the disk are included, solutions exist for γ above a certain minimum value. The strength of the shock decreases with decreasing disk thickness. An effective "α parameter" can be defined, it varies with disk thickness like $(h/r)^{1.5}$. An analogy between the present problem and the case of spherically symmetric accretion is discussed.

064.032   Radiation–driven winds of hot luminous stars. IV. The influence of multi–line effects.
J. Puls.
Astron. Astrophys., Vol. 184, No. 1/2, p. 227 – 248 (1987).
The paper presented here investigates an effect neglected in prior calculations of radiation driven wind models in this series of publications – the influence of multi–line effects. These effects arise through the possibility that photons can be processed more than once in different transitions when propagating through the wind and are essential for hot stars as both the line density and the line strengths are large in the UV. Three different methods – the Sobolev–, comoving frame and Monte–Carlo–approach – have been used to investigate the processes as a whole and some approximations. The influence of these effects is calculated for O–star–models. It turns out that the treatment of the line force in the author's prior calculations overestimates this quantity and that the actual line force is only of order 70% to 90% compared to the former results. With the modified line force, both mass loss rate and terminal velocity are reduced in selfconsistent wind models (ζ Pup, τ Sco) compared with the results of Pauldrach (1987).

064.033   Physical properties of stellar wind flow from hot stars.
Eh. Ya. Vil'koviskij.
Astrofizika, Tom 27, Vyp. 1, p. 41 – 51 (1987). In Russian. English translation in Astrophysics, Vol. 27, No. 1.
On the basis of an analysis of the radiation–driven flow equations, taking into account the "kinetic" heat, the principal characteristics of stellar winds from OB stars are considered.

064.034   Accretion of matter on degenerate hydrogen–helium very low–mass dwarfs.
A. V. Fedorova, V. S. Imshennik.
Nauchn. Inf., Vyp. 63, p. 145 – 151 (1987). In Russian.
Accretion of matter on degenerate hydrogen–helium stars of initial masses 0.005 and 0.01 $M_\odot$ is investigated. Calculations show that in the case when deuterium is present in the accreted matter, deuterium burning with strong screening results in rapid expansion of the accreting star. In the course of subsequent evolution the radius of the accreting star is much more than equilibrium one. In the case when deuterium is absent in the accreted matter, the expansion takes place on the later stage of evolution due to hydrogen burning with strong screening. Before hydrogen burning the accretion results in very strong contraction of the star.

064.035   Rotating stellar atmospheres.
J. P. Cassinelli.
Physics of Be stars, p. 106 – 122 (1987). – See Abstr. 012.030 (IAU Colloq. No. 92).
The effects of rapid rotation on the emergent energy distribution, line profiles, atmospheric motions and polarization are discussed. A simplified explanation of some of the effects is presented. Theoretical models for the intrinsic polarization of

Be stars are critically discussed. It is stressed that polarization is a powerful diagnostic for determining the asymmetrical structure of the outer atmospheres of the Be stars.

064.036 Rotationally–enhanced stellar winds.
J. M. Marlborough.
Physics of Be stars, p. 316 – 338 (1987). – See Abstr. 012.030 (IAU Colloq. No. 92).
The rotationally–enhanced stellar wind model for Be stars represents one attempt to understand many of the phenomena related to Be stars in terms of a stellar wind distorted and perhaps also enhanced by rapid stellar rotation. This review concentrates exclusively on this particular approach.

064.037 The spheroidal/ellipsoidal, variable mass–loss, decelerated Be star model.
V. Doazan.
Physics of Be stars, p. 384 – 410 (1987). – See Abstr. 012.030 (IAU Colloq. No. 92).
The author presents the basic observational characteristics that any Be star model must be able to represent in a self–consistent way. Then, he discusses the basic thermodynamic conditions underlying Be star modelling, and defines, in a quantitative way, the radial sequence of regions which describes the entire Be star atmosphere.

064.038 Magnetic–loop model for Be stars.
A. B. Underhill.
Physics of Be stars, p. 411 – 427 (1987). – See Abstr. 012.030 (IAU Colloq. No. 92).
The magnetic–loop model for a stellar mantle is described and it is shown how this model may be used to understand the differences in spectrum and change of spectrum which are observed for main–sequence stars having effective temperatures in the range from 2.5×10^4 to 3.0×10^4K. It is suggested that the difference in spectrum between a shell star such as Gamma Cas when it is in an active state and a Wolf–Rayet star may be due to a difference in environment.

064.039 Profiles of emission lines from rotating disks.
D. A. Brown.
Physics of Be stars, p. 428 – 430 (1987). – See Abstr. 012.030 (IAU Colloq. No. 92).
Profiles of emission lines contain key information on the dynamics, the physical condition, and the geometry of the cool envelopes of Be stars. The basic rotating model is reconsidered and profile formation studied from a physical point of view.

064.040 Discrete components in OB and Be stars: the shocking truth?
P. K. Barker.
Physics of Be stars, p. 431 – 433 (1987). – See Abstr. 012.030 (IAU Colloq. No. 92).
Preliminary calculations of the absorption profiles produced if multiple shells are ejected from the photosphere into a non–monotonic wind, suggest that this geometrical model can reproduce a wide variety of simple and complex profiles observed in the discrete components in OB and Be stars.

064.041 Consistent spherical NLTE–models for Be–stars.
P. Höflich.
Physics of Be stars, p. 434 – 436 (1987). – See Abstr. 012.030 (IAU Colloq. No. 92).
Spherical NLTE–model photospheres surrounded by envelopes are calculated in order to interpret classical Be–stars. The existence of the H II–region can be understood by such models. The dependence of the calculated spectra on parameters corresponds to the observations. Applications to special objects show good agreement between the predicted and observed spectra in the visual and infrared wavelength range.

064.042 A rotating, magnetic, radiation–driven wind model applied to Be stars.
C. H. Poe, D. B. Friend.
Physics of Be stars, p. 437 – 439 (1987). – See Abstr. 012.030 (IAU Colloq. No. 92).
With their rotating, magnetic, radiation–driven wind model, Friend & MacGregor (1984) found that rapid rotation and an open magnetic field could enhance the mass loss rate and terminal velocity in an O star wind. The purpose of this paper is to see if this model could help explain the winds from Be stars.

064.043 Mass accretion processes in magnetic fields: formation of quasi–Keplerian discs.
O. Kaburaki.
Mon. Not. R. Astron. Soc., Vol. 229, No. 2, p. 165 – 182 (1987).
An axisymmetric, steady–state solution is obtained for geometrically thin accretion discs in external magnetic fields. The main features of this solution are as follows. The azimuthal velocity of a disc plasma is somewhat reduced from the Keplerian value owing to the pressure effect enhanced by the inclusion of a magnetic field. The magnetic stress takes the place of viscous stress in the standard disc model, and extracts angular momentum from the disc. About a half of the gravitational energy is released in the disc, through the Joule dissipation and the work done against the pressure force. The vertical flow and current are also included in the calculation. Analogously to the parameter α in the standard model, the author's solution contains essentially one parameter $\varDelta$ which specifies the size of the electrical resistivity.

064.044 NLTE models for cocoon stars.
P. Höflich, R. Wehrse.
Astron. Astrophys., Vol. 185, No. 1/2, p. 107 – 116 (1987).
The authors calculate NLTE model photospheres surrounded by spherical envelopes with density distribution $\varrho \sim r^{-n}$ in order to interpret cocoon stars similar to the Becklin–Neugebauer object. Both parts are assumed to be in radiative equilibrium. With these models the existence of extended H II regions around stars with luminosities of B main sequence stars can be understood as a consequence of photoionizations by Balmer continuum photons, bound–bound collisions, large optical depths in hydrogen lines due to small escape probabilities of photons, and high photon densities in the optically thin continua. For several relatively well observed objects the authors derive the atmospheric and shell parameters. The observed line fluxes and line ratios are in general well reproduced. Calculated and observed radio fluxes are consistent.

064.045 The dynamical instability of a rotating cylinder as a model for a Keplerian disk.
T. Hanawa.
Astron. Astrophys., Vol. 185, No. 1/2, p. 160 – 164 (1987).
Dynamical instabilities are investigated for a cylinder whose angular velocity $\varOmega$ is proportional to $r^{-1.5}$ where r is the radial distance. The eigenfunctions and the growth rates of the instability are computed. The cylinder is found to be unstable against non–axisymmetric perturbations. The rotation law adopted here is that of a Keplerian disk, i.e., a geometrically thin disk rotating around a compact object. The results imply that such disks are unstable for all ratios of the outer to the inner radius in excess of some small value. For the cylinder considered here, the critical value was 1.4. The amplitude of the tangential displacement of the eigenfunctions is large at the corotation point where the mean velocity is equal to the pattern velocity of the perturbation. Thus, it is concluded that this instability is due to the corotation resonance.

064.046 Line blanketed model atmospheres of Ap–stars. VI. HD 221568.
K. Stepień, H. Muthsam.
Astron. Astrophys., Vol. 185, No. 1/2, p. 225 – 228 (1987).
Visual and UV spectroscopic observations of HD 221568 together with line blanketed model atmospheres show that the star has an effective temperature of 10300K if $\log g = 3.7$ is

assumed. The temperature may be lower by about 100K in the phase of reddest $B-V$. Models with variable chemical composition are not able to reproduce observed light variations, particularly a large decrease of flux between the Balmer jump and about 5000 Å observed in the "red" phase. It is suggested that this decrease may be due to numerous lines of rare earth elements not included in the computations. To explain observed light variations a small decrease of effective temperature in the "red" phase accompanied by an increase of the visible disk and increased line blanketing are suggested.

064.047 Models for stellar coronae: thin coronae with radiative forces.
A. G. Hearn.
Astron. Astrophys., Vol. 185, No. 1/2, p. 247 – 252 (1987).

Models are calculated for small coronae heated by saw tooth waves with radiative forces acting in the cool region above the corona. The radiative forces are introduced in a parameterized form. The mass loss rates obtained in the models are orders of magnitude lower than the mass loss rates observed in OB supergiants. Attempts to produce models with higher mass loss rates failed.

064.048 Kinematic structure of OH/IR stars.
J. Sun, S. Kwok.
Astron. Astrophys., Vol. 185, No. 1/2, p. 258 – 266 (1987).

A kinematic model is constructed for 1612 MHz OH maser emssion in OH/IR stars. The spatial distributions of OH maser intensity are calculated from a model of spherically–symmetric uniformly–expanding circumstellar shell. By comparing VLA/VLBI maps of OH/IR stars with model results, the acceptable range of combination of physical parameters $\dot{M}/V_e$, $(n_{H2})_{max}$, $(n_{H2})_{min}$, $f_{OH} = [n_{OH}]/[n_{H2}]$ are derived. The theoretical relations between OH shell radius R_0 and mass loss rate $\dot{M}$ and between OH maser luminosity L_{OH} and $\dot{M}$ are also obtained. These relations are in good agreement with empirical relations established by Bowers et al. (1983) and Baud et al. (1983). The ranges of $(n_{H2})_{max}$ and $(n_{OH})_{min}$ under different $\dot{M}$ required for operating saturated 1612 MHz OH masers are also discussed. The authors find that the OH emission phase can last over 1000 years after the termination of the asymptotic giant branch and many protoplanetary nebulae may have the characteristics of OH/IR stars.

064.049 Model study of wavelength–dependent limb–darkening and radii of M–type giants and supergiants.
M. Scholz, Y. Takeda.
Astron. Astrophys., Vol. 186, No. 1/2, p. 200 – 212 (1987).

Wavelength–dependent center–to–limb variations are investigated from model photospheres of superluminous M type stars. Normal giants and supergiants with fairly compact photospheres have smooth limb–darkening curves which, however, are not always well approximated by a linear limb–darkening law. Center–to–limb variation in low–mass, high–luminosity non–Miras with spherically extended photospheres and in Miras may show a complicated behaviour in diverse filters used by observers. The interpretation of a measured monochromatic radius and its correlation to the stellar radius defining the effective temperature is not trivial. Selected observed lunar occultation and speckle interferometry radii are discussed.

064.050 Improved NLTE profiles of He II lines in hot stars including their overlap with hydrogen.
A. Herrero.
Astron. Astrophys., Vol. 186, No. 1/2, p. 231 – 240 (1987).

The Accelerated Lambda Iteration method is used to obtain better theoretical profiles of He II in hot stars by considering more levels and Stark broadening in the statistical equilibrium equations. It is found that an appropriate He II model atom must consist of 14 NLTE levels and that Stark broadening must be included for all lines with upper level $n_{up} \leqslant 7$. The new profiles are deeper than those given by former calculations and can modify the spectroscopical analysis when using high quality observational data. With the same method, the problem of the overlapping of hydrogen and ionized helium lines is solved in a

quantitative way for the first time. It is found that this overlapping has no importance in stellar atmospheres. It cannot explain, even partially, the $\lambda 4686$ emission, nor has it influence over the optical profiles of both ions. Thus, the line formation calculations of H and He II may be performed separately, as has been done until now.

064.051 Shock formation from the nonlinear evolution of instabilities in line–driven stellar winds.
S. P. Owocki, J. I. Castor, G. B. Rybicki.
Bull. Am. Astron. Soc., Vol. 19, No. 2, p. 702 (1987). Abstract. – See Abstr. 010.061.

064.052 The radiation–hydrodynamics of expanding atmospheres.
D. C. Abbott.
Bull. Am. Astron. Soc., Vol. 19, No. 2, p. 705 (1987). Abstract. – See Abstr. 010.061.

064.053 Opacity sampled, spherically extended model atmospheres for cool stars.
J. A. Brown, H. R. Johnson, D. R. Alexander, R. Wehrse.
Bull. Am. Astron. Soc., Vol. 19, No. 2, p. 705 (1987). Abstract. – See Abstr. 010.061.

064.054 Demonstration of the Kelvin–Helmholtz instability in accretion disks.
J. E. Tohline, I. Hachisu.
Bull. Am. Astron. Soc., Vol. 19, No. 2, p. 727 (1987). Abstract. – See Abstr. 010.061.

064.055 Evolutionary helium and CNO anomalies in the atmospheres and winds of massive hot stars.
N. R. Walborn.
Space Telesc. Sci. Inst., Prepr. Ser., No. 202, 12 pp. (1987). To appear in IAU Colloq. No. 108.

064.056 Accretion disks in symbiotic stars.
W. J. Duschl.
MPA Rep., No. 309, 12 pp. (1987). To appear in IAU Colloq. 103.

064.057 Reconnection–driven oscillations in dwarf nova disks.
T. Tajima, D. Gilden.
Astrophys. J., Vol. 320, No. 2, p. 741 – 745 (1987).

A class of oscillations observed during eruption of dwarf novae has been interpreted as oscillations of the accretion disks in these systems. These oscillations are quasi–periodic with coherence times typically between three and 15 cycles. The authors show that magnetic field reconnection at high magnetic Reynolds number ($R_m > 10^6$) can drive disk oscillations. The expected stochastic geometry of disk magnetic fields could naturally produce the observed phase incoherency.

064.058 Polarized radiation from extended magnetic polar caps.
G. Chanmugam, J. Frank.
Astrophys. J., Vol. 320, No. 2, p. 746 – 755 (1987).

X–ray studies suggest that in DQ Her binaries the accretion flow lands on a large polar cap. The authors have calculated, for the first time, the cyclotron emission from such extended regions taking into account the magnetic field distribution (assumed dipolar) and self–eclipses. The method is based on the additivity of Stokes parameters for incoherent beams. The results are qualitatively consistent with the polarimetric observations of the DQ Her binary 3A 0729 + 103 provided that the white dwarf has a sufficiently strong field (~ 10 MG) and supports the suggestion that DQ Her binaries have large polar caps.

064.059 A new formulation of the starspot model, and the consequences of starspot structure.
J. D. Dorren.
Astrophys. J., Vol. 320, No. 2, p. 756 – 767 (1987).

The author presents a new formulation of the starspot model for circular spots, derived by direct surface integration over the spots. The resulting expression explicitly exhibits the dependence on the two independent parameters of the problem: α, the angular spot radius, and β, the angle between the surface normal at the spot center and the line of sight. This formulation is extended to include the possibility of spots with concentric circular umbral and penumbral regions and is then employed to investigate the consequences of such a starspot structure.

064.060 Stellar mass loss and atmospheric instability.
C. de Jager, H. Nieuwenhuijzen.
Working Group Stars with Extended Atmospheres, Prepr., No. 104, 12 pp. (1987).

064.061 Turbulence–driven atmospheric instability and large–scale motions in super– and hypergiants.
B. Boer, J. Carpay, A. de Koter, C. de Jager, H. Nieuwenhuijzen, A. Piters, F. Spaan.
Working Group Stars with Extended Atmospheres, Prepr., No. 104, 2 pp. (1987).

064.062 Radiative scattering in ion lines as the main cause of formation and heating of the stellar winds of hot stars.
Eh. Ya. Vil'koviskij.
Tr. Astrofiz. Inst. Alma–Ata, Tom 48, p. 62 – 66 (1987). In Russian.

064.063 On the theory of microstructure of the condensation layer in the atmosphere of a white dwarf.
V. G. Zubko.
Inst. Teor. Fiz. Akad. Nauk USSR, Prepr., No. 52R, p. 1 – 17 (1987). In Russian. Abstr. in Ref. Zh., 51. Astron. 11.51.544 (1987).

064.064 The photosphere of T Tauri stars.
N. Calvet, Z. Marín.
Rev. Mex. Astron. Astrofis., Vol. 14, No. 1, p. 353 – 358 (1987). – See Abstr. 012.042.
The authors have calculated the absorption spectrum in the wavelength interval 4880 – 5025 Å for a set of models with $T_{eff} = 4000K$ and log g = 3.5 and a chromospheric temperature rise. These models are considered as representative of the atmosphere of T Tauri stars. The position of the temperature minimum goes from 0.1 to 2 gr cm^{-2} in the models. Populations are in LTE and the two level plus continuum approximation is used for the source function. The authors compare the calculated spectra with those observed in the program of rotational velocity determination by Vogel and Kuhi (1981).

064.065 Modifications and erosion of circumstellar matter by wind particles.
V. Pironello.
Nucl. Instrum. Methods Phys. Res., Sect. B, Vol. B19–20, Part 2, p. 959 – 962 (1987). Abstr. in Phys. Abstr., Vol. 90, No. 1310, Entry 94861 (1987).

064.066 Episodic mass loss in late–type stars due to acoustic wave packets.
M. Cuntz.
Astron. Astrophys., Vol. 188, No. 1, p. L5 – L8 (1987).
Non–linear time–dependent acoustic wave calculations have been undertaken in an atmosphere model similar to α Boo. The model was chosen to represent conditions in a red giant beyond the Linsky–Haisch division line. It is considered that the wave period changes stochastically in the short period range. The distribution of the periods is assumed as Gaussian around 1.4×10^4s. The calculations predict a time–averaged period of 1.27×10^5s which is in excellent agreement with observations. It is found that stochastically recurring supersonic flows are generated which lead to episodic mass loss.

064.067 Singly ionized iron as a diagnostic of stellar envelopes. I. The methods.
M. Friedjung, G. Muratorio.
Astron. Astrophys., Vol. 188, No. 1, p. 100 – 108 (1987).
Emission and absorption lines of Fe II formed outside LTE frequently dominate the optical and ultraviolet spectra of luminous blue stars and other astrophysical objects. A new method of emission line analysis is proposed with the aim of easily deriving physical information about the emitting region. In spite of the simple assumptions made, it is possible with this method to conclude that for some stars absorption and emission do not take place in the same medium and to postulate the presence of a disk–like structure around these stars.

064.068 Thermal–convective instability of a composite plasma in a stellar atmosphere with rotation and magnetic field.
K. C. Sharma.
Astrophys. Space Sci., Vol. 137, No. 2, p. 297 – 302 (1987).
Thermal–convective instability of a composite plasma in a stellar atmosphere, when the effects of a uniform rotation and a uniform magnetic field are included simultaneously, is discussed. It is found that the criterion for monotonic instability is the same as in the absence or presence (separately) of these two effects.

064.069 Evidence of the connection between internal magnetic fields and chromospheric activity in late–type stars.
B. Montesinos, M. J. Fernández–Figueroa, E. de Castro.
Mon. Not. R. Astron. Soc., Vol. 229, No. 4, p. 627 – 641 (1987).
The authors try to establish a connection between the dynamo effect, which operates in the convection zone of late–type stars, and the chromospheric activity observed in these objects. They estimate in a theoretical way, for a sample of main–sequence active stars with spectral types G and K, the magnetic field generated in a region close to the base of the convection zone and the filling factor (fraction of stellar surface covered by strong magnetic fields). The analysis of the behaviour both of the filling factor and the activity indicator R(Ca II + Mg II) against several stellar parameters ($B–V$ and the Rossby number) supports the existence of a link between the interior processes and the surface activity.

064.070 Magnetically collimated winds from accretion disks.
T. Sakurai.
Publ. Astron. Soc. Jpn., Vol. 39, No. 5, p. 821 – 835 (1987).
A steady, axisymmetric wind from a magnetized accretion disk is studied by applying a numerical scheme developed for the stellar wind modeling (Sakurai, 1985). As in the magnetic stellar winds, the collimation of the wind toward the rotation axis is seen in the wind from a disk. The driver of the collimation is the toroidal magnetic field which develops in the wind due to the rotation of the disk. Magnetic winds from rotating disks can therefore naturally lead to the formation of collimated jets, which are found in star–forming regions and in extragalactic radio sources.

064.071 Radiatively driven winds from magnetic, fast–rotating stars: Wolf–Rayet stars?
S. Nerney, S. T. Suess.
Astrophys. J., Vol. 321, No. 1, p. 355 – 369 (1987).
An analytical procedure is developed to solve the magnetohydrodynamic equations for a stellar wind in the strong magnetic field, optically thick limit for hot stars. The slow–mode, Alfvén, and fast–mode critical points are modified by the radiation terms in the force equation. The magnetic field and a radiation parameter specify the terminal wind velocity. High rotation rates and a modified slow mode critical point close to the stellar surface determine the high mass–loss rates. Wolf–Rayet stars are modeled with 1000 G fields but require stellar rotational velocities approaching breakup values.

064.072 Equilibrium temperature of ice grains formed around a star as a function of stellar parameters.
J. F. Crifo.
J. Phys. Colloq., Vol. 48, No. C–1, p. 587 – 592 (1987). Abstr. in Phys. Abstr., Vol. 90, No. 1314, Entry 120380 (1987).

064.073 Polarization line radiative transfer in the atmospheres of magnetic white dwarfs.
K. N. Nagendra.
Astron. Nachr., Vol. 308, No. 5, p. 303 – 312 (1987).
Some physical mechanisms which can affect the Zeeman line profiles of magnetic white dwarfs are studied. The pure absorption polarization transfer equation is solved including these mechanisms. The broadening of lines in magnetic white dwarfs is briefly discussed.

064.074 Mechanisms for mass loss from cool stars.
M. Morris.
Publ. Astron. Soc. Pac., Vol. 99, No. 621, p. 1115 – 1122 (1987).
= Invited paper presented at the Symposium on cool stars and the structure of galaxies, at the 99th Annual Scientific Meeting of the Astronomical Society of the Pacific, at Pomona College, Claremont, California, July 1987.
The mechanisms believed responsible for the loss of mass from cool, red giant stars are reviewed. The current picture involves a two–step process wherein pulsations act to levitate matter well above the photosphere to the point at which the gas is sufficiently cool for dust grains to form. Radiation pressure on the dust then drives the matter to infinity. Whereas this model is applicable to spherically symmetric mass loss, the outflowing matter in many mass–losing systems displays a pronounced bipolarity, implying axial symmetry on the large scale. A secondary star appears to be responsible for the geometry of such systems. A new scenario involving two winds is presented to describe how the bipolar geometry might be produced.

064.075 Molecules and the structure of spherically extended model atmospheres.
J. A. Brown, H. R. Johnson, D. R. Alexander, R. Wehrse.
Publ. Astron. Soc. Pac., Vol. 99, No. 621, p. 1146 (1987). Abstract. – See Abstr. 010.281.

064.076 Wave energy in white dwarf atmospheres. I. Magnetohydrodynamic energy spectra for homogeneous DB and layered DA stars.
Z. E. Musielak.
Astrophys. J., Vol. 322, No. 1, p. 234 – 247 (1987).
The generation and propagation of the magnetohydrodynamic (MHD) waves in white dwarf atmospheres is considered. The MHD waves are treated in a one–dimensional approach, and radiative as well as viscous damping is taken into account. Theoretical expressions for acoustic and magnetic energy fluxes are obtained as a function of wave frequency, magnetic field, and multipole coefficients. The results of the calculations are presented as the MHD energy spectra obtained for layered DA and homogeneous DB white dwarfs, in a wide range of effective temperatures and masses.

064.077 Selective mass loss, abundance anomalies, and helium–rich stars.
G. Michaud, J. Dupuis, G. Fontaine, T. Montmerle.
Astrophys. J., Vol. 322, No. 1, p. 302 – 314 (1987).
In the presence of mass loss, the He abundance in the line–forming region is modified by the chemical separation that takes place not only in the atmospheric region but also in the wind and in the envelope. The authors discuss here diffusion in the wind, the atmosphere, and the envelope simultaneously in order to obtain constraints on the hydrodynamics of mass loss. It is shown that to explain the He enrichment by separation in the atmosphere in stars of both $T_{eff} = 20,000$K and $T_{eff} = 25,000$K requires that the mass–loss rate decreases when T_{eff} increases. A magnetic field seems to be required to reduce the mass–loss rate where the separation occurs. It is suggested that most of the mass loss occurs at the poles, while the chemical separation leading to

He enrichment occurs at the magnetic equator where the magnetic field reduces mass loss.

064.078 Spectral flux from low–density photospheres: low–temperature results.
S. Hershkowitz, R. V. Wagoner.
Astrophys. J., Vol. 322, No. 2, p. 967 – 975 (1987).
Previous studies of low–density, hydrogen–dominated, quasi-static photospheres (such as those of type II supernovae) are extended to effective temperatures $T_e = 4000 – 6000$K. These temperatures include those at which hydrogen recombines at the densities investigated, corresponding to scale heights $\Delta R = 10^{14} – 10^{16}$cm. Results are presented for the emergent spectral flux (required to obtain supernova distances via the Baade method).

064.079 Effect of Balmer continuum absorption on line emission from Be stars.
K. M. V. Apparao, S. P. Tarafdar.
Astrophys. J., Vol. 322, No. 2, p. 976 – 980 (1987).
The authors show that the observed Balmer line emission from Be stars cannot be accounted for by the ionized region produced by the Lyman continuum of the star. They calculate the enhancement of the ionized region due to absorption of the Balmer continuum from the star and find that the line emission still cannot be accounted for in the later types. A source of EUV photons is needed to produce the required ionization to account for the observed line emission.

064.080 OH and CH continuous opacity in solar and stellar atmospheres.
R. L. Kurucz, E. F. van Dishoeck, S. P. Tarafdar.
Astrophys. J., Vol. 322, No. 2, p. 992 – 998 (1987).
Continuous absorption cross sections of OH and CH have been computed for the temperature range 1000K to 9000K. Both OH and CH produce significant ultraviolet opacity in the Sun and cool stars. CH is also significant in the visible at 400 nm.

064.081 Photochemistry and molecular ions in oxygen–rich circumstellar envelopes.
G. A. Mamon, A. E. Glassgold, A. Omont.
Astrophys. J., Vol. 323, No. 1, p. 306 – 315 (1987).
A theory for the ionization of the circumstellar envelopes around O–rich red giants is developed from the photochemical model. The main source of ionization is photoionization of H_2O, OH, and C by the interstellar UV radiation field, supplemented by cosmic–ray ionization of hydrogen. Significant amounts of H_3O^+ and HCO^+ are produced, with peak abundances $\sim 10^{-7}$ at intermediate distances from the star. Although H_3O^+ may be difficult to detect with current instrumentation, HCO^+ is probably detectable in nearby O–rich envelopes with large millimeter–wave telescopes.

064.082 Formation of chromospheric lines in cool dwarf stars.
L. E. Cram, M. S. Giampapa.
Astrophys. J., Vol. 323, No. 1, p. 316 – 324 (1987).
The authors have developed a simple theory of the formation of Hα and Ca II K in dwarf M stars which relates the observed variations in the emission strenghts of Ca II K and the equivalent widths of Hα to physical differences in the underlying chromospheric structure of the stars. The theory is used to elucidate aspects of the available observational data of Hα and Ca II K lines in late–type dwarfs.

064.083 Cosmic–ray–modified stellar winds. I Solution topologies and singularities.
C. M. Ko, G. M. Webb.
Astrophys. J., Vol. 323, No. 2, p. 657 – 671 (1987).
A two–fluid hydrodynamical model describing the modification of a stellar wind flow owing to its interaction with Galactic cosmic rays is investigated. The cosmic rays are coupled to the stellar wind by scattering with hydromagnetic waves or irregularities propagating in the backgrounud flow. A one–fluid polytropic model is used to describe the thermal stellar wind gas. The

cosmic rays are considered to be a hot, low–density gas with negligible mass flux, but with a significant pressure and energy flux. The solutions are shown to possess a line of singularities in (r,u,P_c)–space, where r, u, and P_c denote the radial distance from the star, the radial wind flow speed, and the cosmic–ray pressure, respectively. Astrophysical applications are briefly discussed.

Zvezdnye atmosfery (*Stellar atmospheres*).
See Abstr. 003.005.

The M–type stars.
See Abstr. 003.075.

Opacity Project: astrophysical and fusion applications.
See Abstr. 013.024.

Quasistatic absorption coefficient of two–component gases.
See Abstr. 022.074.

Doppler imaging.
See Abstr. 036.003.

Chemical abundances.
See Abstr. 061.035.

Magnetic braking in weakly ionized media.
See Abstr. 062.082.

Axisymmetric accretion flows in astrophysics.
See Abstr. 062.092.

Self–similar stong shock waves with thermal radiation in an atmosphere of exponentially–decreasing density.
See Abstr. 062.098.

Numerical observations of dynamic behavior in two–dimensional compressible convection.
See Abstr. 062.142.

Transfer of resonant line photons in spherically accelerating envelopes.
See Abstr. 063.009.

Multidimensional radiative transfer in stratified atmospheres. V. Energy transport by radiation.
See Abstr. 063.018.

Model atmospheres for type I supernovae: curvature effects.
See Abstr. 063.040.

The Poynting–Robertson effect of a rotating star.
See Abstr. 063.052.

Line formation in expanding atmospheres: multi–level calculations using approximate lambda operators.
See Abstr. 063.085.

Stellar atmospheres in non–LTE: model construction and line formation calculations using approximate lambda operators.
See Abstr. 063.086.

Reexamination of the metal contribution to astrophysical opacity.
See Abstr. 063.111.

The thermodynamics of anisotropic radiating systems.
See Abstr. 063.119.

Physical constraints on variable Eddington factors for the gray problem in radiative transfer.
See Abstr. 063.120.

Grids of evolutionary models of massive stars with mass loss and overshooting. Properties of Wolf–Rayet stars sensitive to over-shooting.
See Abstr. 065.015.

Initial–final mass relation for low and intermediate mass stars.
See Abstr. 065.016.

The formation of the first generation stars. II. Maximum masses of stars.
See Abstr. 065.034.

Stellar accretion of matter possessing angular momentum.
See Abstr. 065.094.

Nuclear runaways in a C/O white dwarf accreting H–rich material possessing angular momentum.
See Abstr. 065.095.

Low mass stars with mass loss and low–luminosity carbon star formation.
See Abstr. 065.097.

The Wolf–Rayet mass–luminosity relation.
See Abstr. 065.107.

Evolution of a 0.7 $M_\odot$ red giant.
See Abstr. 065.118.

Thick accretion discs: the role of self–gravity.
See Abstr. 067.085.

Model atmospheres of near–Eddington limit X–ray bursters.
See Abstr. 067.178.

Convection, magnetic fields, and line asymmetry in the sun and stars.
See Abstr. 071.001.

LTE modeling of inhomogeneous chromospheric structure using high–resolution limb observations.
See Abstr. 073.065.

Stellar wind – observations and theories. Part II. Theoretical models of the solar wind.
See Abstr. 074.065.

Constraints for models of Be stars derived from UV and IRAS observations.
See Abstr. 112.018.

Infrared photometry of late–type Wolf–Rayet stars.
See Abstr. 112.019.

A hot, low mass–loss rate inner envelope in IRC + 10216.
See Abstr. 112.049.

Optical properties of circumstellar silicates in the visible and the near–infrared.
See Abstr. 112.051.

Stellarer Massenverlust durch Sternwinde.
See Abstr. 112.058.

A study of UV spectra of ζ Aur/VV Cep stars. X. Mass–loss of α Sco A from high–resolution IUE spectra of α Sco B.
See Abstr. 112.063.

Dust shells around Miras and OH/IR stars: interpretation of IRAS and other infrared measurements.
See Abstr. 112.123.

Stellar wind – observations and theories. Part III. Mass outflow mechanisms from late type giants and supergiant stars.
See Abstr. 112.166.

Mass loss from evolved stars. VII. OH maser shell radii and mass–loss rates for OH/IR stars.
See Abstr. 112.190.

On the colours of Am stars.
See Abstr. 113.010.

Silicon absorption in UV spectra of ApSi stars.
See Abstr. 114.006.

Upper limit to the boron abundance in the Population II star HD 140283.
See Abstr. 114.016.

Wolf–Rayet stars.
See Abstr. 114.029.

Balmer jumps in Ap stars.
See Abstr. 114.033.

On the Balmer discontinuity of B Ia supergiants.
See Abstr. 114.056.

The diffusion of gallium in main–sequence peculiar stars.
See Abstr. 114.060.

On the evolutionary status of Mu Leonis.
See Abstr. 114.073.

The interacting binary β Lyr. II. Non–LTE model analysis and evolutionary conclusions.
See Abstr. 117.011.

A new approach to symbiotic stars.
See Abstr. 117.029.

The spectral evolution of dwarf nova outbursts.
See Abstr. 117.166.

Light element abundances in the atmosphere of the principal component of the binary system v Sgr.
See Abstr. 117.190.

The Roche coordinates in non–synchronous binaries.
See Abstr. 117.236.

IUE observations of the dwarf nova HL Canis Majoris and the winds of cataclysmic variables.
See Abstr. 117.321.

Synoptic studies of T Tauri stars.
See Abstr. 121.001.

Time–dependent bow shocks and Herbig–Haro objects.
See Abstr. 121.034.

Spectral energy distributions of T Tauri stars: disk flaring and limits on accretion.
See Abstr. 121.064.

Long period variables and stellar mass loss.
See Abstr. 122.002.

Energetics of activity of flare stars and the Sun: a synergetical approach.
See Abstr. 122.043.

An investigation of Cepheid variable stars using hydrostatic model atmospheres.
See Abstr. 122.053.

Computed ultraviolet spectra for SN 1987a.
See Abstr. 125.209.

Supernova 1987A: LTE line strengths as a guide to line identifications.
See Abstr. 125.297.

Massive disk formation resulting from the collision of a main–sequence star with a white dwarf in a globular cluster core.
See Abstr. 151.028.

065 Stellar Structure and Evolution

065.001 **Modal selection in stellar pulsators. II. Application to RR Lyrae models.**
J. R. Buchler, G. Kovács.
Astrophys. J., Vol. 318, No. 1, p. 232 – 247 (1987).
    The authors subject the numerically produced dynamical evolution of several RR Lyrae models from various initial conditions to a time–dependent Fourier analysis. This generates the temporal behavior of the amplitudes and phases of the few long–lived transient modes in addition to those of the ultimately surviving mode. It is shown that the amplitude equation formalism of Buchler and Goupil gives a remarkably good description of the observed behavior of the models.

065.002 **The formation and evolution of tidal binary systems.**
    S. L. W. McMillan, P. N. McDermott, R. E. Taam.
Astrophys. J., Vol. 318, No. 1, p. 261 – 277 (1987).
    The spectra of the nonradial oscillations induced in polytropes ($n = 3$ and $n = 3/2$) and low–mass Population II main–sequence stars by close encounters with other stars are investigated in the context of binary star formation in dense stellar systems. The

energy dissipated in the tidal interaction process, the associated capture cross sections, and the subsequent stellar response to the tidal heating are determined. It is found that the critical periastron separation for capture from typical globular cluster orbits ranges from around 3.3 stellar radii for 0.4 $M_\odot$ main–sequence stars to less than two stellar radii for an evolved 0.8 $M_\odot$ star. When tidal heating produces luminosities exceeding about 100 times the nuclear luminosity, the stellar envelope becomes significantly distended, possibly leading to contact between the binary components.

065.003 **Neutrinos from gravitational collapse.**
    R. Mayle, J. R. Wilson, D. N. Schramm.
Astrophys. J., Vol. 318, No. 1, p. 288 – 306 (1987).
    Detailed calculations are made of the neutrino spectra emitted during gravitational collapse events (Type II supernova?). Those aspects of the neutrino signal which are relatively independent of the collapse model and those aspects which are sensitive to model details are discussed. The easier to detect high–energy tail of the emitted neutrinos has been calculated using the Boltzmann

equation. The harder to detect electron antineutrino background from historical supernova might be enhanced by matter oscillation of higher energy mu and tau neutrinos to electron antineutrinos.

065.004 **On the collapse of $8-10 \, M_\odot$ stars due to electron capture.**
S. Miyaji, K. Nomoto.
Astrophys. J., Vol. 318, No. 1, p. 307 – 315 (1987).
In the final stages of evolution, $8-10 \, M_\odot$ stars undergo electron captures on ^{24}Mg and ^{20}Ne. The authors have investigated the density at which electron capture can ignite oxygen burning within the uncertainties of the semiconvection theory. It is found that oxygen burning is not ignited by electron captures on ^{24}Mg and ^{24}Na at a central density of $\sim 4 \times 10^9 g \, cm^{-3}$ but by capture on ^{20}Ne at $\sim 1 \times 10^{10} g \, cm^{-3}$. This density is still high enough to induce a collapse by electron capture rather than a total explosion of the core.

065.005 **Spin–up and mixing in accreting white dwarfs.**
M. Livio, J. W. Truran.
Astrophys. J., Vol. 318, No. 1, p. 316 – 325 (1987).
The authors demonstrate that existing theories of mixing in accreting white dwarfs encounter difficulties when confronted with observations of enrichments in nova ejecta. They present arguments, which suggest that angular momentum transport from the accreted material to the white dwarf is more efficient than previously thought. This should lead to matter spreading over the entire white dwarf surface, as well as inward mixing. When efficient transfer of angular momentum is taken into account, the gross features of nova outbursts can be reproduced, with the runaway occurring in a mixed layer.

065.006 **Subluminous Type I supernovae: their theoretical rate in our Galaxy and in ellipticals.**
A. Tornambè, F. Matteucci.
Astrophys. J., Lett. Ed., Vol. 318, No. 1, p. L25 – L28 (1987).
The rate of subluminous Type I SNs (Type Ib) is computed by means of galactic evolutionary models and compared with that of classical Type I SNs (Type Ia) in our Galaxy and in ellipticals. The assumed progenitors for Type Ib SNs are binary systems consisting of a C–O degenerate dwarf plus a nondegenerate He star, whereas those for Type Ia are binary systems made of two C–O degenerate dwarfs. The present time rate of Type Ib SNs in the solar neighborhood is predicted to be almost identical to the present time rate of Type Ia, in agreement with the observational estimates. Furthermore, it is shown that the Type Ib SN rate in ellipticals goes to zero at $\sim 10^9 yr$ after galaxy formation time, after undergoing a very sharp peak at around $0.05 \times 10^9 yr$.

065.007 **Inhibition of degeneracy by intense magnetic fields: derivation and astrophysical application.**
R. L. Ingraham, J. M. Wilkes.
Astrophys. Space Sci., Vol. 135, No. 1, p. 87 – 104 (1987).
Strong magnetic fields inhibit degeneracy in Fermi gases, that is, they postpone degeneracy to higher densities or lower temperatures. The authors derive this principle, virtually unknown in the physical and astrophysical literature, for the case of an ideal Dirac electron gas. They consider an application to hypothetical degenerate stellar objects with arbitrarily strong fields to see what effects the changed equations of state would lead to. One result is that the luminosity–temperature relation of the star is changed: the luminosity is reduced for given mass and interior temperature.

065.008 **Emden–Chandrasekhar axisymmetric, rigid–body rotating polytropes. IV. Exact configurations for $n = 5$.**
R. Caimmi.
Astrophys. Space Sci., Vol. 135, No. 2, p. 347 – 364 (1987).
In connection with the basic theory reported in a previous paper for EC1 (rigidly rotating) polytropes, exact configurations are defined as configurations for which the equilibrium equation has solutions which are infinitely close to some analytical

function and the related gravitational potential coincides, in fact, with the gravitational potential due to mass distribution, at any point not outside the system. The author restricts to the special case $n = 5$ and divides the related polytropes into two components, a massive body where each mass element has a finite (polytropic) distance from the centre, and a massless atmosphere where each mass element has an infinite (polytropic) distance from the centre. In the special case $n = 0$ it is shown that a particular configuration, the spheroidal one, is an exact configuration and evidence is given that spheroidal configurations are the most stable among all the allowed (axisymmetric) configurations. It is also pointed out that EC1 polytropes with $n = 0$ and incompressible Maclaurin spheroids belong to different sequences, even if they exhibit some common features.

065.009 **Another look at the R–modes.**
C. Koen.
Astrophys. Space Sci., Vol. 135, No. 2, p. 393 – 398 (1987).
Possible oscillation modes for a rotating star are listed. The only assumption made is that oscillations are adiabatic and that rotation is uniform. It is found that two modes not present for non–rotating stars are possible. Oscillation frequencies of these modes are rather different from those given in the literature for so–called r–modes.

065.010 **Evolution stellaire et supernovae.**
G. Meynet.
Orion, 45. Jahrg., Nr. 221, p. 126 – 129 (1987).

065.011 **The Chandrasekhar theory of stellar collapse as the limit of quantum mechanics.**
E. H. Lieb, H.–T. Yau.
Commun. Math. Phys., Vol. 112, No. 1, p. 147 – 174 (1987).
Starting with a "relativistic" Schrödinger Hamiltonian for neutral gravitating particles, the authors prove that as the particle number $N \rightarrow \infty$ and the gravitation constant $G \rightarrow 0$ one obtains the well known semiclassical theory for the ground state of stars.

065.012 **The value of the a/m ratio in binary systems.**
F. de Felice, L. Sigalotti.
Gravitational collapse and relativity, p. 377 – 386 (1986). – See Abstr. 012.006.
The authors compare the a/m ratios of observed binary systems with those deduced from theoretical models of star formation from originally rotating clouds. It is argued that the ratio a/m can be a good indicator of the physical plausibility of the models.

065.013 **Evolutionary constraints for young stellar clusters. I. The luminosity function of H–burning stars.**
E. Brocato, V. Castellani.
Astron. Astrophys., Vol. 182, No. 1, p. 36 – 40 (1987).
The authors performed theoretical simulations of the sequence of hot luminous stars in young stellar clusters formed by main-sequence stars plus H–shell burning stars after the phase of overall contraction (OC). They find that the location of the OC phase is revealed by an abrupt variation in the slope of the star luminosity function. They discuss this occurrence in the frame of open questions about H–burning evolution of intermediate mass stars, showing that the shape of the luminosity function should be able to discriminate whether or not convective core overshooting is efficient in these stars.

065.014 **Evolution of massive stars without convective core overshooting.**
D. Vanbeveren.
Astron. Astrophys., Vol. 182, No. 2, p. 207 – 218 (1987).
The purpose of this paper is to compare evolutionary calculations for massive stars ($M \geqslant 20 \, M_\odot$) without convective core overshooting, and observations of massive stars. The calculations were performed with an improved evolutionary code compared to earlier computations for both single stars and close binaries.

065.015   Grids of evolutionary models of massive stars with mass loss and overshooting. Properties of Wolf–Rayet stars sensitive to overshooting.
A. Maeder, G. Meynet.
Astron. Astrophys., Vol. 182, No. 2, p. 243 – 263 (1987).

Drawing the lessons of previous works about overshooting from convective cores and mass loss, the authors try to establish a reference grid of massive star models. The evolution of stars of initial mass 120, 85, 60, 40, 25, 20, and 15 $M_\odot$ is followed from the zero age sequence to the end of the C–burning phase with up-to–date physical ingredients. The various model outputs are analysed: internal structure, tracks in the HR diagram, lifetimes, t_{He}/t_H ratios, isochrones, ages of O–stars, evolution of surface C/N and O/N ratios. The authors also examine the various predicted properties of WR stars.

065.016   Initial–final mass relation for low and intermediate mass stars.
Yu. L. Frantsman.
Astrophysics, Vol. 25, No. 3, p. 655 – 660 (1987). English translation of 42.065.141.

065.017   Erratum: "Time–dependent models of magnetic stars – IV. Perpendicular axes" [Mon. Not. R. Astron. Soc., Vol. 226, No. 2, p. 281 – 296 (1987)].
D. Moss.
Mon. Not. R. Astron. Soc., Vol. 227, No. 2, p. 543 (1987). See Abstr. 43.065.049.

065.018   Magnetic activity in pre–main sequence stars.
R. J. Tayler.
Mon. Not. R. Astron. Soc., Vol. 227, No. 3, p. 553 – 561 (1987).

There is a much greater spread of surface activity in pre–main-sequence stars, which are otherwise similar, than in main-sequence stars. It is suggested that some pre–main–sequence stars may contain "fossil" magnetic fields in their outer layers which are stronger than any dynamo field would be. If surface activity is related to the magnetic field, the stars would exhibit stronger activity than those without significant "fossil" fields. Before the star reaches the main sequence, the "fossil" field in the convection zone may be affected by ohmic and turbulent decay and reduce to the dynamo value. If this is the case, it can explain the smaller spread of surface activity in main–sequence stars.

065.019   Hydrodynamic simulations of white dwarf–massive main–sequence star collisions in dense galactic nuclei.
O. Regev, M. M. Shara.
Mon. Not. R. Astron. Soc., Vol. 227, No. 4, p. 967 – 973 (1987).

The authors have carried out two–dimensional hydrodynamic simulations of a 0.6 $M_\odot$ white dwarf (WD) colliding head–on with a 10 $M_\odot$ main–sequence star (MSS) at 2000 and 6000 km s^{-1}, respectively. The nuclear energy generated during each collision is small compared to the kinetic energy of the MSS relative to the WD. The WD compresses accreted MSS material around itself. This compressed matter acts as a hard barrier to deflect further infalling matter. The redirection of the MSS kinetic energy leads to significant disruption of the MSS in both collisions in less than an hour. High–speed WD–MSS collisions with small impact parameters should yield the same result, while collisions with large impact parameter ($\gtrsim R_{MSS}/2$) should do considerably less damage to the MSS.

065.020   Equation of state of hot dense matter.
M. Lassaut, H. Flocard, P. Bonche, P. H. Heenen, E. Suraud.
Astron. Astrophys., Vol. 183, No. 1, p. L3 – L6 (1987).

With the Thomas–Fermi method, the authors determine the equation of state of hot dense matter at nuclear subsaturation densities. They investigate several geometries with nucleons organized in crystals structures of nuclei, rods, slabs, tubes and bubbles. They calculate the adiabatic index relevant for the first stage of the collapse of medium mass stars.

065.021   Theory of non–radial pulsations in massive early–type stars.
Y. Osaki.
Instabilities in luminous early type stars, p. 39 – 53 (1987). – See Abstr. 012.012.

The present theory of non–radial pulsations in massive early–type stars is reviewed. The author first describes the basic property of nonradial pulsations in stars. Variation in model-property with stellar evolution is then discussed with the help of the so–called propagation diagram. Problems of vibrational instabilities, effects of rotation, possible consequence of existence of non–axisymmetric waves on the stellar atmosphere are briefly discussed.

065.022   Theory of vibrational instabilities in luminous early type stars.
I. Appenzeller.
Instabilities in luminous early type stars, p. 55 – 72 (1987). – See Abstr. 012.012.

Because of the high temperature sensitivity of thermonuclear reactions massive blue stars are expected to be vibrationally unstable above a certain critical mass M_{crit} which depends on the evolutionary state. For ZAMS stars the most accurate calculations resulted in M_{crit} values between ~ 90 $M_\odot$ and 440 $M_\odot$. The origin of this discrepancy is unclear at present. Although the pulsations may enhance the mass loss from very massive stars, for ZAMS stars radiative driving forces will probably always dominate the mass loss rate.

065.023   Instabilities due to convection and rotation.
J.–P. Zahn.
Instabilities in luminous early type stars, p. 143 – 157 (1987). – See Abstr. 012.012.

In early–type stars, efficient thermal convection arises only in the central region, but even a convective core may be the cause of various instabilities. Some of the instabilities which are associated with convection are discussed, with the possible consequences on the behavior of the surface layers. The instabilities due to rotation are briefly reviewed. Some of them are likely to generate a mild turbulence, whose nature and efficiency are still a matter of conjecture.

065.024   Non radially pulsating Wolf–Rayet stars.
A. Noels, R. Scuflaire.
Instabilities in luminous early type stars, p. 213 – 216 (1987). – See Abstr. 012.012.

Some non radial modes can be amplified in H–burning shell models, with periods in the range of the observed periods. Such unstable models can only represent WN stars, and maybe even only late type WN stars, as they must still have enough hydrogen in the external layers but this is still in agreement with the observations, showing such variations mostly in WN stars.

065.025   The effects of neutrino transport on the collapse of iron stellar cores.
E. S. Myra, S. A. Bludman, Y. Hoffman, I. Lichtenstadt, N. Sack, K. A. Van Riper.
Astrophys. J., Vol. 318, No. 2, p. 744 – 759 (1987).

A multigroup flux–limited diffusion approximation to neu-trino transport, correct to first order in material velocities, is described. This scheme is incorporated into a hydrodynamic computer simulation using the LLPR equation of state in order to calculate stellar collapse for a variety of initial conditions and input physics. The authors report here on calculations using iron cores of medium and high mass. They also study the role of neutrino–electron scattering in determining core deleptonization.

065.026   Thresholds for rapid mass transfer in binary systems. I. Polytropic models.
M. S. Hjellming, R. F. Webbink.
Astrophys. J., Vol. 318, No. 2, p. 794 – 808 (1987).

The stability of a lobe–filling star in close binary systems against mass loss on a dynamical time scale is determined by the adiabatic response of that star to mass loss. The adiabatic properties (radii as functions of remnant mass) of various families of polytropic models are explored. Mass loss is parameterized by a variation of the central pressure, the result of which is calculated by a reformulation of the Lane–Emden equation in Lagrangian (mass) coordinates. The limitations of these models, and their relevance to binary evolution (particularly common envelope evolution) are discussed.

065.027 Deflagrating white dwarfs and the statistical properties of Type I supernovae.
J. R. Graham.
Astrophys. J., Lett. Ed., Vol. 318, No. 2, p. L47 – L50 (1987).

It is generally supposed that the luminosity of a Type I supernova (SN I) is powered by the decay of the radioactive isotope ^{56}Ni, which is synthesized and ejected when a white dwarf explodes. The class of models in which thermonuclear ignition leads to the formation of a turbulent burning front and total disruption of the white dwarf is well favored because of the excellent agreement between theory and observations. A careful examination of the distribution of radioactive material near the center of the explosion suggests how deflagrations may encompass the observed variation in the properties of SN I.

065.028 Convection and the mechanism of Type II supernovae.
A. Burrows.
Astrophys. J., Lett. Ed., Vol. 318, No. 2, p. L57 – L61 (1987).

The author presents some results from a study of the effects of convection on the early life of a neutron star and the mechanism of Type II supernovae. It is found that entropy–driven convection can enhance the neutrino luminosities in the quasi–static postcollapse stage of the core of a massive star by at times an order of magnitude. Furthermore, the neutrinospheric temperatures and, hence, the energy of the emitted neutrinos are increased by up to $\sim 50\%$.

065.029 Thermonuclear processes in accreting white dwarfs (novae, symbiotic stars, and type–I supernovae).
E. V. Ergma.
Astrophys. Space Phys. Rev., Vol. 5, p. 181 – 224 (1987). – See Abstr. 003.002.

Contents: 1. Introduction. 2. Evolutionary formation of white dwarfs. 3. Numerical calculations. 4. Thermal flashes with finite amplitude in shell sources (semianalytic approximation): The accretion phase. Thermal flash. 5. Accreting white dwarfs as presupernovae: Accretion onto a hydrogen dwarf. Accretion onto a helium dwarf. Accretion onto a carbon–oxygen dwarf. Accretion onto an O–Ne–Mg white dwarf. 6. Formation of dust in circumstellar envelopes. 7. Nucleosynthesis and type–I supernovae: Nucleosynthesis in hydrogen, helium and C–O supernovae (central burning). Nucleosynthesis in double–detonation supernova. 8. Conclusions; observations and theory for symbiotic stars, and the slow novae RR Tel, RT Ser, and PU Vul (?). For the Russian original see 42.065.106.

065.030 The numerical determination of the eigenfrequencies of a rotating star.
C. Koen.
Astrophys. Space Sci., Vol. 136, No. 1, p. 91 – 99 (1987).

A technique for finding the oscillation spectrum of a rotating star based on the solution of two simultaneous ordinary differential equations is described. Results are presented for some g–modes of the polytrope of index 2. It appears that the functional dependence of frequency on rotational velocity is different for different azimuthal wave numbers m.

065.031 A discrete–dynamic study of pulsating stars.
M. Takeuti.
Astrophys. Space Sci., Vol. 136, No. 1, p. 129 – 132 (1987).

A sequence of successive kinetic energy maxima at expanding phases of a pulsating star is considered. Transition from strictly periodic stellar pulsation to irregular one can be explained by the change in the rate of dissipation.

065.032 On the pre–main–sequence evolution of stars with mass accretion.
S. K. Bhattacharjee.
Mon. Not. R. Astron. Soc., Vol. 228, No. 2, p. 289 – 292 (1987).

The evolutionary path for a pre–main–sequence star accreting mass has been constructed by an analytical method by assuming a polytropic stellar model with no nuclear energy source. The result is found to be in fairly good agreement with that of Bhattacharjee & Williams (1980) obtained by detailed numerical calculation. It is also found that the energy supplied to the star by the accreted material is mostly gravitational. The thermal energy content of the on–falling matter has no significant effect on the evolution.

065.033 Very low mass stars.
J. Liebert, R. G. Probst.
Annu. Rev. Astron. Astrophys., Vol. 25, p. 473 – 519 (1987). – See Abstr. 003.003.

The authors discuss several theoretical and observational topics involved in discovering and analyzing very low mass stellar objects below about 0.3 $M_\odot$, as well as their likely extension into the substellar range. They outline recent theoretical work on low–mass stellar interiors and atmospheres, the determination of the hydrogen–burning mass limit, important dynamical evidence bearing on the expected numbers of such objects, and the expectations for such objects from star–formation theory. The emphasis of the discussion on brown dwarfs is on the properties of substellar objects near the stellar mass limit. Observational techniques used to discover and analyze very low mass objects are summarized. The authors compare the stellar parameters derived from observations with the theory and discuss the chromospheric and coronal activity with emphasis on how these properties might be used to derive stellar ages and population parameters. They conclude with a discussion of the luminosity and mass functions for very low mass disk and halo stars in the solar neighborhood.

065.034 The formation of the first generation stars. II. Maximum masses of stars.
Yu. I. Izotov.
Kinematika Fiz. Nebesn. Tel, Tom 3, No. 4, p. 30 – 39 (1987). In Russian. English translation in Kinematics Phys. Celest. Bodies.

The dynamics of the protostar envelopes cooled by molecular hydrogen are studied. The main mechanisms limiting the matter accretion on the star core are analyzed. Dependence of the maximum mass of a star with primordial chemical composition on initial abundance of molecular hydrogen is determined.

065.035 Instability of radial oscillations of Wolf–Rayet star models.
H. Kirbiyik.
Astrophys. Space Sci., Vol. 136, No. 2, p. 321 – 330 (1987).

By taking some evolutionary models of an initially 60 $M_\odot$–star, their radial vibrational stabilities have been investigated. These models, evolving with mass loss, are in the advanced stages of their evolution, and burn He in their cores. Calculations have been performed for the first and second harmonics as well as for the fundamental mode; some of these models were found to be vibrationally unstable in the fundamental mode.

065.036 Estimate of the stage of carbon burning in the stellar CNO–cycle.
B. M. Kuzhevskij, B. B. Makhanov.
Mater. nauchn. konf. mol. uchenykh fiz. Fak. Kazan. Univ., Alma–Ata, p. 64 – 68 (1987). In Russian. Abstr. in Ref. Zh., 51. Astron., 8.51.355 (1987).

065.037 The influence of the helium abundance on the convection threshold in surface layers of rotating stars.
A. Z. Dolginov, A. V. Klyachkin.
Pis'ma Astron. Zh., Tom 13, No. 9, p. 764 – 772 (1987). In Russian. English translation in Sov. Astron. Lett., Vol. 13.

The criterion is obtained for the convective instability in a spherical hydrogen–helium shell taking into account nonadiabatic effects and rotation. It is shown that both the decreasing and increasing of the temperature gradient necessary for the convection is possible in dependence on the rotational velocity and helium number density.

065.038   The fate of accreting white dwarfs: Type I supernovae vs. collapse.
K. Nomoto.
Prog. Part. Nucl. Phys., Vol. 17, p. 249 – 266 (1986). – See Abstr. 012.024.
The final fate of accreting $C+O$ white dwarfs is either thermonuclear explosion or collapse, if the white dwarf mass grows to the Chandrasekhar mass. The author discusses how the fate depends on the initial mass, age, composition of the white dwarf and the mass accretion rate. Relatively fast accretion leads to a carbon deflagration at low central density that gives rise to a Type Ia supernova. Slower accretion induces a helium detonation that could be observed as a Type Ib supernova.

065.039   Late stages of massive star evolution and nucleosynthesis.
K. Nomoto, M.–A. Hashimoto.
Prog. Part. Nucl. Phys., Vol. 17, p. 267 – 285 (1986). – See Abstr. 012.024.
The evolution of massive stars in the mass range of $8 - 25\ M_\odot$ is reviewed. The effect of electron degeneracy on the gravothermal nature of stars is discussed. Depending on the stellar mass, the stars form three types of cores, namely, non–degenerate, semi–degenerate, and strongly degenerate cores. The evolution for these cases is quite distinct from each other and leads to the three different types of final fate.

065.040   Late–time neutrino heating and energetics of stalled shocks in type II supernovae.
A. Ray, K. Kar.
Astrophys. J., Vol. 319, No. 1, p. 143 – 148 (1987).
The hydrodynamic shock formed outside the homologously collapsing core of a type II supernova progenitor seems to stall for certain mass ranges in many numerical calculations. A way to revive the stalled shock through late–time neutrino heating was proposed by Bethe and Wilson. The authors calculate through semianalytic models the extent of neutrino heating self–consistently with the changes in nuclear equilibrium behind the shockfront. It is found that whether the shock is revived sufficiently or not depends critically on the total neutrino luminosity and to a lesser extent on the neutrino–sphere radius.

065.041   The core mass–radius relation for giants: a new test of stellar evolution theory.
P. C. Joss, S. Rappaport, W. Lewis.
Astrophys. J., Vol. 319, No. 1, p. 180 – 187 (1987).
Theoretical studies of advanced stages of stellar evolution indicate that there is a nearly unique functional relationship between the radius of a giant and the mass of its core. A single-valued, continuous, and smooth relationship can be fitted both to stars on the first red giant branch and to asymptotic giant–branch stars, with total masses up to $\sim 2\ M_\odot$. The authors note that the measurable properties of systems containing degenerate dwarfs can be used as a direct test of this core mass–radius relation if the final stages of the loss of the envelope of the progenitor giant occurred via stable critical lobe overflow. Once critical lobe overflow has ceased and the remnant core of the giant has become a degenerate dwarf, the system parameters no longer change significantly. The examination of six diverse binaries shows that such a process probably took place in five of them.

065.042   Lower bounds on the masses of rapidly rotating white dwarfs.
G. Chanmugam, M. Rao, J. E. Tohline.
Astrophys. J., Vol. 319, No. 1, p. 188 – 191 (1987).
The rotational stability of rapidly rotating white dwarfs is considered for several equations of state. Lower bounds on the masses of the white dwarfs are deduced. It is also shown that if the period of rotation is less than ~ 14 s, the composition of the white dwarf must consist of ions heavier than He.

065.043   Statistical mechanics of partially ionized stellar plasmas: the Planck–Larkin partition function, polarization shifts, and simulations of optical spectra.
W. Däppen, L. Anderson, D. Mihalas.
Astrophys. J., Vol. 319, No. 1, p. 195 – 206 (1987).
The authors discuss a recent controversy about the Planck–Larkin partition function, and present optical simulations of high–quality spectra from laboratory hydrogen plasmas using several partition function formalisms. A Planck–Larkin cancellation may still have its place in equations of state that are based on quantum–statistical many–body theory. However, experimental evidence shows that it is inconsistent to use the Planck–Larkin partition function as the internal partition function in simple models of reacting gases. The authors also address the question of plasma polarization shifts of bound–state energies.

065.044   Application of time–dependent Fourier analysis to nonlinear pulsational stellar models.
K. Kovács, J. R. Buchler, C. G. Davis.
Astrophys. J., Vol. 319, No. 1, p. 247 – 259 (1987).
The variation of the modal content of the numerical hydrodynamical integration with various initializations of nonlinear RR Lyrae models is studied with the aid of a time–dependent Fourier analysis. The authors discuss the sensitivity of the derived temporal behavior of the amplitudes and phases to the parameters of the method. New light is shed on many very interesting details of mode interaction during the transient phases, not seen before, and a better assessment of the final state of the amplitude evolution results.

065.045   Second–overtone models of RR Lyrae stars.
R. B. Stothers.
Astrophys. J., Vol. 319, No. 1, p. 260 – 263 (1987).
Linear models of RR Lyrae stars pulsating in the second overtone indicate that this mode is self–excited only for rather low luminosities if the masses are normal. An unstable second-overtone model has been followed up to limiting amplitude by using a nonlinear hydrodynamical computer program. Deep splitting of the main light peak in the theoretical light curve occurs but does not resemble any observed feature. Moreover, the distribution of periods of RR Lyrae stars in globular clusters can be accounted for adequately by fundamental–mode and first–overtone pulsation. It is concluded that second–overtone pulsators probably do not exist among RR Lyrae stars.

065.046   Evolutionary sequences for horizontal–branch stars.
A. V. Sweigart.
Astrophys. J., Suppl. Ser., Vol. 65, No. 1, p. 95 – 135 (1987).
A new grid of 120 horizontal–branch evolutionary sequences including semiconvection has been constructed for helium abundances of 0.20, 0.25, and 0.30 and heavy–element abundances of 0.0001, 0.001, and 0.01. Each of these canonical sequences was evolved until the central helium abundance reached 0.05. The present computations include the change in the envelope helium abundance caused by the first dredge–up and use the canonical value of the core mass for each composition. Extensive tabulations of the numerical results are provided for application to globular cluster observations.

065.047   The stage of spontaneous flame propagation in supernovae.
S. I. Blinnikov, A. M. Khokhlov.
Pis'ma Astron. Zh., Tom 13, No. 10, p. 868 – 878 (1987). In Russian. English translation in Sov. Astron. Lett., Vol. 13.
The stage of spontaneous flame propagation in a carbon–oxygen stellar core is considered. An estimate of the critical temperature of the runaway is obtained. The structure and evolution of the spontaneous burning front is studied analytically and numerically for the isentropic temperature profile. At the final stage of spontaneous flame propagation the pressure, the density and the velocity of matter reach the Chapman–Jouguet detonation values.

065.048 A hydrodynamic pulsation model under subharmonic resonance.
T. Aikawa, O. Bruegman.
Astrophys. Space Sci., Vol. 137, No. 1, p. 115 – 127 (1987).
The authors found a hydrodynamic pulsation model of yellow supergiants under subharmonic resonance. It is confirmed that the feature of pulsation by the resonance is long–lived in the model by performing the hydrodynamical simulation for a long time.

065.049 Evolution of stellar binaries formed by tidal capture.
A. Ray, A. K. Kembhavi, H. M. Antia.
Astron. Astrophys., Vol. 184, No. 1/2, p. 164 – 172 (1987).
Two–body tidal capture, as proposed by Fabian et al., is the favoured mechanism for the formation of X–ray binaries in globular clusters. The authors consider here the tidal capture formation and subsequent evolution of a system consisting of a neutron star and a low mass main sequence star. They obtain the amount of tidal energy deposited during the first and later close passages, and the radial distribution of this energy. Going further, they examine the effects of the viscous dissipation of the tidal energy on the structure of the low–mass star and on the binary system. The tidal energy is thermalized on a timescale of 10^4yr. The consequent high tidal luminosity causes the star to expand and overflow its Roche lobe, resulting in the formation of a common envelope. This makes the stellar core and the neutron star spiral towards each other because of the frictional drag. The system may evolve into any of the following configurations: an X–ray binary, a detached binary, a neutron star surrounded by a massive accretion disk and a cloud of matter, or a Thorne–Zytkow object.

065.050 An evolutionary scenario for the formation of highly eccentric Be/X–ray binaries.
G. M. H. J. Habets.
Astron. Astrophys., Vol. 184, No. 1/2, p. 209 – 214 (1987).
Previous evolutionary scenarios for the formation of Be/X–ray binaries from moderately massive close binaries explain the observed range of orbital periods (P_{orb}) but not the observed large orbital eccentricities e unless an asymmetric supernova (SN) explosion is assumed. As an alternative, the author suggests a scenario in which highly eccentric Be/X–ray binaries are formed from very massive close binaries in a symmetric SN explosion after highly non–conservative evolution. This scenario may apply to systems with $P_{orb} \lesssim 50$ d and $e \gtrsim 0.3$ and, in particular, to the X–ray binary A 0538–668. The age and runaway velocity of A 0538–668 as derived from the alternative scenario are discussed. It is concluded that the system is in any case old ($\gtrsim 2 – 3 \times 10^7$yr). This agrees with the derived ages from the pulse period and magnetic field strength of the neutron star in A 0538–668. The derived runaway velocity in any evolutionary scenario is so large (since e is so large) that A 0538–668 is unlikely to be a member of the O–association against which it is projected in the LMC.

065.051 White dwarfs: the connection with the parents' masses.
P. R. Amnuehl', O. Kh. Gusejnov, Kh. I. Novruzova, Yu. S. Rustamov.
Astrofizika, Tom 27, Vyp. 1, p. 53 – 64 (1987). In Russian. English translation in Astrophysics, Vol. 27, No. 1.
The dependence between the mass of white dwarfs and that of their main sequence parent star is determined. The birth–rates of white dwarfs with different masses and the death–rate of the main sequence stars are interconsistent.

065.052 Calculation of a semiconvective zone with diffusion.
E. I. Staritsyn.
Nauchn. Inf., Vyp. 63, p. 97 – 104 (1987). In Russian.
Diffusion of turbulence elements in the semiconvective region of massive stars ($M > 10\ M_\odot$) is discussed. The formula for the diffusion coefficient and method of the hydrogen distribution definition is presented. This method takes into account physical conditions in semiconvective regions and their evolution changes.

065.053 Rotating stellar interiors.
R. C. Smith.
Physics of Be stars, p. 123 – 146 (1987). – See Abstr. 012.030 (IAU Colloq. No. 92).
In this article the author concentrates on the effects of rotation on the internal structure of a star, relating the results wherever possible to the specific problem of the Be stars. He therefore considers only stars with a convective core and a radiative envelope (i.e. with $M > 1.5\ M_\odot$.

065.054 The evolution of rapidly rotating B stars.
R. C. Smith.
Physics of Be stars, p. 486 – 499 (1987). – See Abstr. 012.030 (IAU Colloq. No. 92).
Be stars are located in or near the main–sequence band for non–rotating stars. Although this stage of evolution is relatively well understood, there are two main effects that make it impossible to say whether all Be stars are in the same stage of evolution and, if so, what that stage is. One effect is the spread in observed magnitude and colour as a result of rotation. The other effect is that there are uncertainties in the theoretical evolutionary tracks because the amount of convective overshooting is unclear. A useful way forward may be to try to understand individual stars in as much detail as possible.

065.055 Cooling of white dwarfs with account of non–equilibrium beta processes.
G. S. Bisnovatyj–Kogan.
Pis'ma Astron. Zh., Tom 13, No. 11, p. 1014 – 1018 (1987). In Russian. English translation in Sov. Astron. Lett., Vol. 13.
It is shown that non–equilibrium heating of a white dwarf during two–step neutronization and formation of a small, but finite core of the new phase is the main source of energy when $T < 0.1\ \Theta_{DC}$ and the Coulomb cristal becames degenerate.

065.056 On the theory of nuclear matter.
R. M. Avakyan, A. V. Sarkisyan.
Astrophysics, Vol. 26, No. 1, p. 77 – 82 (1987). English translation of 43.065.012.

065.057 Conditions for the formation of massive stars.
M. G. Wolfire, J. P. Cassinelli.
Astrophys. J., Vol. 319, No. 2, p. 850 – 867 (1987).
Upper limits on the masses of stars that can form are reinvestigated, and models for the inflow of matter through cocoons around stars of 60, 100, and 200 $M_\odot$ are calculated. Radiative forces on dust play a crucial role in halting an accretion inflow. Limits are derived on the dust–to–gas ratios and mass inflow rates that will permit inflow onto very massive stars. The inflows require mass accretion rates of $\sim 10^{-3} M_\odot \text{yr}^{-1}$ or more for there to be sufficiently strong ram pressure to overcome the strong radiative deceleration that occurs at the inner boundary of the shell. Inflow can occur if intermediate–sized grains ($0.05 – 0.25\ \mu$m) are missing from the initial gas/dust mixture.

065.058 A dynamical instability model for the period gap of cataclysmic variable binary systems.
R. E. Taam, P. N. McDermott.
Astrophys. J., Lett. Ed., Vol. 319, No. 2, p. L83 – L87 (1987).
A dynamical instability in a convective secondary is proposed as an explanation of the period gap of cataclysmic variable systems. The authors demonstrate that if the secondary component detaches from its Roche lobe when it is fully convective, then at the next onset of mass transfer a dynamical instability may ensue. It is shown that such a system at orbital period near 3 hr will be unstable to mass transfer and that stability will be regained at longer orbital periods when the system becomes detached. If the time scale for orbital decay by angular momentum losses associated with gravitational radiation and/or a magnetically coupled stellar wind is shorter than the Hubble time the system will then resume mass transfer at periods near 2 hr.

065.059   Interaction of nuclear reactions with weak oscillations in stars.
V. G. Gavryusev.
Mater. Konf. mol. uchenykh NII yader. fiz. MGU, Moskva 3 – 4 iyunya 1985. MGU, Moskva, p. 57 – 62 (1987). In Russian. Abstr. in Ref. Zh., 51. Astron., 9.51.458 (1987).

065.060   Collapse of low–mass "iron" cores of stars and supernovae.
S. I. Blinnikov, M. A. Rudzskij.
Particles and cosmology, p. 119 – 123 (1987). In Russian. Abstr. in Ref. Zh., 51. Astron., 9.51.474 (1987). – See Abstr. 012.034.

065.061   Roxburgh's criterion for convective overshooting.
N. H. Baker, R. Kuhfuß.
Astron. Astrophys., Vol. 185, No. 1/2, p. 117 – 120 (1987). See Abstr. 21.065.052.
An integral relation developed by Roxburgh has been used by several authors to estimate the extent of 'overshooting' from stellar convection zones. The authors show that Roxburgh's relation rests on a set of inconsistent approximations. It does not prove that there must be overshooting and should not be used in stellar structure calculations.

065.062   Evolutionary models for R CrB stars.
A. Weiss.
Astron. Astrophys., Vol. 185, No. 1/2, p. 165 – 177 (1987).
Stellar models with carbon–enriched helium mixtures have been evolved from different starting models (e.g. homogeneous models or such having a carbon–oxygen core). This kind of chemical composition is found in variable stars of type R CrB, extreme helium stars and hydrogen deficient carbon (HdC) stars. It has been investigated, for which stellar mass it is possible to evolve the models into the region of observed effective temperatures and luminosities for R CrBs. The resulting evolutionary tracks cover the whole region of observed R CrB stars and are also applicable as models for extreme helium and hydrogen deficient carbon stars.

065.063   Linear nonadiabatic pulsations of R CrB models.
A. Weiss.
Astron. Astrophys., Vol. 185, No. 1/2, p. 178 – 188 (1987).
Linear nonadiabatic radial pulsations of R CrB models have been examined. The model parameters for the calculations, i.e. mass, chemical composition, luminosity and effective temperature, were taken from the evolutionary models of Weiss (1987). Despite the difficulties in interpreting the results, imposed by the extreme nonadiabaticity of the pulsations, the author finds that most of the R CrB models possess at least one unstable mode. The periods as well as the blue edges of the instability strips agree with observational data. The implications of the results for questions concerning R CrB stars, such as origin and mass loss, are also discussed.

065.064   Theoretical expressions for evolutionary period changes in non–radially pulsating stars.
P. Bruggen, P. Smeyers.
Astron. Astrophys., Vol. 186, No. 1/2, p. 170 – 176 (1987).
The two–time–variable approach is used to study the equations governing linear pulsations, radial or non–radial, of a star evolving on its Helmholtz–Kelvin time scale. It is assumed that the isentropic approximation of the period of the pulsation mode considered is much shorter than the star's Helmholtz–Kelvin time scale. An integral expression for the rate of evolutionary change of the isentropic approximation of the pulsation period is derived and is shown to be determined by the isentropic approximation of the eigenfunctions and by the rates of change of three properties of the mass layers inside the evolving star: the radial distance to the center, the isentropic sound velocity, and the density. A decrease or an increase of the isentropic sound velocity in a mass layer contributes to an increase or a decrease, respectively, of the period. A tentative approximation for the rates of evolutionary change of the isentropic approximations of periods of high–order non–radial g–modes is derived under the restriction that the region of the star that predominates in determining these periods is not located close to the surface.

065.065   Merger of components in intermediate mass close binaries.
A. V. Tutukov, L. R. Yungelson (*L. R. Yungel'son*).
Comments Astrophys., Vol. 12, No. 2, p. 51 – 65 (1987).
The analysis of evolution of close binary stars (CBS) with moderate initial masses of components ($M_2 < M_1 \lesssim 10\,M_\odot$) leads to the inference that a considerable part of their components may merge in a Hubble time due to the loss of orbital angular momentum. A double star thus becomes a single one. The authors discuss possible means for the merger to occur and its consequences.

065.066   Are classical novae and dwarf novae the same systems?
M. Livio.
Comments Astrophys., Vol. 12, No. 2, p. 87 – 97 (1987).
Dwarf novae (DN) and classical novae (CN) both involve a semi–detached cataclysmic binary with a white dwarf primary and a Roche lobe filling secondary. The question is, are these *the same* systems? Namely, do the same systems undergo a cyclic evolution in which they appear sometimes as DN and sometimes as CN. The author presents here the main reasons for considering this possibility.

065.067   Collapse of $9\,M_\odot$ stars.
E. Baron, J. Cooperstein, S. Kahana.
Astrophys. J., Vol. 320, No. 1, p. 300 – 303 (1987).
The authors report on general relativistic hydrodynamical calculations of the collapse of O + Ne + Mg cores of a $9\,M_\odot$ star ($2.2\,M_\odot$ He core). Collapse is induced by rapid electron captures as the O + Ne + Mg is burned to nuclear statistical equilibrium. It is found that the hydrodynamic shock stalls before reaching the edge of the O + Ne + Mg core and thereby fails to produce a successful supernova explosion by the direct mechanism. The authors find no enhancement in the shock energy due to nuclear burning.

065.068   Collapsing white dwarfs.
E. Baron, J. Cooperstein, S. Kahana, K. Nomoto.
Astrophys. J., Vol. 320, No. 1, p. 304 – 307 (1987).
The authors present the results of the hydrodynamic collapse of an accreting C + O white dwarf. Collapse is induced by electron captures in the iron core behind a conductive deflagration front. The shock wave produced by the hydrodynamic bounce of the iron core stalls at about 115 km, and thus a neutron star formed in such a model would be formed as an optically quiet event.

065.069   Period doubling bifurcations and chaos in W Virginis models.
J. R. Buchler, G. Kovács.
Astrophys. J., Lett. Ed., Vol. 320, No. 1, p. L57 – L62 (1987).
The numerical hydrodynamical study of a sequence of models is shown to exhibit a typical series of period doubling bifurcations and a transition to chaos as a control parameter, the effective temperature, is varied. Depending on their behavior these models would be classified as W Vir, RV Tau, or semiregular.

065.070   Deuterium burning and the stellar birthline.
S. W. Stahler.
Bull. Am. Astron. Soc., Vol. 19, No. 2, p. 711 (1987). Abstract. – See Abstr. 010.061.

065.071   Fission of rapidly rotating protostars.
H. A. Williams, J. E. Tohline.
Bull. Am. Astron. Soc., Vol. 19, No. 2, p. 727 (1987). Abstract. – See Abstr. 010.061.

065.072 **A numerical study of shear flow instability in rotating protostar models.**
R. H. Durisen, X. Yang.
Bull. Am. Astron. Soc., Vol. 19, No. 2, p. 727 (1987). Abstract. – See Abstr. 010.061.

065.073 **Theoretical models of the lower main sequence.**
B. Dorman, L. A. Nelson, W. Y. Chau.
Bull. Am. Astron. Soc., Vol. 19, No. 2, p. 756 (1987). Abstract. – See Abstr. 010.061.

065.074 **Density discontinuity g–modes.**
P. N. McDermott.
Bull. Am. Astron. Soc., Vol. 19, No. 2, p. 757 (1987). Abstract. – See Abstr. 010.061.

065.075 **Weight functions in adiabatic stellar pulsations. I. Radially symmetric motion.**
W. D. Pesnell.
Publ. Astron. Soc. Pac., Vol. 99, No. 619, p. 975 – 985 (1987).
Through the use of two classes of simple stellar models, the author illustrates a weight function for radial pulsations that can be interpreted in terms of two types of potential energy. The traditional weight function is related to the present one via an integration by parts, and it is argued that both formulations should be examined to minimize possible errors that can arise from neglecting several surface terms. The present form provides a different interpretation of radial oscillations of stars with small central condensations. The author also follows the radial fundamental of a highly centrally condensed polytrope through the dynamic instability at $\gamma = 4/3$ and demonstrates that the envelope is not involved in the instability.

065.076 **Evolution of very low mass stars and brown dwarfs. II. The Population II.**
F. D'Antona.
Astrophys. J., Vol. 320, No. 2, p. 653 – 662 (1987).
This paper addresses a number of problems connected with the evolution of low–mass ($M \leqslant 0.8\,M_\odot$) and very low mass ($M \geqslant 0.07\,M_\odot$) stars having metal contents much smaller than the solar metal content. Two main chemical compositions bracketing the Population II regime are considered: helium and metal contents $Y = 0.23$, $Z = 10^{-4}$ (Cox and Stewart opacities) and $Y = 0.25$, $Z = 10^{-3}$ (Alexander, Johnson, and Rypma opacities). For these compositions, the hydrogen burning minimum mass is evaluated. It is confirmed that the atmospheric opacities are the main factor influencing the minimum main–sequence luminosity. Population II luminosity functions (LFs) are constructed and applied to the interpretation of globular cluster main–sequence LFs.

065.077 **On the origin of the solar system s–process abundances.**
R. A. Malaney, A. I. Boothroyd.
Astrophys. J., Vol. 320, No. 2, p. 866 – 870 (1987).
In the search for the origin of the solar system s–process abundances much attention has been focused on the intershell zones of thermally pulsing asymptotic giant branch (AGB) stars. It has recently been suggested that, relative to the poor fits obtained from intermediate–mass AGB models, low–mass AGB models may result in much better fits to the observed solar system abundances. Using new data, the s–process enhancements occurring in the intershell zones of low–mass AGB stars are calculated. A nonsolar distribution of s–process abundances is reported for all realistic AGB models studied. Other possible astrophysical sites for the origin of the solar system s–process abundances are discussed.

065.078 **On some group properties of Newtonian static star structure equations.**
M. Biesiada, Z. Golda, M. Szydlowski.
J. Phys. A, Vol. 20, No. 6, p. 1313 – 1321 (1987). Abstr. in Phys. Abstr., Vol. 90, No. 1309, Entry 88380 (1987).

065.079 **Ne–O–F cycle reactions in stellar interiors.**
Proc. Natl. Acad. Sci. India, Sect. A, Vol. 55, Part 4, p. 409 – 411 (1985). Abstr. in Phys. Abstr., Vol. 90, No. 1309, Entry 88385 (1987).

065.080 **Stellar convection theory.**
D.–r. Xiong.
Publ. Purple Mt. Obs., Vol. 6, No. 2, p. 101 – 126 (1987). In Chinese.
The mixing–length and statistical theories of convection are reviewed. The dynamic behavior of turbulent convection is studied. In the authors' statistical theory, a complete set of simultaneous equations of auto–correlation and cross–correlation functions of turbulent velocity, turbulent temperature and turbulent concentration of chemical elements is derived from the dynamic equations of fluid under several very general simplifications and approximations. The local theory of time–dependent convection available for both radial and nonradial oscillations of stars and non–local theory of quasi–steady convection are discussed. The theoretical calculations of oscillations of variable stars, the evolution of massive stars and the structure of the solar convective zone are compared with the observations.

065.081 **Collapse of white dwarfs in low mass binary systems.**
J. Isern, R. Canal, E. García–Berro, M. Hernanz, J. Labay.
Rev. Mex. Astron. Astrofis., Vol. 14, No. 1, p. 260 – 264 (1987). – See Abstr. 012.042.
The present work shows that massive, initially cold white dwarfs can collapse nonexplosively if they accrete mass at a rate greater than $1.0 \times 10^{-7}\,M_\odot$ per year.

065.082 **Uncertainties in the determination of the upper mass limit for zero–age main sequence stars.**
J. Klapp, N. Langer, K. J. Fricke.
Rev. Mex. Astron. Astrofis., Vol. 14, No. 1, p. 265 – 270 (1987). – See Abstr. 012.042.
In a recent investigation Klapp et al. (1987) obtained a critical mass of 440 $M_\odot$ for the overstability of very massive extreme population I stars at the main sequence. In this work the authors investigate the dependence of Klapp et al. (1987) results upon the program input physics. They find that stars in the 100 – 500 $M_\odot$ range are marginally stable (or unstable) and that this mass range should be considered as a transition region from stability to overstability of very massive stars.

065.083 **Analytical results connecting stellar structure parameters and neutrino fluxes.**
H. J. Haubold, A. M. Mathai.
Ann. Phys. (Leipzig), Vol. 44, No. 2, p. 103 – 116 (1987). Abstr. in Phys. Abstr., Vol. 90, No. 1310, Entry 94850 (1987).

065.084 **On the spherical symmetry of static perfect fluid spacetimes and the positive–mass theorem.**
A. K. M. Masood–ul–Alam.
Classical Quantum Gravity, Vol. 4, No. 3, p. 625 – 633 (1987). Abstr. in Phys. Abstr., Vol. 90, No. 1312, Entry 107808 (1987).

065.085 **Calculation of relativistic model stars using Regge calculus.**
J. Porter.
Classical Quantum Gravity, Vol. 4, No. 3, p. 651 – 661 (1987). Abstr. in Phys. Abstr., Vol. 90, No. 1312, Entry 107809 (1987).

065.086 **Axion bounds from stellar evolution.**
G. G. Raffelt, D. S. P. Dearborn.
ESO Conf. Workshop Proc., No. 27, p. 179 – 184 (1987). – See Abstr. 012.051.
Number counts of "clump giants" in open clusters determine the lifetime of helium burning stars. These observations can be used to constrain the possibility of "exotic cooling" of stars through the emission of light, weakly interacting particles. The authors derive new bounds on the interaction of "invisible axions" with photons.

065.087   **Horizontal branch stars: theoretical expectations.**
A. Tornambè.
ESO Conf. Workshop Proc., No. 27, p. 307 – 319 (1987). – See Abstr. 012.051.

The author recalls the theoretical properties of population II horizontal branch (HB) stars and discusses the efforts made in the last ten years to obtain as more reliable as possible HB star models. He then presents new HB models where the overabundance of oxygen and other elements, relative to iron, is accounted for. Implications of these models for the interpretation of some observational properties of the HBs of galactic globular clusters are discussed.

065.088   **Planetary systems and their central stars.**
A. V. Tutukov.
Astron. Vestn., Tom 21, No. 4, p. 299 – 301 (1987). In Russian. – See Abstr. 012.052.

065.089   **Planetary systems as ultimate or by–product of binary star formation.**
Eh. M. Drobyshevskij.
Astron. Vestn., Tom 21, No. 4, p. 313 – 317 (1987). In Russian. – See Abstr. 012.052.

065.090   **Numerical modelling of the compression of rotating clouds.**
I. V. Igumenshchev, B. M. Shustov.
Astron. Vestn., Tom 21, No. 4, p. 318 – 320 (1987). In Russian. – See Abstr. 012.052.

065.091   **Some embarrassments in current treatments of convective overshooting.**
A. Renzini.
Astron. Astrophys., Vol. 188, No. 1, p. 49 – 54 (1987).

In this paper several versions of the Mixing Length Theory (MLT) approach to overshooting from convective cores are analysed, and it is argued that a confusion between local and non–local quantities introduces a fatal physical inconsistency, which invalidates the corresponding overshooting algorithms used in stellar evolutionary calculations. It is further argued that the MLT is unlikely to provide an adequate description of convective overshooting, and possible alternatives are briefly discussed.

065.092   **Stellar structure with a Boltzmann factor: an extension to chemically heterogeneous H–He–stars.**
W. Verschueren, D. K. Callebaut.
Astrophys. Space Sci., Vol. 137, No. 2, p. 257 – 266 (1987).

The authors present an attempt for an extension of the modified Boltzmann model, which was introduced by Callebaut et al. (1982) as an improvement of the polytropic models, to the case of chemically–heterogeneous stars in equilibrium, containing H and He, by proposing a density profile of the form $\varrho = C_1 T^N \exp(-\mu_H m(\phi-\phi_*)/kT) + C_2 T^N \exp(-\mu_{He} m(\phi-\phi_*)/kT)$. Analytical properties are derived and numerical as well as analytical arguments are presented for the conclusion that this hypothesis for a density profile imposes an almost constant chemical profile to the model as a whole, thereby making it in this form unsuited for the study of heterogeneous stars. A comparison is made with the former Boltzmann model in the homogeneous limit.

065.093   **Similarity solutions for a stellar line explosion.**
S. N. Ojha, O. Nath.
Astrophys. Space Sci., Vol. 138, No. 1, p. 49 – 59 (1987).

Similarity solutions for line explosion in a non–uniform self–gravitating medium including the effects of magnetic field radiation flux and neglecting the radiation pressure and energy are investigated. Gas is assumed to be grey and opaque and the shock to be transparent and isothermal.

065.094   **Stellar accretion of matter possessing angular momentum.**
G. S. Kutter, W. M. Sparks.
Astrophys. J., Vol. 321, No. 1, p. 386 – 393 (1987).

The authors develop the theory of accretion onto stars of matter possessing angular momentum, including the shear forces between the accreting matter and the star (assumed to be nonrotating), distribution of accreted matter and its angular momentum by turbulent mixing, thermalizing of rotational kinetic energy, and loss of energy by horizontal expansion.

065.095   **Nuclear runaways in a C/O white dwarf accreting H–rich material possessing angular momentum.**
W. M. Sparks, G. S. Kutter.
Astrophys. J., Vol. 321, No. 1, p. 394 – 403 (1987).

The authors present the results of numerical calculations of the accretion of H–rich material onto 1 $M_\odot$ C/O white dwarfs and the resulting nuclear runaway. The accreting material arrives on the white dwarfs with angular momentum in the range of 10% – 100% of the Keplerian value and at rates of 10^{-11}– $10^{-8} M_\odot \mathrm{yr}^{-1}$. The matter is assumed to arrive from an accretion disk and to mix by shear turbulence into the white dwarfs' surface layers. Maximum nuclear burning rates in the range $0.38 – 6.9 \times 10^{15}\mathrm{ergs\ g}^{-1}\mathrm{s}^{-1}$ are reached, but no nova–like mass ejection results. So far, all models of nova outbursts lying between the two limiting cases of radial accretion and accretion with Keplerian angular momentum have failed to obtain a sufficiently strong nuclear runaway for nova–like mass ejection. Possible explanations of this failure are considered.

065.096   **Gamma ray lines from novae.**
M. D. Leising.
Diss. Abstr. Int., Sect. B, Vol. 48, No. 5, p. 1384–B (1987). Thesis, Rice University, 119 pp. (1987). Order No. DA8718741.

065.097   **Low mass stars with mass loss and low–luminosity carbon star formation.**
A. I. Boothroyd.
Diss. Abstr. Int., Sect. B, Vol. 48, No. 6, p. 1710–B (1987). Thesis, California Institute of Technology, 271 pp. (1987). Order No. DA8719650.

065.098   **The production of ^{26}Al in supermassive stars and the gamma–ray line flux from the Galactic center.**
W. Hillebrandt, F.–K. Thielemann, N. Langer.
Astrophys. J., Vol. 321, No. 2, p. 761 – 767 (1987).

The evolution through contraction and explosion of a high–metallicity ($Z = 0.04$) supermassive star of $5 \times 10^5 M_\odot$ is computed. It is shown that in the inner 20% of the star a significant fraction of the preexisting magnesium is converted into ^{26}Al leading to the production of about 50 $M_\odot$ of ^{26}Al. This amount is sufficient to explain the observed 1.8 MeV γ–ray line flux from the Galactic center if such a supermassive star exploded there within the last 2 million years.

065.099   **Monte Carlo simulations of radio pulsars and their progenitors.**
R. J. Dewey, J. M. Cordes.
Astrophys. J., Vol. 321, No. 2, p. 780 – 798 (1987).

The authors combine models for the evolution of main–sequence stars with an initial mass function and orbital period distribution to simulate the formation of neutron stars. It is assumed that isolated stars in the range of $8 – 20 M_\odot$ form neutron stars directly in supernovae. Among the effects included in the simulation are (1) mass exchange due to Roche lobe overflow; (2) mass loss from a system in a contact binary stage; (3) changes in a system's orbit caused by mass loss or exchange; (4) the swelling of the secondary as it accretes matter; (5) changes in a system's orbit due to sudden mass ejection in a supernova; and (6) the effect of supernova shell impact on a companion star. It is found that it is difficult to reproduce both the observed velocity distribution of radio pulsars and the observed incidence of binary pulsars. The authors believe it is unlikely that all pulsars are formed in close binary systems and that asymmetries in the supernova explosion, or processes that accelerate a pulsar soon after its birth, are important in determining pulsar velocities.

065.100 The s–process nucleosynthesis of barium stars.
R. A. Malaney.
Astrophys. J., Vol. 321, No. 2, p. 832 – 845 (1987).

In order to analyze the observations of enhanced s–process elements in barium stars, new nuclear reaction network calculations of the heavy–element synthesis occurring in thermally pulsing asymptotic giant branch (AGB) stars are presented. In addition, the s–process abundances from single neutron exposure calculations are presented. It is found that a mass transfer mechanism from an AGB star, in which the ^{22}Ne source operates, cannot be responsible for the s–process abundances observed in the classical barium stars HR 774 and ζ Cap, and in the mild barium star o Vir. Low neutron density single exposures of ~ 1.0 mb^{-1}, on the other hand, are shown to result in good agreement with the barium star observations. The most plausible site for such single exposures still seems to be the helium core flash.

065.101 The dynamical oscillation and propulsion of magnetic fields in the convective zone of a star. II. Thermal shadows.
E. N. Parker.
Astrophys. J., Vol. 321, No. 2, p. 984 – 1008 (1987).

This paper is a preliminary theoretical investigation of the dynamics of thermal shadows formed in the convective zone of a star around an insulating obstacle such as a horizontal band of intense magnetic field. It is shown that the depth of the shadow on the cool side of the obstacle (and the pileup of heat on the hot side) is determined principally by the width of the obstacle multiplied by the temperature gradient. The nonlinear convective heat transport reduces the depth of the shadow by $\sim 20\%$. The horizontal temperature gradients within the shadow set up convective rolls. The broad bands of azimuthal field in the convective zone of the Sun produce thermal shadows pressing fields as strong as 10^4G downward against the bottom of the convective zone in opposition to the magnetic buoyancy.

065.102 The dynamical oscillation and propulsion of magnetic fields in the convective zone of a star. III. Accumulation of heat and the onset of the Rayleigh–Taylor instability.
E. N. Parker.
Astrophys. J., Vol. 321, No. 2, p. 1009 – 1030 (1987).

This paper studies the time–dependent accumulation of heat beneath a thermal barrier simulating a broad band of intense azimuthal field in the lower convective zone of the Sun, or other star. The accumulating heat causes the gas density beneath the field to decline to the point that the gas is less dense than the gas within the magnetic field. The resulting Rayleigh–Taylor instability causes tongues of heated gas to penetrate upward through the field. The author suggests that the tongues of hot gas take on the nature of thermal plumes, extending all the way to the surface, where they provide the emerging magnetic fields that give rise to the activity of the Sun.

065.103 Neutrino pair energy deposition in supernovae.
J. Cooperstein, L. J. van den Horn, E. Baron.
Astrophys. J., Lett. Ed., Vol. 321, No. 2, p. L129 – L132 (1987).

The authors investigate whether the process $v\bar{v} \rightarrow e^+ e^-$ can increase the energy of a delayed Type II supernova explosion significantly, and they find it cannot.

065.104 Synchronization–induced period gaps and ultra–short periods in magnetic cataclysmic binaries.
D. Q. Lamb, F. Melia.
Astrophys. J., Lett. Ed., Vol. 321, No. 2, p. L133 – L137 (1987).

When synchronization occurs in magnetic cataclysmic variables, the spin angular momentum of the white dwarf primary is injected directly into the binary system by the magnetohydrodynamic torque. The authors show that this process brings the system out of contact and produces a new kind of period gap. If synchronization occurs when the orbital period $P_b \lesssim 1.5$ hr, the secondary is degenerate when contact is reestablished and the orbital period can be as short as 25 minutes. Thus binaries with hydrogen–rich companion stars can form with periods much shorter than the usual minimum period of ~ 72 minutes.

065.105 Dynamical mass transfer in cataclysmic binaries.
F. Melia, D. Q. Lamb.
Astrophys. J., Lett. Ed., Vol. 321, No. 2, p. L139 – L143 (1987).

The authors explore the consequences of the hypothesis that temporary storage of orbital angular momentum in an accretion disk does lead to dynamical mass transfer in cataclysmic binaries and does not lead to coalescence. They find that systems coming into contact longward of the classical period gap undergo dynamical mass transfer. In some cases, systems undergo a series of dynamical mass transfer episodes. These episodes typically terminate in stable mass transfer at periods shortward of the usual period gap, i.e., $P_b < 2$ hr. Few, if any, systems are expected longward of the usual period gap, i.e., $P_b > 3$ hr, and the origin of the observed long–period systems is left unanswered.

065.106 Asymptotic approximations for higher order nonradial oscillations of a spherically symmetric star.
P. Smeyers, M. Tassoul.
Astrophys. J., Suppl. Ser., Vol. 65, No. 3, p. 429 – 449 (1987).

Higher order asymptotic approximations for p– and g–modes of high radial orders and low degrees of a spherically symmetric star are developed by starting from second–order differential equations that are expressed in a single dependent variable. This is unlike Tassoul's treatment, which uses different variables in the two members of the starting equations. The ratio of the amplitude of the transverse component of the Lagrangian displacement to the amplitude of the radial component is shown to be of the order of the reciprocal of the eigenfrequency in a very large domain of a star.

065.107 The Wolf–Rayet mass–luminosity relation.
M. Beech.
Astrophys. Space Sci., Vol. 139, No. 1, p. 149 – 153 (1987).

The author considers the mass–luminosity relation proposed for the Wolf–Rayet stars on the basis of detailed numerical models. It is found that the linear form of this relation can be explained in a straightforward manner and is due to essentially three factors, (1) the WR stars are approximately chemically homogeneous, (2) the stars evolve under constant radiation pressure, and (3) the stars evolve with high mass loss.

065.108 The Pomeau–Manneville intermittent transition to chaos in hydrodynamic pulsation models.
T. Aikawa.
Astrophys. Space Sci., Vol. 139, No. 2, p. 281 – 293 (1987).

The author performed hydrodynamic simulations for a series of stellar models of luminosity $\log (L/L_\odot) = 3.505$, and $T_e = 5300$K with the range of the mass $1.4 M_\odot \leqslant M \leqslant 1.5 M_\odot$. With decreasing the mass, he confirms a transition from limit cycles to irregular oscillations. The nature of the transition is finally specified by examining the dissipation of pulsational kinetic energies in limit cycle models, when pulsations start with larger amplitudes than their limiting pulsations.

065.109 A note on the linear stellar model.
J. W. Anderson, A. M. Mathai, H. J. Haubold.
Astron. Nachr., Vol. 308, No. 5, p. 313 – 318 (1987).

For a purely gaseous self–gravitating stellar configuration with linear matter density distribution the total power generated by nuclear reactions is considered. The analytic connection between physical parameters of the macroscopic and microscopic levels of the stellar equilibrium configuration is revealed.

065.110 The construction of a difference grid in calculations of stellar evolution.
A. B. Men'shchikov.
Nauchn. Inf., Vyp. 59, p. 124 – 127 (1986). In Russian.

A simple way for the construction of a difference grid in calculations of stellar structure and evolution is presented. The application of a similar approach to the choice of a time step is briefly described.

065.111   Wolf–Rayet star models and radial vibrations.
H. Kirbiyik.
Doga Turk Fiz. Astrofiz. Derg., Vol. 11, No. 2, p. 173 – 179 (1987). Abstr. in Phys. Abstr., Vol. 91, No. 1320, Entry 10560 (1988).

065.112   Convection in supernova theory.
H. A. Bethe, G. E. Brown, J. Cooperstein.
Astrophys. J., Vol. 322, No. 1, p. 201 – 205 (1987).
Convection takes place behind the shock front if and when the strength of the shock declines. However, the convection does not strengthen the shock but weakens it. It is therefore useless for promoting an early explosion (at times of order 10 ms). At late times (several tenths of a second), convection may take place behind the neutrino sphere and may increase the strength of the late shock caused by neutrino heating.

065.113   Evolution of 8–10 $M_\odot$ stars toward electron capture supernovae. II. Collapse of an O + Ne + Mg core.
K. Nomoto.
Astrophys. J., Vol. 322, No. 1, p. 206 – 214 (1987).
A helium core of an 8.8 $M_\odot$ star (initial core mass of 2.2 $M_\odot$) is evolved from the helium–burning stage through the early stage of collapse of an O + Ne + Mg core. When the mass interior to helium–burning shell reaches 1.38 $M_\odot$, electron captures induce the rapid contraction of the core. Entropy production associated with electron capture on ^{20}Ne as well as gravitational contraction leads to an ignition of the oxygen deflagration that incinerates the core material into nuclear statistical equilibrium. The star will become a Type II supernova to leave a neutron star less massive than 1.3 $M_\odot$. The effects of oxygen burning upon the hydrodynamic behavior of collapse and the possible nucleosynthetic yields are discussed.

065.114   The evolutionary status of 4U 1820–30.
S. Rappaport, L. A. Nelson, C. P. Ma, P. C. Joss.
Astrophys. J., Vol. 322, No. 2, p. 842 – 851 (1987).
The authors calculate the evolution of an ultracompact binary wherein mass transfer is driven by the emission of gravitational radiation, with emphasis on a systematic study of how the rate of change in orbital period depends on the various system parameters. The authors derive analytic scaling laws describing how the mass of the secondary, the mass transfer rate, and $\dot{P}$ depend on the binary system parameters, and they apply these simple relations to the 4U 1820–30 system, located in the globular cluster NGC 6624. The authors then carry out detailed numerical evolutionary calculations utilizing Zapolsky–Salpeter (1969) models for cold stars, but allowing for the possibility that the secondary has not yet reached a completely degenerate configuration. Finally, the results are used to set significant constraints on the deviation from a completely degenerate configuration of the secondary, the systemic mass–loss parameters, and the mass of the neutron star in the 4U 1820–30 system.

065.115   On synchronization in early–type binaries.
J.–L. Tassoul.
Astrophys. J., Vol. 322, No. 2, p. 856 – 861 (1987).
Tidal friction alone does not explain the general trend toward synchronism in the early–type binaries. In this paper the author presents a purely hydrodynamical mechanism which tends to synchronize the axial and orbital motions in the components of a double star. This very effective process involves a large–scale meridional flow, superposed on the motion around the rotation axis of the tidally distorted component. These transient, mechanically driven currents – which are much faster than the steady, thermally driven Eddington–Vogt currents – cease to exist as soon as synchronization has been achieved in the star. The application of the mechanism to Am stars in binaries and to RS CVn systems is discussed.

065.116   Magnetic activity, tides, and orbital period changes in close binaries.
J. H. Applegate, J. Patterson.
Astrophys. J., Lett. Ed., Vol. 322, No. 2, p. L99 – L102 (1987).
The authors propose that a variable quadrupole moment produced by magnetic activity in the outer convection zone of one of the stars in a close binary is responsible for the orbital period changes which occur on a time scale of order 10 yr. In this model magnetic activity modulates the orbital period, superposed on any long–term period changes that may occur. The model predicts that the orbital period changes should be periodic with the period of the activity cycle.

065.117   A rigorous examination of the Chandrasekhar theory of stellar collapse.
E. H. Lieb, H.–T. Yau.
Astrophys. J., Vol. 323, No. 1, p. 140 – 144 (1987).
Some of the results of a rigorous analysis of the Chandrasekhar semiclassical theory of stellar collapse are presented. They are of two kinds. The first concerns the Chandrasekhar equation itself; the authors prove the uniqueness of the solution and also prove that the solution has certain properties not noted before. The second is a derivation of the Chandrasekhar equation from quantum mechanics without making *a priori* assumptions about the smallness of correlation effects. A parallel derivation is made for stars composed of bosons; the resulting equation is quite different from the Chandrasekhar equation, for it is of the Hartree type and involves density gradients.

065.118   Evolution of a 0.7 $M_\odot$ red giant.
A. Harpaz, A. Kovetz, G. Shaviv.
Astrophys. J., Vol. 323, No. 1, p. 154 – 158 (1987).
The evolution of a 0.7 $M_\odot$ red giant is followed along the red giant branch, with mass loss included according to Reimers' formula. The red giant star completes its evolution by totally exhausting its envelope, without igniting the helium in the core. At this point, the core mass, which turns into a white dwarf, is 0.43 $M_\odot$. The consequences of such an evolutionary path are discussed, together with the constraints that these results impose on the mass–loss rate during this evolutionary phase.

065.119   Positron annihilation gamma rays from novae.
M. D. Leising, D. D. Clayton.
Astrophys. J., Vol. 323, No. 1, p. 159 – 169 (1987).
The potential for observing annihilation gamma rays from novae is investigated. These gamma rays would result from the annihilation of positrons emitted by β^+–unstable nuclei produced near the peak of the runaway and carried by rapid convection to the surface of the nova envelope. Simple models serve as input into investigations of the fate of the emitted positrons and transfer of the resulting gamma–ray photons. The resulting estimates suggest that nearby Galactic fast novae could yield detectable fluxes of electron–positron annihilation gamma rays produced by the decay of ^{13}N and ^{18}F.

065.120   Three–dimensional hydrodynamical simulations of stellar collisions. I. Equal–mass main–sequence stars.
W. Benz, J. G. Hills.
Astrophys. J., Vol. 323, No. 2, p. 614 – 628 (1987).
Fully three–dimensional calculations of collisions between identical stars, treated as polytropes of index $n = 1.5$, are presented. Mechanisms of mass loss operating under various encounter conditions are studied in detail. This work is directly applicable to collisions between two lower–main–sequence stars, between two white dwarfs, and between two very massive upper–main–sequence stars whose internal structure is dominated by large convective cores. The results should be applicable to collisions between main–sequence stars in globular clusters and galactic nuclei and to the coalescence of massive stars bound gravitationally in a fragmentation hierarchical structure.

Beta–decay rates of highly ionized heavy atoms in stellar interiors.
See Abstr. 002.075.

The M–type stars.
See Abstr. 003.075.

Dark stars: the evolution of an idea.
See Abstr. 004.042.

The origin and evolution of neutron stars. Proceedings of the 125th IAU Symposium, held in Nanjing, China, 26 – 30 May 1986.
See Abstr. 012.001.

Asteroseismology and SYNOP.
See Abstr. 013.001.

Simple theory of stellar structure. Research with Professor Hayashi.
See Abstr. 013.012.

Order–of–magnitude "theory" of stellar structure.
See Abstr. 014.062.

Search for neutrino bursts from gravitational stellar collapses.
See Abstr. 034.050.

On a possibility of recording a stellar collapse from the interaction of neutrinos and antineutrinos with carbon–12.
See Abstr. 061.009.

Binary stars as sources of iron and of s–process isotopes.
See Abstr. 061.014.

Neutrino transport in stellar matter.
See Abstr. 061.022.

Radio–dating the Galaxy.
See Abstr. 061.033.

Thorium in G–dwarf stars as a chronometer for the Galaxy.
See Abstr. 061.034.

Radiative transitions in ^{27}Al and their relevance to the ^{27}Al$(p,\gamma)^{27}$Si reaction under stellar conditions.
See Abstr. 061.061.

Laboratory limits on solar axions from an ultralow–background germanium spectrometer.
See Abstr. 061.068.

Stellar reaction rate of ^{14}N$(p,\gamma)^{15}$O and hydrogen burning in massive stars.
See Abstr. 061.091.

Astrophysical aspects of the ^{25}Mg$(p,\gamma)^{26}$Al reaction.
See Abstr. 061.092.

Astrophysical S(E) factor of ^{3}He$(^3$He, 2p$)^4$He at solar energies.
See Abstr. 061.093.

Current status of nuclear astrophysics.
See Abstr. 061.094.

Stellar krypton cross sections at kT = 25 and 52 keV.
See Abstr. 061.108.

Neutrino transport in relativity.
See Abstr. 061.112.

Low–energy cross sections for ^{11}B$(p, 3\alpha)$.
See Abstr. 061.119.

Paraphotons and axions: similarities in stellar emission and detection.
See Abstr. 061.123.

Neutrinos.
See Abstr. 061.125.

Resonant neutrino oscillations and stellar collapse.
See Abstr. 061.154.

The stability of differentially rotating self–gravitating gas clouds. II. Polytropic configurations.
See Abstr. 062.010.

Implicit 2D–radiation hydrodynamics.
See Abstr. 062.125.

A composite stellar model of geostrophic flows. I.
See Abstr. 062.140.

Hydrodynamics in curvilinear coordinates.
See Abstr. 062.143.

Nonlinear magnetohydrodynamic waves in a steady zonal circulation for a shallow fluid shell on the surface of a rotating sphere.
See Abstr. 062.156.

Fragmentation of rapidly rotating gas clouds. I. A universal criterion for fragmentation.
See Abstr. 062.161.

A new example of the Papaloizou–Pringle instability.
See Abstr. 062.162.

Effect of ion correlations on free–free opacities.
See Abstr. 063.034.

Reexamination of the metal contribution to astrophysical opacity.
See Abstr. 063.111.

Mass–loss of globular cluster red giants. A semi–empirical estimation.
See Abstr. 064.030.

NLTE models for cocoon stars.
See Abstr. 064.044.

Gravitational radiation from rotating gravitational collapse.
See Abstr. 066.020.

Rapidly rotating general relativistic polytropes.
See Abstr. 066.026.

Propagation of a shock wave in a radiating spherically symmetric distribution of matter.
See Abstr. 066.067.

Gravitational radiation from realistic relativistic stars: odd–parity fluid perturbations.
See Abstr. 066.099.

Stellar resonant oscillations coupled to gravitational waves.
See Abstr. 066.104.

Realistic exact solutions of Einstein's field equations.
See Abstr. 066.146.

Numerical relativistic gravitational collapse with spatial time slices.
See Abstr. 066.153.

The birth of neutron stars.
See Abstr. 067.012.

Numerical experiments and neutron star formation.
See Abstr. 067.013.

Neutron star formation in theoretical supernovae – low mass stars and white dwarfs.
See Abstr. 067.014.

Neutron–star models in Ni's theory of gravity.
See Abstr. 067.039.

Core collapse and formation of a rapidly rotating neutron star.
See Abstr. 067.064.

The effect of heating from the boundary layer on accretion models for novae and other compact objects.
See Abstr. 067.129.

Curious, quark–like and metastable neutron stars.
See Abstr. 067.132.

A comparison between mass–losing and standard solar models.
See Abstr. 080.033.

Hydrogen–helium–diffusion in solar models.
See Abstr. 080.048.

Be stars as nonradial pulsators.
See Abstr. 112.104.

Wolf–Rayet stars.
See Abstr. 114.029.

On the absence of young white dwarf companions to five technetium stars.
See Abstr. 114.030.

Additional late–type stars with technetium.
See Abstr. 114.031.

Wolf–Rayet stars: current status and major questions.
See Abstr. 114.057.

Neutron–exposure variations in MS and S stars, and the implications for s–process nucleosynthesis.
See Abstr. 114.078.

Does the slope of the initial mass function change for low–mass stars?
See Abstr. 115.009.

Wolf–Rayet, the storm–stars.
See Abstr. 115.015.

Rotation and magnetic activity in main–sequence stars.
See Abstr. 116.021.

The Pleiades rapid rotators: evidence for an evolutionary sequence.
See Abstr. 116.025.

The beat frequency model for QPOs.
See Abstr. 117.004.

A new approach to symbiotic stars.
See Abstr. 117.029.

An evolutionary scenario for the black hole binary A 0620–00.
See Abstr. 117.054.

Spin–down of the white dwarf in the intermediate polar V1223 Sgr/4U 1849–31.
See Abstr. 117.077.

Basic characteristics of evolved binaries according to computations of evolutionary models of close binary systems. I. Catalogue of evolutionary models of close binary systems.
See Abstr. 117.123.

Evolutionary models for Be X–ray binaries.
See Abstr. 117.138.

Binary system parameters and the hibernation model of cataclysmic variables.
See Abstr. 117.145.

The spectral evolution of dwarf nova outbursts.
See Abstr. 117.166.

On OB–type close binary stars.
See Abstr. 117.192.

The Roche coordinates in non–synchronous binaries.
See Abstr. 117.236.

Precession and the long–time magnetic variables.
See Abstr. 117.272.

The period distribution and evolution of short–period cataclysmic variables.
See Abstr. 117.319.

Time–dependent bow shocks and Herbig–Haro objects.
See Abstr. 121.034.

Radio and infrared properties of young stars.
See Abstr. 121.038.

Long period variables and stellar mass loss.
See Abstr. 122.002.

Hydrodynamic models for the short–period, classical Cepheid, SU Cas.
See Abstr. 122.009.

The complex structure of Cas A. Consistent model calculations.
See Abstr. 125.027.

Supernovae and supernova remnants.
See Abstr. 125.031.

Type II supernova models: nucleosynthesis and isotopic anomalies.
See Abstr. 125.035.

A critical study of Type II supernovae: equations of state and general relativity.
See Abstr. 125.036.

Supernova explosion in a very massive star.
See Abstr. 125.037.

Prospects for gamma–ray line observations of individual supernovae.
See Abstr. 125.086.

Supernova 1987A in the Large Magellanic Cloud: the explosion of a $\sim 20\ M_\odot$ star which has experienced mass loss?
See Abstr. 125.215.

Neutrinos from SN 1987A.
See Abstr. 125.217.

Supernova theory and Supernova 1987A.
See Abstr. 125.221.

Light–curve models for supernova SN1987A in the Large Magellanic Cloud.
See Abstr. 125.235.

An interacting binary model for SN 1987A.
See Abstr. 125.236.

Prospects for observations of nucleosynthetic gamma–ray lines and continuum from SN 1987A.
See Abstr. 125.248.

Supernova 1987A in the Large Magellanic Cloud: possible s–process enhancements in the progenitor.
See Abstr. 125.296.

The birthrate and initial spin period of single radio pulsars.
See Abstr. 126.054.

The initial–final mass relation: galactic disk and Magellanic Clouds.
See Abstr. 126.082.

Star formation in molecular clouds: observation and theory.
See Abstr. 131.066.

Protostellar formation in rotating interstellar clouds. VI. Nonuniform initial conditions.
See Abstr. 131.077.

Hydrogen absorption line profiles of ionising star clusters.
See Abstr. 131.202.

The crucial role of cooling in the making of molecular clouds and stars.
See Abstr. 131.303.

The evolution of central stars of planetary nebulae: pop. I vs. pop. II.
See Abstr. 134.047.

Massive disk formation resulting from the collision of a main-sequence star with a white dwarf in a globular cluster core.
See Abstr. 151.028.

Weiße Zwerge in Sternhaufen: ein Schlüssel zur Bestimmung der Anfangs–Endmassen–Beziehung in der Sternentwicklung.
See Abstr. 153.004.

Theoretical stellar luminosity functions and the ages and compositions of globular clusters.
See Abstr. 154.002.

The chicken came first.
See Abstr. 154.016.

Globular clusters and stellar evolution.
See Abstr. 154.058.

A flux–limited sample of Galactic carbon stars.
See Abstr. 155.152.

Integrated properties of star clusters in the LMC.
See Abstr. 156.045.

The development of the red giant branch in intermediate age clusters of the Magellanic Clouds.
See Abstr. 156.047.

Synthetic clusters in the Large Magellanic Cloud.
See Abstr. 156.048.

The evolution of asymptotic giant branch stars in the Magellanic Clouds. III. The problem of intermediate–mass stars.
See Abstr. 156.054.

The chemical evolution of galaxies.
See Abstr. 157.105.

The ratio numbers of carbon and oxygen stars in galaxies.
See Abstr. 157.122.

Chemical evolution of elliptical galaxies.
See Abstr. 157.208.

A model of spectrophotometric evolution for high–redshift galaxies.
See Abstr. 157.210.

Pregalactic–primordial low–mass stars.
See Abstr. 161.066.

066 Relativistic Astrophysics, Gravitation Theory

066.001 Why is G the least precisely known physical constant?
C. C. Speake, G. T. Gillies.
Z. Naturforsch., A, Vol. 42, No. 7, p. 663 – 669 (1987).

CODATA has recently published its readjustment of the fundamental physical constants and assigns a relative precision of 128×10^{-6} to G, the Newtonian constant of gravitation. Given that most of the other constants in physics have relative precisions of $\sim 10^{-6}$ or better, the author examines the reasons why the value for G remains so imprecise: The role of G in physics in general is considered and the most recent experimental determinations are examined. Constraints are given for perturbing effects in G measurements and a key result is that horizontal ground movements must be taken more carefully into account in future more precise terrestrial experiments.

066.002 Gravitationslinsen.
U. Borgeest, R. Kayser, S. Refsdal.
Phys. Bl., 43. Jahrg., Heft 7, p. 219 – 226 (1987).

066.003 Gravitational field mass.
R. Penrose.
Gravitational collapse and relativity, p. 43 – 59 (1986). – See Abstr. 012.006.

The author's definition for the mass–momentum/angular momentum surrounded by a spacelike 2–surface with S^2 topology is presented, this definition being motivated by some ideas from twistor theory in relation to linearized gravitational theory. The status of this definition is examined in relation to many examples which have been worked out. The reason for introducing a slight modification to the original definition is also presented.

066.004 On the propagation problem in gravitational radiation theory.
T. Damour.
Gravitational collapse and relativity, p. 63 – 73 (1986). – See Abstr. 012.006.

This paper discusses the relative merits of analytical and numerical methods in dealing with gravitational radiation problems. Analytical methods are needed, on the one hand to provide boundary conditions to numerical codes, and on the other hand to relate the gravitational field at the outer edge of the grid with the asymptotic outgoing wave form. The author presents an explicit formula which may be useful in computing the gravitational wave form emitted during the three–dimensional collapse of a star.

066.005   The new formula for gravitational radiation energy loss and its applications.
F. I. Cooperstock, P. H. Lim.
Gravitational collapse and relativity, p. 74 – 80 (1986). – See Abstr. 012.006.
The authors describe and discuss the essential features which led to the development of a new radiation formula. The results of its application to the problem of fall in a line and results to date on its application to the circular rotating binary problem are discussed.

066.006   The strong field point particle limit and the quadrupole formula in the binary pulsar system.
T. Futamase.
Gravitational collapse and relativity, p. 81 – 87 (1986). – See Abstr. 012.006.
The author constructs an asymptotic approximation in general relativity to a binary pulsar system by taking a point–particle limit along a sequence of spacetimes. The near–zone calculation is performed up to radiation reaction order. It is found that the rate of loss of Newtonian energy for the orbital motion exactly balances the energy carried by the radiation in the far zone, and is given by the quadrupole formula. The equation of spin precession is also calculated from the near zone metric and the same result as for weak gravity is obtained.

066.007   Surface integrals associated with the canonical formalism on a null surface.
J. N. Goldberg.
Gravitational collapse and relativity, p. 88 – 95 (1986). – See Abstr. 012.006.
The author describes an approach to quantum gravity based on the study of the gravitational radiation field in asymptotically flat space–times.

066.008   Positive energy and Poincaré gauge theory.
D.–C. Chern, J. M. Nester.
Gravitational collapse and relativity, p. 106 – 113 (1986). – See Abstr. 012.006.
Some alternative theories of gravity fit the experimental evidence just as well as Einstein's theory. Such is the case for certain Poincaré gauge theories. Hence, there is a need for a strong theoretical test. The requirement of positive total energy, recently established for Einstein's theory, promises to fulfill this need. Preliminary results indicate that this test will eliminate almost all of the otherwise viable Poincaré gauge theories.

066.009   Computer–aided classification of exact solutions in general relativity.
M. A. H. MacCallum.
Gravitational collapse and relativity, p. 127 – 140 (1986). – See Abstr. 012.006.
The theory and practical implementation of the procedure for providing a unique local invariant characterisation of a spacetime is discussed: this enables one to study whether two solutions of Einstein's field equations locally describe the same spacetime.

066.010   Derivation of Nutku–Halil colliding plane wave solution from isotropic Kasner metric using double–Harrison transformation.
F. J. Ernst.
Gravitational collapse and relativity, p. 141 – 149 (1986). – See Abstr. 012.006.
The author presents a derivation of the Nutku–Halil colliding wave solution from the isotropic Kasner metric using a double–Harrison (Bäcklund) transformation. The method when applied to other seed metrics is expected to supply useful candidates for new colliding wave solutions. All calculations were carried out using a symbolic evaluation program designed for manipulating differential forms on MS–DOS microcomputers.

066.011   Analysis of a directional singularity.
S. M. Scott.
Gravitational collapse and relativity, p. 158 – 175 (1986). – See Abstr. 012.006.
An indepth analysis of the directional singularity of the Curzon solution is presented. New properties of geodesics and other curves lying in the spatial sections t = constant are derived. These lead to the construction of a new compactified coordinate system in which the real singularity of the Curzon solution appears as a ring with finite radius. A new region of spacelike infinity is also revealed on the "other side" of this ring, and is approached by geodesics threading through the ring.

066.012   On multipole moments in general relativity.
C. Hoenselaers.
Gravitational collapse and relativity, p. 176 – 184 (1986). – See Abstr. 012.006.

066.013   Frame bundles: covariance, invariant forms and exterior differential systems for Ricci–flat solutions.
F. B. Estabrook.
Gravitational collapse and relativity, p. 185 – 196 (1986). – See Abstr. 012.006.

066.014   Solvable lattice models and the elliptic theta functions.
M. Jimbo, T. Miwa, M. Okado.
Gravitational collapse and relativity, p. 197 – 203 (1986). – See Abstr. 012.006.

066.015   Some linear prolongations of the vacuum gravitational field equations.
M. Gürses.
Gravitational collapse and relativity, p. 204 – 210 (1986). – See Abstr. 012.006.
Two types of linear prolongations of the vacuum gravitational field equations are given. The first one is shown to produce non-trivial Bäcklund transformations and the second one is given in a form that resembles the prolongation of the self–dual Yang–Mills and of the sigma–model field equations.

066.016   Clifford algebras in general relativity.
F. J. Chinea.
Gravitational collapse and relativity, p. 211 – 220 (1986). – See Abstr. 012.006.

066.017   Exact solutions, numerical relativity and gravitational radiation.
J. Winicour.
Gravitational collapse and relativity, p. 221 – 230 (1986). – See Abstr. 012.006.
Contents: 1. Introduction. 2. Oppenheimer–Snyder collapse. 3. Boost–rotation symmetric solutions.

066.018   Bianchi type–I solutions in Brans–Dicke theory with torsion.
S.–W. Kim, B. H. Cho.
Gravitational collapse and relativity, p. 231 – 238 (1986). – See Abstr. 012.006.
A modified Brans–Dicke theory is investigated in Bianchi type–I space–time models with isotropic dust. It is shown that the scalar field and the spin cannot exist simultaneously.

066.019   An interpretation of the Hermitian theory of relativity.
S. Antoci.
Gravitational collapse and relativity, p. 239 – 246 (1986). – See Abstr. 012.006.
This paper demonstrates that an interpretation of the Hermitian theory of relativity is possible, where a unique tensor field represents both electromagnetic and strong interactions.

066.020 **Gravitational radiation from rotating gravitational collapse.**
R. F. Stark, T. Piran.
Gravitational collapse and relativity, p. 249 – 275 (1986). – See Abstr. 012.006.
The authors present results for the waveform, polarization and efficiency of the gravitational radiation emission from the collapse to a black hole of rotating relativistic polytropic stars. The dependence of black hole formation on the star's angular momentum, and the hydrodynamics of the collapse, are also discussed. These results have been obtained using a fully general relativistic computer code that evolves rotating axisymmetric configurations and directly computes their gravitational radiation emission.

066.021 **Gravitational radiation in the 3–D processes such as fragmentation and collision of fluids.**
S. M. Miyama, M. Nagasawa, T. Nakamura.
Gravitational collapse and relativity, p. 282 – 294 (1986). – See Abstr. 012.006.
Gravitational radiation from a non–axisymmetric collapse of a rotating fluid or a non–head–on collision between two fluids is calculated by using a three–dimensional, time–dependent, Newtonian hydrodynamical code. The fluid is treated in an isothermal or an adiabatic ($\Gamma = 5/3$) approximation. Gravitational radiation is computed by using Landau–Lifshitz's formula as a small perturbation. For a non–axisymmetric collapse leading to fragmentation of the fluid, the efficiency $\Delta E_{GR}/Mc^2$ is much larger than for an axisymmetric collapse.

066.022 **Gravitational radiation and 3D numerical relativity.**
T. Nakamura.
Gravitational collapse and relativity, p. 295 – 312 (1986). – See Abstr. 012.006.
Recent research activities on numerical relativity are reviewed. Main topics discussed are 2D rotating collapse, phase cancellation effects and perturbation calculation of the gravitational radiation from a particle falling into a black hole. New numerical results on 3D time evolution of pure gravitational waves are also presented.

066.023 **Numerical quantum gravity.**
J. B. Hartle.
Gravitational collapse and relativity, p. 329 – 338 (1986). – See Abstr. 012.006.
Potential and actual applications of numerical simulations to issues in quantum gravity are discussed and compared with numerical simulations in the classical theory.

066.024 **Regge calculus in numerical relativity.**
M. R. Dubal.
Gravitational collapse and relativity, p. 339 – 349 (1986). – See Abstr. 012.006.
Regge calculus offers a fundamentally different approach to numerical relativity from that of finite differencing of partial differential equations. As a test it is shown that, in a $3+1$ form with a velocity potential representation of the hydrodynamics, this approach can model astrophysically interesting spacetimes quite accurately. The dynamical collapse of polytropic spheres is studied in particular.

066.025 **Hubble's constant from gravitational wave observations.**
B. F. Schutz.
Gravitational collapse and relativity, p. 350 – 368 (1986). – See Abstr. 012.006.
This paper demonstrates that by observing the gravitational radiation from an ultracompact binary consisting of two neutron stars the absolute distance of the system can be derived. It is suggested that these observations may lead to a reliable method for determining the Hubble constant.

066.026 **Rapidly rotating general relativistic polytropes.**
H. Komatsu, Y. Eriguchi, I. Hachisu.
Gravitational collapse and relativity, p. 369 – 376 (1986). – See Abstr. 012.006.
The authors have generalized the self–consistent field method used in the computation of Newtonian rotating star to general relativistic polytropes. The remarkable result obtained is that the mass increase due to rotation is much less (less than 1%) for differential rotation than for uniform rotation.

066.027 **Are extra dimensions visible near the event horizon? Black holes coupled to scalar fields.**
A. Tomimatsu.
Gravitational collapse and relativity, p. 417 – 425 (1986). – See Abstr. 012.006.
The author studies in the framework of the Kaluza–Klein theory some properties of the spherically symmetric gravitational field coupled to a massless scalar field. He then discusses the final state of gravitational collapse. The static field with a naked singularity is unstable and generates a large disturbance of the scalar waves. It is suggested that the extra space can enlarge with the lapse of time to a visible size near the disturbed black hole.

066.028 **The Einstein gravity as an attractor in higher–dimensional theories.**
K.–i. Maeda.
Gravitational collapse and relativity, p. 426 – 433 (1986). – See Abstr. 012.006.
In higher–dimensional theories with a suitable potential, the 4–dim Einstein gravity is always realized as an attractor in the dynamical system if the 3–space is expanding. On the other hand, if the 3–space is contracting, the potential minimum is no longer an attractor. The space–time leaves it, leading to the blowing up of the internal space and/or to a Jordan–Brans–Dicke gravity.

066.029 **Can mini–superspace quantization be justified?**
K. V. Kuchař, M. P. Ryan Jr.
Gravitational collapse and relativity, p. 451 – 464 (1986). – See Abstr. 012.006.
The authors show in simple examples taken from ordinary quantum mechanics and field theory that, under certain conditions, a mini–superspace model can be regarded as an approximation to the full quantum theory.

066.030 **Quantum effects in non static black hole space–times.**
R. Balbinot.
Gravitational collapse and relativity, p. 482 – 490 (1986). – See Abstr. 012.006.
Quantum fields propagating in two–dimensional black hole space–times are discussed. The expectation values of the renormalized stress energy tensor are computed on the event horizon of the hole. The effects induced by the quantized fields are then examined.

066.031 **Nonzero effect of the Eötvös experiment?**
Y. Fujii.
Gravitational collapse and relativity, p. 491 – 499 (1986). – See Abstr. 012.006.
A critical review of a recent claim by Fischbach et al. on the nonzero effect of the Eötvös experiment is presented together with a brief sketch of the theoretical implications of a possible finite–range force.

066.032 **On detecting stochastic background gravitational radiation with terrestrial detectors.**
P. F. Michelson.
Mon. Not. R. Astron. Soc., Vol. 227, No. 4, p. 933 – 941 (1987).
Several mechanisms for generating stochastic gravitational radiation have been postulated that lead to radiation in the frequency range accessible to terrestrial detectors ($\sim 1 - 10^3$Hz). The signal–to–noise ratio in a cross–correlation experiment with two detectors that have arbitrary relative angular orientation and separation is discussed.

066.033   **A search for sinusoidal gravitational radiation in the period range 30 – 2000 seconds.**
J. W. Armstrong, F. B. Estabrook, H. D. Wahlquist.
Astrophys. J., Vol. 318, No. 2, p. 536 – 541 (1987).

The Doppler tracking time series of a distant spacecraft can be used to search for very low frequency gravitational radiation. Earth and the spacecraft act as free test masses with the radio tracking system continuously measuring their relative dimensionless velocity, $\Delta v/c$. An incident gravitational wave produces small perturbations in this Doppler tracking record. The authors report here a search for sinusoidal gravitational radiation using Pioneer 11 tracking data taken in the spring of 1983. For the period range 30 – 2000 s the search yielded results consistent with noise only; the one standard deviation sensitivity threshold for the amplitude of sinusoidal fractional frequency fluctuations was 1.5×10^{-14}.

066.034   **Strings from gravity.**
G.'t Hooft.
Phys. Scr., Vol. T15, p. 143 – 150 (1987). – See Abstr. 012.014.

The quantum properties of black holes are compared with those of elementary and composite particles. It is desirable to search for a theory of black holes in which quantum mechanical "information" is not drained away by the horizon, but such a theory requires a drastically new approach in formulating general coordinate transformations with horizons in a quantum theory. It is subsequently shown that a closed string with string tension $T = 1/8\pi G$ reproduces in a remarkable way the horizon fluctuations so that a new geometric interpretation of strings is suggested.

066.035   **On "radiating Vaidya metric imbedded in de Sitter space".**
D. Bhattacharyya, D. Ray.
Gen. Relativ. Gravitation, Vol. 19, No. 7, p. 651 – 654 (1987).

Mallett has generalized the Vaidya metric as well as the de Sitter metric to obtain what he called the "Vaidya–Schwarzschild–de Sitter metric" and has obtained the condition under which this metric represents a pure radiation field. The present note obtains the complete first integral of that equation as well as some particular solutions.

066.036   **Pseudotensors in asymptotically curvilinear coordinates.**
E. Nahmad–Achar, B. F. Schutz.
Gen. Relativ. Gravitation, Vol. 19, No. 7, p. 655 – 663 (1987).

The authors show how to calculate pseudotensor–based conserved quantities for isolated systems in general relativity, in a way which allows an arbitrary asymptotic behavior of the coordinate system used. The method allows the asymptotic evaluation of energy, momentum, and angular momentum in any coordinate system. The authors carry out the calculation for the Schutz–Sorkin gravitational Noether operator.

066.037   **Proposal for an interpretation of the Hermitian theory of relativity.**
S. Antoci.
Gen. Relativ. Gravitation, Vol. 19, No. 7, p. 665 – 679 (1987).

The equilibrium conditions for charges and currents, apparent in exact solutions of the field equations, lead one to regard the Hermitian theory of relativity as the theory of a field endowed with two sources: electromagnetic and color four–currents.

066.038   **Reversible evolution of charged ergoregions.**
K. Kokkotas, N. Spyrou.
Gen. Relativ. Gravitation, Vol. 19, No. 7, p. 681 – 691 (1987).

The reversible evolution of a charged rotating ergoregion, due to the injection into it of particles with mass–energy and angular momentum, is studied systematically. As in the uncharged case, a bulge always forms on the outer boundary of the ergoregion due to the latter's angular momentum. The behavior of the bulge's position, relative to the black hole's rotation axis and equatorial plane, is studied, on the basis of the cosmic censorship hypothesis, during the ergoregion's reversible evolution.

066.039   **Some real and complex solutions of Einstein's equations.**
D. C. Robinson.
Gen. Relativ. Gravitation, Vol. 19, No. 7, p. 693 – 698 (1987).

The construction of some real solutions of Eintein's vacuum field equations from certain half flat holomorphic metrics is described.

066.040   **Ghost neutrino fields in flat space–time.**
G. F. Torres del Castillo.
Gen. Relativ. Gravitation, Vol. 19, No. 7, p. 699 – 705 (1987).

The Weyl neutrino equation is integrated in flat space–time assuming that the energy–momentum tensor of the neutrino field vanishes. It is shown that the flux vector of the neutrino field is tangent to a twist–free and shear–free congruence of null geodesics, which is a special Robinson congruence and constitutes a geometrical representation of a null twistor.

066.041   **On the universality of Einstein equations.**
A. Jakubiec, J. Kijowski.
Gen. Relativ. Gravitation, Vol. 19, No. 7, p. 719 – 727 (1987).

It is proved that a Lagrangian field theory based on a linear connection in space–time is equivalent to Eintein's general relativity interacting with additional matter fields.

066.042   **Exact solution of a static charged sphere in general relativity.**
X. Wang.
Gen. Relativ. Gravitation, Vol. 19, No. 7, p. 729 – 737 (1987).

An exact solution of the field equations of general relativity is obtained for a static, spherically symmetric distribution of charge and mass. Their physical properties are studied in some detail.

066.043   **Wormhole solutions in the Einstein–Yang–Mills–Higgs system: solution of first–order equations for $G = SU(2)$.**
F. Degen.
Gen. Relativ. Gravitation, Vol. 19, No. 7, p. 739 – 761 (1987).

For an $SU(2)$ Einstein–Yang–Mills–Higgs model the author studies the extreme wormhole solutions. He uses an iterative method based on expansion in the radial distance N from the boundary of the hole. Here the nontrivial solutions of the first–order equations are presented. They give useful information about existing extremal wormholes. A nonabelian solution exists if the value of the Higgs scalar at the horizon is equal to the Planck mass and if the magnetic charge b and the electric charge e of the hole satisfy $b = 1/e$.

066.044   **Sectional curvature in general relativity.**
G. S. Hall, A. D. Rendall.
Gen. Relativ. Gravitation, Vol. 19, No. 8, p. 771 – 789 (1987).

This paper gives a detailed account of the sectional curvature function in general relativity. Some recent results are rederived by more systematic methods and some new results are obtained. Symmetries of the sectional curvature are also considered, as is the topological structure of the space of sectional curvature functions of all Lorentz metrics on a given four–dimensional manifold.

066.045   **General relativistic electromagnetic mass models of neutral spherically symmetric systems.**
J. Ponce de León.
Gen. Relativ. Gravitation, Vol. 19, No. 8, p. 797 – 807 (1987).

The recent work of Grøn concerning charged analogues of Florides' class of solutions is discussed and generalized. The properties of this kind of model are investigated. In particular it is shown that the ratio m/r as well as the acceleration of gravity are maximum inside the body rather than at the boundary. Some exact solutions of the Einstein–Maxwell equations illustrating these properties are presented. The solutions are matched continuously to the exterior Schwarzschild solution and they represent electromagnetic mass models of neutral systems.

066.046 Gravitational field with toroidal topology.
B. W. Stewart, L. Witten, D. Papadopoulos.
Gen. Relativ. Gravitation, Vol. 19, No. 8, p. 827 – 839 (1987).

The Weyl formalism for static, axially symmetric solutions to Einstein's equations is employed to examine solutions of Einstein's equations with toroidal topology. The authors investigate a particular family of solutions that contain the Schwarzschild, Curzon, Bach–Weyl, and γ metric (also known as the Zipoy–Voorhees metric) as special cases.

066.047 The verification of Killing tensor components for metrics in general relativity using the computer algebra system SHEEP.
G. C. Joly.
Gen. Relativ. Gravitation, Vol. 19, No. 8, p. 841 – 845 (1987).

The author reports on a program, written in the computer algebra system SHEEP, for verifying the components of Killing tensors and conformal Killing tensors. Some examples are given, including the components of the Killing tensor admitted by the Kerr metric.

066.048 The gravitational field in the wave zone. I. The radiating terms of the metric tensor.
S. Persides.
Gen. Relativ. Gravitation, Vol. 19, No. 9, p. 847 – 870 (1987).

The author considers an asymptotically flat space–time generated by a perfect fluid source of compact spatial support. Using the de Donder gauge conditions, the Einstein equations are reduced to a new form of Poisson–type equations. A formal iterative scheme is set up to solve these equations by expanding the components of the metric tensor in powers of c^{-1}.

066.049 The gravitational field in the wave zone. II. Consequences of asymptotic structure.
S. Persides.
Gen. Relativ. Gravitation, Vol. 19, No. 9, p. 871 – 886 (1987).

The author considers an asymptotically flat and empty space–time generated by a bounded source of perfect fluid. A simplified expression is derived for the coefficient of r^{-1} of the metric tensor after an expansion in powers of c^{-1}. The result is a very simple expression for the dominant term of $g_{\alpha\beta,0}$ in the radiation zone in terms of the quadrupole moment of the source. The author calculates in the framework of full general relativity the radiated energy per unit time and proves that the first term is identical with the quadrupole radiation as given by the linearized version of general relativity.

066.050 A class of spherically symmetric solutions with conformal Killing vectors.
C. C. Dyer, G. C. McVittie, L. M. Oattes.
Gen. Relativ. Gravitation, Vol. 19, No. 9, p. 887 – 898 (1987).

Spherically symmetric solutions with a conformal Killing vector in the (r, t) surface allow the null geodesics to be found with relative ease. Knowledge of the null geodesics is essential to calculating the optical properties of a solution via the optical scalar equations. Solutions of this type may be useful for the treatment of the optical properties of an inhomogeneous universe. The authors address the question of whether the large class of spherically symmetric solutions found by McVittie possess conformal symmetry.

066.051 On the conformal transformations of metrics in Rosen's bimetric gravitation theory.
I. Lukačević.
Gen. Relativ. Gravitation, Vol. 19, No. 9, p. 907 – 916 (1987).

The conformal transformations of the background metric are considered and the transformed expressions of the gravitational energy and of the gravitational field equations are obtained. The conditions of conservation of the nongravitational energy are formulated with respect to the transformed metric.

066.052 A family of rotating disks as sources of the Kerr metric.
V. H. Hamity, W. Lamberti.
Gen. Relativ. Gravitation, Vol. 19, No. 9, p. 917 – 925 (1987).

A method of constructing a family of regular rotating disks as sources of the Kerr metric is discussed. The algebraic type of the energy–momentum tensor is analyzed, and it is found that none of the disks satisfies the dominant energy condition and, in some cases, even the weak energy condition is violated.

066.053 Gravitational field of the quantized electromagnetic plane wave.
T. Padmanabhan.
Gen. Relativ. Gravitation, Vol. 19, No. 9, p. 927 – 938 (1987).

The quantum and classical descriptions of an electro–magnetic field are connected by the correspondence principle. The author considers the electromagnetic field as a source for gravity and compares the metrics due to a classical and quantized electromagnetic field. The quantization of the source demands the quantization of gravity.

066.054 Bianchi type V perfect fluid models with source–free electromagnetic fields.
B. K. Nayak, G. B. Bhuyan.
Gen. Relativ. Gravitation, Vol. 19, No. 9, p. 939 – 948 (1987).

The Einstein–Maxwell field equations characterizing a non-tilted Bianchi type V perfect fluid model with source–free electromagnetic field are solved exactly in the nonlocally rotationally symmetric case. It is found that these equations admit one and only one exact solution, expressible, however, in terms of two arbitrary functions.

066.055 Consistency of field equations in "self–creation" cosmologies.
C. H. Brans.
Gen. Relativ. Gravitation, Vol. 19, No. 9, p. 949 – 952 (1987).

It is pointed out that the field equations in one of Barber's two "self–creation" cosmologies are not only in disagreement with experiment, but are actually inconsistent, in general. The construction of consistent general relativistic field equations involving field variables, without invoking Lagrangian techniques, requires careful checking that sufficient functional freedom has been provided so as to produce a consistent set of equations.

066.056 Possible effects on the sun and close binary systems from background gravitational radiation with period 160 min.
A. G. Kosovichev.
Bull. Crimean Astrophys. Obs., Vol. 75, p. 30 – 35 (1987). English translation of 42.066.157.

066.057 Birkhoff–type theorem in the scale–covariant theory of gravitation.
D. R. K. Reddy, R. Venkateswarlu.
Astrophys. Space Sci., Vol. 136, No. 1, p. 191 – 194 (1987).

It is shown that an analog of Birkhoff's theorem of general relativity exists in the scale–covariant theory of gravitation when the gauge function which occurs in the theory is independent of time.

066.058 An analytically soluble problem in fully nonlinear statistical gravitational lensing.
P. Schneider.
Astrophys. J., Vol. 319, No. 1, p. 9 – 13 (1987).

The amplification probability distribution $p(I)dI$ for a point source behind a random star field which acts as the deflector exhibits a I^{-3} behavior for large amplification, as can be shown from the universality of the lens equation near critical lines. In this paper it is shown that the amplitude of the I^{-3} tail can be derived exactly for arbitrary mass distribution of the stars, surface mass density of stars and smoothly distributed matter, and large–scale shear.

066.059   Gravity waves: a progress report.
V. Trimble.
Sky Telesc., Vol. 74, No. 4, p. 364 – 369 (1987).

066.060   In pursuit of the fifth force.
E. Iacopini.
Nature, Vol. 328, No. 6131, p. 578 – 579 (1987).
This note describes recent and future experimental searches for a short–range departure from Newton's law of gravity and for deviations from g–universality. Positive results of these searches would imply the existence of a fifth fundamental interaction modifying Newtonian gravity at intermediate ranges.

066.061   Experiments on gravitation.
A. H. Cook.
Three hundred years of gravitation, p. 49 – 79 (1987). – See Abstr. 003.012.
Contents: 1. Introduction. 2. Theoretical framework. 3. The inverse square law. 4. The weak equivalence principle. 5. Preferred frames and locations. 6. Additional deviations from general relativity. 7. The measurement of the constant of gravitation. 8. Conclusion.

066.062   Experimental gravitation from Newton's *Principia* to Einstein's general relativity.
C. M. Will.
Three hundred years of gravitation, p. 80 – 127 (1987). – See Abstr. 003.012.
Contents: 1. Introduction. 2. Newtonian gravitation and general relativity: commonalities and contrasts. 3. Testing the fundamentals of gravitation theory. 4. Testing the applications of gravitation theory. 5. Beyond the tercentenary. Appendix: The parametrized post–Newtonian formalism.

066.063   The problem of motion in Newtonian and Einsteinian gravity.
T. Damour.
Three hundred years of gravitation, p. 128 – 198 (1987). – See Abstr. 003.012.
1. Introduction. 2. The N–extended–body problem in Newtonian gravity. 3. The external and the internal problems of motion (Newtonian case). 4. The effacement of internal structure in the external problem. 5. The effacement of external structure in the internal problem. 6. Newton and the strong principle of equivalence. 7. Solving the Newtonian problems of motion. 8. The N–extended–body problem in Einsteinian gravity. 9. Approximation methods. 10. The post–Newtonian approximation methods. 11. The post–Minkowskian approximation methods. 12. Singular perturbation methods. 13. The external and the internal problems of motion (Einsteinian case). 14. The effacement of internal structure in the external problem (Einsteinian case). 15. The problem of gravitational radiation damping and the relativistic Laplace effect. 16. Conclusion.

066.064   Gravitational radiation.
K. S. Thorne.
Three hundred years of gravitation, p. 330 – 458 (1987). – See Abstr. 003.012.
Contents: 1. Introduction. 2. The physical and mathematical description of a gravitational wave. 3. The generation and propagation of gravitational waves. 4. Astrophysical sources of gravitational waves. 5. Detection of gravitational waves. 6. Conclusion.

066.065   Gravitational interactions of cosmic strings.
A. Vilenkin.
Three hundred years of gravitation, p. 499 – 523 (1987). – See Abstr. 003.012.
Contents: 1. Introduction. 2. Some properties of strings. 3. Nambu action. 4. Gauge conditions. 5. Oscillating loops. 6. Waves on infinite strings. 7. Gravitational field of a straight string. 8. Gravitational field of a straight string (continued). 9. Physics in conical space. 10. Other metrics with straight strings. 11. Gravitational field of oscillating loops. 12. Grav-

itational radiation from loops. 13. Gravitational rocket effect. 14. Gravitational field of waves on infinite strings.

066.066   Gravitational wave astronomy – potential and possible realisation.
J. Hough, B. J. Meers, G. P. Newton, N. A. Robertson, H. Ward, B. F. Schutz, I. F. Corbett, R. W. P. Drever.
Vistas Astron., Vol. 30, Part 2, p. 109 – 134 (1987).
There are a number of interesting astrophysical sources of gravitational waves including coalescing compact binary star systems, stellar collapses and rotating neutron stars, and to detect all of these is likely to require a strain sensitivity better than 10^{-22} over a bandwidth of a few hundred Hz at frequencies at or below 1 kHz. To achieve such sensitivity requires considerable experimental ingenuity; however work in a number of laboratories suggests that such performance should be attainable using laser interferometry between freely suspended masses separated by a distance of the order of a kilometre. This paper includes a review of possible sources and outlines methods of detection currently being developed or planned, with particular emphasis on long baseline laser interferometers.

066.067   Propagation of a shock wave in a radiating spherically symmetric distribution of matter.
L. Herrera, L. Núñez.
Astrophys. J., Vol. 319, No. 2, p. 868 – 884 (1987).
A method used to study the evolution of radiating spheres reported by L. Herrera, J. Jiménez, and G. Ruggeri in 1980 is extended to the case in which the sphere is divided in two regions by a shock wave front. The equations of state at both sides of the shock are different, and the solutions are matched on it via the Rankine–Hugoniot conditions. The outer region metric is matched with a Vaidya solution on the boundary surface of the sphere. The results are discussed in the light of recent work on gravitational collapse and supernovae.

066.068   Coincidence probabilities for a network of interferometric detectors of gravitational waves.
M. Tinto.
Mon. Not. R. Astron. Soc., Vol. 229, No. 2, p. 315 – 332 (1987).
The coincidence probabilities for a network of Laser Interferometric Gravitational–Wave detectors, observing bursts from coalescing binary systems containing neutron stars or black holes, are calculated by developing the theoretical results deduced in a previous paper.

066.069   On some physical properties and the stability of an exact model for a relativistic star.
H. Knutsen.
NORDITA Prepr., No. 87/9A, 39 pp. (1987).
The physical properties of an exact model for a relativistic fluid sphere found independently by Adler, Kuchowicz and Adams/ Cohen has been investigated in great detail. The author finds that the most important parameter is the degree of compressibility. If this is less than 0.82 the fluid sphere is stable. Then the weak energy condition is fulfilled, and the speed of sound is less than the speed of light. In this case the adiabatic index is also increasing outwards. However, the degree of compressibility must be less than 0.58 for the strong energy condition to hold everywhere inside the sphere.

066.070   On the instability of Buchdahl's model for a gaseous relativistic star.
H. Knutsen.
NORDITA Prepr., No. 87/12A, 16 pp. (1987).
It is shown that Buchdahl's exact solution for a gaseous relativistic star is unstable.

066.071   An exact model for a gaseous regular bouncing sphere in general relativity.
H. Knutsen.
NORDITA Prepr., No. 87/20A, 12 pp. (1987).
An exact model for a relativistic "gaseous" sphere (i.e. the density vanishes at the outer boundary of the nonstatic sphere

together with the pressure) is given. The model has a bounce, i.e. the collapsing sphere comes to rest when the boundary is still outside the Schwarzschild radius of the matter sphere. Then there is a macroscopic bounce, and the matter of the expanding sphere is spread all over the universe. This bouncing solution of Einstein's field equations is physically valid at any moment. The mass function is positive, and the circumference is an increasing function of radial coordinate. This solution may represent an easily surveyable model for a supernova explosion where the explosion is so violent that no remnant whatsoever is left.

066.072 **On the stability and physical properties for an exact relativistic model for a superdense star.**
H. Knutsen.
NORDITA Prepr., No. 87/21A, 28 pp. (1987).
An exact solution of Einstein's field equations for a static spherically symmetric perfect fluid is investigated in detail. It is shown that the model is stable with respect to infinitesimal radial pulsations. The author also finds that the adiabatic speed of sound is smaller than the speed of light everywhere inside the fluid sphere if and only if the radius of the sphere is larger than 1.46 times its Schwarzschild radius. The necessary and sufficient criterion for the sound speed to be decreasing outwards close to the center is given, but if this criterion is fulfilled the fluid must necessarily be supraluminal somewhere. It is further found that the strong energy condition is fulfilled everywhere if it is fulfilled at the origin, and the ratio of the pressure and the density is decreasing outwards. This necessarily yields that the temperature is decreasing outwards.

066.073 **Possible evidence for weak violation of special relativity.**
F. Winterberg.
Z. Naturforsch., A, Vol. 42a, No. 11, p. 1374 – 1375 (1987).

066.074 **An electromagnetic field in the Plebanski space–time.**
M. Ahmed.
Gen. Relativ. Gravitation, Vol. 19, No. 10, p. 953 – 959 (1987).
The electromagnetic wave equation in the Plebanski space–time has been reduced to a one–dimensional wave equation with real short–range potential. It is found that the superradiance phenomenon occurs.

066.075 **Quasi–regular singularities based on null planes.**
R. V. Bruno, L. C. Shepley, G. F. R. Ellis.
Gen. Relativ. Gravitation, Vol. 19, No. 10, p. 973 – 982 (1987).
The authors examine quasi–regular singularities that take the form of invariant two–dimensional null planes in Minkowski space–time, thus extending earlier studies of "conical" singularities based on timelike and spacelike planes. The result is described in terms of a deficit parameter. The form of the Riemann curvature tensor at the singularity is also examined.

066.076 **On the exterior gravitational field of a mass with a multipole moment.**
H. Quevedo.
Gen. Relativ. Gravitation, Vol. 19, No. 10, p. 1013 – 1023 (1987).
Recently, Gutsunaev and Manko presented a procedure for obtaining new static axisymmetric solutions of Einstein's vacuum field equations from a known one. The author shows that this procedure is based on the property that the derivatives of a harmonic function are harmonic. The special case of a metric with mass and quadrupole moment is investigated and compared with the Erez–Rosen metric.

066.077 **Distinguishing properties of causality conditions.**
I. Rácz.
Gen. Relativ. Gravitation, Vol. 19, No. 10, p. 1025 – 1031 (1987).
The author shows that certain causality conditions, such as causality, (future–, past–) distinguishing, strong causality, and stable causality have the common feature that there exists a map from the space–time manifold M to the power set $P(M)$ which is injective if and only if the right causality condition holds on M.

066.078 **Quantum effects in a homogeneous dust cloud collapse.**
P. S. Joshi, S. S. Joshi.
Gen. Relativ. Gravitation, Vol. 19, No. 10, p. 1033 – 1042 (1987).
The status of a classical space–time singularity, when quantum effects are taken into account, has remained a matter of intense interest ever since the epoch–making paper of DeWitt on quantum gravity. The authors examine the evolution of quantum fluctuations in the vicinity of the singularity arising out of the classical collapse of a homogeneous dust cloud. The traditional operator approach to quantum theory is used. It is shown that the quantum uncertainty diverges in the limit of approach to the classically singular epoch and that nonsingular, nonclassical states can occur with finite probability.

066.079 **Møller was right.**
J. Novotný.
Gen. Relativ. Gravitation, Vol. 19, No. 10, p. 1043 – 1052 (1987).
Direct calculation proves that the total energy–momentum vector derived from the Møller energy–momentum complex does not transform like a free 4–vector with respect to the Lorentz transformation. This conforms with the conclusion formulated by Møller himself, but it contradicts the result of the critical analysis of Kovacs.

066.080 **Geometry of space–time singularities and their stability.**
D. Canarutto.
General relativity and gravitational physics, p. 3 – 17 (1987). – See Abstr. 012.038.
The nature of space–time singularities and the problem of their stability are discussed. Some recent results on this subject are reviewed.

066.081 **Smoothing out spacetime geometry.**
M. Carfora, A. Marzuoli.
General relativity and gravitational physics, p. 19 – 34 (1987). – See Abstr. 012.038.
The authors discuss some results in differential geometry of Riemannian manifolds which may allow a significant extension of the smoothing out procedure recently introduced in relativistic cosmology.

066.082 **Solutions of the Einstein equations with data at past infinity.**
F. Cagnac, Y. Choquet–Bruhat, N. Noutcheguene.
General relativity and gravitational physics, p. 35 – 54 (1987). – See Abstr. 012.038.
Many studies in general relativity suppose the existence of solutions of Einstein equations for an infinite time in the past, which tend to a stationary solution at past infinity. In this paper the authors give a proof of the existence, for an infinite proper time, of infinitely many solutions of the system of Einstein equations coupled with conservative sources.

066.083 **Self–gravitating toroidal figures of equilibrium around a rotating black hole: the numerical method.**
A. Lanza.
General relativity and gravitational physics, p. 67 – 82 (1987). – See Abstr. 012.038.
The multigrid method is being applied to solve numerically the stationary and axisymmetric Einstein equations for self–gravitating tori rotating around rapidly rotating black holes. Preliminary results for two test problems, necessary steps for constructing the numerical code, are presented.

066.084 **Exact solutions and the meaning of the Hermitian theory of relativity.**
S. Antoci.
General relativity and gravitational physics, p. 83 – 88 (1987). – See Abstr. 012.038.
Singular external currents are allowed for at the right–hand–sides of two field equations of Hermitian relativity; then particular solutions enable the theory to account for both electrodynamic and strong interactions.

066.085 The possible role of dissipative structures in gravitation theory and cosmology.
E. Pessa.
General relativity and gravitational physics, p. 95 – 99 (1987). – See Abstr. 012.038.

066.086 On the embedding approach to general relativity: a status report.
V. Tapia.
General relativity and gravitational physics, p. 105 – 109 (1987). – See Abstr. 012.038.

The author considers four–dimensional space–time as locally and isometrically embedded in a ten–dimensional flat space with global Euclidean topology. The space–time is obtained by a dimensional reduction of the extra six dimensions and its metric is isometrically induced by the embedding.

066.087 Comments on the canonical formalism for general relativity.
T. Regge.
General relativity and gravitational physics, p. 241 – 243 (1987). – See Abstr. 012.038.

066.088 Detection of low frequency gravitational waves with interplanetary spacecraft.
B. Bertotti.
General relativity and gravitational physics, p. 419 – 430 (1987). – See Abstr. 012.038.

This is a report on the gravitational wave experiment to be carried out with the spacecraft ULYSSES. After outlining the underlying theory, the author reviews the current progress in the experiment preparation and the long range perspective for gravitational wave astronomy in the low frequency band.

066.089 A new laser ranged satellite, LAGEOS 3, to measure the gravitomagnetic field.
I. Ciufolini.
General relativity and gravitational physics, p. 431 – 438 (1987). – See Abstr. 012.038.

The author describes a project to measure the Lense–Thirring effect with a new geodynamic satellite of the LAGEOS type.

066.090 Feasibility of detection of curl h_0 gravitational fields.
M. Bonaldi, M. Cerdonio, P. Falferi, G. Prodi, C. Rovelli, S. Vitale.
General relativity and gravitational physics, p. 497 – 501 (1987). – See Abstr. 012.038.

066.091 Coincidence probability for gravitational wave detection using laser interferometers.
M. Tinto.
General relativity and gravitational physics, p. 503 – 507 (1987). – See Abstr. 012.038.

066.092 4U 1820–30 as a potential test of the nonsymmetric gravitational theory of Moffat.
T. P. Krisher.
Astrophys. J., Lett. Ed., Vol. 320, No. 1, p. L47 – L50 (1987).

Recent observations of the X–ray burst source 4U 1820–30 have revealed a 685 s modulation of the luminosity. The author discusses how this system could provide a stringent test of the nonsymmetric gravitational theory (NGT) of Moffat, provided the observed periodicity is due to orbital motion of a binary system. The possible orbital period change predicted by general relativity may be detectable in this system.

066.093 Statistical treatment of fluctuations in the gravitational focusing of light due to stellar masses within a gravitational lens.
S. Deguchi, W. D. Watson.
Phys. Rev. Lett., Vol. 59, No. 24, p. 2814 – 2817 (1987).

When light from small, distant sources in the Universe is gravitationally focused by an intervening galaxy, the gravitational lens can be influenced by the granularity of the matter distribution which is caused by the stellar (or other compact) masses in the galaxy. A largely analytic, statistical calculation for a gravitational lens due to a collection of compact masses – valid for sources of finite size and for large (as well as small) "optical depths" for the lens – is developed to treat fluctuations in the light caused by such "microfocusing" effects. Previous treatments have been either numerical simulations of the Monte Carlo type or limited to single–star (i.e., low–optical–depth) effects.

066.094 Borehole measurement of the Newtonian gravitational constant.
A. T. Hsui.
Science, Vol. 237, No. 4817, p. 881 – 883 (1987).

It has been reported that the geophysically determined Newtonian gravitational constant is consistently larger than the laboratory value by 1 to 2% on the basis of gravity measurements in Australian mines. This discrepancy may have strong implications for the physics of gravitation. To test whether similar results can be observed in a different geological environment, gravity measurements in a Michigan borehole have been examined. Although these results cannot be taken as conclusive, owing to the large uncertainties involved in mass determination on a geophysical scale, these measurements are generally consistent with those of the Australian experiment.

066.095 Tests of general relativity using radar observations of the inner planets.
E. V. Pit'eva.
Byull. Inst. Teor. Astron., Tom 15, No. 9, p. 538 – 543 (1986). In Russian.

Radar observations of the inner planets made during 1961 – 1982 have been used for tests of general relativity. Computations have been performed using three theories of motion of the planets: JPL Ephemerides (DE–200), ITA relativistic theory and Newtonian theory. The results are as follows: the correction to the secular shift of Mercury's perihelion $\Delta\pi_1 = -0''.15 \pm 0''.12$ per century; parameters of post–Newtonian formalism are $(2+2\gamma-\beta)/3 = 0.995 \pm 0.003$, $(1+\gamma)/2 = 0.944 \pm 0.380$, $(\gamma+2\beta)/3 = 0.942 \pm 0.095$, what quite agrees with general relativity. The positive secular variation of the gravitational constant $\dot{G}/G = (4.0 \pm 0.8) \times 10^{-11} \text{yr}^{-1}$ has been found.

066.096 Microstructure of space–time in quantum gravity.
A. Ashtekar.
MPA Rep., No. 318, 13 pp. (1987). To appear in the proceedings of the eighth Workshop on Grand Unification.

066.097 Gravitational lensing by isothermal spheres with finite core radii: galaxies and dark matter.
G. Hinshaw, L. M. Krauss.
Astrophys. J., Vol. 320, No. 2, p. 468 – 476 (1987).

The authors calculate the gravitational lensing properties of an isothermal sphere in which the singularity at the origin is replaced by a finite core and contrast this with the usual singular case. This can significantly alter the statistics of the lensing of distant quasars by both the known distribution of intervening galaxies and also by a possible distribution of "dark" galaxies, as might be predicted in biased galaxy formation models. For example, it is found that the mean angular separation predicted to result from galactic lenses can be much larger than previously estimated.

066.098 Probing gravity to the second post–Newtonian order and to one part in 10^7 using the spin axis of the Sun.
K. Nordtvedt.
Astrophys. J., Vol. 320, No. 2, p. 871 – 874 (1987).

The close alignment of the Sun's spin axis with the solar system's planetary angular–momentum vector after 5 billion yr implies that the post–Newtonian gravitational interaction is Lorentz invariant to high precision. A metric gravity first post–Newtonian order (PPN) coefficient α_2 is constrained to be of order 10^{-7} or less, and second post–Newtonian order $(1/c^4)$ aspects of the gravitational interaction are modestly constrained by the present–day solar system configuration. Observations related to possible precession of the spin axes of fast pulsars also are relevant to this precision test of the Lorentz invariance of gravity.

066.099 Gravitational radiation from realistic relativistic stars: odd–parity fluid perturbations.
E. Seidel, T. Moore.
Phys. Rev. D, Vol. 35, No. 8, p. 2287 – 2296 (1987). Abstr. in Phys. Abstr., Vol. 90, No. 1309, Entry 80534 (1987).

066.100 A smooth oscillating cosmological solution.
N. Deruelle, J. Madore.
Phys. Lett. B, Vol. 186, No. 1, p. 25 – 28 (1987). Abstr. in Phys. Abstr., Vol. 90, No. 1309, Entry 80540 (1987).

066.101 Is the compactified vacuum semiclassically unstable?
K. Maeda.
Phys. Lett. B, Vol. 186, No. 1, p. 33 – 37 (1987). Abstr. in Phys. Abstr., Vol. 90, No. 1309, Entry 80541 (1987).

066.102 Path–integral and operator formalism im quantum gravity.
H. Arisue, T. Fujiwara, M. Kato, K. Ogawa.
Phys. Rev. D, Vol. 35, No. 8, p. 2309 – 2314 (1987). Abstr. in Phys. Abstr., Vol. 90, No. 1309, Entry 80569 (1987).

066.103 Dirac equation in Bianchi I metrics.
L. P. Chimento, M. S. Mollerach.
Phys. Lett. A, Vol. 121, No. 1, p. 7 – 10 (1987). Abstr. in Phys. Abstr., Vol. 90, No. 1309, Entry 80989 (1987).

066.104 Stellar resonant oscillations coupled to gravitational waves.
Y. Kojima.
Prog. Theor. Phys., Vol. 77, No. 2, p. 297 – 309 (1987). Abstr. in Phys. Abstr., Vol. 90, No. 1309, Entry 88386 (1987).

066.105 Arcs from gravitational lensing.
S. A. Grossman, R. Narayan.
Prepr. Steward Obs., No. 756, 10 pp. (1987). Submitted to Astrophys. J., Lett. Ed.

066.106 On the application of a hydrodynamic analogy to the general theory of relativity.
M. M. Abdil'din.
Tr. Astrofiz. Inst. Alma–Ata, Tom 47, p. 113 – 114 (1987). In Russian.

066.107 Dynamics of gravitating spinning masses in a chronometric reference system.
L. M. Chechin.
Tr. Astrofiz. Inst. Alma–Ata, Tom 47, p. 115 – 120 (1987). In Russian.

066.108 Thermal relaxation time of a mixture of relativistic electrons and neutrinos.
M. A. Herrera, S. Hacyan.
Rev. Mex. Astron. Astrofis., Vol. 14, No. 1, p. 47 – 51 (1987). – See Abstr. 012.042.
The interaction between the components of a relativistic binary mixture is studied by means of a fully covariant formalism. Assuming both components to differ slightly in temperature, an application of the relativistic Boltzmann equation yields general expressions for the energy transfer rate and for the relaxation time of the system. The resulting relation is then applied to a mixture of relativistic electrons and neutrinos to obtain numerical values of its relaxation time.

066.109 Ultimas noticias sobre el sistema de lente gravitacional 0957 + 561.
E. E. Falco.
Rev. Mex. Astron. Astrofis., Vol. 14, No. 1, p. 127 – 133 (1987). – See Abstr. 012.042.
The gravitational lens system 0957 + 561 has been studied in detail. VLBI observations have allowed the author to determine detailed models for the lens. The models set useful limits on the value of Hubble's constant. The author is thus beginning to develop the applications of gravitational lenses as cosmological tools.

066.110 The gravity of cosmic loops.
T. Vachaspati.
Gen. Relativ. Gravitation, Vol. 19, No. 11, p. 1053 – 1058 (1987).
It is demonstrated that the gravitational field of a loop of cosmic string can repel particles. This leads to a very interesting pattern of accretion of matter around a cosmic loop.

066.111 Characterization of standard clocks by means of light rays and freely falling particles.
V. Perlick.
Gen. Relativ. Gravitation, Vol. 19, No. 11, p. 1059 – 1073 (1987).
A mathematical characterization of standard clocks (i.e., clocks measuring proper time) is presented, which yields an experimental method to test whether or not a given clock is a standard clock. The only tools needed are light rays and freely falling particles. As the underlying space–time model the author uses a Weyl manifold (instead of a Lorentz manifold, which is the usual model of general relativity).

066.112 Vorticity and isotropic singularities.
S. W. Goode.
Gen. Relativ. Gravitation, Vol. 19, No. 11, p. 1075 – 1082 (1987).
This paper proves that if a solution of the Einstein field equations with perfect fluid source and γ law equation of state $[p = (\gamma-1)\mu]$ admits an isotropic singularity, then necessarily the fluid is irrotational. This shows the essential equivalence of seemingly distinct concepts of quasi–isotropic singularities and Friedmann–like singularities.

066.113 Causal measurability in chronological spaces.
L. B. Szabados.
Gen. Relativ. Gravitation, Vol. 19, No. 11, p. 1091 – 1100 (1987).
It is shown that the causal structure determines a volume measurability up to sets of zero measure. In space–time manifolds this causal measurability, apart from sets of zero measure, agrees with the a priori four–dimensional Lebesgue measurability, provided the strong causality condition holds.

066.114 The Doppler response to gravitational waves from a binary star source.
H. Wahlquist.
Gen. Relativ. Gravitation, Vol. 19, No. 11, p. 1101 – 1113 (1987).
The equations are developed for spacecraft Doppler detection of periodic gravitational waves from a single binary star source. Graphical examples are included to indicate the great variety of Doppler signals which can be generated by these systems.

066.115 A Lagrangian formulation of high–spin fields coupled to gravity.
A. Al–Saad, I. M. Benn, R. W. Tucker.
Gen. Relativ. Gravitation, Vol. 19, No. 11, p. 1115 – 1125 (1987).
A coupling of high (half–integer) spin fields to gravity is discussed in terms of symmetric tensor–valued spinors for which restrictive integrability constraints are avoided. The theory is generated from a spin–invariant action which provides a means of computing the associated stress tensor. Explicit no–ghost solutions are presented.

066.116   The quality factor of approximate solutions of Einstein's equations.
L. Bel.
Gen. Relativ. Gravitation, Vol. 19, No. 11, p. 1127 – 1130 (1987).
  The author defines the quality factor of arbitrary metrics to be used as a measure of their approximation to a solution of Einstein's vacuum equations.

066.117   Comment on conformally Ricci–flat perfect fluids of Petrov–type N.
N. Van den Bergh.
Gen. Relativ. Gravitation, Vol. 19, No. 11, p. 1131 – 1135 (1987).
  It is shown that vacuum solutions of the Einstein field equations, which are of Petrov–type N, cannot be conformally transformed into nonvacuum perfect fluid space–times.

066.118   Three–dimensional space–times.
  G. S. Hall, T. Morgan, Z. Perjés.
Gen. Relativ. Gravitation, Vol. 19, No. 11, p. 1137 – 1147 (1987).
  A real version of the Newman–Penrose formalism is developed for $(2+1)$–dimensional space–times. The complete algebraic classification of the (Ricci) curvature is given. The field equations of Deser, Jackiw, and Templeton, expressing balance between the Einstein and Bach tensors, are reformulated in triad terms.

066.119   A note on a paper by Massa and Pagani.
  S. B. Edgar.
Gen. Relativ. Gravitation, Vol. 19, No. 11, p. 1149 – 1150 (1987).
  Massa and Pagani (38.066.164) have given a neat refutal to the conjecture that the Riemann tensor is derivable from a tensor potential. In this note it is shown that the crucial equation in their argument can be obtained quite naturally and easily in ordinary tensor notation.

066.120   Coalescing binaries – probe of the Universe.
  A. Krolak, B. F. Schutz.
Gen. Relativ. Gravitation, Vol. 19, No. 12, p. 1163 – 1171 (1987).
  At present, coalescing binary systems containing neutron stars or black holes are thought to be the most likely sources of gravitational waves to be detected by long baseline interferometers being currently designed. In this essay the authors calculate the characteristics of the signal from a coalescing binary to the first post–Newtonian order. The authors also consider the effects of the expansion of the universe on the signal. They show that observations of these signals will provide a wealth of astrophysical information, e.g. determination of the Hubble constant, new rungs on the cosmic distance ladder, information about the mass distribution in the Universe, highly accurate tests of general relativity, and constraints on neutron–star equations of state.

066.121   On the gravitational field of a plane plate in general relativity.
J. Novotný, J. Kučera, J. Horský.
Gen. Relativ. Gravitation, Vol. 19, No. 12, p. 1195 – 1201 (1987).
  It is shown that a static solution of the Einstein equations inside an infinite plate of an ideal liquid with continuous metric coefficients and their first derivatives cannot have a plane of mirror symmetry. As a consequence, the boundaries of the plate are joined with qualitatively different vacuum solutions on both sides of the plate.

066.122   General solution of Liouville's equation in Robertson–Walker space–times.
S. D. Maharaj, R. Maartens.
Gen. Relativ. Gravitation, Vol. 19, No. 12, p. 1217 – 1222 (1987).
  The authors obtain the general solution of Liouville's equation in Robertson–Walker space–times, based on Killing vector constants of the motion and a new time–dependent first integral. For photons the solution is generated by conformal Killing vectors.

066.123   Invariant solutions of Liouville's equation in Robertson–Walker space–times.
R. Maartens, S. D. Maharaj.
Gen. Relativ. Gravitation, Vol. 19, No. 12, p. 1223 – 1234 (1987).
  The authors find all solutions of Liouville's equation in Robertson–Walker space–times that are either spatially homogeneous or isotropic or both. Some of these solutions depend on constants of motion that are not generated by Killing vectors. The authors indicate how these solutions may be used to find Einstein–Liouville solutions.

066.124   The connection between general observers and Lanczos potential.
M. Novello, A. L. Velloso.
Gen. Relativ. Gravitation, Vol. 19, No. 12, p. 1251 – 1265 (1987).
  The authors present some algorithms to find the explicit form of the Lanczos potential in an arbitrary geometry.

066.125   Erratum: "A Poincaré gauge theory of gravity" [Gen. Relativ. Gravitation, Vol. 18, No. 10, p. 995 – 1018 (1986)].
T. Kawai.
Gen. Relativ. Gravitation, Vol. 19, No. 12, p. 1285 (1987). See Abstr. 42.066.058.

066.126   Axisymmetric expanding universe with viscous fluid and heat flow.
L. K. Patel, S. S. Koppar.
Phys. Lett. A, Vol. 121, No. 6, p. 267 – 268 (1987). Abstr. in Phys. Abstr., Vol. 90, No. 1310, Entry 89058 (1987).

066.127   Spacetime as a membrane in higher dimensions.
  G. W. Gibbons, D. L. Wiltshire.
Nucl. Phys. B, Part. Phys., Vol. B287, No. 4, p. 717 – 742 (1987). Abstr. in Phys. Abstr., Vol. 90, No. 1310, Entry 89080 (1987).

066.128   Testing Newtonian gravity in space.
  A. M. Nobili, A. Milani, P. Farinella.
Phys. Lett. A, Vol. 120, No. 9, p. 437 – 441 (1987). Abstr. in Phys. Abstr., Vol. 90, No. 1310, Entry 89103 (1987).

066.129   On Newton's Earth–Moon–test of 1665/66.
  G. Richter.
Ann. Phys. (Leipzig), Vol. 44, No. 3, p. 176 – 197 (1987). In German. Abstr. in Phys. Abstr., Vol. 90, No. 1310, Entry 94641 (1987).

066.130   A homogeneous perfect fluid cosmological model in general relativity.
S. Narain.
Acta Phys. Pol., Ser. B, Vol. B18, No. 5, p. 407 – 409 (1987). Abstr. in Phys. Abstr., Vol. 90, No. 1311, Entry 95412 (1987).

066.131   Analysis of 18 months of data of the GEOGRAV experiment.
S. Frasca, M. Gabellieri, G. V. Pallottino.
Nuovo Cimento C, Vol. 10C, Ser. 1, No. 1, p. 1 – 26 (1987). Abstr. in Phys. Abstr., Vol. 90, No. 1311, Entry 95450 (1987).

066.132   Gravity makes a spectacle of itself.
  E. Falco.
New Sci., Vol. 113, No. 1552, p. 46 – 49 (1987). Abstr. in Phys. Abstr., Vol. 90, No. 1311, Entry 101677 (1987).

066.133   Critique of the theory of a cosmic potential.
  H. H. Soleng.
Astrophys. Space Sci., Vol. 137, No. 2, p. 403 – 406 (1987).
  The use of a cosmic potential in relativistic cosmology is criticized. It is pointed out that the energetic closure of the Universe follows from general relativity and from quantized superspace cosmology without the introduction of the cosmic potential.

066.134   **The general relativistic {2 + 2} formulation of magneto-hydrodynamics.**
S. M. Carioli.
Astrophys. Space Sci., Vol. 137, No. 2, p. 407 – 411 (1987).
In this paper the author shows how the general relativistic treatment of magnetohydrodynamics may find a natural and powerful tool in the {2 + 2} approach for the solution of Einstein equations.

066.135   **Experimental verification of gravitational interaction of bodies immersed in fluids.**
B. Vybíral.
Astrophys. Space Sci., Vol. 138, No. 1, p. 87 – 98 (1987).
This article deals with the experimental verification of the generalized Newton's gravitational law, formulated by Z. Horák. According to this law, the gravitational force between two resting homogeneous bodies immersed in resting homogeneous fluids is dependent on the densities ϱ_1, ϱ_2 of both the bodies and densities ϱ_1', ϱ_2' of both the fluids: furthermore, $\mathbf{F}' = (1-\varrho_1'/\varrho_1)(1-\varrho_2'/\varrho_2)\mathbf{F} = h_1 h_2 \mathbf{F}$, where $\mathbf{F}$ is the force between the bodies in a vacuum and h_1, h_2 are the density factors. The aim of the experimental verification of the law was to determine the density factors by exploring the phenomenon that is influenced by the gravitational interaction of the bodies immersed in different fluids. With the relative error of 0.4%, the validity of Horák's gravitational law was proved.

066.136   **Self–gravitating fluid in a conformally–flat space–time.**
D. R. K. Reddy.
Astrophys. Space Sci., Vol. 138, No. 1, p. 121 – 125 (1987).
The Einstein field equations for an irrotational perfect fluid with pressure p, equal to energy density are studied when the space–time is conformally flat. The coordinate transformation to co–moving coordinates is discussed. The energy and Hawking–Penrose inequalities are studied. Static and non–static solutions of the field equations are obtained. It is interesting to note that in the static case the only spherically–symmetric conformally flat solution for self–gravitating fluid is simply the empty flat space–time of general relativity.

066.137   **True "gravitational lens" effect for cylindrical deflectors.**
M. J. Bazin, L. de Freitas.
Astrophys. Space Sci., Vol. 138, No. 2, p. 381 – 386 (1987).
The authors show that gravitational light deflection is truly equivalent to the optical effect of a prism and a thin diverging lens when the massive deflector has cylindrical symmetry. They then investigate the possibility of cosmological measurements analogous to those proposed by Refsdal in the spherically–symmetric case.

066.138   **On the absence of gravitational forces in general relativity.**
H. H. Soleng.
Astrophys. Space Sci., Vol. 138, No. 2, p. 419 – 420 (1987).
The interpretation of Einstein's general theory of relativity is discussed. The concept of gravitational forces is shown to be redundant in general relativity.

066.139   **Partition function and energy density of a scalar field at finite temperature in Robertson–Walker spacetime.**
M.–y. Zhou, L.–f. Chen.
Sci. Sin., Ser. A, Vol. 30, No. 12, p. 1292 – 1306 (1987).
The authors extend the Euclidean path integral method, which is very successful in static spacetimes, to a dynamic spacetime, namely the Robertson–Walker spacetime, and take a scalar field as an example to study the effect of the evolution of the spacetime on the finite temperature field theory.

066.140   **Cosmological applications of singular hypersurfaces in general relativity.**
P. Laguna–Castillo.
Diss. Abstr. Int., Sect. B, Vol. 48, No. 5, p. 1384–B (1987). Thesis, The University of Texas at Austin, 200 pp. (1987). Order No. DA8717462.

066.141   **Relativistic stellar pulsations.**
L. S. Finn.
Diss. Abstr. Int., Sect. B, Vol. 48, No. 6, p. 1710–B (1987). Thesis, California Institute of Technology, 252 pp. (1987). Order No. DA8719665.

066.142   **Maclaurin π–ellipsoid in parametrized post–Newtonian formalism.**
N. P. Bondarenko.
Kiev. Politekh. Inst., Kiev, 15 pp. (1987). In Russian. Abstr. in Ref. Zh., 51. Astron., 12.51.155 (1987).

066.143   **Diffraction of surface gravitational waves in a system of rotating bodies.**
V. V. Yakovlev, A. V. Pyatetskij.
Dokl. Akad. Nauk USSR, A, No. 8, p. 36 – 39 (1987). In Russian. Abstr. in Ref. Zh., 51. Astron., 12.51.156 (1987).

066.144   **Lorentz invariance as a dynamic symmetry.**
F. Winterberg.
Z. Naturforsch., A, Vol. 42a, No. 12, p. 1428 – 1442 (1987).
If all the forces of nature can be reduced to those which follow from a linear combination of a scalar and vector potential, as in electrodynamics, Lorentz invariance can be derived as a dynamic symmetry. All that has to be done is to assume that there is an all pervading substratum or ether, transmitting those forces through space, and that all physical bodies actually observed are held together by those forces.

066.145   **A new self–similar space–time.**
L. K. Chi.
J. Math. Phys., Vol. 28, No. 7, p. 1539 – 1540 (1987). Abstr. in Phys. Abstr., Vol. 90, No. 1313, Entry 108327 (1987).

066.146   **Realistic exact solutions of Einstein's field equations.**
D. C. Durgapal.
Classical and quantum aspects of gravitation, p. 15 – 22 (1986). Abstr. in Phys. Abstr., Vol. 90, No. 1314, Entry 115864 (1987). – See Abstr. 012.065.

066.147   **Killing approximation for vacuum and thermal stress–energy tensor in static space–times.**
V. P. Frolov, A. I. Zel'nikov.
Phys. Rev. D, Vol. 35, No. 10, p. 3031 – 3044 (1987). Abstr. in Phys. Abstr., Vol. 90, No. 1314, Entry 115885 (1987).

066.148   **Astrophysically significant solutions and their derivation.**
K. C. Das.
Classical and quantum aspects of gravitation, p. 73 – 85 (1986). Abstr. in Phys. Abstr., Vol. 90, No. 1314, Entry 120275 (1987). – See Abstr. 012.065.

066.149   **Gravitational imaging by isolated elliptical potential wells. I. Cross sections.**
R. D. Blandford, C. S. Kochanek.
Astrophys. J., Vol. 321, No. 2, p. 658 – 675 (1987).
The imaging properties of isolated gravitational lenses are investigated numerically using a model two–dimensional potential parametrized by its hardness, core radius, and ellipticity. Cross sections for creating image arrangements with specific characteristics are presented and analyzed. The authors present the results in the form of rough "rules of thumb" that express the variation of image arrangements with potential shape. The distribution of an important observable, the arrival time differences for the different images, is also discussed.

066.150 **Gravitational imaging by isolated elliptical potential wells. II. Probability distributions.**
C. S. Kochanek, R. D. Blandford.
Astrophys. J., Vol. 321, No. 2, p. 676 – 685 (1987).

Optical depths for the magnification and positions of multiple quasar images created by a cosmological distribution of isolated elliptical potential wells are computed. Introducing a core radius into a singular potential well can reduce the optical depth for multiple imaging by a large factor. Strong lenses predominantly produce three images (of which two are opposed) at low magnification and five images at high magnification when amplification bias may become important. Marginal lenses produce mainly three comparatively bright images, but with relatively small probability. As an application, the authors estimate a probability $\sim 10^{-6}$ that a given galaxy at $z_L \sim 0.04$ will lens a background quasar (e.g., 2237+0305).

066.151 **The marginal gravitational lensing.**
I. Kovner.
Astrophys. J., Vol. 321, No. 2, p. 686 – 705 (1987).

Four types of gravitational lenses are investigated here. Their common feature is "marginality": one type has a pair of *merging images*, another is a lens *barely able* to split images, the third is a lens made of a galaxy superposed on a cluster at a site where merging images produced by the *cluster alone* could have been, the fourth is made of a galaxy superposed on a *cluster almost able* to split images on its own. The marginal gravitational lenses are important if either or both of the luminosity and density bias hypotheses are true. The first hypothesis assumes a very steep quasar luminosity function, the second assumes a very steep falloff of the distribution of clusters in surface mass densities. A crude model statistics, based on available statistics of clusters and quasars gives an estimate for the uncertainties in numbers and characteristics of lenses of large separations, on the basis of the presently available data.

066.152 **Why ultrarelativistic numerical hydrodynamics is difficult.**
M. L. Norman, K.–H. A. Winkler.
Astrophysical radiation hydrodynamics, p. 449 – 475 (1986). – See Abstr. 012.069.

The authors address the numerical difficulties associated with solving the equations of ultrarelativistic hydrodynamics in a flat background spacetime, i.e. special relativistic hydrodynamics, although their findings have relevance to numerical techniques for studying general relativistic fluid flows as well.

066.153 **Numerical relativistic gravitational collapse with spatial time slices.**
C. R. Evans, L. L. Smarr, J. R. Wilson.
Astrophysical radiation hydrodynamics, p. 491 – 529 (1986). – See Abstr. 012.069.

The authors present a general approach for solving the Einstein equations of general relativity coupled to hydrodynamics. This coupled nonlinear system of partial differential equations naturally split into equations of constraint and of evolution. The constraints can be solved for the longitudinal pieces of the gravitational field which generalizes the Newtonian scalar potential. The evolution equations determine the generation and propagation of gravitational radiation, a feature not found in Newtonian gravity. The general framework is then specialized to the axisymmetric case in which no rotation occurs. A computer code was written using finite difference techniques to solve these equations.

066.154 **The characteristic initial value problem in general relativity.**
J. M. Stewart.
Astrophysical radiation hydrodynamics, p. 531 – 568 (1986). – See Abstr. 012.069.

A new approach to the numerical solution of Einstein's Equations based on the characteristic initial value problem is outlined. Because this approach requires integration through caustics and up to singularities, this topic is described and some novel finite–differencing techniques are introduced for integrating up to singularities of non–linear differential equations. Finally some preliminary results from this new approach are presented.

066.155 **Magnetic moments of astrophysical objects as a consequence of general relativity.**
A. Widom, D. V. Ahluwalia.
Chin. J. Phys., Vol. 25, No. 1, p. 23 – 26 (1987). Abstr. in Phys. Abstr., Vol. 90, No. 1315, Entry 120883 (1987).

066.156 **Newtonian analogue of force and motion of a free particle in the gravitational field of Kerr–de Sitter space–time.**
T. Singh, R. S. Srivastava.
Astrophys. Space Sci., Vol. 139, No. 2, p. 263 – 273 (1987).

The Newtonian analogue of force for the Kerr–de Sitter metric has been investigated. To the first–order of approximation, a component of the force vector corresponds to the Newtonian gravitational force and a cosmic force arising due to the cosmological constant Λ. In the higher order of approximation, the relativistic correction terms due to rotation and the presence of Λ are obtained. In the second part of the paper the motion of a freely–falling body has been investigated. It is found that plane orbits are not possible. Also a radial fall is not possible and there is a rotational drag on the particle.

066.157 **Soliton solutions to the Einstein gravitational field equations in the presence of a spherically–symmetric static background field.**
P. C. W. Fung, F. Z. Tao.
Astrophys. Space Sci., Vol. 139, No. 2, p. 311 – 320 (1987).

The authors have found two sets of new solutions to Einstein's equation of gravity in the presence of a spherically–symmetrical gravitational background, like the Earth. The transverse and longitudinal components of the metric tensor representing the gravity waves are all soliton solutions, propagating towards the origin of the Earth. If the situation is considered where the static background field is absent, the solutions still remain soliton–like in nature. The difference between the authors' result and Einstein's is attributed to the two approximations taken previously – weak field and "harmonic condition".

066.158 **Jedinstvena teorija polja u fizici R. Boškovića.**
B. Jovanović.
Vasiona, Année 35, No. 2, p. 52 – 54 (1987). In Croatian.

In the light of contemporary physics knowledge the Bošković ideas on the force's law and generalized force's are analyzed, connected with the foundation of unifying fields.

066.159 **Comment on "Mercury's precession according to special relativity" [Am. J. Phys., Vol. 54, No. 3, p. 245 – 247 (1986)] (and reply).**
P. C. Peters, T. Phipps Jr.
Am. J. Phys., Vol. 55, No. 8, p. 757 – 759 (1987). Abstr. in Phys. Abstr., Vol. 90, No. 1316, Entry 128225 (1987). – See Abstr. 41.066.225.

066.160 **The deflection of light by the Sun due to three–space curvature.**
J. G. Ellingson.
Am. J. Phys., Vol. 55, No. 8, p. 759 – 760 (1987). Abstr. in Phys. Abstr., Vol. 90, No. 1316, Entry 128226 (1987).

066.161 **Graviton and topology contributions to self–consistent cosmology.**
M. A. Castagnino, J. P. Paz, N. Sanchez.
Phys. Lett. B, Vol. 193, No. 1, p. 13 – 22 (1987). Abstr. in Phys. Abstr., Vol. 90, No. 1316, Entry 128414 (1987).

066.162   **A rotating mass in a Gödel universe with an electromagnetic field.**
L. K. Patel, S. S. Koppar.
Acta Phys. Hung., Vol. 61, No. 3 – 4, p. 363 – 367 (1987). Abstr. in Phys. Abstr., Vol. 90, No. 1317, Entry 134583 (1987).

066.163   **Static axisymmetric discs and gravitational collapse.**
A. Chamorro, R. Gregory, J. M. Stewart.
Proc. R. Soc. London, Ser. A, Vol. 413, No. 1844, p. 251 – 262 (1987). Abstr. in Phys. Abstr., Vol. 90, No. 1317, Entry 134592 (1987).

066.164   **A test of the equivalence principle with polarised light.**
B. Lesche, M. L. Bedran.
Rev. Bras. Fis., Vol. 17, No. 1, p. 100 – 108 (1987). Abstr. in Phys. Abstr., Vol. 90, No. 1317, Entry 134593 (1987).

066.165   **Could a dilaton solve the cosmological constant problem?**
J. Ellis, N. C. Tsamis, M. Voloshin.
Phys. Lett. B, Vol. 194, No. 2, p. 291 – 296 (1987). Abstr. in Phys. Abstr., Vol. 90, No. 1318, Entry 140584 (1987).

066.166   **Optimum positioning of attracting masses in the Heyl–type determination of the Newtonian gravitational constant.**
C. C. Speake.
Metrologia, Vol. 24, No. 2, p. 97 – 99 (1987). Abstr. in Phys. Abstr., Vol. 90, No. 1318, Entry 140657 (1987).

066.167   **On Geroch's limit of space–times and its relation to a new topology in the space of Lie groups.**
H.–J. Schmidt.
J. Math. Phys., Vol. 28, No. 8, p. 1928 – 1936 (1987). Abstr. in Phys. Abstr., Vol. 90, No. 1318, Entry 146124 (1987).

066.168   **A new finite element method for spherically symmetric relativistic collapse.**
P. J. Mann.
J. Comput. Phys., Vol. 72, No. 2, p. 467 – 485 (1987). Abstr. in Phys. Abstr., Vol. 91, No. 1319, Entry 155 (1988).

066.169   **Boundary conditions and the cosmological constant.**
J. Wudka.
Phys. Rev. D, Vol. 36, No. 4, p. 1036 – 1040 (1987). Abstr. in Phys. Abstr., Vol. 91, No. 1319, Entry 4521 (1988).

066.170   **Exact model for a gaseous regular bounding sphere in general relativity.**
H. Knutsen.
Int. J. Theor. Phys., Vol. 26, No. 9, p. 895 – 900 (1987). Abstr. in Phys. Abstr., Vol. 91, No. 1320, Entry 4774 (1988).

066.171   **Chaos in Kaluza–Klein models.**
Y. Elskens, M. Henneaux.
Classical Quantum Gravity, Vol. 4, No. 5, p. L161 – L167 (1987). Abstr. in Phys. Abstr., Vol. 91, No. 1320, Entry 4784 (1988).

066.172   **Kaluza–Klein theories.**
D. Bailin, A. Love.
Rep. Prog. Phys., Vol. 50, No. 9, p. 1087 – 1170 (1987). Abstr. in Phys. Abstr., Vol. 91, No. 1320, Entry 4799 (1988).

066.173   **Non–Schrödinger forces and pilot waves in quantum cosmology.**
F. J. Tipler.
Classical Quantum Gravity, Vol. 4, No. 5, p. L189 – L195 (1987). Abstr. in Phys. Abstr., Vol. 91, No. 1320, Entry 4805 (1988).

066.174   **Wormholes in space–time.**
I. Moss.
Nature, Vol. 330, No. 6145, p. 210 (1987).
The author reports on the theory of wormholes, gives their quantum description and an application to cosmology – wormholes can connect parts of the Universe to separate, self–contained pieces of space. The possible classification of black holes as special wormholes is considered. The author reports on Hawking's new theory of wormhole excitations, and though this is not yet a full quantum theory of wormholes it seems likely that the best opportunity to test the theory is with the Higgs particle.

066.175   **A galactic gravitational lens as the ultimate astronomical telescope.**
B. McBreen, L. Metcalfe.
Nature, Vol. 330, No. 6146, p. 348 – 350 (1987).
The authors point out that the caustic surfaces of galactic gravitational lenses are powerful telescopes, with point–source intensifications varying with frequency from 10^5 for radio waves to 10^8 for X rays. The diffraction limit to the angular resolution varies from 10^{-11} to 10^{-17}arc s from radio waves to X rays and exceeds the resolution of man–made telescopes by many orders of magnitude. The signature of a point source crossing the caustic is a burst of radiation modulated in time by a characteristic diffraction pattern. The microcaustic surfaces generated by stars in the lensing galaxy are also powerful telescopes that yield large intensifications of small sources crossing the microcaustic. Decaying cosmic strings could be an important source of cosmic rays with energies above 10^{10}GeV.

066.176   **Experiment and the general theory of gravitation.**
C. Klimčík, J. Niederle.
Vesmír, Vol. 66, No. 8, p. 437 – 446 (1987). In Czech.

066.177   **Does gravity decrease with time?**
P. Andrle, V. Marvanová.
Vesmír, Vol. 66, No. 10, p. 569 – 571 (1987). In Czech.

066.178   **A possible semisecular variation in orbital period of binary pulsar system PSR 1913+16 and Lorentz invariance of gravity.**
K. Nordtvedt Jr.
Astrophys. J., Vol. 322, No. 1, p. 288 – 290 (1987).
If the post–Newtonian gravitational equations of motion exhibit a "preferred inertial frame", a possible observational consequence is that orbital systems whose center of mass moves relative to the preferred frame will have semisecular time variation in the orbital periods – the semisecular period being the period of precession of orbital periastron location. A contribution to the binary pulsar PSR 1913+16 orbital period rate of change above the level of observational accuracy would result for a parameterized post–Newtonian (PPN) coeffient α_1. The fast precession rate of the binary pulsar system's periastron (4.2 arc degrees yr^{-1}) should permit distinguishing this α_1 effect from the secular gravitational radiation reaction effect.

066.179   **New mechanism of appearance of proper redshifts in spectra of compact objects.**
M. F. Khodyachikh.
Astrofizika, Tom 27, Vyp. 2, p. 359 – 370 (1987). In Russian. English translation in Astrophysics, Vol. 27, No. 2.
It is supposed that the interaction between particles by means of "heavy quanta" is realized alongside with the gravitational interaction. This interaction may be reduced to the gravitational interaction of particles whose masses are changed on account of additional interaction. The mass of the smaller particle decreases, and the sum of particle masses remains constant. The mass of an electron in the medium decreases for augmentation of the quantity of matter in its neighbourhood at distances of the order of the interaction radius. This must bring about an increase of the wavelengths of spectral lines. An increase of the mean radial velocity of stars has been revealed in the central, most denser zone of the nucleus of the star cluster M3 relative to stars outside the nucleus at 3.9 ± 1.3 km s^{-1}, which is in agreement with the

prediction of the theory qualitatively. The mechanism has been used for the explanation of quasar redshifts and proper redshifts of galaxies revealed by Arp.

066.180 Degenerate dwarf binaries as promising, detectable sources of gravitational radiation.
C. R. Evans, I. Iben Jr, L. Smarr.
Astrophys. J., Vol. 323, No. 1, p. 129 – 139 (1987).

The authors discuss the implications for gravitational wave astronomy of a recently hypothesized population of double degenerate dwarf binaries (DDBs). In theoretical studies of the stellar evolution of intermediate–mass binary systems, remnant binaries containing two degenerate dwarfs are a frequent outcome, with a Galactic birth rate perhaps as high as $v_T \approx 0.15 \text{ yr}^{-1}$. Observationally there is as yet only circumstantial evidence for a large population of DDBs. The observational difficulties in detecting precontact binaries is formidable since there is little to distinguish these from single degenerate dwarfs, and it is most probable to find these at their widest separations. If DDBs remain viable theoretical candidates for producing SNe I, then it may be that future low–frequency gravitational wave astronomy observations will provide important constraints on supernova theories. The authors discuss one such low–frequency gravitational wave detection scheme employing space–based laser interferometers. The proposed sensitivity is adequate to detect the presence of a large population of DDBs as well as to detect individual sources.

Gravitation: Raum–Zeit–Struktur und Wechselwirkung.
See Abstr. 003.055.

Osnovy relyativisticheskoj teorii gravitatsii (*Foundations of a relativistic theory of gravitation*).
See Abstr. 003.091.

Spinors and space–time. Vol. 1: Two–spinor calculus and relativistic fields.
See Abstr. 003.110.

The evolution of relativity.
See Abstr. 003.115.

Vsemirnoe tyagotenie i teorii prostranstva i vremeni (*Gravitation and the theory of space and time*).
See Abstr. 003.148.

Newton's *Principia*.
See Abstr. 004.040.

Newtonianism and today's physics.
See Abstr. 004.041.

Dark stars: the evolution of an idea.
See Abstr. 004.042.

Trajectoires et impasses de la solution de Schwarzschild.
See Abstr. 004.061.

String theory, quantum cosmology and quantum gravity, integrable and conformal invariant theories. Proceedings of a colloquium held at Paris and Meudon, France, September 1986.
See Abstr. 012.083.

Superstrings, unified theories and cosmology. Proceedings of a workshop held at Trieste, Italy, June 1986.
See Abstr. 012.084.

The Runge–Lenz vector and Einstein perihelion precession.
See Abstr. 014.033.

Stellar sky as seen from the vicinity of a black hole.
See Abstr. 014.035.

The use of amorphous metallic ribbon as a torsion spring for the measurement of the gravitational constant.
See Abstr. 014.036.

A close look at a canonical semiclassical derivation of the gravitational redshift.
See Abstr. 014.042.

A student experiment for accurate measurements of the Newtonian gravitational constant.
See Abstr. 014.060.

Automatic recording for the Cavendish balance.
See Abstr. 014.063.

The Rutherford cross section and the perihelion shift of Mercury with the Runge–Lenz vector.
See Abstr. 014.065.

The analytical operations system GRATOS. Part 1: conceptions and possibilities.
See Abstr. 021.013.

The seismic noise challenge to the detection of low frequency gravitational waves.
See Abstr. 034.051.

Operation of the cryogenic detector of the Rome group at CERN.
See Abstr. 034.052.

Resonant–mass detectors of gravitational radiation.
See Abstr. 034.053.

Research of Q values of some materials for gravitational wave antennae at low temperatures.
See Abstr. 034.112.

The two–body problem in the (truncated) PPN theory.
See Abstr. 042.007.

The two–body problem in general relativity.
See Abstr. 042.013.

Coordinate invariance of the Schwarzschild shift of the pericenter.
See Abstr. 042.020.

Translational and rotational motion of a test rigid body in relativistic celestial mechanics.
See Abstr. 042.026.

The Keplerian metric of space–time.
See Abstr. 042.058.

Tetrad description of astrometric reference systems.
See Abstr. 043.001.

Strings and superspace.
See Abstr. 061.002.

Higher curvature supergravity and superstrings.
See Abstr. 061.003.

Gravity from strings.
See Abstr. 061.004.

Mass and dual mass: a gravitational analogue of the Dirac quantization rule and its physical implications.
See Abstr. 061.005.

Superstrings.
See Abstr. 061.007.

Newton, quantum theory and reality.
See Abstr. 061.036.

Superstring unification.
See Abstr. 061.037.

Covariant description of canonical formalism in geometrical theories.
See Abstr. 061.038.

A unified approach to the gauging of space–time and internal symmetries.
See Abstr. 061.039.

Curvature and torsion as induced quantum effects.
See Abstr. 061.046.

Recent developments in supergravity theories.
See Abstr. 061.047.

Renormalization group techniques in lattice gravity.
See Abstr. 061.057.

Covariant canonical formalism for polynomial supergravity in any dimension.
See Abstr. 061.058.

Prospects for fifth force fade.
See Abstr. 061.098.

Neutrino transport in relativity.
See Abstr. 061.112.

On the thermodynamics of a relativistic Fermi–gas.
See Abstr. 062.059.

On hydrodynamics of radiatively driven flows.
See Abstr. 062.065.

Non–linear stability of a 3–dimensional spatially symmetric solution of the relativistic Poisson Vlasov equation.
See Abstr. 062.066.

Gravitational stability of a torus in the presence of a magnetic field.
See Abstr. 062.085.

Influence of a magnetic field on the gravitational stability of a cylinder.
See Abstr. 062.086.

On the spherical symmetry of static perfect fluid spacetimes and the positive–mass theorem.
See Abstr. 065.084.

Calculation of relativistic model stars using Regge calculus.
See Abstr. 065.085.

The spin–torsion coupling precession of spin and its effects on single pulses of pulsars.
See Abstr. 067.029.

Neutron–star models in Ni's theory of gravity.
See Abstr. 067.039.

An intuitive understanding of black–hole evaporation by viewing it in terms of more familiar quantum–field effects in flat spacetime.
See Abstr. 067.040.

A charged analogue of the Vaidya–Tikekar solution.
See Abstr. 067.042.

The third law.
See Abstr. 067.049.

Black holes and monopoles.
See Abstr. 067.050.

3–D relativistic hydrodynamics around a Kerr black hole.
See Abstr. 067.051.

Apparent horizon of initial data for black hole–collision.
See Abstr. 067.052.

Accretion of charged matter by collapsing objects.
See Abstr. 067.068.

The thermodynamics of a system containing two black holes and black–body radiation.
See Abstr. 067.103.

Polar orbits in the Kerr space–time.
See Abstr. 067.105.

The orbit of Pluto and the cosmological constant.
See Abstr. 101.027.

Influence of a periodic gravitational wave on the parameters of a binary system.
See Abstr. 117.121.

Influence of the spectrum's shape on the effect of the "relativistic searchlight" in close binaries.
See Abstr. 117.144.

Spectrum of gravitational radiation of binary systems.
See Abstr. 117.288.

The binary pulsar: gravity waves exist.
See Abstr. 126.098.

Computer simulations of relativistic star clusters.
See Abstr. 151.013.

Simulations of axisymmetric, Newtonian star clusters: prelude to 2 + 1 general relativistic computations.
See Abstr. 151.027.

The collapse of dense star clusters to supermassive black holes: binaries and gravitational radiation.
See Abstr. 151.130.

Concerning the limit on the mean mass distribution of galaxies from their gravitational lens effect.
See Abstr. 157.315.

Spectroscopy of the extranuclear line–emitting regions associated with the gravitational lens system 2016 + 112.
See Abstr. 158.067.

The triple radio source 0023 + 171: a candidate for a dark gravitational lens.
See Abstr. 158.325.

Statistical gravitational lensing: influence of compact objects on the number counts of quasars.
See Abstr. 159.017.

Observations of the new gravitational lens system UM673 = Q0142–100.
See Abstr. 159.076.

Quasar microlensing and dark matter.
See Abstr. 159.085.

Further data on the blue ring–like structure in A 370.
See Abstr. 160.060.

Extraction of cosmological information from multiimage gravitational lenses.
See Abstr. 161.004.

An anisotropic cosmological model in a scalar–tensor theory of gravitation.
See Abstr. 161.009.

Gauge–invariant cosmological density perturbations.
See Abstr. 161.018.

Higher dimensional cosmology.
See Abstr. 161.019.

Higher dimensional solutions of modified Einstein equations with curvature square terms.
See Abstr. 161.020.

Who decides boundary conditions for the wave function of the universe?
See Abstr. 161.021.

Cosmological solutions of the Einstein equation with the backreaction effect of quantized conformally invariant fields.
See Abstr. 161.022.

Inhomogeneous generalizations of the Robertson–Walker cosmological models.
See Abstr. 161.023.

The density matrix of the universe.
See Abstr. 161.053.

Eternally existing self–reproducing inflationary universe.
See Abstr. 161.054.

Uncertainty principle and the horizon size of our universe.
See Abstr. 161.057.

Thermodynamics and general relativity could determine the geometry of the universe.
See Abstr. 161.058.

A note on variable–G cosmologies.
See Abstr. 161.067.

Robertson–Walker–type universes with conformally–invariant scalar field.
See Abstr. 161.084.

Anisotropic viscous–fluid cosmological model.
See Abstr. 161.085.

The Bianchi type–V solution in the scale–covariant theory.
See Abstr. 161.087.

Robertson–Walker Lyttleton–Bondi universe with cosmological constant.
See Abstr. 161.088.

Cosmological aspects of superstring models.
See Abstr. 161.107.

The singularities in quantum cosmology.
See Abstr. 161.118.

Gravitational lensing effect on the fluctuations of the cosmic background radiation.
See Abstr. 161.122.

Inflation and quantum cosmology.
See Abstr. 161.134.

Quantum cosmology.
See Abstr. 161.135.

Dynamics of anisotropic cosmological model with ultrarelativistic matter, magnetic field, and fluxes of free isotropic particles.
See Abstr. 161.168.

Nonminimal gravitational coupling: the spectrum of cosmic solutions.
See Abstr. 161.169.

Inflation in a completely anisotropic Einstein–Cartan cosmological model.
See Abstr. 161.171.

Cosmology with twice compactified internal space and higher–order gravitational Lagrangian.
See Abstr. 161.172.

Generation of gravitational waves by the anisotropic phases in the early universe.
See Abstr. 161.199.

Vacuum Friedmann cosmological models in Dunn's scalar–tensor theory of gravitation.
See Abstr. 161.214.

Exact cosmological solutions of gravitational theories.
See Abstr. 161.219.

The constancy of physics.
See Abstr. 161.233.

String theory and quantum cosmology.
See Abstr. 161.265.

Cosmology and extra dimensions.
See Abstr. 161.266.

Is the space–time dimension 11 distinguished in cosmology?
See Abstr. 161.267.

A new class of spherically symmetric interior solution with cosmological constant Λ.
See Abstr. 161.268.

Kaluza–Klein cosmology: Friedmann models with phenomenological matter.
See Abstr. 161.271.

A dust–filled Kantowski–Sachs Universe with $\Lambda > 0$.
See Abstr. 161.288.

A note on vacuum self–creation cosmological models.
See Abstr. 161.304.

Inhomogeneous viscous fluid cosmological model with electromagnetic field in general relativity.
See Abstr. 161.305.

The maximum density attained at the non–linear stage of gravitational instability in a collisionless medium with thermal velocities.
See Abstr. 161.354.

Self–creation cosmological solutions.
See Abstr. 161.373.

A gravitationally non–degenerate cosmological model with expanding and shearing viscous fluid in general relativity.
See Abstr. 161.374.

High–temperature quantum effects in multidimensional mixmaster models.
See Abstr. 161.386.

Quantum cosmology of superstrings.
See Abstr. 161.387.

Dynamical neutralization of the cosmological constant.
See Abstr. 161.398.

The wave functions and the effective action in quantum cosmology: covariant loop expansion.
See Abstr. 161.399.

Ergodic theory of the mixmaster universe in higher space–time dimensions. II.
See Abstr. 161.415.

Ergodic theory of the mixmaster model in higher space–time dimensions.
See Abstr. 161.418.

Symmetry behavior in curved spacetime: finite–size effect and dimensional reduction.
See Abstr. 161.433.

New LRS perfect–fluid cosmological models.
See Abstr. 161.435.

Cosmology from nothing.
See Abstr. 161.444.

Entropy and cosmology.
See Abstr. 161.445.

067 Astrophysics of Compact Objects (Neutron Stars, Black Holes)

067.001 Pair production in intense electromagnetic fields of pulsars.
T. Y. Shi.
The origin and evolution of neutron stars, p. 61 (1987). Abstract. – See Abstr. 012.001 (IAU Symp. No. 125).

067.002 A modified pulsar model, Green function, period distribution.
L. F. Chen, S. R. Liang.
The origin and evolution of neutron stars, p. 62 (1987). Abstract. – See Abstr. 012.001 (IAU Symp. No. 125).

067.003 Neutron star coupling to its environment.
M. Salvati, F. Pacini.
The origin and evolution of neutron stars, p. 79 – 89 (1987). – See Abstr. 012.001 (IAU Symp. No. 125).

The authors review facts, myths and theories related to the formation of neutron stars and their coupling to the environment.

067.004 On the origin of neutron stars in globular clusters.
J. E. Grindlay.
The origin and evolution of neutron stars, p. 173 – 185 (1987). – See Abstr. 012.001 (IAU Symp. No. 125).

The formation of neutron stars in globular clusters is discussed in light of a number of recent results and, in particular, studies of the origin and evolution of the high luminosity X-ray binaries found in globular clusters. The author argues that the neutron stars most probably arise from the accretion–induced collapse of white dwarfs in compact binary systems, themselves detectable as low luminosity cluster X-ray sources.

067.005 Accretion onto magnetized neutron stars: polar cap flow and centrifugally driven winds.
J. Arons.
The origin and evolution of neutron stars, p. 207 – 225 (1987). – See Abstr. 012.001 (IAU Symp. No. 125).

Some basic concepts of accretion onto the polar caps of magnetized neutron stars are reviewed. Preliminary results of new, multidimensional, time–dependent calculations of polar cap flow are outlined, and are used to suggest the possible observability of fluctuations in the X-ray intensity of accretion powered pulsars on time scales of 10 – 100 msec. The possible relevance of such fluctuations to quasi–periodic oscillations is suggested. Basic concepts of the interaction between a disk and the magnetosphere of a neutron star are also discussed. Some recent work on the disk–magnetosphere interaction is outlined, leading

to the suggestion that a neutron star can lose angular momentum by driving some or all of the mass in the disk off as a centrifugally driven wind. The relevance of such mass loss to the orbital evolution of the binary is pointed out.

067.006 Line radiation from accreting magnetized neutron stars.
D. Y. Wang.
The origin and evolution of neutron stars, p. 227 – 232 (1987). – See Abstr. 012.001 (IAU Symp. No. 125).

A new non–thermal model on cyclotron line radiation from accreting magnetized neutron stars have been reviewed. Above the magnetic polar cap of accreting neutron stars, the maser instability may be excited by a part of non–thermal electrons, and the radiation with the frequency near electron cyclotron frequency and its second harmonic may be emitted as well. According to this model, the intensity and energy of line radiation will vary with pulsation phase of X-ray pulsars. These phenomena have been observed from Her X–1.

067.007 Polar cap accretion onto magnetized neutron stars: an analytic solution.
J. Arons, R. I. Klein.
The origin and evolution of neutron stars, p. 245 (1987). Abstract. – See Abstr. 012.001 (IAU Symp. No. 125).

067.008 Radiation gas dynamics of polar cap accretion onto magnetized neutron stars.
R. I. Klein, J. Arons.
The origin and evolution of neutron stars, p. 246 (1987). Abstract. – See Abstr. 012.001 (IAU Symp. No. 125).

067.009 An inverse Compton scattering model for the spectra of X–ray pulsars.
G. J. Qiao, X. J. Wu, H. Chen, X. Y. Xia.
The origin and evolution of neutron stars, p. 248 (1987). Abstract. – See Abstr. 012.001 (IAU Symp. No. 125).

067.010 The X–ray radiation mechanism of the compact (neutron) binary stars.
Y. Tong, G. Fang, X.–j. Mao.
The origin and evolution of neutron stars, p. 249 (1987). Abstract. – See Abstr. 012.001 (IAU Symp. No. 125).

067.011 Model atmospheres for X–ray bursting neutron stars.
R. A. London, R. E. Taam, W. M. Howard.
The origin and evolution of neutron stars, p. 251 (1987). Abstract. – See Abstr. 012.001 (IAU Symp. No. 125).

067.012 The birth of neutron stars.
    S. É. Woosley.
The origin and evolution of neutron stars, p. 255 – 272 (1987). –
See Abstr. 012.001 (IAU Symp. No. 125).
    Presupernova models of massive stars are discussed and their
explosion by either the "core bounce" or neutrino energy
transport mechanism briefly reviewed. Special consideration is
given to those attributes of the stellar evolution and explosion
that might influence the properties of the neutron star remnant:
its mass, rotation rate, magnetic field, and "kick" velocity.

067.013 Numerical experiments and neutron star formation.
    W. D. Arnett.
The origin and evolution of neutron stars, p. 273 – 280 (1987). –
See Abstr. 012.001 (IAU Symp. No. 125).
    Theory and numerical experiments on neutron star formation
are critically reviewed. Several new numerical experiments are
summarized, and the importance of advection is discussed.

**067.014 Neutron star formation in theoretical supernovae – low
    mass stars and white dwarfs.**
K. Nomoto.
The origin and evolution of neutron stars, p. 281 – 303 (1987). –
See Abstr. 012.001 (IAU Symp. No. 125).
    The presupernova evolution of stars that form semi–degener-
ate or strongly degenerate $O + Ne + Mg$ cores is discussed. For
the $10 – 13\,M_\odot$ stars, behavior of off–center neon flashes is
crucial. The $8 – 10\,M_\odot$ stars do not ignite neon and eventually
collapse due to electron captures. Properties of supernova
explosions and neutron stars expected from these low mass
progenitors are compared with the Crab nebula. The author also
examines the conditions for which neutron stars form from
accretion–induced collapse of white dwarfs in close binary
systems.

**067.015 On the long term stability of neutron star magnetic
    fields.**
D. Eichler, Z.–z. Wang.
The origin and evolution of neutron stars, p. 375 (1987).
Abstract. – See Abstr. 012.001 (IAU Symp. No. 125).

**067.016 Tortion influence on the magnetic field of rotating
    neutron stars.**
C. N. Zhang, J. G. Shi, F. P. Chen.
The origin and evolution of neutron stars, p. 379 (1987).
Abstract. – See Abstr. 012.001 (IAU Symp. No. 125).

**067.017 Monte Carlo simulations of radio pulsars and their
    progenitors.**
R. J. Dewey, J. M. Cordes.
The origin and evolution of neutron stars, p. 408 (1987).
Abstract. – See Abstr. 012.001 (IAU Symp. No. 125).

067.018 Neutron stars interiors.
    C. Alcock.
The origin and evolution of neutron stars, p. 413 – 424 (1987). –
See Abstr. 012.001 (IAU Symp. No. 125).
    The current status of the theory of neutron star interiors is
reviewed. The various phases of matter that might exist are
discussed and their relation to other areas of astrophysics briefly
noted. Observational distinction between the models is not
simply obtained; analysis of pulsar post–glitch relation, of
neutron star precession and neutron star cooling are promising in
this regard.

067.019 Two types of pulsars.
    J. H. Huang.
The origin and evolution of neutron stars, p. 425 – 437 (1987). –
See Abstr. 012.001 (IAU Symp. No. 125).
    To sort out the whole sample of pulsars with measured P and Ṗ
into two types has much to do with the origin and evolution of
neutron stars. The author has discussed a variety of pulsar
properties, including their radio emission mechanism, space
velocities, interior structures and evolutionary modes. The fact

that different type of pulsars does have quite different properties
indicates that the processes to create neutron stars may have two
distinct types, say Type II supernova explosion and the collapse
of accreting white dwarfs. The evolutionary mode for the
author's Type I pulsars provides such a key link between binary
pulsars and X–ray binary pulsars that may be proposed to be a
self–consistent scenario for binary pulsars, X–ray binary pulsars,
fast pulsars as well as Type I pulsars.

**067.020 Neutron star cooling: critical test of dense matter
    physics.**
N. Itoh.
The origin and evolution of neutron stars, p. 439 – 446 (1987). –
See Abstr. 012.001 (IAU Symp. No. 125).
    Recent developments in the standard theory of neutron star
cooling is critically reviewed. Emphasis is placed on the recent
developments in the calculations of thermal conductivity and
neutrino energy loss rates.

**067.021 Neutron stars formed from supernova explosion and
    quark matter.**
Y. C. Li, X. J. Kong, C. W. Wei, Y. Z. Ge.
The origin and evolution of neutron stars, p. 447 (1987).
Abstract. – See Abstr. 012.001 (IAU Symp. No. 125).

067.022 The problem of solidification in neutron stars.
    W. H. Huang, S. H. Gao.
The origin and evolution of neutron stars, p. 448 (1987).
Abstract. – See Abstr. 012.001 (IAU Symp. No. 125).

067.023 Interior structures for two types of pulsars.
    J.–H. Huang, Z.–G. Deng, X.–Y. Xia.
The origin and evolution of neutron stars, p. 449 (1987).
Abstract. – See Abstr. 012.001 (IAU Symp. No. 125).

**067.024 Modes of energy loss from isolated magnetized neutron
    star.**
S. Shibata.
The origin and evolution of neutron stars, p. 450 (1987).
Abstract. – See Abstr. 012.001 (IAU Symp. No. 125).

**067.025 Effects of magnetization on structure parameters of
    neutron stars.**
Q. M. Wang, S. H. Gao.
The origin and evolution of neutron stars, p. 451 (1987).
Abstract. – See Abstr. 012.001 (IAU Symp. No. 125).

**067.026 Synchro–curvature radiation and magnetic pair produc-
    tion of relativistic electrons in strong curved magnetic
fields.**
S.–Y. An.
The origin and evolution of neutron stars, p. 452 (1987).
Abstract. – See Abstr. 012.001 (IAU Symp. No. 125).

**067.027 Inverse Compton scattering in strong magnetic fields:
    applied to the radiation mechanism of PSR 0531 + 21.**
H. Chen, X. J. Wu, G. J. Qiao.
The origin and evolution of neutron stars, p. 453 (1987).
Abstract. – See Abstr. 012.001 (IAU Symp. No. 125).

067.028 The nutations of neutron stars and core–crust coupling.
    C. R. Gwinn.
The origin and evolution of neutron stars, p. 454 (1987).
Abstract. – See Abstr. 012.001 (IAU Symp. No. 125).

**067.029 The spin–torsion coupling precession of spin and its
    effects on single pulses of pulsars.**
Y. Zhang.
The origin and evolution of neutron stars, p. 455 (1987).
Abstract. – See Abstr. 012.001 (IAU Symp. No. 125).

067.030   **Neutron star cooling and the Vela pulsar.**
K. Nomoto, S. Tsuruta.
The origin and evolution of neutron stars, p. 456 (1987).
Abstract. – See Abstr. 012.001 (IAU Symp. No. 125).

067.031   **Model atmospheres for cooling neutron stars.**
R. W. Romani, R. D. Blandford, L. Hernquist.
The origin and evolution of neutron stars, p. 459 (1987).
Abstract. – See Abstr. 012.001 (IAU Symp. No. 125).

067.032   **The neutron stars with magnetic charges.**
Y.–J. Wang, Q.–H. Peng.
The origin and evolution of neutron stars, p. 460 (1987).
Abstract. – See Abstr. 012.001 (IAU Symp. No. 125).

067.033   **The nonlinear dispersion relation and the relationship of the forming soliton area to the evolution of pulsars.**
X.–j. Mao, Y. Tong.
The origin and evolution of neutron stars, p. 461 (1987).
Abstract. – See Abstr. 012.001 (IAU Symp. No. 125).

067.034   **The origin of the enhanced dissipation "α" in accretion discs and its relation to gamma bursts.**
S. A. Colgate.
The origin and evolution of neutron stars, p. 546 (1987).
Abstract. – See Abstr. 012.001 (IAU Symp. No. 125).

067.035   **Cosmic ray particle acceleration in pulsar magnetospheres.**
K. O. Thielheim.
The origin and evolution of neutron stars, p. 555 (1987).
Abstract. – See Abstr. 012.001 (IAU Symp. No. 125).

067.036   **Where neutron stars come from, how neutron stars evolve, and where neutron stars go.**
L. Woltjer.
The origin and evolution of neutron stars, p. 559 – 562 (1987). – See Abstr. 012.001 (IAU Symp. No. 125).
The author briefly reviews the current knowledge on neutron stars.

067.037   **Nonradial g–mode oscillations in X–ray bursting neutron stars.**
P. N. McDermott, R. E. Taam.
Astrophys. J., Vol. 318, No. 1, p. 278 – 287 (1987).
The oscillation spectrum of nonradial g–modes in X–ray bursting neutron stars has been studied. The pulsation periods are found to be sensitive to the envelope temperature and range from ~ 15 ms to ~ 50 ms for the $l = 1$ g_1–mode during the X–ray burst. From a quasi–adiabatic stability analysis it is likely that a spectrum of l–pole g–modes is unstable due to the ε–mechanism associated with rapid alpha captures. As the thermal structure of the envelope of the neutron star changes on time scales less than 0.2 s during the rise of the X–ray burst, the oscillations are expected to be quasi–coherent during this phase. The pulsations are short–lived (~ 1 s) and are most likely to be seen in the immediate vicinity of the burst peak.

067.038   **On the observability of the magnetic precession of the black hole accretion disks.**
A. N. Aliev, D. V. Galt'sov.
Astrophys. Space Sci., Vol. 135, No. 1, p. 81 – 86 (1987).
The precession of a rotating charged black hole in an external magnetic field is considered. The possibility of observing the effect in black hole systems with magnetized accretion disks is discussed.

067.039   **Neutron–star models in Ni's theory of gravity.**
J. P. Sharma.
Astrophys. Space Sci., Vol. 135, No. 2, p. 409 – 415 (1987).
The author has studied the structure of massive neutron–star models with polytropic indices $n = 0, 1$, and 2 in Ni's theory of gravity in post–Newtonian approximation. Solutions of the structure equation are shown.

067.040   **An intuitive understanding of black–hole evaporation by viewing it in terms of more familiar quantum–field effects in flat spacetime.**
R. M. Nugayev (*R. M. Nugaev*).
Z. Naturforsch., A, Vol. 42, No. 7, p. 657 – 662 (1987).
The article is aimed at an intuitive understanding of the recently explored deep connections between thermal physics, quantum field theory and general relativity. The physical effects involved in particle creation by a black hole are viewed in terms of more familiar quantum–field effects in flat spacetime. Black hole evaporation is investigated in terms of temperature correction to the Casimir effect. The application of the Casimir effect results and those for accelerated mirrors reveals that a black hole should produce the blackbody radiation at a temperature that exactly coincides with Hawking's result.

067.041   **Particle creation by a black hole as a consequence of the Casimir effect.**
R. M. Nugayev (*R. M. Nugaev*).
Commun. Math. Phys., Vol. 111, No. 4, p. 579 – 592 (1987).
Particle creation by a black hole is investigated in terms of temperature corrections to the Casimir effect. The reduction of the Hawking effect to more familiar effects observed in the laboratory enables one to reveal the mechanism of particle creation.

067.042   **A charged analogue of the Vaidya–Tikekar solution.**
L. K. Patel, S. S. Koppar.
Aust. J. Phys., Vol. 40, No. 3, p. 441 – 447 (1987).
The authors present an interior solution of the Einstein–Maxwell equations for a charged static fluid sphere. The physical 3–space $t = $ constant of the solution is spheroidal. The solution is interpreted as an exact relativistic model for a charged superdense star.

067.043   **The birth of neutron stars and black holes.**
A. Burrows.
Phys. Today, Vol. 40, No. 9, p. 28 – 37 (1987).
The core of a massive star that lives for 10 million years collapses within one second, initiating a series of some of the most exotic and extreme events that occur in the universe.

067.044   **Global trapped oscillations of relativistic accretion disks.**
A. T. Okazaki, S. Kato, J. Fukue.
Publ. Astron. Soc. Jpn., Vol. 39, No. 3, p. 457 – 473 (1987).
Axially symmetric, isothermal pulsations of a geometrically thin disk rotating around a nonrotating relativistic object are investigated. The disk is assumed to be isothermal in the vertical direction and to have a constant vertical scale height. It is found that there exist global pulsation modes trapped in an inner region of the disk. Eigenfrequencies of the modes are in the range of 0 to $\varkappa_{max}$, where $\varkappa_{max}$ is the maximum value of an epicyclic frequency in the disk. For an oscillation to be trapped in the disk, the eigenfunctions must have at least one node in the vertical direction. The results suggest that the presence of global trapped oscillation modes is a general characteristic of thin relativistic disks. The oscillation period obtained is briefly compared with observations.

067.045   **Energy spectrum at infinity emitted from a rapidly rotating neutron star.**
I. Asaoka, R. Hoshi.
Publ. Astron. Soc. Jpn., Vol. 39, No. 3, p. 475 – 484 (1987).
The energy spectrum at infinity is calculated for a rapidly rotating neutron star, which emits blackbody radiation in its local Lorentz frame. The dragging of the inertial frame due to rotation considerably deforms the original blackbody spectrum, if the neutron star has a large mass–to–radius ratio and is rotating with a period less than ~ 0.5 ms (for a distant observer). When the rotation period is increased to as long as ~ 1 ms, the energy spectrum at infinity can be expressed by a blackbody. However, the transverse Doppler effect cannot be neglected, although the gravitational redshift is the major process in deforming the spectral shape.

067.046   **Erratum: "X-ray spectra and atmospheric structures of bursting neutron stars" [Publ. Astron. Soc. Jpn., Vol. 39, No. 2, p. 287 – 308 (1987)].**
T. Ebisuzaki.
Publ. Astron. Soc. Jpn., Vol. 39, No. 3, p. 539 (1987). See Abstr. 43.067.075.

067.047   **Magnetohydrodynamics of plasma in the crust of a neutron star.**
D. M. Sedrakyan, A. K. Avetisyan.
Astrofizika, Tom 26, Vyp. 3, p. 489 – 500 (1987). In Russian. English translation in Astrophysics, Vol. 26, No. 3, p. 295 – 302 (1987).
Considering that ions form the Boltzmann liquid and electrons the relativistic degenerated gas, it is shown that the plasma in the crust of a neutron star is the Lorentz one. Magnetohydrodynamic equations are written and conditions of their reliability for such plasma are discussed. The kinetic coefficients for this plasma are calculated and it is shown, that for $Z \geqslant 27$ they mainly depend on the plasma density. Also, it is shown that when $\varrho \gtrsim 3 \times 10^8 \mathrm{g/cm^3}$ the kinetic coefficients are not affected by the magnetic field of the magnitude of $B \lesssim 10^{12}$ gauss. The numerical values of these coefficients when $3 \times 10^8 \mathrm{g/cm^3} \leqslant \varrho \leqslant 2 \times 10^{14} \mathrm{g/cm^3}$ are presented.

067.048   **The membrane paradigm and black–hole thermodynamics.**
K. S. Thorne.
Gravitational collapse and relativity, p. 13 – 19 (1986). – See Abstr. 012.006.
A brief overview is given of the theoretical underpinnings of the membrane paradigm for black–hole physics. Then those underpinnings are used to elucidate the paradigm's view that the laws of black–hole thermodynamics (including the statistical origin of black–hole entropy) are just a special case of the laws of thermodynamics for an ordinary, rotating, thermal reservoir.

067.049   **The third law.**
W. Israel.
Gravitational collapse and relativity, p. 96 – 105 (1986). – See Abstr. 012.006.
The third law of black hole mechanics is here established in the following form: no continuous process in which the stress–energy tensor of matter stays bounded and satisfies the weak energy condition in a neighbourhood of the apparent horizon can reduce the surface gravity of a black hole to zero within a finite advanced time.

067.050   **Black holes and monopoles.**
M. Yamazaki.
Gravitational collapse and relativity, p. 150 – 157 (1986). – See Abstr. 012.006.
The expression on the gravitational field made by lining up of N Kerr black holes is manifestly symmetric with respect to any permutation of N black holes. This fact may help the attack on the problem of finding the N Yang–Mills–Higgs monopoles solutions of total charge N having finite separation among the locations.

067.051   **3–D relativistic hydrodynamics around a Kerr black hole.**
J.-A. Marck, T. Piran.
Gravitational collapse and relativity, p. 276 – 281 (1986). – See Abstr. 012.006.
The authors present a numerical code for solving the relativistic hydrodynamic equations around a Kerr black hole (expressed in the Boyer–Lindquist coordinates). With an appropriate choice of the variables they remove the horizon singularity and the pseudo–singularity associated to "spherical–like" coordinate system at the polar axis $\sin \theta = 0$. The hydrodynamic equations are solved using (pseudo) spectral methods with Chebychev polynomials expansion in the r– and θ–direction and Fourier series in the φ–direction and explicit integration in time.

067.052   **Apparent horizon of initial data for black hole–collision.**
K.–i. Oohara.
Gravitational collapse and relativity, p. 313 – 319 (1986). – See Abstr. 012.006.
The apparent horizon of three dimensional initial data for a black hole–collision is examined. The author gives a method of determining the apparent horizon in three dimensional numerical relativity using the $(3+1)$–formalism of the Einstein equation.

067.053   **Gravitational radiation from a Kerr black hole.**
Y. Kojima.
Gravitational collapse and relativity, p. 320 – 328 (1986). – See Abstr. 012.006.
Perturbation calculations are used to study the gravitational radiation emitted when a test particle is falling into a Kerr black hole.

067.054   **The incompressibility of hot, neutron–rich nuclear matter.**
X. Vinas, M. Barranco, J. Treiner, S. Stringari.
Astron. Astrophys., Vol. 182, No. 1, p. L34 – L36 (1987).
The authors have carried out Hartree–Fock calculations of the nuclear matter incompressibility K at finite temperatures. These calculations cover the range of proton concentrations $Y_e = Z/A$ from symmetric matter, i.e. $Y_e = 0.5$, to $Y_e = 0.25$ and temperatures T from 0 to 8–10 MeV; the authors investigate the role of the effective mass of the interaction. For values of T and Y_e relevant for stellar collapse calculations, they find that K decreases by a factor 2 to 3 with respect to its value at saturation of symmetric matter. An analytical equation of state is proposed valid up to densities around 1.5 times the nuclear matter saturation density.

067.055   **Electric fields generated by a rotating neutron star in a vacuum with allowance for GTR effects.**
A. G. Muslimov, A. I. Tsygan.
Sov. Astron., Vol. 30, No. 5, p. 567 – 570 (1986). English translation of 42.067.041.

067.056   **Ejection of the photosphere of a neutron star with a strong magnetic field.**
I. G. Mitrofanov.
Sov. Astron., Vol. 30, No. 5, p. 571 – 573 (1986). English translation of 42.067.042.

067.057   **Optical and γ–ray bursts on neutron stars.**
G. S. Bisnovatyj–Kogan, A. F. Illarionov.
Sov. Astron., Vol. 30, No. 5, p. 582 – 588 (1986). English translation of 42.067.043.

067.058   **Energy release in pulsars due to the motion of vortices.**
D. M. Sedrakyan.
Astrophysics, Vol. 25, No. 2, p. 539 – 542 (1987). English translation of 42.067.131.

067.059   **Paramagnetic effects in a superconducting neutron star.**
K. M. Shakhabasyan.
Astrophysics, Vol. 25, No. 3, p. 665 – 671 (1987). English translation of 42.067.164.

067.060   **g–modes in zero–temperature neutron stars.**
L. S. Finn.
Mon. Not. R. Astron. Soc., Vol. 227, No. 2, p. 265 – 293 (1987).
It is well known that in an isentropic, perfect–fluid star, chemical inhomogeneities perturb some g–modes away from zero–frequency. Here this phenomenon is studied for the idealized case of zero–temperature, perfect–fluid neutron stars.

067.061   **Stabilization of non–axisymmetric instabilities in a rotating flow by accretion on to a central black hole.**
O. M. Blaes.
Mon. Not. R. Astron. Soc., Vol. 227, No. 4, p. 975 – 992 (1987).
The normal modes of a sequence of two–dimensional flows around a black hole are investigated. When the flow is one of

pure rotation, numerous non–axisymmetric dynamical instabilities are found which are similar to those discovered by previous authors. When the inner edge of the flow crosses the critical 'cusp' radius, a purely rotating equilibrium flow is no longer possible and the fluid in the inner regions flows into the hole transonically. All the unstable modes found for the purely rotating flow are quickly stabilized by this process. The relevance of these results to the dynamical stability of accreting thick tori is discussed.

067.062 Disc accretion by magnetized neutron stars: a reassessment of the torque.
Y.–M. Wang.
Astron. Astrophys., Vol. 183, No. 2, p. 257 – 264 (1987).

The torque exerted on a magnetized neutron star undergoing steady, axisymmetric disc accretion is re–evaluated. As in the widely adopted model of Ghosh and Lamb, the magnetospheric field is assumed to thread the disc both inside and outside the radius of corotation, yielding opposing contributions to the net torque from the forward– and backward–swept field lines. It is shown, however, that the particular pitch distribution postulated by Ghosh and Lamb is not consistent, since the resulting magnetic pressure would disrupt their disc beyond the corotation point. By taking the winding rate proportional to the poloidal rather than the toroidal component of the field, the author obtains a more realistic behaviour for the magnetic stresses generated by the rotational shear. He then discusses the results in the light of period–change measurements for several binary X–ray pulsars.

067.063 Power spectra of quasi–periodic oscillations in luminous X–ray stars.
N. Shibazaki, F. K. Lamb.
Astrophys. J., Vol. 318, No. 2, p. 767 – 785 (1987).

Quasi–periodic oscillations (QPOs) with frequencies in the range 5–50 Hz have recently been discovered in the intensity of more than half a dozen luminous X–ray stars. Random process models of the X–ray intensity time series provide a useful framework for analyzing the constraints imposed on physical models of QPO sources by measurements of the power–density spectra of their X–ray time series. Here, the authors use this framework to study the effects on the power spectrum of several physical phenomena that may occur in the beat–frequency model of accreting neutron stars.

067.064 Core collapse and formation of a rapidly rotating neutron star.
N. V. Ardelyan, G. S. Bisnovatyj–Kogan, Yu. P. Popov, S. V. Chernigovskij.
Astron. Zh., Tom 64, Vyp. 4, p. 761 – 772 (1987). In Russian. English translation in Sov. Astron., Vol. 31, No. 4.

Two–dimensional stellar collapse is investigated, beginning from the uniformly rotating white dwarf near the stability border up to the formation of an equilibrium differentially rotating neutron star. It is supposed that amplification of the magnetic field in a differentially rotating neutron star because of twisting of magnetic field lines leads to the magneto–rotational supernova explosion.

067.065 Joint evolution of normal and magnetized compact stars in low–mass binary systems. Analytical description of the degenerate component evolution.
V. M. Lipunov, K. A. Postnov.
Astron. Zh., Tom 64, Vyp. 4, p. 773 – 784 (1987). In Russian. English translation in Sov. Astron., Vol. 31, No. 4.

The evolution of the magnetized degenerate component (white dwarf, neutron star) in a low–mass binary system is described. The possible evolutionary paths of such stars in the $\lg P - \lg (\dot{M}/\mu^2)$ diagram are considered qualitatively.

067.066 Coherent scalar–field oscillations forming compact astrophysical objects.
I. I. Tkachev.
Sov. Astron. Lett., Vol. 12, No. 5, p. 305 – 308 (1987). English translation of 42.067.020.

067.067 A black hole as a particle–collision agent.
V. I. Dokuchaev.
Sov. Astron. Lett., Vol. 12, No. 5, p. 322 – 325 (1987). English translation of 42.067.025.

067.068 Accretion of charged matter by collapsing objects.
U. S. Pandey.
Astrophys. Space Sci., Vol. 136, No. 1, p. 195 – 199 (1987).

Hydrodynamical equations of accreting charged matter for the radial flow are presented in the Schwarzschild background space–time. The product δ of entropy–density S and charge–density inverse ε^{-1} is constant during motion. The author also demonstrates the occurrence of a critical region where δ and the enthalpy density h become equal in magnitude. Application of the model to stellar X–ray sources is also remarked.

067.069 Neutron star magnetic field decay: flux expulsion from the superconducting interior.
P. B. Jones.
Mon. Not. R. Astron. Soc., Vol. 228, No. 2, p. 513 – 520 (1987).

The viscous force acting on a moving vortex in a proton type II superconductor has been calculated by application of Chambers' representation of the Boltzmann equation. Equilibrium between the buoyancy, Magnus and viscous forces defines a vortex drift velocity such that magnetic flux is expelled from the superconducting interior to the solid crust in less than 10 Myr, a time many orders of magnitude smaller than the previous consensus. Ohmic dissipation in the crust may then be the cause of the 5 Myr time constant seen in the decay of radio pulsar dipole moments. The longer time constant ($\gtrsim 10^3$Myr), for which some evidence exists, could be explained by a small central volume of a nonsuperconducting phase, such as a normal Fermi baryon liquid.

067.070 Gravitational energy for spherically symmetric configurations.
E. Nahmad–Achar.
Mon. Not. R. Astron. Soc., Vol. 228, No. 3, p. 51P – 53P (1987).

Lynden–Bell & Katz (1985) showed by physical arguments that the gravitational field energy of a Schwarzschild black hole should lie entirely outside its horizon. Here the author shows that one can also give physical arguments by which the total energy should lie within the hole's horizon.

067.071 Observations of accretion instabilities at super Eddington accretion rates.
M. A. Abramowicz.
Variability of galactic and extragalactic X–ray sources, p. 137 – 150 (1987). – See Abstr. 012.019.

The author discusses the observational evidence that accretion disks around black holes and neutron stars with accretion rates roughly between 10^{-2} and 10^3 times the Eddington rate show remarkably more thermal variability than disks with accretion rates above or below this instability strip. A theoretical explanation is given.

067.072 Quasi–normal modes of a black hole.
N. Panchapakesan, B. Majumdar.
Astrophys. Space Sci., Vol. 136, No. 2, p. 251 – 257 (1987).

The quasi–normal modes of a black hole are calculated requiring that the ingoing amplitudes vanish at infinity. The amplitude is obtained by making the solutions valid close to and far away from the horizon. The normal modes are obtained in the form of polynomial solutions.

067.073   General relativistic effects on collimation of a jet.
R. C. Kapoor.
Astrophysical jets and their engines, p. 245 – 246 (1987). – See Abstr. 012.021.

067.074   Photon absorption and splitting in the magnetospheres of neutron star gamma–ray bursters.
I. G. Mitrofanov, A. S. Pozanenko, V. Sh. Dolidze, C. Barat, K. Hurley, M. Niel, G. Vedrenne.
Sov. Astron., Vol. 30, No. 6, p. 659 – 663 (1987). English translation of 42.067.149.

067.075   Accretion disc models around compact objects.
U. S. Pandey.
Astrophys. Space Sci., Vol. 137, No. 1, p. 77 – 84 (1987).
The author presents a survey of accretion disc models around compact objects – in particular the accretion onto white dwarfs, neutron stars, and black holes. He discusses both the thin disc as well as thick disc models and also the feasibility where either of these can be applied in the astrophysical systems. The crucial role of magnetic field in facilitating the formation of accretion discs in neutron stars is indicated. The prime significance of accretion discs in the generation of soft and hard X–rays is also discussed. Thick disc models are found to explain the observations of active galactic nuclei and also collimated and persistent jets in some of the radio sources.

067.076   Hydromagnetic flows from rapidly rotating compact objects. II. The relativistic axisymmetric jet equilibrium.
M. Camenzind.
Astron. Astrophys., Vol. 184, No. 1/2, p. 341 – 360 (1987).
The author develops the theory and the numerical methods for the treatment of the relativistic and axisymmetric magnetohydrodynamic jet equilibrium. This leads to a nonlinear system consisting of a quasilinear partial differential equation for the magnetic flux function and an algebraic wind equation for the poloidal velocity of the plasma. The relativistic Grad–Schlüter–Shafranov equation is solved with the method of finite elements on a grid scaled by the light cylinder radius. Since the grid has to follow the shape of the Alfvèn surface, the nodal points have to be redistributed iteratively. The hot wind equation for adiabatic ion flows is transformed to a polynomial of degree 16 in the poloidal velocity. The author discusses the values of the equilibrium parameters for galactic and extragalactic objects. Preliminary results of the modelling of low density plasma flows are presented for the case of a rapidly rotating magnetosphere of a compact object in interaction with an accretion disk. The asymptotic plasma flow can reach Lorentz factors in the range $1 < \gamma \leqslant 5$.

067.077   Relativistic corrections to the gravitational radiation from rapidly rotating magnetized neutron stars.
V. P. Tsvetkov, A. N. Tsirulev.
Astron. Zh., Tom 64, Vyp. 5, p. 1117 – 1120 (1987). In Russian. English translation in Sov. Astron., Vol. 31, No. 5.
A method for calculating relativistic corrections to the total intensity of gravitational radiation from rapidly rotating neutron stars in the model of an oblique rotator is developed. The concrete values of these corrections are calculated and their analysis for various parameters of neutron stars is made.

067.078   Limits on hadronic cosmic ray production by young pulsars.
D. Eichler, J. R. Letaw.
Nature, Vol. 328, No. 6133, p. 783 – 784 (1987).
Since the discovery of supernova 1987A several authors have noted that it may provide an excellent opportunity to observe the cosmic ray output of a young pulsar through the "beam dump" that the supernova ejecta provide. It has been suggested that neutrino emission from p–p collisions is possible. In this letter the authors argue that the cosmic abundances of Li, Be and B set significant constraints on the comsic ray proton production in the young ($t \leqslant 1$ yr) remnant, and, in particular, rule out neutrinos

from shock–accelerated protons in the ejecta at currently detectable levels.

067.079   Astrophysical black holes.
R. D. Blandford.
Three hundred years of gravitation, p. 277 – 329 (1987). – See Abstr. 003.012.
Contents: 1. Introduction. 2. Black holes in astrophysics. 3. Observational evidence for black holes. 4. Future prospects.

067.080   Pair production and Compton scattering in compact sources and comparison to observations of active galactic nuclei.
A. P. Lightman, A. A. Zdziarski.
Astrophys. J., Vol. 319, No. 2, p. 643 – 661 (1987).
The authors determine the steady state particle and photon distributions under the assumption of continuous injection of relativistic electrons and low–energy photons throughout a spherical volume. Through inverse Compton scattering, the primary electrons produce γ–rays, which then produce electron–positron pairs. The pairs constitute a secondary electron injection, and the process continues, with each generation of pairs cooling down to subrelativistic energies before thermalizing and annihilating. The authors compute detailed emergent spectra for a wide range of parameters, and compare the results to the data from active galactic nuclei.

067.081   Gamma–ray pulsar model.
J. M. Cohen, E. Mustafa.
Astrophys. J., Vol. 319, No. 2, p. 930 – 938 (1987).
Extremely high energy bursts of pulsed γ–rays have recently been detected from several binary X–ray pulsars and young radio pulsars. The authors propose that curvature synchrotron radiation from electrons (accelerated along the open magnetic field lines) is a possible origin of this emission. The electric and magnetic fields in this model are computed in the near zone (close to the magnetic poles), and the effects of space charge limited flow and pair creation are considered. It is found that the measured photon energy of the emission from most of these pulsars is consistent with the model.

067.082   On the theory of radio radiation of pulsars (Review).
V. S. Beskin, A. V. Gurevich, Ya. N. Istomin.
Izv. Vyssh. Uchebn. Zaved., Radiofiz., Tom 30, No. 2, p. 161 – 186 (1987). In Russian. Abstr. in Ref. Zh., 51. Astron., 9.51.717 (1987).

067.083   Neutron star precession and the dynamics of the superfluid interior.
A. Alpar, H. Ögelman.
Astron. Astrophys., Vol. 185, No. 1/2, p. 196 – 202 (1987).
The authors consider the dynamics of the superfluid interior of a precessing neutron star. They formulate Euler's equations with the current models of internal torques and find that the superfluid interior of the star has steady states in which the interior follows the crust's precession. Using the current understanding of the neutron star interior based on observations of radio pulsars, the authors calculate the internal torques and energy dissipation rates. They show that the torques do not exceed those available in the Her X–1/HZ Her binary system. They discuss the implications of interpreting the 35 day cycle turn–on fluctuations of Her X–1 in terms of fluctuating internal pinning torques and show that this would lead to a large energy dissipation that is not compatible with the observations.

067.084   Thick accretion disks: theory vs. observations.
M. A. Abramowicz, M. Calvani, P. Madau.
Comments Astrophys., Vol. 12, No. 2, p. 67 – 85 (1987).
The authors briefly discuss the current state of research on geometrically thick accretion disks which are believed to form when the accretion rate is supercritical. They then review those galactic and extragalactic objects (SS433, QPO sources, active galactic nuclei) where supercritical accretion seems to occur, or

for which thick accretion disks have been invoked to explain the observations.

067.085 Thick accretion discs: the role of self–gravity.
G. Bodo, A. Curir.
General relativity and gravitational physics, p. 399 – 404 (1987). – See Abstr. 012.038.

The authors present some preliminary results concerning the structure of self–gravitating accretion discs. Self–gravity is introduced in a fully self–consistent way and the resulting equations are solved numerically using the self–consistent field method. The main results are the possibility of solutions which do not exist when self–gravity is neglected, the possibility of accretion in a purely newtonian potential and a limitation on the size of the discs imposed by self–gravity.

067.086 How big are supermassive black holes formed from the collapse of dense star clusters?
C. S. Kochanek, S. L. Shapiro, S. A. Teukolsky.
Astrophys. J., Vol. 320, No. 1, p. 73 – 84 (1987).

If a star cluster becomes sufficiently relativistic in its center, it is unstable to catastrophic collapse to a black hole. In general, the final state consists of a central massive black hole surrounded by a halo of orbiting stars. The authors present a method for estimating how much mass goes into the black hole and how much is left in the halo following collapse. They find that extreme core–halo clusters, such as those arising from the gravothermal catastrophe, can produce black holes with masses substantially larger than the core mass. This may be crucial for explaining the birth of quasars and active galactic nuclei.

067.087 On two–stream instability in pulsar magnetospheres.
V. V. Usov.
Astrophys. J., Vol. 320, No. 1, p. 333 – 335 (1987).

It has been shown in earlier work that electron–positron pair creation near a pulsar may be a very nonstationary process. In that case the outflowing electron–positron plasma gathers into separate clouds. The clouds move along magnetic field lines and disperse as they go farther from the pulsar. At a distance of $\sim 10^8$cm from the pulsar surface the high–energy particles of a given cloud catch up with the low–energy particles that belong to the cloud going ahead of it. In this region of a pulsar magnetosphere the energy distribution of plasma particles is two–humped, and a two–stream instability may develop.

067.088 An estimate of the event rate for neutron star binary coalescence and its implications for gravitational wave detection.
R. J. Dewey.
Bull. Am. Astron. Soc., Vol. 19, No. 2, p. 684 (1987). Abstract. – See Abstr. 010.061.

067.089 Accretion onto the polar caps of magnetized neutron stars: an analytic solution.
J. Arons, R. I. Klein.
Bull. Am. Astron. Soc., Vol. 19, No. 2, p. 755 – 756 (1987). Abstract. – See Abstr. 010.061.

067.090 Thermal radiation of neutron stars.
G. G. Arutyunyan, V. V. Papoyan, G. S. Saakyan.
Astrophysics, Vol. 26, No. 2, p. 155 – 161 (1987). English translation of 43.067.072.

067.091 The interaction of a magnetic neutron star with an accretion disc.
C. G. Campbell.
Mon. Not. R. Astron. Soc., Vol. 229, No. 3, p. 405 – 422 (1987).

The interaction of a magnetic neutron star with an accretion disk is considered, in the axisymmetric case. Stellar poloidal field penetrates the disc and acts as a source of toroidal field, generated by material shears. For typical neutron star magnetic moments the back reaction of the field on the main part of the disc is small. However, in the inner parts of the disc magnetic stresses become comparable to material stresses and the angular velocity of matter tends towards that of the star. Magnetic field solutions are found for two forms of diffusion coefficient and for two angular velocity distributions. These solutions are used to derive the magnetic torque on the star as a function of its rotation rate. Magnetic and accretion torques are compared and expressions are found for the time evolution of the stellar angular velocity and the equilibrium period at which the torques cancel.

067.092 Ohmic decay of crustal neutron–star magnetic fields.
Y. Sang, G. Chanmugam.
Space Telesc. Sci. Inst., Prepr. Ser., No. 211, 9 pp. (1987). To appear in Astrophys. J., Lett. Ed.

067.093 Thermische Strahlung von Neutronensternen.
M. E. Schaaf.
MPE Rep., No. 203, 255 pp. (1987). Dissertation, Tech. Univ. München.

067.094 Particle acceleration in a flow accreting through shock waves.
H. C. Spruit.
MPA Rep., No. 304, 19 pp. (1987). Submitted to Astron. Astrophys.

067.095 Accretion disks in soft potential wells.
W. J. Duschl.
MPA Rep., No. 306, 12 pp. (1987). To appear in the proceedings of the 10th European Regional Meeting of the IAU, Prague, 24 – 29 August 1987.

067.096 Dynamical instability of a Keplerian disk; dependence on azimuthal wave number, March number, and the size of the disk.
T. Hanawa.
MPA Rep., No. 307, 28 pp. (1987). To appear in Astron. Astrophys.

067.097 Accretion disks in the centers of galaxies.
W. J. Duschl.
MPA Rep., No. 315, 10 pp. (1987). To appear in the proceedings of the "2nd workshop on high energy astrophysics", Ringberg Castle, 13 – 18 July 1987.

067.098 Electron injection by relativistic protons in active galactic nuclei.
M. Sikora, J. G. Kirk, M. C. Begelman, P. Schneider.
Astrophys. J., Lett. Ed., Vol. 320, No. 2, p. L81 – L85 (1987).

The authors make the case that efficient acceleration of protons by strong shocks in active galactic nuclei will automatically channel most of the available energy into relativistic electrons and positrons. Thus, direct acceleration of electrons need not be efficient. It is shown that, for the most energetic protons likely to be accelerated in AGNs, p–p collisions are less important than collisions of protons with soft photons. Pairs produced by p–γ collisions in this scenario have much higher energies than the maximum energies attainable by Fermi acceleration of electrons. They are energetic enough to initiate and sustain a pair cascade even when synchrotron radiation is the primary cooling process.

067.099 Primitive black hole generated from unstable Minkowski spacetime.
Z. Zhao.
Chin. Phys. Lett., Vol. 3, No. 11, p. 527 – 528 (1986). Abstr. in Phys. Abstr., Vol. 90, No. 1309, Entry 88494 (1987).

067.100 Particle production in de Sitter spacetime.
T. Mishima, A. Nakayama.
Prog. Theor. Phys., Vol. 77, No. 2, p. 218 – 222 (1987). Abstr. in Phys. Abstr., Vol. 90, No. 1309, Entry 88496 (1987).

067.101   Exact solutions of distribution of the density of material in a relativistic polytrope with a compact core.
Y. Zhou, G.–z. Xie, X.–g. Wang, X.–s. Tian, L. Chen.
Publ. Yunnan Obs., No. 1, p. 1 – 6 (1987). In Chinese.

The authors have re–discussed in detail the distribution of the density of material in a relativistic polytrope with a compact core. They solved the Euler equation and got some exact analytical solutions. The results show that the distribution of the density of material in a relativistic polytrope with a compact core is inversely proportional to the cube of radius r and it is different from the isothermal gas sphere.

067.102   Convection, type II supernovae, and the early evolution of neutron stars.
A. Burrows, J. M. Lattimer.
Prepr. Steward Obs., No. 769, 26 pp. (1987). To appear in Phys. Rep.

067.103   The thermodynamics of a system containing two black holes and black–body radiation.
N. T. Bishop, P. T. Landsberg.
Gen. Relativ. Gravitation, Vol. 19, No. 11, p. 1083 – 1090 (1987).

The problem of two black holes in a box separated by a partition is considered. Each black hole is in thermal equilibrium with its appropriate black–body radiation. The separating partition is removed and the possible final states of the combined system are investigated. Limits on the amount of gravitational wave energy that can be produced are calculated.

067.104   Are there topological black–hole solitons in string theory?
P. O. Mazur.
Gen. Relativ. Gravitation, Vol. 19, No. 12, p. 1173 – 1180 (1987).

The author points out that the celebrated Hawking effect of quantum instability of black holes seems to be a purely semiclassical but nonperturbative effect in string theory. Studying quantum dynamics of strings in the gravitational background of black holes the author finds classical instability due to emission of massless string excitations. The topology of a black hole seems to play a fundamental role in developing the string theory classical instability due to the effect of sigma model instantons. It is argued that string theory allows for a qualitative description of black holes with very small masses and it predicts topological solitons with a quantized spectrum of masses.

067.105   Polar orbits in the Kerr space–time.
E. Stoghianidis, D. Tsoubelis.
Gen. Relativ. Gravitation, Vol. 19, No. 12, p. 1235 – 1249 (1987).

The motion of test particles in polar orbit about the source of the Kerr field of gravity is studied using Carter's first integrals for timelike geodesics in the Kerr space–time. Expressions giving the angular coordinates of such particles as functions of the radial one are derived, both for the case of a rotating black hole as well as for that of a naked singularity.

067.106   Dilaton fields and event horizon.
T. Koikawa, M. Yoshimura.
Phys. Lett. B, Vol. 189, No. 1 – 2, p. 29 – 33 (1987). Abstr. in Phys. Abstr., Vol. 90, No. 1311, Entry 95416 (1987).

067.107   Pulsar drift and rotational collapse.
A. Qadir, M. Rafique, A. W. Siddiqui.
Chin. Phys. Lett., Vol. 4, No. 4, p. 177 – 180 (1987). Abstr. in Phys. Abstr., Vol. 90, No. 1311, Entry 101834 (1987).

067.108   The return currents in pulsar models.
J.–l. Tan, S.–r. Liang.
Chin. Phys. Lett., Vol. 4, No. 5, p. 225 – 228 (1987). Abstr. in Phys. Abstr., Vol. 90, No. 1311, Entry 101835 (1987).

067.109   Soliton stars and the critical masses of black holes.
T. D. Lee.
Phys. Rev. D, Vol. 35, No. 12, p. 3637 – 3639 (1987). Abstr. in Phys. Abstr., Vol. 90, No. 1312, Entry 102240 (1987).

067.110   Mini–soliton stars.
R. Friedberg, T. D. Lee, Y. Pang.
Phys. Rev. D, Vol. 35, No. 12, p. 3640 – 3657 (1987). Abstr. in Phys. Abstr., Vol. 90, No. 1312, Entry 102241 (1987).

067.111   Scalar soliton stars and black holes.
R. Friedberg, T. D. Lee, Y. Pang.
Phys. Rev. D, Vol. 35, No. 12, p. 3658 – 3677 (1987). Abstr. in Phys. Abstr., Vol. 90, No. 1312, Entry 102242 (1987).

067.112   Fermion soliton stars and black holes.
T. D. Lee, Y. Pang.
Phys. Rev. D, Vol. 35, No. 12, p. 3678 – 3694 (1987). Abstr. in Phys. Abstr., Vol. 90, No. 1312, Entry 102243 (1987).

067.113   Some Kerr–like cosmological solutions of the Einstein–Maxwell equations.
L. K. Patel, S. R. Yadav.
J. Math. Phys. Sci., Vol. 21, No. 2, p. 167 – 188 (1987). Abstr. in Phys. Abstr., Vol. 90, No. 1312, Entry 102260 (1987).

067.114   Quantisation of scalar and vector fields inside the cosmological event horizon and its application to the Hawking effect.
A. Higuchi.
Classical Quantum Gravity, Vol. 4, No. 3, p. 721 – 740 (1987). Abstr. in Phys. Abstr., Vol. 90, No. 1312, Entry 102273 (1987).

067.115   The intrinsic character of a black hole.
M. Ludvigsen.
Classical Quantum Gravity, Vol. 4, No. 3, p. 619 – 623 (1987). Abstr. in Phys. Abstr., Vol. 90, No. 1312, Entry 107889 (1987).

067.116   An axionic laser in the center of a galaxy?
I. I. Tkachev.
Phys. Lett. B, Vol. 191, No. 1 – 2, p. 41 – 45 (1987). Abstr. in Phys. Abstr., Vol. 90, No. 1312, Entry 107979 (1987).

067.117   Boson instability of charged black holes.
A. B. Gaina, I. M. Ternov.
Sov. Astron. Lett., Vol. 12, No. 6, p. 394 – 396 (1986). English translation of 42.067.064.

067.118   The influence of external magnetic fields on the structure of thin accretion disks.
U. Anzer, G. Börner, E. Meyer–Hofmeister.
Astron. Astrophys., Vol. 188, No. 1, p. 85 – 88 (1987).

The internal structure of thin accretion disks is studied for the case where strong outside magnetic fields interact with these disks. Such configurations are expected to occur in close binary X–ray pulsars such as Her X–1. The authors find that the magnetic pressure exerted on the disk has two main effects: the vertical disk structure shows a density increase towards the interface between disk and magnetosphere; and the thickness of the disk decreases more rapidly as one approaches the inner edge of the disk. The second effect gives a stronger irradiation of disk surface close to the inner edge and this results in a higher surface temperature.

067.119   Akkretion und kompakte Quellen.
H. G. Paul, G. Rüdiger.
Sterne, 63. Band, Heft 4, p. 213 – 224 (1987).

067.120   Nonexistence theorems for Yang–Mills fields and harmonic maps in the Schwarzschild spacetime.
H.–s. Hu.
Lett. Math. Phys., Vol. 14, No. 3, p. 253 – 262 (1987).

The nonexistence of static solutions to pure Yang–Mills equations and nonconstant harmonic maps defined on the Schwarzschild spacetime outside the black hole ($r > 2\,M$) is considered. Nonexistence theorems for pure Yang–Mills equations and harmonic maps in the region $r \geqslant 5\,M$ and $r \geqslant 3\,M$ are obtained, respectively.

067.121 Nonexistence theorems for Yang–Mills fields and harmonic maps in the Schwarzschild spacetime (II).
H. S. Hu, S. Y. Wu.
Lett. Math. Phys., Vol. 14, No. 4, p. 343 – 351 (1987).

The authors continue the study of the nonexistence of static pure Yang–Mills fields and harmonic maps defined on the Schwarzschild spacetime outside the black hole. Both the conditions on the regions and on the energy density are improved.

067.122 Grand Unified Models.
R. D. Blandford.
Superluminal radio sources, p. 310 – 327 (1987). – See Abstr. 012.053.

The author summarizes the ideas of this symposium to superluminal radio sources and reviews ideas about "Grand Unified Models", that is to say attempts to place compact radio sources within a general interpretive framework for galactic nuclear activity.

067.123 Doubly periodic thermal relaxation oscillations on an accreting neutron star.
M. Yasutomi.
Publ. Astron. Soc. Jpn., Vol. 39, No. 5, p. 769 – 779 (1987).

Thermal relaxation oscillations with double periods are found to occur in a two–zone model for the neutron star which accretes matter even at a constant rate. The model consists of a nuclear fuel zone and an ash zone left after a shell flash. The thermal oscillation tends to a limit cycle with small amplitude and short period when the accretion rate is large, and to that with large amplitude and long period when the accretion rate is small. For the intermediate range of the accretion rate, the above two cycles occur alternately. If small perturbations are allowed to the model, the recurrence of a shell flash becomes even aperiodic.

067.124 X–ray polarizations from accreting strongly magnetized neutron stars: case studies for the X–ray pulsars 4U 1626–67 and Hercules X–1.
T. Kii.
Publ. Astron. Soc. Jpn., Vol. 39, No. 5, p. 781 – 800 (1987).

The linear polarization of X–rays from an accreting neutron star with a strong magnetic field of $10^{12} – 10^{13}$G is calculated to investigate a possibility of observing the polarization of X–rays. Two binary X–ray pulsars, 4U 1626–67 and Her X–1, are considered as examples. The polarization degree is expected to be larger than 10% for some pulse phases. The result suggests a feasibility of detecting the X–ray polarization from X–ray binary pulsars with techniques available at present.

067.125 Formation of a millisecond pulsar in a globular cluster.
R. W. Romani, S. R. Kulkarni, R. D. Blandford.
Nature, Vol. 329, No. 6137, p. 309 – 310 (1987).

The discovery of an isolated 3–millisecond pulsar PSR 1821–24 in the globular cluster M28 has important implications for the genesis of such objects. The authors suggest a model appropriate to the cluster site, namely disruption of an expanded tidal capture binary by the perturbations of field stars. They also attempt to relate the millisecond pulsar formation rate to that of the cluster X–ray sources.

067.126 Origin of millisecond pulsars.
F. C. Michel.
Nature, Vol. 329, No. 6137, p. 310 – 311 (1987).

The recent discovery of a 3–millisecond pulsar in the globular cluster M28 suggests that a few per cent of type I supernovae form neutron stars. If the contact–binary white dwarf model for type Is is invoked, it seems probable that millisecond pulsars of low magnetic field would be formed in binary systems which might survive the supernova event.

067.127 Formation of isolated millisecond pulsars in globular clusters.
F. Verbunt, E. P. J. van den Heuvel, J. van Paradijs, S. A. Rappaport.
Nature, Vol. 329, No. 6137, p. 312 – 314 (1987).

Millisecond pulsars are probably 'recycled' old pulsars which have been spun–up through accretion from a binary stellar companion. In a globular cluster such a binary is likely to have formed in a tidally dissipative collision between an old neutron star and a low–mass field star from the cluster. The authors also explore the possibility that accretion from a massive disk is able directly to spin up the neutron star; in this case, however, the predicted number of millisecond pulsars in globular clusters would be ten times the number of bright low–mass X–ray binaries. In the model where the neutron star is spun in a binary the predicted number of millisecond pulsars is significantly smaller. If the millisecond pulsar in M28 has a magnetic moment similar to that of the other three known millisecond pulsars, the epoch of spin–up must have terminated $\lesssim 8 \times 10^8$yr ago.

067.128 On nonthermal models for active galactic nuclei.
D. L. Band.
Astrophys. J., Vol. 321, No. 1, p. 80 – 93 (1987).

In the nonthermal model for radio–quiet active galactic nuclei (AGNs) proposed in 1986 by Band and Grindlay, the flat hard X–ray power–law spectrum results from the scattering of a luminous ultraviolet source by the relativistic electrons that also produce the infrared and soft X–ray continuum by synchrotron emission. In the current work, the quantitative model is shown to be relatively insensitive to more complicated, realistic geometries. The effects of photon–photon pair production within this model are studied. The model's parameters are consistent with a nonthermal source 10 Schwarzschild radii around a massive black hole with a magnetic field in approximate equipartition with the photon flux from the accretion disk.

067.129 The effect of heating from the boundary layer on accretion models for novae and other compact objects.
G. Shaviv, S. Starrfield.
Astrophys. J., Lett. Ed., Vol. 321, No. 1, p. L51 – L53 (1987).

The authors investigate the assumption commonly made in accretion onto compact objects, namely that the matter lands "softly" on the surface. They find that (1) this assumption is not justified for currently believed parameters for accretion rates and (2) the effects of deviations from this assumption on the thermonuclear runaway are profound.

067.130 Neutron stars observations as astrophysical probes.
R. W. Romani.
Diss. Abstr. Int., Sect. B, Vol. 48, No. 6, p. 1711–B – 1712–B (1987). Thesis, California Institute of Technology, 203 pp. (1987). Order No. DA8719698.

067.131 Black holes and gravitational collapse.
Ts. Radoslavova.
Priroda (NRB), Tom 36, No. 3, p. 3 – 7 (1987). In Bulgarian. From Ref. Zh., 51. Astron., 12.51.160 (1987).

067.132 Curious, quark–like and metastable neutron stars.
M. I. Krivoruchenko.
Pis'ma Zh. Ehksp. Teor. Fiz., Tom 46, No. 1, p. 5 – 8 (1987). In Russian. Abstr. in Ref. Zh., 51. Astron., 12.51.572 (1987).

067.133 Radiation anisotropy of the accretion disc around a black hole, heating of the optical component and parameters of the binary system LMC X–3.
N. G. Bochkarev, R. A. Syunyaev, T. S. Khruzina, A. M. Cherepashchuk, N. I. Shakura.
Inst. Kosm. Issled. Akad. Nauk SSSR, Prepr., No. 1245, p. 1 – 41 (1987). In Russian. Abstr. in Ref. Zh., 51. Astron., 12.51.753 (1987).

067.134   Electron outflow in axisymmetric pulsar magnetospheres. II.

R. R. Burman.

Aust. J. Phys., Vol. 40, No. 5, p. 687 – 703 (1987).

In the axisymmetric pulsar magnetosphere model of Mestel et al. (1985), electrons, following injection with non–negligible speeds from the stellar surface, flow with moderate acceleration, and with poloidal motion that is closely tied to poloidal magnetic field lines, before reaching a limiting surface, near which rapid acceleration occurs. The formalism introduced by Mestel et al. for the description of the outflow is applied in an extended version which fully incorporates γ_0, the emission Lorentz factor of the particles. This treatment removes the singularity of γ_0 at the stellar poles that occurred in the earlier work.

067.135   Sustained magnetic fields in binary millisecond pulsars.

G. Chanmugam, K. Brecher.

Nature, Vol. 329, No. 6141, p. 696 – 698 (1987).

If correct, the fraction of pulsars which are binary millisecond pulsars is too large by a factor of > 100. This severe discrepancy is removed if, as the authors propose here, the magnetic fields of neutron stars do not decay either in such binaries or in general. They also show that, if such neutron stars are formed from the accretion–induced magnetic flux and angular momentum–conserving collapse of white dwarfs, most of them are likely to have been born, and remain, spinning rapidly and to have weak magnetic fields, in agreement with observations of binary millisecond pulsars and low–mass X–ray binaries.

067.136   Self–gravitating accretion disks in active galactic nuclei.

I. Shlosman, M. C. Begelman.

Nature, Vol. 329, No. 6142, p. 810 – 812 (1987).

Accretion onto supermassive black holes (SBHs) is widely believed to be responsible for the phenomena of active galactic nuclei (AGN). It is not known whether the accretion flow is fuelled by mass loss from the dense cluster of stars surrounding the SBH, or enters the nucleus from the galactic interstellar medium. Here the authors adopt the latter view, assuming that most of the incoming fuel forms a thin accretion disk at distances of $\sim 10 - 10^3 \mathrm{pc}$ from the SBH. Such a disk must be vertically self–gravitating. They analyse some of its thermal and dynamical properties, and conclude that its energetics is likely to be dominated by backscattered AGN radiation. They also discuss the conditions under which Jeans fragmentation of such a disk into weakly interacting cloudlets can be avoided, and give a necessary condition for disk fragmentation to occur.

067.137   Separability of the Killing–Maxwell system underlying the generalized angular momentum constant in the Kerr–Newman black hole metrics.

B. Carter.

J. Math. Phys., Vol. 28, No. 7, p. 1535 – 1538 (1987). Abstr. in Phys. Abstr., Vol. 90, No. 1313, Entry 108339 (1987).

067.138   On the solution of the Tolman–Oppenheimer–Volkov equation with the ultrarelativistic equation of state.

T. Toimela.

J. Math. Phys., Vol. 28, No. 7, p. 1541 – 1543 (1987). Abstr. in Phys. Abstr., Vol. 90, No. 1313, Entry 115494 (1987).

067.139   On the conversion of neutron stars into strange stars.

A. V. Olinto.

Phys. Lett. B, Vol. 192, No. 1 – 2, p. 71 – 75 (1987). Abstr. in Phys. Abstr., Vol. 90, No. 1313, Entry 115495 (1987).

067.140   Mining energy from a rotating black hole in a magnetic field by the Penrose process.

N. Dadhich.

Classical and quantum aspects of gravitation, p. 63 – 72 (1986). Abstr. in Phys. Abstr., Vol. 90, No. 1314, Entry 120643 (1987). – See Abstr. 012.065.

067.141   Quantum energy–momentum tensor in space–time with time–like–Killing vector.

V. P. Frolov, A. I. Zel'nikov.

Phys. Lett. B, Vol. 193, No. 2 – 3, p. 171 – 174 (1987). Abstr. in Phys. Abstr., Vol. 90, No. 1314, Entry 120666 (1987).

067.142   A numerical simulation of the non–stationary electron acceleration in a pulsar magnetosphere.

Yu. A. Rylov.

Astron. Zh., Tom 64, Vyp. 6, p. 1220 – 1232 (1987). In Russian. English translation in Sov. Astron., Vol. 31, No. 6.

The energy of electrons accelerated stationarily inside the outflow channel is calculated in the framework of a globally self–consistent pulsar magnetosphere model. The stationary acceleration is shown to be almost always unstable. A numerical simulation of non–stationary electron acceleration is produced. It is shown that by the non–stationary acceleration the electron bunches are formed and the energy consumed by electrons inside the outflow channel increases by a few orders of magnitude as compared with that by the stationary acceleration. The energy consumed inside the outflow channel occurs to be sufficient for the pulsar radio emission energetics explanation, the radiopulsar phenomenon being explained as a thermal radio emission of the gas consisting of electron bunches. Such common properties of the pulsar radio emission as the high brightness temperature, the sharp radio emission directivity and the characteristic turn–over of the radio emission spectrum at the frequency of the order of $10^8 \mathrm{Hz}$ find a natural explanation in the framework of this model.

067.143   Dissociative equilibrium of the H_2^+ molecular ion in the magnetic field of a neutron star.

V. K. Khersonskij.

Astron. Zh., Tom 64, Vyp. 6, p. 1233 – 1242 (1987). In Russian. English translation in Sov. Astron., Vol. 31, No. 6.

The equation of dissociative equilibrium for the molecular ion H_2^+ in a strong magnetic field $B = 10^{12} - 10^{13} \mathrm{G}$ is discussed. The equilibrium calculations are performed for the temperature region $T \lesssim 2 \times 10^6 \mathrm{K}$. Ionization and dissociation balance is considered for the molecular ions H_2^+ and hydrogen atoms at total proton number desity $N_p = 10^{20} - 10^{22} \mathrm{cm}^{-3}$. It is shown that when $T \lesssim 2 \times 10^5 \mathrm{K}$, the major component of such a medium may be neutral hydrogen atoms. The number density of H_2^+ admixture can amount to $10^{-7} - 10^{-4}$ of the total proton density and depends on the magnetic field intensity and N_p. Different spin modifications of H_2^+ in the total density of hydrogen molecular ions are also considered.

067.144   Electromagnetic jets from compact objects.

F. C. Michel.

Astrophys. J., Vol. 321, No. 2, p. 714 – 720 (1987).

The author examines the possibility that at least some astrophysical jets are initially electromagnetic in origin. Subsequent pick–up of ionization would convert such electromagnetic jets into hydrodynamic jets. In such a model, relativistic outflow is formed into highly collimated beams simply through the interaction with the surrounding medium. The overall properties of such jets are largely determined by a single dimensionless parameter: the characteristic electrostatic potential drop rewritten as a particle Lorentz factor: $\sigma \equiv e \Delta\phi / mc^2$.

067.145   A Kruskal–like model with finite density.

C. Hellaby.

Classical Quantum Gravity, Vol. 4, No. 3, p. 635 – 650 (1987). Abstr. in Phys. Abstr., Vol. 90, No. 1312, Entry 102225 (1987).

067.146   Angular momentum transport by star–gas interaction and structure of accretion disks.

F. Hagio.

Publ. Astron. Soc. Jpn., Vol. 39, No. 6, p. 887 – 893 (1987).

The structure of a gaseous disk around a massive black hole (mass $M_H = 10^{6-8} M_\odot$) embedded in a dense stellar system is studied. It is supposed that the disk is formed by the transport of angular momentum through the star–gas interaction. The author assumes that the density n_c and velocity dispersion v_c of stars around the black hole are given by the Bahcall–Wolf (1976) distribution. The results are discussed. The radiation generated by the star–gas interaction seems sufficient to explain the luminosity of Seyferts and QSOs.

067.147   Magnetic fields of old neutron stars.
   V. V. Usov.
Astron. Tsirk., No. 1469, p. 1 – 2 (1986). In Russian.
   It is shown that during the evolution of a neutron star its magnetic field B first decays with the time t (B ~ exp(–t/τ), where τ ≅ 10^6years), and then becomes quasi–stationary. The non–decaying magnetic field of a neutron star is generated by a degenerate electron gas which is in the Landau–orbital–ferromagnetism state.

067.148   Sigma model calculations of neutron–rich nuclear matter.
   M. Prakash, T. L. Ainsworth.
Phys. Rev. C, Vol. 36, No. 1, p. 346 – 353 (1987). Abstr. in Phys. Abstr., Vol. 90, No. 1315, Entry 121688 (1987).

067.149   Strong–field point–particle limit and the equations of motion in the binary pulsar.
   T. Futamase.
Phys. Rev. D, Vol. 36, No. 2, p. 321 – 329 (1987). Abstr. in Phys. Abstr., Vol. 90, No. 1315, Entry 127937 (1987).

067.150   Role of hyperons and pions in neutron stars and supernova.
   N. K. Glendenning.
Z. Phys., A, Vol. 327, No. 3, p. 295 – 300 (1987). Abstr. in Phys. Abstr., Vol. 90, No. 1315, Entry 127939 (1987).

067.151   Thermodynamics of higher dimensional black holes.
   F. S. Accetta, M. Gleiser.
Ann. Phys. (N.Y.), Vol. 176, No. 2, p. 278 – 300 (1987). Abstr. in Phys. Abstr., Vol. 90, No. 1315, Entry 127940 (1987).

067.152   Can particle creation by a black hole be described in terms of more familiar laboratory processes?
   R. M. Nugayev (*R. M. Nugaev*).
Int. J. Theor. Phys., Vol. 26, No. 5, p. 407 – 428 (1987). Abstr. in Phys. Abstr., Vol. 90, No. 1315, Entry 127941 (1987).

067.153   Instability of higher dimensional de Sitter space.
   P. F. Gonzalez–Diaz.
Phys. Lett. B, Vol. 191, No. 3, p. 263 – 266 (1987). Abstr. in Phys. Abstr., Vol. 90, No. 1315, Entry 127942 (1987).

067.154   Radial oscillations of warm cores in neutron stars.
   J. M. Marti, J. A. Miralles, J. M. Ibáñez.
Astrophys. Space Sci., Vol. 139, No. 1, p. 93 – 101 (1987).
   The authors have solved the relativistic equations for the radial oscillations of warm cores in neutron stars by assuming a given law for the interior distribution of temperature – resulting from the condition of relativistic thermal equilibrium – and focussed on the properties of the fundamental modes. The results establish well–defined regions of stability in a diagram "central temperature versus central density".

067.155   Spectrum of dust heated by thick accretion disks of pregalactic black holes.
   G.–z. Xie, M.–x. Bao, W.–x. Gu, R.–w. Lu, Y. Zhang.
Chin. Astron. Astrophys., Vol. 11, No. 3, p. 234 – 236 (1987). English translation of Acta Astron. Sin., Vol. 28, No. 2, p. 168 – 172 (1987).
   The authors discuss the spectrum of dust heated by accretion disks of supermassive objects (SMOs) and the effect of the re-radiation on the cosmic microwave background. The main results are: (1) Whether or not there is Lyman–alpha cutoff, the dust radiation in the case of SMOs is very strong and will appreciatively distort the microwave background. (2) In spite of (1), the dust radiation spectrum is different when Lyman–α cutoff is present

and when it is absent, and the two cases can be distinguished observationally.

067.156   White holes and their thermodynamics.
   Y.–x. Gui.
Chin. Astron. Astrophys., Vol. 11, No. 4, p. 275 – 281 (1987). English translation of Acta Astrophys. Sin., Vol. 7, No. 3, p. 177 – 188 (1987).
   The author discusses the thermodynamics of white holes and gives their thermodynamical laws. He points out that there is a flaw in the view that white holes must necessarily violate the Second Law of Thermodynamics. If the burst of white holes obeys the Second Law, then this will give certain constraints on the burst, and may provide an explanation of the time of burst of lagging–cores.

067.157   The variation of the angle between the magnetic and rotation axes in a collapsing star.
   J.–l. Zhang, R.–s. Gong.
Chin. Astron. Astrophys., Vol. 11, No. 4, p. 297 – 298 (1987). English translation of Acta Astrophys. Sin., Vol. 7, No. 3, p. 165 – 168 (1987).
   The authors derive a formula for the variation of the magnetic axis in a collapsing oblique rotator. Their results show that the case of the magnetic axis coincident with the rotation axis is unstable, and the angle between the two axis will grow on any small perturbation.

067.158   Charged rotating black hole from five dimensional point of view.
   V. P. Frolov, A. I. Zel'nikov, U. Bleyer.
Ann. Phys. (Leipzig), Vol. 44, No. 5, p. 371 – 377 (1987). Abstr. in Phys. Abstr., Vol. 90, No. 1316, Entry 134266 (1987).

067.159   A simplified derivation of stimulated emission black holes.
   R. D. Sorkin.
Classical Quantum Gravity, Vol. 4, No. 4, p. L149 – L155 (1987). Abstr. in Phys. Abstr., Vol. 90, No. 1316, Entry 134267 (1987).

067.160   Quantum black holes and Planck's constant.
   D. K. Ross.
Classical Quantum Gravity, Vol. 4, No. 4, p. 995 – 1001 (1987). Abstr. in Phys. Abstr., Vol. 90, No. 1316, Entry 134268 (1987).

067.161   From strange matter to strange stars.
   P. Haensel.
Acta Phys. Pol., Ser. B, Vol. B18, No. 8, p. 739 – 757 (1987). Abstr. in Phys. Abstr., Vol. 90, No. 1317, Entry 135238 (1987).

067.162   Stars of bosons with non–minimal energy–momentum tensor.
   J. J. van der Bij, M. Gleiser.
Phys. Lett. B, Vol. 194, No. 4, p. 482 – 486 (1987). Abstr. in Phys. Abstr., Vol. 90, No. 1317, Entry 140116 (1987).

067.163   Neutron stars in the early Universe.
   C. Gimmi, B. Mauro.
Acta Phys. Hung., Vol. 61, No. 3 – 4, p. 375 – 377 (1987). Abstr. in Phys. Abstr., Vol. 90, No. 1317, Entry 140146 (1987).

067.164   Mass, dual mass, and gravitational entropy.
   A. Magnon.
J. Math. Phys., Vol. 28, No. 9, p. 2149 – 2154 (1987). Abstr. in Phys. Abstr., Vol. 90, No. 1317, Entry 140150 (1987).

067.165   A mechanism for pulsar drift.
   A. Qadir.
Chin. Phys. Lett., Vol. 4, No. 7, p. 289 – 292 (1987). Abstr. in Phys. Abstr., Vol. 90, No. 1318, Entry 146018 (1987).

067.166   **Origin of Hawking radiation.**
P. Hajicek.
Phys. Rev. D, Vol. 36, No. 4, p. 1065 – 1079 (1987). Abstr. in
Phys. Abstr., Vol. 91, No. 1319, Entry 4277 (1988).

067.167   **Neutron tori and the origin of r–process elements.**
C. J. Hogan, J. H. Applegate.
Nature, Vol. 330, No. 6145, p. 236 – 238 (1987).
If an accretion disk gets hotter than a few MeV, nuclei in the
infalling matter are dissociated into their constituent neutrons
and protons. Neutrons released by dissociation of matter falling
at high accretion rates into a black hole or neutron star
accumulate in a dense "neutron torus". Matter in accretion disks
around compact objects may thus provide ideal conditions for
classical rapid or r–process nucleosynthesis. Systems in which
accretion powers an outflow from a region near the compact
object might thereby enrich the interstellar medium in r–process
elements. The r–process mechanism considered was inspired by
some ideas for manufacturing deuterium in pregalactic accretion
disks.

067.168   **Early neutron stars and quark matter.**
Y. Li, X. Kong, C. Wei, Y. Ge.
Acta Astrophys. Sin., Vol. 7, No. 4, p. 266 – 272 (1987). In
Chinese.
There may exist quark matter inside the early hot neutron
stars. By means of the general method used by G. Baym (1981)
and S. A. Chin (1978) the authors obtain the pressures and
densities of the neutron matter – quark matter phase transition at
different temperatures and compare these densities with the
central densities of the stable hot neutron stars at each
temperature.

067.169   **Axion bremsstrahlung in dense stars.**
M. Nakagawa, Y. Kohyama, N. Itoh.
Astrophys. J., Vol. 322, No. 1, p. 291 – 295 (1987).
The energy loss rate due to axion bremsstrahlung in dense stars
such as white dwarfs or neutron stars is calculated by taking into
account the ionic correlation effects accurately. It is found that
the ionic correlation effects suppress the axion bremsstrahlung
rate typically by a factor 2–50.

067.170   **X–ray irradiated accretion disks and bimodal states.**
H. Inoue, R. Hoshi.
Astrophys. J., Vol. 322, No. 1, p. 320 – 323 (1987).
X–ray irradiation plays an important role in the structure of an
outer accretion disk. The authors examine the outer portion of an
accretion disk, carefully taking into account X–ray irradiation
within the context of α–viscosity. Bimodal states observed in the
X–ray sources Cyg X–1 and GX 339–4 are explained naturally in
terms of an accretion disk irradiated by X–rays originating at the
inner portion of the disk.

067.171   **The spectra of accretion disks and their application to
low–mass X–ray binaries.**
R. E. Taam, P. Mészáros.
Astrophys. J., Vol. 322, No. 1, p. 329 – 341 (1987).
The continuous spectra emitted by accretion disks in low–mass
X–ray binaries and, in particular, in X–ray bursters in their
quiescent state as well as in their burst active state is studied in the
context of the α disk model. Specific attention is focused on the
detailed electron–scattering–dominated spectrum (in contrast to
multiple blackbodies) of the accretion disk in circumstances
where the disk is irradiated from regions corresponding to the
entire neutron star surface or from an accretion belt (or boundary
layer).

067.172   **Erratum: "Neutrino–pair bremsstrahlung in dense stars.
IV. Phonon contributions in the crystalline lattice
phase." [Astrophys. J., Vol. 285, No. 1, p. 304 – 311 (1984)].**
N. Itoh, Y. Kohyama, N. Matsumoto, M. Seki.
Astrophys. J., Vol. 322, No. 1, p. 584 (1987). See Abstr.
38.067.098.

067.173   **A pulsar emission model: observational tests.**
F. C. Michel.
Astrophys. J., Vol. 322, No. 2, p. 822 – 830 (1987).
The author reexamines the idea of pulsar emission in the form
of coherent curvature radiation from particle "bunches" following curved magnetic field lines near the neutron star, where the
field lines are assumed not to be purely dipolar. The radiation is
assumed to be concentrated on relatively narrow isolated flux
tubes which generate wide individual fan beams. A given pulsar
may rotate into view several such beams, leading to complex
pulse profiles. Radiation is *not* from a "hollow cone" but rather
concentrated near the low latitude edge of the magnetic polar
cap, which is tipped at some angle to the spin axis. Observable
consequences of this pulsar emission model are outlined.

067.174   **The effect of decay of the amplitude of oscillation on
random process models for QPO X–ray stars.**
N. Shibazaki, R. F. Elsner, M. C. Weisskopf.
Astrophys. J., Vol. 322, No. 2, p. 831 – 837 (1987).
Random process models provide a useful mathematical
framework for analysis of the quasi–periodic oscillations (QPO)
recently discovered in some X–ray binaries. The authors examine
the effects on the power spectrum of the decay of the amplitude of
oscillation that is expected in some physical models for the QPO
phenomena. The resulting changes in the power spectrum depend
on the ratio of the decay time of the amplitude of oscillation to
the lifetime of the shot envelope. The results of this analysis are
applied to the beat–frequency modulated accretion model.

067.175   **New insights from a global view of X–ray bursts.**
I. Fushiki, D. Q. Lamb.
Astrophys. J., Lett. Ed., Vol. 323, No. 1, p. L55 – L60 (1987).
The authors show that the behavior of nuclear burning on
accreting neutron stars depends on the mass accretion rate per
unit area $\dot{\sigma}$, the temperature T at the bottom of the accreted
matter, and the column density of accreted matter σ, and is thus
intrinsically a three–dimensional problem. They therefore introduce the concepts of a nuclear fuel surface S and an ignition
surface S^{ign} in $(\dot{\sigma}, T, \sigma)$–space. The authors show that certain cuts
taken through the ignition surface in $(\dot{\sigma}, T, \sigma)$–space correspond to
particular physical situations. However, observed sources need
not lie on any of these cuts. They do still lie on the ignition surface
S^{ign} in $(\dot{\sigma}, T, \sigma)$–space; thus the calculations provide a complete
framework for interpreting the observations.

067.176   **Ohmic decay of crustal neutron star magnetic fields.**
Y. Sang, G. Chanmugam.
Astrophys. J., Lett. Ed., Vol. 323, No. 1, p. L61 – L64 (1987).
Calculations of ohmic decay of dipolar magnetic fields which
are created so that they are initially confined to the crust are
presented for the first time. It is shown that the field does not
decay exponentially as has been assumed in most analyses of
pulsar observations and that, if the field occupies the entire crust,
it decays by less than a factor of order 100 in the Hubble time.
Fields that are confined to the outer crust ($\varrho \leqslant 10^{11} \text{g cm}^{-3}$)
decay too quickly, and again nonexponentially, in conflict with
observations. Thus it is difficult to justify, at present, any of the
models for field decay which have been proposed. The authors'
results, therefore, suggest that neutron star magnetic fields may
not decay significantly.

067.177   **Pair–creation effects in accretion–shock models of active
galactic nuclei.**
J. M. Blondin, A. Königl.
Astrophys. J., Vol. 323, No. 2, p. 451 – 455 (1987).
The authors explore the possibility that the copious production
of $e^+ e^-$ pairs that is expected to occur in active galactic nuclei
with large compactness parameters could have an important
effect on the structure of the accretion shock that might form
near the central black hole in these sources. The basic effect lies in
the fact that, if the pairs are well coupled to the inflowing plasma,
then they could give rise to a substantial decrease in the effective
value of the Eddington luminosity within the shock. As a result,
the radiation generated in the shock could play a significant role

in the deceleration and thermalization of the flow even when the measured luminosity from the source is much below the nominal Eddington value.

067.178 Model atmospheres of near–Eddington limit X–ray bursters.
A. Babul, B. Paczyński.
Astrophys. J., Vol. 323, No. 2, p. 582 – 591 (1987).

The authors have developed some simple, plane–parallel models of atmospheres of X–ray bursters that are very close to the Eddington limit. The dominant opacity source is assumed to be incoherent Thomson scattering. At large optical depths one has LTE, while at small optical depths electron scattering is coherent. The authors develop models with luminosities up to $L = 0.9999 L_E$. The spectral (i.e. color) temperature in the extreme models is more than twice the effective temperature.

067.179 Radiation–hydrodynamic calculation of sub–Eddington accretion disks.
G. E. Eggum, F. V. Coroniti, J. I. Katz.
Astrophys. J., Vol. 323, No. 2, p. 634 – 646 (1987).

The authors report the results of self–consistent, azimuthally symmetric, radiation–hydrodynamic calculations of subcritical accretion disks about a Newtonian pseudo–black hole. Energy generation is described by a kinematic viscosity law and a modified α–disk model. A disk with constant kinematic viscosity settles to a nearly steady stae which approximates a thin disk solution. Its unstable vertical distribution of entropy leads to mild subsonic correction. The α–disk is unstable, as expected, and collapses to a thin cold sheet with low accretion rate and low luminosity.

067.180 On the radius of neutron stars.
P. Mészáros, H. Riffert.
Astrophys. J., Lett. Ed., Vol. 323, No. 2, p. L127 – L130 (1987).

The authors discuss recent calculations of general relativistic effects in the beaming, spectrum, and pulse properties of accreting neutron stars. Some possible models for X–ray pulsars and QPOs are analyzed, which indicate that current observational and theoretical requirements can be explained with a value of the radius smaller than about two Schwarzschild radii.

067.181 Neutron evaporation from the surface of neutron stars.
K. Tennakone.
J. Natl. Sci. Council Sri Lanka, Vol. 13, No. 2, p. 107 – 113 (1985).

It is shown that a newly born neutron star could cool more rapidly by neutron evaporation than through neutrino emission. The mass loss due to neutron evaporation is estimated.

Dark stars: the evolution of an idea.
See Abstr. 004.042.

Accretion research: how it started.
See Abstr. 004.063.

The origin and evolution of neutron stars. Proceedings of the 125th IAU Symposium, held in Nanjing, China, 26 – 30 May 1986.
See Abstr. 012.001.

Stellar sky as seen from the vicinity of a black hole.
See Abstr. 014.035.

The black hole as a gravitational "lens".
See Abstr. 014.041.

Behind the event horizon.
See Abstr. 014.056.

Hydrogen molecule ion in the magnetic field of a neutron star. Probabilities of vibration–rotation transitions.
See Abstr. 022.110.

Observations of neutron stars planned by the High Speed Photometer team using Space Telescope.
See Abstr. 051.001.

Nuclear matter under extreme conditions.
See Abstr. 061.013.

Current gradient–driven linear and nonlinear electromagnetic waves in a magnetized electron–positron plasma.
See Abstr. 062.003.

The effect of electron screening on thermonuclear reaction rates.
See Abstr. 062.008.

The central engine.
See Abstr. 062.039.

X–ray pulsars: accretion flow deceleration.
See Abstr. 062.068.

Energetic particle acceleration in spherically symmetric accretion flows and shocks.
See Abstr. 062.081.

Two–photon annihilation radiation in strong magnetic field: the case of small longitudinal velocities of electrons and positrons.
See Abstr. 062.096.

Shocked relativistic magnetohydrodynamic flows with application to pulsar winds.
See Abstr. 062.102.

Hydrodynamics in curvilinear coordinates.
See Abstr. 062.143.

Solitons in strongly magnetized electron–positron plasmas and pulsar microstructure.
See Abstr. 062.151.

On the relativistic theory of Alfvén solitons.
See Abstr. 062.154.

A simple model to account for the effects of plasma screening on thermonuclear reaction rate.
See Abstr. 062.155.

The formation of iron features in the cooling spectrum of X–ray bursts.
See Abstr. 063.025.

Variability from pair atmospheres.
See Abstr. 063.031.

Quantized synchrotron radiation in strong magnetic fields.
See Abstr. 063.046.

X–ray spectra from the boundary–layer region of a (disk) accreting weakly magnetized neutron star.
See Abstr. 063.056.

Radiation pressure on the plasma above degenerate stars with a strong magnetic field.
See Abstr. 063.112.

Inverse Comptonization by one–dimensional relativistic electrons.
See Abstr. 063.125.

Effects of electron scattering on the oscillations of an X–ray source.
See Abstr. 063.126.

Neutrinos from gravitational collapse.
See Abstr. 065.003.

The Chandrasekhar theory of stellar collapse as the limit of quantum mechanics.
See Abstr. 065.011.

The value of the a/m ratio in binary systems.
See Abstr. 065.012.

Equation of state of hot dense matter.
See Abstr. 065.020.

The effects of neutrino transport on the collapse of iron stellar cores.
See Abstr. 065.025.

Convection and the mechanism of Type II supernovae.
See Abstr. 065.028.

Thermonuclear processes in accreting white dwarfs (novae, symbiotic stars, and type–I supernovae).
See Abstr. 065.029.

Evolution of stellar binaries formed by tidal capture.
See Abstr. 065.049.

An evolutionary scenario for the formation of highly eccentric Be/X–ray binaries.
See Abstr. 065.050.

Collapse of 9 $M_\odot$ stars.
See Abstr. 065.067.

Collapsing white dwarfs.
See Abstr. 065.068.

Density discontinuity g–modes.
See Abstr. 065.074.

Monte Carlo simulations of radio pulsars and their progenitors.
See Abstr. 065.099.

The evolutionary status of 4U 1820–30.
See Abstr. 065.114.

A rigorous examination of the Chandrasekhar theory of stellar collapse.
See Abstr. 065.117.

Gravitational radiation from rotating gravitational collapse.
See Abstr. 066.020.

Gravitational radiation and 3D numerical relativity.
See Abstr. 066.022.

Hubble's constant from gravitational wave observations.
See Abstr. 066.025.

Rapidly rotating general relativistic polytropes.
See Abstr. 066.026.

Are extra dimensions visible near the event horizon? Black holes coupled to scalar fields.
See Abstr. 066.027.

Quantum effects in non static black hole space–times.
See Abstr. 066.030.

Strings from gravity.
See Abstr. 066.034.

Reversible evolution of charged ergoregions.
See Abstr. 066.038.

Wormhole solutions in the Einstein–Yang–Mills–Higgs system: solution of first–order equations for $G = SU(2)$.
See Abstr. 066.043.

On some physical properties and the stability of an exact model for a relativistic star.
See Abstr. 066.069.

On the instability of Buchdahl's model for a gaseous relativistic star.
See Abstr. 066.070.

On the stability and physical properties for an exact relativistic model for a superdense star.
See Abstr. 066.072.

Self–gravitating toroidal figures of equilibrium around a rotating black hole: the numerical method.
See Abstr. 066.083.

4U 1820–30 as a potential test of the nonsymmetric gravitational theory of Moffat.
See Abstr. 066.092.

Coalescing binaries – probe of the Universe.
See Abstr. 066.120.

Realistic exact solutions of Einstein's field equations.
See Abstr. 066.146.

Killing approximation for vacuum and thermal stress–energy tensor in static space–times.
See Abstr. 066.147.

Astrophysically significant solutions and their derivation.
See Abstr. 066.148.

Wormholes in space–time.
See Abstr. 066.174.

A possible semisecular variation in orbital period of binary pulsar system PSR 1913+16 and Lorentz invariance of gravity.
See Abstr. 066.178.

Degenerate dwarf binaries as promising, detectable sources of gravitational radiation.
See Abstr. 066.180.

Neutron density and neutron source determination in barium stars.
See Abstr. 114.110.

Hardness ratio in evolving low–mass X–ray binary systems.
See Abstr. 117.001.

Quasi–periodic oscillations in low–mass X–ray binaries.
See Abstr. 117.002.

The beat frequency model for QPOs.
See Abstr. 117.004.

What type of binary system is Cygnus X–3?
See Abstr. 117.006.

Black hole candidates in X–ray binaries.
See Abstr. 117.021.

Disk formation at the magnetosphere of wind–fed pulsars: application to Vela X–1.
See Abstr. 117.030.

Optical and UV spectroscopy of the black hole binary candidate LMC X–1.
See Abstr. 117.047.

An evolutionary scenario for the black hole binary A 0620–00.
See Abstr. 117.054.

Evidence for black holes in X–ray binary systems.
See Abstr. 117.094.

A theory of soft X–ray transients.
See Abstr. 117.096.

The relation between optical and X–ray flux variations of the black–hole candidate LMC X–3.
See Abstr. 117.118.

EXOSAT observations of 4U/MXB 1636–53: on the relation between the amount of accreted fuel and the strength of an X–ray burst.
See Abstr. 117.147.

On the theory of type I X–ray bursts: the energetics of bursts and the nuclear fuel reservoir in the envelope.
See Abstr. 117.148.

Hard spectral components in soft X–ray transients.
See Abstr. 117.159.

A classification of fast quasi–periodic X–ray oscillators: is 6 Hz a fundamental frequency?
See Abstr. 117.162.

Model of binary stellar systems containing a black hole for explaining the properties of some cepheids in the IC 1613 galaxy.
See Abstr. 117.170.

The origin of the ultra–compact binary 4U 1820–30.
See Abstr. 117.186.

Deterministic chaos in accreting systems: analysis of the X–ray variability of Hercules X–1.
See Abstr. 117.206.

Fluctuation of wind–driven accretion in X–ray binaries.
See Abstr. 117.241.

On the nonthermal radio emission of double stars with relativistic components.
See Abstr. 117.247.

The high–energy X–ray spectrum of black hole candidate GX 339–4 during a transition.
See Abstr. 117.280.

Constraints on models of Cygnus X–3 from high–energy gamma–ray absorption at source.
See Abstr. 117.318.

Neutron stars in twelve supernova remnants.
See Abstr. 125.002.

The disruption of a light neutron star in an ultra–close binary and the second neutrino burst from SN 1987A.
See Abstr. 125.241.

The nature of the companion of SN 1987A.
See Abstr. 125.244.

A possible explanation of the second neutrino burst from SN 1987A.
See Abstr. 125.307.

SN 1987A supernova: a black–hole precursor?
See Abstr. 125.312.

Particle acceleration and production of energetic photons in SN 1987A.
See Abstr. 125.314.

Interstellar scintillations and neutron star kinematics.
See Abstr. 126.004.

Toward an empirical theory of pulsar emission.
See Abstr. 126.011.

Radio emission mechanisms for two types of pulsars.
See Abstr. 126.012.

Pulse asymmetry of millisecond pulsars.
See Abstr. 126.013.

Pulsar polarization limiting radii and the evolution of pulsar beams.
See Abstr. 126.014.

Triplicity of pulsar profiles and orthogonal polarization modes.
See Abstr. 126.016.

The effects of inverse Compton scattering on the pulsars' radiation.
See Abstr. 126.017.

The galactic pulsar population and neutron star birth.
See Abstr. 126.021.

The progenitors of pulsars.
See Abstr. 126.022.

The structure of pulsar nebulae.
See Abstr. 126.024.

Accreting neutron stars.
See Abstr. 126.026.

On the evolution of magnetic inclination with age.
See Abstr. 126.027.

Timescale for the decay of magnetic fields of pulsars.
See Abstr. 126.028.

Binary pulsars: observations and implications.
See Abstr. 126.029.

Millisecond pulsar formation and evolution.
See Abstr. 126.030.

Secondary components of binary pulsars & magnetic field decay in neutron stars.
See Abstr. 126.031.

Constraints to possible progenitor systems of PSR 1831–00.
See Abstr. 126.032.

Geodetic precession in binary pulsars.
See Abstr. 126.033.

Einstein Observatory limits on neutron star surface temperatures.
See Abstr. 126.034.

Thermal radiation from a radio pulsar: PSR 1055–52.
See Abstr. 126.035.

The luminosity decay of radio pulsars and some related matters.
See Abstr. 126.038.

Soft X–ray observations of the radio pulsar PSR 1055–52.
See Abstr. 126.039.

Characteristic times of pulsars and their ages.
See Abstr. 126.042.

Studies of isolated neutron stars, pulsars and pulsar–driven nebulae with the Advanced X–Ray Astrophysics Facility (AXAF).
See Abstr. 126.053.

The birthrate and initial spin period of single radio pulsars.
See Abstr. 126.054.

Optical and X–ray radiation from fast pulsars: effects of duty cycle and spectral shape.
See Abstr. 126.084.

Outer magnetospheric fluctuations and pulsar timing noise.
See Abstr. 126.087.

Could glitches inducing magnetospheric fluctuations produce low–frequency pulsar timing noise?
See Abstr. 126.088.

Observations of X–ray burst sources.
See Abstr. 142.001.

X–ray bursting neutron stars.
See Abstr. 142.007.

X–ray irradiated accretion disk and bimodal states.
See Abstr. 142.008.

X–ray spectra and atmospheric structures of bursting neutron stars.
See Abstr. 142.009.

Optical variability of the black hole candidate GX 339–4 (X 1659–487, V821 Ara) – limits on periodic modulation.
See Abstr. 142.016.

The complex cross–spectra of Cygnus X–2 and GX 5–1.
See Abstr. 142.021.

Coherence effects on quasi–periodic oscillations from galactic X–ray sources.
See Abstr. 142.034.

EXOSAT observations of a giant X–ray burst in XB 1905+ 000.
See Abstr. 142.035.

Multipeaked X–ray bursts from 4U/MXB 1636–53: evidence against burst–induced accretion disk coronae.
See Abstr. 142.043.

Some constraints on neutron star properties from gamma ray burster observations.
See Abstr. 143.003.

Theory of gamma ray bursters.
See Abstr. 143.004.

HEAO–1 observations of gamma–ray bursts.
See Abstr. 143.041.

High energy cosmic rays from young neutron stars.
See Abstr. 144.001.

Very high energy emission in accretion onto compact objects.
See Abstr. 144.078.

Origin and acceleration of 10^{20}eV cosmic–ray protons.
See Abstr. 144.085.

Computer simulations of relativistic star clusters.
See Abstr. 151.013.

Relativistic collapse of a uniform star cluster.
See Abstr. 151.044.

Dynamics in the centres of triaxial elliptical galaxies.
See Abstr. 151.052.

The collapse of dense star clusters to supermassive black holes: binaries and gravitational radiation.
See Abstr. 151.130.

The statistical specific angular momentum of gas within a star cluster.
See Abstr. 151.132.

A new kind of stellar orbit in a galactic potential.
See Abstr. 151.138.

The globular cluster population of X–ray binaries.
See Abstr. 154.001.

The central object: some comments and speculations.
See Abstr. 155.048.

Further "loss of weight" by a black hole at the galactic center.
See Abstr. 155.071.

Discovery of a pulsating component in a decametric frequency range suggesting a rotating super black hole at the Galactic Center as a source.
See Abstr. 155.151.

Population of accreting neutron stars in external galaxies.
See Abstr. 157.001.

The evidence for and against the existence of supermassive black holes in E galaxies.
See Abstr. 157.126.

Cores of early–type galaxies.
See Abstr. 157.148.

Supermassive black holes in galaxy nuclei.
See Abstr. 157.250.

Improved accretion disk models of continuum emission from active galactic nuclei.
See Abstr. 158.113.

The central power source in active galaxies.
See Abstr. 158.169.

A nonthermal model for the XUV, soft X–ray emission of AGN and QSOs.
See Abstr. 158.173.

Quasar as a superstar with magnetic monopoles.
See Abstr. 159.034.

Constraints on quasar accretion disks from the optical/ultraviolet/ soft X–ray big bump.
See Abstr. 159.116.

High–amplitude peaks of density perturbations and primordial black holes formation in the dust–like universe.
See Abstr. 161.059.

Does thermodynamics require our cosmos to undergo a series of contraction/expansion cycles?
See Abstr. 161.097.

Fundamental physics and dark matter.
See Abstr. 161.145.

The monopole problem and the primordial black hole problem in the inflationary universe.
See Abstr. 161.216.

Can Planck–mass relics of evaporating black holes close the Universe?
See Abstr. 161.316.

No hair theorem for the Universe.
See Abstr. 161.351.

Cosmology from nothing.
See Abstr. 161.444.

Entropy and cosmology.
See Abstr. 161.445.

Sun

071 Photosphere, Spectrum

071.001 **Convection, magnetic fields, and line asymmetry in the sun and stars.**
W. Livingston, Y.–R. Huang.
The SHIRSOG Workshop, p. 1 – 4 (1986). – See Abstr. 012.003.

071.002 **A photometric search for solar giant convection cells.**
W.–H. Chiang, L. D. Petro, P. V. Foukal.
Sol. Phys., Vol. 110, No. 1, p. 129 – 138 (1987).
The authors limit the photometric contrast of solar giant convection cells using λ525.6 nm continuum images obtained on 15 days in May 1985. The r.m.s. of the giant cell intensity pattern must be less than or equal to the observed r.m.s. on spatial scales 80 to 240 Mm which is 0.023% or, equivalently, 0.33K. However, the spatial scale and time–scale dependence of the variance demonstrate that giant cells are not the source of the observed variance. Consequently, a tighter constraint on the r.m.s. of the giant cell pattern may be placed, namely 0.016% or 0.23K. This limit is consistent with temperature perturbations estimated from recent nonlinear simulations of global–scale solar convection. The authors use this limit on the r.m.s. of the giant cell pattern to estimate that the contribution of giant cells to the fluctuation of the solar irradiance on a one–month time–scale is less than 3×10^{-5}s.

071.003 **Asymmetry and variations of solar limb darkening along the diameter defined by diurnal motion in April 1981.**
H. Neckel, D. Labs.
Sol. Phys., Vol. 110, No. 1, p. 139 – 170 (1987).
Unexpected asymmetries and variations, which showed up in the first, preliminary reductions of new limb–darkening observations made in June 1986, near the present minimum of solar activity, stimulated a re–analysis of the limb–darkening observations made in April 1981 at Kitt Peak (Neckel and Labs, 1984). The results seem to indicate rather definitely that the intensity distribution across the disk varies at all observed wavelengths between 3300 and 6600 Å with amplitudes in the order of 1 – 2% and time–scales from minutes to hours. Asymmetries in the intensity distribution with respect to the disk center are a frequent phenomenon. There can be no doubt, that also the absolute disk center intensity undergoes variations with comparable size and modulation. Examples for widely differing limb darkening curves are given.

071.004 **The gradient of the small–scale velocity fluctuation in the solar atmosphere.**
A. Nesis, W. Mattig, K. H. Fleig, E. Wiehr.
Astron. Astrophys., Vol. 182, No. 1, p. L5 – L7 (1987).
First observations with the Gregory–Coudé–Telescope at Izaña (Tenerife) are used to study the small–scale velocity fluctuations in the solar atmosphere. The results confirm that they decrease steeply with height in the photosphere. The results also demonstrate the high quality of the observatory in Izaña.

071.005 **Observations of 5–minute oscillations in the brightness of the solar photosphere.**
L. V. Didkovskij, V. A. Kotov.
Bull. Crimean Astrophys. Obs., Vol. 74, p. 123 – 132 (1987). English translation of 41.071.036.

071.006 **Identification and classification of Fe I lines. Solar spectrum.**
A. G. Gasanalizade.
Soln. Dannye, Byull., 1986, No. 11, p. 63 – 68 (1987). In Russian.
From a comparison of laboratory and solar wavelengths many solar lines have been newly identified as Fe I. The total number of solar lines is 60, of which 32 weak lines are unblended. In order to investigate the solar redshift some new unblended (which are more dubious) lines have been omitted from the statistical analysis.

071.007 **Variations of the fine structure of Fraunhofer lines during flares.**
N. N. Kondrashova.
Soln. Dannye, Byull., 1986, No. 11, p. 69 – 75 (1987). In Russian.
An analysis is made of the fine structure of Fraunhofer lines observed during the flares of 8 September 1978 and 7 October 1979. Variations of the line profiles are found to be different. It may be due to an increase of the magnetic field in the active region on September 1978 and a downward transport of material during the 7 October 1979 flares.

071.008 **Depths of the intensity formation in the sodium D–lines at different positions on the solar disc.**
B. T. Babij, M. I. Stodilka.
Soln. Dannye, Byull., 1986, No. 11, p. 80 – 85 (1987). In Russian.
Depths of the intensity formation in sodium D–lines at various distance of line centers at different positions on the solar disc are determined. It is shown that only very deep parts of line cores are formed in the chromosphere. The rest radiation outside the cores ($\Delta\lambda \geqslant 0.1$ Å) has a photospheric nature.

071.009 **Temporal variations of the He I 10830.3 Å line on the Sun.**
G. F. Sitnik, L. M. Kozlova, M. I. Divlekeev, G. A. Porfir'eva.
Soln. Dannye, Byull., 1986, No. 12, p. 67 – 73 (1987). In Russian.
Using photoelectric infrared spectra obtained during several years in the periods of increase, maximum and decrease of solar activity a temporal behaviour of the He I 10830.3 Å line was investigated. Observations were made at the Sternberg Astronomical Institute with three solar instruments: at Kushchino, Moscow and by the Tyanshanskaya expedition near Alma–Ata. It is shown that at maximum solar activity the helium chromosphere noticeably increases both in quiet regions and in facular fields on the solar disc.

071.010 **Influence of the solar atmosphere inhomogeneities on the determination of equivalent widths of Fraunhofer lines and the chemical element abundances.**
B. T. Babij, M. M. Koval'chuk, P. A. Olijnyk, R. E. Rykalyuk.
Soln. Dannye, Byull., 1987, No. 2, p. 66 – 69 (1987). In Russian.
The influence is investigated of a six–stream solar atmospheric model on the equivalent widths of some iron and oxygen lines and abundances determination. It is shown that the abundances determined using inhomogeneous models is almost equal to that determined using homogeneous models.

071.011   **Solar emission lines revisited: extended study of magnesium.**
E. S. Chang.
Phys. Scr., Vol. 35, No. 6, p. 792 – 797 (1987).

Edléns polarization formula is found to fit the high–*l* solar infrared emission lines attributed to Mg I to the accuracy of the high–precision observations. For several previously unidentified weak lines, assignments are made, a few to Mg I. With some revision of laboratory energy levels, a more precise ionization limit for Mg I is determined to be 61671.056 (10)cm^{-1}. The fine structure of the two strongest lines at 811 and 818 cm^{-1} is found to consist of 4 components spaced about 0.001 cm^{-1} apart with nearly equal intensities. Their behaviour in the penumbra magnetic field is well described by the Paschen–Back effect.

071.012   **Fine structure of the velocity field of the photosphere. Power spectrum and coherence.**
V. E. Efremov, R. N. Ikhsanov, M. V. Kushnir.
Soln. Dannye, Byull., 1987, No. 3, p. 87 – 94 (1987). In Russian.

Power spectra and coherence spectra are calculated on the basis of the measurement of radial velocities for six lines and a continuous spectrum using spectrograms with the resolution of about 0".5. Radial velocities vary slowly with height in the middle and upper photosphere. A correlation is observed between brightness and velocity up to 250 – 350 km for certain discrete scales. Gradual disappearance of the correlation with height takes place beginning with small elements.

071.013   **On a method for theoretical calculation of Fraunhofer lines.**
N. S. Petrova.
Soln. Dannye, Byull., 1987, No. 4, p. 57 – 62 (1987). In Russian.

A comparison of the theoretical calculation computer programs (LTE) which are operated by a logically identical scheme showed that the difference in the initial parameters of a model solar atmosphere and the incomplete coincidence of the physical constants led to a discrepancy in the final result by about 2%. The residual discrepancy in the results, attaining 1 – 1.8%, should be regarded as due to the different methods and, accordingly, to different formulae for calculating the continuous absorption coefficient and the damping constant.

071.014   **Solar limb darkening accounting for absorption lines.**
S. N. Osipov.
Kinematika Fiz. Nebesn. Tel, Tom 3, No. 5, p. 57 – 64 (1987). In Russian. English translation in Kinematics Phys. Celest. Bodies.

Solar limb darkening data in the integral (lines + continuum) spectrum are obtained for 111 spectral intervals (width of 5 nm) by the drift–curve method. The results obtained in Kiev ($\lambda\lambda$355 – 670 nm) and on peak Terskol ($\lambda\lambda$317.5 – 362.5 nm) agree well. Corrections for atmospheric transparency fluctuations, for scattered light in the spectrograph and for atmospheric and instrumental scattered light and seeing are included in the reductions. The results are presented in tabular form.

071.015   **Magnetosensitive Fe I lines and FeH lines in the $\lambda\lambda$525.0 – 525.9 nm solar spectrum region.**
G. A. Porfir'eva.
Kinematika Fiz. Nebesn. Tel, Tom 3, No. 5, p. 87 – 90 (1987). In Russian. English translation in Kinematics Phys. Celest. Bodies.

The magnetosensitive Fe I 525.022, 525.065, 525.347 nm lines are shown to be blended by FeH lines. The FeH lines are seen in the photospheric spectrum and noticeably increase in the umbra spectrum. In the 525.00 – 525.95 nm wavelength region of the Liège solar atlas (1973) many weak lines not indicated in Rowland's Table (1966) are revealed. Their wavelengths are approximately evaluated, 18 "new" lines are identified with FeH.

071.016   **Erratum: "Observations of a solar latitude–dependent limb brightness variation" [Astrophys. J., Vol. 290, No. 2, p. 758 – 764 (1985)].**
J. R. Kuhn, K. G. Libbrecht, R. H. Dicke.
Astrophys. J., Vol. 319, No. 2, p. 1010 (1987). See Abstr. 39.071.013.

071.017   **Polarimetry in the Mg II h and k lines.**
W. Henze, J. O. Stenflo.
Sol. Phys., Vol. 111, No. 2, p. 243 – 254 (1987).

The Ultraviolet Spectrometer and Polarimeter (UVSP) on the SMM satellite has been used to record the linear polarization profile across the Mg II h and k lines, including its center–to–limb variation. Linear polarization with an orientation of the electric vector parallel to the solar limb is detected in the line wings on the short wavelength side of the k line and on the long wavelength side of the h line. The predicted negative polarization (electric vector perpendicular to the limb) between the h and k lines is however not confirmed by the observations. The authors have examined possible explanations of such a discrepancy between theory and observations.

071.018   **The solar platinum content.**
N. H. Youssef, N. M. Khalil.
Astron. Astrophys., Vol. 186, No. 1/2, p. 333 – 334 (1987).

Using the spectral synthesis method with the Holweger–Müller solar model atmosphere, the solar abundance of platinum is derived from an LTE analysis of the photospheric spectrum. The recent value of the oscillator strength, deduced from lifetime measurements is used. The authors obtain $A_{Pt} = 1.74$ in the usual scale where log $A_H = 12.00$.

071.019   **Investigation of the He I λ10830 line in the spectrum of the sun as a star during 1981 – 1984.**
Z. A. Shcherbakova, A. G. Shcherbakov.
Izv. Krymskoj Astrofiz. Obs., Tom 76, p. 98 – 102 (1987). In Russian. English translation in Bull. Crimean Astrophys. Obs., Vol. 76.

Sixty nine spectrograms of the whole solar disk have been obtained in the 10830 Å region during 1981 – 1984. The spectrograms were taken with a high speed infrared film. Dispersion 0.83 Å/mm. The analysis of the He I λ10830 line shows the correlation between the intensity of helium line and the 11 year cycle of solar activity. But the line half width is not connected with the activity cycle. There probably exists intensity line variability on time scale near to the solar rotational period.

071.020   **On the solar mercury abundance.**
L. S. Lyubimkov, N. G. Zalaletdinova.
Izv. Krymskoj Astrofiz. Obs., Tom 76, p. 102 – 110 (1987). In Russian. English translation in Bull. Crimean Astrophys. Obs., Vol. 76.

Using the method of synthetic spectra three spectral sections containing Hg I λ5460.74, λ4358.34 and λ4046.56 lines were investigated. A good agreement of the latest data obtained by different authors for the oscillator strengths of these lines has been revealed. The calculations were made for two models of the solar atmosphere, and the results were compared with observed spectra.

071.021   **The line–of–sight photospheric velocities and Hα chromospheric fine structure in the active region SD No. 135.**
L. G. Kartashova, I. E. Kozhevatov, E. Kh. Kulikova,
M. B. Ogir', N. N. Stepanyan.
Izv. Krymskoj Astrofiz. Obs., Tom 76, p. 110 – 119 (1987). In Russian. English translation in Bull. Crimean Astrophys. Obs., Vol. 76.

Hα–films and line– of–sight photospheric velocity maps registered by integral spectrometer for the developing active region SD No. 135 (June, 1984) were investigated and the results are presented.

071.022   **Differential observations of photospheric brightness oscillations of the Sun.**
L. V. Didkovskij, V. A. Kotov.
Izv. Krymskoj Astrofiz. Obs., Tom 76, p. 119 – 138 (1987). In Russian. English translation in Bull. Crimean Astrophys. Obs., Vol. 76.

A photodiode array illuminated by the pin–hole image of the Sun is used to measure relative (center–to–limb) brightness variations of the photosphere at 0.82 μm, associated with global oscillations of the Sun. The observations performed in 1983 and 1984 during 86 days yielded about 448 hours of data on differential brightness oscillations. These data were analysed in the frequency range from 55 to 165 μHz (100 to 300 min in period). The power spectra obtained vary significantly from season to season. The data show the existence of global oscillations with the well-known 160 min period. The mean brightness amplitude of the data for this 160-min mode is found to be $\sim 5 \times 10^{-5}$ in units of average solar brightness. It corresponds to $\sim (0.2 - 2.0) \times 10^{-6}$ for a harmonic amplitude of the 160-min variations in the solar irradiance in the 0.82 μm spectral range.

071.023 Observations of far–infrared solar continuum variations due to compression waves.

C. Lindsey, E. E. Becklin, F. Q. Orrall, M. W. Werner, T. R. Roellig, G. Kopp.
Bull. Am. Astron. Soc., Vol. 19, No. 2, p. 741 (1987). Abstract. – See Abstr. 010.061.

071.024 The solar boron abundance.

R. C. M. Learner, C. J. Harris.
Astrophys. J., Vol. 320, No. 2, p. 926 – 927 (1987).
The concern that the measurement of the solar boron abundance (log $A_B = 2.60 \pm 0.30$) is affected by an unassigned iron line is shown to be unfounded.

071.025 Experimental observations of the spectroheliogram of the solar D_2 line.

J.-y. Xuan, Q. Zhang.
Publ. Yunnan Obs., No. 1, p. 111 – 113 (1987). In Chinese.
The observations showed that the spectroheliogram of the solar D_2 ($\lambda 5890$ Å) line can provide some characteristics of an active object in a level under the chromosphere, but the resolutions of space and time of the photography was low. It should be better to use a higher collective receiver as a radiative receiver system.

071.026 The effects of departure from the local thermodynamical equilibrium in the solar Fraunhofer spectrum. Oxygen.

N. G. Shchukina.
Kinematika Fiz. Nebesn. Tel, Tom 3, No. 6, p. 36 – 45 (1987). In Russian. English translation in Kinematics Phys. Celest. Bodies.
The non–LTE formation of the O I lines $\lambda\lambda$ 130.2 – 130.6, 135.6 – 135.9, 777.2 – 777.5, 844.6 nm observable in the spectrum of quiet regions on the solar disk is discussed. The departure coefficients for twelve levels of O I and O II are obtained. The non–LTE center–to–limb profiles and equivalent widths of the lines are evaluated and compared with LTE–case and with observations. The influence of radiative and collisional processes, hydrogen radiation of Lyman–β and Lyman–continuum is analysed. The non–LTE effects for lines of O I are higher than the errors of precise solar observations and must be taken into account.

071.027 Temporal variations of solar spectral line profiles induced by the 5–minute photospheric oscillation.

M. T. Gomez, C. Marmolino, G. Roberti, G. Severino.
Astron. Astrophys., Vol. 188, No. 1, p. 169 – 177 (1987).
The authors simulated the variations induced by the 5 min photospheric oscillation on the line profiles. They found that a phase lag of the order of 150 degree between temperature and velocity wave perturbations can explain the observed differences between the oscillations of the line flanks at residual intensity levels $I/I_c < 0.7$. Such a phase relation in the 5 min oscillation differs from that of the adiabatic case in which the temperature and pressure fluctuations are 90 degrees out of phase with respect to the velocity. It is shown that a simple model of radiative damping in the solar photosphere can produce the required phase lag between temperature and velocity. Finally, it is also shown that the granulation can affect differentially the oscillations of the line flanks. This effect, however, does not fit the observed behaviour of the flank oscillations.

071.028 Bright threads in the inner wing of solar Ca II K line.

Z. Suemoto, E. Hiei, Y. Nakagomi.
Sol. Phys., Vol. 112, No. 1, p. 59 – 66 (1987).
On spectrograms of the K line at quiet regions of the Sun, bright threads visible in the real continuum due to the granulations are also seen in the outer wing as far as $|\Delta\lambda| \sim 3$ Å from the line centre. At the inner wing (3 Å $\gtrsim |\Delta\lambda| \gtrsim 0.5$ Å) bright threads are also seen, but their spatial distribution is different from the former ones. The threads at the inner wing appear at intergranular regions, and many of them are seen inside the supergranulation. Their size and number density are about the same as those of the granulation. These facts reflect that the penetration of the granular high temperature layer stops at a certain height in the photosphere, and that the intergranular bright threads at the inner wing are due to a hotter temperature layer, located at a considerably higher photospheric layer than the granulation.

Theoretical emission line ratios for Si XIII compared to solar observations.
See Abstr. 022.121.

System of energy levels and wavelengths of 26757 lines of neutral iron Fe I (1622 – 99948 Å). Part 1: Introduction, energy levels, Grotrian diagram and description.
See Abstr. 022.123.

Preliminary data obtained with the integrating spectrometer mounted on the solar tower telescope BST–2 of the Crimean Astrophysical Observatory.
See Abstr. 034.011.

Study of sizes, brightnesses and dynamics of solar facular points.
See Abstr. 036.117.

On the central part of the Hα–line profile in the gas–dynamic model of a flare.
See Abstr. 063.014.

Effects of acoustic and gravity waves on the curve–of–growth.
See Abstr. 063.098.

Motions around a decaying sunspot.
See Abstr. 072.074.

A determination of the ^{3}He/H ratio in the solar photosphere from flare gamma–ray line observations.
See Abstr. 073.043.

Observational maps of the moments of strong line profiles on the solar disk.
See Abstr. 073.086.

Comparison of computed fluxes for Fe X and Fe XIV lines with observed values at 1980 eclipse.
See Abstr. 074.024.

The magnetic and velocity structure adjacent to solar active regions.
See Abstr. 075.001.

Investigation of the magnetic field distribution in the undisturbed atmosphere of the sun.
See Abstr. 075.024.

Evolution of network magnetic fields of solar quiet regions.
See Abstr. 075.028.

Waves in the solar photosphere.
See Abstr. 080.012.

Evidence of global circulation currents from solar–limb temperature variations.
See Abstr. 080.030.

072 Sunspots, Faculae, Activity Cycles, Solar Patrol

072.001 A catalogue of sunspot observations from 165 BC to AD 1684.
A. D. Wittmann, Z. T. Xu.
Astron. Astrophys., Suppl. Ser., Vol. 70, No. 1, p. 83 – 94 (1987).

The authors have compiled a new catalogue of sunspot observations covering the period 165 BC to AD 1684 by updating and merging previously published catalogues and by adding a substantial amount of new data. The catalogue is in machine-readable form, the total number of entries being 235. Epoch analyses of the data have been made with regard to (a) the usefulness of naked–eye observable sunspots as tracers of the maximum epochs, and (b) the long–term phase behaviour of the sunspot cycle.

072.002 Torsional oscillations and the solar cycle.
H. B. Snodgrass.
Sol. Phys., Vol. 110, No. 1, p. 35 – 49 (1987). – See Abstr. 012.005.

Both the net torsional pattern and its derivative, the shear oscillation, are studied in relation to the solar activity cycle using data collected at Mount Wilson from 1967 – 1986. The shear zones run from pole to equator, clearly indicating that the cycle begins at the poles. Total transit takes roughly 18 years, and the active zones emerge to span the zones of shear enhancement after the latter have reached sunspot latitudes. This 18–yr transit time is seen as the proper duration of the cycle: successive cycles begin roughly 11 years apart and thus overlap. The polar origin of the torsional pattern is found to be phenomenologically connected with variations in the polar field amplitude. In a general discussion of torsional oscillations and their role in the solar cycle, the "net pattern" and "$k = 2$ wave" interpretations of the torsional phenomenon are contrasted and reasons for preferring the net pattern are presented. A model is proposed in which the torsional oscillations are the surface signature of an azimuthal convective–roll pattern. This model could provide the original Babcock model with a mechanism for trapping and further amplifying the toroidal field.

072.003 Sunspot cycle variations of ensemble–averaged active regions.
J. K. Lawrence.
Sol. Phys., Vol. 110, No. 1, p. 73 – 79 (1987). – See Abstr. 012.005.

The author examines published sunspot and calcium plage areas for 1620 solar active regions between 1974 and 1985. With these data he studies the properties of ensemble–averaged active regions. The average sunspot area per region, the average plage to sunspot area ratio, and the average plage intensity of regions all vary significantly with the sunspot cycle and in correlation with one another. The results suggest the existence of some energetic connection between active region sunspot areas and plage intensities. Further, if energy balance between sunspot luminosity deficits and facular luminosity excesses holds, then standard models relating these quantities to sunspot and plage areas will have to be modified. Overall energy balance can neither be established nor ruled out.

072.004 One possible explanation of Maunder's minimum of sunspots.
M. Kopecký, G. V. Kuklin.
Bull. Astron. Inst. Czech., Vol. 38, No. 4, p. 193 – 200 (1987).

The influence of effective resolving power of sunspot groups observations in the 17th century in connection with the visibility function and distribution of sunspot groups according to their importance enables the authors to explain the low level of solar activity observed at that time. This result is even strengthened if Maunder's minimum coincided with both the minimum of the super–long cycle of the frequency of origin of sunspot groups and the minimum of the 80–year period of importance of sunspot groups. A real decrease of the level of solar activity was not so large as to necessitate an introduction of the hypothesis of turning of mechanisms of solar activity. The actual cause was a substantial qualitative change in the phenomena of solar activity which was even strengthened by the observation conditions of that time in connection with the visibility function.

072.005 Quelques résultats d'observations du 21e cycle d'activité solaire.
D. Yvergneaux.
Ciel Terre, Vol. 103, No. 4, p. 103 – 106 (1987).

072.006 L'activité solaire.
M.–J. Martres, G. Zlicaric.
Astronomie, Vol. 101, p. 451, 504 – 505, 556 – 557, 607 – 609, 659 – 661 (1987).

072.007 The interpretation of oscillations in sunspot umbrae.
Y. D. Žugžda (*Yu. D. Zhugzhda*), V. Locāns, J. Staude.
Astron. Nachr., Vol. 308, No. 4, p. 257 – 269 (1987).

The authors review possibilities for an interpretation of oscillations observed in several period bands (3 min., 5 min., 20 min.) and at different heights in sunspot umbrae. At subphotospheric depths two independent resonators are acting: A resonator for slow, quasi–transverse waves can explain the lifetimes of bright umbral dots ($\gtrsim 20$ sec.), while a resonator for fast (acoustic), quasi–longitudinal waves could result in the 5–min. oscillations. The acoustic resonator strongly couples with the slow–mode longitudinal resonator at photospheric and chromospheric heights, the latter produces the resonance peaks in the 3–min. period band. The whole scheme of resonance levels generalizes and corroborates a chromospheric resonator model earlier proposed by the present authors. Comparisons with alternative models and recent measurements show that the present model most naturally explains the majority of observed data.

072.008 Velocity field and flare activity in a sunspot group.
S. I. Gopasyuk, L. F. Lazareva.
Bull. Crimean Astrophys. Obs., Vol. 74, p. 80 – 85 (1987). English translation of 41.072.064.

072.009 Large–scale structure of the solar magnetic field and distribution of sunspot groups with an anomalous inclination of the bipolar axis to the equator.
E. V. Ivanov.
Soln. Dannye, Byull., 1986, No. 11, p. 52 – 56 (1987). In Russian.

Bipolar sunspot groups with an anomalous inclination of bipolar axes to the equator ($\varphi \geqslant |15°|$) have been chosen for an analysis of their distribution with respect to the boundaries of large–scale magnetic field structures on the Sun for the period 1966 – 1979.

072.010 Sunspot groups of different types of the Mt. Wilson classification and solar cycles.
Yu. I. Vitinskij.
Soln. Dannye, Byull., 1986, No. 11, p. 85 – 89 (1987). In Russian.

Latitude–time diagrams are analysed for α, β, $\beta\gamma$, γ sunspot groups (Mt. Wilson classification) in solar cycles 16, 17, 18 for the entire solar disc and separately for the Northern and Southern solar hemispheres. It is shown that deviations from the Schwabe–Wolf and Spörer laws and N–S asymmetry seem to be mainly due to $\beta\gamma$ and γ sunspot groups.

072.011 The early stage of the AR 135 (S.D.) evolution (June 1984).
B. A. Ioshpa, L. I. Starkova.
Soln. Dannye, Byull., 1986, No. 11, p. 90 – 94 (1987). In Russian.

The early evolutional stage of magnetic and velocity fields of the active region with the sunspot group N 135 according "Solar

Data" (AR 135, S.D.) was studied. The AR is characterized by very rapid emergence of magnetic fluxes. From 22 to 23 June 1984 the magnetic fluxes were increased more than by a factor 2. It is found that before the emergence of a new magnetic flux there was an anomalous relationship between the magnetic intensity on two height levels: the fields at the temperature minimum level were higher than the fields in the photosphere. The velocity structure in the Hα–flare of 23 June is presented.

072.012   **On the spatial homogeneity of oscillatory motions in a sunspot's umbra.**
V. S. Loskutnikov, N. V. Muslakova, R. F. Khisamov.
Soln. Dannye, Byull., 1986, No. 12, p. 46 – 52 (1987). In Russian.
   Fluctuations of velocity and intensity at three points of the umbra and one point of a sunspot's penumbra are studied using spectrograms of H Ca II lines. It is shown that notwithstanding differences in forms of the power spectra at different parts of the umbra, one can detect an oscillation mode observed at all the umbra points. The nature of the observed oscillatory motions in a sunspot's umbra is discussed.

072.013   **A study of the velocity field of polar faculae.**
Yu. A. Nagovitsyn, E. Yu. Nagovitsyna.
Soln. Dannye, Byull., 1986, No. 12, p. 52 – 56 (1987). In Russian.
   The longitudinal $\partial L/\partial t$ and latitudinal $\partial \varphi/\partial t$ components of the velocity field of polar faculae were studied using the method of a highly accurate determination of heliographic coordinates "Helicor". It is confirmed that "polar" faculae are observed all over the latitude interval. Their differential rotation satisfies the law $\partial L/\partial t = (14.40 \pm 0.57) - (4.94 \pm 0.82) \sin^2 \varphi$. The meridional motion of "polar" faculae is determined by two flows, directed to the pole and the equator, with the line of divergence near the latitude $\varphi = 45°$.

072.014   **On the mean latitude of the polar faculae zone.**
E. I. Khusainov, V. F. Chistyakov.
Soln. Dannye, Byull., 1986, No. 12, p. 62 – 67 (1987). In Russian.
   With the help of photoheliograms for the period 1981 – 1985 a poleward drift of the polar faculae has been discovered. The mean drift velocities were +1°/year in the north and +1.3°/year in the south. Moreover, oscillations of the middle latitudes were found. They have a quasi–period of nearly 2 years and an amplitude of nearly 4°.

072.015   **On some regularities in the distribution of sunspots on the solar surface at different phases of an 11 year solar cycle.**
E. V. Miletskij.
Soln. Dannye, Byull., 1986, No. 12, p. 73 – 77 (1987). In Russian.
   Distributions of the longitudinal component of the sunspot index have been calculated for the solar surface. On the basis of an analysis of these distributions made for different phases of an 11 year cycle (from N 15 to N 19), a conclusion is made on the presence of long–lived zones of high and low activity on the Sun. Some qualitative regularities of the evolution of the form and sizes of these zones have been found. These regularities depend on the phase and the number of the cycle. A supposition is made that the regularities found may be caused by long–period oscillatory processes in the structure of the subphotospheric magnetic fields.

072.016   **Sources of sunspot activity and solar flares.**
Yu. M. Slonim, K. F. Kuleshova.
Soln. Dannye, Byull., 1987, No. 1, p. 77 – 87 (1987). In Russian.
   Twenty five sunspot groups are studied, where many large flares of ≥2B occurred.

072.017   **Forecast of the Wolf number in the maximum of the 22–nd solar cycle.**
A. I. Ol'.
Soln. Dannye, Byull., 1987, No. 2, p. 58 – 59 (1987). In Russian.
   According to the forecast the value of the annual average Wolf number for the maximum of cycle No. 22 is expected to be 150 ± 11.

072.018   **A determination of solar activity fluctuations of the Sun treated as exceeding the low–frequency background.**
A. V. Mordvinov.
Soln. Dannye, Byull., 1987, No. 2, p. 81 – 86 (1987). In Russian.
   An effective, simple procedure for the determination of solar activity fluctuations is proposed. The fluctuations are treated as a high–frequency positive addition to the low–frequency component that represents an 11–year cycle. The proposed recursive procedure, named the background filter, performs a special kind of numerical filtering. The transfer function is determined with the aid of spectral relationships between the filter input and output. A sinusoid with varying frequency was taken as an input process. General recommendations for application of the new filter were suggested.

072.019   **The contrast of faculae near the solar limb.**
H. Wang, H. Zirin.
Sol. Phys., Vol. 110, No. 2, p. 281 – 293 (1987).
   The authors have measured the contrast of solar faculae near the limb on direct digital video images made with the 65 cm vacuum reflector at the Big Bear Solar Observatory. They used six broad band filters with different wavelengths from red to violet. The range of heliocentric angle covered in their measurements is $0.05 < \mu = \cos \theta < 0.4$ ($\theta = 87° - 66°$). About 300 images were measured from observations made during the summers of 1983 and 1985. Over 20,000 faculae were measured. By averaging the contrasts of faculae and plotting them vs heliocentric angle, the authors found that contrast increases monotonically towards the limb for the shorter wavelengths; for longer wavelengths, contrast has a tendency to peak around $\mu = 0.15$, and then decrease towards the extreme limb. The contrast increases as wavelength decreases.

072.020   **Frequency modulation and stochastic variability of the Elatina varve record: a proxy for solar cyclicity?**
C. P. Sonett, G. E. Williams.
Sol. Phys., Vol. 110, No. 2, p. 397 – 410 (1987).
   The authors report a statistical analysis of a sequence of laminae, interpretable as periglacial varves from the late Precambrian (~680 million years old) Elatina Formation in South Australia. These "varves" contain an apparent solar record that is much longer and more detailed than any available from other sources. The rock record discloses a sequence of varve–cycles, empirically similar to solar activity cycles, which displays deterministic and stochastic elements in the succession of varve–cycle periods as well as modulation of varve–cycle amplitudes.

072.021   **Polfackeln – Neuland für den Amateursonnenbeobachter. 2. Teil.**
D. Brauckhoff, M. Delfs, H. Stetter.
Sonne, Jahrg. 11, Nr. 42, p. 42 – 43 (1987).

072.022   **Sonnenfleckenminimum im August 1986.**
K. Reinsch.
Sonne, Jahrg. 11, Nr. 43, p. 72 (1987).

072.023   **Das Sonnenfleckenminimum 1986.**
H. U. Keller.
Sonne, Jahrg. 11, Nr. 43, p. 72 – 73 (1987).

072.024   **Polfackeln – Neuland für den Amateursonnenbeobachter. 3. Teil.**
D. Brauckhoff, M. Delfs, H. Stetter.
Sonne, Jahrg. 11, Nr. 43, p. 79 (1987).

072.025   **Sonnenfleckenrelativzahlen des SONNE–Netzes und des S.I.D.C. für Mai 1987.**
Sterne Weltraum, 26. Jahrg., Nr. 7 – 8, p. 416 (1987).

072.026   **Sonnenfleckenrelativzahlen des SONNE–Netzes und des S.I.D.C. für Juni und Juli 1987.**
Sterne Weltraum, 26. Jahrg., Nr. 9, p. 496 (1987).

072.027 **A.L.P.O. solar section observations for rotations 1761 – 1770 (1985 April 17 to 1986 January 14).**
R. E. Hill.
Strolling Astron., Vol. 32, Nos. 3 – 4, p. 73 – 78 (1987).

072.028 **Sonnenfleckenrelativzahlen des SONNE–Netzes und des S.I.D.C. für August 1987.**
Sterne Weltraum, 26. Jahrg., Nr. 10, p. 563 (1987).

072.029 **Ne IX emission–line ratios in solar active regions.**
F. P. Keenan, D. L. McKenzie, S. M. McCann, A. E. Kingston.
Astrophys. J., Vol. 318, No. 2, p. 926 – 929 (1987).
Recent calculations of electron excitation rates for transitions in Ne IX are used to derive several theoretical density–sensitive as well as temperature–sensitive line ratios. Electron temperatures deduced from observed ratios obtained by the SOLEX B spectrometer on the *P*78–1 satellite for nonflaring solar active regions are in excellent agreement, with discrepancies that lie within the observational uncertainties.

072.030 **Stokes profile analysis and vector magnetic fields. III. Extended temperature minima of sunspot umbrae as inferred from Stokes profiles of Mg I λ4571.**
B. W. Lites, A. Skumanich, D. E. Rees, G. A. Murphy, M. Carlsson.
Astrophys. J., Vol. 318, No. 2, p. 930 – 939 (1987).
Observed Stokes profiles of Mg I λ4571 are analyzed as a diagnostic of the magnetic field and thermal structure at the temperature minimum of sunspot umbrae. Multilevel non–LTE transfer calculations of the Mg I–II–III excitation and ionization balance in model umbral atmospheres show (1) Mg I to be far less ionized in sunspot umbrae than in the quiet Sun, and (2) LTE excitation of λ4571. Umbral atmospheres with extended temperature minima (i.e., chromospheric temperature rise deferred to a smaller column mass) are suggested. Implications for chromospheric heating mechanisms and the utility of this line for solar vector magnetic field measurements are discussed.

072.031 **Periodicities of the flare occurrence rate in solar cycle 19.**
T. Bai.
Astrophys. J., Lett. Ed., Vol. 318, No. 2, p. L85 – L91 (1987).
The occurrence rate of major flares during solar cycle 19 has been analyzed. A periodicity of 51^d is found, which is one–third of the period found from the flare rates of solar cycles 20 and 21 by various authors. The statistical significance of the periodicity is estimated to be at 99.85% confidence level. This periodicity is thought to be related to the 153^d periodicity. The template of the phase diagram is well described by a constant plus a sine function.

072.032 **Peculiarities of radio radiation of flare–active sunspot groups of December 1982.**
V. E. Abramov–Maksimov, N. G. Peterova.
Problems of solar flares, p. 100 – 106, 205 (1986). In Russian. Abstr. in Ref. Zh., 51. Astron., 7.51.325 (1987). – See Abstr. 012.018.

072.033 **Magnetic fields of flare–active sunspot groups according to observations of the phenomenon of sign change in circular polarization.**
N. G. Peterova, R. E. Rodriges.
Problems of solar flares, p. 83 – 88, 203 (1986). In Russian. Abstr. in Ref. Zh., 51. Astron., 7.51.327 (1987). – See Abstr. 012.018.

072.034 **On the distribution of sunspot cycle periods.**
R. M. Wilson.
J. Geophys. Res., Vol. 92, No. A9, p. 10,101 – 10,104 (1987).
A comparison is made between the observed distribution of sunspot cycle periods and distributions based on uniform, normal, and bimodal distributions. The bimodal distribution, composed of short– and long–period cycles, each normally distributed about its respective mean, is found to best describe the observed distribution. Compared to the normal distribution for the most reliably determined cycles (cycles 8 – 20), the bimodal distribution has a residual (sum of squares of differences) that is about 86% smaller. Means for short– and long–period cycles are estimated to be 122 ± 4 months and 140 ± 5 months, respectively.

072.035 **Long–periodic oscillations in sunspot group No. 199, 1984.**
E. Yu. Nagovitsyna.
Soln. Dannye, Byull., 1987, No. 3, p. 58 – 62 (1987). In Russian.
The results are given of a study of long–periodic oscillations of the sunspot group axis, rotation of sunspots and velocities of proper motions for sunspot group S.D. No. 199, 1984.

072.036 **On the relationship between polar faculae, X–ray bright points and ephemeral active regions on the sun.**
V. I. Makarov, V. V. Makarova.
Soln. Dannye, Byull., 1987, No. 3, p. 62 – 70 (1987). In Russian.
A study was made on the basis of an analysis of photoheliograms of the Kislovodsk Station of the Pulkovo Observatory (during the minima of solar activity in 1964, 1976, 1985). Under the assumption of a homogeneous longitudinal distribution it is found that about 900 polar faculae are observed simultaneously over the entire solar disc. Through a 14 day analysis of solar X–ray observations it has been discovered that polar faculae coincide in coordinates with X–ray bright points (XBP) in 66% of the cases. On certain days the coincidence attains 85%. A conclusion is made that polar faculae, EAR and XBP seem to be a particular type of solar activity, differing from sunspot activity. This difference is discussed and the importance of polar faculae studies in understanding the global process of solar activity is stressed.

072.037 **Jahresauswertung des SONNE–Beobachternetzes 1986.**
K. Reinsch, V. Gericke, M. Schwab.
Sterne Weltraum, 26. Jahrg., Nr. 11, p. 638 – 641 (1987).

072.038 **"Butterfly" diagram for polar faculae and sunspots during 1940 – 1985.**
V. I. Makarov, V. V. Makarova, K. R. Sivaraman.
Soln. Dannye, Byull., 1987, No. 4, p. 62 – 65 (1987). In Russian.
Data on the time–latitude distribution of polar faculae and sunspots for 1940 – 1985 are given. They characterize two waves of the toroidal component of the magnetic field of the global process of solar activity.

072.039 **Parameters of sunspot magnetic fields as determined using superpenumbral structure.**
M. M. Molodenskij, L. I. Starkova, B. P. Filippov.
Soln. Dannye, Byull., 1987, No. 4, p. 70 – 77 (1987). In Russian. With 4 figures.
A field of the fibrils directions has been calculated for a penumbral and superpenumbral region of sunspots possessing their own field which have "a charge", dipole and quadrupole momentum and a longitudinal current. For a number of spots turning around curves are constructed and the data corresponding to the parameters of spots are obtained. The relative size of the azimuthal component characterizing the current in a spot is of the order 0.2 – 0.3.

072.040 **Some kinematic characteristics of sunspot proper motions.**
J. P. Doval, M. A. Cid.
Soln. Dannye, Byull., 1987, No. 4, p. 87 – 95 (1987). In Russian.
Some kinematic characteristics of proper motions of sunspots in 41 groups, observed during 1972 – 1976 are studied. The observed distribution curve of the absolute values of velocities significantly differs from the normal distribution. For the first time a method of mean velocities has been used. It is shown that 63% of 45 umbrae manifest variations in acceleration for a period of several weeks. It is supposed that sunspot proper motions may be associated with the plasma motion in supergranules.

072.041　Some data on sunspots for 1979 – 1985 from catalogues of solar activity.
R. S. Gnevysheva.
Soln. Dannye, Byull., 1987, No. 5, p. 70 – 81 (1987). In Russian.
　　Data on total areas and relative Wolf numbers of sunspots, and a list of large sunspot groups for 1979 – 1985 are given. The data are planned to be published in the Catalog of Solar Activity for 1979 – 1985, the last issue of which was printed for 1978.

072.042　On characteristics of fluctuations of solar activity during the last years of solar cycle 21.
Yu. I. Vitinskij.
Soln. Dannye, Byull., 1987, No. 5, p. 91 – 95 (1987). In Russian.
　　Values of the fluctuation index and indices of "perturbations" for Wolf numbers and density of solar radio emission flux at 2800 MHz during 1984 – 1986 are given. On the basis of their comparison with the corresponding data for the last three years of solar cycles 18, 19 and 20 a conclusion is made on manifestation of the secular and 22 year cycles in characteristics of fluctuations.

072.043　The Sun's spots and flares.
　　D. M. Rust.
The solar wind and the Earth, p. 2 – 17 (1987). – See Abstr. 003.007.
　　Recently, a NASA satellite has not only gathered much new information about sunspots and flares, but it also has had the unique experience of being repaired in outer space by astronauts. What has been accomplished with the telescopes of the satellite observatory, and what do solar physicists hope to achieve with it in the near future?
Contents: Introduction. The repair mission. Research highlights. Gamma rays. X rays. Ultraviolet rays. Mass ejection. The solar constant.

072.044　Sunspot numbers: April 1987.
　　Sky Telesc., Vol. 74, No. 1, p. 107 (1987).

072.045　Sunspot numbers: May 1987.
　　Sky Telesc., Vol. 74, No. 2, p. 221 (1987).

072.046　Sunspot numbers: June 1987.
　　Sky Telesc., Vol. 74, No. 3, p. 333 (1987).

072.047　Sunspot numbers: July 1987.
　　Sky Telesc., Vol. 74, No. 4, p. 445 (1987).

072.048　Sunspot numbers: August 1987.
　　Sky Telesc., Vol. 74, No. 5, p. 563 (1987).

072.049　White light sunspot observations from the Solar Optical Universal Polarimeter on Spacelab–2.
R. A. Shine, A. M. Title, T. D. Tarbell, K. P. Topka.
Science, Vol. 238, No. 4831, p. 1264 – 1267 (1987).
　　The Solar Optical Universal Polarimeter (SOUP) flew on the space shuttle Challenger as part of the Spacelab–2 mission from 29 July to 6 August 1985. Because of electronic and thermal problems, only the white–light film data were scientifically useful. However, the high resolution (approximately 0.5 arc sec or 350 km) and, more importantly, the stability and freedom from variable atmospheric distortion of these data have provided an unprecedented opportunity to study the dynamics of the solar photosphere by viewing and analyzing movie sequences.

072.050　Periodicities in the sunspot cycle.
　　P. A. M. Berry.
Vistas Astron., Vol. 30, Part 2, p. 97 – 108 (1987).
　　Statistical analyses of sunspot counts have indicated the presence of a number of different periodicities with wavelengths ranging from days to centuries; however, conflicting values are obtained for the frequencies of these variations. This paper presents an outline of existing methods and results, together with an independent statistical examination of the Waldmeier sunspot numbers. Analysis of the data did not provide evidence for the existence of short–period variations as previously reported. Some evidence was found supporting the hypothesis that long–term effects may modulate the amplitude of the eleven–year cycle; uncertainty about the accuracy of early sunspot data precluded firm identification of such effects.

072.051　On the prospect of using butterfly diagrams to predict cycle minimum.
R. M. Wilson.
Sol. Phys., Vol. 111, No. 2, p. 255 – 265 (1987).
　　On the basis of butterfly diagrams for the period 1874 – present (covering late cycle 11 through late cycle 21), features are identified that may be useful for predicting the beginning and the length of a solar cycle, as well as the discernment of turning points in the period–growth dichotomy.

072.052　A.L.P.O. Solar Section observations for rotations 1771 – 1783 (1986 Jan 14 to 1987 Jan 3).
R. E. Hill.
Strolling Astron., Vol. 32, Nos. 5 – 6, p. 131 – 135 (1987).

072.053　Generation and structure of the electric currents in a flaring activity complex.
J. C. Hénoux, B. V. Somov.
Astron. Astrophys., Vol. 185, No. 1/2, p. 306 – 314 (1987).
　　Assuming the existence of organised photospheric velocity fields in an active region, the authors study the creation of photospheric and coronal D.C. currents by vortex and radial motions. Vortex motions leading to opposite temporal evolutions of the magnetic flux in old and new centres of activity create two systems of coronal currents flowing along the lines of force that, in the active region complex, connect the old centre of activity to the new. The magnetic energy stored in these currents is higher than the energy of the largest flares. The effect of radial photospheric motions is also discussed. Two systems of coronal currents can be distinguished. They are located inside two different magnetic cells, and interact along the separator which is the intersection of the magnetic line surfaces that separate the magnetic cells produced by the photospheric magnetic sources. Due to source motions and electric current evolution, the separator may be the location of reconnection of the magnetic field.

072.054　Velocity oscillations in a sunspot penumbra.
　　A. P. Kramynin, V. S. Loskutnikov, R. F. Khisamov.
Vladivostok, Sev.–Vost. Kompleks NII, Prepr., 13 pp. (1987). In Russian. Abstr. in Ref. Zh., 51. Astron., 10.51.493 (1987).

072.055　Torsional oscillations of sunspots.
　　S. I. Gopasyuk, G. V. Lyamova.
Izv. Krymskoj Astrofiz. Obs., Tom 77, p. 17 – 24 (1987). In Russian. English translation in Bull. Crimean Astrophys. Obs., Vol. 77.
　　Using the photoheliograms obtained in 1970 – 1982 at the Crimean Astrophysical Observatory according to the program «sluzhba solntsa» the sunspots rotation has been studied. Single spots and spots with the simple penumbrae and umbrae in complex groups have been selected. It has been shown that torsional oscillations of sunspots are not unique, but rather a characteristic state. The spectrum of oscillations in periods and amplitudes is rather wide. On the average the amplitude of oscillations increases with the period. The authors conclude that during the minimum of solar activity the magnetic flux tube that formed sunspots rised to the surface of the sun from not very deep layers. With the approach of maximum activity of the Sun the rise of magnetic flux tubes envolves more and more layers. This in turn leads to the conclusion that during maximum solar activity the convective motions have higher velocities and the dimensions of convective elements enlarge.

072.056 Model of a sunspot and of a flare above a sunspot.
Eh. A. Baranovskij, Z. A. Shcherbakova.
Izv. Krymskoj Astrofiz. Obs., Tom 77, p. 25 – 31 (1987). In Russian. English translation in Bull. Crimean Astrophys. Obs., Vol. 77.

The profiles of Ca II lines H, K, $\lambda 8498$, $\lambda 8542$, $\lambda 8662$ are obtained for two sunspots. The wings of these lines profiles are used for umbra model photosphere calculations and for derivations of the opacity enhancement in the continuum (3900 – 4000 Å). The additional opacity is found to be 2 – 4 times the opacity caused by H⁻ ions. The umbra model chromosphere is calculated on the basis of the central parts of the Ca II lines and the lines H_α, H_β, H_δ. The flare model chromosphere for two flares, situated above the spots, is also derived.

072.057 Study of spectral line asymmetry in sunspots.
A. Ludmány.
Publ. Debrecen Heliophys. Obs., Vol. 6, No. 1, p. 5 – 10 (1986).

The previously found inverse line asymmetry is studied in sunspots near to the solar disc center.

072.058 Two–component modelling of sunspot spectral line profiles.
A. Ludmány.
Publ. Debrecen Heliophys. Obs., Vol. 6, No. 1, p. 11 – 23 (1986).

A two–component model is discussed as a possible interpretation of the negative asymmetry of spectral line $\lambda 5714$ Ti I in sunspots.

072.059 Study of solar Hα–line profiles by means of filtergrams.
T. Baranyi.
Publ. Debrecen Heliophys. Obs., Vol. 6, No. 1, p. 25 – 37 (1986).

The shape of the Hα–profile has been determined in 15 points of a solar active region on the basis of monochromatic images. A short overview is given about the developed version of the procedure published earlier.

072.060 Manila Observatory, Solar Division. Solar Maps and Activity. 1987 February – April.
Sol. Maps Act. (1987).

072.061 A preliminary discussion of measuring accuracy of sunspot magnetic fields at Yunnan Observatory.
T. Luan.
Publ. Yunnan Obs., No. 2, p. 34 – 40 (1987). In Chinese.

072.062 Faraday rotation and measurements of magnetic fields in sunspots.
S.–h. Ye, J.–h. Jin.
Publ. Purple Mt. Obs., Vol. 6, No. 3, p. 227 – 232 (1987). In Chinese.

In order to clarify the three–dimensional configuration of sunspot magnetic fields and in particular the orientation of filaments and fibrils it is necessary to determine the direction of transverse fields. Observations can provide only the azimuthal angle (ϕ) of the plane of polarization. Under the action of the Faraday rotation ϕ may differ significantly from the azimuth of the transverse field (χ). In order to get the intrinsic value of χ from the measured ϕ, the authors study the relation between these two angles.

072.063 A catalogue of non–telescopic sunspot observations from 165 BC to AD 1684.
A. D. Wittmann, Z. T. Xu.
Publ. Purple Mt. Obs., Vol. 6, No. 3, p. 233 – 285 (1987).

The authors have compiled a new catalogue of non–telescopic sunspot observations covering the period 165 BC to AD 1684 by updating and merging previously published catalogues and by adding a substantial amount of new data. The catalogue is in machine–readable form, the total number of entries being 235. Epoch analyses of the data have been made.

072.064 On the factor k of sunspot relative–numbers at the Tokyo Astronomical Observatory.
T. Natori, M. Irie.
Tokyo Astron. Obs. Rep., Vol. 21, No. 1, p. 120 – 126 (1987). In Japanese.

072.065 Structure of solar active regions from VLA and RATAN–600 observations in July 1982. Part 1: AR 3804.
Sh. B. Akhmedov, V. M. Bogod, V. N. Borovik, R. F. Willson, G. B. Gel'frejkh, V. N. Dikij, A. N. Korzhavin, K. R. Lang, Z. E. Petrov.
Astrofiz. Issled. Izv. Spets. Astrofiz. Obs., Tom 25, p. 105 – 134 (1987). In Russian. English translation in Bull. Spec. Astrophys. Obs. – North Caucasus, Vol. 25.

The results of combined RATAN–600 and VLA observations of solar active regions in July 1982 are discussed.

072.066 Evaluation of effective global sunspot number.
I. Kuriki, H. Inuki.
Rev. Radio Res. Lab., Vol. 33, No. 166, p. 13 – 19 (1987). In Japanese. Abstr. in Phys. Abstr., Vol. 90, No. 1311, Entry 101758 (1987).

072.067 Sunspot numbers: September 1987.
Sky Telesc., Vol. 74, No. 6, p. 683 (1987).

072.068 Definitive Sonnenfleckenrelativzahlen für 1986.
A. Koeckelenbergh.
Sterne, 63. Band, Heft 4, p. 235 (1987).

072.069 Sonnenfleckenrelativzahlen des SONNE–Netzes und des S.I.D.C. für Oktober 1987.
Sterne Weltraum, 26. Jahrg., Nr. 12, p. 703 (1987).

072.070 A comparison of the oscillations in sunspot umbrae, penumbrae, and the surrounding photosphere.
H. Balthasar, G. Küveler, E. Wiehr.
Sol. Phys., Vol. 112, No. 1, p. 37 – 48 (1987).

Time series of the nonsplit Fe I 7090 Å line have been observed in several sunspots with a 100 × 100 diode array corresponding to 48 arc sec times 1.39 Å. The spatial behaviour of Doppler motions along one fixed slit position has been studied as a function of time. Former results are confirmed, that the power in the five minute range decreases from the photosphere to the umbra, where, however, values still well above the noise level are measured. Regarding the penumbra, the power tends to exhibit a maximum at locations where the line–of–sight component of a radial horizontal field should be maximal. This indicates that the direction of the oscillatory velocities might be influenced by the magnetic field or the Evershed flow. No significant power is found in the 3 min range.

072.071 Umbral dots: a case of penetrative convection between sunspot fragments.
J. I. García de la Rosa.
Sol. Phys., Vol. 112, No. 1, p. 49 – 58 (1987).

A study of the observations made on the development and, in some cases, even the decay of 15 large active regions is presented. It is shown that the mature spots result from the subphotospherically controlled attraction of several large fragments of $1 – 2 \times 10^{21}$ Mx, which are themselves made of smaller elements. The fragments are more stable structures than the spots they constitute; and usually survive after the spot decay. In the process of coalescence of fragments to form the spot, the fusion is never complete and properly exposed photographs reveal light bridges or saddle–like distributions of umbral dots in the interstices between fragments. These are also the regions along which the break up of the spot occurs.

072.072 Microwave emission above steady and moving sunspots.
F. Chiuderi–Drago, C. Alissandrakis, M. Hagyard.
Sol. Phys., Vol. 112, No. 1, p. 89 – 105 (1987).

Two–dimensional maps of radio brightness temperature and polarization, computed assuming thermal emission with free–free and gyroresonance absorption, are compared with observations of active region 2502, performed at Westerbork at $\lambda = 6.16$ cm during a period of 3 days in June 1980. A very good agreement is found with the model used. A strong radio source, associated with a new–born moving sunspot, cannot be ascribed to thermal emission. It is suggested that this source may be due to synchrotron radiation by mildly relativistic electrons accelerated by resistive instabilities occuring in the evolving magnetic configuration.

072.073 A statistical study of the geometrical Wilson effect.
M. Collados, J. C. Del Toro Iniesta, M. Vazquez.
Sol. Phys., Vol. 112, No. 2, p. 281 – 293 (1987).

An analysis has been carried out of the centre–to–limb variation of the apparent sizes of the umbra, penumbra and whole spot. The umbral size decreases with height. The authors interpreted this result in terms of the penumbral geometrical height scale. A value of 230 km, which is larger than that of the photosphere or the umbra, explains the observed decrease. An intrinsic asymmetry in the penumbra of old sunspots has also been found, the western penumbra being slightly shorter than the rest of the penumbra. This explains why the inverse Wilson effect is present, preferentially, in the western hemisphere, as found in previous investigations. A comparison with other works is also made.

072.074 Motions around a decaying sunspot.
E. Muller, B. Mena.
Sol. Phys., Vol. 112, No. 2, p. 295 – 303 (1987). = Contrib. Kwasan Hida Obs., Univ. Kyoto.

The authors have measured the motion of facular points and granules in the same region near a decaying sunspot. It is found that both features move away across the moat surrounding the sunspot. The mean speed of facular points is larger than that of granules: 0.65 km s^{-1} and 0.4 km s^{-1}, respectively. These results are consistent with previous measurements of the speed of bright network features and moving magnetic fields, as well as of non–magnetic photospherical material. They support models in which a decaying sunspot is at the center of a supergranule, whose horizontal motions sweep out granules and magnetic flux tubes associated to the facular points. It is also found that granules are dragged by supergranular motions away from the moat.

072.075 The Faraday rotation of sunspots.
S.–h. Ye, J.–h. Jin.
Sol. Phys., Vol. 112, No. 2, p. 305 – 312 (1987).

The numerical solutions of the Unno–Beckers equations for the magneto–sensitive line Fe I $\lambda 5250.216$ are used to demonstrate the importance and role of Faraday rotation in sunspot magnetic fields and to study the influence of this effect on the measurements of the azimuth of the transverse field. The authors propose a method to determine the intrinsic direction of the transverse field with the observed azimuthal angle of the plane of linear polarization.

072.076 The 17–month periodicity of sunspot activity.
M. Akioka, J. Kubota, M. Suzuki, K. Ichimoto, I. Tohmura.
Sol. Phys., Vol. 112, No. 2, p. 313 – 316 (1987).

A statistical study of sunspot activity during 1969 to 1986 was carried out by using the number of sunspot groups and their areas. The authors found a 17–month periodicity, which is consistent with the 500–day periodicity of flare occurrence (Ichimoto et al., 1985).

072.077 Provisional sunspot numbers for June – November 1987.
Yamamoto Circ., Nos. 2087, 2098, 2092, 2093, 2097, 2100 (1987). In Japanese.

072.078 Asymmetric pulses of the Sun in the antiphase of activity.
T. S. Razmadze.
Astron. Tsirk., No. 1487, p. 5 – 7 (1987). In Russian.

The 11–year cyclicity in the change of the module of the north–south asymmetry of the sunspot areas is revealed. On the basis of the high positive connection between the sunspot areas and its magnetic fields the analogous conclusions, concerning the magnetic asymmetry of the Sun, are drawn.

072.079 The visibility function and its effect on the observed characteristics of sunspot groups. 6. Various types of diagrams derived from Minnaert's classical diagram.
M. Kopecký, G. V. Kuklin.
Bull. Astron. Inst. Czech., Vol. 38, No. 6, p. 344 – 350 (1987). With plates 5 – 8.

The purpose of this study is to develop Minnaert's diagram in time into a three–dimensional model (3D model). If this 3D model is projected into the three basic spatial planes, one plane displays Minnaert's classical diagram, the second the diagram of observational conditions of sunspot groups, and the third a "diagram of boundaries of observed sunspot group areas".

072.080 Daily maps of the sun and of magnetic fields of sunspots.
Soln. Dannye, Byull., 1986, 1987, Nos. 11, 12, Nos. 1 – 5 (1987). In Russian.

072.081 Fuzzy sets. Classification of solar active regions and flare prediction.
X.–z. Liu, W. Li.
Chin. Astron. Astrophys., Vol. 11, No. 3, p. 186 – 190 (1987). English translation of 43.072.067.

072.082 Erratum: "A two cavity model for umbral oscillations" [J. Korean Astron. Soc., Vol 20, No. 1, p. 27 – 47 (1987)].
J. W. Lee, H. S. Yun.
J. Korean Astron. Soc., Vol. 20, No. 2, p. 95 (1987). See Abstr. 43.072.102.

072.083 The vertical gradient of sunspot magnetic fields based on a force–free field model.
Y. Lin, F. Wu.
Acta Astrophys. Sin., Vol. 7, No. 4, p. 312 – 316 (1987). In Chinese.

The vertical gradient of the longitudinal magnetic field above the preceeding sunspot of the active region Boulder 2744 on 23 October 1980 was calculated based on the model of constant α force–free field. In connection with the longitudinal magnetic field of the chromosphere–corona transition region observed in the spectral line C IV 1548, the effective height of the C IV 1548 emission region was estimated. These results are quite different from those obtained by Hagyard et al. (1983) for the same active region based on a model of a potential magnetic field. This shows that the adoption of a model of a potential field or a model of a force–free field will lead to very different results in some cases.

072.084 Relative sunspot numbers for June – December 1987.
Am. Assoc. Variable Star Obs. Sol. Bull., Vol. 43, Nos. 6 – 12 (1987).

072.085 Simultaneous SMM flat crystal spectrometer and Very Large Array observations of solar active regions.
K. R. Lang, R. F. Willson, K. L. Smith, K. T. Strong.
Astrophys. J., Vol. 322, No. 2, p. 1035 – 1043 (1987).

The authors compare high–resolution images of the quiescent emission from two solar active regions at 20 cm (VLA) and soft X–ray (SMM FCS) wavelengths. There are regions where the X–ray coronal loops have been completely imaged at 20 cm wavelength. The X–ray data were used to infer average electron temperatures, T_e, and average electron densities for the X–ray emitting plasma. The 20 cm brightness temperatures, T_B, were always less than T_e. The results are interpreted with a model involving thermal bremsstrahlung and thermal gyroresonance radiation from the coronal plasma.

072.086   Solar active region physical parameters inferred from a thermal cyclotron line and soft X–ray spectral lines.
K. R. Lang, R. F. Willson, K. L. Smith, K. T. Strong.
Astrophys. J., Vol. 322, No. 2, p. 1044 – 1051 (1987).

The authors present simultaneous high–resolution observations of coronal loops at 20 cm wavelength with the VLA and at soft X–ray wavelengths with the SMM FCS. The results show that 20 cm VLA maps can image X–ray coronal loops. The X–ray spectral lines are used to infer an average electron temperature of about $2.6 \pm 0.1 \times 10^6$K and an average electron density of $3.1 \pm 0.3 \times 10^9 \text{cm}^{-3}$. These parameters are used to show that the layers emitting 20 cm radiation can be optically thick to either thermal bremsstrahlung or thermal gyroresonance radiation. The detection of a linelike feature in the radio spectrum indicates that gyroresonance absorption exceeds free–free absorption. This feature is attributed to a thermal cyclotron line. The X–ray values for T_e and N_e are combined with plausible values for gyroresonant optical depth and the magnetic scale height to show that the 20 cm radiation is probably at the fourth harmonic of the gyrofrequency.

072.087   On the large–scale dynamics and magnetic structure of solar active regions.
J. A. Klimchuk.
Astrophys. J., Vol. 323, No. 1, p. 368 – 379 (1987). With plates 3 – 8.

The author has studied sets of carefully coaligned C IV Dopplergrams, photospheric magnetograms, and Hα filtergrams to infer the flow properties of active regions and the relationship of these flows to the active region magnetic fields. The combined data show that active regions can be naturally divided into three basic parts: strong field regions, weak field corridors between strong fields of opposite polarity, and surrounding weak field areas. An idealized topological model in which the vertical fields of strong field regions diverge very rapidly with height to become essentially horizontal in the adjacent low–lying areas is proposed. The picture is similar to canopy structures.

072.088   On the centre–to–limb variation of some visible and infrared atomic hydrogen lines in three facular models.
K. R. Bondal, V. P. Gaur.
Bull. Astron. Soc. India, Vol. 15, Nos. 2 – 3, p. 65 – 69 (1987).

The centre–to–limb behaviour of the residual intensities in the line wings of the Balmer, Paschen, and Brackett series lines (alpha to delta) in three facular models at four wing points has been investigated with a view to distinguish amongst the models.

072.089   Sunspot observations for the year 1982.
N. Doğan.
Commun. Fac. Sci. Univ. Ankara, Sér. A3, Tome 31 – 32, No. 1, p. 1 – 38 (1983/84).

Sunspot groups seen between solar rotations 1717 and 1729, their development and the distribution of these groups over northern and southern latitudes are given. The number of spot groups by type in each rotation is also presented.

Korotkoperiodnye tsikly solnechnoj aktivnosti (*Short–period cycles of solar activity*).
See Abstr. 003.140.

Old sunspot records.
See Abstr. 004.109.

Solar Cycle Workshop. A review.
See Abstr. 011.001.

Einführung in die Sonnenbeobachtung. I.
See Abstr. 014.009.

Einführung in die Sonnenbeobachtung. II.
See Abstr. 014.054.

Stokes profile analysis and vector magnetic fields. I. Inversion of photospheric lines.
See Abstr. 063.109.

The polarization properties of model sunspots: the broad–band polarization signature of the Schlüter–Temesvary representation.
See Abstr. 063.110.

The line–of–sight photospheric velocities and Hα chromospheric fine structure in the active region SD No. 135.
See Abstr. 071.021.

Time variation of the flare index during the 21st solar cycle.
See Abstr. 073.005.

Velocity field and flare activity in a sunspot group.
See Abstr. 073.032.

Observational peculiarities of flare behaviour in sunspot umbrae.
See Abstr. 073.033.

Rotation of individual background magnetic field components during the formation of the white–light flare region of April 1984 (NOAA 4474).
See Abstr. 073.092.

Solar cycle evolution of solar wind speed structure between 1973 and 1985 observed with the interplanetary scintillation method.
See Abstr. 074.012.

Solar cycle invariance in solar wind proton temperature relationships.
See Abstr. 074.042.

Behaviour of the coronal intensity in the line 5303 Å and latitude zonal structure of the magnetic field. Period 1944 – 1974.
See Abstr. 074.043.

Actividad de los hoyos coronales solares de baja latitud.
See Abstr. 074.058.

Properties of solar coronal active regions deduced from X–ray line spectra.
See Abstr. 074.080.

Is there a weak mixed polarity background field? Theoretical arguments.
See Abstr. 075.006.

Solar activity indices and periods of general magnetic field inversion of the Sun.
See Abstr. 075.012.

The origin of rigidly rotating magnetic field patterns on the Sun.
See Abstr. 075.016.

Activity waves of the solar magnetic field.
See Abstr. 075.017.

Magnetic waves of solar activity.
See Abstr. 075.018.

The system of electric currents and magnetic field structure in active regions.
See Abstr. 075.022.

On the factor of Zeeman splitting of the Sr I λ4607.3 Å line.
See Abstr. 075.023.

Solar soft X–ray pulsations.
See Abstr. 076.002.

Parameters of solar radio plages and their relations to other active region characteristics.
See Abstr. 077.033.

10.7 cm solar radio flux and the magnetic complexity of active regions.
See Abstr. 077.036.

Radio spectrum of a solar active region.
See Abstr. 077.041.

S–component model computations applied to mean mm–wave measurements.
See Abstr. 077.049.

Studies of some aspects of solar proton events and related phenomena.
See Abstr. 078.001.

Distribution of solar proton events over a solar activity cycle.
See Abstr. 078.002.

A prediction of the major proton event for the first time in cycle 22.
See Abstr. 078.008.

Prolonged minima and the 179–yr cycle of the solar inertial motion.
See Abstr. 080.005.

Rotation of solar high–latitude regions in 1971 – 1978.
See Abstr. 080.008.

Differential rotation of polar regions of the solar atmosphere.
See Abstr. 080.009.

Variations of solar irradiance due to magnetic activity.
See Abstr. 080.017.

Acoustic absorption by sunspots.
See Abstr. 080.024.

Is solar neutrino capture rate correlated with sunspot number?
See Abstr. 080.038.

The 160 minutes period, internal rotation and 11–year cycle of the Sun: evidence for a relationship?
See Abstr. 080.043.

The interaction of solar p–modes with a sunspot. II. Simple theoretical models.
See Abstr. 080.049.

Rigid rotation and localisation of sunspots.
See Abstr. 080.061.

On the proposed associations of solar neutrino flux with solar particles, cosmic rays, and the solar activity cycle.
See Abstr. 080.062.

Neutrinos and sunspots.
See Abstr. 080.074.

Geomagnetic and solar data. February 1987.
See Abstr. 084.008.

Geomagnetic and solar data. March 1987.
See Abstr. 084.009.

Geomagnetic and solar data, April 1987.
See Abstr. 084.010.

Geomagnetic and solar data. May 1987.
See Abstr. 084.016.

Geomagnetic and solar data. July 1987.
See Abstr. 084.046.

Geomagnetic and solar data. August 1987.
See Abstr. 084.073.

Some solar cycle phenomena related to the geomagnetic activity from 1868 to 1980. III. Quiet–days, fluctuating activity or the solar equatorial belt as the main origin of the solar wind flowing in the ecliptic plane.
See Abstr. 085.003.

Statistical analysis of the geomagnetic disturbing index Ap with some solar activities.
See Abstr. 085.013.

Solar activity and heliosphere–wide cosmic ray modulation in mid–1982.
See Abstr. 144.009.

Solar cycle modulation of galactic protons and electrons.
See Abstr. 144.010.

073 Chromosphere, Flares, Prominences

073.001 **Analysis of ultraviolet and X–ray observations of three homologous solar flares from SMM.**
C.–C. Cheng, R. Pallavicini.
Astrophys. J., Vol. 318, No. 1, p. 459 – 473 (1987).
The authors present the study of three homologous flares observed in the UV–lines of Fe XXI and O V and in X–rays from the Solar Maximum Mission. The homology of the flares was best recognized in Fe XXI and soft X–ray emissions, whose spatial emission patterns were very similar for the three flares. The three flares shared many of the same loop footpoints, which were located in O V bright kernels associated with hard X–ray bursts. Although there was strong spatial homology, the temporal evolution in UV and X–ray emissions varied from flare to flare.

073.002 **Particle acceleration in solar flares.**
R. Ramaty, M. A. Forman.
NASA Conf. Publ., NASA CP–2464, p. 47 – 73 (1987). – See Abstr. 012.004.
The most direct signatures of particle acceleration in flares are energetic particles detected in interplanetary space and in the Earth's atmosphere, and gamma rays, neutrons, hard X–rays, and radio emissions produced by the energetic particles in the solar atmosphere. The authors review the stochastic and shock acceleration theories in flares, and discuss the implications of observations on particle energy spectra, particle confinement and escape, multiple acceleration phases, particle anisotropies, and solar atmospheric abundances.

073.003   Hydrogen emission from moving solar prominences.
P. Heinzel, B. Rompolt.
Sol. Phys., Vol. 110, No. 1, p. 171 – 189 (1987).

Brightness variations of the lines arising from a five–level hydrogen model atom, depending upon prominence velocities, have been investigated using a combination of two non–LTE techniques. The importance of the Doppler brightening and/or Doppler dimming effects is demonstrated for the lines of the Lyman and Balmer series.

073.004   Effects of impulsively heated electrons in solar flares.
M. Karlický.
Bull. Astron. Inst. Czech., Vol. 38, No. 4, p. 201 – 206 (1987).

Using a one–dimensional particle code, the evolution of impulsively heated electrons in the flare loop is examined. The processes are studied in two regimes: with and without the electric current. One run was made for the system with the constant electric current, simulating the large inductance of the current loop. It was shown that impulsive electron heating leads to the formation of a double layer. Moreover, this heating can trigger the ion–acoustic instability in the current system of a flare loop. Both these effects increase the flare energy release. The generated electrostatic high–frequency waves can be observed as a radio burst.

073.005   Time variation of the flare index during the 21st solar cycle.
T. Ataç.
Astrophys. Space Sci., Vol. 135, No. 1, p. 201 – 205 (1987).

The flare index during the solar activity cycle 21, was calculated for every day from 1 January 1976 to 31 March 1986. The authors have determined that the chromospheric activity for the whole disc was significantly higher during the 21st solar cycle, as compared to the preceding cycle. They have also found out that the activity began and reached its maximum earlier on the northern hemisphere. Finally, they determined that the time variation of the flare indices for the eastern and western hemispheres are in good agreement.

073.006   Stabilizing effect of a transverse magnetic field in current sheets of solar flares.
A. I. Verneta, B. V. Somov.
Pis'ma Astron. Zh., Tom 13, No. 8, p. 716 – 722 (1987). In Russian. English translation in Sov. Astron. Lett., Vol. 13.

Quasi–steady high–temperature current sheets are an energy source during the "main" or "hot" phase of solar flares. Such sheets are shown to be stabilized with respect to the tearing instability by a small transverse component of a magnetic field existing in the sheets.

073.007   Similarities in the structure of the chromosphere and corona.
R. A. Gulyaev.
Sov. Astron., Vol. 30, No. 5, p. 607 – 608 (1986). English translation of 42.073.032.

073.008   Solar–flare grouping in time.
M. B. Ogir'.
Bull. Crimean Astrophys. Obs., Vol. 74, p. 86 – 93 (1987). English translation of 41.073.096.

073.009   Characteristic time dependence of the structure of some chromospheric formations.
L. G. Kartashova.
Bull. Crimean Astrophys. Obs., Vol. 74, p. 94 – 103 (1987). English translation of 41.073.097.

073.010   Comparison of the total energy release in the optical and hard X–ray ranges of solar flares (Catalog).
V. M. Zenchenko, A. V. D'yachkov, L. N. Kurochka, V. G. Kurt, E. V. Yurovitskaya.
Soln. Dannye, Byull., 1987, No. 2, p. 48 – 57 (1987). In Russian.

An instrument is briefly described which measures hard X–ray radiation with an energy $E_x > 50$ keV from solar flares. The measurements were made on board the Venera 13, 14 spacecraft. These flares were used as a basis for a catalog where the total energy flux in the optical E_{opt} and hard X–ray E_{Hx} ranges of solar flares is compared. The procedure of E_{Hx} and E_{opt} calculations is described.

073.011   Local rigid rotation and the emergence of Active Centres.
Z. Mouradian, M. J. Martres, I. Soru–Escaut, L. Gesztelyi.
Astron. Astrophys., Vol. 183, No. 1, p. 129 – 134 (1987).

Deviations from the mean solar rotation rule of filaments are frequent. They correspond to real physical behaviour. Using filaments as tracers, limited solar areas of rigid rotation are observed: these are "pivot points" around which the filaments rotate during two or more successive rotations. It is shown that a relation exists between the pivot points and Active Centres: many Active Centres appear close to a pivot point whatever their latitude, whereas filaments which reveal the classical differential rotation do not generate any Active Centres. The magnetic flux emergence is associated in time with apparent reduced speed of the filament tilt about the pivot point. Some examples show that the relation exists using a reduced time scale of several days as well. The new Active Centre may destroy or displace the associated pivot point.

073.012   Fine structures in solar filaments. I. Observations and thermal stability.
P. Démoulin, M. A. Raadu, J. M. Malherbe, B. Schmieder.
Astron. Astrophys., Vol. 183, No. 1, p. 142 – 150 (1987).

Limb observations of quiescent prominences show very fine structures of less than one arcsecond. $H\alpha$ observations on the disk, made with the Multichannel Subtractive Double Pass (MSDP) spectrograph operating at the Observatoire du Pic du Midi are presented. They show long fine structures in the body of the filament, both in intensity and velocity maps. Many prominence models ignore such fine structures and suppose that the quiescent filament is a uniform body. It has been suggested that in a current–sheet model of a prominence, fine structures may be produced as a consequence of the tearing mode coupled to thermal instability. The authors investigate the role of parallel and orthogonal conduction on the stability of a periodic fine structure. Equilibrium conditions are found and growth rates determined for linear perturbations.

073.013   Microwave radiation from a dense magneto–active plasma.
K.–L. Klein.
Astron. Astrophys., Vol. 183, No. 2, p. 341 – 350 (1987).

The author presents a formalism for the approximate computation of gyrosynchrotron radiation from mildly relativistic electrons. It enables a rapid numerical computation in the case of an isotropic or weakly anisotropic electron population. The author discusses quantitatively a solar flare with a peculiar flux spectrum: The flux increases from long to short millimetre waves, while the emission at centimetric waves, where the bulk of the microwave radiation is generally emitted, is negligible. Various authors considered that such a spectrum is inconsistent with the general interpretation of hard X–ray and microwave bursts in terms of respectively bremsstrahlung and gyrosynchrotron radiation from electrons up to some MeV. It is shown here that the observed microwave spectrum is readily explained by suppression and absorption of gyrosynchrotron radiation, and that the hard X–ray and microwave radiation of this event are consistent with a common distribution of mildly relativistic electrons.

073.014   Formation of the hydrogen spectrum in quiescent prominences: one–dimensional models with standard partial redistribution.
P. Heinzel, P. Gouttebroze, J.–C. Vial.
Astron. Astrophys., Vol. 183, No. 2, p. 351 – 362 (1987).

Departures from complete frequency redistribution (CRD) in hydrogen lines are investigated for quiescent solar prominences, within the frame of one–dimensional, isothermal–isobaric static slab models. Partial redistribution (PRD) effects on the hydrogen

La line are found to be very important not only in the far wings (as expected), but also in the near wings of Lα. Theoretical PRD Lα profiles exhibit strong (symmetrical) peaks which are not present in the CRD case.

073.015 Soft X–ray line profiles in the impulsive phase of electron–heated solar flares.
A. G. Emslie, D. Alexander.
Sol. Phys., Vol. 110, No. 2, p. 295 – 303 (1987).

Using a model of the hydrodynamic response of a flare loop to energy input in the form of a non–thermal electron beam, the authors compute the predicted line profile of the Ca XIX resonance line at 3.177 Å. The results are compared with observations and are shown to generally agree, both qualitatively and quantitatively, with these observations. The model does, however, predict an overall blueshift in the line profile. The authors conclude that electron–heated models of solar flares quite naturally predict the observed features in the Ca XIX line profile, without the need to introduce non–thermal motions or an artificial division of the flare plasma into two distinct components, as suggested by previous authors.

073.016 Hard X–ray emission in electron–heated solar flares – a comparison of nonthermal and thermal contributions.
D. G. Brown, A. G. Emslie.
Sol. Phys., Vol. 110, No. 2, p. 305 – 315 (1987).

The authors calculate the spatial structure of hard X–ray emission during the impulsive phase of electron–heated solar flares. Both direct non–thermal bremsstrahlung and the thermal bremsstrahlung arising from the heated plasma are considered. The results indicate that the spread of non–thermal emission into the upper parts of the loop, through evaporation of the chromospheric target, may be more important than the appearance of a hot thermal source in the corona. The effects of varying the viewing angle to the flare loop, and of finite–size resolution element, are also considered. The results are compared with observations from the Solar Maximum Mission Hard X–Ray Imaging Spectrometer. By contrasting the predicted structures with those predicted by other models of flare energy release, it is found that the electron–heated model provides the most satisfactory agreement with the observations.

073.017 Chromospheric and coronal explosions in solar flares.
C. de Jager.
Sol. Phys., Vol. 110, No. 2, p. 337 – 342 (1987).

The phenomenon described as a coronal explosion results directly from a chromospheric explosion. These two phenomena always occur together. They are the manifestations of the impulsive phase explosions in solar flares. These explosive processes occur during and immediately after the onset of the impulsive phase of flares. A previously presented model, describing the relation between the two kinds of explosions, appears to be able to explain qualitatively, and in many cases also quantitatively, the observations relevant to these explosive processes.

073.018 Hα diagnostics of (post)–flare loops based on narrow–band filtergram observations.
P. Heinzel, M. Karlický.
Sol. Phys., Vol. 110, No. 2, p. 343 – 357 (1987).

Using narrow–band Hα filtergrams, the authors develop a quantitative non–LTE approach to determine the physical conditions prevailing at the tops of (post)–flare loops observed against the solar disc. At temperatures 10,000 – 15,000K, the tops of flare loops turn to emission at Hα line center when the gas pressure P_g reaches 1 dyn cm^{-2} and should be clearly visible for $P_g \gtrsim 3$ dyn cm^{-2}, independently of the loop diameter. This situation corresponds to the electron density of the order 10^{12}cm^{-3}. The contrast of flare–loops (in projection on the disc) at Hα line center is mainly the function of P_g, while in the line wings (Hα $\pm$ 1 Å) the loop can be visible in absorption or emission only when rather strong microturbulence is present or for unrealistically high gas pressures. Finally, the authors briefly discuss their diagnostical results in frame of the latest (post)–flare loop model.

073.019 Characteristics of the expansion associated with eruptive prominences.
R. G. Athay, B. C. Low, B. Rompolt.
Sol. Phys., Vol. 110, No. 2, p. 359 – 379 (1987).

Gradients of Hα and electron scattering intensities derived from instantaneous radial distributions of erupting prominence material observed at several solar radii by Illing and Athay (1986) are often markedly smaller than those inferred by comparing the intensities observed near several radii to average prominence intensities observed near the limb. The authors show that gradients derived by following individual features in their outward progression with time yield values that are consistent with limb observations and that usually exceed the values obtained from instantaneous distributions. They conclude from the diversity of observed gradients that the prominence eruption cannot be described by a self–similar expansion in which the expansion velocity is a function of radius and time only. However, one cannot rule out possible self–similar solutions that allow the expansion velocity to be a function of angular direction.

073.020 Rapid motions of filaments in solar active regions.
M. M. Molodenskij, B. P. Filippov.
Astron. Zh., Tom 64, Vyp. 4, p. 825 – 834 (1987). In Russian. English translation in Sov. Astron., Vol. 31, No. 4.

Static magnetic field distribution above the photosphere surface is given, as a rule, by the solution of the potential theory problem called Neuman problem. The results are applied to the filament "activating" phenomenon of November 2, 1980. It is shown that filament motion can be described by the drift, and that there exists correspondence between the observed distribution of the flare brightness and the isolines of the Pointing vector z–component.

073.021 Concerning the relationship between the filament appearance and magnetic field changes in an active region.
V. P. Maksimov, L. V. Ermakova.
Astron. Zh., Tom 64, Vyp. 4, p. 841 – 849 (1987). In Russian. English translation in Sov. Astron., Vol. 31, No. 4.

A study has been made of the dynamical behaviour of the longitudinal magnetic field gradient distribution in the nearest vicinity of the polarity inversion line. It is shown that a necessary condition for a filament to appear within an active region is the displacement of grad $H_\parallel$ distribution toward lower gradient values as well as the homogeneity of the polarity inversion line with respect to the magnetic field. Also, maximum values of grad $H_\parallel$ must not exceed a certain limiting value.

073.022 Once more on the interpretation of the results of the July 31, 1981 solar eclipse.
G. P. Apushkinskij, O. B. Smirnova.
Astron. Zh., Tom 64, Vyp. 4, p. 875 – 876 (1987). In Russian. English translation in Sov. Astron., Vol. 31, No. 4.

The prominence brightness distribution at 1.35 cm wavelength obtained during the solar eclipse coincides better with Ca II line brightness distribution than with Hα one. The prominence was an active not a long–lived one.

073.023 Solar flare layering.
Eh. A. Baranovskij.
Bull. Crimean Astrophys. Obs., Vol. 75, p. 1 – 5 (1987). English translation of 42.073.048.

073.024 Simulating thermal and gasdynamic processes in solar–flare pulse phases.
A. G. Kosovichev.
Bull. Crimean Astrophys. Obs., Vol. 75, p. 6 – 18 (1987). English translation of 42.073.049.

073.025 **Temperature and density variations in flocculi in a developing active region.**
Eh. A. Baranovskij, N. B. Ograpishvili, N. N. Stepanyan.
Bull. Crimean Astrophys. Obs., Vol. 75, p. 43 – 47 (1987).
English translation of 42.073.050.

073.026 **Examining whisker polarization with a Hα filter.**
A. N. Babin, A. N. Koval'.
Bull. Crimean Astrophys. Obs., Vol. 75, p. 48 – 53 (1987).
English translation of 42.073.051.

073.027 **160 min solar flare activity modulation.**
V. A. Kotov, L. S. Levitskij.
Bull. Crimean Astrophys. Obs., Vol. 75, p. 54 – 60 (1987).
English translation of 42.073.052.

073.028 **Acceleration and energization by currents and electric fields.**
A. O. Benz.
Sol. Phys., Vol. 111, No. 1, p. 1 – 18 (1987). – See Abstr. 012.016.
The acceleration of the influential $\lesssim 100$ keV electrons in flares observed in hard X–rays and several radio emissions is unknown. Shock–waves and MHD turbulence, successfully applied to interprete interplanetary energetic particles, have recently been called in question concerning energetic flare electrons and ions. Other possible mechanisms are considered which are closely related to the primary flare energy release. In particular, runaway acceleration by the electric field of the reconnection current sheet, bulk heating by microturbulence, and cross–field ion currents due to bulk motion as a primary result of reconnection are reviewed. All three are likely to occur in some way. Their relative importance cannot be definitively assessed due to the lack of information on non–thermal, low energy protons.

073.029 **Evidence for interacting loop process in a phase of the May 16, 1981 flare.**
B. Vršnak, V. Ruždjak, M. Messerotti, P. Zlobec.
Sol. Phys., Vol. 111, No. 1, p. 23 – 29 (1987). – See Abstr. 012.016.
The behaviour of the flare in the period of enhancement and maximum of hard X–ray, microwave and decimetric type IV continuum is analysed. The elongation of the Hα ribbons and microwave source disclose that the energy release site was shifting through a system of loops with a velocity less than 200 km s^{-1}, and that the energy was carried down the field lines with a velocity of about 1000 km s^{-1}, implying the thermal conduction front mechanism of energy transport. Several processes of energy release are considered and it is concluded that an explanation in terms of succeeding interactions of neighbouring loops, involving fast reconnection of their poloidal components is in best agreement with the observations.

073.030 **Spotless flares and the associated radio continuum emission.**
V. Ruždjak, M. Messerotti, M. Nonino, A. Schroll, B. Vršnak, P. Zlobec.
Sol. Phys., Vol. 111, No. 1, p. 103 – 111 (1987). – See Abstr. 012.016.
The authors studied 24 spotless flares of Hα importance $\geqslant 1$ which occurred during the 21st cycle of solar activity. The spotless flares could be grouped in three categories according to their location and time history of the associated active region. The association of the flares with radio events was based on relative timing and on the flare importances. Weak microwave gradual rise and fall events were frequently recorded during the occurrence of the spotless flares. A few flares from this sample could be associated with impulsive and complex microwave bursts. Only in one case an association of a spotless flare with a significant metric type II/IV event seems to be justified.

073.031 **The 160 min–period with regard to the frequency of occurrence of chromospheric flares.**
V. A. Kotov, L. S. Levitskij.
Problems of solar flares, p. 112 – 118, 207 (1986). In Russian. Abstr. in Ref. Zh., 51. Astron., 7.51.304 (1987). – See Abstr. 012.018.

073.032 **Velocity field and flare activity in a sunspot group.**
S. I. Gopasyuk, L. F. Lazareva.
Problems of solar flares, p. 7 – 18, 198 (1986). In Russian. Abstr. in Ref. Zh., 51. Astron., 7.51.326 (1987). – See Abstr. 012.018.

073.033 **Observational peculiarities of flare behaviour in sunspot umbrae.**
V. P. Maksimov.
Problems of solar flares, p. 126 – 131, 208 (1986). In Russian. Abstr. in Ref. Zh., 51. Astron., 7.51.344 (1987). – See Abstr. 012.018.

073.034 **Investigation of the physical parameters of a solar flare in the optical range.**
V. A. Ostapenko.
Problems of solar flares, p. 107 – 111, 206 (1986). In Russian. Abstr. in Ref. Zh., 51. Astron., 7.51.349 (1987). – See Abstr. 012.018.

073.035 **Solar flares: distribution functions according to different components.**
E. I. Dajbog, V. G. Kurt, Yu. I. Logachev, V. G. Stolpovskij.
Problems of solar flares, p. 169 – 175, 212 (1986). In Russian. Abstr. in Ref. Zh., 51. Astron., 7.51.353 (1987). – See Abstr. 012.018.

073.036 **Magnetic fields and high–temperature plasma in the solar flare of November 5, 1980.**
O Gym Den, B. V. Somov.
Problems of solar flares, p. 19 – 31, 199 (1986). In Russian. Abstr. in Ref. Zh., 51. Astron., 7.51.355 (1987). – See Abstr. 012.018.

073.037 **Observational aspects in investigating the plasma turbulence in solar flares.**
N. M. Firstova.
Problems of solar flares, p. 65 – 82, 202 (1986). In Russian. Abstr. in Ref. Zh., 51. Astron., 7.51.356 (1987). – See Abstr. 012.018.

073.038 **Particle acceleration in pulsed solar flares.**
G. E. Kocharov, G. A. Koval'tsov, P. E. Semukhin, L. G. Kocharov.
Problems of solar flares, p. 176 – 184, 213 (1986). In Russian. Abstr. in Ref. Zh., 51. Astron., 7.51.363 (1987). – See Abstr. 012.018.

073.039 **On a topologic trigger of solar flares.**
V. S. Gorbachev, S. R. Kel'ner, B. V. Somov, A. S. Shvarts.
Problems of solar flares, p. 41 – 64, 201 (1986). In Russian. Abstr. in Ref. Zh., 51. Astron., 7.51.364 (1987). – See Abstr. 012.018.

073.040 **Model representations of the problem of solar flare heating.**
O. P. Shmeleva.
Problems of solar flares, p. 190 – 196, 215 (1986). In Russian. Abstr. in Ref. Zh., 51. Astron., 7.51.369 (1987). – See Abstr. 012.018.

073.041 Information of the variable "flare brightness" and prospects of its application as a criterion for forecasting the flare number in an active region.
N. Lopes Reyes, I. Ferro Ramos.
Soln. Dannye, Byull., 1987, No. 3, p. 71 – 79 (1987). In Russian.

Information of variables characterizing flares is studied using methods of pattern recognition. It is found that the variable of brightness gives more information than the variable of the area. Three rules are formulated for a prediction of the number of flares in an active region.

073.042 Observation of nonthermal energy distributions during the impulsive phase of solar flares.
J. F. Seely, U. Feldman, G. A. Doschek.
Astrophys. J., Vol. 319, No. 1, p. 541 – 554 (1987).

The Fe XXV resonance line and dielectronic satellite intensities have been measured as functions of time for several flares recorded by the Naval Research Laboratory crystal spectrometer flown on the P78–1 spacecraft. The intensity ratios of the Fe XXV resonance line, the Fe XXIV $n = 2$ satellite line j, and the Fe XXIV $n = 3$ satellite line d13 indicate that nonthermal electron energy distributions occur during the impulsive phase of the flares. For all of the flares that were studied, hard X–ray bursts occurred near the time of the nonthermal distributions.

073.043 A determination of the ^{3}He/H ratio in the solar photosphere from flare gamma–ray line observations.
X.–M. Hua, R. E. Lingenfelter.
Astrophys. J., Vol. 319, No. 1, p. 555 – 566 (1987).

The time dependence of the 2.223 MeV gamma–ray line emission from the capture of flare–produced neutrons in the solar photosphere provides a sensitive measure of the ^{3}He abundance there. The authors have made Monte Carlo calculations of this time dependence as a function of both ^{3}He abundance and observing angle for neutrons produced by thick–target interactions of flare–accelerated ions. It is found that the SMM measurements of the time dependence of the 2.223 MeV line emission from the solar flare of 1982 June 3 imply a solar photospheric ^{3}He/H ratio of $(2.3 \pm 1.2) \times 10^{-5}$ at the 90% confidence level.

073.044 Short solar events with evidence of repetitive structures.
U. D. Desai, C. Kouveliotou, C. Barat, K. Hurley, M. Niel, R. Talon, G. Vedrenne, I. V. Estulin (*I. V. Ehstulin*), V. C. Dolidze (*V. Sh. Dolidze*).
Astrophys. J., Vol. 319, No. 1, p. 567 – 573 (1987).

Several short (< 40 s) solar hard X–ray events (> 100 keV) have been observed simultaneously with identical instruments on the Venera 11, 12, 13, and 14 and Prognoz 7 (Franco–Soviet Signe experiments) spacecraft. High time resolution ($\geqslant 2$ ms) data were recorded. The authors present the observations of modulation with a period of 1.6 s for the event on 1978 December 3. They also present evidence for fast time fluctuations from an event on 1979 November 6, and another on 1981 September 6. The authors have used power spectrum analysis, epoch folding, and Monte Carlo simulation to evaluate the statistical significance of persistent time delays between features.

073.045 Estimation of most significant qualitative variables, associated with Z flares.
I. Ferro Ramos.
Soln. Dannye, Byull., 1987, No. 5, p. 88 – 91 (1987). In Russian.

Most significant qualitative variables, associated with Z flares are found using an heuristic correlation method for qualitative variables. All the indices found characterize the importance of flares.

073.046 UV emission in quiescent prominences. I. Neutral oxygen.
N. N. Morozhenko.
Kinematika Fiz. Nebesn. Tel, Tom 3, No. 5, p. 65 – 71 (1987). In Russian. English translation in Kinematics Phys. Celest. Bodies.

Ionization and excitation of neutral oxygen in structurally and physically inhomogeneous models of quiescent prominences are considered. The intensities of O I $\lambda\lambda$130.22, 130.49, 130.60 and 135.56 nm lines are calculated. A comparison with the observations obtained on Skylab permits concluding that the O I lines are excited by radiation in cold components of quiescent prominences.

073.047 Estimating the anomalously large fluences of solar flare protons with tritium data from 1900 to 1952.
H. Nutley, M. D. Voth.
J. Geophys. Res., Vol. 92, No. A10, p. 11,179 – 11,182 (1987).

An analysis is presented for estimating the "anomalously large" fluences from solar flare protons by measuring tritium deposited between 1900 and 1952 in a high–latitude ice core. Most of the tritium from cosmic rays and solar flare protons comes by way of indirect (p, n) and (n, t) reactions rather than from direct (p, t) reactions. Therefore the neutron spectrum produced by protons of the August 1972 flare is estimated first. Then the calculated neutron spectrum is compared with that known from galactic cosmic rays. Finally, the tritium produced by (n, t) reactions is compared for the solar protons and cosmic rays.

073.048 A search for a plasma turbulence in the flare of September 15, 1981, 00:09 UT.
N. M. Firstova, P. Kotrč.
Bull. Astron. Inst. Czech., Vol. 38, No. 5, p. 257 – 261 (1987).

The linear polarization in the Hα and Hβ lines in the flare of 15th Sept. 1981 is studied by a photographic method which enables instrumental polarization to be eliminated. Considerable values of the degree of polarization occur in the cores of the lines, at some points as much as 17 per cent. A blue shift corresponding to the velocities –2.5 km/s and –1.3 km/s has been found in the cores of the Hα and Hβ lines, respectively. The blue shift changes into red at about the full width of the half maximum (FWHM) of the lines.

073.049 Rapid motions of filaments in solar active regions. II
M. M. Molodenskij, B. P. Filippov.
Astron. Zh., Tom 64, Vyp. 5, p. 1079 – 1087 (1987). In Russian. English translation in Sov. Astron., Vol. 31, No. 5.

Many motions of filaments can be described as a drift of plasma in the crossed fields, but there are such motions as well which cannot be explained by the drift only. One must consider that motions having an acceleration higher than that of the gravity on the Sun are the result of the filament's own current. The conditions of the equilibrium and stability of a filament with strong current in an external field were obtained in the framework of the two–dimensional model. The motion of such a filament originating as a result of weak disturbances of the external field was considered. The acceleration of a filament is much higher than that of the gravity. For the vector potential, the general boundary problem determining the magnetic field and the current above the photosphere is formulated.

073.050 Ultraviolet observations of solar fine structure.
K. P. Dere, J.–D. F. Bartoe, G. E. Brueckner, J. W. Cook, D. G. Socker.
Science, Vol. 238, No. 4831, p. 1267 – 1269 (1987).

The High Resolution Telescope and Spectrograph was flown on the Spacelab–2 shuttle mission to perform extended observations of the solar chromosphere and transition zone at high spatial and temporal resolution. Ultraviolet spectroheliograms show the temporal development of macrospicules at the solar limb. The C IV transition zone emission is produced in discrete emission elements that must be composed of exceedingly fine (less than 70 kilometers) subresolution structures.

073.051 The solar chromosphere.
R. W. Noyes, E. H. Avrett.
Spectroscopy of astrophysical plasmas, p. 125 – 164 (1987). – See Abstr. 003.010.

Contents: 1. Introduction: the nature of the solar chromosphere. 2. The mean structure of the chromosphere. 3. The three–dimensional structure of the chromosphere: Plages and network – the role of magnetic fields. Small–scale chromospheric structure. Structure and dynamics of the high chromosphere. Magnetic fields in the chromosphere. 4. Conclusions.

073.052 **Simultaneous VLA observations of a flare at 6 and 20 centimeter wavelengths.**
T. Velusamy, M. R. Kundu, E. J. Schmahl, M. McCabe.
Astrophys. J., Vol. 319, No. 2, p. 984 – 992 (1987). With plate 18.
Using the VLA at 6 and 20 cm wavelengths, the authors have mapped a solar flare on 1980 May 15. The 1B flare, as observed in Hα at Mees Solar Observatory, Mauri, Hawaii, appeared as two sequential flares occurring at different neutral lines. The peaks of the time profile at 20 cm were delayed with respect to the 6 cm counterparts, but they were related to each other and to the Hα activity. The main 20 cm emission appeared to be displaced limbward from the 6 cm burst. If both the 6 and 20 cm emission originated in the same system of loops, the apparent lateral displacement was caused by a height difference of 33,000 km in the sources of emission.

073.053 **The stability of line tied force–free cylindrical arcades: is an active region filament a requirement for a two–ribbon flare?**
A. Hood, U. Anzer.
Sol. Phys., Vol. 111, No. 2, p. 333 – 346 (1987).
The MHD stability of force–free, cylindrical arcades is investigated, including the stabilising effect of photospheric line tying. It is found that a wide variety of fields are stable. This suggests that either a departure from a force free equilibrium or suppression of line tying is necessary if a two–ribbon flare is to be triggered. It is postulated that in both circumstances, the existence of an active region filament is an essential preflare requirement for a two–ribbon flare.

073.054 **Linear polarization of hydrogen Balmer lines in optically thick quiescent prominences. I. Theoretical investigation.**
E. Landi Degl'Innocenti, V. Bommier, S. Sahal–Bréchot.
Astron. Astrophys., Vol. 186, No. 1/2, p. 335 – 353 (1987).
A theoretical investigation is presented for deducing the expected linear polarization of H_α and H_β in quiescent prominences having a non–negligible optical thickness in H_α. The prominence is modelled as an infinite slab of finite thickness standing vertically over the solar surface, while the physical characteristics of the prominence plasma are deduced, in a first approximation, from the conventional prominence models of Heasley and Milkey (1976, 1978).

073.055 **Stability of current–sheet models of quiescent prominences and equatorial disks.**
F. Wu.
Astrophys. J., Vol. 320, No. 1, p. 418 – 425 (1987).
The stability of two–dimensional, infinitely thin solar prominence models is investigated in the MHD approximation. Anzer's stability criteria are used for the vertical plane prominences. An example of stable equilibrium meeting these criteria is given. The stability of curved prominence sheets is investigated. In spherical geometry, sufficient stability criteria are derived for plane disk–shaped current sheets and applied to a simple model for a nonrotating equatorial accretion disk.

073.056 **On the unresolved fine structures of the solar atmosphere. II. The temperature region $2 \times 10^5 - 5 \times 10^5$K.**
U. Feldman.
Astrophys. J., Vol. 320, No. 1, p. 426 – 429 (1987). With plates 9 – 10.
In a previous paper, experimental evidence was presented to argue that the solar plasma in the temperature region 4×10^4K $< T_e < 2 \times 10^5$K occurred primarily in unresolved fine structures that were isolated from the chromosphere and corona. Here, the author extends the argument to include in the same category also the plasma emitted in the temperature range of 2×10^5K $< T_e < 5 \times 10^5$K.

073.057 **Investigations of turbulent and direct motions in solar flares.**
R. D. Bentley, A. Fludra, J. R. Lemen, J. Jakimiec, J. Sylwester.
Bull. Am. Astron. Soc., Vol. 19, No. 2, p. 750 (1987). Abstract. – See Abstr. 010.061.

073.058 **Particle acceleration by MHD waves in solar flares.**
J.–F. de La Beaujardière, E. G. Zweibel.
Bull. Am. Astron. Soc., Vol. 19, No. 2, p. 750 (1987). Abstract. – See Abstr. 010.061.

073.059 **SKYLAB XUV observations of densities and thermal structure in a compact flare.**
K. G. Widing, J. W. Cook.
Bull. Am. Astron. Soc., Vol. 19, No. 2, p. 753 (1987). Abstract. – See Abstr. 010.061.

073.060 **The H, K and IR triplet of Ca II in plages.**
Eh. A. Baranovskij, A. Kuchera (*A. Kučera*), Z. A. Shcherbakova.
Izv. Krymskoj Astrofiz. Obs., Tom 77, p. 3 – 8 (1987). In Russian. English translation in Bull. Crimean Astrophys. Obs., Vol. 77.
The observed profiles of Ca II lines H, K, $\lambda 8498$, $\lambda 8542$, $\lambda 8662$ are obtained and the model chromospheres are derived for plages of different brightness. The calculations involve the program of non–LTE spectra analysis. The density in the bright plages is of an order of 2 higher than that of the undisturbed chromosphere. The models of weak plages do not differ significantly from the undisturbed model chromosphere.

073.061 **Investigation of linear polarization of moustaches.**
A. N. Babin, A. N. Koval'.
Izv. Krymskoj Astrofiz. Obs., Tom 77, p. 9 – 16 (1987). In Russian. English translation in Bull. Crimean Astrophys. Obs., Vol. 77.
An observational technique to estimate linear polarization of solar emission features in the H_α–line is described, permitting to obtain a set of four polarized spectrograms of the same region on the Sun at four positions of analyser differing by 45° during one exposure. The proposed method permits to determine all parameters of linear polarization of emission.

073.062 **On the fibrils system activation connected with flares.**
L. G. Kartashova.
Izv. Krymskoj Astrofiz. Obs., Tom 77, p. 32 – 38 (1987). In Russian. English translation in Bull. Crimean Astrophys. Obs., Vol. 77.
The variations in the structure of some fibrils systems were studied in connection with flares and brightenings. The variations with time of fibrils size and intensity are compared with earlier published data.

073.063 **On the quasi–homologous limb flares observed on 3 August 1981.**
L. Dezsö, A. Fludra, O. Gerlei, J. Jakimiec, Á. Kovács, T. Pettauer.
Heliophys. Obs. Hung. Acad. Sci., Prepr., No. 12, 8 pp. (1986). = Mitt. Sonnenobs. Kanzelhöhe, Nr. 73. Submitted to Adv. Space Res.

073.064 **The intensities and the widths of spectral lines in a loop prominence observed on November 3, 1973.**
J. Kubota, T. Kureizumi.
Contrib. Kwasan Hida Obs., Univ. Kyoto, No. 274, 16 pp. (1986).

073.065 **LTE modeling of inhomogeneous chromospheric structure using high–resolution limb observations.**
C. Lindsey.
Astrophys. J., Vol. 320, No. 2, p. 893 – 897 (1987).
The author discusses local thermodynamic equilibrium (LTE) modeling of rough atmospheres with the intent of applying recent

high–resolution observations of the solar limb in the far–infrared and radio continuum to the modeling of chromospheric spicules. This paper explains how the continuum limb observations can be combined with morphological knowledge of spicule structure to model the physical conditions in chromospheric spicules.

073.066   A solar chromosphere and spicule model based on far–infrared limb observations.
D. Braun, C. Lindsey.
Astrophys. J., Vol. 320, No. 2, p. 898 – 903 (1987).

The authors use techniques developed for LTE radiative transfer problems in a rough atmosphere to compute a model chromosphere containing spicules consistent with high–resolution solar limb observations from 100 μm to 2.6 mm. The model consists of a smooth, plane–parallel temperature minimum region extending from the photosphere to a height of 1000 km and randomly distributed cylindrical spicules above this height. The observed limb brightness profiles are well fitted by spicules with electron temperatures on the order of 7000K.

073.067   Nonlocal thermal transport in solar flares.
J. T. Karpen, C. R. DeVore.
Astrophys. J., Vol. 320, No. 2, p. 904 – 912 (1987).

The authors modeled a flaring solar atmosphere assuming classical thermal transport, locally limited thermal transport, and nonlocal thermal transport. The classical, local, and nonlocal expressions for the heat flux yield significantly different temperature, density, and velocity profiles throughout the rise phase of the flare. The classical and locally limited results are generally comparable, while the nonlocal results diverge markedly from the other two cases. Much higher coronal temperatures are achieved in the nonlocal calculations owing to the combined effects of delocalization and flux limiting. It is concluded that nonlocal thermal transport strongly affects the evolution of a flare event.

073.068   Skylab XUV observations of densities, thermal structure, and mass motions in a compact flare.
K. G. Widing, J. W. Cook.
Astrophys. J., Vol. 320, No. 2, p. 913 – 925 (1987). With plates 27 – 28.

The energetic limb flare of 1973 December 17 was one of the best Skylab examples of a class characterized by small volume, high energy density, and short lifetime. In the present paper the authors have studied the extensive series of slit spectra of this flare obtained with the NRL slit spectrograph on Skylab operating in the 1000 – 1945 Å range. An overview of the flare development and exposure chronology is given. Examples of the flare morphology observed in He II, Ne IV, and Mg VII are contrasted with the morphology observed in Fe XVII. Slit profiles of Fe XXI 1354 Å, O V 1371 Å, and Si IV 1403 Å are presented. The electron density is derived at selected phases from diagnostic line ratios of O IV, O V, S IV, and S V observed in the slit spectra. In the decay phase these densities are combined with the densities derived from high–temperature lines to obtain the density distribution in the flare between 10^5K and 6×10^6K. The arch structure within the $5'' \times 5''$ kernel is resolved and an improved picture of the density spatial structure of the flare kernel is derived.

073.069   Analysis of a large flare on the solar disc on 1984 February 25.
H.–l. Ye, Q.–y. Li, J.–y. Xuan.
Publ. Yunnan Obs., No. 2, p. 41 – 46 (1987). In Chinese.

073.070   White light flare and its associated Moreton wave of 25 April 1984.
M. Miyazawa, K. Yamaguchi.
Tokyo Astron. Obs. Rep., Vol. 21, No. 1, p. 101 – 119 (1987). In Japanese.

073.071   Catalogue of LDE (*Long Duration Event*) flares (January 1969 – March 1986).
A. Antalová.
Contrib. Astron. Obs. Skalnaté Pleso, Vol. 16, p. 79 – 190 (1987).

The catalogue of LDE flares contains data on 646 flares observed in the 20th and on 1029 flares observed in the 21st cycle of solar activity.

073.072   The two–ribbon flare of May 14, 1981.
B. Vršnak, V. Ruždjak, P. Zlobec.
Hvar Obs. Bull., Vol. 10, No. 1, p. 17 – 21 (1986).

The flare behaviour at radio waves, X–rays and in H–alpha is described. The flare geometry is discussed and compared with the model of Heyvaerts et al. (1977).

073.073   Calculating the weakening of intensity of direct solar radiation in a prominence.
V. S. Galishev.
Ural. Univ. Sverdlovsk, 40 pp. (1987). In Russian. Abstr. in Ref. Zh., 51. Astron. 11.51.392 (1987).

073.074   On the possibility of electron detection from the decay of solar neutrons.
E. I. Dajbog, V. G. Stolpovskij.
Pis'ma Astron. Zh., Tom 13, No. 12, p. 1086 – 1093 (1987). In Russian. English translation in Sov. Astron. Lett., Vol. 13.

The authors consider the possibility to detect energetic electrons from the decay of neutrons produced in solar flares. The proposed method allows to estimate the order of magnitude of the total number of neutrons with energies $\geqslant 1$ MeV for flares like the June 3, 1982 and April 24, 1984 flares. The dependence of the integral electron flux per total number of neutrons for various heliolongitudes is estimated.

073.075   Radiation energy estimation and analysis of optical solar flares observed in 1978.
L. N. Kurochka, V. G. Lozitskij, P. Paluš.
Kinematika Fiz. Nebesn. Tel, Tom 3, No. 6, p. 46 – 49, 67 (1987). In Russian. English translation in Kinematics Phys. Celest. Bodies.

Radiation energies of 5508 flares observed in 1978 are estimated. The radiation in all lines and in continua of hydrogen series (beginning with Lyman) with typical parameters of the flares of different intensities is taken into consideration. Mean energy values are obtained for flares of every importance which increase with importance, whereas the number of the flares observed decreases. With the flare energy decrease their number increases at first up to 5×10^{20}J and then fastly decreases. This suggests that the flares on the Sun are independent events among other non–stationary processes.

073.076   Calcium ionization balance and argon/calcium abundance in solar flares.
E. Antonucci, D. Marocchi, A. H. Gabriel, G. A. Doschek.
Astron. Astrophys., Vol. 188, No. 1, p. 159 – 162 (1987).

An earlier analysis of solar flare calcium spectra from XRP and P78–1 aimed at measuring the calcium ionization balance resulted in an ambiguity due to a line blend between the calcium q line and an Ar XVII line. In the present work the calcium line "r" is included in the analysis in order to resolve this problem.

073.077   Impulsive brightenings/velocity transients in quiescent solar prominences. II. Time–varying reversal regions.
G. D. Toot, J. M. Malville.
Sol. Phys., Vol. 112, No. 1, p. 67 – 81 (1987).

In some quiescent prominences, areas are found where the Hα emission profiles are centrally reversed. By combining good spatial, spectral, and temporal resolution, the detailed behavior of these reversal regions has been investigated. The detailed behavior of the Hα line profiles is consistent with these reversal features being true self–reversal of the line, indicating unusually high column masses in these areas. Some models of condensation of coronal material to the prominence state predict temporary regions of high density, perhaps high enough to produce the observed reversal. This implies that reversal features are the result of on–going condensation of coronal material into already formed prominences, a result which impacts models of prominence formation and stability.

073.078 The old calcium plages differential rotation latitudinal profile: its gross and fine structure evolution with the solar cycle.
M. Ternullo.
Sol. Phys., Vol. 112, No. 1, p. 143 – 151 (1987).

This work is a study of the rotational properties of the solar calcium plages, during the time interval 1967 – 1977; only plages older than 4 days have been the object of this research. The author has looked systematically for any significant change occurring during the course of the solar cycle, and any kind of 'anomaly' or fine structure in the differential rotation latitudinal profile. Such a profile undergoes a cyclic transformation.

073.079 The impact of aging on the rotation rate of calcium plages in the years 1967 – 1977.
M. Ternullo.
Sol. Phys., Vol. 112, No. 1, p. 153 – 163 (1987).

An analysis of Ca II spectroheliograms obtained at Catania Astrophysical Observatory throughout the years 1967 – 1977 has been carried out, to throw light on the complex relationship linking the angular rotation rate of Ca plages with their age, as well as with solar cycle phase, and with latitude.

073.080 Nuclear processes and accelerated particles in solar flares.
R. Ramaty, R. J. Murphy.
Space Sci. Rev., Vol. 45, Nos. 3+4, p. 213 – 268 (1987).

Nuclear processes and particle acceleration in solar flares are reviewed. The theory of gamma–ray and neutron production is discussed and results of calculations are compared to gamma–ray, neutron, and charged–particle observations from solar flares. The implications of these comparisons on particle energy spectra, total numbers, anisotropies, electron–to–proton ratios, as well as on acceleration mechanisms and the interaction site, are presented. The information on elemental and isotopic abundances derived from gamma–ray observations is compared to abundances obtained from escaping accelerated particles and other sources.

073.081 The Hα spectral counterparts of solar hard X–ray microflares.
R. C. Canfield, T. R. Metcalf.
Astrophys. J., Vol. 321, No. 1, p. 586 – 592 (1987).

X–ray observations have revealed energetically significant numbers of very small hard X–ray bursts, termed microflares by Lin et al. The authors have simultaneously observed the Hα counterparts of serveral of these microflares. It is found that microflares occur in regions that are also productive of larger flares, suggesting that they may be components of the larger flare event. All but the weakest miroflares show pronounced impulsive–phase red asymmetry in Hα. Their energetics, interpreted using the thick–target nonthermal model of electron transport, implies that these events are substantially underresolved at the authors' spatial resolution and have a true area of $10^{15} - 10^{16} cm^{-2}$.

073.082 Simultaneous 2 and 6 centimeter wavelength observations of a solar flare using the VLA.
M. R. Kundu, T. Velusamy, S. M. White.
Astrophys. J., Vol. 321, No. 1, p. 593 – 605 (1987). With plates 5 – 11.

VLA observations of a solar active region and a flare are discussed. The event was observed at wavelengths of 2 and 6 cm simultaneously. Radio maps prior to the flare delineate the most important magnetic structures in the region. Interaction between these structures apparently led to preheating of plasma above the active region some 30 min prior to the flare. The 2 and 6 cm flare positions were coincident, and the time profiles of the burst at the two wavelengths were almost identical. Emission was probably nonthermal gyrosynchrotron radiation, and the physical conditions in the burst source are derived using this assumption.

073.083 Closed magnetic structures in the chromosphere and in the transition region.
J. M. Malherbe, B. Schmieder, G. Simon, P. Mein, E. Tandberg–Hanssen.
Sol. Phys., Vol. 112, No. 2, p. 233 – 253 (1987).

Using simultaneous observations of the same solar regions in the lines Hα and C IV 1548 Å, the authors have derived schematic models of closed magnetic lines from dynamical constraints. They conclude that the magnetic loops are closed at higher levels above facular than above non–facular regions. This result remains valid whatever the assumed density models are, even by taking into account the 3 min oscillations. The center–to–limb behaviour is well predicted by taking into account the relative opacity in the chromosphere and transition region.

073.084 North–south asymmetry in sudden disappearances of solar prominences.
G. Vizoso, J. L. Ballester.
Sol. Phys., Vol. 112, No. 2, p. 317 – 323 (1987).

The N–S asymmetry in sudden disappearances (SD) of solar prominences during solar cycles 18 – 21 is studied. The N–S SD asymmetry curve is not in phase with the solar cycle and peaks about the time of solar minimum, the asymmetry reverses in sign during the solar maximum, being, this change of sign, coincident with the reversal of the Sun's magnetic dipole. The SD asymmetry curve can be fitted by a sinusoidal function with a period of eleven years. On the other hand, the SD asymmetry curve shows a strong coincidence with the N–S asymmetries presented by other solar activity manifestations as studied by different authors.

073.085 Energetic flare zones on the Sun.
V. K. Verma, M. C. Pande, W. Uddin.
Sol. Phys., Vol. 112, No. 2, p. 341 – 346 (1987).

The authors have studied the latitudinal, longitudinal (northern and southern hemispheric) distributions based on 1737 major flares observed during solar cycles 19 and 20. At least 5 flare zones are present in each hemisphere. In fact these zones seem to produce more than 50% of the total number of energetic fares investigated but occupy only <4% area of the Sun.

073.086 Observational maps of the moments of strong line profiles on the solar disk.
B. Caccin, A. Donati–Falchi, R. Falciani, L. A. Smaldone, G. P. Tozzi.
Sol. Phys., Vol. 112, No. 2, p. 383 – 386 (1987).

Using the method of solar bidimensional spectroscopy based on the Universal Birefringent Filter (UBF), the authors have determined the bidimensional maps of moments of some chromospheric lines. The observational material, referring to a quiet region on the disk center, have been acquired with the UBF of the NSO at Sacramento Peak on Aug. 27, 1985. The authors present the work in progress and the new observational aspects arising from the diagnostic method.

073.087 An empirical basis for the 'alternative interpretation' of Sawyer et al.
P. A. Stahl.
Sol. Phys., Vol. 112, No. 2, p. 387 – 390 (1987).

An analysis of the data of Achong and Stahl (1984) discloses discrete intervals which differ significantly in SID flare incidence and the proportion of complex sunspot groups present. This supports the alternative interpretation of Sawyer et al. (1985) in respect of the strong dependence of SID flare production on complex magnetic class (see Abstr. 40.073.048).

073.088   **Chromospheric and coronal heating due to the radiation and collisional damping of fast magnetosonic surface waves.**
W. Sahyouni, Z. Kiss'ovski, I. Zhelyazkov.
Z. Naturforsch., A, Vol. 42a, No. 12, p. 1443 – 1450 (1987).

The authors study the propagation of surface/pseudosurface modes in the structured solar atmosphere assuming that the surface waves may be able to heat the chromosphere and corona. The wave energy can be dissipated by radiation, ion viscosity, and electron heat conduction. For the solar corona, it is found that the pseudosurface waves, trapped in the coronal loops, dissipate efficiently only if their periods are longer than 200 seconds and only if the background magnetic field is smaller than 5 gauss.

073.089   **The Hα characteristics of hard X–ray bursts.**
Z.-x. Shi, J.-x. Wang.
Chin. J. Space Sci., Vol. 6, No. 4, p. 243 – 251 (1986). In Chinese. Abstr. in Phys. Abstr., Vol. 90, No. 1313, Entry 115342 (1987).

073.090   **Current loop coalescence model of solar flares.**
T. Tajima, J. Sakai, H. Nakajima, T. Kosugi, F. Brunel, M. R. Kundu.
Astrophys. J., Vol. 321, No. 2, p. 1031 – 1048 (1987).

The nonlinear coalescence instability of current–carrying loops provides keys to understanding many of the characteristics of solar flares. In this paper the authors investigate the physical consequences of current coalescence by using computer simulations. The physical signatures of coalescence include time scales, the amount of energy release, the temporal profile of magnetic and kinetic energies of the loop, simultaneous acceleration of particles in various energy spectra, radiation domains, radiation spectra, and various other associated effects. The physical consequences of coalescence are checked against and compared with actual flare observations.

073.091   **Study of a two–ribbon flare by the equidensitometric method.**
M. S. Toropova, V. S. Prokudina, G. F. Sitnik.
Astron. Tsirk., No. 1479, p. 3 – 5 (1987). In Russian.

The fine structure of emission knots are revealed having the form of a chain stretched along the axis of ribbons and which, perhaps, is connected with the existence of high–temperature X–ray arches.

073.092   **Rotation of individual background magnetic field components during the formation of the white–light flare region of April 1984 (NOAA 4474).**
V. Bumba, L. Gesztelyi.
Bull. Astron. Inst. Czech., Vol. 38, No. 6, p. 351 – 355 (1987). With plates 1 – 4.

In the present continuation of their study of processes related to the development of the white–light flare region of April 1984, the authors pay greater attention to the problem of rotation rates of certain components of the background magnetic field, constituting the main patterns of the weak as well as strong fields, from three points of view: as they are demonstrated by the distribution of chromospheric filaments, from point of view of the existence of so–called "pivot points" (Mouradian et al., 1987) and of the rotation of the strongest magnetic flux sources.

073.093   **Two–dimensional distributions of physical parameters in the prominence of 1984 March 23.**
Q.-z. Zhang, C. Fang.
Chin. Astron. Astrophys., Vol. 11, No. 3, p. 215 – 220 (1987). English translation of 43.073.221.

073.094   **The motion of matter in flare loops of the solar disc.**
A.-a. Xu.
Chin. Phys., Vol. 7, No. 2, p. 446 (1987). Abstr. in Phys. Abstr., Vol. 91, No. 1320, Entry 10515 (1988).

073.095   **An investigation on the symmetry of the line profile of a solar prominence. I. The effects of the line of sight velocity on the line profile.**
J. Chen, Y. Lin.
Acta Astrophys. Sin., Vol. 7, No. 3, p. 207 – 219 (1987). In Chinese.

This paper is a theoretical analytic investigation of the line profile of a prominence affected by the line of sight velocity. The analysis shows clearly that the line profile of a prominence can be represented by two terms. One of the terms depends only on the model of the velocity distribution, and is independent of the source function. The other term evaluates the effect of the distribution of the source function upon the line profile. Assuming that the source function is constant, the authors discuss whether the line profile is symmetrical or not using some models of the velocity distribution.

073.096   **An investigation on the symmetry of the line profile of a solar prominence. II. The effects of the source function on the line profile.**
J. Chen, Y. Lin.
Acta Astrophys. Sin., Vol. 7, No. 4, p. 287 – 297 (1987). In Chinese.

As a continuation of paper I, this paper discusses the line profile of a solar prominence effected by variation of the source function and velocity with depth.

073.097   **The loop prominence and its velocity field of Feb. 18, 1984.**
X. Gu, Q. Li.
Acta Astrophys. Sin., Vol. 7, No. 4, p. 298 – 304 (1987). In Chinese. With plates 1 – 2.

Photographic observations of the "post"–flare loop prominence of 1984 Feb. 18 have been made using an Hα–spectrum–spectroheliograph and a chromospheric telescope in Hα and its vicinty. These are presented along with an analysis of the prominence morphology and velocity fields.

073.098   **Energy distribution of 15,000 solar flares.**
L. N. Kurochka.
Sov. Astron., Vol. 31, No. 2, p. 231 – 233 (1987). English translation of 43.073.126.

073.099   **Impulsive Hα diagnostics of electron–beam–heated solar flare model chromospheres.**
R. C. Canfield, K. G. Gayley.
Astrophys. J., Vol. 322, No. 2, p. 999 – 1009 (1987).

The authors have computed time–dependent Hα profiles for the dynamic model atmospheres of Fisher, Canfield, and McClymont simulating the effects of an intense impulsively initiated power–law beam of electrons incident on the chromosphere. The temporal response of Hα arises from three separate physical mechanisms. The fastest variations (typically < 0.1 s) arise from energy imbalance. Slower variations (~ 0.5 s) arise from ionization imbalance. The slowest variations ($\geqslant 1$ s) arise from hydrodynamic effects and are related to the formation of a chromospheric condensation.

073.100   **X–ray observations of solar flares with the Einstein Observatory.**
J. H. M. M. Schmitt, F. R. Harnden Jr., H. Fink.
Astrophys. J., Vol. 322, No. 2, p. 1023 – 1034 (1987).

The authors present the first Einstein Observatory Imaging Proportional Counter observations of solar flares. These flares were detected in scattered X–ray light when the X–ray telescope was pointed at the sunlit Earth. The authors discuss the propagation and scattering of solar X–rays in the Earth's atmosphere in order to deduce the solar X–ray flux incident on top of the atmosphere from their scattered X–ray intensity measurements. After this correction the scattered X–ray data are interpreted as full–disk observations of the Sun obtained with the same instrumentation used for observations of flares on other stars. The results are compared with "known" properties of compact loop flares.

073.101 Microflares in the solar magnetic network.
J. G. Porter, R. L. Moore, E. J. Reichmann,
O. Engvold, K. L. Harvey.
Astrophys. J., Vol. 323, No. 1, p. 380 – 390 (1987).

It is suggested that the events observed by HRTS are microflares in tiny magnetic bipoles (some in cell interiors but most in the magnetic network) and that these same events, when strong enough and frequent enough in some of the larger bipoles, sustain X–ray bright points. In this paper, the authors present new evidence in favor of this hypothesis. Using C IV spectroheliograms in combination with magnetograms and He I λ10,830 spectroheliograms they find that impulsive heating events of the class observed by HRTS are common at small bipoles in the network, both at bipoles corresponding to X–ray bright points and at many weaker bipoles that show no sustained enhanced coronal brightness.

073.102 On the relationship of flare size and particle anisotropy in solar gamma–ray flares.
C. D. Dermer.
Astrophys. J., Vol. 323, No. 2, p. 795 – 798 (1987).

The author considers the effects of radiation directivity on the interpretation of SMM γ–ray flare observations. If the electron anisotropy is independent of flare size, then the size distribution of 300 keV – 1 MeV flares is much flatter than that of X–ray flares. This implies that larger flares have, on average, harder electron spectra. Distributions of electrons strongly peaked downward into the solar photosphere are not consistent with the data.

073.103 Transient ionization and solar flare X–ray spectra.
G. A. Doschek, K. Tanaka.
Astrophys. J., Vol. 323, No. 2, p. 799 – 809 (1987).

In this paper the effects of a transiently ionizing solar flare plasma on the X–ray spectrum of iron between 1.85 and 1.92 Å are considered. The atomic physics of the nonequilibrium spectrum is discussed, and reasons for differences in appearance from ionization equilibrium spectra are explained. The efect of spectral resolution on the ability to detect transient ionization in the iron X–ray spectrum is illustrated by synthetic spectra. A synthetic transiently ionizing spectrum is applied to the interpretation of spectra obtained from the SOX 1 spectrometer on the Japanese Hinotori spacecraft.

The High Altitude Observatory–Lowell Observatory Solar–Stellar Spectrophotometry Project.
See Abstr. 013.003.

Identification of intercombination transitions in Fe XIV and Fe XIII in the spectra of foil–excited ions and solar flares.
See Abstr. 022.056.

Radiative reconnection – a fast start for a flare?
See Abstr. 062.074.

Numerical simulation of a magnetic reconnection process.
See Abstr. 062.090.

Marfes: radiative condensation in tokamak edge plasma.
See Abstr. 062.150.

On the central part of the Hα–line profile in the gas–dynamic model of a flare.
See Abstr. 063.014.

Proton–electron bremsstrahlung polarization: solar flare X–ray and gamma–ray polarization.
See Abstr. 063.122.

A photometric search for solar giant convection cells.
See Abstr. 071.002.

Variations of the fine structure of Fraunhofer lines during flares.
See Abstr. 071.007.

Temporal variations of the He I 10830.3 Å line on the Sun.
See Abstr. 071.009.

The interpretation of oscillations in sunspot umbrae.
See Abstr. 072.007.

Sources of sunspot activity and solar flares.
See Abstr. 072.016.

Stokes profile analysis and vector magnetic fields. III. Extended temperature minima of sunspot umbrae as inferred from Stokes profiles of Mg I λ4571.
See Abstr. 072.030.

Periodicities of the flare occurrence rate in solar cycle 19.
See Abstr. 072.031.

Magnetic fields of flare–active sunspot groups according to observations of the phenomenon of sign change in circular polarization.
See Abstr. 072.033.

The Sun's spots and flares.
See Abstr. 072.043.

Generation and structure of the electric currents in a flaring activity complex.
See Abstr. 072.053.

Model of a sunspot and of a flare above a sunspot.
See Abstr. 072.056.

Fuzzy sets. Classification of solar active regions and flare prediction.
See Abstr. 072.081.

Why coronal loops cannot exist without solar gravity.
See Abstr. 074.001.

Theoretical study of onset conditions for solar eruptive processes.
See Abstr. 074.046.

Three–dimensional structure and velocities of the motion of matter in dark loops of H$_\alpha$ near–limb flares.
See Abstr. 074.063.

Explosive chromospheric instability in hydrodynamic loop flare models: the problem and its cure.
See Abstr. 074.074.

Nonlocal thermal conduction in hydrodynamic loop flare models.
See Abstr. 074.075.

Energy buildup in coronal magnetic flux tubes.
See Abstr. 074.079.

Simulations of the Ca XIX spectral emission from a flaring solar coronal loop. I. Thermal case.
See Abstr. 074.081.

The ejection of helical field structures through the outer corona.
See Abstr. 074.086.

Local magnetic structures on the sun and energetics of solar flares.
See Abstr. 075.011.

Stimulated dissipation of magnetic discontinuities and the origin of solar flares.
See Abstr. 075.020.

Hard X–ray emission processes in solar flares.
See Abstr. 076.003.

Directivity and polarization of hard X–ray emission of solar flares.
See Abstr. 076.005.

Studies on X–ray and gamma–ray bursts associated with the solar limb event of 1981 April 27.
See Abstr. 076.007.

Calculations of X–ray anisotropy for energies > 150 keV and its comparison with PVO/ISEE–3 observations.
See Abstr. 076.008.

X–ray emission of solar flares generated by anisotropic electron beams.
See Abstr. 076.010.

The directivity of high–energy emission from solar flares: Solar Maximum Mission observations.
See Abstr. 076.011.

A high–frequency "zebra"–pattern in radio emission of solar flares.
See Abstr. 077.008.

Microwave sources of solar flares: loop top or foot points?
See Abstr. 077.021.

Electron cyclotron maser emission from solar flares.
See Abstr. 077.024.

Evidence of harmonic microwave radiation during solar flares.
See Abstr. 077.027.

Imaging observations of the evolution of meter–decameter burst emission during a major flare.
See Abstr. 077.037.

Changes of polarization in the dm–m range during the flare of May 16, 1981.
See Abstr. 077.046.

On preflare changes of solar decimeter radio emission fluctuations. 1. Short–period fluctuations.
See Abstr. 077.054.

Solar neutron emissivity during the large flare on 1982 June 3.
See Abstr. 078.003.

Determination of energy spectra of solar electrons under different scenarios in solar flare sources.
See Abstr. 078.009.

A multispacecraft study of the injection and transport of solar energetic particles.
See Abstr. 078.015.

Solar flare neutron and accelerated ion angular distributions.
See Abstr. 078.016.

Nuclear processes in the solar atmosphere and the problem of particle acceleration.
See Abstr. 080.019.

Fast transient X–rays from flare stars and RS CVn binaries.
See Abstr. 112.160.

Energetics of activity of flare stars and the Sun: a synergetical approach.
See Abstr. 122.043.

074 Corona, Solar Wind

074.001   Why coronal loops cannot exist without solar gravity.
I. J. D. Craig, A. N. McClymont.
Astrophys. J., Vol. 318, No. 1, p. 421 – 427 (1987).
The authors develop a simple stability analysis of coronal loops which demonstrates that a loop is unstable to coronal temperature fluctuations unless a sufficiently large pressure perturbation accompanies the disturbance. Such pressure fluctuations are possible only if the chromospheric footpoints of the loop exhibit the property of dynamical stiffness, preventing the corona from expanding under the influence of a temperature increase. The stiffness of the chromosphere is shown to derive from its gravitational stratification and its "spring constant" to depend on the gravitational scale height. Any loop whose coronal length is large compared to the chromospheric scale height must therefore "see" the chromospheric footprints as essentially rigid.

074.002   Rotation of the coronal magnetic field.
J. T. Hoeksema, P. H. Scherrer.
Astrophys. J., Vol. 318, No. 1, p. 428 – 436 (1987).
The coronal magnetic field rotates differently than the photosphere. The field configuration of the corona can be modeled using the observed photospheric field and a potential field model. Correlation of the field patterns at different latitudes with a lag near one solar rotation shows much less differential rotation than observed in the photospheric field. Consideration of longer lags indicates different rotation rates in the northern and southern hemispheres. During solar cycle 21, the southern hemisphere field rotates with a 28.1 day period, while the northern hemisphere rotates faster, having a period of 26.9 days.

074.003   The energy source of the interplanetary medium and the heliosphere.
E. N. Parker.
NASA Conf. Publ., NASA CP–2464, p. 75 – 85 (1987). – See Abstr. 012.004.
The activity of the interplanetary medium arises from occasional transient outbursts of the active corona and, for the most part, from the interaction of fast and slow streams in the solar wind. The basic driver is the heat input to the corona, both transient and steady. The fast streams issue from coronal holes where the heat input may be Alfvén waves with rms fluid velocities of nearly 10^2 km/sec or may be wholly or in part the waves refracted into the hole from neighboring active regions. If the latter, then the character of the wind from the coronal hole depends upon the proximity and vigor of active regions, with significant differences between the polar and low latitude solar wind.

074.004   Particle propagation channels in the solar wind.
K. A. Anderson, W. M. Dougherty.
NASA Conf. Publ., NASA CP–2464, p. 87 – 98 (1987). – See Abstr. 012.004.
The intensities of low energy solar–interplanetary electrons and ions at 1 AU occasionally change in a "square wave" manner. The changes may be increases or decreases and they typically have durations of from one hour to a few hours. In some cases these channels are bounded by discontinuities in the interplanetary field and the plasma properties differ from the surrounding solar wind. In one case solar flare particles were confined to a channel of width 3×10^6 km at Earth. At the Sun

this dimension extrapolates to about 12000 km, a size comparable to small flares.

074.005 The heating and cooling in certain active region loops.
S. Chandra.
Astrophys. Space Sci., Vol. 135, No. 1, p. 195 – 196 (1987). – See Abstr. 40.074.083.

Narain and Kumar (1985) have mixed up various situations which are quite different from each other. In view of that, the conclusions drawn by them may not be reliable.

074.006 Latitudinal dependence of solar wind speed.
C. D. Fry, S.-I. Akasofu.
Planet. Space Sci., Vol. 35, No. 7, p. 913 – 920 (1987).

The authors study the solar–cycle evolution of solar wind bulk speed as a function of source magnetic field strength. Effects of solar transient events are removed. The background solar wind speed follows a function of the form $v_{sw} = v_0 + v_1(1-\cos^n\lambda_m)$, where λ_m is the magnetic latitude, set equal to the arc sine of the normalized field strength. The authors find the value of n is largest at solar minimum, and smallest at solar maximum, while v_0 and v_1 remain fairly constant throughout the solar cycle. This implies the latitudinal gradient in background solar wind speed is steepest at solar minimum and most broad at solar maximum. The lowest and highest background speeds remain fairly constant throughout the solar cycle. The new function is inserted into the improved kinematic code of Hakamada and Akasofu (1982), and solar wind speed and IMF are simulated for two periods in the solar cycle. The authors are able to reproduce observed parameters for specific coronal hole passages.

074.007 A search for forerunner activity associated with coronal mass ejections.
J. T. Karpen, R. A. Howard.
J. Geophys. Res., Vol. 92, No. A7, p. 7227 – 7234 (1987).

The authors have performed a systematic search for forerunners using the white–light coronagraph observations obtained with the Solwind instrument on board the P78–1 satellite. Forty-four bright, well–observed events were selected and analyzed, employing selection criteria and analysis methods similar to those used by Jackson and Hildner (1978). Approximately half of these events either do not exhibit low–density plateaus in front or are questionable (e.g., a frontal plateau only appears intermittently). Based on their analysis of the remaining coronal mass ejections (CMEs), the authors conclude that identification of the forerunner as a distinct entity probably is not warranted, and that the low–density plasma is an integral part of the CME itself.

074.008 A comparison of coronal rotation and interplanetary recurrence during 1964 – 1976.
G. D. Parker.
J. Geophys. Res., Vol. 92, No. A7, p. 7235 – 7240 (1987).

This paper is an observational investigation of the relationships between interplanetary recurrence and the differential rotation of the white light corona during 1964 – 1976.

074.009 Meridional transport of magnetic flux in the solar wind between 1 and 10 AU: a theoretical analysis.
V. J. Pizzo, B. E. Goldstein.
J. Geophys. Res., Vol. 92, No. A7, p. 7241 – 7253 (1987).

Pioneer 10 observations suggest that the mean (longitudinally averaged) solar wind azimuthal field strength, B_ϕ, near the ecliptic plane falls off more rapidly with heliocentric distance than would be expected in a classic Parker expansion, showing a deficit of 10 – 20% (as compared to the projected 1–AU value) by 10 AU. Though this observational interpretation has been challenged by subsequent analyses of Voyager data, it has nevertheless stimulated efforts to explain the inferred deficit on the basis of systematic north–south magnetic pressure gradients generated by the differential spiral wrapping of magnetic field lines in interplanetary space. The authors reexamine this issue from the theoretical perspective using a three–dimensional MHD nonlinear numerical model for steady, corotating flow.

074.010 Generation of solar wind proton tails and double beams by Coulomb collisions.
S. Livi, E. Marsch.
J. Geophys. Res., Vol. 92, No. A7, p. 7255 – 7261 (1987).

A kinetic model is presented for the generation of proton heat flux tails and double beams by Coulomb collisions in the solar wind. The combined action of the large–scale interplanetary magnetic field (mirror effect) and of the collisional scattering (runaway) is shown to be able to produce variously skewed velocity distributions. All intermediate cases between the exospheric and isotropic adiabatic expansion can be simulated as a function of the collisionality of the plasma. Good qualitative agreement is achieved between in situ measurements and model results in the collisional domains of the solar wind.

074.011 Observational evidence for marginal stability of solar wind ion beams.
E. Marsch, S. Livi.
J. Geophys. Res., Vol. 92, No. A7, p. 7263 – 7268 (1987).

Solar wind proton double streams and proton–alpha differential streaming may provide sufficient free energy to drive field–aligned magnetosonic waves unstable. The authors carry out a statistical analysis of this electromagnetic beam instability based on measured Helios ion distributions. Observational evidence is indeed found for unstable distributions, whereby the largest growth rates are obtained in the collisionless high–speed wind. Alpha particles by themselves are not able to excite these waves but mostly tend to stabilize an existing proton double stream configuration by enhancing the cyclotron damping of the main proton population.

074.012 Solar cycle evolution of solar wind speed structure between 1973 and 1985 observed with the interplanetary scintillation method.
M. Kojima, T. Kakinuma.
J. Geophys. Res., Vol. 92, No. A7, p. 7269 – 7279, 7763 – 7765 (1987).

The solar cycle evolution of solar wind speed structure was studied for the years from 1973 to 1985 on the basis of interplanetary scintillation observations using a new method for mapping solar wind speed to the source surface.

074.013 On the He²⁺ to H⁺ temperature ratio in slow solar wind.
R. Hernández, S. Livi, E. Marsch.
J. Geophys. Res., Vol. 92, No. A7, p. 7723 – 7727 (1987).

On the basis of a three–fluid model with energy exchange by Coulomb collisions and heat flux degradation the ratio T of alpha to proton temperatures is investigated. For various relative ion densities and plasma parameters the collisional evolution of T is modeled and compared with Helios in situ measurements. Good agreement between data and theory is found.

074.014 Correction to "Evolution of interstellar pickup ions in the solar wind" [J. Geophys. Res., Vol. 92, No. A2, p. 1067 – 1073 (1987)].
P. A. Isenberg.
J. Geophys. Res., Vol. 92, No. A7, p. 7762 (1987). – See Abstr. 43.074.012.

074.015 Bidirectional solar wind electron heat flux events.
J. T. Gosling, D. N. Baker, S. J. Bame,
W. C. Feldman, R. D. Zwickl, E. J. Smith.
J. Geophys. Res., Vol. 92, No. A8, p. 8519 – 8535 (1987).

Normally the $\gtrsim 80$–eV electrons which carry the solar wind electron heat flux are collimated along the interplanetary magnetic field (IMF) in the direction pointing outward away from the sun. Occasionally, however, collimated fluxes of $\gtrsim 80$–eV electrons are observed traveling both parallel and antiparallel to the IMF. The authors present the results of a survey of such bidirectional electron heat flux events as observed with the plasma and magnetic field experiments aboard ISEE 3 at times when the spacecraft was not magnetically connected to the earth's bow shock. Consistent with previous work they interpret the bidirectional heat flux as evidence for a closed field topology in interplanetary space. Further, the authors suggest that these events are one of the more prominent signatures of coronal mass ejection events in the solar wind at 1 AU.

074.016   The line λ10830 He I as a probe of coronal holes: theoretical aspect.
Zh. A. Pozhalova.
Pis'ma Astron. Zh., Tom 13, No. 7, p. 610 – 615 (1987). In Russian. English translation in Sov. Astron. Lett., Vol. 13.

It is shown that measurements of the central intensity of the He I λ10830 line can be used to numerically estimate the coronal UV flux illuminating the chromosphere. The analysis is based on non–LTE calculations of He I level populations. The VAL–C model atmosphere was used with additional coronal UV–illumination considered as a parameter.

074.017   The theory of magnetic coronal heating.
G. E. Vekstein.
Astron. Astrophys., Vol. 182, No. 2, p. 324 – 328 (1987).

Coronal heating connected with electric currents generated by cellular convective flows on the photosphere is discussed. These current systems represent a potential source of free energy which may be released by the complex reconnection processes. The latter may be treated phenomenologically (Heyvaerts and Priest, 1984) as the relaxation to the minimum energy state in a time scale τ_r. The relaxed state is determined by the constraint of global magnetic helicity conservation in highly conductive plasmas (Taylor, 1974). The model of an array of closely packed flux tubes (Parker, 1983; Browning et al., 1986) is considered, and such an approach allows to express the rate of coronal heating in terms of a photospheric driver.

074.018   Interpretation of polarization observations of forbidden coronal lines.
K. I. Nikol'skaya.
Sov. Astron., Vol. 30, No. 5, p. 603 – 606 (1986). English translation of 42.074.024.

074.019   A new method of analyzing data on the white corona.
O. G. Badalyan, M. A. Livshits.
Sov. Astron., Vol. 30, No. 5, p. 609 – 615 (1986). English translation of 42.074.025.

074.020   Coronal millimeter–wave sources.
I. G. Moiseev, N. S. Nesterov.
Bull. Crimean Astrophys. Obs., Vol. 74, p. 104 – 108 (1987). English translation of 41.074.094.

074.021   Radio emission from coronal and interplanetary shocks.
H. V. Cane.
Radio astronomy from space, p. 283 – 288 (1987). – See Abstr. 012.009.

From the study of low frequency (less than 2 MHz) solar emissions two questions have arisen that will only be conclusively answered if one can obtain sensitive, multi–frequency observations in the frequency range not adequately covered by previous experiments (0.5 – 20 MHz). The first question concerns the relationship between coronal and interplanetary shocks. The second question concerns the relationship between coronal shocks and certain fast drift kilometric bursts.

074.022   On different temperature regions in the inner corona.
V. L. Merzlyakov.
Soln. Dannye, Byull., 1987, No. 2, p. 91 – 94 (1987). In Russian.

A method of determining the temperature and nonthermal velocities in the inner corona is proposed. The coronal lines obtained at the total eclipses are examined by this method. The coronal lines proved to originate in three types of regions with different temperature: $T_1 \cong 1.2 \times 10^6 K$, $T_2 \cong 1.9 \times 10^6 K$, $T_3 \cong 2.7 \times 10^6 K$. The nonthermal velocities are $V_t = 16 – 17$ km/s for all types of regions.

074.023   A model of the solar wind turbulence from radio occultation experiments.
N. A. Armand, A. I. Efimov, O. I. Yakovlev.
Astron. Astrophys., Vol. 183, No. 1, p. 135 – 141 (1987).

A model of the solar wind turbulence is developed using the experimental data of radio occultation experiments carried out with both coherent spacecraft signals and natural radio sources. It is shown that the electron density fluctuation spectrum is a power law with the definite spectral index for each radial distance from the Sun. Another specific feature of this model is a monotonic increase of the spectral index, and both inner and outer scales of turbulence with radial distance. The proposed model is consistent with all measurements obtained with the radio occultation experiments.

074.024   Comparison of computed fluxes for Fe X and Fe XIV lines with observed values at 1980 eclipse.
P. K. Raju, J. Singh.
Sol. Phys., Vol. 110, No. 2, p. 271 – 280 (1987).

Fluxes have been computed for Fe X (6374 Å) and Fe XIV (5303 Å) lines as a function of solar radii and at various coronal temperatures. The electron density derived from the white light corona during the total solar eclipse of 1980 were used in the computations. Fluxes in adjacent continua have also been computed. The computed ratios of line flux to the square of continuum flux at a coronal temperature of $1.6 \times 10^6 K$ show a good fit with the observed values for Fe X line. Further, radiative excitation seems to dominate over collisional excitation beyond 1.3 solar radius.

074.025   Formation mechanism and mathematical description of viscosity and anisotropic pressure in the solar wind.
X.–p. Zhao.
Sci. Sin., Ser. A, Vol. 30, No. 9, p. 955 – 962 (1987).

Starting from the second order moment equation of the proton distribution function in the solar wind the formation mechanism of the anisotropic pressure and the viscosity has been analysed. The effects on the evolution of the pressure anisotropy of the various factors such as the magnetic field, the velocity field, the anisotropic heat flow, the external heating sources, and the Coulomb collisions and wave–particle interactions are investigated.

074.026   Optical coronal polarization and solar dust ring.
S. Isobe, T. Hirayama, N. Baba, N. Miura.
Publ. Astron. Soc. Jpn., Vol. 39, No. 4, p. 667 – 677 (1987).

Observations of the outer solar corona on the Java island were carried out on June 11, 1983, at a 30–km altitude using a B–15 balloon. At 5325, 5965, 7200, and 8015 Å, data on polarizations in a field of 5° × 5° centered nearly on the sun were obtained. An excess of polarization at the four wavelengths was found in the ecliptic plane and at the location of a coronal streamer. High polarization at the coronal streamer is caused mainly by coronal electrons, but dust grains in the region out of the ecliptic plane contribute also in a few percent to the high polarization degree in this streamer. It is confirmed by additional data that there is a peak in the polarization excess in the ecliptic between 4 $R_\odot$ and 5 $R_\odot$. This excess is considered to be due to an enhanced distribution of dust in a ring or a thick wide band around the sun.

074.027   Electron acceleration within coronal loops: a wave–particle process?
D. Le Quéau, A. Roux.
Sol. Phys., Vol. 111, No. 1, p. 19 – 22 (1987). – See Abstr. 012.016.

A simple model is given to explain how electrons could be accelerated within solar coronal loops due to their trapping by quasi–monochromatic whistler waves.

074.028   Theory of radio pulsations in coronal loops.
M. J. Aschwanden.
Sol. Phys., Vol. 111, No. 1, p. 113 – 136 (1987). – See Abstr. 012.016.

"Pulsations" include a wide range of phenomena from strictly sinusoidal oscillations up to quasi–periodic fine structures, observed in the radio, microwave and X–ray frequency range. The various versions of pulsation models are reviewed and classified in three groups according to their driver mechanisms: (1) Magnetic flux tube oscillations (the emissivity of trapped particles is modulated by a standing or propagating MHD wave), (2) cyclic self–organizing systems of plasma instabilities (wave–particle, wave–wave interactions), and (3) modulation of acceleration (acceleration/injection of particles into the source). Observational references illustrate the applicability of the models. In conclusion, discrimination criteria of models are discussed, in order to give a key for interpretation of observations.

074.029   Solar wind transition region from occultation observations within meter wavelengths.
N. A. Lotova, Yu. I. Alekseev, Ya. V. Nagelis.
Kinematika Fiz. Nebesn. Tel, Tom 3, No. 4, p. 70 – 74 (1987). In Russian. English translation in Kinematics Phys. Celest. Bodies.
Occultation experiments using the interplanetary plasma near the Sun have been carried out at meter wavelength $\lambda = 2.92$ m. The radial dependence of the scattering angle $\theta(R)$ displays the predicted phenomenon: the existence of the anomalous enhanced scattering region, stretching from about 16 to 30 $R_\odot$ which is associated with the solar wind transition region. Analysis of the theoretical relations has shown that in the interplanetary medium close to the Sun $(R \lesssim 40 R_\odot)$ both radial profiles of the scattering angle $\theta(R)$ and scintillation index $m(R)$ coincide. A combination of two different modifications of the occultation method extends significantly the possibilities in investigations of the structure and geometry of the solar wind transition region.

074.030   Energetic interplanetary shocks, radio emission, and coronal mass ejections.
H. V. Cane, N. R. Sheeley Jr., R. A. Howard.
J. Geophys. Res., Vol. 92, No. A9, p. 9869 – 9874 (1987).
The interplanetary shocks which generate detectable low-frequency (< 1 MHz) radio emission, represent as a group, the most energetic shocks produced by the sun. For all interplanetary (IP) shocks which generated so–called IP type II events, the authors find, when observations were available, that the associated solar events involved fast (> 500 km/s) coronal mass ejections (CMEs). In comparison with the set of all CMEs detected by the Solwind coronagraph, the CMEs associated with IP type II events are the most massive and energetic. The majority ($> 50\%$) belong to the structural class described by the Solwind researchers as "curved front" or "halo". Evidence presented suggests that these are the same class viewed from a different perspective. The results are consistent with there being a close relationship between interplanetary shocks and fast CMEs.

074.031   An analysis of solar wind fluctuations between 1 and 10 AU.
M. Vellante, A. J. Lazarus.
J. Geophys. Res., Vol. 92, No. A9, p. 9893 – 9900 (1987).
A cross–correlation analysis between different parameters of the solar wind plasma based on Voyager 1 and 2 data from 1 to 9.5 AU evidences, as a predominant feature of the microscale structure of the solar wind, a persistent presence of time periods characterized by high anticorrelation between the magnetic field magnitude and the proton density. These events are present at least 25% of the total time, and their frequency of occurrence progressively increases with increasing heliocentric distance. These results have been satisfactorily interpreted as being due to the presence of nonpropagating structures with internal pressure balance and characteristic time scale of $\lesssim 10$ hours. Alfvénic fluctuations are confirmed to dominate the microscale at $1 – 2$ AU, but they seem to become less important at larger heliocentric distances.

074.032   Velocity of iron ions in the solar wind.
J. Schmid, P. Bochsler, J. Geiss.
J. Geophys. Res., Vol. 92, No. A9, p. 9901 – 9906 (1987).
From a set of approximately 7000 spectra provided by the ISEE 3 ion composition instrument at the time near the maximum of solar cycle 21 the authors have been able to derive

iron velocities and to compare them with velocities of helium ions obtained with the same instrument. They find a strong correlation among these velocities ($r_{cor} = 0.975 \pm 0.001$). Whereas no significant velocity differences between helium and iron are found at low solar wind speeds, it appears that iron tends to flow at a somewhat lower speed ($\Delta v \sim 10$ km/s) in high–speed streams.

074.033   Simultaneous observations of Pc 3 – 4 pulsations in the solar wind and in the earth's magnetosphere.
M. J. Engebretson, L. J. Zanetti, T. A. Potemra,
W. Baumjohann, H. Lühr, M. H. Acuna.
J. Geophys. Res., Vol. 92, No. A9, p. 10,053 – 10,062 (1987).

074.034   On characteristics of the solar corona above a coronal hole.
I. V. Chashej.
Pis'ma Astron. Zh., Tom 13, No. 9, p. 797 – 802 (1987). In Russian. English translation in Sov. Astron. Lett., Vol. 13.
It is shown that more fast than radial divergence of the magnetic field lines results in more strong cooling influence of the plasma flow on the solar corona regime. This explains the lower values of the plasma density and temperature and the higher flow velocity typical of coronal holes.

074.035   Solar transition region and coronal response to heating rate perturbations.
J. T. Mariska.
Astrophys. J., Vol. 319, No. 1, p. 465 – 480 (1987).
Observations of Doppler shifts in UV emission lines formed in the solar transition region show continual plasma downflows and impulsive plasma upflows. Using numerical simulations, the author examines the conjecture that areas of downflowing plasma are the base regions of coronal loops in which the heating is gradually decreasing and that areas of upflowing plasma are the base regions of coronal loops in which the heating rate is gradually increasing. While significant mass motions do develop in the simulations, both the emission measure and the velocity at 10^5 K do not show the characteristics present in UV observations.

074.036   On an analog of the hydrodynamical picture of solar wind flow around comets and the geomagnetosphere during sudden pulses.
N. G. Ptitsyna, Z. A. Kereselidze.
Inst. Zemn. Magn., Ionos. Rasprostr. Radiovoln Akad. Nauk SSSR, Prepr., No. 6/695, p. 3 – 17 (1987). In Russian. Abstr. in Ref. Zh., 62. Issled. Kosm. Prostranstva, 8.62.520 (1987).

074.037   Solar wind and heliosphere.
M. Dryer.
The solar wind and the Earth, p. 20 – 35 (1987). – See Abstr. 003.007.
Contents: Introduction. Solar wind generation. High speed corotating streams. Interplanetary flare–generated shock waves.

074.038   The nature and evolution of magnetohydrodynamic fluctuations in the solar wind: Voyager observations.
D. A. Roberts, L. W. Klein, M. L. Goldstein,
W. H. Matthaeus.
J. Geophys. Res., Vol. 92, No. A10, p. 11,021 – 11,040 (1987).
The magnetic field and plasma data collected by the Voyager spacecraft between 1 and 11 AU are used to study the properties of interplanetary MHD fluctuations and to attempt to answer several related questions about the Alfvénicity of solar wind fluctuations: First, to what extent are magnetic and velocity fluctuations Alfvénic? Second, does the dominant propagation direction of Alfvénic fluctuations evolve with heliocentric distance? Third, is the presence of Alfvénic fluctuations correlated with large–scale structures, such as stream interaction regions? In addition, the authors investigated the contributions of compressive modes to the interplanetary fluctuations.

074.039 Hydromagnetic wave excitation by ionized interstellar hydrogen and helium in the solar wind.
M. A. Lee, W.–H. Ip.
J. Geophys. Res., Vol. 92, No. A10, p. 11,041 – 11,052 (1987).

Interstellar atoms penetrate the heliosphere, are ionized by solar UV radiation or charge exchange with solar wind ions, and are "picked up" by the solar wind onto a ring distribution in velocity space. The ring distribution is unstable to the generation of hydromagnetic waves with growth time scales which are small compared with those of the continual pickup process and convection with the solar wind. First, the growth rates for parallel propagation are derived, presented, and compared with previous work. Then the spatially homogeneous quasi–linear equations describing the subsequent hydromagnetic wave excitation and pickup ion velocity diffusion in pitch angle and energy are presented under the assumption of wave propagation parallel to the ambient magnetic field. Neglecting ion energy changes, analytical expressions for the time–asymptotic wave spectra accompanying the time–asymptotic isotropic ion distribution are derived.

074.040 Do slow shocks precede some coronal mass ejections?
A. J. Hundhausen, T. E. Holzer, B. C. Low.
J. Geophys. Res., Vol. 92, No. A10, p. 11,173 – 11,178 (1987).

The observed speeds of coronal mass ejections are often below the estimated Alfvén speed but above the sound speed for the background solar corona. This suggests that slow magnetohydrodynamic shocks may form as mass ejections sweep through the corona. The authors argue on the basis of the Rankine–Hugoniot relations and the propagation of small–amplitude slow mode waves that the shape of a slow shock front would be flattened (with respect to a sun–centered sphere) or perhaps even concave outward (from the sun) and thus present a very different appearance from the fast coronal shock waves that have been commonly modeled as wrapping around a mass ejection.

074.041 Simulation of January 1 – 7, 1978, events.
J. K. Chao, M. B. Moldwin, S.–I. Akasofu.
J. Geophys. Res., Vol. 92, No. A10, p. 11,183 – 11,188 (1987).

The solar wind disturbances during January 1 – 7, 1978, are reconstructed by a modelling method. The method is useful in reconstructing a very complicated chain of interplanetary events observed by a number of spacecraft.

074.042 Solar cycle invariance in solar wind proton temperature relationships.
R. E. Lopez.
J. Geophys. Res., Vol. 92, No. A10, p. 11,189 – 11,194 (1987).

The relationships between solar wind proton temperature and velocity and between temperature and momentum flux density at 1 AU are examined using National Space Science Data Center IMP 8 solar wind data obtained from late 1984 to early 1985. These relationships are compared with similar ones obtained from a variety of solar wind data spanning 14 years, from 1966 to 1980. It is found that these relationships, particularly the one between temperature and velocity, are very stable over the solar cycles from which these data were drawn. This suggests that the basic physical processes which accelerate and heat the solar wind have remained unchanged for the last 20 years.

074.043 Behaviour of the coronal intensity in the line 5303 Å and latitude zonal structure of the magnetic field. Period 1944 – 1974.
V. I. Makarov, J. L. Leroy, J. C. Noëns.
Astron. Zh., Tom 64, Vyp. 5, p. 1072 – 1078 (1987). In Russian. English translation in Sov. Astron., Vol. 31, No. 5.

A comparison of variations of intensity of the corona using the 5303 Å line (Pic du Midi observations) with the latitude zonal structure of the coronal magnetic field for 1944 – 1974 is made using Hα charts. It is shown that the global process of solar activity originally is to be seen in the polar zone after the polarity reversal and lasts for about 11 years up to the next reversal of the magnetic field of the Sun. A second and more intense manifestation of this process takes place at + 30° to –30° in latitude and

also lasts for about 11 years. This means that the total duration of the global solar activity cycle at all latitudes attains 17 – 18 years.

074.044 Spectroscopy of the solar corona.
J. B. Zirker.
Spectroscopy of astrophysical plasmas, p. 165 – 184 (1987). – See Abstr. 003.010.

Contents: 1. Introduction. 2. Coronal structures. 3. Coronal temperature diagnostics. 4. Electron density. 5. Electron temperature and density in flares. 6. Velocity. 7. Magnetic field. 8. Summary.

074.045 Magnetohydrodynamic modeling of coronal bright points.
W. L. Waldron, D. J. Mullan.
Astrophys. J., Vol. 319, No. 2, p. 971 – 983 (1987).

Coronal bright points are known to be magnetic in nature. The authors have developed a two–dimensional MHD code to investigate the response of a stratified atmosphere to a localized magnetic structure. If a bright point occurs at a region where flux tubes of opposite polarity have encountered by chance, in the presence of stratification, a net upward mass flux results. Upflow velocities of up to one half of the Alfvén speed are produced. The implications of these results for mass outflows from bright points in coronal holes are discussed.

074.046 Theoretical study of onset conditions for solar eruptive processes.
W. Zwingmann.
Sol. Phys., Vol. 111, No. 2, p. 309 – 331 (1987).

The author applies the model of quasistatic equilibrium sequences to describe the time development of magnetic field structures in the plasma of the solar corona, and to determine onset points of a dynamical evolution. The representation of the magnetic field by Euler potentials provides a realistic modeling of the photospheric boundary conditions. He presents a numerical method suited for the computation of magnetohydrodynamic equilibrium states and for analysing their stability against perturbations within ideal MHD. Pressure and magnetic footpoint displacement can be prescribed separately as boundary conditions. The author considers magnetic arcade structures typical for large two–ribbon flares. The results indicate that a finite pressure gradient seems to be essential for the existence of onset points. Furthermore, it is shown that magnetic shear destabilizes for intermediate values, but can have a stabilizing effect for a large amount of shear.

074.047 Meterwave observations of a coronal hole.
Z. Wang, E. J. Schmahl, M. R. Kundu.
Sol. Phys., Vol. 111, No. 2, p. 419 – 428 (1987).

The authors present meterwave maps showing a coronal hole at 30.9, 50.0, and 73.8 MHz using the Clark Lake Radioheliograph in October 1984. The coronal hole seen against the disk at all three frequencies shows interesting similarities to, and significant differences from its optical signatures in He I λ10830 spectroheliograms. Using the model of coronal holes by Dulk et al. (1977) the authors derive the electron density from the radio observations of the brightness temperature.

074.048 Solar wind flow associated with stream–free sector boundaries at 1 AU.
J. H. Sastri.
Sol. Phys., Vol. 111, No. 2, p. 429 – 437 (1987).

The correlations between the plasma characteristics of the solar wind flow in the vicinity (± 12 hr) of stream–free sector boundaries near Earth are examined using the composite data base of interplanetary plasma for the period 1965 – 1980. The author confirms the result of Lopez et el. (1986) of an inverse relationship of the proton temperature (T_p) with the momentum flux density (NV^2) in the low speed wind at 1 AU. The coefficients of lines of best fit to the T_p vs NV^2 (as well as T_p vs V) distribution in our sample are, however, significantly different from those of the undifferentiated sample of low speed wind considered by Lopez et al. such that T_p is, in general, lower than expected. The author finds further that the proton number density (N) varies as the inverse cube of the flow speed (V) indicating an invariance of the kinetic energy flux density (NV^3) relative to velocity structure in the plasma flow around stream-free boundaries. These average relationships, which are unaffected by interplanetary dynamical processes, are suggested to be due to sub–sonic addition of momentum and energy to the solar wind flow from the source structures, namely coronal streamers.

074.049 A numerical study of the nonlinear thermal stability of solar loops.

J. A. Klimchuk, S. K. Antiochos, J. T. Mariska.
Astrophys. J., Vol. 320, No. 1, p. 409 – 417 (1987).

The authors have used a time–dependent numerical model to study the nonlinear thermal stability of static loops of various heights. Under the assumption of uniform energy input, high–lying loops ($z_{apex} \gtrsim 5 \times 10^3$km) must be hot ($T \approx 10^6$K), but lower–lying loops can exist in either a hot or a cool ($T \lesssim 10^5$K) state. The hot state is thermally unstable whenever the loop height is less than about 10^3km. An initially hot atmosphere in low–lying, compact loops evolves to a state best described as an extended chromosphere, with temperatures far below 10^5K. The simulations indicate that high–lying hot loops are stable to all reasonable perturbations.

074.050 Numerical calculations of thermal waves in the solar corona.

A. S. Andreev, A. G. Kosovichev.
Izv. Krymskoj Astrofiz. Obs., Tom 76, p. 186 – 192 (1987). In Russian. English translation in Bull. Crimean Astrophys. Obs., Vol. 76.

A numerical simulation of thermal waves observed in high-temperature coronal plasma during solar flares has been carried out. A heat flux saturation and energy transfer from electrons to ions are shown to be two principal physical factors responsible for thermal waves propagation. The former prevails, mainly, during the first few second of flare energy release, the latter is more significant in the succeding period of time. The dependence of thermal wave velocity on initial thermodynamic state of plasma has been determined. It is found that the mean velocity depends on density rather than on the intitial temperature. The calculations show a good qualitative agreement with the observational data, and make it possible to estimate the parameters of coronal plasma in magnetic arch structures.

074.051 Origin and evolution of fluctuations in the solar wind: Helios observations and Helios–Voyager comparisons.

D. A. Roberts, M. L. Goldstein, L. W. Klein, W. H. Matthaeus.
J. Geophys. Res., Vol. 92, No. A11, p. 12,023 – 12,035 (1987).

Using hour–averaged data from the Helios and Voyager spacecraft, the authors have investigated the origin and evolution of low–frequency interplanetary fluctuations from 0.3 to 20 AU. The results support the view that outward propagating Alfvénic fluctuations are generated near the Sun and that substantial dynamical evolution, probably involving shear–generated nonlinear couplings, is important at all heliocentric distances examined.

074.052 Solar wind iron charge states preceding a driver plasma.

A. B. Galvin, F. M. Ipavich, G. Gloeckler, D. Hovestadt, S. J. Bame, B. Klecker, M. Scholer, B. T. Tsurutani.
J. Geophys. Res., Vol. 92, No. A11, p. 12,069 – 12,081 (1987).

During September 28 and 29, 1978, the ISEE 3 spacecraft observed several distinct types of high–speed solar wind flows when a coronal hole–associated high–speed stream was followed by two interplanetary shocks, one of which was driven by flare ejecta contained in a "magnetic cloud" or interplanetary "plasmoid". Using the University of Maryland/Max–Planck–Institut ultralow energy charge analyzer (ULECA) on ISEE 3, the authors present solar wind Fe and Si/S charge state and Fe density measurements for the different plasma regimes associated with the flare–related shock and combine these measurements with the Los Alamos National Laboratory proton observations to obtain iron/hydrogen density ratios and iron/hydrogen velocity differences. The authors place special emphasis on the postshock shell of turbulent and compressed ambient solar wind and interplanetary magnetic fields constituting the "sheath region" preceding the driver plasma.

074.053 The eastward deflection of fast coronal mass ejecta in interplanetary space.

J. T. Gosling, M. F. Thomsen, S. J. Bame, R. D. Zwickl.
J. Geophys. Res., Vol. 92, No. A11, p. 12399 – 12406 (1987).

Previous work has shown that a bidirectional solar wind electron heat flux is one of the more prominent signatures of a coronal mass ejection event in the solar wind at 1 AU. Using ISEE 3 solar wind electron measurements obtained during 1978 and 1979, the authors have used this signature to identify the fast coronal mass ejecta driving 19 interplanetary shocks. They suggest that the preferred eastward deflection of these fast coronal mass ejecta is a consequence of solar rotation and the spiral geometry of the ambient solar wind.

074.054 The Sun's outer corona at radio wavelengths.

M. R. Kundu.
Indian J. Radio Space Phys., Vol. 16, No. 1, p. 172 – 191 (1987). Abstr. in Phys. Abstr., Vol. 90, No. 1309, Entry 88356 (1987).

074.055 Large–scale changes in the intensity of the green corona and their reflection in the solar wind particle flux. Part 1: The equatorial problem.

L. Kulčár.
Contrib. Astron. Obs. Skalnaté Pleso, Vol. 16, p. 27 – 41 (1987).

The way marked changes of the intensity of the green emission corona from the region of the solar equator are reflected in the solar wind particle flux is studied. The analysis, based on a quantitative as well as qualitative approach, indicates that an increase or decrease in the solar wind particle flux is observed with equal probability after an increase in the intensity of the green corona. However, a marked decrease in the intensity of the green corona is accompanied by an increase in the solar wind particle flux with a probability exceeding 70%.

074.056 MHD waves detected by ICE at distances $\geqslant 28 \times 10^6$km from comet P/Halley: cometary or solar wind origin?

B. T. Tsurutani, A. L. Brinca, E. J. Smith, R. M. Thorne, F. L. Scarf, J. T. Gosling, F. M. Ipavich.
Astron. Astrophys., Vol. 187, No. 1/2, p. 97 – 102 (1987). – See Abstr. 003.013, 43.074.056.

Spectral analyses of the high resolution magnetic field data are employed to determine if there is evidence of cometary heavy ion pickup when ICE was closest to Halley, $\sim 28 \times 10^6$km. No evidence is found for the presence of heavy ion cyclotron waves. However, from this search, two new wave modes are discovered in the solar wind: electromagnetic waves near the proton cyclotron frequency and drift mirror mode "waves". Both modes have scales of $10 – 60$ s (1 to 6 times the proton gyro period) in the spacecraft frame. The possibility of wave generation by cometary hydrogen pickup is explored. Theoretical arguments and further experimental evidence indicates that cometary origin is improbable. The most likely source is instabilities in high speed solar wind streams, presumably associated with stream–stream interactions. VLF electrostatic emissions are found to occur in field minima or at gradients of the drift mirror structures. Possible generation mechanisms of drift mirror mode waves, cyclotron waves and electrostatic waves are discussed.

074.057 On maximum velocity of the solar wind.

I. V. Chashej.
Kratk. Soobshch. Fiz., No. 6, p. 45 – 47 (1987). In Russian. Abstr. in Ref. Zh., 51. Astron. 11.51.438 (1987).

074.058   Actividad de los hoyos coronales solares de baja latitud.
S. Bravo, B. Mendoza, R. Pérez–Enríquez,
J. Valdés–Galicia.
Rev. Mex. Astron. Astrofis., Vol. 14, No. 2, p. 688 – 692 (1987).
– See Abstr. 012.042.

Strong variations of the coronal holes of mid and low latitudes during the ascending and descending phase of the solar cycle have been observed by means of IPS records (the interplanetary scintillation of small diameter radio sources). These variations induce the propagation of disturbances in the interplanetary medium which produce the IPS and are also responsable of cosmic ray modulation. The authors suggest that the 11 years modulation is influenced by such a coronal hole behaviour.

074.059   Numerical analysis of the azimuthal transport of solar particles.
J. Pérez–Peraza, M. Alvarez–Madrigal, F. Rivero.
Rev. Mex. Astron. Astrofis., Vol. 14, No. 2, p. 693 – 699 (1987).
– See Abstr. 012.042.

The authors derive particle energy spectra at the level of the roots of the coronal magnetic field for two specific solar proton events. From the confrontation of these spectra with those derived from interplanetary demodulation of solar proton data at the level of the earth level, the authors are able to infer about the large–scale coronal magnetic field topology.

074.060   About the interest of solar interferometric observations.
J. Heyvaerts.
ESA Spec. Publ., ESA SP–273, p. 11 – 13 (1987). – See Abstr. 012.044.

It is shown that the understanding of some basic solar physical processes calls for high spatial resolution observation on the visible and in EUV lines. These processes are of general astrophysical interest. The scientific return expected from UV line observations of coronal loops with an angular resolution of 0.″01 across the loop is described in some more detail.

074.061   Diagnostics of solar coronal loops at interferometric angular resolution.
B. H. Foing, M. Faucherre, L. Damé.
ESA Spec. Publ., ESA SP–273, p. 217 – 219 (1987). – See Abstr. 012.044.

The advent of very high angular resolution (equivalent to 20 km on the sun) for extreme ultraviolet observations would allow to diagnose the fine structure in density and temperature of solar coronal loops. In the framework of the variety and uncertainties of the existing theoretical models of loops, the high angular resolution is of particular importance to estimate the filling factor of loops by hot and cool material, to measure radial gradients of temperature and density, to observe flows, spatio-temporal evolution due to heating mechanisms and interaction between loops.

074.062   On the equation of state and collision time for a multicomponent, anisotropic solar wind.
E. Marsch, A. K. Richter.
Ann. Geophys., Ser. A, Vol. 5, No. 2, p. 71 – 82 (1987). Abstr. in Phys. Abstr., Vol. 90, No. 1310, Entry 94623 (1987).

074.063   Three–dimensional structure and velocities of the motion of matter in dark loops of H_α near–limb flares.
N. S. Shilova, L. I. Starkova.
Kinematika Fiz. Nebesn. Tel, Tom 3, No. 6, p. 28 – 35 (1987). In Russian. English translation in Kinematics Phys. Celest. Bodies.

Two near–limb flares filtergrams with filter passband at H_α and $H_\alpha \pm 0.2$ nm are analysed. Correction of the loop geometry in accordance with the projection effect gave the possibility to determine the dark flare loop form in its own plane, the orientation of the loop plane with respect to the solar surface and velocities along the loop. The ascent from the solar surface with the following descent along the loop is a typical motion of matter. By means of contrast pro-files the loop electron density N_e and hydrogen density N_1

are calculated: $7 \times 10^{10} \text{cm}^{-3} \leqslant N_e \leqslant 10^{11} \text{cm}^{-3}$, $6 \times 10^9 \text{cm}^{-3} \leqslant N_1 \leqslant 2 \times 10^{11} \text{cm}^{-3}$.

074.064   On the inability of magnetically constricted transition regions to account for the 10^5 to 10^6K plasma in the quiet solar atmosphere.
J. F. Dowdy Jr., A. G. Emslie, R. L. Moore.
Sol. Phys., Vol. 112, No. 2, p. 255 – 279 (1987).

The authors re–examine static models of the atmosphere which incorporate not only conduction and radiation but also the effects of large magnetic constrictions. They find the structure of the plasma depends not only on the magnitude of the constriction but also on the tube's shape. The results show that no model with a constriction of order 100 can simultaneously (a) produce the variation of differential emission measure with temperature derived from measured line intensities and (b) satisfy the ob-served constraint (Reeves, 1976) that EUV emission from below $\approx 7 \times 10^5$K be confined to the supergranular network. This suggests that the bulk of the $10^5 - 10^6$K plasma in the quiet solar atmosphere is magnetically isolated from the corona and heated internally. For any permitted tube shape, constriction factors of order 100 reduce the coronal conductive energy losses to the transition region to a value which is less than a third of the value for an unconstricted field.

074.065   Stellar wind – observations and theories. Part II. Theo-retical models of the solar wind.
M. Krogulec.
Postepy Astron., Tom 34, Zesz. 3, p. 123 – 156 (1986). In Polish.

Theoretical models of the solar wind – Parker's thermally driven, MHD waves driven and Weber and Davis wind model from rotating Sun are described. The influence of the additional sources or sinks of the energy and momentum on the value of mass loss rates Ṁ and terminal velocity v_∞, according to Leer and Holzer results, is presented.

074.066   The analysis of periodic structure of solar wind velocity.
X.–b. Zhang, G.–l. Zhang, G.–z. Qi.
Chin. J. Space Sci., Vol. 6, No. 4, p. 252 – 257 (1986). In Chinese. Abstr. in Phys. Abstr., Vol. 90, No. 1313, Entry 115195 (1987).

074.067   Geometry of the transition region of the solar wind in the 11–year activity cycle.
N. A. Lotova, D. Blūms.
Issled. Solntsa Krasnykh Zvezd, No. 25, p. 13 – 28 (1987). In Russian.

Using the mean solar wind velocity of the year, measured by the interplanetary scintillation far from the Sun, $R \gtrsim 100\, R_\odot$, the structure of the transition region from 1972 till 1982 is found. The structure of the transition region considerably changes with the solar activity.

074.068   Drawing inferences about solar wind acceleration from coronal minor ion observations.
R. Esser, T. E. Holzer, E. Leer.
J. Geophys. Res., Vol. 92, No. A12, p. 13,377 – 13,389 (1987).

A parameter study is designed and carried out to illustrate the physical effects that can be studied through analysis and interpretation of coronal minor ion spectral line observations. It is shown that minor ion line width, together with the coronal Ly α line width and coronal white light observations, can yield important information concerning the transport and dissipation of energy carried outward from the coronal base by hydromag-netic waves. It is concluded that the observation of coronal minor ion spectral lines represents an important component of a concerted observational approach to the solar wind acceleration problem.

074.069   A kinetic study of solar wind mass loading and cometary bow shocks.
N. Omidi, D. Winske.
J. Geophys. Res., Vol. 92, No. A12, p. 13,409 – 13,426 (1987).

A detailed numerical study is conducted to understand the kinetic processes associated with solar wind mass loading due to pickup of cometary ions and the formation of cometary bow shocks. The results are compared and shown to be in qualitative agreement with the recent observations at comets.

074.070   The evolution of a coronal streamer prior to mass ejection.
R. Wolfson, C. Conover, R. M. E. Illing.
J. Geophys. Res., Vol. 92, No. A12, p. 13,641 – 13,646 (1987).
The authors developed a model describing the quasistatic evolution of a coronal helmet streamer as it is inflated with excess mass. They have fit their model to SMM coronagraph/polarimeter observations showing the slow growth of a coronal streamer prior to its disruption in association with an eruptive prominence and mass ejection on August 18, 1980. The results suggest that the early phase of this event is consistent with the quasi–static evolution of the corona in response to the slow addition of mass to the closed–field region of the streamer.

074.071   Evidence for slow mode MHD turbulence in the solar wind: post–slow shock observations at 0.31 AU.
A. K. Richter, A. H. Luttrell.
J. Geophys. Res., Vol. 92, No. A12, p. 13,653 – 13,657 (1987).
First observational evidence is presented for low–frequency slow mode MHD turbulence in the solar wind by employing power and coherence spectral analysis of the Helios high time resolution solar wind plasma and magnetic field observations. This turbulence occurs directly behind an interplanetary slow forward shock wave at 0.31 AU. As the fore–shock plasma conditions are such that steepening of slow waves is favored in comparison to Landau damping, the observations presented here do in addition provide unique support to the theoretical results of Hada and Kennel (1985).

074.072   Correction to "Solar wind proton temperature–velocity relationship" by R. E. Lopez, J. W. Freeman [J. Geophys. Res., Vol. 91, No. A2, p. 1701 – 1705 (1986)] and "The cold solar wind" by J. W. Freeman, R. E. Lopez [J. Geophys. Res., Vol. 90, No. A10, p. 9885 – 9887, 9949 (1985)].
J. Geophys. Res., Vol. 92, No. A12, p. 13,679 (1987). See abstracts 41.074.005 and 40.074.038.

074.073   Propagation of solar energetic particles from the corona into the interplanetary space.
N. N. Volodichev, V. P. Grigor'eva, V. S. Prokudina.
Astron. Tsirk., No. 1481, p. 3 – 5 (1987). In Russian.
The delay time of escape for energetic protons relative to the explosive phases of flares is investigated. It is shown that the moments of escape of energetic protons are close in time to those of shock wave passages or a coronal transient in the corona. It is suggested that the delay of the protons' escape is due to the capture of accelerated particles in a closed magnetic configuration.

074.074   Explosive chromospheric instability in hydrodynamic loop flare models: the problem and its cure.
G. Peres, S. Serio, R. Rosner.
Nuovo Cimento B, Vol. 99B, Ser. 11, No. 1, p. 15 – 28 (1987).
Abstr. in Phys. Abstr., Vol. 90, No. 1315, Entry 127814 (1987).

074.075   Nonlocal thermal conduction in hydrodynamic loop flare models.
G. Peres, R. Rosner, S. Serio.
Nuovo Cimento B, Vol. 99B, Ser. 11, No. 1, p. 29 – 44 (1987).
Abstr. in Phys. Abstr., Vol. 90, No. 1315, Entry 127815 (1987).

074.076   Concentration, plasma inhomogeneities and kinetic energy of the solar wind from radio occultation measurements using Venera 15 and 16.
S. N. Rubtsov, O. I. Yakovlev, A. I. Efimov.
Kosm. Issled., Tom 25, Vyp. 4, p. 620 – 625 (1987). In Russian. English translation in Cosm. Res.

074.077   Critical layers of hydromagnetic gravity inertial waves in the solar helmet streamers and magnetic sectors.
O. M. El Mekki.
Proc. R. Soc. London, Ser. A, Vol. 413, No. 1845, p. 415 – 427 (1987). Abstr. in Phys. Abstr., Vol. 91, No. 1319, Entry 4360 (1988).

074.078   Heating of coronal loops by phase–mixed shear Alfvén waves.
T. E. Abdelatif.
Astrophys. J., Vol. 322, No. 1, p. 494 – 502 (1987).
The author investigates the dissipation of shear Alfvén waves in a coronal loop driven externally by an incident wave in the subcoronal region. The phase mixing of these incident shear Alfvén waves serves as the dissipation mechanism in the corona. The wave solution found by Heyvaerts and Priest for coronal holes is used to compute the total energy deposited in a loop. The energy deposited is shown to depend upon the magnetic diffusivity and viscosity. The energy deposited in a three–layer model is computed for incident waves with periods of 5 minutes or 5 seconds.

074.079   Energy buildup in coronal magnetic flux tubes.
R. S. Steinolfson, T. Tajima.
Astrophys. J., Vol. 322, No. 1, p. 503 – 511 (1987).
The response of the magnetic field in coronal loops to photospheric motion is investigated using a time–dependent, two–dimensional MHD simulation. Starting with an initially uniform field, circular sections of the ends of the loop are slowly rotated to represent the photospheric motion. The field lines at the base are modified by this flow in a manner consistent with the generated electric fields.

074.080   Properties of solar coronal active regions deduced from X–ray line spectra.
D. L. McKenzie.
Astrophys. J., Vol. 322, No. 1, p. 512 – 521 (1987).
Spectra from the SOLEX B RAP spectrometer have been used to analyze the temperature and density structure of over 100 nonflaring solar active regions. Density measurements that used the R ratio of O VII indicated that few regions have electron densities higher than $\sim 3 \times 10^9 \mathrm{cm}^{-3}$. In a few cases, flare–productive regions had measured densities approximately twice this high. Temperature–sensitive line ratios in the helium–like ions O VII, Ne IX, and Mg XI were used to decude the general properties of the differential emission–measure function $B(T)$ for nonflaring regions. $B(T)$ falls off with increasing temperature above a peak temperature that is almost always lower than $T_m(\mathrm{O\ VII}) = 1.8 \times 10^6 \mathrm{K}$.

074.081   Simulations of the Ca XIX spectral emission from a flaring solar coronal loop. I. Thermal case.
E. Antonucci, M. A. Dodero, G. Peres, S. Serio, R. Rosner.
Astrophys. J., Vol. 322, No. 1, p. 522 – 543 (1987).
Simulations of Ca XIX line profiles observed from a flaring coronal loop have been carried out for two impulsive heating models, which assume either (1) a thermal source at the apex of the loop, or (2) thermal sources localized near the footpoints of the loop. The hydrodynamic evolution of the coronal loop in response to this impulsive heating has been calculated by using the Palermo–Harvard numerical model for magnetically confined plasmas. The simulated spectra are compared with observed soft X–ray spectra obtained with the SMM satellite. This comparison clearly favours case (2) as impulsive heating model for the flares.

074.082   On the energy flux of MHD waves emerging into the solar corona.
I. V. Chashej, V. I. Shishov.
Sov. Astron., Vol. 31, No. 2, p. 208 – 210 (1987). English translation of 43.074.060.

074.083   Formation of current layers in magnetic irregularities of the solar wind.
S. I. Vajnshtejn.
Geomagn. Aehron., Tom 27, No. 4, p. 529 – 535 (1987). In Russian. English translation in Geomagn. Aeron.

074.084   Formation of energy and mass fluxes of the solar wind in a model with a wave source.
I. V. Chashej, V. I. Shishov.
Geomagn. Aehron., Tom 27, No. 5, p. 705 – 711 (1987). In Russian. English translation in Geomagn. Aeron.

074.085   Acceleration of charged particles in a nonuniform solar wind stream.
V. N. Lomonosov.
Geomagn. Aehron., Tom 27, No. 6, p. 889 – 892 (1987). In Russian. English translation in Geomagn. Aeron.

074.086   The ejection of helical field structures through the outer corona.
L. L. House, M. A. Berger.
Astrophys. J., Vol. 323, No. 1, p. 406 – 413 (1987). With plates 9 – 12.
The 1980 May 5 coronal mass ejection observed by the SMM coronagraph polarimeter contained an eruptive prominence with apparent helical structure. After a general discussion of the morphology of this event, the authors describe the evolution of the prominence structure as it traversed the outer corona. Particular attention is given to the distribution of magnetic helicity, a measure of helical structure.

074.087   Radio range measurements of coronal electron densities at 13 and 3.6 centimeter wavelengths during the 1985 solar conjunction of Voyager 2.
J. D. Anderson, T. P. Krisher, S. E. Borutzki, M. J. Connally, P. M. Eshe, H. B. Hotz, S. Kinslow, E. R. Kursinski, L. B. Light, S. E. Matousek, K. I. Moyd, D. C. Roth, D. N. Sweetnam, A. H. Taylor, G. L. Tyler, D. L. Gresh, P. A. Rosen.
Astrophys. J., Lett. Ed., Vol. 323, No. 2, p. L141 – L143 (1987).
Radio range measurements were generated by the Deep Space Network at two wavelengths (3.6 and 13 cm) during the solar conjunction of the Voyager 2 spacecraft in 1985 December. The difference in range at the two wavelengths provides a direct measurement of the integrated electron density along the ray path between Earth stations and the spacecraft. Derived electron density profiles on ingress and egress between 7 and 40 solar radii revealed a surprising asymmetry in the radial power–law dependence of the coronal electron density.

Progress with LSM optics for solar observations within the French space program.
See Abstr. 031.006.

The Sakigake/Suisei encounter with comet P/Halley.
See Abstr. 051.042.

Thermal overstability of hydromagnetic surface waves.
See Abstr. 062.016.

Resonance absorption of magnetohydrodynamic surface waves: viscous effects.
See Abstr. 062.083.

Resonant mode conversion of a shear Alfvén wave.
See Abstr. 062.091.

Time dependent MHD models for the cometary magnetosphere.
See Abstr. 062.137.

Cherenkov damping of surface waves.
See Abstr. 063.123.

Cosmic–ray–modified stellar winds. I Solution topologies and singularities.
See Abstr. 064.083.

Generation and structure of the electric currents in a flaring activity complex.
See Abstr. 072.053.

Simultaneous SMM flat crystal spectrometer and Very Large Array observations of solar active regions.
See Abstr. 072.085.

Solar active region physical parameters inferred from a thermal cyclotron line and soft X–ray spectral lines.
See Abstr. 072.086.

Similarities in the structure of the chromosphere and corona.
See Abstr. 073.007.

Chromospheric and coronal explosions in solar flares.
See Abstr. 073.017.

Hα diagnostics of (post)–flare loops based on narrow–band filtergram observations.
See Abstr. 073.018.

Chromospheric and coronal heating due to the radiation and collisional damping of fast magnetosonic surface waves.
See Abstr. 073.088.

Magnetic field geometry in the plasma ejected by flares: geometry at 1 A.U. from the sun.
See Abstr. 075.014.

On the uniqueness of the determination of the coronal potential magnetic field from line–of–sight boundary conditions.
See Abstr. 075.019.

Solar soft X–ray pulsations.
See Abstr. 076.002.

Radio evidence for a magnetic mirror effect on beams of subrelativistic electrons in the solar corona.
See Abstr. 077.031.

On the role of a coronal shock wave for high–energy solar protons.
See Abstr. 078.005.

Some features of solar wind protons, α particles and heavy ions behaviour. The Prognoz 7 and Prognoz 8 experimental results.
See Abstr. 078.012.

Alfvén wave dissipation in the solar atmosphere.
See Abstr. 080.023.

The solar wind–magnetosphere–ionosphere coupling during a magnetic storm.
See Abstr. 083.004.

Solar cosmic ray electrons in the Earth's magnetosphere.
See Abstr. 084.027.

Entry velocity of the solar wind's energy into the earth's magnetosphere.
See Abstr. 084.030.

Model of solar–wind magnetosphere coupling.
See Abstr. 084.060.

Some solar cycle phenomena related to the geomagnetic activity from 1868 to 1980. III. Quiet–days, fluctuating activity or the solar equatorial belt as the main origin of the solar wind flowing in the ecliptic plane.
See Abstr. 085.003.

North–south asymmetry in response of geomagnetic activity to different solar events.
See Abstr. 085.018.

Steady state flow/field model of solar wind interaction with Venus: global implications of local effects.
See Abstr. 093.009.

Characteristics of the Marslike limit of the Venus–solar wind interaction.
See Abstr. 093.012.

Some problems of the solar wind interaction with Venus.
See Abstr. 093.023.

Interaction of mass–loaded solar wind flow with a blunt body.
See Abstr. 093.024.

The role of the hot oxygen corona in the interaction of Venus with solar wind.
See Abstr. 093.057.

Analysis of Martian ionosphere and solar wind electron gas data from the planar retarding potential analyzer on the Viking spacecraft.
See Abstr. 097.061.

Electron impact ionization in the vicinity of comets.
See Abstr. 102.011.

Interactions between cometary plasmas and the solar wind.
See Abstr. 102.027.

Encounters with comets: discoveries and puzzles in cometary plasma physics.
See Abstr. 102.046.

Unusual characteristics of electromagnetic waves excited by cometary newborn ions with large perpendicular energies.
See Abstr. 102.048.

Resonant interactions between cometary ions and low frequency electromagnetic waves.
See Abstr. 102.054.

Alfvén wave plasma turbulence during solar wind–comet interaction.
See Abstr. 102.060.

Charge exchange avalanche at the cometopause.
See Abstr. 102.061.

Magnetohydrodynamic turbulence in the solar wind interacting with a comet.
See Abstr. 103.047.

The pick–up of cometary protons by the solar wind.
See Abstr. 103.465.

Alfvénic turbulence in the solar wind flow during the approach to comet P/Halley.
See Abstr. 103.466.

General features of comet P/Halley: solar wind interaction from plasma measurements.
See Abstr. 103.467.

Waves in the magnetic field and solar wind flow outside the bow shock at comet P/Halley.
See Abstr. 103.468.

Macroscopic perturbations of the IMF by P/Halley as seen by the Giotto magnetometer.
See Abstr. 103.470.

Low–frequency magnetic field fluctuations in comet P/Halley's magnetosheath: Giotto observations.
See Abstr. 103.471.

Fine structure of the magnetic field in comet P/Halley's coma.
See Abstr. 103.472.

Giotto magnetic–field results on the boundaries of the pile–up region and the magnetic cavity.
See Abstr. 103.473.

Spatial distribution of water–group ions near comet P/Halley observed by Suisei.
See Abstr. 103.482.

Description of the main boundaries seen by the Giotto electron experiment inside comet P/Halley–solar wind interaction region.
See Abstr. 103.484.

In–situ observations of a bi–modal ion distribution in the outer coma of comet P/Halley.
See Abstr. 103.485.

Pick–up ions at comet P/Halley's bow shock: observations with the IIS spectrometer on Giotto.
See Abstr. 103.487.

Possible models on disturbances of the plasma tail of comet Halley during the 1985–1986 apparition.
See Abstr. 103.494.

Structure and dynamics of the plasma tail of comet P/Halley. I. Knot event on December 31, 1985.
See Abstr. 103.495.

Structure and dynamics of the plasma tail of comet P/Halley. II. Kink event on January 10 – 11, 1986.
See Abstr. 103.496.

Two disconnection events in comet P/Halley and possible solar causes.
See Abstr. 103.503.

The cause of two plasma–tail disconnection events in comet P/Halley during the ICE–Halley radial period.
See Abstr. 103.505.

Charge exchange of solar wind ions in the coma of comet P/Halley.
See Abstr. 103.510.

Solar wind observation at the time of closest encounter with comet Halley.
See Abstr. 103.624.

Diffusive shock acceleration: comparison of a unified shock model to bow shock observations.
See Abstr. 106.002.

Low–energy protons associated with interplanetary shocks as a coherent population.
See Abstr. 106.006.

Aspects of interplanetary plasma turbulence.
See Abstr. 106.007.

ISEE 3 observations of low–energy proton bidirectional events and their relation to isolated interplanetary magnetic structures.
See Abstr. 106.018.

Numerical evaluation of statistical acceleration for energetic particles in the interplanetary medium at 2.5 and 5.0 AU.
See Abstr. 106.032.

Interplanetary and geophysical effects of a coronal transient.
See Abstr. 106.039.

Cosmic ray fluctuations and dynamical processes in the solar wind.
See Abstr. 144.029.

Variations of cosmic ray density and solar wind with the solar activity cycle.
See Abstr. 144.031.

075 Magnetic Fields

075.001 **The magnetic and velocity structure adjacent to solar active regions.**
R. G. Athay, J. A. Klimchuk.
Astrophys. J., Vol. 318, No. 1, p. 437 – 444 (1987).
Results from a number of earlier papers relating patterns observed in the C IV line at 154.8 nm to photospheric magnetic–field patterns are combined to develop a qualitative model of the magnetic–field geometry outside of the strong field areas of active regions. The motion is assumed to originate at the crests of magnetic arcades and flow downward along field lines, which are assumed to be elliptical in shape with the major axis in the photosphere. It is found that the ratio of the major axis to the minor axis of the ellipse must be less than 2 for fields under 100 G.

075.002 **Does the large–scale solar magnetic field distribution really reflect the convective velocity fields?**
V. Bumba.
Sol. Phys., Vol. 110, No. 1, p. 51 – 57 (1987). – See Abstr. 012.005.
The authors tried to decide whether the typical circular cellular–like features, which are striking during some intervals in the large–scale distribution of weak magnetic fields measured with low resolution, are related to large–scale convective motions. Two scales of such patterns were found and their morphological, kinematical and evolutionary behaviour was estimated. Their slower and overall rotation is also demonstrated in comparison with the rotation of highly averaged sunspot and magnetic fields. It is difficult to explain all the observed characteristics as random, or due to the method of field measurement and map construction used. The change of their magnetic field polarities with the solar polar field reversal is also discussed.

075.003 **The separation velocity of emerging magnetic flux.**
D.–Y. Chou, H. Wang.
Sol. Phys., Vol. 110, No. 1, p. 81 – 99 (1987). – See Abstr. 012.005.
The authors measure the separation velocity of opposite poles from 24 new bipoles on the Sun. They find that the measured velocities range from about 0.2 to 1 km s^{-1}. The fluxes of the bipoles range over more than two orders of magnitude, and the mean field strength and the sizes range over one order of magnitude. The measured separation velocity is not correlated with the flux and the mean field strength of the bipole. The predicted separation velocity is about one order of magnitude higher than those measured, or else the flux tubes are almost vertical at the photosphere. There is no correlation between the measured separation velocity and the theoretical value. The predicted rising velocity is also higher than the vertical velocity near the line of inversion in emerging flux regions observed by other authors.

075.004 **Weak solar fields and their connection to the solar cycle.**
H. Zirin.
Sol. Phys., Vol. 110, No. 1, p. 101 – 107 (1987). – See Abstr. 012.005.
The author discusses the weak solar magnetic fields as studied with the Big Bear Solar Observatory videomagnetograph. Weak fields mean those outside active and unipolar regions. These are found everywhere on the Sun, even where there never have been sunspots. These fields consist of the network and intranetwork elements. The smallest detectable elements at present are 10^{16}Mx.

075.005 **Observation of solar differential rotation with the aid of magnetic tracers.**
V. Bumba, L. Hejna.
Sol. Phys., Vol. 110, No. 1, p. 109 – 113 (1987). – See Abstr. 012.005.
The authors tried to search for the manifestation of differential rotation in the distribution of weak remnants of magnetic fields measured with a very low resolution. They found that, during the periods of low solar activity and in parts of the solar photosphere with smaller density of new magnetic flux sources, it was possible to observe the distribution of magnetic tracers in the form of differential rotation parabolas which increase their curvature from one rotation to the next. The obtained differential rotation rates are not far from those given by highly averaged sunspot data or by the daily magnetic fields. The characteristic differential rotation parabolas as well as specific cellular–like features disturbing their smooth patterns are always formed from fields of one main polarity, the sign of which depends on the phase of the activity cycle.

075.006 **Is there a weak mixed polarity background field? Theoretical arguments.**
H. C. Spruit, A. M. Title, A. A. van Ballegooijen.
Sol. Phys., Vol. 110, No. 1, p. 115 – 128 (1987). – See Abstr. 012.005.
A number of processes associated with the formation of active regions produce "U–loops": fluxtubes having two ends at the photosphere but otherwise still embedded in the convection zone. The mass trapped on the field lines of such loops makes them behave in a qualitative different way from the "omega–loops" that form active regions. It is shown that U–loops will disperse through the convection zone and form a weak (down to a few gauss) field that covers a significant fraction of the solar surface. This field is tentatively identified with the inner–network fields observed at Kitt Peak and Big Bear. The process by which these fields escape through the surface is described; a remarkable property is that it can make active regions fields apparently disappear in situ. The mixed polarity moving magnetic features near sunspots are interpreted as a locally intense form of this disappearance by escape of U–loops.

075.007   On the two types of neutral lines of the large–scale solar magnetic field.
V. I. Makarov, V. P. Mikhajlutsa.
Soln. Dannye, Byull., 1986, No. 12, p. 81 – 86 (1987). In Russian.

Two types of solar magnetic neutral lines have been defined on the criterion of topology of the field above them. The lines above which closed systems of the line–of–force loops are observed that transform into helmet–like formations in the corona are referred to the first type. The neutral lines with deeply open field structures identified with cavities in the filaments and filament channels in the chromosphere are referred to the second type. The even number of the neutral line in the given latitudinal interval beginning with the pole, which separates latitude zonal structures of the magnetic field and the presence of a symmetrically scattered flocculus relative to the neutral line are necessary and sufficient conditions for the existence of a cavity in the chromosphere. The first type of neutral lines always bears an odd number, the second type an even number.

075.008   Global magnetic field of the Sun and solar activity cycle.
V. S. Berdichevskaya.
Astron. Zh., Tom 64, Vyp. 4, p. 835 – 840 (1987). In Russian. English translation in Sov. Astron., Vol. 31, No. 4.

It is shown that, in considering the global magnetic field of the Sun approximately as dipole, the variation of the angle between the rotation axis of the Sun and the dipole axis can be explained by the introduction of the dipole rotation hypothesis. The determined value of the magnetic fields mean velocity for such rotation appears to coincide approximately with the observed value. The dipole rotation hypothesis can also account for coronal configurations, polar magnetic fields sign change and other periodical phenomena. Also, under dipole rotation conditions, the connection between large–scale fields and active region (sunspot) fields (in accordance with Ol's effect) leads to the explanation of Spörer's law.

075.009   Elements and patterns in the solar magnetic field.
C. Zwaan.
Annu. Rev. Astron. Astrophys., Vol. 25, p. 83 – 111 (1987). – See Abstr. 003.003.

This review concentrates on observational studies that may reveal the MHD processes in the solar interior and photosphere that generate and shape the magnetic field. The intrinsically strong magnetic field is contained in seemingly isolated elements, ranging from the thick boundles of flux tubes in sunspots to the hypothetical thin flux fibers. These elements are arranged in the typical patterns observed in active regions and in the magnetic network. The processes of emergence of magnetic flux into the atmosphere and removal of flux from the photosphere are directly related to the magnetic structure and dynamics in the solar convective envelope.

075.010   Structure of the magnetic field and electric streams in an active region.
V. I. Abramenko, S. I. Gopasyuk.
Problems of solar flares, p. 32 – 40, 200 (1986). In Russian. Abstr. in Ref. Zh., 51. Astron., 7.51.334 (1987). – See Abstr. 012.018.

075.011   Local magnetic structures on the sun and energetics of solar flares.
N. N. Kontor, G. P. Lyubimov.
Problems of solar flares, p. 132 – 144, 209 (1986). In Russian. Abstr. in Ref. Zh., 51. Astron., 7.51.365 (1987). – See Abstr. 012.018.

075.012   Solar activity indices and periods of general magnetic field inversion of the Sun.
R. T. Gushchina, L. I. Dorman.
Kosm. luchi, Moskva, No. 24, p. 119 – 128 (1987). In Russian. Abstr. in Ref. Zh., 51. Astron., 7.51.398 (1987).

075.013   Transfer of a large–scale solar magnetic field caused by inhomogeneity of matter density of the convective zone.
V. N. Krivodubskij.
Pis'ma Astron. Zh., Tom 13, No. 9, p. 803 – 810 (1987). In Russian. English translation in Sov. Astron. Lett., Vol. 13.

The transfer velocity of the horizontal large–scale magnetic field in the Sun due to the radial density gradient of the solar turbulent plasma is calculated in a nonlinear approach for three mixing length models of the convective zone. The value of the transfer velocity is changed from 10^4cm/s in the surface layers to 10^2cm/s near the deep layers of the convective zone bottom. Maximum values of the stationary magnetic field in the deep layers of the convective zone are estimated which is conserved under the joint action of the magnetic buoyancy and downward flux of the magnetic field due to the radial gradient of the matter density.

075.014   Magnetic field geometry in the plasma ejected by flares: geometry at 1 A.U. from the sun.
M. S. Bobrov.
Pis'ma Astron. Zh., Tom 13, No. 9, p. 811 – 819 (1987). In Russian. English translation in Sov. Astron. Lett., Vol. 13.

The hodographs of the interplanetary field vector **B** near the Earth during 29 flare–induced geomagnetic disturbances are considered. It is found that **B** either rotates in one diretion, or oscillates, or moves chaotically. The chaotic component is also present in cases of regular rotation or oscillation. These features are interpreted as evidence of the closed, or nearly closed, magnetic field geometry in the plasma ejected by flares. The chaotic component may be due to external forces which cause local deformations of some parts of magnetic loops.

075.015   A potential calculation of multipoles of the solar global magnetic field.
V. P. Mikhajlutsa.
Soln. Dannye, Byull., 1987, No. 3, p. 53 – 57 (1987). In Russian.
A method is proposed for the calculation of multipoles of the solar magnetic field in a potential approximation without using spherical functions. The method is very convenient for an analysis of geometric and physical properties of multipoles of the lowest order at heights $> 1.8 \, R_\odot$.

075.016   The origin of rigidly rotating magnetic field patterns on the Sun.
N. R. Sheeley Jr., A. G. Nash, Y.–M. Wang.
Astrophys. J., Vol. 319, No. 1, p. 481 – 502 (1987).

Using analytical calculations and numerical simulations, the authors show that a meridional component of magnetic–flux transport will offset the shearing effect of differential rotation and give rise to rigidly rotating patterns of large–scale magnetic field. The nonaxisymmetric field attains a striped polarity pattern which rotates rigidly like a barber pole while its individual small–scale flux elements rotate at the differential rate of the latitudes they are crossing. On the Sun, the meridional transport is provided by supergranular diffusion possibly assisted by a small poleward flow. New sources of flux retard this process and exclude the rigid rotation from the sunspot belts until well into the declining phase of the sunspot cycle.

075.017   Activity waves of the solar magnetic field.
V. I. Makarov, A. A. Ruzmajkin, S. V. Starchenko.
Soln. Dannye, Byull., 1987, No. 5, p. 82 – 88 (1987). In Russian.
An asymptotic solution of generation equations for the solar magnetic field is given. A variation in the angular velocity of rotation with depth is assumed on the basis of helioseismic data. Mean helicity is calculated on the basis of mixing length theory. It is shown that three dynamo–waves of the magnetic field are excited.

075.018   Magnetic waves of solar activity.
V. I. Makarov, A. A. Ruzmaikin (*A. A. Ruzmajkin*), S. V. Starchenko.
Sol. Phys., Vol. 111, No. 2, p. 267 – 277 (1987).
An asymptotic solution of generation equations for the solar mean magnetic field is given and studied. The variation of

rotational angular velocity with depth is taken from helioseismo-logical data. Average helicity is prescribed according to the mixing length theory. It is shown that three dynamo waves of the magnetic field are excited. The first wave is generated at the surface layer and concentrates at latitudes of about 60°. Its activity becomes apparent in the poleward migration of the zone of polar faculae formation. The second more powerful wave of the field is excited in the center of the convection zone and its activity shows up in a sunspot cycle. The third wave which is similar to the first wave, is generated at the bottom of the convection zone and attenuates towards the surface. Its activity may appear as a three–fold reversal of the polar magnetic field.

075.019 On the uniqueness of the determination of the coronal potential magnetic field from line–of–sight boundary conditions.
J. J. Aly.
Sol. Phys., Vol. 111, No. 2, p. 287 – 296 (1987).

The author considers a simple model in which the coronal magnetic field **B** is assumed to be potential in the region between the solar surface Γ_0 and an exterior "source–surface" Γ_1 of arbitrary shape. He proves that the boundary value problem that determines **B** from the value B_l of its component on Γ_0 along either $\vec{l} = \hat{\omega}$ (orthoradial direction) or $\vec{l} = \hat{x}$ (fixed direction) has at most one solution. On the other hand, he shows that a solution can exist only if B_l satisfies some "solubility conditions".

075.020 Stimulated dissipation of magnetic discontinuities and the origin of solar flares.
E. N. Parker.
Sol. Phys., Vol. 111, No. 2, p. 297 – 308 (1987).

It is proposed that the principal cause of the confined solar flare is the dissipation of magnetic energy at the many small–scale pre–existing tangential discontinuities in the local bipolar mag-netic field. The discontinuities are a consequence of the continu-ous shuffling and intermixing of the footpoints of the bipolar field by the turbulent photospheric granules. The X–ray corona within the bipolar field is presumed to be a consequence of the continuing dissipation by reconnection at these discontinuities. A flare results when static deformation and/or internal agitation of the field stimulates the onset of rapid reconnection at the many small internal discontinuities. The discontinuities are partially exhausted by the flare, so that the post–flare X–ray emission of that particular loop is substantially below the pre–flare level for a period of some hours while the discontinuities are being rejuvenated.

075.021 Structure of elements of the solar magnetic field.
S. I. Gopasyuk, L. B. Demkina.
Izv. Krymskoj Astrofiz. Obs., Tom 76, p. 138 – 147 (1987). In Russian. English translation in Bull. Crimean Astrophys. Obs., Vol. 76.

The structure of the solar magnetic field elements has been investigated using observational data in Fe I λ5250.2 Å line obtained on a double magnetograph installed on the Tower Solar telescope of the Crimean Astrophysical observatory. It has been found that the magnetic field elements are elongated along the line of solar rotation. Weak magnetic fields of elements are compressed at the westward side. The authors conclude that magnetic field elements of the total magnetic field are moving westward relative to photospheric plasma with a velocity of 50 – 60 m/s. This exceeds the rotation rates of the plasma in the photosphere by 2.5 – 3% as measured by Doppler shift and coincides with the rotation rate of the chromosphere.

075.022 The system of electric currents and magnetic field structure in active regions.
V. I. Abramenko, S. I. Gopasyuk.
Izv. Krymskoj Astrofiz. Obs., Tom 76, p. 147 – 168 (1987). In Russian. English translation in Bull. Crimean Astrophys. Obs., Vol. 76.

Magnetic field structure of two active regions has been studied by comparing the observed magnetic field vector with the computed one in the frame of a potential model. The investiga-tion permitted to reveal the existence of a global electric current (of approximately 2×10^{12}A) distributed over the whole area of the active region. A system of local small–scale–current struc-tures is superimposed over the global current. The energy of the global current magnetic field is 10^{32}erg.

075.023 On the factor of Zeeman splitting of the Sr I λ4607.3 Å line.
M. D. Gusejnov.
Izv. Krymskoj Astrofiz. Obs., Tom 76, p. 169 – 172 (1987). In Russian. English translation in Bull. Crimean Astrophys. Obs., Vol. 76.

The spectral line Sr I λ4607.3 Å is being used in observations of magnetic fields on the sun. According to the theory of atom line splitting in weak magnetic fields, it can be considered as a normal Zeeman triplet with $g_t = 1,00$. Polarized spectrograms of differ-ent spots (9 spectrograms) are obtained by the author with the Echelle spectrograph of the Solar Tower telescope of the Crimean Observatory. Values of magnetic splitting for four lines with known experimental values of the Lande factor are compared with those of the Sr I line. The observed magnetic field splitting of the Sr I λ4607.3 Å line in the spectra of sunspots confirms the theoretical value of Zeeman splitting multiplier. It can be evaluated as $g = 1.00 \pm 0.02$.

075.024 Investigation of the magnetic field distribution in the undisturbed atmosphere of the sun.
D. N. Rachkovskij.
Izv. Krymskoj Astrofiz. Obs., Tom 76, p. 172 – 178 (1987). In Russian. English translation in Bull. Crimean Astrophys. Obs., Vol. 76.

The ratios of pulses observed with a magnetograph in undisturbed solar regions for Fe I lines with Lande factor being 2 or less can be easily explained provided that the magnetic field is plane parallel and concentrated in the upper levels of photo-sphere. Its optical depth in continuum is of the order of 0.04 or 200 km.

075.025 Properties of solar magnetic fluxtubes from only two spectral lines.
S. K. Solanki, C. Keller, J. O. Stenflo.
Astron. Astrophys., Vol. 188, No. 1, p. 183 – 197 (1987).

A procedure requiring only the Stokes V and Q profiles of two spectral lines (Fe I 5250.2 Å and Fe I 5247.1Å) to determine magnetic field strength, velocity and temperature inside solar magnetic fluxtubes, as well as their inclinations and filling factors with a minimum of a priori assumptions is presented. The procedure is then applied to spectra of the two lines obtained at various distances from the solar limb.

075.026 The decay of the large–scale solar magnetic field.
C. R. DeVore.
Sol. Phys., Vol. 112, No. 1, p. 17 – 35 (1987).

The author studies the time–asymptotic behavior of the large–scale photospheric magnetic field by solving analytically for the eigenstates of the transport equation. The nonaxisymmetric eigenstates are barberpole patterns of flux, localized near either the solar equator or the poles, where the rotational shear vanishes. They rotate rigidly at approximately the equatorial or polar rate, respectively, and decay on the geometric mean of the short time–scale for shearing by differential rotation and the long time–scale for dispersal by supergranular diffusion. The author also investigates the effects of meridional convection on the evolution of the field, by including a hypothetical poleward flow in the transport equation. Such a flow contributes to the decay of the nonaxisymmetric patterns on its own, intermediate time–scale, and also hastens the relaxation of the axisymmetric field to a modified dipolar configuration.

075.027 **On the spectrum of small–scale magnetic structures on the Sun.**
N. N. Kontor.
Vestn. Mosk. Univ., Ser. 3. Fiz. Astron., Tom 28, No. 4, p. 81 – 86 (1987). In Russian. Abstr. in Ref. Zh., 51. Astron., 12.51.461 (1987).

075.028 **Evolution of network magnetic fields of solar quiet regions.**
Z.–x. Shi, J.–x. Wang.
Chin. Astron. Astrophys., Vol. 11, No. 3, p. 221 – 228 (1987). English translation of Acta Astron. Sin., Vol. 28, No. 2, p. 111 – 119 (1987).
From a time sequence of magnetograms of a quiet region of 1983 October 14, the evolution of 300 network features was measured. The magnetograms have a spatial resolution of 2 to 3 arcsec and a time resolution of about 2 hr.

075.029 **Rotation of the sun and structure of magnetic elements.**
S. I. Gopasyuk, L. B. Demkina.
Sov. Astron., Vol. 31, No. 2, p. 205 – 207 (1987). English translation of 43.075.010.

Correlation properties of fine–structured magnetic fields.
See Abstr. 062.014.

Propagation of compressive waves through fibril magnetic fields. II. Scattering from a slab of magnetic flux tubes.
See Abstr. 062.019.

Propagation of compressive waves through fibril magnetic fields. III. Waves that propagate along the magnetic field.
See Abstr. 062.020.

Electric current sheet formation in a magnetic field induced by continuous magnetic footpoint displacements.
See Abstr. 062.158.

Stokes profile analysis and vector magnetic fields. I. Inversion of photospheric lines.
See Abstr. 063.109.

The polarization properties of model sunspots: the broad–band polarization signature of the Schlüter–Temesvary representation.
See Abstr. 063.110.

Convection, magnetic fields, and line asymmetry in the sun and stars.
See Abstr. 071.001.

Solar emission lines revisited: extended study of magnesium.
See Abstr. 071.011.

Large–scale structure of the solar magnetic field and distribution of sunspot groups with an anomalous inclination of the bipolar axis to the equator.
See Abstr. 072.009.

The early stage of the AR 135 (S.D.) evolution (June 1984).
See Abstr. 072.011.

Stokes profile analysis and vector magnetic fields. III. Extended temperature minima of sunspot umbrae as inferred from Stokes profiles of Mg I λ4571.
See Abstr. 072.030.

Generation and structure of the electric currents in a flaring activity complex.
See Abstr. 072.053.

On the large–scale dynamics and magnetic structure of solar active regions.
See Abstr. 072.087.

Stabilizing effect of a transverse magnetic field in current sheets of solar flares.
See Abstr. 073.006.

Concerning the relationship between the filament appearance and magnetic field changes in an active region.
See Abstr. 073.021.

Rotation of individual background magnetic field components during the formation of the white–light flare region of April 1984 (NOAA 4474).
See Abstr. 073.092.

Microflares in the solar magnetic network.
See Abstr. 073.101.

Rotation of the coronal magnetic field.
See Abstr. 074.002.

The theory of magnetic coronal heating.
See Abstr. 074.017.

Behaviour of the coronal intensity in the line 5303 Å and latitude zonal structure of the magnetic field. Period 1944 – 1974.
See Abstr. 074.043.

Theoretical study of onset conditions for solar eruptive processes.
See Abstr. 074.046.

Rotation of the Sun and rotation of its general magnetic field.
See Abstr. 080.042.

Solar oscillations and magnetograph signal formation in the presence of differences in brightness of the splitting components.
See Abstr. 080.060.

On a relationship between the interplanetary magnetic field and the large–scale distribution of the solar magnetic field.
See Abstr. 106.008.

076 UV, X, Gamma Radiation

076.001 **Why do stars emit X rays?**
E. N. Parker.
Phys. Today, Vol. 40, No. 7, p. 36 – 42 (1987).
Careful study of our closest star, the Sun, suggests that bundles of twisted magnetic flux tubes extending from subsurface layers may account for the surprising prevalence of X rays from most rather ordinary stars.

076.002 **Solar soft X–ray pulsations.**
R. A. Harrison.
Astron. Astrophys., Vol. 182, No. 2, p. 337 – 347 (1987).
Using data from the Hard X–ray Imaging Spectrometer on the Solar Maximum Mission, soft X–ray (3.5 – 5.5 keV) pulsations are identified, which originate from a compact active region which lies at one footpoint of a large coronal loop. It is believed

that this is the first report of soft X–ray pulsations from the non-flaring sun. The pulsations were of period 24 min and were detected for six hours. The periodicity is thought to be produced by a standing wave or a travelling wave "packet" which exists within the loop. The candidates for the wave are fast or Alfvén MHD modes of Alfvénic surface waves.

076.003 Hard X–ray emission processes in solar flares.
N. Vilmer.
Sol. Phys., Vol. 111, No. 1, p. 207 – 223 (1987). – See Abstr. 012.016.
Solar hard X–ray emission is one of the most direct diagnostics of accelerated particles during solar flares. In this review, the current understanding of hard X–ray emission processes is discussed: first the different emission mechanisms (in particular inverse Compton radiation, energetic ion or electron bremsstrahlung) are presented and the plausibility of each of these mechanisms is discussed. Then, different types of hard X–ray models (thermal or non–thermal, homogeneous or inhomogeneous emission regions) are presented together with the comparison of their predictions with X–ray observations (spectral, spatial and temporal informations – directivity and polarization).

076.004 Hard X–ray and radio emission at the onset of great solar flares.
K.-L. Klein, M. Pick, A. Magun, B. R. Dennis.
Sol. Phys., Vol. 111, No. 1, p. 225 – 233 (1987). – See Abstr. 012.016.
A study of the onset phase of ten great hard X–ray bursts is presented. It is shown from hard X–ray and radio observations in different wavelength ranges that the energization of the electrons proceeds on a global time–scale of some tens of seconds. In nine of the bursts two phases of emission can be distinguished during the onset phase: the pre–flash phase, during which emission up to an energy limit ranging from some tens of keV to 200 keV is observed, followed ten to some tens of seconds later by the flash phase, where the count rate in all detector channels rises simultaneously to within some seconds. For two of the events strong γ–ray line emission is observed and is shown to start close to the onset of the flash phase.

076.005 Directivity and polarization of hard X–ray emission of solar flares.
S. V. Bogovalov, S. R. Kel'ner, Yu. D. Kotov.
Problems of solar flares, p. 156 – 168, 211 (1986). In Russian. Abstr. in Ref. Zh., 51. Astron., 7.51.351 (1987). – See Abstr. 012.018.

076.006 High–resolution spectroscopy of solar X–ray radiation.
I. A. Zhitnik, V. V. Korneev, V. V. Krutov, S. N. Oparin, A. M. Urnov.
Tr. Fiz. Inst. Akad. Nauk SSSR, Tom 179, p. 39 – 59 (1987). In Russian. Abstr. in Ref. Zh., 51. Astron., 9.51.331 (1987).

076.007 Studies on X–ray and gamma–ray bursts associated with the solar limb event of 1981 April 27.
H.–q. Zhang, E. L. Chupp.
Publ. Purple Mt. Obs., Vol. 6, No. 3, p. 295 – 309 (1987).
Using the Hα observations of the Astrophysical Observatory in Catania, Italy and hard X–ray and gamma–ray burst data from the Solar Maximum Mission Gamma–ray Spectrometer, a major eruptive loop prominence was studied during the limb solar flare event of 1981 April 27. Preliminary analysis shows that there seems to exist a second abrupt energy release for this event, almost 20 minutes after the end of the impulsive phase of the flare. This energy release is probably associated with the rapidity in upward motion or activation of the loop prominence. A possible candidate for such a process could be the reconnection of the old magnetic field with a newly emerging magnetic field.

076.008 Calculations of X–ray anisotropy for energies > 150 keV and its comparison with PVO/ISEE–3 observations.
N. Nazir, R. R. Rausaria, P. N. Khosla.
Publ. Astron. Soc. Jpn., Vol. 39, No. 5, p. 761 – 768 (1987).
The evolution of electron energy and angular distributions has been studied at different levels in the solar atmosphere by combining a small–angle analytical treatment with large–angle Monte Carlo calculations for electron energies greater than 500 keV. Using these distributions energy spectra and angular distributions of photons for energies greater than 150 keV have been computed as a function of height. The anisotropy ratio for these photon energies first decreases then increases with decrease in height. The results are compared with the observations of PVO/ISEE–3.

076.009 Study of electron energy and angular distributions and calculations of X–ray, EUV line flux and rise times.
R. Bakaya, S. Peshin, R. R. Rausaria, P. N. Khosla.
J. Astrophys. Astron., Vol. 8, No. 3, p. 263 – 270 (1987).
Evolution of energy and angular distributions of electrons has been studied by combining small–angle analytical treatment with large–angle Monte Carlo calculations as a function of column density for initially monoenergetic and monodirectional electrons. Time evolution of extreme ultraviolet (EUV) spectrum has been studied. The slopes of the curves calculated compare well with the experimentally observed curve.

076.010 X–ray emission of solar flares generated by anisotropic electron beams.
S. V. Bogovalov, S. R. Kel'ner, Yu. D. Kotov.
Astron. Zh., Tom 64, Vyp. 6, p. 1280 – 1290 (1987). In Russian. English translation in Sov. Astron., Vol. 31, No. 6.
For three types of the initial angle distribution of fast electrons, energy spectra, directivity, and polarization of the bremsstrahlung have been computed with an account for multiple scattering and energy losses. The influence of Compton scattering and of photoabsorption on the observed hard X–ray emission of solar flares has been investigated. It is obtained that the photon spectrum index depends not only on the spectrum of electrons but also on the registered energy range and on the angle of view of the flare. In the 10 – 40 keV range the spectrum is softer at the limb than in the solar disc centre; in the 60 – 360 keV the situation is reverse, the spectrum being softer in the solar disc centre.

076.011 The directivity of high–energy emission from solar flares: Solar Maximum Mission observations.
W. T. Vestrand, D. J. Forrest, E. L. Chupp, E. Rieger, G. H. Share.
Astrophys. J., Vol. 322, No. 2, p. 1010 – 1022 (1987).
The data base consisting of flares detected by the gamma–ray spectrometer on board the SMM satellite is used to study the directivity of high–energy radiation. The authors analyze the position distribution of gamma–ray flares and show that one can rule out the hypothesis that the emission is isotropic. This conclusion is reinforced by the observed difference in the spectral index distributions for limb and disk flares. Evidence for center-to–limb spectral variations at both hard X–ray and gamma–ray energies is presented. Further, there is evidence for a strong radiation anisotropy at energies > 10 MeV. Some of the implications of these results are discussed.

X–ray photographs of a solar active region with a multilayer telescope at normal incidence.
See Abstr. 035.040.

Cross sections for production of the 15.10 MeV and other astrophysically significant gamma–ray lines through excitation and spallation of ^{12}C and ^{16}O with protons.
See Abstr. 061.067.

Proton–electron bremsstrahlung polarization: solar flare X–ray and gamma–ray polarization.
See Abstr. 063.122.

Ne IX emission–line ratios in solar active regions.
See Abstr. 072.029.

Analysis of ultraviolet and X–ray observations of three homologous solar flares from SMM.
See Abstr. 073.001.

Comparison of the total energy release in the optical and hard X–ray ranges of solar flares (Catalog).
See Abstr. 073.010.

Microwave radiation from a dense magneto–active plasma.
See Abstr. 073.013.

Soft X–ray line profiles in the impulsive phase of electron–heated solar flares.
See Abstr. 073.015.

Hard X–ray emission in electron–heated solar flares – a comparison of nonthermal and thermal contributions.
See Abstr. 073.016.

Acceleration and energization by currents and electric fields.
See Abstr. 073.028.

Evidence for interacting loop process in a phase of the May 16, 1981 flare.
See Abstr. 073.029.

Observation of nonthermal energy distributions during the impulsive phase of solar flares.
See Abstr. 073.042.

A determination of the ^{3}He/H ratio in the solar photosphere from flare gamma–ray line observations.
See Abstr. 073.043.

Short solar events with evidence of repetitive structures.
See Abstr. 073.044.

Skylab XUV observations of densities, thermal structure, and mass motions in a compact flare.
See Abstr. 073.068.

Nuclear processes and accelerated particles in solar flares.
See Abstr. 073.080.

The Hα spectral counterparts of solar hard X–ray microflares.
See Abstr. 073.081.

The Hα characteristics of hard X–ray bursts.
See Abstr. 073.089.

X–ray observations of solar flares with the Einstein Observatory.
See Abstr. 073.100.

Microflares in the solar magnetic network.
See Abstr. 073.101.

On the relationship of flare size and particle anisotropy in solar gamma–ray flares.
See Abstr. 073.102.

Transient ionization and solar flare X–ray spectra.
See Abstr. 073.103.

On the inability of magnetically constricted transition regions to account for the 10^5 to 10^6K plasma in the quiet solar atmosphere.
See Abstr. 074.064.

Properties of solar coronal active regions deduced from X–ray line spectra.
See Abstr. 074.080.

A high–energy solar flare burst complex and the physical properties of its source region.
See Abstr. 077.010.

Repetition rates of fast pulses in a solar burst observed at mm–waves and hard X–rays.
See Abstr. 077.023.

Study on solar X–ray bursts based on abnormal phenomena of LF wave propagation.
See Abstr. 083.029.

077 Radio, Infrared Radiation

077.001 **Solar brightness temperature at submillimeter wavelengths.**
R. T. Boreiko, T. A. Clark.
Astrophys. J., Vol. 318, No. 1, p. 445 – 450 (1987). = Rothney Astrophys. Obs. Publ., No. 41.
The brightness temperature of the Sun over the spectral range $(20 – 90 \text{ cm}^{-1})$ for which the source depths bracket the temperature minimum region of the solar atmosphere has been determined from calibrated high–resolution spectra obtained at balloon altitudes with a Michelson interferometer. The solar spectral continuum level was determined by utilizing high spectral resolution (0.015 cm^{-1}). Calibration was performed in flight using a 1165K blackbody source. The measured temperature minimum of 4300 $(+130, –200)$K at about 65 cm^{-1} is in close agreement with the prediction of model M of the solar atmosphere proposed by Vernazza, Avrett, and Loeser.

077.002 **Solar Radio Data. 1987 April – December.**
Sol. Radio Data, Part II – IV (1987). In microfiche form.

077.003 **Night–time reception of solar radio bursts over wide bandwidths.**
J. J. Riihimaa.
Earth, Moon, Planets, Vol. 38, No. 3, p. 305 – 308 (1987).
Spectra of night–time solar radio bursts observed over wide bandwiths are described. The bursts covered frequency ranges of 20–30 and 24–33 MHz without displaying any abrupt frequency structures. It seems that a scatter mechanism of some kind is involved in the night–time propagation.

077.004 **On the nature of fluctuations of solar radio noise storms of type I.**
A. A. Gnezdilov, V. V. Fomichev.
Pis'ma Astron. Zh., Tom 13, No. 8, p. 704 – 709 (1987). In Russian. English translation in Sov. Astron. Lett., Vol. 13.
A possible mechanism of fluctuations and quasi–periodic variations of intensity of solar radio noise storms is discussed. The mechanism is based on the development of relaxation oscillations of plasma parameters in adiabatic traps which are the coronal magnetic loops. The power of energetic particle sources

in radio sources is estimated using the observed period of radio emission variations.

077.005   Depression of radio emission of fine structures in the region of transition temperatures of the solar atmosphere.
A. S. Grebinskij.
Pis'ma Astron. Zh., Tom 13, No. 8, p. 710 – 715 (1987). In Russian. English translation in Sov. Astron. Lett., Vol. 13.
It is shown that the matter of the solar atmosphere in the region of transition temperatures $2 \times 10^4 - 2 \times 10^5$K gives practically no contribution to the total radio emission of the quiet sun, if the bulk of this matter is confined in unresolved fine structures. Taking into account the depression of radio emission from this part of the atmosphere gives good agreement between calculated and observed radio spectra in a wide wavelength band (1 – 100 cm).

077.006   Ion–collision broadening of solar lines in the far–infrared and submillimeter spectrum.
D. Hoang–Binh, P. Brault, J. Picart, N. Tran–Minh, O. Vallée.
Astron. Astrophys., Vol. 181, No. 1, p. 134 – 137 (1987).
The authors give new theoretical data for calculating the ion impact broadening of solar far–infrared and submillimeter lines. These are more accurate than, and should be preferred to data extrapolated directly from Griem's (1967) work on radio recombination lines. The variation of the line width as a function of the upper level of the transition has also been investigated. It may be a potentially important diagnostic tool for determining the physical conditions in the solar atmosphere.

077.007   A method of reduction of one–dimensional scans of the Sun.
V. N. Borovik, N. G. Peterova.
Soln. Dannye, Byull., 1987, No. 1, p. 66 – 70 (1987). In Russian.
A method is given for separation of scans taken during solar passage through the diagram of the Large Pulkovo Radio Telescope into B and S components. The authors give here a way of a reduction from antenna temperatures to the units of measurement of the spectral density of an emission flux and a determination of the degree of circular polarization of local sources. The directivity diagram of the Large Pulkovo Radio Telescope necessary for an estimation of the size of local sources is reported. The causes of the errors and their value in determination of different characteristics of local sources are noted.

077.008   A high–frequency "zebra"–pattern in radio emission of solar flares.
O. G. Gontarev, A. P. Classen.
Soln. Dannye, Byull., 1987, No. 1, p. 87 – 93 (1987). In Russian.
The results of observations of the flares on August 1, 1983 and April 24, 1985 made with a spectrograph in the 1 – 2 GHz ranges are discussed. It is shown that a high–frequency "zebra"–pattern with $\Delta f / f \leqslant 0.1$ and strips amplitude $\geqslant 25\%$ of continuous spectra level occurs at a flash phase of these bursts. For interpretation of the "zebra"–pattern the plasma mechanism is attracted. Appearance of strips is explained as a result of interaction of the plasma wave hybrid band and the electron–cyclotron harmonics in a magnetic loop trap. The flare source parameters are discussed.

077.009   On some characteristics of precursors of solar impulsive microwave bursts in the 8 – 12 GHz range.
Yu. V. Tikhomirov, V. M. Fridman, O. A. Shejner.
Soln. Dannye, Byull., 1987, No. 2, p. 70 – 77 (1987). In Russian.
Multi–frequency observations (in the 8 – 12 GHz frequency range) of solar burst's precursors during November 1981 are analysed. It is shown that the step–like increase in total intensity precedes at least 30% subsequent solar flares. The temporal behavior of precursor's spectra and the quasi–periodic component of fluctuations are investigated.

077.010   A high–energy solar flare burst complex and the physical properties of its source region.
C. de Jager, J. Kuijpers, E. Correia, P. Kaufmann.
Sol. Phys., Vol. 110, No. 2, p. 317 – 326 (1987).
The authors discuss a solar flare microwave burst complex, which included a major structure consisting of some 13 spikes of 60 ms FWHM each, observed 21 May, 1984 at 90 GHz (3 mm). It was associated with a simultaneous very hard X–ray burst complex. The authors suggest that the individual spikes of both bursts were caused by the same electron population: the X–bursts by their bremsstrahlung, and the microwave bursts by their gyrosynchrotron emission.

077.011   Fine structures in solar microwave flares.
N. Gopalswamy.
Sol. Phys., Vol. 110, No. 2, p. 327 – 335 (1987).
The pulsed electron acceleration and release from the energy release volume in solar flares implies that there is a possibility of interaction between a group of electrons reflected from the foot of a bipolar flux tube with a newly injected beam. It is shown that interaction can lead to the stoppage of the synchrotron maser instability caused by the loss cone distribution and hence can produce further millisecond fine structures in the solar microwave bursts.

077.012   On the intermediate drift burst model.
G. Mann, M. Karlický, U. Motschmann.
Sol. Phys., Vol. 110, No. 2, p. 381 – 389 (1987).
A modification of the presently existing intermediate drift burst model by Kuijpers (1975) and Bernold (1983) is suggested. It is shown that whistler solitons cannot be responsible for intermediate drift bursts. Here, they are interpreted as the radio signature of localized formstable whistler wave packets propagating along the magnetic field in a coronal loop. In the frame of this modified model, the magnetic field strengths derived from fiber burst data agree with previous estimates by Dulk and McLean (1978).

077.013   Chaotic behaviour of solar radio flux.
L. Romanelli, M. A. Figliola, F. A. Hirsch, S. M. Radicella.
Sol. Phys., Vol. 110, No. 2, p. 391 – 395 (1987).
The presence of a chaotic attractor is investigated in time series of 10.7 cm solar flux. The correlation dimension and the Kolmogorov entropy have been calculated for the time period 1964 – 1984. The values found for the Kolmogorov entropy show that chaos is indeed present. The correlation dimension found for high solar activity is 3.3 and for low solar activity is 4.5, indicating that a low–dimension chaotic attractor is present in the time series analysed.

077.014   Das solare Radio–Minimum: September/Oktober 1986.
K. I. Malde.
Sonne, Jahrg. 11, Nr. 42, p. 38 – 39 (1987).
A precise determination of the time of minimum solar activity using radio flux data of the Algonquin Radio Observatory is presented.

077.015   Active latitudinal zones and the radio–fluctuation power spectrum.
N. S. Nesterov.
Bull. Crimean Astrophys. Obs., Vol. 75, p. 61 – 70 (1987). English translation of 42.077.016.

077.016   Radio–emission fluctuations at 2.25 cm from the quiet sun.
L. I. Tsvetkov.
Bull. Crimean Astrophys. Obs., Vol. 75, p. 71 – 77 (1987). English translation of 42.077.017.

077.017 Quasiregularity in the burst component of solar noise storms.
Yu. F. Yurovskij, O. Alvarez.
Bull. Crimean Astrophys. Obs., Vol. 75, p. 78 – 82 (1987).
English translation of 42.077.018.

077.018 Effects of turbulence development in solar surges.
V. Carbone, G. Einaudi, P. Veltri.
Sol. Phys., Vol. 111, No. 1, p. 31 – 44 (1987). – See Abstr. 012.016.

Several authors have claimed for correlations between surges (dark features) and various kinds of solar emissions (radio, microwave, X–ray). The authors propose a model to explain such correlations, in particular presenting the properties of the instabilities resulting from the coupling between material flow, connected to the appearance of a surge, and magnetic field topology. As a consequence of such instability a turbulent energy cascade to small characteristic lengths grows up. Depending of the relevant parameters of the surge (dark feature), different regimes can be found, producing different levels of electrons acceleration and mass motion deceleration. The authors try to correlate the different developments of the instability with the behavior observed in type I and type III radio bursts related to surges.

077.019 Type II solar radio emission – a self–consistent approach.
G. Thejappa.
Sol. Phys., Vol. 111, No. 1, p. 45 – 51 (1987). – See Abstr. 012.016.

An attempt is made to construct a self–consistent model for type II radio bursts. It is proposed that a majority of the type II shocks are super–critical and the reflected ions from such type II shock fronts are described by the drifted Maxwellian in the upstream and by the Dory–Guest–Harris distribution in the downstream. The low–frequency waves excited by these ions accelerate electrons resonantly along the field lines both in the upstream as well as in the downstream, which are responsible for the lower–frequency and upper–frequency bands in the dynamic spectrum of a type II radio burst. The functional behaviour of the distribution functions of the accelerated electrons is the same in both the cases whereas the number densities of the accelerated electrons in the downstream is smaller than that in the upstream.

077.020 A type IV burst associated with a coronal streamer disruption event.
M. R. Kundu.
Sol. Phys., Vol. 111, No. 1, p. 53 – 57 (1987). – See Abstr. 012.016.

A type IV burst was observed on February 17, 1985 with the Clark Lake Radio Observatory multifrequency radioheliograph operating in the frequency range 20 – 125 MHz. This burst was associated with a coronal streamer disruption event. From two-dimensional images produced at 50 MHz, evidence is shown of a type II burst and a slow moving type IV burst. The observations of the moving type IV burst suggests that a plasmoid containing energetic electrons can result from the disruption of a coronal streamer.

077.021 Microwave sources of solar flares: loop top or foot points?
K. Kai.
Sol. Phys., Vol. 111, No. 1, p. 81 – 87 (1987). – See Abstr. 012.016.

The microwave images of solar flares obtained with the VLA are reviewed from a standpoint that the microwave source is near the top or foot point(s) of a flaring loop. The review is focused on whether extended structure is missed due to the lack of short baseline components leading to an incorrect interpretation of the processed images. The author concludes that at short cm (< 2 cm) there is no conclusive evidence for the source near the loop top whereas at longer cm (6, 20 cm) the source tends to occupy a significant portion of a loop. The observed bipolar structure could be unambiguously interpreted as evidence for the

source at the loop top, only when it is confirmed that a more extended structure has not been missed.

077.022 Beat structure in pulsating type IV solar radio bursts and a possible mechanism.
H.–W. Li, M. Messerotti, P. Zlobec.
Sol. Phys., Vol. 111, No. 1, p. 137 – 142 (1987). – See Abstr. 012.016.

Pulsating type IV solar radio bursts with beat structure are presented and analysed. Based upon the theory of whistler soliton emission the authors interpret the beat structure by the combination of two components with different pulsation frequencies due to radial oscillations of two legs of the magnetic loop. The large depth of pulsation is also explained in this model.

077.023 Repetition rates of fast pulses in a solar burst observed at mm–waves and hard X–rays.
E. Correia, P. Kaufmann.
Sol. Phys., Vol. 111, No. 1, p. 143 – 154 (1987). – See Abstr. 012.016.

The solar burst of 21 May, 1984, 13:26 UT, showed radio spectral emission with a turnover frequency above 90 GHz, well correlated in time with the hard X–ray emission. It consisted of seven major time structures (1 – 3 s in duration), of which each was composed of several fast pulses with rise times between 30 and 60 ms. The spectral indices of the millimeter and hard X–ray emission exhibited sudden changes during each major time structure. The subsecond pulses were nearly in phase at 30 and 90 GHz, but their relative amplitude at 90 GHz ($\approx 50\%$) were considerably larger than at 30 GHz (< 5%). It was also found that the 90 GHz and the 100 keV X–ray fluxes were proportional to the repetition rate of the subsecond pulses, and that the hard X–ray power law index hardens with increasing repetition rate.

077.024 Electron cyclotron maser emission from solar flares.
L. Vlahos.
Sol. Phys., Vol. 111, No. 1, p. 155 – 166 (1987). – See Abstr. 012.016.

A review of the recent results on the theory of electron cyclotron maser (ECM) is presented. Following questions are discussed: (a) What are the characteristics of the linear growth rate of the ECM during solar flares? (b) How does the ECM saturate and what is its efficiency? (c) How does the ECM generated radiation modify the flare environment? Finally outstanding questions in the theory of ECM are reviewed and the theoretical predictions are related to current observations.

077.025 The saturation of electron–cyclotron maser and the time profile of emitted spikes.
H.–W. Li.
Sol. Phys., Vol. 111, No. 1, p. 167 – 173 (1987). – See Abstr. 012.016.

The evolution of the hollow beam distribution of energetic electrons giving rise to ECM instability is investigated and the spatial dispersion term is included in the equation of wave energy. The instability causes the growth of wave energy, while the propagation of waves evacuates the electromagnetic energy from the source region. By analysing these two effects spike–like time profiles of waves are obtained. It is found that the saturation time t_s of ECM emission and the duration of spikes increase with the decrease of the frequency of solar radio spike emission. The approximate expressions of t_s and of the peak wave energy density are derived.

077.026 Harmonic emission and polarization of millisecond radio spikes.
A. O. Benz, M. Güdel.
Sol. Phys., Vol. 111, No. 1, p. 175 – 180 (1987). – See Abstr. 012.016.

The spectral distribution of millisecond radio spikes observed by the Zürich spectrometers in the 200 – 1100 MHz range has been studied. In one event out of a total of 36 the authors found clearly developed harmonic structure. The ratio between the two bands of emission was 1:1.39 ± 0.01. They have also determined the sense of circular polarization of the spike events and compared it to the magnetic polarity of the leading spot of the flaring active region. According to the "Leading Spot Rule" the majority of the events (10 out of 13) were emitted in the ordinary mode.

077.027 Evidence of harmonic microwave radiation during solar flares.
M. Stähli, A. Magun, E. Schanda.
Sol. Phys., Vol. 111, No. 1, p. 181 – 188 (1987). – See Abstr. 012.016.

First observational evidence of harmonic radiation at microwave frequencies during solar bursts is presented for the event of April 28, 1983. The recordings between 3.1 and 19.6 GHz show a typical continuum with a spectral maximum near 5.2 GHz. Superimposed fine structures with durations in the order of some seconds exhibit a very unusual spectral behaviour. Narrow–banded intensity peaks appeared at 5.2 and 11.8 GHz which were barely visible at 3.1, 8.4 and 19.6 GHz. These structures can be interpreted as harmonic emission. Harmonic radio emission can be generated either by plasma radiation, gyroradiation, electron–cyclotron maser or by nonlinear conversion processes. However, all of those mechanisms require extreme assumptions on the source and the ambient plasma in order to account for the observations.

077.028 Low–level decimetric (1.6 GHz) solar burst activity.
H. S. Sawant, J. E. R. Costa, R. H. Trevisan, C. J. B. Lattari, P. Kaufmann.
Sol. Phys., Vol. 111, No. 1, p. 189 – 199 (1987). – See Abstr. 012.016.

Observations of solar bursts at 1.6 GHz were carried out in the month of July 1985 for about two weeks. Five intervals of solar burst activity, each one lasting for a couple of minutes, were observed. Predominantly, two classes of fast bursts were observed: viz: "spike" and "blips". However, some of these bursts were two orders of magnitude less intense than those reported earlier. This investigation suggests that blips probably originate at second harmonic by beam plasma interaction as that of metric type III bursts. Also, low–level ms–spikes with the half power duration in the range of 5 to 20 ms suggest that source sizes be smaller than 50 km if the process of emission is electron–cyclotron maser.

077.029 Direct generation of solar and stellar radio bursts by energetic electron maser.
P. Louarn, D. Le Quéau, A. Roux.
Sol. Phys., Vol. 111, No. 1, p. 201 – 206 (1987). – See Abstr. 012.016.

A fully relativistic electron maser is proposed for the explanation of certain non–thermal solar and stellar radio bursts. This mechanism (maser synchrotron) is based on a gyroresonant interaction between waves and electrons of high energies and uses the free energy contained in an electronic distribution function that peaks for energies around 1 MeV. The synchrotron maser instability appears to be a direct and efficient amplification process for considerably larger physical conditions than the cyclotron maser.

077.030 A statistical study of properties of microwave bursts.
R. E. Rodriges.
Soln. Dannye, Byull., 1987, No. 3, p. 79 – 87 (1987). In Russian.

This study was made on the basis of an analysis of the spectrum of the emission flux density at $\lambda = 1 - 21$ and polarization at $\lambda = 4.5$ cm for ~600 bursts. It is found that the distribution of the polarizaton degree keeps constant in 85% of the cases. It is also shown that the spectrum pattern for most bursts (>95%) does not depend on burst morphology. A generalized inversion curve for bursts is given, on whose basis one can suggest that the region of polarized emission of the burst has a smaller size than that of the corresponding source of the S–component and is displaced relative to the source. The absence of regularities in behaviour of various properties of bursts is discussed.

077.031 Radio evidence for a magnetic mirror effect on beams of subrelativistic electrons in the solar corona.
C. Caroubalos, M. Poquérusse, J.–L. Bougeret, R. Crépel.
Astrophys. J., Vol. 319, No. 1, p. 503 – 513 (1987). With plates 8 – 10.

The authors present new radio spectral observations of the Sun which reveal a magnetic mirror effect on impulsively injected beams of subrelativistic electrons $(0.1 - 0.3\ c)$ in coronal arch structures. The authors report the detection of a third branch following the descending branch of some U–bursts, suggesting the capital letter N on the dynamic spectra. The authors call these events "N–bursts". A detailed analysis of the digital time profiles shows that the duration of the beam emission regularly increased with time along the three successive branches of the N–bursts. It is demonstrated that the N–bursts actually trace reflections on magnetic mirrors near the feet of magnetic arches.

077.032 VLA observations of a solar noise storm.
K. R. Lang, R. F. Willson.
Astrophys. J., Vol. 319, No. 1, p. 514 – 519 (1987).

The authors present the first VLA observations of the Sun at 92 cm wavelength (328 MHz). A solar noise storm, which lasted at least 3 hr, was detected at this wavelength; it consisted of burstlike spikes superposed on a slowly varying background, and both storm components were 95% ± 5% right–hand circularly polarized. A long–duration soft X–ray event preceded the radio radiation by 30 minutes. The 92 cm noise storm was resolved with an angular resolution of 9″ for time intervals as short as 13 s. The observations confirm previously reported trends for a decrease in source size at higher frequencies, but they suggest a hitherto unresolved complexity in source structure.

077.033 Parameters of solar radio plages and their relations to other active region characteristics.
J. Kleczek, B. Ružičková–Topolová.
Bull. Astron. Inst. Czech., Vol. 38, No. 5, p. 262 – 271 (1987).

Radio spectroheliograms for 9.1 cm and 21 cm wavelength and other data on solar active regions published in Solar–Geophysical Data have been used to deduce characteristic parameters of radio plages (670 cases). Their relations to optical and magnetic parameters have been statistically investigated. The distribution functions of the radio characteristics (area, maximum brightness temperature, luminosity) are represented graphically. Correlation coefficients between selected radio and optical parameters of associated active regions are given in tabular form. A conspicuous enhancement of radio characteristic values has been found in the radio plages with underlying spot groups of $\beta\gamma$ and δ magnetic configurations. The results are in agreement with conclusions of many previous papers and complement them in various respects.

077.034 On the interpretation of the frequency spectrum of the millimetre–wave radiation from solar active regions.
E. Ya. Zlotnik.
Astron. Zh., Tom 64, Vyp. 5, p. 1088 – 1093 (1987). In Russian. English translation in Sov. Astron., Vol. 31, No. 5.

Peculiarities of the s–component of millimetre– and short centimetre–wave solar radio emission are considered. Physical conditions in the source region (temperature, chromospheric and coronal densities) which determine the form of the frequency spectrum (a monotonic decrease in intensity with wavelength or a complex spectrum with the minimum intensity at $\lambda \lesssim 2$ cm and increasing intensity towards shorter wavelengths) are studied.

077.035 Confirmation of the existence of the $2^2P_{3/2}$–$2^2S_{1/2}$ (3.05 cm) hydrogen line in the solar radio radiation, and brightness distribution of this line over the solar disk from observations with RATAN–600.
A. F. Dravskikh, Z. V. Dravskikh.
Spets. Astrofiz. Obs. Akad. Nauk SSSR, Prepr., No. 42L (1987). In Russian. Abstr. in Ref. Zh., 51. Astron., 9.51.352 (1987).

077.036   10.7 cm solar radio flux and the magnetic complexity of active regions.
R. M. Wilson, D. Rabin, R. L. Moore.
Sol. Phys., Vol. 111, No. 2, p. 279 – 285 (1987).

During sunspot cycles 20 and 21, the maximum in smoothed 10.7 cm solar radio flux occurred about 1.5 yr after the maximum smoothed sunspot number, whereas during cycles 18 and 19 no lag was observed. Thus, although 10.7 cm radio flux and Zürich sunspot number are highly correlated, they are not interchangeable, especially near solar maximum. The number of sunspots in an active region is one measure of the complexity of the magnetic structure of the region, and the coincidence in the maxima of radio flux and number of sunspots apparently reflects higher radio emission from active regions of greater magnetic complexity. The presence of a lag beween sunspot–number maximum and radio–flux maximum in some cycles but not in others argues that some aspect of the average magnetic complexity near solar maximum must vary from cycle to cycle.

077.037   Imaging observations of the evolution of meter–decameter burst emission during a major flare.
N. Gopalswamy, M. R. Kundu.
Sol. Phys., Vol. 111, No. 2, p. 347 – 363 (1987).

The authors present the results of a study of the evolution of 3 February, 1986 flare at meter–decameter wavelengths using the two dimensional imaging observations made with the Clark Lake multifrequency radioheliograph. The flare was complex and produced various types of meter–decameter bursts. The preflare activity was observed in the form of type III bursts some tens of minutes prior to the impulsive onset. From the positional analysis of the preflare and impulsive phase type III bursts and other measured characteristics the authors discuss the characteristics of energy release and possible magnetic field configurations in the vicinity of energy release region. From positional and temporal studies of the flare continuum and type II burst in relation to the microwave and hard X–ray emissions, they discuss the possible magnetic field structures in which the accelerated particles are confined or along which they propagate. They develop a schematic model of the flaring region.

077.038   Herringbone bursts associated with type II solar radio emission.
I. H. Cairns, R. D. Robinson.
Sol. Phys., Vol. 111, No. 2, p. 365 – 383 (1987).

The authors report detailed observations of the herringbone (HB) fine structure on type II solar radio bursts. Data from the Culgoora radiospectrograph, radiometer and radioheliograph are analyzed. The authors determine the characteristic spectral profiles, frequency drift rates and exciter velocities, fluxes, source sizes, brightness temperatures, and polarizations of individual HB bursts. Correlations between individual bursts within the characteristic groups of bursts and the properties of the associated type II bursts are examined. The data are compatible with HB bursts being radiation at multiples of the plasma frequency generated by electron streams accelerated by the type II shock. The authors conclude that HB bursts are physically distinct phenomena from type II and type III bursts, differing significantly in emission processes and/or source conditions; this conclusion indicates that many of the presently available theoretical ideas for HB bursts are incorrect.

077.039   Solar radio pulsations at 410 MHz.
C. S. Wright, G. J. Nelson.
Sol. Phys., Vol. 111, No. 2, p. 385 – 395 (1987).

On 6 September, 1982 very regular, narrow–band radio pulsations of solar origin were observed on the 410 MHz solar radiometer at the Learmonth Solar Observatory. Initial low–amplitude pulsations with a period of about 3 min gave way to large–amplitude pulsations with a period of about 5 min following a 1B solar flare. Position measurements at 327 MHz with the Culgoora Radioheliograph indicated two sources: a strong, extended source located above a unipolar magnetic region near the centre of the disk and a much weaker source near the west limb. Polarization measurements indicate the burst to be plasma emission. The radio pulsations were unique in their association with both sympathetic radio emission and optical flares at widely different locations. Interpretation of the observations in terms of "sausage" mode standing oscillations in a coronal flux tube leads to an estimate of the magnetic flux density $B = 46$ G at the 400 MHz plasma level. Also a 2.8–fold density increase in the loop after the 1B flare is inferred.

077.040   Characteristics of shock–associated fast–drift kilometric radio bursts.
R. J. MacDowall, R. G. Stone, M. R. Kundu.
Sol. Phys., Vol. 111, No. 2, p. 397 – 418 (1987).

The existence of a class of fast–drift, shock–associated (SA), kilometric radio bursts which occur at the time of metric type II emission and which are not entirely the kilometric continuation of metric type III bursts has been reported previously (Cane et al., 1981). In this paper, the authors establish unambiguous SA event criteria for the purpose of statistically comparing SA events with conventional kilometric type III bursts. They apply these criteria to all long–duration, fast–drift bursts observed by the ISEE–3 spacecraft during a 28–month interval and find that more than 70% of the events satisfying the criteria are associated with the radio signatures of coronal shocks. If a given event in this sample is associated with a metric type II or type IV burst, it is 13 times more likely to satisfy the SA criteria than an event associated only with metric type III activity. Compared with conventional kilometric type III bursts, the characteristics of these SA events are longer duration, higher maximum intensity, and a larger number of components.

077.041   Radio spectrum of a solar active region.
F. Chiuderi–Drago, S.–c. Ji.
Sci. Sin., Ser. A, Vol. 30, No. 11, p. 1199 – 1202 (1987).

The radio spectrum of the A.R. (active region) 2490, in the range of wavelengths between 8 mm and 20 cm, is presented. The A.R. presents a double structure at $\lambda \geqslant 2.8$ cm. It is shown that gyroresonance absorption at $\lambda = 6$ cm, which is the main source of opacity in the portion of the A.R. overhanging the sunspots, plays an important role also in the radio source associated with the Hα plage.

077.042   Method for automated initial reduction of results obtained by observations of the Sun with the RATAN–600 radio telescope and foundation of archives.
S. A. Andrianov.
Glav. Geofiz. Obs. Goskomiteta SSSR Gidrometeorol. Kontr. Prirod. Sredy. Leningrad, 59 pp. (1987). In Russian. Abstr. in Ref. Zh., 51. Astron., 10.51.479 (1987).

077.043   On the enhanced radio brightness of solar polar regions.
I. G. Moiseev, N. S. Nesterov.
Izv. Krymskoj Astrofiz. Obs., Tom 77, p. 83 – 89 (1987). In Russian. English translation in Bull. Crimean Astrophys. Obs., Vol. 77.

The results of observations of solar polar regions made in 1984 – 1985 by means of the 22–meter radio telescope of the Crimean Astrophysical Observatory at wavelengths of 0.32, 0.82, 1.35 and 2.25 cm are discussed.

077.044   Spectral features of solar gradual microwave bursts.
J.–x. Yao.
Publ. Purple Mt. Obs., Vol. 6, No. 3, p. 286 – 294 (1987). In Chinese.

The author presents data and spectral analysis of five solar gradual microwave bursts (GMB's), which are associated with the gradual hard X–ray bursts (GHB's). The durations of GMB's are about tens of minutes and are longer than that of impulsive bursts (5 min.), and the sources of GMB's are high in the corona. Therefore, one may attribute the long durations and spectral index decrease to the high radio sources.

077.045   Solar radio emission, January – December 1982.
Q. Bull. Sol. Act., Vol. 24, Part V, p. 1 – 141 (1987).

077.046   Changes of polarization in the dm–m range during the flare of May 16, 1981.
V. Ruždjak, B. Vršnak, P. Zlobec.
Hvar Obs. Bull., Vol. 10, No. 1, p. 11 – 16 (1986).
The polarization behaviour in the dm–m radio range during the flare of May 16, 1981 is studied and related to other observed phenomena. The evolution of the flare is discussed.

077.047   Source sizes of type III bursts at hectometric wavelengths as determined from ionospheric cutoffs.
R. Schreiber, J. Hanasz.
Astron. Astrophys., Vol. 188, No. 1, p. 178 – 182 (1987).
A new method of source size determination for type III bursts from swept frequency recordings received from spacecraft has been proposed. The method uses the frequency profiles of the ionospheric cutoffs for the ordinary and extraordinary burst radiation. The sizes of the burst sources are deduced from a regression based on a limited sample of events (4 bursts, 19 cutoff profiles). The mean e–folding radii vary from 0.9° at 4.2 MHz to 3.4° at 0.7 MHz, which is in general agreement with determinations by Dulk et al. (1984). The first determinations of source sizes are made for frequencies between 2 and 4.2 MHz.

077.048   A search for local sources of the S–component at decameter wavelengths.
L. L. Bazelyan.
Sol. Phys., Vol. 112, No. 1, p. 107 – 117 (1987).
In an effort to find local sources of the Slowly Varying Component (SVC), an analysis is made of the episodical observations carried out since 1972 during periods of low solar activity at 20 and 25 MHz. No local sources were found. 25 MHz radioheliograms of April 29 and 30, 1976, are presented. The author concludes that at present, in spite of reports of some workers, there is no convincing evidence for the existence of local SVC sources at decameter wavelengths.

077.049   S–component model computations applied to mean mm–wave measurements.
S. Urpo, J. Hildebrandt, A. Krüger.
Sol. Phys., Vol. 112, No. 1, p. 119 – 131 (1987).
Selected mm–wave observations with arc–min resolution of active regions in the central part of the solar disk obtained by the Metsähovi Radio Research Station during the years 1978 – 1984 are discussed from the perspective of recently developed mean S–component models. It is concluded that the mm–wave emission has a considerable (but at present unresolved) spatial fine structure. Basing on model calculations, predictions are made concerning the mm–wave brightness of the sunspot and plage components. Averaging over the whole S–component sources provides a good correspondence between observation and calculation.

077.050   Correlated type III burst emission from distant sources on the sun.
M. R. Kundu, N. Gopalswamy.
Sol. Phys., Vol. 112, No. 1, p. 133 – 142 (1987).
The authors report the observation and interpretation of a correlated type III burst emitted from distant sources on the Sun. The angular separation between the distant sources is as large as 26′ or ~10^6km. There was an active region ~30° behind the limb, and it is believed that the type III burst emission originated from activity in this region. The implications of the locations of the correlated sources with regard to the geometry of the magnetic structures involved in the flare process are discussed.

077.051   Quiet Sun and slowly varying component at meter and decameter wavelengths.
P. Lantos, C. E. Alissandrakis, T. Gergely, M. R. Kundu.
Sol. Phys., Vol. 112, No. 2, p. 325 – 340 (1987).
Comparison of maps of the Sun obtained over the period June 29 to July 8, 1982 at 169 MHz with the Nançay Radioheliograph and at 73.8, 50, and 30.9 MHz with the Clark Lake Radioheliograph shows that the slowly varying component at meter and decameter wavelengths is not always thermal emission.

Weak noise storm continua were the most frequent sources of the slowly varying component at 169 and 73.8 MHz. Most filaments show no radio counterpart on the disk. A streamer has been detected on the disk from 169 to 30.9 MHz with an optimum observability at 50 MHz. The brightest source of the slowly varying component from 73.8 to 30.9 MHz for most of the period was located above an extended coronal hole in a region where a depression was observed at 169 MHz.

077.052   On a sequence of remarkable fine structures in the type IV burst of 24 April, 1985.
H. Aurass, G. P. Chernov, M. Karlický, J. Kurths, G. Mann.
Sol. Phys., Vol. 112, No. 2, p. 347 – 357 (1987).
During the type IV burst on 24 April, 1985 the authors observed at 234 MHz an untypical, strong, nearly six hours lasting continuum emission incorporating several groups of braodband pulsations, zebra patterns, fiber bursts, and a new fine structure phenomenon. The power spectra of the groups of broadband pulsations reveal no simple structure. There is only one common periodic component between 0.3 s and 0.4 s. Slowly drifting chains of narrowband fiber bursts are described as a new fine structure by spectrograms and simultaneously recorded single frequency intensity profiles. A qualitative model of this new fine structure is suggested.

077.053   An example of type IV minute scale pulsations and the standing magnetoacoustic wave model.
H. Aurass, G. Mann.
Sol. Phys., Vol. 112, No. 2, p. 359 – 364 (1987).
An example of m–Dm solar radio pulsations consisting of a mixture of different minute scale periodic components is discussed. The ratios of the observed periods are independent of the observing frequency. They can be reproduced by the pulsation model of standing magnetoacoustic waves in coronal flux tubes driven by trapped protons.

077.054   On preflare changes of solar decimeter radio emission fluctuations. 1. Short–period fluctuations.
E. A. Aver'yanikhina, M. Paupere, G. Ozolinš, M. Eliāss.
Issled. Solntsa Krasnykh Zvezd, No. 25, p. 42 – 55 (1987). In Russian.
Results of an analysis of solar decimetre preflare radio emission fluctuations at 755 and 612 MHz during 1979 – 1984 are considered. Two kinds of preflare changes are detected.

077.055   Bright region at the polar cap of the sun at millimeter wavelengths.
E. Hiei.
Publ. Astron. Soc. Jpn., Vol. 39, No. 6, p. 937 – 940 (1987).
Polar–cap brightenings in both north and south polar regions of the sun were observed. A brightness enhancement by 3 – 7% at 36 GHz (8.3 mm) and no enhancement at 98 GHz (3.1 mm) were reported (Kosugi et al. 1986). Brightening at the polar region is estimated from a model atmosphere (Vernazza and Noyes 1972). The estimated enhancement has turned out to be too weak, and the observed enhancement is not explained by the existing models of the atmospheric structure of the polar region as derived from the EUV observations, and requires a revision of the models.

077.056   Radio emission of the Sun at millimetre wavelengths.
V. G. Nagnibeda, V. V. Piotrovich.
Tr. Astron. Obs., Leningrad, Tom 41, p. 5 – 80 (1987). Uch. Zap. LGU, No. 420, Ser. Mat. Nauk, Vyp. 63. In Russian.
This review article deals with the radio emission originating from different solar atmospheric regions – the quiet solar atmosphere, active regions and solar flares. All experimental data of the quiet Sun brightness temperature at the region of 0.1 – 20 mm wavelength are summarized. The quiet Sun brightness distributions across the disk and values of the solar radio radius are reviewed. The properties of the sources of sunspot–associated active region emission and radio brightness depression associated with Hα–filaments are considered in comparison with observations at centimetre and optical domains. The observational properties of millimetre wave bursts and their correlations

with similar phenomena at other domains are reviewed. Special reference is devoted to nearly 100% correlation impulsive radio bursts with hard X–ray bursts. Existence of the fine temporal structure containing many spikes with time scales up to 10 ms as well as observations of quasi–periodic millisecond oscillations are discussed.

077.057 Observations of kilometer radio bursts for some proton events.
N. N. Volodichev, V. P. Grigor'eva, V. S. Prokudina.
Astron. Tsirk., No. 1481, p. 1 – 3 (1987). In Russian.
Results of the kilometer radio bursts observations at "Prognoz–8" for large chromospheric flares in April 1981 are given. During the events the flux of protons with energies $\varepsilon \geqslant 100$ MeV and $\varepsilon \geqslant 500$ MeV were observed.

Operative presentation of astrophysical results of observations of the Sun at the RATAN–600 radio telescope.
See Abstr. 033.031.

Program design for spectral observations of type I solar radio bursts.
See Abstr. 036.154.

Imaging the Sun at 1.4 GHz.
See Abstr. 036.219.

Plasma emission: a review.
See Abstr. 062.031.

Evolution of the axial electron cyclotron maser instability, with applications to solar microwave spikes.
See Abstr. 063.108.

Observations of far–infrared solar continuum variations due to compression waves.
See Abstr. 071.023.

Peculiarities of radio radiation of flare–active sunspot groups of December 1982.
See Abstr. 072.032.

Structure of solar active regions from VLA and RATAN–600 observations in July 1982. Part 1: AR 3804.
See Abstr. 072.065.

Microwave emission above steady and moving sunspots.
See Abstr. 072.072.

Effects of impulsively heated electrons in solar flares.
See Abstr. 073.004.

Microwave radiation from a dense magneto–active plasma.
See Abstr. 073.013.

Acceleration and energization by currents and electric fields.
See Abstr. 073.028.

Evidence for interacting loop process in a phase of the May 16, 1981 flare.
See Abstr. 073.029.

Spotless flares and the associated radio continuum emission.
See Abstr. 073.030.

Simultaneous VLA observations of a flare at 6 and 20 centimeter wavelengths.
See Abstr. 073.052.

Simultaneous 2 and 6 centimeter wavelength observations of a solar flare using the VLA.
See Abstr. 073.082.

Coronal millimeter–wave sources.
See Abstr. 074.020.

Radio emission from coronal and interplanetary shocks.
See Abstr. 074.021.

Meterwave observations of a coronal hole.
See Abstr. 074.047.

The Sun's outer corona at radio wavelengths.
See Abstr. 074.054.

Hard X–ray and radio emission at the onset of great solar flares.
See Abstr. 076.004.

A method of determining the solar radio radius by scanning.
See Abstr. 080.007.

1980 – 1984 measurements of the solar radio radius at 2.25 cm wavelength.
See Abstr. 080.045.

Analysis of correlation between the total radiation fluxes at 2902 MHz, 3653 MHz and 9375 MHz and the solar constant in 1980.
See Abstr. 080.051.

Aspects of interplanetary plasma turbulence.
See Abstr. 106.007.

078 Cosmic Radiation

078.001 Studies of some aspects of solar proton events and related phenomena.
T. K. Das, T. B. Chakraborty, M. K. Das Gupta.
Bull. Astron. Inst. Czech., Vol. 38, No. 4, p. 206 – 210 (1987).
A statistical investigation has been done of the proton events which occurred during the period from 1955 to 1985. The important results obtained are as follows: (1) the occurrence frequency distribution of proton fluxes obeys a power law, (2) the rate of rise of proton flux varies directly with exponent 1.2; (3) although the occurrence of proton events more or less follows the phase of the solar cycle, they are maximum at the 20th solar cycle peak which is less intensive than those of the 19th and 21st

cycles, (4) N–S and E–W asymmetries display definite latitude and longitude dependences, respectively.

078.002 Distribution of solar proton events over a solar activity cycle.
M. V. Zil', V. G. Mitrikas, V. M. Petrov.
Soln. Dannye, Byull., 1987, No. 1, p. 71 – 76 (1987). In Russian.
On the basis of the data on 406 solar proton events for 1965 – 1984 a significant variation of their occurrence with the 11–year solar cycle has been found. A variation with a period 1.8 ± 0.2 years of frequency of the occurrence of solar proton events is shown. Moreover, in their power spectra peaks have

been detected that correspond to the periods of 1 year, 8 months and 5 months.

078.003 Solar neutron emissivity during the large flare on 1982 June 3.

E. L. Chupp, H. Debrunner, E. Flückiger, D. J. Forrest, F. Golliez, G. Kanbach, W. T. Vestrand, J. Cooper, G. Share.
Astrophys. J., Vol. 318, No. 2, p. 913 – 925 (1987).

Observations made with the gamma–ray spectrometer on the SMM satellite and with the Jungfraujoch neutron monitor are used to determine the directional solar neutron emissivity spectrum from ∼ 100 MeV to ∼ 2 GeV during the solar flare on 1982 June 3. The experimental data require a time–extended emission of the neutrons at the Sun with the majority of the neutrons produced after the impulsive phase. The observations require that the first GeV protons producing the GeV neutrons interacted at the Sun within a time span of at most 16 s implying neutron production at densities $n > 10^{14} \text{cm}^{-3}$.

078.004 Evolution of an active proton complex on the Sun in May 1981.

V. S. Prokudina, G. F. Sitnik.
Problems of solar flares, p. 89 – 99, 204 (1986). In Russian. Abstr. in Ref. Zh., 51. Astron., 7.51.341 (1987). – See Abstr. 012.018.

078.005 On the role of a coronal shock wave for high–energy solar protons.

G. A. Bazilevskaya, A. I. Sladkova.
Problems of solar flares, p. 119 – 125, 208 (1986). In Russian. Abstr. in Ref. Zh., 51. Astron., 7.51.381 (1987). – See Abstr. 012.018.

078.006 Two types of flare events in the electron component of solar cosmic rays.

E. I. Dajbog, V. G. Kurt, V. G. Stolpovskij.
Problems of solar flares, p. 185 – 189, 214 (1986). In Russian. Abstr. in Ref. Zh., 51. Astron., 7.51.392 (1987). – See Abstr. 012.018.

078.007 Spectrum of solar cosmic rays in the source with allowance for their coronal propagation.

M. Alvarez–Madrigal, L. I. Miroshnichenko, J. Pérez–Peraza, F. Rivero–Garduso.
Sov. Astron., Vol. 30, No. 6, p. 690 – 695 (1987). English translation of 42.078.015.

078.008 A prediction of the major proton event for the first time in cycle 22.

Q. Zhang.
Publ. Yunnan Obs., No. 2, p. 31 – 33 (1987). In Chinese.

By analysis of past proton events it is found that the major proton events for the first time appeared 1.9 years after the start of each solar activity cycle.

078.009 Determination of energy spectra of solar electrons under different scenarios in solar flare sources.

A. Gallegos, J. Pérez–Peraza.
Rev. Mex. Astron. Astrofis., Vol. 14, No. 2, p. 700 – 704 (1987). – See Abstr. 012.042.

The energy spectrum of the solar flare electrons is investigated within the frame of three different scenarios. Solutions of the Fokker–Planck equation are employed under the assumption of a short fly time of particles within the source and low density therein (thin–geometry). The obtained results are compared with observational spectra measured at the earth level. This intercomparison allows the authors to infer about the appropriate scenario in each particular solar event.

078.010 Catalogue of energy spectra of solar proton events in 1970 – 1979.

Yu. I. Logachev.
Inst. Zem. Magn. Ionos. Rasporstr. Radiovoln Akad. Nauk SSSR. Moskva, 234 pp. (1986). In Russian. Abstr. in Ref. Zh., 62. Issled. Kosm. Prostranstva, 11.62.373 (1987).

078.011 Interplanetary proton ($0.61 < E_p < 3.41$ MeV) events observed with Pioneer 11, 1973 – 1986 and out to 22.4 AU.

J. A. Van Allen.
Sol. Phys., Vol. 112, No. 1, p. 165 – 179 (1987).

A survey of interplanetary proton events is summarized in graphical and tabular form. Two hundred and sixty–five distinct events are identified. The spectra and intensities of the protons, presumed to be originally of solar origin, are influenced dramatically by propagative and accelerative processes in the interplanetary medium.

078.012 Some features of solar wind protons, α particles and heavy ions behaviour. The Prognoz 7 and Prognoz 8 experimental results.

L. Avanov, N. Borodkova, Z. Nemecek, A. Omeltchenko (A. Omel'chenko), J. Safrankova, A. Skalski, Yu. Yermolaev (Yu. Ermolaev), G. Zastenker.
Czech. J. Phys., Sect. B, Vol. B37, No. 6, p. 759 – 774 (1987). Abstr. in Phys. Abstr., Vol. 90, No. 1314, Entry 120232 (1987).

078.013 Solar cosmic rays in the interplanetary space.

E. V. Gorchakov.
Problems of cosmic ray physics, p. 30 – 50 (1987). In Russian. Abstr. in Ref. Zh., 62. Issled. Kosm. Prostranstva, 12.62.336 (1987). – See Abstr. 003.015.

078.014 The energy spectrum of the solar proton event of February 16, 1984.

L. I. Miroshnichenko, M. O. Sorokin.
Geomagn. Aehron., Tom 27, No. 6, p. 893 – 899 (1987). In Russian. English translation in Geomagn. Aeron.

078.015 A multispacecraft study of the injection and transport of solar energetic particles.

J. Beeck, G. M. Mason, D. C. Hamilton, G. Wibberenz, H. Kunow, D. Hovestadt, B. Klecker.
Astrophys. J., Vol. 322, No. 2, p. 1052 – 1072 (1987).

Two "prompt" solar particle events were observed simultaneously by the Helios and Voyager space probes and the ISEE 1 satellite. During the events, which were on 1977 November 22 and 1977 December 27, the spacecraft instrumentation observed H, He, and (for 1977 November 22) O and Fe flux increases over the energy range ∼0.5 to more than 20 MeV per nucleon. The analysis of the data shows: 1. The time–intensity profiles of these events can be adequately fitted using a standard model of interplanetary propagation including diffusion, convection, and adiabatic energy loss. 2. The acceleration or release times, or both, at the Sun are of extended duration with decay times of ∼4 – 8 hr. 3. The interplanetary scattering mean free path is ∼0.1 AU and increases with particle rigidity.

078.016 Solar flare neutron and accelerated ion angular distributions.

X.–M. Hua, R. E. Lingenfelter.
Astrophys. J., Vol. 323, No. 2, p. 779 – 794 (1987).

The authors made Monte Carlo calculations of the angular and energy distributions of neutrons escaping from the solar atmosphere, expected from thick–target interactions of flare–accelerated ions with a variety of energy spectra and incident angular distributions, ranging from downward beaming to magnetic mirroring. Comparing the calculations with measurements of the neutron flux from the flares of 1980 June 21 at a heliocentric angle of 89° and 1982 June 3 at 72°, the authors found that the bulk of the observed neutron flux was consistent with that expected from neutron production with an essentially δ–function time history following that of the 4 – 7 MeV emission.

Both the angular distribution and the energy spectrum of the accelerated ions were determined.

The α–effect generated by random Alfvén waves in a stratified atmosphere and particle acceleration.
See Abstr. 062.159.

Particle acceleration in solar flares.
See Abstr. 073.002.

Estimating the anomalously large fluences of solar flare protons with tritium data from 1900 to 1952.
See Abstr. 073.047.

Nuclear processes and accelerated particles in solar flares.
See Abstr. 073.080.

Numerical analysis of the azimuthal transport of solar particles.
See Abstr. 074.059.

Propagation of solar energetic particles from the corona into the interplanetary space.
See Abstr. 074.073.

Observations of kilometer radio bursts for some proton events.
See Abstr. 077.057.

Nuclear processes in the solar atmosphere and the problem of particle acceleration.
See Abstr. 080.019.

Spectrum of accelerated particles and neutrino generation in solar flares.
See Abstr. 080.036.

On the proposed associations of solar neutrino flux with solar particles, cosmic rays, and the solar activity cycle.
See Abstr. 080.062.

Solar cosmic ray electrons in the Earth's magnetosphere.
See Abstr. 084.027.

Recurrent increases of proton intensity from Prognoz 7 measurements.
See Abstr. 106.038.

Cosmic rays in the heliosphere.
See Abstr. 144.002.

Cosmic rays and energetic particles in the heliosphere.
See Abstr. 144.021.

079 Solar Eclipses

079.001 **Long duration total solar eclipses in Britain.**
P. Macdonald.
Astron. Now, Vol. 1, No. 4, p. 6 – 11 (1987).
At present, the maximum possible duration for a total solar eclipse is 7 minutes 31 seconds. Eclipses approaching this length occur infrequently and are visible only from equatorial regions. In the latitude of the British Isles, it is possible to witness an eclipse producing more than five minutes' totality, if only rarely.

079.002 **The shadow chasers.**
D. Allen.
1988 Yearbook of Astronomy, p. 135 – 146 (1987). – See Abstr. 003.006.

079.003 **Der Canon der Finsternisse und die Rotationsdauer der Erde.**
K. Bretterbauer.
Vorträge beim Theodor Ritter von Oppolzer–Gedächtnissymposium, p. 37 – 47 (1987).
The significance of solar eclipses for researches on the secular variations of Earth–rotation and the role of Oppolzer's ''Canon of Eclipses'' is discussed.

079.004 **Aktuelle Beobachtungen von Sonnen– und Mondfinsternissen.**
H. Haupt.
Vorträge beim Theodor Ritter von Oppolzer–Gedächtnissymposium, p. 48 – 52 (1987). – See Abstr. 012.073.

Solar eclipse 1988 March 17 – 18

079.101 **The total solar eclipse of March, 1988.**
E. M. Brooks.
Sky Telesc., Vol. 74, No. 1, p. 21 – 22 (1987).

079.102 **Prospects for the March 1988 total solar eclipse.**
J. Anderson.
Astronomy, Vol. 15, No. 8, p. 38 – 42 (1987). Abstr. in Phys. Abstr., Vol. 90, No. 1317, Entry 139890 (1987).

Sonnenfinsternisse auf prähistorischen Kultplätzen durch Felsritzungen dokumentiert.
See Abstr. 004.001.

Canon der verduisteringen.
See Abstr. 004.026.

Once more on the interpretation of the results of the July 31, 1981 solar eclipse.
See Abstr. 073.022.

Comparison of computed fluxes for Fe X and Fe XIV lines with observed values at 1980 eclipse.
See Abstr. 074.024.

The ionospheric effects of the July 31, 1981 solar eclipse.
See Abstr. 083.026.

080 Atmosphere, Figure, Internal Constitution, Neutrinos, Rotation, etc.

080.001 Is the solar oblateness variable? Measurements of 1985.
R. H. Dicke, J. R. Kuhn, K. G. Libbrecht.
Astrophys. J., Vol. 318, No. 1, p. 451 – 458 (1987).

The solar oblateness measured in 1985 is $\Delta r = r_{eq} - r_p = 14.6 \pm 2.2$ arc ms, where the error is only a formal standard deviation assuming normally distributed and uncorrelated errors. The above result is significantly greater than the 1984 value which, in turn, is significantly less than the 1983 and 1966 values. The differences may be physically significant and are consistent with the hypothesis that the oblateness oscillates with the 11.14 yr period of the solar cycle. The data at present only weakly support this hypothesis.

080.002 The dynamo dilemma.
E. N. Parker.
Sol. Phys., Vol. 110, No. 1, p. 11 – 21 (1987). – See Abstr. 012.005.

The recent determination that the angular velocity Ω of the Sun declines downward through the convective zone raises serious questions about the nature of the solar dynamo. The principal qualitative features of the Sun are the azimuthal fields that migrate toward the equator in association with an oscillating poloidal field which reverses at about the time of maximum appearance of bipolar magnetic regions. If Ω decreases downward, or is negligible, the horizontal gradient in Ω produces a dynamo with some of these essential characteristics. There is reason to think that the dynamo is confined to the lower half of the convective zone where α has the opposite sign from the usual ($\alpha > 0$ in the northern hemisphere) producing equatorward migration but reversing the sign of the associated poloidal field. Meridional circulation may play an essential role in shaping the dynamo. At the present time it is essential to measure Ω accurately and determine the nature of the meridional circulation.

080.003 Large–scale motions on the sun: an overview.
R. S. Bogart.
Sol. Phys., Vol. 110, No. 1, p. 23 – 34 (1987). – See Abstr. 012.005.

The history and present status of observations of large–scale velocity fields in the solar atmosphere are reviewed. Observations of the torsional oscillation and of mean meridional circulation suggest a connection of large–scale dynamics with the solar cycle. Significant problems must be solved before Doppler observations can match the precision of tracer measurements, particularly allowing for the effects of changes in line–profile asymmetries and for scattered light. Coordinated observations would establish the reliability of Doppler techniques, but Doppler measurements with precision of order $1 \, m \, s^{-1}$ made in a proper spatial–temporal window appear necessary for the identification of sub–global velocity fields varying with time–scales less than that of the solar cycle.

080.004 Solar rotation and the giant cells.
P. R. Wilson.
Sol. Phys., Vol. 110, No. 1, p. 59 – 71 (1987). – See Abstr. 012.005.

Departures from the mean solar differential rotation rate as a function of latitude, longitude, and epoch of the solar cycle, together with variations in the rotation rate as determined by spectroscopic and tracer measurements are reviewed. It is shown that, if giant convection cells do exist as predicted, real variations in the subsurface rotation rate should occur and that this may be responsible for the observed surface anomalies. In terms of this hypothesis, a simple account is given for the anomalous rotation rates of sunspots. Furthermore, the torsional oscillations are identified as a modulation of the differential rotation produced by a system of toroidal convective rolls generated near the poles and propagating towards the equator.

080.005 Prolonged minima and the 179–yr cycle of the solar inertial motion.
R. W. Fairbridge, J. H. Shirley.
Sol. Phys., Vol. 110, No. 1, p. 191 – 210 (1987).

The authors employ the JPL long ephemeris DE–102 to study the inertial motion of the Sun for the period A.D.760 – 2100. Defining solar orbits with reference to the Sun's successive close approaches to the solar system barycenter, occurring at mean intervals of 19.86 yr, they find simple relationships linking the inertial orientation of the solar orbit and the amplitude of the precessional rotation of the orbit with the occurrence of the principal prolonged solar activity minima of the current millenium (the Wolf, Spörer, and Maunder minima). The progression of the inertial orientation parameter is controlled by the 900–yr "great inequality" of the motion of Jupiter and Saturn, while the precessional rotation parameter is linked with the 179–yr cycle of the solar inertial motion previously identified by Jose (1965). A new prolonged minimum of solar activity may be imminent.

080.006 Long periods in solar diameter variations observed with the solar astrolabe.
F. Laclare.
C. R. Acad. Sci., Sér. II, Tome 305, No. 6, p. 451 – 454 (1987). In French.

A solar astrometry program, started at C.E.R.G.A. Observatory in 1975, covers now 95% of the activity cycle No. 21. An analysis of the complete observational data reveals or confirms oscillations, some of which seem correlated with solar activity. Several long periods, from 50 to 1,000 days with characteristic amplitude 0.10", have been detected, and a negative correlation between diameter measurements and the sunspot cycle seems to exist.

080.007 A method of determining the solar radio radius by scanning.
A. F. Bachurin.
Bull. Crimean Astrophys. Obs., Vol. 74, p. 109 – 112 (1987). English translation of 41.080.102.

080.008 Rotation of solar high–latitude regions in 1971 – 1978.
V. V. Makarova, Yu. A. Solonskij.
Soln. Dannye, Byull., 1986, No. 11, p. 56 – 63 (1987). In Russian.

A 8–year series of daily observations of solar photospheric polar faculae is analysed. A formula is found for the middle sidereal rotation rate ω of the solar high–latitude regions. It is shown that the rotation rates and the differential rotation curves were changing during the period of observations. Time variations of the solar rotation rates received from the polar faculae, the Doppler shifts and the sunspots are compared. It is shown that the authors' results correlate rather well with the Doppler measurements.

080.009 Differential rotation of polar regions of the solar atmosphere.
V. V. Makarova, Yu. A. Solonskij.
Soln. Dannye, Byull., 1986, No. 12, p. 56 – 62 (1987). In Russian.

A 8–year (1971 – 1978) series of daily observations of solar photospheric polar faculae is analysed. It is found that at that time the average sidereal rotation rate ω of the solar high–latitude regions is expressed by the formula $\omega = (13.106 \pm 0.55) - (1.646 \pm 0.51) \sin^2 \varphi - (0.215 \pm 0.058) \sin^4 \varphi$ (degrees/day) (φ is the heliographic latitude). It is shown that the rotation rates and the differential rotation curves were changing during the period of the observations. Time variation of the solar rotation received from the polar faculae, the Doppler shifts and the sunspots are compared. It is shown that the authors' results correlate rather well with the Doppler measurements.

080.010 **Oscillations of the solar convective zone and Hill's spectrum.**
A. I. Khlystov.
Soln. Dannye, Byull., 1987, No. 2, p. 77 – 81 (1987). In Russian.
   The solid wall approximation for the lower boundary of the solar convective zone is shown to be quite adequate. Having used this boundary condition the author derives formulas for calculations of free periods in oscillations of the solar convective zone. One of them approximates Hill's spectrum with a high accuracy and gives also the period nearly 160 minutes. These formulas can be used for analysis of stars pulsations.

080.011 **Evidence for a small, high–Z, iron–like solar core. IV. Sensitivity studies of the five–minute band frequencies to the gravitational perturbation and of the 160–minute period of oscillation to the space mesh.**
C. A. Rouse.
Sol. Phys., Vol. 110, No. 2, p. 211 – 235 (1987).
   Radial and nonradial oscillation equations without and with the gravitation perturbation (with and without the Cowling approximation, CA) are solved numerically using the profile from a more accurate high–Z core (HZC) solar model. This more accurate HZC model was generated with the CRAY X–MP/48 supercomputer at the San Diego Supercomputer Center. Frequencies of oscillation in the five–min band (5MB) and frequencies with period near 160 min are presented in tables and plotted in echelle diagrams. In either case, it is concluded that the first–order, radially–symmetric structure of the model outside the HZC is close to the structure of the real Sun. This is of fundamental importance because a real gas adiabatic temperature gradient (Rouse, 1964, 1971) is used in the outer convective region without free parameters. Other aspects of agreements and differences between radial and nonradial solutions, with CA and without CA are discussed. In particular, the $l = 4, 6, 8$, and 9 g–mode solutions with CA indicate that the observed 160.01 min period may be a common l–mode period of oscillation. More research is proposed.

080.012 **Waves in the solar photosphere.**
R. Stebbins, P. R. Goode.
Sol. Phys., Vol. 110, No. 2, p. 237 – 253 (1987).
   Time–sequences of line profile data have been subjected to a unique analysis which produces an amplitude and phase of the velocity and intensity at several line depths for each time sample and spatial point on the Sun. The data have been filtered to pass only the frequencies and spatial wavenumbers of the 5–min band. Yet, a secondary oscillation emerges, the phase of which propagates downward. Empirical eigenfunctions for velocity and intensity are given, and the kinetic energy flux is computed.

080.013 **Inertial oscillations in the solar convection zone. III. A cylindrical model for nonaxisymmetric oscillations in a superadiabatic gradient.**
P. A. Gilman.
Astrophys. J., Vol. 318, No. 2, p. 904 – 912 (1987).
   The author examines the effects of a superadiabatic gradient and differential rotation upon inertial oscillations that are nonaxisymmetric about the axis of rotation as a generalization of earlier calculations published by Gilman and Guenther in 1985 for a cylindrical model of a stellar convection zone. It is shown that the frequencies of these oscillations are sensitive to both superadiabatic gradient and differential rotation. If such oscillations can be detected on the Sun, then they could be used to measure the degree of superadiabaticity of the convection zone.

080.014 **Power output of a turbulent current sheet on the sun.**
V. M. Bardakov.
Sov. Astron. Lett., Vol. 12, No. 5, p. 330 – 332 (1987). English translation of 42.080.008.

080.015 **Processing photodiode–array solar–oscillation observations.**
A. G. Kosovichev.
Bull. Crimean Astrophys. Obs., Vol. 75, p. 19 – 29 (1987). English translation of 42.080.027.

080.016 **Solution of an inverse helioseismological problem from observations on solar gravitational oscillations.**
A. G. Kosovichev.
Bull. Crimean Astrophys. Obs., Vol. 75, p. 36 – 42 (1987). English translation of 42.080.028.

080.017 **Variations of solar irradiance due to magnetic activity.**
G. A. Chapman.
Annu. Rev. Astron. Astrophys., Vol. 25, p. 633 – 667 (1987). – See Abstr. 003.003.
   The author concentrates on solar luminosity variability due to magnetic activity. The review discusses the use of proxy data – that is, the position and area or some other property that is not in itself a photometric measure of intensity. The direct photometric measurements are reviewed and instrumentation and ongoing observing programs are described. A brief discussion of a few of the key theoretical problems that are posed by the observed short– and long–term solar irradiance variations is given.

080.018 **Upward streams in the solar atmosphere.**
G. P. Lyubimov, N. N. Kontor.
Problems of solar flares, p. 153 – 155, 210 (1986). In Russian. Abstr. in Ref. Zh., 51. Astron., 7.51.366 (1987). – See Abstr. 012.018.

080.019 **Nuclear processes in the solar atmosphere and the problem of particle acceleration.**
G. E. Kocharov.
Itogi Nauki i Tekhniki. Seriya Astronomiya. Tom 32. Astrophysics and space physics, p. 43 – 141 (1987). In Russian. – See Abstr. 003.004.

080.020 **A solar model having a detached unstable 160–min mode.**
Yu. V. Vandakurov.
Pis'ma Astron. Zh., Tom 13, No. 9, p. 789 – 796 (1987). In Russian. English translation in Sov. Astron. Lett., Vol. 13.
   A 4–zone solar model is considered which contains a convective core, a chemically inhomogeneous convectively neutral zone, a radiative zone and a convective envelope. On the core boundary such an abrupt change in the density and molecular weight is supposed to be present due to which a deeply localized 160–min mode turns out to exist. This mode is usually unstable. In this model the heavy element abundance at the bottom of the inhomogeneous neutral zone having a superadiabatic temperature gradient is slightly larger than in the core. A brief discussion of the frequencies of deeply penetrating 5–min modes for the model in question is also given.

080.021 **A revisited standard solar model.**
M. Cassé, S. Cahen, C. Doom.
Neutrinos and the present–day universe, p. 49 – 70 (1986). – See Abstr. 012.025.
   Using the authors' version of the standard solar model and recent estimates of nuclear reaction rates and heavy element abundances, they derive an initial solar helium content $Y = 0.285$ by mass, close to the Orion value, and neutrino counting rates of 7.4 SNU on chlorine and 111 SNU on gallium (using the capture cross sections of Bahcall et al. 1985). From the solar helium abundance and the primordial one, inferred from the observations of metal poor extragalactic nebulae, the authors deduce a galactic helium production from big–bang to the birth of the sun $\Delta Y \cong 0.05$. This value sets interesting constraints on models of the chemical evolution of the Galaxy.

080.022 The gallium solar neutrino experiment GALLEX.
D. Vignaud.
Neutrinos and the present–day universe, p. 71 – 84 (1986). – See Abstr. 012.025.

The GALLEX experiment will measure the solar neutrino flux by counting the ^{71}Ge atoms produced in a 30 ton gallium target. The detector will be set up in the Gran Sasso Underground Laboratory (Italy). The obtained result will allow a better understanding of solar models and enables discrimination between the problems of solar models and possible neutrino oscillations.

080.023 Alfvén wave dissipation in the solar atmosphere.
G. Einaudi, Y. Mok.
Astrophys. J., Vol. 319, No. 1, p. 520 – 530 (1987). = Univ. Calif., Irvine, Tech. Rep. No. 86–23.

Dissipative Alfvén waves in a nonuniform plasma are studied by using a normal mode analysis. Specific magnetic geometries are used to simulate various magnetic structures in the solar atmosphere. The authors have computed the real and imaginary parts of the eigenfrequency for each of the configurations and found that these eigenmodes can deposit a substantial amount of energy to the corona because of their short damping distances.

080.024 Acoustic absorption by sunspots.
D. C. Braun, T. L. Duvall Jr., B. J. LaBonte.
Astrophys. J., Lett. Ed., Vol. 319, No. 1, p. L27 – L31 (1987).

The authors present the initial results of a series of observations designed to probe the nature of sunspots by detecting their influence on high–degree p–mode oscillations in the surrounding photosphere. The analysis decomposes the observed oscillations into radially propagating waves described by Hankel functions in a cylindrical coordinate system centered on the sunspot. From measurements of the differences in power between waves traveling outward and inward it is demonstrated that sunspots appear to absorb as much as 50% of the incoming acoustic waves. For all three sunspots observed, the amount of absorption increases linearly with horizontal wavenumber.

080.025 Magnetic manifestations of solar rotation.
A. A. Ruzmajkin, S. V. Starchenko.
Astron. Zh., Tom 64, Vyp. 5, p. 1057 – 1065 (1987). In Russian. English translation in Sov. Astron., Vol. 31, No. 5.

An analytical solution of the mean–field solar dynamo equations is obtained. The asymptotic method used allows the finding of the basic features of magnetic fields, excited as dynamo–waves under arbitrary angular velocities of rotation. The dynamo–wave migrates along the surfaces close to the isorotational ones. The maximum of its amplitude is shifted from the location of maximum intensity of the sources in the direction of migration. Generally, there are two dynamo–waves on the solar surface, diverging from a certain latitude towards the pole and the equator. The first one is weakened, and the other is more intense. From the observed surface fields, the regions of field generation are localized; the conditions are imposed on the distribution of the rotational angular velocity in these regions, which agree with the modern helioseismology data.

080.026 On the depth dependence of the solar rotation velocity.
A. S. Gadun, R. I. Kostyk, V. A. Sheminova.
Astron. Zh., Tom 64, Vyp. 5, p. 1066 – 1071 (1987). In Russian. English translation in Sov. Astron., Vol. 31, No. 5.

A new method to determine the rotation rate of the solar atmosphere is suggested. In the region from $h = 100$ km to $h = 1100$ km, the rotation velocity is almost depth–independent: $\omega_e(\mu \, \text{rad s}^{-1}) = 2.83 + 1.1 \times 10^{-4} H$ (km) ($\pm 0.06 \pm 1.0$).

080.027 The solar–stellar connection.
M. S. Giampapa.
Sky Telesc., Vol. 74, No. 2, p. 142 – 146 (1987).

080.028 GONG: to see inside our sun.
J. W. Harvey, J. R. Kennedy, J. W. Leibacher.
Sky Telesc., Vol. 74, No. 5, p. 470 – 476 (1987).

080.029 The sun and its neutrinos.
R. L. Mössbauer.
Naturwissenschaften, 24. Jahrg., Heft 11, p. 511 – 519 (1987). In German.

Neutrinos provide instant information on solar fusion reactions, the sources of solar energy. The reasons for the observed shortage by a factor of 3 in the high–energy portion of the solar neutrino flux are unknown at present. The European Gallex Collaboration will perform a solar neutrino experiment aiming at the low–energy portion of the solar neutrino spectrum, which is directly related to the known solar luminosity. Any deficit of solar neutrinos observed in the new experiments would be indicative of neutrino properties such as neutrino masses and neutrino mixing.

080.030 Evidence of global circulation currents from solar–limb temperature variations.
J. R. Kuhn, K. G. Libbrecht, R. H. Dicke.
Nature, Vol. 328, No. 6128, p. 326 – 327 (1987).

The temperature distribution in a turbulent rotating photosphere is non–spherical. The authors report new solar observations that should help to understand the solar global dynamics problem. From about 1,400 h of solar–limb data obtained during the summers of 1983 – 85 they find that the solar–limb temperature variation is not spherically symmetric and is ~ 1K. These results also indicate that the limb temperature departs from its expected $l = 2$ spatial harmonic form and has, at most, a weak dependence on solar cycle.

080.031 Models of motions in the Sun.
R. F. Howard.
Nature, Vol. 328, No. 6132, p. 667 – 668 (1987).

This note comments on the model of sub–surface convective rolls proposed by H. B. Snodgrass and P. R. Wilson (see 080.032) to explain the observed large–scale motions on the solar surface.

080.032 Solar torsional oscillations as a signature of giant cells.
H. B. Snodgrass, P. R. Wilson.
Nature, Vol. 328, No. 6132, p. 696 – 699 (1987).

The existence of giant cells as the fundamental mode of solar convection has long been proposed on theoretical grounds. During one search, using Mount Wilson magnetograph data, Howard and La Bonte discovered a pattern of latitudinal velocity bands that move from the poles towards the equator in synchrony with the sunspot cycle, and they interpreted this pattern as a torsional wave or "oscillation" with wavenumber $k = 2$ hemisphere^{-1}. Here the authors suggest that this signal is not in fact an oscillation but represents a modulation of the mean differential rotation caused by a system of giant convective rolls which start at the poles at 11–yr intervals and migrate to the equator in a period of 18 – 22 yr. Additional evidence for the presence of these rolls is found in the zero offsets in the Mount Wilson data and in latitude variations of the limb temperature.

080.033 A comparison between mass–losing and standard solar models.
J. A. Guzik, L. A. Willson, W. M. Brunish.
Astrophys. J., Vol. 319, No. 2, p. 957 – 965 (1987).

As part of exploring the implications of the main–sequence mass–loss hypothesis of Willson, Bowen, and Struck–Marcell (1987), a short epoch of early main–sequence mass loss is incorporated into solar evolutionary calculations. Three mass–losing solar models, each of initial mass $2 \, M_\odot$, having initial mass–loss rates of several times $10^{-9} M_\odot \text{yr}^{-1}$ that decrease exponentially with e–folding times of a few times 10^8yr are compared with a standard $1 \, M_\odot$ model. The mass–losing models require slightly lower initial ^{4}He abundances, and higher mixing length/pressure scale height ratios than the standard model to attain $1 \, L_\odot$ and $1 \, R_\odot$ at age 4.6 Gyr.

080.034 **Solar oscillations: generation of a *g*–mode by two *p*–modes.**
D. G. Wentzel.
Astrophys. J., Vol. 319, No. 2, p. 966 – 970 (1987).

Three modes of solar oscillations can be coupled resonantly by the nonlinear terms in the equations of motion. A general integral for the coupling rate was derived by Dziembowski. The author evaluates the coupling of two *p*–modes, of nearly identical frequencies, so as to generate a *g*–mode. The coupling occurs primarily in the convection zone. A rather select set of *g*–modes of suitably low order and degree and with weak linear damping may grow, but the modes saturate when surface amplitudes are still unobservably small.

080.035 **Semi–annual variations of the flux of solar neutrinos according to data obtained in the experiment Davis during 1979 – 1982.**
A. I. Veselov, M. I. Vysotskij, V. P. Yurov.
Yader. Fiz., Tom 45, No. 5, p. 1392 – 1402 (1987). In Russian. Abstr. in Ref. Zh., 51. Astron., 9.51.333 (1987).

080.036 **Spectrum of accelerated particles and neutrino generation in solar flares.**
L. I. Miroshnichenko.
Particles and cosmology, p. 142 – 145 (1987). In Russian. Abstr. in Ref. Zh., 51. Astron., 9.51.390 (1987). – See Abstr. 012.034.

080.037 **Neutrino helioseismology.**
Yu. S. Kopysov.
Particles and cosmology, p. 124 – 142 (1987). In Russian. Abstr. in Ref. Zh., 51. Astron., 9.51.462 (1987). – See Abstr. 012.034.

080.038 **Is solar neutrino capture rate correlated with sunspot number?**
J. N. Bahcall, G. B. Field, W. H. Press.
Astrophys. J., Lett. Ed., Vol. 320, No. 1, p. L69 – L73 (1987).

Davis has described an apparently remarkable correlation between sunspot number and the capture rate of solar neutrinos in the ^{37}Cl experiment. Here the authors quantify the statistical significance of the apparent correlation between sunspots and observed neutrino rate. They show that the correlation derives from a small number of experimental points. By a calculation based on standard electroweak theory and well–understood processes for neutrino production, it is shown that a correlation, if real, would be extremely puzzling on energetic grounds alone.

080.039 **Limits on sensitivity of large silicon bolometers for solar neutrino detection.**
C. J. Martoff.
Science, Vol. 237, No. 4814, p. 507 – 509 (1987).

Estimates are given for ultimate limits on background in proposed direct–counting measurements of neutrino scattering from large silicon crystals. Methods of background reduction are discussed. In the best case, the limiting backgrounds due to activities from cosmic–ray spallation of silicon would be less than the expected true event rate for reactor neutrino measurements of coherent neutral–current scattering from silicon nuclei. Considerable reduction of the estimated high–energy backgrounds would be required for a good signal–to–noise ratio in solar neutrino detection.

080.040 **Nonlinear effects of acoustic oscillations in the Sun.**
A. G. Kosovichev.
Izv. Krymskoj Astrofiz. Obs., Tom 76, p. 179 – 185 (1987). In Russian. English translation in Bull. Crimean Astrophys. Obs., Vol. 76.

A finite–difference method for the adiabatic equations of gas dynamics was used to study nonlinear interactions of radial acoustic modes ($p_{10} - p_{30}$) in the Sun. It is shown that nonlinear effects generate a low frequency oscillation and high frequency oscillations whose frequencies are the sums of the acoustic mode frequencies. It has been found that although the period of low frequency oscillation may be close to 160 min, the amplitudes of acoustic modes must be $10^2 - 10^3$ times the observed ones independently of the number of interacting modes and their phases. Thus the author tends to conclude that the 160–min oscillation cannot be explained in terms of nonlinear effects of acoustic mode oscillations in the Sun.

080.041 **The influence of giant–cell convection on eigenfrequencies of solar oscillations.**
E. M. Lavely, T. H. Jordan.
Bull. Am. Astron. Soc., Vol. 19, No. 2, p. 741 (1987). Abstract. – See Abstr. 010.061.

080.042 **Rotation of the Sun and rotation of its general magnetic field.**
V. A. Kotov.
Izv. Krymskoj Astrofiz. Obs., Tom 77, p. 39 – 50 (1987). In Russian. English translation in Bull. Crimean Astrophys. Obs., Vol. 77.

The author analysed the measurements of the solar mean magnetic field (SMMF, 1968 – 1984) and also the time series of polarity of the interplanetary magnetic field (IMF 1926 – 1983). Power spectra of both sets show the presence of two dominant periods of rotation: $26\overset{d}{.}94 \pm 0\overset{d}{.}02$ and $28\overset{d}{.}20 \pm 0\overset{d}{.}02$ (synodic). The first period is interpreted as the most coherent period of rotation of the surface magnetic fields near equator. The other period corresponds to the highest peak in the IMF power spectrum and does not show any splittings due to differential rotation and 22–year cycle. It is conjectured that this $28\overset{d}{.}20$ period reflects rotation of a deeply rooted solar magnetic field which penetrates the radiative zone of the Sun. This tentative conclusion appears to be compatible with recent helioseismological observations. The author notes that the $28\overset{d}{.}20$ period corresponds to rotation rate of the photosphere at heliolatitudes of about 28° where spots usually originate at the onset of each solar cycle.

080.043 **The 160 minutes period, internal rotation and 11–year cycle of the Sun: evidence for a relationship?**
V. A. Kotov, L. S. Levitskij.
Izv. Krymskoj Astrofiz. Obs., Tom 77, p. 51 – 71 (1987). In Russian. English translation in Bull. Crimean Astrophys. Obs., Vol. 77.

In order to specify an exact value of the 160^m period of solar oscillations, the authors analyse a sample of about 19,000 chromospheric flares on the Sun as observed by the world network of solar observatories in 1947 – 1980. For the dominant peak in the 160^m range studied the authors obtain the best period's value of $160\overset{m}{.}01015 \pm 0\overset{m}{.}00008$, which is in excellent agreement with the result of Doppler observations. Some physical relation between fine structure of the 160^m period and solar cycle might exist. The time sequence of flares reveals also a strong peak in the 80^m range. Discrete frequencies of 160^m oscillations (as seen in solar flares) may be dependent on detailed structure of the deep solar interior and therefore may be used to map out the interior rotation rate. It is also pointed out that the model with a rapidly spinning small core may reasonably explain the existence of a slower rotation zone inside the Sun, near $0.3\ R_\odot$, deduced recently by Duvall et al. (1984) from helioseismological observations.

080.044 **On the possibility of rapid rotation of the solar core.**
V. A. Kotov, A. G. Kosovichev.
Izv. Krymskoj Astrofiz. Obs., Tom 77, p. 72 – 78 (1987). In Russian. English translation in Bull. Crimean Astrophys. Obs., Vol. 77.

The authors discuss a conjecture that the central core of the Sun rotates plausibly with very short period near 160^m. The rotational splitting of low degree *p*–modes oscillations was calculated for the standard solar model, but with the inner core of various size spinning with the 160^m period. The results of calculations agree with the actual splitting of *l* = 1, 2 and 3 modes inferred by Duvall et al. (1984) from observations of 5^m–oscillations, if one assumes that the radius of a rapidly spinning core is less than $0.08\ R_\odot$. Such small core contains about 6% of the total solar mass. The resulting gravitational quadrupole moment J_2 of the Sun, $\lesssim 4 \times 10^{-6}$, does not contradict the results of observational data on solar oblateness.

080.045 1980 – 1984 measurements of the solar radio radius at 2.25 cm wavelength.
A. F. Bachurin.
Izv. Krymskoj Astrofiz. Obs., Tom 77, p. 79 – 82 (1987). In Russian. English translation in Bull. Crimean Astrophys. Obs., Vol. 77.

The solar radio radius was measured in 1980 – 1984 in polar directions at 2.25 cm wavelength. Its maximum value was observed in 1981. Radio radius of the Sun differs from optical, measured at the same time, by about 3″. It has been shown that more precise determination of such variations needs more accurate account of the radio emission in polar regions.

080.046 160–min pulsations of the Sun studied by a geophysical method.
V. P. Bobova.
Izv. Krymskoj Astrofiz. Obs., Tom 77, p. 90 – 96 (1987). In Russian. English translation in Bull. Crimean Astrophys. Obs., Vol. 77.

The effect of 160–min pulsations of the Sun found earlier in the geomagnetic AE–index, was investigated in a short time–scale to study the variability of the amplitude and the phase. It was revealed that the mean phase was stable for a long time (1966 – 1979), but that there existed periodical variations with the quasiperiod of about 8.5 months.

080.047 The sun among the stars: what the stars indicate about solar variability.
D. R. Soderblom, S. L. Baliunas.
Space Telesc. Sci. Inst., Prepr. Ser., No. 214, 24 pp. (1987). To appear in NATO advanced research workshop "Secular solar and geomagnetic variations in the last 10,000 years".

080.048 Hydrogen–helium–diffusion in solar models.
J. Wambsganß.
MPA Rep., No. 310, 9 pp. (1987). Submitted to Astron. Astrophys.

080.049 The interaction of solar p–modes with a sunspot. II. Simple theoretical models.
T. E. Abdelatif, J. H. Thomas.
Astrophys. J., Vol. 320, No. 2, p. 884 – 892 (1987).

The authors present two simple theoretical models of the interaction of solar p–modes with a sunspot magnetic flux tube. The sunspot magnetic flux tube is modeled as a uniform magnetic slab or circular magnetic cylinder surrounded by field–free gas. Gravity is neglected, and trapped p–modes are set up by rigid, reflecting horizontal boundaries. Although these models are highly simplified, they do predict qualitatively the two important observed effects reported in Paper I, namely, the horizontal wavelength shift and the selective filtering of p–modes by the sunspot.

080.050 The ground–based spectrophotometry of solar radiation and its application.
L.–s. Zhong, Y. Shan, Z.–l. Qiu.
Publ. Yunnan Obs., No. 1, p. 102 – 110 (1987). In Chinese.

The authors report on the work on the groundbased determination of solar radiation as a function of wavelengths done in the Kunming area, Yunnan Province, China; and the analysis and processing of the 196 groups of data obtained at four time intervals. The solar irradiation curves are given and discussed.

080.051 Analysis of correlation between the total radiation fluxes at 2902 MHz, 3653 MHz and 9375 MHz and the solar constant in 1980.
Y. Ma.
Publ. Yunnan Obs., No. 2, p. 47 – 50 (1987). In Chinese.

080.052 Numerical modelling of formation of zones with increased temperature and flux density in the atmosphere of the Sun.
A. Ya. Bojko.
Vestn. MGU. Vychisl. Mat. Kibernet., No. 3, p. 42 – 46 (1987). In Russian. Abstr. in Ref. Zh., 51. Astron. 11.51.387 (1987).

080.053 Can the solar neutrino problem be the first detected signature of dark matter from the halo of our galaxy? (Candidates for solar cosmions).
G. Gelmini.
The early universe, p. 351 – 360 (1988). – See Abstr. 012.043.

The solar neutrino problem can be solved if the sun has accreted a sufficient number of weakly interacting particles from the dark matter of the galaxy. Dark matter particles with mass and cross–sections in a small range are required. It is an interesting problem of model building to give concrete particle candidates for these particles. The author presents it as an example of the deep interconnection of astrophysics and particle physics necessary to deal with dark matter candidates.

080.054 Dark matter induced neutrinos from the Sun: theory versus experiment.
K.–W. Ng, K. A. Olive, M. Srednicki.
Phys. Lett. B, Vol. 188, No. 1, p. 138 – 142 (1987). Abstr. in Phys. Abstr., Vol. 90, No. 1310, Entry 94616 (1987).

080.055 Limits on the flux of energetic neutrinos from the Sun.
J. M. Losecco, J. C. van der Velde, R. M. Bionta, G. Blewitt, C. B. Bratton, D. Casper, A. Ciocio, B. Cortez, S. Dye, S. Errede, G. Foster, W. Gajewski, K. S. Ganezer, M. Goldhaber, T. J. Haines, T. W. Jones, D. Kielczewska, W. R. Kropp, J. G. Learned, E. Lehmann, J. Matthews, H. S. Park, F. Reines, J. Schultz, S. Seidel, E. Shumard, D. Sinclair, H. W. Sobel, J. L. Stone, L. Sulak, R. Svoboda, G. Thornton, C. Wuest.
Phys. Lett. B, Vol. 188, No. 3, p. 388 – 392 (1987). Abstr. in Phys. Abstr., Vol. 90, No. 1310, Entry 94617 (1987).

080.056 Limits on neutrino oscillation parameters from the chlorine solar–neutrino experiment.
M. Cribier, J. Rich, M. Spiro, D. Vignaud, W. Hampel, B. T. Cleveland.
Phys. Lett. B, Vol. 188, No. 1, p. 168 – 170 (1987). Abstr. in Phys. Abstr., Vol. 90, No. 1310, Entry 94619 (1987).

080.057 The search for solar neutrinos.
L. S. Peak.
Aust. Phys., Vol. 24, No. 2, p. 32 – 35 (1987). Abstr. in Phys. Abstr., Vol. 90, No. 1310, Entry 94844 (1987).

080.058 Solar neutrino spectroscopy.
R. D. Scott.
Resonance ionization spectroscopy 1986, p. 281 (1987). Abstr. in Phys. Abstr., Vol. 90, No. 1310, Entry 94848 (1987). – See Abstr. 012.047.

080.059 Feasibility of a $^{81}Br(v,e^-)^{81}Kr$ solar neutrino experiment.
G. S. Hurst.
Resonance ionization spectroscopy 1986, p. 283 – 288 (1987). Abstr. in Phys. Abstr., Vol. 90, No. 1310, Entry 94849 (1987). – See Abstr. 012.047.

080.060 Solar oscillations and magnetograph signal formation in the presence of differences in brightness of the splitting components.
M. L. Demidov.
Kinematika Fiz. Nebesn. Tel, Tom 3, No. 6, p. 19 – 27 (1987). In Russian. English translation in Kinematics Phys. Celest. Bodies.

The problem concerning magnetograph signal formation with proper account of the possible difference in brightness of the splitting components is considered. The errors which may appear when this difference is ignored are numerically estimated. It is shown that this problem is of particular relevance to the study of small–amplitude phenomena such as global oscillations of the Sun. Besides, a method is suggested which makes it possible to measure (along with differential velocities) differential brightness variations in the same line what is very important in the study of oscillatory processes.

080.061   **Rigid rotation and localisation of sunspots.**
   I. Soru–Escaut, M.–J. Martres, Z. Mouradian.
C. R. Acad. Sci., Sér. II, Tome 305, No. 19, p. 1493 – 1497 (1987). In French.
   A spatial association between a sunspot and solar structures (plages and/or filaments) observed in previous rotations is pointed out. Using all spots (225) observed in 1977 with a diameter larger than 2,000 km, one finds that 25% are located at the tip of a previous filament, 35% at the border of a plage and 17% present the double association. During the evolution, the spot reaches, at its maximum, the location occupied by the features associated to it in the previous rotations. Therefore, this location is in rigid rotation.

080.062   **On the proposed associations of solar neutrino flux with solar particles, cosmic rays, and the solar activity cycle.**
   R. M. Wilson.
Sol. Phys., Vol. 112, No. 1, p. 1 – 15 (1987).
   Solar neutrino flux was found not to correlate strongly with cosmic–ray flux, and the Ap geomagnetic index was found not to correlate strongly with sunspot number.

080.063   **Variation of the solar constant during 1983 and 1984.**
   J. Pap.
Sol. Phys., Vol. 112, No. 1, p. 181 – 193 (1987).
   Measurements of the Nimbus–7/ERB and SMM/ACRIM radiometers indicated several dips in the total solar irradiance in 1983 and in the first part of 1984. The dips in 1983, which should have a real solar origin, were selected according to the peaks of the projected areas of the active sunspot groups above the $2\,\sigma$ error limit of their data set. In the first part of 1984 the sunspot activity was strong and few irradiance dips with relatively large amplitudes were observed. In the second part of 1984 the sunspot activity disappeared and at that time the solar constant only fluctuated around its mean.

080.064   **On spatial filtering of low–degree global oscillations of the Sun.**
   A. L. Balandin, V. M. Grigoryev (*V. M. Grigor'ev*),
   M. L. Demidov.
Sol. Phys., Vol. 112, No. 2, p. 197 – 209 (1987).
   For the spatial filters at the Crimean and Stanford telescopes and at a telescope with which Sun–as–a–star observations were conducted at the geographic south pole of the Earth, the authors have computed the functions of sensitivity (functions of spatial response) to global oscillations of the Sun for degrees $l = 0$ to 21. The results of the calculations are presented for both the vertical and horizontal components of the oscillations and a brief description of the computational technique is given.

080.065   **Chlorine and bromide experiments on recording of solar neutrinos.**
   R. Davis, B. T. Cleveland, J. K. Rowley.
Izv. Akad. Nauk SSSR. Ser. Fiz., Tom 51, No. 6, p. 1228 – 1229 (1987). In Russian. Abstr. in Ref. Zh., 51. Astron., 12.51.457 (1987). – See Abstr. 012.058.

080.066   **On excitation of magnetic line oscillations by a plasma stream.**
   M. P. Ryutova.
Inst. Yader. Fiz. SO Akada. Nauk SSSR, Prepr., No. 19, p. 3 – 9 (1987). In Russian. Abstr. in Ref. Zh., 51. Astron., 12.51.469 (1987).

080.067   **Search for ^{8}B solar neutrinos at KAMIOKANDE–II.**
   K. Hirata, T. Kajita, M. Koshiba, M. Nakahata,
   Y. Oyama, N. Sato, A. Suzuki, M. Takita, Y. Totsuka,
   T. Kifune, T. Suda, K. Takahashi, T. Tanimori, K. Miyano,
   M. Yamada, E. W. Beier, L. R. Feldscher, S. B. Kim,
   A. K. Mann, F. M. Newcomer, R. Van Berg, W. Zhang,
   B. G. Cortez.
The standard model. The supernova 1987A, p. 689 – 700 (1987). – See Abstr. 012.074.
   A search for ^{8}B solar neutrinos in the KAMIOKANDE–II detector is presented. No definite signature of the solar neutrinos has been observed yet. A flux upper limit of $4.3 \times 10^6 v_e/\mathrm{cm}^2/\mathrm{sec}$ on ^{8}B solar neutrino is obtained at 90% confidence level, based on 146 days of detector live time.

080.068   **The radio radius of the Sun at 0.82 and 1.35 cm.**
   N. A. Topchilo.
Tr. Astron. Obs., Leningrad, Tom 41, p. 143 – 155 (1987). Uch. Zap. LGU, No. 420, Ser. Mat. Nauk, Vyp. 63. In Russian.
   A new method of solar radius measurements based on the use of the derivative of scans across opposite sides of the solar limb is proposed. The corrections for the shape and width of the radio telescope diagram are computed. For the observations of 1982 – 1984 the average excess of radio radius above the optical one is $18.8''$ at $\lambda = 1.35$ cm and $12.4''$ at $\lambda = 0.82$ cm.

080.069   **L'eliosismologia. II.**
   L. Paternò.
G. Astron., Vol. 13, N. 1 – 2, p. 2 – 14 (1987).

080.070   **Transition of magnetohydrodynamical waves in the solar atmosphere.**
   L. Nocera.
Advances in turbulence, p. 284 – 290 (1987). Abstr. in Phys. Abstr., Vol. 90, No. 1316, Entry 134220 (1987). – See Abstr. 012.077.

080.071   **Discriminating among scenarios to resolve the solar neutrino puzzle.**
   E. W. Beier, M. L. Cherry, B. Kayser, W. Marciano,
   L. Wolfenstein.
Physics of the superconducting supercollider, p. 673 – 680 (1987). Abstr. in Phys. Abstr., Vol. 91, No. 1319, Entry 4256 (1988). – See Abstr. 012.080.

080.072   **The status of the search for solar neutrinos at KAMIOKANDE II.**
   E. W. Beier.
Physics of the superconducting supercollider, p. 694 – 696 (1987). Abstr. in Phys. Abstr., Vol. 91, No. 1319, Entry 4257 (1988). – See Abstr. 012.080.

080.073   **Energy transport and evaporation of weakly interacting particles in the Sun.**
   M. Nauenberg.
Phys. Rev. D, Vol. 36, No. 4, p. 1080 – 1087 (1987). Abstr. in Phys. Abstr., Vol. 91, No. 1319, Entry 4358 (1988).

080.074   **Neutrinos and sunspots.**
   J. N. Bahcall.
Nature, Vol. 330, No. 6146, p. 318 (1987).
   The discrepancy between the calculated and the observed rate of solar neutrino captures in chlorine is a long–standing problem. The author reports on the inverse correlation of the neutron capture rate with sunspot activity. For an other view see Abstr. 080.075.

080.075   **Atmospheric radioactivity and variations in the solar neutrino flux.**
   A. de la Zerda Lerner, K. O'Brien.
Nature, Vol. 330, No. 6146, p. 353 – 354 (1987).
   Attention has been devoted recently to possible time variations in the neutrino flux data, and some investigators have found an inverse correlation with the 11–year sunspot cycle. Here the

authors present a calculation of the inventory of positron emitters in the Earth's atmosphere that results from hadronic cascades initiated by galactic cosmic rays, and the variation of this source of low–energy neutrinos with the 11–year solar cycle. Their results rule out the possibility that cosmogenic neutrino emitters are responsible for an apparent solar–cycle dependence. As atmospheric secondary–particle decay was previously eliminated as a significant neutrino source, any such variations, if corroborated over the next sunspot cycle, would appear to be caused by phenomena outside the Earth's atmosphere, most likely in the Sun itself.

080.076   How accurate were seventeenth–century measurements of solar diameter?
C. R. O'Dell, A. Van Helden.
Nature, Vol. 330, No. 6149, p. 629 – 631 (1987).

The assumption that the solar diameter is constant over periods that are short compared to the nuclear process has recently been questioned. The authors discuss the accuracy of Jean Picard's micrometer measurements of the solar diameter and his successors' Mercury transit measurements in light of the claim by Ribes et al. that these data provide evidence for a change in the apparent size of the Sun at the end of the Maunder minimum and a difference from the current size. Their evidence suggests that the necessary corrections to the measurement of sizes with early telescopes are larger than Ribes et al. assume. Therefore the authors call into question the conclusion that the Sun rapidly changed in size during this period.

Volny v atmosfere Solntsa (*Oscillations and waves on the Sun*). Third seminar of the working group "Waves in the atmosphere of the Sun", held at Novosibirsk, USSR, 30 March – 1 April 1987.
See Abstr. 012.086.

The DUMAND project.
See Abstr. 013.038.

Observational neutrino astrophysics.
See Abstr. 013.045.

Neutrons, the sun and Canada.
See Abstr. 013.081.

Electrooptic Fabry–Perot filter: development for the study of solar oscillations.
See Abstr. 034.062.

A heavy water detector to resolve the solar neutrino problem.
See Abstr. 034.196.

Solar interferometry with a 4–aperture non–redundant and stabilized network.
See Abstr. 035.080.

Radiative alpha–capture rates leading to $A = 7$ nuclei: applications to the solar neutrino problem and big bang nucleosynthesis.
See Abstr. 061.024.

Bounds on galactic cold dark matter particle candidates and solar axions from a Ge–spectrometer.
See Abstr. 061.026.

Superstring candidates for dark matter.
See Abstr. 061.028.

The effects of resonant neutrino oscillations on the solar neutrino experiments.
See Abstr. 061.030.

Chemical abundances.
See Abstr. 061.035.

Analytic treatments of matter–enhanced solar neutrino oscillations.
See Abstr. 061.064.

Laboratory limits on solar axions from an ultralow–background germanium spectrometer.
See Abstr. 061.068.

"Just so" neutrino oscillations.
See Abstr. 061.080.

E_6 based mechanism for the generation of fermion electric dipole moments: an application to the solar neutrino puzzle.
See Abstr. 061.081.

Neutrino oscillations and the solar–neutrino problem.
See Abstr. 061.088.

Resonant neutrino oscillations.
See Abstr. 061.089.

Astrophysical S(E) factor of ^{3}He(^{3}He, 2p)^{4}He at solar energies.
See Abstr. 061.093.

Weakly interacting massive particle distribution in and evaporation from the Sun.
See Abstr. 061.101.

Permutation symmetry and the fermion mass matrix.
See Abstr. 061.106.

Neutrino oscillations and the Landau–Zener formula.
See Abstr. 061.107.

Opportunities for particle physics: solar neutrinos and superstrings.
See Abstr. 061.111.

Supersymmetric dark matter.
See Abstr. 061.113.

On the non adiabatic neutrino oscillations in matter.
See Abstr. 061.116.

Mikheyev–Smirnov–Wolfenstein enhancement of oscillations as a possible solution to the solar neutrino problem.
See Abstr. 061.118.

Paraphotons and axions: similarities in stellar emission and detection.
See Abstr. 061.123.

A simple solution to the solar neutrino and missing mass problems.
See Abstr. 061.133.

A review of matter oscillations and solar neutrinos.
See Abstr. 061.140.

A possible experimental method for measurement of the neutrino flux from the reaction e + ^{7}Be → ^{7}Li + v_e in the Sun.
See Abstr. 061.141.

Propagation of compressive waves through fibril magnetic fields. II. Scattering from a slab of magnetic flux tubes.
See Abstr. 062.019.

Propagation of compressive waves through fibril magnetic fields. III. Waves that propagate along the magnetic field.
See Abstr. 062.020.

Mode classification and wave propagation in a magnetically structured medium.
See Abstr. 062.026.

On negative viscosity in global circulations.
See Abstr. 062.062.

Hydromagnetic dynamo as source of planetary, solar and galactic magnetism.
See Abstr. 062.071.

Excitation of non–axisymmetrical modes of the mean magnetic field of the Sun.
See Abstr. 062.087.

MHD instabilities of a cylindrical plasma with a realistic energy equation.
See Abstr. 062.106.

A model of an isolated magnetic flux tube in a stratified atmosphere.
See Abstr. 062.136.

The interaction of quasiperpendicular shock waves in a collisionless plasma.
See Abstr. 062.144.

The α–effect generated by random Alfvén waves in a stratified atmosphere and particle acceleration.
See Abstr. 062.159.

Turbulent heat transport and latitudinal temperature variations in rotating convective envelopes.
See Abstr. 064.002.

OH and CH continuous opacity in solar and stellar atmospheres.
See Abstr. 064.080.

On the origin of the solar system s–process abundances.
See Abstr. 065.077.

The dynamical oscillation and propulsion of magnetic fields in the convective zone of a star. II. Thermal shadows.
See Abstr. 065.101.

The dynamical oscillation and propulsion of magnetic fields in the convective zone of a star. III. Accumulation of heat and the onset of the Rayleigh–Taylor instability.
See Abstr. 065.102.

Possible effects on the sun and close binary systems from background gravitational radiation with period 160 min.
See Abstr. 066.056.

Probing gravity to the second post–Newtonian order and to one part in 10^7 using the spin axis of the Sun.
See Abstr. 066.098.

The deflection of light by the Sun due to three–space curvature.
See Abstr. 066.160.

A test of the equivalence principle with polarised light.
See Abstr. 066.164.

The gradient of the small–scale velocity fluctuation in the solar atmosphere.
See Abstr. 071.004.

The solar boron abundance.
See Abstr. 071.024.

Temporal variations of solar spectral line profiles induced by the 5–minute photospheric oscillation.
See Abstr. 071.027.

Torsional oscillations and the solar cycle.
See Abstr. 072.002.

Local rigid rotation and the emergence of Active Centres.
See Abstr. 073.011.

Does the large–scale solar magnetic field distribution really reflect the convective velocity fields?
See Abstr. 075.002.

Observation of solar differential rotation with the aid of magnetic tracers.
See Abstr. 075.005.

The origin of rigidly rotating magnetic field patterns on the Sun.
See Abstr. 075.016.

Rotation of the sun and structure of magnetic elements.
See Abstr. 075.029.

Ion–collision broadening of solar lines in the far–infrared and submillimeter spectrum.
See Abstr. 077.006.

On the origin of magnetic fields of cosmic objects.
See Abstr. 091.033.

Solar oscillations and rotation of close binary systems in the Galaxy: the 160 minute period.
See Abstr. 117.035.

Heliomagnetic dipole moment and daily variation of cosmic rays underground.
See Abstr. 144.025.

Earth

081 Structure, Figure, Gravity, Orbit, etc.

081.001 Search for, and study of, Paleozoic impact ejecta: progress made during the past year.
W. F. Read.
NASA Tech. Memo., NASA TM–89810, p. 439 – 440 (1987). –
See Abstr. 003.001.

081.002 Secular cooling of Earth as a source of intraplate stress.
S. C. Solomon.
NASA Tech. Memo., NASA TM–89810, p. 455 – 457 (1987). –
See Abstr. 003.001.

It is likely that the Earth has been steadily cooling over the last 3 – 4 billion years, and the global contraction that accompanied such cooling would have led to a secular decrease in the radius of curvature of the plates. The author explores here the implications of this global cooling and contraction for the intraplate stress field and the evolution of continental plates.

081.003 A test of the hypothesis that impact–induced fractures are preferred sites for later tectonic activity.
S. C. Solomon, E. D. Duxbury.
NASA Tech. Memo., NASA TM–89810, p. 458 – 460 (1987). –
See Abstr. 003.001.

081.004 Mass extinctions caused by large bolide impacts.
L. W. Alvarez.
Phys. Today, Vol. 40, No. 7, p. 24 – 33 (1987).

Evidence indicates that the collision of Earth and a large piece of Solar System debris such as a meteoroid, asteroid or comet caused the great extinctions of 65 million years ago, leading to the transition from the age of the dinosaurs to the age of the mammals.

081.005 Values of 16th–order harmonics in the geopotential from analysis of resonant orbits.
D. G King–Hele, D. M. C. Walker.
Planet. Space Sci., Vol. 35, No. 7, p. 937 – 946 (1987).

Analysis of the orbits of three new satellites at 16th–order resonance, together with two previously analysed, has given five pairs of values of lumped 16th–order harmonics of odd degree over a range of inclinations between 50° and 92°. These values of the lumped harmonics are used to derive values of individual harmonic coefficients of order 16 and degree 17, 19, 21, 23 and 25. Of these, the values for degrees 17, 19 and 21 have an average standard deviation equivalent to 2 cm in geoid height. The lumped harmonics are also used to test existing comprehensive models of the gravity field: the authors conclude that, for order 16, these models have deficiencies, especially at high latitudes.

081.006 The accuracy of a Goddard TOPEX gravity model as seen by independent resonant data.
C. A. Wagner.
Planet. Space Sci., Vol. 35, No. 8, p. 997 – 1008 (1987). With a correction in Vol. 35, No. 9, p. 1229 (1987).

081.007 Structure of the earth: mantle and core.
T. Lay.
Rev. Geophys., Vol. 25, No. 6, p. 1161 – 1167 (1987). U.S.National Report to IUGG 1983–1986.

The deep interior structure of the earth has been extensively analyzed using a wide variety of seismic phases and techniques during the last four years. Most studies have emphasized quantitative three–dimensional mapping of the lateral velocity heterogeneity of the mantle and core. These aspherical velocity variations are believed to be direct manifestations of thermal and compositional heterogeneity associated with convective process-es. A first generation of global models for the lateral velocity variations in the deep earth has been produced, providing tantalizing images of large scale structures suggestive of a non-steady state thermal convection system.

081.008 Structure of the earth: oceanic crust and uppermost mantle.
J. A. Orcutt.
Rev. Geophys., Vol. 25, No. 6, p. 1177 – 1196 (1987). U.S.National Report to IUGG 1983–1986.

Contents: Rise axis structure: magma chambers. Fracture zone structure: thin oceanic crust. Crustal and uppermost mantle anisotropy. Evolution of the oceanic crust and uppermost mantle. Arctic exploration. Propagation of high frequency $P_n/S_n/ T$ phases: reverberation or scattering. Attenuation. Seafloor noise and topographic scattering. Seafloor and subseafloor receivers and sources. Theoretical seismology including the inversion of data. Seismicity. Multichannel and reflection seis-mology. Seismic refraction studies. Epilogue.

081.009 Mantle convection and the state of the earth's interior.
B. H. Hager, M. Gurnis.
Rev. Geophys., Vol. 25, No. 6, p. 1277 – 1285 (1987). U.S.National Report to IUGG 1983–1986.

During 1983 – 1986, the four year period covered by this review, emphasis in the study of mantle convection shifted away from fluid mechanical analysis of simple systems with uniform material properties and simple geometries, toward analyzing the effects of more complicated, presumably more realistic models.

081.010 An estimate of equatorial gravity from terrestrial and satellite data.
R. H. Rapp.
Geophys. Res. Lett., Vol. 14, No. 7, p. 730 – 732 (1987).

Equatorial gravity has been estimated from terrestrial gravity data and from satellite implied data. The terrestrial estimate was based on the analysis of 43271 $1° \times 1°$ mean free air gravity anomalies. The value estimated was 978032.35 ± 0.5 mgals. An alternate estimate based on the comparison of terrestrial anoma-lies and altimeter derived anomalies yields 978032.6 mgals. Equatorial gravity from space data depends on GM from Lageos and lunar laser data and on a new estimate (6378136.2 ± 0.5 m) of the equatorial radius. The resultant equatorial gravity is 978032.78 ± 0.2 mgals. The difference between the two terrestrial estimates is 0.4 and 0.2 mgals. The agreement is within the accuracy of the two separate estimates. These comparisons can put constraints on parameters of non–Newtonian force models.

081.011 Mantle devolatilization and convection: implications for the thermal history of the earth.
M. J. Jackson, H. N. Pollack.
Geophys. Res. Lett., Vol. 14, No. 7, p. 737 – 740 (1987).
The authors assess the possible effects of mantle devolatilization on terrestrial thermal evolution, by means of a parameterized convection model with volatile–dependent rheology.

081.012 End–Cretaceous mass extinction event: argument for terrestrial causation.
A. Hallam.
Science, Vol. 238, No. 4831, p. 1237 – 1242 (1987).
The end–Cretaceous mass extinctions were not a geologically instantaneous event and were selective in character. These features are incompatible with the original Alvarez hypothesis of their being caused by a single asteroid impact that produced a world–embracing dust cloud with devastating environmental consequences. By analysis of physical and chemical evidence from the stratigraphic record it is shown that a modified extraterrestrial model in which stepwise extinctions resulted from encounter with a comet shower is less plausible than one intrinsic to the earth, involving significant disturbance in the mantle.

081.013 Cosmogenic ^{10}Be in Zaire alluvial diamonds: implications for ^{3}He contents of diamonds.
D. Lal, K. Nishiizumi, J. Klein, R. Middleton, H. Craig.
Nature, Vol. 328, No. 6126, p. 139 – 141 (1987).
To determine the amounts of cosmic–ray produced ("cosmogenic") ^{3}He which could be created in diamonds during their post–eruptive residence in near–surface regions, the authors have measured ^{10}Be in an alluvial diamond sample, using radiochemical and accelerator mass spectrometry (AMS) techniques. The observed ^{10}Be concentration leads to an effective surface exposure of 3.3×10^5yr for these diamonds. This corresponds to a substantial cosmogenic production of ^{3}He in the alluvial diamonds, $\geqslant 29\%$ of the total ^{3}He. It is concluded that the high ^{3}He/^{4}He ratios in at least some diamonds are due to cosmogenic ^{3}He production in near–surface regions during their residence in alluvial or weathering deposits.

081.014 Magnetohydrodynamics of the Earth's core.
D. Gubbins, P. H. Roberts.
Geomagnetism, Vol. 2, p. 1 – 183 (1987). – See Abstr. 003.011.
Contents: 1. The reference state for for the Earth's core. 2. Density stratification. 3. Conduction in solids. 4. Fluids in rotation. 5. Perfect fluids and their application to the theory of the secular variation. 6. Diffusionless waves. 7. Diffusion. 8. Magnetoconvection in rotating systems.

081.015 Origin of the main field: kinematics.
P. H. Roberts, D. Gubbins.
Geomagnetism, Vol. 2, p. 185 – 249 (1987). – See Abstr. 003.011.
Contents: 1. Driving mechanisms for convection and the dynamo. 2. Kinematic dynamo theory. 3. The α–effect. 4. α^2–dynamos. 5. $\alpha\omega$–dynamos.

081.016 Origin of the main field: dynamics.
P. H. Roberts.
Geomagnetism, Vol. 2, p. 251 – 306 (1987). – See Abstr. 003.011.
Contents: 1. Background. 2. Electrodynamic background. 3. Strong–field models. 4. Weak–field models. 5. Two–dimensional models. 6. Conclusion.

081.017 Determination of the external gravity field of the earth and the geoid.
K. H. Ilk, C. Reigber, H.–G. Wenzel.
Dtsch. Geod. Komm. Bayer. Akad. Wiss., Reihe B, Heft Nr. 284, p. 157 – 165 (1987). In German and English.

081.018 Tidal parameters for an inelastic Earth.
V. Dehant.
Phys. Earth Planet. Inter., Vol. 49, Nos. 1 – 2, p. 97 – 116 (1987).
The gravimetric factors for an elliptical uniformly rotating Earth are computed according to a new definition with respect to Wahr's results and considering the mantle inelasticity. The Love numbers h and k are defined in the elliptical Earth case to correspond to the Love spherical definition. They become latitude dependent. The tidal parameters discussed in this paper are computed for an inelastic Earth's mantle. The numerical results show an increase of $\sim 0.4\%$ of the gravimetric factors with respect to the Wahr's values. The Love numbers h and k increase by $\sim 1.4\%$ to 3% with respect to the elastic case.

081.019 Integration of the gravitational motion equations for an elliptical uniformly rotating earth with an inelastic mantle.
V. Dehant.
Phys. Earth Planet. Inter., Vol. 49, Nos. 3 – 4, p. 242 – 258 (1987).
The author presents numerical integration solutions of the deformation equations for an elliptical uniformly rotating earth with an inelastic mantle. Inelasticity is introduced in Wahr's model by using complex rheological parameters in the frequency domain. Many choices for their profile inside the earth are analysed. The author concludes that the cut gaussian form of the stess relaxation time distribution of Zschau for the mantle shear modulus profile, combined with any realistic value for the bulk modulus profile and the elastic profile inside the outer core and inner core are the best choice. The corresponding increase of the tidal gravimetric factor is of the order of one or two tenths of a percent. Mantle inelasticity can thus explain a part of the discrepancy between observations and computations.

081.020 Multiple–taper spectral analysis of terrestrial free oscillations: part I.
J. Park, C. R. Lindberg, D. J. Thomson.
Geophys. J. R. Astron. Soc., Vol. 91, No. 3, p. 755 – 794 (1987).

081.021 Multiple–taper spectral analysis of terrestrial free oscillations: part II.
C. R. Lindberg, J. Park.
Geophys. J. R. Astron. Soc., Vol. 91, No. 3, p. 795 – 836 (1987).

081.022 Combination of temporal changes of gravity, height and potential coefficients for the determination of secular changes of the geoid.
L. E. Sjöberg.
Z. Vermessungswes., Vol. 112, No. 4, p. 167 – 172 (1987). Abstr. in Phys. Abstr., Vol. 90, No. 1311, Entry 101184 (1987).

081.023 Das Aussterben der Dinosaurier: ein kosmisches Ereignis?
H. Hilbrecht.
Sterne Weltraum, 26. Jahrg., Nr. 12, p. 696 – 698 (1987).

081.024 Do slow orbital periodicities appear in the record of Earth's magnetic reversals?
R. B. Stothers.
Geophys. Res. Lett., Vol. 14, No. 11, p. 1087 – 1090 (1987).
Time–series spectral analysis has been performed on the dates of geomagnetic reversals of the last 20 Myr BP and earlier. Possible evidence is found from the presence of high spectral peaks for two very long periodicities, 0.4 Myr and 1.3 Myr, that may be associated with slow variations of the Earth's orbital eccentricity as predicted by Berger. However, statistical significance tests and a number of other arguments do not confirm the two detections.

081.025 Comet showers as a cause of mass extinctions.
P. Hut, W. Alvarez, W. P. Elder, T. Hansen, E. G. Kauffman, G. Keller, E. M. Shoemaker, P. R. Weissman.
Nature, Vol. 329, No. 6135, p. 118 – 126 (1987).
If at least some mass extinctions are caused by impacts, why do they extend over intervals of one to three million years and have a partly stepwise character? The solution may be provided by multiple cometary impacts. Astronomical, geological and palaeontological evidence is consistent with a causal connection

between comet showers, clusters of impact events and stepwise mass extinctions.

081.026 Palaeomagnetic data suggest link between the Archaean–Proterozoic boundary and inner–core nucleation.
C. J. Hale.
Nature, Vol. 329, No. 6136, p. 233 – 237 (1987).

Recent determinations of the geomagnetic palaeointensity are consistent with a sharp increase in the magnitude of the geomagnetic field between 2.7 and 2.1 Gyr. Analysis of published late Archaean and early Proterozoic apparent polar wander paths shows that an abrupt change in the rate and style of continental motion occurred at about 2.5 – 2.6 Gyr ago. If this change is also interpreted to be related to the growth of an inner core, then, an intriguing conjecture arises that the Archaean–Proterozoic transition is related to inner core nucleation.

081.027 Confusion at the boundary.
S. K. Donovan.
Nature, Vol. 329, No. 6137, p. 288 (1987).

This note comments on the Precambrian/Cambrian boundary, at about 570 million years ago, that immediately preceded the great radiation of the first shelly faunas following the genesis of metazoan organisms in the late Precambrian. It has been suggested variously that this boundary does or does not represent a mass extinction.

081.028 Acid rain at the K/T boundary.
P. J. Crutzen.
Nature, Vol. 330, No. 6144, p. 108 – 109 (1987).

The author reports on the Cretaceous/Tertiary (K/T) mass extinction 65 Myr ago and gives explanations of this event.

081.029 The Bruns formula in three dimensions.
A. Dermanis.
Bull. Géod., Vol. 61, No. 4, p. 297 – 309 (1987).

The Bruns formula is generalized to three dimensions with the derivation of equations expressing the height anomaly vector or the geoid undulation vector as a function of the disturbing gravity potential and its spatial derivatives.

081.030 Progress towards a new measurement of the mass of the Earth.
G. T. Gillies, A. Marussi.
Boll. Geofis. Teor. Appl., Vol. 28, No. 111 – 112, p. 193 – 197 (1986). Abstr. in Phys. Abstr., Vol. 90, No. 1315, Entry 127070 (1987).

081.031 Current loops fitted to geomagnetic model spherical harmonic coefficients.
L. R. Alldredge.
J. Geomagn. Geoelectr., Vol. 39, No. 5, p. 271 – 296 (1987). Abstr. in Phys. Abstr., Vol. 90, No. 1316, Entry 133568 (1987).

081.032 Thermal effects of core formation in planet Earth.
U. Schmit.
Gerlands Beitr. Geophys., Vol. 96, No. 3 – 4, p. 239 – 250 (1987). Abstr. in Phys. Abstr., Vol. 90, No. 1317, Entry 139525 (1987).

081.033 A model–Z geodynamo.
S. I. Braginskij, P. H. Roberts.
Geophys. Astrophys. Fluid Dyn., Vol. 38, No. 4, p. 327 – 349 (1987). Abstr. in Phys. Abstr., Vol. 90, No. 1318, Entry 145327 (1987).

081.034 The cometary contribution to the oceans of primitive Earth.
C. F. Chyba.
Nature, Vol. 330, No. 6149, p. 632 – 635 (1987).

Recent compilations of the lunar impact record, combined with the mass–scaling law for crater diameters in the large–body regime, allow an estimate of the total mass incident on the Moon during the period of heavy bombardment. The calculation

provides a lower limit which is independent of cometary flux models, extrapolations of the lunar impact record, or assumptions about the cometary mass spectrum. The results imply that the Earth would have acquired an exogeneous ocean of water between ~ 4.5 and ~ 3.8 Gyr ago if comets comprised $\gtrsim 10\%$ by mass of the impacting population.

081.035 Magnetic field at the core boundary in the nearly symmetric hydromagnetic dynamo Z.
A. P. Anufriev, I. Cupal.
Stud. Geophys. Geod., Vol. 31, No. 1, p. 37 – 42 (1987).

081.036 Definition of the normal gravity field including the constant part of tides.
A. Zeman.
Stud. Geophys. Geod., Vol. 31, No. 2, p. 113 – 120 (1987).

081.037 Secular variations in the curvature of equipotential surfaces and directions of verticals due to geodynamic phenomena.
M. Burša.
Stud. Geophys. Geod., Vol. 31, No. 3, p. 225 – 227 (1987).

Formulae and numerical estimates are given for the non–periodical variations in the curvature of equipotential surfaces, horizontal forces and directions of the vertical, caused by the decrease in the second zonal geopotential harmonic, by the decrease in the angular velocity of the Earth's rotation and by secular polar motion.

081.038 Global geoidal undulations as inferred from the Bruns formula and from inversion of the geopotential series.
Z. Martinec, K. Pěč.
Stud. Geophys. Geod., Vol. 31, No. 3, p. 228 – 251 (1987).

081.039 Improvement of the convergence of an expansion of a planetary potential. Radial derivative of the earth's potential and a gravity anomaly.
M. S. Petrovskaya.
Sov. Astron., Vol. 31, No. 2, p. 213 – 215 (1987). English translation of 43.081.031.

081.040 An improved model of the Earth's gravitational field: *GEM–T1*.
J. G. Marsh, F. J. Lerch, B. H. Putney, D. C. Christodoulidis, T. L. Felsentreger, B. V. Sanchez, D. E. Smith, S. M. Klosko, T. V. Martin, E. C. Pavlis, J. W. Robbins, R. G. Williamson, O. L. Colombo, N. L. Chandler, K. E. Rachlin, G. B. Patel, S. Bhati, D. S. Chinn.
NASA Tech. Memo., NASA TM–4019, 6 + 354 pp. (1987).

This report describes the preliminary gravity model GEM–T1, which is exclusively based upon direct satellite tracking observations. This spherical harmonic model, complete to degree and order 36 is a direct result of the gravity field improvement effort.

Geohistory. Global evolution of the Earth.
See Abstr. 003.106.

Interior structure of the Earth and planets.
See Abstr. 003.145.

Remote sensing from space. Proceedings of Symposium 3, Workshop V and of the COSPAR Interdisciplinary Scientific Commission A (Meeting A2) of the COSPAR Twenty–sixth Plenary Meeting held in Toulouse, France, 30th June – 11th July 1986.
See Abstr. 012.067.

"Solid Earth" panel report.
See Abstr. 013.077.

Tidal disruption of a solid body.
See Abstr. 014.034.

Geological cyclicity and its sources.
See Abstr. 015.007.

Computer simulations of 10–km–diameter asteroid impacts into oceanic and continental sites – preliminary results on atmospheric passage, cratering, and ejecta dynamics.
See Abstr. 021.002.

Transformation of geocentric to geodetic coordinates without approximations.
See Abstr. 021.025.

Radiation chemistry under unconventional conditions: dosimetry and aqueous radiolysis relevant to comet nuclei and early Earth structure.
See Abstr. 022.115.

Relative motion near the triangular libration points in the Earth–Moon system.
See Abstr. 042.054.

Global geodynamics.
See Abstr. 044.009.

Changes in the Earth's rotation and low–degree gravitational field induced by earthquakes.
See Abstr. 044.011.

Snow load effect on the earth's rotation and gravitational field, 1979 – 1985.
See Abstr. 044.013.

Electromagnetic core–mantle coupling and the Earth's rotation.
See Abstr. 044.040.

Secular tidal and nontidal variations in the Earth's rotation.
See Abstr. 044.042.

Non–tidal motions between Mizusawa and Washington, D.C. with the aid of the PZT observations.
See Abstr. 045.007.

Microwave geodesy by interferometric radiometry.
See Abstr. 045.009.

VLBI for geodynamics involving the Hartebeesthoek Radio Astronomy Observatory in South Africa.
See Abstr. 045.014.

Potential of remote sensing for the study of global change. COSPAR report to the International Council of Scientific Unions (ICSU).
See Abstr. 051.018.

The estimation of the influence of tesseral harmonics perturbations on the orbits of some geodetic satellites.
See Abstr. 052.010.

Satellite–to–satellite tracking, satellite gravity gradiometry.
See Abstr. 052.012.

Geopotential orbit variations: applications to error analysis.
See Abstr. 052.017.

Earth gravity field and high satellite orbits.
See Abstr. 052.033.

On the interpretation of the geomagnetic energy spectrum.
See Abstr. 084.039.

Geomagnetic secular variation, core motions and implications for the earth's wobbles.
See Abstr. 084.043.

Annual variation in the surface magnetic field due to magnetospheric current systems.
See Abstr. 084.078.

Crater production on Venus and Earth by asteroid and comet impact.
See Abstr. 093.003.

Secular deceleration of the Moon and of the Earth's rotation and variation in the zonal geopotential harmonic.
See Abstr. 094.028.

Volatiles on Earth and Mars: a comparison.
See Abstr. 097.065.

Can episodic comet showers explain the 30–Myr cyclicity in the terrestrial record?
See Abstr. 107.014.

082 Atmosphere (Refraction, Scintillation, Extinction, Airglow, Site Testing)

082.001 **On possible influence of meteors on the Earth atmosphere.**
J. Rajchl.
Bull. Astron. Inst. Czech., Vol. 38, No. 4, p. 235 – 240 (1987).

In the enlarged system of fireballs and faint meteors as sources and one response in the form of noctilucent clouds (NLC) mutually inter–related by global atmospheric circulation on the two hemispheres of the Earth two types of junction are found: the serial (s) and the parallel (p) type. This s–p pattern helps in determining a response in the form of the so–called enhanced airglow of the night sky (EA) as complementary to the NLC. An essential factor for the origin of NLC proves to be meteoric ions rather than neutral dust, whereas the case of EA should be of opposite nature.

082.002 **Temperature trends in the stratosphere and lower mesosphere of the northern hemisphere.**
C. Varotsos.
Earth, Moon, Planets, Vol. 39, No. 1, p. 93 – 99 (1987).

Rocketsonde–derived temperature fluctuations within the northern hemisphere are examined for the stratosphere and lower mesosphere in seasonal basis for the years 1969 – 78, inclusive. The rocketsonde records presented here are homogeneous because they are mostly based on the Datasone system. It is suggested that stratospheric–lower mesospheric temperature variations are about one order of magnitude larger than those recorded in the literature before. The main feature in all seasons is that the cooling trend has maximum values at low latitudes in the lower mesosphere.

082.003   Site evaluation for the VLT: a status report.
M. Sarazin.
Messenger, No. 49, p. 37 – 39 (1987).

082.004   On the detection and utilization of gravity waves in airglow studies.
C. O. Hines, D. W. Tarasick.
Planet. Space Sci., Vol. 35, No. 7, p. 851 – 866 (1987).
The authors develop theoretical relations between fluctuations of airglow brightness, fluctuations of temperature as revealed by airglow, and the atmospheric gravity waves that are believed to cause these fluctuations.

082.005   Seasonal variability of the OH Meinel bands.
H. Le Texier, S. Solomon, R. R. Garcia.
Planet. Space Sci., Vol. 35, No. 8, p. 977 – 989 (1987).
Seasonal and latitudinal variations of the OH Meinel bands nightglow emission are studied with a two–dimensional dynamical and photochemical coupled numerical model. Comparisons of the model predictions with observations provide an important test for the theoretical understanding of mesospheric transport processes, particularly the relative roles of diffusive and advective transport.

082.006   Comment on the letter "On the influx of small comets into the earth's atmosphere. II. Interpretation" [Geophys. Res. Lett., Vol. 13, No. 4, p. 307 – 310 (1986)].
J. T. Wasson, F. T. Kyte.
Geophys. Res. Lett., Vol. 14, No. 7, p. 779 – 780 (1987). See Abstr. 42.082.058.

082.007   Reply to the comment on the letter "On the influx of small comets into the earth's atmosphere. II. Interpretation" by J. T. Wasson, F. T. Kyte.
L. A. Frank, J. B. Sigwarth, J. D. Craven.
Geophys. Res. Lett., Vol. 14, No. 7, p. 781 – 782 (1987). See Abstr. 082.006.

082.008   Correction to printer's error in "Reply to Cragin et al.", by L. A. Frank, J. B. Sigwarth, and J. D. Craven [Geophys. Res. Lett., Vol. 14, No. 5, p. 577 – 580 (1987)].
Geophys. Res. Lett., Vol. 14, No. 7, p. 784 (1987). See Abstr. 043.082.097.

082.009   Hydrogen Balmer alpha intensity distributions and line profiles from multiple scattering theory using realistic geocoronal models.
D. E. Anderson Jr., R. R. Meier, R. R. Hodges Jr., B. A. Tinsley.
J. Geophys. Res., Vol. 92, No. A7, p. 7619 – 7642 (1987).
The H Balmer α nightglow is investigated by using Monte Carlo models of asymmetric geocoronal atomic hydrogen distributions as input to a radiative transfer model of solar Lyman β radiation in the thermosphere and exosphere. The radiative transfer model includes all orders of scattering, temperature variation with altitude and solar zenith angle, and anisotropic velocity distributions. The influences of multiple scattering of Lyman β radiation and of observing geometry on the H Balmer α intensity and effective temperature are evaluated in detail. Morning and evening hydrogen distributions for minimum, medium, and maximum solar activity are used in calculations of nightglow emission rates and line profiles. For each of the hydrogen models the H Balmer α intensity and effective temperature are displayed as a function of solar depression angle, observation zenith angle, and azimuth relative to the sun, to make possible detailed comparison to observations.

082.010   Kinetic parameter related to sources and sinks of vibrationally excited OH in the nightglow.
I. C. McDade, E. J. Llewellyn.
J. Geophys. Res., Vol. 92, No. A7, p. 7643 – 7650 (1987).
Kinetic parameters related to vibrational deactivation and chemical removal of vibrationally excited OH radicals in the mesosphere are deduced from ground–based measurements of the mean vibrational distribution of the OH Meinel bands in the nightglow. The deduced kinetic parameters should be particularly useful in future Meinel band studies as they have been obtained from an analysis for which there is no assumption about the very uncertain OH radiative lifetimes.

082.011   Effects of atmospheric disturbances on polar mesopause airglow OH emissions.
G. G. Sivjee, R. L. Walterscheid, J. H. Hecht, R. M. Hamwey, G. Schubert, A. B. Christensen.
J. Geophys. Res., Vol. 92, No. A7, p. 7651 – 7656 (1987).
Temporal variations in the absolute intensity (I) and the rotational temperature (T) of the polar winter airglow OH(8, 3) bands were observed over Longyearbyen, Spitsbergen (78.2°N), in December 1984. The measured temporal variations in I and T are used to determine the spectral contents of $(I - \bar{I})/\bar{I}$, $(T - \bar{T})/\bar{T}$, and their ratio for the airglow OH(8, 3) band.

082.012   Comment on "The O$_2$ atmospheric dayglow in the thermosphere" by M. R. Torr, B. Y. Welsh, D. G. Torr [J. Geophys. Res., Vol. 91, No. A4, p. 4561 – 4566 (1986)].
T. G. Slanger, E. J. Llewellyn, I. C. McDade, G. Witt.
J. Geophys. Res., Vol. 92, No. A7, p. 7753 – 7755 (1987). See Abstr. 41.082.014.

082.013   Reply to the comment on "The O$_2$ atmospheric dayglow in the thermosphere" by T. G. Slanger, E. J. Llewellyn, I. C. McDade, G. Witt.
M. R. Torr, J. K. Owens, D. G. Torr.
J. Geophys. Res., Vol. 92, No. A7, p. 7756 – 7760 (1987). See Abstr. 082.012.

082.014   On the global mean structure of the thermosphere.
R. G. Roble, E. C. Ridley, R. E. Dickinson.
J. Geophys. Res., Vol. 92, No. A8, p. 8745 – 8758 (1987).
With a self–consistent model of the global mean structure of the thermosphere, the authors have examined the ionospheric and thermospheric processes that control the structure of the thermosphere for solar minimum and maximum conditions during geomagnetic quiet periods. The model includes the physical and chemical processes believed to be dominant in the thermosphere and ionosphere.

082.015   Atomic hydrogen and solar Lyman α flux deduced from STP 78–1 UV observations.
D. E. Anderson Jr., L. J. Paxton, R. P. McCoy, R. R. Meier, S. Chakrabarti.
J. Geophys. Res., Vol. 92, No. A8, p. 8759 – 8766 (1987).
Simultaneous observations of the Lyman α and 1026 Å airglow by a spectrometer on the STP 78–1 satellite have been analyzed with a spherical radiative transport model to obtain the atomic hydrogen density distribution.

082.016   The O I 3d^{3}D^0–2p^4 3P transition at 1026 Å in the day airglow.
R. R. Meier, D. E. Anderson Jr., L. J. Paxton, R. P. McCoy, S. Chakrabarti.
J. Geophys. Res., Vol. 92, No. A8, p. 8767 – 8773 (1987).
An emission feature at 1026 Å observed at 600 km in the dayglow with a spectrometer on board the STP 78–1 spacecraft cannot be explained by geocoronal hydrogen Lyman β. An accidental resonance of some lines of the O I multiplet at 1026 Å with the broad solar Lyman β line can in fact account for the observed emission rate, based on analysis with a Monte Carlo radiative transport model. Photoelectron excitation of atomic oxygen is negligible compared with resonant scattering. These airglow results support a ratio of 66 to 1 for the solar Lyman α to Lyman β line center fluxes.

082.017 **A dynamical–chemical model of tidally driven fluctuations in the OH nightglow.**
R. L. Walterscheid, G. Schubert.
J. Geophys. Res., Vol. 92, No. A8, p. 8775 – 8780 (1987).

A theory is presented to explain tidally induced oscillations in the emission intensity I and rotational temperature T of the OH nightglow. The theory includes photochemical reactions among H, O, O_3, OH, and HO_2 and the complete dynamics of tides in an isothermal, uniformly rotating atmosphere.

082.018 **On the quality of solar photographs at Pamirs in winter time.**
Kh. I. Abdusamatov, A. G. Zlatopol'skij, G. V. Komissarov, N. N. Lashkin.
Soln. Dannye, Byull., 1986, No. 12, p. 87 – 94 (1987). In Russian.

An attempt has been made to develop quantitative objective criteria for deciding upon the best place for an optical ground solar telescope. Preliminary results of monitoring the quality of the solar image with the use of a quality analyser are discussed. A drastic decrease in good quality and general deterioration of the solar image at Pamirs in winter time, was found.

082.019 **Site–testing in Aniane.**
C.-H. Jahn.
Sonne, Jahrg. 11, Nr. 42, p. 48 – 49 (1987).

082.020 **Das Ozonloch.**
T. Bührke, F. Arnold.
Sterne Weltraum, 26. Jahrg., Nr. 7 – 8, p. 394 – 396 (1987).

082.021 **Low humidity and submillimeter transparency above the Vostok Antarctic station.**
L. P. Burova, V. D. Gromov, N. I. Luk'yanchikova, G. B. Sholomitskij.
Sov. Astron. Lett., Vol. 12, No. 5, p. 339 – 341 (1987). English translation of 42.082.011.

082.022 **Ground–based lidar and atmospheric studies.**
I. S. McDermid.
Surv. Geophys., Vol. 9, No. 2, p. 107 – 122 (1987). Review paper presented at the 14th meeting on "Atmospheric studies by optical methods", held at Cambridge, U.K., August 1986.

This review considers the requirements and possibilities for the development of a ground–based network for long–term observations of the atmosphere. This network would be specifically designed to provide early detection of changes in the composition and structure of the stratosphere. The species and parameters identified as being important and amenable to ground–based measurements are summarized, as are the currently available techniques capable of making the required measurements. Ultraviolet laser remote sensing is identified as the most promising technique for the measurement of ozone and temperature profiles which are considered to have the highest priority for network measurements. The laser techniques, and the research at JPL Table Mountain Observatory, to implement ozone and temperature measurements are discussed in greater detail.

082.023 **Infrared remote sensing of the middle atmosphere from satellites: the Stratospheric and Mesospheric Sounder experiment 1978 – 1983.**
F. W. Taylor.
Surv. Geophys., Vol. 9, No. 2, p. 123 – 148 (1987). Review paper presented at the 14th meeting on "Atmospheric studies by optical methods", held at Cambridge, U.K., August 1986.

Some reasons for the recent scientific interest in the Earth's middle atmosphere are discussed. It is thought that a better understanding of basic processes related to solar and planetary electromagnetic radiation, atmospheric fluid dynamics, and the photochemical behaviour of certain minor constituents, is the key to predicting, and perhaps avoiding, the threat of natural and anthropogenically–induced changes to the atmospheric and surface environment. One of the most powerful experimental techniques available for research into this problem is infrared remote sensing, and a noval example of this was the Stratospheric

and Mesospheric Sounder (SAMS) instrument on the satellite Nimbus 7. This article reviews what has been learned so far using SAMS data, and what remains to be done.

082.024 **Atmospheric and glow images from the Shuttle.**
E. J. Llewellyn, I. C. McDade, M. R. Gale, D. J. W. Kendall, R. L. Gattinger, S. B. Mende, G. R. Swenson, W. S. C. Brooks.
Surv. Geophys., Vol. 9, No. 2, p. 149 – 168 (1987). Review paper presented at the 14th meeting on "Atmospheric studies by optical methods", held at Cambridge, U.K., August 1986.

The use of simple photographic cameras on early Shuttle missions allowed spacecraft glow to be clearly identified, and its potential for the contamination of weak atmospheric emissions to be estimated. Since those early flights the equipment has been extensively modified so that it is now possible to obtain images with a spectral resolution of 0.1 nm. The early Shuttle glow observations are reviewed and the use of spatially scanned filters to obtain spectral results is described. These glow measurements are discussed in terms of some current ideas for vehicle induced glows and it is suggested that the glow intensity may be controlled by the temperature of the glowing surface. An example of an atmospheric image obtained with the interference filter camera is presented and the limitations in the use of such images are discussed.

082.025 **Antarctica – a unique laboratory for atmospheric studies by optical methods.**
M. J. Rycroft.
Surv. Geophys., Vol. 9, No. 2, p. 215 – 229 (1987). Review paper presented at the 14th meeting on "Atmospheric studies by optical methods", held at Cambridge, U.K., August 1986.

Optical methods of studying the atmosphere are one valuable means of investigating atmospheric behaviour at heights ranging from less than 1 km to several hundred km. Some examples are given of results from various optical experiments carried out at Halley, Antarctica ($76°S$, $27°W$; $L = 4.2$), as is a consideration of the results of some complementary experiments. By combining observations made using different techniques, an improved understanding of atmospheric, ionospheric and magnetospheric processes is obtained.

082.026 **Stratospheric water vapor *in–situ* measurements from infra–red Montgolfière.**
F. Goutail, J. P. Pommereau.
Adv. Space Res., Vol. 7, No. 7, p. 111 – 114 (1987). – See Abstr. 012.023.

Stratospheric water vapor has been measured by a Lyman alpha fluorescence technique on board a long duration balloon during a ten days flight. The results demonstrate that such a long period is needed for a complete outgasing of the instrument. Data recorded during the last day of flight at $40°S$ south of Australia in August 1985 provides an upper limit of water vapor mixing ratio at 20 km of 4 ± 1 ppmV.

082.027 **Peculiarities of atmospheric influences on astronomical–geodetical measurements in the Arctic and Antarctic.**
F. D. Zabolotskij.
Sostoyanie i perspektivy razvitiya geod. i kartogr. Mater. Vses. nauchn.-tekh. konf. Moskva, 5 – 7 sent., 1984. Moskva, p. 190 – 192 (1986). In Russian. Abstr. in Ref. Zh., 52. Geod. Aehrosemka, 8.52.100 (1987).

082.028 **The earth's thermosphere.**
R. G. Roble.
The solar wind and the Earth, p. 244 – 264 (1987). – See Abstr. 003.007.

Contents: Introduction. Global mean properties. Circulation. Auroral processes. Circulation at high latitudes. Interactions with other regions.

082.029   Ozone and the stratosphere.
T. Shimazaki.
The solar wind and the Earth, p. 266 – 284 (1987). – See Abstr.
003.007.

082.030   ETON 5: simultaneous rocket measurements of the OH Meinel $\Delta v = 2$ sequence and (8,3) band emission profiles in the nightglow.
I. C. McDade, E. J. Llewellyn, D. P. Murtagh, R. G. H. Greer.
Planet. Space Sci., Vol. 35, No. 9, p. 1137 – 1147 (1987).
Simultaneous rocket measurements of the emission profiles of the OH Meinel (8,3) band and the $\Delta v = 2$ sequence at 1.61 μm are presented and analysed. It is shown that the $v = 8$ level of the hydroxyl radical must suffer significant loss in the mesosphere due to collisions with O_2 and/or N_2. The rate coefficients for this removal process are obtained for certain limiting assumptions about the excitation mechanism and the coefficients are found to be in good agreement with those deduced from an independent analysis of ground–based observations.

082.031   A twilight measurement of the OH(8–3) Meinel band and atmospheric temperature.
D. P. Murtagh, J. Stegman, G. Witt, E. J. Llewellyn, I. C. McDade.
Planet. Space Sci., Vol. 35, No. 9, p. 1149 – 1155 (1987).
A measurement of the OH(8–3) Meinel band under twilight conditions is reported and, in view of the inherent difficulties, a detailed description of the experiment and data reduction procedure is presented. The derived intensity for the (8–3) band is ~100 R, in reasonable agreement with model calculations. The Rayleigh scattered sunlight that is the major source of background emission has been used to determine the temperature profile in the mesosphere. This derived profile is in satisfactory agreement with that expected from the CIRA–72 model for the season.

082.032   Night airglow O_2 (0–1) atmospheric band emission during the northern polar winter.
H. K. Myrabø.
Planet. Space Sci., Vol. 35, No. 10, p. 1275 – 1279 (1987).

082.033   Issues relating to "holes" in the O I 1304 Å far u.v. dayglow.
R. R. Meier.
Planet. Space Sci., Vol. 35, No. 10, p. 1297 – 1299 (1987).
The results from a detailed model of the multiple scattering of atomic oxygen radiation at 1304 Å developed by Meier and Lee (1982) have been used to draw conclusions regarding the interpretation of "holes" observed in the far ultraviolet airglow by the Dynamics Explorer 1 satellite. It is argued that the "hole" phenomenon cannot be the result either of depletions in thermospheric atomic oxygen or of persistent absorbing clouds within the atmosphere.

082.034   Measurements of OH($X^2\pi$) in the stratosphere by high resolution UV spectroscopy.
D. G. Torr, M. R. Torr, W. Swift, J. Fennelly, G. Liu.
Geophys. Res. Lett., Vol. 14, No. 9, p. 937 – 940 (1987).
The authors report the first results obtained using high spectral resolution imaging ultraviolet spectroscopy to observe multiple rotational lines of OH $A^2\Sigma$-$X^2\pi$(0–0) band. A 9.2 Å spectral segment from 3075.8 Å to 3085.0 Å is imaged at 0.08 Å FWHM spectral resolution allowing the simultaneous acquisition of six of the brightest OH resonance fluorescence emission lines. The instrument was flown to an altitude of 40 km on August 25, 1983, and again on June 12, 1986, on scientific balloons from Palestine, Texas. The OH profiles inverted from the limb scans made during these flights are reported here.

082.035   Observations of anomalous refraction at radio wavelengths.
W. J. Altenhoff, J. W. M. Baars, D. Downes, J. E. Wink.
Astron. Astrophys., Vol. 184, No. 1/2, p. 381 – 385 (1987).
Anomalous refraction has been observed at millimeter wavelengths with the IRAM 30–m telescope, and at centimeter wavelengths with the Effelsberg 100–m telescope. During refraction "events", radio sources are displaced by up to 40″, in both azimuth and elevation, for periods up to 30 s of time. More typically, sources are displaced by a few arcsec for a few seconds of time. The most likely explanation is in terms of moist air packets of a size comparable to the aperture, moving through the beam of the telescope. The effect is observed at high and low elevation angles, and does not require a long line of sight through the atmosphere, indicating that the moist air packets are in the lower troposphere. The effect is clearly dependent on meteorological conditions, as temperature and humidity.

082.036   Are fast atmospheric pulsations optical signatures of lightning–induced electron precipitation?
J. LaBelle.
Geophys. Res. Lett., Vol. 14, No. 10, p. 1023 – 1026 (1987).
Fast atmospheric light pulsations (FAP's) consist of millisecond time–scale bursts of light which have been observed at $L = 1.5 – 2.2$ during searches for atmospheric light emissions associated with supernovae. Their statistics of occurrence resemble those of lightning–induced electron precipitation (Trimpi events) observed at somewhat higher L–shells. Here the author proposes that FAP's are in fact optical signatures of LEP events associated with the $\geqslant 2$ MeV electrons of the inner radiation belt ($L \approx 1.4$).

082.037   A means to sharper images.
E. K. Hege.
Nature, Vol. 328, No. 6127, p. 198 (1987).
This note comments on a paper by L. A. Thompson and C. S. Gardner (see abstr. 082.038) discussing the use of an artificial guide star in an attempt to reduce the effects of atmospheric turbulence on telescopic image formation.

082.038   Experiments on laser guide stars at Mauna Kea Observatory for adaptive imaging in astronomy.
L. A. Thompson, C. S. Gardner.
Nature, Vol. 328, No. 6127, p. 229 – 231 (1987).
Atmospheric turbulence severely limits the resolution of ground–based astronomical telescopes. During the past 15 years, adaptive optical systems with electrically deformable mirrors have been developed to compensate for turbulence. Foy and Labeyrie were the first to suggest that lasers could be used to create artificial guide stars that might be suitable in controlling an adaptive imaging system. The authors have identified the key engineering parameters that optimize the performance of a laser–guided imaging system. On the nights of 21 and 22 January 1987, they conducted experiments at the Mauna Kea Observatory on the island of Hawaii to test the feasibility of using a laser to generate an artificial guide star in the mesospheric sodium layer.

082.039   The extinction properties of Saharan dust over La Palma.
D. C. B. Whittet, M. F. Bode, P. Murdin.
Vistas Astron., Vol. 30, Part 2, p. 135 – 144 (1987).
The extinction produced by wind–blown Saharan dust particles in the atmosphere above the Canary Islands has been modelled by means of Mie theory calculations, on the basis of available evidence relating to its composition and size distribution.

082.040   Two theorems of astronomical refraction.
L. S. Yunoshev.
Issled. v obl. izmerenij vremeni i chastoty. Moskva, p. 87 – 93 (1986). In Russian. Abstr. in Ref. Zh., 52. Geod. Aehrosemka, 9.52.69 (1987).

082.041 The nature of the anomalous sky glow connected with the Tunguska meteorite.
V. A. Bronshtehn.
20th All–Union meteoritic conference, p. 167 – 168 (1987). In Russian. Abstr. in Ref. Zh., 51. Astron., 9.51.330 (1987). – See Abstr. 012.017.

082.042 Atmospheric extinction and night–sky brightness at Mauna Kea.
K. Krisciunas, W. Sinton, D. Tholen, A. Tokunaga, W. Golisch, D. Griep, C. Kaminski, C. Impey, C. Christian.
Publ. Astron. Soc. Pac., Vol. 99, No. 618, p. 887 – 894 (1987).

The authors present a summary of atmospheric extinction values obtained at the 4200–m and 2800–m elevations of Mauna Kea over the years 1980 to 1986. The wavelengths of the 17 filters in question range from 0.44 μm to 32 μm. During 1986 the night–sky brightness at the 2800–m level was measured to be V = 21.5, B = 22.3 mag sec^{-2}.

082.043 Precipitable water vapor at Mount Laguna Observatory.
R. J. Angione.
Publ. Astron. Soc. Pac., Vol. 99, No. 618, p. 895 – 898 (1987).

Measurements of precipitable water vapor at Mount Laguna Observatory are presented and discussed. The method, using a filter–wheel radiometer in conjunction with spectrographic calibration, is described. The mean value of precipitable water vapor on photometric days was 5.2 mm.

082.044 Choice of a mathematical model of measurement results when determining the differential astronomical refraction with the photographic method.
V. V. Kirichuk, I. T. Liva.
L'vov, Politekh. Inst., 21 pp. (1987). In Russian. Abstr. in Ref. Zh., 51. Astron., 10.51.114 (1987).

082.045 Determination of the nonspherical terrestrial atmosphere from measurements of the astronomical refraction at different phases of the Moon.
N. A. Vasilenko, M. R. Fedyanin.
Izv. Akad. Nauk SSSR. Fiz. Atmos. Okean, Tom 23, No. 6, p. 616 – 621 (1987). In Russian. Abstr. in Ref. Zh., 51. Astron., 10.51.179 (1987).

082.046 Difficulties in site–testing by differential image motion.
H. M. Martin.
Bull. Am. Astron. Soc., Vol. 19, No. 2, p. 689 (1987). Abstract. – See Abstr. 010.061.

082.047 Sky brightness on Kitt Peak.
C. A. Pilachowski, J. L. Africano, B. Goodrich, W. Binkert.
Bull. Am. Astron. Soc., Vol. 19, No. 2, p. 750 (1987). Abstract. – See Abstr. 010.061.

082.048 Predictions for collisional broadening of far–infrared OH rotational lines of atmospheric interest.
G. Buffa, O. Tarrini, M. Inguscio.
Appl. Opt., Vol. 26, No. 15, p. 3066 – 3068 (1987).
Collisional line shapes for OH in the presence of N_2 are computed using the Anderson theory. The theoretical results compare satisfactorily with the experimental results now available. Predictions for all strong OH far–infrared lines are given together with the dependence of broadening and shift on temperature. The results will be useful in the interpretation of atmospheric or astrophysical spectra.

082.049 Atmospheric phase measurements with the Mark III stellar interferometer.
M. M. Colavita, M. Shao, D. H. Staelin.
Appl. Opt., Vol. 26, No. 19, p. 4106 – 4112 (1987).
The Mark III interferometer is a phase–coherent stellar interferometer designed for astrometry. Operating through the turbulent atmosphere, the instrument is also a sensitive detector of atmospheric phase fluctuations. The effect of phase fluctuations on astrometric accuracy is reviewed, and phase measurements obtained with the instrument at Mt. Wilson using a 12–m base line are presented.

082.050 On the saturation of the refractive index structure function. II. Influence of the correlation length on astronomical 'seeing'.
P. Venkatakrishnan.
Mon. Not. R. Astron. Soc., Vol. 229, No. 3, p. 379 – 382 (1987).
Loss of phase correlation in long–exposure images could occur on much shorter scales than the scale of saturation of the phase structure function. The departure of the phase structure function from the 5/3 power law at large separations leads to better expectation of 'seeing' at longer wavelengths.

082.051 Mesospheric 5577 Å green line and atmospheric motions – Atmosphere Explorer satellite observations.
J. H. Yee, V. J. Abreu.
Planet. Space Sci., Vol. 35, No. 11, p. 1389 – 1395 (1987).
Photometric measurements of the 5577 Å $O(^1S)$ green line mesospheric emission obtained by the Visible Airglow Experiment (VAE) on board the Atmosphere Explorer (AE) satellite have been analyzed. The inverted volume emission rate profiles showed a peak at approximately 96 – 97 km with a half–width of ~8 km. The diurnal variation of the intensity indicates the presence of a wave component with 10 ~ 12 h period, probably of solar semi–diurnal tide. Shorter time scale variations due to the presence of travelling waves were also observed.

082.052 A survey of research on nightglow variability.
R. J. Forsyth, P. C. Wraight.
Planet. Space Sci., Vol. 35, No. 11, p. 1449 – 1461 (1987).
The literature from 1970 to 1987 on nightglow variability, particularly from the mesosphere and lower thermosphere, is surveyed and classified. Earlier work is referenced via reviews. The work is classified both in terms of its coverage of short period, diurnal and long term variations, and also by the observational methods used.

082.053 Geocoronal structure. 2. Inclusion of a magnetic dipolar plasmasphere.
J. Bishop, J. W. Chamberlain.
J. Geophys. Res., Vol. 92, No. A11, p. 12,377 – 12,388 (1987).
Calculations of exospheric quantities (hydrogen atom density, satellite atom fractional density, kinetic temperature, and escape flux) at locations along the Earth–Sun axis in the noon and midnight directions have been extended to incorporate a plasmasphere characterized by a dipolar shape and an empirical temperature profile. A careful discussion of the handling of plasmaspheric charge exchange collisions and solar ionization is included, and the effect on the exospheric kinetic distribution is analyzed in terms of pertinent examples. In addition, the geotail is demonstrated to stem primarily from the imposition of an exopause by radiation pressure dynamics.

082.054 Geocoronal structure. 3. Optically thin, Doppler–broadened line profiles.
J. Bishop, J. W. Chamberlain.
J. Geophys. Res., Vol. 92, No. A11, p. 12389 – 12397 (1987).
Theoretical line profiles, applicable to the analysis of geocoronal Hα profile measurements, are presented for illustrative cases. While retaining a number of simplifications (classical exobase and diffusive equilibrium plasmasphere conditions), distinctive spectral signatures of mechanisms governing the geocorona are isolated. Examining the consequences of solar radiation pressure dynamics is the main point here. In the prototype evaporative case, radiation pressure acts to form narrow profiles via the creation of an extensive quasi–satellite component. Comparison with a simple extension of the earlier analytic theory discloses the influence of an exopause in this regard. The main modifications to evaporative spectral shapes in the geocoronal application, for shadow heights greater than 2 R_E, are discussed.

082.055   Modeling of the O I 989–Å to 1173–Å ratio in the terrestrial dayglow.
G. R. Gladstone, R. Link, S. Chakrabarti, J. C. McConnell.
J. Geophys. Res., Vol. 92, No. A11, p. 12445 – 12450 (1987).

The O I 989–Å and 1173–Å intensities in the mid–latitude dayglow are modeled for the conditions of a January 17, 1985, rocket flight. It is found that the branching ratio required to fit the 1173–Å data is in agreement with the laboratory measurement of $1.3 - 1.5 \times 10^{-4}$. The results also suggest that the electron impact excitation cross section for the $3s'\ ^3D^0$ state in current use may be too large by a factor of $\sim 2 - 3$, in agreement with the most recent laboratory results. If this is indeed the case, then the 1173–Å triplet is by far the most important branch for the 989–Å multiplet.

082.056   Mesospheric temperatures and the OH layer height as derived from ground–based lidar and OH* spectrometry.
U. von Zahn, K. H. Fricke, R. Gerndt, T. Blix.
J. Atmos. Terr. Phys., Vol. 49, Nos. 7/8, p. 863 – 869 (1987).

Night–time mesospheric temperatures were simultaneously determined from the Doppler broadening of the D_2 resonance line of atmospheric sodium excited by a laser and from the rotational distribution of the $P_1(1)$, $P_1(3)$ and $P_1(4)$ lines of the OH(3,1) band by an i.r. spectrometer. Both instruments were located at the Andøya Rocket Range (69°N, 16°E). The mesospheric temperature gradient permits determination of the altitude of the OH* emitting layer from a comparison of the equivalent layer temperatures calculated from the height–resolved Na Doppler temperatures with the observed OH* rotational temperatures. The altitude of the OH* layer maximum is determined with an accuracy of ± 4 km. For 3 nights in January 1986 the OH* emission layer is found near an altitude of 86 km.

082.057   Airglow studies in India.
V. V. Agashe.
Indian J. Radio Space Phys., Vol. 16, No. 1, p. 84 – 101 (1987).
Abstr. in Phys. Abstr., Vol. 90, No. 1309, Entry 87994 (1987).

082.058   Site testing for observations of large–scale phenomena of comet Halley in southern Yunnan.
M.–c. Wu, P.–z. Qiu, T.–l. Qian, J.–m. Yu.
Publ. Yunnan Obs., No. 1, p. 89 – 95 (1987). In Chinese.

082.059   Report of site selection for a millimeter wave radio astronomical station.
W.–h. Li, H.–m. Hu, G.–q. Chen, Q.–z. Shang.
Publ. Yunnan Obs., No. 2, p. 90 – 98 (1987). In Chinese.

The principle of the site selection for a millimetre wave radio astronomical station is briefly described in this paper. The statistical results of calculation for the depositable water vapour contents in the atmosphere above each site and other related meteorological conditions are also presented.

082.060   Image motion as a measure of seeing quality.
H. M. Martin.
Prepr. Steward Obs., No. 759, 43 pp. (1987). To appear in Publ. Astron. Soc. Pac.

082.061   On some conditions of meridian observations of the Sun at the High–Altitude Station near Kislovodsk.
V. A. Varina, G. S. Kosin.
Glav. Astron. Obs. Akad. Nauk SSSR, Leningrad, 7 pp. (1987). In Russian. Abstr. in Ref. Zh., 51. Astron. 11.51.68 (1987).

082.062   ETON 6: a rocket measurement of the O_2 infrared atmospheric (0–0) band in the nightglow.
I. C. McDade, E. J. Llewellyn, R. G. H. Greer, D. P. Murtagh.
Planet. Space Sci., Vol. 35, No. 12, p. 1541 – 1552 (1987).

Co–ordinated rocket measurements of the $O_2(a^1\Delta_g - X^3\Sigma_g^-)$ infrared atmospheric (0–0) band emission profile and the atomic oxygen densities in an undisturbed night–time atmosphere are used to investigate the processes responsible for the excitation of $O_2(a^1\Delta_g)$ in the terrestrial nightglow.

082.063   Estimation of the horizontal component of astronomical refraction in an ellipsoidal atmosphere.
M. R. Fedyanin, N. A. Vasilenko, V. A. Vol'f.
Kinematika Fiz. Nebesn. Tel, Tom 3, No. 6, p. 84 – 88 (1987). In Russian. English translation in Kinematics Phys. Celest. Bodies.

An algorithm of computation of vertical and horizontal components of astronomical refraction for the homogeneous ellipsoidal Earth atmosphere is presented. For the atmosphere ellipsoid with parameters determined from the measurements of the vertical component near the horizon maximum the value of the horizontal refraction component is $0.1''$ for the zenith distance 88°. The real atmosphere is approximated by a three–axial ellipsoid with the atmosphere maximum slope about $0''.3$ at the zenith.

082.064   Day–time seeing statistics at Sacramento Peak Observatory.
P. N. Brandt, H. A. Mauter, R. Smartt.
Astron. Astrophys., Vol. 188, No. 1, p. 163 – 168 (1987).

A method for the photoelectric measurement of angle–of–arrival fluctuations at the solar limb is described, from which Fried's seeing parameter r_0 can be determined. From a set of 2092 measurements, each of 10 s duration, performed on 146 observing days in the period from June 84 to September 86 at the solar vacuum tower telescope of the Sacramento Peak Observatory, a log–normal distribution of the r_0 values gave a median $r_0 = 8.7$ cm (measured at $\lambda = 510$ nm), with a standard deviation $\sigma = 0.25$ in logarithmic units. The results are compared with atmospheric sounding experiment data and longterm day– and night–time seeing statistics obtained at other observatories.

082.065   Preliminary analysis of the turbidimetric characteristics of the Bruncu Spina (Sardinia) radiometer station.
V. Quesada, A. Banni, I. Porceddu, E. Proverbio.
Nuovo Cimento C, Vol. 9C, N. 3, p. 690 – 700 (1986). = Pubbl. Stn. Astron. Int. Latitudine, Carloforte–Cagliari, Nuov. Ser., N. 129.

082.066   Dati meteorologici rilevati presso l'Osservatorio Astronomico di Cagliari (Punta sa Menta) nel 1985.
M. Putzulu.
Circ. Stn. Astron. Int. Latitudine, Carloforte–Cagliari, N. 33 (Ser. C(9)), 26 pp. (1986).

082.067   Dati meteorologici ottica dell'atmosfera radiazione solare rilevati in Cagliari negli anni 1982, 1983, 1984.
M. Serrau.
Circ. Stn. Astron. Int. Latitudine, Carloforte–Cagliari, N. 34 (Ser. C(10)), 72 pp. (1986).

082.068   Osservazioni turbidimetriche effettuate presso l'Osservatorio Astronomico di Cagliari nel 1985 e 1986.
V. Quesada, M. Serrau.
Circ. Stn. Astron. Int. Latitudine, Carloforte–Cagliari, N. 35 (Ser. C(11)), 20 pp. (1987).

082.069   Mechanisms and observations for isotope fractionation of molecular species in planetary atmospheres.
J. A. Kaye.
Rev. Geophys., Vol. 25, No. 8, p. 1609 – 1658 (1987).

Chemical and physical processes which may give rise to isotope fractionation of molecular species in the atmospheres of both Earth and other planets are reviewed, along with observations of isotopically substituted molecules in planetary atmospheres. Mechanisms for production of isotope fractionation considered include escape and effect of isotope substitution on equilibrium constants (including those of phase changes), photolysis rates, and chemical reaction rates.

082.070   **Comparisons of theoretical models and observations of the thermosphere and ionosphere during extremely disturbed geomagnetic conditions during the last solar cycle.**
D. Rees, T. J. Fuller–Rowell.
Adv. Space Res., Vol. 7, No. 8, p. 27 – 38 (1987). – See Abstr. 012.055.

082.071   **An interim reference model for the middle atmosphere water vapor distribution.**
J. M. Russell III.
Adv. Space Res., Vol. 7, No. 9, p. 5 – 18 (1987). – See Abstr. 012.056.
The intent of this paper is to use the Limb Infrared Monitor of the Stratosphere experiment on the Nimbus 7 spacecraft (LIMS) results and other available data to generate an interim water vapor reference model for the middle atmosphere.

082.072   **About a possible reference model for stratospheric NO_2.**
J. P. Naudet, R. J. Thomas, H. K. Roscoe,
J. M. Russell III.
Adv. Space Res., Vol. 7, No. 9, p. 19 – 23 (1987). – See Abstr. 012.056.
After a brief review of observations of the global distribution of stratospheric NO_2, the authors discuss the possibilities of combining different data sets to construct a NO_2 reference model.

082.073   **Proposed reference model for nitric acid.**
J. C. Gille, P. L. Bailey, C. A. Craig.
Adv. Space Res., Vol. 7, No. 9, p. 25 – 35 (1987). – See Abstr. 012.056.
A nearly global set of data on the nitric acid distribution was obtained for seven months by the Limb Infrared Monitor of the Stratosphere (LIMS) experiment of the Nimbus 7 spacecraft. The evaluation of the accuracy, precision and resolution of these data is described, and a description of the major features of the nitric acid distributions is presented.

082.074   **Proposed reference models for ozone.**
G. M. Keating, M. C. Pitts.
Adv. Space Res., Vol. 7, No. 9, p. 37 – 47 (1987). – See Abstr. 012.056.
Ozone reference models are proposed, similar to the Keating and Young (1985) models which were prepared for the new COSPAR International Reference Atmosphere. This paper updates tables provided in the Keating and Young ozone model, giving improved monthly zonal mean total column ozone in 10° latitude increments, improved monthly zonal mean ozone volume mixing ratios (ppmv) from 20 to 0.003 mb in 10° latitude increments, and conversion tables providing ozone vertical structure in other units. Also, a new table is provided giving ozone vertical structure as a function of altitude (from 25 to 80 km), latitude, and month. The models are based on measurements from six contemporary satellite instruments.

082.075   **Proposed reference models for nitrous oxide and methane in the middle atmosphere.**
F. W. Taylor, A. Dudhia, C. D. Rodgers.
Adv. Space Res., Vol. 7, No. 9, p. 49 – 62 (1987). – See Abstr. 012.056.
Data from the Stratospheric and Mesospheric Sounder (SAMS) on the Nimbus 7 satellite, for the three years form January 1979 – December 1981, are used to prepare a reference model for the long–lived trace gases methane (CH_4) and nitrous oxide (N_2O) in the stratosphere.

082.076   **Proposed reference models for CO_2 and halogenated hydrocarbons.**
P. Fabian.
Adv. Space Res., Vol. 7, No. 9, p. 63 – 72 (1987). – See Abstr. 012.056.
The vertical distribution of carbon dioxide, halocarbons and their sink products, HCl and HF, have become available, mainly by means of balloon measurements. This report attempts to combine the available data for presentation of reference models for CO_2, CCl_4, CCl_3F, CCl_2F_2, $CClF_3$, CF_4, $CCl_2F–CClF_2$, $CClF_2–CClF_2$, $CClF_2–CF_3$, $CF_3–CF_3$, CH_3Cl, $CHClF_2$, $CH_3–CCl_3$, $CBrClF_2$, $CBrF_3$, HCl and HF.

082.077   **Background stratospheric aerosol reference model.**
M. P. McCormick, P.–H. Wang.
Adv. Space Res., Vol. 7, No. 9, p. 73 – 80 (1987). – See Abstr. 012.056.
The objective of this paper is to present a proposed reference model for the background stratospheric aerosol based on currently available data from satellite observations.

082.078   **An empirical model of the global total ozone distribution.**
V. I. Bekoryukov, V. N. Glazkov, V. V. Fedorov.
Adv. Space Res., Vol. 7, No. 9, p. 115 – 117 (1987). – See Abstr. 012.056.
On the basis of the World Ozone Network data for 1957 – 1983 in the Northern Hemisphere and a combination of the network data and satellite data (Nimbus 4) in the Southern Hemisphere, global empirical models of the total ozone and its variability are developed using analytical methods of describing meteorological fields.

082.079   **Precision measurements of stratospheric ozone profiles by rocket–borne optical ozonesondes.**
T. Watanabe, T. Ogawa.
Adv. Space Res., Vol. 7, No. 9, p. 123 – 126 (1987). – See Abstr. 012.056.
Sunset observations of the upper stratospheric and mesospheric ozone were made at Uchinoura (31.25°N, 131.08°E) with rocket–borne optical ozonesondes, which consist of multi–color solar ultraviolet radiometers and sun tracking devices. Three ozone density profiles were obtained in this study.

082.080   **Measurements of stratospheric trace constituent distributions from balloon–borne far infrared observations.**
M. M. Abbas, J. Guo, B. Carli, F. Mencaraglia, A. Bonetti, M. Carlotti, I. G. Nolt.
Adv. Space Res., Vol. 7, No. 9, p. 137 – 140 (1987). – See Abstr. 012.056.
Far infrared limb thermal emission spectra obtained from balloon borne measurements made as a part of the Balloon Intercomparison Campaign (BIC) have been analyzed for retrieval of stratospheric trace constituent distributions. The measurements were made with a high resolution Michelson Interferometer and covered the $15 – 180 \, cm^{-1}$ spectral range with an unapodized spectral resolution of $0.0033 \, cm^{-1}$. The retrieved vertical profiles of O_3, H_2O, HDO, HCN, CO and isotopes of O_3 are presented.

082.081   **Interpretation of infrared measurements of the high–latitude thermosphere from a rocket–borne interferometer.**
R. H. Picard, J. R. Winick, R. D. Sharma, A. S. Zachor, P. J. Espy, C. R. Harris.
Adv. Space Res., Vol. 7, No. 10, p. 23 – 30 (1987). – See Abstr. 012.057.
A preliminary analysis of high–resolution infrared spectra of the aurorally dosed lower thermosphere above Poker Flat Research Range, Alaska, obtained by an uplooking cryogenic field–widened interferometer is presented. Both models and spectral–fitting/resolution–enhancement methods are used to discuss the behavior of NO, CO, NO^+, and $CO_2 \, v_3$ vibrational bands in the high–latitude thermosphere.

082.082   **Rocket–borne measurements of atmospheric infrared emissions by spectrometric techniques.**
H. G. Brückelmann, K. U. Grossmann, D. Offermann.
Adv. Space Res., Vol. 7, No. 10, p. 43 – 46 (1987). – See Abstr. 012.057.
As part of the MAP/WINE Campaign 1983/84 a liquid helium cooled infrared grating spectrometer measured night zenith radiances of CO_2, O_3, and H_2O in the mesosphere and lower

thermosphere. From a comparison of the measured spectral radiances with results from LTE radiative transfer calculations atmospheric temperatures and concentration profiles of H_2O and O_3 are determined.

082.083   Rocket observations of the atomic and molecular oxygen emissions in the equatorial region.
H. Takahashi, B. R. Clemesha, Y. Sahai, P. P. Batista, A. Eras, A. H. P. Chaves, B. Rossire, J. R. Daniel.
Adv. Space Res., Vol. 7, No. 10, p. 47 – 50 (1987). – See Abstr. 012.057.
A Brazilian sounding rocket with two airglow photometers and two ionospheric electron density probes, was launched successfully from Natal (5.8°S, 35.2°W), Brazil, on December 11, 1985, at 23:30 GMT. The observed height profiles of the atomic oxygen O I 5577 Å and molecular oxygen atmospheric (0,0) band at 7619 Å emissions are discussed.

082.084   Middle atmosphere models and comparison with Shuttle reentry density data.
K. S. W. Champion.
Adv. Space Res., Vol. 7, No. 10, p. 77 – 82 (1987). – See Abstr. 012.057.
Several middle atmosphere models are reviewed, including a new set of models produced by Groves in 1985. The latter models are based on rocket and rawinsonde in situ measurements and satellite remote sounding temperature data. The models are compared with measurements made with instruments on board U.S. Shuttles during their reentry. Very useful atmospheric density data have been obtained in the altitude region from 50 to 80 km.

082.085   Middle atmosphere temperature measurements as compared to atmospheric models.
D. Offermann, R. Gerndt, R. Küchler.
Adv. Space Res., Vol. 7, No. 10, p. 97 – 104 (1987). – See Abstr. 012.057.
Two large sets of temperature data measured in Europe are compared to atmospheric models that were prepared for the NEW CIRA. The data were taken by rocket flights during the Energy Budget and MAP/WINE Campaigns, and during several years of operation of ground based OH* spectrometers. Monthly mean temperature profiles from 20 – 80 km are available for November, December, and January. They are in very good agreement with the new models.

082.086   Ozone reference models for CIRA.
G. M. Keating, D. F. Young, M. C. Pitts.
Adv. Space Res., Vol. 7, No. 10, p. 105 – 115 (1987). – See Abstr. 012.057.
The ozone reference model which is to be incorporated in the COSPAR International Reference Atmosphere (CIRA) is described and compared with other measurements of the Earth's ozone distribution.

082.087   Atmospheric structure between 80 and 120 km.
J. M. Forbes, G. V. Groves.
Adv. Space Res., Vol. 7, No. 10, p. 135 – 141 (1987). – See Abstr. 012.057.
Recent progress in modelling the temperature, density, and pressure specifications between 80 and 100 km is reported. The data base consists primarily of rocket and incoherent scatter measurements of temperature. An analytic polynomial scheme is described whereby smooth transitions with tabulations below 80 km and the MSIS–83 model above 120 km are obtained.

082.088   A theoretical thermosphere model for CIRA.
D. Rees, T. J. Fuller–Rowell.
Adv. Space Res., Vol. 7, No. 10, p. 185 – 197 (1987). – See Abstr. 012.057.
Theoretical and semi–empirical descriptions of the solar and geomagnetic driving forces affecting the terrestrial mesosphere and thermosphere have been used to generate a series of representative numerical models of the thermosphere, covering a wide range of solar and geomagnetic activity levels, for all seasons. These numerical models are compared with observations, and with the most recent experimental and semi–empirical models of the thermosphere.

082.089   Models of the thermosphere total density for satellite dynamics.
L. Sehnal.
Adv. Space Res., Vol. 7, No. 10, p. 203 – 206 (1987). – See Abstr. 012.057.
Mathematical expressions for models of the upper atmosphere total density distribution and variation, to be used for the analytical determination of drag effects, are developed. The models are computed by a transformation of the other model values of the thermosphere and from the observed data.

082.090   Atmospheric iridium at the South Pole as a measure of the meteoritic component.
G. Tuncel, W. H. Zoller.
Nature, Vol. 329, No. 6141, p. 703 – 705 (1987).
The authors present a determination of average particle–borne Ir concentration in the South Pole atmosphere. The average value suggests that the concentration of extraterrestrial material in the South Pole atmosphere is not large enough to explain the enrichments of anomalously enriched elements; however, meteoritic material contributes significantly to the observed concentrations of Co, Fe and Mn. They estimate an accretion rate for background extraterrestrial material of 11,000 tons annually.

082.091   Algorithms of the determination of air density of the upper atmosphere of the earth.
L. L. Filenko, A. M. Fominov.
Algoritm. Nebesnoj Mekh., No. 51, 34 pp. (1987). In Russian.

082.092   Comparison of tropospheric models used for VLBI data analysis.
S. Fischer, F. Flechtner, J. Campbell, H. Schuh.
Mitt. Geod. Inst. Rheinischen Friedrich–Wilhelms–Univ. Bonn, Nr. 72, p. 69 – 80 (1987).

082.093   The N II 2143 Å emission in the dayglow.
D. D. Cleary, C. A. Barth.
J. Geophys. Res., Vol. 92, No. A12, p. 13,635 – 13,640 (1987).
Observations of the N II 2143 Å emission in the Earth's dayglow were obtained by a rocket–borne spectrograph flown on November 9, 1981 and August 11, 1982. The ultraviolet spectra were analyzed using synthetic spectra to separate the contribution of the 2143 Å emission from the overlapping nitric oxide gamma band. The 2143 Å column emission rate profiles for both flights were quite similar with peak column emission intensities of approximately 500 Rayleighs at an altitude of 160 km. When the observed column emission rate profiles are compared with theoretical calculations, the production efficiency of this emission was found to be 0.10 ± 0.04. The 2143 Å emission varies by more than a factor of 3 over the solar cycle and therefore should be useful as an indicator of the solar EUV flux.

082.094   Satellite–borne measurements of middle–atmosphere temperature.
J. J. Barnett.
Philos. Trans. R. Soc. London, Ser. A, Vol. 323, No. 1575, p. 527 – 544 (1987). – See Abstr. 012.068.
Satellites were first used to measure middle–atmosphere temperatures in the early 1960s. There has been steady progress towards the present position where we have routine observations of the whole stratosphere with 10 – 15 km vertical resolution by operational satellites, and where experimental instruments provide data as high as the mesopause, with 3 km vertical resolution in some cases. The paper outlines the present state of the art and attempts to show how well one can hope to do in the future.

082.095 Satellite–borne measurements of middle–atmosphere composition.
J. M. Russell III, M. P. McCormick.
Philos. Trans. R. Soc. London, Ser. A, Vol. 323, No. 1575, p. 545 – 565 (1987). – See Abstr. 012.068.
A number of satellite experiments have been launched in recent years with the goal of providing fundamental data needed for analysis of photochemistry, radiation, dynamics, and transport processes. Collectively, these experiments have accumulated information on the vertical and horizontal distributions of a host of minor constituents in the middle atmosphere.

082.096 Rocket photometry and the lower–thermospheric oxygen nightglow.
R. G. H. Greer, D. P. Murtagh, I. C. McDade, E. J. Llewellyn, G. Witt.
Philos. Trans. R. Soc. London, Ser. A, Vol. 323, No. 1575, p. 579 – 595 (1987). – See Abstr. 012.068.
Experimental aspects of rocket photometry are briefly discussed in the light of the requirement to measure the nightglow with high accuracy. In relation to measurements of the nightglow green line and the (0,0) atmospheric band system, the development of our understanding of their excitation mechanisms in particular and the associated photochemistry in general is traced from Chapman to the present.

082.097 A progress report on the LEST site testing.
T. A. Darvann, U. Kusoffsky, Z. Li.
LEST Found., Tech. Rep., No. 25, 50 pp. (1987). See Abstr. 012.071.

082.098 Objectives and conditions for the LEST site survey.
O. Engvold.
LEST Found., Tech. Rep., No. 26, p. 7 – 9 (1987). – See Abstr. 012.071.

082.099 Recommendations for the LEST site testing campaign.
E. H. Schröter.
LEST Found., Tech. Rep., No. 26, p. 11 – 14 (1987). – See Abstr. 012.071.

082.100 Identification of optimum sites for daytime and night-time observations at Mauna Kea Observatory.
D. A. Erasmus.
LEST Found., Tech. Rep., No. 26, p. 17 – 24 (1987). – See Abstr. 012.071.
Reprinted from proceedings of the international conference on identification, optimization and protection of optical telescope sites. 22 – 23 May 1986, Flagstaff, Arizona. See Abstr. 43.082.050.

082.101 Night time seeing at observatories in the Canary Islands.
F. J. Fuentes Gandia.
LEST Found., Tech. Rep., No. 26, p. 25 – 31 (1987). – See Abstr. 012.071.

082.102 Site testing for the VLT – instrumentation and preliminary results.
M. Sarazin.
LEST Found., Tech. Rep., No. 26, p. 33 – 39 (1987). – See Abstr. 012.071.

082.103 Day–time seeing statistics at Sacramento Peak Observatory.
P. N. Brandt, H. A. Mauter, R. Smartt.
LEST Found., Tech. Rep., No. 26, p. 41 – 52 (1987). – See Abstr. 012.071.
The results of the extended series of day–time seeing measurements performed at the vacuum tower telescope of Sacramento Peak show that the median r_0 value of 8.7 cm is about a factor of 2 higher than measured at other mountain sites during day–time. Even without correction for zenith distance this median value ranks among the highest measured in longterm *night–time* tests at observatories like Kitt Peak, Flagstaff, Chile, Calar Alto (Spain),

Roque de los Muchachos (Canary Islands) and others – although one should be cautious when comparing results obtained with different measuring techniques.

082.104 Morning minimum of the $C_n{}^2$ at Mt. Sanglok.
M. S. Pekur, I. V. Petenko, E. A. Shurygin.
Astron. Tsirk., No. 1460, p. 4 – 5 (1986). In Russian.

082.105 On the quality of day–time seeing on the Pamirs in winter.
Kh. I. Abdusamatov, A. G. Zlatopol'skij, G. V. Komissarov, N. N. Lashkin.
Astron. Tsirk., No. 1466, p. 1 – 3 (1986). In Russian.
A quantitative criterion is proposed for the estimation of the efficiency of a site for solar observations. According to a preliminary analysis in winter on the Pamirs the situation is essentially worse than in summer–autumn.

082.106 Heterochromatic extinction coefficients and spectral transparency of the atmosphere.
G. A. Terez.
Astron. Tsirk., No. 1470, p. 1 – 3 (1986). In Russian.
It is shown that heterochromatic extinction coefficients for the atmosphere can be obtained from the data of monochromatic observations with an error of order 0.5%.

082.107 Sky conditions at the High–Altitude Station of the Sternberg Institute at Mt. Majdanak in 1978 – 1985.
S. A. Gladyshev, M. G. Shirokova.
Astron. Tsirk., No. 1477, p. 6 – 8 (1987). In Russian.
The average number of photometric nights per year is 212, spectroscopic – 231, the average number of observational hours – 1734.

082.108 Clouds of the twilight (*noctilucent clouds*).
D. McConnell.
Astronomy, Vol. 15, No. 7, p. 42 – 47 (1987). Abstr. in Phys. Abstr., Vol. 90, No. 1315, Entry 127623 (1987).

082.109 6300 Å night airglow in winter.
B. S. Salaria, N. S. Chauhan, H. S. Gurm.
Indian J. Radio Space Phys., Vol. 16, No. 2, p. 221 – 224 (1987). Abstr. in Phys. Abstr., Vol. 90, No. 1315, Entry 127628 (1987).

082.110 Precision in determination of astronomical refraction from aerological data.
K. Kurzyńska.
Astron. Nachr., Vol. 308, No. 5, p. 323 – 328 (1987).
Various errors in determination of the local pure astronomical refraction are evaluated versus the apparent zenith distance z_0. Numerical integration with the step imposed by heights of an aerological sounding brings the error smaller than $0\rlap{.}''01$ until $z_0 = 70°$. For larger zenith distances integration with a more dense step is possible after fitting the data to a five–parameter functional dependence of the refractive index on height. The fitting is simultaneously a good equalization of aerological data reducing considerably their experimental error as well as short–term local fluctuations of the atmosphere. After equalization, the error in the refraction originating from the error in the aerological data is found to approach $0\rlap{.}''01$ at $z_0 = 72°$, $0\rlap{.}''1$ at $z_0 = 82°$, $1''$ at $z_0 = 87°$, and $10''$ at the horizon.

082.111 The atmospheric refraction.
M. O. Suliman.
Nauchn. Inf., Vyp. 58, p. 51 – 65 (1986). In Russian.
A review of the historical development of the study of astronomical refraction, anomalous refraction and the methods of their investigation is given.

082.112 Recognising and photographing nuctilucent clouds.
D. McConnell.
Weather, Vol. 42, No. 6, p. 180 – 184 (1987). Abstr. in Phys. Abstr., Vol. 90, No. 1316, Entry 134044 (1987).

082.113 **A numerical investigation of nonlinear internal gravity waves and their influence on the mean flow.**
S. D. Mobbs.
Ann. Geophys., Ser. A, Vol. 5, No. 4, p. 197 – 208 (1987). Abstr. in Phys. Abstr., Vol. 90, No. 1317, Entry 139820 (1987).

082.114 **The effect of inhomogeneities on the resonant parametric interaction of gravity waves in the atmosphere.**
B. Inhester.
Ann. Geophys., Ser. A, Vol. 5, No. 4, p. 209 – 218 (1987). Abstr. in Phys. Abstr., Vol. 90, No. 1317, Entry 139821 (1987).

082.115 **A simultaneous observation of the height profiles of the night airglow O I 5577 Å, O_2 Herzberg and atmospheric bands.**
T. Ogawa, N. Iwagami, M. Nakamura, M. Takano, H. Tanabe, A. Takechi, A. Miyashita, K. Suzuki.
J. Geomagn. Geoelectr., Vol. 39, No. 4, p. 211 – 228 (1987). Abstr. in Phys. Abstr., Vol. 90, No. 1317, Entry 139828 (1987).

082.116 **Sky brightness on Kitt Peak.**
C. A. Pilachowski, J. Africano, P. Massey, B. D. Goodrich, W. S. Binkert.
Publ. Astron. Soc. Pac., Vol. 99, No. 621, p. 1149 (1987). Abstract. – See Abstr. 010.281.

082.117 **New aspects of the production mechanism for excited atoms, molecules and their ions associated with short–time luminosity variations in the upper atmosphere's layers.**
I. P. Bogdanova, V. I. Yakovleva.
Geomagn. Aehron., Tom 27, No. 5, p. 855 – 857 (1987). In Russian. English translation in Geomagn. Aeron.

082.118 **On the effect of meteor activity in the night–time emission of the middle atmosphere.**
L. M. Fishkova, K. D. Kvavadze.
Geomagn. Aehron., Tom 27, No. 5, p. 858 – 860 (1987). In Russian. English translation in Geomagn. Aeron.

082.119 **Modelling the atomic nitrogen emission in the extreme ultraviolet spectral region.**
E. I. Gol'brajkh, A. V. Gurvich.
Geomagn. Aehron., Tom 27, No. 6, p. 1028 – 1031 (1987). In Russian. English translation in Geomagn. Aeron.

082.120 **Atmospheric precipitable water vapour over Manora Peak, Naini Tal.**
B. C. Bhatt, H. S. Mahra.
Bull. Astron. Soc. India, Vol. 15, Nos. 2 – 3, p. 116 – 120 (1987).
  Precipitable water vapor is one of the variable atmospheric constituents and its short–term and seasonal variations may affect the infrared photometric observations. Form the analysis of the measurements it is concluded that the observatory site at Manora Peak shows a good potential for infrared observations during the period November to March and this is also the period of maximum number of photometric nights at Manora Peak.

History of the Earth's atmosphere.
See Abstr. 003.034.

Theory of planetary atmospheres. An introduction to their physics and chemistry.
See Abstr. 003.037.

Remote sensing from space. Proceedings of Symposium 3, Workshop V and of the COSPAR Interdisciplinary Scientific Commission A (Meeting A2) of the COSPAR Twenty–sixth Plenary Meeting held in Toulouse, France, 30th June – 11th July 1986.
See Abstr. 012.067.

Specification of LEST site survey program 1987 – 88.
See Abstr. 013.076.

Remote sensing of atomic oxygen: some observational difficulties in the use of the forbidden O I λ1173 Å and O I λ1641 Å transitions.
See Abstr. 022.070.

Comments on absolute absorption coefficients of atmospheric water vapor at CO_2 laser wavelengths.
See Abstr. 022.101.

A scheme for measuring differential image motion (Proposal for LEST site testing).
See Abstr. 031.070.

An optical telescope for site testing.
See Abstr. 032.048.

Instrumentation of the initial LEST site testing campaign.
See Abstr. 034.177.

Intercalibration of two solar limb motion meters.
See Abstr. 034.178.

Image selection and binning for improved atmospheric calibration of infrared speckle data.
See Abstr. 036.020.

Ultraviolet detection of very low–surface–brightness objects.
See Abstr. 036.053.

Atmospheric phase measurements with the Mk II and Mk III interferometers at Mt. Wilson.
See Abstr. 036.113.

Two–color method for optical astrometry: theory and preliminary measurements with the Mark III stellar interferometer.
See Abstr. 036.141.

Kalman filtering VLA phase data with a supercomputer.
See Abstr. 036.220.

High resolution interferometric imaging using a large optical telescope.
See Abstr. 036.243.

Infrared differential speckle interferometry with Fourier's spectral selectivity.
See Abstr. 036.244.

Problem of the elimination of the refractional effects in Doppler positioning.
See Abstr. 045.001.

Potential of remote sensing for the study of global change. COSPAR report to the International Council of Scientific Unions (ICSU).
See Abstr. 051.018.

The natural background at Shuttle altitudes.
See Abstr. 051.020.

The extreme and far ultraviolet environment at shuttle altitudes.
See Abstr. 051.026.

Analysis of the orbit of Intercosmos 10 rocket (1973–82B) in its last 15 days.
See Abstr. 052.002.

Isolated eddy models in geophysics.
See Abstr. 062.089.

Self–similar stong shock waves with thermal radiation in an atmosphere of exponentially–decreasing density.
See Abstr. 062.098.

Geophysical and astrophysical turbulence.
See Abstr. 062.141.

Application of an extended ESFT method to calculation of solar heating rates by water vapor absorption.
See Abstr. 063.008.

Effects of parameter variations in the radiative transfer equation: application to the lower stratosphere in the far infrared.
See Abstr. 063.066.

Atmospheric radioactivity and variations in the solar neutrino flux.
See Abstr. 080.075.

Observations of the source and propagation of atmospheric gravity waves.
See Abstr. 084.028.

Volatiles on Earth and Mars: a comparison.
See Abstr. 097.065.

Mineralogy of chondritic interplanetary dust particles.
See Abstr. 106.010.

083 Ionosphere

083.001 Generation of gravity waves within the ionosphere by transient heating during high–power wave propagation.
S. S. De, S. K. Adhikari, S. Bandyopadhyay,
J. Bandyopadhyay.
Astrophys. Space Sci., Vol. 135, No. 2, p. 325 – 334 (1987).

Generation of short–range gravity waves within the ionosphere due to inhomogeneous heating in the presence of space–localized inhomogeneities during high–power radio wave–propagation has been investigated. The magnitude and form of the anticipated atmospheric wave–trains are obtained. The derived expression of electric field within the ionosphere under the stated perturbed condition may be considered to be manifested through Lorentz–force and Joule–dissipation that influence the neutral gas of the atmosphere via collision–mechanism and thereby gravity waves are launched. The expressions for the low–frequency part of the fractional pressure variations have been derived which are applied to the E–region of the ionosphere.

083.002 Satellite data alters view on Earth–space environment.
D. Dooling.
Spaceflight, Vol. 29, Suppl. No. 1, p. 21 – 29 (1987).

The environment between the top of the atmosphere and the edge of deep space has been re–drawn as more dynamic than scientists had expected even as a little as five years ago. How subtle and important this region is was discovered by the two Earth–orbiting Dynamics Explorer satellites. The data that both returned continue to alter our view of the ionosphere and its role in the Earth–space environment.

083.003 Hemispheric models of electric fields and currents in the ionosphere.
X.–p. Zhao, Z. Xiao.
Sci. Sin., Ser. A, Vol. 30, No. 9, p. 963 – 969 (1987).

The basic equations describing the electric fields and currents in the global ionosphere with the effects of both neutral wind and field–aligned currents are derived. Given the height–integrated electric conductivities and the neutral wind, two kinds of models have been developed and the appropriate schemes for numerical solutions proposed.

083.004 The solar wind–magnetosphere–ionosphere coupling during a magnetic storm.
C.–s. Shen.
Acta Geophys. Sin., Vol. 30, No. 4, p. 331 – 340 (1987). In Chinese.

The indices of IMF, ground magnetic field and observations of the ionospheric E–field are used to study the response of the E–field in high and middle latitude ionosphere to variation of the solar wind and the magnetosphere. It is shown that although the electrostatic potential distribution at the magnetopause driven by the solar wind can be directly mapped to the ionosphere along the magnetic field line to the polar cap boundary, the response of the ionospheric E–field at high and middle latitudes is a kind of

relaxation process related to the existence and development of the ring current.

083.005 Estimating the possibilities of semi–empirical modelling of the ionosphere.
V. M. Polyakov, A. I. Agaryshev, M. K. Ivel'skaya,
A. D. Kazmirov, M. A. Koen, S. V. Lopatkin, V. I. Sazhin,
V. E. Sukhodol'skaya.
Ionos. Issled., Moskva, No. 42, p. 5 – 9 (1987). In Russian. Abstr. in Ref. Zh., 62. Issled. Kosm. Prostranstva, 8.62.564 (1987).

083.006 Semi–empirical models of the quiet ionosphere.
M. A. Koen, S. N. Barajshchuk, S. N. Burenkov,
A. D. Kazmirov, D. V. Khazanov.
Ionos. Issled., Moskva, No. 42, p. 10 – 21 (1987). In Russian. Abstr. in Ref. Zh., 62. Issled. Kosm. Prostranstva, 8.62.565 (1987).

083.007 The ionosphere.
A. D. Richmond.
The solar wind and the Earth, p. 124 – 140 (1987). – See Abstr. 003.007.

Contents: Introduction. Ionospheric regions. Ionization processes. Movements of plasma. Electrical conductivity. Global currents and electric fields.

083.008 Spacelab–2 Plasma Depletion Experiments for ionospheric and radio astronomical studies.
M. Mendillo, J. Baumgardner, D. P. Allen, J. Foster, J. Holt,
G. R. A. Ellis, A. Klekociuk, G. Reber.
Science, Vol. 238, No. 4831, p. 1260 – 1264 (1987).

The Spacelab–2 Plasma Depletion Experiments were a series of studies to examine shuttle–induced perturbations in the ionosphere and their application to ground–based radio astronomy. A burn was conducted over a low–frequency radio observatory in Hobart, Australia, to create an "artificial window" for ground–based observations at frequencies normally below the natural ionospheric cutoff (penetration) frequency. The Hobart experiment succeeded in making high–resolution observations at 1.7 megahertz through the induced ionospheric hole.

083.009 A study of the meridional structure of the equatorial red arcs in the night–time ionosphere from the ISIS–II satellite.
V. P. Bhatnagar, G. G. Shepherd.
J. Atmos. Terr. Phys., Vol. 49, No. 10, p. 959 – 973 (1987).

Equatorial 6300 Å arcs observed by the ISIS–II satellite close to the magnetic equator over the African and Asian zones are studied for night–time conditions from 21:00 h to 02:00 h local time in the summer and spring of 1972 – 1974 and 1976, respectively. Case studies of the arcs have been made for quiet geomagnetic conditions and for minor storms.

083.010   **Fifty years of the ionosphere.**
K. Maeda.
J. Radio Res. Lab., Vol. 33, No. 140, p. 103 – 167 (1986). Abstr. in Phys. Abstr., Vol. 90, No. 1309, Entry 88024 (1987).

083.011   **Analysis of variations of the parameters of the F2 region in the semicycle of solar activity.**
D. A. Dzyubanov, I. V. Sokolova, A. F. Kononenko.
Vestn. Khar'kov. Politekh. Inst., No. 248, p. 7 – 10 (1987). In Russian. Abstr. in Ref. Zh., 62. Issled. Kosm. Prostranstva, 11.62.429 (1987).

083.012   **Progress in modeling the ionospheric peak and topside electron density.**
D. Bilitza, K. Rawer, S. Pallaschke, C. M. Rush, N. Matuura, W. R. Hoegy.
Adv. Space Res., Vol. 7, No. 6, p. 5 – 12 (1987). – See Abstr. 012.054.

083.013   **Joint analytical profile of electron density through the whole ionosphere.**
K. Rawer.
Adv. Space Res., Vol. 7, No. 6, p. 25 – 33 (1987). – See Abstr. 012.054.
A fully analytical profile is obtained by suitably combining three height ranges in each of which an independent description by a linear combination of Epstein LAY–functions is arranged. The present IRI description of the topside is directly reformulated in terms of two such functions. Three to four functions will be needed in the middle, up to three in the lower ionosphere (excluding a C–layer). Indications are given, how the different parameters could be determined.

083.014   **An empirical model for the global distributions of density, temperature and effective collision frequency of electrons in the ionosphere.**
Yu. K. Chasovitin, A. V, Shirochkov, A. S. Besprozvannaya, T. L. Gulyaeva, P. F. Denisenko, O. A. Armenskaya, S. E. Ivanova, A. I. Kashirin, N. M. Klyueva, E. A. Koryakina, L. S. Mironova, T. N. Sykilinda, V. B. Shushkova, V. I. Vodolazkin, V. V. Sotsky (*V. V. Sotskij*), N. E. Sheidakov. Adv. Space Res., Vol. 7, No. 6, p. 49 – 52 (1987). – See Abstr. 012.054.
An improved ionospheric global empirical reference model (named RIM–85) is established for the distributions of density (N_e), temperature (T_e) and effective collision frequency (ν_e) of electrons between 65 and 600 km.

083.015   **Electron–density profiles from 100 to 500 km altitude during high–solar activity: a comparison with the IRI.**
G. Johanning, C.–U. Wagner.
Adv. Space Res., Vol. 7, No. 6, p. 61 – 64 (1987). – See Abstr. 012.054.
Simultaneous measurements made at the ionosonde station Juliusruh (GDR) and with the topside sounder onboard satellite Intercosmos–19 have been used to compile electron–density profiles for the altitude range from 100 to 500 km. These profiles have been compared with the new version of the International Reference Ionosphere (IRI–9).

083.016   **Modelling of the lower ionosphere according to the IRI guidelines.**
Y. V. Ramanamurty, K. Rawer.
Adv. Space Res., Vol. 7, No. 6, p. 77 – 82 (1987). – See Abstr. 012.054.

083.017   **Description of the mean behaviour of ionospheric plasma temperatures.**
D. Bilitza.
Adv. Space Res., Vol. 7, No. 6, p. 93 – 98 (1987). – See Abstr. 012.054.
A status report on the empirical modeling of ionospheric electron and ion temperatures is given with special emphasis on the models used in the International Reference Ionosphere (IRI).

083.018   **Ionospheric electron temperature at solar maximum.**
L. H. Brace, R. F. Theis, W. R. Hoegy.
Adv. Space Res., Vol. 7, No. 6, p. 99 – 106 (1987). – See Abstr. 012.054.
Langmuir probe measurements made at solar maximum from the Dynamics Explorer–2 satellite in 1981 and 1982 are employed to examine the latitudinal variation of electron temperature at altitudes between 300 and 400 km and its response to 27 day variations of solar EUV.

083.019   **Goodness of approximation of lower ionosphere parameters given by a theoretical model and by the International Reference Ionosphere (IRI).**
V. A. Vlaskov, N. V. Smirnova, O. F. Ogloblina, P. I. Vellinov (*P. I. Velinov*).
Adv. Space Res., Vol. 7, No. 6, p. 121 – 124 (1987). – See Abstr. 012.054.

083.020   **Banded ion morphology.**
R. A. Frahm.
Diss. Abstr. Int., Sect. B, Vol. 48, No. 5, p. 1384–B (1987). Thesis, Rice University, 236 pp. (1987). Order No. DA8718712.

083.021   **Ionospheric refraction correction in radio astronomy.**
Y. Chai, W.–j. Han.
Chin. J. Space Sci., Vol. 6, No. 4, p. 266 – 271 (1986). In Chinese. Abstr. in Phys. Abstr., Vol. 90, No. 1313, Entry 115140 (1987).

083.022   **Response of the ionosphere to variations of the interplanetary medium parameter – a case study.**
M. D. Fligel', A. S. Besprozvannaya, N. P. Benkova, N. K. Osipov, G. Johanning, C.–U. Wagner.
Gerlands Beitr. Geophys., Vol. 96, No. 3 – 4, p. 341 – 351 (1987). Abstr. in Phys. Abstr., Vol. 90, No. 1317, Entry 139837 (1987).

083.023   **Control of particle precipitation into the middle atmosphere by regular changes of the interplanetary magnetic field.**
J. Bremer.
Gerlands Beitr. Geophys., Vol. 96, No. 3 – 4, p. 362 – 371 (1987). Abstr. in Phys. Abstr., Vol. 90, No. 1317, Entry 139839 (1987).

083.024   **Convection patterns in the polar ionosphere for northward IMF inferred from ground–based magnetometer data.**
B.–H. Ahn, Y. Kamide, S.–I. Akasofu, C.–I. Meng.
J. Geomagn. Geoelectr., Vol. 39, No. 6, p. 313 – 331 (1987). Abstr. in Phys. Abstr., Vol. 90, No. 1317, Entry 139846 (1987).

083.025   **Intensity of VLF and ELF phenomena in the outer ionosphere.**
Ya. I. Likhter, V. I. Larkina, F. Jiříček, P. Tříska.
Stud. Geophys. Geod., Vol. 31, No. 1, p. 85 – 91 (1987).

083.026   **The ionospheric effects of the July 31, 1981 solar eclipse.**
S. I. Musatenko, I. S. Ivchenko, A. D. Avramchuk, E. P. Datsko, O. I. Maksimenko, M. M. Medvedskij, V. I. Moskalyuk.
Geomagn. Aehron., Tom 27, No. 4, p. 553 – 559 (1987). In Russian. English translation in Geomagn. Aeron.

083.027   **Global structure of the ionosphere and its peculiarities at the transition from high to mid–latitudes.**
N. K. Osipov, V. V. Bychkov, G. I. Vasina, T. A. Gerasimova.
Geomagn. Aehron., Tom 27, No. 5, p. 746 – 751 (1987). In Russian. English translation in Geomagn. Aeron.

083.028   **Height distribution of the electron density in the ionospheric D region at mid–latitudes during the winter anomaly.**
V. V. Belikovich, E. A. Benediktov, V. D. Vyakhirev, N. P. Goncharov, L. V. Grishkevich.
Geomagn. Aehron., Tom 27, No. 6, p. 906 – 909 (1987). In Russian. English translation in Geomagn. Aeron.

083.029   **Study on solar X–ray bursts based on abnormal phenomena of LF wave propagation.**
L.–d. Pan.
Acta Astron. Sin., Vol. 28, No. 3, p. 260 – 267 (1987). In Chinese.
   Some aspects concerning the effects of solar X–ray bursts on the ionosphere are analysed. The flux of solar X–ray bursts, the concerned emission measure and the plasma temperature are deduced or estimated from LF (100 kHz) signals. The results closely conform with satellite measurements.

Preliminary observations from the auroral and ionospheric remote sensing imager.
See Abstr. 035.092.

Numerical modelling of three–dimensional plasma cloud evolution in crossed E × B fields.
See Abstr. 062.007.

On the global mean structure of the thermosphere.
See Abstr. 082.014.

Comparisons of theoretical models and observations of the thermosphere and ionosphere during extremely disturbed geomagnetic conditions during the last solar cycle.
See Abstr. 082.070.

Plasma populations in the magnetosphere.
See Abstr. 084.019.

The role of magnetospheric substorms in magnetosphere–ionosphere coupling.
See Abstr. 084.021.

Combined analysis of frequency spectra of cosmic ray intensity, ionospheric parameters and the geomagnetic field strength.
See Abstr. 084.044.

A suggested model for the IRI plasmaspheric distribution.
See Abstr. 084.054.

Effect of double layers on magnetosphere–ionosphere coupling
See Abstr. 084.072.

084   Aurorae, Geomagnetic Field, Magnetosphere

084.001   **The origin of polarity asymmetries in the history of the geomagnetic field.**
E. H. Levy.
NASA Tech. Memo., NASA TM–89810, p. 147 (1987). Abstract. – See Abstr. 003.001.

084.002   **The steady state toroidal magnetic field at the core–mantle boundary.**
S. J. Pearce, E. H. Levy.
NASA Tech. Memo., NASA TM–89810, p. 148 (1987). Abstract. – See Abstr. 003.001.

084.003   **Auroral particles.**
D. S. Evans.
NASA Conf. Publ., NASA CP–2464, p. 19 – 45 (1987). – See Abstr. 012.004.
   Contents:1. Introduction. 2. Progress from World War II to 1970. 3. Theoretical explanations for particle energization near the Earth. 4. Summary.

084.004   **Latitude dependence of auroral frequency in relation to solar–terrestrial and interplanetary parameters.**
Y. Liritzis, B. Petropoulos.
Earth, Moon, Planets, Vol. 39, No. 1, p. 75 – 91 (1987).
   The auroral frequency of occurrences (A) for the 20th solar cycle and for the geomagnetic latitudes 54°–63° N has been investigated in relation to sunspot numbers (R_z), number of flares (F), the solar wind streams derived from the coronal holes (H) and the geomagnetic index (A_p). The relationship between A and the other indices were found to be strongly latitude dependent. At around 57°–58° N, a drastic change in this relationship occurs, and an attempt is made qualitatively to evaluate this latitudinal variation.

084.005   **Satellite observations of O II (7320 – 7330 Å) emission in aurora.**
L. L. Cogger, G. G. Shepherd, M. M. Gogoshev, Ts. P. Dachev, M. G. Gerdjikova.
Planet. Space Sci., Vol. 35, No. 7, p. 845 – 850 (1987).
   Photometric measurements obtained by the IC–Bulgaria–1300 satellite during 1981 and 1982 have been used to study the characteristics of the O II (7320 – 7330 Å) emission in aurora. The $I(7320)/I(4278)$ emission rate ratio is found to increase in a systematic way with increasing $I(5577)/I(4278)$. This is interpreted as arising from a correlation with the atomic oxygen concentration. The $I(7320)/I(4278)$ ratio also changes with electron mean energy, but for typical nightside electron energy spectra the changes are less important than those caused by changes in neutral composition.

084.006   **Latitudinal structures of discrete arcs resulting from viscous interaction between sheared plasma flows.**
T. Yamamoto, N. Hori.
Planet. Space Sci., Vol. 35, No. 8, p. 1009 – 1020 (1987).
   Latitudinal structures of discrete arcs are modelled as a consequence of the quasi–steady magnetosphere–ionosphere coupling involving viscous interaction between sunward and anti–sunward plasma flows in the magnetosphere.

084.007   **An analytical approach to the magnetic field of the Earth's crust.**
H. J. Nolte, M. Siebert.
J. Geophys., Vol. 61, No. 2, p. 69 – 76 (1987).
   A method is introduced that allows calculation of the magnetic crustal field of the Earth from crustal and geomagnetic core field data.

084.008   **Geomagnetic and solar data. February 1987.**
H. E. Coffey.
J. Geophys. Res., Vol. 92, No. A7, p. 7778 (1987).

084.009   **Geomagnetic and solar data. March 1987.**
H. E. Coffey.
J. Geophys. Res., Vol. 92, No. A7, p. 7779 (1987).

084.010   **Geomagnetic and solar data, April 1987.**
H. E. Coffey.
J. Geophys. Res., Vol. 92, No. A8, p. 8831 (1987).

084.011   **Spectrum of variations in the geomagnetic AE index in the range 130 – 200 min as a possible reflection of the spectrum of global oscillations of the Sun.**
V. P. Bobova, B. M. Vladimirskij, M. I. Pudovkin.
Bull. Crimean Astrophys. Obs., Vol. 74, p. 113 – 122 (1987). English translation of 41.084.042.

084.012 The magnetopause as a source of nonthermal continuum radiation.
D. Jones.
Phys. Scr., Vol. 35, No. 6, p. 887 – 894 (1987).

Second only to auroral kilometric radiation as the Earth's most prominent radio emission, magnetospheric nonthermal continuum radiation has been the subject of numerous observational and theoretical investigations. It has been well established that nonthermal continuum radiation which exists in both ordinary (O) and extraordinary (X) electromagnetic modes, results from the frequency smearing of low frequency Terrestrial myriametric radiation (TMR) due to multiple reflections within the magnetospheric cavity; higher frequency TMR penetrates the magnetosheath and propagates away through the solar wind. It is shown that the magnetopause far from being only a passive reflector of TMR is itself an important source region. Data from satellites within the magnetospheric cavity are presented to illustrate that the magnetopause could well be as active a TMR source as the plasmapause.

084.013 Plasma sheet ion composition at various levels of geomagnetic and solar activity.
W. Lennartsson.
Phys. Scr., Vol. 36, No. 2, p. 367 – 371 (1987).

Observations made in recent years by a number of spaceborne ion mass spectrometers have profoundly altered our view of the earth's plasma environment, especially our view of the energetic portion of this plasma, the one with energies of the order of a keV or more. The primary objective is to reaffirm the evidence for a dual origin of the energetic plasma in the magnetosphere. The secondary objective is to highlight some major differences in the statistical properties of different ions, which may have a bearing on the physics of the solar wind–magnetosphere interaction.

084.014 Het poollicht van 20–4–'85. Een reconstruktie.
F. Witte.
Radiant, Jaarg. 8, Nr. 5, p. 98 – 99 (1986).

084.015 Rocket–borne spectroscopic measurements in the ultraviolet aurora: the Lyman–Birge–Hopfield bands.
R. W. Eastes, W. E. Sharp.
J. Geophys. Res., Vol. 92, No. A9, p. 10,095 – 10,100 (1987).

Ultraviolet emissions from Earth's aurora were observed at wavelengths between 1675 and 2075 Å by a sounding rocket payload launched at Churchill, Canada, on March 28, 1980. The emissions from the Lyman–Birge–Hopfield and Vegard–Kaplan bands of N_2 were observed and analyzed to determine the relative populations of the $v' = 0$–6 levels of the $a^1\pi_g$ state and the $v' = 4$, 6, 7, and 8 levels of the $A^3\Sigma_g^+$ state, respectively. The relative population of higher vibrational levels of the $A^3\Sigma_g^+$ state are consistent with direct excitation and cascade. The relative populations of the vibrational levels of the $a^1\pi_g$ state peak at $v' = 2$. Such a distribution has not been observed previously in the aurora.

084.016 Geomagnetic and solar data. May 1987.
H. E. Coffey.
J. Geophys. Res., Vol. 92, No. A9, p. 10,145 (1987).

084.017 The earth as a magnet.
J. C. Cain.
The solar wind and the Earth, p. 56 – 69 (1987). – See Abstr. 003.007.

Contents: Introduction. Dipole and non–dipole field. Spectrum of the field. Secular change. Patterns of change. Theory of earth's field. Magnetic irregularities. Defining the field.

084.018 The magnetosphere.
C. T. Russell.
The solar wind and the Earth, p. 72 – 100 (1987). – See Abstr. 003.007.

Contents: Introduction. Bow shock. The radiation belts. The plasmasphere. Magnetosphere dynamics. Other magnetospheres. The future.

084.019 Plasma populations in the magnetosphere.
J. L. Burch.
The solar wind and the Earth, p. 102 – 122 (1987). – See Abstr. 003.007.

Contents: Introduction. Guide to the plasma regions of the magnetosphere. The plasma sheet. The plasmasphere. Ring current and radiation belts. Auroral plasma. New observations of plasmas in the magnetosphere. Transpolar auroral arcs. Plasmas in the deep geomagnetic tail. Upward acceleration of ionospheric ions. Future directions in magnetospheric plasma research.

084.020 The aurora.
S.–I. Akasofu, Y. Kamide.
The solar wind and the Earth, p. 142 – 159 (1987). – See Abstr. 003.007.

Contents: Introduction. Electron beams and auroral spectra. Auroral discharge circuit. Auroral potential structure. Solar flares, interplanetary magnetic field and the aurora. Auroral substorms. Polar cap arcs. Auroral effects on human life. Summary.

084.021 The role of magnetospheric substorms in magnetosphere–ionosphere coupling.
G. Rostoker.
The solar wind and the Earth, p. 162 – 181 (1987). – See Abstr. 003.007.

Contents: Introduction. Electric current systems as a quantitative measure of magnetosphere–ionosphere coupling. Acceleration of auroral particles and the role of velocity shear. The magnetospheric substorm. A global view of magnetosphere–ionosphere coupling and substorm processes.

084.022 Plasma wave phenomena in the earth's magnetosphere.
H. Fukunishi.
The solar wind and the Earth, p. 184 – 212 (1987). – See Abstr. 003.007.

Contents: Introduction. Types of magnetospheric plasma waves. Acceleration of auroral electrons by the parallel potential drop. Plasma waves generated by auroral electrons. Standing oscillations of the geomagnetic field lines. Plasma waves generated by trapped particles in the radiation belt. Helium ion effect on Pc 1 emissions. Modulation of whistler mode waves by hydromagnetic waves. Triggered emissions and power line radiation. Conclusion.

084.023 Geomagnetic pulsations associated with MHD–waveguide in magnetospheric ducts.
A. V. Buzevich, A. S. Leonovich, V. A. Parkhomov.
Planet. Space Sci., Vol. 35, No. 9, p. 1093 – 1100 (1987).

A theory of waveguide propagation of MHD waves in magnetospheric ducts (detached plasma regions) is presented. It is shown that Alfvénic waveguide modes can be excited due to the resonance interaction with the waveguide FMS modes. The Alfvénic modes penetrate to the ground and lead to electromagnetic field variations, registered as local geomagnetic Pc1–2 pulsations. The problem of the Alfvénic waveguide modes propagation from the magnetosphere to the Earth's surface is considered and a comparison is made of theoretical results with experimental data.

084.024 Criteria of interplanetary parameters causing intense magnetic storms ($D_{st} < -100$ nT).
W. D. Gonzalez, B. T. Tsurutani.
Planet. Space Sci., Vol. 35, No. 9, p. 1101 – 1109 (1987).

Ten intense magnetic storms ($D_{st} < -100$ nT) occurred during the 500 days from 16 August 1978 to 28 December 1979. From analysis of *ISEE–3* field and plasma data, it is found that the interplanetary causes of these storms are long–duration, large and negative (< -10 nT) IMF B_z events, associated with interplanetary duskward–electric fields > 5 mV m^{-1}, that last for intervals > 3 h. Because the authors find a one–to–one relationship between these interplanetary events and intense storms, they suggest that these criteria can, in the future, be used as predictors of intense storms by an inerplanetary monitor such as *ISEE–3*.

084.025 Effect of possible passage through Halley's magnetic tail on geomagnetic activity.
C. T. Russell, J. L. Phillips, J. A. Fedder, J. H. Allen, L. Morris, R. A. Craig.
J. Geophys. Res., Vol. 92, No. A10, p. 11,195 – 11,200 (1987).

During the 1910 apparition, comet Halley transited the Sun on May 19 at 0400 UT. Since Halley was about 0.16 AU closer to the Sun that was the Earth at the time of transit, the solar plasma passing Halley would take about 0.63 day to reach the Earth if it were moving at 440 km/s. In this paper the authors consider whether this passage of the comet between the Earth and the Sun had any discernible effect on the Earth's magnetosphere.

084.026 Interplanetary magnetic field B_y effects on the magnetic field at synchronous orbit.
T. Nagai.
J. Geophys. Res., Vol. 92, No. A10, p. 11,215 – 11,220 (1987).

084.027 Solar cosmic ray electrons in the Earth's magnetosphere.
S. N. Kuznetsov, E. N. Sosnovets, L. V. Tverskaya, K. Kudela.
Bull. Astron. Inst. Czech., Vol. 38, No. 5, p. 276 – 282 (1987).

The electron fluxes on two low–altitude polar orbiting satellites during two events when solar cosmic ray electrons (SCRE) are present in the interplanetary medium are examined with the aim to deduce positions of dayside cusp and plasmasheet projection into low altitudes. SCRE penetration into the magnetosphere has different character for $B_z > 0$ and $B_z < 0$, respectively. For $B_z > 0$ the polar caps are filled by electrons relatively slowly while for $B_z < 0$ the intensity in the whole polar cap is high and has a character of plateau. An empirical formula for dependence of boundary position in kinetic pressure of solar wind is found while its dependence on geodipole tilt angle is negligible. The inconsistency in difference in boundary position in south and north hemisphere with magnetospheric models is stressed. The results illustrate the possibilities of mapping the boundary regions of magnetosphere by energetic electron registration at low altitudes.

084.028 Observations of the source and propagation of atmospheric gravity waves.
G. Crowley, P. J. S. Williams.
Nature, Vol. 328, No. 6127, p. 231 – 233 (1987).

Atmospheric gravity waves play an important role in transporting energy from one region of the atmosphere to another. The authors report on measurements made during the Worldwide Atmospheric Gravity Wave Study in the auroral source region showing that enhanced Joule heating and Lorentz forcing lead to the generation of large–scale thermospheric waves. Both of these source mechanisms depend on the auroral electric field and the intrinsic periodicity of that field determines the periodicity of waves observed an hour later propagating southwards over the UK.

084.029 Observed beaming of terrestrial myriametric radiation.
D. Jones, W. Calvert, D. A. Gurnett, R. L. Huff.
Nature, Vol. 328, No. 6129, p. 391 – 395 (1987).

Observations by the Dynamics Explorer 1 satellite provide validation of the theory that terrestrial myriametric radiation is produced by the linear conversion of electrostatic upper hybrid waves to electromagnetic radiation via a radio window. This theory predicts that the myriametric radiation is beamed relative to the magnetic field lines.

084.030 Entry velocity of the solar wind's energy into the earth's magnetosphere.
M. I. Pudovkin, V. S. Semenov, M. F. Khejn, Kh. Birnat.
Magnitosfer. Issled., Moskva, No. 9, p. 5 – 16 (1987). In Russian. Abstr. in Ref. Zh., 62. Issled. Kosm. Prostranstva, 9.62.653 (1987).

084.031 The intensity of the geomagnetic field at 3.5 Ga: paleointensity results from the Komati Formation, Barberton Mountain Land, South Africa.
C. J. Hale.
Earth Planet. Sci. Lett., Vol. 86, No. 2/4, p. 354 – 364 (1987).

084.032 The secular variation of the geomagnetic field and the shape of the earth.
B. Alessandrini, G. M. Papi, G. de Franceschi, G. P. Gregori.
Phys. Earth Planet. Inter., Vol. 48, Nos. 1 – 2, p. 84 – 114 (1987).

084.033 Optimum use of satellite intensity and vector data in modelling the main geomagnetic field.
F. J. Lowes, J. E. Martin.
Phys. Earth Planet. Inter., Vol. 48, Nos. 3 – 4, p. 183 – 192 (1987).

084.034 Some effects of quiet geomagnetic field changes upon values used for main field modeling.
W. H. Campbell.
Phys. Earth Planet. Inter., Vol. 48, Nos. 3 – 4, p. 193 – 199 (1987).

084.035 Estimating cosmic ray vertical cutoff rigidities as a function of the McIlwain L–parameter for different epochs of the geomagnetic field.
M. A. Shea, D. F. Smart, L. C. Gentile.
Phys. Earth Planet. Inter., Vol. 48, Nos. 3 – 4, p. 200 – 205 (1987).

The McIlwain L–parameter is commonly used to estimate cosmic ray vertical cutoff rigidities. However, in some areas of the world, secular changes in the geomagnetic field between 1955 and 1980 have been large enough to produce significant differences in both the vertical cutoff rigidity and in the L value for specified positions. The authors show that these changes are complimentary, and that the relationship between the L value and the vertical cutoff rigidity is essentially invariant.

084.036 Spherical harmonic modeling of the geomagnetic field using the fast Fourier transform.
J. M. Quinn, G. A. Barrick.
Phys. Earth Planet. Inter., Vol. 48, Nos. 3 – 4, p. 206 – 220 (1987).

084.037 A new method for analysing geomagnetic impulses.
K. A. Whaler.
Phys. Earth Planet. Inter., Vol. 48, Nos. 3 – 4, p. 221 – 240 (1987).

Geomagnetic records from some observatories give a strong indication that there was an extremely rapid, possibly impulsive, change in geomagnetic secular acceleration around 1969, and perhaps at other earlier epochs, which originated in the Earth's core. Controversy exists as to whether the change is indeed impulsive. A new application of a method of data analysis which has been used extensively on palaeomagnetic demagnetisation data, is proposed.

084.038 Geomagnetic field modeling incorporating constraints from frozen–flux electromagnetism.
E. R. Benton, R. H. Estes, R. A. Langel.
Phys. Earth Planet. Inter., Vol. 48, Nos. 3 – 4, p. 241 – 264 (1987).

084.039 On the interpretation of the geomagnetic energy spectrum.
E. R. Benton, L. R. Alldredge.
Phys. Earth Planet. Inter., Vol. 48, Nos. 3 – 4, p. 265 – 278 (1987).

Two recent high–degree magnetic energy spectra, based mostly on MAGSAT data, are compared and found to agree very well out to order and degree $n = 15$, but the spectrum remains somewhat uncertain for higher degrees. The hypothesis that a primary break in the slope of the spectrum, plotted semi–logarithmically, is due to a transition from dominance by core sources to dominance by crustal magnetization is tested. Simple arrays of dipoles and current loops are found whose combined fields fit the spectrum. Two distinctly different ranges of source depth are found to be adequate. Because one range is shallow and the other deep, the hypothesis is supported.

084.040   International geomagnetic reference field: the fourth generation.
D. R. Barraclough.
Phys. Earth Planet. Inter., Vol. 48, Nos. 3–4, p. 279–292 (1987).

In August 1985 the International Association of Geomagnetism and Aeronomy revised the International Geomagnetism Reference Field (IGRF). This is the third revision since the first IGRF was produced in 1968. The revised IGRF now consists of 10 spherical harmonic models of the main geomagnetic field and its secular variation and covers the interval 1945–1990. For the interval 1965–1980 the constituent models are definitive (DGRFs), in the sense that it is not intended to revise them in the future. A brief description of the derivation of the revised IGRF is given, together with a brief review of basic formulae and a set of world contour maps of the geomagnetic elements for 1985.

084.041   Testing recent geomagnetic field models via magnetic flux conservation at the core–mantle boundary.
E. R. Benton, C. V. Voorhies.
Phys. Earth Planet. Inter., Vol. 48, Nos. 3–4, p. 350–357 (1987).

084.042   An intriguing empirical correlation between the Earth's magnetic field and plate motions.
A. K. Goodacre.
Phys. Earth Planet. Inter., Vol. 49, Nos. 1–2, p. 3–5 (1987). = Contrib. Geol. Surv. Canada, No. 10387.

Spatial variations in the intensity of the Earth's magnetic field appear to be related to plate motions with plates tending to move from areas where the intensity of the magnetic field is diminished to where it is enhanced. This empirical correlation may be due to the presence of regions of enhanced heat transfer at the core–mantle boundary.

084.043   Geomagnetic secular variation, core motions and implications for the earth's wobbles.
J. Hinderer, H. Legros, C. Gire, J. L. Le Mouël.
Phys. Earth Planet. Inter., Vol. 49, Nos. 1–2, p. 121–132 (1987).

Motions at the top of the core are known to be responsible for the secular variation of the earth's magnetic field. If this flow is supposed geostrophic, the associated pressure field can have an appropriate geometry to exert a pressure torque upon the elliptical core–mantle boundary and, also, to alter the elastic products of inertia in such a way as to excite the earth's and core's wobbles. The authors consider some schematic excitation functions and the resulting amplitudes of the earth's and core's rotational motions. The proposed mechanism is shown to be efficient for exciting the long–period Markowitz wobble of the rotation axis and also the Chandler wobble if the variations in the pressure field have the right time scales, as indeed suggested by the available secular variation data.

084.044   Combined analysis of frequency spectra of cosmic ray intensity, ionospheric parameters and the geomagnetic field strength.
V. P. Antonova, L. F. Churunova.
Voln. vozmushcheniya v ionos. Alma–Ata, p. 134–142 (1987). In Russian. Abstr. in Ref. Zh., 62. Issled. Kosm. Prostranstva, 10.62.439 (1987).

084.045   Global quantitative models of the geomagnetic field in the cislunar magnetosphere for different disturbance levels.
N. A. Tsyganenko.
Planet. Space Sci., Vol. 35, No. 11, p. 1347–1358 (1987).

A previously proposed model (Tsyganenko and Usmanov, 1982) is further developed, using IMP–A, C, D, E, F, G, H, I, J and HEOS–1, –2 spacecraft measurements made during 1966–1980. The main improvement consists of a considerable extension of the modeling region by adding to the original data set a large number of magnetotail field measurements. Data points used in the present study cover a vast range of distances $4 \lesssim R \lesssim 70\, R_E$ and comprise an unprecedently large number of measurements.

084.046   Geomagnetic and solar data. July 1987.
H. E. Coffey.
J. Geophys. Res., Vol. 92, No. A11, p. 12471 (1987).

084.047   Wave acceleration of auroral electrons.
D. A. Bryant.
Rutherford Appleton Lab., Rep., RAL–87–044, 7 pp. (1987). To appear in 8th ESA symposium on European rocket and balloon programmes and related research, Sunne, Sweden, 17–23 May 1987.

084.048   The Earth's magnetopause and related phenomena.
R. N. Singh.
Indian J. Radio Space Phys., Vol. 16, No. 1, p. 161–171 (1987). Abstr. in Phys. Abstr., Vol. 90, No. 1309, Entry 88040 (1987).

084.049   Irregular, broad–band ELF/VLF emissions and optical aurora at cusp latitudes in the post–noon sector.
A. Egeland, H. Liao, P. E. Sandholt.
Ann. Geophys., Ser. A, Vol. 5, No. 2, p. 89–92 (1987). Abstr. in Phys. Abstr., Vol. 90, No. 1310, Entry 94577 (1987).

084.050   Monochromatic auroral images observed at Syowa Station, in Antarctica.
T. Ono, M. Ejiri, T. Hirasawa.
J. Geomagn. Geoelectr., Vol. 39, No. 2, p. 65–95 (1987). Abstr. in Phys. Abstr., Vol. 90, No. 1311, Entry 101616 (1987).

084.051   Evaluation of the GSFC (9/80) model of the geomagnetic field and determination of local anomaly solutions of observatories in China.
W.–j. Wu, T.–x. Bai, X.–x. Ren.
Acta Geophys. Sin., Vol. 30, No. 2, p. 178–185 (1987). In Chinese.

The GSFC (9/80) model of the geomagnetic field is evaluated on the basis of the data of the observatories in China. The results show that this model is successful. The local anomaly solutions of these observatories are determined. It is suggested to subtract the local anomalies from the annual means for the study of main magnetic field in order to improve the accuracy of data.

084.052   The aurora 1985.
R. J. Livesey.
J. Br. Astron. Assoc., Vol. 98, No. 1, p. 18–22 (1987).

This report summarises observations of the aurora and magnetic field disturbances sent in to the Aurora Section in 1985.

084.053   The northern and southern auroral ovals in response to the IMF By component.
H. Nakai.
Geophys. Res. Lett., Vol. 14, No. 11, p. 1162–1165 (1987).

The auroral boundary in terms of precipitating electrons is used to examine the influence of the By component of the interplanetary magnetic field (IMF) on the size of the auroral oval.

084.054 A suggested model for the IRI plasmaspheric distribution.
M. J. Rycroft, I. R. Jones.
Adv. Space Res., Vol. 7, No. 6, p. 13 – 22 (1987). – See Abstr. 012.054.

A model has been constructed which grafts IRI topside profiles to field–aligned diffusive equilibrium profiles at a "reference level" near 650 km altitude. For different values of geographic latitude, longitude, sunspot number, season and time, the properties of the plasma distribution in the equatorial plane are investigated. These are compared with observations made by satellite (particularly GEOS and ISEE) and also with deductions made from whistler observations.

084.055 Structure of the polar oval from simultaneous observations of the optical emissions and particle precipitations during the period of high solar activity 1981 – 1982.
M. M. Gogoshev, G. G. Shepherd, P. V. Maglova, V. C. Guineva, T. P. Datchev.
Adv. Space Res., Vol. 7, No. 8, p. 7 – 10 (1987). – See Abstr. 012.055.

The observations on board the IC–Bulgaria–1300 satellite, obtained during the period 1981 – 1982, have been analysed. The optical emissions have been measured by the optical photometer EMO–5. The simultaneous precipitating electron and proton fluxes have been measured by the ANEPE spectrometer. The structure of the oval is studied as a function of the magnetic local time and the geomagnetic activity.

084.056 Spectrophotometric observations of the auroras in the far UV range on board the Intercosmos–Bulgaria–1300 satellite during the magnetic storm of 2 – 3 March 1982.
M. M. Gogoshev, S. I. Sargoychev, I. D. Mendev, V. M. Balebanov, L. P. Smirnova, V. S. Bassolo.
Adv. Space Res., Vol. 7, No. 8, p. 11 – 13 (1987). – See Abstr. 012.055.

The storm–time H I(1216 Å), O I(1304 Å), O I(1356 Å) emission lines and NO molecule γ(1,1) band computed from the onboard Intercosmos–Bulgaria–1300 measurements are examined. The auroral particles and ring current development are discussed as possible sources of the observed storm–time intensity increase over the theoretical intensity–solar zenith angle dependencies in the evening–midnight sector.

084.057 Model calculation of hydrogen Balmer emissions under various modes of proton precipitation.
V. Srivastava, V. Singh.
Adv. Space Res., Vol. 7, No. 8, p. 21 – 25 (1987). – See Abstr. 012.055.

A theoretical model is developed to study the H_α and H_β volume emission rates in proton auroras. The H_α and H_β photon yields are deduced by using recent experimental cross–sections. These photon yields are then coupled with the primary proton fluxes to get the altitude profiles of the emissions. The prediction of the altitude profiles are found in good agreement with the auroral measurements. The effects of various modes of proton precipitation on the emission intensities (IH_α and IH_β) are discussed and a comparison is made with earlier theoretical models.

084.058 Radiative transfer effects on aurora enhanced 4.3 micron emission.
J. R. Winick, R. H. Picard, R. D. Sharma, R. A. Joseph, P. P. Wintersteiner.
Adv. Space Res., Vol. 7, No. 10, p. 17 – 21 (1987). – See Abstr. 012.057.

084.059 Conditions for geomagnetically trapping sub–micron lunar ejecta between L = 1.2 and L = 3.1.
L. M. Lodhi.
Diss. Abstr. Int., Sect. B, Vol. 48, No. 5, p. 1384–B (1987). Thesis, Baylor University, 151 pp. (1987). Order No. DA8718339.

084.060 Model of solar–wind magnetosphere coupling.
F. R. Toffoletto.
Diss. Abstr. Int., Sect. B, Vol. 48, No. 5, p. 1385–B – 1386–B (1987). Thesis, Rice University, 92 pp. (1987). Order No. DA8718773.

084.061 Rapid changes and near–stationarity of the geomagnetic field during a polarity reversal.
C. Laj, S. Guitton, C. Kissel.
Nature, Vol. 330, No. 6144, p. 145 – 148 (1987).

The authors report the results of a detailed study of a geomagnetic reversal recorded in Middle Miocene marine clays, with temporal resolution almost comparable with that of the volcanic records and, in their opinion, a more reliable chronological control. The record displays large fluctuations in the rate of directional change and different episodes of collapse and recovery of the field intensity.

084.062 Two hundred years of auroral activity (1780 – 1979).
J.–P. Legrand, P. A. Simon.
Ann. Geophys., Ser. A, Vol. 5, No. 3, p. 161 – 167 (1987). Abstr. in Phys. Abstr., Vol. 90, No. 1313, Entry 115135 (1987).

084.063 A case study of an active aurora observed by monochromatic auroral TV on the ground and particle analyzers on board the EXOS–C satellite.
T. Ono, M. Ejiri, T. Hirasawa, N. Kaya, T. Mukai.
J. Geomagn. Geoelectr., Vol. 39, No. 3, p. 119 – 128 (1987). Abstr. in Phys. Abstr., Vol. 90, No. 1313, Entry 115136 (1987).

084.064 On the location of the stationary reconnection region in the Earth's magnetotail.
J. Büchner, L. M. Zeleny.
Gerlands Beitr. Geophys., Vol. 96, No. 2, p. 179 – 181 (1987). Abstr. in Phys. Abstr., Vol. 90, No. 1313, Entry 115169 (1987).

084.065 A gyroviscous model of the magnetotail current layer and the substorm mechanism.
K. Stasiewicz.
Phys. Fluids, Vol. 30, No. 5, p. 1401 – 1409 (1987). Abstr. in Phys. Abstr., Vol. 90, No. 1313, Entry 115177 (1987).

084.066 The theory of radio windows in planetary magnetospheres.
K. G. Budden, D. Jones.
Proc. R. Soc. London, Ser. A, Vol. 412, No. 1842, p. 1 – 23 (1987). Abstr. in Phys. Abstr., Vol. 90, No. 1313, Entry 115178 (1987).

084.067 Theory of wave polarization of radio waves in magnetospheric cavities.
K. G. Budden, D. Jones.
Proc. R. Soc. London, Ser. A, Vol. 412, No. 1842, p. 25 – 44 (1987). Abstr. in Phys. Abstr., Vol. 90, No. 1313, Entry 115179 (1987).

084.068 Anomalous transport in discrete arcs and simulation of double layers in a model auroral circuit.
R. A. Smith.
Laser Part. Beams, Vol. 5, Part 2, p. 381 – 391 (1987). Abstr. in Phys. Abstr., Vol. 90, No. 1314, Entry 117737 (1987). – See abstracts. 012.064, 43.084.109.

084.069 Double layers above the aurora.
M. Temerin, F. S. Mozer.
Laser Part. Beams, Vol. 5, Part 2, p. 203 – 210 (1987). Abstr. in Phys. Abstr., Vol. 90, No. 1314, Entry 120194 (1987). – See abstracts 012.064, 43.084.112.

084.070 Weak double layers in the auroral ionosphere.
M. K. Hudson, T. L. Crystal, W. Lotko, C. Barnes.
Laser Part. Beams, Vol. 5, Part 2, p. 295 – 313 (1987). Abstr. in Phys. Abstr., Vol. 90, No. 1314, Entry 120204 (1987). – See abstracts 012.064, 43.084.110.

084.071   Conditions for double layers in the Earth's magneto-sphere and perhaps in other astrophysical objects.
L. R. Lyons.
Laser Part. Beams, Vol. 5, Part 2, p. 191 – 196 (1987). Abstr. in Phys. Abstr., Vol. 90, No. 1314, Entry 120217 (1987). – See abstracts 012.064, 43.084.111.

084.072   Effect of double layers on magnetosphere–ionosphere coupling
R. L. Lysak, M. K. Hudson.
Laser Part. Beams, Vol. 5, Part 2, p. 351 – 366 (1987). Abstr. in Phys. Abstr., Vol. 90, No. 1314, Entry 120218 (1987). – See abstracts 012.064, 43.084.108.

084.073   Geomagnetic and solar data. August 1987.
H. E. Coffey.
J. Geophys. Res., Vol. 92, No. A12, p. 13,680 (1987).

084.074   Drift wave instability in the plasmasphere.
G. Renuka, E. Sreevalsan.
Indian J. Radio Space Phys., Vol. 16, No. 2, p. 232 – 235 (1987). Abstr. in Phys. Abstr., Vol. 90, No. 1315, Entry 127642 (1987).

084.075   Radiation belts of the Earth.
Yu. I. Logachev, B. A. Tverskoj.
Problems of cosmic ray physics, p. 5 – 29 (1987). In Russian. Abstr. in Ref. Zh., 62. Issled. Kosm. Prostranstva, 12.62.381 (1987). – See Abstr. 003.015.

084.076   Modelling the distribution of energetic ions in the radiation belts of the earth.
M. F. Goryainov, M. I. Panasyuk, V. V. Senkevich.
Kosm. Issled., Tom 25, Vyp. 4, p. 556 – 565 (1987). In Russian. English translation in Cosm. Res.

084.077   Dynamics of aurora radiation in the far ultraviolet region during the magnetic storm on 2 – 3 March 1982.
V. M. Balebanov, M. M. Gogoshev, L. P. Smirnova,
V. S. Bassolo, S. I. Sargojchev, I. D. Mendev.
Kosm. Issled., Tom 25, Vyp. 4, p. 566 – 576 (1987). In Russian. English translation in Cosm. Res.

084.078   Annual variation in the surface magnetic field due to magnetospheric current systems.
G. K. Mukherjee, R. Rajaram.
Ann. Geophys., Ser. A, Vol. 5, No. 4, p. 239 – 246 (1987). Abstr. in Phys. Abstr., Vol. 90, No. 1317, Entry 139442 (1987).

084.079   Understanding models of the geomagnetic field by Fourier analysis.
Z. Wang.
J. Geomagn. Geoelectr., Vol. 39, No. 6, p. 333 – 347 (1987). Abstr. in Phys. Abstr., Vol. 90, No. 1317, Entry 139450 (1987).

084.080   Observational evidences for a new approach to an old problem: magnetotail stability and substorm onset caused by chaotic electron dynamics.
J. Büchner, L. M. Zelenyi.
Gerlands Beitr. Geophys., Vol. 96, No. 3 – 4, p. 326 – 330 (1987). Abstr. in Phys. Abstr., Vol. 90, No. 1317, Entry 139860 (1987).

084.081   Theory and observations of Alfvén solitons in the finite beta magnetospheric plasma.
V. L. Patel, B. Dasgupta.
Physica D, Vol. 27D, No. 3, p. 387 – 398 (1987). Abstr. in Phys. Abstr., Vol. 90, No. 1317, Entry 139865 (1987).

084.082   Simultaneous observation of fundamental and second harmonic radio emission from the terrestrial foreshock.
D. Burgess, C. C. Harvey, J.–L. Steinberg, C. Lacombe.
Nature, Vol. 330, No. 6150, p. 732 – 735 (1987).
The authors present measurements made with the sounder experiment aboard ISEE–1 and the radio experiment on ISEE–3

of simultaneous fundamental and second harmonic electromagnetic emission from the Earth's foreshock.

084.083   The aurora of July 12 and 13, 1982.
L. Křivský.
Říše hvězd, Vol. 68, No. 8, p. 151 – 153 (1987). In Czech.

084.084   On the possibilities of observing ballistic plasma wave processes in the magnetosphere.
E. D. Poezd, V. N. Oraevskij, V. Fiala, P. Tříska.
Stud. Geophys. Geod., Vol. 31, No. 2, p. 197 – 213 (1987).

084.085   Coupling between auroral kilometric radiation and auroral radio noise based on wave–particle resonance interaction.
E. D. Poezd, V. Fiala, P. Tříska.
Stud. Geophys. Geod., Vol. 31, No. 1, p. 73 – 84 (1987).

084.086   Variations of the undisturbed level of the mid–latitude geomagnetic field in the solar cycle and in dependence on the intensity of ionospheric and magnetospheric current systems.
B. M. Kuznetsov, N. G. Ptitsyna.
Geomagn. Aehron., Tom 27, No. 4, p. 620 – 624 (1987). In Russian. English translation in Geomagn. Aeron.

084.087   The night–time trend in the average brightness of discrete aurora forms.
N. I. Dzyubenko, V. I. Dzyubenko.
Geomagn. Aehron., Tom 27, No. 5, p. 848 – 850 (1987). In Russian. English translation in Geomagn. Aeron.

084.088   11–year variations of the main characteristics of the geomagnetic field.
Yu. D. Kalinin, T. S. Rozanova.
Geomagn. Aehron., Tom 27, No. 5, p. 868 – 869 (1987). In Russian. English translation in Geomagn. Aeron.

084.089   The mean statistical structure of the magnetospheric tail as given by satellite data.
N. A. Tsyganenko.
Geomagn. Aehron., Tom 27, No. 6, p. 987 – 993 (1987). In Russian. English translation in Geomagn. Aeron.

Ancient Chinese auroral records: interpretation problems and methods.
See Abstr. 004.084.

Preliminary observations from the auroral and ionospheric remote sensing imager.
See Abstr. 035.092.

Auroral images from space: imagery, spectroscopy, and photometry.
See Abstr. 035.093.

The Polar BEAR magnetic field experiment.
See Abstr. 035.094.

International Geophysical Calendar 1988.
See Abstr. 046.022.

Active plasma experiments.
See Abstr. 051.034.

Adiabatic consideration of the motion of charged particles in superposed dipole and uniform magnetic fields.
See Abstr. 062.021.

Plasma turbulence in space.
See Abstr. 062.032.

The guidance of oblique whistler mode waves along magnetospheric field lines.
See Abstr. 062.034.

Hydromagnetic dynamo as source of planetary, solar and galactic magnetism.
See Abstr. 062.071.

Some aspects of shock–wave research.
See Abstr. 062.084.

Numerical simulation of a magnetic reconnection process.
See Abstr. 062.090.

An amended magnetohydrodynamic equation which predicts field–aligned current sheets.
See Abstr. 062.095.

Electrostatic waves at frequencies close to the harmonics of electron gyrofrequency.
See Abstr. 062.097.

Some aspects of double layer formation in a plasma constrained by a magnetic mirror.
See Abstr. 062.116.

Particle simulation of auroral double layers.
See Abstr. 062.119.

Weak compressive ion acoustic double layers in presence of negative ions on the auroral field lines.
See Abstr. 062.134.

Simultaneous observations of Pc 3 – 4 pulsations in the solar wind and in the earth's magnetosphere.
See Abstr. 074.033.

Magnetohydrodynamics of the Earth's core.
See Abstr. 081.014.

Origin of the main field: kinematics.
See Abstr. 081.015.

Origin of the main field: dynamics.
See Abstr. 081.016.

Palaeomagnetic data suggest link between the Archaean–Proterozoic boundary and inner–core nucleation.
See Abstr. 081.026.

Current loops fitted to geomagnetic model spherical harmonic coefficients.
See Abstr. 081.031.

A model–Z geodynamo.
See Abstr. 081.033.

Antarctica – a unique laboratory for atmospheric studies by optical methods.
See Abstr. 082.025.

The solar wind–magnetosphere–ionosphere coupling during a magnetic storm.
See Abstr. 083.004.

Banded ion morphology.
See Abstr. 083.020.

Some solar cycle phenomena related to the geomagnetic activity from 1868 to 1980. III. Quiet–days, fluctuating activity or the solar equatorial belt as the main origin of the solar wind flowing in the ecliptic plane.
See Abstr. 085.003.

Statistical analysis of the geomagnetic disturbing index Ap with some solar activities.
See Abstr. 085.013.

On the origin of magnetic fields of the earth and celestial bodies.
See Abstr. 091.034.

A comparison of the Mercury and Earth magnetospheres: electron measurements and substorm time scales.
See Abstr. 092.011.

Diffusive shock acceleration: comparison of a unified shock model to bow shock observations.
See Abstr. 106.002.

Interplanetary and geophysical effects of a coronal transient.
See Abstr. 106.039.

085 Solar-terrestrial Relations

085.001 Equatorial middle atmospheric temperature response to solar activity, 1971 – 1982, from spectral analyses.
S. Devanarayanan, K. Mohanakumar.
Planet. Space Sci., Vol. 35, No. 7, p. 959 – 964 (1987).
Rocket temperature data of the middle atmosphere (20 – 80 km) over the rocket launching station, THUMBA (8°N, 77°E) during the period 1971 – 1982 are used. Auto– and cross–spectral analyses are made with these temperature data and the sunspot number data during the same period, at various altitude levels.

085.002 The solar wind–magnetosphere–ionosphere current–voltage relationship.
J. A. Fedder, J. G. Lyon.
Geophys. Res. Lett., Vol. 14, No. 8, p. 880 – 883 (1987).
The nature of the solar wind–magnetosphere–ionosphere (SW–M–I) coupling has been a subject of intense study and scientific interest. The authors report results from a numerical simulation of the SW–M–I system which shed light on the physics and behavior of the controlling processes.

085.003 Some solar cycle phenomena related to the geomagnetic activity from 1868 to 1980. III. Quiet–days, fluctuating activity or the solar equatorial belt as the main origin of the solar wind flowing in the ecliptic plane.
P. A. Simon, J. P. Legrand.
Astron. Astrophys., Vol. 182, No. 2, p. 329 – 336 (1987).
The quiet–days form the largest set of geomagnetic activity indices while the fluctuating activity is the most frequent form of storminess. Both categories of geomagnetic phenomena are associated with the flow, in the ecliptic plane, of low velocity wind for the quiet–days and moderate velocity wind for the fluctuating activity, respectively. Therefore their solar origin takes place in the narrow "equatorial belt" scanned twice a year by the ecliptic plane. The authors write down a series of conclusions coming from experimental evidences relating the occurrence in the "equatorial belt" of bipolar structures of high density, identified either with coronal streamers of with the heliosheet, with the flow in the ecliptic plane of a slow wind. These conclusions concern the distribution of the wind velocity on both sides of the associated neutral sheet as well as the cyclical

evolution of its shape, size and stability. Finally the authors discover an inverse ratio relation in between the thickness of the heliosheet at the sunspot minimum and the following peak of the sunspot activity.

085.004   Interacting regions of solar corpuscular streams and geomagnetic activity.
G. I. Ol'.
Soln. Dannye, Byull., 1986, No. 11, p. 75 – 80 (1987). In Russian.
   An increase in geomagnetic activity is shown to be associated with the appearance of high–speed corpuscular streams near the Earth rather than the formation of interacting regions there. When fast isolated streams are observed near the Earth gradual-commencement storms are found to occur more frequently while the storms with sudden commencement are observed when fast corpuscular streams and interacting regions occur simultaneously.

085.005   Solar activity related variability of volcanic eruptions for 300 years.
E. G. Kravchuk.
Soln. Dannye, Byull., 1986, No. 12, p. 78 – 81 (1987). In Russian.
   Time structure of great volcanic eruptions for the recent three centuries is considered. The frequency of volcanic eruptions increases near the minima of secular cycles in solar activity. The frequency structure of volcanic eruptions includes a change in the long–period component whose length is close in an 11–year cycle at the solar activity maximum. When the solar activity is low the component is observed to grow up to 14 – 18 years. The supposed nature of the cycles mentioned is discussed.

085.006   On the discovery of the solar–troposphere effect in the Southern Hemisphere.
Eh. R. Mustel', V. E. Chertoprud, N. B. Mulyukova.
Astron. Zh., Tom 64, Vyp. 4, p. 873 – 875 (1987). In Russian. English translation in Sov. Astron., Vol. 31, No. 4.
   The solar–troposphere effect is discovered in the Southern Hemisphere as well as in the Northern one. At the fourth–fifth day after the entering of the Earth into the solar corpuscular streams, an increase of instability of the troposphere is observed in the moderate latitudes of the Southern Hemisphere.

085.007   Solar sources of geomagnetic storms.
V. P. Mikhajlutsa, M. N. Gnevyshev.
Soln. Dannye, Byull., 1987, No. 4, p. 65 – 69 (1987). In Russian.
   Magnetospheric storms are caused by a low–speed solar wind, transferring the polarity division line of the solar global magnetic field to interplanetary space. The sector boundary of the interplanetary magnetic field is formed either above the system of closed lines–of–force of the coronal field (then it is observed as a ray in the corona), or above an open neutral line (then the ray in the corona is absent, but there is a cavity in the chromosphere and the corona).

085.008   Investigation of the dependence of the tropopause's pressure on solar activity: the case of the North Atlantic.
B. A. Sandler.
Nauchn. Inf., Vyp. 59, p. 128 – 138 (1986). In Russian.
   A possible influence of solar flares on the transition region between troposphere and stratosphere (tropopause) above the North Atlantic and adjacent regions is investigated by the method of superposed epochs. The nine flare–induced winter geomagnetic storms during the period 1961 – 1970 were analyzed.

085.009   Sunspot–weather correlation found.
R. A. Kerr.
Science, Vol. 238, No. 4826, p. 479 – 480 (1987).
   A stunningly strong correlation between the sunspot cycle and weather has been found; will it persist and what, if any, physical connection is responsible?

085.010   Investigation of the connection of ionospheric disturbances and cosmic ray variations with fast solar wind streams.
A. G. Zusmanovich, O. N. Kryakunova, L. F. Churunova.
Voln. vozmushcheniya v ionos. Alma–Ata, p. 127 – 134 (1987). In Russian. Abstr. in Ref. Zh., 62. Issled. Kosm. Prostranstva, 10.62.438 (1987).

085.011   On the role of wave and corpuscular radiation of the sun in geomagnetic field variations.
A. V. Vinitskij, V. V. Kazantseva.
Magadan, Sev.–Vost. Kompleks NII, Prepr., 18 pp. (1987). In Russian. Abstr. in Ref. Zh., 62. Issled. Kosm. Prostranstva, 10.62.440 (1987).

085.012   Relationship between sudden ionospheric disturbances in the D layer and solar microwave bursts.
S.–q. Wang, L.–d. Pan.
Publ. Shaanxi Astron. Obs., Vol. 9, No. 1, p. 5 – 10 (1986). In Chinese.
   The relationship between the sudden phase anamolies (SPA) and the solar microwave bursts are analysed and it leads the results that the best correlation exists between the SPA events and the solar microwave bursts at 3 – 8 cm band, and as for the different types of solar microwave bursts, the complex type has the largest probability of accompaning the SPA events.

085.013   Statistical analysis of the geomagnetic disturbing index Ap with some solar activities.
Q. Zhang.
Publ. Yunnan Obs., No. 2, p. 25 – 30 (1987). In Chinese.
   In order to have predictions of the geomagnetic disturbances caused by solar activities, a statistical analysis is made for the correlation of the daily planetary index Ap with some solar activities by means of superposed epochs. The obtained results are useful for the medium range (4 – 6 day) forecasts of geomagnetic disturbances.

085.014   Investigation of a connection between sodium emission and solar activity.
M. P. Korobejnikova.
Izv. Akad. Nauk TurkSSR. Ser. Fiz.–Tekh., Khim. Geol. Nauk, No. 3, p. 89 – 90 (1987). In Russian. Abstr. in Ref. Zh., 51. Astron. 11.51.457 (1987).

085.015   Solar cycle variations in the annual mean values of the geomagnetic components of observatory data.
T. Yukutake, J. C. Cain.
J. Geomagn. Geoelectr., Vol. 39, No. 1, p. 19 – 46 (1987). Abstr. in Phys. Abstr., Vol. 90, No. 1312, Entry 107342 (1987).

085.016   Response of geomagnetic pulsations to solar eclipse.
N. A. Zolotukhina, V. A. Parkhomov.
Issled. Geomagn. Aehron. Fiz. Solntsa, Moskva, No. 78, p. 40 – 48 (1987). In Russian. Abstr. in Ref. Zh., 62. Issled. Kosm. Prostranstva, 11.62.417 (1987).

085.017   146–day signal in the geomagnetic field – a probable association with the periodicity of the solar flare index.
D. R. K. Rao, B. D. Kadam.
Proc. Indian Acad. Sci., Earth Planet. Sci., Vol. 96, No. 2, p. 119 – 122 (1987).
   Results on the spectral analysis using geomagnetic fields at three low latitude stations and the planetary magnetic activity index have shown peaks in the power densities in a broad band centred around 146–day period. This periodic behaviour appears to be close to that shown by the solar flare activity index for the same interval. It is suggested that the geoeffectiveness of the flare activity signal in different phases of the solar cycle can be better worked out using long series of ground–based geomagnetic data.

085.018 North–south asymmetry in response of geomagnetic activity to different solar events.
G. K. Rangarajan.
Proc. Indian Acad. Sci., Earth Planet. Sci., Vol. 96, No. 2, p. 195 – 202 (1987).

Study of the response of geomagnetic activity to five different kinds of solar events reveals that an average north–south asymmetry of about 15% exists which diminishes with enhanced geomagnetic activity. The response of the geomagnetic field is quite significant only when high speed solar wind streams in association with the sector boundary of the interplanetary magnetic field or solar proton streams near the sector boundary sweep past the earth.

085.019 Influence of solar flares on the state of the ionosphere.
S. I. Didenko, V. T. Ustinovich, Yu. I. Podyachij, N. A. Smaglo.
Vestn. Khar'k. Politekh. Inst., No. 248, p. 14 – 16 (1987). In Russian. Abstr. in Ref. Zh., 51. Astron., 12.51.552 (1987).

085.020 Forecast of abnormally high solar activity in the next century and of its possible negative effects on the Earth.
M. Kopecký.
Stud. Geophys. Geod., Vol. 31, No. 1, p. 43 – 59 (1987).

085.021 Manifestation of solar activity in the polar–cap troposphere.
R. V. Smirnov, Eh. V. Kononovich, V. V. Afanas'ev.
Sov. Astron., Vol. 31, No. 2, p. 227 – 228 (1987). English translation of 43.085.006.

085.022 The effect of solar wind wave conductivity on the convection in the polar cap.
B. V. Rezhenov.
Geomagn. Aehron., Tom 27, No. 4, p. 616 – 619 (1987). In Russian. English translation in Geomagn. Aeron.

085.023 Spatial distribution of geomagnetic field oscillations associated with a solar flare.
Yu. V. Gutop, V. M. Sorokin.
Geomagn. Aehron., Tom 27, No. 6, p. 977 – 981 (1987). In Russian. English translation in Geomagn. Aeron.

Inter–agency coordination of solar–terrestrial science projects.
See Abstr. 051.003.

Latitude dependence of auroral frequency in relation to solar–terrestrial and interplanetary parameters.
See Abstr. 084.004.

Planetary System

091 Physics and Dynamics of the Planetary System

091.001 **Shapes of small satellites.**
P. Thomas.
NASA Tech. Memo., NASA TM–89810, p. 35 – 36 (1987). – See Abstr. 003.001.

091.002 **Investigations of the surfaces and interiors of outer planet satellites.**
G. Schubert.
NASA Tech. Memo., NASA TM–89810, p. 40 – 42 (1987). – See Abstr. 003.001.

091.003 **Failure strength of icy lithospheres.**
M. P. Golombek, W. B. Banerdt.
NASA Tech. Memo., NASA TM–89810, p. 43 – 45 (1987). – See Abstr. 003.001.
In this paper the authors will introduce lithospheric strengths derived from friction on pre–existing fractures and ductile flow laws, derive these relations for icy lithospheres, show that the tensile strength of intact ice under applicable conditions is actually an order of magnitude stronger than widely assumed, and demonstrate that this strength is everywhere greater than that required to initiate frictional sliding on pre–existing fractures and faults.

091.004 **Studies of outer planet satellites, Mercury, and Uranus.**
W. B. McKinnon, P. M. Schenk.
NASA Tech. Memo., NASA TM–89810, p. 46 – 48 (1987). – See Abstr. 003.001.

091.005 **Investigations of planetary ring phenomena.**
J. A. Burns.
NASA Tech. Memo., NASA TM–89810, p. 52 – 54 (1987). – See Abstr. 003.001.

091.006 **Solar system dynamics.**
J. Wisdom.
NASA Tech. Memo., NASA TM–89810, p. 117 – 119 (1987). – See Abstr. 003.001.
I. Rotational dynamics of irregularly shaped satellites. II. Origin of the Kirkwood gaps.

091.007 **Orbital resonances, unusual configurations and exotic rotation states among planetary satellites.**
S. J. Peale.
NASA Tech. Memo., NASA TM–89810, p. 120 (1987). Abstract. – See Abstr. 003.001.

091.008 **Dynamics of satellites, asteroids, and rings.**
S. F. Dermott.
NASA Tech. Memo., NASA TM–89810, p. 121 – 123 (1987). – See Abstr. 003.001.

091.009 **Planetary ring dynamics and morphology.**
J. N. Cuzzi, R. H. Durisen, F. H. Shu.
NASA Tech. Memo., NASA TM–89810, p. 125 – 127 (1987). – See Abstr. 003.001.

091.010 **Bidirectional reflectance properties of planetary surface materials.**
B. Buratti, W. Smythe, R. Nelson, V. Gharakhani, B. Hapke.
NASA Tech. Memo., NASA TM–89810, p. 201 – 204 (1987). – See Abstr. 003.001.

091.011 **On the sputter alteration of regoliths of outer solar system bodies.**
B. Hapke.
NASA Tech. Memo., NASA TM–89810, p. 220 (1987). Abstract. – See Abstr. 003.001.
For the full paper see 41.091.045.

091.012 **Solid sulfur in vacuum: sublimation effects on surface microtexture, color, and spectral reflectance, and applications to planetary surfaces.**
D. B. Nash.
NASA Tech. Memo., NASA TM–89810, p. 227 – 228 (1987). – See Abstr. 003.001.

091.013 **The 1984 Mauna Loa eruption and planetary geology.**
H. J. Moore.
NASA Tech. Memo., NASA TM–89810, p. 351 – 353 (1987). – See Abstr. 003.001.

091.014 **Impact crater scaling laws.**
K. A. Holsapple.
NASA Tech. Memo., NASA TM–89810, p. 392 – 393 (1987). – See Abstr. 003.001.
This is a progress report on research on impact cratering scaling laws, on numerical studies that were designed to investigate those laws, and on various applications of the scaling laws developed by the author and his colleagues.

091.015 **Impact and cratering processes on asteroids, satellites, and planets.**
C. R. Chapman, D. R. Davis, S. J. Weidenschilling.
NASA Tech. Memo., NASA TM–89810, p. 399 – 401 (1987). – See Abstr. 003.001.

091.016 **Natural fracture systems on planetary surfaces: genetic classification and pattern randomness.**
L. A. Rossbacher.
NASA Tech. Memo., NASA TM–89810, p. 499 – 501 (1987). – See Abstr. 003.001.
One method for classifying natural fracture systems is by fracture genesis. This approach involves the physics of the formation process, and it has been used most frequently in attempts to predict subsurface fractures. This classification system can also be applied to larger fracture systems on any planetary surface.

091.017 **A satellite–asteroid mystery and a possible early flux of scattered C–class asteroids.**
W. K. Hartmann.
Icarus, Vol. 71, No. 1, p. 57 – 68 (1987).
Spectrophotometric measurements of the subset of satellites thought to be of capture origin, orbiting Saturn, Jupiter, and Mars, indicate that they typically (all?) have neutral spectra and

low albedo, indicating spectral class C. This finding is puzzling in light of evidence that class–C objects are native only to the outer half of the asteroid belt. How did C–like objects approach, and get capture by, planets? Probably Jupiter resonances scattered a high flux of C–type objects out of the belt and throughout much of the primordial solar system. Such scattering could occur only at the close of planet accretion when extended atmospheres could affect capture.

091.018   Inelastic collisions in narrow planetary rings.
W. Wiesel.

Icarus, Vol. 71, No. 1, p. 78 – 90 (1987).

It is shown that a particle ring with energy dissipation has an extremum in its energy when all the particles are in the same circular orbit. This extremum is a relative maximum in radial directions, indicating possible radial expansion; but it is a relative minimum in the particle velocity components, indicating a tendency for the velocity distribution to collapse. An N-body model of ring evolution incorporating two–body dynamics, oblateness perturbations, inelastic collisions, and phase averaging is described.

091.019   Magnetospheric environments of outer planet rings: influence of Saturn's axially symmetric magnetic field.
L. L. Hood.

Icarus, Vol. 71, No. 1, p. 115 – 136 (1987). = Arizona Theor. Astrophys. Contrib., No. 87–1.

The Jovian and Uranian rings exist within severe energetic particle and plasma environments where magnetosphere–related losses of small ring particles and surface reflectance alteration by sputtering are likely to be important. In contrast, the main Saturnian rings exist within a zone where magnetopsheric losses and surface alteration effects are negligible, primarily because of solid–body absorption of inwardly diffusing magnetospheric particles. It is shown that solid–body absorption of radially diffusing ions is a much more efficient process in the inner Saturnian magnetosphere than in the inner Jovian and Uranian magnetospheres because of the near axial symmetry of the planetary magnetic field with respect to the rotational equatorial plane.

091.020   Solar system radio astronomy at low frequencies.
M. D. Desch.

Radio astronomy from space, p. 239 – 253 (1987). – See Abstr. 012.009.

In the past nine years, the two Voyager spacecraft have flown past Jupiter and Saturn, and Voyager 2 recently encountered Uranus. Both spacecraft carry a radio astronomy package, the first radio astronomy experiment to travel to the outer planets. This paper provides a brief survey of the major observations, none of which could have been made from the earth.

091.021   Collisions in the Solar System – IV. Cometary impacts upon the planets.
D. Olsson–Steel.

Mon. Not. R. Astron. Soc., Vol. 227, No. 2, p. 501 – 524 (1987).

The collision probability with each of the planets is calculated for all known comets, and hence the mean collision probability for each planet is determined; from the results the mean cratering rates may be estimated if it is assumed that the presently observed samples are diagnostic of the long–term populations.

091.022   The axial–rotation velocity distribution in the solar system and the sun's 160–min oscillations.
V. A. Kotov.

Bull. Crimean Astrophys. Obs., Vol. 75, p. 83 – 103 (1987). English translation of 42.080.029.

091.023   Fast–ion induced chemical evolution in the outer solar system.
G. Strazzulla.

Adv. Space Res., Vol. 7, No. 5, p. 17 – 21 (1987). – See Abstr. 012.022.

The importance of the role of fast ion–bombardment in the chemical evolution of the surfaces of objects in the outer solar system is made clear by laboratory simulation experiments. The author compares the results of these simulations with recent observations by the Giotto and Voyager spacecraft.

091.024   Dark matter in the solar system.
D. P. Cruikshank.

Adv. Space Res., Vol. 7, No. 5, p. 109 – 120 (1987). – See Abstr. 012.022.

Many small bodies in the solar system, including planetary satellites, comets, and asteroids, have a surface component consisting at least in part of a very low albedo (0.06 or less) solid substance of neutral or red color in the wavelength region $0.3 - 2.5\,\mu m$. Laboratory spectra of organic residues from meteorites and mixtures with hydrous silicates suggest that complex mixtures of complex organic molecules (kerogens) together with aqueous alteration products of igneous minerals may be the source of the dark matter that is distributed so widely throughout the solar system.

091.025   Europa, tidally heated oceans, and habitable zones around giant planets.
R. T. Reynolds, C. P. McKay, J. F. Kasting.

Adv. Space Res., Vol. 7, No. 5, p. 125 – 132 (1987). – See Abstr. 012.022.

Tidal dissipation in the satellites of a giant planet may provide sufficient heating to maintain an environment favorable to life on the satellite surface or just below a thin ice layer. In our own solar system, Europa, one of the Galilean satellites of Jupiter, could have a liquid ocean which may occasionally receive sunlight through cracks in the overlying ice shell. In such a case, sufficient solar energy could reach liquid water that organisms similar to those found under Antarctic ice could grow. In other solar systems, larger satellites with more significant heat flow could represent environments that are stable over an order of Aeons and in which life could perhaps evolve. The authors define a zone around a giant planet in which such satellites could exist as a tidally–heated habitable zone.

091.026   Model of the spectral albedo for atmosphereless cosmic bodies.
Yu. G. Shkuratov.

Kinematika Fiz. Nebesn. Tel, Tom 3, No. 5, p. 39 – 46 (1987). In Russian. English translation in Kinematics Phys. Celest. Bodies.

A new geometro–optic model of spectral albedo for powder surfaces is presented. Formally it is a one–dimensional model. However, it takes into account the specificity of three–dimensions by using the average Fresnel coefficients. Description of the model is given. It is shown that the model is more suitable for surfaces with albedo <10% than Stokes–Bodo's one. The laboratory spectrophotometry data confirming the theoretical model are presented. Applications of the model to interpretation of spectral albedo absorption bands are considered.

091.027   Study of some statistical relationships for polarimetric quantities of atmosphereless bodies.
L. O. Kolokolova.

Kinematika Fiz. Nebesn. Tel, Tom 3, No. 5, p. 47 – 51 (1987). In Russian. English translation in Kinematics Phys. Celest. Bodies.

From strict statistical criterions correlations between characteristics of negative polarization branch are studied for asteroids, satellites and laboratory samples. On the ground of the light–scattering theory for surfaces the causes of correlations are analysed. The relationships between polarimetric quantities and albedo as well as between polarimetric quantities themselves are explained. It is shown that some correlations are caused by surface structure variations. The common notion about structure similarity of all atmosphereless bodies is refuted.

091.028 Die Atmosphären der jupiterartigen Planeten und ihrer großen Satelliten.
C. Hänsel.
Sterne, 63. Band, Heft 3, p. 158 – 171 (1987).

091.029 Planetary magnetism.
C. T. Russell.
Geomagnetism, Vol. 2, p. 457 – 523 (1987). – See Abstr. 003.011.
    Contents: 1. Introduction. 2. Mercury. 3. Venus. 4. Mars.
5. Jupiter. 6. Saturn. 7. Uranus. 8. Neptune and Pluto.
9. Concluding remarks.

091.030 Influence of the azimuthal nonsymmetrical magnetic field of a planet on the density of charged particles in the region of a stationary orbit.
S. V. Smolin.
Magnitosfer. Issled., Moskva, No. 9, p. 100 – 109 (1987). In Russian. Abstr. in Ref. Zh., 51. Astron., 9.51.264 (1987).

091.031 Radio propagation experiments in the outer solar system with Voyager.
G. L. Tyler.
Proc. IEEE, Vol. 75, No. 10, p. 1404 – 1431 (1987).
    Microwave telecommunications transmissions from the two Voyager spacecraft are being used to make detailed studies of planetary atmospheres, rings, and magnetic fields in the outer solar system. Studies have been carried out during the spacecraft encounters of Jupiter, Saturn, and Uranus; observations of Neptune's system are planned for August, 1989. The required occultation geometries are obtained either as by–products of the gravity–assist trajectories employed to travel among the planets or, in some cases, by design. Both spacecraft and ground systems are specifically modified and improved to support radio investigations.

091.032 Isotopic abundances: inferences on solar system and planetary evolution.
G. J. Wasserburg.
Earth Planet. Sci. Lett., Vol. 86, No. 2/4, p. 129 – 173 (1987). = Contrib. Div. Geol. Planet. Sci., Calif. Inst. Technol., No. 4479(562).

091.033 On the origin of magnetic fields of cosmic objects.
K. N. Bystrov.
Mosk. Poligr. Inst., Moskva, 11 pp. (1987). In Russian. Abstr. in Ref. Zh., 51. Astron., 10.51.148 (1987).

091.034 On the origin of magnetic fields of the earth and celestial bodies.
A. Z. Dolginov.
Usp. Fiz. Nauk, Tom 152, No. 2, p. 231 – 262 (1987). In Russian. Abstr. in Ref. Zh., 51. Astron., 10.51.149 (1987).

091.035 Correction to "Approximate methods for finding CO_2 15–μm band transmission in planetary atmospheres" [J. Geophys. Res., Vol. 91, No. D11, p. 11851 – 11866 (1986)].
D. Crisp, S. B. Fels, M. D. Schwarzkopf.
J. Geophys. Res., Vol. 92, No. D10, p. 12021 (1987). See Abstr. 42.091.044.

091.036 Ionospheres of outer planets.
K. K. Mahajan.
Indian J. Radio Space Phys., Vol. 16, No. 1, p. 192 – 206 (1987). Abstr. in Phys. Abstr., Vol. 90, No. 1309, Entry 88172 (1987).

091.037 Small satellites.
P. Farinella.
The evolution of the small bodies of the solar system, p. 276 – 300 (1987). – See Abstr. 012.041.
    Contents: 1. Introduction. 2. Shapes and strengths. 3. Collisional evolution. 4. Origins and thermal histories, 5. Dynamical peculiarities. 6. Some individual satellites (Phobos and Deimos, Amalthea, Adrastea, Metis, Mimas and Enceladus,

Hyperion, Phoebe, F–ring shepherding and co–orbital satellites). 7. Open problems and conclusions.

091.038 Thermal effects of insolation propagation into the regoliths of airless bodies.
R. H. Brown, D. L. Matson.
Icarus, Vol. 72, No. 1, p. 84 – 94 (1987).
    The authors have investigated thermal models for planetary surfaces composed of particles that are bright and optically thin in the visual, and dark and opaque in the thermal infrared. The models incorporate the assumption that insolation is absorbed over a finite distance in the regolith, predicting lower daytime and higher nighttime temperatures than those predicted if the insolation were absorbed only at the surface. The magnitude of the effect depends on the scale length for absorption of insolation relative to the diurnal skin depth for thermal diffusion, and can be significant when insolation penetrates to a depth comparable to the diurnal skin depth.

091.039 Urey Prize Lecture: Chaotic dynamics in the solar system.
J. Wisdom.
Icarus, Vol. 72, No. 2, p. 241 – 275 (1987).
    Newton's equations have chaotic solutions as well as regular solutions. There are several physical situations in the solar system where chaotic solutions of Newton's equations play an important role. There are examples of both chaotic rotation and chaotic orbital evolution. Hyperion is currently tumbling chaotically. Many of the other irregularly shaped satellites in the solar system have had chaotic rotations in the past. This episode of chaotic tumbling could have had a significant effect on the orbital histories of these satellites. Chaotic orbital evolution seems to be an essential ingredient in the explanation of the Kirkwood gaps in the distribution of asteroids. Chaotic trajectories at the 3/1 commensurability have the correct properties to provide a dynamical route for the transport of meteoritic material from the asteroid belt to Earth.

091.040 Eventos planetarios periódicos.
    J. D. Flores Gutiérrez.
Rev. Mex. Astron. Astrofis., Vol. 14, No. 2, p. 643 (1987). Abstract. – See Abstr. 012.042.

091.041 A numerical suimulation of planetary rings. II. Monte Carlo model.
J.–M. Petit, M. Hénon.
Astron. Astrophys., Vol. 188, No. 1, p. 198 – 205 (1987).
    The authors consider a numerical model of planetary rings in which particles interact through mutual attraction and inelastic collisions. Arbitrary mass distributions are allowed. This paper describes the basic algorithm. The authors derive also an approximate equation of evolution, and illustrate the method with a simple example: the radial spreading of an isolated ring.

091.042 Erratum: "Planetary fast neutral emission and effets on the solar wind: a cometary exosphere analog" [Astrophys. J., Vol. 310, No. 2, p. 927 – 936 (1986)].
D. D. Barbosa, A. Eviatar.
Astrophys. J., Vol. 321, No. 2, p. 1049 (1987). See Abstr. 42.091.061.

091.043 Diffuse reflection of sunlight by a convex rotating body with regular surface.
A. P. Mosin.
Astron. Tsirk., No. 1459, p. 3 – 5 (1986). In Russian.
    The geometry of diffuse reflection by a rotating body is considered. Formulas giving the possibility to derive the object configuration from observational data are obtained.

091.044 **On "mirror symmetry" in the C–distribution of minor bodies of the solar system.**
A. K. Terent'eva.
Astron. Tsirk., No. 1472, p. 3 – 5 (1986). In Russian.

The C–distribution (C being the Tisserand's constant for Jupiter as a perturbing planet) for long–periodic (1), parabolic (2) and short–periodic (3) comets, asteroids (4), minor meteor streams (5) and large meteor bodies, including meteorites (6) are presented. For 34 meteorites the orbits were revealed using photographic data, no less than 10% of them would be of cometary origin. A mirror symmetry in C–distributions between populations (3) and (4), (5) and (6) relative to the "gap" opening in collinear libration points L_2 and L_3 has been found. A possible interpretation is given in the following paper.

091.045 **On the origin of minor bodies of the solar system.**
A. K. Terent'eva.
Astron. Tsirk., No. 1472, p. 5 – 6 (1986). In Russian. See Abstr. 091.044.

091.046 **Description and behavior of streamlines in planetary rings.**
N. Borderies, P. Y. Longaretti.
Icarus, Vol. 72, No. 3, p. 593 – 603 (1987).

The dynamical behavior of planetary rings is often described in terms of streamline deformations and motions. It is argued here that the standard formula giving the shape of the streamlines involves geometric elements rather than orbital elements, when the oblateness of the planet is taken into account. The authors relate the geometric elements to the orbital elements and derive the differential equations which govern the variations with time of the geometric elements.

091.047 **Inferred heat transfer mechanisms and tectonic development of planets and satellites.**
A. P. W. Hodder.
Earth, Moon, Planets, Vol. 39, No. 3, p. 237 – 241 (1987).

A schematic diagram showing the relative importance of conduction, convection and hotspots as heat transfer mechanisms on planets has been previously described by Solomon and Head (1982). In their construction they assumed that the majority of heat transfer on Earth involved mantle convection (and hence, plate recycling), with Io and Mercury dominated by hotspot and conduction, respectively. This diagram is here quantified and used to deduce the tectonic regime of Jovian and Saturnian satellites.

091.048 **Gravitational scattering in planetary rings.**
F. Spahn.
Earth, Moon, Planets, Vol. 39, No. 3, p. 243 – 249 (1987).

The gravitational influence of a large body (moonlet; satellite) on the radial structure of planetary rings has been calculated numerically. A drastical change of the surface mass density is obtained even after a single scattering process of the ring–particles on a moonlet (satellite). The final surface density shows a significant radial structure, which has been used to estimate radius and mass of satellites embedded in rings of low optical depth (E–ring, Cassini–division, C–ring of Saturn).

091.049 **Colombo's Top.**
J. Henrard, C. Murigande.
Celest. Mech., Vol. 40, Nos. 3 – 4, p. 345 – 366 (1987).

The authors analyse in detail a simple dynamical system proposed by G. Colombo for the description of the rotational state of planets and satellites. They show that the derivatives of the critical areas are simple analytical functions of the parameters of the problem. These quantities are instrumental in computing the probabilities of capture of the precession of the spin axis in resonance with the precession of the orbit.

091.050 **First order planetary perturbations with elliptic functions.**
C. A. Williams, T. van Flandern, E. A. Wright.
Celest. Mech., Vol. 40, Nos. 3 – 4, p. 367 – 391 (1987).

The differential equations of planetary theory are solved analytically to first order for the two–dimensional case, using only Jacobian elliptic functions and the elliptic integrals of the first and second kind. This choice of functions leads to several new features potentially of importance for planetary theory.

091.051 **Chaotic behaviour in the solar system.**
J. Wisdom.
Proc. R. Soc. London, Ser. A, Vol. 413, No. 1844, p. 109 – 129 (1987). Abstr. in Phys. Abstr., Vol. 90, No. 1317, Entry 139892 (1987).

091.052 **Minor bodies, fragmentation and solar system evolution.**
R. A. Mackenzie.
Meteoros (GB), Vol. 17, No. 4, p. 63 – 69 (1987). Abstr. in Phys. Abstr., Vol. 90, No. 1318, Entry 145842 (1987).

091.053 **The changing shape of planetary rings.**
L. W. Esposito.
Astronomy, Vol. 15, No. 9, p. 6 – 17 (1987). Abstr. in Phys. Abstr., Vol. 91, No. 1319, Entry 4306 (1988).

091.054 **Geometrical insolation of a planet.**
O. Godart.
Ann. Soc. Sci. Brux., Sér. I, Vol. 100, No. 2, p. 53 – 85 (1986).

Analytical expressions for geometrical insolation of an area of a spherical planet for any period of time are given in function of the parameters of the planet, inclination and period of rotation, as well as parameters of its orbit. Certain approximations have been envisaged. This paper could be used for numerical evaluations by introducing numerical values of relevant parameters.

MicroVAX–based data management and reduction system for the regional planetary image facilities.
See Abstr. 002.002.

Planetary nomenclature.
See Abstr. 002.003.

Hvordan får planeter og måner navn?
See Abstr. 002.009.

Theory of planetary atmospheres. An introduction to their physics and chemistry.
See Abstr. 003.037.

Planetary landscapes.
See Abstr. 003.065.

Die Planeten des Sonnensystems.
See Abstr. 003.092.

Interior structure of the Earth and planets.
See Abstr. 003.145.

Application of numerical methods to planetary radiowave scattering.
See Abstr. 021.001.

High pressure cosmochemistry of major planetary interiors: laboratory studies of the water–rich region of the system ammonia–water.
See Abstr. 022.002.

Experiments pertaining to the formation and equilibration of planetary cores. I. High–pressure metallization of FeO and implications for the Earth's core.
See Abstr. 022.003.

Experiments pertaining to the formation and equilibration of planetary cores. II. The melting curve of iron to over 100 GPa and of iron sulfide to over 60 GPa.
See Abstr. 022.004.

Studies of the scattering/absorption properties of minerals.
See Abstr. 022.008.

Mass transport by aeolian saltation on Earth, Mars and Venus: the effects of full saltation cloud development and choking.
See Abstr. 022.013.

Development of wind tunnel techniques for the solution of problems in planetary aeolian processes.
See Abstr. 022.014.

Reynolds number effects on surface shear stress patterns around isolated hemispheres.
See Abstr. 022.016.

Determination of surface shear stress with the naphthalene sublimation technique.
See Abstr. 022.017.

Groundwater sapping channels: summary of effects of experiments with varied stratigraphy.
See Abstr. 022.018.

Creep of ice: further studies.
See Abstr. 022.029.

Diode laser measurements of CO line widths at planetary atmospheric temperatures.
See Abstr. 022.064.

Measurements on 4.7 μm CH_3D lines broadened by H_2 and N_2 at temperatures relevant to planetary atmospheres.
See Abstr. 022.066.

Self- and foreign-gas broadening of ethane lines determined from diode laser measurements at 12 μm.
See Abstr. 022.067.

Hydrogen and nitrogen broadening of the lines of C_2H_2 at 14 μm.
See Abstr. 022.068.

Measurement and analysis of the far infrared absorption spectrum of the gaseous mixture H_2–CH_4.
See Abstr. 022.069.

Laboratory experiments in the study of the chemistry of the outer planets.
See Abstr. 022.072.

Optical efficiencies of lightning in planetary atmospheres.
See Abstr. 022.081.

Electrostatic charge on a dust size distribution in a plasma.
See Abstr. 022.100.

Vapour pressure of amorphous H_2O ice and its astrophysical implications.
See Abstr. 022.142.

Static compression of H_2O–ice to 128 GPa (1.28 Mbar).
See Abstr. 022.143.

Cartography of irregularly shaped satellites.
See Abstr. 036.008.

Methods of remote sensing with automatic interplanetary stations in mapping planets.
See Abstr. 036.048.

The radar–glory theory for icy moons with implications for radar mapping.
See Abstr. 036.059.

Right ascensions of major planets and the Sun obtained at the Tashkent meridian circle in 1978 – 1980.
See Abstr. 041.005.

On the confinement of planetary arcs.
See Abstr. 042.010.

Expansion of the disturbing force–function for the study of high–eccentricity librations.
See Abstr. 042.011.

The motion of a satellite in a ring potential.
See Abstr. 042.077.

Computing ephemerides for the Sun, the Moon and the planets based on modern theories.
See Abstr. 046.020.

Adiabatic consideration of the motion of charged particles in superposed dipole and uniform magnetic fields.
See Abstr. 062.021.

On negative viscosity in global circulations.
See Abstr. 062.062.

Some aspects of shock–wave research.
See Abstr. 062.084.

Electrostatic waves at frequencies close to the harmonics of electron gyrofrequency.
See Abstr. 062.097.

A unified treatment of polarized light emerging from a homogeneous plane–parallel atmosphere.
See Abstr. 063.020.

The adding method for multiple scattering calculations of polarized light.
See Abstr. 063.021.

The Poynting–Robertson effect of a rotating star.
See Abstr. 063.052.

Calculation of the radiation field in a homogeneous anisotropically scattering plane–parallel atmosphere.
See Abstr. 063.071.

Tests of general relativity using radar observations of the inner planets.
See Abstr. 066.095.

Mechanisms and observations for isotope fractionation of molecular species in planetary atmospheres.
See Abstr. 082.069.

The theory of radio windows in planetary magnetospheres.
See Abstr. 084.066.

Implications of convection in the moon and the terrestrial planets.
See Abstr. 094.004.

The probability of falling of Apollo–Amor–type interplanetary bodies on the surface of planets.
See Abstr. 104.078.

A method for calculating the trajectories of meteorites in the atmospheres of planets with ablation and change of the form.
See Abstr. 105.018.

Accumulation and migration of the bodies from the zones of giant planets.
See Abstr. 107.015.

Chemical condensation in the outflowing matter from the proto–Sun and its application to meteorites.
See Abstr. 107.022.

Deuterium fractionation in the presolar nebula: kinetic limitations on surface catalysis.
See Abstr. 107.027.

Evolutionary tracks of the terrestrial planets.
See Abstr. 107.036.

On the gravitational instability in thin gaseous Kepler disks.
See Abstr. 107.038.

Distance law and formation of satellite systems.
See Abstr. 107.039.

The IRAS view of the Galaxy and the solar system.
See Abstr. 155.024.

092 Mercury

092.001   **A preliminary analysis of the Mariner 10 color ratio map of Mercury.**
B. Rava, B. Hapke.
NASA Tech. Memo., NASA TM–89810, p. 229 (1987). Abstract. – See Abstr. 003.001.

092.002   **A statistical study of Mercurian crater classes applied to the emplacement of the intercrater plains.**
A. Woronow, K. Love.
NASA Tech. Memo., NASA TM–89810, p. 408 – 410 (1987). – See Abstr. 003.001.
Studies of crater classes have been few in number and disappointing in value. Perhaps this stems from the difficulties in statistically treating a nonstationary, multivariate data set, or from the propensity to treat all crater data phenomenologically rather than statistically. In either case, recent statistical innovations (Aitchison, 1982) afford a new opportunity to reconsider the value of crater classes.

092.003   **An investigation of the motions of the node and perihelion of Mercury.**
N. C. Rana.
Astron. Astrophys., Vol. 181, No. 1, p. 195 – 202 (1987).
The secular motions of the node and the perihelion of Mercury are calculated from numerically integrated ephemerides, and compared to those already existing obtained from analytical studies and observational analysis. The differences in the analytical rates are found to be negligible compared to $0\overset{''}{.}2$ per century, the estimated error of the present computation, which has been numerical. The classical method of determining the secular motions from the observations has been critically examined to find that, because of the geometry of the orbits, the geocentric angular data of low precision ($\pm 1''$) are incapable of giving the apse rate better than $3''$ per century. Only very accurate range measurements (± 1 km or less) over carefully selected portions of the orbit as well as over a long enough period, say one or two decades, can give the apse rate to an accuracy of about $0\overset{''}{.}1$ per century.

092.004   **The mass, gravity field, and ephemeris of Mercury.**
J. D. Anderson, G. Colombo, P. B. Esposito, E. L. Lau, G. B. Trager.
Icarus, Vol. 71, No. 3, p. 337 – 349 (1987). Paper presented at the Tucson Mercury Conference, held at Tucson, Ariz., USA, 6 – 9 August 1986.
This paper represents a final report on the gravity analysis of radio Doppler and range data generated by the Deep Space Network (DSN) with Mariner 10 during two of its encounters with Mercury in March 1974 and March 1975. A combined least–squares fit to Doppler data from both encounters has resulted in a determination of two second degree gravity harmonics, $J_2 = (6.0 \pm 2.0) \times 10^{-5}$ and $C_{22} = (1.0 \pm 0.5) \times 10^{-5}$ referred to an equatorial radius of 2439 km, plus an indication of a gravity anomaly in the region of closest approach of Mariner 10 to Mercury in March 1975 amounting to a mass deficiency of about $GM = -0.1$ km^3 sec^{-2}. The authors obtain a mass $GM = 22,032.09 \pm 0.91$ km^3 sec^{-2} or a Sun to Mercury mass ratio of $6,023,600 \pm 250$. The corresponding mean density of Mercury is 5.43 ± 0.01 g cm^{-3}. A discussion of the utility of the DSN radio range data obtained with Mariner 10 is included. These data are most applicable to the improvement of the ephemeris of Mercury, in particular the determination of the precession of the perihelion.

092.005   **The chronology of Mercury's geological and geophysical evolution: the vulcanoid hypothesis.**
M. A. Leake, C. R. Chapman, S. J. Weidenschilling, D. R. Davis, R. Greenberg.
Icarus, Vol. 71, No. 3, p. 350 – 375 (1987). Paper presented at the Tucson Mercury Conference, held at Tucson, Ariz., USA, 6 – 9 August 1986.
The authors have investigated whether constraints on Mercury's chronology could be relaxed by effects of a Mercury–specific bombarding population of planetesimals interior to its orbit, encountering the planet only occasionally due to secular perturbations. Such "vulcanoids" could have been a significant source of early cratering. However, those in orbits that can cross Mercury's are depleted by mutual collisions in $\lesssim 1$ Gyr, and can provide at most a modest extension of the period of heavy bombardment. Further inside Mercury's orbit, lower collisional velocities might allow survival of vulcanoids to the present. The authors report on a search for such bodies and on observational limits to such a population. They also review evidence that Mercury's intercrater plains are of volcanic origin and mainly predate Caloris, and that scarp formation (and global contraction) mainly postdates Caloris and has continued to recent times.

092.006   **Mercurian crater–filling classes constrain the emplacement process of the intercrater plains material.**
A. Woronow, K. M. Love.
Icarus, Vol. 71, No. 3, p. 376 – 385 (1987). Paper presented at the Tucson Mercury Conference, held at Tucson, Ariz., USA, 6 – 9 August 1986.
A multivariate analysis of crater–filling classes for craters larger than 10 km diameter on Mercury constrains the process of intercrater plains emplacement to have been one that affected both the intercrater plains and the densely cratered terrain in a similar manner. Any emplacement process considerably restricted in time, such as ejecta from a single basin or an episode of volcanic eruptions that was brief compared to the time span of the late heavy bombardment, violates the observations.

092.007 Mercury: wavelength and longitude dependence of polarization.
T. Gehrels, R. Landau, G. V. Coyne.
Icarus, Vol. 71, No. 3, p. 386 – 396 (1987). Paper presented at the Tucson Mercury Conference, held at Tucson, Ariz., USA, 6 – 9 August 1986.

Linear polarization was observed on the integrated disk of Mercury with seven filters between 0.3 and 1.0 μm, and between 53° and 130° phase angle. Polarization–time variations were found and they are mostly explained by longitude dependence through variation in brightness or other properties over the surface. The polarization–wavelength dependence is flatter than for the Moon, and the polarization–albedo relation also differs, indicating a difference in surface composition and/or texture.

092.008 An analysis of the Mariner 10 color ratio map of Mercury.
B. Rava, B. Hapke.
Icarus, Vol. 71, No. 3, p. 397 – 429 (1987). Paper presented at the Tucson Mercury Conference, held at Tucson, Ariz., USA, 6 – 9 August 1986.

The results of a geological analysis of the Mariner 10 orange/UV color ratio map of Mercury (B. Hapke et al., 1980) are given. Certain errors that occurred in reproducing the published version of the 1980 map are pointed out. The relationships between color and terrain are distinctly nonlunar.

092.009 Magnetosphere, exosphere, and surface of Mercury.
A. F. Cheng, R. E. Johnson, S. M. Krimigis, L. J. Lanzerotti.
Icarus, Vol. 71, No. 3, p. 430 – 440 (1987). Paper presented at the Tucson Mercury Conference, held at Tucson, Ariz., USA, 6 – 9 August 1986.

The discovery of an atomic sodium exosphere at Mercury raises the question of whether Mercury, like Io at Jupiter, can maintain a heavy ion magnetosphere. The authors suggest that it does, and that heavy ions (mainly Na^+) from the exosphere are typically accelerated to keV energies and make important or dominant contributions to the mass (~ 300 g sec^{-1}) and energy ($\sim 3 \times 10^9$W) budgets of the magnetosphere. The sodium supply to the exosphere is largely from within Mercury itself, with external sources like meteoroid infall and the solar wind being relatively unimportant. Therefore Mercury is in the process of losing its semivolatiles. Photosputtering dominates charged particle sputtering and can maintain an adequate rate of Na ejection from the surface.

092.010 Dynamics of electrons and heavy ions in Mercury's magnetosphere.
W.-H. Ip.
Icarus, Vol. 71, No. 3, p. 441 – 447 (1987). Paper presented at the Tucson Mercury Conference, held at Tucson, Ariz., USA, 6 – 9 August 1986.

The possibility exists that plasma–surface interaction could be essential in forming a "metallic" magnetosphere at Mercury. It is thus of interest to examine some of the dynamical properties of the exospheric ions in the magnetospheric environment. The author considers the basic configuration of Mercury's magnetosphere as a scaled–down model of the Earth's magnetosphere. In particular, the global electrostatic potential distribution and particle drift pattern are reviewed from the point of view of comparative study of these two magnetospheric systems. The finite gyroradius effect on the trajectories of heavy ions is examined. A few examples are used to illustrate the potential importance of direct electric field acceleration of the Na^+ ions on the loss and recirculation of the exospheric sodium and potassium atoms.

092.011 A comparison of the Mercury and Earth magnetospheres: electron measurements and substorm time scales.
S. P. Christon.
Icarus, Vol. 71, No. 3, p. 448 – 471 (1987). Paper presented at the Tucson Mercury Conference, held at Tucson, Ariz., USA, 6 – 9 August 1986.

This paper contains both a review of relevant information from the first Mariner 10 encounter with Mercury (Mercury 1) and analyses which utilize some aspects of the Mercury 1 measurements. These data (about a half hour in length), along with the Mercury 3 data (16 March 1975), may be the most important magnetospheric data yet returned from a planetary mission, solely because they depict a magnetosphere which appears more similar to Earth's than any other in the solar system. The author shows the similarity extends from microscopic (quiet–time plasmas) to macroscopic (substorm activity) levels. The analysis sections deal with the Mercury 1 measurements in some detail. The last section develops a perspective of the importance of the similarity through comparison of substorm time scales in the Earth and Mercury magnetospheres.

092.012 Variation of sodium on Mercury with solar radiation pressure.
A. E. Potter, T. H. Morgan.
Icarus, Vol. 71, No. 3, p. 472 – 477 (1987). Paper presented at the Tucson Mercury Conference, held at Tucson, Ariz., USA, 6 – 9 August 1986.

Sodiums atoms in the atmosphere of Mercury can be accelerated by solar radiation pressure, and several authors have suggested that radiation pressure could sweep sodium off the planet. As a consequence, the sodium abundance might be expected to decrease as the radiation pressure increases. The authors have measured the average sodium abundance over a range of solar radiation pressures and found that the sodium abundance does decrease with increasing radiation pressure. Possible explanations for the observed variation are (1) that radiation pressure sweeps away transient high–velocity sodium atoms generated upon meteoric material impacts, thus reducing the supply rate of sodium, or (2) that the accommodation coefficient of sodium for surface interactions is less than unity, so that radiation pressure can effectively push sodium to the dark side of the planet, where it cannot be detected by scattered sunlight.

092.013 On the atmosphere of Mercury.
R. I. Kiladze.
Soobshch. Akad. Nauk GSSR, Tom 126, No. 3, p. 529 – 531 (1987). In Russian. Abstr. in Ref. Zh., 51. Astron., 10.51.167 (1987).

092.014 Radio–interferometric imaging of the subsurface emissions from the planet Mercury.
J. O. Burns, G. R. Gisler, J. E. Borovsky, D. N. Baker, M. Zeilik.
Nature, Vol. 329, No. 6136, p. 224 – 226 (1987).

The VLA was used to make the first centimetre–wavelength interferometric observations that resolve Mercury's disk. The authors have mapped the distribution of total and polarized intensities from the planet's subsurface layers. They report the first detection of a hot pole along the hermean equator, which they model as black–body reradiation from preferential diurnal heating. These observations appear to rule out any internal sources of heat within Mercury. Polarized emission from the limb of the planet, which is understood in terms of the dielectric properties of the hermean surface is found.

092.015 Heat conduction limits on calorimetric effects at Mercury due to solar wind–magnetosphere interactions.
V. M. Vasyliunas.
J. Geophys. Res., Vol. 92, No. A12, p. 13,658 – 13,660 (1987).

Heat produced at the surface of Mercury by the highly time–variable precipitation of charged particles need not be radiated away immediately but can also be conducted into the interior of the planet. For a given precipitated energy flux density, the rise of surface temperature can be computed taking both heat conduction and radiation into account. When the energy input varies on time scales shorter than a characteristic period, estimated to be one (terrestrial) day for conditions typical of Mercury's dark side, heat conduction dominates over radiation and the predicted surface temperature rise becomes negligibly small.

092.016 **Mercury's magnetospheric irradiation effect on the surface.**
W.-H. Ip.
Geophys. Res. Lett., Vol. 14, No. 12, p. 1191 – 1194 (1987).

Using numerical model calculations, the author investigates the transient heating effect from magnetospheric charged particle precipitation as proposed by Baker et al. (1987). Under extreme circumstances, the transient temperature increase may reach 1K if the heating duration lasts more than one minute and the energy influx reaches 10^4ergs cm^{-2}s^{-1}. Otherwise, the expected temperature rise should be substantially smaller ($\Delta T < 0.05$K). From this point of view, optical observations capable of differentiating various degrees of particle irradiation effect might be more appropriate to pinpoint signature of Mercury's auroral zones.

092.017 **The formation of Mercury's smooth plains.**
W. S. Kiefer, B. C. Murray.
Icarus, Vol. 72, No. 3, p. 477 – 491 (1987). = Calif. Inst. Technol., Div. Geol. Planet. Contrib. No. 4417.

There has been extensive debate about whether Mercury's smooth plains are volcanic features or impact ejecta deposits. The authors present new indirect evidence which supports a volcanic origin for two different smooth plains units. In Borealis Planitia, stratigraphic relations indicate at least two distinct stages of smooth plains formation. At least one of these stages must have had a volcanic origin. In the Hilly and Lineated Terrain, Petrarch and several other anomalously shallow craters apparently have been volcanically filled. Areally extensive smooth plains volcanism evidently occurred at these two widely separated areas on Mercury. These results, combined with work by other researchers on the circum–Caloris plains and the Tolstoj basin, show that smooth plains volcanism was a global process on Mercury.

The Rutherford cross section and the perihelion shift of Mercury with the Runge–Lenz vector.
See Abstr. 014.065.

Determination of right ascensions of the Sun, Venus and Mercury in Tashkent in the period 1981 – 1983.
See Abstr. 041.006.

Right ascensions of the Sun, Mercury and Venus observed at the transit instrument in Nikolaev in 1985.
See Abstr. 041.020.

Comment on "Mercury's precession according to special relativity" [Am. J. Phys., Vol. 54, No. 3, p. 245 – 247 (1986)] (and reply).
See Abstr. 066.159.

Basin–ring spacing on the Moon, Mercury, and Mars.
See Abstr. 094.023.

The investigation of intersecting craters on the Moon, Mars and Mercury.
See Abstr. 094.041.

Appulses and occultations of SAO stars by Mercury: 1987 – 1995.
See Abstr. 096.008.

093 Venus

093.001 **Gravity data analysis.**
W. L. Sjogren.
NASA Tech. Memo., NASA TM–89810, p. 144 – 146 (1987). – See Abstr. 003.001.

093.002 **Oxidized basalts on the surface of Venus: compositional implications of measured spectral properties.**
C. M. Pieters, W. Patterson, S. Pratt, J. W. Head, J. Garvin.
NASA Tech. Memo., NASA TM–89810, p. 165 – 166 (1987). – See Abstr. 003.001.

093.003 **Crater production on Venus and Earth by asteroid and comet impact.**
E. M. Shoemaker, R. F. Wolfe.
NASA Tech. Memo., NASA TM–89810, p. 402 – 404 (1987). – See Abstr. 003.001.

The authors have carried out new calculations of the collision probabilities of asteroids and comets with Venus based on the orbits of the known Venus–crossing asteroids and comets. For comparison, they recalculated asteroid and comet collision probabilities and cratering rates on the Earth and Moon and normalized the estimated cratering rates on Venus to those of the Earth.

093.004 **The surface age of Venus: applying the terrestrial cratering rate.**
G. G. Schaber, E. M. Shoemaker, R. C. Kozak.
NASA Tech. Memo., NASA TM–89810, p. 405 – 407 (1987). – See Abstr. 003.001.

The population of Venusian craters having suspected impact–crater morphology has been reported from 115×10^6km^2 of the northern hemisphere of the planet. They estimated the average age of the surface to be approximately 1 b.y. (± 0.5 b.y.) on the basis of lunar crater–production curves corrected for Venus. Such an old average age is somewhat difficult to reconcile with the similarity in size and mass of Venus and Earth and with Earth's high heat flow and crustal resurfacing rate.

093.005 **Comments on the tectonism of Venus.**
R. C. Kozak, G. G. Schaber.
NASA Tech. Memo., NASA TM–89810, p. 443 – 445 (1987). – See Abstr. 003.001.

093.006 **Lithospheric structure on Venus from tectonic modelling of compressional features.**
W. B. Banerdt, M. P. Golombek.
NASA Tech. Memo., NASA TM–89810, p. 446 – 448 (1987). – See Abstr. 003.001.

093.007 **Venusian tectonics: convective coupling to the lithosphere?**
R. J. Phillips.
NASA Tech. Memo., NASA TM–89810, p. 449 – 451 (1987). – See Abstr. 003.001.

093.008 **Bilateral symmetry across Aphrodite Terra.**
L. S. Crumpler, J. W. Head, D. B. Campbell.
NASA Tech. Memo., NASA TM–89810, p. 452 – 454 (1987). – See Abstr. 003.001.

093.009 **Steady state flow/field model of solar wind interaction with Venus: global implications of local effects.**
P. A. Cloutier, H. A. Taylor Jr., J. E. McGary.
J. Geophys. Res., Vol. 92, No. A7, p. 7289 – 7307 (1987).

The structure and variability of ion concentrations, velocities, and electric and magnetic fields in the ionosphere of Venus result from a combination of physical and chemical processes driven by the interaction of the solar wind with the atmosphere of Venus. In this paper the authors describe these processes and formulate a model describing their interrelationships in a self–consistent manner. They show that the consequences of this model are in agreement with Pioneer Venus observations in the Venus ionosphere.

093.010 Frequency functions of Venus nightside ion densities.
W. C. Knudsen.
J. Geophys. Res., Vol. 92, No. A7, p. 7308 – 7316 (1987). With a correction in Vol. 92, No. A11, p. 12469 (1987).

The variability of the nightside Venus ionospheric density and the cause or causes of the variability have been subjects of substantial interest since the first measurements of the nightside ionosphere. The purpose of the study reported in this paper is to define the variability of the total and major ion densities by deriving an experimental frequency function for these quantities and, to a limited extent, to investigate the relationship between the frequency functions required or suggested by the present ideas of the chemistry and ionization sources responsible for the nightside ionosphere.

093.011 A one–dimensional time–dependent model of the magnetized ionosphere of Venus.
H. Shinagawa, T. E. Cravens, A. F. Nagy.
J. Geophys. Res., Vol. 92, No. A7, p. 7317 – 7330 (1987).

The behavior and time evolution of the large–scale magnetic fields and ionospheric plasma of the dayside Venus ionosphere were studied using a one–dimensional model. The coupled continuity, momentum, and Maxwell's equations were solved simultaneously for three ions (O^+, O_2^+, H^+) and the magnetic field. The CO_2^+ ion was included photochemically. The calculated magnetic field profiles are in good agreement with observations made by the Pioneer Venus orbiter magnetometer. Good agreement was also obtained between the calculated and observed plasma densities for altitudes below 250 km including the electron density "ledge" near 190 km in magnetized ionospheres. However, the significant disagreement between the calculated and observed plasma densities at high altitudes suggests that under magnetized conditions, horizontal transport processes become important in removing the plasma and/or the magnetic field even in the subsolar region.

093.012 Characteristics of the Marslike limit of the Venus–solar wind interaction.
J. G. Luhmann, C. T. Russell, F. L. Scarf, L. H. Brace, W. C. Knudsen.
J. Geophys. Res., Vol. 92, No. A8, p. 8545 – 8557 (1987).

Many authors have already noted that Mars' interaction with the solar wind may be like Venus', e.g., that of a supermagnetosonic plasma flowing past an effectively unmagnetized body having an ionosphere. However, at Mars, the incident solar wind dynamic pressure usually exceeds the peak ionospheric plasma pressure, while at Venus this condition prevails only when the incident dynamic pressure is extraordinarily high. With the aim of predicting what might be expected at Mars, this study examines the subset of Pioneer Venus Orbiter observations obtained during intervals of extremely high solar wind dynamic pressure.

093.013 Color inhomogeneity on Venus: preliminary analysis of Venera 13, 14 television imagery.
Yu. G. Shkuratov, M. A. Kreslavskij, A. T. Bazilevskij.
Sov. Astron. Lett., Vol. 12, No. 5, p. 332 – 334 (1987). English translation of 42.093.015.

093.014 Ring–shaped formations on the Venus plains as witnesses of its geological history.
O. V. Nikolaeva, L. B. Ronka, A. T. Bazilevskij.
Geokhimiya, No. 5, p. 579 – 589 (1986). In Russian. Abstr. in Ref. Zh., 51. Astron., 7.51.165 (1987).

093.015 Mass spectrometer for investigation of the chemical composition of the Venus atmosphere.
Eh. P. Sheretov.
Metod. apparatura analit. veshchestva dlya kosm. issled., Ryazan', p. 5 – 12 (1986). In Russian. Abstr. in Ref. Zh., 62. Issled. Kosm. Prostranstva, 7.62.174 (1987).

093.016 Numerical experiments with the model "four–day circulation" of the Venus atmosphere.
V. G. Vasin.
Mat. zadachi priklad. aehron., Moskva, p. 155 – 168 (1987). In Russian. Abstr. in Ref. Zh., 62. Issled. Kosm. Prostranstva, 7.62.519 (1987).

093.017 Radar images of Venus (Results of Venera–15 and Venera–16 space probes).
Yu. N. Aleksandrov, A. T. Bazilevskij, V. A. Kotel'nikov, G. M. Petrov, O. N. Rzhiga, A. I. Sidorenko.
Itogi Nauki i Tekhniki. Seriya Astronomiya. Tom 32. Astrophysics and space physics, p. 201 – 234 (1987). In Russian. – See Abstr. 003.004.

093.018 Venus "lightning" signals reinterpreted as in situ plasma noise.
H. A. Taylor Jr., P. A. Cloutier, Z. Zheng.
J. Geophys. Res., Vol. 92, No. A9, p. 9907 – 9919 (1987).

The authors have examined details of the Pioneer Venus Orbiter (PVO) measurements in the nightside ionosphere of Venus for evidence bearing on the interpretation that 100–Hz electric field noise bursts provide evidence for lightning in the lower atmosphere. They find that the altitude and propagating characteristics of the electric field transients plus their association with and localization within ion density troughs are consistent with the interpretation that the 100–Hz noise is the signature of local plasma perturbations which generate plasma waves near the PVO.

093.019 Finite Larmor radius effect on ion pickup at Venus.
J. L. Phillips, J. G. Luhmann, C. T. Russell, K. R. Moore.
J. Geophys. Res., Vol. 92, No. A9, p. 9920 – 9930 (1987).

The interaction of the solar wind with Venus is influenced by the pickup of newly born exospheric oxygen ions by the convecting magnetosheath plasma. The flow and field configuration of the magnetosheath plasma, together with the large gyroradius of the pickup ions, cause mass loading to occur preferentially on one side of the magnetosheath. The observed hemispherical asymmetry in the magnetic field in the near–planet magnetosheath, attributed to this pickup process, is confirmed by direct observation of the picked–up planetary particles. Test particle calculations show that a current system created by ion pickup has the appropriate location and magnitude to account for the magnetic field asymmetry. The results indicate that a fluid treatment of the Venus mass–loading problem is not entirely appropriate; a hybrid or kinetic model is necessary to incorporate the finite Larmor radius of the pickup particles which produces the observed asymmetry.

093.020 Magnetotails at unmagnetized bodies: comparison of comet Giacobini–Zinner and Venus.
D. J. McComas, J. T. Gosling, C. T. Russell, J. A. Slavin.
J. Geophys. Res., Vol. 92, No. A9, p. 10,111 – 10,117 (1987).

Both comet Giacobini–Zinner (G–Z) and Venus have magnetotails consisting of draped interplanetary magnetic field lines. This field line draping is caused by a velocity shear between regions of greater flow speeds away from the bodies and lesser flow speeds near to the bodies. Data obtained within the Venus magnetotail by the Pioneer Venus Orbiter and within the G–Z tail by the International Cometary Explorer traversal of G–Z have previously been combined with stress balance considerations to infer many of the physical characteristics of these two magnetotails. In the present paper the authors compare and contrast these physical characteristics and thereby examine those aspects of the interactions with the solar wind and draped magnetotail forming processes which are common at the two bodies, and those which are different.

093.021 The face of Venus.
O. N. Rzhiga, Yu. N. Aleksandrov, A. I. Sidorenko, A. T. Bazilevskij.
Nauka v SSSR, No. 2, p. 30 – 38 (1987). In Russian. Abstr. in Ref. Zh., 51. Astron., 8.51.125 (1987).

093.022 On the chemical composition of the atmosphere and surface of Venus.
V. I. Chesnokov.
Fiz.–Tekh. Inst. Akad. Nauk SSSR, Prepr., No. 1099, p. 3 – 9 (1987). In Russian. Abstr. in Ref. Zh., 62. Issled. Kosm. Prostranstva, 8.62.461 (1987).

093.023 Some problems of the solar wind interaction with Venus.
T. K. Breus, A. M. Krymskii (*A. M. Krymskij*).
Planet. Space Sci., Vol. 35, No. 9, p. 1213 – 1219 (1987).
The aim of this paper is to analyze the effect of solar wind mass–loading due to hot–oxygen Venus corona photoionization on the plasma flow parameters in the nose part of the magnetosheath and the flow stability, taking into consideration the axial symmetry of the flow.

093.024 Interaction of mass–loaded solar wind flow with a blunt body.
T. K. Breus, A. M. Krymskii (*A. M. Krymskij*), V. Ya. Mitnitskii (*V. Ya. Mitnitskij*).
Planet. Space Sci., Vol. 35, No. 9, p. 1221 – 1227 (1987).
The aim of this paper is the numerical modeling of the solar wind interaction with Venus taking into account the mass loading effect due to the photoionization of the Venus neutral oxygen corona. The analysis has shown that this effect unambiguously explains the number of peculiarities of the SW–Venus interaction pattern that could not be quantitatively explained before, namely the shock front position, and the characteristics of the SW flow and magnetic field in the Venus ionosheath observed from experiments onboard of Venera–9 and –10 and Pioneer–Venus spacecraft.

093.025 Laying bare Venus' dark secrets.
D. A. Allen.
Sky Telesc., Vol. 74, No. 4, p. 350 – 353 (1987).

093.026 Comment on "On the response of ionospheric magnetization to solar wind dynamic pressure from Pioneer Venus measurements" by J. Kar, K. K. Mahajan [Geophys. Res. Lett., Vol. 14, No. 5, p. 507 – 510 (1987)].
J. L. Phillips.
Geophys. Res. Lett., Vol. 14, No. 10, p. 1068 – 1069 (1987). – See Abstr. 43.093.055.

093.027 Reply to the comment on "On the response of ionospheric magnetization to solar wind dynamic pressure from Pioneer Venus measurements" by J. L. Phillips.
J. Kar, K. K. Mahajan.
Geophys. Res. Lett., Vol. 14, No. 10, p. 1070 – 1071 (1987). – See Abstr. 093.026.

093.028 Correction to "On the response of ionospheric magnetisation to solar wind dynamic pressure from Pioneer Venus measurements" [Geophys. Res. Lett., Vol. 14, No. 5, p. 507 – 510 (1987)].
J. Kar, K. K. Mahajan.
Geophys. Res. Lett., Vol. 14, No. 10, p. 1072 (1987). – See Abstr. 43.093.055.

093.029 Waves in the atmosphere of Venus.
P. J. Gierasch.
Nature, Vol. 328, No. 6130, p. 510 – 512 (1987).
The recent Soviet/French instrumented balloon experiments in the Venus atmosphere have shown that turbulent vertical motions with amplitudes of several metres per second exist within the principal cloud deck at about 50 km elevation. The motions are particularly intense over the Aphrodite mountains. The author points out that a patch of turbulence stationary over a surface feature can generate a wake–like pattern of horizontally propagating waves which can explain the general visual appearance of Venus, and which can produce a momentum exchange from mid–latitudes to low latitudes, helping to maintain the Venus atmospheric rotation.

093.030 Making a radar map of Venus.
Yu. N. Aleksandrov, V. M. Dubrovin, A. I. Zakharov, V. A. Kotel'nikov, A. A. Krymov, G. M. Petrov, O. N. Rzhiga, A. I. Sidorenko, V. P. Sinilo, G. A. Sokolov.
Probl. sovrem. radiotekh. i ehlektron. Moskva, p. 46 – 69 (1987). In Russian. Abstr. in Ref. Zh., 62. Issled. Kosm. Prostranstva, 9.62.535 (1987).

093.031 Analysis of distribution of impact craters on the Venus surface (according to photos of Venera 15 and 16).
V. P. Kryuchkov.
Izv. Akad. Nauk SSSR. Ser. Geol., No. 6, p. 75 – 83 (1987). In Russian. Abstr. in Ref. Zh., 62. Issled. Kosm. Prostranstva, 9.62.537 (1987).

093.032 Determination of the ages of craters on Venus.
W. K. Hartmann.
Izv. Akad. Nauk SSSR. Ser. Geol., No. 6, p. 67 – 74 (1987). In Russian. Abstr. in Ref. Zh., 62. Issled. Kosm. Prostranstva, 9.62.538 (1987).

093.033 On the chemical composition of the atmosphere and surface of Venus.
V . I. Chesnokov.
Fiz.–Tekhnol. Inst. Akad. Nauk SSSR, Prepr., No. 1099, p. 3 – 9 (1987). In Russian. Abstr. in Ref. Zh., 51. Astron., 9.51.231 (1987).

093.034 Hydrogen dissipation from the Venus atmosphere.
M. I. Pudovkin, I. V. Golovchanskaya.
Magnitosfer. Issled., Moskva, No. 9, p. 94 – 99 (1987). In Russian. Abstr. in Ref. Zh., 51. Astron., 9.51.238 (1987).

093.035 The 1983 – 84 and 1985 – 86 western (morning) apparitions of Venus: visual and photographic observations.
J. L. Benton Jr.
Strolling Astron., Vol. 32, Nos. 5 – 6, p. 93 – 101 (1987).
Visual and photographic observations of the 1983 – 84 and 1985 – 86 western (morning) apparitions of the planet Venus are summarized, emphasizing the sources of data and the instruments employed in the observation of this planet. There is a statistical analysis of the categories of features on the apparent surface, seen or suspected in visual wavelengths, for both observing periods.

093.036 Was Venus wet? Deuterium reconsidered.
D. H. Grinspoon.
Science, Vol. 238, No. 4834, p. 1702 – 1704 (1987).
The ratio of deuterium to hydrogen on Venus has been accepted as proof of a wetter, more Earth–like past on that planet. However, the present–day water abundance and the nonthermal hydrogen escape flux on Venus imply that hydrogen is in a steady state and that a hydrogen source, most likely cometary infall, is present. An alternative interpretation of the D/H ratio is offered, in which the measured value is consistent with a steady–state evolution over the age of the solar system. No past water excess is required to explain the isotopic data.

093.037 Compilation of a preliminary catalogue of coordinates of reference points in the region of the Maxwell Mountains on Venus.
E. G. Belen'kij, L. M. Kadnichanskaya, T. A. Kamenetskaya, S. E. Reshetova.
Mater. 18 nauchn.–tekh. konf. mol. uchenykh i spets. TsNII geod., aehrosemki i kartogr. Moskva, 28 – 29 apr., 1986. Moskva, p. 167 – 183 (1987). In Russian. Abstr. in Ref. Zh., 52. Geod. Aehrosemka, 10.52.220 (1987).

093.038 Peculiarities of the image of the relief and problems of generalization when mapping the Venus with a 1:5,000,000 scale taking the region of the Maxwell Mountains as an example.
I. B. Korobova.
Mater. 18 nauchn.–tekh. konf. mol. uchenykh i spets. TsNII geod., aehrosemki i kartogr. Moskva, 28 – 29 apr., 1986. Moskva, p. 249 – 254 (1987). In Russian. Abstr. in Ref. Zh., 52. Geod. Aehrosemka, 10.52.221 (1987).

093.039 On mechanisms of formation of the cloud layer of the Venus atmosphere.
Yu. V. Zhulanov, L. M. Mukhin, D. F. Nenarokov.
Dokl. Akad. Nauk SSSR. Ser. Mat. Fiz., Tom 295, No. 2, p. 330 – 334 (1987). In Russian. Abstr. in Ref. Zh., 62. Issled. Kosm. Prostranstva, 10.62.375 (1987).

093.040 Spectra of particle sizes of the cloud layer of the Venus atmosphere (experiment VEGA).
Yu. V. Zhulanov, L. M. Mukhin, D. F. Nenarokov, A. A. Lushnikov.
Dokl. Akad. Nauk SSSR. Ser. Mat. Fiz., Tom 295, No. 1, p. 67 – 70 (1987). In Russian. Abstr. in Ref. Zh., 62. Issled. Kosm. Prostranstva, 10.62.376 (1987).

093.041 Distribution of whistler mode bursts at Venus.
F. L. Scarf, K. F. Jordan, C. T. Russell.
J. Geophys. Res., Vol. 92, No. A11, p. 12407 – 12411 (1987).
Several thousand impulsive whistler mode noise bursts were detected by the Pioneer Venus wave instrument during the first 10 seasons with nightside traversals at low altitudes. The authors study the occurrence rate for these events as a function of spacecraft altitude and position.

093.042 Venus gravity: a harmonic analysis.
B. G. Bills, W. S. Kiefer, R. L. Jones.
J. Geophys. Res., Vol. 92, No. B10, p. 10335 – 10351 (1987).
An improved model of Venusian global gravity has been obtained by fitting an eighteenth–degree and eighteenth–order spherical harmonic series to 78 orbital arcs of high altitude (950 – 1350 km at periapsis) tracking data and 351 orbital arcs of lower–altitude (150 – 200 km at periapsis) data from the Pioneer Venus Orbiter.

093.043 Impact craters of Venus: a continuation of the analysis of data from the Venera 15 and 16 spacecraft.
A. T. Basilevsky (A. T. Bazilevskij), B. A. Ivanov, G. A. Burba, I. M. Chernaya, V. P. Kryuchkov, O. V. Nikolaeva, D. B. Campbell, L. B. Ronca.
J. Geophys. Res., Vol. 92, No. B12, p. 12869 – 12901 (1987).

093.044 An approximate analytical method for calculating tides in the atmosphere of Venus.
S. B. Fels.
J. Atmos. Sci., Vol. 43, No. 23, p. 2757 – 2772 (1986). Abstr. in Phys. Abstr., Vol. 90, No. 1310, Entry 94680 (1987).

093.045 Dissipative tides: application to Venus' lower atmosphere.
C. Covey, R. L. Walterscheid, G. Schubert.
J. Atmos. Sci., Vol. 43, No. 24, p. 3273 – 3278 (1986). Abstr. in Phys. Abstr., Vol. 90, No. 1311, Entry 101712 (1987).

093.046 Geological–morphological description of the Laima Tessera, Tellus Regio and Leda Planitia (photomap of the Venus surface, Sheet B–13).
A. L. Sukhanov, N. N. Bobina, G. A. Burba, Yu. S. Tyuflin, M. V. Ostrovskij, L. S. Ledovskaya, V. A. Kotel'nikov, O. N. Rzhiga, A. I. Sidorenko, Yu. N. Aleksandrov, G. M. Petrov, A. I. Zakharov, A. P. Krivtsov.
Astron. Vestn., Tom 21, No. 3, p. 195 – 206 (1987). In Russian. English translation in Sol. Syst. Res.
The central part of the area is occupied by Leda Planitia. Its formation may be explained as the result of plume ascent from planetary interiors and its lateral spreading. Systems of areal dislocations of several directions ("parquet") around Leda Planitia seem to be originated from the material moved downslope as viscous flows (Laima and Dekla Tesserae, Tellus Regio). Belts of ridges and furrows (Kamari Dorsa and others) were formed along the edges of the "parquet" areas above the astenospheric currents moving downwards.

093.047 Geological–morphological description of Tellus Regio (photomap of the Venus surface, Sheet B–24).
M. S. Markov, V. P. Shashkina, G. A. Burba, Yu. S. Tyuflin, M. V. Ostrovskij, V. A. Kotel'nikov, O. N. Rzhiga, G. M. Petrov, Yu. N. Aleksandrov, A. I. Sidorenko, V. P. Sinilo, N. V. Rodionova.
Astron. Vestn., Tom 21, No. 4, p. 286 – 297 (1987). In Russian. English translation in Sol. Syst. Res.
The systems of areal dislocations of several directions ("parquet") at the highland of Tellus Regio are located in the central and eastern parts of the area. The eastern border of the area is occupied with Niobe Planitia, and western part of the area contains Leda Planitia, which have a number of linear topographic uplifts elongated in NE direction (Mardezh–ava Dorsa).

093.048 Patterns of spatial distribution of craters in the northern hemisphere of Venus.
A. E. Altynov.
Astron. Vestn., Tom 21, No. 4, p. 335 – 339 (1987). In Russian. English translation in Sol. Syst. Res.

093.049 The reversal of the spin of Venus.
J. R. Dormand, J. McCue.
J. Br. Astron. Assoc., Vol. 98, No. 1, p. 23 – 25 (1987).
A collision between an asteroid–sized body and Venus during the final stage of planetary formation may have been responsible for the reversal of the spin of Venus.

093.050 Solar cycle changes in the ionization sources of the nightside Venus ionosphere.
W. C. Knudsen, A. J. Kliore, R. C. Whitten.
J. Geophys. Res., Vol. 92, No. A12, p. 13,391 – 13,398 (1987).
The authors argue, using new and previous experimental evidence, that the height–integrated flux of O^+ ions transported from the dayside Venus ionosphere into the nightside ionosphere is strongly reduced (shut off) at solar cycle minimum and that the electron impact source becomes the predominant nightside ionization source during this phase of the solar cycle.

093.051 The $^{12}C/^{13}C$ and $^{16}O/^{18}O$ ratios in the atmosphere of Venus from high–resolution 10–μm spectroscopy.
B. Bézard, J. P. Baluteau, A. Marten, N. Coron.
Icarus, Vol. 72, No. 3, p. 623 – 634 (1987).
Spectra of Venus in the 925– to 980–cm^{-1} spectral range were recorded in January 1985 at a resolution of 0.06 cm^{-1}. Several lines from the $v_3 - v_1$ bands of $^{13}CO_2$ and $^{12}C^{16}O^{18}O$ were observed for the first time. Synthetic spectra, which include absorption from CO_2 bands and from sulfuric acid clouds, are compared to the observations. Taking into account measurement noise as well as systematic errors, the analysis yields $^{12}C/^{13}C = 86 \pm 12$ and $^{16}O/^{18}O = 500 \pm 80$, in agreement with the terrestrial ratios. The results are consistent with previous ground–based near–infrared studies and with in situ mass spectrometer measurements.

093.052 Morphologic and gravimetric investigations of Bell and Eisila Regiones on Venus.
P. Janle, D. Jannsen, A. T. Basilevsky (*A. T. Bazilevskij*).
Earth, Moon, Planets, Vol. 39, No. 3, p. 251 – 273 (1987). = Contrib. Nr. 343, Inst. Geophys. Univ. Kiel, F.R. Germany.

The aim of this paper is to calculate crustal density models for Bell and Eisila Regiones and to compare the results with the morphology derived from the Pioneer Venus Orbiter and the Venera 15 and 16 missions. The isostatic state of the two regions is investigated. The role of these two features for the tectonics of Venus is discussed. The discussion is concentrated on Bell Regio because of the lack of high resolution radar pictures for Eisila Regio.

093.053 Tidal variations in the rotation of Venus.
C.–z. Zhang, M. Shen.
Bull. Astron. Inst. Czech., Vol. 38, No. 6, p. 325 – 328 (1987).

By using the recently reported Stokes coefficients C_{22} and S_{22} in the Venusian gravitational field expansion, the periodic variations of the Venusian spin rate, $\Delta\omega_1$, generated by the torque of solar gravitation, and $\Delta\omega_2$, generated by the variation of the principal moment of inertia C, c_{33}, are estimated. Whether the gravitational torque due to the coupling between the permanently deformed equatorial zone of Venus and the Earth may control the state of Venus' rotation is also discussed.

093.054 Compilation of a preliminary catalogue of coordinates of reference points in the region of the Maxwellian Mountains on Venus.
E. G. Belen'kij, L. M. Kadnichanskaya, T. A. Kamenetskaya, S. E. Reshetova.
Mater. 18 nauchn.–tekh. konf. mol. uchenykh i spets. TsNII Geod., Aehrosemki, Kartogr., Moskva, 28 – 29 apr., 1986. Moskva, p. 167 – 183 (1987). In Russian. Abstr. in Ref. Zh., 62. Issled. Kosm. Prostranstva, 12.62.307 (1987).

093.055 Peculiarities of the representation of the relief and problems of generalization when mapping Venus with the scale 1:5000000 taking the Maxwellian Mountains as an example.
I. B. Korobova.
Mater. 18 nauchn.–tekh. konf. mol. uchenykh i spets. TsNII Geod., Aehrosemki, Kartogr., Moskva, 28 – 29 apr., 1986. Moskva, p. 249 – 254 (1987). In Russian. Abstr. in Ref. Zh., 62. Issled. Kosm. Prostranstva, 12.62.308 (1987).

093.056 Hydrogen dissipation from the Venus atmosphere.
M. I. Pudovkin, I. V. Golovchanskaya.
Magnitosfer. Issled., Moskva, No. 9, p. 94 – 99 (1987). In Russian. Abstr. in Ref. Zh., 62. Issled. Kosm. Prostranstva, 12.62.309 (1987).

093.057 The role of the hot oxygen corona in the interaction of Venus with solar wind.
T. K. Breus, A. M. Krymskij, V. Ya. Mitnitskij, T. Gombosi, A. Nagy.
Kosm. Issled., Tom 25, Vyp. 4, p. 626 – 634 (1987). In Russian. English translation in Cosm. Res.

093.058 The automatic stations Vega 1 and Vega 2. Operation of the reentry vehicles in the Venus atmosphere.
V. A. Deryugin, V. P. Dolgopolov, V. P. Karyagin, V. M. Kovtunenko, R. S. Kremnev, K. M. Pichkhadze, G. N. Rogovskij, R. Z. Sagdeev.
Kosm. Issled., Tom 25, Vyp. 5, p. 649 – 654 (1987). In Russian. English translation in Cosm. Res.

093.059 Thermal structure of the Venus atmosphere from measurement results of Vega 2 instruments.
V. M. Linkin, J. Blamont, S. I. Devyatkin, S. P. Ignatova, V. V. Kerzhanovich, A. N. Lipatov, C. Malique, B. I. Stadnyk, Ya. V. Sanotskij, P. G. Stolyarchuk, A. V. Terterashvili, G. A. Frank, L. I. Khlyustova.
Kosm. Issled., Tom 25, Vyp. 5, p. 659 – 672 (1987). In Russian. English translation in Cosm. Res.

093.060 Vega 1 and Vega 2: vertical profiles of the wind velocity from Doppler measurement data aboard the reentry vehicles.
V. V. Kerzhanovich, N. M. Antsibor, R. V. Bakit'ko, V. P. Lysov, E. P. Molotov, V. I. Puchkov, A. I. Rutkovskij, V. F. Tikhonov, K. M. Pichkhadze.
Kosm. Issled., Tom 25, Vyp. 5, p. 673 – 677 (1987). In Russian. English translation in Cosm. Res.

093.061 Water vapour distribution in the middle and lower Venus atmosphere.
Yu. A. Surkov, O. P. Shcheglov, M. L. Ryvkin, D. M. Shejnin, N. A. Davydov.
Kosm. Issled., Tom 25, Vyp. 5, p. 678 – 690 (1987). In Russian. English translation in Cosm. Res.

093.062 Investigation of UV absorption in the Venus atmosphere with the Vega 1 and Vega 2 reentry vehicles.
J.–L. Bertaux, A. P. Ehkonomov, B. Mege, V. I. Moroz, A. Abergel, V. I. Gnedykh, A. V. Grigor'ev, B. E. Moshkin, A. Hauchecorne, J.–P. Pommereau, S. B. Sergeeva.
Kosm. Issled., Tom 25, Vyp. 5, p. 691 – 706 (1987). In Russian. English translation in Cosm. Res.

093.063 The vertical structure of the Venus cloud layer at the landing sites of Vega 1 and Vega 2.
V. I. Gnedykh, L. V. Zasova, V. I. Moroz, B. E. Moshkin, A. P. Ehkonomov.
Kosm. Issled., Tom 25, Vyp. 5, p. 707 – 714 (1987). In Russian. English translation in Cosm. Res.

093.064 Gas–chromatographic analysis of products of thermal reactions of aerosol in the Venus cloud layer on board Vega 1 and Vega 2.
N. V. Porshnev, L. M. Mukhin, B. G. Gel'man, D. F. Nenarokov, V. A. Rotin, A. V. D'yachkov, V. B. Bondarev.
Kosm. Issled., Tom 25, Vyp. 5, p. 715 – 720 (1987). In Russian. English translation in Cosm. Res.

093.065 Radiometric analysis of aerosol in the Venus clouds with the automatic interplanetary stations Vega 1 and Vega 2.
B. M. Andrejchikov, I. K. Akhmetshin, B. N. Korchuganov, L. M. Mukhin, B. I. Ogorodnikov, I. V. Petryanov, V. I. Skitovich.
Kosm. Issled., Tom 25, Vyp. 5, p. 721 – 736 (1987). In Russian. English translation in Cosm. Res.

093.066 Chemical composition and structure of Venus clouds from results of X–ray radiometric experiments made with the Vega 1 and Vega 2 automatic interplanetary stations.
B. M. Andrejchikov.
Kosm. Issled., Tom 25, Vyp. 5, p. 737 – 743 (1987). In Russian. English translation in Cosm. Res.

093.067 Determination of the chemical composition of Venus cloud aerosols with the "Malachite" mass–spectroscopy instrument aboard Vega 1.
Yu. A. Surkov, V. F. Ivanova, A. N. Pudov, Eh. P. Sheretov, B. I. Kolotilin, M. P. Safonov, R. Thomas, G. Israël, J. Lespagnol, D. Imbault, A. Hauser, D. Caramelle.
Kosm. Issled., Tom 25, Vyp. 5, p. 744 – 750 (1987). In Russian. English translation in Cosm. Res.

093.068 Element composition of the Venus rocks in the north–eastern part of Aphrodite Terra (from Vega 2 data).
Yu. A. Surkov, L. P. Moskaleva, V. P. Kharyukova, A. D. Dudin, G. G. Smirnov, S. E. Zajtseva, A. N. Tikhomirov, O. S. Manvelyan.
Kosm. Issled., Tom 25, Vyp. 5, p. 751 – 761 (1987). In Russian. English translation in Cosm. Res.

093.069   The content of natural radio active elements in Venus rocks according to data of the Vega 1 and Vega 2 stations.
Yu. A. Surkov, F. F. Kirnozov, V. N. Glazov,
A. G. Dunchenko, L. P. Tatsij.
Kosm. Issled., Tom 25, Vyp. 5, p. 762 – 767 (1987). In Russian. English translation in Cosm. Res.

093.070   Superrotation induced by critical–level absorption of gravity waves on Venus: an assessment.
A. Y. Hou, B. F. Farrell.
J. Atmos. Sci., Vol. 44, No. 7, p. 1049 – 1061 (1987). Abstr. in Phys. Abstr., Vol. 90, No. 1318, Entry 145857 (1987).

093.071   Precession and long period nutation of Venus.
C.–z. Zhang.
Acta Astron. Sin., Vol. 28, No. 3, p. 242 – 249 (1987). In Chinese.
With the inertial frame defined by the ecliptic and dynamical equinoxes for the epoch J2000.0, a simplified theory of precession and nutation of Venus has been developed. The precession rate and the long period nutations have been evaluated in terms of the recent dynamical parameters of Venus. A precession of $-7828''.6144$ cy^{-1} has been found.

Vénus dévoilée. Voyage autour d'une planète.
See Abstr. 003.029.

Another reports for observation of Venus transit by Avicenna and its effects on ancient astronomy.
See Abstr. 004.053.

Aeolian abrasion on Venus: preliminary results from the Venus simulator.
See Abstr. 022.015.

Determination of $O_2(a^1 \Delta g)$ and $O_2(b^1 \Sigma^+ g)$ yields in the reaction $O + ClO \rightarrow Cl + O_2$: implications for photochemistry in the atmosphere of Venus.
See Abstr. 022.078.

A compilation system for Venus radar mission (Magellan).
See Abstr. 035.001.

Methods of remote sensing with automatic interplanetary stations in mapping planets.
See Abstr. 036.048.

Processing of Venus radar images with the analyzer "Madzhiskan–2".
See Abstr. 036.049.

Determination of right ascensions of the Sun, Venus and Mercury in Tashkent in the period 1981 – 1983.
See Abstr. 041.006.

Right ascensions of the Sun, Mercury and Venus observed at the transit instrument in Nikolaev in 1985.
See Abstr. 041.020.

Weltraummissionen zu Venus und Mars. Teil 1: Missionen zur Venus.
See Abstr. 051.016.

The Magellan altimetry experiment.
See Abstr. 051.050.

Scientific results of the Vega mission.
See Abstr. 051.063.

Volatiles on Earth and Mars: a comparison.
See Abstr. 097.065.

Analysis of planetary evolution with emphasis on differentiation and dynamics.
See Abstr. 107.004.

094 Moon

094.001   Dynamical constraints on the origin of the moon.
A. P. Boss, S. J. Peale.
NASA Tech. Memo., NASA TM–89810, p. 115 (1987). Abstract. – See Abstr. 003.001.

094.002   Lunar science from lunar laser ranging.
J. G. Williams, X X Newhall, J. O. Dickey.
NASA Tech. Memo., NASA TM–89810, p. 116 (1987). Abstract. – See Abstr. 003.001.

094.003   Lunar magnetization concentrations (Magcons) antipodal to young large impact basins.
R. P. Lin, K. A. Anderson, L. L. Hood.
NASA Tech. Memo., NASA TM–89810, p. 149 – 150 (1987). – See Abstr. 003.001.

094.004   Implications of convection in the moon and the terrestrial planets.
D. L. Turcotte.
NASA Tech. Memo., NASA TM–89810, p. 151 (1987). – See Abstr. 003.001.

094.005   Global petrologic variations on the moon: a ternary–diagram approach.
P. A. Davis, P. D. Spudis.
NASA Tech. Memo., NASA TM–89810, p. 152 – 154 (1987). – See Abstr. 003.001.

094.006   A chemical and petrological model of the lunar crust.
P. D. Spudis, P. A. Davis.
NASA Tech. Memo., NASA TM–89810, p. 155 – 157 (1987). – See Abstr. 003.001.

094.007   Compositional information for the moon: some characteristics of current near–IR spectra (telescopic and laboratory).
C. M. Pieters.
NASA Tech. Memo., NASA TM–89810, p. 208 – 209 (1987). – See Abstr. 003.001.

094.008   Compositional stratigraphy of crustal material from near–infrared spectra.
C. M. Pieters.
NASA Tech. Memo., NASA TM–89810, p. 210 (1987). Abstract. – See Abstr. 003.001.

094.009   Preliminary results of spectral reflectance studies of Tycho crater.
B. R. Hawke, C. R. Coombs, P. G. Lucey, J. F. Bell,
R. Jaumann, G. Neukum, C. M. Pieters.
NASA Tech. Memo., NASA TM–89810, p. 211 – 213 (1987). – See Abstr. 003.001.
Spectral studies of lunar craters and their ejecta deposits provide valuable information about the composition of the lunar

crust as well as insights into the impact cratering process. Materials from a variety of depths have been excavated by these impacts and deposited in a systematic manner around the craters. This paper presents the preliminary analysis and interpretation of near–infrared spectra obtained for both the interior and exterior deposits associated with Tycho crater.

094.010 Preliminary results of geologic and remote sensing studies of Rima Mozart.
C. R. Coombs, B. R. Hawke.
NASA Tech. Memo., NASA TM–89810, p. 214 – 216 (1987). – See Abstr. 003.001.

094.011 Photometric properties of lunar terrains derived from Hapke's equation.
P. Helfenstein, J. Veverka.
NASA Tech. Memo., NASA TM–89810, p. 217 – 218 (1987). – See Abstr. 003.001.

094.012 New very high resolution radar studies of the moon.
P. J. Mouginis–Mark, B. Campbell.
NASA Tech. Memo., NASA TM–89810, p. 237 (1987). – See Abstr. 003.001.

094.013 High resolution radar map of the moon.
T. W. Thompson.
NASA Tech. Memo., NASA TM–89810, p. 238 – 239 (1987). – See Abstr. 003.001.

Previous radar mappings of the moon at 70 cm wavelength in the late 1960s by Thompson (1974) have been replaced with a new set of observations conducted between 1981 and 1984 using the 430 MHz radar at the Arecibo Observatory, Puerto Rico.

094.014 Landform identification – lunar radar images.
H. J. Moore, T. W. Thompson.
NASA Tech. Memo., NASA TM–89810, p. 240 – 242 (1987). – See Abstr. 003.001.

Three sets of polarized radar–echo images of the moon are being examined to establish the relation between radar resolution and landform–identification resolution. The results of the study should be valuable to those planning to acquire or interpret radar images of the earth or other planetary bodies.

094.015 Physiographic constraints on the origin of lunar wrinkle ridges.
M. P. Golombek, B. J. Franklin.
NASA Tech. Memo., NASA TM–89810, p. 461 – 463 (1987). – See Abstr. 003.001.

094.016 Lithospheric loading and tectonics of the lunar irregular maria.
J. L. Hall, S. C. Solomon.
NASA Tech. Memo., NASA TM–89810, p. 464 – 466 (1987). – See Abstr. 003.001.

In this paper the authors test the hypothesis that the irregular lunar maria share with the circular maria the processes leading to the formation of associated graben and mare ridge systems. They apply a formulation of the lithospheric flexure problem that accounts for the variable distribution of basalt loads for the irregular maria. On the basis of the tectonic structures and geologic history of each of the major irregular maria, as well as models for the distribution of mare basalt for each region, the authors compare the predicted stress fields with the distribution of tectonic features for a range of assumed values of the thickness of the elastic lithosphere. They then compare the best fitting values for lithospheric thickness beneath the irregular maria with those previously inferred for the circular maria.

094.017 A unified lunar control network.
M. E. Davies, T. R. Colvin, D. L. Meyer.
NASA Tech. Memo., NASA TM–89810, p. 537 – 538 (1987). – See Abstr. 003.001.

At this time, control on the Moon is composed of a number of independent regional networks. These networks frequently have different origins but never have common ties, even in overlapping areas. The objective of the unified network program is to tie the regional networks into a single consistent planetwide control network.

094.018 Ontstond de maan bij een kosmische botsing?
G. Beekman.
Zenit, Jaarg. 14, Nr. 7/8, p. 254 (1987).

094.019 Magnetic field and remanent magnetization effects of basin–forming impacts on the moon.
L. L. Hood.
Geophys. Res. Lett., Vol. 14, No. 8, p. 844 – 847 (1987).

Maps of the distribution of lunar surface magnetic fields have shown that the largest observed concentrations of lunar crustal magnetization occur antipodal to 4 relatively young large impact basins: Imbrium, Orientale, Serenitatis, and Crisium. A model is proposed for the formation of these magnetization concentrations (or "magcons") in which the partially ionized vapor cloud produced in a hypervelocity (> 10 km s^{-1}) basin–forming impact expands around the moon forcing a pre–existing ambient magnetic field to be concentrated for a brief (< 1 day) time period in the antipodal zone. The model implies that basin–forming impacts have played a major role in determining the large–scale distribution of crustal magnetization detectable from lunar orbit.

094.020 On the relative and absolute ages of seven lunar front face basins. I. From viscosity arguments.
R. B. Baldwin.
Icarus, Vol. 71, No. 1, p. 1 – 18 (1987).

The seven basins, Orientale, Imbrium, Crisium, Nectaris, Humorum, Serenitatis and an unnamed basin between Werner and the Altai ring show rims whose absolute and relative heights are correlated with the sharpness and crispness of the features. On the assumpton that the decline in average outer rim height, not scarp height, measures the age of the basin and also that the decline represents a hot creep of rocks of very high viscosity, absolute ages were derived. Basins were found to increase in age in the sequence listed above, with a range from about 3.82 to 4.30×10^9 years. The average or effective viscosity of the surface layers down to whatever level was involved in the creep was calculated as increasing from 9.46×10^{24} poises at about 4.30×10^9 years to 1.86×10^{30} poises at present.

094.021 On the relative and absolute ages of seven lunar front face basins. II. From crater counts.
R. B. Baldwin.
Icarus, Vol. 71, No. 1, p. 19 – 29 (1987). = Oliver Tech. Publ., No. 133.

From counts of postbasin craters larger than 30 km in diameter, lying within or near to seven giant front face lunar basins, relative ages for the basins may be obtained. These relative ages correlate well with absolute basin ages found from viscosity arguments in R. B. Baldwin (1987). From crater counts the basins are in the following sequence of increasing relative age: Orientale, Imbrium, Crisium, Serenitatis, Nectaris, Humorum, and the unnamed basin lying between Werner and the Altai ring. The absolute ages from Baldwin (1987) range from 3.80 to 4.30×10^9 years while a correlation with the relative ages of this paper yields a range of 3.79 to 4.27×10^9 years.

094.022 Lunar domes – first observed by Johann Hieronymus Schroeter.
J. H. Phillips.
Strolling Astron., Vol. 32, Nos. 3 – 4, p. 67 – 73 (1987).

094.023 Basin–ring spacing on the Moon, Mercury, and Mars.
R. J. Pike, P. D. Spudis.
Earth, Moon, Planets, Vol. 39, No. 2, p. 129 – 194 (1987).

Radial spacing between concentric rings of impact basins that lack central peaks is statistically similar and nonrandom on the Moon, Mercury, and Mars, both inside and outside the main ring. One spacing interval, $(2.0 \pm 0.3)^{0.5} D$, or an integer multiple of it, dominates most basin rings. Three analytical approaches yield similar results from 296 remapped or newly mapped rings of 67 multi–ringed basins: least–squares of rank–grouped rings, least–squares of rank and ring diameter for each basin, and averaged ratios of adjacent rings. The statistically constant and target–invariant spacing of so many rings suggests that this characteristic may constrain formational models of impact basins on the terrestrial planets. It is suggested that some type of wave mechanism dominates the location, but not necessarily the formation, of basin rings. The waves may be standing, rather than travelling.

094.024 Peculiarities in the distribution of lunar craters with different states of preservation.
Zh. F. Rodionova, V. V. Shevchenko, A. A. Karlov,
T. F. Smolyakova.
20th All–Union meteoritic conference, p. 56 – 58 (1987). In Russian. Abstr. in Ref. Zh., 51. Astron., 7.51.203; 62. Issled. Kosm. Prostranstva, 7.62.495 (1987). – See Abstr. 012.017.

094.025 The role of evaporation and dissipation processes in the formation of the moon.
O. I. Yakovlev, O. M. Markova, B. M. Manzon.
Geokhimiya, No. 4, p. 467 – 482 (1987). In Russian. Abstr. in Ref. Zh., 51. Astron., 8.51.170 (1987).

094.026 The lunar samples difference from the natural lunar surface.
O. I. Kvaratskheliya, L. O. Kolokolova.
Kinematika Fiz. Nebesn. Tel, Tom 3, No. 5, p. 90 – 92 (1987). In Russian. English translation in Kinematics Phys. Celest. Bodies.

A statistical analysis is made for the polarimetric measurements of lunar surface regions and of lunar samples. These measurements are performed at Abastumani Astrophysical Observatory. The dependence of polarization degree on phase angle is investigated. It is shown that almost all polarimetric properties of lunar samples differ from analogous properties of the lunar surface. The accepted opinion, that polarimetric properties of the Moon and of the lunar powder samples are identical, is a consequence of an incorrect method of polarimetric values comparison.

094.027 A study of the efficiency of some inversion techniques applied to a simple model of the Moon.
A. I. Denis.
Astron. Astrophys., Vol. 184, No. 1/2, p. 373 – 380 (1987).

The author discusses some aspects of several commonly used inversion techniques: the Backus–Gilbert method, the Moore–Penrose inversion, and the singular value decomposition. He shows that a very close connection exists between the generalized Moore–Penrose inverse and the solution obtained by singular value decomposition. The degree of non–uniqueness in inverting the free oscillation data is studied numerically on simple lunar models. The parameters which are searched for by inversion are density and shear modulus. An algorithm based on the singular value decomposition inversion scheme is found to handle in a straightforward fashion both inconsistent and rank–deficient data within a certain tolerance.

094.028 Secular deceleration of the Moon and of the Earth's rotation and variation in the zonal geopotential harmonic.
M. Burša.
Bull. Astron. Inst. Czech., Vol. 38, No. 5, p. 309 – 313 (1987).

The angular momentum balance of the Earth–Moon–Sun system has been refined using the most recent value of the decrease in the Moon's mean motion obtained by lunar laser ranging. It has been proved that the decrease in the second zonal geopotential harmonic J_2, detected by LAGEOS LR, fits the angular momentum budget well. The dynamics at the Earth's core–mantle boundary is assumed to be responsible for dJ_2/dt.

094.029 On the stability of rotation of the Moon according to Cassini's laws.
A. A. Khentov.
Astron. Zh., Tom 64, Vyp. 5, p. 1105 – 1113 (1987). In Russian. English translation in Sov. Astron., Vol. 31, No. 5.

The problem of rotation of the Moon is considered with account for the gravitational torques and small tidal friction as well as the evolution of the orbit. Using Cassini's empirical laws, it is proved that when the central ellipsoid of inertia of the Moon is close enough to a sphere in a close neighbourhood of rotation, there exists a solution for the complete system of equations, which is stable at an unlimited interval of time. Estimations are obtained of admissible initial and possible current deviations from the rotation according to Cassini's laws.

094.030 Lunar paleomagnetism.
M. Fuller, S. M. Cisowski.
Geomagnetism, Vol. 2, p. 307 – 455 (1987). – See Abstr. 003.011.

Contents: 1. Introduction. 2. The Apollo–returned lunar samples. 3. Lunar–rock magnetism. 4. Paleomagnetic record of Apollo samples. 5. Present lunar magnetic fields. 6. The origin of lunar magnetism. 7. Conclusions and directions for future research.

094.031 The moon's ancient magnetism.
S. K. Runcorn.
Sci. Am., Vol. 257, No. 6, p. 34 – 42 (1987).

The moon is now a dead body, but it seems once to have generated its own magnetic field. Since then it has been shifted with respect to its spin axis – perhaps by collisions with moon–orbiting satellites.

094.032 Morphological catalogue of lunar craters.
V. V. Shevchenko.
Izdatel'stvo Moskovskogo Gosudarstvennogo Universiteta, Moskva, 173 pp. (1987). In Russian. Abstr. in Ref. Zh., 52. Geod. Aehrosemka, 9.52.160 (1987).

094.033 Het ontstaan van de maan.
J. van Diggelen.
Zenit, 14. Jaarg., Nr. 12, p. 413 – 419 (1987).

094.034 Names of lunar formations.
Heavens, Vol. 68, No. 11, p. 334 – 335 (1987). In Japanese.

094.035 Recent comet impacts on the Moon: the evidence from remote–sensing studies.
J. F. Bell, B. R. Hawke.
Publ. Astron. Soc. Pac., Vol. 99, No. 618, p. 862 – 867 (1987).

Multispectral images and near–infrared reflectance spectra were obtained for the Reiner Gamma Formation, a swirl–like albedo and magnetic anomaly in the western portion of the lunar near side, as well as for nearby geologic units in Oceanus Procellarum. The data indicate that the Reiner Gamma swirls were formed by the disruption and alteration of the normal lunar regolith and support the hypothesis that Reiner Gamma and other lunar swirls are the imprints of relatively recent comet impacts on the Moon.

094.036 Further investigations into lunar palaeointensity determinations.
S. K. Chowdhary, D. W. Collinson, A. Stephenson,
S. K. Runcorn.
Phys. Earth Planet. Inter., Vol. 49, Nos. 1 – 2, p. 133 – 141 (1987).

094.037 Numerical model of the Moon's physical libration.
G. I. Eroshkin.
Byull. Inst. Teor. Astron., Tom 15, No. 9, p. 511 – 518 (1986). In Russian.

Six independent variables determining the magnitude of the Moon's angular momentum vector and its orientation with respect to the lunar equator and a fixed ecliptic are taken as components of the physical libration of the Moon. A set of six ordinary differential equations of the first order describing the Moon's physical libration is constructed. All harmonics up to the fourth order of the lunar gravitational potential and those of the second order in the figure–figure part of the Earth–Moon mutual gravitational potential are taken into account in the expansion of the Moon–Earth–Sun mutual gravitational potential in spherical functions. The fundamental ephemeris DE200/LE200 provides the selenocentric coordinates both of the Earth and the Sun. The numerical integration of the constructed set of equations is carried out by means of Everhart's method over the 80 year span from JD 2440400. The present numerical model is compared with Eckhardt's semianalytic theory of the Moon's physical libration.

094.038 Main results of spectropolarimetric observations of 100 areas on the lunar surface.
O. I. Kvaratskheliya.
Soobshch. Akad. Nauk GSSR, Tom 126, No. 2, p. 313 – 316 (1987). In Russian. Abstr. in Ref. Zh., 51. Astron., 10.51.187; 62. Issled. Kosm. Prostranstva 10.62.366 (1987).

094.039 Photometric properties of lunar terrains derived from Hapke's equation.
P. Helfenstein, J. Veverka.
Icarus, Vol. 72, No. 2, p. 342 – 357 (1987).
The primary objective of the present investigation is to derive Hapke parameters for the lunar surface from both disk–integrated and disk–resolved photometric data and to examine the mutual consistency of values derived for three principal albedo classes of lunar terrains. Fundamental photometric properties such as geometric albedo and phase integral are evaluated and compared to the results of previous investigators. Finally, the geological significance of results is considered.

094.040 Approximation of the lunar potential by a set of point masses using the method of one–dimensional optimization.
P. M. Zazulyak, A. L. Tserklevich.
Geod., Kartogr. Aehrofotosemka, L'vov, No. 46, p. 40 – 46 (1987). In Russian. Abstr. in Ref. Zh., 52. Geod. Aehrosemka, 11.52.190 (1987).

094.041 The investigation of intersecting craters on the Moon, Mars and Mercury.
T. P. Skobeleva.
Astron. Vestn., Tom 21, No. 3, p. 221 – 224 (1987). In Russian. English translation in Sol. Syst. Res.
Eleven thousand of intersecting craters were investigated on the Moon, Mars and Mercury. A statistic analysis was made, the morphological characteristics were considered. The rose–diagrams of directions of azimuths between the centres of intersecting craters were drawn. The intersecting craters are distributed practically evenly in all directions thus indicating the accidental distribution of these craters on the surfaces of the Moon, Mars and Mercury.

094.042 Impacts that magnetize.
M. Fuller.
Nature, Vol. 329, No. 6141, p. 674 – 675 (1987).
The author gives comments to the following questions on lunar magnetism: Did the Moon once have a molten core that acted like a dynamo, or could transient fields at the surface of the Moon account for the magnetization?

094.043 Geology and petrology of the Apollo 15 Landing Site: past, present and future understanding.
P. D. Spudis, G. Ryder.
EOS Trans. Am. Geophys. Union, Vol. 66, No. 27 – 53, 7 pp. (1985). Abstr. in Phys. Abstr., Vol. 90, No. 1313, Entry 115263 (1987).

094.044 Preliminary results of small phase–angle spectropolarimetric measurements of the Moon.
A. L. Gural'chuk, N. V. Opanasenko, Yu. G. Shkuratov.
Astron. Tsirk., No. 1462, p. 6 – 8 (1986). In Russian.

094.045 The tidal evolution of the Earth–Moon system.
M. Burša.
Bull. Astron. Inst. Czech., Vol. 38, No. 6, p. 321 – 325 (1987).
On the basis of the most recent laser ranging data and the tidal evolution theory of the Earth–Moon system, several dynamic parameters of the system are refined: (1) the theoretical relation between the tidal deceleration of the Moon and of the Earth's rotation; (2) the tidal decrease in the angular velocity of the Earth's rotation generated separately by the Moon and by the Sun; (3) the total secular decrease in the angular velocity of the Earth's rotation.

094.046 Catalogue of 10484 heights in the limb zone of the Moon taking into account A. A. Yakovkin's second model of the figure of the Moon (monograph).
A. A. Nefed'ev, V. S. Borovskikh, Yu. A. Nefed'ev.
Kazan. Inzh.–Stroit. Inst., Kazan', 137 pp. (1987). In Russian. Abstr. in Ref. Zh., 62. Issled. Kosm. Prostranstva, 12.62.296 (1987).

094.047 Distance of the Moon 2,5 billion years ago.
Z. Urban.
Říše hvězd, Vol. 68, No. 7, p. 135 – 136 (1987). In Czech.

094.048 On using the moon as a source with a standard intensity in the 0.1 – 30 cm wavelength range.
V. D. Krotikov, S. A. Pelyushenko.
Sov. Astron., Vol. 31, No. 2, p. 216 – 219 (1987). English translation of 43.094.021.

094.049 Physical libration of the moon in longitude.
A. S. Dubrovskij, Yu. A. Chikanov.
Sov. Astron., Vol. 31, No. 2, p. 223 – 225 (1987). English translation of 43.094.022.

Atlas of reflectance spectra of terrestrial, lunar and meteoritic powders and frosts from 92 to 1800 nm.
See Abstr. 002.001.

Radio astronomy on the moon.
See Abstr. 013.017.

Programme for the calculation of apparent places of the Moon.
See Abstr. 021.027.

Spectral properties of plagioclase and pyroxene mixtures and the interpretation of lunar soil spectra.
See Abstr. 022.133.

On the representation of stationary time series by the sum of harmonic terms.
See Abstr. 042.036.

Computing ephemerides for the Sun, the Moon and the planets based on modern theories.
See Abstr. 046.020.

On Newton's Earth–Moon–test of 1665/66.
See Abstr. 066.129.

The cometary contribution to the oceans of primitive Earth.
See Abstr. 081.034.

Conditions for geomagnetically trapping sub–micron lunar ejecta between L = 1.2 and L = 3.1.
See Abstr. 084.059.

Isotopic abundances: inferences on solar system and planetary evolution.
See Abstr. 091.032.

Gravity data analysis.
See Abstr. 093.001.

Crater production on Venus and Earth by asteroid and comet impact.
See Abstr. 093.003.

Cometesimals in the inner Solar System.
See Abstr. 102.087.

Collisional and dynamical processes in Moon and planet formation.
See Abstr. 107.008.

The origin of the Moon and the single–impact hypothesis, II.
See Abstr. 107.013.

095 Lunar Eclipses

095.001   **Saros et périodicité des éclipses.**
L. Piotin.
Obs. Trav., No. 11, p. 53 – 60 (1987).
   L'auteur tente de calculer la durée du Saros et applique le principe utilisé à un essai de prévision de la périodicité des éclipses de Lune.

095.002   **Éclipses totales de lune. Graphique des vitesses de prises de vue.**
H. Arioli.
Astronomie, Vol. 101, p. 497 – 500 (1987).

Der Canon der Finsternisse und die Rotationsdauer der Erde.
See Abstr. 079.003.

Aktuelle Beobachtungen von Sonnen– und Mondfinsternissen.
See Abstr. 079.004.

096 Lunar and Planetary Occultations

096.001   **The September 13th Pleiades passage.**
D. W. Dunham.
Occultation Newsl., Vol. 4, No. 5, p. 99 – 101 (1987).

096.002   **Grazing occultations.**
D. Stockbauer.
Occultation Newsl., Vol. 4, No. 5, p. 103 – 104 (1987).

096.003   **Graze predictions and XZ catalog improvements.**
D. W. Dunham.
Occultation Newsl., Vol. 4, No. 5, p. 104 – 105 (1987).

096.004   **Explanation of lunar occultation observation reductions.**
D. Büttner.
Occultation Newsl., Vol. 4, No. 5, p. 105 – 106 (1987).

096.005   **Asteroidal, cometary, and Jovian occultations during 1987.**
D. W. Dunham.
Occultation Newsl., Vol. 4, No. 5, p. 113 – 134 (1987).

096.006   **Grazing occultation reports.**
B. Fraser.
Mon. Notes Astron. Soc. S. Afr., Vol. 46, Nos. 7 – 8, p. 103 – 110 (1987).

096.007   **Lunar occultations: from conjecture to results.**
N. M. White.
Vistas Astron., Vol. 30, Part 1, p. 13 – 25 (1987). – See Abstr. 012.007.
   An historical sketch of the lunar occultation phenomenon and its astronomical uses is given. A brief review of the parameters affecting an occultation diffraction pattern is followed by chronological descriptions of occultation instrumentation, pre-telescopic observations, the lunar atmosphere debate, timings and angular resolution.

096.008   **Appulses and occultations of SAO stars by Mercury: 1987 – 1995.**
D. J. Mink.
Icarus, Vol. 71, No. 3, p. 478 – 481 (1987). Paper presented at the Tucson Mercury Conference, held at Tucson, Ariz., USA, 6 – 9 August 1986.
   Predictions are presented for 78 close approaches by Mercury to stars in the SAO catalog between 1 January 1987 and 31 December 1995. No easily observable occultations were found, but 7 occultations of stars with visual magnitudes of 7 or fainter were found. These events are presented for the use of observers searching for material in the vicinity of Mercury and for those who need close bright stars for the improvement of ground–based images of Mercury.

096.009   **De bijzondere Plejadenbedekking van 27 januari 1988.**
A. Gerritsen, H. Bulder.
Zenit, 14. Jaarg., Nr. 11, p. 365 – 367 (1987).

096.010   **Finite–parametric models in inverse problems of astrophysics. Lunar occultations of stars.**
A. V. Goncharskij, S. Yu. Romanov, V. V. Stepanov, A. M. Cherepashchuk.
Sov. Astron., Vol. 30, No. 6, p. 710 – 716 (1987). English translation of 42.096.008.

096.011 Occultations of stars by solar system objects. VII. Occultations of catalog stars by asteroids in 1988 and 1989.
L. H. Wasserman, E. Bowell, R. L. Millis.
Astron. J., Vol. 94, No. 5, p. 1364 – 1372 (1987).

Predictions are given for 102 occultations of stars by minor planets occurring in 1988 and 1989. The predictions are based on a computerized comparison of asteroid ephemerides with star positions given in eight major astrometric catalogs. The search is complete for all numbered asteroids whose orbits are accurately known and that reach an angular diameter of at least 0.08 arcsec during the search years. Preliminary information sufficient for planning is given for each occultation.

096.012 Occultation by Neptune.
IAU Circ., No. 4418 (1987).

096.013 Eastern hemisphere grazing occultation supplement for 1988.
D. W. Dunham.
International Occultation Timing Association, 10 pp. (1987).

096.014 Occultations of stars by major and minor planets – 1988.
J. E. Carroll.
Joseph E. Carroll, 4261 Queens Way, Minnetonka, MN 55345, USA. 2 pp. (1987).

Upon request the author provides a list of appulses of solar-system objects with stars which should be observable from an observer's specific location during 1988.

096.015 Occultation by minor planets: observation program for 1988.
R. Boninsegna.
Roland Boninsegna, Dourbes, Belgium, 12 pp. (1987).

A catalog of stellar angular diameters measured by lunar occultation.
See Abstr. 002.020.

Let's report occultation timings on diskettes.
See Abstr. 036.010.

Projet d'observation des occultations par la Pleine Lune.
See Abstr. 036.195.

Model study of wavelength–dependent limb–darkening and radii of M–type giants and supergiants.
See Abstr. 064.049.

The size, shape, density, and albedo of Ceres from its occultation of BD + 8°471.
See Abstr. 098.086.

The occultation of KME 17 by Uranus and its rings.
See Abstr. 101.022.

Stellar occultation probes of the Uranian rings at 0.1 and 2.2 μm: a comparison of Voyager UVS and earth–based results.
See Abstr. 101.026.

Photoelectric observations of lunar occultations.
See Abstr. 115.003.

Lunar occultations at La Silla.
See Abstr. 115.017.

Photoelectric observations of lunar occultations of stars. Angular diameter of the carbon star Y Tauri and its physical characteristics.
See Abstr. 115.020.

The possibility of detection of effects of linear polarization of radiation in lunar occultation observations of stars.
See Abstr. 116.063.

Don't be afraid to observe the next series of lunar occultations of the galactic center, 1986 – 1989!
See Abstr. 155.069.

The January 87 galactic centre occultation observed with the UKIRT IR camera, IRCAM.
See Abstr. 155.154.

097 Mars, Mars Satellites

097.001 The case for a wet, warm climate on early Mars.
J. B. Pollack, J. F. Kasting, S. M. Richardson, K. Poliakoff.
NASA Tech. Memo., NASA TM–89810, p. 160 (1987). Abstract. – See Abstr. 003.001.

097.002 The role of regolith adsorption in the transition from early to late Mars climate.
F. P. Fanale, S. E. Postawko, A. P. Zent, J. R. Salvail.
NASA Tech. Memo., NASA TM–89810, p. 161 – 162 (1987). – See Abstr. 003.001.

097.003 Variations of Martian surface albedo: evidence for yearly dust deposition and removal.
P. R. Christensen.
NASA Tech. Memo., NASA TM–89810, p. 167 – 169 (1987). – See Abstr. 003.001.

097.004 On the spectral reflectance properties of materials exposed at the Viking landing sites.
E. Guinness, R. Arvidson, M. Dale–Bannister, R. Singer, E. Bruckenthal.
NASA Tech. Memo., NASA TM–89810, p. 170 – 172 (1987). – See Abstr. 003.001.

097.005 Spectral mixture modeling: further analysis of rock and soil types at the Viking Lander sites.
J. B. Adams, M. O. Smith.
NASA Tech. Memo., NASA TM–89810, p. 173 – 174 (1987). – See Abstr. 003.001.

097.006 Hisingerite and iddingsite on Mars: degradation of iron–rich basalts.
R. G. Burns.
NASA Tech. Memo., NASA TM–89810, p. 175 (1987). – See Abstr. 003.001.

097.007 Gossans on Mars: spectral features attributed to jarosite.
R. G. Burns.
NASA Tech. Memo., NASA TM–89810, p. 176 – 177 (1987). – See Abstr. 003.001.

097.008 Characterization of surficial units on Mars using Viking Orbiter multispectral image and thermal data.
M. A. Presley, R. E. Arvidson, P. R. Christensen.
NASA Tech. Memo., NASA TM–89810, p. 178 – 180 (1987). – See Abstr. 003.001.

097.009 **CO$_2$: adsorption on palagonite and the Martian regolith.**
A. P. Zent, F. P. Fanale, S. E. Postawko.
NASA Tech. Memo., NASA TM–89810, p. 181 – 183 (1987). –
See Abstr. 003.001.

097.010 **Investigation of Martian H$_2$O and CO$_2$ via gamma–ray spectroscopy.**
S. W. Squyres, L. G. Evans.
NASA Tech. Memo., NASA TM–89810, p. 184 – 186 (1987). –
See Abstr. 003.001.

097.011 **Goldstone radar observations of Mars: the 1986 opposition.**
T. W. Thompson.
NASA Tech. Memo., NASA TM–89810, p. 243 – 244 (1987). –
See Abstr. 003.001.

097.012 **A diffuse radar scattering model from Martian surface rocks.**
W. M. Calvin, B. M. Jakosky, P. R. Christensen.
NASA Tech. Memo., NASA TM–89810, p. 245 – 247 (1987). –
See Abstr. 003.001.
Remote sensing of Mars has been done with a variety of instrumentation at a variety of wavelengths. Recently, a surface rock distribution map from –60° to +60° latitude has been generated by Christensen (1986). The authors' objective is to model the diffuse component of radar reflection based on this surface distribution of rocks.

097.013 **Mars: seasonally variable radar reflectivity.**
L. E. Roth, R. S. Saunders, T. W. Thompson.
NASA Tech. Memo., NASA TM–89810, p. 248 – 250 (1987). –
See Abstr. 003.001.
Since reflectivity is a quantity characteristic of a given target at a particular viewing geometry, the same (temporally unchanging) target examined by radar at different occasions should have the same reflectivity. Zisk and Mouginis–Mark (1980) noted that the average reflectivities in the Goldstone Mars Data increased as the planet's S hemisphere passed from the late spring into early summer. The authors have examined the same data set and confirmed the presence in the data, of the phenomenon of the apparent seasonal variability of radar reflectivity.

097.014 **Bright sand/dark dust: the identification of active sand surfaces on the Earth and Mars.**
H. G. Blount II, R. Greeley, P. R. Christensen, R. Arvidson.
NASA Tech. Memo., NASA TM–89810, p. 257 – 258 (1987). –
See Abstr. 003.001.
Aeolian features are common on Mars. Although wind streaks are known to be active features, controversey has arisen regarding the activity of sand sheets and dunes, particularly in the north polar erg. Active sand bodies are here defined as those with saltation surfaces on which wind–induced ripple marks and/or slipfaces are observable in field outcrop. Inactive sands are those surfaces which do not show evidence of wind movement. The reasons for the albedo difference between active and inactive sands are partly explained by a textural analysis of the two terrains.

097.015 **Regional sources and sinks of dust on Mars: Viking observations of Cerberus, Solis Planum, and Syrtis Major.**
S. W. Lee.
NASA Tech. Memo., NASA TM–89810, p. 259 – 260 (1987). –
See Abstr. 003.001.
Seasonal variability of classical martian albedo features has long been noted by terrestrial observers. A study of seasonal variations of albedo features in the Cerberus, Solis Planum, and Syrtis Major regions has been based on Viking Orbiter data obtained over more than one complete Martian year. Contour maps of Lambert albedo and single–point thermal inertia have been constructed from the Infrared Thermal Mapper experiment data, and Orbiter images have been used to determine the pattern

and variability of regional winds (inferred from wind streak orientations).

097.016 **High resolution thermal infrared mapping of Martian channels.**
R. A. Craddock, R. Greeley, P. R. Christensen.
NASA Tech. Memo., NASA TM–89810, p. 261 – 263 (1987). –
See Abstr. 003.001.
Morphologic studies have not yielded a unique interpretation for channel formation because of the ambiguity between geomorphologic features and their similarities of one mode of formation with another. Viking Infrared thermal mapper high resolution (2 to 5 km) data have been compiled and compared to Viking Visual imaging subsystem data and available 1:5M geologic maps for several Martian channels in an effort to determine the surface characteristics and the processes active during and after the formation of these channels.

097.017 **Mars: morphology of southern hemisphere intracrater dunefields.**
N. Lancaster, R. Greeley.
NASA Tech. Memo., NASA TM–89810, p. 264 – 265 (1987). –
See Abstr. 003.001.

097.018 **Timing of frost deposition on Martian dunes: a clue to properties of dune particles?**
P. Thomas.
NASA Tech. Memo., NASA TM–89810, p. 266 – 267 (1987). –
See Abstr. 003.001.
The nature of the particles in Martian dunes has been a source of controversy since the discovery of dune forms on Mars. Their color and high thermal inertias have suggested that the dark dunes are made of medium to coarse grained, minimally weathered basaltic or similar materials. The authors has now mapped the occurrences of dune brightening in the fall and has done the quantitative photometry needed to confirm that the dunes are the sites of early frost deposition.

097.019 **Eolian saltation on Mars.**
D. J. MacKinnon.
NASA Tech. Memo., NASA TM–89810, p. 271 – 273 (1987). –
See Abstr. 003.001.
At the high wind speeds necessary for saltation on Mars, individual sand grains assume relatively long low angle trajectories compared with those on the Earth. The author has developed a computer model in order to examine the effects of both the wind field and surface collisions on the Martian saltation cloud; results are summarized here.

097.020 **Fluvial valleys on Martian volcanoes.**
V. R. Baker, V. C. Gulick.
NASA Tech. Memo., NASA TM–89810, p. 294 – 296 (1987). –
See Abstr. 003.001.
Channels and valleys have been known on the Martian volcanoes since their discovery by the Mariner 9 mission. Their analysis has generally centered on interpretation of possible origins by fluvial, lava, or viscous flows (debris, lahar, etc.). As summarized by Baker (1982), fluvial and related degradational processes (sapping) produce landforms strikingly similar to those observed on some Martian volcanoes. However, the possible fluvial dissection of Martian volcanoes has received scant attention in comparison to that afforded outflow, runoff, and fretted channels.

097.021 **Possible origin of some channels on Alba Patera, Mars.**
S. E. Postawko, P. Mouginis–Mark.
NASA Tech. Memo., NASA TM–89810, p. 303 – 304 (1987). –
See Abstr. 003.001.
Several alternative models have been proposed for the origin and mode of formation of channels and valley networks on martian volcanoes, notably Hecates Tholus, Ceraunius Tholus, Alba Patera. As part of a continuing program to better understand the eruptive history of the young volcanic centers on Mars, the authors have identified numerous channels on the

flanks of Alba Patera that resemble the channels on Hecates. As a result, they are exploring the possibility that some of the small channels on the flanks of Alba Patera may be fluvial in origin, and are examining potential water sources and modes of formation.

097.022 Volatile reservoirs below the surface of the Elysium Region of Mars: geomorphic evidence.
E. H. Christiansen, J. A. Hopler.
NASA Tech. Memo., NASA TM–89810, p. 307 – 309 (1987). – See Abstr. 003.001.

The Elysium volcanic province contains a variety of geomorphic evidence for the existence of large volatile reservoirs of subsurface volatiles. Study of these landforms yields insight into the distribution and size of these reservoirs and how they interact with the surface environment and will ultimately place constraints on the geometry, constitution, origin, time of formation, and temporal evolution of these important components of the martian crust.

097.023 Evidence for glaciation in Elysium.
D. M. Anderson.
NASA Tech. Memo., NASA TM–89810, p. 310 – 312 (1987). – See Abstr. 003.001.

Evidence for the existence of permafrost and the surface modification due to frost effects and the presence of ice on Mars dates from early observations. The Viking landers I and II confirmed the presence of water in the regolith and the periodic occurrence of frost at the surface of Mars. Similarities have been pointed out between a number of streamlined Martian channel features and similar streamlined landforms created by Antarctic ice sheet movements.

097.024 Water and ice on Mars: evidence from Valley Marineris.
B. K. Lucchitta.
NASA Tech. Memo., NASA TM–89810, p. 313 – 315 (1987). – See Abstr. 003.001.

097.025 Formation of the layered deposits in the Valles Marineris, Mars.
S. S. Nedell, S. W. Squyres.
NASA Tech. Memo., NASA TM–89810, p. 316 – 318 (1987). – See Abstr. 003.001.

The authors examine in detail the conditions necessary for the existence of ice–covered martian paleolakes, and consider mechanisms for sediment deposition in them.

097.026 Geomorphic evidence for ancient seas in West Deuteronilus Mensae, Mars–I: Regional geomorphology.
T. J. Parker, D. M. Schneeberger, D. C. Pieri, R. S. Saunders.
NASA Tech. Memo., NASA TM–89810, p. 319 – 321 (1987). – See Abstr. 003.001.

097.027 Geomorphic evidence for ancient seas in West Deuteronilus Mensae, Mars–II: From very high resolution Viking Orbiter images.
T. J. Parker, D. M. Schneeberger, D. C. Pieri, R. S. Saunders.
NASA Tech. Memo., NASA TM–89810, p. 322 – 324 (1987). – See Abstr. 003.001.

097.028 Explosive volcanic deposits on Mars: preliminary investigations.
D. A. Crown, L. A. Leshin, R. Greeley.
NASA Tech. Memo., NASA TM–89810, p. 327 – 329 (1987). – See Abstr. 003.001.

Although many investigators have considered explosive volcanism on Mars, currently much of the evidence for extensive deposits is tenuous, and the occurrence of large–scale explosive volcanism remains controversial. The present studies are intended to address the question in more detail with the aid of new and previously unused sources of information.

097.029 Pseudocraters as indicators of ground ice on Mars.
H. Frey.
NASA Tech. Memo., NASA TM–89810, p. 330 – 332 (1987). – See Abstr. 003.001.

097.030 Eruptive history of the Elysium volcanic province of Mars.
K. L. Tanaka, D. H. Scott.
NASA Tech. Memo., NASA TM–89810, p. 333 – 335 (1987). – See Abstr. 003.001.

New geologic mapping of the Elysium volcanic province and crater counts provide a basis for describing its overall eruptive history.

097.031 Late–stage flood lavas in the Elysium region, Mars.
J. B. Plescia.
NASA Tech. Memo., NASA TM–89810, p. 336 – 338 (1987). – See Abstr. 003.001.

097.032 Relative ages of lava flows at Alba Patera, Mars.
D. M. Schneeberger, D. C. Pieri.
NASA Tech. Memo., NASA TM–89810, p. 339 – 341 (1987). – See Abstr. 003.001.

097.033 Martian volcanism: festoon–like ridges on terrestrial basalt flows and implications for Mars.
E. Theilig, R. Greeley.
NASA Tech. Memo., NASA TM–89810, p. 342 – 344 (1987). – See Abstr. 003.001.

097.034 Revision of the Martian relative age chronology.
N. G. Barlow.
NASA Tech. Memo., NASA TM–89810, p. 413 – 415 (1987). – See Abstr. 003.001.

This study provides a more detailed chronology than currently exists in the literature and creates some changes to the currently accepted geological evolutionary sequence of Mars.

097.035 Early changes in gradation styles and rates on Mars.
P. H. Schultz, D. Britt.
NASA Tech. Memo., NASA TM–89810, p. 416 – 417 (1987). – See Abstr. 003.001.

The history of an active surface–atmosphere exchange of volatiles on Mars is recorded in the ancient cratered terrains. Large impact basins and craters provide a means to document this process and any changes in style with time.

097.036 Cratering and obliteration history of the south polar region of Mars.
J. J. Plaut, R. E. Arvidson, E. A. Guinness, R. Kahn.
NASA Tech. Memo., NASA TM–89810, p. 418 – 419 (1987). – See Abstr. 003.001.

The diverse assemblage of geological units in the south polar region of Mars implies a complex geological history, with exogenic processes strongly modulated by climatic phenomena. In this study results are presented from a comprehensive analysis of crater size frequency distributions compiled from Viking Orbiter images of south polar terrains.

097.037 Martian rampart craters: morphologic clues for the physical state of the target at time of impact.
P. J. Mouginis–Mark.
NASA Tech. Memo., NASA TM–89810, p. 420 – 422 (1987). – See Abstr. 003.001.

097.038 Crater ejecta morphology and the presence of water on Mars.
P. H. Schultz.
NASA Tech. Memo., NASA TM–89810, p. 423 – 425 (1987). – See Abstr. 003.001.

097.039 Thermal inertia characteristics of the Martian crater Curie.
V. M. Horner, J. R. Zimbelman.
NASA Tech. Memo., NASA TM–89810, p. 426 – 428 (1987). – See Abstr. 003.001.

097.040 Characteristic structures of the highland boundary on Mars: evidence against a single mega–impact event?
A. M. Semeniuk, H. Frey.
NASA Tech. Memo., NASA TM–89810, p. 467 – 468 (1987). – See Abstr. 003.001.
Wilhelms and Squyres have suggested that an early mega–impact event might explain the fundamental crustal dichotomy on Mars. Detailed morphological mapping of the structures which characterize the boundary between the cratered highlands and northern plains does not support this idea: the distribution of these features along and especially away from the boundary is more consistent with a larger number of smaller but overlapping impacts.

097.041 The Martian crustal dichotomy: product of accretion and not a specific event?
H. Frey, R. A. Schultz, T. A. Maxwell.
NASA Tech. Memo., NASA TM–89810, p. 469 – 471 (1987). – See Abstr. 003.001.
Attempts to explain the fundamental crustal dichotomy on Mars range from purely endogenic to extreme exogenic processes but to date no satisfactory theory has evolved which is generally accepted. Wilhelms and Squyres call on a single mega–impact event, essentially an instantaneous rearrangement of the crustal structure (topography and lithospheric thickness). Wise et al. (1979) prefer an internal mechanism: a period of vigorous convection subcrustally erodes the northern one third of Mars, causing foundering and isostatic lowering of that part of Mars. In this paper the authors review the evidence for each of these two extreme theories, conclude there is little to recommend or require either, and suggest an alternative.

097.042 Implications of Viking color data for evolution of the Amenthes region, Mars.
T. A. Maxwell.
NASA Tech. Memo., NASA TM–89810, p. 472 – 473 (1987). – See Abstr. 003.001.
Using higher resolution individual Viking frames, the authors identified two spectral types of plateaus in the Amenthes–Aeolis regions. The morphologic characteristics of these plateaus do not differ, but spectrally, one group is similar to the cratered terrain, whereas the other is intermediate between the cratered terrain and smooth plains. Such compositional indications are being used to map the extent of scarp retreat of cratered terrain boundary, and to infer its former position.

097.043 Timing of ancient extensional tectonic features on Mars.
R. Wichman, P. H. Schultz.
NASA Tech. Memo., NASA TM–89810, p. 474 – 475 (1987). – See Abstr. 003.001.
Although numerous studies have delineated the Tharsis and post–Tharsis volcanic/tectonic history on Mars, only a few attempts have examined the earlier epochs. Tanaka previously reported a method for determining the areal density for narrow elongate or linear areas. The authors' modification uses a discrete unbinned count where an observed arithmetic mean is calculated without an assumption about crater production. This expression has been successfully tested for arbitrary lines drawn through martian plains units with different crater ages.

097.044 Stress history of the Tharsis region, Mars.
R. A. Francis.
NASA Tech. Memo., NASA TM–89810, p. 476 – 477 (1987). – See Abstr. 003.001.
The events that led to the formation of the Tharsis region continue to generate debate. Three geophysical models for the formation of Tharsis are now in general contention and each of these models has been used to predict a characteristic stress–field.

Each of these models has been used by its proponents to predict some of the features observed in the Tharsis region but none accurately accounts for all of the fracture features observed. To constrain the origin of Tharsis, as opposed to its later history, one should look for the oldest fractures related to Tharsis and compare these to the predictions made by the models.

097.045 History and morphology of faulting in the Noctis Labyrinthus–Claritas Fossae region of Mars.
K. L. Tanaka, P. A. Davis.
NASA Tech. Memo., NASA TM–89810, p. 478 – 480 (1987). – See Abstr. 003.001.

097.046 Thin and thick–skinned deformation in the Tharsis region of Mars.
T. R. Watters.
NASA Tech. Memo., NASA TM–89810, p. 481 – 483 (1987). – See Abstr. 003.001.

097.047 Flexurally–resisted uplift of the Tharsis province, Mars.
R. J. Phillips, N. H. Sleep.
NASA Tech. Memo., NASA TM–89810, p. 484 – 486 (1987). – See Abstr. 003.001.

097.048 Tectonic domains in the eastern hemisphere of Mars.
T. A. Maxwell, C. E. Leff.
NASA Tech. Memo., NASA TM–89810, p. 487 – 488 (1987). – See Abstr. 003.001.

097.049 Age of fracturing and mesa development in the Elysium area, northern Martian plains.
G. E. McGill.
NASA Tech. Memo., NASA TM–89810, p. 489 – 491 (1987). – See Abstr. 003.001.
This abstract summarizes progress on one aspect of a long–range geological study intended to constrain hypotheses for the dichotomy by tracing the history of the northern plains from the most recent events backward – essentially the same approach used to understand old events in Earth history.

097.050 Geometry and relative age of large patterned fractures in southern Acidalia Planitia, Mars.
M. C. Borrello.
NASA Tech. Memo., NASA TM–89810, p. 492 – 494 (1987). – See Abstr. 003.001.

097.051 Origin of fractures, Martian polygonal terrain.
L. S. Hills.
NASA Tech. Memo., NASA TM–89810, p. 495 – 496 (1987). – See Abstr. 003.001.

097.052 Curvilinear ridges and related features in southwest Cydonia Mensae, Mars.
T. J. Parker, D. M. Schneeberger, D. C. Pieri, R. S. Saunders.
NASA Tech. Memo., NASA TM–89810, p. 502 – 504 (1987). – See Abstr. 003.001.

097.053 Progress in compilation of the 1:2,000,000–scale topographic map series of Mars.
S. S. C. Wu, R. Jordan, F. J. Schafer.
NASA Tech. Memo., NASA TM–89810, p. 530 (1987). Abstract. – See Abstr. 003.001.

097.054 Mars digital terrain model.
S. S. C. Wu, A.–E. Howington.
NASA Tech. Memo., NASA TM–89810, p. 531 – 533 (1987). – See Abstr. 003.001.

097.055 Digital cartography of Mars.
R. M. Batson.
NASA Tech. Memo., NASA TM–89810, p. 534 – 535 (1987). – See Abstr. 003.001.
A medium–resolution Digital Image Model (DIM) of Mars is being compiled. A DIM is a mosaic of radiometrically corrected,

photometrically modelled spacecraft images displaying accurate reflectance properties at uniform resolution, and geometrically tied to the best available control.

097.056   The control network of Mars: October 1986.
M. E. Davies.
NASA Tech. Memo., NASA TM–89810, p. 536 (1987). – See Abstr. 003.001.

097.057   Martian terrains.
H. Masursky, M. G. Chapman, P. A. Davis, A. L. Dial Jr., M. E. Strobell.
NASA Tech. Memo., NASA TM–89810, p. 545 – 547 (1987). – See Abstr. 003.001.

Terrain studies of candidate landing sites for a future rover/sample–return mission to Mars are being conducted to evaluate the geologic and trafficability aspects of each site.

097.058   The oblateness effect on the mean seasonal daily insolations at the Martian surface during global dust storms.
E. Van Hemelrijck.
Earth, Moon, Planets, Vol. 38, No. 3, p. 209 – 216 (1987). With a correction in Vol. 39, No. 3, p. 291 (1987).

The combined effect of global dust storms and the oblateness on the mean seasonal daily insolations at the Martian surface is investigated. Due to the flattening, the mean summertime insolation is increased at equatorial and low latitudes, decreased at mid– and high latitudes. In winter, the mean daily insolations are decreased over the entire latitudinal interval; the maximum values are found at polar region latitudes.

097.059   Ancient dorsa–related stresses of the Tharsis region on Mars.
J. T. Raitala.
Earth, Moon, Planets, Vol. 38, No. 3, p. 285 – 297 (1987).

Topographic information, surface structures and construction of the Martian Tharsis bulge are used to estimate the previous stresses across the low–lying peripheral margins of the crustal blocks in terms of simple compensation models. Hot mantle activity, crustal roots, isostasy, and late–stage extensive lithosphere thickening together with volcanic building have been in combined response to the high–elevated Tharsis bulge.

097.060   Observations de Mars en 1984 et 1986: apparition des prémices de la couverture polaire et formation de la couverture hivernale.
S. Ebisawa, A. Dollfus.
Astronomie, Vol. 101, p. 403 – 410 (1987).

097.061   Analysis of Martian ionosphere and solar wind electron gas data from the planar retarding potential analyzer on the Viking spacecraft.
G. P. Mantas, W. B. Hanson.
J. Geophys. Res., Vol. 92, No. A8, p. 8559 – 8569 (1987).

Approximate expressions for the electron current collected by a planar retarding potential analyzer (RPA) mounted on a moving, conducting, charged spacecraft are derived. They are utilized for the analysis of electron current data obtained by the RPAs on the Viking spacecraft in the ionosphere of Mars and in the disturbed and undisturbed solar wind near this planet.

097.062   Mars: VLA observations of the Northern Hemisphere and the North Polar Region at wavelengths of 2 and 6 cm.
D. J. Rudy, D. O. Muhleman, G. L. Berge, B. M. Jakosky, P. R. Christensen.
Icarus, Vol. 71, No. 1, p. 159 – 177 (1987).

Observations of Mars at wavelengths of 2 and 6 cm were made using the VLA in its A configuration. Whole–disk brightness temperatures were estimated to be $193.2 \pm 1.0K$ at 2 cm and $191.2 \pm 0.6K$ at 6 cm. Calculations yielded a whole–disk effective dielectric constant of 2.34 ± 0.05, which implied a subsurface density of $1.24 \pm 0.11 \, g \, cm^{-3}$ at 2 cm. The same calculations at 6 cm yielded an effective density of $1.45 \pm 0.10 \, g \, cm^{-3}$ and dielectric constant of 2.70 ± 0.10. From the mapped data these parameters were also estimated as a function of latitude between latitudes of 15°S and 60°N. In addition the radio absorption length of the subsurface was estimated.

097.063   A brief report on the 1986 perihelic apparition of Mars.
J. D. Beish.
Strolling Astron., Vol. 32, Nos. 3 – 4, p. 79 (1987).

097.064   The case for a wet, warm climate on early Mars.
J. B. Pollack, J. F. Kasting, S. M. Richardson, K. Poliakoff.
Icarus, Vol. 71, No. 2, p. 203 – 224 (1987). Paper presented at the "Symposium on Mars: evolution of its climate and atmosphere", Washington, D.C., July 1986.

Theoretical arguments are presented in support of the idea that Mars possessed a dense CO_2 atmosphere and a wet, warm climate early in its history.

097.065   Volatiles on Earth and Mars: a comparison.
G. Dreibus, H. Wänke.
Icarus, Vol. 71, No. 2, p. 225 – 240 (1987). Paper presented at the "Symposium on Mars: evolution of its climate and atmosphere", Washington, D.C., July 1986.

Based on new and previous determinations of halogens in SNC meteorites, the bulk concentrations of halogens in the parent body, which is thought to be Mars, are estimated. The two–component model for the formation of terrestrial planets as proposed by A. E. Ringwood and H. Wänke is further substantiated. It is postulated that the present atmospheres of Venus, Earth, and Mars were formed by degassing the interiors of the planets. It is also postulated that the large differences in the amounts of primordial rare gases in the atmospheres of Venus, Earth, and Mars are due mainly to different loss factors. Except for gaseous species, Mars is found to be richer in volatile (halogens) and moderately volatile elements than the Earth.

097.066   Carbon dioxide: adsorption on palagonite and partitioning in the Martian regolith.
A. P. Zent, F. P. Fanale, S. E. Postawko.
Icarus, Vol. 71, No. 2, p. 241 – 249 (1987). Paper presented at the "Symposium on Mars: evolution of its climate and atmosphere", Washington, D.C., July 1986.

The authors report new CO_2 adsorption measurements on palagonites. These results are used together with earlier results on basalt and nontronite adsorption to derive a "generic" relationship which is valid to within a factor of 3 for likely mixtures of basalt and weathering products of basalt. The relationship involves only T, P (CO_2), and the specific surface area, and is relatively insensitive to mineralogy. It is used to predict the distribution and exchange of CO_2 on Mars.

097.067   Comets, volcanism, the salt–rich regolith, and cycling of volatiles on Mars.
B. C. Clark.
Icarus, Vol. 71, No. 2, p. 250 – 256 (1987). Paper presented at the "Symposium on Mars: evolution of its climate and atmosphere", Washington, D.C., July 1986.

Estimates of the total inventory of the volatile elements C, H, O, and N on Mars, based upon atmospheric gas tracers, vary by a factor of 25 among different authors. Accretion of comets as the source of volatiles can account for less than 5% of the actual inventory, assuming the chondritic S/Cl abundance ratio in comets and a Martian outgassing ratio for these two elements no lower than for the estimated excess volatile inventory on Earth. Sulfate salt formation with the igneous minerals in the regolith can be a major sink for H_2O, but first will recycle C and N incorporated in carbonate and nitrate minerals back to the atmosphere. Extrusive and shallow intrusive volcanism, at a persistent but decreasing rate, can interfere with this volatile recycling by irreversibly masking much of the incorporated inventory, resulting in the inevitable evolution to a relatively volatile–poor environment at the outermost, observable surface of Mars.

097.068   Spatial resolution and the geologic interpretation of Martian morphology: implications for subsurface volatiles.
J. R. Zimbelman.
Icarus, Vol. 71, No. 2, p. 257 – 267 (1987). Paper presented at the "Symposium on Mars: evolution of its climate and atmosphere", Washington, D.C., July 1986.

Viking Orbiter images with a spatial resolution > 50 m/pixel provide valuable information about the tectonic and stratigraphic relationships of surface terrain units, but this resolution cannot provide strong constraints on the relative importance of geologic processes that have acted upon the surface. Images with a spatial resolution < 10 m/pixel reveal the presence of features, such as small mounds interpreted to be aeolian dunes, that are significant to a complete evaluation of the surface history but which may be invisible at > 50 m/pixel resolution. Landforms proposed as indicators of subsurface volatiles will only be as useful as the degree to which their geologic history can be assessed from the available image resolution.

097.069   Water or ice in the Martian regolith?: clues from rampart craters seen at very high resolution.
P. J. Mouginis–Mark.
Icarus, Vol. 71, No. 2, p. 268 – 286 (1987). Paper presented at the "Symposium on Mars: evolution of its climate and atmosphere", Washington, D.C., July 1986.

Very high resolution Viking Orbiter images (8 – 17 m per pixel) have been used to investigate the morphology of Martian rampart crater ejecta blankets and the crater interiors, with the objective of identifying the fluidizing medium for the ejecta and the physical properties of the target rock. The occurrence of well-preserved, small–scale pressure ridges and scour marks, evidence for subsidence around isolated buried blocks in partially eroded ejecta lobes, and the stability of crater walls and distal ramparts argue for ground ice being the dominant state for volatiles within the target rocks at the time of impact. Rare examples of channels (190 – 650 m wide) on the surfaces of ejecta blankets, and on the inner walls of the crater Cerulli, indicate that in some instances liquid water was incorporated into the ejecta during its emplacement.

097.070   Ring furrows: inversion of topography in Martian highland terrains.
R. A. De Hon.
Icarus, Vol. 71, No. 2, p. 287 – 297 (1987). Paper presented at the "Symposium on Mars: evolution of its climate and atmosphere", Washington, D.C., July 1986.

Ring furrows are flat–floored trenches, circular in plan view, surrounding a central, flat–topped, mesa or plateau. The internal plateau is about the same elevation or lower than the plain outside the ring. The outer wall is often breached by valley drainage or opened to low, degraded surfaces. Related landforms range from ring furrows with fractured central plateaus to those with isolated circular mesas without depressed rings. Ring furrows are superposed on many types of materials, but they are most common on cratered plateau–type materials which are interpreted as volcanic flows overlying ancient cratered terrain. Most rings occur in or near regions of fretted terrain. Ring furrows are formed by preferential removal of the exposed rims of partially buried craters. Ground ice decay and sapping followed by fluvial erosion are proposed for removal of the least resistant rim materials.

097.071   Interannual variability of Mars' south polar cap.
P. B. James, K. M. Malolepszy, L. J. Martin.
Icarus, Vol. 71, No. 2, p. 298 – 305 (1987). Paper presented at the "Symposium on Mars: evolution of its climate and atmosphere", Washington, D.C., July 1986.

Telescopic data on the twentieth–century regressions of Mars' south polar cap have been reexamined for evidence of interannual variability. Several regressions, particularly that of 1956, are found to differ significantly from the mean. The possibility of correlations with major dust storm is explored.

097.072   Martian north polar cap and circumpolar clouds: 1975 – 1980 telescopic observations.
P. B. James, M. Pierce, L. J. Martin.
Icarus, Vol. 71, No. 2, p. 306 – 312 (1987). Paper presented at the "Symposium on Mars: evolution of its climate and atmosphere", Washington, D.C., July 1986.

International Planetary Patrol pictures from 1975 – 1980 have been used to derive partial regression curves for Mars' north polar cap during the 3 Martian years included in the study. The results confirm the existence of a surface cap during late winter and early spring which extends to at least 55°N lat and the mid-spring plateau in cap recession noted by previous studies. Comparison of images aquired using different filters reveals that a substantial increase in circumpolar clouds accompanies the resumption of cap sublimation at the end of the plateau period.

097.073   Dynamical modeling of a planetary wave mechanism for a Martian polar warming.
J. R. Barnes, J. L. Hollingsworth.
Icarus, Vol. 71, No. 2, p. 313 – 334 (1987). Paper presented at the "Symposium on Mars: evolution of its climate and atmosphere", Washington, D.C., July 1986.

A dynamical mechanism for the Martian (atmospheric) polar warming observed by the Viking IRTM during the 1977 winter solstice dust storm (and a similar one possibly observed by Mariner 9 in 1971) is proposed, and investigated using a simplified nonlinear model. The model is of a type previously used to successfully simulate the essential aspects of terrestrial sudden stratospheric warmings. The dynamical mechanism is, in part, very similar fundamentally to that responsible for these warmings, involving planetary–scale waves. A number of numerical experiments have been conducted to assess the basic variability of such a mechanism for the Martian polar warming and to examine its sensitivity to several factors.

097.074   The inventory and distribution of water on Mars.
M. H. Carr.
Adv. Space Res., Vol. 7, No. 5, p. 85 – 94 (1987). – See Abstr. 012.022.

The role of water in the evolution of the martian surface remains one of the most puzzling aspects of martian geology. In this paper the volatile inventory is assessed from geologic evidence.

097.075   Strategies for the future exploration of Mars.
H. P. Klein.
Adv. Space Res., Vol. 7, No. 5, p. 95 – 98 (1987). – See Abstr. 012.022.

While the Viking mission yielded a wealth of scientific information about Mars, many intriguing new questions were raised about its chemical and physical environment. For many reasons, Mars thus continues to be an object of intense scientific interest. In addition, many scenarios for the further exploration of that planet have been advanced in recent years, and these is also keen public interest in future Mars missions. In looking ahead, one important aspect in planning a strategy for the exploration of Mars is whether or not to assume that Mars is a dead planet and also whether it is capable of supporting the growth of terrestrial organisms. Three very different mission strategies are presented, depending upon how these assumptions are made.

097.076   Deflection of Mars from the hydrostatic equilibrium figure and its interpretation.
Yu. A. Tarakanov, V. P. Trubitsyn, N. Sh. Kambarov, V. A. Prikhod'ko.
Pis'ma Astron. Zh., Tom 13, No. 9, p. 824 – 829 (1987). In Russian. English translation in Sov. Astron. Lett., Vol. 13.

The second and fourth hydrostatic coefficients were obtained by means of the principle of adjustment of Stoke's constants and measured precession constant of Mars. The sources of the gravitational anomalies were calculated relative to the hydrostatic spheroid with flattening of 1/185. The sources of the anomalies are located at the lithosphere–mantle boundary. The thickness of the lithosphere is estimated to be about 400 km.

097.077   Future exploration of Mars.
G. E. Hunt.
1988 Yearbook of Astronomy, p. 147 – 155 (1987). – See Abstr. 003.006.

097.078   Modeling the spectral dependence of the albedos of Phobos and Deimos.
Yu. G. Shkuratov, N. P. Stadnikova, S. N. Yarmolenko.
Sov. Astron., Vol. 30, No. 6, p. 698 – 701 (1987). English translation of 42.097.044.

097.079   Martian satellites.
V. N. Zharkov, A. V. Kozenko.
Priroda, No. 9, p. 4 – 11 (1987). In Russian.

097.080   Phobos: crevices or grooves?
V. P. Belov.
Priroda, No. 9, p. 12 – 17 (1987). In Russian.

097.081   Radiometry of Deimos.
G. J. Veeder, D. L. Matson, E. F. Tedesco, L. A. Lebofsky, J. C. Gradie.
Astron. J., Vol. 94, No. 5, p. 1361 – 1363 (1987).
The authors report ground–based infrared photometry of Deimos at 4.8, 10, and 20 μm. The observed fluxes are significantly brighter than predicted by the "standard" thermal model. Recent recalibrations that modify the model beam pattern of the infrared emission are marginally consistent with the observations at 10 and 20, but not at 4.8 μm.

097.082   The meteorology of Mars – Part III.
J. D. Beish, D. C. Parker.
Strolling Astron., Vol. 32, Nos. 5 – 6, p. 101 – 114 (1987).
This paper is the final part of the A.L.P.O. Mars Section Reports on the meteorology of Mars. Final statistical analyses are presented along with tables and graphs of the Martian meteorology observed during the apparitions from 1964 – 66 through 1983 – 85. Tables are presented in order to be used as test cases for predicting the Martian meteorology observed during the 1985 – 87 and 1987 – 89 apparitions.

097.083   Possible tornado–like tracks on Mars.
J. A. Grant, P. H. Schultz.
Science, Vol. 237, No. 4817, p. 883 – 885 (1987).
Distinct atmospheric conditions suggest that dark, ephemeral, filamentary lineations on the martian surface may be formed during the passage of intense atmospheric vortices.

097.084   Relative ages and the geologic evolution of martian terrain units.
N. G. Barlow.
Diss. Abstr. Int., Sect. B, Vol. 48, No. 2, p. 472–B (1987). Thesis, The University of Arizona, 139 pp. (1987). Order No. DA8711625.

097.085   Dayside temperatures of the Martian upper atmosphere.
H. Bittner, K. H. Fricke.
J. Geophys. Res., Vol. 92, No. A11, p. 12,045 – 12,055 (1987).
One central problem in understanding the Martian upper atmosphere is the poor correlation between exospheric temperatures and the energy input from the Sun in the EUV and UV. Turbulence heats the atmosphere by dissipation of turbulent energy and cools it by downward heat transport. A time–variable turbulence may introduce a stochastic component in addition to the solar–driven, regular variation of the exospheric temperatures. To investigate the possible range of temperatures on the basis of this assumption, the authors develop a one–dimensional mean–dayside model of the energy balance of the Martian upper atmosphere. With plausible assumptions on the range of the eddy

diffusion coefficient, they find a stochastic component of ± 63K for the exospheric temperatures. The comparison of observed data with the results of their model yields a best value for the efficiency of the heating by absorption of solar ultraviolet radiation of 0.145 ± 0.05.

097.086   Polar basal melting on Mars.
S. M. Clifford.
J. Geophys. Res., Vol. 92, No. B9, p. 9135 – 9152 (1987).
The potential importance of basal melting on Mars is illustrated through the discussion of four examples: (1) the origin of the major polar reentrants, (2) the removal and storage of an ancient Martian ice sheet, (3) the mass balance of the polar terrains, and (4) the possibility of basal melting at temperate latitudes. This analysis suggests that the process of basal melting may play a key role in understanding the evolution of the Martian polar terrains and the long–term climatic behavior of water on Mars.

097.087   Investigation of Martian H_2O and CO_2 via orbital gamma ray spectroscopy.
L. G. Evans, S. W. Squyres.
J. Geophys. Res., Vol. 92, No. B9, p. 9153 – 9167 (1987).
The authors examine ways in which orbital gamma ray data can be used to address a number of important scientific questions regarding H_2O and CO_2 on Mars.

097.088   Generation of liquid water on Mars through the melting of a dusty snowpack.
G. D. Clow.
Icarus, Vol. 72, No. 1, p. 95 – 127 (1987).
The possibility that snowmelt could have provided liquid water for valley network formation early in the history of Mars is investigated using an optical–thermal model developed for dusty snowpacks at temperate latitudes. The heating of the postulated snow is assumed to be driven primarily by the absorption of solar radiation during clear sky conditions. Radiative heating rates are predicted as a function of depth and are shown to be sensitive to the dust concentration and the size of the ice grains while the thermal conductivity is controlled by temperature, atmospheric pressure, and bulk density.

097.089   Valles Marineris, Mars: wet debris flows and ground ice.
B. K. Lucchitta.
Icarus, Vol. 72, No. 2, p. 411 – 429 (1987).
The author first discusses landslides and present observations suggesting that they were wet when emplaced; second, the walls of Valles Marineris are discussed and it is explained why they may have been underlain by mixtures of rock and ice. A final part of this reports addresses the temporal aspects of both landslide emplacement and the presence of ground ice in the equatorial area of Mars.

097.090   Infrasound generation under radiative–mechanical synchronism. I. Terminator's variant in the system Sun–Martian atmosphere.
V. P. Vasil'ev.
Kinematika Fiz. Nebesn. Tel, Tom 3, No. 6, p. 3 – 9 (1987). In Russian. English translation in Kinematics Phys. Celest. Bodies.
It is shown that in the Martian atmosphere conditions for a speed–resonance infrasonic terminator wave generation of thermoelastic nature are realized. In accordance with the observational data these waves arise in the troposphere at planetographic latitudes of $\sim 10 - 30°$ (depending on the season) with the basic–period of about 3 min and characteristic wavelength of about 40 km.

097.091   Mars: high resolution VLA observations at wavelengths of 2 and 6 cm and derived properties.
D. J. Rudy.
Diss. Abstr. Int., Sect. B, Vol. 48, No. 6, p. 1712–B (1987). Thesis, California Institute of Technology, 156 pp. (1987). Order No. DA8719700.

097.092 **Mars: geological regions and crater population.**
M. A. Kazennov.
Astron. Tsirk., No. 1472, p. 1 – 2 (1986). In Russian.

The population of craters of four stages of sharpness have been investigated in five geological regions of Mars. An analysis of correlation between albedo, altitude of surface and density of craters has been made. It is supposed that the dark material in different geological regions has various composition; for example, volcanic lava, aeolian dunes or ancient primary material.

097.093 **Mars: north–polar atmospheric warming during dust storms.**
B. M. Jakosky, T. Z. Martin.
Icarus, Vol. 72, No. 3, p. 528 – 534 (1987).

Measurements of winter polar atmospheric temperatures on Mars are presented for the two 1977 dust storms observed by Viking. These data show a strong time and spatial dependence of the polar warming and provide important boundary conditions for dynamical models of the phenomenon.

097.094 **Tectonics of Tharsis dorsa on Mars.**
J. Raitala.
Earth, Moon, Planets, Vol. 39, No. 3, p. 275 – 289 (1987).

The tectonics of the Tharsis and adjoining areas is considered to be associated wih the convection in the Martian mantle. Convection and mantle plume have been responsible for the primary uplift and volcanism of the Tharsis area. The radial compressional forces generated by the tendency for downslope movement of surface strata, vertical volcanic intrusions and traction of mantle spreading beneath Tharsis were transmitted through the lithosphere to form peripheral mare ridge zones. The locations of mare ridges were thus mainly controlled by the Tharsis–radial compression. The load–induced stresses then contributed on further ridge formation over an extended period of time by the isostatic readjustment which was responsible for long–term stresses in the adjoining areas. Extrusions, changes in internal temperature and possible phase changes may also have caused changes in mantle volume giving rise to additional compressional forces and crustal deformations.

097.095 **Volcanic input to the atmosphere from Alba Patera on Mars.**
L. Wilson, P. J. Mouginis–Mark.
Nature, Vol. 330, No. 6146, p. 354 – 357 (1987).

The authors present what they believe to be the first estimates for the amount of water vapour and/or carbon dioxide released, together with the corresponding release rates, from specific volcanic deposits on a relatively young martian volcanic construct (Alba Patera).

Martian geomorphology and its relation to subsurface volatiles. Proceedings of a MECA special session at the XVII. Lunar and Planetary Science Conference, held at Houston, TX, USA, 17 March 1986 (Abstracts of papers).
See Abstr. 012.085.

Reflectance spectra of mafic silicates and phyllosilicates from 0.6 to 4.6 μm.
See Abstr. 022.006.

Wind ripples in low density atmospheres.
See Abstr. 022.012.

Non–equilibrium freezing of water–ice in sandy basaltic regoliths and implications for fluidized debris flows on Mars.
See Abstr. 022.019.

Undercooled water in basaltic regoliths and implications for fluidized debris flows on Mars.
See Abstr. 022.134.

The Mars Observer Camera.
See Abstr. 035.002.

Enhanced Landsat images of Antarctica and planetary exploration.
See Abstr. 036.009.

The Phobos project.
See Abstr. 051.009.

Balloons for Mars missions.
See Abstr. 051.064.

Small satellites.
See Abstr. 091.037.

Colombo's Top.
See Abstr. 091.049.

Characteristics of the Marslike limit of the Venus–solar wind interaction.
See Abstr. 093.012.

Basin–ring spacing on the Moon, Mercury, and Mars.
See Abstr. 094.023.

The investigation of intersecting craters on the Moon, Mars and Mercury.
See Abstr. 094.041.

Calcium carbonate and calcium sulfate in Martian meteorite EETA 79001.
See Abstr. 105.003.

Martian meteorites are arriving.
See Abstr. 105.036.

The large crater origin of SNC meteorites.
See Abstr. 105.037.

098 Minor Planets

098.001 Observations on the magnitude–frequency distribution of earth–crossing asteroids.
E. M. Shoemaker, C. S. Shoemaker.
NASA Tech. Memo., NASA TM–89810, p. 72 – 74 (1987). – See Abstr. 003.001.

098.002 Asteroid families, dynamics, and astrometry.
J. G. Williams, J. Gibson.
NASA Tech. Memo., NASA TM–89810, p. 75 – 76 (1987). – See Abstr. 003.001.

098.003 1986 DA and 1986 EB: M–class asteroids in near–earth orbits.
J. Gradie, E. Tedesco.
NASA Tech. Memo., NASA TM–89810, p. 77 – 79 (1987). – See Abstr. 003.001.

098.004 Evolution of the inner asteroid belt: paradigms and paradoxes from spectral studies.
M. J. Gaffey.
NASA Tech. Memo., NASA TM–89810, p. 80 (1987). – See Abstr. 003.001.

098.005 Meteorite spectroscopy and characterization of asteroid surface materials.
M. J. Gaffey.
NASA Tech. Memo., NASA TM–89810, p. 81 – 83 (1987). – See Abstr. 003.001.

098.006 Asteroid lightcurve inversion.
S. J. Ostro, R. Connelly.
NASA Tech. Memo., NASA TM–89810, p. 90 – 91 (1987). – See Abstr. 003.001.

098.007 Physical studies of asteroids XVI: photoelectric photometry of 17 asteroids.
C.-I. Lagerkvist, G. Hahn, P. Magnusson, H. Rickman.
Astron. Astrophys., Suppl. Ser., Vol. 70, No. 1, p. 21 – 32 (1987).
The authors present photoelectric photometry of 17 asteroids observed during the period 1979 – 1986. Composite lightcurves have been derived for 76, 236, 439, 516 and 591, and at least a reasonable single–lightcurve coverage has been obtained for the majority of the remaining ones. Reference is made to Tholen's (1984) taxonomic classification and a brief discussion of light-curve statistics in this framework is given. Differences between the spin rates and amplitudes of the C, S and M classes are considered and previous conclusions regarding the peculiar characteristics of M asteroids are confirmed. An indication is found that C and S asteroids have a particularly slow spin when situated near the 3/1 or 5/2 Kirkwood gaps.

098.008 Reports of asteroidal appulses and occultations.
J. Stamm.
Occultation Newsl., Vol. 4, No. 5, p. 106 – 108 (1987).

098.009 More information on two asteroidal appulses.
R. Boninsegna.
Occultation Newsl., Vol. 4, No. 5, p. 108 (1987).

098.010 Planetoide voor Bert van Sprang.
Zenit, Jaarg. 14, Nr. 7/8, p. 255 (1987).

098.011 V + I photoelectric photometry of asteroids 4 Vesta and 9 Metis.
F. J. Melillo.
Minor Planet Bull., Vol. 14, No. 3, p. 21 – 23 (1987).
V + I photometry observations of the asteroids 4 Vesta and 9 Metis were made from North Valley Stream Observatory during the period of Oct. 7 to Dec. 5, 1986 UT. For both asteroids V + I lightcurves help to determine the rotation period more accurately and to detect any albedo variations.

098.012 General report of position observations by the ALPO Minor Planets Section for the year 1986.
F. Pilcher.
Minor Planet Bull., Vol. 14, No. 3, p. 23 – 28 (1987).
Observations of positions of minor planets by members of the Minor Planets Section in calendar year 1986 are summarized, along with significant discrepancies from the predicted magnitudes noted for a few planets.

098.013 Orbital statistics of Apollo–Amors.
C. J. Cunningham.
Minor Planet Bull., Vol. 14, No. 3, p. 29 – 31 (1987).
A survey of the orbital elements of the Apollo–Amors is presented. Their distribution in q–a space is shown, followed by a histogram of their Jacobi constants. Apollo–Amors possibly of cometary orgin are identified on the basis of Kresak's P criterion.

098.014 Filling the gaps: missing longitude coverage for large rapidly rotating asteroids.
D. R. Davis, R. P. Binzel.
Minor Planet Bull., Vol. 14, No. 3, p. 31 – 33 (1987).
In modelling asteroid shapes from lightcurves, observations are needed from a range of ecliptic longitudes. Ecliptic longitude ranges in which there are no published lightcurves for selected target asteroids are tabulated.

098.015 Comments on the new magnitude system for asteroids.
G. R. Harvey.
Minor Planet Bull., Vol. 14, No. 3, p. 34 – 35 (1987).
Comments on the new magnitude system for asteroids are presented. The 1998 opposition of (2) Pallas is explored with various slope parameters. Personal experience with the system is noted.

098.016 Index to unpublished lightcurve data.
R. P. Binzel.
Minor Planet Bull., Vol. 14, No. 4, p. 39 – 41 (1987).
An index to unpublished asteroid lightcurve data is given. The purpose of this list is to foster collaboration among observers so that improved results can be obtained.

098.017 Photoelectric photometry of asteroids 5 Astraea and 22 Kalliope.
F. J. Melillo.
Minor Planet Bull., Vol. 14, No. 4, p. 42 – 43 (1987).
Photoelectric photometry observations of the asteroids 5 Astraea and 22 Kalliope were made from North Valley Stream Observatory in February 1987 and December 1986. Astraea and Kalliope displayed lightcurve amplitudes of 0.20 and 0.16 magnitude, respectively.

098.018 On the rotational period of the asteroid (4) Vesta.
A. N. Aksenov, Yu. A. Egorov, V. G. Tejfel',
G. A. Kharitonova.
Pis'ma Astron. Zh., Tom 13, No. 7, p. 616 – 620 (1987). In Russian. English translation in Sov. Astron. Lett., Vol. 13.
The electrophotometric observations of Vesta were carried out in September 1986 during seven successive nights within the time intervals from 4.5 to 5.3 hours. From these data the reliable mean light curves were plotted separately for even and odd 5^h342 cycles. A comparison of these light curves has shown their coincidence in all details as well as in their amplitudes. It gives a good reason to suppose that the true period of the Vesta rotation is equal to the period of brightness variations (5^h342) but not twice as much.

098.019 Coordinated observations of asteroids 1219 Britta and 1972 Yi Xing.
R. P. Binzel, A. L. Cochran, E. S. Barker, D. J. Tholen,
A. Barucci, M. Di Martino, R. Greenberg,
S. J. Weidenschilling, C. R. Chapman, D. R. Davis.
Icarus, Vol. 71, No. 1, p. 148 – 158 (1987).

Britta's sidereal rotation period is found to be 5.57497 ± 0.00013 hr and its rotation is retrograde. The lightcurve amplitude ranged from 0.60 to 0.70 mag, depending on phase angle. Britta can be classified as an S–type asteroid based on its measured spectra and albedo. The absolute magnitude and slope parameter derived from the lightcurve maxima are $H_0 = 11.67 \pm 0.03$ and $G_0 = 0.30 \pm 0.04$. A 0.002 mag deg^{-1} phase reddening in B–V was also measured. 1972 Yi Xing was less well observed but a unique synodic period of 14.183 ± 0.003 hr was determined. The observed lightcurve amplitude was 0.18 mag. Five-color measurements are consistent with an S–type classification. For an assumed slope parameter $G = 0.25$, Yi Xing's (lightcurve maximum) absolute magnitude $H_0 = 13.32 \pm 0.01$.

098.020 Radar astrometry of near–earth asteroids.
D. K. Yeomans, S. J. Ostro, P. W. Chodas.
Astron. J., Vol. 94, No. 1, p. 189 – 200 (1987).

In an effort to assess the extent to which radar observations can improve the accuracy of near–Earth asteroid ephemerides, an uncertainty analysis has been conducted using four asteroids with different histories of optical and radar observations.

098.021 Averaging the elliptic asteroidal problem near a first–order resonance.
S. Ferraz–Mello.
Astron. J., Vol. 94, No. 1, p. 208 – 212 (1987).

A derivation of a completely integrable dynamical system that represent the averaged motion of an asteroid moving in a first–order resonance with Jupiter is presented. In this system Jupiter lies on an elliptic orbit and the model is more suited for the analysis of the gravitational phenomena in the asteroidal belt than classical integrable models imposing a circular motion on Jupiter. The model reproduces the coupling of oscillations found in numerical calculations and may be used as an intermediate orbit in averaging methods, as well as in studying the motion of resonant asteroids and the motion of small planetary satellites or ring particles involved in first–order resonance with a large satellite.

098.022 Rotation and variability of the large C–type asteroid 375 Ursula.
H. J. Schober.
Astron. Astrophys., Vol. 183, No. 1, p. 151 – 155 (1987).

The large C–type asteroid 375 Ursula with a diameter of 216 ± 10 km (Millis et al., 1984) was observed photo–electrically in UBV in August 1981. It is one of the very few asteroids with a diameter larger than 200 km for which no rotation rate was known. From three consecutive observing nights the rotation period of $P = 16^h83 \pm 0^h07$ corresponding to $0^d701 \pm 0^d003$ was derived. The resulting light curve shows double mode wave structure with primary and secondary extrema and a total amplitude of 0.17 mag. This proves that 375 Ursula is not totally spherical as suggested by Millis et al. (1984). Absolute magnitude and colors are derived to be $V(1,0) = 7.63 \pm 0.01$ (for the primary maximum), $B-V = 0.66 \pm 0.05$ and $U-B = 0.29 \pm 0.06$.

098.023 Spectrometry of the minor planets: the $0.44 - 0.56\,\mu$ spectrum of Juno.
L. F. Golubeva.
Sov. Astron. Lett., Vol. 12, No. 5, p. 335 – 336 (1987). English translation of 42.098.008.

098.024 Minor planets discoveries at ESO, La Silla, in February 1985. Error simulation.
H. Debehogne.
Acta Astron., Vol. 37, No. 1, p. 115 – 140 (1987).

This paper contains the positions of 6 discoveries and observations of numbered and one unnumbered asteroids, some of them being observed for orbit improvement and theoretical work. Moreover, the author gives the dependences and the residuals of the reference stars for future studies, for future improvements, and for error effect studies.

098.025 Pole coordinates and phase dependence of brightness of the asteroid (21) Lutetia.
D. F. Lupishko, F. P. Velichko, I. N. Bel'skaya,
V. G. Shevchenko.
Kinematika Fiz. Nebesn. Tel, Tom 3, No. 5, p. 36 – 38 (1987). In Russian. English translation in Kinematics Phys. Celest. Bodies.

Photometric observations of Lutetia in 1985 are presented. Using these data as well as observations in 1981 and 1983 the coordinates of the pole of the asteroid $\lambda_{01} = 42°$; $\beta_{01} = 40°$ or $\lambda_{02} = 223°$; $\beta_{02} = 48°$, the semiaxes ratios $1.36{:}1.09{:}1$ and the magnitude phase dependence for phase angles ranging from 1.6 to 25° are determined.

098.026 Spectrometry of asteroids. (4) Vesta: $0.48 - 0.55\,\mu$ spectral region.
L. F. Golubeva, D. G. Pogosbekov, D. I. Shestopalov.
Sov. Astron., Vol. 30, No. 6, p. 696 – 697 (1987). English translation of 42.098.075.

098.027 Dynamical relations between asteroids, meteorites and Apollo–Amor objects.
G. W. Wetherill.
Philos. Trans. R. Soc. London, Ser. A, Vol. 323, No. 1572, p. 323 – 337 (1987). – See Abstr. 012.026.

This paper describes recent work that seeks to identify distinct portions of the asteroid belt as source regions for the most abundant type of stony meteorite (ordinary chondrites) and the majority of differentiated stony meteorites. Many of the Earth–approaching Apollo–Amor objects of 1 km in diameter are members of the same "collision hierarchy" of asteroidal debris produced in these same regions.

098.028 Statistisches zu Entdeckern von Kleinplaneten.
J. Jahn.
KPM, Jahrg. 2, Nr. 6, p. 51 (1987).

098.029 Asteroid 5025 P–L, comet 1967 II Rudnicki, and the Taurid meteoroid complex.
D. Olsson–Steel.
Observatory, Vol. 107, No. 1079, p. 157 – 160 (1987).

It is shown that the Taurid system of small interplanetary bodies in low–inclination, short period, high–eccentricity orbits contains in addition a Jupiter–crossing Apollo asteroid (5025 P–L) and a long–period comet (1967 II Rudnicki), the existence of which may give important clues to the origin and evolution of the overall complex.

098.030 Identifications and identification changes of minor planets.
Minor Planet Circ., Nos. 11887, 12025, 12165 – 12166, 12253, 12360, 12494 (1987).

098.031 Ephemerides of minor planets and comets.
Minor Planet Circ., Nos. 12018 – 12024, 12144 – 12164, 12211 – 12252, 12326 – 12358, 12460 – 12492, 12585 – 12624 (1987).

098.032 Orbital elements of one–opposition minor planets.
Minor Planet Circ., Nos. 11981 – 11982, 12116 – 12118, 12186 – 12187, 12301 – 12302, 12431 – 12432, 12535 – 12560 (1987).

098.033 New names of minor planets.
Minor Planet Circ., Nos. 12012 – 12018, 12208 – 12211, 12456 – 12459 (1987).

098.034 Critical list of minor planets.
Minor Planet Circ., Nos. 12493 – 12494 (1987).

098.035 Index to orbital elements of comets and minor planets.
Minor Planet Circ., Nos. 12026 – 12031 (1987).
Index concerning all elements published in Minor Planet Circ., Nos. 11000 – 12012.

098.036 Observations of minor planets.
Minor Planet Circ., Nos. 11896 – 11981, 12034 – 12116, 12170 – 12186, 12257 – 12301, 12365 – 12431, 12500 – 12535 (1987).
Observations made at the following stations are published: Belgrade, Brorfelde, Bulgarian Natl. Obs., Burlington remote site, Calar Alto, Campinas, Caussols, Chamberlin Obs. Field Stn., Chorzow, Crimean Astrophys. Obs., Eldagsen, ESO, Floirac, Geisei, Goethe Link, Haute Provence, Hoher List, JCPM Sapporo Stn., Kavalur, Kitami, Kitt Peak (Steward Obs.), Kleť, Kushiro, Lowell Obs., Lowell Obs. Anderson Mesa Stn., Mauna Kea, Mount John, Oak Ridge, Oishi, Ojima, Okutama, Oss. Chaonis, Palomar, Perth, Poznan, Reintal, San Fernando, San Vittore, Saratov, Sendai Obs. Ayashi Stn., Shizuoka, Siding Spring, St. Etienne, Tautenburg, Toyota, Turku, Uenohara, Valencia, Victoria (Climenhaga Obs.), Yatsugatake–Kobuchizawa, Yebes, Zimmerwald.

098.037 Orbital elements of numbered minor planets.
Minor Planet Circ., Nos. 11887 – 12624 (1987).
The numbered minor planets are listed according to their definitive number. Newly numbered objects are indicated by an asterisk. The names of the orbit computers are given behind the respective M.P.C. numbers:
(1) 12187 E. Goffin; (7) 11982 L. D. Schmadel; (12), (17) 12187 E. Goffin; (20) 11982 L. D. Schmadel; (22), (25), (42) 12188 E. Goffin; (44) 11982 L. D. Schmadel; (48), (52), (53) 12188, (56), (57), (58), (62) 12189, (65) 12302 E. Goffin; (70) 12118 L. D. Schmadel; (75), (78) 12189, (89), (90), (95), (103), (104), (105) 12190, (112), (114), (121), (134) 12191, (135), (139), (141), (169), (175) 12303, (192) 12432, (198) 12303, (206), (209), (211), (256), (276), (318) 12304 E. Goffin; (363) 12560 L. D. Schmadel; (372), (423), (426), (453), (466), (479) 12305, (516), (566), (586), (623) 12306, (632), (702) 12432, (751) 12433 E. Goffin; (801) 12119 L. D. Schmadel; (841) 12433 E. Goffin; (847) 12119 L. D. Schmadel; (871) 12307 D. W. E. Green; (872), (899), (940), (961) 12433, (1049) 12561, (1118), (1152), (1200), (1209), (1262), (1332) 12434 E. Goffin; (1350) 12200 D. W. E. Green; (1406), (1425), (1449) 12119, (1492), (1496), (1500) 12120 L. D. Schmadel; (1507) 12306, (1516) 12435 E. Goffin; (1522), (1523), (1536) 12120, (1609), (1614), (1625), (1675), (1681) 12121 L. D. Schmadel; (1682) 12435, (1685) 12191 E. Goffin; (1713) 12121, (1825) 11983 L. D. Schmadel; (1849) 12435, (1864) 12191, (1866), (1915), (1916), (1917) 12192, (1927), (1935), (1940) 11983, (1943), (1980) 12192 E. Goffin; (1981) 12194 D. K. Yeomans, M. S. Keesey; (1984) 12435, (1988) 11983, (1990) 12435, (2009), (2014), (2016) 11984, (2039) 12435, (2040) 12436, (2051) 11984, (2059), (2061), (2063) 12193, (2064) 12306, (2071) 12436, (2079), (2085) 11984, (2093) 12436, (2112) 11985, (2119) 12307, (2126) 12436, (2129) 11985, (2135) 12193, (2139) 12436, (2154), (2156) 11985, (2164) 12307, (2165), (2186) 11985, (2190), (2199) 11986, (2200) 12436, (2204) 11986, (2207), (2212) 12193 E. Goffin; (2223) 12308 D. W. E. Green; (2225) 12437, (2231) 11986, (2233) 12437, (2256), (2257) 11986, (2260) 11987, (2276), (2279) 12437, (2280) 12561, (2285) 11987, (2303) 11987, (2340) 12194, (2354), (2359), (2360) 11987, (2364) 11988, (2372) 12437 E. Goffin; (2394), (2396) 12194, (2416) 12195 S. Nakano; (2424) 11983 L. D. Schmadel, M. Gressmann; (2456) 12194 E. Goffin; (2475), (2476), (2477), (2478) 12195 S. Nakano; (2488) 12561, (2496) 11988 E. Goffin; (2497) 12195 S. Nakano; (2498) 11988 E. Goffin; (2499) 12196 S. Nakano; (2520) 12561 E. Goffin; (2528), (2529), (2530), (2531), (2532) 12196 S. Nakano; (2542) 12561, (2551) 11988, (2560) 12561, (2562), (2567) 12562 E. Goffin; (2583), (2584), (2585), (2586), (2587), (2589) 12197, (2590) 12198 S. Nakano; (2593) 12562 E. Goffin; (2616), (2617), (2618) 12198 S. Nakano; (2619), (2631), (2643) 12562, (2644) 11988, (2674) 12563, (2677) 11988 E. Goffin; (2722) 11992 C. M. Bardwell; (2726) 11989, (2729), (2733) 12563, (2754) 11989, (2757), (2773), (2798) 12563, (2818), (2823) 11989, (2828), (2839), (2860) 12564, (2868), (2872) 11989, (2874), (2875), (2886)

12564, (2909), (2917) 12565, (2920) 11990, (2952), (2961), (2966), (2983), (2999) 12565, (2999) 11990 E. Goffin; (3005) 12200 D. W. E. Green; (3009), (3010) 11990, (3029) 12566, (3057) 11990, (3121), (3124), (3162) 12566, (3192) 11990, (3375) 12566, (3631)* 11991 E. Goffin; (3632)*–(3640)* 11992–11995 C. M. Bardwell; (3641)*–(3643)* 11998–11999 S. Nakano; (3644)*–(3647)* 12002–12003 T. Kobayashi; (3648)*–(3652)* 12005–12007 H. Oishi; (3653)* 12008 T. Urata; (3654)*–(3655)* 12009–12010 B. G. Marsden.
For a continuation of this list see Abstr. 098.038.

098.038 Orbital elements of numbered minor planets.
Minor Planet Circ., Nos. 11887 – 12624 (1987).
This list is a continuation of Abstr. 098.037. The numbered minor planets are listed according to their definitive number. Newly numbered objects are indicated by an asterisk. The names of the orbit computers are given behind the respective M.P.C. numbers:
(3656)*–(3658)* 12125–12126 D. W. E. Green; (3659)*–(3664)* 12128–12130 S. Nakano; (3665)*–(3673)* 12136–12139 B. G. Marsden; (3674)*–(3677)* 12140–12142 C. M. Bardwell; (3678)*–(3680)* 12198–12199 S. Nakano; (3681)* 12201 D. W. E. Green; (3682)* 12201–12202 B. G. Marsden; (3683)* 12204 W. Landgraf; (3684)* 12208 H. Oishi; (3685)*, (3686)* 12308 D. W. E. Green; (3687)*–(3694)* 12309–12312 B. G. Marsden; (3695)*–(3701)* 12314–12316 C. M. Bardwell; (3702)*–(3706)* 12318–12320 S. Nakano; (3707)*, (3708)* 12322–12323 H. Oishi; (3709)* 12438 C. M. Bardwell; (3710)*–(3713)* 12440–12442 B. G. Marsden; (3714)* 12446 S. Nakano; (3715)*, (3716)* 12566–12567 D. W. E. Green; (3717)*, (3718)* 12577–12578 S. Nakano; (3719)*, (3720)* 12580–12581 T. Kobayashi.

098.039 Orbital elements of unnumbered minor planets.
Minor Planet Circ., Nos. 11887 – 12624 (1987).
The unnumbered minor planets are sorted by their provisional designation. The names of the orbit computers are given behind the respective M.P.C. numbers:
[1928 UF] 12142 C. M. Bardwell; [1931 TC$_2$] 12578, [1931 TC$_4$] 12447 S. Nakano; [1935 SP$_1$] 12442 B. G. Marsden; [1938 GG] 11999 S. Nakano; [1940 RG] 12442 B. G. Marsden; [1940 WA] 12437 E. Goffin; [1949 PV] 12454 T. Kobayashi; [1950 JB] 11999 S. Nakano; [1953 UD] 12316 C. M. Bardwell; [1959 LM] 12139, [1961 CX] 12312 B. G. Marsden; [1962 RN] 12204–12205 T. Kobayashi; [1966 TP] 12447 S. Nakano; [1967 UQ] 12581 T. Kobayashi; [1968 HP], [1968 OC$_1$] 12450 K. Hurukawa; [1970 OB] 12456 W. Landgraf; [1970 WC] 12450–12451 K. Hurukawa; [1971 OH] 12323–12324 T. Kobayashi; [1971 RA] 12142 C. M. Bardwell; [1971 SX$_3$] 12007 H. Oishi; [1971 UQ] 12442, [1972 RF] 12312 B. G. Marsden; [1972 RQ], [1972 YR] 12324 T. Kobayashi; [1973 SM] 12451 K. Hurukawa; [1973 SA$_2$] 12320, [1974 QT$_1$] 12003, [1974 RG$_1$] 12004 T. Kobayashi; [1974 SF] 12447 S. Nakano; [1974 SR$_1$], [1975 UF] 12004 T. Kobayashi; [1975 XH] 12199 S. Nakano; [1975 XJ] 11991 E. Goffin; [1976 GJ$_1$] 12199 S. Nakano; [1976 GO$_3$] 12122 H. Oishi; [1976 GQ$_6$] 12143 C. M. Bardwell; [1976 SD$_3$] 12451 K. Hurukawa; [1976 UG$_{15}$] 12438 C. M. Bardwell; [1977 AL$_1$] 12447, [1977 AZ$_1$] 12448, [1977 CD] 12320–12321 S. Nakano; [1977 CZ] 12438 C. M. Bardwell; [1977 DN$_4$] 12451 K. Hurukawa; [1977 EO] 11999 S. Nakano; [1977 EK$_1$], [1977 EM$_1$] 12004, [1977 EG$_7$] 12581 T. Kobayashi; [1977 KL$_1$] 12324–12325 A. Lowe; [1977 QD$_3$] 12005 T. Kobayashi; [1977 QH$_4$] 12143 C. M. Bardwell; [1977 RF$_2$] 12202 B. G. Marsden; [1977 RJ$_6$] 12567–12568 C. M. Bardwell; [1977 RR$_6$] 12123–12124 T. Kobayashi; [1977 RY$_6$], [1977 RD$_7$] 12568, [1977 RO$_7$] 12568–12569, [1977 RR$_7$], [1977 SS$_2$], [1977 SD$_3$] 12569, [1977 SG$_3$] 12570 C. M. Bardwell; [1977 TQ$_6$] 12578, [1977 TG$_7$] 12578–12579 S. Nakano; [1978 OK] 11995 C. M. Bardwell; [1978 PY$_2$], [1978 PL$_4$], [1978 RG$_1$] 12443 B. G. Marsden; [1978 RE$_3$] 12452 K. Hurukawa; [1978 SH$_1$] 12325 T. A. Vinogradova; [1978 SQ$_4$] 11995 C. M. Bardwell; [1978 SP$_6$] 12131 S. Nakano; [1978 TB$_2$] 12326 T. A. Vinogradova; [1978 TP$_6$] 12325 A. Lowe; [1978 UU$_1$] 12203 C. M. Bardwell; [1978 VN] 12008 T. Urata; [1978 VS$_5$] 12579 S. Nakano; [1978 VH$_8$] 12576 B. G. Marsden; [1978 VL$_{11}$]

11995 C. M. Bardwell; [1978 XQ] 12131 S. Nakano; [1979 HG$_5$] 12205 T. Kobayashi; [1979 ML] 12202 B. G. Marsden; [1979 MK$_1$] 12131, [1979 ME$_9$] 11999–12000 S. Nakano; [1979 QM$_1$] 11996 C. M. Bardwell; [1979 QW$_3$] 11991 E. Goffin; [1979 SJ] 12143 C. M. Bardwell; [1979 SS] 12312, [1979 SU$_9$] 12010 B. G. Marsden; [1979 WE$_2$] 12438–12439 C. M. Bardwell; [1980 FO$_3$] 12000 S. Nakano; [1980 FH$_5$] 12126 D. W. E. Green; [1980 OG] 12576, [1981 ET] 12443–12444 B. G. Marsden; [1981 EF$_2$] 12321 S. Nakano; [1981 EK$_5$] 12576 B. G. Marsden; [1981 EC$_{13}$] 12143 C. M. Bardwell; [1981 EP$_{20}$] 12452 K. Hurukawa; [1981 EA$_{26}$] 12444, [1981 ES$_{32}$] 12313, [1981 EO$_{34}$] 12444, [1981 ER$_{35}$], [1981 EG$_{39}$] 12577, [1981 FD] 12010, [1981 FQ] 12010–12011 B. G. Marsden; [1981 GD] 12126–12127 D. W. E. Green; [1981 JE$_3$] 12122 H. Oishi; [1981 PK] 12205 T. Kobayashi; [1981 QF] 12208, [1981 QH$_2$] 12122, [1981 QY$_2$] 12452–12453, [1981 QA$_3$] 12323 H. Oishi; [1981 RQ] 12205 T. Kobayashi; [1981 RD$_2$] 12444–12445 B. G. Marsden; [1981 RU$_3$] 12323 H. Oishi; [1981 RD$_5$] 12313 B. G. Marsden; [1981 SE$_2$] 12325 G. R. Kastel'; [1981 SJ$_7$] 12453 H. Oishi; [1982 BP$_2$] 12308–12309 D. W. E. Green; [1982 DY$_1$] 12321 S. Nakano; [1982 DV$_2$] 12585 K. Ichikawa.

098.040   Orbital elements of unnumbered minor planets.
Minor Planet Circ., Nos. 11887 – 12624 (1987).

This list is a continuation of Abstr. 098.039. The unnumbered minor planets are sorted by their provisional designation. The names of the orbit computers are given behind the respective M.P.C. numbers:

[1982 FZ$_1$] 12131–12132 S. Nakano; [1982 SF] 12011 B. G. Marsden; [1982 SU$_1$] 12132 S. Nakano; [1982 TT] 12445 B. G. Marsden; [1982 UQ$_5$] 12007 K. Ichikawa; [1982 UC$_{11}$] 12452 K. Hurukawa; [1982 XV], [1982 XQ$_1$] 12000 S. Nakano; [1983 AJ] 12570 C. M. Bardwell; [1983 AY] 12453 H. Oishi; [1983 AF$_2$] 12570 C. M. Bardwell; [1983 LM] 12321 S. Nakano; [1983 RT$_3$] 12317, [1983 RC$_4$] 12203 C. M. Bardwell; [1983 TW$_1$] 12454 T. Kobayashi; [1984 DA] 11996 C. M. Bardwell; [1984 HK$_1$], [1984 JA$_1$] 12001, [1984 JP$_1$] 12579 S. Nakano; [1984 QS] 12455 T. Kobayashi; [1984 SR$_1$] 12001, [1984 SX$_5$] 12579 S. Nakano; [1984 UX$_2$] 12202 B. G. Marsden; [1984 WM$_1$] 12205–12206 T. Kobayashi; [1984 YH$_1$] 12580 S. Nakano; [1985 AE] 12005 T. Kobayashi; [1985 FZ$_1$] 12144 C. M. Bardwell; [1985 PO] 12580 S. Nakano; [1985 RW] 11996 C. M. Bardwell; [1985 RE$_4$] 12200 S. Nakano; [1985 TE$_3$] 12570, [1985 UL] 11996–11997 C. M. Bardwell; [1985 UT$_4$] 12326 T. A. Vinogradova;   [1985 UY$_4$],   [1985 UB$_5$]   12317 C. M. Bardwell;   [1985 UG$_5$]   12321–12322   S. Nakano; [1985 VK$_2$] 12317–12318 C. M. Bardwell; [1985 YP] 12011–12012 B. G. Marsden; [1986 CL$_1$] 12318 C. M. Bardwell; [1986 EM$_2$] 12140 B. G. Marsden; [1986 EN$_4$] 12132 S. Nakano; [1986 EZ$_4$] 12581–12582, [1986 EE$_5$] 12455 T. Kobayashi; [1986 GU] 12203 C. M. Bardwell; [1986 GZ] 12313 B. G. Marsden; [1986 JT] 12439, [1986 PA] 11997 C. M. Bardwell; [1986 PJ$_4$] 12132, [1986 QL] 12132–12133, [1986 QB$_1$] 12133, [1986 QL$_1$] 12133–12134, [1986 QP$_1$] 12134 S. Nakano; [1986 QV$_2$] 12206 T. Kobayashi; [1986 QZ$_2$] 12134, [1986 QA$_3$] 12134–12135 S. Nakano; [1986 QB$_3$] 12206 T. Kobayashi; [1986 QM$_3$], [1986 QN$_3$] 12127 D. W. E. Green; [1986 QX$_3$] 12207 T. Kobayashi; [1986 VV$_6$], [1986 VW$_6$] 12584 H. Oishi; [1986 WB$_1$] 12001 S. Nakano; [1986 XH] 12005, [1987 BB$_2$] 12207 T. Kobayashi; [1987 DE] 11997 C. M. Bardwell; [1987 DJ] 12001–12002, [1987 DF$_1$], [1987 FF$_1$] 12002 S. Nakano; [1987 GG] 11997, [1987 KF] 11998 C. M. Bardwell; [1987 MK] 12322 S. Nakano; [1987 OA] 12204 C. M. Bardwell; [1987 OM] 12207 T. Kobayashi; [1987 OQ] 12322 S. Nakano; [1987 PA] 12204 C. M. Bardwell; [1987 PB] 12203, [1987 QA] 12313, 12445, [1987 QB] 12314 B. G. Marsden; [1987 QC] 12448 S. Nakano; [1987 QX], [1987 QF$_7$] 12439 C. M. Bardwell; [1987 RG] 12448, [1987 RJ] 12448–12449 S. Nakano; [1987 SB] 12439 C. M. Bardwell; [1987 SE] 12582, [1987 SG], [1987 SJ] 12455, [1987 SK] 12456 T. Kobayashi; [1987 SL] 12445 B. G. Marsden; [1987 SV] 12449 S. Nakano; [1987 SY] 12440 C. M. Bardwell; [1987 SN$_1$] 12449 S. Nakano; [1987 SW$_1$] 12560 L. D. Schmadel; [1987 SB$_2$] 12456 T. Kobayashi; [1987 SV$_2$] 12449 S. Nakano; [1987 SF$_3$] 12440

C. M. Bardwell; [1987 SW$_3$] 12449–12450, [1987 SE$_4$] 12450 S. Nakano; [1987 UA] 12440 C. M. Bardwell; [1987 UJ] 12580 S. Nakano; [1987 UK], [1987 UQ$_1$] 12582 T. Kobayashi; [2024 P–L] 12585 H. Oishi; [2121 P–L] 12570–12571 C. M. Bardwell; [2142 P–L] 12582–12583 T. Kobayashi; [2208 P–L], [2574 P–L] 12571, [3108 P–L] 12571–12572 C. M. Bardwell; [4153 P–L] 12585 H. Oishi; [4665 P–L] 12583 T. Kobayashi; [4831 P–L] 12572 C. M. Bardwell; [5568 P–L] 12583 T. Kobayashi; [6047 P–L] 12208 H. Oishi; [6092 P–L] 12144 C. M. Bardwell; [6568 P–L], [6575 P–L] 12583 T. Kobayashi; [6608 P–L] 12572 C. M. Bardwell; [7604 P–L], [7618 P–L] 12584 T. Kobayashi; [2041 T–3] 12572–12573, [2141 T–3], [2321 T–3] 12573, [2402 T–3] 12573–12574, [2480 T–3], [2672 T–3], [3134 T–3] 12574, [5142 T–3] 12575 C. M. Bardwell.

098.041   Normal places for Pallas 1802 – 1978.
W. Landgraf.
Astron. Astrophys., Suppl. Ser., Vol. 71, No. 2, p. 197 – 199 (1987).

101 normal places for Pallas were determined from about eight hundred individual observations between 1802 and 1978.

098.042   Photopolarimetry of asteroids and meteorites as a method of investigating the microstructure of their surfaces.
Yu. G. Shkuratov, L. A. Akimov, L. Ya. Melkumova.
20th All–Union meteoritic conference, p. 179 – 180 (1987). In Russian. Abstr. in Ref. Zh., 51. Astron., 9.51.247 (1987). – See Abstr. 012.017.

098.043   On some peculiarities of spectra of dark asteroids and meteorites in the ultraviolet region.
Yu. G. Shkuratov.
20th All–Union meteoritic conference, p. 177 – 179 (1987). In Russian. Abstr. in Ref. Zh., 51. Astron., 9.51.248 (1987). – See Abstr. 012.017.

098.044   Photographic positions of seven minor planets.
J. Bem, B. Szczodrowska–Kozar.
Acta Astron., Vol. 37, No. 2, p. 189 – 191 (1987).

Topocentric positions of minor planets Herculina, Eleonora, Massalia, Nysa, Sappho, Euphrosyne and Leto, observed in the years 1982 – 83 in the Wrocław Astronomical Observatory, are given.

098.045   Positions of minor planets and comets obtained at the Chorzów Observatory.
I. Włodarczyk.
Acta Astron., Vol. 37, No. 2, p. 193 – 195 (1987).

The paper contains results of photographic observations of minor planets and comet Halley made in the years 1977 – 1986.

098.046   Photometric properties of the minor planets: observations of (4) Vesta in 1985.
A. J. Hollis.
J. Br. Astron. Assoc., Vol. 97, No. 6, p. 350 – 352 (1987).

098.047   No satellites of asteroids.
R. A. Kerr.
Science, Vol. 237, No. 4812, p. 250 (1987).

098.048   Organic matter on asteroid 130 Elektra.
D. P. Cruikshank, R. H. Brown.
Science, Vol. 238, No. 4824, p. 183 – 184 (1987).

Infrared absorption spectra of a low–albedo water–rich asteroid appear to show a weak 3.4–micrometer carbon–hydrogen stretching mode band, which suggests the presence of hydrocarbons on asteroid 130 Elektra. The organic extract from the primitive carbonaceous chondritic Murchison meteorite shows similar spectral bands.

098.049 **The use of crossing–point observations in reference frames and for the refinement of minor planet masses.**
A. L. Whipple, R. L. Duncombe, P. D. Hemenway.
Bull. Am. Astron. Soc., Vol. 19, No. 2, p. 742 (1987). Abstract. – See Abstr. 010.061.

098.050 **On the simple models for the interpretation of centimeter–wavelength radio observations of asteroids.**
W. J. Webster Jr.
Publ. Astron. Soc. Pac., Vol. 99, No. 619, p. 1009 – 1013 (1987).
The predictions of the two–layer models used to interpret radio spectra of the asteroids have been examined. Although the predictions are in accord with physical expectations, careful attention to the circumstances of the observations is essential. Some simple improvements to the models based on studies of the lunar radio emission by Keihm and using radar measurements of asteroid surface characteristics by Ostro and colleagues are suggested.

098.051 **Introduction: history, definitions, nomenclature (*of small bodies of the solar system*).**
B. G. Marsden.
The evolution of the small bodies of the solar system, p. 3 – 9 (1987). – See Abstr. 012.041.

098.052 **The systems of interplanetary objects.**
Ĺ. Kresák.
The evolution of the small bodies of the solar system, p. 10 – 32 (1987). – See Abstr. 012.041.
Contents: 1. Introduction. 2. The system of asteroids. 3. The system of comets. 4. The system of meteoroids.

098.053 **Dynamics of asteroids.**
H. Scholl.
The evolution of the small bodies of the solar system, p. 53 – 78 (1987). – See Abstr. 012.041.
Contents: 1. Introduction. 2. The observational data. 3. Resonances. 4. Hirayama families. 5. Earth–crossing asteroids 6. Interrelation between asteroids and comets.

098.054 **Physics, chemistry and collisional evolution of the asteroids.**
R. Chapman.
The evolution of the small bodies of the solar system, p. 79 – 90 (1987). – See Abstr. 012.041.
Contents: 1. Introduction. 2. Physical properties (mass and density, diameters, shapes and spins, asteroidal satellites). 3. Compositions (C types, D types, S types, other types). 4. Collisional evolution.

098.055 **Physical and statistical interpretations of asteroid light-curves.**
V. Zappalà.
The evolution of the small bodies of the solar system, p. 91 – 110 (1987). – See Abstr. 012.041.
Contents: 1. Introduction. 2. Lightcurves and rotational properties. 3. Surface topography and morphology. 4. Rotation axes. 5. Binary asteroids. 6. Statistical analyses. 7. Collisional evolution. 8. Conclusions.

098.056 **Properties of asteroids from observations and laboratory simulations.**
M. Fulchignoni.
The evolution of the small bodies of the solar system, p. 111 – 130 (1987). – See Abstr. 012.041.
Contents: 1. Photometric data (asteroid lightcurves, colorimetry, phase function). 2. Reflectance data (ultraviolet reflectance spectra, visible and infra–red reflectance spectra, infra–red reflectance spectra, radiometry, taxonomy). 3. Other observational methods (occultation observations, radar observations, speckle interferometry, polarimetry). 4. Analytical/numerical simulations (synthetic lightcurves, the scattering effects, the convex–profile inversion method). 5. Laboratory simulations (photometric studies, impact experiments).

098.057 **Physical properties of asteroids from radar observations.**
S. J. Ostro.
The evolution of the small bodies of the solar system, p. 131 – 146 (1987). – See Abstr. 012.041.
Contents: 1. Introduction. 2. Observational techniques and derived quantities. 3. Asteroid surface characteristics. 4. Radar constraints on asteroid shapes. 5. Radar constraints on asteroid pole directions. 6. Conclusion.

098.058 **Near–Earth asteroid searches: status and prospects.**
E. F. Helin.
The evolution of the small bodies of the solar system, p. 147 – 156 (1987). – See Abstr. 012.041.
Contents: 1. Introduction. 2. Search requirements. 3. Recent near–Earth asteroid searches. 4. International search networks. 5. CCD Schmidts. 6. Conclusion.

098.059 **A photoelectric survey of 130 asteroids.**
R. P. Binzel.
Icarus, Vol. 72, No. 1, p. 135 – 208 (1987).
Photoelectric observations of 130 asteroids were obtained at the University of Texas McDonald Observatory and Cerro Tololo Inter–American Observatory on 150 nights between November 1982 and February 1985. The program objects were primarily sampled from six distinct regions in the asteroid belt to illuminate two specific problems: the role of collisions in clearing the 3/1 resonance Kirkwood gap and the formation and evolution of the Eos and Koronis families. Over 400 individual lightcurves were obtained which when combined allowed rotational periods and lightcurve amplitudes to be derived. $B–V$ and $U–B$ colors were measured for a majority of the objects and absolute magnitudes, H, have been computed.

098.060 **Palomar–Leiden minor planets: proper elements, frequency distributions, belt boundaries, and family memberships.**
J. G. Williams, J. E. Hierath.
Icarus, Vol. 72, No. 2, p. 276 – 303 (1987).
Proper elements have been calculated for 1227 higher accuracy orbits from the Palomar–Leiden Survey of faint minor planets. Tabulations are given for the special orbits: Earth and deep Mars crossers, Trojans, Hildas, and one 2:1 librator. The frequency distributions of the proper semimajor axis, eccentricity, sine inclination, longitudes of perihelion and node plus their rates, and the closest distances of approach to Mars and Jupiter are displayed as histograms and discussed.

098.061 **Classification of asteroids using G–mode analysis.**
M. A. Barucci, M. T. Capria, A. Coradini, M. Fulchignoni.
Icarus, Vol. 72, No. 2, p. 304 – 324 (1987).
A revised version of the G–mode multivariate statistics has been used to classify the 438 asteroids for which the eight–color photometric data and IRAS albedo are available. At a confidence level of 99.7%, seven taxonomic units of asteroids are separated by the method, while with higher values of the confidence level no separation occurs in the adopted sample. Decreasing the confidence level (i.e., accepting a higher probability of a wrong decision in classifying the asteroids), the authors obtain a more detailed grouping, which results in a successive subdivision of the first units found. At a confidence level of 97.5%, two groups are added to the original ones: C asteroids are in fact subdivided on the basis of their albedo into three different units. In total, 18 groups of objects can be distinguished.

098.062 **Rotation properties of four L5 Trojan asteroids from CCD photometry.**
L. M. French.
Icarus, Vol. 72, No. 2, p. 325 – 341 (1987).
CCD observations made with the Cerro Tololo 0.9–m telescope of four L5 Trojan asteroids during June and July of 1986 are presented. Partial lightcurves for (1208) Troilus and (1867) Deiphobus suggest long periods ($P > 24$ hr). Complete lightcurves were obtained for two asteroids, (1173) Anchises

$(P = 11.6095 \pm 0.0036$ hr, $\qquad \Delta m = 0.57 \pm 0.01$ mag) and (2674) Pandarus $(P = 8.4803 \pm 0.0019,$ $\Delta m = 0.58 \pm 0.01)$. The large lightcurve amplitudes are unique. The author suggests that the Trojans may represent a primordial asteroid population, with shapes not significantly altered since the formation of the solar system.

098.063   Results of photographic position observations of the minor planets Parthenope, Pallas and Juno in 1984 – 1985 at the Astronomical Observatory of the Kharkov State University.
P. P. Pavlenko, L. S. Pavlenko.
Red. Zh. Kinematika Fiz. Nebesn. Tel, Kiev, 7 pp. (1987). In Russian. Abstr. in Ref. Zh., 51. Astron. 11.51.71 (1987).

098.064   Die Entdeckung des kleinen Planeten (3338) Richter.
F. Börngen.
Sterne, 63. Band, Heft 4, p. 230 – 234 (1987).

098.065   Il pianetino (3344) Modena scoperto il 15 maggio 1982 dagli astrofili dell'Osservatorio S. Vittore (Bologna).
E. Colombini.
Astronomia, N. 3 – 4, p. 33 – 35 (1987).

098.066   Photometric properties of the minor planets: observations of (15) Eunomia in 1972, 1981 and 1985.
A. J. Hollis.
J. Br. Astron. Assoc., Vol. 98, No. 1, p. 14 – 17 (1987).
(15) Eunomia was observed visually in 1972 and 1981 and visually and photoelectrically in 1985. A method of estimating the rotation period by folding visual observations is described.

098.067   Observations of selected minor planets in Nikolaev in the years 1976 – 1978. Communication 15.
V. I. Voronenko, G. K. Gorel', L. A. Gudkova, F. F. Kalikhevich.
Glav. Astron. Obs. Akad. Nauk SSSR, Leningrad, 93 pp. (1987). In Russian. Abstr. in Ref. Zh., 51. Astron., 12.51.84 (1987).

098.068   Les occultations stellaires par les astéroides.
J. Lecacheux, R. Boninsegna.
Ann. Phys. (Paris), Vol. 12, Colloq. No. 1, p. 225 – 233 (1987). – See Abstr. 012.061.
Stellar occultations by the asteroids are also fast phenomena. This type of observations may bring us valuable informations mainly on the shape of these bodies. This paper intends to present the european network of observers necessary to obtain interesting results and the different possibility of observations.

098.069   Prédictions de "dernière minute" des occultations astéroidales.
G. Dourneau, M. Rapaport, Y. Réquième.
Ann. Phys. (Paris), Vol. 12, Colloq. No. 1, p. 235 – 237 (1987). – See Abstr. 012.061.
Predictions of the stellar occultations by asteroids have an accuracy of only 0″.5 to 1″.0 that corresponds to 1000 km on the location, on Earth, of the area of visibility. Using the meridian circle, Bordeaux Observatory is able to improve this accuracy. "Last minute" predictions may reach a precision of 0″.05, i.e. 50 km on Earth. This is made possible with observations of the asteroid and the star a month before the phenomenon.

098.070   1959 LM = 1987 MB.
IAU Circ., No. 4416 (1987).

098.071   1928 UF.
IAU Circ., No. 4431 (1987).

098.072   1987 OA.
IAU Circ., Nos. 4436, 4441, 4446 (1987).

098.073   1987 PA.
IAU Circ., No. 4437 (1987).

098.074   1987 QB.
IAU Circ., Nos. 4451, 4455 (1987).

098.075   1987 QA.
IAU Circ., No. 4454 (1987).

098.076   1987 SL.
IAU Circ., Nos. 4465, 4466, 4469 (1987).

098.077   1987 SY.
IAU Circ., Nos. 4465, 4467, 4472 (1987).

098.078   1987 SJ$_3$.
IAU Circ., Nos. 4467, 4469 (1987).

098.079   1987 SF$_3$.
IAU Circ., No. 4477 (1987).

098.080   1987 UA.
IAU Circ., No. 4478 (1987).

098.081   1982 XB.
IAU Circ., Nos. 4495, 4499 (1987).

098.082   (3607) Naniwa.
Yamamoto Circ., No. 2088 (1987). In Japanese.

098.083   (3200) Phaethon.
Yamamoto Circ., No. 2094 (1987). In Japanese.

098.084   Apollo asteroid 1987 SY and the δ–Leonid meteor shower.
D. Olsson–Steel.
WGN, Vol. 15, Nr. 6, p. 179 – 180 (1987).
The recently–discovered Apollo–type asteroid 1987 SY has a theoretical radiant which is very close to that of the δ–Leonid shower, and the asteroid and stream orbits are similar: it is therefore suggested that this asteroid may be the parent of the shower.

098.085   Catalogue of orbital elements of 3516 minor planets numbered by November 16, 1987, on magnetic tape.
Yu. V. Batrakov, V. A. Shor.
Bull. Inf. Cent. Données Stellaires, No. 33, p. 131 – 134 (1987).
The catalogue on magnetic tape containing the orbital elements of 3516 permanently numbered minor planets and some connected information is described.

098.086   The size, shape, density, and albedo of Ceres from its occultation of BD + 8°471.
R. L. Millis, L. H. Wasserman, O. G. Franz, R. A. Nye, R. C. Oliver, T. J. Kreidl, S. E. Jones, W. Hubbard, L. Lebofsky, R. Goff, R. Marcialis, M. Sykes, J. Frecker, D. Hunten, B. Zellner, H. Reitsema, G. Schneider, E. Dunham, J. Klavetter, K. Meech, T. Oswalt, J. Rafert, E. Strother, J. Smith, H. Povenmire, B. Jones, D. Kornbluh, L. Reed, K. Izor, M. F. A'Hearn, R. Schnurr, W. Osborn, D. Parker, W. T. Douglas, J. D. Beish, A. R. Klemola, M. Rios, A. Sanchez, J. Piironen, M. Mooney, R. S. Ireland, D. Leibow.
Icarus, Vol. 72, No. 3, p. 507 – 518 (1987).
The occultation of BD + 8°471 by Ceres on 13 November 1984 was observed photoelectrically at 13 sites in Mexico, Florida, and the Caribbean. These observations indicate that Ceres is an oblate spheroid having an equatorial radius of 479.6 ± 2.4 km and a polar radius of 453.4 ± 4.5 km. The mean density of this minor planet is 2.7 g/cm^3 $\pm$ 5%, and its visual geometric albedo is 0.073. While the surface appears globally to be in hydrostatic equilibrium, firm evidence of real limb irregularities is seen in the data.

098.087   A simple analytical model for the secular resonance v_6 in the asteroidal belt.

M. Yoshikawa.

Celest. Mech., Vol. 40, Nos. 3 – 4, p. 233 – 272 (1987).

The dynamics of the secular resonance v_6 is investigated by a simple analytical model, in which the third degree terms of the eccentricity and inclination are taken into account. The eccentricity variations of asteroids located near this resonance are represented clearly by the diagrams of equi–Hamiltonian curves on the plane of $\tilde{\omega}–\tilde{\omega}_S$ versus e ($\tilde{\omega},\tilde{\omega}_S$: the longitude of perihelion of asteroids and Saturn, e: the eccentricity of asteroids). These diagrams predict that the eccentricity of these asteroids suffers a large increase or decrease, and that the secular resonance argument $\tilde{\omega}–\tilde{\omega}_S$ librates about 0° and 180°. In order to confirm these predictions, numerical integrations are carried out over one million years.

098.088   The role of gravitation and nuclear forces in the formation of asteroids and comets.

B. A. Vorontsov–Vel'yaminov.

Komet. Tsirk., No. 370 (1987). In Russian.

098.089   1987 – a favourable year for observations of the unusual asteroid 3200 (Phaethon).

G. O. Ryabova.

Komet. Tsirk., No. 371 (1987). In Russian.

Neue Namen von Kleinplaneten.
See Abstr. 002.028.

Asteroid reference catalogue.
See Abstr. 002.068.

Amphitrite and the astronomy of Regent's Park.
See Abstr. 004.037.

Perturbation computations and numerical modelling experiments.
See Abstr. 021.011.

Optical evolution of laboratory–produced organics: applications to Phoebe, Iapetus, outer belt asteroids and cometary nuclei.
See Abstr. 022.079.

Expansion of the disturbing force–function for the study of high–eccentricity librations.
See Abstr. 042.011.

Evolution of asteroid orbits in the case of second– and third–order commensurabilities.
See Abstr. 042.027.

Some instabilities in the 3/1 resonant problem.
See Abstr. 042.051.

Sôbre algumas técnicas de perturbação utilizadas no problema ressonante 3/1.
See Abstr. 042.052.

A first order analytical elliptic planar model for 3/1 resonance problem.
See Abstr. 042.053.

Relaxation–chaos phenomena in celestial mechanics. I. On wisdom's model for the 3/1 Kirkwood gap.
See Abstr. 042.057.

The Comet Rendezvous Asteroid Flyby mission.
See Abstr. 051.056.

Low–thrust navigation to comets or asteroids.
See Abstr. 052.001.

Mathematical modelling of light scattering by surfaces in the polarimetry of atmosphereless cosmic bodies.
See Abstr. 063.053.

A satellite–asteroid mystery and a possible early flux of scattered C–class asteroids.
See Abstr. 091.017.

Study of some statistical relationships for polarimetric quantities of atmosphereless bodies.
See Abstr. 091.027.

Asteroidal, cometary, and Jovian occultations during 1987.
See Abstr. 096.005.

Occultations of stars by solar system objects. VII. Occultations of catalog stars by asteroids in 1988 and 1989.
See Abstr. 096.011.

Occultations of stars by major and minor planets – 1988.
See Abstr. 096.014.

Occultation by minor planets: observation program for 1988.
See Abstr. 096.015.

Les applications astrométriques de la photométrie bidimensionnelle.
See Abstr. 099.066.

A review of cometary sciences.
See Abstr. 102.030.

Aging of comets and their evolution into asteroids.
See Abstr. 102.044.

Periodic comet Smirnova–Chernykh 1967 XV = 1978 NA_6 = 1981 UH_{18}.
See Abstr. 103.256.

Periodic comet Urata–Niijima (1986o) = 1986 TD_4 = 1986 WP_5.
See Abstr. 103.961.

Distribution of parent bodies of meteorites of different generations according to their sizes.
See Abstr. 104.047.

Unity of small bodies of the solar system.
See Abstr. 104.061.

Geminid fireballs and the peculiar asteroid 3200 Phaethon.
See Abstr. 104.067.

A search for spectral alteration effects in chondritic gas–rich breccias.
See Abstr. 105.002.

Photometry and polarimetry of possible meteorite and terrestrial analogs of M–asteroids.
See Abstr. 105.032.

Visual albedo of meteorites and asteroids.
See Abstr. 105.033.

Four–colour photometry of meteorites and asteroids.
See Abstr. 105.034.

Accumulation of the planets.
See Abstr. 107.001.

Meteorites, asteroids and comets – a study of their interrelationship.
See Abstr. 107.016.

099 Jupiter, Jupiter Satellites

099.001   SO₂ on Io: a thermodynamic perspective.
A. P. Zent, F. P. Fanale.
NASA Tech. Memo., NASA TM–89810, p. 49 – 51 (1987). – See Abstr. 003.001.

099.002   Charged particle modification of surfaces in the outer solar system.
R. E. Johnson.
NASA Tech. Memo., NASA TM–89810, p. 221 – 222 (1987). Abstract. – See Abstr. 003.001.

099.003   Io: comparison of photometric scans produced by the Minnaert and Hapke functions.
D. P. Simonelli, J. Veverka.
NASA Tech. Memo., NASA TM–89810, p. 223 – 225 (1987). – See Abstr. 003.001.
The authors' work with Io data indicates that the empirical Minnaert function, while not a perfect model of real photometric behaviour, does provide a very useful parameterization of limb darkening at phase angles out to 90°, and is especially useful near opposition.

099.004   Sulfur–oxygen processes on Io.
R. M. Nelson, W. D. Smythe.
NASA Tech. Memo., NASA TM–89810, p. 226 (1987). – See Abstr. 003.001.

099.005   Dome craters on Ganymede.
J. M. Moore, M. C. Malin.
NASA Tech. Memo., NASA TM–89810, p. 429 – 431 (1987). – See Abstr. 003.001.

099.006   Non–Newtonian ice rheology and the retention of craters on Ganymede.
P. J. Thomas, G. Schubert.
NASA Tech. Memo., NASA TM–89810, p. 432 – 434 (1987). – See Abstr. 003.001.

099.007   Ejecta types on Ganymede and Callisto.
V. M. Horner, R. Greeley.
NASA Tech. Memo., NASA TM–89810, p. 435 – 437 (1987). – See Abstr. 003.001.

099.008   Local–scale stratigraphy of grooved terrain on Ganymede.
S. L. Murchie, J. W. Head, P. Helfenstein, J. B. Plescia.
NASA Tech. Memo., NASA TM–89810, p. 505 – 507 (1987). – See Abstr. 003.001.

099.009   Stratigraphy of the south polar region of Ganymede.
R. A. De Hon.
NASA Tech. Memo., NASA TM–89810, p. 508 – 510 (1987). – See Abstr. 003.001.

099.010   Geology of Io.
R. Greeley, R. A. Craddock, D. A. Crown, L. A. Leshin, G. G. Schaber.
NASA Tech. Memo., NASA TM–89810, p. 519 – 521 (1987). – See Abstr. 003.001.

099.011   Large scale topography of Io.
R. W. Gaskell, S. P. Synnott.
NASA Tech. Memo., NASA TM–89810, p. 522 – 523 (1987). – See Abstr. 003.001.

099.012   The Galilean satellite geological mapping program, 1986.
B. K. Lucchitta.
NASA Tech. Memo., NASA TM–89810, p. 524 (1987). Abstract. – See Abstr. 003.001.

099.013   Voyager cartography.
R. M. Batson, E. M. Lee, K. F. Mullins.
NASA Tech. Memo., NASA TM–89810, p. 527 (1987). – See Abstr. 003.001.

099.014   The control networks fo the satellites of Jupiter.
M. E. Davies.
NASA Tech. Memo., NASA TM–89810, p. 539 (1987). Abstract. – See Abstr. 003.001.

099.015   Bilan de la campagne "PHÉMU 85" (*phénomènes mutuels*): la participation des amateurs.
L. Bergeal.
Astronomie, Vol. 101, p. 487 – 496 (1987).

099.016   H₃⁺ in the Jovian ionosphere.
J. C. McConnell, T. Majeed.
J. Geophys. Res., Vol. 92, No. A8, p. 8570 – 8578 (1987).
Recent measurements and theoretical calculations suggest that H_3^+ ions in their ground vibrational level recombine much more slowly than in the vibrational levels $v \geqslant 3$. Using these results as a guide, the authors have investigated the impact of modified H_3^+ chemistry on the structure of the Jovian ionosphere for both a solar EUV ionization source and a high–altitude ionization source, associated with the electroglow.

099.017   The emission of narrow–band Jovian kilometric radiation.
S. F. Fung, K. Papadopoulos.
J. Geophys. Res., Vol. 92, No. A8, p. 8579 – 8593 (1987).
A model based on the nonlinear coupling of electrostatic plasma waves is proposed to explain the emission of the narrow–band Jovian kilometric radiation observed by the Voyager spacecraft. A possible excitation mechanism for the electrostatic waves is also discussed.

099.018   Volcanic control of the Io atmosphere and neutral and plasma torus.
A. Eviatar.
J. Geophys. Res., Vol. 92, No. A8, p. 8800 – 8804 (1987).
Rate equations for the column density of Io's atmosphere and the number density of neutral atoms and ions in the torus are formulated in terms of surface sublimation and venting of volcanic gases, sputtering efficiency, thermal escape, and the plasma processes that take place in the torus which included electron impact ionization, charge exchange, and transport. In the limit of low volcanic activity, estimates of the densities can be obtained, and the baseline configuration that would exist in the absence of volcanic activity is found.

099.019   Voyager and Nançay observations of the Jovian radio–emission at different frequencies: solar wind effect and source extent.
F. Genova, P. Zarka, C. H. Barrow.
Astron. Astrophys., Vol. 182, No. 1, p. 159 – 162 (1987).
Voyager observations of the Jovian radio emission at hectometer– and decameter–wavelengths are well correlated. On the other hand, the probability of observing Io–independent Jovian decametric emission from Nançay is shown to be highly variable. These variations correspond well to fluctuations of hectometric activity, influenced by the solar wind, observed by Voyager. This implies that the hectometric and non–Io decametric events originate from the same source regions, at high latitudes in the Jovian magnetosphere, and that the altitude extent of the source covers several planetary radii, like the source of Saturn kilometric radiation. No effect is found for the Io–dependent emission, which is consistent with a source close to the Io field line. Spacecraft low–frequency and ground–based high frequency observations of the Jovian radio–emission are thus shown to be complementary.

099.020 **Theoretical studies of the faint features in the $S_0(0)$ line of H_2 observed in the Voyager IRIS mission.**
J. Schaefer.
Astron. Astrophys., Vol. 182, No. 2, p. L40 – L42 (1987).

The faint features of the $S_0(0)$ line of H_2 observed at the Jovian atmosphere in the Voyager IRIS mission have been made subject of thorough theoretical studies. The presently best available non-spherical interaction potential of H_2–H_2 and very recently improved ab initio results of the induced dipole moment have been used for obtaining absorption spectra which are quantitatively reliable. Some significant corrections of previously published theoretical results have been found and are discussed in this letter.

099.021 **Synoptic report for the 1981 – 82 apparition of the planet Jupiter.**
P. K. Mackal, R. Néel.
Strolling Astron., Vol. 32, Nos. 3 – 4, p. 53 – 67 (1987).

099.022 **Compositional anomaly of Ganymede and Callisto among the ice satellites as inferred from impact crater morphology.**
K. D. Trego.
Earth, Moon, Planets, Vol. 39, No. 2, p. 195 – 196 (1987).

Differences in crater morphology between the Jovian and Saturnian–Uranian ice satellites implies a weaker surface strength for Ganymede and Callisto and thus a more concentrated composition of water. This compositional anomaly among the ice satellites is apparently due to a more complete migration of heavy material toward the inner part of the pre–planetary disc of the Jovian system than occurred in the discs of the Saturnian and Uranian systems.

099.023 **Interpretation of polarimetric observations of Jupiter.**
M. I. Mishchenko.
Inst. Teor. Fiz. Akad. Nauk USSR, Prepr., No. 22R, 24 pp. (1987). In Russian. Abstr. in Ref. Zh., 51. Astron., 7.51.181 (1987).

099.024 **Radial force balance within Jupiter's dayside magnetosphere.**
B. H. Mauk, S. M. Krimigis.
J. Geophys. Res., Vol. 92, No. A9, p. 9931 – 9941 (1987).

A local field stress technique, developed previously in a study of the Saturnian magnetosphere, is introduced to the problem of determining the radial force balance characteristics of Jupiter's magnetosphere. The authors begin by estimating the near-equatorial, radial magnetic force densities using the data obtained by Voyager 1 principally on the dayside (inbound) portion of its trajectory ($7.4 \leqslant R(R_J) \leqslant 42$). Using the low-energy charged particle data ($\gtrsim 30$ keV) and other published data they then explore ways in which the field forces might be balanced. Comparing present results with the results of a previous study, the authors note that contrary to common expectations, Saturn rather than Jupiter is unique in having the corotation centrifugal forces dominate over other sources of radial particle force in some regions of the middle (ring current) equatorial magnetosphere.

099.025 **Detection of a hot plasma component within the core regions of Jupiter's distant magnetotail.**
E. C. Sittler Jr., R. P. Lepping, B. H. Mauk, S. M. Krimigis.
J. Geophys. Res., Vol. 92, No. A9, p. 9943 – 9948 (1987).

The authors have combined the Voyager 2 low–energy plasma data from the Plasma Science Experiment (PLS) and the magnetic field data from the Magnetometer Experiment (MAG) with the Voyager 2 Low Energy Charged Particle Experiment (LECP) ion data ($E > 28$ keV) for the previously described distant magnetotail observations ($5000 < R < 9000\ R_J$). They show for the first time a definite enhancement of LECP fluxes within the core regions where the PLS densities and magnetic field pressure are lower than surrounding regions, indicating that this hot tenuous plasma is present within the core regions. In general there is a strong anticorrelation between PLS density and

LECP fluxes, while a less pronounced anticorrelation between magnetic field pressure and LECP fluxes is observed. Estimates of LECP pressures suggest that this hot plasma will provide the previously described missing pressure in the core if heavy ions dominate the ion composition.

099.026 **Io's interaction with the plasma torus: a self–consistent model.**
D. A. Wolf–Gladrow, F. M. Neubauer, M. Lussem.
J. Geophys. Res., Vol. 92, No. A9, p. 9949 – 9961 (1987).

The interaction between Io and the plasma torus is simulated by a three–dimensional numerical model which allows the calculation of electric fields, current density distributions, and magnetic fields in Io's vicinity and in the Alfvén wings. The authors show results for different up–stream plasma conditions and atmosphere/ionosphere models and discuss consequences of variations in neutral density, Alfvénic Mach number, and other parameters.

099.027 **The interaction of Io's Alfvén waves with the Jovian magnetosphere.**
A. N. Wright.
J. Geophys. Res., Vol. 92, No. A9, p. 9963 – 9970 (1987).

The author studies, numerically, the propagation of Alfvén waves through the Io torus. It is shown that the WKB limit is not valid for Io's Alfvén waves and that the waves strongly interact with the torus. The effect of this interaction will be to produce more complicated wave patterns behind Io, and the author discusses the possible new features one may expect.

099.028 **Cross–polarized interferometry of a Jovian decametric radio storm.**
J. A. Phillips, T. D. Carr, W. B. Greenman, J. Levy.
J. Geophys. Res., Vol. 92, No. A9, p. 9971 – 9977 (1987).

Observations of an 18–MHz Io–A storm have been conducted using a 46 km cross–polarized interferometer at the University of Florida. The pre– and postdetection correlation properties of LH and RH elliptically polarized L bursts have been studied. Fringe visibility measurements suggest that two discrete and independent sources are sometimes simultaneously active, the polarization sense of one being LH and that of the other RH. A scintillation analysis supports the conclusion that two or more sources were active in such cases.

099.029 **Z mode radiation in Jupiter's magnetosphere.**
C. F. Kennel, R. F. Chen, S. L. Moses, W. S. Kurth, F. V. Coroniti, F. L. Scarf, F. F. Chen.
J. Geophys. Res., Vol. 92, No. A9, p. 9978 – 9996 (1987).

The paper presents initial results of a survey of Voyager plasma wave instrument wide–band frames that exhibit a narrow–band emission below the low–frequency cutoff of the continuum band. These waves were first reported by Gurnett et al. (1980), and the authors' analysis of over 400 wide–band frames has enabled them to identify them as the Z mode.

099.030 **Detection of neutral oxygen and sulfur emissions near Io using *IUE*.**
G. E. Ballester, H. W. Moos, P. D. Feldman, D. F. Strobel, M. E. Summers, J.–L. Bertaux, T. E. Skinner, M. C. Festou, J. H. Lieske.
Astrophys. J., Lett. Ed., Vol. 319, No. 1, p. L33 – L38 (1987).

Two very long *IUE* exposures obtained with precise control of the spacecraft pointing show emissions due to neutral oxygen and sulfur near Io. The emitting region is centered on Io and is less than five Io diameters across. The two spectra obtained near eastern and western elongation in 1986 July and October, respectively, are similar, independent of whether the downstream or upstream side of the torus plasma flow past Io was viewed. If the spectra were produced solely by electron impact on O and S, the effective electron temperature is low (<2 eV), and large O and S densities are required. Electron impact on SO_2 could contribute substantially to the observed spectra.

099.031   Vibrationally excited molecular hydrogen in the upper atmosphere of Jupiter.
T. E. Cravens.
J. Geophys. Res., Vol. 92, No. A10, p. 11,083 – 11,100 (1987).
Experiments on the Voyager 1 and 2 spacecraft and observations made by the International Ultraviolet Explorer (IUE) have provided evidence for the existence of energetic particle precipitation into the upper atmosphere of Jupiter from the magnetosphere. This auroral precipitation has been shown to generate large ionization and dissociation rates, to excite auroral emissions, and also to vibrationally excite molecular hydrogen. A theoretical model of vibrationally excited H_2 in the upper atmosphere of Jupiter is presented in this paper. Models are considered for both the auroral region and also for lower latitudes, where H_2 is vibrationally excited owing to processes associated with the absorption of solar ultraviolet radiation.

099.032   Comment on "Periodic amplitude variations in Jovian continuum radiation" by W. S. Kurth, D. A. Gurnett, F. L. Scarf [J. Geophys. Res., Vol. 91, No. A12, p. 13,523 – 13,530 (1986)].
D. D. Barbosa.
J. Geophys. Res., Vol. 92, No. A10, p. 11,269 – 11,272 (1987). See Abstr. 42.099.069.

099.033   Reply to the comment on "Periodic amplitude variations in Jovian continuum radiation" by D. D. Barbosa.
W. S. Kurth, D. A. Gurnett, F. L. Scarf.
J. Geophys. Res., Vol. 92, No. A10, p. 11,273 – 11,276 (1987). See Abstr. 099.032.

099.034   Correlating east–west asymmetries in the Jovian magnetosphere and the Io sodium cloud.
W. H. Smyth, M. R. Combi.
Geophys. Res. Lett., Vol. 14, No. 9, p. 973 – 976 (1987).
Three–east–west intensity asymmetries in the Jupiter system that would appear to an Earth observer in optical S^+, extreme ultraviolet S^{++} and optical Na emissions are reviewed.

099.035   Corotation lag limit on mass–loss rate from Io.
T. S. Huang, G. L. Siscoe.
Astrophys. J., Vol. 319, No. 2, p. 1003 – 1009 (1987).
Considering rapid escape of H_2O from Io during an early hot evolutionary epoch, the authors construct an H_2O plasma torus by balancing dissociation and ionization products against centrifugally driven diffusion, including for the first time the effects of corotation lag resulting from mass loading. An ultimate mass loading limit of $1.3 \times 10^7 kg\,s^{-1}$ is derived. Connecting this limit with the variations of Io's temperature in its early evolution epoch gives an estimate of the upper limit on the total mass loss from Io, $\sim 3.0 \times 10^{20} kg$ (for high–opacity nebula) and $\sim 8.9 \times 10^{20} kg$ (for low–opacity nebula). These limits correspond to eroding 8 km and 22 km of H_2O from the surface.

099.036   Polarimetric investigation of the Jovian satellite Callisto.
R. A. Chigladze.
Tr. Tbil. Univ., Tom 264, p. 257 – 266 (1986). In Russian. Abstr. in Ref. Zh., 51. Astron., 9.51.273 (1987).

099.037   Near IR imaging of Io.
R. R. Howell.
Interferometric imaging in astronomy, p. 251 – 252 (1987). – See Abstr. 012.035.

099.038   Galilean satellite eclipse timings: 1985/86 report.
J. E. Westfall.
Strolling Astron., Vol. 32, Nos. 5 – 6, p. 114 – 128 (1987).

099.039   Eclipse measurements of Io's sodium atmosphere.
N. M. Schneider, D. M. Hunten, W. K. Wells, L. M. Trafton.
Science, Vol. 238, No. 4823, p. 55 – 58 (1987).
The satellites of Jupiter eclipsed each other in 1985, and these events allowed an unusual measurement of the sodium in Io's extended atmosphere. Europa was used as a mirror to look back through the Io atmosphere at the sun. The measured column abundances suggest that the atmosphere is collisionally thin above 700 kilometers and may be collisionally thin to the surface. The sodium radial profile above 700 kilometers resembles a 1500 kelvin exosphere with a surface density near 2×10^4 sodium atoms per cubic centimeter, but a complete explanation of the dynamics requires a more complex nonthermal model: the calculated loss rates suggest that the atmosphere is being replaced on a time scale of hours.

099.040   The Jupiter–Io connection: an Alfvén engine in space.
J. W. Belcher.
Science, Vol. 238, No. 4824, p. 170 – 176 (1987).
Io continuously generates an Alfvén wing that carries two billion kilowatts of power into the jovian ionosphere. Concurrently, Io is acted upon by a $J \times B$ force tending to propel it out of the jovian system. The energy source for these processes is the rotation of Jupiter. This unusual planet–satellite coupling serves as an archetype for the interaction of a large moving conductor with a magnetized plasma, a problem of general space and astrophysical interest.

099.041   Voyager photometry of Io.
D. P. Simonelli.
Diss. Abstr. Int., Sect. B, Vol. 48, No. 1, p. 162–B (1987). Thesis, Cornell University, 405 pp. (1987). Order No. DA8708989.

099.042   The surfaces of Europa, Ganymede, and Callisto: an investigation using Voyager IRIS thermal infrared spectra.
J. R. Spencer.
Diss. Abstr. Int., Sect. B, Vol. 48, No. 3, p. 793–B – 794–B (1987). Thesis, The University of Arizona, 228 pp. (1987). Order No. DA8712912.

099.043   Observations of large–amplitude MHD waves in Jupiter's foreshock in connection with a quasi–perpendicular shock structure.
M. B. Bavassano–Cattaneo, G. Moreno, M. T. Scotto, M. Acuña.
J. Geophys. Res., Vol. 92, No. A11, p. 12413 – 12418 (1987).
Plasma and magnetic field observations performed on board the Voyager 2 spacecraft have been used to investigate Jupiter's foreshock. Large–amplitude waves have been detected in association with the quasi–perpendicular structure of the Jovian bow shock, thus proving that the upstream turbulence is not a characteristic signature of the quasi–parallel shock.

099.044   Ice–covered water volcanism on Ganymede.
M. L. Allison, S. M. Clifford.
J. Geophys. Res., Vol. 92, No. B8, p. 7865 – 7876 (1987).
Eruption of liquid H_2O magmas along extensional fractures and graben–bounding normal faults may have played a critical role in the development of Ganymede's grooved terrain. The resurfacing potential of a water magma is dependent on a variety of factors, including the areal extent of the source region, the rate of discharge, the thickness of the flow, and the time that it takes the flow to completely freeze to its base. In this paper the thermal evolution of such a flow is considered in detail.

099.045   Erratum: "The contraction of Io's orbit" [Astron. J., Vol. 92, No. 1, p. 199 – 202 (1986)].
S. J. Goldstein Jr., K. C. Jacobs.
Astron. J., Vol. 94, No. 6, p. 1694 (1987). See Abstract 42.099.050.

099.046 Rings around planets.
J. A. Burns.
The evolution of the small bodies of the solar system, p. 301 – 307 (1987). – See Abstr. 012.041.
1. Introduction. 2. Rings around Jupiter. 3. Rings around Uranus. 3. Ringlike arc around Neptune, 5. Rings around Saturn. 6. Concluding remarks.

099.047 Ring geometry on Ganymede and Callisto.
P. M. Schenk, W. B. McKinnon.
Icarus, Vol. 72, No. 1, p. 209 – 234 (1987).
The authors assess the degree of circular symmetry inherent to multiringed impact structures on icy satellites by mapping the well–preserved multiring Valhalla (and Asgard) structure(s) on Callisto. With the aid of improved Ganymede coordinate control, they remap the furrow system and analyze its regional and overall circularity, both intrinsically and in comparison with Valhalla. The degree of alignment of furrows in Marius and Galileo Regio is determined, and in addition, at least three other multiring systems of varying sizes are identified. Based on furrow geometry and occurrence, the authors then reexamine large impact and alternate origins for the furrow system and evaluate the hypothesis of lateral motion within Ganymede's lithosphere.

099.048 Spectral geometric albedos of the Galilean satellites from 0.24 to 0.34 micrometers: observations with the International Ultraviolet Explorer.
R. M. Nelson, A. L. Lane, D. L. Matson, G. J. Veeder, B. J. Buratti, E. F. Tedesco.
Icarus, Vol. 72, No. 2, p. 358 – 380 (1987).
The authors present the results of an 8–year program of spectrophotometry of the Galilean satellites of Jupiter that was undertaken using the International Ultraviolet Explorer (IUE) Spacecraft. The ultraviolet geometric albedos of all four satellites are low. This is consistent with the hypothesis that sulfurous materials escaping from the surface of Io are being distributed by magnetospheric processes to the surfaces of the other three objects.

099.049 Variability of ethane on Jupiter.
T. Kostiuk, F. Espenak, M. J. Mumma, D. Deming, D. Zipoy.
Icarus, Vol. 72, No. 2, p. 394 – 410 (1987).
Abundances and spatial distributions of ethane in Jupiter's stratosphere were obtained from ultrahigh–resolution $(\lambda/\Delta\lambda \sim 10^6)$ spectra of individual C_2H_6 emission lines in the ν_9 band near 12 μm. The accuracy of the retrieved C_2H_6 mole fractions was evaluated in the context of varying stratospheric temperature profiles and C_2H_6 altitude distributions. A twofold uncertainty in the accuracy of the obtained abundances is possible. A mean equatorial value for the C_2H_6 mole fraction of $2.8 \pm 0.6 \times 10^{-6}$ was retrieved. Significant variability in the ethane line emission and retrieved mole fractions was found near the footprint of Io's flux tube and within the auroral regions. An increase in the ethane emission and abundance is obtained near the south polar region, relative to equatorial and northern latitudes. A significant decrease in ethane emission and abundance was observed in April 1983 near the known "hot spot".

099.050 Three excitation bands of whistlers in Jupiter's radiation belts.
P. A. Bespalov.
Pis'ma Astron. Zh., Tom 13, No. 12, p. 1094 – 1099 (1987). In Russian. English translation in Sov. Astron. Lett., Vol. 13.
Instability of Jupiter's radiation belts to the excitation of whistler waves at cyclotron resonance is studied. It is shown that for the experimentally observed dumb–bell–shaped distributions of energetic electrons the instability is possible in three spectral bands. One is below the relativistic gyrofrequency, ω_B/γ. The other two are centred round the half of the nonrelativistic gyrofrequency ($\omega_B/2$).

099.051 Photoelectric photometry of some mutual events of Galilean satellites in 1985.
F. J. Melillo.
I.A.P.P.P. Commun., No. 30, p. 15 – 19 (1987).

099.052 Erratum: "La planète Jupiter en 1984" [Obs. Trav., No. 10, p. 5 – 28 (1987)].
R. Neel.
Obs. Trav., No. 12, p. 60 (1987). See Abstr. 43.099.063.

099.053 Jupiter without satellites.
J. Meeus.
J. Br. Astron. Assoc., Vol. 98, No. 1, p. 35 – 37 (1987).
The list is given of all appearances of Jupiter without satellites from AD 1900 to 2100.

099.054 Les observations astrométriques des satellites galiléens.
J.–E. Arlot.
Ann. Phys. (Paris), Vol. 12, Colloq. No. 1, p. 7 – 11 (1987). – See Abstr. 012.061.
The author discusses the different methods of astrometric observations of the Galilean satellites. Two main types of observations are possible: observations of phenomena, and observations of positions.

099.055 Le bilan de la campagne.
J.–E. Arlot.
Ann. Phys. (Paris), Vol. 12, Colloq. No. 1, p. 13 – 17 (1987). – See Abstr. 012.061.
Many phenomena were observable in the 1985 – 1986 period. This paper is a report on the results obtained in French observatories, and in some other countries. On 105 possible phenomena, 82 have been observed, which lead to 200 light curves recorded.

099.056 Les observations de phénomènes mutuels faites à l'OHP et au Chiran.
W. Thuillot, J.–E. Arlot, V. D'Ambrosio, J. Lecacheux, C. Ruatti.
Ann. Phys. (Paris), Vol. 12, Colloq. No. 1, p. 59 – 63 (1987). – See Abstr. 012.061.
This paper gives the list of observations made at the "Observatoire de Haute Provence" and at the "Mont Chiran". 12 observers obtained 25 "good" light–curves using the 80 cm telescope (OHP) and the 1 m telescope (Chiran). Photoelectric photometers were used.

099.057 Les observations à Grenade (Espagne).
J.–C. Valtier, J.–M. Le Contel, J.–P. Sareyan.
Ann. Phys. (Paris), Vol. 12, Colloq. No. 1, p. 65 – 66 (1987). – See Abstr. 012.061.
A team of observers of short period variable stars used their telescope and photometer for the successful observations of some mutual phenomena. The observations were made at Granada Observatory (Spain).

099.058 Les phénomènes mutuels à Teramo.
R. Burchi, A. Di Paolantonio.
Ann. Phys. (Paris), Vol. 12, Colloq. No. 1, p. 67 – 71 (1987). – See Abstr. 012.061.
Twenty three mutual phenomena were observed at Teramo Observatory (Italy) between June and December 1985. The particulars of these observations are given.

099.059 Photométrie photoélectrique des phénomènes mutuels à l'Observatoire Astrophysique de Catane.
V. D'Ambrosio, C. Blanco.
Ann. Phys. (Paris), Vol. 12, Colloq. No. 1, p. 73 – 77 (1987). – See Abstr. 012.061.
This paper reports nine mutual phenomena observed at Catania Observatory using the 91 cm Cassegrain telescope. A comparison of observations and predictions is shown.

099.060   1985 mutual phenomena of Jupiter's satellites observed in Brazil.
J. Barroso Jr., F. J. Jablonski, G. R. Quast.
Ann. Phys. (Paris), Vol. 12, Colloq. No. 1, p. 109 – 115 (1987). –
See Abstr. 012.061.

099.061   Les observations photographiques des phénomènes mutuels.
F. Colas.
Ann. Phys. (Paris), Vol. 12, Colloq. No. 1, p. 117 – 122 (1987). –
See Abstr. 012.061.
Some mutual phenomena have been observed using the photographic method: only few observations have been made but it seems that this method may be used by amateur astronomers who have not other material. The accuracy of such observations is not yet of good quality but may be easily improved.

099.062   Les observations effectuées par le "GEOS".
S. Ferrand.
Ann. Phys. (Paris), Vol. 12, Colloq. No. 1, p. 123 – 126 (1987). –
See Abstr. 012.061.
The observations described in this paper were made by amateur astronomers belonging to the GEOS network of observers. Light–curves were obtained either by the visual method or using photoelectric photometers.

099.063   Modèle de chute en magnitude au cours d'un phénomène mutuel des satellites de Jupiter.
B. Morando.
Ann. Phys. (Paris), Vol. 12, Colloq. No. 1, p. 139 – 146 (1987). –
See Abstr. 012.061.
A method to compute a model for the drop in magnitude during a mutual occultation or a mutual eclipse is given. This method allows the scattering law to be chosen and takes into account the darkening of the solar limb. Phase effects, formerly thought to be negligible, might give measurable discrepancies.

099.064   Les albédos des satellites galiléens.
J. Lecacheux.
Ann. Phys. (Paris), Vol. 12, Colloq. No. 1, p. 147 – 156 (1987). –
See Abstr. 012.061.
This paper gives information about the albedos of the Galilean Satellites. Since the observations of the mutual events are of photometrical type and very precise, it is necessary to consider, even for astrometrical use, the albedos. A good value for the magnitude drop during a mutual event may so be obtained.

099.065   Influence de la non–uniformité de brillance.
J. F. Ferrier.
Ann. Phys. (Paris), Vol. 12, Colloq. No. 1, p. 157 – 165 (1987). –
See Abstr. 012.061.
This paper tries to explain why some light–curves are not symmetrical. A model is built to fit a light–curve obtained at ESO in 1979. The variations of the albedo depending on the location on the surfaces of the satellites may modify the shape and the magnitude drop of the observed light–curves. Such a study is made in order to improve the accuracy of the (O–C) to be calculated.

099.066   Les applications astrométriques de la photométrie bidimensionnelle.
W. Thuillot, P. Laques.
Ann. Phys. (Paris), Vol. 12, Colloq. No. 1, p. 195 – 199 (1987). –
See Abstr. 012.061.
Bidimensional photometry has been extensively used for the first time on the observations of the mutual phenomena of the Galilean Satellites of Jupiter. The advantages of this technique are examined and its application to other types of phenomena such as stellar occultations by asteroids is advocated.

099.067   Rapprochements et phénomènes des satellites galiléens.
W. Thuillot.
Ann. Phys. (Paris), Vol. 12, Colloq. No. 1, p. 211 – 216 (1987). –
See Abstr. 012.061.
The modern techniques of observations used during the PHEMU campaign should be applied to other types of phenomena such as close approaches of two satellites or the classical phenomena of the Galilean satellites. This could give a great amount of precise data useful for the improvement of the knowledge of the motions.

099.068   Voids in Jovian magnetosphere revisited: evidence of spacecraft charging.
K. K. Khurana, M. G. Kivelson, T. P. Armstrong, R. J. Walker.
J. Geophys. Res., Vol. 92, No. A12, p. 13,399 – 13,408 (1987).
The authors have put forward a new interpretation of cold plasma voids observed by Voyager 2 near the orbit of Ganymede in terms of spacecraft charging. They believe that the proposed charging yields reasonably consistent interpretation of plasma and field observations.

099.069   On an explanation of the variability of temperature contrasts in the solar eclipse zone on Jupiter.
V. P. Vasil'ev, V. M. Litvinov.
Astron. Tsirk., No. 1461, p. 1 – 2 (1986). In Russian.
It is suggested that the appearance of an additional thermal flow from Jovian solar eclipse regions is connected with sound wave generation that leads to an increase of the atmosphere's infrared transparency.

099.070   Photoelectric observations of a mutual occultation of Jovian satellites.
M. B. Bogdanov.
Astron. Tsirk., No. 1466, p. 7 – 8 (1986). In Russian.
The light curves of the mutual occultation of the Jovian satellites Europa and Ganymede observed on 1985 August 27 are given.

099.071   On the intensity of methane and ammonia molecular bands in the spectrum of Jupiter.
Sh. M. Namazov.
Astron. Tsirk., No. 1470, p. 3 – 4 (1986). In Russian.
From observations of Jupiter carried out in June, 1982 it is found that the intensity of methane and ammonia molecular absorption bands changes both during one night (20 – 30%) and from night to night (15 – 20%). It is supposed that the essential changes of intensity are mainly the result of changes of the physical conditions in the cloud layer.

099.072   Electron capture decay in Jovian planets.
R. R. Zito, D. Schiferl.
Icarus, Vol. 72, No. 3, p. 647 – 649 (1987).
In Jovian planets and stars, where interior pressures may be comparable to or even exceed 45 Mbar, extra interior heat may be generated by pressure–accelerated electron capture decay of certain isotopes. These isotopes include ^{40}K, ^{50}V, ^{123}Te, ^{138}La, ^{26}Al, and ^{36}Cl.

099.073   Cloud chemistry on Jupiter.
B. E. Carlson, M. J. Prather, W. B. Rossow.
Astrophys. J., Vol. 322, No. 1, p. 559 – 572 (1987).
Chemical equilibrium models have been used to interpret observations and constrain models of Jupiter. The authors reexamine the chemical reactions controlling the composition of the cloud–forming region from 10 to 0.1 bar and the thermodynamic data used in these models. Particular attention is focussed on reactions involving H_2S and NH_3. Best agreement with the observations is found for 2 times solar abundances of N, O, and S relative to H.

099.074 **An explanation for characteristics of HOM radiation from Jupiter.**
P.–s. Chen, T. D. Carr.
Acta Astron. Sin., Vol. 28, No. 3, p. 268 – 273 (1987). In Chinese.
 An interpretation of the observed HOM radiation from Jupiter based on magnetic field properties of a tilted dipole is presented.

Les satellites galiléens de Jupiter de 1610 à 1985: observations et mouvements.
See Abstr. 004.111.

Observations of industrial sulfur flows and implications for Io.
See Abstr. 022.030.

Intensities and H_2–broadened half–widths of germane lines around 4.7 μm at temperatures relevant to Jupiter's atmosphere.
See Abstr. 022.065.

Sulfur in vacuum: sublimation effects on frozen melts, and applications to Io's surface and torus.
See Abstr. 022.107.

Laboratory measurements of the microwave opacity of gaseous ammonia (NH_3) under simulated conditions for the Jovian atmosphere.
See Abstr. 022.108.

Electric discharge synthesis of HCN in simulated Jovian atmospheres.
See Abstr. 022.109.

Rate constant for the reaction of atomic oxygen with phosphine at 298K.
See Abstr. 022.135.

Observations PHEMU85 à l'Observatoire de Bordeaux.
See Abstr. 034.126.

Un photomètre adapté aux phénomènes mutuels.
See Abstr. 034.127.

A mid–infrared cryogenic echelle spectrometer.
See Abstr. 034.169.

The radar–glory theory for icy moons with implications for radar mapping.
See Abstr. 036.059.

Observations vidéo de phénomènes mutuels faites à Meudon, au Pic du Midi et à Nice.
See Abstr. 036.189.

Observations effectuées par le "GEA" (Group d'Estudis Astronomics).
See Abstr. 036.190.

La campagne PHEMU85 a l'ESO.
See Abstr. 036.191.

Les informations à extraire de l'observation des phénomènes mutuels et leur utilisation.
See Abstr. 036.193.

Long–term numerical integrations and synthetic theories for the motion of the outer planets.
See Abstr. 042.008.

Isolated eddy models in geophysics.
See Abstr. 062.089.

Magnetospheric environments of outer planet rings: influence of Saturn's axially symmetric magnetic field.
See Abstr. 091.019.

Europa, tidally heated oceans, and habitable zones around giant planets.
See Abstr. 091.025.

Small satellites.
See Abstr. 091.037.

A simple two–layer model for the Uranus satellites.
See Abstr. 101.053.

Gas flow in the solar nebula leading to the formation of Jupiter.
See Abstr. 107.011.

An evolutionary framework for the Jovian and Saturnian satellites.
See Abstr. 107.037.

100 Saturn, Saturn Satellites

100.001 **Accretional heating of the satellites of Saturn and Uranus.**
S. W. Squyres, R. T. Reynolds, A. L. Summers, F. Shung.
NASA Tech. Memo., NASA TM–89810, p. 37 – 39 (1987). – See Abstr. 003.001.

100.002 **Subcentimeter–size particle distribution functions in planetary rings from Voyager radio and photopolarimeter occultation data.**
H. A. Zebker, G. L. Tyler, E. A. Marouf.
NASA Tech. Memo., NASA TM–89810, p. 55 – 56 (1987). – See Abstr. 003.001.

100.003 **Observational studies of Saturn's rings.**
C. C. Porco.
NASA Tech. Memo., NASA TM–89810, p. 57 – 58 (1987). – See Abstr. 003.001.

100.004 **The production of "braids" in Saturn's F ring.**
J. J. Lissauer, S. J. Peale.
NASA Tech. Memo., NASA TM–89810, p. 59 (1987). Abstract. – See Abstr. 003.001.

100.005 **Hamiltonian theory of nonlinear waves in planetary rings.**
G. R. Stewart.
NASA Tech. Memo., NASA TM–89810, p. 124 (1987). Abstract. – See Abstr. 003.001.

100.006 **Dynamical studies of Saturn's rings.**
P. D. Nicholson, C. C. Porco.
NASA Tech. Memo., NASA TM–89810, p. 128 – 129 (1987). – See Abstr. 003.001.

100.007   Studies of early intense cratering and possible saturation effects.
W. K. Hartmann.
NASA Tech. Memo., NASA TM–89810, p. 411 – 412 (1987). – See Abstr. 003.001.
Crater counts on Rhea were completed, to be compared and combined with independent counts by Steve Squyres; the material is now being integrated into a study of cratering on Rhea and other Saturn satellites by Squyres, Lissauer, and Hartmann. Special attention is being paid to lighting and other effects on the apparent changes in crater density from one region to another.

100.008   The effect of Saturn's rings on the upper–boundary insolation of its atmosphere.
E. Van Hemelrijck.
Earth, Moon, Planets, Vol. 38, No. 3, p. 217 – 235 (1987). With a correction in Vol. 39, No. 3, p. 291 (1987).
The daily solar radiation incident at the top of Saturn's atmosphere, taking into account both the oblateness of the planet and the shadow of the ring system, is calculated. It is found that the decrease of the daily insolation in winter is important near the solstices up to mid–latitudes and in the neighborhood of the equinoxes for equatorial and low latitudes. The combined effect of Saturn's rings and its flattening on the mean winter and annual daily insolations is also studied. The numerical results show that the mean wintertime insolation falls gradually in the (0 – 20°) latitude region to a peak value of about 50%. Beyond 20° the loss of insolation decreases and from approximately 45° up to polar region latitudes the decrease reaches a practically constant level of 35%. The mean annual daily insolation is maximally reduced by about 20% at localities of 20°.

100.009   Saturn 1985.
A. W. Heath.
J. Br. Astron. Assoc., Vol. 97, No. 5, p. 263 – 266 (1987).

100.010   Interpretation of electrostatic noise observed by Voyager 1 in Titan's wake.
T. Z. Ma, D. A. Gurnett, C. K. Goertz.
J. Geophys. Res., Vol. 92, No. A8, p. 8595 – 8602 (1987).
During the Voyager 1 spacecraft flyby of Titan on November 12, 1980, an intense band of low–frequency electric field noise was observed during the inbound wake crossing. This analysis shows that the noise is generated by a beam–plasma interaction between the corotating magnetosphere of Saturn and newly created ions from the atmosphere of Titan. The analysis is based on plasma and wave measurements from Voyager 1 and reasonable assumptions. The results agree quite well with the observation.

100.011   Scattering properties of a moonlet (satellite) embedded in a particle ring: application to the rings of Saturn.
F. Spahn.
Icarus, Vol. 71, No. 1, p. 69 – 77 (1987).
The gravitational influence of moonlets or satellites on the radial structure of the rings of Saturn has been calculated numerically. A drastic change in the surface mass density is obtained even after a single scattering process of the ring particles on a moonlet (satellite). The final surface density shows a significant radial structure, which has been used to estimate the radius and the mass of moonlets or satellites embedded in rings of low optical depth (E ring, Cassini division, C ring).

100.012   Corrections to the theory of the orbit of Saturn's satellite Hyperion.
D. B. Taylor, A. T. Sinclair, P. J. Message.
Astron. Astrophys., Vol. 181, No. 2, p. 383 – 390 (1987).
A spectral analysis has been made of the residuals obtained from fitting Woltjer's theory of Hyperion to the numerical integration of Sinclair and Taylor (1985) extended to ± 25 yr from epoch. The major correction found was to the phase of the libration argument and is probably due to a slightly inaccurate value of the frequency of this argument. The new theory now includes all perturbations corresponding to an observed displacement of 0."07 or more at the opposition distance of 8.5 AU. The new amplitudes of the long–period perturbations are in broad agreement with those obtained from recent analytical work by Message. The new theory fitted to observations from 1967 to 1983 gave a root–mean–square residual of 0."28 compared to 0."44 for the fit of the old theory given in Taylor (1984) to observations in the same period.

100.013   How primitive are the gases in Titan's atmosphere?
T. Owen.
Adv. Space Res., Vol. 7, No. 5, p. 51 – 54 (1987). – See Abstr. 012.022.

100.014   Titan's atmospheric interaction with its plasma environment.
W.–H. Ip.
Adv. Space Res., Vol. 7, No. 5, p. 55 – 64 (1987). – See Abstr. 012.022.
Physical processes involved in Titan's atmospheric interaction with the Saturnian magnetosphere and the solar wind are considered with special attention to the ionization coupling of Titan's extended hydrogen (and nitrogen) gas torus with the magnetospheric plasma. The effect of direct plasma interaction with the neutral atmosphere on the ionospheric transport and chemistry is also discussed in comparison with the Venus–type interaction process.

100.015   Titan's ionosphere and atmospheric evolution.
S. J. Bauer.
Adv. Space Res., Vol. 7, No. 5, p. 65 – 69 (1987). – See Abstr. 012.022.
The ionosphere of Titan is expected to consist of hydrocarbon and nitrile ions resulting from the ionization of the predominant constituent N_2 and chemical reactions with the minor constituent methane. Impact ionization of N_2 and dissociative recombination of the resulting ions lead to energetic neutrals which can escape from Titan's exosphere. Interaction of the solar wind, particularly during a T–Tauri phase, with Titan's ionosphere when it lies outside Saturn's magnetosphere will lead to removal of atmospheric species from Titan. Estimates of atmospheric mass loss due to these processes and its significance for the evolution of Titan's atmosphere are presented.

100.016   Organic chemistry in the oceans of Titan.
F. Raulin.
Adv. Space Res., Vol. 7, No. 5, p. 71 – 81 (1987). – See Abstr. 012.022.
On Titan, most of the organics present in the atmosphere must condense in the lower stratosphere and be solid near the surface, except methane, ethane, propane, propene and 1–butene which must be liquid and could form oceans containing large fractions of dissolved N_2. Chemical evolution on Titan must have followed a way very different from the terrestrial one, involving physical chemical processes in a cryogenic apolar solvent mainly composed of CH_4–C_2H_6–N_2, in place of organic chemistry in water. Systematic study of the volumic mass and solubility of organics in such a cryogenic mixture of various compositions, at 94K, is presented, using thermodynamic modelling. The results suggest that the oceans of Titan could be free of any "icebergs" of organic compounds. These oceans could be very rich in dissolved organics, with relatively high concentrations, in the range $1 - 10^{-6}$M. In addition, the concentration of several of the organic solutes should be constant, buffered by a bottom layer of the corresponding compound in the solid phase.

100.017   Limits on the extent of Saturn's hydrogen cloud.
J. D. Richardson, A. Eviatar.
Geophys. Res. Lett., Vol. 14, No. 10, p. 999 – 1,002 (1987).
The authors model the effects of a dense hydrogen cloud on the plasma in the inner magnetosphere of Saturn, and conclude that observed plasma properties are not consistent with the presence of such a cloud. Therefore the source of the neutral hydrogen cloud around Saturn must be Titan rather than Saturn's

atmosphere, and plasma sources and dynamics in the outer magnetosphere are predominantly the result of satellite–magnetosphere interactions.

100.018 The nature of Saturn's atmospheric Great White Spots.
  A. Sanchez–Lavega, E. Battaner.
Astron. Astrophys., Vol. 185, No. 1/2, p. 315 – 326 (1987).
  Saturn's Great White Spots (GWS) are the biggest asymmetric cloud systems observable in the atmosphere of this planet, which have been detected in only four occasions in 1876, 1903, 1933 and 1960. In this paper the authors present new data about GWS, extending their previous work on morphology and kinematics (Sanchez–Lavega, 1982), and including a photometric analysis of the calibrated photographs of the GWS obtained in 1933 in four colors, which allows to evaluate the approximate pressure level attained by the GWS's cloud top. Then the authors present a buoyant parcel model which is a first approach to the origin of Saturn's GWS, and which is similar to that employed by Stoker (1986) to explain the posssible actuation of a trigger mechanism in the emergence of these unusual events and the importance for dynamical studies of a long–term survey in the visible of Saturn's cloud systems.

100.019 Spokes in Saturn's B ring: dynamical and physical properties deduced from Voyager Saturn ring images.
  R. E. Eplee Jr.
Diss. Abstr. Int., Sect. B, Vol. 48, No. 3, p. 792–B (1987). Thesis, The University of Arizona, 160 pp. (1987). Order No. DA8712872.

100.020 A partially collisional model of the Titan hydrogen torus.
  D. A. Hilton.
Diss. Abstr. Int., Sect. B, Vol. 48, No. 3, p. 792–B (1987). Thesis, The University of Arizona, 203 pp. (1987). Order No. DA8712879.

100.021 Numerical theory of motion of Titan and Hyperion.
  T. K. Nikol'skaya.
Byull. Inst. Teor. Astron., Tom 15, No. 9, p. 534 – 537 (1986). In Russian.
  A description of a numerical theory of the motions of Saturn VI and VII (Titan and Hyperion) is given. The theory is compared with astrometric observations of these satellites made during 1967 – 1981. The mean–squares errors of representation of the observational data after improvement of the orbital elements are 0."45 for Titan and 0."58 for Hyperion. The osculating elements as well as the accuracy estimates for the initial date integration, i.e. 1972, October 10.0 JD 2441600.5 are given.

100.022 Titan's atomic nitrogen torus: inferred properties and consequences for the Saturnian aurora.
  D. D. Barbosa.
Icarus, Vol. 72, No. 1, p. 53 – 61 (1987).
  This paper follows up the lead suggested by Barbosa and Eviatar (1986) that Titanogenic nitrogen ions are a key component of the magnetospheric particle populations and can account for the energetics of the Saturnian aurora without undue assumptions. Nitrogen atoms resulting from electron impact dissociations of N_2 (Strobel and Shemansky 1982) escape from Titan and form a large doughnut–shaped ring around the satellite's orbit that is cospatial with the McDonough–Brice (1973) hydrogen cloud. Processes attendant to the ionization and pickup of nitrogen ions include the production of a warm kiloelectronvolt electron population and the excitation of the UV aurora by particle precipitation from the outer magnetosphere.

100.023 Estimated impact shock production of N_2 and organic compounds on early Titan.
  T. D. Jones, J. S. Lewis.
Icarus, Vol. 72, No. 2, p. 381 – 393 (1987).
  The authors present the results of a chemical model that calculates the equilibrium composition of a Titan gas mixture subjected to shock heating and rapid quenching, with the shock energy derived from the infall of meteoritic and cometary debris. Early bombardment of Titan by circum–Saturnian and heliocentric projectiles delivered energy to the atmosphere in amounts dwarfing the current input from ultraviolet radiation and any existing lightning source. The resulting N_2 and organic production would have had major effects on the early atmospheric and surface evolution of Titan, converting enough primordial NH_3 to produce the observed N_2–rich atmosphere, and depositing an organic layer hundreds of meters thick.

100.024 Eccentric features in Saturn's outer C ring.
  C. C. Porco, P. D. Nicholson.
Icarus, Vol. 72, No. 2, p. 437 – 467 (1987).
  A systematic search has been made for as yet unrecognized eccentric and inclined features in Saturn's outer C ring. The radii of all sharp–edge features in the outer C ring were measured in Voyager data consisting of six high–resolution images, the Photopolarimeter occultation data, and the Radio Science $\lambda 3.6$–cm occultation data corrected for the effects of diffraction. Besides the well–known Maxwell ringlet at 87,491 km (1.450 RS), whose eccentric shape and kinematics have already been studied, two other narrow ringlets at 88,716 km (1.470 RS), and 90,171 km (1.495 RS) have been found to be demonstrably eccentric. The former has a mean width of ~ 16 km and is located within a gap ~ 30 km wide. The latter has a mean width of ~ 62 km and is only partially isolated: its outer edge is defined by a gap ~ 15 km wide.

100.025 An update of nitrile photochemistry of Titan.
  Y. L. Yung.
Icarus, Vol. 72, No. 2, p. 468 – 472 (1987). = Contrib. No. 4390, Div. Geol. Planet. Sci., Calif. Inst. Technol., Pasadena, USA.
  New chemical schemes leading to the formation of cyanogen (C_2N_2) and dicyanoacetylene (C_4N_2) in the upper atmosphere of Titan are proposed and examined in light of recent laboratory kinetics experiments and Voyager observations.

100.026 First order perturbation in the Enceladus–Dione system.
  J. da Silva Bevilacqua, W. Sessin.
Rev. Mex. Astron. Astrofis., Vol. 14, No. 2, p. 627 – 630 (1987). – See Abstr. 012.042.
  A new orbit for the pair of Saturn's satellites Enceladus and Dione, is constructed based on the intermediate solution obtained by Salgado and Sessin (1985). Secular terms are aggregated to the Hamiltonian when all first order perturbations are taken into account. Consequently, the periods of the new orbit generated by this Hamiltonian are better determined.

100.027 Seasonal changes display in the Saturn atmosphere.
  A. P. Vid'machenko.
Kinematika Fiz. Nebesn. Tel, Tom 3, No. 6, p. 10 – 12 (1987). In Russian. English translation in Kinematics Phys. Celest. Bodies.
  By investigating the methane absorption distribution ($\lambda = 0.619 \mu m$) across the Saturn disk it is shown that it is possible to study the seasonal processes in planetary atmospheres for short time intervals.

100.028 Tidal flows in Saturn's rings and in thin disks in binary stellar systems.
  E. V. Smel'chakova.
Astron. Zh., Tom 64, Vyp. 6, p. 1317 – 1320 (1987). In Russian. English translation in Sov. Astron., Vol. 31, No. 6.
  The expression for the tidal flow velocities in a thin disk revolving around the component of a binary system are obtained. The tidal velocities produced by the satellites and the Sun in Saturn's rings are calculated. The tidal flow velocities in the disk–

shaped shells around the stars in not very close binaries are calculated similarly.

100.029   Predictions of the electrical conductivity and charging of the aerosols in Titan's atmosphere.
W. J. Borucki, Z. Levin, R. C. Whitten, R. G. Keesee, L. A. Capone, A. L. Summers, O. B. Toon, J. Dubach.
Icarus, Vol. 72, No. 3, p. 604 – 622 (1987).

The ionization caused by galactic cosmic rays and electron precipitation causes the formation of positive ions and electrons in Titan's atmosphere. Because the thermal velocity of electrons is higher than that of positive ions and ion clusters, the aerosols will become negatively charged. The lack of electrophilic molecular species results in large concentrations of free electrons throughout the lower atmosphere. These free electrons increase the conductivity of the lower atmosphere above values typical for the terrestrial atmosphere and cause additional absoprtion of radio waves propagating through the lower atmosphere. Because the aerosol particles are highly charged, coagulation is inhibited, particles are smaller, and settling speeds are reduced. Therefore the atmosphere contains large numbers of particles at all altitudes and has a much higher optical depth than it would without particle charging.

100.030   Comparison of the results of observations of the Saturn system in 1975 with theories of motion.
S. V. Tolbin.
Sov. Astron., Vol. 31, No. 2, p. 211 – 212 (1987). English translation of 43.100.007.

Experimental studies on the impact properties of water ice.
See Abstr. 022.001.

Measurements of the dielectric constants for planetary volatiles.
See Abstr. 022.011.

Trapping of gases by water ice and implications for icy bodies.
See Abstr. 022.071.

Optical evolution of laboratory–produced organics: applications to Phoebe, Iapetus, outer belt asteroids and cometary nuclei.
See Abstr. 022.079.

Rate constant for the reaction of atomic oxygen with phosphine at 298K.
See Abstr. 022.135.

Clustering reactions of H_2CN^+ ions with HCN.
See Abstr. 022.138.

A mid–infrared cryogenic echelle spectrometer.
See Abstr. 034.169.

Long–term numerical integrations and synthetic theories for the motion of the outer planets.
See Abstr. 042.008.

Magnetospheric environments of outer planet rings: influence of Saturn's axially symmetric magnetic field.
See Abstr. 091.019.

Small satellites.
See Abstr. 091.037.

Charged particle modification of surfaces in the outer solar system.
See Abstr. 099.002.

Voyager cartography.
See Abstr. 099.013.

Radial force balance within Jupiter's dayside magnetosphere.
See Abstr. 099.024.

Rings around planets.
See Abstr. 099.046.

Electron capture decay in Jovian planets.
See Abstr. 099.072.

A simple two–layer model for the Uranus satellites.
See Abstr. 101.053.

An evolutionary framework for the Jovian and Saturnian satellites.
See Abstr. 107.037.

101   Uranus, Neptune, Pluto, Transplutonian Planets

101.001   The cratering record at Uranus: implications for satellite evolution and the origin of impacting objects.
R. G. Strom.
NASA Tech. Memo., NASA TM–89810, p. 3 – 5 (1987). – See Abstr. 003.001.

101.002   Crater morphology and morphometry on the Uranian satellites.
S. K. Croft.
NASA Tech. Memo., NASA TM–89810, p. 6 – 8 (1987). – See Abstr. 003.001.

101.003   Cratering history of Miranda.
J. B. Plescia, J. M. Boyce.
NASA Tech. Memo., NASA TM–89810, p. 9 – 11 (1987). – See Abstr. 003.001.

101.004   The viscosity of Miranda.
P. J. Thomas, R. T. Reynolds, S. W. Squyres, P. M. Cassen.
NASA Tech. Memo., NASA TM–89810, p. 12 – 14 (1987). – See Abstr. 003.001.

101.005   Mechanical and thermal properties of planetologically important ices.
S. K. Croft.
NASA Tech. Memo., NASA TM–89810, p. 15 – 17 (1987). – See Abstr. 003.001.

101.006   Could Ariel have been heated by tidal friction?
S. J. Peale.
NASA Tech. Memo., NASA TM–89810, p. 18 (1987). Abstract. – See Abstr. 003.001.

101.007   Geology and cratering history of Ariel.
J. B. Plescia, J. M. Boyce.
NASA Tech. Memo., NASA TM–89810, p. 19 – 21 (1987). – See Abstr. 003.001.

101.008   On the lack of commensurabilities in the mean motions of the satellites of Uranus and the resurfacing of Ariel.
S. J. Peale.
NASA Tech. Memo., NASA TM–89810, p. 22 (1987). Abstract. – See Abstr. 003.001.

101.009 Orbital dynamics of the Uranian satellites based on Voyager data.
J. B. Plescia.
NASA Tech. Memo., NASA TM–89810, p. 23 – 25 (1987). – See Abstr. 003.001.

101.010 Why no orbital resonances among the satellites of Uranus?
S. J. Peale.
NASA Tech. Memo., NASA TM–89810, p. 26 (1987). Abstract. – See Abstr. 003.001.

101.011 Voyager observations of 1985U1.
P. Thomas, J. Veverka.
NASA Tech. Memo., NASA TM–89810, p. 27 – 29 (1987). – See Abstr. 003.001.

101.012 Kinematics and dynamics of the Uranian rings.
R. G. French.
NASA Tech. Memo., NASA TM–89810, p. 30 – 32 (1987). – See Abstr. 003.001.

101.013 Uranus and Neptune: questions and possible answers.
R. T. Reynolds, M. Podolak.
NASA Tech. Memo., NASA TM–89810, p. 109 – 111 (1987). – See Abstr. 003.001.

101.014 Properties of planetary fluids at high pressure and temperature.
W. J. Nellis, D. C. Hamilton, N. C. Holmes, H. B. Radousky, F. H. Ree, M. Ross, D. A. Young, M. Nicol.
NASA Tech. Memo., NASA TM–89810, p. 141 – 143 (1987). – See Abstr. 003.001.

101.015 Aspects of Voyager photogrammetry.
S. S. C. Wu, F. J. Schafer, R. Jordan, A.-E. Howington.
NASA Tech. Memo., NASA TM–89810, p. 528 – 529 (1987). – See Abstr. 003.001.

101.016 Magnetospheres of the outer planets.
J. A. Van Allen.
NASA Conf. Publ., NASA CP-2464, p. 1 – 17 (1987). – See Abstr. 012.004.
Contents: 1. Introduction. 2. Conditions for the existence of a planetary magnetosphere. 3. Source of magnetospheric particles. 4. In situ energization of particles. 5. The magnetosphere of Uranus. 6. The magnetosphere of Neptune. 7. Conclusion.

101.017 Position observations of the five greatest Uranian satellites and comparison with theory.
C. H. Veiga, R. Vieira Martins, C. Veillet, D. Lazzaro.
Astron. Astrophys., Suppl. Ser., Vol. 70, No. 3, p. 325 – 334 (1987).
Positions of the Uranian satellites from 264 photographic plates obtained at the Brazilian 1.6 m reflector are given for the oppositions of 1982, 1983, 1984 and 1985. These positions were compared with those theoretically calculated from orbital parameters determined by Veillet and Jacobson. The O–C (observed minus calculated) residuals had standard deviation of the order 0".1, which is compatible with the orbital parameters used.

101.018 Occurrence of central peak craters on the Uranian satellites: implications for surface structure and comparison with the Jovian and Saturnian systems.
K. D. Trego.
Earth, Moon, Planets, Vol. 38, No. 3, p. 299 – 303 (1987).
Craters with central peaks occur on the Uranian satellites Ariel, Umbriel, Titania, and Oberon; but do not occur on Miranda. The inelastic surface of Miranda is apparently due to the heavy tectonic reworking of its surface. A theory of expansion/contraction is proposed to explain the tectonic history of Miranda. The existence of central peak craters on the four largest satellites of Uranus implies that they have surface strengths similar to those of the Saturnian satellites and silicate bodies of the inner solar system which all have central peak craters. The absence of central peak craters on Miranda implies that it has an inelastic surface similar to those of the Jovian ice satellites Ganymede and Callisto whose surfaces do not contain central peak craters.

101.019 The rings of Uranus.
J. N. Cuzzi, L. W. Esposito.
Sci. Am., Vol. 257, No. 1, p. 42 – 48 (1987).
Why are the rings of Uranus so narrow and dark? Findings from the Voyager 2 encounter suggest that the austere ring system may be only a fleeting stage in a continuing saga of creation and destruction.

101.020 Pluto and Charon, a mysterious couple.
M. Lachièze-Rey.
Recherche, Vol. 18, No. 191, p. 1118 – 1119 (1987). In French.

101.021 The Uranian bow shock: Voyager 2 inbound observations of a high Mach number shock.
F. Bagenal, J. W. Belcher, E. C. Sittler Jr., R. P. Lepping.
J. Geophys. Res., Vol. 92, No. A8, p. 8603 – 8612 (1987).
The Voyager 2 magnetometer and plasma detector measured a high Mach number (fast magnetosonic number $M_{MS} > 15$), high β (~ 3) bow shock on the dayside of the Uranian magnetosphere. Although the average conditions on either side of the shock are consistent with the Rankine–Hugoniot (MHD) relations for a stationary, quasi–perpendicular shock, the data revealed both detailed structure in the transition region as well as considerable variability in the downstream magnetosheath plasma. The bulk plasma parameters and the magnetic field exhibited some of the characteristics of a supercritical shock: an overshoot followed by damped oscillations downstream, consistent with recent theoretical models of high Mach number quasi–perpendicular shocks.

101.022 The occultation of KME 17 by Uranus and its rings.
J. L. Elliot, I. S. Glass, R. G. French, J. A. Kangas.
Icarus, Vol. 71, No. 1, p. 91 – 102 (1987).
The occultation of KME 17 by Uranus and its rings was observed with the 1.8–m telescope at SAAO through a K filter with an InSb detector on 25 March 1983. Immersion of the nine main rings and the emersion of rings, 5, 4, α, and β were recorded in the nonchopping mode. A diffracted square–well model was fitted to the data, and the midtime, width, and equivalent depth were determined for each profile. The profile model also includes the diameter of the occulted star as a free parameter. The immersion and emersion atmospheric events give mean temperatures of 166 ± 15 and 149 ± 15 K within the pressure altitude range 1 – 10 μbar. Photometry in the JHKL bands for KME 17 and other stars previously occulted by Uranian rings is presented.

101.023 The near–infrared phase curve of the Uranian rings.
T. M. Herbst, M. F. Skrutskie, P. D. Nicholson.
Icarus, Vol. 71, No. 1, p. 103 – 114 (1987).
The authors performed K–band observations of the Uranian rings during the 1985 opposition in order to determine if a measurable opposition effect occurs. The investigation, carried out over a period of several months and over a range of 0.021° to 2.5° in solar phase angle, revealed an opposition surge of approximately 0.2 mag. This result is interpreted in terms of two simple models: mutual shadowing and surface microstructure. If entirely due to interparticle shadowing, the brightening implies a ring thickness inconsistent with surface mass densities derived from dynamical considerations. If the surge results from surface effects, the ring particles have dark surfaces with a porosity comparable to that of terrestrial snow.

101.024 The control networks of the satellites of Uranus.
M. E. Davies, T. R. Colvin, F. Y. Katayama, P. C. Thomas.
Icarus, Vol. 71, No. 1, p. 137 – 147 (1987).
Control networks of the five large satellites of Uranus have been established photogrammetrically from pictures taken by the

Voyager 2 spacecraft. The control networks cover the illuminated southern hemisphere of each satellite. Coordinates are listed for 103 points on Miranda, 52 points on Ariel, 43 points on Umbriel, 46 points on Titania, and 34 points on Oberon; some points are identified on the U.S. Geological Survey maps of these satellites. Miranda is ellipsoidal in shape with radii of 241, 235, and 232 km. Mean radii are 579 km for Ariel, 586 km for Umbriel, 790 km for Titania, and 762 km for Oberon.

101.025   **A semi–analytical solution for the eccentricities and longitudes of the pericenter of the Uranian satellites.**
D. Lazzaro, S. Ferraz–Mello, R. Vieira Martins.
Astron. Astrophys., Vol. 182, No. 1, p. 150 – 158 (1987). With a correction in Vol. 186, No. 1/2, p. 360 (1987).
Semi–analytical solutions for the variations in time of the eccentricities and longitudes of the pericenter are developed as part of a complete semi–analytical theory for the motion of the Uranian satellites. All the long and short period variations of these elements, up to the order of 10^{-5}, are considered. The results show that the oscillations around the mean value obtained by other authors through the analysis of the observations of Uranus' system, are significant. A discussion on the importance of the determination of osculating orbital elements, in order to obtain integration constants consistent with the real movement of Uranian satellites, is presented.

101.026   **Stellar occultation probes of the Uranian rings at 0.1 and 2.2 µm: a comparison of Voyager UVS and earth–based results.**
J. B. Holberg, P. D. Nicholson, R. G. French, J. L. Elliot.
Astron. J., Vol. 94, No. 1, p. 178 – 188 (1987).
Stellar–occultation observations of the Uranian ring system obtained by the Voyager 2 ultraviolet spectrometer, at an effective wavelength of 0.11 µm, are compared with previous 2.2 µm Earth–based occultation data.

101.027   **The orbit of Pluto and the cosmological constant.**
M. D. Roberts.
Mon. Not. R. Astron. Soc., Vol. 228, No. 2, p. 401 – 405 (1987).
The effect of a non–zero cosmological constant in central orbit problems is analysed. The analysis is applied to the orbit of Pluto, and it is found that the cosmological constant would have to be about 12 orders of magnitude bigger than the upper bound deduced from cosmological considerations in order for it to explain the irregularities in Pluto's orbit.

101.028   ***IRAS* observations of the Pluto–Charon system.**
H. H. Aumann, R. G. Walker.
Astron. J., Vol. 94, No. 4, p. 1088 – 1091 (1987).
High–signal–to–noise–ratio observations of the Pluto–Charon system at 25, 60, and 100 µm, using *IRAS* are combined with visual–magnitude and mutual–eclipse constraints to evaluate thermal models of Pluto and Charon. The most likely model for Charon is the standard asteroid model, typical for the icy Galilean and Saturnian satellites. Charon models with a significant atmosphere can be ruled out. Based on currently available radius and albedo constraints, no significant numerical distinction is possible between Pluto models ranging from isothermal spheres with surface emissivity between 0.4 and 0.9. Concerns regarding the viability of an emissivity as low as 0.4 favor the higher–emissivity models. The globally uniform surface temperature of Pluto may thus at present be as low as 45K, with a methane column abundance of 6.7 cm atm. The most likely models are centered on radii of 1180 and 747 km and albedos of 0.47 and 0.26 for Pluto and Charon, respectively.

101.029   **Pluto and Charon: the dance goes on.**
J. K. Beatty.
Sky Telesc., Vol. 74, No. 3, p. 248 – 251 (1987).

101.030   **Neue Daten von den Uranusringen.**
D. Stoll, H. Tiersch.
Astron. Raumfahrt, 25. Jahrg., Heft 5, p. 138 – 141 (1987).

101.031   **Charon: the frosty satellite.**
P. Moore.
Astron. Now, Vol. 1, No. 5, p. 40 – 42 (1987).

101.032   **Warming of Miranda during chaotic rotation.**
R. Marcialis, R. Greenberg.
Nature, Vol. 328, No. 6127, p. 227 – 229 (1987).
Miranda, a satellite of Uranus, appears in Voyager images to have had an active geological history, seemingly characterized by relaxation of very large–scale topography or the sinking of large blocks of material accreted by the satellite, with associated extrusion, eruption, and flow of the icy material on the body. The low temperature of Miranda ($<100K$) would preclude such mobilization of water ice. Here the authors offer an explanation: topography or anomalous blocks themselves may once have provided the conditions for heating and softening the supporting material, even pure ice, by permitting chaotic rotation of the satellite.

101.033   **Spectrophotometry of Pluto–Charon mutual events: individual spectra of Pluto and Charon.**
S. R. Sawyer, E. S. Barker, A. L. Cochran, W. D. Cochran.
Science, Vol. 238, No. 4833, p. 1560 – 1563 (1987).
Time–resolved spectra of the 3 March and 4 April 1987 mutual events of Pluto and its satellite Charon were obtained with spectral coverage from 5,500 to 10,000 angstroms with 25 angstrom spectral resolution. Charon has a featureless reflectance spectrum, with no evidence of methane absorption. Charon's reflectance appears neutral in color and corresponds to a geometric albedo of ~0.37 at 6000 angstroms. The Pluto reflectance spectrum displays methane absorption bands at 7300, 7900, 8400, 8600, and 8900 angstroms and is red in color, with a geometric albedo of ~0.56 at 6000 angstroms.

101.034   **Comparison of Bretagnon's VSOP 82 theory with observations of Neptune.**
R. S. Gomes, S. Ferraz–Mello.
Astron. Astrophys., Vol. 185, No. 1/2, p. 327 – 331 (1987).
A comparison of the observations of Neptune in the period 1846 – 1982 with Bretagnon's theory VSOP 82 is presented. The yearly averages of the residuals in longitude and latitude are discussed by Keplerian and spectral analysis, leading to small corrections to orbital elements and to significant periodicities.

101.035   **Improved orbital and physical parameters for the Pluto–Charon system.**
D. J. Tholen, M. W. Buie, R. P. Binzel, M. L. Frueh.
Science, Vol. 237, No. 4814, p. 512 – 514 (1987).
Analysis of the observations of several Pluto–Charon occultation and transit events in 1985 and 1986 has provided a more detailed knowledge of the system. The sum of the radii of Pluto and Charon is 1786 ± 19 kilometers, but the individual radii are somewhat more poorly determined; Pluto is 1145 ± 46 kilometers in radius and Charon is 642 ± 34 kilometers in radius. The mean density of the system is 1.84 ± 0.19 grams per cubic centimeter implying that more than half of the mass is due to rock. Charon appears to have hemispheres of two different colors, the Pluto–facing side being neutral in color and the opposite hemisphere being a reddish color similar to Pluto.

101.036   **IRAS Serendipitous Survey observations of Pluto and Charon.**
M. V. Sykes, R. M. Cutri, L. A. Lebofsky, R. P. Binzel.
Science, Vol. 237, No. 4820, p. 1336 – 1340 (1987).
On 16 August 1983 the Infrared Astronomical Satellite made two separate pointed observations of Pluto and its moon Charon. Detections were made at 60 and 100 micrometers with color–corrected flux densities of 581 ± 58 and 721 ± 123 millijanskys, respectively. Pluto is best described as having a dark equatorial band, and brighter polar caps of methane ice extending to $\pm 45°$ latitude, at most. An upper limit of approximately 9 meter–amagats is placed on the column abundance of a methane atmosphere on Pluto, which is comparable to recent upper limits based on independent ground–based spectroscopy.

101.037   The surface composition of Charon: tentative identification of water ice.
R. L. Marcialis, G. H. Rieke, L. A. Lebofsky.
Science, Vol. 237, No. 4820, p. 1349 – 1351 (1987).

The 3 March 1987 Charon occultation by Pluto was observed in the infrared at 1.5, 1.7, 2.0, and 2.35 micrometers. Subtraction of fluxes measured between second and third contacts from measurements made before and after the event has yielded individual spectral signatures for each body at these wavelengths. Charon's surface appears depleted in methane relative to Pluto. Constancy of flux at 2.0 micrometers throughout the event shows that Charon is effectively black at this wavelength, which is centered on a very strong water absorption band. Thus, the measurements suggest the existence of water ice on Pluto's moon.

101.038   Circumstances for Pluto–Charon mutual events in 1988.
D. J. Tholen, M. W. Buie, C. E. Swift.
Astron. J., Vol. 94, No. 6, p. 1681 – 1685 (1987).

Circumstances are tabulated for 89 Pluto–Charon mutual events occurring during the 1988 opposition. All superior events will be total, as in 1987, but only slightly more than half of the inferior events will be total. 1988 is the last year during which total events will occur. Two new stars have been selected as comparison stars for events occurring before opposition in 1988. Two transformation stars have also been selected so that observers can determine the necessary color terms to convert their instrumental magnitudes to the standard system.

101.039   IRAS serendipitous survey observations of Pluto and Charon.
M. V. Sykes, R. M. Cutri, L. A. Lebofsky, R. P. Binzel.
Prepr. Steward Obs., No. 752, 25 pp. (1987). To appear in Science.

101.040   An atmospheric rotation period of Neptune determined from methane–band imaging.
H. B. Hammel, M. W. Buie.
Icarus, Vol. 72, No. 1, p. 62 – 68 (1987).

Direct imaging of Neptune through an 8900–Å methane–band filter with the University of Hawaii 2.24–m telescope at Mauna Kea Observatory shows discrete atmospheric cloud features. A rotation period of 17.86 ± 0.02 hr is derived from the observations of two transits of a bright feature in the southern hemisphere during May and June 1986. This period is consistent with earlier observations of cloud motion on Neptune. The imaging also shows that bright features in Neptune's northern hemisphere seen as recently as in 1983 by earlier investigations have disappeared, markedly changing the overall distribution of reflected light from the planetary disk.

101.041   The colors of the Uranian rings.
C. C. Porco, J. N. Cuzzi, M. E. Ockert, R. J. Terrile.
Icarus, Vol. 72, No. 1, p. 69 – 78 (1987).

Observations of the Uranian rings were made in several color filters by the Voyager Imaging Science experiment in January 1986 for the purpose of determining the color of the rings. The results of the analysis are consistent with the α, β, η, γ, δ, and ε rings being very dark, with flat spectra throughout the visible, and are comparable to the latest Voyager results showing a lack of color for the Uranian satellites. The general lack of color in the ring/satellite system of Uranus is remarkably different than the case of the distinctly reddish systems of Jupiter and Saturn. The unique combination of low absolute reflectivity and flat spectrum which characterizes the Uranian rings supports the concept that the Uranian ring material is compositionally distinct from either the Si– and S–rich Jovian ring and inner satellites, or the water–ice–rich rings and inner satellites of Saturn. The candidate which best matches the low brightness and flat spectrum of the Uranian rings is carbon.

101.042   Voyager observations of 1985U1.
P. Thomas, J. Veverka, T. V. Johnson, R. H. Brown.
Icarus, Vol. 72, No. 1, p. 79 – 83 (1987).

1985U1 is an irregularly shaped satellite about 75 km in average radius that orbits Uranus betweeen the ε ring and Miranda. It is the only one of the 10 satellites of Uranus discovered by Voyager that was resolved in sufficient detail to permit determinations of the size, shape, and photometric properties. The albedo (0.07 – 0.09) is considerably lower than those of the large Uranian satellites and slightly higher than that of Saturn's Phoebe. The satellite's opposition magnitude should be about $+20.5$ in the V filter.

101.043   Observations of the satellites of Uranus.
R. Vieira–Martins, C. H. Veiga, D. Lazzaro.
Rev. Mex. Astron. Astrofis., Vol. 14, No. 2, p. 642 (1987). Abstract. – See Abstr. 012.042.

101.044   GUST 86. An analytical ephemeris of the Uranian satellites.
J. Laskar, R. A. Jacobson.
Astron. Astrophys., Vol. 188, No. 1, p. 212 – 224 (1987).

The General Uranus Satellite Theory GUST (Laskar, 1986) is used for the construction of an analytical ephemeris for the Uranian satellites. The theory is fitted against Earth–based observations from 1911 to 1986 and all radio and optical data obtained during Voyager encounter with Uranus. Earth–based observations alone allow the determination of masses which are within 15% of the values determined by the Uranus flyby. The analysis of all the observations confirms the values of the masses obtained during the encounter (Stone and Miner, 1986) and gives a complete set of dynamical parameters for the analytical theory. The authors obtain an analytical ephemeris, GUST 86, with an estimated precision of about 100 km with respect to Uranus.

101.045   Pluto–Charon: implications of Bouet's empirical relation.
S. A. Stern.
J. R. Astron. Soc. Can., Vol. 81, No. 5, p. 157 – 159 (1987).

The recently discussed Bouet relation between the mass and rotation frequency of a planet, and the masses and resolution frequencies of its satellites is discussed as it relates to the Pluto–Charon system. The empirical Bouet framework fails in this case unless there exists an undiscovered satellite or ring interior to Charon.

101.046   Uranus und sein Ringsystem.
M. Reichstein.
Astron. Sch., 24. Jahrg., Heft 6, p. 123 – 127 (1987).

101.047   The role of aqueous chemistry in determining the composition and cloud structure of the upper troposphere on Uranus.
B. E. Carlson, M. J. Prather, W. B. Rossow.
Astrophys. J., Lett. Ed., Vol. 321, No. 1, p. L97 – L101 (1987).

Aqueous chemistry on Uranus affects the atmopsheric abundances of NH_3 and H_2S below the methane cloud base. The authors present a complete thermochemical equilibrium model for the $H_2O–NH_3–H_2S$ system. Inclusion of H_2S increases the aqueous removal of NH_3 to 20% – 30%, but aqueous chemistry alone cannot for the depletion of NH_3 in the 150 – 200K region of the atmosphere required to fit microwave observations. Formation of NH_4SH clouds can account for the oberved depletion provided the H_2S/NH_3 ratio is enhanced by a factor of 4 relative to solar.

101.048   L'observation des phénomènes mutuels "Pluton–Charon".
J.–F. Le Borgne.
Ann. Phys. (Paris), Vol. 12, Colloq. No. 1, p. 217 – 219 (1987). – See Abstr. 012.061.

Mutual phenomena "Pluto–Charon" occur until 1989 and it is important to make observations since next phenomena will occur 124 years from now. Predictions of those events are given.

101.049   Water frost on Charon.
M. W. Buie, D. P. Cruikshank, L. A. Lebofsky,
E. F. Tedesco.
Nature, Vol. 329, No. 6139, p. 522 – 523 (1987).

The authors present new spectra of the Pluto–Charon system taken just before and during a total eclipse of the satellite. From these data they have extracted the spectrum of the satellite, Charon, which reveals the spectral signature of water ice. There is no evidence for any methane or ammonia frost on the surface of Charon. This observation places important constraints on the composition and origin of this planetary system.

101.050   Estimate of perturbing accelerations in the motion of the second Neptune satellite (Nereid).
E. V. Alfimova, I. A. Gerasimov.
Astron. Tsirk., No. 1470, p. 7 – 8 (1986). In Russian.

It is shown that main perturbations of Nereid's motion are caused by the Sun and Triton.

101.051   Absorption bands in the Uranus spectrum in the $\lambda\lambda 4000 – 5000$ Å region.
A. A. Atai, N. A. Gusejnov.
Astron. Tsirk., No. 1475, p. 1 – 3 (1987). In Russian.

New (unidentified) depressions were revealed in the spectrum of Uranus.

101.052   Oblateness, radius, and mean stratospheric temperature of Neptune from the 1985 August 20 occultation.
W. B. Hubbard, P. D. Nicholson, E. Lellouch, B. Sicardy,
A. Brahic, F. Vilas, P. Bouchet, R. A. McLaren, R. L. Millis,
L. H. Wasserman, J. H. Elias, K. Matthews, J. D. McGill,
C. Perrier.
Icarus, Vol. 72, No. 3, p. 635 – 646 (1987).

The occultation of a bright ($K \sim 6$) infrared star by Neptune revealed a central flash at two stations and provided accurate measurements of the limb position at these and several additional stations. The authors have fitted this data ensemble with a general model of an oblate atmosphere to deduce the oblateness e and equatorial radius a_0 of Neptune at the 1 μbar pressure level, and the position angle p_n of the projected spin axis. The results are $e = 0.0209 \pm 0.0014$, $a_0 = 25269 \pm 10$ km, $p_n = 20.1° \pm 1°$. As an alternative to the methane absorption model proposed by Lellouch et al. (1986), the authors explain an observed reduction in the central flash intensity by a decrease in temperature from 150 to 135K as the pressure rises from 1 to 400 μbar. Implications of the oblateness results for Neptune interior models are briefly discussed.

101.053   A simple two–layer model for the Uranus satellites.
J. Leliwa–Kopystynski.
Earth, Moon, Planets, Vol. 39, No. 3, p. 215 – 223 (1987).

A two–layer model of a satellite interior with a rocky core with a density $3 – 3.4$ g cm^{-3} and with a H$_2$O mantle with a density $0.94 – 1.2$ g cm^{-3} is applied for the icy satellites. The case of Mimas is discussed separately. A comparison of the results with these obtained for more complicated models as applied for Jupiter and Saturn icy satellites has been carried out. This comparison shows that the two–layer model offers a reasonable approximation and, therefore, it can be applied for the satellites of Uranus. The author obtained the dimensionless core radii $0.55 – 0.74$, $0.45 – 0.68$, $0.59 – 0.67$, $0.55 – 0.65$, and dimensionless core masses $0.42 – 0.72$, $0.26 – 0.63$, $0.47 – 0.61$, $0.41 – 0.57$, for Ariel, Umbriel, Titania, and Oberon, respectively.

101.054   Rings of Uranus.
P. Koubský.
Říše hvězd, Vol. 68, No. 7, p. 128 – 131 (1987). In Czech.

Long–term numerical integrations and synthetic theories for the motion of the outer planets.
See Abstr. 042.008.

On the confinement of planetary arcs.
See Abstr. 042.010.

Magnetospheric environments of outer planet rings: influence of Saturn's axially symmetric magnetic field.
See Abstr. 091.019.

Small satellites.
See Abstr. 091.037.

First order planetary perturbations with elliptic functions.
See Abstr. 091.050.

Occultation by Neptune.
See Abstr. 096.012.

Rings around planets.
See Abstr. 099.046.

Electron capture decay in Jovian planets.
See Abstr. 099.072.

Accretional heating of the satellites of Saturn and Uranus.
See Abstr. 100.001.

102   Comets (Origin, Structure, Atmospheres, Dynamics)

102.001   Comet thermal modeling.
P. R. Weissman, H. H. Kieffer.
NASA Tech. Memo., NASA TM–89810, p. 66 – 68 (1987). – See Abstr. 003.001.

102.002   The loss and depth of CO$_2$ ice in comet nuclei.
F. P. Fanale, J. R. Salvail.
NASA Tech. Memo., NASA TM–89810, p. 69 (1987). Abstract. – See Abstr. 003.001.

102.003   Dynamics of long period comets.
P. R. Weissman.
NASA Tech. Memo., NASA TM–89810, p. 70 – 71 (1987). – See Abstr. 003.001.

102.004   The distribution of short–period comet magnitudes.
J. R. Donnison.
Earth, Moon, Planets, Vol. 38, No. 3, p. 263 – 272 (1987).

Short–period comets with $P \leqslant 15$ yr represent one of the most complete comet samples. The magnitude distribution of these comets was analysed using a maximum likelihood method. The brightness (magnitude) index for the comets with $H_{10} \leqslant 11$ mag was estimated together with the large sample errors and found to be 0.62 ± 0.09. It was found that comets satisfying the combined criteria $P \leqslant 15$ yr, $H_{10} \leqslant 11$ mag, $q < 1.5$ AU probably represent the most complete set of comets available. The brightness index of this sample estimated by maximum likelihood was 0.69 ± 0.14. This translates into a mass distribution index s of 1.69 ± 0.14 indicating that most of the mass is contained in a few of the larger comets rather than spread throughout the smaller ones.

102.005   **Light scattering in cometary dust comae.**
G. Herman, H. Salo.
Earth, Moon, Planets, Vol. 39, No. 1, p. 51 – 74 (1987).
Detailed single and multiple scattering calculations were carried out for a spherically symmetric cometary atmosphere irradiated by a plane–parallel source. Using simplifying assumptions in the single scattering approximation, analytical expressions were derived for the total flux impinging the cometary nucleus, which was shown to be a decreasing function of the coma opacity.

102.006   **The lifetimes of comets. II. Their disappearance.**
Ľ. Kresák.
Int. Comet Q., Vol. 9, No. 2, p. 59 – 61 (1987).

102.007   **La diversité des comètes.**
A. H. Delsemme.
Ciel, Vol. 49, p. 273 – 278 (1987).

102.008   **The influence of strong hydromagnetic turbulence on newborn cometary ions.**
C. P. Price, C. S. Wu.
Geophys. Res. Lett., Vol. 14, No. 8, p. 856 – 859 (1987).
By means of a test particle model the authors have studied the time evolution of the velocity distribution function of newborn cometary ions in the presence of strong hydromagnetic turbulence. The test particle model employs a realistic spectrum of hydromagnetic turbulence corresponding to the observations of the International Cometary Explorer spacecraft at comet Giacobini–Zinner. It is found that pitch–angle scattering processes can rapidly result in the formation of a shell distribution which has been observed near comets Giacobini–Zinner and Halley.

102.009   **Numerical simulation of the generation of turbulence from cometary ion pick–up.**
M. L. Goldstein, D. A. Roberts, W. H. Matthaeus.
Geophys. Res. Lett., Vol. 14, No. 8, p. 860 – 863 (1987).
Observations of magnetic field fluctuations near comet Halley have revealed a rapid development of a Kolmogoroff–like turbulence spectrum extending from below 10^{-2}Hz to above 0.1 Hz. Spectra obtained far from the comet have a strong peak in power near the Doppler–shifted ion–cyclotron frequency of singly ionized water. Closer to the comet, the spectrum at higher frequencies is enhanced in power level over the background solar wind spectrum by approximately an order of magnitude. The authors solved the equations of incompressible MHD using a two–dimensional 256×256 mode spectral method code to simulate this spectral evolution as an inertial range turbulent cascade. Both the time scale and the increase in power level of the turbulence seen in the simulation are in accord with the Giotto observations.

102.010   **A new model of cometary ionospheres.**
A. Körösmezey, T. E. Cravens, T. I. Gombosi,
A. F. Nagy, D. A. Mendis, K. Szegö, B. E. Gribov,
R. Z. Sagdeev, V. D. Shapiro, V. I. Shevchenko.
J. Geophys. Res., Vol. 92, No. A7, p. 7331 – 7340 (1987).
The coupled continuity, momentum, and energy equations were solved for ionospheric conditions appropriate for comet Halley at 1 AU. The numerical scheme used is such that any shock transition appears naturally in the solution and no a priori assumptions are necessary. Solutions were obtained for a number of different assumptions concerning electron heating rates, but all showed that the electron temperatures increase rapidly and significantly at a distance from the nucleus where collisional electron–neutral cooling becomes unimportant. This temperature increase is accompanied by a sharp increase in both the plasma pressure and its associated polarization electric field, causing the supersonic plasma flow to go subsonic. It is not clear at this time whether or not this sonic transition is accompanied by a shock.

102.011   **Electron impact ionization in the vicinity of comets.**
T. E. Cravens, J. U. Kozyra, A. F. Nagy,
T. I. Gombosi, M. Kurtz.
J. Geophys. Res., Vol. 92, No. A7, p. 7341 – 7353 (1987).
Electron distributions measured in the vicinity of comets Halley and Giacobini–Zinner by instruments on the VEGA and ICE spacecraft, respectively, are used to calculate electron impact ionization frequencies. Ionization by electrons is of comparable importance to photoionization in the magnetosheaths of Comets Halley and Giacobini–Zinner. The ionization frequency in the inner part (radial distance $\approx 10^4$km) of the cometary plasma region of comet Halley is several times greater than the photoionization value. Tables of ionization frequencies as functions of electron temperature are presented for H_2O, CO_2, CO, O, N_2, and H.

102.012   **Energy diffusion of pickup ions upstream of comets.**
P. A. Isenberg.
J. Geophys. Res., Vol. 92, No. A8, p. 8795 – 8799 (1987).
The author presents a steady state model of pickup ion energization upstream of a cometary bow wave in order to investigate the effects of quasi–linear energy diffusion in the turbulence there. The model assumes that the ions are immediately isotropized at pickup, and it includes the effects of adiabatic acceleration in the slowing solar wind and of continual pickup of ions as the comet is approached. By taking all physical quantities to fall off as power laws with distance from the comet, an analytical expression is obtained for the distribution function of pickup ions in the reference frame moving with the solar wind. To illustrate the application of this model, the author compares the model results to the observations of pickup ions at comet Giacobini–Zinner.

102.013   **Time–dependent injection of Oort cloud comets into earth–crossing orbits.**
J. A. Fernández, W.–H. Ip.
Icarus, Vol. 71, No. 1, p. 46 – 56 (1987).
The effect of close stellar encounters in modulating the influx rate of Oort cloud comets is investigated. In particular, it is shown that comet showers intense enough to be reflected in crater statistics can be produced at intervals of 80 million years or so, provided we are dealing with an Oort cloud consisting of a heavy core of comets. The authors have also performed numerical simulations of the time evolution of comet showers or bursts.

102.014   **Sources of cometary radicals and their jets: gases or grains.**
M. R. Combi.
Icarus, Vol. 71, No. 1, p. 178 – 191 (1987).
The recent discovery of CN and C_2 gas jets in comet Halley has led to basic speculation as to their physical source mechanism. A basic quantitative study of the photosputtering of CHON grains and the spatial evolution of trace gas jets is presented. Two possible single sources, a parent gas and CHON grains, for both the jet and the background gas, are also investigated.

102.015   **Photoprocessing of H_2S in interstellar grain mantles as an explanation for S_2 in comets.**
R. J. A. Grim, J. M. Greenberg.
Astron. Astrophys., Vol. 181, No. 1, p. 155 – 168 (1987).
The observation of S_2 in comet IRAS–Araki–Alcock 1983 VII has raised critical questions regarding whether the precometary formation of this molecule occurred in the gas phase or in the solid interstellar ices. The implications of these results on the origin of comets are interpreted using laboratory analogue experiments on S_2 formation. It is shown that ultraviolet photolysis of H_2S embedded in dirty ices composed of molecules such as H_2O, CO and CH_4 converts the H_2S molecules into S_2 and other sulfur bearing species. From these experiments, using the cosmic abundance of sulfur relative to oxygen, it is calculated that the $[S_2]/[H_2O]$ ratio in interstellar grains lies between 2×10^{-4} and 1.4×10^{-3}, consistent with the observed ratio $[S_2]/[OH] = 5 \times 10^{-4}$ in IRAS–Araki–Alcock 1983 VII. It is argued that S_2 cannot be simply used as a discriminant for the

comet formation temperature, unless one properly takes into account the aggregation time scales. The experiments suggest that for comet formation times of 10^6yr temperatures well under 100K are indeed required so that comets could have been formed no closer than the Uranus–Neptune region.

102.016   **A model for the excitation of water in comets.**
D. Bockelée–Morvan.
Astron. Astrophys., Vol. 181, No. 1, p. 169 – 181 (1987).
The vibrational and rotational excitation of the H_2O molecule in cometary atmospheres is investigated using a model which includes infrared vibrational pumping by the solar radiation flux, thermal excitation by collisions and radiation trapping in the rotational and rovibrational lines. Collisional effects are treated using a non–isothermal and a cold kinetic temperature profile. The escape probability formalism is used for solving the radiative transfer problem. Steady state is not assumed in the rate equations and the H_2O rotational population distribution is computed as a function of distance to nucleus. The rotational line intensities are evaluated for an unresolved coma observed at a distance of 1 AU. The model fails to explain the $6_{16} - 5_{23}$ H_2O radio line tentatively detected in two comets.

102.017   **Structure of the Oort cometary cloud.**
L. S. Marochnik, G. B. Sholomitskij.
Sov. Astron., Vol. 30, No. 5, p. 615 – 621 (1986). English translation of 42.102.010.

102.018   **Fountain model and the rotation of cometary nuclei.**
D.–h. Chen, X.–t. Zheng.
Sci. Sin., Ser. A, Vol. 30, No. 8, p. 860 – 867 (1987).
The authors extend Eddington's fountain model in consideration of the rotation of cometary nucleus, derive the corresponding formulas for calculating the apparent density distribution of neutral molecules in cometary atmospheres and prove that the distribution of the apparent density in comets depends upon the position of the observer with respect to the comet.

102.019   **Comets and their composition.**
H. Spinrad.
Annu. Rev. Astron. Astrophys., Vol. 25, p. 231 – 269 (1987). – See Abstr. 003.003.
This review discusses the origin theories of comets, then moves to empirical studies of cometary nuclei from the ground and from space. The author also reviews gas production rates in cometary comae, the cometary ionosphere, and future expectations.

102.020   **Laboratory modelling of ion–molecular cluster formation in cometary nuclei.**
N. M. Khashilov, Sh. Sh. Shoekubov.
Dokl. Akad. Nauk TadzhSSR, Tom 29, No. 7, p. 402 – 405 (1986). In Russian. Abstr. in Ref. Zh., 51. Astron., 7.51.205 (1987).

102.021   **On numerical modelling of thermal processes in cometary nuclei.**
A. V. Reshetov, Yu. V. Skorov.
Mat. zadachi priklad. aehron., Moskva, p. 169 – 177 (1987). In Russian. Abstr. in Ref. Zh., 62. Issled. Kosm. Prostranstva, 7.62.561 (1987).

102.022   **Method for determining the temperature of the inner coma of a comet.**
D. V. Bisikalo, S. V. Repin, V. S. Strel'nitskij.
Mat. zadachi priklad. aehron., Moskva, p. 178 – 198 (1987). In Russian. Abstr. in Ref. Zh., 62. Issled. Kosm. Prostranstva, 7.62.562 (1987).

102.023   **Taking into account nonequilibrium processes in modelling the spectral luminosity of a cometary coma.**
G. I. Smievskaya, A. E. Korolev, M. M. Maurakh.
Mat. zadachi priklad. aehron., Moskva, p. 210 – 234 (1987). In Russian. Abstr. in Ref. Zh., 62. Issled. Kosm. Prostranstva, 7.62.563 (1987).

102.024   **Chemical interactions of solar wind with cometary nuclei and comae.**
K. Roessler.
Adv. Space Res., Vol. 7, No. 5, p. 7 – 11 (1987). – See Abstr. 012.022.
This paper restricts itself to the discussion of the effects of solar radiation on an actual short period comet such as P/Halley and the probability of hot chemical processes in the icy surfaces of nucleus or ejected grains and the gas phases of coma and tails.

102.025   **Radiation dosimetry and chemistry of a cometary nucleus.**
I. G. Draganić, M. P. Ryan Jr., Z. D. Draganić.
Adv. Space Res., Vol. 7, No. 5, p. 13 – 16 (1987). – See Abstr. 012.022.
The authors discuss the contributions of cosmic rays and imbedded radionuclides to the absorbed doses and the chemistry in a cometary nucleus. Some of recently made observations during the close encounter with comet Halley are considered in the light of radiation dosimetry and chemistry.

102.026   **Comets and life.**
J. Oró, J. M. Berry.
Adv. Space Res., Vol. 7, No. 5, p. 23 – 32 (1987). – See Abstr. 012.022.
The amount of cometary carbon–containing matter captured by the Earth, as calculated by different authors, is several times larger than the total amount of organic matter present in the biosphere (10^{18}g). The major classes of reactions which were probably involved in the formation of key biochemical compounds are discussed. The authors' tentative conclusions are that: (1) comets played a predominant role in the emergence of life on our planet, and (2) they are the cosmic connection with extraterrestrial life.

102.027   **Interactions between cometary plasmas and the solar wind.**
Z.–y. Pu.
Acta Geophys. Sin., Vol. 30, No. 4, p. 433 – 440 (1987). In Chinese.
The author provides an overview on both major features of interactions between cometary plasmas and the solar wind and the associated MHD models. He presents a brief account of preliminary observation results obtained from the encounters of both ICE with comet Giacobini–Zinner in November, 1985, and 6 spacecraft (Vega 1, 2, Suisei, Sakigake, Giotto and ICE) with comet Halley in March and April 1986.

102.028   **On the dynamics of short–period comets.**
V. V. Emel'yanenko.
Kinematika Fiz. Nebesn. Tel, Tom 3, No. 5, p. 52 – 56 (1987). In Russian. English translation in Kinematics Phys. Celest. Bodies.
The evolution of the observed short–period comets is studied numerically on time intervals reaching 5×10^4years. The peculiarities of comet's libration and stochastic motion are considered. The comets which do not have close approaches to Jupiter near the contemporary epoch are indicated.

102.029   **Possibility of observing the Oort cometary cloud in the far–infrared region.**
L. S. Marochnik, G. B. Sholomitskij.
Sov. Astron., Vol. 30, No. 6, p. 702 – 710 (1987). English translation of 42.102.066.

102.030   **A review of cometary sciences.**
F. L. Whipple.
Philos. Trans. R. Soc. London, Ser. A, Vol. 323, No. 1572, p. 339 – 347 (1987). – See Abstr. 012.026.
This paper presents an elementary description of comets and their nature. It deals with the contribution from comets to the solid particles that produce the Zodiacal light and discusses the possibility that some comets in short–period orbits may degenerate into asteroids. A section of the paper deals with the possible

nature of comets and their origin such that some might become superficially indistinguishable from asteroids.

102.031   The nature of comets.
J. C. Brandt.
Philos. Trans. R. Soc. London, Ser. A, Vol. 323, No. 1572, p. 437 – 446 (1987). With 4 plates. – See Abstr. 012.026.

The vast scientific campaign associated with the 1986 return of Halley's comet has greatly improved and expanded our knowledge of comets. An overview of the first results is presented with emphasis on the large–scale structure, the chemistry, and the nucleus.

102.032   Motion of long–period comets in the perturbing field of the Galaxy. Irregular forces.
V. A. Antonov, Z. P. Todriya.
Astron. Zh., Tom 64, Vyp. 5, p. 1094 – 1104 (1987). In Russian. English translation in Sov. Astron., Vol. 31, No. 5.

Perturbing influence of the galactic irregular forces upon the motion of long–period comets is estimated more accurately, assuming that stellar masses and randomly directed stellar velocities are respectively equal in magnitude. The cumulative effect is shown to perturb cometary orbits more intensely than mean single close encounters with stars and giant interstellar clouds. Comparison of the roles played by regular and irregular forces indicates that in a detailed investigation of cometary orbit evolution, both types of perturbations should be taken into account.

102.033   Radiogenic heating of comets by ^{26}Al and implications for their time of formation.
D. Prialnik, A. Bar–Nun, M. Podolak.
Astrophys. J., Vol. 319, No. 2, p. 993 – 1002 (1987).

The effect of radiogenic heating on the thermal evolution of spherical icy bodies with radii 1 km $< R <$ 100 km was investigated. The radioisotopes considered were ^{26}Al, ^{40}K, ^{232}Th, ^{235}U, and ^{238}U. The main object of this study is to examine the conditions under which the transition temperature from amorphous into cubic ice (T_a = 137K) would be reached. It is shown that the influence of the short–lived radionuclide ^{26}Al dominates the effect of other radioactive species for bodies of radii up to $\sim$50 km. Consequently, if one requires comets to retain their ice in amorphous form, as suggested by observations, an *upper limit* of $\sim 4 \times 10^{-9}$ is obtained for the initial ^{26}Al abundance in *comets*, a factor of 100 lower than that of the inclusions in the Allende meteorite. A *lower limit* for the formation time of comets may thus be derived.

102.034   Cometary models: excitation of molecules at radio wavelengths and thermodynamics of the coma.
J. Crovisier.
Cometary radio astronomy, p. 45 – 58 (1987). – See Abstr. 012.033.

The author presents models for molecular excitation under physical conditions of cometary atmospheres. A review of cometary thermodynamical models is also given, and the relations between such models and cometary molecular observations are discussed.

102.035   Taking into account nonequilibrium processes in modelling the spectral luminosity of a cometary coma.
G. I. Zmievskaya, A. E. Korolev, M. M. Maurakh.
Mat. zadachi prikl. aehron., Moskva, p. 210 – 234 (1987). In Russian. Abstr. in Ref. Zh., 51. Astron., 9.51.290 (1987).

102.036   Kometastronomin på nya spår, I. En utvärdering efter Halley–passagen.
H. Rickman.
Astron. Tidsskr., Årg. 20, Nr. 4, p. 145 – 174 (1987).

102.037   Numerical investigation of the photochemistry of an H$_2$O–dominant cometary atmosphere.
M. Ya. Marov, V. I. Shematovich.
Inst. Prikl. Mat. Akad. Nauk SSSR, Prepr., No. 90, p. 3 – 26 (1987). In Russian. Abstr. in Ref. Zh., 51. Astron., 10.51.224 (1987).

102.038   The formation and extent of the solar system comet cloud.
M. Duncan, T. Quinn, S. Tremaine.
Bull. Am. Astron. Soc., Vol. 19, No. 2, p. 742 (1987). Abstract. – See Abstr. 010.061.

102.039   The frequency and intensity of comet showers from the Oort cloud.
J. Heisler, S. Tremaine, C. Alcock.
Bull. Am. Astron. Soc., Vol. 19, No. 2, p. 751 (1987). Abstract. – See Abstr. 010.061.

102.040   Stochasticity in the Kepler problem and a model of possible dynamics of comets in the Oort cloud.
R. Z. Sagdeev, G. M. Zaslavskij.
Nuovo Cimento B, Vol. 97B, Ser. 11, No. 2, p. 119 – 130 (1987). Abstr. in Phys. Abstr., Vol. 90, No. 1309, Entry 88209 (1987).

102.041   Comets, and comet Halley 1982 – 1985.
J. Rahe, H. Fechtig, R. L. Newburn Jr.
The evolution of the small bodies of the solar system, p. 159 – 167 (1987). – See Abstr. 012.041.

Contents: 1. Introduction. 2. Orbital properties of comets. 3. Physical properties of comets. 4. The International Halley Watch. 5. The Inter–Agency Consultative Group. 6. The comet Halley Armada. 7. Comet Halley.

102.042   The Oort cloud and its galactic environment.
H. Scholl.
The evolution of the small bodies of the solar system, p. 168 – 183 (1987). – See Abstr. 012.041.

Contents: 1. Introduction. 2. The galactic environment of the Sun. 3. The perturbational effects of passing stars on comets in the Oort cloud. 4. Monte Carlo simulation models for the dynamical evolution of the Oort cloud. 5. The perturbations of giant molecular clouds and of Nemesis on the Oort cloud. 6. A modified view of the Oort cloud.

102.043   Dynamics of comets.
B. G. Marsden.
The evolution of the small bodies of the solar system, p. 184 – 190 (1987). – See Abstr. 012.041.

102.044   Aging of comets and their evolution into asteroids.
Ĺ. Kresák.
The evolution of the small bodies of the solar system, p. 202 – 216 (1987). – See Abstr. 012.041.

Contents: 1. Introduction. 2. The aging processes in comets. 3. The lifetimes of comets. 4. The end fates of comets.

102.045   On the origin of comets.
V. S. Safronov.
The evolution of the small bodies of the solar system, p. 217 – 226 (1987). – See Abstr. 012.041.

Contents. 1. The hypotheses on the origin of comets. 2. The formation of the cometary cloud in the process of accumulation of giant planets.

102.046   Encounters with comets: discoveries and puzzles in cometary plasma physics.
A. A. Galeev.
Astron. Astrophys., Vol. 187, No. 1/2, p. 12 – 20 (1987). – See Abstr. 003.013, 43.102.021.

The encounters of various cometary probes with comets led to the discovery of a number of phenomena that were not envisaged by the existing theories based mainly on the magnetohydrodynamic description of the solar wind interaction with cometary plasma: the generation of strong magnetohydrodynamic turbulence by the loading of solar wind by cometary ions, the discovery of thin plasma structures in an inner coma that correlate with the bursts of plasma waves and the anomalously thin boundaries in a solar wind flow near comets that indicate the importance of collective plasma processes for the formation of these structures and boundaries. The paper reviews the observations of these phenomena from spacecraft and the kinetic theories of cometary plasma that were developed recently and that are able to provide a natural explanation of observed puzzling phenomena.

102.047   Fluid simulation of comet P/Halley's ionosphere.
K. Baumgärtel, K. Sauer.
Astron. Astrophys., Vol. 187, No. 1/2, p. 307 – 310 (1987). – See Abstr. 003.013, 43.102.033.

The formation of a magnetic cavity around the cometary nucleus is studied in a non–stationary quasi one–dimensional fluid treatment. Characteristic features of spatial density, velocity and magnetic field profiles observed on comet P/Halley by the Giotto spacecraft can be reproduced. The observed relative density maximum ahead of the cavity (pile–up region) may be interpreted as a consequence of reduced recombination.

102.048   Unusual characteristics of electromagnetic waves excited by cometary newborn ions with large perpendicular energies.
A. L. Brinca, B. T. Tsurutani.
Astron. Astrophys., Vol. 187, No. 1/2, p. 311 – 319 (1987). – See Abstr. 003.013.

Solution of the linear kinetic dispersion equation shows that cometary new–born ions with large perpendicular energies (large initial angles between the solar wind velocity and the IMF) can excite a wave mode with rest frame frequencies of the order of the heavy ion cyclotron frequency which exhibits maximum growth slightly away ($6° - 15°$) from parallel propagation, together with high mass density compression ratios and almost linear polarization. The association of these properties is relevant to space observations of MHD–like waves at comets and in planetary foreshocks.

102.049   Radiation formation of a non–volatile comet crust.
R. E. Johnson, J. F. Cooper, L. J. Lanzerotti, G. Strazzulla.
Astron. Astrophys., Vol. 187, No. 1/2, p. 889 – 892 (1987). – See Abstr. 003.013, 43.102.023.

Ion irradiation of the outer meters of a cometary surface can produce new molecular species in the solid state. These species segregate in an irreversible way into a non–volatile residue and new very volatile species. The latter are ejected directly or lost when the comet enters the inner solar system. Therefore, a comet exposed to background particle radiations in the Oort cloud obtains an outer web of non–volatile material which will lead to the formation of a substantial "crust". When a new comet enters the inner solar system there will be early activity, inital fizzures in the crust and the break–off of unstable pieces of the crust, due to warming of subsurface species. If this comet enters a periodic orbit in the inner solar system the remaining mantle should be continuously hardened due, primarily, to thermal processing. There will also be permanently active regions on such a comet which were initially shaped from the cosmic ray radiation when the comet was in the Oort cloud or which subsequently lost their crust.

102.050   High–order librations of Halley–type comets.
A. Carusi, L. Kresák, E. Perozzi, G. B. Valsecchi.
Astron. Astrophys., Vol. 187, No. 1/2, p. 899 – 905 (1987). – See Abstr. 003.013, 43.102.028.

Halley–type comets, usually referred to as those with revolution periods of 20 to 200 yr, can be dynamically defined by the values of their Tisserand invariants with respect to Jupiter, which account for the processes that brought these comets into the present orbits. Long–term integrations of their motion show oscillations of their heliocentric elements when the comets are far from the Sun and the centre of their motion is the barycentre of

the solar system. The shift from barycentric to heliocentric motion is also responsible for the major changes of orbital elements within a single revolution which, in the case of direct orbits, may follow regular libration patterns with typical periods of 350 – 400 yr. This is the reason for the concentration of periods around 70 yr. Some examples of this behavior and a short discussion of its importance for the dynamical history of Halley–type comets are presented.

102.051   Galactic tides affect the Oort cloud: an observational confirmation.
A. H. Delsemme.
Astron. Astrophys., Vol. 187, No. 1/2, p. 913 – 918 (1987). – See Abstr. 003.013, 43.102.027.

There are 152 known original orbits of comets with a period larger than ten thousand years. Their aphelia are shown here to avoid three zones of the celestial sphere, namely the two galactic polar caps and a strip along the galactic equator. Such an axial symmetry of the aphelia distribution in respect to the plane of the Galaxy cannot be explained by the traditional mechanism proposed to bring the Oort cloud comets into visibility – namely perturbations from random stellar passages. It cannot be explained either by any other random mechanism – like the passage of molecular clouds, a more recent variation of the previous idea. Finally, none of the known observational biases could introduce such a specific symmetry with respect to the Galaxy. Standing in contrast, the observed symmetry is predicted by the "vertical" galactic tide, namely the differential attraction, on the sun and on each comet of the Oort cloud respectively, that comes from the mass distribution within the thick galactic disk.

102.052   The evolution of topography on a comet.
J. E. Colwell, B. M. Jakosky.
Icarus, Vol. 72, No. 1, p. 128 – 134 (1987).

The authors have developed a simple model of an infinite cylindrical trench on a comet. The energy balance equation has been modified to include physical processes which are relevant with topography present, and includes shadowing, radiative heating from the opposing walls, and the condensation energy of sublimed gas molecules striking the walls instead of escaping to space. The model is designed to indicate the general course of the evolution of topography on a comet, and is not intended as a complete model of a cometary nucleus. By running the model for trenches of different depths and at different solar distances, the authors draw conclusions about the evolution of topography through an orbit and the consequences this evolution has on the evolution of the nucleus as a whole over the course of several perihelion passages.

102.053   On the hypothesis of eruption of comets from the satellites of Saturn, Uranus and Neptune.
M. V. Nikolaeva, V. P. Tomanov.
Tr. Astrofiz. Inst. Alma–Ata, Tom 48, p. 149 – 156 (1987). In Russian.

102.054   Resonant interactions between cometary ions and low frequency electromagnetic waves.
R. M. Thorne, B. T. Tsurutani.
Planet. Space Sci., Vol. 35, No. 12, p. 1501 – 1511 (1987).

The authors explore the conditions for resonance between cometary pick–up ions and parallel propagating electromagnetic waves. A model ring–beam distribution for the pick–up H_2O^+ ions is adopted which allows a direct comparison of the source of free energy for growth from either the beam or the gyrating ring in the limit near marginal stability. Under average solar wind conditions in the inner solar system, the gyrating ring provides the dominant contribution to wave growth. The authors consider this the most likely mechanism to account for the interior MHD waves observed by satellites over an extended spatial region surrounding comets Giacobini–Zinner and Halley.

102.055 Particle acceleration in cometary processes of magnetic reconnection.
J. Pérez–Peraza, M. Alvarez–Madrigal, A. Sánchez.
Rev. Mex. Astron. Astrofís., Vol. 14, No. 2, p. 682 – 687 (1987). – See Abstr. 012.042.

The source of high energy charged particles in comets may be found in the impulsive reconnection process associated to disconnection events of comet tails, when it crosses through an interplanetary sector boundary, or in a long lived process associated with the activation of the tail current sheet by interplanetary or flare shock waves. In order to probe the nature of the reconnection process the authors investigate the energy spectrum of the accelerated particles and they make predictions about the characteristics of photon emission fluxes produced by electron capture of the energetic ions during their interaction with cometary plasmas and neutral matter.

102.056 Solution of linear equations of the orbital motion of a comet with account for non–gravitational effects.
E. N. Polyakhova.
Astron. Vestn., Tom 21, No. 3, p. 233 – 241 (1987). In Russian. English translation in Sol. Syst. Res.

The classical Meshcherskij equation was integrated for the case of the orbital motion of a comet as a point with variable mass taking two non–gravitational effects into account: the "parashute" effect (solar radiation pressure in the nucleus–tail system) and the reactive acceleration effect (gas ejection from the cometary nucleus). The solution of this equation allows to estimate the contribution of these effects to the secular perturbations in orbital elements and can be useful for the prediction of orbital motions of comets.

102.057 Numerical modelling of the evolution of Oort's Cloud.
E. I. Dolgopolova, L. S. Marochnik, D. A. Usikov.
Astron. Vestn., Tom 21, No. 4, p. 331 – 334 (1987). In Russian. – See Abstr. 012.052.

102.058 On the stable multi–belt interplanetary reservoirs of cometary bodies.
Yu. K. Gulak.
Kinematika Fiz. Nebesn. Tel, Tom 3, No. 6, p. 13 – 18 (1987). In Russian. English translation in Kinematics Phys. Celest. Bodies.

The results of computer simulation of cometary orbit evolution are compared with the positions of comets calculated on the basis of the author's statistical theory of stable isoenergetic complexes. The data obtained permit suggesting that the complexes and the multi–belt interplanetary reservoirs of comets are allied (or the same) formations of stable structures of the solar system.

102.059 Oort's Cloud.
L. S. Marochnik, D. A. Usikov, E. I. Dolgopolova.
Priroda, No. 12, p. 36 – 45 (1987). In Russian.

102.060 Alfvén wave plasma turbulence during solar wind–comet interaction.
F. Verheest.
Astrophys. Space Sci., Vol. 138, No. 1, p. 209 – 215 (1987).

The large differences in drift velocities between the solar wind protons and the picked–up ions of cometary origin cause the Alfvén waves to become unstable and generate turbulence. A self–consistent treatment of such instabilities has to take into account that these cometary ions affect the solar wind plasma in a decisive way. With the help of a previously developed formalism one finds the correct Alfvén instability criterion, which is here nondispersive, in contrast to recent calculations where the cometary ions are treated as a low–density, high–speed, and non–neutral beam through an otherwise undisturbed solar wind. The true bulk speed of the combined solar wind plus cometary ion plasma clearly shows the mass–loading and deceleration of the solar wind near the cometary nucleus, indicating a bow shock. The instability criterion is also used to determine the region upstream where the Alfvén waves can be unstable, based upon recent observations near comet Halley.

102.061 Charge exchange avalanche at the cometopause.
T. I. Gombosi.
Geophys. Res. Lett., Vol. 14, No. 11, p. 1174 – 1177 (1987).

A sharp ($\leqslant 10^4$ km thick) transition from a solar wind proton dominated flow to a plasma population primarily consisting of relatively cold cometary heavy ions has been observed at a cometocentric distance of about 1.6×10^5 km by the VEGA and GIOTTO missions. This boundary (the cometopause) was thought to be related to charge transfer processes, but its location and thickness are inconsistent with conventionally estimated ion – neutral coupling boundaries. In this paper a two–fluid model is used to investigate the major physical processes at the cometopause.

102.062 Studying the evolution of the exterior zone of Oort's cometary cloud by numerical modelling.
V. M. Chepurova, S. L. Shershkina.
Astron. Tsirk., No. 1478, p. 5 – 8 (1987). In Russian.

Differential equations of cometary motions in the Oort's cloud are studied by the Runge–Kutta numerical integration method taking into account different kinds of perturbations. It is shown that (1) the gravitational field of the Galaxy increases the dimensions of Oort's cloud, but at the same time it presses this cloud. (2) The exterior zone of Oort's cloud is destroyed after one encounter of the Solar System with any star or interstellar H_2– cloud.

102.063 Diversity and similarity of comets.
A. H. Delsemme.
ESA Spec. Publ., ESA SP–278, p. 19 – 30 (1987). – See Abstr. 012.075.

Although comets appear superficially very different, it is suspected that they were originally very similar. In particular, the elemental abundances of all pristine comets are likely to be "primitive", that is in solar abundance ratios for all elements including C, N, O, S, but with the exception of H (and assumedly He and Ne) that are severely depleted. The solid phase was originally in very fine grains, typically 0.1 μm or less, eventually sintered into larger grain clusters. The volatile phase contains H, C, N, O, S molecules frozen in the pores of the grain clusters; cosmic ray plus solar irradiation change the volatile to refractory ratio of the crust. Differences in dust tails and in plasma tails, in photometry between young and old comets and in the variable carbon depletion of the gas phase seem to be induced by the decay processes.

102.064 The absolute total magnitudes of periodic comets and their variations.
Ľ. Kresák, M. Kresáková.
ESA Spec. Publ., ESA SP–278, p. 37 – 42 (1987). – See Abstr. 012.075.

Absolute total magnitudes of all periodic comets are determined. Since the available photometric data are strongly biased by instrumental effects (time–dependent and increasing with decreasing brightness), the list is based on the extreme magnitudes recorded during individual comet apparitions. Brightness dependence on the inverse fourth power of the heliocentric distance is assumed, and empirical corrections are applied to lower values of apparent brightness. The absolute magnitude distribution is found to be substantially different for the Jupiter family and for the Halley–type comets. There are only five statistically significant cases of progressive fading.

102.065 Cometary magnitude distribution: the tabulated data.
D. W. Hughes.
ESA Spec. Publ., ESA SP–278, p. 43 – 48 (1987). – See Abstr. 012.075.

The numerical distribution of cometary absolute magnitudes reveals information as to the way in which cometary numbers vary as a function of the radius, activity and mass of the cometary nucleus. In this paper comets are divided into three groups according to their orbital periods. Tables are provided listing all the known absolute magnitudes. Graphs are presented showing how the cumulative number of comets, N, varies as a function of their absolute magnitude, H_{10} (the apparent magnitude a comet would have if it were 1 AU from both Sun and Earth).

102.066 The comet flux and the 15 Myr galactic cycle.
S. V. M. Clube.
ESA Spec. Publ., ESA SP–278, p. 49 – 53 (1987). – See Abstr. 012.075.

The near parabolic flux of comets fluctuates in response to variations of the tidal radius of the Oort cloud in the presence of stars and molecular clouds. Variations are of two kinds, strong and weak, respectively giving rise to stochastic (showers) and cyclic (background) flux components. The latter includes a 15 Myr cycle due to the Sun's vertical motion through layers of zero tide on either side of the galactic plane produced by the distribution of discrete clouds in an otherwise plane stratified disc. Since the comet flux modulates the course of evolution by perturbing the boundary conditions at the core, outer mantle and atmosphere of the Earth, it follows that the otherwise unattributed 15 Myr cycle in the H–reversal record signifies ultimate galactic control.

102.067 The relationship between low activity comets and more active comets.
A. L. Cochran, J. R. Green, E. S. Barker.
ESA Spec. Publ., ESA SP–278, p. 151 – 156 (1987). – See Abstr. 012.075.

Previously, the authors have published a study of the abundance correlations of CN, C_3, and C_2 among a set of relatively active comets. They showed that there is a relatively strong correlation of C_3 and C_2 to CN. In this paper, the authors have extended this work to a group of low activity comets. They show that these low activity comets are intrinsically different from the "normal" comets in that they are depleted in C_2 and C_3 relative to the observed CN, or that they are without any emissions at small heliocentric distances.

102.068 The photochemistry of some possible cometary CN parent species.
J. B. Halpern.
ESA Spec. Publ., ESA SP–278, p. 159 – 162 (1987). – See Abstr. 012.075.

Radio observations of comet Halley have detected HCN in amounts of the order needed to account for the CN emission. However, the character of that emission cannot be accounted for by assuming that HCN is the parent species. A number of other precursor species have been considered. A necessary starting point for such a consideration is an understanding of the photochemistry of these molecules. This paper summarizes recent photochemical studies on the cyanoacetylenes, HC_3N and C_4N_2 and acetonitrile.

102.069 Photodissociation lifetimes of CH and CD radicals in comets.
P. D. Singh, A. Dalgarno.
ESA Spec. Publ., ESA SP–278, p. 177 – 180 (1987). – See Abstr. 012.075.

The rates of photodissociation of the CH radical from absorption of solar radiation are calculated to lie between $4.16 \times 10^{-3} s^{-1}$ and $2.67 \times 10^{-2} s^{-1}$ for heliocentric velocities between -42 and $42 \, km \, s^{-1}$ at 1 AU from the sun. For CD, the corresponding rates lie between 4.33×10^{-3} and $1.31 \times 10^{-2} s^{-1}$. The CH and CD radicals may be detectable in comets through absorption lines of the $C^2\Sigma^+ - X^2\Pi_r$ electronic bands near 3143.2 Å and 3142.1 Å, respectively.

102.070 The role of water in the thermal balance of the coma.
D. Bockelée–Morvan, J. Crovisier.
ESA Spec. Publ., ESA SP–278, p. 235 – 240 (1987). – See Abstr. 012.075.

Water has a dominant role in the thermodynamics of cometary atmospheres. Its photolysis is an important heating source. Its strong rotational transitions may allow radiative cooling, or, just the opposite, heating by absorbing the local radiation field. By coupling in a consistent model water excitation and gas thermal balance, the authors have attempted such an evaluation and tried to predict the kinetic temperature and expansion velocity of the coma. These predictions are in good agreement with recent observations of comet P/Halley.

102.071 A model of the anisotropic structure of the neutral gas coma of a comet.
N. I. Kömle, W.–H. Ip.
ESA Spec. Publ., ESA SP–278, p. 247 – 254 (1987). – See Abstr. 012.075.

Many observations during the recent comet Halley return indicated a marked asymmetry between the sunward– and the tailward–directed part of the neutral gas coma. In this paper the authors present some new model calculations simulating the observed asymmetries from a gasdynamical point of view. In particular it is discussed under which conditions outgassing anisotropies at the nucleus could be maintained out to large distances.

102.072 Anisotropic emission from comets: fans versus jets. I. Concept and modeling.
Z. Sekanina.
ESA Spec. Publ., ESA SP–278, p. 315 – 322 (1987). – See Abstr. 012.075.

Distinct jets – linear, spiral, and halo–like – displayed by a number of comets are examples of continuous ansiotropic ejection of dust (and, in some instances, of gas) from discrete emission sources on the sunlit side of the rotating nucleus. So are the broad, virtually featureless fans exhibited, in particular, by many short–period comets. Computer–generated images of these features illustrate the diversity of patterns that can be explained by varying the position of the spin axis, rotation period, location and extent of the active area, and the ejecta's dynamical constants for a given Sun–Earth–comet geometry. It is shown that an emission fan is generated by rotation of a high–latitude source about the spin axis of the nucleus with a relatively high obliquity, but that the fan's formation is rather insensitive to the rotation period.

102.073 Similarity and diversity of the polarization of comets.
J.–M. Perrin, P. L. Lamy.
ESA Spec. Publ., ESA SP–278, p. 411 – 415 (1987). – See Abstr. 012.075.

The authors present an analysis of the polarization measurements of the dust coma of 10 comets including P/Halley. Considerable attention is given to the proper separation of the dust from the gas emphasizing the value of the narrow–band and spectropolarimetric data. Different polarization curves are deduced implying unambiguous differences in the physical properties of the cometary dust.

102.074 Porosity and fractality of a cometary nucleus.
B. Lang.
ESA Spec. Publ., ESA SP–278, p. 483 – 486 (1987). – See Abstr. 012.075.

Porosity is a quantity resulting from the difference between the apparent density as received from astronomical optical measurements on a cometary nucleus and that attributed to the cometary solids. The calculation of the latter is made per analogy to the available data on mineral species believed to occur in cometary nuclei.

102.075 Crustal models of the evolving comet nucleus.
M. K. Wallis, N. C. Wickramasinghe.
ESA Spec. Publ., ESA SP–278, p. 495 – 499 (1987). – See Abstr. 012.075.

An evolving inhomogeneous surface and structured surface layers seem necessary to explain comets' behaviour. The comet missions have validated the two–phase surface model of Shul'man, with the modification that loose grains are consolidated via physico–chemical bonds and gas deposition into a coherent crust.

102.076 Diversity of comets, synergic temperature effects and the stability of organic compounds in icy nuclei.
K. Roessler, B. Nebeling.
ESA Spec. Publ., ESA SP–278, p. 509 – 513 (1987). – See Abstr. 012.075.
This paper concentrates on the interaction of temperature dependent processes, in particular chemical effects, induced by reaching the inner solar system and along the passages.

102.077 On the interpretation of the CH cometary spectrum.
C. Arpigny, C. J. Zeippen, M. Klutz, P. Magain, D. Hutsemékers.
ESA Spec. Publ., ESA SP–278, p. 607 – 612 (1987). – See Abstr. 012.075.
The authors describe a procedure set up to compute cometary spectra taking into account both radiative processes and collisional effects. Such a programme has been applied to the CH radical and the results are presented briefly.

102.078 The role of hydrogen thermalization in shaping the Lyman–α coma.
M. R. Combi, W. H. Smyth.
ESA Spec. Publ., ESA SP–278, p. 621 – 626 (1987). – See Abstr. 012.075.
Two important and interrelated aspects of the thermalization of cometary hydrogen have been addressed using a three-dimensional Monte Carlo particle trajectory model. These are the photochemical heating of the coma and the velocity distribution of cometary hydrogen leaving the inner coma that produces the extended Lyman–α coma. A wide variety of observed cometary conditions have been examined which range from productive comets like Kohoutek at very small heliocentric distances to intrinsically small comets like Giacobini–Zinner at 1 AU.

102.079 Interaction mechanisms of comets with the zodiacal dust cloud.
S. Ibadov.
ESA Spec. Publ., ESA SP–278, p. 655 – 656 (1987). – See Abstr. 012.075.
The comparison of realizability conditions of meteor–like and explosion–type interaction mechanisms of comets with the zodiacal dust cloud shows that in dusty comets of Halley's type the explosion–type mechanism is the dominant one.

102.080 The mass loss rates of periodic comets.
Ľ. Kresák, M. Kresáková.
ESA Spec. Publ., ESA SP–278, p. 739 – 744 (1987). – See Abstr. 012.075.
The mass loss rates integrated over the last two centuries are determined for all the known periodic comets of $q < 3$ AU, on the basis of their absolute total brightness and orbital evolution. They cover a range from 10^8 to 10^{12} kg/century and undergo appreciable temporal variations.

102.081 A note on comets and chemical evolution of the Galaxy.
V. Vanysek.
ESA Spec. Publ., ESA SP–278, p. 745 – 746 (1987). – See Abstr. 012.075.
A possible influence of comets on the growth of heavy elements abundance suggested several years ago by Tinsely and Cameron is shortly discussed. It is shown that this effect will be significant only if the formation of comets produces about 10^{28}kg cometary material per one solar mass which is equivalent to about 10^{14} cometary nuclei of average size per one solar mass.

102.082 The loss and depth of CO_2 ice in comet nuclei.
F. P. Fanale, J. R. Salvail.
Icarus, Vol. 72, No. 3, p. 535 – 554 (1987).
An analytical model has been developed to simulate the chemical differentiation of a homogeneous, initially unmantled cometary nucleus composed of water ice, putative unclathrated CO_2 ice, and silicate dust in specified proportions. The model includes the effects of nucleus rotation, arbitrary orientation of the rotation axis, latitude, heat conduction into the interior of the nucleus, restriction of CO_2 gas outflow by the water ice and dust layer, and the use of thermal conductivities for both amorphous and crystalline water ice as appropriate features. The model also accounts for the erosion of the water ice surface.

102.083 Axisymmetric dusty gas jet in the inner coma of a comet. II. The case of isolated jets.
Y. Kitamura.
Icarus, Vol. 72, No. 3, p. 555 – 567 (1987).
The behavior of isolated pure and dusty gas jets ejected from an active spot on the sunlit side of the nucleus surface is hydrodynamically investigated in the inner coma of an H_2O–dominated comet that is assumed to have no ambient ejection of the gas and dust from the dust–covered surface except the active spot. Steady–state solutions of the expanding jets are obtained by numerically solving the axisymmetric, time–dependent, coupled hydrodynamic equations of H_2O gas and the dust in polar coordinates.

102.084 Nonlinear propagation of Alfvén waves in cometary plasmas.
G. S. Lakhina, P. K. Shukla.
Astrophys. Space Sci., Vol. 139, No. 2, p. 275 – 279 (1987).
Large–amplitude Alfvén waves propagating along the guide magnetic field in a three–component plasma are shown to be spatially localized due to their nonlinear interaction with nonresonant electrostatic density fluctuations. A new class of subsonic Alfvén soliton solutions are found to exist in the three–component plasma. The Alfvén solitons can be relevant in explaining the properties of hydromagnetic turbulence near the comets.

102.085 Numerical modelling of circumcometary quasi–parallel shock waves.
A. S. Lipatov, I. N. Syrovatskij.
Kosm. Issled., Tom 25, Vyp. 6, p. 952 – 957 (1987). In Russian. English translation in Cosm. Res.

102.086 Encounters of the second kind.
P. D. Feldman.
Nature, Vol. 330, No. 6148, p. 518 – 519 (1987).
The author reports on miniature comets with unusual properties in the inner Solar System. See also Abstr. 102.087.

102.087 Cometesimals in the inner Solar System.
T. M. Donahue, T. I. Gombosi, B. R. Sandel.
Nature, Vol. 330, No. 6148, p. 548 – 550 (1987).
The authors have examined spectra obtained by the ultraviolet spectrometer on the Voyager 2 spacecraft between 1 and 2.5 AU and have found evidence for a very large number of 'cometesimals' with radii between a few metres and a few tens of metres in the neighbourhood of the Earth. They propose that these cometesimals are ice–coated, porous, low density refractory boulders that may be the building blocks of ordinary comet nuclei. They show that the cometesimals required to produce the observed Lyman–α emission can also account for all the lunar craters with diameters between 200 m and 1500 m produced during the past 3200 million years at sites such as Mare Tranquillitatis.

102.088 On the distribution of absolute magnitudes of comets according to the perihelion distance.
A. S. Guliev, D. G. Abbasov.
Komet. Tsirk., No. 372 (1987). In Russian.

102.089 On the possible excitation of internal gravitational waves in cometary atmospheres.
Yu. N. Redkoborodyj.
Komet. Tsirk., No. 373 (1987). In Russian.

102.090 On the difference in the photometric indices of new and periodic comets.
V. A. Dranevich, I. S. Lizunkova.
Komet. Tsirk., No. 374 (1987). In Russian.

102.091   On the probable structure of cometary ice.
V. A. Dranevich.
Komet. Tsirk., No. 374 (1987). In Russian.

102.092   Estimate of the ejection velocity of cometary dust according to the length of the terrestrial path in meteor streams.
G. V. Andreev.
Komet. Tsirk., No. 374 (1987). In Russian.

102.093   On the longitude distribution of aphelion distances of periodic comets.
A. S. Guliev.
Komet. Tsirk., No. 376 (1987). In Russian.

Periodic nature of cometary motions as known to Indian astronomers before eleventh century A.D.
See Abstr. 004.091.

Classification of comets recorded in ancient China (300 BC – 1800 AD) and statistics of type–I tails.
See Abstr. 004.119.

Perturbation computations and numerical modelling experiments.
See Abstr. 021.011.

Laboratory studies on cometary molecules.
See Abstr. 022.036.

Optical evolution of laboratory–produced organics: applications to Phoebe, Iapetus, outer belt asteroids and cometary nuclei.
See Abstr. 022.079.

On the applicability of the Sobolev approximation to the calculation of profiles and integral intensities of emission lines in optically thick slowly expanding media.
See Abstr. 022.080.

Interstellar problems and matrix solutions.
See Abstr. 022.104.

Radiation chemistry under unconventional conditions: dosimetry and aqueous radiolysis relevant to comet nuclei and early Earth structure.
See Abstr. 022.115.

Simulation of cometary nuclei.
See Abstr. 022.128.

Different trapping mechanisms of gases by water ice and their relevance for comet nuclei.
See Abstr. 022.129.

Laboratory amorphous carbon: a possible analogue of cometary dust.
See Abstr. 022.130.

Physical–mechanical properties of matrixes on the comet nuclei surface models.
See Abstr. 022.131.

Sublimation characteristics of H_2O comet nucleus with CO_2 impurities.
See Abstr. 022.132.

Thermally activated release of stored chemical energy in cryogenic media.
See Abstr. 022.141.

The cometary cloud in the solar system and the Resibois–Prigogine singular invariants of motion.
See Abstr. 042.079.

Rosetta: a mission to sample the nucleus of a comet.
See Abstr. 051.058.

Low–thrust navigation to comets or asteroids.
See Abstr. 052.001.

Cometary gas and plasma flow with detailed chemistry.
See Abstr. 062.080.

Time dependent MHD models for the cometary magnetosphere.
See Abstr. 062.137.

A kinetic study of solar wind mass loading and cometary bow shocks.
See Abstr. 074.069.

Comet showers as a cause of mass extinctions.
See Abstr. 081.025.

The cometary contribution to the oceans of primitive Earth.
See Abstr. 081.034.

Collisions in the Solar System – IV. Cometary impacts upon the planets.
See Abstr. 091.021.

Recent comet impacts on the Moon: the evidence from remote–sensing studies.
See Abstr. 094.035.

Introduction: history, definitions, nomenclature (*of small bodies of the solar system*).
See Abstr. 098.051.

The systems of interplanetary objects.
See Abstr. 098.052.

Dynamics of asteroids.
See Abstr. 098.053.

The role of gravitation and nuclear forces in the formation of asteroids and comets.
See Abstr. 098.088.

Electrostatic charging and fragmentation of dust near P/Giacobini–Zinner and P/Halley.
See Abstr. 103.035.

Estimates of masses, volumes and densities of short–period comet nuclei.
See Abstr. 103.043.

Study of the isotopic features of Swan bands in comets.
See Abstr. 103.102.

Characteristics of cometary picked–up ions in a global model of Giacobini–Zinner.
See Abstr. 103.121.

A model of the nongravitational motion of comet P/Encke.
See Abstr. 103.175.

Comet Halley: a carrier of interstellar dust chemical evolution.
See Abstr. 103.422.

Cometary MHD and chemistry.
See Abstr. 103.515.

The 2.7 μm water band of comet P/Halley: interpretation of observations by an excitation model.
See Abstr. 103.529.

Improved gas–kinetic treatment of cometary water sublimation and recondensation: application to comet P/Halley.
See Abstr. 103.532.

The cometary nucleus: current concepts.
See Abstr. 103.588.

The evolution of the mass distribution of cometary particles.
See Abstr. 103.666.

Consequences of the size determination of P/Halley by space probes on the scale of sizes of cometary nuclei.
See Abstr. 103.688.

Three approaches to the modelling of comet Halley's motion over long periods of time.
See Abstr. 103.690.

Anisotropic emission from comets: fans versus jets. II. Periodic comet Tempel 2.
See Abstr. 103.931.

On orbits and associations of meteor streams with comets P/Halley and P/Encke.
See Abstr. 104.096.

Local and exotic components of primitive meteorites, and their origin.
See Abstr. 105.020.

Interplanetary dust dynamics. II. Poynting–Robertson drag and planetary perturbations on cometary dust.
See Abstr. 106.036.

Interplanetary dust dynamics. III. Dust released from P/Encke: distribution with respect to the zodiacal cloud.
See Abstr. 106.037.

Can episodic comet showers explain the 30–Myr cyclicity in the terrestrial record?
See Abstr. 107.014.

Meteorites, asteroids and comets – a study of their interrelationship.
See Abstr. 107.016.

The formation and extent of the solar system comet cloud.
See Abstr. 107.021.

Interstellar molecules.
See Abstr. 131.086.

103 Comets (Individual Objects)

103.001 Recent news and research concerning comets.
D. W. E. Green.
Int. Comet Q., Vol. 9, No. 2, p. 58, 92 – 95 (1987).

103.002 Periodic comets for the visual observer in 1987.
A. Hale.
Int. Comet Q., Vol. 9, No. 2, p. 62 – 64 (1987).

103.003 Tabulation of comet observations.
Int. Comet Q., Vol. 9, No. 2, p. 65 – 92 (1987).
Concerning comets: 1910 II P/Halley, 1985 XII Shoemaker, 1985 XVII Hartley–Good, 1985 XIX Thiele, 1985n P/Boethin, 1985q P/Wirtanen, 1986a P/Shoemaker 3, 1986e P/Machholz, 1986h P/Schwassmann–Wachmann 2, 1986i Churyumov–Solo-dovnikov, 1986*l* Wilson, 1986n Sorrells, 1986o P/Urata–Niijima, 1986p P/Lovas 2, 1987a Levy, 1987b P/Wiseman–Skiff, 1987c Nishikawa–Takamizawa–Tago, 1987d Terasako, P/Schwass-mann–Wachmann 1.

103.004 Die Kometen Thiele (1985m) und P/Boethin (1985n).
G. Wagner.
KPM, Jahrg. 2, No. 5, p. 6 – 9 (1987).

103.005 Sulphur compounds in cometary *IUE* spectra.
K. S. Krishna Swamy, M. K. Wallis.
Mon. Not. R. Astron. Soc., Vol. 228, No. 2, p. 305 – 312 (1987).
Bands of S_2 have been identified in several comets including Černis, Bowell, Churyumov–Gerasimenko, Encke, etc., covering a wide range in heliocentric distances. S_2 seems to have been present in comet Halley even from 1985 early December when it was at 1.5 AU. The authors have also tentatively identified bands of the *B–X* systems of the molecule SO in IRAS–Araki–Alcock in the 2200 – 2400 Å region. These bands appear to be present in comets Borrelly, Stephan–Oterma and Meier. Emissions of S^+ ions, tentatively SH and several bands of CS, apart from the commonly observed ones, are evident in the spectra of comet Halley of 1986 March. Sulphur compounds are more prevalent than had been realized. Assessments of cometary dust are confused by S_2 emissions. The S_2 of comet Černis at 3.3 AU

means that the molecule is not bound in H_2O–ice, so postulated formation processes via cosmic ray or UV irradiation of ice are doubtful.

103.006 Comet Digest.
J. E. Bortle.
Sky Telesc., Vol. 74, No. 1, p. 108 (1987).
Concerning comets: 1982i P/Halley, 1986*l* Wilson, 1986m P/Grigg–Skjellerup, 1986n Sorrells, 1987c Nishikawa–Takami-zawa–Tago, 1987i P/Klemola, P/Denning–Fujikawa, P/Borrelly.

103.007 Comet Digest.
J. E. Bortle.
Sky Telesc., Vol. 74, No. 2, p. 217 – 218 (1987).
Concerning comets: 1986*l* Wilson, 1986m P/Grigg–Skjellerup, 1986n Sorrells, 1987i P/Klemola, 1987m P/Brooks 2, 1987p P/Borrelly, P/Schwassmann–Wachmann 1, P/Denning–Fujik-awa, P/Encke.

103.008 Comet Digest.
J. E. Bortle.
Sky Telesc., Vol. 74, No. 3, p. 329 – 330 (1987).
Concerning comets: 1957 V Mrkos, 1987i P/Klemola, 1987m P/Brooks 2, 1987p P/Borrelly, P/Encke.

103.009 Comet Digest.
J. E. Bortle.
Sky Telesc., Vol. 74, No. 4, p. 442 (1987).
Concerning comets: 1986*l* Wilson, 1986n Sorrells, 1987i P/Klemola, 1987m P/Brooks 2, 1987p P/Borrelly, 1987s Brad-field, 1987u Rudenko.

103.010 Comet Digest.
J. E. Bortle.
Sky Telesc., Vol. 74, No. 5, p. 562 – 563 (1987).
Concerning comets: 1987p P/Borrelly, 1987s Bradfield, 1987u Rudenko.

103.011   Erläuterungen zum VdS–Archiv.
   J. Jahn.
KPM, Jahrg. 2, Nr. 6, p. 16 – 35 (1987).
   Concerning comets: 1982i P/Halley, 1985m Thiele, 1985n P/Boethin, 1986n Sorrells, 1987c Nishikawa–Takamizawa–Tago, 1987d Terasako.

103.012   The criteria for cometary remarkability.
   D. W. Hughes.
Vistas Astron., Vol. 30, Part 2, p. 145 – 152 (1987).
   The history of astronomy has, from time to time, been enlivened by the appearance of a 'remarkable comet'. But what characteristics place a comet in this category, and are their numbers increasing or decreasing? To answer these questions the author investigates 'great' comets between 1066 and 1843 to study what these comets had in common and considers if any of the 'great' comets that appeared in the last two centuries should join the remarkable category.

103.013   Cometary plasma – review.
   J. C. Brandt, M. B. Niedner Jr.
Cometary radio astronomy, p. 1 – 18 (1987). – See Abstr. 012.033.
   An overview of large–scale phenomena is presented based on results of spacecraft probing of comets Halley and Giacobini–Zinner and on worldwide submissions to the Large–Scale Phenomena Discipline Specialist Team of the International Halley Watch. Examples of tail phenomena and science are presented with emphasis on the observed Disconnection Events. The archive of this material will clearly be very valuable for studying the solar–wind/comet interaction during the 1985 – 1986 apparition of Halley's comet. Estimates of the expected yield from the Network are included.

103.014   Radio continuum observations – thermal emission.
   W. J. Altenhoff.
Cometary radio astronomy, p. 25 – 26 (1987). – See Abstr. 012.033.
   The author reports on the possible origin of cometary radio radiation.

103.015   A search for HCN and other parent molecules in comets P/Giacobini–Zinner 1984e and P/Halley 1982i.
   D. Bockelée–Morvan, J. Crovisier, D. Despois, T. Forveille, E. Gérard, J. Schraml, C. Thum.
Cometary radio astronomy, p. 59 – 63 (1987). – See Abstr. 012.033.
   The authors searched for the rotational lines of several potential parent molecules in comets P/Giacobini–Zinner 1984e and P/Halley 1982i with the IRAM 30–m radio telescope. The $J = 1 \rightarrow 0$ lines of HCN were unambiguously detected in P/Halley in November – December 1985 and February 1986. Upper limits were obtained for HCN in P/Giacobini–Zinner, and for HC_3N, CH_3CN, OCS and CO in P/Halley. The paper is only a summary of the results.

103.016   Searches for parent molecules at MPIfR.
   M. K. Bird, W. K. Huchtmeier, A. von Kapp–herr, J. Schmidt, C. M. Walmsley.
Cometary radio astronomy, p. 85 – 89 (1987). – See Abstr. 012.033.
   Searches for radio line emission/absorption from comet Halley and comet Giacobini–Zinner were made at the 100–m Effelsberg Telescope in late 1985/early 1986. No detections were made for the K–band lines of water and ammonia, the ground state rotational transition of formaldehyde, or the excited lambda–doublet transitions of OH. Strong emission lines of the ground state OH molecule were observed in late January 1986.

103.017   The OH radio lines in comets: a review.
   E. Gerard.
Cometary radio astronomy, p. 91 – 99 (1987). – See Abstr. 012.033.
   Contents: Detection of the OH radical. OH excitation. OH radiation transfer. Mapping the OH coma. OH line profiles. Detection of weak cometary signals: SYMCOMET.

103.018   Observations of OH at 18 cm in comets P/Halley 1982i, P/Giacobini–Zinner 1984e, C/Hartley–Good 1985l and C/Thiele 1985m.
   E. Gérard, D. Bockelée–Morvan, G. Bourgois, P. Colom, J. Crovisier.
Cometary radio astronomy, p. 125 – 133 (1987). – See Abstr. 012.033.
   The 1667 MHz and 1665 MHz transitions of the OH radical were recently monitored in 4 comets with the Nançay radio telescope. These monitorings are a contribution to the International Halley Watch. A preliminary report of these observations is given here.

103.019   Comet notes. XI. 1987's comet harvest.
   D. H. Levy.
Strolling Astron., Vol. 32, Nos. 5 – 6, p. 129 (1987).

103.020   Recent news and research concerning comets.
   D. W. E. Green.
Int. Comet Q., Vol. 9, No. 3, p. 99 (1987).
   Concerning comets: 1987n P/Harrington, 1987o Shoemaker, 1987p P/Borrelly, 1987q P/Russell 2, P/Schwassmann–Wachmann 1, 1985 VIII Machholz.

103.021   Tabulation of comet observations.
   Int. Comet Q., Vol. 9, No. 3, p. 99 – 135 (1987).

103.022   1988 Comet Handbook.
   S. Nakano, with an introduction by D. W. E. Green.
Int. Comet. Q., Vol. 9, No. 5, 56 pp. (1987).

103.023   Identifications of minor planets with comets.
   Minor Planet Circ., No. 12025 (1987).

103.024   Observations of comets.
   Minor Planet Circ., Nos. 11887 – 11896, 12031 – 12034, 12166 – 12170, 12253 – 12257, 12360 – 12365, 12494 – 12500 (1987).
   Observations made at the following stations are published: Alma–Ata, Assah, Baldone, Bosscha Obs., Boyeros Obs., Burlington remote site, Burwash, Cambridge, Canberra, Cape, Carter Obs. Black Birch Stn., Caussols, Cerro El Roble, Cerro Tololo, Chamberlin Obs. Field Stn., Chorzow, Crimea, Eastfield, Engelhardt Obs., Engelhardt Obs. Zelenchukskaya Stn., Geisei, Gissar, Golosseevo–Kiev, Hatamae, Heidelberg, JCPM Sapporo Stn., Kambah, Kharkov, Kiev, Kitab, Kitami, Kitt Peak, Kitt Peak (Univ. Ariz.), Kiyosato, Klet, Kourovskaya, Kushiro, Lowell Obs. Anderson Mesa Stn., Mount John, Muro, Nikolaev, Oak Ridge, Odessa, Odessa–Mayaki, Oishi Astron. Obs., Okutama, Ordubad, Palomar, Perth, Rixeyville, Sapporo Sci. Cent., Sendai Obs. Ayashi Stn., Shizuoka, Siding Spring, Smolyan, Stakenbridge, Tokyo Obs. Kiso Stn., Uenohara, Uppsala South. Stn., YGCO Hoshikawa and Nagano Stn., Zvenigorod.
1974 II   P/Schwassmann–Wachmann 1,   1977 V   P/Kopff, 1978 XX P/Haneda–Campos, 1982 I Bowell, 1982 VIII P/Churyumov–Gerasimenko, 1982i P/Halley, 1984 III P/Hartley–IRAS, 1984 V P/Smirnova–Chernykh, 1984 VI P/Encke, 1984 VIII P/Clark, 1984 XIII Austin, 1984 XV Shoemaker, 1984 XIX P/Neujmin 1, 1984 XXI P/Arend–Rigaux, 1985 II Shoemaker, 1985 IV P/Gehrels 3, 1985 VIII Machholz, 1985 XII Shoemaker, 1985 XIII P/Giacobini–Zinner, 1985 XIV Hartley, 1985 XVI P/Ciffréo, 1985 XIX Thiele, 1986b Shoemaker, 1986g P/Forbes, 1986i Churyumov–Solodovnikov, 1986k P/Kohoutek, 1986l Wilson, 1986m P/Grigg–Skjellerup, 1986n Sorrells, 1986q P/du Toit–Hartley, 1987a Levy, 1987b P/Wiseman–Skiff, 1987c Nishikawa–Takamizawa, Tago, 1987e P/Wild 3, 1987g

P/Tempel 2, 1987h P/Howell, 1987i P/Klemola, 1987j Torres, 1987*l* P/Reinmuth 2, 1987m P/Brooks 2, 1987n P/Harrington, 1987o Shoemaker, 1987p P/Borrelly, 1987q P/Russell 2, 1987r P/Reinmuth 1, 1987s Bradfield, 1987t P/Jackson–Neujmin, 1987u Rudenko, 1987v P/Gehrels 1, 1987w P/Helin, 1987x P/West–Kohoutek–Ikemura, 1987y Levy, 1987z P/Shoemaker–Holt, 1987a₁ P/Mueller, 1987b₁ McNaught, 1987c₁ P/Longmore, 1987d₁ Ichimura, 1987e₁ P/Tempel 1, 1987f₁ Furuyama.

103.025 Orbital elements of comets.
Minor Planet Circ., Nos. 11887 – 12624 (1987).

The comets are listed according to their Roman numeral designation or preliminary designation. The names of the authors are given behind the respective M.P.C. numbers.

1977 XII P/Sanguin, 1977 XIII P/Tritton, 1978 XI P/Wild 2 12125 D. W. E. Green; 1978 XXIII P/Clark 12128 S. Nakano; 1978 XXV P/Tuttle–Giacobini–Kresák 12135 B. G. Marsden; 1979 VIII P/Schwassmann–Wachmann 3 12122 W. Landgraf; 1980 V P/Lovas 1 12135 B. G. Marsden; 1981 XVII P/Gehrels 2 12124 D. W. E. Green; 1981 XX P/Kearns–Kwee 12123 T. Kobayashi; 1982 III P/Peters–Hartley 12128 S. Nakano; 1982 IX P/Russell 3 12136 B. G. Marsden; 1983 IV P/Pons–Winnecke 12123 T. Kobayashi; 1983 IX P/du Toit–Neujmin–Delporte 12135 B. G. Marsden; 1983 XIII P/Kopff, 1983 XVIII P/Johnson 12123 T. Kobayashi; 1984 I P/Russell 4, 1984 II P/Taylor 12136 B. G. Marsden; 1984 VI P/Encke 12577 S. Nakano; 1984 XV Shoemaker 12307 D. W. E. Green; 1985 III P/Honda–Mrkos–Pajdušáková 12128 S. Nakano; 1985 VIII Machholz 12009 B. G. Marsden; 1985 XII Shoemaker 12124, 1985 XIV Hartley 12307, 1986b Shoemaker 12307 D. W. E. Green; 1986i Churyumov–Solodovnikov 12453, 1986*l* Wilson, 1986n Sorrells 12454 T. Kobayashi; 1986o P/Urata–Niijima 12128 S. Nakano; 1986p P/Lovas 2, 1987a Levy, 1987b P/Wiseman–Skiff 12124 D. W. E. Green; 1987c Nishikawa–Takamizawa–Tago 12009 B. G. Marsden; 1987d Terasako, 1987e P/Wild 3 12200, 1987j Torres, 1987o Shoemaker 12008 D. W. E. Green; 1987s Bradfield 12201, 12309, 12440, 1987u Rudenko 12201, 12309 B. G. Marsden, 12446 D. W. E. Green; 1987v P/Gehrels 1 12201, 1987w P/Helin 12309, 12440 B. G. Marsden; 1987x P/West–Kohoutek–Ikemura 12454 T. Kobayashi; 1987y Levy 12445 D. W. E. Green, 12575 B. G. Marsden; 1987z P/Shoemaker–Holt 12446 D. W. E. Green, 12576 B. G. Marsden; 1987a₁ P/Mueller, 1987b₁ McNaught 12446 D. W. E. Green, 12575, 1987d₁ Ichimura, 1987f₁ Furuyama 12575 B. G. Marsden.

103.026 Ephemerides of comets.
Minor Planet Circ., Nos. 12144 – 12148, 12211 – 12212, 12326 – 12330, 12460 – 12463, 12585 – 12589 (1987).

Concerning ephemerides of the following comets:
1981 VIII P/West–Kohoutek–Ikemura, 1982 X P/Gunn, 1982i P/Halley, 1986h P/Schwassmann–Wachmann 2, 1986j P/Comas Solá, 1986k P/Kohoutek, 1986*l* Wilson, 1987c Nishikawa–Takamizawa–Tago, 1987f P/Bus, 1987g P/Tempel 2, 1987j Torres, 1987o Shoemaker, 1987s Bradfield, 1987u Rudenko, 1987v P/Gehrels 1, 1987w P/Helin, 1987y Levy, 1987z P/Shoemaker–Holt, 1987a₁ P/Mueller, 1987b₁ McNaught, 1987c₁ P/Longmore, 1987d₁ Ichimura, 1987e₁ P/Tempel 1, 1987f₁ Furuyama.

103.027 High–resolution imaging studies of the near–nucleus regions of comets.
B. A. Goldberg, R. J. Bamberg, I. Halliday, B. A. McIntosh, G. C. L. Aikman, J. A. Slavin, F. L. Scarf, A. F. Cook.
Bull. Am. Astron. Soc., Vol. 19, No. 2, p. 742 (1987). Abstract. – See Abstr. 010.061.

103.028 Photoelectric photometry of comets in the system of standard IHW filters and the special case of comet P/Halley.
B. Stecklum, W. Pfau, M. Hesse.
Astron. Nachr., Vol. 308, No. 4, p. 239 – 246 (1987). = Mitt. Univ.–Sternw. Jena, Nr. 176.

Magnitudes of comets P/Giacobini–Zinner (1984e), P/Halley (1982i), P/Hartley–Good (1985*l*), and Thiele (1985m) in the bandpasses of the standard IHW comet filters are presented. For comet P/Halley production rates for CN, C_3, C_2, and solids were derived. For the gaseous components these show a strong dependence on heliocentric distance. The dependence is less steep for the solids which may be due to relatively pronounced backscattering properties in case of comet P/Halley. During one night (1985 Dec. 22/23) intensity profiles along three sections through the coma of comet P/Halley were measured. Compared with theoretical profiles they show a global anisotropy of the coma and possibly local structure.

103.029 Observations of cometary plasma–wave phenomena.
F. L. Scarf, F. V. Coroniti, C. F. Kennel, D. A. Gurnett, W.–H. Ip, E. J. Smith.
Astron. Astrophys., Vol. 187, No. 1/2, p. 109 – 116 (1987). – See Abstr. 003.013, 43.103.008.

The ICE plasma wave investigation utilized very long electric antennas (90 meters tip–to–tip) and a very high sensitivity magnetic search coil to obtain: (1) significant local information on plasma physics phenomena occurring in the distant pickup regions of comet Giacobini–Zinner and comet P/Halley, and (2) information on the processes that developed in the coma and tail of Giacobini–Zinner. Since ICE traversed cometary regions that complemented those sampled by Vega and Sakigake, it is important to compare observations from the three missions that carried dedicated wave instruments. Here the authors summarize ICE plasma wave measurements associated with both comet encounters and relate the very high sensitivity ICE observations to corresponding measurements from the other Halley space-craft.

103.030 Plasma structures in comets P/Halley and Giacobini–Zinner.
J. C. Brandt, M. B. Niedner Jr.
Astron. Astrophys., Vol. 187, No. 1/2, p. 281 – 286 (1987). – See Abstr. 003.013, 43.103.419.

An overview of large–scale phenomena is presented based on results of spacecraft probing of comets Halley and Giacobini–Zinner and on worldwide submissions to the Large–Scale Phenomena Discipline Specialist Team of the International Halley Watch. Examples of tail phenomena and science are presented with emphasis on observed disconnection events. The archive of this material will clearly be very valuable for studying the comet/solar–wind interaction during the 1985 – 1986 apparition of Halley's comet.

103.031 Resolution of the [O I] + NH₂ blend in comet P/Halley.
C. Arpigny, P. Magain, J. Manfroid, F. Dossin, A. C. Danks, D. L. Lambert.
Astron. Astrophys., Vol. 187, No. 1/2, p. 485 – 488 (1987). – See Abstr. 003.013, 43.103.571.

Spectra of comet P/Halley have been taken at very high resolution (0.015 nm) with a view to evaluating the contamination of the forbidden oxygen doublet at 630 nm by NH₂ features at low resolution. Comparison is made with a few other bright comets. Appreciable variations are found not only in the NH₂/[O I] ratio but also in the relative intensities of the various NH₂ emissions themselves. The authors comment upon the accuracy obtained on the oxygen abundance derived from the [O I] lines. Mapping of these emissions over the coma is required in order to correctly remove any important NH₂ contribution at low resolution. This should in addition provide information bearing upon the origin of the oxygen atoms in the ^1D level.

103.032 Polarimetry of comet P/Halley: continuum versus molecular bands.
J. F. Le Borgne, J. L. Leroy, J. Arnaud.
Astron. Astrophys., Vol. 187, No. 1/2, p. 526 – 530 (1987). – See Abstr. 003.013, 43.103.016.

Measurements of the linear polarization of molecular bands in comets P/Halley and Hartley–Good are presented. The measured bands are the C2 Swan (5140 Å), the CN (3880 Å), the C3 (4060 Å) and the OH (3090 Å). The phase angle dependence of the polarization of C2, C3 and CN is in good agreement with theory. However the polarization of OH is found 5 times smaller than expected. The direction of polarization of C2, C3 and OH is usually perpendicular to the scattering plane, though some measurements are parallel to it. For CN half the measures show significant deviations from the perpendicular to the scattering plane.

103.033 Observations of comet P/Halley at minimum phase angle.
K. J. Meech, D. C. Jewitt.
Astron. Astrophys., Vol. 187, No. 1/2, p. 585 – 593 (1987). – See Abstr. 003.013, 43.103.497.

The authors present time–series spectrophotometric observations of comet P/Halley obtained over 8 nights in 1985 November in near backscattering geometry (phase angle $1° \leqslant \alpha \leqslant 10°$). They have used the spectra to search for enhanced backscattering of the type reported earlier in comets P/Ashbrook–Jackson, Bowell (1982 I), and P/Stephan–Oterma and to study the time dependence of the molecular band and continuum emissions. No opposition surge in brightness greater $\approx 20\%$ was observed in P/Halley; instead, the brightening is consistent with a small linear phase coefficient $\beta = 0.02 \pm 0.01$ mag/deg in the observed phase angle range. The phase functions for P/Halley and the other comets are steeper than the phase function of the interplanetary dust. The comet phase functions are also unlike those of the low albedo asteroids. Two continuum brightness outbursts observed on UT 1985 November 15 and 21 are discussed.

103.034 Albedo maps of comets P/Halley and P/Giacobini–Zinner.
H. B. Hammel, C. M. Telesco, H. Campins, R. Decher, A. D. Storrs, D. P. Cruikshank.
Astron. Astrophys., Vol. 187, No. 1/2, p. 665 – 668 (1987). – See Abstr. 003.013, 43.103.019.

Near–simultaneous infrared and visual images of P/Halley are combined to create a map of the spatial variation of average albedo (which is generally a factor of four larger than the geometric albedo defined by Hanner et al., 1981). The map is compared with a similar map of P/Giacobini–Zinner (P/G–Z) obtained previously. P/G–Z showed a minimum in albedo of 0.07 with an increase by a factor of 2 over an angular distance of 30″. The lowest albedos in P/G–Z were offset from the nucleus in the anti–sunward direction, coincident with a dust tail observed in the infrared. The P/Halley albedos are higher than those found for P/G–Z, ranging from 0.2 – 0.4, but both comets have relatively low albedo in the anti–sunward direction (along the tail). The authors attribute the albedo distribution to large, dark, fluffy grains confined to the orbital plane close to the nucleus. The high albedo values in P/Halley may be due to enhanced flux in the visual image because of the comet's very small phase angle.

103.035 Electrostatic charging and fragmentation of dust near P/Giacobini–Zinner and P/Halley.
H. Boehnhardt, H. Fechtig.
Astron. Astrophys., Vol. 187, No. 1/2, p. 824 – 828 (1987). – See Abstr. 003.013.

In the coma of a comet, the dust grains emitted by the cometary nucleus are embedded in a surrounding plasma and exposed to the sunlight. Several effects (impinging charged plasma particles, photo–, secondary, thermionic, and field emission of electrons, sputtering of ions) might cause the electrostatic charging and possibly the fragmentation of some of these grains. Using in–situ–measurements of the comets P/Giacobini–Zinner and P/Halley, the electrostatic charging and fragmentation of spherical dust particles of carbon and silicate in the cometary comae is investigated theoretically. The results are compared with in–situ–measurements of dust in the coma of P/Halley.

103.036 Dormant phases in the aging of periodic comets.
L. Kresák.
Astron. Astrophys., Vol. 187, No. 1/2, p. 906 – 908 (1987). – See Abstr. 003.013, 43.103.024.

Implications of the observations of P/Halley for the problem of comet aging are discussed. It is suggested that the active lifetimes of short–period comets are intermitted by dormant phases during which they remain completely, or almost completely, inactive. Such phases are tentatively identified for several short–period comets by the absence of observations of their prediscovery apparitions at which they should have been very bright and favourably placed.

103.037 Comet Digest.
J. E. Bortle.
Sky Telesc., Vol. 74, No. 6, p. 681 (1987).

Concerning comets: 1987p P/Borrelly, 1987s Bradfield, 1987u Rudenko.

103.038 Organic chemicals in comets.
T. R. Geballe.
Nature, Vol. 329, No. 6140, p. 583 (1987).

The author reports on organic molecules detected in comet P/Halley and comet 1986l Wilson, see also Abstr. 103.213.

103.039 A comparison between wave observations performed in the environments of comets Halley and Giacobini–Zinner.
R. Grard, F. Scarf, J. G. Trotignon, M. Mogilevsky (M. Mogilevskij).
ESA Spec. Publ., ESA SP–278, p. 97 – 105 (1987). – See Abstr. 012.075.

The American spacecraft ICE flew through the tail of Comet Giacobini–Zinner at 7800 km from its nucleus on 11 September 1985. The two Soviet probes VEGA intercepted Comet Halley at comparable distances from its nucleus (8890 km and 8030 km) on 6 and 9 March 1986, but on the sunward side. The scope of this study is limited to a comparison between the results obtained with the plasma wave instrument flown on ICE and those collected by one of the two wave analysers carried by VEGA–1 and VEGA–2.

103.040 Recent observations of the OH radio lines in comets.
P. Colom, D. Bockelée–Morvan, G. Bourgois, J. Crovisier, E. Gérard.
ESA Spec. Publ., ESA SP–278, p. 241 – 245 (1987). – See Abstr. 012.075.

As part of their continuing program of observations of the OH 18–cm lines in bright comets with the Nançay radio telescope, the authors have recently monitored these lines in comets P/Halley 1982i and C/Wilson 1986l. They present an analysis of the time variations of the signal and an attempt to find out periodic variability. They discuss the variation of the line shapes with heliocentric distance and gas production rate.

103.041 Objective–prism spectra of the comets Honda (1968e), Hartley–Good (1985l) and Halley.
A. Florsch, J. Marcout, A. Laval, G. Traversa.
ESA Spec. Publ., ESA SP–278, p. 289 – 292 (1987). – See Abstr. 012.075.

Several spectra were taken of the comets Honda (1968e), Hartley–Good (1985l) and Halley with a Fehrenbach Objective–prism and can be compared in the field ranging from 380 nm to 500 nm. The main images of the coma in these limits are those given by C_2, C_3 and CN. No fundamental differences appear by means of this method.

103.042 The striae in the dust tails of great comets – a formal comparison of the goodness of fit by various theories, and a new proposal.
H.–E. Fröhlich, P. Notni, W. Thänert.
ESA Spec. Publ., ESA SP–278, p. 385 – 389 (1987). – See Abstr. 012.075.

The motion of the striae in the dust tails of comets Mrkos 1957 V and West 1976 VI is checked whether it corresponds to the predictions of the high–speed ejection theory or to a special two–step kinematical sequence. The authors find that only the two–step kinematics is able to explain all striae in both comets.

103.043 **Estimates of masses, volumes and densities of short–period comet nuclei.**
H. Rickman, L. Kamél, M. C. Festou, C. Froeschlé.
ESA Spec. Publ., ESA SP–278, p. 471 – 481 (1987). – See Abstr. 012.075.
An estimate of masses of 29 short–period comets is presented. By comparing nongravitational parameters with the jet forces expected from the gas production curves, an allowed mass range could be calculated for each comet. Nothing definite can yet be said about the densities of individual comets, except for P/Halley. The authors' approach, rather, leads to a statistical picture which supports the idea of cometary nuclei as highly porous objects.

103.044 **On the forbidden red lines of oxygen in comets.**
C. Arpigny, J. Manfroid, P. Magain, R. Haefner.
ESA Spec. Publ., ESA SP–278, p. 571 – 576 (1987). – See Abstr. 012.075.
The main purpose of this work is to examine the applicability of the hypothesis that previous authors had to make, for lack of accurate quantitative data, to disentangle the 630 blend and derive the [OI] flux from low–resolution spectra. The authors have presented evidence, in particular from very high resolution observations of two recent comets, showing that this assumption although appropriate in some cases, may be at fault in others.

103.045 **Determination of the content of CO^+ and H_2O^+ ions in the tail of comet Giacobini–Zinner and comet Halley.**
V. Ivanova, K. Jockers, V. Shkodrov.
ESA Spec. Publ., ESA SP–278, p. 577 – 581 (1987). – See Abstr. 012.075.
Objective prism spectra of comets Giacobini–Zinner and Halley were obtained with the Schmidt telescope of the Bulgarian National Observatory through a special double bandpass filter. Using extrafocal images of UBV standard stars taken immediately before or after the cometary exposures through the same filter, the plates have been put on absolute scale and the CO^+ and H_2O^+ content of the comet tail has been derived.

103.046 **Comparison of Halley's monochromatic coma with others.**
Z.–w. Hu.
ESA Spec. Publ., ESA SP–278, p. 583 – 589 (1987). – See Abstr. 012.075.
From November 1986 to April 1987, observations of Halley's comet with narrow band filters were made in China. Some similarities and differences of Halley's comet and others can be found from their comparison and are discussed. Finally, some conclusions on the similarity and the difference among comets are summarized, and observational techniques are reviewed briefly.

103.047 **Magnetohydrodynamic turbulence in the solar wind interacting with a comet.**
A. A. Galeev, A. N. Polyudov, R. Z. Sagdeev, K. Szegö, V. D. Shapiro, V. I. Shevchenko.
Zh. Ehksp. Teor. Fiz., Tom 92, No. 6, p. 2090 – 2105 (1987). In Russian. Abstr. in Ref. Zh., 62. Issled. Kosm. Prostranstva, 12.62.347 (1987).

103.048 **Photometric parameters of comets of 1981.**
D. A. Andrienko, A. V. Karpenko.
Komet. Tsirk., No. 371 (1987). In Russian.

103.049 **Definitive designations of comets of 1985.**
Komet. Tsirk., No. 377 (1987). In Russian.

103.050 **Evolution of new short–period comets.**
N. A. Belyaev, K. P. Ivanovskaya.
Komet. Tsirk., No. 378 (1987). In Russian.

103.051 **Photometric parameters of comets.**
Komet. Tsirk., No. 380 (1987). In Russian.
Concerning comets: 1967 X P/Tempel 2, 1969 IX Tago–Sato–Kosaka, 1970 XV Abe, 1974 III Bradfield.

103.052 **Results of observations of comets.**
Komet. Tsirk., No. 380 (1987). In Russian.
Concerning comets: 1984m Schaumasse, 1984t Levy–Rudenko, 1985m Thiele, 1985p Ciffreo, 1982i Halley, 1986*l* Wilson, 1986n Sorrells.

103.053 **On the possible identity of comets 1770 II and 1979 X.**
V. V. Radzievskij, A. V. Artem'ev, A. A. Brun, A. V. Gobetskij, L. N. Kokurina, I. N. Nekhlebova.
Komet. Tsirk., No. 381 (1987). In Russian.

103.054 **Predictive ephemerides for selected one–apparition periodic comets 1987 – 2000.**
C.. Townsend, J. Rogers.
Available from Charles Townsend, 3521 San Juan Ave., Oxnard, CA 93033, USA. 64 pp. Price US$ 6.00 (1987).
Review in Strolling Astron., Vol. 32, Nos. 5 – 6, p. 136, 1987 (*D. Machholz*).

Comet 1976 VI West

103.101 **The dust grains in the coma of Comet West.**
K. S. Krishna Swamy, G. A. Shah.
Earth, Moon, Planets, Vol. 38, p. 273 – 283 (1987).
The observed variation of reddening as a function of the heliocentric distance and the spatial variation of reddening within the coma of Comet West in the visual wavelength range have been considered to infer the properties of the cometary dust grains. The relevant model incorporates the variation in the size distribution function as well as the composition of the spherical grains. As a by–product, the model also satisfies the observed polarization and albedo for Comet West.

103.102 **Study of the isotopic features of Swan bands in comets.**
K. S. Krishna Swamy.
Astron. Astrophys., Vol. 187, No. 1/2, p. 388 – 390 (1987). – See Abstr. 003.013, 43.103.025.
It is shown from a detailed statistical equilibrium calculation of the $^{12}C^{13}C$ molecule that the interpretation of the observed intensities of Swan bands of the normal and the isotopic molecule of C_2 in terms of the abundance ratio of ^{12}C and ^{13}C is a reasonable one. The synthetic profile of some isotopic features in the (0,0) Swan band is compared with the observed profiles for comet West.

Periodic comet Takamizawa

103.106 **Periodic comet Takamizawa 1984 VII.**
OAA Comput. Sect. Circ., NK 506, 3 pp. (1987).

Comet 1983 VII IRAS–Araki–Alcock

103.111 **The rotation of comet 1983 VII IRAS–Araki–Alcock.**
J.–i. Watanabe.
Publ. Astron. Soc. Jpn., Vol. 39, No. 3, p. 485 – 503 (1987).
Sequential photographic observations of the coma of comet 1983 VII IRAS–Araki–Alcock were carried out with a 50–cm Schmidt telescope at the Dodaira Observatory from May 7 through May 11, 1983. Applying a new image–processing method to 19 plates, the author finds time variation of strong asymmetry of the coma. This result suggests that the surface of the rotating nucleus is fairly inhomogeneous and has some active regions. The rotation period of the nucleus is found to be between 18 hr and 170 hr and the obliquity of rotation axis to the orbital plane is between 64° and 80°, from an analysis of the author's results with an "active region" model.

103.112 **An examination of possible solar wind sources for a sudden brightening of comet IRAS–Araki–Alcock.**
C. T. Russell, J. G. Luhmann, D. N. Baker.
Geophys. Res. Lett., Vol. 14, No. 10, p. 991 – 994 (1987).
Possible solar wind sources for a sudden global brightening of comet IRAS–Araki–Alcock are examined. No increases in solar

wind momentum flux, solar energetic particles or solar activity occurred coincident with these brightenings. The only change in the solar wind coincident with the brightenings was a rotation of the interplanetary magnetic field to a more flow–aligned state. If this rotation did not lead to the cometary brightening, the brightening must have been intrinsic to the comet.

Periodic comet Hartley 2

103.116   **Periodic comet Hartley 2 1985 V.**
OAA Comput. Sect. Circ., NK 507, 1 p. (1987).

Periodic comet Giacobini–Zinner

103.121   **Characteristics of cometary picked–up ions in a global model of Giacobini–Zinner.**
C. D. Kimmel, J. G. Luhmann, J. L. Phillips, J. A. Fedder.
J. Geophys. Res., Vol. 92, No. A8, p. 8536 – 8544 (1987).
Energetic ions observed during the International Cometary Explorer (ICE) spacecraft flyby of comet Giacobini–Zinner provide information about both the constitution of comets and the plasma physical processes associated with their interaction with the solar wind. In this investigation the details of ion "pickup", in the limit where small–scale fluctuations in the plasma and magnetic field are neglected, are modeled by following the motion of a large number of initially cold, heavy (mass 18) ions in a global magnetohydrodynamic model of the local plasma and magnetic field. The results indicate how the background or macroscopic velocity and magnetic field structure of the comet can affect the average spatial and spectral characteristics of the observed cometary ions.

103.122   **Observations of comet Giacobini–Zinner at the 1.667 GHz OH line.**
J. Galt.
Astron. J., Vol. 94, No. 1, p. 174 – 177 (1987).
Comet Giacobini–Zinner was observed during an 18 day interval just prior to its perihelion passage. The 1.667 GHz OH line appeared in absorption, as expected, but showed large intensity fluctuations from day to day. Comparisons are made with subsequent observations of comet Halley.

103.123   **ICE observations of comet Giacobini–Zinner.**
S. W. H. Cowley.
Philos. Trans. R. Soc. London, Ser. A, Vol. 323, No. 1572, p. 405 – 420 (1987). – See Abstr. 012.026.
The first spacecraft encounter with a comet took place on 11 September 1985 when the International Cometary Explorer spacecraft passed through the tail of comet Giacobini–Zinner at a distance of 7800 km from the nucleus. It provided the first definitive in–situ information concerning the interaction of a cometary atmosphere with the flowing solar–wind plasma, and the results of initial analyses are reviewed. An unexpected feature of the interaction was the extreme levels of field and plasma turbulence, and broadband wave activity observed in the region of mass–loaded flow.

103.124   **Observations of energetic water–group ions at comet Giacobini–Zinner: implications for ion acceleration processes.**
I. G. Richardson, S. W. H. Cowley, R. J. Hynds, C. Tranquille, T. R. Sanderson, K.–P. Wenzel, P. W. Daly.
Planet. Space Sci., Vol. 35, No. 10, p. 1323 – 1345 (1987).
In this paper the primary focus of interest has been on the form of the energy spectrum of water–group pick–up ions in the rest–frame energy range from a few tens to a few hundred kiloelec-tronvolts in the region within $\sim 3 \times 10^5$ km of comet Giacobini–Zinner, as determined by the EPAS instrument on the ICE spacecraft. The ultimate aim of the study is to contribute to an understanding of the processes which scatter and accelerate these ions in the vicinity of the comet by comparing the observed spectral forms with those derived from theory. As a result of this

analysis the authors have also been able to study the variations of ion isotropy, intensity and spectral hardness with distance from the comet which also bear upon this topic.

103.125   **Steepened magnetosonic waves at comet Giacobini–Zinner.**
B. T. Tsurutani, R. M. Thorne, E. J. Smith, J. T. Gosling, H. Matsumoto.
J. Geophys. Res., Vol. 92, No. A10, p. 11,074 – 11,082 (1987).
The authors examine intense hydromagnetic waves at comet Giacobini–Zinner to investigate the mode and direction of wave propagation and thereby provide important constraints on potential mechanisms for wave origin in the vicinity of the comet.

103.126   **Infrared images of comets. I. P/Giacobini–Zinner (1985e).**
T. L. Hayward, G. L. Grasdalen.
Astron. J., Vol. 94, No. 5, p. 1339 – 1349 (1987). With plates 116 – 118.
The authors present near– and mid–infrared images of periodic comet Giacobini–Zinner taken during the summer of 1985. The near–infrared H and K observations covered five nights from 27 June to 13 September, including 11 September, the date of the *ICE* encounter with the comet. The comet was most active for the mid–August observations; then it faded by mid–September to approximately the absolute brightness observed in June. On 15 and 17 August the authors' maps show jetlike structure near the nucleus, and on all dates the radial surface brightness falls off as $\sim \varrho^{-1.5}$. The 10 μm maps, taken 31 August and 13 September, are of low signal–to–noise, but confirm the decrease in brightness by mid–September. The existence of jets suggests that the coma production rate is variable, which may account for the steep surface–brightness gradient.

103.127   **Energy spectra of energetic ions in the vicinity of comet P/Giacobini–Zinner.**
I. G. Richardson, S. W. H. Cowley, V. Moore, K. Staines, R. J. Hynds, T. R. Sanderson, K.–P. Wenzel, P. W. Daly.
Astron. Astrophys., Vol. 187, No. 1/2, p. 276 – 280 (1987). – See Abstr. 003.013, 43.103.131.
The energy spectrum of water–group pick–up ions with energies > 35 keV observed during the encounter of comet Giacobini–Zinner by the ICE spacecraft is examined using data from the EPAS experiment, combined with published data from the ULECA sensor. In the plasma frame the combined spectra consist mainly of a steeply–falling tail extending to energies of a few hundred keV, but there is also evidence for flattening at speeds somewhat below the local pick–up speed. The latter feature is most evident in the solar wind outside the mass–loaded region of slowed flow, where the local pick–up speed is sufficiently fast that it falls within the range of the combined EPAS and ULECA data. Above the local pick–up speed the ion distribution function is well approximated by a decreasing exponential of ion speed. The characteristic speed of this distribution increases as the comet is approached. It is demon-strated that EPAS data can give detailed information on this portion of the spectrum with 32 s time resolution.

103.128   **Photometric observations of comet P/Giacobini–Zinner.**
D. G. Schleicher, R. L. Millis, P. V. Birch.
Astron. Astrophys., Vol. 187, No. 1/2, p. 531 – 538 (1987). – See Abstr. 003.013.
Narrowband filter photometry of comet P/Giacobini–Zinner was obtained on 15 nights at Lowell Observatory in the interval from 15 June to 20 November 1985. The emission bands of OH, NH, CN, C_3, and C_2 were isolated, along with continuum points at 3650 Å and 4845 Å. All observed species showed pronounced asymmetry about perihelion, with pre–perihelion abundances being greater by more than a factor of 2 than post–perihelion abundances over the heliocentric distance range of 1.0 – 1.5 AU. The characteristics of the asymmetry were significantly different than observed in previous apparitions, with the peak production rates occurring in late July/early August. The unusually low C_2/CN abundance ratio seen in previous apparitions is confirmed

by the present observations. Furthermore, the OH data demonstrate that the C_2 in Giacobini–Zinner (along with C_3 and NH) is depleted when compared to water (the dominant species), rather than CN being unusually overabundant.

103.129   Narrow–band photometry of comet P/Giacobini–Zinner at the time of ICE encounter.
C. Sterken, J. Manfroid.
ESA Spec. Publ., ESA SP–278, p. 191 – 194 (1987). – See Abstr. 012.075.
The authors present narrow–band multi–diaphragm photometry of comet Giacobini–Zinner, obtained at Wise Observatory (Israel) shortly after the International Cometary Explorer passed behind the comet's nucleus. Derived fluxes are presented, and Haser model production rates are given for each night data.

103.130   Gas coma of comet Giacobini–Zinner: emission from grains.
M. K. Wallis, N. P. Meredith, D. Rees.
ESA Spec. Publ., ESA SP–278, p. 283 – 287 (1987). – See Abstr. 012.075.
Comet Giacobini–Zinner was observed from 26 August to 12 September 1985 from La Palma Observatory using a sensitive IPD attached to the 1 m JKT and 0.2 m Meade telescopes, with narrow band and IHW filters. Emissions were recorded to the edge of the field of view of the smaller telescope at 300,000 km radius. The images in CN, C_2 antd OI showed roughly circular isophotes, interpreted as probably due to daughter molecules emitted from parent molecules or grains having lower source speeds, rather than emission from day and night sides of the nucleus.

103.131   Ground–based constraints on the H_2O^+/CO^+ abundance ratio and dust impact rate in comet P/Giacobini–Zinner: comparison with the ICE spacecraft results.
R. M. Wagner, B. L. Lutz, S. Wyckoff.
Astrophys. J., Vol. 322, No. 1, p. 544 – 548 (1987).
The results of ground–based spectrophotometry of the comet P/Giacobini–Zinner (1984e) near perihelion and the time of the *International Cometary Explorer (ICE)* encounter are presented. The authors find that the abundance ratio $[H_2O^+/CO^+]$ is 1.2 ± 0.6 compared with the *ICE* result of approximately 5. Analysis of the dust production as measured by the optical continuum predicts a dust impact rate on the S antenna of the *ICE* of $\leqslant 0.01\ s^{-1}$ at closest approach. This upper limits is consistent with the *ICE* results.

103.132   Position observations of comet Giacobini–Zinner at the Astronomical Observatory of the Ural University.
Komet. Tsirk., No. 373 (1987). In Russian.

Periodic comet Schuster

103.141   Periodic comet Schuster 1978 I.
OAA Comput. Sect. Circ., NK 508, 2 pp. (1987).

Periodic comet Kowal 2

103.146   Periodic comet Kowal 2 1979 II.
OAA Comput. Sect. Circ., NK 509, 1 p. (1987).

Periodic comet Kowal–Mrkos

103.151   Periodic comet Kowal–Mrkos 1984 X.
OAA Comput. Sect. Circ., NK 510, 1 p. (1987).

Periodic comet Shoemaker 1

103.156   Periodic comet Shoemaker 1 1984 XVI.
OAA Comput. Sect. Circ., NK 511, 2 pp. (1987).

Periodic comet Arend–Rigaux

103.161   The nucleus of comet P/Arend–Rigaux.
G. J. Veeder, M. S. Hanner, D. J. Tholen.
Astron. J., Vol. 94, No. 1, p. 169 – 173 (1987). With plate 20.
The authors report simultaneous broadband visual and infrared photometry of comet P/Arend–Rigaux. Its *VJHK* colors are similar to other periodic comets and the class D asteroids. The visual flux decreased by 0.5 mag within 2 hr, while the thermal flux at 10 μm continued to decrease by 0.7 mag over 3 hr. These observations are consistent with thermal models of a nucleus with a geometric albedo of 0.03 and equivalent radii of 5.1 and 3.8 km.

Periodic comet Swift–Gehrels

103.166   Periodic comet Swift–Gehrels 1972 VII = 1981 XIX.
OAA Comput. Sect. Circ., NK 512, 2 pp. (1987).

Periodic comet Encke

103.171   Linkage of the 53 observed perihelion times of the periodic comet Encke.
G. Sitarski.
Acta Astron., Vol. 37, No. 1, p. 99 – 113 (1987).
To search the nongravitational effects in the comet's motion, the 53 perihelion times in the period 1786 – 1984 were used as observational data. Formulae for differential coefficients were derived to correct the parameters of motion by the least squares method using in the observational equations the values of mean anomaly computed for the observed perihelion times. The nongravitational effects were examined in two ways: (1) as a secular change $\dot{a}$ of the semi–major axis of the comet's orbit, (2) as the Marsden's nongravitational parameters A_1 and A_2. It was found that in both cases the nongravitational effects are similar functions of time. A successful linkage of 66 positional observations from 1977 – 84 with 51 perihelion times from 1977 – 86 also was made.

103.172   Periodic comet Encke.
IAU Circ., Nos. 4425, 4447, 4503 (1987).

103.173   Periodic comet Encke.
Yamamoto Circ., No. 2088 (1987). In Japanese.

103.174   P/comet Encke.
Br. Astron. Assoc. Circ., No. 670 (1987).

103.175   A model of the nongravitational motion of comet P/Encke.
G. Sitarski.
ESA Spec. Publ., ESA SP–278, p. 751 – 752 (1987). – See Abstr. 012.075.
Motion of the comet over the last two centuries was investigated assuming the nongravitational effects as a daily change of the semi–major axis of the comet's orbit. It was found that a variability of the nongravitational effects could be approximated by a sinusoidal function of time. This model of the comet's motion allows to represent the observed perihelion times to within 0.025 of a day.

103.176   Periodic comet Encke 1974 V = 1977 XI = 1980 XI = 1984 VI.
OAA Comput. Sect. Circ., NK 519, 4 pp. (1987).

Comet 1983 V Sugano–Saigusa–Fujikawa

103.181   Comet Sugano–Saigusa–Fujikawa (1983 V) – a small, puzzling comet.
M. S. Hanner, R. L. Newburn, H. Spinrad, G. J. Veeder.
Astron. J., Vol. 94, No. 4, p. 1081 – 1087 (1987).
Spectroscopic and infrared observations of comet Sugano–Saigusa–Fujikawa (1983 V) were obtained during its close

approach to the Earth on 11 – 14 June 1983. The [O I] production rates of $1.8 \pm 0.9 \times 10^{26}$ atoms/s observed on 12.3 June and $7 \pm 3.5 \times 10^{26}$ atoms/s on 13.4 June lead to derived water–production rates of 3×10^{27} mol/s on 12 June and 1.1×10^{28} mol/s on 13 June. The abundances of the minor species NH_2, CN, C_2, and C_3 are unusually low relative to [O I]. The upper limit to the average nuclear radius from the infrared and visual photometry on 12 – 13 June (assuming that the entire signal came from the nucleus) is ~ 370 m. The dust/gas mass ratio was < 0.01 on June 12 and < 0.005 on June 13.

Periodic comet Kowal–Vavrova

103.186   **Periodic comet Kowal–Vavrova 1983 III.**
OAA Comput. Sect. Circ., NK 514, 1 p. (1987).

Comet 1982 I Bowell

103.191   **Comet Bowell at record heliocentric distance.**
K. J. Meech, D. Jewitt.
Nature, Vol. 328, No. 6130, p. 506 – 509 (1987).
The authors present new observations of comet Bowell at the record heliocentric distance $R = 13.6$ AU. An extended coma is present, the size of which is consistent with the same slow expansion rate $v \approx 1$ m s^{-1} detected around perihelion (March 1982 and $R \approx 3.36$ AU). The cross–section of the solid grains within the central 10 arc s of the coma has decreased by over an order of magnitude since 1980 – 1984, which indicates that the coma production is declining. The coma at $R \leqslant 10$ AU may be formed by sublimation of CO_2 or an ice of similar volatility from the nucleus.

Periodic comet Chernykh

103.196   **Periodic comet Chernykh 1978 IV.**
OAA Comput. Sect. Circ., NK 515, 6 pp. (1987).

Periodic comet Wolf–Harrington

103.201   **Investigation of motion of comet P/Wolf–Harrington.**
S. Szutowicz.
Acta Astron., Vol. 37, No. 2, p. 179 – 187 (1987).
The motion of the comet was investigated using 181 observations made in the period 1924 – 1985 during its seven apparitions. The existence of nongravitational effects in the comet's motion either as a daily change of the semi–major axis of the orbit á and or as the Marsden's nongravitational function $A_i g(r)$ was taken into account. Linking three and four successive apparitions of the comet a variation of the transverse component A_2 with time was searched. The ephemeris for the next return of the comet in 1991 was computed on the basis of the orbital elements as determined by linkage of the last three apparitions.

Periodic comet Harrington–Abell

103.206   **Periodic comet Harrington–Abell 1969 III = 1976 VIII = 1983 XVII.**
OAA Comput. Sect. Circ., NK 516, 2 pp. (1987).

Comet 1986*l* Wilson

103.211   **Postperihelion activity in comet Wilson 1986*l*.**
G. C. L. Aikman.
Bull. Am. Astron. Soc., Vol. 19, No. 2, p. 751 (1987). Abstract. – See Abstr. 010.061.

103.212   **CCD observations of comet Wilson at the ESO 1 m telescope with a focal reducer.**
K. Jockers, E. H. Geyer.
Messenger, No. 50, p. 48 – 50 (1987).

103.213   **Discovery of organic grains in comet Wilson.**
D. A. Allen, D. T. Wickramasinghe.
Nature, Vol. 329, No. 6140, p. 615 – 616 (1987).
Organic materials have been shown to be a major component of the solid grains in long path lengths through the Galaxy. Their identification was through the C–H bond stretching vibration transition near 3.4 μm, which in grains is seen as a broad, weakly structured absorption feature on the radiation of bright background sources. A similar spectral feature was seen in emission in comet Halley. The surface appears to have been heavily processed by long exposure to sunlight, and it remains unclear whether the organic material represented a release of pristine grains from within the nucleus, or the stripping from the surface of the processed material. Here the authors report the detection of a similar, but distinct, emission feature in comet Wilson, a comet probably making its first visit to the solar vicinity.

103.214   **Comet Wilson (1986*l*).**
IAU Circ., Nos. 4439, 4445, 4463, 4491, 4503 (1987).

103.215   **Comet Wilson (1986*l*).**
Yamamoto Circ., Nos. 2089, 2091 (1987). In Japanese.

103.216   **Observations of comet Wilson (1986*l*).**
Komet. Tsirk., Nos. 375, 379 (1987). In Russian.

Periodic comet Kopff

103.221   **Nongravitational motion of comet P/Kopff during 1958 – 1983.**
H. Rickman, G. Sitarski, B. Todorovic–Juchniewicz.
Astron. Astrophys., Vol. 188, No. 1, p. 206 – 211 (1987).
225 astrometric observations from the last five apparitions of comet Kopff were collected in order to investigate the orbital motion of the comet. Values of all three nongravitational parameters (A_1, A_2 and A_3) were found for three linkages of three consecutive apparitions, and by ascribing these to the respective mid–moments the authors observe a regular change with time for A_2 and A_3. Further they computed thermal models for the nucleus of comet Kopff in order to find the physical interpretations of the orbital results. These models considered an oblate spheroidal nucleus, and it was found that an equatorial radius of 3.0 km with 75% of the surface area in a state of free sublimation of H_2O ice gives a reasonable account of the light curves observed at recent apparitions.

Periodic comet Wirtanen

103.226   **Periodic comet Wirtanen 1961 IV = 1967 XIV = 1974 XI = 1985q.**
OAA Comput. Sect. Circ., NK 517, 2 pp. (1987).

Comet 1987c Nishikawa–Takamizawa–Tago

103.231   **Comet Nishikawa–Takamizawa–Tago (1987c).**
IAU Circ., Nos. 4414, 4424 (1987).

103.232   **Comet Nishikawa–Takamizawa–Tago (1987c).**
Yamamoto Circ., Nos. 2087, 2088 (1987). In Japanese.

103.233   **Comet 1987c and the ε Geminids.**
Br. Astron. Assoc. Circ., No. 671 (1987).

103.234   **Observations of comet Nishikawa (1987c) in Odessa.**
Komet. Tsirk., No. 379 (1987). In Russian.

Periodic comet van Biesbroeck

103.236 **Periodic comet van Biesbroeck 1954 IV = 1966 III = 1978 XXIV.**
OAA Comput. Sect. Circ., NK 518, 5 pp. (1987).

Periodic comet Russell 2

103.241 **Periodic comet Russell 2 (1987q).**
IAU Circ., No. 4415 (1987).

103.242 **Periodic comet Russell 2 (1987q).**
Yamamoto Circ., No. 2087 (1987). In Japanese.

103.243 **Periodic comet Russell 2 1980 III = 1987q.**
OAA Comput. Sect. Circ., NK 522, 1 p. (1987).

103.244 **Rediscovery of short–period comet Russell 2 (1987q).**
Komet. Tsirk., No. 370 (1987). In Russian.

Periodic comet Shoemaker 2

103.246 **Periodic comet Shoemaker 2 1984 XVIII.**
OAA Comput. Sect. Circ., NK 513, 1 p. (1987).

Periodic comet Howell

103.251 **Periodic comet Howell (1987h).**
IAU Circ., Nos. 4415, 4420, 4450 (1987).

Periodic comet Smirnova–Chernykh

103.256 **Periodic comet Smirnova–Chernykh 1967 XV = 1978 NA$_6$ = 1981 UH$_{18}$.**
OAA Comput. Sect. Circ., NK 520, 4 pp. (1987).

Periodic comet Reinmuth 2

103.261 **Periodic comet Reinmuth 2 (1987*l*).**
IAU Circ., No. 4417 (1987).

Periodic comet du Toit–Hartley

103.266 **Periodic comet du Toit–Hartley 1982 II = 1986q.**
OAA Comput. Sect. Circ., NK 521, 2 pp. (1987).

Periodic comet Brooks 2

103.271 **Periodic comet Brooks 2 (1987m).**
IAU Circ., Nos. 4417, 4451, 4472, 4488 (1987).

Periodic comet Hartley–IRAS

103.276 **Photographic astrometry of periodic comet Hartley–IRAS.**
Komet. Tsirk., No. 371 (1987). In Russian.

Comet 1986n Sorrells

103.281 **Comet Sorrells (1986n).**
IAU Circ., Nos. 4418, 4432, 4462, 4500 (1987).

103.282 **Comet Sorrells (1986n).**
Yamamoto Circ., No. 2087 (1987). In Japanese.

103.283 **Photographic observations of comet Sorrells (1986n).**
Komet. Tsirk., No. 376 (1987). In Russian.

Comet 1985 XVII Hartley–Good

103.286 **Photographic observations of comet Hartley–Good (1985*l*) in Alma–Ata.**
Komet. Tsirk., No. 371 (1987). In Russian.

Comet 1987j Torres

103.291 **Comet Torres (1987j).**
IAU Circ., No. 4421 (1987).

Periodic comet Stephan–Oterma

103.296 **The ring–like structure in the head of comet Stephan–Oterma.**
V. P. Tarashchuk, V. D. D'yakonova.
Komet. Tsirk., No. 378 (1987). In Russian.

Periodic comet Grigg–Skjellerup

103.301 **Periodic comet Grigg–Skjellerup (1986m).**
IAU Circ., Nos. 4421, 4450 (1987).

103.302 **Periodic comet Grigg–Skjellerup (1986m).**
Yamamoto Circ., No. 2087 (1987). In Japanese.

Comet 1965 VIII Ikeya–Seki

103.306 **On the problem of radial structures of the tail of comet Ikeya–Seki (1965f).**
Sh. N. Sabitov, O. V. Chumak.
Komet. Tsirk., No. 378 (1987). In Russian.

Periodic comet Klemola

103.311 **Periodic comet Klemola (1987i).**
IAU Circ., Nos. 4422, 4433, 4462 (1987).

103.312 **Periodic comet Klemola (1987i).**
Yamamoto Circ., No. 2089 (1987). In Japanese.

103.313 **Periodic comet Klemola 1987i.**
Br. Astron. Assoc. Circ., No. 669 (1987).

Comet 1987o Shoemaker

103.321 **Comet Shoemaker (1987o).**
IAU Circ., No. 4423 (1987).

Periodic comet Reinmuth 1

103.331 **Periodic comet Reinmuth 1 (1987r).**
IAU Circ., No. 4424 (1987).

103.332 **Periodic comet Reinmuth 1 (1987r).**
Yamamoto Circ., No. 2088 (1987). In Japanese.

103.333 **Rediscovery of short–period comet Reinmuth 1 (1987r).**
Komet. Tsirk., No. 371 (1987). In Russian.

Periodic comet Borrelly

103.341 **Periodic comet Borrelly (1987p).**
IAU Circ., Nos. 4426, 4439, 4453, 4480, 4492, 4501, 4512, 4520 (1987).

103.342 **Periodic comet Borrelly (1987p).**
Yamamoto Circ., Nos. 2088, 2093 (1987). In Japanese.

103.343 **Rediscovery of short–period comet Borrelly (1987p).**
Komet. Tsirk., No. 370 (1987). In Russian.

Comet 1987s Bradfield

103.351 **Comet Bradfield (1987s).**
IAU Circ., Nos. 4431, 4433, 4434, 4437, 4438, 4442, 4448, 4458, 4460, 4466, 4477, 4483, 4487, 4496, 4503, 4509, 4513, 4518 (1987).

103.352 **Comet Bradfield (1987s).**
Yamamoto Circ., Nos. 2089 – 2094, 2096 – 2098, 2100 (1987). In Japanese.

103.353 **Comet Bradfield 1987s.**
Br. Astron. Assoc. Circ., Nos. 670 – 672, 674 (1987).

103.354 **New comet Bradfield (1987s).**
Komet. Tsirk., Nos. 372, 374, 380, 381 (1987). In Russian.

Periodic comet Jackson–Neujmin

103.361 **Periodic comet Jackson–Neujmin (1987t).**
IAU Circ., No. 4438 (1987).

103.362 **Periodic comet Jackson–Neujmin (1987t).**
Yamamoto Circ., No. 2091 (1987). In Japanese.

103.363 **Rediscovery of short–period comet Jackson–Neujmin (1987t).**
Komet. Tsirk., No. 373 (1987). In Russian.

Comet 1987u Rudenko

103.371 **Comet Rudenko (1987u).**
IAU Circ., Nos. 4440 – 4443, 4448, 4458, 4462, 4479, 4488, 4520 (1987).

103.372 **Comet Rudenko (1987u).**
Yamamoto Circ., Nos. 2090, 2091, 2093, 2096, 2097 (1987). In Japanese.

103.373 **Comet Rudenko 1987u.**
Br. Astron. Assoc. Circ., Nos. 671, 672 (1987).

103.374 **New comet Rudenko (1987u).**
Komet. Tsirk., Nos. 374, 380, 381 (1987). In Russian.

Periodic comet Gehrels 1

103.381 **Periodic comet Gehrels 1 (1987v).**
IAU Circ., No. 4444 (1987).

103.382 **Periodic comet Gehrels 1 (1987v).**
Yamamoto Circ., No. 2092 (1987). In Japanese.

103.383 **Rediscovery of short–period comet Gehrels 1 (1987v).**
Komet. Tsirk., No. 374 (1987). In Russian.

Periodic comet Helin

103.391 **Periodic comet Helin (1987w).**
IAU Circ., Nos. 4448, 4449 (1987).

103.392 **Periodic comet Helin (1987w).**
Yamamoto Circ., Nos. 2092, 2093 (1987). In Japanese.

103.393 **New short–period comet Helin (1987w).**
Komet. Tsirk., No. 375 (1987). In Russian.

Periodic comet Halley

103.401 **Temperatures and minimum thickness of the inactive surface layer of comet Halley.**
F. P. Fanale, J. R. Salvail.
NASA Tech. Memo., NASA TM–89810, p. 65 (1987). Abstract. – See Abstr. 003.001.

103.402 **Photographic observations of comet Halley (1982i).**
S. C. Joshi, J. B. Srivastava, C. D. Kandpal, T. D. Padalia, B. B. Sanwal, U. S. Chaubey, M. Singh.
Earth, Moon, Planets, Vol. 38, No. 3, p. 249 – 252 (1987).
Positions of comet Halley have been measured from the photographic plates taken at Cassegrain focus of the 104–cm telescope of Uttar Pradesh State Observatory.

103.403 **A cometary aurora.**
D. A. Mendis.
Earth, Moon, Planets, Vol. 39, No. 1, p. 17 – 20 (1987).
It is proposed that the cometary analog of a terrestrial aurora was responsible for the enhanced fluxes of suprathermal (keV) electrons and associated plasma waves observed in the cometosheath of comet Halley during its VEGA 2 encounter. The non–detection of such suprathermal electron fluxes during the GIOTTO encounter is ascribed to the much quieter solar wind conditions at that time.

103.404 **The chemical composition of the dust of comet P/Halley as measured by "PUMA" on board VEGA–1.**
F. R. Krueger, J. Kissel.
Naturwissenschaften, 74. Jahrg., Heft 7, p. 312 – 316 (1987).
The chemical composition, mass, and density distributions of the dust particles of comet P/Halley were measured by an impact mass spectrometer on board the spacecraft VEGA–1. The fluffy dust particles consist of an anorganic, as a mean chondritic, fraction and an organic one composed of highly unsaturated hydrocarbons, as analyzed by means of atomic and molecular ions produced on impact.

103.405 **Plasma transport near the magnetic cavity surrounding comet Halley.**
G. Haerendel.
Geophys. Res. Lett., Vol. 14, No. 7, p. 673 – 676 (1987).
The dominant forces resisting the transport of magnetic field into the inner coma of a comet are ion mass loading from and friction with the expanding neutral atmosphere. A magnetic cavity is thereby created. Close to it the frictional force is most important. Careful interpretation of the magnetic field profile measured during the Giotto flyby of comet P/Halley reveals the existence of an inward directed component of plasma flow of a few km/s, which drops to zero at the boundary of the cavity. The energy transferred from the neutral gas to the plasma by friction and mass loading is responsible for the strongly elevated ion temperatures outside the magnetic cavity. Fitting of the observed magnetic profile and ion temperature distribution yields quantitative determinations of some crucial parameters of the coma.

103.406 An estimate of the mass and density of the nucleus of comet Halley.
R. Z. Sagdeev, P. E. Ehl'yasberg, V. I. Moroz.
Pis'ma Astron. Zh., Tom 13, No. 7, p. 621 – 629 (1987). In Russian. English translation in Sov. Astron. Lett., Vol. 13.

The mass of the nucleus of comet Halley is evaluated from the lag between successive perihelion passages and "jet propulsion" effect created by the mass outflow from the nucleus. A mass of $\sim 3 \times 10^{11}$t was found for the nominal model of the outflow. This value combined with the volume defined by the Vega-1 and Vega-2 TV data lead to the average density of the nucleus 0.6 g cm^{-3}.

103.407 The use of comet Halley observations of the current appearance for providing space projects.
Yu. F. Kolyuka, S. M. Kudryavtsev, V. P. Tarasov, V. F. Tikhonov.
Pis'ma Astron. Zh., Tom 13, No. 7, p. 630 – 637 (1987). In Russian. English translation in Sov. Astron. Lett., Vol. 13.

The used observations of comet Halley in the current appearance for the improvement of the comet's orbit for purposes of the projects "Vega" and "Pathfinder" are described. The data on comet observations from different observatories and spacecraft "Vega" are analyzed. The improved orbit parameters are given.

103.408 Observation by Pioneer 7 of He$^+$ in the distant coma of Halley's comet.
J. D. Mihalov, H. R. Collard, D. S. Intriligator, A. Barnes.
Icarus, Vol. 71, No. 1, p. 192 – 197 (1987).

During the approach of Pioneer 7 to a distance of 12.1×10^6km from the nucleus of comet P/Halley on March 20, 1986, features were observed in plasma analyzer data that the authors interpret as He$^+$ produced by charge exchange of solar wind He^{++} with neutral cometary material. The maximum flux of the He$^+$ was observed several hours after the closest approach of the spacecraft of the nucleus, and is unexpectedly large. Remarkable large discontinuous flux changes were also observed.

103.409 One–micron CN observations of comet Halley.
G. S. Rossano, R. J. Rudy, R. W. Russell, R. C. Puetter, S. C. Chapman.
Publ. Astron. Soc. Pac., Vol. 99, No. 620, p. 1099 – 1101 (1987).

Three spectra of comet Halley in the 0.86 to 1.37–micron wavelength region have been obtained. These spectra were taken at similar geocentric and heliocentric distances but on opposite sides of perihelion, with data taken at 02:40 UT on 1985 December 4, 07:00 UT on 1986 April 28, and 05:30 UT on 1986 April 29.

103.410 A possible neck–line structure in the dust tail of Comet Halley.
M. Fulle.
Astron. Astrophys., Vol. 181, No. 1, p. L13 – L14 (1987).

The author proposes an interpretation of the anomalous dust structures observed in the dust tail of Comet Halley during April, May and June 1986. Sekanina et al. (1986) related these features to the rotation of the comet nucleus. The author considers the model of the neck–line structures (Pansecchi et al., 1987), which necessarily take place in every dust tail and which can be detected under proper geometrical circumstances (verified during the observations). This interpretation is able to explain the observations within the errors of measurement avoiding some strong hypotheses which are necessary in the Sekanina's interpretation.

103.411 Die Lichtkurve von P/Halley: Eine Auswertung aus Amateurbeobachtungen.
R. J. Bouma.
KPM, Jahrg. 2, No. 5, p. 30 – 37 (1987).

103.412 Das neue Bild vom Kern des Kometen Halley. Beobachtet mit der Halley–Mehrfarben–Kamera.
H. U. Keller.
Sterne Weltraum, 26. Jahrg., Nr. 9, p. 478 – 482 (1987).

103.413 La comète de Halley: vers un premier bilan des observations.
T. Encrenaz.
J. Astron. Fr., No. 30, p. 3 – 5 (1987).

103.414 First in situ observations of energetic particles near comet Halley.
A. J. Somogyi, K. I. Gringauz, K. Szegö, L. Szabó, G. Kozma, A. P. Remizov, J. Erö, I. N. Klimenko, I. T. Szücs, M. I. Verigin, J. Windberg, T. E. Cravens, A. V. D'yachkov, G. Erdös, M. Faragó, T. I. Gombosi, K. Kecskeméty, E. Keppler, A. Kondor, T. Kovács, Yu. I. Logachev, L. Lohonyai, R. Marsden, R. Redl, A. K. Richter, V. G. Stolpovskij, J. Szabó, I. Szentpétery, A. Szepesváry, M. Tátrallyay, A. Varga, G. A. Vladimirova, K. P. Wenzel, A. Zarándy.
Sov. Astron. Lett., Vol. 12, No. 5, p. 277 – 279 (1987). English translation of 42.103.622.

103.415 First in situ plasma and neutral–gas measurements near comet Halley: preliminary Vega results.
K. I. Gringauz, T. I. Gombosi, A. P. Remizov, I. Apáthy, T. Szemerey, L. I. Denshchikova, A. V. D'yachkov, E. Keppler, I. N. Klimenko, A. K. Richter, A. J. Somogyi, K. Szegö, S. Szendrö, M. Tátrallyay, A. Varga, M. I. Verigin, G. A. Vladimirova.
Sov. Astron. Lett., Vol. 12, No. 5, p. 279 – 282 (1987). English translation of 42.103.623.

103.416 Dual–frequency Vega radio sounding of comet Halley.
N. A. Savich, V. I. Altunin, V. E. Andreev, Yu. F. Basos, N.–E. E. Boguslavskaya, O. N. Doroshchuk, A. L. Gavrik, M. M. Kruchkovich, V. I. Lyaskovskaya, E. P. Molotov, I. M. Morskoj, A. S. Nabatov, M. A. Ovsyannikova, V. V. Seleznev, A. S. Sheverdyaev, K. G. Sukhanov, A. S. Vyshlov.
Sov. Astron. Lett., Vol. 12, No. 5, p. 283 – 286 (1987). English translation of 42.103.624.

103.417 Vega observations of electric fields and plasma in the comet Halley environment.
R. Grard, C. Beghin, M. Mogilevskij, V. Formisano, Yu. Mikhajlov, O. Molchanov, A. Pedersen, J.–G. Trotignon.
Sov. Astron. Lett., Vol. 12, No. 5, p. 286 – 288 (1987). English translation of 42.103.625.

103.418 ELF plasma waves near comet Halley.
S. I. Klimov, S. P. Savin, Ya. N. Aleksevich, G. Avanesova, V. M. Balebanov, M. A. Balikhin, A. A. Galeev, B. Gribov, M. N. Nozdrachev, V. N. Smirnov, A. Yu. Sokolov, O. L. Vajsberg, P. Oberc, Z. Krawczyk, S. Grzedzielski, J. Juchniewicz, K. Nowak, D. Orlowski, G. Parfianowicz, Z. Zbyszyński, Ja. Vojta, P. Triska.
Sov. Astron. Lett., Vol. 12, No. 5, p. 288 – 291 (1987). English translation of 42.103.626.

103.419 A model for the coma of comet Halley, based on the Astron ultraviolet spectrophotometry.
A. A. Boyarchuk, V. P. Grinin, P. P. Petrov, A. I. Shejkhet, A. M. Zvereva.
Sov. Astron. Lett., Vol. 12, No. 5, p. 291 – 296 (1987). English translation of 42.103.627.

103.420 Energy distribution for comet Halley.
M. Singh, J. P. Chaturvedi.
Earth, Moon, Planets, Vol. 39, No. 2, p. 197 – 201 (1987).

Thermal energy distribution of comet Halley have been obtained by using the data available from the Indian astronomical ephemeris. The magnitude vs wavelength plots have also been enumerated and discussed.

103.421   Preliminary analysis of the element abundance of comet Halley's dust component.
L. M. Mukhin, R. Z. Sagdeev, E. N. Evlanov, O. F. Prilutskij.
20th All–Union meteoritic conference, p. 155 (1987). In Russian.
Abstr. in Ref. Zh., 51. Astron., 7.51.213; 62. Issled. Kosm.
Prostranstva, 7.62.578 (1987). – See Abstr. 012.017.

103.422   Comet Halley: a carrier of interstellar dust chemical evolution.
J. M. Greenberg.
Adv. Space Res., Vol. 7, No. 5, p. 33 – 44 (1987). – See Abstr. 012.022.
The following key questions are discussed in this paper: are comets born as aggregated interstellar dust and what evidence do we have that comet nuclei maintain or do not maintain their initial composition and structure?

103.423   Wissenschaftler definieren das Bild des Kometen Halley.
M. J. Schmidt.
Orion, 45. Jahrg., Nr. 222, p. 163 – 173 (1987).

103.424   Physical processes near the cometopause according to in–situ plasma, magnetic field and wave measurements on board the "Vega-2" spacecraft.
M. I. Verigin, A. A. Galeev, R. Grard, K. I. Gringauz,
E. G. Eroshenko, S. I. Klimov, M. Eh. Mogilevskij,
A. P. Remizov, W. Riedler, R. Z. Sagdeev, S. P. Savin,
K. Szegö, A. Yu. Sokolov, M. Tátrallyay, K. Schwingenschuh.
Pis'ma Astron. Zh., Tom 13, No. 10, p. 907 – 916 (1987). In Russian. English translation in Sov. Astron. Lett., Vol. 13.
The physical processes near the cometopause are analyzed on the basis of simultaneous plasma, magnetic field and plasma wave data measured on board the "Vega-2" spacecraft. Large–scale magnetic field and plasma flow variations, indicating the fire–hose instability development, lead to the pitch angle scattering and deceleration of proton flow. As a result, the charge exchange becomes an efficient process for the increasing loading of the plasma flow. The enhanced plasma wave intensity in lower hybrid frequency range results in suprathermal electron acceleration and in the high–frequency oblique Langmuir wave excitation.

103.425   The history of Halley's comet.
D. W. Hughes.
Philos. Trans. R. Soc. London, Ser. A, Vol. 323, No. 1572, p. 349 – 367 (1987). – See Abstr. 012.026.
The history of Halley's comet can be approached in three ways. First one can start with the origin of the comet and follow it until one finally arrives at its present position as a typical middle–aged comet with a large associated meteoroid stream. Secondly one can retreat from four and a half thousand million years of history to a mere two thousand years and can start with the first known record made of Halley's comet by mankind. In the third approach one can wait until the comet actually received its name and then one can chart its highlights. This paper reviews all three approaches.

103.426   Earth–based observations of comet Halley dust and gas.
A. J. Meadows.
Philos. Trans. R. Soc. London, Ser. A, Vol. 323, No. 1572, p. 369 – 379 (1987). With 1 plate. – See Abstr. 012.026.
Very extensive ground–based observations of P/Halley have now been made, both to provide a standard cometary archive and to help interpret data from the Halley space probes. A number of new results are reported, of which the most important is probably the major role played by sporadic activity of the nucleus in the development of the comet.

103.427   Giotto observations of comet Halley dust.
J. A. M. McDonnell, J. C. Zarnecki, R. E. Olearczyk,
S. C. Chakaveh, G. S. A. Pankiewicz, S. T. Evans.
Philos. Trans. R. Soc. London, Ser. A, Vol. 323, No. 1572, p. 381 – 395 (1987). – See Abstr. 012.026.
Measurements of the dust environment of comet P/Halley by the dust experiments on the Giotto and Vega spacecraft are discussed, as well as some aspects of the dust modelling that was carried out before the spacecraft encounters with the comet. These data have highlighted certain shortcomings in the models. The mass–loss rate is determined and implications for the past history of the comet are discussed. It is also shown how the dust–impact data from the spacecraft experiments can be used to derive information on the distribution of emission over the surface of the nucleus.

103.428   Comet Halley: the gas composition derived from space missions.
T. Encrenaz.
Philos. Trans. R. Soc. London, Ser. A, Vol. 323, No. 1572, p. 397 – 404 (1987). – See Abstr. 012.026.
Important results have been obtained by the Vega and Giotto missions concerning comet Halley's gas composition. Water vapour and carbon dioxide have been identified. In addition, there is evidence for the presence of hydrocarbons and/or carbonaceous material in large amounts in the immediate vicinity of the nucleus.

103.429   The spectrum of comet P/Halley from 3.0 to 4.0 μm.
A. C. Danks, T. Encrenaz, P. Bouchet, T. Le Bertre,
A. Chalabaev.
Astron. Astrophys., Vol. 184, No. 1/2, p. 329 – 332 (1987).
Infrared observations of comet Halley have been obtained in the spectral range 3.0 – 4.0 μm, using a CVF spectrophotometer with a resolving power of 80. Observations carried out on 1986 March 28 show a number of emission features superimposed on a 300K blackbody continuum. The main feature is asymmetric and centered at 3.375 μm; another strong feature is seen at 3.5 μm. Observations made on 1985 Dec. 30 show no significant emission at 3.375 μm. The broad feature observed on March 28 is most likely due to the vibrational mode of the H–C bond in a large molecule. If the emission is due to gaseous hydrocarbon molecules, then their production rate is of the order of $10^{28} s^{-1}$. The 3.5 μm feature, and possibly the weaker 3.58 μm feature could be due to H_2CO, with a mixing ratio of a few percent with respect to H_2O.

103.430   Halley, premier bilan.
J. Manfroid.
Ciel, Vol. 49, p. 356 – 361 (1987).

103.431   Ion energetics in the inner coma of comet Halley.
T. E. Cravens.
Geophys. Res. Lett., Vol. 14, No. 10, p. 983 – 986 (1987).
The cometary plasma in the magnetic barrier just outside the diamagnetic cavity which surrounds the nucleus of comet Halley is virtually stagnant. The outflowing neutral gas exerts an outward ion–neutral drag force on this plasma, which balances the inward magnetic pressure gradient force in the vicinity of the contact surface. The cometary ions are frictionally heated due to the relative motion of the ion and neutral gases. The ion flow velocity must have a few km/s non–radial component in order to explain the ion temperatures measured by the ion mass spectrometer on Giotto.

103.432   Acceleration of cometary plasma in the vicinity of comet Halley associated with an interplanetary magnetic field polarity change.
M. I. Verigin, W. I. Axford, K. I. Gringauz, A. K. Richter.
Geophys. Res. Lett., Vol. 14, No. 10, p. 987 – 990 (1987).
Based on the ion plasma and magnetic field observations of Vega-1 near its closest approach to comet Halley a self–consistent scenario is developed according to which the observed magnetic field topology, the observed burst of ions at energies

200 – 600 eV, and the observed directional dependence of the flow of these ions leads to the conclusion that these burst–particles are cometary ions which have been accelerated by the process of merging of magnetic field lines of opposite polarity.

103.433 The variation of protons, alpha particles, and the magnetic field across the bow shock of comet Halley.
M. Neugebauer, F. M. Neubauer, H. Balsiger, S. A. Fuselier, B. E. Goldstein, R. Goldstein, F. Mariani, H. Rosenbauer, R. Schwenn, E. G. Shelley.
Geophys. Res. Lett., Vol. 14, No. 10, p. 995 – 998 (1987).
Data from the Ion Mass Spectrometer and the magnetometer on the Giotto spacecraft are used to examine the structure of the inbound crossing of the comet Halley bow shock on March 13, 1986.

103.434 Organic dust in comet Halley.
F. Hoyle, N. C. Wickramasinghe.
Nature, Vol. 328, No. 6126, p. 117 (1987).

103.435 Reply to "Organic dust in comet Halley" by F. Hoyle and N. C. Wickramasinghe.
J. Kissel, F. R. Krueger.
Nature, Vol. 328, No. 6126, p. 117 (1987).

103.436 Catalogue of astrometric observations of comet P/Halley at its apparition 1909 – 1911.
S. Röser.
Astron. Astrophys., Suppl. Ser., Vol. 71, No. 2, p. 363 – 395 (1987).
This catalogue contains a collection of astrometric observations of comet P/Halley at its apparition 1909 – 1911. More than 2300 visual astrometric observations from 41 observatories and the photographic observations from Greenwich have been re-reduced using the best presently available data on positions and proper motions of reference stars. A detailed publication of these data is chosen to enable an up–dating of the catalogue in the future, when positions and proper motions of reference stars will be improved. The tables are available upon request on magnetic tape.

103.437 Photographing comet Halley.
G. A. Avanesov, B. S. Novikov.
Nauka v SSSR, No. 3, p. 58 – 65 (1987). In Russian. Abstr. in Ref. Zh., 62. Issled. Kosm. Prostranstva, 9.62.614 (1987).

103.438 Near–nucleus optical observations of P/Halley.
S. M. Larson.
Cometary radio astronomy, p. 19 – 24 (1987). – See Abstr. 012.033.
The Near–Nucleus Studies Net of the International Halley Watch has obtained an extensive series of high resolution optical images of P/Halley during its most active phases in 1985 – 86 which may be useful in interpreting radio observations of comet Halley. They often show coma structure resulting from anisotropic emission of dust and gas from the inhomogeneous nucleus. Images were obtained in broadband spectral regions to study dust coma morphology, and in medium to narrow spectral bands to isolate the principal emissions of CN, C_3, C_2, CO^+ and H_2O^+. The goals and methods of near–nucleus studies are discussed and recent studies of 1910 images are briefly reviewed. The role of dust jets and cometary activity in P/Halley is discussed and several examples of anisotropic emission of dust during the current apparition are shown.

103.439 Radio observations of comet Halley at 22 and 43 GHz.
E. Scalise Jr., J. L. Monteiro do Vale, J. W. S. Vilas Boas, Z. Abraham, L. C. L. Botti, C. E. Tateyama, A. C. O. Cancoro, P. Kaufmann, L. F. Del Ciampo.
Cometary radio astronomy, p. 27 – 30 (1987). – See Abstr. 012.033.
Radio continuum and water molecular line observations of comet Halley were carried out at Itapetinga Radio Observatory,

and for the first time radio continuum emission from the comet was detected at 22 and 43 GHz. It was prominent when the comet was approaching the Earth, after perihelion passage. The signal was time variable. The observed flux densities can be explained as the thermal emission from the ice grains that form the halo, with a radius of 1000 km and grain temperature of 300K. No line emission was detected at 22 GHz.

103.440 A search for continuum radiation from comet P/Halley.
S. Hoban, S. Baum.
Cometary radio astronomy, p. 31 – 33 (1987). – See Abstr. 012.033.
The authors report briefly on observations of comet P/Halley on 1985 November 17 with the VLA. They found an upper limit to the 2–cm flux density of 1×10^{-4} Jy. This result is consistent with the detections of comet Halley by Altenhoff et al. (1986) at 1.3 and 3.5 mm if the millimeter emission comes from submillimeter sized icy grains or from many small refractory grains.

103.441 Scintillations of 4 radio sources occulted by the plasma tail of comet Halley.
S. Ananthakrishnan, P. K. Manoharan, V. R. Venugopal.
Cometary radio astronomy, p. 35 – 39 (1987). – See Abstr. 012.033.
During the passage of comet Kohoutek in 1974, Ananthakrishnan et al. (1975) had reported the successful detection of increased scintillation when the plasma tail of the comet occulted the compact radio source 2025-15. Similar observations have been recently reported by Alurkar et al. (1986) and Slee et al. (1986) on comet Halley. The authors present extensive series of carefully controlled measurements on comet Halley during Feb – April, 1986, where they show that there was no unambiguous increase in the scintillation by the plasma tail of comet Halley.

103.442 A continuum observation of P/Halley at 843 MHz.
J. D. Biggs, J. E. Reynolds.
Cometary radio astronomy, p. 41 – 43 (1987). – See Abstr. 012.033.
P/Halley was observed with the Molonglo Observatory Synthesis Telescope in order to determine its continuum emission at 843 MHz. In the absence of a significant detection, the authors present upper limits to emission from the nucleus and near–tail regions of P/Halley.

103.443 Observations of HCN in comet Halley.
F. P. Schloerb, W. M. Kinzel, D. A. Swade, W. M. Irvine.
Cometary radio astronomy, p. 65 – 73 (1987). – See Abstr. 012.033.
The authors discuss their observations of the HCN J = 1–0 rotational transition at 3.4 mm wavelength in comet P/Halley. The data were obtained during a total of 56 individual observing sessions between 18 November 1985 and 11 May 1986 and represent the first time that a cometary parent molecule has been so extensively monitored. The HCN production rate is well correlated with the total visual magnitude of the comet, indicating that HCN follows the overall gas production. There is also evidence of time variability and variations in the HCN hyperfine ratios from their LTE values. Spectra obtained by binning the HCN data with heliocentric distance show that the HCN line width, and thus the parent outflow velocity, increases with decreasing heliocentric distance.

103.444 Observations of J = 1–0 HCN in comet P/Halley at Onsala Space Observatory (OSO).
A. Winnberg, L. Ekelund, A. Ekelund.
Cometary radio astronomy, p. 75 (1987). Abstract. – See Abstr. 012.033.

103.445 A search for CO^+ in comet P/Halley at millimeter wavelengths.
S. Baum, S. Hoban.
Cometary radio astronomy, p. 77 (1987). Summary. – See Abstr. 012.033.

103.446 Search for molecules in comet Halley at millimeter wavelengths.
D. A. Swade, F. P. Schloerb, W. M. Irvine, W. M. Kinzel.
Cometary radio astronomy, p. 79 – 83 (1987). – See Abstr. 012.033.

An unsuccessfull search for the possible CN parent molecules HNC, HC_3N, and CH_3CN has been conducted in comet Halley at millimeter wavelengths with the Five College Radio Astronomy Observatory 13.7 m radio telescope. Upper limits to the beam averaged column densities and production rates of these molecules are calculated. The maximum relative abundances $HNC/HCN = 0.3$, $HC_3N/HCN = 0.4$, and $CH_3CN/HCN = 0.8$, indicate that these three molecules are not a major source of the CN radical observed in optical and ultraviolet spectroscopy. An upper limit for the beam averaged column density for the formyl ion, HCO^+, of $<1 \times 10^{11} cm^{-2}$ is also obtained. These limits provide important constraints for chemical models of the coma.

103.447 OH radio observations of comet Halley from the southern hemisphere.
I. F. Mirabel, V. Boriakoff, E. Bajaja, E. M. Arnal,
J. C. Cersosimo, F. R. Colomb, M. C. Martin, J. Mazzaro,
R. Morras, W. G. L. Pöppel, A. M. Silva.
Cometary radio astronomy, p. 101 – 107 (1987). – See Abstr. 012.033.

Post–perihelion observations of the 1667 MHz OH line in comet Halley are reported. Using a 30–m antenna of the Instituto Argentino de Radioastronomia, P/Halley was detected from March 22 to April 30, 1986. Strong absorption by the OH in the coma was observed during occultation of galactic background radio sources. The daily monitoring has revealed outbursts with variations in the content of gas of up to a factor of 2.7 in 24 – 28 hours. Three OH bursts lasting six days each were observed. The observations are consistent with discrete and anisotropic release of gas toward the coma, and with the estimated lifetimes of the H_2O and OH.

103.448 Comet Halley hydroxyl observations at 18 cm during the transits of the Galactic plane and Centaurus A.
M. J. Gaylard.
Cometary radio astronomy, p. 109 – 119 (1987). – See Abstr. 012.033.

The 1665 and 1667 MHz OH lines from comet Halley were seen in absorption during the transits of the Galactic plane and Centaurus A. Peak line flux densities of −13.4 Jy and −4.0 Jy were observed at 1667 MHz during the two transits. The mean line width was $2500 m s^{-1}$, but the line profile showed rapid variations in shape. The profiles are consistent with parent H_2O and daughter OH molecules in the coma having similar velocities of about $1000 m s^{-1}$, with small velocity dispersions of about $100 m s^{-1}$, and a clumpy distribution.

103.449 Monitoring the 1.667 GHz OH line in comet Halley.
J. Galt.
Cometary radio astronomy, p. 121 (1987). Abstract. Submitted to Astron. J. – See Abstr. 012.033.

103.450 OH observations at Onsala Space Observatory (OSO).
A. Winnberg, L. Ekelund, C. Andersson.
Cometary radio astronomy, p. 123 (1987). Abstract. – See Abstr. 012.033.

103.451 Radio OH observations of P/Halley with the NRAO 43–m telescope.
M. J. Claussen, F. P. Schloerb.
Cometary radio astronomy, p. 135 – 141 (1987). – See Abstr. 012.033.
The authors summarize the results of an OH–observing campaign at the NRAO 43–m telescope.

103.452 Models of the OH 18–cm line profiles of comet Halley.
L. Tacconi–Garman, F. P. Schloerb.
Cometary radio astronomy, p. 143 – 148 (1987). – See Abstr. 012.033.

In an attempt to better understand the kinematics of the cometary coma the authors have begun modeling the OH 18–cm line profiles observed during the recent apparition of comet Halley. There are obviously many physical properties which are responsible for producing the observed line profiles. Here the authors report the effect that the outflow of parents, specifically the parent outflow velocity, has on the mean velocity and the line width of the emergent profiles.

103.453 OH observations of comet P/Halley.
B. M. Lewis, J. M. Cordes, Y. Terzian,
A. Donati–Falchi, G. Tofani.
Cometary radio astronomy, p. 149 – 157 (1987). – See Abstr. 012.033.

The OH microwave lines were monitored in comet Halley, to look for temporal variations in flux, velocity and width. While flux variations are common, no significant variations in velocity, polarization or line–ratios occurred during the approach to perihelion. Some mapping information was also obtained, which on two occasions showed that the nuclear flux was less than that at off–axis points.

103.454 The first radio images of OH emission from comet Halley.
I. de Pater, P. Palmer, L. E. Snyder.
Cometary radio astronomy, p. 159 – 162 (1987). – See Abstr. 012.033.

The authors observed the 1667.359 MHz transition of OH in comet Halley on IAT days 1985 November 13 and 16 and 1986 January 4 and 6. The results are shown with some brief comments on the interpretation.

103.455 Evidence for chain molecules enriched in carbon, hydrogen, and oxygen in comet Halley.
D. L. Mitchell, R. P. Lin, K. A. Anderson, C. W. Carlson,
D. W. Curtis, A. Korth, H. Rème, J. A. Sauvaud, C. d'Uston,
D. A. Mendis.
Science, Vol. 237, No. 4815, p. 626 – 628 (1987).

In situ measurements of the composition and spatial distribution of heavy thermal positive ions in the coma of comet Halley were made. Above 50 atomic mass units an ordered series of mass peaks centered at 61, 75, 91, and 105 atomic mass units were observed. These observations suggest the presence of chain molecules that are enriched in carbon, oxygen, and hydrogen, such as polyoxymethylene (polymerized formaldehyde), in comet Halley.

103.456 First polymer in space identified in comet Halley.
W. F. Huebner.
Science, Vol. 237, No. 4815, p. 628 – 630 (1987).

The heavy–ion mass spectrum obtained in the inner coma of comet Halley with the PICCA instrument on the Giotto spacecraft has been examined. Short polymer chains of polyoxymethylene and their decay products are identified as the source for the spectrum with six mass peaks between about 45 and 120 atomic mass units. The properties of polyoxymethylene are consistent with many of the unexpected observations in the coma.

103.457 Seasonal brightness variations on Halley's comet: spring time for Halley.
P. R. Weissman.
Bull. Am. Astron. Soc., Vol. 19, No. 2, p. 741 – 742 (1987). Abstract. – See Abstr. 010.061.

103.458 Solar Maximum Mission (SMM) coronagraph/polarimeter (C/P) observations of comet Halley.
R. J. Oliversen, J. M. Hollis, N. B. Niedner.
Bull. Am. Astron. Soc., Vol. 19, No. 2, p. 750 – 751 (1987). Abstract. – See Abstr. 010.061.

103.459   Infrared investigation of water in Halley's comet.
H. A. Weaver, M. J. Mumma, H. P. Larson.
Space Telesc. Sci. Inst., Prepr. Ser., No. 179, 18 pp. (1987). To
appear in Astron. Astrophys.

103.460   Infrared monitoring of Halley's comet.
D. Lorenzetti, A. Moneti, R. Stanga, F. Strafella.
ESO Sci. Prepr., No. 522, 23 pp. (1987). Submitted to Astron.
Astrophys.

103.461   Polyoxymethylene in comet Halley.
W. F. Huebner, D. C. Boice, C. M. Sharp.
Astrophys. J., Lett. Ed., Vol. 320, No. 2, p. L149 – L152 (1987).
The RPA2–Positive Ion Cluster Composition Analyser on the
Giotto spacecraft detected five distinct groups of mass abundance peaks with regular spacing of about 15 amu and masses up
to about 105 amu. Starting with the first peak at about 45 amu,
these peaks decrease in intensity with increasing mass. Within
their half–width they are in good agreement with dissociation
products of formaldehyde polymer, also known as polyoxymethylene (POM). POM has been produced in the laboratory at low
temperatures by irradiating formaldehyde with bremsstrahlung
from 5 MeV electrons and gamma rays from ^{60}Co. It is suggested
that cosmic radiation formed POM on grains that were aggregated into comet nuclei or formed POM on the surface layers of
nuclei in the Oort cloud.

**103.462   Observations of large–scale phenomena during the 1985/
86 apparition of comet Halley at Yunnan Observatory.**
P.–z. Qiu, M.–c. Wu, Y. Zhang, C.–d. Wang, S.–c. Wang.
Publ. Yunnan Obs., No. 1, p. 96 – 101 (1987). In Chinese.

**103.463   Test observation of the 21–cm wavelength continuos
radiation of Halley's comet.**
Q.–z. Shang, S.–q. Lu, Y.–f. Gong, H.–m. Hu, S.–b. Shi,
J.–s. Wang.
Publ. Yunnan Obs., No. 2, p. 78 – 82 (1987). In Chinese.

**103.464   Observations of comet Halley (1982i) at Dodaira
Observatory.**
H. Shibasaki, Y. Iizuka, M. Noguchi.
Tokyo Astron. Bull., Second Ser., No. 280, p. 3251 – 3266
(1987).

103.465   The pick–up of cometary protons by the solar wind.
M. Neugebauer, A. J. Lazarus, K. Altwegg,
H. Balsiger, B. E. Goldstein, R. Goldstein, F. M. Neubauer,
H. Rosenbauer, R. Schwenn, E. G. Shelley, E. Ungstrup.
Astron. Astrophys., Vol. 187, No. 1/2, p. 21 – 24 (1987). – See
Abstr. 003.013, 43.103.414.
The HERS detector of the Ion Mass Spectrometer on the
Giotto spacecraft measured the 3–dimensional distribution of
picked–up cometary protons over a distance of ~8 million km
upstream of the bow shock of comet P/Halley. The protons were
observed to be elastically scattered out of their original cycloidal
trajectories such that they were nonuniformly distributed over a
spherical shell in velocity space. The shell radius (relative to its
expected radius) and thickness increased as the bow shock was
approached. Down–stream of the shock, the cometary protons
could not be distinguished from the heated solar wind protons.

**103.466   Alfvénic turbulence in the solar wind flow during the
approach to comet P/Halley.**
A. D. Johnstone, A. J. Coates, J. Heath, M. F. Thomsen,
B. Wilken, K. Jockers, V. Formisano, E. Amata,
J. D. Winningham, H. Borg, D. A. Bryant.
Astron. Astrophys., Vol. 187, No. 1/2, p. 25 – 32 (1987). – See
Abstr. 003.013, 43.103.415.
The Johnstone Plasma Analyser on Giotto operated almost
continuously for two days before the encounter with comet
Halley. Throughout this period it made observations of the
proton and alpha particle distributions in the solar wind with
8 seconds time resolution. As the comet was approached,
fluctuations were observed in all the primary bulk parameters, i.e.
density, temperature and flow velocity, of both directions at

levels above the usual solar wind turbulence. The authors present
a survey of the data from a distance of 5×10^6km from the
nucleus up to the cometary foreshock at 1.4×10^6km.

**103.467   General features of comet P/Halley: solar wind interac-
tion from plasma measurements.**
H. Rème, J. A. Sauvaud, C. d'Uston, A. Cros, K. A. Anderson,
C. W. Carlson, D. W. Curtis, R. P. Lin, A. Korth,
A. K. Richter, D. A. Mendis.
Astron. Astrophys., Vol. 187, No. 1/2, p. 33 – 38 (1987). – See
Abstr. 003.013, 43.103.416.
The Giotto RPA–COPERNIC plasma experiment identified
several regions in which the solar wind interaction with the
cometary plasma displayed characteristic features: (1) Beginning
~4.6×10^6km from the comet there is an upstream region with
sporadic connection to the comet; (2) An electron foreshock is
present up to 2.5×10^5km away from the bow shock; (3) A bow
shock is detected at 1.15×10^6km; (4) Between the bow shock and
the cometopause the outer regions can be devided into 3 parts;
(5) A cometopause is found at ~1.35×10^5km and a density
decrease is detected at ~4.5×10^4km from the comet. The
detailed plasma features associated with these regions are
successively described.

**103.468   Waves in the magnetic field and solar wind flow outside
the bow shock at comet P/Halley.**
A. Johnstone, K. Glassmeier, M. Acuna, H. Borg, D. Bryant,
A. Coates, V. Formisano, J. Heath, F. Mariani, G. Musmann,
F. Neubauer, M. Thomsen, B. Wilken, J. Winningham.
Astron. Astrophys., Vol. 187, No. 1/2, p. 47 – 54 (1987). – See
Abstr. 003.013, 43.103.453.
The existence of hydromagnetic waves in the mass–loaded
solar wind upstream from the bow shock of a comet is well–
established both for comet Giacobini–Zinner and for comet
Halley. Whereas previous reports have been concerned either
with the magnetic field observations or with plasma observations,
here the authors combine observations of the magnetic field with
the solar wind proton and alpha particle distributions. This
allows the three possible modes of propagation for these waves to
be separated. The magnetic component is predominantly transverse to the magnetic field and linearly polarised. The flow vector
also has a substantial amount of power parallel to the magnetic
field. An examination of the pressure variations shows that slow
magnetosonic waves are more common than the fast mode.

103.469   Solar wind flow through the comet P/Halley bow shock.
A. J. Coates, A. D. Johnstone, M. F. Thomsen,
V. Formisano, E. Amata, B. Wilken, K. Jockers,
J. D. Winningham, H. Borg, D. A. Bryant.
Astron. Astrophys., Vol. 187, No. 1/2, p. 55 – 60 (1987). – See
Abstr. 003.013, 43.103.451.
The bow shock of a comet is formed by the interaction of three
different particle populations; solar wind ions, cometary ions and
electrons. The authors follow the behaviour of the solar wind
protons through the Giotto inbound shock crossing at comet
P/Halley. A foreshock boundary is seen at ~1.4×10^6km where the
level of solar wind fluctuations increases substantially. The shock
itself is seen some 2.5×10^5km closer. It is a complex structure
with high–amplitude waves in the cometary ion foot of the shock
preceding a permanent drop in the solar wind speed. The width of
the shock structure is ~40,000 km. There is evidence for a further
excursion back into the cometary ion foot after the initial shock
crossing. The cometary ion density is inferred from the solar wind
speed changes and is compared with measurements.

**103.470   Macroscopic perturbations of the IMF by P/Halley as
seen by the Giotto magnetometer.**
J. Raeder, F. M. Neubauer, N. F. Ness, L. F. Burlaga.
Astron. Astrophys., Vol. 187, No. 1/2, p. 61 – 64 (1987). – See
Abstr. 003.013, 43.103.589.
Giotto magnetic field data have been used to analyze the
macroscopic field structure in the vicinity of comet P/Halley.
During the Giotto flyby at P/Halley the IMF showed quite a
stable "away" polarity. Draping of magnetic field lines is clearly
observed along the outbound leg of the trajectory. Inside the
magnetic pile–up region the field reverses its polarity several

times. A symmetry of oppositely magnetized sheets with respect to the nucleus is found and can be explained in terms of convected IMF features.

103.471 Low–frequency magnetic field fluctuations in comet P/Halley's magnetosheath: Giotto observations.

K. H. Glaßmeier, F. M. Neubauer, M. H. Acuña, F. Mariani.
Astron. Astrophys., Vol. 187, No. 1/2, p. 65 – 68 (1987). – See Abstr. 003.013, 43.103.588.

The interaction region between comets and the solar wind is characterized by large amplitude, low frequency magnetic field fluctuations, both within the upstream region as well as in the magnetosheath. Magnetosheath observations of the magnetic field experiment onboard Giotto indicate values of $\delta B/B \cong O(1)$. Power spectral peaks appear at frequencies of $\cong 10$ mHz with the spectrum exhibiting a power law dependence with an exponent of the order 2. Radial variation of the fluctuation level does not clearly increase with decreasing distance from the cometary nucleus as observed by the magnetometer onboard Vega–1 and as expected from quasilinear theory. The entrance into the cometary bow shock is furthermore characterized by an order of magnitude increase of the fluctuation level, both of the in– and outbound pass of Giotto.

103.472 Fine structure of the magnetic field in comet P/Halley's coma.

Y. G. Yeroshenko (E. G. Eroshenko), V. A. Styashkin (V. A. Styazhkin), W. Riedler, K. Schwingenschuh, C. T. Russell.
Astron. Astrophys., Vol. 187, No. 1/2, p. 69 – 72 (1987). – See Abstr. 003.013, 43.103.437.

During the encounter with comet P/Halley the magnetic field measurements aboard Vega–1 and Vega–2 have revealed short magnetic field depressions (MFD) at times of a generally rather smooth rise or fall of the magnetic field. A maximum MFD magnitude of 40 nT and a maximum duration of 50 s were observed at distances of $50 - 70 \times 10^3$km. Most likely these events are due to spatial plasma irregularities inside the coma region and not to temporal field variations. The observed effects may be caused by cometary plasma rays.

103.473 Giotto magnetic–field results on the boundaries of the pile–up region and the magnetic cavity.

F. M. Neubauer.
Astron. Astrophys., Vol. 187, No. 1/2, p. 73 – 79 (1987). – See Abstr. 003.013, 43.103.417.

The application of the minimum–variance–technique to the magnetic field measurements obtained during the Giotto encounter at comet P/Halley has led to the following results on the inner plasma boundaries in the magnetoplasma surrounding the comet: (1) Modelling the boundary of the magnetic pile–up region firstly by a half–sphere cylinder combination and secondly by a paraboloid with the axes of both surfaces aligned with the aberrated solar wind has led to stand–off distances towards the aberrated solar wind of 143,000 km and 215,000 km, respectively. (2) The ionopause, i.e. the outer boundary of the magnetic cavity region, had either a smooth surface of radius 4,600 km displaced from the nucleus by 1,500 km or, alternatively, it had no average displacement but strong ripples aligned with the aberrated solar wind direction and a wavelength of at least 200 km. Combinations are also possible.

103.474 Identification of boundaries in the cometary environment from ac electric field measurements.

M. Mogilevskij (M. Mogilevskij), Y. Mikhailov (Yu. Mikhajlov), O. Molchanov, R. Grard, A. Pedersen, J. G. Trotignon, C. Béghin, V. Formisano, V. Shapiro, V. Shevchenko.
Astron. Astrophys., Vol. 187, No. 1/2, p. 80 – 82 (1987). – See Abstr. 003.013, 43.103.628.

Electric fields are measured with the APV–V experiment in the frequency range 8 Hz – 300 kHz. The field amplitude increases significantly, first at a distance of 2×10^5km, then at distances of $1.2 - 1.5 \times 10^5$km and $5 - 7 \times 10^4$km from the nucleus. These phenomena have been observed both on Vega–1 and Vega–2. The electric field measurements are compared with data obtained

from dust and plasma experiments; possible mechanisms responsible for the existence of these boundaries are discussed.

103.475 Dust observations of comet P/Halley by the plasma–wave analyser.

J. G. Trotignon, C. Béghin, R. Grard, A. Pedersen, V. Formisano, M. Mogilevsky (M. Mogilevskij), Y. Mikhailov (Yu. Mikhajlov).
Astron. Astrophys., Vol. 187, No. 1/2, p. 83 – 88 (1987). – See Abstr. 003.013, 43.103.622.

This paper reports the electric field observations made with the high–frequency plasma–wave analyser, during the Vega spacecraft encounters with comet Halley on 1986, March 6 and 9. Electric fields are measured in the bandwidth 0 – 300 kHz with an antenna made of two spheres separated by a distance of 11 m. As the nucleus is approached, the average amplitude and bandwidth of the AC electric field increase dramatically; at frequencies above 1 kHz, the power law index of the electric field spectrum is about -3.4 ($E^2 \sim f^{-3.4}$), a value close to those reported elsewhere for random electric pulses induced by dust particles. Moreover, a strong correlation is found between spikes in the quasi–static electric field and small–scale variations in dust count rates for masses less than 10^{-12}g. Finally, the authors identify several large scale gradients in the AC electric field intensity within 140,000 km from the nucleus which are believed to be associated with enhancements of dust jet activity. The locations of these gradients are used to estimate the apparent rotation period of the cometary nucleus and the radial velocity of the small dust particles.

103.476 Comparative study of the low–frequency waves near comet P/Halley during the Vega–1 and Vega–2 flybys.

S. Savin, G. Avanesova, M. Balikhin, D. Wozniak, P. Wronowski, S. Klimov, Z. Krawczyk, M. Nozdrachev, D. Orlowski, A. Sokolov, J. Juchniewicz.
Astron. Astrophys., Vol. 187, No. 1/2, p. 89 – 93 (1987). – See Abstr. 003.013, 43.103.626.

Magnetosonic waves in the critical ionization velocity events in the inner coma of comet P/Halley are studied using the data of the low frequency wave experiments aboard Vega–1 and –2 spacecraft. Three boundaries downflow the bow shock are also found: "secondary wave" at the distance near 7×10^5km from the comet; outer boundary of the mantle; cometopause. Some common features and differences between these two flybys are discussed.

103.477 Plasma flow in the cometosheath of P/Halley during the encounter of Suisei.

S. Takahashi, T. Terasawa, T. Mukai, M. Kitayama, W. Miyake, K. Hirao.
Astron. Astrophys., Vol. 187, No. 1/2, p. 94 – 96 (1987). – See Abstr. 003.013.

The authors estimate proton and alpha particle bulk parameters within the cometosheath. They observed that the density of proton was almost constant (~ 200 cm^{-3}) except near the bow shock and that of alpha particle was almost constant (~ 7 cm^{-3}) in the cometosheath. Temperature of protons seems to decrease from bow shock to about 2×10^5km, and increase thereafter slightly, but that of alpha particle seems to be almost constant. This decrease is likely caused by the cooling effect accompanying with the charge exchange process as well as with the mass addition of cold pickup ions. On the contrary, the observed temperature of alpha particles stayed almost constant throughout the cometosheath. This is because there is no effective cooling process for alpha particles within the cometosheath.

103.478 Plasma–tail activity at the time of the Vega encounters.

M. B. Niedner Jr., K. Schwingenschuh.
Astron. Astrophys., Vol. 187, No. 1/2, p. 103 – 108 (1987). – See Abstr. 003.013, 43.103.624.

Physical associations are sought between plasma–tail activity seen in ground–based imagery and near–comet, interplanetary magnetic field (IMF) measurements obtained by the Vega spacecraft. Emphasis is placed on Disconnection Events (DE's) and on testing the sector boundary/frontside magnetic reconnection model (Niedner and Brandt, 1978) of their origin. Strong support for the model comes from the spectacular DE whose onset, on 1986 March 7 – 8, is strongly correlated with a reversal of the comet's magnetic barrier observed by Vega–1 and Vega–2 at the encounters on March 6.3 and 9.3 UT. Additional work is needed to determine the possible role of solar wind plasma effects in producing the DE. Relationships between other occurrences of plasma–tail activity and Halley Armada spacecraft measurements remain to be explored.

103.479 **Hydromagnetic waves associated with cometary water group ions: Sakigake observation.**
K. Yumoto, T. Saito, T. Nakagawa.
Astron. Astrophys., Vol. 187, No. 1/2, p. 117 – 120 (1987). – See Abstr. 003.013, 43.103.448.

Long–period hydromagnetic waves near the local "water group" ion cyclotron frequency were detected at the closest approach of Sakigake to comet P/Halley. These transverse waves having $\sim 2nT$ peak–to–peak amplitude propagated approximately parallel to the ambient field. The necessary cometary pickup–ion density of "water group" ions for excitation of these waves at 7×10^6 km upstream from the comet can be inferred to be of the order of 10^{-3} cm^{-3}.

103.480 **Plasma properties from the upstream region to the cometopause of comet P/Halley: Vega observations.**
M. I. Verigin, K. I. Gringauz, A. K. Richter, T. I. Gombosi, A. P. Remizov, K. Szegö, I. Apáthy, I. Szemerey, M. Tátrallyay, L. A. Lezhen.
Astron. Astrophys., Vol. 187, No. 1/2, p. 121 – 124 (1987). – See Abstr. 003.013, 43.103.433.

Based on the Plasmag–1 plasma measurements on board Vega–1 and –2, evidence is provided for the deceleration upstream, for the heating at and for the thermalization and deceleration behind the bow shock of comet Halley. In the cometosheath region two separate ion populations are observed: the first one consists of cometary ions being picked up in the vicinity of the point of observation; the energy of these ions coming from the solar direction decreases much faster than the energy of the solar wind ions. The second one consists of cometary ions being picked up by the solar wind far away from the point of observation. Considerable oscillations in the plasma flow direction occur in the cometosheath region.

103.481 **Observations of heavy energetic ions far upstream from comet P/Halley.**
T. R. Sanderson, K.–P. Wenzel, P. W. Daly, S. W. H. Cowley, R. J. Hynds, I. G. Richardson, E. J. Smith, S. J. Bame, R. D. Zwickl.
Astron. Astrophys., Vol. 187, No. 1/2, p. 125 – 128 (1987). – See Abstr. 003.013, 43.103.427.

On March 25, 1986, the ICE spacecraft came within 28×10^6 km of the nucleus of comet Halley. For several days around this time, bursts of heavy ions were observed by the ICE energetic ion experiment. These bursts were observed only during periods when the solar wind velocity was considerably higher than its nominal value. The authors examine the characteristics of these ions, in particular their anisotropies. Using the well–known formulae for transformation of distributions from the solar wind frame of reference to the spacecraft frame, they examine the angular distributions expected from either protons, or heavy ions from the water group, and show that the measurements are consistent with heavy ions, and not with protons. They discuss other sources of heavy ions and conclude that the most likely source of these ions is comet Halley.

103.482 **Spatial distribution of water–group ions near comet P/Halley observed by Suisei.**
T. Mukai, W. Miyake, T. Terasawa, M. Kitayama, K. Hirao.
Astron. Astrophys., Vol. 187, No. 1/2, p. 129 – 131 (1987). – See Abstr. 003.013, 43.103.422.

The spatial distribution of water–group ions observed by Suisei during the encounter with comet P/Halley is discussed in terms of the pickup model. The observed result is most consistently interpreted when we take the water production rate of 2×10^{30} s^{-1}. The model calculation shows the inbound–

outbound asymmetry in the density of cometary ions in good agreement with the observation. It is suggested that this asymmetry is mainly caused by the difference of orbital conditions between the inbound and the outbound observations in reference to the comet.

103.483 **An interpretation of the ion pile–up region outside the ionospheric contact surface.**
W.–H. Ip, R. Schwenn, H. Rosenbauer, H. Balsiger, M. Neugebauer, E. G. Shelley.
Astron. Astrophys., Vol. 187, No. 1/2, p. 132 – 136 (1987). – See Abstr. 003.013, 43.103.443.

An analysis interrelating some of the published results of plasma measurements by the Giotto spacecraft in the inner coma of comet P/Halley suggests that the formation of the plasma pile–up region at a distance of 10,000 km – 30,000 km may be the result of accumulation of ion density in a stagnant flow region with a sharp electron temperature transition forming a recombination front at about 10,000 km.

103.484 **Description of the main boundaries seen by the Giotto electron experiment inside comet P/Halley–solar wind interaction region.**
C. d'Uston, H. Rème, J. A. Sauvaud, A. Cros, K. A. Anderson, C. W. Carlson, D. Curtis, R. P. Lin, A. Korth, A. K. Richter, A. Mendis.
Astron. Astrophys., Vol. 187, No. 1/2, p. 137 – 140 (1987). – See Abstr. 003.013, 43.103.423.

The Giotto electron–plasma experiment identified several boundaries inside the comet Halley–solar wind region. Two of them are particularly interesting; they separate very different plasma regimes on quite sharp length scales and furthermore their existence was not foreseen in theoretical models. They are the limit between the transition region and the sheath detected at 550,000 km from the nucleus, and the cometopause detected at 135,000 km from the nucleus. Characteristic features of the electron distribution functions and of the macroscopic parameters of the electron plasma through these main boundaries are described.

103.485 **In–situ observations of a bi–modal ion distribution in the outer coma of comet P/Halley.**
M. F. Thomsen, W. C. Feldman, B. Wilken, K. Jockers, W. Stüdemann, A. D. Johnstone, A. Coates, V. Formisano, E. Amata, J. D. Winningham, H. Borg, D. Bryant, M. K. Wallis.
Astron. Astrophys., Vol. 187, No. 1/2, p. 141 – 148 (1987). – See Abstr. 003.013.

Observations obtained by the Johnstone Plasma Analyzer on the Giotto fly–by of comet Halley showed a fairly sudden decrease in the count rate of energetic (~ 30 keV) water–group ions inside about 5×10^5 km from the nucleus. This decrease was accompanied by the appearance of a new water–group ion population at slightly lower energies (< 10 keV). Close inspection reveals that this lower–energy peak was also present somewhat earlier in the post–shock flow but only became prominent near the sudden transition just described. It is shown that the observed bimodal ion distribution is well explained in terms of the velocity history of the accreting solar wind flow in the outer coma. The decline in count rate of the energetic pick–up distribution is due to a relatively sudden slowing of the bulk flow there and not to a loss of particles. Hence charge–exchange cooling of the flow is probably not important at these distances from the nucleus. Finally, the observations suggest that pitch–angle scattering is fairly efficient at least after the bow shock, but that energy diffusion is probably not very efficient.

103.486 **The composition and radial dependence of cometary ions in the coma of P/Halley.**
A. Korth, A. K. Richter, D. A. Mendis, K. A. Anderson, C. W. Carlson, D. W. Curtis, R. P. Lin, D. L. Mitchell, H. Rème, J. A. Sauvaud, C. d'Uston.
Astron. Astrophys., Vol. 187, No. 1/2, p. 149 – 152 (1987). – See Abstr. 003.013, 43.103.439.

The heavy ion analyzer, RPA2–PICCA, on board the Giotto spacecraft detected an increase in the densities of cometary ions within a cometocentric distance of 150,000 km. The composition of cometary ions changed dramatically as the comet's nucleus was approached, but it was clearly dominated by the water group. The second and third most abundant ions identified were associated with the CO– or S–group and the CO_2–group, respectively. Ions of larger atomic mass units were also present closer to the comet, and they possibly correspond to sulphur compounds and/or various hydrocarbons. Radial profiles of various groups of heavy ions and certain abundance ratios are presented. The peak density for all mass groups was detected at a cometocentric distance of 11,000 km. A distinct boundary, where the ion velocity and temperature dropped significantly, was identified at about 27,000 km.

103.487   Pick–up ions at comet P/Halley's bow shock: observations with the IIS spectrometer on Giotto.

B. Wilken, A. Johnstone, A. Coates, H. Borg, E. Amata,
V. Formisano, K. Jockers, H. Rosenbauer, W. Stüdemann,
M. F. Thomsen, J. D. Winningham.
Astron. Astrophys., Vol. 187, No. 1/2, p. 153 – 159 (1987). – See Abstr. 003.013.

Gaseous material expanding from the nucleus of comet Halley into space form the neutral coma around the comet. Ionisation in the solar UV radiation removes particles from the coma and injects them into the solar wind plasma. These freshly created ions are accelerated by the interplanetary electric field on cycloidal trajectories with gyrocenters moving with the speed of the magnetic field lines. In the solar wind frame of reference these particles move along the magnetic field lines with a fixed pitchangle. Pitchangle scattering and energy diffusion reduce quickly the initial energy anisotropy which is associated with the narrow pick–up structures. First observations of heavy cometary pick–up ions (water group ions) at the bowshock are presented. The evolution of the distribution function in the vicinity of the shock and radial density profiles are discussed.

103.488   Ion temperature and flow profiles in comet P/Halley's close environment.

R. Schwenn, W.–H. Ip, H. Rosenbauer, H. Balsiger, F. Bühler,
R. Goldstein, A. Meier, E. G. Shelley.
Astron. Astrophys., Vol. 187, No. 1/2, p. 160 – 162 (1987). – See Abstr. 003.013, 43.103.444.

The HIS instrument of the ion mass spectrometer on board the Giotto spacecraft identified the contact surface at 4800 km distance from the comet nucleus. This boundary is clearly seen by a drastic drop in the temperatures of different ion species from about 2000K outside to values as low as 300K inside. Also, inside the contact surface an outflow speed of $>1\,km\,s^{-1}$ was measured, in contrast to a value around zero right outside. The authors discuss how these numbers might be affected by a potential charge–up of the spacecraft. Outside the contact surface, the ion temperature rises gradually with increasing distance. Between 9000 and 10000 km distance the ion density increases by a factor of 4. This "ion pile–up" is not yet explained uniquely. Between 25000 and 28000 km distance there is again a rather abrupt jump to significantly higher temperatures and higher outflow speeds.

103.489   The composition and dynamics of cometary ions in the outer coma of comet P/Halley.

H. Balsiger, K. Altwegg, F. Bühler, S. A. Fuselier, J. Geiss,
B. E. Goldstein, R. Goldstein, W. T. Huntress, W.–H. Ip,
A. J. Lazarus, A. Meier, M. Neugebauer, U. Rettenmund,
H. Rosenbauer, R. Schwenn, E. G. Shelley, E. Ungstrup,
D. T. Young.
Astron. Astrophys., Vol. 187, No. 1/2, p. 163 – 168 (1987). – See Abstr. 003.013, 43.103.426.

During its flyby at comet Halley, the Giotto spacecraft encountered high densities of cometary ions inside approximately 200,000 km from the nucleus. Their properties changed drastically as the comet was approached. The authors present density profiles of solar wind alpha particles and of the major cometary ions as obtained by the IMS–HERS sensor between 340,000 km and 60,000 km from the nucleus. Typical mass spectra at various

distances are presented and angular and velocity distributions of the cometary ions are discussed.

103.490   Expansion velocity and temperatures of gas and ions measured in the coma of comet P/Halley.

P. Lämmerzahl, D. Krankowsky, R. R. Hodges,
U. Stubbemann, J. Woweries, I. Herrwerth, J. J. Berthelier,
J. M. Illiano, P. Eberhardt, U. Dolder, W. Schulte,
J. H. Hoffman.
Astron. Astrophys., Vol. 187, No. 1/2, p. 169 – 173 (1987). – See Abstr. 003.013, 43.103.435.

In situ measurements of flow velocities and temperature in the inner coma were obtained from ram energy spectra of molecules and ions observed by the NMS experiment onboard the Giotto probe to comet Halley. Radial flow speed was (800 ± 50) m s^{-1} at distances from the nucleus between 1000 km and 4000 km. Whereas ions became stagnant just outside the contact surface (located near 4500 km), the outflow of water vapour even accelerated and was above 1 km s^{-1} at 30000 km. Ion temperatures were around 200K inside the contact surface where the plasma is collisionally coupled to the gas; at this boundary ion temperature rises by about 1000K and then further increases to reach about 5000K at 15000 km. Gas temperature remains below 300 – 500K over this range of distances. A near zero volt spacecraft potential was inferred.

103.491   Giotto–IMS observations of ion–flow velocities and temperatures outside the magnetic cavity of comet P/Halley.

B. E. Goldstein, M. Neugebauer, H. Balsiger, J. Drake,
S. A. Fuselier, R. Goldstein, W.–H. Ip, U. Rettenmund,
H. Rosenbauer, R. Schwenn, E. G. Shelley.
Astron. Astrophys., Vol. 187, No. 1/2, p. 174 – 178 (1987). – See Abstr. 003.013, 43.103.445.

Fluid parameters for He^{2+} ions obtained from the Giotto Ion Mass Spectrometer HERS sensor are presented; also investigated are proton densities and velocities and thermal speeds of protons, alpha particles, and heavy ions in the hour before closest approach. A bow shock transition lasting ten minutes is observed. A region of enhanced He^{2+} ion densities and velocity, and decreased temperature is observed from 20:26 to 21:45. Sharp decreases in the proton density are observed at 23:30 and at 23:41; there are also sharp drops in alpha particle density and temperature at 23:30. There is a relative flow velocity between alpha particles and oxygen ions of approximately 20 km s^{-1} during a period roughly from 22:55 to 23:10; by 23:30 the difference in flow velocity is less than the experimental uncertainties.

103.492   Spatial distribution of heavy ions in comet P/Halley's coma.

O. L. Vaisberg (O. L. Vajsberg), G. Zastenker, V. Smirnov,
B. Khazanov, A. Omelchenko (A. N. Omel'chenko),
A. Fedorov, D. Zakharov.
Astron. Astrophys., Vol. 187, No. 1/2, p. 183 – 190 (1987). – See Abstr. 003.013, 43.103.436.

Radial profiles of heavy cometary ions at large solar zenith angles were obtained with the BD–3 plasma detector. The relatively sharp (thickness ~8000 km) boundary at ~1.5 × 10^5km of the region where cometary ions dominate, may result from a violation of the frozen–in condition of solar wind protons along the field lines. The density of the heavy ions mantle varies approximately as R^{-2}. A large–scale density enhancement was observed during the inbound crossing at 65,000 – 25,000 km on March 6, 1986. Many small–scale (250 – 1,000 km) density enhancements were observed. A local maximum of ~4 × 10^3cm^{-3} at ≈1.2 × 10^4km was observed on inbound and on outbound trajectories of Vega–1 which shows the permanent existence of this pile–up feature. Data on the bow shock crossing and on the mantle density distribution are discussed and compared with the data of other experiments.

103.493 Quasi–periodic features and the radial distribution of cometary ions in the cometary plasma region of comet P/Halley.
K. I. Gringauz, M. I. Verigin, A. K. Richter, T. I. Gombosi, K. Szegö, M. Tátrallyay, A. P. Remizov, I. Apáthy.
Astron. Astrophys., Vol. 187, No. 1/2, p. 191 – 194 (1987). – See Abstr. 003.013, 43.103.425.

Based on the measurements of the ion electrostatic analyzer CRA of the Plasmag–1 instrument on board Vega–2, which was oriented along the spacecraft–comet relative velocity direction, the authors present observations (1) of quasi–periodic features in the intensity of cometary ions occurring inside the cometopause, (2) of the mass composition of the cometary ions in the mass range of 1 to 100 atomic mass units, and (3) of the radial dependence of the density of certain, well–defined groups of cometary ions.

103.494 Possible models on disturbances of the plasma tail of comet Halley during the 1985–1986 apparition.
T. Saito, K. Saito, T. Aoki, K. Yumoto.
Astron. Astrophys., Vol. 187, No. 1/2, p. 201 – 208 (1987). – See Abstr. 003.013, 43.103.587.

More than 500 photographs of the plasma tail of comet P/Halley taken from the ground during this apparition are surveyed to study the solar wind–plasma tail interaction. The main disturbances of the plasma tail are tabulated, classifying them as outstanding streamers, rays, condensations helices, arcades, kinks, or disconnection events. Based on the photographs, a standard three–dimensional model of the cometary magnetosphere is proposed. The classified plasma tail disturbances are explained further by possible interaction models taking the IMF observation by Sakigake into account.

103.495 Structure and dynamics of the plasma tail of comet P/Halley. I. Knot event on December 31, 1985.
T. Saito, K. Yumoto, K. Hirao, S. Minami, K. Saito, E. Smith.
Astron. Astrophys., Vol. 187, No. 1/2, p. 209 – 214 (1987). – See Abstr. 003.013, 43.103.586.

Many photographs of an outstanding disconnection event (DE)–like knot in the plasma tail of comet Halley were taken by many Japanese astronomers on December 31, 1985. According to the solar wind magnetic field observed by Sakigake, comet Halley encountered not a sector boundary, but a high–speed stream on that day, demanding an explanation for the DE–like event, different from the sector boundary crossing model. The purpose of the present paper is to analyze 20 photographs taken by Japanese astronomers to examine the dynamic pressure model that was proposed in the authors' previous paper to explain the mechanism of the knot event appeared in the plasma tail of the comet in New Year Eve.

103.496 Structure and dynamics of the plasma tail of comet P/Halley. II. Kink event on January 10 – 11, 1986.
K. Tomita, T. Saito, S. Minami.
Astron. Astrophys., Vol. 187, No. 1/2, p. 215 – 219 (1987). – See Abstr. 003.013.

An outstanding DE–like kink event occurred in the plasma tail of comet P/Halley on January 10 – 11, 1986. The kink was traced for 33 h in these days by surveying 24 photographs of the comet. The images of the kink on eleven representative photographs out of the 24 were transcribed on the Palomar Sky Survey Atlas to investigate its motion and formation. In order to explain the kink event, the following two problems are discussed: (1) how the observed relation of the nucleus–kink distance versus time is explained and (2) how the kink is formed.

103.497 Hot ions observed by the Giotto ion mass spectrometer at the comet P/Halley contact surface.
R. Goldstein, D. T. Young, H. Balsiger, F. Buehler, B. E. Goldstein, M. Neugebauer, H. Rosenbauer, R. Schwenn, E. G. Shelley.
Astron. Astrophys., Vol. 187, No. 1/2, p. 220 – 224 (1987). – See Abstr. 003.013, 43.103.441.

Just inside the contact surface (~ 4700 km) the High Energy Range Spectrometer (HERS) sensor of the Giotto Ion Mass

Spectrometer (IMS) detected a sudden, intense burst of ions that lasted until the HERS sensor ceased transmitting data at a distance of ~ 3000 km from comet P/Halley. During this brief interval ions with $M/Q = 1, 2, 12, 14, 16 - 19, 24$, and 28 were observed. The heavier ions appear in two populations (in the S/C frame): (1) a very low energy, almost omnidirectional distribution, and (2) a more energetic ($\sim$ ram speed) population coming from the ram direction. In this paper the authors investigate the possible origin(s) of these two populations. In particular, they discuss whether the low energy ions belong to the natural Halley environment or are generated at the spacecraft by dust and gas bombardment.

103.498 Photographic observations of tail–formation activities of comet P/Halley in November 1985.
Z.–l. Liu.
Astron. Astrophys., Vol. 187, No. 1/2, p. 225 – 228 (1987). – See Abstr. 003.013, 43.103.432.

On November 12, 1985 the plasma tail of comet P/Halley made its first appearance during the latest apparition. Subsequently the comet displayed a great variety of tail activities. These included structures of different shapes in different bands even on the same night. The activities in the U and B bands usually preceded those in the V and R bands and were more prominent. According to the morphologies and lengths of the structures which appeared in the B and U bands they were plasma tails rather than dust jets. The deviation of the tail from the antisolar direction seemed to show a general tendency of decrease as the comet got closer and closer to the sun.

103.499 The outburst of comet P/Halley on December 12, 1985.
J. Watanabe, H. Kawakami, K. Tomita, H. Kinoshita, T. Nakamura, Y. Kozai.
Astron. Astrophys., Vol. 187, No. 1/2, p. 229 – 232 (1987). – See Abstr. 003.013, 43.103.600.

A strong outburst with a southward jet was found during 9^h07^m to 12^h29^m UT on December 12, 1985 while the authors were observing the near nucleus region of comet P/Halley. The projected expansion velocity is estimated to be 0.3 ± 0.1 km s^{-1} and the dust number density is 9 times as large as that in the coma background, assuming the jet cone angle is 30°. As the width of the jet in the C_2 emission band is 10% broader than that in white light, it can be concluded that the C_2 molecules are created from some parent body in the ejected matter.

103.500 Structure and dynamics of plasma–tail condensations of comet P/Halley 1986 and inferences on the structure and activity of the cometary nucleus.
W. E. Celnik, T. Schmidt–Kaler.
Astron. Astrophys., Vol. 187, No. 1/2, p. 233 – 248 (1987). – See Abstr. 003.013.

Comet P/Halley has been observed photographically by Bochum observers from 1986 Feb. 17, to Apr. 17. Coma, dust, and plasma tail were recorded with high resolution and wide–field cameras. First results on these matters were published by Celnik et al. (1986), Celnik (1986), and Schlosser et al. (1986). The present paper presents results from a deeper and more refined analysis of the plasma tail dynamics during the observation period. The "Bochum Halley Project" is described in more detail, the instrumentation and the full data base of the whole photographic campaign are summarized. The method of extracting useful data from the photographic images is described. The results concerning plasma tail, neutral coma, and cometary nucleus are presented and some physical conclusions are drawn.

103.501 Observations of the coma of comet P/Halley and the outburst of 1986 March 24 – 25 (UT).
T. W. Rettig, J. R. Kern, R. Ruchti, B. Baumbaugh, A. E. Baumbaugh, K. L. Knickerbocker, J. Dawe.
Astron. Astrophys., Vol. 187, No. 1/2, p. 249 – 255 (1987). – See Abstr. 003.013, 43.103.573.

Observations using a fast CCD and custom built Video Data Acquisition System at the Nasmyth focus of the 2.3 meter telescope at Siding Spring Observatory, Australia produced digitized spectra and images of comet Halley. On March 25 (UT), an overall brightening and anti–sunward jet was recorded continuously over several hours and is shown to be consistent with the Giotto measurements. Preliminary analysis of the spectral enhancement of March 24 is presented. The time resolution system allowed a spectral range from 3800 Å to 7000 Å to be analyzed every 1/2 s. Indications of unexplained rapid variations in several molecular emission band persist and are noted.

103.502 Observations of ions in comet P/Halley with a focal reducer.

K. Jockers, E. H. Geyer, H. Rosenbauer, A. Hänel.
Astron. Astrophys., Vol. 187, No. 1/2, p. 256 – 260 (1987). – See Abstr. 003.013, 43.103.421.

Comet Halley was observed with the ESO 1 m telescope and a dedicated focal reducer instrumentation. Narrow–band images of the comet in the light of the CO_2^+ ion at 367.4 nm were obtained for the first time. On these images, the cometary plasma can be observed as close as 10^4km to the nucleus. No plasma envelopes are visible. Multislit spectra show emissions of OH^+, CO_2^+, CO^+, N_2^+ and CH^+. They indicate the presence of CO^+ ions in a large halo around the comet, not only in the tail streamers.

103.503 Two disconnection events in comet P/Halley and possible solar causes.

H. Lundstedt, P. Magnusson.
Astron. Astrophys., Vol. 187, No. 1/2, p. 261 – 263 (1987). – See Abstr. 003.013, 43.103.431.

Disconnection events in comet P/Halley in April 1986 are discussed. Observations were made both from La Palma, Canary Islands and Siding Spring, Australia. Different solar phenomena, such as solar flares, coronal mass ejections, coronal holes and the coronal neutral line, are examined as possible causes of the disconnection events. The solar wind counterpart of the coronal holes and the coronal neutral lines, the so called high speed plasma streams and the heliospheric current sheet, are the most often suggested causes. In their two cases the authors found the heliospheric current sheet to be the most likely cause, but also a high speed plasma stream might have caused the DE of April 12.

103.504 Activity of the plasma tail of comet P/Halley in March, 1986.

M.–c. Wu, P.–z. Qiu.
Astron. Astrophys., Vol. 187, No. 1/2, p. 264 – 266 (1987). – See Abstr. 003.013, 43.103.581.

A photographic sequence of Comet Halley was taken at the Simao Observation Station of the Yunnan Observatory from March 4 to 22, recording the large–scale activity of the plasma tail. The various phenomena are described in the present article, and it is suggested from the analysis that at least two disconnection events occurred on March 8 – 9 and 21, respectively. The observed results are in accordance with those obtained by the observers in other countries. Finally, the process of the activity of the plasma tail and the correlation between the activity of the plasma tail and the solar activity are preliminary discussed.

103.505 The cause of two plasma–tail disconnection events in comet P/Halley during the ICE–Halley radial period.

J. W. Brosius, G. D. Holman, M. B. Niedner, J. C. Brandt, J. A. Slavin, E. J. Smith, R. D. Zwickl, S. J. Bame.
Astron. Astrophys., Vol. 187, No. 1/2, p. 267 – 275 (1987). – See Abstr. 003.013.

The authors have examined ICE magnetometer and electron plasma data for possible causes of the plasma tail disconnection events (DE's) which were observed in Halley's comet on 1986 March 20 – 22 and April 11 – 12. They attribute the DE of March 20 – 22 to an interplanetary magnetic field polarity reversal, and the DE of April 11 – 12 to either a compression region in the solar wind, an interplanetary magnetic field polarity reversal, or a combination of the two.

103.506 Analysis of the electron measurements from the Plasmag–1 experiment on board Vega 2 in the vicinity of comet P/Halley.

K. I. Gringauz, A. P. Remizov, M. I. Verigin, A. K. Richter, M. Tàtrallyay, K. Szegö, I. N. Klimenko, I. Apàthy, T. I. Gombosi, T. Szemerey.
Astron. Astrophys., Vol. 187, No. 1/2, p. 287 – 289 (1987). – See Abstr. 003.013, 43.103.438.

Measurements of electron spectra, as obtained by the Plasmag–1 experiment on board Vega 2 in the vicinity of comet Halley, are presented. It is shown that the temperature for thermal electrons gradually decreases when the comet is approached from about 4×10^5km behind the cometary bow shock to about 2×10^5km at the cometopause. In the region inside the cometopause a fast increase in the flux of energetic electrons of about 1 keV energy is observed. Various possibilities are discussed regarding the differences in the electron spectra measured by Vega 2 and Giotto, respectively.

103.507 The upstream region, foreshock and bow shock wave at comet P/Halley from plasma electron measurements.

K. A. Anderson, C. W. Carlson, D. W. Curtis, R. P. Lin, H. Rème, J. A. Sauvaud, C. d'Uston, A. Korth, A. K. Richter, D. A. Mendis.
Astron. Astrophys., Vol. 187, No. 1/2, p. 290 – 292 (1987). – See Abstr. 003.013, 43.103.450.

The authors survey the plasma electron parameters from a distance of 2.7×10^6km from the comet nucleus to the bow shock wave at about 1.1×10^6km and somewhat beyond. They describe the features of the electron foreshock lying outside the shock to a distance of about 230,000 km. It is a region of intense solar wind–comet plasma interaction in which energetic electrons are prominent. Several spikes of electrons whose energies extend to about 2.5 keV appear in front of the shock. These energetic electrons may have been accelerated in the same way electrons are accelerated at the Earth's bow shock to energies of 1 to 10 keV. The direction of the electron bulk flow direction changes abruptly between 1920 and 1922 UT, and the flow speed begins a sharp decline at the same time. The authors believe the spacecraft has entered the bow shock wave sometime between 1920 and 1922 UT. They found that the electron density variations at P/ Halley are very much smaller than those reported at Giacobini–Zinner.

103.508 Stochastic Fermi acceleration of ions in the pre–shock region of comet P/Halley.

B. E. Gribov, K. Kecskeméty, R. Z. Sagdeev, V. D. Shapiro, V. I. Shevchenko, A. J. Somogyi, K. Szegö, G. Erdös, E. G. Eroshenko, K. I. Gringauz, E. Keppler, R. G. Marsden, A. P. Remizov, A. K. Richter, W. Riedler, K. Schwingenschuh, K.–P. Wenzel.
Astron. Astrophys., Vol. 187, No. 1/2, p. 293 – 296 (1987). – See Abstr. 003.013, 43.103.452.

Energetic cometary ion fluxes measured between 2.5×10^6km and 10^6km from the cometary nucleus along the inbound trajectory of s/c VEGA–1 are used to derive the temperatures of the ion distributions in the solar wind frame. The increase of the temperature is modelled by the temperature change derived from a Fokker–Planck type equation with a source term and a stochastic acceleration term. The temperature increase predicted by theory is about 3 keV, higher than the observed one ($\cong 1.4$ keV). The difference may be due to the approximations applied. The second order Fermi mechanism is thus capable of producing the temperature increase observed.

103.509 Measurements of low energy electrons and spacecraft potentials near comet P/Halley.

A. Pedersen, R. Grard, J. G. Trotignon, C. Beghin, Y. Mikhailov (*Yu. Mikhajlov*), M. Mogilevsky (*M. Mogilevskij*).
Astron. Astrophys., Vol. 187, No. 1/2, p. 297 – 303 (1987). – See Abstr. 003.013, 43.103.625.

Data collected with the Langmuir probes and the electric field sensors on Vega–1 and Vega–2 show that the spacecraft potentials were a few volts positive in the outer coma and less positive near the closest approach. This leads to the conclusion that the energy of the secondary electrons generated by gas and dust impacts was lower than assumed in previous model work. Steep increases of the electron density were observed at a distance of 150,000 km from the nucleus, in the vicinity of the cometopause.

103.510   Charge exchange of solar wind ions in the coma of comet P/Halley.
E. G. Shelley, S. A. Fuselier, H. Balsiger, J. F. Drake, J. Geiss, B. E. Goldstein, R. Goldstein, W.–H. Ip, A. J. Lazarus, M. Neugebauer.
Astron. Astrophys., Vol. 187, No. 1/2, p. 304 – 306 (1987). – See Abstr. 003.013, 43.103.455.

In addition to solar wind pick–up of cometary ions, charge exchange between the solar wind ions and cometary neutrals is, potentially, an important comet–solar wind interaction closer to the nucleus. One monitor of the charge exchange rate is the relative abundances of solar wind He^{2+} and He^+. Radial profiles of these two ions measured by the Giotto IMS/HERS instrument on the inbound trajectory to comet Halley are presented. Results indicate that charge exchange may be an important solar wind loss process in the comet Halley coma at distances $\sim 10^5$km from the nucleus.

103.511   Far–ultraviolet objective spectra of comet P/Halley from sounding rockets.
C. B. Opal, R. P. McCoy, G. R. Carruthers.
Astron. Astrophys., Vol. 187, No. 1/2, p. 320 – 324 (1987). – See Abstr. 003.013, 43.103.477.

A sounding rocket payload to obtain far ultraviolet spectral images of comet P/Halley was launched on 1986 February 24 and again on 1986 March 13. The second flight occurred 13 hours prior to the Giotto spacecraft flyby of the comet. The payload included an objective grating spectrograph, consisting of an f/1.5 microchannel plate intensified electrographic Schmidt camera pointed at a plane grating. The instrument field of view was 12° with a spatial resolution of 1' and a point source wavelength resolution of 0.1 nm. High quality images of the oxygen coma at 130.4 nm, the carbon coma at 156.1 nm and 165.7 nm, and the sulphur coma at 181.4 nm were obtained. The images have been analyzed using a radial outflow model to derive production rates for these species.

103.512   IUE observations of comet P/Halley: evolution of the ultraviolet spectrum between September 1985 and July 1986.
P. D. Feldman, M. C. Festou, M. F. A'Hearn, C. Arpigny, P. S. Butterworth, C. B. Cosmovici, A. C. Danks, R. Gilmozzi, W. M. Jackson, L. A. McFadden, P. Patriarchi, D. G. Schleicher, G. P. Tozzi, M. K. Wallis, H. A. Weaver, T. N. Woods.
Astron. Astrophys., Vol. 187, No. 1/2, p. 325 – 328 (1987). – See Abstr. 003.013, 43.103.460.

A complete characterization of the UV spectrum of the comet was obtained which allows to derive coma abundances and to study the light emission mechanisms of the observed species. IUE observations at the time of the Giotto encounter provide a unique opportunity to compare the in situ measurements with remote observations of the principal coma species. The activity of the nucleus was highly variable, particularly at the end of December 1985 and during March and April 1986. The production rates of OH, CS and dust were derived for the entire period of the observations. The total water loss rate for this period is estimated to be 3×10^8 metric tons.

103.513   Some diatomic molecules from comet P/Halley's UV spectra near spacecraft flybys.
M. K. Wallis, K. S. Krishna Swamy.
Astron. Astrophys., Vol. 187, No. 1/2, p. 329 – 332 (1987). – See Abstr. 003.013, 43.103.558.

The abundances of diatomic molecules seen in the UV both give clues to the parent compounds and help unravel the gas and ion mass spectrometry. From IUE spectra of comet Halley, the authors find upper limits on SH and CS^+ column densities, and estimates of probable NO, S_2, and SO. In particular, judged from the 226 nm γ-band, NO was relatively abundant at $2 - 8 \times 10^{13}$cm^{-2} on March 9 – 14. The production rate of S_2 was around 10^{27} molecules s^{-1} at that time, but both showed day–to–day variability by 2 – 3 times.

103.514   Activity of comet P/Halley 23 – 25 March 1986: IUE observations.
L. A. McFadden, M. F. A'Hearn, P. D. Feldman, E. E. Roettger, D. M. Edsall, P. S. Butterworth.
Astron. Astrophys., Vol. 187, No. 1/2, p. 333 – 338 (1987). – See Abstr. 003.013.

A large but gradual increase in the brightness of the emission bands of comet Halley and thus in the coma abundance of OH, CS, CO_2^+ and dust was observed with the International Ultraviolet Explorer (IUE) satellite from 23 to 25 March 1986. This brightness change was also monitored by the Fine Error Sensor (FES) tracking and acquisition camera at a higher temporal resolution than that of the spectrophotometric measurements. The amplitude of the increase varied among the species depending on the lifetime of the parent molecule. By comparing the FES lightcurve with the optical lightcurve measured in the light of C_2 by Millis and Schleicher (1986) it can be shown that this increase in brightness corresponds to part of the periodic lightcurve of the comet. It is conclued that the activity reported here is controlled primarily by photodissociation and ionization of molecules ejected from an active area of the nucleus that rotates into the field of view of the sun approximately every 7.4 days.

103.515   Cometary MHD and chemistry.
R. Wegmann, H. U. Schmidt, W. F. Huebner, D. C. Boice.
Astron. Astrophys., Vol. 187, No. 1/2, p. 339 – 350 (1987). – See Abstr. 003.013, 43.103.684.

The authors' magnetohydrodynamical and chemical comet–coma model has been applied to describe and analyze the plasma flow, the magnetic field and the ion abundances in comet P/Halley in a consistent manner. The authors assume the volatile composition to consist of 80% water and 20% carbon–, nitrogen–, oxygen–, and sulfur–compounds. The radius of the nucleus is 3.36 km.

103.516   Atomic hydrogen production rates for comet P/Halley from observations with Dynamics Explorer 1.
J. D. Craven, L. A. Frank.
Astron. Astrophys., Vol. 187, No. 1/2, p. 351 – 356 (1987). – See Abstr. 003.013.

The distribution of atomic hydrogen surrounding comet Halley is observed in resonantly scattered solar Lyman–α radiation with the imaging photometer for vacuum–ultraviolet wavelengths on the Earth–orbiting spacecraft Dynamics Explorer 1. Measurements are made of the total Lyman–α flux at Earth due to the cometary neutral hydrogen distribution and the hydrogen production rate determined as a function of heliocentric distance, r. Corrections are made for the finite field–of–view of the photometer and for the ~ 700-R background at the spacecraft arising from the presence of geocoronal, interplanetary and galactic hydrogen.

103.517   The spectrum of P/Halley's coma obtained with an objective–prism.
A. Florsch, J. Marcout, G. Traversa.
Astron. Astrophys., Vol. 187, No. 1/2, p. 357 – 359 (1987). – See Abstr. 003.013, 43.103.567.

Spectra of comet P/Halley were taken with an objective prism at the Haute–Provence Observatory in November, December 1985 and January 1986. They lead to approximate values of the size of the coma for C_2, C_3 and CN and to relative surface brightness for these molecules.

103.518   Comet P/Halley neutral gas density profile along the Vega–1 trajectory measured by the Neutral Gas Experiment.

C. C. Curtis, C. Y. Fan, K. C. Hsieh, D. M. Hunten, W.–H. Ip, E. Keppler, A. K. Richter, G. Umlauft, V. V. Afonin, A. V. Dyachkov (*A. V. D'yachkov*), J. Erö Jr., A. J. Somogyi.

Astron. Astrophys., Vol. 187, No. 1/2, p. 360 – 362 (1987). – See Abstr. 003.013, 43.103.471.

Two complementary gas analyzers comprised the Nautral Gas Experiment on the two Vega spacecraft sent to comet Halley. Although no significant mass spectra were obtained, data returned from the Vega-1 experiment have permitted the determination of the total neutral gas density profile along the spacecraft trajectory during the comet encounter. Discounting small fluctuations, the FIS instrument measured a density profile which varied approximately as the inverse radial distance squared. Data from the EIS instrument yielded a series of calibration points; e.g. the neutral density at a radial distance of 10^5km was about 10^4cm^{-3}. The combined data provided a calibrated total density profile, and implied a neutral production rate of 10^{30}molecules/second (assuming an outflow velocity of 1 km s^{-1}) during the hours prior to the spacecraft encounter.

103.519   Low–resolution maps of comet P/Halley in principal atomic and molecular species.

M. R. Kidger, J. A. Acosta, F. Garzón, M. Prieto, R. Gómez.

Astron. Astrophys., Vol. 187, No. 1/2, p. 363 – 368 (1987). – See Abstr. 003.013, 43.103.621.

The authors present maps of a 3′ × 2′ region of the inner coma of comet P/Halley, at 10″ resolution, in CN (3883 Å); C_2 (5100 Å); C_3 (4040 Å); CN (4216 Å); NH_2 (6107 Å); [O I] (6300 Å) and CH (4322 Å). The maps were prepared from multi–position absolute spectrophotometry of comet Halley, with 3.6 Å resolution, in the wavelength range 3700 – 7598 Å, using the 2.5–m Isaac Newton Telescope in La Palma, Spain. This data represents a sub–set of a larger database of mapping at a range of heliocentric distances.

103.520   Pioneer Venus measurements of H, O, and C production in comet P/Halley near perihelion.

A. I. F. Stewart.

Astron. Astrophys., Vol. 187, No. 1/2, p. 369 – 374 (1987). – See Abstr. 003.013.

The Pioneer Venus Orbiter Ultraviolet Spectrometer measured the production of H, C, and O in comet Halley from late December 1985 to early March 1986. Water production rates obtained from the H data rose from 3.3×10^{29}s^{-1} at 1 AU inbound to 1.2×10^{30}s^{-1} at perihelion; there was a further increase to 1.6×10^{30}s^{-1} followed by a slow decline to 1.4×10^{30}s^{-1} at the time of the Vega-1 encounter. The H measurements show a periodic behavior that imply variations in the water production rate having structure within a 7.4–day sidereal period, in phase with similar variations seen in radical and dust production in March and April. Average H:O:C ratios of 1.0:0.7:0.07 were measured from 10 to 26 days after perihelion.

103.521   Anisotropy of the neutral gas distribution of comet P/Halley deduced from NGE/Vega 1 measurements.

K. C. Hsieh, C. C. Curtis, C. Y. Fan, D. M. Hunten, W.–H. Ip, E. Keppler, A. K. Richter, G. Umlauft, V. V. Afonin, J. Erö Jr., A. J. Somogyi.

Astron. Astrophys., Vol. 187, No. 1/2, p. 375 – 379 (1987). – See Abstr. 003.013, 43.103.476.

The density of neutral gas near comet Halley was measured by the Neutral Gas Experiment on the Vega 1 spacecraft. Plots of gas density vs. distance to the comet nucleus show an asymmetry between the inbound and the outbound legs of the spacecraft flyby. A multidimensional analysis is presented. It permits gas densities detected at distances between 8.9×10^3 and 7.2×10^4km to be mapped back to source regions on or near the comet nucleus, thus providing information on the spatio–temporal properties of the source of neutral gas in comet Halley. Sample results of the mapping–back process using an outflow velocity of

0.9 km s^{-1}, scale lengths of 2.9×10^5km and 7.0×10^4km, and the faster rotation period of 2.09 days are presented.

103.522   The atomic carbon distribution in the coma of comet P/Halley.

T. N. Woods, P. D. Feldman, K. F. Dymond.

Astron. Astrophys., Vol. 187, No. 1/2, p. 380 – 384 (1987). – See Abstr. 003.013, 43.103.478.

The radial distributions of CO, O I, C I, and C II emissions in the coma of comet Halley were measured by a long–slit far ultraviolet spectrograph aboard a sounding rocket on 26 February and 13 March 1986. While radial outflow models of CO can match the observed CO profiles at cometocentric distances $> 10^4$km, the observed carbon distribution is not consistent with the model, suggesting an additional source of atomic carbon in the inner coma. The in situ plasma measurements from the Vega and Giotto spacecraft suggest a possible additional source from the dissociative recombination of electron–impact ionized CO in the inner coma.

103.523   Carbon–isotope ratio in PUMA 1 spectra of P/Halley dust.

M. Šolc, V. Vanýsek, J. Kissel.

Astron. Astrophys., Vol. 187, No. 1/2, p. 385 – 387 (1987). – See Abstr. 003.013, 43.103.011.

Carbon isotopes ^{12}C and ^{13}C have yet been identified both in cometary gas and dust. Although the determination from optical C_2 and CN emissions from ground–based observations is not precise, the ratio ^{12}C/^{13}C is roughly the same as the mean ratio derived from PUMA 1 mass spectra of cometary grains. This value both in volatile and refractory material does not differ much from the nominal solar system value 90. However, the collection of 38 PUMA 1 mode 0 spectra shows surprisingly variable 12/13 amu lines ratio within 3 decades, increasing with carbon abundance in the grains. Possible instrumental effects are discussed.

103.524   Kinematic properties of the neutral gas outflow from comet P/Halley.

H. P. Larson, M. J. Mumma, H. A. Weaver.

Astron. Astrophys., Vol. 187, No. 1/2, p. 391 – 397 (1987). – See Abstr. 003.013, 43.103.462.

High resolution airborne infrared spectra of H_2O in comet Halley were used to characterize the velocity field in the neutral gas outflow from the nuclear surface. The observed H_2O emission appeared to be the partially isotropized sum of gas production from numerous active areas primarily on the sunlit side of the nucleus. The spatial distribution and expansion velocity of the H_2O molecules were deduced from the positions, width, and shape of the H_2O lines. These line profiles demonstrated that the H_2O distribution was anisotropic in the circular field–of–view of approximately 10^4km radius centered on the nucleus.

103.525   The spectrum of comet P/Halley between 0.9 and 2.5 microns.

J. P. Maillard, J. Crovisier, T. Encrenaz, M. Combes.

Astron. Astrophys., Vol. 187, No. 1/2, p. 398 – 404 (1987). – See Abstr. 003.013, 43.103.466.

High resolution Fourier–transform spectra of comet P/Halley have been recorded at the CFH 3.60 m telescope (Hawaii) on 1985 December 20 – 23, and 1986 April 28 – 30. The spectral interval ranges from 3800 to 12000 cm^{-1} (0.83 to 2.6 μm) and the spectral resolution is equal to 10 cm^{-1}. Dust temperatures of 300 ± 40K and 240 ± 40K are derived for 1985 December 20 – 23, and 1986 April 28 – 30, respectively, from the slope of the thermal component. Bands of the CN red system are observed at 1.46 μm (1–2), 1.10 μm (0–0), 0.94 μm (2–1), and 0.92 μm (1–0). In addition, an emission feature centered at 2.44 μm (4100 cm^{-1}) is detected. The authors tentatively identify this feature as the $v_1 + v_3 - 2v_2$ band of H_2O. Ground–based monitoring of this feature on future bright comets may be a direct method for measuring their H_2O production rate.

103.526 Anisotropic non–stationary gas flow dynamics in the coma of comet P/Halley.
N. I. Kömle, W.-H. Ip.
Astron. Astrophys., Vol. 187, No. 1/2, p. 405 – 410 (1987). – See Abstr. 003.013, 43.103.493.
One of the major results from the Giotto and Vega observations was that the gas outflow from the nucleus should be far from being spherically symmetric. The authors present results from numerical model calculations which address two particular outflow patterns that may occur at comet Halley: (1) steady state outflow dominated by a strong gas source at the subsolar point and (2) steady state gas flow caused by a ring–like outgassing region. In both cases it is found that the anisotropies introduced at the comet's surface are maintained out to large distances, i.e. several thousand kilometers. In addition to the steady state solutions the authors discuss the transient stages of the gas flow immediately after onset or offset of the strong outgassing at the subsolar point.

103.527 Infrared investigation of water in comet P/Halley.
H. A. Weaver, M. J. Mumma, H. P. Larson.
Astron. Astrophys., Vol. 187, No. 1/2, p. 411 – 418 (1987). – See Abstr. 003.013, 43.103.461.
The authors used the University of Arizona Fourier transform spectrometer (FTS; Davis et al., 1980) to obtain high resolution spectra of the (001–000) and (011–010) bands of H_2O near $\lambda \sim 2.65\ \mu m$. This paper summarizes the results of this investigation. The authors describe in detail how the observed H_2O excitation differs from their original expectations, and analyze the differences between the pre- and post–perihelion data. New discussion is presented on the observed short–term temporal variations, and further evidence is introduced which supports the reality of the previously reported large pre- to post–perihelion asymmetry in water production at $R \sim 1$ AU. Regarding the latter point, the authors have compared their findings to those from UV and radio observations of OH, and to Suisei and Pioneer Venus observations of H. Finally, they also comment on possible future directions for high resolution IR spectroscopy of comets.

103.528 The ortho–para ratio of water vapor in comet P/Halley.
M. J. Mumma, H. A. Weaver, H. P. Larson.
Astron. Astrophys., Vol. 187, No. 1/2, p. 419 – 424 (1987). – See Abstr. 003.013, 43.103.463.
The abundance ratio of ortho–H_2O to para–H_2O is shown to be an invariant in the cometary coma, its dependence on temperature in thermal equilibrium is given, and the nuclear-spin–temperature is defined. The relation between the physical temperature of the cometary ice and the nuclear-spin–temperature is discussed, and the prospects for using the observed ortho–para ratio to infer properties of the cometary nucleus are explored. The ortho–para ratio in Halley's comet, derived from high resolution infrared spectra of the ν_3–band near 2.7 μm wavelength, is given. On UT 1985 December 24.1, it was 2.73 ± 0.17, and on UT 1986 March 22.7/26.7, it was 3.23 ± 0.37. The nuclear-spin–temperature was 35K (+9K, –5K) pre–perihelion, and >40K post–perihelion, at the 67% confidence limit.

103.529 The 2.7 μm water band of comet P/Halley: interpretation of observations by an excitation model.
D. Bockelée–Morvan, J. Crovisier.
Astron. Astrophys., Vol. 187, No. 1/2, p. 425 – 430 (1987). – See Abstr. 003.013, 43.102.029.
In order to interpret infrared observations of parent molecules in comets, the authors have developed an excitation model which takes into account excitation by collisions, pumping of the fundamental bands of vibration by the infrared solar radiation field, and radiation trapping in the rotational and rovibrational lines. This model allows to derive infrared synthetic spectra for various species. It is applied to interpret recent infrared observations of water in comet P/Halley. The saturation of the 2.7 μm water band observed by the IKS instrument on Vega is evaluated. The relative water line intensities observed by the KAO in March 1986 are well fitted by the model, when a coma kinetic temperature of $\sim 60K$ is assumed in the $r \gtrsim 1000$ km region.

103.530 Curves of growth of emission lines in cometary spectra. Implications for H_2O and OH bands of comet P/Halley.
V. A. Krasnopolsky (*V. A. Krasnopol'skij*), A. Y. Tkachuk (*A. Yu. Tkachuk*).
Astron. Astrophys., Vol. 187, No. 1/2, p. 431 – 434 (1987). – See Abstr. 003.013, 43.103.484.
Curves of growth of emission lines are calculated for both parent and secondary species of comets assuming a Haser model distribution. Variations of these curves when varying input parameters is discussed. The technique developed is applied to the measurements of H_2O 2.7 and 1.38 μm bands by the Vega spacecraft and to OH 3090 Å band measurements by IUE. The evaluation of $10^{30}s^{-1}$ for the H_2O total production rate on 1986 March 9 is obtained by this technique.

103.531 The D/H ratio in water from comet P/Halley.
P. Eberhardt, U. Dolder, W. Schulte, D. Krankowsky, P. Lämmerzahl, J. H. Hoffman, R. R. Hodges, J. J. Berthelier, J. M. Illiano.
Astron. Astrophys., Vol. 187, No. 1/2, p. 435 – 437 (1987). – See Abstr. 003.013, 43.103.496.
The neutral gas mass spectrometer on the Giotto spacecraft made neutral and ion composition measurements with a high mass resolution. From a preliminary evaluation of the ion data within the contact surface the authors obtain for the water in Halley an ^{18}O abundance of $^{18}O/^{16}O = 0.0023 \pm 0.0006$ and for the D/H ratio $0.6 \times 10^{-4} \leqslant$ D/H $\leqslant 4.8 \times 10^{-4}$. The cometary ^{18}O abundance is within error limits identical to the terrestrial one. The D/H ratio in Halley is definitely not compatible with the D/H in gaseous hydrogen of the protosolar nebula or the Jovian and Saturnian atmospheres. It is in the range observed for hydrogen in solar system objects which acquired their hydrogen as part of volatile molecules, e.g. as ices.

103.532 Improved gas–kinetic treatment of cometary water sublimation and recondensation: application to comet P/Halley.
J. F. Crifo.
Astron. Astrophys., Vol. 187, No. 1/2, p. 438 – 450 (1987). – See Abstr. 003.013, 43.103.495.
Two effects associated with the solid–gas phase transition assumed to control cometary activity, i.e. nucleus ice sublimation, and homogeneous recondensation in the fluid expansion region are reconsidered using appropriate gasdynamic methods.

103.533 Detection of OH rotational emission from comet P/Halley in the far–infrared.
G. J. Stacey, J. B. Lugten, R. Genzel.
Astron. Astrophys., Vol. 187, No. 1/2, p. 451 – 454 (1987). – See Abstr. 003.013, 43.103.467.
The authors detected the $^2\Pi_{3/2}$ ($J = 5/2 \to 3/2$; $+ \to -$ parity) transition of OH in comet Halley at 119.44 μm and placed an upper limit of the line intensity of the $- \to +$ parity transition at 119.23 μm. The ratio of these lines is consistent with radiative pumping and inconsistent with photochemical pumping of the OH rotational levels. These far–infrared lines probe the inner regions of the comet coma where OH is produced through photodissociation of a parent (presumably H_2O) molecule. Hence, these lines complement the 18 cm radio measurements which are sensitive to the outer regions of the comet coma. The OH production rate on May 3, 1986 was $\sim 3.6 \times 10^{29}$ molecules s^{-1}.

103.534 18–cm wavelength radio monitoring of the OH radical in comet P/Halley 1982i.
E. Gérard, D. Bockelée–Morvan, G. Bourgois, P. Colom, J. Crovisier.
Astron. Astrophys., Vol. 187, No. 1/2, p. 455 – 461 (1987). – See Abstr. 003.013, 43.103.502.
The 1667 MHz and 1665 MHz transitions of the OH radical were detected and monitored nearly daily in comet Halley 1982i with the Nançay radio telescope. The observations cover the period July 1985 to July 1986 and provide a unique test for the excitation mechanisms of the OH molecule and the kinematic

models of the OH coma. The heliocentric velocity range was fully observed between −26.8 and 26.8 km s⁻¹ and 6 line reversals were covered. The OH–parent production rate of P/Halley was measured between 3.4 AU (pre–perihelion) and 2.8 AU from the Sun (post–perihelion). The authors find new effects like significant short–term variations of the OH–parent production rate in a few days, large variations of the half power profile width and anomalous line shapes.

103.535   10.7 GHz continuum observations of comet P/Halley.
A. Falchi, L. Gagliardi, F. Palagi, G. Tofani,
G. Comoretto.
Astron. Astrophys., Vol. 187, No. 1/2, p. 462 – 464 (1987). – See Abstr. 003.013, 43.103.577.

A search for continuum emission at 10.7 GHz from Halley's comet has been performed with the Bologna 32 m radio telescope near the first comet perigee. Owing to the expected low fluxes the beam switching ON–OFF technique was adopted for the observations. Sky background level was determined by subtraction of instrumental and sky long term drifts using a least square fit method. Long time integrations (> 3000 s) were obtained for 7 days. The results indicate 2 detections greater than 3σ level with fluxes of ∼14 mJy.

103.536   Rotational structure of the (2,0) Phillips band of C₂ in comet P/Halley.
I. Appenzeller, G. Münch.
Astron. Astrophys., Vol. 187, No. 1/2, p. 465 – 468 (1987). – See Abstr. 003.013.

The results of the study of a digital echellogram of the nuclear region of comet Halley, containing the (2,0) Phillips band of C_2 fully resolved, are presented. The intensities of individual lines, measured from the R–branch head up to $R36$, $Q30$, and $P24$, indicate a very high degree of rotational excitation. The population of the levels from where the lines arise, when approximated by a Boltzmann distribution, is found to be characterized by a temperature higher than that obtained from the Swan bands in other comets. An absolute calibration of the line fluxes has been obtained from narrow band CCD–imagery, to make feasible their comparison with measurements in the Swan system, probably carried out by other observers of comet Halley.

103.537   OH radio observations of comet P/Halley.
F. P. Schloerb, M. J. Claussen, L. Tacconi–Garman.
Astron. Astrophys., Vol. 187, No. 1/2, p. 469 – 474 (1987). – See Abstr. 003.013, 43.103.501.

The authors describe the first results of their group's efforts to monitor the 18–cm OH transitions from comet Halley with the NRAO 43 m telescope. Since this telescope is equipped with the world's most sensitive 18–cm receiver, the authors' effort was largely directed toward achieving several high signal–to–noise ratio spectra for the purpose of modeling the kinematics of OH in the coma. In this initial work, they emphasize both the long–term average behavior of the OH production rate, as monitored by the OH radio lines, and the OH kinematics derived from model fitting of the observed line profiles.

103.538   Observations of HCN in comet P/Halley.
F. P. Schloerb, W. M. Kinzel, D. A. Swade,
W. M. Irvine.
Astron. Astrophys., Vol. 187, No. 1/2, p. 475 – 480 (1987). – See Abstr. 003.013, 43.103.500.

The authors present observations of the HCN J = 1–0 rotational transition at 3.4 mm wavelength in comet P/Halley. The data were obtained during a total of 56 individual observing sessions between November 1985 and May 1986. The HCN production rate is well correlated with the total visual magnitude of the comet, and comparison of the HCN production to the total gas production of the comet indicates that it is a relatively minor constituent with 0.1% the abundance of H_2O. Comparison of HCN and CN production suggests that HCN is a major parent molecule of CN. HCN spectra obtained by binning the data with heliocentric distance show that the line width, and thus the parent

outflow velocity, increases with decreasing heliocentric distance, and that there is a tendency for the lines to be blue shifted due to anisotropic outgassing from the nucleus. Finally, there is evidence of day–to–day time variability in the total HCN emission and in the hyperfine ratios.

103.539   The CO and N₂ abundance in comet P/Halley.
P. Eberhardt, D. Krankowsky, W. Schulte,
U. Dolder, P. Lämmerzahl, J. J. Berthelier, J. Woweries,
U. Stubbemann, R. R. Hodges, J. H. Hoffman, J. M. Illiano.
Astron. Astrophys., Vol. 187, No. 1/2, p. 481 – 484 (1987). – See Abstr. 003.013, 43.103.469.

A preliminary evaluation of the mass 28 amu/e signal observed in the neutral mode of the Giotto neutral gas mass spectrometer (NMS) is presented. At 1000 km from the nucleus the authors obtain for the CO number density $n(CO)/n(H_2O) \leqslant 0.07$. The production rate of CO as parent molecule directly from the nucleus is thus less than 7% of the H_2O production rate. However, CO is also produced from an extended source in the inner coma ($R < 20,000$ km) and at 20,000 km from the nucleus the authors obtain for the total equivalent CO production rate $0.05 \leqslant Q(CO)/Q(H_2O) \leqslant 0.15$. For N_2 an upper limit $Q(N_2)/Q(H_2O) \leqslant 0.1$ is derived. No parent molecule for the CO could be identified which would be in agreement with the NMS measurements. It is proposed that CO or a very short–lived parent is released in the coma from cometary dust grains, such as the "CHON" particles.

103.540   A rotational–state population analysis of the high–resolution IUE observation of CS emission in comet P/Halley.
M. G. Prisant, W. M. Jackson.
Astron. Astrophys., Vol. 187, No. 1/2, p. 489 – 496 (1987). – See Abstr. 003.013.

The high resolution spectrum of comet P/Halley has been measured in the region of the CS emission. CS in comet Halley is produced by photodissociation of CS_2 and subsequently re–excited by broad band solar radiation. An inversion procedure is introduced to extract the rotational population distribution from the dispersed fluorescence spectrum. The fluorescence profile from the R branch of the $v' = 0$ to $v'' = 0$ is analyzed using this method. The extracted rotational distribution shows a peak at $J'' = 1$ superimposed on a background extending to $J'' = 29$. The distribution is interpreted by considering radiative and collisional relaxation processes in the comet. Intensity anomalies in the wavelength region corresponding to P and Q branch emission are considered.

103.541   Search for methane in comet P/Halley.
S. Drapatz, H. P. Larson, D. S. Davis.
Astron. Astrophys., Vol. 187, No. 1/2, p. 497 – 501 (1987). – See Abstr. 003.013, 43.103.464.

A search for gaseous neutral CH_4 has been conducted in the coma of comet Halley on March 20, 1986 using a Fourier Transform Spectrometer on the NASA Kuiper Airborne Observatory. In the field–of–view (41″ diam. corresponding to 24,000 km) the CH_4 rotational temperature is equal to the kinetic temperature of the coma molecules due to collisions in most of the region and is frozen in at the outer edge. The spectrum of the v_3 band is due to fluorescence with the R(3) to R(6) lines being most suitable for Earth–based observations. Measured upper (3σ) limits of ∼4×10^{-19} W cm⁻² for several lines lead to a (5σ) upper limit of 4×10^{28} CH₄ molecules per second. This value corresponds to ∼4% of the water vapor production rate at the time of observation.

103.542   Evidence for methane and ammonia in the coma of comet P/Halley.
M. Allen, M. Delitsky, W. Huntress, Y. Yung, W.–H. Ip,
R. Schwenn, H. Rosenbauer, E. Shelley, H. Balsiger, J. Geiss.
Astron. Astrophys., Vol. 187, No. 1/2, p. 502 – 512 (1987). – See Abstr. 003.013.

Methane and ammonia abundances in the coma of Halley are derived from Giotto IMS data using an Eulerian model of chemical and physical processes inside the contact surface to simulate Giotto HIS ion mass spectral data for mass–to–charge ratios (m/q) from 15 to 19. The ratio $(m/q) = 19/18$ as a function of distance from the nucleus is not reproduced by a model for a pure water coma. It is necessary to include the presence of NH_3, and uniquely NH_3, in coma gases in order to explain the data. A ratio of production rates $Q(NH_3)/Q(H_2O) = 0.01 - 0.02$ results in model values approximating the Giotto data. Methane is identified as the most probable source of the distinct peak at $m/q = 15$. The observations are fit best with $Q(CH_4)/Q(H_2O) = 0.02$. The chemical composition of the comet nucleus implied by these production rates is unlike that of the outer planets. On the other hand, there are also significant differences from observations of gas phase interstellar material.

103.543 Detection of parent molecules in comet P/Halley from the IKS–Vega experiment.

V. I. Moroz, M. Combes, J. P. Bibring, N. Coron, J. Crovisier, T. Encrenaz, J. F. Crifo, N. Sanko, A. V. Grigoryev (*A. V. Grigor'ev*), D. Bockelée–Morvan, R. Gispert, Y. V. Nikolsky (*Yu. V. Nikol'skij*), C. Emerich, J. M. Lamarre, F. Rocard, V. A. Krasnopolsky (*V. A. Krasnopol'skij*), T. Owen.
Astron. Astrophys., Vol. 187, No. 1/2, p. 513 – 518 (1987). – See Abstr. 003.013, 43.103.465.

Several parent molecules have been detected in comet P/Halley using the $2.5 - 5\,\mu m$ channel of the IKS experiment aboard Vega 1. The spectral resolving power was about 40 at 2.5 μm and 70 at 5 μm. The ν_3 band of H_2O has been observed at 2.7 μm, with an intensity corresponding to a production rate of about $10^{30}\,s^{-1}$. The ν_3 Band of CO_2 has been detected for the first time at 4.3 μm; the derived CO_2/H_2O ratio is 2×10^{-2}. In addition, there are emission features between 3.3 μm and 3.7 μm, also observed for the first time in a comet, which are attributed to CH–bearing molecules. Their mixing ratio relative to H_2O is on the order of a few percent. The ν_1 band of H_2CO might be present at 3.6 μm, but this interpretation is tentative. A weaker feature also appears at 2.8 μm; it could be tentatively identified as the OH (1–0) band. The CO/H_2O ratio is lower than 0.2.

103.544 Detection of a new emission band at 2.8 μm in comet P/Halley.

A. T. Tokunaga, T. Nagata, R. G. Smith.
Astron. Astrophys., Vol. 187, No. 1/2, p. 519 – 522 (1987). – See Abstr. 003.013.

A 2.8– to 3.7 μm spectrum of comet Halley taken in May 1986 is presented. The spectrum shows (1) the 3.36 μm emission feature that most likely arises from carbonaceous material, (2) a previously unknown emission band at 2.8 – 2.9 μm, and (3) no evidence for the 3 μm water–ice band either in reflection or absorption. The 2.8– to 2.9 μm emission feature does not arise from infrared fluorescence from water or OH. Since LTE emission from water is unlikely, the nature of this feature is uncertain.

103.545 Photometry of P/Halley (1982i).

C. Sterken, J. Manfroid, C. Arpigny.
Astron. Astrophys., Vol. 187, No. 1/2, p. 523 – 525 (1987). – See Abstr. 003.013, 43.103.481.

The authors report the results of photoelectric observations carried out during several weeks between 1985 September 12, and 1986 June 24. All measurements were made using the standard IAU/IHW cometary filters. Large amplitude night–to–night changes are superimposed on the general trend of slow brightness variation and support a $7\overset{d}{.}4$ period. The changes are seen as smooth variations with a range up to 0.2 mag on time scales of several hours.

103.546 Circular polarization near the nucleus of comet P/Halley.

K. Metz, R. Haefner.
Astron. Astrophys., Vol. 187, No. 1/2, p. 539 – 542 (1987). – See Abstr. 003.013.

Circular polarization measurements at different positions near the nucleus of comet P/Halley have been carried out for phase angles normally producing a high linear polarization. The observed circular polarization degree up to 2% was highly variable but never changed its sign.

103.547 Spectrophotometry of comet P/Halley. I. Flux, column density and emission gradients within the coma in the emission bands and the continuum.

K. R. Sivaraman, G. S. D. Babu, B. S. Shylaja, R. Rajamohan.
Astron. Astrophys., Vol. 187, No. 1/2, p. 543 – 550 (1987). – See Abstr. 003.013, 43.103.487.

The authors made spectral scans of comet P/Halley using the scanner at the Cassegrain focus (f/13) of the 102 cm reflector at Kavalur. The scans cover generally the wavelength region from 3900 to 6200 Å and are at 40 Å resolution and on a few occasions in the blue region at 20 Å resolution. The f/13 beam provided an image of the coma with a scale of $15\overset{''}{.}5$ mm^{-1}. The authors have obtained scans on several consecutive nights in March and April, 1986 with a $25\overset{''}{.}9$ entrance aperture of the scanner at discrete and systematically displaced locations within the coma in the anti-sunward direction as well as normal to it about the position of the nucleus. They have derived the brightness profiles of the neutrals and dust within the coma and have discussed the variations of the spatial gradients of the profiles.

103.548 Spectrophotometry of comet P/Halley at wavelengths 275 – 710 nm from Vega 2.

G. Moreels, J. Clairemidi, J. P. Parisot, J. M. Zucconi, J. L. Bertaux, J. E. Blamont, M. Hersé, V. A. Krasnopolsky (*V. A. Krasnopol'skij*), V. I. Moroz, A. A. Krys'ko, A. Yu. Tkachuk, M. Gogoshev, T. Gogosheva, R. Werner, S. Spasov.
Astron. Astrophys., Vol. 187, No. 1/2, p. 551 – 559 (1987). – See Abstr. 003.013, 43.103.482.

Spectrophotometric observations of the inner coma of comet Halley in the range 275 – 710 nm were conducted from Vega 2 on March 8, 9, 10, 11, 1986. During the encounter session, when the impact parameter was minimum, p = 120 km, the gaseous emissions were still present, superimposed on an intense continuum. In the inner coma, C_2 and C_3 emissions steadily increase when the impact parameter decreases. The emissions of OH at 309 nm and CN in its band head at 388 nm show a plateau between 1500 and 600 km. The decrease in OH emission when the impact parameter is less than 500 km is interpreted in terms of high optical thickness of water vapor at Lyman–alpha. Pixel–to–pixel ratios of spectra show that the color of dust continuum presents small fluctuations which might be produced by a discontinuous emission of icy grains.

103.549 The visual brightness behavior of P/Halley during 1981 – 1987.

D. W. E. Green, C. S. Morris.
Astron. Astrophys., Vol. 187, No. 1/2, p. 560 – 568 (1987). – See Abstr. 003.013, 43.103.506.

The overall behavior of the total visual coma brightness of P/Halley during 1981 – 1987 is discussed. Observations used in this study include ~ 75 published V magnitudes obtained with large telescopes (usually with CCDs) prior to July 1985 and ~ 3500 total magnitudes estimated visually with smaller telescopes during the period July 1985 – February 1987. The authors discuss the observed macroscopic light curve and provide preliminary least–square fits to the traditional power–law formula. While the heliocentric brightness variation with respect to heliocentric distance did not follow expectations closely, P/Halley's maximum brightness during 1986 Feb. – Apr. fell somewhere between the most pessimistic and the most optimistic forecasts. The authors also review brightness variations on smaller timescales (hours, days), noting that several definite outbursts in total visual magnitude occurred during Oct. 1985 – May 1986 which can be correlated with outbursts observed at other wavelengths.

103.550 The spectral behavior of P/Halley at large heliocentric distance in light of the Giotto/Vega results.

M. J. S. Belton, H. Spinrad, P. A. Wehinger, S. Wyckoff, D. K. Yeomans.
Astron. Astrophys., Vol. 187, No. 1/2, p. 569 – 574 (1987). – See Abstr. 003.013.

The authors describe preperihelion spectroscopic observations of P/Halley while it was beyond 5.6 AU, the reduction procedures they followed, and derive V magnitudes and $(g-r)$ and

(V–R) colour indices. They outline two possible interpretations of these observations. The first, which ascribes part of the light in the spectra to reflection from the nucleus and part to reflection from grains in the developing coma, depends on an earlier interpretation of a 54–h periodicity in the comet's brightness in terms of nuclear rotation. Uncertainties in this interpretation due to the discovery of a second photometric periodicity near 7.4 days and evidence of sporadic activity at large heliocentric distances lead the authors to consider a second possible interpretation in which coma particulates maybe entirely responsible for the spectrophotometric behavior.

103.551 Periodicities in the light curve of P/Halley and the rotation of its nucleus.

M. C. Festou, P. Drossart, J. Lecacheux, T. Encrenaz, F. Puel, J. L. Kohl–Moreira.
Astron. Astrophys., Vol. 187, No. 1/2, p. 575 – 580 (1987). – See Abstr. 003.013.

Early photometric measurements of comet P/Halley exhibit a photometric periodicity of 7.307 ± 0.005 days. Brightness variations occur together with the periodic release of solid particles during bursts of activity or activity increases of the nucleus. The mechanical behavior of the comet nucleus can be described as that of an axisymmetric ellipsoid. In the torque free approximation it is found that the nucleus rotates about its long axis in 14.614 days. A free precession about an axis tilted by 84° to the rotation axis occurs in 2.088 days. Due to the peculiar geometry, the true photometric period is 14.6 days with two halves nearly identical, thus mimicking a periodicity of 7.3 days. The permanence of the rotation during at least half the apparition of the comet indicates that no internal damping of the nucleus rotation exists and that the nucleus must then be rigid.

103.552 Photometry of comet P/Halley at near post–perihelion phases.

T. Neckel, G. Münch.
Astron. Astrophys., Vol. 187, No. 1/2, p. 581 – 584 (1987). – See Abstr. 003.013.

Photometric measurements of Halley's comet at phases around its recent perigee are presented. Narrow band filters isolating the C_2 Swan band emission and a neighboring continuum, besides the standard UBV, have been used with 4 nucleus centered field stops subtending angles between 41″ and 1050″. The brightness fluctuations detected, especially conspicuous in C_2 and the nearby continuum, indicate a period of 7.3 days. Some evidence for the double peaked nature of the periodic variation is marginally present.

103.553 Chinese observations of comet P/Halley in China and abroad.

S. M. Gong, G. J. Wu, P. S. Chen, X. F. Zhang, S. S. Sun.
Astron. Astrophys., Vol. 187, No. 1/2, p. 594 – 600 (1987). – See Abstr. 003.013, 43.103.598.

A brief account of Chinese observations of Halley's comet in China and abroad together with some preliminary results is given.

103.554 Thermal infrared imaging of comet P/Halley.

H. Campins, C. M. Telesco, R. Decher, B. D. Ramsey.
Astron. Astrophys., Vol. 187, No. 1/2, p. 601 – 604 (1987). – See Abstr. 003.013, 43.103.517.

Thermal infrared images ($\lambda = 10.8$, 12.8 and 19.2 μm) of comet P/Halley were obtained on 1986 March 28.7, 29.7 and 30.7 UT using the NASA–MSFC 20–pixel bolometer array at the NASA Infrared Telescope Facility. A preliminary analysis of the observations shows the following results: (1) the presence of spatial structure indicative of dust ejection occurring mainly on the sunward side of the nucleus; (2) a dust coma deficient in large particles (radius $\gtrsim 100$ μm) compared with that of comet P/Giacobini–Zinner; (3) temporal variability which is consistent with an extrapolation of a quasi–sinusoidal lightcurve observed by IUE on 1986 March 23, 24, and 25; (4) a maximum, at the

nuclear condensation, of the color temperature and the strength of the silicate emission feature.

103.555 Low resolution mapping of comet P/Halley in the near–infrared.

C. Lázaro, F. Garzón, M. J. Arévalo.
Astron. Astrophys., Vol. 187, No. 1/2, p. 605 – 608 (1987). – See Abstr. 003.013, 43.103.620.

The authors present low resolution maps, of a $4' \times 0\rlap{.}'5$ region, of comet Halley, in photometric bands J and K, a central scan in the H band and $JHKL$ photometry taken with the 1.5 m Sánchez–Magro Telescope of the Instituto de Astrofisica de Canarias. They obtain central and integrated magnitudes and colors and discuss its variation with distance from the nucleus. The black–body color temperature of the grains is found to be ~ 545K, significantly greater than the black–body equilibrium temperature (228K). The profile of the scans seem to follow an r^{-2} density law in the dust distribution, although the central core has a much steeper density law.

103.556 Infrared monitoring of comet P/Halley.

D. Lorenzetti, A. Moneti, R. Stanga, F. Strafella.
Astron. Astrophys., Vol. 187, No. 1/2, p. 609 – 615 (1987). – See Abstr. 003.013, 43.103.619.

The authors present the results of an infrared monitoring program of comet P/Halley. An analysis of the JHK data indicates that Halley was typically somewhat redder than the Sun ($J–H = 0.48$, $H–K = 0.18$) and that its color did not change throughout the monitoring period, implying that the properties of the dust in the coma remained constant. Also, with the exception of November 4 and March 21, when Halley was presumably undergoing an outburst, the multiaperture data are consistent with an r^{-2} density distribution of scattering dust in the coma. Furthermore, the flux from Halley in its quiescent phase increased as $R^{-4.5}$, where R is the heliocentric distance. The 10 μm silicate feature in emission was detected at about the 2σ level on 4 and 7 November, and from the spectral energy distribution at that time an albedo of 0.05 for the scattering dust in the coma is deduced.

103.557 Airborne and groundbased spectrophotometry of comet P/Halley from 5 – 13 micrometers.

J. D. Bregman, H. Campins, F. C. Witteborn, D. H. Wooden, D. M. Rank, L. J. Allamandola, M. Cohen, A. G. G. M. Tielens.
Astron. Astrophys., Vol. 187, No. 1/2, p. 616 – 620 (1987). – See Abstr. 003.013, 43.103.523.

Spectrophotometry from 5 – 10 μm ($\Delta\lambda/\lambda \approx 0.02$) of comet Halley was obtained from the Kuiper Airborne Observatory on 1985 December 12.1 and 1986 April 8.6 and 10.5 UT. 8 – 13 μm data were obtained on 17.2 December 1985 from the Nickel Telescope at Lick Observatory. The spectra show a strong broad emission band at 10 μm and a weak feature at 6.8 μm. The authors do not confirm the strong 7.5 μm emission feature observed by the Vega 1 spacecraft. Color temperatures significantly higher than the equilibrium blackbody temperature indicate that small particles are abundant in the coma. Significant spatial and temporal variations in the spectrum have been observed and show trends similar to those observed by the spacecraft and from the ground. Temporal variability of the silicate emission relative to the 5 – 8 μm continuum suggests that there are at least two physically separate components of the dust.

103.558 The near–infrared polarization and color of comet P/Halley.

T. Y. Brooke, R. F. Knacke, R. R. Joyce.
Astron. Astrophys., Vol. 187, No. 1/2, p. 621 – 624 (1987). – See Abstr. 003.013, 43.103.516.

The linear polarization and color of comet P/Halley in broadband filters $J(1.25$ μm), $H(1.65$ μm), and $K(2.2$ μm) were measured at phase angles $9\rlap{.}°4 \leqslant \beta \leqslant 65\rlap{.}°3$. Comet Halley exhibited a red color characteristic of comet dust at these wavelengths with no clear variation with β. The JHK polarizations reach 25% at the largest phase angles with a reversal at $\beta = 20°$ and negative polarizations of a few per cent at smaller β. The polarization increases with wavelength at the larger phase angles. Mie theory calculations of the polarization from a grain size distribution based on measurements by the Giotto spacecraft indicate that no single component fits the data; a two–component model is needed. The two components that give the best fit correspond loosely to a "dirty" silicate and a moderately absorbing material. The model gives fair agreement with both the polarization and color of the Halley dust. The average albedo of the grains in the model is 0.02.

103.559 The 3.2 – 3.5 μm emission features in comet P/Halley: spectral identifications and similarities.

R. F. Knacke, T. Y. Brooke, R. R. Joyce.
Astron. Astrophys., Vol. 187, No. 1/2, p. 625 – 628 (1987). – See Abstr. 003.013, 43.103.518.

Gound–based spectroscopic observations of emission features near 3.3 μm in comet Halley are reported. At a resolution of $\Delta\lambda/\lambda = 0.012$ the emissions consist of several components between 3.2 μm and 3.6 μm, probably superpositions of C–H group bands. The grains responsible for the $3 - 20$ μm thermal emission do not appear to be the site of the band emitter, but either molecular or small particle emission is possible. The band shape differs from those of interstellar emitting substances in having strong features longward of 3.3 μm bands. This extra structure may be due to alkyl side–chains or to compounds characteristic of low formation temperatures. Emission at 3.52 ± 0.02 μm could be evidence for oxygen containing molecules, possibly formaldehyde.

103.560 Airborne spectrophotometry of P/Halley from 16 to 30 microns.

T. Herter, H. Campins, G. E. Gull.
Astron. Astrophys., Vol. 187, No. 1/2, p. 629 – 631 (1987). – See Abstr. 003.013, 43.103.522.

The authors report on the first medium resolution spectrophotometric observations of a comet ever obtained in the 16– to 30–μm region. Comet P/Halley was observed using the Cornell University 7–channel spectrometer ($\Delta\lambda/\lambda \sim 0.02$) on board the Kuiper Airborne Observatory on 1985 December 14.2. Measurements were made centered on the nuclear condensation. These observations show a complex spectrum which could not be fit with simple models. The 20 μm silicate feature, if present, is very weak; and a relatively narrow ($\Delta\lambda \sim 0.5$ μm), strong feature centered at approximately 28.4 μm is observed and attributed to dust emission.

103.561 Photometry of comet P/Halley from 40 to 160 microns.

H. Campins, M. Joy, P. M. Harvey, D. F. Lester, H. B. Ellis Jr.
Astron. Astrophys., Vol. 187, No. 1/2, p. 632 – 634 (1987). – See Abstr. 003.013, 43.103.520.

Far infrared photometry of comet P/Halley was obtained from NASA's Kuiper Airborne Observatory 1986 March 15 – 16. A preliminary analysis of the observations shows the following results: (1) The brightness of the comet decreased by nearly a factor of two on the second day, in phase with the variability observed in blue light by IUE during the same time period. (2) The energy distribution varied marginally with a shallower (cooler) spectrum on the second day. (3) Observations centered 30″ and 60″ away from the nuclear condensation show lower $50 - 100$ μm color temperatures on both dates. This is consistent with spacecraft observations which indicate a higher proportion of large particles away from the nucleus.

103.562 Airborne spectrophotometry of P/Halley from 20 to 65 microns.

W. Glaccum, S. H. Moseley, H. Campins, R. F. Loewenstein.
Astron. Astrophys., Vol. 187, No. 1/2, p. 635 – 638 (1987). – See Abstr. 003.013, 43.103.521.

The authors report on simultaneous 20 – 65 μm spectrometry and 100 μm photometry of P/Halley obtained on board the Kuiper Airborne Observatory in 1985 December and 1986 April.

Spectra with resolution $\lambda/\Delta\lambda \cong 20 - 40$ were obtained with the NASA/Goddard 24–channel grating spectrometer. Measurements were made on the nucleus as well as 5 points along and perpendicular to the sun–tail direction. The comet spectrum is smooth and contains no strong ($>6\%$) spectral features. The color temperature of the dust is higher than that expected for a rapidly rotating blackbody and varies over time scales as short as 2 days. The coma is brigther at these wavelengths on the sunward side than on the tailward side, but the color temperature does not vary within one arcmin of the nucleus. The spectrum is fitted with a power law distribution of dark, featureless, spherical grains.

103.563 Comet P/Halley near–nucleus phenomena in 1986.

S. Larson, Z. Sekanina, D. Levy, S. Tapia, M. Senay.
Astron. Astrophys., Vol. 187, No. 1/2, p. 639 – 644 (1987). – See Abstr. 003.013, 43.103.526.

Computer enhanced high resolution groundbased images taken during Halley's most active phases in 1985–86 exhibit all of the coma structures seen in 1910 photographs. Improved detector technology compensated for the poor observing circumstances of the current apparition yielding a wealth of complex coma structure. This structure resulted mostly from discrete dust jet activity producing spirals, arcs, and jagged–edged envelopes. Several episodes of the "multiple nucleus" and antisunward jet phenomenon were also observed. The variation in dust jet activity during the Vega and Giotto flybys is easily seen on the groundbased images. The authors' images of jets in the 388 nm CN (0,0) band show that they usually do not correlate very well with visible dust jets. As in 1910, the observed expansion velocities and jet curvature indicate a rotation period near 2.2 days. Because the short period dominates the shape of the jets, detailed jet modeling will be required to detect longer term rotation components such as the proposed 7.4 day period.

103.564 The sunward spike of Halley's comet.

Z. Sekanina, S. M. Larson, G. Emerson, E. F. Helin, R. E. Schmidt.
Astron. Astrophys., Vol. 187, No. 1/2, p. 645 – 649 (1987). – See Abstr. 003.013, 43.103.532.

On wide–field photographs from the late April till early June 1986, comet Halley is seen to display a spike–like extension in the general direction of the Sun, projecting to distances of about 700,000 km from the nucleus. The spike was composed of dust and its enormous sunward extent (compared to other dust features) suggests an anomalously high ratio of particle ejection velocity to solar radiation pressure. The grains are either dielectric or slightly absorbing, $\ll 0.1$ μm in size, and undetected optically from Earth except when it is located in or very near a plane of their concentration. The only plane to which these grains ejected from Halley's wobbling nucleus can possibly be confined for long is the plane normal to the comet's angular momentum vector. This concept is applied to interpret the spike observations.

103.565 Complex refractive index of grain material deduced from the visible polarimetry of comet P/Halley.

T. Mukai, S. Mukai, S. Kikuchi.
Astron. Astrophys., Vol. 187, No. 1/2, p. 650 – 652 (1987). – See Abstr. 003.013, 43.103.513.

Visible polarimetry of comet Halley over the phase angle α (sun–comet–earth angle) from $1°6$ to $64°4$ has shown negative values of the polarization p in $\alpha \leqslant 20°$ and linearly increase of p with $\alpha > 20°$. In addition, a slight increase of p with wavelength λ is likely to exist. By applying the Mie calculations to explain both α– and λ–dependences of p, the authors have found that it's likely for the cometary dust to consist of the material with the average complex refractive index (m = n–ik), i.e. (n, k) = (1.392, 0.024) at $\lambda = 0.365$ μm, (1.387, 0.031) at $\lambda = 0.484$ μm, (1.385, 0.035) at $\lambda = 0.62$ μm and (1.383, 0.038) at $\lambda = 0.73$ μm, for grains assumed to have the size spectrum reported by the Vega missions.

103.566 Infrared emission from P/Halley's dust coma during March 1986.

M. S. Hanner, A. T. Tokunaga, W. F. Golisch, D. M. Griep, C. D. Kaminski.

Astron. Astrophys., Vol. 187, No. 1/2, p. 653 – 660 (1987). – See Abstr. 003.013.

2 to 20 μm photometry of the inner dust coma of comet Halley was obtained at the NASA IRTF on March 6.85, 12.8, 13.75, 17.7, and 24.8. Positions offset 10″ were measured as well as the central brightness. The strength of the 10 μm emission feature was observed to vary with location in the coma. The infrared emission is in general agreement with the dust size distribution measured from the Vega and Giotto spacecraft. March 6.8, 17.7, and 24.8 correspond to strong dust jet activity. The strength of the 10 μm silicate emission is shown to be a sensitive indicator of grain size and thus of jet activity. Dust production rate on March 13.75, 6 h before Giotto encounter, was 10^7g s^{-1}.

103.567 The dust tail of comet P/Halley in April 1986.

P. L. Lamy, H. Pedersen, R. Vio.

Astron. Astrophys., Vol. 187, No. 1/2, p. 661 – 664 (1987). – See Abstr. 003.013, 43.103.515.

Images of the dust tail of comet Halley were obtained with the ESO wide–field CCD camera during April 1986 with the Johnson B, V, and R filters in polarized light. An analysis of selected images taken on April 6 and 11, processed for photometric analysis, is presented. Results are given for the absolute brightness, the color, the polarization and its variation with wavelength.

103.568 Polarimetry of grains in the coma of P/Halley. I. Observations.

A. Dollfus, J.–L. Suchail.

Astron. Astrophys., Vol. 187, No. 1/2, p. 669 – 688 (1987). – See Abstr. 003.013.

With the new photopolarimeter that the authors built at the Observatoire de Paris, Section de Meudon, the authors analysed the polarization of the light in different areas over the coma of P/Halley, from October to December 1985 with the 100 cm telescope of Meudon, and then in greater detail in April 1986 with the 152 cm ESO telescope at La Silla, Chile. The data include 164 determinations of the linear polarization Q/I, 114 measurements of U/I and 74 values of the circular polarization V/I. The linear polarization is expressed as a function of the phase angle, separately for the dust recently ejected, for the oldest dust having travelled over at least 3000 km and for the bright envelope around the nucleus. Wavelength dependences are given. Deviations of the azimuth of polarization and circular polarizations are mapped over the coma.

103.569 Polarimetry of comet P/Halley.

S. Kikuchi, Y. Mikami, T. Mukai, S. Mukai, J. H. Hough.

Astron. Astrophys., Vol. 187, No. 1/2, p. 689 – 692 (1987). – See Abstr. 003.013.

The linear polarization P of comet P/Halley was observed simultaneously at eight wavelengths ranging from 0.36 to 0.90 μm, for phase angles α (sun–comet–earth angle) between 2° and 65°. In addition, observations in the J and H bands were made for $\alpha = 30°$. In the continuum bands, the authors have obtained (1) negative polarization for $\alpha \lesssim 20°$, (2) a quasi–linear increase of the polarization level with phase angle for $20° < \alpha < 50°$, and (3) a slight increase of P with wavelength from the visible through to the near infrared.

103.570 Dust in comet P/Halley from Vega observations.

E. P. Mazets, R. Z. Sagdeev, R. L. Aptekar', S. V. Golenetskii (*S. V. Golenetskij*), Yu. A. Guryan, A. V. D'yachkov, V. N. Ilyinskii (*V. N. Il'inskij*), V. N. Panov, G. G. Petrov, A. V. Savvin, I. A. Sokolov, D. D. Frederiks, N. G. Khavenson, V. D. Shapiro, V. I. Shevchenko.

Astron. Astrophys., Vol. 187, No. 1/2, p. 699 – 706 (1987). – See Abstr. 003.013, 43.103.507.

Direct measurements of the dust particle spatial and mass distributions in comet Halley have been carried out on Vega spacecraft over the mass range $\sim 10^{-16} - 10^{-6}$g with the SP–2

dust particle detector consisting of a set of acoustic and impact plasma sensors. These measurements have revealed a large–scale, time–variable structure of the dust coma, the position and characteristics of its boundaries, and a strongly pronounced angular directivity of dust emission from the cometary nucleus. An unexpectedly high concentration of tiny dust particles with masses below 10^{-14}g and strong systematic variations of the shape of the particle mass distribution as a function of distance to the cometary nucleus have been observed. The mean dust production rate by the nucleus has been estimated at $(10 - 13) \times 10^6$g s^{-1} (Vega 1) and $(5 - 7) \times 10^6$g s^{-1} (Vega 2).

103.571 Properties of dust in comet P/Halley measured by the Vega–2 three–channel spectrometer.

V. A. Krasnopolsky (*V. A. Krasnopol'skij*), V. I. Moroz, A. A. Krys'ko, A. Yu. Tkachuk, G. Moreels, J. Clairemidi, J. P. Parisot, M. Gogoshev, T. Gogosheva.

Astron. Astrophys., Vol. 187, No. 1/2, p. 707 – 711 (1987). – See Abstr. 003.013.

Optical properties of cometary dust exhibit no substantial variations at distances from 100 to 15,000 km from the nucleus. Dust ejection can be approximated by cos φ where φ is the angle with the sun–nucleus axis. Phase functions of dust scattering at 10 wavelengths from 328 nm to 1.4 μm are obtained and discussed. Dust reflectivity is nearly constant in the range 0.3 μm to 1.8 μm. Vega–2 in situ measurements are analysed and compared with optical data. Phase functions calculated for Mie and Fresnel spheres fail to reproduce adequately the obtained data. More promising seems to be the approximation of rough spheres. Dust density was evaluated as 0.3 g cm^{-3} from comparison of the in situ data with the authors' measurements. Dust velocities are approximated by the expression $V_g/V_d = 1 + (\varrho r/b)^{1/2}$ with $b = 4$ g μm cm^{-3} obtained in the measurements. Total dust production rates equals 10 ton s^{-1} and $Q_d/Q_g \approx 0.3$ on 1986 March 9.

103.572 Dust distribution of comet P/Halley's inner coma determined from the Giotto Radio–Science Experiment.

P. Edenhofer, M. K. Bird, J. P. Brenkle, H. Buschert, E. R. Kursinski, N. A. Mottinger, H. Porsche, C. T. Stelzried, H. Volland.

Astron. Astrophys., Vol. 187, No. 1/2, p. 712 – 718 (1987). – See Abstr. 003.013, 43.103.535.

Measurements of the Giotto Radio–Science Experiment, consisting of the Doppler frequency shift and the intensity level of the X–band downlink signal of the Giotto spacecraft during Halley encounter, are analysed and interpreted. Continuous radio–science data were recorded throughout the encounter. The Doppler shift observed over a time interval of about 100 s is attributed to a drag effect in the cometary atmosphere causing a deceleration of the spacecraft. The total change of velocity of Giotto was 23.05 ± 0.05 cm s^{-1}. The cometary mass fluence and mean mass density along the Giotto trajectory was found to be 1.2×10^{-4}kg m^{-2} and 1.8×10^{-11}kg m^{-3}, respectively. Within a time interval ± 45 s about closest approach, at least four well–established, sharply confined dust jets are distinguished, for which the characteristic geometries and production rates can be inferred. The total dust production rate is estimated to be 1.5×10^3kg s^{-1}. The comet's jet structure derived from these Doppler data is shown to correlate with measurements on board Giotto.

103.573 The dust distribution within the inner coma of comet P/Halley 1982i: encounter by Giotto's impact detectors.

J. A. M. McDonnell, W. M. Alexander, W. M. Burton, E. Bussoletti, G. C. Evans, S. T. Evans, J. G. Firth, R. J. L. Grard, S. F. Green, E. Grün, M. S. Hanner, D. W. Hughes, E. Igenbergs, J. Kissel, H. Kuczera, B. A. Lindblad, Y. Langevin, J.–C. Mandeville, S. Nappo, G. S. A. Pankiewicz, C. H. Perry, G. H. Schwehm, Z. Sekanina, T. J. Stevenson, R. F. Turner, U. Weishaupt, M. K. Wallis, J. C. Zarnecki.

Astron. Astrophys., Vol. 187, No. 1/2, p. 719 – 741 (1987). – See Abstr. 003.013.

Analysis of the data from Giotto's Dust Impact Detection System experiment (DIDSY) is presented. These data represent measurement of the size of dust grains incident on the Giotto dust shield along its trajectory through the coma of comet P/Halley on 1986 March 13/14. First detection occurred at some 287,000 km distance from the nucleus on the inbound leg; the majority of the DIDSY subsystems remained operational after closest approach (604 km) yielding the last detection at about 202,000 km from the nucleus. In order to improve the data coverage (and especially for the smallest grains, to approximately 10^{-19}kg particle mass), data from the PIA instrument has been combined with DIDSY data. For the close encounter period (-5 min to $+5$ min), the cumulative mass distribution function has been investigated.

103.574 The dust coma of comet P/Halley: measurements on the Vega–1 and Vega–2 spacecraft.

J. A. Simpson, D. Rabinowitz, A. J. Tuzzolino, L. V. Ksanfomaliti, R. Z. Sagdeev.
Astron. Astrophys., Vol. 187, No. 1/2, p. 742 – 752 (1987). – See Abstr. 003.013, 43.103.508.

The spatial, temporal and mass distributions of coma dust particles for masses $> 10^{-13}$g have been confirmed. The authors have analyzed the inverse square dependence ($\propto R^{-2}$) of flux with distance R from the nucleus and show that there are well-defined envelope boundaries for the measured range of masses. Regions of enhanced fluxes above the R^{-2} 'baselines' inside these boundaries are identified as dust jets. Of special interest is the giant flux enhancement near closest approach for Vega–1. The authors have discovered particles with a wide range of masses arriving in clusters or 'packets' (i.e., a non-Poisson distribution of dust particles) throughout the regions beyond the envelope boundaries – namely, the fringe regions of the dust coma. These observations suggest the emission from the nucleus of large conglomerates of small particles which gently disintegrate as they travel outward to account for the 10^{-13}g particles observed beyond the envelope boundaries. Also, dust fluences over the missions have been determined.

103.575 Spatial and mass distribution of low–mass dust particles ($m < 10^{-10}$g) in comet P/Halley's coma.

O. L. Vaisberg (*O. L. Vajsberg*), V. Smirnov, A. Omelchenko (*A. N. Omel'chenko*), L. Gorn, M. Iovlev.
Astron. Astrophys., Vol. 187, No. 1/2, p. 753 – 760 (1987). – See Abstr. 003.013, 43.103.509.

The spatial distribution of cometary dust particles in the mass range $\sim 10^{-10}$ to $\sim 3 \times 10^{-17}$g was obtained with the plasma impact detectors SP–1 on Vega–1 and Vega–2. The mass spectrum continues to rise in the range $10^{-14} - 3 \times 10^{-17}$g. The evolution of the mass spectrum with distance suggests radiation pressure separation and some other effects. Several dust shells are easily distinguished by spatial gradients and by different properties of the grains. The spatial mass dispersion in the prominent jet observed on Vega–2 is used to evaluate the rotation of the nucleus and to calculate the velocity dispersion of the dust.

103.576 First statistical analysis of 5000 mass spectra of cometary grains obtained by PUMA 1 (Vega 1) and PIA (Giotto) impact ionization mass spectrometers in the compressed modes.

Y. Langevin, J. Kissel, J.–L. Bertaux, E. Chassefière.
Astron. Astrophys., Vol. 187, No. 1/2, p. 761 – 766 (1987). – See Abstr. 003.013.

Impact ionization mass spectrometers have been flown on board the space vehicles Giotto (instrument PIA), Vega 1 and Vega 2 (instruments PUMA 1 and 2). All three instruments obtained a large collection of mass spectra of individual cometary dust particles in the mass range $3 \times 10^{-16} - 3 \times 10^{-10}$g. Spectra from PUMA 2 are still being recovered from data transmission problems, and only results from PUMA 1 and PIA are presented. A large fraction of particles ($\sim 30\%$) are dominated by light elements (H, C, O, and N, or "CHON"). Approximately 35% of particles can be interpreted as minerals with a compositional range similar to that encountered in carbonaceous chondrites. The remaining 35% appear to be mixtures in varying amounts of these two major components. Statistically significant variations have been observed as a function of time.

103.577 Comet P/Halley: implications of the mass distribution function for the photopolarimetric properties of the dust coma.

P. L. Lamy, E. Grün, J. M. Perrin.
Astron. Astrophys., Vol. 187, No. 1/2, p. 767 – 773 (1987). – See Abstr. 003.013.

The dust particle fluences measured aboard the Vega (SP–1 and SP–2 impact sensors) and Giotto probes (DIDSY impact sensors) are analyzed to obtain the differential size distribution function of the dust in the coma of comet Halley. The brightness integral is then calculated for perfectly spherical (Mie scattering) and rough grains of various compositions. It is shown that the photopolarimetric observations rule out the dominating presence of weakly obsorbing silicates such as olivine but are compatible with rough moderately absorbing silicate grains having a density decreasing with radius and with rough graphite grains. A mixture of the two types gives in fact the best agreement with the observations.

103.578 Systematics of the "CHON" and other light–element particle populations in comet P/Halley.

B. C. Clark, L. W. Mason, J. Kissel.
Astron. Astrophys., Vol. 187, No. 1/2, p. 779 – 784 (1987). – See Abstr. 003.013, 43.103.617.

Based on chemical signatures measured by the PIA experiment during the Giotto flyby of comet P/Halley, a number of particle classifications have been designated. In addition to silicate–like grains and particles of mixed (cosmic) composition, there appear to be several light–element rich populations, including the CHON, (H, C), (H, C, O), and (H, C, N) particle types. In some cases, these compositional classes are further distinguished by differences in mass distributions, a density indicator, and variations in relative abundance within the coma. These particle populations are evidence for chemical heterogeneity in the surface of the cometary nucleus. Some particles found mainly in the inner coma may be volatile icy grains. Most of the N of the comet may be found in up to three different populations of grains; one or more of these may be responsible for the observation of cyanojets.

103.579 Secondary electron emission induced by gas and dust impacts on Giotto, Vega–1 and Vega–2 in the environment of comet P/Halley.

R. J. L. Grard, J. A. M. McDonnell, E. Grün, K. I. Gringauz.
Astron. Astrophys., Vol. 187, No. 1/2, p. 785 – 788 (1987). – See Abstr. 003.013, 43.103.614.

Giotto, Vega–1 and Vega–2 have been bombarded by a flow of gas and dust particles during their flyby of comet Halley. The emission of secondary electrons and sputtered ions caused by the largest impact velocities (70 – 80 km s^{-1}) perturbed the plasma density in the spacecraft vicinity and was a possible source of interference for electric field and plasma measurements. Identical impact plasma detectors were mounted on the three space probes; the saturation currents of secondary electrons emitted from gold targets were measured once per second. The results obtained during the three flybys are presented and compared. Information about the gas density profile and nucleus gas production rate can be derived from the measurements.

103.580 Dust environment of comet P/Halley: a review.

Z. Sekanina.
Astron. Astrophys., Vol. 187, No. 1/2, p. 789 – 795 (1987). – See Abstr. 003.013, 43.103.525.

Recent major advances in the understanding of comet Halley's dust environment are highlighted and a wide range of related problems reviewed. Addressed individually are selected spacecraft results, namely the dust composition, the detection of subfemtogram grains, and the particle–mass distribution; jets, the dust–emission pattern, and the existence of discrete sources on the nucleus surface; CN and C_2 jets; large–scale phenomena, including an antitail, streamers, and a sunward spike; infrared observations, highlighted by recent detections of the 3.4 μm emission feature, first applications of array detectors, and rare far–infrared data; the chemical modification of low–temperature ices by UV and cosmic–ray irradiation processes, their laboratory simulations, and applications to the evolution of comets; the "interstellar–connection" models of comet origin; and the formation of a mantle made of refractory organic material on the nucleus surface.

103.581   Evaporating grains in P/Halley's coma.
M. K. Wallis, R. Rabilizirov, N. C. Wickramasinghe.
Astron. Astrophys., Vol. 187, No. 1/2, p. 801 – 806 (1987). – See Abstr. 003.013, 43.103.537.

Evidence is presented for the hypothesis that the organic grains streaming from a cometary nucleus lose substantial volatile constituents within 10^5s. To develop the hypothesis, the authors study grain properties appropriate to both the comet–probe encounters at 0.8 – 0.9 AU and to comae observed at 4 – 5 AU. The authors adopt kerogen as a prototype organic complex having a range of bond energies. Temperatures and evaporation rates of kerogen grains are calculated using Arrhenius relations and Mie–Güttler radiative scattering theory with complex refractive index as measured for two cases: organic material of biological origin and UV–processed methane ice ("tholin"). The 10 μm IR feature can be explained by grains of an organo–siliceous material or organic mantle on a siliceous core, a few μm in size and about 20 kcal/mole bond strength.

103.582   Comet P/Halley's nucleus and its activity.
H. U. Keller, W. A. Delamere, W. F. Huebner, H. J. Reitsema, H. U. Schmidt, F. L. Whipple, K. Wilhelm, W. Curdt, R. Kramm, N. Thomas, C. Arpigny, C. Barbieri, R. M. Bonnet, S. Cazes, M. Coradini, C. B. Cosmovici, D. W. Hughes, C. Jamar, D. Malaise, K. Schmidt, W. K. H. Schmidt, P. Seige.
Astron. Astrophys., Vol. 187, No. 1/2, p. 807 – 823 (1987). – See Abstr. 003.013.

The Halley Multicolour Camera on board ESA's Giotto spacecraft observed the nucleus of comet P/Halley and its environment and returned more than 2000 images. The observations are summarized, their calibration is described, the status of the analysis and the results are discussed. Topographic features on the nuclear surface and areas of activity are identified. The optical thickness of the dust produced in jet–like features is estimated. The impact and constraints of the observations on cometary nucleus models are discussed.

103.583   Evolution of comet P/Halley in early March 1986 as observed from Vega pictures.
A. Abergel, J. L. Bertaux.
Astron. Astrophys., Vol. 187, No. 1/2, p. 829 – 834 (1987). – See Abstr. 003.013.

From 4 to 10 March 1986, the internal dust coma of comet P/Halley was observed every day with Vega–1 and 2 narrow angle cameras of TVS system. The distribution of dust was monitored within 30,000 km from the nucleus. Strong variations are observed from one day to the other, indicating corresponding variations of the nucleus activity. These variations affect both the overall dust number density and the shape of its distribution around the nucleus. Dust intensities are given in absolute units of brightness.

103.584   The spatial distribution of dust jets during the Vega 2 flyby.
R. Z. Sagdeev, B. Smith, K. Szegö, S. Larson, I. Tóth, E. Merényi, G. A. Avanesov, V. A. Krasikov, V. A. Shamis, V. I. Tarnapolski (*V. I. Tarnopol'skij*).
Astron. Astrophys., Vol. 187, No. 1/2, p. 835 – 838 (1987). – See Abstr. 003.013, 43.103.543.

One important physical characteristic of comet P/Halley is dust jet activity, seen from Earth as changing coma patterns. At spacecraft resolution it becomes possible to precisely locate the sources of active jets on an irregular shaped nucleus. The data set employed in this analysis consists of 11 full–format (512 × 512 px) Vega–2 images taken over a 15 min interval starting approximately 370 s before closest approach. The distance to the nucleus was always less than 45,000 km and

became as close as 8030 km. Most of the active zones formed at least one quasi–linear structure on the surface. The authors found evidence of the activity on the dark hemisphere, at least one of the faint but active jet sources is there.

103.585   Temperature and size of the nucleus of comet P/Halley deduced from IKS infrared Vega 1 measurements.
C. Emerich, J. M. Lamarre, V. I. Moroz, M. Combes, N. F. San'ko, Y. V. Nikolsky (*Yu. V. Nikol'skij*), F. Rocard, R. Gispert, N. Coron, J. P. Bibring, T. Encrenaz, J. Crovisier.
Astron. Astrophys., Vol. 187, No. 1/2, p. 839 – 842 (1987). – See Abstr. 003.013, 43.103.555.

The infrared thermal emission of the nucleus of Halley's comet was measured by the imaging channel of the IKS spectrometer in two wavelengths bands (7 – 10 μm and 9 – 14 μm) and in two perpendicular directions of analysis. The effective dimensions of the infrared emissive region in these directions are estimated from a simulation of the flyby configuration, and compared to observations made in the visible range. A color temperature larger than 360K was deduced using on–ground preflight calibrations. A temperature of the same order is obtained independently, comparing the absolute value of the measured fluxes with model simulations. Such temperatures, significantly higher than the sublimating temperature of ices, and near the blackbody temperatures at 0.8 AU, suggest the existence of an essentially inactive dark crust covering a significant part of the surface of the nucleus.

103.586   Fine dust structures in the emission of comet P/Halley observed by the Halley Multicolour Camera on board Giotto.
N. Thomas, H. U. Keller.
Astron. Astrophys., Vol. 187, No. 1/2, p. 843 – 846 (1987). – See Abstr. 003.013.

Halley Multicolour Camera observations on board the ESA Giotto spacecraft provided images of comet P/Halley with resolution 4 orders of magnitude better than from the ground in March 1986. The fine structure seen in the emission from the cometary surface has led to some confusion over the nomenclature of jets and jet–like features streaming from the nucleus. Emission features are classified according to directions with respect to the sun and to positions with respect to the nucleus. In total, 17 fine structures (filaments) are identified, more than doubling the number presented earlier (Keller et al., 1986).

103.587   Detailed analysis of a surface feature on comet P/Halley.
G. Schwarz, H. Craubner, A. Delamere, M. Göbel, M. Gonano, W. F. Huebner, H. U. Keller, R. Kramm, E. Mikusch, H. Reitsema, F. L. Whipple, K. Wilhelm.
Astron. Astrophys., Vol. 187, No. 1/2, p. 847 – 851 (1987). – See Abstr. 003.013, 43.103.553.

The surface of the nucleus of comet P/Halley is visible an many images taken by the Halley Multicolour Camera on board the Giotto spacecraft. A number of structural features like "mountains" or "surface dips" are clearly discernible. This paper outlines the methods that can be applied to analyze a surface feature with the aim of obtaining its characteristics. A relatively large craterlike surface feature is selceted for detailed analysis and an attempt is made to determine its size and shape as well as to assess the attainable accuracy. The surface topography is computed using a photoclinometric approach, while dimensions are determined based on apparent object sizes and an estimate of aspect angle.

103.588   The cometary nucleus: current concepts.
F. L. Whipple.
Astron. Astrophys., Vol. 187, No. 1/2, p. 852 – 858 (1987). – See Abstr. 003.013.

The Vega and Giotto space missions have established invaluable check points with regard to comet nuclei. Striking is the low geometric albedo of 0.044 deduced from the directly observed dimensions of the nucleus and the brightness at large solar distances. The dimensions, the observed nongravitational motions and the jet activity combine to suggest an extremely low density for the nucleus of Halley's comet – in the range of 0.1 to 0.5 g cm^{-3}. The gross composition of the nucleus appears to conform roughly to abundances of the elements and isotope ratios expected if comets were frozen out of a Solar–System mix, supporting theories of cometary origin concurrent with the Sun, planets and asteroids. The extremely low albedo and the prevalent "CHON" composition of the dust, the newly observed dust particles of mass less than 10^{-17}g, and evidence of formation at low temperatures support accumulation processes for comets from material like interstellar grains imbedded in a proto–planetary nebula.

103.589 Composition measurements and the history of cometary matter.
J. Geiss.
Astron. Astrophys., Vol. 187, No. 1/2, p. 859 – 866 (1987). – See Abstr. 003.013.

The author presents estimates for elemental abundances in the material emitted by Halley's comet. They are based on in situ measurements on the gas and dust in the coma. It appears that Halley's nucleus is not depleted in the volatile elements C and O, whereas N is deficient. The abundance data show that a large fraction of the material in Halley's nucleus condensed at very low temperature. Comparison of the results published so far by the Giotto and Vega groups on molecular, elemental and isotopic abundances with data from meteorites, meteors, and the interstellar gas confirm that comets are regular members of the solar system which have preserved the original characteristics of the condensed and accreted matter better than other bodies in this system.

103.590 Modeling Halley before and after the encounters.
N. Divine, R. L. Newburn Jr.
Astron. Astrophys., Vol. 187, No. 1/2, p. 867 – 872 (1987). – See Abstr. 003.013.

Numerical models developed prior to the 1986 spacecraft flybys at comet P/Halley described anticipated dust and gas environments. Predicted values for gas production, for dynamic ranges of gas and dust experiments, and for likely spacecraft effects were matched during the flybys well within expected margins; in particular the dust–impact–generated attitude disturbance experienced by Giotto several seconds before closest approach was within the envelope of expectations from the models. The authors have identified three major areas in which 1986 results for Halley might be used to improve the models, namely (1) gas production values from Earth–based and in–situ observations, (2) dust flux and fluence values in the coma from three spacecraft, and (3) dust size distributions from the in–situ data which require the presence of numerous particles at masses less than 10^{-17}kg.

103.591 Post–perihelion brightening of comet P/Halley: springtime for Halley.
P. R. Weissman.
Astron. Astrophys., Vol. 187, No. 1/2, p. 873 – 878 (1987). – See Abstr. 003.013, 43.103.637.

Increased brightness and gas production rates for Halley's comet after perihelion can be explained as a result of seasonal effects on an obliquely rotating nucleus. The highly eccentric cometary orbit causes a rapid change in solar declination as the comet rounds perihelion, resulting in drastic changes in the insolation reaching the northern and southern hemispheres of the nucleus. The rapid heating of the northern hemisphere post–perihelion likely results in substantial cracking of the non–volatile surface crust due to thermal stresses, exposing areas of fresh volatile ices. The orientation of the triaxial ellipsoid nucleus may also play a role in exposing more surface area to continuous sunlight and sublimation after perihelion. Post–perihelion brightening models based on heat flow and storage in sub–surface layers of the nucleus pre–perihelion are likely not viable because of the low thermal conductivities of porous, low density cometary surface materials.

103.592 P/Halley dust characteristics: a comparison between Orionid and Eta Aquarid meteor observations and those from the flyby spacecraft.
D. W. Hughes.
Astron. Astrophys., Vol. 187, No. 1/2, p. 879 – 888 (1987). – See Abstr. 003.013.

Dust from P/Halley was detected by the Giotto and Vega spacecraft in mid March 1986 and can also be observed from Earth every May and October in the form of the Eta Aquarid and Orionid meteor showers. Unfortunately, the two techniques measure dust in dramatically different places (with respect to the parent comet) and over non–overlapping mass ranges. This paper compares and contrasts the fluxes and mass distribution indices obtained.

103.593 Astrometric positions of comet P/Halley.
C. Barbieri, A. Kranjc, M. Scardia, G. Cremonese.
Astron. Astrophys., Vol. 187, No. 1/2, p. 893 – 895 (1987). – See Abstr. 003.013, 43.103.603.

One hundred astrometric positions of comet P/Halley (mostly in the pre–perihelion phase) have been derived from images taken at the Asiago and Merate Observatories. These positions have been used to determine a set of orbital elements for the comet that, although in good agreement with those of IHW Orbit Nr. 45, give smaller residuals.

103.594 A new approach to investigations of the long–term motion of comet P/Halley.
G. Sitarski, K. Ziolkowski.
Astron. Astrophys., Vol. 187, No. 1/2, p. 896 – 898 (1987). – See Abstr. 003.013, 43.103.607.

A new method, based on the moments of perihelion passages as observational data to the orbit improvement, was used to link all the observed apparitions of comet Halley in 240 BC – 1986 by one system of orbital elements. Preliminary results are presented. The problem of secular variations in the non–gravitational parameters of the comet's motion, influencing the results of the orbit improvement, is discussed. It is shwon that a successful modeling of the motion of comet Halley over the last two millennia is also possible if a variability of the non–gravitational effects with time is taken into account.

103.595 The dynamical lifetime of comet P/Halley.
D. I. Olsson–Steel.
Astron. Astrophys., Vol. 187, No. 1/2, p. 909 – 912 (1987). – See Abstr. 003.013.

The stability of the orbit of Halley's comet under close encounters with the planets is investigated, and it is found that the orbital changes due to such encounters expected over a long time–base are of the same order as the effects of more distant planetary perturbations and non–gravitational effects. For the present orbit the half–life of the comet against ejection from the solar system as a result of a close encounter is 68,000 apparitions; the half–life against planetary impacts or catastrophic disruption in a close approach is about a million apparitions: dynamical loss is therefore unlikely. For a past lifetime of 2000 – 3000 perihelion passages there is a probability of about 50% of a single large deflection into an orbit of ($q < 0.5$ or > 0.7 AU), ($a < 15$ or > 25 AU), or ($i < 150°$ or $> 170°$). During this period the perihelion distance has most probably remained between 0.5 and 0.7 AU, so that physical models based upon sensibly constant rates of material loss from the surface through its lifetime appear to be justified.

103.596 Maps of the emission of CN and other molecular species in the spectrum of comet Halley.
J. A. Acosta, M. Kidger, M. Prieto, J. Cepa, C. Muñoz Tuñón.
Rev. Mex. Astron. Astrofis., Vol. 14, No. 2, p. 644 – 650 (1987). – See Abstr. 012.042.

The authors present isophotal maps of the distribution of the flux emitted by transitions of various molecular species in the spectrum of comet Halley at a heliocentric distance of 1.92 A.U., after perihelion. This contribution is a part of a more general program, which aims to study the production rates of a number of species and its variation with heliocentric distance.

103.597 Observaciones fotometricas en bandas moleculares del cometa Halley.
J. J. Clariá, R. F. Sisteró, E. Lapasset.
Rev. Mex. Astron. Astrofis., Vol. 14, No. 2, p. 651 – 659 (1987). – See Abstr. 012.042.
Results from narrow–band photoelectric photometry of comet Halley after its perihelion passage are presented. The largest fluxes measured successively correspond to the molecules of C_2, CN, and C_3. The CN emission is typically 70% of the one of C_2, while the one of C_3 is about 20%. The observations show that: (1) There is no detectable contribution of molecular ions CO^+ during the period in consideration. (2) Between March 15 and 22 of 1986 there is a decrease in the proportion of molecules of C_2, CN, and C_3, with respect to the presence of dust indicated by the continuum.

103.598 Earth based observations of comet Halley.
J. F. Barral, M. R. Berumen, M. R. Herrera, A. Quintero, A. Pani, R. Peniche, J. H. Peña.
Rev. Mex. Astron. Astrofis., Vol. 14, No. 2, p. 660 – 663 (1987). – See Abstr. 012.042.
Direct photographs and low resolution spectra of comet Halley were obtained, which have been digitalized by means of a PDS.

103.599 Strip photometry of comet Halley.
J. R. Ducati, T. S. Bergmann, C. M. Bevilacqua, C. Bonatto, R. L. Cavalcanti, R. D. D. Costa, H. A. Dottori, L. Girardi, D. Hadjimichef, S. O. Kepler, S. H. B. Livi, M. G. Pastoriza, J. F. Santos, A. Schmidt, M. F. S. Schröder.
Rev. Mex. Astron. Astrofis., Vol. 14, No. 2, p. 664 – 667 (1987). – See Abstr. 012.042.

103.600 Espectroscopia del cometa Halley.
O. Naranjo, F. Fuenmayor, I. Ferrín, P. Bulka, C. Mendoza.
Rev. Mex. Astron. Astrofis., Vol. 14, No. 2, p. 668 (1987). Abstract. – See Abstr. 012.042.

103.601 Procesamiento digital de imágenes del cometa Halley.
I. Ferrín, F. Fuenmayor, O. Naranjo, P. Bulka, C. Mendoza.
Rev. Mex. Astron. Astrofis., Vol. 14, No. 2, p. 669 (1987). Abstract. – See Abstr. 012.042.

103.602 Radio–observaciones del OH en la coma del cometa Halley desde el hemisferio sur.
A. M. Silva, E. Bajaja, R. Morras, J. C. Cersosimo, M. C. Martin, E. M. Arnal, W. G. L. Pöppel, F. R. Colomb, J. Mazzaro, J. C. Olalde.
Rev. Mex. Astron. Astrofis., Vol. 14, No. 2, p. 670 (1987). Abstract. – See Abstr. 012.042.

103.603 Observações do cometa de Halley no contínuo de 22 GHz e 44 GHz, e na raia maser de vapor de H_2O em 22.2 GHz.
Z. Abraham, E. Scalise Jr., L. C. L. Botti, A. C. O. Cancoro, J. L. M. do Vale, L. F. del Ciampo, C. E. Tateyama, J. W. S. Vilas Boas, J. L. Homor, P. Kaufmann.
Rev. Mex. Astron. Astrofis., Vol. 14, No. 2, p. 671 (1987). Abstract. – See Abstr. 012.042.

103.604 Collisionless tearing instability as the cause of an observed structure in the plasma tail of comet Halley.
J. Galindo Trejo.
Rev. Mex. Astron. Astrofis., Vol. 14, No. 2, p. 672 – 677 (1987). – See Abstr. 012.042.
A wavy structure in the near tail of comet Halley was observed in Tonantzintla on 12 April 1986. Recent spacecraft encounters with comets have confirmed the Alfvén field line draping model of cometary tails. On the basis of this result, the author discusses the origin of the structure by considering an equilibrium which

consists of a neutral current sheet with a non–vanishing amount of magnetic flux through it. The author proposes that the observed structure is a consequence of the onset of a collisionless tearing instability in the above mentioned plasma equilibrium. Such a magnetic reconnection process could be triggered by an interplanetary shock.

103.605 The P/Halley's comet encountering the interplanetary medium.
E. Chavira.
Rev. Mex. Astron. Astrofis., Vol. 14, No. 2, p. 678 – 681 (1987). – See Abstr. 012.042.

103.606 Halley's Comet: a smooth, dark, dirty snowball.
D. Hughes.
New Sci., Vol. 113, No. 1546, p. 50 – 55 (1987). Abstr. in Phys. Abstr., Vol. 90, No. 1311, Entry 101738 (1987).

103.607 Hydroxyl–line observations in Halley's comet.
N. V. Bystrova, G. S. Golubchin, I. V. Gosachinskij, A. S. Morozov, E. K. Nizhel'skaya, N. A. Yudaeva.
Pis'ma Astron. Zh., Tom 13, No. 12, p. 1100 – 1103 (1987). In Russian. English translation in Sov. Astron. Lett., Vol. 13.
The results of observations of the hydroxyl line at the frequency of 1667 MHz in Halley's comet are reported. Observations were made in November and December 1985 and in March 1986 with the RATAN–600 radio telescope. The emission and absorption lines were found at rather high radial velocities relative to the comet – from $+5$ to $+25$ km/s.

103.608 Observation of star SAO 128409 occultation by Halley's comet.
M. A. Eritsyan, L. G. Akhverdyan.
Kinematika Fiz. Nebesn. Tel, Tom 3, No. 6, p. 89 – 90 (1987). In Russian. English translation in Kinematics Phys. Celest. Bodies.
An occultation of star SAO 128409 by Halley's comet was observed at Byurakan on December 9, 1985. The star passed in $12''$ from the photometric nucleus of the comet. The maximum extinction of the star because of occultation was approximately 1^m which corresponds to the optical depth of $\tau = 0.9$.

103.609 Burning as the main cause of comet Halley's activity.
Eh. M. Drobyshevskij.
Fiz.–Tekh. Inst. Akad. Nauk SSSR, Prepr., No. 1132, p. 3 – 24 (1987). In Russian. Abstr. in Ref. Zh., 62. Issled. Kosm. Prostranstva, 11.62.363 (1987).

103.610 An interpretation of the break in comet Halley's tail.
A. M. Obukhov, A. V. Danilov, M. V. Kurganskij.
Sov. Astron. Lett., Vol. 12, No. 6, p. 393 – 394 (1986). English translation of 42.103.656.

103.611 Surface microstructure of the nucleus of comet P/Halley.
G. A. Steigmann, M. B. Dodsworth.
Observatory, Vol. 107, No. 1081, p. 263 – 267 (1987).
Polarimetric analysis of the light reflected from a range of carbon–dominated materials has been undertaken. The data obtained have been compared with published work carried out on Halley's Comet as it approached perihelion. Results strongly suggest that the nucleus of comet P/Halley is covered with a thick crust of carbon–dominated material, having an albedo of 0.032 and existing in an extremely finely divided and roughened state.

103.612 Que nous a appris la comète de Halley?
J. Manfroid.
Ciel Terre, Vol. 103, No. 6, p. 135 – 142 (1987).
Comet Halley is already back in the outer solar system, forgotten by most of us. But the cometary specialists are still busy with the wealth of data gathered during a few months of frantic activity, peaking with the spacecraft fly–bys. The author summarizes here some of the new results which have been obtained up to now about P/Halley.

103.613   **Cometary organics but no evidence for bacteria.**
C. Chyba, C. Sagan.
Nature, Vol. 329, No. 6136, p. 208 (1987).
F. Hoyle and N. Wickramasinghe compare the $3-4 \mu m$ spectrum of Comet Halley with the predictions of a bacterial model. No evidence for the existence of cometary bacteria is shown.

103.614   **Hydrodynamical model of gas outflow from the comet's nucleus (dust coma).**
A. V. Kolesnichenko, M. Ya. Marov, Yu. V. Skorov.
Inst. Prikl. Mat. Akad. Nauk SSSR, Prepr., No. 113, p. 3 – 23 (1987). In Russian. Abstr. in Ref. Zh., 51. Astron., 12.51.248 (1987).

103.615   **On some periodicities in the motion of comet Halley (relative to the mass center).**
V. A. Sarychev, V. V. Sazonov, V. V. Savchenko, V. I. Tarnopol'skij.
Inst. Prikl. Mat. Akad. Nauk SSSR, Prepr., No. 133, p. 1 – 23 (1987). In Russian. Abstr. in Ref. Zh., 51. Astron., 12.51.252 (1987).

103.616   **Investigations of the motion of Halley's comet. Part II.**
K. Ziołkowski.
Postepy Astron., Tom 34, Zesz. 4, p. 249 – 268 (1986). In Polish.
The review of works devoted to the determination of orbit and the investigation of the long–term motion of Halley's comet is presented in this part. The first efforts to link some apparitions of the comet by one system of orbital elements as well as the attempts to recognize the past orbit by means of numerical integrations of the equations of the comet's motion are surveyed. The comparison of results obtained by different authors is discussed.

103.617   **A Halley watch from Eastern Island.**
W. Liller.
Mercury, Vol. 16, No. 5, p. 130 – 137 (1987).

103.618   **Occultation observations of compact radio sources through comet Halley's plasma tail.**
S. Ananthakrishnan, P. K. Manoharan, V. R. Venugopal.
Nature, Vol. 329, No. 6141, p. 698 – 700 (1987).
Radio sources suitable for studying the plasma tail of a comet must have a significant fraction of their flux density in the sub-arc second structure, so a set of 30 sources lying within $2°$ of the path of comet Halley and with a flux density of $S_{408} \gtrsim 0.8$ Jy were chosen from the Molonglo Catalogue and interplanetary scintillation observations of them were made at 327 MHz. From these, four were found to be suitable for cometary scintillation observations. The authors monitored nearly simultaneously other scintillating sources outside the tail for comparison and, in contrast with earlier claims of positive detection, found no significant increase in the level of turbulence that could be attributed to the plasma tail.

103.619   **Plasma and high energy charged particles measurements in the vicinity of Halley's comet.**
G. Erdös, T. Gombosi, K. Kecskemety, A. Somogyi, M. Tatrallyay, A. Varga.
Meres Autom., Vol. 35, No. 3, p. 85 – 90 (1987). In Hungarian. Abstr. in Phys. Abstr., Vol. 90, No. 1313, Entry 115310 (1987).

103.620   **Periodic comet Halley (1982i).**
IAU Circ., Nos. 4496, 4512 (1987).

103.621   **Periodic comet Halley (1982i).**
Yamamoto Circ., No. 2099 (1987). In Japanese.

103.622   **Near–infrared spectrophotometric observations of periodic comet Halley (1982i).**
H. Suto, T. Maihara, K. Mizutani, T. Yamamoto, J. A. Thomas.
Publ. Astron. Soc. Jpn., Vol. 39, No. 6, p. 925 – 932 (1987).
Periodic comet Halley (1982i) was observed spectrophotometrically in selected wavelength ranges between 2.0 and 3.5 μm with spectral resolutions of 170 to 270 on 1985 November 28 and 1986 January 3 and March 21 – 25. Although no molecular emission feature has been detected in the spectra observed, upper limits for the production rates of CO and CH_4 have been obtained. A notable day–to–day variation of the near–infrared brightness of the inner coma was observed, consistent with an apparent periodicity of 2.4 d in dust producing activity. The dust production rate of about 2 to $5 \times 10^6 g\,s^{-1}$ has also been derived by assuming the mean grain radius of 1 μm.

103.623   **How typical is Halley's comet?**
P. R. Weissman.
ESA Spec. Publ., ESA SP–278, p. 31 – 36 (1987). – See Abstr. 012.075.
In the coming years there will be a natural tendency to generalize the results from the Comet Halley spacecraft encounters and observations in 1985–86 to all comets, both long and short–period. Therefore, it is important to consider whether or not Halley is a truly representative comet. The comet's orbit is quite unusual for a short–period (SP) comet, characterized by a very small perihelion distance and a retrograde inclination. Additionally, the comet is much brighter than most SP comets, and is even relatively bright for a long–period comet. Although Halley's orbit crosses the orbits of seven of the nine planets, it can currently make close approaches to only Venus, the Earth, and Mars, planets likely too small to have played a major role in capturing Halley to a short–period orbit. This, and other evidence, suggests that Halley has been in its current orbit for many returns.

103.624   **Solar wind observation at the time of closest encounter with comet Halley.**
K. I. Oyama, T. Abe.
ESA Spec. Publ., ESA SP–278, p. 57 – 61 (1987). – See Abstr. 012.075.
On the 10th and 12th of March 1986, Japan's interplanetary spacecraft, Sakigake traversed the front of comet Halley. Small fluctuation of solar wind velocity due to H_2O^+ ions was first detected at the closest approach. However maximum fluctuation was detected between 22h30m and 24h00m on the 11th of March. The time corrresponds to the closest approach to the trajectory of comet Halley along which the comet went through. This might suggest that H_2O^+ still remains along the pass of comet Halley.

103.625   **Variability of comet Halley's coma: VEGA–1 and VEGA–2 magnetic field observations.**
K. Schwingenschuh, W. Riedler, H. I. M. Lichtenegger, J. L. Phillips, J. G. Luhmann, C. T. Russell, J. A. Fedder, A. Somogyi, Ye. G. Yeroshenko (E. G. Eroshenko).
ESA Spec. Publ., ESA SP–278, p. 63 – 67 (1987). – See Abstr. 012.075.
The differences of the VEGA–1 and VEGA–2 magnetic field investigations near comet Halley are discussed. The different draping structures of the magnetic field observed by VEGA–1 and VEGA–2 are caused by variations of the solar wind parameters.

103.626   **Calculation of the shape of the contact surface at comet Halley.**
Z.–J. Wu.
ESA Spec. Publ., ESA SP–278, p. 69 – 73 (1987). – See Abstr. 012.075.
Shapes of the contact surface of comet Halley are calculated according to a model in which the position of the contact surface is determined by a balance between the magnetic field pressure and the ion–neutral drag. Shapes of the contact surface under various conditions show some similar features: a round head and

a finite tail. The dimension of the contact surface increases as the comet approaches the sun.

103.627   The foreshock region upstream from the comet Halley bow shock.
S. A. Fuselier, K. A. Anderson, H. Balsiger, K. H. Glassmeier, B. E. Goldstein, M. Neugebauer, H. Rosenbauer, E. G. Shelley.
ESA Spec. Publ., ESA SP–278, p. 77 – 82 (1987). – See Abstr. 012.075.

Increased magnetic field fluctuations in the comet Halley upstream region are correlated with magnetic connection to the bow shock. No backstreaming proton population has been conclusively identified for the times of magnetic connection.

103.628   Physical processes in the vicinity of the cometopause interpreted on the basis of plasma, magnetic field and plasma wave data measured on board the VEGA–2 spacecraft.
A. A. Galeev, K. I. Gringauz, S. I. Klimov, A. P. Remizov, R. Z. Sagdeev, S. P. Savin, A. Yu. Sokolov, M. I. Verigin, K. Szegő, M. Tátrallyay, R. Grard, Ye. G. Yeroshenko (*E. G. Eroshenko*), M. Mogilevsky (*M. Mogilevskij*), W. Riedler, K. Schwingenschuh.
ESA Spec. Publ., ESA SP–278, p. 83 – 87 (1987). – See Abstr. 012.075.

The authors attempt to give a complex approach to the phenomena occuring at the cometopause on the basis of plasma, magnetic field and plasma wave data simultaneously recorded on board the VEGA–2 spacecraft.

103.629   Some new features of plasma wave phenomena at Halley: APV–N observations.
P. Oberc, W. Parzydło, P. Koperski, D. Orłowski, S. Klimov.
ESA Spec. Publ., ESA SP–278, p. 89 – 95 (1987). – See Abstr. 012.075.

The authors have first presented some specific plasma wave features of the two fine boundaries observed by both Vega spacecraft within the bow shocks. These boundaries have not been predicted theoretically and the question what types of plasma discontinuities they represent seems to be largely open.

103.630   Correlation between magnetic and electric field fluctuations observed during flyby near Halley's comet.
Y. Mikhailov (*Yu. Mikhajlov*), M. Moguilevsky (*M. Mogilevskij*), O. Molchanov, E. G. Eroshenko, R. Grard, A. Pedersen, C. Beghin, J.–G. Trotignon, V. Formisano, K. Schwingenschuh.
ESA Spec. Publ., ESA SP–278, p. 109 – 112 (1987). – See Abstr. 012.075.

Comparison is made among the results of measuring the electric and magnetic fields and the plasma density fluctuations in 0.001 – 0.5 Hz frequency band after traversing shock wave during flyby near Halley's comet. The spectra of electric fields, density, and magnetic field B_z–component are of the same shape and exhibit a peak at 1.329 mHz. The electric–to–magnetic field ratio is proportional to the Alfven velocity, thereby confirming the MHD nature of the examined waves.

103.631   Magnetic cavity formation at comet Halley and at the AMPTE Li release.
K. Sauer, K. Baumgärtel.
ESA Spec. Publ., ESA SP–278, p. 113 – 118 (1987). – See Abstr. 012.075.

Main features of the cavity formation around comets and at AMPTE neutral gas releases can be described by a quasi one–dimensional fluid model. Especially, some useful informations about the global plasma flow dynamics and the magnetic field structure have been obtained by this strong coupling approach.

103.632   Plasma phenomena in the vicinity of the closest approach of VEGA–1, 2 spacecraft to the comet Halley nucleus.
V. M. Balebanov, K. I. Gringauz, M. I. Verigin.
ESA Spec. Publ., ESA SP–278, p. 119 – 124 (1987). – See Abstr. 012.075.

The plasma phenomena observed on board VEGA–1,2 spacecraft inside the cometary plasma region are summarized and their nature is discussed. The observed quasi–periodic modulation of the heavy ion density in this region could be explained by the development of the large–scale instability connected with the cyclotron resonance of ions with Alfven oscillations.

103.633   Energy spectra of pick–up ions recorded during the encounter of Giotto with comet Halley.
S. McKenna–Lawlor, B. Wilken, P. Daly, W.–H. Ip, E. Kirsch, A. Coates, A. Johnstone, A. Thompson, D. O'Sullivan, K.–P. Wenzel.
ESA Spec. Publ., ESA SP–278, p. 133 – 137 (1987). – See Abstr. 012.075.

In the present paper, ion data in the range 2 – 280 keV, obtained by combining the particle measurements made aboard Giotto by the Energetic Particle Detector and by the Implanted Ion Spectrometer, March 13 – 14, 1986 are considered. The results are compared with model particle spectra computed on the assumption of adiabatic compression in the cometary accretion flow and of varying degrees of stochastic acceleration from Alfven scattering.

103.634   Stochastic acceleration of cometary ions: the effects of composition and source strength distributions.
W.–H. Ip, W. I. Axford.
ESA Spec. Publ., ESA SP–278, p. 139 – 144 (1987). – See Abstr. 012.075.

For comet Halley, the major chemical components of the volatile ice have been determined to be H_2O, CO and CO_2 by remote–sensing methods and by spacecraft in–situ measurements. The authors evaluate the relative importance of these different species in contributing to the fast neutral population. They focus therefore on the atomic fragments of H_2O, CO and CO_2 molecules, i.e., carbon and oxygen atoms. Some simple model calculations based on the stochastic acceleration process are discussed to investigate the effect of fast neutral source on the ion energy spectra at large distances from the comet.

103.635   Observation of energetic particles (E > 30 keV) by the Giotto experiment EPA in the magnetic cavity of comet Halley.
E. Kirsch, S. McKenna–Lawlor, D. O'Sullivan, A. Thompson, P. W. Daly.
ESA Spec. Publ., ESA SP–278, p. 145 – 148 (1987). – See Abstr. 012.075.

The energetic Particle Analyzer EPA/EPONA onboard the Giotto S/C detected on 14 March 1986 in the magnetic cavity of comet Halley energetic ions with energies > 3.5 MeV as well as electrons > 300 keV. Between the forward and backward viewing telescopes an anisotropy ratio in the count rates of ~2:1 was observed which results from energetic ions since neutral particles should not be detectable. It is suggested that ions and electrons are accelerated by field line merging processes which seem to be possible at the sunward side of the pile up region.

103.636   Detection of parent molecules in the 2.5 to 5 µm spectrum of comet Halley.
J. Crovisier, V. I. Moroz, T. Encrenaz, M. Combes, A. V. Grigoriev (*A. V. Grigor'ev*), N. F. Sanko, J. F. Crifo, J. P. Bibring, N. Coron, D. Bockelée–Morvan, R. Gispert, C. Emerich, J. M. Lamarre, F. Rocard, V. A. Krasnopolsky (*V. A. Krasnopol'skij*), T. Owen.
ESA Spec. Publ., ESA SP–278, p. 157 – 158 (1987). Abstract. – See Abstr. 012.075.

103.637 Evidence for first polymer in comet Halley: polyoxymethylene.
W. F. Huebner, D. C. Boice, C. M. Sharp, A. Korth, R. P. Lin, D. L. Mitchell, H. Reme.
ESA Spec. Publ., ESA SP–278, p. 163 – 167 (1987). – See Abstr. 012.075.
The RPA2–Positive Ion Cluster Composition Analyser on the Giotto spacecraft detected five mass peaks with regular spacing of about 15 amu up to about 120 amu. Starting at about 45 amu, the peaks decrease in intensity with increasing mass. Within their half–width they are in good agreement with dissociation products of formaldehyde polymer (POM). The authors suggest a production sequence in which cosmic radiation formed POM from water and carbon monoxide on grains that were aggregated into cometesimals.

103.638 Polarization of molecular fluorescence bands in comets: recent observations and interpretation.
J. F. Le Borgne, J. Crovisier.
ESA Spec. Publ., ESA SP–278, p. 171 – 175 (1987). – See Abstr. 012.075.
The authors review the theory of polarization of fluorescence emission and its application to cometary molecules. Special consideration is given to OH: it is predicted that the linear polarization of its near–UV band depends upon the comet heliocentric velocity and is small, due to the presence of both directions of polarization in individual lines. They compare these theoretical results with recent polarization obser...vations of the CN, C_2, C_3 and OH bands in P/Halley.

103.639 Photometry of P/Halley (1982i): what we have learned is useful for other comets?
F. A. Catalano, G. A. Baratta, G. Strazzulla.
ESA Spec. Publ., ESA SP–278, p. 181 – 184 (1987). – See Abstr. 012.075.
Photometric observations of P/Halley obtained at the Catania (Italy) observatory, before and after its passage at perihelion are presented. Gas (CN, C_2, C_3) and dust production rates and their variations with the solar distance are evaluated. It is shown that the results as a whole are conciliable with a nucleus whose sublimation is regulated by a dark (polymer–like) material. The ejected dust seems also to contain some absorbing material. The comparison of production rates of CN with those of HCN (taken from recent literature) supports, in the authors' opinion, the hypothesis that HCN is a major parent species of CN.

103.640 Water vapor and hydroxyl distributions in the inner coma of comet Halley measured by the VEGA 2 three channel spectrometer TKS.
V. A. Krasnopolsky (V. A. Krasnopol'skij), A. Yu. Tkachuk, G. Moreels, M. Gogoshev.
ESA Spec. Publ., ESA SP–278, p. 185 – 190 (1987). – See Abstr. 012.075.
Measurements of the H_2O 1.38 μm band on March 9, 1986, demonstrate that water vapor is ejected in a cone of $(40 \pm 10)°$ in the sunward direction with a total production rate of $(8 \pm 2) \times 10^{29} s^{-1}$. Water vapor jets are seen up to 1000 – 2000 km. Gas and dust velocities in the observed jet are 0.85 ± 0.15 and 0.55 ± 0.05 km/s, respectively. Measurements of the OH 309 nm band are reproduced by a model.

103.641 CCD–observations of gas and dust jets in the coma of comet Halley.
C. B. Cosmovici, P. Mack, G. Schwarz, A. Craubner, W.–H. Ip.
ESA Spec. Publ., ESA SP–278, p. 195 – 207 (1987). – See Abstr. 012.075.
The authors report on CCD–observations of comet Halley at the 192 cm SAAO–Telescope (Sutherland, South Africa) using the IHW filters for near nucleus studies and special narrow band filters. A "ring masking technique" was applied in order to visualize the gas and dust jet behavior during five days of observation around the GIOTTO fly–by.

103.642 Photoelectric scans of comet Halley.
C. Sterken, J. Manfroid, D. Hutsemékers, C. Arpigny.
ESA Spec. Publ., ESA SP–278, p. 211 – 215 (1987). – See Abstr. 012.075.
The authors compare radial photometric scans of comet Halley obtained in April 1986 at the European Southern Observatory, La Silla, Chile, with theoretical profiles derived from Haser's model.

103.643 Spatial distribution of $O(^1D)$ from comet Halley.
F. L. Roesler, F. Scherb, K. Magee–Sauer, R. J. Reynolds, J. Harlander, R. J. Oliversen.
ESA Spec. Publ., ESA SP–278, p. 217 – 222 (1987). – See Abstr. 012.075.
Images of comet Halley in [O I] 6300 Å emission obtained using the Wisconsin 150 mm Fabry–Perot spectrometer in the imaging mode have been combined with spectra taken in the scanning mode to deduce the distribution of cometary $O(^1D)$ within an approximately 10 arc min field of view centered on the comet head. The results are modeled to provide photodestruction scale lengths for cometary H_2O and OH, the predominant parents of $O(^1D)$.

103.644 Distribution of neutral gas molecules at large distances from Halley's comet.
G. Erdös, K. Kecskeméty.
ESA Spec. Publ., ESA SP–278, p. 223 – 227 (1987). – See Abstr. 012.075.
Ballistic trajectories of neutral gas molecules around Halley's comet are traced in the cometocentric frame up to large distances from the comet. It is shown that the density of gas significantly deviates from what would be expected if the orbital motion of comet were disregarded. The trajectory calculation suggests that some observations, cayrried out far from Halley's comet, are inconsistent with low outflow velocity.

103.645 Halley's environments observed by the Japanese Suisei spacecraft.
M. Shimizu.
ESA Spec. Publ., ESA SP–278, p. 229 – 232 (1987). – See Abstr. 012.075.
The scientific results on the atomic and ionic environments of comet Halley obtained from the observation of the Suisei spacecraft are reviewed, emphasizing the following topics: determination of rotational period 52.9 ± 0.1 hours, detection of organic hydrogen atom and related biological discussion, and ion density observation in the plasma environment.

103.646 Profiles of the major gas emissions and distribution of dust particles in the inner coma ($0–10^4$ km) of comet Halley.
J. P. Parisot, J. M. Zucconi, M. C. Festou, G. Moreels, J. Clairemidi, J. L. Bertaux, J. E. Blamont, M. Hersé, V. A. Krasnopolsky (V. A. Krasnopol'skij), V. I. Moroz, A. Yu. Tkachuk, M. Gogoshev, T. Gogosheva.
ESA Spec. Publ., ESA SP–278, p. 255 – 262 (1987). – See Abstr. 012.075.
Spectroscopic observations of the inner coma of comet Halley in the visible spectral range (275 – 710 nm) were obtained on March 9 1986 by the TKS spectrometer placed on board the Vega 2 spacecraft. In the inner part of the coma 370 spectra were recorded with a spatial resolution and a temporal resolution never achieved so far from the ground. Density profiles of the dust and the major gases emitting in the optical window were obtained and are analysed and compared with cometary coma models in order to derive the parameters describing the expansion and destruction of both the parent and the daughter species.

103.647   Analysis of the coma outburst of comet Halley March 24 – 25 (UT), 1986.
T. W. Rettig, J. R. Kern, R. Ruchti, B. Baumbaugh,
A. E. Baumbaugh, K. L. Knickerbocker, J. Dawe.
ESA Spec. Publ., ESA SP–278, p. 265 – 269 (1987). – See Abstr. 012.075.
Observations using a fast CCD and custom built Video Data Acquisition System at the Nasmyth focus of the 2.3 meter telescope at Siding Spring Observatory, Australia produced digitized images of comet Halley. Nightly variations in inner coma brightness and shape are noted. On March 25 (UT), an overall brightening and anti–sunward jet was recorded continuously over several hours and is shown to be consistent with the increased activity recorded just prior to these observations.

103.648   Dust coma morphology of comet Halley during the Giotto encounter.
W.–H. Ip, C. B. Cosmovici, P. Mack, H. Craubner,
G. Schwarz.
ESA Spec. Publ., ESA SP–278, p. 271 – 275 (1987). – See Abstr. 012.075.
The CCD images of comet Halley obtained at the South African Astronomical Observatory during the time interval of Giotto encounter on March 14, 1986, were analysed by applying image enhancement techniques. The anisotropic structures of the dust coma so revealed provide additional information of the dust emission pattern.

103.649   Cometary MHD and chemistry: application to Halley.
R. Wegmann, H. U. Schmidt, W. F. Huebner,
D. C. Boice.
ESA Spec. Publ., ESA SP–278, p. 277 – 282 (1987). – See Abstr. 012.075.
Th authors' MHD and chemical comet coma model has been applied to describe and analyze the plasma flow, the magnetic field and the ion abundances in comet Halley in a consistent manner. The physics and chemistry of the coma are modeled. The authors resolve the contact surface (magnetopause). A comparison of the results is made with the data from the HIS ion mass spectrometer.

103.650   Color fluctuations in the inner coma of comet Halley as measured from Vega 2.
J. Clairemidi, G. Moreels, J.–L. Bertaux, M. Hersé,
V. A. Krasnopolsky (*V. A. Krasnopol'skij*), V. I. Moroz,
M. Gogoshev, T. Gogosheva.
ESA Spec. Publ., ESA SP–278, p. 293 – 301 (1987). – See Abstr. 012.075.
A detailed analysis of color variations in the inner coma was made by using the 390 spectra in the 275 – 710 nm range transmitted by the three–channel spectrometer of the Vega 2 spacecraft during the 34 m–period preceding closest approach. The analysis is based on two differential data processing methods. The results show that the pixel–to–pixel ratios of spectra, which are proportional to the ratios of albedos, noticeably vary from a place in the field of view to another. Then the color indexes are used to draw charts where the regions of higher blue or red index are depicted. These regions show a certain degree of spatial periodicity. The regions of higher blue color index appear well correlated with the presence of CO_2^+ emission.

103.651   New properties of cometary dust.
E. Grün, L. Massonne, G. Schwehm.
ESA Spec. Publ., ESA SP–278, p. 305 – 314 (1987). – See Abstr. 012.075.
Previously unknown phenomena involving dust and new properties of cometary matter have been observed by the space probes to comet Halley as well as by Earth–based observations. Much more refined modeling is necessary to interpret the new data. The observation of the coma boundaries for particles in the mass range 10^{-16}g to 10^{-10}g is mostly compatible with optically absorbing particles. The observed variability of the dust flux of a factor 3 to 5 is due to spatially and timely varying dust emission

on the nuclear surface. Because of the flat shape of the cumulative mass distribution the total dust mass production is critically dependent on the mass of the largest particle emitted from the nucleus which escaped the detection by the space probes.

103.652   Comet P/Halley's near–nucleus jet activity.
N. Thomas, H. U. Keller.
ESA Spec. Publ., ESA SP–278, p. 337 – 342 (1987). – See Abstr. 012.075.
The Halley Multicolour Camera (HMC) on board the European spacecraft, Giotto, returned high resolution images of the nuclear and near–nuclear regions of comet P/Halley in March 1986. The images transmitted in the seconds prior to closest approach allow a detailed study of an active region down to a resolution of ~ 50 m per pixel. Variations in activity within the main emission have been observed and results are presented which indicate deviations from a r^{-1} law to a less steep profile near the source region. The results obtained provide new constraints on models of cometary jet activity.

103.653   Dust photometry in the near nucleus region of comet Halley from VEGA–2 observations.
I. Tóth, K. Szegö, A. Kondor.
ESA Spec. Publ., ESA SP–278, p. 343 – 347 (1987). – See Abstr. 012.075.
The images obtained by VEGA–2 spacecraft during the encounter with Halley's comet show interesting jet structures. A semi–empirical method has been developed and presented in this paper for the special near nucleus conditions to estimate the dust and gas contribution to brightness distribution. Some of the jets have been analysed using this modified Haser model.

103.654   Dynamics of charged submicron grains in Halley's coma.
M. K. Wallis, M. H. A. Hassan.
ESA Spec. Publ., ESA SP–278, p. 351 – 353 (1987). – See Abstr. 012.075.
Two characteristics of the smallest (femtogramme) grains in comet Halley's coma are ascribed to electrical forces on them, namely, their penetration to large distances with any remnants of 'envelopes' eliminated, and the dispersion with grain size of dust jets in the inner coma. The authors show this quantitatively for the data from Vega–2, when the magnetic field was at a suitable orientation.

103.655   Composition of cometary dust particles.
P. Hsiung, J. Kissel.
ESA Spec. Publ., ESA SP–278, p. 355 – 358 (1987). – See Abstr. 012.075.
The dust impact analyser on board the Giotto spacecraft provide more than 3000 mass spectra of cometary dust grains during its flyby of Halley's comet. About 120 of them were transmitted in mode 0 which gives direct access to the mass numbers. A detailed analysis of these spectra has been done including the composition of single grains and a global elemental composition. The variation of the elemental composition with the grain's mass clearly shows a dependence between C, N, and H on one side, and between Si and O on the other side.

103.656   Isotopic composition of Halley dust.
M. Šolc, V. Vanýsek, J. Kissel.
ESA Spec. Publ., ESA SP–278, p. 359 – 362 (1987). – See Abstr. 012.075.
On the basis of PUMA 1 mode 0 mass spectra of Halley solid grains isotopic patterns have been found for C, O, Mg, S, Fe (and Cl, probably not of cometary origin). While isotopic abundances calculated for O, Mg, S, Cl and Fe are near to the terrestrial values, carbon isotope ratio $^{12}C/^{13}C$ varies from about 30 to several hundreds. Temporal and local carbon isotope ratio dependence is given along the path of Vega 1.

103.657 Cometary evolution: clues from chondritic interplanetary dust particles.
F. J. M. Rietmeijer, I. D. R. Mackinnon.
ESA Spec. Publ., ESA SP–278, p. 363 – 367 (1987). – See Abstr. 012.075.
Chondritic interplanetary dust particles have properties consistent with properties expected for cometary dust. The complex and varied mineralogy of these particles may be indicative of mineral alteration processes that occur in comet nuclei. Depending on the thermal budget of a comet, the upper few metres of nucleus material may maintain temperatures within regimes of hydrocryogenic ($\sim 200 - 273K$) and low–temperature aqueous ($274 - 400K$) alteration. Thus, layer silicates, carbonates and sulfates may be important components of cometary dust and correspondingly are common constituents of chondritic interplanetary dust particles.

103.658 On the interpretation of the 3 μm emission feature in the spectrum of comet Halley: abundances in comet Halley and in interstellar matter.
T. Encrenaz, J. L. Puget, J. P. Bibring, M. Combes, J. Crovisier, C. Emerich, L. d'Hendecourt, F. Rocard.
ESA Spec. Publ., ESA SP–278, p. 369 – 376 (1987). – See Abstr. 012.075.
By comparing the nature and the interpretation of the 3 μm feature in interstellar spectra and in the spectrum of comet Halley, the authors have derived a possible interpretation and an excitation mechanism in comet Halley, which has been used to estimate the abundance of hydrocarbons in P/Halley. Then the authors have used these numbers to derive relative abundances of H, O, C and N in P/Halley, and they have compared them with cosmic and interstellar dust abundances. This comparison first indicates that important fractions of carbon (40%) and nitrogen (more than 50%) are trapped in cometary grains.

103.659 P/Halley dust coma: grains or rocks?
S. F. Green, J. A. M. McDonnell, C. H. Perry, S. Nappo, J. C. Zarnecki.
ESA Spec. Publ., ESA SP–278, p. 379 – 384 (1987). – See Abstr. 012.075.
The Giotto Dust Impact Detection System and Particle Impact Analyser measured the dust flux in the coma of P/Halley for masses $>10^{-19}$kg. The mass distribution is discussed with emphasis on the large particles in the mg range.

103.660 The origin of low mass particles within and beyond the dust coma envelopes of comet Halley.
J. A. Simpson, D. Rabinowitz, A. J. Tuzzolino, L. V. Ksanfomality (*L. V. Ksanfomaliti*), R. Z. Sagdeev.
ESA Spec. Publ., ESA SP–278, p. 391 – 397 (1987). – See Abstr. 012.075.
Measurements from the Dust Counter and Mass Analyzer instruments on the VEGA–1 and VEGA–2 spacecraft have revealed unexpected fluxes of low mass ($\geqslant 10^{-13}$g) dust particles at very great distances from the nucleus ($\sim 3 \times 10^5$ to 6×10^5km). These particles are detected in clusters (~ 10 second duration), preceded and followed by relatively long time intervals during which no dust is detected. The authors show that this cluster phenomenon also occurs inside the envelope boundaries.

103.661 Optical and hydrodynamic implications of comet Halley dust size distribution.
J. F. Crifo.
ESA Spec. Publ., ESA SP–278, p. 399 – 408 (1987). – See Abstr. 012.075.
With the help of a specially developed radiative hydrodynamic code, the author examines the compatibility between (1) the dust fluences recorded by detectors on board the "Giotto" and "Vega" flyby spacecraft, (2) the rate of deceleration of the Giotto spacecraft during encounter, and (3) a set of selected pre– and post–perihelion spectra of the comet extending from near to far infrared wavelengths.

103.662 Compatibility of the in–situ mass distribution with photopolarimetric observations of comet Halley.
P. L. Lamy, E. Grün, J. M. Perrin.
ESA Spec. Publ., ESA SP–278, p. 409 – 410 (1987). – See Abstr. 012.075. The full paper will appear in Astron. Astrophys.

103.663 Cometary grains and interplanetary dust: a polarimetric sensing.
A. Dollfus.
ESA Spec. Publ., ESA SP–278, p. 419 – 421 (1987). – See Abstr. 012.075.
Physical properties of the solid grains ejected from the nucleus of comet P/Halley have been sensed by optical telescopic polarimetry. In the coma, the presence is indicated of a large number of fluffy aggregates made of very dark small particles. Such type of flakes are subsequently subjected to the Poynting–Robertson effect and they are progressively transported in the inner part of the Solar System.

103.664 The dust around the nucleus of comet Halley.
P. Bastien, L. Drissen, F. Ménard, N. St–Louis, R. Nadeau.
ESA Spec. Publ., ESA SP–278, p. 423 – 426 (1987). – See Abstr. 012.075.
The dust in the coma of P/Halley has been studied by polarimetry in narrow band filters. The authors discuss observations obtained before and after perihelion. In the continuum, the polarization was mostly independent of wavelength before perihelion, but after perihelion, it decreased slowly from the red to the blue. A depolarization was clearly observed in the emission bands.

103.665 Variation of grain properties at the dust outbursts.
T. Mukai, S. Mukai, S. Kikuchi.
ESA Spec. Publ., ESA SP–278, p. 427 – 430 (1987). – See Abstr. 012.075.
The authors examine the simultaneous appearance of higher polarization and brighter scattered light in comets by considering a variation of grain properties during the dust outbursts.

103.666 The evolution of the mass distribution of cometary particles.
A. Hajduk, I. Kapišinský.
ESA Spec. Publ., ESA SP–278, p. 441 – 444 (1987). – See Abstr. 012.075.
A model for the evolution of the mass distribution of particles ejected from comet Halley is proposed on the basis of results of space experiments and ground based observations of Halley meteor showers. It is shown that the mass index of dust particles varies in the evolutionary process, following the diffusion of particles in the interplanetary medium.

103.667 The nucleus of comet Halley.
H. U. Keller.
ESA Spec. Publ., ESA SP–278, p. 447 – 454 (1987). – See Abstr. 012.075.
This contribution briefly summarizes the various aspects contributing to our understanding of the nature of cometary nuclei. It then focuses on the results of the imaging experiments and in particular on those of the Halley Multicolour Camera on board the ESA Giotto spacecraft.

103.668 Comparison of active regions on the nucleus of comet Halley.
H. J. Reitsema, W. A. Delamere, F. L. Whipple.
ESA Spec. Publ., ESA SP–278, p. 455 – 459 (1987). – See Abstr. 012.075.
The images of the nucleus of comet Halley returned by the GIOTTO spacecraft reveal a number of active regions on the surface, one of which is near the expected location of the rotation axis. This feature is larger and brighter than other source regions, suggesting that the mechanism which drives this source is also different. At this active region near the rotation pole the sun is circumpolar for a significant portion of the solar encounter.

Continuous solar insolation causes heating of the nucleus to greater depths and results in the unique properties of this source region.

103.669 Reconstruction of the orientation and shape of the nucleus of comet Halley.
K. Szegö, A. Kondor, I. Tóth, R. Z. Sagdeev, K. Wilhelm, H. U. Keller.
ESA Spec. Publ., ESA SP–278, p. 463 – 469 (1987). – See Abstr. 012.075.
In this paper a constructive modelling technique is developed for the analysis of the imaging data of VEGA and GIOTTO. As a first result the position of the small and big ends of the nucleus is obtained unambiguously for the different encounters.

103.670 Properties of the nucleus of P/Halley.
D. Möhlmann, M. Danz, H. Börner.
ESA Spec. Publ., ESA SP–278, p. 487 – 492 (1987). – See Abstr. 012.075.
The reality of surface structures of P/Halley, as they can be seen on processed VEGA images, is critically discussed. As a result of these reinvestigations the reality of these large–scale structures has been confirmed. Based on the assumption that these surface structures are related to large–scale internal structures, it is concluded that cometary nuclei have grown by accretion of large km–sized "building blocks".

103.671 Plasma tail of comet Halley in January 1986.
E. M. Pittich, J. Zvolánková, D. Kubáček.
ESA Spec. Publ., ESA SP–278, p. 549 – 553 (1987). – See Abstr. 012.075.
A set of large–scale phenomena exposures of comet Halley between January 5 and 10, 1986 is investigated. Within this period a disruption of the comet's plasma tail occured. The elbow of the disrupted plasma tail observed on January 10 moved from the nucleus at a mean radial velocity of 67 ± 4 km/s. This disconnection event is studied in relation to the solar wind flow and the interplanetary magnetic field.

103.672 Structure of interplanetary magnetic field during the approach of P/Halley and associated solar activity phenomena.
B. Valníček.
ESA Spec. Publ., ESA SP–278, p. 555 – 557 (1987). – See Abstr. 012.075.
Interplanetary magnetic field during the period December 1985 – April 1986 was very chaotic. Therefore, its systematic study is very difficult. An analysis of the occurrence of disconnection phenomena in the tail of P/Halley demonstrates narrow association to the coronal holes and sunspot groups. Therefore it can be said that not only the crossing of neutral sheet can cause disconnection events, but also the interaction with solar wind streamers.

103.673 The evolution of CN emission in Halley's comet with heliocentric distance.
M. Prieto, C. García, F. Garzón, M. Kidger.
ESA Spec. Publ., ESA SP–278, p. 561 – 565 (1987). – See Abstr. 012.075.
The authors have made observations of the spectrum of comet Halley on various dates between September 1985 and December 1986, obtaining multipositional long–slit absolute spectrophotometry of the inner region of the coma in the visible range. They present a study of the emission in the 3883 Å line of the (0–0) transition of CN. The observed CN surface brightness spatial distribution is used to obtain the distribution of molecules and is fitted to Haser's model to obtain scale lengths.

103.674 Study of the ionic and neutral species in the coma of comet Halley with an image intensifier camera.
T. Chandrasekhar, C. D. Prasad, J. N. Desai, N. M. Ashok, R. Gupta.
ESA Spec. Publ., ESA SP–278, p. 567 – 570 (1987). – See Abstr. 012.075.
Imaging Fabry–Perot spectroscopy in the lines of hydrogen Balmer alpha and C_2 and IHW filter photography on comet Halley carried out with a small telescope and an image intensifier camera are described. A special event seen in H_2O^+ emission and in the H alpha interferogram on March 13, 1986 is detailed.

103.675 P/Halley activity in November 1985.
T. Bonev, V. Shkodrov, V. Ivanova, V. Boneva.
ESA Spec. Publ., ESA SP–278, p. 591 – 594 (1987). – See Abstr. 012.075.
U– and B–plates of 12/13 Nov. 1985, taken on the 50/70–cm Schmidt and the 2–m RCC telescope of the Rozhen National Astronomical Observatory, show an ejection from the inner coma region towards the south. On 15/16 Nov. on U– and B–plates again, a strong ejection was recorded, extended to the north–northeast. At the same time in the R–band only a slight deformation of the coma can be seen. To explain these observations a model is proposed interpreting them as a high velocity ejection of matter off the rotating and precessing comet nucleus.

103.676 Outburst of Halley's comet.
P. S. Chen, Y. K. Chen, H. Gao, L. Li, Y. Zhang, J. Yang, M. Mei.
ESA Spec. Publ., ESA SP–278, p. 595 – 597 (1987). – See Abstr. 012.075.
During Mar. 20 – Apr. 3 of 1986 a photometric monitoring for Halley's comet at optical and near infrared regions was made at Yunnan Observatory in China. It was shown that Halley's comet had a remarkable outburst after Mar. 20 but before Mar. 26 (perhaps on Mar. 22). Characteristics of variations in all observed wavelengths and possible explanations for some phenomena during the outburst are demonstrated.

103.677 Comet P/Halley: visual magnitude estimates and gas production.
D. Fischer, S. Hüttemeister.
ESA Spec. Publ., ESA SP–278, p. 599 – 605 (1987). – See Abstr. 012.075.
More visual magnitude estimates have been made of comet P/Halley than of any other comet in history. This unique data base allows to work out systematic sources of error and to remove them by a rigid selection procedure. From the remaining sample brightness laws of two different types are deduced. Several studies of molecular production rates and their changes during the 1985–1986 apparition of P/Halley are compared with these laws in order to obtain formulae transforming visual magnitude estimates directly into production rates.

103.678 Spectrophotometric observations of comet Halley.
J. A. de Freitas Pacheco, S. J. C. Landaberry, P. D. Singh.
ESA Spec. Publ., ESA SP–278, p. 627 – 631 (1987). – See Abstr. 012.075.
Spectrophotometric observations of comet Halley in the wavelength range (3600 – 7000) Å during the period October 15, 1985 and April 9, 1986 are presented. The observed fluxes of $O(^1D)$, CN(388 nm) and C_2(516 nm) are analysed in the framework of Haser's model and production rates of these cometary species are determined.

103.679 Nature of comet Halley's organic grains inferred from infrared spectra.
M. K. Wallis, R. Rabilizirov, N. C. Wickramasinghe, S. Al–Mufti.
ESA Spec. Publ., ESA SP–278, p. 635 – 637 (1987). – See Abstr. 012.075.
The infrared radiation from a distribution of organic grains in Halley's coma is compared with astronomical spectra, concentrating on the $3 - 4 \mu m$ region which discriminates between

compositions. Grain temperatures appropriate to free space are calculated for non–evaporating and chemically charred material making up grains of various sizes. The particle size distribution found by the comet probes is used together with Mie theory to calculate the scattered light and thermal emission.

103.680 The dust tail of comet Halley.
M. Fulle, C. Barbieri, G. Cremonese.
ESA Spec. Publ., ESA SP–278, p. 639 – 644 (1987). – See Abstr. 012.075.
This work concerns the photometric analysis of 10 wide field CCD images (702 nm and RG 830 filters) taken at ESO between 24–2 and 5–4–1986. Derived isophote fields were used as input for the inverse approach to the Finson–Probstein method for interpreting cometary dust tails.

103.681 Determination of the aberration angle of Halley's comet (1982i).
L.–s. Yan, Z.–L. Ma, Y. Wang, M. Zhang, L.–x. Dai.
ESA Spec. Publ., ESA SP–278, p. 653 – 654 (1987). – See Abstr. 012.075.
To determine the value of aberration angle ε, 21 plates taken with 15 cm astrograph at Xinjing province, northwest of China from 30 Dec. 1985 to 5 Jan. 1986 have been chosen. They are just at the interval of two DE. Calculations show that the angle was wobbled about 3°.

103.682 On the periodic southward jets of comet Halley before perihelion passage.
J.–i. Watanabe, H. Kawakami, K. Tomita, K. Takagishi, H. Kinoshita, T. Nakamura, Y. Kozai.
ESA Spec. Publ., ESA SP–278, p. 657 – 663 (1987). – See Abstr. 012.075.
The authors report that the period of southward jets observed before perihelion passage is about 7 – 8 days. This period is determined by the morphological analysis of near–nucleus images of comet Halley.

103.683 Dust tail streamers and Halley's nucleus rotation.
K. Beisser, H. Boehnhardt.
ESA Spec. Publ., ESA SP–278, p. 665 – 670 (1987). – See Abstr. 012.075.
An overview of the various appearances of the dust tail of Comet Halley during its 1986 return is outlined in a series of synchrone–syndyne–plots. Using ESO–Schmidt–telescope images of Comet Halley a clear indication of the rotation period was found in the dust streamers (= synchrones) of the tail.

103.684 Near–infrared imaging of comet Halley: discovery of a color gradient in the inner coma.
M. J. Rieke, H. Campins.
ESA Spec. Publ., ESA SP–278, p. 671 – 675 (1987). – See Abstr. 012.075.
Near–infrared images of comet Halley were obtained in the standard J, H and K bandpasses, on 1985 Nov. 3.5 with a HgCdTe camera at the U. of Arizona 1.54 m Telescope. A well defined gradient in the J–H and H–K colors within ~ 5,000 km of the nucleus has been discovered with the bluest colors at the photocenter. Surface brightness profiles steeper than the "canonical" $1/\varrho$ are observed in the same region. A preliminary analysis indicates that the color gradient and the brightness profiles can both be explained by the presence of volatile (dirty ice) grains in the inner coma.

103.685 Linear and circular polarization of comet Halley light.
N. G. Beskrovnaja (N. G. Beskrovnaya), N. A. Silant'ev, N. N. Kiselev, G. P. Chernova.
ESA Spec. Publ., ESA SP–278, p. 681 – 685 (1987). – See Abstr. 012.075.
Polarimetry of comet Halley revealed two new observational facts: gradual turning of the polarization plane relative to the scattering plane in the inversion angle vicinity (α_0) and the circular polarization. The polarization plane turning has no explanation in terms of Mie scattering theory.

103.686 The light curve of P/Halley and the rotation of its nucleus.
M. C. Festou, P. Drossart, J. Lecacheux, T. Encrenaz, F. Puel, J. L. Kohl–Moreira.
ESA Spec. Publ., ESA SP–278, p. 695 – 701 (1987). For the full paper see Abstr. 44.103.551. – See Abstr. 012.075.

103.687 Temperature of the nucleus of comet Halley.
C. Emerich, R. Gispert, J. M. Lamarre, N. Coron, M. Combes, J. Crovisier, T. Encrenaz, F. Rocard, J. P. Bibring, V. I. Moroz, N. F. Sanko (N. F. San'ko), Yu. V. Nikolsky (Yu. V. Nikol'skij).
ESA Spec. Publ., ESA SP–278, p. 703 – 706 (1987). – See Abstr. 012.075.
The infrared thermal radiation of the nucleus of Halley's comet at 0.8 AU from the Sun was measured by the imaging channel of the IKS–VEGA 1 spectrophotometer in two wavelengths bands (7 – 10 μm and 9 – 14 μm). The surface temperature of the nucleus is shown to be subject to high gradients relative to the Sun direction. These observations suggest that a large part of the Halley's comet nucleus is covered by a refractory, dark and inactive crust.

103.688 Consequences of the size determination of P/Halley by space probes on the scale of sizes of cometary nuclei.
J. Svoreň.
ESA Spec. Publ., ESA SP–278, p. 707 – 712 (1987). – See Abstr. 012.075.
Photometric data referring to heliocentric distances beyond 2.5 AU on the pre–perihelion arc of the orbit of P/Halley are used. On the basis of the estimate of a projected area of about 100 km² and of the brightness 24.1^m at the extreme distance, the albedo p = 0.055 has been derived. This value of albedo is used to calculate the new values of radii of short–period comets. One of the possibilities, to explain the small difference between sizes of long–period and short–period comets, is that the mass loss causes the decrease of density of a nucleus.

103.689 The splitting of comet Halley 1986.
D.–h. Chen, Z.–l. Liu, J.–q. Zheng, L.–s. Yan, L.–z. Liu, Z.–x. Wu, X.–h. Zhou, A. C. Gilmore.
ESA Spec. Publ., ESA SP–278, p. 721 – 724 (1987). – See Abstr. 012.075.
Photographs, taken on March 25, 1986, show clearly that the nucleus of comet Halley has split into two parts: a primary nucleus and a secondary nucleus. It is of considerable interest to note that the secondary nucleus has its own coma and jet.

103.690 Three approaches to the modelling of comet Halley's motion over long periods of time.
K. Ziołkowski.
ESA Spec. Publ., ESA SP–278, p. 747 – 750 (1987). – See Abstr. 012.075.
Diversity of the existing models of the motion of Halley's comet makes uncertain our knowledge of the comet's dynamic behavior over long periods of time. In order to better recognize the problem it is worth to make a thorough study of different approaches to the investigations of the long–term motion of the comet.

103.691 Brightness distribution in the coma of comet Halley (1982i).
S. C. Joshi, B. B. Sanwal.
Earth, Moon, Planets, Vol. 39, No. 3, p. 203 – 205 (1987).
Photographic photometry on a photograph of comet Halley taken on March 15, 1986, through a Schott GG 385 filter on Kodak 103a–O plate has been carried out. The average magnitude per square mm of the coma at various distances from the nucleus have been estimated. The total integrated magnitude within two arc minutes of the nucleus has been estimated to be 5.5 magnitude.

103.692 Evidence for the nucleus rotation in streamer patterns of comet Halley's dust tail.
K. Beisser, H. Boehnhardt.
Astrophys. Space Sci., Vol. 139, No. 1, p. 5 – 12 (1987).

A generalized Finson–Probstein formalism was used to analyse the streamer patterns in the dust tail of comet Halley as observed between 22 February and 11 March, 1986. A periodic appearance of a pattern of double streamers was found in the dust tail, indicating a time interval in the emission of this structure of about 6.5 days, if interpreted as synchrones. The time interval between the emission of the two components of each double streamer is about 2.5 days. The results are discussed with respect to the rotation of Halley's nucleus.

103.693 On some periodicities in the motion of comet Halley relative to the mass center.
V. A. Sarychev, V. V. Sazonov, V. V. Savchenko, V. I. Tarnopol'skij.
Kosm. Issled., Tom 25, Vyp. 5, p. 768 – 780 (1987). In Russian. English translation in Cosm. Res.

103.694 Results of the IKS experiment. 1. Infrared radiation of parent molecules in comet Halley.
V. I. Moroz, M. Combes, A. V. Grigor'ev, J. F. Crifo, T. Encrenaz, J. Crovisier, J. P. Bibring, N. F. San'ko, N. Coron, D. Bockeleé–Morvan, Yu. V. Nikol'skij, R. Gispert, J. M. Lamarre, V. A. Krasnopol'skij, C. Emerich, F. Rocard, T. Owen.
Kosm. Issled., Tom 25, Vyp. 5, p. 781 – 792 (1987). In Russian. English translation in Cosm. Res.

103.695 Results of the IKS experiment. 2. Infrared radiometry of the nucleus of comet Halley.
Yu. V. Nikol'skij, C. Emerich, J. M. Lamarre, N. F. San'ko, M. Combes, V. I. Moroz, J. Crovisier, T. Encrenaz, F. Rocard, A. V. Grigor'ev, R. Gispert, J. P. Bibring, A. V. Kiselev, N. Coron.
Kosm. Issled., Tom 25, Vyp. 5, p. 793 – 809 (1987). In Russian. English translation in Cosm. Res.

103.696 Analysis of the $\lambda 3.35\ \mu m$ hydrocarbon band in the spectrum of comet Halley from data of the IKS installed aboard the Vega 1 spacecraft.
A. V. Grigor'ev.
Kosm. Issled., Tom 25, Vyp. 5, p. 810 – 814 (1987). In Russian. English translation in Cosm. Res.

103.697 Photometric characteristics of the nucleus and inner coma of comet Halley.
R. Z. Sagdeev, G. A. Avanesov, B. S. Zhukov, I. V. Zenkin, Ya. L. Ziman, V. I. Moroz, V. I. Tarnopol'skij.
Kosm. Issled., Tom 25, Vyp. 6, p. 831 – 839 (1987). In Russian. English translation in Cosm. Res.

103.698 The dust envelope of comet Halley according to data obtained with PUMA.
R. Z. Sagdeev, E. N. Evlanov, B. V. Zubkov, N. P. Kutyreva, V. N. Khromov, V. D. Shapiro, O. F. Prilutskij, M. N. Fomenkova.
Kosm. Issled., Tom 25, Vyp. 6, p. 840 – 848 (1987). In Russian. English translation in Cosm. Res.

103.699 Analysis of the composition of dust particles of comet Halley according to measurements with the PUMA mass spectrometer.
R. Z. Sagdeev, E. N. Evlanov, B. V. Zubkov, L. M. Mukhin, O. F. Prilutskij, M. N. Fomenkova, V. N. Khromov.
Kosm. Issled., Tom 25, Vyp. 6, p. 849 – 855 (1987). In Russian. English translation in Cosm. Res.

103.700 Classification of dust particles of comet Halley according to data obtained with the PUMA–2 mass spectrometer.
R. Z. Sagdeev, L. M. Mukhin, E. N. Evlanov, B. V. Zubkov, O. F. Prilutskij, M. N. Fomenkova.
Kosm. Issled., Tom 25, Vyp. 6, p. 856 – 859 (1987). In Russian. English translation in Cosm. Res.

103.701 Recording of dust particles in the nucleus environment of comet Halley with "Foton".
S. I. Anisimov, V. P. Karyagin, V. M. Kovtunenko, A. B. Konstantinov, R. S. Kremnev, V. A. Kudryashov, Yu. A. Osip'yan, Yu. A. Ryzhov, S. B. Svirshchevskij, A. Z. Strukov, A. V. Terterashvili, V. E. Fortov.
Kosm. Issled., Tom 25, Vyp. 6, p. 860 – 866 (1987). In Russian. English translation in Cosm. Res.

103.702 Mass spectrum and spatial distribution of dust in comet Halley's head according to data obtained with the SP–1 detector on board Vega 1 and Vega 2.
O. L. Vajsberg, V. N. Smirnov, L. S. Gorn, M. V. Iovlev.
Kosm. Issled., Tom 25, Vyp. 6, p. 867 – 883 (1987). In Russian. English translation in Cosm. Res.

103.703 Tomographic reconstruction of the inner coma of comet Halley from TV–information data of Vega 1.
A. Yu. Kogan, V. N. Khejfets.
Kosm. Issled., Tom 25, Vyp. 6, p. 884 – 894 (1987). In Russian. English translation in Cosm. Res.

103.704 Measurement of neutral particle concentration in comet Halley's environment with PLAZMAG–1 on board Vega 1 and Vega 2.
A. P. Remizov, M. I. Verigin, K. I. Gringauz, I. Apathy, I. Szemerey, T. Gombosi, A. K. Richter.
Kosm. Issled., Tom 25, Vyp. 6, p. 895 – 899 (1987). In Russian. English translation in Cosm. Res.

103.705 Position and structure of the shock wave near comet Halley according to Vega 1 and Vega 2 measurements.
A. A. Galeev, B. Eh. Gribov, T. Gombosi, K. I. Gringauz, S. I. Klimov, P. Oberts, A. P. Remizov, W. Riedler, R. Z. Sagdeev, S. P. Savin, A. Yu. Sokolov, V. D. Shapiro, V. I. Shevchenko, K. Szegö, M. I. Verigin, E. G. Eroshenko.
Kosm. Issled., Tom 25, Vyp. 6, p. 900 – 906 (1987). In Russian. English translation in Cosm. Res.

103.706 Peculiarities of the plasma transition region of comet Halley (cometary sheath) according to Vega 1 and Vega 2 data.
M. I. Verigin, K. I. Gringauz, A. Richter, T. Gombosi, A. P. Remizov, K. Szegö, I. Apathy, I. Szemerey, M. Tátrallyay, L. A. Lezhen.
Kosm. Issled., Tom 25, Vyp. 6, p. 907 – 913 (1987). In Russian. English translation in Cosm. Res.

103.707 The region of cometary ions in comet Halley's head according to Vega 2 data.
K. I. Gringauz, M. I. Verigin, A. Richter, T. Gombosi, K. Szegö, M. Tátrallyay, A. P. Remizov, I. Apathy.
Kosm. Issled., Tom 25, Vyp. 6, p. 914 – 920 (1987). In Russian. English translation in Cosm. Res.

103.708 On the possible experimental detection of cometary plasma acceleration connected with the variation of the magnetic field's direction according to PLAZMAG–1 data installed aboard Vega 1.
M. I. Verigin, I. Axford, K. I. Gringauz, A. Richter.
Kosm. Issled., Tom 25, Vyp. 6, p. 921 – 926 (1987). In Russian. English translation in Cosm. Res.

103.709 **The electron component of plasma in comet Halley's environment according to data of measurements made with PLAZMAG–1 aboard Vega 2.**
K. I. Gringauz, A. P. Remizov, M. I. Verigin, A. Richter,
M. Tátrallyay, K. Szegö, I. N. Klimenko, I. Apathy,
T. Gombosi, I. Szemerey.
Kosm. Issled., Tom 25, Vyp. 6, p. 927 – 931 (1987). In Russian.
English translation in Cosm. Res.

103.710 **Measurements of energetic cometary ions in the solar wind before the shock wave of comet Halley.**
K. Kecskeméty, T. Cravens, V. V. Afonin, A. Varga,
K.–P. Wenzel, M. I. Verigin, Lu Gan, T. Gombosi,
K. I. Gringauz, E. G. Eroshenko, E. Keppler, I. N. Klimenko,
R. Marsden, A. Nagy, A. P. Remizov, W. Riedler, A. Richter,
K. Szegö, M. Tátrallyay, K. Schwingenschuh, A. Somogyi,
G. Erdös.
Kosm. Issled., Tom 25, Vyp. 6, p. 932 – 942 (1987). In Russian.
English translation in Cosm. Res.

103.711 **Investigations of plasma waves aboard the automatic interplanetary station Vega 1.**
S. P. Klimov, Z. Krawczyk, V. E. Korepanov, S. P. Savin,
V. M. Balebanov, G. B. Simonenko, K. Nowak,
Ya. N. Aleksevich, A. Yu. Sokolov.
Kosm. Issled., Tom 25, Vyp. 6, p. 943 – 951 (1987). In Russian.
English translation in Cosm. Res.

103.712 **Possible nuclear splitting of comet Halley in March 1986.**
D.–h. Chen, Z.–l. Liu, J.–q. Zheng, L.–s. Yan.
Chin. Astron. Astrophys., Vol. 11, No. 4, p. 307 – 308 (1987).
English translation of 43.103.698.

103.713 **An explosion on comet Halley.**
P.–s. Chen, Y.–k. Chen, L. Lu, H. Gao, J. Yang,
M. Mei.
Chin. Astron. Astrophys., Vol. 11, No. 4, p. 307 – 308 (1987).
English translation of 43.103.697.

103.714 **Infrared photometry and spectrophotometry of comet Halley.**
P.–s. Chen, H. Gao, Y. Zhang, Y.–k. Chen, J. Yang, M. Mei.
Chin. Astron. Astrophys., Vol. 11, No. 4, p. 309 – 311 (1987).
English translation of Acta Astron. Sin., Vol. 28, No. 2, p. 211 – 214 (1987).
The authors report on photometric and spectrophotometric measurements of comet Halley in the near infrared. The observations confirmed the explosion found in a preceding article and showed the post–explosion behaviour of the comet was different in the visible and the infrared. Water emission lines at 1.4 cm and 1.9 cm were found.

103.715 **Near nuclear region of comet Halley based on the imaging results of the VEGA mission.**
R. Z. Sagdeev, K. Szegö.
Report KFKI–1987–35/C, Hungarian Acad. Sci., Budapest, Hungary, 27 pp. (1987). Abstr. in Phys. Abstr., Vol. 90, No. 1317, Entry 139984 (1987).

103.716 **Observing comet Halley's near–nucleus features.**
S. Larson, D. H. Levy.
Astronomy, Vol. 15, No. 9, p. 90 – 95 (1987). Abstr. in Phys. Abstr., Vol. 91, No. 1319, Entry 4316 (1988).

103.717 **Infrared emission by organic grains in the coma of comet Halley.**
C. Chyba, C. Sagan.
Nature, Vol. 330, No. 6146, p. 350 – 353 (1987).
The authors discuss transmission spectra of organic residues produced by laboratory irradiation of candidate cometary ices. The laboratory synthesis of solid organic residue from irradiated low–occupancy methane ice clathrate simulates the radiation processing experienced by comet Halley. The transmission

spectrum of this residue, convolved with a simple two–component thermal emission model fits the 3.4 μm feature, provides optical depths in excellent agreement with those observationally determined, and accounts for the absence of features at longer wavelengths.

103.718 **How does the nucleus of the comet Halley rotate?**
L. Kresák.
Kozmos, Vol. 18, No. 6, p. 194 – 195 (1987). In Slovak.

103.719 **Look–back for comet Halley.**
V. Vanýsek.
Říše hvězd, Vol. 68, No. 11, p. 210 – 214 (1987). In Czech.

103.720 **Observation of the OH 1667 MHz line toward comet Halley from the southern hemisphere.**
E. Bajaja, R. Morras, W. G. L. Pöppel, J. C. Cersosimo,
M. C. Martín, J. Mazzaro, J. C. Olalde, A. Silva, M. E. Arnal,
F. R. Colomb, I. F. Mirabel, V. Boriakoff.
Astrophys. J., Vol. 322, No. 1, p. 549 – 558 (1987).
Halley's comet has been tracked for observing at the Instituto Argentino de Radioastronomía (IAR) in the 1667 MHz line of OH, 4 hr a day, from 1986 February 6 to May 1. It has been detected in absorption from March 24 to April 23. The total number of OH molecules has been estimated for each day from the observed profiles. The average of the OH production rate derived from these observations is $(4.6 \pm 0.2) \times 10^{29}$ molecules s^{-1}.

103.721 **Observations of comet Halley.**
Komet. Tsirk., Nos. 372, 374 – 377, 379 – 381 (1987).
In Russian.

Periodic comet West–Kohoutek–Ikemura

103.801 **Periodic comet West–Kohoutek–Ikemura (1987x).**
IAU Circ., Nos. 4456, 4464 (1987).

103.802 **Periodic comet West–Kohoutek–Ikemura (1987x).**
Yamamoto Circ., No. 2093 (1987). In Japanese.

103.803 **Short–period comet West–Kohoutek–Ikemura (1987x).**
Komet. Tsirk., No. 375 (1987). In Russian.

Periodic comet Kohoutek

103.811 **Periodic comet Kohoutek (1986k).**
IAU Circ., Nos. 4461, 4492, 4503 (1987).

Comet 1987y Levy

103.821 **Comet Levy (1987y).**
IAU Circ., Nos. 4468, 4469, 4471, 4474, 4476, 4487 (1987).

103.822 **Comet Levy (1987y).**
Yamamoto Circ., Nos. 2093, 2094, 2096 (1987). In Japanese.

103.823 **Comet Levy 1987y.**
Br. Astron. Assoc. Circ., No. 672 (1987).

103.824 **New comet Levy (1987y).**
Komet. Tsirk., No. 375 (1987). In Russian.

Periodic comet Schwassmann–Wachmann 1

103.831 **Periodic comet Schwassmann–Wachmann 1.**
IAU Circ., Nos. 4471, 4503 (1987).

103.832 **Periodic comet Schwassmann–Wachmann 1.**
Yamamoto Circ., No. 2094 (1987). In Japanese.

103.833 **Comet P/Schwassmann–Wachmann 1.**
Br. Astron. Assoc. Circ., No. 672 (1987).

Periodic comet Shoemaker–Holt

103.841 **Periodic comet Shoemaker–Holt (1987z).**
IAU Circ., Nos. 4472, 4479, 4508 (1987).

103.842 **Periodic comet Shoemaker–Holt (1987z).**
Yamamoto Circ., Nos. 2095 – 2097 (1987). In Japanese.

103.843 **New short–period comet Shoemaker–Holt (1987z).**
Komet. Tsirk., No. 376 (1987). In Russian.

Periodic comet Mueller

103.851 **Periodic comet Mueller (1987a$_1$).**
IAU Circ., Nos. 4472, 4480 (1987).

103.852 **Periodic comet Mueller (1987a$_1$).**
Yamamoto Circ., Nos. 2095, 2096 (1987). In Japanese.

103.853 **New short–period comet Mueller (1987a$_1$).**
Komet. Tsirk., No. 376 (1987). In Russian.

Comet 1987b$_1$ McNaught

103.861 **Comet McNaught (1987b$_1$).**
IAU Circ., Nos. 4473 – 4475, 4519 (1987).

103.862 **Comet McNaught (1987b$_1$).**
Yamamoto Circ., Nos. 2095 – 2097 (1987). In Japanese.

103.863 **New comet McNaught (1987b$_1$).**
Komet. Tsirk., Nos. 377, 378 (1987). In Russian.

Periodic comet Harrington

103.871 **Periodic comet Harrington (1987n).**
IAU Circ., No. 4487 (1987).

Periodic comet Longmore

103.881 **Periodic comet Longmore (1987c$_1$).**
IAU Circ., No. 4493 (1987).

103.882 **Periodic comet Longmore (1987c$_1$).**
Yamamoto Circ., No. 2099 (1987). In Japanese.

103.883 **Periodic comet Longmore 1974 XIV = 1981 XVI.**
OAA Comput. Sect. Circ., NK 493, 1 p. (1987).

103.884 **Rediscovery of short–period comet Longmore (1987c$_1$).**
Komet. Tsirk., No. 378 (1987). In Russian.

Comet 1987d$_1$ Ichimura

103.891 **Comet Ichimura (1987d$_1$).**
IAU Circ., Nos. 4494, 4497, 4498, 4504, 4505, 4507, 4512 (1987).

103.892 **Comet Ichimura (1987d$_1$).**
Yamamoto Circ., Nos. 2098, 2099 (1987). In Japanese.

103.893 **Comet Ichimura (1987d$_1$).**
Br. Astron. Assoc. Circ., No. 674 (1987).

103.894 **New comet Ichimura (1987d$_1$).**
Komet. Tsirk., No. 378 (1987). In Russian.

Periodic comet Tempel 1

103.901 **Periodic comet Tempel 1 (1987e$_1$).**
IAU Circ., No. 4498 (1987).

103.902 **Periodic comet Tempel 1 (1987e$_1$).**
Yamamoto Circ., No. 2099 (1987). In Japanese.

103.903 **Rediscovery of short–period comet Tempel 1 (1987e$_1$).**
Komet. Tsirk., No. 379 (1987). In Russian.

Comet 1987f$_1$ Furuyama

103.911 **Comet Furuyama (1987f$_1$).**
IAU Circ., Nos. 4499, 4502, 4506, 4520 (1987).

103.912 **Comet Furuyama (1987f$_1$).**
Yamamoto Circ., Nos. 2098 – 2100 (1987). In Japanese.

103.913 **Comet Furuyama (1987f$_1$).**
Br. Astron. Assoc. Circ., No. 674 (1987).

103.914 **New comet Furuyama (1987f$_1$).**
Komet. Tsirk., Nos. 380, 381 (1987). In Russian.

Comet 1987g$_1$ Jensen–Shoemaker

103.921 **Comet Jensen–Shoemaker (1987g$_1$).**
IAU Circ., Nos. 4503, 4505, 4521 (1987).

103.922 **Comet Jensen–Shoemaker (1987g$_1$).**
Yamamoto Circ., Nos. 2099, 2100 (1987). In Japanese.

103.923 **New comet Shoemaker (1987g$_1$).**
Komet. Tsirk., No. 381 (1987). In Russian.

Periodic comet Tempel 2

103.931 **Anisotropic emission from comets: fans versus jets. II. Periodic comet Tempel 2.**
Z. Sekanina.
ESA Spec. Publ., ESA SP–278, p. 323 – 336 (1987). – See Abstr. 012.075.

Application of the concept of emission cone to P/Tempel 2 has resulted in complicated sets of constraints on the position of the spin axis, especially for common solutions over longer periods of time. A successful search was made for an axial position which satisfies all constraints from the apparitions 1920 – 1967 and some from 1899 and 1983. Thus, the observations are consistent with the assumption of the spin axis invariable between at least 1920 and 1967, even though a variety of solutions involving a precessing nucleus could be equally satisfactory.

Periodic comet Finlay

103.936   **Periodic   comet   Finlay   1960 VIII =   1967 IX =
1974 X = 1981 XII.**
OAA Comput. Sect. Circ., NK 491, 2 pp. (1987).

Periodic comet Gehrels 3

103.941   **Periodic comet Gehrels 3 1977 VII = 1985 IV.**
OAA Comput. Sect. Circ., NK 494, 2 pp. (1987).

Periodic comet Machholz

103.946   **Periodic comet Machholz (1986e).**
OAA Comput. Sect. Circ., NK 495, 3 pp. (1987).

Periodic comet Shajn–Schaldach

103.951   **Periodic comet Shajn–Schaldach 1971 IX = 1979 I =
1985i.**
OAA Comput. Sect. Circ., NK 496, 2 pp. (1987).

Periodic comet Singer Brewster

103.956   **Periodic comet Singer Brewster (1986d).**
OAA Comput. Sect. Circ., NK 497, 2 pp. (1987).

Periodic comet Urata–Niijima

103.961   **Periodic comet Urata–Niijima (1986o) = 1986 TD$_4$ =
1986 WP$_5$.**
OAA Comput. Sect. Circ., NK 498, 3 pp. (1987).

Periodic comet Kowal 1

103.966   **Periodic comet Kowal 1 1977 III.**
OAA Comput. Sect. Circ., NK 499, 1 p. (1987).

Periodic comet Russell 1

103.971   **Periodic comet Russell 1 1979 V = 1985 IX.**
OAA Comput. Sect. Circ., NK 500, 1 p. (1987).

Periodic comet Skiff–Kosai

103.976   **Periodic comet Skiff–Kosai 1976 XVI.**
OAA Comput. Sect. Circ., NK 501, 1 p. (1987).

Periodic comet Clark

103.981   **Periodic   comet   Clark   1973 V =   1978 XXIII =
1984 VIII.**
OAA Comput. Sect. Circ., NK 502, 3 pp. (1987).

Periodic comet Peters–Hartley

103.986   **Periodic comet Peters–Hartley 1982 III.**
OAA Comput. Sect. Circ., NK 503, 1 p. (1987).

Periodic comet Haneda–Campos

103.991   **Periodic comet Haneda–Campos 1978 XX.**
OAA Comput. Sect. Circ., NK 504, 2 pp. (1987).

Periodic comet Hartley 1

103.996   **Periodic comet Hartley 1 1985 VII.**
OAA Comput. Sect. Circ., NK 505, 1 p. (1987).

**A catalog of precise reference star positions for the astrometry
network of the international comet P/Halley campaign.**
See Abstr. 002.032.

Atlas of comet Halley 1910 II.
See Abstr. 002.033.

Begegnung mit Halley.
See Abstr. 003.082.

**New information on comet P/Halley as depicted by Giotto di
Bondone and other Western artists.**
See Abstr. 004.056.

**Tales from the first international Halley watch (1755 – 59):
2. Thomas Stevenson of Barbados and his two–comet theory.**
See Abstr. 004.113.

**Comet notes: X. The Fourth American Workshop on Cometary
Astronomy.**
See Abstr. 011.010.

**Diffuse matter in the solar system: comet Halley and other studies.
Proceedings of a Royal Society Discussion Meeting, held at
London, UK, 21 – 22 May 1986.**
See Abstr. 012.026.

**Cometary radio astronomy. Proceedings of an NRAO Workshop
held at the National Radio Astronomy Observatory, Green Bank,
WV, USA, 24 – 26 September 1986.**
See Abstr. 012.033.

Trapping of gases by water ice and implications for icy bodies.
See Abstr. 022.071.

Fluorescent and collisional excitation in diatomic molecules.
See Abstr. 022.073.

**Electronic spectroscopy and relaxation of some molecular cations
of cometary interest.**
See Abstr. 022.105.

Charging of dust particles in comets and in interplanetary space.
See Abstr. 022.106.

**Laboratory amorphous carbon: a possible analogue of cometary
dust.**
See Abstr. 022.130.

**The control circuit of a telescope for the observation of Halley's
comet.**
See Abstr. 034.089.

The prediction of the four–axis tracking of Halley's comet.
See Abstr. 034.090.

Infrared imaging with JPL's linear array camera.
See Abstr. 034.155.

**The dependence of mass resolution and sensitivity of the PUMA
instrument on the energy spread of ions produced by hypervelocity
impacts.**
See Abstr. 035.067.

**Calibration of the DIDSY–IPM dust detector and application to
other impact ionisation detectors on board the P/Halley probes.**
See Abstr. 035.068.

An attempt to evaluate the structure of cometary dust particles.
See Abstr. 035.069.

Hungarian results in the exploration of Halley's comet.
See Abstr. 035.096.

The Giotto magnetometer experiment.
See Abstr. 035.109.

The Giotto electron plasma experiment.
See Abstr. 035.110.

The Giotto dust impact detection system.
See Abstr. 035.111.

The ion mass spectrometer on Giotto.
See Abstr. 035.112.

The heavy ion analyser PICCA for the comet Halley fly–by with Giotto.
See Abstr. 035.113.

The Giotto three–dimensional positive ion analyser.
See Abstr. 035.114.

The Halley multicolour camera.
See Abstr. 035.115.

The lightweight energetic particle detector EPONA and its performance on Giotto.
See Abstr. 035.116.

The Giotto implanted ion spectrometer (IIS): physics and techniques of detection.
See Abstr. 035.117.

Image restoration using the point spread function of the Halley Multicolour Camera.
See Abstr. 036.016.

Image processing on VEGA pictures.
See Abstr. 036.216.

Methods and algorithms for reducing television images of the nucleus of comet Halley. Some results.
See Abstr. 036.234.

Measurement technique for the Giotto radio science experiment.
See Abstr. 036.237.

Active plasma experiments.
See Abstr. 051.034.

The Sakigake/Suisei encounter with comet P/Halley.
See Abstr. 051.042.

Japanese Halley's Comet probes.
See Abstr. 051.047.

The Comet Rendezvous Asteroid Flyby mission.
See Abstr. 051.056.

The Giotto extended mission.
See Abstr. 051.057.

Retrieving samples from comet nuclei.
See Abstr. 051.059.

Proposition for a mission to comet P/Schwassmann–Wachmann 1.
See Abstr. 051.060.

Scientific results of the Vega mission.
See Abstr. 051.063.

The Giotto mission to comet Halley.
See Abstr. 051.067.

On the necessity of the space mission to comet Encke.
See Abstr. 051.070.

Alternative cometary targets for the Giotto extended mission.
See Abstr. 052.026.

The automatic stations Vega 1 and Vega 2. Operation of the instruments at the encounter with comet Halley.
See Abstr. 053.013.

On an analog of the hydrodynamical picture of solar wind flow around comets and the geomagnetosphere during sudden pulses.
See Abstr. 074.036.

MHD waves detected by ICE at distances $\geqslant 28 \times 10^6$ km from comet P/Halley: cometary or solar wind origin?
See Abstr. 074.056.

A kinetic study of solar wind mass loading and cometary bow shocks.
See Abstr. 074.069.

Effect of possible passage through Halley's magnetic tail on geomagnetic activity.
See Abstr. 084.025.

Fast–ion induced chemical evolution in the outer solar system.
See Abstr. 091.023.

Magnetotails at unmagnetized bodies: comparison of comet Giacobini–Zinner and Venus.
See Abstr. 093.020.

Asteroid 5025 P–L, comet 1967 II Rudnicki, and the Taurid meteoroid complex.
See Abstr. 098.029.

Ephemerides of minor planets and comets.
See Abstr. 098.031.

Index to orbital elements of comets and minor planets.
See Abstr. 098.035.

Positions of minor planets and comets obtained at the Chorzów Observatory.
See Abstr. 098.045.

The distribution of short–period comet magnitudes.
See Abstr. 102.004.

The influence of strong hydromagnetic turbulence on newborn cometary ions.
See Abstr. 102.008.

Numerical simulation of the generation of turbulence from cometary ion pick–up.
See Abstr. 102.009.

Electron impact ionization in the vicinity of comets.
See Abstr. 102.011.

Energy diffusion of pickup ions upstream of comets.
See Abstr. 102.012.

Photoprocessing of H_2S in interstellar grain mantles as an explanation for S_2 in comets.
See Abstr. 102.015.

A model for the excitation of water in comets.
See Abstr. 102.016.

Comets and their composition.
See Abstr. 102.019.

Chemical interactions of solar wind with cometary nuclei and comae.
See Abstr. 102.024.

Radiation dosimetry and chemistry of a cometary nucleus.
See Abstr. 102.025.

The nature of comets.
See Abstr. 102.031.

Comets, and comet Halley 1982 – 1985.
See Abstr. 102.041.

Fluid simulation of comet P/Halley's ionosphere.
See Abstr. 102.047.

Unusual characteristics of electromagnetic waves excited by cometary newborn ions with large perpendicular energies.
See Abstr. 102.048.

High–order librations of Halley–type comets.
See Abstr. 102.050.

Galactic tides affect the Oort cloud: an observational confirmation.
See Abstr. 102.051.

Alfvén wave plasma turbulence during solar wind–comet interaction.
See Abstr. 102.060.

The role of water in the thermal balance of the coma.
See Abstr. 102.070.

A model of the anisotropic structure of the neutral gas coma of a comet.
See Abstr. 102.071.

Anisotropic emission from comets: fans versus jets. I. Concept and modeling.
See Abstr. 102.072.

On the interpretation of the CH cometary spectrum.
See Abstr. 102.077.

The role of hydrogen thermalization in shaping the Lyman–α coma.
See Abstr. 102.078.

Comet Nishikawa–Takamizawa–Tago (1987c) and the Epsilon Geminid meteor shower.
See Abstr. 104.033.

Estimate of the mass of the meteor complex generated by comet Encke.
See Abstr. 104.053.

Unity of small bodies of the solar system.
See Abstr. 104.061.

The P/Halley meteor showers in 1985 – 1986.
See Abstr. 104.070.

The spectra of meteors from Halley's comet.
See Abstr. 104.071.

Meteoroids from comet P/Halley. The comet's mass production and age.
See Abstr. 104.072.

The 1985 return of the Giacobinid meteor stream.
See Abstr. 104.073.

The meteor stream associated with comet P/Grigg–Skjellerup.
See Abstr. 104.074.

Meteor contribution by short–period comets.
See Abstr. 104.075.

Associations between ancient comets and meteor showers.
See Abstr. 104.076.

Revealing a genetic connection between the α–Capricornid meteor stream and comets.
See Abstr. 104.079.

On orbits and associations of meteor streams with comets P/Halley and P/Encke.
See Abstr. 104.096.

The structure of the Halley meteor stream and the orbital history of the comet.
See Abstr. 104.097.

On the formation of meteor showers of comet Halley.
See Abstr. 104.099.

Meteoroids from comet Bennett 1970 II.
See Abstr. 106.009.

Particles injected by comets.
See Abstr. 106.035.

Interplanetary dust dynamics. II. Poynting–Robertson drag and planetary perturbations on cometary dust.
See Abstr. 106.036.

Interplanetary dust dynamics. III. Dust released from P/Encke: distribution with respect to the zodiacal cloud.
See Abstr. 106.037.

Search for the primitive (*material in the solar system*).
See Abstr. 107.035.

The effect of comet ion tails on radio source scintillation.
See Abstr. 141.013.

104 Meteors, Meteor Streams

104.001   European Network fireballs photographed in 1978.
Z. Ceplecha, P. Spurný, J. Boček, M. Nováková,
G. Polnitzky, V. Porubčan, T. Kirsten, J. Kiko.
Bull. Astron. Inst. Czech., Vol. 38, No. 4, p. 211 – 222 (1987).
Geometric, dynamic, photometric, and orbital data on all 15 fireballs photographed within the European Network in the year 1978 are given. A few remarks on the measuring and reduction procedures are included and different methods of least–squares solutions for velocities and decelerations are briefly described. No meteorite impacts were predicted for any of these 15 fireballs due to insignificant terminal masses.

104.002   Geometric, dynamic, orbital and photometric data on meteoroids from photographic fireball networks.
Z. Ceplecha.
Bull. Astron. Inst. Czech., Vol. 38, No. 4, p. 222 – 234 (1987).
Methods and procedures of computing geometric, dynamic, orbital and photometric data from multi–station photographic records of fireballs are presented and corresponding mathematical formulae are put forth. These methods and procedures are currently used in evaluating photographic records from the European Fireball Network.

104.003   Teleskopische Meteorbeobachtung.
B. Koch, M. Nolle.
KPM, Jahrg. 2, No. 5, p. 10 – 19 (1987).

104.004   Prospects for an enhanced ε–Geminid shower in 1987.
D. Olsson–Steel.
WGN, Vol. 15, Nr. 4, p. 109 – 111 (1987).
It appears likely that comet Nishikawa–Takamizawa–Tago, a long–period comet which passed perihelion in March 1987, is the parent of the ε–Geminid meteor shower. If this comet is in fact the parent, then there is a good chance of a strong meteor shower or even a storm in October 1987.

104.005   Reflection duration determination: an experiment.
C. Steyaert.
WGN, Vol. 15, Nr. 4, p. 114 – 116 (1987).
The results of an experiment are discussed that was set up to determine the accuracy of reflection duration estimates in radio meteor observing.

104.006   The relation between visual magnitude and echo duration.
J. Van Wassenhove.
WGN, Vol. 15, Nr. 4, p. 116 – 118 (1987).
The relationship between the echo duration and the visual magnitude of a simultaneous radio–visual meteor is examined. A linear relationship between the visual magnitude and the logarithm of the echo duration seems to fit rather well.

104.007   On the Perseids of 1986.
P. Brown.
WGN, Vol. 15, Nr. 4, p. 119 – 129 (1987).
In 1986 members of the Alberta Meteor Group, Canada, observed the Perseids. In all 651 Perseids and 517 sporadic meteors were recorded. An analysis of the magnitude distribution was made. The average particle size is found to gradually increase from the beginning of the activity to 3 – 5 days before maximum. Thereafter the particle size falls off.

104.008   Meteor streams of the southern hemisphere: The ϰ–Pavonid meteor stream in 1986. The γ–Normid meteor stream in 1986. The δ–Pavonid meteor stream in 1986.
J. Wood.
WGN, Vol. 15, Nr. 4, p. 129 – 132 (1987).

104.009   Finland: Geminids and Ursids in 1986.
T. Hankamäki.
WGN, Vol. 15, Nr. 4, p. 133 (1987).

104.010   Norway: annual report 1986.
T. E. Hillestad.
WGN, Vol. 15, Nr. 4, p. 134 (1987).
In 1986, 14 members of the Norwegian Meteor Section saw over 6000 meteors during 48 nights. A brief account of these observations is given.

104.011   VVS–Meteor Section. Radio subsection: Annual report 1986.
J. Van Wassenhove.
WGN, Vol. 15, No. 5, p. 145 – 153 (1987).

104.012   Comet 1944 I and the Nov Monocerotids.
B. A. Lindblad.
WGN, Vol. 15, No. 5, p. 154 (1987).
The association between comet van Gent–Peltier–Daimaca (1944 I) and the November Monocerotids is investigated. This association is finally confirmed by visual observations in 1985.

104.013   On the Perseid meteor stream. I. Magnitude data.
P. Roggemans.
WGN, Vol. 15, No. 5, p. 155 – 162 (1987).
Visual observations yielded 36876 magnitude estimates of which 11320 were sporadics and 21148 were Perseids. Sufficient numbers of meteor estimates were available for the Perseids from July 29 – 30 until August 16, inclusive, to study mass distribution variations from day to day. A remarkable increment in faint Perseids was found for August 8 – 9, 1986. The 1986 Perseids were remarkably strong in faint meteors during the whole activity period.

104.014   The Perseids 1985 in the Soviet Union.
V. V. Martynenko, A. S. Levina.
WGN, Vol. 15, No. 5, p. 163 – 164 (1987).

104.015   Quadrantids 1987 in Finland.
T. Hankamäki.
WGN, Vol. 15, No. 5, p. 165 – 166 (1987).

104.016   Quadrantids 1987 in Norway.
T. E. Hillestad.
WGN, Vol. 15, No. 5, p. 167 (1987).

104.017   Quadrantids 1987 in Spain.
J.–M. Trigo–Campoy.
WGN, Vol. 15, No. 5, p. 167 – 168 (1987).

104.018   Quadrantids 1987 in Japan.
M. Koseki.
WGN, Vol. 15, No. 5, p. 168 – 171 (1987).

104.019   Quadrantids 1987 in Belgium.
J. Van Wassenhove.
WGN, Vol. 15, No. 5, p. 171 – 173 (1987).

104.020   De vuurbol van 12 Augustus '86 21^{h}03^{m}23^{s}UT.
H. Betlem, M. de Lignie.
Radiant, Jaarg. 8, Nr. 5, p. 94 – 95 (1986).

104.021   De telescopische Draconiden 1985.
F. Witte.
Radiant, Jaarg. 8, Nr. 5, p. 102 (1986).

104.022 **Vuurbol simultaan!**
H. Betlem.
Radiant, Jaarg. 8, Nr. 6, p. 108 – 109 (1986).
On November 9th 1986 a very bright sporadic meteor was photographed by two Dutch stations of the European Network (EN). The fireball travelled a 43 km trajectory in two seconds time. Orbital and trajectory data are presented here.

104.023 **EN (*European Network*) 13.08.86 Puimichel.**
H. Betlem, M. de Lignie.
Radiant, Jaarg. 8, Nr. 6, p. 112 – 114 (1986).
On August 13th 1985 $1^h04^m00^s$UT an extremely bright sporadic fireball was photographed by DMS observers Klaas Jobse and Koen Miskotte, working with a double station set of cameras near the village of Puimichel in southern France. Orbital and trajectory data are presented here.

104.024 **De Hohenlangenbeck meteoriet. De baan in de ruimte.**
P. Jenniskens, H. Betlem, J. Rendtel.
Radiant, Jaarg. 8, Nr. 6, p. 118 – 119 (1986).

104.025 **De vorm van de Perseiden curve.**
P. Jenniskens.
Radiant, Jaarg. 8, Nr. 6, p. 120 – 126 (1986).

104.026 **Baanelementen: een meteoorbaan om de zon.**
P. Jenniskens, M. de Lignie.
Radiant, Jaarg. 9, Nr. 1, p. 10 – 17 (1987).

104.027 **De vuurbol van 23 September boven West Europa.**
W. van Utrecht.
Radiant, Jaarg. 9, Nr. 3, p. 35 – 37 (1987).

104.028 **Simultaanopnamen 1986. Baanelementen en trajekten.**
H. Betlem, M. de Lignie, C. ter Kuile.
Radiant, Jaarg. 9, Nr. 3, p. 43 – 52 (1987).

104.029 **Telescopic observation of Draconids in 1985.**
J. Hollan, M. Vorel, V. Znojil.
Bull. Astron. Inst. Czech., Vol. 38, No. 6, p. 376 (1987).
An upper limit of activity index of Draconids outside the maximum in 1985 is given.

104.030 **Visuele resultaten van de Zomerakties 1986.**
R. Veltman.
Radiant, Jaarg. 9, Nr. 4, p. 63 – 72 (1987).

104.031 **Determination of the "age" of meteor streams.**
G. V. Andreev.
Komet. Tsirk., No. 376 (1987). In Russian.

104.032 **The determination of shower meteor parameters from single station observations.**
A. G. Duffy, R. L. Hawkes, J. Jones.
Mon. Not. R. Astron. Soc., Vol. 228, No. 1, p. 55 – 75 (1987).
An effective method for identifying shower meteors and determining atmospheric trajectories from single station observations has been developed and tested with computer–simulated and low light level television meteor data. The apparent velocity and calculated height information are used to reduce significantly the number of test radiants in the shower identification algorithm. With typical data the technique permits isolation of shower meteors from a sporadic background when the ratio of the number of meteors from a given shower to the total sample size is as low as 1/20 (for a sample size of about 80). The technique has been applied to 3.2 hr of low light level television data (limiting stellar sensitivity approximately +8.5 mag) obtained at a station in Benmiller, Ontario, Canada during the 1984 Orionid shower.

104.033 **Comet Nishikawa–Takamizawa–Tago (1987c) and the Epsilon Geminid meteor shower.**
D. Olsson–Steel.
Mon. Not. R. Astron. Soc., Vol. 228, No. 2, p. 23P – 28P (1987).
Comet Nishikawa–Takamizawa–Tago (1987c) appears to be an excellent candidate as parent of the Epsilon Geminid meteor shower. A comparison of its orbital parameters and theoretical meteor radiant with the observed characteristics of the Epsilon Geminids, along with the fact that the Earth makes its closest approach to the comet's path in October when the shower is seen and a more distant approach in July when no shower has been detected, indicates that 1987c is more likely to be the parent than a previously proposed comet (1964 VIII Ikeya).

104.034 **Modelling the interaction between meteor bodies and the earth's atmosphere and surface.**
V. P. Korobejnikov, P. I. Chushkin, L. V. Shurshalov.
20th All–Union meteoritic conference, p. 157 (1987). In Russian. Abstr. in Ref. Zh., 51. Astron., 7.51.235 (1987). – See Abstr. 012.017.

104.035 **The influence of superradiation of the surface and of the nonuniform volatilization on the ablation and change in the form of a meteor body.**
Eh. Z. Apshtejn.
20th All–Union meteoritic conference, p. 160 – 161 (1987). In Russian. Abstr. in Ref. Zh., 51. Astron., 7.51.236 (1987). – See Abstr. 012.017.

104.036 **One of the possible models of thermal disruption of meteor and meteoritic bodies.**
V. G. Kruchinenko.
20th All–Union meteoritic conference, p. 161 – 162 (1987). In Russian. Abstr. in Ref. Zh., 51. Astron., 7.51.237; 62. Issled. Kosm. Prostranstva, 7.62.579 (1987). – See Abstr. 012.017.

104.037 **Physical parameters of cosmic bodies according to results of investigations on meteors and bolides.**
V. V. Kalenichenko.
20th All–Union meteoritic conference, p. 163 (1987). In Russian. Abstr. in Ref. Zh., 51. Astron., 7.51.238 (1987). – See Abstr. 012.017.

104.038 **On the possibility of meteor swarm origin at a collision of minor planets.**
G. V. Andreev.
20th All–Union meteoritic conference, p. 169 – 170 (1987). In Russian. Abstr. in Ref. Zh., 51. Astron., 7.51.239 (1987). – See Abstr. 012.017.

104.039 **Investigation of the orbit of the Leonid meteor stream from 1660 to 1799.**
E. D. Kondrat'eva.
Kazan. Univ. Kazan', 9 pp. (1987). In Russian. Abstr. in Ref. Zh., 51. Astron., 7.51.240 (1987).

104.040 **On the problem of measuring the deceleration of meteor particles with radio methods.**
N. S. Andrianov.
Meteor. rasprostr. radiovoln, Kazan', No. 20, p. 9 – 15 (1987). In Russian. Abstr. in Ref. Zh., 51. Astron., 7.51.241 (1987).

104.041 **Interpretation of light curves of bright meteors taking fragmentation into account.**
P. B. Babadzhanov, G. G. Novikov, N. A. Konovalova, A. V. Blokhin.
Dokl. Akad. Nauk TadzhSSR, Tom 29, No. 9, p. 523 – 526 (1986). In Russian. Abstr. in Ref. Zh., 51. Astron., 7.51.242 (1987).

104.042 Numerical modelling and calculation of radiative thermal fluxes in interactions between meteor bodies and the earth's atmosphere.
Eh. Z. Apshtejn, V. I. Sakharov.
20th All–Union meteoritic conference, p. 159 – 160 (1987). In Russian. Abstr. in Ref. Zh., 51. Astron., 7.51.244 (1987). – See Abstr. 012.017.

104.043 Calculating the coefficient for the transition from the observed number of meteors to the density of a falling stream for different models of meteor stream formation.
K. V. Kostylev, T. K. Filimonova.
Meteor. rasprostr. radiovoln, Kazan', No. 20, p. 3 – 9 (1987). In Russian. Abstr. in Ref. Zh., 51. Astron., 7.51.245 (1987).

104.044 On the possibility of forming of the Geminid meteor stream during the collision of asteroids.
G. O. Ryabova.
20th All–Union meteoritic conference, p. 168 (1987). In Russian. Abstr. in Ref. Zh., 51. Astron., 7.51.246; 62. Issled. Kosm. Prostranstva, 7.62.580 (1987). – See Abstr. 012.017.

104.045 The Chulym bolide.
D. F. Anfinogenov, L. I. Budaeva, V. G. Fast, N. P. Fast.
20th All–Union meteoritic conference, p. 163 – 165 (1987). In Russian. Abstr. in Ref. Zh., 51. Astron., 7.51.249 (1987). – See Abstr. 012.017.

104.046 Orbits of meteorites and their interaction with small bodies.
A. K. Terent'eva.
20th All–Union meteoritic conference, p. 152 – 153 (1987). In Russian. Abstr. in Ref. Zh., 51. Astron., 7.51.259 (1987). – See Abstr. 012.017.

104.047 Distribution of parent bodies of meteorites of different generations according to their sizes.
A. M. Kazantsev, L. M. Sherbaum.
20th All–Union meteoritic conference, p. 171 (1987). In Russian. Abstr. in Ref. Zh., 51. Astron., 7.51.264 (1987). – See Abstr. 012.017.

104.048 Secular perturbations of the orbital elements and cosmic age of bodies generating bolides.
E. N. Kramer, I. S. Shestaka.
20th All–Union meteoritic conference, p. 170 – 171 (1987). In Russian. Abstr. in Ref. Zh., 51. Astron., 8.51.59 (1987). – See Abstr. 012.017.

104.049 Acceleration of macroparticles to high velocities.
V. M. Titov, G. A. Shvetsov.
Din. splosh. sredy, Novosibirsk, No. 78, p. 128 – 136 (1986). In Russian. Abstr. in Ref. Zh., 51. Astron., 8.51.226 (1987).

104.050 Ursiden 1986.
P. Jenniskens.
Radiant, Jaarg. 9, Nr. 6, p. 101 – 102 (1987).

104.051 Boötiden 1987: 771 visuele meteoren.
R. Veltman.
Radiant, Jaarg. 9, Nr. 2, p. 18 – 20 (1987).

104.052 Statistical analysis of Chinese historical records of fireballs.
X.–l. Liu.
Chin. Astron. Astrophys., Vol. 11, No. 4, p. 312 – 319 (1987). English translation of Acta Astrophys. Sin., Vol. 7, No. 3, p. 220 – 229 (1987).

104.053 Estimate of the mass of the meteor complex generated by comet Encke.
V. G. Kruchinenko.
Komet. Tsirk., No. 375 (1987). In Russian.

104.054 Wanneer meteoren waarnemen?
P. Jenniskens.
Radiant, Jaarg. 9, Nr. 2, p. 30 – 33 (1987).

104.055 Lyrid meteor shower of 1982: enhanced activity observed at Ottawa, Canada.
V. Porubčan, B. A. McIntosh.
Bull. Astron. Inst. Czech., Vol. 38, No. 5, p. 313 – 317 (1987).
Radar observations of the Lyrid meteor shower in 1982 at Ottawa, Canada, (Springhill Meteor Observatory) showed enhanced activity on April 22 peaking at 06:49 UT (solar longitude 31.380°, equinox 1950). The duration of the storm was 22 minutes between half–maximum points, and 50 minutes to quarter maximum. Small particles predominated. A search for non–random pairing of observed particles gave no positive result.

104.056 No evidence for the Giacobinid meteor shower in 1986.
M. Šimek.
Bull. Astron. Inst. Czech., Vol. 38, No. 5, p. 317 – 318 (1987).
Ferrin's presumption of Giacobinid meteor shower occurrence between October 8.8 and 9.4 UT, 1986 was not confirmed from radar observation at the Ondřejov Observatory.

104.057 Meteors, meteorites and cosmic dust.
V. N. Lebedinets.
Priroda, No. 9, p. 34 – 36 (1987). In Russian.

104.058 Orioniden 1987.
K. Jobse, M. Olie.
Radiant, Jaarg. 9, Nr. 6, p. 98 – 100 (1987).

104.059 Meteor streams of comet Encke.
P. B. Babadzhanov, Yu. V. Obrubov, N. Makhmudov.
Komet. Tsirk., No. 373 (1987). In Russian.

104.060 Geminiden 1986 DMS (*Dutch Meteor Society*).
P. Jenniskens.
Radiant, Jaarg. 9, Nr. 6, p. 103 – 105 (1987).

104.061 Unity of small bodies of the solar system.
A. K. Terent'eva.
Priroda, No. 9, p. 36 – 39 (1987). In Russian.

104.062 Vuurbollen in april.
F. Bettonvil.
Zenit, 14. Jaarg., Nr. 12, p. 406 – 407 (1987).

104.063 Observing meteors. XI.
D. H. Levy.
Strolling Astron., Vol. 32, Nos. 5 – 6, p. 129 – 130 (1987).

104.064 On mathematical modelling of deceleration of meteor fragmentation.
V. N. Lebedinets.
Tr. Inst. Ehksp. Meteorol. Goskomgidrometa, No. 19/125, p. 3 – 16 (1987). In Russian. Abstr. in Ref. Zh., 51. Astron., 10.51.266 (1987).

104.065 Quasi–continuous fragmentation of meteor bodies taking deceleration into account.
P. B. Babadzhanov, G. G. Novikov, V. N. Lebedinets, A. V. Blokhin.
Tr. Inst. Ehksp. Meteorol. Goskomgidrometa, No. 19/125, p. 16 – 21 (1987). In Russian. Abstr. in Ref. Zh., 51. Astron., 10.51.267 (1987).

104.066 Influence of mass distribution of fragments on light curves and tails of meteors.
G. G. Novikov, V. N. Lebedinets, A. V. Blokhin.
Tr. Inst. Ehksp. Meteorol. Goskomgidrometa, No. 19/125, p. 22 – 25 (1987). In Russian. Abstr. in Ref. Zh., 51. Astron., 10.51.268 (1987).

104.067 Geminid fireballs and the peculiar asteroid 3200 Phaethon.
I. Halliday.
Bull. Am. Astron. Soc., Vol. 19, No. 2, p. 751 (1987). Abstract. – See Abstr. 010.061.

104.068 Light curves of faint meteors.
R. L. Hawkes, P. E. Hitchcock, J. D. D. Fyfe, A. G. Duffy, J. Jones.
Bull. Am. Astron. Soc., Vol. 19, No. 2, p. 751 (1987). Abstract. – See Abstr. 010.061.

104.069 Physics and orbits of meteoroids.
B. A. Lindblad.
The evolution of the small bodies of the solar system, p. 229 – 251 (1987). – See Abstr. 012.041.
Contents: 1. Introduction (meteoroids, meteor magnitude, micrometeroids, meteor showers, sporadic meteors, meteor streams, interstellar meteors, results of visual programs, orbits). 2. Meteoroid orbits (meteor photography, definition of a meteor stream, velocity–elongation diagram, short–period meteor streams, long–period meteor streams, distribution of $1/a$, radio orbits). 3. Meteor theory (luminosity equation, luminous efficiency, meteor magnitudes, shape factor of a meteoroid, drag equation, fragmentation). 4. Meteoroid densities (mean meteoroid density, individual meteoroid densities). 5. Evolution of short–period meteor streams.

104.070 The P/Halley meteor showers in 1985 – 1986.
M. Hajduková, A. Hajduk, G. Cevolani, C. Formiggini.
Astron. Astrophys., Vol. 187, No. 1/2, p. 919 – 920 (1987). – See Abstr. 003.013, 43.104.014.
The main characteristics of the meteor streams associated with comet P/Halley, Orionids and Eta Aquarids, have been deduced from simultaneous radar observations at Ondřejov (Czechoslovakia) and Budrio (Italy), during the shower periods in 1985 and 1986. The flux of particles with masses m $\geq 10^{-6}$kg was found to be F $= 1.5 \times 10^{-12}$m^{-2}s^{-1} for the central parts of the streams. This value is less than one third of the maximum values observed within the stream. The differential mass index, s $= 1.85$, derived for the last three returns of the showers, is compared with the results from 1980 – 1983, s $= 2.2$. The difference suggests the presence of two populations of particles in the stream.

104.071 The spectra of meteors from Halley's comet.
I. Halliday.
Astron. Astrophys., Vol. 187, No. 1/2, p. 921 – 924 (1987). – See Abstr. 003.013, 43.104.013.
The spectra of 16 bright Orionid meteors are studied in the spectral range from 3440 to 8700 Å. The brightest members of the group have peak absolute magnitudes near –8, indicating masses of a few grams and providing clear evidence that large particles from comet Halley survive to encounter the Earth's atmosphere. Spectral lines from six neutral atomic species and three singly-ionized species are observed. The Orionid spectra are very similar to Perseid meteor spectra, suggesting any differences in abundances in the parent comets appear to be marginal for the elements observed.

104.072 Meteoroids from comet P/Halley. The comet's mass production and age.
A. Hajduk.
Astron. Astrophys., Vol. 187, No. 1/2, p. 925 – 927 (1987). – See Abstr. 003.013, 43.104.012.
The flux of meteoroid size particles ejected from comet P/Halley in the mass range from 10^{-9} to 10^{-1}kg as observed by radar and optical techniques is compared with the flux of dust particles in the mass range from 10^{-20} to 10^{-8}kg as measured by space probes. The theory of the orbital motion of the comet and of the shell model of the meteor stream give within the observational errors of particle fluxes an initial mass of the comet of about 3.2×10^{15}kg and an effective diameter of 21 km. An estimated age of about 2300 revolutions is derived.

104.073 The 1985 return of the Giacobinid meteor stream.
B. A. Lindblad.
Astron. Astrophys., Vol. 187, No. 1/2, p. 928 – 930 (1987). – See Abstr. 003.013, 43.104.010.
Radar observations of the 1985 return of the Giacobinid meteor shower was carried out at the Onsala Space Observatory in Sweden. The observations showed that the shower was of approximately 3^h duration, rising steeply to a maximum at 09^{h}35^m UT on 1985 October 8. The meteor shower thus showed the same short period characteristics as the intense showers of 1933 and 1946. Peak meteor echo rates of 75 occurred in the ten-minute interval 09^{h}30^m – 09^{h}40^m UT. The corresponding shower zenithal rate at maximum after removal of background activity was 484 echoes per hour. The longitude of the descending node of the shower at maximum was 194°55 (equinox 1950). The plane of the central part of the 1985 shower was displayed by 0°15 in longitude with respect to the orbital plane of the comet.

104.074 The meteor stream associated with comet P/Grigg–Skjellerup.
B. A. Lindblad.
Astron. Astrophys., Vol. 187, No. 1/2, p. 931 – 932 (1987). – See Abstr. 003.013, 43.104.017.
A 1964 Jupiter approach of P/Grigg–Skjellerup perturbed the comet's orbit so that very close approaches to the Earth's orbit now occur. A recently observed southern hemisphere meteor shower, the σ Puppids, is associated with P/Grigg–Skjellerup. Observations of this meteor shower now provide us with a unique opportunity to observe the birth and evolution of a meteoroid stream.

104.075 Meteor contribution by short–period comets.
J. Štohl.
Astron. Astrophys., Vol. 187, No. 1/2, p. 933 – 934 (1987). – See Abstr. 003.013, 43.104.009.
The contribution of P/Halley and other comets to the meteor streams are discussed. Particular attention is given to the meteoric production by P/Encke as the potential major contributor to the meteoric complex in the inner part of the solar system. The density and total mass of the meteoric stream associated with P/Encke is derived, taking into account the whole complex both of the showers and of the very broad and diffuse stream of seemingly sporadic meteoroids, produced by P/Encke in its long history via its showers.

104.076 Associations between ancient comets and meteor showers.
M. Kresáková.
Astron. Astrophys., Vol. 187, No. 1/2, p. 935 – 936 (1987). – See Abstr. 003.013, 43.104.015.
Association criteria for comets and meteor showers are reviewed. Their application to the records in the Far East chronicles suggests three or four such associations, the most probable pairs being the comets and meteor showers of 461, 647, 684, 1063, and 1539. Most of them seem to be of low inclination and short revolution period. The number of meteor events associated with comet Halley implies its uniqueness in meteoroid production.

104.077 Model of an explosion of a meteor body.
L. V. Shurshalov.
Izv. Akad. Nauk SSSR. Mekh. Zhidkosti, No. 3, p. 138 – 144 (1987). In Russian. Abstr. in Ref. Zh., 51. Astron. 11.51.267 (1987).

104.078 The probability of falling of Apollo–Amor–type interplanetary bodies on the surface of planets.
E. N. Kramer, I. S. Shestaka.
Astron. Vestn., Tom 21, No. 3, p. 225 – 232 (1987). In Russian. English translation in Sol. Syst. Res.
The influence of secular gravitational perturbations of orbits of the Apollo–Amor–type interplanetary bodies on their collision probabilities with Mercury, Venus, Earth and Mars has been studied. New formulae for calculating the probability of the

interplanetary bodies falling upon the surfaces of these planets have been obtained. A new method is suggested for the estimation of meteor orbits' cosmic weight.

104.079   Revealing a genetic connection between the α–Capricornid meteor stream and comets.
N. V. Kulikova.
Astron. Vestn., Tom 21, No. 3, p. 242 – 250 (1987). In Russian. English translation in Sol. Syst. Res.
To gain a clearer insight into the origin of the meteor stream of α–Capricornids the matter ejection from the nuclei of the comets 1948 XII, 1954 III and 1770 I at any point of cometary orbits with r ⩽ 2 AU was simulated using the method of statistical tests. It is shown that the stream is most likely to be related to the comet Honda–Mrkos–Pajdušakova.

104.080   On the peculiarities of meteoroid density distributions in different meteor streams.
V. N. Lebedinets.
Astron. Vestn., Tom 21, No. 3, p. 262 – 271 (1987). In Russian. English translation in Sol. Syst. Res.
Density and fragmentation energy of meteoroids for 98 faint photographic meteors from 17 meteor streams were evaluated by mathematical simulation of quasi–continuous fragmentation and deceleration of meteoroids. The result is of a great interest for cometary physics and cosmogony of the solar system.

104.081   Le brillant météore du 23 septembre 1986 (2ᵉpartie).
J. Sauval.
Ciel Terre, Vol. 103, No. 6, p. 149 – 153 (1987).

104.082   Perseidi 1986: osservazioni del massimo di attività.
E. Stomeo.
Astronomia, N. 3 – 4, p. 36 – 37 (1987).

104.083   The activity of the Orionid meteor stream in 1985.
G. H. Spalding.
J. Br. Astron. Assoc., Vol. 98, No. 1, p. 26 – 34 (1987).
World–wide visual observations of the 1985 Orionid meteor stream, made as part of the International Halley Watch campaign, have been analysed. The level of activity of the Orionids was similar to that in other years.

104.084   Numerical modelling of the flight and explosion of natural cosmic bodies.
V. P. Korobejnikov, P. I. Chushkin, L. V. Shurshalov.
Probl. Prikl. Mat. Informat., Moskva,, p. 33 – 47 (1987). In Russian. Abstr. in Ref. Zh., 51. Astron., 12.51.356 (1987).

104.085   The minimum recorded electron density of a meteor train.
A. A. Gajdaev.
Geomagn. Aehron., Tom 27, No. 6, p. 1009 – 1011 (1987). In Russian. English translation in Geomagn. Aeron.

104.086   The mineralogical density of meteor bodies.
V. V. Benyukh.
Komet. Tsirk., No. 371 (1987). In Russian.

104.087   The ZHR profile of the 1985 Perseids, recomputed.
P. Roggemans.
WGN, Vol. 15, Nr. 6, p. 181 – 183 (1987).
The use of a zenith distance correction with a zenith exponent ε = 1.0 introduced short duration features on the ZHR profile, initially explained as the result of undercorrection. A new attempt to use another zenith distance correction with ε = 1.5 smoothed only some of the short duration features, but other fluctuations became more pronounced on the ZHR profile.

104.088   Nightly ZHR and HR variations for European Perseid results 1985.
P. Roggemans.
WGN, Vol. 15, Nr. 6, p. 183 – 187 (1987).
European Perseid ZHR's of 1985 enabled to consider the ZHR variation night by night. Results support the use of a zenith exponent ε > 1.0, while remaining post–maximum submaxima and depressions in the hourly rates are most likely due to a fine inner–core layered structure of the Perseid meteor stream.

104.089   The ε–Geminids 1987 in southern France.
P. Roggemans.
WGN, Vol. 15, Nr. 6, p. 188 – 189 (1987).

104.090   Radio work in connection with the ε–Geminids.
J. Van Wassenhove.
WGN, Vol. 15, Nr. 6, p. 189 – 190 (1987).

104.091   More radio observations of the Ursids 1986.
I. Reimann.
WGN, Vol. 15, Nr. 6, p. 191 – 192 (1987).
Some as yet unpublished radio observations of the Ursids in 1986 are presented. They support the time of maximum which can be deduced from the visual observations in Norway.

104.092   The Taurids 1986 by the Dutch group "Delphinus".
K. Miskotte, B. Rispens.
WGN, Vol. 15, Nr. 6, p. 195 – 197 (1987).

104.093   Observational results 1986 – 87 from Florida, USA.
N. W. McLeod III.
WGN, Vol. 15, Nr. 6, p. 198 – 200 (1987).
A short description is given of observations of the Perseids 1986, the Lyrids 1987 and the η–Aquarids 1987. Much attention is paid to the role played by the perception of the observer.

104.094   The η–Aquarids from California, USA.
R. Lunsford.
WGN, Vol. 15, Nr. 6, p. 200 – 201 (1987).

104.095   Canadian summer 1987 results.
P. Brown.
WGN, Vol. 15, Nr. 6, p. 201 – 205 (1987).
During the summer 1987, the author observed both in Southern France and in Northern Alberta, Canada. His results are presented and discussed.

104.096   On orbits and associations of meteor streams with comets P/Halley and P/Encke.
V. Porubčan, J. Štohl.
ESA Spec. Publ., ESA SP–278, p. 435 – 440 (1987). – See Abstr. 012.075.
Mean orbits of meteor streams associated with comet P/Encke, together with the dispersion of individual orbits and changes of the orbital elements across the streams are evaluated on the basis of precise photographic observations. The results are compared and interrelated to those obtained for some other streams, especially those associated with comet P/Halley. Reality and origin of the whole meteoric complex, including the diffuse "sporadic stream", associated with comet P/Encke is discussed.

104.097   The structure of the Halley meteor stream and the orbital history of the comet.
M. Hajduková.
ESA Spec. Publ., ESA SP–278, p. 645 – 647 (1987). – See Abstr. 012.075.
Long base observations of the Orionid and Eta Aquarid meteor showers carried out in Czechoslovakia and simultaneously at other observatories show structural features which can be related to the orbital history of the parent comet. There are different possibilities how to fit the observational data with theories of the motion of particles ejected from the comet.

104.098   **The structure of the Giacobinid 1985 meteor shower from radar observations in Dushanbe and Ondřejov.**
R. P. Chebotarev, M. Šimek.
Bull. Astron. Inst. Czech., Vol. 38, No. 6, p. 362–367 (1987).

Results from the joint investigation of the Giacobinid 1985 meteor shower from radar observations in Dushanbe and Ondřejov are presented. The main attention was focused on the fine structure of the shower in the overdense echo–range. A strong concentration of echoes in the range $0.5 \text{ s} < T_D < 6 \text{ s}$ was found. The maximum flux of the shower particles occurred at $9^h48^m \pm 6^m$UT, October 8.

104.099   **On the formation of meteor showers of comet Halley.**
P. B. Babadzhanov, Yu. V. Obrubov, A. N. Pushkarev, A. Hajduk.
Bull. Astron. Inst. Czech., Vol. 38, No. 6, p. 367–371 (1987).

Orbits of test particles, ejected from the nucleus of Halley's comet at its perihelion passage in 1910 with different velocities are studied for the next passages of the comet up to 2134 taking into consideration perturbations from all planets. Some characteristics of the stream formation are presented. The calculations show that the return of the comet to its perihelion cannot produce an immediate influence on the activity of its meteor showers.

104.100   **Concentration of small particles in Orionids.**
V. Znojil, J. Hollan, A. Hajduk.
Bull. Astron. Inst. Czech., Vol. 38, No. 6, p. 372–375 (1987).

Results of telescopic observations of meteors carried out during the Orionid shower period in 1985 are given. The magnitude distribution function derived at different solar longitudes shows an increased concentration of small particles in the youngest belt of the stream.

Meteory i ikh nablyudenie (*Meteors and their observations*).
See Abstr. 003.024.

Meteory, meteority, meteoroidy (*Meteors, meteorites, meteoroids*).
See Abstr. 003.033.

A mysterious woodcut.
See Abstr. 004.032.

The Roman fireball of 76 BC.
See Abstr. 004.038.

The Chinese "candle star" of 76 BC.
See Abstr. 004.039.

On the trail of meteor trains.
See Abstr. 004.069.

Een TV–systeem voor meteoren.
See Abstr. 034.021.

Meteorstrombeobachtung: Amateurprogramm.
See Abstr. 036.012.

Die Bestimmung der ZHR bei Meteorbeobachtungen. Teil 1: Die Grenzgrößenkorrektur.
See Abstr. 036.026.

Photography: see what you will get.
See Abstr. 036.027.

Visuele radiantbepaling.
See Abstr. 036.034.

GNOMPLOT – ein Hilfsprogramm für Meteorbeobachter.
See Abstr. 036.062.

FORWARD: extension for multiple stations.
See Abstr. 036.209.

On possible influence of meteors on the Earth atmosphere.
See Abstr. 082.001.

On the effect of meteor activity in the night–time emission of the middle atmosphere.
See Abstr. 082.118.

Asteroid 5025 P–L, comet 1967 II Rudnicki, and the Taurid meteoroid complex.
See Abstr. 098.029.

Introduction: history, definitions, nomenclature (*of small bodies of the solar system*).
See Abstr. 098.051.

The systems of interplanetary objects.
See Abstr. 098.052.

Apollo asteroid 1987 SY and the δ–Leonid meteor shower.
See Abstr. 098.084.

Estimate of the ejection velocity of cometary dust according to the length of the terrestrial path in meteor streams.
See Abstr. 102.092.

Comet 1987c and the ε Geminids.
See Abstr. 103.233.

P/Halley dust characteristics: a comparison between Orionid and Eta Aquarid meteor observations and those from the flyby spacecraft.
See Abstr. 103.592.

Comparison of characteristics of meteorites and meteor bodies.
See Abstr. 105.013.

A method for calculating the trajectories of meteorites in the atmospheres of planets with ablation and change of the form.
See Abstr. 105.018.

The Innisfree meteorite: dynamical history of the orbit; possible family of meteor bodies.
See Abstr. 105.045.

105 Meteorites, Meteorite Craters

105.001 Source of the optical red–slope in iron–rich meteorites.
D. T. Britt, C. M. Pieters, P. H. Schultz.
NASA Tech. Memo., NASA TM–89810, p. 84 (1987). – See Abstr. 003.001.

105.002 A search for spectral alteration effects in chondritic gas–rich breccias.
J. F. Bell, K. Keil.
NASA Tech. Memo., NASA TM–89810, p. 85 – 86 (1987). – See Abstr. 003.001.

105.003 Calcium carbonate and calcium sulfate in Martian meteorite EETA 79001.
J. L. Gooding, S. J. Wentworth.
NASA Tech. Memo., NASA TM–89810, p. 158 – 159 (1987). – See Abstr. 003.001.

105.004 Shatter cones in Illinois: evidence for meteoritic impacts at Glasford and Des Plaines.
J. F. McHone, M. L. Sargent, W. J. Nelson.
NASA Tech. Memo., NASA TM–89810, p. 438 (1987). – See Abstr. 003.001.

105.005 Micrométéorites et débris orbitaux.
S. Koutchmy.
Astronomie, Vol. 101, p. 426 – 430 (1987).

105.006 Melting experiment on a model chondritic mantle composition at 25 GPa.
E. Ohtani, H. Sawamoto.
Geophys. Res. Lett., Vol. 14, No. 7, p. 733 – 736 (1987).
The authors have conducted melting experiments on a chondritic mantle composition in the five–component system CaO–FeO–MgO–Al_2O_3–SiO_2 at 25 GPa. The liquidus phase at this pressure is majorite garnet, and the second crystallizing phase is Mg–silicate perovskite or magnesiowustite; these phases seem to be consumed into the melt at almost the same temperature. The present experiments indicate that majorite garnet fractionation might have occurred in the chondritic primitive mantle, if the partial molten zone extended to a depth equivalent to the upper part of the lower mantle in the early stages of terrestrial evolution.

105.007 Depth distribution of cosmogenic nuclides in boring core samples of Jilin meteorite and its cosmic ray irradiation history.
Z.–y. Ouyang, C.–y. Fan, W.–x. Yi, X.–b. Wang, F. Begemann, T. Kersten, G. Heusser, E. Pernicka.
Sci. Sin., Ser. A, Vol. 30, No. 8, p. 885 – 896 (1987).
Two boring cores were sampled from the Jilin No. 1 meteorite in such a way that they were kept parallel or perpendicular to the surface of the 1st stage parent body, but across the center of the 2nd stage parent body. Cosmogenic nuclides (^{3}He, ^{20,21,22}Ne, ^{22}Na, ^{26}Al, ^{53}Mn, ^{60}Co), radiogenic gases and trapped noble gases in the two cores have been studied in detail, which has confirmed the two–staged cosmic ray irradiation history proposed by the authors as being typical of the Jilin meteorite, and also justified their previous models regarding the ages, shapes and sizes of the two parent bodies related to the two stages as well as the emplacement of various samples.

105.008 Origin and evolution of the ureilite parent magmas: multi–stage igneous activity on a large parent body.
C. A. Goodrich, J. H. Jones, J. L. Berkley.
Geochim. Cosmochim. Acta, Vol. 51, No. 9, p. 2255 – 2273 (1987).
The authors present new electron microprobe data for minor elements in olivine and pigeonite in a suite of eight ureilites. Their purpose in obtaining these data was to see if minor element – major element trends among the ureilites provide evidence for fractional crystallization or partial melting, and whether they support the importance of reduction or oxidation. In order to obtain a self–consistent set of high precision data, they used a high beam current and long counting times.

105.009 Ureilites: trace element clues to their origin.
M.–J. Janssens, J. Hertogen, R. Wolf, M. Ebihara, E. Anders.
Geochim. Cosmochim. Acta, Vol. 51, No. 9, p. 2275 – 2283 (1987).
To reexamine the origin of ureilites, the authors have analyzed by RNAA two hand–picked vein separates from Haverö and Kenna, as well as a bulk sample of Kenna – the ureilite richest in siderophiles.

105.010 Aqueous alteration in carbonaceous chondrites: mass balance constraints on matrix mineralogy.
H. Y. McSween Jr.
Geochim. Cosmochim. Acta, Vol. 51, No. 9, p. 2469 – 2477 (1987).
Although a rather detailed picture of the identities, structures, and compositions of secondary minerals in chondrite matrices is beginning to emerge from various studies, no quantitative information is available on the relative proportions of these phases or how these proportions may change during progressive alteration. The purpose of this paper is to provide that information. Mass balance calculations using the compositions of secondary phases and of bulk matrices will be used to estimate relative mineral proportions.

105.011 Isotopic characterisation of kerogen–like material in the Murchison carbonaceous chondrite.
J. F. Kerridge, S. Chang, R. Shipp.
Geochim. Cosmochim. Acta, Vol. 51, No. 9, p. 2527 – 2540 (1987).
Isotopic data for C, H and N in acid–resistant residues from carbonaceous chondrites show substantial variability during stepwise pyrolysis and/or combustion. After subtraction of contributions due apparently to inorganic C grains, of probably circumstellar origin, considerable isotopic variability remains, attributable to the kerogen–like organic fraction. That variability may be interpreted in terms of three or four distinct components, based on C, H and N isotopes. The different isotopic components are tentatively identified in terms of specific chemical/structural moieties within the kerogen–like material. This combination of chemical, structural and isotopic information suggests a complex history for meteoritic organic matter.

105.012 Geochemical investigations to explain iodine–overabundances in Antarctic meteorites.
K. G. Heumann, M. Gall, H. Weiss.
Geochim. Cosmochim. Acta, Vol. 51, No. 9, p. 2541 – 2547 (1987).
Iodine, bromine, and chlorine concentrations were determined in different Antarctic meteorite specimens (eucrites, high–iron and low–iron chondrites) with isotope dilution mass spectrometry. In all Antarctic meteorites I–overabundances have been analysed compared with the concentrations for non–Antarctic meteorites of the same class. By analysing different types of Antarctic rocks, a significant decrease of the I concentration – but not of the Br and Cl concentration – was determined from the surfaces to the centers of the rocks. This shows that atmospheric I compounds interact with the surfaces of Antarctic rocks and, therefore, with those of Antarctic meteorites as well. The authors construct a preliminary hypothesis for a geochemical I cycle in Antarctica, taking into consideration long–distance and short–distance transportation of different I compounds from the coast to inland Antarctica.

105.013 Comparison of characteristics of meteorites and meteor bodies.
V. N. Lebedinets.
20th All–Union meteoritic conference, p. 153 – 154 (1987). In Russian. Abstr. in Ref. Zh., 51. Astron., 7.51.234 (1987). – See Abstr. 012.017.

105.014 The Sverdlovsk chondrite.
V. A. Koroteev, G. V. Pal'gueva, V. A. Vilisov, I. S. Chashukhin, I. A. Yudin, V. N. Loginov.
20th All–Union meteoritic conference, p. 150 (1987). In Russian. Abstr. in Ref. Zh., 51. Astron., 7.51.257 (1987). – See Abstr. 012.017.

105.015 New meteorites of the Trans–Ural.
A. M. Dymkin, V. I. Glazyrina, I. S. Chashukhin, G. V. Pal'gueva, I. A. Yudin, V. N. Loginov, V. N. Savel'ev.
20th All–Union meteoritic conference, p. 149 (1987). In Russian. Abstr. in Ref. Zh., 51. Astron., 7.51.258 (1987). – See Abstr. 012.017.

105.016 The luminosity of meteorites: theory and comparison with observations.
A. I. Vislyj, V. P. Stulov.
20th All–Union meteoritic conference, p. 159 (1987). In Russian. Abstr. in Ref. Zh., 51. Astron., 7.51.260 (1987). – See Abstr. 012.017.

105.017 Methods for determining preatmospheric sizes of meteorites.
G. K. Ustinov, V. A. Alekseev, A. K. Lavrukhina.
20th All–Union meteoritic conference, p. 112 – 113 (1987). In Russian. Abstr. in Ref. Zh., 51. Astron., 7.51.261 (1987). – See Abstr. 012.017.

105.018 A method for calculating the trajectories of meteorites in the atmospheres of planets with ablation and change of the form.
G. I. Petrov, A. I. Vislyj, V. N. Mirskij, V. P. Stulov.
20th All–Union meteoritic conference, p. 158 (1987). In Russian. Abstr. in Ref. Zh., 51. Astron., 7.51.262 (1987). – See Abstr. 012.017.

105.019 Statistics of meteorite fall.
V. A. Alekseev.
20th All–Union meteoritic conference, p. 172 – 173 (1987). In Russian. Abstr. in Ref. Zh., 51. Astron., 7.51.263 (1987). – See Abstr. 012.017.

105.020 Local and exotic components of primitive meteorites, and their origin.
E. Anders.
Philos. Trans. R. Soc. London, Ser. A, Vol. 323, No. 1572, p. 287 – 304 (1987). – See Abstr. 012.026.

All meteoritic matter originally came from outside the solar system and thus is exotic. However, much of this material was reprocessed in the early solar system and became isotopically homogeneous. Such commonplace material is properly called "local", leaving the term "exotic" for material that still retains an anomalous isotopic signature. Qualitatively, a similar picture may be expected for comets and so primitive meteorites are a good frame of reference for interpretation of cometary data. The author reviews the properties and classification of chondrites and their local and exotic components.

105.021 Stable isotope measurements of meteorites and cosmic dust grains.
C. T. Pillinger.
Philos. Trans. R. Soc. London, Ser. A, Vol. 323, No. 1572, p. 313 – 322 (1987). – See Abstr. 012.026.

It is now established that dust probably produced in novae and red giants can be located in primitive meteorites and the latest state of knowledge in respect of such components is reviewed.

Nitrogen isotopic measurements have been helpful in distinguishing another form of dust that is carbonaceous but does not have a distinctive ^{13}C abundance. Likewise they suggest a non–carbonaceous material present in the meteorite Bencubbin could be a relief of supernovae outbursts. None of the components seen in meteorites can be detected in deep–sea spheres or stratospheric grains to provide a link between interstellar matter and comets. Until now interstellar dust has been the realm of observing astronomers and theoretians; stable isotope measurements are responsible for recognizing a material which it should be possible to isolate and study in the laboratory.

105.022 On the cometary origin of chondrites.
V. A. Alekseev.
Priroda, No. 9, p. 39 – 41 (1987). In Russian.

105.023 Organic matter in meteorites.
O. V. Nikolaeva.
Priroda, No. 9, p. 41 – 43 (1987). In Russian.

105.024 What geologists find in meteorites.
V. L. Masajtis.
Priroda, No. 9, p. 43 – 45 (1987). In Russian.

105.025 Copper and nickel partitioning in iron meteorites.
S. R. Sutton, J. S. Delaney, J. V. Smith, M. Prinz.
Geochim. Cosmochim. Acta, Vol. 51, No. 10, p. 2653 – 2662 (1987).

The classification of iron meteorites on the basis of their trace element abundances has proven to be important in constraining the origins of these materials. In magmatic iron groups, the bulk Cu content of the meteorites is inversely related to the bulk Ni content, whereas in non–magmatic groups, such as the IAB and IIICD groups, Cu can be either chalcophile or siderophile. This paper reports measurements of the distribution of Cu and Ni between troilite and metal from nine meteorites representing groups IAB, IIB, IIIA, IIICD and IVA and quenched spherules from the Canyon Diablo impact crater (Meteor Crater, Arizona). The observed relationships between partitioning and bulk meteorite properties are discussed.

105.026 Composition and mineralogy of refractory–metal–rich assemblages from a Ca, Al–rich inclusion in the Allende meteorite.
A. Bischoff, H. Palme.
Geochim. Cosmochim. Acta, Vol. 51, No. 10, p. 2733 – 2748 (1987).

The authors have separated four refractory metal–rich samples (10 μg to 190 μg) from a single Ca, Al–inclusion of the Allende meteorite. Chemical analyses were performed by INAA. From the analysed samples polished sections were prepared. Mineral phases from six large Fremdlinge and the surrounding inclusion were analysed by EDS.

105.027 The oxidation state of iron in tektite glass.
R. F. Fudali, M. D. Dyar, D. L. Griscom, H. D. Schreiber.
Geochim. Cosmochim. Acta, Vol. 51, No. 10, p. 2749 – 2756 (1987).

Fe^{3+}/Fe^{2+} ratios in tektite glass have the potential for constraining the $f_{O_2}-T$ conditions associated with tektite melts prior to quenching. But, even discounting older analyses, values of Fe^{3+}/Fe^{2+} in the literature range from zero to 0.82. The authors have analyzed five tektites from populations that have been reported to have Fe^{3+}/Fe^{2+} ratios both of zero and also a range of ratios up to 0.67. The range they found of 0.02 – 0.12 is in excellent agreement with the literature values that the authors consider to be consistently most reliable and they believe that most if not all Fe^{3+}/Fe^{2+} ratios in tektites fall within this range.

105.028   Meteoroid sonic shock–wave–generated seismic signals observed at a seismic array.
F. M. Anglin, R. A. W. Haddon.
Nature, Vol. 328, No. 6131, p. 607 – 609 (1987).
At approximately 04:20 local time (10:20 UT) on 19 September 1986, a local resident observed a meteor in the sky above Yellowknife, North–west Territory, Canada. The authors subsequently found seismic signals, recorded at the Yellowknife seismic array in North–west Territory, Canada, that coincided with the visual observation of the meteor in the sky overhead. The signals are shown to be consistent with their having been generated by the shock wave created by the associated meteoroid as it passed through the atmosphere overhead.

105.029   Secrets of black dust revealed.
I. D. R. Mackinnon.
Nature, Vol. 328, No. 6132, p. 670 – 671 (1987).
Commenting on a new study of Greenland cryoconite by M. Maurette et al. (see 105.030), this note reviews recent research on the total flux and origin of micrometeorites.

105.030   Characteristics and mass distribution of extraterrestrial dust from the Greenland ice cap.
M. Maurette, C. Jéhanno, E. Robin, C. Hammer.
Nature, Vol. 328, No. 6132, p. 699 – 702 (1987).
The authors present analyses of extraterrestrial grains extracted from six samples of cryoconite (black dust) collected from the melt zone of the Greenland ice cap. In addition to families of grains never reported before, the authors have found a surprisingly high abundance of unmelted chondritic fragments, in which the non–volatile component of the parent bodies has been well preserved. The mass distribution of the grains is very similar to that of the micrometeorite flux at 1 AU, indicating that most of the grains are micrometeorites, and not ablation products of larger meteorites.

105.031   Magnetism of meteorites.
S. M. Cisowski.
Geomagnetism, Vol. 2, p. 525 – 560 (1987). – See Abstr. 003.011.
Contents: 1. Introduction. 2. Magnetic classification of meteorites. 3. Magnetic properties of meteorites. 4. Paleointensity studies on meteorites. 5. Summary.

105.032   Photometry and polarimetry of possible meteorite and terrestrial analogs of M–asteroids.
I. N. Bel'skaya, D. F. Lupishko.
20th All–Union meteoritic conference, p. 180 – 182 (1987). In Russian. Abstr. in Ref. Zh., 51. Astron., 9.51.249 (1987). – See Abstr. 012.017.

105.033   Visual albedo of meteorites and asteroids.
V. V. Titov, V. Ya. Vokhmentsev,
V. D. Kolomenskij, D. I. Shestopalov, L. F. Golubeva.
20th All–Union meteoritic conference, p. 182 – 183 (1987). In Russian. Abstr. in Ref. Zh., 51. Astron., 9.51.250 (1987). – See Abstr. 012.017.

105.034   Four–colour photometry of meteorites and asteroids.
D. I. Shestopalov, L. F. Golubeva,
V. Ya. Vokhmentsev, V. D. Kolomenskij, V. V. Titov.
20th All–Union meteoritic conference, p. 183 – 184 (1987). In Russian. Abstr. in Ref. Zh., 51. Astron., 9.51.329 (1987). – See Abstr. 012.017.

105.035   Fine–grained aggregates in L3 chondrites.
S. Watanabe, M. Kitamura, N. Morimoto.
Earth Planet. Sci. Lett., Vol. 86, No. 2/4, p. 205 – 213 (1987).
The textures and chemical compositions of the constituent minerals of the fine–grained aggregates (FGA's) of L3 chondrites were studied by the backscattered electron image technique, electron probe microanalysis, and transmission electron microscopy. Bulk compositions of the FGA's are within the range of those of chondrules, so some chondrules probably were produced by complete melting of the same precursor materials as those of the FGA's.

105.036   Martian meteorites are arriving.
R. A. Kerr.
Science, Vol. 237, No. 4816, p. 721 – 723 (1987).
The eight SNC meteorites found on earth are probably from Mars, most researchers now agree, but how they ever got off their home planet remains a question.

105.037   The large crater origin of SNC meteorites.
A. M. Vickery, H. J. Melosh.
Science, Vol. 237, No. 4816, p. 738 – 743 (1987).
A large body of evidence strongly suggests that the shergottite, nakhlite, and Chassigny (SNC) meteorites are from Mars. Various mechanisms for the ejection of large rocks at martian escape velocity (5 kilometers per second) have been investigated, but none has proved wholly satisfactory. This article examines a number of possible ejection and cosmic–ray exposure histories to determine which is most plausible. SNC meteorites were probably ejected from a very large crater (> 100 kilometers in diameter) about 200 million years ago, and cosmic–ray exposure of the recovered meteorites was initiated after collisional fragmentation of the original ejecta in space at much later times (0.5 to 10 million years ago).

105.038   Fission–track dating of Haughton Astrobleme and included biota, Devon Island, Canada.
G. Omar, K. R. Johnson, L. J. Hickey, P. B. Robertson,
M. R. Dawson, C. W. Barnosky.
Science, Vol. 237, No. 4822, p. 1603 – 1605 (1987).

105.039 ^{41}K isotope anomaly in meteoritic matter.
A. A. Ul'yanov.
Geokhimiya, No. 7, p. 1037 – 1040 (1987). In Russian. Abstr. in Ref. Zh., 51. Astron., 10.51.300 (1987).

105.040   UV and vacuum UV spectra of organic extract from Yamato carbonaceous chondrites.
S. Yabushita, T. Inagaki, E. T. Arakawa, K. Wada.
Mon. Not. R. Astron. Soc., Vol. 229, No. 3, p. 45p – 48p (1987).
Earlier UV measurements of the absorbance of an organic extract from the Murchison carbonaceous chondrite showed a shoulder close to 220 nm, which is sometimes interpreted as evidence which supports the organic nature of interstellar grains. Here the authors present UV and vacuum UV absorbance spectra of organic extracts from the Yamato carbonaceous chondrites. The most conspicuous peak is found at $\lambda = 190$ nm and only a weak shoulder is visible at 220 nm. The result indicates that organic dust grains are not likely to be carriers of the well–known 220 nm peak of the IS extinction curve.

105.041   Erratum: "Detection of a meteorite 'stream': observations of a second meteorite fall from the orbit of the Innisfree chondrite" [Icarus, Vol. 69, No. 3, p. 550 – 556 (1987)].
I. Halliday.
Icarus, Vol. 72, No. 1, p. 239 (1987). See Abstr. 43.105.199.

105.042   Formation of crater Siljan.
K. B. Bruberg.
Mekh.: Novoe v zarubezh. nauke, Moskva, No. 42, p. 235 – 272 (1987). In Russian. Abstr. in Ref. Zh., 51. Astron. 11.51.340 (1987).

105.043   Trace element contents of primitive meteorites; a test of solar system abundance smoothness.
D. S. Woolum, D. S. Burnett, T. M. Benjamin, P. S. Z. Rogers,
C. J. Duffy, C. J. Maggiore.
Nucl. Instrum. Methods Phys. Res., Sect. B, Vol. B22, No. 1 – 3, p. 376 – 379 (1987). Abstr. in Phys. Abstr., Vol. 90, No. 1310, Entry 94832 (1987).

105.044 On the use of a nuclear microprobe for trace element analysis in meteorites and cosmic dust.
R. D. Vis, C. C. A. H. van der Stap, D. Heymann.
Nucl. Instrum. Methods Phys. Res., Sect. B, Vol. B22, No. 1 – 3, p. 380 – 385 (1987). Abstr. in Phys. Abstr., Vol. 90, No. 1310, Entry 94833 (1987).

105.045 The Innisfree meteorite: dynamical history of the orbit; possible family of meteor bodies.
I. V. Galibina, A. K. Terent'eva.
Astron. Vestn., Tom 21, No. 3, p. 251 – 261 (1987). In Russian. English translation in Sol. Syst. Res.
Evolution of the Innisfree meteorite orbit caused by secular perturbations is studied over the time interval of 500000 yrs (from the current epoch backwards). Calculations are made by the Gauss–Halphen–Gorjatschew method taking into account perturbations from the four outer planets – Jupiter, Saturn, Uranus and Neptune. In the above mentioned time interval the meteorite orbit has undergone no essential transformations. The Innisfree orbit intersected in 91 cases the Earth orbit and in 94 – the Mars orbit. A system of small and large meteor bodies (producing ordinary meteors and fireballs) which may be genetically related to the Innisfree meteorite has been found, i.e. there probably exists an Innisfree family of meteor bodies.

105.046 Compositional evidence regarding the origins of rims on Semarkona chondrules.
J. N. Grossman, J. T. Wasson.
Geochim. Cosmochim. Acta, Vol. 51, No. 11, p. 3003 – 3011 (1987).
Six of seven chondrule rims have enhanced contents of siderophiles and chalcophiles relative to chondrule interiors. The authors suspect that this indicates that during chondrule formation, metal/sulfide melts migrated to the exterior of the chondrule, and later reheating caused it to spread out into fine-grained rim material.

105.047 Chemical and physical studies of type 3 chondrites. VIII. Thermoluminescence and metamorphism in the CO chondrites.
B. D. Keck, D. W. G. Sears.
Geochim. Cosmochim. Acta, Vol. 51, No. 11, p. 3013 – 3021 (1987).
In view of the mineralogical similarity of the CO and type 3 ordinary chondrites, and their potentially similar metamorphic history which caused glass devitrification in both types of meteorites, the authors have undertaken a study of the TL properties of the CO chondrite class in the hope that the technique may contribute to a quantitative understanding of the metamorphic history of this class.

105.048 Variable $^{190}Os/^{184}Os$ ratio in acid residues of iron meteorites.
P. S. Goel.
Proc. Indian Acad. Sci., Earth Planet. Sci., Vol. 96, No. 2, p. 81 – 102 (1987).
In residual materials obtained on dissolution of iron meteorites in 2M H_2SO_4, the ratio of $^{190}Os/^{184}Os$ has been measured by radiochemical neutron activation analysis. Most residues have a normal isotopic ratio (to within $\pm 2\%$). However, in some residues both positive and negative deviations in the isotopic ratio are seen. The most spectacular deviations are in the insoluble fragments (nuggets) from Sikhote Alin iron meteorite where the $^{190}Os/^{184}Os$ ratio is about 50% of the normal value. The new results confirm our earlier observations that iron meteorites contain pre–solar grains.

105.049 Sulla caduta di meteoriti, la loro ricerca e classificazione.
M. Eltri.
Astronomia, N. 3 – 4, p. 19 – 26 (1987).
The incidence of falling meteoritic objects and the problems regarding the research and identification of the specimens are

briefly examined. A simple and simplified mineralogical classification is included together with details on how to distinguish meteorites from terrestrial rocks.

105.050 Isotopically anomalous ^{196}Hg and ^{202}Hg in Antarctic achondrites.
S. Jovanovic, G. W. Reed Jr.
Geophys. Res. Lett., Vol. 14, No. 11, p. 1127 – 1130 (1987).

105.051 Study on chemical compositions of Bo Xian chondrite, Anhui [China].
B.–x. Li, J.–f. Gu, X. Zhang.
Chin. J. Space Sci., Vol. 6, No. 4, p. 321 – 324 (1986). In Chinese. Abstr. in Phys. Abstr., Vol. 90, No. 1313, Entry 115327 (1987).

105.052 New types of spherules from Antarctica: meteoritic impact origin?
Y. Tazawa, Y. Fujii.
Geophys. Res. Lett., Vol. 14, No. 12, p. 1199 – 1202 (1987).
Spherules collected from Antarctic ice have been studied by using instrumental neutron activation analysis, energy dispersive X–ray spectrometry and X–ray diffraction photography. Peculiar spherules, Ca–Ti–rich (perovskite) type (CTS) and Fe–Cr–Ni–rich type (FCN), were found in the Mizuho ice core at depths of 32 to 33.5 m. In the Allan Hills bare ice, only a "chondritic" type without depletion of Au and S (CAS) was recognized. All the results combine to suggest that CTS and FCN may be droplets strewn by the impact of a huge meteorite, and CAS must be debris from one of the chondrites that fell on the source region of the Allan Hills bare ice and survive terrestrial alterations.

105.053 Die Meteoritensammlung E. F. F. Chladnis.
G. Hoppe.
Sterne, 63. Band, Heft 6, p. 315 – 329 (1987).

105.054 The Bouvante eucrite. Chemistry, petrology and mineralogy.
M. Christophe–Michel–Levy, M. Bourot–Denise, H. Palme, B. Spettel, H. Wänke.
Bull. Mineral. (France), Vol. 110, No. 4, p. 449 – 458 (1987). In French. Abstr. in Phys. Abstr., Vol. 90, No. 1316, Entry 134199 (1987).

105.055 Mössbauer study of shock–induced effects in the ordered alloy $Fe_{50}Ni_{50}$ in meteorites.
R. B. Scorzelli, I. S. Azevedo, J. Danon, M. A. Meyers.
J. Phys. F, Vol. 17, No. 9, p. 1993 – 1997 (1987). Abstr. in Phys. Abstr., Vol. 90, No. 1318, Entry 145902 (1987).

105.056 Isotopic lead investigations on Bo County chondrite of Anhui Province, China.
Z.–c. Peng.
Chin. Phys., Vol. 7, No. 2, p. 450 – 454 (1987). Abstr. in Phys. Abstr., Vol. 91, No. 1320, Entry 10510 (1988).

105.057 Evidence for interstellar SiC in the Murray carbonaceous meteorite.
T. Bernatowicz, G. Fraundorf, T. Ming, E. Anders, B. Wopenka, E. Zinner, P. Fraundorf.
Nature, Vol. 330, No. 6150, p. 728 – 730 (1987).
The authors have identified silicon carbide in two separates from the Murray carbonaceous chondrite. The SiC is present in the form of crystalline grains, 0.1 – 1 µm in size. The anomalous isotopic composition of its carbon, nitrogen and silicon indicates a pre–solar origin. An additional silicon– and oxygen–rich phase shows large isotopic anomalies in nitrogen and silicon, also associated with a pre–solar origin.

105.058 Large isotopic anomalies of Si, C, N and noble gases in interstellar silicon carbide from the Murray meteorite.
E. Zinner, T. Ming, E. Anders.
Nature, Vol. 330, No. 6150, p. 730 – 732 (1987).
Primitive meteorites contain several noble gas components with anomalous isotopic compositions which imply that they –

and their solid 'carrier' phases – are of exotic, pre–solar origin. The authors found that minor fractions of the Murray meteorite contain two minerals not previously seen in meteorites: silicon carbide and an amorphous Si–O phase. They report ion microprobe analyses of these phases which reveal very large isotopic anomalies in silicon, nitrogen and carbon, exceeding the highest anomalies previously measured by factors of up to ~50. It is concluded that these phases are circumstellar grains from carbon–rich stars, whose chemical inertness allowed them to survive in exceptionally well–preserved form.

105.059 ^{16}O excesses in Murchison and Murray hibonites: a case against a late supernova injection origin of isotopic anomalies in O, Mg, Ca, and Ti.
A. J. Fahey, J. N. Goswami, K. D. McKeegan, E. K. Zinner.
Astrophys. J., Lett. Ed., Vol. 323, No. 1, p. L91 – L95 (1987).

The oxygen isotopic compositions of seven individual hibonite samples from the carbonaceous chondrites Murchison and Murray have been measured by means of ion microprobe mass spectrometry. All seven samples show large ^{16}O excesses relative to terrestrial oxygen. These hibonites have the largest ^{48}Ca and ^{50}Ti isotopic anomalies found to date; thus there is no intrinsic relationship between anomalies of a nucleosynthetic origin and isotopic mass fractionation effects. There is no correlation between the oxygen isotopic compositions, the initial ^{26}Al/^{27}Al ratios, the Ca and Ti isotopic abundances, or the rare earth element patterns of individual hibonite grains. In particular, the large ^{16}O excess seen in the Murchison grain BB–5 argues against a late injection of exotic material from a nearby supernova as a source for the isotopic anomalies.

Atlas of reflectance spectra of terrestrial, lunar and meteoritic powders and frosts from 92 to 1800 nm.
See Abstr. 002.001.

Meteory, meteority, meteoroidy (Meteors, meteorites, meteoroids).
See Abstr. 003.033.

Meteoritnye kratery na Zemle (Meteorite craters on the Earth).
See Abstr. 003.079.

Meteority Ukrainy (Meteorites of the Ukraine).
See Abstr. 003.127.

The Bernardino Rivadavia meteorite collection.
See Abstr. 009.028.

Horizons of present–day meteoritics.
See Abstr. 011.015.

The XXth All–Union meteoritic conference held in Tallinn, 10 – 12 February 1987.
See Abstr. 011.026.

20th All–Union meteoritic conference, held in Tallinn, 10 – 12 February 1987.
See Abstr. 012.017.

Application of RIMS to the study of noble gases in meteorites.
See Abstr. 022.114.

Mass–independent isotopic fractionation in nonadiabatic molecular collisions.
See Abstr. 022.117.

Grain formation through the nucleation process in astrophysical environments. II. Nucleation and grain growth accompanied by chemical reaction.
See Abstr. 022.136.

Nucleosynthesis contributions to the solar nebula.
See Abstr. 061.025.

Nuclear structure of ^{49}Ca above 5 MeV excitation from n + ^{48}Ca and astrophysics for 30 keV neutrons.
See Abstr. 061.066.

The nature of the anomalous sky glow connected with the Tunguska meteorite.
See Abstr. 082.041.

Atmospheric iridium at the South Pole as a measure of the meteoritic component.
See Abstr. 082.090.

Isotopic abundances: inferences on solar system and planetary evolution.
See Abstr. 091.032.

Meteorite spectroscopy and characterization of asteroid surface materials.
See Abstr. 098.005.

Dynamical relations between asteroids, meteorites and Apollo–Amor objects.
See Abstr. 098.027.

Photopolarimetry of asteroids and meteorites as a method of investigating the microstructure of their surfaces.
See Abstr. 098.042.

On some peculiarities of spectra of dark asteroids and meteorites in the ultraviolet region.
See Abstr. 098.043.

One of the possible models of thermal disruption of meteor and meteoritic bodies.
See Abstr. 104.036.

Orbits of meteorites and their interaction with small bodies.
See Abstr. 104.046.

Distribution of parent bodies of meteorites of different generations according to their sizes.
See Abstr. 104.047.

Meteors, meteorites and cosmic dust.
See Abstr. 104.057.

Analytical electron microscopy of fine–grained phases in primitive interplanetary dust particles and carbonaceous chondrites.
See Abstr. 106.001.

Interstellar polycyclic aromatic hydrocarbons and carbon in interplanetary dust particles and meteorites.
See Abstr. 106.021.

Magnetic flares in the protoplanetary nebula and the origin of meteorite chondrules.
See Abstr. 107.009.

Meteorites, asteroids and comets – a study of their interrelationship.
See Abstr. 107.016.

Isotope abundance anomalies and the early solar system: (some) facts and (some) implications.
See Abstr. 107.017.

Search for prototerrestrial matter.
See Abstr. 107.019.

Meteorites – witnesses of the evolution of the protoplanetary disk.
See Abstr. 107.020.

Chemical condensation in the outflowing matter from the proto–Sun and its application to meteorites.
See Abstr. 107.022.

What the solar nebula was like.
See Abstr. 107.041.

Shock processing of interstellar dust: diamonds in the sky.
See Abstr. 131.130.

106 Interplanetary Matter, Interplanetary Magnetic Field, Zodiacal Light

106.001 Analytical electron microscopy of fine–grained phases in primitive interplanetary dust particles and carbonaceous chondrites.
I. D. R. Mackinnon, F. J. M. Rietmeijer, D. S. McKay.
NASA Tech. Memo., NASA TM–89810, p. 87 – 89 (1987). – See Abstr. 003.001.

106.002 Diffusive shock acceleration: comparison of a unified shock model to bow shock observations.
D. C. Ellison, E. Möbius.
Astrophys. J., Vol. 318, No. 1, p. 474 – 484 (1987).
Recent AMPTE/IRM observations of diffuse ions seen up-stream from the Earth's bow shock when the interplanetary magnetic field was nearly parallel to the solar wind direction are compared with a previously published collisionless, quasi-parallel shock model (Ellison; Ellison and Eichler). These observations, which provide the proton spectrum over all of velocity space, give a direct measure of the efficiency of shock acceleration and show how thermal solar wind ions are injected into the diffusive shock acceleration mechanism.

106.003 Studies of the interplanetary magnetic field: IMP's (*Interplanetary Monitoring Platforms*) to Voyager.
N. F. Ness.
NASA Conf. Publ., NASA CP–2464, p. 99 – 121 (1987). – See Abstr. 012.004.
Contents: 1. Introduction. 2. Early IMP period: 1963 – 1967. 3. Sector structure of the interplanetary magnetic field. 4. Cosmic ray modulation. 5. Summary.

106.004 Doppler shifts in zodiacal light.
D. C. Hirschi, D. B. Beard.
Planet. Space Sci., Vol. 35, No. 8, p. 1021 – 1027 (1987).
Doppler shifts in zodiacal light are calculated for various eccentricities, dust sizes, and assumptions about radiation pressure. The purpose is to determine what effects in spectra might be observed which would enhance our understanding of the origin and lifetimes of interplanetary dust.

106.005 Magnetic helicity of the IMF and the solar modulation of cosmic rays.
J. W. Bieber, P. Evenson, W. H. Matthaeus.
Geophys. Res. Lett., Vol. 14, No. 8, p. 864 – 867 (1987).
Using interplanetary magnetic field data acquired at 1 AU, the authors show that the distribution of magnetic helicity in the heliosphere is asymmetric about the current sheet, in accord with recent theoretical predictions. Such an asymmetric distribution of helicity can in principle explain the sensitivity of the solar modulation of cosmic rays to the sun's magnetic polarity, as observed in charge sign dependent modulation, Jovian electrons, and anomalous ^{4}He.

106.006 Low–energy protons associated with interplanetary shocks as a coherent population.
B. Sanahuja, V. Domingo.
J. Geophys. Res., Vol. 92, No. A7, p. 7280 – 7288 (1987).
The authors investigate the flow pattern of low–energy protons (35 – 1600 keV) associated with interplanetary shocks observed by ISEE–3 between August 1978 and April 1980. The analysis of the shape of the distribution function in the solar wind frame and its temporal evolution indicates that the low–energy protons can behave as a coherent, independent population of particles in the solar wind. Ahead of the shock this population propagates along the magnetic field in the same direction as the solar wind flow, while after the passage of the perturbed region associated with the shock, it propagates in the opposite sense. The behavior of the flow pattern of this population through the shock front is discussed for the 17 largest events observed in this period.

106.007 Aspects of interplanetary plasma turbulence.
L. M. Celnikier, L. Muschietti, M. V. Goldman.
Astron. Astrophys., Vol. 181, No. 1, p. 138 – 154 (1987).
Data acquired in the free solar wind by the ISEE 1 and ISEE 2 propagation experiment were used to obtain the power spectra of fluctuations in electron density; a maximum entropy technique has allowed full exploitation of the intrinsic precision of the in situ measurements. The results have led to a new determination of the anisotropy of density fluctuations in the solar wind, which is in conflict with certain expectations based on the propagation of Langmuir waves, but is in agreement with multistation interplanetary scintillation measurements. The authors have also investigated the relation between the fluctuation level and the mean density, and present tentative evidence for a correlation between the relative fluctuation level and proton temperature.

106.008 On a relationship between the interplanetary magnetic field and the large–scale distribution of the solar magnetic field.
E. V. Ivanov.
Soln. Dannye, Byull., 1987, No. 1, p. 59 – 66 (1987). In Russian.
The data on the IMF and space–time distribution of sunspot groups with an inclination of the bipolar axis $\varphi \geqslant |15°|$ for 1966 – 1979 have been used to reveal a correspondence of two IMF components with rotation periods of $\sim 27.2 – 27.3$ and $\sim 28 – 29$ days, respectively, to 2 components of the large–scale solar magnetic field with a similar period. Histograms showing the distribution of sunspot groups with $\varphi \geqslant |15°|$ relative to IMP sector boundaries are discussed.

106.009 Meteoroids from comet Bennett 1970 II.
M. Fulle.
Astron. Astrophys., Vol. 183, No. 2, p. 392 – 396 (1987).
The author considers the problem of the contribution of dust from long–period comets to the zodiacal cloud. He derives a new general expression, which takes into account the dust ejection velocity from the inner coma, for the diameter of the smallest particles injected into bounded orbits. For the case of ejection at zero–velocity the author confirms previous results (Burns et al., 1979) which state that no dust can be injected into the interplanetary cloud. But the consideration of an ejection velocity different from zero shows that this conclusion, suggested also by Jambor (1976) and Sekanina (1977), is not realistic. The application to the results of the inversion of the Finson–Probstein functional (Fulle, 1986) for the long–period comet Bennett 1970 II, shows that a significant fraction of the dust mass is injected into bounded orbits before the perihelion, whereas almost the whole dust is lost after. The extrapolation of the results to all long–period comets gives an isotropic source of meteoroids larger than ≈ 0.4 mm with an input mass rate of $(0.4–1.5) \times 10^6 \mathrm{g\,s^{-1}}$, quite far from the $\approx 10^7 \mathrm{g\,s^{-1}}$ required by the collisional models of the zodiacal cloud (Grün et al., 1985).

106.010 Mineralogy of chondritic interplanetary dust particles.
I. D. R. Mackinnon, F. J. M. Rietmeijer.
Rev. Geophys., Vol. 25, No. 7, p. 1527 – 1553 (1987).
Contents: Introduction. Sample collection contamination and preparation. Discussion: Mineral observations. Classification and comparison with carbonaceous chondrites. Sources of interplanetary dust. Solar system history: processes. Conclusions.

106.011 Radial and latitudinal variations of the interplanetary magnetic field.
L. W. Klein, L. F. Burlaga, N. F. Ness.
J. Geophys. Res., Vol. 92, No. A9, p. 9885 – 9892 (1987).
This paper presents observations of the radial and latitudinal variations of the interplanetary magnetic field measured by the Voyager 1 (V1) and Voyager 2 (V2) spacecraft from mid–1977 to

mid–1985. The data extend from 1 to 20 AU and from –5° to 26° in heliographic latitude. Data obtained at 1 AU are used to separate temporal variations from radial variations, and plasma measurements from V2 are used to consider the effect of temporal variations in the bulk speed. The observations are compared with the predictions of Parker's (1958, 1963) model.

106.012   Spectral signatures of jumps and turbulence in interplanetary speed and magnetic field data.
D. A. Roberts, M. L. Goldstein.
J. Geophys. Res., Vol. 92, No. A9, p. 10,105 – 10,110 (1987).

106.013   On the theory of radiation of plasmoids generated by collisions of cometary and zodiacal dust particles.
S. Ibadov.
Dokl. Akad. Nauk TadzhSSR, Tom 29, No. 10, p. 587 – 590 (1986). In Russian. Abstr. in Ref. Zh., 51. Astron., 8.51.223 (1987).

106.014   North–south asymmetric propagation of flare–associated interplanetary shock waves.
F.–s. Wei, X.–d. Deng.
Acta Geophys. Sin., Vol. 30, No. 5, p. 443 – 449 (1987). In Chinese.

The possible model of a flare–associated shock wave propagating in the meridian plane has been deduced from the north–south asymmetry in cosmic ray Forbush decreases, by using the dimensionless jump conditions of MHD. The results show that the propagation of a flare–associated shock wave in the meridian plane exhibits a north–south asymmetry relative to the flare normal, and the direction of the fastest propagation velocity tends to the heliospherical current sheet. This physical model is consistent with the results from interplanetary scintillation observations and could explain the north–south asymmetrical effects in the flare–associated geomagnetic disturbances and cosmic ray Forbush decreases.

106.015   Interplanetary matter and geomagnetic activity under stationary conditions.
A. A. Danilov.
Geofiz. issled. na shirotakh avroral. zony. Yakutsk, p. 20 – 41 (1986). In Russian. Abstr. in Ref. Zh., 62. Issled. Kosm. Prostranstva, 8.62.562 (1987).

106.016   Morphological, chemical and mineralogical studies of cosmic dust.
D. E. Brownlee.
Philos. Trans. R. Soc. London, Ser. A, Vol. 323, No. 1572, p. 305 – 311 (1987). – See Abstr. 012.026.

Significant numbers of 5 μm – 1 mm particles of interplanetary dust have been collected and subjected to laboratory analysis. The extraterrestrial origin of selected samples has been established by detection of space–exposure effects. The collected samples should contain both cometary and asteroidal particles. The elemental composition of the majority of particles is similar to CI and CM chondrites. Most particles can be grouped into one of two general classes, those that contain hydrated minerals and those that are anhydrous.

106.017   The origin of dust in the solar system.
S. V. M. Clube.
Philos. Trans. R. Soc. London, Ser. A, Vol. 323, No. 1572, p. 421 – 436 (1987). – See Abstr. 012.026.

The assumption that the zodiacal cloud is a predominantly meteoritic rather than a meteoroidal complex is questioned. On the basis of (1) the observed exposure ages of interplanetary dust particles collected from the stratosphere, (2) the compressive strength of the commonest fireballs, (3) the existence of a broad ecliptic stream centred on the Taurids and (4) the observation of substantial short–lived meteoroid swarms therein, a suitably consistent replenishment model is constructed in which the zodiacal cloud appears to derive from a now defunct large comet that arrived in an Earth–crossing orbit ca. 10 – 100 ka ago.

106.018   ISEE 3 observations of low–energy proton bidirectional events and their relation to isolated interplanetary magnetic structures.
R. G. Marsden, T. R. Sanderson, C. Tranquille, K.–P. Wenzel, E. J. Smith.
J. Geophys. Res., Vol. 92, No. A10, p. 11,009 – 11,019 (1987).

Bidirectional, field–aligned flows of low–energy particles in interplanetary space have been proposed as a possible signature of large–scale, looplike magnetic structures. The authors present the results of a survey of bidirectional anisotropy observations using data from the low–energy proton and magnetometer experiments on board ISEE 3 covering a 45–month period corresponding to the last solar maximum.

106.019   Asymmetry of the heliosphere.
S. T. Suess, D. H. Hathaway, A. J. Dessler.
Geophys. Res. Lett., Vol. 14, No. 9, p. 977 – 980 (1987).

The outflowing solar wind interacts with the local interstellar medium to form the heliospheric cavity within which the solar wind is supersonic. Because the interstellar medium is moving with respect to the Sun, and because the solar wind has a latitude dependence, the heliosphere is asymmetric. The purpose in this paper is to describe some of the important physical processes and considerations that determine the character of the asymmetry of the heliosphere and to present quantitative estimates of its magnitude.

106.020   The structure of ULF waves produced by a tethered satellite system.
A. N. Wright.
Astron. Astrophys., Vol. 186, No. 1/2, p. 354 – 358 (1987).

The Alfvén waves produced by a tethered satellite system (TSS) are modelled within the MHD approximation. Expressions for the current, velocity, electric and magnetic fields are derived using the general formalism given by Wright and Southwood (1987) to describe Stationary Alfvénic Structures. The current closure in the Alfvén wave is found to have some novel features. The subsidiary current system that flows in a plane perpendicular to the wave structure has a quadrupolar character, in contrast to the dipolar magnetic field and flow perturbations. The topology of the perpendicular current has not been understood in previous TSS models. The worked example that the author presents compliments the work of Wright and Southwood (1987) who anticipated that the polar order of the perpendicular currents would be twice that of the magnetic field and plasma flow perturbations.

106.021   Interstellar polycyclic aromatic hydrocarbons and carbon in interplanetary dust particles and meteorites.
L. J. Allamandola, S. A. Sandford, B. Wopenka.
Science, Vol. 237, No. 4810, p. 56 – 59 (1987).

Both interplanetary dust particles (IDPs) and meteorites may contain material that is similar to polycyclic aromatic hydrocarbons (PAHs). The Raman spectra of IDPs and meteorites show features that are similar in position and relative strength to interstellar infrared emission features that have been attributed to vibrational transitions in free, molecular–sized PAHs. The observations suggest that some of the carbonaceous material in IDPs and meteorites may have been produced in circumstellar dust shells and only slightly modified in interstellar space.

106.022   Refractory interplanetary dust particles.
M. E. Zolensky.
Science, Vol. 237, No. 4821, p. 1466 – 1468 (1987).

Criteria are described by which refractory interplanetary dust particles (IDPs) can be differentiated from the products of spacecraft debris. These criteria have been used to discover and characterize IDPs that are composed predominantly of refractory phases. Two of these particles contain hibonite, perovskite, spinel, refractory glass, and a melilite; only hibonite was identified within a third. The grain size for all particles ranges from 0.05 to 1 micrometer, so that they are much finer grained than the refractory calcium– and aluminum–rich inclusions in meteorites. The glass–containing refractory IDPs may be primitive nebular condensates that never completely crystallized and thus have been preserved extant.

106.023 Oxygen isotopes in refractory stratospheric particles: proof of extraterrestrial origin.
K. D. McKeegan.
Science, Vol. 237, No. 4821, p. 1468 – 1471 (1987).

The oxygen and magnesium isotopic compositions of five individual particles that were collected from the stratosphere and that bear refractory minerals were measured by secondary ion mass spectrometry. Four of the particles exhibit excesses of oxygen–16 similar to those observed in anhydrous mineral phases of carbonaceous chondrites and thus are extraterrestrial. Magnesium in the four extraterrestrial particles is isotopically normal. It is unlikely that these particles are derived from carbonaceous chondrites and thus such particles probably represent a new type of collected extraterrestrial material.

106.024 Techniques of zodiacal–light modelling.
T. Kelsall.
Bull. Am. Astron. Soc., Vol. 19, No. 2, p. 750 (1987). Abstract. – See Abstr. 010.061.

106.025 Enhanced low energy (1 MeV) ion fluxes in the outer heliosphere.
R. E. Gold, L. J. Lanzerotti, C. G. Maclennan.
Planet. Space Sci., Vol. 35, No. 11, p. 1359 – 1366 (1987).

The Low Energy Charged Particle (LECP) experiment on the Voyager 2 spacecraft in the outer heliosphere (>10 a.u.) has observed several occasions when there was a peak in the interplanetary ion spectra for ions of energies $\sim 0.5 – 1.0$ MeV). Such enhancements can last for several days, suggesting that at these times particles of these energies dominate the low energy cosmic population in this region of the heliosphere. Two specific cases are discussed. The enhancements seem to be associated with the passage of transient interplanetary shock events, with the ion anisotropies generally showing outflow. The most straight-forward explanation for the observations seems to involve only a propagation effect of ions from the inner to the outer solar system. This conclusion is supported by simple modeling of the propagation of an event observed at 1 a.u. to the spacecraft at ~ 12 a.u.

106.026 Statistical properties of interplanetary fluctuations behind the bow shock.
F. Wei, H. Du, Y. C. Whang.
Planet. Space Sci., Vol. 35, No. 11, p. 1419 – 1428 (1987).

Snell's transmission model can determine all modes of refracted waves if the incident angle is between the two critical angles. However, the model cannot be used to study transmission of waves at all incident angles. Assuming that interplanetary fluctuations impinge on the bow shock from all directions, an alternative model is introduced in this paper to simulate their interaction.

106.027 Inversion of the zodiacal infrared brightness integral.
S. S. Hong, I. K. Um.
Astrophys. J., Vol. 320, No. 2, p. 928 – 935 (1987).

An inversion method is developed for the brightness integral of zodiacal thermal emission and is applied to recent infrared observations by Murdock and Price. It is found that the dependence on the heliocentric distance r of the volumetric absorption cross section, $n(r)\sigma_{abs}(r;\lambda)$, of interplanetary dust particles at IR wavelengths cannot be described by the usual power–law relation which is frequently used in the visible. The discrepancy between IR emission and visible scattering indicates that the mean absorption cross section of interplanetary particles at IR wavelengths varies significantly with heliocentric distance.

106.028 The motion of interplanetary dust.
J. A. Burns.
The evolution of the small bodies of the solar system, p. 252 – 275 (1987). – See Abstr. 012.041.

Contents: 1. Introduction. 2. Derivation of radiation pressure and Poynting–Robertson drag (perfectly absorbing particles, scattering particles, comparison of radiation pressure to gravity, values of radiation forces). 3. Orbital consequences (radiation pressure effects on heliocentric orbits, radiation pressure effects on planetocentric orbits, Poynting–Robertson drag effects on heliocentric orbits, Poynting–Robertson collapse of planetocentric orbits). 4. The Yarkovsky effect. 5. Planetary perturbations. 6. Dynamical models of the IRAS dust bands. 7. Conclusions.

106.029 Brightness contribution of zodiacal dust along the line of sight in and out of the ecliptic plane and in the F–corona.
N. Y. Misconi, E. T. Rusk.
Planet. Space Sci., Vol. 35, No. 12, p. 1571 – 1574 (1987).

Model calculations are used to determine the location of interplanetary dust particles that contribute most of the brightness of the zodiacal light as seen from Earth, in and out of the ecliptic plane and in the F–corona. It is found that as one observes in increasing ecliptic latitude, the distance to the Earth decreases for dust contributing equal fractions to the line–of–sight brightness.

106.030 Radiative braking and resonance pattern in the motion of interplanetary dust.
A. Yu. Kogan.
Pis'ma Astron. Zh., Tom 13, No. 12, p. 1080 – 1085 (1987). In Russian. English translation in Sov. Astron. Lett., Vol. 13.

Dissipation due to the Poynting–Robertson effect generates resonance structures in the plane model of the Sun–Jupiter–dust system with non–interacting dust particles. Their parameters are computed. The particles captured in the fixed resonance are formed in a synodical orbit, Jupiter being enclosed in one of its loops. Particle motions are synchronized in a manner making the form render itself continuously rotating about the Sun following Jupiter.

106.031 Filtering of the local interstellar medium at the heliopause.
S. Bleszynski.
Ir. Astron. J., Vol. 18, No. 2, p. 88 (1987). Abstract. For the full paper see 43.106.036. – See Abstr. 012.036.

106.032 Numerical evaluation of statistical acceleration for energetic particles in the interplanetary medium at 2.5 and 5.0 AU.
X. Moussas, J. J. Quenby, J. F. Valdes–Galicia.
Sol. Phys., Vol. 112, No. 2, p. 365 – 382 (1987).

Numerical integration of particle trajectories is performed to evaluate the statistical acceleration coefficients D_{TT} for 1 to 100 MeV protons in a solar wind corotating interaction region (CIR) seen at 2.5 and 5.0 AU. Acceleration is followed in the solar wind reference frame and is due to random wave–particle interactions and to random drift motion in moderate scale field gradients. Comparison is made between the time constant for statistical acceleration within this CIR and estimates for diffuse shock acceleration and adiabatic deceleration. All three time constants are of the same order, but deceleration is faster than shock acceleration which in turn is faster than statistical acceleration.

106.033 Oblique propagating ion solitary waves in a magnetized plasma.
L.–t. Song.
Chin. J. Space Sci., Vol. 6, No. 4, p. 258 – 265 (1986). In Chinese. Abstr. in Phys. Abstr., Vol. 90, No. 1313, Entry 115196 (1987).

106.034   Large–scale fluctuations between 13 AU and 25 AU and their effects on cosmic rays.
L. F. Burlaga, N. F. Ness, F. B. McDonald.
J. Geophys. Res., Vol. 92, No. A12, p. 13,647 – 13,652 (1987).
The purpose of this paper is to investigate the temporal and latitudinal variation of large–scale fluctuations in the magnetic field strength observed by Voyagers 1 and 2 during the years 1984 and 1985 and to examine their effects on the cosmic ray intensity profile.

106.035   Particles injected by comets.
S. F. Singer.
ESA Spec. Publ., ESA SP–278, p. 433 – 434 (1987). – See Abstr. 012.075.
Submicron sized particles are released by comets into interplanetary space. The author has determined the total mass and lifetime of these particles and deduced their electric potential. Space debris, which, in principle, could affect the observations, is shown to be unimportant.

106.036   Interplanetary dust dynamics. II. Poynting–Robertson drag and planetary perturbations on cometary dust.
B. Å. S. Gustafson, N. Y. Misconi, E. T. Rusk.
Icarus, Vol. 72, No. 3, p. 568 – 581 (1987).
The aim of this second paper in a series is to deduce an empirical expression, in analytic form, for the average rate at which 30–μm sized cometary dust particles spiral toward the Sun under the influence of the Poynting–Robertson effect combined with corpuscular drag and planetary perturbations. This study is based on numerical calculations of trajectories corresponding to over 200 particles released from comet P/Encke.

106.037   Interplanetary dust dynamics. III. Dust released from P/Encke: distribution with respect to the zodiacal cloud.
B. Å. S. Gustafson, N. Y. Misconi, E. T. Rusk.
Icarus, Vol. 72, No. 3, p. 582 – 592 (1987).
Numerical simulations of the trajectories of over 200 30–μm–radius dust particles released by Comet P/Encke were designed to study the evolution and redistribution of orbital elements as the dust particles spiral in toward the Sun. The hypothesis that short–period comets may have contributed in a major way to the zodiacal cloud is compatible with the results. The study is directly relevant to, and supports, Whipple's suggestion that Comet P/Encke may have been a major source to the zodiacal cloud.

106.038   Recurrent increases of proton intensity from Prognoz 7 measurements.
O. R. Grigoryan, M. A. Zel'dovich, Yu. I. Logachev, V. G. Stolpovskij.
Kosm. Issled., Tom 25, Vyp. 4, p. 592 – 598 (1987). In Russian. English translation in Cosm. Res.

106.039   Interplanetary and geophysical effects of a coronal transient.
P. E. Sandholt.
Ann. Geophys., Ser. A, Vol. 5, No. 4, p. 219 – 229 (1987). Abstr. in Phys. Abstr., Vol. 90, No. 1317, Entry 139876 (1987).

106.040   Influence of the IMF sector boundaries on cosmic rays and tropospheric vorticity.
J. Laštovička.
Stud. Geophys. Geod., Vol. 31, No. 2, p. 213 – 218 (1987).

106.041   East–west asymmetry of interplanetary plasma disturbances propagating from the Sun.
V. I. Vlasov.
Geomagn. Aehron., Tom 27, No. 4, p. 657 – 659 (1987). In Russian. English translation in Geomagn. Aeron.

106.042   Reciprocal space–time localization of the sector structure of the interplanetary magnetic field and the large–scale field of the Sun.
V. N. Obridko, B. D. Shel'ting.
Geomagn. Aehron., Tom 27, No. 4, p. 660 – 662 (1987). In Russian. English translation in Geomagn. Aeron.

Modulyatsiya kosmicheskikh luchej v solnechnoj sisteme (*Cosmic ray modulation in the solar system*).
See Abstr. 003.147.

Diffuse matter in the solar system: comet Halley and other studies. Proceedings of a Royal Society Discussion Meeting, held at London, UK, 21 – 22 May 1986.
See Abstr. 012.026.

Charging of dust particles in comets and in interplanetary space.
See Abstr. 022.106.

The interaction of quasiperpendicular shock waves in a collisionless plasma.
See Abstr. 062.144.

Structural features and scattering properties of dust particles.
See Abstr. 063.037.

The energy source of the interplanetary medium and the heliosphere.
See Abstr. 074.003.

Particle propagation channels in the solar wind.
See Abstr. 074.004.

A comparison of coronal rotation and interplanetary recurrence during 1964 – 1976.
See Abstr. 074.008.

Meridional transport of magnetic flux in the solar wind between 1 and 10 AU: a theoretical analysis.
See Abstr. 074.009.

Solar cycle evolution of solar wind speed structure between 1973 and 1985 observed with the interplanetary scintillation method.
See Abstr. 074.012.

Bidirectional solar wind electron heat flux events.
See Abstr. 074.015.

Radio emission from coronal and interplanetary shocks.
See Abstr. 074.021.

Optical coronal polarization and solar dust ring.
See Abstr. 074.026.

Solar wind transition region from occultation observations within meter wavelengths.
See Abstr. 074.029.

Energetic interplanetary shocks, radio emission, and coronal mass ejections.
See Abstr. 074.030.

Solar wind and heliosphere.
See Abstr. 074.037.

Simulation of January 1 – 7, 1978, events.
See Abstr. 074.041.

The eastward deflection of fast coronal mass ejecta in interplanetary space.
See Abstr. 074.053.

The analysis of periodic structure of solar wind velocity.
See Abstr. 074.066.

Radio range measurements of coronal electron densities at 13 and 3.6 centimeter wavelengths during the 1985 solar conjunction of Voyager 2.
See Abstr. 074.087.

Interplanetary proton $(0.61 < E_p < 3.41$ MeV) events observed with Pioneer 11, 1973 – 1986 and out to 22.4 AU.
See Abstr. 078.011.

A multispacecraft study of the injection and transport of solar energetic particles.
See Abstr. 078.015.

Rotation of the Sun and rotation of its general magnetic field.
See Abstr. 080.042.

Response of the ionosphere to variations of the interplanetary medium parameter – a case study.
See Abstr. 083.022.

Control of particle precipitation into the middle atmosphere by regular changes of the interplanetary magnetic field.
See Abstr. 083.023.

Convection patterns in the polar ionosphere for northward IMF inferred from ground–based magnetometer data.
See Abstr. 083.024.

Criteria of interplanetary parameters causing intense magnetic storms $(D_{st} < -100$ nT).
See Abstr. 084.024.

Interplanetary magnetic field B_y effects on the magnetic field at synchronous orbit.
See Abstr. 084.026.

The northern and southern auroral ovals in response to the IMF By component.
See Abstr. 084.053.

Conditions for geomagnetically trapping sub–micron lunar ejecta between $L = 1.2$ and $L = 3.1$.
See Abstr. 084.059.

Model of solar–wind magnetosphere coupling.
See Abstr. 084.060.

Solar sources of geomagnetic storms.
See Abstr. 085.007.

North–south asymmetry in response of geomagnetic activity to different solar events.
See Abstr. 085.018.

A review of cometary sciences.
See Abstr. 102.030.

Unusual characteristics of electromagnetic waves excited by cometary newborn ions with large perpendicular energies.
See Abstr. 102.048.

Interaction mechanisms of comets with the zodiacal dust cloud.
See Abstr. 102.079.

The pick–up of cometary protons by the solar wind.
See Abstr. 103.465.

Alfvénic turbulence in the solar wind flow during the approach to comet P/Halley.
See Abstr. 103.466.

Waves in the magnetic field and solar wind flow outside the bow shock at comet P/Halley.
See Abstr. 103.468.

Macroscopic perturbations of the IMF by P/Halley as seen by the Giotto magnetometer.
See Abstr. 103.470.

The cause of two plasma–tail disconnection events in comet P/Halley during the ICE–Halley radial period.
See Abstr. 103.505.

Cometary evolution: clues from chondritic interplanetary dust particles.
See Abstr. 103.657.

Cometary grains and interplanetary dust: a polarimetric sensing.
See Abstr. 103.663.

Plasma tail of comet Halley in January 1986.
See Abstr. 103.671.

Structure of interplanetary magnetic field during the approach of P/Halley and associated solar activity phenomena.
See Abstr. 103.672.

Stable isotope measurements of meteorites and cosmic dust grains.
See Abstr. 105.021.

On the use of a nuclear microprobe for trace element analysis in meteorites and cosmic dust.
See Abstr. 105.044.

Interstellar scattering and resolution limitations.
See Abstr. 131.036.

Penetration of charged interstellar dust into the solar system.
See Abstr. 131.039.

On the inhomogeneous stucture of the interplanetary and interstellar plasma under the conditions of a developed ion–acoustic turbulence.
See Abstr. 131.157.

Solar activity and heliosphere–wide cosmic ray modulation in mid–1982.
See Abstr. 144.009.

Solar cycle modulation of galactic protons and electrons.
See Abstr. 144.010.

High–energy cosmic ray intensity waves.
See Abstr. 144.011.

A survey of the cosmic ray diurnal variation during 1973 – 1979 – I. Persistence of solar diurnal variation.
See Abstr. 144.022.

A survey of the cosmic ray diurnal variation during 1973 – 1979 – II. Application of diffusion – convection model to diurnal anisotropy data.
See Abstr. 144.023.

Solar modulation of galactic antiprotons.
See Abstr. 144.024.

Heliomagnetic dipole moment and daily variation of cosmic rays underground.
See Abstr. 144.025.

Flares of cosmic radiation due to particle acceleration by the solar wind.
See Abstr. 144.030.

107 Cosmogony

107.001 Accumulation of the planets.
G. W. Wetherill.
NASA Tech. Memo., NASA TM–89810, p. 92 – 94 (1987). – See Abstr. 003.001.

107.002 Protostellar disks and the primitive solar nebula.
P. M. Cassen, J. B. Pollack, T. Bunch, O. Hubickyj, P. Moins, C. Yuan.
NASA Tech. Memo., NASA TM–89810, p. 97 – 99 (1987). – See Abstr. 003.001.

107.003 Timescales for planetary accretion and the structure of the protoplanetary disk.
J. J. Lissauer.
NASA Tech. Memo., NASA TM–89810, p. 100 (1987). Abstract. – See Abstr. 003.001.

107.004 Analysis of planetary evolution with emphasis on differentiation and dynamics.
W. M. Kaula, W. I. Newman.
NASA Tech. Memo., NASA TM–89810, p. 102 – 104 (1987). – See Abstr. 003.001.

107.005 Evolution of planetesimal velocities.
G. R. Stewart, G. W. Wetherill.
NASA Tech. Memo., NASA TM–89810, p. 105 (1987). Abstract. – See Abstr. 003.001.

107.006 Accumulation of solid bodies in the solar nebula.
S. J. Weidenschilling, D. R. Davis.
NASA Tech. Memo., NASA TM–89810, p. 106 – 107 (1987). – See Abstr. 003.001.

SM107.007 A scaling law for accretion zone sizes.
Y. Greenzweig, J. J. Lissauer.
NASA Tech. Memo., NASA TM–89810, p. 108 (1987). Abstract. – See Abstr. 003.001.

107.008 Collisional and dynamical processes in Moon and planet formation.
C. R. Chapman, D. R. Davis, S. J. Weidenschilling, W. K. Hartmann, D. Spaute.
NASA Tech. Memo., NASA TM–89810, p. 112 – 114 (1987). – See Abstr. 003.001.

107.009 Magnetic flares in the protoplanetary nebula and the origin of meteorite chondrules.
E. H. Levy, S. Araki.
NASA Tech. Memo., NASA TM–89810, p. 147 (1987). Abstract. – See Abstr. 003.001.

107.010 Generation of a dynamo magnetic field in a protoplanetary accretion disk.
T. Stepinski, E. H. Levy.
NASA Tech. Memo., NASA TM–89810, p. 148 (1987). Abstract. – See Abstr. 003.001.

107.011 Gas flow in the solar nebula leading to the formation of Jupiter.
M. Sekiya, S. M. Miyama, C. Hayashi.
Earth, Moon, Planets, Vol. 39, No. 1, p. 1 – 15 (1987).
Three–dimensional gas flow in the solar nebula, which is subject to the gravity of the Sun and proto–Jupiter, is numerically calculated by using a three–dimensional hydrodynamic code – the so–called smoothed–particle method. The flow is circulating around the Sun as well as falling into a potential well of proto–Jupiter. The results for various masses of proto–Jupiter show that (1) the e–folding growth time of proto–Jupiter by accretion of the nebular gas is as short as about 300 years in stages where the mass of proto–Jupiter is 0.2 ~ 0.5 times the present Jovian mass, and

that (2) proto–Jupiter begins to push away the nebular gas from the orbit of proto–Juiter and form a gap around the orbit, when its mass is about 0.7 times the present Jovian mass. It is possible that this pushing–away process determined the present Jovian mass.

107.012 Planetesimal rotations induced by collisions.
S. Clairemidi.
Earth, Moon, Planets, Vol. 39, No. 1, p. 21 – 36 (1987).
Collision events between celestial bodies on prograde orbits in a Keplerian disc produce an excess of prograde rotations because of the geometry of the shocks. Circular orbits produce prograde rotations in 80% of the cases. With a simple collisional law based on some friction of the surfaces, the rotation increment is found to be roughly proportional to $R^{-1.36}$ on a circular path of radius R. Nearly tangent eccentric orbits induce larger prograde rotations but reduce the collisions frequencies. Crossing orbits generate both prograde and retrograde rotations with an excess of prograde ones.

107.013 The origin of the Moon and the single–impact hypothesis, II.
W. Benz, W. L. Slattery, A. G. W. Cameron.
Icarus, Vol. 71, No. 1, p. 30 – 45 (1987).
This paper is the second paper devoted to the numerical study of planetary collisions as a possible scenario for forming the Moon. The authors present a series of nine simulations of a collision between the protoearth and an impactor of various sizes. The mass ratio between the protoearth and the impactor ranged from 0.1 to 0.25. The authors were able to model both planets with iron cores, having modified their smoothed particle hydrodynamics code to allow the inclusion of up to 10 different material types.

107.014 Can episodic comet showers explain the 30–Myr cyclicity in the terrestrial record?
M. E. Bailey, D. A. Wilkinson, A. W. Wolfendale.
Mon. Not. R. Astron. Soc., Vol. 227, No. 4, p. 863 – 885 (1987).
It is shown that comet showers with a mean interval of $\cong 30$ Myr cannot be produced using perturbations of the Oort cloud by known stars or molecular clouds unless the currently accepted mass density of these bodies is much too low. This applies whether or not the showers are periodic. It is conceivable that hypothetical dark material might produce a sufficient rate of strong, close encounters, but the existence of bodies with the required masses and mass density seems most unlikely. If there is indeed an apparent 30–Myr periodicity in the mass extinction and geological records, the authors argue that astronomically induced processes are unlikely to be the primary cause.

107.015 Accumulation and migration of the bodies from the zones of giant planets.
S. I. Ipatov.
Earth, Moon, Planets, Vol. 39, No. 2, p. 101 – 128 (1987).
Within the model of solid–body accumulation of planets (or their nuclei) the accumulation and migration of bodies from the feeding zones of the giant planets are investigated. The investigation is based on results of computer simulation of evolving disks which initially consisted of hundreds of particles moving about the Sun and coagulating under collisions. In this paper the characteristics of an initial protoplanetary circumsolar cloud, the body migration in the forming solar system, the planet orbit evolution, the formation of the beyond–Neptune belt and asteroid belts between the giant planet orbits are considered. The results obtained confirm many analytical estimates earlier made by V. S. Safronov and his colleagues.

107.016   Meteorites, asteroids and comets – a study of their interrelationship.
S. Ramadurai.
Adv. Space Res., Vol. 7, No. 5, p. 121 – 124 (1987). – See Abstr. 012.022.

The origin of the three kinds of extra–terrestrial bodies, meteorites, asteroids, and comets can be traced to the origin of solar system. From a study of their properties, a generic relationship of the form interstellar gas and dust condensing in the primitive solar nebula leading to the formation of comets, the comets getting degassed and ablated to become asteroids and the asteroids on fragmentation due to collisions becoming meteorites, is advocated in the paper. The isotopic abundances of carbon are suggested as the key to unlock the mystery of this relationship.

107.017   Isotope abundance anomalies and the early solar system: (some) facts and (some) implications.
F. Begemann.
Prog. Part. Nucl. Phys., Vol. 17, p. 349 – 364 (1986). – See Abstr. 012.024.

Variations have been found in the isotopic composition of a number of chemical elements which appear to reflect primordial variations in space and/or time of the composition of the protosolar nebula. Data for xenon, oxygen, titanium and evidence for the former existence of now extinct ^{244}Pu and ^{26}Al are discussed in some detail as well as implications from the presence of ^{26}Al as an effective heat source in the early history of the solar system.

107.018   Stars and planets.
A. V. Tutukov.
Nauchn. Inf., Vyp. 63, p. 161 – 188 (1987). In Russian.

The formation of planetary systems around stars is discussed. Possibilities of the discovery of planetary systems by traditional methods are analysed. The influence of the central star's evolution on the fate of the planetary system is studied. The role of cosmic factors in the earth's climate evolution is discussed.

107.019   Search for prototerrestrial matter.
L. K. Levskij.
Priroda, No. 9, p. 30 – 31 (1987). In Russian.

107.020   Meteorites – witnesses of the evolution of the protoplanetary disk.
T. V. Ruzmajkina.
Priroda, No. 9, p. 32 – 34 (1987). In Russian.

107.021   The formation and extent of the solar system comet cloud.
M. Duncan, T. Quinn, S. Tremaine.
Astron. J., Vol. 94, No. 5, p. 1330 – 1338 (1987).

The authors present the results of numerical simulations of the origin and evolution of the solar system comet cloud. They assume that comets formed in the outer planetary region and that their orbits evolved to their current state through planetary perturbations, stellar encounters, and the galactic tide. The evolution is followed using a hybrid integration scheme which directly integrates the regularized equations of motion for cometary orbits with large semimajor axes, while solving an energy–diffusion equation for more tightly bound orbits. Stellar encounters are introduced via a Monte Carlo approach using the impulse approximation. The results of the simulations are discussed in detail.

107.022   Chemical condensation in the outflowing matter from the proto–Sun and its application to meteorites.
B. M. P. Trivedi.
Astrophys. J., Vol. 320, No. 1, p. 430 – 436 (1987).

For the past 20 years, cosmochemists have been interpreting meteoritic mineralogy in terms of chemical condensation in a hot and monotonically cooling solar nebula. However, there is some question as to whether the solar nebula was ever hot enough (> 2000K) at the site of meteorite formation, 2 – 4 AU from the Sun. The author proposes an alternative model, which is based on the observations of nebular disks around young solar–type stars in their T Tauri phase and the evidence that these disks formed from the material condensed from the outflowing gases from these stars. Consequently, it is suggested that chemical condensation took place in the outflowing matter from the proto–Sun and that the solar nebula formed from this condensed material. The duration of the T Tauri phase is ~10^{6}yr. During this period, the gas and the condensed dust would accumulate in the protosolar cavity to form the solar nebula. The characteristics of this nebula and the chemical condensation sequence in it are discussed.

107.023   The structure and evolution of the solar nebula.
S. P. Ruden.
Diss. Abstr. Int., Sect. B, Vol. 48, No. 3, p. 793–B (1987). Thesis, University of California, Santa Cruz, 158 pp. (1986). Order No. DA8709135.

107.024   Vortex model of the solar system.
V. P. Sivtsov.
Tom. Politekh. Inst., Tomsk, 35 pp. (1987). In Russian. Abstr. in Ref. Zh., 51. Astron., 10.51.147 (1987).

107.025   Accretion disk model of the primordial solar nebula.
D. N. C. Lin.
Bull. Am. Astron. Soc., Vol. 19, No. 2, p. 711 (1987). Abstract. – See Abstr. 010.061.

107.026   Formation of planetesimals.
A. Coradini, G. Magni, C. Federico.
The evolution of the small bodies of the solar system, p. 33 – 49 (1987). – See Abstr. 012.041.

Contents: 1. Introduction. 2. Protosolar–cloud properties and disk formation. 3. Disk structure. 4. Coagulation of grains. 5. Gravitational instabilities. 6. Relations between planetesimals and primitive bodies.

107.027   Deuterium fractionation in the presolar nebula: kinetic limitations on surface catalysis.
D. H. Grinspoon, J. S. Lewis.
Icarus, Vol. 72, No. 2, p. 430 – 436 (1987).

Models assuming low temperature equilibrium fractionation have previously been employed to explain the range of D/H values observed in the solar system and to make predictions of deuterium enhancement in the outer planets. While the reaction rates of the homogeneous partitioning reactions are prohibitively low at nebular temperatures, it has been suggested that catalysis on grains could shorten equilibration times sufficiently. This idea is quantitatively tested here. It is found that under highly idealized conditions – the full cosmic abundance of nickel available for catalysis in pure 5–μm grains – the equilibration time constant becomes greater than the lifetime of the nebula at temperatures lower than 560K. Even this firm lower limit is not cold enough to allow strong fractionation. Speculations are offered on alternative explanations for the distribution of hydrogen isotopes in the solar system.

107.028   Solid–body accumulation of terrestrial planets.
S. I. Ipatov.
Astron. Vestn., Tom 21, No. 3, p. 207 – 215 (1987). In Russian. English translation in Sol. Syst. Res.

The evolution of spatial discs of gravitating bodies coagulating under their collisions permit a solid–body mechanism of accumulation of terrestrial planets. The number of large ($\geqslant 0.05$ m$_s$) planetary embryos might surpass the real number of these planets. During the process of accumulation the average eccentricity of bodies' orbits might exceed 0.2. Relatively large eccentricities of Mercury and Mars might be caused by the influence of bodies entering in the terrestrial planets' zone from the giant planets' zone.

107.029   Evolution of the dust component of the circumsolar protoplanetary disk.
V. S. Safronov.
Astron. Vestn., Tom 21, No. 3, p. 216 – 220 (1987). In Russian. English translation in Sol. Syst. Res.

Settling of dust particles to the central plane of the non–turbulent gas–dust circumsolar disk is discussed. Gravitational instability in the dust subdisk is considered. Dispersion equation for radial perturbations is obtained. Critical values of the density and wavelength λ_c are found which give maximum masses of dust condensations 10^{22} and 10^{24}g in the zones of Jupiter and Neptune respectively. Due to negligible small thermal velocities of particles in the gas all modes with $\lambda < \lambda_c$ are unstable, shortwave perturbations growing faster than longwave ones. Earlier formation of small dust condensations removes difficulties with obtaining kilometersize bodies, necessary for formation of the Oort cometary cloud.

107.030   Formation of the protoplanetary disc.
T. V. Ruzmajkina.
Astron. Vestn., Tom 21, No. 4, p. 303 – 308 (1987). In Russian. – See Abstr. 012.052.

107.031   Formation of planets and satellites.
A. V. Vityazev, G. V. Pechernikova.
Astron. Vestn., Tom 21, No. 4, p. 322 – 324 (1987). In Russian. – See Abstr. 012.052.

107.032   The thermal regime of the protoplanetary disc.
A. B. Makalkin.
Astron. Vestn., Tom 21, No. 4, p. 324 – 327 (1987). In Russian. – See Abstr. 012.052.

107.033   Improvement of the model of solid matter formation in the protoplanetary disc.
M. N. Izakov.
Astron. Vestn., Tom 21, No. 4, p. 327 – 331 (1987). In Russian. – See Abstr. 012.052.

107.034   A collapse model of the turbulent presolar nebula.
W. M. Tscharnuter.
Astron. Astrophys., Vol. 188, No. 1, p. 55 – 73 (1987).

A sequence of axially–symmetric collapse models for studying the formation of a presolar–nebula–like object is presented. As an initial configuration an interstellar cloud fragment containing a total mass $M = 1.2\,M_\odot$ within a sphere of radius $R = 4.06 \times 10^{15}$cm has been adopted. The main result of the calculations is the formation of a fairly *stable* stellar core – the precursor of the early Sun in a solar nebula – which takes shape after the completion of the second collapse and a subsequent violent redistribution of angular momentum.

107.035   Search for the primitive (*material in the solar system*).
R. Berry.
Astronomy, Vol. 15, No. 6, p. 6 – 22 (1987). Abstr. in Phys. Abstr., Vol. 90, No. 1313, Entry 115261 (1987).

107.036   Evolutionary tracks of the terrestrial planets.
T. Matsui, Y. Abe.
Earth, Moon, Planets, Vol. 39, No. 3, p. 207 – 214 (1987).

On the basis of the model proposed by Matsui and Abe, the authors show that two major factors – distance from the Sun and the efficiency of retention of accretional energy – control the early evolution of the terrestrial planets. A diagram of accretional energy versus the optical depth of a proto–atmosphere provides a means to follow the evolutionary track of surface temperature of the terrestrial planets and an explanation for why the third planet in our solar system is an "aqua"–planet.

107.037   An evolutionary framework for the Jovian and Saturnian satellites.
R. J. Stevenson.
Earth, Moon, Planets, Vol. 39, No. 3, p. 225 – 236 (1987).

The position of the satellite within the protonebula, the influence of the parent planet, particularly the relative effects of tidal (gravitational) as opposed to radiogenic (internal) heat generating processes, as well as the type of ice, exert a control on the evolutionary histories of the Jovian and Saturnian satellites. Two thermal–drive models are proposed based on: an expression for externally derived gravitational influences between two bodies; and internal heat generation via radiogenic decay (expressed by surface area/volume ratio). Both parameters, for the Galilean satellites, are plotted against an inferred product of tectonic processes – the age of the surface terrain. From these diagrams, the tectonic evolutionary state of the more distant Saturnian system are predicted. These moons are fitted into an evolutionary framework for the Solar System.

107.038   On the gravitational instability in thin gaseous Kepler disks.
G. Rüdiger, R. Tschäpe.
Astron. Nachr., Vol. 308, No. 6, p. 349 – 358 (1987).

The idea that the Titius–Bode law reflects an unstable mode of a self–gravitational instability in very thin Keplerian disks makes a careful discussion of the Poisson equation especially necessary. Due to the planetary distances in the solar system ($\delta r/r \cong 0.5$) the well–known short–wave approximation is not appropriate for definite assertions. The authors use a simple series expansion of the relation between the radial and vertical wave numbers of the disturbances which is additionally valid for medium–scale and non–zonal modes. The numerical solution of the dispersion relation reveals an extra unstable branch for wave–lengths of rings and spirals two orders of magnitudes larger than those already known. Though the authors are not yet able to consider modes long enough for application to the planetary system, they feel the existence of the medium–wave instability ($\delta r/r \cong 0.1$) to be a serious challenge for a better, i.e. non–local theory.

107.039   Distance law and formation of satellite systems.
Z.–w. Hu, Z.–x. Chen.
Astron. Nachr., Vol. 308, No. 6, p. 359 – 362 (1987).

In the solar system satellite systems of Jupiter, Saturn and Uranus are typical ones. The distribution of the semi–major axis of satellite orbits in each system may be expressed by an empirical formula corresponding to the Titius–Bode law. The authors found that it can be written as $a_n = B' \cdot B^n$, where B' and B are constants. Values of B' and B depend on formation conditions of each system. Satellites should be formed in the gas–satellitesimal disk around a planet and by aggregation of satellitesimals. The gas is the major component in the disk and its damping effect must play an important role in the process of aggregation of satellitesimals. It may be proved that radial small perturbation in the disk can cause the gravitational instability and the formation of gaseous rings with increased density, where satellitesimals can easy aggregat into satellites.

107.040   Supernova as a stimulator of the creation of the solar system.
Z. D. Dohčević.
Vasiona, Année 35, No. 2, p. 44 – 51 (1987). In Croatian.

A review of hypotheses about the formation of the solar system is presented including the supernova triggering hypothesis.

107.041   What the solar nebula was like.
J. Maddox.
Nature, Vol. 330, No. 6150, p. 691 (1987).

The author reports on two papers (see abstracts 105.057, 105.058) which are a pointer to the ways in which it may be possible to learn how the material from which the Solar System formed was itself produced.

Geohistory. Global evolution of the Earth.
See Abstr. 003.106.

Seminar on "Astrophysical problems of planetary cosmogony", held by the Astronomical Council of the USSR of Sciences, April 1, 1987.
See Abstr. 012.052.

High–explosive simulation of supernovae.
See Abstr. 022.089.

Nucleosynthesis contributions to the solar nebula.
See Abstr. 061.025.

Equilibrium temperature of ice grains formed around a star as a function of stellar parameters.
See Abstr. 064.072.

Planetary systems and their central stars.
See Abstr. 065.088.

Comet showers as a cause of mass extinctions.
See Abstr. 081.025.

A satellite–asteroid mystery and a possible early flux of scattered C–class asteroids.
See Abstr. 091.017.

The chronology of Mercury's geological and geophysical evolution: the vulcanoid hypothesis.
See Abstr. 092.005.

Estimated impact shock production of N_2 and organic compounds on early Titan.
See Abstr. 100.023.

Possibility of observing the Oort cometary cloud in the far–infrared region.
See Abstr. 102.029.

Radiogenic heating of comets by ^{26}Al and implications for their time of formation.
See Abstr. 102.033.

The formation and extent of the solar system comet cloud.
See Abstr. 102.038.

On the stable multi–belt interplanetary reservoirs of cometary bodies.
See Abstr. 102.058.

Oort's Cloud.
See Abstr. 102.059.

Composition measurements and the history of cometary matter.
See Abstr. 103.589.

On the peculiarities of meteoroid density distributions in different meteor streams.
See Abstr. 104.080.

Local and exotic components of primitive meteorites, and their origin.
See Abstr. 105.020.

Trace element contents of primitive meteorites; a test of solar system abundance smoothness.
See Abstr. 105.043.

Evidence for interstellar SiC in the Murray carbonaceous meteorite.
See Abstr. 105.057.

Large isotopic anomalies of Si, C, N and noble gases in interstellar silicon carbide from the Murray meteorite.
See Abstr. 105.058.

^{16}O excesses in Murchison and Murray hibonites: a case against a late supernova injection origin of isotopic anomalies in O, Mg, Ca, and Ti.
See Abstr. 105.059.

Mineralogy of chondritic interplanetary dust particles.
See Abstr. 106.010.

The origin of dust in the solar system.
See Abstr. 106.017.

Stars

111 Parallaxes, Proper Motions, Radial Velocities, Space Motions, Distances

111.001 **Systematic and external errors of trigonometric parallaxes.**
L. A. Breakiron.
Astron. Astrophys., Suppl. Ser., Vol. 70, No. 2, p. 157 – 188 (1987).

The trigonometric parallaxes for 3035 stars in the General Catalogue of Trigonometric Stellar Parallaxes (Jenkins, 1952) and its 1963 Supplement are compared with the corresponding spectroscopic parallaxes in order to determine the systematic and external errors of the former. Significant dependences of both were found upon observatory and, depending on observatory, upon right ascension, declination, zenith distance, galactic latitude, trigonometric and spectroscopic parallax, internal trigonometric parallax error, absolute and apparent magnitude, and color. Many of these correlations are found to be primarily statistical, rather than instrumental, in origin, including the magnitude error in the Allegheny parallaxes. No evidence is found for variations with sector opening, but some evidence is found for a change of systematic error with time.

111.002 **Radial velocities in three fields along the southern galactic equator.**
J. Denoyelle.
Astron. Astrophys., Suppl. Ser., Vol. 70, No. 3, p. 373 – 388 (1987).

A list of radial velocities for 764 stars is given for three fields in the Vela–Carina region of the Galaxy. They were obtained from GPO–plates taken at La Silla and reduced following Fehrenbach's method. Slit–spectra were collected with the 152 cm–spectrographic telescope at La Silla, to derive an accurate radial velocity for a sufficient number of calibration stars: out of the 29 stars, 26 had no formerly published value. The global motions of 10 to 14 km/s can be considered as normal on the basis of galactic rotation. Some stars however show high velocities. The case of HD 81471 (Sp type A7 Iab) suggest that a detailed study of this star is to be recommended.

111.003 **The rebirth of stellar radial–velocity research.**
D. W. Latham.
Vistas Astron., Vol. 30, Part 1, p. 77 – 82 (1987).

The author gives a brief overview of the areas of astronomical research where modern stellar radial–velocity observations are having an impact. He discusses in more detail the recent progress in two areas of interest to David S. Evans: the establishment of radial–velocity standard stars and the determination of distances to pulsating variable stars.

111.004 **Lick Northern Proper Motion program. I. Goals, organization, and methods.**
A. R. Klemola, B. F. Jones, R. B. Hanson.
Astron. J., Vol. 94, No. 2, p. 501 – 515 (1987). = Lick Obs. Bull. No. 1068.

The Lick Northern Proper Motion (NPM) program determines absolute proper motions, measured with respect to an extragalactic reference frame, for selected stars in the blue magnitude range 9 – 18, north of declination –23°. Positions and photographic photometry are also measured. The final NPM catalog will contain some 300000 stars, whose motions are referred to some 70000 galaxies. This paper, the first in a series presenting the NPM results, describes the history and goals of the NPM program, the photographic survey, the selection of stars and galaxies, the plate measurements, the reduction procedures for positions, proper motions, and photometry, and the tests of the accuracy and precision of the NPM data.

111.005 **Lick Northern Proper Motion program. II. Solar motion and galactic rotation.**
R. B. Hanson.
Astron. J., Vol. 94, No. 2, p. 409 – 415 (1987). = Lick Obs. Bull., No. 1069.

The Lick Northern Proper Motion (NPM) program has been described in Paper I. The NPM reductions for absolute proper motions and positions have been completed for 617 NPM fields, almost completely covering the sky more than ~10° from the galactic plane between declinations –3° and +68°. To test the accuracy of the NPM proper motions, and to begin their use in galactic structure studies, the author has done solutions for galactic rotation and solar motion. He finds: (1) $A = +11.31 \pm 1.06$ and $B = -13.91 \pm 0.92$ km s^{-1}kpc^{-1}, consistent with a nearly flat galactic–rotation curve with a local circular velocity near 200 km s^{-1}; (2) solar apex locations near the standard apex for low galactic latitudes, but trending strongly toward the direction of galactic rotation for high galactic latitudes; (3) external error estimates for the Lick proper motions: 0".2 century^{-1} for the absolute zero point in a typical NPM field, and 0".06 $\pm$ 0".02 century^{-1} for the overall systematic zero–point error.

111.006 **Computations of radial velocities for O– and B–type stars.**
M. Missana.
Astrophys. Space Sci., Vol. 136, No. 1, p. 167 – 170 (1987).

The author gives the computations of radial velocities and generalized Compton effects in the spectra of the O stars HD 108, HD 151804, HD 152408, HD 148937, and of the B stars HD 41117, HD 152236, Wray 977, and in the spectrum of YY Orionis.

111.007 **Photographic astrometry of binary and proper–motion stars. III.**
W. D. Heintz.
Astron. J., Vol. 94, No. 4, p. 1077 – 1080 (1987).

65 parallaxes and 13 binary star mass ratios are listed and discussed.

111.008 **Radial velocities in selected B–G stars.**
P. C. Frisch.
Astrophys. J., Suppl. Ser., Vol. 65, No. 2, p. 313 – 317 (1987).

Radial velocities and stellar line–broadening parameters are given for about 100 B–G stars in selected regions of space near the Sun. The resulting accuracy in radial velocities is ± 3 km s^{-1} for sharp–lined stars.

111.009 Proper motion of the compact, nonthermal radio source in the galactic center.
D. C. Backer, R. A. Sramek.
AIP Conf. Proc., No. 155, p. 163 – 165 (1987). – See Abstr. 012.029.

Observations with the VLA for four epochs from 1981 to 1985 have resulted in the first two–axis measurement of the proper motion of the compact, nonthermal radio source in the galactic center, Sgr A*. The observed motion is consistent with that expected for an object in the galactic center whose peculiar velocity is no larger than $40\,r$ km s^{-1}, where r is the distance to the galactic center in units of 8.5 kpc. If there is a massive black hole in the galactic center, then these observations suggest that the radio emission must be coming from a synchrotron corona surrounding the hole. The converse is not true: these observations, by themselves and at their present level of accuracy, do not require the existence of a massive black hole.

111.010 Erratum: "List of radial velocities of 258 stars near Alpha Persei" [Astron. Astrophys., Suppl. Ser., Vol. 68, No. 3, p. 515 – 520 (1987)].
C. Fehrenbach, R. Burnage, J. Figuière, G. Traverse, C. Agniel.
Astron. Astrophys., Suppl. Ser., Vol. 71, No. 1, p. 185 (1987). In French. See Abstr. 43.111.005.

111.011 Mesures de vitesses radiales. I. Accompagnement au sol du programme d'observation du satellite Hipparcos.
C. Fehrenbach, R. Burnage, M. Duflot, A. Peton, L. Rolland, V. Genty, C. Mannone.
Astron. Astrophys., Suppl. Ser., Vol. 71, No. 2, p. 263 – 274 (1987).

The authors have published a catalogue of radial velocities (RV) of 272 stars generally brighter than $m_{pg} = 8.5$, chosen for their possible membership to the not yet available Hipparcos Input catalogue. They developed an automatic spectral classification from the registered spectra available for RV measurements.

111.012 Mesures de vitesses radiales. II. Accompagnement au sol du programme d'observation du satellite Hipparcos.
C. Fehrenbach, M. Duflot, R. Burnage, C. Mannone, A. Peton, V. Genty.
Astron. Astrophys., Suppl. Ser., Vol. 71, No. 2, p. 275 – 295 (1987).

The authors give 446 new radial velocity (RV) results obtained by the Fehrenbach (1947) objective prism method. The measuring technique is now automatic. For each field, the results are preceded by a list of RV from the literature, which have been used for calibration.

111.013 Parallax calibration of the population II main sequence.
T. E. Lutz, R. B. Hanson, W. F. van Altena.
Bull. Am. Astron. Soc., Vol. 19, No. 2, p. 675 – 676 (1987). Abstract. – See Abstr. 010.061.

111.014 An atlas of identification charts for Luyten common proper motion stars with probable degenerate components.
T. D. Oswalt, P. M. Hintzen, W. J. Luyten.
Bull. Am. Astron. Soc., Vol. 19, No. 2, p. 707 (1987). Abstract. – See Abstr. 010.061.

111.015 Vitesses radiales pour le programme HIPPARCOS. II.
C. Fehrenbach, M. Duflot, R. Burnage, C. Mannone, A. Peton, V. Genty.
Obs. Haute Provence, Pré–Publ., No. 23, 50 pp. (1987). Submitted to Astron. Astrophys., Suppl. Ser.

111.016 The Tautenburg field of the programme of studying the main meridional section of the Galaxy.
E. Schilbach.
Tartu Astrofüüs. Obs. Teated, Nr. 84, p. 8 – 17 (1987).

The astrometric observational programme for the determination of proper motions with reference to galaxies for 5000 stars in 17 fields in the main meridional section of the Galaxy is presented. The proper motions will be derived on plates taken with the Tautenburg 2 m–Schmidt telescope. For each field 2 or 3 first–epoch plates obtained before 1970 are available. Preliminary estimates to the mean error of an individual proper motion have yielded about $\pm 0\rlap{.}''6$ per century both for the bright $(10^m - 12^m)$ and for the faint $(16^m - 18^m)$ stars.

111.017 The galactic orbit of the remarkable high–velocity wide binary LDS 519.
C. Allen, M. A. Martos, A. Poveda.
Rev. Mex. Astron. Astrofis., Vol. 14, No. 1, p. 213 – 222 (1987). – See Abstr. 012.042.

The nature of the high–velocity, nearby, wide binary LDS 519 is discussed, and arguments are given against a possible extragalactic origin for this object. Its space velocity is shown to be lower than the escape velocity if the galaxy has a massive halo; therefore, a galactic origin for this binary is suggested. The galactic orbit of LDS 519 is numerically integrated using a realistic model for the galactic mass distribution. On the basis of the information provided by the orbit a possible scenario for the origin of LDS 519 is proposed.

111.018 The galactic orbits of nearby, high–velocity stars.
C. Allen, M. A. Martos.
Rev. Mex. Astron. Astrofis., Vol. 14, No. 1, p. 224 (1987). Abstract. – See Abstr. 012.042.

111.019 Discovery of very low luminosity stars.
M. T. Ruiz, J. Maza.
Rev. Mex. Astron. Astrofis., Vol. 14, No. 1, p. 381 – 384 (1987). – See Abstr. 012.042.

During the blinking process of plates taken for a supernova search, a group of faint stars ($m_{ph} \gtrsim 14$) with high proper motions ($\mu \gtrsim 0\rlap{.}''7$/year) were discovered. The authors present spectrophotometry and proper motion determinations for three of them, showing that ER2 and ER6 are red dwarfs at distances of 6 pc and 20 pc respectively. The featureless spectra of ER8 can be fitted by a black body at 3500K and corresponds to a cold degenerate star with a luminosity $L \cong 5 \times 10^{-5} L_\odot$ at about 13 pc from us.

111.020 An astrometric search for a stellar companion to the Sun.
S. Perlmutter.
Diss. Abstr. Int., Sect. B, Vol. 48, No. 5, p. 1385–B (1987). Thesis, University of California, Berkeley, 71 pp. (1986). Order No. DA8718121.

111.021 Discussion of trigonometric parallaxes determined along right ascension and declination.
J.–l. Zhao, Y.–p. He.
Chin. Astron. Astrophys., Vol. 11, No. 4, p. 269 – 274 (1987). English translation of Acta Astron. Sin., Vol. 28, No. 2, p. 182 – 189 (1987).

The authors discuss the data of trigonometric parallaxes determined along α and δ published by the US Naval Observatory and Van Vleck Observatory. They found that the absolute difference $\Delta\pi = \pi_x - \pi_y$ shows a very similar periodic variation with α for both observations. This variation is shown to be due to errors in π_y and to be related to the current practice of maximising the parallax factor for parallax determination in right ascension.

111.022 The proper motion of the AGK3 variable stars. V: Z Del and NSV 12844.
C. E. Lopez.
Inf. Bull. Variable Stars, No. 3100, 2 pp. (1987).

On the general catalogue of stellar proper motions with respect to galaxies in the areas of the Galaxy main meridional section.
See Abstr. 002.022.

On systematic errors of proper motions of stars contained in the catalogue of the USSR Time Service.
See Abstr. 002.034.

The Third Catalogue of Nearby Stars with special emphasis on wide binaries.
See Abstr. 002.064.

The formation of an astrometric data bank in the framework of the complex programme of studying the main meridional section of the Galaxy (MEGA).
See Abstr. 002.072.

On a special compiled catalogue of absolute proper motions of stars.
See Abstr. 002.073.

The magnetic tape version of the LDS catalogue.
See Abstr. 002.092.

Catalogue of determinations of metallicity, velocity components and orbital elements of F5–K5 dwarfs in the neighbourhood of 80 parsecs from the Sun.
See Abstr. 002.101.

Catalogue of determinations of [Fe/H], velocity components and orbital elements of F stars.
See Abstr. 002.102.

Catalogue of determinations of [Fe/H], distances, velocity components and orbital elements of F stars of the southern Galactic pole.
See Abstr. 002.103.

The calibration problem. III. First–order solution for mean absolute magnitude and dispersion.
See Abstr. 036.017.

A survey of proper–motion stars. II. Extracting metallicities from high–resolution, low–S/N spectra.
See Abstr. 036.054.

Direct reduction of astrometric data in terms of centroids and variances.
See Abstr. 036.163.

The calibration problem. IV. The Lutz–Kelker correction.
See Abstr. 036.178.

An improvement of the method of plate constants for the determination of stellar trigonometric parallaxes.
See Abstr. 036.248.

Precise proper motions and positions of stars from a combination of fundamental catalogues with the HIPPARCOS catalogue.
See Abstr. 041.009.

Optical positions and proper motions of selected radio stars.
See Abstr. 041.010.

The nature of the F str $\lambda 4077$ stars.
See Abstr. 113.030.

Kinematics and age of the low–mass X–ray binaries.
See Abstr. 117.165.

Orbital elements for double stars of Population II. The high–velocity system COD –48°1741.
See Abstr. 117.234.

Binary star statistics: the mass ratio distribution for very wide systems.
See Abstr. 118.033.

Radial velocities of the components of triple stars.
See Abstr. 118.034.

Candidates for spectroscopic binaries found in the Mount Wilson Halo–Mapping Program.
See Abstr. 120.009.

Radial velocity study of the Be star ζ Tau (HR 1910).
See Abstr. 120.019.

The variable star HD 79889.
See Abstr. 122.045.

Membership in the young cluster Trumpler 37.
See Abstr. 153.005.

The kinematical and binary properties of association and field O stars.
See Abstr. 155.007.

The visually brightest early–type supergiants in the spiral galaxies NGC 2403, M81, and M101.
See Abstr. 157.203.

112 Stellar Environments (Chromospheres, Coronae, Stellar Winds, Shells, Masers, etc.)

112.001 Circumstellar envelopes and mass loss in OB stars.
C. T. Bolton.
The SHIRSOG Workshop, p. 66 – 70 (1986). – See Abstr. 012.003.

There are a number of clearly defined scientific questions related to the atmospheric structure and stellar winds of OB stars which can only be addressed by long and nearly continuous series of spectroscopic observations. Ideally these should be carried out in conjunction with optical photometric and polarimetric and UV spectroscopic observing programs. The improved understanding of OB stars that will result from these observations will have an important impact on our ideas about the structure and evolution of OB stars, supernovae, chemical evolution of galaxies, star formation, and the structure of the interstellar medium.

112.002 The infrared emission–line spectrum of γ Cassiopeiae.
F. Hamann, M. Simon.
Astrophys. J., Vol. 318, No. 1, p. 356 – 362 (1987).

The near–IR spectrum of γ Cas contains emission lines of H I, He I, and Mg II. No lines of low–excitation species, such as are found in cool and dense environments, are detected. The observed Brα and Brγ profiles were double–peaked, with FWHM ≈ 260 km s^{-1}. The IR hydrogen line fluxes indicate that these lines are formed in a small, dense, optically thick region where the density of ionized gas declines sharply with distance from the star. Both the line profiles and fluxes are shown to be inconsistent with the predictions of standard stellar wind theory, but are in qualitative agreement with a rotating disk model.

112.003 Long–term and mid–term spectroscopic variations of the Be–shell star HD 184279 (V1294 Aql). I. Observational data.
D. Ballereau, J. Chauville.
Astron. Astrophys., Suppl. Ser., Vol. 70, No. 2, p. 229 – 245 (1987).

The Be–shell star HD 184279 showed in the past important changes through B and Be–shell phases. Long–term and mid–term spectroscopic variations between 1971 and 1984 have been studied from a set of spectra with a dispersion of 12.3 Å mm^{-1} in the blue spectral range. Emission and absorption lines have been measured and traced. During this period, the spectrum of this star showed permanent emission lines on the first terms of Balmer series and shell absorption lines on hydrogen, He I, Ca II, Fe III and sometimes Mg II, Fe II, Si II. The main results are outlined.

112.004 Spectral features of the B2e star EW Lac before and during the variable shell phase.
A. M. Hubert, M. Floquet, J. Chauville, M. T. Chambon.
Astron. Astrophys., Suppl. Ser., Vol. 70, No. 3, p. 443 – 464 (1987).

On the basis of an important observational material in the visual range covering the period of 1960 – 1984, a study of spectral characteristics (circumstellar lines) of the B2e shell star EW Lac, before and during the variable shell phase, has been carried out. The observations are consistent with the two–zone structure model previously proposed by Hubert–Delplace et al. (1983): a cool equatorial disk and an expanding, rather hot, atmosphere seen at high latitude.

112.005 Stellar chromosphere – its energy balance and line formation.
T. Watanabe.
Astron. Her., Vol. 80, No. 7, p. 192 – 196 (1987). In Japanese.

112.006 The rapid photometric variability of some Ae/Be–star envelopes.
M. A. Pogodin.
Pis'ma Astron. Zh., Tom 13, No. 8, p. 695 – 703 (1987). In Russian. English translation in Sov. Astron. Lett., Vol. 13.

The results of a narrowband photometry of some Ae/Be stars of different types are presented. Some indications to variability are found for emission in Hα and Hβ lines and shortward Balmer limit with characteristic times from several minutes to several days. Some mechanisms are analysed which may be responsible for the observed variability.

112.007 HD 213985: a hot post–AGB star in the galactic halo.
C. Waelkens, L. B. F. M. Waters, A. Cassatella, T. Le Bertre, H. J. G. L. M. Lamers.
Astron. Astrophys., Vol. 181, No. 1, p. L5 – L8 (1987).

HD 213985 is a 9th–magnitude blue star situated at a galactic latitude $b = -57°$. The authors' measurements of this star in the Geneva photometric system show that it is a large–amplitude, long–term, photometric variable. Visual and UV spectra indicate a supergiant character, and the IRAS data reveal strong dust components around the star. A 220 nm absorption feature of circumstellar origin is observed. It is concluded that HD 213985 is a low–mass object evolving from the AGB to the left of the HR diagram. The remarkable similarity of this star with the enigmatic object HR 4049 makes it likely that the latter is also a low–mass star.

112.008 Sulfur in IRC + 10216.
J. Cernicharo, M. Guélin, H. Hein, C. Kahane.
Astron. Astrophys., Vol. 181, No. 1, p. L9 – L12 (1987).

The authors report the detection of 12 lines of C_2S and 11 lines of C_3S in IRC +10216. These lines, which sample a large range of rotational energies, yield accurate values of these species' column densities. The abundances of C_2S and C_3S are compared to those of other sulfur molecules and carbon chains. C_2S and C_3S are a factor of ≈ 40 less abundant than CS, but about as abundant as C_5H and C_6H, two carbon chain radicals of similar weight. The detection of these molecules and of H_2S shows that sulfur chemistry in circumstellar envelopes is more interesting than previously thought.

112.009 Polarization investigations in four peculiar supergiants with high IR excess.
U. C. Joshi, M. R. Deshpande, A. K. Sen, A. Kulshrestha.
Astron. Astrophys., Vol. 181, No. 1, p. 31 – 33 (1987).

First polarimetric measurements of four peculiar supergiants (HD 89353, HD 163506, HD 161796 and HD 101584) with high IR excess have been carried out. These investigations indicate the presence of large dust grains ($>0.3\,\mu m$) in the dust shells, generally uniformly distributed, around the stars. HD 89353 shows large changes in polarization angle with wavelength, which can be attributed to the presence of two dust shells containing, respectively, large ($\sim 0.8\,\mu m$) and small ($\sim 0.05\,\mu m$) dust particles.

112.010 B and A type stars with unexpectedly large colour excesses at IRAS wavelengths.
J. Coté.
Astron. Astrophys., Vol. 181, No. 1, p. 77 – 84 (1987).

The authors present IRAS observations at 12, 25, 60 and 100 μm of 18 bright stars of spectral type A, 6 of spectral type B, with strongly deviating IR fluxes at those wavelengths. Eight of these stars were already discussed in previous studies. Tables with the IRAS magnitudes, the derived colour excesses and some stellar parameters are presented. For each star the reliability of the observations is discussed. The excess IR fluxes are interpreted either in terms of thermal radiation from circumstellar dust, or in terms of free–free radiation from hot, ionized, circumstellar gas. Dust temperatures are typically in the range between 60 and 130K. Three B–type stars, HR 2501, ϕ Sgr and HR 7739, were found to be candidates for Be stars.

112.011   The extended radio emission of P Cygni.
J. W. M. Baars, H. J. Wendker.
Astron. Astrophys., Vol. 181, No. 1, p. 210 – 212 (1987).

New observations of the P Cyg area with the 100 m–RT at 4.8 GHz and with the WSRT at 1420 MHz are presented. The authors confirm the existence of extended emission around P Cyg. Contrary to earlier speculation, it is of thermal origin. New information on its spatial structure could not be derived. A list of point sources in the field is also presented.

112.012   The peculiar emission–line supergiant HD 37836.
O. Stahl, B. Wolf.
Astron. Astrophys., Vol. 181, No. 2, p. 293 – 299 (1987).

The luminous emission–line star HD 37836 of the LMC has been studied using high–dispersion spectra in the UV and in the optical. In addition, IUE low–dispersion spectra and *UBVRIJHKL* photometry have been used to study the energy distribution. The emission–line spectrum of the star is very peculiar. Its UV absorption–line spectrum is more normal and indicates that HD 37836 is a late O type supergiant. Specifically, strong stellar–wind lines with an edge velocity of ≈ 2400 km s^{-1} have been found. The most likely explanation for the variety of velocities which are found appears to be a disk surrounding the star. It is shown that the continuum energy distribution is in agreement with the presence of a disk. The two LMC stars HDE 269445 and Hen S131 have similar energy distributions and emission–line spectra and may thus represent other cases of luminous, hot supergiants surrounded by gaseous disks.

112.013   The UV high resolution spectrum of A–type supergiants.
A. Talavera, A. I. Gomez de Castro.
Astron. Astrophys., Vol. 181, No. 2, p. 300 – 314 (1987).

A study of all the A–supergiants observed with IUE at high resolution shows that they can be divided into two groups: stars which show evidence of wind and mass–loss, and stars which do not show these characteristics. In the first group the resonance lines of Mg II, Fe II, Si II, C II are asymmetric, showing variable shortward shifted components. In the second group, formed by the faintest stars, these lines are symmetric. The velocity of the most shortward shifted component observed in the Fe II lines and the terminal velocity remain constant in most of the stars. However, the intermediate shortward shifted components of Fe II lines vary with time, indicating the ejection of shells from time to time or the existence of CIR (corotating interaction regions). The terminal velocity decreases with increasing escape velocity which is not in agreement with the radiation–driven–wind theory.

112.014   IRAS observations of CP stars.
R. Kroll.
Astron. Astrophys., Vol. 181, No. 2, p. 315 – 322 (1987).

In order to resolve the presence of circumstellar matter, the IRAS point sources catalogue was scanned for chemically peculiar (CP) stars, for which near IR photometry is accessible. 40 CP stars are found. The flux distribution is equal to that of normal upper main sequence stars for CP2 and CP3 type stars, no flux excesses are reported. Two CP4 stars display circumstellar dust. This shows that the CP4 class contains also stars with Be characteristics.

112.015   Echelle and spectropolarimetric observations of the η Carinae nebulosity.
J. Meaburn, R. D. Wolstencroft, J. R. Walsh.
Astron. Astrophys., Vol. 181, No. 2, p. 333 – 342 (1987).

Spatially resolved observations have been made with the Manchester echelle spectrometer of the Hα and [N II] line profiles from the Homunculus and outer shell of the η Carinae nebulae. These observations have been combined with new spectropolarimetric measurements over the line profiles and of the scattered continuum light from the Homunculus as well as with previous observations of proper motions.

112.016   Vibrationally excited CS in IRC + 10216.
B. E. Turner.
Astron. Astrophys., Vol. 182, No. 1, p. L15 – L18 (1987).

Vibrationally excited ($v = 1$) CS has been detected via its $J = 2–1$ and $5–4$ transitions in the circumstellar shell IRC +10216. Its properties are quite different than those of vibrationally excited SiO and HCN, the only other such species so far detected in IRC +10216. Negative results for CO $v = 1$ an CN $v = 1$ are consistent with current understanding of the IRC +10216 envelope. The author reports the detection of vibrationally excited HCN (0,2,0) in Ori (KL), which quantitatively confirms that vibrationally excited HCN in the hot core is excited solely by IR radiation from IRc 2. The author analyzes his negative results for $v = 1$, CS, CO, and CN in terms of this picture, and finds that their hot core abundances are not greatly enhanced, unlike those of HCN, NH$_3$, H$_2$O which seem to be the products of grain disruption.

112.017   Long–term variability of the far–UV high velocity components in γ Cas (1978 – 1986).
V. Doazan, L. Rusconi, G. Sedmak, R. N. Thomas, B. Bourdonneau.
Astron. Astrophys., Vol. 182, No. 1, p. L25 – L28 (1987).

A synthesis of the authors' long term observations of γ Cas in the visual and in the far UV, combined with IUE archived spectra, shows that: (1) The occurrence of the high velocity components observed in the Si IV, C IV, and N V resonance lines exhibits a long term variability pattern, which is associated with the cyclic V/R variations of the Balmer emission–lines: they are frequently observed when V/R > 1, while they are absent/rarely present when V/R < 1. (2) During the present V/R > 1 half–cycle, the observed components do not fit the column density vs. velocity correlation derived by Henrichs et al. (1983). γ Cas is one more example of a Be star which exhibits associated far UV and visual long term variability patterns.

112.018   Constraints for models of Be stars derived from UV and IRAS observations.
H. J. G. L. M. Lamers, L. B. F. M. Waters.
Astron. Astrophys., Vol. 182, No. 1, p. 80 – 90 (1987).

The observations of UV line profiles and IR excess of 10 Be stars are interpreted quantitatively in order to derive constraints for models of Be stars. Two models are considered: a disc model in which the UV lines are formed in the high velocity polar region and the IR excess is formed in the high density equatorial disc, and a circumstellar shell model where the UV lines are formed in a spherically symmetric wind and the IR excess is formed in a circumstellar shell which surrounds the wind region. The UV and IR observations can be explained by both models. The results of the studies of the polarization and the Balmer emission profiles in Be stars, strongly support the disc model and do not agree with the circumstellar shell model. The large contrast in mass flux and density between the polar region and the disc region cannot be explained by radiation driven winds of rapidly rotating stars. Some other mechanism which enhances the mass loss in the equatorial region is required.

112.019   Infrared photometry of late–type Wolf–Rayet stars.
P. M. Williams, K. A. van der Hucht, P. S. Thé.
Astron. Astrophys., Vol. 182, No. 1, p. 91 – 106 (1987).

The authors present infrared photometry of 41 Population I Wolf–Rayet stars, including all known WC 8–10 and WN 9–10 stars, and the emission line star LSS 4005 originally thought to define the WN 11 subtype in a study of the incidence of circumstellar dust associated with these objects. Circumstellar dust emission was observed from five of the ten WC 8 stars, 14 of the 17 WC 9 stars, the only WC 10 star, LSS 4005 and probably the only WN 10 star. It is demonstrated that the circumstellar dust is being formed in the winds. Models of the shells, assuming spherical symmetry and an inverse square law dust density distribution, are presented and the size of that for WR 104 agrees well with that measured by speckle interferometry.

112.020 The relation between the visual polarisation and UV narrow absorption lines in irregular Be star variations.
J. C. Brown, H. F. Henrichs.
Astron. Astrophys., Vol. 182, No. 1, p. 107–114 (1987).

Using recent coordinated UV and optical data on ω Ori (Hayes and Guinan, 1984, Sonneborn et al., 1987), and the polarimetric theory of Brown and McLean (1977), geometric, kinematic and physical constraints are established for non–steady mass loss processes associated with irregular Be star variations.

112.021 New doublets in IRC + 10216: vibrationally excited C_4H?
M. Guélin, J. Cernicharo, S. Navarro, D. R. Woodward, C. A. Gottlieb, P. Thaddeus.
Astron. Astrophys., Vol. 182, No. 2, p. L37–L39 (1987).

The authors have discovered in the envelope of the carbon star IRC + 10216 a new series of doublets arising from a reactive species in a $^2\Sigma$ or ^2A state. This new molecule is also observed in the laboratory in an acetylene discharge and it is therefore probably a hydrocarbon radical. Its rotation and centrifugal distortion constants are quite close to those of ground state C_4H suggesting that the authors are observing C_4H in an excited state, vibrational or electronic.

112.022 Z CMa resolved at near infrared wavelengths: one more piece to the puzzle.
C. Leinert, M. Haas.
Astron. Astrophys., Vol. 182, No. 2, p. L47–L50 (1987).

Near infrared speckle observations have shown Z CMa to be extended in E–W direction. The full width at half maximum is about 0.10 arcsec (115 AE) at 1.65 μm, and apparently decreases with increasing wavelength as does the fractional brightness of the extended component. The width is less by about a factor of two in N–S direction. Although the object is not fully resolved, which makes an unambiguous interpretation difficult, the discussion of the data suggests that the infrared radiation is scattered light coming from a dusty disk seen approximately edge–on. Then a high albedo of the particles and a special geometry appear to be required to fit the considerable infrared scattering cross–sections implied in such a picture to the moderate extinction observed in the visible.

112.023 The optical counterpart of the X–ray transient EXO 2030 + 375.
C. Motch, E. Janot–Pacheco.
Astron. Astrophys., Vol. 182, No. 2, p. L55–L58 (1987).

The authors present CCD photometric observations of the error circle of the X–ray source EXO 2030 + 375. The X–ray position and the optical and infrared colours allow the identification of the transient source with a highly reddened (E(B–V) = 3.8) early type star. The optical candidate has photometric characteristics consistent with a Be/X–ray system located at a distance of 2 – 7 kpc.

112.024 Direct imagery of circumstellar shells around Ofpe/WN9 stars in the Galaxy and in the LMC.
O. Stahl.
Astron. Astrophys., Vol. 182, No. 2, p. 229–236 (1987).

The galactic Ofpe/WN9 stars AG Car and He3–519 and the LMC objects S 61, R 84, R 99, and R 127 have been observed with narrow–band imagery. Both galactic stars are surrounded by ring nebulae. Nebulae around the LMC objects have previously been suspected from spectroscopy. The author re–solved the nebulosities around S 61 and R 127, while he could not resolve the emission from R 84 and R 99. The linear sizes of the resolved objects are of the order of $\leqslant 1$ pc, their kinematic age is of the order of 10^4yr and their ionized masses are of the order of $1 M_\odot$, i.e. they are larger and more massive than a typical classical planetary nebula. They are also clearly different from classical planetary nebulae with respect to their nebular spectra

and the luminosity of the central stars. It is suggested that the nebulae have been ejected during violent S Dor outbursts of the central stars.

112.025 Chemical composition of the atmosphere of K giant α Ser.
M. E. Boyarchuk, M. Ya. Orlov, A. V. Shavrina.
Astrophysics, Vol. 25, No. 3, p. 603–605 (1987). English translation of 42.112.163.

112.026 Achernar.
L. A. Balona, C. A. Engelbrecht, F. Marang.
Mon. Not. R. Astron. Soc., Vol. 227, No. 1, p. 123–133 (1987).

The authors present results of intensive photometry over two seasons for the bright Be star Achernar. Some medium–dispersion spectroscopy simultaneous with the photometry is also presented. The variability of the star in both radial velocity and light has a period of 1.26 day. The form of the light curve is complex and probably changes with time. There also seems to be a weak but significant variation of the strength of the Mg II λ4881 line with the same period. The authors discuss these observations in the light of the non–radial oscillation, rotational modulation and binary hypotheses.

112.027 A photometric survey of the bright southern Be stars.
C. Stagg.
Mon. Not. R. Astron. Soc., Vol. 227, No. 1, p. 213–240 (1987).

Repeated UBV photometric measurements were made of the 86 bright Be stars south of declination –20°, and a network of comparison stars was set up. From a statistical study of the differential photometry it was found that short– or intermediate–term variability seems to be occurring in about half of the Be stars, and to be more evident in the stars of earlier spectral type. It was also possible to identify 11 individual short– or intermediate–term variables. Four of these (all of early B spectral type) appear to exhibit significant variability on a time–scale of a day or less. More intensive observations of one of these stars, 28ω CMa, indicate short–term variations consistent with the published spectroscopic period of 1.37 day.

112.028 Infrared studies of Eta Carinae – I. Spectroscopy and a composite dust model.
G. Robinson, R. M. Mitchell, D. K. Aitken, G. P. Briggs, P. F. Roche.
Mon. Not. R. Astron. Soc., Vol. 227, No. 2, p. 535–542 (1987).

The authors report spectral observations of η Carinae between 8 and 13 μm which compare the central peak of the homunculus with its periphery. The spectra not only show a lower effective temperature for the outer regions but also that the grains here have a different emissivity function. This has been shown to be consistent with the presence of large grains (Mitchell & Robinson, 1986) and here the modelling is extended to that of a spherical core containing small grains ($a \cong 0.2$ μm) and a disc of larger grains ($a \cong 2.0$ μm); the model predictions are consistent with observations.

112.029 IR excesses in Be stars.
A. Chokshi, M. Cohen.
Astron. J., Vol. 94, No. 1, p. 123–128 (1987).

Infrared excesses are empirically derived for a sample of bright Be stars based on IRAS data and previously published near–infrared photometry. The excesses are compared with published Hα luminosities and are found to be correlated. The data are compared with the theoretically expected relation between the sum of free–free and free–bound continua and the Hα recombination–line intensity from a common envelope. The inclusion of temperature and optical–depth effects is necessary to produce agreement between observations and theory. No significant correlation is detected between $L_{IR}/L_{H\alpha}$ ratios and rotational velocities.

112.030   Metals in IRC + 10216: detection of NaCl, AlCl, and KCl, and tentative detection of AlF.

J. Cernicharo, M. Guélin.
Astron. Astrophys., Vol. 183, No. 1, p. L10 – L12 (1987).

The authors report the first detection of metal halides in IRC + 10216. The millimetre–wave line profiles suggest that these species are concentrated in the inner circumstellar envelope. The abundances derived for NaCl, AlCl, KCl, and tentatively for AlF, are in the range $10^{12} - 10^{14} cm^{-2}$ and are $10^6 - 10^8$ times lower than the abundance of H_2. They agree with the chemical equilibrium abundances calculated by Tsuji (1973) for a carbon-rich stellar atmosphere with a temperature 1200 – 1500K. The lines of NaCl and AlCl, observed with the IRAM 30m telescope, are strong enough to allow the detection of these species' rare ^{37}Cl isotopes. The derived $^{35}Cl/^{37}Cl$ isotopic ratio (2.3 ± 0.5) is consistent with the terrestrial elemental isotopic ratio.

112.031   CO(J = 1–0) observations of bright carbon stars.

H. Olofsson, K. Eriksson, B. Gustafsson.
Astron. Astrophys., Vol. 183, No. 2, p. L13 – L16 (1987).

The authors have surveyed a sample of bright N–type stars, with recent estimates of chemical composition, in the CO(J = 1–0) line. Almost all stars were detected. The mass loss rate is well correlated with a far–IR excess measure, and the gas–to–dust mass ratio, estimated to be 350 ± 200, seems relatively constant for this sample of stars. The mass loss rate appears to be dependent on the effective temperature, the carbon excess, and the $^{12}C/^{13}C$–ratio of the central star. In particular, the peculiar ^{13}C–rich stars have comparatively low mass loss rates. A weak dependence of gas expansion velocity of the circumstellar envelope on the carbon excess may exist.

112.032   Detection of vibrationally excited SiS in IRC + 10216.

B. E. Turner.
Astron. Astrophys., Vol. 183, No. 2, p. L23 – L26 (1987).

Vibrationally excited (v = 1) SiS has been detected via its J = 12–11, 13–12, and 14–13 transitions in the circumstellar shell IRC + 10216. The unusually narrow linewidths indicate that the SiS v = 1 emission arises in the innermost, accelerating zone of the envelope. Brightness temperatures are therefore $\gtrsim 600K$, and are shown to be consistent with thermal excitation (via IR radiation) of the v = 1 state, although maser activity is also possible. While the observed lack of v = 1 emission from CO and SiO is shown to be consistent with current estimates of CO and SiO abundances, analysis of v = 1 SiS and CS emission, which show contrasting characteristics, indicates that the SiS abundance must decrease significantly outside the inner 600K core, while the CS abundance must increase outside the inner core, even more sharply.

112.033   Chemical modelling of molecular sources. V. IRC + 10216.

L. A. M. Nejad, T. J. Millar.
Astron. Astrophys., Vol. 183, No. 2, p. 279 – 286 (1987).

The authors present detailed results of a chemical kinetic model of the outer envelope ($10^{16} cm$ to $10^{18} cm$) of the carbon-rich star IRC + 10216. The chemistry is driven by a combination of cosmic–ray ionization and ultraviolet radiation and, starting from 7 parent molecules injected into the envelope, the authors find that a complex chemistry ensues. Ion–molecule reactions can efficiently build hydrocarbon species and account for the observed abundances of CH_3CN and HNC. Reactions involving CO may lead to observable abundances of oxygen–bearing molecules such as C_3O, CH_2CO and HCO^+.

112.034   Different regions of line formation in the envelope of the early emission line star HD 190073.

A. E. Ringuelet, M. Rovira, L. Cidale, J. Sahade.
Astron. Astrophys., Vol. 183, No. 2, p. 287 – 294 (1987).

A description is presented of the spectral features that characterize the spectrum of HD 190073 both in the photographic region from about 3600 Å to about 6600 Å, and in the IUE ultraviolet, from about 1150 to about 3200 Å. A number of different types of profiles can be distinguished and this seems to imply that many different "broad" regions of line formation coexist in the extended envelope of the star, including regions with densities differing in several orders of magnitude. A very interesting feature in the spectrum of HD 190073 is the presence of relatively narrow emissions at 3685 Å, 3760 Å and at Hα which suggests a similarity between a compact planetary nebula – NGC 7027 – and this A star with infrared excess.

112.035   Wind activity in 66 Oph (B2 IVe) from 1984 – 1987.

C. A. Grady, O. L. Lupie, G. Sonneborn, C.–C. Wu, G. J. Peters, K. S. Bjorkman.
Be Star Newsl., No. 16, p. 5 – 6 (1987).

112.036   Radial velocity variations of Zeta Tau in 1978 – 1986.

Y.–l. Guo, W.–s. Gao.
Be Star Newsl., No. 16, p. 7 – 8 (1987).

112.037   The unstable shell of ζ Tauri.

W.–s. Gao, Y.–l. Guo.
Be Star Newsl., No. 16, p. 9 – 12 (1987).

112.038   Giant outbursts of the Eta Carinae – P Cygni type.

K. Davidson.
Instabilities in luminous early type stars, p. 127 – 141 (1987). – See Abstr. 012.012.

Certain very massive stars, including η Car and P Cyg, give observational hints that their most crucial mass loss occurs in giant eruptions. During such an event, the mass loss rate can be of the order of 0.1 $M_\odot yr^{-1}$ for several years. We do not yet understand this phenomenon, partly because most relevant theoretical work has been devoted instead to more familiar, less dramatic stellar winds.

112.039   The unstable O6.5f?p star HD 148937 and its interstellar environment.

C. Leitherer, C. Chavarria–K.
Instabilities in luminous early type stars, p. 209 – 210 (1987). – See Abstr. 012.012.

112.040   Episodic distortion and dust formation in the wind of WR 140.

P. M. Williams, K. A. van der Hucht, H. van der Woerd, W. M. Wamsteker, T. R. Geballe, C. D. Garmany, A. M. T. Pollock.
Instabilities in luminous early type stars, p. 221 – 226 (1987). – See Abstr. 012.012.

In 1985 April, the WC7 + abs star WR 140 = HD 193793 was observed to have brightened by over two mag. in the infrared owing to the formation of a new dust shell. The growth and evolution of the shell was monitored by infrared observations using UKIRT during the remainder of 1985. Examination of infrared photometry of this star since 1979 and previously published data indicate that the dust formation occurs at intervals of 7.9 years. Phasing the published radial velocities of the absorption line component with this period confirms that it is a member of an eccentric binary system having periastron passage shortly before dust formation.

112.041   Photometric variability of Wolf–Rayet stars.

A. F. J. Moffat, M. M. Shara, R. Lamontagne.
Instabilities in luminous early type stars, p. 227 – 230 (1987). – See Abstr. 012.012.

A potentially important factor in the initiation and acceleration of the winds in WR stars may be related to instabilities in the core or in the wind. To explore this possibility, the authors are in the process of making extensive surveys of complete samples of WR stars using broadband, optical photometry during contiguous–night observing runs spread over several weeks at a time.

112.042 **Improved mass loss rates for WC7 – 9 stars: their effect on the Wolf–Rayet stellar wind momentum problem.**
J. P. Cassinelli, K. A. van der Hucht.
Instabilities in luminous early type stars, p. 231 – 235 (1987). – See Abstr. 012.012.

Applying recent surface abundance determinations for Wolf–Rayet stars of notably the WC sequence, the effect of model atmosphere energy distributions on the ionization and mass loss rates is investigated. The authors conclude that these stars lose more mass than previously considered, which makes it even more difficult to explain WR stellar winds by radiation pressure alone.

112.043 **An extended nebulosity surrounding the S Dor variable R 127.**
I. Appenzeller, B. Wolf, O. Stahl.
Instabilities in luminous early type stars, p. 241 – 244 (1987). – See Abstr. 012.012.

New high resolution spectrograms of R 127 show the presence of an extended expanding gaseous nebula around this high–mass loss S Dor variable.

112.044 **The LMC – S Dor variable R 71: an IRAS point source.**
B. Wolf, F.–J. Zickgraf.
Instabilities in luminous early type stars, p. 245 – 248 (1987). – See Abstr. 012.012.

The S Dor variable R 71 of the LMC was found to be an IRAS point source. Its flux densities at 12, 25, and 60 μ are explained to be due to the radiation of an extended (d $\approx$ 8000 stellar radii) dust envelope of very low temperature. The total amount of matter (i.e. gas and dust) of the envelope is estimated to M $\approx 3 \times 10^{-2} M_\odot$.

112.045 **Eclipse spectrum of the LMC P Cyg star R 81.**
O. Stahl, B. Wolf, F.–J. Zickgraf.
Instabilities in luminous early type stars, p. 249 – 252 (1987). – See Abstr. 012.012.

R 81 is a particularly interesting P Cygni star since its absolute luminosity is reliably known (due to its membership to the LMC) and it offers the possibility to derive the mass of a P Cygni star. For this purpose phase–dependent spectroscopic observations of high resolution and high S/N ratio are badly needed.

112.046 **The light– and colour variation of Eta Carinae for the years 1983 – 1986 in the VBLUW system.**
A. M. van Genderen, P. S. Thé.
Instabilities in luminous early type stars, p. 253 – 255 (1987). – See Abstr. 012.012.

New VBLUW photometry of the peculiar variable star Eta Carinae, was made in the interval 1983 to 1986, to extend the detailed light– and colour curves for the epoch 1974 to 1983. A new light maximum has been observed in 1986, with a height similar to the maximum of 1981/1982.

112.047 **Do superluminous stars really explode?**
R. Viotti.
Instabilities in luminous early type stars, p. 257 – 261 (1987). – See Abstr. 012.012.

The author discusses three rather well documented cases of superluminous stars: Eta Carinae, P Cygni and AG Carinae, which are well representative of the uppermost part of the H–R diagram. It is concluded that the observed historical variations of these objects occurred at almost constant bolometric magnitude, i.e. the use of the term outburst is not appropriate. The implications for other cases, including the Hubble–Sandage variables, are discussed.

112.048 **The X–ray emission of Tau Sco, B0 V, and the problems posed for embedded shock models.**
J. P. Cassinelli.
Instabilities in luminous early type stars, p. 273 – 277 (1987). – See Abstr. 012.012.

Main sequence OB–stars are useful for testing X–ray source models, because the mass loss rates are low as compared to those of the OB supergiants and the Of stars hence the winds are optically thin at nearly all frequencies. Tau Sco is a very well studied B0 V star. The discovery of the resonance doublet of O VI (1030 Å) in its Copernicus satellite spectrum led to the recognition of the now well known "superionization" problem. This led to the prediction that O and B stars would be X–ray sources.

112.049 **A hot, low mass–loss rate inner envelope in IRC + 10216.**
R. Sahai.
Astrophys. J., Vol. 318, No. 2, p. 809 – 822 (1987).

The author has modeled CO $J = 3$–2 and $J = 6$–5, and new high spatial resolution, $J = 1$–0, observations of the late–type carbon–rich Mira, IRC + 10216, in addition to previous $J = 1$–0 and $J = 2$–1 data in order to constrain the mass loss and gas kinetic temperature in the inner regions of its molecular envelope. It is found that, in the $\sim 1''$–$6''$ radial region, the kinetic temperature distribution is significantly higher than that determined from the Kwan and Linke thermodynamic model and the mass–loss rate is significantly smaller than in the extended envelope. The lower mass–loss rate provides a natural explanation for the higher temperature in the inner envelope. The inner envelope probably represents a phase of reduced mass loss in the evolution of IRC + 10216.

112.050 **The distribution of maser emission in OH/IR stars.**
A. D. Welty, J. D. Fix, R. L. Mutel.
Astrophys. J., Vol. 318, No. 2, p. 852 – 867 (1987).

The authors have mapped, at three epochs over a 2.5 yr period of time, the 1612 MHz OH emission from five OH/IR stars. Although the stars were observed at very different phases in the radio light curve of each, there were no remarkable changes in the appearance of the maps. The maps have been used to produce star–centered surface maps of the distribution of maser emission from each star. The surface maps generally are sparsely filled with OH emission and are dominated by relatively few (~ 10) major clumps of emission.

112.051 **Optical properties of circumstellar silicates in the visible and the near–infrared.**
B. Pégourié.
Astrophys. Space Sci., Vol. 136, No. 1, p. 133 – 148 (1987).

Previous results of modelling the infrared emission of a sample of 23 O–rich giants and supergiants are used to deduce the optical properties of circumstellar silicates in the visible and near–infrared. This leads to a unique set of optical properties for the grain material which permits to reproduce the circumstellar emission of such stars. Planck mean efficiencies both for emissivity and radiation pressure are presented in the temperature and grain radius ranges of interest for computations of a red giant environment.

112.052 **The stellar winds of early–type stars in the Small Magellanic Cloud.**
R. K. Prinja.
Mon. Not. R. Astron. Soc., Vol. 228, No. 1, p. 173 – 192 (1987).

Results are presented from a study of the stellar winds of early–type stars in the Small Magellanic Cloud, based on low resolution *IUE* spectra. Well developed P Cygni doublets observed in 25 main–sequence, giant, and supergiant stars, are fitted using model profiles, allowing for the *IUE* low–resolution instrumental profile. Some characteristics of mass–loss in these stars are derived, including terminal velocities, stellar wind densities, and the products of mass–loss rate and ionization fraction. These results are compared with the observed stellar wind properties of galactic O stars of similar spectral types and luminosities.

112.053 **Radio detection of the Be star ψ Persei.**
A. R. Taylor, L. B. F. M. Waters,
H. J. G. L. M. Lamers, P. Persi, K. S. Bjorkman.
Mon. Not. R. Astron. Soc., Vol. 228, No. 3, p. 811 – 817 (1987).

The authors present the results of a VLA survey for radio emission from a sample of Be stars having strong excess emission in the *IRAS* bands. Radio emission was detected from one star,

ψ Persei. The radio flux density, together with nearly simultaneous ultraviolet and near–infrared observations, argue strongly against a spherically symmetric model for the circumstellar envelope, thus favouring equatorial disc models. In general it is found that the spectral index of the excess emission increases between the radio and far–infrared, suggesting either increased acceleration of the outflow at large distances from the star, partial recombination of the outer regions of the disc or divergence of the equatorial disc at large radii.

112.054 Hα echelle spectroscopy of Be stars: an atlas.
D. Ballereau, M. Alvarez, J. Chauville, R. Michel.
Rev. Mex. Astron. Astrofis., Vol. 15, No. 1, p. 29 – 52 (1987).

From the analysis of 78 spectroscopic plates of the Hα emission line of 15 Be and shell stars, known to be variable from spectroscopic and photometric studies, the authors present an atlas of the Hα line. They emphasize the spectroscopic changes of the line to obtain criteria to further study the short–term variability of these interesting objects. The authors can easily detect the long–term variability comparing their Hα line observations to previous work. From this study, the authors found that all the stars observed show variability on a time scale from 2 to 14 days. On a shorter time scale, Hα line variability is present in HD 37202, HD 184279, HD 200120, HD 217050, HD 224559, HD 174638 and HD 183656. The authors include in this atlas 3 spectra of the Hα absorption line of 2 β CMa stars observed during the campain.

112.055 Multiband photometry (8 – 13 μm) of Herbig Ae/Be stars.
F. Berrilli, D. Lorenzetti, P. Saraceno, F. Strafella.
Mon. Not. R. Astron. Soc., Vol. 228, No. 4, p. 833 – 842 (1987).

New multifilter 8 – 13 μm observations of 14 Herbig Ae/Be stars are presented. Silicate features are found in emission or absorption; a simple model to fit the obtained data provides an evaluation of the dust shell temperature and of the optical depth τ_{10}. These parameters are used both to define the circumstellar dust shell and to discuss the applicability of the free–free hypothesis to explain the near–infrared spectrum (J,H,K,L' and M).

112.056 OH/IR stars without OH masers: nondetection statistics.
B. M. Lewis, J. Eder, Y. Terzian.
Astron. J., Vol. 94, No. 4, p. 1025 – 1034 (1987).

Potential OH/IR stars are readily recognized from their far–infrared colors. These are confirmed as OH/IR stars by detecting their 1612 MHz masers. Color selection from the sources listed in the *IRAS Point Source Catalog* therefore provides an efficient means for identifying new objects. However, this approach is not so efficient with weak sources ($S_{25} < 2$ Jy). Color–selected sources with $S_{25} > 2$ Jy that are not associated with OH masers have the same color–color range, the same IR flux range, and a similar galactic distribution to the detections. These sources are explicable as normal OH/IR stars in which the OH maser is being suppressed. This would occur when a star with a circumstellar envelope also has a companion star that is able to disrupt the velocity coherence of the normal masing shell. The detection to nondetection ratio then provides a tool for studying the duplicity of these stars over the whole Galaxy.

112.057 An investigation of stellar coronae with AXAF.
J. L. Linsky.
Astrophys. Lett. Commun., Vol. 26, Nos. 1 – 2, p. 21 – 34 (1987).
– See Abstr. 012.020.

The author lists some of the fundamental questions of stellar coronal physics and indicates to what extent the different instruments on AXAF will be able to answer these questions. First, AXAF will obtain very rich spectra of bright sources with which one may compute detailed models of their coronae. Second, AXAF will obtain time–resolved spectra during the eclipses of two prototypical systems, AR Lacertae and YY Geminorum, which can be used to identify and model individual geometrical structures (i.e. active regions) on these stars. Third, AXAF will obtain time–resolved spectra during stellar flares with which one may infer the changes in the flaring plasma temperatures as a function of time.

112.058 Stellarer Massenverlust durch Sternwinde.
C. Friedemann.
Astron. Sch., 24. Jahrg., Heft 5, p. 98 – 101 (1987).

112.059 Abundances in red giant stars: carbon and oxygen isotopes in carbon–rich molecular envelopes.
P. G. Wannier, R. Sahai.
Astrophys. J., Vol. 319, No. 1, p. 367 – 382 (1987).

Millimeter–wave observations have been made of isotopically substituted CO toward the envelopes of 11 carbon–rich stars. In every case, ^{13}CO was detected and model calculations were used to estimate the ^{12}C/^{13}C abundance ratio. C^{17}O was detected toward three, and possibly four, envelopes. The C^{18}O variant was detected in two envelopes. As with other classes of red giant stars, the carbon–rich giants seem to be significantly, though variably, enriched in ^{17}O.

112.060 Broad–band imaging of the Beta Pictoris circumstellar disk.
F. Paresce, C. Burrows.
Astrophys. J., Lett. Ed., Vol. 319, No. 1, p. L23 – L25 (1987). With plates L2 – L3.

The β Pictoris circumstellar disk is imaged at subarcsecond resolution between 5″ and 15″ from the star in the B, V, R and I_c bands. The shape of the emitted spectrum is indistinguishable within statistical uncertainty from that of β Pictoris itself. The surface brightness of the disk in the B, V, R and I_c band at 6″ from the star is 16.6, 16.3, 15.9 and 15.9 mag per square arcsecond, respectively. The results are consistent with the emission mechanism being scattering of starlight and the bulk of the scattering particles having sizes much greater than 1 μm.

112.061 Further observations of the He I 10830 Å chromospheric line in stars.
D. L. Lambert.
Astrophys. J., Suppl. Ser., Vol. 65, No. 2, p. 255 – 271 (1987).

Observations of the chromospheric He I triplet at 10830 Å are presented for late–type stars. O'Brien and Lambert's suggestion that the triplet was broad, shallow, and constant in K giants to the left of the coronal dividing line is confirmed and extended to 8 years for β Gem. Other giants and supergiants show a variable 10830 Å feature. This may be rotationally modulated for Arcturus with a period of about 233 days.

112.062 A study of the massive O–type binary Iota Orionis.
D. J. Stickland, C. D. Pike, C. Lloyd, I. D. Howarth.
Astron. Astrophys., Vol. 184, No. 1/2, p. 185 – 192 (1987).

Iota Orionis is an O9 III + B1 III binary with a highly eccentric orbit. Using coudé spectra the authors have updated its ephemeris and established a value for the apsidal motion. Combining the orbital elements, the astrophysical data (the primary is an O9 III standard star in the Orion Association), and photometry through a grazing eclipse, they have determined the dimensions of the system with, they believe, some precision. This shows that at periastron, the centres of the stars are separated only by about 1.5 times the sum of the radii, which suggests that strong tidal forces will be generated and enhanced mass loss may be expected. The authors have searched for evidence of this in a sequence of high resolution IUE spectra taken through periastron passage; some perturbations of the wind lines of C IV, N V and Si IV are seen but it is not clear that they are phase–related. Sharp components of the wind lines appear to have decoupled from the orbital motion.

112.063 A study of UV spectra of ζ Aur/VV Cep stars. X. Mass–loss of α Sco A from high–resolution IUE spectra of α Sco B.
H.–J. Hagen, K. Hempe, D. Reimers.
Astron. Astrophys., Vol. 184, No. 1/2, p. 256 – 262 (1987).

The authors redetermine the rate of mass–loss of the M supergiant α Sco by means of circumstellar lines observed with IUE in the ultraviolet spectrum of its B star companion. They first show (1) that nearly all UV lines, on which previous mass–loss determinations (v. d. Hucht et al., 1978; Bernat, 1982) were based, are blends with interstellar lines and (2) that the remaining 'pure' circumstellar lines have P Cyg profiles for which radiative transfer had been treated incorrectly. This explains both the discrepancy between mass–loss rates from UV lines and optical Ti II lines (Kudritzki and Reimers, 1978) and the large scatter (factor 30) among rates from individual ions (Bernat, 1982). By application of correct line transfer calculations to P Cyg type lines of Si II, O I and Fe II and Ti II (optical) lines the authors obtain a mass–loss rate of $\sim 1 \times 10^{-6} M_\odot$/yr with an uncertainty of a factor of at least 2 either way, in excellent agreement with the values from optical lines (Kudritzki and Reimers, 1978) and from radio emission of the H II region (Hjellming and Newell, 1983).

112.064 Studies in Be–star variability. 2. Analysis of published radial velocities of six bright emission–line stars.
P. Harmanec.
Bull. Astron. Inst. Czech., Vol. 38, No. 5, p. 283 – 296 (1987).

The published radial velocities of six emission–line stars, o Cas, HR 1763, HR 2370, HR 2577, 7 Oph and HD 105056, are collected and analyzed. No satisfactory period could be found for the velocity variations of o Cas and it is concluded that the observed velocity changes probably arise from the superposition of long–term and rapid velocity variations. Tentative periods are suggested for the other five stars, and it is argued that all of them can be spectroscopic binaries. However, the evidence for real periodicity is not fully conclusive in either case and new series of observations are urgently needed.

112.065 P Cygni: a hypergiant well worth watching.
M. De Groot.
J. Am. Assoc. Variable Star Obs., Vol. 16, No. 1, p. 12 – 18 (1987).

The hypergiant P Cygni is discussed, and an invitation for the collaborative observation of this interesting star is issued.

112.066 Diamond formation in carbon–star atmospheres.
T. J. Wdowiak.
Nature, Vol. 328, No. 6129, p. 385 (1987).

112.067 The use of terms and definitions in the study of Be stars.
G. W. Collins II.
Physics of Be stars, p. 3 – 21 (1987). – See Abstr. 012.030 (IAU Colloq. No. 92).

The author examines the use and misuse of some astronomical terminology as it is commonly found in the literature. A basic definition of the Be phenomena is suggested and other stellar characteristics whose interpretation may change when used for non–spherical stars, is discussed. Special attention is paid to a number of terms whose semantic nature is misleading when applied to the phenomena they are intended to represent. The use of model–dependent terms is discussed and some comments are offered which are intended to improve the clarity of communication within the subject.

112.068 Observations of rapid variability in Be stars.
J. R. Percy.
Physics of Be stars, p. 49 – 65 (1987). – See Abstr. 012.030 (IAU Colloq. No. 92).

Photometric and/or spectroscopic variability on time scales of approximately 0.2 to 2 days has been observed in over 40 Be stars, and is suspected in many more. This paper reviews the observational aspects of this phenomenon: both surveys and studies of individual objects.

112.069 First measurements of γ Cassiopeiae's hydrogen envelope.
P. Granes, C. Thom, F. Vakili.
Physics of Be stars, p. 66 – 67 (1987). – See Abstr. 012.030 (IAU Colloq. No. 92).

112.070 Short term photometric and spectroscopic variability of a sample of Be stars.
M. Alvarez, R. Michel.
Physics of Be stars, p. 74 – 77 (1987). – See Abstr. 012.030 (IAU Colloq. No. 92).

Previous studies of variable Be stars show that the short–term variability is a common, but not well understood phenomena. The authors choose some interesting Be objects to study their behaviour both in spectroscopy and photometry. They are reporting some preliminary results of their observations.

112.071 Is HR 9070 really pulsating?
J. P. Sareyan, M. Alvarez, J. Chauville,
J. M. Le Contel, R. Michel, D. Ballereau.
Physics of Be stars, p. 78 – 81 (1987). – See Abstr. 012.030 (IAU Colloq. No. 92).

112.072 Simultaneous spectroscopy and polarimetry of Be stars.
D. R. Gies, M. McDavid.
Physics of Be stars, p. 84 – 86 (1987). – See Abstr. 012.030 (IAU Colloq. No. 92).

The authors describe an initial attempt to search for simultaneous variations in continuum polarization, Hα emission, and the He I $\lambda6678$ photospheric absorption line in Be stars in order to investigate correlated changes on short timescales.

112.073 The short–period photometric variability of Be stars.
L. A. Balona, C. A. Engelbrecht.
Physics of Be stars, p. 87 – 89 (1987). – See Abstr. 012.030 (IAU Colloq. No. 92).

The fact that some Be stars have periodic light variations with time scales of the order of one day is well established. The explanation which seems to have gained the most acceptance is nonradial pulsation. It is of great importance to ascertain the proportion of Be stars which are short–period variables and to determine the link between the variations and the emission–line phase. The authors have started a photometric project to investigate these problems. In this note they present some preliminary results and tentative conclusions.

112.074 Some important results from two photometric campaigns on short term variability in Be stars.
C. R. Stagg.
Physics of Be stars, p. 90 – 92 (1987). – See Abstr. 012.030 (IAU Colloq. No. 92).

In 1983 and 1984 a programme of UBV observations was carried out to set up a network of standard stars, identify short term variables, and analyse the variability of the 86 bright ($V \gtrsim 6.5$) southern ($\delta < -20°$) Be stars.

112.075 Short–term spectroscopic variations of the Be star 11 Cam.
J. Chauville, M. Alvarez, J. P. Sareyan, D. Ballereau,
R. Michel.
Physics of Be stars, p. 93 – 94 (1987). – See Abstr. 012.030 (IAU Colloq. No. 92).

112.076 A transient shock–wave in the shell of the Be star HD 184279.
D. Ballereau, J. Chauville.
Physics of Be stars, p. 95 – 98 (1987). – See Abstr. 012.030 (IAU Colloq. No. 92).

112.077 Rapid variation of shell line parameters in ζ Tau.
M. Bossi, G. Guerrero, M. Scardia.
Physics of Be stars, p. 99 – 100 (1987). – See Abstr. 012.030 (IAU Colloq. No. 92).

112.078   Photometric and H–alpha variability in some Be stars.
S. Catalano, G. Umana.
Physics of Be stars, p. 101 – 103 (1987). – See Abstr. 012.030 (IAU Colloq. No. 92).
The authors present a preliminary report of photometric and spectrographic observations of Be stars with the aim of studying short–term variations which could eventually be interpreted in terms of surface inhomogeneities.

112.079   Optical emission–line spectra of Be stars.
J. Dachs.
Physics of Be stars, p. 149 – 171 (1987). – See Abstr. 012.030 (IAU Colloq. No. 92).
Recent observational work on spectra obtained at optical and near–infrared wavelengths is reviewed for "ordinary" (non–supergiant non–peculiar) Be and Be–shell stars, with particular emphasis on comparison between high–resolution spectral measurements and current Be star models.

112.080   Spectral energy distribution and interstellar reddening.
L. Houziaux, J. Manfroid.
Physics of Be stars, p. 172 – 191 (1987). – See Abstr. 012.030 (IAU Colloq. No. 92).
The authors review the recent results on flux distribution of Be stars from 320 to 850 nm. It appears that data are available for only 10 to 15% of the objects recognized as "normal" Be stars. The main current problems appear to be the variability of the continuous radiation, and the correction for interstellar reddening. Various attempts to solve these matters are discussed.

112.081   Structure of the envelope of EW Lac in 1971 – 1984.
T. Kogure, M. Suzuki.
Physics of Be stars, p. 192 – 194 (1987). – See Abstr. 012.030 (IAU Colloq. No. 92).

112.082   Observational constraints on cool disk models of Be stars.
A. M. Hubert, M. Floquet, J. Chauville, M. T. Chambon.
Physics of Be stars, p. 195 – 196 (1987). – See Abstr. 012.030 (IAU Colloq. No. 92).

112.083   Long–term polarization changes of 88 Her.
J. Arsenijević, S. Jankov, G. Djurašević.
Physics of Be stars, p. 200 – 201 (1987). – See Abstr. 012.030 (IAU Colloq. No. 92).
Linear optical polarization of 88 Her has been measured in V spectral region during the period 1974 – 1985. The mean annual values of the intrinsic polarization parameters are presented. The polarization percentage changes from 0.15% (1976) to 0.56% (1979). Small values of the polarization percentage correspond to the period when the envelope effect is negligible. The maximum of polarization percentage has been found during the early period of strong shell phase. The polarization position angle varies between 53 and 83 degrees.

112.084   Red and infrared photometry of Be stars.
G. C. Kilambi, P. Vivekananda Rao, M. B. K. Sarma.
Physics of Be stars, p. 202 – 205 (1987). – See Abstr. 012.030 (IAU Colloq. No. 92).

112.085   Spectral energy distribution of Be stars.
D. Kaiser, R. W. Hanuschik.
Physics of Be stars, p. 206 – 207 (1987). – See Abstr. 012.030 (IAU Colloq. No. 92).

112.086   High–resolution emission–line spectroscopy of Be stars: comparison of Hα and weak emission lines.
R. W. Hanuschik.
Physics of Be stars, p. 208 – 210 (1987). – See Abstr. 012.030 (IAU Colloq. No. 92).
For a number of bright southern Be stars, high–resolution, high S/N spectroscopy has been performed for Hα, Si II λ6371 and four Fe II emission lines. With the exception of four stars, both Hα and weak emission line profiles exhibit resolved double–peak structure. Peak separation increases with stellar projected rotational velocity. In many cases, Hα profiles (but not Fe II lines) show resolved fine–structure (inflections in the flanks) indicating a two–component structure of the disk. Envelope models with arbitrary density and rotational velocity laws have been fitted to the Fe II line profiles with satisfactory results.

112.087   Absolute spectrophotometry of Be stars.
T. Stiff, S. Jeffers.
Physics of Be stars, p. 211 – 213 (1987). – See Abstr. 012.030 (IAU Colloq. No. 92).
Absolute spectrophotometric data have been obtained for a sample of Be stars. The data have been corrected for differential atmospheric extinction, instrumental response, and interstellar extinction and calibrated in terms of absolute flux. Absolute Hα and Hβ fluxes have been determined.

112.088   Interstellar absorption of some Be stars.
D. Briot, J. Zorec.
Physics of Be stars, p. 214 – 216 (1987). – See Abstr. 012.030 (IAU Colloq. No. 92).
The aim of this work is to test the dereddening method from the 2200 Å bump for "classical" Be stars, in determining the interstellar absorption in the region of the Be star, using the surrounding normal stars.

112.089   An optical study of southern Be stars.
F. Giovannelli, C. Rossi, A. A. Vittone.
Physics of Be stars, p. 217 – 218 (1987). – See Abstr. 012.030 (IAU Colloq. No. 92).

112.090   Infrared observations of Be stars.
H. J. G. L. M. Lamers.
Physics of Be stars, p. 219 – 236 (1987). – See Abstr. 012.030 (IAU Colloq. No. 92).
The author discusses the IRAS observations of the Be stars and the correlations between the IR excess and the stellar parameters. He describes the quantitative interpretation of the IR excess and the results in terms of characteristics of the circumstellar material. He also describes some of the major problems of the IR studies of Be stars and possible future investigations.

112.091   An infrared study of southern Be stars: ground–based and IRAS observations.
P. Persi, M. Ferrari–Toniolo.
Physics of Be stars, p. 242 – 244 (1987). – See Abstr. 012.030 (IAU Colloq. No. 92).
Infrared observations represent a powerful tool in understanding the physical conditions of the circumstellar matter around Be stars. For this purpose, the authors have undertaken an IR study of 21 southern Be stars including JHKLM photometry, CVF spectra at resolution R = 100 around the IR H I emission lines Brγ, Brα, and Pfγ, and IRAS data at 12 and 25 microns taken from the Point Sources Catalogue.

112.092   The relation between mass loss and luminosity for Be stars.
L. B. F. M. Waters, H. J. G. L. M. Lamers, J. Coté.
Physics of Be stars, p. 245 – 249 (1987). – See Abstr. 012.030 (IAU Colloq. No. 92).
The mass loss rates of a large sample of Be stars derived from the UV and the IR are compared. The IR rates were derived using a simple equatorial disc model, and are typically a factor 100 larger than those derived from the UV. In terms of mass fluxes, the mass flux in the polar regions (derived from the UV observations) is about a factor 10^3 smaller than the mass flux in the equatorial regions. The dependence of $\dot{M}_{IR}$ and $\dot{M}_{UV}$ on stellar luminosity is studied. It is shown that $\dot{M}_{IR}$ depends weaker on L than $\dot{M}_{UV}$. This suggests that two different mechanisms are responsible for the mass flux in polar and equatorial regions.

112.093 **High–energy phenomena in Be stars.**
T. P. Snow.
Physics of Be stars, p. 250 – 262 (1987). – See Abstr. 012.030
(IAU Colloq. No. 92).
  When Be stars began to be observed in the ultraviolet it was
found that many of them have stellar winds, and furthermore
that excess ionization was also characteristic of these winds, as it
was in the more luminous O and B stars. This excess ionization,
which today is often called "superionization", along with the
associated stellar wind phenomena, forms the basis for this
review. Here the term "high–energy phenomena" is taken to
mean any phenomena requiring energies higher than can be
derived from radiative equilibrium with the stellar photosphere.
Hence this paper discusses superionization, stellar winds and all
their characteristics, and high–energy (X–ray) emission.

112.094 **Fe II and Fe III lines as a diagnostic of the physical
conditions in the atmospheres of Be stars.**
G. B. Baratta, M. Friedjung, G. Muratorio, R. Viotti.
Physics of Be stars, p. 263 – 264 (1987). – See Abstr. 012.030
(IAU Colloq. No. 92).
  Fe II lines are frequently observed in the optical spectra of
early–type emission–line stars. Space observations of their
ultraviolet spectra led to the discovery of a considerable line
absorption due to ionized metal lines formed in the stellar
envelopes. The authors describe some examples and show that
spectral synthesis is required especially for the analysis of the low
resolution spectra, and may provide information about the
structure of the envelopes and the mass loss rates. They also
discuss the Fe III lines in the UV spectra of a Herbig Be star and
of β Lyrae.

112.095 **Narrow absorption components in the UV spectra of
HD 110432 (B1 IIIe).**
R. K. Prinja, H. F. Henrichs.
Physics of Be stars, p. 265 – 267 (1987). – See Abstr. 012.030
(IAU Colloq. No. 92).
  The authors present high resolution IUE ultraviolet observa-
tions of multiple narrow absorption components seen in the
Si III, N V, Si IV, and C IV profiles of the B1 IIIe star,
HD 110432. Spectra taken during March 1986, spanning
~ 11 days, are modelled using line profile fits. Central velocities
and column densities of the discrete features are derived.

112.096 **The evidence for aspect–dependent winds in Be stars.**
C. A. Grady, K. S. Bjorkman, T. P. Snow.
Physics of Be stars, p. 268 – 270 (1987). – See Abstr. 012.030
(IAU Colloq. No. 92).
  The authors present the results of a survey of stellar winds in 62
Be and 43 normal B stars in spectral types B0 to B5 and
luminosity classes V–III. They find that in general the wind
absorption seen in the resonance lines of C IV and Si IV in
Be stars is the result of the superposed absorption from multiple
shortward–shifted discrete absorption components.

112.097 **IUE spectra of the Be star HD 174237.**
S. P. Tarafdar.
Physics of Be stars, p. 271 (1987). Abstract. – See Abstr. 012.030
(IAU Colloq. No. 92).

112.098 **The flux distributions of Be stars in the far–UV.**
R. Stalio, R. S. Polidan, G. J. Peters.
Physics of Be stars, p. 272 – 273 (1987). Abstract. – See Abstr.
012.030 (IAU Colloq. No. 92).

112.099 **Voyager observations of Zeta Tau.**
T. E. Carone, R. S. Polidan.
Physics of Be stars, p. 274 – 277 (1987). – See Abstr. 012.030
(IAU Colloq. No. 92).

112.100 **Long–termed variability in the far ultraviolet flux of
Be stars.**
G. J. Peters, R. S. Polidan.
Physics of Be stars, p. 278 – 281 (1987). – See Abstr. 012.030
(IAU Colloq. No. 92).
  To investigate the nature of the photometric variability in
Be stars, the authors have been monitoring the FUV flux of
selected objects with the ultraviolet spectrometers on the
Voyager 1 and 2 spacecrafts, and report the initial results from
these observations in this paper.

112.101 **Short–term photometric variability in Be stars in the
far–ultraviolet – a preliminary report.**
R. S. Polidan, G. J. Peters.
Physics of Be stars, p. 282 – 285 (1987). – See Abstr. 012.030
(IAU Colloq. No. 92).
  In this paper the authors report preliminary results on the first
search for short term photometric variations in the far–UV in
Be stars. Results for μ Cen and 28 Cyg indicate no significant
variations.

112.102 **Ubiquitous C IV 1550 variability in B and Be stars.**
G. Sonneborn, M. P. Garhart, C. A. Grady.
Physics of Be stars, p. 286 – 288 (1987). – See Abstr. 012.030
(IAU Colloq. No. 92).
  Studies of line profile variability of the ultraviolet 1550 Å
resonance transitions of C IV in Be stars have prompted an
investigation into the short– and long–term behavior of the C IV
lines in other types of B stars. The authors present examples of
two well–studied Be stars, Omega Orionis and 66 Ophiuchi, and
two non–Be stars, Beta Cephei and the standard star Zeta
Cassiopeiae.

112.103 **HD 166596: a silicon star with Hα emission?**
R. Hirata, D. N. Dawanas, G. Jasniewicz.
Physics of Be stars, p. 289 – 290 (1987). – See Abstr. 012.030
(IAU Colloq. No. 92).

112.104 **Be stars as nonradial pulsators.**
D. Baade.
Physics of Be stars, p. 361 – 383 (1987). – See Abstr. 012.030
(IAU Colloq. No. 92).
  The observational status of the ubiquitous rapid variability of
Be stars is summarized. The most comprehensive interpretation
is obtained with a traveling velocity field and associated
temperature variations. But neither the available observations
nor theoretical predictions presently allow an unambiguous
mode determination of these nonradial pulsations. In addition to
rapid rotation, an NRP mode of low azimuthal order, $m \approx 2$,
seems another prerequisite for a B star becoming a Be star. The
amplitudes of these modes are variable and have been observed
to decrease with some delay after an outburst. Mechanisms for
the transfer of energy to the atmosphere and its transformation
into the kinetic energy of an outburst are discussed.

112.105 **Ultraviolet observations of CX Draconis.**
P. Koubský, J. Horn, P. Harmanec, G. J. Peters,
R. S. Polidan, P. K. Barker.
Physics of Be stars, p. 443 – 444 (1987). – See Abstr. 012.030
(IAU Colloq. No. 92).

112.106 **o And photometry, polarimetry and a tentative model of
the light variability.**
P. Harmanec, K. Oláh, H. Božić, P. Hadrava, J. Horn,
P. Koubský, S. Kříž, N. H. Minikunov (*N. Kh. Minikulov*),
M. Muminović, J. R. Percy, A. G. Skcherbakov
(*A. G. Shcherbakov*), M. Stupar, A. E. Tarasov.
Physics of Be stars, p. 456 – 459 (1987). – See Abstr. 012.030
(IAU Colloq. No. 92).

112.107 **Nonradial pulsations and the Be phenomenon.**
G. D. Penrod.
Physics of Be stars, p. 463 (1987). Abstract. – See Abstr. 012.030
(IAU Colloq. No. 92).

112.108   Spectral transients in the line profiles of λ Eridani.
M. A. Smith, D. R. Gies, G. D. Penrod.
Physics of Be stars, p. 464 – 467 (1987). – See Abstr. 012.030 (IAU Colloq. No. 92).
The authors describe a number of transient spectral features that they have observed in the He I λ6678 line profile of the Be star λ Eri, that are qualitatively different from the travelling bump pattern of a single mode or from beating between two well defined modes.

112.109   Galactic distribution, kinematics, locations in clusters and H–R diagrams, and duplicity of Be stars.
H. A. Abt.
Physics of Be stars, p. 470 – 485 (1987). – See Abstr. 012.030 (IAU Colloq. No. 92).

112.110   Synthetic uvby–β photometry of HD 12856 and HD 13890.
A. V. Torres, C. D. Garmany.
Physics of Be stars, p. 519 – 522 (1987). – See Abstr. 012.030 (IAU Colloq. No. 92).
Strömgren and Hβ colors have been measured from spectrophotometric observations of two Be stars without published photometry in Per OB1: HD 12856 (B0 pe) and HD 13890 (B1 III:pe). Stellar parameters and improved spectral types are then derived from the color indices and the calibrations of Jakobsen (1986). These are compared with the parameters of normal B stars and they are used to estimate the evolutionary status of the stars.

112.111   Spectral variability of Beta Pictoris and the search for nearby proto–planetary systems.
F. C. Bruhweiler, Y. Kondo.
Physics of Be stars, p. 526 – 528 (1987). – See Abstr. 012.030 (IAU Colloq. No. 92).

112.112   A photometric study of Herbig Ae/Be stars.
E. E. Mendoza V.
Physics of Be stars, p. 529 – 531 (1987). – See Abstr. 012.030 (IAU Colloq. No. 92).
The author has selected the $\alpha(16)\Lambda(9)$ (Mendoza, 1979), and the JHKL–photometric systems, in order to study some photometric properties of Be stars.

112.113   Reflections on Be stars and the Be phenomenon.
M. J. Plavec.
Physics of Be stars, p. 553 – 557 (1987). – See Abstr. 012.030 (IAU Colloq. No. 92).

112.114   Spectroscopy of circumstellar shells.
B. Zuckerman.
Spectroscopy of astrophysical plasmas, p. 185 – 209 (1987). – See Abstr. 003.010.
Contents: 1. Overview. 2. Observations of gas and dust grains: Ultraviolet. Optical. Infrared. Microwave. 3. Analysis and interpretation: Physical conditions. Chemical composition. 4. Some near future developments.

112.115   The line Hα in chromospheric diagnostics.
T. A. Kipper.
Astrophysics, Vol. 26, No. 1, p. 35 – 39 (1987). English translation of 43.112.016.

112.116   Hot carbon stars: more about V348 Sgr.
L. Houziaux, P. Bouchet, A. Heck, J. Manfroid.
Q. J. R. Astron. Soc., Vol. 28, No. 3, p. 231 – 238 (1987). – See Abstr. 012.037.
New observational techniques (sensitive receivers, access to ultraviolet and infrared spectroscopy), as well as recent theoretical investigations on advanced stages of stellar evolution focus attention again on hydrogen–poor and helium–rich objects known as "carbon stars". The case of the hot peculiar variable V348 Sgr is discussed. New visible and infrared observations are presented, but the exact nature of the star remains much of an enigma.

112.117   Giant grains around protostars.
M. E. Bailey.
Q. J. R. Astron. Soc., Vol. 28, No. 3, p. 242 – 247 (1987). – See Abstr. 012.037.
Powerful stellar winds and outflows from young stars produce dense decelerating shells which are an excellent environment for the growth of large interstellar grains by coagulation. Observations of dense clouds around star–forming regions should be interpreted within a framework allowing the possiblity that such regions may contain an anomalous population of large interstellar grains.

112.118   Variable dust emission from Wolf–Rayet stars.
P. M. Williams, K. A. van der Hucht, P. S. Thé.
Q. J. R. Astron. Soc., Vol. 28, No. 3, p. 248 – 253 (1987). – See Abstr. 012.037.
An infrared photometric survey of late type Wolf–Rayet stars shows little evidence for secular variation in mass loss or dust formation rates apart from the three known variables: WR 48a (Danks' WC 9 star), WR 137 (HD 192641) and WR 140 (HD 193793, the prototype). In 1985, WR 140 again brightened sharply in the infrared owing to the formation of circumstellar dust. The relation of this to the previous dust condensation episode strongly supported suggestions that these episodes were periodic ($P = 7.9$ yr). This leads to confirmation that the system is a binary, and provides a framework for the interpretation of the radio and X–ray variations.

112.119   On the output of energy, gas and dust by Wolf–Rayet stellar winds.
K. A. van der Hucht, P. M. Williams, P. S. Thé.
Q. J. R. Astron. Soc., Vol. 28, No. 3, p. 254 – 260 (1987). – See Abstr. 012.037.
The influence of Wolf–Rayet stars on the ambient interstellar medium and on heavy element enrichment of the Galaxy is reviewed. Their strong stellar winds provide energy, gas and dust in huge quantities, and can explain abundances and isotope ratios found in supernovae, cosmic rays and meteorites.

112.120   IRAS observations of Be stars. II. Far–IR characteristics and mass loss rates.
L. B. F. M. Waters, J. Coté, H. J. G. L. M. Lamers.
Astron. Astrophys., Vol. 185, No. 1/2, p. 206 – 224 (1987).
Observations of bright, classical Be stars detected at 12, 25 and 60 μ by the Infrared Astronomical Satellite are presented. The far–IR energy distribution of the Be stars is studied in terms of the slope of the spectrum and the IR excess as a function of wavelength. The far–IR colours of Be stars indicate the presence of fully ionized circumstellar material, and no indication of dust. The IR excess as a function of wavelength is interpreted in terms of a simple equatorial disc model. The analysis yields information about the density structure of the circumstellar discs.

112.121   Chromospheric Mg II h and k emissions free of interstellar contamination: velocity structure in late–type dwarfs and giants.
G. Vladilo, P. Molaro, L. Crivellari, B. H. Foing, J. E. Beckman, R. Genova.
Astron. Astrophys., Vol. 185, No. 1/2, p. 233 – 246 (1987).
The authors have used high resolution IUE spectra from their own studies and from the archive to examine the Mg II h and k chromospheric emission cores of a sample of late–type dwarfs and giants. Sharp photospheric absorptions were used to provide a velocity rest–frame with respect to each stellar photosphere with the IUE–limited precision of ± 4 km s^{-1}. The knowledge of the kinematics of the local interstellar medium (LISM) could then be used to identify cases where either the cores or the wings, or in best circumstances both features of the chromospheric lines were uncontaminated by LISM absorption. The authors derive, using

only LISM–free emission wings, accurate Wilson–Bappu relations for both the *h* and *k* line, characterized by a slope higher than in previous determinations.

112.122 The Beta Pictoris circumstellar disk. V. Time variations of the Ca II–K line.

R. Ferlet, L. M. Hobbs, A. Vidal–Madjar.
Astron. Astrophys., Vol. 185, No. 1/2, p. 267 – 270 (1987).

Observations of the Ca II–K line profile in *β* Pictoris (A5 V) at different epochs show drastic changes on time scales of months, days and hours. At the center of the stellar line, a relatively stable narrow absorption feature is present while, on its red wing, absorptions can appear and disappear. These changes could be related either to a "classical" shell very close to the star or to the much more extended circumstellar dust disk present around *β* Pic. In this last case, they may be related to cometary–like objects falling on the star.

112.123 Dust shells around Miras and OH/IR stars: interpretation of IRAS and other infrared measurements.

P. J. Bedijn.
Astron. Astrophys., Vol. 186, No. 1/2, p. 136 – 152 (1987).

It is shown that the locus of optical M–type Miras and OH/IR variables in the IRAS two–color diagram can be modelled by a sequence of single–shell dust shells of varying optical depth (mass loss rate) with inner radii such that the dust temperature at the inner boundary is the same for all models. The dependence of a theoretical two–color sequence on model parameters is investigated. Also, fits to the spectrum of OH 26.5 + 0.6 are given which might indicate an observable increase of the mass loss rate for this source during the past evolution. It is demonstrated that the position of the non–variable sources in the IRAS two–color diagram can be understood in terms of a sequence of dust shells, whose inner radii are progressively larger than those of the models mentioned above. It is suggested that these sources therefore represent an evolutionary phase shortly after the cessation of pulsation and of mass loss of the central star and before the formation of a planetary nebula. A model fit to the spectrum of OH 17.7–2.0 indicates that the mass loss does not stop "abruptly".

112.124 Additional constraints on cool–disk models of Be stars based on long observational sequences in the visual range.

A. M. Hubert, M. Floquet, M. T. Chambon.
Astron. Astrophys., Vol. 186, No. 1/2, p. 213 – 222 (1987).

The authors have examined the main common spectral properties of the 3 well–known early type ($\sim B2$–$B4$) V/R variable shell stars EW Lac, ζ Tau, 48 Lib, before and during their variable shell phase characterized by long RV (radial velocity) and V/R (intensity ratio of violet to red emission) cycles. They then recall the two current cool envelope models which could explain the V/R or R/V cycles (in particular the $V/R > 1$ phase or positive R/V phase): (1) the non–axisymmetrical elliptical disk, and (2) the rotating–pulsating axisymmetrical envelope (or disk) model. The authors present the spectroscopic observations deduced from their comparison of the 3 V/R variable shell stars EW Lac, ζ Tau, 48 Lib which are in favour of or against these two models. They conclude that the 3 stars present signatures of both a geometrical model (elongated disk with apsidal motion) and a physical model (variable radial flux).

112.125 An analysis of the emission features of the IRAS low–resolution spectra of carbon stars.

Y. Baron, M. de Muizon, R. Papoular, B. Pégourié.
Astron. Astrophys., Vol. 186, No. 1/2, p. 271 – 279 (1987).

The Low–Resolution Spectra (LRS) data in class 4n reveal a considerable wealth of features and a remarkable variety of behaviours of the mid–IR spectra of C–stars. The main thrust of this work is on the SiC feature at 11.3 μm, whose shape and intensity were studied in detail. The feature becomes stronger and narrower as the temperature of the underlying continuum increases. The narrowest features are quite similar in shape to the mass absorption spectrum of laboratory α–SiC. A new weak feature at 11.7 μm may be due to solid SiC_2 or a–C:H. When the whole range (8 – 23 μm) of the LRS is considered, evidence is found for crystalline as well as amorphous carbon in the emitting dust; the proportion of the former with respect to the latter increases with the dust temperature. The "unidentified" 7.7 and 8.6 μm emission bands are also detected at the highest temperatures. A large number of weak SiC features are also found in LRS class 1n; the proportion of C–rich to O–rich giants in the LRS is hence increased to an estimated 60%.

112.126 Highly ionized stellar winds in Be stars: the evidence for aspect dependence.

C. A. Grady, K. S. Bjorkman, T. P. Snow.
Astrophys. J., Vol. 320, No. 1, p. 376 – 397 (1987).

The authors present the results of an ultraviolet survey of stellar winds in 62 Be and 43 normal B stars covering spectral types B0.5 – B5 and luminosity classes V – III. They find that the wind absorption seen in the resonance lines of C IV, Si IV, and Si III in Be stars is often the result of blended absorption from multiple shortward–shifted discrete absorption components. There is evidence of a threshold in $v \sin i$ for the presence of strong and highly variable winds in Be stars. Wind absorption in the form of shortward–shifted discrete components is observed only for Be stars having $v \sin i \geqslant 150$ km s^{-1}. The recent discovery that many, if not all, Be stars appear to be nonradial pulsators, and the observation that the pulsation characteristics of these stars change at $v \sin i = 150$ km s^{-1} suggests that nonradial pulsation may be important in generating the Be phenomenon. The authors also present evidence that the high–velocity and highly ionized stellar wind observed in some, but not all, moderate to high $v \sin i$ Be stars is a function of latitude.

112.127 Diffraction–limited imaging of Alpha Orionis.

E. K. Hege, J. C. Christou, J. C. Hebden, A. Eckart.
Bull. Am. Astron. Soc., Vol. 19, No. 2, p. 702 (1987). Abstract. – See Abstr. 010.061.

112.128 Hα as a tracer of stellar mass loss in metal–poor galaxies.

C. Leitherer.
Bull. Am. Astron. Soc., Vol. 19, No. 2, p. 702 – 703 (1987). Abstract. – See Abstr. 010.061.

112.129 An analysis of X–ray fluxes of O–type stars.

T. Chlebowski.
Bull. Am. Astron. Soc., Vol. 19, No. 2, p. 703 (1987). Abstract. – See Abstr. 010.061.

112.130 Stellar winds from O stars in the Small Magellanic Cloud.

C. D. Garmany, E. L. Fitzpatrick.
Bull. Am. Astron. Soc., Vol. 19, No. 2, p. 703 – 704 (1987). Abstract. – See Abstr. 010.061.

112.131 Pleione shell phase fading fast.

R. F. Garrison.
Bull. Am. Astron. Soc., Vol. 19, No. 2, p. 704 – 705 (1987). Abstract. – See Abstr. 010.061.

112.132 VLA observations of 1612 and 1665 MHz OH masers in the circumstellar envelope of U Orionis.

P. F. Bowers, K. J. Johnston.
Bull. Am. Astron. Soc., Vol. 19, No. 2, p. 705 – 706 (1987). Abstract. – See Abstr. 010.061.

112.133 Circumstellar disks and mass outflows associated with young stars.

S. E. Strom.
Bull. Am. Astron. Soc., Vol. 19, No. 2, p. 710 – 711 (1987). Abstract. – See Abstr. 010.061.

112.134 Red supergiant infrared shells and the galactic metallicity gradient.
J. E. Pesce, R. E. Stencel, W. Hagen Bauer.
Bull. Am. Astron. Soc., Vol. 19, No. 2, p. 724 (1987). Abstract. – See Abstr. 010.061.

112.135 Emission from SiO, SiS, and HC₃N in IRC + 10216: the appearance and disappearance of molecules in the outflow.
L. Likkel, M. Morris, C. Masson, A. Wootten.
Bull. Am. Astron. Soc., Vol. 19, No. 2, p. 755 (1987). Abstract. – See Abstr. 010.061.

112.136 Infrared spectroscopy of radio luminous OH/IR stars.
T. J. Jones, A. R. Hyland, J. D. Fix, M. L. Cobb.
Bull. Am. Astron. Soc., Vol. 19, No. 2, p. 760 (1987). Abstract. – See Abstr. 010.061.

112.137 IRAS spectroscopy of carbon stars.
J. H. Goebel.
Bull. Am. Astron. Soc., Vol. 19, No. 2, p. 760 (1987). Abstract. – See Abstr. 010.061.

112.138 Highly ionized stellar winds in Be stars: C IV in B6 – B9.5e stars.
C. A. Grady, K. S. Bjorkman, G. Sonneborn, S. N. Shore, T. P. Snow.
Bull. Am. Astron. Soc., Vol. 19, No. 2, p. 761 – 762 (1987). Abstract. – See Abstr. 010.061.

112.139 The 1985 – 1986 spectroscopic observations of γ Cassiopeiae. Analysis of H_α and H_β emission lines.
T. S. Galkina.
Izv. Krymskoj Astrofiz. Obs., Tom 77, p. 115 – 125 (1987). In Russian. English translation in Bull. Crimean Astrophys. Obs., Vol. 77.

The results of H_α and H_β-emission line observations in the spectrum of the Be star γ Cas obtained from September 1985 till February 1986 are presented. The H_α and H_β emission line profiles reveal variations of the multicomponential structure in September – November 1985 to a double–component emission (by H_β) and an one–component (by H_α). Different parameters of the H_α and H_β emission lines are defined.

112.140 Variations in the envelope of the shell star HD 50845.
J. Sahade, A. E. Ringuelet, N. Rotstein.
Publ. Astron. Soc. Pac., Vol. 99, No. 619, p. 971 – 974 (1987).

Spectrographic observations of the shell star HD 50845 made in December 1984 have disclosed variations in the appearance of Hα and a decrease in the intensity of the strongest metastable and resonance lines of Fe I and Ca I, with respect to observations made with the same equipment ten months earlier. The underlying star seems to be a velocity variable F5 V object.

112.141 Detection of X–ray emission from the young low–mass star Rossiter 137B.
O. Vilhu, J. L. Linsky.
Publ. Astron. Soc. Pac., Vol. 99, No. 620, p. 1071 – 1075 (1987).

The authors have detected Rst 137B, a close M dwarf companion to the active K star HD 36705, from a High Resolution Image in the Einstein Archive. The X–ray surface fluxes (0.2 – 4 keV) from both stars are close to the empirical saturation level, $F_X/F_{bol} \approx 10^{-3}$, defined by rapid rotators and very young stars. This supports the earlier results of the youthfulness of the system. Rst 137B is one of the latest spectral types and thus lowest–mass pre–main–sequence stars yet detected as an X–ray source.

112.142 Dust disks around low mass main sequence stars.
R. D. Wolstencroft, H. J. Walker.
Edinb. Astron. Prepr., No. 19/87, 25 pp. (1987). To appear in Philos. Trans. R. Soc. London, Ser. A.

112.143 Time–resolved high–resolution spectroscopy of an Hα outburst of μ Cen (B2 IV Ve).
D. Baade, J. Dachs, R. van de Weygaert, F. Steeman.
ESO Sci. Prepr., No. 551, 35 pp. (1987). To appear in Astron. Astrophys.

112.144 CO observations of IRAS sources with 11.3 micron silicon carbide dust features.
D. A. Leahy, S. Kwok, R. A. Arquilla.
Astrophys. J., Vol. 320, No. 2, p. 825 – 841 (1987). = Publ. Rothney Astrophys. Obs., No. 44.

The authors report the detection of CO emission from seven late–type stars selected from the sources observed by the IRAS Low Resolution Spectrometer to have 11 μm SiC features. All these sources have no existing catalog associations and are typically 10 times weaker at 12 μm than previously detected radio carbon stars. They are probably excellent candidates for hitherto unidentified carbon stars. Comparison of these objects with IRC + 10216 suggests that the distance to IRC + 10216 may have been overestimated in the past.

112.145 Time variability of Gamma Cassiopeiae in X–rays.
F. Frontera, D. Dal Fiume, N. R. Robba, G. Manzo, S. Re, E. Costa.
Astrophys. J., Lett. Ed., Vol. 320, No. 2, p. L127 – L131 (1987).

A 22,000 s observation with EXOSAT of the X–ray source 4U 0053 + 60/Gamma Cassiopeiae reveals oscillations in the X–ray flux with a period of about 6000 s. The source also exhibits chaotic variability with time scales down to a few tens of seconds. This observation strengthens the hypothesis that the X–ray source is a neutron star in a wide orbit around the Be star γ Cas.

112.146 The Beta Pictoris circumstellar disk. VI. Evidence for material falling to the star.
A. M. Lagrange–Henri, A. Vidal–Madjar, R. Ferlet.
Inst. Astrophys. Paris, Pré–Publ., No. 193, 31 pp. (1987). To appear in Astron. Astrophys.

112.147 Spectropolarimetry as a probe of the structure of the envelopes of Wolf–Rayet stars.
G. D. Schmidt.
Prepr. Steward Obs., No. 770, 14 pp. (1987). To appear in the proceedings of the Vatican workshop on circumstellar polarization, June 1987.

112.148 Spectroscopy of southern Be stars 1984 – 1987.
R. I. Mennickent, N. Vogt.
Astrophys. Prepr. Ser., No. 18, 37 pp. (1987). To appear in Astron. Astrophys., Suppl. Ser.

112.149 Spectrometric patrol of the Hα and Hβ emission lines in the spectrum of the Be star φ Per.
A. Kh. Mamatkazina.
Tr. Astrofiz. Inst. Alma–Ata, Tom 48, p. 67 – 75 (1987). In Russian.

112.150 Spectroscopy and photometry of some intrinsic B variable stars.
M. Alvarez, D. Ballereau, J. P. Sareyan, J. Chauville, R. Michel, J. M. Le Contel.
Rev. Mex. Astron. Astrofis., Vol. 14, No. 1, p. 315 – 322 (1987). – See Abstr. 012.042.

From recent studies of the variability of Be stars the authors report some preliminary results found.

112.151 Spectroscopic monitoring of southern Be stars.
L. H. Barrera, N. Vogt.
Rev. Mex. Astron. Astrofis., Vol. 14, No. 1, p. 323 – 329 (1987). – See Abstr. 012.042.

The authors present first results from their long–term spectroscopic observations of Be stars. 36 bright Be stars (V ≲ 6ᵐ) are regularly being monitored for variations in radial velocity, line profiles and equivalent widths (3900 – 5100 Å, 20 Å/mm at Hγ).

A spectrophotometric analysis of 35 spectrograms of μ Cen (HR 5193 = HD 120324) is discussed in detail.

112.152   Photometric observations of the Be/X–ray binary system 2S0114 + 650 = LSI + 65°010.
L. Corral, G. Koenigsberger.
Rev. Mex. Astron. Astrofis., Vol. 14, No. 1, p. 330 – 335 (1987).
– See Abstr. 012.042.
An interpretation for the X-ray variability of the Be/X–ray source 2S0114 + 650 is presented in the context of preliminary optical photometric observations and the orbital solution for the binary system derived by Crampton et al. (1985).

112.153   Changes in the Be stars in a Cassiopeia–Perseus region.
B. Iriarte, E. Chavira, O. Cardona.
Rev. Mex. Astron. Astrofis., Vol. 14, No. 1, p. 336 – 337 (1987).
– See Abstr. 012.042.
The authors started a new survey similar to the one carried out in Tonantzintla in 1952 on Be stars, with the purpose to find changes in thirty years. They obtained the material with the Tonantzintla Schmidt camera. They found changes of around 13% in the Cassiopeia stars, and 9% in the Perseus area in the intensity of Hα.

112.154   An observational study of the Herbig Ae star VV Serpentis and R–stars associated with its dark cloud.
C. Chavarría–K., J. Ocegueda, E. de Lara, U. Finkenzeller, E. E. Mendoza.
Rev. Mex. Astron. Astrofis., Vol. 14, No. 1, p. 338 (1987). Abstract. – See Abstr. 012.042.

112.155   Observaciones con IUE e IRAS de gigantes y supergigantes rojas.
W. Hagen, K. G. Carpenter, R. E. Stencel.
Rev. Mex. Astron. Astrofis., Vol. 14, No. 1, p. 380 (1987). Abstract. – See Abstr. 012.042.

112.156   On the masses of the circumstellar dust disks around the YSO's R Mon and R50.
L. Carrasco, I. Cruz–González, G. L. Grasdalen.
Rev. Mex. Astron. Astrofis., Vol. 14, No. 2, p. 518 (1987). Abstract. – See Abstr. 012.042.

112.157   The ultraviolet spectra of cool star chromospheres; excitation processes and plasma diagnostics.
C. Jordan.
Phys. Scr., Vol. T17, p. 104 – 108 (1987). – See Abstr. 012.045.
The advent of ultraviolet observations from space has led to rapid advances in our knowledge of the conditions in the chromospheres which surround cool stars. In contrast to the solar chromosphere and corona, where electron temperatures reach $T_e > 10^6$K, observations with IUE have shown that stars with low surface gravities have only cool chromospheres with $T_e \lesssim 2 \times 10^4$K. The gas density regime is also lower and the excitation of many of the ultraviolet lines is by radiative processes rather than by ion/electron collisions as in the sun. In particular, higher opacities lead to multiple photon scattering and line leakage in semi-forbidden transitions. Quantitative analyses of such spectra are at an early stage and a wide variety of atomic rate coefficients need to be known more accurately.

112.158   Betelgeuse.
D. H. Levy, P. Jedicke.
Astronomy, Vol. 15, No. 4, 1 pp. (1987). Abstr. in Phys. Abstr., Vol. 90, No. 1311, Entry 101776 (1987).

112.159   Observational phenomena of protoplanetary discs around young stars.
G. M. Rudnitskij.
Astron. Vestn., Tom 21, No. 4, p. 311 – 313 (1987). In Russian. – See Abstr. 012.052.

112.160   Fast transient X–rays from flare stars and RS CVn binaries.
A. R. Rao, M. N. Vahia.
Astron. Astrophys., Vol. 188, No. 1, p. 109 – 113 (1987).
The authors have studied the fast transient X-ray (FTX) observations of the Ariel V satellite. They find that the FTX have characteristics very similar to the stellar flares detected in flare stars and RS CVn binaries by other satellites. It is found that, of the possible candidate objects, only the flare stars and RS CVn binaries can be associated with the Ariel V observations. 11 new flare stars and RS CVn binaries are associated with the FTX. This brings the total number of identifications with the flare stars and RS CVn binaries to 17. The authors further study the flare properties and correlate the peak X–ray luminosity of these Ariel V sources with the bolometric luminosity of the candidate stars. They discuss a solar flare model and show that the observed correlation can be explained under the assumption of constant temperature loops of binary sizes.

112.161   Amorphous carbon around carbon stars.
V. Orofino, L. Colangeli, E. Bussoletti, F. Strafella.
Astrophys. Space Sci., Vol. 138, No. 1, p. 127 – 140 (1987).
The authors present the results of a simplified model to determine the flux emerging from dust envelopes around cool stars. The model proposed holds under the hypotheses of negligible scattering effects and spherical geometry of the dust cloud. The aim of this work is to compare the effects of a graphitic or amorphous composition of the carbon grains in the envelopes. To do this the authors have used, for the first time, experimental extinction data obtained in the laboratory for submicron amorphous carbon particles. The model has been used to fit the FIR spectral trend of 78 optically thin sources and to reproduce the full spectra of two of the most IR luminous optically–thick sources: CIT 6 and IRC + 10216.

112.162   P Cygni: ein Überriese, den man beobachten sollte.
M. de Groot.
BAV Rundbrief, 36. Jahrg., Nr. 4, p. 196 – 201 (1987).

112.163   Hα observations of supergiant stars: R Puppis and ϱ Cassiopeiae.
J. R. Sowell, D. J. Bord.
Observatory, Vol. 107, No. 1081, p. 259 – 263 (1987).

112.164   IRAS observations of the giant shell surrounding λ Orionis.
D. Malone, B. McBreen, G. G. Fazio.
Ir. Astron. J., Vol. 18, No. 2, p. 91 – 94 (1987). – See Abstr. 012.036.
A large shell of far infrared emission surrounds the star λ Orionis. The shell has fragmented into a number of regions of enhanced emission. It has been shown that for some of these sources internal heat sources are required to explain the observed luminosity. The mechanism whereby the expanding shell has fragmented into clumps of about equal size and separation can be accounted for by a Rayleigh–Taylor instability.

112.165   Émission maser de la molécule SiO provenant des variables à longue période.
V. Bujarrabal.
Bull. Assoc. Fr. Obs. Etoiles Variables, No. 42, p. 12 – 17 (1987).

112.166   Stellar wind – observations and theories. Part III. Mass outflow mechanisms from late type giants and supergiant stars.
M. Krogulec.
Postepy Astron., Tom 34, Zesz. 4, p. 269 – 290 (1986). In Polish.
Lα radiation pressure, absorption of the photospheric radiation by circumstellar dust grains and Alfvén waves pressure as possible mechanisms of mass outflow from cool giants and supergiant stars are presented.

112.167   Evidence for strong magnetic fields in the inner envelopes of late–type stars.
R. Barvainis, G. McIntosh, C. R. Predmore.
Nature, Vol. 329, No. 6140, p. 613 – 615 (1987).

The authors report sensitive measurements of circular polarization in SiO masers associated with late–type stars, which reveal significantly polarized features at the several per cent level in five of six stars observed. Under the Zeeman splitting hypothesis, these measurements imply magnetic fields in the masing gas in the range 10 – 100 G. The masers associated with IRc 2 in the Orion–KL region were not detected in circular polarization.

112.168   X–ray emission and structure of coronae of active late–type dwarfs.
M. M. Katsova, O. G. Badalyan, M. A. Livshits.
Astron. Zh., Tom 64, Vyp. 6, p. 1243 – 1255 (1987). In Russian.
English translation in Sov. Astron., Vol. 31, No. 6.

A method to determine the density in stellar coronae from X-ray data is suggested. Its application to 45 G, K, and M main-sequence stars shows that the density at the corona base increases by the transition from G stars to M0 – M3 stars, and drops thereafter. The rising–density branch is connected with the intensification of solar–type activity processes on G – K stars. The properties of the X–ray emission of active M3 – M6 stars, its rapid variability and its association with flare activity and with nonthermal radio emission apparently evidences that heating of the coronae of these stars is due to numerous processes of the subflare type.

112.169   The Na I lines as an indicator of chromospheres of late–type stars.
N. A. Sakhibullin.
Astron. Zh., Tom 64, Vyp. 6, p. 1269 – 1279 (1987). In Russian.
English translation in Sov. Astron., Vol. 31, No. 6.

This paper presents results of a theoretical analysis of the Na I line formation to determine the possible existence of chromospheres of late–type stars. The method of the analysis is based on the solution of the equation of radiative transfer coupled to statistical equilibrium equations. The analysis is performed for stars with effective temperatures in the range 6500 – 3300K. It is shown that Na I resonance lines are collisionally dominated, so they may give information about the temperature distribution near the temperature minimum for the stars later than K0.

112.170   V348 Sagittarii.
IAU Circ., No. 4432 (1987).

112.171   Que se passe–t–il autour de l'étoile Bêta Pictoris?
A.–M. Lagrange–Henri, R. Ferlet, A. Vidal–Madjar.
Astronomie, Vol. 101, p. 581 – 587 (1987).

112.172   The rich molecular spectrum and the rapid outflow of OH 231.8 + 4.2.
M. Morris, S. Guilloteau, R. Lucas, A. Omont.
Astrophys. J., Vol. 321, No. 2, p. 888 – 897 (1987).

Emission lines of eight molecules were observed in the circumstellar outflow of the bipolar nebula OH 231.8 + 4.2. Five of these molecules, including HNC, CS, and NH_3, were previously undetected in this object, and two of these – HCO^+ and OCS – are reported for the first time in any post–main-sequence circumstellar outflow. Of the 13 molecules now known in this outflow, five contain sulfur. A discussion of the circumstellar chemistry in OH 231.8 + 4.2 is presented. Shocks may be important for producing the sulfur molecules with their observed abundances. Line profiles are used to discuss the kinematics of the outflow. An examination of the latitude dependence of the outflow velocity and mass–loss rate is presented.

112.173   Extinction and abundance properties of Alpha Scorpii circumstellar grains.
T. P. Snow Jr., R. H. Buss Jr., D. P. Gilra, J. P. Swings.
Astrophys. J., Vol. 321, No. 2, p. 921 – 936 (1987).

The authors have investigated the ultraviolet extinction and element depletions in the circumstellar envelope of the M supergiant binary system, α Sco, by using IUE, Copernicus, and ANS data of the B dwarf companion, α Sco B, which can be seen in absorption through the cool–star envelope. After assessing the interstellar, circumstellar, and atmospheric column densities toward α Sco B, the authors estimated the circumstellar reddening, the circumstellar depletions, and the circumstellar spectral extinction toward α Sco B. It is found that the circumstellar grains in this oxygen–rich environment are larger than 0.05 μm and siliceous in composition, yet contain less depleted elements than the diffuse–interstellar grains in the Scorpius region or the general Galactic diffuse–interstellar grains.

112.174   Airborne spectrophotometry of Eta Carinae from 4.5 to 7.5 microns and a model for source morphology.
R. W. Russell, D. K. Lynch, J. A. Hackwell, R. J. Rudy, G. S. Rossano, M. W. Castelaz.
Astrophys. J., Vol. 321, No. 2, p. 937 – 942 (1987).

Spectrophotometric observations of η Car between 4.5 and 7.5 μm show a featureless thermal–like spectrum with no fine-structure lines or broad emission or absorption features. The color temperature of the spectrum is approximately 375K. Silicate emission at 10.5 μm is observed which exceeds the extrapolated gray–body curve by a factor of about 2. High spatial resolution maps at 3.5, 4.8, and 10 μm obtained from the ground are used to discuss the dust distribution and temperature structure, and to present a model for general source morphology.

112.175   Possible dishevelled structure of the atmospheres of WR stars.
I. I. Antokhin, A. F. Kholtygin, A. M. Cherepashchuk.
Astron. Tsirk., No. 1460, p. 1 – 3 (1986). In Russian.

Ionization state calculations of WR atmospheres have been carried out which demonstrate that the simultaneous co-existence of He^+ and He^{++} (and other) zones of ionization is possible only in a very narrow range of T_{eff}: $\Delta T_{eff} \approx 2000$K. This result is in contradiction with observations. The contradiction can be removed assuming a dishevelled structure of WR atmospheres consisting of dense "clouds" and low–density "intercloud" matter.

112.176   Evidence for the effect of the interstellar far UV–radiation on circumstellar shells.
M. Szymczak.
Astrophys. Space Sci., Vol. 139, No. 1, p. 63 – 73 (1987).

The interstellar UV field at 1565 Å is calculated around nearby OH/IR sources. The front–back asymmetry observed in the 1612 MHz maser line profile is well correlated with anisotropy of the interstellar UV flux. For some sources the spatial positions of the 1612 MHz masers are confined to the position angles for which stronger UV radiation occurs. These facts strongly support the theory of the photoproduction of OH from H_2O induced by ambient interstellar UV photons penetrating the circumstellar shell. A simple model of the 1612 MHz maser with OH photoproduction suggests that the influence of the UV field on the observed maser profiles is governed by the mass loss rate and the relative abundances of OH and H_2O molecules.

112.177   Circumstellar disks and rings – observational results.
H. Elsässer.
Astron. Nachr., Vol. 308, No. 5, p. 285 – 291 (1987).

Recent advances in the observation of star–forming regions at visual, infrared and radio wavelengths have demonstrated that disks or rings of dust and molecules are suspected or even seen to exist around newly borne stars. They might be the prestage of a planetary system. Typical objects of this kind which are discussed in some detail are the bipolar nebula S 106 and the molecular cloud surrounding it as well as the CO outflow source in the dark cloud L 1551 with its central infrared star IRS 5.

112.178   Arecibo OH survey of OH/IR stars identified by IRAS colors.
J. Eder, B. M. Lewis, Y. Terzian.
Publ. Astron. Soc. Pac., Vol. 99, No. 621, p. 1147 (1987).
Abstract. – See Abstr. 010.281.

112.179   Red supergiant infrared shells and the galactic metallicity gradient.
R. E. Stencel, J. E. Pesce, W. H. Bauer.
Publ. Astron. Soc. Pac., Vol. 99, No. 621, p. 1150 (1987).
Abstract. – See Abstr. 010.281.

112.180   Carbon grains in the envelope of IRC + 10216.
P. G. Martin, C. Rogers.
Astrophys. J., Vol. 322, No. 1, p. 374 – 392 (1987).
IRC + 10216 is a well–observed carbon star with copious mass loss, surrounded by an expanding envelope of dust and molecular gas. The authors have studied the dust component using radiative transfer calculations by the half–range moment method. The models are sensitive to the choice of grain properties, basically size and composition. Two types of carbon material were considered: "amorphous carbon" and graphite. Small particles of amorphous carbon (radius $\approx 0.05 - 0.1 \, \mu$m) seem most suitable. The authors conclude that graphite is not a major component of carbon–rich stardust. SiC is also not a major constituent material. These models provide information on the spectrum, the surface brightness, the polarization, and the relative importance of direct starlight, scattered light, and thermal emission. The authors assess the extent to which observations show that a spherically symmetric density distribution is an oversimplification.

112.181   Recurrent episodic mass loss in a B2e star: 66 Ophiuchi, 1982 – 1985.
C. A. Grady, G. Sonneborn, C.–C. Wu, H. F. Henrichs.
Astrophys. J., Suppl. Ser., Vol. 65, No. 4, p. 673 – 694 (1987).
High–dispersion IUE spectra of 66 Ophiuchi (B2 IV–Ve) have revealed the existence of recurrent, episodic variation in the highly ionized stellar wind. Significant profile variations are seen on time scales spanning over 2 orders of magnitude, ranging from small–amplitude changes detected on time scales of hours to transitions from minimal–wind absorption to strong–wind absorption, which take months. Comparison of the IUE data obtained during 1982 and published linear polarimetry suggests that little correlation exists between the highly ionized wind properties and linear polarization episodes.

112.182   Variable shell strength of Pleione (BU Tau).
P. S. Goraya, N. S. Tur, B. S. Rautela.
Inf. Bull. Variable Stars, No. 3052, 4 pp. (1987).

112.183   Recent onset of an outburst in μ Centauri.
K. K. Ghosh, C. Velu, K. Kuppuswamy,
K. Jaykumar, M. J. Rosario.
Inf. Bull. Variable Stars, No. 3056, 4 pp. (1987).

112.184   Hydrogen emissions on low–dispersion spectrograms of B–stars.
T. Radoslavova.
Inf. Bull. Variable Stars, No. 3074, 2 pp. (1987).

112.185   Rapid variation in Hα emission of γ Cassiopeiae.
S. C. Joshi, R. K. Srivastava, J. B. Srivastava.
Inf. Bull. Variable Stars, No. 3116, 4 pp. (1987).

112.186   On the continued immaculateness of the Be star Mu Centauri.
D. Baade.
Inf. Bull. Variable Stars, No. 3124, 6 pp. (1987).

112.187   Kinematics of the circumstellar gas of HL Tauri and R Monocerotis.
A. I. Sargent, S. Beckwith.
Astrophys. J., Vol. 323, No. 1, p. 294 – 305 (1987).
The authors present interferometric observations of ^{13}CO emission from HL Tau and R Mon. The data permit quantitative determinations of the column density, mass, and dimensions of the circumstellar gas around each object. Simultaneous measurements of the 2.7 mm continuum emission allow an investigation of the circumstellar dust at high spatial resolution, furnishing a secondary probe of the circumstellar mass. Kinematic analysis of the weak, extended, molecular emission detected around both objects is consistent with the hypothesis that the gas orbits the stars in bound disks.

112.188   Water maser emission toward V788 Cygni, a carbon star.
Y. Nakada, H. Izumiura, T. Onaka, O. Hashimoto, N. Ukita,
S. Deguchi, T. Tanabé.
Astrophys. J., Lett. Ed., Vol. 323, No. 1, p. L77 – L80 (1987).
Radio emission lines of H_2O, SiO, HCN, and CO have been searched in the circumstellar envelopes of carbon stars which exhibit the 10 μm silicate feature in the IRAS low–resolution spectra. Water maser emission at 22 GHz has been detected in V778 Cyg. No other emission has been detected in the carbon stars BM Gem, V778 Cyg, and EU And. Two competing interpretations for this curious association of the silicate feature and the water maser with the carbon star, the transient stage hypothesis and the binary one, are examined based upon the present observations. Five other carbon stars and two S stars were also observed for comparison.

112.189   Detection of H^{13}CN maser emission from the ground state toward Y Canum Venaticorum.
H. Izumiura, N. Ukita, R. Kawabe, N. Kaifu, T. Tsuji,
W. Unno, K. Koyama.
Astrophys. J., Lett. Ed., Vol. 323, No. 1, p. L81 – L85 (1987).
The authors have detected unusually strong molecular emission of the H^{13}CN $J = 1$–0 transition in the ground vibrational state, comparable to that of the H^{12}CN, from the carbon star Y Canum Venaticorum. The emission has a spike component with a full width at half–maximum of 1.7 km s^{-1}. The intensity of the spike has shown a significant time variation in only 1 month. These observational facts suggest that the spike component of the H^{13}CN emission is a maser emission.

112.190   Mass loss from evolved stars. VII. OH maser shell radii and mass–loss rates for OH/IR stars.
N. Netzer, G. R. Knapp.
Astrophys. J., Vol. 323, No. 2, p. 734 – 748 (1987).
The equations of photodissociation and radiative transfer are solved for the photochemical chain $H_2O \rightarrow OH + H \rightarrow O + H + H$ in the circumstellar envelopes produced by steady mass loss from cool evolved stars. The solution is performed for a large range of mass–loss rates, outflow velocities, and interstellar UV radiation field densities. Comparison with observations shows that, over a wide range of mass–loss rates, the radius at which the OH density peaks coincides with the radius defined by maser emission in the 1612 MHz line. This allows the use of these models, together with observations of OH maser shell radii and wind outflow velocities, to measure mass–loss rates. Mass–loss rates are derived for 42 stars.

112.191   High–excitation SiO masers in evolved stars.
P. R. Jewell, D. F. Dickinson, L. E. Snyder,
D. P. Clemens.
Astrophys. J., Vol. 323, No. 2, p. 749 – 755 (1987).
The authors report the detection of SiO maser lines from highly excited rotational states, including the $v = 1$, $J = 6$–5 transition in R Leo – the highest rotational transition detected to date. Maser emission from the $v = 1$, $J = 5$–4 transition appears to be fairly common in evolved stars, with detectability mainly determined by telescope sensitivity. VX Sgr and possibly

VY CMa show $v = 2$, $J = 5 - 4$ emission and $v = 3$, $J = 5 - 4$ emission may be present in VX Sgr. The observations of this study do not support theories advocating a cascade of photons down the entire SiO rotational ladder.

112.192   H₂O masers in circumstellar envelopes.

A. P. Lane, K. J. Johnston, P. F. Bowers,
J. H. Spencer, P. J. Diamond.
Astrophys. J., Vol. 323, No. 2, p. 756 – 765 (1987).

The spatial structure of H_2O maser emission at 22.235 GHz for 12 late–type variable stars has been measured with a resolution of 0".07. The H_2O maser regions of Mira variables range from 9 or less to 108 AU in total extent, while those of supergiant long–period variables are 300 – 720 AU in extent. The size of the H_2O maser regions of giants and supergiants is correlated with the stellar mass–loss rate. The Mira variable IK Tau exhibits ~ 31 maser features. The spatial distribution of these features appears to be constistent with two spherical shells of radii 25 and 59 AU, both expanding at a velocity of 15.5 km s^{-1} from the star.

112.193   Detection of a new Be star – γ Lup.

K. K. Ghosh, K. Kuppuswamy, K. Jaykumar,
M. J. Rosario, C. Velu.
Inf. Bull. Variable Stars, No. 3057, 3 pp. (1987).

Valinhos 2.2 micron survey of the southern galactic plane. II. Near–IR photometry, IRAS identifications and nature of the sources.
See Abstr. 002.029.

The M–type stars.
See Abstr. 003.075.

The High Altitude Observatory–Lowell Observatory Solar–Stellar Spectrophotometry Project.
See Abstr. 013.003.

Panel discussion of future research on Be stars.
See Abstr. 013.043.

Extinction coefficients of carbon grains in far–infrared.
See Abstr. 022.102.

Laboratory microwave spectroscopy of the vibrational satellites for the v_7 and $2v_7$ states of C_4H and their astronomical identification.
See Abstr. 022.152.

High spatial resolution imaging with linear arrays.
See Abstr. 034.163.

Near infrared spectroscopy with the array spectrometer of the Anglo–Australian telescope.
See Abstr. 034.166.

Recent astronomical results obtained with the AFGL ten micron Array Spectrometer.
See Abstr. 034.170.

Observational requirements for a synoptic spectroscopic study of nonradial pulsations in OB stars.
See Abstr. 036.002.

Fourier removal of stripe artifacts in *IRAS* images.
See Abstr. 036.055.

Infrared speckle interferometry on Calar Alto.
See Abstr. 036.119.

Imaging of low mass binary companions and circumstellar disks.
See Abstr. 036.120.

Two dimensional high angular resolution infrared imaging of circumstellar shells.
See Abstr. 036.121.

10 μm imaging of circumstellar dust.
See Abstr. 036.134.

Fourier removal of stripe artifacts in IRAS images.
See Abstr. 036.146.

Sobolev type line profile in case of non radial wind density perturbations.
See Abstr. 063.022.

Structure and kinematics of stellar wind bubbles.
See Abstr. 064.005.

Radiation driven winds of hot luminous stars. III. Detailed statistical equilibrium calculations for hydrogen to zinc.
See Abstr. 064.007.

Stellar instabilities in the upper part of the Hertzsprung–Russell diagram.
See Abstr. 064.011.

The influence of photospheric turbulence on stellar mass loss.
See Abstr. 064.012.

Nonlinear dynamics of instabilities in line–driven stellar winds.
See Abstr. 064.013.

On the possibility of interpretation of Mg II profiles in hybrid stars.
See Abstr. 064.019.

Analysis of the atmospheres of K–giants.
See Abstr. 064.023.

Ionization equilibrium and accuracy of determining the chemical abundance of the atmospheres of K–giants.
See Abstr. 064.024.

Dust formation in stellar winds.
See Abstr. 064.029.

Radiation–driven winds of hot luminous stars. IV. The influence of multi–line effects.
See Abstr. 064.032.

Rotating stellar atmospheres.
See Abstr. 064.035.

Rotationally–enhanced stellar winds.
See Abstr. 064.036.

The spheroidal/ellipsoidal, variable mass–loss, decelerated Be star model.
See Abstr. 064.037.

Magnetic–loop model for Be stars.
See Abstr. 064.038.

Profiles of emission lines from rotating disks.
See Abstr. 064.039.

Discrete components in OB and Be stars: the shocking truth?
See Abstr. 064.040.

Consistent spherical NLTE–models for Be–stars.
See Abstr. 064.041.

A rotating, magnetic, radiation–driven wind model applied to Be stars.
See Abstr. 064.042.

Kinematic structure of OH/IR stars.
See Abstr. 064.048.

The radiation–hydrodynamics of expanding atmospheres.
See Abstr. 064.052.

Evolutionary helium and CNO anomalies in the atmospheres and winds of massive hot stars.
See Abstr. 064.055.

Radiatively driven winds from magnetic, fast–rotating stars: Wolf–Rayet stars?
See Abstr. 064.071.

Equilibrium temperature of ice grains formed around a star as a function of stellar parameters.
See Abstr. 064.072.

Selective mass loss, abundance anomalies, and helium–rich stars.
See Abstr. 064.077. ·

Photochemistry and molecular ions in oxygen–rich circumstellar envelopes.
See Abstr. 064.081.

The evolution of rapidly rotating B stars.
See Abstr. 065.054.

Shell stars in the Geneva photometric system.
See Abstr. 113.021.

The short–period photometric variability of four Be stars.
See Abstr. 113.022.

Recent Hα observations at Kitt Peak National Observatory.
See Abstr. 114.019.

Sudden flashes/outbursts in X Oph.
See Abstr. 114.020.

Near infrared spectra of 103 bright Be stars.
See Abstr. 114.046.

Near infrared spectra of southern Be stars.
See Abstr. 114.047.

On the technetium abundance determination in the atmospheres of red giants.
See Abstr. 114.049.

Analysis of the chemical composition of the atmosphere of θ Vir.
See Abstr. 114.050.

Determination of the physical parameters of the atmospheres of B and A stars in the spectral interval near the Balmer discontinuity.
See Abstr. 114.051.

Wolf–Rayet stars: current status and major questions.
See Abstr. 114.057.

High–dispersion spectroscopy of the Of/WN stars R 84 and S 61 of the Large Magellanic Cloud.
See Abstr. 114.059.

TiO molecular emission in the spectrum of VY Canis Majoris.
See Abstr. 114.062.

Chemical composition of the atmospheres of the K0–giant stars β Gem and γ^1 Leo
See Abstr. 114.065.

Chemical composition of the atmosphere of the "metallic–line" star Sirius.
See Abstr. 114.066.

Spectrophotometry of northern Wolf–Rayet stars: a sample of WN4, WN5 and WN6 stars.
See Abstr. 114.091.

A spectrographic study of two Of stars: HD 163758 and HD 117797.
See Abstr. 114.093.

Resonance line–profiles in galactic disk UV–bright stars.
See Abstr. 114.095.

The chemical composition of the atmospheres of the K giants 9 Bootis and ϱ Bootis.
See Abstr. 114.107.

Continuum energy dissipation of 48 Persei.
See Abstr. 114.119.

Rate of mass–loss in the Hertzsprung–Russell diagram.
See Abstr. 115.005.

Fundamental parameters of the underlying Be stars.
See Abstr. 115.010.

The Einstein view of the Wolf–Rayet stars.
See Abstr. 115.012.

Wolf–Rayet, the storm–stars.
See Abstr. 115.015.

On the polarization of Herbig Ae/Be stars.
See Abstr. 116.003.

Discovery of flare activity on BD + 3°4138 B.
See Abstr. 116.004.

Atmospheric activity in red dwarf stars.
See Abstr. 116.005.

The unknown, remembered gate: present and future observations of active chromosphere stars.
See Abstr. 116.006.

Recent spectacular activity in the Be star HR 2855 (FY CMa).
See Abstr. 116.013.

Observations of the carbon star CIT 6 in the CO and HCN lines.
See Abstr. 116.016.

An IUE survey of activity in red giants and supergiants.
See Abstr. 116.024.

Magnetic fields in Be stars?
See Abstr. 116.027.

The magnetic Be star Sigma Orionis E.
See Abstr. 116.029.

Double periodicity in Be stars.
See Abstr. 116.030.

A radio–continuum survey of the coolest M and C giants.
See Abstr. 116.033.

The magnetosphere and stellar wind of the helium–strong star, HD 184927.
See Abstr. 116.043.

On the variable elliptical polarization of Alpha Orionis.
See Abstr. 116.049.

High angular resolution speckle imaging of α Ori.
See Abstr. 116.050.

Activity in early F–type stars in the Hyades.
See Abstr. 116.053.

Spectrophotometric investigation of the Be star Merope.
See Abstr. 116.056.

A model for the intrinsic linear polarization of cool giant and supergiant stars.
See Abstr. 116.058.

Further VLA observations of hydrogen deficient stars.
See Abstr. 116.062.

Chromospheric–coronal activity at saturated levels.
See Abstr. 116.064.

Polarization variability among Wolf–Rayet stars. I. Linear polarization of a complete sample of southern Galactic WC stars.
See Abstr. 116.071.

Polarization variability among Wolf–Rayet stars. II. Linear polarization of a complete sample of southern galactic WN stars.
See Abstr. 116.072.

The interacting binary β Lyr. II. Non–LTE model analysis and evolutionary conclusions.
See Abstr. 117.011.

Model light curves of close binary systems with circumstellar envelopes.
See Abstr. 117.017.

Polarization and infrared colors of symbiotic stars.
See Abstr. 117.026.

On the nature of 623 + 71: a cataclysmic binary surrounded by a bow–shock–like emission nebula.
See Abstr. 117.027.

A search for non–stellar contributions to the optical and near–IR flux of RS CVn binaries. I. The cases of TY Pyx, UV Psc, RU Cnc and VV Mon.
See Abstr. 117.056.

Winds in collision. III. Modeling the interaction nebulae of eruptive symbiotics.
See Abstr. 117.058.

Changes in the relative intensities of the V and R components of the Hα emission line and in the radial velocity derived from it for X Persei.
See Abstr. 117.069.

Thermal radiation of the corona in binary X–ray sources.
See Abstr. 117.085.

Chemical composition of the atmosphere of the main component of the υ Sgr binary system.
See Abstr. 117.104.

X–ray observations of B–emission stars.
See Abstr. 117.128.

The Be/X–ray system HDE 245770/A0535 + 26 in an active phase.
See Abstr. 117.129.

Emission line variability in the Be star X–ray binaries 4U1258–61 and 4U2206 + 54.
See Abstr. 117.130.

Emission–line stars as interacting binaries.
See Abstr. 117.131.

Be binary systems with a cool companion: are they interacting?
See Abstr. 117.134.

Interacting binaries as Be stars.
See Abstr. 117.135.

Pole–on cataclysmic variables as Be stars.
See Abstr. 117.136.

Observations and evolutionary scenario for Be/X–ray binaries.
See Abstr. 117.137.

Evolutionary models for Be X–ray binaries.
See Abstr. 117.138.

Effect of compact objects near Be stars.
See Abstr. 117.139.

Photometry and spectroscopy of the O–type variable HD 167971.
See Abstr. 117.156.

Rotational modulation and flares on RS CVn and BY Dra stars. IV. The spatially resolved chromosphere of AR Lacertae.
See Abstr. 117.163.

Circumstellar matter in the AU Monocerotis system.
See Abstr. 117.181.

Rapid emission–line variations in the UV spectrum of the Wolf–Rayet system HD 90657.
See Abstr. 117.194.

Extended dust emission around R Aquarii observed with the JPL IR array camera.
See Abstr. 117.262.

Have circumstellar envelopes been detected around nearby M–dwarfs?
See Abstr. 118.004.

Prospects of infrared prospecting for planets.
See Abstr. 118.011.

Eclipsing binary stars.
See Abstr. 119.001.

Light curve variations of the eclipsing binary V367 Cygni.
See Abstr. 119.003.

The *BVJK* light curves of the short–period eclipsing binary CG Cygni.
See Abstr. 119.008.

Photometry and spectroscopy of the eclipsing P Cygni star R 81 in the Large Magellanic Cloud.
See Abstr. 119.021.

The peculiar Be star V644 Mon = HD 51480 as an interacting binary.
See Abstr. 119.030.

Radial velocity study of the Be star ζ Tau (HR 1910).
See Abstr. 120.019.

Synoptic studies of T Tauri stars.
See Abstr. 121.001.

Rotational modulation of the wind of the PMS star AB Aur: new observations in C IV and Mg II.
See Abstr. 121.003.

The strange "spots" on the T Tauri star RY Lupi.
See Abstr. 121.010.

Observations of infrared emission lines and radio continuum emission from pre–main–sequence objects.
See Abstr. 121.030.

Radio and infrared properties of young stars.
See Abstr. 121.038.

On the infrared emission of the exciting star of the Herbig–Haro objects 1 and 2.
See Abstr. 121.045.

Long period variables and stellar mass loss.
See Abstr. 122.002.

Shape of the visual light curve and detection of a 1.35 cm H_2O line in single M Miras.
See Abstr. 122.013.

The rapidly oscillating Ap stars as a test of stellar chromospheric heating mechanisms.
See Abstr. 122.014.

Mechanism of excitation of chromospheric emission of pulsating stars of the type of δ Scuti.
See Abstr. 122.016.

Variations in luminous blue variables.
See Abstr. 122.026.

Energy dissipation of high–frequency radial oscillations as possible source for heating of the outer layers of the atmosphere of τ Pegasi.
See Abstr. 122.037.

Extreme hydrogen–deficient stars.
See Abstr. 122.049.

The light curve of V441 Herculis.
See Abstr. 122.074.

Doppler imaging of variable early–type stars.
See Abstr. 122.080.

Nonradial pulsations and the Be phenomenon.
See Abstr. 122.081.

The onset of a type–II supernova explosion in a circumstellar envelope created by intense stellar wind from a presupernova.
See Abstr. 125.078.

Maps of millimeter wave emission from three galactic star–forming regions.
See Abstr. 131.007.

C_6H: astronomical study of its fine and hyperfine structure.
See Abstr. 131.025.

Searches for interstellar and circumstellar metal oxides and chlorides.
See Abstr. 131.031.

Interstellar molecules.
See Abstr. 131.086.

Water vapor masers associated with young visible stars.
See Abstr. 131.152.

Variability of the H_2O maser emission at 1.35 cm wavelength. I. Observational results.
See Abstr. 131.219.

La nebulosa cometaria 1548C27.
See Abstr. 131.226.

RCW 114: another case of H I bubble related to a WR star.
See Abstr. 131.229.

Magnetic fields in the regions of possible formation of planetary systems (from observations of H_2O masers).
See Abstr. 131.233.

H_2O maser outburst – protoplanetary rings.
See Abstr. 131.234.

EXOSAT observations of the ring nebula NGC 6888 and HD 192163.
See Abstr. 132.002.

Further observations of the peculiar galactic radio source BG 2107 + 49.
See Abstr. 132.004.

Systematic identification of IRAS point sources.
See Abstr. 133.001.

A study of the silicate emission features of the IRAS low resolution spectra.
See Abstr. 133.006.

Ground–based infrared observations of variable IRAS sources as candidates for late asymptotic giant branch stars.
See Abstr. 133.022.

Direct imaging at 12 microns of the star–forming region W51 IRS 2.
See Abstr. 133.024.

The kinematical structure of the bipolar planetary nebula 19W 32.
See Abstr. 134.024.

The ultraviolet spectra of central stars of planetary nebulae.
See Abstr. 134.026.

Models for the wind of the central star of NGC 6543.
See Abstr. 134.053.

Be stars as members of open clusters.
See Abstr. 153.016.

On the percentage of Be stars in galactic open clusters.
See Abstr. 153.017.

Variable Be stars in h and χ Persei.
See Abstr. 153.018.

Photoelectric search for CP2–stars in open clusters. XII. Alpha Persei, Praesepe and NGC 7243.
See Abstr. 153.022.

The IRAS view of the Galaxy and the solar system.
See Abstr. 155.024.

OH/IR stars and Galactic structure.
See Abstr. 155.164.

Properties of evolved mass–losing stars in the Milky Way and variations in the interstellar dust composition.
See Abstr. 155.173.

Are there stellar winds in the hot stars of M31?
See Abstr. 157.344.

113 Photometric Properties

113.001 Photometric variability of some CP stars.
A. Heck, G. Mathys, J. Manfroid.
Astron. Astrophys., Suppl. Ser., Vol. 70, No. 1, p. 33 – 48 (1987).
The photometric data relative to twenty–three southern CP stars have been re–analysed. New or improved parameters of the lightcurves are presented for eleven of them.

113.002 An objective–prism survey for Hα–emission–line stars of a field in Puppis.
B. Pettersson.
Astron. Astrophys., Suppl. Ser., Vol. 70, No. 1, p. 69 – 75 (1987).
52 new Hα–emission–line objects have been found in an objective–prism survey covering 27 square degrees in the north–western quadrant of the Gum Nebula, centered north of the H II region RCW 19 (at $l = 254°$, $b = 0°$). Iris photometry of the stars and photoelectric photometry for calibration of the iris photometry are presented. Accurate coordinates are given.

113.003 Erratum: "Photometric variability of Ap and He–weak stars in clusters and associations. II" [Astron. Astrophys., Suppl. Ser., Vol. 69, No. 3, p. 371 – 396 (1987)].
P. North.
Astron. Astrophys., Suppl. Ser., Vol. 70, No. 1, p. 141 (1987). See Abstr. 43.113.043.

113.004 Evidence for no short time scale photometric variations in the Bp–Si star HD 92664.
C. Mégessier, P. North.
Astron. Astrophys., Suppl. Ser., Vol. 70, No. 2, p. 247 – 255 (1987).
The Bp–Si star HD 92664 has been observed in Geneva photometry during seven nights in order to detect possible periodicities of about one hour or more in the vicinity of the phase of maximum magnetic field intensity. No periodic variation was found with periods shorter than 2 hours. Frequency analysis shows some peaks near 2 and 3 hours, but they are shown to be due to atmospheric perturbations. It is concluded that HD 92664 exhibits no periodic variations shorter than a few hours with an amplitude larger than 0.75 millimagnitude. The rotational period is confirmed and the corresponding light curve is shown for the magnitudes of the Geneva system and for the $\Delta(V1–G)$ peculiarity parameter.

113.005 Infrared observations of metal–deficient stars.
S. Arribas, C. Martinez Roger.
Astron. Astrophys., Suppl. Ser., Vol. 70, No. 3, p. 303 – 310 (1987).
Infrared magnitudes in the J, H, K and L bands for 64 metal–deficient stars spanning a wide range in effective temperature, luminosity and metal content are presented. An accuracy of 0.02 magnitude is obtained for the JHK bands and 0.03 for the L filter. Infrared–infrared and optical–infrared colour–colour diagrams are discussed and compared with the mean intrinsic tracks for Population I stars. It is concluded that infrared colours are not notably dependent on metallicity, with some exceptions for the reddest giants, which can be also interpreted by residual effects in the transformation equations between different systems. The authors also discuss briefly the near infrared photometric system of the Observatorio del Teide, from the Instituto de Astrofisica de Canarias.

113.006 *UBVRI* photometry of FKSZ stars. I.
G. Carrasco, P. Loyola.
Astron. Astrophys., Suppl. Ser., Vol. 70, No. 3, p. 369 – 372 (1987).
UBVRI photometry for 179 stars of the Faint Fundamental Stars Catalogue (FKSZ) in the declination zone $0° > \delta > –30°$ is presented.

113.007 La photométrie astronomique et son évolution au cours du 20ᵉ siècle.
M. Golay.
Astronomie, Vol. 101, p. 461 – 472 (1987).

113.008 The missing opacity and the temperature calibration of solar–type stars.
P. Magain.
Astron. Astrophys., Vol. 181, No. 2, p. 323 – 327 (1987).
A new temperature calibration of some colour indices is established for F and G dwarfs. It is based on the infrared flux method and extends the work of Saxner and Hammerbäck to the Population II stars. The disagreement between the empirical colours and some model predictions is discussed and showed to be most naturally interpreted in terms of the so–called "missing opacity".

113.009 Infrared photometry of FG Sge.
O. G. Taranova.
Astrophysics, Vol. 25, No. 3, p. 616 – 620 (1987). English translation of 42.113.050.

113.010 On the colours of Am stars.
J. B. Lester.
Mon. Not. R. Astron. Soc., Vol. 227, No. 1, p. 135 – 141 (1987).
The Am stars are observed to have excess flux at $\lambda^{-1} = 2.09 \, \mu m^{-1}$ ($\lambda = 4785$ Å) compared with the predicted fluxes of line blanketed model stellar atmospheres or with the fluxes of normal stars. This excess flux affects both the Strömgren b–magnitude and the β index so that the effective temperature of an Am star derived from these measures is erroneously high. The use of the contaminated b–magnitude in the construction of the c_1 and m_1 indices leads to surface gravities and metallicities that are too large. The best way to avoid these hazards is to use the full energy distribution. Synthetic spectra of the blue spectral region show that the excess flux is not due to elements, such as calcium and scandium, that are deficient in Am stars. It is suggested that the enhanced brightness near $2.09 \, \mu m^{-1}$ might be due to a change in the ionization balance of the iron–peak elements.

113.011 Hα photometry of dwarf K and M stars.
W. Herbst, A. C. Layden.
Astron. J., Vol. 94, No. 1, p. 150 – 157 (1987).
Hα photometry of 188 dM and dK stars has been obtained with the Perkin telescope at Van Vleck Observatory. The photometric index, $N–W$, has been calibrated using 56 stars with equivalent widths determined spectroscopically by Stauffer and Hartmann (1986). An external standard error of 0.08 Å is found for the equivalent width (EW) measurements of non–emission stars; it may be higher for those with emission lines. The distribution of stars in the $N–W$, $R–I$ plane is discussed. Most lie along a single "main sequence" of absorption which slopes up towards 0 EW with increasing $R–I$. Eighteen dKe or dMe stars are found, all clearly separated from the absorption sequence. The absorption EWs are large compared with expectation based on photospheric models, indicating either that the models are in error or that chromospheres are present in essentially all late–type dwarfs. Three stars (Gl 182, Gl 873AB, and Gl 890) were found to vary significantly in $N–W$. All are emission stars, and the variability can probably be attributed to flares.

113.012 Observations at the Corralitos Observatory. II.
E. M. Halbedel.
Be Star Newsl., No. 16, p. 13 – 14 (1987).

113.013 **A photoelectric *BVRI* sequence in the field of the globular cluster NGC 6121 (M4).**
G. Glementini.
Publ. Astron. Soc. Pac., Vol. 99, No. 617, p. 637 – 641 (1987).
A photoelectric *BVRI* sequence of eleven stars in the field of the galactic globular cluster NGC 6121 (M4) is presented. Their *BVRI* magnitudes are in the Cousins photometric system. A comparison is made with previous data collected on these stars.

113.014 **The Cousins and Kron *VRI* systems.**
M. S. Bessell, E. W. Weis.
Publ. Astron. Soc. Pac., Vol. 99, No. 617, p. 642 – 644 (1987).
Cubic transformations have been derived between colors measured in the Cousins and the Kron systems using measurements of common stars with spectral types from A0 V to M6 V. Polynomial fits are also derived for the mean main sequence for late–type dwarfs.

113.015 **Photometry of faint blue stars – VII. More southern stars.**
D. Kilkenny.
Mon. Not. R. Astron. Soc., Vol. 228, No. 3, p. 713 – 719 (1987).
Photoelectric *uvby* photometry is presented for 103 southern stars, mostly from the LB and PB catalogues. Using photometric criteria, the stars appear to be a mixture of hot subdwarfs, horizontal–branch and population II objects. Four high–latitude hot stars which are too red for their c_1 indices are suggested to be possible binary systems. Observations of metal–weak secondary standards indicate that the *uvby* photometry of the population II stars is close to the standard system.

113.016 **The photometric variability of the extremely hydrogen–deficient stars KS Persei and LS II + 33°5.**
K. Morrison, G. P. H. Willingale.
Mon. Not. R. Astron. Soc., Vol. 228, No. 3, p. 819 – 825 (1987).
Strömgren photometry of the extremely hydrogen–deficient binary KS Persei and the extremely hydrogen–deficient star (extreme helium star) LS II + 33°5 is presented. The observations of KS Persei suggest light and colour variations on a time–scale of ~5 day in addition to the longer ~30 day light variation previously reported. The suspected variability in LS II + 33°5 is confirmed with maximum changes in brightness over 3 – 4 day.

113.017 **On the (*B–V*) colors of the bright stars.**
G. T. Davidson, E. S. Claflin, B. M. Haisch.
Astron. J., Vol. 94, No. 3, p. 771 – 791 (1987).
The Bright Star Catalogue provides (*B–V*) colors for 8769 stars of visual magnitude less than magnitude 6.5. MK spectral classifications are available for most of those stars; the number for which MK classifications are given or for which a luminosity class can be interpreted or assumed with high probability is 8458. When stars of uncertain spectral class, including variables, are excluded, that number is reduced to 7651. The color distributions were examined for selected MK subclasses to determine whether the distributions can be understood in terms of the expected causes of color variability, including observational errors, the intrinsic variability within a subclass, and reddening by interstellar dust. To facilitate a quantitative analysis, a new statistical model for reddening by interstellar dust clouds was developed.

113.018 **Precision photometric monitoring of southern variable Wolf–Rayet stars with a comment on the overall continuum variability of WR stars.**
R. Lamontagne, A. F. J. Moffat.
Astron. J., Vol. 94, No. 4, p. 1008 – 1024 (1987).
The authors report on the photometric monitoring in the visible continuum of eight southern Wolf–Rayet (WR) stars that are known or suspected to show large variations. Combining these data with previous photometry of similar quality for other WR stars, they find that: (1) Nearly all WR + O binaries show phase–dependent light variations with a generally broad shallow minimum centered at phase zero when the WR star passes closest to the observer. (2) Late subtype WN and WC stars tend to vary with greater amplitude than the hotter subtypes.(3) If the binary

frequency is normal among WN8 stars, some of their strong variations might be due to interaction with a companion in some cases. (4) A few hotter WN stars without O companions show relatively large variations, possibly related to the presence of a compact companion or some other perturbation.

113.019 **On the relation between b–y and V–I$_c$.**
A. W. J. Cousins.
Mon. Notes Astron. Soc. S. Afr., Vol. 46, Nos. 9 – 10, p. 144 – 146 (1987).
There is a tight relationship between b–y and V–I$_c$ for unreddened stars with b–y < 0.45 (s.d. $\pm$ 0^{m}005 for E region stars). It is less well defined for the later spectral types. $E_{b-y} = 0.59\ E_{V-I}$, and this has important effects for hot stars. Balmer absorption and emission have small effects. Duplicity has to be considered for the cooler stars.

113.020 **Infrared photometry of two carbon stars.**
O. G. Taranova.
Astrofizika, Tom 27, Vyp. 1, p. 29 – 40 (1987). In Russian. English translation in Astrophysics, Vol. 27, No. 1.
The JHKLM observations of the carbon stars UV Aur and RW LMi are analysed. From the integrated flux in 0.5 – 5 μm range, the effective temperature in the maximum brightness is found to be 2700K, the luminosity 6.5 × 10^{37}erg/s and the radius ~600 $R_\odot$ (at the star's distance 1.35 kpc). The observed IR brightness variations of UV Aur are analysed in the framework of the models of a variable star and of a constant star with a dust shell. The asymmetry of the phase dependence of the energy distribution is noted in the spectrum of UV Aur in the range 0.7 – 5 μm. The measurements of the RW LMi in the range of 1.25 – 5 μm correspond to the sum of radiation of two black body sources with $T \cong 1200$K and 650K, the angular dimensions of these sources are 0".013 and 0".21 respectively.

113.021 **Shell stars in the Geneva photometric system.**
B. Hauck.
Physics of Be stars, p. 523 – 525 (1987). – See Abstr. 012.030 (IAU Colloq. No. 92).

113.022 **The short–period photometric variability of four Be stars.**
L. A. Balona, F. Marang, P. Monderen, A. Reitermann, F.–J. Zickgraf.
Astron. Astrophys., Suppl. Ser., Vol. 71, No. 1, p. 11 – 24 (1987).
The authors present results of intensive photometric monitoring of the Be stars λ Eri, ω Ori, 27 CMa and 28 CMa for two seasons. They confirm the periodicity of λ Eri previously obtained from radial velocity observations, but the amplitude is very low. They find ω Ori and 27 CMa to be new short–period variables. The former has a well–defined double–wave light curve of large amplitude. The authors could not detect any definite periodicity in 28 CMa, but the star was very active over a time scale of a few days.

113.023 **GALAXY and the Galaxy. The RGO selected area proper motion survey. I. Photometric sequences in selected areas.**
N. Reid, D. L. King, R. W. Argyle.
Astron. Astrophys., Suppl. Ser., Vol. 71, No. 2, p. 397 – 410 (1987).
B, *V*, and in some cases Cousins *I* photoelectric observations are presented for 119 stars in 11 of Kapteyn's Selected Areas. Magnitudes extend down to $V \lesssim 15.7$ and $0.1 \lesssim (B–V) \lesssim 1.6$.

113.024 **Instrumental effects and the Strömgren photometric system.**
J. Manfroid, C. Sterken.
Astron. Astrophys., Suppl. Ser., Vol. 71, No. 3, p. 539 – 555 (1987).
The authors present simultaneous observations obtained with various *uvby* photometers in order to check the validity of the usual colour transformation methods. The results agree with previous theoretical investigations and show that substantial

errors arise from the non–conformity of passbands. The need for a more strictly defined standard system and for closely matching instrumental systems is stressed.

113.025   UBV photometry of stars whose positions are accurately known. V.
T. Oja.
Astron. Astrophys., Suppl. Ser., Vol. 71, No. 3, p. 561 – 564 (1987).
UBV photometry is given for about 560 stars of the AGK3R and NPZT catalogues between BD declinations 11° and 23°.

113.026   Photometry of faint blue stars – VIII. Photometry in UKST areas 345 and 788/9.
D. Kilkenny.
Mon. Not. R. Astron. Soc., Vol. 229, No. 2, p. 345 – 354 (1987).
Strömgren *uvby* photoelectric photometry is presented for 130 stars brighter than about 16 mag and selected from UKST objective–prism plates to be not later than type F0. The very low dispersion prism plates appear to be well–suited to searches for faint blue stars and the completeness is probably as good as attained by other methods. The relatively high surface densities of Ahb stars down to at least 16 mag could make these objects very useful for studies of the Galaxy and its halo.

113.027   Which photometric period for WR 16?
J. Manfroid, E. Gosset, J. M. Vreux.
Astron. Astrophys., Vol. 185, No. 1/2, p. L7 – L9 (1987).
Analysis of new and already published photometric observations shows that the Wolf–Rayet star WR 16 = HD 86161 is not a binary system (WR + compact, $P = 5.365$ d) as claimed by Moffat and Niemela (1982). Instead it is shown that the simultaneous presence of two periodicities (about 1.3 and 2.5 days) is necessary to account for the observed variations during at least one observing run. The variability of the B9p star HD 86199, previously used as a comparison for a photometric study of WR 16, is established.

113.028   A high precision photometric investigation of the micro–variations of Wolf–Rayet stars.
A. M. van Genderen, K. A. van der Hucht, W. J. G. Steemers.
Astron. Astrophys., Vol. 185, No. 1/2, p. 131 – 146 (1987).
Walraven *VBLUW* photometric monitoring of 7 Wolf–Rayet stars made in March/April 1986 is presented and discussed. The sample consists of 4 real or suspected single line spectroscopic binaries and 2 single stars, all of spectral type WN, and 1 star of spectral type WC. The light and colour variations, caused by binary modulation and/or by an intrinsic random mechanism are shown and discussed.

113.029   A comparison between two–dimensional classifications from Vilnius photometry and those on the MK system.
C. J. Corbally, R. P. Boyle.
Astron. Astrophys., Vol. 186, No. 1/2, p. 114 – 119 (1987).
A feasibility study (Smriglio et al., 1986) for a planned galactic structure survey has used seven–colour photographic photometry in the Vilnius system to determine automatically two-dimensional stellar classifications for a region in Lyra. Twenty-three of these predicted classifications are now tested against MK classifications from digital slit spectra. Stars later than G0 agreed to 79% in type and to 86% in luminosity class. However, in the A to F region the predicted types were much earlier than the MK types. To investigate this failure, the photographic photometry was compared with photoelectric Vilnius photometry (Janulis, 1986) for the 28 stars common to both photometric studies. Save for a small (0.08) shift in the (U–P) index, the photographic results followed the photoelectric ones within the expected errors, and the photometric classes, determined respectively by the automatic and a non–automatic method, also agreed well. However, a new, non–automatic determination of the photometric classes for seven problem stars revealed that the derived classifications are sensitive to which diagrams relating photometric indices to spectral class are favoured.

113.030   The nature of the F str λ4077 stars.
P. North.
Astron. Astrophys., Vol. 186, No. 1/2, p. 191 – 199 (1987).
Geneva photometry of seven stars classified F str λ4077 by Bidelman (1981) is presented. Published *uvbyβ* measurements of five additional stars of this type are used as well to look for a possible link with the classical Am or Ap stars. It is shown that these stars exhibit a whole range of metallicities, the latter being correlated with effective temperature. The existing proper motions point to a Population I membership, although one star seems to have a rather high velocity. It is suggested that this new type of peculiarity may be interpreted in the framework of the diffusion theory and that at least some F str λ4077 stars may represent an intermediate stage between Am and λ Bootis stars, although they are generally cooler.

113.031   Near–infrared excesses of barium stars.
J. Hakkila, B. J. McNamara.
Astron. Astrophys., Vol. 186, No. 1/2, p. 255 – 260 (1987).
A sample of thirty one barium stars are found to exhibit weak ($< 0^m.1$) infrared excesses at 1.6 μm and 4.8 μm. There is a possible correlation of the excess at 1.6 μm with the barium intensity (defined by the abnormal strength of the Ba II line at λ4554). Two mechanisms for the production of these excesses are discussed: (1) redistribution of energy from the Bond–Neff depression into the infrared, and (2) thermal emission caused by mass transfer in a binary system. The current observations do not allow one to conclusively differentiate between the two models, but the correlation of infrared excess at 1.6 μm with suppressed optical flux is suggestive of energy redistribution.

113.032   Theoretical two–colour diagrams in the broad–band UBVRIJHKLM–system for different sources of radiation.
N. Kh. Minikulov.
Izv. Krymskoj Astrofiz. Obs., Tom 76, p. 43 – 51 (1987). In Russian. English translation in Bull. Crimean Astrophys. Obs., Vol. 76.
A program to compute colorimetric characteristics of continuous and linear radiation of hydrogen gas in broad–band UBVRIJHKLM–system has been developed. Colour indices were calculated for different I_e, I_*, N_e, W, and $\log \beta_{12}$. The most reliable diagram to separate gaseous radiation from that of dust has been searched for using different two–colour diagrams. The following systems appeared to be the most effective: $(U–B, V–R)$, $(V–R, B–V)$, $(B–V, H–K)$ and $(H–K, K–L)$.

113.033   U–B and B–V–diagram for hydrogen gas with 10^6cm^{-3} electron density.
T. S. Belyakina, N. Kh. Minikulov.
Izv. Krymskoj Astrofiz. Obs., Tom 76, p. 51 – 54 (1987). In Russian. English translation in Bull. Crimean Astrophys. Obs., Vol. 76.
The U–B and B–V colour indices have been calculated for hydrogen gas with $n_e = 10^6 \text{cm}^{-3}$, electron temperature $T_e = 1.5 \times 10^4$K, 2×10^4K, the exciting star temperature $T_* = 3 \times 10^4$K, 5×10^4K, 10^5K, $\beta_{12} = 1$, 10^{-1}, 10^{-2} and dilution factor $W = 10^{-4}, 10^{-6}, 10^{-8}$. Results of computations were used for the analysis of UBV–observations of some selected symbiotic stars.

113.034   Determination of UBV–magnitudes based on data of spectrophotometric catalogues.
V. I. Burnashev.
Izv. Krymskoj Astrofiz. Obs., Tom 76, p. 70 – 76 (1987). In Russian. English translation in Bull. Crimean Astrophys. Obs., Vol. 76.
The coefficients of relations between the observed and calculated UBV–magnitudes based on the data in a compiled catalogue were determined. The mean standard deviations of the calculated values of V, B–V, U–B are $\pm 0^m.04$; $\pm 0^m.02$; $\pm 0^m.05$ correspondingly. The discrepancy between the coefficients derived by different authors is analyzed.

113.035 **Four–color CCD observations of BHB stars.**
A. G. D. Philip.
Bull. Am. Astron. Soc., Vol. 19, No. 2, p. 678 (1987). Abstract. –
See Abstr. 010.061.

113.036 **Spectroscopic tests of two faint photometric surveys.**
C. J. Corbally.
Bull. Am. Astron. Soc., Vol. 19, No. 2, p. 680 (1987). Abstract. –
See Abstr. 010.061.

113.037 **The intrinsic colors of OB supergiants in the Large Magellanic Cloud.**
E. L. Fitzpatrick.
Bull. Am. Astron. Soc., Vol. 19, No. 2, p. 704 (1987). Abstract. –
See Abstr. 010.061.

113.038 **Stellar photometry in crowded fields.**
P. B. Stetson.
Bull. Am. Astron. Soc., Vol. 19, No. 2, p. 745 (1987). Abstract. –
See Abstr. 010.061.

113.039 **Photoelectric photometry of M–type stars.**
J. F. Lahulla.
Publ. Astron. Soc. Pac., Vol. 99, No. 619, p. 998 – 1002 (1987).
UBVRI photoelectric photometry has been obtained for 110 M–type stars contained in the Caltech Two–Micron Survey and the revised AFGL Catalog.

113.040 **UBV(RI)$_c$ photometry for Ca II emission stars. 1. Observations at Sutherland.**
T. Lloyd Evans, M. C. J. Koen.
S.Afr. Astron. Obs., Circ., No. 11, p. 21 – 56 (1987).
Extensive photometry on the UBV(RI)$_c$ system is given for 55 stars with strong Ca II H and K emission. The UBV observations were made relative to the standard stars of Lloyd Evans, Koen & Hultzer (1983) while the (RI)$_c$ observations were made relative to E–region standards over greater angular distances. The data are plotted in phase diagrams.

113.041 **UBV(RI)$_c$ photometry for Ca II emission stars. 2. Observations at Mt. John University Observatory and at Mt. Stromlo.**
A. Collier Cameron.
S.Afr. Astron. Obs., Circ., No. 11, p. 57 – 71 (1987).
UBR(RI)$_c$ data are given for 17 stars with strong Ca II H and K emission. Techniques of observation and reduction are described briefly.

113.042 **UBVRI photometry of HD 36705 (AB Dor).**
T. Lloyd Evans.
S.Afr. Astron. Obs., Circ., No. 11, p. 73 – 82 (1987).
A total of 695 sets of UBVRI observations of HD 36705, made in the period 5 – 15 December 1984, are presented. New photometric values for the two local comparison stars HD 36269 and HD 36316 are given.

113.043 **Revised UBV photometry of Magellanic Cloud sequences.**
A. W. J. Cousins.
S.Afr. Astron. Obs., Circ., No. 11, p. 89 – 92 (1987).
The old photometry of UBV secondary standards in the Magellanic Clouds (Cousins 1970) has been revised with the help of new measurements made at Sutherland.

113.044 **Secondary standards for the Strömgren uvby system.**
A. W. J. Cousins.
S.Afr. Astron. Obs., Circ., No. 11, p. 93 – 120 (1987).
Observations of 158 E region stars have been made in the Strömgren system, using the 46 cm reflector at Cape Town. They are mostly brighter than eighth magnitude and are intended for use as secondary standards for the four–color system. The results are given in tables.

113.045 **GALAXY and the Galaxy. The RGO Selected Area proper motion survey. I: Photometric sequences in Selected Areas.**
N. Reid, D. L. King, R. W. Argyle.
R. Greenwich Obs., Prepr., No. 56, 23 pp. (1987). To appear in Astron. Astrophys. Suppl. Ser.

113.046 **21 Com – photometry at Hβ.**
J. Zverko.
Contrib. Astron. Obs. Skalnaté Pleso, Vol. 16, p. 7 – 15 (1987).
The narrow band Hβ photometric observations of the CP2 star 21 Com obtained on 8 nights during 1978 – 1979 are presented. A tentative period analysis confirmed the existence of short–time variations and, in addition to those hitherto published, the author's data indicate a period of about 1.5 hour. As to long periodicity, the data fit the best value P_{rot} = 1.83736 d, which is as much as the seventh estimation of the rotational period of this star.

113.047 **VBLUW–photometry of the two F0 Ib supergiants HD 80404 and HD 90853, and the K5 Ib supergiant HD 85891 with its associated cluster.**
S. E. van der Wal, A. M. van Genderen.
Working Group Stars with Extended Atmospheres, Prepr., No. 103, 20 pp. (1987). To appear in Astron. Astrophys.

113.048 **Photometric classification of early type stars with hydrogen emission lines.**
E. E. Mendoza V.
Rev. Mex. Astron. Astrofis., Vol. 14, No. 1, p. 310 – 314 (1987). – See Abstr. 012.042.
This study is based upon 145 stars, from O4 to A1, whose spectra have or had at some time, one or more hydrogen lines in emission. They have been observed in two photometric systems: (1) Johnson's HKL, (2) Mendoza's α(16)Λ(9). The stars are divided in 5 working groups: (1) Wolf–Rayet stars, (2) Supergiant stars, (3) Classical Be stars, (4) Peculiar Be stars, (5) Herbig Ae/Be stars. The photometric characteristics of each group are discussed. An outline of classification is given.

113.049 **A preliminary report on photometric measurements of the helium line λ10830 Å.**
E. E. Mendoza V.
Rev. Mex. Astron. Astrofis., Vol. 14, No. 1, p. 385 – 389 (1987). – See Abstr. 012.042.
This study is based upon 89 stars taken from the Bright Star Catalog to measure the total absorption of the He I line at λ10830 Å. It is defined a $\psi(25)$ index in the same fashion as the α(16) and Λ(9) indices. The results are compared with Zirin's equivalent widths, and with the α(16)Λ(9) photometric system.

113.050 **Strömgren photometry of high–velocity stars; metal abundances and ages.**
W. J. Schuster, P. E. Nissen.
ESO Conf. Workshop Proc., No. 27, p. 141 – 146 (1987). – See Abstr. 012.051.
uvby–β photometry has been obtained for 711 stars. The majority of stars were selected from various catalogues to have a space velocity with respect to the local standard of rest greater than 80 km/s. About 100 stars were selected on the basis of a spectral classification that indicates very weak metal lines. Only dwarfs and subgiants in the spectral range A5 to K5 were included. From internal and external comparisons the mean error of b–y is estimated to be $\pm 0^m005$ or less and the mean errors of V, m_1, c_1 and β are $\pm 0^m008$ or less. A catalogue with the data is ready for publication. On the basis of stars estimated to be closer than 100 pc, the intrinsic colour index $(b-y)_0$ has been calibrated as a function of β, m_1 and c_1. From this relation the colour excess E(b–y) has been computed for all stars with $\beta > 2^m55$.

113.051 Observations of the Pleiades member Hz II–625 (V811 Tau) with a simultaneous two–star *UBVR*–photometer.
A. K. Magnitskij.
Astron. Zh., Tom 64, Vyp. 6, p. 1323 – 1326 (1987). In Russian.
English translation in Sov. Astron., Vol. 31, No. 6.

Observations with the simultaneous two–star *UBVR*–photometer showed changing of the light curve of the K–type dwarf Hz II–625, one of the 11 late–type variable stars known in the Pleiades cluster.

113.052 La photométrie astronomique et son évolution au cours du 20ᵉ siècle (2).
M. Golay.
Astronomie, Vol. 101, p. 511 – 520 (1987).

113.053 V–photometry of 6 Cas.
V. Ya. Solov'ev.
Astron. Tsirk., No. 1462, p. 5 (1986). In Russian.

113.054 Infrared observations of WY Gem and ZZ CMi.
O. G. Taranova.
Astron. Tsirk., No. 1467, p. 7 – 8 (1986). In Russian.

Infrared photometry of WY Gem and ZZ CMi has shown that the mean value of the colour index (J–K) for WY Gem corresponds to M4 I. No excess radiation is observed. Variability of WY Gem over the observational period did not exceed 0^m20 in all filters. The angular diameter of WY Gem is $\sim 0\overset{\prime\prime}{.}0035$. The mean colour index for ZZ CMi corresponds to M5 III. In the M band (4.7 μm), excess emission possibly exists. The angular diameter of ZZ CMi is $\sim 0\overset{\prime\prime}{.}0024$.

113.055 Obtaining continuous spectra of stars of the NGC 6913 cluster.
R. M. Raznik.
Astron. Tsirk., No. 1471, p. 3 – 4 (1986). In Russian.

Photometric observations of star clusters are more simple, if the comparison star and the investigated stars belong to the same cluster and are of close spectral subclasses. The reddening of the star No. 3 of the cluster NGC 6913 in comparison with the star No. 2 is interpreted by the difference in orientations of the circumstellar dust disks.

113.056 The search for BY Dra–type variability in red dwarf stars.
G. Sh. Rojzman.
Astron. Tsirk., No. 1484, p. 1 – 3 (1987). In Russian.

The results of UBVR photometry for nine red dwarf stars in June – August 1982 are presented.

113.057 The photometric UBVR system of the 50/70 cm Schmidt telescope of the National Astronomical Observatory Rozhen.
M. K. Tsvetkov, T. B. Georgiev, B. P. Bilkina,
A. G. Tsvetkova, E. H. Semkov.
Dokl. Bolg. Akad. Nauk, Tome 40, No. 5, p. 9 – 12 (1987).

This paper is part of the programme investigations on the quality and parameters of the Schmidt telescope of the NAO Rozhen in operation. For the photometric test, open clusters with electrophotometrically measured UBVR magnitudes for about 20 – 40 stars in the range of 9 – 15 magnitudes were used as standard stars.

113.058 Calibration of physical quantities in the DDO photometric system.
H. Sung, S.–W. Lee.
J. Korean Astron. Soc., Vol. 20, No. 2, p. 63 – 94 (1987).

Using many homogeneous data of DDO and UBV colors for all luminosity classes and physical quantities known from spectroscopic observations, new calibration schemes with DDO photometric parameters are presented for metal abundance, effective temperature and surface gravity of stars, and an intrinsic color relation is derived for the reddening estimate.

113.059 Cool star automated photometry.
R. M. Genet, L. J. Boyd, S. L. Baliunas, D. S. Hall.
Publ. Astron. Soc. Pac., Vol. 99, No. 621, p. 1147 – 1148 (1987).
Abstract. – See Abstr. 010.281.

***UBV* photoelectric catalogue (1986). II Analysis of the data.**
See Abstr. 002.030.

UBV photoelectric photometry catalogue (1986). I. The original data (magnetic tape).
See Abstr. 002.031.

A catalogue of seven–colour photometry of 752 stars in the Vilnius system for a region of two square degrees in the direction of the globular cluster M56.
See Abstr. 002.090.

A catalog of bright *uvbyβ* standard stars.
See Abstr. 002.104.

The High Altitude Observatory–Lowell Observatory Solar–Stellar Spectrophotometry Project.
See Abstr. 013.003.

The solar–stellar connection at low spectral resolution.
See Abstr. 013.005.

The Automatic Photoelectric Telescope Service.
See Abstr. 013.031.

Doppler imaging.
See Abstr. 036.003.

Inverse photometric problem for spotted stars.
See Abstr. 036.180.

COLOUR program for data processing of UBVR photometry of variable stars.
See Abstr. 036.215.

The rapid photometric variability of some Ae/Be–star envelopes.
See Abstr. 112.006.

HD 213985: a hot post–AGB star in the galactic halo.
See Abstr. 112.007.

B and A type stars with unexpectedly large colour excesses at IRAS wavelengths.
See Abstr. 112.010.

The peculiar emission–line supergiant HD 37836.
See Abstr. 112.012.

IRAS observations of CP stars.
See Abstr. 112.014.

Infrared photometry of late–type Wolf–Rayet stars.
See Abstr. 112.019.

Hot carbon stars: more about V348 Sgr.
See Abstr. 112.116.

Extinction and abundance properties of Alpha Scorpii circumstellar grains.
See Abstr. 112.173.

MK$_m$ spectral classifications for 37 weak–line stars and their testing of a Strömgren photometry survey.
See Abstr. 114.015.

Studies of some diatomic molecules in the F– and early G–type dwarf stars.
See Abstr. 114.032.

New M–type supergiants in the southern Milky Way.
See Abstr. 114.096.

Spectral synthesis in the ultraviolet. I. Far–ultraviolet stellar library.
See Abstr. 114.116.

The early A type stars: refined MK classification, confrontation with Strömgren photometry, and the effects of rotation.
See Abstr. 114.125.

The helium abundance and luminosity of the main–sequence F stars as derived from uvbyβ photometry.
See Abstr. 115.006.

A diagram for photometric recognition of barium CH, and R–types stars.
See Abstr. 115.018.

The rapidly rotating spotted red dwarf flare star Gliese 890.
See Abstr. 116.012.

Activity in late–type dwarfs. I. Walraven and Johnson photometry of flares and spot variations on Gl 867 A (= FK Aqr) in 1979.
See Abstr. 116.035.

Simultaneous multicolour photometry of OY Carinae during quiescence.
See Abstr. 117.025.

The 67–min X–ray period of EX Hydrae observed with the EINSTEIN observatory.
See Abstr. 117.055.

The spectral classification of the cool components of symbiotic stars.
See Abstr. 117.066.

A uvbyβ survey of northern–hemisphere active binaries. I. The observations.
See Abstr. 117.143.

Photometry and spectroscopy of the O–type variable HD 167971.
See Abstr. 117.156.

Orbital elements for double stars of Population II. The high–velocity system COD –48°1741.
See Abstr. 117.234.

Photometry and spectroscopy of the eclipsing P Cygni star R 81 in the Large Magellanic Cloud.
See Abstr. 119.021.

The strange "spots" on the T Tauri star RY Lupi.
See Abstr. 121.010.

A study of the nature of the Hα–emission stars LkHα 112, 115, 118, and 119.
See Abstr. 121.018.

The pulsation modes of CO Aur.
See Abstr. 122.010.

The variable star HD 79889.
See Abstr. 122.045.

Photometric behaviour of FG Sagittae in 1985.
See Abstr. 122.108.

EUV photometry of DA white dwarfs with EXOSAT.
See Abstr. 126.061.

Study of Fernie's photometric catalogue of supergiants.
See Abstr. 131.276.

Ultraviolet interstellar extinction toward NGC 6530 and the intrinsic energy distribution of 9 Sagittarii and HD 165052.
See Abstr. 131.305.

Observations of stars in H II regions: UBVRI photometry.
See Abstr. 132.011.

The LMC H II regions N11C and E and their stellar contents.
See Abstr. 132.013.

Optical photometry of high latitude 12 μm IRAS sources.
See Abstr. 133.010.

Seven–colour photoelectric photometry of bright stars in the open cluster IC 4756.
See Abstr. 153.015.

Photoelectric search for CP2–stars in open clusters. X. NGC 2232, NGC 2343, Cr 140, and Tr 10.
See Abstr. 153.020.

Photoelectric search for CP2–stars in open clusters. XI. NGC 3532 and NGC 5662.
See Abstr. 153.021.

Photoelectric search for CP2–stars in open clusters. XII. Alpha Persei, Praesepe and NGC 7243.
See Abstr. 153.022.

Strömgren photometry of open clusters. III. NGC 2323, NGC 5662.
See Abstr. 153.023.

DDO photometry of giants in the open cluster NGC 2660.
See Abstr. 153.032.

Photoelectric search for CP2–stars in open clusters. X: NGC 2232, NGC 2343, Cr 140, and Tr 10.
See Abstr. 153.036.

Photoelectric B, V observations of stars in Omega Centauri.
See Abstr. 154.048.

The distribution of OB stars and dust in a Milky Way field at $(l, b) = (335°, 0°)$.
See Abstr. 155.084.

On the distribution of colors for stars in the ninth to fifteenth magnitude range: statistics and implications for galactic structure.
See Abstr. 155.112.

An analysis of the Yale Bright Star Catalog.
See Abstr. 155.113.

Surface brightnesses from standard stars with wide–field cameras.
See Abstr. 157.237.

The stellar content and morphology of the dwarf irregular galaxy Holmberg IX.
See Abstr. 157.304.

114 Spectra, Temperatures, Chemical Composition, etc.

114.001   Optical region elemental abundance analyses of B and A stars. VII. The metallic–lined star 32 Aquarii.
D. Kocer, C. Bolcal, E. Inelmen, S. J. Adelman.
Astron. Astrophys., Suppl. Ser., Vol. 70, No. 1, p. 49 – 56 (1987).

An abundance analysis using photographic region spectrograms and fully line–blanketed model atmospheres has been performed for the metallic–lined (Am) star 32 Aquarii consistent with previous papers of this series. Its pattern of abundance anomalies is not identical with those of the hot Am stars although there are definite similarities. Changes of up to order 1 dex are found by comparison with the previous analysis by Smith. Slightly better agreement is found with the recent analyses of singly–ionized rare earth lines by Magazzu and Cowley. Results are also presented for a model selected on the basis of photometric indices according to the formula of Moon and Dworetsky.

114.002   Isotope ratio of europium in Procyon.
K.–i. Kato.
Publ. Astron. Soc. Jpn., Vol. 39, No. 3, p. 517 – 520 (1987).

The Eu II line at 4129.7 Å in the spectrum of Procyon (α CMi, F5 IV–V) is analyzed by the method of spectrum synthesis to derive the isotope ratio of europium. It is found that the observed profile is well reproduced by the computed profile when the author takes the isotope ratio of $\varepsilon(Eu^{153})/\varepsilon(Eu^{151}) = (35 \pm 15)/(65 \pm 15)$. This is in fairly good agreement with the solar value obtained from the same line in the solar spectrum.

114.003   HD 16523 and HD 17638 – possible Wolf–Rayet stars of type WO.
D. N. Rustamov, A. M. Cherepashchuk.
Pis'ma Astron. Zh., Tom 13, No. 8, p. 680 – 685 (1987). In Russian. English translation in Sov. Astron. Lett., Vol. 13.

Using spectrograms obtained with a dispersion of 75 Å/mm with the 2 m telescope of the Shemakha Astrophysical Observatory the spectral subtype of the WR–type stars HD 16523 and HD 17638 are defined. The spectral subtypes are uncertain. The stars may be considered as WO 5 stars.

114.004   New carbon stars in selected regions of the Milky Way.
M. G. Nikolashvili.
Astrofizika, Tom 26, Vyp. 3, p. 559 – 561 (1987). In Russian. English translation in Astrophysics, Vol. 26, No. 3. – With 3 plates.

Thirty–three new carbon stars are revealed by a low–dispersion (1250 Å/mm near Hγ) spectral survey of selected regions of the Milky Way.

114.005   An upper limit on p–mode amplitudes in β Hyi.
S. Frandsen.
Astron. Astrophys., Vol. 181, No. 2, p. 289 – 292 (1987).

An attempt has been made to use a high resolution spectrograph (the ESO CES) fed by an optical fiber to look for stellar oscillations in β Hyi. A ratio R is formed between the flux in the line cores divided by the flux in the continuum. This ratio should show oscillations at a level higher than the oscillations in the total luminosity, and at the same time be insensitive to scintillation. Due to detector problems only an upper limit to the amplitudes was found corresponding to a few times 10^{-5} in the luminosity.

114.006   Silicon absorption in UV spectra of ApSi stars.
M.–C. Artru, T. Lanz.
Astron. Astrophys., Vol. 182, No. 2, p. 273 – 284 (1987).

The authors investigate the role of silicon UV absorption occurring in the atmosphere of early–type stars. A detailed analysis of the $\lambda\lambda 1250 – 1850$ Å range is presented. High–resolution spectra are obtained from the IUE database for typical silicon stars and for a few normal main–sequence stars. The strongest absorption in these spectra are the resonance lines of Si II, and three large features ($\lambda 1400$, $\lambda 1560$ and $\lambda 1770$ Å) which are strongly enhanced in the spectra of ApSi stars. For the hottest silicon stars, a discontinuity at 1310 Å is observed. The atomic data are carefully reviewed for Si II to perform spectrum synthesis calculations showing out more precisely the silicon absorption.

114.007   Computation of synthetic stellar and solar spectra.
L. S. Lyubimkov.
Bull. Crimean Astrophys. Obs., Vol. 74, p. 1 – 11 (1987). English translation of 41.114.103.

114.008   Variability of Hα emission in the spectrum of the Be star 28 Cyg.
A. E. Tarasov, A. G. Shcherbakov.
Bull. Crimean Astrophys. Obs., Vol. 74, p. 17 – 20 (1987). English translation of 41.114.104.

114.009   Comparative spectral characteristics of the three Be stars ϕ Persei, θ Coronae Borealis, and X Persei.
T. S. Galkina.
Bull. Crimean Astrophys. Obs., Vol. 74, p. 21 – 31 (1987). English translation of 41.114.105.

114.010   Spectrophotometry of ζ Tau using 1983 – 1984 observations. I. Variability of the spectrum.
T. M. Rachkovskaya, Ch. M. Nasibova.
Bull. Crimean Astrophys. Obs., Vol. 74, p. 32 – 39 (1987). English translation of 41.114.106.

114.011   Determination of the uranium abundance in the Ap star 73 Dra using the line U II $\lambda 3859.58$.
I. Kh. Iliev, L. S. Lyubimkov, I. S. Savanov.
Astrophysics, Vol. 25, No. 2, p. 491 – 498 (1987). English translation of 42.114.088.

114.012   Intensities of the resonance lines of neutral and ionized magnesium in stellar spectra.
V. G. Cholakyan.
Astrophysics, Vol. 25, No. 2, p. 531 – 538 (1987). English translation of 42.114.089.

114.013   Observations of M dwarfs beyond 2.2 μm.
G. Berriman, N. Reid.
Mon. Not. R. Astron. Soc., Vol. 227, No. 2, p. 315 – 329 (1987).

This paper presents the first systematic spectroscopic observations of M dwarfs beyond 2.2 μm. The coolest dwarfs show strong water absorption in the 3 μm window, and beyond 4 μm, the energy distributions of all the stars fall slightly less steeply than the Rayleigh–Jeans tail of a blackbody. Spectra between 1 and 4 μm are essential in deriving accurate luminosities of M dwarfs, and possibly in deriving accurate effective temperatures too. New values reported here are not in general well explained by theoretical models of hydrogen burning stars. This is especially true for those cooler than 3000K: in the HR diagram they lie closer to brown dwarfs, in contrast to recent results based only on photometry. This result revives the old issues of the nature of the coolest stars and of the accuracy of the theoretical models.

114.014   Mass to line–strength relations from *IUE* spectra of early–type stars.
K. Sekiguchi, K. S. Anderson.
Astron. J., Vol. 94, No. 1, p. 129 – 136 (1987).

An extensive survey of the 1200 – 1900 Å region in OB type stellar spectra obtained from the *IUE* archives has been used to establish line–strength versus spectral–type relationships. One hundred and sixty–three low–resolution *IUE* spectra of 124 well–classified O3 to B5 stars were examined. The authors confirm earlier results that the equivalent width of Si IV (λ 1400) and C IV (λ 1550) features are well correlated with optical spectral type. The equivalent width/spectral type/luminosity class relations for these features are then established.

114.015 MK_m spectral classifications for 37 weak–line stars and their testing of a Strömgren photometry survey.
C. J. Corbally.
Astron. J., Vol. 94, No. 1, p. 161 – 168 (1987).

Houk's extension of the MK system has been continued as an interim expediency for classifying stars with metal abundances greater or less than the solar abundance. These classifications ("MK_m") were related to the initial results of a *uvby* survey for intermediate and extreme Population II stars at high galactic latitudes and with $13 < V < 16$ mag. The survey is found to have a 91% success rate of identifying metal–weak stars. The photometrically derived [Fe/H] values are used to calibrate the metallicity indices produced by the classifications, yielding [Fe/H] $= 0.13 \, \Delta_{met-H} - 0.26$.

114.016   Upper limit to the boron abundance in the Population II star HD 140283.
P. Molaro.
Astron. Astrophys., Vol. 183, No. 2, p. 241 – 246 (1987).

A high resolution IUE spectrum of the Population II star HD 140283 has been analyzed to search for the B I 2496 Å resonance doublet. No features attributable to this element have been found. An upper limit for the B abundance of $B/H < 10^{-11}$, i.e. more than one order of magnitude less than the solar value has been derived. The presence of Li in the atmosphere of HD 140283 rules out significant depletion for the more refractory B, and it is argued that the protostellar nebula was substantially free of B. This value favours the idea of a gradual increase in B during the Galaxy's life, but a more abrupt B synthesis is still possible if HD 140283 were one of the first stars formed in the Galaxy. The Li/B observed ratio in HD 140283 is more than 25 times that predicted by the spallation theory of high energy GCR with CNO nuclei, ruling out a spallogenic origin for the lithium observed in Population II stars and supporting its primordial origin.

114.017   Physical parameters for Population II stars.
C. Cacciari, M. L. Malagnini, C. Morossi, L. Rossi.
Astron. Astrophys., Vol. 183, No. 2, p. 314 – 318 (1987).

The observed energy distributions from the ultraviolet to the near–infrared for a number of Population II stars have been collected from the literature (Christensen, 1978; Cacciari, 1985) and have been compared with the Kurucz (1979) model atmospheres computed for different [M/H] ratios. Effective temperatures, surface gravities and apparent angular diameters have been obtained for those stars falling in the T_{eff} range covered by the Kurucz models. For each of these stars the results refer to that [M/H] value which permits the best reproduction of the observations according to a weighted LMS error criterion. The quality of the derived parameters is discussed, and comparisons with previous results cited in the literature are presented.

114.018   Abundance anomalies in stars: atomic physics at play.
G. Michaud.
Phys. Scr., Vol. 36, No. 1, p. 112 – 121 (1987).

Abundance anomalies are obtained in a large fraction of both evolved and main sequence stars. They involve most chemical elements and abundances can vary by many orders of magnitude. In this paper the author concentrates on abundance anomalies observed on non–magnetic stars with $6000 < T_{eff} < 10000K$. Some 20% of the stars in that T_{eff} range are involved. In them the separation goes on below the visible region. It is shown that, to obtain accurate enough radiative accelerations, the f values are needed to an accuracy of 30% for the important lines of all states of ionization, up to the 18th for Fe, for instance. An analysis of the needed atomic data is done in detail.

114.019   Recent Hα observations at Kitt Peak National Observatory.
G. J. Peters.
Be Star Newsl., No. 16, p. 15 – 16 (1987).

114.020   Sudden flashes/outbursts in X Oph.
K. K. Ghosh, K. Jayakumar, K. Kuppuswamy.
Be Star Newsl., No. 16, p. 18 – 20 (1987).

114.021   Studies of hot B subdwarfs. V. Continuing investigation of the C, N, and Si abundance patterns in the atmospheres of sdB stars.
R. Lamontagne, F. Wesemael, G. Fontaine.
Astrophys. J., Vol. 318, No. 2, p. 844 – 851 (1987).

High–dispersion ultraviolet observations of two hot, hydrogen–rich subdwarfs have been obtained with the IUE observatory. These data are analyzed with model atmosphere techniques, and element abundances for C, N, and Si are derived. Both PG 0342 + 026 ($T_e = 21,800K$) and PG 1104 + 243 ($T_e = 27,200K$) display abundances of these ions consistent with trends observed previously in generally hotter sdB and sdOB stars.

114.022   A search for $^{14}C^{16}O$ in the atmospheres of evolved stars.
M. J. Harris, D. L. Lambert.
Astrophys. J., Vol. 318, No. 2, p. 868 – 872 (1987).

No evidence for the presence of $^{14}C^{16}O$ was found in the 2.4 μm spectra of 7 MS and S stars and 19 carbon stars. Lower limits on the ratio $^{12}C/^{14}C$ ranging from 5000 to 28,500 were determined. In the framework of the thermal pulse model of s–process nucleosynthesis in asymptotic giant branch stars, these lower limits may be used to constrain models of the $^{13}C(\alpha, n)^{16}O$ neutron source.

114.023   The investigation of the spectral–variable star HD 148112. The spectrophotometry of lines.
I. A. Aslanov.
Astron. Zh., Tom 64, Vyp. 4, p. 815 – 824 (1987). In Russian. English translation in Sov. Astron., Vol. 31, No. 4.

The measurements of the radial velocities, equivalent widths and line component half–widths of Cr, Eu, Fe, Si, Sr, Gd allowed to reveal the presence of some regions on the surface of the star HD 148112 with high concentration of these elements. Brightness minimum in B and V coincides in phase with that of equivalent widths of the above–mentioned element lines. The number of spots for all elements is equal to six, however, their coordinates differ slightly. The investigation of the Balmer hydrogen lines showed that lg $(N_{02}H)$ and lg n_e change with the rotation period. The short–period variations were studied and a period of 2^h5 has been found.

114.024   Heavy elements in the ultraviolet spectra of Ap stars.
A. B. Severnyj.
Astrophys. Space Phys. Rev., Vol. 5, p. 241 – 260 (1987). – See Abstr. 003.002.

Heavy element abundances of Ap stars were determined using ultraviolet spectra obtained from the photoelectric scanner on the Astron space telescope. The procedure by which the observed spectral data were obtained for the individual sections of the spectrum scanned is described. The observed data were analyzed using model atmospheres and known oscillator strengths. In this way, a lead abundance ~ 100 times solar, and a tungsten abundance of more than (10^2times solar were obtained in $\varkappa$ CnC. Anomalous abundances (10^5solar) were also obtained for uranium and thorium in 73 Dra, where the abundance of lead and tungsten is about the same as in $\varkappa$ Cnc. The lead abundance found in several stars rules out the s–process as a possible source for the formation of this element from the iron–peak elements. For the Russian original see 42.114.093.

114.025   An analysis of HD 43819 by wavelength–coincidence statistics.
A. A. Poli, D. J. Bord, C. R. Cowley.
Publ. Astron. Soc. Pac., Vol. 99, No. 617, p. 623 – 628 (1987).

An element identification study has been performed on the sharp–lined Si star HD 43819 using the method of wavelength–coincidence statistics. The results are compared to a study by Adelman who used traditional methods. The data generally confirm the previous investigation but also provide evidence to support the identifications of Dy III, Gd II, Nd III, and Mo II.

114.026 Recognition and classification of strong–CN giants.
P. C. Keenan, S. B. Yorka, O. C. Wilson.
Publ. Astron. Soc. Pac., Vol. 99, No. 617, p. 629 – 636 (1987).

The stars designated as super–metal–rich by Spinrad and Taylor in 1969 can be readily recognized as a distinct group by the CN bands, which are stronger than expected for the luminosities indicated by the usual line ratios. The authors have classified on the Revised MK System 50 of these stars, assigning positive CN indices in recognition of the characteristic excesses of carbon and nitrogen in their atmospheres, as compared to normal Population I stars.

114.027 Elemental abundance analyses with coadded Dominion Astrophysical Observatory spectrograms – II. The mercury–manganese stars 53 Tauri, Mu Leporis and Kappa Cancri.
S. J. Adelman.
Mon. Not. R. Astron. Soc., Vol. 228, No. 3, p. 573 – 594 (1987).

Elemental abundance analyses based on the coaddition of at least 10 2.4 Å mm^{-1} IIa–O Dominion Astrophysical Observatory spectrograms have been performed for three mercury–manganese stars, 53 Tauri, Mu Leporis, and Kappa Cancri. These fine analyses show a greater degree of internal consistency than previous studies based on lower signal–to–noise data. Lines as weak as of order 3 mÅ are employed in these studies and lines of atomic species not previously identified have been discovered.

114.028 Erratum: "The relative merits of seven photometric temperature indices for disk–population K giants" [Astron. J., Vol. 93, No. 5, p. 1253 – 1263 (1987)].
B. J. Taylor, S. B. Johnson, M. D. Joner.
Astron. J., Vol. 94, No. 3, p. 803 (1987). See Abstr. 43.114.055.

114.029 Wolf–Rayet stars.
D. C. Abbott, P. S. Conti.
Annu. Rev. Astron. Astrophys., Vol. 25, p. 113 – 150 (1987). – See Abstr. 003.003.

The authors consider the properties and evolutionary status of W–R stars both from the observable data and from the theoretical models used to describe the stellar structure, atmospheres and winds. Ideally, they should like to describe the characteristics of luminosity, effective temperature, chemical composition, radius, mass, and (for stars such as these) the mass–loss rates and other stellar wind parameters. In practice, however, not all of these are well known given the complex physical phenomena present in W–R stars. The authors thus proceed from the better–known quantities, such as the absolute visual magnitudes and line strengths, through those that can be reasonably inferred from the existing data with very simplified models, such as the composition and mass–loss rates, to more controversial issues, such as the luminosities and effective temperature scale.

114.030 On the absence of young white dwarf companions to five technetium stars.
V. V. Smith, D. L. Lambert.
Astron. J., Vol. 94, No. 4, p. 977 – 980 (1987).

A search for hot companions to five stars of type MS and S has been carried out using the International Ultraviolet Explorer (IUE) satellite. No hot companions were detected for the MS stars HR 85, 4647, 6702, and 8062, and the S star HR 8714. Limits on the luminosities of possible white dwarf companions provide lower limits of $2 - 5 \times 10^8$yr to the ages of any degenerate companions. All five stars exhibit strong Tc I lines, and the presence of technetium, with a half–life of 2.1×10^5yr, signifies recent nucleosynthesis. The limits on the ages of possible white dwarf companions that are $\geqslant 1000$ half–lives of Tc exclude the possibility that the s–process elemental enhancements seen in these MS and S stars resulted from mass transfer from a more highly evolved companion (as is probably the mechanism by which barium stars are created). These MS and S stars represent a sample of true thermally pulsing asymptotic giant–branch stars.

114.031 Additional late–type stars with technetium.
S. J. Little, I. R. Little–Marenin, W. Hagen Bauer.
Astron. J., Vol. 94, No. 4, p. 981 – 995 (1987).

The authors present the results of a survey of 279 late–type giants and supergiants for the spectral lines of the radioactive element technetium (Tc I) at 4297, 4262, and 4238 Å. They reach the following conclusions: (1) the presence of Tc correlates very strongly with the existence of light variability; (2) evolutionary MS stars show Tc and spectroscopic MS stars do not show Tc; (3) single S stars show Tc; (4) SC stars show Tc; (5) about 75% of the C stars show Tc; (6) Ba II stars do not show Tc. The findings are compatible with the predictions from stellar evolution theory that Tc, along with other s–process elements and carbon, is mixed with the surface materials after helium shell flashing episodes. The presence of Tc is a very sensitive indicator of the third dredge–up and can be detected in the spectrum before enhancements of other s–process elements are measurable.

114.032 Studies of some diatomic molecules in the F– and early G–type dwarf stars.
M. Singh, J. P. Chaturvedi.
Astrophys. Space Sci., Vol. 136, No. 2, p. 231 – 246 (1987).

A critical analysis of CH, NH, OH, C_2, and CN molecules/ radicals has been made in twenty–four F– and early G–type dwarfs at different effective temperatures as well as in newly constructed model atmospheres. Molecular indices of bandheads of $A-X$ systems of CH, NH, OH, C_2, and CN have been obtained by using the data available in the literature (thirteen–colour and eight–colour photometry). Besides, some interesting plots of the molecular indices vs θ_{eff}, molecular abundances and molecular indices vs dissociation energy, reduced equivalent widths and FCF's vs dissociation energy for respective molecules have also been enumerated.

114.033 Balmer jumps in Ap stars.
Yu. V. Glagolevskij, G. P. Topil'skaya.
Stellar atmospheres, p. 132 – 134 (1987). In Russian. Abstr. in Ref. Zh., 51. Astron., 8.51.391 (1987). – See Abstr. 003.005.

114.034 Scales of effective temperatures of hot stars.
E. V. Ruban.
Stellar atmospheres, p. 88 – 95 (1987). In Russian. Abstr. in Ref. Zh., 51. Astron., 8.51.394 (1987). – See Abstr. 003.005.

114.035 Synthetic low–resolution spectra of K and M giants.
N. S. Komarov.
Stellar atmospheres, p. 48 – 53 (1987). In Russian. Abstr. in Ref. Zh., 51. Astron., 8.51.395 (1987). – See Abstr. 003.005.

114.036 Fundamental characteristics of late–type stars.
N. S. Komarov, V. D. Motrich.
Stellar atmospheres, p. 30 – 40 (1987). In Russian. Abstr. in Ref. Zh., 51. Astron., 8.51.396 (1987). – See Abstr. 003.005.

114.037 Spectrophotometric standards. Absolute calibration of the energy distribution of 15 stars in the 311 μm – 750 μm range.
Eh. I. Terez.
Red. Zh. Kinematika Fiz. Nebesn. Tel. Akad. Nauk USSR, Kiev, 29 pp. (1987). In Russian. Abstr. in Ref. Zh., 51. Astron., 8.51.397 (1987).

114.038 Chemical composition of the atmosphere of the Hg–Mn star ✕ Cancri.
V. M. Dobrichev, D. V. Rajkova, T. A. Ryabchikova.
Stellar atmospheres, p. 119 – 123 (1987). In Russian. Abstr. in Ref. Zh., 51. Astron., 8.51.405 (1987). – See Abstr. 003.005.

114.039 Atmospheric abundances of classical metallic–line A stars: the visual spectral region.
M. C. Lane, J. B. Lester.
Astrophys. J., Suppl. Ser., Vol. 65, No. 1, p. 137 – 160 (1987).

An abundance study of six classical Am stars finds that calcium and scandium are systematically deficient, and that the elements heavier than zinc are enhanced by up to 1.0 dex. The iron–peak elements, however, have essentially solar abundances that do not vary with T_{eff}. The microturbulent velocities range between 3 and 4 km s^{-1}. A comparison with the predictions of diffusion theory finds agreement for some elements if a small mass loss is present, but other predictions are contradicted for all calculated values of the mass loss.

114.040 IUE observations of the broad continuum feature at 1400 Å in the silicon and related stars.
S. N. Shore, D. N. Brown.
Astron. Astrophys., Vol. 184, No. 1/2, p. 219 – 226 (1987).

The upper main sequence chemically peculiar stars have previously been studied spectrophotometrically using low resolution TD–1 spectra. These show that the broad continuum feature at 1400 Å is a useful indicator of Si anomaly. In this paper the authors extend this to IUE low resolution spectra and show that this feature is indeed well correlated with Si and also with optical chemical peculiarity indices. The authors further demonstrate its utility in selecting silicon stars. This feature is likely due to autoionization of Si II and is the best available discriminator of the overabundance of a key element in the Ap stars.

114.041 Absolute spectrophotometry of 18 carbon stars. III.
I. Eglitis.
Nauchn. Inf., Vyp. 63, p. 46 – 59 (1987). In Russian.

The results of spectrophotometric investigations of 18 carbon stars in the range $\lambda3225 – 8000$ Å are given. The graphical energy distributions and for some objects the fluxes are given. Some correlations of the spectrophotometric indices, obtained on the basis of scans of carbon stars are examined. The similarity of the atmospheric structure of the stars of subclasses R9–N and group J is pointed out.

114.042 The B stars: beacons of the skies.
J. B. Kaler.
Sky Telesc., Vol. 74, No. 2, p. 147 – 150 (1987).

114.043 The spectacular O stars.
J. B. Kaler.
Sky Telesc., Vol. 74, No. 5, p. 464 – 469 (1987).

114.044 Platinum and bismuth in HR 465.
C. R. Cowley.
Observatory, Vol. 107, No. 1080, p. 188 – 194 (1987).

114.045 About the spectral classification of Be stars.
A. Feinstein, H. Tignanelli.
Physics of Be stars, p. 72 – 73 (1987). – See Abstr. 012.030 (IAU Colloq. No. 92).

114.046 Near infrared spectra of 103 bright Be stars.
Y. Andrillat.
Physics of Be stars, p. 237 – 238 (1987). – See Abstr. 012.030 (IAU Colloq. No. 92).

114.047 Near infrared spectra of southern Be stars.
L. Pastori.
Physics of Be stars, p. 239 – 241 (1987). – See Abstr. 012.030 (IAU Colloq. No. 92).

This paper presents general results for southern Be stars obtained from more than 100 spectrograms in the region $\lambda\lambda7750 – 9000$ Å; all the available southern Be stars (right ascension between 4^h and 17^h) listed in the Bright Star Catalogue (Hoffleit and Jaschek, 1982) were observed.

114.048 High resolution observations of stars in the peculiar globular cluster ω Cen.
M. Spite, S. Huille, P. François, F. Spite.
Astron. Astrophys., Suppl. Ser., Vol. 71, No. 3, p. 591 – 596 (1987).

The abundances of the elements in the atmosphere of 6 stars belonging to the peculiar globular cluster ω Centauri are discussed in François et al. (1987). This paper presents the observational material and, line by line, the equivalent widths and the deduced abundances.

114.049 On the technetium abundance determination in the atmospheres of red giants.
T. Kipper.
Pis'ma Astron. Zh., Tom 13, No. 11, p. 1019 – 1022 (1987). In Russian. English translation in Sov. Astron. Lett., Vol. 13.

Technetium abundances in the atmospheres of S– and C–stars are estimated by comparing the synthetic and observed spectra for the Tc I $\lambda5924.47$ region. It was found that reliable estimates are possible only for the pure S–stars and SC–stars. For the late carbon stars only the upper limit for the Tc abundance can be found. For TX Psc (C6.5) this was found to be lg $\varepsilon_{TC} = 0.5$.

114.050 Analysis of the chemical composition of the atmosphere of θ Vir.
V. M. Dobrichev, T. A. Ryabchikova, D. V. Rajkova.
Astrophysics, Vol. 26, No. 1, p. 31 – 34 (1987). English translation of 43.114.016.

114.051 Determination of the physical parameters of the atmospheres of B and A stars in the spectral interval near the Balmer discontinuity.
S. A. Gulyaev, V. E. Panchuk, V. V. Pleshakov, S. G. Pyatkes.
Bull. Spec. Astrophys. Obs. – North Caucasus, Vol. 22, p. 1 – 10 (1987). English translation of 41.114.075.

114.052 Chemical composition of K–giants in the Hyades.
T. V. Mishenina, V. E. Panchuk, N. S. Komarov.
Bull. Spec. Astrophys. Obs. – North Caucasus, Vol. 22, p. 11 – 14 (1987). English translation of 41.114.076.

114.053 Effective temperatures of chemically peculiar stars.
Yu. V. Glagolevskij, N. M. Chunakova.
Bull. Spec. Astrophys. Obs. – North Caucasus, Vol. 22, p. 35 – 44 (1987). English translation of 41.114.077.

114.054 A system of standard wavelengths for white supergiants.
V. M. Dobrichev, E. L. Chentsov,
Z. V. Shkhagosheva.
Bull. Spec. Astrophys. Obs. – North Caucasus, Vol. 22, p. 50 – 53 (1987). English translation of 41.114.078.

114.055 Spectral types for objects in the Kiso survey. III. Data for 102 stars.
G. Wegner, R. K. McMahan, F. I. Boley.
Astron. J., Vol. 94, No. 5, p. 1271 – 1279 (1987).

Spectroscopy and spectral types for 102 ultraviolet–excess objects found in the Kiso Schmidt camera survey are reported. The spectra were observed with the McGraw–Hill 1.3 m telescope at 8 Å resolution and cover the interval $\lambda\lambda4000 – 7200$. Descriptions of some of the more peculiar objects found in this sample are given and include nine subdwarfs, 23 definite DA and one DB white dwarfs, 14 quasars and emission–line objects, and a new cataclysmic variable possibly resembling V Sge.

114.056 On the Balmer discontinuity of B Ia supergiants.
J. Zorec, R. Mercado–Ibañez.
C. R. Acad. Sci., Sér. II, Tome 305, No. 16, p. 1279 – 1284 (1987). In French.

The authors show that the disagreement between the observed Balmer discontinuities of B Ia supergiants and those predicted by the classical models of stellar atmospheres, is a function of the effective temperature. Some attempts at explaining to this disagreement are briefly discussed.

114.057 Wolf–Rayet stars: current status and major questions.
A. J. Willis.
Q. J. R. Astron. Soc., Vol. 28, No. 3, p. 217 – 224 (1987). – See Abstr. 012.037.

Population I WR stars are believed to be the evolved He–burning descendants of initially more massive O–star progenitors in which mass loss has significantly affected the overall evolution, thereby revealing interior CNO–burning (WN stars) and He–burning (WC stars) products. This review outlines the current status of knowledge of WR stars and the uncertainties in basic WR parameters.

114.058 Microturbulence in the upper photosphere of α Persei (F5 Ib) derived from ultraviolet spectral observations.
F. H. P. Spaan, C. de Jager, H. Nieuwenhuijzen, Y. Kondo.
Astron. Astrophys., Vol. 185, No. 1/2, p. 229 – 232 (1987).

High–resolution ultraviolet spectra of the moderate supergiant α Per (F5 Ib) were studied to determine the dynamic state of its upper photosphere. It was found that the line–of–sight microturbulent velocity component ζ_u in the region of origin of the UV spectrum (which is 73,000 km above that of formation of the visual spectrum) is about 5 km s^{-1}, and is slightly smaller than the value derived from the visual spectrum. This is ascribed to dissipation of mechanical energy between the higher and lower layers where, respectively, the ultraviolet and visual light lines originate. Between these two levels, which are one scale height apart, the mechanical energy flux decreases to about 0.3 of its photospheric value. The consequent value for the (outward directed) turbulent acceleration g_t is 24 cm s^{-2}, more than one half the observationally determined effective acceleration of gravity.

114.059 High–dispersion spectroscopy of the Of/WN stars R 84 and S 61 of the Large Magellanic Cloud.
B. Wolf, O. Stahl, W. Seifert.
Astron. Astrophys., Vol. 186, No. 1/2, p. 182 – 190 (1987).

The Of/WN stars R 84 and S 61 of the LMC have been studied with high–dispersion spectroscopy in the optical and satellite–UV range. The high resolution and high S/N spectra in the optical range are particularly distinguished by strong emission lines of H, He I and [N II]. The comparison of R 84 with S 61 shows that the peculiar emission line spectrum is not caused by the previously found late type companion of R 84. It is found that the UV spectra of both stars closely resemble those of late O–supergiants but all absorption lines are violet–shifted. The absorption lines are stronger than in normal O–type stars. The UV–resonance lines indicate low terminal wind velocities of ≈900 km s^{-1} only. Unlike to normal O–type stars the Al III–resonance lines also show pronounced P Cygni profiles with an even lower edge velocity. The mass loss rates are comparable to rates found in normal luminous hot stars. It is suggested that the Of/WN transition type stars are the hotter counterparts of the early B–type P Cygni stars.

114.060 The diffusion of gallium in main–sequence peculiar stars.
G. Alecian, M.–C. Artru.
Astron. Astrophys., Vol. 186, No. 1/2, p. 223 – 230 (1987).

New observations of the gallium abundance in main–sequence peculiar stars are now available in the literature. The present paper analyses these new data in the framework of the diffusion processes. The radiative accelerations of gallium are computed, with up–to–date gallium line data, in various physical conditions of stable main–sequence star atmospheres. Zero mass loss is assumed. The role of the 1400 Å absorption feature, attributed to Si II autoionization, is investigated and appears of minor importance. The main observed trends (the anti–correlation between the gallium and silicon overabundances in Ap stars and the smaller gallium enhancement for lower effective temperature) may be explained by the calculations on the gallium diffusion: the authors propose here that the presence of a magnetic field modifies strongly the gallium accumulation.

114.061 Spectral types and their uses.
P. C. Keenan.
Publ. Astron. Soc. Pac., Vol. 99, No. 618, p. 713 – 723 (1987).

The principles involved in setting up any viable system of spectral classification are discussed. Their application is illustrated by reviewing the essential features of the original MK system of two dimensions and its extension to additional dimensions, particular attention being paid to the Revised MK System as applied to stars of types G0 through M8. The domain of wavelengths within which stars at a given temperature can be classified on a consistent system is defined. The application of spectral types to the preparation of observing programs, to statistical studies of the distribution of selected groups of stars, and to the direct estimation of physical properties of stars is discussed briefly.

114.062 TiO molecular emission in the spectrum of VY Canis Majoris.
J. G. Phillips, S. P. Davis.
Publ. Astron. Soc. Pac., Vol. 99, No. 618, p. 839 – 841 (1987).

Relative line strengths in the rotational structure of the (0–0) band of the γ' system of TiO seen in emission in the spectrum of VY CMa yields a rotational temperature close to 600K. This is intermediate between the temperatures found by Herbig in 1962 and 1966 – 380K and 820K, respectively. It is also in reasonable accord with the theoretical conclusion of Schwartz that the outer boundary of the circumstellar envelope is 300K.

114.063 Addendum: "The abundance of gallium in B–type chemically peculiar stars" [Astrophys. J., Vol. 304, No. 1, p. 425 – 435 (1986)].
M. Takada–Hidai, K. Sadakane, J. Jugaku.
Astrophys. J., Vol. 320, No. 1, p. 437 (1987). See Abstr. 41.114.100.

114.064 A younger universe is seen in the stars.
M. M. Waldrop.
Science, Vol. 237, No. 4813, p. 361 – 362 (1987).

A tricky observation of radioactive thorium in nearby stars leads to a surprisingly young age for the galaxy and for the universe.

114.065 Chemical composition of the atmospheres of the K0–giant stars β Gem and γ^1 Leo
M. E. Boyarchuk, I. S. Savanov.
Izv. Krymskoj Astrofiz. Obs., Tom 76, p. 21 – 36 (1987). In Russian. English translation in Bull. Crimean Astrophys. Obs., Vol. 76.

On the basis of high dispersion spectrograms the K0 III stars β Gem and γ^1 Leo are studied. Using the model–atmosphere method their parameters and chemical composition are obtained. It is suggested that the analysis based on weak and moderately strong lines is insensitive to the temperature structure of the high photosphere. The differences in abundances of the investigated stars and the Sun are discussed.

114.066 Chemical composition of the atmosphere of the "metallic–line" star Sirius.
I. S. Savanov.
Izv. Krymskoj Astrofiz. Obs., Tom 76, p. 37 – 40 (1987). In Russian. English translation in Bull. Crimean Astrophys. Obs., Vol. 76.

Using observational data and the model atmosphere method the chemical composition of the "metallic–line" star Sirius A is found. The abundances of 26 elements have been obtained with T_{eff} = 10150K, lg g = 4.3 and ξ_t = 2.0 km/s. The established chemical composition is compared with the results for several other Am stars.

114.067 Lithium in the Alpha Per cluster.
S. Balachandran, D. L. Lambert, J. R. Stauffer.
Bull. Am. Astron. Soc., Vol. 19, No. 2, p. 702 (1987). Abstract. – See Abstr. 010.061.

114.068 Red stars in the equatorial selected areas.
T. H. Robertson, T. M. Jordan.
Bull. Am. Astron. Soc., Vol. 19, No. 2, p. 703 (1987). Abstract. – See Abstr. 010.061.

114.069 The spectra of extra–galactic Wolf–Rayet stars.
P. Massey, P. S. Conti, T. E. Armandroff.
Bull. Am. Astron. Soc., Vol. 19, No. 2, p. 703 (1987). Abstract. – See Abstr. 010.061.

114.070 A roadmap to the ultraviolet spectrum of cool, non–coronal stars.
K. G. Carpenter, R. E. Stencel, A. Brown.
Bull. Am. Astron. Soc., Vol. 19, No. 2, p. 705 (1987). Abstract. – See Abstr. 010.061.

114.071 Effective temperatures and gravities of S Doradus–like stars in the Large Magellanic Cloud.
B. Bohannan, J. B. Doggett.
Bull. Am. Astron. Soc., Vol. 19, No. 2, p. 706 (1987). Abstract. – See Abstr. 010.061.

114.072 The chemical abundances of Alpha Centauri A.
T. Meylan, I. Furenlid.
Bull. Am. Astron. Soc., Vol. 19, No. 2, p. 761 (1987). Abstract. – See Abstr. 010.061.

114.073 On the evolutionary status of Mu Leonis.
M. J. Harris, D. L. Lambert, V. V. Smith.
Publ. Astron. Soc. Pac., Vol. 99, No. 619, p. 1003 – 1008 (1987).
The abundances of carbon, nitrogen, and oxygen and the carbon and oxygen isotopic ratios have been measured in the CN–strong giant star μ Leo from CO and CN lines measured from a new $2 - 2.5\ \mu$m spectrum and published equivalent widths of C_2 and [O I] lines. The relative abundances of ^{12}C, ^{14}N, and ^{16}O, along with the ^{12}C/^{13}C ratio, indicate that μ Leo has undergone the first red–giant dredge–up.

114.074 Lithium and rotation in the Hyades F dwarfs.
A. M. Boesgaard.
Publ. Astron. Soc. Pac., Vol. 99, No. 620, p. 1067 – 1070 (1987).
Data on Li/H abundances have been assembled for 32 Hyades F dwarfs from high signal–to–noise, high–spectral–resolution observations. The middle F stars show a regular pattern in the Li–temperature profile. This Li–temperature profile is compared with the rotation–temperature profile based on $v \sin i$ data from Kraft. A relationship between the Li abundance and $v \sin i$ may be present for the stars in the Li dip.

114.075 High–latitude A – F supergiants.
H. E. Bond, R. E. Luck.
Space Telesc. Sci. Inst., Prepr. Ser., No. 182, p. 9 – 12 (1987). Paper presented at IAU Colloq. No. 95.

114.076 Chemical compositions of post–AGB stars.
H. E. Bond, R. E. Luck.
Space Telesc. Sci. Inst., Prepr. Ser., No. 183, p. 2 – 5 (1987). Paper presented at ESO Workshop on stellar evolution and dynamics in the outer halo of the Galaxy, April 1987.

114.077 The chemical composition of the extreme halo stars. I. Blue spectra of 20 dwarfs.
P. Magain.
ESO Sci. Prepr., No. 546, 67 pp. (1987). Submitted to Astron. Astrophys.

114.078 Neutron–exposure variations in MS and S stars, and the implications for s–process nucleosynthesis.
V. V. Smith, D. L. Lambert, A. McWilliam.
Astrophys. J., Vol. 320, No. 2, p. 862 – 865 (1987).
High–resolution, near–infrared spectra have been obtained for 34 stars of types MS and S along with 14 M giants used as comparison stars. Equivalent width measurements of spectral lines arising from the heavy s–process species La II and Nd II, relative to the lighter s–process species Y II, indicate a large range of neutron exposures in the MS and S stars. The indicated range of mean neutron exposures is approximately $\tau_0 \approx 0.1 - 0.6\ \mathrm{mb}^{-1}$. The solar system s–process distribution is characterized by $\tau_0 = 0.26\ \mathrm{mb}^{-1}$. The overall enhancement of s–process elements observed in the MS and S stars indicates that they are substantial contributors to the Galactic production of s–process nuclei.

114.079 Spectrum analysis of the extremely metal poor carbon dwarf star G 77–61.
H. Gass, J. Liebert, R. Wehrse.
Prepr. Steward Obs., No. 745, 25 pp. (1987). To appear in Astron. Astrophys.

114.080 Chemically peculiar star π^1Boo – stellar parameters and v sin i.
M. Zboril, J. Žižňovský, J. Zverko.
Contrib. Astron. Obs. Skalnaté Pleso, Vol. 16, p. 47 – 60 (1987).
Using the authors' own spectroscopic and published photometric observations, the atmospheric and stellar parameters of the Hg–Mn star π^1Boo were determined. The synthetic spectrum was computed for a selected part of its spectrum and compared with the observed. The spectral lines were identified and equivalent widths measured.

114.081 Radial velocities of the Be star Kappa Draconis in the years 1983 – 1984.
L. C. Iliev, K. Juza.
Contrib. Astron. Obs. Skalnaté Pleso, Vol. 16, p. 61 – 68 (1987).
In studying the problem of the reality of the variations of the radial velocity of the Be star $\varkappa$ Draconis, the authors determined the radial velocities of this star by measuring 17 spectrograms obtained at the National Astronomical Observatory (NAO) in Rozen. They found no convincing proof of these variations in the years 1983 and 1984.

114.082 The dynamical state of the atmosphere of the supergiant Alpha Cygni (A2 Iae) derived from high–resolution ultraviolet spectra.
B. Boer, C. de Jager, H. Nieuwenhuijzen.
Working Group Stars with Extended Atmospheres, Prepr., No. 104, 4 pp. (1987).

114.083 Quantitative spectral peculiarity indexes of CP stars of the upper main sequence.
I. M. Kopylov.
Astrofiz. Issled. Izv. Spets. Astrofiz. Obs., Tom 24, p. 44 – 67 (1987). In Russian. English translation in Bull. Spec. Astrophys. Obs. – North Caucasus.
The quantitative spectral peculiarity indexes P are determined for 160 chemically peculiar stars of the upper main sequence within the spectral type interval B2 – A9 ($22000 > T_e > 7500$K) from the selected ion lines Si II, Mn II, Sr II, Cr II and Eu II. Dependences for each of these indexes upon the spectral type Sp (effective temperature T_e) are studied for the used sample of CP stars.

114.084 Spectroscopic study of metal–poor F and G dwarfs. 1. Iron spectrum analysis.
I. F. Bikmaev.
Astrofiz. Issled. Izv. Spets. Astrofiz. Obs., Tom 25, p. 3 – 12 (1987). In Russian. English translation in Bull. Spec. Astrophys. Obs. – North Caucasus, Vol. 25.
Tasks and problems of a spectroscopic study of metal–poor F and G dwarfs are discussed. Methodical questions of determination of iron abundance [Fe/H] in the atmospheres of these stars are considered. It is shown that the accuracy of determination is Δ[Fe/H] $= \pm 0.2 - 0.3$ and is limited not by quality of photographic observations but by uncertainties in the atmospheric parameters. Iron abundance has been determined for 9 F and G dwarfs, for the first time for 4 of them it was determined spectroscopically. A comparison with the results of other authors shows a good agreement.

114.085 Balmer discontinuities of chemically peculiar stars.
Yu. V. Glagolevskij, G. P. Topil'skaya.
Astrofiz. Issled. Izv. Spets. Astrofiz. Obs., Tom 25, p. 13 – 23 (1987). In Russian. English translation in Bull. Spec. Astrophys. Obs. – North Caucasus, Vol. 25.
Balmer discontinuities (D) were determined for more than 110 chemically peculiar stars (CP) on the basis of the authors' measurements and literature data. A comparison of D for the

normal and CP stars confirms the data, obtained earlier, about the smaller values of D relative to the normal ones. A dependence of δD upon the surface magnetic field B_S was found. Models of D values are discussed.

114.086   Formation of spectra of hot stars near the Balmer jump.
S. A. Gulyaev, O. G. Patimova, V. V. Pleshakov.
Astrofiz. Issled. Izv. Spets. Astrofiz. Obs., Tom 25, p. 55 – 59 (1987). In Russian. English translation in Bull. Spec. Astrophys. Obs. – North Caucasus, Vol. 25.

The spectra of B and A stars in the spectral region near the Balmer jump are computed. New limiting main quantum numbers are used for calculation with Kurucz program. "Balmer indices" as a function of the star atmospheres' parameters are calculated. They can be used in the "Balmer indices" method – the new method of spectral classification of hot stars.

114.087   Determination of physical characteristics of atmospheres of B and A stars by the Balmer indices method.
S. A. Gulyaev, V. E. Panchuk, S. G. Pyatkes.
Astrofiz. Issled. Izv. Spets. Astrofiz. Obs., Tom 25, p. 60 – 67 (1987). In Russian. English translation in Bull. Spec. Astrophys. Obs. – North Caucasus, Vol. 25.

Spectra (28 Å/mm) have been obtained with the telescopes BTA and Zeiss–600 and spectrophotometry has been performed for 57 bright stars of B and A spectral classes. With the help of a new method of Balmer indices effective temperatures T_e and gravities g in the atmospheres of these stars have been determined. The internal accuracy of the method is high, the effective temperature determinations agree with the calibration by Kopylov. Systematic discrepancy in $\lg g$ found by the method of Balmer indices (from the features of the continuous spectrum in the region of the Balmer jump), and from the equivalent widths of the first members of the Balmer series is revealed. Possible reasons for this discrepancy are discussed.

114.088   Spectral variability of Wolf–Rayet–type stars.
S. V. Marchenko.
Inst. Teor. Fiz. Akad. Nauk USSR, Prepr., No. 55R, p. 1 – 25 (1987). In Russian. Abstr. in Ref. Zh., 51. Astron. 11.51.573 (1987).

114.089   UV line profile variations in Wolf–Rayet stars.
G. Koenigsberger, L. H. Auer.
Rev. Mex. Astron. Astrofis., Vol. 14, No. 1, p. 271 – 276 (1987). – See Abstr. 012.042.

IUE observations of line–profile variability in Wolf–Rayet stars are presented. Different mechanisms are discussed which are expected to produce these variations.

114.090   Identifying lines in the IUE spectrum of the Wolf–Rayet star HD 193077.
G. Koenigsberger, L. H. Auer.
Rev. Mex. Astron. Astrofis., Vol. 14, No. 1, p. 277 – 283 (1987). – See Abstr. 012.042.

The IUE spectrum of HD 193077 (WN6+O+?) is rich in lines. The authors present probable identifications for all emission lines which are evident in the 1240 – 1800 Å wavelength range, and they discuss several aspects of an intriguing narrow absorption line spectrum.

114.091   Spectrophotometry of northern Wolf–Rayet stars: a sample of WN4, WN5 and WN6 stars.
J. F. Barral, G. F. Bisiacchi, C. Firmani, G. Koenigsberger, J. Wampler.
Rev. Mex. Astron. Astrofis., Vol. 14, No. 1, p. 284 – 286 (1987). – See Abstr. 012.042.

Results of the measurement of emission line equivalent widths in WNE stars are presented which suggest that the WN6–B stars have the most massive winds among the WN stars.

114.092   An atlas of optical spectrophotometry of southern Wolf–Rayet carbon stars.
A. V. Torres, P. Massey.
Rev. Mex. Astron. Astrofis., Vol. 14, No. 1, p. 287 – 292 (1987). – See Abstr. 012.042.

The authors present a homogeneous set of optical spectrophotometric observations (3360 – 7350 Å) at moderate resolution (~ 10 Å) of almost all southern WC stars in the Galaxy and the LMC. The data are presented in the form of spectral tracings (in magnitude units) arranged by subtype, with no correction for interstellar reddening. A montage of the prototype stars of each spectral class is also shown. Line identifications are given.

114.093   A spectrographic study of two Of stars: HD 163758 and HD 117797.
D. O. Gómez, V. S. Niemela.
Rev. Mex. Astron. Astrofis., Vol. 14, No. 1, p. 293 – 299 (1987). – See Abstr. 012.042.

The authors have performed a spectrographic study of two Of type stars, namely HD 163758 and HD 117797. Radial velocities were measured and an estimate of the rates of mass loss was derived by comparison of the observed profiles of Hα and He II $\lambda 4686$ with the theoretical ones calculated by Klein and Castor (1978). The mass loss rate for both stars is $2 - 8 \times 10^{-6} M_\odot/y$. Relatively strong C lines are observed in the spectra of both stars, in addition to the spectral lines corresponding to Of type stars.

114.094   Line structures in the IUE spectra of γ^2 Velorum.
E. Brandi, O. E. Ferrer, J. Sahade.
Rev. Mex. Astron. Astrofis., Vol. 14, No. 1, p. 300 (1987). Abstract. – See Abstr. 012.042.

114.095   Resonance line–profiles in galactic disk UV–bright stars.
L. Carrasco, R. Costero, R. Stalio.
Rev. Mex. Astron. Astrofis., Vol. 14, No. 1, p. 301 – 309 (1987). – See Abstr. 012.042.

The authors have made a comparative analysis of UV resonance line–profiles in O–type stars members of young clusters and OB associations, with those of hot stars located away from sites of recent star formation (including "runaway" stars). The resonance line–profiles are found to be generally dominated by stellar winds that appear to depend mainly on the surface gravity and temperature of the star, and not on its mass. The use of only the largest stellar wind velocity detectable in the resonance lines as a stellar population indicator, is disputed.

114.096   New M–type supergiants in the southern Milky Way.
R. F. Wing, D. J. MacConnell, E. Costa.
Rev. Mex. Astron. Astrofis., Vol. 14, No. 1, p. 362 – 366 (1987). – See Abstr. 012.042.

An infrared objective–prism survey of the southern Milky Way from longitude 210° to 320° has revealed a large number of likely supergiants of late type. Follow–up observations by narrow–band photometry and CCD spectroscopy have confirmed that more than 100 of the candidate stars are indeed supergiants of types K and M, increasing very substantially the number of such stars known in this part of the sky. Since distances can be estimated for these stars from the spectroscopic and photometric data, the new supergiants should help to improve our knowledge of the spiral–arm structure of this part of the Galaxy.

114.097   CCD spectroscopy of new M–type supergiants in the southern Milky Way.
D. J. MacConnell, R. F. Wing, E. Costa.
Rev. Mex. Astron. Astrofis., Vol. 14, No. 1, p. 367 – 374 (1987). – See Abstr. 012.042.

The authors obtained CCD spectra of possible new M supergiants identified on low–dispersion, I–N objective–prism plates taken near the southern galactic plane. The spectra cover the range 6400 – 8800 Å with a resolution of 8 Å. The spectra of 180 stars have been observed at least once, and the spectra of about 40 late–type, luminous MK standards have been obtained as well. The standards are used to calibrate equivalent widths of the Ca II triplet as a function of luminosity and temperature, and this is the principal criterion used to classify the CCD spectra. Analysis of this material indicates that there are >60 new K/M supergiants, some very heavily reddened, and a few new S stars in the sample.

114.098   Chemical compositions of post–AGB stars.
H. E. Bond, R. E. Luck.
ESO Conf. Workshop Proc., No. 27, p. 103 – 106 (1987). – See Abstr. 012.051.

The low [Fe/H] ratios seen in high–latitude supergiants (and in RV Tau and W Vir variables) show that these stars belong to an old population. Therefore, they are most probably low–mass post–AGB stars, rather than normal Population I supergiants. The CNO abundances in these post–AGB stars clearly indicate the presence of hydrogen–burning products at the stellar surfaces. Both the high–latitude supergiants and the RV Tau/W Vir variables show systematically low s–process abundances, suggesting that the stars were originally of very low [Fe/H] and that their atmospheres are now significantly hydrogen–deficient.

114.099   HD 213985: a hot post–AGB star in the galactic halo.
C. Waelkens, L. B. F. M. Waters, A. Cassatella, T. Le Bertre, H. J. G. L. M. Lamers.
ESO Conf. Workshop Proc., No. 27, p. 107 (1987). Abstract. The full paper will be published in Astron. Astrophys. Lett. – See Abstr. 012.051.

114.100   The halo metallicity distribution function.
J. B. Laird, M. P. Rupin, B. W. Carney, D. W. Latham.
ESO Conf. Workshop Proc., No. 27, p. 135 – 139 (1987). – See Abstr. 012.051.

A photometric and spectroscopic survey of over 900 Lowell proper motion stars has recently been completed by Carney and Latham. Metallicities of most of these stars have been measured from the low signal–to–noise spectra used for the radial velocities. A sample of true halo stars has been isolated from the survey by choosing stars which have retrograde orbits. Their metallicity distribution function is compared to that of globular clusters, and both are compared to a theoretical distribution based on a simple model with mass loss. The cluster distribution function is somewhat different, having relatively fewer objects at both the metal–poor and metal–rich ends. The differences are not significant.

114.101   The "odd–even" effect in metal–deficient dwarfs.
P. François.
ESO Conf. Workshop Proc., No. 27, p. 147 – 151 (1987). – See Abstr. 012.051.

The analysis of cosmic abundances shows that neutron–rich species (odd elements) are less abundant than other species (even elements). This is usually called the "odd–even" effect. It has been predicted that this effect should be enhanced in metal–deficient stars. The author presents a determination of the "odd–even" effect enhancement in metal deficient stars by means of the ratios [Ag/Mg] and [Na/Mg].

114.102   Manganese abundances in metal poor stars.
R. G. Gratton.
ESO Conf. Workshop Proc., No. 27, p. 153 – 158 (1987). – See Abstr. 012.051.

The possible presence of an enhanced odd–even effect among iron group elements, indicating an underabundance of Mn in metal poor stars, has been suggested several years ago by Helfer et al. (1959). The author analyzed for Mn abundance a sample of 13 metal poor giants observed some years ago with CASPEC. These spectra have already been studied for abundances of several elements (from Na to Ni). They consist of blue (3700 – 4450 Å) and yellow (5000 – 6000 Å) spectra and have S/N ratios from 80 to more than 200 (generally the blue spectra having a lower S/N).

114.103   Boron abundance in the pop II star HD 140283.
P. Molaro.
ESO Conf. Workshop Proc., No. 27, p. 159 – 164 (1987). – See Abstr. 012.051.

A high resolution IUE spectrum of the very metal deficient star HD 140283 has been analyzed to derive the abundance of boron. B appears deficient by at least a factor of 40 compared to the solar value. This is the first measurement of B abundance in a pop II star and, except for the Sun, the only performed on a solar–type star. The present value of B appears to exclude the possibility that some or all of the Li abundances in the pop II stars could be accounted for by spallation processes induced by high energy cosmic rays or by a yet hypothetical component of suprathermal particles.

114.104 ^{9}Be abundances in halo dwarfs: implications for early nucleosynthesis of light elements.
C. Abia, P. Molaro, R. Rebolo, J. Beckman.
ESO Conf. Workshop Proc., No. 27, p. 165 – 171 (1987). – See Abstr. 012.051.

The spectral range of ^{9}Be II resonance doublet at 3130 Å was observed in six highly metal poor dwarfs of known ^{7}Li abundance. In three of the stars the authors detected ^{9}Be with abundances from 1.0×10^{-12} to 2.5×10^{-12}. These are the first reported detections of ^{9}Be in such metal deficient stars. Taken together with literature ^{9}Be abundances in more metal rich objects, these low values rule out models of galactic evolution with an early burst of ^{9}Be production.

114.105   Carbon star surveys in the Magellanic Clouds and the dwarf spheroidals in the galactic halo.
M. Azzopardi.
ESO Conf. Workshop Proc., No. 27, p. 191 – 200 (1987). – See Abstr. 012.051.

114.106   Detection of carbon stars in two independent surveys.
M. F. McCarthy.
ESO Conf. Workshop Proc., No. 27, p. 203 – 205 (1987). – See Abstr. 012.051.

114.107   The chemical composition of the atmospheres of the K giants 9 Bootis and ϱ Bootis.
M. E. Boyarchuk, M. Ya. Orlov, A. V. Shavrina.
Kinematika Fiz. Nebesn. Tel, Tom 3, No. 6, p. 59 – 61 (1987). In Russian. English translation in Kinematics Phys. Celest. Bodies.

The abundances of chemical elements in the atmospheres of K giants 9 Boo and ϱ Boo are determined from high–dispersion spectrograms by the method of model atmospheres.

114.108   Helium abundance in the atmospheres of B stars in open clusters.
V. G. Klochkova, V. E. Panchuk.
Sov. Astron. Lett., Vol. 12, No. 6, p. 387 – 390 (1986). English translation of 42.114.053.

114.109   A new word on the spectral classification of stars.
E. K. Kharadze, R. A. Bartaya.
Astrophys. Space Sci., Vol. 137, No. 2, p. 211 – 215 (1987).

With the deepening of our knowledge on the stellar Universe, the problem is set forth concerning a further development of the spectral classification of stars – the actual means for investigation of the realm of stars. A new word on the spectral classification of stars – the idea of the "MK Process" – has been coined by Morgan; it marks a new stage in the study of our Universe.

114.110   Neutron density and neutron source determination in barium stars.
R. A. Malaney.
Messenger, No. 50, p. 37 – 40 (1987).

114.111   Identification of lines in the satellite ultraviolet: the spectrum of Tau Scorpii.
C. R. Cowley, D. R. Merritt.
Astrophys. J., Vol. 321, No. 1, p. 553 – 559 (1987).
The method of wavelength coincidence statistics (WCS) is applied to the ultraviolet spectra of τ Sco obtained by the IUE. No identifications of elements heavier than Zn were made. With the exception of Zn IV, all identifications had been previously made in an exhaustive study of Copernicus spectra by Rogerson and Ewell. A comparison of the identifications made in the two studies is in very good agreement for those spectra rich in lines, where the WCS parameters indicate the strong presence of the species. Useful information is added by WCS in cases where species are only weakly, or arguably, present.

114.112   A comprehensive exploration of late–type stars in the far–ultraviolet.
J. O. Bennett.
Diss. Abstr. Int., Sect. B, Vol. 48, No. 5, p. 1383–B (1987). Thesis, University of Colorado at Boulder, 206 pp. (1987). Order No. DA8716238.

114.113   Continuum flux depressions of chemically peculiar stars and their connection with the magnetic field.
V. S. Lebedev.
Astron. Zh., Tom 64, Vyp. 6, p. 1308 – 1311 (1987). In Russian. English translation in Sov. Astron., Vol. 31, No. 6.
Existence is suspected of two groups of chemically peculiar stars with different character of dependence between magnetic fields and the amount of flux depression at λ5200 Å. The stars with weak dependence have significantly greater values of radial velocities, while their other observational characteristics do not differ from those of the main group of stars. Hypotheses on the origin of flux depressions are discussed.

114.114   New carbon stars at the galactic longitude of 82°.
A. Alksnis, Z. Alksne, I. Platais, V. Ozolina.
Issled. Solntsa Krasnykh Zvezd, No. 25, p. 5 – 12 (1987). In Russian.
Objective prism infrared plates taken with the Schmidt telescope of the Radioastrophysical Observatory in a 4°5 wide and 25° long zone perpendicular to the galactic equator and centered on the galactic longitude $l = 82°$ were used for carbon stars' search. 16 new carbon stars were found and one earlier suspected carbon star confirmed. Positions of the new carbon stars BC 256 – BC 272 were measured on direct plates and their equatorial and galactic coordinates determined. 17 carbon stars are identified on finding charts.

114.115   Zirconium stars in open clusters.
U. Dzĕrvitis.
Issled. Solntsa Krasnykh Zvezd, No. 25, p. 29 – 41 (1987). In Russian.
Positions of all known zirconium stars and open galactic clusters have been compared and cases searched when the star at the sky is situated inside three radii of the cluster. In such cases, if there were corresponding photometric data, the absolute magnitudes of the zirconium stars were determined assuming the membership to the clusters.

114.116   Spectral synthesis in the ultraviolet. I. Far–ultraviolet stellar library.
M. N. Fanelli, R. W. O'Connell, T. X. Thuan.
Astrophys. J., Vol. 321, No. 2, p. 768 – 779 (1987).
A library of mean stellar energy distributions for use in population synthesis of the ultraviolet spectra of active star–forming galaxies is derived from the ultraviolet spectrophotometry in the IUE Spectral Atlas. The spectra extend from 1230 to 1930 Å with a resolution of 6 Å. The library contains eight main–sequence groups from O3 to A7 V, four giant groups from O5 to B9 III, and three supergiant groups from O9 to A0 I. Several continuum and spectral line indices are computed, and their usefulness as temperature and luminosity discriminants is discussed.

114.117   Barium abundances in a sample of red giants.
M. Cornide, J. L. Fernandez–Villacanas.
Bull. Etoiles Tardives Spectre Particulier, No. 5, p. 3 – 4 (1987).

114.118   The star ν Cyg as a regional spectrophotometric standard in the Cygnus region.
I. S. Shityuk, E. A. Glushkova.
Astron. Tsirk., No. 1470, p. 5 – 6 (1986). In Russian.
The data on the energy distribution E(λ) erg/cm³sec for ν Cyg chosen as a "primary" regional standard for the Cygnus region are presented.

114.119   Continuum energy dissipation of 48 Persei.
N. S. Tur, P. S. Goraya, U. S. Chaubey.
Astrophys. Space Sci., Vol. 139, No. 2, p. 257 – 262 (1987).
The continuum energy distribution data of 48 Per are presented in the optical region (λλ3200 to 8000 Å). The continuum data in the ultraviolet and infrared wavelength range are also gathered to study the behaviour over the wider wavelength base–line.

114.120   A model atmosphere analysis of HD 25823.
C. Bolcal, D. Kocer, A. Duzgelen.
Astrophys. Space Sci., Vol. 139, No. 2, p. 295 – 304 (1987).
An abundance analysis using photographic region spectrograms and fully line–blanketed model atmospheres has been performed for the chemically peculiar star HD 25823. Model parameters were determined from the Hγ profile and photometric data. Its pattern of abundance shows that the heavier elements in HD 25823 are overabundant.

114.121   Surface distribution according to the spectral type of the stars in the region of the Cygnus OB4 association.
Ts. R. Radoslavova.
Dokl. Bolg. Akad. Nauk, Tome 40, No. 1, p. 7 – 9 (1987).

114.122   MK classifications of stars in Cygnus.
C.–c. Huang.
Chin. Astron. Astrophys., Vol. 11, No. 3, p. 229 – 233 (1987). English translation of Acta Astron. Sin., Vol. 28, No. 2, p. 120 – 126 (1987).
MK spectral classifications are given for 24 stars of types B and A in Fehrenbach's selected areas No. 4 for radial velocity investigations. A spectral atlas of the standard stars is given, and the main criteria for classification are described. Many of the target stars have broad lines and one of them, HD 192605 is especially interesting, having a large Balmer progression of radial velocity.

114.123   Low–dispersion spectra of six Me stars.
Y.–w. Huang, J.–y. Liu, X.–c. Feng.
Chin. Astron. Astrophys., Vol. 11, No. 4, p. 337 – 342 (1987). English translation of 43.114.098.

114.124   The third spectrum of erbium in HR 465.
C. R. Cowley, M. Greenberg.
Publ. Astron. Soc. Pac., Vol. 99, No. 621, p. 1201 – 1205 (1987).
The authors discuss the strong presence of Er III in coudé spectra of the chemically peculiar (CP) A star HR 465 at the time of rare–earth maximum. Er III is only a marginal detection on IUE images of HR 465. A very crude abundance calculation, using an assumed oscillator strength for a "typical" line gives a value for erbium some two orders of magnitude below earlier results based on Er II. An incidental identification, Dy III in γ Equ, is briefly discussed.

114.125   The early A type stars: refined MK classification, confrontation with Strömgren photometry, and the effects of rotation.
R. O. Gray, R. F. Garrison.
Astrophys. J., Suppl. Ser., Vol. 65, No. 4, p. 581 – 602 (1987).
The authors refine the MK classification system for the early A type stars and introduce broad–lined (high $v \sin i$) standards. The resulting classifications of 372 early A type stars have a greater degree of precision and are free from systematic effects

with rotational line broadening. The confrontation of these classifications with Strömgren photometry reveals photometric effects of rotation on the $b - y$ and β indices. The effect on the β index is compared with theory.

114.126 Non–LTE analysis of carbon lines in hot–star spectra. III. The C II resonance lines and $\lambda4267.2$ in B stars.
N. A. Sakhibullin.
Sov. Astron., Vol. 31, No. 2, p. 151 – 158 (1987). English translation of 43.114.035.

114.127 Emissions observed on objective–prism spectrograms.
T. Radoslavova.
Inf. Bull. Variable Stars, No. 3047, 2 pp. (1987).

114.128 A remark on TV Cygni.
W. Wenzel.
Inf. Bull. Variable Stars, No. 3113, 1 p. (1987).

114.129 Gamma Lupi does not appear to be a Be star.
D. Baade.
Inf. Bull. Variable Stars, No. 3123, 4 pp. (1987).

114.130 The "Alternative X" question: what causes the strong molecular bands and damping lines in very strong lined (super–metal–rich) stars?
B. J. Taylor, S. B. Johnson.
Astrophys. J., Vol. 322, No. 2, p. 930 – 948 (1987).
This paper considers absorption features in nearby, very strong lined (VSL) G and K stars. These are the stars which traditionally have been called "super–metal–rich" (SMR). The data include measurements made by Spinrad and Taylor, more recent data of the same kind, and a temperature index (Cousins $R - I$). The other indices adopted measure narrow–band blanketing, broad–band blanketing, and strong absorption features of CN, Ca I, CH, Mg I, and Na I. For dwarfs, data for Fe I and Ca II are also analyzed. It is found that with two exceptions (CH in giants, Ca II in dwarfs), G and K VSL stars have enhancements in all these features. For disk G and K stars in general, there is a positive correlation between the temperature–corrected strengths of most indices. A feature is enhanced in VSL stars if and only if it participates in this correlation. The VSL stars are therefore simply strong–featured examples of disk stars. For non–VSL stars, stronger (weaker) features are associated with higher (lower) overall metallicities. If this correlation applies also to VSL stars, they are super–metal–rich. The authors discuss possibilities which might explain the VSL enhancements without appeal to supermetallicity. None of these "Alternatives X" are clearly satisfactory, but the existence for an Alternative X is not ruled out.

An atlas of optical spectrophotometry of Wolf–Rayet carbon and oxygen stars.
See Abstr. 002.081.

Catalogue of determinations of metallicity, velocity components and orbital elements of F5–K5 dwarfs in the neighbourhood of 80 parsecs from the Sun.
See Abstr. 002.101.

Catalogue of determinations of [Fe/H], velocity components and orbital elements of F stars.
See Abstr. 002.102.

The M–type stars.
See Abstr. 003.075.

The High Altitude Observatory–Lowell Observatory Solar–Stellar Spectrophotometry Project.
See Abstr. 013.003.

The solar–stellar connection at low spectral resolution.
See Abstr. 013.005.

Laboratory study of the rotational spectrum of vibrationally excited C_2H.
See Abstr. 022.088.

Doppler imaging.
See Abstr. 036.003.

A survey of proper–motion stars. II. Extracting metallicities from high–resolution, low–S/N spectra.
See Abstr. 036.054.

Speckle imaging at CfA.
See Abstr. 036.135.

The effect of C/O ratio on the Blackwell–Shallis method of determining stellar temperatures for cool evolved stars.
See Abstr. 036.247.

An HIPPARCOS astrometric programme for spectral classification standard stars.
See Abstr. 041.012.

Thorium in G–dwarf stars as a chronometer for the Galaxy.
See Abstr. 061.034.

Th/Nd abundance ratio in the surfaces of G–dwarfs.
See Abstr. 061.099.

Reply to "Th/Nd abundance ratio in the surfaces of G–dwarfs" by D. D. Clayton.
See Abstr. 061.100.

The wings of the calcium infrared triplet lines in solar–type stars.
See Abstr. 064.003.

An extension to the wavelength coincidence statistics for spectral line identification.
See Abstr. 064.004.

Line blanketed model atmospheres of Ap–stars. VI. HD 221568.
See Abstr. 064.046.

Evolutionary helium and CNO anomalies in the atmospheres and winds of massive hot stars.
See Abstr. 064.055.

Selective mass loss, abundance anomalies, and helium–rich stars.
See Abstr. 064.077.

The s–process nucleosynthesis of barium stars.
See Abstr. 065.100.

Convection, magnetic fields, and line asymmetry in the sun and stars.
See Abstr. 071.001.

Computations of radial velocities for O– and B–type stars.
See Abstr. 111.006.

Radial velocities in selected B–G stars.
See Abstr. 111.008.

Mesures de vitesses radiales. I. Accompagnement au sol du programme d'observation du satellite Hipparcos.
See Abstr. 111.011.

Mesures de vitesses radiales. II. Accompagnement au sol du programme d'observation du satellite Hipparcos.
See Abstr. 111.012.

Long–term and mid–term spectroscopic variations of the Be–shell star HD 184279 (V1294 Aql). I. Observational data.
See Abstr. 112.003.

Spectral features of the B2e star EW Lac before and during the variable shell phase.
See Abstr. 112.004.

HD 213985: a hot post–AGB star in the galactic halo.
See Abstr. 112.007.

The peculiar emission–line supergiant HD 37836.
See Abstr. 112.012.

The UV high resolution spectrum of A–type supergiants.
See Abstr. 112.013.

IRAS observations of CP stars.
See Abstr. 112.014.

Long–term variability of the far–UV high velocity components in γ Cas (1978 – 1986).
See Abstr. 112.017.

Chemical composition of the atmosphere of K giant α Ser.
See Abstr. 112.025.

Metals in IRC + 10216: detection of NaCl, AlCl, and KCl, and tentative detection of AlF.
See Abstr. 112.030.

CO(J = 1–0) observations of bright carbon stars.
See Abstr. 112.031.

Chemical modelling of molecular sources. V. IRC + 10216.
See Abstr. 112.033.

Different regions of line formation in the envelope of the early emission line star HD 190073.
See Abstr. 112.034.

Abundances in red giant stars: carbon and oxygen isotopes in carbon–rich molecular envelopes.
See Abstr. 112.059.

Hot carbon stars: more about V348 Sgr.
See Abstr. 112.116.

Chromospheric Mg II h and k emissions free of interstellar contamination: velocity structure in late–type dwarfs and giants.
See Abstr. 112.121.

An analysis of the emission features of the IRAS low–resolution spectra of carbon stars.
See Abstr. 112.125.

Spectroscopy of southern Be stars 1984 – 1987.
See Abstr. 112.148.

The ultraviolet spectra of cool star chromospheres; excitation processes and plasma diagnostics.
See Abstr. 112.157.

Hydrogen emissions on low–dispersion spectrograms of B–stars.
See Abstr. 112.184.

The missing opacity and the temperature calibration of solar–type stars.
See Abstr. 113.008.

On the colours of Am stars.
See Abstr. 113.010.

Spectroscopic tests of two faint photometric surveys.
See Abstr. 113.036.

Photometric classification of early type stars with hydrogen emission lines.
See Abstr. 113.048.

The helium abundance and luminosity of the main–sequence F stars as derived from uvbyβ photometry.
See Abstr. 115.006.

Fundamental parameters of the underlying Be stars.
See Abstr. 115.010.

Accurate angular diameters and effective temperatures for eleven giants cooler than K0 by Michelson interferometry.
See Abstr. 115.016.

Photoelectric observations of lunar occultations of stars. Angular diameter of the carbon star Y Tauri and its physical characteristics.
See Abstr. 115.020.

Narrow–band photometry of the magnetic variable β CrB.
See Abstr. 116.007.

The rapidly rotating spotted red dwarf flare star Gliese 890.
See Abstr. 116.012.

An analysis of the photospheric line profiles in F, G, and K supergiants.
See Abstr. 116.066.

The interacting binary β Lyr. II. Non–LTE model analysis and evolutionary conclusions.
See Abstr. 117.011.

CPD –71°172, a new binary with a hot subdwarf.
See Abstr. 117.015.

The 67–min X–ray period of EX Hydrae observed with the EINSTEIN observatory.
See Abstr. 117.055.

Photometry and spectroscopy of the O–type variable HD 167971.
See Abstr. 117.156.

A deep, Doppler–compensated IUE SWP echellogram of the K0 primary of HR 1099.
See Abstr. 117.173.

Stars of type MS with evidence of white dwarf companions.
See Abstr. 117.178.

Light element abundances in the atmosphere of the principal component of the binary system υ Sgr.
See Abstr. 117.190.

The anomalous ultraviolet spectrum of the AM Her star H 0538 + 608.
See Abstr. 117.235.

A survey study of energy distribution in component stars of Algol–type binary systems.
See Abstr. 117.245.

Spectral investigation of the atmospheres of eclipsing binaries in the stage of mass exchange.
See Abstr. 119.017.

Photometry and spectroscopy of the eclipsing P Cygni star R 81 in the Large Magellanic Cloud.
See Abstr. 119.021.

Atmospheric eclipses in the SMC Wolf–Rayet eclipsing binary HD 5980: the heavy versus the light metal abundance.
See Abstr. 119.063.

The strange "spots" on the T Tauri star RY Lupi.
See Abstr. 121.010.

A study of the nature of the Hα–emission stars LkHα 112, 115, 118, and 119.
See Abstr. 121.018.

Atmospheric parameters and iron abundance in T Tauri from lines of moderate intensity.
See Abstr. 121.055.

Flare stars.
See Abstr. 122.001.

Rotational modulation and flares on RS CVn and BY Dra–type stars. V. EXOSAT and IUE observations of a flare on EQ Pegasi.
See Abstr. 122.011.

The rapidly oscillating Ap stars as a test of stellar chromospheric heating mechanisms.
See Abstr. 122.014.

Rapid variability of the Hα line in the spectrum of τ Pegasi.
See Abstr. 122.039.

Discovery of a magnetic DA white dwarf with distinct Hβ and Hα Zeeman triplets.
See Abstr. 126.045.

EUV photometry of DA white dwarfs with EXOSAT.
See Abstr. 126.061.

A search for interstellar NaH and MgH in diffuse clouds.
See Abstr. 131.241.

Hydrogen and deuterium in the local interstellar medium.
See Abstr. 131.256.

Galactic interstellar abundance surveys with IUE and IRAS.
See Abstr. 131.257.

Ultraviolet interstellar extinction toward NGC 6530 and the intrinsic energy distribution of 9 Sagittarii and HD 165052.
See Abstr. 131.305.

The LMC H II regions N11C and E and their stellar contents.
See Abstr. 132.013.

Ground–based infrared observations of variable IRAS sources as candidates for late asymptotic giant branch stars.
See Abstr. 133.022.

Optical spectroscopy of IRAS sources with infrared emission bands: IRAS 21282 + 5050 and the diffuse interstellar bands.
See Abstr. 133.023.

NGC 40: IUE observations of the nucleus.
See Abstr. 134.005.

Nebular spectrum of CPD –56°8032 and He2–113.
See Abstr. 134.023.

Proto–planetary nebulae: models and IRAS observations.
See Abstr. 134.029.

Lithium in the Coma star cluster.
See Abstr. 153.047.

The calcium abundance in NGC 1851.
See Abstr. 154.043.

A spectroscopic study of HB stars in the galactic globular cluster NGC 6752.
See Abstr. 154.069.

The metallicity distribution of the extreme halo population.
See Abstr. 155.078.

The distribution of OB stars and dust in a Milky Way field at $(l, b) = (335°, 0°)$.
See Abstr. 155.084.

M giants in Baade's window: infrared colors, luminosities, and implications for the stellar content of E and S0 galaxies.
See Abstr. 155.088.

Low–dispersion spectral sky survey to find faint carbon stars. I. Region $115° \leqslant l \leqslant 130°, -5° \leqslant b \leqslant +5°$.
See Abstr. 155.107.

The chemical composition of field halo stars.
See Abstr. 155.134.

The chemical composition of peculiar field stars in the galactic halo.
See Abstr. 155.135.

Isotopic ratios in the halo.
See Abstr. 155.136.

Serendipitous discovery of two distant halo stars of A–spectral type.
See Abstr. 155.138.

The early chemical evolution of the halo and population III.
See Abstr. 155.144.

A low–dispersion spectral sky survey for revealing faint carbon stars. II. The region $130° \leqslant l \leqslant 145°, -5° \leqslant b \leqslant +5°$.
See Abstr. 155.167.

Spectral types of bright stars in the Small Magellanic Cloud Wing.
See Abstr. 156.005.

Spectral types of bright stars in the north–east arm of the Small Magellanic Cloud.
See Abstr. 156.006.

C, M, and S stars in the Magellanic Clouds.
See Abstr. 156.037.

Preliminary abundances in three field supergiants of the SMC.
See Abstr. 156.039.

On the chemical abundance of the SMC stars.
See Abstr. 156.040.

The carbon star population of the Small Magellanic Cloud.
See Abstr. 156.043.

Spectrophotometric abundances of red Magellanic Cloud super-giants.
See Abstr. 156.044.

The spectra of extragalactic Wolf–Rayet stars.
See Abstr. 157.261.

115 Luminosities, Masses, Diameters, HR and other Diagrams

115.001 HR 4049: an old low–mass star disguised as a young massive supergiant.
C. Waelkens, H. Lamers, R. Waters.
Messenger, No. 49, p. 29 – 32 (1987).

115.002 On the slope of the mass function for stars of small masses.
O. Yu. Malkov.
Astrofizika, Tom 26, Vyp. 3, p. 477 – 487 (1987). In Russian.
English translation in Astrophysics, Vol. 26, No. 3, p. 288 – 294 (1987).
It has been shown that the usual method of construction of the initial mass function (IMF) involves a number of uncertainties which can essentially change the IMF slope at small masses ($m < m_\odot$). The influence of the mass–luminosity relation, BC–scale and luminosity function variations on the IMF form is studied. The effect of unresolved binaries is also discussed. Some quantitative estimates are made and it is shown that the slope at small masses can vary in wide range.

115.003 Photoelectric observations of lunar occultations.
B. Stecklum.
Astron. J., Vol. 94, No. 1, p. 201 – 207 (1987).
The results of a program to record photoelectrically lunar occultations at Jena Observatory are described. The analysis of ten observations is reported. Angular–diameter estimates are given for 97 Tau, ε Gem, and μ Psc. The quintuple nature of the star 63 Gem is confirmed.

115.004 The upper HR diagram – an observational overview.
R. M. Humphreys.
Instabilities in luminous early type stars, p. 3 – 22 (1987). – See Abstr. 012.012.
The author reviews the upper HR diagram and the observational uncertainties in its determination. She also discusses the upper luminosity boundary defined by the most luminous hot and cool stars, and the observational evidence for high mass loss rates and instabilities in these stars' atmospheres.

115.005 Rate of mass–loss in the Hertzsprung–Russell diagram.
C. de Jager, H. Nieuwenhuijzen,
K. A. van der Hucht.
Instabilities in luminous early type stars, p. 205 – 207 (1987). – See Abstr. 012.012.
The authors have collected literature data on rates of mass–loss $\dot{M}$ for 264 O through M–type stars and a number of stars of other types. It appears possible to develop $\log(-\dot{M})$ into a series of Chebychev polynomia of a the first kind in $\log T_{eff}$ and $\log(L/L_\odot)$ and their cross products.

115.006 The helium abundance and luminosity of the main–sequence F stars as derived from uvbyβ photometry.
Yu. G. Shevelev, A. A. Suchkov, V. A. Marsakov.
Astron. Zh., Tom 64, Vyp. 4, p. 747 – 754 (1987). In Russian.
English translation in Sov. Astron., Vol. 31, No. 4.
The relation between the metallicity of F stars and their loci on the c_1,β diagram is investigated for a sample of 2000 galactic-disk F stars having uvbyβ data. It is found that the lower envelope and the mean main sequence for stars with larger metallicities lie lower than those for the stars with lesser metallicities, though larger metallicity requires them to lie higher. The discrepancy is suggested to indicate the larger helium abundance in more metal–rich stars. The required difference in the Y value is found to be $\Delta Y \approx 0.1$ if the disk stars constitute two discrete metallicity groups with $z_1 \approx 0.01$ and $z_2 \approx 0.02$. The authors conclude that the c_1,β diagram for F stars gives evidence in favour of the discreteness of the actual metallicity distribution. It is shown that with increasing metallicity the helium abundance runs as $\Delta Y \approx 10\,\Delta Z$ (in contrast to the widely used relation $\Delta Y \approx 3\,\Delta Z$).

115.007 On the characteristics of the high–luminosity OB stars in the M33 galaxy.
P. Z. Kunchev, G. R. Ivanov.
Dokl. Bolg. Akad. Nauk, Tome 39, No. 12, p. 5 – 8 (1986).

115.008 Stellar angular dimensions obtained by the lunar occultation technique.
E. M. Trunkovskij.
Pis'ma Astron. Zh., Tom 13, No. 10, p. 900 – 906 (1987). In Russian. English translation in Sov. Astron. Lett., Vol. 13.
The photoelectric observations of the lunar occultations of 25 stars are obtained in 1982 at the 48–cm reflector of the High–Altitude Observatory of the Sternberg Astronomical Institute near Alma-Ata. The values of stellar angular dimensions for 13 stars or close stellar systems and some other parameters are determined from the analysis of the diffraction patterns registered with resolution time 1 ms. The occultation of SAO 79361 was observed during the total lunar eclipse at 9 January 1982. In the case of the star HR 1997 (the well known astrometric and spectral binary ADS 4392), a duplicity is evidently seen on the trace; this allows the evaluation of a vector separation of $\varrho_x \cong 0\rlap{.}{''}065$. In some cases the values of angular diameters obtained by single–star model reduction are considerably larger than the reasonable indirect evaluations; the author suggests that these stars are close binaries or multiple systems. A table of results including geometric occultation time determinations (with accuracy of 1 – 5 ms) is presented.

115.009 Does the slope of the initial mass function change for low–mass stars?
O. Yu. Malkov.
Nauchn. Inf., Vyp. 63, p. 19 – 35 (1987). In Russian.
The author investigates the influence of the mass–luminosity relation, the luminosity function and of reasons changing their shape on the initial mass function of low–mass stars. It is shown that the uncertainties of the BC–scale and effects of unresolved binaries can essentially influence the slope of the resulting IMF. Various theoretical mass–luminosity relations available in literature also lead to different initial mass functions. Chemical composition only weakly influences the slope of the IMF till it remains in the typical galactic disk population limits.

115.010 Fundamental parameters of the underlying Be stars.
A. Slettebak.
Physics of Be stars, p. 24 – 37 (1987). – See Abstr. 012.030 (IAU Colloq. No. 92).
The various methods for determining masses, radii, luminosities, effective temperatures, spectral types, and rotational velocities of the underlying Be stars are reviewed, and representative values listed for each.

115.011 A search for brown dwarfs and late M dwarfs in the Hyades and the Pleiades.
B. Zuckerman, E. E. Becklin.
Astrophys. J., Lett. Ed., Vol. 319, No. 2, p. L99 – L102 (1987).
The authors measured the J and K colors of 14 white dwarfs that are believed to be single stars and members of either the Hyades or Pleiades clusters or the Hyades supercluster and they found no indication of any excess 2.2 μm emission above that expected from the white dwarf. Based on recently published theoretical cooling curves for brown dwarfs, one can rule out the existence of any cool companion stars with masses greater than approximately 0.03 $M_\odot$ within a radius of 6'' of eight white dwarfs in the Hyades cluster and greater than approximately 0.015 $M_\odot$ toward the single white dwarf in the Pleiades.

115.012 The Einstein view of the Wolf–Rayet stars.
A. M. T. Pollock.
Astrophys. J., Vol. 320, No. 1, p. 283 – 295 (1987).
A uniform analysis of all 48 Wolf–Rayet stars observed with the IPC of the *Einstein Observatory* shows that their X-ray

luminosities cover a range of more than two orders of magnitude. Most of the brightest stars are either also sources of nonthermal radio radiation, in which case the measurements interpreted as Compton scattering of photospheric radiation by relativistic electrons imply surface magnetic fields of up to a few hundred gauss, or massive binary systems where the X–rays could come from colliding winds. The single stars and many of the stars that have been proposed as low–mass binaries are generally among the faintest objects.

115.013   The absolute magnitude of the barium stars.
J. Hakkila.
Bull. Am. Astron. Soc., Vol. 19, No. 2, p. 702 (1987). Abstract. – See Abstr. 010.061.

115.014   Late–type supergiant absolute magnitudes from UV binaries.
S. B. Parsons, T. B. Ake.
Bull. Am. Astron. Soc., Vol. 19, No. 2, p. 708 (1987). Abstract. – See Abstr. 010.061.

115.015   Wolf–Rayet, the storm–stars.
N. Reeves.
Ciel Espace, No. 215, p. 43 – 46 (1987). In French. Abstr. in Phys. Abstr., Vol. 90, No. 1312, Entry 107852 (1987).

115.016   Accurate angular diameters and effective temperatures for eleven giants cooler than K0 by Michelson interferometry.
G. P. Di Benedetto, Y. Rabbia.
Astron. Astrophys., Vol. 188, No. 1, p. 114 – 124 (1987).
Accurate angular diameters of cool bright giants have been measured at 2.2 μm by the I2T interferometer, operating at the CERGA observatory. The derived effective temperatures have been adopted as a high–quality reference sample to improve both the empirical effective temperature scale for K–M giants and the semiempirical estimations of T_{eff}. In the colour diagrams, the new calibration shows evidence of a fine structure, arising in the transition from K to M spectral–type, with fairly different behaviours for the two branches. It is strengthened that the K–scale does not require to be revised, being in excellent agreement with results from model atmospheres. The M–scale should be lowered towards the color temperature, approaching the M0 spectral–type; this appears quite consistent with a recent theoretical investigation.

115.017   Lunar occultations at La Silla.
A. Richichi.
Messenger, No. 50, p. 6 – 8 (1987).

115.018   A diagram for photometric recognition of barium CH, and R–types stars.
V. Straizys, J. Sleivyte.
Bull. Etoiles Tardives Spectre Particulier, No. 5, p. 1 – 2 (1987).

115.019   Colour–magnitude calibrations for red dwarfs.
H. Jahreiß.
Bull. Inf. Cent. Données Stellaires, No. 33, p. 51 – 57 (1987).
A new calibration procedure is presented which appears to be easily applicable to the derivation of new colour–magnitude relations. The introduction of the biweight provides an objective method for the handling of outliers.

115.020   Photoelectric observations of lunar occultations of stars. Angular diameter of the carbon star Y Tauri and its physical characteristics.
E. M. Trunkovskij.
Sov. Astron., Vol. 31, No. 2, p. 195 – 205 (1987). English translation of 43.115.008.

A catalog of stellar angular diameters measured by lunar occultation.
See Abstr. 002.020.

Catalogue of determinations of [Fe/H], distances, velocity components and orbital elements of F stars of the southern Galactic pole.
See Abstr. 002.103.

Status of the Mark III interferometer.
See Abstr. 032.018.

Initial stellar diameter measurements with the Mark III interferometer.
See Abstr. 034.058.

The calibration problem. III. First–order solution for mean absolute magnitude and dispersion.
See Abstr. 036.017.

The CERGA small interferometer.
See Abstr. 036.098.

The calibration problem. IV. The Lutz–Kelker correction.
See Abstr. 036.178.

Model study of wavelength–dependent limb–darkening and radii of M–type giants and supergiants.
See Abstr. 064.049.

Very low mass stars.
See Abstr. 065.033.

The core mass–radius relation for giants: a new test of stellar evolution theory.
See Abstr. 065.041.

Photographic astrometry of binary and proper–motion stars. III.
See Abstr. 111.007.

Red and infrared photometry of Be stars.
See Abstr. 112.084.

Absolute spectrophotometry of Be stars.
See Abstr. 112.087.

Near–infrared excesses of barium stars.
See Abstr. 113.031.

Observations of M dwarfs beyond 2.2 μm.
See Abstr. 114.013.

Wolf–Rayet stars.
See Abstr. 114.029.

HD 213985: a hot post–AGB star in the galactic halo.
See Abstr. 114.099.

The halo metallicity distribution function.
See Abstr. 114.100.

Rotation and magnetic activity in main–sequence stars.
See Abstr. 116.021.

On the parameters of the system Cyg X–1.
See Abstr. 117.065.

Gamma Persei – not overmassive but overluminous.
See Abstr. 118.008.

Direct infrared observations of the very low mass object Gliese 623B.
See Abstr. 118.015.

A dim star with bright prospects.
See Abstr. 118.027.

Excess infrared radiation from a white dwarf – an orbiting brown dwarf?
See Abstr. 118.028.

Epsilon Persei and Mu Centauri as single–periodic rapid variables.
See Abstr. 122.130.

Observational studies of Cepheids. VI. Period–radius relations.
See Abstr. 122.135.

Dynamical mass determinations for the white dwarf components of HZ 9 and Case 1.
See Abstr. 126.052.

Properties of blue stragglers in young OB associations.
See Abstr. 152.001.

Be stars as members of open clusters.
See Abstr. 153.016.

Mass and age distributions of stars in young open clusters.
See Abstr. 153.053.

The chicken came first.
See Abstr. 154.016.

Horizontal branch morphology in the log Te–log g plane: looking for multiple populations.
See Abstr. 154.041.

Determination of space distribution parameters for different stellar groups in the Hertzsprung–Russell diagram.
See Abstr. 155.015.

M giants in Baade's window: infrared colors, luminosities, and implications for the stellar content of E and S0 galaxies.
See Abstr. 155.088.

On the distribution of colors for stars in the ninth to fifteenth magnitude range: statistics and implications for galactic structure.
See Abstr. 155.112.

An analysis of the Yale Bright Star Catalog.
See Abstr. 155.113.

Some properties of the galactic halo from NLTT proper motion stars.
See Abstr. 155.140.

A flux–limited sample of Galactic carbon stars.
See Abstr. 155.152.

Properties of evolved mass–losing stars in the Milky Way and variations in the interstellar dust composition.
See Abstr. 155.173.

The population of Large Magellanic Cloud field stars in a remote southwestern area.
See Abstr. 156.016.

Physical parameters for 12 planetary nebulae and their central stars in the Magellanic Clouds.
See Abstr. 156.018.

C, M, and S stars in the Magellanic Clouds.
See Abstr. 156.037.

HR diagrams and luminosity functions of the Magellanic Cloud field populations.
See Abstr. 156.038.

The evolution of asymptotic giant branch stars in the Magellanic Clouds. III. The problem of intermediate–mass stars.
See Abstr. 156.054.

The stellar populations of Shapley Constellation III.
See Abstr. 156.059.

The use of the brightest blue supergiants as distance indicators – further complications; or the brightest "stars" are not always stars.
See Abstr. 157.084.

The ratio numbers of carbon and oxygen stars in galaxies.
See Abstr. 157.122.

Wolf–Rayet stars in the Andromeda galaxy.
See Abstr. 157.221.

On the luminosity of Hubble–Sandage objects in the M31 galaxy.
See Abstr. 157.284.

Are there stellar winds in the hot stars of M31?
See Abstr. 157.344.

116 Rotation, Magnetic Fields, Activity, Polarization, Radio Radiation

116.001 HD 187474: the first results of surface magnetic field measurements.
P. Didelon.
Messenger, No. 49, p. 5 – 10 (1987).
This preliminary study of Resolved Zeeman Pattern allowed the determination of the surface magnetic field strength of HD 187474, and its mean inclination on the line of sight at the time of observation.

116.002 Phase dependent polarization variations of southern galactic WR + O systems.
A. F. J. Moffat, W. Seggewiss.
Messenger, No. 49, p. 26 – 29 (1987).

116.003 On the polarization of Herbig Ae/Be stars.
N. N. Petrova, V. S. Shevchenko.
Pis'ma Astron. Zh., Tom 13, No. 8, p. 686 – 694 (1987). In Russian. English translation in Sov. Astron. Lett., Vol. 13.
Results of multicolor UBVRI polarimetry of 14 Herbig Ae/Be stars including 7 stars for which observations of polarization have been made for the first time are presented. 6 bright Herbig Ae/Be stars which belong to the star formation region connected with IC 5070 show polarization from 1 to 4.5 per cent with similar θ ($\sim 180°$). The polarimetric variability of BD +46°3471, BD +65°1637, HD 200775 and LkHα 234 is confirmed. Mechanisms of polarization in Herbig Ae/Be stars in circumstellar formations are discussed.

116.004 Discovery of flare activity on BD + 3°4138 B.
B. R. Pettersen, S. L. Hawley.
Astron. Astrophys., Vol. 181, No. 2, p. 402 – 403 (1987).
The dM1.5e star BD +3°4138 B is reported to show flare activity.

116.005 Atmospheric activity in red dwarf stars.
B. R. Pettersen.
Vistas Astron., Vol. 30, Part 1, p. 41 – 51 (1987). – See Abstr. 012.007.
Radiation losses from quiescent and flaring regions of red dwarf atmospheres are compared and found to be equally important. The flare emission on all wavelengths approaches 1% of the stellar bolometric luminosity for active dMe stars, but is 100 times less for some dM stars. The discriminating parameter is flare frequency, which may vary with stellar mass, age, rotation rate or other quantities. Radiation losses from chromospheric emission lines are larger than losses from transition region lines. H I is the most important element in late dMe stars, while Ca II dominates the chromospheric radiation loss from early dMe and dKe stars. The corona is less significant in dKe stars, but dominates over chromospheric radiation losses in dMe stars. A sharp drop in coronal importance is seen near dM 5e, where stars become fully convective. All forms of radiation losses have a common cause, probably to be found in the convective zone. The magnetic field may be the instrument that brings mechanical energy into the outer atmosphere, where it is radiated into space.

116.006 The unknown, remembered gate: present and future observations of active chromosphere stars.
B. W. Bopp.
Vistas Astron., Vol. 30, Part 1, p. 53 – 66 (1987).
The optical behavior of active chromosphere stars is reviewed, with particular reference to determination of the parameters (size, temperature, location, magnetic field strength) that characterize a stellar active region. Three areas of ground–based observational technique are highlighted for their contributions and future promise. Doppler imaging, the reconstruction of the surface brightness distributions of spotted stars by analysis of distortions in spectral line profiles, promises to yield information on location and time evolution of active regions. Magnetic field measurements, especially those utilizing infrared FTS data, will in the future provide complementary data that will permit investigations of the field evolution of spots. Lastly, asteroseismology, utilizing either very high precision radial velocity or intensity measures, will provide input on internal structures of stars. Since parameters such as stellar convection zone depth are crucial input information for modeling the stellar dynamo, asteroseismology will aid the refinement of theoretical models of stellar activity.

116.007 Narrow–band photometry of the magnetic variable β CrB.
V. I. Burnashev, V. P. Malanushenko, N. S. Polosukhina.
Bull. Crimean Astrophys. Obs., Vol. 74, p. 40 – 45 (1987). English translation of 41.116.076.

116.008 Measurement of the magnetic field of β CrB using the $\lambda4254.33$ Cr I line.
S. I. Plachinda.
Bull. Crimean Astrophys. Obs., Vol. 74, p. 46 – 47 (1987). English translation of 41.116.048.

116.009 Observations of the polarization of the radiation of R–association stars.
L. A. Pavlova, F. K. Rspaev.
Astrophysics, Vol. 25, No. 3, p. 620 – 625 (1987). English translation of 42.116.114.

116.010 A very intense, long–lasting radio flare on HD 32918.
O. B. Slee, G. J. Nelson, R. T. Stewart, A. E. Wright, D. L. Jauncey, L. H. Heisler, J. D. Bunton, A. E. Vaughan, M. I. Large, W. L. Peters, S. G. Ryan.
Mon. Not. R. Astron. Soc., Vol. 227, No. 2, p. 467 – 479 (1987).
The authors report the brightest microwave flare yet detected from an active chromosphere star. The K2 III star HD 32918 emitted 6×10^{12}W Hz^{-1} at 8.4 GHz near the flare peak, and flare emission was still detectable three weeks later. The episode consisted of at least three separate flares with successively decreasing amplitudes. The 8.4–GHz emission was essentially unpolarized while the radio spectrum varied as $v^{1.2}$ near the flare peak and as $v^{0.6}$ about three weeks later. At the flare peak the estimated brightness temperature for the 8.4–GHz source, assuming a size of one stellar diameter, was $\sim 3 \times 10^{10}$K. The observed flare parameters can be explained by gyrosynchrotron emission from an optically thick source. A spherically symmetrical source model in which both the magnetic field strength and the density of mildly relativistic electrons decrease outwards can explain the measured radio spectra, brightness temperatures and circular polarization.

116.011 Axial inclination and differential rotation for 19 rapidly rotating stars.
T. R. Stoeckley, W. Buscombe.
Mon. Not. R. Astron. Soc., Vol. 227, No. 4, p. 801 – 813 (1987).
The widths and shapes of He I 4471 and Mg II 4481 absorption lines in spectra of 19 rapidly rotating B–type main–sequence stars were analysed to estimate the axial inclination and the presence or absence of differential rotation. The data are multiple, superimposed photographic coudé spectra. The calculations use non–LTE intensity profiles summed over the disc of a gravity darkened Roche model star. Results show a scatter of aspects from equator–on to pole–on, and a strong indication of a negative differential rotation parameter, suggesting that the angular velocity increases toward the poles.

116.012 The rapidly rotating spotted red dwarf flare star Gliese 890.
B. R. Pettersen, D. L. Lambert, J. Tomkin, W. H. Sandmann, H. Lin.
Astron. Astrophys., Vol. 183, No. 1, p. 66 – 72 (1987).

The authors present the results of spectroscopic, photometric, and polarimetric observations of the spotted rapidly rotating M–dwarf Gliese 890. Broad spectral lines and starspot modulation of the star's luminosity with a period of 0.4312 days imply a rotational velocity of 70 km s^{-1} with the rotation axis inclined 60°–90° to the line of sight. Flare activity is reported for the first time on this dM1.5e red dwarf. The authors discuss briefly some scenarios to explain why this spotted flare star rotates so unusually rapidly.

116.013 Recent spectacular activity in the Be star HR 2855 (FY CMa).
G. J. Peters.
Be Star Newsl., No. 16, p. 17 (1987).

116.014 Variability of Wolf–Rayet stars in linear polarization.
A. F. J. Moffat, P. Bastien, L. Drissen, N. St–Louis, C. Robert.
Instabilities in luminous early type stars, p. 237 – 240 (1987). – See Abstr. 012.012.

The main reasons for observing variations in linear polarization among WR stars are: (1) to look for inhomogeneities and asymmetries in the winds and (2) to determine some basic parameters of WR binary systems. The authors have been acquiring extensive, repeated observations in linear polarization for a large sample of WR stars, including the eight bright Cygnus stars, the 14 brightest WR stars in the south and several selected close binaries in the north.

116.015 Polarization of radiation of early–type stars.
V. M. Loskutov.
Astron. Zh., Tom 64, Vyp. 4, p. 755 – 760 (1987). In Russian. English translation in Sov. Astron., Vol. 31, No. 4.

The degree of polarization is computed by the Feautrier method for early–type stars radiation. The model atmospheres for these stars presented by Kurucz are used. The results of calculation are given.

116.016 Observations of the carbon star CIT 6 in the CO and HCN lines.
I. I. Zinchenko, A. G. Kislyakov, A. A. Krasil'nikov, Eh. P. Kukina, A. V. Lapinov, L. E. Pirogov.
Astron. Zh., Tom 64, Vyp. 4, p. 870 – 872 (1987). In Russian. English translation in Sov. Astron., Vol. 31, No. 4.

The results of observations of emission of the carbon star CIT 6 in the $J = 1-0$ spectral lines of CO and HCN are presented. The observations have been carried out in 1984 – 1985 at the RT–22 radio telescope of the Crimean Astropysical Observatory. Discrepancies of the obtained spectra with the results of some earlier observations are found; this points to a possible variability of CIT 6 molecular radio emission.

116.017 A frequency solution to the light variations in the rapidly oscillating Ap star HD 6532: evidence for an $l = 1$ oblique dipole oscillation.
D. W. Kurtz, M. S. Cropper.
Mon. Not. R. Astron. Soc., Vol. 228, No. 1, p. 125 – 139 (1987).

The authors present 90 hr of new high–speed photometric observations of HD 6532. A frequency analysis of these new data shows that the light variations are explained by an equally spaced frequency triplet which is split by exactly the rotation frequency derived from the mean light variations by Kurtz & Marang (1987). The authors also show that the first harmonic of the central frequency of that triplet is present in the data. The data are consistent with the hypothesis that HD 6532 is pulsating in an oblique dipole mode.

116.018 The determination of the rotational periods of rapidly oscillating Ap stars from their mean light variations – I. The rapidly oscillating Ap star, HD 6532; the F2 IV/V star, HD 6491; and the He–weak B star, HD 5737 (HR 280, α Scl).
D. W. Kurtz, F. Marang.
Mon. Not. R. Astron. Soc., Vol. 228, No. 1, p. 141 – 158 (1987).

The authors have obtained *UBVRI* mean light observations of HD 6532, HD 6491 and HD 5737 with respect to the F5 IV comparison star HD 7259. A frequency analysis of the HD 6532 data shows that it has a rotation period of $P_{rot} = 1.9455 \pm 0.0019$ day. The authors have discovered that HD 6491 has sinusoidal light variations with a period of either 0.82841 ± 0.00011 day, or twice that period, or, slightly less likely, 4.868 ± 0.004 day. No unequivocal variation was found in HD 5737 although a period of 58.8 day is consistent with the data and previous magnetic observations.

116.019 Chromospherically active stars. III. HD 26337 = EI Eri: an RS CVn candidate for the Doppler–imaging technique.
F. C. Fekel, R. Quigley, K. Gillies, J. L. Africano.
Astron. J., Vol. 94, No. 3, p. 726 – 730 (1987).

The variable star HD 26337 = EI Eri is a chromospherically active G5 IV single–lined spectroscopic binary with a period of 1.94722 days. It has moderate strength Ca II H and K emission and strong ultraviolet emission features, while Hα is a weak absorption feature that is variable in strength. The inclination of the system is 46° ± 12°, and the unseen secondary is most likely a late K or early M dwarf. The $v \sin i$ of the primary is 50 ± 3 km s^{-1}, resulting in a minimum radius of 1.9 ± 0.1 $R_\odot$. A mass ratio of $M_1/M_2 \geq 2.6$ and a mass for the primary of at least 1.4 $M_\odot$ is found. HD 26337 has a moderate lithium abundance $\log n(\mathrm{Li}) = 1.75 - 2.0$. It is suggested that many chromospherically active stars may still have moderate lithium abundances because they have evolved from late A or early F type main–sequence stars.

116.020 The sn stars: magnetically controlled stellar winds among the helium–weak stars.
S. N. Shore, D. N. Brown, G. Sonneborn.
Astron. J., Vol. 94, No. 3, p. 737 – 750 (1987).

The authors report observations of magnetically controlled stellar mass outflows in three helium–weak sn stars: HD 21699 = HR 1063, HD 5737 = α Scl, and HD 79158 = 36 Lyn. *IUE* observations show that the C IV resonance doublet is variable on the rotational timescale but that there are no other strong–spectrum variations in the UV. Magnetic fields, which reverse sign on the rotational timescale, are present in all three stars. The authors interpret this phenomenology in terms of jet–like mass loss above the magnetic poles and discuss these objects in the context of a general survey of the C IV and Si IV profiles of other more typical helium–weak stars.

116.021 Rotation and magnetic activity in main–sequence stars.
L. W. Hartmann, R. W. Noyes.
Annu. Rev. Astron. Astrophys., Vol. 25, p. 271 – 301 (1987). – See Abstr. 003.003.

The authors discuss recent observational results on how rotation interacts with convection to produce stellar magnetic fields. They concentrate on the activity of main–sequence stars. They sketch the evolution of stellar rotation from pre–main–sequence to main–sequence phases. Some implications of this rotational evolution for internal velocity fields, and therefore for the dynamo processes, are noted. Surface manifestations of magnetic activity, thought to be generated by the magnetic dynamo acting within rotating, convecting stars, and how this activity depends on mass, rotation, and age are discussed. The authors summarize the present state of our understanding of rotation and magnetic activity in main–sequence stars and outline some promising areas for future work.

116.022   On the observable magnetic fields of the chemically peculiar stars.
D. Moss.
Mon. Not. R. Astron. Soc., Vol. 228, No. 4, p. 993 – 1000 (1987).

Observable effective and surface magnetic fields are calculated for a number of non–axisymmetric field configurations with structures suggested by theoretical considerations. For a number of cases investigated the effective field variations are not very different to those of displaced dipoles, although distortions in the surface field curves are more readily seen. The anomalous effective field variations of HD 37776 are discussed briefly.

116.023   Investigation of the radial velocity of γ Equ with a high time resolution.
V. D. Bychkov.
Pis'ma Astron. Zh., Tom 13, No. 9, p. 773 – 779 (1987). In Russian. English translation in Sov. Astron. Lett., Vol. 13.

Variation of radial velocity of the cool Ap star γ Equ = HD 201601 was investigated with a time resolution of 2 – 4 minutes. On the basis of the measurement of 161 lines, variations of radial velocity with amplitude up to 2.8 km/s are registered, but periodicity in the time range 5 to 35 minutes is not detected.

116.024   An IUE survey of activity in red giants and supergiants.
I. Oznovich, D. M. Gibson.
Astrophys. J., Vol. 319, No. 1, p. 383 – 391 (1987).

The authors have examined chromospheric and transition region line activity in apparently single red giants and supergiants using the IUE archives. Low–resolution, large–aperture spectra (mostly short–wavelength) were used to search for variations of emission–line fluxes in time. An automatic search procedure was applied to ~120 images of 26 stars taken over a period of 7 yr (1978 – 1984). Four stars showed UV emission–line flux variations. α Aqr, β Peg, and σ Oph showed a single enhanced–emission event in all detectable emission lines. γ Aql exhibited an increase in the flux level of the O I (1641 Å) line in mid–1981 with no comparable change in any other lines.

116.025   The Pleiades rapid rotators: evidence for an evolutionary sequence.
R. P. Butler, R. D. Cohen, D. K. Duncan, G. W. Marcy.
Astrophys. J., Lett. Ed., Vol. 319, No. 1, p. L19 – L22 (1987).

Four rapidly rotating early–K dwarfs in the Pleiades are shown to contain an order of magnitude more Li than four slow rotators of the same spectral type, as would be expected if they were systematically younger. This supports the idea that late–type stars first arrive on the main sequence with $V_{rot} \gtrsim 100$ km s^{-1}, that they spin down to $V_{rot} \lesssim 10$ km s^{-1} in $10^7 – 10^8$ yr, and that the Pleiades lower main sequence shows such an age spread.

116.026   Rapid, low–level X–ray variability in active late–type dwarfs.
C. W. Ambruster, S. Sciortino, L. Golub.
Astrophys. J., Suppl. Ser., Vol. 65, No. 2, p. 273 – 305 (1987).

The authors report the detection of rapid (a few hundred to >1000 s) quiescent variability, at $\geq 99\%$ significance, in 24 of 34 Einstein IPC observations of 19 active K and M dwarfs. The IPC light curve and a plot of the variability analysis (using a new, statistically rigorous form of χ^2 analysis) are given for each observation. The quiescent fluctuations appear not to fall on an extension of the $N(S)$ curve for stellar X–ray flares. The authors discuss solar analogs and the recent suggestions that low–level flaring heats quiescent X–ray coronae.

116.027   Magnetic fields in Be stars?
P. K. Barker.
Physics of Be stars, p. 38 – 48 (1987). – See Abstr. 012.030 (IAU Colloq. No. 92).

No mean longitudinal or toroidal magnetic fields have yet been detected on any classical Be star. Models of stellar winds and circumstellar envelopes around magnetic Be stars are not appreciably constrained by present observed upper limits on field strength. A few magnetic Be stars do exist among the helium strong stars, but these objects show spectral phenomenology which is unmistakably distinct from that shown by every other object known as a Be star.

116.028   Differential rotation in B and Be stars.
J. Zorec, L. Divan, R. Mochkovitch, A. Garcia.
Physics of Be stars, p. 68 – 71 (1987). – See Abstr. 012.030 (IAU Colloq. No. 92).

The authors present two kinds of results which show that differential rotation is a highly probable phenomenon in B and Be stars. These results show that angular momentum is an important parameter which must be taken into account for interpreting the observed parameters of B and Be stars and for studying their structure.

116.029   The magnetic Be star Sigma Orionis E.
C. T. Bolton, A. W. Fullerton, D. Bohlender, J. D. Landstreet, D. R. Gies.
Physics of Be stars, p. 82 – 83 (1987). – See Abstr. 012.030 (IAU Colloq. No. 92).

Over the past two years, the authors have obtained high resolution high signal/noise (S/N) spectra of the magnetic Be star σ Ori E. These spectra, which cover the spectral regions 399 – 417.5 and 440 – 458.5 nm and the Hα line and have typical S/N > 200 and spectral resolution $\cong 0.02$ nm, were obtained at a variety of rotational phases in order to study the magnetic field structure, the distribution of elements in the photosphere, and the effects of the magnetic field on the emission envelope.

116.030   Double periodicity in Be stars.
D. Clarke, P. A. McGale.
Physics of Be stars, p. 197 – 199 (1987). – See Abstr. 012.030 (IAU Colloq. No. 92).

Photometric periodicities with two unequal maxima and minima reported for Be stars are shown to result from a fundamental and overtone associated with rotation of an atmospheric bulge. It is the form of the scattering function of the free electron that produces the double periodicity; the stellar geometry controls the amplitudes of the two components. The form of the light curves is consistent with the oblique rotator model.

116.031   Possible fine structure of the magnetic field of the chemically peculiar stars α^2 CVn and β CrB.
I. I. Romanyuk.
Bull. Spec. Astrophys. Obs. – North Caucasus, Vol. 22, p. 22 – 34 (1987). English translation of 41.116.031.

116.032   The determination of the rotational periods of rapidly oscillating Ap stars from their mean light variations. II. HR 1217 (HD 24712).
D. W. Kurtz, F. Marang.
Mon. Not. R. Astron. Soc., Vol. 229, No. 2, p. 285 – 293 (1987).

The authors have obtained new *UBVRI* mean light observations of HR 1217 in 1986 and 1987. A frequency analysis of the new *B* observations along with previous *B* observations, Strömgren *b* observations and Eu II line strength observations leads unambiguously to a revised rotational ephemeris. All of the non–magnetic observations can be fitted with this ephemeris to the accuracy of the data. On the other hand, the magnetic measurements have slightly, but significantly, different times of maxima; magnetic maximum does not occur at exactly the same time as the extrema in the photometric mean light measurements or the Eu II line strength measurements. The phase shift between the photometric and line strengths observations and the magnetic observations is 1.00 ± 0.35 day or 0.080 ± 0.028 periods. The amplitude of the mean light variations decreases monotonically as a function of wavelength for the five bandpasses studied. There is an indication that the phase of the mean light variations may increase as a function of wavelength.

116.033   A radio–continuum survey of the coolest M and C giants.
S. A. Drake, J. L. Linsky, M. Elitzur.
Astron. J., Vol. 94, No. 5, p. 1280 – 1290 (1987).

The authors present the results of a sensitive VLA continuum survey of 22 cool M and C type giants and supergiants, including nine carbon stars, one S type star, and 12 M stars. The purpose of the survey was to probe the physical properties of the partially ionized, expanding chromospheres of the coolest luminous stars. Of the 22 stars observed at 6 cm, none were detected directly, although extended emission was detected near NML Cyg and OH 26.5+0.6, and a point source was detected near AFGL 865. Of the three stars observed at 2 cm, R Aql (M6.5e–9e) was detected as a 0.54 ± 0.17 mJy source, and a point source was detected 7″ from R Cas (M6e–9e) and may be physically associated with the star. These data imply small upper limits for the ionized–mass–loss rates and 2 – 6 cm spectral indices that are significantly steeper than the 0.6 value predicted by the "standard" stellar–wind model. The nondetection of o Ceti (Mira) at both 2 and 6 cm, despite a previous 6 cm detection, supports the idea that the radio–continuum emission of these stars may be variable, perhaps due either to flares or to the passage of pulsation–generated shock waves through their outer atmospheres.

116.034   Flux density and polarization observations of Hipparcos radio stars.
J. M. Paredes, R. Estalella, A. Rius.
Astron. Astrophys., Vol. 186, No. 1/2, p. 177 – 181 (1987).

Simultaneous flux density measurements with the 64 m antenna of the Madrid Deep Space Communication Complex (DSCC) at a wavelength of 13.1 cm and 3.6 cm, or of right and left circular polarization at 3.6 cm of a group of radio stars from the Hipparcos Input Catalogue, are reported. The total number of stars observed is 33, while the number of stars detected is 9. The polarization measurements of RS CVn systems are in agreement with a gyrosynchrotron emission mechanism. The observations of LS I +61°303 confirm the period emission at 8.4 GHz.

116.035   Activity in late–type dwarfs. I. Walraven and Johnson photometry of flares and spot variations on Gl 867 A (= FK Aqr) in 1979.
P. B. Byrne, E. Black, P. S. Thé.
Astron. Astrophys., Vol. 186, No. 1/2, p. 261 – 267 (1987).

The authors present optical photometry of flares and spot variations on the flare/BY Draconis star Gl 867 A (= FK Aqr). These observations are made in the Walraven intermediate passband system. Based on this and other material already published a new period determination has been made. The result suggests that the system may not be corotating as suggested earlier. In this regard it is similar to the prototype spotted star, BY Dra itself. Time–integrated rates of energy lost by optical flaring are derived and compared with previously published results. No evidence of season–to–season changes are evident. The advantages of the Walraven photometric system for the study of stellar flares are discussed and the nature of the optical flare light is examined in the light of measurements made in this system.

116.036   Activity in late–type dwarfs. II. Flares and spot variations on Gl 867 A (= FK Aqr) in 1981.
P. B. Byrne, J. G. Doyle.
Astron. Astrophys., Vol. 186, No. 1/2, p. 268 – 270 (1987).

The authors present optical photometry of flares and spot variations on the flare/BY Draconis star Gl 867 A (= FK Aqr). Time–integrated rates of energy lost by optical flaring are derived and compared with previously published results. No evidence of season–to–season changes are evident. No spot variations were detected but there is an unusually large scatter in the mean light curve.

116.037   Determination of the unique period for the Ap star HD 184905.
E. W. Burke Jr., S. K. Thompson.
Publ. Astron. Soc. Pac., Vol. 99, No. 618, p. 852 – 853 (1987).

This Ap star was discovered to be variable by Burke, Rolland, and Boy in 1970 with the period of 1.855 days. Morrison and Wolff in 1971 reported their data were better fit by a period of 2.17 days. Since that time different observers preferred one or the other of these periods. These observations show the unique period of variation of this star to be 1.84532 ± 0.00004 days.

116.038   Rotational velocities of low–mass stars in the Hyades.
J. R. Stauffer, L. W. Hartmann, D. W. Latham.
Astrophys. J., Lett. Ed., Vol. 320, No. 1, p. L51 – L55 (1987).

The authors have used high–resolution echelle spectra to estimate rotational velocities for K and M dwarfs in the Hyades. All of the K dwarfs have rotational velocities less than the instrumental limit of 10 km s^{-1}. Most of the M dwarfs with $(R-I)_K > 1.0$ have detectable rotational velocities, with $10 < v \sin i < 20$ km s^{-1}. Combining these data with results from the younger Pleiades and α Persei clusters, the authors find that G dwarfs spin down from ~ 100 km s^{-1} to ~ 10 km s^{-1} in less than 4×10^7 yr, whereas M dwarfs take an order of magnitude longer to spin down.

116.039   A slow flare of BY Draconis May 23 – 25, 1984.
P. F. Chugajnov.
Izv. Krymskoj Astrofiz. Obs., Tom 76, p. 54 – 62 (1987). In Russian. English translation in Bull. Crimean Astrophys. Obs., Vol. 76.

Results of photoelectric observations of a slow flare of BY Dra which were obtained in 5 bands UBVRI with a high accuracy are presented. It is shown that the continuous emission originates mainly in an opaque region with a temperature of about 10^4K and a concentration $n_H = 10^{15}$ cm^{-3}. This opaque region exists only near the flare maximum. The screening of a part of the star disc by the opaque region may give rise to conspicuous effects, especially in the I band. An examination of a complex light curve of the flare leads to the conclusion on the presence of oscillations with periods from 1 to 20 min. The possible mechanism of these oscillations is discussed.

116.040   A VLA survey of BY Draconis variables.
J.–P. Caillault, S. Drake, D. Florkowski.
Bull. Am. Astron. Soc., Vol. 19, No. 2, p. 678 (1987). Abstract. – See Abstr. 010.061.

116.041   The spatial distribution of magnetic fields on Xi Bootis A.
S. H. Saar, J. Huovelin, M. S. Giampapa.
Bull. Am. Astron. Soc., Vol. 19, No. 2, p. 703 (1987). Abstract. – See Abstr. 010.061.

116.042   Abundance and magnetic field geometries of helium strong and helium weak stars.
D. A. Bohlender, J. D. Landstreet.
Bull. Am. Astron. Soc., Vol. 19, No. 2, p. 704 (1987). Abstract. – See Abstr. 010.061.

116.043   The magnetosphere and stellar wind of the helium–strong star, HD 184927.
D. N. Brown, P. K. Barker, C. T. Bolton, S. N. Shore.
Bull. Am. Astron. Soc., Vol. 19, No. 2, p. 705 (1987). Abstract. – See Abstr. 010.061.

116.044   VLA observations of rapid 6 cm flux variations in α Ori.
R. E. Stencel, J. Bookbinder, S. A. Drake, T. Simon, J. L. Linsky, D. Florkowski.
Bull. Am. Astron. Soc., Vol. 19, No. 2, p. 706 (1987). Abstract. – See Abstr. 010.061.

116.045   Simultaneous EXOSAT and VLA observations of active cool binaries VW Cep and XY Leo. A flare in VW Cep.
O. Vilhu, J.–P. Caillault, J. Heise.
Bull. Am. Astron. Soc., Vol. 19, No. 2, p. 706 (1987). Abstract. – See Abstr. 010.061.

116.046 Ultraviolet stellar polarimetry from the Solar Maximum Mission satellite.
W. Henze Jr., B. E. Woodgate.
Bull. Am. Astron. Soc., Vol. 19, No. 2, p. 724 (1987). Abstract. – See Abstr. 010.061.

116.047 Synoptic high–sensitivity polarimetry of very bright stars.
J. C. Kemp, G. D. Henson, D. J. Kraus, M. H. Dunaway.
Bull. Am. Astron. Soc., Vol. 19, No. 2, p. 752 (1987). Abstract. – See Abstr. 010.061.

116.048 Discovery of optical circular polarization in Lambda Andromedae.
G. D. Henson, J. C. Kemp, D. J. Kraus, M. H. Dunaway.
Bull. Am. Astron. Soc., Vol. 19, No. 2, p. 752 (1987). Abstract. – See Abstr. 010.061.

116.049 On the variable elliptical polarization of Alpha Orionis.
B. D. Holenstein.
Bull. Am. Astron. Soc., Vol. 19, No. 2, p. 754 – 755 (1987). Abstract. – See Abstr. 010.061.

116.050 High angular resolution speckle imaging of α Ori.
M. Karovska, P. Nisenson, R. Noyes.
Bull. Am. Astron. Soc., Vol. 19, No. 2, p. 755 (1987). Abstract. – See Abstr. 010.061.

116.051 The first evidence for starspot modulation on fully convective M dwarfs.
C. W. Ambruster.
Bull. Am. Astron. Soc., Vol. 19, No. 2, p. 756 (1987). Abstract. – See Abstr. 010.061.

116.052 Detection of radial velocity variations in the rapidly oscillating Ap star HR 1217.
J. M. Matthews, W. H. Wehlau, G. A. H. Walker, S. Yang.
Bull. Am. Astron. Soc., Vol. 19, No. 2, p. 762 (1987). Abstract. – See Abstr. 010.061.

116.053 Activity in early F–type stars in the Hyades.
S. C. Wolff, J. N. Heasley.
Publ. Astron. Soc. Pac., Vol. 99, No. 619, p. 957 – 960 (1987).
Observations of 14 main–sequence stars in the Hyades show that stellar activity in these stars is closely similar to that seen in early F–type field stars. The onset of activity occurs at about $(B–V) = 0^m30$, and the activity level in stars bluer than $(B–V) = 0^m45$ does not correlate with rotation. There is no evidence of a decline of activity with increasing age during the rather narrow time interval spanned by the main–sequence lifetimes of early F–type stars.

116.054 Rotation of pre–main sequence stars from high S/N spectroscopy.
J. Bouvier.
Inst. Astrophys. Paris, Pré–Publ., No. 197, 3 pp. (1987). To appear in IAU Symp. No. 132.

116.055 The magnetic field of the CP star 21 Per.
V. G. El'kin, Yu. V. Glagolevskij, I. I. Romanyuk.
Astrofiz. Issled. Izv. Spets. Astrofiz. Obs., Tom 25, p. 24 – 27 (1987). In Russian. English translation in Bull. Spec. Astrophys. Obs. – North Caucasus, Vol. 25.
The effective field B_e of 21 Per is measured in different phases of the rotation period using the Zeeman spectrograms obtained on the 6–meter telescope with a reciprocal dispersion of 9 Å/mm. A sinusoid plotted from these data using the least squares method shows variations of B_e from –150 up to +280 Gs. The identical sinusoid plotted using the results reported by Babcock shows field variations from –1270 up to +1350 Gs. Measurements made by Preston show the lack of the field in general. The analysis of all the available data showed that there are probably secular variations of the field for this star with a possible period of the order of 40 – 50 years.

116.056 Spectrophotometric investigation of the Be star Merope.
L. M. Sapargalieva.
Tr. Astrofiz. Inst. Alma–Ata, Tom 48, p. 76 – 83 (1987). In Russian.

116.057 Observations of Pleiades spotted solar–type stars.
A. K. Magnitskij.
Pis'ma Astron. Zh., Tom 13, No. 12, p. 1071 – 1079 (1987). In Russian. English translation in Sov. Astron. Lett., Vol. 13.
The September – October 1986 observations discovered periodic light variations in three solar–type stars in the Pleiades cluster: Hz 296 (0.8 $M_\odot$), Hz 152 (0.91 $M_\odot$) and Hz 739 (1.15 $M_\odot$). Periods and amplitudes are accordingly 2^d53 and 0^m11, 4^d12 and 0^m07, 2^d70 and 0^m05. Considerable light variations of these stars in the Pleiades are due to the rotation of spotted stars. Contrast spots of solar–type stars likely exist when stars are young and rapidly rotate.

116.058 A model for the intrinsic linear polarization of cool giant and supergiant stars.
J. A. Marcondes–Machado.
Astron. Astrophys., Vol. 188, No. 1, p. 131 – 136 (1987).
A model for the intrinsic linear polarization of cool giant and supergiant stars is developed. The following assumptions are made: (1) the polarization is due to circumstellar super–paramagnetic grains aligned by a stellar magnetic field; (2) grain size and magnetic field geometry vary with distance from the star; (3) variations in the star effective radius with wavelength cause changes in the degree and position angle of the linear polarization. Observations of μ Cep (Coyne and Magalhães, 1979) and σ Cet (McLean and Coyne, 1978) are then used as a test of the model, which reproduces the wide variety of features observed in the continuum and in the molecular bands and atomic lines.

116.059 A microwave survey of southern active stars.
O. B. Slee, G. J. Nelson, R. T. Stewart, A. E. Wright, J. L. Innis, S. G. Ryan, A. E. Vaughan.
Mon. Not. R. Astron. Soc., Vol. 229, No. 4, p. 659 – 677 (1987).
The authors report the results of a survey of 153 active–chromosphere stars made with the Parkes 64–m telescope at 5.0/8.4 GHz from 1981 to 1987. Most of the stars were observed on at least 10 separate days in order to better establish flare rate and luminosity. The sample included RS CVn and Algol–type binaries, Ca II emitters with as yet unknown stellar content and AM Her–type cataclysmic variables. For some of the strongest stellar flares the authors measured the circular polarization at 8.4 GHz and obtained crude radio spectra involving frequencies in the range $0.843 < \nu < 22$ GHz.

116.060 The activity, variability, and rotation of lower main–sequence Hyades stars.
R. R. Radick, D. T. Thompson, G. W. Lockwood, D. K. Duncan, W. E. Baggett.
Astrophys. J., Vol. 321, No. 1, p. 459 – 472 (1987).
High–precision differential b, y photometric observations of 24 Hyades stars, spectral types F4 V to K8 V, were made at Lowell Observatory during late 1983 and early 1984. All 18 stars of spectral type F8 V or later in this sample were variable, confirming that low–level photometric variability is a ubiquitous and persistent characteristic of Hyades stars later than spectral type F8 V. Contemporaneous Ca II H + K emission flux measurements for five of the 24 stars, spectral types F8 V to G2 V, showed that H + K emission flux varies inversely with respect to photometric brightness on both long–term and rotational time scales. Hyades stars with measured rotation periods conform reasonably well to the "Rossby relation," which relates chromospheric activity to the ratio of rotation period and convective turnover time scale.

116.061 Observations and analysis of the photospheric magnetic fields on dwarf G, K and M stars.
S. H. Saar.
Diss. Abstr. Int., Sect. B, Vol. 48, No. 5, p. 1385–B (1987). Thesis, University of Colorado at Boulder, 240 pp. (1987). Order No. DA8716295.

116.062   Further VLA observations of hydrogen deficient stars.
N. K. Rao, V. R. Venugopal, A. R. Patnaik.
J. Astrophys. Astron., Vol. 8, No. 3, p. 227 – 230 (1987).

VLA observations at 6 cm have been obtained for three hydrogen–deficient objects υ Sgr, V348 Sgr, and Abell 58. Abell 58 was also observed at 2 cm. Only upper limits to the flux density could be set for these sources. The upper limit for 6 cm flux density of V348 Sgr sets an upper limit to its reddening as $E(B-V) \leqslant 0.65$. The hydrogen deficient planetary nebula A 58 shows much lower radio flux than expected from the infrared–radio flux density relationship of planetary nebulae.

116.063   The possibility of detection of effects of linear polarization of radiation in lunar occultation observations of stars.
M. B. Bogdanov.
Astron. Zh., Tom 64, Vyp. 6, p. 1300 – 1307 (1987). In Russian. English translation in Sov. Astron., Vol. 31, No. 6.

A method is presented to investigate the distribution of linear polarization of radiation across a stellar disk from an analysis of lunar occultation curves observed simultaneously with two polarizers which are aligned perpendicular and parallel to the direction of the lunar limb motion. The strip distributions of brightness and the occultation curves observed for these senses of polarization are calculated for plane–parallel and spherically symmetric extended models of a Rayleigh scattering atmosphere. It is shown that polarization effects can be observed only for the stars with quite extended atmospheres. These effects are independent of atmospheric scintillation of the star and of the influence of lunar limb irregularities.

116.064   Chromospheric–coronal activity at saturated levels.
O. Vilhu, F. M. Walter.
Astrophys. J., Vol. 321, No. 2, p. 958 – 966 (1987).

The upper bounds in the activity–color relations for cool dwarfs are defined by the rapid rotators in synchronous binaries. The two components of a late–type contact binary have identical chromospheres and transition regions, as deduced from many IUE emission–line light curves and from high–resolution IUE spectra of W UMa. Due to rotational smearing, the Mg II emission strength of F–type contact binaries have heretofore remained unknown. The authors estimate the Mg II fluxes in these stars by measuring a 4 Å passband centered on the Mg II k line in high–resolution IUE spectra, calibrating the continuum flux using spectra of narrow–lined stars. The authors discuss these measurements and their relation to other published data on Mg II, C IV, and X–ray emission.

116.065   On the possible nature of RS CVn–type star spot activity.
V. P. Vasil'ev.
Astron. Tsirk., No. 1473, p. 5 – 7 (1986). In Russian.

The apparently symmetric star spots in RS CVn systems are connected with speed–resonance generation of infrasonic waves by a thermal terminator moving through the asynchronically rotating stellar atmosphere.

116.066   An analysis of the photospheric line profiles in F, G, and K supergiants.
D. F. Gray, C. G. Toner.
Astrophys. J., Vol. 322, No. 1, p. 360 – 367 (1987).

The authors have measured the spectral–line broadening for 30 F, G, and K Ib supergiants. Fourier analysis for macroturbulence and rotation shows (1) macroturbulent velocities larger than but compatible with values found for lower luminosity stars and (2) rotation results that cannot be fully explained. The authors suggest three plausible scenarios, but favor the one in which angular momentum of the surface "shell" is conserved during the evolutionary changes experienced by these stars.

116.067   Magnetic fields of chemically peculiar stars of different ages.
Yu. V. Glagolevskij, V. G. Klochkova, I. M. Kopylov.
Sov. Astron., Vol. 31, No. 2, p. 188 – 191 (1987). English translation of 43.116.010.

116.068   Photographic observations of KR Aur in 1985/1986.
S. Fujino, M. Huruhata, T. Saito, M. Wakuda.
Variable Star Bull., No. 1, p. 1 (1987).

116.069   Variability of NSV 01715.
M. Huruhata.
Inf. Bull. Variable Stars, No. 3071, 2 pp. (1987).

116.070   BV light curves of BY Dra in 1986.
G. Cutispoto, G. Leto, I. Pagano, G. Santagati, R. Ventura.
Inf. Bull. Variable Stars, No. 3102, 4 pp. (1987).

116.071   Polarization variability among Wolf–Rayet stars. I. Linear polarization of a complete sample of southern Galactic WC stars.
N. St.–Louis, L. Drissen, A. F. J. Moffat, P. Bastien, S. Tapia.
Astrophys. J., Vol. 322, No. 2, p. 870 – 887 (1987).

This paper presents extensive time–dependent linear polarization observations of the seven known southern Wolf–Rayet stars of the carbon sequence brighter than magnitude $b = 9$. The stars can be grouped into three categories: (1) three single–line stars, for which the observed stochastic variability in polarization decreases dramatically for stars with faster winds; (2) two short–period WC + O V binaries which show double–wave orbital modulation; and (3) two relatively long–period WC + O I binaries in which the effects of nonradial pulsations of the bright supergiants are important compared to the binary modulation.

116.072   Polarization variability among Wolf–Rayet stars. II. Linear polarization of a complete sample of southern galactic WN stars.
L. Drissen, N. St.–Louis, A. F. J. Moffat, P. Bastien.
Astrophys. J., Vol. 322, No. 2, p. 888 – 901 (1987).

The authors present linear polarization data for the six brightest southern Wolf–Rayet (W–R) stars of the nitrogen sequence (WN). They are all of the cool WN 7, WN 8 subclasses, and all show intrinsic, apparently random, variations with amplitudes ranging from $\Delta P = 0.15\%$ to 0.6%. The variations are probably caused by blobs of dense plasma being ejected in the wind. Combining these data with similar results for the seven brightest southern WC stars, the authors find a general anticorrelation between the terminal velocity of the wind and the amplitude of the polarimetric variations.

116.073   The discovery of nonthermal radio emission from magnetic Bp–Ap stars.
S. A. Drake, D. C. Abbott, T. S. Bastian, J. H. Bieging, E. Churchwell, G. Dulk, J. L. Linsky.
Astrophys. J., Vol. 322, No. 2, p. 902 – 908 (1987).

In a VLA survey of chemically peculiar B– and A–type stars with strong magnetic fields, the authors have detected five of the 34 stars observed as 6 cm continuum sources. Three of the detections are helium–strong early Bp stars, and two are helium weak, silicon–strong stars with spectral types near A0p. Three–frequency observations indicate that the helium–strong Bp stars are variable nonthermal sources. The emission cannot arise from a stellar wind, but is consistent with gyrosynchrotron emission from continuously injected, mildly relativistic particles trapped in the magnetosphere.

116.074   Magnetic field measurements of helium–strong stars.
D. A. Bohlender, D. N. Brown, J. D. Landstreet, I. B. Thompson.
Astrophys. J., Vol. 323, No. 1, p. 325 – 337 (1987).

The authors present results of a continuing program of photoelectric magnetic field measurements of helium–strong stars. Nine of the 11 helium–strong stars so far observed for circular polarization in the wings of Hβ have shown evidence of large, ordered magnetic fields. The discovery of two new magnetic helium–strong stars and observations of seven other members of the class are reported. They also discuss new measurements of the helium–weak star HD 175362. Magnetic curves have been obtained using the He I λ5876 line for four stars. After revising the measured $v \sin i$ values of three of the helium–strong stars, it is demonstrated that the $v \sin i$ distribution for the helium–strong stars is similar to the distribution for normal early B stars.

The High Altitude Observatory–Lowell Observatory Solar–Stellar Spectrophotometry Project.
See Abstr. 013.003.

The solar–stellar connection at low spectral resolution.
See Abstr. 013.005.

The detection of magnetic fields on late–type stars: progress, problems, and future needs.
See Abstr. 036.004.

The significance of stellar magnetic field measurements obtained with the photographic technique: the spurious magnetic field of the supergiant Canopus.
See Abstr. 036.045.

Inverse photometric problem for spotted stars.
See Abstr. 036.180.

Doppler images of rotating stars using maximum entropy image reconstruction.
See Abstr. 036.185.

Observations of radio stars and radio reference stars with the eight–inch transit circle at Flagstaff.
See Abstr. 041.024.

A new formulation of the starspot model, and the consequences of starspot structure.
See Abstr. 064.059.

Evidence of the connection between internal magnetic fields and chromospheric activity in late–type stars.
See Abstr. 064.069.

Radiatively driven winds from magnetic, fast–rotating stars: Wolf–Rayet stars?
See Abstr. 064.071.

Rotating stellar interiors.
See Abstr. 065.053.

Direct generation of solar and stellar radio bursts by energetic electron maser.
See Abstr. 077.029.

The solar–stellar connection.
See Abstr. 080.027.

The sun among the stars: what the stars indicate about solar variability.
See Abstr. 080.047.

Polarization investigations in four peculiar supergiants with high IR excess.
See Abstr. 112.009.

The extended radio emission of P Cygni.
See Abstr. 112.011.

The relation between the visual polarisation and UV narrow absorption lines in irregular Be star variations.
See Abstr. 112.020.

Radio detection of the Be star ψ Persei.
See Abstr. 112.053.

Observations of rapid variability in Be stars.
See Abstr. 112.068.

Simultaneous spectroscopy and polarimetry of Be stars.
See Abstr. 112.072.

Long–term polarization changes of 88 Her.
See Abstr. 112.083.

***o* And photometry, polarimetry and a tentative model of the light variability.**
See Abstr. 112.106.

Emission from SiO, SiS, and HC$_3$N in IRC + 10216: the appearance and disappearance of molecules in the outflow.
See Abstr. 112.135.

Detection of X–ray emission from the young low–mass star Rossiter 137B.
See Abstr. 112.141.

Fast transient X–rays from flare stars and RS CVn binaries.
See Abstr. 112.160.

Evidence for strong magnetic fields in the inner envelopes of late–type stars.
See Abstr. 112.167.

X–ray emission and structure of coronae of active late–type dwarfs.
See Abstr. 112.168.

Hα photometry of dwarf K and M stars.
See Abstr. 113.011.

Precision photometric monitoring of southern variable Wolf–Rayet stars with a comment on the overall continuum variability of WR stars.
See Abstr. 113.018.

Lithium and rotation in the Hyades F dwarfs.
See Abstr. 114.074.

Chemically peculiar star π^1Boo – stellar parameters and $v \sin i$.
See Abstr. 114.080.

Continuum flux depressions of chemically peculiar stars and their connection with the magnetic field.
See Abstr. 114.113.

The early A type stars: refined MK classification, confrontation with Strömgren photometry, and the effects of rotation.
See Abstr. 114.125.

Fundamental parameters of the underlying Be stars.
See Abstr. 115.010.

BV observations of W UMa–type binaries: CK Boo, BI CVn, and AH Vir.
See Abstr. 117.013.

Polarization and infrared colors of symbiotic stars
See Abstr. 117.026.

The magnetic field strength in the emission line region of the AM Her system EF Eridani (= 2A 0311–277).
See Abstr. 117.053.

Is there period–activity correlation in RS CVn–type binary systems?
See Abstr. 117.072.

Direct photometric problem for spotted stars.
See Abstr. 117.101.

CCD echelle observations of the active RS CVn system II Pegasi.
See Abstr. 117.110.

Correlation between bolometric and X–ray luminosities in RS CVn–type binaries.
See Abstr. 117.117.

An analysis of the light curves of short–period RS Canum Venaticorum stars: starspots and fundamental properties.
See Abstr. 117.146.

Simultaneous five–colour (*UBVRI*) polarimetry of EF Eri.
See Abstr. 117.161.

The ultraviolet variability of AY Ceti.
See Abstr. 117.193.

The energy distributions of magnetic variables: the effects of inhomogeneous accretion columns.
See Abstr. 117.196.

Radio flares from AE Aquarii: a low–power analog to Cygnus X–3?
See Abstr. 117.198.

UX Arietis.
See Abstr. 117.249.

A flare event on the long–period RS Canum Venaticorum system IM Pegasi.
See Abstr. 117.282.

Photometry during the end–phase of a spot cycle in II Peg.
See Abstr. 117.294.

II Pegasi: evidence of evolving star spot regions.
See Abstr. 117.307.

Light curve variations in ER Vulpeculae.
See Abstr. 119.016.

Epsilon Aurigae: pulsations and the post–eclipse (1984 – 1987) polarization and light curves.
See Abstr. 119.034.

Narrow–band photoelectrical observations of the eclipsing binary CQ Cep in 1975 – 1976 and their possible interpretation.
See Abstr. 119.043.

Spectroscopy of the rapidly rotating K star HD 36705.
See Abstr. 121.040.

Rotational modulation and flares on RS CVn and BY Dra–type stars. V. EXOSAT and IUE observations of a flare on EQ Pegasi.
See Abstr. 122.011.

The rapidly oscillating Ap stars as a test of stellar chromospheric heating mechanisms.
See Abstr. 122.014.

Spectropolarimetry of R Coronae Borealis during 1986 – 87.
See Abstr. 122.063.

Discovery of high–degree nonradial oscillations in rapidly rotating δ Scuti stars?
See Abstr. 122.083.

The flare energy spectrum of EV Lac.
See Abstr. 122.096.

Time–evolution of colours of two EV Lac flares.
See Abstr. 122.122.

Discovery of a magnetic DA white dwarf with distinct Hβ and Hα Zeeman triplets.
See Abstr. 126.045.

The rotational velocities of white dwarfs. II.
See Abstr. 126.064.

Hydrodynamical collapse of interstellar clouds. IV. Ionization degree and ambipolar diffusion.
See Abstr. 131.117.

To *B* or not to *B*?
See Abstr. 131.260.

Continuum versus line polarization at the center of the Orion nebula.
See Abstr. 132.018.

VLA observations of radio sources near Wolf 359.
See Abstr. 141.014.

Radio–frequency observations of galactic X–ray sources. II.
See Abstr. 142.033.

The distribution of rotational velocities for low–mass stars in the Pleiades.
See Abstr. 153.001.

117 Close Binaries (Observations, Theory)

117.001 Hardness ratio in evolving low–mass X–ray binary systems.
J. Shaham, M. Tavani.
The origin and evolution of neutron stars, p. 199 (1987).
Abstract. – See Abstr. 012.001 (IAU Symp. No. 125).

117.002 Quasi–periodic oscillations in low–mass X–ray binaries.
M. van der Klis.
The origin and evolution of neutron stars, p. 321 – 331 (1987). – See Abstr. 012.001 (IAU Symp. No. 125).

The properties of the rapid, persistent quasi–periodic oscillations (QPO) discovered with EXOSAT in the X–ray flux of at least 7 bright low–mass X–ray binaries are described. Particular attention is given to the various relations observed between QPO frequency and X–ray intensity, the link between QPO and the low–frequency noise in the X–ray intensity and the bimodal properties of in particular Sco X–1, GX 5–1 and Cyg X–2. The merits of the hypothesis that the QPO indicate the presence of a neutron star with a magnetosphere are considered.

117.003 Observations of quasi–periodic oscillations in Cyg X–2.
G. Hasinger.
The origin and evolution of neutron stars, p. 333 – 345 (1987). – See Abstr. 012.001 (IAU Symp. No. 125).

EXOSAT observations of Cyg X–2 reveal two types of QPO's (quasi–periodic oscillations) which are associated with different spectral behaviour. QPO's in the range 18 – 55 Hz with a hard spectrum are found in all observations where the source is on the "horizontal branch" in an hardness–versus–intensity diagram. On the "vertical branch", however, QPO's with a soft spectrum are found at a stable frequency of 5.6 Hz. A multitude of QPO frequency–intensity relations, found in different "horizontal branch" observations, coalesce to an almost unique relation when the QPO frequency is correlated with spectral hardness. All results are consistent with a model where QPO's represent the Kepler frequency at the edge of the neutron star magnetosphere.

117.004 The beat frequency model for QPOs.
J. Shaham.
The origin and evolution of neutron stars, p. 347 – 361 (1987). – See Abstr. 012.001 (IAU Symp. No. 125).

There are to date reports of quasi–periodic–oscillation (QPO) observations in some twelve X–ray source, of which at least seven are low mass X–ray binaries (van der Klis 1987). They constitute a formidable zoo of phenomena with so much variety that they, at times, do not at all even seem amenable to a single model. The author discusses the beat frequency model which seems to him to be by far the prime model for at least some of the QPOs.

117.005 Quasi–periodic oscillations. The Rapid Burster (MXB 1730–335). Models vs. observations – a brief review.
W. H. G. Lewin.
The origin and evolution of neutron stars, p. 363 – 374 (1987). – See Abstr. 012.001 (IAU Symp. No. 125).

The salient features of quasi–periodic oscillations (QPO) observed in type 2 bursts and in the persistent emission from the Rapid Burster are discussed. In addition, a brief review is given of the models that have recently been proposed to explain high–frequency QPO observed in several bright low–mass X–ray binaries. The mechanism(s) of the QPO are not yet known, it is not even known whether they are magnetospheric in origin. However, some of the proposed ideas could well be relevant to the various rather complex aspects of the QPO. It is likely that more than one mechanism is at work.

117.006 What type of binary system is Cygnus X–3?
J. M. Bonnet–Bidaud.
The origin and evolution of neutron stars, p. 550 (1987).
Abstract. – See Abstr. 012.001 (IAU Symp. No. 125).

117.007 On the phases of period variation of SS433.
W. Q. Luo.
The origin and evolution of neutron stars, p. 551 (1987).
Abstract. – See Abstr. 012.001 (IAU Symp. No. 125).

117.008 Production and interaction of high energy neutrinos in close X–ray binaries.
A. K. Harding, J. J. Barnard, F. W. Stecker, T. K. Gaisser.
The origin and evolution of neutron stars, p. 552 (1987).
Abstract. – See Abstr. 012.001 (IAU Symp. No. 125).

117.009 Absorption features from the accretion column in E 1405–451.
D. T. Wickramasinghe, I. R. Tuohy, N. Visvanathan.
Astrophys. J., Vol. 318, No. 1, p. 326 – 336 (1987).

The authors present phase–resolved circular spectropolarimetric observations of E 1405–451 (V834 Cen) covering the wavelength region 3500 – 7800 Å obtained at the AAT. The data show broad absorption features in a strongly polarized cyclotron–continuum present only near maximum light. Cyclotron harmonic and Zeeman line interpretations are investigated, and it is shown that the stronger features are consistent with a nonphotospheric Zeeman interpretation corresponding to a field range of 20 – 25 MG.

117.010 Radio emission from the nova–like variable AC Cancri and the symbiotic variable AG Draconis.
M. V. Torbett, B. Campbell.
Astrophys. J., Lett. Ed., Vol. 318, No. 1, p. L29 – L32 (1987).

Radio emission at 6 cm has been detected from the nova–like cataclysmic variable AC Cnc and the symbiotic variable AG Dra. The AC Cnc observation constitutes the first radio detection in this class of objects. The AG Dra source is probably resolved and appears to show asymmetric, extended structure. The radio emission can best be explained by thermal bremsstrahlung.

117.011 The interacting binary β Lyr. II. Non–LTE model analysis and evolutionary conclusions.
D. L. Dimitrov.
Bull. Astron. Inst. Czech., Vol. 38, No. 4, p. 240 – 253 (1987).

Results of a model atmosphere analysis and non–LTE treatment of He I lines in the optical spectrum of β Lyr primary as well as calculations of theoretical spectra are presented and discussed. Among the major conclusions are the following: (1) As a direct consequence of the CNO cycle, a significant helium overabundance in the atmosphere is found, He/H = 1.0; nitrogen is also overabundant. (2) A good agreement between observed and synthetic spectra is obtained for line profiles of other elements – Mg, Si, S, Ti, Cr, Fe and Ni, assuming normal abundances; the same is valid also for Ne, allowing for departures from LTE. (3) The appearance of some resonance lines suggests influence from a circumstellar envelope; here, an approximate non–LTE approach (also for some Si II lines) gave better agreement between observed and predicted profiles. These results are consistent over large parts of the (optical) spectrum.

117.012 The nature of classical symbiotic stars.
B. F. Yudin.
Astrophys. Space Sci., Vol. 135, No. 1, p. 143 – 156 (1987).

The results of the optical and infrared observations of classical symbiotic stars Z And, CI Cyg, BF Cyg, AG Dra, AX Per, V443 Her, and YY Her, are summarized. It is shown that the hot component of most classical symbiotic stars is a hot subdwarf and not a main–sequence star with an accretion disc. The energy source of its outbursts is the gravitational energy of the matter accreted from the cool component's surface. The cool component is a red giant filling the Roche lobe and having class II luminosity. It is probable that among classical symbiotic stars there are – in an insignificant quantity – systems in which the hot component is a main–sequence star with an accretion disc.

117.013   BV observations of W UMa–type binaries: CK Boo, BI CVn, and AH Vir.
O. Demircan.
Astrophys. Space Sci., Vol. 135, No. 1, p. 169 – 174 (1987).

Photoelectric BV observations of W UMa–type binaries CK Boo, BI CVn, and AH Vir are presented, and light curve variations and shifts of minima are discussed in the context of stellar activity and mass transfer. The O–C increments in the three systems were found to be 0.08, 0.80, and 0.60 s per revolution, respectively. The new ephemeris for the future observations were also found.

117.014   The supercycles of SU Ursae Majoris.
J. E. Isles.
J. Br. Astron. Assoc., Vol. 97, No. 5, p. 274 – 276 (1987).

Normal outbursts of SU UMa become more energetic as a supermaximum approaches. The cycle–averaged outburst flux is lowest immediately after a supermaximum, but begins increasing straight away.

117.015   CPD –71°172, a new binary with a hot subdwarf.
M. Viton, D. Burgarella, A. Cassatella, L. Prévot.
Messenger, No. 49, p. 2 – 3 (1987).

117.016   A comparison of characteristics of dwarf novae, nova–like stars and polars. A new criterion of search for polars.
N. F. Vojkhanskaya.
Pis'ma Astron. Zh., Tom 13, No. 7, p. 597 – 602 (1987). In Russian. English translation in Sov. Astron. Lett., Vol. 13.

The mean characteristics of binary systems: dwarf novae, nova–like stars and polars are compared. A new method of sampling the candidates for polars according to the typical spectral features is suggested.

117.017   Model light curves of close binary systems with circumstellar envelopes.
I. Pustyl'nik, L. Einasto.
Pis'ma Astron. Zh., Tom 13, No. 7, p. 603 – 609 (1987). In Russian. English translation in Sov. Astron. Lett., Vol. 13.

The technique and the results of calculations of model light curves for Algol–type binaries are described, which account for the influence of a semi–transparent circumstellar gas filling the critical Roche lobe. It is shown that the combined effect of screening of stellar radiation by gas and the contribution of the circumstellar shell to the total luminosity of a binary will lead to an appreciable broadening of the secondary minimum and distortion of the measured ratio of the luminosities of the components. The neglecting of these effects results in an overestimate of both, the luminosity and the radii of subgiant secondaries.

117.018   Parameters of the SS433 jets. Interpretation of the results.
S. N. Fabrika, N. V. Borisov.
Pis'ma Astron. Zh., Tom 13, No. 8, p. 663 – 670 (1987). In Russian. English translation in Sov. Astron. Lett., Vol. 13.

Mass loss rate in the SS433 jets is estimated to be $\dot{M}_j \approx 5 \times 10^{-7} M_\odot/y$. The parameters of the Hα clouds in the jet: dimensions and density, filling factor are presented. Gas in the clouds is heated by the hard radiation of the disk with $\varepsilon \geqslant 0.5$ keV. This radiation should be strongly collimated along the jets, the opening angle of the radiation cone is $\theta_c < 13°$. Arguments for $\theta_c \approx 6°$ are presented. The possible radiation cone structure is discussed.

117.019   Symbiotic stars observed from the IRAS satellite.
L. Luud, T. Tuvikene.
Astrofizika, Tom 26, Vyp. 3, p. 457 – 468 (1987). In Russian. English translation in Astrophysics, Vol. 26, No. 3, p. 276 – 283 (1987).

Symbiotic stars according to Allen's catalogue have been checked for coincidence with the IRAS far–infrared sources. 72 symbiotic and possible symbiotic stars have been identified

with the IRAS sources. A catalogue of identified stars and energy distributions of representative stars are given. It turns out that the dust in symbiotic stars is a more spread phenomenon than it was believed before. Almost 40% of systems are the dusty ones. Among objects with dust temperature some tens of K have been found. It is shown that the only useful two–color diagram is $(K–m_{12}) – (m_{12}–m_{25})$.

117.020   DQ Herculis' orbital period variations according to 1982 – 1986 observations.
E. S. Dmitrienko.
Astrofizika, Tom 26, Vyp. 3, p. 561 – 566 (1987). In Russian. English translation in Astrophysics, Vol. 26, No. 3.

The eclipses of the primary component by the secondary one in DQ Herculis (N Her 1934) were analyzed by the light curves obtained by photometric observations in 1982 – 1986. It is suspected that the value O–C varies with time by the sinusoidal law with a period of about five years. Even being confirmed by further observations, any hypothesis – either the presence of the third component in the system, or the rotation of apsid line – seems inadequate to explain the phenomenon because the secondary minimum is not to be seen on the *UBVRI* light curve.

117.021   Black hole candidates in X–ray binaries.
S. Hayakawa.
Gravitational collapse and relativity, p. 20 – 42 (1986). – See Abstr. 012.006.

A number of X–ray binaries have been thought to consist of black holes on account of characteristic features such as soft spectra accompanied with high energy tails, rapid variabilities without regular periods, and bipolar jets associated with radio lobes. Observational results are presented which show that many of neutron star binaries also show similar features. The most convincing signature of the black hole is therefore considered to be its mass exceeding the neutron star mass limit. Black hole candidates with this signature have been found for Cyg X–1, LMC X–1, A 0620–00, and SS433 from their binary motions.

117.022   On the mechanism of the "noisar" phenomenon in magnetic close binary systems.
I. L. Andronov.
Astron. Nachr., Vol. 308, No. 4, p. 229 – 234 (1987).

The mechanisms of the "noisar" phenomenon in AM Herculis–type stars are discussed. In an accretion column above the surface of the magnetized degenerate star the instability of some types may be excited, such as axi–symmetrical quasi–periodical penetration of the low–density "bulks" from the column axis to the outer parts; "boiling" with "bulks" moving inside or outside the column; "tornados" with low–density regions rapidly rotating around the column axis; "switchings" of the accretion from one half of the "polar cap" to another and vice versa. The oscillations of different plasma clots ("spaghetti") may interfer causing flux changes as well. Such "noisar" oscillations appear in different regions of the accretion column, so hard and soft X–ray fluxes might not have correlation in their variability. The observations are in qualitative agreement with the models.

117.023   V361 Lyrae: an exotic binary system with a "hot spot" between the components?
I. L. Andronov, G. A. Richter.
Astron. Nachr., Vol. 308, No. 4, p. 235 – 238 (1987).

A phenomenological model for V361 Lyr is proposed. Probably it is a binary system which consists of a mass accreting primary star with mass of about $M_1 \approx 0.81\ M_\odot$ and radius $R_1 \approx (6.1 \pm 0.4) \times 10^{10}$ cm and a mass–losing secondary with about $M_2 \approx 0.77\ M_\odot$ and $R_2 \approx 5.8 \times 10^{10}$ cm. The secondary fills its Roche lobe, but the primary is something smaller than this lobe, contrary to the models of W UMa–type systems. So the hot spot appears in the atmosphere of the primary, but not in a disk, like in cataclysmic variables. The luminosity of the hot spot, $L = (6 – 15) \times 10^{32}$ erg/s, is large enough to be the main emission source of the system in visible light. So phenomenologically the object may be somewhat between W UMa–type stars and cataclysmic variables.

117.024 Discovery of 2 – 3 s quasi–periodic oscillations in EF Eri.

S. Larsson.

Astron. Astrophys., Vol. 181, No. 1, p. L15 – L17 (1987).

The AM Her system EF Eri is shown to exhibit optical 2 – 3 second quasi–periodic oscillations. This is the third AM Her system for which such oscillations have been detected. The discovery of oscillations in EF Eri, which is believed to contain a high mass ($\gtrsim 1\,M_\odot$) white dwarf, may pose problems for oscillating shock models. The author points out the similarity in magnetic field orientation for the three systems with known oscillation, which could be an indication that the oscillations are instead associated with the magnetic field.

117.025 Simultaneous multicolour photometry of OY Carinae during quiescence.

R. Schoembs, H. Dreier, H. Barwig.

Astron. Astrophys., Vol. 181, No. 1, p. 50 – 56 (1987).

Using a new multi channel photometer, high time resolved (2s) photometric data of OY Carinae have been obtained simultaneously in 4 colours. OY Car was in quiescent state during the observing period of 10 consecutive nights. A high precision orbital light curve (25 averaged orbital periods) was used to derive system parameters without involving spectroscopic data. Improved masses $m_1 = 0.89\,M_\odot$, $m_2 = 0.1\,M_\odot$ and radii $r_1 = 8.16 \times 10^8$cm, $r_2 = 10^{10}$cm were determined. The primary fits the mass radius relation for white dwarfs within 5%. The radius of the accretion disk determined directly from its eclipse contact times, amounts to 58% of its Roche radius. Colour variations during eclipse confirm the existence of a hot corona above the accretion disk. The orbital period was further improved.

117.026 Polarization and infrared colors of symbiotic stars.

R. E. Schulte–Ladbeck, A. M. Magalhães.

Astron. Astrophys., Vol. 181, No. 1, p. 213 – 216 (1987).

The authors present new results of an ongoing survey of the optical linear polarization properties of symbiotic stars. Their aim is to obtain a statistically significant sample to study the nature and geometrical arrangement of the circumstellar matter and its relationship to the stellar components. It is concluded that dust scattering in the asymmetric circumstellar environment of symbiotic systems does play a role in producing the polarization, at least in D–type objects. In S–type systems, part of the observed intrinsic polarization might originate in photospheric scattering.

117.027 On the nature of 623 + 71: a cataclysmic binary surrounded by a bow–shock–like emission nebula.

J. Krautter, U. Klaas, G. Radons.

Astron. Astrophys., Vol. 181, No. 2, p. 373 – 377 (1987).

Direct CCD images and long–slit spectrograms of the peculiar system 623 + 71 have been obtained. 623 + 71 is a cataclysmic binary surrounded by a faint emission nebula which has a bow–shock–like structure. The dimensions of the nebula are about 105" both in the east–west and south–north directions. Electron densities in the nebula are between 100 and 250 cm^{-3}. Emission line ratios show that shock wave heating and photoionization contribute to the excitation of 623 + 71 as old nova shell which has been ejected during an eruption some time ago. The observations do not support an interpretation of 623 + 71 as old nova shell which has been ejected during an eruption some time ago. A planetary nebula nature would have important implications for the evolution of cataclysmic binaries. Evidence for such a nature is the strong morphological similarity with the peculiar planetary nebula Abell 35. A third explanation is that the nebula was formed by an interaction of a stellar wind with the surrounding interstellar medium.

117.028 Soft X–ray transients in quiescence: observations of Aql X–1 and Cen X–4.

J. van Paradijs, F. Verbunt, R. A. Shafer, K. A. Arnaud.

Astron. Astrophys., Vol. 182, No. 1, p. 47 – 50 (1987).

The authors observed the soft X–ray transients Aql X–1 and Cen X–4 during quiescence with EXOSAT. For Aql X–1 they obtain a 3σ upper limit to the X–ray luminosity L_x (in the energy range between 0.5 and 4.5 keV) of $< 3.0 \times 10^{33}$erg s^{-1}, Cen X–4 was detected at L_x (0.5 – 4.5 keV) $\cong$ 1.5 to 4.2×10^{33}erg s^{-1}. An optical spectrum of Cen X–4 obtained two years after its 1979 outburst shows a K3 V continuum with Balmer emission lines. Combining their EXOSAT observations with a previous EINSTEIN detection the authors argue that the X–ray luminosity of Cen X–4 is due to continued accretion, and not to thermal radiation of a cooling neutron star.

117.029 A new approach to symbiotic stars.

H. Nussbaumer, M. Vogel.

Astron. Astrophys., Vol. 182, No. 1, p. 51 – 62 (1987).

The authors assume symbiotic stars to be double star systems. They show with detailed model calculations that a typical outbreak of a symbiotic star can be induced by an increase in the mass–loss of the cool star. The onset of that stellar wind will provide a fast increasing target for the photons emitted by the hot star. The cool stellar wind is ionized by the radiation of the hot companion. Depending on the temperature of the hot star, the mass–loss, and the stellar separation, the emission of the lowly and medium ionized atoms may originate at various regions between the cool and the hot star. Differences in the combination of mass–loss, stellar separation, and radiation temperature may also explain the observationally established division of symbiotic stars into s–type and d–type. The authors calculate the evolution of M_B and M_V as a function of the mass–loss of the cool star. The calculated line profiles show periodic variations in intensity and wavelength position.

117.030 Disk formation at the magnetosphere of wind–fed pulsars: application to Vela X–1.

G. Börner, S. Hayakawa, F. Nagase, U. Anzer.

Astron. Astrophys., Vol. 182, No. 1, p. 63 – 70 (1987).

The disk formation in a wind–fed, rotating, accreting neutron star is considered in view of the fact that the angular momentum transfer by wind accretion is too small to account for large rates of spin–up and spin–down as observed. Under the assumption that the closed magnetosphere extends to the corotation radius it is shown that a disk will form through the interaction of the spherically accreting matter with the rotating magnetosphere. Long term spin–up and spin–down episodes then become possible due to the difference between the accretion torque transmitted by the disk and the deceleration by the accretion at the magnetosphere surface. Large rates of short–term pulse period change are attributed to the change in the accretion torque transmitted by the disk. Application of this model to the X–ray source Vela X–1 requires a large magnetic moment $\mu \sim 10^{32}$ Gauss cm^3, but once the high value for the magnetic field (10^{13} Gauss $\lesssim B \lesssim 10^{14}$ Gauss) is accepted, a variety of observations fit well into the picture.

117.031 Discovery of soft X–ray oscillations in VW Hydri.

H. van der Woerd, J. Heise, F. Paerels, K. Beuermann, M. van der Klis, C. Motch, J. van Paradijs.

Astron. Astrophys., Vol. 182, No. 2, p. 219 – 228 (1987).

The authors report the discovery of short–period oscillations in the soft X–ray flux, emitted by the dwarf nova VW Hydri during superoutburst. A modulation of 15% with a period of 14.06 ± 0.02 s was detected near the end of the November 1983 superoutburst. The oscillation was observed for two hours and is coherent to within the limits of observation ($Q > 2500$). A less coherent oscillation was detected near the maximum of the October 1984 superoutburst. This oscillation shows erratic period changes between 14.2 and 14.4 s. The pulsed emission probably has a harder spectrum than the non–pulsed emission. Upper limits to the amplitude of oscillations during other observations of VW Hyi during normal outburst, superoutburst and quiescent indicate that these 14 s oscillations are a transient effect. The properties, and origin of the soft X–ray oscillations in dwarf novae are discussed.

117.032   A new, distant dwarf nova: 2138–453.
    M. R. S. Hawkins, P. Véron.
Astron. Astrophys., Vol. 182, No. 2, p. 271 – 272 (1987).
    The authors report on the discovery of a distant dwarf nova, located at about 3.0 kpc below the galactic plane. They comment on the frequency of such objects.

117.033   Some observational manifestations of the operation of a clock mechanism in the Her X–1/HZ Her system.
    E. A. Karitskaya, N. G. Bochkarev, Yu. N. Gnedin.
Sov. Astron., Vol. 30, No. 5, p. 592 – 598 (1986). English translation of 42.117.068.

117.034   On the possibility of detecting close binary degenerate dwarfs.
    A. V. Tutukov, L. R. Yungel'son.
Sov. Astron., Vol. 30, No. 5, p. 598 – 600 (1986). English translation of 42.117.069.

117.035   Solar oscillations and rotation of close binary systems in the Galaxy: the 160 minute period.
    V. A. Kotov.
Bull. Crimean Astrophys. Obs., Vol. 74, p. 65 – 79 (1987). English translation of 41.117.210.

117.036   Observations of the symbiotic stars AS 296, AS 360, and AS 338 during outbursts.
    V. F. Esipov, A. P. Ipatov, B. F. Yudin.
Astrophysics, Vol. 25, No. 2, p. 487 – 491 (1987). English translation of 42.117.229.

117.037   The symbiotic star AG Dra.
    A. P. Ipatov, B. F. Yudin.
Astrophysics, Vol. 25, No. 3, p. 605 – 612 (1987). English translation of 42.117.298.

117.038   Spectral observations of AG Dra in 1983.
    Z. A. Ismailov, Kh. M. Mikajlov.
Astrophysics, Vol. 25, No. 3, p. 612 – 615 (1987). English translation of 42.117.299.

117.039   Absolute magnitudes of cataclysmic variables.
    B. Warner.
Mon. Not. R. Astron. Soc., Vol. 227, No. 1, p. 23 – 73 (1987).
    Absolute magnitudes M_v of the accretion discs (with allowance for effects of inclination) of stars in the principal classes of cataclysmic variable stars are derived from a variety of techniques. For dwarf novae, whose distances are found mostly from infrared observations of their secondaries, a tight relationship is found between M_v(max) at maximum of outburst and orbital period: M_v(max) = 5.64 − 0.259 P(hr). Use of this equation provides M_v(min) at minimum light for further dwarf novae without known distances. For many of these stars it is possible also to derive M_v(mean), averaged over the outburst cycle. Analysis of this extensive data set discloses that M_V(min) and M_V(mean) are functions of orbital period P and mean time T_n between normal outbursts. An amplitude −T_n(Kukarkin-Parenago) relationship is found to exist for the dwarf novae. A scenario is given in which recent suggestions on cyclical evolution through classical novae, dwarf novae and nova–like variables are shown to agree qualitatively with the systematics found in this study.

117.040   Mass transfer rates and the soft X–ray excess in AM Herculis binaries.
    A. R. King, M. G. Watson.
Mon. Not. R. Astron. Soc., Vol. 227, No. 1, p. 205 – 211 (1987).
    The authors consider the mass transfer rates implied for the AM Her systems by their hard X–ray fluxes alone, assuming that no soft X–ray excess exists. Distance estimates suggest that these rates are at least an order of magnitude lower than is compatible with currently favoured pictures of the secular evolution of cataclysmic variables, and would present severe difficulties for any evolutionary scheme, as the implied mass transfer time–scales would exceed a Hubble time. By contrast, the soft excesses already directly measured for some systems imply transfer rates in rough agreement with those expected.

117.041   Tidal heating in close binary stellar systems.
    M. Rieutord, S. Bonazzola.
Mon. Not. R. Astron. Soc., Vol. 227, No. 2, p. 295 – 314 (1987).
    Tidal heating of a low–mass star in a close binary system, resulting from the conjugate effect of angular momentum loss and tidal action, is investigated via a detailed study of the flow inside the secondary.

117.042   Photometry and spectroscopy of the interacting binary white dwarf V803 Cen (AE–1).
    D. O'Donoghue, J. W. Menzies, P. W. Hill.
Mon. Not. R. Astron. Soc., Vol. 227, No. 2, p. 347 – 359 (1987).
    High–speed photometry of the helium variable V803 Cen (AE–1) is presented which shows rapid oscillation at 1611 and 175 s. The 1611–s period has a complex and variable harmonic structure. Both periods remained coherent over four days when the system was at ∼13.4 in B. After the system faded to $B \sim 15.4$, the coherency of the 1611–s period seemed to be breaking down and a significant period change in the 175–s oscillation was found. A spectrum of the system at $B \sim 15.4$ showed evidence for weak He I emission. These properties establish that V803 Cen is the same kind of object as PG 1346 + 082. The overall properties of the two stars are considered with those of AM CVn and GP Com and are found to be consistent with the interacting binary white dwarf model.

117.043   Contact and near–contact binary systems – VII. EZ Hydrae, AD Phoenicis and RS Columbae.
    T. M. McFarlane, R. W. Hilditch.
Mon. Not. R. Astron. Soc., Vol. 227, No. 2, p. 381 – 402 (1987).
    $BVRI_c$ photometry is presented for three late–type contact binaries: EZ Hydrae, AD Phoenicis and RS Columbae. EZ Hydrae is identified as a W–type system of orbital period 0.45 day, while AD Phoenicis and RS Columbae are probably A–type systems, with orbital periods of 0.38 and 0.67 day, respectively. Although a value for the mass ratio of EZ Hydrae had been obtained from spectroscopic observations, no photometric solution could be obtained because of severe 'disturbances' in its light curve. On the other hand, photometric solutions were obtained for AD Phoenicis and RS Columbae, but these were very insensitive to mass ratio, which tended toward the unlikely value of unity.

117.044   RZ Pyxidis: an early–type marginal contact binary.
    S. A. Bell, G. J. Malcolm.
Mon. Not. R. Astron. Soc., Vol. 227, No. 2, p. 481 – 500 (1987).
    The first modern photometric and spectroscopic study of the early–type binary RZ Pyx is presented. The analysis suggests that the system is in a marginal contact configuration with components on the zero–age main sequence. The masses and radii of the two components are found to be (5.3 ± 0.4) and (4.3 ± 0.2) solar masses and (2.61 ± 0.10) and (2.44 ± 0.06) solar radii, respectively. The age estimate for the system is less than 2×10^6 yr. The system may have evolved into contact within this period of time or have arrived on the main sequence in a contact configuration.

117.045   An EXOSAT observation of 1.5 orbital cycles of the 0.7 day short–period RS CVn system ER Vul.
    N. E. White, J. L. Culhane, A. N. Parmar, M. A. Sweeney.
Mon. Not. R. Astron. Soc., Vol. 227, No. 3, p. 545 – 551 (1987).
    Two EXOSAT observations of the short–period 0.7 day eclipsing RS CVn binary system ER Vul are reported. The first observation in 1984 October lasted for 3 hr, while the second in 1985 May covered 1.5 orbital cycles. No strong orbital modulation of the X–ray flux was evident, nor were any deep eclipses seen. The spectrum measured in the 0.05 – 6 keV band was well

fitted by a two–component thermal plasma model with temperatures of 6 and 40 million degrees. The failure to detect any strong orbital modulation indicates for both temperature components either loop heights larger than a stellar radius, or more compact loops that are uniformly distributed as a function of longitude.

117.046 Tidal resonances in binary star systems.
M. E. Alexander.
Mon. Not. R. Astron. Soc., Vol. 227, No. 4, p. 843 – 861 (1987).

The Hamiltonian describing the coupled system of orbital motion and oscillations of a non–rotating star is derived and then transformed to action–angle variables. The averaged Hamiltonian for a single or multiple resonance or near resonance is obtained and constants of motion derived. The conditions for resonance are derived in detail for the case of small orbital eccentricity. The equations of motion are numerically integrated to demonstrate the evolution through tidal capture and partial circularization of the elliptical orbit. The extension of the analysis to a rotating star is discussed, and the possibility of resonance capture due to evolution by tidal dissipation is briefly considered.

117.047 Optical and UV spectroscopy of the black hole binary candidate LMC X–1.
J. B. Hutchings, D. Crampton, A. P. Cowley, L. Bianchi, I. B. Thompson.
Astron. J., Vol. 94, No. 2, p. 340 – 344 (1987).

The authors present both further optical spectroscopy of the binary star identified with LMC X–1, obtained between 1983 and 1985, and a series of IUE UV spectra taken during a 5 day interval in 1984. The optical data are used to refine the orbital period to 4.2288 days, and improved orbital parameters are derived. The most probable component masses are approximately 20 $M_\odot$ for the primary and near 6 $M_\odot$ (for the X–ray star), suggesting that the latter may be a black hole. The UV spectra show very weak, low–velocity stellar–wind lines. The authors suggest that much of the surrounding medium is highly ionized by the X–ray flux. The 'nonwind' UV spectral lines and the UV continuum temperature are consistent with the optical data, indicating a late O type star of $M_{bol} = -8.5$.

117.048 Outburst and quiescence observations of the dwarf nova V101 in the globular cluster M5.
M. M. Shara, A. F. J. Moffat, M. Potter.
Astron. J., Vol. 94, No. 2, p. 357 – 359 (1987). With plates 27 – 28.

The authors have observed the dwarf nova V101 towards the globular cluster M5 through most of an eruption, and during quiescence. Assuming it to be a member of the cluster, V101 reached $M_B \cong +4.3$ at maximum and $M_B \sim +7.6$ at minimum, in good agreement with determinations for field dwarf novae. The outburst rise and fall times and total duration imply a long orbital period (~ 11 hr) and possibly an evolved secondary. This hypothesis is consistent with the red continuum of V101 seen in the spectrum of Margon et al. (1981).

117.049 The light variation of AG Draconis in its quiet state.
J. B. Kaler.
Astron. J., Vol. 94, No. 2, p. 437 – 451 (1987).

Light curves of the yellow symbiotic star AG Draconis are constructed in 11 intermediate and narrow wavelength bands from the near ultraviolet to the far red, from observations made on 80 nights between 17 March 1977 and 18 October 1980, while the star was in its normal state. The curves from Strömgren $uvby$, $\lambda 4428$, and at wide and narrow Hβ are carried through the entire 1311 day observing interval, which encompasses 2.4 cycles. Red data at $\lambda 6503$ and Hα, and regular He II observations at $\lambda 4686$, were added later and run through the last 1224, 905, and 592 days (2.2, 1.6, and 1.1 cycles), respectively. In addition, $\lambda 8200$ was sporadically monitored. The observations were terminated by the November 1980 outburst.

117.050 Photometry of the 1980 outburst of AG Draconis.
J. B. Kaler, C. A. Stoehr, W. I. Hartkopf, R. A. Shaw, K. Hufford, E. C. Olson, A. Shankar, J. P. Hickey, K. B. Kwitter.
Astron. J., Vol. 94, No. 2, p. 452 – 462 (1987).

The authors present detailed 11–color (continuum and line) photometry of the November 1980 outburst of AG Draconis, which increased the brightness of the star by two magnitudes in the ultraviolet and one in the visual. The star was observed intensively for the first two weeks after discovery, was monitored at a slower rate for the next eight months, and checked sporadically through 1986. The Strömgren y data were used to calibrate the AAVSO light curve, which fills in the gaps in the photoelectric record. By June 1982, after 900 days (1.6 normal cycles), the star had returned to its quiet state. Following an uneventful 550 day period, the AAVSO light curve then shows two smaller additional bursts.

117.051 Ultraviolet and optical observations of AG Draconis.
J. H. Lutz, T. E. Lutz, J. D. Dull, D. D. Kolb.
Astron. J., Vol. 94, No. 2, p. 463 – 483 (1987).

Ultraviolet and optical observations have been obtained of the yellow symbiotic star AG Draconis at irregular intervals between 1975 and 1984. The series of observations with the International Ultraviolet Explorer satellite includes data from before, during, and after the 1980/81 outburst. Due to recent calculations of several new atomic–line ratios, particularly for high–density conditions, the authors are able to use some of the data to estimate physical parameters such as electron density. They do not have sufficient data to present a comprehensive model of AG Dra, but they discuss how their observations compare to predictions of various models.

117.052 Variability of the intrinsic polarization of the supergiant binary VV Cephei.
R. J. Pfeiffer, R. H. Koch.
Astron. J., Vol. 94, No. 2, p. 484 – 500 (1987).

Polarization observations in u, b, g, and r bandpasses have been obtained for VV Cep from before the 1977 primary eclipse to past the 1985 secondary conjunction. These are mostly linear measurements, but a few uncalibrated circular observations have been obtained also. The data show a mean polarization variation locked to Keplerian phase. In addition, there are many transients from the mean. The sporadic activity increases towards periastron, and includes a large polarization pulse that is interpreted to signal the supergiant's overflow of its Roche lobe or the ejection of a relatively large amount of material from the M star, through the L_1 point toward the hot component. The mean trend of the polarization activity has been successfully reproduced by a multiparameter model for the scattering geometry of the system.

117.053 The magnetic field strength in the emission line region of the AM Her system EF Eridani (= 2A 0311–277).
W. Seifert, R. Östreicher, G. Wunner, H. Ruder.
Astron. Astrophys., Vol. 183, No. 1, p. L1 – L2 (1987).

If the feature around $\lambda 6540$ in the circular polarization spectrum of EF Eri (seen only at polarization phase 0.45) is interpreted in terms of a Zeeman split σ^-–component of Hα emission, at least a significant fraction of the Hα flux originates in a typical magnetic field strength of 1.5 ± 0.3 MG. Assuming a typical polar field strength of 30 MG and a dipolar field structure, this implies a height of the emission region of about 1 – 2 white dwarf radii.

117.054 An evolutionary scenario for the black hole binary A 0620–00.
M. de Kool, E. P. J. van den Heuvel, E. Pylyser.
Astron. Astrophys., Vol. 183, No. 1, p. 47 – 52 (1987).

The authors present an evolutionary scenario for the black hole binary A 0620–00, which starts from an initial configuration consisting of a massive star and a low–mass companion in a very wide orbit. By using the presently observed system parameters and following the evolution backward in time, the authors derive an upper limit to the initial mass of the companion of 2 $M_\odot$, a relation between the present mass of the black hole and the mass of its main–sequence progenitor, which, for the most likely black hole mass of $7\,M_\odot$, yields a progenitor mass between 27 and $46\,M_\odot$ and a lower limit on the initial orbital period of 240 days.

117.055   The 67–min X–ray period of EX Hydrae observed with the EINSTEIN observatory.

J. Heise, R. Mewe, A. Kruszewski, T. Chlebowski.
Astron. Astrophys., Vol. 183, No. 1, p. 73 – 82 (1987).

The cataclysmic variable EX Hydrae has been observed in X–rays with the HRI, IPC and MPC of the EINSTEIN observatory on two occasions. The X–ray spectra indicate multiple spectral components. A constant hard X–ray flux (with $kT \approx 8$ keV) is found in the high–energy channels of the MPC. In the lower energy channels of the MPC and in the IPC a variable, softer component is observed with a period of 67 min. The modulation is approximately in phase with the stable 67–min modulation in the optical brightness. The IPC X–ray spectrum changes slightly with the 67 min phase. The X–ray observations of EX Hya, recently interpreted as an intermediate polar, exhibit many features characteristic for AM Her–type stars. The authors discuss the AM Her–type X–ray properties of EX Hya and the possible absence of a standard accretion disc in this system.

117.056   A search for non–stellar contributions to the optical and near–IR flux of RS CVn binaries. I. The cases of TY Pyx, UV Psc, RU Cnc and VV Mon.

M. Busso, F. Scaltriti, P. Persi, M. Robberto, G. Silvestro.
Astron. Astrophys., Vol. 183, No. 1, p. 83 – 90 (1987).

UBVRIJHK light curves of four active binaries of the RS CVn class are presented together with coordinated spectrophotometric observations in the near infrared (1.4–2.5 μm). Low amplitude time variations in the optical filters are clearly present. They are interpreted as effects of dark spots partially covering the stellar surface. However, the whole spectral distributions (from U to K) of three out of four studied systems (with the exception of TY Pyx) cannot be simply explained by a combination of stellar spectra and of spot–related photospheric perturbations. Moreover, at least in two cases, the data clearly show infrared excesses that cannot be simply explained by classification errors, and seem to require the presence of circumstellar matter. A preliminary analysis suggests that, at least in two systems, thermal radiation from a thin dust shell is likely to explain the data.

117.057   Doppler–effect modulation of the observed radiation flux from ultracompact binary stars.

N. I. Shakura, K. A. Postnov.
Astron. Astrophys., Vol. 183, No. 2, p. L21 – L22 (1987).

The observed radiation flux from ultra–short period binaries must be modulated due to Doppler effect. Different modulation amplitudes of integral flux should be seen depending on the geometry of the emitting object (a disc or a quasi–spherical source). When the source is observed in a narrow frequency band, Doppler modulation will be dependent on the shape of the spectrum also. The modulation will be more pronounced in the exponential tail of the spectrum.

117.058   Winds in collision. III. Modeling the interaction nebulae of eruptive symbiotics.

T. Girard, L. A. Willson.
Astron. Astrophys., Vol. 183, No. 2, p. 247 – 256 (1987).

In Papers I and II, observations of HM Sge and V1016 Cyg were interpreted in terms of two colliding stellar winds in an interacting binary. In this paper the colliding wind model is investigated in more detail. The authors present numerical models which describe the gross geometric and kinematic properties, as well as the time development, of the interaction nebula which forms in the general two colliding wind system. These numerical results are then used to interpret the geometric model of V1016 Cyg derived in Paper I. The authors also discuss the system HM Sge in light of the numerical results.

117.059   A new observing campaign for HR 6469.

R. Wasson.
I.A.P.P.P. Commun., No. 29, p. 21 – 22 (1987).

117.060   HD 185510: a chromospherically active binary with eclipses of a hot subdwarf companion.

L. A. Balona, T. Lloyd Evans, T. Simon, G. Sonneborn.
I.A.P.P.P. Commun., No. 29, p. 29 – 32 (1987).

117.061   Der symbiotische Stern CH Cyg. BAV–Beobachtungen 1972 – 1986.

A. Thomas.
Sterne Weltraum, 26. Jahrg., Nr. 9, p. 498 – 500 (1987).

117.062   Observed variations in Wolf–Rayet stars.

J.–M. Vreux.
Instabilities in luminous early type stars, p. 81 – 97 (1987). – See Abstr. 012.012.

Observed variations in WR stars are reviewed with a special emphasis on the ones which could be intrinsic and periodic.

117.063   Outbursts of dwarf novae.

S. Mineshige.
Astron. Her., Vol. 80, No. 10, p. 284 – 289 (1987). In Japanese.

117.064   On the character of SS433 spectral variability with the 6–day period (June – August, 1981).

I. M. Kopylov, R. N. Kumajgorodskaya, N. N. Somov, T. A. Somova, S. N. Fabrika.
Astron. Zh., Tom 64, Vyp. 4, p. 785 – 802 (1987). In Russian. English translation in Sov. Astron., Vol. 31, No. 4.

New spectral data ($\sim$280 spectra) for the object SS433 obtained during 35 nights in the summer of 1981 with the 1000–channel spectrophotometer of the 6–meter telescope within the range $\lambda\lambda 5300 – 7900$ Å with a resolution of $\sim$4 Å are analysed. The behaviour of hydrogen and helium relativistic line profiles with time is studied and the following parameters for them are determined: equivalent widths W_λ, central intensities R_c, halfwidths $\Delta\lambda(1/2)$ and radial velocities z. A detailed study has been given to the radial velocity variations in the main components of relativistic hydrogen lines Hα and Hβ with a 6–day (nutation) period.

117.065   On the parameters of the system Cyg X–1.

V. V. Sokolov.
Astron. Zh., Tom 64, Vyp. 4, p. 803 – 814 (1987). In Russian. English translation in Sov. Astron., Vol. 31, No. 4.

Estimations are given of the mass of the supergiant HDE 226868 in the system Cyg X–1 which are made on the basis of interpretation of spectroscopic observations by means of the model–atmosphere method. The mass turns out to be $M_* = (16 \pm 3)\,M_\odot$. Provided the zone of formation of the emission He II $\lambda 4686$ Å must be localized near the supergiant surface, the least value of inclination of the system's orbit plane is estimated: $i \gtrsim 35°$. The possibility is pointed out of variability of equivalent widths of He II emission $\lambda 4686$ Å, which is connected with eclipse of the "spot". It is noted that for large angles ($i \gtrsim 35°$) of orbit inclination, the mass of the degenerate star in the system Cyg X–1 does not exceed $10\,M_\odot$.

117.066   The spectral classification of the cool components of symbiotic stars.

O. G. Taranova, B. F. Yudin.
Astron. Zh., Tom 64, Vyp. 4, p. 867 – 870 (1987). In Russian. English translation in Sov. Astron., Vol. 31, No. 4.

From the results of photometric observations of some symbiotic stars, the $(R–J)$, $(J–K)$ and $(K–L)$ colours of their cool components have been determined. It was found that the $(R–J)$ colours correspond to later spectral types than the $(J–K)$ colours. Classifying the cool components in accordance to their $(J–K)$ colours in the IR range, the IR excesses should be observed in the L filter, as a rule.

117.067 *UBVRI* **photometry and polarimetry for the optical component of the A 0535+26 = V725 Tau X–ray source.**
N. I. Shakhovskaya, N. M. Shakhovskoj, N. G. Beskrovnaya.
Bull. Crimean Astrophys. Obs., Vol. 75, p. 110 – 123 (1987).
English translation of 42.117.167.

117.068 *UBV* **photometry of the symbiotic stars Z And and AG Peg.**
T. S. Belyakina.
Bull. Crimean Astrophys. Obs., Vol. 75, p. 124 – 129 (1987).
English translation of 42.117.168.

117.069 **Changes in the relative intensities of the V and R components of the Hα emission line and in the radial velocity derived from it for X Persei.**
T. S. Galkina.
Bull. Crimean Astrophys. Obs., Vol. 75, p. 151 – 153 (1987).
English translation of 42.117.274.

117.070 **Symbiotic Miras.**
P. A. Whitelock.
Publ. Astron. Soc. Pac., Vol. 99, No. 617, p. 573 – 591 (1987).
This paper concerns interacting binary systems involving Mira variables. Twenty–six objects which potentially fall into this category are identified and observations of them covering the spectral regions from X–ray to radio are reviewed. Particular emphasis is given to near–infrared observations which are pertinent to establishing the presence of a Mira variable and also to new far–infrared data from IRAS. The majority of the objects under consideration have been classified as symbiotic stars. It is shown how our knowledge of normal Miras can contribute to the understanding of the evolutionary condition and luminosities of these binary Miras. Distances are derived for those objects with measured pulsation periods. The significance of the relatively long pulsation periods shown by these objects is also discussed.

117.071 **Perturbed orbital elements of close binary systems due to tidal lag in longitude.**
B. Zafiropoulos.
Astrophys. Space Sci., Vol. 136, No. 1, p. 149 – 165 (1987).
This paper deals with the perturbations which tidal lag in longitude can produce to the orbital elements of a close binary system. The expressions obtained for the six elements of the orbit have been presented as functions of the unperturbed true anomaly, measured from the periastron. The study includes the effects produced by the second, third, and fourth tidal harmonic distortions. In order to save space these extremely lengthy equations are given in the compact form of summations, by means of Hansen coefficients. Various recurrence relations, which hold good for Hansen coefficients, are also presented. Finally, this paper includes a second–order approximation only for the secular terms of first–order approximation.

117.072 **Is there period–activity correlation in RS CVn–type binary systems?**
O. Demircan.
Astrophys. Space Sci., Vol. 136, No. 1, p. 201 – 205 (1987).
The rotation period–activity luminosity and the rotation period–surface activity flux correlations for the components of RS CVn–type binaries are discussed. It is argued that such correlations, although they are required by dynamo generated magnetic fields, do not represent the dominant parameter in determining the level of activity. Evidences are given that these correlations may essentially be induced by the period–radius dependence for the components of RS CVn–type binaries, and in addition, this result may well be valid for all the late–type stars (single or binary components).

117.073 **Iron emission line from low–mass X–ray binaries.**
T. Hirano, S. Hayakawa, F. Nagase, K. Masai, K. Mitsuda.
Publ. Astron. Soc. Jpn., Vol. 39, No. 4, p. 619 – 644 (1987).
By observing the X–ray spectra of ten low–mass binary X–ray sources, the authors detected a line emission feature around 6 – 7 keV from six X–ray sources. Most of the line–center energies determined are consistent with their weighted average of 6.66 ± 0.05 keV, and the equivalent widths lie between 20 and 60 eV. For the remaining sources, only a 90% confidence upper limit of 20 – 30 eV was obtained for the equivalent widths. The 6.7–keV iron line observed is ascribed to the Kα radiative transitions of heliumlike ions of iron, due mainly to the recombination process taking place in the accretion disk coronae with temperatures of $\sim 10^6$K at radii of $\sim 10^9$cm.

117.074 **Spin–down of the white dwarf in the DQ Herculis system FO Aquarii (H2215–086).**
A. W. Shafter, J. D. Macry.
Mon. Not. R. Astron. Soc., Vol. 228, No. 1, p. 193 – 202 (1987).
High–speed photometry of FO Aqr (H2215–086) has been obtained in order to update the ephemerides for the orbital and 21 min periodicities. The analysis yields a refined orbital period $P = 0.168018(1)$ day with no evidence for a period change. The data show that the period of the 21 min optical modulation is increasing as reported by Pakull & Beuermann (1987). This period change is interpreted as spin–down of the white dwarf. The authors find that $\dot{P} = (8.0 \pm 1.1) \times 10^{-11}$. This result yields an ephemeris for the times of maxima of the 21 min optical modulation. The authors conclude by discussing the implications of their $\dot{P}$ measurement for constraining the magnetic field of the white dwarf.

117.075 **Observations of X–ray emission from the RS CVn binary HD 155555 and the detection of a nearby serendipitous source.**
M. A. Barstow.
Mon. Not. R. Astron. Soc., Vol. 228, No. 1, p. 251 – 257 (1987).
Results of the EXOSAT observation of the RS CVn binary HD 155555 are reported. A new source EXO 171224–6653.9 has been discovered, lying only 40 arcsec away from HD 155555. Broad band X–ray photometry allows constraints to be placed on the plasma temperature $(5 \times 10^5 - 1.3 \times 10^6$K) and emission measure $(4.5 \times 10^{51} - 8 \times 10^{51}cm^{-3})$ of HD 155555. When compared with earlier X–ray observations, the combined luminosity of the two sources can be seen to have decreased by an order of magnitude. Several explanations of this are discussed. A likely optical counterpart $(m_j = 12.6)$ to EXO 171224–6653.9 has been found. The X–ray to optical luminosity ratio indicates that it is an accretion driven source, probably a cataclysmic variable or high mass X–ray binary.

117.076 **Simultaneous observations of the X–ray and optical eclipse of SS433 and their implications.**
G. C. Stewart, M. G. Watson, M. Matsuoka, W. Brinkmann, J. Jugaku, K. Takagishi, T. Omodaka, J. C. Kemp, G. D. Kenson, D. J. Kraus, T. Mazeh, E. M. Leibowitz.
Mon. Not. R. Astron. Soc., Vol. 228, No. 2, p. 293 – 303 (1987).
The authors report on the results of coordinated X–ray and optical observations of SS433 made over one binary cycle. The optical light curve shows SS433 to be in a typical state, with two unequal minima separated by 0.5 in phase. The X–ray observations show reduced intensity coincident with the deeper, primary optical minimum. The authors interpret this result, which was suggested in the earlier observations of Grindlay et al. (1984) as the eclipse of the X–ray emitting region by the optical star, thus confirming the optical minimum as the occultation of an accretion disc by the mass–losing star. The characteristics of the X–ray eclipse, when coupled with optical data and the discovery of Watson et al. (1986), that the X–ray emission is from the jets, then constrain the dimensions of all components of the system.

117.077   Spin–down of the white dwarf in the intermediate polar V1223 Sgr/4U 1849–31.
S. van Amerongen, T. Augusteijn, J. van Paradijs.
Mon. Not. R. Astron. Soc., Vol. 228, No. 2, p. 377 – 388 (1987).

From a set of 10 times of maximum light of the 13.2 min pulsational light curve of the intermediate polar V1223 Sgr the authors find that the white dwarf rotation period increases on a time–scale $P/\dot{P} = (1.1 \pm 0.1) \times 10^6 \mathrm{yr}$. V1223 Sgr is the fifth intermediate polar for which a change of the white dwarf spin rate has been measured. The authors derive estimates of the white dwarf magnetic fields of these five intermediate polars. They find that for some of these systems the white dwarf spin rate is not close to its equilibrium value. The authors suggest that the rather rapid spin–up and spin–down time–scales, observed for intermediate polars, are the result of substantial variations in the mass accretion rate on time–scales less than $10^5 - 10^6 \mathrm{yr}$.

117.078   Polarimetry of BL Hyi (H 0139–68) in high and low states.
M. Cropper.
Mon. Not. R. Astron. Soc., Vol. 228, No. 2, p. 389 – 399 (1987).

Observations of BL Hyi in the bright and faint states are presented. The orbital ephemeris has been updated. The orbital behaviour of the polarization indicates that the accreting pole lies $\sim 27°$ from the rotation axis ($\beta \cong 153°$) on the far side of the line–of–sight and that we view the system from slightly above the orbital plane ($\iota \cong 70°$). The bright phase duration is longer than expected and is caused either by the accretion shock being substantially higher above the white dwarf than in the other AM Her systems where this has been determined or by accretion at a second pole. Both possibilities are discussed in the paper.

117.079   Time resolved optical spectroscopy of the eclipsing intermediate polar EX Hydrae.
C. Hellier, K. O. Mason, S. R. Rosen, F. A. Córdova.
Mon. Not. R. Astron. Soc., Vol. 228, No. 2, p. 463 – 481 (1987).

Time resolved spectroscopy of the eclipsing intermediate polar EX Hya has been obtained. The authors confirm that the dominant variation in the spectral line flux occurs with the 67–min white dwarf rotation period and is in phase with the 67–min optical and X–ray continuum variation. Further, they find that the pulse fraction is greatest in the wings of the emission lines and that there is a variation in the relative strengths of the blue and red line wings on the 67–min period. No evidence is found for an occultation of the line flux corresponding to the narrow continuum eclipse that recurs every 98–min orbital cycle. The authors do, however, find changes in the line profiles that extend for ~ 0.15 of the orbital cycle, either side of the photometric eclipse. These they interpret as a progressive occultation of a rotating accretion disc by the companion star. The emission line data are interpreted in terms of three components: double peaked emission from a disc, an S–wave component caused by a bright region on the outside of the disc, and emission pulsed with the 67–min period which originates close to the white dwarf.

117.080   A new analysis of the infrared eclipses of the ultra–short period dwarf nova OY Carinae.
G. Berriman.
Mon. Not. R. Astron. Soc., Vol. 228, No. 3, p. 729 – 743 (1987).

The paper investigates what can be learned about the eclipsed components of OY Car – the disc, the white dwarf and bright spot – from the morphology of its infrared light curve.

117.081   LTE models of the emission lines of the dwarf nova Z Cha.
T. R. Marsh.
Mon. Not. R. Astron. Soc., Vol. 228, No. 3, p. 779 – 796 (1987).

The authors compare the Balmer line profiles of Z Cha with Local Thermodynamic Equilibrium (LTE) models of the disc. Even with the inclusion of Stark broadening, the theoretical line profiles are too narrow in comparison to the observations. Stark broadening has little effect on the LTE models because the disc becomes opaque in the inner regions where Stark broadening would have most effect. The effects of the velocity shear in the disc are important at all radii in the disc. Optically thin discs produce more optical flux in their outer parts than do opaque discs of the same effective temperature. Despite this, the author's models cannot fit the observed eclipse depth of Z Cha in quiescence which is unusually shallow. He concludes that the outer disc is $\geqslant 3$ times brighter than expected relative to the inner disc.

117.082   On the nature of the central object in Z Cha.
J. H. Wood.
Mon. Not. R. Astron. Soc., Vol. 228, No. 3, p. 797 – 803 (1987).

Observational evidence, in quiescence and outburst, regarding the nature of the object at the centre of the accretion disc in the eclipsing dwarf nova Z Cha is examined. At quiescence the most likely central object is a white dwarf, luminous over most of its surface, unocculted by the accretion disc, with no boundary layer contributing significantly to the light at visual wavelengths. This result is in disagreement with the results of Smak (1986), who, in order to model the observed behaviour of the central object through an outburst, proposed that in quiescence it is just a narrow equatorial boundary layer. The author shows, however, that the behaviour during outburst is consistent with their central object if, during the outburst, an extended toroidal boundary layer develops and the inner regions of the accretion disc are optically thick.

117.083   Six years of photometry of the RS CVn binary EI Eri = HD 26337.
D. S. Hall, S. A. G. Osborn, E. R. Seufert, L. J. Boyd, R. M. Genet, R. E. Fried.
Astron. J., Vol. 94, No. 3, p. 723 – 725 (1987).

Six years of differential UBV photoelectric photometry of this G5 IV SB1 RS CVn binary is presented. Periodogram analysis yields a photometric period of either $1^{\mathrm{d}}95$ or $2^{\mathrm{d}}04$, but it is shown that the latter is an alias. There were season–to–season variations in the photometric period (by $\sim 1\%$), the light–curve amplitude ($0^{\mathrm{m}}07 - 0^{\mathrm{m}}20$), and the mean light level (by $\sim 10\%$) not correlated with each other. The mean value $P(\mathrm{phtm.}) = 1^{\mathrm{d}}945 \pm 0^{\mathrm{d}}005$ is so close to $P(\mathrm{orb.}) = 1^{\mathrm{d}}9472$ that the authors conclude the G5 IV star is rotating synchronously.

117.084   Studies of period variation in close binary systems.
S. L. Lipari, R. F. Sistero.
Astron. J., Vol. 94, No. 3, p. 792 – 802 (1987).

New photoelectric observations and period studies for the systems V758 Cen, CT Eri, EZ Hya, TY Men, V502 Oph, and V1010 Oph are presented. TZ Boo was also studied from all published minima and by means of those derived by the authors from photoelectric data in the literature. Five new cases of period variation were detected, one confirmed and another one suspected. Probable causes of variability are discussed.

117.085   Thermal radiation of the corona in binary X–ray sources.
O. A. Tsiopa.
Glav. Astron. Obs. Akad. Nauk SSSR. Leningrad, 10 pp. (1987).
In Russian. Abstr. in Ref. Zh., 51. Astron., 7.51.594 (1987).

117.086   Hα region spectroscopy of the RS CVn system HR 5110.
Z. Eker, L. R. Doherty.
Mon. Not. R. Astron. Soc., Vol. 228, No. 4, p. 869 – 881 (1987).

High–resolution (0.3 Å) échelle spectra covering all orbital phases of the bright RS CVn binary HR 5110 have been obtained for the Hα region. Lines from the cool secondary (other than Ca II H and K) have been detected for the first time, providing new mass and light ratios: $M_2/M_1 = 0.54$ and $F_2/F_1 = 0.20$. Effective temperatures of 6750 and 4700K for the primary and secondary, respectively, have been estimated from the line spectrum. A composite comparison spectrum was subtracted from the weak Hα absorption profile, revealing strong, variable Hα emission with two components: a narrow and steady emission peak and a variable and extremely broad component. Stark broadening in a dense chromosphere of the K star can explain the broad component, while the narrow peak appears to come from an unusually bright Hα core in the F star.

117.087 **Symbiotic binaries: I. Spectrophotometry of AX Persei.**
J. Mikołajewska, T. Iijima.
Acta Astron., Vol. 37, No. 1, p. 17 – 28 (1987).

Secular and eclipse variations of optical emission lines during almost three orbital cycles of the symbiotic star AX Per are presented. The permitted lines show pronounced but nontotal eclipse effects while forbidden lines (i.e. [O III], [Ne III], [Fe VII]) do not show such effects. The data are discussed in terms of physical conditions and geometry of the line formation region. The possible presence of the reflection of a hot star light from a red–giant companion is considered.

117.088 **Statistical analysis of dwarf nova outbursts.**
A. Gicger.
Acta Astron., Vol. 37, No. 1, p. 29 – 39 (1987).

Correlations between maximum brightness, outburst width, lengths of preceding and following intervals have been studied for 14 dwarf novae. Significant correlations ($\varrho \geqslant 0.4$) occur only in 16 per cent of cases. Global correlations have also been studied between mean photometric parameters and binary system parameters using a sample including over 30 objects. The most interesting result is the strong correlation ($\varrho = +0.94$) between the orbital period and the outburst duration. It implies that the quantity $\alpha(z_0/r)^2$ is approximately constant for all dwarf novae.

117.089 **UBV light variation and orbital elements of W Gruis.**
M. A. Cerruti.
Acta Astron., Vol. 37, No. 1, p. 41 – 51 (1987).

In each of the *UBV* band–passes 2440 observations define for the first time the photoelectric light variation of the binary system W Gru. A consistent set of elements is derived in the framework of Russell's classic and Napier's synthesis models. Absolute dimensions are derived. The system seems to be a well–detached binary consisting of two normal, not interacting, F5 IV components.

117.090 **Phase–dependent variations of Hα equivalent width in VW Cep.**
W. Herbst, L. E. Bishel.
Astron. J., Vol. 94, No. 4, p. 1051 – 1054 (1987).

Hα photometry of the W UMa type eclipsing binary VW Cep has been obtained on four nights in 1985 at Van Vleck Observatory. A phase–dependent variation of Hα equivalent width (EW) has been discovered. The absorption EW decreases by a small (0.1 – 0.2 Å) but significant amount during both eclipses. No significant difference in the amplitude of the effect is seen between primary and secondary eclipse. The most likely interpretation is that the photospheric absorption–line EW decreases during eclipse. It is also possible that an emission component to the line is seen more prominently during both eclipses, resulting in a net decrease in the absorption EW. No evidence for plage–like concentrations of Hα flux on either component of the binary is found.

117.091 **Time–resolved IUE studies of cataclysmic variables. I. Eclipsing systems IP Peg, PG 1030 + 590, and V1315 Aql.**
P. Szkody.
Astron. J., Vol. 94, No. 4, p. 1055 – 1061 (1987).

IUE time–resolved spectra of the high–inclination cataclysmic variables IP Peg, PG 1030 + 590, and V1315 Aql are analyzed in order to determine the characteristics of the disks, hotspots, and white dwarfs. The UV continuum flux distributions are generally flatter than systems of low inclination and high mass–transfer rate, and the white dwarfs/inner disks appear to be relatively cool (15000 – 19000K) for their orbital periods, possibly because the boundary layers are blocked from view. The continuum fluxes increase at spot phases, with the spot providing the dominant flux in IP Peg. The spot temperatures range from hot (20000K) in IP Peg, and perhaps in PG 1030 + 590, to cool (11000K) in V1315 Aql. The C IV emission lines show slightly larger decreases at spot phases than during eclipse, which implies an extended stream area.

117.092 **A 685 second orbital period from the X–ray source 4U 1820–30 in the globular cluster NGC 6624.**
L. Stella.
Variability of galactic and extragalactic X–ray sources, p. 157 – 164 (1987). – See Abstr. 012.019.

The author reports the discovery of a 685 s modulation of the X–ray flux of 4U 1820–30 and argues that it represents the orbital period of the system, the first determined in a globular cluster source and the shortest of any known binary.

117.093 **X–ray spectral formation in LMXRB.**
N. E. White.
Variability of galactic and extragalactic X–ray sources, p. 165 – 173 (1987). – See Abstr. 012.019.

The spectral properties of the non–pulsing low mass X–ray binaries, LMXRB, are reviewed using recent EXOSAT observations. Various models for the inner region of an accretion disk and its interaction with a neutron star are considered.

117.094 **Evidence for black holes in X–ray binary systems.**
S. A. Ilovaisky.
Variability of galactic and extragalactic X–ray sources, p. 175 – 184 (1987). – See Abstr. 012.019.

The author reviews the various criteria used to distinguish black holes from neutron stars in X–ray binaries. X–ray signatures previously thought to be exclusively characteristic of disk accretion onto black holes have been also found in systems containing neutron stars. The compact source's mass, derived from an optical determination of the mass function, appears the only believable signature.

117.095 **The contribution of QPO to our understanding of low–mass X–ray binaries.**
M. van der Klis.
Variability of galactic and extragalactic X–ray sources, p. 185 – 192 (1987). – See Abstr. 012.019.

The status of QPO (quasi–periodic oscillations) as a probe into the structure of LMXB is briefly reviewed. Although no consensus has been reached yet about the correct model, the observed relations between QPO properties and other source characteristics allow some preliminary conclusions to be drawn.

117.096 **A theory of soft X–ray transients.**
A. R. King, J. M. Hameury, J. P. Lasota.
Variability of galactic and extragalactic X–ray sources, p. 193 – 200 (1987). – See Abstr. 012.019.

The authors review the properties of a model for soft X–ray transients and its evolutionary consequences.

117.097 **Millisecond X–ray binary pulsars.**
K. S. Wood, J. P. Norris, P. Hertz, P. F. Michelson.
Variability of galactic and extragalactic X–ray sources, p. 213 – 216 (1987). – See Abstr. 012.019.

Millisecond binary X–ray pulsars are expected theoretically, but remain undetected. The authors describe a novel search technique which applies a grid of quadratic time transformations to effect coherence recovery of smeared pulsar signals. They present upper limits on pulsed fluxes from the binary X–ray systems Cyg X–2, Cyg X–3, and 1820–30, and discuss how to improve future searches.

117.098 **X–ray emission from low–mass X–ray binaries.**
M. Czerny.
Variability of galactic and extragalactic X–ray sources, p. 217 – 220 (1987). – See Abstr. 012.019.

Qualitative arguments, as well as the impact of an Eddington burst on persistent emission from 4U 1636–53, suggest that X–ray emission from low–mass X–ray binaries is produced in optically thick accretion disks, boundary layers, and optically thin hot coronas.

117.099   Periodicities of low mass X–ray binaries from HEAO–1.
P. Hertz, K. S. Wood.
Variability of galactic and extragalactic X–ray sources, p. 221 – 224 (1987). – See Abstr. 012.019.
The authors have undertaken a study of the variability of all X–ray sources brighter than 10 UFU in the 1H catalog. They have discovered periodic behavior for three low mass X–ray binaries: a 4.2 hour period in GX9+9, a 10.5 hour period in Sco X–2, and a 19.8 hour period in GX17+2. They confirm the 21.2 hour period reported by EXOSAT for Nor X–1, and present the first X–ray detection of the 8.5 hour optical period in 4U2129+12, the X–ray source in the core of M15.

117.100   BR Lupi: a new SU UMa–star.
D. O'Donoghue.
Astrophys. Space Sci., Vol. 136, No. 2, p. 247 – 250 (1987).
From the presence of superhumps in its outburst light curve, the dwarf nova BR Lup is shown to be a member of the SU UMa sub–class of cataclysmic variables. Its orbital period is estimated to be 0.0789 d.

117.101   Direct photometric problem for spotted stars.
D. P. Kjurkchieva.
Astrophys. Space Sci., Vol. 136, No. 2, p. 289 – 297 (1987).
In the framework of the direct photometric problem the author obtained an analytical representation of the light variations of binary systems of stars, one of which having spots on its surface. The photometric effect is due to the mutual eclipses and to the rotation of the spotted component. These expressions can be used for the solution of the inverse problem for spotted stars.

117.102   Jets in X–ray binaries.
H. Ögelman.
Astrophysical jets and their engines, p. 67 – 90 (1987). – See Abstr. 012.021.
The three galactic X–ray binaries that show evidence for jets, Sco X–1, Cyg X–3, and SS433 are examined. SS433 is the only source that shows rich jet features in all wavelengths from radio to gamma rays; the jets of Sco X–1 and Cyg X–3 are only evident in the radio region. Observational features for the three systems, as X–ray binaries, are summarised and compared. A possible common model for the central engine, involving a fast spinning neutron star with weak surface field ($B_s \lesssim 10^9$ gauss), is outlined.

117.103   Physical characteristics of the peculiar object V1016 Cygni.
Ya. T. Blagodyr, V. V. Golovatyj.
Stellar atmospheres, p. 170 – 178 (1987). In Russian. Abstr. in Ref. Zh., 51. Astron., 8.51.463 (1987). – See Abstr. 003.005.

117.104   Chemical composition of the atmosphere of the main component of the υ Sgr binary system.
V. V. Leushin, G. P. Topil'skaya.
Stellar atmospheres, p. 112 – 114 (1987). In Russian. Abstr. in Ref. Zh., 51. Astron., 8.51.621 (1987). – See Abstr. 003.005.

117.105   Evolutionary variation of the carbon abundance in binaries.
V. V. Leushin.
Stellar atmospheres, p. 115 – 118 (1987). In Russian. Abstr. in Ref. Zh., 51. Astron., 8.51.623 (1987). – See Abstr. 003.005.

117.106   Spectral characteristics of matter overflow in the close binary systems V448 Cygni and XZ Cephei.
L. V. Glazunova.
Stellar atmospheres, p. 104 – 111 (1987). In Russian. Abstr. in Ref. Zh., 51. Astron., 8.51.625 (1987). – See Abstr. 003.005.

117.107   The evolutionary status of MXB 1820–30 and other short–period low–mass X–ray binaries.
A. V. Tutukov, A. V. Fedorova, E. V. Ergma, L. R. Yungel'son.
Pis'ma Astron. Zh., Tom 13, No. 9, p. 780 – 788 (1987). In Russian. English translation in Sov. Astron. Lett., Vol. 13.
The evolutionary status of secondaries in X–ray binaries with orbital periods $P < 1^h$ is discussed. Most probably they are degenerate hydrogen–helium dwarfs, which are remnants of stars, that had filled their Roche lobes in the core hydrogen exhaustion stage. However, they may be as well nondegenerate helium stars or helium or carbon–oxygen degenerate dwarfs. The absence of bright X–ray sources with $P = 2^h - 4^h$ can be explained by an interruption of mass exchange, caused by the switch–off of the magnetic stellar wind. At $P < 2^h$ accretion onto a rapidly spinning weakly magnetized neutron star is impossible due to "propeller" action. Only systems that evolve to ultra–short periods could become bright X–ray sources, because they have high enough values of $\dot{M}$ at $P \lesssim 1^h$.

117.108   RW Comae Berenices. III. Light curve solution and absolute parameters.
E. F. Milone, R. E. Wilson, B. J. Hrivnak.
Astrophys. J., Vol. 319, No. 1, p. 325 – 333 (1987). = Publ. Rothney Astrophys. Obs., No. 43.
UBV light curves of the W–type W UMa system RW Com have been modeled with the Wilson–Devinney synthetic light curve program and solution elements have been recovered despite the presence of a large O'Connell effect in the light curves. The system is a W–type W UMa system in which the smaller and less massive component is the hotter one. The masses of the components are 0.20 ± 0.03 and 0.56 ± 0.06 solar masses; the bolometric magnitudes are 6.76 ± 0.24 and 5.97 ± 0.26, for the hotter and cooler components, respectively. The separation of centers is 1.48 ± 0.01 solar radii and the ratio of radii about 0.6. The components are shown to be in shallow contact.

117.109   Reexamination of the *SAS* 2 Cygnus X–3 data.
C. E. Fichtel, D. J. Thompson, R. C. Lamb.
Astrophys. J., Vol. 319, No. 1, p. 362 – 366 (1987).
Recent observations of Cygnus X–3 have shown marked variability of the radiation on short time scales. In particular, the bursts lasting on the order of 10 minutes, seen in both the infrared and very high energy ($>10^{11}$ eV) gamma–ray regions, have stimulated a reanalysis of the 1973 March 6 to 13 *SAS* 2 high–energy gamma–ray data. Although a clear periodicity in the $E > 35$ MeV gamma radiation is observed at the 4.79 hr period seen in X–rays, there is no evidence for major variations of the radiation from one day to the next and no statistically significant evidence for bursts on the 10 minute time scale.

117.110   CCD echelle observations of the active RS CVn system II Pegasi.
D. P. Huenemoerder, L. W. Ramsey.
Astrophys. J., Vol. 319, No. 1, p. 392 – 402 (1987).
Optical spectra were obtained of II Peg on eight different nights in 1984 and 1985 to assess the strength and variability of surface activity indicators in this very active RS CVn system. These cross–dispersed echelle spectra covered the range from 390 nm to 900 nm at a resolution of 12,000. Emission was seen in the first four Balmer lines, in the Ca II infrared triplet, Ca II H lines, and in one observation, in He I D_3. The ratio of energy emitted in the Hα line to that in Hβ is similar to that in solar prominences, except during enhancements when the ratio decreases toward values more typical of solar flares.

117.111   First unambiguous X–ray detection of R Aquarii.
R. Viotti, L. Piro, M. Friedjung, A. Cassatella.
Astrophys. J., Lett. Ed., Vol. 319, No. 1, p. L7 – L11 (1987). With plate L1.
The authors report coordinated ultraviolet and X–ray observations of the symbiotic Mira R Aqr obtained with *IUE* and *EXOSAT* in 1985 June and December at phases 0.0 and 0.5 of the Mira light curve. They have detected with the Low–Energy

EXOSAT detector a weak X–ray flux of 5.4 and 4.6×10^{-3} counts s^{-1} at the two epochs, respectively. The source is probably soft, with a 0.2 – 1 keV luminosity of $0.8 - 1.0 \times 10^{30}$erg s^{-1}. An *IUE* image of the jet obtained in 1985 December shows high-temperature lines of N V and He II stronger than in an exposure of the star. It is argued that the X–rays are mostly emitted from the jet. Possible radiation mechanisms of the X–ray source in R Aqr are considered.

117.112   WBVR and Hα–photometry of SS433.

A. A. Aslanov, V. G. Kornilov, N. A. Lipunova, A. M. Cherepashchuk.
Pis'ma Astron. Zh., Tom 13, No. 10, p. 879 – 885 (1987). In Russian. English translation in Sov. Astron. Lett., Vol. 13.

A new WBVR and Hα–photometry of SS433 is given. Observations were carried out with the 2–meter reflector of the Shemakha Astrophysical Observatory at the dates corresponding to the phases $\psi = 0$ and $\psi = 0.35$ of the precessional 164^{d}–period. The ratio of depths for the primary and secondary minima in W and V on the orbital light–curves at a precessional phase $\psi = 0$ is such that it cannot be explained in terms of a simple geometrical model of the accretion disk eclipses. These W and V depths can be understood taking into account the "reradiation" of the accretion disk radiation by the "normal" star. The orbital light–curves obtained at $\psi = 0$ and $\psi = 0.35$ are compared with those observed by Cherepashchuk et al. (1982) at $\psi = 0.5$. The observational data are discussed in comparison with model simulations for SS433.

117.113   The cyclic variations of dwarf nova UU Aquilae outbursts.

L. I. Shakun.
Kinematika Fiz. Nebesn. Tel, Tom 3, No. 5, p. 72 – 77 (1987). In Russian. English translation in Kinematics Phys. Celest. Bodies.

On the basis of the literature data and the author's own observations the linear elements with $P_0 = 54.136^d$ describing the behaviour of outbursts of UU Aql from 1938 to 1986 are determined. The cyclic change of the time intervals between the outbursts with the cycle length $\Pi = 4697^d \approx 89\,P_0$ is accompanied by alterations of brightness at the moment of the outburst. Two star states with a mean cycle length between the outbursts $P_{01} = 49.617^d$ and $P_{02} = 59.498^d$ (periodically alternating) are indicated.

117.114   Theoretical profiles of emission lines of Wolf–Rayet stars with low–mass satellites.

I. I. Antokhin.
Sov. Astron., Vol. 30, No. 6, p. 680 – 689 (1987). English translation of 42.117.262.

117.115   Photoelectric observations and the wave minimum of RS CVn.

R. K. Srivastava.
Astrophys. Space Sci., Vol. 137, No. 1, p. 63 – 72 (1987).

B, V observations of the eclipsing binary RS CVn have been presented. A dip around $0^{p}1$ appears to be a wave minimum which fits well in the 'wave minimum phase–time' relation, but deviates from the 'wave amplitude–time' relation, derived for RS CVn. Either, the 'wave amplitude–time' relation requires a modification or the amplitude of the wave minimum appears masked by the intrinsic variability of one of the components or by the Sun–spot activity of the system. The colour exhibits variation. The secondary component appears active.

117.116   Gravitational radiation and spiralling time of close binary systems.

T. D. Padalia.
Astrophys. Space Sci., Vol. 137, No. 1, p. 191 – 194 (1987).

Power–output by gravitational radiation (P_B) and spiral time (τ_0) for individual systems of sixteen eclipsing binary stars have been evaluated, and a relation between P_B and τ_0 obtained.

117.117   Correlation between bolometric and X–ray luminosities in RS CVn–type binaries.

O. Demircan.
Astrophys. Space Sci., Vol. 137, No. 1, p. 195 – 199 (1987).

Re–analysis of the X–ray emission from RS CVn–type binaries has revealed that – contrary to the claims by many authors – their coronal activity is not independent of bolometric luminosity (and thus mass). It becomes clear that activity in late–type stars is also powered – just as in the case of early–type stars – mostly by photospheric radiation.

117.118   The relation between optical and X–ray flux variations of the black–hole candidate LMC X–3.

J. van Paradijs, M. van der Klis, T. Augusteijn, P. Charles, R. H. D. Corbet, S. Ilovaisky, L. Maraschi, C. Motch, M. Pakull, A. P. Smale, A. Treves, S. van Amerongen.
Astron. Astrophys., Vol. 184, No. 1/2, p. 201 – 208 (1987).

The authors present the results of V–band photometry of the optical counterpart of the black–hole candidate LMC X–3. The observations were made during a three–week period in December 1984, using CCD cameras at the Danish 1.5 m, and MPI 2.2 m telescopes at ESO, and the 1.0 m telescope at SAAO. In addition to an orbital brightness modulation of 0.15 mag LMC X–3 showed a secular brightness increase of ∼0.3 mag. Evidence is found that either the non–orbital variations are not smooth on a time scale of days, or the shape of the orbital light curve is not constant (or both). During the period of the optical observations LMC X–3 was observed five times with EXOSAT. The long–term optical brightness variation is correlated with the X–ray flux. If the X–ray flux does not show orbital variations, the observed long–term correlated X–ray/optical brightness variations of LMC X–3 can be described by a model comprising a constant secondary star, with $V \sim 17.5$ and B–$V \sim$ –0.20, and a rather cool ($T < 15,000$K) accretion disk of variable brightness, radiating through reprocessing of X rays.

117.119   The radiation parameters of the X–ray binary A 0535 + 26 = HDE 245770.

V. M. Larionov.
Astrofizika, Tom 27, Vyp. 1, p. 19 – 27 (1987). In Russian. English translation in Astrophysics, Vol. 27, No. 1.

An analysis of Shakhovskaya et al.'s observations of the X–ray binary A 0535 + 26 = HDE 245770 made it possible to distinguish in its radiation the two components connected with the visible star (O9 III) and the accretion disc around the neutron star. The interstellar polarization parameters are in accordance with Serkowski's formula and the observations of field stars. The IR and optical variability can be explained in terms of variable accretion disc radiation. The intrinsic polarization parameters obtained can be used to predict, in the model proposed, the directions of the polarization vectors in the IR and X–ray bands.

117.120   Gas flow in close binary systems of low mass stars.

L. N. Ivanov.
Astrofizika, Tom 27, Vyp. 1, p. 159 – 195 (1987). In Russian. English translation in Astrophysics, Vol. 27, No. 1.

Contents: Introduction. The observational events. The problem of the nonsynchronous rotation. The stationary mass loss from the component of CBS. The nonstationary one. Gas flow through the interstellar space. The accretion of the gas. The models of the outbursts of the dwarf novae.

117.121   Influence of a periodic gravitational wave on the parameters of a binary system.

R. A. Kochkin, Yu. G. Sbytov.
Astron. Zh., Tom 64, Vyp. 5, p. 1030 – 1036 (1987). In Russian. English translation in Sov. Astron., Vol. 31, No. 5.

The method of averaging is used in analysing secular terms in the perturbed parameters of a binary system which is subject to the action of a periodical gravitational wave under resonance conditions. The formulae for secular variation of the eccentricity, semimajor axis and pericenter position are derived. For two values of the ratio of the gravitational wave frequency ω to that of the orbital motion ω_0, $\omega/\omega_0 = 1, 3$, the secular terms in the pericenter shifts depend on the eccentricity e as e^{-1}, resulting in a possibility of increasing the effect as e decreases.

117.122   Photometric investigations of SS433. The 1979 – 1986 observational results.
S. A. Gladyshev, V. P. Goranskij, A. M. Cherepashchuk.
Astron. Zh., Tom 64, Vyp. 5, p. 1037 – 1056 (1987). In Russian. English translation in Sov. Astron., Vol. 31, No. 5.

300 photographic and photoelectric observations of SS433 obtained at the Sternberg Astronomical Institute in 1982 – 1986 are given. The photometric parameters and their principal phase relations with orbital and precession periods are re–examined using all published data, the precession and orbital periods having been defined more precisely. The accordance of modern theoretical claims and models with multicolour photometry and spectroscopy was under analysis. Some observational divergences from the classical model of a close binary system with a thick accretion disc were picked out. Eclipse modelling has shown that the contribution of the "normal" star to the total light of the system does not exceed 20%.

117.123   Basic characteristics of evolved binaries according to computations of evolutionary models of close binary systems. I. Catalogue of evolutionary models of close binary systems.
Z. T. Krajcheva.
Nauchn. Inf., Vyp. 63, p. 105 – 144 (1987). In Russian.

The catalogue contains evolutionary models of 165 close binary systems whose primary components fill their Roche lobes during core hydrogen burning, core hydrogen exhaustion, shell hydrogen burning, core helium burning, and then exhaustion of helium in the core. Information about parameters of systems, luminosities and radii of components before mass transfer and at the end of different phases and after mass transfer is given. Information of durations of different phases of mass transfer and basic assumptions of computations are presented, too.

117.124   X–ray system with very short periods.
E. V. Ergma.
Priroda, No. 7, p. 103 – 104 (1987). In Russian.

117.125   The frequency of outbursts in SS Aurigae.
L. M. Cook.
J. Am. Assoc. Variable Star Obs., Vol. 16, No. 1, p. 8 – 11 (1987).

SS Aurigae is a U Geminorum–type dwarf nova that displays both narrow and wide outbursts. Narrow outbursts predominate during intervals of frequent outbursts and wide outbursts predominate during intervals of less frequent outbursts. The outbursts are found to occur not strictly at random but with intervals of more or less frequent outbursts.

117.126   SS433 continues to perplex.
B. Margon.
Nature, Vol. 328, No. 6128, p. 293 – 294 (1987).

117.127   Evolving radio structure of the binary star SS433 at a resolution of 15 marc s.
R. C. Vermeulen, R. T. Schilizzi, V. Icke, I. Fejes, R. E. Spencer.
Nature, Vol. 328, No. 6128, p. 309 – 313 (1987).

A comprehensive series of VLBI observations of SS433, carried out over an 11–day period on the European VLBI Network at 5 GHz, reveals the motion and evolution of a series of faint "blobs" ejected from the system. These blobs are observed to brighten at a substantial distance from the binary star.

117.128   X–ray observations of B–emission stars.
E. P. J. van den Heuvel, S. Rappaport.
Physics of Be stars, p. 291 – 308 (1987). – See Abstr. 012.030 (IAU Colloq. No. 92).

In this review the authors summarize the present state of knowledge regarding X–ray emission from Be stars. As to soft X–ray emission, which is thought to be characteristic of stellar

coronae or colliding stellar winds, little information has been obtained thus far. The only Be star for which a soft X–ray flux has been detected is ζ Ophiuchi. All other information regarding the X–ray emission from the vicinity of Be stars concerns hard X–rays (typically $1 - 20$ keV), which are characteristic of the pulsating Be/X–ray binaries. The detected hard X–ray luminosities of these systems are all $\geqslant 10^{33}$ergs/s. This review is confined to these systems.

117.129   The Be/X–ray system HDE 245770/A0535 + 26 in an active phase.
C. Bartolini, M. Burger, E. L. van Dessel, F. Giovannelli, A. Guarnieri, C. de Loore, A. Piccioni.
Physics of Be stars, p. 309 – 310 (1987). – See Abstr. 012.030 (IAU Colloq. No. 92).

The authors report on the sudden appearance and disappearance of (mainly) Fe II emission lines in the Be/X–ray system HDE 245770/A0535 + 26, about 24 days before an X–ray outburst was observed.

117.130   Emission line variability in the Be star X–ray binaries 4U1258–61 and 4U2206 + 54.
R. H. D. Corbet.
Physics of Be stars, p. 311 – 313 (1987). – See Abstr. 012.030 (IAU Colloq. No. 92).

Results of programs to monitor Hα emission line variability in two Be star X–ray binaries are presented. These systems provide a means of investigating the influence of a binary companion on the circumstellar envelope of a Be star.

117.131   Emission–line stars as interacting binaries.
P. Harmanec.
Physics of Be stars, p. 339 – 360 (1987). – See Abstr. 012.030 (IAU Colloq. No. 92).

A general binary hypothesis of the emission–line phenomenon in early–type stars was published eleven years ago. Since then, new observational techniques and theoretical concepts have led to a number of exciting findings which have made the emission–line phenomenon even more puzzling and challenging than before. It is therefore of interest to ask whether the binary hypothesis can withstand all these developments or not.

117.132   He I lines in the Be + K binary KX And.
S. Štefl.
Physics of Be stars, p. 440 – 442 (1987). – See Abstr. 012.030 (IAU Colloq. No. 92).

KX And (HD 218393) is a peculiar interacting binary with the period of about 38.9 days. This study is a first attempt to model the system as a double–line binary. The main results of the He I–line analysis are summarized.

117.133   The far UV spectrum of the binary system AX Mon.
E. Danezis.
Physics of Be stars, p. 445 – 447 (1987). – See Abstr. 012.030 (IAU Colloq. No. 92).

The author describes the main features of the far UV spectrum of HD 45910 observed with IUE at phase 0.5. The far UV spectrum of AX Mon presents a large range of ionization stages going from C I, O I, N I, Mg I to Si IV, C IV, N V.

117.134   Be binary systems with a cool companion: are they interacting?
M. Floquet, A. M. Hubert, J. P. Maillard, J. Chauville.
Physics of Be stars, p. 448 – 450 (1987). – See Abstr. 012.030 (IAU Colloq. No. 92).

High resolution spectrograms of the two well known Be binary systems ζ Tau and KX And were obtained in the IR ($2 - 2.5$ μm) with the Fourier Transform Spectrometer of the CFHT. They do not reveal evidence of a cool giant companion.

117.135   **Interacting binaries as Be stars.**
M. J. Plavec.
Physics of Be stars, p. 451 – 455 (1987). – See Abstr. 012.030
(IAU Colloq. No. 92).

The author has discovered high–ionization emission lines of
N V, C IV, Si IV, Fe III, etc. in the ultraviolet spectra of totally
eclipsing Algols. They probably originate in circumstellar turbulent regions at fairly high electron temperatures, of the order of
100000K. They are not detectable in most non–eclipsing systems,
but may be there and may play an important role in the dynamics
of accretion and mass outflow from the systems.

117.136   **Pole–on cataclysmic variables as Be stars.**
R. F. Garrison.
Physics of Be stars, p. 460 – 462 (1987). – See Abstr. 012.030
(IAU Colloq. No. 92).

117.137   **Observations and evolutionary scenario for Be/X–ray
binaries.**
G. M. H. J. Habets.
Physics of Be stars, p. 509 – 513 (1987). – See Abstr. 012.030
(IAU Colloq. No. 92).

117.138   **Evolutionary models for Be X–ray binaries.**
C. de Loore, C. H. B. Sybesma.
Physics of Be stars, p. 514 – 515 (1987). – See Abstr. 012.030
(IAU Colloq. No. 92).

The aim of this paper is to make an analysis of possible
sequences of massive binary evolution calculated with various
assumptions, and to select plausible scenarios.

117.139   **Effect of compact objects near Be stars.**
K. M. V. Apparao, S. P. Tarafdar.
Physics of Be stars, p. 516 – 518 (1987). – See Abstr. 012.030
(IAU Colloq. No. 92).

117.140   **Speckle interferometric measurements of binary
stars: IV.**
A. Blazit, D. Bonneau, R. Foy.
Astron. Astrophys., Suppl. Ser., Vol. 71, No. 1, p. 57 – 62 (1987).

The authors report speckle interferometric observations of
interferometric binaries, close visual double stars and nearby
stars suspected to be binaries with the 3.6–meter Canada–
France–Hawaii telescope. As a part of this program, the stars
Gl 616.2 and Gl 831 are clearly resolved as binary for the first
time and Gl 793.1 appears to be marginally resolved. Gl 747.2
and Gl 866 are confirmed as double stars.

117.141   **The Fe II emission in the UV spectrum of CH Cyg.**
C. Marsi, P. L. Selvelli.
Astron. Astrophys., Suppl. Ser., Vol. 71, No. 1, p. 153 – 162
(1987).

Fe II emission lines are present in all UV spectra of CH Cyg,
with a great increase in number and intensity since 1985. The
authors give a list of emission intensities and equivalent widths
for the prominent Fe II lines. Optical depth effects are clearly
evident from a comparison of the relative emission intensities
within couples of multiplets which share the same upper term.
The anomalously high intensity of some emissions seems to be
due to a selective fluorescence mechanism originating from Ly α.

117.142   **Ultraviolet observations of cataclysmic variables: the
IUE archive.**
F. Verbunt.
Astron. Astrophys., Suppl. Ser., Vol. 71, No. 2, p. 339 – 361
(1987).

A comparative study of the ultraviolet properties of cataclysmic variables is made on the basis of the IUE data archive. The
reddening to 51 systems is determined from the 2200 Å feature.
The spectral flux distribution of cataclysmic variables does not
depend on system type, orbital period or (for dwarf novae)
average length of interval between outburst maxima; it does not
depend strongly on inclination. The exception is formed by the
DQ Her systems, which may have relatively low fluxes at short

wavelengths as compared to the other systems. All systems for
which observations are available show variability on a time scale
of hours. In two nova–like variables in a low state and in five
dwarf novae in quiescence there is evidence that most of the
ultraviolet flux at short wavelengths is due to the white dwarf.
Disk model spectra are calculated to show that the slope of the
spectrum depends strongly on the white dwarf mass. Hence
determinations of the mass transfer rate from the spectral slope
are subject to large errors in cases where the white dwarf mass is
not known.

117.143   **A _uvbyβ_ survey of northern–hemisphere active binaries.
I. The observations.**
V. Reglero, A. Giménez, E. de Castro,
M. J. Fernandez–Figueroa.
Astron. Astrophys., Suppl. Ser., Vol. 71, No. 3, p. 421 – 429
(1987).

A recent detailed calibration of the _uvby_ photometric system
for late–type stars has opened the possibility of studying
photometric effects of stellar activity in the RS CVn group of
binaries. A _uvbyβ_ photometric study of 72 northern–hemisphere
binary systems with active chromospheres has been carried out at
the Roque de los Muchachos Observatory (La Palma, Canary
Islands) with a People's photometer attached to the JKT 1.0 m
telescope. The standardized colours and V magnitudes are
presented as a function of the orbital phase, along with an
indication of their accuracy. A detailed discussion of atmospheric
extinction, characteristics of the photometric system and transformation coefficients is also included.

117.144   **Influence of the spectrum's shape on the effect of the
"relativistic searchlight" in close binaries.**
N. G. Bochkarev.
Pis'ma Astron. Zh., Tom 13, No. 11, p. 1007 – 1013 (1987). In
Russian. English translation in Sov. Astron. Lett., Vol. 13.

The shape of a radiating object's spectrum and angular
distribution of radiation intensity influence significantly the
value of "relativistic searchlight" effect: modulation of observed
spectral density radiation flux from a short–period stellar binary
due to Doppler effect and aberration of light can be changed by
the factor 2 – 4 in comparison to observation of bolometric flux.
The effect can exceed 1% for binaries with semiamplitude radial
velocity $K \gtrsim 150$ km/s. For the optical radiation of A 0620–00
the effect is about 1.5%, and for the radiation of the low–massive
component of Geminga it can reach near 10%.

117.145   **Binary system parameters and the hibernation model of
cataclysmic variables.**
M. Livio, M. M. Shara.
Astrophys. J., Vol. 319, No. 2, p. 819 – 826 (1987).

The "hibernation" model, in which nova systems spend most
of the time between eruptions in a state of low mass transfer rate,
is examined. The authors determine which binary systems are
more likely to undergo hibernation. The predictions of the
hibernation scenario are shown to be consistent with available
observational data. The authors indicate how the hibernation
scenario provides links between classical novae, dwarf novae, and
novalike variables, all of which represent different stages in the
cyclic evolution of the same systems.

117.146   **An analysis of the light curves of short–period
RS Canum Venaticorum stars: starspots and fundamen-
tal properties.**
E. Budding, M. Zeilik.
Astrophys. J., Vol. 319, No. 2, p. 827 – 835 (1987).

The authors perform an analysis of selected light curves for the
short–period RS CVn group: UV Psc, XY UMa, RT And,
SV Cam, BH Vir, ER Vul, WY Cnc, and CG Cyg. The authors
optimize the photometric fitting parameters for the "distorted"
light curves in order to derive the maculation wave for each
system. A dark circular spot model is then fitted to the
maculation wave to infer the longitudes and sizes of one or two
spot groups presumed to account for these effects. New optimal
solutions are derived, which give the geometrical, orbital, and

physical parameters for the stars of these systems, which fall on the lower end of the main sequence.

117.147   EXOSAT observations of 4U/MXB 1636–53: on the relation between the amount of accreted fuel and the strength of an X–ray burst.

W. H. G. Lewin, W. Penninx, J. van Paradijs, E. Damen, M. Sztajno, J. Trümper, M. van der Klis.
Astrophys. J., Vol. 319, No. 2, p. 893 – 901 (1987).

During EXOSAT observations of 4U/MXB 1636–53 in 1985 August, three bursts were observed in 6 hr and 24 bursts in 79 hr of uninterrupted observing. The persistent X–ray flux varied by about a factor of 2.4, the burst intervals by a factor of 24 (from 35 minutes to 14 hr), and the integrated burst fluxes (the burst fluences), and burst peak fluxes both approximately by a factor of 6. Very globally, the burst fluence is approximately linearly proportional to the interval since the preceding burst. In the context of the thermonuclear flash models, all bursts suffer from nuclear energy losses due to stable hydrogen burning between the bursts. This provides a mechanism in which the percentage of "lost" energy increases with increasing burst intervals.

117.148   On the theory of type I X–ray bursts: the energetics of bursts and the nuclear fuel reservoir in the envelope.

M. Y. Fujimoto, M. Sztajno, W. H. G. Lewin, J. van Paradijs.
Astrophys. J., Vol. 319, No. 2, p. 902 – 915 (1987).

A comparison is made between the observed properties of type I X–ray bursts from 4U/MXB 1636–53 and those of models of thermonuclear flashes on accreting neutron stars. Possible ways are discussed to explain variations in the burst recurrence properties without an apparent correlation with the accretion rate, including the rapid succession of bursts at intervals shorter than ~10 minutes. An ignition mechanism of the bursts is proposed in terms of elemental mixing and dissipative heating associated with hydrodynamical instabilities in the neutron star envelope caused by angular momentum carried inward by accreted gas. The energy of the weak bursts is shown to be available in the outer layers where accreted fuel can survive thermonuclear flashes because of the relatively low pressures.

117.149   2–D hydrodynamical models of the stream–disk interaction in cataclysmic binaries.

M. Różyczka, A. Schwarzenberg–Czerny.
Acta Astron., Vol. 37, No. 2, p. 141 – 162 (1987).

Two–dimensional hydrodynamical models of the collision region between the stream and the disk in a cataclysmic binary are presented. The adiabatic collision may result in significant mass loss from the system and produce large turbulent regions in outer parts of the disk. In the more realistic case, radiative cooling of the collision region leads to a semicontinuous injection of mass into the disk in the form of dense blobs of gas. In both cases the hot regions obtained as a result of the collision are over 1/6 of the circumference of the disk long and almost 40% of its radius deep, deserving to be renamed "hot stripes".

117.150   The 1985 May superoutburst of the dwarf nova OY Carinae. I. Optical and infrared photometry.

T. Naylor, P. A. Charles, B. J. M. Hassall, G. T. Bath, G. Berriman, B. Warner, J. Bailey, K. Reinsch.
Mon. Not. R. Astron. Soc., Vol. 229, No. 2, p. 183 – 202 (1987).

The authors present optical and infrared photometry from the 1985 May superoutburst of OY Carinae, and estimate the temperature and area of the superhump region to be approximately 8000K and ~10^{20}cm^2 respectively. There is evidence for extended vertical structure in the disc, possibly analogous to that which causes X–ray dips seen in low mass X–ray binaries. The authors show that the size of the O–C variations of the eclipse timings are significantly smaller than previously thought.

117.151   Wavelength dependence of superhumps in VW Hyi.

S. van Amerongen, H. Bovenschen, J. van Paradijs.
Mon. Not. R. Astron. Soc., Vol. 229, No. 2, p. 245 – 251 (1987).

The authors present results of five–colour photometric observations of the SU UMa system VW Hyi, made on six nights

during the November 1984 superoutburst. The light curve is dominated by superhump variations, whose amplitude in all passbands decreases with time. The superhump light curve depends strongly on wavelength. In particular it appears that the light curves in different passbands are mutually shifted: the larger the wavelength is, the more the light curve is delayed.

117.152   Light and colour variations in the extremely hydrogen–deficient binary CPD –58°2721.

K. Morrison, P. W. Hill, C. S. Jeffery, F. Marang, J. Spencer Jones.
Mon. Not. R. Astron. Soc., Vol. 229, No. 2, p. 269 – 284 (1987).

Strömgren and UBVRI photometry of the extremely hydrogen–deficient binary star CPD –58°2721 (LSS 1922) is presented. Light and colour variations are present with amplitudes 0.17 mag in V and 0.07 mag in (U–B) and (u–b). A detailed frequency analysis of the complex light curve was carried out and possible interpretations of the light curve are discussed. The lack of any obvious evidence for eclipse–related variations during this interval suggests that CPD –58°2721 is not an eclipsing binary. Its photometric behaviour is similar to that of another hydrogen–deficient binary v Sgr for which variations on a timescale of 20 day have been reported recently.

117.153   Observations of SS 433 at 2695 and 8085 MHz, 1979 – 1985.

R. L. Fiedler, K. J. Johnston, J. H. Spencer, E. B. Waltman, D. R. Florkowski, D. N. Matsakis, F. J. Josties, P. E. Angerhofer, W. J. Klepczynski, D. D. McCarthy.
Astron. J., Vol. 94, No. 5, p. 1244 – 1250 (1987).

The authors present flux densities of SS 433 at 2695 and 8085 MHz spanning the period July 1979 – December 1985. The variations may be characterized as a clustering of flare events separated by periods of quiescent emission. A harmonic analysis reveals no significant periodicities in either the entire 7 yr data set or in subsets corresponding to active and quiescent time frames. There is, however, broadband power from roughly 50 to 200 days which fades during 1985. Moreover, the authors find that the temporal power distribution of SS 433 at both frequencies strongly resembles $1/f$ noise. Finally, a comparison with 408 MHz data (Bonsignori–Facondi et al., 1986) shows variations in the light curves to lag those at 2695 and 8085 MHz by approximately 4 days.

117.154   Visual and infrared photometry of the ultrashort–period dwarf nova HT Cassiopeiae.

G. Berriman, S. Kenyon, C. Boyle.
Astron. J., Vol. 94, No. 5, p. 1291 – 1298 (1987).

This paper presents simultaneous visual (V) and infrared (H) photometry of the eclipsing dwarf nova binary HT Cas during two eclipses in October 1985, and infrared (JHK) photometry out of eclipse in January 1986. These are the first infrared observations of this object. They require that the red dwarf is fainter than $F_v = 0.5$ mJy at $H(H = 15.7)$, and that the system is at least 215 pc away, for a red dwarf radius of $R_R = 0.15\,R_\odot$. At this distance, the white dwarf has a brightness temperature at V of $T_b = 26000$K. The accretion disk around the white dwarf is highly variable from epoch to epoch: it was 70% brighter at H in January 1986 than in October 1985. Such variability in a quiescent system is unprecedented. The disk consisted of optically thin material in January, when it had a kinetic temperature of $T_{kin} = 10000 – 20000$K, and was most likely optically thin in October too.

117.155   U, B, V, R photometry and light–curve solution for WY Cancri.

S. A. Naftilan.
Astron. J., Vol. 94, No. 5, p. 1327 – 1329 (1987).

New light curves obtained in UBV and R are presented for the short–period RS CVn–like binary WY Cancri. The system exhibits a distorted light curve at phases near second quadrature, and on one night excess light was seen at these phases. A light–curve–synthesis solution yields parameters that generally agree with the earlier transit model of Awadella and Budding (1979), although the derived temperature of the secondary is higher and its limb darkening lower. The recent observations fail to indicate any period change.

117.156 Photometry and spectroscopy of the O–type variable HD 167971.
C. Leitherer, D. Forbes, A. C. Gilmore, J. Hearnshaw, G. Klare, J. Krautter, H. Mandel, O. Stahl, W. Strupat, B. Wolf, F.–J. Zickgraf, E. Zirbel.
Astron. Astrophys., Vol. 185, No. 1/2, p. 121 – 130 (1987).

The authors present photometric and spectroscopic observations of the O–type variable HD 167971. From a period analysis of the visual light–curve they find $P = 3.3213$ days. Combining the photometric and spectroscopic data they derive a consistent model for the system which can be understood in terms of a close O–type eclipsing binary and a third, more distant companion of spectral type O. The latter star is the most luminous component of the system and dominates the optical and the UV spectrum. The mass loss rate of this star is derived from the Hα– and UV–line profiles. The authors find $\dot{M} \cong 2 \times 10^{-6} M_\odot \mathrm{yr}^{-1}$ and $v_\infty = 3100$ km s^{-1}.

117.157 CCD photometry of V926 Sco, the optical counterpart of the X–ray burst source 4U/MXB 1735–44.
S. van Amerongen, H. Pedersen, J. van Paradijs.
Astron. Astrophys., Vol. 185, No. 1/2, p. 147 – 149 (1987).

The authors present the results of V–band CCD photometry of V926 Sco, the optical counterpart of the X–ray burst source 4U/MXB 1735–44. A period search in the variations of the V magnitude of V926 Sco confirms the period found recently by Corbet et al. (1986). An ephemeris for the time of maximum light of the orbital light curve is given, covering the period 1984 – 1986.

117.158 Light–curve analysis of the W Serpentis objects W Crucis and RX Cassiopeiae.
W. Strupat.
Astron. Astrophys., Vol. 185, No. 1/2, p. 150 – 154 (1987).

A light–curve analysis of the W Serpentis objects RX Cassiopeiae and W Crucis based on the Wilson–Devinney approach is presented. Orbital parameters and absolute dimensions of these interacting giant and supergiant binaries are derived.

117.159 Hard spectral components in soft X–ray transients.
A. R. King, J. P. Lasota.
Astron. Astrophys., Vol. 185, No. 1/2, p. 155 – 159 (1987).

The authors show that at low luminosities ($\leqslant 10^{35}$erg^{-1}) boundary–layer accretion onto a non–magnetic neutron star is likely to produce an ion–supported corona around the star. The electron temperature in the corona is $\geqslant 10^8$K, and gives rise to a hard X–ray spectrum.

117.160 Five–colour (UBVRI) polarimetry of H 0139–68 = BL Hydri.
V. Piirola, A. Reiz, G. V. Coyne.
Astron. Astrophys., Vol. 185, No. 1/2, p. 189 – 195 (1987).

The authors present observations of linear and circular polarization obtained simultaneously in five colour bands during a low activity state of the AM Her type magnetic binary H 0139–68 in November 1984.

117.161 Simultaneous five–colour (UBVRI) polarimetry of EF Eri.
V. Piirola, A. Reiz, G. V. Coyne.
Astron. Astrophys., Vol. 186, No. 1/2, p. 120 – 128 (1987).

The authors' simultaneous multicolor observations have revealed new features in the wavelength and orbital phase dependence of the linear and circular polarization of EF Eri. These have been interpreted in terms of cyclotron emission from two discrete regions at largely differing ($\sim 80°$) latitudes but having the same magnetic field polarity. This implies a near-equatorial emission region deviating from the dipole field geometry. The strong ($30° – 35°$) rotation of position angle as a

function of wavelength from U to I could be explained by Faraday rotation. Magnetic field estimates from the ratio of linear to circular polarization give stronger fields for the shorter wavelengths than those found in the red and infrared.

117.162 A classification of fast quasi–periodic X–ray oscillators: is 6 Hz a fundamental frequency?
G. Hasinger.
Astron. Astrophys., Vol. 186, No. 1/2, p. 153 – 158 (1987).

In a number of low–mass X–ray binaries different types of quasi–periodic oscillations (QPOs) are strictly related to different spectral states. It is demonstrated that nearly all observed QPOs can be grouped into two major classes: intensity–dependent QPOs with frequencies of 20 – 55 Hz associated with strong red noise occur on the "horizontal" spectral branch; the majority of QPOs, however, is found during "normal branch" behavior with very weak red noise and rather constant frequencies clustering around 6 Hz. The fact that the QPO frequency in this mode is stable and almost the same between different systems suggests a mechanism that can be described by fundamental physical properties of the neutron star and it's close environments. It is suggested here that on the normal branch the X–ray flux is modulated by a resonant density wave in the Eddington–limited accretion flow above the neutron star surface. This model is able to explain both the low value and the constancy of the QPO frequency as well as the absence of red noise in the normal branch.

117.163 Rotational modulation and flares on RS CVn and BY Dra stars. IV. The spatially resolved chromosphere of AR Lacertae.
F. M. Walter, J. E. Neff, D. M. Gibson, J. L. Linsky, M. Rodonò, D. E. Gary, C. J. Butler.
Astron. Astrophys., Vol. 186, No. 1/2, p. 241 – 254 (1987).

The authors observed the RS CVn system AR Lacertae systematically over an orbital period with the International Ultraviolet Explorer in October 1983. Contemporaneous radio observations were obtained at the Very Large Array. The spectra of the Mg II k emission line were analyzed using a Doppler imaging technique. In this way, the authors identified three discrete regions of emission in the outer atmosphere of the K star – two "plages" and a chromospheric brightening that was related to a radio flare. The widths of the plage profiles indicate that the two plages together cover about 2% of the visible stellar hemisphere, and their $v \sin i$ values indicate that they lie close to the equator of the K star. The Mg II k surface flux in the plages is about five times the mean Mg II k surface flux of the K star. The authors then used the far–ultraviolet spectra obtained at the eclipse phases to separate the individual contributions of the two stars and the plage and flare regions in order to estimate their line surface fluxes.

117.164 B and V light curves of TY Bootis.
R. G. Samec, B. B. Bookmyer.
Publ. Astron. Soc. Pac., Vol. 99, No. 618, p. 842 – 848 (1987).

The short–period eclipsing binary system TY Boo was observed on five consecutive nights. The observations covering the eclipse portions of the light curves yielded five epochs of minimum light. A period study gives no indication of the cyclic period variation indicated by Szafraniec. The light curves, defined by 348 observations with the B filter and 353 with the V filter, are of the W Ursae Majoris type and are symmetric. The system becomes redder during each eclipse with a momentary dereddening at phase 0.0. The depths of the eclipse curves are about the same as those in the light curves published by Carr. However, the system appears to have undergone a slight reddening.

117.165 Kinematics and age of the low–mass X–ray binaries.
A. P. Cowley, J. B. Hutchings, D. Crampton, F. D. A. Hartwick.
Astrophys. J., Vol. 320, No. 1, p. 296 – 299 (1987).

Radial velocities have been measured and collected from the literature for the low–mass X–ray binaries (LMXB) in order to study their kinematic properties. They show a large velocity dispersion (~ 110 km s^{-1}) suggesting they belong to a very old stellar population. The galactic rotation obtained from this group lies between that observed for the metal–rich globular clusters and that of the metal–poor globular clusters which lie within 8 kpc from the galactic center. Both the galactic distribution and kinematics of the LMXB suggest an age of $\sim 15 \times 10^9$yr.

117.166 The spectral evolution of dwarf nova outbursts.
J. K. Cannizzo, S. J. Kenyon.
Astrophys. J., Vol. 320, No. 1, p. 319 – 332 (1987).
The authors present calculations of the spectral evolution of disk instability models for dwarf nova outbursts. Observed stellar spectra are used to model the radiation emitted by optically thick annuli within the disk. It is found that eruptions in disks with low background accretion rates begin near the white dwarf and quickly spread to larger disk radii. The continuum flux rises simultaneously at all wavelengths in these "inside–out" outbursts. Outbursts in disks with higher mass–transfer rates commence near the outer edge of the disk. The optical continuum flux rises before the UV continuum by $\sim 0.1 – 0.3$ day for the parameters chosen in this study, which is smaller than the $\sim 0.5 – 1.0$ day delays observed in several dwarf novae.

117.167 Rapid optical flaring in MXB 1735–444 and an optical burst from GX 17 + 2.
J. N. Imamura, T. Y. Steiman–Cameron, J. Middleditch.
Astrophys. J., Lett. Ed., Vol. 320, No. 1, p. L41 – L45 (1987).
The authors report the results of high–speed white–light photometry of five low–mass X–ray binaries. They find that MXB 1735–444 was in a previously unreported state. It showed rapid optical flaring on time scales of $\leqslant 10$ s to ~ 20 minutes over the course of a 2 hr observation. During the bursts, the optical luminosity increased by up to a factor of 5. Qualitatively similar long type II X–ray bursts have been observed from the rapid burster, MXB 1730–335, and long X–ray bursts of undetermined type have been observed from GX 17 + 2. The authors also report the detection of an optical burst from the source GX 17 + 2.

117.168 On light curves variability of DQ Her from 1954 to 1985.
E. S. Dmitrienko.
Izv. Krymskoj Astrofiz. Obs., Tom 76, p. 62 – 70 (1987). In Russian. English translation in Bull. Crimean Astrophys. Obs., Vol. 76.
The variations of light curve shapes obtained by photometrical observations of DQ Her in 1954 – 1985 have been analyzed. The comparison of the light curves showed that the light variations of DQ Her in the period from 1954 to 1985 were caused by its primary component luminosity variations. All the light curves can be classified into three types according to maximal noneclipsing light in 0.40 – 0.60 phases, its excess of the light in 0.2 phase and the height of the shoulder appearing prior to the eclipse. Each state of the power variations in the primary component is characterized by one of the three types of light curves. A possible similarity between luminosity variations from maximal to minimal in DQ Her and developing a normal outburst into a superburst for dwarf novae of SU UMa type has been also considered.

117.169 Spectrophotometric analysis of some unresolved double stars. II. 52 Per, 58 Per, 5 Lac.
V. I. Burnashev.
Izv. Krymskoj Astrofiz. Obs., Tom 76, p. 76 – 80 (1987). In Russian. English translation in Bull. Crimean Astrophys. Obs., Vol. 76.
Spectroscopic observations of three unresolved double stars were carried out during 1979 – 1982. Spectral types of the components and their luminosities were determined, while their masses were estimated approximately.

117.170 Model of binary stellar systems containing a black hole for explaining the properties of some cepheids in the IC 1613 galaxy.
V. V. Mityanok.
Red. Zh. Izv. Vuzov. Fiz., Tomsk, 12 pp. (1987). In Russian. Abstr. in Ref. Zh., 51. Astron., 10.51.765 (1987).

117.171 The jets in SS433.
W. Brinkmann, S. Massaglia.
Bull. Am. Astron. Soc., Vol. 19, No. 2, p. 692 (1987). Abstract. – See Abstr. 010.061.

117.172 Gamma–ray temporal and spectral variability of Cygnus X–1.
J. C. Ling, W. A. Mahoney, W. A. Wheaton, A. S. Jacobson.
Bull. Am. Astron. Soc., Vol. 19, No. 2, p. 694 (1987). Abstract. – See Abstr. 010.061.

117.173 A deep, Doppler–compensated IUE SWP echellogram of the K0 primary of HR 1099.
J. O. Bennett, T. R. Ayres, E. Jensen, O. Engvold.
Bull. Am. Astron. Soc., Vol. 19, No. 2, p. 706 (1987). Abstract. – See Abstr. 010.061.

117.174 IR emission from X–ray binaries: IRAS observations.
H. A. Smith, J. Beall, K. Wood.
Bull. Am. Astron. Soc., Vol. 19, No. 2, p. 708 (1987). Abstract. – See Abstr. 010.061.

117.175 HD 26337 = EI Eri: an RS CVn candidate for the Doppler imaging technique.
F. C. Fekel, R. Quigley, K. Gillies, J. L. Africano.
Bull. Am. Astron. Soc., Vol. 19, No. 2, p. 708 (1987). Abstract. – See Abstr. 010.061.

117.176 HD 17433 (VY Ari): a young chromospherically active binary.
B. W. Bopp, S. Saar, P. A. Feldman, R. Dempsey, M. Allen, C. Armbruster, S. P. Barden.
Bull. Am. Astron. Soc., Vol. 19, No. 2, p. 709 (1987). Abstract. – See Abstr. 010.061.

117.177 Luminous accretion disks in binary stars.
R. S. Polidan.
Bull. Am. Astron. Soc., Vol. 19, No. 2, p. 709 (1987). Abstract. – See Abstr. 010.061.

117.178 Stars of type MS with evidence of white dwarf companions.
B. F. Peery Jr.
Bull. Am. Astron. Soc., Vol. 19, No. 2, p. 709 – 710 (1987). Abstract. – See Abstr. 010.061.

117.179 Polarization of light from accretion disks in cataclysmic variables.
F. H. Cheng, G. Shields, D. N. C. Lin, J. E. Pringle.
Bull. Am. Astron. Soc., Vol. 19, No. 2, p. 710 (1987). Abstract. – See Abstr. 010.061.

117.180 V444 Cygni, WN5 + O6, revisited.
A. B. Underhill, S. Yang, G. M. Hill.
Bull. Am. Astron. Soc., Vol. 19, No. 2, p. 710 (1987). Abstract. – See Abstr. 010.061.

117.181 Circumstellar matter in the AU Monocerotis system.
G. J. Peters.
Bull. Am. Astron. Soc., Vol. 19, No. 2, p. 713 (1987). Abstract. – See Abstr. 010.061.

117.182 Grossly non–synchronous rotation in the RS CVn–type binary HD 181809.
D. S. Hall, L. Pazzi.
Bull. Am. Astron. Soc., Vol. 19, No. 2, p. 713 – 714 (1987). Abstract. – See Abstr. 010.061.

117.183 **A polarimetric, photometric and spectrophotometric study of the very massive close binary DH Cephei.**
M. F. Corcoran.
Bull. Am. Astron. Soc., Vol. 19, No. 2, p. 714 (1987). Abstract. – See Abstr. 010.061.

117.184 **Studies of low–mass X–ray binaries.**
M. R. Garcia.
Bull. Am. Astron. Soc., Vol. 19, No. 2, p. 720 (1987). Abstract. – See Abstr. 010.061.

117.185 **EXOSAT X–ray absorption spectra from the binary system 4U 1700–37/HD 153919.**
F. Haberl, N. E. White, M. Gottwald.
Bull. Am. Astron. Soc., Vol. 19, No. 2, p. 721 (1987). Abstract. – See Abstr. 010.061.

117.186 **The origin of the ultra–compact binary 4U 1820–30.**
S. Miyaji, I. Hachisu, H. Saio.
Bull. Am. Astron. Soc., Vol. 19, No. 2, p. 721 (1987). Abstract. – See Abstr. 010.061.

117.187 **IUE and optical observations of the symbiotic star/ nebula He2–104.**
J. H. Lutz.
Bull. Am. Astron. Soc., Vol. 19, No. 2, p. 753 (1987). Abstract. – See Abstr. 010.061.

117.188 **A rapid change of the Her X–1 pulse profile and high state duration.**
Y. Soong, D. E. Gruber, R. E. Rothschild.
Bull. Am. Astron. Soc., Vol. 19, No. 2, p. 756 (1987). Abstract. – See Abstr. 010.061.

117.189 **An analysis of the chemical composition and other parameters of the star π Sgr as a binary system.**
L. S. Lyubimkov, Z. A. Samedov.
Izv. Krymskoj Astrofiz. Obs., Tom 77, p. 97 – 114 (1987). In Russian. English translation in Bull. Crimean Astrophys. Obs., Vol. 77.
A method for investigation of the combined spectrum of binary system components based on model atmospheres is proposed and applied to determine the fundamental parameters and chemical composition of the A and B component of π Sgr: T_{eff} = 7300K, log g = 2.6 for component A, T_{eff} = 6200K and log g = 2.3 for component B. The chemical composition of both stars appears to be close to the solar one, whereas the standard analysis of π Sgr as a single star leads to common underabundance of elements and especially low abundance of Ca and Sc. The distance to π Sgr (d = 180 – 195 pc) and the lower limit of the orbital period in this binary system ($P \geqslant 20$ years) are estimated. It is shown that the lines of A and B components in the combined spectrum of π Sgr can be separated by no more than 0.4 Å provided this separation is observed in different phases.

117.190 **Light element abundances in the atmosphere of the principal component of the binary system υ Sgr.**
V. V. Leushin, G. P. Topil'skaya.
Astrophysics, Vol. 26, No. 2, p. 117 – 125 (1987). English translation of 43.117.232.

117.191 **Numerical calculations of mass transfer flow in semi– detached binary systems.**
D. A. Edwards, J. E. Pringle.
Mon. Not. R. Astron. Soc., Vol. 229, No. 3, p. 383 – 394 (1987).
The authors calculate numerically the details of the mass transfer flow near the inner Lagrangian point in a semi–detached binary system. They use a polytropic equation of state with n = 3/2. They calculate the dependence of the mass transfer rate on the degree to which the star over–fills its Roche lobe, and find good agreement with previous analytic estimates. The authors calculate the variation of mass transfer rate which occurs if the binary system has a small eccentricity, and use this to cast doubt

on the model for superhumps in dwarf novae proposed by Papaloizou and Pringle (1979).

117.192 **On OB–type close binary stars.**
R. W. Hilditch, S. A. Bell.
Mon. Not. R. Astron. Soc., Vol. 229, No. 3, p. 529 – 538 (1987).
The authors present a compilation of masses, radii, effective temperatures and luminosities for the components of 31 binary systems with spectral types in the range O – B5. Comparisons of the data on the 16 detached systems in the sample with main-sequence models for individual stars provides further observational confirmation of the recent (1981 – 86) conclusions from theoretical studies that convective overshooting and stellar–wind mass loss must be included in evolutionary models. A convenient empirical mass–luminosity relationship for the mass range 3.5 – 20 $M_\odot$ is also determined from these detached components. Of the remaining systems, eight are in semi–detached configurations and seven are in (mostly) marginal–contact states. Comparisons of these observational data with evolutionary models for binary stars published by Sybesma are made and it is concluded that all the systems in this sample have evolved or will evolve through a case A mass–transfer process.

117.193 **The ultraviolet variability of AY Ceti.**
T. Simon, G. Sonneborn.
Astron. J., Vol. 94, No. 6, p. 1657 – 1663 (1987).
AY Cet is a single–line binary comprised of a spotted G5 III primary and a white dwarf secondary. The authors obtained a series of UV spectra with the *IUE* satellite on five different dates covering a substantial part of the optical cycle of the primary star. They found no evidence that the continuum or the Lyα absorption line of the secondary star varied. There were significant changes in the strengths of the UV emission lines, but the variations were only weakly correlated with either the orbital phase of the binary or the rotational phase of the primary. The UV emission lines were especially strong near maximum visual brightness at a time when the starspot(s) on the primary was least visible. The authors attribute the enhanced line emission to a flare event on the primary, most likely at a high–latitude site close to the pole of this star. The UV radiative losses of this flare were comparable with those of flares previously observed on the RS CVn variables λ And and HR 1099.

117.194 **Rapid emission–line variations in the UV spectrum of the Wolf–Rayet system HD 90657.**
G. Koenigsberger, L. H. Auer.
Publ. Astron. Soc. Pac., Vol. 99, No. 620, p. 1080 – 1083 (1987).
The authors report IUE observations of the Wolf-Rayet binary system HD 90657 which indicate that a major change in the properties of the W–R wind occurred on time scales of less than 90 minutes. These are very rapid variations when compared to those resulting from atmospheric eclipse effects, which are also observed in this system. The rapid change is attributed to wind instabilities, possibly induced by pulsations of the underlying helium–burning core.

117.195 **Hamuy's blue variable star in Orion.**
H. E. Bond, A. D. Grauer, D. Burnstein, R. O. Marzke.
Publ. Astron. Soc. Pac., Vol. 99, No. 620, p. 1097 – 1098 (1987).
A 14th–magnitude blue variable star in Orion, serendipitously discovered by M. Hamuy, is shown to be a new member of the class of nova–like (UX UMa–type) variables on the basis of spectroscopic and high–speed photometric observations.

117.196 **The energy distributions of magnetic variables: the effects of inhomogeneous accretion columns.**
H. S. Stockman, A. F. Lubenow.
Space Telesc. Sci. Inst., Prepr. Ser., No. 177, 8 pp. (1987). To appear in IAU Colloq. No. 93.

117.197   The radial velocity curve and peculiar TiO distribution of the red secondary star in Z Chamaeleontis.
R. A. Wade, K. Horne.
Space Telesc. Sci. Inst., Prepr. Ser., No. 194, 33 pp. (1987). To appear in Astrophys. J.

117.198   Radio flares from AE Aquarii: a low–power analog to Cygnus X–3?
T. S. Bastian, G. A. Dulk, G. Chanmugam.
Space Telesc. Sci. Inst., Prepr. Ser., No. 196, 17 pp. (1987). To appear in Astrophys. J.

117.199   A model of the symbiotic star RX Puppis.
D. A. Allen, A. E. Wright.
Anglo–Aust. Obs., Prepr., No. 222, 28 pp. (1987). To appear in Mon. Not. R. Astron. Soc.

117.200   Hamuy's blue variable star in Orion.
H. E. Bond, A. D. Grauer, D. Burstein, R. O. Marzke.
Space Telesc. Sci. Inst., Prepr. Ser., No. 199, 6 pp. (1987). To appear in Publ. Astron. Soc. Pac.

117.201   LTE models of the emission lines of the dwarf nova Z Cha.
T. R. Marsh.
R. Greenwich Obs., Prepr., No. 55, 28 pp. (1987). To appear in Mon. Not. R. Astron. Soc.

117.202   Relative C, N, O abundances in red giants, planetary nebulae, novae and symbiotic stars.
H. Nussbaumer, H. Schild, H. M. Schmid, M. Vogel.
R. Greenwich Obs., Prepr., No. 67, 12 pp. (1987). To appear in Astron. Astrophys.

117.203   Der 35–Tage–Zyklus von Hercules X–1.
P. Kahabka.
MPE Rep., No. 204, 139 pp. (1987).

117.204   Pre–main sequence binaries.
B. Reipurth.
ESO Sci. Prepr., No. 548, 14 pp. (1987). Paper presented at the NATO ASI meeting "Formation and evolution of low mass stars", Viano do Castelo, Portugal, October 1987.

117.205   Time–resolved spectroscopy of the cataclysmic variable V426 Ophiuchi.
F. V. Hessman.
Max–Planck–Inst. Astron. Heidelb., Prepr., 54 pp. (1987). Submitted to Astron. Astrophys.

117.206   Deterministic chaos in accreting systems: analysis of the X–ray variability of Hercules X–1.
W. Voges, H. Atmanspacher, H. Scheingraber.
Astrophys. J., Vol. 320, No. 2, p. 794–802 (1987).
The irregular X–ray variability of Her X–1 has been investigated in different phases of the 1.7 day orbital period of the binary system Her X–1/HZ Her. The analyzed data have been sampled by the *EXOSAT* satellite during two main–on states of the 35 day cycle of the system. Using new system theoretical methods, the attractors of different processes have been reconstructed which are responsible for the temporal evolution of the observed X–ray radiation. For the emission of X–ray radiation from Her X–1, an attractor has been established which is of intriguingly low fractal dimension. This fact provides evidence for a non–stochastic, deterministic chaotic origin of the irregular variability of the emitted X–ray radiation.

117.207   New physics from Cygnus X–3.
J. Collins, F. Olness.
Phys. Lett. B, Vol. 187, No. 3–4, p. 376–380 (1987). Abstr. in Phys. Abstr., Vol. 90, No. 1309, Entry 88539 (1987).

117.208   AG Dra a symbiotic star with an uncommon cool component.
M. Friedjung.
Inst. Astrophys. Paris, Pré–Publ., No. 205, 6 pp. (1987). To appear in IAU Colloq. No. 103.

117.209   Concluding remarks. Presented at IAU Colloquium No. 103 "The symbiotic phenomenon", Torun, Poland, 18–21 August 1987.
M. Friedjung.
Inst. Astrophys. Paris, Pré–Publ., No. 206, 6 pp. (1987).

117.210   The infrared spectrum of the symbiotic star CI Cyg at phase 0.5.
M. Bensammar, M. Friedjung, N. Letourneur, J. P. Maillard.
Inst. Astrophys. Paris, Pré–Publ., No. 207, 11 pp. (1987). To appear in Astron. Astrophys.

117.211   The gravity–darkening of highly distorted stars in close binary systems. IV. Practical analysis of secondary components filling the Roche lobe in semi–detached systems.
M. Kitamura, Y. Nakamura.
Ann. Tokyo Astron. Obs., Second Ser., Vol. 21, No. 4, p. 387–397 (1987).
From a quantitative analysis of the observed photometric ellipticity effect, the exponent of gravity–darkening has been empirically determined for the secondary components filling the Roche lobe in nine well–understood semi–detached close binary systems. In the analysis, the exponent of gravity–darkening for the main–sequence primaries with spectral types of B1 V – A2 V has been assumed as the unity. The result of the present analysis indicates that the empirical values of the exponent deduced for the secondary components are significantly greater than the unity. Such greater values of the exponent for the secondaries could not be reconciled by any adjustment of the physical elements used as the input parameters within the extent of reduction errors.

117.212   The current observational outlook on AM Herculis variables.
G. D. Schmidt.
Prepr. Steward Obs., No. 771, 22 pp. (1987). To appear in the proceedings of the Vatican workshop on circumstellar polarization, June 1987.

117.213   Some remarks on the dwarf nova V1504 Cyg.
G. A. Richter.
Mitt. Veränderliche Sterne, Band 11, Heft 2, p. 35–37 (1987).

117.214   Photographische UBV–Beobachtungen an dem Polar AM Her aus dem Jahre 1986.
W. Götz.
Mitt. Veränderliche Sterne, Band 11, Heft 2, p. 47–50 (1987).
In supplementing and completing the previous list of observations 156 photographic UBV observations from 58 nights covering the time interval between 1986 February 11 and 1986 December 5 are given. Most of the plates were obtained in B.

117.215   Maxima von SU Ursae Majoris ab 1940.
E. Splittgerber.
Mitt. Veränderliche Sterne, Band 11, Heft 3, p. 54–60 (1987).
A list of maxima data, mainly derived from published individual observations and from estimates on Sky Patrol plates, is given in order to make possible the continuation of a statistical treatment of the eruptions of this prototype.

117.216   Zur Periodenänderung des W–UMa–Sterns GU Ori.
R. Steiner–Sohn.
Mitt. Veränderliche Sterne, Band 11, Heft 3, p. 61–64 (1987).

117.217   A turbulent region in CH Cyg?
A. Skopal.
Contrib. Astron. Obs. Skalnaté Pleso, Vol. 16, p. 69 – 77 (1987).
Photoelectric U, B, V observations of the symbiotic eclipsing binary CH Cygni made in 1983 – 1986 are published. A gradual decrease in the amplitudes of brightness variations was observed after the activity maximum in 1982 until December 1984. On Oct. 6 – 7, 1985, no rapid changes in brightness were observed at the time of the full eclipse of the hot component. After the eclipse, at the end of 1985, a sudden increase in the amplitudes of the star's brightness fluctuations was observed, especially in the U colour. The possibility of observing the turbulent region of the accretion disk is discussed.

117.218   Roche lobe in eccentric orbits.
P. Hadrava.
Hvar Obs. Bull., Vol. 10, No. 1, p. 1 – 10 (1986).
The shape of stars distorted by both rotation and tidal force in binaries with eccentric orbits is studied in approximation of Roche potential. A particular attention is devoted to the influence of response of component's structure to variations of tidal force.

117.219   Long–time and orbital behaviour of Hα in the spectrum of HDE 226868 (Cygnus X–1).
O. Eh. Aab.
Astrofiz. Issled. Izv. Spets. Astrofiz. Obs., Tom 25, p. 28 – 40 (1987). In Russian. English translation in Bull. Spec. Astrophys. Obs. – North Caucasus, Vol. 25.
The Hα line is studied in the spectrum of the supergiant HDE 226868 – optical component of the X–ray source Cygnus X–1 from spectrograms obtained during five years with the 6–meter telescope. Position and photometrical line parameters are measured and reduced. A long–time and orbital variability of the Hα intensity is shown. The radial velocity curves and the calculated orbital curves testify the formation of Hα emission mainly in the mass–loss atmosphere of the supergiant.

117.220   Spectral and photoelectrical observations of SS433 in the main minimum.
V. P. Goranskij, I. M. Kopylov, V. Yu. Rakhimov, N. V. Borisov, L. V. Bychkova, S. N. Fabrika, G. P. Chernova.
Soobshch. Spets. Astrofiz. Obs., Vyp. 52, p. 5 – 50 (1987). In Russian.
The results of spectral and photoelectrical observations of SS433, carried out from the end of May through the beginning of June, 1986 are presented. The observations were made near the eclipse orbital phase of the accretion disk by main system component (Min I), and in the precession phase of maximal turn of the disk to the observer. The spectral observations were carried out in $\lambda\lambda3500 – 6770$, spectral resolution $\Delta\lambda = 3$ Å. The photoelectrical observations in U, B, V, R, I, and Hα. The results are presented in tables and diagrams which show the brightness variations of SS433 in different filters and the general character of spectrum variations in the observations.

117.221   On the problem of variation of periods of binary systems.
R. K. Mukhametkalieva.
Tr. Astrofiz. Inst. Alma-Ata, Tom 47, p. 51 – 59 (1987). In Russian.

117.222   Influence of viscosity on tidal streams in binary systems.
A. Z. Dolginov, E. V. Smel'chakova.
Fiz.–Tekh. Inst. Akad. Nauk SSSR, Prepr., No. 1124, p. 3 – 33 (1987). In Russian. Abstr. in Ref. Zh., 51. Astron. 11.51.510 (1987).

117.223   Evolution of the axial rotation and orbital motion in binary systems.
A. Z. Dolginov, E. V. Smel'chakova.
Fiz.–Tekh. Inst. Akad. Nauk SSSR, Prepr., No. 1123, p. 3 – 35 (1987). In Russian. Abstr. in Ref. Zh., 51. Astron. 11.51.632 (1987).

117.224   Espectrofotometria de las estrellas simbióticas He2–417, He2–467 y He2–468.
S. Navarro, R. Costero, A. Serrano P. G., L. Carrasco.
Rev. Mex. Astron. Astrofis., Vol. 14, No. 1, p. 339 – 343 (1987). – See Abstr. 012.042.
A spectrophotometric study of He2–417, 467 and 468 is made. Their spectra show mainly emission lines of H I, He I and He II, as well as absorption bands characteristic of late–type stars. No forbidden lines were detected in the spectra of the objects. The authors report the line intensity ratios relative to Hβ and compare them with those previously reported in the literature. Some line ratios have varied in He2–417 and 467, the two objects with previous observations.

117.225   Study of the light curve of LSI + 61°303.
J.–M. Paredes.
Rev. Mex. Astron. Astrofis., Vol. 14, No. 1, p. 395 – 400 (1987). – See Abstr. 012.042.
A model based on deformations of the primary star by a compact star in an eccentric system is presented for explaining the optical variability observed in LSI + 61°303.

117.226   A photometric analysis of the massive contact binary BR Muscae.
E. Lapasset, M. N. Gómez, J. J. Clariá.
Rev. Mex. Astron. Astrofis., Vol. 14, No. 2, p. 402 – 409 (1987). – See Abstr. 012.042.
The UBV photometry of the early–type contact binary BR Muscae is analysed by means of the differential correction method of Wilson and Devinney. The best solutions found correspond to mass–ratio values close to unity. The remaining parameters are well defined and practically independent of the mass–ratio. General considerations on the hot contact binaries are presented and discussed.

117.227   Spectroscopic study of the binary system HD 153919.
G. R. Solivella, J. Menzies, Y. Kondo, J. Sahade.
Rev. Mex. Astron. Astrofis., Vol. 14, No. 2, p. 417 (1987). Abstract. – See Abstr. 012.042.

117.228   Looking for new cataclysmic variables.
J. E. Steiner, D. Cieslinskí, F. J. Jablonski.
Rev. Mex. Astron. Astrofis., Vol. 14, No. 2, p. 436 (1987). Abstract. – See Abstr. 012.042.

117.229   The history of a strange variable star; AM Herculis.
M. Verdenet.
Ciel Espace, No. 215, p. 47 (1987). In French. Abstr. in Phys. Abstr., Vol. 90, No. 1312, Entry 107900 (1987).

117.230   On the symbiotic nature of the emission–line object K4–45.
A. Yu. Shchelkanova.
Pis'ma Astron. Zh., Tom 13, No. 12, p. 1061 – 1064 (1987). In Russian. English translation in Sov. Astron. Lett., Vol. 13.
The symbiotic nature of the emission–line object K4–45 (96 + 1°1) is revealed on the basis of spectral observations. The visible component has a spectral type no later than M3. The interstellar absorption for the object is more than 3ᵐ2.

117.231   The 1984 – 1985 optical fade and radio brightening of CH Cygni.
L. S. Luud, T. Tomov, J. A. Vennik, L. Leedjarv.
Sov. Astron. Lett., Vol. 12, No. 6, p. 364 – 367 (1986). English translation of 42.117.098.

117.232   On the origin of the radio flare in CH Cygni.
B. F. Yudin.
Sov. Astron. Lett., Vol. 12, No. 6, p. 368 – 369 (1986). English translation of 42.117.099.

117.233 Spectrophotometry of AG Pegasi.
A. P. Ipatov, B. F. Yudin.
Sov. Astron. Lett., Vol. 12, No. 6, p. 390 – 392 (1986). English translation of 42.117.109.

117.234 Orbital elements for double stars of Population II. The high–velocity system COD –48°1741.
H. Lindgren, A. Ardeberg, E. Zuiderwijk.
Astron. Astrophys., Vol. 188, No. 1, p. 39 – 45 (1987).
Radial–velocity data for COD –48°1741 are analyzed. The object has been identified as a high–velocity binary system. High–precision radial–velocity data have been obtained with a CORAVEL scanner. From 45 individual measurements of radial velocity, orbital elements have been determined. The system velocity is $+307.9$ km s^{-1} and the orbital period 7.56 days. The object has been measured photometrically in the $uvby$ and UBV systems. A search for eclipses showed no significant evidence of variability. The $uvby$ data indicate the main component to be a low–luminosity star of early G type with a metallicity of [Fe/H] $= -1.5$. Further astrophysical and space–velocity data are derived and discussed.

117.235 The anomalous ultraviolet spectrum of the AM Her star H 0538 + 608.
J. M. Bonnet–Bidaud, M. Mouchet.
Astron. Astrophys., Vol. 188, No. 1, p. 89 – 94 (1987).
The X–ray source H 0538 + 608, recently identified as an AM Her system, was observed with the IUE satellite and discovered to have a unique ultraviolet spectrum. The strong resonance emission lines commonly observed in the AM Her sources are present but with very unusual intensities. Such an anomaly is not satisfactorily explained in terms of orbital variability or differences in the ionization structure of the emitting region in the context of the present available photoionization models. A non–solar chemical composition such as produced in nova–type outbursts cannot be excluded as a possible explanation. If confirmed this would imply that nova outbursts can occur in strongly magnetized systems.

117.236 The Roche coordinates in non–synchronous binaries.
P. Hadrava.
Astrophys. Space Sci., Vol. 138, No. 1, p. 61 – 69 (1987).
The Cayley–Darboux problem for the Roche model of binaries is reinvestigated. Generalised Roche coordinates are then defined and calculated in the form of power series of potential for the general case of non–synchronous binaries with eccentric orbits.

117.237 Kataklysmische Veränderliche. I. Übersicht über die Erscheinungsformen.
G. A. Richter.
Sterne, 63. Band, Heft 5, p. 275 – 282 (1987).

117.238 Symbiotische Sterne – rätselhafte Außenseiter in unserem Milchstraßensystem?
R. Luthardt.
Sterne, 63. Band, Heft 5, p. 292 – 301 (1987).

117.239 Giant spots on II Peg in late 1986.
P. B. Byrne, F. Marang.
Ir. Astron. J., Vol. 18, No. 2, p. 84 – 87 (1987). – See Abstr. 012.036.
During September 1986 a light curve was recorded for II Peg which was of an unprecedentedly large amplitude. To establish the duration of this unusual phenomenon a second set of photometric observations was made which established that the curve was essentially the same a full two months later. The purpose of this paper is to report the preliminary results of the analysis of these observations and to outline briefly their implications for the starspot hypothesis.

117.240 Long–term cycles in cosmic X–ray sources.
W. C. Priedhorsky, S. S. Holt.
Space Sci. Rev., Vol. 45, Nos. 3 + 4, p. 291 – 348 (1987).
Most of what we know about galactic X–ray binaries comes from time variation, particularly periodic variations corresponding to neutron star rotation, and binary motion. Longer cycles or quasi–cycles are much harder to observe because of the shortage of instrumentation suitable for long–term monitoring. Nonetheless, cycles with periods up to a few years have been seen in several galactic binaries. The review summarizes present knowledge about cycles in this time domain.

117.241 Fluctuation of wind–driven accretion in X–ray binaries.
C. Ho, J. Arons.
Astrophys. J., Vol. 321, No. 1, p. 404 – 417 (1987).
The authors study the stability and fluctuation of the mass transfer onto a binary X–ray source fed by the primary's radiation–driven wind and regulated by the ionization feedback of the X–rays. Overstability is found with oscillation period typically 10^2s, and the growth time scale approximately equals the wind transit time. Such propagating waves will lead to asymmetry in mass density and velocity and to specific angular momentum greater than previously estimated for the steady state. Depending on the detailed angular momentum dissipation mechanism, there may be instantaneous disk formation which leads to large spin–up or spin–down torque. Random episodes of spin–up and spin–down are equally likely, which may contribute to the observed timing behavior of Vela X–1.

117.242 The primary orbit and the absorption lines of HDE 226868 (Cygnus X–1).
Z. Ninkov, G. A. H. Walker, S. Yang.
Astrophys. J., Vol. 321, No. 1, p. 425 – 437 (1987).
From Reticon spectra of ~ 1 Å resolution taken between 1980 and 1984, the radial velocity curve of HDE 226868 is found to be characteristic of a single–line spectroscopic binary with $K = 75.0 \pm 1.0$ km s^{-1} and $e = 0.0$. Using all available data, a period of 5.59964 ± 0.00001 days is found. An absolute magnitude of -6.5 ± 0.2 is derived for the primary from the equivalent width of Hγ which is consistent with the spectral classification of O9.7 Iab. Assuming 20 $M_\odot$ as a reasonable estimate for the mass of the primary implies a mass of 10 ± 1 $M_\odot$ for the secondary.

117.243 The He II $\lambda4686$ and Hα emission lines of Cygnus X–1.
Z. Ninkov, G. A. H. Walker, S. Yang.
Astrophys. J., Vol. 321, No. 1, p. 438 – 446 (1987).
The results of a long–term (1980 – 1984) monitoring program of the He II $\lambda4686$ and Hα emission lines of the massive X–ray binary Cygnus X–1 are presented. The radial velocities of the He II $\lambda4686$ emission profiles can be fitted to a smooth sinusoid with little scatter and with no significant eccentricity. There is no significant variation in the K amplitude estimated at different epochs which implies a stable origin for the emission. The data support the model published by Friend and Castor in 1982 in which the primary star almost fills its Roche lobe and has an enhanced mass flow toward the secondary.

117.244 The large–scale radio structure of R Aquarii.
J. M. Hollis, M. Kafatos, A. G. Michalitsianos, R. J. Oliversen, F. Yusef–Zadeh.
Astrophys. J., Lett. Ed., Vol. 321, No. 1, p. L55 – L59 (1987).
Radio continuum observations of the R Aquarii symbiotic star system, using the VLA at 6 cm wavelength, reveal a large–scale $\sim 2'$ structure engulfing the binary which has long been known to have a similar optical nebula. This optical/radio nebula possesses $\sim 4 \times 10^{42}$ ergs of kinetic energy which is typical of a recurrent nova outburst. Moreover, a cluster of a dozen additional 6 cm radio sources were observed in proximity to R Aquarii; the nature of these sources is briefly discussed.

117.245   A survey study of energy distribution in component stars of Algol–type binary systems.
J. J. Dobias.
Diss. Abstr. Int., Sect. B, Vol. 48, No. 6, p. 1710–B (1987). Thesis, University of California, Los Angeles, 334 pp. (1987). Order No. DA8719941.

117.246   Photoelectric photometry of white dwarf eclipsing binary V471 Tauri.
C. Ibanoglu, Z. Tunca, S. Evren.
Doğa Turk Fiz. Astrofiz. Dergisi (Turkey), Vol. 11, No. 1, p. 157–166 (1987). In Turkish. Abstr. in Phys. Abstr., Vol. 90, No. 1313, Entry 115521 (1987).

117.247   On the nonthermal radio emission of double stars with relativistic components.
V. M. Lipunov, M. E. Prokhorov.
Astron. Zh., Tom 64, Vyp. 6, p. 1189–1198 (1987). In Russian. English translation in Sov. Astron., Vol. 31, No. 6.

The article is devoted to an analysis of possible manifestations of ejecting neutron stars in double systems with normal companions. For the first time, the problem of the propagation of radio emission from a point source orbiting in the stellar wind of the optical star is solved accounting for refraction. In the framework of this model, an attempt is made to explain the recently detected nonthermal radio emission from some OB stars.

117.248   VY Aquarii.
IAU Circ., Nos. 4413–4415, 4418, 4427 (1987).

117.249   UX Arietis.
IAU Circ., No. 4424 (1987).

117.250   AM Herculis.
IAU Circ., No. 4425 (1987).

117.251   CAL 83: a puzzling X–ray source in the Large Magellanic Cloud.
D. Crampton, A. P. Cowley, J. B. Hutchings, P. C. Schmidtke, I. B. Thompson, J. Liebert.
Astrophys. J., Vol. 321, No. 2, p. 745–754 (1987).

Spectroscopic observations of the X–ray point source no. 83 in the Columbia Astrophysics Laboratory Einstein survey of the LMC have been accumulated over 4 yr. The optical spectrum shows no stellar absorption features, but only emission lines typical of an accretion disk. Radial velocity measurements reveal a small velocity variation modulated at 0^d93, which appears to be the orbital period. No strong constraints can be put on the mass of the collapsed object, but its companion must be a low–mass, evolved star. Evidence for a precessing disk with a possible period of 69 days is presented.

117.252   The eclipses of cataclysmic variables. II. U Geminorum.
E.–H. Zhang, E. L. Robinson.
Astrophys. J., Vol. 321, No. 2, p. 813–821 (1987).

U Gem is an eclipsing dwarf nova with an orbital period of 4^h15^m. The authors have obtained high–speed, multicolor photometric observations of U Gem in its quiescent state. A light synthesis program is used to derive the properties of U Gem from its eclipses. The authors find $i = 69°7 \pm 0.7$, $M_1 = 1.12 \pm 0.13\, M_\odot$, and $M_2 = 0.53 \pm 0.06\, M_\odot$. The radial temperature distribution across the accretion disk in U Gem shows that the disk is a hollow ring around the white dwarf with $R_{out} = (0.30 \pm 0.04)$ and $R_{in} = (0.12 \pm 0.05)a$, where a is the separation of the two stars. The temperature of the ring is 4800 ± 300K.

117.253   The inner beams of SS 433.
J. D. Romney, R. T. Schilizzi, I. Fejes, R. E. Spencer.
Astrophys. J., Vol. 321, No. 2, p. 822–831 (1987).

The authors report VLBI observations of SS 433 at eight epochs in 1980 and 1981. The images derived are in excellent agreement with predictions of the "kinematic model" which posits ballistic motions at mildly relativistic velocity along a cone

of directions swept out in a 164 day cycle. By fitting some of the results to the predicted trajectories, the authors derive a distance to SS 433 of 5.0 ± 0.5 kpc. The maps exhibit numerous discrete features which appear to be ejected from the center of SS 433 in association with flares in the flux density.

117.254   Photometric study of V361 Lyr: a "hot spot" between the components of a binary system?
G. A. Richter, I. L. Andronov.
Mitt. Veränderliche Sterne, Band 11, Heft 1, p. 27–33 (1986).

The variable star V361 Lyr was investigated on 328 Sonneberg plates, taken between 1963 and 1985. Times of extrema were derived for 7 seasons of observations, and an upper limit of $|\dot{P}/P| < 8 \times 10^{-8}\mathrm{yr}^{-1}$ was derived, four times more than the theoretically predicted value. A model is proposed, intermediate between the models of W Ursae Maioris and cataclysmic variables.

117.255   GX 9 + 9.
IAU Circ., No. 4478 (1987).

117.256   V818 Scorpii.
IAU Circ., Nos. 4485, 4489 (1987).

117.257   CH Cygni.
IAU Circ., No. 4491 (1987).

117.258   EXO 023432–5232.3.
IAU Circ., No. 4491 (1987).

117.259   TV Columbae.
IAU Circ., No. 4508 (1987).

117.260   VY Aquarii.
Yamamoto Circ., No. 2087 (1987). In Japanese.

117.261   VY Aquarii.
Br. Astron. Assoc. Circ., Nos. 669, 670 (1987).

117.262   Extended dust emission around R Aquarii observed with the JPL IR array camera.
H. E. Schwarz, C. Aspin, M. Hanner, J. Zarnecki.
Infrared astronomy with arrays, p. 312–315 (1987). – See Abstr. 012.070.

The authors present the discovery of spatially extended IR emission at 3.45 μm associated with the R Aquarii system. The emission extends over 30″ and its morphology is essentially circularly symmetrical about the central object. This contrasts with the complex structure previously observed in the UV, optical and radio wavebands. This result indicates the presence of dust at distances of up to 0.02 pc from the symbiotic star.

117.263   Kataklysmische Veränderliche. II. Modellvorstellungen und Zustandsgrößen.
G. A. Richter.
Sterne, 63. Band, Heft 6, p. 334–343 (1987).

117.264   Photometric behaviour of AM Her in 1984–85.
V. P. Smykov, L. I. Shakun.
Astron. Tsirk., No. 1461, p. 3–4 (1986). In Russian.

117.265   RX Cassiopeiae. New elements.
D. Ya. Martynov.
Astron. Tsirk., No. 1463, p. 3–4 (1986). In Russian.

A new quadratic formula for minima of RX Cas is determined from observations 1983–1986.

117.266   Parameters of the hot components of symbiotic stars.
A. P. Ipatov, B. F. Yudin.
Astron. Tsirk., No. 1471, p. 4–6 (1986). In Russian.

Temperatures and bolometric fluxes were estimated for the hot components of several symbiotic stars. These results are based on two models of the hot source. The first model uses a black body and the second one – the main sequence star as a source of hot radiation.

117.267 Search for periods of infrared variability of symbiotic stars.
O. G. Taranova.
Astron. Tsirk., No. 1473, p. 7 – 8 (1986). In Russian.

For ten symbiotic stars the variability periods are found to be $20 – 50^d$.

117.268 Parameters of the X–ray binary system A0620–00 ≡ V616 Mon.
E. A. Karitskaya, N. G. Bochkarev.
Astron. Tsirk., No. 1477, p. 1 – 3 (1987). In Russian.

The area of possible parameter values of A0620–00 = V616 Mon in the model of a tidally distorted K–star with partial eclipse by the outer dark part of the accretion disk has been obtained. The main conclusions are: the 2nd component has the mass $3\,M_\odot \leqslant M_x \leqslant 5.3\,M_\odot$, which confirms the possible existence of a black hole in this system; the radius of the accretion disk R_d is variable, the disk often almost fills the Roche lobe. The system inclination is $75° \leqslant i \leqslant 83°$.

117.269 Periodic light variations of the yellow symbiotic star He2–467.
V. P. Arkhipova, R. I. Noskova.
Astron. Tsirk., No. 1478, p. 1 – 3 (1987). In Russian.

The light variations of the yellow, symbiotic star He2–467 were found to be periodic, $P \cong 500^d$. The light amplitude depends essentially on the wavelengths. In the U–band it is about $1^m.8$. He2–467 may be a binary system containing gas in any form. The visibility of the gaseous component varies with the period shown.

117.270 Photographic observations of HZ Herculis in 1985.
V. P. Smykov, L. I. Shakun.
Astron. Tsirk., No. 1479, p. 1 – 3 (1987). In Russian.

The HZ Her photographic light curve is considered. In spite of the anomalous behaviour of the X–ray emission marked changes of the reflection effect did not appear. A narrow secondary minimum of $0^d.07$ width is observed and the increase of brightness on the phases $0.07 – 0.11$ of the orbital period, independent of the phase of the 35–day cycle.

117.271 CH Cygni.
L. N. Berdnikov.
Astron. Tsirk., No. 1485, p. 8 (1987). In Russian.

In summer 1986 photoelectric observations of CH Cyg show the deep minimum, which began in 1984, to be continuing.

117.272 Precession and the long–time magnetic variables.
H. Lehmann.
Astron. Nachr., Vol. 308, No. 6, p. 333 – 341 (1987).

The model of forced precession of a star gravitationally influenced by a companion is tested for a small group of stars having time–scales of magnetic changes in the order of some years up to some decades. Results show that the observed secular periods cannot be reduced to a common time–scale of regular precession. The main difficulty is the evident discrepancy between the oblateness of the star required by the model and that attainable from rotational or magnetic flattening. Only the observed behaviour of the star 52 Her could be interpreted as a precessional motion.

117.273 Photoelectric photometry of V541 Cassiopeiae.
R.–x. Zhang, J.–t. Zhang, Q.–s. Li, D.–s. Zhai.
Chin. Astron. Astrophys., Vol. 11, No. 3, p. 237 – 243 (1987). English translation of Acta Astron. Sin., Vol. 28, No. 2, p. 131 – 138 (1987).

The authors give the results of two–colour photoelectric photometry of V541 Cas. Using the Wilson–Devinney method, they obtained a photometric solution which showed the system to be a detached binary, with a mass ratio 2.083, and the less massive component filling its Roche lobe more closely.

117.274 Polarization measurement and study of thirty RS CVn stars.
X.–f. Liu, H.–s. Tan.
Chin. Astron. Astrophys., Vol. 11, No. 3, p. 244 – 262 (1987). English translation of Acta Astron. Sin., Vol. 28, No. 2, p. 139 – 159 (1987).

The authors made polarization measurement on 30 RS CVn systems in July – December 1984, of which 26 systems were measured for the first time. The results are given. They show that for most RS CVn systems, the optical linear polarization is weak, generally below 0.45%, averaging about 0.20%.

117.275 Cataclysmic variables. III. AC Cancri.
E.–h. Zhang.
Chin. Astron. Astrophys., Vol. 11, No. 4, p. 320 – 327 (1987). English translation of Acta Astrophys. Sin., Vol. 7, No. 4, p. 245 – 254 (1987).

AC Cnc is a nova–like, eclipsing binary of period 7^h13^m. A photometric solution gives inclination $i = 74.5° \pm 0.8°$, mass of white dwarf $M_1 = 0.74 \pm 0.07\,M_\odot$, mass of the late–type companion, $M_2 = 0.97 \pm 0.8\,M_\odot$. Rate of mass transfer from the late–type star to the white dwarf is $\sim 7 \times 10^{-9}\,M_\odot/\text{yr}$. The distance of AC Cnc is $\sim 500 \pm 100$ pc.

117.276 Cygnus X–3: is there convincing evidence for new physics?
M. L. Cherry.
Physics of the superconducting supercollider, p. 640 – 654 (1987). Abstr. in Phys. Abstr., Vol. 91, No. 1320, Entry 10620 (1988). – See Abstr. 012.080.

117.277 Infrared radiation of RS CVn systems.
G. Wang, J.–y. Hu, Z.–y. Qian, X. Zhou.
Chin. Astron. Astrophys., Vol. 11, No. 4, p. 328 – 336 (1987). English translation of Acta Astrophys. Sin., Vol. 7, No. 4, p. 255 – 265 (1987).

The authors give the results of J, H, K photometry on 41 RS CVn systems and the data on 40 RS CVn systems identified with IRAS point sources. For those systems in which the components have individual spectral types, the authors discuss their infrared color excess. They found only very few systems with infrared excesses. For systen having fluxes at 12 and 25 μm and K magnitudes, the authors plotted a color–color diagram and found the majority of points to be lying close to the black body line.

117.278 The search for the elusive companion of EG Andromedae.
J. E. Pesce, R. E. Stencel, N. A. Oliversen.
Publ. Astron. Soc. Pac., Vol. 99, No. 621, p. 1178 – 1183 (1987).

The authors report observations at opposite quadratures of the interacting symbiotic binary EG And (HD 4174, Period $= 470^d$). After correcting for absolute motion at the system, it appears, surprisingly, that many of the nebular lines arise from material that moves with the red giant star. This fact is used to interpret the observed complex line profiles of C IV and He II in the object.

117.279 A discussion of the Roche model.
L. Liu.
Acta Astrophys. Sin., Vol. 7, No. 3, p. 169 – 176 (1987). In Chinese.

In this paper, the variation of Roche's critical equipotential surface (namely the surface of zero velocity corresponding to the inner Lagrangian point, the constant of Jacobian integral being $C = C_2$) in some close binary systems is discussed.

117.280 The high–energy X–ray spectrum of black hole candidate GX 339–4 during a transition.
J. F. Dolan, C. J. Crannell, B. R. Dennis, L. E. Orwig.
Astrophys. J., Vol. 322, No. 1, p. 324 – 328 (1987).

The X–ray emitting system GX 339–4 contains one of the prime candidates for a stellar mass–sized black hole. The authors report here the first observations of the $E > 20$ keV spectrum of

GX 339–4 during a transition between luminosity states. The hard spectral state is the lower luminosity state of the system. GX 339–4 has a power–law spectrum above 20 keV which pivots during transitions between distinct luminosity states.

117.281 Search for X–ray emission from the radio lobes of Scorpius X–1.
B. Geldzahler, P. Hertz.
Astrophys. J., Vol. 322, No. 1, p. 342 – 348 (1987).

Images obtained with the low–energy imaging telescope on board EXOSAT have been searched for X–ray emission from the radio lobes of Sco X–1. No significant additional X–ray flux from the radio lobes can be detected above the background. The 3 σ upper limit is less than 0.7 μJy for the northeast radio lobe and less than 1.0 μJy for the southwest radio lobe. This eliminates the embedded source model of Kundt and Gopal–Krishna as a viable model of the radio emission.

117.282 A flare event on the long–period RS Canum Venaticorum system IM Pegasi.
D. L. Buzasi, L. W. Ramsey, D. P. Huenemoerder.
Astrophys. J., Vol. 322, No. 1, p. 353 – 359 (1987).

During a monitoring program of the visible and ultraviolet spectra of the long–period RS CVn system IM Pegasi a flare event was detected in the short–wavelength region with IUE. This low–resolution spectrum showed enhancements of up to a factor of 5 in some emission lines. Visible spectra obtained some hours earlier failed to show any indication of this event. Emission fluxes of both the quiescent state and flare event were used to construct models of the density and temperature variation with height. These models reveal a downward shift of the transition region during the flare.

117.283 The discovery of near–infrared polarized cyclotron emission in the intermediate polar BG Canis Minoris.
S. C. West, G. Berriman, G. D. Schmidt.
Astrophys. J., Lett. Ed., Vol. 322, No. 1, p. L35 – L39 (1987).

BG CMi is the only intermediate polar known to be magnetic. New observations reported here reveal that its wavelength dependence of circular polarization increases rapidly into the infrared, from $V_I = -0.25\% \pm 0.06\%$ in the optical red to $V_J = -1.74\% \pm 0.26\%$ (1.25 μm). This dramatic rise, which may even continue to a level of –4% at H (1.5 μm), is the first conclusive identification of cyclotron emission in an intermediate polar, and confirms that the long–held basic model of a magnetic accreting white dwarf is correct. The field in the cyclotron emission region of BG CMi is likely in the range ~5 – 10 MG.

117.284 The ultraviolet spectrum of Beta Lyrae.
P. A. Mazzali.
Astrophys. J., Suppl. Ser., Vol. 65, No. 4, p. 695 – 720 (1987).

A complete list of line identifications for the high–dispersion UV spectrum of β Lyrae observed with the IUE (1225 – 3125 Å) is presented. The main spectral features are P Cygni profiles of resonance lines of high ionization states typical of stellar winds, some of which are combined with a broad emission feature due probably to an accretion disk; P Cygni profiles due to moderately ionized iron group atoms (mostly Fe III, Ni III); and absorption lines of elements in lower ionization stages. The absorption spectrum is very rich in Fe II lines.

117.285 On the equilibrium of Hercules X–1.
V. M. Lipunov.
Sov. Astron., Vol. 31, No. 2, p. 167 – 169 (1987). English translation of 43.117.083.

117.286 Long–period optical variability of the Cygnus X–1 system.
J. C. Kemp, E. A. Karitskaya, M. I. Kumsiashvili,
V. M. Lyutyj, T. S. Khruzina, A. M. Cherepashchuk.
Sov. Astron., Vol. 31, No. 2, p. 170 – 179 (1987). English translation of 43.117.084.

117.287 Parameters of the X–ray binary system SK 160 = 4U 0115–73 (SMC X–1).
T. S. Khruzina, A. M. Cherepashchuk.
Sov. Astron., Vol. 31, No. 2, p. 180 – 187 (1987). English translation of 43.117.085.

117.288 Spectrum of gravitational radiation of binary systems.
V. M. Lipunov, K. A. Postnov.
Sov. Astron., Vol. 31, No. 2, p. 228 – 230 (1987). English translation of 43.117.086.

117.289 On a model of the symbiotic star AG Dra.
Kh. Mikajlov, L. Luud.
Astrofizika, Tom 27, Vyp. 2, p. 219 – 230 (1987). In Russian. English translation in Astrophysics, Vol. 27, No. 2.

The radial velocities of the He II λ4686 line from the Shemakha Observatory indicate that AG Dra is a spectroscopic binary whose spectra of both components are observable. The parameters of the system have been determined. A model of a white dwarf accreting K giant's wind is proposed. The authors have adopted a gK star with the mass loss rate $5 \times 10^{-8} M_{\odot}$/yr and a white dwarf with an approximate mass of $1 - 1.2 \, M_{\odot}$ as the most adequate case. The ultraviolet flux is radiated by a small disk and the X–ray flux from the polar areas of the disk.

117.290 Quasi–periodic light variations of dwarf nova SS Aurigae at quiescence.
G. G. Tovmasyan.
Astrofizika, Tom 27, Vyp. 2, p. 231 – 236 (1987). In Russian. English translation in Astrophysics, Vol. 27, No. 2.

The results of $UBVRI$ photoelectric observations of dwarf nova SS Aur are presented. It has been shown that there are quasi–periodic variations in the light curve of the order of 20 – 30 minutes at quiescence. These variations have significant amplitudes at shorter wavelengths. At U and B bands the amplitudes of variations have been $\gtrsim 0^{m}.5$ and did not depend on the general luminosity of the system. It has been suggested that disk instability or shock waves at hot spots are responsible for the observed quasi–periodic light variations.

117.291 RS Canum Venaticorum: note on the minimum reported in this Bulletin.
R. Diethelm.
BBSAG Bull., No. 84, p. 6 (1987).

117.292 Photoelectric observations of EX Hydrae during the 1986 July – August outburst.
I. Bond, R. V. Freeth, B. F. Marino, W. S. G. Walker.
Inf. Bull. Variable Stars, No. 3037, 3 pp. (1987).

117.293 1986 light curve of V711 Tau.
M. V. Mekkaden.
Inf. Bull. Variable Stars, No. 3042, 3 pp. (1987).

117.294 Photometry during the end–phase of a spot cycle in II Peg.
M. V. Mekkaden.
Inf. Bull. Variable Stars, No. 3043, 3 pp. (1987).

117.295 Optical behaviour of the X–ray source EXO 020528 + 1454.8 in the season 1986/87.
W. Götz.
Inf. Bull. Variable Stars, No. 3045, 3 pp. (1987).

117.296 Request for VRI photometry of the RS CVn type binary HD 26337 during 15. through 30. December 1987.
K. G. Strassmeier.
Inf. Bull. Variable Stars, No. 3049, 3 pp. (1987).

117.297 B, V photometry of ER Vulpeculae.
M. C. Akan, C. Ibanoglu, Z. Tunca, S. Evren,
V. Keskin.
Inf. Bull. Variable Stars, No. 3059, 3 pp. (1987).

117.298 The 1986 light curves of UV Piscium.
V. Keskin, M. C. Akan, C. Ibanoglu, Z. Tunca,
S. Evren.
Inf. Bull. Variable Stars, No. 3060, 2 pp. (1987).

117.299 Optical behaviour of AT Cancri in the season 1986/87.
W. Götz.
Inf. Bull. Variable Stars, No. 3066, 3 pp. (1987).

117.300 Optical behaviour of the polar ST Leonis Minoris = CW 1103 + 254 in the season 1986/87.
W. Götz.
Inf. Bull. Variable Stars, No. 3067, 3 pp. (1987).

117.301 Photometric observations of the AM Herculis system MR Serpentis = PG 1550 + 191.
C. La Dous, R. Schoembs.
Inf. Bull. Variable Stars, No. 3068, 4 pp. (1987).

117.302 Photoelectric observations of the symbiotic binary EG Andromedae.
R. Luthardt.
Inf. Bull. Variable Stars, No. 3075, 2 pp. (1987).

117.303 Let's forget DO Dra.
J. Patterson, N. Eisenman.
Inf. Bull. Variable Stars, No. 3079, 3 pp. (1987).

117.304 New UBV photoelectric observations of GZ And.
X.-f. Liu, J. Yang, H.-s. Tan.
Inf. Bull. Variable Stars, No. 3080, 3 pp. (1987).

117.305 Long–term light–curve of the cataclysmic binary V425 Cassiopeiae.
W. Wenzel.
Inf. Bull. Variable Stars, No. 3086, 2 pp. (1987).

117.306 Light changes in the primary eclipse of RT Lac.
V. Keskin, S. Evren, C. Ibanoglu, Z. Tunca,
M. C. Akan.
Inf. Bull. Variable Stars, No. 3087, 3 pp. (1987).

117.307 II Pegasi: evidence of evolving star spot regions.
P. T. Boyd, K. R. Garlow, E. F. Guinan,
G. P. McCook, J. P. McMullin, S. W. Wacker.
Inf. Bull. Variable Stars, No. 3089, 7 pp. (1987).

117.308 On the period of the W UMa star NSV 12040.
M. A. Seeds.
Inf. Bull. Variable Stars, No. 3090, 3 pp. (1987).

117.309 Additional photometric data for the X–ray source TT Arietis during 1985 – 1986.
Z. Kraicheva, A. Antov, V. Genkov.
Inf. Bull. Variable Stars, No. 3093, 4 pp. (1987).

117.310 V471 Tauri eclipse timings.
J. J. Eitter.
Inf. Bull. Variable Stars, No. 3095, 2 pp. (1987).

117.311 Near simultaneous polarimetry and phase–resolved spectroscopy of the AM Her system H0538 + 608.
P. A. Mason, J. W. Liebert, G. D. Schmidt.
Inf. Bull. Variable Stars, No. 3104, 2 pp. (1987).

117.312 AY Lyrae superoutburst photometry.
M. Szymanski, A. Udalski.
Inf. Bull. Variable Stars, No. 3105, 2 pp. (1987).

117.313 1987 amplitude changes of II Peg light curve.
J. A. Cano, R. Casas, C. Gallart, J. M. Gomez,
E. Jariod, M. Peracaula.
Inf. Bull. Variable Stars, No. 3107, 2 pp. (1987).

117.314 1986 BV light curves of BH Virginis.
M. J. Arévalo, B. Robayna, J. J. Fuensalida,
D. K. Bedford.
Inf. Bull. Variable Stars, No. 3117, 3 pp. (1987).

117.315 Photometric observations on SZ Psc.
K. Thompson.
Inf. Bull. Variable Stars, No. 3119, 2 pp. (1987).

117.316 HD 30861, a new ellipsoidal variable.
W. Verschueren, H. Hensberge, H. Schneider,
K. Pavlovski.
Inf. Bull. Variable Stars, No. 3120, 2 pp. (1987).

117.317 Optical behaviour of the polar AM Her in 1987.
W. Götz.
Inf. Bull. Variable Stars, No. 3126, 2 pp. (1987).

117.318 Constraints on models of Cygnus X–3 from high–energy gamma–ray absorption at source.
R. J. Protheroe, T. Stanev.
Astrophys. J., Vol. 322, No. 2, p. 838 – 841 (1987).
   Very high energy gamma rays have been observed from several X–ray binaries. Using Cygnus X–3 as an example, the authors discuss the interaction of high–energy gamma rays with thermal radiation from the source. They conclude that this interaction must play a significant part in the modulation of the observed IR and X–ray fluxes. Furthermore, the detection of high–energy gamma–ray radiation from Cygnus X–3 restricts the dimensions and the temperatures of the hot plasma at and around the system.

117.319 The period distribution and evolution of short–period cataclysmic variables.
B. Warner, M. Livio.
Astrophys. J., Lett. Ed., Vol. 322, No. 2, p. L95 – L98 (1987).
   The period distribution of cataclysmic variables below the orbital period gap shows an intriguing feature: the SU UMa stars and the AM Her stars separately show a tendency to cluster, but the period ranges favored by one class are avoided by the other. The authors show that this is a statistically significant effect and consider possible explanations without successfully identifying a physical cause.

117.320 On the stability of the 13.2 minute oscillation of V1223 Sagittarii.
F. Jablonski, J. E. Steiner.
Astrophys. J., Vol. 323, No. 2, p. 672 – 677 (1987).
   Partial results of a photometric program covering seven consecutive observational seasons on the intermediate polar V1223 Sgr are reported here. It is found that the period of the 13.2 minute oscillation of V1223 Sgr increased with the time derivative $\dot{P} = +2.8 \times 10^{-11}$ since its discovery. The authors also obtain an improved ephemeris for predicting the times of maximum light of the orbital modulation.

117.321 IUE observations of the dwarf nova HL Canis Majoris and the winds of cataclysmic variables.
C. W. Mauche, J. C. Raymond.
Astrophys. J., Vol. 323, No. 2, p. 690 – 713 (1987).
   In order to investigate the nature of the winds of cataclysmic variables the authors present an observational and theoretical study of the P Cygni profiles of these systems, giving particular attention to the profiles of the dwarf nova HL CMa. After presenting the IUE observations of HL CMa, the authors give the results of a synthetic spectral code, which is used to assist in the interpretation of the ultraviolet data and to place constraints on the mass accretion rate through the disk during outburst. Theoretical P Cygni profiles are developed, and profiles are presented for various inclinations, velocity laws, and mass–loss rates under the assumptions of a spherically symmetric wind originating near the white dwarf and of a constant ionization fraction in the wind. As the results concerning the mass–loss rate and the character of the P Cygni profiles depend intimately on the true ionization structure of the wind, the authors consider a

model of a shocked wind for cataclysmic variables analogous to current models of the winds of early–type stars.

117.322 Discovery of radio emission from AE Aquarii.
J. A. Bookbinder, D. Q. Lamb.
Astrophys. J., Lett. Ed., Vol. 323, No. 2, p. L131 – L135 (1987).

Using the VLA, the authors have searched for radiation at 1.4 and 4.9 GHz from six DQ Her–type cataclysmic variables: AE Aqr, FO Aqr (H2215–086), AO Psc (H2252–035), BG CMi (3A 0729 + 103), TV Col, and EX Hya. The authors report the discovery of variable radio emission from AE Aqr, with flux densities of 8–16 mJy at 4.9 GHz and 3–5 mJy at 1.4 GHz. The radio emission from AE Aqr is best explained by synchrotron emission from mildly relativistic electrons. The energy source for the radio radiation is briefly discussed.

117.323 Non–conservative evolution of close binaries: evolutionary computations for a case A binary.
H.–l. Cheng, R.–q. Huang.
Acta Astron. Sin., Vol. 28, No. 2, p. 127 – 130 (1987). In Chinese.

Detailed evolutionary computations for a case A binary of $8.5 + 4.5\,M_\odot$ are described usind the method proposed by Huang and Xie (1986). Changes in the mass loss and angular momentum loss coefficients in the process of mass transfer are discussed.

117.324 Evolutionary state of contact binaries.
Q.–y. Liu.
Acta Astron. Sin., Vol. 28, No. 3, p. 274 – 281 (1987). In Chinese.

A new method for estimating the evolutionary state of contact binary systems from observations is described. Investigations of 38 systems show that all the A subtype systems and some of the W subtype systems are evolved. Some of these systems are not in contact at zero–age.

Catalogue of cataclysmic binaries, low–mass X–ray binaries and related objects (fourth edition).
See Abstr. 002.004.

An atlas and catalogue of northern dwarf novae.
See Abstr. 002.005.

Atlas of time–resolved spectrophotometry of cataclysmic variables.
See Abstr. 002.080.

Katalog orbital'nykh ehlementov, mass i svetimostej tesnykh dvojnykh zvezd. (Catalogue of orbital elements, masses and luminosities of close binaries).
See Abstr. 002.109.

Preliminary results on atmospheric neutrinos and Cygnus X–3 in the Fréjus detector.
See Abstr. 034.033.

Rotationally–perturbed orbital elements of close binary systems.
See Abstr. 042.023.

Perturbations in close binary systems produced by the tides lagging in latitude.
See Abstr. 042.030.

Symmetric motions in the Equatorial Magnetic–Binary problem.
See Abstr. 042.067.

Dynamical effects of large–scale magnetic fields in jets.
See Abstr. 062.046.

Energy dependent delay measurements of quasi–periodic oscillations in low–mass X–ray binaries.
See Abstr. 063.026.

Formation of X–ray spectra in close binary systems. Reflection effects.
See Abstr. 063.044.

Effects of electron scattering on the oscillations of an X–ray source.
See Abstr. 063.126.

Astrophysical jets in SS433.
See Abstr. 064.016.

Accretion of matter on degenerate hydrogen–helium very low–mass dwarfs.
See Abstr. 064.034.

Accretion disks in symbiotic stars.
See Abstr. 064.056.

Reconnection–driven oscillations in dwarf nova disks.
See Abstr. 064.057.

Polarized radiation from extended magnetic polar caps.
See Abstr. 064.058.

A new formulation of the starspot model, and the consequences of starspot structure.
See Abstr. 064.059.

The formation and evolution of tidal binary systems.
See Abstr. 065.002.

Spin–up and mixing in accreting white dwarfs.
See Abstr. 065.005.

Evolution of massive stars without convective core overshooting.
See Abstr. 065.014.

Thresholds for rapid mass transfer in binary systems. I. Polytropic models.
See Abstr. 065.026.

Thermonuclear processes in accreting white dwarfs (novae, symbiotic stars, and type–I supernovae).
See Abstr. 065.029.

The core mass–radius relation for giants: a new test of stellar evolution theory.
See Abstr. 065.041.

Evolution of stellar binaries formed by tidal capture.
See Abstr. 065.049.

An evolutionary scenario for the formation of highly eccentric Be/X–ray binaries.
See Abstr. 065.050.

A dynamical instability model for the period gap of cataclysmic variable binary systems.
See Abstr. 065.058.

Merger of components in intermediate mass close binaries.
See Abstr. 065.065.

Are classical novae and dwarf novae the same systems?
See Abstr. 065.066.

Collapse of white dwarfs in low mass binary systems.
See Abstr. 065.081.

Planetary systems as ultimate or by–product of binary star formation.
See Abstr. 065.089.

Monte Carlo simulations of radio pulsars and their progenitors.
See Abstr. 065.099.

Synchronization–induced period gaps and ultra–short periods in magnetic cataclysmic binaries.
See Abstr. 065.104.

Dynamical mass transfer in cataclysmic binaries.
See Abstr. 065.105.

The evolutionary status of 4U 1820–30.
See Abstr. 065.114.

On synchronization in early–type binaries.
See Abstr. 065.115.

Magnetic activity, tides, and orbital period changes in close binaries.
See Abstr. 065.116.

Three–dimensional hydrodynamical simulations of stellar collisions. I. Equal–mass main–sequence stars.
See Abstr. 065.120.

Possible effects on the sun and close binary systems from background gravitational radiation with period 160 min.
See Abstr. 066.056.

4U 1820–30 as a potential test of the nonsymmetric gravitational theory of Moffat.
See Abstr. 066.092.

The Doppler response to gravitational waves from a binary star source.
See Abstr. 066.114.

A possible semisecular variation in orbital period of binary pulsar system PSR 1913 + 16 and Lorentz invariance of gravity.
See Abstr. 066.178.

Degenerate dwarf binaries as promising, detectable sources of gravitational radiation.
See Abstr. 066.180.

On the origin of neutron stars in globular clusters.
See Abstr. 067.004.

The X–ray radiation mechanism of the compact (neutron) binary stars.
See Abstr. 067.010.

Neutron star formation in theoretical supernovae – low mass stars and white dwarfs.
See Abstr. 067.014.

Nonradial g–mode oscillations in X–ray bursting neutron stars.
See Abstr. 067.037.

Disc accretion by magnetized neutron stars: a reassessment of the torque.
See Abstr. 067.062.

Power spectra of quasi–periodic oscillations in luminous X–ray stars.
See Abstr. 067.063.

Joint evolution of normal and magnetized compact stars in low–mass binary systems. Analytical description of the degenerate component evolution.
See Abstr. 067.065.

Astrophysical black holes.
See Abstr. 067.079.

Gamma–ray pulsar model.
See Abstr. 067.081.

Neutron star precession and the dynamics of the superfluid interior.
See Abstr. 067.083.

Thick accretion disks: theory vs. observations.
See Abstr. 067.084.

An estimate of the event rate for neutron star binary coalescence and its implications for gravitational wave detection.
See Abstr. 067.088.

The influence of external magnetic fields on the structure of thin accretion disks.
See Abstr. 067.118.

X–ray polarizations from accreting strongly magnetized neutron stars: case studies for the X–ray pulsars 4U 1626–67 and Hercules X–1.
See Abstr. 067.124.

Origin of millisecond pulsars.
See Abstr. 067.126.

Formation of isolated millisecond pulsars in globular clusters.
See Abstr. 067.127.

Radiation anisotropy of the accretion disc around a black hole, heating of the optical component and parameters of the binary system LMC X–3.
See Abstr. 067.133.

Sustained magnetic fields in binary millisecond pulsars.
See Abstr. 067.135.

Strong–field point–particle limit and the equations of motion in the binary pulsar.
See Abstr. 067.149.

X–ray irradiated accretion disks and bimodal states.
See Abstr. 067.170.

The spectra of accretion disks and their application to low–mass X–ray binaries.
See Abstr. 067.171.

Tidal flows in Saturn's rings and in thin disks in binary stellar systems.
See Abstr. 100.028.

An atlas of identification charts for Luyten common proper motion stars with probable degenerate components.
See Abstr. 111.014.

Galactic distribution, kinematics, locations in clusters and H–R diagrams, and duplicity of Be stars.
See Abstr. 112.109.

Variable dust emission from Wolf–Rayet stars.
See Abstr. 112.118.

On the output of energy, gas and dust by Wolf–Rayet stellar winds.
See Abstr. 112.119.

Time variability of Gamma Cassiopeiae in X–rays.
See Abstr. 112.145.

Photometric observations of the Be/X–ray binary system 2S0114 + 650 = LSI + 65°010.
See Abstr. 112.152.

Fast transient X–rays from flare stars and RS CVn binaries.
See Abstr. 112.160.

The photometric variability of the extremely hydrogen–deficient stars KS Persei and LS II + 33°5.
See Abstr. 113.016.

Precision photometric monitoring of southern variable Wolf–Rayet stars with a comment on the overall continuum variability of WR stars.
See Abstr. 113.018.

Which photometric period for WR 16?
See Abstr. 113.027.

Near–infrared excesses of barium stars.
See Abstr. 113.031.

$U–B$ and $B–V$–diagram for hydrogen gas with $10^6 cm^{-3}$ electron density.
See Abstr. 113.033.

The search for BY Dra–type variability in red dwarf stars.
See Abstr. 113.056.

Stellar angular dimensions obtained by the lunar occultation technique.
See Abstr. 115.008.

The Einstein view of the Wolf–Rayet stars.
See Abstr. 115.012.

Late–type supergiant absolute magnitudes from UV binaries.
See Abstr. 115.014.

Phase dependent polarization variations of southern galactic WR + O systems.
See Abstr. 116.002.

Chromospherically active stars. III. HD 26337 = EI Eri: an RS CVn candidate for the Doppler–imaging technique.
See Abstr. 116.019.

Flux density and polarization observations of Hipparcos radio stars.
See Abstr. 116.034.

Simultaneous EXOSAT and VLA observations of active cool binaries VW Cep and XY Leo. A flare in VW Cep.
See Abstr. 116.045.

Discovery of optical circular polarization in Lambda Andromedae.
See Abstr. 116.048.

A microwave survey of southern active stars.
See Abstr. 116.059.

Further VLA observations of hydrogen deficient stars.
See Abstr. 116.062.

Chromospheric–coronal activity at saturated levels.
See Abstr. 116.064.

On the possible nature of RS CVn–type star spot activity.
See Abstr. 116.065.

Polarization variability among Wolf–Rayet stars. I. Linear polarization of a complete sample of southern Galactic WC stars.
See Abstr. 116.071.

Polarization variability among Wolf–Rayet stars. II. Linear polarization of a complete sample of southern galactic WN stars.
See Abstr. 116.072.

Multiple stars: anathemas or friends?
See Abstr. 118.005.

Up–to–date parameters of the eclipsing triple system IU Aur.
See Abstr. 118.007.

RS CVn–binary RW Com: a possible three–body system.
See Abstr. 118.032.

Eclipsing binary stars.
See Abstr. 119.001.

A photometric study of the eclipsing binary 68 u Her.
See Abstr. 119.007.

FS Lupi: a contact binary in poor thermal contact.
See Abstr. 119.010.

The new light curves and period study of AK Herculis.
See Abstr. 119.012.

Photometric solutions and the rate of mass transfer in U Sagittae.
See Abstr. 119.015.

The distortions in the light curves of MM Herculis.
See Abstr. 119.020.

The peculiar Be star V644 Mon = HD 51480 as an interacting binary.
See Abstr. 119.030.

Ephemerides of eclipsing binaries among cataclysmic variables for the year 1988.
See Abstr. 119.038.

The light curve analysis of MM Herculis.
See Abstr. 119.046.

The division of eclipsing binary stars according to the stages of filling up the Roche lobe.
See Abstr. 119.059.

The light curve variations in UV Piscium.
See Abstr. 119.060.

The light curve analysis of SZ Piscium.
See Abstr. 119.061.

Distribution of mass ratios in spectroscopic binaries.
See Abstr. 120.001.

Low–mass star formation in the high galactic latitude dark cloud L 1642.
See Abstr. 121.002.

Fast scanning slit near IR photometry of T Tauri stars in close binaries: Elias 22 and Chamaeleon.
See Abstr. 121.042.

The 160 min period in RR Lyr stars in globular clusters and highly evolved close binary systems.
See Abstr. 122.028.

The reclassification of the supposed dwarf nova V1285 Cygni as a semiregular variable.
See Abstr. 122.050.

The outbursts of classical and recurrent novae.
See Abstr. 124.004.

Orbital eccentricity in classical novae.
See Abstr. 124.005.

What does an erupting nova do to its red dwarf companion?
See Abstr. 124.009.

RZ Leonis.
See Abstr. 124.012.

An ancient planetary nebula surrounding the old nova GK Persei.
See Abstr. 124.231.

High–resolution radio observations of W50, the remnant associated with SS 433.
See Abstr. 125.065.

Binary systems as supernova progenitors (some frequency estimates).
See Abstr. 125.067.

An interacting binary model for SN 1987A.
See Abstr. 125.236.

The disruption of a light neutron star in an ultra–close binary and the second neutrino burst from SN 1987A.
See Abstr. 125.241.

The nature of the companion of SN 1987A.
See Abstr. 125.244.

Accreting neutron stars.
See Abstr. 126.026.

Dynamical mass determinations for the white dwarf components of HZ 9 and Case 1.
See Abstr. 126.052.

Millisecond X–ray binary pulsars.
See Abstr. 126.070.

An upper limit to the space density of short–period, noninteracting binary white dwarfs.
See Abstr. 126.095.

The (C III λ1909/Si III λ1892) ratio as a diagnostic for planetary nebulae and symbiotic stars.
See Abstr. 134.018.

Close–binary central stars of planetary nebulae.
See Abstr. 134.035.

PC 11: symbiotic star or planetary nebula?
See Abstr. 134.039.

Herculis X–1: results and interpretation.
See Abstr. 142.003.

A hard X–ray observation of Cyg X–1 in 1985.
See Abstr. 142.004.

The complex cross–spectra of Cygnus X–2 and GX 5–1.
See Abstr. 142.021.

A rapid change of the Hercules X–1 pulse profile and high–state duration.
See Abstr. 142.022.

Radio–frequency observations of galactic X–ray sources. II.
See Abstr. 142.033.

EXOSAT observations of a giant X–ray burst in XB 1905+000.
See Abstr. 142.035.

The discovery of 7 and 24 – 28 Hertz quasi–periodic oscillations in the X–ray flux of GX 17+2 (4U 1813–14).
See Abstr. 142.042.

Multipeaked X–ray bursts from 4U/MXB 1636–53: evidence against burst–induced accretion disk coronae.
See Abstr. 142.043.

The X–ray sources of the galactic bulge, X–ray bursters and X–ray sources in globular clusters.
See Abstr. 142.044.

X–ray pulsars.
See Abstr. 142.045.

EXO 033319–2554.2.
See Abstr. 142.047.

0.65–second oscillation at the peak of an X–ray burst from X 1608–522.
See Abstr. 142.049.

EXOSAT observations of 4U 1705–44: type I bursts and persistent emission.
See Abstr. 142.052.

The properties of bursts with short recurrence times from the transient X–ray source EXO 0748–676.
See Abstr. 142.053.

Neutron stars and gamma–rays.
See Abstr. 143.001.

Cygnus X–3 and other ultra–high–energy γ–ray sources.
See Abstr. 143.005.

Possible explanation of the γ–ray light curve and time variability in Cygnus X–3.
See Abstr. 143.020.

Very high energy gamma–ray binary stars.
See Abstr. 143.025.

A search for TeV gamma rays from the X–ray binary 1E 2259 + 586.
See Abstr. 143.031.

Long–term gamma–ray spectral variability of Cygnus X–1.
See Abstr. 143.043.

Cygnus X–3 observation in gamma–ray energy range $> 10^{14}$eV.
See Abstr. 143.045.

Sporadic and periodic 10 – 1000 TeV gamma rays from Cygnus X–3.
See Abstr. 143.050.

Underground search for muons correlated with Cygnus X–3.
See Abstr. 144.063.

Supersymmetric cosmic accelerators: fluxes at Earth and companion stability.
See Abstr. 144.066.

Radiation from Cygnus X–3 and the related muon puzzle – an explanation.
See Abstr. 144.071.

Mass limits on particles from pulsed sources: how reliable are they?
See Abstr. 144.077.

The tidally circularized binaries in open clusters: a new clock for age determination.
See Abstr. 153.030.

The globular cluster population of X–ray binaries.
See Abstr. 154.001.

X–ray and UV observations of ω Centauri with EXOSAT.
See Abstr. 154.008.

White dwarfs and cataclysmic variables in the globular cluster M71.
See Abstr. 154.032.

The low–luminosity X–ray sources in Omega Centauri.
See Abstr. 154.076.

The galactic distribution of Wolf–Rayet stars.
See Abstr. 155.009.

Binarias masivas en las Nubes de Magallanes.
See Abstr. 156.036.

Population of accreting neutron stars in external galaxies.
See Abstr. 157.001.

The spatial distribution and population of novae in M31.
See Abstr. 157.081.

Activity in galaxy nuclei and the SS433 phenomenon.
See Abstr. 158.057.

118 Visual Binaries, Multiple Stars, Astrometric Binaries

118.001 Measurements of visual double stars made at Pic du Midi and at Nice.
P. Couteau.
Astron. Astrophys., Suppl. Ser., Vol. 70, No. 2, p. 193 – 199 (1987). In French.

Table I gives 50 visual measures of 37 very close binaries observed with the 2–m telescope at Pic du Midi. Table II gives 601 measures of 245 close binaries observed with the 74–cm and 50–cm refractors at Nice. All measurements were made by micrometer with illuminated wires. At the 2–m telescope, binaries as close as 0.″07 are separated, using magnifying powers from 2500 to 5000.

118.002 Visual binary systems in the solar neighbourhood: the mass–ratio distribution.
M. A. Giannuzzi.
Astrophys. Space Sci., Vol. 135, No. 2, p. 245 – 252 (1987).

Nearby visual binaries, with both components on the main sequence, have been considered in order to obtain information about the distribution of their mass ratios. These systems have their primary components ranging from A0 to G9. The data have been corrected for selection effects and the differences ΔV of the visual magnitudes have been transformed into mass–ratio values. The frequency distribution of the mass ratios appears to be bimodal, with a peak around unity and a maximum at about 0.25. It is suggested that this feature may be indicative of different mechanisms of formation for wide binaries.

118.003 Micrometer measurements of visual double stars (3[rd]list).
M. Scardia.
Astron. Nachr., Vol. 308, No. 4, p. 271 – 281 (1987). In French.

The results of 521 micrometric measurements of 140 binary stars are given. The measurements have been made at Brera–Merate and Collurania (Teramo) observatories during the period February 1984 – December 1985 with two refractors of 23 cm and 40 cm, respectively.

118.004 Have circumstellar envelopes been detected around nearby M–dwarfs?
J.–M. Mariotti, C. Perrier, F. Lacombe.
Astron. Astrophys., Vol. 182, No. 1, p. L11 – L14 (1987).

In recent papers, two groups report the discovery around nearby M dwarfs of halos at 2.2 μm, interpreted as scattering of the central starlight by dusty envelopes. The authors' observations of the same stars by infrared speckle interferometry do not reveal any significant amount of extended emission: both stars appear as point sources in the limit of accuracy currently reached by infrared speckle observations. The authors compare the available data and discuss the discrepancies in terms of variability of the transfer function.

118.005 Multiple stars: anathemas or friends?
F. C. Fekel.
Vistas Astron., Vol. 30, Part 1, p. 69 – 76 (1987).

Multiple systems are generally divided into two groups: Trapezium systems and hierarchical systems. Trapezium systems consist of three or more stars whose separations are roughly equal. These systems are expected to be dynamically unstable. In hierarchical systems successive separations increase by large factors. The multiplicity of such hierarchies may be detected by a number of different techniques including astrometry, spectroscopy, photometry and speckle interferometry. Photometrically, an eclipsing system orbiting a third star will have periodic changes in eclipse timings as a result of light travel time effects. This manifests itself as an apparent change of the eclipse pair. Such nodal precession may also be detected spectroscopically as a change in the semiamplitude of the orbit. Also, unexpectedly large and/or systematic radial velocity residuals often indicate a change in the center–of–mass velocity of the short period system due to orbital motion about a third component. Observational examples of such occurrences are discussed. Finally, preliminary theoretical work on multiple star stability and formation is discussed and compared in some cases with observations.

118.006 Dynamical states of close triple stars.
Zh. P. Anosova.
Astrophysics, Vol. 25, No. 2, p. 524 – 530 (1987). English translation of 42.118.036.

118.007 Up–to–date parameters of the eclipsing triple system IU Aur.
P. Mayer, H. Drechsel.
Astron. Astrophys., Vol. 183, No. 1, p. 61 – 65 (1987).

IU Aur is a triple system consisting of an eclipsing binary orbiting around a third stellar component. Due to the changing inclination angle of the eclipsing binary orbital plane with respect to the line of sight caused by the nodal line rotation, the amplitudes of primary and secondary minima are currently increasing. UBV light curves obtained during different epochs (including new observations made between 1983 and 1985) have been solved with the Wilson–Devinney appraoch. The inclination has been increasing with a rate of 0.42°/year during the last twenty years and will reach 90° at about 1990. Accurate system parameters are derived.

118.008 Gamma Persei – not overmassive but overluminous.
D. M. Popper, H. A. McAlister.
Astron. J., Vol. 94, No. 3, p. 700 – 711 (1987).

Measurement and analysis of the set of Michigan spectrograms of the 14.″6 binary γ Per shows that the masses of the A3 and G8 III stars are 2.0 $M_\odot$ and 3.0 $M_\odot$ rather than the abnormally large values for the types found by McLaughlin (1948), 2.8 and 4.9. The decreases are primarily due to an upward revision of the large orbital eccentricity. Speckle interferometric observations of high quality covering nine years with the components resolved are analyzed. The well–determined parallax, 0.″014, along with Bahng's (1958) evaluation of the magnitude difference between the components, leads to absolute magnitudes M_V of $+0.3$ and -1.1 for the A star and G giant, respectively, values more than a magnitude more luminous than "standard" values for the spectral types. Thus, each star appears to be in a state of rapid evolution, a situation not permitted by evolutionary theory for stars of such different mass if they have a common origin.

118.009 Micrometer observations of double stars and new pairs. XIII.
W. D. Heintz.
Astrophys. J., Suppl. Ser., Vol. 65, No. 1, p. 161 – 174 (1987).
Measures of 1300 pairs, including 117 new doubles, were obtained at Swarthmore and Cerro Tololo in the time 1984.77 – 1987.06.

118.010 Statistical investigation of visual binaries according to data of the IDS catalogue.
S. V. Vereshchagin, Z. T. Krajcheva, E. I. Popova,
A. V. Tutukov, L. R. Yungel'son.
Nauchn. Inf., Vyp. 63, p. 3 – 18 (1987). In Russian.
A statistical study of the IDS catalogue visual binaries is performed. For samples of physical pairs contained in the catalogue (FIDS – 48168 stars) and visual binaries of luminosity class V (FIDS V – 2811 stars) the authors derive distributions over observed parameters and study the correlations between them. For FIDS V sample stars they estimate the masses and mass ratios of components, as well as large semiaxes of orbits, using spectrum–mass, and spectrum–absolute magnitude relations.

118.011 Prospects of infrared prospecting for planets.
H. H. Aumann.
Nature, Vol. 328, No. 6127, p. 208 (1987).

118.012 Micrometer measurements of visual double stars obtained at the Nice and Pic du Midi Observatories.
J. F. Ling.
Astron. Astrophys., Suppl. Ser., Vol. 71, No. 1, p. 115 – 118 (1987). In French.
The author reports on 158 micrometer measurements of 62 visual double stars obtained at the Nice and Pic du Midi Observatories.

118.013 Orbital elements of 26 double stars.
P. Baize.
Astron. Astrophys., Suppl. Ser., Vol. 71, No. 1, p. 177 – 184 (1987). In French.
The orbital elements of twenty–six binaries are given (5 new, 21 revised), with the measures and their residuals, the dynamical parallax, the mass, the ephemeris, and, in Notes, the discussion of the results and other features.

118.014 Orbits of six visual binary stars.
P. Couteau.
Astron. Astrophys., Suppl. Ser., Vol. 71, No. 3, p. 569 – 574 (1987). In French.
Elements of the pairs COU 79, Φ 342, ADS 5726, COU 292, ADS 15487, COU 542 have been determined with recent interferometric and visual observations. The pairs COU 79 and Φ 342 are discussed.

118.015 Direct infrared observations of the very low mass object Gliese 623B.
D. W. McCarthy Jr., T. J. Henry.
Astrophys. J., Lett. Ed., Vol. 319, No. 2, p. L93 – L98 (1987).
A low–mass, probably stellar, companion to the nearby star Gliese 623 has been detected interferometrically at 1.25, 1.65, and 2.2 μm, and its motion has been monitored over more than one full revolution cycle. The interferometry, together with available astrometric and radial velocity measurements, yields improved

orbital elements and the first photometric measurements and mass determinations of the component stars. The indicated luminosity ($M_{bol} = 12.3$) and mass ($\sim 0.09\, M_\odot$) of Gl 623B appear inconsistent with stellar models which predict a rapid decline of luminosity with decreasing mass ($\leqslant 0.2\, M_\odot$) for stars of age $\geqslant 2$ Gyr.

118.016 Astrometric–spectroscopic binaries. II. Gamma Geminorum.
K. W. Kamper, W. R. Beardsley.
Astron. J., Vol. 94, No. 5, p. 1302 – 1308 (1987).
Using astrometric observations from the Allegheny, Yale, and Van Vleck Observatories combined with the 447 radial velocities available to them, the authors have derived a definitive astrometric–spectroscopic orbit for the bright A0 IV star γ Geminorum. The orbital period is 12.632 yr, but the unusually high eccentricity, 0.896, causes most of the rise from minimum to maximum velocity to occur over an interval of only two months. In addition, the authors find a well–determined trigonometric parallax of 0.″0326 ± 0.″0025 and a provisional orbital parallax of 0.″0300 ± 0.″0026.

118.017 ICCD speckle observations of binary stars. III. A survey for duplicity among high–velocity stars.
P. K. Lu, P. Demarque, W. van Altena, H. McAlister,
W. Hartkopf.
Astron. J., Vol. 94, No. 5, p. 1318 – 1326 (1987).
A survey program to identify binary candidates among high–velocity dwarf stars using the GSU speckle camera has been carried out. The purposes of this study are: (1) to determine the binary frequency of the halo population to provide information on the star–formation processes in the galactic halo; and (2) to eventually derive the orbital elements of the newly discovered binaries. In this paper, the authors report speckle interferometry data that have been obtained and analyzed for a sample of 182 stars. Based on these data, ten stars are found to be binary. The authors find that their data are compatible with a total frequency for high–velocity long–period doubles as large as for low–velocity stars. Distances have been estimated for the ten binary stars using their spectroscopic parallaxes and visual magnitudes. Of these ten stars, all are within 100 pc of the Sun and eight have linear separations <20 AU. Using the mass–luminosity relation and assuming circular orbits, four stars are found to have periods less than 20 yr. These ten candidates will be monitored to determine their orbital elements.

118.018 Wide stellar systems of the Trapezium type.
G. N. Salukvadze, G. Sh. Dzhavakhishvili.
Soobshch. Akad. Nauk GSSR, Tom 126, No. 1, p. 69 – 72 (1987). In Russian. Abstr. in Ref. Zh., 51. Astron., 10.51.747 (1987).

118.019 A search for sub–stellar companions around late–type dwarfs.
V. Lindsay, G. W. Marcy, K. Wilson, D. Moore.
Bull. Am. Astron. Soc., Vol. 19, No. 2, p. 714 (1987). Abstract. – See Abstr. 010.061.

118.020 A search for brown dwarf or planetary–mass companions to solar–type stars with high precision radial velocities.
B. Campbell, G. A. H. Walker, S. Yang.
Bull. Am. Astron. Soc., Vol. 19, No. 2, p. 762 (1987). Abstract. – See Abstr. 010.061.

118.021 The red–dwarf binary Σ2398.
W. D. Heintz.
Publ. Astron. Soc. Pac., Vol. 99, No. 620, p. 1084 – 1088 (1987).
Photographic data over 69 yr give a parallax of 0.″289 ± 0.″001. From a revised orbit ($P = 408$ yr, $a = 13.″88$) component masses of 0.36 and 0.30 solar masses are found. There are no periodic effects indicative of a third body.

118.022   Observations of binaries Σ 73, Σ 346AB and Kpr 23 by the speckle technique.
Y.-h. Qiu, R.-n. Lu, P. Qian.
Publ. Yunnan Obs., No. 2, p. 51 – 54 (1987). In Chinese.

118.023   Spectroscopic orbital elements and photometry of the multiple system Epsilon Hydrae.
G. A. Bakos, J. Tremko.
Contrib. Astron. Obs. Skalnaté Pleso, Vol. 16, p. 17 – 26 (1987).
In combination with older plate material, the recently obtained high–dispersion spectrograms have been used to derive the orbital elements of the AB pair of this multiple system. The new orbital elements have been compared with those already published. In addition, over the period of seven years a two–color photometry of the system has been conducted and light variations of about 0.1 mag have been observed. The derived period is 72 days.

118.024   Orbits of eleven binary stars.
A. A. Tokovinin.
Pis'ma Astron. Zh., Tom 13, No. 12, p. 1065 – 1070 (1987). In Russian. English translation in Sov. Astron. Lett., Vol. 13.
The orbits of HR 3880 = 19 Leo, ADS 7952 and Cou 1145 have been computed for the first time. The previously known orbits of ADS 7662, ADS 10624, ADS 13777 and Cou 542 were recalculated, and orbits of Cou 79, ADS 9744, ADS 10092, ADS 14893 have been corrected in order to eliminate unacceptably large discrepancy with recent interferometric observations. Dynamical parallaxes and masses are derived.

118.025   Mesures d'étoiles doubles visuelles (Nice, 1986).
J. Le Beau.
Obs. Trav., No. 12, p. 37 – 44 (1987).

118.026   The Alpha Centauri system.
D. R. Soderblom.
Mercury, Vol. 16, No. 5, p. 138 – 140 (1987).

118.027   A dim star with bright prospects.
D. Lindley.
Nature, Vol. 330, No. 6144, p. 105 (1987).
The author reports on brown dwarfs, especially on the faint companion to the white dwarf Giclas 29–38 (see also Abstr. 118.028), their evolution and their contribution to the total mass of the Galaxy.

118.028   Excess infrared radiation from a white dwarf – an orbiting brown dwarf?
B. Zuckerman, E. E. Becklin.
Nature, Vol. 330, No. 6144, p. 138 – 140 (1987).
The authors have discovered that the white dwarf star Giclas 29–38 appears to emit substantial radiation at wavelengths between 2 and 5 μm, far in excess of that expected from an extrapolation of the visual and near–infrared spectrum of the star. The characteristics are similar to those that have been calculated for substellar objects called brown dwarfs. The most natural interpretation of these observations is that there is a substellar, somewhat Jupiter–like brown dwarf in orbit around G29–38.

118.029   Stars and planetary systems.
A. V. Tutukov.
Astron. Zh., Tom 64, Vyp. 6, p. 1264 – 1268 (1987). In Russian. English translation in Sov. Astron., Vol. 31, No. 6.
The study of the distribution of single and double stars over the specific angular momentum gives the possibility of estimating probable limits for the latter, necessary for planetary system formation. The analysis of traditional methods of the giant Jupiter–type planets search around of close (0.1 – 0.4) $M_\odot$ stars leads to the conclusion about the relatively high efficiency of photometry for that aim.

118.030   Was dürfen Doppelsterne?
P. Brosche.
Sterne, 63. Band, Heft 6, p. 330 – 333 (1987).

118.031   Photoelectric UBV photometry of the components of triple stars.
Zh. P. Anosova, S. V. Sudakov.
Tr. Astron. Obs., Leningrad, Tom 41, p. 80 – 96 (1987). Uch. Zap. LGU, No. 420, Ser. Mat. Nauk, Vyp. 63. In Russian.
Photoelectric UBV photometry has been obtained for components of 30 bright and nearby ($V \lesssim 10^m$, $r \lesssim 100$ pc) triple stars for purposes of finding physically linked triple systems, their distances and the subsequent investigation of their dynamical evolution. For every component colour excesses E_{B-V} and E_{U-B} are determined and photometric parallaxes π_{ph} are calculated.

118.032   RS CVn–binary RW Com: a possible three–body system.
R. K. Srivastava.
Astrophys. Space Sci., Vol. 139, No. 2, p. 373 – 387 (1987).
A first detailed period study of the eclipsing RS CVn–binary system RW Com is presented. A new period ($P = 0^d2373455$) based on 223 minima is given. The O–C diagrams of RW Com have been presented for the first time. Period changes in different portions of the O–C diagram have been estimated. The total change in period ($\Delta P/P$) ranges from 5.5×10^{-7} to 6.4×10^{-6}. Thus, ΔP ranges from 1.3×10^{-7}d to 1.5×10^{-6}d. It is suspected that RW Com could be a three–body system. The period of variation due to a third body appears to be nearly 16 years.

118.033   Binary star statistics: the mass ratio distribution for very wide systems.
V. Trimble.
Astron. Nachr., Vol. 308, No. 6, p. 343 – 347 (1987).
The distribution of mass ratios for a sample of common proper motion (CPM) binaries is determined and compared with that of 798 visual binaries (VB's) studied earlier, in hopes of answering the question: Can the member stars of these systems have been drawn at random from the normal initial mass function for single stars? The observed distributions peak strongly toward $q = 1.0$ for both kinds of systems, but less strongly for the CPM's than for the VB's. Due allowance having been made for assorted observational selection effects, it seems quite probable that the CPM's represent the observed part of a population drawn at random from the normal IMF, while the VB's are much more difficult to interpret that way and could, perhaps, result from a formation mechanism that somewhat favors systems with roughly equal components.

118.034   Radial velocities of the components of triple stars.
Zh. P. Anosova, V. N. Sementsov, A. A. Tokovinin.
Sov. Astron., Vol. 31, No. 2, p. 220 – 221 (1987). English translation of 43.118.015.

118.035   Orbites nouvelles.
Circ. Inf., No. 103 (1987).

118.036   Etoiles doubles nouvelles.
P. Couteau, G. M. Popovic.
Circ. Inf., No. 103 (1987).

118.037   Variability of the double star HD 41824 and its multiplicity.
Z. Kviz, M. Mayor, F. Rufener.
Inf. Bull. Variable Stars, No. 3091, 4 pp. (1987).

The Third Catalogue of Nearby Stars with special emphasis on wide binaries.
See Abstr. 002.064.

The catalogue of Trapezium–Type Multiple Systems in machine–readable form.
See Abstr. 002.089.

The magnetic tape version of the LDS catalogue.
See Abstr. 002.092.

The future of high angular resolution astronomy: seeing the unseen.
See Abstr. 013.014.

Speckle–interferometry on the AZT–8 telescope at the Astronomical Observatory of the Kharkov University.
See Abstr. 034.029.

The circumstellar imaging telescope – direct detection of extra–solar planets.
See Abstr. 035.043.

In search of other planetary systems.
See Abstr. 036.015.

Infrared speckle interferometry on Calar Alto.
See Abstr. 036.119.

Imaging of low mass binary companions and circumstellar disks.
See Abstr. 036.120.

Very low mass stars.
See Abstr. 065.033.

Planetary systems and their central stars.
See Abstr. 065.088.

Planetary systems as ultimate or by–product of binary star formation.
See Abstr. 065.089.

Stars and planets.
See Abstr. 107.018.

Photographic astrometry of binary and proper–motion stars. III.
See Abstr. 111.007.

The galactic orbit of the remarkable high–velocity wide binary LDS 519.
See Abstr. 111.017.

A study of the massive O–type binary Iota Orionis.
See Abstr. 112.062.

A study of UV spectra of ζ Aur/VV Cep stars. X. Mass–loss of α Sco A from high–resolution IUE spectra of α Sco B.
See Abstr. 112.063.

Detection of X–ray emission from the young low–mass star Rossiter 137B.
See Abstr. 112.141.

Stellar angular dimensions obtained by the lunar occultation technique.
See Abstr. 115.008.

A search for brown dwarfs and late M dwarfs in the Hyades and the Pleiades.
See Abstr. 115.011.

Discovery of flare activity on BD + 3°4138 B.
See Abstr. 116.004.

Activity in late–type dwarfs. I. Walraven and Johnson photometry of flares and spot variations on Gl 867 A (= FK Aqr) in 1979.
See Abstr. 116.035.

Activity in late–type dwarfs. II. Flares and spot variations on Gl 867 A (= FK Aqr) in 1981.
See Abstr. 116.036.

Speckle interferometric measurements of binary stars: IV.
See Abstr. 117.140.

The double system HD 135421.
See Abstr. 119.002.

Search for low–mass objects. II.
See Abstr. 126.044.

51 Tauri and the Hyades distance modulus.
See Abstr. 153.029.

Small clusters or multiple stars?
See Abstr. 153.045.

The binary frequency of high–velocity field dwarfs as obtained with CCD measures.
See Abstr. 155.014.

119 Eclipsing Binaries

119.001 Eclipsing binary stars.
C. T. Bolton.
The SHIRSOG Workshop, p. 71–76 (1986). – See Abstr. 012.003.

Synoptic observations of binary stars are an essential tool for learning about the distribution and motions of circumstellar material in these systems. These observations are the only hope for obtaining the empirical estimates of mass and angular momentum loss rates in semidetached binaries which are required before further progress can be made in understanding mass exchange evolution of close binary stars.

119.002 The double system HD 135421.
P. Rovithis, H. Rovithis–Livaniou.
Astron. Astrophys., Suppl. Ser., Vol. 70, No. 1, p. 63–68 (1987).

The double visual system HD 135421 ≡ ADS 9537 consists of the eclipsing binaries BV and BW Dra. New B and V light curves for BW Dra obtained during 1982 are presented. The light curves for both members of the double visual system ADS 9537 are analysed using Frequency Domain techniques. New elements for the system are given.

119.003 Light curve variations of the eclipsing binary V367 Cygni.
M. C. Akan.
Astrophys. Space Sci., Vol. 135, No. 1, p. 157–167 (1987).

The long–period eclipsing binary star V367 Cygni has been observed photoelectrically in two colours, B and V, in 1984, 1985, and 1986. These new light curves of the system have been discussed and compared for the light–variability with the earlier ones presented by Heiser (1962). Using some of the previously published photoelectric light curves and the present ones, several primary minima times have been derived to calculate the light elements. Any attempt to obtain a photometric solution of the binary is complicated by the peculiar nature of the light curve caused by the presence of the circumstellar matter in the system. Despite this difficulty, however, some approaches are being carried out to solve the light curves which are briefly discussed.

119.004 Period changes in EI Cephei.
R. K. Srivastava.
Astrophys. Space Sci., Vol. 135, No. 2, p. 229–235 (1987).

A new period ($P = 8^{d}439422$) of the eclipsing system EI Cephei has been given, which is based on all available times of minima. Periods using Strohmeier's (1958) epoch have also been presented for the observations given by other investigators. A period based on only photoelectric minima comes out to be $8^{d}439336$, which is lesser than the earlier periods given in the literature. O–C diagrams of EI Cephei have been presented for the first time, and period variations have been estimated in different portions of the O–C diagram of EI Cephei. Strong period changes have occurred around the years 1959 and 1965. The total change in period ($\Delta P/P$) ranges from 6.7×10^{-5} to 4.3×10^{-4}. The existence of a third body in the system could not be confirmed.

119.005 Orbital elements of the eclipsing binary ζ Phe.
A. Nikolov, K. Maslev.
Astrophys. Space Sci., Vol. 135, No. 2, p. 283–286 (1987).

The elements of the orbit in the cases of elliptic orbit and the nonspherical form of the components were obtained on the basis of the published photoelectric and spectroscopic observations of the eclipsing binary system ζ Phe.

119.006 Infrared photometry of UV Piscium.
E. Antonopoulou.
Astrophys. Space Sci., Vol. 135, No. 2, p. 335–345 (1987).

The eclipsing binary system UV Psc was observed in the near–infrared (JHK), and wave–like distortions have been observed similar to those in the visible. The out–of–eclipse observations have shown a very large amount of scatter. This cannot be attributed to the existence of spots on one of the components, which can explain the wave–like distortion of the light curves, but probably to the intrinsic variability of one of them.

119.007 A photometric study of the eclipsing binary 68 u Her.
S. R. Jabbar, N. L. Jabir, H. A. Fleyeh.
Astrophys. Space Sci., Vol. 135, No. 2, p. 377–388 (1987).

The aim of this paper is to present and study the light curves (B and V) of the eclipsing binary 68 u Her from the photometric observations made at Al–Battani Observatory. New photometric elements have been obtained by a Fourier analysis of the light changes in the frequency–domain.

119.008 The BVJK light curves of the short–period eclipsing binary CG Cygni.
D. K. Bedford, J. J. Fuensalida, M. J. Arévalo.
Astron. Astrophys., Vol. 182, No. 2, p. 264–270 (1987).

BVJK light curves of CG Cygni show an apparent infrared excess ≲0.2 mag out of eclipse, assuming the spectral types of the components to be G9.5 V + K3 V. The depths of the eclipses also show an excess equal (within the errors) to that at quadrature. The source of these excesses is most probably a circum–system shell of cool material; it cannot be material at the inner Lagrangian point. A subsidiary third minimum is present in at least the J light curve; it suggests the presence of material localised in the orbital plane, but not at the Lagrangian point.

119.009 Spectrophotometric analysis of the primary component of V380 Cyg.
V. V. Leushin, G. P. Topil'skaya.
Astrophysics, Vol. 25, No. 2, p. 503–513 (1987). English translation of 42.119.078.

119.010 FS Lupi: a contact binary in poor thermal contact.
L. Milano, G. Russo, A. Terzan.
Astron. Astrophys., Vol. 183, No. 2, p. 265–273 (1987).

The UBV photoelectric lightcurves of the eclipsing binary FS Lupi, observed at ESO, La Silla, during 1982–1984, have been analyzed with a Roche–model–based lightcurve synthesis method. The system, alternatively classified in the past as an EA or EB system, results an A–type contact binary with a relatively large degree of overcontact (33%) and with a large difference in the temperatures of the components (1120K), which explains the EB–type light curve. The asymmetries and the large scatter of the observed data, when investigated in terms of deviations from the (synthetic) mean light curve, show a systematic trend which has been studied with three different methods, searching for periodicity.

119.011 The IUE spectrum of the early–type eclipsing binary μ¹Scorpii.
J. Sahade, L. G. García.
Publ. Astron. Soc. Pac., Vol. 99, No. 617, p. 617–622 (1987).

The ultraviolet spectrum of the system μ^{1}Sco at two phases of the orbital cycle is described. Evidence for nonthermal sources and for wind in the system is discussed.

119.012 The new light curves and period study of AK Herculis.
Z. Tunca, V. Keskin, M. C. Akan, S. Evren, C. Ibanoğlu.
Astrophys. Space Sci., Vol. 136, No. 1, p. 63–76 (1987).

B and V observations of the W Ursae Majoris–type eclipsing variable system AK Her were made on five nights at the Ege University Observatory. Several times of minima were obtained during the observations and the new light elements were calculated. The light-time period was found to be about 75.72 years. The light curve of the system appears to change in each cycle for both colours. The secondary minimum of the system seems to be a total eclipse with a duration of about $42^{m}5$.

119.013 Properties of the main–sequence eclipsing binary V442 Cygni.
C. H. Lacy, M. L. Frueh.
Astron. J., Vol. 94, No. 3, p. 712 – 722 (1987).

Radial velocities, line profiles, and V, R photometry of V442 Cyg (F1 + F2) are analyzed to obtain the fundamental properties of this detached main–sequence binary. The minima in the light curve (0.56 and 0.49 mag) are deep enough for definitive photometric orbits, despite the relatively small number of observations within eclipses. Comparison of the properties of these stars with theoretical models of stellar evolution shows a good match at an age of 1.3×10^9 yr and a composition of $x = 0.7$, $z = 0.03$. The primary is very near the end of its core–hydrogen–burning lifetime.

119.014 The surface–brightness anomaly in eclipsing binaries.
C. H. Lacy, M. L. Frueh, A. E. Turner.
Astron. J., Vol. 94, No. 4, p. 1035 – 1042 (1987).

Surface–brighness ratios of the eclipsing binaries V477 Cyg, TX Her, CM Lac, RR Lyn, and EE Peg are determined from new V, R light curves. Along with the previously determined values for IQ Per, these ratios are used as probes for the $B–V$ and $V–R$ surface–brightness calibrations in a region of low data density. An anomaly is found in the sense that if the primaries are assumed to lie on the calibration curve, then the secondaries are less bright than expected, or if the secondaries are assumed to lie on the calibration curve, then the primaries are brighter than expected. It does not seem possible at this time to find an unambiguous interpretation of this anomaly.

119.015 Photometric solutions and the rate of mass transfer in U Sagittae.
E. C. Olson.
Astron. J., Vol. 94, No. 4, p. 1043 – 1050 (1987).

Five–color intermediate–band photoelectric observations of the totally eclipsing Algol binary U Sagittae were obtained from 1978 to 1985. Photometric solutions are made using WINK and the Wilson–Devinney programs. The photometric mass ratio agrees well with the value found recently by Van Hamme and Wilson (1986). Small depressions in the light curves between phases 0.80 and 0.95 were produced by light extinction in a mass–transferring stream. A new method using the theory of stream structure by Lubow and Shu (1975, 1976) is used to find $M \lesssim 2 \times 10^{-7} M_\odot$ per year. An attempt is made at a quantitative interpretation of small brightness changes of the cool star in this system.

119.016 Light curve variations in ER Vulpeculae.
C. Ibanoğlu, S. Evren, Z. Tunca.
Astrophys. Space Sci., Vol. 136, No. 2, p. 225 – 229 (1987).

Photoelectric observations of the peculiar eclipsing variable ER Vul were obtained in blue and yellow light, in the 1981 and 1982 observing seasons. The light curves suffer to change in short time–intervals. The wave–like distortion superimposed on the light curves is clearly seen, but sometimes there is no indication about its existence. The migration period has been estimated roughly about eight months. Moreover, small–amplitude light fluctuations in the light curves are noticeable. These variations seem to occur randomly. When the IUE and optical observations are taken into consideration together it is strongly suggested that both of the components in the system ER Vul are too active.

119.017 Spectral investigation of the atmospheres of eclipsing binaries in the stage of mass exchange.
V. G. Karetnikov.
Stellar atmospheres, p. 96 – 103 (1987). In Russian. Abstr. in Ref. Zh., 51. Astron., 8.51.587 (1987). – See Abstr. 003.005.

119.018 V677 Centauri: overcontact and completely eclipsing but A–type or W–type?
P. M. Kilmartin, D. H. Bradstreet, R. H. Koch.
Astrophys. J., Vol. 319, No. 1, p. 334 – 339 (1987).

The first photoelectric light curves of the contact binary V677 Cen show the system to be a completely eclipsing one. The

mass and radius ratios between the component stars are far from unity. In distinction to most binaries of comparably short period, V677 Cen shows no temperature difference between the member stars, at least at the time of the present light curves. Main–sequence radius and mass values for the member stars are considered and rejected. An interpretation that describes it as a nearly limiting A type contact binary is also shown to have difficulties.

119.019 Physical characteristics of the stellar atmospheres of the eclipsing binary system TX Ursae Majoris.
V. G. Karetnikov, V. V. Kovtyukh.
Sov. Astron., Vol. 30, No. 6, p. 675 – 679 (1987). English translation of 42.119.085.

119.020 The distortions in the light curves of MM Herculis.
S. Evren.
Astrophys. Space Sci., Vol. 137, No. 1, p. 151 – 166 (1987).

The RS CVn–type eclipsing binary MM Her was observed photoelectrically in B and V colours. The light curves obtained in 1984 and 1985 are presented. It was found that the depths of the primary minima are decreased from 1983 to 1985. However, the amplitude of the wave–like distortion outside the eclipses was detected to increase since 1976. The period of migration was determined to be about 3.57 ± 0.08 years.

119.021 Photometry and spectroscopy of the eclipsing P Cygni star R 81 in the Large Magellanic Cloud.
O. Stahl, B. Wolf, F.–J. Zickgraf.
Astron. Astrophys., Vol. 184, No. 1/2, p. 193 – 200 (1987).

IUE high resolution data of the LMC P Cyg star R 81 confirm that the wind characteristics of R 81 and P Cyg are very similar. Photometric monitoring from 1982 to 1986 revealed several brightness minima with a period of 74.59 days. It is concluded that R 81 is an eclipsing binary with this period. From the photometry and the period the authors derive a preliminary mass of $\approx 33\ M_\odot$. Interestingly, the scatter around the mean light curve is only about 0.05 mag, i.e. considerably less than the variations usually quoted for P Cyg stars. Since the period of R 81 has been found only after extensive monitoring with high precision within Sterken's long–term photometry of variables program, this finding may imply that irregular variations reported for P Cyg stars on the basis of less complete data sets may in fact be due to periodic variations.

119.022 The strange case of Beta Lyrae.
J. Tomkin, D. L. Lambert.
Sky Telesc., Vol. 74, No. 4, p. 354 – 357 (1987).

119.023 Period changes in eclipsing binary stars observed by amateurs.
A. Mallama.
J. Am. Assoc. Variable Star Obs., Vol. 16, No. 1, p. 4 – 7 (1987).

Recent period changes in UU And, SW Cyg, WW Cyg, TY Del, TW Dra, TU Her, Z Per, and Y Psc have been revealed by visual times of minima obtained over the past two decades. New ephemerides are reported for these stars.

119.024 Photoelectric radial velocities of eclipsing binaries. V. Orbital elements of V643 Ori.
M. Imbert.
Astron. Astrophys., Suppl. Ser., Vol. 71, No. 1, p. 69 – 73 (1987). In French.

A spectroscopic orbit has been determined for the faint double lined eclipsing binary V643 Ori. The radial velocities were obtained with the Coravel radial velocity spectrometer. Using 35 Coravel measurements for the primary component and 31 for the secondary, the first spectroscopic orbit was calculated, yielding accurate elements. Considerations about spectral type, duration of main eclipse and consecutive loss of light and other parameters deduced from Coravel peaks, lead the author to adopt for this system a model in agreement with the whole spectroscopic and photometric data. The system so appears to be a relatively unusual association of two giant K stars.

119.025   BV photometry of β Lyrae in 1979 and 1981.
Z. Aslan, E. Derman, S. Engin, N. Yilmaz.
Astron. Astrophys., Suppl. Ser., Vol. 71, No. 3, p. 597 – 601 (1987).

Photoelectric observations in *BV* made in 1979 and 1981, together with three determinations of minima, are given. Both in the light and colour curves, changes of slope near the phases of the beginning and end of eclipses are noted. The colour curve is suspected to have two red–dips at phases 0.39 and 0.60. In the light curve there is a random scatter, of about $\pm 0^m03$ in *V*, which is attributed to irregular changes in the light of β Lyr.

119.026   *UBV* photometry of eclipsing binaries in the LMC.
T. J. Davidge.
Astron. J., Vol. 94, No. 5, p. 1169 – 1177 (1987).

UBV photoelectric observations are presented for three eclipsing binaries in the LMC: HV 2241, HV 2765, and HV 5943. Prior to solving the light curves, the photoelectric data were corrected for crowding by studying models, based on CCD observations, of the fields around the variables. The brightnesses of contaminating stars, both near the variables and in the areas selected for sky measurements, were determined by simulating aperture photometry on the modeled star fields. The light curves were solved with the Wilson–Devinney program by making a systematic search of parameter space. Preliminary absolute dimensions are computed for the components, and the evolutionary status of the systems is discussed.

119.027   Photometry of long–period Algol binaries. III. The accretion disk and mass transfer in RZ Ophiuchi.
E. C. Olson.
Astron. J., Vol. 94, No. 5, p. 1309 – 1317 (1987).

Five–color photometric observations of RZ Oph have been obtained from 1981 through 1986, in an effort to deduce the properties of the accretion disk in this long–period Algol. The partial eclipse of the disk by the cool star, and the partial occultation of the cool star by the disk, have both been observed in some detail. A simple gravitationally stratified model of the disk accounts very well for the emitted flux.

119.028   Symbiotic eclipsing binary star CI Cyg. The cold component variability.
T. S. Belyakina.
Izv. Krymskoj Astrofiz. Obs., Tom 76, p. 40 – 43 (1987). In Russian. English translation in Bull. Crimean Astrophys. Obs., Vol. 76.

It has been shown that the cold component of the CI Cyg eclipsing binary system is a variable star. The amplitude of its brightness variations is about $\Delta V \leqslant 0^m4$ within the time interval of 40 – 60 days. This component being the red giant M4 III can be attributed as SR Type variability according to GCVS classification.

119.029   A search for accretion shock in 31 Cygni.
I. A. Ahmad.
Bull. Am. Astron. Soc., Vol. 19, No. 2, p. 707 (1987). Abstract. – See Abstr. 010.061.

119.030   The peculiar Be star V644 Mon = HD 51480 as an interacting binary.
J. H. Kastner, M. J. Plavec.
Bull. Am. Astron. Soc., Vol. 19, No. 2, p. 708 (1987). Abstract. – See Abstr. 010.061.

119.031   Non–stellar ultraviolet radiation in Algol–type binaries TT Hydrae and RY Geminorum.
M. J. Plavec.
Bull. Am. Astron. Soc., Vol. 19, No. 2, p. 708 – 709 (1987). Abstract. – See Abstr. 010.061.

119.032   AK Canis Minoris.
R. M. Robb, J. D. Moffatt.
Bull. Am. Astron. Soc., Vol. 19, No. 2, p. 709 (1987). Abstract. – See Abstr. 010.061.

119.033   Light curve solutions for five eclipsing binaries in the Magellanic Clouds.
T. J. Davidge.
Bull. Am. Astron. Soc., Vol. 19, No. 2, p. 714 (1987). Abstract. – See Abstr. 010.061.

119.034   Epsilon Aurigae: pulsations and the post–eclipse (1984 – 1987) polarization and light curves.
D. J. Kraus, J. C. Kemp, G. D. Henson, M. H. Dunaway, J. L. Hopkins, P. C. Schmidtke.
Bull. Am. Astron. Soc., Vol. 19, No. 2, p. 752 (1987). Abstract. – See Abstr. 010.061.

119.035   Properties of the main–sequence eclipsing binary AY Camelopardalis.
C. H. Lacy.
Astron. J., Vol. 94, No. 6, p. 1670 – 1672 (1987).

Radial velocities and line profiles of AY Cam are analyzed, and the results are combined with previous light–curve analyses to yield absolute properties of this binary. Comparison of the properties of these stars with theoretical models of stellar evolution indicates that the primary component is significantly beyond the point of core–hydrogen exhaustion and the secondary is near the point of core–hydrogen exhaustion.

119.036   Properties of the main–sequence eclipsing binary KP Aquilae.
C. H. Lacy.
Astron. J., Vol. 94, No. 6, p. 1673 – 1674 (1987).

Radial velocities and line profiles of KP Aql (F0 + F0) are analyzed, and the results are combined with those of the light–curve analysis of Ibanoglu and Gülmen to yield absolute properties of this main–sequence binary. Comparison of the properties of these stars with theoretical models of stellar evolution indicates an age of approximately 8×10^8 yr.

119.037   Ephemerides of eclipsing binaries for the year 1988.
E. Danielkiewicz–Krośniak,
M. Kurpińska–Winiarska.
Rocznik Astronomiczny Obserwatorium Krakowskiego 1988, p. 1 – 110 (1987). – See Abstr. 046.024.

119.038   Ephemerides of eclipsing binaries among cataclysmic variables for the year 1988.
J. M. Kreiner.
Rocznik Astronomiczny Obserwatorium Krakowskiego 1988, p. 111 – 114 (1987). – See Abstr. 046.024.

119.039   Elemente des Algolsterns LU Persei.
H. Geßner.
Mitt. Veränderliche Sterne, Band 11, Heft 2, p. 38 (1987).

119.040   Neue Elemente des Algolsterns BE Persei.
H. Geßner.
Mitt. Veränderliche Sterne, Band 11, Heft 2, p. 38 – 39 (1987).

119.041   Research on the eclipsing system V366 Cyg.
J. M. Kreiner, J. Tremko.
Contrib. Astron. Obs. Skalnaté Pleso, Vol. 16, p. 191 – 206 (1987).

Photoelectric observations in the B spectral range were made of the eclipsing system V366 Cyg at the Astronomical Observatory in Skalnaté Pleso. The published minimum epochs and the newly derived minimum epochs were used to study changes of period. A secular variation of the period was found and its possible cause is discussed.

119.042   Photoelectric photometry of the eclipse of Epsilon Aurigae.
D. Chochol, J. Žižňovský.
Contrib. Astron. Obs. Skalnaté Pleso, Vol. 16, p. 207 – 211 (1987).

The U, B, V photometric observations of ε Aur during the ingress and totality phases obtained at the Skalnaté Pleso

Observatory in 1982–1983 are published. The mid–eclipse brightening is confirmed. The observations suggest that the structure of the eclipsing dusty disk is not homogeneous.

119.043   Narrow–band photoelectrical observations of the eclipsing binary CQ Cep in 1975–1976 and their possible interpretation.
T. A. Kartasheva.
Astrofiz. Issled. Izv. Spets. Astrofiz. Obs., Tom 24, p. 35–43 (1987). In Russian. English translation in Bull. Spec. Astrophys. Obs. – North Caucasus.
Detailed results of narrow–band photoelectrical observations of the eclipsing binary CQ Cephei carried out in 1975–1976 are presented. An assumption is made that the high activity of the star in 1975 was caused by the strong matter outflow from the WR–component and frequent expulsions of the outer layers of the WR–envelope.

119.044   LZ Her: an eclipsing binary star.
R. Garrido, T. Gómez, R. Peniche, J. H. Peña.
Rev. Mex. Astron. Astrofis., Vol. 14, No. 2, p. 416 (1987). Abstract. – See Abstr. 012.042.

119.045   Hα profile variations in Epsilon Aurigae.
A. Arellano Ferro, L. Parrao, A. Giménez, M. López Arroyo, D. D. Sasselov.
Rev. Mex. Astron. Astrofis., Vol. 14, No. 2, p. 418 (1987). Abstract. – See Abstr. 012.042.

119.046   The light curve analysis of MM Herculis.
S. Evren.
Astrophys. Space Sci., Vol. 137, No. 2, p. 357–371 (1987).
The photoelectric light curves of MM Her obtained in 1983 and 1984 by Evren (1985, 1987) were analyzed by two different methods. Firstly, the effects of the wave–like distortions on the observations were removed from the observed magnitudes by obtaining its mathematical expression. The remaining light curves were analyzed by using Wood's approach. Later, the light curves of the same years were treated by the method of Wilson–Devinney and distortions seen in the light curves were thought to be explained by locating the spots on the surface of the cooler component. The results obtained by two different approaches are in good agreement.

119.047   An analysis of the light changes of the eclipsing binary XY Ceti in the frequency–domain.
R. K. Srivastava.
Astrophys. Space Sci., Vol. 138, No. 1, p. 197–207 (1987).
The revised geometrical elements of the eclipsing binary XY Ceti have been obtained by the method of Fourier analysis of the light changes in the frequency–domain, which was developed by Kopal (1979). These have been compared with the earlier (Srivastava and Padalia, 1975) results obtained by employing Russell and Merrill's (1952) method. The revised absolute dimensions of XY Ceti have been obtained using the spectroscopic elements given by Popper (1971), and the newly derived geometrical elements. The Roche radii have been derived to discuss the evolution of the system. The secondary component lies reasonably near to the Main Sequence, while the primary component falls above it. The evolutionary discussion indicates that the system is a detached one.

119.048   Apsidal motion in the eclipsing binary system OX Cas.
Kh. F. Khaliullin, V. S. Kozyreva, S. E. Leontiev.
Astrophys. Space Sci., Vol. 138, No. 2, p. 361–368 (1987).
Multi–colour $WBVR$ photoelectric observations of the eclipsing binary OX Cas were carried out. The photometric elements, absolute parameters and the angular rate of the apsidal motion ($\dot{\omega} = 9.1$ deg yr^{-1}) were obtained. The apsidal parameter k_2 derived for this system is by 15–25% smaller than the theoretical parameter k_2.

119.049   Der Bedeckungsveränderliche DV Cephei.
P. Frank, D. Lichtenknecker.
BAV Mitt., Nr. 47, 6 pp. (1987).
The authors present first light–elements for the EA–type eclipsing binary DV Cep and a light–curve based on extensive photographic photometry.

119.050   Helligkeitsfluktuationen von V388 Cyg im Hauptminimum ebenfalls beobachtet.
W. Braune.
BAV Rundbrief, 36. Jahrg., Nr. 4, p. 164–165 (1987).

119.051   Minimum von 32 Cygni beobachtet.
D. Böhme.
BAV Rundbrief, 36. Jahrg., Nr. 4, p. 166–168 (1987).

119.052   Algol – noch immer ein rätselhafter Stern.
S. Rössiger.
Sterne, 63. Band, Heft 5, p. 283–291 (1987).

119.053   Spectrophotometry of Epsilon Aurigae.
D. T. Thompson, B. L. Lutz, G. W. Lockwood, J. R. Sowell.
Astrophys. J., Vol. 321, No. 1, p. 450–458 (1987).
The authors have made 4 Å and 8 Å resolution spectrophotometric observations, 3295–8880 Å, of the peculiar eclipsing binary ε Aurigae during the 1982–1984 eclipse. They determine a reddening of $E(B-V) = 0.30$ mag and a spectrophotometric spectral type in the MK region of F3–F4 Ia. The behavior of several absorption lines during eclipse is described. The observations suggest the presence of neutral sodium, potassium, hydrogen, and the molecule CH in the eclipsing body.

119.054   Spectral investigations of the eclipsing binary star TX UMa.
V. G. Karetnikov, V. V. Kovtyukh.
Astron. Zh., Tom 64, Vyp. 6, p. 1256–1263 (1987). In Russian. English translation in Sov. Astron., Vol. 31, No. 6.
Spectral characteristics of the eclipsing binary star TX UMa were studied from 15 diffraction spectrograms obtained with reciprocal dispersions 9 and 37 Å/mm. Motions in the circumstellar gas structures and rotations of stars were determined from the discovered emission and absorption components of contours of the lines H$_\beta$, H$_\gamma$, K Ca II and Mg II $\lambda 4481$ Å. Electron concentrations in the envelope of the system and in the atmosphere of the main star were computed. Parameters of the stars were corrected and the value of the mass loss rate of the system was found to be 1.7×10^{-6} solar mass per year.

119.055   Absolute magnitude results from the Copenhagen eclipsing binary programme.
J. Andersen.
Bull. Inf. Cent. Données Stellaires, No. 33, p. 59–61 (1987).
Since about 1971, a programme of coordinated photometric and spectroscopic observations of selected double–lined eclipsing binary systems has been carried out. The main aim of the programme is to use suitably detached binary systems as a tool to determine very accurately the masses, radii, and temperatures of stars which can be assumed to be representative of normal, single stars.

119.056   Infrared photometry of TX UMa.
O. G. Taranova.
Astron. Tsirk., No. 1462, p. 4 (1986). In Russian.

119.057   Unusual change in the BM Ori spectrum.
N. Z. Ismailov.
Astron. Tsirk., No. 1466, p. 3–5 (1986). In Russian.
The analysis of spectral variablity of BM Ori on JD 2446459.28 is given. The observed continuum corresponds to a spectral class F0–F5. The spectrum suggests a high variability of the line He I $\lambda 4921$ Å and strengthening of metal emission lines. Such a non–stationarity is proposed to be connected with the ejection of matter from a hot companion in a binary system.

119.058 The third body in the eclipsing binary system V889 Aql.
Kh. F. Khaliullin, A. I. Khaliullina.
Astron. Tsirk., No. 1485, p. 4 – 5 (1987). In Russian.
Multicolour UBV photoelectric observations of the eclipsing binary system V889 Aql have been carried out. Photometric elements of the orbit and absolute characteristics of the components have been determined. A third light is found to be 18.5%; it corresponds to a main–sequence star of 1.8 M$_\odot$.

119.059 The division of eclipsing binary stars according to the stages of filling up the Roche lobe.
V. G. Karetnikov.
Astron. Tsirk., No. 1485, p. 6 – 7 (1987). In Russian.
The stage has been determined of filling up the Roche lobe with stars of 246 eclipsing binary systems of Svechnikov and Bessonova Catalogue. Graphical comparison of the values estimated has permitted the division of the objects according to their types and the description of two sequences with empirical formulae. It is indicated that in this way the types of eclipsing systems may be specified.

119.060 The light curve variations in UV Piscium.
C. Ibanoğlu.
Astrophys. Space Sci., Vol. 139, No. 1, p. 139 – 147 (1987).
Two–colour photoelectric observations of the short–period RS CVn–type eclipsing binary UV Psc have been made between 1981 – 1986 at the Ege University Observatory. The present work deals with the light curve variations of the binary obtained in the observing seasons of 1981, 1982, and 1984. The shape of the light curve, the depths of the minima, and the total brightness of the system seemed to change in the course of time that covers three years of observing time. The value of the migration period of the wave–like distortion was roughly estimated to be between 1.5 and 2.0 years.

119.061 The light curve analysis of SZ Piscium.
G. A. Bakos, J. Tremko.
Bull. Astron. Inst. Czech., Vol. 38, No. 6, p. 356 – 362 (1987).
From 16 epochs of primary minima of SZ Psc the authors have determined an improved period of the system. The O–C residuals exhibit a sine wave modulation with a period of 45.2 years. Analysing the light curves of 1957 – 1984 for the distortion wave the authors have derived a period of twelve years with a semiamplitude of 0.04 of the light intensity of the system. In addition, a migration of the distortion wave, reported by other authors, has been followed and a period of 1.5 years has been derived.

119.062 UBV photoelectric photometry of the 1976 – 1978 eclipse of VV Cephei.
Y.–w. Huang, Z.–h. Guo.
Chin. Astron. Astrophys., Vol. 11, No. 3, p. 207 – 214 (1987). English translation of 43.119.094.

119.063 Atmospheric eclipses in the SMC Wolf–Rayet eclipsing binary HD 5980: the heavy versus the light metal abundance.
G. Koenigsberger, A. F. J. Moffat, L. H. Auer.
Astrophys. J., Lett. Ed., Vol. 322, No. 1, p. L41 – L44 (1987).
Phase–dependent variations in the *IUE* spectra of the SMC 19.6 day eclipsing binary system HD 5980 (WN4 + O7 I:) are presented. The effects due to selective atmospheric eclipse at the N V, N IV, He II, and C IV lines are clearly present as in Galactic counterparts. However, the variations in the wavelength band $\lambda\lambda 1350 – 1490$ Å, prominent in Galactic WN binary systems, are virtually absent in HD 5980. This is most likely a result from the lower heavy element abundances (specifically iron) in the SMC.

119.064 Spectral investigation of the eclipsing binary star RY Geminorum.
V. G. Karetnikov, E. V. Menchenkova.
Sov. Astron., Vol. 31, No. 2, p. 192 – 195 (1987). English translation of 43.119.031.

119.065 Observations of the eclipse of IP Peg in 1985.
S. Akita, S. Fujino, T. Kato.
Variable Star Bull., No. 2, p. 5 (1987).

119.066 Observations of eclipse of IP Peg in 1986.
S. Fujino.
Variable Star Bull., No. 2, p. 6 (1987).

119.067 117th list of minima of eclipsing binaries.
BBSAG Bull., No. 84, p. 1 – 5 (1987).

119.068 The amplitude of DK Herculis.
A. Paschke.
BBSAG Bull., No. 84, p. 5 (1987).

119.069 V418 Aquilae: the duration and brightness of the totality.
K. Locher.
BBSAG Bull., No. 84, p. 6 (1987).

119.070 NSV 11987 Draconis: the more accurate period.
K. Locher.
BBSAG Bull., No. 84, p. 6 (1987).

119.071 118th list of minima of eclipsing binaries.
BBSAG Bull., No. 85, p. 1 – 4 (1987).

119.072 The correct period of DO Andromedae.
K. Locher.
BBSAG Bull., No. 85, p. 5 (1987).

119.073 The correct period of EY Pegasi.
K. Locher.
BBSAG Bull., No. 85, p. 5 – 6 (1987).

119.074 Times of minimum light for U Cephei in 1986.
F. Sato, A. Nishimura.
Inf. Bull. Variable Stars, No. 3040, 2 pp. (1987).

119.075 UBV photometry of Epsilon Aurigae outside eclipse.
F. Sato, A. Nishimura.
Inf. Bull. Variable Stars, No. 3041, 2 pp. (1987).

119.076 PX Cep: a new large amplitude eclipsing binary.
R. Boninsegna.
Inf. Bull. Variable Stars, No. 3048, 4 pp. (1987).

119.077 B, V, R, I observations of CE Leonis.
R. G. Samec, B. B. Bookmyer.
Inf. Bull. Variable Stars, No. 3053, 4 pp. (1987).

119.078 HD 185510: a calcium–emission binary with eclipses of a subdwarf companion.
L. A. Balona, T. Lloyd Evans, T. Simon, G. Sonneborn.
Inf. Bull. Variable Stars, No. 3061, 4 pp. (1987).

119.079 Time of minimum determination of the eclipsing binary V1143 Cygni.
E. F. Guinan, S.–I. Najafi, F. Zamani–Noor, P. T. Boyd, S. M. Carroll.
Inf. Bull. Variable Stars, No. 3070, 3 pp. (1987).

119.080 Photoelectric minima of eclipsing binaries.
E. Pohl, M. C. Akan, C. Ibanoglu, C. Sezer, N. Güdür.
Inf. Bull. Variable Stars, No. 3078, 8 pp. (1987).

119.081 The eclipse of the long period eclipsing binary 32 Cygni in 1987.
D. Böhme.
Inf. Bull. Variable Stars, No. 3083, 1 p. (1987).

119.082 **New observed times of minima for EE Aqr, RV Gru and RW PsA.**
E. Covino, L. Milano, D. de Martino, A. A. Vittone.
Inf. Bull. Variable Stars, No. 3088, 2 pp. (1987).

119.083 **The 1986 observations and the period study of CN Andromedae.**
S. Evren, C. Ibanoglu, Z. Tunca, M. C. Akan, V. Keskin.
Inf. Bull. Variable Stars, No. 3109, 4 pp. (1987).

119.084 **The nature of the variability of 42 Per.**
A. H. Batten.
Inf. Bull. Variable Stars, No. 3111, 3 pp. (1987).

119.085 **32 Cyg: UBV photometry of eclipse in 1987.**
A. Dolžan.
Inf. Bull. Variable Stars, No. 3112, 2 pp. (1987).

119.086 **New minima times and light elements for Sigma Aquilae.**
D. B. Williams.
Inf. Bull. Variable Stars, No. 3118, 3 pp. (1987).

119.087 **Photoelectric times of minima of four eclipsing binaries.**
T. Hegedüs.
Inf. Bull. Variable Stars, No. 3125, 2 pp. (1987).

Katalog orbital'nykh ehlementov, mass i svetimostej tesnykh dvojnykh zvezd. (Catalogue of orbital elements, masses and luminosities of close binaries).
See Abstr. 002.109.

Timing binary star eclipses.
See Abstr. 036.028.

Spectral variability of Wolf–Rayet–type stars.
See Abstr. 114.088.

Phase dependent polarization variations of southern galactic WR + O systems.
See Abstr. 116.002.

BV light curves of BY Dra in 1986.
See Abstr. 116.070.

DQ Herculis' orbital period variations according to 1982 – 1986 observations.
See Abstr. 117.020.

An EXOSAT observation of 1.5 orbital cycles of the 0.7 day short–period RS CVn system ER Vul.
See Abstr. 117.045.

Variability of the intrinsic polarization of the supergiant binary VV Cephei.
See Abstr. 117.052.

HD 185510: a chromospherically active binary with eclipses of a hot subdwarf companion.
See Abstr. 117.060.

Is there period–activity correlation in RS CVn–type binary systems?
See Abstr. 117.072.

Time resolved optical spectroscopy of the eclipsing intermediate polar EX Hydrae.
See Abstr. 117.079.

A new analysis of the infrared eclipses of the ultra–short period dwarf nova OY Carinae.
See Abstr. 117.080.

Phase–dependent variations of Hα equivalent width in VW Cep.
See Abstr. 117.090.

Time–resolved IUE studies of cataclysmic variables. I. Eclipsing systems IP Peg, PG 1030 + 590, and V1315 Aql.
See Abstr. 117.091.

RW Comae Berenices. III. Light curve solution and absolute parameters.
See Abstr. 117.108.

Photoelectric observations and the wave minimum of RS CVn.
See Abstr. 117.115.

Gravitational radiation and spiralling time of close binary systems.
See Abstr. 117.116.

Photometry and spectroscopy of the O–type variable HD 167971.
See Abstr. 117.156.

A turbulent region in CH Cyg?
See Abstr. 117.217.

The eclipses of cataclysmic variables. II. U Geminorum.
See Abstr. 117.252.

The ultraviolet spectrum of Beta Lyrae.
See Abstr. 117.284.

Up–to–date parameters of the eclipsing triple system IU Aur.
See Abstr. 118.007.

RS CVn–binary RW Com: a possible three–body system.
See Abstr. 118.032.

Statistical investigation of chemically peculiar stars. V. Spectroscopic binary stars.
See Abstr. 120.012.

Verbesserte Periode von V445 Cassiopeiae.
See Abstr. 122.101.

The galactic foreground reddening in the direction of the nearby Triangulum galaxy M33.
See Abstr. 131.046.

Veränderliche Sterne in offenen Sternhaufen.
See Abstr. 153.044.

120 Spectroscopic Binaries

120.001 Distribution of mass ratios in spectroscopic binaries.
J. L. Halbwachs.
Astron. Astrophys., Vol. 183, No. 2, p. 234 – 240 (1987).
205 main–sequence spectroscopic binaries were selected in order to derive the distribution of mass ratios of unevolved close binaries.

120.002 Spectroscopic binary orbits from photoelectric radial velocities. Paper 75: BD + 28°413.
R. F. Griffin.
Observatory, Vol. 107, No. 1079, p. 154 – 156 (1987).

120.003 Spectroscopic binary orbits from photoelectric radial velocities. Paper 76: HD 8997.
R. F. Griffin.
Observatory, Vol. 107, No. 1080, p. 194 – 200 (1987).
HD 8997 is a double–lined spectroscopic binary with components of types K1 V and K6 V in an 11–day orbit with a small but definitely non–zero eccentricity.

120.004 Spectroscopic binary orbits from ultraviolet radial velocities. Paper 3: δ Orionis.
A. S. Harvey, D. J. Stickland, I. D. Howarth, E. J. Zuiderwijk.
Observatory, Vol. 107, No. 1080, p. 205 – 210 (1987).
The purpose of the present contribution is to add a modern set of orbital elements to those derived from data of some antiquity in the pursuit of a value for the apsidal motion constant for the bright, hot supergiant δ Orionis.

120.005 Nonvariability of the radial velocity of Eta Cassiopeiae A.
R. S. McMillan, P. H. Smith.
Publ. Astron. Soc. Pac., Vol. 99, No. 618, p. 849 – 851 (1987).
Twenty–eight measurements of the radial velocity of the G0 V star η Cas A on 14 nights between 1986 December 11 and 1987 February 19 UT (inclusive) were made with an uncertainty per observation between ± 10 and ± 20 m s^{-1}. The velocity of the star never exceeded 30 m s^{-1} from the mean, even though the star was monitored on three occasions for three consecutive nights. The standard deviation of all 28 measurements is ± 13.7 m s^{-1}.

120.006 A premain–sequence double–lined binary system near the Trapezium.
L. A. Marschall, R. D. Mathieu.
Bull. Am. Astron. Soc., Vol. 19, No. 2, p. 707 (1987). Abstract. – See Abstr. 010.061.

120.007 The spectroscopic orbits of U Ophiuchi and QS Aquilae.
D. Holmgren.
Bull. Am. Astron. Soc., Vol. 19, No. 2, p. 709 (1987). Abstract. – See Abstr. 010.061.

120.008 The double–lined eccentric spectroscopic binary Beta Ari.
J. Tomkin, H. Tran.
Astron. J., Vol. 94, No. 6, p. 1664 – 1669 (1987).
Low–noise, near–infrared, Reticon spectra reveal β Ari as a double–lined spectroscopic binary. The secondary velocities provide a well–defined velocity curve with $K = 66.6$ km s^{-1} and $e = 0.88$. This eccentricity represents a slight reduction of the eccentricity determinations from earlier analyses of the primary velocity curve. The difference between the A5 V primary's minimum mass $M_1 \sin^3 i = 0.87 \, M_\odot$ and its estimated actual mass indicates an orbital inclination $i \approx 50°$, ruling out the possibility of eclipses. The spectroscopic results, parallax ($\pi = 0\overset{''}{.}074$), and estimated orbital inclination require a maximum separation in the apparent orbit of $0\overset{''}{.}084$. Thus, although β Ari has not been resolved by speckle interferometry, this day cannot be far off.

120.009 Candidates for spectroscopic binaries found in the Mount Wilson Halo–Mapping Program.
G. Fouts.
Publ. Astron. Soc. Pac., Vol. 99, No. 619, p. 986 – 997 (1987).
Forty–six candidates for spectroscopic binaries have been found from the four–year radial–velocity data base of the Mount Wilson Halo–Mapping Project. HD 16031 was added to the candidate list when two external velocity data bases were compared. Six of the candidates have previous orbital elements in the literature, to which the present data have been added, and new elements calculated. Comparison of the Mount Wilson velocities with the velocity survey of Carney and Latham (1987) shows a velocity dispersion in the total 309 star overlap sample of $\sigma = 6.4$ km s^{-1}. When the 101 stars which Carney and Latham flag as suspected velocity variables are extracted from this overlap sample and compared to the Fouts and Sandage (1986) data, a dispersion of $\sigma = 8.0$ km s^{-1} is found in the residuals. This suggests that if, indeed, the majority of the Carney–Latham 101 suspects are binaries, they will probably be found to have very small velocity amplitudes.

120.010 Radial velocities of calcium emission stars. 1. Observations at Sutherland.
L. A. Balona.
S. Afr. Astron. Obs., Circ., No. 11, p. 1 – 12 (1987).
The author presents radial velocities of 53 bright late–type stars with emission in the H and K lines of Ca II, which is indicative of a high level of chromospheric activity. Most of the stars are spectroscopic binaries for which orbital elements are presented.

120.011 Radial velocities of calcium emission stars. 2. Observations at Mt. Stromlo and Siding Spring Observatories and at Mt. John University Observatory.
A. Collier Cameron.
S. Afr. Astron. Obs., Circ., No. 11, p. 13 – 19 (1987).
The author presents radial velocities of 17 bright, late–type stars with strong emission cores in the Ca II H and K lines. Most of the stars are spectroscopic binaries. Orbital elements derived from these observations combined with observations made at SAAO are given by Balona (44.120.010).

120.012 Statistical investigation of chemically peculiar stars. V. Spectroscopic binary stars.
V. S. Lebedev.
Astrofiz. Issled. Izv. Spets. Astrofiz. Obs., Tom 25, p. 41 – 54 (1987). In Russian. English translation in Bull. Spec. Astrophys. Obs. – North Caucasus, Vol. 25.
The observable parameters of orbits for 83 spectroscopic binary stars with chemically peculiar components are reported. Distribution functions of the principal parameters are found. A comparison with a control sample of a normal spectroscopic binary star showed that distribution of Ap stars in periods, eccentricities and amplitudes of radial velocities differ from those of normal stars. Physical characteristics of stars, inclination angles of orbital planes to the line of sight, absolute sizes of orbits, the sizes of Roche lobes and angular orbital moments are estimated. It is found that the inclination angles of rotation axes of Ap components and those of their orbits do not correlate with each other. For 4 systems a motion of line of apsides with velocities of $(3 – 70)$E–4 grad/day is suspected.

120.013 Spectroscopic binary frequency among CNO stars.
H. Levato, S. Malaroda, B. García, N. Morrell.
Rev. Mex. Astron. Astrofis., Vol. 14, No. 2, p. 401 (1987). – See Abstr. 012.042.
Radial velocity variations are analyzed through a sample of 35 OB stars with CH anomalies. Bolton and Rogers' proposal (1978) is confirmed in the sense that the OBN stars appear preferably in short–period binary systems, in contrast to OBC stars.

120.014   Photometric variability of the binary HD 1826.
S. F. González, A. Rolland, A. Giménez,
P. López de Coca, R. Garrido, M. A. Hobart, J. H. Peña.
Rev. Mex. Astron. Astrofis., Vol. 14, No. 2, p. 410 – 413 (1987).
– See Abstr. 012.042.

Photoelectric two–color measurements of the single–lined spectroscopic binary HD 1826 are presented showing its photometric variability. The period of the light variations has been found to be in excellent agreement with the orbital period derived from the radial velocity measurements. An estimation of the amplitude of the variations of 0.03 mag has been obtained roughly independent of color. The system appears to be an ellipsoidal variable and the available preliminary results are discussed in light of the most probable binary model.

120.015   Spectroscopic binary orbits from photoelectric radial velocities. Paper 77: HD 90442.
R. F. Griffin.
Observatory, Vol. 107, No. 1081, p. 248 – 251 (1987).

120.016   Photometric search for eclipses in the spectroscopic binary HD 186943 containing a WR star.
N. A. Lipunova.
Astron. Tsirk., No. 1463, p. 1 – 2 (1986). In Russian.

120.017   On the initial epochs of spectroscopic binary systems.
M. I. Lavrov.
Astron. Tsirk., No. 1474, p. 4 – 6 (1987). In Russian.

The moments of the upper conjunction T_c are proposed as initial epochs for calculations of the orbital elements of spectroscopic binary systems instead of the traditional epochs T_n, T_{Ω}, T_o. The arguments for such substitution are discussed.

120.018   A new version of the method for determining the orbital elements of spectroscopic binary systems.
M. I. Lavrov.
Astron. Tsirk., No. 1474, p. 6 – 8 (1987). In Russian.

120.019   Radial velocity study of the Be star ζ Tau (HR 1910).
M. M. Jarad.
Astrophys. Space Sci., Vol. 139, No. 1, p. 83 – 91 (1987).

Extensive radial velocity investigation has been presented in this paper for the bright Be star HR 1910, based upon thirty–three spectrograms. The analysis of these radial velocity measurements has demonstrated clearly the binary nature of the star and new orbital elements were determined. A brief discussion of the previous work on this system appears in the relevant sections.

120.020   The O–type spectroscopic binary system HD 165921.
N. D. Morrison, V. S. Niemela.
Publ. Astron. Soc. Pac., Vol. 99, No. 621, p. 1149 (1987).
Abstract. – See Abstr. 010.281.

Studies in Be-star variability. 2. Analysis of published radial velocities of six bright emission–line stars.
See Abstr. 112.064.

A high precision photometric investigation of the micro–variations of Wolf–Rayet stars.
See Abstr. 113.028.

UBV(RI)$_c$ photometry for Ca II emission stars. 1. Observations at Sutherland.
See Abstr. 113.040.

UBV(RI)$_c$ photometry for Ca II emission stars. 2. Observations at Mt. John University Observatory and at Mt. Stromlo.
See Abstr. 113.041.

Activity in late–type dwarfs. I. Walraven and Johnson photometry of flares and spot variations on Gl 867 A (= FK Aqr) in 1979.
See Abstr. 116.035.

Activity in late–type dwarfs. II. Flares and spot variations on Gl 867 A (= FK Aqr) in 1981.
See Abstr. 116.036.

HD 26337 = EI Eri: an RS CVn candidate for the Doppler imaging technique.
See Abstr. 117.175.

On a model of the symbiotic star AG Dra.
See Abstr. 117.289.

Spectroscopic orbital elements and photometry of the multiple system Epsilon Hydrae.
See Abstr. 118.023.

Light curve variations in ER Vulpeculae.
See Abstr. 119.016.

Photoelectric radial velocities of eclipsing binaries. V. Orbital elements of V643 Ori.
See Abstr. 119.024.

The orbit of the classical Cepheid U Aquilae.
See Abstr. 122.030.

The orbit of the cepheid AW Per.
See Abstr. 122.058.

A new orbit for the binary cepheid S Sagittae.
See Abstr. 122.059.

The binary cepheid DL Cas and the open cluster NGC 129.
See Abstr. 153.008.

Binary frequencies in the open cluster IC 2602.
See Abstr. 153.039.

Spectroscopic binary stars in globular clusters.
See Abstr. 154.034.

The binary frequency of high–velocity field dwarfs as obtained with CCD measures.
See Abstr. 155.014.

121 Early-stage Stars (T Tauri Stars, Herbig-Haro Objects, etc.)

121.001 Synoptic studies of T Tauri stars.
M. S. Giampapa.
The SHIRSOG Workshop, p. 53 – 59 (1986). – See Abstr. 012.003.

The synoptic study of pre–main sequence stars at non–redundant wavelengths throughout the electromagnetic spectrum can elucidate critical problems in stellar evolution. Among these problems are: (1) The process of star formation in molecular clouds. (2) The redistribution of angular momentum as cloud collapse occurs. (3) The evolution of stellar magnetic activity up to the ZAMS. (4) The origin of chromospheres and coronae, including the operative heating mechanisms in the atmosphere and the associated mass flows. These processes may ultimately influence the structure of the star forming region and thereby play a role in regulating star formation rates.

121.002 Low–mass star formation in the high galactic latitude dark cloud L 1642.
G. Sandell, B. Reipurth, G. Gahm.
Astron. Astrophys., Vol. 181, No. 2, p. 283 – 288 (1987).

The authors have carried out CCD imaging, spectroscopy and infrared photometry in a study of young stars in the high galactic latitude cloud L 1642. Two infrared sources, L 1642–1 and L 1642–2, both associated with IRAS point sources are identified with faint nebulous binary stars. L 1642–1A is spectroscopically identified as a K7 IV T Tauri star, obscured by $\sim 2^m$ of visual extinction. The secondary becomes quite bright in the far–red, and may possibly dominate the infrared emission. L 1642–2A is associated with a small compact reflection nebula. Its secondary is also very red and might be responsible for the far–infrared emission. Both binaries have very low luminosity, $\lesssim 0.5\,L_{\odot}$, if a distance of 100 pc is adopted. Both secondaries appear very active, L 1642–1B has Brγ emission, while L 1642–2B exhibits molecular hydrogen emission in the infrared. L 1642 is one of the very nearest star forming clouds.

121.003 Rotational modulation of the wind of the PMS star AB Aur: new observations in C IV and Mg II.
C. Catala, F. Praderie, P. Felenbok.
Astron. Astrophys., Vol. 182, No. 1, p. 115 – 119 (1987).

In continuation of a time monitoring program devoted to short term spectral variability in Herbig pre–main–sequence stars, the authors present new results obtained for the star AB Aur. On a one day time scale, the variability of the blue edge velocity of the Mg II 2795 Å line is confirmed, and that of the C IV 1548 Å line established. Since the observations cover only 6 days and since the phase coverage is not tight enough, the new data can only support the periodic variability of V_s found previously. Periods between 40 and 50 hr are possible. Scenarios for the wind model of AB Aur are discussed.

121.004 Molecular hydrogen emission in Herbig–Haro complexes. II. The high latitude nebulosities HH 52/53/54.
G. Sandell, W. J. Zealey, P. M. Williams, K. N. R. Taylor, J. V. Storey.
Astron. Astrophys., Vol. 182, No. 2, p. 237 – 242 (1987).

The authors present deep gunn r and gunn z CCD images of HH 52, 53 and 54 and high spatial resolution H_2 maps covering HH 52, HH 53 and the bright HH 54 nebulosity. Clumpy H_2 emission is seen in all three HH objects with a generally similar distribution to that of the visible emission. The morphology is discussed in terms of existing models of bipolar outflows. HH 53 is found to have a jet like appearance reminiscent of those described by Mundt. HH 54 appears bipolar and morphologically similar to the HH 1/HH 2 system. Spatial differences between the H_2 peaks and visible emission peaks are particularly pronounced in HH 54. These are interpreted as being caused by differential absorption occurring in shocked cloudlets in the stellar outflow. Possible exciting sources have been identified for HH 52 and HH 53 but in the case of HH 54 no obvious source is present.

121.005 Herbig–Haro objects in the vicinity of NGC 2023.
D. F. Malin, K. Ogura, J. R. Walsh.
Mon. Not. R. Astron. Soc., Vol. 227, No. 2, p. 361 – 372 (1987).

Two groups of Herbig–Haro objects have been discovered in the vicinity of the reflection nebula NGC 2023 and the Horsehead nebula (Barnard 33). Prime focus photographs, objective prism and low–dispersion spectra, and high–resolution line profiles have been obtained for the HH objects and proper motions have been derived for the brighter knots. Low–dispersion spectra of two possible exciting stars for the HH objects have also been obtained. The HH group closer to NGC 2023 shows a spatial distribution and proper motions suggesting a nearby faint star with an IR excess and strong Hα emission as the probable exciting source. However, no strong candidate was found for the exciting star of the second group of HH objects which are situated near V615 Ori and the base of the Horsehead nebula.

121.006 Optical polarization studies of Herbig Haro objects – I. The HH57 region.
S. M. Scarrott, T. M. Gledhill, R. F. Warren–Smith.
Mon. Not. R. Astron. Soc., Vol. 227, No. 3, p. 701 – 704 (1987).

An optical linear polarization map of the nebulosity associated with the Herbig Haro object HH57 is presented. The HH complex is illuminated and excited by a recently recognized FU Ori type star within the nebulosity. The possibility of the "interstellar nozzle" mechanism playing a part in the formation of HH57 is considered.

121.007 Follow–up Zeeman observations of the T Tauri star RU Lup.
R. M. Johnstone, M. V. Penston.
Mon. Not. R. Astron. Soc., Vol. 227, No. 3, p. 797 – 800 (1987).

The authors have obtained further Zeeman observations of the extreme T Tauri star RU Lup. They fail to confirm the suggestion of the presence of a magnetic field obtained in earlier work. A 3σ upper limit of 494 G is obtained for the mean longitudinal field strength if the present and previous data are averaged. A 3σ upper limit of 0.12 per cent is set for the average circular polarization around 5100 Å.

121.008 Optical polarization studies of Herbig Haro objects – II. The HH 24 and SSCV 140 nebulosities.
S. M. Scarrott, T. M. Gledhill, R. F. Warren–Smith.
Mon. Not. R. Astron. Soc., Vol. 227, No. 4, p. 1065 – 1071 (1987).

The authors present optical linear polarization data for the Herbig Haro object HH 24 and nearby nebulous star SSCV 140 which show that these systems are independent reflection nebulae illuminated by individual central sources. The authors tentatively identify the source of a bipolar molecular outflow in the region.

121.009 Photometric variations of Orion population stars. V. A search for periodicities.
W. Herbst, J. F. Booth, D. L. Koret, G. V. Zajtseva, N. I. Shakhovskaya, F. J. Vrba, E. Covino, L. Terranegra, A. Vittone, D. Hoff, L. Kelsey, R. Lines, W. Barksdale.
Astron. J., Vol. 94, No. 1, p. 137 – 149 (1987).

An intensive monitoring campaign aimed at detecting rotation periods in a set of bright T Tauri and Herbig Ae/Be stars was undertaken during the 1985/86 Milky Way season. In addition to the rotation period found for T Tau, on the basis of these data (Herbst et al. 1986). The authors find a probable period for SU Aur of 1.55 or 2.73 ± 0.03 days, the ambiguity arising because of a sampling frequency of (1 day)$^{-1}$, and a possible period for RY Tau of 5.6 days. Z CMa displayed a remarkably regular variation of peculiar form on a time scale of 60 days, which could be its rotation period. This study suggests that periodicity due to rotational modulation may be present in all T Tauri light curves. Other aspects of the variability of these stars are also discussed, and absolute photometry is given for comparison and check stars.

121.010 The strange "spots" on the T Tauri star RY Lupi.
R. Liseau, K. P. Lindroos, C. Fischerström.
Astron. Astrophys., Vol. 183, No. 2, p. 274–278 (1987).

Simultaneous photometric and spectroscopic observations of the suspected spotted T Tauri star RY Lup are reported. The time coverage was more than a full cycle of the star's fundamental of the periodic variations. The temporal behaviour of the fluxes of the chromospheric emission in the $H\alpha$, $H\beta$ and Ca II infrared triplet lines is anticorrelated with the broad band variations as expected within the spot theory. In contrast, the predicted correlation between brightness and color variation is very weak, if existent at all. The stellar photospheric absorption spectrum does apparently not change over the range of more than three magnitudes in the visual. The application of spot models to the photometry leads to solutions which do not agree with the spectroscopic results. The star spot hypothesis would imply luminosity changes ($\Delta M_{bol} > 1$ mag) on time scales as short as ~2 day. Variable circumstellar dust opacities have probably to be invoked in order to explain the brightness variations of RY Lup.

121.011 The atmospheres of T Tauri stars. I. High–resolution calibrated observations of moderately active stars.
U. Finkenzeller, G. Basri.
Astrophys. J., Vol. 318, No. 2, p. 823–843 (1987). With plate 8.

Calibrated optical spectra from 3700 to 8700 Å with high resolution and signal–to–noise ratio have been obtained quasi-simultaneously for three pairs of cool low– and intermediate-activity southern T Tauri stars and one G2 T Tauri star with well–defined photospheric spectra. Together with observations of spectral standards, these are used to obtain spectral types, reddening corrections, radial velocities, $v \sin i$, and in combination with near–IR and IRAS data, photospheric and systemic luminosities. From positions in the H–R diagram, the authors determine the mass, radius, and surface gravity for each object. Surface fluxes in the emission lines are discussed. For all targets, ratio or difference plots versus appropriate standards are analyzed and show that important features of T Tauri spectra are clearly chromospheric.

121.012 The photospheric spectrum of T Tau.
S. A. Korotin, V. I. Krasnobabtsev.
Bull. Crimean Astrophys. Obs., Vol. 75, p. 154–161 (1987). English translation of 42.121.044.

121.013 Optical polarization studies of Herbig Haro objects – III. HH 100 in Corona Australis.
S. M. Scarrott, T. M. Gledhill, R. F. Warren–Smith, R. D. Wolstencroft.
Mon. Not. R. Astron. Soc., Vol. 228, No. 2, p. 533–536 (1987).

Optical linear polarization data for the HH 100 nebulosity are presented. The authors are unable to confirm that HH 100 IRS is the sole source of excitation and illumination of the nebula and they suggest that there may be an additional and/or alternative source deeply embedded within the nebula itself.

121.014 Narrowband imaging of the Herbig–Haro object HH 46/47.
A. C. Raga, M. Mateo.
Astron. J., Vol. 94, No. 3, p. 684–699 (1987). With plate 74.

The authors have obtained narrowband, CCD images of the HH 46/47 system in the light of the $H\alpha$, [N II] 6583, [S II] 6717, and [S II] 6731 emission lines. The images include HH 46, HH 47B, and HH 47A. The authors have carried out a calibration for these images that allows them to calculate line ratios, and then use these line ratios as diagnostics of the physical conditions in the radiating gas. The study shows that the bright condensation HH 47A has a higher electron density and a lower excitation spectrum than the jet that joins this condensation to the central source. This result does not agree with the observations of other morphologically similar Herbig–Haro objects.

121.015 T Tauri stars, pre–T Tauri stars, and stellar jets.
M. Cohen.
Astrophysical jets and their engines, p. 91–102 (1987). – See Abstr. 012.021.

The properties of the pre–main–sequence T Tauri stars are described in relation to the likely properties of their predecessors. These predecessors, the "pre–T Tauri stars", are thought to be the stars responsible for exciting Herbig–Haro objects. Pre–T Tauri stars are capable of generating "stellar jets" by which bipolar flows carry material out into the ambient medium out of which low–mass stars form. Manifestations of these bipolar flows are: strings of optical Herbig–Haro objects; radio jets; large flattened dusty structures orthogonal to the flows. The author describes the observations (optical, infrared, far–infrared and radio continuum) that provide evidence for these components.

121.016 Das HH–34–System: Prototyp stellarer Jets.
T. Bührke, R. Mundt.
Sterne Weltraum, 26. Jahrg., Nr. 11, p. 616–620 (1987).

121.017 Jets from young stars: CCD imaging, long–slit spectroscopy, and interpretation of existing data.
R. Mundt, E. W. Brugel, T. Bührke.
Astrophys. J., Vol. 319, No. 1, p. 275–303 (1987). With plate 2.

High–velocity jets and collimated outflows are now recognized as phenomena commonly associated with young stars. The authors discuss new CCD imaging of five objects and in particular spatially resolved spectroscopy of eight highly collimated flows. They have discovered three new jets, and they show that several previously known Herbig–Haro objects have extended bow–shock–like structures. In most of the latter cases a jet is pointing from the star toward the bow–shock apex. Using a data base of ~20 known jets, the authors have compiled a detailed list of observational criteria describing these jets. Based on this compilation, a physical description and interpretation of jets from pre–main–sequence stars is presented.

121.018 A study of the nature of the $H\alpha$–emission stars LkHα 112, 115, 118, and 119.
B. Boesono, P. S. Thé, H. R. E. Tjin A Djie.
Astrophys. Space Sci., Vol. 137, No. 1, p. 167–181 (1987).

The present study of the nature of the stars LkHα 112, LkHα 115, LkHα 118, and LkHα 119 is based on low–dispersion IDS spectra and photometric measurements in the wavelength range between 0.33 and 3.8 μm. These stars are located in the direction of the extremely young open cluster NGC 6530 (Walker, 1957). The purpose of this study is, in general, to know whether they belong to the group of intermediate mass pre–main sequence objects, also known as Herbig Ae/Be stars, and, in particular, what special characteristics they possess. The result is as follows. The stars are very young; probably only LkHα 112 and LkHα 115 are members of the above–mentioned class of objects. The membership of LkHα 118 and LkHα 119 in this group is doubtful.

121.019 The kinematic structure of the HH 24 complex derived from high–resolution spectroscopy.
J. Solf.
Astron. Astrophys., Vol. 184, No. 1/2, p. 322–328 (1987).

High–resolution long–slit spectra obtained from various positions within the HH 24 complex are presented. The radial velocity field deduced from the spatially resolved emission line features indicates the presence of a bipolar system of highly collimated mass flow (jets) with velocities exceeding 200 km s^{-1}. The bipolar structure presents a remarkable degree of symmetry and consists of the elongated nebular component HH 24C and a new weak component, designated HH 24F, which has been identified by its high positive radial velocity. HH 24F is elongated ($2'' \times 15''$) resembling its counterpart HH 24C in both shape and orientation. The infrared source SSV 63 is located near the midpoint on the line joining HH 24C and HH 24F, suggesting that the bipolar jets are emanating from that source. Hence SSV 63 is probably a young stellar object and the main source of excitation of the HH 24 complex. Some evidence is presented that HH 24A traces the region where the head of the jet (HH 24F) interacts with a condensation in the ambient material.

121.020 Observational study of Fuors. I. On the light curve of V1057 Cygni.
M. A. Ibragimov, V. S. Shevchenko.
Astrofizika, Tom 27, Vyp. 1, p. 5 – 17 (1987). With one plate. In Russian. English translation in Astrophysics, Vol. 27, No. 1.

The photographic magnitudes m_{pg}, m_{pv}, m_{pr} of the Fuor V1057 Cygni for July 1968 – August 1970 and photoelectric UBVRI' observations made in July 1978 – December 1985 are presented. In the epoch of light increase and light maximum 26 estimations of m_{pg}, 20 m_{pv} and 3 m_{pr} for V1057 Cygni have been obtained. The upper limit of the color–index V–1 < 3^{m}5 in the preoutburst epoch is estimated. A periodic component with the period of about 12 days and amplitude of about 0^{m}1 V was derived from the light curves in B, V, R for 300 observational nights. From 1982 to 1986 the rate of brightness decrease did not exceed 0^{m}2 V while during four previous years the light decrease was ΔV > 0^{m}4.

121.021 The T Tauri stars.
G. Herbig.
J. Am. Assoc. Variable Star Obs., Vol. 16, No. 1, p. 1 – 3 (1987).

The T Tauri stars are described, and several interesting examples of this class of star in relation to the work of the AAVSO are discussed.

121.022 Anomalous distribution of axial inclinations of T Tauri stars.
W. B. Weaver.
Astrophys. J., Lett. Ed., Vol. 319, No. 2, p. L89 – L92 (1987).

Axial inclinations and their associated errors are derived for 20 T Tauri stars. The distribution of these inclinations is shown to be significantly different than that expected if the inclinations were distributed at random. A suggestion for current observing programs is made.

121.023 An X–ray survey for pre–main–sequence stars in the Taurus–Auriga and Perseus molecular cloud complexes.
E. D. Feigelson, J. M. Jackson, R. D. Mathieu, P. C. Myers, F. M. Walter.
Astron. J., Vol. 94, No. 5, p. 1251 – 1259 (1987). With plates 105 – 109.

Seventy–five fields from the Einstein X–Ray Observatory IPC detector between 3^h and 5^h right ascension and 15° and 35° declination are examined to search for X–ray–emitting low–mass pre–main–sequence (PMS) stars. Six such stars were previously found; they appear to be similar to T Tauri stars but without dense circumstellar envelopes or winds. The authors present here finding charts for 59 X–ray sources that may be PMS stars. While some are likely to be spurious X–ray sources, chance coincidences with unrelated stars, or non–PMS stellar sources such as RS CVn type binaries, approximately half are probably X–ray–selected PMS stars. Detailed study of the candidates will be reported elsewhere.

121.024 Proper motions of Herbig–Haro objects. VIII. The region of NGC 2068.
B. F. Jones, M. Cohen, P. A. Wehinger, T. Gehren.
Astron. J., Vol. 94, No. 5, p. 1260 – 1270 (1987). With plates 110 – 112 = Lick Obs. Bull. No. 1081.

The authors have determined proper motions and have obtained long–slit low–resolution spectroscopy for many of the Herbig–Haro (HH) objects in the region of NGC 2068. The HH 24 complex alone contains two dynamically distinct bipolar "jets". The authors believe there are three exciting stars in this region, two within the bounds of HH 24: one (SSV 63) is responsible for HH 24 knots CEA and the objects HH 19 and HH 20 to the northwest of HH 24, and possibly for HH 27 to the south; a second for HH 24 knots G1 – G3 and probably for HH 23. A third star (probably SSV 59) is responsible for HH 25 and HH 26, which also define a bipolar flow. The proper motions

and radial velocities suggest that objects have undergone both acceleration in the vicinity of the exciting stars, and deceleration a parsec away.

121.025 M20–04: a T Tauri star.
M. T. Ruiz, J. Maza, L. E. Gonzalez, M. Wischnjewsky.
Astron. J., Vol. 94, No. 5, p. 1299 – 1301 (1987). With plates 113 – 115.

Using spectrophotometry, the authors have classified M20–04 as a T Tauri star, based on the presence of strong chromospheric lines in emission. The object, found in an objective–prism survey, has the photospheric absorption spectrum of an M3 type star, with a large UV excess. The authors estimate the distance to be 170 pc, suggesting that M20–04 could be a member of the Upper Scorpius OB2 association.

121.026 Astrophysical results on young stars and active objects.
C. Perrier, A. Chelli, H. Zinnecker.
Interferometric imaging in astronomy, p. 247 – 250 (1987). – See Abstr. 012.035.

Several young stars in the Ophiuchus dark cloud were observed in July 1985 and May 1986 with the ESO standard infrared specklegraph.

121.027 Structure of a possible circumbinary disk around T Tauri.
D. A. Weintraub, C. R. Masson, B. Zuckerman.
Astrophys. J., Vol. 320, No. 1, p. 336 – 343 (1987).

Interferometric measurements of ^{12}CO and ^{13}CO ($J = 1 \rightarrow 0$) emission toward T Tau reveal two concentrations of molecular gas offset by several arcseconds from the star. The concentrations could represent material bound in Keplerian orbits lying nearly in the plane of the sky around T Tau and/or its infrared companion. Representative masses for these clumps are $\sim 10^{-3} M_\odot$. In this binary system, the bulk of the gaseous material seems to be in a distant orbit around the close (100 AU) binary pair. There is no evidence for a dense concentration of gas at the stellar positions themselves.

121.028 Milliarcsecond resolution infrared observations of young stars in Taurus and Ophiuchus.
M. Simon, R. R. Howell, A. J. Longmore, B. A. Wilking, D. M. Peterson, W.–P. Chen.
Astrophys. J., Vol. 320, No. 1, p. 344 – 355 (1987).

The authors report K band lunar occultation observations of 18 stars in the Taurus and Ophiuchus star–forming regions. Four of the systems, HQ Tau, FF Tau, and SR 12 and ROX 31 in Ophiuchus, are binaries. Their separations, as observed in the projection along the directions of their occultations, range from ~ 5 to 186 milliarcseconds. SR 12 was also observed by speckle interferometry in the J, H, and K bands. It is shown that SR 12 is a ~ 0".30 binary system whose components are late–type stars still approaching the main sequence. Extended structure has been discovered in the sources Elias 29 and YLW 16A in Ophiuchus.

121.029 A possible protostar near HH 7–11.
E. N. Grossman, C. R. Masson, A. I. Sargent, N. Z. Scoville, S. Scott, D. P. Woody.
Astrophys. J., Vol. 320, No. 1, p. 356 – 363 (1987).

The authors report millimeter wavelength, aperture synthesis observations of the bipolar outflow region associated with Herbig–Haro objects HH 7–11. At 98 and 110 GHz, the thermal dust continuum emission is resolved into a binary source, with a separation of approximately 5000 AU, oriented normal to the outflow axis. One source coincides with the embedded near–infrared source SVS 13, previously identified as the outflow center, while the companion to the southwest has not been detected at other wavelengths. The apparent lack of a near–IR luminosity source suggests that the companion may be a true protostar.

**121.030   Observations of infrared emission lines and radio contin-
uum emission from pre–main–sequence objects.**
N. J. Evans II, R. M. Levreault, S. Beckwith, M. Skrutskie.
Astrophys. J., Vol. 320, No. 1, p. 364 – 375 (1987).

Moderate–resolution ($\lambda/\Delta\lambda \sim 800 - 1000$) observations of the
Brα, Brγ, PFγ, and H$_2$ $v = 1 \rightarrow 0$ S(1) emission lines are presented
for 16 pre–main–sequence stars, most of which are low–mass
($M < 5\,M_\odot$) objects. Brackett line emission was detected in nine
of 16 objects. H$_2$ emission was confirmed in T Tauri and
probably detected in MWC 1080 and L1551 IRS 5. Radio
continuum observations at 6 cm are also presented. These results
are compared to the predictions of existing models for stellar
winds. The results are also compared to the data on molecular
outflows.

**121.031   Observations and bowshock models of Herbig–Haro
objects.**
P. M. Hartigan.
Diss. Abstr. Int., Sect. B, Vol. 48, No. 2, p. 472–B (1987). Thesis,
The University of Arizona, 273 pp. (1987). Order No.
DA8711632.

**121.032   Radio emission from pre–main sequence stars in Corona
Australis.**
A. Brown.
Bull. Am. Astron. Soc., Vol. 19, No. 2, p. 728 (1987). Abstract. –
See Abstr. 010.061.

121.033   A catalog of Mg II emission line fluxes for T Tauri stars.
C. L. Imhoff, G. Basri, M. S. Giampapa.
Bull. Am. Astron. Soc., Vol. 19, No. 2, p. 728 (1987). Abstract. –
See Abstr. 010.061.

121.034   Time–dependent bow shocks and Herbig–Haro objects.
A. C. Raga, K. H. Böhm, M. Mateo, J. Solf.
Bull. Am. Astron. Soc., Vol. 19, No. 2, p. 728 (1987). Abstract. –
See Abstr. 010.061.

**121.035   Rotation modulation of the ultraviolet spectrum of
BP Tau.**
T. Simon, F. J. Vrba, W. Herbst.
Bull. Am. Astron. Soc., Vol. 19, No. 2, p. 741 (1987). Abstract. –
See Abstr. 010.061.

**121.036   An unexpected distribution of axial inclinations of
T Tauri stars.**
W. B. Weaver.
Bull. Am. Astron. Soc., Vol. 19, No. 2, p. 741 (1987). Abstract. –
See Abstr. 010.061.

**121.037   The magnetic field geometry in the vicinity of
HH 7–11/HH 12 and HH 33/HH 40.**
M. H. Heyer, S. E. Strom, K. M. Strom.
Astron. J., Vol. 94, No. 6, p. 1653 – 1656 (1987). With plates
137 – 138.

The small–scale structure of the interstellar magnetic field in
the vicinity of the Herbig–Haro objects HH 7–11/HH 12 and
HH 33/40 is derived from CCD imaging polarimetry of stars
nearby to these young stellar object driven outflows. The outflow
associated with SVS 13/HH 7–11 lies approximately parallel to
the local field direction. The outflow mapped by HH 33/HH 40 is
also well aligned with the local magnetic field. However, the
orientation of the outflow associated with HH 12 located 2' to the
north of HH 7–11 is offset from the inferred magnetic field
direction by 60°. These results taken together with those of
previous studies suggest (1) magnetic fields are directly or
indirectly responsible for the observed orientation of outflows
associated with newborn stars in the majority of cases but (2) in
some cases, the flow properties may depend upon other
characteristics of the parent cloud core and, in particular, the
ionization.

121.038   Radio and infrared properties of young stars.
N. Panagia.
Space Telesc. Sci. Inst., Prepr. Ser., No. 203, 28 pp. (1987). Paper
presented at the NATO ASI "Galactic and extragalactic star
formation", Whistler, B.C., Canada, June 1987.

121.039   Flows and jets from young stars.
R. Mundt.
Max–Planck–Inst. Astron. Heidelb., Prepr., 23 pp. (1987). Paper
presented at the NATO Advanced Study Institute on "Forma-
tion and evolution of low mass stars", Viano do Castelo,
Portugal, 21 September – 2 October 1987.

121.040   Spectroscopy of the rapidly rotating K star HD 36705.
O. Vilhu, B. Gustafsson, B. Edvardsson.
Astrophys. J., Vol. 320, No. 2, p. 850 – 861 (1987).

Spectroscopic observations of HD 36705, performed with the
CAT/CES telescope at ESO 1984 December, are reported. The
authors derive 100 ± 5 km s^{-1} and 29 ± 2 km s^{-1} for the $v \sin i$
and the radial velocity, respectively. The central wavelength of
the broad Hα emission is varying, possibly regularly, as a
function of rotation phase with an amplitude of 16 km s^{-1}. The
authors derive T_{eff}, log g, the metal content and the lithium
abundance of the star. They propose HD 36705 to be a
contracting star with a mass in the interval 1–2 solar masses and
an age of $10^6 - 3 \times 10^7$yr.

**121.041   Water masers associated with low–mass stars: a survey
of the Rho Ophiuchi infrared cluster.**
B. A. Wilking, M. J. Claussen.
Astrophys. J., Lett. Ed., Vol. 320, No. 2, p. L133 – L137 (1987).

The authors report the detection of H$_2$O masers toward two
extremely young pre–main–sequence objects embedded in the
nearby Rho Ophiuchi molecular cloud complex. These embedded sources (YLW 16 and IRAS 16293–2422) are among the
lowest luminosity objects known to be associated with maser
emission (16 and 27 $L_\odot$). The masers exhibit multiple velocity
components which varied substantially in intensity over a period
of 9 months and which are spatially unresolved (< 70 A.U.). The
authors suggest a model whereby the maser activity in both
sources arises from the interaction of a strong stellar wind with
circumstellar material.

**121.042   Fast scanning slit near IR photometry of T Tauri stars in
close binaries: Elias 22 and Chamaeleon.**
L. Carrasco, A. Chelli, H. Zinnecker, I. Cruz–González,
C. Perrier.
Rev. Mex. Astron. Astrofis., Vol. 14, No. 1, p. 359 (1987).
Abstract. – See Abstr. 012.042.

121.043   Isolated post–T Tauri stars.
G. R. Quast, C. A. P. C. O. Torres, R. de La Reza,
G. F. P. Mello.
Rev. Mex. Astron. Astrofis., Vol. 14, No. 1, p. 360 (1987).
Abstract. – See Abstr. 012.042.

121.044   V4046 Sgr – an isolated binary post T Tauri star.
C. A. P. C. O. Torres, G. R. Quast, R. de La Reza,
G. F. P. Mello.
Rev. Mex. Astron. Astrofis., Vol. 14, No. 1, p. 361 (1987).
Abstract. – See Abstr. 012.042.

**121.045   On the infrared emission of the exciting star of the
Herbig–Haro objects 1 and 2.**
M. Tapia, M. Roth, L. Carrasco, M. T. Ruiz.
Rev. Mex. Astron. Astrofis., Vol. 14, No. 2, p. 517 (1987).
Abstract. – See Abstr. 012.042.

**121.046   Speckle observations of the ice feature in the young
double source Serpens SVS 20.**
C. Eiroa, C. Leinert.
Astron. Astrophys., Vol. 188, No. 1, p. 46 – 48 (1987).

The authors present one–dimensional near–infrared speckle
observations in the ice feature of the low–luminosity infrared

double source Serpens SVS 20. The depth of the 3.1 μm feature appears to be the same for both components of SVS 20. This suggests that the ice–carrying grains belong to the material distributed on a larger scale within the Serpens molecular cloud and are not confined to circumstellar environments. The available constraints on luminosity and temperature indicate that both stars of the system SVS 20 have not yet reached the T Tauri stage and may still be embedded in circumstellar dust shells.

121.047   T–Tauri–Sterne.
W. Wenzel.
Sterne, 63. Band, Heft 5, p. 264 – 274 (1987).

121.048   Optical imaging and polarization mapping of the variable nebulosity associated with the PMS star PV Cephei.
T. M. Gledhill, R. F. Warren–Smith, S. M. Scarrott.
Mon. Not. R. Astron. Soc., Vol. 229, No. 4, p. 643 – 652 (1987).
The authors present observations of the variable optical nebulosity associated with the T Tauri star PV Cephei, spanning the period 1981 August – 1985 October. Polarization mapping and direct imaging show that this object has a biconical structure and is located within an asymmetric distribution of both matter and magnetic field.

121.049   Forbidden line and Hα profiles in T Tauri star spectra: a probe of anisotropic mass outflows and circumstellar disks.
S. Edwards, S. Cabrit, S. E. Strom, I. Heyer, K. M. Strom, E. Anderson.
Astrophys. J., Vol. 321, No. 1, p. 473 – 495 (1987).
The authors report the results of a high–resolution spectroscopic study of 10 T Tauri stars (TTS) and two Herbig emission stars. Red echelle spectra include the lines of [O I] $\lambda6300$, [N II] $\lambda6584$, [S II] $\lambda\lambda6716$, 6731, as well as Hα. The forbidden lines display a continuous progression of profile types, from single–peaked profiles with maxima near the stellar rest velocity and broad (150 km s^{-1}) blue wings, to double–peaked profiles with blueshifted maxima displaced by as much as 200 km s^{-1} from the second peak near the stellar rest velocity. It is found that the observed range of profile types is best explained by a wind with a latitude dependent velocity field, such that the wind velocity is higher at the poles than at the equator. The authors also present estimates of the average densities (10^4cm^{-3}), sizes (tens of AU) for the TTS forbidden emission regions, and compute mass–loss rates (10^{-7} to $10^{-9}M_\odot$yr^{-1}). The absence of redshifted forbidden line emission demands that the receding part of the mass ouflows must be obscured; an optically opaque circumstellar disk seems a logical candidate for such a screen. IRAS far–infrared fluxes are used to estimate the disk sizes on the assumption that they are optically thick at 100 μm. The implied masses of these circumstellar disks range from 0.01 to 0.1 $M_\odot$.

121.050   The 2 micron spectrum of L1551 IRS 5.
J. S. Carr, P. M. Harvey, D. F. Lester.
Astrophys. J., Lett. Ed., Vol. 321, No. 1, p. L71 – L74 (1987).
Two micron spectroscopy of L1551 IRS 5 is presented. The main result is the presence of strong absorption in the first–overtone bands of CO. From the equivalent width of the CO bands, a spectral type of K2 III is derived. The CO bands and marginal evidence for H$_2$O vapor absorption provide additional evidence that IRS 5 may be a member of the FU Orionis class of pre–main–sequence stars. Emission lines of H$_2$ are detected, suggesting the presence of shocked molecular gas in the vicinity of IRS 5. A marginal detection of the hydrogen Brγ line was also made.

121.051   Observations of jets from low–luminosity stars. II. SVS 12, HH 30, near HL Tauri, and HH 34.
M. Cohen, B. F. Jones.
Astrophys. J., Vol. 321, No. 2, p. 846 – 854 (1987). = Lick Obs. Bull., No. 1071.
The authors have obtained spatially resolved spectral scans for four Herbig–Haro jets emanating from low–luminosity pre–main–sequence stars. There appears to be a general tendency for the excitation and electron density to diminish along these jets. For three jets, the electron density dependence is close to r^{-1}. HH 30's scattered stellar continuum showed much weaker Fe II lines in 1985 than in 1979, indicative of variable stellar activity. The most distant knot in HH 34's jet has the lowest excitation of any known HH object.

121.052   Studies of young stellar objects with IRCAM at UKIRT.
H. Zinnecker, I. S. McLean, I. M. Coulson.
Infrared astronomy with arrays, p. 291 – 294 (1987). – See Abstr. 012.070.
The authors report the discovery of an embedded NIR binary object in the IRAS source L1495 in the Taurus dark cloud, and show and describe observations of other candidate pre–main–sequence objects in the Rho Ophiucus dark cloud.

121.053   Quick time variability of the Hα line in the star DI Cep.
N. Z. Ismailov.
Astron. Tsirk., No. 1457, p. 6 – 8 (1986). In Russian.
Rapid photometric variability of the T Tau–type star DI Cep has been observed in Hα on time–scales 30 s and 1.5 min with amplitudes $0^m.10 - 0^m.15$ and $0^m.25$, respectively.

121.054   On a group of Ina variable stars.
V. I. Kardopolov.
Astron. Tsirk., No. 1464, p. 1 – 3 (1986). In Russian.
The brightness–color index diagrams of V586 Ori, BF Ori and RR Tau Ina variable stars are examined. The photoelectric data confirmed that the stars may really belong to a homogeneous group of extremely young stellar objects.

121.055   Atmospheric parameters and iron abundance in T Tauri from lines of moderate intensity.
Yu. V. Borisov, A. N. Krasnobaev.
Astron. Tsirk., No. 1468, p. 1 – 3 (1986). In Russian.

121.056   On some properties of the spectral variability of the T Tau–type star DI Cep.
T. I. Timoshenko.
Astron. Tsirk., No. 1471, p. 6 – 8 (1986). In Russian.
Wavelengths of some 2000 features were measured for ten DI Cep spectrograms, and some of them were identified. The spectrum shows noticeable variability. Absorption components due to photospheric absorption appear in the region of H and K Ca II lines. The intensity of absorption by CH molecules was estimated by depression depth of quasi–continuum level in the region of $\lambda4275$ Å and was compared with the brightness in the V pass–band. Molecular absorption increases with decreasing brightness in V.

121.057   On the YY Ori–type activity.
V. I. Kardopolov.
Astron. Tsirk., No. 1481, p. 7 – 8 (1987). In Russian.
The dependence of U–B, B–V, V–R and V–I colors on V magnitude are examined for the RW Aur, CO Ori and SU Aur irregular variable stars. The same systematic trends in the color variations have been found. CO Ori and SU Aur are probably YY Ori type stars.

121.058   A recent change in the brightness and form of the HH46 reflection nebula.
J. A. Graham.
Publ. Astron. Soc. Pac., Vol. 99, No. 621, p. 1174 – 1177 (1987).
A several–fold increase in the surface brightness of the scattered light component of the southern Herbig–Haro object HH46 took place between January 1984 and May 1986. A more subtle change in the shape of the nebula also occurred in that a small region, formerly only prominent in the infrared, is now the brightest part of the reflection nebula in visible red light. The change is due either to a flaring of the adjacent but still invisible young stellar object or to a partial clearing of the dust which still surrounds it. In May 1987, the nebula still appeared bright suggesting that the change may be a long–term one.

121.059 High spectral resolution infrared observations of V1057 Cygni.
L. Hartmann, S. J. Kenyon.
Astrophys. J., Vol. 322, No. 1, p. 393 – 398 (1987).

The authors report high–resolution near–infrared spectra which confirm a key prediction of their accretion disk model for V1057 Cygni. In this model, the outbursts of V1057 Cyg and other FU Ori objects are caused by rapid accretion from a circumstellar disk nebula onto a pre–main–sequence star. Infrared spectra obtained with the KPNO 4 m FTS show that V1057 Cygni does rotate more slowly at 2.3 μm than at 6000 Å, by an amount quantitatively consistent with the simple disk models. The absence of any radial velocity variations in either the infrared or optical spectral regions make it very unlikely that the accretion disk is fed by a companion star but support the view that the accreted material arises from a remnant disk of protostellar material.

121.060 Near–infrared H$_2$ emission from Herbig–Haro objects. I. A survey of low–excitation objects.
R. D. Schwartz, M. Cohen, P. M. Williams.
Astrophys. J., Vol. 322, No. 1, p. 403 – 411 (1987).

A survey for H$_2$ 1–0 S(1) emission in 16 Herbig–Haro (HH) objects and three exciting stars for HH objects is reported. Eleven HH objects which show low–excitation optical spectra exhibit H$_2$ emission. One object (HH 43) is more than twice as bright as any previously reported HH object. In addition, spectra in the range 1.6 – 2.55 μm are reported for HH 43 and HH 120, and a 2.0 – 2.55 μm spectrum is presented for HH 26. The spectra yield estimates of the H$_2$ density and temperature ranges in these objects. Models which may account for the combined ultraviolet, optical, and near–IR spectra of HHs are briefly analyzed.

121.061 Radio emission from pre–main–sequence stars in Corona Australis.
A. Brown.
Astrophys. J., Lett. Ed., Vol. 322, No. 1, p. L31 – L34 (1987).

The central region of the Corona Australis molecular cloud surrounding the stars R and TY CrA has been studied using the VLA at 6 cm. Eleven radio sources are detected including five associated with pre–main–sequence objects. The most striking is associated with the near–IR source IRS 7 and shows a complex structure comprising two strong pointlike sources positioned either side of the deeply embedded IR source and two extended lobes of radio emission. The other detected sources include the massive pre–main–sequence star TY CrA, the near–IR sources IRS 1 and IRS 5, and the Herbig–Haro object HH 101.

121.062 Photometric behaviour of DR Tauri in the season 1986/ 87.
W. Götz.
Inf. Bull. Variable Stars, No. 3084, 2 pp. (1987).

121.063 Time–dependent bow shocks and the condensation structure of Herbig–Haro objects.
A. C. Raga, K. H. Böhm.
Astrophys. J., Vol. 323, No. 1, p. 193 – 210 (1987).

The authors show that an interpretation of the whole Herbig–Haro object as a single, time–dependent bow shock provides a natural explanation for the occurrence of condensations with different proper motions. To this effect, the authors have developed time–dependent, axisymmetric, nonadiabatic bow shock models from which they obtain predictions for spatially resolved Hα intensity maps, and then they compare these predictions qualitatively with observations of a few Herbig–Haro objects.

121.064 Spectral energy distributions of T Tauri stars: disk flaring and limits on accretion.
S. J. Kenyon, L. Hartmann.
Astrophys. J., Vol. 323, No. 2, p. 714 – 733 (1987).

The authors analyze spectral energy distributions of T Tauri stars (TTS) to place limits on disk accretion in this early phase of

stellar evolution. The results reinforce the conclusion of Adams, Lada, and Shu that much of the infrared excess emission arises from reprocessing of stellar radiation by a dusty circumstellar disk. Observational constraints on disk geometries and disk masses are derived. The relative importance of accretion and solar–type chromospheric activity in T Tauri stars is assessed. The present analysis indicates that disk accretion in the T Tauri phase does not modify stellar evolution significantly and that the angular momentum of accreted material can be lost in a stellar wind.

121.065 HL Tauri: a site for planet formation?
S. Beckwith, A. Sargent.
Mercury, Vol. 16, No. 6, p. 178 – 181 (1987).

The sources and their models.
See Abstr. 062.038.

A model for the thermal radio continuum emission produced by a shock wave and its application to the Herbig–Haro objects 1 and 2.
See Abstr. 063.079.

The photosphere of T Tauri stars.
See Abstr. 064.064.

Magnetic activity in pre–main sequence stars.
See Abstr. 065.018.

Fission of rapidly rotating protostars.
See Abstr. 065.071.

A numerical study of shear flow instability in rotating protostar models.
See Abstr. 065.072.

Accretion disk model of the primordial solar nebula.
See Abstr. 107.025.

Z CMa resolved at near infrared wavelengths: one more piece to the puzzle.
See Abstr. 112.022.

Multiband photometry (8 – 13 μm) of Herbig Ae/Be stars.
See Abstr. 112.055.

A photometric study of Herbig Ae/Be stars.
See Abstr. 112.112.

Giant grains around protostars.
See Abstr. 112.117.

Detection of X–ray emission from the young low–mass star Rossiter 137B.
See Abstr. 112.141.

Circumstellar disks and rings – observational results.
See Abstr. 112.177.

Kinematics of the circumstellar gas of HL Tauri and R Monocerotis.
See Abstr. 112.187.

An objective–prism survey for Hα–emission–line stars of a field in Puppis.
See Abstr. 113.002.

On the polarization of Herbig Ae/Be stars.
See Abstr. 116.003.

Rotation of pre–main sequence stars from high S/N spectroscopy.
See Abstr. 116.054.

Pre–main sequence binaries.
See Abstr. 117.204.

Flares of Orion Population variables in the association Taurus T3.
See Abstr. 122.114.

Low–mass stars in the Orion region.
See Abstr. 131.021.

An Hα emission–line survey of the ϱ Ophiuchi dark cloud complex.
See Abstr. 131.045.

OH outflows in star–forming regions.
See Abstr. 131.054.

Three new H_2O masers located near Herbig–Haro–like nebulosities.
See Abstr. 131.064.

Star formation in molecular clouds: observation and theory.
See Abstr. 131.066.

Large–scale radio brightness distribution of the ionized gas in the Canis Major OB1 association.
See Abstr. 131.084.

Dense cores in dark clouds.
See Abstr. 131.095.

High–resolution images of the L1551 bipolar outflow: evidence for an expanding, accelerated shell.
See Abstr. 131.128.

Water vapor masers associated with young visible stars.
See Abstr. 131.152.

A search for T Tauri stars in high–latitude molecular clouds.
See Abstr. 131.167.

A new cometary nebula in Cygnus.
See Abstr. 131.212.

The structure of the molecular outflow near SSV 13 and HH 7–11 in the NGC 1333 region.
See Abstr. 131.217.

A survey of IRAS point sources in Taurus for high–velocity molecular gas.
See Abstr. 131.249.

To B or not to B?
See Abstr. 131.260.

Interaction of the high–density gas with the bipolar outflow in Cepheus A.
See Abstr. 131.264.

Investigation of the interstellar absorption in the region of the two young stars DI Cep and AS 501.
See Abstr. 131.277.

Radio recombination lines from fast shocks in molecular clouds, with application to bipolar flows.
See Abstr. 131.297.

Fluorescent excitation of interstellar H_2.
See Abstr. 131.299.

Temporal changes of the IRS 5 jet in L1551.
See Abstr. 131.301.

The EINSTEIN survey of the young stars in the Orion Nebula.
See Abstr. 132.022.

Identification of new young stellar objects associated with *IRAS* point sources. I. The southern galactic plane.
See Abstr. 133.005.

Near–infrared and optical observations of *IRAS* sources in and near dense cores.
See Abstr. 133.008.

122 Intrinsic Variables (Pulsating Variables, Spectrum Variables, etc.)

122.001 Flare stars.
B. W. Bopp.
The SHIRSOG Workshop, p. 60 – 65 (1986). – See Abstr. 012.003.

The observational needs for a program on optical spectroscopy of stellar flares are as follows: (1) The program will need a dedicated telescope. (2) The program will need to observe faint targets. The best stars for flare observations are those of spectral type dM3 or later. This corresponds to objects of tenth magnitude or fainter. (3) The program will require high time resolution observations. Time resolution of a few seconds or less should be aimed for. (4) The program will require "modest" spectral resolution. (5) Lastly, there will be a very great need for the optical observations to be coordinated with those at other wavelengths.

122.002 Long period variables and stellar mass loss.
G. Wallerstein.
The SHIRSOG Workshop, p. 77 – 79 (1986). – See Abstr. 012.003.

122.003 FY Aquilae and the gamma–ray burst event of 1979 March 31.
D. Hartmann, R. W. Pogge.
Astrophys. J., Vol. 318, No. 1, p. 363 – 369 (1987). = Lick Obs. Bull., No. 1056. With plate 4.

The authors present deep CCD images, spectrophotometry, and astrometry of the variable star FY Aquilae, which is located within the error box of the 1979 March 31 gamma–ray burst event. Deep red CCD images reveal that FY Aql is surrounded by a faint, extended nebulosity centered on the variable star. Spectrophotometry of FY Aql and its nebulosity suggests that it is an M4e III Mira variable star embedded in an extended reflection nebula. No compelling observational evidence suggests that this system is the optical counterpart of the 1979 March 31 gamma–ray burst.

122.004 Observations de la variable TT Aql 1982 – 1986.
M. Dumont.
Obs. Trav., No. 11, p. 5 – 16 (1987).

122.005 The semiregular variable RX Bootis.
M. D. Taylor.
J. Br. Astron. Assoc., Vol. 97, No. 5, p. 277 – 279 (1987).

Observations of RX Boo in 1968 – 84 indicate an extreme visual range 7.5 – 8.9. A period of 160 d is apparent from inspection of the light curve, and is confirmed by Fourier analysis which also indicates a secondary period of 302 d.

122.006 Programme Hipparcos: courbes de lumière.
Bull. Assoc. Fr. Obs. Etoiles Variables, No. 41, p. 8 – 9 (1987).

122.007 Les étoiles variables β CMa.
E. Chapellier.
Bull. Assoc. Fr. Obs. Etoiles Variables, No. 41, p. 10 – 16 (1987).

122.008 Discussion of results of polarization observations of cepheids.
T. A. Polyakova.
Astrofizika, Tom 26, Vyp. 3, p. 469 – 475 (1987). In Russian. English translation in Astrophysics, Vol. 26, No. 3, p. 283 – 287 (1987).

The character of polarization of cepheids S Sge, SU Cyg, W Gem, ζ Gem and SU Cas is like that of the earlier investigated cepheids. The polarization diagram for RR Lyr constructed from observations of Piirola probably shows small interstellar polarization which is present in this star light. The sizes of "rosettes" of polarization diagrams for 19 cepheids decrease with decrement of the errors of observations, probably till errors of $\pm 0.04 – 0.03\%$ and "rosette" sizes of $\sim 0.2\%$.

122.009 Hydrodynamic models for the short–period, classical Cepheid, SU Cas.
T. Aikawa, E. Antonello, N. R. Simon.
Astron. Astrophys., Vol. 181, No. 1, p. 25 – 30 (1987).

Hydrodynamic models are constructed for the 1.95–day, classical Cepheid, SU Cas. The models cover three luminosities (479, 955 and 2239 $L_\odot$) and a range of temperatures (6200K $\leqslant T_e \leqslant$ 6500K), in line with observational and theoretical results from the work of previous authors. The authors discuss the pulsation of SU Cas in the light of stellar evolution calculations, and summarize the evidence regarding its pulsation mode. They conclude that SU Cas is almost certainly not an F–mode pulsator, although a completely satisfactory overtone model has yet to be constructed.

122.010 The pulsation modes of CO Aur.
J. Babel, G. Burki.
Astron. Astrophys., Vol. 181, No. 1, p. 34 – 40 (1987).

Photoelectric measurements in radial velocity (spectrophotometer CORAVEL) and photometry (Geneva system) of the peculiar double–mode cepheid CO Aur, confirm the two periods $P_1 = 1\overset{d}{.}78$ and $P_2 = 1\overset{d}{.}43$. A mean radius of $43 \pm 13\,R_\odot$ is derived by applying the Baade–Wesselink method. A complete discussion of the pulsation is given: CO Aur is radially pulsating in two overtone modes, probably the 1st and 2nd ones.

122.011 Rotational modulation and flares on RS CVn and BY Dra–type stars. V. EXOSAT and IUE observations of a flare on EQ Pegasi.
B. M. Haisch, C. J. Butler, J. G. Doyle, M. Rodono.
Astron. Astrophys., Vol. 181, No. 1, p. 96 – 102 (1987).

The authors have obtained simultaneous time–trailed ultraviolet spectra and a soft X–ray light curve during a flare on the binary dMe star EQ Peg AB. This is the first time to the authors' knowledge that there has been time–resolved ultraviolet spectroscopy simultaneous with a measured X–ray flare on a star. The IUE ultraviolet spectra were obtained by the long–wavelength (2000 – 3200 Å) spectrograph, and thus manifest primarily the chromospheric response to the coronal flare energy release and to the presence of an overlying hot coronal flare plasma.

122.012 HD 37819 ≡ V356 Aur, a double–mode δ Sct star with an unusual period ratio.
E. Poretti, L. Mantegazza, E. Antonello.
Astron. Astrophys., Vol. 181, No. 2, p. 273 – 282 (1987).

An analysis of new BV measures made at the Osservatorio Astronomico di Merate shows that HD 37819 ≡ V356 Aur is a double–mode pulsator with $P_1 = 0.189$ d and $P_2 = 0.156$ d. A new discussion of the photometric and radial velocity measures of Burki and Major (1981) strengthens this conclusion. The application of the Balona–Stobie method (1978) and the unusual ratio between the two periods (0.826) suggest that the pulsation modes thus excited could be non–radial: other observational parameters, however, show that this hypothesis should be accepted with great caution.

122.013 Shape of the visual light curve and detection of a 1.35 cm H₂O line in single M Miras.
M. S. Vardya.
Astron. Astrophys., Vol. 182, No. 1, p. 75 – 79 (1987).

The probability of detecting a 1.35 cm H_2O vapor line from single M Mira variables has been found to depend on the actual shape of the visual light curve. Following the classification scheme of Ludendorff (1928), the probability of detection of H_2O is highest for α–class light curves, reduces drastically for β–class, and is almost nil for γ–class. Similar tendency is exhibited by the mean luminosity of H_2O in the three classes as well.

122.014   The rapidly oscillating Ap stars as a test of stellar chromospheric heating mechanisms.
S. N. Shore, D. N. Brown, G. Sonneborn, D. M. Gibson.
Astron. Astrophys., Vol. 182, No. 2, p. 285 – 289 (1987).

The discovery of a new class of magnetic stars, the rapidly pulsating Ap stars, provides an unprecedented tool for the study of Alfvénic heating in the outer atmospheres of main sequence stars. In this note, the authors point out that detection of rotationally modulated chromospheric emission line fluxes, especially the Ca II, Mg II, and C IV lines, can provide a novel method for investigating the role of Alfvénic heating in stellar atmospheres.

122.015   Atmospheres of δ Sct pulsating variables. IV. Abundance of thorium in 20 CVn, 28 And, V644 Her, and δ Del.
L. S. Lyubimkov, T. M. Rachkovskaya.
Bull. Crimean Astrophys. Obs., Vol. 74, p. 12 – 16 (1987). English translation of 41.122.115.

122.016   Mechanism of excitation of chromospheric emission of pulsating stars of the type of δ Scuti.
G. A. Garbuzov, S. M. Andrievskij.
Astrophysics, Vol. 25, No. 2, p. 498 – 503 (1987). English translation of 42.122.155.

122.017   Color characteristics of irregular variable stars.
G. V. Zajtseva.
Astrophysics, Vol. 25, No. 3, p. 626 – 634 (1987). English translation of 42.122.182.

122.018   Optical and infrared observations of RV Tauri stars.
M. J. Goldsmith, A. Evans, J. S. Albinson, M. F. Bode.
Mon. Not. R. Astron. Soc., Vol. 227, No. 1, p. 143 – 159 (1987).

The authors present optical and IR photometry of RV Tauri stars, much of which was obtained nearly simultaneously over the wavelength range $0.36 - 10 \,\mu m$. From the dereddened optical–IR flux distributions they deduce stellar and (where appropriate) dust shell parameters. The deduced stellar colours are in good agreement with those inferred from optical spectroscopy. The Planck mean optical depths of the dusty RV Tauri stars range from 0.07 to 0.63, the corresponding circumstellar contributions to $E(B-V)$ ranging from ~ 0.02 to ~ 0.2. The observations show that the flux distributions of oxygen– and carbon–rich types are indistinguishable photometrically, that some of these objects may have multiple dust shells and hence that dust production in RV Tauri stars probably occurs sporadically rather than continuously.

122.019   Erratum: "Mode identification of Beta Cephei stars in NGC 3293" [Mon. Not. R. Astron. Soc., Vol. 223, No. 1, p. 189 – 206 (1986)].
C. A. Engelbrecht.
Mon. Not. R. Astron. Soc., Vol. 227, No. 2, p. 543 (1987). See Abstr. 42.122.085.

122.020   Non–radial pulsations in the extreme helium star HD 160641.
A. E. Lynas–Gray, D. Kilkenny, I. Skillen, C. S. Jeffery.
Mon. Not. R. Astron. Soc., Vol. 227, No. 4, p. 1073 – 1087 (1987).

Mean periods of 0.35 ± 0.01, 0.71 ± 0.05, 1.12 ± 0.13 and 1.77 ± 0.34 day are identified from separately analysed photometric observations (obtained in 1979 and 1982) of the extreme helium star HD 160641. Radial velocities were obtained simultaneously with photometry on three nights in 1982. Optical variations are not accompanied by detectable colour changes, and radial velocities are not characteristic of radial pulsation. HD 160641 is therefore regarded as a non–radial pulsator. Radial velocity and light variations are tentatively interpreted as

$l = 4$ mode pulsation, corresponding to a Wesselink radius of $7.9 \pm 1.0 \, R_\odot$. The consequent luminosity of $\log(L/L_\odot) = 4.7 \pm 0.3$ would be consistent with Schönberner's evolution model for a 1 $M_\odot$ extreme helium star.

122.021   The enigmatic object Variable A in M33.
R. M. Humphreys, T. J. Jones, R. D. Gehrz.
Astron. J., Vol. 94, No. 2, p. 315 – 323 (1987).

Variable A is a very luminous ($M_{bol} \cong -9.5$ mag), highly unstable star in M33. In 1950, it was one of the visibly brightest stars in M33, with the spectrum of a very luminous F supergiant. It then rapidly declined in brightness by 3.5 mag, becoming faint and red after slowly increasing in brightness during the previous 50 yr. In this paper, the authors report current spectroscopy and visual and infrared photometry. Variable A is still faint and red and now has the spectrum of an M supergiant! It also has a large infrared excess and is today as bright as 10 μm as it was at its visual maximum in 1950. The authors discuss its energy distribution, luminosities, and mass–loss rate. They present a possible explanation for Variable A's bizarre behavior.

122.022   Frequency analysis of the short–period Cepheid EU Tauri.
W. P. Gieren, J. M. Matthews.
Astron. J., Vol. 94, No. 2, p. 431 – 436 (1987).

New photometric and radial–velocity observations of the short–period Cepheid EU Tauri show systematic deviations from a smooth curve when plotted in phase according to the accepted period of 2.1025 days. These deviations suggest that the star may be multiperiodic, or now have a single period which differs from the older value. Frequency analysis of the existing data was undertaken to investigate these possibilities.

122.023   Erneute Periodenzunahme bei RR Leonis.
E. Wunder.
BAV Rundbrief, 36. Jahrg., Nr. 3, p. 109 – 115 (1987).

122.024   Observations and interpretations of stellar pulsations.
G. Burki.
Instabilities in luminous early type stars, p. 23 – 38 (1987). – See Abstr. 012.012.

The three most useful methods for the identification of the stellar pulsation modes are described. Some of the most significant examples are given for each method, taken among the Cepheid, RR Lyrae, β Cephei and δ Scuti stars. For two classes the complex variability of stars is described in detail: the 53 Persei stars and the supergiants. In these two classes non–radial modes have been identified and the complete mode identification has still to be largely improved.

122.025   Observed variations in O and Of stars.
D. Baade.
Instabilities in luminous early type stars, p. 73 – 79 (1987). – See Abstr. 012.012.

122.026   Variations in luminous blue variables.
H. J. G. L. M. Lamers.
Instabilities in luminous early type stars, p. 99 – 125 (1987). – See Abstr. 012.012.

The photometric and spectroscopic variability of luminous blue variables is discussed and compared with predictions. The photometric variations occur on all timescales from weeks to centuries. There are three kinds of photometric variations: large variations with $\Delta V \gtrsim 3^m$, due to large eruptions; moderate variations of $\Delta V \sim 1^m$, due to shell ejections; and microvariations with $\Delta V \sim 0^m_. 1$ probably due to pulsation. If the large eruptions are recurrent, the time of recurrence is of the order of centuries. There are three kinds of spectral variations: changes of the spectral type; large variations of the shapes of the line profiles; small variations of absorption components superimposed on the overall line profiles. The first two variations are due to large variations in the mass loss, and are accompanied by moderate photometric variations of $\Delta V \sim 1^m$.

122.027 Two comments on the β Cephei variable 12 Lacertae.
M. Jerzykiewicz.
Instabilities in luminous early type stars, p. 211 – 212 (1987).
See Abstr. 012.012.

According to a popular notion, the dominant pulsation mode in the β Cephei variables is radial. In the case of 12 Lacertae there is no evidence for long–term variations of the pulsation amplitudes such as those that were recently discovered in several other β Cephei stars.

122.028 The 160 min period in RR Lyr stars in globular clusters and highly evolved close binary systems.
V. A. Kotov.
Bull. Crimean Astrophys. Obs., Vol. 75, p. 104 – 109 (1987).
English translation of 42.122.074.

122.029 Atmospheres of δ Sct pulsating variables, Part 5. Spectrum analysis for δ Sct (HR 7020).
T. M. Rachkovskaya.
Bull. Crimean Astrophys. Obs., Vol. 75, p. 137 – 142 (1987).
English translation of 42.122.110.

122.030 The orbit of the classical Cepheid U Aquilae.
D. L. Welch, N. R. Evans, R. W. Lyons,
H. C. Harris, T. G. Barnes III, M. H. Slovak, T. J. Moffett.
Publ. Astron. Soc. Pac., Vol. 99, No. 617, p. 610 – 616 (1987).

A total of 62 new radial–velocity observations of the classical Cepheid U Aql have been obtained during the interval 1969 – 86. The authors present the first determination of a spectroscopic binary orbit for this star. The orbital elements derived from both new and published velocities are given. *IUE* observations reported by Böhm–Vitense and Proffitt (1985) indicate the presence of an early–type main–sequence companion. The orbital elements combined with estimates of the companion mass result in upper limits for the mass of the Cepheid in the range $6.4 – 8.8\ M_\odot$. The possibility of spatially resolving the system using interferometric techniques is discussed.

122.031 A photometric study of BL Herculis.
A. L. Alexander, M. D. Joner, D. H. McNamara.
Publ. Astron. Soc. Pac., Vol. 99, No. 617, p. 645 – 653 (1987).

The aims of the authors' investigations are to determine the absolute magnitude, radius, chemical composition, mean temperature, surface gravity, and mass of the prototype variable BL Her. To meet these aims they have secured very extensive $uvby\beta$ photometry of the variable.

122.032 Radial modes excited in the Delta Scuti star 44 Tau.
P. López de Coca, A. Rolland, R. Garrido,
E. Rodríguez.
Rev. Mex. Astron. Astrofis., Vol. 15, No. 1, p. 59 – 62 (1987).

B–band photoelectric photometry and Fourier analysis of the δ Scuti type variable 44 Tau are presented. It appears that 44 Tau pulsates in the first and second overtones with probably the fundamental mode also excited. The physical parameters are redetermined.

122.033 Cepheids as distance indicators.
M. W. Feast, A. R. Walker.
Annu. Rev. Astron. Astrophys., Vol. 25, p. 345 – 375 (1987). –
See Abstr. 003.003.

The crucial importance of Cepheids as distance indicators in both galactic and extragalactic research has long been recognized. At the present time, for instance, they appear to be the most satisfactory indicators of relative and absolute distances to nearby galaxies and to be the foundation for extensions of the distance scale to cosmologically important distances. The authors have attempted in this review to give a detailed assessment of the current situation, taking into account the very considerable improvements on both the observational and the theoretical side since earlier studies.

122.034 The stability of period in DD Lacertae.
T. Ciurla.
Acta Astron., Vol. 37, No. 1, p. 53 – 62 (1987).

334 observed maxima of brightness and velocities were redetermined from the observation curves since 1914 up to 1983. No secular variations of the period larger than 0.08 s/century were found, but there is an indication for duplicity of the star with an orbital period of about 44 years.

122.035 A photometric analysis of BW Vulpeculae.
J. H. Peña, R. Peniche, A. Arellano Ferro,
M. Rios Berumen, M. Rios Herrera.
Acta Astron., Vol. 37, No. 1, p. 63 – 77 (1987).

New times of maximum light have been determined and combined with all the previous times available in the literature to analyze the period variation of BW Vul. A period increase rate of 2.5 s/century is found. To study a possible cause of these variations, the radial velocities available over the last sixty years were analyzed. Significant systemic velocity variations were found; it is suggested that such variations may be produced by photospheric oscillations about a dynamic equilibrium point, as a consequence of the existence of more than one overtone acting simultaneously.

122.036 A new scale of classical Cepheid color excesses.
J. D. Fernie.
Astron. J., Vol. 94, No. 4, p. 1003 – 1007 (1987).

The $uvby\beta$ photometry of Feltz and McNamara (1980) for 41 classical Cepheids has been combined with theoretical colors derived by Lester, Gray, and Kurucz (1986) using theoretical gravities and temperatures from Cox (1980) to yield color excesses. The latter can be derived from each of $m1$, $c1$, and β, and on average the three results agree well if solar metallicities and evolutionary masses are assumed; different metallicities or masses degrade the agreement considerably. Comparison with five other major Cepheid–excess studies shows satisfactory agreement with the present results being at the midpoint of the range among the other studies.

122.037 Energy dissipation of high–frequency radial oscillations as possible source for heating of the outer layers of the atmosphere of τ Pegasi.
S. M. Andrievskij.
Stellar atmospheres, p. 144 – 152 (1987). In Russian. Abstr. in Ref. Zh., 51. Astron., 8.51.393 (1987). – See Abstr. 003.005.

122.038 Physical conditions in the atmosphere of a variable RR–Lyrae star.
Yu. S. Romanov.
Stellar atmospheres, p. 124 – 131 (1987). In Russian. Abstr. in Ref. Zh., 51. Astron., 8.51.429 (1987). – See Abstr. 003.005.

122.039 Rapid variability of the Hα line in the spectrum of τ Pegasi.
S. M. Andrievskij, G. A. Garbuzov, V. P. Malanushenko.
Stellar atmospheres, p. 135 – 143 (1987). In Russian. Abstr. in Ref. Zh., 51. Astron., 8.51.437 (1987). – See Abstr. 003.005.

122.040 Spectroscopy of the optical photosphere of the hot pulsating star BW Vulpeculae.
I. Furenlid, A. Young, T. Meylan, C. Haag, G. Crinklaw.
Astrophys. J., Vol. 319, No. 1, p. 264 – 274 (1987).

The authors present spectroscopic observations of BW Vul which were made at high spectral and temporal resolution while preserving a high signal–to–noise ratio. The authors present a synchronous, narrow–band, photometric light curve which establishes unambiguously the temporal relationship between structures in the light curve and spectroscopic line behavior. They further derive the systemic radial velocity and present a radial velocity curve in the rest frame of the star which represents the actual kinematics of the optical photosphere, and which implies that the pulsation results in a symmetrical expansion and contraction of the stellar atmosphere by a total of about 7% of the stellar radius.

122.041   Observational studies of Cepheids. V. Radial velocities of bright Cepheids.
T. G. Barnes III, T. J. Moffett, M. H. Slovak.
Astrophys. J., Suppl. Ser., Vol. 65, No. 2, p. 307 – 312 (1987).
Over 600 radial velocities are given for 24 bright Cepheids accessible from McDonald Observatory. The accuracy of a single observation of unit weight is typically ± 3.8 km s^{-1}. These velocities are contemporaneous with previously published *BVRI* photometry.

122.042   The infrared variability of FG Sagittae in 1985 – 1986.
O. G. Taranova.
Pis'ma Astron. Zh., Tom 13, No. 10, p. 891 – 893 (1987). In Russian. English translation in Sov. Astron. Lett., Vol. 13.
From an analysis of 115-day variations of the IR brightness of FG Sge, it follows that these variations are mainly due to stellar photospheric temperature variations. From the energy distribution in the spectrum of FG Sge in the range $1.25 – 3.5 \mu m$, the star is classified as G2 I and F5 I near the minimum and maximum of the 115-day variations respectively. The luminosity of FG Sge at the maximum of the 115-day cycle is almost by a factor of 2 larger than at the minimum. The radius variations do not exceed 10%. By the character of the IR brightness variations FG Sge is at present similar to the variables of the δ Cep of SRd types.

122.043   Energetics of activity of flare stars and the Sun: a synergetical approach.
R. E. Gershberg, Eh. I. Mogilevskij, V. N. Obridko.
Kinematika Fiz. Nebesn. Tel, Tom 3, No. 5, p. 3 – 17 (1987). In Russian. English translation in Kinematics Phys. Celest. Bodies.
The analysis of the energy spectra of flares on flare stars and on the Sun shows that the maximum energy of optical radiation of stellar flares is close to 10^{29}J and of solar flares is several units of 10^{25}J; taken into account the electromagnetic emission in other wavelength ranges, losses on particle acceleration and hydrodynamic losses, it is concluded that the total energy released in such flares must exceed the above estimates by an order of magnitude. Difficulties of the standard model of current sheets in providing a high energy release in powerful solar and stellar flares are discussed and the necessity of a synergetical approach to analyse structural and kinematic properties of convective zones in the stars with a magnetic field is substantiated. The model of soliton gas as a mechanism of rapid transport of large portions of magnetic energy from deep convection layers into the atmosphere is suggested and discussed in brief. This mechanism seems to be able to provide the energetics of the most powerful solar and stellar flares.

122.044   Hα line in the spectrum of the unique Cepheid V473 Lyr.
S. M. Andrievskij, G. A. Garbuzov.
Kinematika Fiz. Nebesn. Tel, Tom 3, No. 5, p. 94 – 96 (1987). In Russian. English translation in Kinematics Phys. Celest. Bodies.
Five spectrograms of HR 7308 = V473 Lyr are obtained in the range of Hα (dispersion 4 nm/mm) at the maximum light with a high time resolution. Variations of the Hα profile in the spectrum of HR 7308 of classical Cepheids and of beat Cepheids are compared. It is concluded that rapid variations in the Hα profile observed in HR 7308 are not characteristic of Cepheids. The qualitative similarity of these variations with the spectral manifestation of nonradial pulsations is accentuated.

122.045   The variable star HD 79889.
T. Oja.
Astron. Astrophys., Vol. 184, No. 1/2, p. 215 – 218 (1987).
The star HD 79889 is shown to be a short–period ($P = 0.096$ days), high–amplitude ($0^{m}4$ in V) "dwarf cepheid" variable, very probably of the δ Sct type.

122.046   Hydrogen ionization front dynamics in type–II Cepheid envelopes.
A. B. Fokin.
Nauchn. Inf., Vyp. 63, p. 87 – 96 (1987). In Russian.
The motion of gas near the hydrogen ionization front in population II Cepheids is investigated. It is shown that a discontinuity of velocity appears on the recombination front, transforming after the front stopping into two shocks. A relation between the initial amplitudes of the shocks and the effective mass of hydrogen ionization zone is obtained. These recombination shocks form a shoulder on the light curve, as well as on the radial velocity curve.

122.047   GY Cygni – Erforschung eines veränderlichen Sterns.
D. Böhme.
Astron. Raumfahrt, 25. Jahrg., Heft 5, p. 153 – 155 (1987).

122.048   Spectroscopic investigation of AC Herculis. 1. Parameters of the atmosphere and chemical composition.
Yu. V. Borisov, V. E. Panchuk.
Bull. Spec. Astrophys. Obs. – North Caucasus, Vol. 22, p. 15 – 21 (1987). English translation of 41.122.101.

122.049   Extreme hydrogen–deficient stars.
P. W. Hill.
Q. J. R. Astron. Soc., Vol. 28, No. 3, p. 225 – 230 (1987). – See Abstr. 012.037.
Some properties and problems of the extreme hydrogen–deficient stars are discussed. These stars comprise the R CrB variables, the hydrogen–deficient carbon stars and the extreme helium stars. Also discussed are the hydrogen–deficient binaries and the pulsating helium star V652 Her.

122.050   The reclassification of the supposed dwarf nova V1285 Cygni as a semiregular variable.
A. Bruch, R. Aniol, B. Cunow.
Astron. Astrophys., Vol. 185, No. 1/2, p. 203 – 205 (1987).
V1285 Cyg is classified in the General Catalogue of Variable Stars as a possible dwarf nova of Z Cam type. New photographic photometry and spectrographic observations presented here lead to a reclassifaction of V1285 Cyg as a semiregular variable of type SRB.

122.051   The period of BW Vulpeculae.
D. van der Linden, C. Sterken.
Astron. Astrophys., Vol. 186, No. 1/2, p. 129 – 135 (1987).
The data of the 1982 BW Vul observing campaign are examined for periodicity. A precise redetermination of the period yields $P = 0^{d}2010425 \pm 0^{d}0000007$. The presence of other periodicities with amplitudes larger than $0^{m}01$ is excluded. The different shapes of the stillstand phases do not repeat themselves in a periodic way, as far as periods longer than two days are concerned. An independent analysis of the period variation is carried out using historic and new times of minimum light.

122.052   Cepheids in the Magellanic Clouds. II. Search for double mode Cepheids in the LMC.
G. K. Andreasen.
Astron. Astrophys., Vol. 186, No. 1/2, p. 159 – 169 (1987).
The author uses photographic B and V observations published by Martin et al. (1981) and Wayman et al. (1984) for 244 Cepheids in the Large Magellanic Cloud to search for double mode pulsators. Constructing power spectra by means of simultaneous least squares fits with two periods, he selects about 10 candidates for double mode pulsation. Two of these stars: DV 14 and HV 2345 with fundamental periods $\Pi_0 = 3.298$ and 4.743 days consistently show the pulsation behaviour expected for double mode pulsators in three checks. They have period ratios $\cong 0.70$, and, therefore, seem to be the first found members of the group of double mode Cepheids expected in the Magellanic Clouds.

122.053   An investigation of Cepheid variable stars using hydrostatic model atmospheres.
R. B. Hindsley.
Diss. Abstr. Int., Sect. B, Vol. 48, No. 1, p. 161–B (1987). Thesis, University of Maryland College Park, 204 pp. (1986). Order No. DA8709078.

122.054 Atmospheres of Delta Scuti pulsating variables. VI. Analysis of the spectrum of HD 127986.
T. M. Rachkovskaya.
Izv. Krymskoj Astrofiz. Obs., Tom 76, p. 3 – 10 (1987). In Russian. English translation in Bull. Crimean Astrophys. Obs., Vol. 76.

Using the spectrograms with dispersion 8 and 12 Å/mm a model atmosphere analysis of the pulsating star HD 127986 has been carried out. The values of effective temperature $T_{eff} = 6150 \pm 200$K, surface gravity $\log g = 3.3 \pm 0.2$ and microturbulent velocity $\xi_t = 3.8 \pm 0.5$ km/s were determined. The abundance of 20 elements in the atmosphere of HD 127986 were determined. Chemical composition of HD 127986 differs from the solar one, and is the same as for the pulsating variables δ Del, 44 Tau and V644 Her. The mass, the radius, the luminosity and the age of HD 127986 is found.

122.055 The oscillation period of Delta Scuti stars close to 160 minutes.
V. A. Kotov.
Izv. Krymskoj Astrofiz. Obs., Tom 76, p. 10 – 20 (1987). In Russian. English translation in Bull. Crimean Astrophys. Obs., Vol. 76.

The so–called «resonance power spectrum» (or «commensurability spectrum») computed for 217 δ Sct stars shows that the dominant (the most commensurate) period for the total set of oscillation periods of these stars equals to 162.2 ± 2.8 min. This value coincides fairly well, within the limits of error, with the famous 160–min period of global oscillations of the Sun. The finding appears to favour the nonlinear mechanism of resonant interaction between different modes of stellar oscillations, with the 160–min period being the most characteristic (resonant) one.

122.056 Nonradial oscillations in δ Scuti stars.
G. A. H. Walker, S. Yang, G. G. Fahlman.
Bull. Am. Astron. Soc., Vol. 19, No. 2, p. 707 (1987). Abstract. – See Abstr. 010.061.

122.057 A spectroscopic search for binaries among southern hemisphere long period variable stars.
R. B. Culver, P. A. Ianna.
Bull. Am. Astron. Soc., Vol. 19, No. 2, p. 709 (1987). Abstract. – See Abstr. 010.061.

122.058 The orbit of the cepheid AW Per.
D. L. Welch, N. R. Evans.
Bull. Am. Astron. Soc., Vol. 19, No. 2, p. 710 (1987). Abstract. – See Abstr. 010.061.

122.059 A new orbit for the binary cepheid S Sagittae.
M. H. Slovak, T. G. Barnes.
Bull. Am. Astron. Soc., Vol. 19, No. 2, p. 753 (1987). Abstract. – See Abstr. 010.061.

122.060 A database of RR Lyr star radial velocities.
D. J. Westpfahl, D. L. Welch.
Bull. Am. Astron. Soc., Vol. 19, No. 2, p. 753 (1987). Abstract. – See Abstr. 010.061.

122.061 The Cepheid surface–brightness relation and the slope of the P–L relation.
T. G. Barnes, T. J. Moffett, W. H. Jefferys, S. L. Hawley.
Bull. Am. Astron. Soc., Vol. 19, No. 2, p. 754 (1987). Abstract. – See Abstr. 010.061.

122.062 Distances to galactic cepheids.
T. J. Moffett.
Bull. Am. Astron. Soc., Vol. 19, No. 2, p. 754 (1987). Abstract. – See Abstr. 010.061.

122.063 Spectropolarimetry of R Coronae Borealis during 1986 – 87.
G. C. Clayton, S. A. Stanford, M. R. Meade, B. A. Whitney, M. A. Murison, C. M. Anderson, M. A. Nook, K. H. Nordsieck.
Bull. Am. Astron. Soc., Vol. 19, No. 2, p. 754 (1987). Abstract. – See Abstr. 010.061.

122.064 Analysis of visual observations of cepheid variable stars.
J. R. Percy, K. M. Marcus.
Bull. Am. Astron. Soc., Vol. 19, No. 2, p. 754 (1987). Abstract. – See Abstr. 010.061.

122.065 Shock wave phenomena in β Cephei stars.
R. A. Crowe, D. Gillet.
Bull. Am. Astron. Soc., Vol. 19, No. 2, p. 754 (1987). Abstract. – See Abstr. 010.061.

122.066 RU Piscium: a double–mode RR Lyrae star?
C. Mendes de Oliveira, J. M. Nemec.
Bull. Am. Astron. Soc., Vol. 19, No. 2, p. 755 (1987). Abstract. – See Abstr. 010.061.

122.067 RR Lyrae light curves: another look.
N. R. Simon.
Bull. Am. Astron. Soc., Vol. 19, No. 2, p. 762 (1987). Abstract. – See Abstr. 010.061.

122.068 Multisite observations of UU Herculis: 1985 results.
D. Sasselov, E. Zsoldos, J. D. Fernie, A. Arellano Ferro.
Publ. Astron. Soc. Pac., Vol. 99, No. 619, p. 967 – 970 (1987).

The results of a joint effort to monitor photometrically the remarkable variable star UU Her from four observatories in Europe and North America are presented. UU Her is known to exhibit two distinct pulsational modes. During 1984 – 85 the amplitude of the pulsations increased and there seemed to be a change in the $V, (B-V)$ phase shift from negative in 1984 to positive in 1985. Meanwhile, there was no discernible change in the pulsational period or the shape of the light curve. High–dispersion spectra of UU Her from 3500 Å to 6900 Å were also obtained. There are no obvious traces of mass loss or shell phenomena in its optical spectrum. The status of UU Her remains unclear.

122.069 A check on EU Tauri.
J. D. Fernie.
Publ. Astron. Soc. Pac., Vol. 99, No. 620, p. 1093 – 1096 (1987).

New $uvby\beta RI$ photometry has been obtained to check the suggestion that the two–day classical Cepheid EU Tau had changed its period slightly and also showed a secondary period of half the primary period. The present photometry does not support this suggestion. The earlier period is confirmed and improved, and there is no evidence of a secondary period. The view that EU Tau is an overtone pulsator is supported.

122.070 Ephemerides of RR Lyrae–type variables for the year 1988.
B. N. Firmaniuk (*B. N. Firmanyuk*), J. M. Kreiner.
Rocznik Astronomiczny Obserwatorium Krakowskiego 1988, p. 115 – 132 (1987). – See Abstr. 046.024.

122.071 Multivariability of RS Bootis.
S. Kanyó.
Commun. Konkoly Obs., No. 87, p. 1 – 37 (1986). ISBN 963–8361–24–7.

3420 photoelectric observations obtained between 1971 and 1978 are reported. The long period variation obtained by Oosterhoff has been confirmed and its change defined ($P_1 = 533$ days). A shorter cycle of 62 days has been found superimposed on the long period. A secular change in the pulsation period has been found ($\beta = 10^{-11}$ day cycle^{-1}). Detached O–C_2 residuals give a curve of a sinusoidal form which, in a binary system, could be due to an orbit period motion of 70 years.

122.072 Photoelectric observations of SZ Lyncis.
M. A. Soliman, M. A. Hamdy, B. Szeidl, L. Szabados.
Commun. Konkoly Obs., No. 88, p. 39 – 56 (1986). ISBN 963–8361–25–5.

1018 photoelectric observations of SZ Lyn obtained at Kottamia and Konkoly Observatories between 1984 and 1986 are reported. The long–period variation in time of maximum light discovered by van Genderen was investigated and the value of its period ($P_B = 1173.5$ days = 3.213 years) and semiamplitude have been improved, and a secular change in the pulsation period has been found.

122.073 Period changes of RR Lyrae stars. II. TW Her, VZ Her, AV Peg and TU UMa.
B. Szeidl, K. Oláh, A. Mizser.
Commun. Konkoly Obs., No. 89, p. 57 – 110 (1986). ISBN 963–8361–26–3.

Photographic and photoelectric observations obtained at Konkoly Observatory during the past 35 years are presented. Using all available observations the O–C diagrams of TW Her, VZ Her, AV Peg and TU UMa are constructed. The period of TW Her is constant whereas that of VZ Her, AV Peg and TU UMa is changing.

122.074 The light curve of V441 Herculis.
G. A. Bakos.
Commun. Konkoly Obs., No. 90, p. 111 – 119 (1987). ISBN 963–8361–27–1.

The light variations of the F–type supergiant V441 Her have been followed for a period of five years in the B and V pass bands. A pulsation period of 63.9 days has been derived. The residuals show a harmonic distribution with a period of about 1500 days. Arguments are presented in favour of a circumstellar envelope around the F–type star and an infrared companion to this star.

122.075 A search for pulsating stars similar to PG 1159–035 and K1–16.
A. D. Grauer, H. E. Bond, J. Liebert, T. A. Fleming, R. F. Green.
Space Telesc. Sci. Inst., Prepr. Ser., No. 185, 30 pp. (1987). To appear in Astrophys. J.

122.076 Optical and infrared observations of the carbon Mira R For. Dust shell modelling as a function of phase.
T. Le Bertre.
ESO Sci. Prepr., No. 527, 30 pp. (1987). To appear in Astron. Astrophys.

122.077 Observations of RR Lyrae variables in Baade's window.
P. Mack, A. R. Walker.
S.Afr. Astron. Obs., Circ., No. 11, p. 121 – 124 (1987).

The authors present photometry for 13 short period variable stars in the NGC 6522 Baade window field. An analysis of the results has been made by Walker & Mack (1986). The observations were made with the CCD camera on the SAAO 1.0–m telescope using B, V and I filters. A table contains the heliocentric Julian dates of observation and the corresponding V, B, and I magnitudes.

122.078 Notes on Cepheid variables in galaxies more distant than the Magellanic Clouds.
A. R. Walker.
S.Afr. Astron. Obs., Circ., No. 11, p. 125 – 130 (1987).

In order to estimate the distances to the galaxies, notes and references are given for galaxies more distant than the Magellanic Clouds, which are known to contain Cepheid variables.

122.079 Notes on galactic Cepheids useful for determining the period–luminosity and the period–luminosity–colour zeropoints.
A. R. Walker.
S.Afr. Astron. Obs., Circ., No. 11, p. 131 – 139 (1987).

Notes and references are given for the galactic Cepheid variables which lie in clusters and associations. Some other stars that have at times been thought to be in clusters are also noted, together with a few stars which may become useful calibrators when more data pertaining to them are obtained.

122.080 Doppler imaging of variable early–type stars.
D. Baade.
ESO Sci. Prepr., No. 529, p. 9 – 13 (1987). Paper presented at IAU Colloq. No. 132.

122.081 Nonradial pulsations and the Be phenomenon.
D. Baade.
ESO Sci. Prepr., No. 529, p. 3 – 7 (1987). Paper presented at IAU Colloq. No. 132.

122.082 The nonradial oscillations of ε Persei. II. Nonlinear characteristics.
M. A. Smith, A. W. Fullerton, J. R. Percy.
Astrophys. J., Vol. 320, No. 2, p. 768 – 793 (1987).

During five nights in 1984 November the authors obtained over 300 high quality CCD exposures of the Si III $\lambda4552$–4574 lines in ε Persei (B0.7 III), a star with remarkably large line profile variations. Detailed line profile models have been computed for a sample of over 60 of them, and these show that the variations are described well in the framework of nonradial pulsation (NRP) by two modes, a dominant one described by $-m = l = 4$ (mean period, 3.85 ± 0.02 hr) and a secondary one by $-m = l = 6$ (2.25 ± 0.3 hr). This study provides also a definition of the nonlinear properties of NRP waves, which can show extreme variations in amplitude both in time and around the wave pattern. The analysis suggests a simple picture in which a wave becomes marginally supersonic in the stellar atmosphere, develops an increasingly nonsinusoidal waveform, and partially loses its coherence by interfering with itself as it circuits the star.

122.083 Discovery of high–degree nonradial oscillations in rapidly rotating δ Scuti stars?
G. A. H. Walker, S. Yang, G. G. Fahlman.
Astrophys. J., Lett. Ed., Vol. 320, No. 2, p. L139 – L143 (1987).

The authors obtained time–resolved spectral series covering a single photometric period in the region of $\lambda4500$ for four rapidly rotating (~ 120 km s^{-1}) δ Scuti stars. In all four cases the line profiles show the unmistakable signature of traveling subfeatures already seen in Spica and ζ Ophiuchi, where they have been attributed to high–degree nonradial pulsations. On the simplifying assumption that the subfeatures are periodic and can be ascribed to a nonradial, spherical harmonic motion with $l = |m|$, and that stellar rotation dominates the acceleration, the authors calculate values of $|m| = 8, 14, 16, 16$ for 21 Mon, $\varkappa^2$ Boo, υ UMa, o^1 Eri, respectively.

122.084 Visuelle Beobachtung des halbregelmäßigen Veränderlichen RR CrB durch den AKV.
D. Böhme.
Mitt. Veränderliche Sterne, Band 11, Heft 2, p. 46 – 47 (1987).

122.085 CM Ursae Majoris – ein RR–Lyrae–Stern.
G. Hacke.
Mitt. Veränderliche Sterne, Band 11, Heft 3, p. 51 – 53 (1987).
CM UMa is an RR Lyr type variable with a period of $0^{d}59$.

122.086 Balmer emission profiles in radially pulsating stars: the case of the double Hα emission.
D. Gillet.
Obs. Haute Provence, Pré–Publ., No. 24, 7 pp. (1987). To appear in IAU Symp. No. 132.

122.087   Bump, hump and shock waves in the RR Lyrae stars: X Ari and RR Lyr.
D. Gillet, R. A. Crowe.
Obs. Haute Provence, Pré–Publ., No. 28, 29 pp. (1987). Submitted to Astron. Astrophys.

122.088   Variacion en la amplitud de la curva de luz de la variable del tipo β CMa, δ Cet.
S. González, J. P. Sareyan, R. Garrido, A. Delgado, E. Chapellier.
Rev. Mex. Astron. Astrofis., Vol. 14, No. 1, p. 391 – 394 (1987). – See Abstr. 012.042.
The light variation amplitudes of the β CMa variable δ Cet are increasing at a rate consistent with that observed in radial velocity, which means that the pulsation keeps its characteristics through amplitude variation or modulation. Excellent fit of the observations with the most recent ephemerides shows that δ Cet could eventually be a 13 days period binary.

122.089   Photometric study of the Delta Scuti star Delta Serpentis.
A. Rolland, P. López de Coca, M. A. Hobart, J. H. Peña, L. Parrao.
Rev. Mex. Astron. Astrofis., Vol. 14, No. 2, p. 419 (1987). – See Abstr. 012.042.
The results of the analysis for determining the frequencies of pulsation of the variable star Delta Ser are presented.

122.090   Variaciones fotométricas en la estrella B Iota Herculis.
E. Chapellier, J. M. Le Contel, J. C. Valtier, S. González–Bedolla, D. Ducatel, P. J. Morel, J. P. Sareyan, I. Geiger, P. Antonelli.
Rev. Mex. Astron. Astrofis., Vol. 14, No. 2, p. 424 – 428 (1987). – See Abstr. 012.042.
Spectrographic and photometric observations of the B Type variable star ι Herculis are presented. The relation of ι Herculis with the β CMa group is discussed. The authors confirm the existence of β CMa–like variations outside the instability strip towards lower temperatures.

122.091   Period variations in SX Phe stars: CY Aqr, DY Peg and HD 94033.
J. H. Peña, R. Peniche, S. F. González, M. A. Hobart.
Rev. Mex. Astron. Astrofis., Vol. 14, No. 2, p. 429 – 430 (1987). – See Abstr. 012.042.
An analysis of the times of maximum light of three SX Phe stars was carried out. The results support a monotonous decrement of the period which is consistent with the theoretical models of pre–white–dwarfs of 0.2 $M_\odot$.

122.092   Spectroscopic studies of the southern Cepheid β Dor.
S. Giridhar.
Rev. Mex. Astron. Astrofis., Vol. 14, No. 2, p. 431 – 434 (1987). – See Abstr. 012.042.
Fine analysis of the southern Cepheid β Dor has been conducted at different phases with the aim to derive atmospheric parameters at those phases. Atmospheric abundances have been derived for Fe group elements and s–process elements. The derived parameters of β Dor are compared with earlier estimates.

122.093   The luminosities of the binary Cepheids SU Cyg, SU Cas and W Sgr.
N. R. Evans, A. Arellano Ferro.
Rev. Mex. Astron. Astrofis., Vol. 14, No. 2, p. 435 (1987). Abstract. – See Abstr. 012.042.

122.094   A study of the RR Lyrae in Plaut's field 3.
T. Wesselink, R. S. Le Poole, J. Lub.
ESO Conf. Workshop Proc., No. 27, p. 185 – 189 (1987). – See Abstr. 012.051.

122.095   RR Lyrae variables in galactic globular clusters: observational properties and open problems.
F. Caputo.
ESO Conf. Workshop Proc., No. 27, p. 321 – 330 (1987). – See Abstr. 012.051.
The main observational properties of globular cluster RR Lyrae variables are presented and discussed. For some apparent disagreement between theory and observation a solution is given as suggested from the analysis of RR Lyrae variables in the "anomalous" cluster ω Centauri.

122.096   The flare energy spectrum of EV Lac.
L. N. Mavridis, S. Avgoloupis.
Astron. Astrophys., Vol. 188, No. 1, p. 95 – 99 (1987).
A study is made of the flare energy spectrum of the flare star EV Lac for the period 1974 – 79, based on the homogeneous series of observational data obtained at the Stephanion Observatory. The spectrum varies significantly during the 5–year activity cycle of the star. Analogous variation is also observed in the time–averaged rate of flare energy release of the star, which becomes at least twice as high during the years of maximum flare activity than during the rest of the cycle. Arguments are presented which show that the validity of the power law for the flare energy spectrum should be considered with due caution.

122.097   Verbesserung der Elemente für RS Bootis und XZ Cygni.
E. Wunder.
BAV Rundbrief, 36. Jahrg., Nr. 4, p. 157 – 163 (1987).

122.098   Klassische Pulsationsveränderliche.
G. Hacke.
Sterne, 63. Band, Heft 5, p. 302 – 310 (1987).

122.099   Langzeitverhalten des Mirasterns T Herculis.
E. Pfitzner.
Mitt. Veränderliche Sterne, Band 11, Heft 1, p. 18 (1986).

122.100   Photoelektrische Beobachtung von R CrB im Maximum.
D. Böhme.
Mitt. Veränderliche Sterne, Band 11, Heft 1, p. 25 (1986).

122.101   Verbesserte Periode von V445 Cassiopeiae.
K. Häußler.
Mitt. Veränderliche Sterne, Band 11, Heft 1, p. 25 – 26 (1986).

122.102   Photoelektrische Beobachtung von W Boo.
D. Böhme.
Mitt. Veränderliche Sterne, Band 11, Heft 1, p. 34 (1986).

122.103   RV Andromedae.
IAU Circ., No. 4489 (1987).

122.104   V482 Cygni.
IAU Circ., Nos. 4511, 4515 (1987).

122.105   V482 Cyg.
Yamamoto Circ., No. 2100 (1987). In Japanese.

122.106   Polarimeric investigation of five cepheids.
T. A. Polyakova.
Tr. Astron. Obs., Leningrad, Tom 41, p. 137 – 143 (1987). Uch. Zap. LGU, No. 420, Ser. Mat. Nauk, Vyp. 63. In Russian.
Results of polarimetric observations of SZ Cas, S Sge, RT Aur, RX Cam, and X Cyg in 1983, 1984 are presented. The $p_x p_y$ diagrams and the change of their intrinsic polarization with phase are discussed.

122.107   Two–color photoelectric observations of the δ Sct variable V1208 Aquilae.
N. L. Magalashvili, Ya. I. Kumsishvili.
Astron. Tsirk., No. 1459, p. 7 – 8 (1986). In Russian.
Photoelectric BV–observations of the δ Sct star V1208 Aql were carried out during 1984 – 1985. The light amplitude and the form of the light curves are changing with time.

122.108   Photometric behaviour of FG Sagittae in 1985.
V. P. Arkhipova, G. V. Zajtseva, R. I. Noskova.
Astron. Tsirk., No. 1463, p. 6 – 8 (1986). In Russian.
The photoelectric UBV–observations of FG Sge carried out during 46 nights in 1985 are discussed. The prolonged reddening of the star was strongly reduced due to the possible stop of the star cooling.

122.109   Results of an investigation of light curves of Mira–type variable stars.
I. A. Klyus.
Astron. Tsirk., No. 1464, p. 5 – 6 (1986). In Russian.
It is found that in some Mira–type variable stars the process of light variation represents the superposition of purely accidental and deterministic periodic processes. This phenomenon can be accounted for by the action of two mechanisms: that of local fluctuations of surface brightness and that providing periodic light variation of the whole star.

122.110   New variable star in the region of the M56 globular cluster.
B. L. Shaganyan.
Astron. Tsirk., No. 1487, p. 7 – 8 (1987). In Russian.
A new variable star possibly of SRa–type with the cycle length of 59^d9 was discovered. Its coordinates are α = $19^h01^m42^s$, δ = + 30°05′.1 (1900).

122.111   Photoelectric observation of AD Canis Minoris and the change in its period.
S.–y. Jiang.
Chin. Astron. Astrophys., Vol. 11, No. 4, p. 343 – 347 (1987). English translation of 43.122.122.

122.112   Galactic classical Cepheids: gamma velocities.
T. J. Moffett, T. G. Barnes III.
Publ. Astron. Soc. Pac., Vol. 99, No. 621, p. 1206 – 1208 (1987).
The γ velocities for 61 galactic classical Cepheids have been determined from recent homogeneous radial–velocity measurements combined with older published velocities. The results confirm those of Caldwell and Coulson (1987), who used a largely different data set and an independent analysis. This makes it unlikely that the net blueshift of 3 ± 1 km s^{-1} found by Caldwell and Coulson for galactic Cepheids is a result of approximations used in determining the γ velocities. The Cepheids Y Sct and Z Lac are found to be binary star candidates.

122.113   Fourier coefficients of the RR Lyrae variables in NGC 6171.
R. F. Stellingwerf, R. J. Dickens.
Astrophys. J., Vol. 322, No. 1, p. 133 – 141 (1987).
The authors present the set of Fourier coefficients of the 20 RR Lyrae stars in the globular cluster NGC 6171, based on the observations of Dickens. The Fourier phases and amplitudes are shown to vary smoothly and correlate well with two simple quantities: the skewness of the light curve (determines the amplitudes), and the acuteness, or narrowness, of the light curve (determines the phases). The fundamental and overtone pulsators follow a smooth relation as these quantities vary, but a change in slope is apparent when the mode changes.

122.114   Flares of Orion Population variables in the association Taurus T3.
A. S. Khodzhaev.
Astrofizika, Tom 27, Vyp. 2, p. 207 – 217 (1987). In Russian. English translation in Astrophysics, Vol. 27, No. 2.
Thirteen new flare stars, proved to be irregular variables of the Orion Population, were discovered. Seventeen flares on these stars were detected for about 750 hours of the effective observing time. A great variety and multiplicity and various dynamics of flare energy release processes are shown. The existence of flare stars with some properties typical for both of the T Tauri and UV Ceti stars simultaneously indicates an intimate relation between the above mentioned types of young nonstable stars. The population of flare stars in the Taurus Dark Cloud region is apparently as young as in Orion and Monoceros.

122.115   Period shortning of SS CMa.
M. Koshiro.
Variable Star Bull., No. 1, p. 2 – 3 (1987).

122.116   Periodicity of V1152 Cygni.
M. Huruhata.
Variable Star Bull., No. 2, p. 8 (1987).

122.117   A revised ephemeris for the RRab variable BB Bootis.
P. Louis.
GEOS Circ., RR 10, 4 pp. (1987).
The author obtains the following revised ephemeris: Max: Hel. J. D. 2440872.715 + 0.4727524 E.

122.118   Instabilities in the light curve of the RR Lyrae star SS Leo.
I. Skillen, J. A. Fernley, D. Kilkenny.
Inf. Bull. Variable Stars, No. 3036, 2 pp. (1987).

122.119   Flares of HD 97766.
Z.–s. Zhang, Y.–l. Li, X.–h. Wang.
Inf. Bull. Variable Stars, No. 3050, 3 pp. (1987).

122.120   UBV photometric observations of DY Pegasi.
H. A. Mahdy.
Inf. Bull. Variable Stars, No. 3055, 5 pp. (1987).

122.121   McNaught's variable at 7^h23^m –03° is a Mira star.
W. Wenzel.
Inf. Bull. Variable Stars, No. 3063, 1 p. (1987).

122.122   Time–evolution of colours of two EV Lac flares.
K. P. Panov, T. Korhonen.
Inf. Bull. Variable Stars, No. 3064, 4 pp. (1987).

122.123   The optical behaviour of KR Aurigae in the season 1986/87.
W. Götz.
Inf. Bull. Variable Stars, No. 3065, 2 pp. (1987).

122.124   Flares on AD Leo in 1973 and 1983.
R. B. Herr, D. B. Opie.
Inf. Bull. Variable Stars, No. 3069, 8 pp. (1987).

122.125   Colorimetric observations of Y Ori.
N. D. Melikian (*N. D. Melikyan*),
R. Sh. Natsvlishvili, M. della Valle.
Inf. Bull. Variable Stars, No. 3072, 3 pp. (1987).

122.126   Important parameters of four small–amplitude cepheids from uvby photometry.
A. Arellano Ferro, S. Giridhar, L. Parrao.
Inf. Bull. Variable Stars, No. 3077, 4 pp. (1987).

122.127   UBV photometry of the R Coronae Borealis star RY Sgr – 1984 to 1986.
W. A. Lawson, P. M. Kilmartin, A. C. Gilmore, M. Clark.
Inf. Bull. Variable Stars, No. 3085, 4 pp. (1987).

122.128 HD 27503 = LB 3345, a new Delta Scuti star.
D. Kilkenny.
Inf. Bull. Variable Stars, No. 3092, 2 pp. (1987).

122.129 Some unpublished photometric observations of AC Herculis.
M. Santangelo.
Inf. Bull. Variable Stars, No. 3094, 2 pp. (1987).

122.130 Epsilon Persei and Mu Centauri as single–periodic rapid variables.
P. Harmanec.
Inf. Bull. Variable Stars, No. 3097, 4 pp. (1987).

122.131 Pulsation periods in Delta Serpentis.
P. López de Coca, A. Rolland.
Inf. Bull. Variable Stars, No. 3108, 1 p. (1987).

122.132 Photoelectric observations of R CrB.
D. Böhme.
Inf. Bull. Variable Stars, No. 3115, 1 p. (1987).

122.133 More about FY Aql and GRBS 790331.
R. Hudec.
Inf. Bull. Variable Stars, No. 3121, 2 pp. (1987).

122.134 A search for pulsating stars similar to PG 1159–035 and K1–16.
A. D. Grauer, H. E. Bond, J. Liebert, T. A. Fleming,
R. F. Green.
Astrophys. J., Vol. 323, No. 1, p. 271 – 279 (1987).
High–speed photometric observations have been obtained for 15 stars sharing some of the properties of the class of hot, hydrogen–deficient pulsators represented by PG 1159–035 and the nucleus of the planetary nebula K1–16. With the exception of PG 0122 + 200, the stars showed no evidence of periodic variability on time scales of 60 s up to as long as 1 – 2 hr; typical amplitude limits are a few millimagnitudes. Also presented are *UBV* photometry and optical spectroscopy for several of the stars. The authors discuss what is known about the temperatures, gravities, luminosities, and compositions of the program stars and assess what has been learned about the location of the instability strip in the HR diagram.

122.135 Observational studies of Cepheids. VI. Period–radius relations.
T. J. Moffett, T. G. Barnes III.
Astrophys. J., Vol. 323, No. 1, p. 280 – 287 (1987).
Using the visual surface brightness technique, the authors have determined the radii of 63 classical Cepheids covering a period range of 3 to 45 days. The resulting period–radius relation is in reasonable agreement with the theoretical period–radius relation for Cepheids. After reviewing all of the modern Baade–Wesselink type solutions, the authors find that they bracket the theoretical period–radius relation. These results cast doubt on the validity of the beat/bump period–radius relation and suggest that the distance and/or temperature scales for classical Cepheids may need adjustment.

Machine–readable version of the Tonantzintla catalogue of the Pleiades flare stars.
See Abstr. 002.091.

Activité de l'A.F.O.E.V. en 1986.
See Abstr. 010.101.

Techniques for observing stellar oscillations.
See Abstr. 036.001.

Observational requirements for a synoptic spectroscopic study of nonradial pulsations in OB stars.
See Abstr. 036.002.

On the inversion of the Baade–Wesselink technique.
See Abstr. 036.124.

Modal selection in stellar pulsators. II. Application to RR Lyrae models.
See Abstr. 065.001.

Theory of non–radial pulsations in massive early–type stars.
See Abstr. 065.021.

Non radially pulsating Wolf–Rayet stars.
See Abstr. 065.024.

Application of time–dependent Fourier analysis to nonlinear pulsational stellar models.
See Abstr. 065.044.

Second–overtone models of RR Lyrae stars.
See Abstr. 065.045.

Evolutionary models for R CrB stars.
See Abstr. 065.062.

Linear nonadiabatic pulsations of R CrB models.
See Abstr. 065.063.

Period doubling bifurcations and chaos in W Virginis models.
See Abstr. 065.069.

Weight functions in adiabatic stellar pulsations. I. Radially symmetric motion.
See Abstr. 065.075.

Relativistic stellar pulsations.
See Abstr. 066.141.

Energy distribution of 15,000 solar flares.
See Abstr. 073.098.

HD 213985: a hot post–AGB star in the galactic halo.
See Abstr. 112.007.

Long–term variability of the far–UV high velocity components in γ Cas (1978 – 1986).
See Abstr. 112.017.

An extended nebulosity surrounding the S Dor variable R 127.
See Abstr. 112.043.

The LMC – S Dor variable R 71: an IRAS point source.
See Abstr. 112.044.

The light– and colour variation of Eta Carinae for the years 1983 – 1986 in the VBLUW system.
See Abstr. 112.046.

Do superluminous stars really explode?
See Abstr. 112.047.

Hα echelle spectroscopy of Be stars: an atlas.
See Abstr. 112.054.

Studies in Be–star variability. 2. Analysis of published radial velocities of six bright emission–line stars.
See Abstr. 112.064.

Observations of rapid variability in Be stars.
See Abstr. 112.068.

Is HR 9070 really pulsating?
See Abstr. 112.071.

Simultaneous spectroscopy and polarimetry of Be stars.
See Abstr. 112.072.

The short–period photometric variability of Be stars.
See Abstr. 112.073.

Some important results from two photometric campaigns on short term variability in Be stars.
See Abstr. 112.074.

Be stars as nonradial pulsators.
See Abstr. 112.104.

Nonradial pulsations and the Be phenomenon.
See Abstr. 112.107.

Dust shells around Miras and OH/IR stars: interpretation of IRAS and other infrared measurements.
See Abstr. 112.123.

Spectroscopy and photometry of some intrinsic B variable stars.
See Abstr. 112.150.

Spectroscopic monitoring of southern Be stars.
See Abstr. 112.151.

Fast transient X–rays from flare stars and RS CVn binaries.
See Abstr. 112.160.

Émission maser de la molécule SiO provenant des variables à longue période.
See Abstr. 112.165.

H_2O masers in circumstellar envelopes.
See Abstr. 112.192.

Which photometric period for WR 16?
See Abstr. 113.027.

A high precision photometric investigation of the micro–variations of Wolf-Rayet stars.
See Abstr. 113.028.

The search for BY Dra–type variability in red dwarf stars.
See Abstr. 113.056.

An upper limit on p–mode amplitudes in β Hyi.
See Abstr. 114.005.

Chemical compositions of post–AGB stars.
See Abstr. 114.098.

Low–dispersion spectra of six Me stars.
See Abstr. 114.123.

A frequency solution to the light variations in the rapidly oscillating Ap star HD 6532: evidence for an $l = 1$ oblique dipole oscillation.
See Abstr. 116.017.

Rapid, low–level X–ray variability in active late–type dwarfs.
See Abstr. 116.026.

Observations and analysis of the photospheric magnetic fields on dwarf G, K and M stars.
See Abstr. 116.061.

BV light curves of BY Dra in 1986.
See Abstr. 116.070.

Observed variations in Wolf–Rayet stars.
See Abstr. 117.062.

Symbiotic Miras.
See Abstr. 117.070.

Model of binary stellar systems containing a black hole for explaining the properties of some cepheids in the IC 1613 galaxy.
See Abstr. 117.170.

Photometric study of V361 Lyr: a "hot spot" between the components of a binary system?
See Abstr. 117.254.

PG 0122 + 200: a new member of the GW Virginis (PG 1159–035) class of extremely hot pulsating white dwarfs.
See Abstr. 126.090.

Systematic identification of IRAS point sources.
See Abstr. 133.001.

A study of the silicate emission features of the IRAS low resolution spectra.
See Abstr. 133.006.

VLA observations of radio sources near Wolf 359.
See Abstr. 141.014.

New flare stars in the region of Mon I association.
See Abstr. 152.004.

Membership of Cepheids and red giants in 8 open clusters: NGC 129, 6067, 6087, 6649, 6664, IC 4725, Ly 6, Ru 79.
See Abstr. 153.003.

The binary cepheid DL Cas and the open cluster NGC 129.
See Abstr. 153.008.

CCD photometry of galactic clusters containing Cepheid variables. V. Ruprecht 79.
See Abstr. 153.011.

A photometric study of short period variable stars in open clusters.
See Abstr. 153.040.

The distance and reddening of the open cluster NGC 7790 and the luminosity of its three Cepheids.
See Abstr. 153.042.

Veränderliche Sterne in offenen Sternhaufen.
See Abstr. 153.044.

On the role of flare stars in young open clusters.
See Abstr. 153.056.

H–R diagram of the Pleiades' flare stars.
See Abstr. 153.059.

The distance to M5 from its RR Lyrae variables.
See Abstr. 154.003.

The Oosterhoff dichotomy revisited. I. The ranking of RR Lyrae periods versus metallicity.
See Abstr. 154.017.

The calcium abundance in NGC 1851.
See Abstr. 154.043.

Distribution of classical cepheids in the galactic plane.
See Abstr. 155.018.

Distance moduli and structure of the Magellanic Clouds from near–infrared photometry of classical Cepheids.
See Abstr. 156.055.

Near–infrared observations of Cepheids: the distance to NGC 300.
See Abstr. 157.216.

A search for RR Lyrae variables in NGC 147.
See Abstr. 157.262.

The long–period variable stars of M33.
See Abstr. 157.331.

The 160–minutes period in extragalactic objects: the LMC (RR Lyr stars) and NGC 4151.
See Abstr. 158.222.

123 Variable Stars (Surveys, Lists of Observations, Charts, etc.)

123.001 Beobachtungsergebnisse der Berliner Arbeitsgemein-schaft für Veränderliche Sterne e.V. (BAV).
W. Braune, J. Hübscher.
BAV Mitt., Nr. 46, 16 pp. (1987).
   This 20th compilation of BAV results contains 678 observed minima and maxima on 261 variable stars including 16 photo-electric results.

123.002 Tableaux des observations reçues à l'AFOEV pendant les mois d'avril, mai et juin 1987.
Bull. Assoc. Fr. Obs. Etoiles Variables, No. 41, p. 18 – 46 (1987).

123.003 Expected number of new variable stars by TYCHO photometry with HIPPARCOS.
H. Mauder, E. Høg.
Astron. Astrophys., Vol. 185, No. 1/2, p. 349 – 353 (1987).
   Empirical detection probabilities and statistics of known variable stars lead to the conclusion, that several thousand new variables may be found in the TYCHO photometric survey with HIPPARCOS. In the present paper predicted numbers of detectable variables of several types are derived, based on the 3rd edition of the General Catalogue of Variable Stars by Kukarkin et al. (1969), if not stated otherwise.

123.004 Die veränderlichen Sterne der nördlichen Milchstraße. Teil XVIII.
H. Geßner, E. Splittgerber, H. Busch, K. Häußler.
Veröff. Sternw. Sonneberg, Band 10, Heft 2, p. 169 – 254 (1986).
   The paper comprises informations for the fields γ Aquilae, β Aurigae and 2 Lacertae (including parts of 3 Lacertae and ϱ Cygni). The material presented is not quite homogeneous.

123.005 Neuer Veränderlicher S 10921 Persei.
H. Geßner.
Mitt. Veränderliche Sterne, Band 11, Heft 2, p. 39 (1987).

123.006 Verbesserte Elemente von vier Veränderlichen in Per-seus.
H. Geßner.
Mitt. Veränderliche Sterne, Band 11, Heft 3, p. 65 – 67 (1987).

123.007 Bearbeitung von 17 Veränderlichen (Feld γ Aquilae, Teil IV).
H. Geßner.
Mitt. Veränderliche Sterne, Band 11, Heft 3, p. 67 – 72 (1987).

123.008 Tableaux des observations reçues à l'AFOEV pendant les mois de juillet, août et septembre 1987.
Bull. Assoc. Fr. Obs. Etoiles Variables, No. 42, p. 23 – 64 (1987).

123.009 NSV 6708.
IAU Circ., Nos. 4419, 4420, 4429 (1987).

123.010 Veränderliche im Feld β Trianguli.
L. Meinunger.
Mitt. Veränderliche Sterne, Band 11, Heft 1, p. 1 – 18 (1986).
   On plates of a field around β Trianguli taken with the 40/160 and 40/195 cm astrographs of Sonneberg, 11 new variable stars have been found. For these objects and some other variable stars the type of variability, light curves, coordinates, and charts are given.

123.011 Beobachtungsergebnisse des Arbeitskreises "Veränder-liche Sterne" im Kulturbund der DDR (Teil XIII).
Mitt. Veränderliche Sterne, Band 11, Heft 1, p. 19 – 24 (1986).

123.012 Neuer Veränderlicher S 10920 Persei.
H. Geßner.
Mitt. Veränderliche Sterne, Band 11, Heft 1, p. 33 (1986).

123.013 Supplement to the catalogue of RR Lyrae stars.
B. N. Firmanyuk, V. P. Bezdenezhnyj,
V. G. Derevyagin, L. E. Lysova.
Astron. Tsirk., No. 1462, p. 3 – 4 (1986). In Russian.

123.014 List of variable stars discovered in the USSR and preliminary SVS designations.
Astron. Tsirk., No. 1476, p. 6 – 8 (1987). In Russian.

123.015 A new red variable in Cygnus.
M. Huruhata.
Variable Star Bull., No. 2, p. 6 – 7 (1987).

123.016 The highly probable constancy of IS Geminorum.
F. Fumagalli.
GEOS Circ., SR 10, 10 pp. (1987).
   The analysis of visual estimates shows that IS Gem, classified as an SRd by the GCVS, does not reveal any significant variation of brightness greater than 0.1 mag. It therefore cannot be considered as an SRd and, with a high probability, is not a variable star.

123.017 Predicted dates of maxima and minima of long period variables for 1988.
J. A. Mattei.
Am. Assoc. Variable Star Obs. Bull., No. 51, 14 pp. (1987).

123.018 Observations of dwarf novae, novae, irregular and unusual variables for June – November 1987.
Am. Assoc. Variable Star Obs. Circ., Nos. 201 – 206 (1987).

123.019 **Some new possible variable stars II.**
E. M. Halbedel.
Inf. Bull. Variable Stars, No. 3039, 2 pp. (1987).

123.020 **A possible new binary star in Sagittarius.**
J. H. Pena, J. Campos, R. Peniche.
Inf. Bull. Variable Stars, No. 3054, 4 pp. (1987).

123.021 **The 68ᵗʰ name–list of variable stars.**
P. S. Kholopov, N. N. Samus', E. V. Kazarovets,
N. N. Kireeva.
Inf. Bull. Variable Stars, No. 3058, 30 pp. (1987).

123.022 **New Hα–emission stars in the regions NGC 7000, IC 5068 and IC 5070.**
N. D. Melikian (*N. D. Melikyan*), V. S. Shevchenko,
S. Ju. Melnikov (*S. Yu. Melnikov*).
Inf. Bull. Variable Stars, No. 3073, 3 pp. (1987).

123.023 **Photometric study of the blue variables IW, IZ and IO Andromedae.**
L. Meinunger, I. L. Andronov.
Inf. Bull. Variable Stars, No. 3081, 3 pp. (1987).

123.024 **Identification of variable stars in Sagittarius.**
M. Morel.
Inf. Bull. Variable Stars, No. 3096, 3 pp. (1987).

123.025 **BD + 43°3749 recovered.**
G. Welin.
Inf. Bull. Variable Stars, No. 3098, 1 p. (1987).

123.026 **H–alpha variables in the field of NGC 7000, IC 5068 and IC 5070.**
L. G. Balazs, N. D. Melikyan, S. Yu. Melnikov,
V. S. Shevchenko.
Inf. Bull. Variable Stars, No. 3099, 4 pp. (1987).

123.027 **Flare star observations in the Pleiades–region.**
J. Kelemen.
Inf. Bull. Variable Stars, No. 3103, 6 pp. (1987).

123.028 **A new red variable star.**
M. Kun.
Inf. Bull. Variable Stars, No. 3106, 2 pp. (1987).

123.029 **A new variable star in Cassiopeia.**
J. L. Sedano, E. Rodriguez, P. López de Coca.
Inf. Bull. Variable Stars, No. 3122, 2 pp. (1987).

Status of the Perseus optical flasher.
See Abstr. 013.023.

117th list of minima of eclipsing binaries.
See Abstr. 119.067.

118th list of minima of eclipsing binaries.
See Abstr. 119.071.

Photoelectric minima of eclipsing binaries.
See Abstr. 119.080.

Radio pulses from the Perseus flasher?
See Abstr. 141.012.

The Perseus Flasher and satellite glints.
See Abstr. 143.026.

Evidence from meteor patrol photographs for a nonastronomical origin of the reported optical flashes in Perseus.
See Abstr. 143.034.

Optical flashes from a γ–burster?
See Abstr. 143.046.

11 new variables in globular cluster M5.
See Abstr. 154.080.

124 Novae

124.001 *UBV* **photometry of novae.**
S. van den Bergh, P. F. Younger.
Astron. Astrophys., Suppl. Ser., Vol. 70, No. 1, p. 125 – 140 (1987).
A literature search has been made for all *UBV* photometry of novae. These data were then used to construct *V* lightcurves and *B–V* and *U–B* color curves for Galactic novae. Reddening values have been used to recalibrate the absolute magnitudes of novae. A comparison of 6 photoelectrically observed novae in the Galaxy with those in M31 yields a distance modulus $(m-M)_B = 24.08 \pm 0.25$ for the Andromeda nebula.

124.002 **Radio survey of classical novae.**
M. F. Bode, E. R. Seaquist, A. Evans.
Mon. Not. R. Astron. Soc., Vol. 228, No. 1, p. 217 – 227 (1987).
The authors report the first high–sensitivity radio survey of classical novae. A total of 26 objects were observed with the VLA at either 1.465 or 4.885 GHz. Of these, only two recent novae were detected, and the emission is probably thermal bremsstrahlung. The results suggest that the non–thermal emission of the old nova GK Per is not a common characteristic of novae. The authors further conclude that classical novae are not a major contributor to the cosmic ray electron flux.

124.003 **Models for the remnants of recurrent novae – II. Dynamical effect of radiative heat loss.**
T. J. O'Brien, F. D. Kahn.
Mon. Not. R. Astron. Soc., Vol. 228, No. 1, p. 277 – 287 (1987).
A simple model for the remnant of a recurrent nova was proposed in Paper I (Bode and Kahn, 1985) of this series. One of the conclusions of that paper was that cooling seriously affects the dynamics of such objects at an early stage, some two months in the case of RS Ophiuchi (1985). However, Paper I did not explicitly include the effects of radiative cooling in its treatment of the dynamics. In this paper the authors develop a set of linearized equations, with the cooling parameter q as a first order quantity. The effects of heat loss produce their most pronounced changes in the centre of the remnant so that eventually the hot material is confined to a shell behind the outer shock.

124.004 **The outbursts of classical and recurrent novae.**
M. Livio.
Variability of galactic and extragalactic X–ray sources, p. 201 – 212 (1987). – See Abstr. 012.019.
The "hibernation" scenario for classical novae is discussed. In this model, classical nova systems spend most of the time between eruptions in a state of very low mass transfer rate. The predictions of this model are compared with available observational material. Models for the outbursts of recurrent novae are examined. General properties of outbursts powered by thermonuclear runaways and by accretion events are outlined.

124.005 **Orbital eccentricity in classical novae.**
D. A. Edwards, J. E. Pringle.
Nature, Vol. 328, No. 6130, p. 505 (1987). = Contrib. Lick Obs., No. 445.
The authors consider the effect on the orbital parameters of a classical nova of the ejection of mass during the nova explosion. The most easily observable consequence is the generation of a small eccentricity in the orbit which leads to a luminosity modulation at a period just longer than the orbital period. Observation of such an effect, would have implications not just for interpreting the dynamics of the explosion but also for measuring the secular effect of tidal interaction after the outburst.

124.006 **A search for gamma–ray lines from novae.**
M. D. Leising, G. H. Share, R. L. Kinzer,
E. L. Chupp, D. J. Forrest, E. Rieger.
Bull. Am. Astron. Soc., Vol. 19, No. 2, p. 694 (1987). Abstract. – See Abstr. 010.061.

124.007 **Dust formation in novae.**
·D. Johnson, M. W. Friedlander, J. I. Katz.
Bull. Am. Astron. Soc., Vol. 19, No. 2, p. 724 (1987). Abstract. – See Abstr. 010.061.

124.008 *IRAS* **additional observations of classical novae.**
C. M. Callus, A. Evans, J. S. Albinson,
R. M. Mitchell, M. F. Bode, R. F. Jameson, A. R. King, M. Sherrington.
Mon. Not. R. Astron. Soc., Vol. 229, No. 3, p. 539 – 548 (1987).
Additional (pointed) observations of classical novae, obtained with the Infrared Astronomy Satellite (IRAS), are presented. The data include both old novae and novae in eruption. In the latter case, the IRAS data complement and confirm the results of ground–based infrared observations; while in the former case the IRAS data provide useful constraints on the long–term evolution of the material ejected in nova eruptions.

124.009 **What does an erupting nova do to its red dwarf companion?**
A. Kovetz, D. Prialnik, M. M. Shara.
Space Telesc. Sci. Inst., Prepr. Ser., No. 197, 16 pp. (1987). To appear in Astrophys. J.

124.010 **The large intractable nova shells.**
H. W. Duerbeck.
Messenger, No. 50, p. 8 – 11 (1987).

124.011 **Probable nova in Sagittarius.**
IAU Circ., No. 4482 (1987).

124.012 **RZ Leonis.**
IAU Circ., Nos. 4504, 4506 (1987).

124.013 **Probable novae in M31.**
IAU Circ., Nos. 4515, 4516 (1987).

124.014 **Probable nova in Sagittarius.**
Yamamoto Circ., No. 2097 (1987). In Japanese.

124.015 **Recent observations of novae.**
Br. Astron. Assoc. Circ., Nos. 669, 670, 672 (1987).
Concerning: Nova And 1986, nova Cyg 1986, nova Her 1987, nova Sgr 1987, and nova Vul 1984 No. 2.

124.016 **Positions and proper motions of three novae.**
M. Geffert, A. Barteldrees, B.–C. Kämper, A. Peters,
H.–P. Schmitz, H. Weiland, D. Warnke.
Inf. Bull. Variable Stars, No. 3038, 4 pp. (1987).

124.017 **Spectrum of the recent novae: nova Andromedae 1986 and nova Centauri 1986.**
B. N. Ashoka.
Inf. Bull. Variable Stars, No. 3062, 2 pp. (1987).

124.018 **On the possible nova of Solovyov.**
B. E. Schaefer.
Inf. Bull. Variable Stars, No. 3082, 2 pp. (1987).

124.019 **Positions and finding charts of nova Herculis 1987 and nova Sagittarii 1987.**
H. W. Duerbeck.
Inf. Bull. Variable Stars, No. 3110, 3 pp. (1987).

124.020 **Additional remark on Solovyov's so–called nova Aquilae 1949.**
W. Wenzel.
Inf. Bull. Variable Stars, No. 3114, 1 p. (1987).

Nova Vulpeculae 1984 No. 2 = QU Vulpeculae

124.101 **The unusual radio outburst of Nova Vulpeculae 1984 No. 2.**
A. R. Taylor, E. R. Seaquist, J. M. Hollis, S. R. Pottasch.
Astron. Astrophys., Vol. 183, No. 1, p. 38 – 46 (1987).

Multi–frequency radio monitoring of Nova Vulpeculae 1984 No. 2 has revealed a unique radio light curve, exhibiting a strong outburst which precedes the appearance of normal radio emission from the principal ejecta of the nova by at least 100 days. The early emission is extremely optically–thick and has a brightness temperature in excess of 10^5K, suggestive of either a non–thermal mechanism or free–free emission from shocked gas. A model is discussed in which the radiation is produced by a strong shock propagating outward through the principal ejecta of the nova, as a result of an interaction with a later, high velocity wind from the central source. It is shown that the general features of the radio light curve can be explained by the presence of a central wind with a mass loss rate of $\sim 10^{-5} M_\odot \mathrm{yr}^{-1}$, lasting for a period of 200 to 300 days after the optical outburst. The authors also present the first radio map of nova ejecta shortly after outburst. Comparison of the angular expansion rate of the ejection velocity implies a distance of 3.6 kpc.

124.102 **Nova Vulpeculae 1984 No. 2 (QU Vulpeculae).**
IAU Circ., Nos. 4447, 4461, 4498, 4520 (1987).

Nova Aquilae 1982

124.121 **Nova Aquilae 1982.**
M. A. J. Snijders, T. J. Batt, P. F. Roche, M. J. Seaton, D. C. Morton, T. A. T. Spoelstra, J. C. Blades.
Mon. Not. R. Astron. Soc., Vol. 228, No. 2, p. 329 – 376 (1987).

The paper gives results obtained from observations of Nova Aquilae 1982 made with the International Ultraviolet Explorer satellite, the Anglo–Australian Telescope and the Westerbork Synthesis Radio Telescope, and it discusses the interpretation of these results together with those obtained by other observers.

Nova Vulpeculae 1984 No. 1 = PW Vulpeculae

124.141 **The nova PW Vulpeculae: the spectrum and *UBV*–magnitudes in 1986.**
E. A. Kolotilov, R. I. Noskova.
Pis'ma Astron. Zh., Tom 13, No. 10, p. 886 – 890 (1987). In Russian. English translation in Sov. Astron. Lett., Vol. 13.

The results of spectral and *UBV*–observations of PW Vul (nova Vulpeculae 1984 No. 1) carried out in August–September 1986 are presented. During the observations the star brightness was by $\Delta V \approx 8^{\mathrm{m}}2$ below maximum and the nova was at the nebular stage, as its spectral characteristics confirmed (the brightest emission lines were [O III] an [N II]). The PW Vul track on the two–colour diagram $(U–B)–(B–V)$ from the premaximal to the nebular stage is considered. It is possible that the shell thrown out by the nova had inhomogeneous structure like "the equatorial ring + two polar caps". The authors suggest that the cause of the IR–radiation excess is the primordial dust heated by nova burst.

124.142 **Photographische Beobachtungen der Nova Vul 1984 I = PW Vul.**
G. Hacke.
Mitt. Veränderliche Sterne, Band 11, Heft 2, p. 40 – 45 (1987).

Observations on photographic plates of the Sonneberg Observatory and the Sternberg State Institute, Moscow, are presented. A chart, a list of the comparison stars with their brightness, the list of the observations, and a lightcurve are given.

124.143 **Nova PW Vulpeculae: photometry and spectroscopy in 1984 – 1985.**
E. A. Kolotilov, R. I. Noskova.
Sov. Astron. Lett., Vol. 12, No. 6, p. 370 – 372 (1986). English translation of 42.124.142.

124.144 **PW Vulpeculae.**
IAU Circ., No. 4453 (1987).

124.145 **Visual observations of PW Vul.**
M. Watanabe.
Variable Star Bull., No. 1, p. 3 – 4 (1987).

Nova 1986 No. 32 in M31

124.161 **Spectroscopic observations of nova 1986 No. 32 in M31.**
A. P. Cowley, S. G. Starrfield.
Publ. Astron. Soc. Pac., Vol. 99, No. 618, p. 854 – 857 (1987).

A nova near the nucleus in M31 (1986 No. 32) has been observed spectroscopiclly approximately seven weeks after discovery. Narrow–emission lines principally of hydrogen and Fe II, were detected. The spectrum differs from that of previously observed M31 novae both in the width of the emission lines and in the steepness of the Balmer decrement. The spectral appearance and linewidths suggest that this was a very slow nova. The authors failed to detect a second nova (1986 No. 29) which apparently declined more rapidly.

Nova Cygni 1986 = V1819 Cygni

124.181 **Spectropolarimetry of Nova Cygni 1986.**
B. A. Whitney, G. C. Clayton, M. A. Murison, M. A. Nook, C. M. Anderson, K. H. Nordsieck, M. R. Meade.
Bull. Am. Astron. Soc., Vol. 19, No. 2, p. 753 (1987). Abstract. – See Abstr. 010.061.

124.182 **Nova Cygni 1986 (V1819 Cygni).**
IAU Circ., Nos. 4419, 4435, 4449, 4457, 4496 (1987).

Nova Cygni 1975 = V1500 Cygni

124.201 **Photometry of Nova V1500 Cygni eleven years after outburst.**
J. Kaluzny, I. Semeniuk.
Space Telesc. Sci. Inst., Prepr. Ser., No. 180, 7 pp. (1987). To appear in Acta Astron., Vol. 37.

124.202 **V1500 Cygni.**
IAU Circ., Nos. 4413, 4415, 4458 (1987).

124.203 **Pre–outburst light curve of nova Cygni 1975 (V1500 Cyg).**
R. A. Wade.
Inf. Bull. Variable Stars, No. 3101, 2 pp. (1987).

Nova Ophiuchi 1977 = H 1705–25

124.211 **On the nature of X–ray emission from Nova Ophiuchi (H1705–25).**
V. A. Krol'.
Astrofizika, Tom 27, Vyp. 2, p. 237 – 244 (1987). In Russian. English translation in Astrophysics, Vol. 27, No. 2.

Explanation of the nature of the X–ray emission from the transient source nova Ophiuchi (H1705–25) based on the mechanism of thermal bremsstrahlung from a thin gas layer in a stellar envelope heated by a strong shock wave in the nova outburst has been given. An analytical expression for the observed spectrum that has time–dependent intensity and spectral distribution is obtained. The calculated spectra agree well with the observed data.

Nova Puppis 1942 = CP Puppis

124.221   **Orbital elements of the exnova CP Puppis.**
H. W. Duerbeck, W. C. Seitter, R. Duemmler.
Mon. Not. R. Astron. Soc., Vol. 229, No. 4, p. 653 – 657 (1987).
Radial velocity measurements of the exnova CP Pup yield an orbital amplitude of 92 km s^{-1} and a period of 0.0614215 day, which is shorter than the photometric periods determined by Warner. The mass of the secondary is 0.14 $M_\odot$, that of the primary is below 0.86 $M_\odot$, perhaps as low as 0.12 $M_\odot$.

Nova Persei 1901 = GK Persei

124.231   **An ancient planetary nebula surrounding the old nova GK Persei.**
M. F. Bode, E. R. Seaquist, D. A. Frail, J. A. Roberts,
D. C. B. Whittet, A. Evans, J. S. Albinson.
Nature, Vol. 329, No. 6139, p. 519 – 521 (1987).
As part of a continuing programme of observations of the old nova GK Persei, the authors examined IRAS images in the region of the nova, and discovered extended emission in the far infrared. These observations are interpreted in terms of an ancient planetary nebula ejected from the central binary system. If this interpretation is correct, then several unique phenomena associated with GK Per may be explained. In addition, GK Per would then provide valuable clues to the evolutionary status of classical novae, and the later stages of planetary nebula formation.

Nova Andromedae 1986 = OS Andromedae

124.241   **Nova Andromedae 1986 (OS Andromedae).**
IAU Circ., Nos. 4419, 4452, 4485, 4513 (1987).

124.242   **UBV photometry of nova Andromedae 1986.**
G. Milani, A. Tonello, G. Favero.
Inf. Bull. Variable Stars, No. 3046, 2 pp. (1987).

Nova Herculis 1987

124.251   **Nova Herculis 1987.**
IAU Circ., Nos. 4423, 4425, 4430, 4433, 4459, 4487 (1987).

Recurrent nova V394 Coronae Austrinae

124.261   **V394 Coronae Austrinae.**
IAU Circ., Nos. 4428 – 4430, 4445 (1987).

124.262   **V394 CrA.**
Yamamoto Circ., Nos. 2088, 2090 (1987). In Japanese.

124.263   **V394 CrA (nova).**
Br. Astron. Assoc. Circ., No. 670 (1987).

Nova Sagittarii 1987

124.271   **Nova Sagittarii 1987.**
IAU Circ., No. 4428 (1987).

Nova in LMC

124.281   **Nova in the Large Magellanic Cloud.**
IAU Circ., Nos. 4453, 4456, 4459, 4468 (1987).

Nova Vulpeculae 1979 = PU Vulpeculae

124.291   **PU Vulpeculae.**
IAU Circ., No. 4474 (1987).

Nova Vulpeculae 1987

124.301   **Nova Vulpeculae 1987.**
IAU Circ., Nos. 4488, 4489, 4492, 4493, 4501, 4504, 4509, 4511, 4517 (1987).

124.302   **Nova Vulpeculae 1987.**
Yamamoto Circ., Nos. 2097, 2099 (1987). In Japanese.

Nova Centauri 1986 = V842 Centauri

124.311   **V842 Centauri (nova Centauri 1986).**
IAU Circ., No. 4519 (1987).

Recurrent nova WZ Sagittae

124.321   **The 1978 outburst of the recurrent nova WZ Sagittae.**
X.–l. Hao, B. Mei.
Chin. Astron. Astrophys., Vol. 11, No. 4, p. 288 – 290 (1987).
English translation of Acta Astrophys. Sin., Vol. 7, No. 3, p. 189 – 194 (1987).
The authors give results of photographic observations of the recurrent nova WZ Sge during its 1978 outburst. They discuss the observed properties and calculate the total energy released during the outburst to be about 1.07×10^{40} erg and estimate that the next outburst will occur around the year 2011.

Nova Herculis 1963 = V533 Herculis

124.331   **Photographic photometry of the old nova V533 Her (= Nova Her 1963).**
X. Hao, B. Mei.
Acta Astrophys. Sin., Vol. 7, No. 4, p. 320 – 324 (1987). In Chinese.
Photographic observing results of V533 Her from June 1980 to November 1982 are given in this paper. These results indicate that: (1) During the observing period the old nova declined approximately to the pre–nova level at $m_{pg} \cong 15$. But the brightness fluctuation is unstable. (2) From the light–curve of V533 Her the authors have found that a long–term brightness fluctuation existed with a range of more than one magnitude ($\Delta m \cong 1\overset{m}{.}5$). (3) Short–term fluctuations ($\Delta m \cong 0\overset{m}{.}2$) also existed during one night of the observations.

Near infrared spectroscopy with the array spectrometer of the Anglo–Australian telescope.
See Abstr. 034.166.

Spin–up and mixing in accreting white dwarfs.
See Abstr. 065.005.

Thermonuclear processes in accreting white dwarfs (novae, symbiotic stars, and type–I supernovae).
See Abstr. 065.029.

Are classical novae and dwarf novae the same systems?
See Abstr. 065.066.

Stellar accretion of matter possessing angular momentum.
See Abstr. 065.094.

Nuclear runaways in a C/O white dwarf accreting H–rich material possessing angular momentum.
See Abstr. 065.095.

Gamma ray lines from novae.
See Abstr. 065.096.

Positron annihilation gamma rays from novae.
See Abstr. 065.119.

The effect of heating from the boundary layer on accretion models for novae and other compact objects.
See Abstr. 067.129.

DQ Herculis' orbital period variations according to 1982 – 1986 observations.
See Abstr. 117.020.

Binary system parameters and the hibernation model of cataclysmic variables.
See Abstr. 117.145.

On light curves variability of DQ Her from 1954 to 1985.
See Abstr. 117.168.

Relative C, N, O abundances in red giants, planetary nebulae, novae and symbiotic stars.
See Abstr. 117.202.

Parameters of the X–ray binary system A0620–00 ≡ V616 Mon.
See Abstr. 117.268.

Ancient Guest Stars as harbingers of neutron star formation.
See Abstr. 125.011.

The luminosity oscillation of X–ray bursters and recurrent novae.
See Abstr. 142.040.

Search for the optical variability of the X–ray source E 2000 + 223.
See Abstr. 142.050.

The rate of nova production in the Galaxy.
See Abstr. 155.017.

The spatial distribution and population of novae in M31.
See Abstr. 157.081.

Observations of novae in the Virgo cluster.
See Abstr. 160.019.

125 Supernovae, Supernova Remnants

125.001 Crab–like supernova remnants.
R. H. Becker.
The origin and evolution of neutron stars, p. 91 – 97 (1987). – See Abstr. 012.001 (IAU Symp. No. 125).
 Crab–like SNR are markers for recently formed pulsars. The current catalog of 15 objects allows a direct measure of the space distribution of pulsars, the beaming factor of pulsars, and the energetics at young pulsars. The 15 Crab–like SNR are equally divided between those with surrounding shells and those without. No other properties appear to correlate with the presence or lack of a shell. The ratio of 10 to 1 between all remnants and Crab–like remnants has important implications for the formation of pulsars.

125.002 Neutron stars in twelve supernova remnants.
F. D. Seward.
The origin and evolution of neutron stars, p. 99 – 108 (1987). – See Abstr. 012.001 (IAU Symp. No. 125).
 X–ray observations of selected SNR are summarized. Five contain internal spinning neutron stars – four isolated and one in a binary system. Another seven contain central unresolved sources or bright nebulae. Observations of these nebulae, probably due to synchrotron emission, are used to estimate characteristics of the unseen pulsars.

125.003 High–resolution radio observations of the Crablike supernova remnant 3C 58.
S. P. Reynolds, H. D. Aller.
The origin and evolution of neutron stars, p. 123 (1987). Abstract. – See Abstr. 012.001 (IAU Symp. No. 125).

125.004 Supernova remnants with radio jets.
M. J. Kesteven, J. L. Caswell, R. S. Roger,
D. K. Milne, R. F. Haynes, K. J. Wellington.
The origin and evolution of neutron stars, p. 125 (1987). Abstract. – See Abstr. 012.001 (IAU Symp. No. 125).

125.005 The evolution of young supernova remnants.
J. R. Dickel, E. M. Jones, J. A. Eilek.
The origin and evolution of neutron stars, p. 126 (1987). Abstract. – See Abstr. 012.001 (IAU Symp. No. 125).

125.006 On the origin of Kepler's supernova remnant.
R. Bandiera.
The origin and evolution of neutron stars, p. 127 (1987). Abstract. – See Abstr. 012.001 (IAU Symp. No. 125).

125.007 The X–ray spectra of Tycho and SN 1006.
K. Koyama.
The origin and evolution of neutron stars, p. 128 (1987). Abstract. – See Abstr. 012.001 (IAU Symp. No. 125).

125.008 The galactic sources G 5.4–1.2 and G 5.27–0.90.
J. L. Caswell, M. J. Kesteven, R. F. Haynes,
D. K. Milne, M. M. Komesaroff, R. T. Stewart, S. G. Wilson.
The origin and evolution of neutron stars, p. 129 (1987). Abstract. – See Abstr. 012.001 (IAU Symp. No. 125).

125.009 Supernova remnants in the Magellanic Clouds observed at the Molonglo Observatory.
A. J. Turtle.
The origin and evolution of neutron stars, p. 130 (1987). Abstract. – See Abstr. 012.001 (IAU Symp. No. 125).

125.010 A statistical study of the correlation between galactic SNRs and spiral arms.
Z. W. Li, J. C. Wheeler, F. N. Bash.
The origin and evolution of neutron stars, p. 131 (1987). Abstract. – See Abstr. 012.001 (IAU Symp. No. 125).

125.011 Ancient Guest Stars as harbingers of neutron star formation.
Z.–R. Wang.
The origin and evolution of neutron stars, p. 305 – 318 (1987). – See Abstr. 012.001 (IAU Symp. No. 125).
 The well–known AD 1006, 1054, 1572, and 1604 were described as "Guest Stars" by Chinese, Japanese and Korean. In most cases, it might thus be possible to expect a Guest Star to be a term for supernova or nova. There are a lot of records concerning ancient Guest Stars in Chinese historical books. Two catalogues were compiled by Xi (1955) and Xi and Bo (1965, 1966) that listed 90 probable novae or supernovae observed between 1400 BC and AD 1700. Clark and Stephenson (1977), Ho (1962) and Kanda (1935) collected more or less similar records. Among all the

historical records more than 80% are from China. The discussion presented in this paper is based on them.

125.012   Forbidden coronal iron–line emission in the Puppis A shock front: the effect of inhomogeneities.
R. G. Teske, R. Petre.
Astrophys. J., Vol. 318, No. 1, p. 370 – 378 (1987).
    The authors have obtained CCD images of the shock front at the eastern rim of Puppis A in [Fe X] $\lambda 6374$ and [Fe XIV] $\lambda 5303$ and have compared the optical data to Einstein high resolution image soft X–ray data. Optical and X–ray data are consistent in showing a nearly flat gradient of ionization temperature behind the shock. Scans of surface brightness across the shock in the optical lines were compared to surface brightness predicted by idealized Sedov models. The authors were unable to match both the red and green line scans by a simple, single–component model, and have ascribed the failure to the presence of the density inhomogeneities.

125.013   Does apparent spatial coincidence of pulsars and super-novae mean generic association?
C. E. Akujor.
Astrophys. Space Sci., Vol. 135, No. 1, p. 187 – 190 (1987).
    The probability of pulsars and supernovae being found close to each other on the celestial sphere by chance is calculated. It is found that this corresponds to the number of pulsars found close to supernovae in recent investigations. Thus, the author concludes that given the inaccuracy of distances involved, the apparent spatial coincidence of pulsars and supernovae does not generally imply generic associations.

125.014   G 109.1–1.0: a supernova remnant interacting with a molecular cloud.
K.–i. Tatematsu.
Astron. Her., Vol. 80, No. 8, p. 220 – 222 (1987). In Japanese.

125.015   A redetermination of the X–ray spectrum of SN 1006 and excess diffuse emission from the Lupus region.
K. Koyama, H. Tsunemi, R. H. Becker, J. P. Hughes.
Publ. Astron. Soc. Jpn., Vol. 39, No. 3, p. 437 – 445 (1987).
    X–rays from SN 1006 and from the adjacent Lupus region were separately observed with the Tenma gas scintillation proportional counters. The spectrum of the local excess emission from the Lupus region can be consistently fitted with either a thin thermal bremsstrahlung spectrum with a temperature of 7.5 ± 2.6 keV or a power–law spectrum with a photon index of 2.1 ± 0.1. The X–ray emission from SN 1006, after subtraction of this local excess, has a spectrum which can be described as a power–law spectrum with a photon index of 3.3 ± 0.1 or a thin thermal bremsstrahlung spectrum with a temperature of 1.9 ± 0.1 keV which is much softer than the previously reported spectrum. No significant iron line emission was observed in the SN 1006 spectrum.

125.016   Radioactive mechanism for the luminosity during the late stage of type II supernovae and the ionization of hydrogen.
N. N. Chugaj.
Pis'ma Astron. Zh., Tom 13, No. 8, p. 671 – 679 (1987). In Russian. English translation in Sov. Astron. Lett., Vol. 13.
    The ionization of hydrogen in the type II supernova SN 1970g at the late stage ($t \approx 270$ days) is studied assuming that the supernova luminosity is due to the radioactive decay $^{56}\text{Co} \rightarrow {}^{56}\text{Fe}$. The ionization balance of the hydrogen is calculated. It is shown that (a) nonthermal ionization itself is inadequate; (b) when thermal ionization being taken into account the required ionization degree is reached at $T_e \approx 7000$K; (c) main process of the hydrogen ionization is the photoionization from the level $n = 2$ by Mg II $\lambda 2798$ quanta. The author predicts the intensity of the emission line of Mg II $\lambda 2798$ in the late stage spectra of the SN II comparable with that of the Hα line.

125.017   Properties of supernova remnants at known distances. II. The effect of ambient density on number–diameter relations.
E. M. Berkhuijsen.
Astron. Astrophys., Vol. 181, No. 2, p. 398 – 401 (1987).
    The diameter distributions of SNRs at known distances were statistically analysed. The observed exponent of about $+1$ of the cumulative N–D relations could result from a random distribution of diameters. Such a distribution may be largely ascribed to variations in the ambient density. The diameters were corrected for variations in the ambient density (using the results of Paper I) and the resulting cumulative N–D relations then have exponents considerably larger than $+1$. However, the statistical significance of these exponents does not yet permit definite conclusions on the average expansion law of SNRs.

125.018   Non–spherical supernova remnants. IV. Sequential explosions in OB associations.
G. Tenorio–Tagle, P. Bodenheimer, M. Różyczka.
Astron. Astrophys., Vol. 182, No. 1, p. 120 – 126 (1987).
    Multi–supernova remnants, driven by sequential supernova explosions in OB associations are modelled by means of two-dimensional hydrodynamical calculations. It is shown that due to the Rayleigh–Taylor instability the remnants quickly evolve into highly irregular structures. A critical evaluation of the multi-supernova model as an explanation for supershells is given.

125.019   Evolution of supernova remnants with a central pulsar.
T. A. Lozinskaya.
Sov. Astron., Vol. 30, No. 5, p. 542 – 548 (1986). English translation of 42.125.043.

125.020   Evolution of the Cassiopeia A radio spectrum: evidence for cyclic variability.
A. P. Barabanov, V. P. Ivanov, K. S. Stankevich, S. P. Stolyarov.
Sov. Astron., Vol. 30, No. 5, p. 549 – 555 (1986). English translation of 42.125.044.

125.021   Stratification of optical emission in adiabatic supernova remnants.
K. V. Bychkov, T. G. Sitnik, O. V. Fedorova.
Sov. Astron., Vol. 30, No. 5, p. 555 – 559 (1986). English translation of 42.125.045.

125.022   Observations of X–ray emission of the Crab Nebula and the pulsar NP 0532 on the Astron automatic station.
V. G. Kurt, M. S. Burgin, I. M. Golynskaya, L. S. Gurin, A. V. D'yachkov, V. M. Zenchenko, I. F. Kopaeva, T. A. Mizyakina, V. I. Rubanovskaya, N. A. Savel'eva, V. A. Sklyankin, A. S. Smirnov, V. M. Shamolin, E. Yu. Shafer, E. K. Sheffer.
Sov. Astron., Vol. 30, No. 5, p. 560 – 562 (1986). English translation of 42.125.046.

125.023   Possible binary nature of peculiar type I supernovae; is the satellite a red supergiant?
N. N. Chugaj.
Sov. Astron., Vol. 30, No. 5, p. 563 – 567 (1986). English translation of 42.125.047.

125.024   Neutral hydrogen in the neighborhood of the supernova remnant W50.
I. V. Gosachinskij, V. K. Khersonskij.
Astrophysics, Vol. 25, No. 2, p. 518 – 523 (1987). English translation of 42.125.083.

125.025   Chemical composition of the filaments in the Crab nebula. II. Dispersion of the chemical abundances in the filaments.
V. V. Golovatyj, V. I. Pronik.
Astrophysics, Vol. 25, No. 2, p. 542 – 551 (1987). English translation of 42.125.084.

125.026 Observations at 408 MHz of the supernova remnant HB 3 (G132.6 + 1.5).
T. L. Landecker, J. F. Vaneldik, P. E. Dewdney, D. Routledge.
Astron. J., Vol. 94, No. 1, p. 111 – 122 (1987).

Observations are presented of the supernova remnant HB 3 (G132.6 + 1.5) made at 408 MHz with an angular resolution of 3.5×4.0 arcmin. The flux density at 408 MHz is 75 ± 15 Jy, and the spectral index is 0.60 ± 0.04. The observations show the full extent of the SNR, now seen to be 90×123 arcmin (EW $\times$ NS). The 408 MHz map is compared with X–ray, optical, and CO data. The SNR is a shell remnant with a distorted outline, which suggests that it is in a late stage of its evolution where its development is influenced by the irregularities of the interstellar medium. There is no strong evidence of interaction with the adjacent H II region W3. However, the SNR appears to be immediately behind a CO cloud, and may possibly be interacting with it. A distance of 2.7 ± 0.3 kpc can be assigned to HB 3, making its physical size 71×97 pc. The 408 MHz observations also include the H II region W4.

125.027 The complex structure of Cas A. Consistent model calculations.
M. Contini.
Astron. Astrophys., Vol. 183, No. 1, p. 53 – 60 (1987).

Radiation emissions in the different wavebands from the Cas A supernova remnant are discussed and explained in the light of consistent model calculations. The calculations indicate preshock densities $n_0 \cong 5 \, \mathrm{cm}^{-3}$ and shock velocities, $v_s \sim 5000 \, \mathrm{km \, s^{-1}}$ in the hard X–ray emission region, $n_0 \sim 10$–$50 \, \mathrm{cm}^{-3}$ and $v_s \cong 1000$–$4000 \, \mathrm{km \, s^{-1}}$ in the fast moving knots and $n_0 \cong 100$–$300 \, \mathrm{cm}^{-3}$ and $v_s < 500 \, \mathrm{km \, s^{-1}}$ in the quasi stationary flocculi (QSF). A strong magnetic field $B_0 \cong (10^{-5}$–$10^{-4} \mathrm{g})$ is confirmed in the QSF region. Soft X–rays are created by the reverse shock wave propagating into the circumstellar region and in the accelerated QSF. The intensity of the radio emission produced in the shock fronts by Fermi mechanisms depends on the same parameters. Therefore, a large scale correlation between X–rays, optical and radio emissions is justified while correlations on smaller scales are meaningless, due to the inhomogeneous nature of the interstellar medium.

125.028 Barrel–shaped supernova remnants.
M. J. Kesteven, J. L. Caswell.
Astron. Astrophys., Vol. 183, No. 1, p. 118 – 128 (1987).

The authors suggest that in many supernova remnants the three–dimensional distribution of emission is barrel–shaped, confined to the "staves", and with little or no emissivity in the "end–caps". Two excellent examples are examined in detail. Estimates are given of the wall thickness, and limits are set on the emissivity of the end–caps. Maps of 70 well–resolved remnants are used to examine the incidence of the barrel–shaped morphology. From the statistics it is argued that the majority of supernova remnants fall into the barrel category. Some mechanisms that might be responsible for this appearance are suggested.

125.029 Spectra of Type I supernovae.
D. Branch.
Phys. Scr., Vol. 35, No. 6, p. 787 – 791 (1987).

The optical–ultraviolet spectrum of a classical Type I supernova (Type Ia) appears to be formed in a pure heavy–element plasma. During the first weeks after the explosion the spectrum forms in the outer, high–velocity layers of the ejected material, where the composition is a mixture of intermediate–mass elements from carbon to calcium. At later times the spectrum forms in deeper, slower material consisting of a time–dependent mixture of iron and cobalt, resulting from the decay of radioactive $^{56}\mathrm{Ni}$ that was synthesized by nuclear fusion at the time of the explosion.

125.030 On spectrophotometric temperatures of supernova envelopes.
A. K. Kolesov.
Vestn. Leningr. Univ., Mat. Mekh. Astron., No. 1, p. 86 – 102 (1987). In Russian. Abstr. in Ref. Zh., 51. Astron., 7.51.582 (1987).

125.031 Supernovae and supernova remnants.
S. I. Blinnikov, T. A. Lozinskaya, N. N. Chugaj.
Itogi Nauki i Tekhniki. Seriya Astronomiya. Tom 32. Astrophysics and space physics, p. 142 – 200 (1987). In Russian. – See Abstr. 003.004.

125.032 Einstein X–ray observations of the supernova remnant HB21.
D. A. Leahy.
Mon. Not. R. Astron. Soc., Vol. 228, No. 4, p. 907 – 913 (1987).

In radio, HB21 is a large ($\sim 2^\circ$ diameter) supernova remnant. It has been observed with incomplete coverage by five Einstein IPC fields. Here is reported the first detection of X–ray emission from HB21. The 0.2 – 4 keV X–ray image of HB21 does not correlate well with radio maps. HB21 has an X–ray temperature of $7(+6, -2) \times 10^6 \mathrm{K}$ and a luminosity of $1.9(+7.6, -1.2) \times 10^{34} \mathrm{erg \, s^{-1}}$. A Sedov model applied to HB21 gives too large a temperature and shock velocity. Supernova remnant models for expansion into a 3–component ISM can fit HB21, from which an age of 8000 – 15000 yr is inferred.

125.033 Spectrophotometry of the Crab Nebula as a whole.
K. Davidson.
Astron. J., Vol. 94, No. 4, p. 964 – 971 (1987).

A spatial scanning technique has been used to observe the visual–wavelength spectrum integrated over the Crab Nebula. The emission–line fluxes are useful for estimates of the nebular mass and chemical composition, while limits are placed on a possible emission–line halo around the Crab. One result is quite unexpected: Estimates of the [O III]/continuum flux ratio during the past 25 yr, including that presented here, are mutually discrepant, and one possible explanation is that the visual continuum of the Crab may be changing rapidly. This would have serious implications for models of the nonthermal SNR and pulsar.

125.034 Helium–rich supernovas.
J. C. Wheeler, R. P. Harkness.
Sci. Am., Vol. 257, No. 5, p. 50 – 58 (1987).

Computer modeling suggests that helium–rich supernovas result when the denuded core of a massive star collapses. These "peculiar" supernovas are close cousins of the 1987 bright event.

125.035 Type II supernova models: nucleosynthesis and isotopic anomalies.
W. Hillebrandt.
Prog. Part. Nucl. Phys., Vol. 17, p. 215 – 230 (1986). – See Abstr. 012.024.

Core–collapse models of type II supernovae are discussed. The conditions for prompt explosions caused by a hydrodynamic shock wave as well as the prospects of delayed explosions triggered by neutrino heating are investigated. Predictions of nucleosynthetic yields from various models are presented and are compared with observations.

125.036 A critical study of Type II supernovae: equations of state and general relativity.
S. Kahana.
Prog. Part. Nucl. Phys., Vol. 17, p. 231 – 247 (1986). – See Abstr. 012.024.

The relevance of relativistic gravitation and of the properties of nuclear matter at high density to supernova explosions is examined in detail. The existing empirical knowledge on the nuclear equation of state at densities greater than saturation, extracted from analysis of heavy ion collisions and from the breathing mode in heavy nuclei, is also considered. Particulars of the prompt explosions recently obtained theoretically by Baron, Cooperstein, and Kahana are presented.

125.037 Supernova explosion in a very massive star.
M. F. El Eid, N. Langer.
Prog. Part. Nucl. Phys., Vol. 17, p. 295 – 304 (1986). – See Abstr. 012.024.

The authors describe the final evolution of a 100 solar mass star following an evolutionary scenario during which the star evolves from a Wolf–Rayet stage through the electron–positron pair creation supernova. They find that the star is completely disrupted by explosive oxygen burning, and this type of explosion as a possible scenario for the Cassiopeia A remnant. This scenario seems to be also applicable to the supernova 1985f according to the recent observations of this object.

125.038 *IRAS* observations of collisionally heated dust in Large Magellanic Cloud supernova remnants.
J. R. Graham, A. Evans, J. S. Albinson, M. F. Bode, W. P. S. Meikle.
Astrophys. J., Vol. 319, No. 1, p. 126 – 135 (1987).

IRAS additional observations are presented which show that luminous ($10^4 - 10^5 L_\odot$) far–infrared sources are associated with the LMC supernova remnants N63A, N49, N49B, and N186D. Comparison of the infrared and X-ray data shows that a substantial fraction of the infrared emission from three of the remnants can be accounted for by collisionally heated dust. The evolution of a supernova remnant in a two–phase interstellar medium is considered, and it is concluded that dust cooling will not dominate supernova remnant evolution.

125.039 A scanning modulation collimator observation of the high–energy X-ray source in the Crab Nebula.
R. M. Pelling, W. S. Paciesas, L. E. Peterson, K. Makishima, M. Oda, Y. Ogawara, S. Miyamoto.
Astrophys. J., Vol. 319, No. 1, p. 416 – 425 (1987).

The authors have synthesized two–dimensional maps of the 22 – 64 keV emission from the Crab Nebula with an angular resolution of 15″. The maps are generated by application of a Maximum Entropy Method operating on a series of one–dimensional scans obtained with a balloon–borne modulation collimator telescope. The authors have, for the first time, measured the two–dimensional size, shape and orientation of the hard X–ray nebula relative to the pulsar.

125.040 Difference between the light curves of type–Ia and Ib supernovae.
D. Yu. Tsvetkov, N. N. Chugaj.
Astron. Tsirk., No. 1465, p. 3 – 4 (1986). In Russian.

The authors have compared B light curves for SN Ia and SN Ib and found that the fall of B magnitude from the maximum down to the inflection point for SN Ib is on average 0.44 mag. smaller than for SN Ia.

125.041 Influence of radioactivity on the development of a type II supernova outburst.
Eh. K. Grasberg, D. K. Nadezhin.
Sov. Astron., Vol. 30, No. 6, p. 670 – 674 (1987). English translation of 42.125.092.

125.042 On the spectrum of the low–frequency radio emission of Cassiopeia A.
E. N. Vinyajkin, V. A. Nikonov, A. F. Tarasov, Yu. V. Tokarev, M. A. Yurishchev.
Astron. Zh., Tom 64, Vyp. 5, p. 987 – 991 (1987). In Russian. English translation in Sov. Astron., Vol. 31, No. 5.

Flux densities of the radio emission of Cassiopeia A have been measured at the frequencies of 5.6 MHz and 8.9 MHz. Possible causes for the low–frequency spectrum cut–off are considered, and it is concluded that the main factor is absorption of radio emission in the fossil zone of ionized hydrogen around the supernova remnant.

125.043 On the anisotropy of Crab Nebula's synchrotron L_c–emission.
V. V. Golovatyj.
Astrofizika, Tom 27, Vyp. 1, p. 65 – 77 (1987). In Russian. English translation in Astrophysics, Vol. 27, No. 1.

Values of spectral index α of Crab Nebula's synchrotron L_c–emission are determined from spectra of 39 filaments. The significant difference of these values is derived and investigated. This difference is shown to be real and gives evidence of the anisotropy of the nebula's synchrotron L_c–spectra.

125.044 Visually discovered extra–galactic supernovae.
C. E. Spratt.
J. Am. Assoc. Variable Star Obs., Vol. 16, No. 1, p. 19 – 22 (1987).

This paper is intended as a supplement to the author's recent paper (Spratt 43.125.036) entitled, "A Checklist of Supernovae in the NGC and IC Galaxies Through 1985", and is a listing of those extra–galactic supernovae discovered visually.

125.045 32 GHz radio continuum observations of four shell–type supernova remnants.
H. W. Morsi, W. Reich.
Astron. Astrophys., Suppl. Ser., Vol. 71, No. 2, p. 189 – 195 (1987).

32 GHz maps with the Effelsberg 100–m telescope of the four supernova remnants G11.2–0.3, G29.7–0.3, G41.1–0.3 and G43.3–0.2 are presented. The radio spectra of these sources are discussed. G11.2–0.3 seems to belong to the class of supernova remnants showing both shell–type and plerionic characteristics.

125.046 Neutral hydrogen around some young supernova remnants.
I. V. Gosachinskij, V. K. Khersonskij.
Astrophysics, Vol. 26, No. 1, p. 40 – 44 (1987). English translation of 43.125.016.

125.047 On the origin of Kepler's supernova remnant.
R. Bandiera.
Astrophys. J., Vol. 319, No. 2, p. 885 – 892 (1987).

This paper presents a scenario for Kepler's supernova remnant whereby its emission is mostly due to the interaction of the blast wave with dense circumstellar matter, whose distribution is in turn determined by the interaction with a diffuse interstellar medium. The kind of observed asymmetry is easily explained if Kepler's supernova progenitor was a runaway object, subject to strong mass loss. A mass loss with $\dot{M}w \sim 5 \times 10^{-4}(M_\odot \text{yr}^{-1})\,(\text{km s}^{-1})$ is estimated for the progenitor. The runaway nature of the progenitor is discussed, and similarities with known runaway objects are presented.

125.048 Pulsar–like emission from the supernova remnant CTB 80.
R. G. Strom.
Astrophys. J., Lett. Ed., Vol. 319, No. 2, p. L103 – L107 (1987). With plates L7 – L8.

A compact radio source has been found immersed in the flat–spectrum central component of the peculiar supernova remnant CTB 80. The object's apparent steep radio spectrum, moderate degree of polarization, and near–coincidence with an unresolved X–ray source are consistent with a pulsar/neutron star which provides energy to power the flat–spectrum component. Its eccentric location and other evidence suggest a fast–moving object interacting with its environment.

125.049 Giant–scale supernova remnants. The role of differential galactic rotation and the formation of molecular clouds.
G. Tenorio–Tagle, J. Palouš.
Astron. Astrophys., Vol. 186, No. 1/2, p. 287 – 294 (1987).

The evolution of remnants produced by the total supernova power from an evolved OB association in a differentially rotating galactic disk is presented. The calculations at 5 kpc and 10 kpc from the galactic center lead to column densities across the remnant shell, or across sections of the remnants, which eventually exceed the opacity criterion $N_{opacity} = 10^{21}Z_\odot/Z$ cm^{-2} (Franco and Cox, 1986) and thus form molecular clouds. The resultant clouds have masses larger than $10^5 M_\odot$, dimensions of several hundred parsecs and a separation larger than 1 kpc. In contrast, at 20 kpc from the galactic center the opacity criterion is never fulfilled.

125.050 **IRAS observations of supernova remnants: a comparison between their infrared and X-ray cooling rates.**
E. Dwek, R. Petre, A. Szymkowiak, W. L. Rice.
Astrophys. J., Lett. Ed., Vol. 320, No. 1, p. L27 – L33 (1987).
IRAS infrared observations of several Galactic supernova remnants (SNR), including the brightest X-ray remnants, reveal that infrared emission from collisionally heated dust is the dominant cooling mechanism during much of their lifetimes. The authors compare the infrared fluxes from selected Galactic SNR observed by IRAS with their existing X-ray fluxes to derive an infrared-to-X-ray flux ratio. This ratio is a measure of the relative importance of remnant cooling by gas-grain collisions to its cooling by atomic processes, and for all the observed remnants the ratio is significantly greater than unity. The observed infrared-to-X-ray flux ratio is also compared to theoretical models of SNR cooling mechanisms.

125.051 **A new particle acceleration mechanism in Crab Nebula.**
H. Karimabadi, K. Papadopoulos, C. R. Menyuk.
Bull. Am. Astron. Soc., Vol. 19, No. 2, p. 684 (1987). Abstract. – See Abstr. 010.061.

125.052 **Ionization structure of the X-ray emitting plasma in Puppis A.**
K. F. Fischbach, C. R. Canizares, T. H. Markert, J. M. Coyne.
Bull. Am. Astron. Soc., Vol. 19, No. 2, p. 721 (1987). Abstract. – See Abstr. 010.061.

125.053 **Multiwaveband observations of the supernova remnant HB 9.**
D. A. Leahy, R. S. Roger.
Bull. Am. Astron. Soc., Vol. 19, No. 2, p. 729 (1987). Abstract. – See Abstr. 010.061.

125.054 **A CO (J = 1–0) survey of five supernova remnants at $l = 70° – 110°$.**
K. Tatematsu, Y. Fukui.
Bull. Am. Astron. Soc., Vol. 19, No. 2, p. 729 (1987). Abstract. – See Abstr. 010.061.

125.055 **A new shell SNR in the direction of the young pulsar PSR 1930+22.**
D. Routledge, J. F. Vaneldik.
Bull. Am. Astron. Soc., Vol. 19, No. 2, p. 729 (1987). Abstract. – See Abstr. 010.061.

125.056 **Early supernova expansion with flux-limited heat conduction.**
D. L. Band, R. A. Chevalier.
Bull. Am. Astron. Soc., Vol. 19, No. 2, p. 729 (1987). Abstract. – See Abstr. 010.061.

125.057 **Morphological studies of the supernova remnant E0102.2–72.2.**
J. P. Hughes, R. P. Kirshner, P. F. Winkler.
Bull. Am. Astron. Soc., Vol. 19, No. 2, p. 730 (1987). Abstract. – See Abstr. 010.061.

125.058 **"Sweeping" spectroscopy of the Crab Nebula.**
K. Davidson.
Bull. Am. Astron. Soc., Vol. 19, No. 2, p. 736 (1987). Abstract. – See Abstr. 010.061.

125.059 **Echelle spectroscopy of M33 supernova remnants.**
W. P. Blair, Y.–H. Chu, R. C. Kennicutt.
Bull. Am. Astron. Soc., Vol. 19, No. 2, p. 737 (1987). Abstract. – See Abstr. 010.061.

125.060 **Neutrino transport during the core bounce phase of a type II supernova explosion.**
P. M. Giovanoni, D. C. Ellison.
Bull. Am. Astron. Soc., Vol. 19, No. 2, p. 740 (1987). Abstract. – See Abstr. 010.061.

125.061 **X– and γ–ray spectra from supernovae.**
P. Pinto.
Bull. Am. Astron. Soc., Vol. 19, No. 2, p. 756 – 757 (1987). Abstract. – See Abstr. 010.061.

125.062 **Theoretical light curves of "type IIb" supernovae.**
L. Ensman, S. E. Woosley.
Bull. Am. Astron. Soc., Vol. 19, No. 2, p. 757 (1987). Abstract. – See Abstr. 010.061.

125.063 **The morphology and dynamics of a multi–lobed supernova remnant in the LMC (DEM 34a, N11L).**
J. Meaburn.
Mon. Not. R. Astron. Soc., Vol. 229, No. 3, p. 457 – 468 (1987). With 3 pp. plates.
Photon–counting imagery of the 16×24 pc supernova remnant, DEM 34a, has revealed a secondary lobe projecting from the bright ring of emission line filaments. Spatially resolved, long–slit echelle spectra of the Hα and [N II] emission lines show that complex motions are occurring within this remnant. Multiple lobes expanding with velocities of up to 350 km s^{-1} are present. The total Hα flux of DEM 34a is measured as 2.1×10^{-12}erg s^{-1}cm^{-2} (uncorrected for interstellar extinction) and the Hα/[N II] 6584+6548 Å and Hα/[S II] 6731+6716 Å brightness ratios as 3.2 and 1.1 respectively. These values are consistent with a 71 km s^{-1} velocity for the radiative shock which ionizes this supernova remnant.

125.064 **High–resolution X–ray and radio images of the galactic SNR G39.2–0.3.**
R. H. Becker, D. J. Helfand.
Astron. J., Vol. 94, No. 6, p. 1629 – 1632 (1987).
Images of the galactic supernova remnant G39.2–0.3 at radio, infrared, and X–ray wavelengths are presented. The 6 and 20 cm VLA maps reveal a shell–brightened source with substantial emission in the central regions and a rather high degree of linear polarization throughout; no significant spectral variations over the face of the remnant are apparent. The Einstein X–ray image also shows emission from both the limb and the central region, while the IRAS data show no flux that can be unambiguously associated with the remnant, although an uncataloged small H II region is seen adjacent to the shell. The authors conclude that, contrary to previous suggestions, there is no evidence for a Crab–like component in G39.2–0.3, and that the central emission plateau is most likely simply a shell filament which happens to fall across the center of the remnant.

125.065 **High–resolution radio observations of W50, the remnant associated with SS 433.**
R. Elston, S. Baum.
Astron. J., Vol. 94, No. 6, p. 1633 – 1640 (1987). With plate 128.
The authors have used the NRAO VLA to observe the radio source W50 with 30″ and 1′ resolutions at 20 cm. They find that the overall structure of W50 is well delineated by filamentary structure. W50 possesses a nearly circular shell reminiscent of an evolved supernova remnant such as CTB1 and the Cygnus Loop. The most unusual features of W50 are the edge–brightened radio ears which extend beyond the shell. The authors suggest that the ears are the result of the interaction of the ram pressure of the collimated jets of SS 433 with a supernova remnant shell. At 30″ resolution the western ear of W50 is depolarized, suggesting that W50 and S74 are at a common distance of about 3 kpc.

125.066 **Infrared spectroscopy of supernova remnants.**
A. F. M. Moorwood, E. Olivia, I. J. Danziger.
ESO Sci. Prepr., No. 516, 6 pp. (1987). Paper presented at IAU Colloq. No. 101.

125.067 Binary systems as supernova progenitors (some frequency estimates).
A. Tornambè, F. Matteucci, I. Iben Jr., K. Nomoto.
ESO Sci. Prepr., No. 523, 9 pp. (1987). Paper presented at the 4th workshop on "Nuclear physics", Ringberg Castle, F.R. Germany, April 1987.

125.068 On the progenitors of type II supernovae in M83.
M. Rosa, O.–G. Richter.
ESO Sci. Prepr., No. 525, 29 pp. (1987). To appear in Astron. Astrophys.

125.069 A search for supernovae in Shapley–Ames galaxies.
R. D. McClure, S. van den Bergh, R. Evans.
Publ. Dom. Astrophys. Obs., Vol. 16, No. 16, p. 281 – 296 (1987).
The present paper lists surveillance times for 748 Shapley–Ames galaxies that were regularly searched for supernovae between November of 1980 and October of 1985. Entries are given in a table for assumed search limits of V = 14.0, V = 14.5 and V = 15.0. During the 5–year period Evans discovered 11 supernovae in this galaxy sample.

125.070 Multi–frequency radio observations of supernova remnants in the range between $l = 85°$ and $l = 135°$.
S. A. Trushkin, V. V. Vitkovskij, N. A. Nizhel'skij.
Astrofiz. Issled. Izv. Spets. Astrofiz. Obs., Tom 25, p. 84 – 104 (1987). In Russian. English translation in Bull. Spec. Astrophys. Obs. – North Caucasus, Vol. 25.
The results of the multi–frequency radio observations of fourteen supernova remnants (SNRs) made with the RATAN–600 radio telescope in 1985 are reported. Flux densities of these SNRs are measured at least at two wavelengths: 31.3 and 7.6 cm. The radio spectra of some SNRs are precised. The one–dimensional brightness distributions of all SNRs are given. A comparison of the observational data with a simple model of a uniform, optically thin, spherical shell is made. Radio sources detected inside or near the SNRs are listed. The spectra of some point sources are discussed.

125.071 Type I supernova models.
R. Canal, J. Isern, J. Labay.
Rev. Mex. Astron. Astrofis., Vol. 14, No. 1, p. 245 – 251 (1987). – See Abstr. 012.042.
The authors briefly describe the characteristics of Type I supernova outbursts and present the theoretical models so far advanced to explain them. They especially insist on models based on the thermonuclear explosion of a white dwarf in a close binary system, even regarding the recent division of Type I supernovae into the Ia and Ib subtypes. Together with models assuming explosive thermonuclear burning in a fluid interior, the authors consider in some detail those based on partially solid interiors. They finally discuss models that incorporate nonthermonuclear energy contributions, suggested in order to explain Type Ib outbursts.

125.072 Radiative transfer in type I supernovae atmospheres.
J. Isern, R. López, E. Simonneau.
Rev. Mex. Astron. Astrofis., Vol. 14, No. 1, p. 252 – 259 (1987). – See Abstr. 012.042.
As the only direct information concerning the physics and the triggering mechanism of supernova explosions comes from the spectrophotometry of the emitted radiation, it is worthwhile to put considerable effort on the understanding of the radiation transfer in the supernovae envelopes in order to set constraints on the theoretical models of such explosions. In this paper the authors analyze the role played by the layers curvature on the radiative transfer.

125.073 Computational problems in supernova simulations.
E. Müller.
Comput. Phys. Commun., Vol. 44, No. 3, p. 271 – 277 (1987). Abstr. in Phys. Abstr., Vol. 90, No. 1312, Entry 107872 (1987). – See Abstr. 012.049.

125.074 Radio polarization of the semicircular supernova remnant G 109.1–1.0.
K. Tatematsu, Y. Fukui, M. Nakano, T. Iwata.
Publ. Astron. Soc. Jpn., Vol. 39, No. 5, p. 755 – 760 (1987).
The first detection of linear polarization of the semicircular supernova remnant G 109.1–1.0 is reported. The observations were made at 10 GHz with the Nobeyama 45–m telescope. The degree of polarization typically ranges from several to 30%. The distribution of the projected magnetic field cannot be simply explained as tangential to the shell of the supernova remnant, but is rather consistent with the general galactic magnetic field.

125.075 Experimental investigations of the radio radiation of the Crab nebula.
A. P. Barabanov, V. P. Ivanov, K. S. Stankevich, S. P. Stolyarov.
Nauchn.–Issled. Radiofiz. Inst. Gor'kij, Prepr., No. 234, p. 3 – 25 (1987). In Russian. Abstr. in Ref. Zh., 51. Astron., 12.51.887 (1987).

125.076 The Cygnus Loop: an older supernova remnant.
W. Straka.
Mercury, Vol. 16, No. 5, p. 150 – 154 (1987).

125.077 The neutral hydrogen distribution in the vicinity of the supernova remnant 3C 396.
I. V. Gosachinskij, V. K. Khersonskij.
Astron. Zh., Tom 64, Vyp. 6, p. 1184 – 1188 (1987). In Russian. English translation in Sov. Astron., Vol. 31, No. 6.
Observations of neutral hydrogen in the vicinity of 3C 396, which were performed at the RATAN–600 radio telescope with the resolution $2' \times 130' \times 6.3$ km/s, show that the H I distribution is similar to a clumpy envelope with the diameter 310 pc and mass $1.9 \times 10^4 M_\odot$. It is obtained that the most probable mechanism for envelope formation is the outflow of stellar wind during the star evolution along the main sequence. The estimated mass of the star is about 19 $M_\odot$, spectral class is O8.5. The mass which was ejected in the explosion is estimated (16 $M_\odot$).

125.078 The onset of a type–II supernova explosion in a circumstellar envelope created by intense stellar wind from a presupernova.
Eh. K. Grasberg, D. K. Nadezhin.
Astron. Zh., Tom 64, Vyp. 6, p. 1199 – 1209 (1987). In Russian. English translation in Sov. Astron., Vol. 31, No. 6.
The influence of extended circumstellar envelopes on photometric properties and dynamics of type–II supernovae is investigated. The typical stellar wind of late giants and supergiants (with $C_\varrho \lesssim 10^{14}$ g/cm) is shown to have virtually no influence on supernova light curves. If the mass loss from the presupernova is more intense, the shape of the light–curve plateau changes. For circumstellar envelopes of very high density ($C_\varrho \gtrsim 10^{17}$ g/cm), the near–maximum dome may appear on the light curve. The interaction of the supernova envelope with the stellar wind leads to formation of a thin (spatially but not optically!) and dense spherical layer in which true absorption dominates Thomson scattering.

125.079 Supernovae 1987J and 1987K.
IAU Circ., Nos. 4426 – 4428 (1987).

125.080 Supernovae (1987J and 1987K).
Br. Astron. Assoc. Circ., No. 670 (1987).

125.081 Automatic search for supernovae.
V. Porubčan.
Kozmos, Vol. 18, No. 4, p. 116 (1987). In Slovak.

125.082 How does the supernova explode?
T. Fabini, K. Rosová.
Kozmos, Vol. 18, No. 4, p. 117 – 121 (1987). In Slovak.

125.083 How do supernovae explode?
P. Hadrava.
Říše hvězd, Vol. 68, No. 12, p. 228 – 231 (1987). In Czech.

125.084 Why do supernovae explode? (II).
Z. Mikulášek.
Říše hvězd, Vol. 68, No. 12, p. 225 – 227 (1987). In Czech.

125.085 X–ray observations of the supernova remnant N103B in the Large Magellanic Cloud.
K. P. Singh, N. J. Westergaard, H. W. Schnopper,
D. J. Helfand.
Astrophys. J., Vol. 322, No. 1, p. 80 – 87 (1987).
The X–ray spectrum of the supernova remnant N103B, as measured with the ME + LE experiment aboard *EXOSAT* and the IPC experiment aboard the *Einstein Observatory*, clearly indicates a predominantly thermal source ($kT \sim 1.2$ keV) of X–ray emission. From the observed temperature, high X–ray luminosity, and small size of this SNR the authors derive an age of ~600 to 1200 yr and a high density inside and surrounding the remnant.

125.086 Prospects for gamma–ray line observations of individual supernovae.
N. Gehrels, M. Leventhal, C. J. MacCallum.
Astrophys. J., Vol. 322, No. 1, p. 215 – 233 (1987).
The authors have studied the γ–ray line emission from individual Type I and II supernovae using numerical simulations and photon propagation codes to predict flux levels and line shapes. A record of historical supernovae has been compiled to determine the discovery rate of extragalactic events close enough for line detections with current and planned γ–ray spectrometers. For both Type I and II supernovae, the γ–ray lines with the highest flux from an individual event are the 0.847 and 1.238 MeV lines from the ^{56}Ni $\rightarrow$ ^{56}Co $\rightarrow$ ^{56}Fe decay chain.

125.087 Low–frequency radio emission of the Tycho supernova remnant, 3C 10.
E. N. Vinyajkin, Yu. V. Volodin, R. D. Dagkesamanskij,
K. P. Sokolov.
Sov. Astron., Vol. 31, No. 2, p. 141 – 145 (1987). English translation of 43.125.029.

125.088 Experiments on the evolution of the Crab Nebula radio spectrum.
A. P. Barabanov, V. P. Ivanov, I. A. Malyshev,
K. S. Stankevich, S. P. Stolyarov.
Sov. Astron., Vol. 31, No. 2, p. 145 – 150 (1987). English translation of 43.125.030.

125.089 Einstein IPC imaging and spectral observations of the supernova remnant HB 9.
D. A. Leahy.
Astrophys. J., Vol. 322, No. 2, p. 917 – 921 (1987).
In radio wavelengths, HB 9 is a large ($\sim 2°$ diameter) supernova remnant. The central area of HB 9 has been observed by four Einstein IPC fields. A soft X–ray image and results from analysis of X–ray spectra from the IPC observations are presented. HB 9 has spatial variations in X–ray temperature from ~0.4 to ~1.2 keV. For a distance of 1.1 kpc, the 0.2 – 4 keV luminosity is 5×10^{34} ergs s^{-1}. Supernova remnant models for expansion into a three–component ISM yield an age of ~14,000 yr.

125.090 The supernova rate in Shapley–Ames galaxies.
S. van den Bergh, R. D. McClure, R. Evans.
Astrophys. J., Vol. 323, No. 1, p. 44 – 53 (1987).
During a 5 yr period R. Evans has carried out a visual search for supernovae in 1017 bright galaxies. The subsample comprising 748 Shapley–Ames galaxies surveyed by him constitutes a homogeneous data base that is suitable for a determination of the supernova rate. For the supernova rates recently derived by Tammann one would have expected 54 supernovae to have been observed in the sample. In fact, Evans found only 11. The authors conclude from their observations that Tammann's supernova

rates are probably too high by a factor of ~3. They also find that the majority of Galactic supernova remnants must have been produced by massive progenitors.

125.091 Light echoes: supernovae 1987A and 1986G.
B. E. Schaefer.
Astrophys. J., Lett. Ed., Vol. 323, No. 1, p. L47 – L49 (1987).
The sudden brilliance of a supernova eruption will be reflected on surrounding dust grains to create a phantom nebula. The author presents a series of calculations in which the apparent brightness of this "light echo" is predicted for a variety of situations where the dust is part of the interstellar medium (ISM). He finds that the supernova 1987A will have a very bright echo off the ISM. At a time of 400 days past maximum, the SN 1986G is found to be 2.7 mag brighter than would be predicted by an extrapolation of its light curve. This unique property has an easy explanation as a light echo off the dust in the dust lane of Cen A.

125.092 Light echoes: Type II supernovae.
B. E. Schaefer.
Astrophys. J., Lett. Ed., Vol. 323, No. 1, p. L51 – L54 (1987).
The author extends the calculations of Chevalier so as to produce realistic light curves of SNs where a dust reflection echo off a circumstellar shell is present. He examines the morphology of Type II SN light and concludes that light echo effects dominate the late–time brightness of most Type II SNs. Three predictions for testing this conclusion are presented.

Supernova 1987A in LMC

125.201 Supernova 1987a: laatste nieuws.
M. Drummen.
Zenit, Jaarg. 14, Nr. 7/8, p. 264 – 265 (1987).

125.202 La supernova 1987A: courbe de lumière.
Bull. Assoc. Fr. Obs. Etoiles Variables, No. 41, p. 7 (1987).

125.203 Die Supernova SN1987A in der Großen Magellanschen Wolke.
A. Weiß, J. Wambsganß.
Phys. Bl., 43. Jahrg., Heft 9, p. 371 – 372 (1987).

125.204 Bang: the supernova of 1987.
D. Helfand.
Phys. Today, Vol. 40, No. 8, Part 1, p. 24 – 32 (1987).

125.205 Time profile of the neutrino burst from SN 1987A in the Large Magellanic Cloud.
H. Suzuki, K. Sato.
Publ. Astron. Soc. Jpn., Vol. 39, No. 3, p. 521 – 528 (1987).
Using the data of the neutrino burst from SN 1987A detected by Kamiokande and IMB the authors investigated the time evolution of the neutrino temperature, the luminosity, and the radius of neutrinosphere.

125.206 X–ray emission from SN 1987A.
H. Itoh, S. Hayakawa, K. Masai, K. Nomoto.
Publ. Astron. Soc. Jpn., Vol. 39, No. 3, p. 529 – 537 (1987).
X–ray emission from SN 1987A will be enhanced substantially when the blast shock hits the circumstellar matter arising from the mass ejection in the red–supergiant stage of the progenitor. The enhanced X–ray emission may be detected with the satellite Ginga shortly, depending on the mass–loss history of the progenitor.

125.207 Hydrodynamical models of supernova SN 1987A in the LMC.
Eh. K. Grasberg, V. S. Imshennik, D. K. Nadezhin,
V. P. Utrobin.
Pis'ma Astron. Zh., Tom 13, No. 7, p. 547 – 553 (1987). In Russian. English translation in Sov. Astron. Lett., Vol. 13.

It is shown that the properties of SN 1987A in the LMC can be described well by hydrodynamical models of explosions of compact massive stars. In accordance with these models, the mass of the expelled envelope, M, the presupernova radius, R, and the total energy of explosion, E, are evaluated for SN 1987A to be $\sim 16\,M_\odot$, $\sim 30\,R_\odot$, and $\sim 3 \times 10^{51}$ erg respectively. The progenitor of supernova remnant Cas A may be considered as the prototype of the SN 1987A in our own Galaxy. In other galaxies, this subtype of supernovae can be represented by SN 1948B in NGC 6946. If the energy of explosion transfers from the collapsed core of the star to its envelope with a characteristic time less than 1 hour, then delay of $\Delta t \sim 3$ hours between the neutrino pulse and the steep rise of optical luminosity of SN 1987A does not contradict to the scenario of explosions of compact massive stars.

125.208 The Hα velocity structure during the first month of SN 1987a in the LMC.
R. W. Hanuschik, J. Dachs.
Astron. Astrophys., Vol. 182, No. 1, p. L29 – L30 (1987).

The authors present first results of a spectroscopic study of the P Cyg–type Hα profiles of SN 1987a in the LMC, covering the period February 25.1 until March 29.2. During that time, the terminal velocity of the absorption trough decreased from –31000 to –11600 km/s, and the velocity of the minimum of the absorption trough from –18000 to –7400 km/s. The peak emission velocity was around –4000 km/s at the end of February, around 0 km/s between March 10 and 25 and around –400 km/s afterwards. A weak, later pronounced peak in the blueward emission wing developed on March 17.4; a bump in the redward emission wing has been visible since March 19.

125.209 Computed ultraviolet spectra for SN 1987a.
L. B. Lucy.
Astron. Astrophys., Vol. 182, No. 1, p. L31 – L33 (1987).

Monte Carlo calculations of the early UV spectra of supernova 1987A are presented. These confirm the crucial role of line blocking in the differentially expanding atmosphere in explaining the unusual and rapidly changing appearance of the IUE spectra as well as providing identifications for several major features. Consistent with the expectation of a relatively massive progenitor, the largely successful reproduction of the UV spectra is achieved with the present day LMC metallicity, $Z_\odot/2.75$.

125.210 Constraints on the interpretation of the neutrino experiments by the optical observations of SN 1987a.
E. J. Wampler, J. W. Truran, L. B. Lucy, P. Höflich,
W. Hillebrandt.
Astron. Astrophys., Vol. 182, No. 2, p. L51 – L54 (1987).

The authors' aim here is to show that the brief time interval, $\delta t = 1.1 \times 10^4$ sec, between the detection of the neutrinos by the Kamiokande and IMB experiments and the observation of the supernova at $V = 6.4$ mag. cannot be easily understood if the Kamiokande–IMB neutrino detection signals the beginning of the supernova's envelope expansion.

125.211 Spectroscopic and photometric observations of SN 1987a: the first 50 days.
J. W. Menzies, R. M. Catchpole, G. van Vuuren, H. Winkler,
C. D. Laney, P. A. Whitelock, A. W. J. Cousins, B. S. Carter,
F. Marang, T. H. H. Lloyd Evans, G. Roberts, D. Kilkenny,
J. Spencer Jones, K. Sekiguchi, A. P. Fairall,
R. D. Wolstencroft.
Mon. Not. R. Astron. Soc., Vol. 227, No. 3, p. 39P – 49P (1987).

The authors present spectroscopic and photometric observations of SN 1987a in the Large Magellanic Cloud made at the Sutherland field station of SAAO and in Cape Town during the first 50 days after its discovery was announced.

125.212 Could there be terrestrial signatures of the EUV pulse from Supernova 1987A?
H. Ögelman, H. Böhringer, S. Buchert, S. Cakir, J. LaBelle,
R. A. Treumann.
Astron. Astrophys., Vol. 183, No. 2, p. L27 – L29 (1987).

Models for the emergence of the shock wave from the envelope of a type II supernova indicate that an ultraviolet to soft X-ray (EUV) pulsar should accompany the break–out. The spectrum and the time profile of this pulse is sensitive to the energetics of the explosion as well as the envelope structure of the progenitor. For the case of SN 1987A, the expected EUV pulse should be delayed about one hour with respect to the collapse time, and should have an effective temperature around 10^6 K, a total energy of 10^{46} to 10^{48} ergs, and a duration of several hundred seconds. This burst would have given rise to an energy flux of $0.01 – 1.0$ erg cm^{-2} at the earth leading to an enhancement of the electron concentration in the E–layer of the ionosphere by up to 7×10^3 cm^{-3}. A detection of this enhancement by experiments monitoring ionospheric parameters could give indirect information on the otherwise unobservable EUV pulse.

125.213 SN 1987a: Zwei Monate Beobachtungen.
K. S. de Boer, T. Richtler.
Sterne Weltraum, 26. Jahrg., Nr. 7 – 8, p. 388 – 393 (1987).

125.214 Observations de la supernova SN 1987A avec UFT à bord du satellite ASTRON.
A. A. Boyarchuk, R. E. Gershberg, A. M. Zvereva,
P. P. Petrov, A. A. Severny, A. V. Terebizh, C. T. Hua,
A. I. Sheikhet (A. I. Shejkhet).
J. Astron. Fr., No. 30, p. 2 (1987).

125.215 Supernova 1987A in the Large Magellanic Cloud: the explosion of a $\sim 20\,M_\odot$ star which has experienced mass loss?
S. E. Woosley, P. A. Pinto, P. G. Martin, T. A. Weaver.
Astrophys. J., Vol. 318, No. 2, p. 664 – 673 (1987). = Lick Obs. Bull., No. 1067.

The recently discoverd Type II supernova in the LMC is interpreted as the explosion of a massive star having a helium core mass $\sim 6\,M_\odot$, although mass loss might have significantly reduced this value. Theoretical UBV light curves are presented, and the expected fluxes of γ-ray lines from ^{56}Co, ^{57}Co, and ^{44}Ti are estimated. The necessity of a compact progenitor is emphasized.

125.216 Calculated gamma–ray line fluxes from the Type II supernova 1987A.
K. W. Chan, R. E. Lingenfelter.
Astrophys. J., Lett. Ed., Vol. 318, No. 2, p. L51 – L55 (1987).

The authors have calculated the time–dependent flux in the 847 keV gamma–ray line from the decay of ^{56}Co that might be expected from the Type II supernova 1987A in the Large Magellanic Cloud. They find that for a wide range of assumed ^{56}Co and supernova ejecta masses this line should be detectable by planned gamma–ray observations with flux sensitivities of about 10^{-4} photons cm^{-2}s^{-1}.

125.217 Neutrinos from SN 1987A.
A. Burrows, J. M. Lattimer.
Astrophys. J., Lett. Ed., Vol. 318, No. 2, p. L63 – L68 (1987).

The detections by the Kamiokande II and IMB collaborations of the neutrinos from the supernova SN 1987A have provided an opportunity to probe deeply into a collapsed core and watch the birth of a neutron star. The authors compare model calculations of the neutrino emissions following core collapse with these data and obtain reasonable agreement for the total energy, average neutrino energy, and burst duration. The simultaneous observation of both 10 MeV neutrinos and second, not millisecond, characteristic times indicates that the neutrinos do indeed diffuse out of the core.

125.218   Observations with "Astron": supernova 1987A in the Large Magellanic Cloud.
A. A. Boyarchuk, R. E. Gershberg, A. M. Zvereva, P. P. Petrov, A. B. Severnyj, A. V. Terebizh, C. T. Hua, A. I. Shejkhet.
Pis'ma Astron. Zh., Tom 13, No. 9, p. 739 – 743 (1987). In Russian. English translation in Sov. Astron. Lett., Vol. 13.

The results of observations of the ultraviolet spectrum of supernova 1987A that have been carried out from 4 to 12 March 1987 with the Astrophysical station "Astron" are presented.

125.219   Gamma radiation expected from the supernova 1987A in the Large Magellanic Cloud.
O. S. Bartunov, S. I. Blinnikov, L. V. Levakhina, D. K. Nadezhin.
Pis'ma Astron. Zh., Tom 13, No. 9, p. 744 – 750 (1987). In Russian. English translation in Sov. Astron. Lett., Vol. 13.

The authors present the results of Monte–Carlo simulations of expected photon spectra in the range 0.1 – 3.5 MeV emerging from SN 1987A which are born either in the ^{56}Co decay or by a young pulsar. The influence of the variations of the parameters of the expanding envelope on the results is demonstrated.

125.220   Die Supernova 1987A: Beobachtungen und Interpretationen.
R. Wehrse.
Sterne Weltraum, 26. Jahrg., Nr. 11, p. 612 – 614 (1987).

125.221   Supernova theory and Supernova 1987A.
W. D. Arnett.
Astrophys. J., Vol. 319, No. 1, p. 136 – 142 (1987).

The implications of Supernova 1987A are examined and compared with theoretical expectations. The relatively low luminosity follows from the progenitor having lower than solar abundances. The explosion energy is $(1.0 - 2.0) \times 10^{51}$ ergs for a 15 $M_\odot$ star. The neutrino signal is strikingly that expected from recent models of the core collapse process with advective overturn.

125.222   Supernova!
P. Moore.
1988 Yearbook of Astronomy, p. 176 – 180 (1987). – See Abstr. 003.006.

125.223   Observations of the supernova in the Large Magellanic Cloud.
A. P. Vid'machenko.
Kinematika Fiz. Nebesn. Tel, Tom 3, No. 5, p. 84 – 85 (1987). In Russian. English translation in Kinematics Phys. Celest. Bodies.

Spectrograms of the Supernova 1987A in the Large Magellanic Cloud obtained at the Tarija Observatory (Bolivia) during March 5 – 23, 1987 in the spectral region of 390 – 640 nm with resolution 0.5 – 5 nm are presented. The estimates of the magnitudes V are also given.

125.224   Spectroscopic and photometric observations of SN 1987a – II. Days 51 to 134.
R. M. Catchpole, J. W. Menzies, A. S. Monk, W. F. Wargau, D. Pollacco, B. S. Carter, P. A. Whitelock, F. Marang, C. D. Laney, L. A. Balona, M. W. Feast, T. H. H. Lloyd Evans, K. Sekiguchi, J. D. Laing, D. M. Kilkenny, J. Spencer Jones, G. Roberts, A. W. J. Cousins, G. van Vuuren, H. Winkler.
Mon. Not. R. Astron. Soc., Vol. 229, No. 1, p. 15P – 25P (1987).

The authors present spectroscopic and photometric observations of SN 1987a in the Large Magellanic Cloud for days 51–134 after the Kamiokande–II neutrino event. During this period both the bolometric flux from the supernova and the apparent angular radius of the photosphere reached a maximum and subsequently declined. Some discussion is included based on a comparison between the derived fluxes and those predicted by selected theoretical models.

125.225   The light curve of SN 1987A.
R. Schaeffer, M. Cassé, R. Mochkovitch, S. Cahen.
Astron. Astrophys., Vol. 184, No. 1/2, p. L1 – L4 (1987).

The authors use a semi–analytical model of supernova light curves that extends Arnett's scheme to include the effect of recombination of hydrogen, helium and heavy elements in the expanding ejecta. Introducing in the model the physical parameters of Sanduleak –69°202 and a plausible composition derived from presupernova models, the salient characteristics of the light curve of SN 1987A are reasonably reproduced over 100 days, without invoking feeding by radioactive ^{56}Ni or by a newly born pulsar.

125.226   Neutrinos from supernova 1987a.
V. S. Berezinskij.
Priroda, No. 8, p. 95 – 97 (1987). In Russian.

125.227   SN 1987A: Watching and waiting.
R. A. Schorn.
Sky Telesc., Vol. 74, No. 1, p. 14 – 15 (1987).

125.228   Supernova 1987A's fading glory.
R. A. Schorn.
Sky Telesc., Vol. 74, No. 3, p. 258 – 259 (1987).

125.229   Supernova 1987A after 200 days.
R. A. Schorn.
Sky Telesc., Vol. 74, No. 5, p. 477 – 479 (1987).

125.230   Supernova 1987A.
D. Alloin, E. Schatzman.
Recherche, Vol. 18, No. 194, p. 1494 – 1502 (1987). In French.

125.231   Supernova 1987A: some answers – but more questions.
P. Murdin.
Astron. Now, Vol. 1, No. 5, p. 8 – 16 (1987).

125.232   Circumstellar matter and the nature of the SN1987A progenitor star.
R. A. Chevalier, C. Fransson.
Nature, Vol. 328, No. 6125, p. 44 – 45 (1987).

The radio observations of the supernova SN1987A can be interpreted in terms of its interaction with circumstellar matter. The early turn–on of the radio emission implies a relatively low density circumstellar medium, with $\dot{M}/V_w = 8.8 \times 10^{-6} M_\odot \mathrm{yr}^{-1}$ per (550 km s^{-1}) where $\dot{M}$ is the mass loss rate from the progenitor star and V_w is the wind velocity. The optical properties of the supernova imply that the progenitor star had a smaller radius than that of a typical type II supernova progenitor. The authors predict the thermal X–ray luminosity of the supernova and note that it is below the current upper limit.

125.233   A relativistic jet from SN1987A?
M. J. Rees.
Nature, Vol. 328, No. 6127, p. 207 (1987).

125.234   The progenitor of SN1987A.
R. Gilmozzi, A. Cassatella, J. Clavel, C. Fransson, R. Gonzalez, C. Gry, N. Panagia, A. Talavera, W. Wamsteker.
Nature, Vol. 328, No. 6128, p. 318 – 320 (1987).

The autors use spatial and spectroscopic information from IUE spectra in the range 1,150 – 1,950 Å to demonstrate that Sk –69°202, the star coinciding positionally with the LMC supernova has disappeared from sight. Two weaker sources, named star 2 and star 3 in the astrometric analysis of the image of Sk –69°202 before the supernova outburst, are still present in the ultraviolet spectra. The authors isolate their spectra and give a spectral classification. They conclude that Sk –69°202 is the progenitor of the LMC supernova.

125.235 Light–curve models for supernova SN1987A in the Large Magellanic Cloud.
T. Shigeyama, K. Nomoto, M. Hashimoto, D. Sugimoto.
Nature, Vol. 328, No. 6128, p. 320 – 323 (1987).

By computing the propagation of the shock wave, the subsequent expansion of ejected matter and the optical light curve, the authors find that the early light curve of SN 1987A can be accounted for by the diffusive release of energy deposited by the shock wave, that the progenitor's radius should be as small as $\sim(1-3) \times 10^{12}$cm, and that the explosion energy per unit mass should be relatively large, $\sim(2-3) \times 10^{51}$erg for the ejected matter of $7-10\ M_\odot$. The light curve after eight days is well reproduced by using a model with constant energy input. The energy input may be due to the activity of an embedded pulsar.

125.236 An interacting binary model for SN 1987A.
A. C. Fabian, M. J. Rees, E. P. J. van den Heuvel, J. van Paradijs.
Nature, Vol. 328, No. 6128, p. 323 – 324 (1987).

The recent supernova in the LMC, SN 1987A, is positionally coincident with the B3 I supergiant star, Sk –69°202. Ultraviolet observations of this region made with the IUE satellite after the explosion now show this star to have disappeared. Initial doubts that a blue supergiant could be a supernova progenitor may be overcome by calculations of the evolution of massive stars at reduced metallicity, as expected in the Magellanic Clouds. The authors present an alternative model for SN 1987A in which the progenitor was a close binary companion of Sk –69°202.

125.237 Regularly pulsed neutrinos from supernova SN 1987A?
M. Harwit, P. L. Biermann, H. Meyer, I. M. Wasserman.
Nature, Vol. 328, No. 6130, p. 503 – 504 (1987).

The neutrino events from SN 1987A observed with the Kamiokande and IMB experiments appear to show evidence of a period (P) of approximately 8.9 milliseconds. While the statistical significance of this period is marginal in each individual experiment, the two sets of data are compatible. The authors remark on a few consequences of interest should the periodicity be confirmed by observations. Interpreting the apparent period $P \approx 8.9$ ms as a rotation of a compact object would imply that: (1) neutrino emission is anisotropic; (2) the neutrino mass, averaged over all flavours observed is less than $0.2\ \mathrm{eV}/c^2$.

125.238 The multiple star system Sk–69°202.
S. van den Bergh.
Nature, Vol. 328, No. 6133, p. 768 (1987).

125.239 Cosmic rays in the shell of SN 1987A and its gamma–radiation.
V. S. Berezinskij, V. L. Ginzburg.
Pis'ma Astron. Zh., Tom 13, No. 11, p. 931 – 944 (1987). In Russian. English translation in Sov. Astron. Lett., Vol. 13.

The search for cosmic rays in the SN 1987A shell with the help of observations of gamma, neutrino and neutron radiation is discussed. The highest sensitivity to the presence of cosmic rays in the SN shell is shown to have gamma–radiation with $E_\gamma \gtrsim 1$ TeV observed with ground–based Cherenkov telescopes. It allows to detect cosmic rays in the shell if the power of their generation (CR luminosity) is at least $L_p \sim 10^{39}$erg/s. These observations can solve the question whether the acceleration inside the SN shells is the main mechanism of CR acceleration in the Galaxy.

125.240 Expected X–ray radiation spectra of supernova 1987A. Monte–Carlo computations.
S. A. Grebenev, R. A. Syunyaev.
Pis'ma Astron. Zh., Tom 13, No. 11, p. 945 – 963 (1987). In Russian. English translation in Sov. Astron. Lett., Vol. 13.

X–ray spectra of radiation escaping the supernova envelope are computed using the Monte–Carlo techniques. Two types of hard photon sources are considered: (a) the gamma–rays of $0.1\ M_\odot$ radioactive ^{56}Co decay and (b) the young energetic pulsar. Dominant elementary processes leading to the formation of the spectra are Comptonization and photoabsorption of the

electrons of the K–shells of heavy elements. The evolution of X–ray spectra with time is presented. The models with strongly perturbed density distribution (the rarefied phase with dense plasma clouds imbedded), with strong stratification of heavy elements' abundance, with toroidal geometry of the expanding envelope were computed.

125.241 The disruption of a light neutron star in an ultra–close binary and the second neutrino burst from SN 1987A.
L. Stella, A. Treves.
Astron. Astrophys., Vol. 185, No. 1/2, p. L5 – L6 (1987).

In correspondence to SN 1987A two neutrino bursts were reported separated by ~4.7 hr. Assuming that both of them are real, the authors propose a picture where the first supernova explosion generates a very close binary system ($P_{orb} \sim 0.2$ s) consisting of two collapsed objects, a neutron star plus a neutron star or a black hole. The evolution of the system is dominated by gravitational radiation energy losses, and after a few hours the lighter component fills its Roche lobe. Then matter starts to be transferred at a huge rate and on a timescale of $\ll 1$ s most of the mass of the lighter neutron star is expelled from its Roche lobe and accreted by the primary. In this process an energy of $\sim10^{53}$ergs is released, which may explain the second neutrino burst.

125.242 Deconvolution of a pre–outburst picture of SN 1987A.
S. R. Heap, D. J. Lindler.
Astron. Astrophys., Vol. 185, No. 1/2, p. L10 – L12 (1987).

The authors have applied the block iterative method (Young 1971) of algebraic image restoration to a photographic plate of the field around SN 1987A obtained four years before outburst by Chu at the CTIO 4–meter telescope. By setting appropriate constraints for the solution on a pixel–by–pixel basis, they have restored the image of a starfield centered on Sk –69°202, the precursor to the supernova. It is found that this star has two companions, which others (e.g. Walborn et al. 1987, West et al. 1987) have also noted and designated as Star 2 and Star 3. The authors also find marginal evidence for Star 4, a weak source at $\theta = 222°$, r = 1".9. Correction for the contributions of these companions yields an estimate of the magnitude for Sk –69°202 alone of V = 12.37.

125.243 The modulation of neutrinos from SN 1987A during stellar collapse.
O. C. de Jager.
Astron. Astrophys., Vol. 185, No. 1/2, p. L13 – L15 (1987).

The time separation between the neutrino events from SN 1987A measured by the KAMIOKANDE–II and IMB detectors are examined for deviations from exponentiality: the significance of the existence of an $\cong7$s delay of neutrinos in the KAMIOKANDE data is 3.1σ while the IMB data shows no significant deviation from exponentiality.

125.244 The nature of the companion of SN 1987A.
I. Goldman.
Astron. Astrophys., Vol. 186, No. 1/2, p. L3 – L4 (1987).

A model for the companion of SN 1987A, observed recently by optical speckle interferometry, is proposed. It is suggested that the observed relativistic velocity and the high luminosity can be naturally explained within a framework of a rotating collapse, producing two neutron stars in a binary system which decays due to emission of gravitational radiation. This leads eventually to the tidal disruption of the less massive member and the ejection of a fraction of the disrupted star of mass $\lesssim 0.1\ M_\odot$ at a relativistic velocity. In the present model there is no need for any energy source separate from the companion, as is the case with models proposed so far.

125.245 The interaction of the UV burst of Supernova 1987A with a nearby cloud: a possible explanation of the speckle images.
W. Hillebrandt, P. Höflich, H. U. Schmidt, J. W. Truran.
Astron. Astrophys., Vol. 186, No. 1/2, p. L9 – L10 (1987).

Five weeks after the explosion of Supernova 1987A in the LMC, speckle interferometric observations gave evidence for a second object having a velocity of a fraction of the velocity of light and a brightness of about a tenth of the luminosity of the supernova at that time. The possibility is discussed that this second object may reflect the interaction of the early UV burst of the supernova with a nearby dense and cool protostellar cloud. If this interpretation is correct, the cloud should show up as a very bright X–ray source about one year after the explosion.

125.246   The neutrino burst from Supernova 1987A: a search for periodicities.
D. Fischer.
Astron. Astrophys., Vol. 186, No. 1/2, p. L11 – L13 (1987).

The neutrino showers from Supernova 1987A detected by the Kamiokande II and IMB experiments are checked systematically for periodicities between 5 and 15 msec, in order to assess the recent hypothesis that an 8.9 msec period may exist in both data sets (Harwit et al., 1987). Many more fits are found for each detector's sequence of events but none stands out sharply, including the 8.9 msec result. The multitude of mediocre period fits turns out to be rather typical for events distributed randomly.

125.247   Detection of a very bright source close to the LMC supernova SN 1987A.
P. Nisenson, C. Papaliolios, M. Karovska, R. Noyes.
Astrophys. J., Lett. Ed., Vol. 320, No. 1, p. L15 – L18 (1987). With plate L1.

High angular resolution observations of the supernova SN 1987A have revealed a bright source separated from the SN by approximately 60 mas with a magnitude difference of 2.7 at 656 nm (Hα). Speckle imaging techniques were applied to data recorded with the CfA two–dimensional photon counting detector on the CTIO 4 m telescope on March 25 and April 2. The nature of this object is as yet unknown, though it is almost certainly a phenomenon related to the SN.

125.248   Prospects for observations of nucleosynthetic gamma–ray lines and continuum from SN 1987A.
N. Gehrels, C. J. MacCallum, M. Leventhal.
Astrophys. J., Lett. Ed., Vol. 320, No. 1, p. L19 – L22 (1987).

Expected flux levels for nucleosynthetic γ–ray line and continuum emissions from SN 1987A are calculated for several models. The dominant line emission is from freshly synthesized ^{56}Ni and its decay daughters, and the continuum is from Compton scattering of line photons. For a 15 $M_\odot$ Type II model, the light curve for the 0.847 MeV γ–ray line peaks in 1988 September at 3×10^{-4} photons cm^{-2}s^{-1}. This is detectable only by new, ultrasensitive balloon–borne spectrometers. For models with substantial mass loss from their envelopes, the peak is in early 1988 at $\sim 10^{-2}$ photons cm^{-2}s^{-1}, which is detectable at high significance levels by all current instruments.

125.249   Blackbody emissivity of supernova 1987A and the extragalactic distance scale.
D. Branch.
Astrophys. J., Lett. Ed., Vol. 320, No. 1, p. L23 – L25 (1987).

An application of the Baade method of supernova distance determination to SN 1987A in the Large Magellanic Cloud yields a distance of 55 ± 5 kpc. The good agreement with the known distance to the LMC means that the absolute emissivity of the supernova photosphere was like that of a blackbody. This result lends weight to a supernova–based extragalactic distance scale, $H_0 \approx 60$ km s^{-1}Mpc^{-1}.

125.250   Physical implications of the Kamioka observation of neutrinos from supernova 1987A.
J. Arafune, M. Fukugita.
Phys. Rev. Lett., Vol. 59, No. 3, p. 367 – 369 (1987).

It is shown that the neutrino events associated with SN 1987A are characterized by the first energetic ($E_\nu \sim 25$ MeV) electron–capture ν_e pulse with quite a high source neutrino flux $[L_{\nu,e} \cong (1 - 2) \times 10^{53}$ erg], and by the subsequent thermal pair neutrinos with a modest average energy $\langle E_\nu \rangle \cong 7$ MeV and also with a reasonable source neutrino flux $L_\nu \cong 3 \times 10^{53}$ erg. Some significant constraints on particle physics are also derived.

125.251   Type II supernovae from prompt explosions.
E. Baron, H. A. Bethe, G. E. Brown, J. Cooperstein, S. Kahana.
Phys. Rev. Lett., Vol. 59, No. 6, p. 736 – 739 (1987).

Evidence is cited that supernova 1987A involved a large explosion energy, $\cong (2 - 3) \times 10^{51}$ erg. Such a large explosion energy has not come from delayed shocks to date, nor is it likely to. Improved physics in the presupernova evolution, especially the inclusion of Coulomb interactions, has brought the iron–core mass down by $\gtrsim 0.1$ $M_\odot$ in the 13 $M_\odot$ star which has recently been evolved. The authors find that supernova explosion energies up to 3×10^{51} erg can be obtained by the prompt–explosion mechanism, provided that a somewhat soft equation of state is used at supranuclear densities.

125.252   Erratum: "Energetic (> 1 GeV) neutrinos as a probe of acceleration in the new supernova" [Phys. Rev. Lett., Vol. 58, No. 16, p. 1695 – 1697 (1987)].
T. K. Gaisser, T. Stanev.
Phys. Rev. Lett., Vol. 59, No. 7, p. 844 (1987). See Abstr. 43.125.235.

125.253   Neutrinos from SN 1987A and current models of stellar–core collapse.
S. W. Bruenn.
Phys. Rev. Lett., Vol. 59, No. 8, p. 938 – 941 (1987).

The neutrino signatures obtained from extended hydrodynamic calculations of stellar–core collapse are described and compared with the data of Hirata et al. and Bionta et.al. The comparisons suggest that SN 1987A resulted from the collapse of a small core, and marginally suggest the delayed–shock mechanism.

125.254   Constraints on the lifetime of massive neutrinos from SN 1987A.
A. Dar, S. Dado.
Phys. Rev. Lett., Vol. 59, No. 20, p. 2368 – 2370 (1987).

The absence of a detectable γ–ray emission from SN 1987A following the neutrino burst from this source has closed the window left by accelerator experiments and by previous astrophysical and cosmological observations for the existence of heavy neutrinos whose mass is less than 100 MeV and which are mixed with ν_e.

125.255   Search for high–energy neutrinos from SN 1987A: first six months.
Y. Oyama, K. Hirata, T. Kajita, M. Koshiba, M. Nakahata, N. Sato, A. Suzuki, M. Takita, Y. Totsuka, T. Kifune, T. Suda, K. Nakamura, K. Takahashi, T. Tanimori, K. Miyano, M. Yamada, E. W. Beier, L. R. Feldscher, S. B. Kim, A. K. Mann, F. M. Newcomer, R. Van Berg, W. Zhang, B. G. Cortez.
Phys. Rev. Lett., Vol. 59, No. 22, p. 2604 – 2606 (1987).

Upward–going muons with energy greater than 1.7 GeV produced in the nearby rock by neutrinos from SN 1987A were searched for in a 2140–ton underground water Cherenkov detector, Kamiokande II. No upward–going muons from the direction of SN 1987A were found from 23 February 1987 to 1 September 1987; the 90%–confidence–level flux limit is 1.2×10^{-13} cm^{-2}s^{-1} for an energy threshold of 1.7 GeV.

125.256   Supernova 1987A: a mysterious stranger.
M. M. Waldrop.
Science, Vol. 237, No. 4810, p. 25 – 26 (1987).

A computerized technique for high–resolution imaging suggests that the supernova has a companion.

125.257   A simple model for neutrino cooling of the Large Magellanic Cloud supernova.
D. N. Spergel, T. Piran, A. Loeb, J. Goodman, J. N. Bahcall.
Science, Vol. 237, No. 4821, p. 1471 – 1473 (1987).

A simplified analytic model of a cooling hot neutron star, motivated by detailed computer calculations, describes well the neutrinos detected from the recent supernova in the Large

Magellanic Cloud. The observations do not require explanations that invoke exotic physics or complicated astrophysics. The parameters in this simple model are not severely constrained.

125.258 **Supernova 1987A on center stage.**
M. M. Waldrop.
Science, Vol. 238, No. 4830, p. 1038 – 1041 (1987).

125.259 **Supernova Shelton 1987A: the discovery and the news via short–wave radio from last night.**
R. F. Garrison.
Bull. Am. Astron. Soc., Vol. 19, No. 2, p. 722 (1987). Abstract. – See Abstr. 010.061.

125.260 **Optical and infrared observations of SN 1987A from Cerro Tololo Inter–American Observatory.**
M. M. Phillips.
Bull. Am. Astron. Soc., Vol. 19, No. 2, p. 722 (1987). Abstract. – See Abstr. 010.061.

125.261 **Ultraviolet observations of supernova 1987A.**
R. P. Kirshner.
Bull. Am. Astron. Soc., Vol. 19, No. 2, p. 722 (1987). Abstract. – See Abstr. 010.061.

125.262 **Optical spectroscopy of supernova 1987A – the first week.**
M. D. Gregg, R. A. Kimble, A. F. Davidsen.
Bull. Am. Astron. Soc., Vol. 19, No. 2, p. 734 (1987). Abstract. – See Abstr. 010.061.

125.263 **Radio emission from SN 1987A in the Large Magellanic Cloud.**
R. N. Manchester, D. L. Jauncey, M. J. Kesteven, R. P. Norris, M. C. Storey, A. J. Turtle, D. Campbell–Wilson, J. D. Bunton, J. E. Reynolds.
Bull. Am. Astron. Soc., Vol. 19, No. 2, p. 734 (1987). Abstract. – See Abstr. 010.061.

125.264 **Spatially–resolved far ultraviolet spectroscopy of the field around supernova 1987A.**
G. Sonneborn, R. P. Kirshner.
Bull. Am. Astron. Soc., Vol. 19, No. 2, p. 734 (1987). Abstract. – See Abstr. 010.061.

125.265 **The ultraviolet interstellar line spectrum toward the supernova 1987A.**
F. C. Bruhweiler.
Bull. Am. Astron. Soc., Vol. 19, No. 2, p. 734 (1987). Abstract. – See Abstr. 010.061.

125.266 **The interstellar medium towards SN 1987A.**
A. K. Dupree, R. P. Kirshner, G. Nassiopoulos, J. C. Raymond, G. Sonneborn.
Bull. Am. Astron. Soc., Vol. 19, No. 2, p. 734 (1987). Abstract. – See Abstr. 010.061.

125.267 **Search for prompt gamma–ray emission from SN 1987A.**
E. L. Chupp, W. T. Vestrand, A. Ghosh, D. J. Forrest, G. H. Share, C. Reppin.
Bull. Am. Astron. Soc., Vol. 19, No. 2, p. 734 – 735 (1987). Abstract. – See Abstr. 010.061.

125.268 **A search for gamma–ray lines from SN 1987A in the LMC.**
S. M. Matz, G. H. Share, M. D. Leising, W. R. Purcell, E. L. Chupp, W. T. Vestrand, A. Ghosh, C. Reppin.
Bull. Am. Astron. Soc., Vol. 19, No. 2, p. 735 (1987). Abstract. – See Abstr. 010.061.

125.269 **Gamma–rays from supernova 1987A.**
K. Brecher.
Bull. Am. Astron. Soc., Vol. 19, No. 2, p. 735 (1987). Abstract. – See Abstr. 010.061.

125.270 **Infrared prognosis for SN 1987A.**
E. Dwek.
Bull. Am. Astron. Soc., Vol. 19, No. 2, p. 735 (1987). Abstract. – See Abstr. 010.061.

125.271 **X–ray spectrometers for observing SN 1987A.**
F. Marshall, E. Boldt, S. Holt, R. Kelley, R. Mushotzky, R. Petre, P. Serlemitsos, J. Swank, A. Szymkowiak.
Bull. Am. Astron. Soc., Vol. 19, No. 2, p. 735 (1987). Abstract. – See Abstr. 010.061.

125.272 **X–ray emission from supernova explosions.**
J. H. Beall, K. S. Wood.
Bull. Am. Astron. Soc., Vol. 19, No. 2, p. 736 (1987). Abstract. – See Abstr. 010.061.

125.273 **Inside supernova 1987A.**
R. McCray, J. M. Shull, P. Sutherland.
Bull. Am. Astron. Soc., Vol. 19, No. 2, p. 736 (1987). Abstract. – See Abstr. 010.061.

125.274 **Comprehensive model light curves for SN 1987A.**
A. Fu, W. D. Arnett.
Bull. Am. Astron. Soc., Vol. 19, No. 2, p. 739 (1987). Abstract. – See Abstr. 010.061.

125.275 **Atmosphere models for SN 1987A.**
R. P. Harkness, J. C. Wheeler.
Bull. Am. Astron. Soc., Vol. 19, No. 2, p. 740 (1987). Abstract. – See Abstr. 010.061.

125.276 **Supernova 1987A: a critical test of the Baade method of distance determination.**
R. V. Wagoner, E. V. Linder, S. Hershkowitz.
Bull. Am. Astron. Soc., Vol. 19, No. 2, p. 740 (1987). Abstract. – See Abstr. 010.061.

125.277 **Theoretical models for supernova 1987A.**
B. A. Fryxell, W. D. Arnett.
Bull. Am. Astron. Soc., Vol. 19, No. 2, p. 740 (1987). Abstract. – See Abstr. 010.061.

125.278 **Theoretical models for SN 1987A.**
S. E. Woosley.
Bull. Am. Astron. Soc., Vol. 19, No. 2, p. 740 (1987). Abstract. – See Abstr. 010.061.

125.279 **SN 1987A and a most probable value of the neutrino mass.**
H.–Y. Chiu, Y. Kondo, K. L. Chan.
Bull. Am. Astron. Soc., Vol. 19, No. 2, p. 740 (1987). Abstract. – See Abstr. 010.061.

125.280 **IUE low–dispersion spectra of supernova 1987A.**
D. M. Crenshaw, G. Sonneborn, R. P. Kirshner.
Bull. Am. Astron. Soc., Vol. 19, No. 2, p. 752 (1987). Abstract. – See Abstr. 010.061.

125.281 **Rapid ionization of the environment of SN 1987A.**
A. C. Raga.
Astron. J., Vol. 94, No. 6, p. 1578 – 1584 (1987).
It has been suggested by some authors that IUE observations of the supernova SN 1987A show the presence of a strong component of the interstellar C IV 1550 and Si IV 1393 absorption lines at a velocity that approximately corresponds to the velocity of the LMC. It is possible that this component might come from originally neutral (or at least not very highly ionized) gas which has been photoionized by the initially very strong

ionizing radiation field of the supernova. Theoretical considerations of this scenario lead to the study of fast (with velocities $\sim c$) ionization fronts. The author shows that for reasonable model parameters it is possible to obtain considerably large C IV column densities, in agreement with the IUE observations. On the other hand, the models do not so easily predict the large Si IV column densities that are also obtained from the IUE observations. The author finds that only models in which the interstellar medium surrounding SN 1987A is initially composed of already ionized hydrogen and helium predict substantial Si IV column densities. This result provides an interesting prediction of the ionization state of the environment of the presupernova star.

125.282   Early optical spectroscopy of supernova 1987A.
J. A. Tyson, P. C. Boeshaar.
Publ. Astron. Soc. Pac., Vol. 99, No. 619, p. 905 – 914 (1987).

CCD spectroscopy of SN 1987A from 1987 February 24 – 28 is presented. Spectra with 10 Å resolution and signal–to–noise ratio in excess of 200, over a range 3900 Å to 9000 Å, are reduced to flux relative to the continuum at 8000 Å. Spectra as a function of time and nightly difference spectra are obtained. Spectra as a function of angular distance from SN 1987A are also obtained, showing the adjacent nebular emission of the LMC. There is no evidence for a circumstellar shell. These data are most consistent with an unusually low envelope mass progenitor.

125.283   Photometry of supernova 1987A.
A. H. Jarrett.
Inf. Bull. Variable Stars, No. 3076, 2 pp. (1987).

125.284   Observations of the supernova 1987A at Boyden Observatory.
A. H. Jarrett.
Boyden Obs., Occas. Publ., No. 4, 5 pp. (1987).

125.285   The composite image of Sanduleak –69°202, candidate precursor to supernova 1987A in the Large Magellanic Cloud.
N. R. Walborn, B. M. Lasker, V. G. Laidler, Y.–H. Chu.
Space Telesc. Sci. Inst., Prepr. Ser., No. 193, 13 pp. (1987). To appear in Astrophys. J., Lett. Ed.

125.286   Ultraviolet photometry and the energetics of SN 1987A.
N. Panagia.
Space Telesc. Sci. Inst., Prepr. Ser., No. 210, 8 pp. (1987). To appear in the proceedings of the ESO workshop on "SN 1987A", July 1987.

125.287   VLBI observations of supernova 1987A.
I. Shapiro, N. Bartel, R. Preston, D. Jones, A. Kemball, G. Nicolson, D. Jauncey, J. Reynolds, A. Whitney, A. Rogers, R. Phillips, T. Clark, D. Robertson.
Cent. Astrophys., Prepr. Ser., No. 2553, 2 pp. (1987). To appear in IAU Symp. No. 129.

125.288   Spectropolarimetry of SN 1987A: observations up to 1987 July 8.
M. Cropper, J. Bailey, J. McCowage, R. D. Cannon, W. J. Couch, J. R. Walsh, J. O. Straede, F. Freeman.
Anglo–Aust. Obs., Prepr., No. 218, 40 pp. (1987). Submitted to Mon. Not. R. Astron. Soc.

125.289   On the interaction of the UV burst of supernova 1987A with a nearby cloud: a possible explanation of the speckle images.
W. Hillebrandt, P. Höflich, H. U. Schmidt, J. W. Truran.
MPA Rep., No. 312, 4 pp. (1987). Submitted to Astron. Astrophys.

125.290   The progenitor of SN 1987A: spatially resolved ultraviolet spectroscopy of the supernova field.
G. Sonneborn, B. Altner, R. P. Kirshner.
Astrophys. J., Lett. Ed., Vol. 323, No. 1, p. L35 – L39 (1987).

Careful deconvolution of spatially resolved *IUE* ultraviolet spectra shows that only two of the three stars detected in plate material before 1987 are now present near the site of SN 1987A. The separation, magnitudes, and spectra of these two are consistent with their identification as star 2 and star 3 of the Sanduleak –69°202 trio. The clear implication is that star 1, the 12th mag B3 I star, has disappeared, and it may be identified as the progenitor of SN 1987A.

125.291   Very low upper limits on the strength of interstellar lithium lines toward SN 1987A.
D. Baade, P. Magain.
ESO Sci. Prepr., No. 541, 17 pp. (1987). To appear in Astron. Astrophys.

125.292   SN 1987A: observational results obtained at ESO.
I. J. Danziger, P. Bouchet, R. A. E. Fosbury, C. Gouiffes, L. B. Lucy, A. F. M. Moorwood, E. Oliva, F. Rufener.
ESO Sci. Prepr., No. 554, 14 pp. (1987). Paper presented at the 4th George Mason University Workshop in Astrophysics "SN 1987A in LMC", Fairfax, Va., USA, October 1987.

125.293   Modelling the atmosphere of SN 1987A.
L. B. Lucy.
ESO Sci. Prepr., No. 557, 12 pp. (1987). Paper presented at the 4th George Mason University Workshop in Astrophysics "SN 1987A in LMC", Fairfax, Va., USA, October 1987.

125.294   Supernova 1987A in the Large Magellanic Cloud: initial observations at Cerro Tololo.
V. M. Blanco, B. Gregory, M. Hamuy, S. R. Heathcote, M. M. Phillips, N. B. Suntzeff, D. M. Terndrup, A. R. Walker, R. E. Williams, M. G. Pastoriza, T. Storchi–Bergmann, J. Matthews.
Astrophys. J., Vol. 320, No. 2, p. 589 – 596 (1987).

Optical and infrared observations of SN 1987A in the Large Magellanic Cloud (LMC) covering the first 7 weeks after discovery are presented. Over this period, the spectra were dominated by strong P Cygni emission lines of hydrogen, making this supernova a Type II event. Nevertheless, nearly all aspects of the behavior of SN 1987A have been unusual. The optical spectral and color evolution, while closely resembling that of a normal "plateau" Type II supernova, took place at a rate that was ~10 times faster than normal. Measurements of several presupernova plates and an objective–prism spectrum of the suspected progenitor, Sanduleak –69°202, rule out the presence of a previously undetected luminous red supergiant, but do suggest the existence of a less luminous red star.

125.295   Ultraviolet observations of SN 1987A.
R. P. Kirshner, G. Sonneborn, D. M. Crenshaw, G. E. Nassiopoulos.
Astrophys. J., Vol. 320, No. 2, p. 602 – 608 (1987).

Ultraviolet observations of the supernova in the Large Magellanic Cloud, SN 1987A, were carried out with the IUE. The first observations were obtained at 1987 February 24.80, 14 hr after the discovery. The earliest data show that the UV flux from the supernova was already declining while the optical flux was still rising. The UV spectrum at these epochs consists of broad features associated with the supernova atmosphere punctuated by sharp interstellar absorptions. The long–wavelength ultraviolet resembles the spectrum of a SN I, which is attributed to the absence of a circumstellar envelope and the presence of line absorption.

125.296 Supernova 1987A in the Large Magellanic Cloud: possible s–process enhancements in the progenitor.
R. E. Williams.
Astrophys. J., Lett. Ed., Vol. 320, No. 2, p. L117 – L120 (1987).

Absorption lines which appear in the spectrum of SN 1987A are identified as Ba II transitions which are also observed to be strong in classical barium stars. The presence of these lines and similar Sr II transitions argues for s–process element enhancements in the supernova progenitor. The same lines may also be present in some type I supernovae, suggesting the possibility that Ba II stars could be progenitors for some SN I outbursts.

125.297 Supernova 1987A: LTE line strengths as a guide to line identifications.
D. Branch.
Astrophys. J., Lett. Ed., Vol. 320, No. 2, p. L121 – L125 (1987).

Optical depths of spectral lines at the photosphere of SN 1987A are calculated on the basis of a simple model and are found to be in good qualitative agreement with observation. The calculated optical depths will serve as a guide to detailed studies of line identifications in the optical, infrared, and ultraviolet spectra of SN 1987A during its early photospheric phases.

125.298 Neutrinos from SN 1987A: theory versus observations.
A. Burrows.
Prepr. Steward Obs., No. 749, 7 pp. (1987). = Ariz. Theor. Astrophys. Prepr. #87–24. To be published in the proceedings on the ESO Workshop on the SN 1987A, held in Garching bei München, F.R. Germany, 6 – 8 July 1987.

125.299 Peering into the abyss: the neutrinos from SN 1987A.
A. Burrows.
Prepr. Steward Obs., No. 766, 11 pp. (1987). = Ariz. Theor. Astrophys. Prepr., #87–34. Presented at the George Mason Workshop on SN 1987A, Fairfax, VA, USA, 12 – 14 October 1987.

125.300 IR speckle observations of SN 1987a.
C. Perrier, A. A. Chalabaev, J. M. Mariotti, P. Bouchet.
Obs. Haute Provence, Pré–Publ., No. 26, 9 pp. (1987). To appear in ESO Workshop on SN 1987A.

125.301 IR speckle–interferometry of SN 1987A.
A. A. Chalabaev, C. Perrier, J. M. Mariotti.
Obs. Haute Provence, Pré–Publ., No. 29, 7 pp. (1987).

125.302 A tentative approach to the second neutrino burst of SN 1987A.
H. J. Haubold, B. Kaempfer, A. V. Senatorov, D. N. Voskresenski.
Inst. Astrophys. Paris, Pré–Publ., No. 207a, 9 pp. (1987). To be published in Astron. Astrophys.

125.303 On the event observed in the Mont Blanc Underground Neutrino Observatory during the occurrence of supernova 1987a.
M. Aglietta, G. Badino, G. Bologna, C. Castagnoli, A. Castellina, V. L. Dadykin, W. Fulgione, P. Galeotti, F. F. Kalchukov (*F. F. Khalchukov*), V. B. Kortchaguin (*V. B. Korchagin*), P. V. Kortchaguin (P. V. Korchagin), A. S. Malguin (*A. S. Mal'gin*), V. G. Ryassny (*V. G. Ryasnyj*), O. G. Ryazhskaya, O. Saavedra, V. P. Talochkin, G. Trinchero, S. Vernetto, G. T. Zatsepin, V. F. Yakushev.
Europhys. Lett., Vol. 3, No. 12, p. 1315 – 1320 (1987). Abstr. in Phys. Abstr., Vol. 90, No. 1312, Entry 107873 (1987).

125.304 Comments on the two events observed in neutrino detectors during the supernova 1987A outburst.
M. Aglietta, G. Badino, G. Bologna, C. Castagnoli, A. Castellina, V. L. Dadykin, W. Fulgione, P. Galeotti, F. F. Kalchukov (*F. F. Khalchukov*), V. B. Kortchaguin (*V. B. Korchagin*), P. V. Kortchaguin (*P. V. Korchagin*), A. S. Malguin (*A. S. Mal'gin*), V. G. Ryassny (*V. G. Ryasnyj*), O. G. Ryazhskaya, O. Saavedra, V. P. Talochkin, G. Trinchero, S. Vernetto, G. T. Zatsepin, V. F. Yakushev.
Europhys. Lett., Vol. 3, No. 12, p. 1321 – 1324 (1987). Abstr. in Phys. Abstr., Vol. 90, No. 1312, Entry 107874 (1987).

125.305 Detection of hard X–ray emission from SN 1987A. Preliminary results from the Kvant module on the Mir space station.
R. A. Syunyaev, A. Kaniovskij, V. Efremov, M. Gil'fanov, E. Churazov, S. Grebenev, A. Kuznetsov, A. Melioranskij, N. Yamburenko, S. Yunin, D. Stepanov, I. Chulkov, N. Pappe, M. Boyarskij, E. Gavrilova, V. Loznikov, A. Prudkoglyad, V. Rodin, C. Reppin, W. Pietsch, J. Engelhauser, J. Trümper, W. Voges, E. Kendziorra, M. Bezler, R. Staubert, A. C. Brinkman, J. Heise, W. A. Mels, R. Jager, G. K. Skinner, O. Al–Emam, T. G. Patterson, A. Willmore.
Pis'ma Astron. Zh., Tom 13, No. 12, p. 1027 – 1041 (1987). In Russian. English translation in Sov. Astron. Lett., Vol. 13.

1987 August 10, hard X–ray emission from the region of supernova 1987A was detected. A set of observations up to September 15 allowed to localize a hard X–ray source (its position coincides with the supernova's with accuracy 10 arcmin) and to determine its spectrum in the 20 – 300 keV band. Upper flux limits in the 3 – 20 keV and 300 – 1000 keV spectral bands were obtained. Upper limits are set on the emission measure of hot gas, heated by the shock wave, on the density of the gas before the front of the shock wave and the mass loss rate several tenth of years before the star explosion. The theoretical models are in good agreement with the observational data.

125.306 Expected X–ray emission from SN 1987A. Analytical consideration.
S. A. Grebenev, R. A. Syunyaev.
Pis'ma Astron. Zh., Tom 13, No. 12, p. 1042 – 1054 (1987). In Russian. English translation in Sov. Astron. Lett., Vol. 13.

In terms of diffusion (for the space) and Fokker–Planck (for photon energy) approximations a fast converging series describing the supernova envelope hard X–ray emission ($h\nu < 1$ MeV) was obtained assuming the central source of hard photons of any type. Compton scattering on cold electrons follwed by the recoil effect and photoabsorption are the main processes which determine the spectrum's shape. An opportunity of envelope geometry investigation during the observations of direct emission γ–quanta is also considered.

125.307 A possible explanation of the second neutrino burst from SN 1987A.
D. N. Voskresenskij (*D. N. Voskresenskij*), A. V. Senatorov, B. Kaempfer, H. J. Haubold.
Astrophys. Space Sci., Vol. 138, No. 2, p. 421 – 424 (1987).

It is argued that the neutrino bursts registered on February 23.316 UT, 1987 signalized the transition of a fresh–borne neutron star into a superdense state. The neutron star is supposed to be formed approximately five hours before February 23.12 UT in the supernova SN 1987A in the Large Magellanic Cloud.

125.308 Die Supernova 1987A.
B. Stecklum.
Sterne, 63. Band, Heft 4, p. 203 – 212 (1987).

125.309 Observations of the LMC supernova spectrum.
M. Parkinson.
Aust. J. Astron., Vol. 2, No. 2, p. 49 – 53 (1987).

Observations of the spectrum of SN 1987a in the LMC at a dispersion on the film plane of ~10^2Å mm^{-1} with a Unitron direct–vision star spectroscope fitted to a Celestron C8 are reported. On the evening of May 5 when SN 1987a was about m_v

3.0, at least ten broad absorption regions in addition to the broad hydrogen alpha emission line with blueward flanking P Cygni signature were seen clearly. Visual observations are consistent with spectra of SN 1987a published in the literature.

125.310 **A 1–5 μm infrared spectrum of SN 1987A.**
E. Oliva, A. F. M. Moorwood, I. J. Danziger.
Messenger, No. 50, p. 18 – 20 (1987).

125.311 **Electric charge of the neutrinos from SN 1987A.**
G. Barbiellini, G. Cocconi.
Nature, Vol. 329, No. 6134, p. 21 – 22 (1987).

125.312 **SN 1987A supernova: a black–hole precursor?**
S. Nussinov, I. Goldman, G. Alexander,
Y. Aharonov.
Nature, Vol. 329, No. 6135, p. 134 – 135 (1987).

The authors show that an identification of the first two neutrino events from SN 1987A in the Kamiokande data as v_e prompt events requires a mass, $M \gtrsim 6\,M_\odot$, which implies a black–hole formation. This is also consistent with the probable identification of the progenitor as a B3 I supergiant of $\sim 25\,M_\odot$.

125.313 **Photon–photon pair production and the opacity of SN 1987A to TeV and PeV γ–rays.**
R. J. Protheroe.
Nature, Vol. 329, No. 6135, p. 135 – 138 (1987).

Supernovae have long been considered as likely sites of cosmic–ray acceleration. Interaction of newly accelerated cosmic–ray nuclei with target material within the expanding supernova is expected to produce an observable flux from SN 1987A of very high–energy and possibly ultra high–energy γ–rays in the TeV and PeV ranges. The presence of intense infrared emission from the supernova itself will, however, make some regions of SN 1987A opaque to TeV and PeV γ–rays due to pair–production interactions. The author discusses the question of photon–photon pair–production interactions and calculates from which regions of SN 1987A may be observed TeV and PeV γ–rays.

125.314 **Particle acceleration and production of energetic photons in SN 1987A.**
T. K. Gaisser, A. Harding, T. Stanev.
Nature, Vol. 329, No. 6137, p. 314 – 316 (1987).

Young supernova remnants are likely to be bright sources of energetic photons and neutrinos through the collision of particles accelerated inside the remnant. If $> 10^{39}$ erg s^{-1} in protons above 10 TeV is injected into the target region, TeV photons from SN 1987A could be observable with present detectors. Synchrotron X–rays and γ–rays up to 10 MeV generated by accelerated electrons, may well also be detectable. The authors discuss a pulsar model for acceleration of particles and find that it would produce observable signals if the spin period of the pulsar is ⩽ 10 ms.

125.315 **The polarization spectrum of supernova 1987A interpreted in terms of shape asymmetry.**
D. J. Jeffery.
Nature, Vol. 329, No. 6138, p. 419 – 421 (1987).

Polarimetry on the type II supernova 1987A on 6 and 7 March 1987 showed variation in polarization across line profiles. This polarization structure is interpreted as arising from an asymmetric, homologously expanding, scattering atmosphere surrounding an asymmetric continuum–producing photosphere. Sobolev–method radiative transfer calculations with axisymmetric oblate ellipsoidal models have been carried out to fit the observed data.

125.316 **Modelling of the radio burst from SN 1987A.**
M. C. Storey, R. N. Manchester.
Nature, Vol. 329, No. 6138, p. 421 – 423 (1987).

The authors propose that the radio emission mechanism for SN 1987A is the same as that proposed for previous type II supernovae, namely, synchrotron emission from a thin shell near the edge of the expanding ejecta. The differences are mainly due to the absence of a thick circumstellar envelope around the progenitor star, Sk –69°202, at least in the line of sight to the supernova. They discuss several possible models for the emission and absorption processes and set limits on the density of a circumstellar envelope.

125.317 **The composite image of Sanduleak –69°202, candidate precursor to supernova 1987A in the Large Magellanic Cloud.**
N. R. Walborn, B. M. Lasker, V. G. Laidler, Y.–H. Chu.
Astrophys. J., Lett. Ed., Vol. 321, No. 1, p. L41 – L44 (1987).
With plates L7 – L10.

The image of Sk –69°202 has been analyzed on the basis of eight CTIO 4 m prime–focus plates obtained in 1974 – 1983, covering wavelengths from blue through near–infrared. Density differences and intensity syntheses based upon reference stars from the same plates are presented. The 12th mag blue supergiant (star 1) has two companions with V magnitudes, position angles, and separations 15.3, 315°, 3″(star 2) and 15.7, 115°, 1″.5(star 3), respectively. Both companions appear to be early–type stars; there is no evidence for a very red star in the system.

125.318 **Interstellar calcium towards supernova 1987A in the Large Magellanic Cloud.**
P. Magain.
Nature, Vol. 329, No. 6140, p. 606 – 608 (1987).

The author presents high–resolution and high signal–to–noise spectra of interstellar calcium towards SN 1987A in LMC. They provide new results for the Ca I spectrum not only at velocities corresponding to our Galaxy and the LMC, but also at intermediate velocities. These spectra, that allow to estimate the ionization balance in these interstellar clouds, provide some clues about their physical state and location. In particular, the components between 150 and 200 km s^{-1} show a much lower ionization degree than other components. This may be due to recent compression of the gas by a shock. This interpretation would require the corresponding clouds to be located inside the LMC, indicating that at least some intermediate velocity components are not of halo origin.

125.319 **Speckle interferometric observations of supernova 1987A and of a bright associated source.**
W. P. S. Meikle, S. J. Matcher, B. L. Morgan.
Nature, Vol. 329, No. 6140, p. 608 – 611 (1987).

The authors report optical speckle interferometric observations of SN1987A in LMC on days 38, 47 and 50 after the explosion on 23 February 1987. On day 30 after the explosion a CfA group discovered a second source lying at a small angular displacement from the supernova. On day 50 the authors observed a second source at 6,585 Å. Its offset from the supernova was 0.074 ± 0.008 arc s and it was ~ 3 mag fainter than the supernova at 6,585 Å at the same epoch. Such a bright source was not present before the explosion, implying that it was caused by the supernova. The source the authors observed was similar in magnitude and position angle to the feature discovered by the CfA group.

125.320 **Neutrino spectroscopy of supernova 1987A.**
L. M. Krauss.
Nature, Vol. 329, No. 6141, p. 689 – 694 (1987).

Two of the four reported observations of neutrino bursts associated with supernova 1987A are consistent with each other and with the theoretical expectations of supernova neutrino luminosities. Testing a range of models for properties of the incident neutrino signal, together with calculation of interaction rates in the chlorine solar neutrino detector, helps to provide constraints on the mass and number of light neutrinos and raises the possibility of vacuum neutrino oscillations.

125.321 **Cosmic rays and gamma radiation from the shell of SN 1987A.**
V. S. Berezinsky (*V. S. Berezinskij*), V. L. Ginzburg.
Nature, Vol. 329, No. 6142, p. 807 – 809 (1987).

High–energy gamma–ray observations, the authors show, can discover cosmic rays in the SN 1987A shell if they are produced inside the shell with luminosity down to $L_p \sim 10^{39} \text{erg s}^{-1}$. This can support or reject a very wide class of the models of cosmic ray production by supernovae. They argue that such measurements for SN 1987A will be possible during the next 1 – 2 years.

125.322   Jets in supernova 1987A?
T. Piran, T. Nakamura.
Nature, Vol. 330, No. 6143, p. 28 (1987).

Speckle observations of SN 1987A reveal a secondary light source, 2 – 3 magnitudes dimmer than the primary and 0.057 ± 0.014 arcs s away from it. The authors suggest that the observed secondary object is a relativistic jet and they compare their model with those of other authors.

125.323   The neutrino emission of SN 1987A.
R. Schaeffer, Y. Declais, S. Jullian.
Nature, Vol. 330, No. 6144, p. 142 – 144 (1987).

Several groups reported neutrino events, a few hours before the onset of the optical signal of the supernova 1987A in LMC. The authors show that the assumption that the the Mont Blanc (or Baksan) observation is due to the formation of a neutron star leads to excessive energy requirements. If confirmed, these two detections would invalidate any of the presently proposed explanations for the SN 1987A event. The Kamioka and IMB observations, on the other hand, match perfectly the standard theory and it is suggested that they were the tracers of the neutron star formation. They can then be used to constrain the number of neutrino flavours to be less than four.

125.324   A burst of discovery. The first days of supernova 1987A.
R. Talcott.
Astronomy, Vol. 15, No. 6, p. 90 – 95 (1987). Abstr. in Phys. Abstr., Vol. 90, No. 1313, Entry 115473 (1987).

125.325   A limit on the neutrino–neutrino scattering cross section from the supernova.
A. Manohar.
Phys. Lett. B, Vol. 192, No. 1 – 2, p. 217 – 218 (1987). Abstr. in Phys. Abstr., Vol. 90, No. 1314, Entry 120481 (1987).

125.326   Supernova 1987A in the Large Magellanic Cloud.
IAU Circ., Nos. 4413, 4414, 4417, 4419, 4421, 4423, 4426, 4427, 4431, 4432, 4435, 4438, 4440, 4445, 4447, 4448, 4450, 4452, 4453, 4456, 4457, 4463, 4466, 4468, 4474, 4481, 4482, 4484 – 4486, 4488, 4494, 4500, 4506, 4510, 4514, 4515, 4518, 4521 (1987).

125.327   La supernova 1987A dans le Grand Nuage de Magellan (suite).
M. Dennefeld.
Astronomie, Vol. 101, p. 563 – 565 (1987).

125.328   Supernova 1987A in LMC.
Yamamoto Circ., Nos. 2088, 2091, 2092 (1987). In Japanese.

125.329   Observations of the supernova 1987A in the Large Magellanic Cloud.
J. Lequeux.
The standard model. The supernova 1987A, p. 707 – 715 (1987).
– See Abstr. 012.074.

Observations of the supernova 1987A in the Large Magellanic Cloud up to mid–may 1987 are summarized: light curve and colors, optical, UV and IR spectrum, nature of the progenitor, interstellar lines in the spectrum. It is shown that this is a somewhat atypical Type II supernova resulting from the explosion of a massive star, but definitive conclusions on the exact nature of its progenitor will have to wait for some time.

125.330   Neutrino observations from supernova 1987a.
M. Aglietta, G. Badino, G. Bologna, C. Castagnoli, A. Castellina, V. L. Dadykin, W. Fulgione, P. Galeotti, F. F. Kalchukov, V. B. Kortchaguin (*V. B. Korchagin*), P. V. Kortchaguin (*P. V. Korchagin*), A. S. Malguin (*A. S. Mal'gin*), V. G. Ryassny (*V. G. Ryasnyj*), O. G. Ryazhskaya, O. Saavedra, V. P. Talochkin, G. Trinchero, S. Vernetto, G. T. Zatsepin, V. F. Yakushev.
The standard model. The supernova 1987A, p. 717 – 725 (1987).
– See Abstr. 012.074.

The authors discuss here the characteristics of the event detected in the Mont Blanc Underground Neutrino Observatory on February 23, 1987, consisting of 5 interactions recorded during 7 sec. The measured energies of the 5 pulses, the duration of the burst, and the advance of the detection time in comparison with the first optical observations give evidence that the event can be explained in terms of detection of neutrinos emitted during the stellar collapse in the Large Magellanic Cloud.

125.331   Observation of a neutrino burst from the supernova SN 1987A.
K. Hirata, T. Kajita, M. Koshiba, M. Nakahata, Y. Oyama, N. Sato, A. Suzuki, M. Takita, Y. Totsuka, T. Kifune, T. Suda, K. Takahashi, T. Tanimori, K. Miyano, M. Yamada, E. W. Beier, L. R. Feldscher, S. B. Kim, A. K. Mann, F. M. Newcomer, R. Van Berg, W. Zhang, B. G. Cortez.
The standard model. The supernova 1987A, p. 727 – 734 (1987).
– See Abstr. 012.074.

A neutrino burst was observed in the KAMIOKANDE–II detector on 23 February, 7:35:35 UT (± 1 minute) during a time interval of 13 seconds. The signal consisted of 11 electron events of energy 7.5 to 36 MeV, of which the first 2 point back to the Large Magellanic Cloud with angles 18° ± 18° and 15° ± 27°.

125.332   Observation of a neutrino burst from supernova SN 1987A.
R. M. Bionta, G. Blewitt, C. B. Bratton, D. Casper, A. Ciocio, R. Claus, M. Crouch, S. T. Dye, S. Errede, G. W. Foster, W. Gajewski, K. S. Ganezer, M. Goldhaber, T. J. Haines, T. W. Jones, D. Kielczewska, W. R. Kropp, J. G. Learned, J. M. LoSecco, J. Matthews, R. Miller, M. Mudan, H. S. Park, L. R. Price, F. Reines, J. Schultz, S. Seidel, E. Shumard, D. Sinclair, H. W. Sobel, J. L. Stone, L. Sulak, R. Svoboda, G. Thornton, J. C. van der Velde, C. Wuest.
The standard model. The supernova 1987A, p. 735 – 738 (1987).
– See Abstr. 012.074.

A burst of eight neutrino events preceding the optical detection of the supernova in the Large Magellanic Cloud has been observed in a large underground water Cherenkov detector. The events span an interval of 6 s and have visible energies in the range 20 – 40 MeV.

125.333   On possible detection of a neutrino burst on February 23, 1987 at the Baksan underground scintillation telescope.
E. N. Alexeyev (*E. N. Alekseev*), L. N. Alexeyeva (*L. N. Alekseeva*), I. V. Krivosheina, V. I. Volchenko.
The standard model. The supernova 1987A, p. 739 – 744 (1987).
– See Abstr. 012.074.

On February 23, 1987 an event was found at the Baksan underground scintillation telescope at the time indicated by the KAMIOKANDE II detector. The event at 7:36:11 ± 2 sec UT consists of 5 signals with energies of 12 – 33 MeV during 9 seconds from which the first 3 in 1.7 seconds.

125.334   SN 1987A. Theory.
R. Schaeffer.
The standard model. The supernova 1987A, p. 745 – 755 (1987).
– See Abstr. 012.074.

SN 1987A was unique in many aspects. The most striking, undoubtedly, is its low luminosity, nearly two orders of magnitude below the expectations based on supernovae currently observed in external galaxies. Other peculiarities can, however, be noticed. The author will discuss how these strange events fit with the theoretical models of supernova explosions, how they differ in some cases, and try to evaluate the degree of certainty – or uncertainty – of our present knowledge on how these extremely powerful star explosions occur.

125.335 MSW effect and (anti)neutrinos from SN 1987A.
P. O. Lagage, M. Cribier, J. Rich, D. Vignaud.
The standard model. The supernova 1987A, p. 757 – 759 (1987). – See Abstr. 012.074.

The MSW effect is calculated for ν_e, ν_μ and $\bar\nu_e$, $\bar\nu_\mu$ from the supernova SN 1987A in the Large Magellanic Cloud.

125.336 Neutrino mass determinations from the supernova SN 1987A bursts.
M. Roos.
The standard model. The supernova 1987A, p. 761 – 764 (1987). – See Abstr. 012.074.

The time and energy coordinates of the neutrino signals are fitted to kinematics. Under the most conservative hypothesis the electron neutrino mass is found to be <5 eV.

125.337 MSW effect and (anti) neutrinos from SN 1987A.
P. O. Lagage, M. Cribier, J. Rich, D. Vignaud.
Phys. Lett. B, Vol. 193, No. 1, p. 127 – 130 (1987). Abstr. in Phys. Abstr., Vol. 90, No. 1315, Entry 127927 (1987).

125.338 May a supernova bang twice?
A. de Rujula.
Phys. Lett. B, Vol. 193, No. 4, p. 514 – 524 (1987). Abstr. in Phys. Abstr., Vol. 90, No. 1315, Entry 127928 (1987).

125.339 Constraints on light particles from supernova SN 1987A.
J. Ellis, K. A. Olive.
Phys. Lett. B, Vol. 193, No. 4, p. 525 – 530 (1987). Abstr. in Phys. Abstr., Vol. 90, No. 1315, Entry 127929 (1987).

125.340 The LMC supernova (SN 1987A) as a probe for the outcome of stellar collapse.
H.-J. Haubold, S. Gottlöber, J. P. Mücket, V. Müller, B. Kämpfer.
Astron. Nachr., Vol. 308, No. 6, p. 329 – 331 (1987).

Implications from the available information on the supernova SN 1987A are discussed for supernova models. The authors derive an upper bound of 10 – 25 eV for the neutrino rest mass.

125.341 Supernova 1987A.
J. Skuljan.
Vasiona, Année 35, No. 2, p. 42 – 44 (1987). In Croatian.

A short presentation of the supernova in the Large Magellanic Cloud including results of neutrino detections is given.

125.342 Surprising Supernova 1987A.
N. I. Shakura.
Zemlya Vselennaya, No. 6, p. 54 – 55 (1987). In Russian.

125.343 Neutrino observations from supernova 1987A.
M. Aglietta, G. Badino, G. Bologna, C. Castagnoli, A. Castellina, V. L. Dadykin, W. Fulgione, P. Galeotti, F. F. Kalchukov (*F. F. Khalchukov*), V. B. Kortchaguin (*V. B. Korchagin*), P. V. Kortchaguin (*P. V. Korchagin*), A. S. Malguin (*A. S. Mal'gin*), V. G. Ryassny (*V. G. Ryasnyj*), O. G. Ryazhskaya, O. Saavedra, V. P. Talochkin, G. Trinchero, S. Vernetto, G. T. Zatsepin, V. F. Yakushev.
Helv. Phys. Acta, Vol. 60, No. 5 – 6, p. 619 – 628 (1987). Abstr. in Phys. Abstr., Vol. 90, No. 1317, Entry 140126 (1987).

125.344 Information on neutrino masses obtained from supernova 1987A data.
Z.-m. Chen, L. Gou, Z.-q. Ma, X.-j. Zhou.
Chin. Phys. Lett., Vol. 4, No. 9, p. 421 – 423 (1987). Abstr. in Phys. Abstr., Vol. 91, No. 1319, Entry 4419 (1988).

125.345 Neutrino mass and supernova 1987A.
L.-z. Fang.
Chin. Phys. Lett., Vol. 4, No. 9, p. 424 – 426 (1987). Abstr. in Phys. Abstr., Vol. 91, No. 1319, Entry 4420 (1988).

125.346 Neutrinos from supernova 1987A.
S. H. Kahana, J. Cooperstein, E. Baron.
Phys. Lett. B, Vol. 196, No. 3, p. 259 – 266 (1987). Abstr. in Phys. Abstr., Vol. 91, No. 1319, Entry 4427 (1988).

125.347 Total energy of the neutrino burst from the supernova 1987A and the mass of the neutron star just born.
K. Sato, H. Suzuki.
Phys. Lett. B, Vol. 196, No. 3, p. 267 – 272 (1987). Abstr. in Phys. Abstr., Vol. 91, No. 1319, Entry 4428 (1988).

125.348 Severe restrictions of neutrino masses and mixing angles from SN 1987A.
D. Notzold.
Phys. Lett. B, Vol. 196, No. 3, p. 315 – 320 (1987). Abstr. in Phys. Abstr., Vol. 91, No. 1319, Entry 4429 (1988).

125.349 Discovery of hard X–ray emission from supernova 1987A.
R. Sunyaev (*R. A. Syunyaev*), A. Kaniovsky, V. Efremov, M. Gil'fanov, E. Churazov, S. Grebenev, A. Kuznetsov, A. Melioranskiy (*A. S. Melioranskij*), N. Yamburenko, S. Yunin, D. Stepanov, I. Chulkov, N. Pappe, M. Boyarskiy (*M. N. Boyarskij*), E. Gavrilova, V. Loznikov, A. Prudkoglyad, V. Rodin, C. Reppin, W. Pietsch, J. Engelhauser, J. Trümper, W. Voges, E. Kendziorra, M. Bezler, R. Staubert, A. C. Brinkman, J. Heise, W. A. Mels, R. Jager, G. K. Skinner, O. Al-Emam, T. G. Patterson, A. P. Willmore.
Nature, Vol. 330, No. 6145, p. 227 – 229 (1987).

The authors report the discovery of hard X–rays from the region of the supernova SN 1987A in the Large Magellanic Cloud. The measured spectrum extends from 20 keV to 300 keV and is extremely hard, having a photon power law index of ~1.4. At lower energies the spectrum becomes even flatter and there is indication of a cutoff between 10 and 25 keV. The luminosity over the energy range 20 – 300 keV is ~2×10^{38} erg s^{-1} (assuming a distance of 55 kpc). The error box for the hard source has a (2σ) radius of 10 arc min and contains SN 1987A. The results have important implications for current models of SN 1987A.

125.350 Discovery of an unusual hard X–ray source in the region of supernova 1987A.
T. Dotani, K. Hayashida, H. Inoue, M. Itoh, K. Koyama, F. Makino, K. Mitsuda, T. Murakami, M. Oda, Y. Ogawara, S. Takano, Y. Tanaka, A. Yoshida, K. Makishima, T. Ohashi, N. Kawai, M. Matsuoka, R. Hoshi, S. Hayakawa, T. Kii, H. Kunieda, F. Nagase, Y. Tawara, I. Hatsukade, S. Kitamoto, S. Miyamoto, H. Tsunemi, K. Yamashita, M. Nakagawa, M. Yamauchi, M. J. L. Turner, K. A. Pounds, H. D. Thomas, G. C. Stewart, A. M. Cruise, B. E. Patchett, D. H. Reading.
Nature, Vol. 330, No. 6145, p. 230 – 231 (1987).

The authors have discovered a new hard X–ray source near supernova 1987A, in the Large Magellanic Cloud, from the X–ray astronomy satellite Ginga. The present error box of $0.2° \times 0.3°$ includes the supernova. The energy spectrum is very hard above 10 keV and unusual for any of the known classes of X–ray sources. The source intensity increased steadily throughout July, August and early September, but an observation in late September revealed no further increase. The positional agreement, a steady brightening and an unusual hard spectrum support the identification of the source as SN 1987A.

125.351 A search for soft X–rays from supernova 1987A.
B. Aschenbach, U. G. Briel, E. Pfeffermann, H. Bräuninger, H. Hippmann, J. Trümper.
Nature, Vol. 330, No. 6145, p. 232 – 233 (1987).

The authors searched for soft X–ray emission (0.2 – 2.1 keV) from SN 1987A in LMC by means of a sounding rocket–borne experiment on 24 August. No positive signal was detected, and an upper limit for the luminosity of 1.5×10^{36}erg s^{-1}, between 0.2 – 2.1 keV at a 95% confidence level has been derived.

125.352 X–rays expected from supernova 1987A compared with the source discovered by the Ginga satellite.
M. Itoh, S. Kumagai, T. Shigeyama, K. Nomoto, J. Nishimura.
Nature, Vol. 330, No. 6145, p. 233 – 235 (1987).

The authors have constructed hydrodynamic models of SN 1987A and calculated the optical light curve and the X–ray light curve and spectrum from ^{56}Co decay. They compare these models with the observations made with the X–ray astronomy satellite Ginga. X–ray emission from radioactive decay may qualitatively account for the hard component of X–rays from SN 1987A observed by Ginga, if ^{56}Co has been mixed into the outer layers. The soft component of the spectrum cannot be explained by the radioactive model and must be due to another mechanism.

125.353 Thermal X–ray emission from supernova 1987A.
K. Masai, S. Hayakawa, H. Itoh, K. Nomoto.
Nature, Vol. 330, No. 6145, p. 235 – 236 (1987). With a correction in Vol. 330, No. 6148, p. 588 (1987).

A new hard–X–ray source has been discovered in the region of SN 1987A by the X–ray astronomy satellite Ginga. The authors show that the observed X–ray spectrum can be accounted for by a superposition of thermal X–rays and Compton degradation of γ–rays, and discuss some constraints on physical quantities related to the mass–loss history of the progenitor.

125.354 Hard X–rays imply more to come.
D. D. Clayton.
Nature, Vol. 330, No. 6147, p. 423 – 424 (1987).

The author reports on the recent detection of hard X–radiation from the supernova 1987A in LMC, and the expectations that γ–ray lines from radioactive ^{56}Co may soon be detected. Hard X–rays and γ–rays from supernovae have never been observed. Astrophysicists have hoped that they will be seen, so that the details will help explain the origin of the elements.

125.355 Making the most of SN 1987A.
T. P. Walker.
Nature, Vol. 330, No. 6149, p. 609 – 610 (1987).

The author reports on the physics derived from the energy budget of SN 1987A in LMC and on the physics derived from the statistics of 19 neutrino events.

125.356 Supernova 1987A in the Large Magellanic Cloud.
T. Fabini, D. Kubáček.
Kozmos, Vol. 18, No. 4, p. 112 – 115 (1987). In Slovak.

125.357 A supernova after four centuries.
M. Wolf.
Vesmír, Vol. 66, No. 8, p. 425 – 428 (1987). In Czech.

125.358 Supernova of the century.
J. Grygar.
Říše hvězd, Vol. 68, No. 7, p. 132 – 134 (1987). In Czech.

125.359 Late emission from SN 1987A.
C. Fransson, R. A. Chevalier.
Astrophys. J., Lett. Ed., Vol. 322, No. 1, p. L15 – L20 (1987).

Assuming that the progenitor star of SN 1987A had an initial mass in the 15 – 25 $M_\odot$ range and had undergone considerable mass loss, the authors model the late emission from the mantle gas under the assumption of energy input by γ–rays from ^{56}Co decays. An emission–line spectrum of low–ionization species is produced at optical through infrared wavelengths. At an age of ~ 700 days an infrared catastrophe occurs and the gas temperature drops to a few hundred K; most of the radiation is then emitted in infrared fine–structure lines. The γ–ray line intensity peaks in the 15 $M_\odot$ model at an age of 270 days. Radiation from a possible central pulsar nebula will be absorbed by the mantle gas; optical depth unity in the 15 $M_\odot$ model at 10 keV is not

reached for 18 yr. The radio emission from the pulsar is likely to be absorbed by the ejecta for several years.

125.360 VRI photometry of LMC SN 1987A.
K. E. Rangarajan, K. Jayakumar, M. Appakutty, H. D. Sheriff.
Inf. Bull. Variable Stars, No. 3051, 1 pp. (1987).

125.361 The ionization effects of shock breakout in SN 1987A.
M. A. Dopita, S. J. Meatheringham, P. Nulsen, P. R. Wood.
Astrophys. J., Lett. Ed., Vol. 322, No. 2, p. L85 – L89 (1987).

The epoch of shock breakout in SN 1987A was almost certainly associated with the production of a pulse of UV photons with a characteristic temperature of order 10^5K and a duration of 2 – 4 hr. The authors propose that this pulse has the characteristics required to ionize the precursor stellar wind, temporarily ionize any nearby remnants of the red giant wind, and can ionize the surrounding interstellar medium out to distances of several parsecs for several thousand years.

Supernova 1984a in NGC 4419

125.501 Observations of supernova 1984a in NGC 4419.
G. N. Kimeridze, D. Yu. Tsvetkov.
Astrophysics, Vol. 25, No. 2, p. 513 – 518 (1987). English translation of 42.125.401.

Supernova 1986J in NGC 891

125.521 Observations of SN 1986J in NGC 891.
M. P. Rupen, J. H. van Gorkom, G. R. Knapp, J. E. Gunn, D. P. Schneider.
Astron. J., Vol. 94, No. 1, p. 61 – 70 (1987). With plate 4.

Supernova SN 1986J was discovered at radio wavelengths with the VLA. The object is currently strongest at $\lambda 6$ cm, and probably turned on at $\lambda 6$ cm in early 1984 and at $\lambda 20$ cm one or two years later. H I absorption–line observations show that the SN is about 7.7 kpc from the center and within 170 pc of the plane of NGC 891. SN 1986J is, in the radio, both the brightest and most luminous supernova yet found. The supernova has been identified optically with a 20th magnitude (September 1986) point source in NGC 891, and was approximately 1 mag brighter in January 1984. The spectrum is dominated by emission lines; H I, He I, [O I], and probably Fe I and Fe II are seen. The Balmer decrement is very large, with an Hα/Hβ flux ratio of ~ 60. The spectrum is similar to those seen in the late phases of Type II supernovae; however, the small linewidth (FWHM of ~ 1000 km s^{-1}) suggests that SN 1986J is actually Type V.

125.522 Circumstellar interaction and a pulsar nebula in the supernova 1986j.
R. A. Chevalier.
Nature, Vol. 329, No. 6140, p. 611 – 612 (1987).

The supernova SN 1986j in NGC 891 was first discovered as a radio source in August 1986. The author shows that the circumstellar interaction model for radio supernovae provides an adequate fit to the existing data. A shell–like source structure is expected. The model together with VLBI observations of the radio source, implies a shock velocity of $\sim 10^4$km s^{-1} and a pre–supernova mass loss rate of $\sim 10^{-4} M_\odot$yr^{-1} for a wind velocity of 10 km s^{-1}. The time behaviour of the optical flux and its high luminosity suggest that a central pulsar nebula is the energy source. The radiated energy from the pulsar neula is absorbed by expanding, helium–rich material extending out to a velocity of ~ 700 km s^{-1} or a radius of 8×10^{15}cm on September 1986. The radius of the pulsar nebula at that time is $\sim 5 \times 10^{15}$cm.

125.523 Supernova 1986J in NGC 891.
IAU Circ., No. 4423 (1987).

Supernova 1985F in NGC 4618

125.541 Light curve of the peculiar supernova 1985f in NGC 4618.
D. Yu. Tsvetkov.
Sov. Astron. Lett., Vol. 12, No. 5, p. 328 – 329 (1987). English translation of 42.125.221.

Supernova 1986G in NGC 5128

125.561 The Type Ia supernova 1986G in NGC 5128: optical photometry and spectra.
M. M. Phillips, A. C. Phillips, S. R. Heathcote, V. M. Blanco, D. Geisler, D. Hamilton, N. B. Suntzeff, F. J. Jablonski, J. E. Steiner, A. P. Cowley, P. Schmidtke, S. Wyckoff, J. B. Hutchings, J. Tonry, M. A. Strauss, J. R. Thorstensen, W. Honey, J. Maza, M. T. Ruiz, A. U. Landolt, A. Uomoto, R. M. Rich, J. E. Grindlay, H. Cohn, H. A. Smith, J. H. Lutz, R. J. Lavery, A. Saha.
Publ. Astron. Soc. Pac., Vol. 99, No. 617, p. 592 – 605 (1987).
Optical light curves and spectra of the Type Ia supernova 1986G in NGC 5128 (Centaurus A) are presented. Although the spectral evolution closely resembled that of the more common "slower" photometric classes of Type Ia supernovae, subtle differences in the maximum–light spectra were detected. The expansion velocity of the photosphere of SN 1986G decreased rapidly at early phases. SN 1986G appears to have been heavily obscured by the dust lane of NGC 5128. This circumstances accounts for the strong interstellar–absorption lines of Ca II H and K and Na I D observed in the spectra as well as for features that the authors identify with the diffuse interstellar bands. SN 1986G provides graphic confirmation of the existence of intrinsic differences in the optical light curves and spectroscopic properties of Type Ia supernovae. Consequently, these objects must be used with considerable caution as cosmological standard candles. The authors derive a relative distance of $D_{NGC\ 5128}/D_{NGC\ 5055} = 0.39 \pm 0.04$.

125.562 Echelle spectroscopy of SN 1986G in NGC 5128: complex Na absorption and interstellar diffuse bands.
R. M. Rich.
Astron. J., Vol. 94, No. 3, p. 651 – 656 (1987).
Spectroscopy at 0.3 Å resolution of the Type Ia supernova 1986G in NGC 5128 has resolved the Na absorption due to NGC 5128 into four components with velocities of 374, 412, 453, and 483 km s^{-1}. A fit to the line profiles shows each component to have a column density of $\approx 10^{13}$cm^{-2}. The absorption is found to agree well with the velocity structure of the ionized gas in the disk of NGC 5128, and is interpreted to arise from a line of sight that passes through the entire disk of NGC 5128. Interstellar diffuse bands at $\lambda\lambda 5780.4, 5797.0, 6283.9,$ and 6613.6 Å are also positively identified as arising in the dust lane.

125.563 Supernova 1986G in NGC 5128.
IAU Circ., No. 4421 (1987).

Supernova 1985P in NGC 1433

125.581 Spectroscopy and photometry of a type II supernova 1985p in NGC 1433.
A. A. Chalabaev, S. Cristiani.
Obs. Haute Provence, Pré–Publ., No. 25, 7 pp. (1987). To appear in ESO Workshop on SN 1987A.

Supernova 1983n in NGC 5236

125.601 Forbidden lines of Si I: an alternate interpretation for an infrared emission feature in the spectrum of SN 1983N (M83).
E. Oliva.
Astrophys. J., Lett. Ed., Vol. 321, No. 1, p. L45 – L49 (1987).
An alternate interpretation for the IR spectral feature observed in SN 1983N is presented and analyzed. The low-resolution, CVF spectrum published by Graham et al. in 1986 around 1.644 μm, originally interpreted in terms of [Fe II] lines, can be fitted assuming pure [Si I] emission (1.6454 μm and 1.6068 μm lines) obtaining a result as good as that of a synthetic [Fe II] spectrum. The mass of silicon inferred from the observed flux of [Si I] 1.6454 μm is consistent with carbon deflagration models.

Supernova 1987I in IC 4963

125.621 Supernova 1987I in IC 4963.
IAU Circ., No. 4417 (1987).

125.622 Supernova in IC 4963 (1987I).
Yamamoto Circ., No. 2087 (1987). In Japanese.

Supernova 1987L in NGC 2336

125.641 Supernova 1987L in NGC 2336.
IAU Circ., Nos. 4441, 4445 – 4447, 4483 (1987).

125.642 Supernova in NGC 2336 (1987L).
Yamamoto Circ., No. 2090 (1987). In Japanese.

125.643 Supernova 1987L.
Br. Astron. Assoc. Circ., Nos. 671, 672 (1987).

Supernova 1987M in NGC 2715

125.661 Supernova 1987M in NGC 2715.
IAU Circ., Nos. 4451, 4459, 4470, 4518 (1987).

125.662 Supernova 1987M in NGC 2715.
Yamamoto Circ., No. 2092 (1987). In Japanese.

Supernova 1987N in NGC 7606

125.681 Supernova 1987N in NGC 7606.
IAU Circ., Nos. 4511, 4513 – 4516 (1987).

125.682 Supernova 1987N in NGC 7606.
Yamamoto Circ., No. 2100 (1987). In Japanese.

Supernova 1987O in UGC 4060

125.701 Supernova 1987O in UGC 4060.
IAU Circ., No. 4521 (1987).

Supernova 1987J in ESO 601–G26

125.721 Supernova 1987J in ESO 601–26.
Yamamoto Circ., No. 2088 (1987). In Japanese.

Supernova 1987K in NGC 4651

125.741 Supernova 1987K in NGC 4651.
Yamamoto Circ., No. 2088 (1987). In Japanese.

Supernova 1984l in NGC 991

125.761 The light curves of type Ib supernovae: SN 1984l in NGC 991.
D. Yu. Tsvetkov.
Pis'ma Astron. Zh., Tom 13, No. 10, p. 894 – 899 (1987). In Russian. English translation in Sov. Astron. Lett., Vol. 13.
Photographic observations of SN 1984l are reported. Parameters of light curves are derived. The light and colour curves for

types Ib, Ia and II supernovae are compared. It is shown that type Ib supernovae have differently shaped light curves.

125.762   Forbidden lines [O I] in the spectrum of the type Ib supernova SN 1984l at an early stage and oxygen mass.
N. N. Chugaj.
Astron. Tsirk., No. 1469, p. 3 – 4 (1986). In Russian.

The author identified the forbidden line [O I] $\lambda5557$ and $\lambda6300$ in the spectrum of SN 1984l at $t \approx t_m + 60^d$. He derived an electron temperature $T_e \approx 6600K$, the neutral oxygen mass $M(O\ I) \approx 1.5\ M_\odot$, and the total oxygen mass $M(O) \approx 3\ M_\odot$.

Supernova 1984h in MCG 8–15–47

125.781   Photometric observations of supernova 1984h in MCG 8–15–47.
N. V. Metlova.
Astron. Tsirk., No. 1460, p. 6 – 8 (1986). In Russian.

Supernova 1987D in UGC 7370

125.801   Optical observations of SN 1987D in UGC 7370.
D. P. Schneider, J. R. Mould, A. C. Porter, M. Schmidt, G. D. Bothun, J. E. Gunn.
Publ. Astron. Soc. Pac., Vol. 99, No. 621, p. 1167 – 1173 (1987).

The authors present optical photometry and spectroscopy of SN 1987D, a Type Ia supernova in the nearby galaxy UGC 7370, during the period immediately subsequent to its discovery. The distance modulus of the galaxy, independently determined by the redshift and the Tully–Fisher relation, is 32.3 ± 1.0. The maximum brightness of SN 1987D was $V \approx 13.2$ and occured on 1987 April 18 ± 4 days; the absolute luminosity is ≈ 0.6 mag brighter than the average for SNe Ia. The spectrum of SN 1987D is typical of SNe Ia, but the expansion velocity (9600 km s^{-1}) is considerably lower than average.

Supernova 1986A in NGC 3367

125.821   First results with a transmission echelle grating on the ESO Faint Object Spectrograph: observations of the SN 1986a in NGC 3367 and of the nucleus of the galaxy.
H. Dekker, S. D'Odorico, R. Arsenault.
ESO Sci. Prepr., No. 518, 32 pp. (1987). Submitted to Astron. Astrophys.

The Shang dynasty's supernova.
See Abstr. 004.022.

Guest stars: historical supernovae and remnants.
See Abstr. 004.110.

Die Supernova 1987A – das astronomische Jahrhundertereignis. Neues vom ESO–Workshop in Garching, BRD.
See Abstr. 011.003.

Supernova 1987A. A summary of the ESO Workshop held from 6 – 8 July 1987.
See Abstr. 011.005.

Supernova 1987A: The parent and its environs.
See Abstr. 011.033.

Observational neutrino astrophysics.
See Abstr. 013.045.

High–explosive simulation of supernovae.
See Abstr. 022.089.

Astrophysical and terrestrial neutrinos in supernova detectors.
See Abstr. 034.032.

IRSPEC: design, performance and first scientific results.
See Abstr. 034.165.

All Sky Supernova and Transient Explorer (ASTRE).
See Abstr. 035.023.

The MIT spectroscopy investigation of AXAF and the study of supernova remnants.
See Abstr. 035.028.

Neutrino emission by supernova explosions.
See Abstr. 061.021.

Neutrino transport in stellar matter.
See Abstr. 061.022.

How reliable are neutrino mass limits derived from SN1987A?
See Abstr. 061.090.

Neutrino mass and mixing constrained from the LMC supernova burst.
See Abstr. 061.131.

Resonant neutrino oscillations and the neutrino signature of supernovae.
See Abstr. 061.132.

Effects of matter oscillations on supernova neutrino flux.
See Abstr. 061.135.

Resonant neutrino oscillations and stellar collapse.
See Abstr. 061.154.

The initial neutrino events from supernovae as evidence for matter versus antimatter.
See Abstr. 061.155.

Mass loaded astronomical flows – III. The structure of supernova remnants and the local soft X–ray background.
See Abstr. 062.027.

The jets.
See Abstr. 062.040.

Dynamical models of radiative shocks – III. Spectra.
See Abstr. 063.017.

Model atmospheres for type I supernovae: curvature effects.
See Abstr. 063.040.

Formation of multicomponent blends in envelopes of supernovae. I. Effects of nonlocal radiative coupling.
See Abstr. 063.060.

Predicted [Ni II] infrared line strengths in the Crab Nebula and IC 443.
See Abstr. 063.107.

The infrared diagnostic of a dusty plasma with applications to supernova remnants.
See Abstr. 063.118.

Spectral flux from low–density photospheres: low–temperature results.
See Abstr. 064.078.

Neutrinos from gravitational collapse.
See Abstr. 065.003.

Subluminous Type I supernovae: their theoretical rate in our Galaxy and in ellipticals.
See Abstr. 065.006.

Evolution stellaire et supernovae.
See Abstr. 065.010.

Deflagrating white dwarfs and the statistical properties of Type I supernovae.
See Abstr. 065.027.

Convection and the mechanism of Type II supernovae.
See Abstr. 065.028.

Thermonuclear processes in accreting white dwarfs (novae, symbiotic stars, and type–I supernovae).
See Abstr. 065.029.

The fate of accreting white dwarfs: Type I supernovae vs. collapse.
See Abstr. 065.038.

Late–time neutrino heating and energetics of stalled shocks in type II supernovae.
See Abstr. 065.040.

The stage of spontaneous flame propagation in supernovae.
See Abstr. 065.047.

Neutrino pair energy deposition in supernovae.
See Abstr. 065.103.

Convection in supernova theory.
See Abstr. 065.112.

Evolution of 8–10 $M_\odot$ stars toward electron capture supernovae. II. Collapse of an O + Ne + Mg core.
See Abstr. 065.113.

Propagation of a shock wave in a radiating spherically symmetric distribution of matter.
See Abstr. 066.067.

An exact model for a gaseous regular bouncing sphere in general relativity.
See Abstr. 066.071.

Neutron star coupling to its environment.
See Abstr. 067.003.

Neutron star cooling: critical test of dense matter physics.
See Abstr. 067.020.

Neutron stars formed from supernova explosion and quark matter.
See Abstr. 067.021.

Limits on hadronic cosmic ray production by young pulsars.
See Abstr. 067.078.

Convection, type II supernovae, and the early evolution of neutron stars.
See Abstr. 067.102.

Role of hyperons and pions in neutron stars and supernova.
See Abstr. 067.150.

Distribution of two types of pulsars in comparison with that of SNRs.
See Abstr. 126.023.

The structure of pulsar nebulae.
See Abstr. 126.024.

Einstein Observatory limits on neutron star surface temperatures.
See Abstr. 126.034.

The low–frequency compact source in the Crab nebula.
See Abstr. 126.047.

Studies of isolated neutron stars, pulsars and pulsar–driven nebulae with the Advanced X–Ray Astrophysics Facility (AXAF).
See Abstr. 126.053.

Detections of diffuse interstellar bands toward the SN 1987A in the Large Magellanic Cloud.
See Abstr. 131.032.

Supernovae and the interstellar medium.
See Abstr. 131.091.

Detection of interstellar CH and CH$^+$ towards SN 1987A.
See Abstr. 131.110.

Molecular clouds in the vicinity of the semicircular supernova remnant G 109.1–1.0.
See Abstr. 131.112.

Coronal interstellar gas and supernova remnants.
See Abstr. 131.123.

IC 443: the interaction of a supernova remnant with a molecular cloud.
See Abstr. 131.135.

Gravitational and dynamical instabilities of a decelerating plane parallel slab of finite thickness.
See Abstr. 131.182.

The interstellar medium toward SN 1987A.
See Abstr. 131.207.

VLA and low–frequency VLBI observations of the radio source 0503 + 467: austere constraints on interstellar scattering in two media.
See Abstr. 131.304.

The identification of galactic radio sources based on a comparison of radio–continuum and infrared emission.
See Abstr. 141.009.

Determinations of distances to radio sources with VLBI.
See Abstr. 141.015.

The catalogue of H I absorption line profiles in the spectra of galactic radio sources. I. Integral profiles.
See Abstr. 141.017.

Gamma–ray observations of the Crab region using a coded–aperture telescope.
See Abstr. 143.040.

Upper limits to the $(10^{13} - 10^{14})$eV gamma–ray fluxes from the Crab Nebula and pulsar.
See Abstr. 143.044.

High energy cosmic rays from young neutron stars.
See Abstr. 144.001.

A new aspect of galactic ridge X–ray emission – SNRs in a tenuous medium?
See Abstr. 155.001.

Thermal and nonthermal emission structures in the galactic center region.
See Abstr. 155.020.

H I in the vicinity of the supernova remnant VRO 42.05.01 (G 166.0 + 4.3)
See Abstr. 155.094.

The galactic thermal pressure gradient.
See Abstr. 155.128.

Chemical evolution of elliptical galaxies.
See Abstr. 157.208.

Supernovae blast waves in proto–dwarf galaxies.
See Abstr. 157.236.

Detached supernovae from undetected dwarf galaxies: expected rates from star formation models.
See Abstr. 157.241.

A radio search for Crab–like nebulae in M33.
See Abstr. 157.340.

Interstellar polarization in the dust lane of Centaurus A (NGC 5128).
See Abstr. 158.062.

VLBI observations of 23 hot spots in the starburst galaxy M82.
See Abstr. 158.349.

126   Degenerate Stars, White Dwarfs, Pulsars

126.001   Pulsar surveys.
  R. N. Manchester.
The origin and evolution of neutron stars, p. 3 – 12 (1987). – See Abstr. 012.001 (IAU Symp. No. 125).
  The current situation regarding pulsar surveys is briefly reviewed. Most of the known pulsars have been found by major radio surveys that are unbiased in the sense of more–or–less uniformly covering a given area of sky. The results from two recent such surveys, the Green Bank 390 MHz survey of Stokes et al. (1985) and the Jodrell Bank 1400 MHz survey of Clifton and Lyne (1986) are compared. Conclusions are drawn regarding the effect of observing frequency on the results of pulsar surveys and the galactic distribution of pulsars and interstellar electrons.

126.002   Millisecond pulsar surveys.
  D. C. Backer.
The origin and evolution of neutron stars, p. 13 – 21 (1987). – See Abstr. 012.001 (IAU Symp. No. 125).
  In 1982 a new class of pulsars was defined by the discovery of a star with a millisecond rotation period, 1.6 ms. The rapid spin of these pulsars is attributed to mass transfer in a low–mass binary progenitor system. The millisecond pulsars provide precise astrophysical clocks that can be used to improve the solar system ephemeredes and to search for a background of gravitational waves that may have been produced in the early stages of the visible universe. Old and ongoing searches for new millisecond pulsars are described in this paper.

126.003   The kinematics of pulsars.
  A. G. Lyne.
The origin and evolution of neutron stars, p. 23 – 33 (1987). – See Abstr. 012.001 (IAU Symp. No. 125).
  Pulsars have a galactic radial distribution similar to that of many galactic populations such as H II regions, massive stars and supernova remnants. However, they are generally much further from the plane of the Galaxy than these objects. Proper motion measurements show that this is because they are typically moving with high velocities. The measurements also indicate that most pulsars were formed a few million years ago close to the plane, within the normal Population I regions. Some pulsars will escape from the Galaxy, although the majority will end up in a halo population.

126.004   Interstellar scintillations and neutron star kinematics.
  J. M. Cordes.
The origin and evolution of neutron stars, p. 35 – 46 (1987). – See Abstr. 012.001 (IAU Symp. No. 125).
  The interstellar scintillation technique for measuring neutron star speeds is described and results are given for 71 radio pulsars. The mean transverse neutron star speed is 100 km s^{-1} and the distribution extends to 300 km s^{-1}. The transverse speed correlates with the z velocity derived independently using distance from the galactic plane, consistent with most neutron stars having been born near the galactic plane. A correlation of transverse speed with the quantity $P\dot{P} \propto$ (magnetic moment)^2 is a

general property of the neutron star population. Monte Carlo simulations of the progenitors of neutron stars show that the velocity distribution is inconsistent with the disruption of binary systems solely by symmetric supernova explosions.

126.005   A deep search for young pulsars in the galactic plane at 1400 MHz.
  T. R. Clifton, A. W. Jones, A. G. Lyne.
The origin and evolution of neutron stars, p. 47 (1987). Abstract. – See Abstr. 012.001 (IAU Symp. No. 125).

126.006   Progenitors of the local pulsars; lower mass limit and beaming factor.
  A. Blaauw.
The origin and evolution of neutron stars, p. 48 (1987). Abstract. – See Abstr. 012.001 (IAU Symp. No. 125).

126.007   The relation between radio luminosity and magnetic field in rotation–powered pulsars.
  S. Pineault.
The origin and evolution of neutron stars, p. 49 (1987). Abstract. – See Abstr. 012.001 (IAU Symp. No. 125).

126.008   Does the radio luminosity of pulsar grow up in its later stage?
  T. Lu, P. C. Zhu, J. S. Kui.
The origin and evolution of neutron stars, p. 50 (1987). Abstract. – See Abstr. 012.001 (IAU Symp. No. 125).

126.009   A few remarks on the two types of pulsars.
  Z.–G. Deng, J.–H. Huang, X.–Y. Xia.
The origin and evolution of neutron stars, p. 51 (1987). Abstract. – See Abstr. 012.001 (IAU Symp. No. 125).

126.010   The kinematic properties of two pulsar types.
  J.–H. Huang, Z.–G. Deng, X.–Y. Xia.
The origin and evolution of neutron stars, p. 52 (1987). Abstract. – See Abstr. 012.001 (IAU Symp. No. 125).

126.011   Toward an empirical theory of pulsar emission.
  J. M. Rankin.
The origin and evolution of neutron stars, p. 53 (1987). Abstract. – See Abstr. 012.001 (IAU Symp. No. 125).

126.012   Radio emission mechanisms for two types of pulsars.
  J.–H. Huang, Z.–G. Deng, X.–Y. Xia.
The origin and evolution of neutron stars, p. 54 (1987). Abstract. – See Abstr. 012.001 (IAU Symp. No. 125).

126.013   Pulse asymmetry of millisecond pulsars.
  K. Y. Chen, J. Shaham.
The origin and evolution of neutron stars, p. 55 (1987). Abstract. – See Abstr. 012.001 (IAU Symp. No. 125).

126.014 Pulsar polarization limiting radii and the evolution of pulsar beams.
J. J. Barnard.
The origin and evolution of neutron stars, p. 56 (1987). Abstract. – See Abstr. 012.001 (IAU Symp. No. 125).

126.015 Polarization angle swings rediscussed.
W. Sieber.
The origin and evolution of neutron stars, p. 57 (1987). Abstract. – See Abstr. 012.001 (IAU Symp. No. 125).

126.016 Triplicity of pulsar profiles and orthogonal polarization modes.
J. A. Gil.
The origin and evolution of neutron stars, p. 58 (1987). Abstract. – See Abstr. 012.001 (IAU Symp. No. 125).

126.017 The effects of inverse Compton scattering on the pulsars' radiation.
X.-Y. Xia, Z.-G. Deng, G.-J. Qiao, X.-J. Wu, H. Chen.
The origin and evolution of neutron stars, p. 59 (1987). Abstract. – See Abstr. 012.001 (IAU Symp. No. 125).

126.018 The geometric study of drifting subpulses.
X.-J. Wu, G.-X. Deng, H. Chen, X.-Y. Xia.
The origin and evolution of neutron stars, p. 60 (1987). Abstract. – See Abstr. 012.001 (IAU Symp. No. 125).

126.019 A massive glitch in PSR 0355+54.
A. G. Lyne.
The origin and evolution of neutron stars, p. 63 (1987). Abstract. – See Abstr. 012.001 (IAU Symp. No. 125).

126.020 A large timing discontinuity in the Vela pulsar, July 1985.
C. S. Flanagan.
The origin and evolution of neutron stars, p. 64 (1987). Abstract. – See Abstr. 012.001 (IAU Symp. No. 125).

126.021 The galactic pulsar population and neutron star birth.
R. Narayan.
The origin and evolution of neutron stars, p. 67 – 78 (1987). – See Abstr. 012.001 (IAU Symp. No. 125).
The radio pulsars in the Galaxy are found predominantly in the disk, with a scale height of several hundred parsecs. After allowing for pulsar velocities, the data are consistent with the hypothesis that single pulsars form from massive stellar progenitors. The number of active single pulsars in the Galaxy is $\sim 1.5 \times 10^5$, and their birthrate is 1 per ~ 60 yrs. There is some evidence that many single pulsars are born spinning slowly, with initial periods $\sim 0.5 - 1\ s$. This could imply an origin through binary "recycling" followed by orbit disruption, or might suggest that the pre–supernova stellar core efficiently loses angular momentum to the envelope through magnetic coupling. The birthrate of binary radio pulsars, particularly of the millisecond variety, seems to be much larger than previous estimates, and might suggest that these systems do not originate in low mass X–ray binary systems.

126.022 The progenitors of pulsars.
G. Srinivasan, D. Bhattacharya.
The origin and evolution of neutron stars, p. 109 – 119 (1987). – See Abstr. 012.001 (IAU Symp. No. 125).
The progenitors of single as well as binary pulsars are discussed with special emphasis on binary pulsars with low mass companions. Several predictions are made concerning millisecond pulsars.

126.023 Distribution of two types of pulsars in comparison with that of SNRs.
H. Sun, Z. W. Li.
The origin and evolution of neutron stars, p. 121 (1987). Abstract. – See Abstr. 012.001 (IAU Symp. No. 125).

126.024 The structure of pulsar nebulae.
R. A. Chevalier, R. T. Emmering.
The origin and evolution of neutron stars, p. 122 (1987). Abstract. – See Abstr. 012.001 (IAU Symp. No. 125).

126.025 Search for plerions in the direction of two young pulsars.
S. Krishnamohan, D. K. Mohanty, A. R. Patnaik, T. Velusamy.
The origin and evolution of neutron stars, p. 124 (1987). Abstract. – See Abstr. 012.001 (IAU Symp. No. 125).

126.026 Accreting neutron stars.
N. E. White.
The origin and evolution of neutron stars, p. 135 – 148 (1987). – See Abstr. 012.001 (IAU Symp. No. 125).
This paper reviews accreting neutron stars in X-ray binaries, with particular emphasis on how variations in magnetic field strength may be responsible for explaining the spectral and temporal properties observed from the various systems. This includes a review of X-ray pulsars in both low and high mass systems, and a discussion of the spectral properties of the low mass X-ray binaries.

126.027 On the evolution of magnetic inclination with age.
L. S. Li.
The origin and evolution of neutron stars, p. 376 (1987). Abstract. – See Abstr. 012.001 (IAU Symp. No. 125).

126.028 Timescale for the decay of magnetic fields of pulsars.
S. Krishnamohan.
The origin and evolution of neutron stars, p. 377 – 378 (1987). Abstract. – See Abstr. 012.001 (IAU Symp. No. 125).

126.029 Binary pulsars: observations and implications.
J. H. Taylor.
The origin and evolution of neutron stars, p. 383 – 392 (1987). – See Abstr. 012.001 (IAU Symp. No. 125).
The author reviews the known facts concerning binary pulsars, and then briefly discusses some implications for the understanding of the place of neutron stars in stellar evolution.

126.030 Millisecond pulsar formation and evolution.
E. P. J. van den Heuvel.
The origin and evolution of neutron stars, p. 393 – 406 (1987). – See Abstr. 012.001 (IAU Symp. No. 125).
The evolutionary history of binary radio pulsars, including the two millisecond binary pulsars, is reviewed. There are two groups of binary pulsars, the PSR 1913+16–group, which descended from massive X-ray binaries, and the PSR 1953+29–group, which descended from fairly wide low–mass X–ray binaries. The neutron stars in the second group probably formed by the accretion–induced collapse of a massive white dwarf. The companion stars in both groups of systems are expected to be dead stars, i.e. white dwarfs or neutron stars. The large total number of millisecond binary pulsars in the galaxy ($\sim 10^4$), indicates that magnetic fields of neutron stars do not decay below a value of order 10^9G. Possible explanations for this phenomenon are discussed. Coalescence with a close degenerate companion provides a viable model for the formation of the single millisecond pulsar.

126.031 Secondary components of binary pulsars & magnetic field decay in neutron stars.
S. R. Kulkarni.
The origin and evolution of neutron stars, p. 407 (1987). Abstract. – See Abstr. 012.001 (IAU Symp. No. 125).

126.032 Constraints to possible progenitor systems of PSR 1831–00.
E. H. P. Pylyser.
The origin and evolution of neutron stars, p. 409 (1987). Abstract. – See Abstr. 012.001 (IAU Symp. No. 125).

126.033 Geodetic precession in binary pulsars.
M. Bailes.
The origin and evolution of neutron stars, p. 410 (1987).
Abstract. – See Abstr. 012.001 (IAU Symp. No. 125).

126.034 Einstein Observatory limits on neutron star surface temperatures.
F. R. Harnden Jr.
The origin and evolution of neutron stars, p. 457 (1987).
Abstract. – See Abstr. 012.001 (IAU Symp. No. 125).

126.035 Thermal radiation from a radio pulsar: PSR 1055–52.
W. Brinkmann, H. Ögelman.
The origin and evolution of neutron stars, p. 458 (1987).
Abstract. – See Abstr. 012.001 (IAU Symp. No. 125).

126.036 The drifting subpulse phenomenon observed in three pulsars with triple profiles.
T. H. Hankins, A. Wolszczan.
Astrophys. J., Vol. 318, No. 1, p. 410 – 420 (1987).
Three pulsars (PSR 1632+24, 1845–01, and 1918+19) with triple average profile shapes have been found which have subpulses that drift across the full profile window. The emission details can be understood in the context of Rankin's core–conal model if two concentric conal emission regions are postulated with the inner cone producing stronger emission than the outer cone. Unlike all previously known pulsars with triple profiles, these objects show no evidence of core emission.

126.037 The new millisecond pulsar.
P. Murdin.
Astron. Now, Vol. 1, No. 3, p. 34 – 36 (1987).
Radio astronomers at Jodrell Bank have just discovered the first millisecond pulsar to be found in a globular cluster. Its discovery throws new light on how these very rapidly–spinning pulsars are formed.

126.038 The luminosity decay of radio pulsars and some related matters.
A. D. Fokker.
Astron. Astrophys., Vol. 182, No. 1, p. 41 – 46 (1987).
Data contained in the Catalogue prepared by Manchester and Taylor (1981) have been used to derive information on the luminosity decay of radio pulsars. It is found that the ratio 'number of old pulsars/number of young pulsars' decreases steadily with increasing distance. This effect can be understood by adopting a certain shape of the luminosity function.

126.039 Soft X–ray observations of the radio pulsar PSR 1055–52.
W. Brinkmann, H. Ögelman.
Astron. Astrophys., Vol. 182, No. 1, p. 71 – 74 (1987).
The radio pulsar PSR 1055–52 was observed with the EXO-SAT X–ray satellite for a total of 10^5 s. The data can be fitted best with blackbody spectra having temperatures of $5 \times 10^5 \leqslant T_\infty \leqslant 8 \times 10^5$K. The obtained temperatures are consistent with recent models for the initial cooling of neutron stars.

126.040 Detection and investigation of the radio emission of the pulsar PSR 1530+27 at 25 MHz.
Yu. M. Bruk, O. M. Ul'yanov, B. Yu. Ustimenko.
Sov. Astron., Vol. 30, No. 5, p. 574 – 577 (1986). English translation of 42.126.019.

126.041 Correlation of intensity fluctuations of individual pulses of PSR 0329+54 at different longitudes of the average profile.
M. V. Popov.
Sov. Astron., Vol. 30, No. 5, p. 577 – 587 (1986). English translation of 42.126.020.

126.042 Characteristic times of pulsars and their ages.
O. Kh. Gusejnov, I. M. Yusifov.
Astrophysics, Vol. 25, No. 3, p. 660 – 664 (1987). English translation of 42.126.068.

126.043 Decametre wavelength observations of PSR 0834+06 and PSR 1919+21.
P. J. Hall.
Mon. Not. R. Astron. Soc., Vol. 227, No. 1, p. 197 – 204 (1987).
The University of Tasmania decametric array radio telescope has been used to observe two well–known pulsars (PSR 0834+06 and PSR 1919+21) in an attempt to confirm the decametric interpulse emission reported by Soviet investigators. Evidence for such emission was found in a 39–MHz profile of PSR 0834+06, but 47 MHz observations of PSR 1919+21 do not indicate significant emission outside the main pulse window. Detailed low–frequency spectra for the two sources are also given.

126.044 Search for low–mass objects. II.
C. K. Kumar.
Astron. J., Vol. 94, No. 1, p. 158 – 160 (1987).
The K and L magnitudes of 20 white dwarf stars were measured to detect brown dwarf companions. None were discovered.

126.045 Discovery of a magnetic DA white dwarf with distinct Hβ and Hα Zeeman triplets.
H.–J. Hagen, D. Groote, D. Engels, U. Haug, F. Toussaint, D. Reimers.
Astron. Astrophys., Vol. 183, No. 1, p. L7 – L8 (1987).
The authors report the discovery of a further single magnetic white dwarf of type DA with distinct Hα and Hβ Zeeman triplets. HS 1254+3430 ($V \approx 17$) was found as an apparent emission line object on a deep objective prism plate and was subsequently identified spectroscopically. From the Zeeman splitting of Hα and Hβ, the magnetic field of HS 1254+3430 is estimated as 9.5 ± 0.5 MG, and the effective temperature derived from the continuous energy distribution is $T_{eff} = 15 \pm 4 \times 10^3$K. The coordinates are $\alpha = +12^h54^m53^s$, $\delta = +34°30'52''$ (1950).

126.046 The pulsewidth–age relation of radio pulsars.
B. N. Candy, D. G. Blair.
Astron. Astrophys., Vol. 183, No. 2, p. L17 – L20 (1987).
The radio pulsar mean pulsewidth – characteristic age relation shows a minimum between 10^6 and 10^7yr. Such a minimum is predicted by evolutionary models involving magnetic alignment, but not by those involving magnetic field decay. Since the statistical significance of the minimum has been questioned (Krishnamohan, 1985), the authors present a statistical analysis of the data. The pulsewidth distribution is highly skewed, which makes the analysis sensitive to the presence of long pulsewidth pulsars at small characteristic ages. The skewness also accounts for the discrepancy between mean and median pulsewidths.

126.047 The low–frequency compact source in the Crab nebula.
V. P. Bovkun, I. N. Zhuk, Ya. M. Sobolev.
Astron. Zh., Tom 64, Vyp. 4, p. 734 – 741 (1987). In Russian. English translation in Sov. Astron., Vol. 31, No. 4.
The parameters of the low–frequency compact source in the Crab nebula are given. In the decameter–wavelength range its coordinates, $\alpha(1950.0) = 5^h31^m31^s$, 62 ± 0^s22 and $\delta(1950.0) = 21°58'52''$, $8 \pm 3''9$, coincide within the error limits with those of the pulsar PSR 0531+21. The apparent angular size of the compact source varies with the frequency ν as $\theta_0 \propto \nu^{-1.95 \pm 0.25}$ from $3''3$ at 12.6 MHz to $0''18$ at 75 MHz. The flux density of the integral radio emission of the compact source and the pulsar S_ν(Jy) in function of the frequency ν(Hz) is approximated by the expression $S_\nu = 44(1 + 2y^2)^{-1}y^{-5/3}$ where $y = \nu/(1.6 \times 10^8)$. The spectrum of the summary emission of the compact source and the pulsar is explained by coherent curvature radiation of charged clots with effective thickness 60 cm.

126.048 A method to determine the dispersion measure for pulsars radio emission.
Yu. M. Bruk, O. M. Ul'yanov, B. Yu. Ustimenko.
Astron. Zh., Tom 64, Vyp. 4, p. 742 – 746 (1987). In Russian. English translation in Sov. Astron., Vol. 31, No. 4.
A new method to determine the dispersion measure and longitude–frequency parameters of pulsars' radio emission is suggested. This method is based on the correlation analysis of Fourier spectra as a function of the dispersion measure and longitude–frequency parameters for pulsar radio emission. The possibility of the account for a real noise pattern in the frequency domain is considered.

126.049 Pulsar radio pulses: the frequency–dependent dispersion measure and the extra arrival–time delay.
A. D. Kuz'min.
Sov. Astron. Lett., Vol. 12, No. 5, p. 325 – 328 (1987). English translation of 42.126.013.

126.050 The drifting subpulses of PSR 0818–13.
J. D. Biggs, P. M. McCulloch, P. A. Hamilton, R. N. Manchester.
Mon. Not. R. Astron. Soc., Vol. 228, No. 1, p. 119 – 123 (1987).
The drifting subpulses of PSR 0818–13 have been analysed in detail using observations of individual pulses at 645 MHz. The subpulse drift is found to have a pronounced decrease in drift rate near the centre of the integrated pulse profile.

126.051 The period structure and stability of the pulsating white dwarf L 19–2.
D. O'Donoghue, B. Warner.
Mon. Not. R. Astron. Soc., Vol. 228, No. 4, p. 949 – 955 (1987).
Observations made in 1983 – 85 confirm the earlier detection of a slight inequality in the splitting of the frequency triplet near 192 s of the pulsating DA white dwarf L 19–2. If interpreted as second–order splitting of rotationally–perturbed non–radial g-mode pulsations, the measured value, 2.0×10^{-6}mHz, is in rough agreement with outdated theoretical calculations which should be repeated with more realistic white dwarf models. The new observations are also used to set an upper limit of 3.0×10^{-14}s s^{-1} on the rate of period change of the 192 s oscillation corresponding to an evolutionary cooling time–scale longer than 2.0×10^{8}yr.

126.052 Dynamical mass determinations for the white dwarf components of HZ 9 and Case 1.
J. R. Stauffer.
Astron. J., Vol. 94, No. 4, p. 996 – 1002 (1987).
Radial–velocity curves for both components of the short–period, white dwarf M dwarf binary systems Case 1 and HZ 9 have been determined from 1–Å–resolution spectra obtained at the MMT. The new data have been used to estimate masses for the white dwarfs, yielding $M = 0.38\ M_\odot$ for the Case 1 DA and $M = 0.51\ M_\odot$ for the HZ 9 DA, with 1 σ uncertainties of about 0.1 $M_\odot$. Both M dwarfs show significant variability in their chromospheric emission–line strengths. Accretion of the dM wind onto the white dwarfs in these systems should produce detectable metal absorption lines in the UV, as has been observed recently with *IUE* spectra of Feige 24.

126.053 Studies of isolated neutron stars, pulsars and pulsar–driven nebulae with the Advanced X–Ray Astrophysics Facility (AXAF).
A. S. Wilson.
Astrophys. Lett. Commun., Vol. 26, Nos. 1 – 2, p. 99 – 111 (1987). With plates I – III.– See Abstr. 012.020.
The AXAF will signal a revolution in our understanding of neutron stars, pulsars and Crab Nebula–type supernova remnants. The mission will provide vital and detailed observational data relevant to the physics of superdense matter inside neutron stars, to the acceleration and radiation of nonthermal particles in their magnetospheres and to the processes whereby these relativistic particles and magnetic fields fuel the surrounding synchrotron nebulosities on scales of parsecs. Highlights of the results should include (1) detection of thermal surface emission from neutron stars, (2) greatly enlarged samples of pulsars that emit magnetospheric (non–thermal) X–rays, and (3) high resolution (spatial and spectral) maps of the nebulae powered by slowing pulsars. The AXAF will extend these studies to external galaxies.

126.054 The birthrate and initial spin period of single radio pulsars.
R. Narayan.
Astrophys. J., Vol. 319, No. 1, p. 162 – 179 (1987).
A statistical analysis of radio pulsar data is presented. Detailed account is taken of known selection effects in pulsar surveys, including the effects of scatter–broadening and period–dependent beaming factor. The birthrate of pulsars in the Galaxy is estimated to be one pulsar in 56 yr, and the local rate in the solar neighborhood is estimated to be one pulsar in $\sim 9 \times 10^4$yr kpc^{-2}. Many pulsars are born with initial periods as slow as 0.5 s, contrary to the usual belief that initial periods are ~ 10 ms. The pulsars that are born slow typically have high magnetic fields, $\log [B(G)] \gtrsim 12.5$. Pulsars that are born fast usually have low fields. Possible explanations are discussed for the correlation between initial spin period and magnetic field.

126.055 Investigation of the microstructure of the pulsars PSR 0809 + 74, 0950 + 08, and 1133 + 16 in the 67 – 102 MHz frequency range.
M. V. Popov, T. V. Smirnova, V. A. Soglasnov.
Astron. Zh., Tom 64, Vyp. 5, p. 1013 – 1029 (1987). In Russian. English translation in Sov. Astron., Vol. 31, No. 5.
The correlation of the microstructure of PSR 0809 + 74, 0950 + 08 and 1133 + 16 was studied from three–frequency simultaneous observations with 10 μs time resolution. It was shown that there are two kinds of microstructure: the short–time one, which does not correlate within ~ 20 MHz frequency separation, and the long–time structure, which has a high correlation of different frequencies and the time delay of it follows the cold plasma dispersion law with high accuracy. The authors obtained the dispersion measures from the correlated microstructure of these pulsars. The results are discussed on the basis of the hollow–cone model.

126.056 White dwarfs: fossil stars.
S. D. Kawaler.
Sky Telesc., Vol. 74, No. 2, p. 132 – 135 (1987).

126.057 The discovery of a millisecond pulsar in the globular cluster M28.
A. G. Lyne, A. Brinklow, J. Middleditch, S. R. Kulkarni, D. C. Backer, T. R. Clifton.
Nature, Vol. 328, No. 6129, p. 399 – 401 (1987).
The authors report the discovery of a pulsar with a 3 millisecond period in the core of the globular cluster M28. The existence of a millisecond pulsar in such a cluster, where there are frequent interactions between cluster stars, provides strong confirmation of the theory that the high rotation rate derives from accretion in a binary system. The pulsar is now isolated, having no binary companion.

126.058 Ultraviolet and visual spectroscopy of DB white dwarfs.
G. Wegner, E. P. Nelan.
Astrophys. J., Vol. 319, No. 2, p. 916 – 929 (1987).
Visual wavelength and *IUE* ultraviolet spectroscopy and model atmospheres of DB white dwarfs are reported. A search for spectral lines due to elements other than helium, and in particular carbon and hydrogen, was made. In no case is there a positive detection of carbon from these data, and upper limits relative to helium by number are derived in the range of C:He $< 10^{-5}$ to 10^{-7} for the 15 DB stars with ultraviolet spectra in the temperature range 11,000K $< T_{\rm eff} < 23,000$K. Additional new DBA stars are reported. Weak hydrogen Balmer lines are observed in the spectra of five of the 27 objects.

126.059   A glitch in the Crab Pulsar.
     A. G. Lyne, R. S. Pritchard.
Mon. Not. R. Astron. Soc., Vol. 229, No. 2, p. 223 – 226 (1987).

Until last year the pulsar in the Crab Nebula, PSR 0531 + 21, had suffered two major glitches, or jumps in rotation rate, since it was discovered in 1968. The first, in 1969, involved a fractional change in period of $\Delta P/P \approx 10^{-8}$ while the second, in 1975, was much larger with $\Delta P/P \approx 4 \times 10^{-8}$. Observations were not made until several days after each event. This paper reports a third glitch, in 1986 August, detected at Jodrell Bank apparently within 1 hr of the event. This is the first occasion on which the recovery from a glitch in the Crab Pulsar has been observed in detail. The recovery in rotation rate is close to a simple exponential and can be understood in terms of a two–component model of the neutron star.

126.060   Spectral behaviour of pulse width in pulsars.
     O. B. Slee, A. D. Bobra, S. K. Alurkar.
Aust. J. Phys., Vol. 40, No. 4, p. 557 – 586 (1987).

The profiles of 24 pulsars have been measured with the Culgoora circular array at 80 and 160 MHz. The low–frequency profile widths have been combined with most of the published higher frequency widths to generate their pulse–widths spectra.

126.061   EUV photometry of DA white dwarfs with EXOSAT.
     S. Jordan, D. Koester, C. Wulf–Mathies, H. Brunner.
Astron. Astrophys., Vol. 185, No. 1/2, p. 253 – 257 (1987).

The authors present EXOSAT EUV observations for 9 white dwarfs of spectral type DA. These are analysed together with other observational data from the literature using theoretical model atmospheres. For 8 objects He/H abundance ratios and interstellar hydrogen column densities are obtained. The data for HZ 43 seem to contradict recent results of other authors; possible explanations for the discrepancies are discussed. The results show that higher He/H ratios as well as a larger range of observed values are found in the hotter objects with $T_{eff} > 50,000$K, whereas the cooler objects ($T_{eff} < 45,000$K) have remarkably similar He/H ratios.

126.062   Soft X–ray imaging observations of the 39 millisecond pulsar PSR 1951 + 32.
H. Ögelman, R. Buccheri.
Astron. Astrophys., Vol. 186, No. 1/2, p. L17 – L19 (1987).

The authors report observations of the newly discovered 39.5 millisecond pulsar PSR 1951 + 32 with the soft X–ray telescope on board the EXOSAT observatory. A point–like source coincident with the position of the pulsar was observed. For a power–law model with photon index of –2, a column density of 6×10^{21}cm^{-2}, and a distance of 2.5 kpc, the observed count rate corresponds to a luminosity of 1.5×10^{34}erg s^{-1} in the energy range 0.03 – 2.4 keV. For a blackbody model, surface temperatures in the range $(1.1 - 2.1) \times 10^6$K would be required to account for the observed count rate. The observed events from the source were searched at the extrapolated radio period for pulsations. A weak signal at 97% confidence level was found, indicating that the X–ray flux from PSR 1951 + 32 can be all pulsed.

126.063   Neutral hydrogen absorption measurements of ten pulsars and the electron density in the galactic plane.
J. M. Weisberg, J. M. Rankin, V. Boriakoff.
Astron. Astrophys., Vol. 186, No. 1/2, p. 307 – 311 (1987).

The authors present H I absorption spectra of ten pulsars measured with the Arecibo 305–m telescope. Kinematic distance limits are set for six of them. Electron densities along the lines of sight are determined from the pulsars' dispersion measures and distances. The results are consistent with typical mean electron densities lying within a factor of two of 0.025 cm^{-3} along kiloparsec–scale paths in the galactic plane.

126.064   The rotational velocities of white dwarfs. II.
     C. A. Pilachowski, R. W. Milkey.
Publ. Astron. Soc. Pac., Vol. 99, No. 618, p. 836 – 838 (1987).

The authors have extended their recent work on the rotational velocities of white dwarfs by analysis of the sharp core of the Hα line to a larger sample of stars with the goal of eventually obtaining the distribution of rotational velocities and thereby improving the understanding of the question of retention of angular momentum in the later stages of evolution of these stars. There are no stars in the sample which show $v \sin i > 60$ km s^{-1}.

126.065   The rotationally modulated Zeeman spectrum at nearly 10^9Gauss of the white dwarf PG 1031 + 234.
W. B. Latter, G. D. Schmidt, R. F. Green.
Astrophys. J., Vol. 320, No. 1, p. 308 – 314 (1987).

The authors evaluate the field strength and morphology on the surface of the white dwarf PG 1031 + 234 by means of high–quality, phase–resolved CCD spectrophotometry and spectral modeling of the rotationally modulated Zeeman features. It is confirmed that PG 1031 + 234 possesses the strongest field yet detected on a white dwarf, with regions on the surface spanning the range ~ 200 to nearly 1000 MG (MG = 10^6G). It is also clear that the field pattern is more complex than a simple dipole; the spectroscopic data are best mimicked by a field pattern containing a slightly offset global component of polar field strength ~ 500 MG together with a localized magnetic "spot" whose central field approaches 1000 MG.

126.066   Interstellar interferometry of the pulsar PSR 1237 + 25.
     A. Wolszczan, J. M. Cordes.
Astrophys. J., Lett. Ed., Vol. 320, No. 1, p. L35 – L39 (1987). With plate L2.

An episode of exceptionally prominent double imaging of the pulsar PSR 1237 + 25 caused by refraction in the interstellar medium has been detected with the Arecibo telescope at 430 MHz and used to resolve the pulsar magnetosphere for the first time. Multiple imaging causes quasi–periodic modulation of the pulsar intensity in time and frequency that represents the fringe pattern of an "interstellar interferometer" with a baseline of the order of 1 AU. From the detected fringe phase shift across the pulse profile of PSR 1237 + 25, the authors have estimated a typical transverse separation between the emitting regions to be $\sim 10^8$cm. This estimate depends on the assumed distance to the refracting region and may be much smaller. It is shown that the observed fringe shifts and their dependence on pulse phase are inconsistent with a simple magnetic dipole model of pulsar emission.

126.067   Millisecond pulsar PSR 1937 + 21: a highly stable clock.
     L. A. Rawley, J. H. Taylor, M. M. Davis,
D. W. Allan.
Science, Vol. 238, No. 4828, p. 761 – 765 (1987).

The stable rotation and sharp radio pulses of PSR 1937 + 21 make this pulsar a clock whose long–term frequency stability approaches and may exceed that of the best atomic clocks. The pulsar's frequency stability is at least as good as 6×10^{-14} for averaging times longer than 4 months. A firm upper limit of 7×10^{-36}gram per cubic centimeter for the energy density of a cosmic background of gravitational radiation at frequencies of about 0.23 cycle per year is estimated. This limit corresponds to approximately 4×10^{-7} of the density required to close the universe.

126.068   The frequency 10 Hz in the distribution of radiopulsar periods.
V. A. Kotov, B. M. Vladimirskij.
Izv. Krymskoj Astrofiz. Obs., Tom 76, p. 93 – 98 (1987). In Russian. English translation in Bull. Crimean Astrophys. Obs., Vol. 76.

A special statistical procedure, based on computation of the so–called "commensurability spectrum", being applied to the distribution of periods of 330 pulsars revealed the dominant frequency $v_0 = 10.05 \pm 0.07$ Hz (period 0.0995 ± 0.0007 s) of a quasiperiodic modulation of the distribution. This modulation means the presence of a statistically significant, at about 3.3 to 4.1–sigma confidence level, deficit of a number of PSR with frequencies $v \approx v_0/n$, where n is integer. The effect appears to be strongly pronounced for PSR with periods from 0.4 to about 1.6 s and can not be ascribed to some unknown observational selection effect.

126.069   **VLBI observations of pulsar interstellar scattering.**
C. R. Gwinn, N. H. Bartel, J. M. Cordes, A. Wolszczan, R. Mutel.
Bull. Am. Astron. Soc., Vol. 19, No. 2, p. 723 (1987). Abstract. – See Abstr. 010.061.

126.070   **Millisecond X–ray binary pulsars.**
K. S. Wood, J. P. Norris, P. Hertz, P. F. Michelson.
Bull. Am. Astron. Soc., Vol. 19, No. 2, p. 752 – 753 (1987). Abstract. – See Abstr. 010.061.

126.071   **Composition and Einstein redshift for the white dwarf van Maanen 2.**
G. L. Hammond.
Bull. Am. Astron. Soc., Vol. 19, No. 2, p. 755 (1987). Abstract. – See Abstr. 010.061.

126.072   **Ultraviolet IUE spectroscopy of the two magnetic white dwarfs BPM25114 and K813–14.**
G. Wegner.
Bull. Am. Astron. Soc., Vol. 19, No. 2, p. 756 (1987). Abstract. – See Abstr. 010.061.

126.073   **Analysis of Balmer profiles from hot DA white dwarfs.**
K. Kidder, J. B. Holberg, J. Liebert, F. Wesemael.
Bull. Am. Astron. Soc., Vol. 19, No. 2, p. 761 (1987). Abstract. – See Abstr. 010.061.

126.074   **Observations of extremely hot DA white dwarfs.**
J. B. Holberg, J. Liebert, F. Wesemael.
Bull. Am. Astron. Soc., Vol. 19, No. 2, p. 761 (1987). Abstract. – See Abstr. 010.061.

126.075   **PG 0122 + 200: a new member of the GW Vir (PG 1159–035) class of extremely hot pulsating white dwarfs.**
H. E. Bond, A. D. Grauer.
Space Telesc. Sci. Inst., Prepr. Ser., No. 198, 11 pp. (1987). To appear in Astrophys. J., Lett. Ed.

126.076   **Sustained magnetic fields in binary millisecond pulsars.**
G. Chanmugam, K. Brecher.
Space Telesc. Sci. Inst., Prepr. Ser., No. 213, 6 pp. (1987). To appear in Nature.

126.077   **White dwarfs and planetary central stars.**
J. Liebert.
Prepr. Steward Obs., No. 768, 10 pp. (1987). To appear in IAU Symp. 131.

126.078   **On the (lack of) proper motion of the Vela pulsar.**
P. A. Caraveo, G. F. Bignami, G. Vacanti.
ESA Spec. Publ., ESA SP–273, p. 151 – 154 (1987). – See Abstr. 012.044.
Optical data taken in Jan. 87 from La Silla compared to the discovery plate of 1975.2 show no proper motion for PSR 0833–45, while a very significant one was expected if the pulsar originated in the center of the Vela SNR, so far associated with it. If such an association is to be kept, either an extreme asymmetry of the SNR is required, or both objects are much older than so far thought. Both alternatives appear not easy, thus possibly casting doubts on the reality of this classic SNR/pulsar association.

126.079   **Does the radio luminosity of pulsar grow up in its later stage?**
T. Lu, P.–c. Zhu, J.–s. Kui.
Chin. Phys. Lett., Vol. 4, No. 2, p. 81 – 84 (1987). Abstr. in Phys. Abstr., Vol. 90, No. 1310, Entry 94936 (1987).

126.080   **White dwarfs in globular clusters: a search in M71.**
H. B. Richer, G. G. Fahlman.
ESO Conf. Workshop Proc., No. 27, p. 299 – 306 (1987). – See Abstr. 012.051.
The inner 6′ × 4′ of the globular cluster M71 was imaged in U, B, and V at CFHT. Stars fainter than 25th magnitude are recorded in all 3 colours. A sequence of faint blue stars was detected, but it is somewhat fainter than expected for DA white dwarfs. Also seen is a sequence of very blue objects which are much too luminous to be single white dwarfs. Their colours and magnitudes are consistent with their identification as cluster cataclysmic variables.

126.081   **Decameter observations of PSR 0809 + 74: average profiles and dispersion measure.**
Yu. M. Bruk, B. Yu. Ustimenko, M. V. Popov, V. A. Soglasnov, A. Yu. Novikov.
Sov. Astron. Lett., Vol. 12, No. 6, p. 381 – 383 (1986). English translation of 42.126.029.

126.082   **The initial–final mass relation: galactic disk and Magellanic Clouds.**
V. Weidemann.
Astron. Astrophys., Vol. 188, No. 1, p. 74 – 84 (1987).
The author studies the question as to how different degrees of core overshoot influence the determination of initial masses for given cluster white dwarfs and demonstrates its strong influence on the upper limiting mass for white dwarf production. He applies the results concerning the dependence of initial masses on overshoot to LMC and SMC clusters, using updated material for cluster AGB maximum luminosities in order to derive new $M_i - M_f$ relations, at the same time considering the influence of changes in the distance modulus. The author describes the new method for the derivation of the $M_i - M_f$ relation by evaluation of LMC/AGB luminosity functions and discusses critically the methods and the reliability of the results.

126.083   **New millisecond pulsar in an unusual environment.**
D. J. Helfand.
Nature, Vol. 329, No. 6137, p. 285 – 286 (1987).
The detection of the fourth millisecond pulsar was of interest because of its very special location. PSR 1821–24 is spinning at 327 Hz, is situated in the core of the globular cluster M28. As such it provides new information of the origin and evolution of these objects as well as offering the first chance for an *in situ* probe of globular cluster dynamics. Theorists are proposing explanations for the peculiar properties of the pulsar see abstracts 067.125 – 067.127.

126.084   **Optical and X–ray radiation from fast pulsars: effects of duty cycle and spectral shape.**
F. Pacini, M. Salvati.
Astrophys. J., Vol. 321, No. 1, p. 447 – 449 (1987).
The optical luminosity of PSR 0540 is considerably stronger than what one would have predicted in a simple model developed earlier where the pulses are synchrotron radiation by secondary electrons near the light cylinder. This discrepancy can be eliminated if one incorporates into the model the effects of the large duty cycle and the spectral properties of PSR 0540. The authors also show that the same model can provide a reasonable fit to the observed X–ray fluxes from fast pulsars.

126.085   **PSR 0042–735.**
IAU Circ., No. 4422 (1987).

126.086   **PSR 1951 + 32.**
IAU Circ., Nos. 4422, 4426, 4429, 4492, 4507 (1987).

126.087   Outer magnetospheric fluctuations and pulsar timing noise.
K. S. Cheng.
Astrophys. J., Vol. 321, No. 2, p. 799 – 804 (1987).

Cheng, Ho, and Ruderman have proposed an outer magnetosphere gap model of fast–spinning pulsars to explain the energetic radiation from the Crab and Vela pulsars. It is shown that short time fluctuations in the size of gaps which cause rapid variations in the current braking torque are possible. Such rapid variations in torque can be approximated by a series of shot noises which automatically give the correct noise spectrum of the Crab pulsar. A generalization to other pulsars is also considered. By comparing observational data and theoretical models, the results are found to be consistent with the stiff equation of state as suggested by the vortex creep model of the neutron star interior.

126.088   Could glitches inducing magnetospheric fluctuations produce low–frequency pulsar timing noise?
K. S. Cheng.
Astrophys. J., Vol. 321, No. 2, p. 805 – 812 (1987).

The pulsar timing–noise power spectra of the frequency derivative for those pulsars with composite spectra, namely, "blue noise" (or "white noise") in high frequencies and "red noise" in low frequencies, could be caused by two related mechanisms. The high–frequency component results from the small–scale internal superfluid unpinning (microglitches). The low–frequency component could possibly come from the sudden change of current braking torque, which is perturbed by the microglitches. The comparison between the theoretical results and the data is discussed. A good fit to the observational data can be obtained if the vortex creep is indeed the dominant heating mechanism inside the pulsars.

126.089   Further observations of the peculiar hot helium–rich degenerate KPD 0005 + 5106.
R. A. Downes, E. M. Sion, J. Liebert, J. B. Holberg.
Astrophys. J., Vol. 321, No. 2, p. 943 – 951 (1987).

A variety of new ultraviolet and optical spectroscopic observations of the peculiar, hot, helium–rich degenerate KPD 0005 + 5106 are presented. While the expected lines of helium appear, the only detected transitions from a heavier element assignable to the stellar photosphere is the N V 1240 Å doublet, from which an atmospheric abundance is derived. However, hydrogen and some probable CNO ions are detected in emission. The overall energy distribution is consistent with a $T_{eff} \sim 80,000K$. The detected nitrogen is most likely due to selective radiative acceleration processes. Except for He II, the emission lines most likely originate above the stellar photosphere, and may be indicative of a recent episode of mass loss.

126.090   PG 0122 + 200: a new member of the GW Virginis (PG 1159–035) class of extremely hot pulsating white dwarfs.
H. E. Bond, A. D. Grauer.
Astrophys. J., Lett. Ed., Vol. 321, No. 2, p. L123 – L128 (1987).

High–speed photometry reveals that the hot ($\sim 100,000K$) hydrogen–deficient white dwarf PG 0122 + 200 is a low–amplitude pulsating variable. The dominant periodicities are in the range 402 – 489 s, and appear to be due to nonradial g–modes. PG 0122 + 200 thus becomes the fifth known member of the PG 1159–035 (GW Vir) class, which includes three other PG white dwarfs and the central star of the planetary nebula K1–16.

126.091   Millisecond pulsar in M4.
IAU Circ., No. 4470 (1987).

126.092   PSR 0656 + 14.
IAU Circ., No. 4490 (1987).

126.093   Two searches for pulsars using a Cyber 205.
A. G. Lyne.
NRAO Workshop, No. 15, p. 133 – 136 (1988). – See Abstr. 012.076.

The 400 known pulsars do not reflect the true population of pulsars in the Galaxy because of observational biases associated with past surveys. In particular, those with short period or high dispersion measure (implying large distances) are under–represented. Moreover, none of the major surveys to date have had any sensitivity at millisecond periods. Until recently even the Crab pulsar with a period as large as 33 ms had not been detected in any of these. The chief reason for this has been the limited computer power available to pulsar astronomers. In 1983, the University of Manchester Regional Computer Centre in England took delivery of a Cyber 205 and this opened up the possibility of correcting the deficiencies of the earlier surveys.

126.094   Polarized γ–rays from Vela.
A. J. Dean.
Nature, Vol. 330, No. 6148, p. 524 (1987).

126.095   An upper limit to the space density of short–period, noninteracting binary white dwarfs.
E. L. Robinson, A. W. Shafter.
Astrophys. J., Vol. 322, No. 1, p. 296 – 301 (1987).

The authors have searched for short–period, detached binary stars in which both stars are white dwarfs by looking for radial velocity variations in spectroscopically identified DA and DB white dwarfs. The search was sensitive to binaries with orbital periods between 30 s and 3 hr and, within that range, the authors would have detected roughly 90% of all binaries. They observed 44 stars without finding any binaries. The fraction of white dwarfs that are binaries is less than 1/20 with a 90% probability and less than 1/37 with a 70% probability. The space density of binary white dwarfs is too low to account for the rate of Type I supernovae in the Galaxy.

126.096   Determination of photospheric helium abundances for the hot white dwarfs LB 1663 and CD –38°10980 from EXOSAT soft X–ray photometry.
F. B. S. Paerels, J. Heise, S. M. Kahn, R. D. Rogers.
Astrophys. J., Vol. 322, No. 1, p. 315 – 319 (1987).

The authors present a measurement of the photospheric helium abundance of LB 1663 and CD –38°10980, based on three–color soft X–ray photometry obtained with the low–energy imaging telescopes on the EXOSAT observatory, using constraints of T_e and absolute flux levels derived from optical and UV observations. The authors find for CD –38°10980 that $n(He)/n(H) < 5 \times 10^{-6}$, almost independent of the other spectral parameters. For LB 1663 the ratio is $n(He)/n(H) = (1-2) \times 10^{-4}$.

126.097   Gravitational redshift and mass–radius relation in white dwarfs.
D. Koester.
Astrophys. J., Vol. 322, No. 2, p. 852 – 855 (1987).

This paper presents gravitational redshifts obtained from observations of Hα in nine DA white dwarfs, which are members of wide binaries or common proper motion pairs. Within the error bars, the data follow the theoretical zero–temperature relation for the pure carbon models of Hamada and Salpeter. Due to the remaining relatively large errors in radii and the intrinsically narrow mass distribution of white dwarfs around a mean mass of 0.58 $M_\odot$ the results do not provide an empirical proof for the shape of the relation. Assuming the validity of this relation, it is possible to derive very accurate masses.

126.098   The binary pulsar: gravity waves exist.
C. Will.
Mercury, Vol. 16, No. 6, p. 162 – 173 (1987).

126.099   L'énigme de Sirius.
J. Manfroid, A. Heck.
Ciel Espace, No. 215, p. 30 – 33 (1987).

A catalog of spectroscopically identified white dwarfs.
See Abstr. 002.108.

Chto takoe pul'sary?
See Abstr. 003.041.

Wide bandwidth signal processor for removing dispersion distortion from pulsar radio signals.
See Abstr. 034.074.

Plasma turbulence in space.
See Abstr. 062.032.

Shocked relativistic magnetohydrodynamic flows with application to pulsar winds.
See Abstr. 062.102.

Gamma ray production by comptonization of electron beams in pulsar magnetospheres.
See Abstr. 063.055.

On the theory of microstructure of the condensation layer in the atmosphere of a white dwarf.
See Abstr. 064.063.

Polarization line radiative transfer in the atmospheres of magnetic white dwarfs.
See Abstr. 064.073.

Wave energy in white dwarf atmospheres. I. Magnetohydrodynamic energy spectra for homogeneous DB and layered DA stars.
See Abstr. 064.076.

Spin–up and mixing in accreting white dwarfs.
See Abstr. 065.005.

Deflagrating white dwarfs and the statistical properties of Type I supernovae.
See Abstr. 065.027.

Thermonuclear processes in accreting white dwarfs (novae, symbiotic stars, and type–I supernovae).
See Abstr. 065.029.

The fate of accreting white dwarfs: Type I supernovae vs. collapse.
See Abstr. 065.038.

The core mass–radius relation for giants: a new test of stellar evolution theory.
See Abstr. 065.041.

Lower bounds on the masses of rapidly rotating white dwarfs.
See Abstr. 065.042.

White dwarfs: the connection with the parents' masses.
See Abstr. 065.051.

Cooling of white dwarfs with account of non–equilibrium beta processes.
See Abstr. 065.055.

Collapsing white dwarfs.
See Abstr. 065.068.

Stellar accretion of matter possessing angular momentum.
See Abstr. 065.094.

Nuclear runaways in a C/O white dwarf accreting H–rich material possessing angular momentum.
See Abstr. 065.095.

Gamma ray lines from novae.
See Abstr. 065.096.

Monte Carlo simulations of radio pulsars and their progenitors.
See Abstr. 065.099.

The strong field point particle limit and the quadrupole formula in the binary pulsar system.
See Abstr. 066.006.

A possible semisecular variation in orbital period of binary pulsar system PSR 1913 + 16 and Lorentz invariance of gravity.
See Abstr. 066.178.

Neutron star coupling to its environment.
See Abstr. 067.003.

Line radiation from accreting magnetized neutron stars.
See Abstr. 067.006.

Neutron star formation in theoretical supernovae – low mass stars and white dwarfs.
See Abstr. 067.014.

Two types of pulsars.
See Abstr. 067.019.

Neutron star cooling: critical test of dense matter physics.
See Abstr. 067.020.

Inverse Compton scattering in strong magnetic fields: applied to the radiation mechanism of PSR 0531 + 21.
See Abstr. 067.027.

The spin–torsion coupling precession of spin and its effects on single pulses of pulsars.
See Abstr. 067.029.

Neutron star cooling and the Vela pulsar.
See Abstr. 067.030.

Where neutron stars come from, how neutron stars evolve, and where neutron stars go.
See Abstr. 067.036.

Limits on hadronic cosmic ray production by young pulsars.
See Abstr. 067.078.

Gamma–ray pulsar model.
See Abstr. 067.081.

On the theory of radio radiation of pulsars (Review).
See Abstr. 067.082.

Ohmic decay of crustal neutron–star magnetic fields.
See Abstr. 067.092.

Formation of a millisecond pulsar in a globular cluster.
See Abstr. 067.125.

Origin of millisecond pulsars.
See Abstr. 067.126.

Formation of isolated millisecond pulsars in globular clusters.
See Abstr. 067.127.

The effect of heating from the boundary layer on accretion models for novae and other compact objects.
See Abstr. 067.129.

Neutron stars observations as astrophysical probes.
See Abstr. 067.130.

Sustained magnetic fields in binary millisecond pulsars.
See Abstr. 067.135.

A mechanism for pulsar drift.
See Abstr. 067.165.

A pulsar emission model: observational tests.
See Abstr. 067.173.

Abundance anomalies in stars: atomic physics at play.
See Abstr. 114.018.

The B stars: beacons of the skies.
See Abstr. 114.042.

Spectral types for objects in the Kiso survey. III. Data for 102 stars.
See Abstr. 114.055.

Spectrum analysis of the extremely metal poor carbon dwarf star G 77–61.
See Abstr. 114.079.

A search for brown dwarfs and late M dwarfs in the Hyades and the Pleiades.
See Abstr. 115.011.

Absorption features from the accretion column in E 1405–451.
See Abstr. 117.009.

Black hole candidates in X–ray binaries.
See Abstr. 117.021.

Simultaneous multicolour photometry of OY Carinae during quiescence.
See Abstr. 117.025.

On the possibility of detecting close binary degenerate dwarfs.
See Abstr. 117.034.

The symbiotic star AG Dra.
See Abstr. 117.037.

Spin–down of the white dwarf in the intermediate polar V1223 Sgr/4U 1849–31.
See Abstr. 117.077.

Millisecond X–ray binary pulsars.
See Abstr. 117.097.

The evolutionary status of MXB 1820–30 and other short–period low–mass X–ray binaries.
See Abstr. 117.107.

On a model of the symbiotic star AG Dra.
See Abstr. 117.289.

IUE observations of the dwarf nova HL Canis Majoris and the winds of cataclysmic variables.
See Abstr. 117.321.

Excess infrared radiation from a white dwarf – an orbiting brown dwarf?
See Abstr. 118.028.

A search for pulsating stars similar to PG 1159–035 and K1–16.
See Abstr. 122.075.

A search for pulsating stars similar to PG 1159–035 and K1–16.
See Abstr. 122.134.

Crab–like supernova remnants.
See Abstr. 125.001.

Neutron stars in twelve supernova remnants.
See Abstr. 125.002.

The galactic sources G 5.4–1.2 and G 5.27–0.90.
See Abstr. 125.008.

Supernova remnants in the Magellanic Clouds observed at the Molonglo Observatory.
See Abstr. 125.009.

Ancient Guest Stars as harbingers of neutron star formation.
See Abstr. 125.011.

Does apparent spatial coincidence of pulsars and supernovae mean generic association?
See Abstr. 125.013.

Evolution of supernova remnants with a central pulsar.
See Abstr. 125.019.

Observations of X–ray emission of the Crab Nebula and the pulsar NP 0532 on the Astron automatic station.
See Abstr. 125.022.

Pulsar–like emission from the supernova remnant CTB 80.
See Abstr. 125.048.

A new shell SNR in the direction of the young pulsar PSR 1930 + 22.
See Abstr. 125.055.

Binary systems as supernova progenitors (some frequency estimates).
See Abstr. 125.067.

Type I supernova models.
See Abstr. 125.071.

The light curve of SN 1987A.
See Abstr. 125.225.

Circumstellar interaction and a pulsar nebula in the supernova 1986j.
See Abstr. 125.522.

The influence of the interstellar scattering on the pulsar mean pulse shape and apparent angular size: general analytical representation.
See Abstr. 131.116.

Results of the timing analyses of X–ray pulsars observed by Hakucho and Tenma.
See Abstr. 142.002.

Herculis X–1: results and interpretation.
See Abstr. 142.003.

The X–ray transient EXO 2030 + 375.
See Abstr. 142.005.

A rapid change of the Hercules X–1 pulse profile and high–state duration.
See Abstr. 142.022.

Neutron stars and gamma–rays.
See Abstr. 143.001.

Upper limits to the $(10^{13} – 10^{14})$eV gamma–ray fluxes from the Crab Nebula and pulsar.
See Abstr. 143.044.

White dwarfs in Omega Centauri?
See Abstr. 154.025.

White dwarfs and cataclysmic variables in the globular cluster M71.
See Abstr. 154.032.

Quasar and stellar objects in the Byurakan Surveys.
See Abstr. 159.040.

Interstellar Matter, Nebulae

131 Interstellar Matter (Molecular Clouds, Reflection Nebulae, etc.), Star Formation

131.001 High mass star formation in the Galaxy.
N. Z. Scoville, J. C. Good.
NASA Conf. Publ., NASA CP–2466, p. 3 – 20 (1987). – See Abstr. 012.002.
The galactic distributions of H I, H_2 and H II regions are reviewed in order to elucidate the high mass star formation occurring in galactic spiral arms and in active galactic nuclei.

131.002 Masses, luminosities, and dynamics of galactic molecular clouds.
P. M. Solomon, A. R. Rivolo, T. J. Mooney, J. W. Barrett, L. J. Sage.
NASA Conf. Publ., NASA CP–2466, p. 37 – 59 (1987). – See Abstr. 012.002.
The authors describe and analyze the masses, luminosities, dynamics and distribution of molecular clouds, primarily giant molecular clouds in the Milky Way. They compare the star formation activity in isolated and interacting galaxies with that of galactic molecular clouds. The interacting galaxies have (IR luminosity/molecular mass) ratios substantially higher than any galactic molecular cloud.

131.003 IRAS colors of VLA identified objects in the Galaxy.
M. Fich, S. Terebey.
NASA Conf. Publ., NASA CP–2466, p. 63 – 66 (1987). – See Abstr. 012.002.
IRAS sources found within 4 degrees of $l = 125°$, $b = 2°$ on the 3rd HCON 60 μ Sky Brightness Images have been observed at the VLA. The intent of this project was to identify regions where massive stars are forming by looking for small areas of radio continuum emission. The IRAS sources could be divided into three groups by their IRAS 12 μ/25 μ and 60 μ/100 μ colors. The group identified with star forming regions contained essentially all of the objects with extended radio emission. In all of these cases the extended radio emission showed a morphology consistent with the identification of these objects as H II regions. The conclusion that may be drawn from this project is that star formation regions can be distinguished from other objects by their infrared colors.

131.004 Star formation in Carina OB1: observations of a giant molecular cloud associated with the η Carinae nebula.
D. A. Grabelsky, R. S. Cohen, P. Thaddeus.
NASA Conf. Publ., NASA CP–2466, p. 67 – 70 (1987). – See Abstr. 012.002.
A giant molecular cloud associated with the η Carinae nebula has been fully mapped in CO. The cloud complex has a mass of roughly $7 \times 10^5 M_\odot$ and extends about 140 pc along the Galactic plane, with the giant Carina H II region situated at one end of the complex. Clear evidence of interaction between the H II region and the molecular cloud is found in the relative motions of the ionized gas, the molecular gas, and the dust; simple energy and momentum considerations suggest that the H II region is responsible for the observed motion of a cloud fragment. The molecular cloud complex appears to be the parent material of the entire Car OB1 association which, in addition to the young clusters in the Carina nebula, includes the generally older clusters NGC 3324, NGC 3293, and IC 2581. The authors estimate the overall star formation efficiency in the cloud complex to be ~ 0.02.

131.005 Star forming regions of the southern Galaxy.
T. B. H. Kuiper, J. B. Whiteoak, J. W. Fowler.
NASA Conf. Publ., NASA CP–2466, p. 71 – 74 (1987). – See Abstr. 012.002.
A catalogue of southern dust cloud properties is being compiled to aid in the planning and analysis of radio spectral line surveys in the southern hemisphere. Ultimately, images of dust temperature and column density will be produced. For the interim, a list of the 60 and 100 μm fluxes has been prepared for the cores and adjacent backgrounds of 65 prominent dust clouds. Dust temperatures and column densities have been derived.

131.006 On the redistribution of OB star luminosity and the warming of nearby molecular clouds.
D. Leisawitz.
NASA Conf. Publ., NASA CP–2466, p. 75 – 78 (1987). – See Abstr. 012.002.
IRAS observations of the neighborhoods of six outer–Galaxy H II regions were combined with CO observations to show that most of the far infrared (FIR) luminosity from within $\sim 25 – 75$ pc of the ionizing stars is contributed by dust in molecular clouds, not by dust in the low–density ionized gas. Dust associated with the clouds is warmed by absorption of UV and visible light from the cluster of stars responsible for the ionization. Most ($\gtrsim 70\%$) of the OB cluster starlight is not absorbed locally.

131.007 Maps of millimeter wave emission from three galactic star–forming regions.
M. Barsony.
NASA Conf. Publ., NASA CP–2466, p. 79 – 82 (1987). – See Abstr. 012.002.
In order to investigate the gas dynamics around young stellar objects, the author has mapped three sources which exhibit supersonic velocities in the 115 GHz, J = 1–0 transition of CO (Bally and Lada 1983). The maps, made with the Owens Valley Radio Observatory Millimeter Interferometer, are the highest spatial resolution ($5'' \times 5''$) images currently available of milli-meter–wave continuum and line emission from the sources S106, S87, and LkHα 101. Observations were made in the CS (J = 2–1) and ^{13}CO (J = 1–0) transitions. The observations indicate that the ionized stellar wind is sweeping up ambient molecular gas. The molecular gas is found adjacent to the outer edges of the ionized winds, which originate in embedded infrared sources. The author infers that the outflowing ionized winds are channeled by the surrounding dense, neutral gas.

131.008 Shock heated dust in L1551: L(IR) > 20 solar luminosities.
F. O. Clark, R. J. Laureijs, G. Chlewicki, C. Y. Zhang, W. van Oosterom, D. Kester.
NASA Conf. Publ., NASA CP–2466, p. 83 – 86 (1987). – See Abstr. 012.002.
The infrared bolometric luminosity of the extended emission from the L1551 flow exceeds 20 solar luminosities. Ultraviolet

radiation from the shock associated with the flow appears to heat the dust requiring shock temperatures from 10,000 to 90,000K in L1551, velocities of ~50 km/s near the end of the flow, and a minimum mechanical luminosity of ~40 solar luminosities. The total energy requirement of the infrared emission over a 10,000 year lifetime is $10^{(46-47)}$ ergs, two orders of magnitude higher than previous estimates for L1551. Infrared radiation offers a new method of probing interstellar shocks, by sampling the ultraviolet halo surrounding the shock. At least one current model for bipolar flows is capable of meeting the energetic requirements.

131.009   Mechanisms for the circular polarization of astrophysical OH masers in star–forming regions and the inferred magnetic fields.
S. Deguchi, W. D. Watson.
NASA Conf. Publ., NASA CP–2466, p. 87 – 90 (1987). – See Abstr. 012.002.

Results of further calculations to explore the cause for the circular polarization of astrophysical OH masers in regions of star–formation are presented. New calculations are given for both the non–linear, Zeeman overlap mechanism and the Cook mechanism. The authors' previous result that magnetic field strengths of a few milligauss or greater are required still survives.

131.010   The large scale gas and dust distribution in the Galaxy: implications for star formation.
T. J. Sodroski, E. Dwek, M. G. Hauser, F. J. Kerr.
NASA Conf. Publ., NASA CP–2466, p. 99 – 102 (1987). – See Abstr. 012.002.

The authors present IRAS observations of the diffuse infrared (IR) emission from the galactic plane at wavelengths of 60 and 100 μm and derive the total far infrared intensity and its longitudinal variation in the disk. They linearly decompose the longitudinal profiles of the 60 and 100 μm emission into three components that are associated with the molecular (H_2), neutral (H I), and ionized (H II) phases in the interstellar medium (ISM), and derive the relevant dust properties (i.e. temperature, IR luminosity per hydrogen mass, total IR luminosity) in each phase. Implications of the findings for various models of the diffuse IR emission and for star formation in the galactic disk are discussed.

131.011   A correlation between the IRAS infrared cirrus at 60 or 100 μm and neutral atomic hydrogen in the outer Galaxy.
S. Terebey, M. Fich.
NASA Conf. Publ., NASA CP–2466, p. 103 – 105 (1987). – See Abstr. 012.002.

The authors find a linear correlation between the infrared cirrus at 100 or 60 μm and neutral atomic hydrogen near the galactic plane. IRAS Sky Brightness images were compared to the 0.5° resolution Weaver–Williams H I survey in two regions of the outer Galaxy near $l = 125°$ and $l = 215°$. The dust temperature inferred is nearly uniform and in reasonable agreement with theoretical predictions of thermal dust emission.

131.012   The Arp Ring: galactic or extragalactic?
J. A. Abolins, W. L. Rice.
NASA Conf. Publ., NASA CP–2466, p. 107 – 110 (1987). – See Abstr. 012.002.

The Arp Ring is a faint, loop–like structure around the northern end of M81 which becomes apparent only on deep optical photographs of the galaxy. The nature of the ring and its proximity to M81 are uncertain. Is it simply foreground structure – part of our own galaxy, or is it within the M81 system? IRAS maps of the region show a far–infrared counterpart of the ring. The new infrared data are compared with previous optical and radio observations to try to ascertain its physical nature. The poor correlation found between the common infrared/optical structure and the distribution of extragalactic neutral hydrogen, and the fact that its infrared properties are indistinguishable from those of nearby galactic cirrus, imply that the Arp Ring is simply a ring structure in the galactic cirrus.

131.013   Infrared properties of dust grains derived from IRAS observations.
G. Chlewicki, R. J. Laureijs, F. O. Clark, P. R. Wesselius.
NASA Conf. Publ., NASA CP–2466, p. 113 – 116 (1987). – See Abstr. 012.002.

The paper presents the analysis of several diffuse interstellar clouds observed by IRAS. The 60/100 μm flux ratios appear to be nearly constant in clouds with up to 1^m of visual extinction at the centre. Observations of a highly regular cloud in Chamaeleon show that the 12/100 μm ratio peaks at an intermediate radial distance and declines towards the centre of the cloud. These observations indicate that non–equilibrium emission accounts only for the 12 and 25 μm bands; strong emission observed at the 60 μm band is probably due to equilibrium thermal radiation. The correlation of 12 μm emission with a red excess observed for a high latitude cloud, L 1780, is shown to be consistent with the assumption that both features are due to fluorescence by the same molecular species.

131.014   IRAS observations of the Pleiades.
P. Cox, A. Leene.
NASA Conf. Publ., NASA CP–2466, p. 117 – 121 (1987). – See Abstr. 012.002.

IRAS observations of the Pleiades region are reported. The data show large flux densities at 12 μm and 25 μm, extended over the optical nebulosity. This strong excess emission, implying temperatures of a few hundred degrees Kelvin, indicates a population of very small grains in the Pleiades. It is suggested that these grains are similar to the small grains needed to explain the surface brightness measurements made in the ultraviolet.

131.015   A simple theory of bimodal star formation.
R. F. G. Wyse, J. Silk.
NASA Conf. Publ., NASA CP–2466, p. 339 – 342 (1987). – See Abstr. 012.002.

The authors present a model of bimodal star formation, wherein massive stars form in giant molecular clouds (GMC), at a rate regulated by supernova energy feedback through the interstellar medium, the heat input also ensuring that the initial mass function (IMF) remains skewed towards massive stars. The low mass stars form at a constant rate. The formation of the GMC is governed by the dynamics of the host galaxy through the rotation curve and potential perturbations such as a spiral density wave. The characteristic masses, relative normalisations and rates of formation of the massive and low mass modes of star formation may be tightly constrained by the requirements of the chemical evolution in the solar neighborhood. The authors obtain good fits to the age metallicity relation and the metallicity structure of thin disk and spheroid stars only for a narrow range of these parameters.

131.016   Star formation and dynamics in starburst nuclei.
C. A. Norman.
NASA Conf. Publ., NASA CP–2466, p. 395 – 400 (1987). – See Abstr. 012.002.

A simple model is presented for gas inflow through a disk galaxy driven by interacting galaxies through the action of a non–axisymmetric disturbance acting on the disk whose gas is modelled as an ensemble of gas clouds. Cloud collisions, as well as being a vital process in forcing gas inflow to the centre of the disk, are also assumed to generate massive stars. This ever increasing rate of gas flow toward the centre of the galaxy and the associated rapid increase in cloud collisions leads to a centrally concentrated starburst.

131.017   Cloud fluid models of gas dynamics and star formation in galaxies.
C. Struck–Marcell, J. M. Scalo, P. N. Appleton.
NASA Conf. Publ., NASA CP–2466, p. 435 – 460 (1987). – See Abstr. 012.002.

The large dynamic range of star formation in galaxies, and the apparently complex environmental influences involved in triggering or suppressing star formation, challenge our understanding. The key to this understanding may be the detailed study of simple physical models for the dominant nonlinear interactions in interstellar cloud fluid systems. The authors describe one such model, a generalized Oort model cloud fluid, and explore two simple applications of it. The first of these is the relaxation of an isolated volume of cloud fluid following a disturbance. The second application is to the modeling of colliding ring galaxies.

131.018   Formation of giant molecular clouds in global spiral structures: the role of orbital dynamics and cloud–cloud collisions.
W. W. Roberts Jr., G. R. Stewart.
NASA Tech. Memo., NASA TM–89810, p. 101 (1987). Abstract. – See Abstr. 003.001.

131.019   Models of molecular cloud cores. III. A multitransition study of H₂CO.
L. G. Mundy, N. J. Evans II, R. L. Snell, P. F. Goldsmith.
Astrophys. J., Vol. 318, No. 1, p. 392–409 (1987).

The dense cores of the M17, S140, and NGC 2024 molecular clouds have been observed in six transitions of ortho–H_2CO. The data have been fitted to spherical LVG models for the excitation to determine densities and ortho–H_2CO column densities. The derived densities (typically $10^6 cm^{-3}$) agree well with those found from studies of CS and $C^{34}S$. The excitation analysis of H_2CO lines and the use of various H_2CO lines as probes of density are both reexamined.

131.020   Erratum: "Optical properties of interstellar graphite and silicate grains" [Astrophys. J., Vol. 285, No. 1, p. 89–108 (1984)].
B. T. Draine, H. M. Lee.
Astrophys. J., Vol. 318, No. 1, p. 485 (1987). See Abstr. 38.131.159.

131.021   Low–mass stars in the Orion region.
S. Isobe.
Astrophys. Space Sci., Vol. 135, No. 2, p. 237–244 (1987).

Recent observational evidence shows that low–mass stars were firstly formed in molecular clouds, and that, at a later stage when massive stars were formed, the formation rate of low–mass stars was still high in the Orion nebula (this paper) but decreases rapidly in NGC 2264 (Adams et al., 1983). This difference is probably caused by the effects of mass ejection and luminous radiation from the stars which were born in the previous period. In this paper, the author discusses examples of low–mass stars in order to find a relationship between the age and location of stars.

131.022   Spectral radio astronomical observations in the 2–4 mm wavelength range.
I. I. Zinchenko, A. B. Burov, V. F. Vdovin, V. N. Voronov, V. M. Demkin, A. G. Kislyakov, A. A. Krasil'nikov, A. V. Lapinov, L. E. Pirogov, V. N. Shanin, V. M. Yurkov.
Pis'ma Astron. Zh., Tom 13, No. 7, p. 582–588 (1987). In Russian. English translation in Sov. Astron. Lett., Vol. 13.

A research program of observations of interstellar clouds in continuum and molecular lines has been carried out at the Institute of Applied Physics of the Academy of Sciences of the USSR. With the help of computer controlled radiometrical equipment the authors have made observations at the 22–meter radiotelescope of the Crimean Astrophysical Observatory in the 2–4 mm wavelength range. The equipment includes a cooled radiometer with the noise temperature of 400–450K (DSB). A survey of a large number of clouds in the $J = 1$–0 HCN line was made. Some sources were observed in the $J = 1$–0 CO, HCO⁺ and H¹³CN lines.

131.023   Complex investigation of the star formation region in the Orion constellation.
V. M. Pashinskij.
Pis'ma Astron. Zh., Tom 13, No. 8, p. 654–662 (1987). In Russian. English translation in Sov. Astron. Lett., Vol. 13.

The proper motions of 54 young stars in the Ori OB1 cluster has been specified. The contraction of the cluster in the time scale of $\approx 3 \times 10^6$years is discovered, and the cluster rotation with the

period of $\approx 6 \times 10^6$years is confirmed as well. The cluster velocity ≈ -10 km/s is determined from spatial velocities of 26 stars. This means that the cluster was situated in the symmetry plane of the Galaxy some 1.4×10^7years ago, that coincides with the oldest subcluster age of the association.

131.024   Populations of the rotational levels of molecules in clouds with large redshifts.
I. E. Val'tts, V. K. Khersonskij.
Astrofizika, Tom 26, Vyp. 3, p. 501–509 (1987). In Russian. English translation in Astrophysics, Vol. 26, No. 3, p. 302–307 (1987).

The effects of the background radiation field on the populations of CO–molecule rotational levels in molecular clouds of galaxies with large redshifts are considered. The kinetic temperature and gas concentration regions are taken to be typical for molecular clouds of our Galaxy: $10^2 \lesssim N(H_2)(cm^{-3}) \lesssim 10^6$, $20 \lesssim T_k(K) \lesssim 100$. The effect of the background radiation field has been studied in the redshift interval $1 \lesssim z \lesssim 5$. It has been shown that the effect is essential even for $N(H_2) < 10^4 cm^{-3}$ and is predominant for $N(H_2) = 10^2 - 10^3 cm^{-3}$.

131.025   C₆H: astronomical study of its fine and hyperfine structure.
J. Cernicharo, M. Guélin, K. M. Menten, C. M. Walmsley.
Astron. Astrophys., Vol. 181, No. 1, p. L1–L4 (1987).

The authors report the detection of the $^2\Pi_{1/2}$ fine structure state of C_6H the new radical discovered by Guélin et al. (1986, 1987) and Suzuki et al. (1986), as well as further observations of its $^2\Pi_{3/2}$ state. The new data, which consist of 50 lines, allows the accurate determination of the radical's rotation and distortion constants as well as its main fine and hyperfine parameters. The authors estimate the C_6H column density to be $3 \times 10^{14} cm^{-2}$ in IRC + 10216 and $10^{13} cm^{-2}$ in TMC1. The C_6H/C_5H abundance ratios are 2 and 3, respectively, in these two sources.

131.026   CO and NH₃ detection of the cone in NGC 2264.
L. P. Pagani, Nguyen–Q–Rieu.
Astron. Astrophys., Vol. 181, No. 1, p. 112–118 (1987).

The authors have detected in CO($J = 1$–0) and in NH_3(1, 1), the Bok globule in the cone head of NGC 2264 which has escaped so far the previous radio observations of this region. The authors also have evidence for the detection of the whole cone from a large scale map of the region. With the help of a simple LTE model, they have derived an estimate of the density ($\sim 10^4$molecules cm^{-3}), mass ($\sim 16 M_\odot$) and kinetic temperature ($\sim 20K$) of the globule. The fact that its LSR velocity, ~ 5 km s⁻¹, is close to the velocity of one of the background nebula features supports the idea that the cone is related to the reflection nebula. Its relatively high kinetic temperature can also be explained by the vicinity of the well–known infrared source embedded in the nebula.

131.027   Magnetic field strengths in molecular clouds.
R. M. Crutcher, I. Kazès, T. H. Troland.
Astron. Astrophys., Vol. 181, No. 1, p. 119–126 (1987).

The authors report additional observations made with the Nançay radio telescope of the Zeeman effect in the 1665 and 1667 MHz absorption lines of OH toward four molecular cloud positions. Magnetic fields determined from these data for each cloud are: S88 B, $+69 \pm 5 \mu G$; W49 B, $+21 \pm 5 \mu G$; W40, $-14.0 \pm 2.6 \mu G$; W22 B, $-32 \pm 9 \mu G$. The authors present as a working hypothesis the suggestion that in cool dust clouds and in warm molecular clouds the total magnetic field strengths are of order 30 and 120 μG, respectively. These detections of the OH Zeeman effect add to a growing body of evidence that magnetic field strengths increase with density in interstellar clouds and that magnetic fields are sufficiently strong to be important in the evolution of dense clouds and in the star formation process.

131.028 The influence of shape on the temperature of small graphite grains.
G. Chlewicki.
Astron. Astrophys., Vol. 181, No. 1, p. 127 – 133 (1987).

The anisotropic optical properties of graphite result in a strong dependence of the temperatures of small graphitic grains ($a \approx 0.02\,\mu$m) on their shape. The temperature attained in the diffuse interstellar medium increases from 20.4K for spheres to 27K for oblate spheroids with axial ratios close to 5. A population of graphite discs is therefore proposed as an explanation for the unexpectedly high temperatures (~ 25K) derived from IRAS observations of diffuse medium dust. The semi–metallic character of graphite is analyzed and it is shown that bulk infrared properties, modified only to account for surface scattering of charge carriers, cannot be applied to particles smaller than 0.01 μm.

131.029 Observations of cold dust in S 106.
P. G. Mezger, R. Chini, E. Kreysa, J. Wink.
Astron. Astrophys., Vol. 182, No. 1, p. 127 – 136 (1987).

The authors mapped the bipolar nebula S 106 in the directon of its exciting star IRS 4. They observed cold dust ($T_d \sim 18$K) emission from a bar–like feature of size $33'' \times$ ($\leqslant 7''$) which is approximately centered on IRS 4. Two models can explain the observed characteristics of S 106 and the dust bar. In the geometrical model the dust bar is located behind S 106 and may represent either the cloud core out of which the exciting star IRS 4 of S 106 has formed or it is a dense condensation of cold dust which by chance coincides spatially with S 106. In the physical model, which the authors prefer, the dust bar represents a massive disk surrounding IRS 4 in its equatorial plane. To get agreement with optical and submm dust observations the excitation cross section of dust in visual, i.e. the ratio A_V/N_H, would have to be decreased by a factor of ~ 20 relative to the standard ratio. While formation of mantles of molecular ices can account for some of this, most of the required decrease of A_V/N_H would probably have to be explained by formation of larger grains due to coagulation.

131.030 A multilevel study of ammonia in star forming regions. II. G 34.3 + 0.2, a new "hot core".
C. Henkel, T. L. Wilson, R. Mauersberger.
Astron. Astrophys., Vol. 182, No. 1, p. 137 – 142 (1987).

The detection of fifteen NH_3 inversion transitions, ranging from 20 to 940K above the ground state, is reported from the molecular cloud G 34.3 + 0.2. It is shown that G 34.3 + 0.2 has one of the largest NH_3 column densities so far determined. The authors find a kinetic temperature, T_{kin}, of 225 ± 75K, an NH_3 column density, $N(NH_3)$, of $10^{18.5 \pm 0.2}$, and an H_2 column density, $N(H_2)$, of $10^{23.6 \pm 0.1}$cm^{-2}. The space densities are $n(NH_3) = 10^{1.6 \pm 0.2}$ and $n(H_2) = 10^{7.2 \pm 0.2}$cm^{-3}; the fractional abundance of NH_3 is $[NH_3]/[H_2] = 10^{-5.6 \pm 0.3}$ and the size 0.04 ± 0.01 pc. All of these values are similar or even larger than those determined from Orion KL. The source, presumably heated by the associated ultracompact H II region, may be optically thick in the IR range up to $\sim 100\,\mu$m.

131.031 Searches for interstellar and circumstellar metal oxides and chlorides.
T. J. Millar, J. Elldér, Å. Hjalmarson, H. Olofsson.
Astron. Astrophys., Vol. 182, No. 1, p. 143 – 149 (1987).

The authors have made sensitive searches for millimetre wave emission lines from MgO, TiO, ClO and CCl in interstellar clouds and circumstellar envelopes. Although several lines were detected in the course of the observations, none of the above molecules were found. The authors discuss this result and conclude that it is consistent with the current knowledge of interstellar and circumstellar chemistry, elemental depletions and the composition of cosmic dust particles. It is possible that metallic atoms may be present in gas–phase molecules other than those for which the authors have searched and it is suggested that, in particular, SiH, and possibly MgH, may be reasonably abundant in interstellar clouds.

131.032 Detections of diffuse interstellar bands toward the SN 1987A in the Large Magellanic Cloud.
G. Vladilo, L. Crivellari, P. Molaro, J. E. Beckman.
Astron. Astrophys., Vol. 182, No. 2, p. L59 – L62 (1987).

The authors report the results of high resolution and high S/N observations of diffuse interstellar bands (DIBs) towards SN 1987A in the LMC. For the first time they clearly detect the LMC component of the 5780, 5797 and 6284 Å DIBs, and possibly that of 5778 and 6269 Å. Moreover, they observe the Galactic components of the 5778, 5780, 5797, 6283, 6376 and 6379 Å bands, along a line of sight with very low reddening [E(B–V) $\cong$ 0.07]. The LMC components are generally weaker than the Galactic ones, while the opposite is observed for ISNa I and K I. The band strengths are compared with the total color excess towards the supernova.

131.033 Deuterated water in Orion–KL and NGC 7538.
C. Henkel, R. Mauersberger, T. L. Wilson,
L. E. Snyder, K. M. Menten, J. G. A. Wouterloot.
Astron. Astrophys., Vol. 182, No. 2, p. 299 – 304 (1987).

The first detections of the 3_{21}–4_{14} and 5_{32}–5_{33} lines of HDO are reported: both lines are observed in emission toward Orion–KL with velocities typical of the "Hot Core" and the "Compact Ridge", while the 3_{21}–4_{14} line is seen in absorption toward NGC 7538–IRS1. The presence of highly excited HDO in these two sources suggests that the observed lines form in a hot and dense molecular environment, where the kinetic temperature exceeds 100K. The Orion data are combined with other line results to derive optical depths, excitation temperature, column densities, and relative abundance. An upper limit to the [HDO]/[H_2O] ratio is given which is inconsistent with gas phase ion–molecule or gas phase shock formation schemes. It is suggested that large amounts of HDO are released from grain mantles when the kinetic temperature exceeds 100K. The data from NGC 7538 are used to obtain a direct estimate of the optical depth, which indicates deviations from thermal equilibrium.

131.034 Bursts of H_2O maser emission.
L. I. Matveenko.
Sov. Astron., Vol. 30, No. 5, p. 589 – 591 (1986). English translation of 42.131.075.

131.035 Observations of diffuse and planetary nebulae on the Astron astrophysical station.
V. I. Pronik, P. P. Petrov.
Sov. Astron., Vol. 30, No. 5, p. 601 – 603 (1986). English translation of 42.131.076.

131.036 Interstellar scattering and resolution limitations.
B. Dennison.
Radio astronomy from space, p. 137 – 143 (1987). – See Abstr. 012.009.

Density irregularities in both the interplanetary medium and the ionized component of the interstellar medium scatter radio waves, resulting in limitations on the achievable resolution. With the Earth–space baselines now planned, it will be possible to search directly for interstellar refraction, which is suspected of modulating the fluxes of background sources.

131.037 H_2O masers and the cosmic distance scale.
M. J. Reid, J. M. Moran, C. R. Gwinn.
Radio astronomy from space, p. 145 – 151 (1987). – See Abstr. 012.009.

VLBI proper–motion studies of H_2O maser sources have revealed energetic expanding flows and provided a means to determine distances across the Milky Way. These studies measure the relative positions of several tens of maser spots in one source over epochs typically spanning months to years. Ground based VLBI observations may be extended to H_2O masers in galaxies less than ≈ 1 Mpc away. The increased angular resolution and dynamic range afforded by space–based VLBI may allow measurement of distances to galaxies out to ≈ 10 Mpc.

131.038 Interstellar electrons and low–frequency radio observations.

S. R. Kulkarni.

Radio astronomy from space, p. 277 – 281 (1987). – See Abstr. 012.009.

The author summarizes the present understanding of the distribution and physical conditions of interstellar electrons. He then discusses key issues for which low–frequency ($\lesssim 10$ MHz) observations provide either unique data or supplement other observations. These issues are (1) measurement of the interstellar low–energy cosmic ray flux, (2) mapping the large–scale structure of the warm, diffuse, ionized medium and (3) identification of the phase responsible for the ISS phenomena.

131.039 Penetration of charged interstellar dust into the solar system.

M. K. Wallis.

Mon. Not. R. Astron. Soc., Vol. 227, No. 2, p. 331 – 339 (1987).

The electric and magnetic fields associated with the solar wind prevent interstellar submicron dust from reaching the neighbourhood of the Earth, primarily because the charged dust grains have large gyroradii and move down the electric field. During passage through an interstellar cloud, the degree of heliosphere contraction (solar wind restriction) depends on the helium gas component, as accelerated in the Sun's gravitational field. For cloud densities exceeding 100 cm^{-3}, the dust component reaches the Earth with little depletion, at least on the upstream part of the Earth's orbit. Formulae are developed for grain densities in the Sun's gravitationally focused wake, as limited by the electromagnetic scattering.

131.040 Energetic N^+ ions in the interstellar medium.

J. H. Yee, S. Lepp, A. Dalgarno.

Mon. Not. R. Astron. Soc., Vol. 227, No. 2, p. 461 – 466 (1987).

Energetic N^+ ions are produced in interstellar clouds by reactions of He^+ ions and N_2 with sufficient energy to overcome the exoergicity of the reaction of N^+ with H_2 and thereby from NH^+ ions. The reacting N^+ ions also slow down by elastic collisions with the ambient H_2 molecules. The equilibrium energy distributions of the N^+ ions in a gas with temperature between 10 and 70K are derived and the effective reaction rate coefficients are calculated. At 30K, the rate coefficient is 1.6×10^{-10}cm^{3}s^{-1}.

131.041 Erratum: "Infrared polarimetry of the reflection nebula near L1551 IRS 5" [Mon. Not. R. Astron. Soc., Vol. 223, No. 1, p. 7P – 11P (1986)].

T. Nagata, T. Yamashita, S. Sato, H. Suzuki, J. H. Hough, R. Garden, I. Gatley.

Mon. Not. R. Astron. Soc., Vol. 227, No. 2, p. 543 (1987). See Abstr. 42.131.169.

131.042 Magnetic fields and star formation: evidence from imaging polarimetry of the Serpens reflection nebula.

R. F. Warren-Smith, P. W. Draper, S. M. Scarrott.

Mon. Not. R. Astron. Soc., Vol. 227, No. 3, p. 749 – 771 (1987).

The authors present CCD polarization mapping and low light level imaging of the Serpens Nebula; an optical bipolar nebula associated with a pre–main–sequence object in Serpens. The authors conduct a detailed analysis of the results to obtain as much information as possible about the structure of the surrounding cloud and its magnetic field configuration. This information is then compared with theoretical models of star formation.

131.043 Theoretical studies of interstellar molecular shocks – VI. The formation of molecules containing two or three carbon atoms.

G. Pineau des Forêts, D. R. Flower, T. W. Hartquist, T. J. Millar.

Mon. Not. R. Astron. Soc., Vol. 227, No. 4, p. 993 – 1011 (1987).

The authors discuss the formation of 'heavy' carbon–bearing molecules in MHD shocks propagating in interstellar clouds. It is shown that substantial column densities of species such as C_2, C_2H, C_2H_2, C_3, C_3H and C_3H_2 can be produced in media where the transverse magnetic field strength is sufficiently large and the ultraviolet radiation field sufficiently attenuated to give rise to extended acceleration zones ('magnetic precursors'). They suggest that the broad emission lines of cyclic C_3H_2, observed by Matthews & Irvine (1985) towards W49, may be formed in a MHD shock.

131.044 IRAS observations of southern molecular clouds.

T. B. H. Kuiper, J. B. Whiteoak, J. W. Fowler, W. Rice.

Mon. Not. R. Astron. Soc., Vol. 227, No. 4, p. 1013 – 1020 (1987).

A report is given of a project to use IRAS Band 3 (60 μm) and Band 4 (100 μm) observations to investigate the far–infrared properties of southern galactic molecular clouds. A method by which dust temperature and total gas column density can be estimated is presented. Results are tabulated for 65 prominent southern far–infrared sources. The dust temperatures are closely grouped between 30 and 50K, while the column densities range between 2×10^{20} and 10^{22}cm^{-2}. Maps of dust temperature and gas column density have been generated for two fields containing far–infrared sources to illustrate the effectiveness of this form of presentation.

131.045 An Hα emission–line survey of the ϱ Ophiuchi dark cloud complex.

B. A. Wilking, R. D. Schwartz, J. H. Blackwell.

Astron. J., Vol. 94, No. 1, p. 106 – 110 (1987). With plates 14 – 19.

An analysis of two Hα objective–prism plates covering about 40 square degrees and including most of the nearby ϱ Ophiuchi cloud complex is presented. These data provide a high–resolution, completely sampled survey of stars in the T Tauri phase of pre–main–sequence evolution in the lower–extinction regions of the dark cloud. Detailed finder charts and accurate positions for 86 sources of definite or probable Hα emission are presented, making this survey useful for future studies of global star formations in the ϱ Oph clouds. In addition, the authors have found the first Herbig–Haro object candidates in this area. The correspondence of the emission–line stars with the densest molecular gas and known T Tauri stars, X–ray sources, and *IRAS* point sources suggest that most of the stars in the sample are indeed young stellar objects.

131.046 The galactic foreground reddening in the direction of the nearby Triangulum galaxy M33.

S. B. Johnson, M. D. Joner.

Astron. J., Vol. 94, No. 2, p. 324 – 339 (1987).

Observations of 314 stars with the $uvby\beta$ photometric system are used to establish the reddening in the direction of the nearby Triangulum galaxy M33. Color excesses are obtained for 151 A and F type stars in a 1° radius centered around the galaxy. A color excess of $E(b-y) = 0.057$ mag ($E(B-V) = 0.077$ mag) is obtained for a distance modulus $m-M = 10$. A secondary result of this investigation is the discovery of six suspected variable stars, as well as ten observations of V Triangulum, a β Lyrae eclipsing variable.

131.047 The fragmentation of molecular clouds. II. Gravitational stability of low–mass molecular cloud cores.

J.–P. Chièze, G. Pineau des Forêts.

Astron. Astrophys., Vol. 183, No. 1, p. 98 – 108 (1987).

In this study, the authors restrict themselves to low mass molecular cloud cores found in Taurus–like clouds unaltered by bursts of OB star formation. So far, complications such as magnetic field or turbulence are ignored. The main result of this study is to provide the basis of an initial mass function of individual clumps on the verge of star formation. High mass pre star–forming cores are predicted to be found at the edges of 200 – 500 $M_\odot$ clumps, while the inner regions embody the typical cores with $M \leqslant 1\ M_\odot$. Growing observational evidences for low mass star clusters in evolved clouds (with present OB star formation) give further impulse to an evolutionary sketch for molecular clouds and time stretching bimodal star formation.

131.048   How abundant are complex interstellar molecules?
T. J. Millar, C. M. Leung, E. Herbst.
Astron. Astrophys., Vol. 183, No. 1, p. 109 – 117 (1987).

In order to resolve discrepancies between the predictions of complex molecule abundances in the dense interstellar cloud chemical models of Herbst and Leung and of Millar and Nejad, the authors have performed a number of calculations by augmenting the smaller Millar and Nejad model with reactions thought to be critical in determining the abundances of complex species, but not yet studied in the laboratory. The augmented model of Millar and Nejad is found to be in excellent agreement with the model of Herbst and Leung. The authors suggest some laboratory studies which would remove much of the current uncertainty in gas phase predictions of molecular abundances in dense interstellar clouds.

131.049   The IRAS cirrus and the diffuse ultraviolet background.
P. Jakobsen, J. S. de Vries, F. Paresce.
Astron. Astrophys., Vol. 183, No. 2, p. 335 – 340 (1987).

The authors show that a correlation exists at high and intermediate galactic latitudes between the diffuse infrared background intensity at 100 μm as measured by IRAS and the diffuse background intensity in the far ultraviolet. The slope of this correlation is in reasonable quantitative agreement with the expected ratio between scattered and thermal emission from high latitude low albedo dust grains illuminated and heated by the integrated interstellar radiation field. The existence of this correlation thus provides strong observational support for both the dust back–scattering explanation for the known correlation between diffuse UV background and line–of–sight 21 cm column density, and the closely related radiatively heated grain emission model for the IRAS cirrus phenomenon.

131.050   Atomic and molecular diagnostics of the interstellar medium.
E. Roueff.
Phys. Scr., Vol. 36, No. 2, p. 319 – 322 (1987).

Ever since molecular species have been discovered in space in the 30's and early 40's by the optical identification of CH, CH$^+$ and CN in absorption towards nearby hot stars, the question of molecule formation has accompanied the observational efforts. The purpose of this paper is to point out presently existing observational constraints and the limits they may cast on our knowledge of the interstellar medium. The need for reliable atomic and molecular data will be emphasized with some specific examples.

131.051   The structure of time–dependent interstellar shocks and grain destruction in the interstellar medium.
C. F. McKee, D. J. Hollenbach, C. G. Seab, A. G. G. M. Tielens.
Astrophys. J., Vol. 318, No. 2, p. 674 – 701 (1987).

Interstellar shocks are the dominant destruction mechanism for refractory interstellar dust. This paper presents several improvements in the theory of the structure of such shocks, focusing on J–shocks. The most significant improvement is the generalization of the theory to weakly time–dependent shocks, in which the driving pressure may vary on a time scale that is long compared to the sound travel time across the shock. Second, the grains are treated as a separate two–dimensional fluid which can exchange mass and energy with the gas. Third, the authors develop an analytic theory for the interaction of a blast wave with an interstellar cloud. Finally, the authors present a more accurate treatment of the charge on the grains gyrating in the shock, including photoelectric emission, secondary electron emission, field emission, and ion and electron sticking probabilities.

131.052   Hydrodynamical processes in the Draco molecular cloud.
S. F. Odenwald, L. J. Rickard.
Astrophys. J., Vol. 318, No. 2, p. 702 – 711 (1987). With plate 5 – 7.

IRAS 100 μm images of the Draco cloud show several cloud components with cometary plumes of material extending 7–10 pc. The structure is consistent with low Reynolds number hydrodynamics ($R < 50$) and is suggestive of a molecular cloud shedding material in a plume as it falls onto the galactic plane from the halo region. A subsequent study of the 100 μm images from the entire IRAS survey has revealed a total of 14 additional cometlike objects with $|b^{II}| > 15°$.

131.053   Temperature and density structure of the collapsing core of G 10.6–0.4.
E. R. Keto, P. T. P. Ho, A. D. Haschick.
Astrophys. J., Vol. 318, No. 2, p. 712 – 728 (1987).

The authors report on high–resolution aperture synthesis observations of the molecular gas surrounding the ultracompact H II region G 10.6–0.4 in the $(J, K) = (1, 1)$ and $(J, K) = (3, 3)$ inversion lines of NH$_3$. Previous observations have suggested that the molecular gas around this source is gravitationally collapsing onto the central star or stars. The new data improve the identifications of the dynamical features. Maps of the NH$_3$ rotational temperature and column density across the collapsing region show a hot, centrally condensed molecular core 0.1 pc in diameter with $n(H_2) = 4 \times 10^6 cm^{-3}$. The H II region embedded in the condensed core is heating the surrounding gas. The data indicate that a rotational motion is superposed on the infall velocity of the cloud. The velocity gradients are consistent with Keplerian rotation. The rotational velocity at 0.05 pc is about the same as the infall velocity at this radius indicating that the infalling gas may be spiraling in sharply toward the H II region.

131.054   OH outflows in star–forming regions.
I. F. Mirabel, A. Ruiz, L. F. Rodríguez, J. Cantó.
Astrophys. J., Vol. 318, No. 2, p. 729 – 737 (1987).

The results from a survey for high–velocity OH in molecular outflows in star–forming regions are reported. High–velocity OH was detected in absorption in nine of these regions. The outflows show similar anisotropic angular distribution as the redshifted and blueshifted CO. The OH transitions are markedly subthermal ($T_{ex} < 3.8K$). The absorbing OH appears to trace gas with higher velocities and lower densities than does the CO and, in some cases, provides information on the structure of the outflows at larger distances from the central source.

131.055   Fluorescent molecular hydrogen emission from the reflection nebula NGC 2023.
I. Gatley, T. Hasegawa, H. Suzuki, R. Garden, P. Brand, J. Lightfoot, W. Glencross, H. Okuda, T. Nagata.
Astrophys. J., Lett. Ed., Vol. 318, No. 2, p. L73 – L76 (1987). With plates L1 – L3.

The near–infrared spectrum of NGC 2023 is shown to be rich with emission lines of vibrationally excited molecular hydrogen. The relative intensities of these lines are in excellent agreement with theoretical predictions for ultraviolet pumped fluorescence. A map of NGC 2023 in the $v = 1$–0 $S(1)$ line of H$_2$ shows that the fluorescent emission originates from a thin shell, of radius ~ 0.3 pc, which surrounds the central star of the reflection nebula. A map of NGC 2023 in $^{12}CO(J = 1$–$0)$ reveals a bright rim of emission lying immediately outside the brightest portion of the fluorescent shell.

131.056   Level population and para/ortho ratio of fluorescent H$_2$ in NGC 2023.
T. Hasegawa, I. Gatley, R. P. Garden, P. W. J. L. Brand, M. Ohishi, M. Hayashi, N. Kaifu.
Astrophys. J., Lett. Ed., Vol. 318, No. 2, p. L77 – L80 (1987).

Observations of vibrationally excited H$_2$ in 10 lines at 2.03 – 2.39 μm toward the reflection nebula NGC 2023 are presented and compared with the corresponding data in Orion– KL. Level populations of H$_2$ in NGC 2023 are characterized by a high vibrational temperature ($T_v > 3600K$) and a low rotational temperature ($T_r \approx 900$ – $1500K$). This confirms the identification of the H$_2$ emission as due to UV excitation followed by fluorescence. The observed para/ortho ratio of H$_2$ is consistently larger than 1:3 and is estimated to be in the range 1 : (1.4 – 2.0).

131.057 Para/ortho abundance ratio of molecular hydrogen in NGC 2023.
K. Takayanagi, K. Sakimoto, K. Onda.
Astrophys. J., Lett. Ed., Vol. 318, No. 2, p. L81 – L84 (1987).

Recent infrared emission–line observations by Hasegawa et al. (131.056) clearly indicate that the para/ortho abundance ratio of H_2 in the reflection nebulae NGC 2023 is quite different from 1:3. A simplified steady state problem is solved to determine the vibrational–rotational level population. In addition to radiative processes, the H_2 formation on grains and the gas–phase para/ortho conversion processes are taken into account. A good fit to the observed line intensity ratio is obtained.

131.058 Electron density of local interstellar medium, based on the Voyager heliospheric–shock observations.
V. B. Baranov.
Sov. Astron. Lett., Vol. 12, No. 5, p. 300 – 302 (1987). English translation of 42.131.038.

131.059 Instability of interaction network for interstellar gas and interstellar diffusive energy in the shear field.
M. Fujimoto, T. Mizuno.
Publ. Astron. Soc. Jpn., Vol. 39, No. 4, p. 605 – 618 (1987).

A model network for interaction between interstellar gas and interstellar diffusive energy is considered in the shear field. Local linearized equations are derived around the equilibrium states which are realized when no shear field exists. A wavy perturbation is followed by employing the WKB method. It is concluded that the shear field brings about various unstable waves depending on their configuration. A great variety of observed dark and luminous pattern in spiral galaxies could be understood as related to these waves.

131.060 HCO⁺ survey of unassociated compact molecular clouds in the *IRAS* Point Source Catalog.
P. J. Richards, L. T. Little, M. Toriseva, B. D. Heaton.
Mon. Not. R. Astron. Soc., Vol. 228, No. 1, p. 43 – 54 (1987).

Criteria for discriminating between compact molecular clouds (CMCs) and other sources in the *IRAS* catalogue are established by a colour comparison of such sources with infrared sources associated with non–stellar masers. Using these criteria, candidate CMCs were selected which are unassociated (i.e. not given an "association" in the *IRAS* catalogue) with known H II regions or molecular clouds. The candidate sources were searched for HCO^+ emission to test whether they are indeed dense molecular clouds; 24 detections out of 34 were obtained, thus verifying the selection procedure.

131.061 Chemical evidence for ambipolar diffusion in interstellar shocks: a suggested observational test.
G. Pineau des Forêts, D. R. Flower.
Mon. Not. R. Astron. Soc., Vol. 228, No. 1, p. 1P – 4P (1987).

Ambipolar diffusion (ion–neutral velocity drift) is a process which has important consequences for the structure of shocks in the interstellar gas and for the chemical processes occurring in such shocks. The authors suggest a chemical test of ambipolar diffusion which should also yield information on the local magnetic field strength.

131.062 An optical counterpart of the infrared bipolar nebula NGC 6334 IRS V.
R. D. Wolstencroft, S. M. Scarrott, R. F. Warren–Smith.
Mon. Not. R. Astron. Soc., Vol. 228, No. 3, p. 805 – 809 (1987).

The authors have discovered a faint optical counterpart of the infrared bipolar nebula NGC 6334–V. It coincides with the infrared source IRS V–2 and is strongly polarized (38 per cent). An optical polarization map in the immediate vicinity of the source indicates that IRS V–3 is the illuminating source. This, the most luminous of the four sources lying along the ridge or axis of the nebula, is very probably the origin of the bipolar outflow.

131.063 An optical polarization study of the Chamaeleon IR nebula.
S. M. Scarrott, R. F. Warren–Smith, R. D. Wolstencroft, H. Zinnecker.
Mon. Not. R. Astron. Soc., Vol. 228, No. 3, p. 827 – 831 (1987).

An optical linear polarization map shows that the Chamaeleon infrared nebula is a bipolar reflection nebula illuminated by a central object located at the position of a known IR source. The source is surrounded by an extensive disc of gas and dust and the data indicate that a magnetic field is concentrated in the disc and remains detectable throughout the lobes of the nebula.

131.064 Three new H_2O masers located near Herbig–Haro–like nebulosities.
A. L. Gyulbudaghian (*A. L. Gyul'budagyan*), L. F. Rodríguez, E. Mendoza–Torres.
Rev. Mex. Astron. Astrofis., Vol. 15, No. 1, p. 53 – 57 (1987).

The authors surveyed 77 Herbig–Haro–like (HHL) nebulosities found at Byurakan for water–vapor maser emission. This list includes the 37 GGD objects reported by Gyulbudaghian, Glushkov and Denisyuk (1978). The authors detected 11 masers, of which 3 are new sources. These new sources are HHL5 (= G1–4), HHL50 (= GGD20) and HHL73 (= G2–11). HHL5 is associated with the infrared source CRL437. There is a strong correlation in HHL objects between the association with H_2O masers and bright far–infrared sources.

131.065 Behind the H II region Sharpless 217: the envelope of the diffuse molecular cloud at G159.1 + 3.3.
J. P. Vallée.
Astron. J., Vol. 94, No. 3, p. 679 – 683 (1987).

Observations of the methylidyne (CH) molecule were made toward the envelope at G159.1 + 3.3 of the molecular cloud located behind the H II region Sharpless 217, at (1950) R.A. = $04^h55^m00^s$, Dec. = $+47°56'$. Some of the envelope parameters derived from these CH line observations include a column density of CH of $2 \times 10^{13} cm^{-2}$, a line optical depth of –0.001, and a microturbulent velocity of 3.4 km s^{-1}. These observed data are also consistent with a blister–type model for the H II region, roughly facing the earth, coming off the envelope of the molecular cloud.

131.066 Star formation in molecular clouds: observation and theory.
F. H. Shu, F. C. Adams, S. Lizano.
Annu. Rev. Astron. Astrophys., Vol. 25, p. 23 – 81 (1987). – See Abstr. 003.003.

Star formation in galaxies is a complex process, which spans roughly 12 orders of magnitude in both mass and linear scale (10^{11} to $10^{-1} M_\odot$, 10^{23} to 10^{11} cm) and involves many diverse physical phenomena. This review concentrates on those aspects of the problem that occur on the scale of a giant molecular cloud (roughly, $10^6 M_\odot$ and 10^{20} cm) and smaller. The authors concern themselves with the problems of present–day star formation, as posed by observations of starforming regions in our own Galaxy and in other spiral galaxies.

131.067 The local interstellar medium.
D. P. Cox, R. J. Reynolds.
Annu. Rev. Astron. Astrophys., Vol. 25, p. 303 – 344 (1987). – See Abstr. 003.003.

Contents: Glossy overview. Questions. Properties of the Local Bubble. Properties of the local clouds. Cloud existence in the Local Bubble.

131.068 Structure of the OH maser source in W33A.
V. E. Velikhov, D. Graham.
Inst. Kosm. Issled. Akad Nauk SSSR, Prepr., No. 1204, 27 pp. (1987). In Russian. Abstr. in Ref. Zh., 51. Astron., 7.51.668 (1987).

131.069 Models of dark clouds and globules.
L. N. Arshutkin, I. G. Kolesnik.
Kinematika Fiz. Nebesn. Tel, Tom 3, No. 4, p. 40 – 44 (1987). In Russian. English translation in Kinematics Phys. Celest. Bodies.

Isothermal models of dark clouds supported in equilibrium by turbulent pressure have been constructed. A power–law relation $v_t \propto r^v$ is proposed. Global parameters of the models agree well with the observations. For $v = 0.5$ the models agree with observations best of all. It is shown that the external pressure for the models is to be within $10^4 - 10^5 \text{K cm}^{-3}$. This model is valid for clouds with mass up to $\approx 100\, M_\odot$.

131.070 Structure and physical parameters of Bok globules.
L. N. Arshutkin.
Kinematika Fiz. Nebesn. Tel, Tom 3, No. 4, p. 45 – 51 (1987). In Russian. English translation in Kinematics Phys. Celest. Bodies.

Physical structure of globules supported in equilibrium by the turbulence pressure is investigated. The isothermal models are calculated for four clouds: B163, B335, L1407, L1551. It is shown that the density distribution outward from the centre is roughly proportional to r^{-2}, but at the inner core the density is nearly constant. In the central regions the volume density is $\sim 10^5 \text{cm}^{-3}$. The formula for cloud mass computation through the density at the edge of the cloud is obtained. The $^{13}\text{CO}/\text{H}_2$ ratios (10^{-6} on the average) are determined by comparing the model column densities of H_2 and the observed column densities of ^{13}CO. The distances to the globules are refined by fitting the model parameters with observations.

131.071 Steady models of radiatively modified conductively driven evaporation from interstellar clouds.
H. Böhringer, T. W. Hartquist.
Mon. Not. R. Astron. Soc., Vol. 228, No. 4, p. 915 – 931 (1987).

The non–equilibrium ionization structures and radiative loss rates in and the dynamical structures of conductive interfaces between spherical interstellar clouds and the hot gas in which they are embedded were calculated self–consistently. Mass loss rates and the column densities of several observable ions were calculated for interfaces between clouds having a range of radii and being embedded in media with temperatures of 5×10^5 and 10^6K. The column densities of O^{5+}, N^{4+}, and C^{3+} in the interfaces are in general compatible with existing ultraviolet absorption data. If the extreme ultraviolet background emission originates in conductive interfaces around nearby clouds, the pressure of the local interstellar medium must be about $10^{-11} \text{erg cm}^{-3}$; the high pressure conductive interfaces do not produce the observed far ultraviolet emission lines.

131.072 The formation of discrete high velocity molecular features.
T. W. Hartquist, J. E. Dyson.
Mon. Not. R. Astron. Soc., Vol. 228, No. 4, p. 957 – 961 (1987).

The authors argue that clumps embedded in a flowing diffuse medium will be dissipated before ram pressure accelerates them substantially. Molecular hydrogen can be accelerated to high speeds by passing through a slow shock leading a shell at the edge of a wind–driven bubble if the density in the ambient medium drops rapidly enough to allow the shell to accelerate subsequently. The shell will be subject to the Rayleigh–Taylor instability which will drive transonic turbulence. The existence of high velocity discrete features in and the magnitude of the linewidth of the H_2 emission from CRL 618 are explained with this acceleration mechanism. High velocity water masers may be formed in a similar fashion, but not Herbig–Haro objects.

131.073 Spectroscopy of the Kleinmann–Low nebula: scattering in a solid absorption band.
R. F. Knacke, S. M. McCorkle.
Astron. J., Vol. 94, No. 4, p. 972 – 976 (1987).

The authors report spectroscopic observations (2.4 – 3.6 μm) of BN, IRc 2, 3, and 4, and three scattering locations in the KL reflection nebula. A previous report (Knacke et al., 1982) of a 2.97 μm spectral feature in the BN object is not confirmed in the new data. The 2.97 μm feature is observed in sources in the

KL nebula, and the spectrum is distorted by a nearby hydrogen line. All the spectra are dominated by absorption along the radiation path, making scattering effects difficult to separate. Scattering could broaden the 3.1 μm interstellar–ice feature, but the effects appear to be small. Except for a long–wavelength wing, the spectra can be modeled reasonably well with core–mantle, silicate–water–ice grains. The wing position and intensity indicate bands of C–H groups or ammonia–ice mixtures.

131.074 Distribution of hydroxyl clouds in the Galaxy.
L. V. Yurevich.
Inst. Teor. Fiz. Akad. Nauk USSR, Prepr., No. 25 R, 23 pp. (1987). In Russian. Abstr. in Ref. Zh., 51. Astron., 8.51.733 (1987).

131.075 Gas content and the intensity of star formation in cluster galaxies.
A. V. Zasov.
Pis'ma Astron. Zh., Tom 13, No. 9, p. 757 – 763 (1987). In Russian. English translation in Sov. Astron. Lett., Vol. 13.

A depletion of H I observed in many disk galaxies in clusters may account for a more intense and efficient star formation process in these galaxies; this is enhanced by compression of interstellar gas layers under the pressure of hot surrounding intergalactic gas in the cluster.

131.076 Dynamische Bedingungen in Molekülwolken.
T. Henning, B. Stecklum.
Sterne Weltraum, 26. Jahrg., Nr. 11, p. 624 – 626, 628 (1987).

131.077 Protostellar formation in rotating interstellar clouds. VI. Nonuniform initial conditions.
A. P. Boss.
Astrophys. J., Vol. 319, No. 1, p. 149 – 161 (1987).

The collapse and fragmentation of rotating protostellar clouds is explored, starting from nonuniform density and nonuniform rotation initial conditions. The numerical models include three–dimensional hydrodynamics, self–gravity, detailed equations of state, and radiative transfer in the Eddington approximation. The calculations show that: (1) whether binary fragmentation occurs during the first dynamic collapse phase depends strongly on the initial density profile; (2) clouds with exponential density profiles are only somewhat more resistant to fragmentation than uniform density clouds; (3) previous scenarios about hierarchical fragmentation are still possible, although the lower bound on the minimum protostellar mass for Population I stars should be increased to $\sim 0.02\, M_\odot$.

131.078 1300 micron continuum and C^{18}O line mapping of the giant molecular cloud cores in Orion, W49, and W51.
F. P. Schloerb, R. L. Snell, P. R. Schwartz.
Astrophys. J., Vol. 319, No. 1, p. 426 – 435 (1987). = Five College Radio Astron. Obs. Contrib. No. 625.

The authors present observations of the 1300 μm continuum emission and the C^{18}O spectral–line emission from three well–studied giant molecular cloud cores: Orion, W49, and W51. The observations provide a means to examine the consistency of these two methods to trace the column density structure of molecular clouds. There is a good general correlation between the 1300 μm continuum, which traces the column density of dust, and the $\text{C}^{18}\text{O}\, J = 2 \rightarrow 1$ line emission, which traces the column density of molecular gas, when the effects of source temperature are taken into consideration. Moreover, nominal values for the gas and dust abundances and the dust properties reproduce the observed continuum–to–line ratios.

131.079 CCD observations of diffuse interstellar bands in reflection nebulae.
K. Josafatsson, T. P. Snow.
Astrophys. J., Vol. 319, No. 1, p. 436 – 445 (1987).

The authors have obtained CCD spectra of 59 stars in order to measure diffuse band strengths between 5690 and 5870 Å. Most (37) of the stars are in reflection nebulae, and the rest are relatively unreddened comparison stars. Six diffuse bands are measurable in these data: 5705, 5778, 5780, 5797, 5844, and 5849 Å. This paper studies the correlation between the bands in pairs, as well as with $E(B-V)$ and the 2175 Å extinction bump. As to the nature of the absorbers, the authors favor a collection of some large, perhaps previously undetected molecules such as the polycyclic aromatic hydrocarbons that are believed to be the source of the IR emission bands at micron wavelengths.

131.080   CO maps of the OMC–1 outflow.
C. R. Masson, K. Y. Lo, T. G. Phillips, A. I. Sargent, N. Z. Scoville, D. P. Woody.
Astrophys. J., Vol. 319, No. 1, p. 446 – 455 (1987).
CO and ^{13}CO emission from OMC–1 have been mapped with a resolution of $6'' \times 9''$. These maps of the hot core and low-velocity flow suggest that essentially all the material near IRc 2 is taking part in a turbulent outflow, the densest clumps of which form the hot core. The high–velocity flow is weakly bipolar and is offset to the north of IRc 2. Analysis of energy sources for the outflow suggests that it must be driven close to the surface of IRc 2, rather than by an orbiting disk.

131.081   A study of the ground–state hydroxyl maser emission associated with 11 regions of star formation.
R. A. Gaume, R. L. Mutel.
Astrophys. J., Suppl. Ser., Vol. 65, No. 2, p. 193 – 253 (1987).
The authors present VLA maps of the OH and 15 GHz continuum emission associated with 11 star–formation regions. They find 20% of the maser clusters not closely associated with continuum sources. The overall distribution of maser features that are associated with H II emission is strongly peaked near the edge of the regions, but the probability of observing OH masers is highest toward their center. Five of the H II regions are of the cometary class. The masers are projected onto the "upstream" edge of these regions.

131.082   On the induced star formation.
Yu. A. Shchekinov.
Pis'ma Astron. Zh., Tom 13, No. 10, p. 862 – 867 (1987). In Russian. English translation in Sov. Astron. Lett., Vol. 13.
The process of induced star formation is described by diffusion–reaction like equation of flame propagation. The possibility of existence of a steady–state star formation wave is shown and the velocity of the wave is found. The characteristics of star formation in our Galaxy are used to estimate this velocity: $u \lesssim 8$ km/s.

131.083   On the intermediate mass spectrum asymptotics in the system of coagulating particles.
G. V. Pechernikova.
Kinematika Fiz. Nebesn. Tel, Tom 3, No. 5, p. 85 – 87 (1987). In Russian. English translation in Kinematics Phys. Celest. Bodies.
The asymptotic power solution $n(m) \propto m^{-q}$ of the coagulation equation with the kernel $A \propto (m^{\alpha} + m'^{\alpha})(m^{\beta} + m'^{\beta})$ is obtained. The exponent q is a function of the parameters α and β.

131.084   Large–scale radio brightness distribution of the ionized gas in the Canis Major OB1 association.
T. B. Pyatunina, Yu. M. Taraskin.
Sov. Astron., Vol. 30, No. 6, p. 648 – 655 (1987). English translation of 42.131.328.

131.085   Two phases of the interstellar medium: the electron density distribution in the galactic disk.
N. Ya. Shapirovskaya, A. A. Bocharov.
Sov. Astron., Vol. 30, No. 6, p. 655 – 658 (1987). English translation of 42.131.329.

131.086   Interstellar molecules.
D. Smith.
Philos. Trans. R. Soc. London, Ser. A, Vol. 323, No. 1572, p. 269 – 286 (1987). – See Abstr. 012.026.
A current list of interstellar molecules is given together with a few other molecular species that have so far been detected only in circumstellar shells. Also listed are those interstellar species that contain rare isotopes of several elements. The gas phase ion chemistry is outlined via which the observed molecules are synthesized, and the process by which enrichment of the rare isotopes occurs in some interstellar molecules is described. Reference is also made briefly to some very recent work interstellar ion chemistry . A list of the atomic and molecular species that have been detected in cometary atmospheres is given and attention is drawn to the similarities and differences between interstellar and cometary molecules.

131.087   Formation and evolution of the largest cloud complexes in spiral galaxies.
B. G. Elmegreen.
Physical processes in interstellar clouds, p. 1 – 12 (1987). – See Abstr. 012.027.
Atomic and molecular cloud complexes containing $10^7 M_\odot$ may form by gravitational instabilities in the ambient galactic gas. A recent calculation gives the observed masses, separations and densities for these clouds, and the estimated formation time is reasonably short. Destabilization from magnetism and spiral density waves are discussed. The Parker instability apparently plays a secondary role in giant cloud formation. Clouds that contain much less than $10^7 M_\odot$ may form by magnetic agglomeration of smaller clouds or by fragmentation of larger clouds.

131.088   Diffuse interstellar gas.
C. Heiles, S. R. Kulkarni.
Physical processes in interstellar clouds, p. 13 – 33 (1987). – See Abstr. 012.027.

131.089   The structure of molecular clouds.
L. Blitz.
Physical processes in interstellar clouds, p. 35 – 58 (1987). – See Abstr. 012.027.
The structure and properties of small, local molecular clouds and giant molecular clouds as typified by the Rosette Molecular Cloud are reviewed.

131.090   Structure and physics of cool giant molecular complexes.
E. Falgarone, M. Pérault.
Physical processes in interstellar clouds, p. 59 – 73 (1987). – See Abstr. 012.027.
The interstellar medium is observed to be fragmented at all scales ranging from that of the supercloud complexes to that of the protostellar cores, forming a hierarchy which is not self-similar. In the solar neighbourhood, the self-similarity breaks down at a salient scale which we call that of "clouds": their mass is several 100 $M_\odot$, their size a few parsecs. The clouds, building blocks of the hierarchy, are supported against gravity by supersonic but subalfvénic turbulence. The authors show that this turbulence can be fed, via magnetohydrodynamic waves, over several 10^7 years, by a pumping of the orbital kinetic energy of the other clouds within the same complex. They also show that local gravitational instabilities may develop within the gravitationally stable clouds and eventually form dense protostellar cores.

131.091   Supernovae and the interstellar medium.
R. McCray.
Physical processes in interstellar clouds, p. 95 – 104 (1987). – See Abstr. 012.027.
Repeated supernovae from an OB association will, in a few $\times 10^7$ yr, create a cavity of coronal gas in the interstellar medium, with radius > 100 pc, surrounded by a dense expanding shell of cool interstellar gas. Such a cavity will likely burst through the gas layer of a disk galaxy. Such holes and "supershells" have been observed in optical and H I radio emission maps of the Milky Way and other nearby galaxies. The gas swept up in the supershell is likely to become gravitationally unstable, providing a mechanism for propagating star formation that may be particularly effective in irregular galaxies.

131.092   Energy dissipation in magnetic cloud complexes.
B. G. Elmegreen.
Physical processes in interstellar clouds, p. 105 – 114 (1987). – See Abstr. 012.027.

A previous calculation of the energy dissipation rate for magnetic cloud complexes is summarized. Emphasis is placed on the possible importance of magnetic entanglements and purely magnetic interactions between clouds. Significant dissipation should occur in several internal crossing times. Dissipation is dominated by cloud–cloud collisions and Alfvén wave propagation away from the cloud. Cloud complexes much older than several crossing times should have a core–halo structure, or, possibly, an R^{-2} density profile, because the kinetic energy loss should be associated with an increased gravitational self–binding. Magnetically–constrained turbulence is also discussed.

131.093   The large–scale motion of the ISM and the interaction with the system of stars.
W. H. Kegel.
Physical processes in interstellar clouds, p. 115 – 124 (1987). – See Abstr. 012.027.

An equation is derived describing the large–scale motion of the interstellar gas in the presence of small–scale fluctuations, which are considered to be due to the gravitational interactions of the ISM with the system of stars. The author discusses in some detail the friction term describing the momentum transfer between the interstellar gas and the system of stars due to small–scale fluctuations. This term is non–zero only if the fluctuation field is anisotropic and if there are phase differences between the density fluctuations of the gas and the potential fluctuations due to the stars. The importance of dissipative processes and of chemical reactions in this context is pointed out.

131.094   Fluctuations in the ISM due to the gravitational interaction with the system of stars.
W. H. Kegel.
Physical processes in interstellar clouds, p. 125 – 136 (1987). – See Abstr. 012.027.

The gravitational interaction between stars and the interstellar gas leads to the excitation of velocity and density fluctuations which may be substantial even for stable modes. Expressions for these fluctuations are derived in the framework of a linearized theory. The relevance of the results for a number of astrophysical problems is pointed out.

131.095   Dense cores in dark clouds.
G. A. Fuller, P. C. Myers.
Physical processes in interstellar clouds, p. 137 – 160 (1987). – See Abstr. 012.027.

The properties of cores in nearby dark clouds are reviewed and a summary of the results of NH_3 observations is presented. Map sizes and line widths of NH_3 line emission are similar to $C^{18}O$ and HC_3N but are smaller than those of CS, even though $n_{critical}(CS) > n_{critical}(NH_3)$. Cores with stars have larger line widths and velocity gradients than cores without stars. There is evidence for interaction between stars and cores. It is shown that cores evolve on timescales comparable to the free fall time of $\sim 2 \times 10^5$ years. Sources detected by IRAS in cores have indirect evidence for circumstellar disks. Some recent theoretical models of the spectra of these objects are described and an extension of these models to account for the infrared spectra of T Tauri stars is outlined.

131.096   Molecular cloud temperature and density determinations and what they teach us.
C. M. Walmsley.
Physical processes in interstellar clouds, p. 161 – 171 (1987). – See Abstr. 012.027.

A review is given of current methods for determining temperature and density in molecular clouds. In the light of the temperature estimates, a brief summary is given of heating and cooling processes. Possible explanations of the high (>60K) temperatures measured in the galactic centre clouds are also given. The advantages and drawbacks of current large velocity gradient density estimates are discussed.

131.097   Formation and heating of molecular cloud cores.
S. Lizano, F. H. Shu.
Physical processes in interstellar clouds, p. 173 – 193 (1987). – See Abstr. 012.027.

The authors review the problem of the structure of molecular clouds and its relation to star formation. They discuss why magnetic fields represent the most reasonable candidate for supporting molecular clouds. This picture naturally leads to a theory for the origin of molecular cloud cores, and a conception of "bimodal star formation ", whereby the modes of formation of high–mass stars and low–mass stars are viewed as being fundamentally different on some mechanistic level. Of the two modes of star formation, the authors examine the low–mass mode in greater detail because of the greater completeness of the developed theory.

131.098   Fragmentation and turbulence in molecular clouds.
L. G. Stenholm.
Physical processes in interstellar clouds, p. 195 – 203 (1987). – See Abstr. 012.027.

Observational data on cloud fragmentation is reviewed. The author concludes that 10% of the cloud mass is contained in fragments. The fluctuation spectrum of the velocity field suggests the presence of some generating mechanism. Many have been suggested, but no one has a clear advantage. Detailed radiative transfer modelling can be used to discriminate between the suggested mechanisms.

131.099   Observational constraints on cloud physics.
L. G. Stenholm.
Physical processes in interstellar clouds, p. 205 – 217 (1987). – See Abstr. 012.027.

A short critical review is given on the physical conditions in molecular clouds. A reasonable description of dark clouds can be summarized as follows: (1) They are in viral equilibrium. (2) The temperature is higher at the edge of the clouds. (3) The density structure is close to $r^{-1.5}$. (4) Molecular abundances are a function of the position within the cloud. (5) Magnetic fields play a major role. (6). The velocity field is due to waves or some other type of small scale motions. (7) Star formation is present in many clouds; these young stars produce bipolar outflows. The physical conditions of hotter molecular clouds are quite uncertain.

131.100   Chemical processes in the interstellar gas.
A. Dalgarno.
Physical processes in interstellar clouds, p. 219 – 239 (1987). – See Abstr. 012.027.

The physical and chemical processes by which molecules are formed and destroyed in the gas phase in conditions characteristic of interstellar clouds are reviewed with an emphasis on recent developments. Some examples of their application to the chemistry of diffuse and dense clouds are briefly discussed.

131.101   The abundance of interstellar CO.
E. F. van Dishoeck, J. H. Black.
Physical processes in interstellar clouds, p. 241 – 274 (1987). – See Abstr. 012.027.

The current understanding of the abundance of CO in interstellar clouds is reviewed from both observational and theoretical points of view. In special circumstances, the CO and H_2 column densities can be measured directly by ultraviolet and infrared absorption line techniques. Indirect methods invoke mean relations between CO line intensities and extinction, diffuse far–infrared flux, and diffuse gamma ray flux, or assume that the clouds are in virial equilibrium. The problems associated with these indirect methods are considered, and the range of applicability of the empirical relations is investigated.

131.102   Infrared emission from interstellar PAHs.
L. J. Allamandola, A. G. G. M. Tielens, J. R. Barker.
Physical processes in interstellar clouds, p. 305 – 331 (1987). – See Abstr. 012.027.

The mid–infrared interstellar emission spectrum with features at 3050, 1610, 1300, 1150, and 885 cm^{-1} (3.28, 6.2, 7.7, 8.7 and 11.3 microns), spectroscopic details and continuum are discussed in terms of the polycyclic aromatic hydrocarbon (PAH) hypothesis. This hypothesis is based on the similarity between the interstellar emission spectrum with the infrared absorption and Raman Spectra of PAHs and soots (collections of PAHs). The fundamental vibrations of PAHs and PAH–like species which determine the IR and Raman properties are discussed. Interstellar IR band emission is due to relaxation from highly vibrationally excited PAHs which have been excited by ultraviolet photons. The excitation/emission process is described in general and the IR fluorescence from one PAH, chrysene, is traced in detail.

131.103   Evolution of interstellar dust.
A. G. G. M. Tielens, L. J. Allamandola.
Physical processes in interstellar clouds, p. 333 – 376 (1987). – See Abstr. 012.027.

This paper presents a review of our current knowledge of interstellar dust. The composition of the interstellar dust is discussed. The global evolution of interstellar dust is described and some current pressing problems are discussed, including dust formation in the outflow from stars and in the interstellar medium, dust destruction by shock, and the interrelationship of interstellar and interplanetary grains. Infrared spectroscopy, the foundation of such analyses, is described with the emphasis on simple molecular mixtures. This is applied to the observations of interstellar icy grain mantles. A discussion of recent KAO observations in the 5 – 8 μm region is given. This is followed by a critical assessment of the molecular identifications for the infrared absorption features observed in protostellar objects.

131.104   The role of dust in interstellar chemistry.
D. A. Williams.
Physical processes in interstellar clouds, p. 377 – 388 (1987). – See Abstr. 012.027.

The effects of grains on interstellar chemistry are explored. Grains catalyze the formation of H$_2$ which is the key to interstellar chemistry. They may also lead to the formation of other molecules. Grains accrete mantles of molecules. In times of $\sim 10^6$ years, all molecules will be removed from the gas unless a mechanism returns this material, and it is proposed that low mass star formation is this mechanism.

131.105   The effects of dust on the ionization structures and dynamics in magnetized clouds.
O. Havnes, T. W. Hartquist, W. Pilipp.
Physical processes in interstellar clouds, p. 389 – 412 (1987). – See Abstr. 012.027.

Most dust grains in interstellar clouds are charged, and in dense dark clouds, small grains may carry a substantial fraction of the negative charge. The charged dust grains are important sites of recombination in dark dense clouds and play a major role in establishing the ionization structure in them. Under certain conditions, the grains are well coupled to the magnetic field. When they are, and the fractional ionization is low, the grain–neutral friction affects significantly the damping of small amplitude waves, the propagation of shocks, and the ambipolar diffusion of the magnetic field. In high temperature shocks in clouds, grain sputtering can have important chemical effects.

131.106   MHD shock waves in diffuse molecular clouds.
B. T. Draine.
Physical processes in interstellar clouds, p. 423 – 428 (1987). – See Abstr. 012.027.

The theory of low–velocity ($v_s \leqslant 30$ km/s) MHD shock waves in diffuse molecular gas is reviewed. The multifluid nature of the gas dynamics is emphasized, and the important microphysics and chemistry is mentioned. Recent attempts to use shock waves to account for the abundance of interstellar CH$^+$ are discussed, and

the required shock frequency and global energy dissipation rates are commented on. Recent and current developments in this area are mentioned.

131.107   Interstellar magnetic fields.
C. Heiles.
Physical processes in interstellar clouds, p. 429 – 452 (1987). – See Abstr. 012.027.

Methods to observe magnetic fields are considered and results for external galaxies are reviewed; the author concludes that most results are questionable. Next, the Galactic field is reviewed. The large–scale field decreases slowly with Galactic radius and z, and has a strength ~ 4 μG near the Sun. It is a roughly circular field, which may reverse one or more times inside the Solar circle. The local value of the uniform component is $\gtrsim 1.6$ μG. The Galactic field is not uniform. A few "magnetic bubbles" of diameter ~ 100 pc stand out quite prominently, and statistical analyses show that the nonuniform component of the field is at least as strong as the uniform component. Finally, observations of the field on small scales are reviewed.

131.108   Star formation in magnetic interstellar clouds: I. Interplay between theory and observations.
T. C. Mouschovias.
Physical processes in interstellar clouds, p. 453 – 489 (1987). – See Abstr. 012.027.

The author summarizes the recent interplay between theoretical calculations and observations of interstellar magnetic fields, and the current understanding of the role of magnetic fields in (and a scenario for) star formation. This includes the relation between the magnetic field strength and the gas density in self-gravitating clouds; support of molecular clouds against self-gravity; rotation of clouds and fragments and magnetic braking; molecular line widths and hydromagnetic waves versus supersonic turbulence; core–envelope separation in molecular clouds and the inefficiency of star formation; ambipolar diffusion in cloud cores and thermalization of line widths. A scenario for star formation (binary, single, and planetary) is given which accounts properly for the redistribution of angular momentum by magnetic braking and of magnetic flux by ambipolar diffusion in clouds and fragments. Key problems remaining unsolved are emphasized.

131.109   Star formation in magnetic interstellar clouds: II. Basic theory. How to manufacture stars by studying waves on strings.
T. C. Mouschovias.
Physical processes in interstellar clouds, p. 491 – 552 (1987). – See Abstr. 012.027.

Magnetic braking and ambipolar diffusion are unavoidable physical processes in interstellar clouds and play a key, perhaps even crucial, role in star formation. The author summarizes those calculations on magnetic braking and ambipolar diffusion which led to the proposal of the scenario for star formation described in the accompanying paper. New insights into the underlying physics of the two processes are provided, and further contact with observations is made.

131.110   Detection of interstellar CH and CH$^+$ towards SN 1987A.
P. Magain, D. Gillet.
Astron. Astrophys., Vol. 184, No. 1/2, p. L5 – L6 (1987).

The authors report the detection of interstellar CH and CH$^+$ towards supernova 1987A. For both of these molecules, one component is detected at a heliocentric velocity of about 280 km/s, corresponding to material inside the LMC. Some implications of the authors' results with regards to the H$_2$–to–dust ratio are briefly discussed.

131.111   IRAS and optical observations of the high–latitude dust cloud Lynds 1642.
R. J. Laureijs, K. Mattila, G. Schnur.
Astron. Astrophys., Vol. 184, No. 1/2, p. 269 – 278 (1987).

The authors have studied the high latitude dark cloud Lynds 1642 ($l = 210°8$, $b = -36°7$), by comparing the IRAS surface brightness measurements with photoelectric and photographic surface brightness observations and with extinction.

131.112 Molecular clouds in the vicinity of the semicircular supernova remnant G 109.1–1.0.
K. Tatematsu, Y. Fukui, M. Nakano, T. Kogure, H. Ogawa, K. Kawabata.
Astron. Astrophys., Vol. 184, No. 1/2, p. 279 – 283 (1987).

The area of the semicircular supernova remnant (SNR) G 109.1–1.0 has been observed in CO ($J = 1$–0) and ^{13}CO ($J = 1$–0) lines. The SNR is found to be in contact with a CO cloud, and it is clear that the existence of the molecular cloud is responsible for the semicircular shape of the SNR. It is found that an armlike CO ridge ($=$CO arm) shows an apparent anticorrelation with the X–ray distribution: the CO arm appears to be surrounded by the curling X–ray feature called the "X–ray jet". If the CO arm exists in front of the SNR, the X–ray emission would suffer absorption by the CO arm. The molecular column density of the CO arm is found to be large enough to explain the appearance of the "X–ray jet" in terms of absorption. Alternatively, the CO arm may substantially exclude the X–ray emitting hot gas, causing relative depression of the X–ray emission. In either case, the CO arm is an important contributor to the formation of the "X–ray jet" feature.

131.113 Molecular line observations of the H II region G 34.3 + 0.2.
N. Matthews, L. T. Little, G. H. Macdonald, M. Andersson, S. R. Davies, P. W. Riley, W. R. F. Dent, D. Vizard.
Astron. Astrophys., Vol. 184, No. 1/2, p. 284 – 290 (1987).

The molecular cloud surrounding the H II region G 34.3 + 0.2 has been extensively mapped in the $J = 1$→0, $J = 2$→1 transitions of ^{12}CO and ^{13}CO and in the $J = 1$→0, $J = 3$→2 transitions of HCO$^+$. Interpretation of the emission is complicated by a deep self absorption feature observed in all spectra and attributed to a tenuous foreground cloud which is moving towards the core of the cloud. The mass of the cloud is estimated from ^{13}CO observations to be greater than $10^3 M_\odot$. A comparison of the $J = 1$→0 and $J = 3$→2 HCO$^+$ line intensities, together with the observed similarity of the HCO$^+$ and ^{13}CO maps, indicate the existence of dense condensations (molecular hydrogen density $\sim 10^6$cm^{-3}) in the core. A blue shifted high velocity molecular outflow is observed in ^{13}CO and ^{12}CO to the NW of the H II region. It is suggested that the driving force for the outflow is a stellar wind which is also responsible for the bipolar appearance of the H II region.

131.114 Formaldehyde absorption and visual extinction in the dark cloud L 1709 in the ϱ Ophiuchi region.
Y. K. Minn, J. M. Greenberg.
Astron. Astrophys., Vol. 184, No. 1/2, p. 315 – 321 (1987).

The authors have mapped the 4.83 GHz H$_2$CO line in the dark cloud L 1709 in the ϱ Oph complex and compared it with the distribution of visual extinction derived from star counts. The contours of the H$_2$CO and the shape of the dark cloud show a remarkable similarity. The H$_2$CO column density and visual extinction have a positive linear correlation with a slope of 0.9 × 10^{13}. The threshold value of visual extinction from which H$_2$CO can be detected in this cloud is $A_v > 0.9$ mag. There is a suggestion of a turnover of the relation between the H$_2$CO column density and A_v at high A_v. Although the limited amount of data does not allow to draw any definite conclusions as to the cause, depletion on grains is a possibility. The volume density and the total mass of the cloud are derived. The radial velocity distribution indicates the presence of independently moving subclouds.

131.115 Gas–and–dust complex NGC 7822 + S 171 connected with the Cep OB4 association.
T. A. Lozinskaya, T. G. Sitnik, M. S. Toropova.
Astron. Zh., Tom 64, Vyp. 5, p. 939 – 955 (1987). In Russian.
English translation in Sov. Astron., Vol. 31, No. 5.

The results of investigations of the gas–and–dust complex connected with the young association Cep OB4 and the cluster Be 59 are presented. Monochromatic photographs of the bright emission nebulae S 171 (W1) and NGC 7822 are obtained. The expansion velocity of a faint extended shell around W1 and Cep OB4 is confirmed to be equal to $30 - 40$ km s^{-1}. IR observations are presented. A second "inner" shell is revealed. The numerous molecular and dust clouds, emission nebulae and two faint extended shells in this region lead to the conclusion that they form a single complex associated with Cep OB4 and the young cluster Be 59. The shell–type morphology of the complex seems to be created by the stellar wind of Cep OB4 and Be 59. The expansion velocity could be explained by a supernova explosion about 3×10^5 years ago or by a short–lived (or invisible) source of a strong stellar wind.

131.116 The influence of the interstellar scattering on the pulsar mean pulse shape and apparent angular size: general analytical representation.
A. A. Bocharov.
Astron. Zh., Tom 64, Vyp. 5, p. 1004 – 1012 (1987). In Russian.
English translation in Sov. Astron., Vol. 31, No. 5.

The influence of scattering in the interstellar medium on the mean pulse profile and apparent angular size of pulsar radiation is considered. The general solution of the equation for the two-frequency mutual coherence function is obtained for a point source in statistically anisotropic and inhomogeneous medium with arbitrary spatial irregularities. The corresponding analytical expressions for pulse spread function and angular spectrum of scattered radiation are presented. The special case of an initially plane wave is also considered.

131.117 Hydrodynamical collapse of interstellar clouds. IV. Ionization degree and ambipolar diffusion.
A. E. Dudorov, Yu. V. Sazonov.
Nauchn. Inf., Vyp. 63, p. 68 – 86 (1987). In Russian.

The evolution of the magnetic field and the hydrodynamic collapse of protostellar clouds is investigated with reference to ambipolar diffusion and decay of the magnetic field. The ionization of collapsing clouds by XR and CR is considered numerically. The evaporation of dust and thermal ionization of metals, hydrogen and helium is estimated. It is shown that the ionization fraction in collapsing clouds can be expressed as a power–law function of neutral gas density, $x \sim n^{-q}$ with $q = 0.3 - 0.5$ in case of radioactive recombination and with $q = 1.0 - 2.2$ for recombination on grains. The evaporation of dust grains and thermal ionization removes the decay of plasma, determining the minimal ionization, $x \approx 10^{-15} - 10^{-17}$ the large exponent q leads to great velocity of ambipolar diffusion, that decreases the intensity of magnetic field $10^2 - 10^3$ times in respect to the freezing magnetic field. The decay of the magnetic field develops slowly and has small effect on the magnetic field of protostars.

131.118 In the shadow of the Horsehead.
D. Malin.
Sky Telesc., Vol. 74, No. 3, p. 253 – 257 (1987).

131.119 On the determination of electron density in diffuse clouds.
S. P. Tarafdar, K. S. Krishna Swamy.
Observatory, Vol. 107, No. 1079, p. 161 – 163 (1987).

It is shown that the inclusion of the wavelength dependence of attenuation for radiation density gives better agreement in the determined electron densities from various ratios of neutral and ionic species in diffuse clouds.

131.120 Refraction it the Galaxy.
A. Hewish.
Nature, Vol. 328, No. 6128, p. 290 (1987).

This note comments on a paper by R. W. Romani, R. D. Blandford and J. M. Cordes (see abstr. 131.121) discussing the nature of interstellar scintillation of extragalactic radio sources.

131.121 Radio caustics from localized interstellar medium plasma structures.
R. W. Romani, R. D. Blandford, J. M. Cordes.
Nature, Vol. 328, No. 6128, p. 324 – 326 (1987).

In a study of 36 extragalactic radio sources observed over 7 years with the Green Bank Interferometer at 2.7 GHz and 8.1 GHz, Fiedler et al. have detected unusual variations in the light curves of several sources. The most dramatic event, found in the flux record of the quasar 0954+658, showed large modulations at both frequencies. Fiedler et al. argue cogently that these variations are unlikely to be intrinsic to the sources and that refractive scintillation in the interstellar medium is probably responsible. The authors present a detailed interpretation of the event in 0954+658, propose possible sites for the refracting clouds and suggest some future observations.

131.122 Observational constraints on interstellar diamonds.
J. H. Hecht.
Nature, Vol. 328, No. 6133, p. 765 (1987).

131.123 Coronal interstellar gas and supernova remnants.
R. A. McCray.
Spectroscopy of astrophysical plasmas, p. 255 – 278 (1987). – See Abstr. 003.010.

Contents: 1. Introduction. 2. Thermally ionized gas: coronal approximation. 3. Photoionized gas: nebular approximation. 4. Thermal conduction. 5. Supernova remnants, interstellar bubbles, and superbubbles.

131.124 Diffuse interstellar clouds.
J. H. Black.
Spectroscopy of astrophysical plasmas, p. 279 – 301 (1987). – See Abstr. 003.010.

Contents: 1. Introduction. 2. Observations: extinction. 3. Observations: atomic and molecular absorption lines. 4. Observations: radio lines. 5. Theory and interpretation. 6. Summary.

131.125 Self–regulating star formation and disk structure.
M. A. Dopita.
Nearly normal galaxies. From the Planck time to the present, p. 144 – 153 (1987). – See Abstr. 012.031.

The model of star–formation proposed in this paper offers a physical model for bimodal star formation which appears to be capable of explaining, in a general way, many of the physical and chemical characteristics of disk galaxies. Applied to the solar neighbourhood, such a model can account for age/metallicity relationships, the increase the O/Fe ratio at low metallicity, the paucity of metal–poor G and K dwarf stars, the missing mass in the disk and, possibly, the existence of a metal poor thick disk. For other galaxies, it accounts for constant w–velocity dispersion of the gas, the relationship between gas content and specific rates of star formation, the surface brightness/metallicity relationship and for the shallow radial gradients in both star formation rates and H I content.

131.126 Molecular and atomic clouds associated with infrared cirrus in Ursa Major.
H. W. de Vries, A. Heithausen, P. Thaddeus.
Astrophys. J., Vol. 319, No. 2, p. 723 – 729 (1987).

Observations of CO and H I revealed that in Ursa Major the high–latitude far–infrared "cirrus" emission discovered by IRAS comes from molecular and atomic clouds. These clouds differ considerably from the large clouds in the Galactic plane. On the assumption of a constant gas–to–dust ratio, it is argued that the cirrus emission in Ursa Major is a good mass tracer. The $N(H_2)/W_{CO}$ ratio derived for those diffuse clouds, $(0.5\pm0.3)\times10^{20}K^{-1}km^{-1}s\,cm^{-2}$, is significantly lower than the ratio applicable to Galactic plane surveys.

131.127 Mass, luminosity, and line width relations of Galactic molecular clouds.
P. M. Solomon, A. R. Rivolo, J. Barrett, A. Yahil.
Astrophys. J., Vol. 319, No. 2, p. 730 – 741 (1987).

The authors present measurements of the velocity line width, size, virial mass, and CO luminosity for 273 molecular clouds in the Galactic disk between longitudes of 8° and 90°. These are obtained from three–dimensional data in the Massachusetts–Stony Brook CO Galactic Plane Survey. It is shown that the molecular clouds are in or near virial equilibrium and are not confined by pressure equilibrium with a warm or hot phase of interstellar matter. The velocity line width is proportional to the 0.5 power of the size, $\sigma_v \propto S^{0.5}$. A tight relationship, over four orders of magnitude, is found between the cloud dynamical mass, as measured by the virial theorem, and the CO luminosity $M \propto (L_{CO})^{0.81}$. The cloud CO luminosity is $L_{CO} \propto \sigma_v^5$.

131.128 High–resolution images of the L1551 bipolar outflow: evidence for an expanding, accelerated shell.
G. H. Moriarty–Schieven, R. L. Snell, S. E. Strom, F. P. Schloerb, K. M. Strom, G. L. Grasdalen.
Astrophys. J., Vol. 319, No. 2, p. 742 – 753 (1987).

Approximately one–half of the L1551 bipolar outflow was mapped in the $J = 1–0$ transition of ^{12}CO using the FCRAO 14 m telescope. The data were reconstructed using a maximum entropy algorithm to obtain images of the high–velociy gas with an angular resolution of $\sim 20''$. The outflow exhibits a striking shell–like structure with the lowest velocity outflowing gas found along the limb of the outflow, and the highest velocity outflowing gas found along the axis of the outflow. The data can be modeled by molecular material located in a thin expanding shell which is accelerating away from IRS 5.

131.129 The magnetic field in star–forming large globules.
K.–W. Hodapp.
Astrophys. J., Vol. 319, No. 2, p. 842 – 849 (1987).

The magnetic field in the three large star–forming globules L810, B335, and ESO 210–6A has been mapped by near–infrared polarimetry of background stars. L810 and B335 show bent magnetic field lines, while in ESO 210–6A, the field lines point away from the center of the Gum nebula. These magnetic field structures are interpreted in a model based on the interaction of a globule with an expanding interstellar shell.

131.130 Shock processing of interstellar dust: diamonds in the sky.
A. G. G. M. Tielens, C. G. Seab, D. J. Hollenbach, C. F. McKee.
Astrophys. J., Lett. Ed., Vol. 319, No. 2, p. L109 – L113 (1987).

The authors have studied the processing of interstellar dust grains by strong shock waves, with the emphasis on the effects of grain–grain collisions. Such collisions provide the high pressures required to transform interstellar graphite and amorphous carbon grains into diamonds. They calculate that about 5% of the C is expected to be in the form of 5 – 100 Å diamonds in the interstellar medium. These results support the suggested interstellar origin for the recently discovered small meteoritic diamonds by providing a feasible interstellar formation mechanism.

131.131 Interstellar extinction and the composition of dust grains.
W. W. Duley.
Mon. Not. R. Astron. Soc., Vol. 229, No. 2, p. 203 – 212 (1987).

Laboratory data on the properties of amorphous carbon and silicate materials are used to develop a new model for interstellar dust.

131.132 Interstellar extinction correlations.
A. P. Jones, W. W. Duley, D. A. Williams.
Mon. Not. R. Astron. Soc., Vol. 229, No. 2, p. 213 – 221 (1987).

The authors have shown that a core mantle grain model, in which the carbon is accreted onto small silicate cores in the form of hydrogenated amorphous carbon (HAC) (Duley et al., 1987), can qualitatively explain the observed extinction correlations.

The bulk of the extinction arises in the small core mantle particles, the silicate cores are responsible for the UV extinction bump at 2200 Å and some for UV absorption, whilst the HAC mantles produce the major part of the visual to far UV extinction (~60 per cent). This clearly explains the correlation between the visual extinction and the UV extinction bump, both features arising in the same particles. The poorer correlation between the visual and the far UV extinction is explained by the range of compositional variations in HAC and the corresponding changes to the visual and UV optical properties.

131.133   Star formation in W49A: gravitational collapse of the molecular cloud core toward a ring of massive stars.
W. J. Welch, J. W. Dreher, J. M. Jackson, S. Terebey, S. N. Vogel.
Science, Vol. 238, No. 4833, p. 1550 – 1555 (1987).

High–resolution molecular line and continuum radio images from the Hat Creek Radio Observatory and the Very Large Array suggest that the core of the W49A star–forming region is undergoing gravitational collapse. The radio continuum shows a 2–parsec ring of at least ten distinct ultracompact H II regions, each associated with at least one O star. The ring is a region of large–scale, organized massive star formation. The molecular cloud core is believed to be collapsing toward the center of the ring.

131.134   Embedded young stellar objects in W75N: interactions with the ambient molecular cloud.
T. J. T. Moore.
Q. J. R. Astron. Soc., Vol. 28, No. 3, p. 264 – 268 (1987). – See Abstr. 012.037.

Young stellar sources in the early stages of evolution interact with the interstellar medium on all scales from a few AU to tens of parsecs. These varied interactions are illustrated by the observations that have been made of the source W75N, at wavelengths from the near infrared to the radio.

131.135   IC 443: the interaction of a supernova remnant with a molecular cloud.
M. Burton.
Q. J. R. Astron. Soc., Vol. 28, No. 3, p. 269 – 276 (1987). – See Abstr. 012.037.

The SNR IC 443 provides an excellent laboratory for studying the interaction of a supernova shock wave with a molecular cloud. Observations of shocked molecular hydrogen emission are reported, and compared with those of other shocked molecular species and with high–velocity atomic hydrogen. It is concluded that the shock is partially dissociative, leaving behind it shocked molecular gas and accelerated atomic gas, but little ionized gas. A model for IC 443 is presented, involving the SN explosion having taken place within the remains of a molecular disk left over from the process of formation of the star which exploded.

131.136   The interstellar 2200 Å feature.
D. J. Carnochan.
Q. J. R. Astron. Soc., Vol. 28, No. 3, p. 284 – 288 (1987). – See Abstr. 012.037.

The cause of the interstellar 2200 Å feature is still unknown, over 20 years after its discovery. The feature varies independently of the rest of the extinction curve implying that a different physical mechanism is at work. The suggestion is made here that the 2200 Å feature might be caused by charge transfer between neutral hydrogen and singly ionized iron, silicon and magnesium.

131.137   Interstellar extinction in the Pleiades.
B. N. G. Guthrie.
Q. J. R. Astron. Soc., Vol. 28, No. 3, p. 289 – 293 (1987). – See Abstr. 012.037.

The different values of the ratio of total to selective absorption R given in the literature for the Pleiades are reviewed, and some new estimates of this ratio are obtained. The results indicate a value $R = 3.2$, which is close to the normal value for the interstellar medium. The reddening of HD 23512 and fainter

stars nearby is due mainly to a small molecular cloud which is moving into the cluster.

131.138   The continuing story of the diffuse interstellar bands.
G. E. Bromage.
Q. J. R. Astron. Soc., Vol. 28, No. 3, p. 294 – 297 (1987). – See Abstr. 012.037.

A brief review is presented of the long and eventful history of the (still unidentified) diffuse interstellar absorption features. Some key contributions to this continuing story are chosen and tabulated, including both observational and theoretical milestones over the last 50 years.

131.139   Magnetic fields and the optical polarization of star formation regions.
R. F. Warren–Smith.
Q. J. R. Astron. Soc., Vol. 28, No. 3, p. 298 – 302 (1987). – See Abstr. 012.037.

It is shown how optical polarization mapping of star formation regions can provide evidence about the magnetic field distribution established within collapsing protostellar clouds. It is suggested that the morphology of bipolar nebulae associated with pre–main–sequence stars results from an interaction between outflowing material and the magnetic field and density distribution established within the surrounding cloud material during the formation of the star.

131.140   Infrared spectroscopy of interstellar dust.
D. C. B. Whittet.
Q. J. R. Astron. Soc., Vol. 28, No. 3, p. 303 – 311 (1987). – See Abstr. 012.037.

Infrared spectroscopy provides a valuable technique for investigating the composition and evolution of interstellar grain material in a variety of astrophysical environments. This paper briefly reviews proposed identifications for the observed features, and discusses some recent observations of the Taurus dark cloud and the galactic centre region.

131.141   Variations in ultraviolet extinction: effect of polarization revisited.
J. M. Greenberg, G. Chlewicki.
Q. J. R. Astron. Soc., Vol. 28, No. 3, p. 312 – 322 (1987). – See Abstr. 012.037.

The alignment of the particles responsible for the polarization and visual extinction is shown to provide a basis for changing the saturation level of the ultraviolet extinction without changing the particle sizes. If the particles are well aligned, it is predicted that there should be significantly lower extinction in the ultraviolet relative to the visible for stars viewed perpendicular to magnetic–field lines (maximum polarization) as compared with those viewed across the field lines. Preliminary evidence for such an effect is noted in Carina.

131.142   Extinction in external galaxies.
G. I. Thompson.
Q. J. R. Astron. Soc., Vol. 28, No. 3, p. 323 – 327 (1987). – See Abstr. 012.037.

It is suggested that there are only two general extinction laws, the galactic type and the linear SMC type. An attempt is made to show that the LMC extinction law is a composite of these two. The result is inconclusive due to the sensitivity of the constants of fit to the empirical data.

131.143   Dust in external galaxies: diffuse bands and the albedo.
D. H. Morgan.
Q. J. R. Astron. Soc., Vol. 28, No. 3, p. 328 – 333 (1987). – See Abstr. 012.037.

The importance of detecting the 4430 Å diffuse absorption band in the Magellanic Clouds is discussed. Recent observations of the band in the spectra of reddened SMC stars are described. The albedo of the interstellar dust near 30 Doradûs is estimated by comparing ultraviolet extinction laws constructed from small– and large–aperture observations.

131.144   Echelle observations of interstellar bubbles.
C. A. Clayton.
Q. J. R. Astron. Soc., Vol. 28, No. 3, p. 334–337 (1987). – See Abstr. 012.037.

Clear observational evidence of expanding interstellar bubbles has been obtained using the dedicated Manchester Echelle Spectrometers. The kinematic nature of these structures is found to be generally consistent with a stellar wind–driven bubble interpretation although some of the features seen cannot be explained in these terms.

131.145   VLA observations of the 6 cm and 2 cm lines of H_2CO in the direction of W3(OH).
H. R. Dickel, W. M. Goss.
Astron. Astrophys., Vol. 185, No. 1/2, p. 271–282 (1987).

Formaldehyde absorption at 2 cm towards the radio continuum source W3(OH) has been mapped with the Very Large Array. The angular resolution of $0.''5$ and the velocity resolution of 0.76 km s^{-1} match those of the previous observations of the 6 cm transition. There are pronounced opacity variations across the source. The density of molecular hydrogen $n(H_2)$ and the formaldehyde column density per unit velocity interval $N_v(H_2CO)$ have been determined as a function of velocity and position across the source. The results support the toroidal model for W3(OH) proposed by Guilloteau et al. (1983).

131.146   NGC 2264: a molecular line study.
E. Krügel, R. Güsten, A. Schulz, C. Thum.
Astron. Astrophys., Vol. 185, No. 1/2, p. 283–290 (1987).

The authors present an extended ammonia map of the central region of the molecular cloud associated with NGC 2264 together with CO measurements over a wide spectral range of the vicinity of the embedded star IRS 1. The ammonia cloud consists of a northern and a southern part each with a mass of about 500 M$_\odot$; the southern one is divided into several subclouds with different velocities and temperatures. The heating of these clouds comes from IRS 1 and embedded low–luminosity stars. There is no cogent evidence for rotation anywhere. On a smaller scale, the CO gas around IRS 1 breaks up into cloudlets with rapidly varying kinematics. The authors do not detect a bipolar high–velocity outflow from IRS 1 at their sensitivity level.

131.147   Variations in UV extinction in galactic associations and perpendicular to the galactic plane.
E. Kiszkurno–Koziej, J. Lequeux.
Astron. Astrophys., Vol. 185, No. 1/2, p. 291–296 (1987).

For about 1200 O – B4 stars observed by ANS, normalized UV extinction parameters, E(bump)$/E(B–V)$ and $E(1550 – 1800)/E(B–V)$, have been calculated together with the distance from the galactic plane, z. A slight correlation of UV extinction parameters with $|z|$ appears: at high $|z|$ low values of E(bump)$/E(B–V)$ and high values of $E(1550 – 1800)/E(B–V)$ are more frequent. Shock waves and/or radiation fields may be responsible for this effect. The same parameters have been studied in 30 OB associations. Association–to–association variations of UV extinction are confirmed, which are not well correlated with galactic longitude. The open cluster Tr 37 is found to exhibit an exceptional behaviour.

131.148   The vicinity of Omicron Per.
R. Bachiller, J. Cernicharo, P. Goldsmith, A. Omont.
Astron. Astrophys., Vol. 185, No. 1/2, p. 297–305 (1987).

The authors have mapped the region around the B1 III star o Per (HD 23180) in the emission of the $J = 1$–0 and $J = 2$–1 lines of ^{12}CO and ^{13}CO with ≈ 1 arcmin resolution. The molecular cloud morphology suggests the presence of a warm hole near the position of the star. It is tempting to think that o Per is the cause of this molecular gap. However, the observed heating of the cloud edge, as well as optical, infrared (IRAS) and carbon recombination line data, all suggest that the local ultraviolet field is enhanced only by a factor $10 – 100$. Therefore o Per must be located at least a few parsecs away from the cloud, and the precise alignment of the observed hole with the star is probably mainly due to chance. Furthermore, other neighbouring OB stars

(mainly BD $+31°643$, ζ Per and X Per) could contribute appreciably (even more than o Per) to the ultraviolet enhancement which seems extend over several degrees on the north–east edge of the Perseus local molecular cloud.

131.149   The spectral hallmark of a contracting protostellar fragment.
G. Anglada, L. F. Rodríguez, J. Cantó, R. Estalella, R. López.
Astron. Astrophys., Vol. 186, No. 1/2, p. 280–286 (1987).

There is no clear observational evidence for the contracting phase that should characterize star formation. In this paper the authors calculate, analytically, the expected CO wing line emission profile from a 1 $M_\odot$ contracting protostellar fragment. The wings have a power law dependence with radial velocity with an index of –3, and the blue wing is about two times stronger than the red wing. This characteristic asymmetric profile can be detected after integration times of tens of hours with the new large radio telescopes for the millimeter waves (Nobeyama and Pico Veleta). Detection of this spectral "hallmark" may provide the observational confirmation of the contracting phase in star formation.

131.150   Interstellar clouds: morphological information from projected shapes.
M. David, W. Verschueren.
Astron. Astrophys., Vol. 186, No. 1/2, p. 295–302 (1987).

The observed image of an interstellar cloud has in general an extremely irregular shape. Nevertheless one can attribute a minor and major dimension (or "axis") to it and construct the distribution of minor to major axis ratios for a given sample of observed cloud images. The authors propose a statistical procedure to analyse such distributions and show that rough morphological features of the clouds, which can be expressed as correlations between their three dimensions, may in fact be detected.

131.151   Infrared radiation of very small dust grains in the Rho Ophiuchi region.
C. Ryter, J. L. Puget, M. Pérault.
Astron. Astrophys., Vol. 186, No. 1/2, p. 312–318 (1987).

The IRAS Sky–flux maps have been used to study the colours of interstellar dust in the Rho Oph region. The 12 μm and 60 μm band intensities relative to the dust emission at longer wavelengths have been obtained in the south–east extension of the molecular cloud and in the surrounding of the stars σ Sco, τ Sco, and HD 147889. It is found that the 60 μm to 100 μm intensity ratio increases gradually with the starlight radiation density heating the dust, as expected. In the reflection nebula NGC 2023, the 12 μm flux is shown to be entirely produced by the "aromatic infrared features" (namely at $\lambda = 7.7, 8.6, 11.3 \mu$m) encompassed in the band, which is thus interpreted as the tracer of the source of the features, believed to be polycyclic aromatic hydrocarbons (PAHs). A strong 12 μm intensity is detected almost everywhere. However, in a radiation field of order 100 eV cm^{-3}, the 12 μm emission is depressed by a factor of order five, which is attributed to a destruction of the polycyclic aromatic hydrocarbons. It is concluded that the particles should in general be made of about 20 to 40 carbon atoms.

131.152   Water vapor masers associated with young visible stars.
L. F. Rodríguez, A. D. Haschick, J. M. Torrelles, P. C. Myers.
Astron. Astrophys., Vol. 186, No. 1/2, p. 319–321 (1987).

The authors report VLA positions for the H_2O masers associated with the young visible stars PV Cep and V1057 Cyg. Using these data as well as published results on LkHα 234 and V645 Cyg, they conclude that H_2O maser spots can be produced and pumped at relatively large distances ($\sim 10^{16}$cm) from the visible exciting star. Remote pumping of H_2O masers is known to occur in the case of luminous, heavily obscured stars and the results suggest that this process also exists for intermediate luminosity ($10^2 – 10^3 L_\odot$) young stars.

131.153   **H I superclouds in the inner Galaxy.**
B. G. Elmegreen, D. M. Elmegreen.
Astrophys. J., Vol. 320, No. 1, p. 182–198 (1987). With plates
3–8.
  Atomic hydrogen clouds in the Weaver–Williams 21 cm survey
are found in the vicinity of the largest molecular cloud complexes
in the first Galactic quadrant. The atomic masses are estimated in
several ways. The clouds are found to contain between 10^6 and
$4 \times 10^7 M_\odot$ of atoms at an average density of $\sim 9\ \mathrm{cm}^{-3}$. They
appear gravitationally bound, their molecular mass fractions
decrease from 70% to 5% with increasing distance from the
galactic center, and they are located in the spiral arms.

131.154   **Star–forming loops in the *IRAS* sky images.**
P. R. Schwartz.
Astrophys. J., Vol. 320, No. 1, p. 258–265 (1987).
  Loops containing diffuse and discrete emission are a feature of
the *IRAS* sky images. Some of these loops are limb–brightened
shells resulting from supernovae or stellar winds acting on the
interstellar medium. Secondary star formation appears to have
occurred at the surface of these shells. A significant proportion of
the early–type stars in the solar neighborhood appear to have
formed in stellar loops.

131.155   **Formaldehyde observations of molecular clouds.**
E. J. Wadiak.
Diss. Abstr. Int., Sect. B, Vol. 48, No. 1, p. 162–B (1987). Thesis,
University of Virginia, 134 pp. (1986). Order No. DA8705679.

131.156   **CCD observations of the spatial structure of the
        hydrogen Balmer alpha (Hα) diffuse galactic back-
ground.**
J. V. Brinkmann.
Diss. Abstr. Int., Sect. B, Vol. 48, No. 3, p. 792–B (1987). Thesis,
The University of Wisconsin–Madison, 211 pp. (1987). Order
No. DA8711015.

131.157   **On the inhomogeneous stucture of the interplanetary and
        interstellar plasma under the conditions of a developed
ion–acoustic turbulence.**
L. G. Genkin.
Red. Zh. Izv. Vuzov. Radiofiz., Gor'kij, 7 pp. (1987). In Russian.
Abstr. in Ref. Zh., 51. Astron., 10.51.137 (1987).

131.158   **Magnetic fields in diffuse H I.**
C. Heiles.
Bull. Am. Astron. Soc., Vol. 19, No. 2, p. 691 (1987). Abstract. –
See Abstr. 010.061.

131.159   **It's everywhere, it's everywhere: or, why do we know so
        little about galactic H I?**
F. J. Lockman.
Bull. Am. Astron. Soc., Vol. 19, No. 2, p. 691 (1987). Abstract. –
See Abstr. 010.061.

131.160   **Observations of galactic H I with arcminute resolution.**
L. A. Higgs.
Bull. Am. Astron. Soc., Vol. 19, No. 2, p. 691 (1987). Abstract. –
See Abstr. 010.061.

131.161   **The Boston University – Arecibo galactic H I survey: the
        CO connection.**
T. M. Bania.
Bull. Am. Astron. Soc., Vol. 19, No. 2, p. 692 (1987). Abstract. –
See Abstr. 010.061.

131.162   **Thermodynamics and small scale structure of the atomic
        gas.**
J. M. Dickey.
Bull. Am. Astron. Soc., Vol. 19, No. 2, p. 692 (1987). Abstract. –
See Abstr. 010.061.

131.163   **The large scale distribution of the atomic phases.**
S. R. Kulkarni.
Bull. Am. Astron. Soc., Vol. 19, No. 2, p. 692 (1987). Abstract. –
See Abstr. 010.061.

131.164   **Supershells**
R. McCray.
Bull. Am. Astron. Soc., Vol. 19, No. 2, p. 696–697 (1987).
Abstract. – See Abstr. 010.061.

131.165   **Theory of star formation.**
P. Bodenheimer.
Bull. Am. Astron. Soc., Vol. 19, No. 2, p. 698 (1987). Abstract. –
See Abstr. 010.061.

131.166   **Infrared polarization in molecular clouds.**
P. G. Martin.
Bull. Am. Astron. Soc., Vol. 19, No. 2, p. 710 (1987). Abstract. –
See Abstr. 010.061.

131.167   **A search for T Tauri stars in high–latitude molecular
        clouds.**
L. Magnani, J.–P. Caillault, L. Armus, I. J. Reilly.
Bull. Am. Astron. Soc., Vol. 19, No. 2, p. 716 (1987). Abstract. –
See Abstr. 010.061.

131.168   **Far–infrared spectroscopy of the DR 21 star formation
        region.**
A. P. Lane, M. R. Haas, D. J. Hollenbach, E. F. Erickson.
Bull. Am. Astron. Soc., Vol. 19, No. 2, p. 716–717 (1987).
Abstract. – See Abstr. 010.061.

131.169   **The diffuse interstellar features at 5780 and 5797 Å in
        star formation regions.**
G. Wallerstein, J. A. Cardelli.
Bull. Am. Astron. Soc., Vol. 19, No. 2, p. 717 (1987). Abstract. –
See Abstr. 010.061.

131.170   **Percolation of ionizing photons in an inhomogenous
        medium.**
D. Van Buren.
Bull. Am. Astron. Soc., Vol. 19, No. 2, p. 723 (1987). Abstract. –
See Abstr. 010.061.

131.171   **Numerical models of superbubble dynamics.**
M.–M. MacLow, M. L. Norman, R. McCray.
Bull. Am. Astron. Soc., Vol. 19, No. 2, p. 723 (1987). Abstract. –
See Abstr. 010.061.

131.172   **Stability of radiative shocks with time dependent cool-
        ing.**
T. J. Gaetz.
Bull. Am. Astron. Soc., Vol. 19, No. 2, p. 723 (1987). Abstract. –
See Abstr. 010.061.

131.173   **Probing the possibility of a $^{12}C/^{13}C$ abundance gradient
        from observations of interstellar CH$^+$.**
I. Hawkins.
Bull. Am. Astron. Soc., Vol. 19, No. 2, p. 723–724 (1987).
Abstract. – See Abstr. 010.061.

131.174   **The diffuse ultraviolet background and interstellar dust.**
M. Hurwitz, S. Bowyer, C. Martin.
Bull. Am. Astron. Soc., Vol. 19, No. 2, p. 724 (1987). Abstract. –
See Abstr. 010.061.

131.175   **A study of C_3H_2 in galactic sources.**
S. C. Madden, H. E. Matthews, T. L. Wilson,
W. M. Irvine.
Bull. Am. Astron. Soc., Vol. 19, No. 2, p. 725 (1987). Abstract. –
See Abstr. 010.061.

131.176   **Ammonia observations of NGC 6334.**
T. B. H. Kuiper, W. L. Peters, J. R. Forster,
F. F. Gardner, J. B. Whiteoak.
Bull. Am. Astron. Soc., Vol. 19, No. 2, p. 725 (1987). Abstract. –
See Abstr. 010.061.

131.177   **Possible outflow sources in nearby molecular clouds.**
J. T. Armstrong, L. Haikala, G. Winnewisser.
Bull. Am. Astron. Soc., Vol. 19, No. 2, p. 725 (1987). Abstract. –
See Abstr. 010.061.

131.178   **Physical conditions in molecular outflows.**
R. M. Levreault.
Bull. Am. Astron. Soc., Vol. 19, No. 2, p. 725 – 726 (1987).
Abstract. – See Abstr. 010.061.

131.179   **IRAS and radio observations of the remarkable bipolar object Lynds 1592/93.**
J. Fischer, L. J. Rickard, A. A. Stark.
Bull. Am. Astron. Soc., Vol. 19, No. 2, p. 726 (1987). Abstract. –
See Abstr. 010.061.

131.180   **Wide–field mapping of ^{13}CO emission from molecular clouds.**
J. Bally, A. A. Stark, R. W. Wilson, W. D. Langer.
Bull. Am. Astron. Soc., Vol. 19, No. 2, p. 726 (1987). Abstract. –
See Abstr. 010.061.

131.181   **A   rotating   dark   cloud   associated   with IRAS 22291 + 7458.**
P. J. Benson, M. A. McDuffie, A. Goodman, P. C. Myers.
Bull. Am. Astron. Soc., Vol. 19, No. 2, p. 726 (1987). Abstract. –
See Abstr. 010.061.

131.182   **Gravitational and dynamical instabilities of a decelerating plane parallel slab of finite thickness.**
G. M. Voit, J. M. Shull.
Bull. Am. Astron. Soc., Vol. 19, No. 2, p. 729 (1987). Abstract. –
See Abstr. 010.061.

131.183   **H I in the dark cloud L134.**
P. P. van der Werf, W. M. Goss, P. A. Vanden Bout.
Bull. Am. Astron. Soc., Vol. 19, No. 2, p. 736 (1987). Abstract. –
See Abstr. 010.061.

131.184   **Kinematics of H I emission and absorption in the Taurus molecular complex.**
W. L. H. Shuter, R. L. Dickman, C. Klatt.
Bull. Am. Astron. Soc., Vol. 19, No. 2, p. 737 (1987). Abstract. –
See Abstr. 010.061.

131.185   **A face–on view of the first galactic quadrant in giant molecular clouds.**
A. R. Rivolo, P. M. Solomon.
Bull. Am. Astron. Soc., Vol. 19, No. 2, p. 758 (1987). Abstract. –
See Abstr. 010.061.

131.186   **Nagoya CO survey of star forming regions.**
Y. Fukui.
Bull. Am. Astron. Soc., Vol. 19, No. 2, p. 758 (1987). Abstract. –
See Abstr. 010.061.

131.187   **Outflows in the Ophiuchi dark cloud.**
C. J. Lada, C. K. Walker, E. Young, M. Margulis.
Bull. Am. Astron. Soc., Vol. 19, No. 2, p. 758 (1987). Abstract. –
See Abstr. 010.061.

131.188   **A spectacular outflow in the Monoceros OB1 molecular cloud.**
M. Margulis, C. J. Lada, T. Hasegawa, S. Hayashi,
M. Hayashi, N. Kaifu, I. Gatley.
Bull. Am. Astron. Soc., Vol. 19, No. 2, p. 758 – 759 (1987).
Abstract. – See Abstr. 010.061.

131.189   **Molecular outflows that are not driven by internal energy sources.**
L. Blitz, L. Magnani, A. Wandel.
Bull. Am. Astron. Soc., Vol. 19, No. 2, p. 759 (1987). Abstract. –
See Abstr. 010.061.

131.190   **The CO J = 2→1 emission from the interstellar gas toward Zeta Ophiuchi.**
S. R. Federman, R. M. Crutcher.
Bull. Am. Astron. Soc., Vol. 19, No. 2, p. 759 (1987). Abstract. –
See Abstr. 010.061.

131.191   **Radio observations of carbon monoxide towards Zeta Ophiuchi.**
A. E. Glassgold, W. D. Langer, R. W. Wilson.
Bull. Am. Astron. Soc., Vol. 19, No. 2, p. 759 (1987). Abstract. –
See Abstr. 010.061.

131.192   **Star formation in galactic molecular clouds.**
T. J. Mooney, P. M. Solomon.
Bull. Am. Astron. Soc., Vol. 19, No. 2, p. 759 (1987). Abstract. –
See Abstr. 010.061.

131.193   **Infrared emission from the Pleiades nebulae and the cirrus clouds.**
L. Luan, M. W. Werner, M. W. Castelaz, E. Dwek,
M. G. Hauser, K. Sellgren.
Bull. Am. Astron. Soc., Vol. 19, No. 2, p. 760 (1987). Abstract. –
See Abstr. 010.061.

131.194   **Ultraviolet interstellar extinction towards NGC 6530 and the intrinsic continua of two early O–type stars.**
A. V. Torres.
Bull. Am. Astron. Soc., Vol. 19, No. 2, p. 760 – 761 (1987).
Abstract. – See Abstr. 010.061.

131.195   **Chromium oxide in the Red Rectangle.**
P. D. Bennett.
Bull. Am. Astron. Soc., Vol. 19, No. 2, p. 761 (1987). Abstract. –
See Abstr. 010.061.

131.196   **An efficient gas phase synthesis for interstellar PN.**
T. J. Millar, A. Bennett, E. Herbst.
Mon. Not. R. Astron. Soc., Vol. 229, No. 3, p. 41p – 44p (1987).
The authors have extended previous models of phosphorus chemistry in dense interstellar clouds to include a gas phase synthesis for the newly observed interstellar molecule PN (phosphorus nitride). A time–dependent calculation shows that the abundance of PN and upper limits to the abundances of other phosphorus–bearing molecules are reproduced without the need to invoke grain disruption or high–temperature reactions.

131.197   **The variability of UV extinction–curve shapes and its impact upon dereddened UV energy distributions.**
D. Massa.
Astron. J., Vol. 94, No. 6, p. 1675 – 1680 (1987).
The global variability of the shapes of UV extinction curves and the errors these variations introduce into the UV continua of reddened objects, which are corrected with a mean UV extinction curve, are studied. The author concentrates on the errors affecting the dereddened 1500 Å V color. The magnitude–limited survey provided by the ANS satellite data is employed in the analysis.

131.198   **Recent massive star formation in 30 Doradus.**
N. R. Walborn, J. C. Blades.
Space Telesc. Sci. Inst., Prepr. Ser., No. 200, 7 pp. (1987). To appear in Astrophys. J., Lett. Ed.

131.199   Structure of galaxies and star formation: workshop summary.
C. A. Norman.
Space Telesc. Sci. Inst., Prepr. Ser., No. 222, p. 9 – 16 (1987). To appear in Proc. 10th Europ. Regional Meeting of the IAU "Structure of galaxies and star formation", Prague, August 1987.

131.200   Mainline OH masers near young H II regions: a correlation with IRAS far–infrared flux density.
T. J. T. Moore, R. J. Cohen, C. M. Mountain.
Edinb. Astron. Prepr., No. 23/87, 18 pp. (1987). To appear in Mon. Not. R. Astron. Soc.

131.201   Radio and (sub)millimeter observations of the initial conditions for star formation.
Å. Hjalmarson, P. Friberg.
Onsala Space Obs., Prepr., No. 87:57, 29 pp. (1987). To appear in Formation and evolution of low mass stars.

131.202   Hydrogen absorption line profiles of ionising star clusters.
A. I. Díaz.
R. Greenwich Obs., Prepr., No. 63, 43 pp. (1987). To appear in Mon. Not. R. Astron. Soc.

131.203   The contamination of cometary globules by the ejecta of nearby massive stars.
J.–P. Arcoragi.
MPA Rep., No. 305, 6 pp. (1987).

131.204   Interstellar CH toward ζ Ophiuchi.
E. Palazzi, N. Mandolesi, P. Crane.
ESO Sci. Prepr., No. 538, 16 pp. (1987). To appear in Astrophys. J.

131.205   Detection of new, high excitation, emission lines of H_2 in the 2.0 – 2.4 μm spectrum of the Orion nebula.
E. Oliva, A. F. M. Moorwood.
ESO Sci. Prepr., No. 549, 17 pp. (1987). To appear in Astron. Astrophys.

131.206   A new cometary nebula in Cygnus.
T. Neckel, H. J. Staude.
Max–Planck–Inst. Astron. Heidelb., Prepr., 19 pp. (1987). To appear in Astrophys. J., Lett. Ed.

131.207   The interstellar medium toward SN 1987A.
A. K. Dupree, R. P. Kirshner, G. E. Nassiopoulos, J. C. Raymond, G. Sonneborn.
Astrophys. J., Vol. 320, No. 2, p. 597 – 601 (1987).
High–resolution spectra of SN 1987A obtained with the IUE show absorption features from the disk and halo of our Galaxy and the Large Magellanic Cloud. A rich variety of elements and ionization stages is found, including allowed resonance lines of abundant elements with ionization states ranging from C I and O I to Al III, Si IV and C IV. The Si IV and C IV absorption at LMC velocities is unusually strong toward the supernova as compared to most other stars in the LMC and SMC. The origin of these lines is discussed.

131.208   Cosmic–ray–induced photodestruction of interstellar molecules in dense clouds.
A. Sternberg, A. Dalgarno, S. Lepp.
Astrophys. J., Vol. 320, No. 2, p. 676 – 682 (1987).
The ultraviolet spectrum of radiation generated by cosmic rays inside dense molecular clouds is presented, and the resulting rates of photodissociation for a variety of interstellar molecules are estimated. The effects of this radiation on the chemistry of dense molecular clouds are discussed, and it is argued that the cosmic–ray–induced photons will significantly inhibit the production of complex molecular species.

131.209   Collisional charging of interstellar grains.
B. T. Draine, B. Sutin.
Astrophys. J., Vol. 320, No. 2, p. 803 – 817 (1987).
Collisional charging of small interstellar particles is reexamined, including effects due to electrostatic polarization of the grain by the electric field of an approaching charged particle. Energy–dependent capture cross sections are derived for spherical particles, and these cross sections are convolved with thermal velocity distributions to obtain rate coefficients for electron and ion collisions with spherical grains of arbitrary size and charge state. Approximate formulae are obtained for both the average grain charge and the average rate of collisions with ions.

131.210   Long–wavelength absorption by fractal dust grains.
E. L. Wright.
Astrophys. J., Vol. 320, No. 2, p. 818 – 824 (1987).
The far–infrared and submillimeter absorption properties of random aggregates of conducting spheres have been studied as a possible source for the long–wavelength absorption of interstellar dust grains. The complex shapes of random aggregates have various fractal dimensions, depending on the process that forms them. The absorption cross section σ versus frequency ν of these fractal grains can follow a power law $\sigma \propto \nu^x$ with $0.6 < \alpha < 1.4$ for two decades in frequency. The long wavelength absorption versus frequency is affected more by the shape of a conducting grain than by its composition.

131.211   The local interstellar medium. VII. The local interstellar wind and interstellar material in front of the nearby star α Ophiuchi.
P. C. Frisch, D. G. York, J. R. Fowler.
Astrophys. J., Vol. 320, No. 2, p. 842 – 849 (1987). With plate 26.
IUE observations of Mg I $\lambda2852.127$ are used to search for warm interstellar gas in the direction of α Oph. Mg I should be a sensitive tracer of warm gas because of dielectronic recombination. However, the authors show that the observed Mg I absorption feature is dominated by material associated with a wisp of relatively cool H I found at the position of α Oph. This H I wisp is also the site of the observed Ca II, Na I, and Ti II absorption lines. If it is assumed that two absorbing regions are present in front of α Oph, one cool ($T \sim 500$K) and one warm ($T \sim 10,000$K), then together these regions fill half or less of the space in front of α Oph. Any gas filling the remaining space must be either hot ($T \gg 10^4$K) or low density ($n \lesssim 10^{-3}$cm^{-3}).

131.212   A new cometary nebula in Cygnus.
T. Neckel, H. J. Staude.
Astrophys. J., Lett. Ed., Vol. 320, No. 2, p. L145 – L148 (1987). With plates L3 – L4.
The authors present CCD images, surface polarimetry, and long–slit spectrograms of a hitherto unknown cometary reflection nebula associated with a dense dust cloud. A bright, compact Herbig–Haro object is embedded in its brightest part. The highly reddened illuminating star of about $3 – 5\ M_\odot$ is located near the apex of the nebula; it emits a collimated bipolar high–velocity flow whose blueshifted component feeds the Herbig–Haro object. The redshifted component can be traced toward the interior of the dark cloud, where the density exceeds 10^5cm^{-3}.

131.213   Erratum: "IRAS 18059–3211: optically known as 'Gomez's Hamburger'" [Astrophys. J., Lett. Ed., Vol. 316, No. 1, p. L21 – L24 (1987)].
M. T. Ruiz, V. Blanco, J. Maza, S. Heathcote, A. Phillips, K. Kawara, C. Anguita, M. Hamuy, A. Gomez.
Astrophys. J., Lett. Ed., Vol. 320, No. 2, p. L157 (1987). See Abstr. 43.131.384.

131.214   Application of an acousto–optical TeO_2 detector in radio astronomy.
J.–s. Wang.
Publ. Yunnan Obs., No. 2, p. 71 – 77 (1987). In Chinese.
A radio astronomical high–resolution acousto–optic spectrograph has been designed and constructed by means of an off–axis slow shear wave TeO_2 acousto–optic detector. In conjunction

with a 4 m millimeter wave radio telescope in Australia, this acousto–optic spectrograph has detected some CO molecular lines with variant shapes from the southern dark clouds. The profile and details of these molecular lines were displayed clearly.

131.215 A new way for presenting the data of a molecular spectral line survey in our Galaxy and an example of its application.
D.–l. Xiang.
Publ. Purple Mt. Obs., Vol. 6, No. 3, p. 215 – 226 (1987).

A new method is developed for presenting the data of molecular spectral line surveys in our Galaxy, to give azimuth information to some extent and improve radial distribution diagrams conventionally adopted in the existing literature.

131.216 On the importance of outflows for molecular clouds and star formation.
C. J. Lada.
Prepr. Steward Obs., No. 751, 20 pp. (1987). Invited review presented at the NATO Advanced Study Institute on "Galactic and extragalactic star formation", June 21 – July 4, 1987, Whistler, Canada.

131.217 The structure of the molecular outflow near SSV 13 and HH 7–11 in the NGC 1333 region.
R. Liseau, G. Sandell, L. B. G. Knee.
Stockholms Obs., Prepr., No. 1987:2, 31 pp. (1987). To appear in Astron. Astrophys.

131.218 Extinction trends at high galactic latitudes as shown by extragalactic extinction indicators.
P. Teerikorpi, S. Haarala.
Rep. Ser., Dep. Phys. Sci., Univ. Turku, No. FTL–R130, 10 pp. (1987). = Turku Univ. Obs., Informo, No. 124. ISBN 951–642–965–3. Submitted to Astrophys. Space Sci.

131.219 Variability of the H_2O maser emission at 1.35 cm wavelength. I. Observational results.
L. Eh. Abramyan, A. P. Venger, I. V. Gosachinskij,
R. A. Kandalyan, R. M. Martirosyan, F. S. Nazaretyan,
V. A. Sanamyan, N. A. Yudaeva.
Astrofiz. Issled. Izv. Spets. Astrofiz. Obs., Tom 24, p. 85 – 92 (1987). In Russian. English translation in Bull. Spec. Astrophys. Obs. – North Caucasus.

The results of observations of 45 H_2O line sources made with the RATAN–600 radio telescope from May 1981 till June 1985 are presented. For 13 sources their line profiles are given, for others – upper limits of their fluxes.

131.220 The NGC 2024 nebula and its exciting stars.
K. G. Dzhakusheva.
Tr. Astrofiz. Inst. Alma–Ata, Tom 48, p. 44 – 49 (1987). In Russian.

131.221 The structure of magnetic fields in R–associations.
L. A. Pavlova.
Tr. Astrofiz. Inst. Alma–Ata, Tom 48, p. 50 – 61 (1987). In Russian.

131.222 The origin of the Orion and Monoceros molecular cloud complexes.
J. Franco, G. Tenorio–Tagle, P. Bodenheimer, M. Rozyczka,
I. F. Mirabel.
Rev. Mex. Astron. Astrofis., Vol. 14, No. 1, p. 240 (1987). Abstract. – See Abstr. 012.042.

131.223 Collisions of high–velocity clouds with a galactic disk.
G. Tenorio–Tagle, P. Bodenheimer, M. Rozyczka,
J. Franco.
Rev. Mex. Astron. Astrofis., Vol. 14, No. 1, p. 241 (1987). Abstract. – See Abstr. 012.042.

131.224 The Orion radio zoo: PIGS (*partially ionized globules*), DEERS (*deeply embedded energetic radio sources*) and FOXES (*fluctuating optical and X–ray emitting sources*).
G. Garay.
Rev. Mex. Astron. Astrofis., Vol. 14, No. 2, p. 489 – 505 (1987). – See Abstr. 012.042.

Recent VLA radio observations of a $\sim 3' \times 3'$ region of the Orion Nebula, centered near the core of the KL nebula, revealing the presence of thirty–five ultracompact radio sources are discussed. Twenty–five of the radio sources are clustered near $\theta^1 C$ Orionis, the most luminous star of the Trapezium cluster, and have optical counterparts. Most of these objects are probably neutral condensations surrounded by ionized envelopes that are excited by $\theta^1 C$. A suggestion for the sequential formation of the Orion radio zoo species is made, as is for the triggering mechanism.

131.225 Variabilidade temporal de fontes maser de vapor d'água associadas ás regiões H II.
B. S. Filho, E. Scalise Jr.
Rev. Mex. Astron. Astrofis., Vol. 14, No. 2, p. 513 (1987). Abstract. – See Abstr. 012.042.

131.226 La nebulosa cometaria 1548C27.
J. M. Vilchez, A. Mampaso.
Rev. Mex. Astron. Astrofis., Vol. 14, No. 2, p. 515 – 516 (1987). – See Abstr. 012.042.

New optical and infrared spectroscopic and photometric observations of the cometary nebula 1548C27 have shown that a young star loosing mass at high velocity (hundreds of km s^{-1}) is the source of energy of the nebula. This source is suggested from its infrared luminosity, to be a Herbig Ae–Be type star surrounded by a dusty envelope radiating at a range of temperatures between 60 and 900K.

131.227 Heating of molecular cloud cores.
S. Lizano, F. H. Shu.
Rev. Mex. Astron. Astrofis., Vol. 14, No. 2, p. 587 – 594 (1987). – See Abstr. 012.042.

The authors examine the role of ambipolar diffusion in heating molecular cloud cores. Following Shu (1983) they calculate the quasistatic evolution of a plane–parallel self–gravitating slab of slightly ionized gas due to the ambipolar diffusion of the field. The temperature in each point of the cloud is computed, taking into account the most important heating and cooling processes. The results can partially explain the thermal differences between the cores which form low and high mass stars.

131.228 Estudio de algunas nubes de alta velocidad del hemisferio sur.
E. Bajaja, C. Cappa de Nicolau, M. C. Martin, R. Morras,
C. A. Olano, W. G. L. Pöppel.
Rev. Mex. Astron. Astrofis., Vol. 14, No. 2, p. 604 – 610 (1987). – See Abstr. 012.042.

Several of the new high–velocity clouds detected in the southern survey by Bajaja et al. (1985) have been observed with better velocity resolution of 2 km s^{-1} and improved spatial grids (generally $0.5° \times 0.5°$). The authors present some of the main results.

131.229 RCW 114: another case of H I bubble related to a WR star.
C. Cappa de Nicolau, V. S. Niemela, G. Dubner, E. M. Arnal.
Rev. Mex. Astron. Astrofis., Vol. 14, No. 2, p. 611 (1987). – See Abstr. 012.042.

131.230 The small scale interstellar dust distribution.
A. Clocchiatti, H. G. Marraco.
Rev. Mex. Astron. Astrofis., Vol. 14, No. 2, p. 613 (1987). Abstract. – See Abstr. 012.042.

131.231 Carbon chemistry of the interstellar medium.
A. C. Danks.
Rev. Mex. Astron. Astrofis., Vol. 14, No. 2, p. 614 – 623 (1987).
– See Abstr. 012.042.
Molecular hydrogen combines with carbon to produce a host
of molecules in the interstellar medium. In diffuse clouds
($A_v < 2$ mag) a number of diatomics can be studied in the visible
ie., CN, CH, CH^+ and C_2.

131.232 Deuterium hyperfine structure in interstellar C_3HD.
M. B. Bell, J. K. G. Watson, P. A. Feldman,
H. E. Matthews, S. C. Madden, W. M. Irvine.
Chem. Phys. Lett., Vol. 136, No. 6, p. 588 – 592 (1987). Abstr. in
Phys. Abstr., Vol. 90, No. 1311, Entry 101891 (1987).

**131.233 Magnetic fields in the regions of possible formation of
planetary systems (from observations of H_2O masers).**
V. S. Strel'nitskij.
Astron. Vestn., Tom 21, No. 4, p. 301 – 303 (1987). In Russian. –
See Abstr. 012.052.

131.234 H_2O maser outburst – protoplanetary rings.
L. I. Matveenko.
Astron. Vestn., Tom 21, No. 4, p. 308 – 311 (1987). In Russian. –
See Abstr. 012.052.

**131.235 Peculiarities of star formation in the Galaxy near the
corotation region.**
L. S. Marochnik, L. M. Mukhin.
Astron. Vestn., Tom 21, No. 4, p. 320 – 322 (1987). In Russian. –
See Abstr. 012.052.

**131.236 Superclouds' giant molecular core formation and origin
of the supersonic turbulence.**
I. G. Kolesnik.
Kinematika Fiz. Nebesn. Tel, Tom 3, No. 6, p. 50 – 58 (1987). In
Russian. English translation in Kinematics Phys. Celest. Bodies.
The physical mechanism of massive turbulent core formation
in superclouds is suggested. It is shown that in a self–gravitating
cloud having a central density which exceeds a definite value
(called the level density) a cooling region appears that turns into a
low–temperature dense core. The conditions for a dense core
formation are appropriate for superclouds with a mass above $5 \times
10^6 M_\odot$ only. The dense core parameters are estimated from
derived analytical expressions: $(1 – 3) \times 10^5 M_\odot$, $100 – 150$ pc.
The original subsonic turbulent motions become supersonic in
the cold core. The turbulence maintains the core in a quasistatic
equilibrium. It is shown that turbulent cores of superclouds after
cooling relaxation acquire parameters typical of the giant
molecular clouds. The nature of the large star forming regions,
giant H II regions, and giant star complexes is discussed on the
basis of these results.

**131.237 Structure of the W3/W4 star formation region.
III. Interstellar extinction and space distribution of dust
clouds in the region of the emission nebula IC 1795.**
L. N. Kolesnik.
Kinematika Fiz. Nebesn. Tel, Tom 3, No. 6, p. 62 – 67 (1987). In
Russian. English translation in Kinematics Phys. Celest. Bodies.
The catalogue of B, V magnitudes and spectral classes of O–B–
A stars in the field around the emission nebula IC 1795 was used
for the investigation of the distribution of dust clouds along the
line of sight by colour excesses of stars. The W3 dusty molecular
cloud complex is located at the distance 2.2 ± 0.2 kpc. The total
visual absorption from the Sun to W3 is between 3.8 ± 0.12^m and
6.7 ± 0.18^m, the extinction in the foreground dust clouds
($r \leqslant 0.9$ kpc) is 1.9 ± 0.15^m. Therefore, the intrinsic foreground
corrected extinction for the region around the emission nebula
IC 1795 ranges from 1.9 ± 0.15^m to 4.8 ± 0.18^m. The eastern and
southern diffuse parts of the H II region of IC 1795 are almost
free from the associated dust matter, the nebula is located at the
edge of the W3 molecular cloud.

131.238 The Orion A helium abundance.
A. P. Tsivilev, A. A. Ershov, G. T. Smirnov,
R. L. Sorochenko.
Sov. Astron. Lett., Vol. 12, No. 6, p. 355 – 359 (1986). English
translation of 42.131.134.

**131.239 Structure of three molecular clouds from HC_3N radio–
line observations.**
S. A. Kolotovkina, R. L. Sorochenko, A. M. Tolmachev.
Sov. Astron. Lett., Vol. 12, No. 6, p. 377 – 380 (1986). English
translation of 42.131.149.

**131.240 Rotational equilibrium of C_2 in diffuse interstellar
clouds. I. Static model: the case of ζ Ophiuchi.**
J. Le Bourlot, E. Roueff, Y. Viala.
Astron. Astrophys., Vol. 188, No. 1, p. 137 – 144 (1987).
A detailed calculation of the statistical equilibrium of C_2 under
conditions prevailing in diffuse interstellar clouds is presented.
New molecular data are used which allow for the consideration of
each individual transition; cascade coefficients are built to make
the computation tractable. The model is applied to the line of
sight toward ζ Oph. A class of models is found, characterised by
the strength of the adopted radiation field, which accounts for
both the excitation of the lower levels of H_2 and most levels of C_2.

**131.241 A search for interstellar NaH and MgH in diffuse
clouds.**
J. Czarny, P. Felenbok, E. Roueff.
Astron. Astrophys., Vol. 188, No. 1, p. 155 – 158 (1987).
The authors report very high signal–to–noise observations in
the wavelength range where interstellar MgH and NaH absorption lines could be observed. They improve the detected
equivalent width limit by a factor over 6 for NaH and 100 for
MgH on the line of sight to ζ Oph. The new upper limit derived
for the column density of MgH is below the one expected if grain
formation processes are efficient.

**131.242 Behind the vdB102 reflection nebula: a study of a
compact molecular cloud's envelope at G 355.5 + 20.9.**
J. P. Vallée.
Astrophys. Space Sci., Vol. 137, No. 2, p. 217 – 224 (1987).
Line observations of the methylidyne (CH) molecule were
performed at Algonquin, toward the reflection nebula vdB102.
An analysis of the molecular cloud behind vdB102 yielded several
envelope parameters, notably a CH column density of $1 \times
10^{13} cm^{-2}$, a microturbulent velocity of 1.4 km s^{-1}, and a total
space density of 1300 cm^{-3}. These observed data are consistent
with a stationary reflection nebula roughly facing the earth,
located on the near side of the surface of a compact molecular
cloud.

131.243 Local interstellar medium.
N. G. Bochkarev.
Astrophys. Space Sci., Vol. 138, No. 2, p. 229 – 302 (1987).
This paper reviews and analyses various observational data
about the local interstellar medium (LISM) – a volume with a
radius of about 200 pc near the Sun. There are collected radio,
IR, optical, UV, and X–ray observations of the ISM and data on
the Sco–Cen association. All available information confirms
Weaver's (1979) conclusions that the Sun is located near an edge
of a giant cavern with a radius of about 180 pc and the cavern
center coincides with the Sco–Cen associated center. Several
arguments are given to show that the bright spots of soft X–rays
($130 – 284$ eV) near the galactic poles are produced by an
interaction of stellar winds with outer edge of the local cloud near
the ends of 'the patch of polarization'. Lyman continuum
radiation from Sco–Cen stars was shown to be probably the main
source of ionization of extended H II regions of low density in the
LISM.

131.244 Interstellare Chemie.
H. Kohl.
Sterne Weltraum, 26. Jahrg., Nr. 12, p. 684 – 689 (1987).

131.245   Internal dynamics of the Gum nebula.
M. Srinivasan, S. R. Pottasch, K. C. Sahu,
J.-C. Pecker.
Messenger, No. 50, p. 11 – 14 (1987).

131.246   Fragmentation of a gas cloud with orthogonal rotation and magnetic axes.
C. G. Campbell, L. Mestel.
Mon. Not. R. Astron. Soc., Vol. 229, No. 4, p. 549 – 572 (1987).
The authors study the evolution of a rotating, cool, self-gravitating gas cloud, permeated by flux from the local galactic magnetic field, and with the magnetic and rotation axes mutually orthogonal.

131.247   MERLIN and VLA observations of the star–forming region G35.2–0.7N.
G. C. Brebner, B. Heaton, R. J. Cohen, S. R. Davies.
Mon. Not. R. Astron. Soc., Vol. 229, No. 4, p. 679 – 689 (1987).
MERLIN observations are presented of OH 1665 MHz masers associated with the bipolar outflow source G35.2–0.7N. The masers are located in a region of expansion near the centre of the molecular disc previously detected in NH_3 and CS observations. The distribution of OH masers has the same elongation and position angle as the NH_3 disc, but on a scale nearly 100 times smaller. VLA maps of the $NH_3(3,3)$ emission from the disc are also presented. The OH masers coincide with a newly–detected compact H II region which lies midway between the two previously known H II regions in G35.2–0.7N, close to the geometrical centre of the disc and the bipolar outflow. The central star is the most likely source of the bipolar outflow. The radio continuum data imply a star of type B0.5 ZAMS, which is consistent with earlier estimates from infrared observations.

131.248   Geometric configuration around GL 2591: an infrared reflection nebula and a CS molecular disk.
T. Yamashita, S. Sato, M. Tamura, H. Suzuki, N. Kaifu,
T. Takano, C. M. Mountain, T. J. T. Moore, I. Gatley,
J. H. Hough.
Publ. Astron. Soc. Jpn., Vol. 39, No. 5, p. 809 – 820 (1987).
The authors present observations of K band ($\lambda = 2.2 \mu m$) polarization and CS $J = 2$–1 and 1–0 emission around GL 2591. The polarization map reveals and infrared reflection nebula extended $(2' \times 1')$ to both east and west. The maps of CS lines delineate a dense disk $(50'' \times 80'')$. The infrared reflection nebula and the bipolar CO flow are parallel to each other, and perpendicular to the CS disk. The density and the mass of the disk are derived to be $1 \times 10^5 cm^{-3}$ and $1000 M_\odot$, respectively. The velocity gradient in the CS emission is nearly perpendicular to the major axis of the disk; this can be interpreted either as rotation about an axis tilted from the disk axis, or as contraction of the disk toward the central source.

131.249   A survey of IRAS point sources in Taurus for high–velocity molecular gas.
M. H. Heyer, R. L. Snell, P. F. Goldsmith, P. C. Myers.
Astrophys. J., Vol. 321, No. 1, p. 370 – 382 (1987).
From a sample of 30 far–infrared point sources, selected from the IRAS Point Source Catalog, the authors have searched for evidence of mass outflow using the ^{12}CO $J = 1$–0 emission to trace the accelerated ambient material. They find evidence for mass outflow toward three far–infrared point sources, two of which are identified as T Tauri stars based on far–infrared colors. The three outflows are dominated by red wing emission, indicative of monopolar outflows. There is also evidence for the large–scale expansion of the B18 cloud from a detailed ^{13}CO map, presumably driven by previous energetic winds of the T Tauri stars now present in the cloud.

131.250   The abundance of H_3^+ ions in dense interstellar clouds.
S. Lepp, A. Dalgarno, A. Sternberg.
Astrophys. J., Vol. 321, No. 1, p. 383 – 385 (1987).
Several simple relationships are written down which express the abundances of C^+, He^+, H_3^+, OH, and H_2O in dense interstellar clouds as functions of the cosmic–ray ionization rate.

It is argued that the primary source of OH is photodissociation of H_2O by cosmic–ray induced photons. The abundance of H_3^+ is derived from observations of OH and predicted to be typically about $2 \times 10^{-5} cm^{-3}$, independent of density. Lower limits of respectively $4 \times 10^{-18} s^{-1}$ and $8 \times 10^{-18} s^{-1}$ are obtained for the cosmic–ray ionization rates in the clouds L134 and B335.

131.251   Interpretation of rotationally excited far–infrared OH emission in Orion–KL.
G. J. Melnick, R. Genzel, J. B. Lugten.
Astrophys. J., Vol. 321, No. 1, p. 530 – 542 (1987).
The authors have observed the $^2\Pi_{1/2}$ OH 163 μm $J = 3/2 \rightarrow 1/2$ rotational transitions in Orion–KL and have also set an upper limit to the line strength of the $^2\Pi_{1/2}$ OH 56 μm $J = 9/2 \rightarrow 7/2$ doublet in this source. The authors analyze the 163 μm line intensities, along with the previously measured $^2\Pi_{3/2}$ 119 and 84 μm rotational line emission in terms of a shock model and of alternate radiative excitation models.

131.252   Detection of interstellar PN: the first identified phosphorus compound in the interstellar medium.
B. E. Turner, J. Bally.
Astrophys. J., Lett. Ed., Vol. 321, No. 1, p. L75 – L79 (1987).
The authors have identified interstellar PN in Ori (KL), W51M, and Sgr B2, by observations of its $J = 2$–1, 3–2, and 5–4 rotational transitions. PN appears to occur in regions of relatively high excitation, with fractional abundances of 1(–11) to 1(–10). Ion–molecule gas phase reactions involving phosphorus at low temperatures cannot account for the observed PN abundances. Other processes, such as high–temperature gas phase reactions (unlikely) or grain disruption seem necessary to explain interstellar PN.

131.253   Detection of interstellar PN: the first phosphorus–bearing species observed in molecular clouds.
L. M. Ziurys.
Astrophys. J., Lett. Ed., Vol. 321, No. 1, p. L81 – L85 (1987).
Phosphorus nitride (PN) has been detected in the interstellar medium. The $J = 2$–1, 3–2, 5–4, and 6–5 rotational lines of this species have been observed toward Orion–KL, and the $J = 2$–1 transition in Sgr B2 and W51. The PN line profiles in Orion indicate that the molecule's emission arises from the "plateau" region associated with the outflow from IRc2. The column densities imply a fractional abundance for PN in the Orion "plateau" of ~ 1–4×10^{-10}. Such a large abundance for phosphorus nitride is not predicted by quiescent cloud ion–molecule chemistry and suggests that high–temperature processes are responsible for the synthesis of PN in the KL outflow.

131.254   Interstellar amorphous carbon.
E. Bussoletti, L. Colangeli, V. Orofino.
Astrophys. J., Lett. Ed., Vol. 321, No. 1, p. L87 – L90 (1987).
The authors discuss amorphous carbon grains as possible candidates for cosmic dust. They have used experimental data on amorphous carbon to construct an "interstellar amorphous carbon", IAC. Extrapolations of laboratory data show that a size distribution with a 10 Å average radius has an extinction efficiency with the peak at 2200 Å satisfactorily matching in shape the interstellar hump. The peak to visual extinction ratio is $\Gamma = 3.8$. IAC requires no more than 18% of the available carbon to produce the hump.

131.255   Ions in grain mantles: the 4.62 micron absorption by OCN^- in W33A.
R. J. A. Grim, J. M. Greenberg.
Astrophys. J., Lett. Ed., Vol. 321, No. 1, p. L91 – L96 (1987).
The 4.62 μm absorption feature in W33A has been uniquely identified with the presence of OCN^- ions in grain mantles. An extensive experimental study of the peak position and absorption strength of this band produced by ultraviolet photolysis of various isotopically labeled ices provided the strict criteria needed to eliminate all conceivable molecular candidates other than the cyanate ion OCN^-. The experiments provide further proof for the ultraviolet photoprocessing of interstellar mantles

leading not only to the formation of radicals and more complex molecules but to molecular ions and salts as well.

131.256   Hydrogen and deuterium in the local interstellar medium.
J. N. Murthy.
Diss. Abstr. Int., Sect. B, Vol. 48, No. 5, p. 1385–B (1987). Thesis, The Johns Hopkins University, 125 pp. (1987). Order No. DA8716635.

131.257   Galactic interstellar abundance surveys with IUE and IRAS.
M. E. Van Steenberg.
Diss. Abstr. Int., Sect. B, Vol. 48, No. 5, p. 1386–B (1987). Thesis, University of Colorado at Boulder, 174 pp. (1987). Order No. DA8716312.

131.258   Investigation of the interstellar absorption in the direction of Kapteyn Selected Areas.
M. S. Kazanasmas, L. A. Zavershneva.
Odes. Univ., Odessa, 13 pp. (1987). In Russian. Abstr. in Ref. Zh., 51. Astron., 12.51.827 (1987).

131.259   The molecules in interstellar medium. Part II. Molecular clouds.
J. Kobus.
Postepy Astron., Tom 34, Zesz. 3, p. 157–179 (1986). In Polish.
The structure and physical conditions in molecular clouds are considered. The molecular cloud in Orion Nebula is discussed in details.

131.260   To B or not to B?
A. P. Boss.
Nature, Vol. 329, No. 6139, p. 486–487 (1987).
The problem of star formation lies in the fact that both gravitationally bound, dense molecular clouds and newly-formed pre–main–sequence stars can be observed, but precious little can be seen in between. What are the intermediate objects, and what physical processes dominate their evolution? The author reports on these two questions, specifically the importance of magnetic fields for star formation and on the chromospheric activity in pre–main sequence stars.

131.261   Water vapor maser in Scutum.
IAU Circ., No. 4414 (1987).

131.262   The magnetic evolution of the Taurus molecular clouds. I. Large–scale properties.
M. H. Heyer, F. J. Vrba, R. L. Snell, F. P. Schloerb, S. E. Strom, P. F. Goldsmith, K. M. Strom.
Astrophys. J., Vol. 321, No. 2, p. 855–876 (1987).
The authors have made finely sampled maps of the ^{13}CO $J = 1$–0 emission from five dark clouds within the Taurus molecular cloud complex and obtained polarization measurements of the optical emission from background and embedded stars to determine the direction of the interstellar magnetic field towards these regions. The clouds have flattened morphologies with the direction of their minor axis parallel to the direction of the magnetic field as expected if the lateral contraction of the gas is inhibited by magnetic pressures. In addition, each cloud appears to be rotating about an axis parallel to the magnetic field direction.

131.263   Are the high–velocity molecular flows really clumpy?
J. Cantó, L. F. Rodríguez, G. Anglada.
Astrophys. J., Vol. 321, No. 2, p. 877–883 (1987).
The authors calculate the line intensities of the low rotational transitions of the ^{12}CO and ^{13}CO molecules for regions with temperature gradients. The relative line intensities are found to be very different from those expected from an isothermal region at any temperature. The results are applied to the standard analysis of CO data from molecular clouds. It is shown that the inferred clumpiness of regions showing peculiar line intensities in the CO may only be an artifact of the assumption of uniform

temperature adopted in the analysis of the data. In particular, this is likely for the high–velocity molecular flows frequently found in regions of current star formation, where strong temperature gradients are expected to be present.

131.264   Interaction of the high–density gas with the bipolar outflow in Cepheus A.
J. M. Torrelles, P. T. P. Ho, L. F. Rodríguez, J. Cantó, J. M. Moran.
Astrophys. J., Vol. 321, No. 2, p. 884–887 (1987).
The authors present the measurements of the CS $J = 1\rightarrow0$ rotational transition line toward Cepheus A made with the 37 m antenna of the Haystack Observatory. The analysis of the CS velocity field suggests that the high–density molecular gas ($n[H_2] \gtrsim 10^4 cm^{-3}$) is disturbed by the stellar wind. A comparison between the CS and NH_3 emission shows that the high–density gas traced by the CS is more disturbed by the stellar wind pressure than that traced by the NH_3.

131.265   A detailed investigation of proposed gas–phase syntheses of ammonia in dense interstellar clouds.
E. Herbst, D. J. DeFrees, A. D. McLean.
Astrophys. J., Vol. 321, No. 2, p. 898–906 (1987).
The initial reactions in two proposed gas–phase syntheses of interstellar ammonia have been studied in detail. The rate of the slightly endothermic reaction $N^+(H_2) \rightarrow NH^+(H)$ under interstellar conditions has been reinvestigated under thermal and nonthermal (translational excitation) conditions based on low-temperature laboratory data. The exothermic reaction $N(H_3^+) \rightarrow NH_2^+(H)$ has been studied via ab initio quantum chemical methods and found to possess a large activation energy barrier. Consequently, this latter reaction does not take place appreciably in interstellar clouds.

131.266   Molecule formation in quasar broad–line cloud gas.
T. Kallman, S. Lepp, P. Giovannoni.
Astrophys. J., Vol. 321, No. 2, p. 907–911 (1987).
Models for the broad–line emitting clouds of quasars typically assume that the clouds have column densities of at most $10^{23} cm^{-2}$. The authors examine the consequences of relaxing this assumption and show that: (1) at slightly larger column densities the gas may cool to $\sim 10^3 K$ as a result of molecule formation; (2) in much of the molecule–forming region the temperature may have either of two values, $\sim 10^3 K$ or 6000–8000K; (3) the strengths of most observable optical lines, including C II] $\lambda2326$ and Fe II lines, are unaffected by such large column densities; (4) lines from low ionization species such as Na I are readily formed at large column densities.

131.267   The unusual dust–scattering properties in the reflection nebula CED 201.
A. N. Witt, R. C. Bohlin, T. P. Stecher, S. M. Graff.
Astrophys. J., Vol. 321, No. 2, p. 912–920 (1987).
Spectrophotometric data available from ground–based and *IUE* observations of the reflection nebula CED 201 and its illuminating star BD +69°1231 are analyzed to deduce the most likely star–nebula geometry and the wavelength dependence of the scattering characteristics of the nebular grains. The wavelength dependence of the optical depth and of the phase function asymmetry in the visible suggests the existence of a narrow size distribution of grains, skewed toward larger than normal particle sizes. The cloud producing CED 201 is a small isolated molecular cloud with a density $n_H > 10^3 cm^{-3}$; the association with the star BD +69°1231 is the result of a random encounter.

131.268   High–resolution ultraviolet observations of interstellar lines toward ζ Persei observed with the balloon–borne ultraviolet stellar spectrometer.
T. P. Snow, H. J. G. L. M. Lamers, C. L. Joseph.
Astrophys. J., Vol. 321, No. 2, p. 952–957 (1987).
The balloon–borne ultraviolet stellar spectrometer payload has been used to obtain high–resolution ($\lambda/\Delta\lambda = 80,000$) data on interstellar absorption lines toward ζ Per. The only lines clearly present in the 2150–2450 region were several Fe II features, which

show double structure. The two velocity components were sufficiently well separated that it was possible to construct separate curves of growth to derive the Fe II column densities for the individual components. A realistic two–component curve of growth for the line of sight to ζ Per was derived, which was then used to reanalyze existing ultraviolet data from Copernicus. The results support the viewpoint that the general diffuse interstellar medium has a nearly constant pattern of depletions.

131.269   Widespread strong methanol masers near H II regions.
R. P. Norris, J. L. Caswell, F. F. Gardner, K. J. Wellington.
Astrophys. J., Lett. Ed., Vol. 321, No. 2, p. L159 – L162 (1987).

A strong methanol maser transition has recently been discovered by Batrla and colleagues. Here the authors report a search of southern sources for masers in this transition. Methanol masers were found in 25 of the 106 star formation regions searched, predominantly those with associated OH masers. A number of other types of object were searched, including OH/IR stars and OH/H_2O megamaser galaxies, but no methanol masers were detected in any of these.

131.270   Observations of the detailed structure and velocity field in the CO bipolar flows associated with L1551 IRS–5.
Y. Uchida, N. Kaifu, K. Shibata, S. S. Hayashi, T. Hasegawa, H. Hamatake.
Publ. Astron. Soc. Jpn., Vol. 39, No. 6, p. 907 – 924 (1987).

The detailed structure and velocity field in the L1551 CO bipolar flows were observed by using the 45–m millimetric–wave telescope at Nobeyama. The observations were made in the 115 GHz ^{12}CO $J = 1$–0 line with spatial and spectral resolution of 18″ and 250 kHz (0.65 km s^{-1} in velocity), respectively. It was revealed as the result that the bipolar flow lobes have a clear hollow cylindrical structure and that both lobes are likely to be spinning with a velocity of $1 – 2$ km s^{-1}. The longitudinal velocity of the flow increases with distance along the axis up to 0.15 pc from IRS–5, the central object. These characteristics coincide well with those predicted by the magnetodynaymic theory proposed by Uchida and Shibata (1985), and indicate the essential importance of the magnetic field in producing such flows. It is also suggested that the angular momentum loss due to the magnetodynamic process is important in the star formation itself.

131.271   High spatial resolution near infrared multi–band imaging of star formation regions.
J. Rayner, I. McLean.
Infrared astronomy with arrays, p. 272 – 280 (1987). – See Abstr. 012.070.

The authors discuss near IR imaging and imaging polarimetry of the star formation regions, GL437 and OMC–2.

131.272   IR imaging of bipolar nebulae/outflows.
C. Aspin, I. S. McLean, J. T. Rayner.
Infrared astronomy with arrays, p. 281 – 285 (1987). – See Abstr. 012.070.

The authors present new infrared images of the bipolar nebulae R Mon and NGC 2261, S106, and OH 0739–14 together with IR imaging polarimetry of the BN/KL region of Orion. Remarkable new structure is seen in R Mon/NGC 2261 where IR streamers and the circumstellar disk are seen for the first time.

131.273   Near–infrared imaging of bipolar outflows.
M. Shure, W. J. Forrest, J. L. Pipher, C. E. Woodward.
Infrared astronomy with arrays, p. 287 – 290 (1987). – See Abstr. 012.070.

The authors have used the University of Rochester's 32×32 InSb/Si–CCD array for 1 to 5 μm imaging of bipolar flows. Images of AFGL 2591, S 140, and Cepheus A show a remarkable amount of structure at the arcsecond scale. In the cases of AFGL 2591 and S 140 the authors have discovered ''bubbles'' on one side of the outflow. On the other end of the evolutionary track is the bipolar nebula associated with

OH 0739–14 which is an outflow from an M9 III giant constrained by a circumstellar disk. The images show the extent of the disk and fine structure in the outflow.

131.274   High resolution images of AFGL 2688 (Egg nebula) with the NASA/GSFC array camera.
D. Jaye, R. Tresch–Fienberg, G. G. Fazio, D. Y. Gezari, G. M. Lamb, P. K. Shu, W. F. Hoffmann, C. McCreight.
Infrared astronomy with arrays, p. 295 – 298 (1987). – See Abstr. 012.070.

High spatial resolution photometric imagery, using the NASA/GSFC 10 μm array camera, of the infrared source at the center of AFGL 2688 (Egg nebulae) at 8.3, 9.8, 11.2 and 12.4 μm is used to investigate its morphology. The infrared source appears as a centrally peaked ellipsoid with its major axis parallel to the axis of the visible nebulosity. Maps of the distribution of color temperature and dust opacity in the source, as derived from the images, suggest no evidence for the dust toroid model, but instead indicate that the source consists of a central star surrounded by a thin dust shell.

131.275   Spectroscopy of the 3 micron emission features with the CGAS (Cooled–Grating Array Spectrometer).
T. Nagata, A. T. Tokunaga, K. Sellgren, R. G. Smith, A. Sakata, S. Wada, T. Onaka, Y. Nakada.
Infrared astronomy with arrays, p. 422 – 425 (1987). – See Abstr. 012.070.

The authors present 3.2 – 3.6 μm spectra of IRAS 21282 + 5050, NGC 7027, and HD 44179.

131.276   Study of Fernie's photometric catalogue of supergiants.
A. S. Miroshnichenko.
Astron. Tsirk., No. 1458, p. 6 – 8 (1986). In Russian.

The parameters of the interstellar extinction lines are obtained, using Fernie's catalogue of UBVRI photometry of supergiants.

131.277   Investigation of the interstellar absorption in the region of the two young stars DI Cep and AS 501.
Yu. K. Bergner, A. S. Miroshnichenko, R. V. Yudin, N. Yu. Yutanov, D. B. Mukanov.
Astron. Tsirk., No. 1459, p. 5 – 7 (1986). In Russian.

The relationship between A_V and distance (D) in the small region near the young stars DI Cep and AS 501 is investigated. It has linear character in the range from O.1 – 0.2 to 1.2 – 1.3 kpc. Some estimates of A_V and D for DI Cep and AS 501 are made.

131.278   On a correlation between width and intensity of H_2O maser lines.
V. S. Strel'nitskij.
Astron. Tsirk., No. 1465, p. 1 – 3 (1986). In Russian.

131.279   Rotational transitions of $(CO_2)H^+$.
V. K. Khersonskij.
Astron. Tsirk., No. 1465, p. 5 – 6 (1986). In Russian.

The frequencies and probabilities of rotational transitions of the interstellar molecule $(CO_2)H^+$ are calculated.

131.280   Millimeter wave spectrum of carbene, HCCN.
V. K. Khersonskij.
Astron. Tsirk., No. 1465, p. 7 – 8 (1986). In Russian.

The frequencies and probabilities of rotational transitions of carbene, HCCN, are calculated for most strong lines in the millimeter wave region of spectrum.

131.281   On a formula for the mean light absorption in the Galaxy.
L. P. Osipkov.
Astron. Tsirk., No. 1467, p. 6 (1986). In Russian.

An expression for the mean light absorption in the Galaxy is found on the basis of a density law.

131.282 A study of clumping in the Cepheus OB3 molecular cloud.
J. S. Carr.
Astrophys. J., Vol. 323, No. 1, p. 170 – 178 (1987).
The author has studied the clumpiness of the molecular gas over a portion of the Cep OB3 giant molecular complex by means of a well–sampled map in the 2.6 mm line of $^{13}CO(1-0)$. The observations reveal a complicated structure in both space and velocity with clumping of the gas down to the spatial resolution. A total of 45 separate clumps of molecular gas are identified based upon correlations in space and velocity. The masses, sizes, and velocity dispersions were measured for all clumps. The Cep OB3 molecular complex appears to be in a state of disruption due to past and present massive star formation. Individual clumps of molecular gas which survive the eventual dispersion of a complex could evolve into clouds similar to either local dark clouds or high–latitude molecular clouds.

131.283 An expanding system of molecular clouds surrounding λ Orionis.
R. J. Maddalena, M. Morris.
Astrophys. J., Vol. 323, No. 1, p. 179 – 192 (1987).
Observations of the $J = 1 \rightarrow 0$ line of CO reveal a ring of molecular gas which is 10° in diameter and centered on the H II region S264 and its ionizing O8 star, λ Ori. A kinematical model for the expanding ring has been fitted to the positions and velocities of the observed molecular peaks. A model is presented for the evolution of a flattened or oblate molecular cloud at whose center an O star forms. The parameters of the model compare favorably with the observed characteristics of the λ Ori system, with the exception of the velocity of λ Ori itself. Possible reasons for this discrepancy are noted.

131.284 Hα scans of high–velocity clouds.
R. J. Reynolds.
Astrophys. J., Vol. 323, No. 2, p. 553 – 556 (1987).
Hα spectra were obtained toward six high–velocity clouds with the Wisconsin large–aperture Fabry–Perot spectrometer. No Hα emission was detected from the clouds down to detection limits of $0.2 - 0.6$ R, which correspond to emission measures smaller than $0.6 - 1.7$ cm^{-6}pc. The clouds appear to be located in a medium of low density, $n < (3-9) \times 10^{-2}$cm^{-3}, or high temperature, $T > 10^4$K, or both.

131.285 Maps of dust clouds at 1.3 millimeters associated with bright H II regions.
M. A. Gordon, P. R. Jewell.
Astrophys. J., Vol. 323, No. 2, p. 766 – 778 (1987).
Maps of the dust clouds associated with W3 Main, NGC 1976, NGC 2024, Sgr B2, M17, W49A, and W51A made at 1.3 mm show brightness distributions often very different from those seen in the FIR. These differences imply that the regions of maximum dust temperature generally do not correspond with regions of maximum column density. The authors give flux densities corrected for free–free emission.

131.286 Gravitational collapse in molecular cloud cores around ultracompact H II regions: two candidates.
E. R. Keto, P. T. P. Ho, M. J. Reid.
Astrophys. J., Lett. Ed., Vol. 323, No. 2, p. L117 – L121 (1987).
The authors report on 10″ resolution observations of NH$_3$(1,1) in the molecular cores surrounding two ultracompact H II regions, W3(OH) and G34.3+0.2. Both sources show extended emission components with velocity gradients consistent with rotation and absorption components that are redshifted with respect to the emission, suggesting gravitational infall. The velocity structure in both sources, and the density and temperature structure in W3(OH), appear similar to that around the previously observed H II region, G10.6–0.4, which is well modeled by a centrally condensed, and centrally heated, molecular core gravitationally collapsing onto the H II region.

131.287 The C/CO ratio in dense interstellar clouds.
R. Gredel, S. Lepp, A. Dalgarno.
Astrophys. J., Lett. Ed., Vol. 323, No. 2, p. L137 – L139 (1987).
Cosmic rays generate energetic secondary electrons which in a dense cloud slow by exciting and ionizing molecular hydrogen. The resulting emission spectrum may cause significant photodestruction of carbon monoxide if the emission lines of H$_2$ overlap photodissociating absorption lines of CO. The authors calculate the cosmic–ray photodestruction rate of CO in dense interstellar clouds, and from models of the chemistry they derive the abundance ratio of neutral atomic carbon to carbon monoxide.

131.288 Nonstability of stars in the starforming region around γ Cygni.
K. P. Tsvetkova.
Star clusters and associations, p. 129 – 140 (1986). – See Abstr. 012.081.
A search for and clarification of the morphology, space distribution and connection between different nonstable types of stars, populating the starforming region around γ Cyg as a basis for the understanding of the galactic structure and the process of star formation in this region have been made.

131.289 H–alpha survey of the galactic nebulae NGC 7129 and IC 5146.
E. Semkov, M. Tsvetkov.
Star clusters and associations, p. 141 – 145 (1986). In Russian. – See Abstr. 012.081.
The results of photometric and spectroscopic (with 4° objective prism) studies in the diffuse nebulae NGC 7129 and IC 5146 are discussed. The data for 50 new Hα–emission stars are given.

131.290 Star formation in molecular clouds of intermediate mass.
P. Mazzei.
Astrophys. Space Sci., Vol. 139, No. 1, p. 37 – 61 (1987).
Self–consistent multicomponent models of evolution of the interstellar medium have been computed by extending the scheme of Habe et al. (1981) and adding some processes of star formation in molecular clouds, induced by supersonic collisions. A monochromatic spectrum of the molecular clouds has been adopted with a cloud mass of $10^4 M_\odot$. The consequences of these simplifying assumptions have been discussed and moreover the influence of several parameters (efficiency of star formation, photoionization rate, cloud radius, and mass) and of the initial conditions has been analyzed.

131.291 Are small diamonds thermodynamically stable in the interstellar medium?
J. A. Nuth III.
Astrophys. Space Sci., Vol. 139, No. 1, p. 103 – 109 (1987).
Very small diamond particles (50 Å diameter) are shown to be thermodynamically stable with respect to similar sized graphite particles for reasonable values of the surface free energies of diamond and graphite. Small diamonds are likely to be stable against both thermal evaporation and chemical attack in the general interstellar medium. A few of the consequences of these conclusions are examined.

131.292 Physical parameters of the diffuse molecular cloud envelope at G 180.9+4.1, next to S241.
J. P. Vallée.
Astrophys. Space Sci., Vol. 139, No. 1, p. 129 – 137 (1987).
Methylidyne (CH) line observations were obtained at Algonquin from the diffuse molecular cloud envelope at G 180.9+4.1, sandwiched between the optical H II region S241 and the molecular cloud core at G 180.8+4.0. An analysis of these observations yields several of the envelope parameters, notably a CH column density of 2×10^{13}cm^{-2}, a microturbulent velocity of 2.6 km s^{-1}, and a total space density of 40 cm^{-3}.

131.293 Star formation, luminous stars, and dark matter.
R. B. Larson.
Am. Sci., Vol. 75, No. 4, p. 376 – 385 (1987). Abstr. in Phys. Abstr., Vol. 90, No. 1317, Entry 140033 (1987).

131.294   A new interstellar maser.
R. D. Brown, D. M. Cragg.
Aust. Phys., Vol. 24, No. 8, p. 184 – 188 (1987). Abstr. in Phys. Abstr., Vol. 90, No. 1318, Entry 146075 (1987).

131.295   Interstellar shock waves and ^{10}Be from ice cores.
C. P. Sonett, G. E. Morfill, J. R. Jokipii.
Nature, Vol. 330, No. 6147, p. 458 – 460 (1987).

The anomalously high concentrations of ^{10}Be in Antarctic ice cores, uncorrelated with δ^{18}O, are consistent with an increase in the atmospheric cosmic ray (CR) flux from CR acceleration in propagating interstellar shock waves, which envelop the heliosphere and whose source may be ancient supernovas. That CR variations attributable to interstellar events such as supernova shock waves have so far not been observed in the CR record is a long–standing issue. If the authors' interpretation of the ^{10}Be spikes is correct, it marks the first such observation.

131.296   H₂O line emission from shocked gas.
D. A. Neufeld, G. J. Melnick.
Astrophys. J., Vol. 322, No. 1, p. 266 – 274 (1987).

The authors compute the H_2O emission expected from a hot astrophysical plasma containing water to obtain (1) a general cooling function for water, and (2) the individual H_2O line intensities in the specific case of the shocked gas region in Orion–KL. It is found that for a shocked molecular region there are several hundred H_2O lines with fluxes that exceed 10^{-18}W cm^{-2} into a 1′beam. The authors also obtain an analytic fit to the total cooling due to water as a function of temperature, H_2 density, and H_2O column density.

131.297   Radio recombination lines from fast shocks in molecular clouds, with application to bipolar flows.
C. F. McKee, D. J. Hollenbach.
Astrophys. J., Vol. 322, No. 1, p. 275 – 287 (1987).

Fast dissociative shocks in molecular clouds are shown to produce detectable radio recombination lines and continuum in addition to optical and infrared recombination lines. Impact broadening sets an upper limit to the brightness of radio recombination lines. Detailed numerical models of the shock structure are used to calculate the effective emission measure, the intensity of optically thin recombination lines, and the effective temperature of the emitting region for shocks with velocities between 40 and 150 km s^{-1} propagating in gas of density 10^4cm${}^{-3} \leqslant n_0 \leqslant 10^6$cm^{-3}. The results are applied to shocks in bipolar flows from young stellar objects in molecular clouds. The H51α line observed by Hasegawa and Akabane in 1984 in Orion–KL is interpreted as emission from a shock in a stellar wind at a velocity of about 70 km s^{-1}.

131.298   Single Gaussian curve of growth abundance determinations from ultraviolet interstellar absorption–line data.
A. W. Harris.
Astrophys. J., Vol. 322, No. 1, p. 368 – 373 (1987).

Data sets used in gas–phase abundance determinations of the elements Fe, P, and S, based on the single Gaussian curve of growth method, are examined for evidence that saturation related errors give rise to, or enhance, the apparent correlations of depletion with mean hydrogen volume density, $\bar{n}_H$. It is shown that the same abundance data which exhibit a strong dependence on $\bar{n}_H$ show no tendency to correlate similarly with the strength or optical depth of the absorption lines, in contrast to the behavior expected if the apparent density dependence were due primarily to effects arising from line saturation.

131.299   Fluorescent excitation of interstellar H₂.
J. H. Black, E. F. van Dishoeck.
Astrophys. J., Vol. 322, No. 1, p. 412 – 449 (1987).

Detailed models of interstellar clouds are used to investigate the infrared emission spectrum of H_2 excited by ultraviolet absorption and fluorescence. The populations of all accessible states of H_2 with rotational quantum number $J \leqslant 15$ are computed as functions of depth through the model clouds. A finding list is presented for the stronger infrared lines of H_2, with

wavelengths based on the best available spectroscopic data. A variety of recent observational results are discussed with reference to the theoretical models. The rich H_2 emission spectrum of the reflection nebula NGC 2023 can be successfully reproduced. The role of fluorescent excitation is explored for NGC 7027, the Galactic center, and the "bright bar" region of the Orion Nebula. Fluxes are also predicted for fluorescent lines of H_2 in T Tauri.

131.300   Radio observations of carbon monoxide toward Zeta Ophiuchi: velocity structure, isotopic abundances, and physical properties.
W. D. Langer, A. E. Glassgold, R. W. Wilson.
Astrophys. J., Vol. 322, No. 1, p. 450 – 462 (1987).

The authors present a map of the ^{12}CO (1–0) emission from the region around ζ Oph at high spectral resolution and high signal–to–noise ratio and also report a measurement of ^{12}CO (2–1) and a detection of ^{13}CO along the line of sight to the star. The CO emission toward the star arises from at least four components with peak emission at –2.0, –0.7, 0.0, and +0.6 km s^{-1}, each with intrinsic velocity dispersion less than or ~ 0.2 km s^{-1}. The observations provide new insights into the physical and dynamical properties of the ζ Oph cloud, often used to test models of interstellar clouds. The existence of numerous narrow velocity components can change abundance determinations and influence the radiation transfer in cloud models.

131.301   Temporal changes of the IRS 5 jet in L1551.
T. Neckel, H. J. Staude.
Astrophys. J., Lett. Ed., Vol. 322, No. 1, p. L27 – L30 (1987).

CCD images taken in 1983, 1985, and 1987 show that in the L1551/IRS 5 jet a new knot appeared near to the central star after 1983. The whole system of knots has moved away from the star by about 1″ in 4 yr.

131.302   Laboratory and astronomical detection of the cyclic C₃H radical.
S. Yamamoto, S. Saito, M. Ohishi, H. Suzuki, S.–I. Ishikawa, N. Kaifu, A. Murakami.
Astrophys. J., Lett. Ed., Vol. 322, No. 1, p. L55 – L58 (1987).

The rotational spectrum of the cyclic C_3H radical was detected in the laboratory and in interstellar space. The radical was produced in the laboratory by discharging a mixture of C_2H_2, CO, and He. Forty–nine lines were assigned, and the molecular constants were precisely determined. The astronomical search was made using the 2_{12}–1_{11} transition at 91.5 GHz, and its four fine and hyperfine components were detected toward TMC–1. The column density of cyclic C_3H is estimated to be 6×10^{12}cm^{-2}, which is comparable to that of the linear C_3H.

131.303   The crucial role of cooling in the making of molecular clouds and stars.
J. E. Tohline, P. H. Bodenheimer, D. M. Christodoulou.
Astrophys. J., Vol. 322, No. 2, p. 787 – 794 (1987). = Lick Obs. Bull., No. B1079.

The authors examine the role that velocity or pressure fluctuations in the interstellar medium can play in initiating compression of sub–Jeans mass diffuse clouds. Nonequilibrium energy arguments are used to determine the fluctuation amplitudes that are required to initiate gravitational collapse of low–mass cloud clumps. The analysis reveals that a cloud which cools under compression – i.e., a cloud of gas for which the effective adiabatic exponent Γ is $0 < \Gamma < 1$ – is particularly sensitive to mild disturbances from its environment. Furthermore, the specific energy required to trigger effective compressions in a cooling medium is nearly independent of the cloud mass. This surprising result suggests that, for a given size disturbance in the H I medium, a wide spectrum of cloud masses below the canonical Jeans mass will condense into self–gravitating clumps. The authors propose that mildly nonlinear disturbances play a primary role in the formation of molecular clouds and, in turn, stars.

131.304   VLA and low–frequency VLBI observations of the radio source 0503 + 467: austere constraints on interstellar scattering in two media.
S. R. Spangler, A. L. Fey, J. M. Cordes.
Astrophys. J., Vol. 322, No. 2, p. 909 – 916 (1987).

The radio source 0503 + 467 lies near the Galactic plane and at the edge of the supernova remnant HB 9. It has a spectrum typical of a compact extragalactic radio source. The small angular size of the source makes it an excellent probe of turbulence in two media: the diffuse or "type A" component of interstellar turbulence and a hypothesized region of hydromagnetic turbulence upstream of the supernova remnant. An eight–station VLBI experiment at 326 MHz indicates that a value of 16 mas is most appropriate as an upper limit to the interstellar scattering contribution to the measured angular size. No evidence is seen for shock–associated turbulence upstream of HB 9.

131.305   Ultraviolet interstellar extinction toward NGC 6530 and the intrinsic energy distribution of 9 Sagittarii and HD 165052.
A. V. Torres.
Astrophys. J., Vol. 322, No. 2, p. 949 – 959 (1987).

The ultraviolet interstellar extinction curve is determined toward 12 early B main–sequence stars and one O8 V star in NGC 6530. The extinction is found to be uniform across the cluster. At short wavelengths, the extinction curve of NGC 6530 falls below the average Galactic curve by a full magnitude, indicating that the population of small grains relative to larger grains is lower than normal. The derived extinction curve is used to deredden the observed fluxes of two early O type stars in NGC 6530: 9 Sgr and HD 165052. The intrinsic continua are compared to model atmospheres in the region 1200 – 8200 Å. The LTE, line–blanketed models of Kurucz fit the observations better than the non–LTE, unblanketed models of Mihalas.

131.306   The interstellar clouds toward 3C 154 and 3C 353.
S. R. Federman, N. J. Evans II, R. F. Willson, E. Falgarone, F. Combes, B. M. Scheufele.
Astrophys. J., Vol. 322, No. 2, p. 960 – 966 (1987).

Molecular observations of the interstellar clouds toward the radio sources 3C 154 and 3C 353 were obtained in order to elucidate the physical conditions within the clouds. Maps of ^{12}CO emission in the $J = 1 \rightarrow 0$ and $J = 2 \rightarrow 1$ lines were compared with observations of the ^{13}CO, CH, and OH molecules. The cloud toward 3C 154 appears to have a low extinction, but a relatively high CO abundance. The cloud toward 3C 353 is considerably denser than that toward 3C 154 and may be more like a dark cloud.

131.307   Velocity waves in 21 centimeter self–absorption toward the Taurus Molecular Complex.
W. L. H. Shuter, R. L. Dickman, C. Klatt.
Astrophys. J., Lett. Ed., Vol. 322, No. 2, p. L103 – L108 (1987).

A $10° \times 7°.5$ region toward the Taurus Molecular Complex has been mapped in 21 cm self–absorption at Arecibo Observatory. The authors find large recorrelation amplitudes in the velocity field of absorbing hydrogen in the cloud complex. Scale lengths are typically 16 pc, and velocity modulation amplitudes 1.5 km s^{-1}. The authors interpret these as velocity waves very likely tightly coupled to, or even resonant with, magnetic field line vibrations. These waves dominate the turbulence within the region.

131.308   VLA observations of H_2CO in absorption against the cosmic background radiation: clumps in the S140 molecular cloud.
N. J. Evans II, M. L. Kutner, L. G. Mundy.
Astrophys. J., Vol. 323, No. 1, p. 145 – 153 (1987). With plate 1.

The authors have used the VLA to observe 6 cm H_2CO absorption in the S140 molecular cloud. They conclude that they are observing anomalous absorption (absorption of the cosmic background radiation). The absorption shows structure on scales from 20″ to 4′. It is possible to construct a consistent model in which the absorption arises primarily in the lower density

envelopes of the dense clumps inferred from multitransition studies of this region. The virial theorem applied to the clumps gives a total mass of 560 – 700 $M_\odot$ for this region, with perhaps 40 $M_\odot$ per clump. It is possible that these clumps represent the first stages in the fragmentation of this portion of the cloud.

The Pulkovo sky survey in the galactic coordinate system.
See Abstr. 002.069.

The Pulkovo sky survey in Magellanic and ecliptic coordinate systems.
See Abstr. 002.070.

Galactic and Magellanic polar areas in the Pulkovo H I survey.
See Abstr. 002.071.

Physics of the Galaxy and interstellar matter.
See Abstr. 003.125.

Physical processes in interstellar clouds. Proceedings of a NATO Advanced Study Institute, held at Irsee, F.R. Germany, 18 – 28 August 1986.
See Abstr. 012.027.

Excitation and dissociation of molecular hydrogen in shock waves at interstellar densities.
See Abstr. 022.031.

Collisional excitation of interstellar sulfur dioxide.
See Abstr. 022.032.

Tabulated extinction efficiencies for various types of submicron amorphous carbon grains in the wavelength range 1000 Å – 300 μm.
See Abstr. 022.034.

The dipole moment of C_3H_2.
See Abstr. 022.055.

C–type transitions in methyl formate.
See Abstr. 022.058.

The laboratory millimeter– and submillimeter–wave spectrum of ^{13}C methanol.
See Abstr. 022.060.

Fluorescent and collisional excitation in diatomic molecules.
See Abstr. 022.073.

Collisional excitation of interstellar cyclopropenylidene.
See Abstr. 022.076.

Infrared spectrum of quenched carbonaceous composite (QCC). II. A new identification of the 7.7 and 8.6 micron unidentified infrared emission bands.
See Abstr. 022.091.

Laboratory and astronomical identification of sulfur–containing carbon–chain molecules, CCS and C_3S.
See Abstr. 022.092.

Heating by PAH molecules or small grains.
See Abstr. 022.093.

High resolution vacuum–ultraviolet spectra and photoabsorption cross sections of carbon monoxide.
See Abstr. 022.094.

Statistical phase space theory of ion–polar molecule systems: application to the reaction $H_2O \cdot H_3O^+ \rightarrow H_2O + H_3O^+$.
See Abstr. 022.103.

Interstellar problems and matrix solutions.
See Abstr. 022.104.

Quasi–statistical solution of early homogeneous nucleation phases.
See Abstr. 022.112.

Amorphous ice. A microporous solid: astrophysical implications.
See Abstr. 022.126.

Grain formation through the nucleation process in astrophysical environments. II. Nucleation and grain growth accompanied by chemical reaction.
See Abstr. 022.136.

Cyclic and linear isomers of $C_3H_2^+$ and $C_3H_3^+$. The $C_3H^+ + H_2$ reaction.
See Abstr. 022.137.

Rotational and vibration–rotational intensities of CS isotopes.
See Abstr. 022.139.

Emission bands of AlH ($X^1\Sigma^+$): the 2–0 sequence.
See Abstr. 022.140.

Spectral parameters of the interstellar molecule H_3O^+.
See Abstr. 022.149.

Laboratory detection of HC_3NH^+ by infrared difference frequency laser spectroscopy.
See Abstr. 022.151.

Laboratory microwave spectroscopy of the vibrational satellites for the v_7 and $2v_7$ states of C_4H and their astronomical identification.
See Abstr. 022.152.

Results with the UKIRT infrared camera.
See Abstr. 034.152.

A 32 × 32 infrared 2 – 5 μm HgCdTe/CCD camera. First results.
See Abstr. 034.158.

IRSPEC: design, performance and first scientific results.
See Abstr. 034.165.

Single–Gaussian curve of growth abundance determinations from UV interstellar absorption line data.
See Abstr. 036.147.

Using SIRTF to study extragalactic star formation.
See Abstr. 051.002.

The origin of the light nuclides.
See Abstr. 061.015.

Astrophysical shocks in diffuse gas.
See Abstr. 062.060.

The origin and cosmogonic implications of seed magnetic fields.
See Abstr. 062.064.

The theory of bipolar flows.
See Abstr. 062.076.

Magnetic braking in weakly ionized media.
See Abstr. 062.082.

Oblique magnetohydrodynamic shock waves in molecular clouds.
See Abstr. 062.101.

Quasi–isentropic non–linear waves in thermally unstable gas.
See Abstr. 062.121.

A hybrid jet of beam–type and wind–type models.
See Abstr. 062.123.

Fragmentation of rapidly rotating gas clouds. I. A universal criterion for fragmentation.
See Abstr. 062.161.

Structural features and scattering properties of dust particles.
See Abstr. 063.037.

Determination of the characteristics of interstellar turbulence from data on the diffusion coefficient of cosmic rays.
See Abstr. 063.062.

On pumping the strong water maser sources.
See Abstr. 063.067.

On pumping the strong water maser sources.
See Abstr. 063.121.

Structure and kinematics of stellar wind bubbles.
See Abstr. 064.005.

Conditions for the formation of massive stars.
See Abstr. 065.057.

Hydromagnetic wave excitation by ionized interstellar hydrogen and helium in the solar wind.
See Abstr. 074.039.

Photoprocessing of H_2S in interstellar grain mantles as an explanation for S_2 in comets.
See Abstr. 102.015.

Comet Halley: a carrier of interstellar dust chemical evolution.
See Abstr. 103.422.

On the interpretation of the 3 μm emission feature in the spectrum of comet Halley: abundances in comet Halley and in interstellar matter.
See Abstr. 103.658.

UV and vacuum UV spectra of organic extract from Yamato carbonaceous chondrites.
See Abstr. 105.040.

Evidence for interstellar SiC in the Murray carbonaceous meteorite.
See Abstr. 105.057.

Large isotopic anomalies of Si, C, N and noble gases in interstellar silicon carbide from the Murray meteorite.
See Abstr. 105.058.

Interstellar polycyclic aromatic hydrocarbons and carbon in interplanetary dust particles and meteorites.
See Abstr. 106.021.

A collapse model of the turbulent presolar nebula.
See Abstr. 107.034.

Echelle and spectropolarimetric observations of the η Carinae nebulosity.
See Abstr. 112.015.

Vibrationally excited CS in IRC + 10216.
See Abstr. 112.016.

The unstable O6.5f?p star HD 148937 and its interstellar environment.
See Abstr. 112.039.

Spectral energy distribution and interstellar reddening.
See Abstr. 112.080.

Interstellar absorption of some Be stars.
See Abstr. 112.088.

Evidence for strong magnetic fields in the inner envelopes of late–type stars.
See Abstr. 112.167.

The rich molecular spectrum and the rapid outflow of OH 231.8 + 4.2.
See Abstr. 112.172.

Airborne spectrophotometry of Eta Carinae from 4.5 to 7.5 microns and a model for source morphology.
See Abstr. 112.174.

Evidence for the effect of the interstellar far UV–radiation on circumstellar shells.
See Abstr. 112.176.

Circumstellar disks and rings – observational results.
See Abstr. 112.177.

Carbon grains in the envelope of IRC + 10216.
See Abstr. 112.180.

On the $(B-V)$ colors of the bright stars.
See Abstr. 113.017.

The spectacular O stars.
See Abstr. 114.043.

Observations of the polarization of the radiation of R–association stars.
See Abstr. 116.009.

IUE and optical observations of the symbiotic star/nebula He2–104.
See Abstr. 117.187.

Synoptic studies of T Tauri stars.
See Abstr. 121.001.

Low–mass star formation in the high galactic latitude dark cloud L 1642.
See Abstr. 121.002.

Molecular hydrogen emission in Herbig–Haro complexes. II. The high latitude nebulosities HH 52/53/54.
See Abstr. 121.004.

Herbig–Haro objects in the vicinity of NGC 2023.
See Abstr. 121.005.

Jets from young stars: CCD imaging, long–slit spectroscopy, and interpretation of existing data.
See Abstr. 121.017.

The kinematic structure of the HH 24 complex derived from high–resolution spectroscopy.
See Abstr. 121.019.

An X–ray survey for pre–main–sequence stars in the Taurus–Auriga and Perseus molecular cloud complexes.
See Abstr. 121.023.

A possible protostar near HH 7–11.
See Abstr. 121.029.

Radio emission from pre–main sequence stars in Corona Australis.
See Abstr. 121.032.

The magnetic field geometry in the vicinity of HH 7–11/HH 12 and HH 33/HH 40.
See Abstr. 121.037.

Water masers associated with low–mass stars: a survey of the Rho Ophiuchi infrared cluster.
See Abstr. 121.041.

Speckle observations of the ice feature in the young double source Serpens SVS 20.
See Abstr. 121.046.

Optical imaging and polarization mapping of the variable nebulosity associated with the PMS star PV Cephei.
See Abstr. 121.048.

The 2 micron spectrum of L1551 IRS 5.
See Abstr. 121.050.

Observations of jets from low–luminosity stars. II. SVS 12, HH 30, near HL Tauri, and HH 34.
See Abstr. 121.051.

A recent change in the brightness and form of the HH46 reflection nebula.
See Abstr. 121.058.

High spectral resolution infrared observations of V1057 Cygni.
See Abstr. 121.059.

Near–infrared H_2 emission from Herbig–Haro objects. I. A survey of low–excitation objects.
See Abstr. 121.060.

FY Aquilae and the gamma–ray burst event of 1979 March 31.
See Abstr. 122.003.

G 109.1–1.0: a supernova remnant interacting with a molecular cloud.
See Abstr. 125.014.

***IRAS* observations of collisionally heated dust in Large Magellanic Cloud supernova remnants.**
See Abstr. 125.038.

Giant–scale supernova remnants. The role of differential galactic rotation and the formation of molecular clouds.
See Abstr. 125.049.

A CO (J = 1–0) survey of five supernova remnants at $l = 70° – 110°$.
See Abstr. 125.054.

Light echoes: supernovae 1987A and 1986G.
See Abstr. 125.091.

The interaction of the UV burst of Supernova 1987A with a nearby cloud: a possible explanation of the speckle images.
See Abstr. 125.245.

Very low upper limits on the strength of interstellar lithium lines toward SN 1987A.
See Abstr. 125.291.

Interstellar calcium towards supernova 1987A in the Large Magellanic Cloud.
See Abstr. 125.318.

Echelle spectroscopy of SN 1986G in NGC 5128: complex Na absorption and interstellar diffuse bands.
See Abstr. 125.562.

Neutral hydrogen absorption measurements of ten pulsars and the electron density in the galactic plane.
See Abstr. 126.063.

Interstellar interferometry of the pulsar PSR 1237 + 25.
See Abstr. 126.066.

VLBI observations of pulsar interstellar scattering.
See Abstr. 126.069.

The Boston University – Arecibo galactic H I survey: the CO connection.
See Abstr. 131.161.

OH observations of galactic radio H II regions.
See Abstr. 132.003.

Further observations of the peculiar galactic radio source BG 2107 + 49.
See Abstr. 132.004.

Dust emission and star formation in compact H II regions.
See Abstr. 132.005.

S201: an H II region produced by an ionization front eroding a molecular cloud.
See Abstr. 132.006.

An infrared study of the stellar population in the direction of the Carina nebula: NGC 3372.
See Abstr. 132.007.

The LMC H II regions N11C and E and their stellar contents.
See Abstr. 132.013.

Optical observations of nebulae.
See Abstr. 132.014.

Centimeter and millimeter recombination lines from W3(OH): expansion or champagne flow?
See Abstr. 132.017.

The EINSTEIN survey of the young stars in the Orion Nebula.
See Abstr. 132.022.

The rotating and collapsing molecular envelope of G 34.25 + 0.14, the "cometary" H II region.
See Abstr. 132.023.

The bipolar H II region S 201.
See Abstr. 132.036.

HCO^+ and SO emission associated with the region G34.3 + 0.2.
See Abstr. 132.040.

The Great Carina Nebula: normal or abnormal extinction?
See Abstr. 132.043.

Submillimeter and far–infrared spectroscopy of M17 and S106: UV–heated, quiescent molecular gas?
See Abstr. 132.048.

Measurements of the ^{3}He abundance in the interstellar medium.
See Abstr. 132.049.

Recent massive star formation in 30 Doradus.
See Abstr. 132.050.

What are "cirrus" point sources?
See Abstr. 133.003.

Near–infrared and optical observations of *IRAS* sources in and near dense cores.
See Abstr. 133.008.

High spatial resolution, far–infrared observations of the Cepheus A region.
See Abstr. 133.012.

Search for the near infrared counterparts of compact H II regions and/or H_2O masers.
See Abstr. 133.020.

Optical spectroscopy of IRAS sources with infrared emission bands: IRAS 21282 + 5050 and the diffuse interstellar bands.
See Abstr. 133.023.

Direct imaging at 12 microns of the star–forming region W51 IRS 2.
See Abstr. 133.024.

Galactic confusion and the detection of 1665 and 1667 MHz OH lines in the neighborhood of planetary nebulae.
See Abstr. 134.046.

The effects of caustics on scintillating radio sources.
See Abstr. 141.007.

The catalogue of H I absorption line profiles in the spectra of galactic radio sources. I. Integral profiles.
See Abstr. 141.017.

Nucleosynthesis and astrophysical gamma ray spectroscopy.
See Abstr. 143.010.

On relativistic particle acceleration in molecular clouds.
See Abstr. 144.017.

Diffusion and nuclear fragmentation of cosmic rays in a cloudy interstellar medium.
See Abstr. 144.026.

Effect of shocks on interstellar turbulence and cosmic–ray dynamics.
See Abstr. 144.056.

The evolution of clumpy gas in young elliptical galaxies.
See Abstr. 151.014.

The stability of a self–gravitating uniform spheroid with azimuthal magnetic field. I.
See Abstr. 151.045.

Accretion of gaseous disks of galaxies. I. Influence of dynamical friction of the large–scale H_2 distribution.
See Abstr. 151.110.

Contraction of a cooling gravitating sphere.
See Abstr. 151.134.

Photometry and spectroscopy of stars in the region of a highly reddened cluster in Ara.
See Abstr. 153.002.

Study of interstellar extinction in some young open clusters.
See Abstr. 153.009.

The Pleiades cluster. IV. The visit of a molecular CO cloud.
See Abstr. 153.024.

Interstellar extinction in Trumpler 37. Infrared results.
See Abstr. 153.031.

The galactic thermal pressure gradient.
See Abstr. 155.128.

Origen cinemático de las nubes obscuras de Tauro y de algunos cúmulos galácticos locales.
See Abstr. 155.129.

The impact of the gas reshuffling between galactic disc and halo on stochastic star formation.
See Abstr. 155.142.

Extinction trends at high galactic latitudes as shown by extragalactic extinction indicators.
See Abstr. 155.147.

The molecules in interstellar medium. Part III. Spiral structure of the Galaxy.
See Abstr. 155.150.

The relative amounts of stars and interstellar matter in the local Milky Way.
See Abstr. 155.162.

The overall structure of gas in the Galaxy.
See Abstr. 155.163.

Large–scale Galactic dust morphology and physical conditions from *IRAS* observations.
See Abstr. 155.166.

On stable hydrostatic equilibrium configurations of the Galaxy and implications for its halo.
See Abstr. 155.170.

A composite CO survey of the entire Milky Way.
See Abstr. 155.171.

The linear filaments of the radio arc near the Galactic center.
See Abstr. 155.172.

Properties of evolved mass–losing stars in the Milky Way and variations in the interstellar dust composition.
See Abstr. 155.173.

A view of the Galactic Hα background: $208° \leqslant l \leqslant 218°$; $-2° \leqslant b \leqslant +8°$.
See Abstr. 155.174.

Star formation in the Magellanic Clouds.
See Abstr. 156.001.

IRAS and ground–based observations of star formation regions in the Magellanic Clouds.
See Abstr. 156.002.

Star formation in the Large Magellanic Cloud.
See Abstr. 156.003.

Molecular hydrogen in H II regions in the Magellanic Clouds.
See Abstr. 156.025.

Molecular clouds in the Large Magellanic Cloud.
See Abstr. 156.034.

Star formation in normal galaxies.
See Abstr. 157.002.

Models for infrared emission from IRAS galaxies.
See Abstr. 157.003.

On the origin of the 40 – 120 micron emission of galaxy disks: a comparison with H–alpha fluxes.
See Abstr. 157.004.

Measuring star formation rates in blue galaxies.
See Abstr. 157.005.

CO observations of galaxies with the Nobeyama 45–m telescope.
See Abstr. 157.006.

CO observations of nearby galaxies and the efficiency of star formation.
See Abstr. 157.007.

Submm observations of IRAS galaxies.
See Abstr. 157.008.

Star formation and spiral structure in M81.
See Abstr. 157.010.

The H II regions in M51: radio and optical observations.
See Abstr. 157.011.

IRAS observations of irregular galaxies.
See Abstr. 157.012.

IUE observations of luminous blue star associations in irregular galaxies.
See Abstr. 157.013.

Neutral hydrogen and star formation in irregular galaxies.
See Abstr. 157.014.

Carbon monoxide emission from small galaxies.
See Abstr. 157.015.

Star formation rates as a function of galaxy mass.
See Abstr. 157.017.

Efficient star formation in the bright bar of M83.
See Abstr. 157.019.

Large scale dissociation of molecular gas in the spiral arms of M51.
See Abstr. 157.020.

Structure and kinematics of the molecular spiral arms in M51.
See Abstr. 157.021.

Modelling the IRAS colors of galaxies.
See Abstr. 157.022.

A simple two–component model for the far–infrared emission from galaxies.
See Abstr. 157.023.

Star forming regions in gas–rich S0 galaxies.
See Abstr. 157.025.

The history of gas in spiral galaxies.
See Abstr. 157.026.

Nuclear star formation on 100 parsec scales: 10$''$ resolution radio continuum, H I, and CO observations.
See Abstr. 157.027.

Induced star formation in interacting galaxies.
See Abstr. 157.029.

The frequency of enhanced star formation in interacting and isolated galaxies.
See Abstr. 157.030.

The azimuthal and radial distributions of H I and H$_2$ in NGC 6946.
See Abstr. 157.033.

The correlation between far–IR and radio continuum emission from spiral galaxies.
See Abstr. 157.038.

Radio continuum, far infrared and star formation.
See Abstr. 157.040.

Near–infrared observations of IRAS minisurvey galaxies.
See Abstr. 157.041.

Enhanced star formation – the importance of bars in spiral galaxies.
See Abstr. 157.043.

Dynamics of a hot ($T \sim 10^7$K) gas cloud with volume energy losses.
See Abstr. 157.085.

Stellar associations and complexes in M33.
See Abstr. 157.090.

The H I distribution in clouds within galaxies.
See Abstr. 157.108.

Formation of shells in elliptical galaxies from interstellar gas.
See Abstr. 157.113.

Optical and near–infrared observations of IRAS galaxies. II.
See Abstr. 157.117.

Star formation in colliding and merging galaxies.
See Abstr. 157.130.

Star formation in disks and spheroids.
See Abstr. 157.131.

Star formation in disks: IRAS results.
See Abstr. 157.132.

Dust and gas: overview.
See Abstr. 157.155.

Properties of elliptical galaxies with dust lanes.
See Abstr. 157.157.

The cold interstellar medium in elliptical galaxies – observations of H I and radio continuum emission.
See Abstr. 157.158.

X–rays from elliptical galaxies.
See Abstr. 157.159.

Metal–enhanced galactic winds.
See Abstr. 157.183.

Infrared emission and star formation in early–type galaxies.
See Abstr. 157.202.

The Leo Triplet spiral galaxy NGC 3628. II. VLA observations of the hydroxyl absorption.
See Abstr. 157.219.

Supernovae blast waves in proto–dwarf galaxies.
See Abstr. 157.236.

Detached supernovae from undetected dwarf galaxies: expected rates from star formation models.
See Abstr. 157.241.

NGC 2403: a flocculent galaxy with two principal centres of star formation.
See Abstr. 157.285.

NGC 6946: kinematics of the starburst in the circum–nuclear zone.
See Abstr. 157.286.

Internal extinction in spiral galaxies. Inclination dependence.
See Abstr. 157.288.

Star formation in blue compact galaxies.
See Abstr. 157.296.

The detection of extragalactic methanol.
See Abstr. 157.303.

A new ionization model for galaxies with dominant star–formation regions.
See Abstr. 157.309.

Giant molecular clouds in M31.
See Abstr. 157.316.

On bursts of star formation in gas–stellar systems with accretion.
See Abstr. 157.328.

Interstellar gas in spiral galaxies.
See Abstr. 157.333.

Molecules in galaxies. V. CO observations of flocculent and grand–design spirals.
See Abstr. 157.334.

The CO contents of dwarf irregular galaxies.
See Abstr. 157.341.

Starburst galaxies.
See Abstr. 158.001.

Infrared spectroscopy of star formation in galaxies.
See Abstr. 158.002.

Molecular gas in the starburst nucleus of M82.
See Abstr. 158.003.

Extragalactic infrared spectroscopy.
See Abstr. 158.005.

Starburst–driven superwinds from infrared galaxies.
See Abstr. 158.006.

A dust scattered halo in starburst galaxy M82?
See Abstr. 158.008.

Radio and infrared observations of (almost) one hundred non–Seyfert Markarian galaxies.
See Abstr. 158.015.

The relation between star formation and active nuclei.
See Abstr. 158.018.

Star formation around active galactic nuclei.
See Abstr. 158.021.

Star formation in Seyfert galaxies.
See Abstr. 158.022.

Circumnuclear "starbursts" in Seyfert galaxies.
See Abstr. 158.023.

Spectrophotometry of Brackett lines in very luminous IRAS galaxies.
See Abstr. 158.025.

Evidence for extended IR emission in NGC 2798 and NGC 6240.
See Abstr. 158.026.

Structure in the nucleus of NGC 1068 at 10 microns.
See Abstr. 158.028.

Summary of symposium: low luminosity sources.
See Abstr. 158.033.

High–dispersion spectroscopy of the clumpy irregular galaxies Markarian 297 and 325.
See Abstr. 158.052.

Interstellar polarization in the dust lane of Centaurus A (NGC 5128).
See Abstr. 158.062.

Infrared emission from young stars in the nucleus of M82.
See Abstr. 158.121.

Star formation in the nucleus of the galaxy NGC 5253.
See Abstr. 158.165.

Zw 15107 + 0724 and the family of OH megamasers.
See Abstr. 158.172.

Starbursts and their dynamics.
See Abstr. 158.229.

Starbursts: nature and implications.
See Abstr. 158.232.

Starbursts: nature and environment.
See Abstr. 158.233.

Enhanced star formation in barred spiral galaxies. II: Radio continuum emission.
See Abstr. 158.239.

Recent star formation in interacting galaxies – III. Evidence from mid–infrared photometry.
See Abstr. 158.240.

H_2O maser emission from the nuclei of NGC 253 and M51.
See Abstr. 158.252.

Molecular hydrogen line emission in Seyfert galactic nuclei.
See Abstr. 158.253.

The megamaser galaxy Markarian 273. I. VLA observations of the hydroxyl emission.
See Abstr. 158.313.

Daily observations of compact extragalactic radio sources at 2695 and 8085 MHz, 1979 – 1985.
See Abstr. 158.327.

OH megamasers in high–luminosity *IRAS* galaxies.
See Abstr. 158.345.

Molecules at early epochs. I. A search for CO toward the quasar PHL 61.
See Abstr. 159.090.

The effect of a quasi–stellar object on its host galaxy: dynamical and physical processes in the interstellar medium around a quasi–stellar object.
See Abstr. 159.141.

Present star formation in spirals of the Virgo cluster.
See Abstr. 160.002.

Molecular gas and star formation in H I–deficient Virgo cluster galaxies.
See Abstr. 160.003.

The interstellar/intergalactic medium in front of supernova 1987A.
See Abstr. 160.099.

132 H II Regions, Emission Nebulae

132.001 Near–IR observations of Sharpless regions. I. S269, S271 and S311.
P. Persi, M. Ferrari–Toniolo, K. Shivanandan, L. Spinoglio.
Astron. Astrophys., Suppl. Ser., Vol. 70, No. 3, p. 437 – 442 (1987).

The authors have searched for near–infrared emission from the Sharpless regions S269, S271, S307 and S311. Seven sources were detected scanning areas of 2.5×2.5 around the radio continuum peaks. The source Irs 2 in S269 shows a very steep infrared energy distribution with $K–L = 2.88$, suggesting the presence of a very young object of luminosity $L(IR) = 5.4 \times 10^3 L_\odot$. Young objects could be present also in the complex H II region S311 as the IRAS observations show. In addition, an extended $2.2\,\mu m$ emission (>2 arcmin) has been observed in this region. Finally, the near–IR observations of S271 and S307 are in agreement with the presence of visible OB stars, at the center of the nebulosity. In particular, the near–IR excess observed in S307, has been interpreted in terms of stellar wind with $\dot{M} = 5 \times 10^{-8} M_\odot/yr$. This wind could affect the evolution of the H II/molecular cloud region.

132.002 EXOSAT observations of the ring nebula NGC 6888 and HD 192163.
H. Kähler, T. Ule, H. J. Wendker.
Astrophys. Space Sci., Vol. 135, No. 1, p. 105 – 109 (1987).

A 25 hr exposure of the ring nebula NGC 6888 was obtained with the EXOSAT X–ray Observatory. X–ray emission of the nebula was not found. Taking all instrumental effects and the large nebular area into account, a conservative upper limit of $1 \times 10^{-12} erg\,s^{-1} cm^{-2}$ (0.05 – 2 keV) is derived. This is about an order of magnitude less than predicted from braking the stellar wind of the central star HD 192163. Two point sources were serendipitously found in the field, HD 192163 and HD 192020.

132.003 OH observations of galactic radio H II regions.
M. A. Braz, P. Sivagnanam.
Astron. Astrophys., Vol. 181, No. 1, p. 19 – 24 (1987).

A sample of 22 galactic radio H II regions has been searched for type I OH masers with the Nançay radiotelescope. Nine OH emission sources were observed, 4 of them being new detections. An analysis of the color indices of the IRAS point sources coincident with the radio continuum of selected H II regions is presented. They are typical of IRAS sources in star forming regions. In the $[25–12] \times [60–25]$ diagram, the IRAS point sources in H II regions with and without associated masers have different distributions of color indices. For the latter sources they are anticorrelated.

132.004 Further observations of the peculiar galactic radio source BG 2107+49.
L. A. Higgs, J. P. Vallée, J. S. Albinson, W. Batrla, W. M. Goss.
Astron. Astrophys., Vol. 181, No. 2, p. 351 – 364 (1987).

The peculiar "comet–shaped" radio source BG 2107+49 (G 19.11+1.58) has been observed with the Very Large Array, the Westerbork Synthesis Radio Telescope, and the NRAO 43–m telescope. These observations have determined its radio morphology at 1.4 GHz, and H I absorption data, in combination with H 103α recombination–line observations, have established that the radio source is about 11 kpc from the Sun. The spectral index is flat, confirmed by observations at the Clark Lake Radio Observatory at 57 MHz and recent data obtained at the Dominion Radio Astrophysical Observatory. The radio source consists of a compact "head" H II region with an approximate diameter of 12 pc, out of which two "tails" appear to emanate. No satisfactory model of the H II complex, which has a total mass in excess of 30000 $M_\odot$, has emerged from this investigation, although it is likely that the object is the result of several physical processes, including a stellar–wind shell and possibly secondary star formation.

132.005 Dust emission and star formation in compact H II regions.
R. Chini, E. Krügel, W. Wargau.
Astron. Astrophys., Vol. 181, No. 2, p. 378 – 382 (1987).

$1 – 5\,\mu m$ observations for 66 compact H II regions are presented and combined with existing FIR and submm data from $12 – 1300\,\mu m$. The total spectra are interpreted in terms of dust emission; radiative transfer calculations indicate that in all regions the exciting stars are surrounded by a cavity of very low dust density. The total energy output of the H II regions places them among the most luminous sources in the Galaxy. A strong relation is found between luminosity and gas mass yielding a star formation efficiency of 8%.

132.006 S201: an H II region produced by an ionization front eroding a molecular cloud.
M. Felli, R. M. Hjellming, R. Cesaroni.
Astron. Astrophys., Vol. 182, No. 2, p. 313 – 323 (1987).

VLA observations at 2 and 6 cm with resolution of $\approx 6''$ of the H II region S201 are presented. The radio source has a bright arc–shaped edge on one side and a smoothly decreasing surface brightness distribution on the opposite side. This configuration is interpreted in terms of a three dimensional electron distribution produced by the ionizing radiation of an early type star located outside a spherical molecular cloud. The classical model of a Strömgren sphere H II region is strongly altered by the gas outflow from the ionization front eroding the molecular cloud. Parameters of the model are optimized by matching a surface brightness model to the observed distribution. An O9 ZAMS star placed at a distance of 0.38 pc from a molecular cloud of 0.42 pc radius allows a good fit to the available data. The molecular cloud is eroded at a rate of $6.2 \times 10^{-5} M_\odot yr^{-1}$ and its gas outflow can produce the observed H II region on a time scale of $8 \times 10^5 yr$.

132.007 An infrared study of the stellar population in the direction of the Carina nebula: NGC 3372.
R. G. Smith.
Mon. Not. R. Astron. Soc., Vol. 227, No. 4, p. 943 – 965 (1987).

Surveys at $2\,\mu m$ have been used to identify the brightest members of the stellar population in the direction of the Carina nebula. Subsequent JHK photometry combined with low–resolution $2\,\mu m$ spectra has been used to separate likely members of the Tr14/16 clusters from the field star population. Several heavily reddened new members of the Tr14/16 association are identified, some of which may be at an earlier evolutionary stage compared to the visible cluster populations.

132.008 A 300 pc thermal spur associated with the H II region S54.
P. Müller, K. Reif, W. Reich.
Astron. Astrophys., Vol. 183, No. 2, p. 327 – 334 (1987).

The authors report multifrequency radio continuum observations of a long narrow thermal spur emerging from the large H II region S54 (W35, NGC 6604). The spur runs perpendicular to the galactic plane at $l \cong 18.5$ from $b \cong 2°$ up to $b \cong 8°$. H110α recombination line emission was detected from both the spur and the H II region at the same radial velocity ($v_{lsr} \cong 30\ km\,s^{-1}$), indicating their physical association. At a kinematic distance of about 2.9 kpc the linear extent of the spur is 25 pc by 300 pc. The authors derive a mean electron temperature of $T_e \cong 4000K$ and a continuous decrease of the mean electron density from $N_e \cong 15\ cm^{-3}$ near S54 to $N_e \cong 1 – 2\ cm^{-3}$ at its largest observed z–distance.

132.009 Brackett–γ mapping observation of RCW 38.
K. Mizutani, H. Suto, H. Takami, T. Maihara, R. K. Sood, J. A. Thomas, H. Shibai, H. Okuda.
Mon. Not. R. Astron. Soc., Vol. 228, No. 3, p. 721 – 727 (1987).

A Brackett–γ map has been obtained in the bright southern H II region RCW 38. The equivalent width of Brackett–γ near the centre is fairly constant and consistent with that expected from recombination theory. A small equivalent width in the periphery of the H II region indicates the presence of some additional continuum emission source, which might be explained by the small grain model applied to reflection nebulae. The existence of a cluster of stars around the central exciting object which has so far been identified as IRS2 is suggested by the comparison of the continuum flux with previous multi–aperture data.

132.010 H166α emission from low–density gas associated with southern H II regions.
I. N. Azcárate, J. C. Cersosimo, F. R. Colomb.
Rev. Mex. Astron. Astrofis., Vol. 15, No. 1, p. 3 – 10 (1987).

H166α line and 1.4 GHz continuum observations have been made of the extended ionized gas associated with the following H II regions: NGC 6334, RCW 97, RCW 116, G 316.8–0.1, NGC 3603, and RCW 131. Electron temperatures and other physical parameters were obtained from the observations. Comparisons have been made with results from higher frequency radio recombination line observations. In general the H II regions present a small and compact central region and an extended envelope of low density.

132.011 Observations of stars in H II regions: *UBVRI* photometry.
J. F. Lahulla.
Astron. J., Vol. 94, No. 4, p. 1062 – 1065 (1987). With plates 80 – 81.

UBVRI photoelectric observations are reported for stars in eight H II regions selected from the Sharpless catalog. Distances and color excesses are given shifting in the $(U–B)$ vs $(B–V)$ two–color diagrams and using the MK or the photometric spectral types. The distances and exciting stars are discussed. The *BVRI* photometry confirms that the extinction law is normal towards the stars in the H II regions.

132.012 Hydrogen recombination lines: a model of the temperature and density in Orion A.
T. L. Wilson, B. Jäger.
Astron. Astrophys., Vol. 184, No. 1/2, p. 291 – 299 (1987).

Five point maps in the H 110α and H 139β lines show that at 4' from the continuum peak, the radio recombination lines are formed in Local Thermodynamic Equilibrium. When combined with radio continuum data, these line results give an average electron temperature, T_e, of 6700 ± 500, where the uncertainty is the RMS scatter. Toward the continuum peak, the value of T_e is 8500K. Based on these results and other recent data a model of Orion A is proposed. This consists of 9 face–on cylindrical slabs, placed on axis.

132.013 The LMC H II regions N11C and E and their stellar contents.
M. Heydari–Malayeri, V. S. Niemela, G. Testor.
Astron. Astrophys., Vol. 184, No. 1/2, p. 300 – 314 (1987).

The authors present a detailed investigation of the LMC H II regions N11C and N11E using several observational techniques. They study the physical properties of these nebulae: gas density, excitation, chemical abundances, extinction, etc. They also study the stars associated with these regions. They determine the spectral types for 9 stars and give the B and V photometry for 58 stars. The authors classify the spectrum of the star Sk –66°41 which has been known as one of the most luminous stars in the LMC, as O5 V and show that it is not a single star but a multiple system. Contrary to what has been believed, this star is not the main exciting source of N11C. Sk –66°43, the main exciting star of N11E, is classified as O4–5 V; several galactic late–type stars in this direction are distinguished. Using CCD long–slit spectra, the authors study the excitation and extinction along several directions in N11C and E. Gas is found to be coupled with dust along several directions. The dust may be associated with the molecular cloud detected near N11.

132.014 Optical observations of nebulae.
B. T. Lynds.
Spectroscopy of astrophysical plasmas, p. 1 – 34 (1987). – See Abstr. 003.010.
Contents: 1. Discovery. 2. Reflection nebulae and interstellar grains. 3. Gaseous nebulae. 4. Kinematics.

132.015 Radio observations of H II regions.
R. L. Brown.
Spectroscopy of astrophysical plasmas, p. 35 – 58 (1987). – See Abstr. 003.010.
Contents: 1. Thermal equilibrium. 2. Thermal bremsstrahlung radiation: Observations and interpretation of the bremsstrahlung radiation from H II regions. Bremsstrahlung emission from inhomogeneous nebulae. 3. Radio recombination line emission: Recombination line emission and absorption coefficients. Transfer of recombination line radiation. Approximations to the recombination line intensity. Departures from local thermodynamic equilibrium. Impact broadening. The relative abundance of helium and hydrogen. Radio recombination lines from atomic carbon. Stimulated radio recombination lines from distant galaxies. 4. Radio observations of H II regions in perspective.

132.016 Echelle observations of the spatially resolved kinematics of a region with high–speed motions in M17 (NGC 6618) – II.
J. Meaburn, C. A. Clayton.
Mon. Not. R. Astron. Soc., Vol. 229, No. 2, p. 253 – 268 (1987). With 3 pp. plates.

In Paper I, Clayton et al. (1985) reported the discovery of a "jet" of ionized gas $\cong 2$ arcsec across emerging from a dark area of M17 with approaching speeds up to 115 km s^{-1}. The vicinity of this jet has now been observed in detail in the light of [O III] 5007 Å. A variety of new high–speed phenomena has been discovered in the ionized gas. The most prominent of these are two isolated knots with radial velocities of –97 and –123 km s^{-1} with respect to the systemic. Also a ring of knots 0.43 pc in extent is found with values of $V_{Hel} = -40$ to -80 km s^{-1} ($\bar{V}_{Hel} = +9$ km s^{-1} for M17). One possibility is that these high–speed phenomena are produced by local sources of wind in the approaching edge of an expanding bubble. In this case the isolated high–speed knots could be ionized stellar jets viewed down their axes or individual bullets of material ejected in the break–up of circumstellar clouds. It also remains possible that many if not all of these high–speed phenomena could be the consequence of the powerful high–speed winds from the central OB association of M17.

132.017 Centimeter and millimeter recombination lines from W3(OH): expansion or champagne flow?
T. L. Wilson, R. Mauersberger, J. Brand, F. F. Gardner.
Astron. Astrophys., Vol. 186, No. 1/2, p. L5 – L8 (1987).

The authors have carried out high precision spectral line and continuum measurements of W3(OH) at 1.3 cm and 7 mm and taken a spectrum at 2 mm in order to study the variation of line and continuum parameters. They compare these and previous data with simple models.

132.018 Continuum versus line polarization at the center of the Orion nebula.
J. L. Leroy, J. F. Le Borgne.
Astron. Astrophys., Vol. 186, No. 1/2, p. 322 – 332 (1987).

New polarization observations have been performed at the center of the Orion nebula with a 10″ spatial resolution, through three filters which allow to study separately the continuum near 5300 Å, Hβ, and the [O III] λ5007 Å. The authors provide the first time line emission polarization measurements which are free from the contamination by the continuum. Their Hβ measurements make it possible to determine the intensity and polarization of the atomic continuum which can be subtracted from the primary continuum data. Eventually, they get the true percentage of polarization P_{sca} for the scattered continuum. These data have been compared with the theoretical model by White et al. (1980). The observations yield P_{sca} values which are generally 2 to 3 times greater than the computed figures. The measurements reveal a good similarity of the Hβ and the [O III] polarization maps which rules out possible intrinsic effects. The polarization of the scattered [O III] radiation has been investigated with the help of a basic geometrical model. Finally, the polarization of the Trapezium stars has been measured.

132.019   Far–infrared [O III] fine–structure lines and ionization structure of RCW 38.
H. Takami, T. Maihara, K. Mizutani, H. Okuda, H. Shibai, T. Nakagawa, J. A. Thomas, R. K. Sood, N. Hiromoto, Y. Kobayashi.
Publ. Astron. Soc. Pac., Vol. 99, No. 618, p. 832 – 835 (1987).

A pair of fine–structure lines of doubly ionized oxygen [O III] 51.8 μm and 88.4 μm have been detected in a southern H II region, RCW 38. A model of the density structure of ionized gas was constructed with the measured line ratio, referring to observations of radio continuum and recombination lines. It is found that there are high–density components of $n_e = 50,000$ cm^{-3} in a low–density ambient gas of $n_e = 150$ cm^{-3}. This model gives the abundance of oxygen, $O/H = 8 \times 10^{-4}$, which is in good agreement with the cosmic abundance.

132.020   [S III] in extragalactic H II regions.
D. R. Garnett.
Bull. Am. Astron. Soc., Vol. 19, No. 2, p. 717 (1987). Abstract. – See Abstr. 010.061.

132.021   S266: an extreme example of a strong stellar wind in an H II region?
M. Fich.
Bull. Am. Astron. Soc., Vol. 19, No. 2, p. 717 (1987). Abstract. – See Abstr. 010.061.

132.022   The EINSTEIN survey of the young stars in the Orion Nebula.
S. Zoonematkermani, J.–P. Caillault.
Bull. Am. Astron. Soc., Vol. 19, No. 2, p. 717 (1987). Abstract. – See Abstr. 010.061.

132.023   The rotating and collapsing molecular envelope of G 34.25 + 0.14, the "cometary" H II region.
D. Wood, E. Churchwell, J. Bieging.
Bull. Am. Astron. Soc., Vol. 19, No. 2, p. 726 (1987). Abstract. – See Abstr. 010.061.

132.024   The variable H II regions in Cepheus A.
V. A. Hughes.
Bull. Am. Astron. Soc., Vol. 19, No. 2, p. 727 (1987). Abstract. – See Abstr. 010.061.

132.025   The velocity width–diameter relation in giant extragalactic H II regions.
R. Arsenault.
Bull. Am. Astron. Soc., Vol. 19, No. 2, p. 727 (1987). Abstract. – See Abstr. 010.061.

132.026   Atomic and molecular gas associated with the H II region S170.
R. S. Roger, P. E. Dewdney, W. H. McCutcheon.
Bull. Am. Astron. Soc., Vol. 19, No. 2, p. 727 – 728 (1987). Abstract. – See Abstr. 010.061.

132.027   Size distributions of H II regions in galaxies. II. Spiral galaxies.
P. W. Hodge.
Publ. Astron. Soc. Pac., Vol. 99, No. 619, p. 915 – 920 (1987).

The size distributions of the H II regions in 21 nearby spiral galaxies are determined from Hα image–tube plates. Almost all integral diameter distributions fit an exponential law, as found for irregular galaxies in Paper I. There is a strong correlation between the luminosity of a galaxy and the slope of the diameter distribution. The slope is shallower in the spiral arms and there is

a weak tendency for it to be shallower also in the outer parts of a galaxy.

132.028   The SMC compact blob N81: a detailed multi–wavelength investigation.
M. Heydari–Malayeri, T. Le Bertre, P. Magain.
ESO Sci. Prepr., No. 521, 51 pp. (1987). Submitted to Astron. Astrophys.

132.029   Ionized gas properties of the peculiar southern H II region RCW 34.
M. Heydari–Malayeri.
ESO Sci. Prepr., No. 540, 46 pp. (1987). Submitted to Astron. Astrophys.

132.030   Correlations between integrated parameters and Hα velocity width in giant extragalactic H II regions: a new appraisal.
R. Arsenault, J.–R. Roy.
ESO Sci. Prepr., No. 556, 25 pp. (1987). Submitted to Astron. Astrophys.

132.031   Far–infrared measurements of N/O in H II regions: evidence for enhanced CN process nucleosynthesis in the inner Galaxy.
D. F. Lester, H. L. Dinerstein, M. W. Werner, D. M. Watson, R. Genzel, J. W. V. Storey.
Astrophys. J., Vol. 320, No. 2, p. 573 – 585 (1987).

Measurements of the [N III] 57.3 μm, [O III] 51.8 μm, and 88.4 μm fine–structure emission lines, obtained with the KAO, are presented for 13 Galactic H II regions spanning a large range in galactocentric distance, and also for two H II regions in the 30 Doradus complex of the LMC. Using newly revised atomic constants, N^{++}/O^{++} ionic ratios are derived for each of these nebulae. The authors argue that this ratio is proportional to the elemental ratio N/O, and its differences around the Galaxy reflect differences in nucleosynthetic history. It is found that N/O varies significantly with galactocentric distance. H II regions inside 7 kpc have N/O that is, on the average, a factor of 3 to 4 higher than those at larger radii. Comparison of the inferred N/O with O/H gives evidence for secondary enrichment of nitrogen in the heavily processed regions of the inner Galaxy. This is expected for nitrogen production by quiescent CN processing in stellar envelopes.

132.032   On the emission of internal dust in four emission nebulae.
D. A. Rozhkovskij.
Tr. Astrofiz. Inst. Alma–Ata, Tom 48, p. 3 – 21 (1987). In Russian.

132.033   On the peripheral structure of the Orion Nebula (M42).
K. G. Dzhakusheva.
Tr. Astrofiz. Inst. Alma–Ata, Tom 48, p. 33 – 43 (1987). In Russian.

132.034   Fotométria UBVRI de las regiones S61, S219, S285 y S299.
J. F. Lahulla.
Rev. Mex. Astron. Astrofis., Vol. 14, No. 2, p. 437 – 439 (1987). – See Abstr. 012.042.

UBVRI photoelectric photometry of stars in four H II regions are reported. Distances and color excesses are given.

132.035   Evolution of H II regions.
H. A. Dottori.
Rev. Mex. Astron. Astrofis., Vol. 14, No. 2, p. 463 – 473 (1987). – See Abstr. 012.042.

The influence of the evolution on general properties of the H II regions is analyzed. The problem of the parametricity, the line ratios generally used as metallicity indicators, and the H II regions as distance calibrators are particularly discussed. Existing determinations of age in the LMC, SMC, and BCG are reviewed. An analysis of radio data on VH + and VHe + to determine ages for H II regions of the Milky Way is presented. Finally, it is also discussed the new idea on the link between H II region evolution and nuclear activity in galaxies.

132.036   The bipolar H II region S 201.
A. Mampaso, P. Pismis, J. M. Vilchez, J. P. Phillips.
Rev. Mex. Astron. Astrofis., Vol. 14, No. 2, p. 474 – 478 (1987).
– See Abstr. 012.042.
The authors discuss the nature of S 201 in the light of new optical and infrared observations, in terms of the interaction between an evolved H II region and a nearby molecular cloud. The energetics of the region and the role of star formation in that area are also briefly discussed.

132.037   Nebulae associated with Of stars: S60, S108, and Of #23.
A. I. Miranda, M. Rosado.
Rev. Mex. Astron. Astrofis., Vol. 14, No. 2, p. 479 – 482 (1987).
– See Abstr. 012.042.
The authors obtained Hα photographs and Fabry–Pérot interferograms of three nebulae presumably associated with Of stars in order to check such associations and to obtain the nebular expansion velocities. Preliminary results indicate that two of these nebulae are wind–blown bubbles.

132.038   The emission object Sharpless 207; is it a planetary nebula?
P. Pismis, I. Hasse, A. Quintero.
Rev. Mex. Astron. Astrofis., Vol. 14, No. 2, p. 483 – 484 (1987).
– See Abstr. 012.042.
A discussion of Fabry–Pérot radial velocities of the Hα line at 21 points over the face of S207 shows that it is an H II region and not a PN as some authors believed. The morphology and the velocity data are consistent with its being a disk–line object seen face–on.

132.039   Estudio cinemático de la región H II S 155: algunos resultados.
M. A. Moreno C.
Rev. Mex. Astron. Astrofis., Vol. 14, No. 2, p. 488 (1987). Abstract. – See Abstr. 012.042.

132.040   HCO⁺ and SO emission associated with the region G34.3 + 0.2.
P. Carral, W. J. Welch, M. C. H. Wright.
Rev. Mex. Astron. Astrofis., Vol. 14, No. 2, p. 506 – 511 (1987).
– See Abstr. 012.042.
The authors have used the Hat Creek millimeter interferometer to map the 89.1 GHz and 86.1 GHz lines of HCO^+ and SO in the direction of the H II region G34.3 + 0.2. The spatial distribution and the measured velocity of the HCO^+ core, support the "champagne phase" model for the compact H II region. A broad (~ 20 km s^{-1}) HCO^+ absorption is detected towards the compact H II region. This suggests the presence of a young object close to the H II region that is producing an outflow.

132.041   Radio observations of M17 and the Orion Nebula: the He/H ratio.
M. Peimbert, N. Ukita, T. Hasegawa, J. Jugaku.
Rev. Mex. Astron. Astrofis., Vol. 14, No. 2, p. 512 (1987). Abstract. – See Abstr. 012.042.

132.042   Modêlos teóricos de linhas de recombinação em radio frequências para regiões H II.
Z. Abraham, A. C. O. Cancoro.
Rev. Mex. Astron. Astrofis., Vol. 14, No. 2, p. 603 (1987). Abstract. – See Abstr. 012.042.

132.043   The Great Carina Nebula: normal or abnormal extinction?
M. Roth, M. Tapia, M. T. Ruiz.
Rev. Mex. Astron. Astrofis., Vol. 14, No. 2, p. 612 (1987). Abstract. – See Abstr. 012.042.

132.044   Solar system–sized condensations in the Orion Nebula.
E. Churchwell, M. Felli, D. O. S. Wood, M. Massi.
Astrophys. J., Vol. 321, No. 1, p. 516 – 529 (1987).
The authors report high–resolution, VLA observations at 2 cm of 22 compact radio continuum sources toward the core of the Orion Nebula. This region contains the highest density of compact radio sources known. The sources are clustered mostly in the region of the Trapezium and the KL nebula. The brightest radio source on 1986 April 28 coincided with θ^1A Orionis. The diameters of the compact radio sources range from less than 4 × 10^{14}cm (<27 AU) to $\sim 3.4 \times 10^{15}$cm (~ 230 AU). All six of the optically visible "nebular condensations" observed by Laques and Vidal (1978) have radio counterparts. Two models for these objects are presented: (1) dense, molecular globules (embedded in the diffuse H II region) with thin but dense ionized envelopes; and (2) low–mass stars (~ 1 $M_\odot$) surrounded by an evaporating protostellar accretion disk. In both cases, ionization is external, produced by UV radiation from θ^1C and other Trapezium stars.

132.045   The velocity structure and turbulence at the center of the Orion Nebula.
H. O. Castañeda.
Diss. Abstr. Int., Sect. B, Vol. 48, No. 5, p. 1383–B – 1384–B (1987). Thesis, Rice University, 190 pp. (1987). Order No. DA8718699.

132.046   Cooled grating spectrometer observations of molecular hydrogen in H II regions in the Magellanic Clouds.
J. Koornneef, F. Israel.
Infrared astronomy with arrays, p. 430 – 433 (1987). See Abstr. 012.070.
The authors obtained near–infrared spectrophotometry of Hα emission line objects in the SMC and the LMC. In six objects, H_2 emission was detected. The molecular hydrogen emission may be caused either by shock excitation due to stellar winds from stars embedded in a molecular cloud or by fluorescence of molecular material in the ultraviolet radiation field of the OB stars exciting the H II region. The molecular hydrogen associated with N88 is at least in part shock–excited. The detected molecular material should have densities in excess of 10^3 H_2 per cm^3.

132.047   Spectrophotometric studies of galactic nebulae. XXVI. The compact H II regions S 270 and Bip 7.
Yu. I. Glushkov, Z. V. Karyagina.
Astron. Tsirk., No. 1457, p. 4 – 6 (1986). In Russian.
Spectra of S 270 and Bip 7 have been obtained. From a comparison of optical and radio observations the authors derive values of extinction A_V = 6.̇6 for S 270 and 3.̇7 for Bip 7. The estimates A_V are in disagreement with those of Neckel, Staude, 1984. A peak density of 2100 cm^{-3} for S 270 and 1100 cm^{-3} for Bip 7 is found.

132.048   Submillimeter and far–infrared spectroscopy of M17 and S106: UV–heated, quiescent molecular gas?
A. I. Harris, J. Stutzki, R. Genzel, J. B. Lugten, G. J. Stacey, D. T. Jaffe.
Astrophys. J., Lett. Ed., Vol. 322, No. 1, p. L49 – L54 (1987).
The authors report measurements of 372 μm $J = 7{\rightarrow}6$ and 186 μm $J = 14{\rightarrow}13$ CO line emission toward the interface between the molecular cloud and the H II region in M17, and toward the center of the bipolar nebula S106. The ratio of $14{\rightarrow}13/7{\rightarrow}6$ lines indicates gas temperatures between 200 and 500 K, and hydrogen densities $\geqslant 10^4$cm^{-3}. The $7{\rightarrow}6$ lines are remarkably narrow, with widths of $\sim 5 – 10$ km s^{-1}, identical to those of the cool quiescent molecular cloud cores. The warm quiescent molecular gas is in the interface between the exciting OB stars and the surrounding molecular cloud. Possible heating mechanisms of the warm CO gas are slow shocks and heating by photoelectrons in UV–illuminated, photodissociation regions.

132.049   Measurements of the ³He abundance in the interstellar medium.
T. M. Bania, R. T. Rood, T. L. Wilson.
Astrophys. J., Vol. 323, No. 1, p. 30 – 43 (1987).
An extensive search for emission from the 8.7 GHz hyperfine line of $^3He^+$ in 17 Galactic H II regions has yielded new detections in six regions and significant upper limits for the others. The authors derive the (^{3}He/H) abundance ratios for their sources and find the largest (^{3}He/H) abundance ratio is more than 10 times greater than their smallest limit. This observation gives strong support for source–to–source variations. There is a marginal tendency for sources outside the solar circle to have larger abundances. For the nine sources where $^3He^+$ was detected, the average (^{3}He/H) abundance ranges from 1.2 to 14.7×10^{-5} by number; for sources with limits, the lowest values are about 1×10^{-5}. The authors discuss briefly the utility of $^3He^+$ as a probe of cosmology or the chemical evolution of the Milky Way.

132.050   Recent massive star formation in 30 Doradus.
N. R. Walborn, J. C. Blades.
Astrophys. J., Lett. Ed., Vol. 323, No. 1, p. L65 – L67 (1987). With plate L1.
Two early O–type stars apparently involved in dense nebular knots have been found in the northeast quadrant of the 30 Doradus nebula. These objects may be very young massive stars just emerging from their protostellar cocoons, and it is suggested that this part of 30 Doradus may represent an earlier evolutionary stage than the central region surrounding R136. A brief survey of recent literature on possibly related objects in the Magellanic Clouds suggests the emergence there of a class corresponding to optically observable, very early evolutionary stages of massive stars.

A ³He cooled bolometer array for the IRAM 30 m telescope.
See Abstr. 034.162.

Speckle masking and speckle spectroscopy.
See Abstr. 036.082.

Chemical abundances.
See Abstr. 061.035.

Astrophysical shocks in diffuse gas.
See Abstr. 062.060.

Quasi–isentropic non–linear waves in thermally unstable gas.
See Abstr. 062.121.

Emission–line ratios for O III in gaseous nebulae and a comparison between theory and observation.
See Abstr. 063.035.

Predicted [Ni II] infrared line strengths in the Crab Nebula and IC 443.
See Abstr. 063.107.

Structure and kinematics of stellar wind bubbles.
See Abstr. 064.005.

NLTE models for cocoon stars.
See Abstr. 064.044.

IRAS observations of the giant shell surrounding λ Orionis.
See Abstr. 112.164.

Airborne spectrophotometry of Eta Carinae from 4.5 to 7.5 microns and a model for source morphology.
See Abstr. 112.174.

Circumstellar disks and rings – observational results.
See Abstr. 112.177.

Observations of the polarization of the radiation of R–association stars.
See Abstr. 116.009.

Observations at 408 MHz of the supernova remnant HB 3 (G132.6 + 1.5).
See Abstr. 125.026.

On the progenitors of type II supernovae in M83.
See Abstr. 125.068.

High mass star formation in the Galaxy.
See Abstr. 131.001.

IRAS colors of VLA identified objects in the Galaxy.
See Abstr. 131.003.

Star formation in Carina OB1: observations of a giant molecular cloud associated with the η Carinae nebula.
See Abstr. 131.004.

On the redistribution of OB star luminosity and the warming of nearby molecular clouds.
See Abstr. 131.006.

Observations of cold dust in S 106.
See Abstr. 131.029.

A multilevel study of ammonia in star forming regions. II. G 34.3 + 0.2, a new "hot core".
See Abstr. 131.030.

IRAS observations of southern molecular clouds.
See Abstr. 131.044.

Temperature and density structure of the collapsing core of G 10.6–0.4.
See Abstr. 131.053.

Behind the H II region Sharpless 217: the envelope of the diffuse molecular cloud at G159.1 + 3.3.
See Abstr. 131.065.

1300 micron continuum and $C^{18}O$ line mapping of the giant molecular cloud cores in Orion, W49, and W51.
See Abstr. 131.078.

CO maps of the OMC–1 outflow.
See Abstr. 131.080.

A study of the ground–state hydroxyl maser emission associated with 11 regions of star formation.
See Abstr. 131.081.

Molecular line observations of the H II region G 34.3 + 0.2.
See Abstr. 131.113.

Gas–and–dust complex NGC 7822 + S 171 connected with the Cep OB4 association.
See Abstr. 131.115.

Coronal interstellar gas and supernova remnants.
See Abstr. 131.123.

Echelle observations of interstellar bubbles.
See Abstr. 131.144.

The vicinity of Omicron Per.
See Abstr. 131.148.

Far–infrared spectroscopy of the DR 21 star formation region.
See Abstr. 131.168.

Mainline OH masers near young H II regions: a correlation with IRAS far–infrared flux density.
See Abstr. 131.200.

Variabilidade temporal de fontes maser de vapor d'água associadas ás regiões H II.
See Abstr. 131.225.

Structure of the W3/W4 star formation region. III. Interstellar extinction and space distribution of dust clouds in the region of the emission nebula IC 1795.
See Abstr. 131.237.

The Orion A helium abundance.
See Abstr. 131.238.

Local interstellar medium.
See Abstr. 131.243.

Interpretation of rotationally excited far–infrared OH emission in Orion–KL.
See Abstr. 131.251.

An expanding system of molecular clouds surrounding λ Orionis.
See Abstr. 131.283.

Maps of dust clouds at 1.3 millimeters associated with bright H II regions.
See Abstr. 131.285.

Gravitational collapse in molecular cloud cores around ultracompact H II regions: two candidates.
See Abstr. 131.286.

Physical parameters of the diffuse molecular cloud envelope at G 180.9 + 4.1, next to S241.
See Abstr. 131.292.

High spatial resolution, far–infrared observations of the Cepheus A region.
See Abstr. 133.012.

Broad IR emission lines from NGC 2024 IRS 2.
See Abstr. 133.013.

Search for the near infrared counterparts of compact H II regions and/or H_2O masers.
See Abstr. 133.020.

High spatial resolution observations of dust and gas in NGC 7027 & the M8 Hourglass.
See Abstr. 134.055.

The identification of galactic radio sources based on a comparison of radio–continuum and infrared emission.
See Abstr. 141.009.

Diffuse Einstein X–ray emission from the region of RCW 49, a bright southern emission nebula.
See Abstr. 142.051.

Stockert's chimney – a galactic fountain.
See Abstr. 155.027.

Structural details of the Sagittarius A complex: evidence for a large–scale poloidal magnetic field in the Galactic center region.
See Abstr. 155.114.

The broad–line region at the center of the Galaxy.
See Abstr. 155.115.

Ionization state in and reddening to the center of the Galaxy.
See Abstr. 155.116.

Spiral structure and age distribution of the H II regions in the Milky Way.
See Abstr. 155.125.

A view of the Galactic Hα background: $208° \leqslant l \leqslant 218°$; $-2° \leqslant b \leqslant +8°$.
See Abstr. 155.174.

IRAS and ground–based observations of star formation regions in the Magellanic Clouds.
See Abstr. 156.002.

Infrared emission and excitation in LMC H II regions.
See Abstr. 156.004.

The nature of dust grains in the clouds of Magellan: $8 - 13$ μm spectra of LMC N44A and SMC N88A.
See Abstr. 156.011.

Molecular hydrogen in H II regions in the Magellanic Clouds.
See Abstr. 156.025.

Spectrophotometric observations of very low ionization H II regions in the LMC.
See Abstr. 156.032.

Star formation and spiral structure in M81.
See Abstr. 157.010.

The H II regions in M51: radio and optical observations.
See Abstr. 157.011.

Neutral hydrogen and star formation in irregular galaxies.
See Abstr. 157.014.

Star forming regions in gas–rich S0 galaxies.
See Abstr. 157.025.

The initial mass function in H II galaxies.
See Abstr. 157.031.

Optical and IR luminosity functions of IRAS galaxies.
See Abstr. 157.037.

Spectroscopic analysis of Kiso ultraviolet–excess galaxies.
See Abstr. 157.060.

Les gaz ionisés hors du noyau d'une galaxie spirale ou irrégulière.
See Abstr. 157.078.

Stellar associations and complexes in M33.
See Abstr. 157.090.

Imaging spectrophotometry of a chain of giant H II regions in the galaxy NGC 2997.
See Abstr. 157.097.

Giant H II regions in M81.
See Abstr. 157.107.

Optical and near–infrared observations of IRAS galaxies. II.
See Abstr. 157.117.

Ionized gas in elliptical galaxies.
See Abstr. 157.156.

Nitrogen abundances in the amorphous galaxy NGC 5253.
See Abstr. 157.201.

The brightest "stars" are not always stars.
See Abstr. 157.232.

Giant H II regions as distance indicators.
See Abstr. 157.290.

A new ionization model for galaxies with dominant star–formation regions.
See Abstr. 157.309.

Star formation around active galactic nuclei.
See Abstr. 158.021.

Mark 277 – clumpy irregular galaxy.
See Abstr. 158.127.

H II region abundances in Seyfert galaxies.
See Abstr. 158.153.

Star formation in the nucleus of the galaxy NGC 5253.
See Abstr. 158.165.

Spectral investigation of the peculiar galaxy NGC 6240.
See Abstr. 158.341.

Cosmological H II regions and the photoionization of the intergalactic medium.
See Abstr. 161.355.

133 Infrared Sources

133.001   Systematic identification of IRAS point sources.
A. Savage, R. G. Clowes, H. T. MacGillivray,
R. D. Wolstencroft, S. K. Leggett, P. J. Puxley.
NASA Conf. Publ., NASA CP–2466, p. 537 – 545 (1987). – See Abstr. 012.002.

The authors have identified sources in 44 Schmidt plate areas including 1300 sources and covering 1100 square degrees. The identifications comprise 700 galaxy identifications (field and cluster members) and 600 stellar identificatons. There are also about 40 sources with no obvious identification but which can be most easily explained by cirrus, confusion between two sources or sources just outside the 2 sigma error box.

133.002   A very deep IRAS survey at $l^{II} = 97°$, $b^{II} = +30°$.
P. Hacking, J. R. Houck.
NASA Conf. Publ., NASA CP–2466, p. 547 – 552 (1987). – See Abstr. 012.002.

A deep far–infrared (12 – 100 μm) survey is presented using over one thousand scans made of a 4 – 6 deg^2 field at the north ecliptic pole by the Infrared Astronomical Satellite (IRAS). Point sources from this survey are up to one hundred times fainter than the IRAS point source catalog at 12 and 25 μm, and up to ten times fainter at 60 and 100 μm. The majority of the 12 μm point sources are stars within the Milky Way. The 25 μm sources are composed almost equally of stars and galaxies. About 80 percent of the 60 μm sources correspond to galaxies on Palomar Observatory Sky Survey (POSS) enlargements. The remaining 20 percent are probably galaxies below the POSS detection limit. The differential source counts are presented and compared with theoretical models.

133.003   What are "cirrus" point sources?
C. Heiles, P. J. McCarthy, W. Reach, M. A. Strauss.
NASA Conf. Publ., NASA CP–2466, p. 553 – 558 (1987). – See Abstr. 012.002.

Most "cirrus" point sources are associated with interstellar gas. The authors have isolated a subset of these, together with other sources showing large band 4 to band 3 flux density ratios, that are not associated with interstellar gas. Most of the point sources are associated with diffuse cirrus emission. The sources appear to be distributed randomly on the sky, with the exception of six clusters, one of which is not associated with any known astronomical object. Six sources out of seventeen that were observed for redshifted H I at Arecibo were found to be associated with relatively nondescript external galaxies. Most of the sources do not appear on the Palomar Sky Survey. Deep optical observations of eight fields revealed some fairly distant galaxies, one object with a very peculiar optical spectrum, and several blank fields.

133.004   An infrared–optical study of IRAS point sources in the Virgo region.
S. K. Leggett, R. G. Clowes, M. Kalafi, H. T. MacGillivray,
P. J. Puxley, A. Savage, R. D. Wolstencroft.
Mon. Not. R. Astron. Soc., Vol. 227, No. 3, p. 563 – 588 (1987).

Optical identifications are given for 199 of the 206 point sources detected by the Infrared Astronomical Satellite (IRAS) in a 113 deg^2 area centred on the Virgo Cluster. The identifications are made using four deep IIIa–J plates taken with the 1.2–m UK Schmidt Telescope. Fifty–four of the sources are associated with stars, 113 with optically bright ($B_j < 16$) galaxies, 32 with faint ($B_j > 16$) galaxies, and seven are apparently empty fields, to the plate limit of $B = 22$. This area is affected by infrared cirrus, with which five of the seven empty fields are associated. The authors have created an infrared–optical Virgo galaxy database, complete to about $B = 16$, by combining their data with the catalogue of Binggeli, Sandage & Tammann. The authors find that the infrared properties of the Virgo cluster galaxies are indistinguishable from those of field galaxies at similar redshifts.

133.005   Identification of new young stellar objects associated with IRAS point sources. I. The southern galactic plane.
S. E. Persson, B. Campbell.
Astron. J., Vol. 94, No. 2, p. 416 – 428 (1987). With plates 50 – 53.

The results of a survey of bright IRAS point sources along the southern Milky Way are presented. The survey was designed to search in the near infrared for new examples of young stellar objects, many of whose northern counterparts are responsible for driving bipolar outflows in molecular clouds. Photometry in the J, H, and K bands is given for nearly all of the 113 near–infrared sources that were found to be associated with the candidate IRAS sources, L magnitudes are given for 60, and M magnitudes for 14. Positions accurate to $\leqslant 2''$ are presented for all the objects detected at K. Forty–five new YSO candidates have been found; the remaining IRAS objects are probably field stars suffering heavy extinction. Deep i band (8000 Å) CCD frames are presented for most of the prime YSO candidates. These are used to examine the surroundings of each source for associated stars and nebulosity; at least half of the candidates have associated optical objects. Low–resolution spectrophotometry in the 2.2 μm region of 15 of the primary candidates confirm that none is a compact H II region or late–type star.

133.006   A study of the silicate emission features of the IRAS low resolution spectra.
O. Gal, M. de Muizon, R. Papoular, B. Pégourie.
Astron. Astrophys., Vol. 183, No. 1, p. 29 – 37 (1987).

Using the IRAS catalog of low resolution spectra, the authors have analyzed 1808 silicate emission features, the union of classes 2n and 6n. The sample size is large enough to establish average properties (e.g. energy distributions); correlations between luminosities, excesses, colour and coordinates; histograms and galactic distributions, with a good degree of confidence. Of particular interest are: the detection of the largest silicate excesses ever observed ($\gtrsim 10$) and their location in the bulge of the Galaxy; the progressive strengthening of the 10–μm silicate feature as the objects become fainter (and hence more distant); the tendency for the 10–μm excess to decrease as the galactic latitude increases; the ranges of "effective" dust temperatures (150 to 500K), star colour temperatures (700 to 5000K) and envelope optical thicknesses (0 to ~ 0.5 at 10 μm). This analysis also points to the similarity of dust properties between IRAS objects and previously observed Long Period Variables.

133.007 The IRAS view of the extragalactic sky.
B. T. Soifer, J. R. Houck, G. Neugebauer.
Annu. Rev. Astron. Astrophys., Vol. 25, p. 187 – 230 (1987). – See Abstr. 003.003.
The authors describe what has been learned to date about the extragalactic sky from the IRAS data. They present an overview of the numerical and morphological properties of the IRAS extragalactic observations and a brief discussion of the mechanisms that give rise to radiation at the survey wavelengths. Observations of normal galaxies and of active galactic nuclei are discussed. Many galaxies have been found to have large infrared luminosities. Cosmology and deeper surveys are addressed.

133.008 Near–infrared and optical observations of *IRAS* sources in and near dense cores.
P. C. Myers, G. A. Fuller, R. D. Mathieu, C. A. Beichman, P. J. Benson, R. E. Schild, J. P. Emerson.
Astrophys. J., Vol. 319, No. 1, p. 340 – 357 (1987).
The authors report photometry from 0.4 to 20 μm of 34 *IRAS* point sources associated with dense cores in dark clouds. The main results of this study are: (1) Stars *near* cores tend to be visible T Tauri stars, while stars *in* cores tend to have circumstellar extinction $A_V \sim 30 – 90$ mag and luminosity $\sim 1\,L_\odot$, similar to that of T Tauri stars. (2) The typical highly obscured star is probably accompanied by a luminous structure of substellar temperature, such as a circumstellar disk. (3) In Taurus–Aurigae, stars in cores probably become visible T Tauri stars less than 1×10^5yr after they attain luminosity greater than $\sim 0.1\,L_\odot$. This implies that they are extremely young and may still be accreting.

133.009 A very deep far–infrared survey.
P. B. Hacking.
Diss. Abstr. Int., Sect. B, Vol. 48, No. 1, p. 161–B (1987). Thesis, Cornell University, 219 pp. (1987). Order No. DA8708882.

133.010 Optical photometry of high latitude 12 μm IRAS sources.
B. D. Goodrich, J. L. Africano, W. X. Binkert, S. G. Kleinmann.
Bull. Am. Astron. Soc., Vol. 19, No. 2, p. 680 (1987). Abstract. – See Abstr. 010.061.

133.011 A very deep far–infrared survey.
P. B. Hacking.
Bull. Am. Astron. Soc., Vol. 19, No. 2, p. 711 (1987). Abstract. – See Abstr. 010.061.

133.012 High spatial resolution, far–infrared observations of the Cepheus A region.
H. B. Ellis Jr., D. F. Lester, P. M. Harvey, M. Joy, C. M. Telesco, R. Decher, M. W. Werner.
Bull. Am. Astron. Soc., Vol. 19, No. 2, p. 726 (1987). Abstract. – See Abstr. 010.061.

133.013 Broad IR emission lines from NGC 2024 IRS 2.
T. R. Geballe, H. A. Smith, J. Fischer.
Bull. Am. Astron. Soc., Vol. 19, No. 2, p. 728 (1987). Abstract. – See Abstr. 010.061.

133.014 High spatial resolution far–infrared observations of the star–forming region AFGL 437.
P. S. Parmar, D. F. Lester, P. M. Harvey, H. B. Ellis Jr.
Bull. Am. Astron. Soc., Vol. 19, No. 2, p. 728 (1987). Abstract. – See Abstr. 010.061.

133.015 High spatial resolution infrared imaging of L1551– IRS5: direct observations of its circumstellar envelope.
A. Moneti, W. J. Forrest, J. L. Pipher, C. E. Woodward.
ESO Sci. Prepr., No. 555, 21 pp. (1987). To appear in Astrophys. J.

133.016 Temporal changes of the IRS 5 jet in L 1551.
T. Neckel, H. J. Staude.
Max–Planck–Inst. Astron. Heidelb., Prepr., 12 pp. (1987). Submitted to Astrophys. J., Lett. Ed.

133.017 Structure of the extended emission in the infrared celestial background.
S. D. Price.
Proc. SPIE Int. Soc. Opt. Eng., Vol. 685, p. 167 – 180 (1986). Abstr. in Phys. Abstr., Vol. 90, No. 1309, Entry 88129 (1987). – See Abstr. 012.039.

133.018 CCD and near infrared observations of the IRAS point sources in LD1448 and LD1455.
C.–p. Liu, Z.–y. Qian, H. Kimura, J. Yan.
Publ. Purple Mt. Obs., Vol. 6, No. 3, p. 207 – 214 (1987).
Results of CCD and near infrared observations of the IRAS point sources in LD1448 and LD1455 are discussed. Based on the energy distributions derived from IRAS data and near infrared observations, and other properties, the structure and evolutionary status of these sources are discussed.

133.019 The energy distribution of 630 sources from the Valinhos 2 μm survey.
N. Epchtein, T. Le Bertre, J. R. D. Lépine, P. Marques dos Santos, O. T. Matsuura, E. Picazzio.
Rev. Mex. Astron. Astrofis., Vol. 14, No. 1, p. 375 – 378 (1987). – See Abstr. 012.042.
630 sources detected by the Valinhos 2 μm survey were observed in the JHKLM bands with the ESO 1–m telescope. The objects were also identified with entries of the IRAS point source catalogue. Based on a statistical investigation of their infrared energy distribution, the authors propose a classification of the sources.

133.020 Search for the near infrared counterparts of compact H II regions and/or H$_2$O masers.
R. Carballo, A. Mampaso, N. Epchtein.
Rev. Mex. Astron. Astrofis., Vol. 14, No. 2, p. 485 – 487 (1987). – See Abstr. 012.042.
The authors present the preliminary results of a search for near–infrared counterparts of compact H II regions and/or H$_2$O masers in the Galactic Plane, probably associated with IRAS sources. *J, H, K, L* and *M* photometry of 10 new detections is reported.

133.021 Fuentes infrarrojas en 2.2 μm en la región asociada al globulo Ori–I–2.
I. Cruz–González, L. Carrasco, M. de la Paz Ramos.
Rev. Mex. Astron. Astrofis., Vol. 14, No. 2, p. 519 (1987). Abstract. – See Abstr. 012.042.

133.022 Ground–based infrared observations of variable IRAS sources as candidates for late asymptotic giant branch stars.
S. Kwok, B. J. Hrivnak, R. T. Boreiko.
Astrophys. J., Vol. 321, No. 2, p. 975 – 983 (1987). = Publ. Rothney Astrophys. Obs., No. 47
Analysis of the color distribution of OH/IR stars and IRAS spectra suggests the presence of a well–defined evolutionary sequence which is populated by late asymptotic giant branch (LAGB) stars. The authors report ground–based identification

and infrared photometry of 10 candidates of new LAGB stars. None of the selected sources are found to have optical counterparts, and eight of the 10 show a strong 10 μm silicate absorption feature. The authors suggest that these stars represent an invisible extension of extreme Mira variables and are some of the most evolved stars observed to date.

133.023   Optical spectroscopy of IRAS sources with infrared emission bands: IRAS 21282 + 5050 and the diffuse interstellar bands.
M. Cohen, B. F. Jones.
Astrophys. J., Lett. Ed., Vol. 321, No. 2, p. L151 – L157 (1987).
   Spectroscopy of the starlike optical counterpart to IRAS 21282 + 5050, a source with the "hydrocarbon" infrared emission band spectrum, shows an O7(f)–[WC 11] planetary nebula nucleus suffering an extinction of 5.7 mag. Emission–line widths in the WC spectrum are only ~ 100 km s^{-1}, indicating a very slow stellar wind. Optical diffuse interstellar bands (DIBs) are prominent. Five DIBs are strongly enhanced, namely $\lambda\lambda 5797$, 6196, 6203,6283, and 6613. The presence of circumstellar hydrocarbon molecules may explain both the infrared emission bands and the enhanced DIBs.

133.024   Direct imaging at 12 microns of the star–forming region W51 IRS 2.
J. Bally, J. F. Arens, R. Ball, R. Becker, J. Lacy.
Astrophys. J., Lett. Ed., Vol. 323, No. 1, p. L73 – L76 (1987). With plate L2.
   The authors have made direct imaging observations of the luminous star–forming region W51 IRS 2. The distribution of the 12 μm emission shows a symmetric "dumbbell" in which the central and brightest structure is $\sim 4'' \times 1''$ in size. The observed surface brightness is consistent with optically thick 12 μm emission from dust at radiative equilibrium with a $3 \times 10^6 L_\odot$ central source. These results suggest that W51 IRS 2 is a hot, very dense massive disk surrounding a very luminous and massive protostellar source.

Valinhos 2.2 micron survey of the southern galactic plane. II. Near–IR photometry, IRAS identifications and nature of the sources.
See Abstr. 002.029.

Catalog of Infrared Observations. Part I: Data. Part II: Appendixes.
See Abstr. 002.111.

Infrared astronomy.
See Abstr. 013.007.

Infrared spectrum of quenched carbonaceous composite (QCC). II. A new identification of the 7.7 and 8.6 micron unidentified infrared emission bands.
See Abstr. 022.091.

Results with the UKIRT infrared camera.
See Abstr. 034.152.

Berkeley–Hughes infrared camera.
See Abstr. 034.161.

Deep infrared surveys.
See Abstr. 036.205.

Astrometric position of IRS–7 in the galactic center.
See Abstr. 041.003.

Infrared observations of contaminants from Shuttle flight 51–F.
See Abstr. 051.029.

Vibrationally excited CS in IRC + 10216.
See Abstr. 112.016.

The kinematic structure of the HH 24 complex derived from high–resolution spectroscopy.
See Abstr. 121.019.

Speckle observations of the ice feature in the young double source Serpens SVS 20.
See Abstr. 121.046.

IRAS colors of VLA identified objects in the Galaxy.
See Abstr. 131.003.

Maps of millimeter wave emission from three galactic star–forming regions.
See Abstr. 131.007.

CO and NH$_3$ detection of the cone in NGC 2264.
See Abstr. 131.026.

The IRAS cirrus and the diffuse ultraviolet background.
See Abstr. 131.049.

HCO$^+$ survey of unassociated compact molecular clouds in the *IRAS* Point Source Catalog.
See Abstr. 131.060.

Spectroscopy of the Kleinmann–Low nebula: scattering in a solid absorption band.
See Abstr. 131.073.

Infrared spectroscopy of interstellar dust.
See Abstr. 131.140.

Star–forming loops in the *IRAS* sky images.
See Abstr. 131.154.

A rotating dark cloud associated with IRAS 22291 + 7458.
See Abstr. 131.181.

A survey of IRAS point sources in Taurus for high–velocity molecular gas.
See Abstr. 131.249.

OH observations of galactic radio H II regions.
See Abstr. 132.003.

The IRAS view of the Galaxy and the solar system.
See Abstr. 155.024.

8.3 and 12.4 micron imaging of the galactic center with the Goddard infrared array camera.
See Abstr. 155.065.

Extremely luminous far–infrared sources (ELFS).
See Abstr. 157.028.

Ground–based follow up of IRAS galaxies.
See Abstr. 157.042.

Studies of IRAS sources at high galactic latitudes – III. Luminosity functions at 25, 60 and 100 μm and the correlation of optical and infrared luminosities.
See Abstr. 157.068.

The *IRAS* bright galaxy sample. II. The sample and luminosity function.
See Abstr. 157.220.

Properties of the unusual galaxy PSC 09104 + 4109.
See Abstr. 158.014.

Extragalactic OH megamasers in strong IRAS sources.
See Abstr. 158.016.

Studies of IRAS sources at high galactic latitudes. IV. New redshifts and the spectroscopic properties of IRAS galaxies.
See Abstr. 158.242.

134 Planetary Nebulae

134.001 Multiple–shell planetary nebulae. I. Morphologies and frequency of occurrence.
Y.-H. Chu, G. H. Jacoby, R. Arendt.
Astrophys. J., Suppl. Ser., Vol. 64, No. 3, p. 529 – 544 (1987).
With plates 13 – 30.

The authors have searched for multiple–shell planetary nebulae (MSPNs) in the sample of 126 planetary nebulae (PNs) in the New General Catalogue and the Index Catalogue. After the biases in distance and nebular evolution are corrected, the frequency of multiple–shell events in PNs is found to be ⩾0.5. Among the 41 known MSPNs, there are two major kinds of morphologies: type I has faint, detached outer shells, and type II has bright, attached outer shells. These two types have different ranges of outer diameter, ratio of outer to inner radius, and ratio of outer to inner surface brightness.

134.002 Photometric and spectrophotometric observations of 10 southern planetary nebulae.
R. Louise, A. Macron, G. Pascoli, E. Maurice.
Astron. Astrophys., Suppl. Ser., Vol. 70, No. 2, p. 201 – 227 (1987).

Spectrophotometric observations in H_α, H_β, [N II] $\lambda 6584 - \lambda 6548$ Å, [O III] $\lambda 5007 - \lambda 4959$ Å [S II] $\lambda 6731 - \lambda 6717$ Å and [O I] of 10 southern planetary nebulae were made. Monochromatic images and isophotal maps in H_α, H_β, [O II], [O III] and [S II] lines were obtained with a CCD detector. The ratios [O III] $\lambda 5007$/[O III] $\lambda 4959$ and [N II] $\lambda 6584$/[N II] $\lambda 6548$ have been found constant in agreement with the theory. The authors discuss the case of the central region of IC 418, the ring shape is clearly visible, even in the [O III] line. This planetary nebula can be considered as an inner bright envelope surrounded by an extended diffuse halo.

134.003 Erratum: "The $-33° \leqslant \delta \leqslant -17°$ zone: probing SRC J film copies for planetary nebulae" [Astron. Astrophys., Suppl. Ser., Vol. 69, No. 3, p. 527 – 531 (1987)].
W. Saurer, R. Weinberger.
Astron. Astrophys., Suppl. Ser., Vol. 70, No. 3, p. 531 (1987). See Abstr. 43.134.030.

134.004 Energy distribution in the emission spectra of the nuclei of planetary nebulae beyond the Lyman limit.
V. V. Golovatyj.
Pis'ma Astron. Zh., Tom 13, No. 7, p. 589 – 596 (1987). In Russian. English translation in Sov. Astron. Lett., Vol. 13.

The energy distribution in the spectra of 29 planetary nebulae central stars beyond the Lyman limit is calculated using observed intensities of 4686 He II and H_β lines. For this purpose the equation of energy balance and the Zanstra equation are used. An emission jump in the spectra of some central stars at $\lambda 228$ Å is discovered. The reality of the jump is discussed.

134.005 NGC 40: IUE observations of the nucleus.
L. Bianchi, M. Grewing.
Astron. Astrophys., Vol. 181, No. 1, p. 85 – 95 (1987).

The authors present ultraviolet spectra of the nucleus of NGC 40. By comparison with model atmospheres, and combining other data from the literature, they derive a foreground reddening of $E(B-V) = 0.50$ and a distance for the object of $D = 980$ pc. These lead to the following parameters for the central star: $T_{eff} = 90,000K$, $\log L/L_\odot = 4.4$, $R = 0.66\,R_\odot$, and $M = 1.1\,M_\odot$. The observed properties of the nucleus are discussed in the framework of recent post–AGB evolutionary scenarios and theories of PN formation. They suggest that the progenitor of NGC 40 was a star with mass larger than $6\,M_\odot$.

134.006 Detection of the hydrocarbon ring molecule C_3H_2 in the planetary nebula NGC 7027.
P. Cox, R. Güsten, C. Henkel.
Astron. Astrophys., Vol. 181, No. 2, p. L19 – L22 (1987).

The authors report the detection of the hydrocarbon ring molecule cyclopropenylidene C_3H_2 in the planetary nebula NGC 7027. The ortho and para transitions at 18.3 and 21.6 GHz have been observed. The results are analyzed in the light of statistical equilibrium calculations, allowing, for the first time, constraints to be put on the physical conditions in the NGC 7027 molecular envelope.

134.007 Distribution of I(He II $\lambda 4686$)/I(Hβ) in planetary nebulae and masses of their nuclei.
R. Szczerba.
Astron. Astrophys., Vol. 181, No. 2, p. 365 – 369 (1987).

This paper discusses a new method for testing the theory of evolution of planetary nebula nuclei independent of the distance to the nebulae. From recent models of the evolution of planetary nebula nuclei a theoretical histogram of the ratio of the intensity of He II $\lambda 4686$ to Hβ is calculated. The results are compared with the observations collected for 194 planetaries. Satisfactory agreement between the observational data and the theoretical prediction is attained only if a correlation between PN and NPN mass is introduced. The best–fitting parameter set suggests that less massive central stars have less massive nebulae and that the mass distribution of the nuclei it very narrowly peaked at $0.60\,M_\odot$.

134.008 The 3.3 μm and 3.4 μm emission features in planetary nebulae.
W. Martin.
Astron. Astrophys., Vol. 182, No. 2, p. 290 – 298 (1987).

Absolute fluxes and relative intensities of the 3.3 μm and 3.4 μm feature in 12 planetary nebulae have been derived from low resolution spectroscopy and narrow band photometry. In addition, the near infrared continuum fluxes in excess of the expected recombination spectrum of hydrogen and helium have been determined. The 3.3 μm feature is present in all 12 objects of the sample. In 4 out of 10 planetaries the 3.4 μm feature could not be detected with CVF spectroscopy. A correlation study shows both features to be associated with the 12 μm to 100 μm emission. Only the 3.3 μm feature is correlated to the excess radiation at 3.8 μm and to the gas phase carbon abundance. The oxygen abundance is anticorrelated to the 3.3 μm and the 3.4 μm feature. The intensities of both are approximately inversely proportional to the nebula's distance from the galactic plane. The data are interpreted in terms of the recent hypothesis on polycyclic aromatic hydrocarbons.

134.009 The Type–I planetary nebula Humason 1–2.
F. Sabbadin, E. Cappellaro, M. Turatto.
Astron. Astrophys., Vol. 182, No. 2, p. 305 – 312 (1987).

Direct Hα+[N II] CCD frames and low and high resolution spectroscopy of the compact bright planetary nebula Humason 1–2 were obtained at Asiago Astrophysical Observatory to study the morphology, physical conditions, chemical abundances and expansion velocity field of this object. The nebula consists of a dense equatorial ring of condensations surrounded by fainter, polar material. The overabundances of He and N found in the nebula, its very short dynamical age and the position of the central star in the H–R diagram suggest that the progenitor of Hu1–2 was a massive star and that the present mass of the exciting star is larger than the mean value generally assumed for central stars of planetary nebulae. Some general properties of Type I planetary nebulae (i.e. those which are believed to derive from massive progenitors) are presented and discussed.

134.010 The determination of the masses of Magellanic Cloud planetary nebulae using [O II] doublet ratio electron densities.
M. J. Barlow.
Mon. Not. R. Astron. Soc., Vol. 227, No. 1, p. 161 – 183 (1987).

Spectrophotometric data, including [O II] 3726, 3729 Å doublet ratios, are presented for 32 planetary nebulae (PN) in the Magellanic Clouds. It is argued that the electron densities derived from these ratios provide a much better diagnostic for the determination of nebular masses than previously assumed. The optically thick PN are found to all have electron densities greater than 6000 cm^{-3}, while the optically thin PN all have electron densities below 5000 cm^{-3}. The optically thin PN show a range of only a factor of 2.0 in their derived masses, and have a mean ionized mass of $0.27 \pm 0.06 \, M_\odot$. The absolute Hβ fluxes of the optically thick nebulae show a range of only a factor of 1.8. The application of these results to Galactic PN would yield distances which are generally larger than those previously estimated. A method of distance determination is proposed for optically thin PN that uses integrated nebular [O II] electron densities rather than angular diameters.

134.011   The physical conditions in SwSt 1: the central star and the nebula.
J. A. de Freitas Pacheco, J. G. Veliz.
Mon. Not. R. Astron. Soc., Vol. 227, No. 3, p. 773 – 782 (1987).
The authors present new optical observations on SwSt 1. They estimate a reddening of $E(B-V) = 0.41$ mag and a distance $D = 1.2 \pm 0.2$ kpc to the nebula. From their data the authors derive an electron density of $(1.0 \pm 0.2) \times 10^5 \text{ cm}^{-3}$ and a temperature of 11400 ± 500K. Carbon, nitrogen, oxygen, chlorine and sulphur are underabundant more or less by a factor of 2 with respect to the "cosmic" values. The central star has an effective temperature of 32000K and is losing mass at a rate of $(6-7) \times 10^{-8} M_\odot \text{yr}^{-1}$.

134.012   The classification of planetary nebulae.
M. Faúndez–Abans, W. J. Maciel.
Astron. Astrophys., Vol. 183, No. 2, p. 324 – 326 (1987).
The classification scheme of planetary nebulae proposed by Peimbert is revised, taking into account the observed heavy-element abundances and radial abundance gradients. The so-called type II nebulae are further divided into the subtypes a and b, according to their nitrogen abundance relative to hydrogen.

134.013   Measurement of the He II radiation field in planetary nebulae through Bowen fluorescence.
A. K. Bhatia, S. O. Kastner.
Phys. Scr., Vol. 35, No. 6, p. 778 – 779 (1987).
Excitation of O III by He II is treated for sources over a useful range of densities to give accurate predictions of Bowen/non-Bowen line ratios. These are applied to recent observations of planetary nebulae to show that Bowen excitation increases monotonically with excitation class, and to deduce other important consequences.

134.014   Determination of temperature of planetary nebulae nuclei from the relative intensities of the lines He I/Hβ and He II/Hβ.
V. V. Golovatyj.
Astron. Zh., Tom 64, Vyp. 4, p. 724 – 733 (1987). In Russian. English translation in Sov. Astron., Vol. 31, No. 4.
The results of determination of planetary nebulae nuclei temperatures are presented. The Zanstra method modified for the He admixture was used. Analytical forms and graphic representations convenient for practical temperature determination are presented. T_* values for 51 planetary nebulae nuclei are found. These T_* values agree with the temperatures found by the energy balance method. Validity of nuclei temperature determination by this method is discussed.

134.015   The evolution of planetary nebulae. I. Structures, ionizations, and morphological sequences.
B. Balick.
Astron. J., Vol. 94, No. 3, p. 671 – 678 (1987). With plates 60 – 73.
An atlas of CCD pictures of fifty–one planetary nebulae (PNs) taken in the light of low–, moderate–, and high–ionization emission lines is presented. The shapes of many of the PNs can be organized into an empirical morphological sequence of round, elliptical, bipolar, and butterfly classes. Many PNs, especially the ones of high surface brightness such as NGC 2392, 3242, 6543,

6826, and 7662, show evidence for thin inclusions of anomalously low ionization in a high–ionization substrate. The morphological sequence of PNs is shown to be consistent with the precepts of the shaping of PNs by interacting winds.

134.016   The shapes and shaping of the planetary nebulae IC 3568, NGC 40, and NGC 6543.
B. Balick, C. R. Bignell, R. M. Hjellming, R. Owen.
Astron. J., Vol. 94, No. 4, p. 948 – 957 (1987).
Radio and optical images of three compact planetary nebulae, NGC 6543, IC 3568, and NGC 40, are found to reveal a variety of morphological and ionization signatures of wind shaping and heating. Both NGC 6543 (one of the most morphologically complex planetaries in the sky) and IC 3568 (one of the simplest) exhibit large density changes on size scales corresponding to sound–crossing times of about a century, arguing that the density gradients are sustained by shocks driven by stellar winds. All of the available data show the structure and ionization of IC 3568 to conform to simple hydrodynamic models of nebulae that are shaped by interior stellar winds. NGC 6543 also shows very local changes in optical line ratios that are best understood in the same manner.

134.017   A wind–blown bubble model for NGC 6543.
B. Balick, H. L. Preston.
Astron. J., Vol. 94, No. 4, p. 958 – 963 (1987). With plate 79.
The structure and kinematics of the planetary nebula NGC 6543 suggest an empirical model for the nebula which consists of two pairs of lobes along nearly perpendicular axes. The concentric, crossed elliptical filaments are associated with the loci of lobe contact interfaces. The outline of the lobes predicted by the model is in complete agreement with the observations of the nebula seen in the radio continuum, in the light of Hα, and [O III].

134.018   The (C III $\lambda 1909$/Si III $\lambda 1892$) ratio as a diagnostic for planetary nebulae and symbiotic stars.
W. A. Feibelman, L. H. Aller.
Astrophys. J., Vol. 319, No. 1, p. 407 – 415 (1987).
The authors examined suitable *IUE* archival material on planetary nebulae to determine log R [F($\lambda 1909$ C III)/F($\lambda 1892$ Si III)] as a discriminant for distinguishing planetary nebulae from symbiotic stars and related objects. The mean value of log R for 73 galactic planetaries is 1.4, while that of extragalactic planetaries appears to be slightly lower, and that for symbiotics is 0.3. The lower value of log R for symbiotics is easily understood as a consequence of their higher densities.

134.019   Physical conditions in planetary nebulae.
V. V. Golovatyj, R. T. Dyak, O. S. Yatsyk.
Sov. Astron., Vol. 30, No. 6, p. 663 – 669 (1987). English translation of 42.134.056.

134.020   Spatial spectroscopic diagnostics of planetary nebulae. VI: Numerical properties of the integral kernel.
J. Hekela, P. Plecháč.
Bull. Astron. Inst. Czech., Vol. 38, No. 5, p. 303 – 308 (1987).
The numerical properties of the kernel function of the Fredholm integral equation of the first kind of spatial spectroscopic diagnostics of planetary nebulae are studied and briefly discussed. Samples of some kernel functions are shown as three–dimensional axonometric figures. A simple reason for inherent limitations of the regularization techniques used is given.

134.021   Misclassified planetary nebulae.
A. Acker, M. Chopinet, S. R. Pottasch, B. Stenholm.
Astron. Astrophys., Suppl. Ser., Vol. 71, No. 1, p. 163 – 175 (1987).
The authors present their opinion on 266 objects, taken from catalogues and lists of planetary nebulae. These opinions are based on observations given in the literature, on the vast photometric information collected by the IRAS satellite, as well

as on the observations from the ongoing project of a spectroscopic survey of all planetary nebulae. 199 objects are definitely rejected as planetary nebulae, 63 others are possibly not planetary nebulae.

134.022 Properties of planetary nebulae. I. Nebular parameters and distance scales.
R. Gathier.
Astron. Astrophys., Suppl. Ser., Vol. 71, No. 2, p. 245 – 253 (1987).

A sample of 30 planetary nebulae with individually determined distances is used to determine nebular parameters such as electron density and ionized mass. The majority of the nebulae studied appear to be ionization bounded. The ionized nebular masses range from $< 2 \times 10^{-3}$ to $\gtrsim 5 \times 10^{-1} M_\odot$. The volume filling factor is found to be ~ 0.75. The distribution of ionized mass as a function of nebular radius and electron density indicates that the observed sample is composed of planetary nebulae with a large spread in intrinsic properties of the central stars and nebulae. The individual distances are compared with distances derived using various statistical distance scales. It appears that such scales in general lead to very uncertain distances in individual cases. The statistical scales that on average show agreement with the 30 individual distances are used to estimate a local planetary nebula birthrate of $2.4 \pm 1.3 \times 10^{-3} \text{kpc}^{-3} \text{yr}^{-1}$. It is argued that there is no serious discrepancy with the local birthrate of white dwarfs.

134.023 Nebular spectrum of CPD –56°8032 and He2–113.
N. Kameswara Rao.
Q. J. R. Astron. Soc., Vol. 28, No. 3, p. 261 – 263 (1987). – See Abstr. 012.037.

The electron density, temperature and ion abundances in the nebulae surrounding the two WC 10 stars CPD –56°8032 and He2–113 have been estimated. In CPD –56°8032 sulphur, nitrogen and oxygen seem to have solar abundances, whereas carbon and neon are enhanced. The nebular properties of all the four WC 10 stars He2–113, CPD –56°8032, M4–18 and V348 Sgr are compared.

134.024 The kinematical structure of the bipolar planetary nebula 19W 32.
J. A. López.
Astron. Astrophys., Vol. 186, No. 1/2, p. 303 – 306 (1987).

Long–slit echelle observations obtained in the light of Hα and [N II] are presented for the bipolar planetary nebula 19W 32. The kinematics of this object shows a difference in velocity of 30 km s⁻¹ as measured from the tip of one lobe to the opposite. Whereas the emission lines are single in the lobes they exhibit a complex, multiple structure at the core of the planetary. An expansion velocity of 50 km s⁻¹ is found for the nebular material surrounding the nucleus. The morphology and velocity structure of the planetary indicate that a circumstellar shell is being driven by a stellar wind. The material that surrounds the core seems to have collimated ionized gas in opposite directions forcing it to flow away in a bipolar mode.

134.025 The evolution of planetary nebulae.
B. Balick.
Bull. Am. Astron. Soc., Vol. 19, No. 2, p. 679 (1987). Abstract. – See Abstr. 010.061.

134.026 The ultraviolet spectra of central stars of planetary nebulae.
S. R. Heap.
Bull. Am. Astron. Soc., Vol. 19, No. 2, p. 701 – 702 (1987). Abstract. – See Abstr. 010.061.

134.027 Two compact planetary nebulae of moderate excitation; NGC 6565 (3 – 4.5) and NGC 6644 (8 – 7.2).
L. H. Aller, C. D. Keyes, W. A. Feibelman.
Bull. Am. Astron. Soc., Vol. 19, No. 2, p. 730 (1987). Abstract. – See Abstr. 010.061.

134.028 An atlas of IUE spectra of planetary nebulae and related objects.
N. A. Oliversen, W. A. Feibelman, J. Nichols–Bohlin.
Bull. Am. Astron. Soc., Vol. 19, No. 2, p. 730 (1987). Abstract. – See Abstr. 010.061.

134.029 Proto–planetary nebulae: models and IRAS observations.
K. Volk, S. Kwok.
Bull. Am. Astron. Soc., Vol. 19, No. 2, p. 730 (1987). Abstract. – See Abstr. 010.061.

134.030 Spectrophotometric observations of the planetary nebula NGC 7027: a search for H₃ molecular emission.
G. T. Gussie.
Bull. Am. Astron. Soc., Vol. 19, No. 2, p. 736 (1987). Abstract. – See Abstr. 010.061.

134.031 Beyond the asymptotic giant branch.
S. Kwok, B. J. Hrivnak.
Bull. Am. Astron. Soc., Vol. 19, No. 2, p. 737 (1987). Abstract. – See Abstr. 010.061.

134.032 Pseudoresonance lines of Si III in the spectra of planetary nebulae.
A. G. Egikyan.
Astrophysics, Vol. 26, No. 2, p. 162 – 169 (1987). English translation of 43.134.032.

134.033 The evolution of planetary nebulae. II. Dynamical evolution of elliptical PNs and collimated outflows.
B. Balick, H. I. Preston, V. Icke.
Astron. J., Vol. 94, No. 6, p. 1641 – 1652 (1987). With plates 129 – 136.

The authors present observations mapping the velocity of the gas in Hα and [N II] for the planetary nebulae NGC 40, 2392, 3242, 6543, 6826, 7009, 7354, and 7662. All of these show spatial and kinematic features suggestive of the existence of highly collimated fast outflows, generally characterized by velocities (projected onto the sky) in the range of 10 – 60 km s⁻¹. NGC 2392 is extreme in its outflow velocities of ~170 km s⁻¹. To explain the observations, the authors propose a hydrodynamic mechanism that collimates the gas as it passes outward through a prolate shock front. They adopt the usual fast–wind model, but assume that the slow–wind envelope that surrounds the star is denser near its equator than near the poles. The model is discussed in detail and a comparison with the observations is presented.

134.034 Spectrophotometry of the compact planetary nebulae NGC 6879 and NGC 6881.
J. B. Kaler, P. Pratap, K. B. Kwitter.
Publ. Astron. Soc. Pac., Vol. 99, No. 619, p. 952 – 956 (1987).

The authors have observed the spectra of the two compact planetary nebulae NGC 6879 and NGC 6881 between λ3727 and λ6731 with both the IRS and the IIDS at Kitt Peak and calculate extinction constants, electron densities and temperatures, and a wide variety of ionic abundances. Both nebulae have high densities. NGC 6881, the more highly excited of the two, is significantly enriched in nitrogen and possibly exhibits some elevation in helium.

134.035 Close–binary central stars of planetary nebulae.
H. E. Bond, A. D. Grauer.
Space Telesc. Sci. Inst., Prepr. Ser., No. 182, p. 1 – 8 (1987). Paper presented at IAU Colloq. No. 95.

134.036 The IR morphology of the proto–planetary nebula M2–9.
C. Aspin, I. S. McLean, M. S. Smith.
Edinb. Astron. Prepr., No. 22/87, 20 pp. (1987). Submitted to Astron. Astrophys.

134.037   The collisional excitation of helium in nebulae.
D. Pequignot, J.–P. Baluteau, R. B. Gruenwald.
Obs. Haute Provence, Pré–Publ., No. 22, 19 pp. (1987).

134.038   Classification of planetary nebulae according to age.
L. N. Kondrat'eva.
Tr. Astrofiz. Inst. Alma–Ata, Tom 48, p. 22 – 32 (1987). In Russian.

134.039   PC 11: symbiotic star or planetary nebula?
A. Gutiérrez–Moreno, H. Moreno, G. Cortés.
Rev. Mex. Astron. Astrofis., Vol. 14, No. 1, p. 344 – 352 (1987).
– See Abstr. 012.042.
The authors have made photographic, spectrophotometric and spectroscopic observations of PC 11 (PK 331–5°1). The analysis of the results suggests that it is a young planetary nebula.

134.040   Photographic and spectroscopic observations of southern planetary nebulae.
H. Moreno, A. Gutiérrez–Moreno, C. Torres, E. Wenderoth.
Rev. Mex. Astron. Astrofis., Vol. 14, No. 2, p. 520 – 527 (1987).
– See Abstr. 012.042.
Photographic and spectroscopic observations of a group of southern planetary nebulae and symbiotic stars are presented. The photographs have been taken in the visual region of the spectrum and in the lines [O III] $\lambda 5007$ and Hα. The spectra cover, in general, the region $\lambda\lambda 3400 – 8600$ Å. Some examples of this work are presented.

134.041   Gradientes de abundancias en la nebulosa planetaria A78.
A. Manchado, A. Mampaso, S. R. Pottasch.
Rev. Mex. Astron. Astrofis., Vol. 14, No. 2, p. 528 – 533 (1987).
– See Abstr. 012.042.
New bidimensional spectroscopic observations of A78 allow the authors to determine the He, O, Ne, and N abundances and their variations along the nebula. A clear overabundance of these elements is found in the central region of the nebula, suggesting different nebular ejections of material processed in the stellar interior and dredged–up subsequently to the envelope.

134.042   JHK photometry of compact planetary nebulae.
M. Peña, S. Torres–Peimbert.
Rev. Mex. Astron. Astrofis., Vol. 14, No. 2, p. 534 – 539 (1987).
– See Abstr. 012.042.
The authors present near–infrared photometry of 31 compact planetary nebulae. A correlation of $(J–H)_0$ with He$^+$ abundance is found. Most high density nebulae show infrared excess in H and K showing that they have warmer dust than lower density nebulae.

134.043   Chemical composition of Type I planetary nebulae. Collisional excitation effects on He I line intensities.
M. Peimbert, S. Torres–Peimbert.
Rev. Mex. Astron. Astrofis., Vol. 14, No. 2, p. 540 – 558 (1987).
– See Abstr. 012.042.
The authors present line intensities for thirteen N–rich planetary nebulae. From these line intensities they have determined the collisional effects in the helium line intensities and the chemical abundances of the most abundant elements.

134.044   The velocity structure of the bipolar planetary nebula 19W32.
J. A. López, E. de Lara.
Rev. Mex. Astron. Astrofis., Vol. 14, No. 2, p. 559 (1987).
Abstract. – See Abstr. 012.042.

134.045   VLA observations of twice ionized helium in NGC 6302.
Y. Gómez, L. F. Rodríguez, J. A. García–Barreto.
Rev. Mex. Astron. Astrofis., Vol. 14, No. 2, p. 560 – 566 (1987).
– See Abstr. 012.042.
The authors present VLA observations of the H76α and He$^+$121α radio recombination lines and their adjacent continua from the planetary nebula NGC 6302. They derive an average value of He^{++}/H$^+$ $\cong 0.06 \pm 0.02$ in agreement with the optical determination. The variations in the H76α–to–continuum ratio that are observed across the face of NGC 6302 are discussed in detail.

134.046   Galactic confusion and the detection of 1665 and 1667 MHz OH lines in the neighborhood of planetary nebulae.
K. C. Turner, Y. Terzian, H. Payne.
Rev. Mex. Astron. Astrofis., Vol. 14, No. 2, p. 567 – 572 (1987).
– See Abstr. 012.042.
OH radiation at 1665 and 1667 MHz was detected in association with the planetary nebula VY2–2. OH lines were also detected in the fields of NGC 1514, M1–5, VV 42, K3–3, He2–459, and M3–35. This gas, however, is more naturally identified with material in the solar neighborhood. A brief OH survey near the galactic plane shows similar gas detection statistics.

134.047   The evolution of central stars of planetary nebulae: pop. I vs. pop. II.
D. Schönberner.
ESO Conf. Workshop Proc., No. 27, p. 519 – 529 (1987). – See Abstr. 012.051.

134.048   A comparison of some physical properties of planetary nebulae in the Magellanic Clouds with those in the Galaxy.
S. R. Pottasch.
ESO Conf. Workshop Proc., No. 27, p. 531 – 541 (1987). – See Abstr. 012.051.
The intrinsic brightness of the nebulae is compared and found to extend over a factor of at least 1000 both in the Magellanic Clouds and in the galactic bulge. Some of the faint nebulae are highly evolved, but a large number are young nebulae with low–temperature central stars. The existence of these low–luminosity nuclei is not predicted by current theories. Finally a discussion of the abundances of helium, oxygen and nitrogen is given.

134.049   Planetary nebulae in the galactic halo.
R. E. S. Clegg.
ESO Conf. Workshop Proc., No. 27, p. 543 – 552 (1987). – See Abstr. 012.051.
Only four planetary nebulae (PN) are known to be definitely located in the halo of our Galaxy. One of these, K 648, is in the globular cluster M15. The author summarises briefly the main properties of the nebula, their chemical compositions, facts about the central stars, and results from the IRAS experiment for one object.

134.050   Abundances, nebular masses and central star parameters for Magellanic Cloud planetary nebulae.
M. J. Barlow.
ESO Conf. Workshop Proc., No. 27, p. 553 – 569 (1987). – See Abstr. 012.051.
Elemental abundances have been derived, from optical data, for 71 Magellanic Cloud planetary nebulae. The abundance of helium in the planetary nebulae (PN) is found to be the same as that in H II regions in the SMC and LMC. A helium mass fraction of Y = 0.247 ± 0.009 is consistent with all of the data on nebulae in both clouds. The oxygen and neon abundances in the PN are the same as those previously found for H II regions in each galaxy, but nitrogen is enhanced by 0.8 dex in PN in both galaxies. Central star parameters have been derived for a sample of ten Magellanic Cloud PN and a mean central star mass of 0.58 ± 0.01 M$_\odot$ is derived. It is argued that WC–type central stars are the progenitors of DB white dwarf stars, the other PN central stars evolving into DA white dwarfs.

134.051   The evolution and dynamics of the Magellanic Cloud planetary nebulae.
M. A. Dopita, S. J. Meatheringham, P. R. Wood, H. C. Ford, B. L. Webster, D. H. Morgan.
ESO Conf. Workshop Proc., No. 27, p. 571 – 586 (1987). – See Abstr. 012.051.

The sample of planetary nebulae (PN) in the Magellanic Clouds is ideal to obtain an understanding of stellar evolution in the post–Asymptotic Giant Branch phase. Furthermore, since the radial velocities can be determined to a very high precision, one can investigate the internal dynamics of the Magellanic Clouds as determined by an intermediate/old population tracer. The authors highlight the discoveries made on the basis of their extensive imaging program and (almost) complete survey of the kinematics, internal dynamics and Hβ photometry of this population.

134.052   Stellar wind from the nucleus of the nebula YM 29?
V. P. Arkhipova, T. A. Lozinskaya, E. I. Moskalenko.
Sov. Astron. Lett., Vol. 12, No. 6, p. 373 – 375 (1986). English translation of 42.134.013.

134.053   Models for the wind of the central star of NGC 6543.
L. B. Lucy, M. Perinotto.
Astron. Astrophys., Vol. 188, No. 1, p. 125 – 130 (1987).

The hypothesis of radiative driving is tested for the wind of the central star of NGC 6543. The data available on this bright, well–observed central star do not exclude the hypothesis that radiative driving is the dominant acceleration mechanism in its wind. The use of stellar wind theory to infer stellar parameters may be regarded as a mechanical analogue of the Zanstra method.

134.054   V605 Aquilae – a star and a nebula with no hydrogen.
W. C. Seitter.
Messenger, No. 50, p. 14 – 17 (1987).

134.055   High spatial resolution observations of dust and gas in NGC 7027 & the M8 Hourglass.
C. E. Woodward.
Diss. Abstr. Int., Sect. B, Vol. 48, No. 6, p. 1712–B (1987). Thesis, The University of Rochester, 366 pp. (1987). Order No. DA8718964.

134.056   A spectroscopic survey of 51 planetary nebulae.
L. H. Aller, C. D. Keyes.
Astrophys. J., Suppl. Ser., Vol. 65, No. 3, p. 405 – 428 (1987).

Line intensity data obtained with an image–tube scanner at the 3 m Lick telescope are used to obtain plasma diagnostics and ionic concentrations. With the aid of theoretical models the authors obtain ionization correction factors and find chemical abundances. Abundance patterns of C, N, and O show considerable spread owing to differing nuclear processing rates in progenitor stars. Most objects show enhanced C or N, suggesting that nucleosynthesis products have been mixed with stellar surface layers.

134.057   Spatially resolved images of NGC 7027 in the 3.28 μm and 3.4 μm dust emission features.
C. E. Woodward, J. L. Pipher, M. A. Shure, W. J. Forrest, K. Sellgren, T. Nagata.
Infrared astronomy with arrays, p. 299 – 303 (1987). – See Abstr. 012.070.

The authors present the first spatially resolved ($\sim 0.42''$/pixel), 1% spectral resolution images of the planetary nebula NGC 7027 in the 3.28 μm and 3.4 μm dust emission features, and in the Brα (4.052 μm) hydrogen recombination line. Comparison of the images clearly shows that the 3.28 μm emission is spatially more extended than that of the ionized gas. Analysis of the 3.28 μm feature strength suggests that the majority of the feature emission arises from a shell at least partially outside of the H II region. The spatial distribution of the 3.4 μm feature emission is found to be similar to that of the 3.28 μm feature emission.

134.058   Infrared imagery of two carbon–rich planetary nebulae: NGC 7027 and BD +30°3639.
J. H. Goebel, D. Rank, M. Cohen.
Infrared astronomy with arrays, p. 305 – 311 (1987). – See Abstr. 012.070.

Narrow–band images of the carbon–rich planetary nebulae NGC 7027 and BD +30°3639 have been obtained at wavelengths in the 8 – 14 μm range corresponding to continuum, atomic emission line, and PAH emission band locations. The images indicate that the spatial distributions of emission from the species responsible are different from one another. The excitation levels and temperatures of the different species are also different from one another. With additional spectral information one can deduce the presence of three distinct carbon dust forms, PAH, α:C–H, and graphitic carbon.

134.059   Spectrophotometric studies of galactic nebulae. XXV. 1. The spectrum of the planetary nebula Sh1–89. 2. Variation in the spectrum of the nebula M1–67.
Yu. I. Glushkov, L. N. Kondrat'eva.
Astron. Tsirk., No. 1457, p. 1 – 4 (1986). In Russian.

In the nebula Sh1–89 the ratio I(6583)/I(Hα) changes from 1.0 to 3.2 along the nebula, for the central part Ne (S II) = 2300 cm^{-3}. In the spectra of the central part of M1–67 (in the vicinity $\pm 2''$ of Merrill's star), obtained in 1980 – 1983, the sharp feature of [N II] $\lambda6583$ is strengthened, compared to the spectra obtained in 1977. The authors conclude from this fact that the compact envelope around Merrill's star is scattering and today its density has decreased up to 10^5cm^{-3}.

134.060   Theoretical nebulae models, advantages, limitations, and frustrations.
L. H. Aller.
Publ. Astron. Soc. Pac., Vol. 99, No. 621, p. 1145 (1987). Abstract. – See Abstr. 010.281.

134.061   CCD images of southern hemisphere planetary nebulae.
J. H. Lutz.
Publ. Astron. Soc. Pac., Vol. 99, No. 621, p. 1148 (1987). Abstract. – See Abstr. 010.281.

134.062   Infrared radiation of the planetary nebula NGC 2242.
J. Hu.
Acta Astrophys. Sin., Vol. 7, No. 4, p. 317 – 319 (1987). In Chinese.

NGC 2242 is located in the position uncertainty ellipse of IRAS 06304 + 4448 and is identified as an IRAS point source. The infrared fluxes in 25 and 60 μm are 0.57 and 0.52 Jy respectively. The interstellar extinction method is used to determine the distance of the nebula as 1.4 kpc. The diameter of the nebula is 4.8×10^{17}cm. The dust temperature and the dust mass contained in the nebula are calculated.

134.063   The ionization structure of planetary nebulae. VII. New observations of the Ring nebula.
T. Barker.
Astrophys. J., Vol. 322, No. 2, p. 922 – 929 (1987).

New optical spectrophotometric observations of emission–line intensities have been made in eight positions in the Ring nebula corresponding to those observed previously with the IUE. Abundance determinations from optical lines and UV spectra are generally in good agreement. UV and optical measurements of the C^{++} abundance disagree; this indicates that the excitation of the $\lambda4267$ line is not adequately understood. The logarithmic abundances (relative to H = 12.00) are He = 11.04, O = 9.05, N = 8.36, Ne = 8.26, C = 9.09, and Ar = 6.38. The rather high abundances of O, N, and C, and, to some extent, N, indicate that some mixing of CNO–processed material into the nebular shell may have occurred in the Ring nebula.

134.064   **Spheromak model of planetary nebulae.**
C. L. Bennett.
Astrophys. J., Lett. Ed., Vol. 323, No. 2, p. L123 – L125 (1987).
With plates L4 – L9.
A spheromak model of planetary nebulae is proposed. This model is able to reproduce essentially the entire range of morphologies observed. It also naturally explains hitherto mysterious features of the velocity distributions of the planetaries.

Collisional effects in He I lines and helium abundances in planetary nebulae.
See Abstr. 022.077.

Astronomical observations at 10 micron wavelength with the NASA/GSFC array camera system.
See Abstr. 034.160.

A mid–infrared cryogenic echelle spectrometer.
See Abstr. 034.169.

A high–resolution Fabry–Pérot spectrometer for emission line studies in planetary nebulae and other extended astronomical objects.
See Abstr. 034.186.

Is de centrale ster in de Ringnevel te zien?
See Abstr. 036.013.

Two–dimensional spectrophotometry of planetary nebulae by CCD imaging.
See Abstr. 036.043.

Emission–line ratios for O III in gaseous nebulae and a comparison between theory and observation.
See Abstr. 063.035.

Predicted [Ni II] infrared line strengths in the Crab Nebula and IC 443.
See Abstr. 063.107.

Non–LTE model atmospheres for central stars of planetary nebulae.
See Abstr. 064.017.

Further VLA observations of hydrogen deficient stars.
See Abstr. 116.062.

On the nature of 623 + 71: a cataclysmic binary surrounded by a bow–shock–like emission nebula.
See Abstr. 117.027.

IUE and optical observations of the symbiotic star/nebula He2–104.
See Abstr. 117.187.

Relative C, N, O abundances in red giants, planetary nebulae, novae and symbiotic stars.
See Abstr. 117.202.

The infrared variability of FG Sagittae in 1985 – 1986.
See Abstr. 122.042.

A search for pulsating stars similar to PG 1159–035 and K1–16.
See Abstr. 122.075.

A search for pulsating stars similar to PG 1159–035 and K1–16.
See Abstr. 122.134.

An ancient planetary nebula surrounding the old nova GK Persei.
See Abstr. 124.231.

White dwarfs and planetary central stars.
See Abstr. 126.077.

Observations of diffuse and planetary nebulae on the Astron astrophysical station.
See Abstr. 131.035.

The formation of discrete high velocity molecular features.
See Abstr. 131.072.

Spectroscopy of the 3 micron emission features with the CGAS (*Cooled–Grating Array Spectrometer*).
See Abstr. 131.275.

Optical spectroscopy of IRAS sources with infrared emission bands: IRAS 21282 + 5050 and the diffuse interstellar bands.
See Abstr. 133.023.

Cold atomic gas in the inner Galaxy.
See Abstr. 155.103.

Physical parameters for 12 planetary nebulae and their central stars in the Magellanic Clouds.
See Abstr. 156.018.

Angular diameters and fluxes of Magellanic Cloud planetary nebulae. II. High–speed direct imaging.
See Abstr. 156.019.

The distance to M81 derived from planetary nebulae.
See Abstr. 157.248.

Radio Sources, X-ray Sources, Cosmic Rays

141 Radio Sources (Surveys, etc.)

141.001 Tasmanian low frequency galactic background surveys.
H. V. Cane.
Radio astronomy from space, p. 289 – 292 (1987). – See Abstr. 012.009.

141.002 Optical identifications of radio sources from the Molonglo deep surveys – I. The declination –20° portion of the first deep survey.
G. L. White.
Mon. Not. R. Astron. Soc., Vol. 227, No. 3, p. 607 – 622 (1987).
Optical identifications have been sought for 304 radio sources from the declination –20° region of the first Molonglo deep survey. Optical objects have been measured using the Palomar Sky Survey to an accuracy of 1 arcsec and magnitude estimates are to 0.4 mag. Sixteen QSOs have been confirmed amongst the blue stellar–like objects and spectroscopy is complete to $m_0 = 19.5$. The mean magnitude of the QSOs is 19.2 and the mean redshift is 1.34. A study of the background densities of objects indicates that about half of the remaining blue stellar–like objects are QSOs and about 80 per cent of the galaxies are associated with the radio emission.

141.003 Radio positions and optical identifications for a complete sample of southern flat–spectrum radio sources – I. Region 06ʰ to 18ʰ.
G. L. White, M. J. Batty, J. D. Bunton, D. R. Brown, J. B. Corben.
Mon. Not. R. Astron. Soc., Vol. 227, No. 3, p. 705 – 715 (1987).
Optical identifications have been sought for 73 flat–spectrum radio sources in the region defined by RA 06ʰ00ᵐ to 18ʰ00ᵐ and Dec –80° to –50°. The radio positions are from Fleurs synthesis telescope observations and have uncertainties of ~2 arcsec. The optical positions are with respect to reference stars from the Perth catalogues and have errors of 0.5 arcsec relative to the FK4 frame. There are 10 empty fields, 43 stellar–like objects, probably QSOs, and 21 galaxies. The identifications are complete to a J–magnitude of 22.5 and have been made independently of the colour of the object. The mean magnitude of both the stellar–like objects and the galaxies is $m_J = 19.6$.

141.004 The 6–m telescope programme of optical identification of the radio sources found in the deep surveys with the RATAN–600 radio telescope.
V. V. Vitkovskij, O. P. Zhelenkova, I. D. Karachentsev, Yu. N. Parijskij, N. A. Tikhonov, V. S. Shergin.
Soobshch. Spets. Astrofiz. Obs., Vyp. 53, p. 86 – 88 (1987). – See Abstr. 012.008.
Information is briefly given on the programme and preliminary results of the optical identification with the 6–meter telescope of the radio sources found in the strip of experiment "COLD" (Berlin et al). New details of the optical image of the strong radio source PKS 2128 + 048 are presented. Plans and perspectives of an optical identification programme with new optical devices are given also.

141.005 Radio sources of the deep survey experiment "Kholod" in an interval of right ascensions between $16^h < \alpha < 17^h$, $4^h < \alpha < 5^h$, $0^h < \alpha < 1^h$: catalogs, investigation, methods of observation and reduction.
Yu. N. Parijskij, N. N. Bursov, R. Wielebinski, V. V. Vitkovskij, U. Klein, N. M. Lipovka, V. N. L'vov, N. S. Soboleva, A. V. Temirova.
Spets. Astrofiz. Obs. Akad. Nauk SSSR, Prepr., No. 41L, 37 pp. (1987). In Russian. Abstr. in Ref. Zh., 51. Astron., 8.51.378 (1987).

141.006 Radio sources of the deep sky survey of experiment "Kholod" in the right ascension intervals $16^h < \alpha < 17^h$, $4^h < \alpha < 5^h$, $0^h < \alpha < 1^h$.
Yu. N. Parijskij, N. N. Bursov, R. Wielebinski, V. V. Vitkovskij, U. Klein, N. M. Lipovka, V. N. L'vov, N. S. Soboleva, A. V. Temirova.
Pis'ma Astron. Zh., Tom 13, No. 10, p. 835 – 842 (1987). In Russian. English translation in Sov. Astron. Lett., Vol. 13.
The catalog of radio sources detected in the 7.6 cm RATAN–600 deep sky surveys in the interval $16^h < \alpha < 17^h$, $\delta = \delta_{SS433} \pm 20'$, the coordinates, flux densities and spectral indices are reported. Flux densities have been measured also with the 100–m Effelsberg mirror for some of the RATAN–600 sources at 6.3 and 2.8 cm. Radio spectra of the objects common to the RATAN–600, Effelsberg and Texas surveys are obtained. The overall number of objects is near 100. Comparison of the catalog at 7.6 cm ($16^h < \alpha < 17^h$) with the data of the experiment "Kholod" at 3.9 cm was carried out. An attempt to search for clustering of radio sources in these strips of sky was made.

141.007 The effects of caustics on scintillating radio sources.
J. J. Goodman, R. W. Romani, R. D. Blandford, R. Narayan.
Mon. Not. R. Astron. Soc., Vol. 229, No. 1, p. 73 – 102 (1987).
The authors consider the scintillation properties of compact radio sources for a spectrum of interstellar electron density fluctuations which is a power law over a finite range of spatial frequencies. In particular, if the power law is truncated at an inner scale intermediate between the diffractive scale and the refractive scale, the authors find that there is additional power in the spectrum of the intensity variations at these scales. This power is associated with strong focusing events, or caustics. These events are best described as simple diffraction catastrophes, which are classified and analysed on the basis of geometrical and wave optics, taking account of the strong dispersion of the ISM scattering, which introduces important frequency dependencies.

141.008 A WSRT 21 cm deep survey of two fields in Hercules.
M. J. A. Oort, H. J. van Langevelde.
Astron. Astrophys., Suppl. Ser., Vol. 71, No. 1, p. 25 – 38 (1987).
The authors present a deep 21 cm survey, carried out with the Westerbork Synthesis Radio Telescope (WSRT), of two fields in the constellation of Hercules. A complete sample is defined, containing 116 radio sources with a peak flux above 5 σ, within the –7 dB attenuation radius (0°464). This complete sample is used to determine the 1412 MHz source counts down to 0.45 mJy. The counts from the current sample show the same

small scale structure at ≈ 1 mJy as was found in previous surveys. Finally, a search was made for the variable sources.

141.009   The identification of galactic radio sources based on a comparison of radio–continuum and infrared emission.
E. Fürst, W. Reich, Y. Sofue.
Astron. Astrophys., Suppl. Ser., Vol. 71, No. 1, p. 63 – 67 (1987).

The ratio of the far–infrared to the radio continuum emission is very different for H II regions and supernova remnants. Using the IRAS 60 μm and radio 2.7 GHz emission it is demonstrated that this difference provides an effective tool to distinguish between thermal and non–thermal sources even in highly confused regions in the galactic plane.

141.010   A deep WSRT 21 cm survey down to 0.1 mJy in the Lynx area.
M. J. A. Oort.
Astron. Astrophys., Suppl. Ser., Vol. 71, No. 2, p. 221 – 243 (1987).

The author presents a very deep 21 cm radio survey, carried out with the Westerbork Synthesis Radio Telescope (WSRT), of a field in the Lynx area. In total, 349 sources above the 5 σ peak flux level were detected. This sample is used to determine the 1412 MHz source counts down to 0.1 mJy and to investigate the angular size distributions of radio sources. At low flux densities the source counts agree very well with other surveys. At higher (0.5 – 10 mJy) flux densities, however, the current survey gives consistently lower values for the source counts than previous WSRT surveys.

141.011   Millimeter–wave spectra and variability of bright, compact radio sources.
R. A. Edelson.
Astron. J., Vol. 94, No. 5, p. 1150 – 1155 (1987).

Observations at 2.7 mm and at 1.5 cm were used to study the millimeter spectra and variability of 176 bright, compact radio sources. More than 20% of the flat–spectrum sources, but none of the steep–spectrum sources, were seen to vary at 1.5 cm by at least 30% over ten months. This is consistent with the hypothesis that flat–spectrum sources are compact and possibly beamed, while steep–spectrum sources are not. These data can also be used to choose sources for VLBI observations and for calibration of millimeter–wave observations.

141.012   Radio pulses from the Perseus flasher?
K. C. Turner.
Bull. Am. Astron. Soc., Vol. 19, No. 2, p. 693 – 694 (1987). Abstract. – See Abstr. 010.061.

141.013   The effect of comet ion tails on radio source scintillation.
C. A. Hajivassiliou, P. J. Duffett–Smith.
Mon. Not. R. Astron. Soc., Vol. 229, No. 3, p. 485 – 493 (1987).

It has been reported that strong intensity fluctuations of three radio sources were observed during occultation by the ion tails of comets Kohoutek and Halley. The authors have analysed an extensive data base in an attempt to correlate an increase in scintillation index with occultation by the ion tails of many comets during the period 1978 May – 1981 March. No such correlation was observed. The authors discuss some weakness of the earlier reports and conclude that the reality of the effect remains in doubt.

141.014   VLA observations of radio sources near Wolf 359.
C. P. O'Dea, W. B. McKinnon.
Publ. Astron. Soc. Pac., Vol. 99, No. 620, p. 1039 – 1043 (1987).

This paper reports high–resolution VLA observations at 6 and 20 cm of the region surrounding Wolf 359. The authors present maps of three extended background sources and determine sensitive upper limits to the quiesent flux density of Wolf 359. These upper limits are discussed in the context of quiescent microwave emission from dwarf M stars.

141.015   Determinations of distances to radio sources with VLBI.
N. Bartel.
Cent. Astrophys., Prepr. Ser., No. 2553, 10 pp. (1987). To appear in IAU Symp. No. 129.

141.016   A deep radio and optical survey near the North Galactic Pole. IV. VLA observations and optical identifications of 5C12 sources.
C. R. Benn, G. Grueff, M. Vigotti, J. V. Wall.
R. Greenwich Obs., Prepr., No. 57, 78 + 28 pp. (1987). To appear in Mon. Not. R. Astron. Soc.

141.017   The catalogue of H I absorption line profiles in the spectra of galactic radio sources. I. Integral profiles.
Z. A. Alferova, A. P. Venger, I. V. Gosachinskij, V. G. Grachev, T. M. Egorova, S. R. Zhelenkov, N. P. Komar, V. G. Mogileva, N. F. Ryzhkov.
Astrofiz. Issled. Izv. Spets. Astrofiz. Obs., Tom 24, p. 93 – 107 (1987). In Russian. English translation in Bull. Spec. Astrophys. Obs. – North Caucasus.

H I absorption line profiles for 55 galactic radio sources are presented. This is the first determination for 20 of them. In other 15 cases the authors could eliminate contradictions in previous data.

141.018   A deep radio survey of the North Ecliptic Pole region at 11 cm.
N. Loiseau, W. Reich, R. Wielebinski, W. Münch.
Rev. Mex. Astron. Astrofis., Vol. 14, No. 2, p. 514 (1987). Abstract. – See Abstr. 012.042.

141.019   The Cambridge IPS survey at 81.5 MHz.
A. Purvis, S. J. Tappin, W. G. Rees, A. Hewish, P. J. Duffett–Smith.
Mon. Not. R. Astron. Soc., Vol. 229, No. 4, p. 589 – 619 (1987). With microfiches MN 229/1 – 229/3.

The authors present a catalogue of 1789 radio sources which exhibit interplanetary scintillation (IPS) at 81.5 MHz. The angular diameters of scintillating components in the range 0.2 – 2 arcsec are listed together with values of the scintillating flux density at a solar elongation of 90°. IPS selects those sources which are highly compact, such as pulsars and some unusual extragalactic sources, or those in which energy is being released from active beams in the outer lobes of intrinsically powerful radio galaxies and quasars.

The Green Bank Third (GB3) Survey of extragalactic radio sources at 1400 MHz.
See Abstr. 002.035.

The first optical identifications of radio sources.
See Abstr. 004.125.

A review of decametric radio astronomy: instruments and science.
See Abstr. 013.022.

Experiment Cold: the first deep sky survey with the RATAN–600 radio telescope.
See Abstr. 033.013.

The three–dimensional data–processing techniques in the research survey.
See Abstr. 036.162.

Image analysis of the observation of the DA240 region with a meter–wave synthesis radio telescope.
See Abstr. 036.249.

Declination improvements in surveys with the RATAN–600 radio telescope.
See Abstr. 041.011.

Observations of anomalous refraction at radio wavelengths.
See Abstr. 082.035.

Scintillations of 4 radio sources occulted by the plasma tail of comet Halley.
See Abstr. 103.441.

Occultation observations of compact radio sources through comet Halley's plasma tail.
See Abstr. 103.618.

Proper motion of the compact, nonthermal radio source in the galactic center.
See Abstr. 111.009.

Refraction it the Galaxy.
See Abstr. 131.120.

Radio caustics from localized interstellar medium plasma structures.
See Abstr. 131.121.

The Orion radio zoo: PIGS (*partially ionized globules*), DEERS (*deeply embedded energetic radio sources*) and FOXES (*fluctuating optical and X–ray emitting sources*).
See Abstr. 131.224.

VLA and low–frequency VLBI observations of the radio source 0503+467: austere constraints on interstellar scattering in two media.
See Abstr. 131.304.

Further observations of the peculiar galactic radio source BG 2107+49.
See Abstr. 132.004.

G0.18–0.04: interaction of thermal and nonthermal radio structures in the Arc near the galactic center.
See Abstr. 155.085.

Cold atomic gas in the inner Galaxy.
See Abstr. 155.103.

A radio continuum survey of the galactic plane at 10 GHz.
See Abstr. 155.148.

A 1.6 MHz survey of the galactic background radio emission.
See Abstr. 155.149.

A complete VLA survey of galactic plane sources.
See Abstr. 155.159.

A non–thermal axially symmetric radio wake towards the galactic centre.
See Abstr. 155.160.

Optical identifications and radio morphology of the complete 5 GHz S5 survey.
See Abstr. 158.146.

Flux densities for 72 radio sources in the declination band $\delta = -6°$ at 7.6 cm wavelength.
See Abstr. 158.268.

Investigation of radio objects with continuum optical spectra. Rapid radio variability from observations at the RATAN–600 radio telescope.
See Abstr. 158.269.

Statistical results from the Effelsberg Abell cluster radio survey.
See Abstr. 160.027.

Some results from the Molonglo cluster survey.
See Abstr. 160.028.

"COLD" experimental data as a test on the initial inhomogeneity of the Universe.
See Abstr. 161.046.

142 UV Sources, X-ray Sources, X-ray Background

142.001   **Observations of X–ray burst sources.**
Y. Tanaka.
The origin and evolution of neutron stars, p. 161 – 172 (1987). – See Abstr. 012.001 (IAU Symp. No. 125).
Recent observational results on X–ray bursts and burst sources are reviewed. Two distinct types of bursts, Type I and Type II bursts, are discussed in relation to the mass and radius of neutron stars and to the problems of unusual mass accretion in some burst sources.

142.002   **Results of the timing analyses of X–ray pulsars observed by Hakucho and Tenma.**
F. Nagase.
The origin and evolution of neutron stars, p. 200 (1987). Abstract. – See Abstr. 012.001 (IAU Symp. No. 125).

142.003   **Herculis X–1: results and interpretation.**
P. Durouchoux.
The origin and evolution of neutron stars, p. 201 (1987). Abstract. – See Abstr. 012.001 (IAU Symp. No. 125).

142.004   **A hard X–ray observation of Cyg X–1 in 1985.**
Y. Q. Ma, G. H. Li, C. M. Zhang, Q. Y. Xiao, Y. M. Qian, Z. G. Li, M. Wu, C. J. Dai, Y. D. Gu, X. Y. Zhang, T. P. Li.
The origin and evolution of neutron stars, p. 202 (1987). Abstract. – See Abstr. 012.001 (IAU Symp. No. 125).

142.005   **The X–ray transient EXO 2030+375.**
A. N. Parmar, N. E. White, L. Stella, P. Ferri.
The origin and evolution of neutron stars, p. 203 (1987). Abstract. – See Abstr. 012.001 (IAU Symp. No. 125).

142.006   **Analysis of X–ray spectrum for globular cluster M15 X–ray source.**
F. Z. Cheng, J. E. Grindlay.
The origin and evolution of neutron stars, p. 204 (1987). Abstract. – See Abstr. 012.001 (IAU Symp. No. 125).

142.007   **X–ray bursting neutron stars.**
H. Inoue.
The origin and evolution of neutron stars, p. 233 – 243 (1987). – See Abstr. 012.001 (IAU Symp. No. 125).
The maximum peak luminosity of the X–ray bursts from a burster is most likely interpreted as the Eddington luminosity of a helium–rich envelope surrounding a neutron star. If this interpretation is true, it is possible to obtain a relation between the mass and the radius of the neutron star in terms of the maximum effective temperature of bursts. On the other hand, the most naive understanding of the origin of the 4.1 keV absorption line often detected in X–ray burst spectra gives another relation of the neutron star mass with its radius. By solving two simultaneous equations, the author determines the values of the mass and the radius of the neutron star, respectively. However, the result is

critical to every neutron star model currently considered. The persistent emissions from X–ray bursters are also discussed.

142.008   X–ray irradiated accretion disk and bimodal states.
R. Hoshi, H. Inoue.
The origin and evolution of neutron stars, p. 247 (1987).
Abstract. – See Abstr. 012.001 (IAU Symp. No. 125).

142.009   X–ray spectra and atmospheric structures of bursting neutron stars.
T. Ebisuzaki.
The origin and evolution of neutron stars, p. 250 (1987).
Abstract. – See Abstr. 012.001 (IAU Symp. No. 125).

142.010   The origin of the diffuse X–ray background.
T. T. Hamilton, D. J. Helfand.
Astrophys. J., Vol. 318, No. 1, p. 93 – 102 (1987). With plates 1 – 3.
An analysis of the arcminute scale fluctuations in the extragalactic diffuse X–ray background has been undertaken using the Deep Survey data gathered by the Einstein Observatory. These data provide a direct measure of the relative contributions to the unresolved X–ray background of point sources and effectively diffuse emission. A sophisticated Monte Carlo scheme is used to compare the X–ray images with a variety of models representing point and diffuse source contributions. Results for two classes of models are described: those which determine the flux cutoff beyond which pointlike sources fail to contribute significantly if the $\log N - \log S$ slope is fixed at the measured value of -1.5, and those which set limits on the slope of the $\log N - \log S$ relation below its last measured point at the Deep Survey threshold. In both cases, the known population of X–ray emitting quasars and any reasonable extrapolations therefrom do not comprise the entire X–ray background.

142.011   Discrete X–ray sources and the X–ray background.
R. Giacconi.
NASA Conf. Publ., NASA CP-2464, p. 317 – 338 (1987). – See Abstr. 012.004.
Since the discovery, more than twenty years ago, of a highly uniform X–ray background (XRB) in the 2 – 10 keV range, its nature has not yet been fully explained. It appears clear from the results of ''Einstein'' medium and deep surveys that at least 50 percent of the XRB is due to individual extragalactic sources when their contribution is integrated to $Z = 3$. This includes contribution from quasi stellar objects, active galactic nuclei, galaxies, and clusters of galaxies. The average spectrum of each of the individual contributing sources is softer than that of the observed XRB (power law index $\alpha \approx -0.4$ from 3 to 10 keV). Therefore, the remaining contribution must have a rather hard spectrum of $\alpha \approx 0.0 - 0.2$. It is unlikely that this spectrum can be produced by diffuse processes. Therefore, the remainder of the XRB must be due to individual sources with the appropriate spectrum. This requires either that the spectrum of the already identified sources changes at early epochs or a new class of objects.

142.012   The cosmic X–ray background.
E. A. Boldt.
NASA Conf. Publ., NASA CP-2464, p. 339 – 378 (1987). – See Abstr. 012.004.
Contents: 1. Overview. 2. Role of the HEAO program. 3. Spectroscopy. 4. Isotropy. 5. Outlook.

142.013   Diffuse background X rays and the density of the intergalactic medium.
D. F. Crawford.
Aust. J. Phys., Vol. 40, No. 3, p. 459 – 464 (1987).
This paper shows that a major part of the X–ray background could be due to thermal bremsstrahlung from a static high temperature plasma, where the X rays are subject to a Hubble frequency shift and a geometric factor consistent with a static Einstein cosmology. There is no evolution of density or pressure. The motivation for this model is the suggestion of a new

mechanism that can explain the Hubble shift without requiring an expanding Universe.

142.014   Time lag between hard and soft X–ray photons in QPO sources.
G. M. Stollman, G. Hasinger, W. H. G. Lewin, M. van der Klis, J. van Paradijs.
Mon. Not. R. Astron. Soc., Vol. 227, No. 2, p. 7P – 12P (1987).
The authors show that the delay (of the order of ms) between the quasi–periodic oscillations (QPO), observed from Cyg X–2 and GX 5–1, in a high–energy and a low–energy photon pass-band, can be the result of Compton scattering of photons in a cloud of hot electron gas. This delay depends on the optical depth, electron temperature, and radius of the cloud, and on the spectrum of the input photons. Using Monte Carlo simulations the authors show how, for an assumed input spectrum of photons, the observed delay and X–ray spectrum can be used to constrain the size of the scattering cloud. The authors find that, in addition to scattering of an oscillating signal in a constant cloud, scattering of a constant source of input photons through a cloud with oscillating optical depth can also give rise to both QPO in the emergent photons, and a time lag between high–energy and low–energy photons.

142.015   Spectral variability of 4U1145–619 during X–ray outburst.
M. C. Cook, R. S. Warwick.
Mon. Not. R. Astron. Soc., Vol. 227, No. 3, p. 661 – 676 (1987).
A series of observations of the X–ray pulsar 4U1145–619 were performed by EXOSAT during successive X–ray outbursts on this source in 1984 July and 1985 January. On all occasions the source exhibited a very hard power–law X–ray spectrum (photon spectral index ~1.0) with a high–energy cut–off above ~6 keV. The low–energy cut–off in the X–ray spectrum, in contrast to the other spectral parameters, exhibited significant variations over the observations. Optical spectroscopy of the Be star Hen 715, identified as the optical counterpart of 4U1145–619, was performed during the 1985 January X–ray outburst. The resulting spectra revealed a marked decrease in the strength of the broad blue component of the dominant Hα emission line as the X–ray outburst progressed. There was, however, no evidence for any pulse phase dependency or short–term variability in either the Hα of Hβ emission lines. These X–ray and optical observations are interpreted in terms of a neutron star passing through the circumstellar envelope of its Be companion.

142.016   Optical variability of the black hole candidate GX 339–4 (X 1659–487, V821 Ara) – limits on periodic modulation.
R. H. D. Corbet, J. R. Thorstensen, P. A. Charles, W. B. Honey, A. P. Smale, J. W. Menzies.
Mon. Not. R. Astron. Soc., Vol. 227, No. 4, p. 1055 – 1064 (1987).
The authors present results of extensive CCD optical photometry (over 1000 frames representing ~150 hr of integration time) of the optical counterpart of the X–ray source GX 339–4 obtained during the high (soft) state. The source was seen to be significantly variable. They do not, however, detect any periodic modulation with semi–amplitude greater than ~0.03 mag for periods less than ~0.5 day or semi–amplitude greater than ~0.07 mag for longer periods. An optical spectrum obtained shortly after GX 339–4 made a transition from a 'low' to a 'high' state is also presented and compared with previous results. The width of the He II λ4686 emission line in the spectrum implies that GX 339–4 does not have an unusually low inclination angle and the authors therefore conclude that the orbital period of GX 339–4 is probably longer than ~0.5 day.

142.017   Search for X rays from the region of the Aries–Perseus flasher.
W. H. G. Lewin, J. van Paradijs, E. Damen, F. Jansen, M. L. McCall, P. A. Feldman, K. F. Tapping.
Astron. J., Vol. 94, No. 2, p. 429 – 430 (1987).
The region containing the Perseus flasher (Katz et al., 1986) was observed for 5.4 hr with the EXOSAT observatory. Upper

limits (2σ levels of confidence) to a point source with a steady X-ray flux were $2 \times 10^{-12}\,\mathrm{erg\,cm^{-2}s^{-1}}$ $(0.1-2\,\mathrm{keV})$ and $6 \times 10^{-12}\,\mathrm{erg\,cm^{-2}s^{-1}}$ $(1-20\,\mathrm{keV})$. Upper limits (4σ level of confidence) to X-ray flashes of $\sim 1\,\mathrm{s}$ duration were $4 \times 10^{-9}\,\mathrm{erg\,cm^{-2}s^{-1}}$ $(0.1-2\,\mathrm{keV})$ and $7 \times 10^{-10}\,\mathrm{erg\,cm^{-2}s^{-1}}$ $(1-20\,\mathrm{keV})$.

142.018   **Discrete X-ray sources and the X-ray background.**
R. Giacconi.
Astrophys. Lett. Commun., Vol. 26, Nos. 1–2, p. 7–19 (1987).
– See Abstr. 012.020.

Since the discovery of a highly uniform X-ray background (XRB) in the $2-10\,\mathrm{keV}$ range, its nature has not yet been fully explained. At least 50 per cent of the XRB is due to individual extragalactic sources, if their contribution is integrated to $Z = 3$. The average spectrum of each of the individual contributing sources is softer than that of the observed XRB incidating the remaining contribution must have a rather hard spectrum. The remainder of the XRB must be due to individual sources with the appropriate spectrum. This requires either that the spectrum of the already identified sources changes at early epochs or a new class of objects. AXAF observations will extend survey sensitivity to limiting fluxes of order of $3 \times 10^{-16}\,\mathrm{erg\,cm^{-2}s^{-1}}$, some 50 times fainter than any previous survey. They will have sufficient sensitivity and angular resolution to permit identification and study of these objects.

142.019   **Stalking the black hole in the star garden of the Unicorn.**
J. McClintock.
Mercury, Vol. 16, No. 4, p. 108–109, 123 (1987).

142.020   **A deep optical study of the field of 1E 0630 + 178.**
G. F. Bignami, P. A. Caraveo, J. A. Paul, L. Salotti,
L. Vigroux.
Astrophys. J., Vol. 319, No. 1, p. 358–361 (1987). With plates 3–6.

The authors present further optical/IR investigation of the HRI error–box of the X-ray source 1E 0630+178, proposed as the counterpart of Geminga (2CG 195+04). After recognizing that the previously suggested $m \approx 21$ object is probably excluded by the recent data, the authors present deep CCD data of the region and discuss the two very faint objects which appear compatible with the X-ray position. No evidence of proper motion is seen, contrary to previous reports. Possible interpretations of this peculiar $L_x/L_v > 1000$ X-ray source are then discussed.

142.021   **The complex cross–spectra of Cygnus X–2 and GX 5–1.**
M. van der Klis, G. Hasinger, L. Stella,
A. Langmeier, J. van Paradijs, W. H. G. Lewin.
Astrophys. J., Lett. Ed., Vol. 319, No. 1, p. L13–L18 (1987).

The intensity variations of Cyg X–2 and GX 5–1 are analyzed with a Fourier cross–spectral technique allowing the measurement of the frequency dependence of the time lags between the intensity variations in two X-ray spectral bands. In the $20-40\,\mathrm{Hz}$ frequency range of the quasi–periodic oscillations (QPO), the hard ($\sim 5-18\,\mathrm{keV}$) signal lags the soft ($\sim 1-5\,\mathrm{keV}$) one, while the reverse is true for the associated red noise. The observed time lags range from 0.4 to 8 ms. For similar QPO parameters, the observed lags are much smaller in GX 5–1 than they are in Cyg X–2.

142.022   **A rapid change of the Hercules X–1 pulse profile and high–state duration.**
Y. Soong, D. E. Gruber, R. E. Rothschild.
Astrophys. J., Lett. Ed., Vol. 319, No. 2, p. L77–L81 (1987).

Her X–1 has been observed in the $13-180\,\mathrm{keV}$ energy range by the *HEAO 1* Low–Energy Detectors during selected phases of the 35^{d} on–off cycle. During a pointing observation in 1978 September, the pulse profile was observed to change continuously from its normal shape to an anomalous double–pulsed form. Moreover, at this time the main–on state, normally of duration 10^{d}, was seen to terminate after only 7^{d}. For one observation of the short–on state the observed profile also had an anomalous double–pulsed form.

142.023   **The Einstein Galactic Plane Survey extension: source counts and the X–ray number–flux relation.**
B. J. Geldzahler, P. Hertz.
Bull. Am. Astron. Soc., Vol. 19, No. 2, p. 680–681 (1987). Abstract. – See Abstr. 010.061.

142.024   **Production of the diffuse X–ray background spectrum by active galactic nuclei.**
D. A. Schwartz, W. Tucker.
Bull. Am. Astron. Soc., Vol. 19, No. 2, p. 696 (1987). Abstract. – See Abstr. 010.061.

142.025   **The X–ray spectra of faint extragalactic sources.**
A. Wolter, I. M. Gioia, T. Maccacaro, J. Stocke,
G. Zamorani.
Bull. Am. Astron. Soc., Vol. 19, No. 2, p. 696 (1987). Abstract. – See Abstr. 010.061.

142.026   **The low luminosity X–ray sources in Omega Centauri.**
B. Margon, M. Bolte.
Bull. Am. Astron. Soc., Vol. 19, No. 2, p. 720 (1987). Abstract. – See Abstr. 010.061.

142.027   **An all–sky study of fast X–ray transients.**
A. Connors.
Bull. Am. Astron. Soc., Vol. 19, No. 2, p. 720 (1987). Abstract. – See Abstr. 010.061.

142.028   **Spectral observation of A 0535 + 26 from 2–180 keV by HEAO–1.**
R. E. Rothschild, J. H. Swank.
Bull. Am. Astron. Soc., Vol. 19, No. 2, p. 720–721 (1987). Abstract. – See Abstr. 010.061.

142.029   **Quasi–periodic oscillations with the blackbody energy spectra.**
T. Ebisuzaki, N. Shibazaki, R. Bussard.
Bull. Am. Astron. Soc., Vol. 19, No. 2, p. 721 (1987). Abstract. – See Abstr. 010.061.

142.030   **Observations of the soft X–ray diffuse background below 0.2 keV.**
M. Juda, J. J. Bloch, D. McCammon, W. T. Sanders,
S. L. Snowden.
Bull. Am. Astron. Soc., Vol. 19, No. 2, p. 722 (1987). Abstract. – See Abstr. 010.061.

142.031   **An *EXOSAT* observation of the bursting X–ray transient 4U 1608–52.**
M. Gottwald, L. Stella, N. E. White, P. Barr.
Mon. Not. R. Astron. Soc., Vol. 229, No. 3, p. 395–403 (1987).

An *EXOSAT* observation of the transient X-ray burster 4U 1608–52 detected the source in a low state. The spectrum of the persistent emission is a power law with a photon index of 1.9 similar to that from other X-ray burst sources. During the 5.5 hr exposure a double–peaked type I burst was seen. A detailed spectral analysis shows this to be caused by changes in the colour temperature rather than radius expansion. Power spectra of the time resolved data show increasing power towards low frequencies (red noise) and a broad peak at $\sim 0.3\,\mathrm{Hz}$ with a significance of $\sim 3\sigma$, that may indicate the presence of quasi–periodic oscillations.

142.032   **A systematic search for variability of X–ray sources. II. The low galactic latitude sample.**
S. Mereghetti, B. Garilli.
Astron. J., Vol. 94, No. 6, p. 1616–1628 (1987).

A search for variability on timescales from 1 day to years has been carried out for 120 X-ray sources observed with the *Einstein* Observatory Imaging Proportional Counter. All the sources are at low galactic latitude and most of them have soft X-ray fluxes in the range $10^{-13} - 10^{-12}$erg cm^{-2}s^{-1}. Significant flux variations have been detected in 17 sources. A strong flare, with a flux increase of a factor ~ 100 in ~ 400 s, has been discovered in an optically unidentified source. This is probably a dwarf M flare star or an RS CVn system.

142.033   Radio–frequency observations of galactic X–ray sources. II.
B. J. Geldzahler.
Publ. Astron. Soc. Pac., Vol. 99, No. 620, p. 1036 – 1038 (1987).

Six galactic low–mass X–ray binary stars were surveyed with the VLA at 4.8 GHz to search for objects with morphologies similar to that of Sco X–1. Two objects, 2129 + 470 and 2142 + 380 (Cyg X–2) were detected at the 1 mJy level; their radio emission is thermal. Variability in the quiescent radio state has been established for 0620–003. None of the sources exhibited a radio morphology similar to the triple–source structure of Sco X–1.

142.034   Coherence effects on quasi–periodic oscillations from galactic X–ray sources.
R. F. Elsner, N. Shibazaki, M. C. Weisskopf.
Astrophys. J., Vol. 320, No. 2, p. 527 – 536 (1987).

Shot noise models provide a useful mathematical representation for some physical models for the quasi–periodic oscillations (QPO) recently observed from several Galactic bulge and burst X–ray sources. For such models, correlations between the shot onset times and oscillation phases can have important effects on the expected autocorrelation function and power spectrum. Making use of analytic and Monte Carlo techniques, the authors have examined in detail two physically plausible correlations of this kind, one corresponding to drift and jitter in the oscillation phase and the other to a random walk in the oscillation phase. The effects of coherence on the expected power spectrum for both cases are examined.

142.035   EXOSAT observations of a giant X–ray burst in XB 1905 + 000.
C. Chevalier, S. A. Ilovaisky.
Obs. Haute Provence, Pré–Publ., No. 27, 19 pp. (1987). Submitted to Astron. Astrophys.

142.036   X–ray investigations of the Hercules X–1 pulsar aboard the automatic station Astron in 1983 – 1985.
E. K. Sheffer, I. F. Kopaeva, G. S. Bisnovatyj–Kogan, I. M. Golanskaya, L. S. Gurin, A. V. D'yachkov, V. M. Zenchenko, V. G. Kurt, T. A. Mizyakina, V. A. Sklyankin, A. S. Smirnov, L. G. Titarchuk, V. M. Shamolin, E. Yu. Shafer, F. Giovannelli.
Inst. Kosm. Issled. Akad. Nauk SSSR, Prepr., No. 1252, p. 3 – 65 (1987). In Russian. Abstr. in Ref. Zh., 51. Astron. 11.51.604 (1987).

142.037   The optical identification of celestial X–ray sources.
I. R. Tuohy.
Aust. Phys., Vol. 23, No. 11, p. 278 – 281 (1986). Abstr. in Phys. Abstr., Vol. 90, No. 1311, Entry 101958 (1987).

142.038   On the spectra of X–ray bursters.
R. A. Syunyaev, L. G. Titarchuk.
Sov. Astron. Lett., Vol. 12, No. 6, p. 359 – 364 (1986). English translation of 42.142.007.

142.039   Comptonization and X–ray burster spectra.
I. I. Lapidus, R. A. Syunyaev, L. G. Titarchuk.
Sov. Astron. Lett., Vol. 12, No. 6, p. 383 – 386 (1986). English translation of 42.142.010.

142.040   The luminosity oscillation of X–ray bursters and recurrent novae.
R. Aquilano, M. Castagnino, L. Lara.
Astrophys. Space Sci., Vol. 138, No. 1, p. 41 – 48 (1987).

'Thermo–mechanical' oscillations of a radiating spherically–symmetric shell containing a fermion gas (i.e., oscillation where only mechanical, thermodynamical, and eventually radiation phenomena are taken into account) are studied. With a reasonable choice of the three relevant parameters all other observational data are similar to the ones of X–ray bursters and recurrent novae. Therefore, thermo–mechanical oscillations could play an important role in the oscillation spectrum of these astronomical objects.

142.041   X–ray observations of the extraordinary pulsar 1E 2259 + 586.
K. Koyama, R. Hoshi, F. Nagase.
Publ. Astron. Soc. Jpn., Vol. 39, No. 5, p. 801 – 807 (1987).

The pulse period and X–ray spectrum of 1E 2259 + 586 were observed with the Tenma satellite in October, 1983. The heliocentric pulse period was determined to be 6.978675 ± 0.000010 s, which is about 4×10^{-5}s longer than that obtained with the Einstein Observatory in 1980 and 1981. An averaged spin–down rate is 5×10^{-13}s s^{-1}. The X–ray spectrum was found to be much softer than usual binary X–ray pulsars.

142.042   The discovery of 7 and 24 – 28 Hertz quasi–periodic oscillations in the X–ray flux of GX 17 + 2 (4U 1813–14).
L. Stella, A. N. Parmar, N. E. White.
Astrophys. J., Vol. 321, No. 1, p. 418 – 424 (1987).

The authors report the discovery of quasi–periodic oscillations (QPO) in the X–ray flux of the bright galactic bulge source GX 17 + 2. The relation between the QPO frequency and the source intensity of GX 17 + 2 is bimodal. On 1985 August 20 – 21 the QPO frequency was ~ 7 Hz and showed no significant change associated with $\sim 10\%$ intensity variations. In another observation 25 days later, the frequency varied between 24 Hz and 28 Hz and was strongly correlated with 5% variations in intensity. The power spectra also displayed a significant increase in power toward low frequencies (red noise: RN). The increase saturated at low frequencies, giving characteristic time scales for the RN activity of 34 ms and 350 ms.

142.043   Multipeaked X–ray bursts from 4U/MXB 1636–53: evidence against burst–induced accretion disk coronae.
W. Penninx, J. van Paradijs, W. H. G. Lewin.
Astrophys. J., Lett. Ed., Vol. 321, No. 1, p. L67 – L69 (1987).

It is shown that the model proposed by Melia in 1987, in which the multipeaked profiles of some X–ray bursts from 4U/MXB 1636–53 are the result of scattering by a transient burst–induced accretion–disk corona, is inconsistent with the observed burst properties.

142.044   The X–ray sources of the galactic bulge, X–ray bursters and X–ray sources in globular clusters.
E. N. Ercan.
Doğa Turk Fiz. Astrofiz. Dergisi (*Turkey*), Vol. 11, No. 1, p. 134 – 156 (1987). Abstr. in Phys. Abstr., Vol. 90, No. 1313, Entry 115629 (1987).

142.045   X–ray pulsars.
IAU Circ., No. 4459 (1987).

142.046   4U 0115 + 63.
IAU Circ., No. 4485 (1987).

142.047   EXO 033319–2554.2.
IAU Circ., Nos. 4486, 4517 (1987).

142.048   0535–668.
IAU Circ., No. 4519 (1987).

142.049   0.65–second oscillation at the peak of an X–ray burst from X 1608–522.
T. Murakami, H. Inoue, K. Makishima, R. Hoshi.
Publ. Astron. Soc. Jpn., Vol. 39, No. 6, p. 879 – 886 (1987).

A 0.65–s period oscillation was detected near the peak of an X–ray burst from X 1608–522. The burst exhibits a clear photospheric expansion at the peak flux which is probably at the

Eddington limit of the underlying star. The 0.65–s oscillation was seen in the contracting phase during which the flux stayed at the maximum value. An anticorrelation between counts in the low– and high–energy bands was observed during the oscillation. This can be interpreted as a radial oscillation of the envelope which was radiating at the Eddington luminosity.

142.050 Search for the optical variability of the X–ray source E 2000 + 223.
I. L. Andronov.
Astron. Tsirk., No. 1463, p. 5 (1986). In Russian.

The old nova E 2000 + 223 was studied on 123 plates of the Moscow collection. The object might be variable ($17^m4 - 17^m7$ pg), but no bursts or systematical luminosity decrease were detected.

142.051 Diffuse Einstein X–ray emission from the region of RCW 49, a bright southern emission nebula.
A. Goldwurm, P. A. Caraveo, G. F. Bignami.
Astrophys. J., Vol. 322, No. 1, p. 349 – 352 (1987). With plates 1 – 3.

An X–ray exploration of a region of the Galactic plane in Carina near $l = 284°$ with the Einstein Observatory is presented, consisting of two IPC and one HRI fields. Three new X–ray sources are seen, of which one is clearly extended and can be associated with RCW 49, an H II region/emission nebula, containing a cluster of young stars. The X–ray data could be consistent with a central source of diffuse emission; however, the X–ray flux appears significantly greater than that expected for an object like RCW 49.

142.052 EXOSAT observations of 4U 1705–44: type I bursts and persistent emission.
A. Langmeier, M. Sztajno, G. Hasinger, J. Trümper, M. Gottwald.
Astrophys. J., Vol. 323, No. 1, p. 288 – 293 (1987).

The authors report on a set of four EXOSAT observations of the bright galactic X–ray source 4U 1705–44 during different intensity levels (from 0.7 to 8.3×10^{-9}ergs cm^{-2}s^{-1} in the 2 – 11 keV energy band). They report also on the discovery of type I X–ray bursts and give an overview of the observations with emphasis on the burst activity pattern. They discuss the spectral properties of the persistent emission as well as the bursts and report on the results.

142.053 The properties of bursts with short recurrence times from the transient X–ray source EXO 0748–676.
M. Gottwald, F. Haberl, A. N. Parmar, N. E. White.
Astrophys. J., Vol. 323, No. 2, p. 575 – 581 (1987).

On 1986 January 13/14 *EXOSAT* performed a 11.3 hr observation of the transient X–ray burst source EXO 0748–676. The source was in a low state and emitted 11 type I X–ray bursts. The bursts show a regular pattern with a long recurrence time always followed by a short one. The long intervals lasted ∼130 to 160 minutes while the short cover the range ∼20 – 70 minutes. The total emitted burst energy depends linearly on the recurrence time, but with an offset in energy of ∼0.3×10^{-7}ergs cm^{-2}. The authors suggest that the bursting behavior indicates the survival and reconsumption of nuclear fuel in different bursts.

X–ray astronomy. Selected reprints.
See Abstr. 003.036.

Variability of galactic and extragalactic X–ray sources. Proceedings of an international symposium held at Villa Olmo, Como, Italy, 20 – 22 October 1986.
See Abstr. 012.019.

What do we learn from space? Space science in Japan.
See Abstr. 013.044.

ROSAT, data centers, and X–ray data analysis, a new era?
See Abstr. 013.051.

X–ray astronomical spectroscopy.
See Abstr. 035.003.

The ultraviolet telescope on the Astron satellite.
See Abstr. 035.021.

All Sky Supernova and Transient Explorer (ASTRE).
See Abstr. 035.023.

X–ray spectroscopy of AGN with the AXAF "microcalorimeter".
See Abstr. 035.026.

Low energy X–ray transmission grating spectrometer for AXAF.
See Abstr. 035.027.

The MIT spectroscopy investigation of AXAF and the study of supernova remnants.
See Abstr. 035.028.

A high sensitivity phoswich scintillator X–ray telescope for hard X–ray (20 – 120 keV) astronomy from balloon platform.
See Abstr. 035.032.

Fabrication and flight performance of a large area balloon borne hard X–ray telescope.
See Abstr. 035.033.

X–ray astronomy satellite "Ginga".
See Abstr. 035.034.

All–sky monitors for X–ray astronomy.
See Abstr. 035.089.

Coded aperture imaging in X– and gamma–ray astronomy.
See Abstr. 035.090.

Ultraviolet detection of very low–surface–brightness objects.
See Abstr. 036.053.

Search for flux variability in a sample of X–ray sources dominated by Poisson statistics.
See Abstr. 036.057.

Techniques for the analysis of data from coded–mask X–ray telescopes.
See Abstr. 036.058.

The SAX mission.
See Abstr. 051.019.

Comments about observing conditions for UV astronomy aboard Spacelab.
See Abstr. 051.025.

The extreme and far ultraviolet environment at shuttle altitudes.
See Abstr. 051.026.

Search for ultraviolet Shuttle glow.
See Abstr. 051.028.

X–ray observations from the Space Shuttle.
See Abstr. 051.030.

The Italian satellite for X–ray astronomy (SAX).
See Abstr. 051.038.

Mass loaded astronomical flows – III. The structure of supernova remnants and the local soft X–ray background.
See Abstr. 062.027.

X–ray pulsars: accretion flow deceleration.
See Abstr. 062.068.

Non–thermal pair production in compact X–ray sources: first–order Compton cascades in soft radiation fields.
See Abstr. 063.015.

The formation of iron features in the cooling spectrum of X–ray bursts.
See Abstr. 063.025.

Energy dependent delay measurements of quasi–periodic oscillations in low–mass X–ray binaries.
See Abstr. 063.026.

Compact radio sources as a plasma turbulent reactor. II. General characteristics of electromagnetic radiation spectra.
See Abstr. 063.041.

Effects of electron scattering on the oscillations of an X–ray source.
See Abstr. 063.126.

An evolutionary scenario for the formation of highly eccentric Be/X–ray binaries.
See Abstr. 065.050.

On the origin of neutron stars in globular clusters.
See Abstr. 067.004.

An inverse Compton scattering model for the spectra of X–ray pulsars.
See Abstr. 067.009.

The X–ray radiation mechanism of the compact (neutron) binary stars.
See Abstr. 067.010.

Model atmospheres for X–ray bursting neutron stars.
See Abstr. 067.011.

Nonradial g–mode oscillations in X–ray bursting neutron stars.
See Abstr. 067.037.

Disc accretion by magnetized neutron stars: a reassessment of the torque.
See Abstr. 067.062.

Power spectra of quasi–periodic oscillations in luminous X–ray stars.
See Abstr. 067.063.

Accretion of charged matter by collapsing objects.
See Abstr. 067.068.

Accretion disc models around compact objects.
See Abstr. 067.075.

Neutron star precession and the dynamics of the superfluid interior.
See Abstr. 067.083.

The influence of external magnetic fields on the structure of thin accretion disks.
See Abstr. 067.118.

X–ray polarizations from accreting strongly magnetized neutron stars: case studies for the X–ray pulsars 4U 1626–67 and Hercules X–1.
See Abstr. 067.124.

Radiation anisotropy of the accretion disc around a black hole, heating of the optical component and parameters of the binary system LMC X–3.
See Abstr. 067.133.

The effect of decay of the amplitude of oscillation on random process models for QPO X–ray stars.
See Abstr. 067.174.

New insights from a global view of X–ray bursts.
See Abstr. 067.175.

Model atmospheres of near–Eddington limit X–ray bursters.
See Abstr. 067.178.

Why do stars emit X rays?
See Abstr. 076.001.

The optical counterpart of the X–ray transient EXO 2030 + 375.
See Abstr. 112.023.

An analysis of X–ray fluxes of O–type stars.
See Abstr. 112.129.

Time variability of Gamma Cassiopeiae in X–rays.
See Abstr. 112.145.

Fast transient X–rays from flare stars and RS CVn binaries.
See Abstr. 112.160.

The Einstein view of the Wolf–Rayet stars.
See Abstr. 115.012.

Hardness ratio in evolving low–mass X–ray binary systems.
See Abstr. 117.001.

Quasi–periodic oscillations in low–mass X–ray binaries.
See Abstr. 117.002.

Observations of quasi–periodic oscillations in Cyg X–2.
See Abstr. 117.003.

The beat frequency model for QPOs.
See Abstr. 117.004.

Quasi–periodic oscillations. The Rapid Burster (MXB 1730–335). Models vs. observations – a brief review.
See Abstr. 117.005.

Soft X–ray transients in quiescence: observations of Aql X–1 and Cen X–4.
See Abstr. 117.028.

Disk formation at the magnetosphere of wind–fed pulsars: application to Vela X–1.
See Abstr. 117.030.

Discovery of soft X–ray oscillations in VW Hydri.
See Abstr. 117.031.

An EXOSAT observation of 1.5 orbital cycles of the 0.7 day short–period RS CVn system ER Vul.
See Abstr. 117.045.

An evolutionary scenario for the black hole binary A 0620–00.
See Abstr. 117.054.

The 67–min X–ray period of EX Hydrae observed with the EINSTEIN observatory.
See Abstr. 117.055.

Doppler–effect modulation of the observed radiation flux from ultracompact binary stars.
See Abstr. 117.057.

Iron emission line from low–mass X–ray binaries.
See Abstr. 117.073.

Observations of X–ray emission from the RS CVn binary HD 155555 and the detection of a nearby serendipitous source.
See Abstr. 117.075.

A 685 second orbital period from the X–ray source 4U 1820–30 in the globular cluster NGC 6624.
See Abstr. 117.092.

X–ray spectral formation in LMXRB.
See Abstr. 117.093.

Evidence for black holes in X–ray binary systems.
See Abstr. 117.094.

The contribution of QPO to our understanding of low–mass X–ray binaries.
See Abstr. 117.095.

A theory of soft X–ray transients.
See Abstr. 117.096.

Millisecond X–ray binary pulsars.
See Abstr. 117.097.

X–ray emission from low–mass X–ray binaries.
See Abstr. 117.098.

Periodicities of low mass X–ray binaries from HEAO–1.
See Abstr. 117.099.

The evolutionary status of MXB 1820–30 and other short–period low–mass X–ray binaries.
See Abstr. 117.107.

Correlation between bolometric and X–ray luminosities in RS CVn–type binaries.
See Abstr. 117.117.

The relation between optical and X–ray flux variations of the black–hole candidate LMC X–3.
See Abstr. 117.118.

X–ray system with very short periods.
See Abstr. 117.124.

X–ray observations of B–emission stars.
See Abstr. 117.128.

EXOSAT observations of 4U/MXB 1636–53: on the relation between the amount of accreted fuel and the strength of an X–ray burst.
See Abstr. 117.147.

On the theory of type I X–ray bursts: the energetics of bursts and the nuclear fuel reservoir in the envelope.
See Abstr. 117.148.

CCD photometry of V926 Sco, the optical counterpart of the X–ray burst source 4U/MXB 1735–44.
See Abstr. 117.157.

Hard spectral components in soft X–ray transients.
See Abstr. 117.159.

Five–colour (UBVRI) polarimetry of H 0139–68 = BL Hydri.
See Abstr. 117.160.

A classification of fast quasi–periodic X–ray oscillators: is 6 Hz a fundamental frequency?
See Abstr. 117.162.

Rapid optical flaring in MXB 1735–444 and an optical burst from GX 17 + 2.
See Abstr. 117.167.

Studies of low–mass X–ray binaries.
See Abstr. 117.184.

Der 35–Tage–Zyklus von Hercules X–1.
See Abstr. 117.203.

The anomalous ultraviolet spectrum of the AM Her star H 0538 + 608.
See Abstr. 117.235.

Long–term cycles in cosmic X–ray sources.
See Abstr. 117.240.

CAL 83: a puzzling X–ray source in the Large Magellanic Cloud.
See Abstr. 117.251.

Photographic observations of HZ Herculis in 1985.
See Abstr. 117.270.

Search for X–ray emission from the radio lobes of Scorpius X–1.
See Abstr. 117.281.

On the equilibrium of Hercules X–1.
See Abstr. 117.285.

Optical behaviour of the X–ray source EXO 020528 + 1454.8 in the season 1986/87.
See Abstr. 117.295.

An X–ray survey for pre–main–sequence stars in the Taurus–Auriga and Perseus molecular cloud complexes.
See Abstr. 121.023.

Rotational modulation and flares on RS CVn and BY Dra–type stars. V. EXOSAT and IUE observations of a flare on EQ Pegasi.
See Abstr. 122.011.

High–resolution radio observations of the Crablike supernova remnant 3C 58.
See Abstr. 125.003.

The X–ray spectra of Tycho and SN 1006.
See Abstr. 125.007.

A scanning modulation collimator observation of the high–energy X–ray source in the Crab Nebula.
See Abstr. 125.039.

Discovery of hard X–ray emission from supernova 1987A.
See Abstr. 125.349.

Discovery of an unusual hard X–ray source in the region of supernova 1987A.
See Abstr. 125.350.

A search for soft X–rays from supernova 1987A.
See Abstr. 125.351.

X–rays expected from supernova 1987A compared with the source discovered by the Ginga satellite.
See Abstr. 125.352.

Thermal X–ray emission from supernova 1987A.
See Abstr. 125.353.

Hard X–rays imply more to come.
See Abstr. 125.354.

Accreting neutron stars.
See Abstr. 126.026.

Soft X–ray observations of the radio pulsar PSR 1055–52.
See Abstr. 126.039.

Studies of isolated neutron stars, pulsars and pulsar–driven nebulae with the Advanced X–Ray Astrophysics Facility (AXAF).
See Abstr. 126.053.

Soft X–ray imaging observations of the 39 millisecond pulsar PSR 1951 + 32.
See Abstr. 126.062.

Millisecond X–ray binary pulsars.
See Abstr. 126.070.

PSR 0656 + 14.
See Abstr. 126.092.

The IRAS cirrus and the diffuse ultraviolet background.
See Abstr. 131.049.

Molecular clouds in the vicinity of the semicircular supernova remnant G 109.1–1.0.
See Abstr. 131.112.

Local interstellar medium.
See Abstr. 131.243.

EXOSAT observations of the ring nebula NGC 6888 and HD 192163.
See Abstr. 132.002.

Neutron stars and gamma–rays.
See Abstr. 143.001.

Cygnus X–3 and other ultra–high–energy γ–ray sources.
See Abstr. 143.005.

Synchrotron X–ray haloes around gamma ray sources.
See Abstr. 143.016.

A search for TeV gamma rays from the X–ray binary 1E 2259 + 586.
See Abstr. 143.031.

Low–energy gamma–rays from pulsar GX 1 + 4 – balloon results.
See Abstr. 143.037.

Low–energy gamma–rays from pulsar GX 1 + 4 – balloon results.
See Abstr. 143.039.

Very high energy emission in accretion onto compact objects.
See Abstr. 144.078.

The globular cluster population of X–ray binaries.
See Abstr. 154.001.

X–ray and UV observations of ω Centauri with EXOSAT.
See Abstr. 154.008.

The low–luminosity X–ray sources in Omega Centauri.
See Abstr. 154.076.

A new aspect of galactic ridge X–ray emission – SNRs in a tenuous medium?
See Abstr. 155.001.

Hard X–ray images of the galactic centre.
See Abstr. 155.161.

Population of accreting neutron stars in external galaxies.
See Abstr. 157.001.

Determination of the masses of elliptical galaxies and clusters of galaxies with AXAF.
See Abstr. 157.103.

Hot coronae around early type galaxies.
See Abstr. 157.138.

X–rays from elliptical galaxies.
See Abstr. 157.159.

Mass distributions in elliptical galaxies at large radii.
See Abstr. 157.160.

A dust scattered halo in starburst galaxy M82?
See Abstr. 158.008.

X–ray intensity and spectral variations in the BL Lac sources H 2155–304 and PKS 0548–322.
See Abstr. 158.063.

Radio and X–ray observations of M87.
See Abstr. 158.080.

The variety of short–term variability in AGN.
See Abstr. 158.098.

Soft X–ray variability and spectra of Seyfert galaxies.
See Abstr. 158.099.

Scale invariant variability in NGC 4051.
See Abstr. 158.100.

X–ray and UV variability of MCG 8–11–11 in 1983 – 86.
See Abstr. 158.101.

Rapid X–ray variability in the Seyfert galaxy NGC 4593.
See Abstr. 158.102.

Rapid X–ray variations in the high luminosity Seyfert galaxy III Zw 2.
See Abstr. 158.103.

High spectral variability in E1615 + 061, a soft X–ray Seyfert 1 galaxy.
See Abstr. 158.104.

Fractal time–variability and spectral invariance of the Seyfert galaxy NGC 5506.
See Abstr. 158.105.

X–ray variability of high luminosity AGN.
See Abstr. 158.106.

BL Lacertae objects: a comparison of X–ray variability to variability in different wavelengths.
See Abstr. 158.107.

X–ray spectral variability of the BL Lac object PKS 2155–304 in 1983 – 85.
See Abstr. 158.108.

X–ray variability of BL Lac objects.
See Abstr. 158.109.

Gas, dust and radio emission in elliptical galaxies.
See Abstr. 158.136.

The inverse Compton test for a large sample of compact radio sources.
See Abstr. 158.160.

EXOSAT observations of Cygnus–A.
See Abstr. 158.188.

Hard X–ray observations of the quasar 3C 273.
See Abstr. 159.006.

The X–ray properties of quasars determined from an iterative multiple regression analysis with censored data.
See Abstr. 159.036.

X–ray observations of the Ophiuchus, PKS 0745–191 and Cygnus–A clusters of galaxies.
See Abstr. 160.016.

X–ray observations of clusters: physical implications.
See Abstr. 160.026.

EXOSAT observations of the Virgo cluster.
See Abstr. 160.031.

X–ray observations of distant blue clusters of galaxies.
See Abstr. 160.043.

Cooling flows and AXAF.
See Abstr. 160.057.

The baryon–symmetric domain cosmology and the cosmic X–ray background.
See Abstr. 161.010.

Constraints on possible precursor AGN sources of the cosmic X–ray background.
See Abstr. 161.452.

143 Gamma-ray Sources, Gamma-ray Background

143.001   Neutron stars and gamma–rays.
G. F. Bignami.
The origin and evolution of neutron stars, p. 465 – 475 (1987). – See Abstr. 012.001 (IAU Symp. No. 125).

The author briefly reviews the highlights of the gamma–ray manifestation of the Crab and Vela pulsars, presents a summary of the evidence of UHE/VHE ($10^{11} - 10^{16}$eV) emission from X–ray binaries, gives the latest Cos–B Collaboration results on high–energy ($\gtrsim 100$ MeV) galactic sources, and finally presents an update on the work in progress toward the identification of Geminga.

143.002   A review of gamma–ray burst observations.
W. D. Evans, J. G. Laros.
The origin and evolution of neutron stars, p. 477 – 487 (1987). – See Abstr. 012.001 (IAU Symp. No. 125).

The authors review the known characteristics of gamma bursts and give new observational results on temporal and spectral properties. They suggest that a class of repeating bursters exists that are spectrally harder than X–ray bursters but significantly softer than "classical" gamma bursts. The March 5, 1979, burst may be the prototype of this class of bursters.

143.003   Some constraints on neutron star properties from gamma ray burster observations.
K. Hurley.
The origin and evolution of neutron stars, p. 489 – 500 (1987). – See Abstr. 012.001 (IAU Symp. No. 125).

The results of recent soft X–ray and optical searches for quiescent gamma ray burster counterparts are used to constrain the properties of the neutron stars responsible for bursters. Ages are restricted to the range 2×10^5y and above based on temperature upper limits and theoretical cooling curves, or 10^7y and above if bursters have evolved from pulsars. Velocities are greater than 20 km/s if the neutron stars are unmagnetized. Practically no main sequence star could have escaped detection in the optical/IR searches, so if the neutron stars are in binary systems, the companion is most likely a degenerate, low mass, low temperature object.

143.004   Theory of gamma ray bursters.
G. S. Bisnovatyj–Kogan.
The origin and evolution of neutron stars, p. 501 – 519 (1987). – See Abstr. 012.001 (IAU Symp. No. 125).

Gamma ray bursters are interpreted as nuclear explosions under the surface of neutron stars. The explosions occur after transportation of the matter with non–equilibrium composition during starquakes in outer layers where the matter becomes unstable and explodes as a result of a developing chain reaction. Formation of a shock wave and total energy output are calculated. Parameters of the mighty burst of 5 March 1979 are estimated. Formation of observed lines in different parts of the γ–range and a possible nature of accompanying optical bursts are discussed with the proposed model as a basis.

143.005   Cygnus X–3 and other ultra–high–energy γ–ray sources.
J. J. Barnard.
The origin and evolution of neutron stars, p. 521 – 533 (1987). – See Abstr. 012.001 (IAU Symp. No. 125).

Recently, several binary X–ray sources have been found to be sources of ultra high energy γ–ray emission. Air shower observations indicate photon energies $> \sim 10^{15}$eV. The author reviews the current status of observations from the source Cygnus X–3, and compare this data with that from the sources Hercules X–1, Vela X–1, and LMC X–4. Current theoretical models for the production of γ–rays and the acceleration of high energy particles are discussed and the consequences for the evolution of such systems are examined.

143.006   EXOSAT news on Geminga.
P. A. Caraveo, G. F. Bignami.
The origin and evolution of neutron stars, p. 545 (1987). Abstract. – See Abstr. 012.001 (IAU Symp. No. 125).

143.007   A model for the 1979 March 5 gamma–ray transient.
C. Alcock, E. Farhi, A. Olinto.
The origin and evolution of neutron stars, p. 548 (1987). Abstract. – See Abstr. 012.001 (IAU Symp. No. 125).

143.008   New distance limit for the March 5, 1979 source.
E. P. Liang.
The origin and evolution of neutron stars, p. 549 (1987). Abstract. – See Abstr. 012.001 (IAU Symp. No. 125).

143.009   High energy gamma ray astronomy.
C. E. Fichtel.
NASA Conf. Publ., NASA CP-2464, p. 227 – 274 (1987). – See Abstr. 012.004.

The galactic plane is the dominant feature of the gamma ray sky, the longitudinal and latitudinal distribution being generally correlated with galactic structural features including the spiral arms. Two molecular clouds have already been seen. Two of the three strongest gamma ray sources are pulsars. The Vela pulsar, PSR 0833–45, exhibits two pulses in the gamma ray region, as opposed to one in the radio region, neither of them in phase with the radio pulse. The highly variable X–ray source Cygnus X–3 was seen at one time in the 100 MeV region, and it has also been observed at very high energies ($>10^{11}$eV). Beyond our Galaxy, there is seen a diffuse radiation, whose origin remains uncertain, as well as at least one quasar, 3C 273. Future projects for high energy gamma ray observatories are reviewed.

143.010 Nucleosynthesis and astrophysical gamma ray spectroscopy.
A. S. Jacobson.
NASA Conf. Publ., NASA CP–2464, p. 275 – 294 (1987). – See Abstr. 012.004.

The HEAO–3 gamma ray spectrometer has provided new evidence in the quest for the understanding of complex element formation in the universe with the discovery of ^{26}Al in the interstellar medium. It has demonstrated that the synthesis of intermediate mass nuclei is currently going on in the Galaxy. The flux is peaked near the galactic center and indicates about 3 $M_\odot$ of ^{26}Al in the interstellar medium, with an implied ratio ^{26}Al/^{27}Al of 1×10^{-5}. Several possible source distributions have been studied but the data gathered thus far do not allow discrimination between them.

143.011 Gamma ray transients.
T. L. Cline.
NASA Conf. Publ., NASA CP–2464, p. 295 – 316 (1987). – See Abstr. 012.004.

143.012 Op jacht naar kosmische lichtflitsen.
G. Schilling.
Zenit, 14. Jahrg., Nr. 9, p. 294 – 299 (1987).

143.013 The SIGNE 2 MP9 cosmic gamma–ray burst experiment: preliminary results.
A. V. Kuznetsov, R. A. Syunyaev, O. V. Terekhov, L. A. Yakubtsev, C. Barat, B. Boer, K. Hurley, M. Niel, G. Vedrenne.
Sov. Astron. Lett., Vol. 12, No. 5, p. 311 – 315 (1987). English translation of 42.143.008.

143.014 SIGNE 2 MP9 data for the powerful gamma–ray burst of August 1, 1983.
A. V. Kuznetsov, R. A. Syunyaev, O. V. Terekhov, L. A. Yakubtsev, C. Barat, B. Boer, K. Hurley, M. Niel, G. Vedrenne.
Sov. Astron. Lett., Vol. 12, No. 5, p. 315 – 318 (1987). English translation of 42.143.009.

143.015 The contribution from active galactic nuclei to the diffuse 1 MeV background.
K. N. Yu.
Astrophys. Space Sci., Vol. 136, No. 1, p. 1 – 10 (1987).

A method is proposed to estimate the contribution from Seyfert galaxies and quasars to the diffuse 1 MeV background. First of all, the author calculates the contribution from these active galactic nuclei (AGNs) to the 2 keV background using traditional methods. He then chooses a suitable spectral X–ray index, which is found to be universal, to find the 1 MeV contribution from these AGNs by extrapolation. Results show that quasars generate about 40% of the 1 MeV background while Seyfert galaxies produce more than 40% taking into account the Penrose Compton Scattering gamma ray emissions. These results indicate that the 1 MeV background is likely to be generated by discrete objects.

143.016 Synchrotron X–ray haloes around gamma ray sources.
T. Kifune, M. Sadzinska, J. Wdowczyk, A. W. Wolfendale, A. J. Norton, R. S. Warwick.
Mon. Not. R. Astron. Soc., Vol. 228, No. 1, p. 243 – 250 (1987).

High energy γ–ray sources, epitomized by Cygnus X–3, may well be releasing significant fluxes of 10^{14}eV electrons into their environment and these electrons can give synchrotron X–rays when deflected in the general interstellar magnetic field. A greatly simplified model for this mechanism is described in terms of electron diffusion parameters and the topography of the galactic

magnetic field and predictions are made for the specific case of Cygnus X–3. *EXOSAT* results for the energy range 2 – 6 keV have been examined and "wings" have been identified which could be due to the postulated mechanism.

143.017 Observational properties of gamma–ray bursts.
E. P. Mazets, S. V. Golenetskij.
Itogi Nauki i Tekhniki. Seriya Astronomiya. Tom 32. Astrophysics and space physics, p. 16 – 42 (1987). In Russian. – See Abstr. 003.004.

143.018 High–energy gamma–astronomy.
A. M. Gal'per, B. I. Luchkov.
Chastitsy i kosmol. mater. 3–j Vsesoyuzn. shk., apr. 1985. Moskva, p. 87 – 97 (1987). In Russian. Abstr. in Ref. Zh., 62. Issled. Kosm. Prostranstva, 8.62.439 (1987).

143.019 Gamma–astronomy.
A. Wolfendale.
Priroda, No. 7, p. 66 – 73 (1987). In Russian.

143.020 Possible explanation of the γ–ray light curve and time variability in Cygnus X–3.
R. J. Protheroe, T. Stanev.
Nature, Vol. 328, No. 6126, p. 136 – 139 (1987).

Very high–energy (VHE) and ultra high–energy (UHE) γ rays, at around 1 TeV and 1 PeV respectively, have been observed from Cygnus X–3 predominantly in two distinct regions of orbital phase. The authors investigate whether this can be understood in terms of the accretion disk corona model of White and Holt. In this model, two bulges in the outer rim of the accretion disk are required to fit the observed X–ray light curve and the authors find these could provide target material for the production of the γ rays observed at two distinct phases.

143.021 COS–B upper limit to the >70 MeV gamma–ray flux from a gamma–ray burst event of 1979 November 9.
T. J. Sumner, D. L. Clements, O. R. Williams, G. K. Rochester.
Astron. Astrophys., Suppl. Ser., Vol. 71, No. 3, p. 557 – 560 (1987).

Using the COS–B data base an upper limit of 2×10^{-7}cm^{-2}s^{-1}keV^{-1} has been set to the high energy (70 – 1000 MeV) gamma–ray flux from a gamma–ray burst event of 1979 November 9. This limit is consistent with a power law extension from the lower energy data with a differential index of ≤ -1.39. An upper limit of 5.4×10^{-11}cm^{-2}s^{-1}keV^{-1} to the persistent emission is also obtained.

143.022 Discovery of a source of repeated soft, short gamma–bursts in Sagittarius.
J. L. Atteia, M. Boer, G. Vedrenne, M. Niel, K. Hurley, J. Laros, E. Fenimore, R. Klebesadel, A. V. Kuznetsov, R. A. Syunyaev, O. V. Terekhov, C. Kouveliotou, T. Cline, B. Dennis, U. Desai, L. Orwig.
Pis'ma Astron. Zh., Tom 13, No. 11, p. 987 – 994 (1987). In Russian. English translation in Sov. Astron. Lett., Vol. 13.

Twelve short cosmic gamma–ray bursts were observed from the same source during the period July – December 1983. Results of this source's localization based on "Prognoz–9" and ICE data are presented. The source position is 10° apart from the galactic center. Data on the localization show that the 7 January 1979 gamma–burst also originated from the same source.

143.023 The source of repeated bursts in Sagittarius. Spectra and time histories.
A. V. Kuznetsov, R. A. Syunyaev, O. V. Terekhov, J. Atteia, M. Boer, K. Hurley, M. Niel.
Pis'ma Astron. Zh., Tom 13, No. 11, p. 995 – 1006 (1987). In Russian. English translation in Sov. Astron. Lett., Vol. 13.

Twelve short cosmic gamma–ray bursts were observed from the same source in Sagittarius by the SIGNE–2MP9 experiment. All twelve events were observed also by the American experiment on the ICE satellite. The SIGNE–2MP9 experimental data on the time history and energy spectra of these events, are presented.

143.024   Search for gamma–quanta with energies $> 10^{14}$eV from the Cygnus X–3 source.
V. V. Alekseenko, A. S. Lidvanskij, V. A. Tizengauzen.
Particles and cosmology, p. 98 – 106 (1987). In Russian. Abstr. in Ref. Zh., 51. Astron., 9.51.755 (1987). – See Abstr. 012.034.

143.025   Very high energy gamma–ray binary stars.
R. C. Lamb, T. C. Weekes.
Science, Vol. 238, No. 4833, p. 1528 – 1534 (1987).
The authors discuss the recent discovery of very high energy (VHE, 10^{11} to 10^{14}eV) and ultrahigh energy (UHE, 10^{14}eV and above) gamma rays from some neutron stars in X–ray binary star systems.

143.026   The Perseus Flasher and satellite glints.
B. E. Schaefer, M. Barber, J. J. Brooks,
A. DeForrest, P. D. Maley, N. W. McLeod III, R. McNiel,
A. J. Noymer, A. K. Presnell, R. Schwartz, S. Whitney.
Astrophys. J., Vol. 320, No. 1, p. 398 – 404 (1987).
The Perseus Flasher (PF) is claimed to be an astrophysical source which frequently emits bright optical flashes. These flashes have all been detected by the naked eye, with the exception of one photographed flash for which an accurate position is measured. The authors have found that the PF is *not* an astrophysical source but is instead merely the observation of glints of reflected sunlight from artificial Earth satellites. This conclusion is based on the following facts. (1) A total of 3400 hr of photographic, video, and CCD observations have detected no flashes in or near the small PF error box – despite the claim of one bright flash every 12 hr. (2) Thirteen of the 26 flashes are shown to be nonastrophysical in origin. (3) Both the observational and theoretical glint rates indicate that most, if not all, PF observations are caused by satellite glints.

143.027   Bias and the gamma–ray burst cumulative–number distribution.
M. C. Jennings.
Bull. Am. Astron. Soc., Vol. 19, No. 2, p. 693 (1987). Abstract. – See Abstr. 010.061.

143.028   A new look at an old gamma–ray burst.
J. G. Laros, E. E. Fenimore, R. W. Klebesadel,
S. R. Kane.
Bull. Am. Astron. Soc., Vol. 19, No. 2, p. 693 (1987). Abstract. – See Abstr. 010.061.

143.029   SMM observation of diffuse galactic γ–ray lines.
G. H. Share, R. L. Kinzer, D. C. Messina,
E. L. Chupp, D. J. Forrest, E. Rieger.
Bull. Am. Astron. Soc., Vol. 19, No. 2, p. 694 (1987). Abstract. – See Abstr. 010.061.

143.030   HEAO 3 search for diffuse galactic positron annihilation radiation.
W. A. Mahoney, J. C. Higdon, J. C. Ling, W. A. Wheaton,
A. S. Jacobson.
Bull. Am. Astron. Soc., Vol. 19, No. 2, p. 694 (1987). Abstract. – See Abstr. 010.061.

143.031   A search for TeV gamma rays from the X–ray binary 1E 2259 + 586.
D. D. Weeks, S. Auloong, R. Becker–Szendy, J. Hudson,
L. Kelley, J. G. Learned, G. Sinnis, L. Resvanis, S. Tsamarias,
G. Voulgaris, A. Szentgyorgyi, E. Gotthelf, J. Gaidos,
F. Loeffler, P. Palfrey, G. Sembroski, C. Wilson, U. Camerini,
J. Finley, W. Fry, M. Jaworski, J. Jennings, A. Kenter,
R. Koepsel, M. Lomperski, R. Loveless, R. March,
J. Matthews, R. Morse, D. Reeder, P. Slane.
Bull. Am. Astron. Soc., Vol. 19, No. 2, p. 721 (1987). Abstract. – See Abstr. 010.061.

143.032   Localization, time histories, and energy spectra of a new type of recurrent high–energy transient source.
J.–L. Atteia, M. Boer, K. Hurley, M. Niel, G. Vedrenne,
E. E. Fenimore, R. W. Klebesadel, J. G. Laros,
A. V. Kuznetsov, R. A. Sunyaev (*R. A. Syunyaev*),
O. V. Terekhov, C. Kouveliotou, T. Cline, B. Dennis, U. Desai,
L. Orwig.
Astrophys. J., Lett. Ed., Vol. 320, No. 2, p. L105 – L110 (1987).
The authors report the detection of a recurrent high–energy transient source which is neither a classical X–ray nor a gamma–ray burster, but whose properties are intermediate between the two. The energy spectra of 12 recurrent events are found to be soft, characterized by kT's of 34 – 56 keV. The time histories are short ($\leqslant 128$ ms) with rise and fall times as fast as ~ 10 ms. The source location is a 0.12 deg^2 region about 10° from the Galactic center.

143.033   A new type of repetitive behavior in a high–energy transient.
J. G. Laros, E. E. Fenimore, R. W. Klebesadel, J.–L. Atteia,
M. Boer, K. Hurley, M. Niel, G. Vedrenne, S. R. Kane,
C. Kouveliotou, T. L. Cline, B. R. Dennis, U. D. Desai,
L. E. Orwig, A. V. Kuznetsov, R. A. Sunyaev (*R. A. Syunyaev*),
O. V. Terekhov.
Astrophys. J., Lett. Ed., Vol. 320, No. 2, p. L111 – L115 (1987).
The high–energy burster SGR 1806–20 (GB 790107) has been observed to repeat, on the order of 100 times, between 1978 August 13 and 1986 June 27. Most of the repetitions are in the latter part of 1983. All are clustered on time scales of hours to months. Simple analyses have shown the repetitive behavior to have very little correlation between burst intensity and interval between bursts. The tremendous range in intervals, with seconds to years (but usually $< 10^5$s) separating the bursts of comparable intensity, is amazing. The range in the intensities of the repetitions spans at least a factor of 30. The luminosity function can be approximated by a power law that flattens at low luminosities.

143.034   Evidence from meteor patrol photographs for a nonastronomical origin of the reported optical flashes in Perseus.
I. Halliday, P. A. Feldman, A. T. Blackwell.
Astrophys. J., Lett. Ed., Vol. 320, No. 2, p. L153 – L155 (1987).
The authors have attempted to verify the astronomical reality of the optical flashes reported by Katz et al. in 1986 by using simultaneous archival meteor patrol photographs. They are unable to confirm four of the reported events, including the single photographed flash of 1985 March 19, which are therefore more likely to be of local (i.e., nonastronomical) origin.

143.035   Search for very–high–energy gamma–rays from the galactic disk.
C. Morello, L. Periale, P. Vallania.
Nuovo Cimento C, Vol. 10C, Ser. 1, No. 1, p. 37 – 42 (1987). Abstr. in Phys. Abstr., Vol. 90, No. 1311, Entry 101642 (1987).

143.036   Analysis of the powerful GRB 830801 time structure.
A. V. Kuznetsov, R. A. Sunyaev, O. V. Terekhov,
M. Boer, G. Vedrenne, M. Niel, K. Hurley.
Pis'ma Astron. Zh., Tom 13, No. 12, p. 1055 – 1060 (1987). In Russian. English translation in Sov. Astron. Lett., Vol. 13.
No pulsations or quasi–periodic oscillations observed were found at the 3 – 4% total intensity level in the 33 ms – 3 s period range. Quasi–period 5.9 s is detected. Time history of this event in the different energy ranges is discussed. At the very beginning of the burst the average energy of photons decreases from 600 to 150 keV during 4 s time interval, then it changes slowly.

143.037   Low–energy gamma–rays from pulsar GX 1 + 4 – balloon results.
U. B. Jayanthi, F. Jablonski, J. Braga.
Astrophys. Space Sci., Vol. 137, No. 2, p. 233 – 238 (1987).
The authors present the results from a search of pulsed emission in low–energy gamma–rays from GX 1 + 4 source

observed during zenith transit in a balloon experiment in April, 1982. The observed pulsar period is 120.6 ± 0.2s with pulsed emission flux of $(1.3 \pm 0.4) \times 10^{-5}$ photons cm^{-2}s^{-1}keV^{-1} at an average energy ~ 342 keV. These pulsations, observed at gamma–ray energies perhaps for the first time from any X–ray pulsar, in conjunction with the period determined in X–rays, indicate a spin–down in contrast with the spin–up behaviour observed by others at earlier epochs.

143.038   Search for optical counterparts of gamma–ray burst sources.
J. Greiner, J. Flohrer, W. Wenzel, T. Lehmann.
Astrophys. Space Sci., Vol. 138, No. 1, p. 155 – 171 (1987).

The examination of nearly 9600 archival photographic plates covering six gamma–ray burst error boxes has unveiled several star–like images. Unfortunately, the reality of none of these images can be proved by duplicate plates. A laboratory test was performed to get a more detailed estimation of the number of plate faults. The result emphasizes the necessity of duplicate synchronous plates to exclude plate defects.

143.039   Low–energy gamma–rays from pulsar GX 1 + 4 – balloon results.
U. B. Jayanthi, F. Jablonski, J. Braga.
Astrophys. Space Sci., Vol. 138, No. 1, p. 183 – 189 (1987).

The authors present the results from a search of pulsed emission in low–energy gamma–rays from GX 1 + 4 source observed during zenith transit in a balloon experiment in April 1982. The observed pulsar period is 120.6 ± 0.2 s with pulsed emission flux of $(1.3 \pm 0.4) \times 10^{-5}$ photons cm^{-2}s^{-1}keV^{-1} at an average energy ~ 342 keV. These pulsations, observed at gamma–ray energies perhaps for the first time from any X–ray pulsar, in conjunction with the period determined in X–rays, indicate a spin–down in contrast with the spin–up behaviour observed by others at earlier epochs.

143.040   Gamma–ray observations of the Crab region using a coded–aperture telescope.
M. L. McConnell, P. P. Dunphy, D. J. Forrest, E. L. Chupp, A. Owens.
Astrophys. J., Vol. 321, No. 1, p. 543 – 552 (1987).

The region of the Galactic anticenter, including the Crab Nebula, was observed during a balloon flight of a new γ–ray telescope employing the coded–aperture imaging technique. The observations covered the energy range of 160 keV to 9.3 MeV with an imaging field of view of 15°2 by 22°8. The centroid of the measured response is located with an uncertainty of $\pm 12'$ (68% confidence level). The derived spectrum of the Crab, incident on the top of the atmosphere, is consistent with a single power–law spectrum. The authors also present upper limits to the flux from the nearby binary X–ray source A0535 + 26 and to the diffuse Galactic emission in the Galactic anticenter region.

143.041   HEAO–1 observations of gamma–ray bursts.
G. J. Hueter.
Diss. Abstr. Int., Sect. B, Vol. 48, No. 6, p. 1710–B – 1711–B (1987). Thesis, University of California, San Diego, 318 pp. (1987). Order No. DA8720172.

143.042   Gamma–lines in cosmic sources.
V. L. Prokhin.
Inst. Kosm. Issled. Akad. Nauk SSSR, Prepr., No. 1131 (1987). In Russian. Abstr. in Ref. Zh., 51. Astron., 12.51.764 (1987).

143.043   Long–term gamma–ray spectral variability of Cygnus X–1.
J. C. Ling, W. A. Mahoney, W. A. Wheaton, A. S. Jacobson.
Astrophys. J., Lett. Ed., Vol. 321, No. 2, p. L117 – L122 (1987).

The JPL HEAO 3 Gamma–Ray Spectroscopy experiment monitored the soft gamma–ray (0.05 – 10 MeV) emission of Cygnus X–1 for ~ 90 days each in the fall of 1979 and spring of 1980. The source exhibited a steady increase of the 100 keV emission, by a factor of ~ 3, during the first 70 days of observations. The spectrum obtained during the first 14 days of

observations showed strong γ–ray emission in the 0.4 – 1.5 MeV band. As the 100 keV flux increased, the MeV flux disappeared, while the overall spectrum softened. The implications of these variations for the physical processes in the accretion disk are discussed.

143.044   Upper limits to the $(10^{13} – 10^{14})$eV gamma–ray fluxes from the Crab Nebula and pulsar.
C. Morello, L. Periale, P. Vallania, G. Navarra.
Nuovo Cimento C, Vol. 10C, Ser. 1, No. 2, p. 142 – 150 (1987). Abstr. in Phys. Abstr., Vol. 90, No. 1316, Entry 134321 (1987).

143.045   Cygnus X–3 observation in gamma–ray energy range $> 10^{14}$eV.
V. V. Alexeenko (*V. V. Alekseenko*), A. E. Chudakov, Ya. S. Elensky, N. S. Khaerdinov, A. S. Lidvanskij, N. I. Metlinskij, S. Kh. Ozrokov, V. V. Sklyarov, V. A. Tizengauzen, G. Navarra.
Nuovo Cimento C, Vol. 10C, Ser. 1, No. 2, p. 151 – 161 (1987). Abstr. in Phys. Abstr., Vol. 90, No. 1316, Entry 134322 (1987).

143.046   Optical flashes from a γ–burster?
G. F. Bignami.
Nature, Vol. 330, No. 6146, p. 316 – 317 (1987).

A new contribution to the understanding of the γ–ray burst phenomenon may have been found. René Hudec and co–workers reported the discovery on archival astronomical plates of a recurrent optical flash emitted very close to the position in the sky of a γ–ray burster, itself not identified at other wavelengths.

143.047   Cosmic gamma–bursts.
J. Grygar.
Vesmír, Vol. 66, No. 8, p. 447 – 450 (1987). In Czech.

143.048   Are cosmic gamma–bursts optically detectable?
R. Hudec.
Vesmír, Vol. 66, No. 9, p. 517 – 523 (1987). In Czech.

143.049   SMM hard X–ray observations of the soft gamma–ray repeater 1806–20.
C. Kouveliotou, J. P. Norris, T. L. Cline, B. R. Dennis, U. D. Desai, L. E. Orwig, E. E. Fenimore, R. W. Klebesadel, J. G. Laros, J.–L. Atteia, M. Boer, K. Hurley, M. Niel, G. Vedrenne, A. V. Kuznetsov, R. A. Sunyaev (*R. A. Syunyaev*), O. V. Terekhov.
Astrophys. J., Lett. Ed., Vol. 322, No. 1, p. L21 – L25 (1987).

A new type of gamma–ray source has been recently discovered. Six bursts from this source were recorded with the Hard X–Ray Burst Spectrometer on the Solar Maximum Mission during a highly active phase in 1983. High time resolution measurements of one burst show rise and decay times of less than 5 ms, the fastest yet observed from this source. Time profiles of these events are simple, but indicate low–level emission before and after the main peaks. Two transients in the series show no spectral evolution over their durations of fractions of a second. Bursts from soft gamma–ray repeaters (SGRs) appear to form a separate class of events. The characteristic properites of the three known SGRs are briefly reviewed.

143.050   Sporadic and periodic 10 – 1000 TeV gamma rays from Cygnus X–3.
R. M. Baltrusaitis, G. L. Cassiday, R. Cooper, B. R. Dawson, J. W. Elbert, B. E. Fick, P. R. Gerhardy, K. D. Green, D. F. Liebing, C. P. Lingle, E. C. Loh, P. Sokolsky, P. Sommers, D. Steck.
Astrophys. J., Vol. 323, No. 2, p. 685 – 689 (1987).

During 1985, small air showers from the direction of Cygnus X–3 were observed using the University of Utah Fly's Eye. Useful spectral information was obtained from these showers. The combined data from 1985 June, July, and August show a 3.9 σ excess at 4.8 hr phase 0.65 – 0.70 for showers with energies above 100 TeV. The excess flux, averaged over all phases, is $4.5 \pm 1.2 \times 10^{-13}$cm^{-2}s^{-1}. Evidence was obtained for a sporadic outburst in 1985 June 17 UT. The outburst occurred at various phases of the 4.8 hr Cygnus X–3 period.

Status of the Perseus optical flasher.
See Abstr. 013.023.

An experiment for observing VHE gamma ray sources.
See Abstr. 034.001.

Ultra–high–energy gamma–ray astronomy using atmospheric Cerenkov detectors at large zenith angles.
See Abstr. 034.114.

Coded aperture imaging in X– and gamma–ray astronomy.
See Abstr. 035.090.

Measurements of background gamma radiation on Spacelab 2.
See Abstr. 051.031.

Cross sections for production of the 15.10 MeV and other atrophysically significant gamma–ray lines through excitation and spallation of ^{12}C and ^{16}O with protons.
See Abstr. 061.067.

Inverse Compton model of gamma ray burst spectra.
See Abstr. 063.001.

Red–shifted annihilation lines in thermal synchrotron model of gamma–ray bursts.
See Abstr. 063.033.

The pion boiler as a model of gamma–ray sources.
See Abstr. 063.057.

Positron annihilation at the Galactic center.
See Abstr. 063.124.

Inverse Comptonization by one–dimensional relativistic electrons.
See Abstr. 063.125.

Positron annihilation gamma rays from novae.
See Abstr. 065.119.

The origin of the enhanced dissipation ″α″ in accretion discs and its relation to gamma bursts.
See Abstr. 067.034.

Optical and γ–ray bursts on neutron stars.
See Abstr. 067.057.

Photon absorption and splitting in the magnetospheres of neutron star gamma–ray bursters.
See Abstr. 067.074.

Gamma–ray pulsar model.
See Abstr. 067.081.

What type of binary system is Cygnus X–3?
See Abstr. 117.006.

Reexamination of the *SAS* 2 Cygnus X–3 data.
See Abstr. 117.109.

Gamma–ray temporal and spectral variability of Cygnus X–1.
See Abstr. 117.172.

A rapid change of the Her X–1 pulse profile and high state duration.
See Abstr. 117.188.

Constraints on models of Cygnus X–3 from high–energy gamma–ray absorption at source.
See Abstr. 117.318.

FY Aquilae and the gamma–ray burst event of 1979 March 31.
See Abstr. 122.003.

More about FY Aql and GRBS 790331.
See Abstr. 122.133.

A search for gamma–ray lines from novae.
See Abstr. 124.006.

Ancient Guest Stars as harbingers of neutron star formation.
See Abstr. 125.011.

Prospects for gamma–ray line observations of individual supernovae.
See Abstr. 125.086.

Photon–photon pair production and the opacity of SN 1987A to TeV and PeV γ–rays.
See Abstr. 125.313.

Cosmic rays and gamma radiation from the shell of SN 1987A.
See Abstr. 125.321.

Polarized γ–rays from Vela.
See Abstr. 126.094.

Radio pulses from the Perseus flasher?
See Abstr. 141.012.

Search for X rays from the region of the Aries–Perseus flasher.
See Abstr. 142.017.

A deep optical study of the field of 1E 0630 + 178.
See Abstr. 142.020.

On relativistic particle acceleration in molecular clouds.
See Abstr. 144.017.

Galactic gamma rays and the origin of cosmic–ray positrons in the diffusion model.
See Abstr. 144.053.

Map of the Galactic center region in the 1.8 MeV ^{26}Al gamma–ray line.
See Abstr. 155.013.

Galactic gamma rays and gas tracers.
See Abstr. 155.035.

Gamma rays and the distribution of cosmic rays in the Galaxy.
See Abstr. 155.037.

Galactic positron annihilation radiation.
See Abstr. 155.047.

Positron annihilation in the galactic center: ″Cheshire Cat″ Compton scattering and ″excess continuum″.
See Abstr. 155.072.

144 Cosmic Rays

144.001 High energy cosmic rays from young neutron stars.
S. Miyaji.
The origin and evolution of neutron stars, p. 554 (1987).
Abstract. – See Abstr. 012.001 (IAU Symp. No. 125).

144.002 Cosmic rays in the heliosphere.
W. R. Webber.
NASA Conf. Publ., NASA CP–2464, p. 125 – 154 (1987). – See
Abstr. 012.004.

144.003 Antiprotons in cosmic rays.
V. K. Balasubrahmanyan, J. F. Ormes,
R. E. Streitmatter.
NASA Conf. Publ., NASA CP–2464, p. 155 – 172 (1987). – See
Abstr. 012.004.
This paper summarizes observational results on cosmic ray
antiprotons and reviews theoretical models of the origin of these
antiparticles.

144.004 Measurements of ultraheavy cosmic rays with HEAO–3.
W. R. Binns, M. H. Israel, J. Klarmann,
T. L. Garrard, E. C. Stone, C. J. Waddington.
NASA Conf. Publ., NASA CP–2464, p. 173 – 189 (1987). – See
Abstr. 012.004.
The HEAO–3 Heavy Nuclei Experiment has measured abundances of elements from ^{18}Ar to ^{92}U in the cosmic rays. The
results on the ultraheavy elements, those with atomic number
greater than 30, indicate that the sources of cosmic rays contain a
mixture of r–process and s–process material similar to that found
in the solar system. This result is at variance with previous
indications that the sources are greatly enhanced with freshly
synthesized r–process material.

144.005 Origin and propagation of galactic cosmic rays.
C. J. Cesarsky, J. F. Ormes.
NASA Conf. Publ., NASA CP–2464, p. 191 – 223 (1987). – See
Abstr. 012.004.
Contents: 1. Introduction. 2. General background. 3. Ob-
servations of cosmic rays. 4. Origin and propagation of dif-
ferent cosmic ray species. 5. Acceleration mechanism.
6. Summary.

144.006 Cosmic ray anisotropy above 10^{15}eV.
R. W. Clay.
Aust. J. Phys., Vol. 40, No. 3, p. 423 – 434 (1987).
An examination is made of published data on cosmic ray
anisotropy at energies above about 10^{15}eV. Both amplitude and
phase results are examined in an attempt to assess the confidence
which can be placed in the observations as a whole. It is found
that whilst many published results individually may suggest quite
high confidence levels of real measured anisotropy, the data
taken as a whole are less convincing. Some internal consistency in
the phase results suggests that a real effect may have been
measured but, again, this is not a high confidence level.

**144.007 ^{10}Be in polar ice: data reflect changes in cosmic ray flux
or polar meteorology.**
D. Lal.
Geophys. Res. Lett., Vol. 14, No. 8, p. 785 – 788 (1987).
The author has analyzed the century scale temporal variations
in the ^{10}Be time series in polar ice during the past few millenia
with a view to examine whether the observations are consistent
with changes due to solar modulation of flux. He has theoretically estimated the extrema production rates of ^{10}Be during periods
of low and high solar activity, and compared the amplitude of
this variation with that observed in ice cores. Based on the known
fallout pattern of fission products injected in the atmosphere, the
author estimates whether or not the ^{10}Be concentrations are
indicative of cosmic ray flux changes.

**144.008 Correction to "Latitudinal gradient of energetic parti-
cles in the outer heliosphere during 1985 – 1986"
[J. Geophys. Res., Vol. 92, No. A4, p. 3375 – 3379 (1987)].**
R. B. Decker, S. M. Krimigis, D. Venkatesan.
J. Geophys. Res., Vol. 92, No. A7, p. 7761 (1987). See Abstr.
43.144.025.

**144.009 Solar activity and heliosphere–wide cosmic ray modula-
tion in mid–1982.**
E. W. Cliver, J. D. Mihalov, N. R. Sheeley Jr., R. A. Howard,
M. J. Koomen, R. Schwenn.
J. Geophys. Res., Vol. 92, No. A8, p. 8487 – 8501 (1987).
A major episode of flare activity in June and July of 1982 was
accompanied by a pair of heliosphere–wide cosmic ray modulation events. In each case, a large Forbush decrease at earth was
followed in turn by apparently related decreases at Pioneer 11
and Pioneer 10. The Pioneer spacecraft were separated by ~155°
in ecliptic longitude. The authors reviewed white light coronagraph and near–sun ($\leqslant 1$ AU) satellite data to identify plausible
solar origins of these modulation events.

144.010 Solar cycle modulation of galactic protons and electrons.
J. S. Perko.
J. Geophys. Res., Vol. 92, No. A8, p. 8502 – 8510 (1987).
A solution of the time–dependent, spherically symmetric
cosmic ray transport equation, with a defensible diffusion
coefficient and a natural model for an 11–year cycle of diffusive
scattering disturbances that originate on the sun and travel
through interplanetary space, accounted simultaneously for the
spectral changes of both galactic protons and electrons, for the
time and phase lag of high– versus low–energy protons, and for
the integral radial gradients of protons > 100 MeV over most of
the solar cycle and over large distances in the heliosphere. Each
individual disturbance caused a sudden particle intensity decrease
as it passed a point in space; recovery of intensity began
immediately afterward. The characteristic recovery time at 1 AU
was in the range found in neutron monitor and satellite data,
except that the recovery time constant was rigidity–dependent,
contrary to these same data. Also, spectral changes over
successive solar minima in 1965 and 1977, heretofore linked to
drifts, can be explained as an adjunct to the hysteresis effect.

144.011 High–energy cosmic ray intensity waves.
R. M. Jacklyn, M. L. Duldig, M. A. Pomerantz.
J. Geophys. Res., Vol. 92, No. A8, p. 8511 – 8518 (1987).
A new mode of galactic cosmic ray modulation has been
observed. It manifests itself as waves of isotropic intensity
variations of high–energy (~150 GeV) galactic cosmic rays. The
wave periods are in synchronism with the alternating toward and
away polarity states of the sectored interplanetary magnetic field.
Since data with the requisite precision for observing the
phenomenon first became available in 1982, isotropic intensity
waves have occurred during the latter half of three successive
years. The variational spectrum $\Delta j(p)/j(p) = kp^\gamma$ (where p is
magnetic rigidity) is hard, with $\gamma \approx 0$. Although the largest
Forbush decrease ever recorded occurred while the 1982 episode
was in progress, isotropic intensity waves are not causally related
to that effect. However, the isotropic intensity waves observed
thus far are in phase in the southern hemisphere with much
smaller concurrent variations of the north–south anisotropy. The
new phenomenon bespeaks a difference, not yet identified, in the
plasma regimes on opposite sides of the neutral sheet.

**144.012 Extensive air showers and the nature of charge to neutral
ratio of hadrons at high energies: some comments.**
S. Bhattacharyya.
Astrophys. Space Sci., Vol. 136, No. 1, p. 77 – 82 (1987).
This paper presents a critical analysis and partial modification
of a very recent work by Sinha (1986) on the same topic with the
help of a specific model. The proposed modifications bring closer
agreements with the experimental data and give the basic model

wider applicability. It is, however, emphasised that the suggested modifications do not alter the basic tenets of the model used by both the authors.

144.013 Cosmic ray kinetics in a strong magnetic field with large–scale inhomogeneities.
L. I. Dorman, M. E. Kats, S. F. Nosov, M. Steglik, Yu. I. Fedorov, B. A. Shakhov, A. K. Yukhimuk.
Kosm. luchi, Moskva, No. 24, p. 49 – 62 (1987). In Russian. Abstr. in Ref. Zh., 51. Astron., 7.51.693 (1987).

144.014 Cosmic radiation in the Galaxy. Review.
V. A. Dogel'.
Izv. Vuzov. Radiofiz., Tom 30, No. 2, p. 187 – 207 (1987). In Russian. Abstr. in Ref. Zh., 62. Issled. Kosm. Prostranstva, 7.62.583; 51. Astron., 8.51.692 (1987).

144.015 Compound nuclei in primary cosmic rays as sources of γ–quanta of superhigh energy.
V. V. Balashov, V. L. Korotkikh, I. V. Moskalenko.
Vestn. Mosk. Univ., Ser. 3. Fiz. Astron., Tom 28, No. 2, p. 76 – 78 (1987). In Russian. Abstr. in Ref. Zh., 62. Issled. Kosm. Prostranstva, 7.62.584 (1987).

144.016 Spectral analysis of small–scale fluctuations of cosmic rays from ground–based observations.
O. V. Gulinskij, V. Yu. Belashov, M. E. Kats, I. Ya. Libin, K. Otaola, S. F. Nosov, R. E. Prilutskij, J. Perez–Peraza, M. Steglik, K. F. Yudakhin.
Kosm. luchi, Moskva, No. 24, p. 63 – 87 (1987). In Russian. Abstr. in Ref. Zh., 62. Issled. Kosm. Prostranstva, 7.62.589 (1987).

144.017 On relativistic particle acceleration in molecular clouds.
V. A. Dogiel (*V. A. Dogel'*), A. V. Gurevich, Ya. N. Istomin, K. P. Zybin.
Mon. N. R. Astron. Soc., Vol. 228, No. 4, p. 843 – 868 (1987).
Equations of particle motion are derived for weakly ionized turbulent plasma containing a magnetic field. In this case particle energies change due to fluctuating electromagnetic fields. Particle motion in coordinate and momentum space is shown to be described by the diffusion approximation. The solutions of these equations are obtained for both magnetized and unmagnetized particles. Magnetized particles are shown to be effectively accelerated under certain conditions. The distribution functions have been calculated for the conditions of galactic giant molecular clouds. It is found that the cosmic ray density may be higher inside clouds than in the intercloud medium. The possible influence of this process on gamma–emissivity of the Galaxy, on the chemical composition of cosmic rays, on the flux of galactic antiprotons and positrons and on the galactic radio emission is discussed.

144.018 Mechanisms of origin and dynamics of anisotropy of the gas pressure of galactic cosmic rays.
S. M. Kamoldinov.
Modulyatsiya kosm. luchej v Soln. sisteme, Yakutsk, p. 49 – 54 (1986). In Russian. Abstr. in Ref. Zh., 51. Astron., 8.51.695 (1987).

144.019 Physical and engineering applications for researches of underground cosmic rays.
D.–z. Zhou, Z.–w. Zhou, C.–j. Cheng, F.–s. Ma.
Acta Geophys. Sin., Vol. 30, No. 5, p. 542 – 548 (1987). In Chinese.

144.020 Search for cosmic ray anisotropy with energies above 10^{15} eV.
G. V. Kulikov, A. A. Silaev, V. I. Solov'eva, A. V. Trubitsyn, G. B. Christiansen.
Mater. konf. mol. uchenykh NII yader. fiz. MGU, Moskva, 3 – 4 iyunya, 1985. MGU. Moskva, p. 100 – 104 (1985). In Russian. Abstr. in Ref. Zh., 62. Issled. Kosm. Prostranstva, 8.62.504 (1987).

144.021 Cosmic rays and energetic particles in the heliosphere.
K. Sakurai.
The solar wind and the Earth, p. 38 – 53 (1987). – See Abstr. 003.007.
Contents: Introduction. Galactic cosmic rays in the heliosphere. Cosmic ray modulations. Solar flares and cosmic rays. Energetic particles from the Sun.

144.022 A survey of the cosmic ray diurnal variation during 1973 – 1979 – I. Persistence of solar diurnal variation.
J. F. Riker, H. S. Ahluwalia.
Planet. Space Sci., Vol. 35, No. 9, p. 1111 – 1115 (1987).
The authors find that the solar diurnal variation in cosmic ray intensity is a persistent phenomenon over the years 1973 – 1979. That is, even during solar minimum, conditions in the heliosphere lead to a net flux of cosmic rays from a particular (time–varying) direction in space. If the authors regard the daily fluctuations in the amplitude and phase of the diurnal variation as random perturbations about the mean vector, they are able to quantify the relative magnitude of the random component.

144.023 A survey of the cosmic ray diurnal variation during 1973 – 1979 – II. Application of diffusion – convection model to diurnal anisotropy data.
J. F. Riker, H. S. Ahluwalia.
Planet. Space Sci., Vol. 35, No. 9, p. 1117 – 1122 (1987).
The authors present a comparison of the standard diffusion–convection theory of the cosmic ray transport with the experimental observations at Embudo, NM and Deep River, Canada for the years 1973 – 1979. In particular, they outline a straightforward method for computing the diffusion coefficients and the gradients in the heliosphere, by making use of the data on the diurnal anisotropy of cosmic rays. The conclusions are generally consistent with the published results, but the authors have also found some inconsistencies which need to be clarified.

144.024 Solar modulation of galactic antiprotons.
J. S. Perko.
Astron. Astrophys., Vol. 184, No. 1/2, p. 119 – 121 (1987).
Galactic antiproton data of current interest lie in an energy regime heavily influenced by solar modulation. Correcting for it needs to be done more carefully than it has been in the past. The author applies the well–known "force–field" analytic approximation of the spherically–symmetric, steady–state, cosmic–ray transport equation to account for modulation down to at least 100 MeV. He gives a sample solution which applies to the currently available antiproton data set (1979 – 80), and can be used to accurately modulate any possible interstellar antiproton spectrum. The solution is easily adapted for comparison to future measurements. It also shows that boosting the low–energy (< 600 MeV) side of the interstellar antiproton spectrum will not affect the low–energy spectrum at 1 AU, due to strong adiabatic deceleration during that time. One needs an excess of particles at around 1 GeV to fit the intense low–energy measurement.

144.025 Heliomagnetic dipole moment and daily variation of cosmic rays underground.
K. Nagashima, H. Ueno, K. Fujimoto.
Nature, Vol. 328, No. 6131, p. 600 – 601 (1987).
Since the discovery of Forbush decrease of cosmic–ray intensity, it has become clear that the intensity and its daily variation are subject to various kinds of solar modulation; these are due to the diffusion–convection of cosmic rays in interplanetary magnetic fields, which is controlled by various kinds of solar activity. The authors demonstrate that the amplitude of diurnal variation of cosmic–ray intensity is intimately correlated with the magnitude of magnetic dipole moment of the Sun and that, contrary to the conventional concept of the modulation, such a correlation exists only in the cosmic–ray high–rigidity region.

144.026 Diffusion and nuclear fragmentation of cosmic rays in a cloudy interstellar medium.
J. L. Osborne, V. S. Ptuskin.
Pis'ma Astron. Zh., Tom 13, No. 11, p. 980 – 986 (1987). In Russian. English translation in Sov. Astron. Lett., Vol. 13.

The transport equation for the nuclear component of cosmic rays in the interstellar medium containing randomly distributed giant molecular clouds is obtained. It takes into account the finite transparency of the clouds for diffusing relativistic particles and the screening of the internal cosmic ray sources. The equation is used to improve the standard model of cosmic ray propagation in the Galaxy.

144.027 Dynamical effects of cosmic rays in the interstellar medium.
V. D. Kuznetsov.
Astrophysics, Vol. 26, No. 1, p. 44 – 52 (1987). English translation of 43.144.012.

144.028 A simulation of high–energy cosmic–ray propagation.
M. Honda.
Astrophys. J., Vol. 319, No. 2, p. 836 – 841 (1987).

A simulation of cosmic–ray propagation in the Galactic magnetic fields is presented in the energy region $10^{16} - 10^{18}$eV, where the gyroradius and the scale of the irregularity of the magnetic field are considered to be comparable. The diffusion tensor is calculated as a function of the ratio of the gyroradius (R_g) and the scale of the magnetic field irregularity (L_0). In the region $R_g < L_0$, the stochastic nature of the magnetic field suggested by Jokipii and Parker controls the process. In the region $R_g > L_0$, the process can be understood in terms of the scattering centers and mean free path.

144.029 Cosmic ray fluctuations and dynamical processes in the solar wind.
V. I. Kozlov.
Modulyatsiya kosm. luchej v soln. sisteme, Yakutsk, p. 80 – 95 (1986). In Russian. Abstr. in Ref. Zh., 51. Astron., 9.51.399 (1987).

144.030 Flares of cosmic radiation due to particle acceleration by the solar wind.
A. T. Filippov.
Modulyatsiya kosm. luchej v soln. sisteme, Yakutsk, p. 66 – 79 (1986). In Russian. Abstr. in Ref. Zh., 51. Astron., 9.51.402 (1987).

144.031 Variations of cosmic ray density and solar wind with the solar activity cycle.
N. P. Chirkov.
Modulyatsiya kosm. luchej v soln. sisteme, Yakutsk, p. 3 – 14 (1986). In Russian. Abstr. in Ref. Zh., 51. Astron., 9.51.426 (1987).

144.032 Modulation of galactic cosmic rays in the region of the magnetoheliopause.
I. A. Transkij.
Modulyatsiya kosm. luchej v soln. sisteme, Yakutsk, p. 23 – 34 (1986). In Russian. Abstr. in Ref. Zh., 51. Astron., 9.51.427 (1987).

144.033 Energetic characteristics of the Forbush effect.
I. A. Transkij, V. P. Mamrukova, G. V. Shafer.
Modulyatsiya kosm. luchej v soln. sisteme, Yakutsk, p. 54 – 57 (1986). In Russian. Abstr. in Ref. Zh., 51. Astron., 9.51.428 (1987).

144.034 Cosmic ray streams during flare perturbations.
V. G. Grigor'ev.
Modulyatsiya kosm. luchej v soln. sisteme, Yakutsk, p. 58 – 66 (1986). In Russian. Abstr. in Ref. Zh., 51. Astron., 9.51.429 (1987).

144.035 Measurements of the fragmentation cross sections of relativistic heavy nuclei and their application to cosmic ray propagation.
M. P. Kertzman.
Diss. Abstr. Int., Sect. B, Vol. 48, No. 3, p. 793–B (1987). Thesis, University of Minnesota, 178 pp. (1987). Order No. DA8713090.

144.036 Investigation of three–dimensional galactic cosmic ray streams in high–velocity recurrent solar wind streams.
I. S. Samsonov, Z. N. Samsonova.
Modulyatsiya kosm. luchej v soln. sisteme, Yakutsk, p. 14 – 23 (1986). In Russian. Abstr. in Ref. Zh., 51. Astron., 10.51.517 (1987).

144.037 Spectrum of solar–diurnal variations of cosmic rays.
P. A. Krivoshapkin, N. G. Kravtsov, G. V. Skripin.
Modulyatsiya kosm. luchej v soln. sisteme, Yakutsk, p. 34 – 39 (1986). In Russian. Abstr. in Ref. Zh., 51. Astron., 10.51.518 (1987).

144.038 The second spherical harmonic in cosmic ray distribution.
P. A. Krivoshapkin, V. P. Mamrukova.
Modulyatsiya kosm. luchej v soln. sisteme, Yakutsk, p. 39 – 48 (1986). In Russian. Abstr. in Ref. Zh., 51. Astron., 10.51.519 (1987).

144.039 Antiproton production of propagating cosmic rays under distributed reacceleration.
M. Simon, U. Heinbach, C. Koch.
Astrophys. J., Vol. 320, No. 2, p. 699 – 702 (1987).

The available measurements on the cosmic–ray $\bar{p}/p$ ratio show an excess of antiprotons above predictions derived in the framework of the standard picture of cosmic–ray origin and propagation. The authors calculated the $\bar{p}$ production from collisions of cosmic rays with the interstellar gas under the condition of distributed reacceleration. It could be shown that the calculated $\bar{p}/p$ ratio is enhanced compared to that derived from the "leaky box" model, but it remains difficult to bring it into agreement with the data by reasonable astrophysical assumptions.

144.040 Search for superheavy grand unified magnetic monopoles in cosmic rays.
M. J. Shepko, C. A. Cagliardi, P. J. Green, P. M. McIntyre, T. Meyer, R. E. Tribble, R. C. Webb.
Phys. Rev. D, Vol. 35, No. 9, p. 2917 – 2920 (1987). Abstr. in Phys. Abstr., Vol. 90, No. 1309, Entry 80562 (1987).

144.041 Search for antiprotons in cosmic rays.
A. C. R. Rao, P. C. M. Yock.
Europhys. Lett., Vol. 3, No. 9, p. 1049 – 1052 (1987). Abstr. in Phys. Abstr., Vol. 90, No. 1309, Entry 88059 (1987).

144.042 Solutions of the Fokker–Planck equation for the energy distribution of suprathermal electrons.
J. Pérez–Peraza, A. Gallegos–Cruz.
Rev. Mex. Astron. Astrofis., Vol. 14, No. 2, p. 705 – 713 (1987). – See Abstr. 012.042.

The authors solve the Fokker–Planck equation (continuity equation in the energy space) under several different assumptions, concerning the phenomena associated with acceleration of electrons. The obtained partial solutions allow for the determination of the particle energy spectra.

144.043 Calculation of the cosmic ray perpendicular gradient.
J. Dubinsky, M. Stehlik.
Acta Phys. Slovaca, Vol. 37, No. 3, p. 198 – 200 (1987). Abstr. in Phys. Abstr., Vol. 90, No. 1311, Entry 101645 (1987).

144.044   Cosmic ray fluctuations at rigidities 4 to 180 GV.
G. Benko, G. Erdos, M. E. Katz, S. F. Nosov,
M. Stehlik.
Acta Phys. Slovaca, Vol. 37, No. 3, p. 192 – 197 (1987). Abstr. in
Phys. Abstr., Vol. 90, No. 1311, Entry 101646 (1987).

144.045   Primary cosmic–ray composition in the energy range of 10^{14}eV to 10^{16}eV and high–energy atmospheric cosmic rays observed with emulsion chambers at Mt. Kanbala.
J. R. Ren, H. H. Kuang, A. X. Huo, S. L. Lu, S. Su,
Y. X. Wang, C. R. Wang, M. He, N. J. Zhang, P. Y. Cao,
J. Y. Li, S. Z. Wang, G. Z. Bai, Z. H. Liu, G. J. Li,
G. X. Gang, W. D. Zhou, R. D. He, M. Amenomori,
H. Nanjo, N. Hotta, I. Ohta, K. Mizutani, K. Kasahara,
T. Yuda, M. Shibata, T. Shirai, N. Tateyama, S. Torii,
H. Sugimoto, K. Taira.
Nuovo Cimento C, Vol. 10C, Ser. 1, No. 1, p. 43 – 60 (1987).
Abstr. in Phys. Abstr., Vol. 90, No. 1311, Entry 101648 (1987).

144.046   Photon interaction with matter: is there a threshold at multi–TeV energy?
F. Halzen, P. Hoyer, N. Yamdagni.
Phys. Lett. B, Vol. 190, No. 1 – 2, p. 211 – 216 (1987). Abstr. in
Phys. Abstr., Vol. 90, No. 1311, Entry 101651 (1987).

144.047   On the discovery of very high energy point sources.
F. Halzen.
Probing the standard model, p. 607 – 632 (1987). Abstr. in Phys.
Abstr., Vol. 90, No. 1312, Entry 107675 (1987). – See Abstr.
012.050.

144.048   Cosmic–ray intensity oscillations immediately before and after the GLE on February 16, 1984.
E. Vainikka, M. Lumme, M. Nieminen, E. Riihonen,
H. Arvela, J. J. Torsti, J. Peltonen, E. Valtonen.
Europhys. Lett., Vol. 3, No. 10, p. 1151 – 1153 (1987). Abstr. in
Phys. Abstr., Vol. 90, No. 1312, Entry 107677 (1987).

144.049   Measurement of the energy spectrum of primary cosmic rays.
K.–z. Xu, Z.–h. Zhang, Z.–z. Tang, M.–q. Wang.
Phys. Energ. Fortis Phys. Nucl., Vol. 11, No. 1, p. 1 – 5 (1987). In
Chinese. Abstr. in Phys. Abstr., Vol. 90, No. 1312, Entry 107679
(1987).

144.050   Measurement of the cosmic–ray energy spectrum between 3×10^{15} and 3×10^{16}eV using a photon–density spectrum technique.
A. G. Gregory, J. R. Patterson, R. J. Protheroe.
J. Phys. G, Vol. 13, No. 4, p. 543 – 551 (1987). Abstr. in Phys.
Abstr., Vol. 90, No. 1312, Entry 107680 (1987).

144.051   The energy spectrum and nuclear composition of primary cosmic rays.
S. I. Nikol'skij.
Problems of cosmic ray physics, p. 169 – 185 (1987). In Russian.
Abstr. in Ref. Zh., 62. Issled. Kosm. Prostranstva, 11.62.366
(1987). – See Abstr. 003.015.

144.052   Variations of cosmic ray intensity (terrestrial observations, methods of investigation, theory).
L. I. Dorman.
Problems of cosmic ray physics, p. 65 – 82 (1987). In Russian.
Abstr. in Ref. Zh., 62. Issled. Kosm. Prostranstva, 11.62.375
(1987). – See Abstr. 003.015.

144.053   Galactic gamma rays and the origin of cosmic–ray positrons in the diffusion model.
V. A. Dogel', A. V. Uryson.
Sov. Astron. Lett., Vol. 12, No. 6, p. 348 – 352 (1986). English
translation of 42.144.029.

144.054   Instability in a shock propagating through gas with a cosmic–ray component.
E. G. Berezhko.
Sov. Astron. Lett., Vol. 12, No. 6, p. 352 – 354 (1986). English
translation of 42.144.030.

144.055   Observation of cosmic ray positrons in the region from 5 to 50 GeV.
R. L. Golden, S. A. Stephens, B. G. Mauger, G. D. Badhwar,
R. R. Daniel, S. Horan, J. L. Lacy, J. E. Zipse.
Astron. Astrophys., Vol. 188, No. 1, p. 145 – 154 (1987).
The absolute flux of cosmic ray positrons has been measured
using a balloon–borne magnet–spectrometer. Based on 193
positrons observed from 3.57 GV/c to 50 GV/c rigidity at the
payload, the integral flux about 5 GeV kinetic energy at the top
of the atmosphere is found to be $(0.33 \pm 0.07)e^+/m^2$–str–s. At the
top of the atmosphere the effective energy interval for the
observation is 4.5 to 64 GeV. In this interval the best–fit
differential flux is $16E^{-(3.0 \pm 0.3)}e^+/(m^2$–str–sec–GeV). The quoted
errors do not include a possible 10% systematic uncertainty in
the exposure factor. The ratio $e^+/(e^+ + e^-)$ was found to be
0.069 ± 0.014 above 5 GeV. The authors compare these results
with other observations and theoretical predictions.

144.056   Effect of shocks on interstellar turbulence and cosmic–ray dynamics.
A. M. Bykov, I. N. Toptygin.
Astrophys. Space Sci., Vol. 138, No. 2, p. 341 – 354 (1987).
A theoretical model for the interstellar turbulence is developed.
In this model the fluctuation spectrum is formed due to reflection
of shocks, produced by supernovae, on interstellar clouds. The
spectra of turbulence and the diffusion coefficient of cosmic rays
are derived. It is demonstrated that local enhancements of the
ionization rate by cosmic rays accelerated by supernova shocks
may be responsible for fast renewal of warm ionized envelopes
around cores of standard ISM clouds.

144.057   Eigenschaften der kosmischen Strahlung und ihre Bedeutung für die Hochenergie–Astrophysik.
I. Halm, F. Jansen.
Sterne, 63. Band, Heft 4, p. 195 – 202 (1987).

144.058   Erratum: "Relativistic transport theory for cosmic rays" [Astrophys. J., Vol. 296, No. 2, p. 319 – 330 (1985)].
G. M. Webb.
Astrophys. J., Vol. 321, No. 1, p. 606 (1987). See Abstr.
40.144.027.

144.059   Modulation in three dimensions.
J. R. Jokipii.
Nature, Vol. 330, No. 6144, p. 109 – 110 (1987).
The author reports on solar modulation of cosmic rays.

144.060   The differences in delay times for air showers initiated by 100 TeV gamma rays and protons.
S. Mikocki, J. Linsley, J. Poirier, A. Wrotniak.
J. Phys. G, Vol. 13, No. 5, p. L85 – L91 (1987). Abstr. in Phys.
Abstr., Vol. 90, No. 1313, Entry 115191 (1987).

144.061   Primary cosmic–ray composition at $10^{15} – 10^{17}$eV inferred from extensive air shower simulations.
T. Cheung, P. K. MacKeown.
J. Phys. G, Vol. 13, No. 5, p. 687 – 705 (1987). Abstr. in Phys.
Abstr., Vol. 90, No. 1313, Entry 115192 (1987).

144.062   The muon component of cosmic–ray air showers in the range $10^{17} – 10^{18}$ eV.
R. Armitage, P. R. Blake, W. F. Nash, C. G. Saltmarsh,
A. J. Sephton, C. C. Shelley, R. B. Strutt.
J. Phys. G, Vol. 13, No. 5, p. 707 – 723 (1987). Abstr. in Phys.
Abstr., Vol. 90, No. 1313, Entry 115193 (1987).

144.063 Underground search for muons correlated with Cygnus X–3.
R. M. Bionta, G. Blewitt, C. B. Bratton, D. Casper, A. Ciocio, R. Claus, M. Crouch, S. T. Dye, S. Errede, G. W. Foster, W. Gajewski, K. S. Ganezer, M. Goldhaber, T. J. Haines, T. W. Jones, D. Kielczewska, W. R. Kropp, J. G. Learned, J. M. LeSecco, J. Matthews, H. S. Park, L. R. Price, F. Reines, J. Schulz, S. Seidel, E. Shumard, D. Sinclair, H. W. Sobel, J. L. Stone, L. Sulak, R. Svoboda, G. Thornton, J. C. van der Velde, C. Wuest.
Phys. Rev. D, Vol. 36, No. 1, p. 30 – 36 (1987). Abstr. in Phys. Abstr., Vol. 90, No. 1314, Entry 120647 (1987).

144.064 Short–wavelength compressive instabilities in cosmic ray shocks and heat conduction flows.
G. P. Zank, J. F. McKenzie.
J. Plasma Phys., Vol. 37, Part 3, p. 347 – 361 (1987). Abstr. in Phys. Abstr., Vol. 90, No. 1315, Entry 127654 (1987).

144.065 The interaction of long–wavelength compressive waves with a cosmic ray shock.
G. P. Zank, J. F. McKenzie.
J. Plasma Phys., Vol. 37, Part 3, p. 363 – 372 (1987). Abstr. in Phys. Abstr., Vol. 90, No. 1315, Entry 127655 (1987).

144.066 Supersymmetric cosmic accelerators: fluxes at Earth and companion stability.
J. R. Cudell, F. Halzen.
Phys. Rev. D, Vol. 36, No. 2, p. 346 – 350 (1987). Abstr. in Phys. Abstr., Vol. 90, No. 1315, Entry 127656 (1987).

144.067 A direct measurement of the energy spectrum of cosmic ray muons in the Mont Blanc underground laboratory.
M. Calicchio, C. De Marzo, O. Erriquez, C. Favuzzi, N. Giglietto, E. Nappi, F. Posa, P. Spinelli.
Phys. Lett. B, Vol. 193, No. 1, p. 131 – 134 (1987). Abstr. in Phys. Abstr., Vol. 90, No. 1315, Entry 127662 (1987).

144.068 Investigation of heavy nuclei streams of galactic cosmic radiation from measurements in circumterrestrial AES and OS orbits in 1974 – 1984.
A. M. Marennyj, R. A. Nymmik, A. A. Suslov.
Kosm. Issled., Tom 25, Vyp. 4, p. 577 – 584 (1987). In Russian. English translation in Cosm. Res.

144.069 Galactic propagation model, the modified proton spectrum and an estimation of the positron flux.
S. Bhattacharyya, P. Pal.
Nuovo Cimento C, Vol. 10C, Ser. 1, No. 2, p. 227 – 234 (1987). Abstr. in Phys. Abstr., Vol. 90, No. 1316, Entry 134096 (1987).

144.070 Low–energy muons in extensive air showers.
D. K. Basak, N. Chaudhuri, S. Sarkar, B. Bhattacharya, B. Ghosh.
Nuovo Cimento C, Vol. 10C, Ser. 1, No. 2, p. 169 – 183 (1987). Abstr. in Phys. Abstr., Vol. 90, No. 1316, Entry 134099 (1987).

144.071 Radiation from Cygnus X–3 and the related muon puzzle – an explanation.
S. Bhattacharyya.
Nuovo Cimento C, Vol. 10C, Ser. 1, No. 2, p. 209 – 226 (1987). Abstr. in Phys. Abstr., Vol. 90, No. 1316, Entry 134100 (1987).

144.072 Possibility of simultaneous observation of nucleus fragments and the γ–ray family in the stratosphere.
Y. Niihori, T. Shibata, I. M. Martin, E. H. Shibuya, A. Turtelli Jr.
Phys. Rev. D, Vol. 36, No. 3, p. 783 – 797 (1987). Abstr. in Phys. Abstr., Vol. 90, No. 1316, Entry 134101 (1987).

144.073 Behavior of cosmic rays in the atmosphere at super high energy region.
A. Ohsawa, S. Yamashita.
Prog. Theor. Phys., Vol. 77, No. 6, p. 1411 – 1433 (1987). Abstr. in Phys. Abstr., Vol. 90, No. 1316, Entry 134102 (1987).

144.074 The differences in angular distributions of secondaries from EAS initiated by gammas and protons.
S. Mikocki, J. Poirier.
J. Phys. G, Vol. 13, No. 9, p. L217 – L221 (1987). Abstr. in Phys. Abstr., Vol. 90, No. 1317, Entry 139872 (1987).

144.075 Angular distributions of secondary charged particles in showers initiated by gammas and protons.
J. Poirier, J. Linsley, S. Mikocki.
Phys. Rev. D, Vol. 36, No. 5, p. 1378 – 1380 (1987). Abstr. in Phys. Abstr., Vol. 90, No. 1317, Entry 139873 (1987).

144.076 Electromagnetic–shower development in concrete and the punch–through effect.
S. Mikocki, J. Poirier.
Phys. Rev. D, Vol. 36, No. 5, p. 1381 – 1384 (1987). Abstr. in Phys. Abstr., Vol. 90, No. 1317, Entry 139874 (1987).

144.077 Mass limits on particles from pulsed sources: how reliable are they?
J. R. Cudell, F. Halzen, P. Hoyer.
Phys. Rev. D, Vol. 36, No. 6, p. 1657 – 1661 (1987). Abstr. in Phys. Abstr., Vol. 91, No. 1319, Entry 4254 (1988).

144.078 Very high energy emission in accretion onto compact objects.
J. E. Grindlay.
Physics of the superconducting supercollider, p. 633 – 636 (1987). Abstr. in Phys. Abstr., Vol. 91, No. 1320, Entry 10375 (1988). – See Abstr. 012.080.

144.079 A cosmic–ray ultra high–energy multijet family event.
B.–t. Zou, C.–r. Wang, J.–r. Ren.
Chin. Phys., Vol. 7, No. 1, p. 141 – 147 (1987). Abstr. in Phys. Abstr., Vol. 91, No. 1320, Entry 10376 (1988).

144.080 The local age parameter and the lateral distribution of electrons in extensive air showers.
D. J. van der Walt.
J. Phys. G, Vol. 13, No. 10, p. L247 – L251 (1987). Abstr. in Phys. Abstr., Vol. 91, No. 1320, Entry 10377 (1988).

144.081 Modulation of galactic cosmic rays with a stepwise dependence of the diffusion coefficient on impulse.
S. V. Chalov.
Geomagn. Aehron., Tom 27, No. 6, p. 900 – 905 (1987). In Russian. English translation in Geomagn. Aeron.

144.082 On fluctuations of the cosmic ray intensity with a period of 160 min.
V. P. Antonova, A. G. Zusmanovich.
Geomagn. Aehron., Tom 27, No. 6, p. 1006 – 1008 (1987). In Russian. English translation in Geomagn. Aeron.

144.083 The effect of the asymmetric magnetosphere on the cut–off rigidity of cosmic rays for mid–latitude stations.
O. A. Danilova, M. I. Tyasto.
Geomagn. Aehron., Tom 27, No. 6, p. 1008 – 1009 (1987). In Russian. English translation in Geomagn. Aeron.

144.084 Cosmic–ray elemental abundances from 1 to 10 GeV per amu for boron through nickel.
R. Dwyer, P. Meyer.
Astrophys. J., Vol. 322, No. 2, p. 981 – 991 (1987).
The relative abundances of cosmic–ray nuclei in the charge range boron through nickel ($5 \leqslant Z \leqslant 28$) over the energy range $\sim 1 – 10$ GeV per amu were measured with a balloon–borne detector. The instrument consists of a scintillation and Cerenkov

counter telescope with a multiwire proportional chamber hodoscope. Good charge resolution (≈ 0.2 charge units at iron) and high statistical accuracy have been achieved. The authors use these data to derive the energy dependence of the leakage path length using the leaky box model of propagation and confinement in the Galaxy.

144.085   Origin and acceleration of 10^{20}eV cosmic–ray protons. W. H. Sorrell.
Astrophys. J., Vol. 323, No. 2, p. 647 – 656 (1987).

The author discusses possible extragalactic origins and acceleration of 10^{20}eV cosmic rays on the assumption that the primary particles are mostly protons. It is concluded that no stochastic (Fermi–like) mechanism appears capable of accelerating protons up to 10^{20}eV. Alternative mechanisms based on the Lovelace electric dynamo are examined. It is proposed that gas accretion onto massive black holes in gaseous halos of spiral galaxies in the local supercluster may be the ultimate energy source for accelerating protons to 10^{20}eV.

Problemy fiziki kosmicheskikh luchej. (*Problems of cosmic ray physics*).
See Abstr. 003.015.

Modulyatsiya kosmicheskikh luchej v solnechnoj sisteme (*Cosmic ray modulation in the solar system*).
See Abstr. 003.147.

History of cosmic ray investigations in the USSR.
See Abstr. 013.032.

Main problems of present–day cosmophysics.
See Abstr. 013.061.

Preliminary results on atmospheric neutrinos and Cygnus X–3 in the Fréjus detector.
See Abstr. 034.033.

Cosmic–ray monopole search at IBM–BNL using superconducting induction detectors.
See Abstr. 034.191.

Design and performance of a 0.18 m^2 inductive detector for cosmic magnetic monopoles.
See Abstr. 034.192.

Cosmic ray tests of 7.6 m drift–tube counters and the readout electronics system of the VENUS muon detector.
See Abstr. 034.194.

Monte Carlo simulations of effects due to delta rays in aluminum drift–tube counters.
See Abstr. 034.195.

Development of low cost liquid scintillator counters for cosmic ray experiments.
See Abstr. 034.197.

A combined cosmic ray muon spectrometer and high energy air shower array.
See Abstr. 034.200.

Cosmic ray events observed in Space Telescope Wide Field Camera exposures.
See Abstr. 035.047.

ASTROMAG: A superconducting particle astrophysics magnet facility for the Space Station.
See Abstr. 035.108.

DUMAND: a detector of both neutrinos and gamma rays?
See Abstr. 036.042.

Comparison of methods for determining the centers of extensive air showers.
See Abstr. 036.240.

Measurements of background gamma radiation on Spacelab 2.
See Abstr. 051.031.

The origin of the light nuclides.
See Abstr. 061.015.

Elemental technetium and promethium as cosmic–ray clocks.
See Abstr. 061.023.

Hadron cross sections at ultrahigh energies and unitarity bounds on diffraction.
See Abstr. 061.130.

Hydrodynamical constraints on cosmic–ray acceleration in relativistic shocks.
See Abstr. 062.051.

First–order Fermi acceleration in the two–stream limit.
See Abstr. 062.053.

Energetic particle acceleration in spherically symmetric accretion flows and shocks.
See Abstr. 062.081.

Particle acceleration at astrophysical shocks.
See Abstr. 062.145.

Fermi acceleration at shocks with arbitrary velocity profiles.
See Abstr. 062.160.

Determination of the characteristics of interstellar turbulence from data on the diffusion coefficient of cosmic rays.
See Abstr. 063.062.

Cosmic–ray–modified stellar winds. I Solution topologies and singularities.
See Abstr. 064.083.

Cosmic ray particle acceleration in pulsar magnetospheres.
See Abstr. 067.035.

Solar cosmic rays in the interplanetary space.
See Abstr. 078.013.

Atmospheric radioactivity and variations in the solar neutrino flux.
See Abstr. 080.075.

Cosmogenic ^{10}Be in Zaire alluvial diamonds: implications for ^{3}He contents of diamonds.
See Abstr. 081.013.

Radiation dosimetry and chemistry of a cometary nucleus.
See Abstr. 102.025.

Magnetic helicity of the IMF and the solar modulation of cosmic rays.
See Abstr. 106.005.

Large–scale fluctuations between 13 AU and 25 AU and their effects on cosmic rays.
See Abstr. 106.034.

New physics from Cygnus X–3.
See Abstr. 117.207.

Cygnus X–3: is there convincing evidence for new physics?
See Abstr. 117.276.

Cosmic rays in the shell of SN 1987A and its gamma–radiation.
See Abstr. 125.239.

Photon–photon pair production and the opacity of SN 1987A to TeV and PeV γ–rays.
See Abstr. 125.313.

Cosmic rays and gamma radiation from the shell of SN 1987A.
See Abstr. 125.321.

Cosmic–ray–induced photodestruction of interstellar molecules in dense clouds.
See Abstr. 131.208.

The abundance of H_3^+ ions in dense interstellar clouds.
See Abstr. 131.250.

Interstellar shock waves and ^{10}Be from ice cores.
See Abstr. 131.295.

Gamma–astronomy.
See Abstr. 143.019.

Stockert's chimney – a galactic fountain.
See Abstr. 155.027.

Gamma rays and the distribution of cosmic rays in the Galaxy.
See Abstr. 155.037.

Acceleration of ultra–high–energy cosmic rays (UHECR) in clusters of galaxies.
See Abstr. 160.038.

Ultra–high–energy cosmic rays from superconducting cosmic strings.
See Abstr. 161.427.

Stellar Systems, Galaxy, Extragalactic Objects, Cosmology

151 Stellar Systems (Kinematics, Dynamics)

151.001 **Models of ring galaxies. I. The growth and disruption of clouds in the expanding density wave.**
P. N. Appleton, C. Struck–Marcell.
Astrophys. J., Vol. 318, No. 1, p. 103 – 123 (1987).
Cloud–fluid models are presented of the evolution of a galactic disk which is centrally perturbed by the passage of a low–mass companion down the rotation axis of the galaxy. When the flow of gas clouds into the expanding density wave occurs on a time scale commensurate with the collision–time between individual clouds in the galaxy, cloud interaction rates are driven out of equilibrium. As a result, a strong enhancement of star formation can occur on the leading edge of the density wave. The models also predict large infall velocities behind the expanding ring, leading to a very compressed inner ring. The inner ring usually undergoes a strong burst of star formation.

151.002 **Gravity in one dimension: selective relaxation?**
C. J. Reidl Jr., B. N. Miller.
Astrophys. J., Vol. 318, No. 1, p. 248 – 260 (1987).
The search for relaxation to equilibrium (thermalization) of one–dimensional, self–gravitating systems consisting of N equal mass particles has been attempted by several groups using computer simulations. Ambiguous results were reported from these calculations. The authors have computed time correlation functions in order to estimate the dynamical time for data to become statistically independent within a given system. They have examined five different 100 particle systems containing characteristics applied in previous research. The results show that all but one of the systems have not thermalized, and that the "closeness" of initial states to equilibrium is neither a sufficient nor relevant condition for relaxation.

151.003 **Density distribution corresponding to one potential of elliptical galaxies.**
P. Andrle.
Bull. Astron. Inst. Czech., Vol. 38, No. 4, p. 253 – 255 (1987).
The potential of a spheroid with a quartic disturbing term is assumed. If the disturbing term is not very small the equidensity surfaces corresponding to this potential are ellipsoids with similar strata. This potential could exist in some elliptical galaxies.

151.004 **Systems of colliding bodies in a gravitational field: impact velocity, inelasticity and size distributions.**
S. Clairemidi.
Earth, Moon, Planets, Vol. 38, No. 3, p. 237 – 248 (1987).
The author studies numerically the dynamical evolution of a system of colliding bodies. Collisions are inelastic and the coefficient of rebound is a function of the impact velocity. Various functions are used. The system can reach an equilibrium state characterized by a finite thickness independent of the sense of variation of the function. Within the frame of various distributions of the body sizes a segregation rather than an equipartition of the kinetic energy of random motions is observed.

151.005 **A dynamical study of the frequency of interacting galaxies.**
T. K. Chatterjee.
Astrophys. Space Sci., Vol. 135, No. 1, p. 131 – 141 (1987).
A study of the expected frequency of multiple interacting systems based on their interpretation as the relics of gravitational

interactions between galaxies is conducted, using the impulsive approximation. Results indicate that if the expected frequency of such galaxies is measured with respect to densely populated regions and using the average values of the parameters corresponding to such regions for its determination, then it comes out to be the order of 5 to 6% of spirals, quite compatible with observational values. But, however, if regions of normal density (where collisions are rare) are considered for frequency determinations, the expected frequency goes down by few orders of magnitude. This indicates that stray hyperbolic encounters are too scarce to explain the formation of such galaxies. Hence, most of these interacting pairs must have already been bound doubles.

151.006 **Relative chaos in stellar systems.**
V. G. Gurzadyan, A. A. Kocharyan.
Astrophys. Space Sci., Vol. 135, No. 2, p. 307 – 324 (1987).
Statistical properties of many–dimensional dynamical systems–stellar systems of different types, are investigated by means of new definition of relative chaos based on the estimation of the Ricci curvature in the direction of the velocity of geodesics. Numerical experiment is performed to calculate the Ricci and scalar curvatures for systems with equal total energy. The results of calculations enable one to obtain schematic classification of stellar systems by increasing degree of chaos.

151.007 **Bisymmetric spiral configuration of magnetic fields in spiral galaxies. II. A quasi–global theory.**
M. Fujimoto, T. Sawa.
Publ. Astron. Soc. Jpn., Vol. 39, No. 3, p. 375 – 392 (1987).
This paper is a sequel to the authors' previous work on large–scale magnetic fields in spiral galaxies. A quasi–global dynamo equation is derived and applied for the "bisymmetric" or two–armed spiral magnetic fields in a gaseous slab in differential rotation. The turbulent diffusivity is included in the dissipation terms of the kinematic dynamo equation. The two–spiraled or the bisymmetric field pattern is found to be maintained without being further twisted by the differential rotation, resembling spiral density waves of disk galaxies.

151.008 **Statistical theory of violent relaxation.**
J. Tanekusa.
Publ. Astron. Soc. Jpn., Vol. 39, No. 3, p. 425 – 436 (1987).
The final state of violent relaxation is discussed. The author has made N–body simulations in a one–dimensional sheet model and analyzed the evolution of the system from the standpoint of coarse–grained entropy. It increases in the initial phase of evolution. However, he finds that it does not necessarily reach the maximum value for the given values of the total mass, the total energy, and the fine–grained phase space density despite the Lynden–Bell theory. The difference in the coarse–grained entropies between the final structure thus obtained and the maximum value is so large that the appearance of such a structure seems impossible. This means that the violent relaxation does not lead to the ergodicity.

151.009 **Global instabilities of flat stellar disks: stabilization by a bulgelike component.**
S. Hozumi, T. Fujiwara, M. T. Nishida.
Publ. Astron. Soc. Jpn., Vol. 39, No. 3, p. 447 – 456 (1987).
The effects of a bulgelike component on the global stability of flat stellar disks are investigated against linear two–armed modes.

The global modes are obtained by integrating numerically the linearized collisionless Boltzmann equation as an initial value problem. Model calculations have been made for a series of disks with bulges of different length scales but of the same mass, half that of the disks. The results show that a very compact bulge can reduce disk instabilities remarkably. It is suggested that the hotness of the central part of a disk and inner Lindblad resonances play an important role in stabilization of a disk.

151.010   **About the possible cause of appearance of the barred spiral structure in galaxies.**
S. G. Gestrin, V. M. Kontorovich.
Pis'ma Astron. Zh., Tom 13, No. 8, p. 648 – 653 (1987). In Russian. English translation in Sov. Astron. Lett., Vol. 13.
It is shown that barred spiral structure can appear if rotation curves have an angular velocity break.

151.011   **The formation of galactic barred structure.**
V. G. Guglenko, A. A. Rumyantsev.
Astrofizika, Tom 26, Vyp. 3, p. 421 – 429 (1987). In Russian. English translation in Astrophysics, Vol. 26, No. 3, p. 255 – 260 (1987).
The formation of bars is closely linked to some kind of eruptive process in the central region of a galaxy. The configuration of the bar appears as a result of some purely geometric factors, notably, the stretched central body, which is being imitated by a cylinder, the way the shock wave front crosses the body and the afterward gas falling upon the galactic disk. The bar collision with the gaseous system of the disk is taken into consideration which can lead to the formation of spiral arms.

151.012   **Dirichlet's problem in stellar dynamics. II. Elements of theory of equilibrium figures.**
B. P. Kondrat'ev, E. A. Malkov.
Astrofizika, Tom 26, Vyp. 3, p. 511 – 526 (1987). In Russian. English translation in Astrophysics, Vol. 26, No. 3, p. 308 – 318 (1987).
A hydrodynamical method for the investigation and construction of equilibrium homogeneous collisionless models has been developed. It has been shown that the angular velocity vector and the vorticity vector both do lie in one of the principal planes or are coincident with the principal axis of the ellipsoid. Conditions for the existence of models are determined and characteristics of equilibrium figures are found. (Part one: 42.151.112).

151.013   **Computer simulations of relativistic star clusters.**
S. L. Shapiro, S. A. Teukolsky.
Gravitational collapse and relativity, p. 387 – 413 (1986). – See Abstr. 012.006.
The authors have constructed a new numerical code which solves Einstein's equations for the dynamical evolution of a collisionless gas of particles in general relativity. The computational scheme combines the tools of numerical relativity with those of N–body particle simulation. The Vlasov equation is solved in general relativity by particle simulation and determine the gravitational field using the ADM $3+1$ formalism. Astrophysical applications include the possible origin of quasars and active galactic nuclei via the collapse of dense star clusters to supermassive black holes.

151.014   **The evolution of clumpy gas in young elliptical galaxies.**
R. Kunze, H.-H. Loose, H. W. Yorke.
Astron. Astrophys., Vol. 182, No. 1, p. 1 – 8 (1987).
The authors calculate numerically the mass flow fueled by stellar mass loss at an early epoch during the evolution of an elliptical galaxy. 10^9 years after "birth" of the galaxy the interstellar gas which has accumulated from an (assumed) initial gas–free state flows towards the center and builds up the fuel of any activity in the nucleus of the galaxy. The authors investigate the influence of the random motions of clouds, generated by broken up supernova remnants and planetary nebulae, on the amount of gas which can be stored in the nucleus of a typical (giant) elliptical galaxy. They find that masses up to approximately $10^5 \, M_\odot$ of neutral gas can be stored in the nucleus. The

massive cloud begins to collapse when the dissipation of its kinetic bulk energy sets in rapidly. The result of the collapse can be a supermassive star.

151.015   **Periodic orbits in a triaxial galaxy. III. Their stability.**
H. Robe.
Astron. Astrophys., Vol. 182, No. 2, p. 202 – 206 (1987).
The author considers the stability of periodic orbits in three–dimensional dynamical systems that represent the inner part of a triaxial galaxy. He investigates the stability of various periodic orbits described in the preceding papers.

151.016   **What observations will be needed to develop a theory for the spiral structure in galaxies?**
A. M. Fridman.
Sov. Astron., Vol. 30, No. 5, p. 525 – 532 (1986). English translation of 42.151.030.

151.017   **Collisionless analogs of Riemann S–ellipsoids with halo.**
M. G. Abramyan.
Astrophysics, Vol. 25, No. 2, p. 563 – 569 (1987). English translation of 42.151.088.

151.018   **The role of equipotential and equidensity surfaces for constructing models of galaxies.**
S. A. Kutuzov, L. P. Osipkov.
Astrophysics, Vol. 25, No. 3, p. 671 – 679 (1987). English translation of 42.151.111.

151.019   **Dirichlet problem in stellar dynamics. I. General case of the motion of a collisionless homogeneous gravitating ellipsoid.**
B. P. Kondrat'ev, E. A. Malkov.
Astrophysics, Vol. 25, No. 3, p. 696 – 707 (1987). English translation of 42.151.112.

151.020   **Erratum: "H–functions and mixing in violent relaxation" [Mon. Not. R. Astron. Soc., Vol. 219, No. 2, p. 285 – 297 (1986)].**
S. Tremaine, M. Hénon, D. Lynden–Bell.
Mon. Not. R. Astron. Soc., Vol. 227, No. 2, p. 543 (1987). See Abstr. 41.151.022.

151.021   **On the velocity fields in elliptical galaxies with dark matter.**
Y. Yoshii, H. Saio.
Mon. Not. R. Astron. Soc., Vol. 227, No. 3, p. 677 – 693 (1987).
The tensor virial theorem is used to investigate a dynamical state of an oblate system of luminous matter which is embedded in the more extended dark matter with reasonable density profiles. The authors derived the relation between the ratio of rotation to random velocities v_p/σ and the ellipticity ε_l of a luminous system. If dark matter is distributed almost spherically, then the rotation velocity in the luminous system with isotropic velocity dispersions is larger than in the isolated system with the same ellipticity. Therefore, the anisotropy in velocity dispersions necessary to be consistent with the observed slow rotations of giant ellipticals should be larger than that for the isolated system without dark matter. The authors also derived the relation between the velocity ratio v_p/σ and the ellipticity ε of isolated systems for modified Newtonian dynamics.

151.022   **Self–consistent equilibrium models for the perfect elliptic disc.**
P. Teuben.
Mon. Not. R. Astron. Soc., Vol. 227, No. 4, p. 815 – 841 (1987).
Numerical self–consistent equilibrium models for the perfect elliptic disc have been constructed using Schwarzschild's linear programming method. Two solutions have been found for each model, which minimize and maximize the total angular momentum in the disc. This is a numerical verification of the non–uniqueness of the solutions. Distribution functions in action space have been derived. The maximum angular momentum solutions have been recalculated using a direct method, and have

been compared with the linear programming solutions. Various numerical effects are discussed. Associating stars with all orbits and gas with the closed elliptic orbits, mean streaming and velocity dispersion fields for the star and gas have been derived.

151.023 The two–stream instability in infinite homogenous and uniformly rotating stellar systems.
S. Araki.
Astron. J., Vol. 94, No. 1, p. 99 – 105 (1987).

The author shows that there is no two–stream instability in infinite homogeneous self–gravitating stellar systems with Maxwellian velocity distributions, although it does occur for suitable non–Maxwellian streams. In more realistic models, the effects of finite size and rotation must be taken into account. The author extends the stability analysis to finite, disk–like systems (counterstreaming Kalnajs disks). By comparing the stability diagram of a counterstreaming disk with that of a reference single–stream disk, he can separate the two–stream instability from the Jeans instability inherent in any self–gravitating system. The author finds that instability for any mode except $(n,m) = (3,1)$. Using a WKB dispersion relation, he examines the asymptotic behavior of high–order modes in the stability diagram.

151.024 Spiralstruktur in Sternsystemen. Teil I: Das stochastische Modell.
J. V. Feitzinger, M. Perschke.
Sterne Weltraum, 26. Jahrg., Nr. 9, p. 483 – 487 (1987).

151.025 Advanced computing for science.
P. Hut, G. J. Sussman.
Sci. Am., Vol. 257, No. 4, p. 136 – 144 (1987).

Computational experiments are enriching scientific investigation. They are now becoming as important as theory, observation and laboratory experiments. The authors consider collisions between galaxies.

151.026 Spiralstruktur in Sternsystemen. Teil II: Die Strukturbildung.
J. V. Feitzinger, M. Perschke.
Sterne Weltraum, 26. Jahrg., Nr. 10, p. 556 – 560 (1987).

151.027 Simulations of axisymmetric, Newtonian star clusters: prelude to 2 + 1 general relativistic computations.
S. L. Shapiro, S. A. Teukolsky.
Astrophys. J., Vol. 318, No. 2, p. 542 – 567 (1987).

The authors have previously analyzed the dynamical behavior of spherical clusters in general relativity. They now treat nonspherical systems and allow for rotation and the emission of gravitational waves. The authors construct in this paper an axisymmetric code for solving the Vlasov equation in the Newtonian limit. The code is based on a mean–field particle simulation scheme, suitable for extension to general relativity. The code is tested by reproducing the known evolution of homogeneous spheroids with and without rotation, including the Lin–Mestel–Shu instability. The code is accurate, even when tracking collapse to a singularity. The authors apply the code to investigate the dynamical stability of equilibrium Freeman spheroids. They compare the gravitational collapse of cold, nonrotating, homogeneous spheroids with the collapse of inhomogeneous spheroids.

151.028 Massive disk formation resulting from the collision of a main–sequence star with a white dwarf in a globular cluster core.
N. Soker, O. Regev, M. Livio, M. M. Shara.
Astrophys. J., Vol. 318, No. 2, p. 760 – 766 (1987).

Collisions between a $0.2\,M_\odot$ main–sequence star and a white dwarf, in the core of a globular cluster, have been studied with the aid of a three–dimensional numerical simulation. For impact parameters between one and two main–sequence star radii, a massive disk forms around the white dwarf. If an accretion phase does follow the formation of massive disks, one can expect luminous X–ray (for neutron star accretors) and UV (for white dwarf accretors) sources to be formed this way in globular cluster cores.

151.029 Nonlinear coupling of galactic spiral modes.
M. Tagger, J. F. Sygnet, E. Athanassoula, R. Pellat.
Astrophys. J., Lett. Ed., Vol. 318, No. 2, p. L43 – L46 (1987).

The authors have worked out the nonlinear coupling of modes with different pattern speeds and propose that it explains some salient features of recent galactic simulations. This nonlinear mechanism may provide a new key for understanding the long–term evolution of spirals in galaxies.

151.030 Performance characteristics of tree codes.
L. Hernquist.
Astrophys. J., Suppl. Ser., Vol. 64, No. 4, p. 715 – 734 (1987).

A Fortran implementation of the Barnes–Hut hierarchical tree algorithm is presented and analyzed in the context of the astrophysical N–body problem. The errors introduced into the force calculation as a result of the clustering of distant particles, and their influence on the relaxation time and physical conservation laws, are considered. Tree algorithms should be suitable for studying a wide range of astrophysical phenomena, including potentially both collisional and collisionless systems. As an example, a simulation of the decay of a satellite orbit around a self–gravitating disk with $N = 32,768$, performed with the tree method, is compared with previous calculations.

151.031 Effect of suprathermal particles on gravothermal oscillation.
J. Makino, D. Sugimoto.
Publ. Astron. Soc. Jpn., Vol. 39, No. 4, p. 589 – 603 (1987).

Evolution of a gravitational 1000–body system is calculated to analyze gravothermal oscillation in discrete particle systems. It is found that suprathermal particles are produced as a result of binary hardenings which transfer their energy to the mean field particles very slowly. Therefore, three modes share the total energy. They are: (1) energy of the mean field which controls global configuration of the system; (2) energy of correlations whose development releases binding energies of binaries and triggers expansion; and (3) energy of suprathermal particles which plays a role of an additional energy reservoir acting as a buffer to avoid a sudden and great amount of energy input from the developing correlations to the mean field. It is concluded that the distribution of inwardly decreasing temperature and the associated heat flow towards the central core drive the gravothermal expansion quite similarly as in the gravothermal oscillation of gaseous models.

151.032 Merging instability in groups of galaxies with dark matter.
J. F. Navarro, M. B. Mosconi, D. García Lambas.
Mon. Not. R. Astron. Soc., Vol. 228, No. 2, p. 501 – 511 (1987).

Several N–body experiments were performed in order to analyse the importance of smoothly distributed dark matter in the merging processes in groups of galaxies.

151.033 Close encounter between galaxies – II. Tidal deformation of a disc galaxy stabilized by massive halo.
M. Noguchi.
Mon. Not. R. Astron. Soc., Vol. 228, No. 3, p. 635 – 651 (1987).

A series of disc galaxy models are constructed and their deformation in the tidal interaction is investigated by N–body simulations. Each galaxy model is composed of a self–gravitating disc made up of many particles and a rigid halo component added to stabilize the disc. Each model is characterized by two parameters; the mass fraction of the disc and the mass concentration of the halo toward the centre. It is found that the self–gravitating discs, when perturbed by the tidal force of another galaxy, develop prominent spiral structures not only in the outer region but also in the inner region. This is a remarkable contrast to the case of the massless discs constructed by test particles, in which only the outer part exhibits a spiral structure. Moreover, the spiral structure generally has a bar–like shape in the inner region. Based on these results, the author predicts an overabundance of bars in interacting galaxies.

151.034   The art of N–body building.
J. A. Sellwood.
Annu. Rev. Astron. Astrophys., Vol. 25, p. 151 – 186 (1987). – See Abstr. 003.003.

The author concentrates on techniques for full N–body simulations. The techniques discussed here might apply to the following problems: 1. Internal dynamics of galaxies, including both elliptical and disk systems with or without bulges or halos. 2. Interactions of galaxies with small companions, such as the decay of satellite orbits, formation of shells around elliptical galaxies, and rings in disk galaxies. 3. Interacting galaxies of more nearly equal mass, principally mergers of disk and elliptical systems. 4. Groups of galaxies. 5. Clustering in an expanding universe.

151.035   Dynamical evolution of globular clusters.
R. Elson, P. Hut, S. Inagaki.
Annu. Rev. Astron. Astrophys., Vol. 25, p. 565 – 601 (1987). – See Abstr. 003.003.

The authors concentrate on the theoretical picture of core collapse and subsequent evaporation of globular clusters; this study is currently gaining momentum both from recent theoretical developments and from observations indicating that a significant fraction of globular clusters may already have completed their collapse phase and are now entering the equivalent of a main–sequence stage in stellar evolution. A theoretical discussion of the physical processes that play an important role in the pre– and postcollapse evolution of a globular cluster is given. The authors review various computational approaches to cluster evolution, summarize relevant observational methods and results, and outline directions for future theoretical and observational research.

151.036   On errors of eccentricities of galactic orbits.
S. A. Kutuzov.
Kinematika Fiz. Nebesn. Tel, Tom 3, No. 4, p. 11 – 18 (1987). In Russian. English translation in Kinematics Phys. Celest. Bodies.

Some definitions of eccentricities are considered. Expressions for eccentricities in terms of integrals of motion are found for a suggested model of the galactic gravitational field. Approximate formulae for the error of eccentricity as a function of errors of observable quantities are presented. The formulae connecting various velocity systems are given in a matrix representation as a by–product. On the basis of precise formulae a numerical analysis is carried out to ascertain the influence of errors in heliocentric distances, radial velocities and proper motions on an estimate of eccentricity. It turns out that as a result of errors in proper motions a tendency to overestimation of the values of small eccentricities is observed (the higher, the greater is the distance).

151.037   A numerical model of the triple system with hidden masses.
L. G. Kiseleva.
Kinematika Fiz. Nebesn. Tel, Tom 3, No. 4, p. 67 – 69, 79 (1987). In Russian. English translation in Kinematics Phys. Celest. Bodies.

A numerical model of a gravitating triple system with hidden mass distributed continuously according to the isothermal law is given. It is shown that the presence of hidden mass prevents the contact collisions and coalescence of the components. The problem on domination of slowly rotating stellar systems (of E galaxies–type) in groups with large hidden masses is discussed.

151.038   On the determination of the potential of rotating barred galaxies.
G. T. Omarova, T. S. Kozhanov.
Vestn. Akad. Nauk KazSSR, No. 3, p. 63 – 66 (1987). In Russian. Abstr. in Ref. Zh., 51. Astron., 8.51.702 (1987).

151.039   Shock–wind mechanism of formation of the shell structure in NGC 5128. Numerical modulation.
B. I. Gnatyk, V. A. Krol'.
Inst. Teor. Fiz. Akad. Nauk USSR, Prepr., No. 46 R, p. 3 – 19 (1987). In Russian. Abstr. in Ref. Zh., 51. Astron., 8.51.705 (1987).

151.040   Dynamical instability of a model of M87.
D. Merritt.
Astrophys. J., Vol. 319, No. 1, p. 55 – 60 (1987).

The stability of Newton and Binney's dynamical model of the giant elliptical galaxy M87 is tested with a mean–field N–body code. The model is unstable to the formation of a bar near the center. This result suggests that stable models of M87 with strong velocity anisotropies may have to be nonspherical.

151.041   The ellipsoidal subsystems in barred spirals.
M. G. Abramyan, D. M. Sedrakyan, M. A. Chalabyan.
Sov. Astron., Vol. 30, No. 6, p. 643 – 647 (1987). English translation of 42.151.099.

151.042   Statistical mechanics and equilibrium sequences of ellipticals.
M. Stiavelli, G. Bertin.
Mon. Not. R. Astron. Soc., Vol. 229, No. 1, p. 61 – 71 (1987).

Elliptical galaxies are expected to have undergone incomplete violent relaxation. Here incomplete relaxation is regarded as a process producing a metastable, long–lived state which is stabilized by the approximate conservation of some set of global quantities in addition to the total energy and the number of particles. The final state corresponds to a maximum of the classical (Boltzmann) entropy provided that the proper phase space partition and set of constraints are chosen. Here the authors explore two different ways of implementing these ideas.

151.043   Maximum entropy states and the structure of galaxies.
S. D. M. White, R. Narayan.
Mon. Not. R. Astron. Soc., Vol. 229, No. 1, p. 103 – 117 (1987).

The authors investigate the properties of maximum entropy states of isolated self–gravitating gas spheres of given mass, energy, extent, and maximum phase space density. They show that if consideration is restricted to objects with power law density profiles, the maximum entropy state has a high central concentration and $\varrho \propto r^{-3}$, as found in violent relaxation experiments. However, if the form of the density profile is not constrained, the highest entropy state is a diffuse system superposed on a central spike of infinitesimal mass but finite binding energy. Limiting the maximum phase space density of the system eliminates the central singularity but does not alter the two component nature of the maximum entropy solutions.

151.044   Relativistic collapse of a uniform star cluster.
G. S. Bisnovatyj–Kogan, L. R. Yangurazova.
Astrofizika, Tom 27, Vyp. 1, p. 79 – 89 (1987). In Russian. English translation in Astrophysics, Vol. 27, No. 1.

An integral numerical method is considered for calculating the dynamical stages of the spherical star cluster evolution. The method is based upon the analysis of the individual stellar orbits and their intersections. The collapse of the cluster with a uniform initial density is computed, initial radius is equal to $5 r_g$, the angular momentum does not exeed $2 mcr_g$. For $N = 100$ all particles fall into a black hole and for $N = 200$ two particles remain in remote orbits.

151.045   The stability of a self–gravitating uniform spheroid with azimuthal magnetic field. I.
V. A. Antonov, O. A. Zheleznyak.
Astrofizika, Tom 27, Vyp. 1, p. 111 – 116 (1987). In Russian. English translation in Astrophysics, Vol. 27, No. 1.

The influence of a secondary magnetic field on the stability of a self–gravitating uniform spheroid with respect to deformation that transforms it to a three–axial ellipsoid has been investigated. It has been shown that the azimuthal magnetic field is a stabilizing factor, thanks to the fact that the spheroid can be stable at e > e_cr = 0.95285.

151.046   What should the gradient of the velocity dispersion of gas clouds in galactic discs be?
V. V. Levy, A. G. Morozov.
Astron. Zh., Tom 64, Vyp. 5, p. 919–928 (1987). In Russian. English translation in Sov. Astron., Vol. 31, No. 5.

The dispersion equation is obtained, describing the dynamics of disturbances in a rotating gravitating thin gas disc with arbitrary distributions of the surface density $\sigma_0(r)$, angular velocity $\Omega(r)$, and velocity dispersion c_s of the particles forming the disc (gas clouds in the case of galactic gas subsystems). A new type of instability in such systems, the entropy–gradient instability, is found. It is shown that such a disc is stable under minimum value of c_s, provided the density $\sigma_0(r)$ decreases towards the disc periphery much faster than the velocity dispersion $c_s(r)$ which agrees with the data of observations of gas subsystems of galaxies.

151.047   Mass models for disk and halo components in spiral galaxies.
E. Athanassoula, A. Bosma.
Nearly normal galaxies. From the Planck time to the present, p. 121–126 (1987). – See Abstr. 012.031.

The authors have recently undertaken an analysis of the rotation curves of a large number of galaxies for which both photometric and kinematic data of reasonable quality are available in the literature. They have applied additional dynamical constraints, based on spiral structure theory, to the modelling. They describe here briefly their method and present the main results.

151.048   The structure and evolution of disk galaxies.
R. G. Carlberg.
Nearly normal galaxies. From the Planck time to the present, p. 129–137 (1987). – See Abstr. 012.031.

The structure and evolution of self–gravitating diks are dominated by the spiral waves that provide the principal source of dynamical relaxation. This review emphasizes the quasilinear theory of shearing waves as providing a substantial basis for understanding the properties of the waves occurring in large N–body experiments. The nonlinear consequences of spirals, heating and angular momentum transfer, can be reliably measured in the N–body experiments, and are comparable to estimates made using quasilinear theory. The heating time scale is of order 10 orbital periods, and the angular momentum transfer time is of order 100 orbital periods. Many of the important aspects of disk galaxies are directly modelled in these N–body experiments. Finally, a speculative origin of S0 galaxies is suggested.

151.049   The effects of satellite accretion on disk galaxies.
P. J. Quinn.
Nearly normal galaxies. From the Planck time to the present, p. 138–143 (1987). – See Abstr. 012.031.

Over a Hubble time, the disk of a spiral galaxy and its system of satellites can be severely altered by the action of dynamical friction. Satellites sink rapidly to the center of the disk when their orbits bring them within the optical radius. Those satellites in direct orbits with orbital planes close to that of the disk sink the fastest. The resultant satellite population does not however have as much projected anisotropy as that observed by Holmberg (1969). The orbital energy of the devoured satellites is deposited into z and radial motions of the disk stars. Self gravitating simulations indicate that the z heating produces a noticeable thickening of the disk even for satellite masses of only a few percent of the mass of the disk.

151.050   Halo response to galaxy formation.
J. E. Barnes.
Nearly normal galaxies. From the Planck time to the present, p. 154–159 (1987). – See Abstr. 012.031.

"Flat" rotation curves appear to require correlations between the parameters of the disk and halo of typical galaxies. These correlations can arise naturally when a disk forms by slow accretion within a pre–existing dark halo. As the disk grows, the inner part of the halo responds to the changing gravitational field. The resulting correlation gives a factor of ~ 2 relief from the "fine tuning" problem posed by flat rotation curves. Models with large initial halo core radii ($r^0_c \geqslant 8\alpha^{-1}$) give approximately flat rotation curves over the widest range of disk mass. Such large r^0_c imply that most of the halo orbits contributing to the mass within the optical radius of the galaxy have large radial excursion. Consequently, approximations based on circular orbits fail to correctly describe the halo response, and the disk has little effect on the shape of the halo.

151.051   Orbits.
J. Binney.
Structure and dynamics of elliptical galaxies, p. 229–239 (1987). – See Abstr. 012.032 (IAU Symp. No. 127).

Orbits that respect at least three isolating integrals of motion have very special structures in phase space. The main characteristics of this structure are reviewed, and the concrete examples that are provided by orbits in Stäckel potentials, are discussed. Many orbits in general potentials admit three approximate isolating integrals and closely resemble orbits in Stäckel potentials. If the potential is that of an elliptical galaxy with negligible figure rotation, the overall orbital structure of the potential differs from that of a Stäckel potential only by the presence of a few unimportant families of resonant orbits. However, this elegant picture is shattered by the introduction of non–negligible figure rotation: though substantial regions of phase space may still be occupied by orbits that individually resemble orbits in Stäckel potentials, the overall orbital structure is radically changed by figure rotation.

151.052   Dynamics in the centres of triaxial elliptical galaxies.
O. E. Gerhard.
Structure and dynamics of elliptical galaxies, p. 241–248 (1987). – See Abstr. 012.032 (IAU Symp. No. 127).

Orbits in the inner kpc of a triaxial galaxy are discussed, taking into account the effect of a central density concentration like a massive black hole, a dense stellar nucleus, or a de Vaucouleurs–type cusp.

151.053   Shells and the potential wells of elliptical galaxies.
P. J. Quinn, L. Hernquist.
Structure and dynamics of elliptical galaxies, p. 249–259 (1987). – See Abstr. 012.032 (IAU Symp. No. 127).

A survey of the possible variety of sharp–edged, caustic features that may arise in the collision of galaxies with very different masses and sizes has shown that in general shells are morphologically very complex. It is therefore not easy to determine the history of the collision that produced the shells nor the properties of the galaxies involved. However, a small number of shell galaxies (notably NGC 3923) have a sufficiently simple and orderly shell distribution that the authors believe the shells were formed by a chance very symmetric and simple encounter. In such cases there is a unique opportunity to investigate the potential well of an elliptical galaxy over a large range in radius ($\cong 0.5\, r_e – 20\, r_e$). An analysis of the NGC 3923 shell system has shown that a large amount of dark matter is present ($M_{dark} \cong 40\, M_{luminous}, r < 17\, r_e$).

151.054   Spherical galaxies: methods and models.
D. O. Richstone.
Structure and dynamics of elliptical galaxies, p. 261–269 (1987). – See Abstr. 012.032 (IAU Symp. No. 127).

Over the last 5 years, considerable progress has been made in our ability to construct self–gravitating stellar equilibria. One of these new methods is essentially a variant of Eddington's (1916) method. Two other key approaches are logical extensions of Schwarzschild's Linear Programming method, and can be applied to nonspherical models as well. These methods are reviewed here. The application of these methods to galaxies has yielded a few very interesting results within the last year or two. The methods described here unambiguously establish M/L's for

M87 and M32 within about 30 arc seconds. They strongly support Tonry's contention that the nucleus of M32 contains a large invisible mass, possibly a $10^7 M_\odot$ black hole. They also suggest that observational recovery of the projected velocity distribution might permit the observer to distinguish between a massive halo and an increasingly tangential velocity distribution function.

151.055   Dynamical models for axisymmetric and triaxial galaxies.
T. de Zeeuw.
Structure and dynamics of elliptical galaxies, p. 271 – 290 (1987).
– See Abstr. 012.032 (IAU Symp. No. 127).

Non–spherical dynamical models for galaxies, and the methods for their construction, are reviewed. The theory for two–integral axisymmetric models is reasonably well developed. Stäckel models give considerable insight in the structure of both three–integral axisymmetric models and non–rotating triaxial systems. Triaxial galaxies with appreciable figure rotation require much further study. Applications to elliptical galaxies and the bulges of disk galaxies are discussed.

151.056   N–body simulations of elliptical galaxies.
T. S. van Albada.
Structure and dynamics of elliptical galaxies, p. 291 – 299 (1987).
– See Abstr. 012.032 (IAU Symp. No. 127).

N–body simulations are a useful tool for constructing equilibrium models of elliptical galaxies and for the exploration of their kinematical properties, in particular the tumbling rate of the figure about some axis and the internal streaming. As yet little is known about these, except that there is a large variety of possible equilibrium models. It is easy to make triaxial systems that tumble about the short axis, with internal streaming aligned with the rotation axis of the figure. Attempts to construct systems with figure rotation and internal streaming in opposite directions have not been successful. Use of current simulation codes for detailed studies of particle orbits is limited to several (about 10) dynamical times due to non–physical fluctuations in the force field.

151.057   Stability of elliptical galaxies. Theoretical aspects.
V. L. Polyachenko.
Structure and dynamics of elliptical galaxies, p. 301 – 314 (1987).
– See Abstr. 012.032 (IAU Symp. No. 127).

The main question for each special system of whether it is stable or in a state of dynamical relaxation is answered by asking whether the parameters of the system belong to stable or unstable regions. Determination of the boundaries between these two regions is the main problem for stability theory. The importance of the problem is sufficiently evident when studying various collisionless systems, in particular, elliptical galaxies. It is important to note that the requirement of stability imposes actually very essential limits on permissible stationary models of collisionless gravitating systems as follows from the results here; for details see Fridmann & Polyachenko, 1984.

151.058   Stability of elliptical galaxies. Numerical experiments.
D. Merritt.
Structure and dynamics of elliptical galaxies, p. 315 – 329 (1987).
– See Abstr. 012.032 (IAU Symp. No. 127).

This paper is divided into four parts. The first part summarizes what has been learned so far about the stability and instability of spherical equilibrium models. Unfortunately, nothing definite is known yet about the stability of triaxial models. The second part discusses how dynamical instabilities might be used to constrain the dynamics of particular well–observed galaxies. The third part describes some preliminary work on the question of whether instabilities could have played an active role during galaxy formation. The fourth part presents an efficient new algorithm for testing the stability of spherical and triaxial models.

151.059   A method to determine the intrinsic axial ratios of individual triaxial galaxies.
O. E. Gerhard, M. Vietri.
Structure and dynamics of elliptical galaxies, p. 401 – 402 (1987).
– See Abstr. 012.032 (IAU Symp. No. 127).

The authors present a method, partly geometrical and partly dynamical, to determine the intrinsic axial ratios of a triaxial galaxy. The method can be applied to bulges of spiral galaxies and to ellipticals containing sufficiently extended and kinematically regular gas disks. Required observational inputs are multi–colour surface photometry and the gas velocity field.

151.060   Initial tidal effects on shell formation.
R. A. James, A. Wilkinson.
Structure and dynamics of elliptical galaxies, p. 471 – 472 (1987).
– See Abstr. 012.032 (IAU Symp. No. 127).

Sharp edged features have been observed by Malin and Carter (1980), Schweizer (1980) and others in the envelopes of many elliptical galaxies. The most promising of current models involves the disruption of a low mass galaxy penetrating close to the centre of a considerably larger ellipsoidal galaxy (Schweizer, 1980, Quinn, 1984, Hernquist and Quinn, 1986, Dupraz and Combes, 1986). The release of stars having a range of energies and with a definite relationship between velocity and position at the point of release generates sharp features by the "phase wrapping" mechanism. The authors examine this effect using a self–consistent particle–mesh code to follow the disruption phase in an almost radial encounter. A standard 2–body code continues the integration up to a Hubble time.

151.061   Shells and encounters of disk galaxies with ellipticals.
T. Piran, J. V. Villumsen.
Structure and dynamics of elliptical galaxies, p. 473 – 474 (1987).
– See Abstr. 012.032 (IAU Symp. No. 127).

Faint shells have been found around a large number of ellipticals. Quinn (1984) has suggested that shells form around ellipticals when a small disk galaxy is tidally disrupted in a near radial orbit. The authors test Quinn's model with fully self–consistent 3–D N–body simulations. The target galaxy is a spherical galaxy with a Hubble profile. The intruder is a self–consistent disk–halo system. The disk is exponential with a finite thickness, and the halo is nearly spherical and isothermal. A number of encounters were performed and followed for a few dynamical times after disruption.

151.062   Dynamical friction and orbit circularization.
S. Casertano, E. S. Phinney, J. V. Villumsen.
Structure and dynamics of elliptical galaxies, p. 475 – 476 (1987).
– See Abstr. 012.032 (IAU Symp. No. 127).

The authors study the change in shape of the orbit of a satellite sinking (because of dynamical friction) towards the center of a larger galaxy. The galaxy is assumed spherically symmetric with distribution function either isotropic or predominantly radial. The satellite is a softened point mass. The orbit evolution is studied analytically using the Chandrasekhar (1943) approximation for the dynamical friction drag and the epicyclic approximation for the orbit of the satellite, and numerically, by direct integration of an N–body system. Fully self–consistent spherical models have been used for the galaxy.

151.063   Stochastic stellar orbits and the shapes of elliptical galaxies.
O. E. Gerhard.
Structure and dynamics of elliptical galaxies, p. 477 – 478 (1987).
– See Abstr. 012.032 (IAU Symp. No. 127).

Using Melnikov's method to study the appearance of stochastic orbits in perturbed Stäckel potentials, a correlation is found between the observed shapes of elliptical galaxies and the occurrence of mainly regular orbits. Some other potential perturbations giving rise to large regions of stochastic orbits, on the other hand, appear to be inconsistent with observations.

151.064   Complex instability around the rotation axis of triaxial systems.
L. Martinet, D. Pfenniger.
Structure and dynamics of elliptical galaxies, p. 479 – 480 (1987).
– See Abstr. 012.032 (IAU Symp. No. 127).

The authors examine the general instability at large amplitude of the radial periodic orbits along the rotation axis of bulges, spheroids and other rotating triaxial ellipsoidal systems.

151.065   Stochasticity in models of elliptical galaxies.
D. Pfenniger, S. Udry.
Structure and dynamics of elliptical galaxies, p. 481 – 482 (1987).
– See Abstr. 012.032 (IAU Symp. No. 127).

The authors measure quantitatively the growth of stochasticity due to various perturbations to a model of elliptical galaxy. This is achieved by computing the Liapunov characteristic exponents of randomly selected orbits.

151.066   Natural action–angle variables.
D. N. Spergel.
Structure and dynamics of elliptical galaxies, p. 483 – 484 (1987).
– See Abstr. 012.032 (IAU Symp. No. 127).

Since galaxies are collisionless relaxed systems, actions are an extremely useful tool for understanding their dynamics. There are many potential applications of actions: (1) When orbits in an N–body simulation are characterized by their actions, the six dimensional distribution function, $f(\vec{x},\vec{p})$, can be reduced to a more tractable three dimensional function, $f(J)$. (2) Actions are adiabatic invariants, and thus are useful for studying slowly evolving systems. (3) The spectral decomposition of an orbit can be used to help generate self–consistent galaxy models (Spergel 1987).

151.067   Integrals of motion in an elliptical galaxy model.
A. Wilkinson, T. de Zeeuw.
Structure and dynamics of elliptical galaxies, p. 485 – 486 (1987).
– See Abstr. 012.032 (IAU Symp. No. 127).

The orbital structure in triaxial elliptical galaxies resembles closely that in the general Stäckel potentials, for which the Hamilton–Jacobi equation separates in ellipsoidal coordinates, and all orbits have three exact integrals of motion H, I_2 and I_3, say (de Zeeuw, 1985). Thus, for N–body models of elliptical galaxies, approximate integrals of motion can be found by the fitting of Stäckel potentials.

151.068   Self–consistent models of perfect triaxial galaxies.
T. S. Statler.
Structure and dynamics of elliptical galaxies, p. 487 – 488 (1987).
– See Abstr. 012.032 (IAU Symp. No. 127).

The author has used Schwarzschild's (1979) method to study the variety of self–consistent solutions available to the family of triaxial "perfect ellipsoids" (de Zeeuw, 1985). The time–averaged density at 240 points within the mass model is computed for each of 1065 orbits distributed regularly in phase space and covering all four major orbit families. Numerical solutions are found for all axis ratios investigated.

151.069   Self–consistent oblate–spheroid models.
J. L. Bishop.
Structure and dynamics of elliptical galaxies, p. 489 – 490 (1987).
– See Abstr. 012.032 (IAU Symp. No. 127).

A method for the construction of models of axisymmetric galaxies is presented. Here the author presents the results from the application of this method to the "perfect" oblate–spheroid mass model. A large class of valid self–consistent distribution functions which depend on three isolating integrals of the motion is found. The kinematics of many models are consistent with those observed for elliptical galaxies.

151.070   Self–consistent elliptical disks.
P. Teuben.
Structure and dynamics of elliptical galaxies, p. 491 – 492 (1987).
– See Abstr. 012.032 (IAU Symp. No. 127).

Self–consistent solutions for the perfect elliptic disk have been obtained. Velocity field, dispersion maps and the distribution function in action space have been derived.

151.071   Stellar dynamics of needles.
S. Tremaine, T. de Zeeuw.
Structure and dynamics of elliptical galaxies, p. 493 – 494 (1987).
– See Abstr. 012.032 (IAU Symp. No. 127).

One dimensional "needles" are a limiting case of general triaxial stellar systems. Self–consistent, finite needles can have arbitrary longitudinal density distributions but have a fixed, universal distribution function. All needles are stable to all longitudinal perturbations but neutral to transverse perturbations.

151.072   Formal inversion of the self–consistent problem for triaxial galaxies.
H. Dejonghe.
Structure and dynamics of elliptical galaxies, p. 495 – 496 (1987).
– See Abstr. 012.032 (IAU Symp. No. 127).

151.073   Analytic axisymmetric dynamical models with three integrals of motion.
H. Dejonghe, T. de Zeeuw.
Structure and dynamics of elliptical galaxies, p. 497 – 498 (1987).
– See Abstr. 012.032 (IAU Symp. No. 127).

151.074   Triaxial scale–free models of highly flattened elliptical galaxies with massive halos.
H. F. Levison, D. O. Richstone.
Structure and dynamics of elliptical galaxies, p. 499 – 500 (1987).
– See Abstr. 012.032 (IAU Symp. No. 127).

Two surveys of dynamical models of highly flattened, triaxial elliptical galaxies with isothermal potentials have been constructed. These models were constructed in order to better understand the range of possible observable dynamical properties of triaxial galaxies. All models have been constructed so that they appear as E6 galaxies when seen from their intermediate axes. However, one set of models is nearly oblate; the other is nearly prolate. The models are constructed with massive halos such that $M/L \propto r$. Triaxial models of either shape can be constructed with their projected axes of rotation at any position angle with respect to the major axes of the galaxies. The most surprising result is that in most models, the position angle of maximum observed rotation is not perpendicular to the position angle of zero rotation.

151.075   Box models without cylindrical rotation.
M. Petrou.
Structure and dynamics of elliptical galaxies, p. 501 – 502 (1987).
– See Abstr. 012.032 (IAU Symp. No. 127).

The author describes here a way of constructing models for box–shaped dynamical systems without cylindrical rotation. She argues that such models may arise as a result of the external heating of a disk.

151.076   Dynamical approach to the $R^{1/4}$ law of ellipticals.
G. Bertin, M. Stiavelli.
Structure and dynamics of elliptical galaxies, p. 503 – 504 (1987).
– See Abstr. 012.032 (IAU Symp. No. 127).

The $R^{1/4}$ luminosity law has often been used as an observational constraint in the construction of self–consistent models of ellipticals. In contrast, developing some ideas proposed in the past, the authors investigate certain theoretical arguments that may lead to a dynamical justification of this important empirical law.

151.077 Statistical mechanics and equilibrium sequences of ellipticals.
M. Stiavelli, G. Bertin.
Structure and dynamics of elliptical galaxies, p. 505 – 506 (1987).
– See Abstr. 012.032 (IAU Symp. No. 127).

Elliptical galaxies are expected to have undergone incomplete violent relaxation. Here incomplete relaxation is regarded as a process producing a metastable, long–lived state which is dynamically stabilized by the approximate conservation of one global quantity in addition to the total energy and number of particles.

151.078 On the distribution function of elliptical galaxies.
A. Kashlinsky.
Structure and dynamics of elliptical galaxies, p. 507 – 508 (1987).
– See Abstr. 012.032 (IAU Symp. No. 127).

Distribution functions containing cutoff in energy impose several limitations on systems they describe, e.g., no circular orbits are allowed in the major part of system and spatial boundary is poorly defined. As opposed to these functions, the author presents here distribution function that describes galaxies of finite extent (i.e., truncated in radius). He discusses properties of systems having this distribution function for spherical and axisymmetric cases and compare their surface brightness, iso-photes, and rotation curves with observations of elliptical galaxies. This distribution function can easily be generalized to a triaxial galaxy.

151.079 Towards a self–consistent model of a galaxy.
W. J. L. V. Durodie, F. D. Kahn.
Structure and dynamics of elliptical galaxies, p. 509 – 510 (1987).
– See Abstr. 012.032 (IAU Symp. No. 127).

Using a theoretical model for the functional distribution of stars in phase space in a spherically symmetric galactic system, it is found upon solving the fundamental equations of stellar dynamics that the rotation curves produced by the model are flat for large distances within the system. The properties of the stellar orbits within such systems are investigated and an N–ring axially and equatorially symmetric model for simulating its dynamics is presented. Poisson's equation is solved by expanding density and potential in Legendre polynomials (c.f. van Albada and van Gorkom, 1977). It is helped to follow the time development of such a system under various forces.

151.080 The envelopes of spherical galaxies.
W. Jaffe.
Structure and dynamics of elliptical galaxies, p. 511 – 512 (1987).
– See Abstr. 012.032 (IAU Symp. No. 127).

The light distribution in the envelopes of spherical galaxies seems to be caused by the existence of a break in the energy distribution, $N(E)$ at $E = 0$. This, in turn is probably caused by the escape of positive energy stars.

151.081 Spherical stellar systems: structure and evolution.
L. M. Ozernoy (*L. M. Ozernoj*), V. A. Volodin.
Structure and dynamics of elliptical galaxies, p. 513 – 514 (1987).
– See Abstr. 012.032 (IAU Symp. No. 127).

The authors analyze a class of spherical stellar systems with an isotropic, non–isothermal, one–particle distribution function which is truncated in energy ε as $(\varepsilon_b - \varepsilon)^\nu$ ($\nu > 0$) near the energy of truncating ε_b and is the Maxwell–Boltzmannian at $|\varepsilon| \gg |\varepsilon_b|$. The stellar systems which correspond to this distribution function appear to be isothermal in their central parts and polytropic in the external parts. Some qualitative results which follow from such a "polytropic–isothermal" distribution function (PIDF) are briefly discussed. A further extention of PIDF enables to analyze in a new way the evolution of spherical stellar systems.

151.082 Instability through anisotropy in spherical stellar systems.
P. L. Palmer, J. Papaloizou.
Structure and dynamics of elliptical galaxies, p. 515 – 516 (1987).
– See Abstr. 012.032 (IAU Symp. No. 127).

The linear stability of spherical stellar systems is considered by solving the Vlasov and Poisson equations which yield a matrix eigenvalue problem to determine the growth rate. The authors consider this for purely growing modes in the limit of vanishing growth rate. They show that a large class of anisotropic models are unstable and derive growth rates for the particular example of generalized polytropic models. A simple method for testing the stability of general anisotropic models is presented. The analysis shows that instability occurs even when the degree of anisotropy is very slight.

151.083 Collisional effects on the density profiles of spherical galaxies.
L. A. Aguilar, S. D. M. White.
Structure and dynamics of elliptical galaxies, p. 517 – 518 (1987).
– See Abstr. 012.032 (IAU Symp. No. 127).

The authors use N–body simulations to study the time evolution and the final shape of the density profiles of non–rotating spherical galaxies that have undergone a tidal encounter. They consider models with the de Vaucouleurs and King surface density profiles and with isotropic, tangential and radially biased velocity distributions.

151.084 Cold collapse as a way of making elliptical galaxies.
L. A. Aguilar, D. Merritt, M. Duncan.
Structure and dynamics of elliptical galaxies, p. 519 – 520 (1987).
– See Abstr. 012.032 (IAU Symp. No. 127).

The authors investigate whether dissipationless collapse starting from very cold, non–rotating initial conditions can produce objects resembling real elliptical galaxies. They also study the effect of various initial geometries on the shape of the final object.

151.085 Angular momentum in elliptical galaxies.
R. F. G. Wyse, S. Lizano.
Structure and dynamics of elliptical galaxies, p. 521 – 522 (1987).
– See Abstr. 012.032 (IAU Symp. No. 127).

The authors have used the available published observations of the rotational properties of elliptical galaxies to test theories of galaxy formation which predict an anti–correlation between the angular momentum of a galaxy and its initial overdensity and hence formation epoch. They find that the prediction, at least in its simplest form, is not supported by the data for well studied elliptical galaxies which have a range in rotational support.

151.086 Violent relaxation and mixing in 1–D gravitational systems.
M. Luwel.
Structure and dynamics of elliptical galaxies, p. 523 – 524 (1987).
– See Abstr. 012.032 (IAU Symp. No. 127).

The one dimensional gravitational model consists of N mass sheets with surface density m_i, parallel to the (y, z)–plane and constrained to move along the x–axis under influence of their mutual gravitational force $F_{ij} = -2\pi Gm_im_j \cdot \mathrm{sgn}(x_i-x_j)$. In order to study the evolution of this one–dimensional system, the N Newtonian equations of motion are integrated numerically, using an "exact" double precision algorithm.

151.087 The dynamical evolution of a star cluster.
G. Som Sunder, R. K. Kochhar.
Structure and dynamics of elliptical galaxies, p. 525 – 526 (1987).
– See Abstr. 012.032 (IAU Symp. No. 127).

Using the second order tensor virial equations and the equation for the rate of change of the kinetic energy tensor, the authors follow the dynamical evolution of a star cluster with anisotropic velocity distribution. They show that the cluster executes finite amplitude oscillations both in size and eccentricity. However, unlike in the isotropic case, the amplitude now depends on the initial eccentricity.

151.088 Numerical investigation of the density distribution of stars and the dispersion of velocities in spiral galaxies.
H.–N. Zhou.
Structure and dynamics of elliptical galaxies, p. 527 (1987). – See Abstr. 012.032 (IAU Symp. No. 127).

151.089 Solution of the problem of stability of a stellar system with Emden's density law and a spherical distribution of velocities.
V. A. Antonov.
Structure and dynamics of elliptical galaxies, p. 531 – 548 (1987).
– See Abstr. 012.032 (IAU Symp. No. 127). Original paper published in Vestn. Leningr. Univ., No. 19, p. 96 – 111 (1962).

Applying a criterion previously derived by the author, the stability of stellar systems with an isotropic velocity distribution and Emden's polytropic density law is demonstrated for the exponent $n = 3/2$.

151.090 On the instability of stationary spherical models with purely radial motion.
V. A. Antonov.
Structure and dynamics of elliptical galaxies, p. 549 – 552 (1987).
– See Abstr. 012.032 (IAU Symp. No. 127). Original paper see Abstr. 10.151.067.

Models of spherical, self–gravitating stellar systems in which all the trajectories pass initially near the center are interesting to examine from different perspectives. Such models are compatible not only with the idea of explosive cosmogony, but also with the concept of formation of stars from a gaseous, spherical cloud, if they are not affected during formation by significant peculiar motions and thus "fall" into the center. Another question is whether such a system is capable of remaining stable for a long period of time. The author shows that such a system is unstable with respect to regular forces and must be reconstructed during the course of one period of revolution of an individual star.

151.091 Quadratic integrals of motion and stellar orbits in the absence of axial symmetry of the potential.
G. G. Kuzmin (*G. G. Kuz'min*).
Structure and dynamics of elliptical galaxies, p. 553 – 556 (1987).
– See Abstr. 012.032 (IAU Symp. No. 127). Original paper see Abstr. 10.151.056.

As is well known, in the case of an axially symmetric and time–invariant gravitational potential, if the potential satisfies one particular additional constraint, there exist three isolating integrals of motion: the energy integral, the area integral, and the third integral which is quadratic in the velocities. This work discusses the case in which there exist quadratic integrals in the absence of axial symmetry of the potential. Such a case has already been examined by Eddington (1916), but in their explicit form, the integrals were introduced by Clark (1936).

151.092 Formation of spiral–whirl structure of VK (*velocity–kink*) galaxies at the linear stage of a shear hydrodynamic instability.
P. V. Baev, Yu. N. Makov, A. M. Fridman.
Pis'ma Astron. Zh., Tom 13, No. 11, p. 964 – 972 (1987). In Russian. English translation in Sov. Astron. Lett., Vol. 13.

The numerical solution of the linearized equations which describe perturbations of a gaseous galactic disk with a velocity kink on the rotation curve shows that already on the linear stage of perturbation growth anticyclonic whirls arise simultaneously with spiral surface density waves. The authors determine the dependence of the whirl locations on the instability increment.

151.093 Catastrophe theory and stellar systems.
V. G. Gurzadyan, A. A. Kocharyan, S. G. Matinyan.
Astrophysics, Vol. 26, No. 1, p. 82 – 90 (1987). English translation of 43.151.024.

151.094 Angular momentum from tidal torques.
J. Barnes, G. Efstathiou.
Astrophys. J., Vol. 319, No. 2, p. 575 – 600 (1987).
The authors describe results for the origin of angular momentum of bound objects in large cosmological N–body simulations. Three sets of models are analyzed: one with white–noise initial conditions and two in which the initial conditions have more power on large scales, as predicted in models with cold dark matter (CDM). Statistical analysis of large catalogs of objects shows that the dimensionless spin parameter λ is remarkably

insensitive to the initial perturbation spectrum and has a median value $\lambda_{me} \sim 0.05$. The quantity λ is weakly correlated with mass and internal substructure and is uncorrelated with initial overdensity. In CDM models, groups with high λ–values tend to have nearer neighbors and are more strongly clustered than groups with low λ; these effects are weaker or absent in white–noise models.

151.095 Evolution of globular clusters including a degenerate component.
H. M. Lee.
Astrophys. J., Vol. 319, No. 2, p. 772 – 800 (1987).
Numerical studies of the dynamical evolution of globular clusters using an orbit–averaged isotropic Fokker–Planck code, including binary formation and their interactions with others, are presented. Two different initial stellar mass categories are assumed: main–sequence stars having individual masses m_{MS} and degenerate stars (either white dwarfs or neutron stars) having individual masses m_D such that $m_D/m_{MS} > 1$. Very hard binaries composed of degenerate–normal pairs are allowed to form via the tidal capture process and moderately hard degenerate–degenerate binaries are allowed to form via the three–body process. If the initial degenerate population is sufficiently large, the three–body binaries among degenerate stars eventually provide enough energy to stop the collapse and cause reexpansion. On the other hand, the degenerate–normal pair of tidally captured binaries are a relatively important energy source if the initial population of degenerate stars is small.

151.096 Dynamical effects of successive mergers on the evolution of spherical stellar systems.
H. M. Lee.
Astrophys. J., Vol. 319, No. 2, p. 801 – 818 (1987).
The dynamical effects of high–mass stars formed out of successive mergers among tidally captured binaries on the evolution of spherical stellar systems are investigated. It is assumed that all tidally captured systems become mergers. Stellar evolution is simulated by computing the mean age of the mass group and applying a specific death rate, as a function of mean age. Successive mergers and stellar evolution are efficient for reversing core–collapse in clusters with large N. For stellar systems with $N = 10^{5-6}$, three–body binary heating among high–mass stars provides significant energy to drive the bounce and postcollapse expansion. For small N, three–body binaries among light stars provide most of the energy to drive the postcollapse expansion. Stellar systems with very large $N (\gtrsim 10^7)$ are vulnerable to the "merger instability," which may lead to the formation of a central black hole.

151.097 A reinvestigation of gas response to an ovally deformed gravitational potential.
T. Matsuda, M. Inoue, K. Sawada, E. Shima, K.–i. Wakamatsu.
Mon. Not. R. Astron. Soc., Vol. 229, No. 2, p. 295 – 314 (1987).
The response of gas to an ovally deformed gravitational potential or a weak bar of a disc galaxy is investigated by the implicit Osher upwind scheme with second–order of accuracy in space and a 180×180 polar grid. Flow patterns obtained are sensitive to the pattern speed and the degree of a distortion from the axisymmetric part of the gravitational potential. Tightly wound spiral patterns were obtained even with a non–axisymmetric perturbation whose amplitude was as low as 0.01. Not only trailing arms but also leading arms are obtained in the inner region of model galaxies. When the amplitude of the non–axisymmetric perturbation is 0.05, the authors obtained loosely wound spirals.

151.098 Dynamical friction and shells around elliptical galaxies.
C. Dupraz, F. Combes.
Astron. Astrophys., Vol. 185, No. 1/2, p. L1 – L4 (1987).
The authors compute the effects of dynamical friction on the radial distribution of a shell system around an elliptical galaxy, by using the Chandrasekhar formula. The scenario presented naturally explains the formation of shells down to the centre of

the E–galaxy. It is shown that the positions of the inner shells cannot be used to derive the age of the shell system and that explaining the radial distribution of a shell system does not require the E–galaxy to be surrounded by any dark matter. As a matter of fact, shell systems cannot be used to determine the mass distribution of ellipticals. For instance, the observed shell distribution of NGC 3923 may be accounted for in the frame of a simple model with no dark halo.

151.099   Dynamics and evolution of hot gas in early type galaxies.
M. Loewenstein.
Diss. Abstr. Int., Sect. B, Vol. 48, No. 1, p. 161–B (1987). Thesis, University of California, Santa Cruz, 71 pp. (1986). Order No. DA8709130.

151.100   The effects of H II regions and supernovae in proto–dwarf galaxies.
A. Noriega–Crespo.
Diss. Abstr. Int., Sect. B, Vol. 48, No. 1, p. 162–B (1987). Thesis, University of California, Santa Cruz, 169 pp. (1986). Order No. DA8709132.

151.101   Gravitational accretion of hot dark matter.
Y. H. Dobyns.
Diss. Abstr. Int., Sect. B, Vol. 48, No. 4, p. 1071–B – 1072–B (1987). Thesis, Princeton University, 108 pp. (1987). Order No. DA8714904.

151.102   Kinematics and dynamics of barred spiral galaxies.
M. N. England.
Diss. Abstr. Int., Sect. B, Vol. 48, No. 4, p. 1072–B (1987). Thesis, University of Florida, 324 pp. (1986). Order No. DA8715989.

151.103   N–body simulations of star clusters with initial binaries and the OB runaway stars.
P. J. T. Leonard, M. J. Duncan.
Bull. Am. Astron. Soc., Vol. 19, No. 2, p. 675 (1987). Abstract. – See Abstr. 010.061.

151.104   Two–dimensional orbits in rotating triaxial potentials.
S. J. Ratcliff, D. V. Guerra.
Bull. Am. Astron. Soc., Vol. 19, No. 2, p. 680 (1987). Abstract. – See Abstr. 010.061.

151.105   Dynamical activity at the center of a galaxy.
B. F. Smith, R. H. Miller.
Bull. Am. Astron. Soc., Vol. 19, No. 2, p. 681 (1987). Abstract. – See Abstr. 010.061.

151.106   New developments in the dynamical structure of clusters.
I. R. King.
Bull. Am. Astron. Soc., Vol. 19, No. 2, p. 687 (1987). Abstract. – See Abstr. 010.061.

151.107   Dissipationless collapse and the origin of triaxiality in elliptical galaxies.
L. Aguilar, D. Merritt.
Bull. Am. Astron. Soc., Vol. 19, No. 2, p. 713 (1987). Abstract. – See Abstr. 010.061.

151.108   Two–body relaxation in small star clusters.
S. L. W. McMillan, P. Hut, S. Casertano.
Bull. Am. Astron. Soc., Vol. 19, No. 2, p. 715 (1987). Abstract. – See Abstr. 010.061.

151.109   Stellar systems as dissipative dynamical systems.
V. G. Gurzadyan, A. A. Kocharyan.
Astrophysics, Vol. 26, No. 2, p. 169 – 174 (1987). English translation of 43.151.070.

151.110   Accretion of gaseous disks of galaxies. I. Influence of dynamical friction of the large–scale H$_2$ distribution.
O. K. Sil'chenko, V. M. Lipunov.
Astrophysics, Vol. 26, No. 2, p. 220 – 228 (1987). English translation of 43.151.071.

151.111   Formation and morphology of shell galaxies. II. Non–sperical potentials.
L. Hernquist, P. J. Quinn.
Space Telesc. Sci. Inst., Prepr. Ser., No. 184, 22 pp. (1987).

151.112   Tidal torques and local density maxima.
A. Heavens, J. Peacock.
Edinb. Astron. Prepr., No. 25/87, 32 pp. (1987). To appear in Mon. Not. R. Astron. Soc.

151.113   Hydrodynamics of the interstellar gas in colliding galaxies. I. Head–on collisions.
G. Mair, E. Müller, W. Hillebrandt, C. N. Arnold.
MPA Rep., No. 314, 38 pp. (1987). Submitted to Astron. Astrophys.

151.114   Escapes from stellar systems.
G. Contopoulos.
ESO Sci. Prepr., No. 517, 8 pp. (1987). To appear in IAU Colloq. No. 97.

151.115   Dynamical evolution of star clusters and the possible disruption of globular clusters in M87 by massive black holes from a dark corona.
R. Wielen.
Astron. Rechen–Inst. Heidelb., Prepr. Ser., No. 14, 3 pp. (1987). To appear in the proceedings of the 10th European Regional Astronomy Meeting of the IAU, Prague, August 1987.

151.116   On entropy and stellar systems.
H. Dejonghe.
Astrophys. J., Vol. 320, No. 2, p. 477 – 481 (1987).
The Tolman proof on the increase of entropy in a collisionless stellar system is discussed: the author emphasizes that it does not prove that generalized H–functions are monotonically increasing in time and that it cannot be invoked to justify an arbitrary convex H–function as a basis for a variational principle. On the other hand, there are entropies that can be given a statistical foundation: the Jaynes entropy is introduced and an application is discussed.

151.117   Kinetic equation of finite Hamiltonian systems with integrable mean field.
J. F. Luciani, R. Pellat.
J. Phys., Vol. 48, No. 4, p. 591 – 599 (1987). Abstr. in Phys. Abstr., Vol. 90, No. 1309, Entry 88554 (1987).

151.118   A hydrodynamic description of disk galaxies.
J. F. Sygnet, R. Pellat, M. Tagger.
Phys. Fluids, Vol. 30, No. 4, p. 1052 – 1058 (1987). Abstr. in Phys. Abstr., Vol. 90, No. 1309, Entry 88642 (1987).

151.119   Non linear coupling of galactic spiral modes in disc galaxies.
J. F. Sygnet, M. Tagger, E. Athanassoula, R. Pellat.
Inst. Astrophys. Paris, Pré–Publ., No. 210, 27 pp. (1987). To appear in Mon. Not. R. Astron. Soc.

151.120   Simulations of sinking satellites revisited.
D. Zaritsky, S. D. M. White.
Prepr. Steward Obs., No. 767, 12 pp. (1987). = Ariz. Theor. Astrophys. Prepr. #87–35.

151.121   Evolution of an ensemble of arbitrary orbits in gravitating systems in the presence of an irregular field.
B. S. Sagintaev, O. V. Chumak.
Tr. Astrofiz. Inst. Alma–Ata, Tom 47, p. 83 – 90 (1987). In Russian.

151.122 Model of a collisionless ellipsoid with inclined rotation: stability relative to ellipsoidal perturbations.
B. P. Kondrat'ev, E. A. Malkov.
Tr. Astrofiz. Inst. Alma–Ata, Tom 47, p. 91 – 103 (1987). In Russian.

151.123 Chaotic regions and invariant manifolds in models for elliptical galaxies.
A. Serrano P. G., A. González, J. F. Barral, E. Recillas–Cruz, A. Sarmiento G.
Rev. Mex. Astron. Astrofis., Vol. 14, No. 1, p. 80 (1987).
Abstract. – See Abstr. 012.042.

151.124 A dynamical explanation for the low frequency of ring galaxies.
T. K. Chatterjee.
Rev. Mex. Astron. Astrofis., Vol. 14, No. 1, p. 120 – 126 (1987). – See Abstr. 012.042.

An explanation for the low frequency of ring galaxies is investigated theoretically on the basis of the collisional theory, according to which the ring galaxy is the aftermath of a collision between a disk and another galaxy.

151.125 A dynamical study of the frequency of merging galaxies. I. Mergers involving spherical galaxies in a single-crossing time.
T. K. Chatterjee.
Astrophys. Space Sci., Vol. 137, No. 2, p. 267 – 279 (1987).

A study of the expected frequency of merging galaxies, involving spherical systems, is conducted, using the impulsive approximation. Results indicate that the expected frequency of such galaxies is several orders of magnitude smaller than the observational value, if mergers taking place only in a single crossing time are considered.

151.126 Spiral structure as a standing density–wave packet.
R. Meinel, G. Rüdiger.
Astrophys. Space Sci., Vol. 138, No. 1, p. 147 – 154 (1987).

Spiral density waves without radial group transport are discussed as a possible explanation of persistent global spiral patterns in galaxies without density–wave driving mechanism. The additional demand for a small winding–up rate favours two-armed patterns in most cases.

151.127 Properties of a spherical galaxy with exponential energy distribution.
K. Petrovay.
Astrophys. Space Sci., Vol. 138, No. 2, p. 323 – 332 (1987).

Some analytical relations for the phase space functions of a self-consistent spherical stellar system are derived. The integral constraints on the distribution function by imposing a given $\varrho(r)$ density distribution and $N(E)$ fractional energy distribution are determined. For the case of radially–anisotropic velocity distribution in the $E{\rightarrow}0$ limit the constraint by an exponential $N(E)$ implies that $f(E,J^2)$ tends to zero in the order $(-E)^{3/2}$. This lends analytical support to the use of the Stiavelli and Bertin (1985) distribution function for modeling elliptical galaxies. Maximum phase space density constraint confirms the necessity of high collapse factors to produce such a distribution function. Limits on the steepness of an exponential $N(E)$ for the case when $\varrho(r)$ resembles the emissivity law of ellipticals are also derived.

151.128 Corrigendum: "Unstable modes from galaxy simulations" [Mon. Not. R. Astron. Soc., Vol. 221, No. 1, p. 195 – 212 (1986)].
J. A. Sellwood, E. Athanassoula.
Mon. Not. R. Astron. Soc., Vol. 229, No. 4, p. 707 – 708 (1987). See Abstr. 42.151.012.

151.129 Self–consistent models of perfect triaxial galaxies.
T. S. Statler.
Astrophys. J., Vol. 321, No. 1, p. 113 – 152 (1987).

The self-consistent problem for the triaxial mass model known as the "perfect ellipsoid" is treated numerically. Schwarzschild's method is applied to 21 perfect ellipsoids of various axis ratios, using an unbiased catalog of 1065 orbits to match the density of the mass model at 240 points. The extent of the space of mathematically allowed solutions for each figure is mapped out (in one projection) by using linear programming to maximize linear combinations of the x and z components of the angular momentum. Lucy's iterative method is used to obtain smooth solutions in the interiors of the solution spaces.

151.130 The collapse of dense star clusters to supermassive black holes: binaries and gravitational radiation.
G. D. Quinlan, S. L. Shapiro.
Astrophys. J., Vol. 321, No. 1, p. 199 – 210 (1987).

A dense Newtonian cluster of compact stars (neutron stars or black holes) can evolve to a relativistic state in a Hubble time, at which point its core undergoes catastrophic collapse to a supermassive black hole. The authors construct a simple model for the dynamical evolution of such a cluster up to the onset of catastrophic collapse, incorporating the role of binary formation. Binaries form by dissipative two–body encounters and by nondissipative three–body encounters. Gravitational radiation, the dissipative agent, plays an important role in both processes. The simple model suggests that the dissipative processes significantly accelerate secular core collapse (the "gravothermal catastrophe") and increase the final core mass that can undergo catastrophic collapse.

151.131 Do H–functions always increase during violent relaxation?
S. Sridhar.
J. Astrophys. Astron., Vol. 8, No. 3, p. 257 – 262 (1987).

Recent work on the violent relaxation of collisionless stellar systems has been based on the notion of a wide class of entropy functions. A theorem concerning entropy increase has been proved. The author draws attention to some underlying assumptions that have been ignored in the applications of this theorem to stellar dynamical problems. Once these are taken into account, the use of this theorem is at best heuristic. The author presents a simple counter–example.

151.132 The statistical specific angular momentum of gas within a star cluster.
A. F. Illarionov.
Astron. Zh., Tom 64, Vyp. 6, p. 1176 – 1183 (1987). In Russian. English translation in Sov. Astron., Vol. 31, No. 6.

The value of the specific angular momentum of gas, which is formed and stored within a quasi–spherical, nonrotating cluster of stars losing their material, is evaluated. It is shown that in stationary clusters, the probability that the gas component of a particular cluster has some value of specific angular momentum is described by the Maxwellian distribution function. Mean regular shift and statistical dispersion of this distribution are calculated. The probability of finding a gas disc structure rotating around a black hole, which is possibly situated in the centre of the cluster, is evaluated.

151.133 The dynamics of small groups of galaxies. I. Virialized groups.
G. A. Mamon.
Astrophys. J., Vol. 321, No. 2, p. 622 – 644 (1987).

A numerical code is developed to follow the dynamical evolution of groups of galaxies, starting from virial equilibrium. The code assigns a single particle to each galaxy and to a diffuse intergalactic background, both with appropriately softened potentials, and explicitly incorporates many of the physical processes occuring in groups, such as collisional stripping, tidal stripping from the background mean field, dynamical friction on the background, mergers, and orbital braking. Groups of eight galaxies with surface densities similar to Hickson's compact groups are unstable against rapid galaxy merging. No dense groups of eight galaxies still appear compact (in Hickson's sense) after $1/2 t_{Hubble}$. These "instability" times are insensitive to the uncertainties of the physics used in the code, but are increased with larger group membership, and larger group mass–luminosity ratios.

151.134   Contraction of a cooling gravitating sphere.
L. P. Osipkov, V. N. Starkov.
Astron. Tsirk., No. 1458, p. 4 – 6 (1986). In Russian.
A system describing the collapse of a gravitating gas sphere and consisting of the Lagrange–Jacobi equation and a cooling law of the sphere has been solved numerically.

151.135   The stabilization of multiple open clusters and the relaxation time of the Galaxy.
E. M. Nezhinskij, L. P. Osipkov.
Astron. Tsirk., No. 1474, p. 1 – 2 (1987). In Russian.
The mass necessary for the stabilization of complexes of open star clusters is estimated. This mass corresponds to masses of giant molecular clouds. The relaxation time of stars on such clouds is evaluated.

151.136   The stability of complexes of open star clusters.
E. M. Nezhinskij, L. P. Osipkov.
Astron. Tsirk., No. 1474, p. 3 – 4 (1987). In Russian.
It is shown that complexes of open star clusters cannot be destroyed by external open clusters, by giant molecular clouds and by usual gas clouds.

151.137   On the dynamics of spherical non–stationary stellar systems.
V. M. Danilov, A. P. Ryazanov.
Astron. Tsirk., No. 1487, p. 3 – 4 (1987). In Russian.
The numerical solution of Vlasov's equation for a "water bag" spherical model is proposed. The high–energy stellar groups are formed in a system during stages of contraction.

151.138   A new kind of stellar orbit in a galactic potential.
J. Greiner.
Celest. Mech., Vol. 40, No. 2, p. 171 – 175 (1987).
Numerical integrations of a star's motion in an axisymmetric galactic potential have unveiled a new kind of orbit in the meridional plane. Especially, neutron stars can be expected to move on such orbits.

151.139   Potential and force due to thin disk and spherical galaxies.
T. K. Chatterjee.
Astrophys. Space Sci., Vol. 139, No. 2, p. 243 – 256 (1987).
The determination of the potential and force due to a thin exponential model disk galaxy and a polytropic $n = 4$ model spherical galaxy are conducted in the light of the study of collision dynamics.

151.140   Asymptotic properties of spherically symmetric self–gravitating mass systems for $t \to$.
J. Batt.
Transp. Theory Stat. Phys., Vol. 16, No. 4 – 6, p. 763 – 7798 (1987). Abstr. in Phys. Abstr., Vol. 91, No. 1320, Entry 10621 (1988).

151.141   Bound star clusters from gas clouds with low star formation efficiency.
F. Pinto.
Publ. Astron. Soc. Pac., Vol. 99, No. 621, p. 1161 – 1166 (1987).
The problem of change of radius of a dynamical system which suffers mass loss is addressed in the impulsive approximation. The results are framed into the general scenario of star–cluster formation. It is shown that the expansion factor R_{cl}/R_0 is not only a function of star formation efficiency ε, as in Hills (1980), but also of initial gas turbulence. A system can undergo collapse after sudden gas removal, and star–formation efficiencies smaller than 50% may produce bound systems. It is also shown how some recent findings in the theory of globular cluster formation, and some numerical calculations, strongly suggest the idea that the present generalization is an appropriate means of avoiding inconsistencies or nonphysical situations.

151.142   The vertical structure of galactic disks.
R. G. Carlberg.
Astrophys. J., Vol. 322, No. 1, p. 59 – 63 (1987).
Galactic disks have a nearly universal vertical structure that this paper suggests is substantially a result of internal evolution. Spiral density waves, the collective motions of disk stars and gas, are the dominant large–scale potential perturbations of the disk and are the effective source of heating for random motions in the plane of the galaxy. Spirals couple weakly to the relatively short–period vertical oscillations of stars. However, massive molecular clouds act as scattering centers that redistribute the energy between vertical and horizontal motions. For a fairly wide range of cloud parameters, the scattering is sufficient to produce the observed local velocity ellipsoid.

151.143   Globular cluster evolution in the Galaxy: a global view.
D. F. Chernoff, S. L. Shapiro.
Astrophys. J., Vol. 322, No. 1, p. 113 – 122 (1987).
The authors provide a framework for evaluating the relative importance of several of the most important competing physical effects that influence globular cluster evolution in the Galaxy. In particular, the authors treat (1) tidal heating of clusters in circular orbits about the Galactic center as they pass through the disk, (2) two–body relaxation and evaporation of the cluster across the tidal boundary, and (3) stellar evolution and mass loss by the stars within the cluster. The cluster evolution is traced to core collapse or tidal disruption using a three–parameter (energy, mass, and tidal radius) sequence of King models. The results suggest that if globular clusters are born, tidally limited, in a single burst of star formation, then only the most concentrated clusters survive the initial phase of mass loss by stellar evolution.

151.144   The evolution and final disintegration of spherical stellar systems in a steady galactic tidal field.
H. M. Lee, J. P. Ostriker.
Astrophys. J., Vol. 322, No. 1, p. 123 – 132 (1987).
The authors have followed the dynamical evolution of simplified N–body systems in a steady galactic tidal field using an orbit–averaged Fokker–Planck code through the core collapse and quasi–static reexpansion phases. Assuming spherical symmetry and two mass categories, single stars and tidally captured binaries, and allowing interactions of binaries with singles and other binaries, the authors find that the mass of the cluster decreases linearly during the reexpansion phase until the cluster totally disintegrates in a finite time.

151.145   Binary–galaxy dynamics. I. Noncontact systems.
D. A. Verner, A. D. Chernin.
Sov. Astron., Vol. 31, No. 2, p. 127 – 132 (1987). English translation of 43.151.057.

151.146   Stability of spherical gravitating collisionless systems.
V. L. Polyachenko.
Astrofizika, Tom 27, Vyp. 2, p. 295 – 309 (1987). In Russian. English translation in Astrophysics, Vol. 27, No. 2.
The stability of collisionless star clusters with various character of anisotropy of velocity distribution is investigated with the help of the common method using the reduction procedure (by reducing the stability problem of a spherical system to the analogous problem of perturbations of the simplest form for a corresponding cyclindrical system). For a spherical system immersed into a massive "halo" or containing a large central mass the equations of eigenfunctions and frequencies of oscillations (the integral ones in the simplest case) are derived.

151.147   The stability of a collisionless ellipsoid with oblique rotation.
B. P. Kondrat'ev, E. A. Malkov.
Astrofizika, Tom 27, Vyp. 2, p. 311 – 323 (1987). In Russian. English translation in Astrophysics, Vol. 27, No. 2.
The equations of non–linear oscillations of a collisionless ellipsoid with homogeneous density are derived. The stability of the model of an ellipsoidal stellar system with oblique rotation with respect to ellipsoid–ellipsoid perturbation is investigated.

The region of stability is defined by the determination of the characteristic frequencies of small oscillations the equations of which have been derived through linearization of the equations of non–linear oscillations.

151.148   Axisymmetric shell models with Stäckel potentials.
J. L. Bishop.
Astrophys. J., Vol. 322, No. 2, p. 618 – 631 (1987).

The author presents self-consistent models for axisymmetric stellar systems with Stäckel potentials which are composed of stars on infinitesimally thin short–axis tube orbits ("shell models"). For shell models, self-consistency is expressed as a one–dimensional integral equation. The solution of this integral equation for the distribution of orbits is unique. Self–consistent models are found for the "perfect" oblate spheroid, modified Hubble, modified Plummer, and modified Jaffe mass distributions.

151.149   Models of ring galaxies. II. Extended starbursts.
C. Struck–Marcell, P. N. Appleton.
Astrophys. J., Vol. 323, No. 2, p. 480 – 504 (1987).

Numerical models of the development of star-formation bursts in collisional ring galaxies are presented. The authors concentrate on target disks which have relatively high mean cloud mass and gas density. In such cases, even relatively low mass intruder galaxies are capable of triggering intense star-formation bursts in the density waves. Although the bursts are very short–lived in any individual gas element, pressure effects stimulate neighboring gas elements to burst, which can result in a sustained enhancement in the net star-formation rate.

151.150   Quasi–empiric determination of angular velocity of galactic spiral arms.
B. A. Balázs.
Star clusters and associations, p. 9 – 32 (1986). In Russian. – See Abstr. 012.081.

If the relevant parameters of the component subsystems of a galaxy have been established, including their rotation curves, surface density distributions, radial velocity dispersions and the number of spiral arms, then relying upon the density–wave theory of C. C. Lin one can, at least in principle, determine the geometrical characteristics of the spiral structure with a single free parameter: the angular velocity of the spiral pattern. It is found that on the usual IAU distance scale the angular speed of the galactic spiral pattern lies somewhere between 20 and 25 km s^{-1}/kpc and therefore the solar system is very close to the zone of co-rotation.

151.151   Tidal interaction between a disc galaxy and a spherical galaxy.
S. N. Zafarullah, K. S. Sastry.
Bull. Astron. Soc. India, Vol. 15, Nos. 2 – 3, p. 86 – 97 (1987).

Tidal interaction between a disc and a Plummer model spherical galaxy is considered. Fractional change in the internal energy and merger velocities for the disc–sphere pair is derived under impulsive approximation for various orientations of the disc. It is shown that in the model considered the chance of two spherical galaxies merging in a head–on collision due to tidal capture is greater than that of a disc–sphere pair.

Galactic dynamics.
See Abstr. 003.028.

Dynamical evolution of globular clusters.
See Abstr. 003.130.

Spiral shocks and accretion in discs.
See Abstr. 062.075.

The formation and evolution of tidal binary systems.
See Abstr. 065.002.

Hydrodynamic simulations of white dwarf–massive main–sequence star collisions in dense galactic nuclei.
See Abstr. 065.019.

Three–dimensional hydrodynamical simulations of stellar collisions. I. Equal–mass main–sequence stars.
See Abstr. 065.120.

How big are supermassive black holes formed from the collapse of dense star clusters?
See Abstr. 067.086.

Accretion disks in the centers of galaxies.
See Abstr. 067.097.

Cloud fluid models of gas dynamics and star formation in galaxies.
See Abstr. 131.017.

The large–scale motion of the ISM and the interaction with the system of stars.
See Abstr. 131.093.

Fluctuations in the ISM due to the gravitational interaction with the system of stars.
See Abstr. 131.094.

Supershells
See Abstr. 131.164.

Collisions of high–velocity clouds with a galactic disk.
See Abstr. 131.223.

Studies of dynamical properties of globular clusters. IV. Detailed structure of 47 Tucanae.
See Abstr. 154.051.

Internal dynamics of globular clusters: from our Galaxy to the Magellanic Clouds.
See Abstr. 154.052.

Destruction mechanisms for the globular clusters in the Milky Way.
See Abstr. 154.057.

Internal dynamics of globular clusters: from our galaxy to the Magellanic Clouds.
See Abstr. 154.074.

The structure of young star clusters in the Large Magellanic Cloud.
See Abstr. 154.083.

Collapsed cores and the structural parameters of old Large Magellanic Cloud star clusters.
See Abstr. 154.084.

Round dynamical models of the galactic halo.
See Abstr. 155.012.

A magnetodynamical model for the galactic center lobes.
See Abstr. 155.019.

Disk rotation curves in triaxial potentials.
See Abstr. 155.081.

Distant satellites as probes of our Galaxy's mass distribution.
See Abstr. 155.111.

A hypothesis of compensation of angular momenta and estimation of masses of binary galaxies.
See Abstr. 157.077.

Dynamics of a hot ($T \sim 10^7$K) gas cloud with volume energy losses.
See Abstr. 157.085.

The distribution and kinematics of stars formed from the cooling flow in M87.
See Abstr. 157.112.

A lower limit to the mass of neutral leptons as constituents of the dark halo around galaxies.
See Abstr. 157.116.

The stellar kinematics of elliptical galaxies.
See Abstr. 157.151.

Formation of stellar shells and X–ray coronae around elliptical galaxies.
See Abstr. 157.197.

Evolution of hot galactic flows.
See Abstr. 157.198.

On the thermal instability of galactic and cluster halos.
See Abstr. 157.199.

Origin and evolution of compact elliptical galaxies.
See Abstr. 157.212.

Steady state cooling flow models for normal elliptical galaxies.
See Abstr. 157.217.

Are elliptical galaxies really round?
See Abstr. 157.230.

The formation of the exponential disk in spiral galaxies.
See Abstr. 157.273.

Observable properties of E0 triaxial galaxies: a test for triaxiality.
See Abstr. 157.274.

Stellar dynamics of radio elliptical galaxies.
See Abstr. 158.138.

Galaxies as dynamical probes of cluster structure.
See Abstr. 160.046.

Can cD galaxies be formed in clusters with anisotropic distribution functions?
See Abstr. 160.048.

Comparisons of the spatial distribution of Abell clusters against models with Gaussian initial conditions.
See Abstr. 160.115.

Galaxy formation by gravitational collapse.
See Abstr. 161.002.

The coalescence criterion of self–gravitating gaseous masses.
See Abstr. 161.037.

Dissipationless formation of elliptical galaxies.
See Abstr. 161.156.

Dissipation and the formation of galaxies.
See Abstr. 161.157.

152 Stellar Associations

152.001 Properties of blue stragglers in young OB associations.
G. Mathys.
Astron. Astrophys., Suppl. Ser., Vol. 71, No. 2, p. 201 – 219 (1987).

The properties of the blue stragglers (BS) belonging to the associations Sco OB1, Cyg OB1, Car OB1, Cen OB1, and Per OB1 as well as of the early O stars belonging to Cas OB6 are studied on the basis of an extensive survey of the literature. Eleven of the thirteen BS for which the relevant information exists show an enhanced atmospheric abundance of N, which is interpreted as supporting the view (Maeder, 1987) that BS are probably quasi–homogeneously evolved stars where the products of the CNO cycle have become observable at the surface. The mechanism inducing the turbulent diffusion responsible for the homogenization of the stellar interior does not clearly appear from the existing data. Tidal forces in close binaries may play a role in some cases, but except in the case of Per OB1 there is no convincing evidence that single BS have abnormally large rotational velocities. Finally the scenario of quasi–homogeneous evolution of the BS is not in contradiction with the observed properties of their likely descendents, the Wolf–Rayet stars, which are present in the associations under consideration.

152.002 Surface distribution according to the spectral type of stars in the region of the Cygnus OB4 association.
Ts. S. Radoslavova.
Dokl. Bolg. Akad. Nauk, Tome 40, No. 1, p. 7 – 9 (1987). Abstr. in Ref. Zh., 51. Astron., 10.51.859 (1987).

152.003 Early type stars in the direction of the association Vulpecula OB4.
T. S. Radoslavova.
Star clusters and associations, p. 75 – 82 (1986). – See Abstr. 012.081.

On the basis of observational material obtained on the 70–cm meniscus telescope of the Abastumani Astrophysical Observatory provided with an 8°–objective prism (dispersion 166 Å/mm at Hγ), the early–type stars in a region of 60 square degrees centered in the association Vul OB4 are identified and their spectral classification is performed.

152.004 New flare stars in the region of Mon I association.
E. S. Parsamian (*Eh. S. Parsamyan*), L. Rosino, O. S. Chavusian (*O. S. Chavushyan*).
Star clusters and associations, p. 123 – 127 (1986). – See Abstr. 012.081.

The first flare stars in the region of the Mon I association (NGC 2264) were discovered by Rosino et al.; Wenzel and Haro and Chavira. Spectral and photometric studies of NGC 2264 have shown the region is rich in variable and T Tau type stars giving every reason to expect that new flare stars can be found in this region.

Synthetic uvby–β photometry of HD 12856 and HD 13890.
See Abstr. 112.110.

The spectacular O stars.
See Abstr. 114.043.

Surface distribution according to the spectral type of the stars in the region of the Cygnus OB4 association.
See Abstr. 114.121.

M20–04: a T Tauri star.
See Abstr. 121.025.

Flares of Orion Population variables in the association Taurus T3.
See Abstr. 122.114.

Non–spherical supernova remnants. IV. Sequential explosions in OB associations.
See Abstr. 125.018.

Complex investigation of the star formation region in the Orion constellation.
See Abstr. 131.023.

Large–scale radio brightness distribution of the ionized gas in the Canis Major OB1 association.
See Abstr. 131.084.

Gas–and–dust complex NGC 7822 + S 171 connected with the Cep OB4 association.
See Abstr. 131.115.

Variations in UV extinction in galactic associations and perpendicular to the galactic plane.
See Abstr. 131.147.

Star–forming loops in the *IRAS* sky images.
See Abstr. 131.154.

Local interstellar medium.
See Abstr. 131.243.

A study of clumping in the Cepheus OB3 molecular cloud.
See Abstr. 131.282.

The kinematical and binary properties of association and field O stars.
See Abstr. 155.007.

The distribution of OB stars and dust in a Milky Way field at $(l, b) = (335°, 0°)$.
See Abstr. 155.084.

Star complexes and associations in the Andromeda galaxy.
See Abstr. 157.053.

Stellar associations and complexes in M33.
See Abstr. 157.090.

Associations in M31 and M33.
See Abstr. 157.348.

153 Open Clusters

153.001   The distribution of rotational velocities for low–mass stars in the Pleiades.
J. R. Stauffer, L. W. Hartmann.
Astrophys. J., Vol. 318, No. 1, p. 337 – 355 (1987).
The authors present new observations which extend their previous photometric and spectroscopic studies of late–type members of the Pleiades. The presence of a large number of late–type stars with $v \sin i \approx 100 \text{ km s}^{-1}$, and the absence of comparable rotational velocities in both younger and older groups of stars, indicate that low–mass stars spin up appreciably during pre–main–sequence contraction, and spin down very rapidly once on the main sequence. The authors show that the wide range of angular momenta exhibited by Pleiades K and M dwarfs can result from a plausible spread in initial angular momenta, coupled with initial main–sequence spin–down rates that are only weakly dependent on rotation.

153.002   Photometry and spectroscopy of stars in the region of a highly reddened cluster in Ara.
B. E. Westerlund.
Astron. Astrophys., Suppl. Ser., Vol. 70, No. 3, p. 311 – 324 (1987).
VRI photographic photometry is presented for 258 stars in the region of the cluster Wd1 in Ara together with near–infrared spectrophotometry of the brightest stars. The brightest member stars of the cluster are shown to be of spectral types B2 Ia to M2 Ia, forming a well defined sequence of supergiants with the maximum visual luminosity being reached in classes A2 and G0. One of the stars is of type Be with an extremely extended shell. Weak hydrogen emission can be traced in the spectra of a number of the stars as well as in between the stars. The interstellar absorption is very heavy, reaching $A(V) = 10$ mag; the surrounding field has an average absorption of about 3 mag. A redetermination of the distance of the cluster leads to a value of about 5 kpc as most likely. The age of the cluster is estimated to about seven million years.

153.003   Membership of Cepheids and red giants in 8 open clusters: NGC 129, 6067, 6087, 6649, 6664, IC 4725, Ly 6, Ru 79.
J. C. Mermilliod, M. Mayor, G. Burki.
Astron. Astrophys., Suppl. Ser., Vol. 70, No. 3, p. 389 – 407 (1987).
The membership of 8 cluster Cepheids, DL Cas, EV Sct, V367 Sct, S Nor, TW Nor, V340 Nor, U Sgr and CS Vel, has been examined by comparing their systemic radial velocity with that observed for the red giants in the same clusters. The membership is very probable in 7 cases, the only–exception being CS Vel. Additional observations of main sequence stars are necessary to confirm some of the conclusions. One spectroscopic binary was discovered among the Cepheids (DL Cas) and at least four were among the red giants. Orbits have been determined for two red giants, one in NGC 129 and one in IC 4725. The position within the instability strip of the non variable F5 Ib star in NGC 129 is due to its probable composite (gK + dB) character. The ratio of the number of Cepheids to the number of red giants is less than 1, in good agreement with the prediction of evolutionary models with core overshooting. This study is based on 382 observations of 45 stars, obtained with the radial velocity scanner CORAVEL.

153.004   Weiße Zwerge in Sternhaufen: ein Schlüssel zur Bestimmung der Anfangs–Endmassen–Beziehung in der Sternentwicklung.
V. Weidemann.
Astron. Raumfahrt, 25. Jahrg., Heft 4, p. 98 – 100 (1987).

153.005   Membership in the young cluster Trumpler 37.
L. A. Marschall, W. F. van Altena.
Astron. J., Vol. 94, No. 1, p. 71 – 83 (1987). With plates 5 – 12.
Astrometric positions for 1387 stars and proper motions for a subset of 1135 stars brighter than $V = 15$ in a 1°5 field surrounding the young open cluster Trumpler 37 are presented.

Membership probabilities are presented for those stars with measured proper motions, and 486 stars with probabilities of >80% are identified.

153.006 Washington photometry of open cluster giants: seven metal–poor anticenter clusters.
D. Geisler.
Astron. J., Vol. 94, No. 1, p. 84 – 91 (1987).

Photometry on the revised Washington system is presented for 75 stars in seven open clusters in the Galactic anticenter. The temperature, a luminosity criterion, and two independent abundance indices are derived for each star. The spread in the individual C–M abundances in NGC 2158 and 2204 suggests that CN–strength variations exist among the giants in these clusters, corroborating earlier work. The mean cluster abundances are [A/H] = −1.3±0.3 for NGC 2112, −0.63±0.15 for NGC 2141, −0.88±0.10 for NGC 2158, −0.47±0.10 for NGC 2204, −0.93±0.15 for NGC 2243, −1.2±0.5 for Tombaugh 2, and −0.46±0.10 for NGC 2506. Four of these open clusters are found to have metallicities as low as, or lower than, 47 Tuc. Both NGC 2112 and Tombaugh 2 are exceptional clusters with regards to their age, galactic location, and extreme metal–poorness. A comparison of their respective age–metallicity relations shows that the chemical–enrichment history of the Galactic anticenter clusters was much more like that experienced in the Large Magellanic Cloud than that which occurred in the solar neighborhood disk.

153.007 Photometric study of the southern open cluster IC 2488.
M. Pedreros.
Astron. J., Vol. 94, No. 1, p. 92 – 98 (1987). With plate 13.

A photometric study of the southern open cluster IC 2488 is presented. A distance modulus $V_0-M_V = 10.80 \pm 0.18$ mag and a color excess E(B–V) = 0.26±0.02 are determined for the cluster by fitting the observed main sequence to an empirical isochrone by means of a previously developed computer code. The location of a concentration of presumed red giant stars in the color-magnitude diagrams in conjunction with the empirical isochrones are used to estimate an age of 1.0×10^8yr for the cluster. The cluster membership status of these red stars, along with a possible blue straggler and a nearby planetary nebula, are analyzed.

153.008 The binary cepheid DL Cas and the open cluster NGC 129.
H. C. Harris, D. L. Welch, R. P. Kraft, E. G. Schmidt.
Astron. J., Vol. 94, No. 2, p. 403 – 408 (1987).

The cepheid DL Cas and the nonvariable supergiant Star A are confirmed as members of NGC 129, and both are found to be spectroscopic binaries with orbital periods of a few years. Periods and orbits are determined for both stars. Star A is apparently rotating; tidal interaction with its companion is discussed and then rejected as the probable cause of the rotation. The membership of other stars and the velocity of the cluster are redetermined.

153.009 Study of interstellar extinction in some young open clusters.
R. Sagar.
Mon. Not. R. Astron. Soc., Vol. 228, No. 2, p. 483 – 499 (1987).

Interstellar extinction has been studied in 15 open clusters, based on reliable cluster members and precise observational data. Out of these, 10 show non–uniform extinction across the cluster region. Most of these show random variation of colour excess over the cluster face except NGC 6530 and 6611, where a systematic spatial variation of reddening is observed. The scatter in colour excess does not depend upon the spectral class between O and K. Only in some of the young clusters (age ⩽ 5 × 10⁶yr) the variation of $E(B–V)$ correlates with luminosity and spectral class, in the sense that brighter cluster members are more highly reddened.

153.010 Spectroscopic and photometric observations in Bochum 10 and 11.
M. P. FitzGerald, S. Mehta.
Mon. Not. R. Astron. Soc., Vol. 228, No. 3, p. 545 – 555 (1987).
= Contrib. Univ. Waterloo Obs. No. 132.

Bochum 10, C 1040–588, is a small loose cluster of OB stars with 21 probable members, including 15 OB stars of which 11 are classified O9.5 – B2 and another B9 Iab; further members are unlikely. Bochum 11, C 1045–598, is a small open cluster with 15 known members, of which seven are classified O8 – B2. The authors present new photometry for 16 members and 18 non-members, and MK spectra for 19 members and nine non-members. The properties of the clusters, based on statistical zero-age main sequence fitting and spectroscopic parallaxes are discussed in detail.

153.011 CCD photometry of galactic clusters containing Cepheid variables. V. Ruprecht 79.
A. R. Walker.
Mon. Not. R. Astron. Soc., Vol. 229, No. 1, p. 31 – 40 (1987).

A colour–magnitude diagram is given for the galactic open cluster Ruprecht 79. Of the 161 stars measured, 77 are considered to be cluster members. The absorption–corrected distance modulus is 12.55±0.16 mag, this assumes $R = 3.05$ and $E(B–V) = 0.794$ for the B stars and is referenced to a Pleiades modulus of 5.57 mag. If the Cepheid CS Vel is a cluster member then it has $<M_v> = -3.24$ and $<B_0>-<V_0> = 0.61$.

153.012 A catalogue of some observational data and elements of galactic orbits of open star clusters.
K. A. Barkhatova, S. A. Kutuzov, L. P. Osipkov.
Astron. Zh., Tom 64, Vyp. 5, p. 956 – 964 (1987). In Russian. English translation in Sov. Astron., Vol. 31, No. 5.

The thoroughly selected values of heliocentric distances, absolute proper motions and radial velocities are listed for 69 open clusters. Galactocentric coordinates and velocities of these clusters are given. Elements of their galactic orbits are found by numerical integration of equations of motion.

153.013 Cluster analysis of young open clusters.
A. M. Ehjgenson, O. S. Yatsyk.
Astron. Zh., Tom 64, Vyp. 5, p. 965 – 979 (1987). In Russian. English translation in Sov. Astron., Vol. 31, No. 5.

Cluster analysis methods are used to consider the galactic distribution of 224 open clusters with age up to 10⁸years. The majority of the clusters is shown to enter condensations with characteristic dimensions of a few hundred parsecs. Some condensations, though being undistinguishable by their density from other ones, have such close values of age, integrated colour and radial velocity of their components that it is impossible to explain this by chance. It means that every such condensation is a physical entity and consists of clusters presumably connected by common origin.

153.014 An estimation of the distance of the Hyades using a geometric method.
A. V. Loktin, N. V. Matkin, V. V. Fedorov.
Astron. Zh., Tom 64, Vyp. 5, p. 1114 – 1116 (1987). In Russian. English translation in Sov. Astron., Vol. 31, No. 5.

The distance of the Hyades cluster is evaluated with the use of a geometric method. The value of the distance modulus of that cluster is equal to $3^m42 \pm 0^m10$.

153.015 Seven–colour photoelectric photometry of bright stars in the open cluster IC 4756.
U. Dzĕrvitis.
Nauchn. Inf., Vyp. 63, p. 36 – 45 (1987). In Russian.

Results of a photoelectric photometry in the Vilnius seven-colour photometric system of 45 stars in the open cluster IC 4756 are given. The obtained quantities have been used to determine individual reddenings, spectral classes; and absolute magnitudes of red giants in the cluster. By combining photometrical and kinematical membership criteria it is concluded that 16 G–K giants are members of the cluster.

153.016   Be stars as members of open clusters.
A. Feinstein.
Physics of Be stars, p. 500 – 502 (1987). – See Abstr. 012.030 (IAU Colloq. No. 92).
A list of 124 Be–type stars belonging to 52 open clusters has been compiled. All of them have photometric UBV data, and many of them spectral classification in the MK system. Besides, the cluster's distance modulus and the mean color excess from the member stars are known. Then, the author has computed the absolute magnitude and the intrinsic colors of each Be star.

153.017   On the percentage of Be stars in galactic open clusters.
A. Reitermann, J. Krautter, B. Wolf, B. Baschek.
Physics of Be stars, p. 503 – 504 (1987). – See Abstr. 012.030 (IAU Colloq. No. 92).

153.018   Variable Be stars in h and χ Persei.
C. L. Waelkens, P. Lampens, J. Cuypers, J. Denoyelle, D. Heynderickx, F. Rufener, P. Smeyers.
Physics of Be stars, p. 505 – 508 (1987). – See Abstr. 012.030 (IAU Colloq. No. 92).

153.019   Strömgren photometry of open clusters. II. NGC 3532.
H. Schneider.
Astron. Astrophys., Suppl. Ser., Vol. 71, No. 1, p. 147 – 152 (1987).
Strömgren photometry of B– and A–type stars in the southern open cluster NGC 3532 up to $V = 11.5$ mag is presented. The membership of the observed stars is discussed and a reddening of $E(b-y) = 0.025$ for the cluster is estimated. Also, the occurrence of chemically peculiar early type (CP1 and CP2) stars is examined.

153.020   Photoelectric search for CP2–stars in open clusters. X. NGC 2232, NGC 2343, Cr 140, and Tr 10.
H. Jenkner, H. M. Maitzen.
Astron. Astrophys., Suppl. Ser., Vol. 71, No. 2, p. 255 – 261 (1987).
100 stars in the regions of the four open clusters NGC 2232, NGC 2343, Cr 140, and Tr 10 were measured in the Δa–system (Maitzen, 1976) in order to detect photometric peculiarity indicating CP2–stars. Of the total of 63 cluster member and probable member stars, only one turned out to be photometrically peculiar (NGC 2232–9), while Cr 140–60 is peculiar, but most likely a non–member. Tr 10–19, a spectroscopically peculiar star according to one source (in disagreement with another), appears to be normal in our observations.

153.021   Photoelectric search for CP2–stars in open clusters. XI. NGC 3532 and NGC 5662.
H. M. Maitzen, H. Schneider.
Astron. Astrophys., Suppl. Ser., Vol. 71, No. 3, p. 431 – 440 (1987).
Δa–photometry was carried out for 164 stars in NGC 3532 (the most populous cluster studied so far in this series) and 27 stars in NGC 5662. Surprisingly, only one mildly peculiar star (Koelbloed 232) was identified this way among the roughly hundred cluster members of NGC 3532. Three photometrically peculiar objects were found among the approximately 60 field stars in the cluster area. This is about the average frequency of CP2–stars, in contrast to the pronounced shortage of CP2–stars in the rather old cluster NGC 3532. In NGC 5662 two CP2–objects, MV6 and MV10, were identified. MV6, however, is a probable non–member based on its proper motion data. As in some previous cases again a negative Δa–object was found with reported Be–characteristics (NGC 5662–MV29).

153.022   Photoelectric search for CP2–stars in open clusters. XII. Alpha Persei, Praesepe and NGC 7243.
H. M. Maitzen, K. Pavlovski.
Astron. Astrophys., Suppl. Ser., Vol. 71, No. 3, p. 441 – 448 (1987).
The authors observed 95 stars in the open clusters Alpha Persei, Praesepe and NGC 7243 by Δa–photometry in order to search for the presence of the $\lambda5200$ feature of CP2 (or CP4) stars. Although peculiarity has been claimed for half a dozen stars in α Persei, only the CP4 star (= He–weak) HL 985 was detected among the cluster members. The behaviour of 3 Be/shell stars (HL 861, 1164, 904) concerning the $\lambda5200$ feature is discussed. Praesepe presents a very new and special case: 3 stars, previously classified as Am have Δa–values around 0.020 mag i.e. they exhibit photometric peculiarity like CP2–stars. NGC 7243 exhibits an outstanding high frequency of CP2–stars: two certain Δa–peculiar stars were found: L 370, already known as peculiar, and the newly identified L 114. Two stars are near the detection level: L 58 and 121. According to Geneva photometry L 487 is markedly peculiar.

153.023   Strömgren photometry of open clusters. III. NGC 2323, NGC 5662.
H. Schneider.
Astron. Astrophys., Suppl. Ser., Vol. 71, No. 3, p. 531 – 537 (1987).
Strömgren photometry of B– and A–type stars in the southern open clusters NGC 2323 and NGC 5662 up to V = 11.5 mag is presented. The membership of the observed stars is discussed and a mean reddening of $E(b-y) = 0.186$ and $E(b-y) = 0.234$ for the clusters is estimated, respectively. In both cases variable reddening across the cluster area is found. Furthermore, the occurrence of chemically peculiar early type (CP2) stars is examined.

153.024   The Pleiades cluster. IV. The visit of a molecular CO cloud.
M. Breger.
Astrophys. J., Vol. 319, No. 2, p. 754 – 759 (1987).
The location, size, and mass of the CO molecular cloud seen in the direction of the Pleiades cluster is determined from a study of the polarization and reddening of cluster members and non–members. Arguments are presented against both a foreground and background location of the molecular cloud, so that the cloud should be presently situated inside the cluster. The extinction determinations for cluster members and background stars indicate a mass of $20\,M_\odot$ for the CO cloud visiting the Pleiades cluster.

153.025   uvbyHβ photoelectric and CCD photometry of IC 4651.
B. J. Anthony–Twarog, B. A. Twarog.
Astron. J., Vol. 94, No. 5, p. 1222 – 1236 (1987). With plates 102 – 103.
$uvbyH\beta$ photoelectric and CCD photometry of the old, open southern cluster IC 4651 is presented and analyzed. The combined data lead to the following cluster characteristics: $E(b-y) = 0.064 \pm 0.006$, $[Fe/H] = +0.23 \pm 0.02$, $(m-M)_0 = 9.83 \pm 0.04$, and an age on the system of VandenBerg (1985) of $2.4 \pm 0.2 \times 10^9$yr. The cluster color–magnitude diagram exhibits some scatter, all of which can be explained by binaries, non–members, and a blueward hook due to the hydrogen–exhaustion phase. The cluster turnoff shows marginal evidence for the presence of bimodality found in NGC 752, a younger and more metal–poor cluster. It is speculated that the bimodality is the result of nonstandard evolution near the Böhm–Vitense gap.

153.026   Photometric study of the southern open clusters NGC 5316 and NGC 6124.
M. Pedreros.
Astron. J., Vol. 94, No. 5, p. 1237 – 1243 (1987). With plate 104.
New UBV photoelectric photometry was obtained for the open clusters NGC 5316 and NGC 6124, both containing red giant stars. Reddening–line slopes were estimated for both fields and were used to derive the individual reddenings of cluster stars. A fitting to an evolved main sequence was performed on the reddening–free colors and magnitudes of cluster members. In general, larger color excesses and distance moduli are found in this work than obtained in earlier investigations.

153.027   Polarimetry of stars in Melotte 15.
H. H. Guetter, F. J. Vrba.
Bull. Am. Astron. Soc., Vol. 19, No. 2, p. 677 (1987). Abstract. – See Abstr. 010.061.

153.028   A BV photographic survey of the old open cluster NGC 3680.
B. J. Anthony–Twarog, E. Heim, B. A. Twarog.
Bull. Am. Astron. Soc., Vol. 19, No. 2, p. 677 (1987). Abstract. – See Abstr. 010.061.

153.029   51 Tauri and the Hyades distance modulus.
D. M. Peterson, R. Solensky.
Bull. Am. Astron. Soc., Vol. 19, No. 2, p. 707 – 708 (1987). Abstract. – See Abstr. 010.061.

153.030   The tidally circularized binaries in open clusters: a new clock for age determination.
R. D. Mathieu, T. Mazeh.
Bull. Am. Astron. Soc., Vol. 19, No. 2, p. 714 (1987). Abstract. – See Abstr. 010.061.

153.031   Interstellar extinction in Trumpler 37. Infrared results.
M. Roth.
Bull. Am. Astron. Soc., Vol. 19, No. 2, p. 724 (1987). Abstract. – See Abstr. 010.061.

153.032   DDO photometry of giants in the open cluster NGC 2660.
J. E. Hesser, G. H. Smith.
Publ. Astron. Soc. Pac., Vol. 99, No. 620, p. 1044 – 1049 (1987).
DDO photometry is presented for giants in the open cluster NGC 2660. A reddening of $E(B-V) = 0.35$ is inferred from the observations, as well as a metallicity which is slightly below solar at $[Fe/H] = -0.4$. An apparent distance modulus of $(V-M_v) = 13.2$ is derived for the cluster. Revised DDO colors are also given for five giants in the intermediate–age cluster NGC 2477.

153.033   On the distances to the young open clusters NGC 2244 and NGC 2264.
M. R. Pérez, P. S. Thé, B. E. Westerlund.
Publ. Astron. Soc. Pac., Vol. 99, No. 620, p. 1050 – 1066 (1987).
A new determination of the distances of the young open clusters NGC 2244 and NGC 2264 is presented. It is based on distance moduli for individual OB–type stars in which the influence of the anomalous ratio $(R = A_v/(E(B-V))$ of total to selective extinction, if any, is taken into account. A distance modulus of $11^m11 \pm 0.16$ has been derived for NGC 2244 and of $9^m88 \pm 0.17$ for NGC 2264, corresponding to distances of 1670 ± 125 pc and 950 ± 75 pc, respectively. Cluster color excesses, $E(B-V)$, were found to be 0^m48 for NGC 2244 and 0^m06 for NGC 2264. In NGC 2264 most of the OB sample stars follow an irregular extinction law.

153.034   Radial velocity for the open cluster NGC 2420.
T. Liu, K. A. Janes.
Publ. Astron. Soc. Pac., Vol. 99, No. 620, p. 1076 – 1079 (1987).
The authors present radial velocities for the bright stars, nine of them giants, in the old open cluster NGC 2420 from high-resolution spectra obtained with the 2.1–meter coudé spectrograph at KPNO. The mean cluster velocity is found to be 70.0 ± 1.4 (rms) km s^{-1} based on these velocity measurements. Among the ten stars observed, two have been confirmed to be nonmembers from their radial–velocity data. In addition, two barium stars in the sample have shown variable velocities when compared with previous observations.

153.035   A spectroscopic study of the open cluster NGC 2281.
J. W. Glaspey.
Publ. Astron. Soc. Pac., Vol. 99, No. 620, p. 1089 – 1092 (1987).
Photographic spectra of 16 stars in NGC 2281 have been used to derive a mean $E(B-V) = 0.11$ mag and a true distance modulus of $(m-M)_0 = 8.3$. Radial velocities of most of the same

stars were derived from low–resolution CCD spectra to yield a mean cluster velocity of $+5$ km s^{-1}. Rotational velocities estimated from low–dispersion photographic spectra yield a distribution of the mean $v \sin i$'s as a function of absolute magnitude typical of intermediate–age open clusters. The latter result suggests that stellar evolution effects largely eliminate differences in the cluster mean rotational velocity distributions by the time clusters are between $50 – 100 \times 10^6$ years old.

153.036   Photoelectric search for CP2–stars in open clusters. X: NGC 2232, NGC 2343, Cr 140, and Tr 10.
H. Jenkner, H. M. Maitzen.
Space Telesc. Sci. Inst., Prepr. Ser., No. 186, 16 pp. (1987). To appear in Astron. Astrophys., Suppl. Ser.

153.037   Some results based on existing photometric data in the open cluster NGC 752.
C. Schalén.
Rep. Obs. Lund, No. 20, 14 pp. (1987).
Some existing determinations of magnitudes and colours in the open cluster NGC 752 are compared. On the basis of colours and spectral types, the colour excess at different angular distance from the centre is studied. No reddening effect can be traced.

153.038   On the dynamics of open star clusters in the Galaxy. II.
T. S. Kozhanov.
Tr. Astrofiz. Inst. Alma–Ata, Tom 47, p. 3 – 11 (1987). In Russian.

153.039   Binary frequencies in the open cluster IC 2602.
H. Levato, C. Hernández, N. Morrell, B. García.
Rev. Mex. Astron. Astrofis., Vol. 14, No. 2, p. 414 – 415 (1987). – See Abstr. 012.042.
The authors present radial velocity measures for 25 stars in the field of IC 2602. They discuss the relation between binary contents and average projected axial rotation for cluster's members.

153.040   A photometric study of short period variable stars in open clusters.
R. Peniche, J. H. Peña.
Rev. Mex. Astron. Astrofis., Vol. 14, No. 2, p. 420 – 422 (1987). – See Abstr. 012.042.
The photoelectric photometry of known Delta Scuti variable stars in selected open clusters (Coma, Praesepe, Pleiades, α Per, and NGC 2264) was carried out in order to determine the periods of pulsation. Multicolor uvby–β photometry and a search for new variables in other open clusters (NGC 2539, NGC 6494, NGC 6882–5, NGC 7062, NGC 7063 and NGC 7686) were also performed.

153.041   The open cluster NGC 6193: spectral types, axial rotation and binarity.
M. Arnal, H. Levato, B. García, N. Morrell.
Rev. Mex. Astron. Astrofis., Vol. 14, No. 2, p. 423 (1987). Abstract. – See Abstr. 012.042.

153.042   The distance and reddening of the open cluster NGC 7790 and the luminosity of its three Cepheids.
J. M. Alcalá, A. Arellano Ferro.
Rev. Mex. Astron. Astrofis., Vol. 14, No. 2, p. 440 (1987). Abstract. – See Abstr. 012.042.

153.043   Study of structural and dynamical characteristics of open clusters (OCl). I. Results of star counts in 50 OCl's.
V. M. Danilov, A. F. Seleznev, E. Yu. Gurto, E. A. Lapina.
Kinematika Fiz. Nebesn. Tel, Tom 3, No. 6, p. 77 – 83 (1987). In Russian. English translation in Kinematics Phys. Celest. Bodies.
Star counts in 50 open clusters up to $B = 16^m$ are carried out. Limiting radii and numbers of stars for the OCls are obtained. Parameters of the King distribution for surface star densities in the clusters, relaxation times, and tidal radii of the clusters are also determined. The estimates of concentrations of molecular

clouds in the neighbourhood of the clusters under consideration are obtained by means of the CO molecular emission line catalogue.

153.044   Veränderliche Sterne in offenen Sternhaufen.
W. Götz.
Sterne, 63. Band, Heft 5, p. 254 – 263 (1987).

153.045   Small clusters or multiple stars?
L. O. Lodén.
Ir. Astron. J., Vol. 18, No. 2, p. 95 – 97 (1987). – See Abstr. 012.036.

The classical opinion that multiple star systems and open clusters are entirely separate phenomena has been reconsidered. No absolutely definite conclusions have been drawn but it is shown that from an observational point of view there is no well-defined border line between small open clusters and wide multiple systems.

153.046   A study of faint young open clusters as tracers of spiral features in our Galaxy. Paper 3: Collinder 97 (OCl 506).
G. S. D. Babu.
J. Astrophys. Astron., Vol. 8, No. 3, p. 219 – 226 (1987).

Photoelectric and photographic photometry of twenty–nine stars was done in the field of the open cluster Collinder 97 ≡ OCl 506. Of these stars, a total of twenty–four have been found to be possible members. There is apparently no interstellar extinction in the direction of this cluster. This cluster is situated at a distance of 0.63 ± 0.01 kpc, which is well within the local arm of our Galaxy. The age of this cluster is in the range of 1×10^8 to 5.9×10^8 yr, which puts it in an older age group. Thus, it cannot be specifically considered as a spiral–arm tracer in the study of our Galaxy.

153.047   Lithium in the Coma star cluster.
A. M. Boesgaard.
Astrophys. J., Vol. 321, No. 2, p. 967 – 974 (1987).

The F dwarfs in the Hyades cluster show large deficiencies of Li at the temperatures of the middle F stars. Observations have now been made of Li in the A and F dwarfs of the Coma cluster at high spectral resolution (0.1 Å) and high signal–to–noise ratios (180–450) with the coudé spectrograph of the CFH telescope. Abundances of both Li and Fe were determined by a model atmosphere abundance analysis. The two Coma stars that have temperatures in the middle of the Hyades Li "chasm" are very depleted in Li. The Hyades empirical curve seems to fit the Coma stars. The five Am stars show real differences relative to one another in both Li content and "metallic–line–ness". One Am star shows Li enhanced by a factor of 4. If the Coma cluster "initial" Li content is like that of the two late F stars which show the highest Li, it is log N(Li) = 2.9.

153.048   Absolute magnitude calibration from H–beta photometry of open star clusters.
J.–C. Mermilliod.
Bull. Inf. Cent. Données Stellaires, No. 33, p. 41 – 49 (1987).

The (M_V, H_β) calibrations are briefly reviewed and compared. The H_β data available for each cluster have also been intercompared and the results are given in a tabular form. Most zero point differences are smaller than 0.010 and the dispersion over the mean is comprised between 0.005 and 0.015. This means that erratic errors of several hundredths of magnitude do exist, resulting in large errors on absolute magnitude determination.

153.049   On cluster analysis of open clusters.
A. M. Ehjgenson, O. S. Yatsyk.
Astron. Tsirk., No. 1461, p. 5 – 6 (1986). In Russian.

A centroid method of cluster analysis (taxonomical analysis) is used to consider the galactic distribution of 361 open clusters. 199 clusters are shown to enter 54 taxons (condensations) with characteristic dimensions of several hundred parsecs. Every taxon consists of clusters connected by common origin.

153.050   On classification of open clusters by the centroid method of taxonomical analysis.
A. M. Ehjgenson, O. S. Yatsyk.
Astron. Tsirk., No. 1461, p. 6 – 8 (1986). In Russian.

In space with coordinates being mass, absolute magnitude, integrated colour, and diameter, the clusters form several taxons (classes) which are not isolated. This possibly shows the absence of gaps in clusters formation process.

153.051   Is there "hidden mass" in multiple open clusters?
K. A. Barkhatova, E. M. Nezhinskij, L. P. Osipkov.
Astron. Tsirk., No. 1472, p. 7 – 8 (1986). In Russian.

The change of configurations of multiple open clusters during their motion in the Galaxy has been computed. It was found that at the present time they are the most compact. The existence of the 'hidden mass' of order of $10^5 - 10^6 M_\odot$ (giant molecular clouds?) is necessary for the gravitational stabilization of the system.

153.052   The virial mass of the open cluster M11.
J. Zhao, Y. He.
Acta Astrophys. Sin., Vol. 7, No. 4, p. 273 – 279 (1987). In Chinese.

The authors recalculated the virial mass of the open cluster M11 using the proper motion data of McNamara et al. (1977). The average distance between cluster members is found to be 1.3 pc and the corresponding virial mass of the cluster 5094 $\mu_\odot$, which is in good agreement with the observed mass, 4671 $\mu_\odot$, recently obtained by Mathieu (1984), the virial coefficent being 0.9.

153.053   Mass and age distributions of stars in young open clusters.
U. C. Joshi, V. I. Myakutin, A. E. Piskunov, R. Sagar.
Star clusters and associations, p. 33 (1986). Abstract. Submitted to Mon. Not. R. Astron. Soc. – See Abstr. 012.081.

153.054   Integral magnitudes of star clusters and the initial mass function.
P. V. Baev, N. M. Spasova.
Star clusters and associations, p. 35 – 39 (1986). In Russian. – See Abstr. 012.081.

The theoretical dependences $M_V - \lg t$ for galactic clusters have been calculated. The method used is similar to Searl's one (Searl et al., 1973), but the initial mass function is different. The high luminosity of the clusters at $\lg t = 7$ can be explained by the initial mass function with $\alpha \varepsilon$ (1.5; 2.0) and can be considered as a superposition of evolutionary clusters' tracks with a different number of stars and α.

153.055   Integrated photometric parameters of open clusters.
B. A. Balázs.
Star clusters and associations, p. 41 – 58 (1986). – See Abstr. 012.081.

Integrated I(M_V) absolute magnitudes of 66 open clusters ($\tau < 10^8$a) have been derived using UBV photometric data from the literature and new distances obtained by the author. As anticipated, the integrated intrinsic colour–absolute magnitude relation – apart from a fairly large scatter – may be considered as linear and the linearity holds for the integrated colour–colour diagram as well. The age dependence of the integrated magnitudes turns out to be also linear but – contrary to the findings of some earlier investigations – the I(B–V)$_0$ versus $\lg \tau$ diagram is probably not linear even for clusters not older than 10^8 years.

153.056   On the role of flare stars in young open clusters.
G. Szécsényi–Nagy.
Star clusters and associations, p. 59 – 70 (1986). – See Abstr. 012.081.

Photometric and spectroscopic investigations of cluster member stars of the nearby open clusters provided us a lot of data about their brighter stars. Main sequences of these clusters are well known in the (O, G) spectral type range corresponding to an absolute visual magnitude range of (-6, +6). MS stars of lower luminosity were only occasionally identified in open clusters. The shortage of red dwarf (late dK and dM) stars seems to be substantial but it can be explained if we take into account various observational difficulties.

153.057 Some remarks on the cluster member FG Vulpeculae.
W. Götz.
Star clusters and associations, p. 71 – 74 (1986). – See Abstr. 012.081.

153.058 What percentage of the stars is known in the open clusters – the example of the Pleiades.
G. Szécsényi–Nagy.
Star clusters and associations, p. 101 – 114 (1986). – See Abstr. 012.081.

A statistical study of 3659 stars and 1110 flare-ups in the Pleiades field is given. It is shown that the total number of flare stars in the Pleiades region is very probably greater than 2500. The distribution of flare stars according to the number of their observed (and published) flare-ups fits almost perfectly the one presented here. It was computed by summing up seven various Poisson distributions with different mean frequencies. It is demonstrated too that at least 50 per cent of these flare stars are members of the cluster, consequently the population of the Pleiades must be well over one thousand.

153.059 H–R diagram of the Pleiades' flare stars.
E. S. Parsamian (*Eh. S. Parsamyan*).
Star clusters and associations, p. 115 – 122 (1986). – See Abstr. 012.081.

At the initial stage of this study the spectral class of only 70 flare stars was known. Being unable to obtain slit spectrograms of faint flare stars, the author found it expedient to determine the spectral class of faint flare stars by applying the method of Nassau et al. Thus, the spectral class of 188 stars was determined with a precision up to one or two subclasses.

Corrections and additions to CDS Catalogue 4017 (cross–identificatins for stars in open clusters).
See Abstr. 002.086.

Machine–readable version of the Tonantzintla catalogue of the Pleiades flare stars.
See Abstr. 002.091.

Evolutionary constraints for young stellar clusters. I. The luminosity function of H–burning stars.
See Abstr. 065.013.

Axion bounds from stellar evolution.
See Abstr. 065.086.

Galactic distribution, kinematics, locations in clusters and H–R diagrams, and duplicity of Be stars.
See Abstr. 112.109.

VBLUW–photometry of the two F0 Ib supergiants HD 80404 and HD 90853, and the K5 Ib supergiant HD 85891 with its associated cluster.
See Abstr. 113.047.

Obtaining continuous spectra of stars of the NGC 6913 cluster.
See Abstr. 113.055.

The photometric UBVR system of the 50/70 cm Schmidt telescope of the National Astronomical Observatory Rozhen.
See Abstr. 113.057.

Chemical composition of K–giants in the Hyades.
See Abstr. 114.052.

Lithium in the Alpha Per cluster.
See Abstr. 114.067.

Lithium and rotation in the Hyades F dwarfs.
See Abstr. 114.074.

Zirconium stars in open clusters.
See Abstr. 114.115.

A search for brown dwarfs and late M dwarfs in the Hyades and the Pleiades.
See Abstr. 115.011.

Rotation and magnetic activity in main–sequence stars.
See Abstr. 116.021.

The Pleiades rapid rotators: evidence for an evolutionary sequence.
See Abstr. 116.025.

Rotational velocities of low–mass stars in the Hyades.
See Abstr. 116.038.

Activity in early F–type stars in the Hyades.
See Abstr. 116.053.

Observations of Pleiades spotted solar–type stars.
See Abstr. 116.057.

The activity, variability, and rotation of lower main–sequence Hyades stars.
See Abstr. 116.060.

A polarimetric, photometric and spectrophotometric study of the very massive close binary DH Cephei.
See Abstr. 117.183.

IRAS observations of the Pleiades.
See Abstr. 131.014.

CO and NH_3 detection of the cone in NGC 2264.
See Abstr. 131.026.

Gas–and–dust complex NGC 7822 + S 171 connected with the Cep OB4 association.
See Abstr. 131.115.

Interstellar extinction in the Pleiades.
See Abstr. 131.137.

NGC 2264: a molecular line study.
See Abstr. 131.146.

Variations in UV extinction in galactic associations and perpendicular to the galactic plane.
See Abstr. 131.147.

The small scale interstellar dust distribution.
See Abstr. 131.230.

Ultraviolet interstellar extinction toward NGC 6530 and the intrinsic energy distribution of 9 Sagittarii and HD 165052.
See Abstr. 131.305.

An infrared study of the stellar population in the direction of the Carina nebula: NGC 3372.
See Abstr. 132.007.

N–body simulations of star clusters with initial binaries and the OB runaway stars.
See Abstr. 151.103.

The stabilization of multiple open clusters and the relaxation time of the Galaxy.
See Abstr. 151.135.

The stability of complexes of open star clusters.
See Abstr. 151.136.

Bound star clusters from gas clouds with low star formation efficiency.
See Abstr. 151.141.

Near–infrared spectral properties of star clusters and galactic nuclei.
See Abstr. 154.026.

On classification of star clusters by the taxonomical analysis.
See Abstr. 154.077.

The kinematical and binary properties of association and field O stars.
See Abstr. 155.007.

Distribution of classical cepheids in the galactic plane.
See Abstr. 155.018.

Star clusters and the thickness of the galactic disk as probes of the outer Galaxy.
See Abstr. 155.110.

Origen cinemático de las nubes obscuras de Tauro y de algunos cúmulos galácticos locales.
See Abstr. 155.129.

The local kinematics of open star clusters.
See Abstr. 155.146.

CMDs for the LMC clusters NGC 2249 and NGC 2241.
See Abstr. 156.009.

BVRI photometry of star clusters in the Bok region of the Large Magellanic Cloud.
See Abstr. 156.010.

Clusters of the Small Magellanic Cloud. II. Age distributions.
See Abstr. 156.015.

Some studies of Magellanic Cloud clusters.
See Abstr. 156.017.

CCD photometry of Large Magellanic Cloud clusters. IV. The metal–rich, remote southern cluster LW 79.
See Abstr. 156.028.

NGC 2209: the nature of the dark patch through the HR diagram.
See Abstr. 156.033.

Integrated properties of star clusters in the LMC.
See Abstr. 156.045.

The development of the red giant branch in intermediate age clusters of the Magellanic Clouds.
See Abstr. 156.047.

Global properties of LMC star clusters.
See Abstr. 156.049.

CCD observations of binary clusters in the Magellanic Clouds.
See Abstr. 156.052.

Infrared emission from young stars in the nucleus of M82.
See Abstr. 158.121.

154 Globular Clusters

154.001 The globular cluster population of X–ray binaries.
    F. Verbunt, P. Hut.
The origin and evolution of neutron stars, p. 187 – 197 (1987). – See Abstr. 012.001 (IAU Symp. No. 125).
    The authors discuss formation mechanisms for low–mass X–ray binaries in globular clusters. They apply the most efficient mechanism, tidal capture in close two–body encounters between neutron and main–sequence stars, to the clusters of our galaxy. The observed number of X–ray sources in these can be explained if the birth velocities of neutron stars are higher than estimated from velocity measurements of radio pulsars, or if the initial mass function steepens at high masses. The authors perform a statistical test on the distribution of X–ray sources with respect to the number of close encounters in globular clusters, and find satisfactory agreement between the tidal capture theory and observation, apart from the presence of low–mass X–ray binaries in four clusters with a very low encounter rate: Ter 1, Ter 2, Gr 1 and NGC 6712.

154.002 Theoretical stellar luminosity functions and the ages and compositions of globular clusters.
S. J. Ratcliff.
Astrophys. J., Vol. 318, No. 1, p. 196 – 214 (1987).
    Using recently calculated tables (Los Alamos) of radiative opacities and the latest thermonuclear reaction rates, stellar evolutionary tracks for low–mass $(0.4 - 1.2\ M_\odot)$ Population II stars $(0.0001 \leqslant Z \leqslant 0.01)$ have been calculated for two helium abundances, $Y = 0.2$ and 0.3. Luminosity functions (LFs) for ages $8\ \mathrm{Gyr} \leqslant t \leqslant 20\ \mathrm{Gyr}$ have been constructed from these tracks, and the resultant dependence of LF morphologies on age

and chemical composition has been studied. The observed M13 LF has been compared to these theortical LFs. Assuming a distance modulus of $(m-M)_0 = 14.3$, a metallicity–independent lower bound of 14 Gyr can be placed on the age of M13.

154.003 The distance to M5 from its RR Lyrae variables.
    J. G. Cohen, G. A. Gordon.
Astrophys. J., Vol. 318, No. 1, p. 215 – 231 (1987).
    The authors have applied a variation of the Baade–Wesselink method to four RR Lyrae variables in the moderately metal–poor globular cluster M5. The radial velocity curves are from observations with the 5 m Hale telescope, while the photometry is derived from CCD images through a Johnson B and a near–infrared i filter. The results must be viewed as preliminary thus far due to apparent discrepancies between the photometrically and spectroscopically deduced angular diameter–phase relationships. The best value obtained for the mean of the 4 RR Lyrae stars in M5 is an absorption–corrected intensity mean absolute V magnitude of $+1.05$ mag $(+0.15, -0.25$ mag$)$.

154.004 Photoelectric photometry of globular clusters in the Andromeda nebula. XII.
A. S. Sharov, V. M. Lyutyj, V. F. Esipov.
Pis'ma Astron. Zh., Tom 13, No. 8, p. 643 – 647 (1987). In Russian. English translation in Sov. Astron. Lett., Vol. 13.
    Results of photoelectric UBV observations of 10 globular clusters in the Andromeda nebula M31 carried out with the 2.6–meter telescope are presented. Improved data for 38 globular clusters based on observations of 1983 – 1986 are also given.

154.005   SMC globular clusters: candidates for mass segregation?
M. Kontizas, E. Theodossiou, E. Kontizas.
Mon. Not. R. Astron. Soc., Vol. 227, No. 2, p. 257 – 263 (1987).

A large number of SMC globular clusters of all evolutionary ages have been examined on 1.2–m UK Schmidt Telescope photographic plates of the same colour but different exposure times in order to derive their density profiles by means of star counts. Five clusters out of 32 were found to show differences in the slope of their radial density profiles. The observed differences in the radial density distributions give strong evidence for differences in the radial distribution of stars with various masses, the so–called "segregation effect".

154.006   C–M diagram and luminosity function of the galactic globular cluster NGC 7099. I. Photographic photometry.
G. Piotto, M. Capaccioli, S. Ortolani, L. Rosino, G. Alcaino, W. Liller.
Astron. J., Vol. 94, No. 2, p. 360 – 371 (1987). With plate 29.

The authors present new photographic photometry for ~ 4400 stars ($16.2 < m_v < 21.2$) in the field of the galactic globular cluster NGC 7099 = M30. A C–M diagram and luminosity function are derived from this photometry. The distance modulus is estimated to be $(m–M)_0 = 14.5 \pm 0.5$. A metallicity [Fe/H] $= -1.9 \pm 0.3$ was computed. The best fit of the theoretical isochrones (VandenBerg and Bell 1985) to the observations gives [Fe/H] $= -1.8$, $(m–M)_V = 14.6$, E(B–V) $= 0.02$, and an age of 17 ± 4 Gyr. The results remove the disagreement between previous photometry (Alcaino and Liller 1980) and the theoretical isochrones, noted by VandenBerg (1983).

154.007   Addendum: "High–resolution CCD spectra of stars in globular clusters. III. M4, M13, and M22" [Astron. J., Vol. 93, No. 5, p. 1137 – 1143 (1987)].
G. Wallerstein, E. M. Leep, J. B. Oke.
Astron. J., Vol. 94, No. 2, p. 523 (1987). See Abstr. 43.154.021.

154.008   X–ray and UV observations of ω Centauri with EXOSAT.
L. Koch–Miramond, M. Aurière.
Astron. Astrophys., Vol. 183, No. 1, p. 1 – 8 (1987).

The globular cluster ω Cen has been observed on two occasions, 6 months apart, with the Channel Multiplier Array on board EXOSAT, in the energy band 0.02 keV – 2.5 keV. The aim of the present investigation is to study the diffuse emission as well as the point sources discovered in the X–ray range with the EINSTEIN satellite (Hartwick et al., 1982). A diffuse UV emission due to the unresolved hot horizontal branch stars of ω Cen is observed. The diffuse X–ray source discovered by Hartwick et al. (1982) is not detected but the authors sensitivity threshold gives an upper limit of the flux which is not inconsistent with the EINSTEIN result. Four UV and three X–ray point sources are detected. One X–ray source shows variability of about a factor 4.5 in flux between the two EXOSAT observations. Flux variations for the X–ray sources between the two EXOSAT observations or with respect to the EINSTEIN ones are discussed.

154.009   The galactic globular cluster system: calibration of the ratio $R = N(HB)/N(RGB)$.
F. Caputo, C. Martinez Roger, E. Paez.
Astron. Astrophys., Vol. 183, No. 2, p. 228 – 233 (1987).

On the basis of Synthetic Horizontal Branches computed with Sweigart and Gross (1976) Horizontal Branch evolutionary tracks, and by means of Sweigart and Gross (1978) Red Giant Branch star models, the calibration of the ratio $R = N(HB)/N(RGB)$ in terms of helium abundance (Y), He–core mass (Mc) and Horizontal Branch morphology is derived. The results obtained in the present paper confirm that the estimate for Y is independent of the Horizontal Branch morphology, at least within the observed range of R–values and for clusters containing RR Lyrae variables. As far as the occurrence of enhanced He–core mass is concerned, the resulting estimate for Y increases by $dY/d\Delta M_c = 0.40$. From star counts in some well studied

RR Lyrae rich globular clusters a mean value for Y equal to 0.24 ± 0.01 is obtained.

154.010   Ages of globular clusters of the Galaxy.
O. K. Sil'chenko.
Astron. Zh., Tom 64, Vyp. 4, p. 686 – 695 (1987). In Russian. English translation in Sov. Astron., Vol. 31, No. 4.

Evolutionary modelling of the integrated $B–V$ colour is used to determine ages of 10 globular clusters of our Galaxy. None of the clusters referred to has an age older than 15 billion years, unlike previous estimates using the turn–off luminosity or effective temperature (up to 18 – 20 billion years). It is argued that metal–poor globular clusters ([Fe/H] < -1.2) are 4 – 5 billion years older than metal–rich ones. In the metallicity range $-1.3 - -2.3$ there is no metallicity–age correlation for this globular clusters sample.

154.011   Globular clusters as extragalactic distance indicators: maximum–likelihood methods.
D. A. Hanes, D. G. Whittaker.
Astron. J., Vol. 94, No. 4, p. 906 – 916 (1987).

The authors have explored the use of maximum–likelihood estimation techniques in the use of globular cluster luminosity functions (LFs) as distance indicators. In particular, they have tested size–of–sample effects through the analysis of Monte Carlo simulations of LFs drawn from an assumed universal population like that characterizing the globular clusters in the Local Group. The authors have also considered the effects of the field objects that will contaminate real cluster LFs in remote galaxies and have tested for biases and the attainable precision in derived distances as a function of limiting magnitude (relative to the turnover in the luminosity function). The findings are that maximum–likelihood methods are very robust. Globular clusters are more far–reaching distance indicators than has previously been realized. The authors apply the maximum–likelihood methods to available data and comment upon the implications of the results and the future promise of the method.

154.012   The CM diagram of the nearby globular cluster NGC 6397.
G. Alcaino, R. Buonanno, V. Caloi, V. Castellani, C. E. Corsi, G. Iannicola, W. Liller.
Astron. J., Vol. 94, No. 4, p. 917 – 947 (1987). With plates 77 – 78.

CCD photometry for faint stars in NGC 6397, combined with a digital reinvestigation of the photographic plates originally used by Alcaino and Liller (1980), has been used to obtain statistically significant samples for the various evolutionary phases, down to $V \sim 21$ mag, i.e., more than 5 mag below the turnoff. With a cluster distance modulus of the order of 12.50 mag this implies that main–sequence stars were observed down to $M_V \sim 8.5$ mag, corresponding to $M \sim 0.5 M_\odot$. The authors report evidence for a flattening of the luminosity function for MS stars fainter than $M_V \sim 6$ mag ($M \sim 0.7 M_\odot$), in agreement with previous indications by other authors. It is shown that, to reconcile the cluster age with current estimates for other galactic globular clusters, one needs to assume $Z = 10^{-4}$ and a cluster reddening of E(B–V) ~ 0.20 mag, plus an additional reddening of $\delta(B–V) = 0.04$ mag in the location of theoretical isochrones in the CM diagram.

154.013 BVI CCD photometry of 47 Tucanae.
G. Alcaino, W. Liller.
Astrophys. J., Vol. 319, No. 1, p. 304 – 313 (1987).

From 31 BVI CCD frames obtained with the 1.54 m Danish telescope at La Silla ESO, the authors have constructed V versus $B–V$, V versus $V–I$, and V versus $B–I$ color–magnitude diagrams in a $4' \times 2\overset{.}{.}5$ field of the globular cluster 47 Tuc. Using $Y = 0.2$, [Fe/H] $= -0.49$, $\alpha = 1.65$, a distance modulus $(m–M)_v = 13.2$ and E(B–V) $= 0.04$, the authors deduce from the location of the main–sequence turnoff a consistent age for 47 Tuc in all three color indices of 17×10^9yr. The agreement between observation and theory is less good for the subgiant branches plotted using colors incorporating the *I* band. The good fit between observation and theory with [Fe/H] = −0.49, and the poor fit with [Fe/H] = −0.79, favors the choice of a value near –0.5 as the best metal abundance for 47 Tuc.

154.014 *RICHFLD* photometry of NGC 6723.
D. H. Martins, D. A. Fraquelli.
Astrophys. J., Suppl. Ser., Vol. 65, No. 1, p. 83 – 93 (1987). With plate 1.

New photoelectrically calibrated photographic photometry is presented for the southern globular cluster NGC 6723. Color–magnitude (*C–M*) diagram morphology remains similar to past results. Most cluster stars brighter than *V* = 16.5 are included in the *C–M* diagram. Cluster reddening is estimated to be 0.1 mag, giving $(B–V)_{0,g} = 0.93$ mag and yielding [Fe/H] = −1.0. An upper horizontal branch may have been detected at the red end of the red horizontal branch; a gap in the subgiant branch is also suggested.

154.015 Studies of dynamical properties of globular clusters. III. Anisotropy in ω Centauri.
G. Meylan.
Astron. Astrophys., Vol. 184, No. 1/2, p. 144 – 154 (1987).

A King–Michie dynamical model has been built, based on an assumed form of the phase–space distribution function *f(E,J)* which induces a radial anisotropic velocity dispersion. This model consists of ten different subpopulations representing heavy remnants, white dwarfs and main sequence stars. From fitting to the surface brightness profile already published in the literature and to the mean radial velocities obtained with CORAVEL for 318 individual stars in ω Centauri, it appears that only models with strong anisotropy of the velocity dispersion ($r_a = 2 − 3\,r_c$) agree with the observations. This result is related to the large half–mass relaxation time, which amounts to $20 − 30 \times 10^9$yr, the central relaxation time being of the order of 10^9yr. The mean value of the mass function exponent x is 1.34. The fraction of the total mass in the form of remnants heavier than the white dwarfs equals 0 to 9% of the total mass of the cluster, being inversely proportional to the fraction of white dwarfs. The total mass $M_{tot} = 3.9 \times 10^6 M_\odot$ gives a mass–to–light ratio $M/L_V = 2.9$.

154.016 The chicken came first.
V. Trimble.
Nature, Vol. 328, No. 6130, p. 474 (1987).

This note comments on the work of G. H. Smith and R. D. McClure on metallicities and initial mass functions (IMF) of galactic globular clusters (see Abstr. 43.154.035).

154.017 The Oosterhoff dichotomy revisited. I. The ranking of RR Lyrae periods versus metallicity.
V. Castellani, M. L. Quarta.
Astron. Astrophys., Suppl. Ser., Vol. 71, No. 1, p. 1 – 10 (1987).

The authors collect and display in a graphical form available pulsational data for RR Lyrae variables in galactic globular clusters. The catalogue has been constructed by ranking all the globular clusters following the metallicity scale given by Zinn and West (1984). The authors report evidences for the inhomogeneity of the Oosterhoff I group, which appears to be split into two subgroups, and they show that the low boundary for periods of ab type RR Lyrae varies between the O.I and O.II groups and within the O.I group following a relation with metallicity which is similar to Sandage's (1982) prescription for single stars.

154.018 Deep photometry of globular clusters. X. The cluster GIC 0435–59 in Reticulum.
R. G. Gratton, S. Ortolani.
Astron. Astrophys., Suppl. Ser., Vol. 71, No. 1, p. 131 – 146 (1987).

Deep CCD photometry ($V_{lim} \sim 24.4$) for the globular cluster GIC 0435–59 in the halo of the Large Magellanic Cloud is presented. The quality of the observations is good, and it allows to derive the main features of the colour magnitude diagram and to discuss the luminosity function. The age is in the range 16 to

18 gyr. A number of blue stragglers are probably present. A (coarse) luminosity function is derived; it resembles the M5 one.

154.019 Systematic properties of extragalactic globular clusters.
D. Burstein.
Nearly normal galaxies. From the Planck time to the present, p. 47 – 56 (1987). – See Abstr. 012.031.

The purpose of this review is to explore three of the more implicit assumptions inherited from the knowledge of Galactic globular clusters: (1) On average, the chemical abundance properties of stars are similar in globular clusters of similar overall metal abundance in different galaxies. (2) The globular cluster system in giant spirals like the Milky Way and M31 are the oldest stellar systems in their respective parent galaxies. (3) The globular cluster systems in giant spirals are only formed during the early stages of formation of the parent galaxy.

154.020 Anisotropy of the velocity dispersion in ω Centauri.
G. Meylan.
Structure and dynamics of elliptical galaxies, p. 449 – 450 (1987). – See Abstr. 012.032 (IAU Symp. No. 127).

By far the brightest and the most massive globular cluster in our Galaxy, ω Cen seems to be, in some of its properties, a kind of transition step between dwarf ellipticals and ordinary globular clusters. For this giant cluster, the comparison between observations and King–Michie multi–mass dynamical models appears possible only using models with strong anisotropy in the velocity dispersion.

154.021 Deep CCD photometry of the globular cluster NGC 7099.
M. Bolte.
Astrophys. J., Vol. 319, No. 2, p. 760 – 771 (1987).

B and *V* photometry to *V* ∼ 24 mag of stars in the globular cluster NGC 7099 has been obtained with the CTIO 4 m CCD prime–focus camera. The color–magnitude diagram obtained shows very well defined, narrow cluster sequences with no blue straggler candidates and no indication of a binary star sequence lying above and to the right of the main sequence. Fitting the NGC 7099 main sequence to a Population II main sequence defined by local subdwarfs, the author derives a distance modulus to the cluster of $(m–M)_V = 14.65$. The best fit to model isochrones is for models with enhanced oxygen abundance and an age near 17 Gyr.

154.022 CCD photometry of the globular cluster NGC 288.
M. W. Pound, K. A. Janes, J. N. Heasley.
Astron. J., Vol. 94, No. 5, p. 1185 – 1201 (1987). With plates 99 – 100.

The authors present CCD photometry of NGC 288. The photometry reaches to *V* ∼ 22.5, about 2.5 mag below the turnoff. The color–magnitude diagram is presented and estimates for the cluster's reddening and distance are derived. NGC 288's heavy–metal abundance is deduced from comparisons to other globular clusters with similar metallicity. The data are compared with previous photometry of the cluster and systematic differences are discussed. The fitting of theoretical isochrones to the color–magnitude diagram is described, along with a detailed comparison to NGC 362.

154.023 Stellar content of the cores of metal–poor globular clusters.
J. A. Rose, P. B. Stetson, M. J. Tripicco.
Astron. J., Vol. 94, No. 5, p. 1202 – 1221 (1987). With plate 101.

Photographic image–tube spectra of the cores of 12 globular clusters have been obtained at a wavelength resolution of 2.5 Å in the interval λλ3400 – 4500 Å. The photographic integrated spectra have been supplemented by integrated CCD spectra of a few clusters and by long–slit CCD and 2D–Frutti spectra. It is found that, whereas ten of the 12 clusters have Hδ equivalent widths of similar strength, two clusters, M30 = NGC 7099 and NGC 4147, have abnormally strong Hδ. Moreover, the Ca II H + Hε feature in those two clusters is abnormally strong relative to Ca K when compared with the other clusters. On the other hand, in an index that measures the Balmer discontinuity, neither cluster is found to be unusual. The stellar content in the cores of M30 and NGC 4147 is analyzed via the Hδ profiles and Balmer discontinuity indices in those clusters compared with "normal" clusters, and in conjunction with imaging data for the core of M30 published by Cordoni and Auriere (1984). Finally, the stellar content of the light cusps found in the cores of some globular clusters is examined.

154.024 An interpretation of the line–strength indices in old stellar populations using an evolutionary synthesis approach.
A. Aragón, J. Gorgas, M. Rego.
Astron. Astrophys., Vol. 185, No. 1/2, p. 97 – 101 (1987).

Evolutionary population synthesis models with different metallicities have been computed in order to interpret the observed Mg_2 and $H\beta$, line–strength indices in old stellar populations. These models have been applied to three different cases, with the following results: (1) Galactic globular clusters do not exhibit a significant dispersion in age, and the metallicities of the computed models resemble those estimated for the clusters. The indices for the metal–poor globulars ([Fe/H] < –1) cannot be attained with these models due to the lack of low metallicity evolutionary tracks and stellar spectra libraries. (2) The line–strength gradients observed in the elliptical galaxy NGC 5813 are due, essentially, to intrinsic variations in metallicity, and they cannot be explained just from changes in the remaining parameters of the stellar population. (3) In order to synthesize the M32 indices it is necessary to introduce a star formation elapsed for a long time–scale, the star formation being still significant $\cong 5$ Gyr ago.

154.025 White dwarfs in Omega Centauri?
S. Ortolani, L. Rosino.
Astron. Astrophys., Vol. 185, No. 1/2, p. 102 – 106 (1987).

Deep B and V CCD images, obtained with the ESO 3.6 m telescope in two fields of the globular cluster Omega Centauri have been used to detect faint, blue objects, possible white dwarf candidates. The contamination by field stars and background extragalactic objects is also discussed.

154.026 Near–infrared spectral properties of star clusters and galactic nuclei.
E. Bica, D. Alloin.
Astron. Astrophys., Vol. 186, No. 1/2, p. 49 – 63 (1987).

The authors present near–infrared CCD spectra with 12.5 Å resolution for 30 star clusters having known ages, metallicities and reddenings. They also observed a sample of 62 nuclei in galaxies of all morphological types. They measure the continuum distribution and the equivalent widths of 13 absorption features for star cluster and galactic nuclei. Analysis of the star cluster sample shows that in the near–infrared, metallicity is the dominant parameter. Age effects in blue clusters disturb the relationship for molecular bands. In view of population synthesis of galactic nuclei, the authors present star cluster grid predictions as a function of age and metallicity for 5 metallic windows and the continuum distribution. Strong–lined globular clusters like NGC 6528 have integrated spectra comparable to those of massive galaxies. Consequently, they are important tools for population synthesis. Strong–lined spectra are observed in the central regions of both spiral galaxies and luminous elliptical galaxies, suggesting a comparable metal content. Even very blue galaxies exhibit, in the near–infrared, spectra quite similar to those observed for the rest of the galaxy sample.

154.027 A CCD color–magnitude study of 47 Tucanae.
J. E. Hesser, W. E. Harris, D. A. VandenBerg,
J. W. B. Allwright, P. Shott, P. B. Stetson.
Publ. Astron. Soc. Pac., Vol. 99, No. 618, p. 739 – 808 (1987).

CCD photometry in B and V is reported for 8800 stars in the field of 47 Tuc. The data, reduced through the DAOPHOT code, comprise results from two long–exposure frame pairs in which the photometry extends from the cluster turnoff down to $V \sim 24$, and two other short–exposure fields nearer to the cluster center

which precisely define the giant and horizontal branches. Color–magnitude diagrams are presented, statistically corrected for background contamination from the Small Magellanic Cloud halo, and a composite CMD and luminosity function are constructed for the entire cluster, covering an 11–magnitude range in luminosity.

154.028 Globular cluster photometry.
G. G. Fahlman.
Bull. Am. Astron. Soc., Vol. 19, No. 2, p. 675 (1987). Abstract. – See Abstr. 010.061.

154.029 Theoretical interpretation of globular cluster CCD data.
D. A. VandenBerg.
Bull. Am. Astron. Soc., Vol. 19, No. 2, p. 675 (1987). Abstract. – See Abstr. 010.061.

154.030 Interpretation of the 47 Tuc color–magnitude diagram.
J. E. Hesser, W. E. Harris, D. A. VandenBerg.
Bull. Am. Astron. Soc., Vol. 19, No. 2, p. 676 (1987). Abstract. – See Abstr. 010.061.

154.031 An IRAS survey of globular clusters.
D. K. Lynch, G. S. Rossano, S. C. Chapman.
Bull. Am. Astron. Soc., Vol. 19, No. 2, p. 676 (1987). Abstract. – See Abstr. 010.061.

154.032 White dwarfs and cataclysmic variables in the globular cluster M71.
H. B. Richer, G. G. Fahlman.
Bull. Am. Astron. Soc., Vol. 19, No. 2, p. 676 (1987). Abstract. – See Abstr. 010.061.

154.033 Main–sequence luminosity functions in 4 globular clusters.
M. Bolte.
Bull. Am. Astron. Soc., Vol. 19, No. 2, p. 676 (1987). Abstract. – See Abstr. 010.061.

154.034 Spectroscopic binary stars in globular clusters.
C. P. Pryor, R. D. McClure, J. E. Hesser,
J. M. Fletcher.
Bull. Am. Astron. Soc., Vol. 19, No. 2, p. 676 – 677 (1987). Abstract. – See Abstr. 010.061.

154.035 Intrinsic integrated UBVRI colors of galactic globular clusters.
B. C. Reed, J. E. Hesser, S. J. Shawl.
Bull. Am. Astron. Soc., Vol. 19, No. 2, p. 677 (1987). Abstract. – See Abstr. 010.061.

154.036 New CCD observations of the globular cluster NGC 2419.
D. H. Martins.
Bull. Am. Astron. Soc., Vol. 19, No. 2, p. 677 (1987). Abstract. – See Abstr. 010.061.

154.037 Globular clusters and the galactic halo population.
B. W. Carney.
Bull. Am. Astron. Soc., Vol. 19, No. 2, p. 687 (1987). Abstract. – See Abstr. 010.061.

154.038 The origin of globular clusters.
S. M. Fall.
Bull. Am. Astron. Soc., Vol. 19, No. 2, p. 687 (1987). Abstract. – See Abstr. 010.061.

154.039 Globular cluster systems and galactic halos.
W. E. Harris.
Bull. Am. Astron. Soc., Vol. 19, No. 2, p. 687 (1987). Abstract. – See Abstr. 010.061.

154.040 Abundances of giant branch stars in four southern hemisphere globular clusters.
M. M. Briley, R. A. Bell, R. J. Dickens.
Bull. Am. Astron. Soc., Vol. 19, No. 2, p. 705 (1987). Abstract. – See Abstr. 010.061.

154.041 Horizontal branch morphology in the log Te–log g plane: looking for multiple populations.
D. A. Crocker, R. T. Rood, R. W. O'Connell.
Bull. Am. Astron. Soc., Vol. 19, No. 2, p. 715 (1987). Abstract. – See Abstr. 010.061.

154.042 *BVRI* CCD photometry of Omega Centauri.
  G. Alcaino, W. Liller.
Astron. J., Vol. 94, No. 6, p. 1585 – 1599 (1987). With plates 124 – 127.
  The authors present the first CCD, *BVRI* main–sequence photometry for ω Cen, matched to the new *BVRI* isochrones of VandenBerg and Bell (1985). Their main conclusions are: (1) The main–sequence turnoffs as seen in the several colors are found to be at $V_{TO} = 18.3 \pm 0.15$; the MSTO color indexes are at: $B–V = 0.55 \pm 0.03$, $V–I = 0.73 \pm 0.03$, and $B–I = 1.28 \pm 0.03$. (2) The magnitude difference between the main–sequence turnoff and the horizontal branch is $\Delta M_V = 3.80 \pm 0.15$, as derived from the color–magnitude diagrams. (3) The authors deduce a consistent age for ω Cen in all three color–magnitude diagrams of 17 ± 1.5 Gyr. (4) The large scatter among the main–sequence stars demonstrates that the chemical inhomogeneity of about $|Fe/H| = \sim 1$ dex persists as well in the unevolved stars, hence suggesting the composition to be primordial.

154.043 The calcium abundance in NGC 1851.
  A. W. Rodgers, P. Harding.
Publ. Astron. Soc. Pac., Vol. 99, No. 619, p. 961 – 966 (1987).
  Spectra have been obtained of eight horizontal branch stars including three RR Lyrae stars, in the globular cluster NGC 1851. The spectra were measured to obtain equivalent widths of Ca II K and profiles of the Hδ line. The equivalent width of the interstellar component of the K line is found to be 0.56 Å. Stetson's (1981) photometry of the cluster is used for an independent measurement of the interstellar reddening and his value of $E(B–V) = 0.02 \pm 0.01$ is confirmed. Measurement of the calcium and hydrogen lines leads to a value of $[Ca/H] = -1.0 \pm 0.15$.

154.044 How similar are the globular clusters in different galaxies?
W. E. Harris.
Publ. Astron. Soc. Pac., Vol. 99, No. 620, p. 1031 – 1035 (1987).
  Recent observations of globular clusters in several galaxies show that the mean globular cluster luminosity l_m is nearly independent of parent galaxy luminosity L_G over more than three orders of magnitude in L_G. These results do not depend strongly on the adopted distance scale parameter H_0, and are at least roughly consistent with the Fall–Rees theoretical models for the origin of globular clusters. The implications for modeling of halo formation, and for the use of globular clusters as "standard candles", are briefly discussed.

154.045 IUE observations of the globular cluster No. 3 in the Fornax dwarf spheroidal galaxy.
C. Cacciari, F. Fusi–Pecci, R. J. Zinn.
Space Telesc. Sci. Inst., Prepr. Ser., No. 183, p. 6 – 12 (1987). Paper presented at ESO Workshop on stellar evolution and dynamics in the outer halo of the Galaxy, April 1987.

154.046 The faint luminosity function of M4.
  A. J. Penny, A. Lubenow, R. J. Dickens.
Space Telesc. Sci. Inst., Prepr. Ser., No. 183, p. 13 – 17 (1987). Paper presented at ESO Workshop on stellar evolution and dynamics in the outer halo of the Galaxy, April 1987.

154.047 A spectroscopic study of chemical abundances in the globular cluster Omega Centauri.
S. P. Caldwell, R. J. Dickens.
Rutherford Appleton Lab., Rep., RAL–87–068, 85 pp. (1987).
  Abundances of C, N, Fe and other heavy elements have been determined by fitting synthetic spectra, calculated from model atmospheres, to the observational data. A new method of performing the fitting, consisting of two sets of computer programs which run on Starlink VAX computers, has been developed. By this method four parameters, including microturbulence, can be determined simultaneously from stellar spectra.

154.048 Photoelectric B, V observations of stars in Omega Centauri.
T. G. Hawarden, E. A. Epps Bingham.
S.Afr. Astron. Obs., Circ., No. 11, p. 83 – 87 (1987).
  Photoelectric BV data are presented for 71 stars brighter than V = 15.6 in ω Cen. The observation and reduction techniques are described, and sources of error are discussed. A comparison with previous photometry of 14 of these stars by Cannon and Stobie shows good agreement, apart from a smaller discrepancy in V.

154.049 A catalogue of concentric aperture UBVRI photoelectric photometry of globular clusters.
C. J. Peterson.
Astron. Data Cent. Bull., Vol. 1, No. 4, p. 277 – 284 (1987).
  The catalogue contains 4204 entries of concentric aperture photometry for 118 globular clusters of our Galaxy. The survey of UBVRI studies is complete and has been supplemented with other photometry that is transformable to the standard photometric system; transformation equations are given.

154.050 A preliminary survey of collapsed cores in Magellanic Clouds globular clusters.
G. Meylan, S. Djorgovski.
ESO Sci. Prepr., No. 526, 8 pp. (1987). To appear in Astrophys. J., Lett. Ed.

154.051 Studies of dynamical properties of globular clusters. IV. Detailed structure of 47 Tucanae.
G. Meylan.
ESO Sci. Prepr., No. 528, 26 pp. (1987). To appear in Astron. Astrophys.

154.052 Internal dynamics of globular clusters: from our Galaxy to the Magellanic Clouds.
G. Meylan.
ESO Sci. Prepr., No. 530, 18 pp. (1987). Paper presented at the ESO workshop on "Stellar evolution and dynamics in the outer halo of the Galaxy", Garching, F.R. Germany, April 1987.

154.053 High resolution observations of stars in the peculiar globular cluster ω Cen.
M. Spite, S. Huille, P. Fraïnçois, F. Spite.
ESO Sci. Prepr., No. 542, 15 pp. (1987). Submitted to Astron. Astrophys., Suppl. Ser.

154.054 High resolution study of different groups of stars in the peculiar globular cluster ω Cen.
P. François, M. Spite, F. Spite.
ESO Sci. Prepr., No. 543, 34 pp. (1987). To appear in Astron. Astrophys.

154.055 CCD stellar photometry in the central region of 47 Tuc.
  M. Aurière, S. Ortolani.
ESO Sci. Prepr., No. 553, 32 pp. (1987). Submitted to Astron. Astrophys.

154.056 CCD photometry of globular cluster core structure. I. NGC 6388, NGC 6624, and M15 – one flat core and two cusps.
P. M. Lugger, H. Cohn, J. E. Grindlay, C. D. Bailyn, P. Hertz.
Astrophys. J., Vol. 320, No. 2, p. 482 – 492 (1987).

The authors have obtained *UBVR* CCD frames of the cores of 72 globular clusters using the CTIO 4 m and KPNO No. 1 0.9 m telescopes. A principal goal of this work is to test the prediction that a significant number of Galactic globular clusters have undergone core collapse and should therefore have central surface brightness cusps. The authors present surface brightness profiles for three clusters: one with a normal flat core profile – NGC 6388 – and two with central cusps – NGC 6624 and M15. These profiles have been fitted with both seeing–convolved King models and seeing–convolved power laws. The central surface brightness cusps of NGC 6624 and M15 are discussed in terms of models for core collapse in a multicomponent cluster.

154.057 Destruction mechanisms for the globular clusters in the Milky Way.
L. A. Aguilar, J. P. Ostriker, P. Hut.
Rev. Mex. Astron. Astrofis., Vol. 14, No. 1, p. 227 (1987). Abstract. – See Abstr. 012.042.

154.058 Globular clusters and stellar evolution.
A. Renzini.
ESO Conf. Workshop Proc., No. 27, p. 289 – 298 (1987). – See Abstr. 012.051.

154.059 Color–magnitude diagrams of galactic globular clusters.
R. Buonanno.
ESO Conf. Workshop Proc., No. 27, p. 331 – 340 (1987). – See Abstr. 012.051.

154.060 Globular clusters in the outer halo.
S. Ortolani.
ESO Conf. Workshop Proc., No. 27, p. 341 – 349 (1987). – See Abstr. 012.051.

With a few exceptions the globular cluster system of the outer halo is dominated by peculiar clusters very loose and markedly poor in stars and spread throughout a wide region far from the galactic center. A noticeable difference between the stellar population of the inner and outer halo globular clusters was revealed with the C–M diagrams for distant clusters. Together with the study of the galactic disk extention, the recent discovery of some very distant clusters and improved C–M diagrams now available, permit a more quantitative statistical approach in the definition of an inner edge of the outer halo or a transition zone between two different populations.

154.061 Space motions of galactic globular clusters.
M. Geffert.
ESO Conf. Workshop Proc., No. 27, p. 351 – 353 (1987). – See Abstr. 012.051.

Although the errors of tangential velocity components are normally larger than the errors of radial velocity components, for two reasons a determination of proper motions of globular clusters is valuable for a discussion of their space motions in the Milky Way: a complete orbit of a globular cluster can only be determined with knowledge of its proper motion. Statistical investigations, which use radial velocities alone cannot give informations about the clusters' tangential velocities (with respect to the galactic centre) far outside the solar circle.

154.062 Star formation in proto–globular cluster clouds.
H. Zinnecker, F. Palla.
ESO Conf. Workshop Proc., No. 27, p. 355 – 361 (1987). – See Abstr. 012.051.

After briefly reviewing several globular cluster formation mechanisms the authors try to constrain the initial conditions of star formation from a hot metal–poor protocluster cloud. They favour shock compression and dismiss self–enrichment. Rapid cooling due to molecular hydrogen and a tiny fraction of CO molecules may play a vital role to reduce the Jeans mass from $10^5 M_\odot$ to about 0.5 $M_\odot$ in less than a free–fall time ($\sim 10^6$yr). Globular clusters may never have formed massive stars.

154.063 Ages of globular clusters derived from BVRI CCD photometry.
G. Alcaino, W. Liller.
ESO Conf. Workshop Proc., No. 27, p. 363 – 374 (1987). – See Abstr. 012.051.

The authors have completed BVRI reductions on five clusters: NGC 104 (47 Tuc), NGC 2298, NGC 5139 (Omega Cen), NGC 6121 (M4), and NGC 6362 and give a summary of the cluster parameters.

154.064 Empirical $L–T_{eff}$ versus metallicity sequences for globular cluster giant branches.
C. Martinez Roger, S. Arribas.
ESO Conf. Workshop Proc., No. 27, p. 375 – 380 (1987). – See Abstr. 012.051.

Mean empirical $L–T_{eff}$ versus metallicity sequences for globular cluster giant branches are derived from a coherent set of photometric data. Using the Ridgway et al. (1980) calibration of the (V–K) index with the T_{eff}, and the bolometric correction to the K magnitude, the photometric indexes are transformed into the physical parameters luminosity and effective temperature.

154.065 CCD photometry of the outer halo clusters Pal 4 and NGC 2419.
C. A. Christian, J. N. Heasley.
ESO Conf. Workshop Proc., No. 27, p. 381 – 385 (1987). – See Abstr. 012.051.

CCD photometry of two outer halo globular clusters, Pal 4 and NGC 2419, has been obtained to sample the stellar population to $V = 25$ mag. The color–magnitude diagram of Pal 4, when compared to both other globular clusters and theoretical isochrones suggest that this object is a moderately metal poor cluster with [Fe/H] ~ -1.7 with $(m–M)_0 = 20.1$, and $E(B–V) = 0.02$ mag. The age of the cluster appears to be ~ 15 Gyr, but the cluster does exhibit the red horizontal branch and distinct asymptotic giant branch typical of "second parameter" clusters. Alternatively NGC 2419 appears to be a close analog to M15, with $(m–M)_0 = 20.0$, and $E(B–V) = 0.12$.

154.066 Palomar 12 – a new color–magnitude diagram.
P. B. Stetson, G. H. Smith.
ESO Conf. Workshop Proc., No. 27, p. 387 – 392 (1987). – See Abstr. 012.051.

154.067 New CCD photometry of the globular cluster NGC 7099 = M30.
M. Capaccioli, S. Ortolani, G. Piotto.
ESO Conf. Workshop Proc., No. 27, p. 393 – 400 (1987). – See Abstr. 012.051.

CCD B–band and V–band images of five regions in the globular cluster NGC 7099 = M30 have been secured. The authors present the preliminary C–M diagrams and luminosity functions for two such regions. An age of 16 ± 4 Gyr and and index $x = 1.5$ for the main sequence mass function are derived.

154.068 The faint luminosity function of M4.
A. J. Penny, A. Lubenow, R. J. Dickens.
ESO Conf. Workshop Proc., No. 27, p. 401 – 405 (1987). – See Abstr. 012.051.

Observations of M4 using good seeing, long exposures, and a new reduction package have given the faintest luminosity function yet measured for a globular cluster. The main new result is a sharply increasing luminosity function from Mv = 8 to Mv = 12. There is some evidence for a turn–over at the faint limit.

154.069 A spectroscopic study of HB stars in the galactic globular cluster NGC 6752.
U. Heber, R. P. Kudritzki, V. Caloi, V. Castellani, J. Danziger.
ESO Conf. Workshop Proc., No. 27, p. 407 – 412 (1987). – See Abstr. 012.051.

Visual spectra of 9 horizontal branch stars in the globular cluster NGC 6752 have been obtained with the UCL IPCS detector attached to the B and C spectrograph on the ESO 3.6 m telescope. The spectral resolution is ≈ 3.3 Å. Additionally, IUE spectra have been obtained for six HB stars.

154.070 A halo cluster of the LMC at 10 kpc: NGC 1841.
J. Andersen, A. Blecha, M. F. Walker.
ESO Conf. Workshop Proc., No. 27, p. 473 – 475 (1987). – See Abstr. 012.051.

The authors' conclusions from a comparison of the CMD of NGC 1841 with other Galactic and Magellanic globulars under the assumption that $E(B-V) = 0.15$ may be summarized as follows: (1) NGC 1841 is a very old, metal–poor globular cluster, closely similar to M92, (2) $(B-V)_{0,g} = 0.68$, corresponding to [Fe/H] = -2.1, (3) from a fit to the CMD of M92, the authors find a true distance modulus to NGC 1841 of $(m-M)_0 = 18.1 \pm 0.3$ (the currently favored distance of the LMC, (4) the luminosity discrepancy between the NGC 1841 giants and those of, e.g. M92, is removed.

154.071 An automated search for halo star clusters and Local Group dwarf galaxies.
H. T. MacGillivray, R. K. Bhatia, S. M. Beard, R. J. Dodd.
ESO Conf. Workshop Proc., No. 27, p. 477 – 479 (1987). – See Abstr. 012.051.

The discovery in recent years of hitherto unknown Local Group dwarf galaxies and intergalactic and extreme halo globular star clusters has highlighted the need for a systematic, statistically complete and homogeneous search for such objects over the sky. The authors have initiated a major programme aimed at carrying out such a survey from objective, machine-produced scans of photographic plates. These plates are already being systematically scanned in order to produce a catalogue of objects down to $B \sim 22$ over the whole of the southern sky.

154.072 Globular clusters in dwarf spheroidals.
F. Fusi Pecci.
ESO Conf. Workshop Proc., No. 27, p. 493 – 509 (1987). – See Abstr. 012.051.

Some properties of the cluster system in the Fornax dwarf spheroidal galaxy are briefly reviewed. Particular emphasis has been given to the analysis of a few aspects more strictly related to the study of Population II stellar evolution.

154.073 IUE observations of the globular cluster N.3 in the Fornax dwarf spheroidal galaxy.
C. Cacciari, G. Clementini, F. Fusi Pecci, R. J. Zinn.
ESO Conf. Workshop Proc., No. 27, p. 511 – 516 (1987). – See Abstr. 012.051.

154.074 Internal dynamics of globular clusters: from our galaxy to the Magellanic Clouds.
G. Meylan.
ESO Conf. Workshop Proc., No. 27, p. 665 – 681 (1987). – See Abstr. 012.051.

Contrary to galactic globular clusters, for some of which radial velocities of individual member stars give new observational constraints for the dynamical models, the Magellanic Clouds globular clusters have still to be studied, from a structural and dynamical point of view, essentially from their observed surface brightness profiles alone. In this review, current knowledge of both galactic and Magellanic globular clusters is described.

154.075 New variable stars in the globular cluster M4.
B.–a. Yao.
Messenger, No. 50, p. 33 – 37 (1987).

154.076 The low–luminosity X–ray sources in Omega Centauri.
B. Margon, M. Bolte.
Astrophys. J., Lett. Ed., Vol. 321, No. 1, p. L61 – L65 (1987). With plate L11.

The authors report multicolor CCD photometry of the fields of three X–ray sources in ω Centauri; two have very accurate (10″) EXOSAT X–ray positions, in addition to observations from the Einstein Observatory. The crowding in these X–ray fields is moderate and does not preclude accurate photometry of individual objects. Although the data in the best case reach to $V = 24.5$, corresponding to $M_V = 10.6$ in the cluster (and well beneath the expected quiescent luminosity of cataclysmic variables (CV), the authors find no evidence for any CV optical candidates.

154.077 On classification of star clusters by the taxonomical analysis.
A. M. Ehjgenson, O. S. Yatsyk.
Astron. Tsirk., No. 1464, p. 7 – 8 (1986). In Russian.

The distribution of 57 open and 90 globular clusters in space is considered. The clusters form two large groups. The first group consists of 72 globular clusters, and the second one of 57 open and 15 globular clusters. Conclusion is made that in clusters formation process perhaps there were no substantial gaps.

154.078 Inferences from a color–magnitude diagram for 47 Tucanae.
J. E. Hesser, W. E. Harris, D. A. VandenBerg.
Publ. Astron. Soc. Pac., Vol. 99, No. 621, p. 1148 (1987). Abstract. – See Abstr. 010.281.

154.079 Ages of globular clusters.
C. J. Peterson.
Publ. Astron. Soc. Pac., Vol. 99, No. 621, p. 1153 – 1160 (1987).

Ages have been calculated for 41 globular clusters for which color–magnitude diagram studies allow a determination of the luminosity difference between the horizontal–branch and the main–sequence turnoff point. The data indicate only weak support for a slight difference in age between clusters in the inner halo and those in the outer halo. No correlation between cluster ages and cluster metallicities is found.

154.080 11 new variables in globular cluster M5.
A. Gerashchenko.
Inf. Bull. Variable Stars, No. 3044, 2 pp. (1987).

154.081 Erratum: "Chemical enrichment of halo protoclusters with differing stellar mass functions" [Astrophys. J., Vol. 316, No. 1, p. 206 – 212 (1987)].
G. H. Smith, R. D. McClure.
Astrophys. J., Vol. 322, No. 2, p. 1074 (1987). See Abstr. 43.154.035.

154.082 A preliminary survey of collapsed cores in the Magellanic Clouds' globular clusters.
G. Meylan, S. Djorgovski.
Astrophys. J., Lett. Ed., Vol. 322, No. 2, p. L91 – L94 (1987).

The authors present a preliminary report on a surface photometry survey for collapsed cores in the Magellanic Clouds' globular clusters. Core morphology classifications are given for the 33 globular clusters examined so far. One cluster, NGC 2019, shows definite signs of a collapsed core, and two others, NGC 1774 and NGC 1951, appear as strong candidates. The fraction of collapsed–core clusters appears to be smaller in the Magellanic Clouds than in the Galaxy.

154.083 The structure of young star clusters in the Large Magellanic Cloud.
R. A. W. Elson, S. M. Fall, K. C. Freeman.
Astrophys. J., Vol. 323, No. 1, p. 54 – 78 (1987).

The authors present surface brightness profiles, based on star counts and aperture photometry, for 10 rich young star clusters in the Large Magellanic Cloud. To investigate the dynamical evolution of the clusters, they derive mass–to–light ratios from stellar population models. They consider various ways in which the clusters might have formed, and suggest several explanations for the observed profiles. Expansion of a newly formed cluster either through mass loss or during violent relaxation could lead to the formation of a halo of unbound stars. To examine this possibility, a calculation of the tidal field of the LMC is included. At least some and perhaps all the clusters in the sample extend beyond their eventual tidal radii, with up to 50% of the total masses in unbound halos.

154.084 Collapsed cores and the structural parameters of old Large Magellanic Cloud star clusters.

M. Mateo.

Astrophys. J., Lett. Ed., Vol. 323, No. 1, p. L41 – L45 (1987).

The analysis of the surface photometry of old LMC clusters has revealed (1) most of the clusters can be satisfactorily fitted to standard, single–mass King profiles; (2) two clusters, NGC 2005 and NGC 2019, do not have surface brightess profiles which can be fitted to King models, and, on the basis of this result and comparisons with "collapsed" galactic globular clusters, these two objects appear to be examples of collapsed LMC cluster cores; and (3) the most concentrated clusters are all located near the LMC center.

154.085 Spectroscopy of the λ7699 K I line among globular cluster giants.

B. Campbell, G. H. Smith.

Astrophys. J., Lett. Ed., Vol. 323, No. 1, p. L69 – L72 (1987).

Echelle spectra have been obtained of giants in the globular clusters NGC 6752, M4, and 47 Tucanae to determine whether the Na I and Al I line enhancements found among the CN–rich stars are accompanied by enhancements of the λ7699 K I line. The K I line is found, however, to be of equal strength in the spectra of the CN–rich and CN–poor stars. This result strengthens the hypothesis that Na and Al line variations are due to real abundance differences among the cluster stars.

154.086 Catalogue of the coordinates of 3600 stars in the globular cluster M15 for the period 1896 – 1910.

N. M. Spasova, Kh. N. Nikov.

Star clusters and associations, p. 147 – 149 (1986). In Russian. – See Abstr. 012.081.

Triennial report on globular star cluster research (1984 – 1986).
See Abstr. 002.067.

A catalogue of seven–colour photometry of 752 stars in the Vilnius system for a region of two square degrees in the direction of the globular cluster M56.
See Abstr. 002.090.

Dynamical evolution of globular clusters.
See Abstr. 003.130.

Mass–loss of globular cluster red giants. A semi–empirical estimation.
See Abstr. 064.030.

Evolutionary sequences for horizontal–branch stars.
See Abstr. 065.046.

Evolution of stellar binaries formed by tidal capture.
See Abstr. 065.049.

Evolution of very low mass stars and brown dwarfs. II. The Population II.
See Abstr. 065.076.

Horizontal branch stars: theoretical expectations.
See Abstr. 065.087.

New mechanism of appearance of proper redshifts in spectra of compact objects.
See Abstr. 066.179.

On the origin of neutron stars in globular clusters.
See Abstr. 067.004.

Formation of a millisecond pulsar in a globular cluster.
See Abstr. 067.125.

Origin of millisecond pulsars.
See Abstr. 067.126.

Formation of isolated millisecond pulsars in globular clusters.
See Abstr. 067.127.

Parallax calibration of the population II main sequence.
See Abstr. 111.013.

A photoelectric *BVRI* sequence in the field of the globular cluster NGC 6121 (M4).
See Abstr. 113.013.

Four–color CCD observations of BHB stars.
See Abstr. 113.035.

High resolution observations of stars in the peculiar globular cluster ω Cen.
See Abstr. 114.048.

High–latitude A – F supergiants.
See Abstr. 114.075.

Outburst and quiescence observations of the dwarf nova V101 in the globular cluster M5.
See Abstr. 117.048.

A 685 second orbital period from the X–ray source 4U 1820–30 in the globular cluster NGC 6624.
See Abstr. 117.092.

The 160 min period in RR Lyr stars in globular clusters and highly evolved close binary systems.
See Abstr. 122.028.

RR Lyrae variables in galactic globular clusters: observational properties and open problems.
See Abstr. 122.095.

Fourier coefficients of the RR Lyrae variables in NGC 6171.
See Abstr. 122.113.

The discovery of a millisecond pulsar in the globular cluster M28.
See Abstr. 126.057.

White dwarfs in globular clusters: a search in M71.
See Abstr. 126.080.

New millisecond pulsar in an unusual environment.
See Abstr. 126.083.

Millisecond pulsar in M4.
See Abstr. 126.091.

Analysis of X–ray spectrum for globular cluster M15 X–ray source.
See Abstr. 142.006.

The low luminosity X–ray sources in Omega Centauri.
See Abstr. 142.026.

The X–ray sources of the galactic bulge, X–ray bursters and X–ray sources in globular clusters.
See Abstr. 142.044.

Massive disk formation resulting from the collision of a main–sequence star with a white dwarf in a globular cluster core.
See Abstr. 151.028.

Dynamical evolution of globular clusters.
See Abstr. 151.035.

Evolution of globular clusters including a degenerate component.
See Abstr. 151.095.

Dynamical effects of successive mergers on the evolution of spherical stellar systems.
See Abstr. 151.096.

New developments in the dynamical structure of clusters.
See Abstr. 151.106.

Dynamical evolution of star clusters and the possible disruption of globular clusters in M87 by massive black holes from a dark corona.
See Abstr. 151.115.

Kinetic equation of finite Hamiltonian systems with integrable mean field.
See Abstr. 151.117.

Bound star clusters from gas clouds with low star formation efficiency.
See Abstr. 151.141.

Globular cluster evolution in the Galaxy: a global view.
See Abstr. 151.143.

The evolution and final disintegration of spherical stellar systems in a steady galactic tidal field.
See Abstr. 151.144.

Chemical evolution of the galactic halo.
See Abstr. 155.143.

CMDs for the LMC clusters NGC 2249 and NGC 2241.
See Abstr. 156.009.

Colour–magnitude diagrams of star clusters in the Magellanic Clouds from wide–field electronography. III. NGC 1841.
See Abstr. 156.012.

Distribution of spectral types in the LMC clusters.
See Abstr. 156.013.

Hot stars in young globular clusters of the Magellanic Clouds.
See Abstr. 156.014.

Clusters of the Small Magellanic Cloud. II. Age distributions.
See Abstr. 156.015.

Surface photometry of the old populous star clusters of the Large Magellanic Cloud.
See Abstr. 156.020.

The age–metallicity relations of the Magellanic Clouds as determined from observations of the integrated spectra of star clusters.
See Abstr. 156.021.

Morphology of LMC clusters.
See Abstr. 156.031.

NGC 1850: a merging binary cluster in the LMC?
See Abstr. 156.051.

The age of the Large Magellanic Cloud cluster NGC 2193.
See Abstr. 156.056.

On the luminosity function of elliptical galaxies.
See Abstr. 157.189.

Near–infrared photometry of globular clusters in the outer halo of M31.
See Abstr. 157.206.

Globular cluster systems: comparative evolution of galactic halos.
See Abstr. 157.302.

Spheroidal systems as a one–parameter family of mass at their birth.
See Abstr. 157.305.

Ellipticity of 30 globular clusters in the Andromeda galaxy.
See Abstr. 157.349.

Spectroscopy of the globular clusters in M87.
See Abstr. 158.143.

Globular clusters belonging to galaxy clusters.
See Abstr. 160.015.

Intercambio de cumulos globulares en cumulos de galaxías.
See Abstr. 160.098.

155 Galaxy

155.001   A new aspect of galactic ridge X–ray emission – SNRs in a tenuous medium?
K. Koyama.
The origin and evolution of neutron stars, p. 535 – 543 (1987). – See Abstr. 012.001 (IAU Symp. No. 125).
Thin thermal natures of a plasma temperature of several keV are observed in the diffuse X–ray spectra from the galactic ridge region by the Tenma satellite. Within the constraints imposed by the observational results, possible origins of the ridge X–ray emission are discussed. The author shows that unidentified young supernova remnants in a tenuous medium could be a significant contributor to the ridge emission.

155.002   Diffuse infrared emission of the Galaxy: large scale properties.
J. L. Puget, M. Perault, F. Boulanger, E. Falgarone.
NASA Conf. Publ., NASA CP–2466, p. 21 (1987). Abstract. – See Abstr. 012.002.

155.003   The origin of the diffuse galactic IR/submm emission: revisited after IRAS.
P. Cox, P. G. Mezger.
NASA Conf. Publ., NASA CP–2466, p. 23 – 35 (1987). – See Abstr. 012.002.
In three previous papers the authors have investigated the origin of the diffuse galactic IR/submm emission by fitting model computations to balloon–borne surveys. In this paper they compare the balloon observations with IRAS observations. For the longitude profiles they find good agreement. However, the dust emission observed by IRAS – contrary to balloon observations which show dust emission only within $|b| \leqslant 3°$ – extends all the way to the galactic pole. The authors also compare the galactic dust emission spectrum with the dust emission spectra of external IRAS galaxies.

155.004   Survey of the galactic disc from $l = -150°$ to $l = 82°$ in the submillimeter range.
E. Caux, G. Serra.
NASA Conf. Publ., NASA CP–2466, p. 93 – 96 (1987). – See Abstr. 012.002.
The authors present new results about the emission of the galactic disc from $l = -150°$ to $l = 82°$ in the submm range ($\lambda_{eff} = 380\,\mu m$). Obervations have been made with the AGLAE83 balloon–borne instrument launched from Brazil in November 1983. In–flight calibration of the instrument was made on Jupiter. The longitude profile obtained exhibits diffuse emission all along the disc with bright peaks associated with resolved sources. The averaged galactic spectrum is in agreement with a temperature distribution of the interstellar cold dust.

155.005   Diffuse infrared emission of the Galaxy: large scale properties.
M. Pérault, F. Boulanger, E. Falgarone, J. L. Puget.
NASA Conf. Publ., NASA CP–2466, p. 97 (1987). Abstract. – See Abstr. 012.002.

155.006   Aperture synthesis observations of the circumnuclear ring in the galactic center.
R. Güsten, R. Genzel, M. C. H. Wright, D. T. Jaffe, J. Stutzki, A. I. Harris.
Astrophys. J., Vol. 318, No. 1, p. 124 – 138 (1987).
The authors report 88 GHz aperture synthesis observations of HCN $J = 1 \rightarrow 0$ emission and absorption in the central 5 pc of the Galaxy and single–dish measurements of HCN $1 \rightarrow 0$ and CO $J = 7 \rightarrow 6$ line profiles at galactocentric radius $R = 5.5$ pc. The HCN synthesis data show a highly inclined, clumpy ring of molecular gas surrounding the ionized center of the Galaxy. The ring is the inner edge of a thin disk extending in the HCN aperture synthesis data to about 5 pc and to $\geqslant 7$ pc in lower excitation lines. The molecular gas is dynamically coupled to ionized gas in the central cavity. The dominant large–scale

velocity pattern of the majority of the molecular gas in the inner 5 pc is rotation. No overall radial motion greater than about $20 - 30$ km s^{-1} is apparent. The neutral gas ring may represent a circumnuclear accretion disk which feeds interstellar matter into the central parsec.

155.007   The kinematical and binary properties of association and field O stars.
D. R. Gies.
Astrophys. J., Suppl. Ser., Vol. 64, No. 3, p. 545 – 563 (1987).
A catalog of 195 O–type stars brighter than $V = 8.0$ has been compiled to compare the velocity distribution and binary frequency among cluster and association, field, and runaway stars. The peculiar radial velocity distributions of the field and runaway stars indicate a net motion away from the Galactic plane, and spectroscopic and visual binaries are deficient among these groups relative to association stars. Many field and runaway stars were probably ejected through gravitational encounters with binaries during an early compact epoch in cluster evolution.

155.008   Accretion of gaseous disks of galaxies. II. Taking into account the viscosity effect in the giant molecular clouds' disk.
O. K. Sil'chenko, V. M. Lipunov.
Astrofizika, Tom 26, Vyp. 3, p. 443 – 456 (1987). In Russian. English translation in Astrophysics, Vol. 26, No. 3, p. 267 – 276 (1987).
The evolution of the distribution of giant molecular clouds in galaxies caused by the dynamical friction in a stellar disk and by the collisionless viscosity is calculated. It has been shown that the summary effect of viscosity and dynamical friction can result in forming the ring–like distribution of giant molecular clouds in the Galaxy with the peak on the $6 - 7$ kpc distance from the centre of the Galaxy after a few billion years.

155.009   The galactic distribution of Wolf–Rayet stars.
C. Doom.
Astron. Astrophys., Vol. 182, No. 2, p. L43 – L46 (1987).
The author investigates mathematically the correlation between the distributions of WR and O stars within 2.5 kpc from the sun. He compares the WR distribution to the distribution of O stars more massive than a given mass M. The results show that the distribution of WR stars is the most compatible with the distribution of O stars more massive than $35\,M_{\odot}$, which is slightly smaller than suggested by Conti et al. (1983). Comparison of the WR distribution with the distribution of O stars within given mass ranges suggests that single WR stars are formed from $23 - 25\,M_{\odot}$ onwards, that binary WR stars are formed from all O–type (binary) stars and that WC stars are formed from O stars with masses between 30 and $45\,M_{\odot}$.

155.010 J, H and K maps of the galactic centre region – II. Qualitative aspects of the interstellar absorption.
I. S. Glass, R. M. Catchpole, P. A. Whitelock.
Mon. Not. R. Astron. Soc., Vol. 227, No. 2, p. 373 – 379 (1987).
J (1.25 μm), H (1.65 μm) and K (2.2 μm) maps of a $2° \times 1°(\alpha \times \delta)$ region of the sky including the galactic centre and having a resolution of a few arcsec are presented. While almost the entire J map is dominated by heavy interstellar extinction, those at H and K show progressively more detail of the inner regions of the Galaxy. Even at K, however, much of the map is dominated by dark clouds, some of which show fine filamentary structure. Most of these clouds can be associated with foreground molecular material having low radial velocity. The high–velocity clouds observed at radio wavelengths toward the galactic centre are not in general apparent and are probably embedded within the dense stellar distribution.

155.011 Density normalization of the thick disc of the Galaxy.
D. W. Evans.
Mon. Not. R. Astron. Soc., Vol. 227, No. 2, p. 13P – 17P (1987).
The conflict between the normalization values for the thick disc found by Gilmore & Reid (2 per cent) and Sandage & Fouts (10 per cent) is resolved.

155.012 Round dynamical models of the galactic halo.
J. Sommer–Larsen.
Mon. Not. R. Astron. Soc., Vol. 227, No. 3, p. 21P – 25P (1987). With a correction in Vol. 229, No. 2, p. 355 (1987).
It is shown that round dynamical models can be fitted to available population II star kinematical data provided the models are not restricted to be scale–free.

155.013 Map of the Galactic center region in the 1.8 MeV ^{26}Al gamma–ray line.
P. von Ballmoos, R. Diehl, V. Schönfelder.
Astrophys. J., Vol. 318, No. 2, p. 654 – 663 (1987).
During a balloon flight with the MPI Compton telescope from Uberaba/Brazil, the 1.809 MeV γ–ray line from radioactive ^{26}Al was detected when the Galactic center came into the field of view. The image is consistent with a "point"–source origin at the Galactic center. The statistical significance of the line detection is 3.8 σ, and the point–source line flux is $(6.4 \pm 2.6) \times 10^{-4} \text{cm}^{-2}\text{s}^{-1}$. The statistical accuracy of the measurement is, however, not sufficient to exclude a diffuse origin of the γ–ray line in interstellar space from unresolved candidate sources.

155.014 The binary frequency of high–velocity field dwarfs as obtained with CCD measures.
H. A. Abt, D. W. Willmarth.
Astrophys. J., Vol. 318, No. 2, p. 786 – 793 (1987).
Published studies of high–velocity or weak–lined dwarfs have differed in concluding that the binary frequency is either substantially lower or similar to that of Population I dwarfs. The authors therefore made a new spectroscopic study with 6 times the previous measuring accuracy of 45 high–velocity FG dwarfs. They found five spectroscopic binaries with derived orbital elements, including three previously known. The resulting frequency of spectroscopic binaries with derived orbital elements is $11\% \pm 5\%$, which is low relative to the 20% for Population I dwarfs. In addition, $\sim 9\% \pm 3\%$ of the high–velocity dwarfs have visual companions, compared with 30% for Population I dwarfs.

155.015 Determination of space distribution parameters for different stellar groups in the Hertzsprung–Russell diagram.
E. K. Kharadze, R. A. Bartaya, S. V. Vereshchagin, O. B. Dluzhnevskaya, E. D. Pavlovskaya, A. Eh. Piskunov.
Astron. Zh., Tom 64, Vyp. 4, p. 696 – 707 (1987). In Russian. English translation in Sov. Astron., Vol. 31, No. 4.
Space densities and parameters of z–distribution for ten stellar groups in the Hertzsprung–Russell diagram (A – KIII, A – GIV, A – GV) were determined. Data from the Abastumani two–dimensional spectral classification catalogue, and a statistical procedure of data processing, allowing to take into account uncertainties of observational data as well as the catalogue selection effect, were used. The study of the space distributions allowed to carry out an independent estimation of the luminosity dispersion for different sequences in the Hertzsprung–Russell diagram.

155.016 Synthesis of galactic deuterium in hot accretion disks.
B. V. Vajner.
Sov. Astron. Lett., Vol. 12, No. 5, p. 319 – 321 (1987). English translation of 42.155.014.

155.017 The rate of nova production in the Galaxy.
W. Liller, B. Mayer.
Publ. Astron. Soc. Pac., Vol. 99, No. 617, p. 606 – 609 (1987).
The ongoing PROBLICOM program in the Southern Hemisphere now makes it possible to derive a reliable value for the overall production rate of Galactic novae. The result,

$73 \pm 24 \text{ y}^{-1}$, indicates that the Galaxy outproduces M31 by a factor of two or three. The authors estimate that the rate of supernova ejecta is one and a half orders of magnitude greater than that of novae in the Galaxy.

155.018 Distribution of classical cepheids in the galactic plane.
C. Kim.
Astrophys. Space Sci., Vol. 136, No. 1, p. 101 – 108 (1987).
Major spiral arms were traced from the distribution of long–period classical cepheids on the projected galactic plane. The position of these spiral features have been compared with those from other optical tracers such as H II regions and OB star groups. Also the galactic longitude distribution of classical cepheids and open clusters are compared.

155.019 A magnetodynamical model for the galactic center lobes.
K. Shibata, Y. Uchida.
Publ. Astron. Soc. Jpn., Vol. 39, No. 4, p. 559 – 571 (1987).
The authors present an MHD model for the galactic center lobes (GCL) by using an axisymmetric 2.5–dimensional MHD simulation. According to this model, GCL is a low–energy jet emanating from the H II gas disk extending beyond $r \sim 100$ pc from the galactic center. The model is based on the "sweeping–magnetic–twist" mechanism developed by the authors for the production of cosmic jets. The authors incorporate the realistic gravitational potential suitable for the galactic center region, in which the rotational velocities are approximately constant for $r = 20 – 100$ pc. The difference between the models with this realistic potential and those with the potential due to a point mass is examined in detail. On the basis of the numerical results, the authors present a scenario for the formation of the GCL.

155.020 Thermal and nonthermal emission structures in the galactic center region.
W. Reich, Y. Sofue, E. Fürst.
Publ. Astron. Soc. Jpn., Vol. 39, No. 4, p. 573 – 587 (1987).
By a detailed comparison of background–filtered maps of the 11–cm (2.695–GHz) radio continuum emission and the 60–μm infrared emission for the central $2° \times 3°$ region of the Galaxy the authors were able to separate nonthermal and thermal emission structures. Based on this comparison they found that the "galactic center lobe (GCL)" basically has nonthermal character–istics. Near Sgr A two possibly nonthermal spurlike structures are noted. The nuclear disk shows unusual high infrared emission. A previously unrecognized supernova remnant was identified.

155.021 A southern atlas of galactic hydrogen (The region $0° \leqslant l \leqslant 12°, -3° \geqslant b \geqslant -17°$).
M. L. Franco, W. G. L. Pöppel, E. R. Vieira.
Rev. Mex. Astron. Astrofis., Vol. 15, No. 1, p. 11 – 28 (1987).
The authors present observational data in the 21 cm line of neutral hydrogen, which have been obtained with the 30–m dish of the IAR. The observations cover $0° \leqslant l \leqslant 12°$, $-3° \geqslant b \geqslant -17°$ at intervals of 1° in both l and b. The radial velocity interval extends from -100 to $+100 \text{ km s}^{-1}$ with a kinematical resolution of 2 km s^{-1}.

155.022 Galactic chaos and the circular velocity at the sun.
R. G. Carlberg, K. A. Innanen.
Astron. J., Vol. 94, No. 3, p. 666 – 670 (1987).
Our galaxy has a substantial nucleus that exhibits itself as a strong rise in the rotation curve inside of 2 kpc, peaking around 500 pc. Stars with very eccentric orbits that pass through the nuclear region generally cannot be confined to a flattened disk distribution; instead, they spend most of their time in the halo. In a local sample of high–velocity–dispersion disk stars, there should be an apparent deficiency of stars with very low angular momenta, because they are scattered to much higher scale heights. The deficiency or gap will be centered on the circular velocity as reflected in the motion of the LSR. Using a new galactic model, the authors calibrate the predicted depression in the distribution of tangential velocities, finding a half–width of 40 km s^{-1} with a depth greater than 80%. They present evidence

that the expected deficiency of low–angular–momentum stars does exist in the local stars. The available data favor a scale and model–free circular velocity at the sun in the range 225 – 245 km s^{-1} with a most probable value of 235 km s^{-1}.

155.023   Physical conditions, dynamics, and mass distribution in the center of the Galaxy.
R. Genzel, C. H. Townes.
Annu. Rev. Astron. Astrophys., Vol. 25, p. 377 – 423 (1987). – See Abstr. 003.003.

This review emphasizes recent investigations of the central 10 pc of the Galaxy and conclusions of energetics, dynamics, and mass distribution derived from them. New information on the Galactic center comes from X– and gamma–ray measurements and from infrared and microwave studies, especially from spectroscopy, high–resolution imaging, and interferometry.

155.024   The IRAS view of the Galaxy and the solar system.
C. A. Beichman.
Annu. Rev. Astron. Astrophys., Vol. 25, p. 521 – 563 (1987). – See Abstr. 003.003.

This article concentrates on IRAS results important to Galactic and solar system astronomy. The infrared emission from the celestial sphere is dominated by the zodiac and the Milky Way. The relative importance of these two planes varies with the wavelength: At 12 and 25 μm, interplanetary dust dominates the emission, while at 60 and 100 μm, emission from interstellar dust becomes increasingly important. This review begins with the zodiacal cloud and ends with the study of the Galaxy as a galaxy. Contents: The solar system. Star formation. Stars: photospheres, mass loss, protoplanetary disks, and brown dwarfs. Diffuse Galactic emission.

155.025   The Galactic spheroid and old disk.
K. C. Freeman.
Annu. Rev. Astron. Astrophys., Vol. 25, p. 603 – 632 (1987). – See Abstr. 003.003.

The author discusses some of the major Galactic components, such as the old disk, the thick disk, the metal–weak halo, the globular cluster system, and the Galactic bulge. The emphasis is on the structure, kinematics, and general chemical properties of these components.

155.026   Investigation of galactic rotation by the maximum likelihood method.
I. V. Petrovskaya.
Kinematika Fiz. Nebesn. Tel, Tom 3, No. 4, p. 19 – 25 (1987). In Russian. English translation in Kinematics Phys. Celest. Bodies.

The maximum likelihood method is proposed for investigation of the rotation law of the neutral hydrogen subsystem both for the galactic plane and for the regions outside this plane. The whole 21 cm line profile is used. The calculations of the rotation curve at the galactic plane beyond the solar circle ($z = 0$, $R > R_0$) gave no evidence for a turnover up to the distance of 20 kpc from the galactic centre.

155.027   Stockert's chimney – a galactic fountain.
W. Kundt, P. Müller.
Astrophys. Space Sci., Vol. 136, No. 2, p. 281 – 287 (1987).

A new type of galactic outflow phenomenon has been discovered and published under the name of "thermal spur" by Müller et al. (1987). It is argued that this outflow consists of relativistic pair plasma with an admixture of partially ionized hydrogen and cosmic rays, escaping from a large H II region. This "chimney" serves as a substantial leak for the cosmic rays, i.e. it may be an essential part of the galactic fountain.

155.028   Orbital eccentricity study for the spherical component of our Galaxy.
S. Ninković.
Astrophys. Space Sci., Vol. 136, No. 2, p. 299 – 314 (1987).

An analysis of the data concerning high–velocity stars from Eggen's catalogue aimed at a determination of the approximate slope of the mass function for the spherical component of our

Galaxy, and at estimating the local circular velocity, as well as the local rotation velocity, as by–products, has been performed. The conclusions are that: a linear dependence of the mass on the radius is very likely; the value of the limiting radius is most likely equal to (40 ± 10) kpc; the two local velocities are approximately equal to each other, being both equal to (230 ± 30) km s^{-1}; the local escape velocity appears to be the most likely equal to (520 ± 30) km s^{-1}; the total mass of a corona, obtained in this way, is $(5 \pm 1) \times 10^{11} M_\odot$.

155.029   Our galactic center.
E. Serabyn.
Astrophysical jets and their engines, p. 47 – 65 (1987). – See Abstr. 012.021.

The last few years have produced large and rapid advances in our understanding of the galactic center. New information has come via a variety of methods, including high spatial resolution radio and infrared continuum mapping, and line observations in the micron through centimeter wavelength range. These observations have addressed the distribution and kinematics of both the stellar population and the interstellar medium in the vicinity of the galactic center. This review summarizes the understanding of the central 100 pc of our galaxy which has emerged from this new body of data.

155.030   Determination of the age of the Galaxy with the method of uranium–thorium isotopic ratios.
Yu. S. Lyutostanskij, S. V. Malevannyj, I. V. Panov, V. M. Chechetkin.
Mosk. Inzh.–Fiz. Inst., Prepr., No. 062, p. 3 – 21 (1986). In Russian. Abstr. in Ref. Zh., 51. Astron., 8.51.348 (1987).

155.031   Investigation of the chemical composition of the galactic disk.
V. G. Klochkova, V. E. Panchuk.
Stellar atmospheres, p. 54 – 62 (1987). In Russian. Abstr. in Ref. Zh., 51. Astron., 8.51.369 (1987). – See Abstr. 003.005.

155.032   Determination of the distance of the solar system to the center of the Galaxy.
V. B. Smirnov.
Ufim. Neft. Inst. Ufa, 4 pp. (1987). In Russian. Abstr. in Ref. Zh., 51. Astron., 8.51.729 (1987).

155.033   Galactic center molecular clouds. I. Spatial and spatial–velocity maps.
J. Bally, A. A. Stark, R. W. Wilson, C. Henkel.
Astrophys. J., Suppl. Ser., Vol. 65, No. 1, p. 13 – 82 (1987).

The authors present maps of the distribution of molecular gas lying within 500 pc of the Galactic center. Based on over 12,000 spectra of the 98 GHz carbon monosulfide line, and over 5000 spectra in the 110 GHz ^{13}CO line, the maps show that the molecular clouds in this part of the Galaxy are more than an order of magnitude denser and have higher internal velocity dispersions than clouds in the Galactic disk. Peculiar (noncircular) velocities comparable in magnitude to the circular velocities about the Galactic center are common.

155.034   The spiral structure of our galaxy from observations of the interstellar extinction.
L. A. Urasin.
Pis'ma Astron. Zh., Tom 13, No. 10, p. 850 – 854 (1987). In Russian. English translation in Sov. Astron. Lett., Vol. 13.

A model of the two–arm spiral structure of the Galaxy is constructed from observations of space distribution of the interstellar dust matter. This model is the logarithmic spiral with characteristic angle 6.5°.

155.035   Galactic gamma rays and gas tracers.
A. W. Strong.
Physical processes in interstellar clouds, p. 75 – 80 (1987). – See Abstr. 012.027.

The methodical study of the relation between galactic γ–rays and gas tracers is being pursued with increasing confidence as the quantity and quality of available data increases. Surveys of CO now cover the whole galactic plane with significant latitude extent. The author discusses the principles of the methods being used, and reviews the results so far presented. Further, new surveys of the southern galaxy are included to show the excellent correlation of γ–rays with gas over the whole galaxy. The new data will enable the previous analyses to be significantly improved.

155.036 The distribution of molecular gas in the Galaxy.
J. L. Osborne, M. Parkinson, K. M. Richardson,
A. W. Wolfendale.
Physical processes in interstellar clouds, p. 81 – 87 (1987). – See Abstr. 012.027.
Attention is given to the contemporary arguments concerning the amount, and distribution, of molecular gas in the Galaxy and new results based on analyses of recent infra–red data obtained by the IRAS satellite are given. A value of 1.0 ± 0.3 for the ratio of H_2 to HI in the important range R = 4 – 8 kpc is found. Concentrating on the region near the Galactic centre (R < 0.5 kpc), the resulting value is $(2 - 4) \times 10^7 M_\odot$ and is much lower than estimates by way of CO emission.

155.037 Gamma rays and the distribution of cosmic rays in the Galaxy.
A. W. Wolfendale.
Physical processes in interstellar clouds, p. 89 – 93 (1987). – See Abstr. 012.027.
A brief assessment is given of the question of a gradient of cosmic ray intensity in the Galaxy. It is concluded that there is certainly a gradient in the electron component. Concerning the nuclear component there is less certainty although the author inclines to the view that there is a gradient here too; certainly a consistent picture can be constructed in which both electrons and nuclei have gradients of magnitude which bear a simple relationship one to the other.

155.038 Mass function of stars in the solar neighbourhood.
N. C. Rana.
Astron. Astrophys., Vol. 184, No. 1/2, p. 104 – 118 (1987).
Using a new law of the star formation proposed by Rana and Wilkinson (1986) and some recent data on luminosity function and scale heights of main sequence disc stars in the solar neighbourhood, the mass function of these stars is investigated and compared with those of Scalo (1986) and Larson (1986). It is found that the classical views such as near constancy of star formation rate, an IMF having a simple power law in mass, no appreciable infall of matter from the halo, and no exotic amount of dark remnants or brown dwarfs, are still able to offer a consistent view of the chemical evolution of the solar neighbourhood.

155.039 A comparative study of galactic radial velocity fields.
J. V. Feitzinger, J. Spicker.
Astron. Astrophys., Vol. 184, No. 1/2, p. 122 – 132 (1987).
The authors have demonstrated several ways in which the basic circular velocity field of the Milky Way system can or should be modified. Their choice of parameters is in no way unique, representing only one or two possible sets of modifications. It should primarily serve to pictorially show how the basic radial velocity field changes its appearance as a consequence of the various modifications. The important possibility that the basic circular orbit assumption may be wrong in a large way (e.g., through the occurrence of elliptical orbits) has been ignored completely, so that there is still much work to be done in this field.

155.040 A molecular counterpart to the galactic center arc.
E. Serabyn, R. Güsten.
Astron. Astrophys., Vol. 184, No. 1/2, p. 133 – 143 (1987).
The ionized arched filaments have been found to lie predominantly along the edges of negative velocity molecular clouds. This gas has a density of a few $\times 10^4 cm^{-3}$, a temperature $\gtrsim 100K$, and a mass ~ 2 orders of magnitude higher than the ionized gas mass.

Molecular material with similar velocities also appears to form the northern ionization boundary of Sgr A. If all of this molecular gas is part of one structure, then the arched filaments lie within ~ 40 pc of the galactic center. The spatial distribution of the negative velocity gas suggests a tidally disrupted molecular cloud. Supporting this hypothesis is a rather high volume filling factor ($\gtrsim 0.3$) for the dense molecular gas. The cloud surfaces could be ionized by a few localized pockets of OB stars, or via collisional ionization by nonthermal particles trapped in the magnetic field. Further observations are required to distinguish between these possibilities.

155.041 Parameters of the initial mass function.
S. V. Vereshchagin, N. A. Kiseleva.
Astron. Zh., Tom 64, Vyp. 5, p. 980 – 986 (1987). In Russian. English translation in Sov. Astron., Vol. 31, No. 5.
The solar–neighbourhood initial mass function (IMF) was constructed using data on more than 22×10^3 stars from the Catalogue of Stellar Identifications with additional data. The IMF slope ($\Gamma = -1.30 \pm 0.20$) in the mass range 0.4 to 25 $M_\odot$ was found to be close to Salpeter's value. Dependence of Γ on the sample limiting magnitude, reddening value, the galactic disk age, rate of star formation, as well as on the adopted calibration scales of stellar masses and ages was studied, and corresponding relations were derived. It was shown that Γ is most strongly influenced by the stellar mass scale. Thus, for determination of the reliable IMF, it is necessary to improve the "MK spectrum–mass" (or mass–luminosity) relation.

155.042 The stellar cluster.
D. A. Allen.
AIP Conf. Proc., No. 155, p. 1 – 7 (1987). – See Abstr. 012.029.
The author reviews the present knowledge of the stellar distribution within the inner parsec of the Galaxy.

155.043 Dust emission and the evidence for star formation.
I. Gatley.
AIP Conf. Proc., No. 155, p. 8 – 18 (1987). – See Abstr. 012.029.
The dust distribution and the structure of the galactic center, as well as the nature of the central source, are discussed in the light of infrared and radio observations.

155.044 Atomic and molecular gas in the circumnuclear disk.
R. Güsten.
AIP Conf. Proc., No. 155, p. 19 – 29 (1987). – See Abstr. 012.029.
Observations of numerous atomic and molecular transitions, probing a wide range of excitation conditions, have given a fairly detailed picture of the physical conditions and energetics of the gas in the galactic center, its velocity field as potential probe of the central mass distribution, and the disk's relation to the nucleus.

155.045 The galactic center compact nonthermal radio source.
K. Y. Lo.
AIP Conf. Proc., No. 155, p. 30 – 38 (1987). – See Abstr. 012.029.
The current observational status of Sgr A*, the compact nonthermal radio source at the galactic center, is reviewed. With its unique properties in the Galaxy and being the only unusual object at the center with dimensions approaching the gravitational radius of a $\sim 10^6 M_\odot$ black hole, Sgr A* is the best candidate for marking the location of a massive collapsed object.

155.046 The ionized gas in the galactic center.
T. R. Geballe.
AIP Conf. Proc., No. 155, p. 39 – 50 (1987). – See Abstr. 012.029.
This paper reviews recent observations and interpretations of the structure and kinematics of the ionized gas within 2 parsecs of the galactic center.

155.047 Galactic positron annihilation radiation.
R. Ramaty, R. E. Lingenfelter.
AIP Conf. Proc., No. 155, p. 51 – 61 (1987). – See Abstr. 012.029.
The production of positron–electron pairs in variable, compact and high luminosity hard X–ray and gamma ray sources is now

recognized to be a dominant process shaping the spectra of such objects. In particular, pair production is thought to play a dominant role in the physics of active galactic nuclei. However, the 511 keV line, which is the most unambiguous signature of positron annihilation, was detected only from the central regions of our Galaxy. In this paper the authors review the line and relevant high energy continuum observations and discuss their interpretations.

155.048   The central object: some comments and speculations.
M. J. Rees.
AIP Conf. Proc., No. 155, p. 71 – 78 (1987). – See Abstr. 012.029.
There is strong circumstantial evidence for a $\sim 10^6 M_\odot$ black hole at the Galactic Center. Such an object could account naturally for the unusual compact radio source. Capture and destruction of stars by the hole could lead to directional ejection of some of the debris; this phenomenon might be relevant to the energetics and morphology within the central "cavity".

155.049   The stellar population at the galactic center.
M. J. Lebofsky, G. H. Rieke.
AIP Conf. Proc., No. 155, p. 79 – 82 (1987). – See Abstr. 012.029.
Preliminary results from a 2 μm survey covering an area of 400 square parsecs ($5' \times 5'$) of the galactic center are presented. That a large amount of star formation has occurred recently is confirmed.

155.050   Kinematics of individual stars in the galactic center.
K. Sellgren, D. N. B. Hall, S. G. Kleinmann,
N. Z. Scoville.
AIP Conf. Proc., No. 155, p. 83 – 86 (1987). – See Abstr. 012.029.
The authors have obtained high–resolution 2 μm spectra of six late–type stars within 2 pc of the galactic center. They have derived spectral types, reddenings, and radial velocities for these stars. They find the supergiant density in the galactic center is lower than in previous determinations. The mass distribution derived from the stellar velocities is compared to those derived from gas velocities and from the 2 μm light.

155.051   Stellar kinematics in the central 10 pc of the Galaxy.
M. T. McGinn, K. Sellgren, E. E. Becklin,
D. N. B. Hall, I. Gatley.
AIP Conf. Proc., No. 155, p. 87 – 90 (1987). – See Abstr. 012.029.
Observations of the profile of the 2.3 μm CO 2–0 bandhead in the galactic center are discussed. These large–beam (20″ or 1 pc) observations provide information on the integrated stellar velocity and velocity dispersion in the inner 10 pc of the Galaxy. There is evidence that the velocity dispersion dominates the rotational velocity at all positions within 5 pc of the galactic center, particularly in the inner regions. The authors also note a significant difference between the stellar velocities they measure and the gas velocities measured in previous studies.

155.052   Is there a cusp in the stellar distribution in the galactic center?
G. H. Rieke, M. J. Lebofsky.
AIP Conf. Proc., No. 155, p. 91 – 94 (1987). – See Abstr. 012.029.

155.053   The 18–cm OH distribution in the galactic center torus.
Aa. Sandqvist, R. Karlsson, J. B. Whiteoak,
F. F. Gardner.
AIP Conf. Proc., No. 155, p. 95 – 98 (1987). – See Abstr. 012.029.
The 18–cm OH distribution in the galactic center region near Sgr A has been studied in all four of the 1612, 1665, 1667 and 1720 MHz OH lines using the VLA with 4 arcsec angular resolution and 9 km s^{-1} velocity resolution. Three 1667 MHz OH spectral line absorption maps, at $+51$, $+25$ and -1 km s^{-1}, covering a 4.3 × 4.3 region around Sgr A are presented together with an 18–cm continuum map. In addition, a complete set of velocity maps from $+139$ to -159 km s^{-1}, covering a $3' \times 3'$ region around the galactic center nuclear torus, is presented.

155.054   NH₃ in the molecular ring at the galactic center.
J. M. Jackson, P. T. P. Ho, A. H. Barrett.
AIP Conf. Proc., No. 155, p. 99 – 102 (1987). – See Abstr. 012.029.
Using the VLA, the NH$_3$ (J,K) = (3,3) inversion line has been mapped toward the galactic center in a $2'$ field with 11″ by 8″ angular resolution (~ 0.5 pc at 10 kpc). A number of NH$_3$ condensations coincident with the molecular ring mapped in HCN by Güsten et al. (1986) are detected. Assuming that the ionized gas flows toward the galactic center and that the NH$_3$ emission defines the "mass–weighted" velocity, the authors conclude that the western side of the ring is behind the galactic center while the eastern side is in front of the galactic center.

155.055   Hat Creek aperture synthesis observations of the circum–nuclear ring in the galactic center.
R. Güsten, R. Genzel, M. C. H. Wright, D. T. Jaffe, J. Stutzki,
A. Harris.
AIP Conf. Proc., No. 155, p. 103 – 105 (1987). – See Abstr. 012.029.
The authors have made aperture synthesis observations of HCN J = 1–0 emission and absorption in the central 5 pc of the Galaxy with 2″ spatial and 4 km/s spectral resolution. The resulting maps show a clumpy ring of molecular gas which surrounds the ionized central 2 pc of the Galaxy. Structure and dynamics of the molecular ring are discussed.

155.056   Rotating molecular ring at the galactic center.
N. Kaifu, M. Hayashi, J. Inatani, I. Gatley.
AIP Conf. Proc., No. 155, p. 106 – 109 (1987). – See Abstr. 012.029.
The rotating molecular ring surrounding the galactic center was found in the CS 2–1, HCN 1–0, HCO$^+$ 1–0 and ^{13}CO 1–0 lines. The 19″ – 15″ resolution maps of above 4 molecules show generally similar structure of the ring with the rotation velocity of 100 km/sec and the radial velocity of roughly 50 km/sec. The ionized gas feature fits very well to the edge of the molecular ring in its south part, but seems to sit in the central hole of the molecular ring in the north part. The structure and dynamics of the central 5 pc of the Galaxy are discussed.

155.057   The CO (J = 2–1) distribution in the inner 10 pc of the Galaxy.
Y. Fukui, E. Churchwell.
AIP Conf. Proc., No. 155, p. 110 – 113 (1987). – See Abstr. 012.029.
The authors report observational results of the J = 2–1 CO emission in the $5' \times 5'$ region toward Sgr A*. The fully sampled high angular resolution (30″) data reveal considerable details of the neutral gas "disk" in the central 10 pc of the Galaxy. Although the CO distribution is generally consistent with a tilted rotating disk, the present data indicate that the distribution is significantly asymmetric with respect to Sgr A*. The CO rotation curve derived from the negative velocity CO lobe indicates that the mass in the inner 10 pc increases linearly with radius from 3 pc to 10 pc.

155.058   CS emission from the galactic center ring.
N. J. Evans II.
AIP Conf. Proc., No. 155, p. 114 – 117 (1987). – See Abstr. 012.029.

155.059   Excitation gradient of the molecular gas in the Sgr A circum–nuclear ring.
J. B. Lugten, G. J. Stacey, A. I. Harris, R. Genzel,
C. H. Townes.
AIP Conf. Proc., No. 155, p. 118 – 122 (1987). – See Abstr. 012.029.
The authors present new measurements of CO submillimeter (J = 7→6) and far–infrared (J = 14→13, J = 21→20) lines in the galactic center. The new data, together with earlier CO measurements, are used to derive the physical conditions of the dense, warm molecular gas at different positions in the Sgr A circum–nuclear ring.

155.060 **Mapping of C$^+$ far–infrared emission in the inner Galaxy.**
J. B. Lugten, R. Genzel, M. K. Crawford, C. H. Townes.
AIP Conf. Proc., No. 155, p. 123 – 126 (1987). – See Abstr. 012.029.

The authors have mapped the [C II] 158 μm emission over a region of about 8′ diameter towards Sgr A. The strongest C$^+$ emission comes from a ring of gas centered on Sgr A*. The ring has an inner radius of about 1.7 pc and can be traced in C$^+$ emission out to a least 8 pc from Sgr A*. The ring is inclined about 70° to the line of sight and tilted about –20° with respect to the galactic plane. The velocity field of the C$^+$ is that expected for a tilted, inclined thin ring which is predominantly rotating about the center.

155.061 **Radio emission from Sgr A and its extended halo.**
M. Morris, F. Yusef–Zadeh.
AIP Conf. Proc., No. 155, p. 127 – 132 (1987). – See Abstr. 012.029.

Observations of the Sgr A complex with the VLA reveal that the diffuse structure surrounding Sgr A East and West consists of two extended components. One is a 20–pc–diameter halo which appears to be associated with the nonthermal Sgr A–E shell source, and which is elongated along the galactic plane. The second component is a set of narrow radio protrusions from the Sgr A–W complex which run perpendicular to the galactic plane. These protrusions are discussed in the context of a model of a directed wind. Comparison of low and high–frequency maps reveals that the bulk of Sgr A–W is located in front of Sgr A–E. This supports the notion that Sgr A–E is unrelated to the activity in the galactic nucleus. New, high–resolution images of the Sgr A–W complex show several new features of the inner few parsecs of the galaxy. Three of these are briefly discussed.

155.062 **86 GHz aperture synthesis observations of the galactic center.**
M. C. H. Wright, R. Genzel, R. Güsten, D. T. Jaffe.
AIP Conf. Proc., No. 155, p. 133 – 137 (1987). – See Abstr. 012.029.

The authors present 86 GHz aperture synthesis observations of the radio continuum with 4″ × 8″ angular resolution in the central 5 pc of the Galaxy. From a comparison with maps with comparable resolution at 1.4, 5, 15, and 24 GHz they have derived the distribution of thermal and non–thermal emission.

155.063 **Small scale structure of the galactic center in the far–IR.**
D. F. Lester, M. Joy, P. M. Harvey, H. B. Ellis Jr.
AIP Conf. Proc., No. 155, p. 138 – 141 (1987). – See Abstr. 012.029.

Far infrared slit scans across the galactic center are presented. These scans, made along the galactic plane through IRS 16, show two luminosity peaks 39″ (2 pc) apart. The NE peak is considerably hotter than that to the SW. When image restoration and deconvolution techniques are applied to the data, it is found that the luminosity peaks are surrounded by a region of cold dust emission. The cool dust emission appears to coincide spatially with the molecular ring, while the luminosity peaks are completely contained within the ring.

155.064 **Observations of galactic center gas dynamics with a cryogenic echelle spectrometer.**
J. H. Lacy, D. F. Lester, J. F. Arens, M. C. Peck, S. Gaalema.
AIP Conf. Proc., No. 155, p. 142 – 145 (1987). – See Abstr. 012.029.

Spectra have been obtained of the [Ne II] (12.8 μm) line from the central 12″ × 14″ of the Galaxy with a new cryogenic echelle spectrometer. The central ″bar″ region of Sgr A West was completely mapped with ~2″ and 30 km s^{-1} resolution. The improved sensitivity and simultaneous spectral and spatial sampling resulting from the use of a 10 × 64 element detector array provide new insight into the flows of gas in the galactic center.

155.065 **8.3 and 12.4 micron imaging of the galactic center with the Goddard infrared array camera.**
D. Y. Gezari, R. Tresch–Fienberg, G. G. Fazio, W. F. Hoffmann, I. Gatley, G. Lamb, P. Shu, C. McCreight.
AIP Conf. Proc., No. 155, p. 146 – 152 (1987). – See Abstr. 012.029.

A 30 × 30 arcsec field at the galactic center (1.5 × 1.5 parsec) has been mapped at 8.3 μm and 12.4 μm with high spatial resolution and accurate relative astrometry. A 3σ upper limit of 0.5 Jy/arcsec^2 was set for the strong near infrared source IRS 16. Color temperature and dust opacity distributions derived from the spatially registered 8.3 μm and 12.4 μm images show that the compact infrared sources are essentially density features.

155.066 **Brackett alpha images of the galactic center.**
W. J. Forrest, M. A. Shure, J. L. Pipher, C. E. Woodward.
AIP Conf. Proc., No. 155, p. 153 – 156 (1987). – See Abstr. 012.029.

Images of the central parsec of the galaxy have been obtained in the Brα emission line of hydrogen. The images are, for the most part, very similar to 5 and 15 GHz radio continuum images of this region. Alignment of the Brα with the radio images gives an improved position of the Sgr A* radio point source with respect to the infrared sources: 5.50″S and 0.15″W of IRS 7, a position remarkably devoid of compact objects in the near infrared. This position rules out identification of Sgr A* with IRS 16 Centre or IRS 16 NW. A number of compact Brα sources with no radio counterparts are seen.

155.067 **Preliminary results of the Spacelab–2 Infrared Telescope survey of the galactic plane at 2.4 μm.**
G. J. Melnick, G. G. Fazio, D. G. Koch, G. H. Rieke, E. T. Young, F. J. Low, W. F. Hoffmann, T. N. Gautier.
AIP Conf. Proc., No. 155, p. 157 – 161 (1987). – See Abstr. 012.029.

The authors present preliminary 2.4 μm maps of the galactic center region and the galactic plane between $l = -10°$ and 110° obtained with the Infrared Telescope (IRT) flown aboard the Space Shuttle in 1985. These results are in qualitative agreement with earlier balloon–borne observations which showed the overall 2.4 μm emission to be more widely distributed perpendicular to the plane than is seen in the longer wavelength IRAS survey.

155.068 **The distance to the center of the galaxy.**
J. M. Moran, M. J. Reid, M. H. Schneps, C. R. Gwinn, R. Genzel, D. Downes, B. Rönnäng.
AIP Conf. Proc., No. 155, p. 166 – 167 (1987). – See Abstr. 012.029.

The authors have estimated the distance to the center of the galaxy from the measurements of the proper motions of 24 H$_2$O maser spots in the source Sgr B2–North. A comparison of the transverse angular velocities measured from VLBI observations and the line of sight Doppler velocities gives a distance estimate of 7.1 ± 1.5 kpc.

155.069 **Don't be afraid to observe the next series of lunar occultations of the galactic center, 1986 – 1989!**
Aa. Sandqvist.
AIP Conf. Proc., No. 155, p. 168 – 171 (1987). – See Abstr. 012.029.

A new series of lunar occultations of the galactic center takes place during the period 1986 – 1989. The previous series of galactic center occultations occurred in 1968 – 1970. That series revealed that the continuum radio source, Sgr A, consists of at least two components (now known as Sgr A West and Sgr A East) and that the continuum source is surrounded by a rotating and contracting cloud of dust and molecules. It could be rewarding to observe the new series of occultations at X–ray, infrared and radio wavelengths. Occultation predictions are available from the author upon request.

155.070 A magnetic loop model for activity in the galactic centre.
    J. Heyvaerts, R. E. Pudritz, C. A. Norman.
AIP Conf. Proc., No. 155, p. 176 – 180 (1987). – See Abstr.
012.029.
    The authors propose that radio structures on 0.1 – 100 pc
scales in the galactic centre radio lobe are manifestations of
magnetic activity in a central source object. The observations
indicate that truly one dimensional structures occur and the
authors hold that this is strong evidence for magnetic loops.

**155.071 Further "loss of weight" by a black hole at the galactic
    center.**
L. M. Ozernoy.
AIP Conf. Proc., No. 155, p. 181 – 183 (1987). – See Abstr.
012.029.
    The author summarizes some recent and new arguments which
constrain the mass of a black hole at the galactic center by a value
not exceeding a few hundred solar masses.

**155.072 Positron annihilation in the galactic center: "Cheshire
    Cat" Compton scattering and "excess continuum".**
M. L. Bildsten, W. H. Zurek.
AIP Conf. Proc., No. 155, p. 184 – 187 (1987). – See Abstr.
012.029.
    Two separate observations of the γ–ray spectrum originating
from the galactic center were made by HEAO–3 in the fall of 1979
and in the spring of 1980. The 2γ 511 keV annihilation line flux
decreased by a factor of three over the corresponding six month
period, whereas the excess γ–ray continuum below the 511 keV
line, often interpreted as 3γ decay of orthopositronium, barely
changed. This apparent discrepancy in the temporal behaviour
makes it difficult to associate the bulk of the excess continuum
with the 3γ decay of positronium. The authors show that
Compton scattering of the line and high energy radiation
provides a natural explanation for the surprisingly small change
seen in the excess continuum.

155.073 Molecular gas associated with the galactic center arc.
    E. Serabyn, R. Güsten.
AIP Conf. Proc., No. 155, p. 188 – 189 (1987). – See Abstr.
012.029.

**155.074 Spatial and kinematic structure of the thermal compo-
    nents of the galactic center arc.**
F. Yusef-Zadeh, M. Morris, J. H. van Gorkom.
AIP Conf. Proc., No. 155, p. 190 – 195 (1987). – See Abstr.
012.029.
    High–resolution radio continuum and radio recombination
line observations of two bright segments of the filamentary arc
near the galactic center have been carried out using the VLA. On
the basis of the polarization and recombination line characteristics of these regions, one can clearly identify and distinguish the
thermal and non–thermal features of the arc. These observations
provide strong evidence in support of a physical interaction
between these two components. The dominant components of
thermal emission are centered on G 0.18–0.04 and G 0.1 + 0.08,
which are integral parts of the arc, but appear to be physically
coupled to two distinct molecular clouds – the 50 and –30 km/s
clouds. A complex and highly organized flow of ionized gas is
seen in both sources. The authors consider several possibilities for
the source of ionization in G 0.18–0.04.

**155.075 Evidence for activity at the galactic center based on low
    frequency radio continuum observations.**
N. E. Kassim, W. C. Erickson, T. N. LaRosa.
AIP Conf. Proc., No. 155, p. 196 – 201 (1987). – See Abstr.
012.029.
    Aperture synthesis observations of the galactic center region at
123.0 and 110.6 MHz reveal striking asymmetric steep spectrum
($\alpha \leqslant -0.8$) radio lobes. The northern galactic lobe appears to be a
unique galactic source and cannot be easily classified as an SNR,
extragalactic source, or a foreground object. The southern "lobe"
appears directly linked to the galactic center and has been
identified by Yusef-Zadeh et al. (1986) as a low energy jet

emanating from Sgr–A. The authors present here new 80 MHz
observations of an expanded region around the galactic center
obtained one year after the initial set of observations.

155.076 The gaseous galactic halo.
    B. D. Savage.
Spectroscopy of astrophysical plasmas, p. 210 – 225 (1987). – See
Abstr. 003.010.
    Contents: 1. Introduction. 2. Ultraviolet observations of halo
gas. 3. Interpretations of the observations: Column densities. The
z extent of galactic halo gas. Ionization and temperature.
Densities. Abundances. 4. Origin of galactic halo gas.
5. Implications for extragalactic astronomy: The evolution of
interstellar elemental abundances. Quasar absorption lines.
6. Future prospects.

155.077 The stellar population of nuclear bulge.
    J. A. Frogel.
Nearly normal galaxies. From the Planck time to the present,
p. 4 – 9 (1987). – See Abstr. 012.031.
    Coincident with this workshop Albert Whitford and the
author submitted for publication a description of the results from
the first phase of their study of stars in the Galactic nuclear bulge.
A slightly amended version of the summary section of their paper
is presented.

**155.078 The metallicity distribution of the extreme halo popula-
    tion.**
T. C. Beers.
Nearly normal galaxies. From the Planck time to the present,
p. 41 – 44 (1987). – See Abstr. 012.031.
    The first results from an objective prism survey for extremely
metal–poor stars indicate that there is no apparent dearth of stars
in the metallicity range [Fe/H] $\leqslant$ –2.6, as had been previously
found. Rather, the evidence seems to favor a metallicity
distribution which remains constant down to the lowest measured [Fe/H].

155.079 The stellar population at the Galactic Center.
    M. J. Rieke.
Nearly normal galaxies. From the Planck time to the present,
p. 90 – 95 (1987). – See Abstr. 012.031.
    Because of the complicating factor of 30 magnitudes of
extinction at visible wavelengths, the study of the Galactic Center
has been done at infrared and radio wavelengths. The availability
of area–format detectors in the infrared recently has permitted
surveying areas of the Galactic center that are comparable in size
to those studied in other galaxies. Preliminary results from a 2 μm
survey covering an area of 400 pc^2 ($5' \times 5'$) centered on the
Galactic Center are presented here. That a large amount of star
formation has occurred recently is confirmed.

**155.080 The relative masses of the Milky Way's components, the
    Ostriker–Caldwell approach, and differential rotation
beyond the solar circle.**
P. L. Schechter.
Nearly normal galaxies. From the Planck time to the present,
p. 116 – 120 (1987). – See Abstr. 012.031.
    A model for the distribution of mass within the Galaxy
includes scale lengths, characteristic densities, and shapes of the
mass distributions for each of its components. Among the
various approaches for determining the parameters which
characterize the components, the author finds the one employed
by Ostriker and Caldwell especially illuminating. This paper is,
for the most part, a review of their work, emphasizing the
strengths of their approach and pointing to those places where
new data would considerably improve the accuracy of the derived
parameters.

155.081   Disk rotation curves in triaxial potentials.
O. E. Gerhard, M. Vietri.
Structure and dynamics of elliptical galaxies, p. 399 – 400 (1987).
– See Abstr. 012.032 (IAU Symp. No. 127).

The apparent rotation curve of cold gas in a triaxial potential often differs from that in an axisymmetric system in characteristic ways. The data argue for triaxial bulges in our Galaxy and several others.

155.082   Galactic center positronium fraction: observations and simulations.
B. L. Brown, M. Leventhal.
Astrophys. J., Vol. 319, No. 2, p. 637 – 642 (1987).

Recent papers discussing positron annihilation data from the Galactic center have used several incompatible definitions of "the positronium fraction". These differences are sorted out and the observations presented in a consistent manner. The authors find that the apparent positronium (Ps) fraction measured could indicate the presence of Ps at a level consistent with a model for a neutral medium, derived from laboratory experiments, and other models such as an ionized medium or one where dust plays a major role.

155.083   A shocked–jet model of the Galactic center bridge and the radio arc.
Y. Sofue, M. Fujimoto.
Astrophys. J., Lett. Ed., Vol. 319, No. 2, p. L73 – L76 (1987). With plate L6.

Discussing the observed radio features in the Galactic center, the authors propose a shocked–jet model for the radio bridge that connects Sgr A and the radio arc. The bridge is interpreted as a tilted magnetized jet filled with a supersonic and shocked gas ejected from the Galactic nucleus. The jet is largely bent when it encounters an ambient poloidal magnetic field. A magnetic bow shock is formed ahead of the jet at the interaction surface with the ambient magnetic field and is observed as the radio arc.

155.084   The distribution of OB stars and dust in a Milky Way field at $(l, b) = (335°, 0°)$.
M. P. FitzGerald.
Mon. Not. R. Astron. Soc., Vol. 229, No. 2, p. 227 – 244 (1987). With 3 pp. plates. = Contrib. Univ. Waterloo Obs., No. 139.

New photoelectric UBV photometry for 103 stars and MK spectroscopy for 110 stars are presented in a 21 deg^2 field at $(l, b) = (335°, 0°)$; these are mostly OB stars and brighter than $V = 14.4$. Combined with other observations from the literature, these provide photoelectric UBV and slit or objective prism spectra for 111 of the 112 SLS stars in this region, and photoelectric measurements for a further 84 non–OB stars, some with spectra. Most of the stars may be separated into three groups associated with the spiral structure of the inner galaxy, A = Ara OB1 at 1.34 ± 0.05 kpc (51 members), B at 2.41 ± 0.08 kpc (25 members), and C at 3.69 ± 0.23 kpc (19 members). The dust is distributed in two distinct clouds, one in the local arm at 190 ± 30 pc and the other in an interarm cloud of variable extinction at 690 ± 70 pc. The dust may have the same two–colour reddening properties as found in Cygnus.

155.085   G0.18–0.04: interaction of thermal and nonthermal radio structures in the Arc near the galactic center.
F. Yusef–Zadeh, M. Morris.
Astron. J., Vol. 94, No. 5, p. 1178 – 1184 (1987). With plates 97 – 98.

Using a high–resolution radio–continuum image made at a frequency of 4.8 GHz, the authors discuss the radio source G0.18–0.04 located at the intersection of the galactic center Arc with the galactic plane. In this direction, the radio Arc is composed of a set of linear, nonthermal filamentary structures superimposed on a narrow, sickle–shaped, thermal feature which crosses them. The radio image suggests a physical interaction between the thermal and nonthermal structures. The authors argue that the magnetic field is dynamically important and estimate that its strength is very large, $\sim 10^{-3}$ G. Furthermore, they suggest that the ionization of G0.18–0.04 is caused by collisional ionization rather than ultraviolet photons from massive stars. The discussion presents possible ways in which some of the energy of the relativistic particles or the magnetic field in the nonthermal filaments might be extracted to ionize ambient gas in the galactic plane.

155.086   The distribution of molecular gas in the Galaxy.
M. Parkinson, K. M. Richardson, A. W. Wolfendale.
Q. J. R. Astron. Soc., Vol. 28, No. 3, p. 277 – 283 (1987). – See Abstr. 012.037.

After a brief introduction and description of the history of attempts to determine the distribution of molecular gas in the Galaxy, new results based on an analysis of recent infrared data obtained by the $IRAS$ satellite and from elsewhere are given.

155.087   Erratum: "The mass density in our Galaxy. I. A dynamical model constrained by general star counts" [Astron. Astrophys., Vol. 180, No. 1/2, p. 94 – 110 (1987)].
O. Bienaymé, A. C. Robin, M. Crézé.
Astron. Astrophys., Vol. 186, No. 1/2, p. 359 (1987). See Abstr. 43.155.104.

155.088   M giants in Baade's window: infrared colors, luminosities, and implications for the stellar content of E and S0 galaxies.
J. A. Frogel, A. E. Whitford.
Astrophys. J., Vol. 320, No. 1, p. 199 – 237 (1987).

Infrared $JHKL$, [10], CO, and H_2O observations of 185 M1–9 giants in Baade's window are presented. The majority of these stars constitute an unbiased sample drawn from *complete* surveys for such stars. Information on bolometric luminosities, metallicity distribution functions, blanketing effects, mass loss and circumstellar emission is derived for nonvariable and long–period variable giants. On the whole, the observed characteristics of the bulge M giants are consistent with their being representatives of a population with a metallicity considerably in excess of solar. The luminosities of both the variables and nonvariables are consistent with an age comparable to that of Galactic globular clusters. Using the bulge M giants for the equivalent component of stellar synthesis models of galaxies results in infrared colors and indices which come close to but do not exactly match those of real galaxies. The remaining problem with the models is that stars other than M giants are not of sufficiently high metallicity. Increasing this metallicity results in a model that closely reproduces the energy distribution of E and S0 galaxies from 0.5 to 10 μm.

155.089   The ionization structure of the galactic halo.
N. A. Daod.
Diss. Abstr. Int., Sect. B, Vol. 48, No. 1, p. 160–B (1987). Thesis, The University of Wisconsin–Madison, 88 pp. (1986). Order No. DA8702299.

155.090   A study of the stellar populations in two high Galactic latitude fields with comparison to Galaxy models.
E. D. Friel.
Diss. Abstr. Int., Sect. B, Vol. 48, No. 1, p. 161–B (1987). Thesis, University of California, Santa Cruz, 299 pp. (1986). Order No. DA8709122.

155.091   CCD photometry of the galactic nuclear bulge.
D. M. Terndrup.
Diss. Abstr. Int., Sect. B, Vol. 48, No. 1, p. 162–B (1987). Thesis, University of California, Santa Cruz, 112 pp. (1986). Order No. DA8709126.

155.092   Galactic 21 cm emission in a low column density region of 40 square degrees.
K. Jahoda, F. J. Lockman.
Bull. Am. Astron. Soc., Vol. 19, No. 2, p. 678 (1987). Abstract. – See Abstr. 010.061.

155.093 BU–Arecibo H I galactic survey progress report.
T. A. Kuchar, T. M. Bania.
Bull. Am. Astron. Soc., Vol. 19, No. 2, p. 678 – 679 (1987).
Abstract. – See Abstr. 010.061.

155.094 H I in the vicinity of the supernova remnant VRO 42.05.01 (G 166.0 + 4.3)
T. L. Landecker, J. F. Vaneldik, S. Pineault, D. Routledge.
Bull. Am. Astron. Soc., Vol. 19, No. 2, p. 679 (1987). Abstract. – See Abstr. 010.061.

155.095 A VLA continuum survey of the galactic center region.
F. Yusef–Zadeh, J. Bally.
Bull. Am. Astron. Soc., Vol. 19, No. 2, p. 679 (1987). Abstract. – See Abstr. 010.061.

155.096 A kinematic study at the galactic poles.
K. M. Yoss, C. Neese, W. I. Hartkopf.
Bull. Am. Astron. Soc., Vol. 19, No. 2, p. 679 (1987). Abstract. – See Abstr. 010.061.

155.097 Kinematic model of our galaxy.
K. U. Ratnatunga, J. N. Bahcall, S. Casertano.
Bull. Am. Astron. Soc., Vol. 19, No. 2, p. 679 (1987). Abstract. – See Abstr. 010.061.

155.098 The role of blue stragglers in the stellar population near the sun.
B. A. Twarog, J. C. Shields.
Bull. Am. Astron. Soc., Vol. 19, No. 2, p. 680 (1987). Abstract. – See Abstr. 010.061.

155.099 Galactic disk evolution in the presence of hot interstellar medium.
Z. Wang.
Bull. Am. Astron. Soc., Vol. 19, No. 2, p. 680 (1987). Abstract. – See Abstr. 010.061.

155.100 The production of deuterium during early activity of the galactic nucleus.
L. Ozernoy.
Bull. Am. Astron. Soc., Vol. 19, No. 2, p. 688 (1987). Abstract. – See Abstr. 010.061.

155.101 Aspects of the global distribution of H I in the Milky Way.
W. B. Burton.
Bull. Am. Astron. Soc., Vol. 19, No. 2, p. 691 (1987). Abstract. – See Abstr. 010.061.

155.102 The kinematical and binary properties of association and field O stars.
D. R. Gies.
Bull. Am. Astron. Soc., Vol. 19, No. 2, p. 715 (1987). Abstract. – See Abstr. 010.061.

155.103 Cold atomic gas in the inner Galaxy.
R. Garwood.
Bull. Am. Astron. Soc., Vol. 19, No. 2, p. 737 (1987). Abstract. – See Abstr. 010.061.

155.104 Kinematic and abundance gradients in the galactic disk.
C. L. Neese.
Bull. Am. Astron. Soc., Vol. 19, No. 2, p. 737 (1987). Abstract. – See Abstr. 010.061.

155.105 The galactic bulge as a prototype population for E and S0 galaxies.
D. M. Terndrup, J. A. Frogel.
Bull. Am. Astron. Soc., Vol. 19, No. 2, p. 737 (1987). Abstract. – See Abstr. 010.061.

155.106 The spectral index distribution of Sgr A at arc sec resolution.
W. M. Goss, J. H. van Gorkom, R. D. Ekers, U. J. Schwarz.
Bull. Am. Astron. Soc., Vol. 19, No. 2, p. 738 (1987). Abstract. – See Abstr. 010.061.

155.107 Low–dispersion spectral sky survey to find faint carbon stars. I. Region $115° \leqslant l \leqslant 130°$, $-5° \leqslant b \leqslant +5°$.
M. G. Nikolashvili.
Astrophysics, Vol. 26, No. 2, p. 125 – 135 (1987). English translation of 43.155.079.

155.108 A kinematic and abundance survey at the galactic poles. III.
K. M. Yoss, C. L. Neese, W. I. Hartkopf.
Astron. J., Vol. 94, No. 6, p. 1600 – 1615 (1987).

A knowledge of the velocity dispersion $\sigma_w(z)$ of stars for the z (distance above the galactic plane) component of space velocity (W) as a function of both z and composition is necessary to the understanding of the early history of the galactic halo and disk. In this paper the authors specifically address six aspects of galactic structure in the z direction, all of which are related but manifest themselves differently from the observational standpoint: (1) The chemical gradient in the z direction. (2) The composition and kinematics of the individual components of the galactic model. (3) The "continuity" or "decoupling" betweeen the thick disk and halo (at [Fe/H] ~ -1.00). (4) The $\sigma_w(z)$ gradient. (5) The presence or lack of solar–composition stars in the halo falling within the zone, and the timescale of the initial halo collapse. (6) The possible "imbalance" of the halo, as proposed by Croswell et al. (1987).

155.109 The effect of metal rich infall on galactic chemical evolution.
M. Tosi.
Space Telesc. Sci. Inst., Prepr. Ser., No. 206, 14 pp. (1987). To appear in Astron. Astrophys.

155.110 Star clusters and the thickness of the galactic disk as probes of the outer Galaxy.
R. Wielen, B. Fuchs.
Astron. Rechen–Inst. Heidelb., Prepr. Ser., No. 13, 7 pp. (1987). To appear in the proceedings of "The outer Galaxy", Lecture Notes in Physics, Springer–Verlag (1987).

155.111 Distant satellites as probes of our Galaxy's mass distribution.
B. Little, S. Tremaine.
Astrophys. J., Vol. 320, No. 2, p. 493 – 501 (1987).

The distant satellites of our Galaxy can be used to constrain the distribution of dark matter in the Galactic halo. The authors have devised a new method of statistical analysis based on Bayes's theorem, which directly yields confidence intervals for the mass M of the Galaxy once the eccentricity distribution of the satellites is specified. Assuming an isotropic velocity distribution for 10 objects at distances of $50 - 140$ kpc, the authors find $M = 2.4(+1.3,-0.7) \times 10^{11} M_\odot$, with $M \lesssim 5.2 \times 10^{11} M_\odot$ at the 95% confidence level. This suggests that the Galaxy's massive dark halo extends to $\lesssim 50$ kpc from the Galactic center. Several effects are investigated which could in principle alter this conclusion.

155.112 On the distribution of colors for stars in the ninth to fifteenth magnitude range: statistics and implications for galactic structure.
B. M. Lasker, P. M. Garnavich, A. P. Reynolds.
Astrophys. J., Vol. 320, No. 2, p. 502 – 514 (1987).

The Guide Star Photometric Catalog consists of a set of $\sim 10^4$ stars distributed nearly uniformly over the Galaxy and over the V–magnitude range from 9 to 15; as the selection was made without regard to the stellar colors, these objects are a random sample of the Galactic color distribution. The distribution shows a blue–yellow component consisting primarily of disk dwarfs. In addition, various subsets of the distribution are bimodal, with a second, redder peak consisting primarily of disk giants. The distributions are in approximate agreement with the "standard" Bahcall–Soneira galactic model.

155.113   An analysis of the Yale Bright Star Catalog.
J. N. Bahcall, S. Casertano, K. U. Ratnatunga.
Astrophys. J., Vol. 320, No. 2, p. 515 – 526 (1987).

The authors analyze the direction, color, and apparent-magnitude distributions of a complete sample of 3993 stars with $m_v \leqslant 6.0$ mag and $B-V \geqslant 0.0$ from the Yale Bright Star Catalog (BSC). Many of the sample stars are intrinsically bright disk giants. The total number and overall distributions of the BSC stars (excluding O and B stars) are well described by the standard Galaxy model with two major density components. The spatial distribution of O and B stars shows a density enhancement due to the Gould Belt in a $15°$ region inclined about $20° \pm 3°$ to the Galactic plane. The total number of O and B stars in the Galaxy is estimated to be $\sim 10^7$.

155.114   Structural details of the Sagittarius A complex: evidence for a large–scale poloidal magnetic field in the Galactic center region.
F. Yusef–Zadeh, M. Morris.
Astrophys. J., Vol. 320, No. 2, p. 545 – 561 (1987).

Radio observations of the Sgr A complex at both 6 and 20 cm indicate that the diffuse radio structure surrounding Sgr A East and Sgr A West consists of two large–scale components. The first of these is a 20 pc diameter halo which appears to be associated with the nonthermal Sgr A East shell source and which is elongated along the Galactic plane. This source can be described by a model which represents an explosion taking place in a uniform poloidal magnetic field. The second large–scale component is a set of narrow radio streamers running roughly perpendicular to the Galactic plane. They are probably associated with the thermal source Sgr A West, which is coincident with the Galactic nucleus. Hypotheses which might account for the structure of these streamers are discussed. The data show clearly that the bulk of Sgr A West is located in front of Sgr A East.

155.115   The broad–line region at the center of the Galaxy.
T. R. Geballe, R. Wade, K. Krisciunas, I. Gatley, M. C. Bird.
Astrophys. J., Vol. 320, No. 2, p. 562 – 569 (1987).

The high–velocity wings of the Brα ($4.05 \mu m$) line at the Galactic center have been mapped with a $2\overset{''}{.}5$ beam and at a velocity resolution of 400 km s^{-1}. The broad–line region is spatially resolved, with a characteristic linear dimension of $\sim 3''$. The peak intensity of the high–velocity line emission is coincident with the position of the source IRS 16 Center, and its intensity distribution about that object is approximately symmetric. The observed motions appear to be neither rotational nor due to well–collimated jets. The authors suggest that the broad–line emission either is from more than one compact wind source or is the result of an interaction between an ultrahigh velocity wind and slower moving ionized gas in the "bar".

155.116   Ionization state in and reddening to the center of the Galaxy.
R. Wade, T. R. Geballe, K. Krisciunas, I. Gatley, M. C. Bird.
Astrophys. J., Vol. 320, No. 2, p. 570 – 572 (1987).

Observations of the He I line at 2.06 μm and the Brackett lines of hydrogen at 2.17 and 4.05 μm at a number of positions in the Galactic center are reported. The intensity of the He I line peaks at IRS 16. The ionization state of the gas, as determined by the He I/Brγ line intensity ratio, also appears to be maximum at or near IRS 16. Measurements of the Brackett line intensities show that the reddening toward the central $0\overset{''}{.}5$ is approximately uniform and corresponds to $A_v \approx 27$ mag.

155.117   Sobre a estrutura galáctica na vizinhança do sol.
P. Pinto.
An. Fac. Cienc. Porto, Vol. 65, Fasc. 1 – 4, p. 35 – 49.
= Publ. Obs. Astron. Fac. Cienc. Porto, No. 37.

Prova–se a existência dum plano de simetria da Galáxia, através do estudo de elipsoides de velocidades de vários grupos estalares, depois usado como plano fundamental para a determinação de coordenadas cilíndricas. Partindo–se da equação de Boltzmann, definida em coordenadas cilíndricas e aplicadas a esses grupos estelares, estabelece–se algumas caracteristicas da estrutura galáctica na vizinhança solar, nomeadamente a distribuição do gradiente da densidade estelar.

155.118   A search for cool carbon stars. II. Anticenter region.
H. Maehara, T. Soyano.
Ann. Tokyo Astron. Obs., Second Ser., Vol. 21, No. 4, p. 423 – 435 (1987).

A search for cool carbon stars has been consecutively made using the $4°$ prism spectra of the Kiso 105 cm Schmidt telescope. The area surveyed in this paper is the 200 square–degree area toward the galactic anticenter ($l = 170°$ to $190°$, $b = -14°$ to $+8°$). The authors present the celestial position, V magnitude, and the finding chart of 125 cool carbon stars of which 21 stars are newly discovered. The variability and the distribution of the present sample are briefly discussed referring to the previous catalogues of carbon stars in this sky area.

155.119   The large–scale distribution of molecular gas in the first galactic quadrant.
D. P. Clemens, D. B. Sanders, N. Z. Scoville.
Prepr. Steward Obs., No. 747, 29 pp. (1987). To appear in Astrophys. J.

155.120   Data for the study of galactic rotation at $R < R_0$ using ζ vs. $\sin^2 l$ diagrams.
I. V. Petrovskaya, P. Teerikorpi.
Rep. Ser., Dep. Phys. Sci., Univ. Turku, No. FTL–R137, 5 pp. (1987). = Turku Univ. Obs., Informo, No. 129. ISBN 951–880–048–0.

A set of H I data has been extracted from published observations, suitable for the study of the galactic rotation inside the solar circle using the graphical variant of the Agekyan et al. method.

155.121   Taking into account the possible expansion of the Galaxy. II.
N. K. Bektasova.
Tr. Astrofiz. Inst. Alma–Ata, Tom 47, p. 79 – 82 (1987). In Russian.

155.122   The spectrum of radio radiation of the Galaxy in the 200 – 300 MHz range.
M. E. Miller.
Izv. Vyssh. Uchebn. Zaved., Radiofiz., Tom 30, No. 5, p. 665 – 669 (1987). In Russian. Abstr. in Ref. Zh., 51. Astron. 11.51.742 (1987).

155.123   Galactic kinematics: an axi–symmetric time–depending model with separable potential.
F. Sala.
Rev. Mex. Astron. Astrofis., Vol. 14, No. 1, p. 195 – 205 (1987). – See Abstr. 012.042.

An axi–symmetric time–depending kinematic galactic model with separable potential has been derived under Chandrasekhar hypothesis, being the motions of the stars obtained through the first integrals of the system. The model has been applied to a sample of luminosity class V, F6 – F7 spectral type stars.

155.124   A model of the stellar velocity field.
P. Cortés, F. Sala.
Rev. Mex. Astron. Astrofis., Vol. 14, No. 1, p. 206 – 212 (1987). – See Abstr. 012.042.

A kinematic galactic model has been derived through the hydrodynamical equations, based on the central momenta obtained for a wide sample of stars. The very general form of the stellar density, the galactic potential, the second order pressures and the velocity of the local standard of the rest have been determined.

155.125   Spiral structure and age distribution of the H II regions in the Milky Way.
H. A. Dottori, M. V. Copetti.
Rev. Mex. Astron. Astrofis., Vol. 14, No. 1, p. 223 (1987).
Abstract. – See Abstr. 012.042.

155.126   Study of the galactic structure using different objects as "tracers".
A. F. L. Botti, J. R. D. Lépine.
Rev. Mex. Astron. Astrofis., Vol. 14, No. 1, p. 225 (1987).
Abstract. – See Abstr. 012.042.

155.127   A simple, realistic model of the galactic mass distribution for orbit computations.
C. Allen, M. A. Martos.
Rev. Mex. Astron. Astrofis., Vol. 14, No. 1, p. 226 (1987).
Abstract. – See Abstr. 012.042.

155.128   The galactic thermal pressure gradient.
J. Bohigas, J. González.
Rev. Mex. Astron. Astrofis., Vol. 14, No. 1, p. 228 – 235 (1987). – See Abstr. 012.042.
The [S II] line ratio 6717/6731 observed in large galactic supernova remnants (SNR's), has a negative gradient as a function of galactocentric distance. This gradient may be at least partially due to a negative density gradient in the warm component of the interstellar medium. This hypothesis is verified by galactocentric distribution of maximum sizes of radio SNR's, where the largest objects are seen at greatest distances from the galactic center. A negative thermal pressure gradient is obtained when this density gradient and the temperature gradient obtained from radio observations of H II regions are combined.

155.129   Origen cinemático de las nubes obscuras de Tauro y de algunos cúmulos galácticos locales.
C. A. Olano, W. G. L. Pöppel.
Rev. Mex. Astron. Astrofis., Vol. 14, No. 1, p. 236 – 239 (1987). – See Abstr. 012.042.
The authors present numerical results about the kinematical origin of the dark clouds in Taurus, as well as of several local galactic clusters, among which are the Pleiades and α Per.

155.130   The C/M5+ ratio in the galactic disc.
F. J. Fuenmayor.
Rev. Mex. Astron. Astrofis., Vol. 14, No. 1, p. 379 (1987).
Abstract. – See Abstr. 012.042.

155.131   Estimates of the density of dark matter near the center of the Galaxy.
J. R. Ipser, P. Sikivie.
Phys. Rev. D, Vol. 35, No. 12, p. 3695 – 3704 (1987). Abstr. in Phys. Abstr., Vol. 90, No. 1312, Entry 107964 (1987).

155.132   Introduction: galaxy and its environment.
J. Jaaniste, J. Einasto.
ESO Conf. Workshop Proc., No. 27, p. 3 – 14 (1987). – See Abstr. 012.051.
The authors give a short review of observational data and theoretical considerations which lead to the concept of hypergalaxies: giant galaxies surrounded by massive coronas and systems of companion galaxies and gaseous clouds and streams. The best object to study hypergalaxies is our own Galaxy and its environment. The key problems of this analysis involve better understanding of mutual interactions between the galactic and extragalactic gas, the influence of the environment on star formation, and galaxy formation, chemical and dynamical evolution in general.

155.133   Our Galaxy – a polar ring galaxy?
U. Haud.
ESO Conf. Workshop Proc., No. 27, p. 15 – 27 (1987). – See Abstr. 012.051.
The pattern of the velocity distribution of high–velocity clouds provides evidence for the existence of two large subsamples of HVC's: (a) a subgroup of relatively nearby features; (b) a subgroup at large distances from the Sun. It is shown that existing observational data on HVC's of the second subgroup permit to consider our Galaxy as a weak polar ring galaxy.

155.134   The chemical composition of field halo stars.
B. Gustafsson.
ESO Conf. Workshop Proc., No. 27, p. 33 – 45 (1987). – See Abstr. 012.051.
Recent studies of the variation of relative abundances of elements with over–all metallicity in halo stars are reviewed. The reality of these "trends" and the scatter around them are commented on. The uncertainties in the analyses are illustrated by empirically traced inconsistencies ascribed to departures from LTE.

155.135   The chemical composition of peculiar field stars in the galactic halo.
D. L. Lambert.
ESO Conf. Workshop Proc., No. 27, p. 47 – 80 (1987). – See Abstr. 012.051.
Data on the chemical composition of selected peculiar field halo stars are reviewed. The selection of unevolved stars includes those early–type stars with a Pop. I composition despite their residency and, perhaps, their birth in the halo. The rare N–rich subdwarfs and dwarf stars with anomalous elemental abundance ratios are discussed. Selection of evolved peculiar stars is restricted to the post He–core flash stars on and beyond the horizontal branch. The chemical composition of A–type horizontal branch, RR Lyrae, Pop. II variable (BL Her, W Vir, RV Tau) and A–F supergiant stars is reviewed.

155.136   Isotopic ratios in the halo.
B. Barbuy.
ESO Conf. Workshop Proc., No. 27, p. 81 – 88 (1987). – See Abstr. 012.051.
The possibilities to obtain isotopic ratios from atomic and molecular lines are described. Data concerning isotopic ratio determinations in halo stars are reported; these concern $^6Li{:}^7Li$, $^{12}C{:}^{13}C$ and $^{24}Mg{:}^{25}Mg{:}^{26}Mg$ ratios. The interpretation of data on the carbon isotopes and magnesium isotopes is presented in terms of nucleosynthesis and the chemical evolution of the Galaxy.

155.137   The galactic spheroid: where is Population II?
G. Gilmore, R. F. G. Wyse.
ESO Conf. Workshop Proc., No. 27, p. 89 – 102 (1987). – See Abstr. 012.051.
Population II stars in the solar neighbourhood have previously been identified with the extreme subdwarfs. Spectroscopic and photometric studies of local subdwarfs show these stars to form a metal–poor, round system whose kinematics are dominated by pressure support. This paradox is here resolved by identifying Population II stars in the solar neighbourhood not with the extreme subdwarfs, but with stars in the thick disk component, which has recently been shown to dominate the stellar distribution with a few kpc of the Galactic plane. Some implications of this for disk galaxy formation and evolution are discussed.

155.138   Serendipitous discovery of two distant halo stars of A–spectral type.
L. E. Campusano, A. Gutiérrez.
ESO Conf. Workshop Proc., No. 27, p. 109 – 115 (1987). – See Abstr. 012.051.
The finding of two distant A–stars in the direction of the South Galactic Pole is reported. They are suspected to be blue horizontal branch stars and their derived distances from the plane of the galaxy are 8.7 and 15.4 kpc. However, the more distant one has metallicity and gravity estimates close to 'normal' main–sequence stars, which reformulates the unanswered question about the evolutionary status of this kind of stars.

155.139   Relaxation time for a polar sample of F stars with solar composition.
J. Knude, H. Schnedler Nielsen, M. Winther.
ESO Conf. Workshop Proc., No. 27, p. 117 – 121 (1987). – See Abstr. 012.051.

From a complete sample of polar F stars the authors find that the velocity dispersions σ_U and σ_V evolve after different power laws: $\gamma \approx 1/2$ and $\gamma \approx 1/4$ respectively. This means that the velocity ellipsoid has a continuously changing shape. The ratio σ_V/σ_U is seen to relaxe to the equilibrium value 0.6 only after a considerable period of time amounting to $4 - 5 \times 10^9$ years.

155.140   Some properties of the galactic halo from NLTT proper motion stars.
M. Grenon.
ESO Conf. Workshop Proc., No. 27, p. 123 – 128 (1987). – See Abstr. 012.051.

Investigations on proper motion stars in the solar vicinity are the natural complement to studies in situ of halo stars, essentially because the star space velocities are accurately determined and, through a galactic potential model, their orbital motions in the Galaxy are known. The distinction between the galactic components, halo, bulge and thick disc, appears to depend critically on the kinematics of their constituents. The halo definition, structure, kinematics and metallicity is re–discussed using a not metal–abundance biased stellar sample.

155.141   Nearby visitors from the outer galactic halo.
B. W. Carney, J. B. Laird, D. W. Latham.
ESO Conf. Workshop Proc., No. 27, p. 129 – 134 (1987). – See Abstr. 012.051.

The authors discuss the results of a study of local proper motion stars, specifically those with the most extreme velocities. The upper limit to their velocities in the Galaxy's restframe yield a local value of the escape velocity of about 550 km sec^{-1}. The implied total mass of the Galaxy exceeds that within the solar orbit by a factor of between 6 and 8. The binary fraction is at least 40%.

155.142   The impact of the gas reshuffling between galactic disc and halo on stochastic star formation.
J. Spicker, J. V. Feitzinger.
ESO Conf. Workshop Proc., No. 27, p. 173 – 178 (1987). – See Abstr. 012.051.

From the distribution of the hot and fountain gas and the equivalence of the time scales envolved the authors conclude, that the galactic fountain gas cycle is the dominating self–regulation mechanism of the stochastic star forming activity in the model galaxy. The observations of 'Velocity Active Regions', with the observationally established relation between fountain gas, magnetic fields and active star formation, provide a new insight into the stochastic nature of pattern formation in galaxies.

155.143   Chemical evolution of the galactic halo.
F. Matteucci.
ESO Conf. Workshop Proc., No. 27, p. 609 – 626 (1987). – See Abstr. 012.051.

A model of chemical evolution for the solar neighborhood, predicting abundance ratios of several elements (^{12}C, ^{13}C, ^{14}N, ^{16}O, ^{28}Si and ^{24}Mg) with respect to ^{56}Fe, both in halo and disk field stars, is presented. The predictions for ^{16}O, ^{24}Mg, ^{28}Si and ^{56}Fe reproduce very well the observed patterns. Some problems are still present in predicting the [C/Fe] and [N/Fe] ratios in halo stars. More observations as well as more nucleosynthesis computations concerning C and N are necessary before drawing firm conclusions. It is suggested that the knowledge of elemental ratios in globular clusters and field stars can be used to assess the hypothesis of self–enrichment in globular clusters.

155.144   The early chemical evolution of the halo and population III.
R. Cayrel.
ESO Conf. Workshop Proc., No. 27, p. 627 – 636 (1987). – See Abstr. 012.051.

155.145   Background starlight at the north and south celestial, ecliptic, and galactic poles.
G. Toller, H. Tanabe, J. L. Weinberg.
Astron. Astrophys., Vol. 188, No. 1, p. 24 – 34 (1987).

Isophotes of background starlight brightness (integrated starlight, diffuse galactic light, cosmic light) have been constructed near the celestial, ecliptic, and galactic poles from observations accumulated by the Pioneer 10 imaging photopolarimeter from beyond the asteroid belt. Blue (3950 Å – 4850 Å) and red (5900 Å – 6900 Å) data and the color index at the poles are presented and are compared with previous photometric and star count results for these regions. Estimates of the diffuse galactic light and extragalactic light in these regions are also obtained.

155.146   The local kinematics of open star clusters.
G. Lyngå, J. Palouš.
Astron. Astrophys., Vol. 188, No. 1, p. 35 – 38 (1987).

Data on positions, distances, and radial velocities of 106 open star clusters available from the computer–based catalogue of open clusters prepared by Lyngå have been used in a statistical analysis of the local kinematics. The authors find a velocity field that deviates significantly from circular streaming. The dispersion of the radial velocity residuals increases with age: for the old clusters it is about twice that for young clusters. This may be a further evidence of heating of the galactic disk. The relatively low velocity residuals of the very oldest clusters are explained as a bias due to the disruption of open clusters.

155.147   Extinction trends at high galactic latitudes as shown by extragalactic extinction indicators.
P. Teerikorpi, S. Haarala.
Astrophys. Space Sci., Vol. 137, No. 2, p. 397 – 402 (1987).

In the galactic latitude range $40° < |b| < 80°$ several extinction indicators of extragalactic type show features with similar longitude dependence. Amplitudes of these variations correspond to B–extinction variations of about $0^m.3 - 0^m.4$. The results favour the view that instead of nearly zero extinction, there are significant amounts of dust at high galactic latitudes.

155.148   A radio continuum survey of the galactic plane at 10 GHz.
T. Handa, Y. Sofue, N. Nakai, H. Hirabayashi, M. Inoue.
Publ. Astron. Soc. Jpn., Vol. 39, No. 5, p. 709 – 753 (1987).

A 10–GHz radio continuum survey of the galactic plane in the range $355° \leqslant l \leqslant 56°$, $-1°.5 \leqslant b \leqslant +1°.5$ was made at the Nobeyama Radio Observatory using the 45–m telescope. The half–power beam size of the resultant map is $3'.0$ and the noise level of the resultant map is typically 15 mK in brightness temperature. Thirty–one contour maps and a list of 144 small–diameter sources are presented.

155.149   A 1.6 MHz survey of the galactic background radio emission.
G. R. A. Ellis, M. Mendillo.
Aust. J. Phys., Vol. 40, No. 5, p. 705 – 708 (1987).

1.6 MHz observations have provided the first measurements of the distribution of the galactic radiation with 25° angular resolution. They show a broad absorption feature along the galactic plane generally consistent with that expected from observations at higher frequencies.

155.150   The molecules in interstellar medium. Part III. Spiral structure of the Galaxy.
J. Kobus.
Postepy Astron., Tom 34, Zesz. 4, p. 291 – 305 (1986). In Polish.

Spiral structure of the Galaxy is described on the basis of the observations of the molecular spectral lines. Some informations on the cosmic masers, dust and isotopes of some elements are given.

155.151   Discovery of a pulsating component in a decametric frequency range suggesting a rotating super black hole at the Galactic Center as a source.
H. Oya, M. Iizima, A. Morioka.
Tôhoku Geophys. J., Vol. 30, No. 2 – 4, p. 15 – 49 (1987). Abstr. in Phys. Abstr., Vol. 90, No. 1313, Entry 115607 (1987).

155.152   A flux–limited sample of Galactic carbon stars.
M. J. Claussen, S. G. Kleinmann, R. R. Joyce, M. Jura.
Astrophys. J., Suppl. Ser., Vol. 65, No. 3, p. 385 – 404 (1987).
The authors summarize the infrared properties of a flux–limited sample of Galactic carbon stars taken from the Two Micron Sky Survey (TMSS). Using data from the TMSS and the *IRAS* survey, and assuming that Galactic carbon stars have the same observed narrow range of absolute 2.2 μm magnitudes as carbon stars in the Magellanic Clouds, the authors deduce their space density, scale height, range of mass–loss rates, the total mass return to the Galaxy, the time scale for the carbon star phase of evolution, and the masses of their main–sequence progenitors.

155.153   Sub–arcsec imaging of the galactic centre in the near infrared.
F. Lacombe, P. Léna, D. Rouan.
Infrared astronomy with arrays, p. 316 – 320 (1987). – See Abstr. 012.076.
Using a prototype IR camera, new IR maps of the galactic centre region have been obtained at 2.2 and 3.8 μ with 0.8″ resolution. Following a brief introduction of the observing procedures, the photometric and astrometric capabilities of the camera are discussed, and some results on the IR/Radio correlation, the surface brightness and colour map are presented.

155.154   The January 87 galactic centre occultation observed with the UKIRT IR camera, IRCAM.
I. S. McLean, C. Aspin, A. J. Longmore, R. I. Dixon.
Infrared astronomy with arrays, p. 321 – 325 (1987). – See Abstr. 012.070.
An infrared camera has been used for the first time to obtain images during an occultation of the galactic centre. The observations were made at 2.2 μm with IRCAM on the 3.8 m UKIRT with an effective spatial resolution of ~0.1″. Despite relatively poor seeing conditions, an excellent data set was acquired which records the appearance of all the major sources in the galacic centre region.

155.155   On an investigation of the distribution of neutral hydrogen in the Galaxy.
A. V. Kostryukov, I. V. Petrovskaya.
Tr. Astron. Obs., Leningrad, Tom 41, p. 155 – 162 (1987). Uch. Zap. LGU, No. 420, Ser. Mat. Nauk, Vyp. 63. In Russian.
The effect of the choice of the latitude interval and the latitude step in the directions of the 21 cm line observations on the neutral hydrogen density determination near the galactic plane is investigated. The picture of the H I distribution for the plane $l = 21°$ corroborates the existence of two hydrogen components in the Galaxy. The increasing of the latitude step, Δb leads to the mixing of these components and to the misrepresentation of the results for the distant region when $\Delta b > 0°.5$.

155.156   The expansion law of the neutral hydrogen subsystem in the inner region of the Galaxy.
Yu. N. Malakhova, I. V. Petrovskaya.
Astron. Tsirk., No. 1458, p. 3 – 4 (1986). In Russian.
The expansion velocity components of the individual H I features for distances between 1.7 and 4 kpc from the galactic centre are used for the determiantion of the smoothed expansion velocity as a function of z.

155.157   Eggen's stellar groups – a test for the determination of the rotation velocity of the Galaxy's spiral pattern.
P. I. Korchagin.
Astron. Tsirk., No. 1462, p. 1 – 3 (1986). In Russian.
The dependence of the kinematic age of stellar groups on the value of angular velocity of the galactic spiral pattern is numerically investigated. The change of Ω_p from 13 to 23 km s^{-1}kpc^{-1} increases the kinematic age of stellar groups from 7×10^8 to 4×10^9 years. The existence of old stellar groups ($T \sim 5 \times 10^9$ years) in the neighbourhood of the Sun confirms values of $\Omega_p \sim 23 - 25$ km s^{-1}kpc^{-1}.

155.158   The turn–off point revealed for F stars. The age of the galactic disk.
V. A. Marsakov, A. A. Suchkov, Yu. G. Shevelev.
Astron. Tsirk., No. 1485, p. 1 – 3 (1987). In Russian.
Is it possible to find the main sequence turn–off point of the oldest field stars and to define the age of the galactic disk from it? The authors have used a sample of 5000 F stars within 80 pc with uvby–photometry from B. Hauck, M. Mermilliod (1985). The sample has been divided into eight metallicity groups, within $-0.6 \leqslant$ [Fe/H] $\leqslant +0.3$. They determine the age of the galactic disk just as ages of clusters. The age is 10 (or 6) Gyrs.

155.159   A complete VLA survey of galactic plane sources.
M. Fich.
NRAO Workshop, No. 15, p. 137 – 139 (1988). – See Abstr. 012.076.
This paper describes an attempt to use the VLA to identify radio sources found in a single–dish radio continuum survey. Later work will focus on measuring radial velocities to H II regions in this part of the Galaxy in order to determine kinematic distances and therefore their positions within the Galaxy.

155.160   A non–thermal axially symmetric radio wake towards the galactic centre.
F. Yusef–Zadeh, J. Bally.
Nature, Vol. 330, No. 6147, p. 455 – 458 (1987).
The authors have discovered a highly unusual radio source lying within 1° of the galactic centre whose ″cometary″ morphology suggests that it is a wake produced by a radio source moving supersonically with respect to the ambient interstellar medium. They report the finding of a collimated axisymmetric nonthermal source.

155.161   Hard X–ray images of the galactic centre.
G. K. Skinner, A. P. Willmore, C. J. Eyles, D. Bertram, M. J. Church, P. K. S. Harper, J. R. H. Herring, J. C. M. Peden, A. M. T. Pollock, T. J. Ponman, M. P. Watt.
Nature, Vol. 330, No. 6148, p. 544 – 547 (1987).
The authors report observations made with a coded mask X–ray telescope flown on the Spacelab 2 mission (29 July to 6 August 1985), yielding for the first time images of the galactic centre in high–energy X–rays up to 30 keV. Components detected include a region of diffuse emission 2° in diameter and several new point sources. At the higher energies emission from the nucleus is weak and the region is dominated by one of the surrounding point sources.

155.162   The relative amounts of stars and interstellar matter in the local Milky Way.
M. Jura.
Publ. Astron. Soc. Pac., Vol. 99, No. 621, p. 1123 – 1126 (1987).
= Invited paper presented at the Symposium on cool stars and the structure of galaxies, at the 99th Annual Scientific Meeting of the Astronomical Society of the Pacific, at Pomona College, Claremont, California, July 1987.
The author considers the balance between star formation and mass loss from evolved stars in the region within 1 kpc of the Sun. There is considerably more mass in stars than in the interstellar medium, and more material is being incorporated into new stars than is being returned by evolved stars. In the simplest interpretation of the data, it appears that unless there is some infall of new interstellar gas, the era of substantial star formation

out of interstellar gas will be over in a few (perhaps 3) billion years.

155.163 **The overall structure of gas in the Galaxy.**
G. R. Knapp.
Publ. Astron. Soc. Pac., Vol. 99, No. 621, p. 1134 – 1143 (1987).
= Invited paper presented at the Symposium on cool stars and the structure of galaxies, at the 99th Annual Scientific Meeting of the Astronomical Society of the Pacific, at Pomona College, Claremont, California, July 1987.
Contents: Introduction. The radial distribution of the atomic and molecular gas in the Galaxy. Diffuse molecular clouds. The local rate of mass return to the interstellar medium by evolved stars. The vertical distribution of gas in the Galaxy.

155.164 **OH/IR stars and Galactic structure.**
J. M. Chapman.
Publ. Astron. Soc. Pac., Vol. 99, No. 621, p. 1144 – 1145 (1987).
Abstract. – See Abstr. 010.281.

155.165 **A kinematic and abundance study at the galactic poles.**
K. M. Yoss, C. L. Neese, W. I. Hartkopf.
Publ. Astron. Soc. Pac., Vol. 99, No. 621, p. 1150 – 1151 (1987).
Abstract. – See Abstr. 010.281.

155.166 **Large–scale Galactic dust morphology and physical conditions from *IRAS* observations.**
T. J. Sodroski, E. Dwek, M. G. Hauser, F. J. Kerr.
Astrophys. J., Vol. 322, No. 1, p. 101 – 112 (1987).
The authors present *IRAS* observations of the 60 and 100 μm Galactic plane emission, from which the contribution of the zodiacal light has been subtracted. Assuming an isothermal dust distribution along each line of sight in the Galactic plane, the authors derive temperature, optical depth, and total far–infrared brightness profiles of the emission. Combining these quantities with available ^{12}CO, 5 GHz radio continuum, and H I data, the authors then derive longitude profiles of the dust–to–gas mass ratio, the infrared luminosity per hydrogen mass, and the infrared excess ratio in the Galaxy.

155.167 **A low–dispersion spectral sky survey for revealing faint carbon stars. II. The region $130° \leqslant l \leqslant 145°$, $-5° \leqslant b \leqslant +5°$.**
M. G. Nikolashvili.
Astrofizika, Tom 27, Vyp. 2, p. 197 – 206 (1987). In Russian. With 7 plates. English translation in Astrophysics, Vol. 27, No. 2.
On the basis of the low–dispersion (1250 Å/mm at H_γ) spectral material 122 carbon stars are revealed. 79 out of them are newly detected. The latitude and longitude distribution is uniform and the surface one – accidental. By the "nearest–neighbour" method is has been shown that among carbon stars located at distances smaller than $0°.15$ from the center of open clusters may be their real members.

155.168 **Multivariate statistical analysis of OB stars around h and χ Persei.**
L. G. Balazs, A. T. Garibdzhanyan.
Astrofizika, Tom 27, Vyp. 2, p. 245 – 252 (1987).
A multivariate statistical analysis has been carried out on 49 OB stars having radial velocities and distance moduli in a field around h and χ Persei. A hierarchical cluster analysis, combined with discriminant analysis, has revealed the probable presence of two groups: one belonging to the Perseus spiral arm (mean distance 2100 pc) the other located at a distance of 1150 pc. Comparison of the difference in mean radial velocities of the groups with the values predicted by the formula of Oort also supports this result.

155.169 **Galactic distance scales.**
M. W. Feast.
The Galaxy, p. 1 – 25 (1987). – See Abstr. 43.012.079.
The author reports on a printer's error in his article concerning the values of $R_0 - ...$ in the mean $R_0 = 7.8$ kpc not 7 kpc. See Abstr. 43.155.115.

155.170 **On stable hydrostatic equilibrium configurations of the Galaxy and implications for its halo.**
J. B. G. M. Bloemen.
Astrophys. J., Vol. 322, No. 2, p. 694 – 705 (1987).
Using a variety of observations, it is shown that the gaseous, magnetic field, and cosmic–ray components in the local region of the Galaxy may be in a large–scale hydrostatic equilibrium that is stable against Parker type instabilities. Such an equilibrium sets limits on the characteristics of the gaseous halo. The density of the base of the halo is found to be approximately given by $(0.15 - 0.3)(h_{halo}/kpc)^{-2} cm^{-3}$, where h_{halo} is the exponential scale height which is argued to be $\gtrsim 5$ kpc. The halo gas has to be hot in a stable equilibrium; typically $\sim 10^6$K in the galactic plane and beyond a few kpc from the plane, but could be only about $(2 - 3) \times 10^5$K at $z \approx 1 - 3$ kpc.

155.171 **A composite CO survey of the entire Milky Way.**
T. M. Dame, H. Ungerechts, R. S. Cohen,
E. J. de Geus, I. A. Grenier, J. May, D. C. Murphy,
L.–Å. Nyman, P. Thaddeus.
Astrophys. J., Vol. 322, No. 2, p. 706 – 720 (1987). With plates 5 – 6.
Large–scale CO surveys of the entire Galactic plane and specific nearby clouds have been combined to produce a panorama of the entire Milky Way in molecular clouds at an angular resolution of 1/2°. Covering 10° – 20° in latitude at all longitudes and all or nearly all large, nearby clouds at higher latitude, the composite survey is the only molecular line survey to date with sky coverage and resolution comparable to that of the early 21 cm surveys. The inner Galaxy spiral arms produce, as expected, a thin, intense ridge of emission along the Galactic plane within $\sim 60°$ of the Galactic center. The local emission shows the same large–scale features as the distribution of dark clouds. The survey provides a thorough inventory of large molecular clouds near the Sun. The overall distribution of clouds within 1 kpc is consistent with the Sun lying near the inner edge of a local spiral arm or spur. The half–thickness at half–intensity of the local molecular cloud layer is 87 pc.

155.172 **The linear filaments of the radio arc near the Galactic center.**
F. Yusef–Zadeh, M. Morris.
Astrophys. J., Vol. 322, No. 2, p. 721 – 728 (1987). With plates 7 – 11.
The authors present high–resolution radio continuum VLA images of a segment of the Galactic center arc centered at G 0.16–0.15 at wavelengths of both 6 and 20 cm. The character, multiplicity, and spacing of the filaments is described. The polarization data indicate an organized azimuthal magnetic structure surrounding the longitudinal field lines defining the nonthermal filaments. The authors consider the energy source for the relativistic electrons in the linear filaments, and suggest that their synchrotron losses can be balanced by twisting of the field lines.

155.173 **Properties of evolved mass–losing stars in the Milky Way and variations in the interstellar dust composition.**
H. A. Thronson Jr, W. B. Latter, J. H. Black, J. Bally,
P. Hacking.
Astrophys. J., Vol. 322, No. 2, p. 770 – 786 (1987).
The authors have used *IRAS* data to produce an extensive, flux–limited survey of evolved stars in the Milky Way. They were able to produce a carefully selected collection of carbon–rich stars and a less exclusive sample of oxygen–rich stars for comparison. From these data, the authors are able to study the distribution of both types of stars in the Galaxy and estimate their local densities and total numbers. They use existing studies

of the relations of CO line emission, thermal infrared emission, and mass loss to calibrate the mass–loss rate for stars in the sample and to estimate the corresponding total mass–return rate for intermediate–mass stars in the Milky Way. Finally, the authors estimate birthrates and lifetimes, attempt to identify the main–sequence precursors to the evolved stars, and discuss the role that these two types of stars have in altering the grain composition in the interstellar medium.

155.174 A view of the Galactic Hα background: $208° \leqslant l \leqslant 218°; -2° \leqslant b \leqslant +8°$.
R. J. Reynolds.
Astrophys. J., Vol. 323, No. 1, p. 118 – 128 (1987).

Faint, diffuse Hα emission from the interstellar medium has been mapped at 1° angular resolution and $12 \, \text{km s}^{-1}$ radial velocity resolution within an $11° \times 11°$ region of the Galactic plane. The observations reveal that warm, ionized gas is widespread throughout the Galactic disk and that it has a complex morphology with faint, previously unidentified Hα emission regions superposed on a more diffuse Hα background. This warm ionized gas appears to contain a substantial fraction (perhaps most) of the H^+ in the interstellar medium. The source of this ionization is not understood. The high [S II] $\lambda 6716$/Hα intensity ratios in the background clearly rule out scattered light as a primary source of the emission.

155.175 Il nucleo della Galassia.
L. Gratton.
Atti Accad. Naz. Lincei, Ser. Ottava, Rend., Vol. 78, Fasc. 6, p. 355 – 377 (1985). With 2 plates.

Valinhos 2.2 micron survey of the southern galactic plane. II. Near–IR photometry, IRAS identifications and nature of the sources.
See Abstr. 002.029.

The Pulkovo sky survey in the galactic coordinate system.
See Abstr. 002.069.

The Pulkovo sky survey in Magellanic and ecliptic coordinate systems.
See Abstr. 002.070.

Galactic and Magellanic polar areas in the Pulkovo H I survey.
See Abstr. 002.071.

Physics of the Galaxy and interstellar matter.
See Abstr. 003.125.

Nineteenth–century Italian contributions to galactic theory.
See Abstr. 004.005.

The galactic center. Proceedings of the symposium honoring C. H. Townes, held at Berkeley, CA, USA, 25 October 1986.
See Abstr. 012.029.

A review of decametric radio astronomy: instruments and science.
See Abstr. 013.022.

Infrared imaging with JPL's linear array camera.
See Abstr. 034.155.

Astronomical observations at 10 micron wavelength with the NASA/GSFC array camera system.
See Abstr. 034.160.

Astrometric position of IRS–7 in the galactic center.
See Abstr. 041.003.

X–ray observations from the Space Shuttle.
See Abstr. 051.030.

Radio–dating the Galaxy.
See Abstr. 061.033.

Thorium in G–dwarf stars as a chronometer for the Galaxy.
See Abstr. 061.034.

Implications of bimodal star formation on the chemical evolution of the Galaxy: the evolution of deuterium.
See Abstr. 061.070.

Th/Nd abundance ratio in the surfaces of G–dwarfs.
See Abstr. 061.099.

Reply to "Th/Nd abundance ratio in the surfaces of G–dwarfs" by D. D. Clayton.
See Abstr. 061.100.

A search for strongly interacting dark matter.
See Abstr. 061.122.

Limits on cold dark matter candidates from an ultralow background germanium spectrometer.
See Abstr. 061.147.

Galactic chronology of thorium/neodymium.
See Abstr. 061.149.

Reply to "Galactic chronology of thorium/neodymium" by W. A. Fowler.
See Abstr. 061.150.

Galactic chronology of thorium/neodymium.
See Abstr. 061.151.

Reply to "Galactic chronology of thorium/neodymium" by P. D. Noerdlinger.
See Abstr. 061.152.

Nuclear chronometers from the r–process and the age of the Galaxy.
See Abstr. 061.157.

On point explosion in a nonuniform medium with a symmetry plane.
See Abstr. 062.052.

Hydromagnetic dynamo as source of planetary, solar and galactic magnetism.
See Abstr. 062.071.

Magnetic monopole plasma oscillations and the survival of galactic magnetic fields.
See Abstr. 062.103.

Positron annihilation at the Galactic center.
See Abstr. 063.124.

Subluminous Type I supernovae: their theoretical rate in our Galaxy and in ellipticals.
See Abstr. 065.006.

The production of ^{26}Al in supermassive stars and the gamma–ray line flux from the Galactic center.
See Abstr. 065.098.

Monte Carlo simulations of radio pulsars and their progenitors.
See Abstr. 065.099.

An estimate of the event rate for neutron star binary coalescence and its implications for gravitational wave detection.
See Abstr. 067.088.

Galactic tides affect the Oort cloud: an observational confirmation.
See Abstr. 102.051.

A note on comets and chemical evolution of the Galaxy.
See Abstr. 102.081.

Radial velocities in three fields along the southern galactic equator.
See Abstr. 111.002.

Lick Northern Proper Motion program. II. Solar motion and galactic rotation.
See Abstr. 111.005.

Proper motion of the compact, nonthermal radio source in the galactic center.
See Abstr. 111.009.

The galactic orbit of the remarkable high–velocity wide binary LDS 519.
See Abstr. 111.017.

The galactic orbits of nearby, high–velocity stars.
See Abstr. 111.018.

Galactic distribution, kinematics, locations in clusters and H–R diagrams, and duplicity of Be stars.
See Abstr. 112.109.

An analysis of the emission features of the IRAS low–resolution spectra of carbon stars.
See Abstr. 112.125.

Red supergiant infrared shells and the galactic metallicity gradient.
See Abstr. 112.134.

Arecibo OH survey of OH/IR stars identified by IRAS colors.
See Abstr. 112.178.

Red supergiant infrared shells and the galactic metallicity gradient.
See Abstr. 112.179.

Spectroscopic tests of two faint photometric surveys.
See Abstr. 113.036.

Strömgren photometry of high–velocity stars; metal abundances and ages.
See Abstr. 113.050.

The chemical composition of the extreme halo stars. I. Blue spectra of 20 dwarfs.
See Abstr. 114.077.

New M–type supergiants in the southern Milky Way.
See Abstr. 114.096.

CCD spectroscopy of new M–type supergiants in the southern Milky Way.
See Abstr. 114.097.

Chemical compositions of post–AGB stars.
See Abstr. 114.098.

HD 213985: a hot post–AGB star in the galactic halo.
See Abstr. 114.099.

The halo metallicity distribution function.
See Abstr. 114.100.

Boron abundance in the pop II star HD 140283.
See Abstr. 114.103.

^{9}Be abundances in halo dwarfs: implications for early nucleosynthesis of light elements.
See Abstr. 114.104.

Carbon star surveys in the Magellanic Clouds and the dwarf spheroidals in the galactic halo.
See Abstr. 114.105.

Kinematics and age of the low–mass X–ray binaries.
See Abstr. 117.165.

Orbital elements for double stars of Population II. The high–velocity system COD –48°1741.
See Abstr. 117.234.

A study of the RR Lyrae in Plaut's field 3.
See Abstr. 122.094.

A statistical study of the correlation between galactic SNRs and spiral arms.
See Abstr. 125.010.

The birthrate and initial spin period of single radio pulsars.
See Abstr. 126.054.

Neutral hydrogen absorption measurements of ten pulsars and the electron density in the galactic plane.
See Abstr. 126.063.

The initial–final mass relation: galactic disk and Magellanic Clouds.
See Abstr. 126.082.

High mass star formation in the Galaxy.
See Abstr. 131.001.

The large scale gas and dust distribution in the Galaxy: implications for star formation.
See Abstr. 131.010.

A correlation between the IRAS infrared cirrus at 60 or 100 μm and neutral atomic hydrogen in the outer Galaxy.
See Abstr. 131.011.

The Arp Ring: galactic or extragalactic?
See Abstr. 131.012.

The galactic foreground reddening in the direction of the nearby Triangulum galaxy M33.
See Abstr. 131.046.

Two phases of the interstellar medium: the electron density distribution in the galactic disk.
See Abstr. 131.085.

Diffuse interstellar gas.
See Abstr. 131.088.

Interstellar magnetic fields.
See Abstr. 131.107.

Self–regulating star formation and disk structure.
See Abstr. 131.125.

Mass, luminosity, and line width relations of Galactic molecular clouds.
See Abstr. 131.127.

Dust in external galaxies: diffuse bands and the albedo.
See Abstr. 131.143.

H I superclouds in the inner Galaxy.
See Abstr. 131.153.

CCD observations of the spatial structure of the hydrogen Balmer alpha (Hα) diffuse galactic background.
See Abstr. 131.156.

A face–on view of the first galactic quadrant in giant molecular clouds.
See Abstr. 131.185.

Extinction trends at high galactic latitudes as shown by extragalactic extinction indicators.
See Abstr. 131.218.

Local interstellar medium.
See Abstr. 131.243.

Galactic interstellar abundance surveys with IUE and IRAS.
See Abstr. 131.257.

On a formula for the mean light absorption in the Galaxy.
See Abstr. 131.281.

Hα scans of high–velocity clouds.
See Abstr. 131.284.

Nonstability of stars in the starforming region around γ Cygni.
See Abstr. 131.288.

Star formation, luminous stars, and dark matter.
See Abstr. 131.293.

Far–infrared measurements of N/O in H II regions: evidence for enhanced CN process nucleosynthesis in the inner Galaxy.
See Abstr. 132.031.

Evolution of H II regions.
See Abstr. 132.035.

A very deep far–infrared survey.
See Abstr. 133.011.

Tasmanian low frequency galactic background surveys.
See Abstr. 141.001.

The X–ray sources of the galactic bulge, X–ray bursters and X–ray sources in globular clusters.
See Abstr. 142.044.

Nucleosynthesis and astrophysical gamma ray spectroscopy.
See Abstr. 143.010.

SMM observation of diffuse galactic γ–ray lines.
See Abstr. 143.029.

HEAO 3 search for diffuse galactic positron annihilation radiation.
See Abstr. 143.030.

Search for very–high–energy gamma–rays from the galactic disk.
See Abstr. 143.035.

Origin and propagation of galactic cosmic rays.
See Abstr. 144.005.

Cosmic radiation in the Galaxy. Review.
See Abstr. 144.014.

Mechanisms of origin and dynamics of anisotropy of the gas pressure of galactic cosmic rays.
See Abstr. 144.018.

A simulation of high–energy cosmic–ray propagation.
See Abstr. 144.028.

Galactic gamma rays and the origin of cosmic–ray positrons in the diffusion model.
See Abstr. 144.053.

On errors of eccentricities of galactic orbits.
See Abstr. 151.036.

Accretion of gaseous disks of galaxies. I. Influence of dynamical friction of the large–scale H$_2$ distribution.
See Abstr. 151.110.

The stabilization of multiple open clusters and the relaxation time of the Galaxy.
See Abstr. 151.135.

The stability of complexes of open star clusters.
See Abstr. 151.136.

Globular cluster evolution in the Galaxy: a global view.
See Abstr. 151.143.

Quasi–empiric determination of angular velocity of galactic spiral arms.
See Abstr. 151.150.

Washington photometry of open cluster giants: seven metal–poor anticenter clusters.
See Abstr. 153.006.

On the dynamics of open star clusters in the Galaxy. II.
See Abstr. 153.038.

A study of faint young open clusters as tracers of spiral features in our Galaxy. Paper 3: Collinder 97 (OCl 506).
See Abstr. 153.046.

On cluster analysis of open clusters.
See Abstr. 153.049.

Is there "hidden mass" in multiple open clusters?
See Abstr. 153.051.

Globular clusters and the galactic halo population.
See Abstr. 154.037.

How similar are the globular clusters in different galaxies?
See Abstr. 154.044.

Internal dynamics of globular clusters: from our Galaxy to the Magellanic Clouds.
See Abstr. 154.052.

Globular clusters in the outer halo.
See Abstr. 154.060.

Space motions of galactic globular clusters.
See Abstr. 154.061.

Population of accreting neutron stars in external galaxies.
See Abstr. 157.001.

IRAS observations of three edge–on galaxies.
See Abstr. 157.064.

The chemical evolution of galaxies.
See Abstr. 157.105.

The ratio numbers of carbon and oxygen stars in galaxies.
See Abstr. 157.122.

Dwarf galaxies and dark matter.
See Abstr. 157.135.

The giants of Fornax: LPV's, carbon and M stars.
See Abstr. 157.297.

On the nature of dark matter in the dwarf spheroidal satellites of the Galaxy.
See Abstr. 157.300.

Globular cluster systems: comparative evolution of galactic halos.
See Abstr. 157.302.

Morphology and phase transitions: from cellular automata to galaxies.
See Abstr. 157.312.

The vertical large–scale magnetic fields in spiral galaxies.
See Abstr. 157.318.

The relation between star formation and active nuclei.
See Abstr. 158.018.

Dark matter in clusters?
See Abstr. 160.049.

A search for QSOs to fit a cosmological model with flat, closed spatial sections.
See Abstr. 161.453.

156 Magellanic Clouds

156.001 **Star formation in the Magellanic Clouds.**
J. A. Frogel.
NASA Conf. Publ., NASA CP–2466, p. 161 – 166 (1987). – See Abstr. 012.002.
There is considerable evidence that star formation in the Clouds has been and is proceeding in a manner different from that found in a typical well–ordered spiral galaxy. Star formation in both Clouds appears to have undergone a number of relatively intense bursts. There exist a number of similarities and differences in the current state of star formation in the Magellanic Clouds and the Milky Way. Examination of IRAS sources with ground based telescopes allows identification of highly evolved massive stars with circumstellar shells as well as several types of compact emission line objects.

156.002 **IRAS and ground–based observations of star formation regions in the Magellanic Clouds.**
J. H. Elias, J. A. Frogel.
NASA Conf. Publ., NASA CP–2466, p. 241 – 244 (1987). – See Abstr. 012.002.
The Infrared Astronomical Satellite (IRAS) detected several hundred individual regions of star formation in the Large and Small Magellanic Clouds. The data show that star formation is considerably less active in the SMC than in the LMC. The sizes of the objects range from less than a few arcsec – objects which look like extremely compact H II regions, with little or no extended radio, optical, or infrared emission – to some tens of arcsec across – giant H II regions, of which the largest and brightest is 30 Doradus. There are no obvious differences in the characteristics of the central portions of the LMC and SMC sources; all look like compact Galactic H II regions of similar luminosity.

156.003 **Star formation in the Large Magellanic Cloud.**
T. J. Jones, A. R. Hyland, P. M. Harvey.
NASA Conf. Publ., NASA CP–2466, p. 245 – 246 (1987). – See Abstr. 012.002.
The authors compare star formation in the LMC with the star formation in our galaxy.

156.004 **Infrared emission and excitation in LMC H II regions.**
V. Ungerer, F. Viallefond.
NASA Conf. Publ., NASA CP–2466, p. 247 – 251 (1987). – See Abstr. 012.002.
The infrared excess (IRE) of LMC H II nebulae is found to correlate positively with the temperature of the ambient radiation field or with the He^+/H^+ abundance ratio. This result is discussed in terms of a selective absorption of the photons in the range 504 – 912 Å relative to the He ionizing photons. This interpretation may explain the paradox of finding highly excited nebulae with only relatively moderate equivalent width of their Balmer lines.

156.005 **Spectral types of bright stars in the Small Magellanic Cloud Wing.**
E. Kontizas, D. H. Morgan, A. Dapergolas, M. Kontizas.
Astron. Astrophys., Suppl. Ser., Vol. 70, No. 1, p. 1 – 14 (1987).
Identification charts and spectral classification catalogues for eight areas in the SMC Wing are presented here. Three of the fields are located in the inner Wing, four fields in the outer Wing and one lies between them. Two additional fields near the Wing are studied for comparison. The classification was made by means of binocular microscope examination of film copies of plates taken with 1.2 m Schmidt Telescope. All the classified stars are brighter than $B \sim 18.5$ mag.

156.006 **Spectral types of bright stars in the north–east arm of the Small Magellanic Cloud.**
A. Dapergolas, E. Kontizas, M. Kontizas, D. H. Morgan.
Astron. Astrophys., Suppl. Ser., Vol. 70, No. 1, p. 15 – 19 (1987).
The spectral types of stars in four selected areas, covering a large proportion of the north–east SMC outer arm, are given with their identification charts. Another field outside the arm has also been studied for comparison. The classification was made by means of binocular microscope examination of film copies of plates taken with the UK 1.2 m Schmidt telescope. All the classified stars are brighter than $B \sim 18.5$ mag.

156.007 **Luminous MS stars in the LMC.**
K. Lundgren.
Messenger, No. 49, p. 4 – 5 (1987).

156.008 **Detection of shell–like features in the north–eastern halo of the Small Magellanic Cloud.**
H. Albers, H. T. MacGillivray, S. M. Beard, F. R. Chromey.
Astron. Astrophys., Vol. 182, No. 1, p. L8 – L10 (1987).
The authors have used the COSMOS automatic, high–speed, plate–scanning machine at the Royal Observatory, Edinburgh in order to obtain new and objective star counts down to faint limiting magnitudes (B $\sim$ 22.5, R $\sim$ 21.5) in the region of the Small Magellanic Cloud (SMC). The counts have been analysed using digital unsharp–masking techniques and reveal previously undetected features in the area to the North–East of the Cloud. These features appear to be concentric about the SMC, extending Southwards into the Wing, and may form part of a ring or shell–type structure in this region of the Cloud. The results reported herein invite a possible new interpretation of the "outer arm" of the SMC as part of a similar structure. These shell–like features also suggest an explanation for the shape of the main subcondensations in the Wing of the SMC.

156.009 **CMDs for the LMC clusters NGC 2249 and NGC 2241.**
J. H. Jones.
Astron. J., Vol. 94, No. 2, p. 345 – 356 (1987). With plates 22 – 26.
Color–magnitude diagrams are derived for two LMC clusters, NGC 2249 and NGC 2241. A technique is introduced for

"subtracting" the field–star contribution from the cluster CMDs using CMDs from nearby LMC fields. The possibility of a "gap" in the main sequence of NGC 2249 is discussed, and, by identifying the gap as the point of core hydrogen exhaustion for stars with convective cores, the isochrones are found to be well placed relative to the turnoff. Based on these isochrones, NGC 2249 is thought to be 550 to 700×10^6 years old, depending on the reddening and distance modulus used. NGC 2241, by virtue of its subgiant branch and giant clump and by comparison with NGC 2506, is thought to be between 3 and 4×10^9 years old.

156.010 *BVRI* photometry of star clusters in the Bok region of the Large Magellanic Cloud.
G. Alcaino, W. Liller.
Astron. J., Vol. 94, No. 2, p. 372 – 402 (1987). With plates 30 – 49.

Photographic *BVRI* color–magnitude and color–color diagrams are provided for 14 star clusters situated in the Bok region of the Large Magellanic Cloud. The authors have estimated the reddening for each cluster by fitting the main sequence of the CMDs to the unreddened ZAMS. Ages have been derived from the MS turnoff and from the brightest blue star following the method developed by Hodge (1983), and from fitting the CMDs to the isochrones of Maeder and Mermilliod (1981). All clusters are found to be young, with ages ranging from $17 \pm 6 \times 10^6$ yr for NGC 1858 to $90 \pm 30 \times 10^6$ yr for NGC 1860.

156.011 The nature of dust grains in the clouds of Magellan: 8 – 13 μm spectra of LMC N44A and SMC N88A.
P. F. Roche, D. K. Aitken, C. H. Smith.
Mon. Not. R. Astron. Soc., Vol. 228, No. 1, p. 269 – 275 (1987).

The authors present 8 – 13 μm spectrophotometry of two Magellanic cloud H II regions, LMC N44A and SMC N88A. The former shows a prominent silicate emission feature and resembles many galactic H II regions, whilst the latter shows no evidence of silicate dust. The implications for the composition of dust particles in the clouds and the species responsible for the 2200 Å feature are discussed.

156.012 Colour–magnitude diagrams of star clusters in the Magellanic Clouds from wide–field electronography. III. NGC 1841.
J. Andersen, A. Blecha, M. F. Walker.
Mon. Not. R. Astron. Soc., Vol. 229, No. 1, p. 1 – 13 (1987). = Contrib. Lick Obs., No. 444.

The colour–magnitude diagram (CMD) of NGC 1841, an outlying globular cluster in the Magellanic System, has been constructed from electronographic photometry of 390 stars in the annular field 22 arcsec $< R <$ 72 arcsec to a limit of about $V = 22.5$. The detailed CMD confirms earlier classifications of NGC 1841, as belonging to the oldest and most metal–poor population of the Magellanic System. For an adopted reddening of $E(B-V) = 0.15$, the CMD of NGC 1841 closely matches that of the galactic globular cluster M92, indicating that the two clusters are of similar age and metal abundance, in agreement with previous spectroscopic studies. The fit of the two CMDs yields a distance modulus to NGC 1841 of $(m-M)_0 = 18.1 \pm 0.3$. This places NGC 1841 within the disc system formed by the other old LMC clusters – some 10 kpc from the centre of the LMC itself – in contrast to earlier results placing it well beyond the LMC. It also eliminates the previous discrepancy between the luminosities of its giant stars and similar Galactic globular cluster giants.

156.013 Distribution of spectral types in the LMC clusters.
E. Kontizas, M. Kontizas, E. Xiradaki.
Astron. Astrophys., Suppl. Ser., Vol. 71, No. 3, p. 575 – 589 (1987).

The distribution of spectral types in 42 LMC globular star clusters are presented. Among them 18 kinematic disk clusters are located near or in the LMC bar and 24 remote clusters are situated at the periphery of the LMC at distances larger then 5° from the centre of the bar. Spectral classification of stars brighter than $B \cong 18.5$ mag was carried out on objective prism plates

taken with the 1.2 m U.K. Schmidt Telescope in Australia. All stars located within the tidal radii of the clusters have been classified and stars in adjoining fields to each cluster were also studied. The selected clusters cover all evolutionary ages and the derived spectral type distributions show that the clusters can be divided into five age categories (from $\sim 10^7$ to $> 10^9$ yr). Several clusters were found to contain carbon stars with C/M ratios ranging from 0.07 to 0.4. These ratios were compared with those found for the SMC clusters, the Milky Way and theoretical models.

156.014 Hot stars in young globular clusters of the Magellanic Clouds.
E. Kontizas, M. Kontizas.
Q. J. R. Astron. Soc., Vol. 28, No. 3, p. 239 – 241 (1987). – See Abstr. 012.037.

The ratio R of early– to late–type luminous stars in young populous globular clusters in the Magellanic Clouds has been obtained via spectral classification. It was found that there are very young globular clusters (age $10^7 - 5 \times 10^7$ yr) with a high percentage of hot stars. The observed ratios R for the LMC clusters exhibit values larger than those found for the SMC, indicating that the hot stars in the SMC clusters are fewer.

156.015 Clusters of the Small Magellanic Cloud. II. Age distributions.
P. Hodge.
Publ. Astron. Soc. Pac., Vol. 99, No. 618, p. 724 – 729 (1987).

From a complete catalog of all star clusters in a nine–region survey of the SMC, the author uses measured cluster properties to study the rate of cluster formation and disintegration in various parts of the SMC. The mean diameters of cataloged SMC clusters is 7 pc; this value reduces to 5.8 pc for the regions for which incompleteness is not a problem. The author finds that the SMC is currently forming clusters at the rate of approximately 100 per 10^8 years. The disintegration rate is significantly slower than for clusters in the solar neighborhood. The median age of SMC clusters is 0.9 Gyr.

156.016 The population of Large Magellanic Cloud field stars in a remote southwestern area.
P. Hodge.
Publ. Astron. Soc. Pac., Vol. 99, No. 618, p. 730 – 733 (1987).

The stars in a remote southwestern part of the LMC show that at this position 4.5 kpc from the center of the Cloud the stellar population is dominated by an intermediate population of stars; a true Population II is not detectable in the color–magnitude diagram, which resembles in many respects those for other remote areas of the LMC. The main–sequence luminosity function is steeper than that reported for the LMC bar, which probably means that this remote field has a somewhat older mean age – found to be about 3 Gyrs. Surprisingly, there is no evidence of significant numbers of stars either younger or older than about 2 – 3 Gyrs.

156.017 Some studies of Magellanic Cloud clusters.
P. Hodge, P. Flower.
Publ. Astron. Soc. Pac., Vol. 99, No. 618, p. 734 – 738 (1987).

Four photoelectric sequences in the *UBV* system for areas in the Large and Small Magellanic Clouds are presented. The stars range in brightness from $V = 8.33$ to $V = 19.42$ and in color from $(B-V) = 0.08$ to $(B-V) = +1.80$. Photographic color–magnitude diagrams for seven clusters in the Clouds are derived from CTIO 4–m telescope plates. The data are used to estimate the ages of the clusters, which range from 0.3 to 1.6 billion years.

156.018 Physical parameters for 12 planetary nebulae and their central stars in the Magellanic Clouds.
L. H. Aller, C. D. Keyes, S. P. Maran, T. R. Gull, A. G. Michalitsianos, T. P. Stecher.
Astrophys. J., Vol. 320, No. 1, p. 159 – 177 (1987).

Nebular and central star parameters and elemental abundances of C, N, O, Ne, S, and Ar are presented for 12 planetary nebulae in the Magellanic Clouds, which were observed with the

IUE. The precursors of the observed nebulae may not have been sufficiently massive to synthesize Ne, S, or Ar, which appear to be deficient with respect to their solar abundances by factors of roughly 4 and 5 for the LMC and SMC, respectively. Similarly, O depletion in the LMC and SMC nebulae is significantly greater than in galactic planetaries. The estimated masses of the 12 remnant central stars range from 0.58 to 0.71 $M_\odot$.

156.019 Angular diameters and fluxes of Magellanic Cloud planetary nebulae. II. High–speed direct imaging.

P. R. Wood, S. J. Meatheringham, M. A. Dopita,
D. H. Morgan.
Astrophys. J., Vol. 320, No. 1, p. 178–181 (1987).

Images and fluxes of Magellanic Cloud planetary nebulae have been obtained with a time resolution of 1/60 s. The high time resolution was used to remove translational seeing effects. Angular diameters greater than 0″.7 have been obtained for 20 Magellanic Cloud planetary nebulae, and fluxes in Hβ or [O III] λ5007 or both have been obtained for 80 nebulae. Masses have been derived for 18 nebulae from the angular diameters and Hβ fluxes. The ionized masses of typical Magellanic Cloud planetary nebulae increase with radius R roughly as $R^{3/2}$ to $R^{5/3}$ until $R \approx 0.1$ pc; thereafter, the ionized mass has a maximum value of $\sim 0.5\ M_\odot$. Bolometric luminosities have been estimated for the central stars of those objects for which Hβ fluxes were obtained.

156.020 Surface photometry of the old populous star clusters of the Large Magellanic Cloud.

M. Mateo.
Bull. Am. Astron. Soc., Vol. 19, No. 2, p. 677 (1987). Abstract. – See Abstr. 010.061.

156.021 The age–metallicity relations of the Magellanic Clouds as determined from observations of the integrated spectra of star clusters.

H. A. Smith, L. Searle, A. Manduca.
Bull. Am. Astron. Soc., Vol. 19, No. 2, p. 678 (1987). Abstract. – See Abstr. 010.061.

156.022 Abundances of Large Magellanic Cloud clusters.

D. P. Geisler.
Bull. Am. Astron. Soc., Vol. 19, No. 2, p. 678 (1987). Abstract. – See Abstr. 010.061.

156.023 CCD and photographic photometry of the northeast arm to halo transition in the Small Magellanic Cloud.

S. R. Baird, H. A. Smith, J. A. Graham.
Bull. Am. Astron. Soc., Vol. 19, No. 2, p. 682 (1987). Abstract. – See Abstr. 010.061.

156.024 A comparison of the luminosity functions of the Hα emission–line stars in the Magellanic Clouds.

J. B. Doggett, B. Bohannan.
Bull. Am. Astron. Soc., Vol. 19, No. 2, p. 704 (1987). Abstract. – See Abstr. 010.061.

156.025 Molecular hydrogen in H II regions in the Magellanic Clouds.

F. P. Israel, J. Koornneef.
Space Telesc. Sci. Inst., Prepr. Ser., No. 209, 16 pp. (1987). To appear in Astron. Astrophys.

156.026 Binary star clusters in the Large Magellanic Cloud.

R. K. Bhatia, D. Hatzidimitriou.
Edinb. Astron. Prepr., No. 17/87, 13 pp. (1987). To appear in Mon. Not. R. Astron. Soc.

156.027 Vacuum ultraviolet images of the Large Magellanic Cloud.

A. M. Smith, R. H. Cornett, R. S. Hill.
Astrophys. J., Vol. 320, No. 2, p. 609–625 (1987). With plate 22.

Images with 50″ resolution of the Large Magellanic Cloud (LMC), obtained with sounding–rocket instrumentation in two vacuum ultraviolet (VUV) bandpasses, are presented. The bandpasses are each ~ 200 Å wide and are centered, for hot stars, near 1500 Å and 1900 Å. Photometry was done on the digitized images for all associations in the list of Lucke and Hodge. The authors discuss the results and their relationship to the overall characteristics of star formation in the LMC. They present a simple model for propagating star formation in the LMC whose results closely resemble the distribution of associations as revealed by VUV images.

156.028 CCD photometry of Large Magellanic Cloud clusters. IV. The metal–rich, remote southern cluster LW 79.

M. Mateo, P. Hodge.
Astrophys. J., Vol. 320, No. 2, p. 626–652 (1987). With plates 23–25.

LW 79 is a populous star cluster located in the remote southern part of the Large Magellanic Cloud. The authors present deep Johnson BV CCD photometry of 1230 stars and Cousins R photometry of 614 stars in and around this cluster. The V versus $(B-V)$ color–magnitude (CM) diagram of LW 79 reveals a well-defined main sequence terminating at $V \approx 20.3$ and $(B-V) \approx 0.43$ and a well-populated red giant branch and red giant clump. Detailed comparisons of the CM diagram of LW 79 with those of NGC 752 and NGC 7789 suggest that the age, metallicity [Fe/H], and distance modulus of LW 79 are $\sim 2.0 \times 10^9$ yr, -0.3 ± 0.3, and 18.4 ± 0.2, respectively. These values agree well with the results determined from comparisons of the main-sequence photometry of LW 79 with theoretical isochrones. Comparisons of the evolved stars in LW 79 with appropriate theoretical models suggest that significant uncertainties remain in accurately predicting post–He–flash stages of stellar evolution.

156.029 Chemical and spectrophotometric evolution of the Magellanic Clouds.

B. Rocca–Volmerange.
Inst. Astrophys. Paris, Pré–Publ., No. 201, 15 pp. (1987).

This paper is published in the ESO Workshop on "Stellar evolution and dynamics in the outer halo of the Galaxy", see Abstr. 156.053.

156.030 A catalog of LMC star clusters outside the Hodge–Wright atlas.

E. W. Olszewski, H. C. Harris, R. A. Schommer,
R. W. Canterna.
Prepr. Steward Obs., No. 757, 24 pp. (1987). To appear in Astron. J.

156.031 Morphology of LMC clusters.

A. F. Zepka, H. A. Dottori.
Rev. Mex. Astron. Astrofis., Vol. 14, No. 1, p. 172–177 (1987). – See Abstr. 012.042.

It is well–known that LMC globular clusters have in general high ellipticities in counterpart to the galactic ones. The authors fit ellipses to the isophotal contours of the clusters. They agree with previous works concerning variation of ellipticity. They also found that position angles do vary in some of them. A pattern in the distribution of these parameters was observed. The authors believe that such variations (specially of position angles) may be evidence of the triaxiality of the actual cluster shape.

156.032 Spectrophotometric observations of very low ionization H II regions in the LMC.

M. Peña, M. T. Ruíz, M. Rubio.
Rev. Mex. Astron. Astrofis., Vol. 14, No. 1, p. 178–182 (1987). – See Abstr. 012.042.

Optical spectrophotometric observations of 17 very low ionization H II regions of the LMC are reported. Physical conditions and chemical composition of these objects are derived from the emission line intensities. The average chemical abundances obtained are: log O/H = 8.49 ± 0.08, log N/H = 6.91 ± 0.07 and log S/H = 6.89 ± 0.10. The authors do not find evidence of any composition gradient in the LMC. The H II regions in the vicinity of the detected molecular cloud complexes show higher nebular reddening.

156.033 NGC 2209: the nature of the dark patch through the HR diagram.
H. Dottori, J. Melnick, E. Bica.
Rev. Mex. Astron. Astrofis., Vol. 14, No. 1, p. 183 – 187 (1987).
– See Abstr. 012.042.

The aim of this paper is to present a study of the nature of the dark patch located near the center of the LMC cluster NGC 2209. The authors use a deep HR diagram of the cluster and analyse the photometric properties of a few stars that appear projected on top of the patch. The conclusion is that the patch is an internal feature of the cluster. The authors also discuss the controversial problem of the metallicity and age of the cluster.

156.034 Molecular clouds in the Large Magellanic Cloud.
M. Rubio.
Rev. Mex. Astron. Astrofis., Vol. 14, No. 1, p. 188 – 190 (1987).
– See Abstr. 012.042.

CO emission has been detected from more than 30 regions in the Large Magellanic Cloud. A preliminary analysis suggests that the LMC molecular clouds may have smaller sizes than those found in our Galaxy.

156.035 Fotometria $W_{H\beta}$ y [O III]/$H\beta$ de regiones H II y la historia de la formación estelar reciente en la Pequeña Nube de Magallanes.
M. V. F. Copetti, H. A. Dottori.
Rev. Mex. Astron. Astrofis., Vol. 14, No. 1, p. 191 (1987).
Abstract. – See Abstr. 012.042.

156.036 Binarias masivas en las Nubes de Magallanes.
V. S. Niemela.
Rev. Mex. Astron. Astrofis., Vol. 14, No. 1, p. 390 (1987).
Abstract. – See Abstr. 012.042.

156.037 C, M, and S stars in the Magellanic Clouds.
B. E. Westerlund.
ESO Conf. Workshop Proc., No. 27, p. 207 – 219 (1987). – See Abstr. 012.051.

Spectroscopic and photometric data are analyzed in order to establish the evolution of carbon stars after the transition form the M–star phase. The location of J–type stars in the transition area of the HR–diagram is found to agree with the evolutionary scenario derived from IRAS data. The existence of a number of stellar generations in the Clouds is confirmed. The importance of dividing the 7 – 2 Gyr generation into subgroups is noted.

156.038 HR diagrams and luminosity functions of the Magellanic Cloud field populations.
K. J. Mighell.
ESO Conf. Workshop Proc., No. 27, p. 221 – 232 (1987). – See Abstr. 012.051.

The history of star formation in the Magellanic Clouds can now be read in their field populations. The last decade has seen the first attempts in this direction, and has made clear the complexity of the evolution of these galaxies. This contribution discusses the present state of this work.

156.039 Preliminary abundances in three field supergiants of the SMC.
F. Spite, M. Spite.
ESO Conf. Workshop Proc., No. 27, p. 233 – 241 (1987). – See Abstr. 012.051.

The comparison of the chemical evolution of the Galaxy and of the Magellanic Clouds is obviously an important source of information about the physical processes which take place in these evolutions. Numerous works have been therefore devoted to the determination of the abundances of objects of various ages in the Magellanic Clouds. The authors analysed a few F– and G–type stars in the Magellanic Clouds.

156.040 On the chemical abundance of the SMC stars.
F. Thévenin.
ESO Conf. Workshop Proc., No. 27, p. 243 – 249 (1987). – See Abstr. 012.051.

The main aim of this short review is to summarise our present knowledge on the stellar abundances in the SMC. If one accepts that the comparison of different star populations is a clue to its history, the analysis of star fields and of star clusters would be very informative. The author presents and uses a compilation of the small existing amount of detailed spectroscopic analysis of late–type stars. Then he presents briefly the well established abundances of light elements obtained from the observations of H II regions or of planetary nebulae. A short discussion of the possible errors for these methods of chemical analysis is also given, and the author presents several questions that one can encounter in comparing the two chemical determinations.

156.041 Detection of features in the north–eastern halo of the SMC from deep star counts.
H. Albers, H. T. MacGillivray, S. M. Beard, F. R. Chromey.
ESO Conf. Workshop Proc., No. 27, p. 251 – 257 (1987). – See Abstr. 012.051.

New and objective star counts down to faint limiting magnitudes (B ~ 22.5, R ~ 21.5) in the region of the Small Magellanic Cloud have been obtained. The counts have been analysed using digital unsharp–masking techniques and reveal previously undetected features in the area to the North–East of the Cloud. The features appear to be concentric about the SMC, extending Southwards into the Wing, and may form part of a ring or shell–type structure in this region of the Cloud. The similarity between these newly found arcs and the "outer arm" of the SMC is striking and suggests that the outer arm may be part of a similar shell–type structure.

156.042 Further evidences for a stellar link between the Magellanic Clouds.
S. Demers, M. J. Irwin, W. E. Kunkel.
ESO Conf. Workshop Proc., No. 27, p. 259 – 262 (1987). – See Abstr. 012.051.

Following the discovery of a substantial population of blue stars east of the SMC wing by Irwin et al. (1985), the authors initiated programs of spectroscopic observations and CCD photometry. Their first aims are to determine the distance of these stars and establish their kinematic properties. This paper is a progress report; the authors present only the first series of spectroscopic measures consisting of more than a dozen blue stars. The data, which should yield not only distances but also luminosity functions, have not yet been fully analysed.

156.043 The carbon star population of the Small Magellanic Cloud.
E. Rebeirot, M. Azzopardi, J. Breysacher, B. E. Westerlund.
ESO Conf. Workshop Proc., No. 27, p. 263 – 267 (1987). – See Abstr. 012.051.

The aim of the present survey is to obtain a clear picture of the carbon star distribution in the various regions of the Small Magellanic Cloud and, thus, to detect possible variations in their luminosity functions. It should give a sufficiently detailed sampling for conclusions regarding the evolution of the carbon stars in that galaxy.

156.044 Spectrophotometric abundances of red Magellanic Cloud supergiants.
T. Richtler, W. Seggewiss.
ESO Conf. Workshop Proc., No. 27, p. 269 – 272 (1987). – See Abstr. 012.051.

156.045 Integrated properties of star clusters in the LMC.
C. Chiosi, G. Bertelli, A. Bressan.
ESO Conf. Workshop Proc., No. 27, p. 415 – 441 (1987). – See Abstr. 012.051.

Synthetic HR diagrams and integrated $(B-V)_0$ and $(U-B)_0$ colours of stellar clusters as a function of the age are obtained with the aid of stellar models that incorporate the effects of convective overshoot all over their major evolutionary phases. The resulting theoretical calibrating relationships are presented and some implications for real clusters of different age are outlined. In particular, the authors derive the age and colour distribution functions (number of clusters per age and colour range) for clusters of LMC, and discuss the causes of the "gap" in the relation between (B–V) and cluster type by Searle at al. (1980) and/or in the V versus (B–V) diagram, or equivalently the bimodality of the colour distribution function.

156.046 Washington CCD photometry of Magellanic Cloud clusters.   •
D. Geisler.
ESO Conf. Workshop Proc., No. 27, p. 443 – 452 (1987). – See Abstr. 012.051.

How far away are the Magellanic Clouds? Until three years ago, despite suggestions of a much smaller distance, a consensus had held for a true LMC distance modulus near 18.7. The ability to obtain good photometry several magnitudes down the main sequence allows to derive very accurate distances to Cloud clusters. The very first published CCD color–magnitude diagrams (CMD) of Magellanic Cloud clusters indicated that the distance to the LMC could be as low as 18.2. The techniques used to determine the distance to a cluster with deep main–sequence photometry, including main–sequence fitting, luminosity function fitting, and matching the clump giants' luminosity, all require an independent knowledge of the cluster metal abundance, [Fe/H]. The derived distance is indeed quite sensitive to this parameter. The authors' research has focused on a new technique for obtaining accurate abundances (uncertainties <0.2 dex) for intermediate to old age Magellanic Cloud clusters: Whashington CCD photometry. He reports on results for two such clusters: NGC 2213 and NGC 2162.

156.047 The development of the red giant branch in intermediate age clusters of the Magellanic Clouds.
L. Greggio.
ESO Conf. Workshop Proc., No. 27, p. 453 – 460 (1987). – See Abstr. 012.051.

156.048 Synthetic clusters in the Large Magellanic Cloud.
E. Brocato, V. Castellani.
ESO Conf. Workshop Proc., No. 27, p. 461 – 466 (1987). – See Abstr. 012.051.

The authors report preliminary results of investigation on young star clusters in the Large Magellanic Cloud devoted to the cluster NGC 1866 and concerning the integrated UV properties of LMC clusters.

156.049 Global properties of LMC star clusters.
M. Mateo.
ESO Conf. Workshop Proc., No. 27, p. 467 – 472 (1987). – See Abstr. 012.051.

The star clusters of the Large Magellanic Cloud (LMC) are ideal probes of various aspects of that galaxy's past history. Specifically, LMC clusters have long been known to span a large range in age, are only slightly reddened and lie at a common distance from the sun. Except for the very oldest clusters, it is now possible to obtain main sequence photometry of most LMC clusters and thereby determine reliable ages for these systems. Combined with knowledge of other properties of these objects, one can then directly study a wide diversity of subjects including the chemical evolution of the LMC, the star cluster formation rate in the LMC throughout the past, the evolution of large n–body stellar systems, and changes in the stellar initial mass function with time to name a few. In this paper, the author describes initial results related to some of these topics.

156.050 Integrated magnitudes and colours of clusters in the LMC with the COSMOS machine.
R. K. Bhatia, R. D. Cannon, H. T. MacGillivray.
ESO Conf. Workshop Proc., No. 27, p. 481 – 484 (1987). – See Abstr. 012.051.

The Magellanic Clouds are rich in clusters: the LMC is estimated to have about 5000 clusters while the SMC has about

2000. The study of these clusters is important not only for understanding their formation and evolution, but also for the study of the structure and evolution of the Magellanic Clouds themselves. It is impractical of expect photoelectric magnitudes for such large number of clusters. An attractive way is to use the powerful combination of the COSMOS machine and Schmidt plates. The authors have therefore initiated a major study for obtaining reliable magnitudes and colours for a majority of the clusters in the LMC and the SMC. They discuss the method and present preliminary results.

156.051 NGC 1850: a merging binary cluster in the LMC?
R. K. Bhatia, H. T. MacGillivray.
ESO Conf. Workshop Proc., No. 27, p. 485 – 487 (1987). – See Abstr. 012.051.

NGC 1850 is a young cluster situated in the eastern end of the bar of the LMC. An examination of this area on Schmidt plates shows that this whole area is "active" and there is a nebulosity around this cluster. The authors have made CCD observations of this area and find that the two clusters are nearly co–eval, the smaller one being slightly younger. This supports their contention that the pair has formed from the same cloud. The authors have made estimates of the time of merger of NGC 1850: 8×10^7 yrs. This is consistent with the age of the cluster 3×10^7 years as also with the results of the simulations done by White (1979) for the mergers of binary stellar systems. They estimate that the two clusters will merge in approximately 10^6 years from now.

156.052 CCD observations of binary clusters in the Magellanic Clouds.
R. K. Bhatia, R. D. Cannon, D. Hatzidimitriou.
ESO Conf. Workshop Proc., No. 27, p. 489 – 490 (1987). – See Abstr. 012.051.

The existence of binary clusters in the Large Magellanic Cloud has been discussed recently. A survey of the LMC on UKSTU plates yielded a total of 69 pairs of clusters with a center–to–center separation of less than ~ 18 pc (for a distance modulus of 18.4 for the LMC). Statistical arguments suggest that taking chance line–ups into account, a significant fraction of the pairs found must be physically associated at the 7σ level. A series of observations have been undertaken in order to establish the physical association of the pairs.

156.053 Chemical and spectrophotometric evolution of the Magellanic Clouds.
B. Rocca–Volmerange.
ESO Conf. Workshop Proc., No. 27, p. 637 – 651 (1987). – See Abstr. 012.051.

From chemical models, the author confirms that very deficient galaxies appear very young ($\leqslant 1$ Gyr) but a large uncertainty on age exists for less deficient galaxies. Spectrophotometric models, from far–UV to near–infrared, simultaneously test emissivity of young stars (UV) and old stars (visible and near–infrared). No old population is detected in bursting galaxies (NGC 4449) and more precise estimates of age are given for magellanic galaxies: 7 Gyrs (SMC) to 9 Gyrs (LMC). Compared to these results, the halo of our Galaxy reaches a less deficient metallicity, explained by a more rapid and efficient collapse in the early phases.

156.054 The evolution of asymptotic giant branch stars in the Magellanic Clouds. III. The problem of intermediate–mass stars.
J. Mould, N. Reid.
Astrophys. J., Vol. 321, No. 1, p. 156 – 161 (1987).

Two samples of red giants in the LMC with bolometric magnitudes in the interval (–4, –8) are examined spectroscopically. These samples are deficient in luminous stars ($M_{bol} < -5.5$) with any evidence of third dredge–up. The most likely explanation is that 2 to 5 $M_\odot$ stars spend their thermal pulsing lifetime in a brief Mira phase, and that their evolution is rapidly truncated by depletion of the stellar envelope by mass loss.

156.055 Distance moduli and structure of the Magellanic Clouds from near–infrared photometry of classical Cepheids.
D. L. Welch, R. A. McLaren, B. F. Madore, C. W. McAlary.
Astrophys. J., Vol. 321, No. 1, p. 162 – 185 (1987).

New *JHK* photometry has been obtained for Cepheid variables in the LMC and SMC and has been merged with the existing data of Laney and Stobie. The total data base comprises 494 *JHK* observations of 65 LMC Cepheids asnd 451 *JHK* observations of 94 SMC Cepheids. From the mean magnitudes, period–luminosity and period–color relations have been determined, and true distance moduli of 18.57 ± 0.05 and 18.93 ± 0.05 mag are found for the LMC and SMC, respectively. The line–of–sight distribution of Cepheids in the LMC and SMC is investigated using residuals from the *H* P–L and P–L–C relations. It is found that the SMC is *not* extended beyond its tidal radius unless the mass of the Galaxy is significantly greater than $10^{12} M_\odot$.

156.056 The age of the Large Magellanic Cloud cluster NGC 2193.
G. S. Da Costa, C. R. King, J. R. Mould.
Astrophys. J., Vol. 321, No. 2, p. 735 – 744 (1987). With plate 12.

The age of the LMC star cluster NGC 2193 is estimated as 2.2 ± 0.6 Gyr from CCD photometry that reaches below the cluster main–sequence turnoff. The principal cause of uncertainty in the estimated age is uncertainty in the LMC distance modulus. A metal abundance of [Fe/H] $= -0.5 \pm 0.3$ dex is deduced for NGC 2193 from the color of the cluster giant branch. The *C–M* diagram for the LMC field near NGC 2193 is dominated by a population of old stars. Isochrone fits indicate an age for this old population in excess of 7 Gyr with some indication of ages perhaps as old as 13 Gyr.

156.057 Clarifications of some identifications for the LMC region.
M. Morel.
Bull. Inf. Cent. Données Stellaires, No. 33, p. 77 – 79 (1987).

Identifications and notes were published for a number of Henry Draper (HD) objects in the LMC region (see Abstr. 43.002.065, 43.156.022). Additional notes are given here.

156.058 Confusion in the Wolf–Rayet literature.
W. P. Bidelman, N. Sanduleak.
Bull. Inf. Cent. Données Stellaires, No. 33, p. 119 (1987).

156.059 The stellar populations of Shapley Constellation III.
N. Reid, J. Mould, I. Thompson.
Astrophys. J., Vol. 323, No. 2, p. 433 – 450 (1987).

The authors have constructed a *V–I* color–magnitude diagram for a 0.6 deg^2 field covering part of the star formation region Shapley III in the LMC. The main–sequence stars have a luminosity function exhibiting a pronounced break at $M_V \sim -3$ which is identified with the turnoff of the first star–forming burst. Using this as an age indicator, the authors have compared stellar evolutionary models with the dynamical age estimate determined by Dopita, Mathewson, and Ford and derive the initial luminosity and mass functions.

Near infrared spectroscopy with the array spectrometer of the Anglo–Australian telescope.
See Abstr. 034.166.

Initial–final mass relation for low and intermediate mass stars.
See Abstr. 065.016.

The peculiar emission–line supergiant HD 37836.
See Abstr. 112.012.

Direct imagery of circumstellar shells around Ofpe/WN9 stars in the Galaxy and in the LMC.
See Abstr. 112.024.

An extended nebulosity surrounding the S Dor variable R 127.
See Abstr. 112.043.

The LMC – S Dor variable R 71: an IRAS point source.
See Abstr. 112.044.

Eclipse spectrum of the LMC P Cyg star R 81.
See Abstr. 112.045.

The stellar winds of early–type stars in the Small Magellanic Cloud.
See Abstr. 112.052.

Stellar winds from O stars in the Small Magellanic Cloud.
See Abstr. 112.130.

The intrinsic colors of OB supergiants in the Large Magellanic Cloud.
See Abstr. 113.037.

Revised UBV photometry of Magellanic Cloud sequences.
See Abstr. 113.043.

High–dispersion spectroscopy of the Of/WN stars R 84 and S 61 of the Large Magellanic Cloud.
See Abstr. 114.059.

Effective temperatures and gravities of S Doradus–like stars in the Large Magellanic Cloud.
See Abstr. 114.071.

An atlas of optical spectrophotometry of southern Wolf–Rayet carbon stars.
See Abstr. 114.092.

Carbon star surveys in the Magellanic Clouds and the dwarf spheroidals in the galactic halo.
See Abstr. 114.105.

Detection of carbon stars in two independent surveys.
See Abstr. 114.106.

Optical and UV spectroscopy of the black hole binary candidate LMC X–1.
See Abstr. 117.047.

CAL 83: a puzzling X–ray source in the Large Magellanic Cloud.
See Abstr. 117.251.

Photometry and spectroscopy of the eclipsing P Cygni star R 81 in the Large Magellanic Cloud.
See Abstr. 119.021.

UBV photometry of eclipsing binaries in the LMC.
See Abstr. 119.026.

Light curve solutions for five eclipsing binaries in the Magellanic Clouds.
See Abstr. 119.033.

Cepheids in the Magellanic Clouds. II. Search for double mode Cepheids in the LMC.
See Abstr. 122.052.

Supernova remnants in the Magellanic Clouds observed at the Molonglo Observatory.
See Abstr. 125.009.

IRAS observations of collisionally heated dust in Large Magellanic Cloud supernova remnants.
See Abstr. 125.038.

Morphological studies of the supernova remnant E0102.2–72.2.
See Abstr. 125.057.

The morphology and dynamics of a multi–lobed supernova remnant in the LMC (DEM 34a, N11L).
See Abstr. 125.063.

Infrared spectroscopy of supernova remnants.
See Abstr. 125.066.

X–ray observations of the supernova remnant N103B in the Large Magellanic Cloud.
See Abstr. 125.085.

Time profile of the neutrino burst from SN 1987A in the Large Magellanic Cloud.
See Abstr. 125.205.

X–ray emission from SN 1987A.
See Abstr. 125.206.

Constraints on the interpretation of the neutrino experiments by the optical observations of SN 1987a.
See Abstr. 125.210.

Spectroscopic and photometric observations of SN 1987a: the first 50 days.
See Abstr. 125.211.

Very low upper limits on the strength of interstellar lithium lines toward SN 1987A.
See Abstr. 125.291.

Interstellar calcium towards supernova 1987A in the Large Magellanic Cloud.
See Abstr. 125.318.

The initial–final mass relation: galactic disk and Magellanic Clouds.
See Abstr. 126.082.

Detections of diffuse interstellar bands toward the SN 1987A in the Large Magellanic Cloud.
See Abstr. 131.032.

Detection of interstellar CH and CH^+ towards SN 1987A.
See Abstr. 131.110.

Extinction in external galaxies.
See Abstr. 131.142.

Dust in external galaxies: diffuse bands and the albedo.
See Abstr. 131.143.

Recent massive star formation in 30 Doradus.
See Abstr. 131.198.

The interstellar medium toward SN 1987A.
See Abstr. 131.207.

The LMC H II regions N11C and E and their stellar contents.
See Abstr. 132.013.

The SMC compact blob N81: a detailed multi–wavelength investigation.
See Abstr. 132.028.

Evolution of H II regions.
See Abstr. 132.035.

Cooled grating spectrometer observations of molecular hydrogen in H II regions in the Magellanic Clouds.
See Abstr. 132.046.

Recent massive star formation in 30 Doradus.
See Abstr. 132.050.

The determination of the masses of Magellanic Cloud planetary nebulae using [O II] doublet ratio electron densities.
See Abstr. 134.010.

A comparison of some physical properties of planetary nebulae in the Magellanic Clouds with those in the Galaxy.
See Abstr. 134.048.

Abundances, nebular masses and central star parameters for Magellanic Cloud planetary nebulae.
See Abstr. 134.050.

The evolution and dynamics of the Magellanic Cloud planetary nebulae.
See Abstr. 134.051.

An all–sky study of fast X–ray transients.
See Abstr. 142.027.

SMC globular clusters: candidates for mass segregation?
See Abstr. 154.005.

Deep photometry of globular clusters. X. The cluster GlC 0435–59 in Reticulum.
See Abstr. 154.018.

Systematic properties of extragalactic globular clusters.
See Abstr. 154.019.

A preliminary survey of collapsed cores in Magellanic Clouds globular clusters.
See Abstr. 154.050.

Internal dynamics of globular clusters: from our Galaxy to the Magellanic Clouds.
See Abstr. 154.052.

A halo cluster of the LMC at 10 kpc: NGC 1841.
See Abstr. 154.070.

Internal dynamics of globular clusters: from our galaxy to the Magellanic Clouds.
See Abstr. 154.074.

A preliminary survey of collapsed cores in the Magellanic Clouds' globular clusters.
See Abstr. 154.082.

The structure of young star clusters in the Large Magellanic Cloud.
See Abstr. 154.083.

Collapsed cores and the structural parameters of old Large Magellanic Cloud star clusters.
See Abstr. 154.084.

The ratio numbers of carbon and oxygen stars in galaxies.
See Abstr. 157.122.

Stellar populations in Local Group galaxies.
See Abstr. 157.129.

The 160–minutes period in extragalactic objects: the LMC (RR Lyr stars) and NGC 4151.
See Abstr. 158.222.

The interstellar/intergalactic medium in front of supernova 1987A.
See Abstr. 160.099.

157 Normal Galaxies (Structure, Evolution, Pairs, etc.)

157.001 Population of accreting neutron stars in external galaxies.
G. Trinchieri.
The origin and evolution of neutron stars, p. 149 – 159 (1987). – See Abstr. 012.001 (IAU Symp. No. 125).
The instruments on board the EINSTEIN Observatory (Giacconi et al., 1979) made possible the study of the X–ray emission of normal galaxies, and the nature, distribution and characteristics of the evolved stellar population in other galaxies through its X–ray emission. The author concentrates on the results regarding X–ray sources in binary systems and disregards all other phenomena that have come out as a result of the study of normal galaxies in X–rays.

157.002 Star formation in normal galaxies.
C. G. Wynn–Williams.
NASA Conf. Publ., NASA CP–2466, p. 125 – 131 (1987). – See Abstr. 012.002.
In this review the author is mainly concerned with the ways in which recent infrared observations, particularly by IRAS, have influenced our ideas about star formation in "normal" galaxies.

157.003 Models for infrared emission from IRAS galaxies.
M. Rowan–Robinson.
NASA Conf. Publ., NASA CP–2466, p. 133 – 152 (1987). – See Abstr. 012.002.
Models for the infrared emission from IRAS galaxies by Rowan–Robinson and Crawford, by de Jong and Brink, and by Helou, are reviewed and compared.

157.004 On the origin of the 40 – 120 micron emission of galaxy disks: a comparison with H–alpha fluxes.
C. J. Lonsdale Persson, G. Helou.
NASA Conf. Publ., NASA CP–2466, p. 153 – 160 (1987). – See Abstr. 012.002.
A comparison of 40 – 120 micron IRAS fluxes with published H–alpha and UBV photometry shows that the far infrared emission of galaxy disks consists of at least two components: a warm one associated with OB stars in H II–regions and yound star–forming complexes, and a cooler one from dust in the diffuse, neutral interstellar medium, heated by the more general interstellar radiation field of the old disk population (a "cirrus"–like component). Most spiral galaxies are dominated by emission from the cooler component in this model. A significant fraction of the power for the cool component must originate with non–ionizing stars.

157.005 Measuring star formation rates in blue galaxies.
J. S. Gallagher III, D. A. Hunter.
NASA Conf. Publ., NASA CP–2466, p. 167 – 177 (1987). With a correction on p. 257. – See Abstr. 012.002.
The problems associated with measurements of star formation rates in galaxies are briefly reviewed, and specific models are presented for determinations of current star formation rates from Hα and FIR luminosities. The models are applied to a sample of optically blue irregular galaxies, and the results are discussed in terms of star forming histories. It appears likely that typical irregular galaxies are forming stars at nearly constant rates, although a few examples of systems with enhanced star forming activity are found among H II regions and luminous irregular galaxies.

157.006 CO observations of galaxies with the Nobeyama 45–m telescope.
Y. Sofue, T. Handa, M. Hayashi, N. Nakai.
NASA Conf. Publ., NASA CP–2466, p. 179 – 196 (1987). – See Abstr. 012.002.
High–resolution (15"), filled aperture maps of the CO (J = 1 – 0) line emission have been obtained of several nearby, CO–bright galaxies like M82, M83, IC 342, NGC 891, etc. in order to study star forming activity in these galaxies.

157.007 CO observations of nearby galaxies and the efficiency of star formation.
J. S. Young.
NASA Conf. Publ., NASA CP–2466, p. 197 – 215 (1987). – See Abstr. 012.002.
The author has observed the CO distributions and total molecular content of 160 galaxies using the 14 meter millimeter Telescope of the FCRAO (HPBW = 45"). For the luminous, relatively face–on Sc galaxies, the azimuthally averaged CO distributions are centrally peaked, while for the Sb and Sa galaxies the CO distributions often exhibit central CO holes up to 5 kpc across. None of the Sc galaxies have CO distributions which resemble that in the Milky Way.

157.008 Submm observations of IRAS galaxies.
R. Chini, E. Kreysa, E. Krügel, P. G. Mezger.
NASA Conf. Publ., NASA CP–2466, p. 217 (1987). Abstract. – See Abstr. 012.002.

157.009 Stellar bars and the spatial distribution of infrared luminosity.
N. Devereux.
NASA Conf. Publ., NASA CP–2466, p. 219 – 226 (1987). – See Abstr. 012.002.
New ground–based 10–μm observations of the central region of over 100 infrared luminous galaxies are presented. A first order estimate of the spatial distribution of infrared emission in galaxies is obtained through a combination of ground–based and IRAS data. The galaxies are nearby and primarily noninteracting, permitting an unbiased investigation of correlations with Hubble type. Approximately 40% of the early–type barred galaxies in this sample are associated with enhanced luminosity in the central (~ 1 kpc diameter) region. The underlying luminosity source is attributed to both Seyfert and star formation activity. Late–type spirals are different in that the spatial distribution of infrared emission and the infrared luminosity are not strongly dependent on barred morphology.

157.010 Star formation and spiral structure in M81.
M. Kaufman, F. N. Bash.
NASA Conf. Publ., NASA CP–2466, p. 227 – 233 (1987). – See Abstr. 012.002.
High resolution digitized images of M81 in the radio continuum, Hα, H I, and I band are used to see how well various density wave models agree in detail with observations. The authors find that the observed width of the nonthermal radio arms favors a cloudy version of a density wave model (e.g., the model of Roberts and Hausman). The radial distribution of the set of giant radio H II regions disagrees with the simple expression of Shu and Visser for star formation by a density wave. The observed displacements of the giant radio H II regions from the spiral velocity shock indicate that some revisions in the details of the ballistic particle model of Leisawitz and Bash are necessary.

157.011 The H II regions in M51: radio and optical observations.
J. M. van der Hulst, R. C. Kennicutt Jr.
NASA Conf. Publ., NASA CP–2466, p. 235 – 238 (1987). – See Abstr. 012.002.
High resolution, dual frequency radio observations and calibrated Hα surface photometry of the spiral galaxy M51 are used to determine the physical properties of the 40 brightest H II region complexes. M51 appears to have a normal H II region population when compared with other nearby Sc galaxies for which good data exist. The authors used the radio and Hα data to measure the extinction toward the H II regions. The extinction is very patchy but appears to have a weak trend to become on average smaller toward large galactocentric radii. This trend is consistent with a possible metallicity gradient in M51. The authors compared the radio determined extinctions with Balmer decrement extinctions and found good agreement between the two, contrary to previous studies of M51 and other galaxies.

157.012 **IRAS observations of irregular galaxies.**
D. A. Hunter, W. Rice, J. S. Gallagher III, F. Gillett.
NASA Conf. Publ., NASA CP–2466, p. 253 – 256 (1987). With a
correction on p. 257. – See Abstr. 012.002.

Normal irregular galaxies seem to be unusual in having
vigorous star formation yet lacking the many dark nebulae
typical of spirals. IRAS observations of a large sample of
irregulars are used to explore the dust contents of these galaxies.
Compared to normal spirals, the irregulars generally have higher
L_{IR}/L_B ratios, warmer f(100)/f(60) dust color temperatures, and
lower globally–averaged dust/gas ratios. The relationship between the infrared data and various global optical properties of
the galaxies is discussed.

157.013 **IUE observations of luminous blue star associations in**
irregular galaxies.
S. A. Lamb, D. A. Hunter, J. S. Gallagher III.
NASA Conf. Publ., NASA CP–2466, p. 259 – 262 (1987). – See
Abstr. 012.002.

Two regions of recent star formation in blue irregular galaxies
have been observed with the IUE in the short wavelength, low
dispersion mode. The spectra indicate that the massive star
content is similar in these regions and is best fit by massive stars
formed in a burst and now approximately $2.5 – 3.0 \times 10^6$ years
old.

157.014 **Neutral hydrogen and star formation in irregular gal-**
axies.
E. D. Skillman.
NASA Conf. Publ., NASA CP–2466, p. 263 – 266 (1987). – See
Abstr. 012.002.

VLA and WSRT H I synthesis observations of seven irregular
galaxies are presented. The total H I images of four Local Group
dwarf irregular galaxies and three larger more distant irregular
galaxies are constructed at the identical resolution of 500 pc. All
galaxies studied show an excellent correlation between the H I
surface density and the presence of H II regions. This correlation
is most easily interpreted in terms of a requisite threshold H I
surface density for massive star formation.

157.015 **Carbon monoxide emission from small galaxies.**
H. A. Thronson Jr., J. Bally.
NASA Conf. Publ., NASA CP–2466, p. 267 – 270 (1987). – See
Abstr. 012.002.

The authors have searched for $J = 1 \to 0$ CO emission from 22
galaxies, detecting half, as part of a survey to study star
formation in small– to medium–size galaxies. Although substantial variation in the star formation efficiencies of the sample
galaxies was found, there is no apparent systematic trend with
galaxy size.

157.016 **Characteristics of UGC galaxies detected by IRAS.**
C. J. Lonsdale Persson, W. Rice, G. D. Bothun.
NASA Conf. Publ., NASA CP–2466, p. 273 – 276 (1987). – See
Abstr. 012.002.

IRAS detection rates at 60 μm have been determined for the
Uppsala General Catalogue of Galaxies (Nilson 1973; the UGC).
Late–type spirals, characterised by a "normal" IR/B ratio of
~ 0.6, are detected to a velocity of ~ 6000 km/s for $L_B = L^*$.
Contrary to the situation for IRAS–selected galaxy samples, the
authors find little evidence for a correlation between IR/B and
60/100 μm in this large optically–selected sample. Thus a
significant fraction of the IRAS–measured far–infrared flux from
normal spirals must originate in the diffuse interstellar medium,
heated by the interstellar radiation field. The authors do not find
support for Burstein and Lebofsky's (1986) conclusion that spiral
disks are optically thick in the far–infrared.

157.017 **Star formation rates as a function of galaxy mass.**
W. Romanishin.
NASA Conf. Publ., NASA CP–2466, p. 293 – 296 (1987). – See
Abstr. 012.002.

Several groups have found correlations between the colors and
absolute magnitudes of spiral galaxies. Using optical and/or near

IR (1.6 micron) colors, they find that lower luminosity spirals are
systematically bluer than higher luminosity spirals. The author
has used IRAS far IR luminosities to investigate the suggestion
that one prime cause of these color – absolute magnitude
correlations is a systematic variation with galaxy mass of the
current star formation rate (SFR) per unit mass. To the extent
that the IRAS fluxes actually measure disk SFR, no correlation
of SFR/ unit mass and galaxy mass is found. Other possible
explanations of the color – absolute magnitude correlations are
discussed.

157.018 **Global properties of the nearby spiral M101.**
C. Beichman, F. Boulanger, W. Rice,
C. J. Lonsdale Persson, F. Viallefond.
NASA Conf. Publ., NASA CP–2466, p. 297 – 302 (1987). – See
Abstr. 012.002.

M101 (NGC 5457) is a classic Sc I spiral galaxy located
sufficiently nearby, 6.8 Mpc, (Aaronson, Mould and Huchra
1980) that its structure can be studied even with the coarse
angular resolution of IRAS. This work addresses the global
infrared properties of M101 including the radial dependence of
its infrared emission.

157.019 **Efficient star formation in the bright bar of M83.**
S. D. Lord, S. E. Strom, J. S. Young.
NASA Conf. Publ., NASA CP–2466, p. 303 – 307 (1987). – See
Abstr. 012.002.

The authors have detected the bright molecular bar in M83
standing out as a 100% enhancement of molecular emission with
respect to the off–bar emission at the same radii. They compare
the spatial variations in the star formation efficiency, as traced by
Hα emission and the surface density of the interstellar gas, in M83
and M51. Both the central bar of M83 and the spiral arms of M51
are regions characterized by high massive star formation rates.
For M83, the authors ascribe the fact that both the gas surface
density and the star formation efficiency are high to the
hydrodynamics of the central region.

157.020 **Large scale dissociation of molecular gas in the spiral**
arms of M51.
R. P. J. Tilanus, R. J. Allen.
NASA Conf. Publ., NASA CP–2466, p. 309 – 313 (1987). – See
Abstr. 012.002.

The authors compare the distribution of the atomic and
ionized hydrogen along the inner spiral arms of M51. As is the
case in M83, the location of both these phases of the interstellar
medium with respect to the major dust lanes suggests that
molecular hydrogen is dissociated on kpc scales in active star–
forming regions, and that this dissociation process may strongly
affect the observed morphology of atomic hydrogen in spiral
arms.

157.021 **Structure and kinematics of the molecular spiral arms in**
M51.
G. Rydbeck, Å. Hjalmarson, L. E. B. Johansson,
O. E. H. Rydbeck, T. Wiklind.
NASA Conf. Publ., NASA CP–2466, p. 315 – 317 (1987). – See
Abstr. 012.002.

Mapping of the CO(1–0) emission from the spiral galaxy M51
has been made with the Onsala 20 m antenna (HPBW = 33″).
The observations show that the emission is considerably
enhanced – above the background disk distribution – in spiral
arms which appear to originate as intense ridges of emission
about 1 kpc from the nucleus (assuming a distance of 9.6 kpc).
Inside this region there is virtually no emission. The excess
emission along the arm is broken up into large scale patches of up
to a few kpc in size. This suggests that the "on–arm" molecular
clouds are assembled into giant complexes with hydrogen masses
estimated to be up to $10^8 M_\odot$.

157.022   Modelling the IRAS colors of galaxies.
G. Helou.
NASA Conf. Publ., NASA CP-2466, p. 319 – 322 (1987). – See Abstr. 012.002.

A physical interpretation is proposed for the color–color diagram of galaxies which are powered only by star formation. The colors of each galaxy result from the combination of two components: cirrus–like emission from the neutral disk, and warmer emission from regions directly involved in on–going star formation. This approach to modelling the emission is based on dust properties, but independent evidence for it is found in the relation between the color sequence and the luminosity sequence. Implications of data and interpretation are discussed and possible tests mentioned for the model.

157.023   A simple two–component model for the far–infrared emission from galaxies.
T. de Jong, K. Brink.
NASA Conf. Publ., NASA CP-2466, p. 323 – 328 (1987). – See Abstr. 012.002.

The authors have constructed a simple model to calculate the far–infrared emission of galaxies made up of a disk component containing cool dust heated by the general interstellar radiation field, and of a molecular cloud component containing warm dust heated by recently formed massive stars. This model is fitted to the optical and far–infrared data of 120 Shapley–Ames galaxies and of 20 optically studied mini–survey galaxies, resulting in the determination of blue face–on extinctions and of the total luminosities of recently born massive stars and of disk stars.

157.024   Detection of CO (J = 1–0) in the dwarf elliptical galaxy NGC 185.
T. Wiklind, G. Rydbeck.
NASA Conf. Publ., NASA CP-2466, p. 331 – 332 (1987). – See Abstr. 012.002.

The authors report the first detection of CO(J = 1–0) emission in the dwarf elliptical galaxy NGC 185. The observations were performed with the Onsala 20 m telescope equipped with an SSB tuned Schottky mixer and, for the most recent observations, an SIS mixer. The previously reported tentative detection of CO in NGC 185 by Johnson and Gottesman (1979) does not agree with the authors' results.

157.025   Star forming regions in gas–rich S0 galaxies.
R. W. Pogge, P. B. Eskridge.
NASA Conf. Publ., NASA CP-2466, p. 333 – 336 (1987). – See Abstr. 012.002.

The authors present the first results of an Hα imaging survey of H I rich S0 galaxies, in which they have searched for H II regions and other sources of emission (e.g., nuclear emission). CCD Hα interference filter images have been made of 16 galaxies. Eight of these galaxies show evidence for on–going star formation (H II regions), one has nuclear emission but no H II regions, and the remaining seven have no emission detected within well defined upper limits. A few of the galaxies are found to be clearly not S0's, or peculiar objects atypical of the S0 class. Using simple models the authors have estimated star formation rates (SFRs) and gas depletion times from the observed Hα fluxes. In general, the derived SFRs are much lower than those found in isolated field spiral galaxies (Kennicutt 1983) and the corresponding gas depletion time scales are also longer.

157.026   The history of gas in spiral galaxies.
P. Maloney.
NASA Conf. Publ., NASA CP-2466, p. 343 – 347 (1987). – See Abstr. 012.002.

The author considers a very simple model for the gas in a spiral galaxy, with a specified initial surface density $\sigma(r)$ and angular velocity $\Omega(r)$. A spiral density wave (pattern speed Ω_p) is present. In each encounter with the density wave some fraction f of the gas is permanently lost in the form of low mass stars and stellar remnants. The actual nature of the density–wave trigger is unimportant. Typical results from this simple model, with parameters appropriate to NGC 6946, are shown.

157.027   Nuclear star formation on 100 parsec scales: 10″ resolution radio continuum, H I, and CO observations.
J. L. Turner, P. T. P. Ho, R. N. Martin.
NASA Conf. Publ., NASA CP-2466, p. 383 – 386 (1987). – See Abstr. 012.002.

The authors report on a program of radio line and continuum studies of star formation in nearby (< 10 Mpc) spiral galaxies. The objective is a search for hot gas and peculiar dynamics in spiral nuclei with 10″ to 30″ angular resolution. Vigorous star formation is found to be a common phenomenon in the inner kpc of spirals. Arcsecond–resolution observations of radio continuum emission at 6 and 2 cm have been used to separate the thermal and nonthermal radio components. It is found that thermal and nonthermal emission are wellmixed even on sizescales of 10 pc. To understand the reason for the increased level of star formation activity in spiral nuclei, the authors are studying H I and CO emission in these galaxies. The CO (J = 3→2) transition has been detected in M51, M82, NGC 253, NGC 6946 and IC 342 with $T_a{}^* \sim 0.5 - 2.0$K, at 20″ angular resolution.

157.028   Extremely luminous far–infrared sources (ELFS).
M. Harwit, J. R. Houck, B. T. Soifer, G. G. C. Palumbo.
NASA Conf. Publ., NASA CP-2466, p. 387 – 393 (1987). – See Abstr. 012.002.

The Infrared Astronomical Satellite (IRAS) survey uncovered a class of Extremely Luminous Far–Infrared Sources (ELFS), exhibiting luminosities up to and occasionally exceeding $10^{12}L_\odot$. The authors present arguments to show that sources with luminosities $L \geqslant 3 \times 10^{10}L_\odot$ may represent gas–rich galaxies in collision. The more conventional explanation of these sources as sites of extremely active star formation fails to explain the observed low optical luminosities of ELFS as well as their high infrared excess. In contrast, a collisional model heats gas to a temperature of $\sim 10^6$K where cooling takes place in the extreme ultraviolet. The UV is absorbed by dust and converted into far–infrared radiation (FIR) without generation of appreciable optical luminosity.

157.029   Induced star formation in interacting galaxies.
R. C. Kennicutt, K. A. Roettiger, W. C. Keel, J. M. van der Hulst, E. Hummel.
NASA Conf. Publ., NASA CP-2466, p. 401 – 408 (1987). – See Abstr. 012.002.

The authors have used measurements of H–alpha emission–line fluxes and FIR fluxes in ~100 interacting spirals to investigate the effects of close tidal interactions of the disk and nuclear star formation rates in galaxies. Two samples of interacting spirals were studied, a complete sample of close pairs, and a set of strongly perturbed systems from the Arp atlas. Both the integrated H–alpha luminosites and FIR luminosities are enhanced in the interacting galaxies, indicating that the encounters indeed trigger massive star formation in many cases.

157.030   The frequency of enhanced star formation in interacting and isolated galaxies.
R. M. Cutri.
NASA Conf. Publ., NASA CP-2466, p. 409 (1987). Abstract. – See Abstr. 012.002.

157.031   The initial mass function in H II galaxies.
A. W. Campbell.
NASA Conf. Publ., NASA CP-2466, p. 479 – 482 (1987). – See Abstr. 012.002.

Observation of a large sample of H II galaxies shows that the emission line ratios of the youngest objects change systematically with gaseous oxygen abundance, which the author interprets as resulting from changes in the initial mass function (IMF) of the ionising cluster. Comparison with cluster/nebula models shows that both the slope and the upper mass limit of the cluster IMF vary with abundance. In H II galaxies with oxygen abundance about 1/10 that of Orion, the IMF for massive stars must have a slope which is about a factor of 2 smaller than in the solar neighbourhood.

157.032   VLA continuum observations of barred spiral galaxies.
    J. A. Garcia–Barreto, P. Pismis.
NASA Conf. Publ., NASA CP–2466, p. 483 – 484 (1987).
Abstract. – See Abstr. 012.002.

157.033   The azimuthal and radial distributions of H I and H₂ in NGC 6946.
L. J. Tacconi–Garman, J. S. Young.
NASA Conf. Publ., NASA CP–2466, p. 491 – 495 (1987). – See Abstr. 012.002.

The authors have completed a study of the atomic and molecular components of the ISM in NGC 6946. The distribution of molecular clouds has been determined from a fully sampled CO map of the inner disk (R ≤ 8 kpc) using the 14–meter telescope of the FCRAO (HPBW = 45″). The distribution of atomic gas was derived from VLA observations at 40″ resolution in the D configuration. While the Hα/CO ratio is found to be approximately constant with radius, the CO/H I ratio decreases by a factor of 30 from the center of the galaxy to R = 10 kpc.

157.034   The infrared morphology of galactic centers.
    C. M. Telesco, R. Decher, B. D. Ramsey,
R. D. Wolstencroft, S. K. Leggett.
NASA Conf. Publ., NASA CP–2466, p. 497 – 500 (1987). – See Abstr. 012.002.

The authors present initial results of a program to map the centers of galaxies in the mid–infrared using the NASA–MSFC 20–pixel bolometer array. Maps at 10.8 μm of the galaxies NGC 5236 (M83), NGC 1808, NGC 4536, and NGC 4527 reveal complex emitting regions ranging in size from 500 pc to 2 kpc. The infrared spatial distributions generally resemble those in the visible and radio. In all cases a large fraction of the IRAS 12 μm flux originates in spatial structures prominent in the maps.

157.035   A near–infrared study of the luminous merging galaxies NGC 2623 and Arp 148.
M. Joy, P. M. Harvey.
NASA Conf. Publ., NASA CP–2466, p. 515 (1987). Abstract. – See Abstr. 012.002.

157.036   A redshift survey of IRAS galaxies.
    B. J. Smith, S. G. Kleinmann, J. P. Huchra,
F. J. Low.
NASA Conf. Publ., NASA CP–2466, p. 565 – 568 (1987). – See Abstr. 012.002.

The authors present results from a redshift survey of all 72 galaxies detected by IRAS in Band 3 at flux levels ≥ 2 Jy, and lying in the region $8^h < \alpha < 17^h$, $23.5° < \delta < 32.5°$. The 60 μm luminosities of these galaxies range from 1.4×10^8 L$_0$ to 5.0×10^{11} L$_0$. The luminosity function at the high luminosity end is proportional to L^{-2}, however, the authors observe a flattening at the low luminosity end indicating that a single power law is not a good description of the entire luminosity function. Only three galaxies in the sample have emission line spectra indicative of AGN's, suggesting that, at least in nearby galaxies, unobscured nuclear activity is not a strong contributor to the far–infrared flux.

157.037   Optical and IR luminosity functions of IRAS galaxies.
    J. P. Vader, M. Simon.
NASA Conf. Publ., NASA CP–2466, p. 569 – 572 (1987). – See Abstr. 012.002.

The optical and infrared luminosity functions are determined for a 60 μm flux–limited sample of 68 IRAS galaxies covering a total area of 150 degrees squared. The IR function is in good agreement with that obtained by other authors. The shape of the optical luminosity function is similar to that of optically selected galaxy samples. The integrated light of most objects in the sample have [N II] to Hα line flux ratios characteristic of H II–region galaxies. In the absolute magnitude range M$_J$ = –18, –22 about 14% of late–type galaxies are IRAS galaxies. The apparent

companionship frequency is about twice as large as that for a comparable sample of non–IRAS late–type galaxies.

157.038   The correlation between far–IR and radio continuum emission from spiral galaxies.
J. M. Dickey, R. W. Garwood, G. Helou.
NASA Conf. Publ., NASA CP–2466, p. 575 – 578 (1987). – See Abstr. 012.002.

The authors have observed a sample of 30 galaxies selected for their intense IRAS flux at 60 and 100 μm using the Arecibo telescope at 21 cm to measure the continuum and H I line luminosities. The centimeter–wave continuum correlates very well with the far–infrared flux, with a correlation coefficient as high as that found for other samples, and the same ratio between FIR and radio luminosities. Weaker correlations are seen between the FIR and optical luminosity and between the FIR and radio continuum. There is very little correlation between the FIR and the H I mass deduced from the integral of the 21 cm line. The strength of the radio continuum correlation suggests that there is little contribution to either the radio or FIR from physical processes not affecting both. If they each reflect time integrals of the star formation rate then the time constants must be similar, or the star formation rate must change slowly in these galaxies.

157.039   The radio–far infrared correlation: spiral and blue compact dwarf galaxies opposed.
U. Klein, E. Wunderlich.
NASA Conf. Publ., NASA CP–2466, p. 583 – 588 (1987). – See Abstr. 012.002.

The recently established correlation between radio continuum and far infrared emission in galaxies has been further investigated by comparing normal spiral and blue compact dwarf galaxies. The puzzling result is that the ratio of radio–to–far infrared luminosity and its dispersion is the same for both samples, although their ratios of blue–to–far infrared luminosity, their radio spectral indices and their dust temperatures exhibit markedly different mean values and dispersions.

157.040   Radio continuum, far infrared and star formation.
    R. Wielebinski, E. Wunderlich, U. Klein,
E. Hummel.
NASA Conf. Publ., NASA CP–2466, p. 589 – 593 (1987). – See Abstr. 012.002.

A very tight correlation has been found between the radio emission and the far infrared emission from galaxies. This has been found for various samples of galaxies (de Jong et al., 1985: Helou et al., 1985) and is explained in terms of recent star formation. The tight correlation would imply that the total radio emission (the sum of thermal and synchrotron emission) is a good tracer of star formation.

157.041   Near–infrared observations of IRAS minisurvey galaxies.
D. P. Carico, B. T. Soifer, J. H. Elias, K. Matthews,
G. Neugebauer, C. Beichman, C. J. Lonsdale Persson,
S. E. Persson.
NASA Conf. Publ., NASA CP–2466, p. 601 – 604 (1987). – See Abstr. 012.002.

Near–infrared photometry at J, H, and K has been obtained for 82 galaxies from the IRAS minisurvey. The near–infrared colors of these galaxies cover a larger range in J–H and H–K than do normal field spiral galaxies, and evidence is presented of a tighter correlation between the near– and far–infrared emission in far–infrared–bright galaxies than exists between the far–infrared and the visible emission. These results suggest the presence of dust in far–infrared–bright galaxies, with hot dust emission contributing to the 2.2 μm emission, and extinction by dust affecting both the near–infrared colors and the visible luminosities. In addition, there is some indication that the infrared emission in many of the minisurvey galaxies is coming from a strong nuclear component.

157.042   Ground–based follow up of IRAS galaxies.
    M. Dennefeld, H. Karoji, P. Bouchet, L. Bottinelli,
L. Gouguenheim.
NASA Conf. Publ., NASA CP–2466, p. 605 – 609 (1987). – See
Abstr. 012.002.
    The authors have undertaken optical, near–infrared, radio–
continuum and H I observations of the galaxies identified with
IRAS sources in a few fields roughly of the size of a sky survey
plate. They present results from two fields at galactic latitude
$+27°$ and $+43°$ over a total area of 100 square degrees (see also
Dennefeld et al. 1986). These regions contained 115 IRAS point
sources, out of which 26 were identified with stars and 81 with
faint galaxies, 10 of which were difficult to recognize on the
Schmidt plates. A further 8 sources could not be identified with
any object down to the limit of the Palomar or ESO Sky Survey
Plates. As judged from the Cirrus Flags, at most 3 could be
spurious sources.

**157.043   Enhanced star formation – the importance of bars in
    spiral galaxies.**
P. J. Puxley, T. G. Hawarden, C. M. Mountain, S. K. Leggett.
NASA Conf. Publ., NASA CP–2466, p. 619 – 622 (1987). – See
Abstr. 012.002.
    The authors have found that amongst an IR–luminous subset
of nearby galaxies, nearly all of the systems with IRAS colours
and luminosities indicative of enhanced star formation are
barred. Radio continuum and IR–spectroscopic results support
the hypothesis that this emission originates within the central
2 kpc; possibly in a circumnuclear ring. The authors also find that
outer rings are over–represented amongst these barred systems
and suggest possible reasons for this phenomena.

157.044   The properties of highly luminous IRAS galaxies.
    R. D. Wolstencroft, P. J. Puxley, J. N. Heasley,
S. K. Leggett, A. Savage, H. T. MacGillivray, R. G. Clowes.
NASA Conf. Publ., NASA CP–2466, p. 623 – 627 (1987). – See
Abstr. 012.002.
    From a complete sample of 154 galaxies identified with IRAS
sources in a 304 deg^2 area centered on the South Galactic Pole, a
sub–sample of 58 galaxies with $L_{IR}/L_B > 3$ has been chosen. Low
resolution spectra have been obtained for 30% of the sub–sample
and redshifts and relative emission–line intensities have been
derived. As a class these galaxies are very luminous with
$<L_{IR}> = 2.9 \times 10^{11} L_\odot$ and (L_{IR}) max $= 1.3 \times 10^{12} L_\odot$. CCD
images and JHK photometry have been obtained for many of the
sub–sample. The galaxies are for the most part newly identified
and are optically faint ($16 < B < 21$), with a majority showing
evidence of a recent interaction. Radio continuum observations
of all galaxies of the sub–sample have recently been obtained at
20 cm (VLA) with about 75% being detected in a typical
integration time of about 10 minutes.

157.045   A study of a flux–limited sample of IRAS galaxies.
    B. J. Smith, S. G. Kleinmann, J. P. Huchra,
F. J. Low.
Astrophys. J., Vol. 318, No. 1, p. 161 – 174 (1987).
    The authors present results from a study of all 72 galaxies
detected by IRAS in band 3 at flux levels $\geqslant 2$ Jy and lying in the
region $8^h < \alpha < 17^h$, $23°5 < \delta < 32°5$. Redshifts and accurate
four–colour IRAS photometry were obtained for the entire
sample. The 60 μm luminosity function was constructed from these
data and is compared with previous results for other infrared
flux–limited samples of galaxies. Comparisons between the
selected IRAS galaxies and an optically complete sample taken
from the CfA redshift survey are also presented. It is found that
the space distribution of the two samples differ: the density
enhancement of IRAS galaxies is only $\sim 1/3$ that of the optically
selected galaxies in the core of the Coma Cluster.

157.046   Isophotal diameters of cluster spirals.
    M. E. Cornell, M. Aaronson, G. Bothun, J. Mould.
Astrophys. J., Suppl. Ser., Vol. 64, No. 3, p. 507 – 528 (1987).
    The authors have obtained CCD frames of several hundred
cluster spirals using the KPNO No. 1 0.9 m telescope, and have

produced surface brightness profiles for each galaxy using a
method which fits elliptical isophotes to the data. Isophotal
diameters at B have been derived and compared to Nilson UGC
B diameters. It is found that for small UGC galaxies ($D \lesssim 2'$),
the transformed UGC diameters are overestimated by about
14% in the mean. The authors also reexamine two correlations
based on galaxy diameters, the diameter/21 cm line width
relation, and the blue absolute magnitude/surface brightness
relation. Implications for distance–scale work are discussed.

157.047   Ultraviolet properties of normal galaxies.
    J. Stryczyński.
Astron. Astrophys., Suppl. Ser., Vol. 70, No. 1, p. 115 – 124
(1987).
    The ultraviolet and optical photometric data for 68 normal
galaxies were combined to derive their colour indices. The
analysis of these indices indicates that the galaxies of this sample
belong to two different groups. The spectral differences seem to
be connected to differences in metallicity within the central
regions of the objects. The galaxies of these groups differ also in
their location in space.

157.048   Accurate positions of Zwicky galaxies. II.
    N. Santagata, L. Basso, M. Gottardi,
G. G. C. Palumbo, G. Vettolani.
Astron. Astrophys., Suppl. Ser., Vol. 70, No. 2, p. 189 – 190
(1987). With 1 microfiche.
    Accurate optical positions are given for the 2562 galaxies
contained in 56 fields of the Zwicky Catalogue.

157.049   Accurate positions of Zwicky galaxies. III.
    N. Santagata, L. Basso, M. Gottardi,
G. G. C. Palumbo, G. Vettolani, M. Vigotti.
Astron. Astrophys., Suppl. Ser., Vol. 70, No. 2, p. 191 – 192
(1987). With 1 microfiche.
    Accurate optical positions are given for the 2378 galaxies
contained in 60 fields of the Zwicky Catalogue.

**157.050   Analysis of absorption–line spectra in a sample of
    164 galactic nuclei.**
E. Bica, D. Alloin.
Astron. Astrophys., Suppl. Ser., Vol. 70, No. 2, p. 281 – 301
(1987).
    The authors present spectral observations in the range
3700 Å $< \lambda <$ 8000 Å, of 154 normal galactic nuclei,
2 amorphous galaxies, as well as of 8 intrinsically faint active
nuclei with their visible spectrum dominated by the stellar
component. This sample of galaxies covers a range in luminosity
$-23.3 \leqslant M_B \leqslant -16.6$ and spans morphological types from E
to Sc. The main purpose of this study is to outline the different
stellar populations which can be found within a given interval of
galaxy luminosity versus morphological type. Some results are
given concerning age and metallicity effects.

157.051   Box–shaped galaxies: a complete list.
    R. E. de Souza, S. dos Anjos.
Astron. Astrophys., Suppl. Ser., Vol. 70, No. 3, p. 465 – 480
(1987).
    The authors present a new list of box–shaped galaxies complete
to $B_r = 13.2$ in the whole sky. A discussion based on this material
shows that the observed frequency of lenticulars exhibiting this
phenomenon is consistent with the idea of all of them being edge–
on barred galaxies.

**157.052   The effects of interactions on spiral galaxies. III. A radio
    continuum survey of galactic nuclei at 1.49 GHz.**
E. Hummel, J. M. van der Hulst, W. C. Keel,
R. C. Kennicutt Jr.
Astron. Astrophys., Suppl. Ser., Vol. 70, No. 3, p. 517 – 530
(1987).
    The radio continuum emission from the central region of a
sample of interacting spiral galaxies (92 galaxies of which 60 in a
complete sample) and of a control sample of more isolated spiral
galaxies (94) was observed with the Very Large Array at

1.49 GHz. The angular resolution of the observations is ~1″3 and the detection limits are ~0.6 mJy and ~1.5 mJy for point sources and extended sources with a half power size of 10″ respectively. This survey, in combination with published optical spectroscopy, provides the data for a detailed comparison of the central region in interacting and more isolated spiral galaxies.

157.053   Star complexes and associations in the Andromeda galaxy.
Yu. N. Efremov, G. R. Ivanov, N. S. Nikolov.
Astrophys. Space Sci., Vol. 135, No. 1, p. 119 – 130 (1987).

Large–scale U and B plates obtained with the 2 m Ritchey–Chrétien telescope of the Rozhen Observatory (Bulgaria) were searched for new resolved star groups and for independent delineation of the boundaries of the known ones in M31. The authors detected 210 groups as real O–associations the mean diameter of which is 80 pc. Many of Hodge's open clusters were also reclassified as O–associations. The majority of van den Bergh's OB–associations were recognized as star complexes and their mean diameter is 650 pc. Almost all O–associations are located inside the star complexes. A dozen new star complexes (mainly around the dark lanes between OB78 and OB22) and numerous groups presumably not containing O–stars were found out.

157.054   The brightest blue stars in M33 galaxy as unresolved star groups.
V. K. Golev, G. R. Ivanov, P. Z. Kunchev.
Astrophys. Space Sci., Vol. 135, No. 2, p. 301 – 306 (1987).

The brightest blue stars in the M33 galaxy have peculiar images. A method to resolve them into components was applied. The brightest member stars at $V \sim 15 - 16$ mag were recognized as multiple ones. Some stars up to $V = 17$ mag have multiple structures, too. The image of the brightest blue stars consist of two or more components and probably some of them, in fact, resemble dense stellar groups such of R 136 in 30 Dor.

157.055   Some features of the galactic halo in NGC 891.
F. X. Hu, R. J. Allen, P. C. van der Kruit, J. H. You.
Astrophys. Space Sci., Vol. 135, No. 2, p. 389 – 392 (1987). With a correction in Vol. 139, No. 2, p. 423 (1987).

Main results of the galactic halo in NGC 891 newly obtained from the Palomar (U', J, F)–WSRT (21,6 cm) observation (both smoothed to 20″ × 20″ resolution) and some of the possible explanations have been summarized.

157.056   De Andromedanevel: anatomie van een naburig sterrenstelsel.
R. Walterbos.
Zenit, Jaarg. 14, Nr. 7/8, p. 236 – 246 (1987).

157.057   Observations of the shell galaxy NGC 3923 with EFOSC.
J.–L. Prieur.
Messenger, No. 49, p. 12 – 14 (1987).

157.058   IC 3370: a box–shaped elliptical or S0 galaxy?
B. Jarvis.
Messenger, No. 49, p. 15 – 18 (1987).

157.059   A new distance indicator for spiral galaxies?
E. Giraud.
Messenger, No. 49, p. 20 – 21, 24 (1987).

157.060   Spectroscopic analysis of Kiso ultraviolet–excess galaxies.
H. Maehara, T. Noguchi, B. Takase, T. Handa.
Publ. Astron. Soc. Jpn., Vol. 39, No. 3, p. 393 – 409 (1987).

The authors present spectroscopic properties of 57 ultraviolet–excess galaxies (KUGs), which were selected from the Kiso survey by Takase et al. (1983). About 85% of this sample exhibit conspicuous emission lines similar to galactic nebulae. The radial velocities of the objects have been obtained from their emission lines as accurate as ± 90 km s^{-1}. The absolute magnitudes estimated from the radial velocities indicate that a wide range exists in blue luminosity of irregular galaxies, and that this sample includes less luminous spiral galaxies. Equivalent widths of emission lines have been measured against the local continuum, and a diagram of the emission line ratio [O III] λ 5007/Hβ versus [N II] λ 6584/Hα is applied to classify these objects.

157.061   Internal motions in three dwarf irregular galaxies.
V. P. Arkhipova, R. I. Noskova, O. K. Sil'chenko, A. V. Zasov.
Pis'ma Astron. Zh., Tom 13, No. 7, p. 575 – 581 (1987). In Russian. English translation in Sov. Astron. Lett., Vol. 13.

Internal motions in the three extreme dwarf galaxies VV 499, 558 and 828 – all being the members of nearby groups of galaxies are studied. Each of them is found to rotate. The masses of the galaxies are obtained. All the galaxies under consideration have absolute magnitudes weaker than -14^{m}, diameters of 0.5 – 2.0 kpc and masses smaller than $10^{8} M_{\odot}$. They contain H II regions about 100 pc in diameter and this fact has led to their formal classification as interacting systems. The comparison of integral characteristics of the dwarf galaxies belonging to different groups (M81, M101 and the Local Group) has revealed their significant resemblance.

157.062   Morphology of some new galaxies with UV excess.
V. S. Tamazyan.
Astrofizika, Tom 26, Vyp. 3, p. 411 – 414 (1987). In Russian. English translation in Astrophysics, Vol. 26, No. 3, p. 249 – 252 (1987).

Morphological study of 32 galaxies with UV excess from the third list published by M. Kazarian are carried out using the plates taken on the 2.6–m telescope of the Byurakan Observatory. About two–thirds of the galaxies investigated are spirals. Among 15 Sa – Sc galaxies 7 are barred spirals. Galaxies No. 281, 289, 332 and 338 having an interesting morphological structure are observed.

157.063   On the structure of the low surface brightness dwarf galaxies in the M81 group.
V. E. Karachentseva, I. D. Karachentsev, G. M. Richter, R. von Berlepsch, K. Fritze.
Astron. Nachr., Vol. 308, No. 4, p. 247 – 255 (1987).

A detailed photometry of spheroidal dwarf galaxies in the M81 group has been carried out. The integral characteristics and the structural parameters of the spheroidal dwarfs have been determined. Their luminosity profiles are well fitted to a King law. The investigated spheroidal dwarfs together with the prototypes of the Local Group form a common sequence according to their main parameters. The observational data presented show that dSphs are not linked evolutionarily with normal E and dE galaxies, but probably form a separate branch together with irregular low surface brightness dwarfs.

157.064   IRAS observations of three edge–on galaxies.
R. J. Wainscoat, T. de Jong, P. R. Wesselius.
Astron. Astrophys., Vol. 181, No. 2, p. 225 – 236 (1987).

The authors present and discuss IRAS observations of NGC 891, NGC 4565, and NGC 5907, and hope that these observations will contribute to a better understanding of less inclined galaxies, and in particular of our own Galaxy.

157.065   The metallicity versus luminosity relationship for early–type galaxies.
E. Bica, D. Alloin.
Astron. Astrophys., Vol. 181, No. 2, p. 270 – 272 (1987).

In the equivalent width (W) vs absolute magnitude (M_B) diagram corresponding to the best metallic features in the visible range, CN and Mg + MgH, field giant E and S0 galaxies are shifted towards weaker line–strengths with respect to galaxies belonging to the Virgo and Fornax clusters. The authors find evidence that this is a genuine stellar population effect. For part of the deviating galaxies it results from an aperture effect, galaxies at larger redshifts including more of the metal poor component outside their nucleus. For the rest, the shift is due to a variable content of intermediate age components in the range 4 – 10 Gyr, superimposed on the very old underlying population.

157.066 Galaxy correlations and the luminosity function.
S. Phillipps, T. Shanks.
Mon. Not. R. Astron. Soc., Vol. 227, No. 1, p. 115 – 121 (1987).

Given the simple assumption that any excess of correlated galaxies seen close on the sky to a galaxy of known distance are also at the same distance, one can determine a galaxy luminosity function from the variation in excess number with magnitude. If one averages over many "centre" galaxies this luminosity function has good statistical accuracy at the faint end compared to the usual direct estimation from magnitude–limited surveys since the latter necessarily contain few intrinsically faint objects. The authors confirm the form of the luminosity function found from recent redshift surveys, a Schechter function with $M^*(b_j) = -19.8$, $\alpha = -1$ fitting the data well.

157.067 Erratum: "Studies of dwarf irregular galaxies – I. B and V CCD photometry of the brighter stars in Sextans A" [Mon. Not. R. Astron. Soc., Vol. 224, No. 4, p. 935 – 944 (1987)].
A. R. Walker.
Mon. Not. R. Astron. Soc., Vol. 227, No. 2, p. 543 (1987). See Abstr. 43.157.019.

157.068 Studies of IRAS sources at high galactic latitudes – III. Luminosity functions at 25, 60 and 100 μm and the correlation of optical and infrared luminosities.
M. Rowan–Robinson, G. Helou, D. Walker.
Mon. Not. R. Astron. Soc., Vol. 227, No. 3, p. 589 – 606 (1987).

The authors have carried out a detailed study of a complete sample of IRAS 25–, 60– and 100–μm sources identified with galaxies brighter that 14.5 mag at $b > 60°$. Redshifts are available for virtually all these galaxies. The luminosity functions for galaxies at 25, 60 and 100 μm are given, together with simple model fits which are consistent with the observed source–counts. The optical luminosity function of the sample is consistent with the Gaussian form found for Virgo spirals by Sandage, Bingelli & Tammann.

157.069 Radio continuum observations of nearby galaxies.
J. I. Harnett.
Mon. Not. R. Astron. Soc., Vol. 227, No. 4, p. 887 – 908 (1987).

This paper is the fourth in a series on radio observations of optically bright galaxies with the Molonglo Observatory Synthesis Telescope. Maps and discussions of 16 galaxies are given together with observed and derived radio properties of the complete sample of 37 galaxies. The radial distributions of non–thermal emission in six face–on galaxies are derived. Of these, NGC 1313, 1566, 1672 and M83 have exponential dependences similar to those of the associated blue light distributions. Scale lengths of the non–thermal emission are used to derive limiting values for the magnetic field and relativistic electron scale lengths.

157.070 Discovery of a huge low–surface–brightness galaxy: a protodisk galaxy at low redshift?
G. D. Bothun, C. D. Impey, D. F. Malin, J. R. Mould.
Astron. J., Vol. 94, No. 1, p. 23 – 29 (1987). With plate 1.

The authors report on the accidental discovery of an extremely large, extremely H I–rich low–surface–brightness galaxy located at a redshift of $z = 0.083$. Its nuclear spectrum exhibits broad, low–level emission lines. Surface photometry at V indicates the presence of a bulge component and a very extended disk, with scale length of $\approx 45''$ (55 kpc for $H_0 = 100$) and with central surface brightness of $V(0) \approx 25.5$ mag arcsec^{-2}. The total amount of H I is at least $1.0 \times 10^{11} M_\odot$. This amount of H I is at least 5 times more H I than any spiral galaxy previously observed. If disk formation is a quiescent process, then it is likely that the authors have caught a disk in the process of formation. They also point out that the properties of this disk are likely to be similar to the suspected sources that produce the observed damped Lyα absorption profiles that are so conspicuous at $z \approx 2$.

157.071 The unusual box–shaped elliptical (?) galaxy IC 3370.
B. Jarvis.
Astron. J., Vol. 94, No. 1, p. 30 – 42 (1987). With plates 2 – 3.

A detailed photometric and kinematic study of the strongly box–shaped galaxy IC 3370 is presented. IC 3370 is bright ($M_B = -22.1$) and kinematically more akin to the bulges of disk galaxies than to elliptical galaxies. Moreover, evidence is given to show that IC 3370 is an S0$_{pec}$ seen nearly edge–on and not an elliptical galaxy as originally classified.

157.072 Colors and the evolution of amorphous galaxies.
J. S. Gallagher III, D. A. Hunter.
Astron. J., Vol. 94, No. 1, p. 43 – 53 (1987).

$UBVRI$ and Hα photometric observations are presented for 16 amorphous galaxies and a comparison sample of Magellanic irregular (Im) and Sc spiral galaxies. These data are analyzed in terms of star-formation rates and histories in amorphous galaxies.

157.073 Molecules in galaxies. IV. Molecular and atomic hydrogen in Virgo cluster galaxies.
G. R. Knapp, G. Helou, A. A. Stark.
Astron. J., Vol. 94, No. 1, p. 54 – 60 (1987).

If gas and dust are well mixed in a galaxy, the dust mass should be proportional to the sum of the masses of atomic and molecular gas. This assumption has been used to estimate the mean conversion ratio between the flux in the CO $J = 1$–0 line and the molecular gas mass of a galaxy using a set of observations of the 100 μm continuum flux density, the H I 21 cm line flux and the CO line flux for a sample of spiral galaxies in the Virgo cluster; these galaxies have a wide variation in the ratio of CO to H I line flux. The result, $N(H_2) = 6.3 \pm 3.5 \times 10^{20}$ H$_2$mol cm^{-2}(K km/s)$^{-1}$, is in agreement with values inferred from observations of the molecular interstellar medium in the Galaxy.

157.074 Arp 227: a case for shells without mergers?
J. M. Schombert, J. F. Wallin.
Astron. J., Vol. 94, No. 2, p. 300 – 305 (1987). With plate 21.

Multicolor surface photometry is presented for shell galaxy Arp 227 (NGC 474). The profile shape and colors confirm that this object is an S0. Unlike previous color measurements of blue shells, the shells of this system are red ($\langle B-V \rangle = 0.90 \pm 0.08$), identical within the errors to the color of the parent galaxy ($B-V = 0.94$). Possible origins for the shells are tidally liberated material from the parent galaxy, stripped matter from a nearby companion, or the merger with a low–luminosity elliptical.

157.075 Surface photometry of six Local Group galaxies.
S. M. Kent.
Astron. J., Vol. 94, No. 2, p. 306 – 314 (1987).

Luminosity profiles for six Local Group galaxies are derived based on observations made with a CCD camera and a set of five telescopes. The telescopes range in size from a 105 mm focal-length lens to the MMT, providing fields ranging from 3' to 5°. This instrumentation permits uniform, calibrated photometry to be obtained over a wide range in angular scale with much greater ease than has been possible previously. The galaxies observed were M31, M32, M33, NGC 147, NGC 185, and NGC 205. Total magnitudes are derived for all galaxies except M33, and nuclear magnitudes are derived for the three galaxies that have distinct nuclear components. Mass/light ratios are computed for the bulge in M31 and M32.

157.076 CCD photometry and dynamics of the peculiar galaxy ESO 217–G09.
A. P. Marston.
Astron. Astrophys., Vol. 183, No. 1, p. 21 – 28 (1987).

The peculiar galaxy ESO 217–G09 has been investigated using blue and red CCD frames together with echelle longslit spectra in the Hα/[N II] wavelength range. These have shown the galaxy to be a barred spiral type with similarities to the Large Magellanic Cloud. Star formation is seen to be occurring across the region causing the galaxy to have a patchy morphology. No nucleus is discernible. The galaxy has a solid–body rotation curve and together with an inclination of 45° this enables a mass of $1.9 \times 10^{10} M_\odot$ to be attributed to it. This then gives a mass to blue luminosity ratio which is similar to that of barred spiral galaxies. Estimates of the temperature of H II regions across the galaxies indicates that the increase in the $H\alpha/[N\ II]$ ratio moving out from the galaxy centre is due to a decreasing abundance of nitrogen.

157.077   A hypothesis of compensation of angular momenta and estimation of masses of binary galaxies.
D. A. Verner.
Soobshch. Spets. Astrofiz. Obs., Vyp. 53, p. 89 – 91 (1987). – See Abstr. 012.008.

A study of the relative orientation of orbital and axial momenta in binary galaxies is important as a test for various theories of the origin of galaxy rotation. Adopting the Helou suggestion for galaxies, the author has determined directions of spins for 19 isolated pairs.

157.078   Les gaz ionisés hors du noyau d'une galaxie spirale ou irrégulière.
J. P. Vallée.
J. R. Astron. Soc. Can., Vol. 81, No. 4, p. 128 – 141 (1987).

In the disk of a spiral galaxy or outside of the nucleus of an irregular galaxy, the morphology of hot ionised gaseous regions (H II regions) can take many shapes: champagne or blister, bubble or half–bubble, giant complex, etc. Two basic types of internal turbulences in the H II regions have been proposed: "ordered" (champagne, wind, etc.) and "random" (relative motions, self–gravitation, differential rotation, etc.). Other deductions have been obtained from the observations, regarding the luminosity and the rate of star formation.

157.079   Massive stars in nearby galaxies.
P. Massey, J. Hutchings, L. Bianchi.
Instabilities in luminous early type stars, p. 201 – 204 (1987). – See Abstr. 012.012.

IUE spectra and optical data (MMT blue spectra and CCD photometry) have been obtained for OB supergiants in M31 and M33. UV and visible data yield consistent values for the effective temperature, bolometric magnitudes, and extinction for the stars, which are classified as late O to early B. The UV resonance lines have very low outflow velocities compared to galactic stars, and no P Cygni profiles, suggesting that there may be significant differences in mass loss mechanisms among Local Group galaxies, as a consequence of different metallicities.

157.080   Massive objects in galactic nuclei may be black holes.
P. H. Andersen.
Phys. Today, Vol. 40, No. 10, p. 22 (1987).

157.081   The spatial distribution and population of novae in M31.
R. Ciardullo, H. C. Ford, J. D. Neill, G. H. Jacoby, A. W. Shafter.
Astrophys. J., Vol. 318, No. 2, p. 520 – 530 (1987).

During the past 5 yr, the authors have been conducting an $H\alpha$ survey for novae in the bulge of M31. Here they report the results from a homogeneous subset of these CCD observations and analyze the spatial distribution of the nova population. It is shown that in M31's central bulge the distribution of novae follows that of the light to within $\sim 10''$ of the nucleus. The authors also reanalyze the Hubble–Arp nova sample and conclude that the novae observed in the central $30' \times 15'$ region belong almost exclusively to the bulge population, implying a ratio of the specific nova rates $\varrho_{Bulge}/\varrho_{Disk} \gtrsim 10$. This result is compared with the observed cataclysmic variable distribution in the Galaxy.

157.082   Detection of retrograde gas streaming in the SB0 galaxy NGC 4546.
G. Galletta.
Astrophys. J., Vol. 318, No. 2, p. 531 – 535 (1987).

The author reports spectroscopic observations of the almost edge–on SB0 galaxy NGC 4546 which reveal a striking discordance between the derived emission and absorption–line velocities. The gas clouds show velocities that are similar in amplitude but opposite in direction from the stars. This discordance is seen in observations obtained through slits oriented in a wide range of position angles. Orbits, elongated both along the bar major axis (prograde, stars) and along the bar intermediate axis (retrograde, gas) are found.

157.083   Isophotal diameters of galaxies at high redshifts.
D. W. Weedman, K. L. Williams.
Astrophys. J., Vol. 318, No. 2, p. 585 – 589 (1987).

New ultraviolet observations and archival spectra are presented for the bright disk of the galaxy NGC 4736 to derive the ultraviolet surface brightness profile. This is used to predict the appearance of such a galaxy at high redshift. The profiles in red light of 100 galaxies determined by S. Kent are used with assumed ultraviolet spectra to preduct the isophotal diameters of these galaxies as a function of redshift. The results show that, for ground–based imagery, normal spiral galaxies should be resolvable to redshifts of $z \approx 1$.

157.084   The use of the brightest blue supergiants as distance indicators – further complications; or the brightest "stars" are not always stars.
R. M. Humphreys, M. Aaronson.
Astrophys. J., Lett. Ed., Vol. 318, No. 2, p. L69 – L72 (1987).

The authors report the results of spectroscopic observations of the candidate brightest blue supergiants in three nearby spirals, NGC 2403, M81, and M101. The spectra show that many of these "stars" are actually compact H II regions, blue clusters, or composite, not single stars. While not unexpected, these results illustrate additional problems with using the brightest blue stars, whether variable or not, as distance indicators and potential pitfalls in the studies of stellar populations in increasingly distant galaxies.

157.085   Dynamics of a hot ($T \sim 10^7$ K) gas cloud with volume energy losses.
A. A. Suchkov, V. G. Berman, Yu. N. Mishurov.
Astron. Zh., Tom 64, Vyp. 4, p. 708 – 719 (1987). In Russian. English translation in Sov. Astron., Vol. 31, No. 4.

The dynamics of a hot ($T = 10^6 - 5 \times 10^7$ K) gas cloud with volume energy losses is investigated by numerical integration of gas dynamics equations. The dynamics is governed by a spherically symmetric gravitational field of the cloud and additional "hidden" mass. The cloud mass is taken in the range $M_0 = 10^{10} - 10^{12} M_\odot$, its radius $R_0 = 50 - 200$ kpc, the "hidden" mass $M_v = 10^{11} - 3 \times 10^{13} M_\odot$. The results show that in such systems a structure can develop in the form of a dense compact nucleus with a radius $R_s \ll R_0$, and an extended rarefied hot envelope with a radius $R_X \sim R_0$. Among the models involved are those where the gas cloud is either entirely blown up or entirely collapses. The results are discussed in connection with the formation and early evolution of galaxies, the history of star formation and chemical evolution of galaxies, the origin of hot gas in galaxies and clusters of galaxies.

157.086   102–MHz observations of M33.
V. S. Artyukh, V. G. Malumyan.
Sov. Astron. Lett., Vol. 12, No. 5, p. 309 – 311 (1987). English translation of 42.157.020.

157.087   Spectroscopy and photometry of elliptical galaxies. II. The spectroscopic parameters.
R. L. Davies, D. Burstein, A. Dressler, S. M. Faber, D. Lynden–Bell, R. J. Terlevich, G. Wegner.
Astrophys. J., Suppl. Ser., Vol. 64, No. 4, p. 581 – 600 (1987).

The authors report measurements of radial velocities, velocity dispersions, and magnesium line strength indices for 469 elliptical galaxies. The scatter of repeat measurements indicates an uncertainty of $\pm 10\%$ and ± 0.01 mag for single determinations of σ and Mg_2, respectively. A correction for the change in linear aperture size as a function of distance has been derived, and mean corrected values of σ and Mg_2 are adopted. The galaxies have been assigned to groups by combining the present velocities with those in the redshift catalog of Huchra and coworkers and using the algorithm of Huchra and Geller.

157.088   Spectroscopy and photometry of elliptical galaxies. III. *UBV* aperture photometry, CCD photometry, and magnitude–related parameters.

D. Burstein, R. L. Davies, A. Dressler, S. M. Faber,
R. P. S. Stone, D. Lynden–Bell, R. J. Terlevich, G. Wegner.
Astrophys. J., Suppl. Ser., Vol. 64, No. 4, p. 601 – 642 (1987).

Nearly 2000 new photoelectric BV aperture measurements of 449 elliptical galaxies are presented. These data are combined with an equal number of aperture observations from the literature and new CCD surface photometry of 70 galaxies to derive three magnitude–related parameters for 505 elliptical galaxies: total magnitude, B_T; effective circular diameter, A_e; and a new diameter, D_n, within which the mean corrected surface brightness of an elliptical galaxy is 20.75 B mag arcsec^{-2}.

157.089   Accretion flows in elliptical galaxies.

P. W. Vedder.
NRAO Workshop, No. 16, p. 127 – 133 (1986). – See Abstr. 012.015.

The author has developed a steady state infall model of gas in elliptical galaxies to investigate the properties and structure of the X–ray emitting gas observed in these systems. Models have been computed for galaxies with an external pressure (as might be important for ellipticals in clusters), and for varying supernova heating rates. All the models exhibit cooling flows, with mass accretion rates of $0.1 - 0.5\ M_\odot yr^{-1}$. The author examines a correlation between the radio luminosity and the X–ray luminosity of elliptical galaxies, which in the context of his infall models may suggest that the radio emission arises from nuclear sources that are powered by the gas accretion flow. These radio sources may also be confined effectively by the X–ray emitting gas.

157.090   Stellar associations and complexes in M33.

G. R. Ivanov.
Astrophys. Space Sci., Vol. 136, No. 1, p. 113 – 128 (1987).

About 460 OB associations were selected by a comparison of the *UBV* plates. The *UBV* photographic photometry of 1944 blue stars in the associations was made. The new associations appear like cores within Humphreys and Sandage's associations. Their star content, size distribution, and mean size ≈ 80 pc confirm their identity with the OB associations in the Galaxy and in the Magellanic Clouds. The genuine OB associations form groups of two or more members with a length scale of 250 pc. Their boundaries were delineated independently, but they coincide with the OB associations of Humphreys and Sandage (1980). These groups represent real concentration of blue massive stars with a large age dispersion. The star complexes unify a group of associations, H II regions, and H I peak distribution. Their mean size is 570 pc. The extensive H I clouds with a mean size of 1.2 kpc contain two or more star complexes. The questions related to star formation are briefly discussed.

157.091   The bisymmetric spiral magnetic field in M31.

Y. Sofue, R. Beck.
Publ. Astron. Soc. Jpn., Vol. 39, No. 4, p. 541 – 546 (1987).

The radio polarization data of M31 taken with the 100–m telescope at 2.7 GHz ($\lambda 11$ cm) are reanalyzed to investigate the magnetic field structure. Variation of polarization angle along an azimuthal circle in the galactic plane can be described as a superposition of a single– and a double–periodic variation. This variation is interpreted as due to a Faraday rotation caused by a bisymmetric open spiral (BSS) magnetic field superposed on the predominant ring field. The coexistence of the two modes (ring and BSS) of magnetic field configurations poses a new aspect on

the origin and maintenance of the large–scale magnetic field in a disk galaxy.

157.092   Vertical dust lanes and magnetic field in spiral galaxies.

Y. Sofue.
Publ. Astron. Soc. Jpn., Vol. 39, No. 4, p. 547 – 557 (1987).

Dark filaments (dust lanes) vertically emerging toward the halo from the 3 – 4 kpc molecular ring in the disk plane are found in the spiral galaxies NGC 253 and NGC 7331. The filamentary structure of dust is interpreted as a trace of a vertical magnetic field penetrating the galaxy disk. The vertical field, being twisted by galactic rotation, is suggested to yield an outflow of disk matter into the halo, which the author calls a "magnetic fountain" of galaxy scale. The magnetic fountain is suggested to be a possible formation mechanism of radio halos in these galaxies. A more condensed vertical field in the central region may be related to a mass ejection from the nucleus of NGC 253.

157.093   The origin of the far–infrared flux from spiral galaxies.

S. K. Leggett, P. W. J. L. Brand, C. M. Mountain.
Mon. Not. R. Astron. Soc., Vol. 228, No. 2, p. 11P – 16P (1987).

Studies of the Virgo cluster spiral galaxies have shown them to have the same mid– and far–infrared properties as redshift–limited samples of field galaxies. This is true even of the galaxies in the core of the cluster which have been stripped of neutral hydrogen. These findings appear to disagree with current models of the far–infrared emission from galaxies which conclude that it is dominated by disc emission from cold H I–associated dust. As the molecular content of the Virgo galaxies is normal, the authors suggest that the far–infrared flux from spirals is in fact dominated by emission from dust associated with the more centrally concentrated molecular gas.

157.094   The relative contributions of bulge and disk to the luminosity density of the universe.

P. L. Schechter, A. Dressler.
Astron. J., Vol. 94, No. 3, p. 563 – 570 (1987).

The authors have obtained redshifts for a virtually complete, magnitude–limited sample of galaxies in a set of randomly chosen fields. They use these in conjunction with previously measured bulge–to–disk ratios to determine the dependence of blue bulge–to–disk ratio on absolute magnitude, and the relative contributions of bulge and disk to the mean luminosity density of the universe. Through there is considerable scatter in bulge–to–disk ratio at all absolute magnitudes, the authors find that B/D varies as $0.61 (L/L_*)^{0.28}$. When the different mass–to–light ratios of bulge and disk are taken into account, they find that roughly equal amounts of mass are incorporated in bulges and disks.

157.095   A redshift survey of IRAS galaxies in the Bootes void area.

J. P. Vader, M. Simon.
Astron. J., Vol. 94, No. 3, p. 636 – 639 (1987).

Results of a redshift survey of IRAS galaxies in a 73 deg^2 region covering $\sim 1/6$ of the Bootes void area are presented. None of the 22 newly observed IRAS galaxies are located within this void. The 60 μm luminosity function obtained from these data and those of Vader and Simon (1987) is in good agreement with previous determinations.

157.096   Accurate radial velocities for carbon stars in the Sculptor dwarf spheroidal.

M. Aaronson, E. W. Olszewski.
Astron. J., Vol. 94, No. 3, p. 657 – 665 (1987).

The authors report velocities good to $\sim \pm 1 - 2$ km s^{-1} for seven of the eight known carbon stars in the Sculptor dwarf spheroidal galaxy. The mean velocity for Sculptor, a quantity important for Galactic mass estimates, but having a controversial past history, is found to be 109.2 ± 4.5 km s^{-1}, which agrees well with recent observations of K giants made by Armandroff and Da Costa (1986). The Sculptor mass–to–light ratio derived is roughly double what is found for typical Galactic globulars, a difference that may arise either from the presence of dark matter, or from variation of the mass function.

157.097 Imaging spectrophotometry of a chain of giant H II regions in the galaxy NGC 2997.
J.–R. Roy, J. R. Walsh.
Mon. Not. R. Astron. Soc., Vol. 228, No. 4, p. 883 – 905 (1987).

An area of $120'' \times 16''$ of the northern arm of the spiral galaxy NGC 2997 was scanned with the slit of the RGO spectrograph using the ASPECT system of the Anglo–Australian Telescope. 530 spectra were obtained. Monochromatic images and diagnostic diagrams of line ratios based on spectral lines at $H\alpha$, $H\beta$, [O II] 3727 Å, [O III] 5007 Å, [N II] 6584 Å and [S II] 6717 – 30 Å, and continua at 3580 Å and 5400 Å are constructed. Correlations are found between the absorption A_V and the oxygen abundance index ([O II] + [O III])/$H\beta$ indicating that the hotter stars are imbedded in more dust than the later–type stars. Reddening also arises from external dust; this dust is closely associated with the H II regions, because the stellar continuum is found to be less affected by reddening. Synthesis of the spiral arm stellar continuum was performed and the spectral signature of early–type stars was clearly detected in the H II regions.

157.098 Shell galaxies detected with IRAS.
A. Wilkinson, I. W. A. Browne, R. D. Wolstencroft.
Mon. Not. R. Astron. Soc., Vol. 228, No. 4, p. 933 – 940 (1987).

Shell galaxies which are infrared sources in the Catalog of Galaxies and Quasars in the IRAS Survey are all radio sources of more than 0.6 mJy at 6 cm.

157.099 Multicolor optical imaging of powerful far–infrared galaxies: more evidence for a link between galaxy mergers and far–infrared emission.
L. Armus, T. Heckman, G. Miley.
Astron. J., Vol. 94, No. 4, p. 831 – 846 (1987).

Broadband optical (B and R) imaging results are presented for a sample of 39 powerful far–infrared galaxies detected by $IRAS$. These objects, which typically have $L_{IR} \sim 10^{11} L_\odot - 10^{12} L_\odot$ and $L_{IR}/L_B \sim 10 - 100$, were chosen to exhibit far–IR spectra similar in shape to the prototypical very luminous infrared galaxies Arp 220 and NGC 6240, and to be faint optically (in apparent magnitude). The imaging program has revealed that: (1) At least 70% of the galaxies have very distorted optical morphologies. (2) There is an unusually low occurrence of apparently flattened members. (3) The galaxies are moderately luminous in the optical. (4) The galaxy colors most closely resemble Sb or Sbc galaxies seen at the appropriate redshift. These results demonstrate that the optical properties of NGC 6240 and Arp 220 are typical of powerful far–infrared galaxies, and add to the impressive body of evidence linking the collision/merger of disk galaxies to the production of intense infrared emission.

157.100 The optical luminosity function of a 60 μm flux–limited sample of IRAS galaxies.
J. P. Vader, M. Simon.
Astron. J., Vol. 94, No. 4, p. 854 – 866 (1987).

The aims of this study are to construct the optical luminosity function of $IRAS$ galaxies and to compare various properties of an infrared–selected sample of $IRAS$ galaxies to those of optically selected non–$IRAS$ galaxies.

157.101 Neutral–hydrogen observations and maps of early–type galaxies.
D. Burstein, N. Krumm, E. E. Salpeter.
Astron. J., Vol. 94, No. 4, p. 883 – 898 (1987).

The authors report measurements of neutral hydrogen in and around 12 elliptical, S0, and early–type spiral galaxies, as well as three spirals of later type. The observations were made with the 4 arcmin "flat" beam of the Arecibo Radio Telescope. Ten of the 12 early–type galaxies were detected, and show small H I masses of 4×10^7 to $8 \times 10^8 M_\odot$ ($H_0 = 100$ km/s/Mpc). Very low upper limits of 3×10^6 and $6 \times 10^6 M_\odot$ of H I have been established for the S0 galaxies NGC 4382 and 7332. A faint dwarf galaxy 3 arcmin south of NGC 4382 was detected. Together with the previously detected dwarfs around NGC 2859, this suggests that H I–rich companions to H I–poor S0 galaxies may be common. In contrast to the other galaxies mapped in this program, the H I

distribution in NGC 3773 is tightly concentrated to the center of the galaxy. The measurements are compared to earlier observations, and satisfactory agreement is found.

157.102 Systematics of the 4000 Angstrom break in the spectra of galaxies.
A. Dressler, S. A. Shectman.
Astron. J., Vol. 94, No. 4, p. 899 – 905 (1987).

The authors have measured the strength of the 4000 Å break from moderate–resolution spectra of 950 galaxies in the fields of 12 rich clusters. They find only a weak dependence of the 4000 Å break amplitude on either the total or bulge luminosity of the galaxy. Since the U–V colors of spheroidal systems appear to correlate with absolute magnitude, a correlation attributed to line blanketing, the authors' result indicates that the 4000 Å break amplitude is insensitive to changes in metal abundance, at least over the range characteristic of galactic bulges. On the other hand, the authors show that the break is a sensitive probe of star formation, by correlating the break amplitude with emission and absorption features indicative of recent star formation. This sensitivity to star formation but insensitivity to metal abundance makes the 4000 Å break a powerful diagnostic for studying evolution with observations of galaxies at large lookback times.

157.103 Determination of the masses of elliptical galaxies and clusters of galaxies with AXAF.
R. F. Mushotzky.
Astrophys. Lett. Commun., Vol. 26, Nos. 1 – 2, p. 43 – 60 (1987).
– See Abstr. 012.020.

The proposed program consists of deep, spatially resolved X–ray spectra of clusters of galaxies and elliptical galaxies to be obtained by the imaging spectrometers on AXAF. The purpose of the investigation is to determine the mass and mass distribution of galaxies and clusters of galaxies. This proposal utilizes the unique AXAF features of large bandwidth, large field of view, high angular resolution, long exposure times and the availability of sophisticated instrumentation. These data will give a firm measurement of the masses and mass distributions of individual elliptical galaxies and clusters of galaxies and thus help solve the "missing" mass problem.

157.104 Morphologie und Kinematik elliptischer Galaxien.
R. Bender.
Diss. Naturwiss.–Math. Gesamtfak., Ruprecht–Karls–Univ., Heidelberg, F.R. Germany, 159 pp. (1987).

157.105 The chemical evolution of galaxies.
C. Chiosi.
Prog. Part. Nucl. Phys., Vol. 17, p. 173 – 214 (1986). – See Abstr. 012.024.

After a short introduction summarizing the observational data on chemical abundances in different astrophysical sites, the theoretical framework of chemical evolution is presented followed by a brief discussion on the current ideas on the initial mass function and star formation rate. Chemical models applying to the solar vicinity and galactic disk are given. In particular, the author discusses the history of C, N, O and Fe in the galactic disk and current models for the evolution of disk galaxies in general.

157.106 The red/infrared evolution in galaxies – effect of the stars on the asymptotic giant branch.
A. Chokshi, E. L. Wright.
Astrophys. J., Vol. 319, No. 1, p. 44 – 54 (1987).

The effect of including the asymptotic giant branch (AGB) population in a spectral synthesis model of galaxy evolution is examined. Stars on the AGB are luminous enough and also evolve rapidly enough to affect the evolution of red and infrared colors in galaxies. The validity of using infrared colors as distance indicators to galaxies is then investigated in detail. The authors find that for $z \leqslant 1$ infrared colors of model galaxies behave linearly with redshift.

157.107   Giant H II regions in M81.
M. Kaufman, F. N. Bash, R. C. Kennicutt Jr.,
P. W. Hodge.
Astrophys. J., Vol. 319, No. 1, p. 61 – 75 (1987). With plate 1.

Hα and VLA radio continuum observations at wavelengths of 6 and 20 cm are used to study the distribution of extinction and the distribution of giant radio H II regions along the spiral arms in M81. Radio flux densities, Hα fluxes, and extinction values are obtained for 42 giant H II regions with high surface brightness. Nearly all the giant radio H II regions lie along the spiral arms or the inner H I ring. The radial distribution of the set of giant radio H II regions exhibits a strong maximum at a galactocentric distance $R \approx 300''$ (4.7 kpc if the distance of M81 is 3.3 Mpc). Unless molecular hydrogen in M81 is also concentrated near $R \approx 300''$, these data disagree with Visser's model for star formation associated with a density wave.

157.108   The H I distribution in clouds within galaxies.
E. J. Shaya, S. R. Federman.
Astrophys. J., Vol. 319, No. 1, p. 76 – 83 (1987).

An interpretation is given for the relatively flat distribution of atomic hydrogen as a function of radius within the visible disks of spiral galaxies and the existence of a correlation between H I surface density and morphological class of spiral galaxy. The H I is mostly contained in constant column density envelopes of molecular clouds. The column density of the cloud envelopes depends on the ratio of flux of H_2 dissociating radiation in the ambient medium to envelope gas density. The flat H I profiles imply the gas density and pressure follow the ultraviolet ambient flux on kiloparsec scales throughout a given galaxy.

157.109   Bimodal star formation: constraints from galaxy colors at high redshift.
R. F. G. Wyse, J. Silk.
Astrophys. J., Lett. Ed., Vol. 319, No. 1, p. L1 – L6 (1987).

The authors investigate the possibility that at early epochs the light from elliptical galaxies is dominated by stars with an initial mass function (IMF) which is deficient in low–mass stars, relative to the solar neighborhood. $V–R$ colors for the optical counterparts of 3CR radio sources offer the most severe constraints on the models. The authors obtain reasonable fits to both the blue, high–redshift colors and the redder, low–redshift colors with a model galaxy which forms with initially equal star formation rates in each of two IMF modes – one lacking low–mass stars, and one with stars of all masses. The net effect is that the time-integrated IMF has twice as many high–mass stars as the solar neighborhood IMF, relative to low mass stars.

157.110   Effects of hidden mass and intergalactic medium on the structure of hot galactic coronae.
V. G. Berman, A. A. Suchkov.
Pis'ma Astron. Zh., Tom 13, No. 10, p. 843 – 849 (1987). In Russian. English translation in Sov. Astron. Lett., Vol. 13.

The structure and origin of hot galactic coronae are studied by numerical integration of gas dynamic equations describing coronae. The results of hydrostatic approximation ("cooling flows") are considered in the light of exact (dynamical) solutions. The predictions are made of the corona temperature profile: the temperature would fall outwards if a corona is bound to a galaxy by a massive dark halo, and it would rise outwards if a corona is retained by a hot intergalactic gas (the thermal conductivity is supposed to be suppressed by a magnetic field). This could be checked in forthcoming X–ray observational programs, and the temperature profile could be used then for testing dark halos and intergalactic gas.

157.111   The radio and optical axes of radio elliptical galaxies.
A. E. Sansom, I. J. Danziger, R. D. Ekers,
R. A. E. Fosbury, W. M. Goss, A. S. Monk, P. A. Shaver,
W. B. Sparks, J. V. Wall.
Mon. Not. R. Astron. Soc., Vol. 229, No. 1, p. 15 – 29 (1987).

New VLA and CCD data are incorporated with previous observations to investigate the alignment of radio and optical axes of radio–emitting early–type galaxies. A comparison of observations with simulations modelling the random projection effects of different geometries on the sky shows that there is no strong physical relation of radio axis with a particular stellar axis. This result is true for both triaxial and axially–symmetric galaxies.

157.112   The distribution and kinematics of stars formed from the cooling flow in M87.
D. S. L. Soares, R. H. Sanders.
Mon. Not. R. Astron. Soc., Vol. 229, No. 1, p. 119 – 128 (1987).

The distribution of stars formed from the cooling flow in M87 has been calculated assuming that the system is spherically symmetric, that the stars follow radial orbits, and that the gas inflow rate has not varied over a Hubble time. The authors find that, while such "cooling flow" stars cannot significantly affect the dynamics of M87, they can affect the observed kinematics: a relatively small population of stars on nearly radial orbits can cause a dramatic increase in the line–of–sight velocity dispersion in the central regions. The cooling flow population in M87 can explain in detail the observed radial dependence of stellar velocity dispersion in this galaxy if this population has roughly the same mass function as in the initial galaxy. Thus, the existence of a super–massive black hole in the centre of M87 is not demanded by the observed high central velocity dispersions.

157.113   Formation of shells in elliptical galaxies from interstellar gas.
M. Loewenstein, A. C. Fabian, P. E. J. Nulsen.
Mon. Not. R. Astron. Soc., Vol. 229, No. 1, p. 129 – 141 (1987).

An alternative to the merger model for producing systems of shells around elliptical galaxies is proposed wherein a one–sided or highly asymmetric disturbance in the nuclear or core region leads to a period of enhanced star formation as it propagates through the hot interstellar medium observed in X–rays. If such a disturbance can induce cooler, denser gas co–existing with the hot phase to form observable stars at a rate comparable to the inferred mass–deposition rates in these galaxies, the newly–formed stars will be created with the appropriate phase–space characteristics for the phase–wrapping mechanism invoked in the context of the merger hypothesis to occur. Thus a system of optical shells with the observed wide range of binding energies can be produced from a single disturbance having a wide variety of possible energies and epochs.

157.114   Evidence of a small velocity dispersion for the Carina dwarf spheroidal?
P. J. Godwin, D. Lynden–Bell.
Mon. Not. R. Astron. Soc., Vol. 229, No. 1, p. 7P – 13P (1987).

Three independent determinations of the velocity dispersion of the Carina dwarf spheroidal give $\sigma_c = 6\pm3, 6\pm2, 8\pm4$ km s^{-1} which appear to be in good agreement. However, star by star comparison shows no correlation between the observers as to which stars move more slowly or faster than the mean. After intercomparison the authors conclude that all the observers have overestimated their accuracy by a factor of 2 in km s^{-1} and that the true velocity dispersion of Carina is unknown with best estimates in the range 1.1 – 3.2 km s^{-1}. This would give this low mass dwarf spheroidal a stellar M/L ratio. In dwarf spheroidals and other systems velocity dispersions close to the observational errors should only be believed when the offsets of stellar velocities from the mean as measured by different observing systems show clear correlation. It is not yet clear that the high M/L ratios found in dwarf spheroidals are real.

157.115   IRAS galaxies: no evidence for a cosmological anisotropy.
R. G. Clowes, A. Savage, G. Wang, S. K. Leggett,
H. T. MacGillivray, R. D. Wolstencroft.
Mon. Not. R. Astron. Soc., Vol. 229, No. 1, p. 27P – 30P (1987).

Rowan–Robinson et al. have found that the surface density of IRAS galaxies is 20 per cent greater in the north galactic cap than in the south galactic cap, and consider the disparity to represent an anisotropy of cosmological significance, with its origin in a northern structure of characteristic dimension 50 – 100 h^{-1} Mpc.

In this paper the authors present an independent north–south comparison of IRAS galaxies. They also find the disparity in surface densities, but their optical information shows that it is strongly concentrated in galaxies having optical, angular major axes greater than 130 arcsec, for which the median distance is $\sim 16\,h^{-1}$ Mpc. The authors conclude, therefore, that the origin of the disparity is the local supercluster rather than a structure of much larger dimensions.

157.116 A lower limit to the mass of neutral leptons as constituents of the dark halo around galaxies.
W. Y. Chau, J. Stone.
Astrophys. Space Sci., Vol. 137, No. 1, p. 17–21 (1987).

The authors have found that the structural equations governing the hydrostatic equilibrium of a thermally relaxed, spherically–symmetric neutral lepton system with a density profile that can account for the dark matter distribution around spiral galaxies do not admit any physically reasonable solutions (namely, the temperature should be positive definite and monotonically decreasing with distance) if the lepton mass is $L \lesssim 17\,\mathrm{eV}$.

157.117 Optical and near–infrared observations of IRAS galaxies. II.
A. F. M. Moorwood, M.–P. Véron–Cetty, I. S. Glass.
Astron. Astrophys., Vol. 184, No. 1/2, p. 63–70 (1987).

Optical spectra, CCD images and near infrared (JHK) photometry have been obtained for a further 23 galaxies in the Soifer et al. (1984) IRAS minisurvey sample. As found also for the first 22 galaxies discussed in Paper I (Moorwood et al., 1986), most appear to be relatively normal spiral galaxies exhibiting H II region like emission spectra except for two cases of relatively weak Seyfert activity. The average far infrared luminosity in this sample ($1.3 \times 10^{11} L_\odot$) is, however, a factor of two larger than for the Paper I sample–partly attributable to a larger number of interacting systems. A good correlation found between J–H and the Balmer decrements provides further evidence that reddening in the near infrared is primarily an extinction effect.

157.118 Multi–frequency radio continuum observations of NGC 5236 (M83).
S. Sukumar, U. Klein, R. Gräve.
Astron. Astrophys., Vol. 184, No. 1/2, p. 71–78 (1987).

The spiral galaxy NGC 5236 (M83) has been observed at 327 MHz, 1465 MHz, and 4750 MHz using the OSRT, VLA and Effelsberg 100–m telescope. Maps of the distributions of radio emission are presented, and a separation of total thermal and nonthermal emission based on the derived integrated flux densities is attempted. The nonthermal spectral index is $\alpha_{nt} = -0.80$, the thermal fraction at 4750 MHz is $f_{\mathrm{th}} = 0.2$, and the average spectral index in the observed frequency range is $\langle\alpha\rangle = -0.75 \pm 0.04$. Flux densities and spectral indices for discrete sources lying in the inner region of the galaxy are also given. The large–scale magnetic field structure of M83 is investigated for the first time by evaluating the measurements of linear polarization. A clear overall spiral configuration of the magnetic field is evident with very high degrees of polarization of up to $\sim 60\%$ in the outer spiral arms. A preliminary analysis of the polarization angles indicates a bisymmetric magnetic field, with the field lines following the spiral arms of M83.

157.119 Standard photometric diameters of galaxies. III. Reduction of the diameters in the ESO–B and SGC catalogues to the standard diameter system at the 25 mag arcsec^{-2} brightness level.
G. Paturel, P. Fouqué, A. Lauberts, E. A. Valentijn, H. G. Corwin, G. de Vaucouleurs.
Astron. Astrophys., Vol. 184, No. 1/2, p. 86–92 (1987).

The standard diameters in the D_{25} system derived in Papers I and II, augmented by newly determined diameters on ESO–B plates, are used to derive the transformation formulae needed to reduce the photographic diameters and axis ratios of galaxies in the ESO–B survey and the Southern Galaxy Catalogue (SGC) to the standard system. The approximate limiting isophotes in the

two catalogues and their dependence on morphological type are determined. A time dependence of the limiting isophotal level is detected in the ESO–B lists, similar to that previously found and corrected in the SGC. No significant dependence on mean surface brightness is detected, at least in the small range covered by the standard galaxies.

157.120 The photometric investigation of the interacting system VV 242.
V. P. Reshetnikov.
Astrofizika, Tom 27, Vyp. 1, p. 91–101 (1987). With two plates. In Russian. English translation in Astrophysics, Vol. 27, No. 1.

The results of a detailed UBV–photometry of the interacting system VV 242 (NGC 7253 a, b) are presented. The standard photometric parameters are determined: the total and absolute magnitudes, colour indices and others. The distribution of the surface brightness in various colour bands, the distribution of the colour B–V are given. The galaxies under consideration are shown to be spirals of morphological types Sbc–Sc or SBbc–SBc. All the observational features of the system (deformation of the outlying regions, non–coplanarity of its structure, superassociations) may be connected with the tidal interaction between galaxies.

157.121 Colour and star formation in the outer parts of disks of spiral galaxies.
A. V. Zasov, A. R. Dzhafarov.
Astron. Zh., Tom 64, Vyp. 5, p. 900–909 (1987). In Russian. English translation in Sov. Astron., Vol. 31, No. 5.

Published UBV colours of outer parts of disks of galaxies (beyond the region of clearly defined spiral arms) are analysed. Influence of different factors on the observed colour of a stellar system is discussed. It is shown that star formation is taking place in outer parts of most galaxies considered. Several early–type spirals have colours of external disks which can be accounted for either by a lower metal abundance of stars (in comparison to other galaxies) or by a sharp stop of star formation several billions of years ago.

157.122 The ratio numbers of carbon and oxygen stars in galaxies.
Yu. L. Frantsman.
Nauchn. Inf., Vyp. 63, p. 60–67 (1987). In Russian.

Simulated populations of AGB stellar models were calculated with different assumptions about mass loss, initial chemical composition and dredge–up efficiency. The AGB was divided into early–AGB (E–AGB) and thermally pulsing–AGB (TP–AGB) phases. The numbers of carbon and oxygen stars per 10^6 generated stars and the ratio (N_C/N_M) of these numbers are presented. It is possible to compare these theoretically obtained N_C/N_M ratios with observations only if the luminosity of observed stars is $M_{\mathrm{bol}} < 3\overset{m}{.}5$, otherwise it is necessary to take into account the E–AGB phase. The authors present the ratio N_C/N_M for the E–AGB phase for 612 per 10^6 generated stars. These ratios are presented for solar vicinity and LMC for different laws of mass loss on the AGB. The numbers of carbon stars per unit of mass in LMC and SMC are quite equal. It is possible to explain this fact only if the rapid mass loss starts suddenly, after the star has reached certain luminosity.

157.123 Spirals from order and chaos.
D. H. Smith.
Sky Telesc., Vol. 74, No. 2, p. 136–138 (1987).

157.124 The Malmquist–type bias in galaxy distance determinations.
M. W. Feast.
Observatory, Vol. 107, No. 1080, p. 185–188 (1987).

It is shown that the classical formulation of the distance bias problem by Malmquist does not apply to methods of determining galaxy distances such as the Tully–Fisher relationship, the Faber–Jackson relationship, and the de Vaucouleurs Λ_c index. A more appropriate formulation shows that at large distances the derived moduli will tend to be over–estimates, contrary to the results of the classical Malmquist formulation.

157.125   Precision of velocity estimates in the face–on galaxy UGC 9500.
B. M. Lewis.
Observatory, Vol. 107, No. 1080, p. 201 – 204 (1987).

Observations of neutral hydrogen in the face–on spiral UGC 9500, at a resolution of 2 km s^{-1} and a S/N ~ 200, provide velocity estimates that are reproducible to 0.05 km s^{-1}. UGC 9500 has a $\sigma_V(z) \sim 9.2$ km s^{-1} and an inclination angle to the plane of the sky of $\sim 3°5$.

157.126   The evidence for and against the existence of supermassive black holes in E galaxies.
W. L. W. Sargent.
AIP Conf. Proc., No. 155, p. 62 – 70 (1987). – See Abstr. 012.029.

The author reviews the history of the controversies surrounding the interpretation of the velocity dispersion and light profile in the center of M87. The most recent theoretical work suggests that the highly anisotropic velocity ellipsoid, which had been proposed as an alternative to a central massive object, is unstable to the formation of a bar. The current observations are just consistent with either a central, essentially point, mass or with a massive star cluster. However, the most recent observations of the center of M31 and M32 are difficult to reconcile with anything but a central point mass.

157.127   Warps, galaxies and haloes.
R. D. Davies.
Nature, Vol. 328, No. 6129, p. 382 – 383 (1987).

157.128   The prodigious warp of NGC 4013.
R. Bottema, G. S. Shostak, P. C. van der Kruit.
Nature, Vol. 328, No. 6129, p. 401 – 403 (1987).

Non–planar distortions, or warps, in the outer gas layers of galaxies are frequently observed. H I radio synthesis observations reveal that the edge–on Sbc galaxy NGC 4013 has the largest regular H I warp so far observed. It extends to a large height above the plane of the galaxy, and begins abruptly at just the radius where photometry indicates the end of the luminous disk. Furthermore, at precisely this position, the rotational velocity is seen to drop by 25 km s^{-1}. This implies that the disk–mass distribution suddenly approaches zero at the radius of the warp onset.

157.129   Stellar populations in Local Group galaxies.
L. L. Stryker.
Nearly normal galaxies. From the Planck time to the present, p. 10 – 17 (1987). – See Abstr. 012.031.

This talk discusses a few of the more interesting recent papers under the subject of populations and luminosity and mass functions in the Local Group.

157.130   Star formation in colliding and merging galaxies.
F. Schweizer.
Nearly normal galaxies. From the Planck time to the present, p. 18 – 25 (1987). – See Abstr. 012.031.

Observations of present–day collisions and mergers give information about (1) the spatial distribution of star formation in such events, (2) global processes that govern the formation of molecular gas and – ultimately – of stars; (3) star formation rates and efficiencies, and (4) the initial mass function. This paper reviews recent progress in these areas and proposes that collisions and mergers may trigger the formation not only of stars, but also of globular clusters.

157.131   Star formation in disks and spheroids.
R. B. Larson.
Nearly normal galaxies. From the Planck time to the present, p. 26 – 35 (1987). – See Abstr. 012.031.

A major goal of studies of the stellar populations in galactic disks and spheroids is to understand how these systems formed. In particular, we wish to understand how the observed stars formed, and how the star formation rate and initial mass function (IMF) may have varied with time and location. It is also important to understand the formation of star clusters, since they serve as crucial population tracers. If the dark matter in galactic halos is in faint stars or stellar remnants, its properties, too, need to be understood in terms of star formation processes.

157.132   Star formation in disks: IRAS results.
C. J. Persson.
Nearly normal galaxies. From the Planck time to the present, p. 36 – 40 (1987). – See Abstr. 012.031.

IRAS far infrared fluxes are potentially powerful as a measure of the total massive star formation rate in galaxies, especially if the flux arises in relatively optically thick situations so that the dust grain properties can be ignored to first order. The IRAS far infrared flux of a star forming region can then be converted to a star formation rate if an initial mass function for the heating sources is assumed, and a correction for the flux missed by IRAS is made. The uncertainty in this correction factor is on the order of a factor of two.

157.133   Stellar populations in dwarf spheroidals.
M. Aaronson.
Nearly normal galaxies. From the Planck time to the present, p. 57 – 66 (1987). – See Abstr. 012.031.

The topics covered in this review encompass structural parameters, stellar content (including carbon and variable stars), ages, abundances, mass–to–light ratios, and some thoughts on the origins of the spheroidals.

157.134   Blue compact dwarfs: extreme dwarf irregular galaxies.
T. X. Thuan.
Nearly normal galaxies. From the Planck time to the present, p. 67 – 75 (1987). – See Abstr. 012.031.

In Hubble's classification scheme, dwarf irregular galaxies (dIs) are low luminosity systems ($M_B \gtrsim -16$) which are at the end of the galaxy morphological sequence and which lack both a dominating nucleus and rotational symmetry. The author concentrates here on a class of dIs which are so extreme in these properties that they appear almost stellar in appearance, with no obvious underlying galaxy. The motivation is twofold: (1) it is easier to discuss the physical characteristics of dIs by focussing on their extreme manifestations; and (2) the properties of dIs have been well reviewed elsewhere. The present review of the blue compact dwarf galaxies (BCDs) is intended to be complementary and attempts to contrast the properties of BCDs with those of dIs.

157.135   Dwarf galaxies and dark matter.
S. Tremaine.
Nearly normal galaxies. From the Planck time to the present, p. 76 – 77 (1987). Abstract. – See Abstr. 012.031.

157.136   Photometry and mass modeling of spiral galaxies.
S. Kent.
Nearly normal galaxies. From the Planck time to the present, p. 81 – 89 (1987). – See Abstr. 012.031.

This paper examines several aspects of galaxy photometry and how photometric and kinematic data can be combined to model the mass distribution in spiral galaxies.

157.137   Hot gas evolution in nearly normal elliptical galaxies.
M. Loewenstein, W. G. Mathews.
Nearly normal galaxies. From the Planck time to the present, p. 96 – 108 (1987). – See Abstr. 012.031.

The authors report the results and interpretation of their gas dynamical calculations. The computed models of the evolution of hot gas described here are an extention of their earlier results (Mathews and Loewenstein, 1986).

157.138 Hot coronae around early type galaxies.
C. Jones.
Nearly normal galaxies. From the Planck time to the present, p. 109 – 115 (1987). – See Abstr. 012.031.

The discovery of hot coronae around early type galaxies has opened a rich field of research. Problems related to the origin of gas, its iron enrichment and evolution over time in various environments will be addressable as will problems of the total galaxy mass and the shape of the underlying potential.

157.139 Cores of early–type galaxies: the nature of dwarf spheroidal galaxies.
J. Kormendy.
Nearly normal galaxies. From the Planck time to the present, p. 163 – 174 (1987). – See Abstr. 012.031.

The systematic study of galaxy cores has become possible with high–resolution CCD surface photometry. The author discusses the scale parameters of cores, i.e., the central surface brightness, the core radius at which the surface brightness has fallen by a factor of two, and the central velocity dispersion. Core parameters are correlated: more luminous galaxies have larger core radii, fainter central surface brightnesses, and higher core mass–to–light ratios. These scaling laws provide constraints on theories of galaxy formation. The main purpose of this paper is to discuss the implications of the core parameter relations for the nature and origin of dwarf spheroidal galaxies.

157.140 Global scaling relations for elliptical galaxies and implications for formation.
S. M. Faber, A. Dressler, R. L. Davies, D. Burstein, D. Lynden–Bell, R. Terlevich, G. Wegner.
Nearly normal galaxies. From the Planck time to the present, p. 175 – 183 (1987). – See Abstr. 012.031.

Two recent surveys of elliptical galaxy structural properties are described. E galaxies are seen to populate a planar distribution in the global logarithmic parameter space (R_e, σ_e, I_e). Two–dimensionality implies that the virial theorem is the only tight constraint on E structure. There is an additional, weaker constraint on radius versus mass that was presumably imposed at formation. The best–fitting plane in logarithmic coordinates has the equation $R_e \sim \sigma^{1.35\pm0.07} I_e^{-0.84\pm0.03}$, which implies $(M/L)_e \sim L^{0.24\pm0.04} I_e^{0.00\pm0.06}$. The planar relation can be used to determine distances to E galaxies to an accuracy of $\pm 23\%$. An analogous, parallel plane exists for core properties. Core and gobal M/Ls argree well, implying that ellipticals are mainly baryon dominated within R_e and that M/Ls are stellar. The effects of other variables such as ellipticity, aspect angle, and rotation on the basic planar relation seem to be small.

157.141 Musings concerning the possible significance of surface brightness variations in disk galaxies.
G. Bothun.
Nearly normal galaxies. From the Planck time to the present, p. 184 – 194 (1987). – See Abstr. 012.031.

Disk galaxies do not have the same surface brightness but instead exhibit a rather large range (unlike the case for luminous ellipticals). The origin of these surface brightness variations is unclear but their existence is important in a number of contexts, some of which the author briefly discusses.

157.142 Core properties of elliptical galaxies.
T. R. Lauer.
Nearly normal galaxies. From the Planck time to the present, p. 207 – 210 (1987). – See Abstr. 012.031.

The author confines his discussion to the basic core structure parameters such as core radius r_C and central surface brightness, I_0, and their relationship to global properties of elliptical galaxies. Most of the discussion presented here is based on the analysis of slightly sub–arcsecond resolution CCD photometry of 42 galaxies presented in Lauer (1985).

157.143 Stellar populations in distant galaxies.
G. Bruzual A.
Nearly normal galaxies. From the Planck time to the present, p. 265 – 275 (1987). – See Abstr. 012.031.

The author reviews the problem of the detection of changes in the population content of distant galaxies with respect to the population content of nearby galaxies. The basic assumptions made by most authors working in the field are examined and criticized. The author evaluates the comparisons of theoretical models with observations of distant and nearby galaxies that lead to different lines of evidence, both in favor and against the detection of evolution of the stellar populations dominating the light in distant and nearby galaxies.

157.144 Dynamics of galaxies at large redshift: prospects for the future.
R. G. Kron.
Nearly normal galaxies. From the Planck time to the present, p. 300 – 309 (1987). – See Abstr. 012.031.

This discussion explores the feasibility of obtaining area–resolved dynamical information for galaxies to $z = 1$ from a combination of observations with the Space Telescope and a very large ground–based optical telescope.

157.145 Dark matter in dwarf galaxies.
K. C. Freeman.
Nearly normal galaxies. From the Planck time to the present, p. 317 – 325 (1987). – See Abstr. 012.031.

This review concentrates mainly on the true dwarf galaxies, i.e. those with $M_B > -14$. First the author discusses the dwarf irregulars, and then the dwarf spheroidal systems.

157.146 Dark matter in early–type galaxies.
S. M. Fall.
Nearly normal galaxies. From the Planck time to the present, p. 326 – 331 (1987). – See Abstr. 012.031.

The evidence for dark matter in early–type galaxies comes from polar rings, shells, and X–rays. Unfortunately, only a few galaxies have been studied in detail and the observations are not always free of ambiguity. The author's purpose in this article is to collect together the fragmentary results that are now available.

157.147 General historical introduction (*concerning elliptical galaxies*).
G. de Vaucouleurs.
Structure and dynamics of elliptical galaxies, p. 3 – 16 (1987). – See Abstr. 012.032 (IAU Symp. No. 127).

A brief historical review of the discovery and exploration of elliptical galaxies in the past two centuries is presented. This review does not include the research of the past ten years or so, which other communications are bound to cover in greater detail.

157.148 Cores of early–type galaxies.
J. Kormendy.
Structure and dynamics of elliptical galaxies, p. 17 – 36 (1987). – See Abstr. 012.032 (IAU Symp. No. 127).

Many cores are well resolved in a photometry program with the Canada–France–Hawaii Telescope. Core profile shape correlates with galaxy luminosity L: the brightest galaxies have isothermal profiles; fainter ellipticals and bulges have profiles that do not completely flatten inward into a core. Core parameters are correlated: more luminous galaxies have larger core radii and fainter central surface brightnesses. Large deviations suggest special events: Fornax A has too small and bright a core for its luminosity; it may be the remnant of a merger with a smaller galaxy. A kinematic search shows strong evidence for a central black hole in M31 and weaker evidence in M32 and NGC 3115.

157.149   Isophote shapes.
R. I. Jedrzejewski.
Structure and dynamics of elliptical galaxies, p. 37 – 46 (1987). – See Abstr. 012.032 (IAU Symp. No. 127).

The shapes of the isophotes of elliptical galaxies are discussed. Ellipticity and position angle variations with distance from the centres of galaxies are described, as well as deviations from perfectly elliptical shape. The factors contributing to these deviations, in particular edge–on disks and boxiness, are described and portrayed.

157.150   Distribution of light: outer regions.
M. Capaccioli.
Structure and dynamics of elliptical galaxies, p. 47 – 61 (1987). – See Abstr. 012.032 (IAU Symp. No. 127).

The data base of photometric properties of galaxies has grown enormously in the last few years. In this discussion, much space is given to selected technical aspects. Their complexity explains the small enthusiasm for and the skepticism about surface photometry at faint levels, and also accounts for the poor quality of some of the results. In addition, to a review of the data and their interpretations, empirical fitting formulae, particularly the $r^{1/4}$ law (de Vaucouleurs, 1948), are discussed for the benefits of theorists, commenting on their deviations from observations at large and small scales. Finally, the author speculates about misclassifications and contamination of the elliptical class by lenticulars.

157.151   The stellar kinematics of elliptical galaxies.
R. L. Davies.
Structure and dynamics of elliptical galaxies, p. 63 – 77 (1987). – See Abstr. 012.032 (IAU Symp. No. 127).

The kinematic properties of elliptical galaxies are summarized. New developments are discussed in four areas: (1) the Faber–Jackson relation and the role of second parameters, (2) the luminosity–rotation relation, (3) the figures of elliptical galaxies and (4) the mass–to–light ratio as a function of radius.

157.152   The manifold of elliptical galaxies.
S. Djorgovski.
Structure and dynamics of elliptical galaxies, p. 79 – 88 (1987). – See Abstr. 012.032 (IAU Symp. No. 127).

Global properties of elliptical galaxies, such as the luminosity, radius, projected velocity dispersion, projected luminosity density, etc., form a two-dimensional family. This "fundamental plane" of elliptical galaxies can be defined by the velocity dispersion and mean surface brightness, and its thickness is presently given by the measurement error–bars only. This is indicative of a strong regularity in the process of galaxy formation. However, all morphological parameters which describe the shape of the distribution of light, and reflect dynamical anisotropies of stars, are completely independent from each other, and independent of the fundamental plane. The M/L ratios show only a small intrinsic scatter in a luminosity range spanning some four orders of magnitude; this suggests a constant fraction of the dark matter contribution in elliptical galaxies.

157.153   Properties of cD galaxies.
J. L. Tonry.
Structure and dynamics of elliptical galaxies, p. 89 – 98 (1987). – See Abstr. 012.032 (IAU Symp. No. 127).

cD galaxies are the most luminous galaxies in the universe. They are characterized by a surface brightness profile that falls off more slowly with radius than most elliptical galaxies. In most respects D galaxies are a continuous extrapolation from other ellipticals: their M/L and their colors are comparable to other ellipticals, their inner parts are fitted by an $r^{1/4}$ law, and they follow the same relation between L and σ. On the other hand, their luminosity is too bright to be consistent with the luminosity function of other ellipticals and they are always found at the center of a cluster of other galaxies. Being at the center of a cluster of galaxies often endows D galaxies with a very faint, very extended halo of luminosity and multiple nuclei, but these are more properly associated with the cluster than the D galaxy itself.

The connection between the formation of cD galaxies and the formation of clusters remains a mystery.

157.154   Compact elliptical galaxies.
J.–L. Nieto, P. Prugniel.
Structure and dynamics of elliptical galaxies, p. 99 – 107 (1987). – See Abstr. 012.032 (IAU Symp. No. 127).

The authors summarize the present knowledge on low–mass high–surface brightness elliptical objects near massive galaxies that are often called M32–type or compact elliptical galaxies. The origin of the low mass of these objects is a controversial matter: is it intrinsic to their formation or produced, as classically believed, by tidal stripping from the massive neighbor? The authors present new observational data allowing to define better the characteristics of these objects and a simple theoretical model whose consequences support the idea that the precursors of compact ellipticals are related to the low–mass end of the luminosity function of elliptical galaxies.

157.155   Dust and gas: overview.
F. Schweizer.
Structure and dynamics of elliptical galaxies, p. 109 – 124 (1987). – See Abstr. 012.032 (IAU Symp. No. 127).

Progress in the 50 years since the discovery of ionized gas in NGC 1052 is reviewed. As discovery has proceeded from H II to H I and recently to 10^7K gas, the known amount of gas in ellipticals has increased dramatically: from $10^{3-6}M_\odot$ for H II, to $10^{5-9}M_\odot$ for H I in some 15% of E's, to $10^{9-10}M_\odot$ for the X–ray emitting gas. Although a few ellipticals with dust have long been known, recent CCD surveys have revealed that dust lanes in nearly half of all E's. The cooler gas – as traced by H I, warm H II, and dust – is generally distributed in the form of a disk that often shows an outer warp. There is mounting evidence that many disks consist of gas accreted by mass transfers and mergers, and that some of them may not yet have reached dynamical equilibrium.

157.156   Ionized gas in elliptical galaxies.
E. M. Sadler.
Structure and dynamics of elliptical galaxies, p. 125 – 133 (1987). – See Abstr. 012.032 (IAU Symp. No. 127).

More than half of all nearby elliptical galaxies contain modest amounts ($10^3 - 10^5 M_\odot$) of ionized gas. In bright elliptical galaxies this gas appears to lie in a rotating, kiloparsec–scale central disk, with a spectrum characteristic of non–thermal ionization. Low–luminosity ellipticals have a clumpy gas distribution and the gas in these galaxies is photoionized by young stars.

157.157   Properties of elliptical galaxies with dust lanes.
F. Bertola.
Structure and dynamics of elliptical galaxies, p. 135 – 144 (1987). – See Abstr. 012.032 (IAU Symp. No. 127).

In this paper the author describes the morphological, statistical, kinematical, and photometric properties, of the elliptical galaxies with dust lanes.

157.158   The cold interstellar medium in elliptical galaxies – observations of H I and radio continuum emission.
G. R. Knapp.
Structure and dynamics of elliptical galaxies, p. 145 – 154 (1987). – See Abstr. 012.032 (IAU Symp. No. 127).

About 10% of nearby elliptical galaxies contain H I, with typical values of M(H I) $\sim 5 \times 10^8 M_\odot$ and M(H I)/L$_B$ $\sim$ 0.03 M$_\odot$/L$_\odot$. The H I content is unrelated to the stellar content, suggesting that the H I in early–type galaxies has an external origin and is not produced by mass loss. The H I and stellar kinematics show that the rotation curves of E and S0 galaxies are approximately flat. Large mass–to–light ratios are found for some systems. Comparison with mass models derived from X–ray emission suggests that these may in some cases overestimate the mass. The presence of H I is shown to enhance the likelihood that an E/S0 galaxy has a nuclear radio continuum source, in agreement with models which suggest that the central engine is fuelled by cold gas. Current data suggest that the gas–to–dust ratio for the cold interstellar medium in ellipticals has a value similar to that found in the solar neighborhood.

157.159 X–rays from elliptical galaxies.
A. C. Fabian, P. A. Thomas.
Structure and dynamics of elliptical galaxies, p. 155 – 165 (1987).
– See Abstr. 012.032 (IAU Symp. No. 127).

X–ray observations have shown that early–type galaxies contain a hot interstellar medium. This implies that the galaxies have (1) a low supernova rate; (2) high total gravitational binding masses and (3) continuous star formation. Much of the gas in isolated galaxies is probably due to stellar mass–loss. The details of its behaviour are complex.

157.160 Mass distributions in elliptical galaxies at large radii.
C. L. Sarazin.
Structure and dynamics of elliptical galaxies, p. 179 – 188 (1987).
– See Abstr. 012.032 (IAU Symp. No. 127).

Recently, X–ray observations have shown that elliptical galaxies generally contain large quantities of hot gas. Central dominant cluster ellipticals have even more gas, which they have accreted from the surrounding clusters. The mass distributions in these galaxies can be derived from the condition of hydrostatic equilibrium. M87, the best studied central dominant galaxy, has a massive, dark halo with a total mass of about $4 \times 10^{12} M_\odot$ within a radius of 300 kpc. The total mass–to–light ratio within this radius is at least 150 $M_\odot/L_\odot$. The X–ray observations of normal ellipticals also strongly suggest that they have heavy haloes, although the distribution of the mass is much less certain than in M87.

157.161 Line–strength gradients in early–type galaxies.
J. Gorgas, G. Efstathiou.
Structure and dynamics of elliptical galaxies, p. 189 – 201 (1987).
– See Abstr. 012.032 (IAU Symp. No. 127).

The authors have measured line–strength gradients in a sample of 15 early–type galaxies. The line–strength measures include the Mg_2 index and the equivalent widths of $H\beta$ and two iron blends at 5270 Å and 5335 Å. In most of the galaxies the authors find gradients in the metallic line–strengths. However, the gradients vary markedly from object to object and do not correlate strongly with other parameters such as total luminosity, rotation, etc. A comparison of the line–strengths in the outer parts of these galaxies with galactic globular clusters suggests relatively modest abundance gradients in early–type galaxies.

157.162 Population synthesis of composite systems.
A. Pickles.
Structure and dynamics of elliptical galaxies, p. 203 – 216 (1987).
– See Abstr. 012.032 (IAU Symp. No. 127).

Metallicity and age dispersion among the stars comprising stellar composite systems are discussed, with emphasis on the present and future implications for the technique of population synthesis.

157.163 The intrinsic shapes of elliptical galaxies.
P. L. Schechter.
Structure and dynamics of elliptical galaxies, p. 217 – 228 (1987).
– See Abstr. 012.032 (IAU Symp. No. 127).

Distribution functions for the intrinsic shapes of elliptical galaxies are discussed, starting with the simplest and proceeding to the more complex. A variety of competing "proxy" observables, which can in principle be used to recover at least some of information lost in the projection of a galaxy onto the plane of the sky, are considered.

157.164 Morphological properties of elliptical galaxies.
S. Djorgovski.
Structure and dynamics of elliptical galaxies, p. 377 – 378 (1987).
– See Abstr. 012.032 (IAU Symp. No. 127).

In the poster as presented at the meeting, the author described global morphological properties of elliptical galaxies, based on the data from a CCD surface photometry survey of

~ 200 ellipticals and ~ 50 S0's (Djorgovski 1985). In this brief summary, he emphasizes two points: (1) there is a very weak and very noisy trend of radial shape with luminosity, in the sense that more luminous galaxies are less concentrated, and (2) there is no preference for low–luminosity ellipticals to show boxy isophotes, and they differ in that respect from the bulges.

157.165 Correlations between $r^{1/4}$–law parameters for bulges and elliptical galaxies.
M. Hamabe, J. Kormendy.
Structure and dynamics of elliptical galaxies, p. 379 – 380 (1987).
– See Abstr. 012.032 (IAU Symp. No. 127).

The correlation between the effective radius r_e and surface brightness μ_e for elliptical galaxies is a fundamental scaling law. The purpose of this paper is to rederive the $\mu_e(\log r_e)$ relations for ellipticals and bulges, taking account of the coupling in the errors and using only high–accuracy CCD data. The preliminary conclusions are: (1) The coupled errors are too small to affect significantly the correlation derived for elliptical galaxies. (2) The correlation for bulges is not very different from that for ellipticals, but the galaxy sample is small and the errors in the parameters are large due to the inherent uncertainty in bulge–disk decomposition.

157.166 The core properties of elliptical galaxies.
T. R. Lauer.
Structure and dynamics of elliptical galaxies, p. 381 – 382 (1987).
– See Abstr. 012.032 (IAU Symp. No. 127).

The core structure of elliptical galaxies is determined by two parameters: total luminosity L and central luminosity density ϱ_C. Scatter in ϱ_C at any L may imply dissipational formation of core structure; once this scatter is accounted for, however, the pure L dependence of the core parameters can be isolated and perhaps used as a metric distance indicator.

157.167 Surface photometry of bright ellipticals.
J. M. Schombert.
Structure and dynamics of elliptical galaxies, p. 383 – 384 (1987).
– See Abstr. 012.032 (IAU Symp. No. 127).

Surface brightness profiles of bright ellipticals ($M_V < -17$, $H_0 = 100$) in rich clusters and the field are reduced to structural parameters. Most fitting functions are found to be inadequate in describing the overall shape of profiles with only the $r^{1/4}$ law providing a good match over the range of surface brightness from 19 to 24 mag arcsec^{-2}. The systematic deviations in structure of first–ranked ellipticals from normal ellipticals (i.e. enlarged characteristic radii and shallow profile slopes) are well explained by comparisons to the N–body simulations of merging systems from Duncan, Farouki, and Shapiro (1982).

157.168 Isophotometry of brightest cluster ellipticals.
A. C. Porter, D. P. Schneider, J. G. Hoessel.
Structure and dynamics of elliptical galaxies, p. 385 – 386 (1987).
– See Abstr. 012.032 (IAU Symp. No. 127).

What are the two–dimensional distributions of projected luminosity in the brightest ellipticals in Abell clusters (hereafter called "El"s)? If we treat isophotes as ellipses, how do their properties vary as functions of surface brightness? Are the isophote parameters correlated with a galaxy's global properties? Are they correlated with any properties of the surrounding cluster? Are Els morphologically similar to other cluster or field ellipticals? This paper describes work in progress to the answers to these questions, and illustrates an interesting result.

157.169 Investigating the scatter in the $V_{26} - \log \sigma$ relation.
M. Gregg, A. Dressler.
Structure and dynamics of elliptical galaxies, p. 387 – 388 (1987).
– See Abstr. 012.032 (IAU Symp. No. 127).

The nature and existence of a second parameter needed to characterize the family of normal elliptical galaxies has been much discussed. The need for a second parameter has been demonstrated by the correlation of the residuals from the well–known magnitude–velocity dispersion relation for ellipticals with other observables such as ellipticity, mass–to–light ratio and Mg line strength or metallicity. Here, evidence for a correlation between residuals in the $V_{26} - \log \sigma$ relation and the strength of Hβ is presented, suggesting that variations in the stellar populations in Virgo elliptical cores may be an important secondary parameter.

157.170   On the relation between radius, luminosity and surface brightness in elliptical galaxies.
E. Recillas–Cruz, A. Serrano.
Structure and dynamics of elliptical galaxies, p. 389 – 390 (1987).
– See Abstr. 012.032 (IAU Symp. No. 127).

The authors have analyzed luminosity profiles of E galaxies in six clusters of galaxies. They have found a relationship between radius, luminosity and surface brightness for galaxies in each of the clusters. Moreover, it seems that there is a dependence of the zero point of the relation with environment. This relationship implies that there is not a universal luminosity profile for elliptical galaxies.

157.171   The isophotal structure of elliptical galaxies.
T. B. Williams, B. Bhattacharya.
Structure and dynamics of elliptical galaxies, p. 391 – 392 (1987).
– See Abstr. 012.032 (IAU Symp. No. 127).

The authors have obtained photographic surface photometry for a large sample of elliptical galaxies. The results for 37 galaxies are summarized here.

157.172   Parent structures in E and S0 galaxies?
R. Michard, F. Simien.
Structure and dynamics of elliptical galaxies, p. 393 – 394 (1987).
– See Abstr. 012.032 (IAU Symp. No. 127).

Isophote shape and flattening are being used for studying a similarity between elliptical and lenticular galaxies.

157.173   Intrinsic shapes of elliptical galaxies from a statistical comparison of two different isophotes.
G. Fasano.
Structure and dynamics of elliptical galaxies, p. 395 – 396 (1987).
– See Abstr. 012.032 (IAU Symp. No. 127).

157.174   Deprojection of galaxies: how much can be learned?
G. B. Rybicki.
Structure and dynamics of elliptical galaxies, p. 397 – 398 (1987).
– See Abstr. 012.032 (IAU Symp. No. 127).

A general discussion, based on the "Fourier Slice Theorem", is given for the problem of deprojecting the observed light distribution of galaxies to obtain their intrinsic three dimensional light distribution or "shape". Several results are obtained : (1) A model–independent deprojection of an axially symmetric galaxy is shown to be possible only if the symmetry axis lies in the plane of the sky. (2) A simple criterion is given to test whether two different galaxies can have the same intrinsic shape, based solely on their observed projections. (3) It is shown that a homogeneous class of galaxies can be deprojected using a sufficiently large number of projections of random perspective.

157.175   Settling of gas disks in elliptical galaxies.
T. Y. Steiman–Cameron, R. H. Durisen.
Structure and dynamics of elliptical galaxies, p. 403 – 404 (1987).
– See Abstr. 012.032 (IAU Symp. No. 127).

Prominent gas disks and dust lanes are found in a number of elliptical galaxies. The authors have developed methods for modeling the evolution of such inclined dissipative galactic gas disks and present some of the results of this work.

157.176   Dust in early type galaxies observed at the CFHT.
J. Kormendy, J. Stauffer.
Structure and dynamics of elliptical galaxies, p. 405 – 406 (1987).
– See Abstr. 012.032 (IAU Symp. No. 127).

A program of CCD imaging of early–type galactic nuclei carried out at the Canada–France–Hawaii Telescope reveals dust in an unusually large fraction of the galaxies.

157.177   Detailed surface photometry of the dust–lane elliptical NGC 6702.
E. Davoust, M. Capaccioli, G. Lelièvre, J.–L. Nieto.
Structure and dynamics of elliptical galaxies, p. 407 – 408 (1987).
– See Abstr. 012.032 (IAU Symp. No. 127).

The authors present preliminary results of a detailed photometric study of NGC 6702, from high resolution photographs taken at the Cassegrain focus of the Canada–France–Hawaii telescope. The luminosity distribution of the galaxy follows an $r^{1/4}$ law (re* = $11\overset{''}{.}6$, μ_e* = 22.17). The axis ratio is 0.75 and the position angle of the major axis is 60° and fairly constant. To reveal the structure of the dust, the authors have subtracted from the previous image the model of an elliptical galaxy having the photometric and geometric parameters quoted above.

157.178   Boxy isophotes and dust lanes in bright Virgo ellipticals.
C. Möllenhoff, R. Bender.
Structure and dynamics of elliptical galaxies, p. 409 – 410 (1987).
– See Abstr. 012.032 (IAU Symp. No. 127).

The CCD camera of the Landessternwarte Heidelberg was used at the 1.2 m Calar Alto Telescope for a V,R,I – survey of ~70 dusty and non–dusty elliptical galaxies. The authors report here about morphological studies of ten bright elliptical galaxies in the Virgo cluster.

157.179   The unusual box–shaped elliptical(?) galaxy IC 3370.
B. Jarvis.
Structure and dynamics of elliptical galaxies, p. 411 – 412 (1987).
– See Abstr. 012.032 (IAU Symp. No. 127).

Photometric and kinematic observations are presented for the box–shaped "elliptical" galaxy IC 3370, that show that this galaxy has characteristics more akin to an S0 galaxy than an elliptical and should be classified S0pec.

157.180   AM2020–5050: an elliptical galaxy with an outer ring.
B. Whitmore, D. McElroy, F. Schweizer.
Structure and dynamics of elliptical galaxies, p. 413 – 414 (1987).
– See Abstr. 012.032 (IAU Symp. No. 127).

Photometric and spectroscopic observations show that the inner component of AM2020–5050 is an elliptical galaxy, unlike other polar–ring galaxies which have an S0 disk at the center. A comparison of the central velocity dispersion with the rotational velocity in the ring suggests the presence of a nearly spherical gravitational potential. The inner component has a rapidly rotating core with rotational velocities at 3″ substantially higher than at 8″. Although the optical ring is quite narrow, Hα emission is observed all the way through the center of the galaxy, indicating the presence of an extended gaseous disk.

157.181   TAURUS observations of S0 polar ring galaxies.
R. A. Nicholson, K. Taylor, W. B. Sparks, J. Bland.
Structure and dynamics of elliptical galaxies, p. 415 – 416 (1987).
– See Abstr. 012.032 (IAU Symp. No. 127).

By use of the TAURUS imaging Fabry–Perot interferometer (Taylor & Atherton,1980) the authors have obtained seeing limited two–dimensional velocity, line width and line flux maps of the ionised gas in two polar ring systems NGC 4650A and NGC 2685.

157.182   The distribution and kinematics of neutral hydrogen in NGC 807.
L. L. Dressel.
Structure and dynamics of elliptical galaxies, p. 423 – 424 (1987).
– See Abstr. 012.032 (IAU Symp. No. 127).

The author has detected 21 cm line emission from neutral hydrogen in the giant elliptical galaxy NGC 807 at Arecibo Observatory, and has mapped this emission with the VLA. Unlike the active and dwarf ellipticals that have been mapped thus far, NGC 807 has a fairly regular disk of gas rotating about the apparent optical minor axis. Combined with observations of active ellipticals, this observation suggests that two classes of H I–rich ellipticals may exist: ellipticals which have accreted gas and become active recently, and quiescent ellipticals which have either produced gas internally or accreted it so long ago that it has reached dynamical equilibrium.

157.183   Metal–enhanced galactic winds.
J. P. Vader.
Structure and dynamics of elliptical galaxies, p. 435 – 436 (1987). – See Abstr. 012.032 (IAU Symp. No. 127).

Constraints on supernova–driven galactic winds from elliptical galaxies at the epoch of star formation are investigated. The occurrence of mass loss is found to depend critically on the supernova rate in the case of dwarf galaxies, while the depth of the potential well is the most important constraint for giant ellipticals. The smallest dwarf ellipticals must have evolved from significantly more massive progenitors in order to have sustained a wind that carried away most of their metal production.

157.184   Line–strength gradients in elliptical galaxies.
R. L. Davies, E. M. Sadler.
Structure and dynamics of elliptical galaxies, p. 441 – 442 (1987). – See Abstr. 012.032 (IAU Symp. No. 127).

The authors have measured line–strength indices as a function of radius in several elliptical galaxies. All of them show strong radial gradients in Mg, but much weaker gradients in Fe and Hβ. The isophotes and contours of constant line–strength have the same flattening. More luminous galaxies have shallower gradients, contrary to the prediction of models of dissipative collapse. Most of the galaxies observed show weak central emission which can partially fill the Balmer absorption lines.

157.185   Visual–IR color gradients in elliptical galaxies.
R. F. Peletier, E. A. Valentijn, R. F. Jameson.
Structure and dynamics of elliptical galaxies, p. 443 – 444 (1987). – See Abstr. 012.032 (IAU Symp. No. 127).

Simultaneous measurements for visual and visual–infrared colors provide the means to determine both the average temperature of the giant branch and the turnoff–temperature of the main sequence. This allows to model fractional contributions of different populations, including age– and metallicity–effects. The authors observed NGC 3379, 4278 and 5813 in B, V, J, H and K. All three galaxies show significant color gradients both in visual and visual–infrared colors, in the sense that the galaxies become bluer going outwards.

157.186   Direct IR determination of the stellar luminosity function to 0.2 $M_\odot$ in elliptical galaxies.
G. Gilmore, K. Arnaud.
Structure and dynamics of elliptical galaxies, p. 445 – 446 (1987). – See Abstr. 012.032 (IAU Symp. No. 127).

There is no evidence for any statistically significant differences in the stellar luminosity function for stars with masses greater than 0.15 $M_\odot$ between elliptical galaxies with and without X–ray cooling flows.

157.187   Dynamics of the Fornax dwarf spheroidal galaxy.
G. Paltoglou, K. C. Freeman.
Structure and dynamics of elliptical galaxies, p. 447 – 448 (1987). – See Abstr. 012.032 (IAU Symp. No. 127).

The mass to light ratio of the inner parts of the Fornax system is 3.2 ± 1.1. Its major axis rotation is only 3.4 ± 2.4 km s^{-1} over ± 1 kpc, which argues against its origin as a stripped dwarf irregular.

157.188   The local density and morphology dependence of the galaxy luminosity function.
J. Choloniewski, M. Panek.
Structure and dynamics of elliptical galaxies, p. 457 (1987). – See Abstr. 012.032 (IAU Symp. No. 127).

157.189   On the luminosity function of elliptical galaxies.
V. S. Popov.
Structure and dynamics of elliptical galaxies, p. 461 – 462 (1987). – See Abstr. 012.032 (IAU Symp. No. 127).

157.190   Spectrophotometry of shell galaxies.
W. D. Pence.
Structure and dynamics of elliptical galaxies, p. 463 (1987). – See Abstr. 012.032 (IAU Symp. No. 127).

Low dispersion spectra of two shell galaxies, NGC 3923 and NGC 3051, have been obtained covering the 5300 Å to 10,000 Å spectral range. These long–slit spectra go through the nucleus of each galaxy and also through 12 shells in NGC 3923 and through 3 shells in NGC 3051.

157.191   Two colour CCD photometry of Malin–Carter shell galaxies.
A. Wilkinson, W. B. Sparks, D. Carter, D. A. Malin.
Structure and dynamics of elliptical galaxies, p. 465 – 466 (1987). – See Abstr. 012.032 (IAU Symp. No. 127).

Shells may be the result of the disruption of a small companion in the potential of a much larger galaxy, the disturbance of a disk system in a tidal encounter, an accumulation of resonant stellar orbits or the result of some shock phenomenon in a hot galactic atmosphere. To distinguish between these formation mechanisms, CCD direct images in B and R have been obtained at the Anglo–Australian 3.8 m telescope for 66 of the 74 galaxies in the range $01^h40' < \alpha < 13^h46'$ in the Malin–Carter (1983) catalogue of shell galaxies.

157.192   Shells and dark matter in elliptical galaxies.
L. Hernquist, P. J. Quinn.
Structure and dynamics of elliptical galaxies, p. 467 – 468 (1987). – See Abstr. 012.032 (IAU Symp. No. 127).

A new method for probing the distribution of matter in elliptical galaxies surrounded by shell systems is described. An an illustration the authors have applied this technique to the giant elliptical galaxy NGC 3923. If the potential is modeled as the sum of an $r^{1/4}$ law and a non–singular isothermal halo, then the best fit to the shell number and shell distribution gives a halo mass (within the shell system ~ 100 h^{-1}kpc) ~ 40 times the mass of the elliptical and a halo core radius ~ 3 times the effective radius of the luminous material.

157.193   Shells around tumbling bars: the mass distribution around NGC 3923.
C. Dupraz, F. Combes, J.–L. Prieur.
Structure and dynamics of elliptical galaxies, p. 469 – 470 (1987). – See Abstr. 012.032 (IAU Symp. No. 127).

157.194   CCD photometry of resolved dwarf irregular galaxies. I. Sextans A.
A. Aparicio, J. M. García–Pelayo, M. Moles, J. Melnick.
Astron. Astrophys., Suppl. Ser., Vol. 71, No. 2, p. 297 – 338 (1987).

CCD *UBV* photometry of 2279 stars in Sextans A is presented. Differences with the photometric scales of Sandage and Carlson (1985) and Hoessel et al. (1983) are discussed; recalibration of the Sandage and Carlson data for Cepheids in Sextans A results in a smaller value for the distance modulus, $\mu = 25.6$. A similar value for the distance modulus is obtained from the colour–colour diagram. Colour–colour and colour–magnitude diagrams for the whole galaxy and, for different parts, are presented indicating a general star formation activity throughout Sextans A. The contribution of stars brighter than $M_{pg} = -3$ to the total blue light of the galaxy amounts to 25%, a much higher value than is found for other dwarf irregular galaxies in the Local Group. The integrated birthrate per unit mass of massive stars for the galaxy is found to be much higher than in the Magellanic Clouds, suggesting that the rate in Sextans A was probably smaller in the past. Different slopes for high mass and low mass IMF parts are needed to account for its low surface brightness. The formation of massive stars seems to be favoured for systems of decreasing mass or, given the relation between total mass and metal content, for galaxies of decreasing metallicity.

157.195   A morphological survey of emission line galaxies.
I. Tarrab.
Astron. Astrophys., Suppl. Ser., Vol. 71, No. 3, p. 449 – 463 (1987).

CCD image photometry in four filters (Johnson B, V and Gunn r, i) is given for 11 emission–line galaxies with compact appearance. Surface brightness profiles have been fitted to three functions: de Vaucouleurs, logarithmic and exp $(1/r^2)$ laws. Various types of morphologies and surface brightness profiles are observed. Only three of them turned out to be compact and blue with one or several bright knots. The most compact and probably the youngest one is PHL 293B. The de Vaucouleurs and the logarithmic functions fit equally well the surface brightness of 6 galaxies.

157.196   Study of the radiation emitted by the nuclei of spiral galaxies.
G. M. Tovmasyan, S. A. Akopyan.
Astrophysics, Vol. 26, No. 1, p. 25 – 30 (1987). English translation of 43.158.056.

157.197   Formation of stellar shells and X–ray coronae around elliptical galaxies.
M. Umemura, S. Ikeuchi.
Astrophys. J., Vol. 319, No. 2, p. 601 – 613 (1987).

The authors propose a wind–accretion flow interaction model for the formation of extended multiple stellar shells and hot X–ray coronae around elliptical galaxies. They examine the successive interactions of a hot galactic wind with nonstationary ambient gas, which accretes onto the galaxy. The similarity solution for the interaction of a wind with accreting gas is obtained, and the authors perform the numerical simulations of the successive collisions between a wind and an accretion flow. The dark matter, the time variations of wind luminosities, and the explosive energy release from young stellar components of shells are taken into account. It is shown that multiple stellar shells actually can be produced by such successive collisions.

157.198   Evolution of hot galactic flows.
M. Loewenstein, W. G. Mathews.
Astrophys. J., Vol. 319, No. 2, p. 614 – 631 (1987). = Lick Obs. Bull., No. 1062.

The time–dependent equations describing galactic flows, including detailed models for the evolving source terms, are integrated over a Hubble time for two elliptical galaxies with total masses of 3.1×10^{12} and $8.3 \times 10^{12} M_\odot$, 90% of which resides in extended, nonluminous halos. The "standard" supernova rate of Tammann and a rate 4 times smaller are considered for each galaxy model. The models lead to the development of massive, quasi–hydrostatic, nearly isothermal distributions of gas at $\sim 10^7$K with cooling inflows inside their galactic cores. For the less massive galaxy with the higher supernova rate, however, a low–luminosity supersonic galactic wind develops.

157.199   On the thermal instability of galactic and cluster halos.
A. Malagoli, R. Rosner, G. Bodo.
Astrophys. J., Vol. 319, No. 2, p. 632 – 636 (1987).

The authors present a detailed study of thermal instabilities in cooling flows associated with galaxies and clusters of galaxies. In the case of purely radiation–driven accretion onto a central object such as the cD galaxy M87, they find that the gas is largely subject to overstability, rather than to monotonic instability. If thermal conductivity is taken into account, the flow is stabilized on scales of several kiloparsecs, even if the conductivity is appreciably reduced (e.g., $\sim 1\%$) with respect to the Spitzer value. The authors present numerical solutions of the local dispersion relation for the cooling flow in M87 and discuss the possible consequences of their results for a correct understanding of cooling flows.

157.200   Evidence of an infrared luminosity indicator for galaxies.
E. D. Feigelson, T. Isobe, D. W. Weedman.
Astrophys. J., Lett. Ed., Vol. 319, No. 2, p. L51 – L55 (1987).

To elucidate the nature of infrared–luminous galaxies discovered with the *IRAS* satellite, the authors compare the optical and infrared luminosities of 1161 Markarian galaxies and 2146 "normal" galaxies from the CfA redshift survey. Survival analysis statistical methods that take upper limits fully into account are used. It is found that L_{IR}/L_B is statistically correlated with $L_{60\mu m}$ in both samples, though they differ in the distribution at low luminosities. The derived correlation shows that L_{IR}/L_B provides an indicator for $L_{60\mu m}$. Since galaxies selected in unbiased *IRAS* surveys will have higher L_{IR}/L_B than optically selected galaxies, they are therefore also selected for high $L_{60\mu m}$.

157.201   Nitrogen abundances in the amorphous galaxy NGC 5253.
J. R. Walsh, J.–R. Roy.
Astrophys. J., Lett. Ed., Vol. 319, No. 2, p. L57 – L62 (1987).

The central complex of ionized gas in the amorphous galaxy NGC 5253 was scanned with the imaging spectroscopy system of the AAT. Monochromatic images corresponding to several nebular lines were obtained with a spatial resolution of 2.3×1.3 arcsec^2 (1 pixel). The electron temperature has been determined over 78 pixels using the line ratio [O III] (4959 + 5007)/4363 and hence abundances of O, N, He, and Ne. A region of high values of log N/O (~ -1.0) was found to correspond with the presence of a cluster of Wolf–Rayet stars.

157.202   Infrared emission and star formation in early–type galaxies.
H. A. Thronson Jr., J. Bally.
Astrophys. J., Lett. Ed., Vol. 319, No. 2, p. L63 – L68 (1987).

The authors have used *IRAS* data for elliptical and S0 galaxies in an effort to determine whether star formation is taking place in galaxies that are often thought to be inert. The authors used two–color diagrams to interpret galaxian infrared colors. It is found that one–third of the sample galaxies have infrared colors that are consistent with emission from dusty regions surrounding young stars. Star formation rates in the range $0.1 - 1 \, M_\odot \mathrm{yr}^{-1}$ are estimated for some normal elliptical galaxies, although with significant systematic uncertainty. In general, star–formation rates calculated for these objects are roughly comparable to the mass–loss rate for evolved stars.

157.203   The visually brightest early–type supergiants in the spiral galaxies NGC 2403, M81, and M101.
R. M. Humphreys, M. Aaronson.
Astron. J., Vol. 94, No. 5, p. 1156 – 1168 (1987).

Moderate–resolution spectroscopy of the candidate brightest blue stars in the nearby spirals NGC 2403, M81, and M101 reveals that many are not single stars, but compact H II regions, clusters, and multiple systems. This is an additional and serious limitation on their use as distance indicators. The authors discuss the derivation of the luminosities of the confirmed single blue supergiants, and, combining these results with their previous work on the brightest red supergiants, they give a brief discussion and comparison of the evolution of their most massive stars. The properties of the brightest blue and red stars in NGC 2403 and M81 are basically in agreement with the pattern of massive–star evolution observed in Local Group galaxies. The distance to M101 is still controversial.

157.204   Magnetic fields in spiral galaxies.
R. D. Wolstencroft.
Q. J. R. Astron. Soc., Vol. 28, No. 3, p. 209 – 216 (1987). – See Abstr. 012.037.

The methods and results of radio and optical studies of galactic magnetic fields are examined. Current understanding of galactic dynamos driven by differential rotation and helical motions of ionized gas is briefly discussed.

157.205 Spectroscopic survey of the Case blue and emission line galaxies.
R. Augarde, P. Figon, D. Kunth, F. Sèvre.
Astron. Astrophys., Vol. 185, No. 1/2, p. 4 – 8 (1987).

A spectroscopic follow–up of 53 galaxies from the Case low–dispersion northern sky survey is presented. Emission lines have been found in 42 of them and have been used to measure recession velocities and absolute magnitudes. No AGN have been detected. The present sample contains a dozen low luminosity galaxies with star formation activity but no extreme metal–poor galaxy has been found.

157.206 Near–infrared photometry of globular clusters in the outer halo of M31.
F. Bònoli, F. Delpino, L. Federici, F. Fusi Pecci.
Astron. Astrophys., Vol. 185, No. 1/2, p. 25 – 32 (1987).

New infrared data for 18 globular clusters in the outer halo of M31 are presented. The whole set of available IR data (68 clusters) is still insufficient to get firm conclusions. Nevertheless, it allows to draw the following considerations: (1) Within 20 kpc from the M31 nucleus, there is no evidence for the existence of any metallicity radial gradient. The metal poor clusters measured out of 20 kpc are still too few to be taken as evidence of any gradient at very large galactocentric distances. (2) There is no support for the claimed existence of a statistical relationship between the absolute integrated K magnitudes and the IR colors of globular clusters to be used as possible distance indicator. (3) The analysis and comparison of the $(U–V)_0 – (V–K)_0$ diagrams for galactic and M31 globulars may indicate the existence of significant "second parameter effects" amongst the M31 clusters.

157.207 The initial mass function for massive stars: a comparison between the total Hα and ultraviolet fluxes of a sample of spiral and irregular galaxies.
V. Buat, J. Donas, J. M. Deharveng.
Astron. Astrophys., Vol. 185, No. 1/2, p. 33 – 38 (1987).

Total Hα and far ultraviolet (2000 Å) fluxes are available for 31 late–type galaxies. The Lyman continuum to far UV luminosity ratio, obtained after correction for extinction and loss of ionizing photons, is compared to model predictions established for different choices of the Initial Mass Function (IMF). The IMF for massive stars does not vary much from galaxy to galaxy. The power–law index has a formal dispersion of ± 0.25 in a subset of 17 Sbc, Sc galaxies. The average absolute value of this index is consistent with most of the existing Field Star IMF determinations. Uncertainties prevent any firm conclusion to be reached about a possible trend with morphological types.

157.208 Chemical evolution of elliptical galaxies.
F. Matteucci, A. Tornambè.
Astron. Astrophys., Vol. 185, No. 1/2, p. 51 – 60 (1987).

Models of chemical evolution for elliptical galaxies, with initial masses in the range $10^9 – 2 \times 10^{12} M_\odot$, taking into account the occurrence of galactic winds powered by supernovae (of type I and II) as well as the most up–to–date nucleosynthesis results and theories on supernova progenitors, are presented. Detailed temporal evolution of ^{16}O, ^{24}Mg, ^{28}Si, and ^{56}Fe in the gas and in the stars, as well as type I and II SN rates, have been computed and compared with the observational constraints.

157.209 Neutral hydrogen observations of four dwarf irregular galaxies in the Virgo Cluster.
E. D. Skillman, G. D. Bothun, M. A. Murray, R. H. Warmels.
Astron. Astrophys., Vol. 185, No. 1/2, p. 61 – 76 (1987).

The authors present results from H I synthesis observations and optical surface photometry of four dwarf irregular galaxies within the Virgo Cluster. H I distributions and rotation curves have been calculated for the four galaxies. One galaxy shows strong evidence that it is presently falling into the Virgo Cluster core and being stripped. Overall, the results support the proposal of Bothun et al. (1985) that the dwarf irregular population is presently outside of the dense core. Three of the galaxies suggest evidence for dark matter in small gravitational systems (luminous

mass $\sim 10^8 M_\odot$). From rotation curve analysis the authors derive total M/L ratios of order 5 to 10.

157.210 A model of spectrophotometric evolution for high–redshift galaxies.
B. Guiderdoni, B. Rocca–Volmerange.
Astron. Astrophys., Vol. 186, No. 1/2, p. 1 – 21 (1987).

The authors present a model of spectrophotometric evolution which predicts apparent magnitudes and colors for galaxies of the Hubble sequence at all redshifts. This model accounts for the effects of cosmology and intrinsic evolution which can thereby be separated. It includes the four main phases of stellar evolution (main sequence, giant branch, horizontal branch, and asymptotic giant branch). The library of 30 stellar spectra (from 220 Å to 10,680 Å) which is used for the evolutionary synthesis is basically observational, with a resolution of 10 Å. The contribution of nebular emission and the effects of internal extinction are taken into account. The model gives good fits of far–UV and visible spectrophotometric properties of *nearby* galaxies along the Hubble sequence, by varying the timescale for the conversion of gas into stars, with uniform IMF.

157.211 Gas kinematics in the nucleus of NGC 6946.
C. Muñoz–Tuñon, J. M. Vilchez.
Astron. Astrophys., Vol. 186, No. 1/2, p. 25 – 29 (1987).

A complex kinematical structure is inferred from the analysis of high resolution spectra of the nuclear region of the galaxy NGC 6946. The spatial resolution of the observations allowed the authors to resolve a dynamically inhomogeneous area of approximately 10 arcseconds in diameter, very close to the centre of the galaxy. Within this region the emission lines show non–gaussian profiles which are generally split into at least two main components, differing in velocity by approximately 50 km s^{-1}. These line profiles show an intricate structure in both the Hα and the [N II] lines, indicating that self–absorption effects in Hα can be ruled out. A simple interpretation of the kinematical structure in terms of a rotational velocity curve alone does not appear to be realistic. An interpretation based on stellar winds plus supernova explosions as proposed by Beckman et al. (1986), is preferred rather than a model with purely gravitationally coupled motions. The authors examine this solution in the light of the particularly rich star formation history of NGC 6946.

157.212 Origin and evolution of compact elliptical galaxies.
J.–L. Nieto, P. Prugniel.
Astron. Astrophys., Vol. 186, No. 1/2, p. 30 – 38 (1987).

The authors study some aspects of the evolution of compact M32–type elliptical galaxies (cEs) during their orbital decay onto a massive neighbour. For a typical cE, they find a very slow orbital decay rate: about a few kpc per 10^9 yr. The present mass loss is small, $\sim 1 – 10 M_\odot$/year. The merging of cEs into their massive neighbour should happen within 10^9 and 10^{10} yr. The small mass loss during the decay implies that progenitors of cEs had already a low–mass when they arose in the vicinity of the massive neighbour, either as formed there or after being captured by it. The authors argue that, in the capture case also, the progenitors were formed with a low mass, ranging between a few times 10^9 and a few times $10^{10} M_\odot$. This gives strong support for cEs to be related to the low mass end of the luminosity function of elliptical galaxies: in particular, cEs could not have evolved from massive ellipticals via tidal stripping. Therefore isolated low–mass ellipticals should exist, that resemble cEs.

157.213 Photometry of Zwicky compact galaxies.
M. Moles, J. M. García–Pelayo, G. del Rio, F. Lahulla.
Astron. Astrophys., Vol. 186, No. 1/2, p. 77 – 83 (1987).

UBVRI colors for 80 Zwicky compact galaxies (ZCG) are presented. It is found that those with sharp emission lines have the same color indices as other objects selected in different ways but also showing stellar formation activity, namely the Markarian (non Seyfert) galaxies, the Blue Compact Dwarfs and the Irregulars. For a fraction of them, the reddest, evolutionary models with declining star formation rates can reproduce conveniently the observed colors. The bluest emission line ZCG have colors of star–burst galaxies but only for some extreme cases like IZw207 there is some evidence for the red colors for the presence of an underlying red supergiant population. For this group of objects, the observed colors indicate that very recent and massive starbursts have occurred or that the IMF could be flatter than the Salpeter law. Regarding the ZCG without emission lines and $U–B > 0.4$, their color indices are all like those of normal elliptical galaxies and can be well accounted by evolutionary models with just on initial burst of star formation. The presence of high luminosity stars can account for the compactness in all the galaxies with $U–B < 0.4$ but for the others, the similarity of their colors with those of normal ellipticals but with much higher surface luminosities point to a high stellar density for the origin of their compactness.

157.214   The magnetic field in M51.
R. Beck, U. Klein, R. Wielebinski.
Astron. Astrophys., Vol. 186, No. 1/2, p. 95 – 98 (1987).

Linearly polarized radio continuum emission at 6.3 cm wavelength has been observed in M51 with the Effelsberg 100–m telescope. The combination of these data with the $\lambda 21.2$ cm Westerbork observations yields the distribution of rotation measures. The mean value is $RM = 13 \pm 6\, \text{rad/m}^2$, without significant variations with azimuthal angle in the plane of M51. The magnetic field lines follow the optical spiral arms. As the present data do not give the direction of the magnetic field, the question whether the field structure is axisymmetric or bisymmetric cannot be answered. Recently published optical polarization data agree with the radio data within the eastern and southern quadrants of M51. In these regions the same magnetic field that gives rise to the radio synchrotron emission also aligns the dust grains. In the western quadrant of the galaxy the field lines as derived from radio and optical data seem to diverge by up to 60°.

157.215   A deep photographic search for Wolf–Rayet stars in M33.
P. Massey, P. S. Conti, A. F. J. Moffat, M. M. Shara.
Publ. Astron. Soc. Pac., Vol. 99, No. 618, p. 816 – 831 (1987).

The authors report on their blink survey for Wolf–Rayet stars in M33 using interference filter photography and provide finding charts for the 80 spectroscopically confirmed W–R stars found from this survey and the 35 good survey candidates that are still to be observed.

157.216   Near–infrared observations of Cepheids: the distance to NGC 300.
B. F. Madore, D. L. Welch, C. W. McAlary, R. A. McLaren.
Astrophys. J., Vol. 320, No. 1, p. 26 – 31 (1987).

Near–infrared H–band observations of two long–period Cepheids in the southern spiral galaxy NGC 300 are presented. A formal near–infrared H–band distance modulus of 26.35 ± 0.25 mag is derived (for a true LMC distance modulus of 18.55 mag), corresponding to a distance of 1.9 Mpc. An appraisal of the previously available optical data on the NGC 300 Cepheids is given, and various uncertainties in the fitting procedure to obtain distances are discussed. The present uncertainty in the true modulus of NGC 300 is probably as high as $\pm$ 0.35 mag. Finally, astrometric positions are given for all the known Cepheids in NGC 300.

157.217   Steady state cooling flow models for normal elliptical galaxies.
C. L. Sarazin, R. E. White III.
Astrophys. J., Vol. 320, No. 1, p. 32 – 48 (1987).

Recent X–ray observations show that normal elliptical galaxies contain large quantities of hot gas. The authors present spherically symmetric, steady state inflow models for hot gas in elliptical galaxies, which is assumed to be from stellar mass loss within the galaxy. They calculate the X–ray luminosities, spectra, and surface brightness profiles of a grid of cooling flow models for galaxies spanning a range of optical luminosities. Galaxy models with and without dark halos and with various supernova

and stellar mass–loss rates are considered. In all of the models, the gas density varies approximately as $\varrho \sim r^{-3/2}$.

157.218   Global properties of interacting disk–type galaxies.
H. A. Bushouse.
Astrophys. J., Vol. 320, No. 1, p. 49 – 72 (1987). With plates 1 – 2.

Optical, far–infrared, and radio observations of global properties for a sample of strongly interacting disk–type galaxies are presented. The data consist of narrow–band Hα digital images, 22″ aperture spectrophotometry in the region 3500 – 7200 Å, IRAS far–IR measurements, and 21 cm H I observations. Star formation rates (SFR) are estimated both from Hα and far–IR luminosities. The interacting galaxies have anomalously high IR/Hα ratios, which leads to IR–derived global SFRs that are, on average, a factor of 6 higher than in isolated spiral galaxies. It is shown that the majority of interaction–induced star–formation activity is concentrated near the nuclei of the galaxies. There is no correlation between global SFR and total H I gas content. Gas–depletion time scales are 1.5 – 3 times lower than for isolated spirals.

157.219   The Leo Triplet spiral galaxy NGC 3628. II. VLA observations of the hydroxyl absorption.
J. T. Schmelz, W. A. Baan, A. D. Haschick.
Astrophys. J., Vol. 320, No. 1, p. 145 – 153 (1987).

The complex hydroxyl absorption in NGC 3628 was observed with the VLA–A with an angular resolution of about 1″5 and a spectral resolution of 195 kHz. NGC 3628 is a peculiar, edge–on Sbc galaxy. It is a member of the Leo Triplet and has an extended, nuclear radio continuum. Eight optically thin features were seen at 1667 MHz; all were also strong enough to be detected at the weaker 1665 MHz transition. The line ratios indicate that the level populations are governed by LTE conditions. Several of these features form part of a rotating molecular disk with a radius of $R = 168$ pc. A second structure seems to be expanding away from this disk at a rate of $\sim 135\, \text{km s}^{-1}$.

157.220   The IRAS bright galaxy sample. II. The sample and luminosity function.
B. T. Soifer, D. B. Sanders, B. F. Madore, G. Neugebauer, G. E. Danielson, J. H. Elias, C. J. Lonsdale, W. L. Rice.
Astrophys. J., Vol. 320, No. 1, p. 238 – 257 (1987).

A complete sample of 324 extragalactic objects with 60 μm flux densities greater than 5.4 Jy has been selected from the IRAS catalogs. Only one of these objects can be classified as a Seyfert nucleus; the others are all galaxies. It is found that far–infrared emission is a significant luminosity component in the local universe, representing 25% of the luminosity emitted by stars in the same volume. Above $10^{11} L_\odot$ the infrared luminous galaxies are the dominant population of objects in the universe, being as numerous as the Seyfert galaxies, and more numerous than quasars at higher luminosities. Approximately 60% – 80% of the far–infrared luminosity of the local universe can be attributed, directly or indirectly, to recent or ongoing star formation.

157.221   Wolf–Rayet stars in the Andromeda galaxy.
A. F. J. Moffat, M. M. Shara.
Astrophys. J., Vol. 320, No. 1, p. 266 – 282 (1987).

The authors have completed a survey of M31 for strong–line Wolf–Rayet (W–R) stars, confirming the trends found previously, that (1) M31 is at present about an order of magnitude less active in star formation than the Galaxy, as reflected in the total number of W–R stars; (2) the number ratio of late to early WC stars, WCL/WCE, varies systematically with galactocentric radius as in the Galaxy; (3) most W–R stars lie in the prominent ring of active star formation at $R = 7 – 12$ kpc form the center of M31.

157.222 Investigations of extragalactic hydroxyl.
J. T. Schmelz.
Diss. Abstr. Int., Sect. B, Vol. 48, No. 4, p. 1072–B (1987). Thesis, The Pennsylvania State University, 204 pp. (1987). Order No. DA8714873.

157.223 Dark matter in the dwarf galaxy DDO 125.
K. Ebneter, M. Davis, N. Jeske, M. Stevens.
Bull. Am. Astron. Soc., Vol. 19, No. 2, p. 681 (1987). Abstract. – See Abstr. 010.061.

157.224 High–resolution radio observations of the bubbles and jets in the center of M51.
P. C. Crane, J. M. van der Hulst, H. C. Ford, D. G. Lawrie, G. H. Jacoby.
Bull. Am. Astron. Soc., Vol. 19, No. 2, p. 681 (1987). Abstract. – See Abstr. 010.061.

157.225 Are peanut–shaped galaxies and polar–ring galaxies related?
M. Bell, B. C. Whitmore, P. Quinn.
Bull. Am. Astron. Soc., Vol. 19, No. 2, p. 681 (1987). Abstract. – See Abstr. 010.061.

157.226 Optical and millimeter–wave studies of NGC 2146.
S. G. Kleinmann, J. S. Young, M. J. Claussen, V. C. Rubin, N. Z. Scoville.
Bull. Am. Astron. Soc., Vol. 19, No. 2, p. 681 (1987). Abstract. – See Abstr. 010.061.

157.227 Radio continuum observations of the spiral galaxy NGC 4736.
N. Duric, E. R. Seaquist.
Bull. Am. Astron. Soc., Vol. 19, No. 2, p. 682 (1987). Abstract. – See Abstr. 010.061.

157.228 Dynamical models of oblate elliptical galaxies with de Vaucouleurs' $r^{1/4}$ law profiles.
H. Levison, J. Fillmore.
Bull. Am. Astron. Soc., Vol. 19, No. 2, p. 682 (1987). Abstract. – See Abstr. 010.061.

157.229 Metallicity gradients in elliptical galaxies.
W. A. Baum, B. Thomsen.
Bull. Am. Astron. Soc., Vol. 19, No. 2, p. 682 (1987). Abstract. – See Abstr. 010.061.

157.230 Are elliptical galaxies really round?
J. L. White, C. R. Canizares.
Bull. Am. Astron. Soc., Vol. 19, No. 2, p. 682 (1987). Abstract. – See Abstr. 010.061.

157.231 Molecular gas in "gas–free" galaxies: the S0s.
H. A. Thronson, L. Tacconi, M. A. Greenhouse, J. Kenney, L. Tacconi–Garman, J. Young.
Bull. Am. Astron. Soc., Vol. 19, No. 2, p. 683 (1987). Abstract. – See Abstr. 010.061.

157.232 The brightest "stars" are not always stars.
R. M. Humphreys, M. Aaronson.
Bull. Am. Astron. Soc., Vol. 19, No. 2, p. 683 (1987). Abstract. – See Abstr. 010.061.

157.233 Settling of inclined gas disks in galaxies with flat rotation curves: a self–similar solution.
T. Y. Steiman–Cameron, R. H. Durisen.
Bull. Am. Astron. Soc., Vol. 19, No. 2, p. 683 (1987). Abstract. – See Abstr. 010.061.

157.234 Maffei 1 at two microns.
S. F. Mason, J. S. Price.
Bull. Am. Astron. Soc., Vol. 19, No. 2, p. 683 (1987). Abstract. – See Abstr. 010.061.

157.235 Stars in Baade's field IV of M31.
P. W. Hodge, M. G. Lee.
Bull. Am. Astron. Soc., Vol. 19, No. 2, p. 683 (1987). Abstract. – See Abstr. 010.061.

157.236 Supernovae blast waves in proto–dwarf galaxies.
A. Noriega–Crespo, P. Bodenheimer.
Bull. Am. Astron. Soc., Vol. 19, No. 2, p. 683 (1987). Abstract. – See Abstr. 010.061.

157.237 Surface brightnesses from standard stars with wide–field cameras.
S. Speck, P. D. Usher, D. Klinglesmith III, M. B. Niedner Jr.
Bull. Am. Astron. Soc., Vol. 19, No. 2, p. 684 (1987). Abstract. – See Abstr. 010.061.

157.238 The largest rotation curve ever derived: H I rotation curve of DDO 154.
C. Carignan, K. C. Freeman.
Bull. Am. Astron. Soc., Vol. 19, No. 2, p. 684 (1987). Abstract. – See Abstr. 010.061.

157.239 Motions of the ionized gas in M83.
C. J. Peterson, M. A. Harper.
Bull. Am. Astron. Soc., Vol. 19, No. 2, p. 684 (1987). Abstract. – See Abstr. 010.061.

157.240 A comparison of mass distributions in cluster and field spirals.
D. A. Forbes, B. C. Whitmore.
Bull. Am. Astron. Soc., Vol. 19, No. 2, p. 685 (1987). Abstract. – See Abstr. 010.061.

157.241 Detached supernovae from undetected dwarf galaxies: expected rates from star formation models.
N. D. Tyson.
Bull. Am. Astron. Soc., Vol. 19, No. 2, p. 686 (1987). Abstract. – See Abstr. 010.061.

157.242 The galaxy two–point angular correlation function in the CTI strip.
M. E. Cornell, J. T. McGraw, D. Batuski.
Bull. Am. Astron. Soc., Vol. 19, No. 2, p. 687 – 688 (1987). Abstract. – See Abstr. 010.061.

157.243 Icebergs and crouching giants.
M. Disney, S. Phillipps.
Nature, Vol. 329, No. 6136, p. 203 – 204 (1987).
The recent serendipitous discovery that an apparently insignificant dwarf Malin 1, is in fact a huge, low–surface–brightness galaxy in the background has demonstrated that new populations of galaxies remain to be detected. Our present knowledge of the galaxy population is so biased by a single insidious selection effect that it is entirely possible that Malin 1 is just the first example of a class of such low–surface–brightness giant galaxies that forms a significant constituent of the Universe.

157.244 The 158 μm [C II] line emission from galaxies: the global significance of [C II] emission.
G. J. Stacey, J. B. Lugten, R. Genzel, C. H. Townes.
Bull. Am. Astron. Soc., Vol. 19, No. 2, p. 711 (1987). Abstract. – See Abstr. 010.061.

157.245 IRAS observations of interacting galaxies.
S. A. Lamb, H. A. Bushouse, M. W. Werner, B. F. Smith.
Bull. Am. Astron. Soc., Vol. 19, No. 2, p. 712 (1987). Abstract. – See Abstr. 010.061.

157.246 Migratory star formation in the dwarf irregular galaxies, DDO 53 and GR 8.
R. Ruotsalainen.
Bull. Am. Astron. Soc., Vol. 19, No. 2, p. 712 (1987). Abstract. – See Abstr. 010.061.

157.247   Narrow–band Hα imaging of early–type galaxies.
P. W. Vedder, C. R. Canizares, P. L. Blizzard.
Bull. Am. Astron. Soc., Vol. 19, No. 2, p. 712 (1987). Abstract. – See Abstr. 010.061.

157.248   The distance to M81 derived from planetary nebulae.
G. Jacoby, H. Ford, J. Booth, R. Ciardullo.
Bull. Am. Astron. Soc., Vol. 19, No. 2, p. 712 – 713 (1987). Abstract. – See Abstr. 010.061.

157.249   The amount of dark matter in spiral galaxies in clusters.
B. C. Whitmore, V. C. Rubin, D. A. Forbes.
Bull. Am. Astron. Soc., Vol. 19, No. 2, p. 713 (1987). Abstract. – See Abstr. 010.061.

157.250   Supermassive black holes in galaxy nuclei.
J. Kormendy.
Bull. Am. Astron. Soc., Vol. 19, No. 2, p. 713 (1987). Abstract. – See Abstr. 010.061.

157.251   A new ionization model for galaxies with dominant star formation regions.
R. I. Thompson.
Bull. Am. Astron. Soc., Vol. 19, No. 2, p. 717 (1987). Abstract. – See Abstr. 010.061.

157.252   Extremely low oxygen abundances in the dwarf irregular galaxies GR 8 and Sextans A.
E. D. Skillman.
Bull. Am. Astron. Soc., Vol. 19, No. 2, p. 717 – 718 (1987). Abstract. – See Abstr. 010.061.

157.253   The nitrogen–to–oxygen ratio in the amorphous galaxy NGC 5253.
J.–R. Roy, J. R. Walsh.
Bull. Am. Astron. Soc., Vol. 19, No. 2, p. 718 (1987). Abstract. – See Abstr. 010.061.

157.254   Radio structure of the nuclei of "edge–on" spiral and Seyfert galaxies.
A. Wehrle.
Bull. Am. Astron. Soc., Vol. 19, No. 2, p. 718 (1987). Abstract. – See Abstr. 010.061.

157.255   Radio emission of spiral galaxies in groups of galaxies.
V. G. Malumyan.
Astrophysics, Vol. 26, No. 2, p. 190 – 195 (1987). English translation of 43.157.174.

157.256   Investigation of the anisotropy in the orientations of galaxies of the Uppsala and ESO/Uppsala catalogs.
A. V. Mandzhos, A. Ya. Gregul', I. Yu. Izotova,
V. V. Tel'nyuk–Adamchuk.
Astrophysics, Vol. 26, No. 2, p. 196 – 203 (1987). English translation of 43.157.175.

157.257   Individual and orbital masses of double galaxies.
V. A. Mineva.
Astrophysics, Vol. 26, No. 2, p. 203 – 213 (1987). English translation of 43.157.176.

157.258   The surface brightness of 1550 galaxies in Fornax: automated galaxy surface photometry – II.
S. Phillipps, M. J. Disney, E. J. Kibblewhite,
M. G. M. Cawson.
Mon. Not. R. Astron. Soc., Vol. 229, No. 3, p. 505 – 515 (1987).
    The authors present a survey of a complete sample of galaxies in the region of the Fornax cluster. Measurements with the APM machine are used to derive the observed distribution of galaxy surface brightness for 1550 objects. Corrections for surface brightness dependent selection effects are then made in order to estimate the true distribution.

157.259   CCD surface photometry of the bright elliptical galaxies NGC 720, NGC 1052, and NGC 4697.
R. I. Jedrzejewski, R. L. Davies, G. D. Illingworth.
Astron. J., Vol. 94, No. 6, p. 1508 – 1518 (1987). With plate 119.
    The authors have analyzed R band CCD frames of three elliptical galaxies of similar luminosity and flattening under fairly good seeing conditions and determined the radial variation of surface brightness, ellipticity, position angle, and $\cos(4\theta)$ component, which measures the degree to which the isophotes may be 'boxy' or 'pointed'. They find that although the galaxies have similar global properties, such as luminosity and flattening, the detailed structure, namely core size, isophote twists, and $\cos(4\theta)$ component, differs markedly. They find evidence for a weak disk in NGC 4697, but argue that this does not contribute significantly to the relatively rapid major–axis rotation seen in this object.

157.260   Two–dimensional photographic and CCD photometry of the S0 galaxy NGC 3115.
M. Capaccioli, E. V. Held, J.–L. Nieto.
Astron. J., Vol. 94, No. 6, p. 1519 – 1537 (1987). With plates 120 – 121.
    The authors present a detailed photometric mapping of the edge–on S0 galaxy NGC 3115, based on photographic and CCD images in B light. Comparison with previous photographic and photoelectric studies shows how poor the state–of–the–art of galaxy photometry still is, even in the case of well–known objects. The final light profiles along the principal axes span the interval $15.6 < \mu < 29.4$ B mag arcsec^{-2}, with an overall accuracy better than 0.1 mag between 16.7 and 26 B mag arcsec^{-2}. The galaxian halo is detected to $a > 20'$ ($=58$ kpc, if the distance $\varDelta = 10$ Mpc). The total magnitude, $B_T = 9.75$, and other photometric and geometric standard parameters are derived.

157.261   The spectra of extragalactic Wolf–Rayet stars.
P. Massey, P. S. Conti, T. E. Armandroff.
Astron. J., Vol. 94, No. 6, p. 1538 – 1555 (1987).
    The authors are studying Wolf–Rayet (WR) stars in nearby galaxies in order to measure the massive–star content of these systems and to see what differences might exist in the evolution and properties of their most massive stars. WR candidates have previously been identified in M31, M33, NGC 6822, and IC 1613, and in this paper the authors discuss spectrophotometry of 93 WR stars in these Local Group galaxies, including 27 newly confirmed ones.

157.262   A search for RR Lyrae variables in NGC 147.
A. Saha, J. G. Hoessel.
Astron. J., Vol. 94, No. 6, p. 1556 – 1563 (1987). With plates 122 – 123.
    Repeated deep CCD observations of a field in the Local Group dwarf elliptical galaxy NGC 147 have been compared, and 34 variable stars have been found. Light curves of 13 of these have been obtained. Periods and shapes indicate that nine of the 13 are RR Lyrae stars. The intensity–averaged phase–weighted–mean g magnitude is 25.15 ($+0.17$ or -0.15), which implies, after correction for extinction, that the distance modulus to NGC 147 is 23.85 ($+0.22$ or -0.20), or 589 kpc, as compared to Pritchet and van den Bergh's (1987) distance of 738 kpc for M31.

157.263   The global luminosity function of the Fornax dwarf elliptical galaxy.
P. B. Eskridge.
Astron. J., Vol. 94, No. 6, p. 1564 – 1577 (1987).
    The global luminosity function for the Fornax dwarf elliptical galaxy has been determined from UK Schmidt plates in B and V. The functions do not extend below the horizontal–branch (HB) peak, and thus provide information on only the bright giants. Both functions show evidence of structure in the giant branch. Comparisons are made with luminosity functions (LFs) of globular clusters covering a range in metallicity similar to that found in the Fornax system. These comparisons indicate the old population of Fornax is similar to that of such moderately metal–poor clusters as M5 and M13. They also demonstrate vividly the existence of a significant excess population of very bright stars

over that seen in pure old systems. Comparisons are also made with theoretical luminosity functions of various ages and helium abundances.

157.264 A deep redshift survey of field galaxies.
D. C. Koo, R. G. Kron.
Space Telesc. Sci. Inst., Prepr. Ser., No. 204, p. 2 – 5 (1987). To appear in the proceedings of the conference "Towards understanding galaxies at large redshifts", Reidel, Dordrecht, The Netherlands.

157.265 The origin of the far–infrared flux from spiral galaxies.
S. K. Leggett, P. W. J. L. Brand, C. M. Mountain.
Edinb. Astron. Prepr., No. 12/87, 15 pp. (1987). To appear in Mon. Not. R. Astron. Soc.

157.266 The nature of "box" and "peanut" shaped galactic bulges.
M. A. Shaw.
Edinb. Astron. Prepr., No. 13/87, 37 pp. (1987). To appear in Mon. Not. R. Astron. Soc.

157.267 The evolution of galactic centers. II. Normal and compact galaxies.
W. J. Duschl.
MPA Rep., No. 308, 28 pp. (1987). Submitted to Astron. Astrophys.

157.268 Chemical evolution of elliptical galaxies.
F. Matteucci, A. Tornambè.
ESO Sci. Prepr., No. 515, 43 pp. (1987). To appear in Astron. Astrophys.

157.269 Iron abundance evolution in spiral and elliptical galaxies.
F. Matteucci.
ESO Sci. Prepr., No. 550, 13 pp. (1987).

157.270 How frequent are tidal interactions between galaxies?
J. W. Fried.
Max–Planck–Inst. Astron. Heidelb., Prepr., 14 pp. (1987). Submitted to Astron. Astrophys.

157.271 The structure and evolution of Hoag's object.
F. Schweizer, W. K. Ford Jr., R. Jedrzejewski, R. Giovanelli.
Astrophys. J., Vol. 320, No. 2, p. 454 – 463 (1987). With plates 11 – 13.
The authors present new imaging, photometric, and spectroscopic observations of Hoag's object, a 16th magnitude galaxy consisting of an almost perfectly round core surrounded by a faint, apparently detached ring. The core of this distant system appears to be a normal spheroid with a half–light radius of 3.6 kpc, and an apparent rotation at $r = 2.5$ kpc of $V_{rot} \sin i = 18$ km s^{-1}. The ring is of comparable luminosity, has a mean radius of 23 kpc, is inclined about $19° \pm 5°$ to the plane of the sky, and shows knotty structure and gaseous emission lines indicative of young stars. The ring rotates with $V_{max} = 300(+100,-60)$ km s^{-1}. The authors rule out several earlier hypotheses on the origin of the ring and propose the new hypothesis that Hoag's object owes its structure to a major accretion event at least 2–3 Gyr ago.

157.272 A search for primeval galaxies using a narrow–band imaging technique.
C. J. Pritchet, F. D. A. Hartwick.
Astrophys. J., Vol. 320, No. 2, p. 464 – 467 (1987).
The authors have searched for faint emission–line objects by using a narrow–band imaging technique at wavelengths of $6000 \lesssim \lambda \lesssim 8000$ Å. These observations are used to place constraints on the number density of Lyα emission objects at redshifts $4 \lesssim z \lesssim 6$: either the surface density of primeval galaxies at $z = 5$ with strong Lyα is less than ~ 0.1 arcmin^{-2}, or such objects possess R magnitudes fainter than $+27.5$.

157.273 The formation of the exponential disk in spiral galaxies.
D. N. C. Lin, J. E. Pringle.
Astrophys. J., Lett. Ed., Vol. 320, No. 2, p. L87 – L91 (1987). = Lick Obs. Bull., No. 1080.
The authors propose a mechanism for the formation of the exponential stellar disk distribution in spiral galaxies. It is assumed that the gas which falls in to form the disk behaves like an accretion disk which evolves on a viscous time scale t_v and forms stars on a time scale t_*. The authors demonstrate that if $t_v \approx t_*$ the resulting stellar disk has an exponential distribution independent of the disk rotation law and of the assumed viscosity prescription. They comment briefly on why the viscous and star formation time scales might be related.

157.274 Observable properties of E0 triaxial galaxies: a test for triaxiality.
H. F. Levison.
Astrophys. J., Lett. Ed., Vol. 320, No. 2, p. L93 – L97 (1987).
It is shown that most triaxial galaxies oriented in space to appear as E0 galaxies should have observable rotation velocities. By projecting Levison and Richstone's triaxial models so that they look like E0's, it was found that v/σ can be as large as 1.0. In a separate argument, the observed distribution of axial ratios of ellipticals was used to show that between 3% and 23% of E0 galaxies are intrinsically flattened systems. Thus, if elliptical galaxies are triaxial then some E0 galaxies should have observable rotation velocities.

157.275 Searching at 21 centimeters for galaxies behind the Milky Way.
F. J. Kerr, P. A. Henning.
Astrophys. J., Lett. Ed., Vol. 320, No. 2, p. L99 – L103 (1987).
Known galaxies are rare in the quarter of the sky which is obscured by the Galactic disk. The authors have carried out a pilot study at 21 cm covering about 1900 arbitrarily chosen points in the zone of avoidance and have detected 16 new galaxies, some of whose line profiles have the signature of spirals. For comparison, a similar study was carried out for a blindly chosen set of points in the "clear" region at higher latitudes. The authors discuss the requirements for a fully sampled study of the whole zone of avoidance.

157.276 A model of spectrophotometric evolution for high–redshift galaxies.
B. Guiderdoni, B. Rocca–Volmerange.
Inst. Astrophys. Paris, Pré–Publ., No. 192, 44 pp. (1987). To appear in Astron. Astrophys.

157.277 Surface photometry of dwarf elliptical galaxies.
S.-i. Ichikawa.
Ann. Tokyo Astron. Obs., Second Ser., Vol. 21, No. 4, p. 437 – 484 (1987).
Photographic surface photometry is carried out for 204 galaxies in the central region and on the outskirts of the Virgo cluster and in two comparison fields outside of the Virgo cluster on the basis of the seven plates in the B band taken with the 2.5 m du Pont telescope at the Las Campanas Observatory. The homogeneous quantitative data are presented in tabular and graphical forms, which can be used as a data base for the study of the nature of dwarf elliptical galaxies.

157.278 2 μm spectroscopy of nearby galaxies and evidence for a late phase starburst in NGC 4736.
C. E. Walker, G. H. Rieke.
Prepr. Steward Obs., No. 740, 21 pp. (1987). To appear in Astrophys. J.

157.279 Dark matter in dwarf galaxies.
M. Aaronson, E. W. Olszewski.
Prepr. Steward Obs., No. 746, 12 pp. (1987). To appear in IAU Symp. No. 130.

157.280 Observations of molecular and atomic clouds in M31.
C. J. Lada, M. Margulis, Y. Sofue, N. Nakai, T. Handa.
Prepr. Steward Obs., No. 762, 41 pp. (1987). To appear in Astrophys. J.

157.281 $I_{CO}/N(H_2)$ conversions and molecular gas abundances in spiral an irregular galaxies.
P. Maloney, J. H. Black.
Prepr. Steward Obs., No. 763, 48 pp. (1987). = Ariz. Theor. Astrophys. Prepr., #87–32. To appear in Astrophys. J.

157.282 Diameters of structures of field galaxies.
N. N. Pavlova.
Tr. Astrofiz. Inst. Alma–Ata, Tom 48, p. 84 – 98 (1987). In Russian.

157.283 Mean surface brightnesses of field galaxies.
N. N. Pavlova.
Tr. Astrofiz. Inst. Alma–Ata, Tom 48, p. 99 – 108 (1987). In Russian.

157.284 On the luminosity of Hubble–Sandage objects in the M31 galaxy.
G. R. Ivanov, P. Z. Kunchev.
Dokl. Bolg. Akad. Nauk, Tome 40, No. 1, p. 11 – 12 (1987).

157.285 NGC 2403: a flocculent galaxy with two principal centres of star formation.
J. Beckman, J. Cepa, M. Prieto, C. Muñoz Tuñón.
Rev. Mex. Astron. Astrofis., Vol. 14, No. 1, p. 134 – 143 (1987). – See Abstr. 012.042.
The authors have mapped the nearby flocculent spiral galaxy in the visible U, B and V bands, as well as in the near infrared J, H and K bands, with a linear resolution of 900 pc. The galaxy is found to have two principal current centres of large–scale star formation, signposted by centres of ultraviolet and blue flux. One is in the nucleus, defined by the geometrical centre of the H I emission, and the other at some 1.5 kpc radial distance away. The outer star–forming region is the more intense and the younger of the two, and corresponds to a local peak in the H I surface density. The authors use the colours of the star–forming regions and of the integrated galaxy to make a first order estimate of the stellar population distribution.

157.286 NGC 6946: kinematics of the starburst in the circum–nuclear zone.
C. Muñoz Tuñón, J. Beckman, M. Prieto.
Rev. Mex. Astron. Astrofis., Vol. 14, No. 1, p. 144 – 148 (1987). – See Abstr. 012.042.
Using spectroscopic observations at relatively high resolution ($\Delta\lambda \sim 0.35$ Å) in the region within 30 arcseconds of the nucleus of NGC 6946 the authors have found clear evidence of a strong expansion, with velocities of order 100 km s^{-1}. By comparison with intensity maps in H I (21 cm) H$_2$ (CO 2.6 mm) and Hα taken from the literature the authors derive a picture of a starburst which contains $\gtrsim 10^4$ O V stars. The gas in the most neutral region has been largely used in star formation, while a placental molecular cloud remains around the starburst. The authors propose a mechanism for maintaining the starburst, and using the molecular mass for progressive star formation.

157.287 Nuclear stellar population of NGC 772 (Arp 78).
A. A. Schmidt, H. A. Dottori.
Rev. Mex. Astron. Astrofis., Vol. 14, No. 1, p. 149 – 155 (1987). – See Abstr. 012.042.
The nuclear spectrum ($5'' \times 13''$) of the Sb I Arp's galaxy NGC 772 was used to synthesize the stellar population through the analysis of the absorption features and the continuum. The results show that the flux around 5360 Å is dominated by G and K stars. The last and weak star formation cycle was found with turn–off–point at A0 V. The possible association between two nearby high redshift objects and nuclear phenomena is discussed.

157.288 Internal extinction in spiral galaxies. Inclination dependence.
G. Magris C., G. Bruzual A.
Rev. Mex. Astron. Astrofis., Vol. 14, No. 1, p. 156 (1987). Abstract. – See Abstr. 012.042.

157.289 Near infrared maps of elliptical and S0 galaxies detected by IRAS.
I. Cruz–González, E. Recillas–Cruz, L. Carrasco.
Rev. Mex. Astron. Astrofis., Vol. 14, No. 1, p. 157 (1987). Abstract. – See Abstr. 012.042.

157.290 Giant H II regions as distance indicators.
J. Melnick, R. Terlevich, M. Moles.
Rev. Mex. Astron. Astrofis., Vol. 14, No. 1, p. 158 – 164 (1987). – See Abstr. 012.042.
The correlations between the integrated Hβ luminosities, the velocity widths of the nebular lines and the metallicities of giant H II regions and H II galaxies are demonstrated to provide powerful distance indicators. They are calibrated on a homogeneous sample of giant H II regions with well determined distances and applied to distant H II galaxies to obtain a value of $H_0 = 95 \pm 10$ for the Hubble parameter, consistent with the value obtained by the Tully–Fisher technique. The effect of Malmquist bias and other systematic effects on the H II region method are discussed in detail.

157.291 Physical conditions of H II regions in the spiral galaxy M101 and the pregalactic helium abundance.
M. Peimbert, S. Torres–Peimbert, J. Fierro.
Rev. Mex. Astron. Astrofis., Vol. 14, No. 1, p. 165 (1987). Abstract. – See Abstr. 012.042.

157.292 Chemical evolution of galaxies.
M. Peimbert.
Rev. Mex. Astron. Astrofis., Vol. 14, No. 1, p. 166 (1987). Abstract. – See Abstr. 012.042.

157.293 Small scale structure of polarized emission in M31.
N. Loiseau, E. Hummel, R. Beck, R. Wielebinski.
Rev. Mex. Astron. Astrofis., Vol. 14, No. 1, p. 167 (1987). Abstract. – See Abstr. 012.042.

157.294 Is there any evidence of evolution in the color distribution of galaxies from J = 13 to J = 24 magnitude?
G. Bruzual.
Rev. Mex. Astron. Astrofis., Vol. 14, No. 1, p. 169 (1987). Abstract. – See Abstr. 012.042.

157.295 On the relation between radius, luminosity, surface brightness and galaxian surface density in E galaxies.
E. Recillas–Cruz, A. Serrano P. G.
Rev. Mex. Astron. Astrofis., Vol. 14, No. 1, p. 170 (1987). Abstract. – See Abstr. 012.042.

157.296 Star formation in blue compact galaxies.
C. Firmani, G. F. Bisiacchi.
Rev. Mex. Astron. Astrofis., Vol. 14, No. 1, p. 171 (1987). Abstract. – See Abstr. 012.042.

157.297 The giants of Fornax: LPV's, carbon and M stars.
S. Demers, W. E. Kunkel, M. J. Irwin.
ESO Conf. Workshop Proc., No. 27, p. 275 – 280 (1987). – See Abstr. 012.051.
Fornax is the most massive of the dwarf spheroidal galaxies surrounding the Milky Way. At a distance of some 140 kpc it is, at best located in the outer halo. Presumably because of its large mass, Fornax was able to retain some interstellar gas to lengthen its period of star formation. An intermediate–age population is revealed by the presence of carbon stars found in its quite extended giant branch.

157.298   C–M diagram of the dwarf irregular galaxy UKS 2323.
M. Capaccioli, S. Ortolani, G. Piotto.
ESO Conf. Workshop Proc., No. 27, p. 281 – 284 (1987). – See Abstr. 012.051.

The authors present a CCD photometry of the resolved stars of the dwarf irregular galaxy UKS 2323, in the B and V bands. The C–M diagram indicates that the dominating bright population consists of blue stars, which suggests recent formation activity. The possible presence of an older population is also argued. From the integrated photometry they derive a distance modulus of $(m-M) = 26 \pm 1$.

157.299   Dark matter in dwarf spheroidal galaxies: new particles or stellar remnants?
J. Melnick.
ESO Conf. Workshop Proc., No. 27, p. 589 – 598 (1987). – See Abstr. 012.051.

The high mass to light ratios observed in the smallest, most metal deficient dwarf spheroidal galaxies can be explained if the IMF of high local star formation rate systems (starbursts) depends on chemical composition. Simple single–burst models are presented that reproduce the correlations between M/L ratio and abundance that are observed for elliptical galaxies and for dwarf–spheroidals, and the lack of correlation observed for globular clusters. The models predict that the fraction of massive stellar remnants in globular clusters should correlate with metallicity. The missing mass in a variety of systems ranging from spiral galaxies to galaxy clusters can probably be explained using similar models thus eliminating the need for non–baryonic invisible matter at these scales.

157.300   On the nature of dark matter in the dwarf spheroidal satellites of the Galaxy.
H. Zinnecker.
ESO Conf. Workshop Proc., No. 27, p. 599 – 605 (1987). – See Abstr. 012.051.

It is proposed that the dark matter in Dwarf Spheroidal galaxies in the Local Group resides in very low mass stars. A simple theoretical argument is presented why the IMF of low mass stars ought to be steeper for a metal–poor system. If the Galaxy forms its halo by swallowing dwarf galaxies, the dark matter in this halo might also be hidden in very low mass stars of Pop II composition.

157.301   Formation and chemical evolution of dwarf galaxies.
J. Silk.
ESO Conf. Workshop Proc., No. 27, p. 653 – 663 (1987). – See Abstr. 012.051.

A new theory of dwarf galaxies is outlined in which dE galaxies, including dwarf spheroidals, form prolifically at $z \sim 5$, but are stripped by supernova–driven winds of most of their gas before star formation is completed. The resulting dE's are metal–poor and of low surface brightness. Within the past 10^9years, the gas–rich dwarfs are observed as apparently young dIrrs, characterized by an underlying faint old stellar population, a large dispersion in chemical abundances, and by a metallicity threshold of about one percent of solar. Large galaxies also trap the enriched intergalactic gas, which provides a preenriched but chemically inhomogeneous source of halo gas.

157.302   Globular cluster systems: comparative evolution of galactic halos.
W. E. Harris.
ESO Conf. Workshop Proc., No. 27, p. 683 – 695 (1987). – See Abstr. 012.051.

Globular cluster systems are found in most, if not all, major galaxies and they provide a unique way to compare galactic halos and trace their very earliest stages of formation. In several important respects, these subsystems of the halo resemble each other from one galaxy to another far more closely than their parent galaxies do; therefore, there is a reasonable basis for supposing that they represent some kind of underlying unity in the protogalaxy formation process. Several of these unifying themes are described.

157.303   The detection of extragalactic methanol.
C. Henkel, T. Jacq, R. Mauersberger, K. M. Menten, H. Steppe.
Astron. Astrophys., Vol. 188, No. 1, p. L1 – L4 (1987).

The detection of emission in the 96 GHz 2_K–1_K lines of methanol is reported toward the central regions of NGC 253 and IC 342. A possible detection is also obtained toward NGC 6946, while no emission is seen toward M82. $(CH_3OH)/(H_2)$ abundance ratios appear to be consistent with those determined for galactic sources. The strength of the CH_3OH emission, however, is not found to be correlated with infrared or CO luminosities.

157.304   The stellar content and morphology of the dwarf irregular galaxy Holmberg IX.
U. Hopp, R. E. Schulte–Ladbeck.
Astron. Astrophys., Vol. 188, No. 1, p. 5 – 12 (1987). See Abstr. 43.157.227.

The authors present CCD images in B, R, I, H_α and long–slit spectroscopy of Ho IX, a dwarf irregular galaxy situated close to M81. Photometry of the resolved stars to magnitude limits of $B = 23^m8$, $R = 23^m8$, and $I = 23^m0$ is used to construct colour–magnitude and colour–colour diagrams. The authors also discuss the photometry of the background light distribution of the subthreshold stars. A revised distance modulus of $m-M = 30^m$ is estimated. The blue, high–mass stars are concentrated towards the center of the galaxy and in at least three OB associations. The blue–to–red supergiant ratio indicates a more or less constant star formation rate within the last 10^7yr. It is proposed that Ho IX has just entered a quiescent phase of its star formation due to lack of interstellar material.

157.305   Spheroidal systems as a one–parameter family of mass at their birth.
Y. Yoshii, N. Arimoto.
Astron. Astrophys., Vol. 188, No. 1, p. 13 – 23 (1987).

A supernova–driven wind model for spheroidal systems is constructed using an evolutionary method of population synthesis. An assumed relation between binding energy and mass $\Omega_G \propto M_G^\eta (\eta = 1.45)$ for protoclouds gives a star formation rate (SFR) per unit mass which is proportional to a negative power of mass. With such a SFR and a universal initial mass function, the authors successfully reproduced the structural and chemical properties of elliptical galaxies, dwarf elliptical galaxies, and globular clusters.

157.306   Magnetic fields in spiral galaxies.
Y. Sofue.
Astron. Her., Vol. 80, No. 12, p. 344 – 348 (1987). In Japanese.

157.307   The evolution of galaxies at moderate redshift.
S. J. Lilly.
Mon. Not. R. Astron. Soc., Vol. 229, No. 4, p. 573 – 588 (1987).

Optical and infrared photometric data on 53 galaxies in five clusters at $0.38 < z < 0.58$ are described and analysed to produce the rest–frame $(U–V)$ and $(V–H)$ colours. The 36 red galaxies form a homogeneous population in each cluster. The colours of the 17 blue galaxies clearly distinguish between the normal spirals and the peculiar 'A–type' galaxies found in these Butcher–Oemler clusters and suggest for the latter a substantial intermediate age 1–Gyr population as indicated by optical spectra. It is shown that inclusion of the evolution of the upper Asymptotic Giant Branch, which is usually neglected, into a simple evolutionary model can explain the observed evolution vector in the $(U–V)/(V–H)$ plane. The observed evolution in the elliptical galaxies is sufficiently small that the implied ages of galaxies at $z \sim 0.45$, relative to their present ages, are large enough to pose problems for cosmological models with $q_0 \sim 0.5$ and/or late epochs of galaxy formation.

157.308   The nature of "box" and "peanut" shaped galactic bulges.
M. A. Shaw.
Mon. Not. R. Astron. Soc., Vol. 229, No. 4, p. 691 – 706 (1987).

The author presents the results of a survey of all large, normal edge–on spiral and lenticular galaxies in the *Second Reference*

Catalogue in an attempt to isolate those objects displaying either "box" or "peanut" shaped central bulges. A total of 20($\pm$4) percent of the sample objects display the features of interest. These box/peanuts are found to show a strong tendency towards intermediate morphological types and do not seemingly influence either the photometric or radio properties of the galaxies in which they reside to any greater or lesser degree than do "normal" bulges. The fraction of "boxy" bulges which occupy cluster or group environments is found to be identical to that fraction for the non–box/peanuts in the sample. The implications of these results for current formation theories of such systems are discussed.

157.309 A new ionization model for galaxies with dominant star–formation regions.
R. I. Thompson.
Astrophys. J., Vol. 321, No. 1, p. 153 – 155 (1987).

Recent observations of M83 by Turner, Ho, and Beck (1987) show volume emission measures derived from infrared lines that are higher than those derived from radio continuum fluxes. This paper utilizes the observed relationship between emission measure and luminosity for sources in Galactic star–formation regions along with the mass–luminosity function and the initial mass function to derive a new ionization model for galaxies or regions which are dominated by star formation. The new model is both qualitatively and quantitatively consistent with the M83 results.

157.310 Electrophotometry of narrow double galaxies.
A. N. Tomov, M. T. Todorova.
Dokl. Bolg. Akad. Nauk, Tome 40, No. 7, p. 5 – 8 (1987).

Observations of 21 double galaxies were carried out at the Crimean Astrophysical Observatory. The data listed are processed in a three–colour UBV system.

157.311 Isolated triple galaxies.
N. A. Tomov, A. N. Tomov.
Dokl. Bolg. Akad. Nauk, Tome 40, No. 5, p. 5 – 7 (1987).

Electrophotometry in a three–colour UBV system of triple galaxies is made. The authors treat some correlations between electrophotometric magnitudes and other parameters for 15 triplets.

157.312 Morphology and phase transitions: from cellular automata to galaxies.
L. S. Schulman.
Patterns, defects and microstructures in nonequilibrium systems. Applications in materials science. Proceedings of a NATO Advanced Research Workshop. Austin, TX, USA, 24 – 28 March 1986. Martinus Nijhoff, Dordrecht, The Netherlands., p. 118 – 142 (1987). Abstr. in Phys. Abstr., Vol. 90, No. 1314, Entry 120618 (1987).

157.313 Studies of kinematics of interacting galaxies. Internal motions in 10 nests of galaxies.
V. P. Arkhipova, A. V. Zasov, R. I. Noskova, O. K. Sil'chenko.
Astron. Zh., Tom 64, Vyp. 6, p. 1161 – 1175 (1987). In Russian. English translation in Sov. Astron., Vol. 31, No. 6.

Spectral observations of ten nests of interacting galaxies have been carried out in 1977 – 1985 with the image–tube UAGS spectrograph in the main focus of the 6–m telescope of the Special Astrophysical Observatory. The radial velocities of components were studied, their rotation curves, luminosities and masses are obtained. The objects VV 148, 566, 575 and possibly VV 270 may be single SBd or IBm galaxies with unusually large H II regions, about 1 kpc or larger. VV 125 was found to be an Scd spiral. The other objects have turned out to be interacting pairs or multiple systems. A strong interaction is observed in the pairs VV 273 and VV 564, and in VV 243 as well. Noncircular motions were found in the main component of VV 273 and in VV 564. VV 131 may be a nest of 3 dwarf galaxies.

157.314 The extraordinary CO luminosity of the far–infrared galaxy VII Zw 31: a possible proto–galactic disk?
L. J. Sage, P. M. Solomon.
Astrophys. J., Lett. Ed., Vol. 321, No. 2, p. L103 – L106 (1987).

The authors report observations of CO millimeter wave emission from the ultraluminous far–infrared galaxy VII Zw 31. At $z = 0.054$, this is the most distant galaxy in which CO emission has been reported. The CO luminosity of 1.1×10^{10} K km s^{-1}pc^2 is about 35 times that of the Milky Way and higher than from any other galaxy reported in the literature. The corresponding molecular hydrogen mass is about $5 \times 10^{10} M_\odot$. The ratio of far–infrared luminosity to molecular mass is similar to that in galactic giant molecular clouds with active O star formation such as W51.

157.315 Concerning the limit on the mean mass distribution of galaxies from their gravitational lens effect.
I. Kovner, M. Milgrom.
Astrophys. J., Lett. Ed., Vol. 321, No. 2, p. L113 – L115 (1987).

Tyson and colleagues deduced bounds on the mass distribution in galaxies, from the (unobserved) distortion of background galaxy images by the gravitational field of foreground galaxies. The authors have checked how critical is their assumption that the background galaxies are at an infinite distance. When the distance distribution of the foreground and background galaxies is taken into account, the upper bounds on the rotation velocities are increased (quite considerably) above those published by Tyson and colleagues in 1984.

157.316 Giant molecular clouds in M31.
S. N. Vogel, F. Boulanger, R. Ball.
Astrophys. J., Lett. Ed., Vol. 321, No. 2, p. L145 – L149 (1987). With plate L12.

The authors have mapped CO ($J = 1$–0) emission near the H II regions BA 292 and P 248 in M31 with the Owens Valley interferometer at 7″ (25 pc) resolution. They resolve a cloud with characteristics similar to galactic giant molecular clouds (GMCs) such as Orion, including brightness, line width, size, mass, and the presence of active massive star formation. It is stated that all CO observations of M31 are consistent with an interstellar medium in which most of the molecular gas resides in GMCs.

157.317 No supernova in M31.
IAU Circ., Nos. 4496, 4498 (1987).

157.318 The vertical large–scale magnetic fields in spiral galaxies.
Y. Sofue, M. Fujimoto.
Publ. Astron. Soc. Jpn., Vol. 39, No. 6, p. 843 – 848 (1987).

A primordial–origin hypothesis is presented to explain a large–scale magnetic field vertically penetrating the galactic nuclear disk. Based on this hypothesis it is shown that a vertical component of a fossil of an intergalactic magnetic field trapped to a protogalaxy is condensed toward the galactic center through a secular accretion of the disk gas. The strong vertical field thus condensed in the nuclear disk may account for the vertical straight structures observed in our galactic center.

157.319 Surface photometry of the warping edge–on galaxy NGC 5907.
T. Sasaki.
Publ. Astron. Soc. Jpn., Vol. 39, No. 6, p. 849 – 878 (1987).

Photographic surface photometry of the edge–on Sc galaxy NGC 5907 was performed with a particular intention to reveal the optical warping of the outer disk. The main results are: (1) The truncation of the disk found by van der Kruit (1979) is confirmed. (2) The disk exhibits outward blueing with a color gradient of –0.048 mag kpc^{-1} in the outer part. (3) The galaxy shows the optical warping of its disk in the region at projected distances from the center of 250″ (13.3 kpc) to 450″ (24.0 kpc) with the maximum displacement of 32″ (1.7 kpc). The optical warp seems to deviate from the H I warp. The tidal interaction with companion galaxies is suggested to be relevant to the characteristics (the truncation, the outward blueing, and the warp) of the disk in the galaxy, but the full explanation remains to be solved.

157.320 High spatial resolution infrared imaging of M82.
J. L. Pipher, A. Moneti, W. J. Forrest,
C. E. Woodward, M. A. Shure.
Infrared astronomy with arrays, p. 326 – 329 (1987). – See Abstr. 012.070.

The authors present images of the central regions of M82 at 1.65 μm and 2.23 μm. From these images they identify eight emission peaks, in addition to the nucleus, four of which are located 8–10″ west of the nucleus. The authors have performed astrometry on the nuclear position, and give flux densities for each of the sources.

157.321 Infrared array imaging of the "hot–spot" galaxy NGC 2903.
D. A. Simons, R. W. Capps, K.–W. Hodapp, E. E. Becklin,
C. G. Wynn–Williams.
Infrared astronomy with arrays, p. 337 – 344 (1987). – See Abstr. 012.070.

The JPL/SISEX infrared CCD and Galileo/IfA 500×500 optical CCD were used to image the nuclear region of NGC 2903 in K (2.2 μm) and V (0.55 μm), respectively. The (2.2 μm)/(0.55 μm) flux ratio has been used to map the reddening across the nucleus of this galaxy.

157.322 Observations of luminous IRAS galaxies with an infra-red array.
S. A. Eales, E. E. Becklin, C. G. Wynn–Williams,
K.–W. Hodapp, R. W. Capps, D. A. Simons.
Infrared astronomy with arrays, p. 345 – 349 (1987). – See Abstr. 012.070.

The authors present observations made with the JPL/SISEX SWIR infrared array of two luminous IRAS galaxies. The observations of NGC 6240 showed the nucleus to have different structures at 1.25, 1.65, and 2.2 μm. The observations of the NGC 3690/IC 694 system disclosed a 2.2 μm double source at the centre of NGC 3690.

157.323 Near infrared imaging of interacting galaxies.
G. S. Wright, I. S. McLean.
Infrared astronomy with arrays, p. 355 – 359 (1987). – See Abstr. 012.070.

The UKIRT two dimensional array infrared camera, IRCAM, has been used to obtain near infrared images of galaxies. Images of the interacting systems 'the Antennae' (Arp 244) and NGC 3690/IC 694 (Arp 299) are presented and discussed.

157.324 Infrared images of galaxies.
I. Vauglin, J.–L. Monin, F. Sibille, P. Merlin.
Infrared astronomy with arrays, p. 360 – 363 (1987). – See Abstr. 012.070.

The authors have obtained near–infrared maps of galaxies as well as galactic sources in the K, L' and M bands. The results on the Seyfert galaxies NGC 1068 and NGC 1275 and on the elliptical galaxy NGC 221 are presented.

157.325 Photométrie des galaxies.
P. Prugniel.
Bull. Inf. Cent. Données Stellaires, No. 33, p. 17 – 25 (1987).

The author presents the techniques and the aims of the photometry of nearby galaxies. He discusses some problems linked to these measurements and the possible solutions to these problems.

157.326 Possibility of the compensation effect of angular momenta in binary galaxies.
D. A. Verner.
Astron. Tsirk., No. 1466, p. 5 – 7 (1986). In Russian.

The ranges of possible directions of orbital angular momenta $\vec{K}$ for 19 isolated pairs of galaxies are determined. The probability that most studied pairs have opposite directions of $\vec{K}$ and of the spin of the larger component, $\vec{S}_l$, is a factor of ~ 2 larger than the probability that most pairs have close directions of $\vec{K}$ and $\vec{S}_l$. This indicates the possibility of the effect of compensation of angular momenta in binary galaxies. The possibility of this effect in the frame of various scenarios of origin and evolution of galaxies is briefly discussed.

157.327 On the activity of the nucleus of M33.
G. Ivanov, P. Kunchev, V. Golev, I. Parov.
Astron. Tsirk., No. 1473, p. 3 – 5 (1986). In Russian.

The photographic structure of the M33 nucleus is examined. There is a well seen jet to the south on isodensity UBV maps of the nucleus. The authors consider the past activity of the nucleus as possible star formation trigger mechanism in the central region of the M33 galaxy.

157.328 On bursts of star formation in gas–stellar systems with accretion.
V. I. Korchagin, A. D. Ryabtsev.
Astron. Tsirk., No. 1479, p. 5 – 7 (1987). In Russian.

A two–phase model of interacting molecular clouds and massive stars describing a star formation burst in a galaxy with growing infall rate of gas is investigated. Star formation triggering by massive stars and via cloud–cloud collisions are taken into account. The system demonstrates damped oscillatory behaviour with first maximum being a well–defined burst of star formation.

157.329 Masses of galaxies.
J. S. Gallagher III.
Physics of the superconducting supercollider, p. 631 – 632 (1987). Abstr. in Phys. Abstr., Vol. 91, No. 1320, Entry 10694 (1988). – See Abstr. 012.080.

157.330 Surface photometry of barred galaxies: global structure of barred galaxies.
H. B. Ann, S. W. Lee.
J. Korean Astron. Soc., Vol. 20, No. 2, p. 49 – 62 (1987).

Using detailed two–dimensional surface photometry of 39 galaxies, the observed profiles are decomposed into spheroid, disk and bar components simultaneously. From the analyses of decomposition parameters, the correlations among the three components are investigated to find the global properties of barred galaxies.

157.331 The long–period variable stars of M33.
J. Mould.
Publ. Astron. Soc. Pac., Vol. 99, No. 621, p. 1127 – 1133 (1987). = Invited paper presented at the Symposium on cool stars and the structure of galaxies, at the 99th Annual Scientific Meeting of the Astronomical Society of the Pacific, at Pomona College, Claremont, California, July 1987.

Infrared photometry of long–period variables in M33 shows that the majority of those identified up to now are supergiants. The tip of the asymptotic giant branch may also be visible, but the present data are not adequate to decide the matter. Calibration photometry presented here confirms that the distance of M33 is 890 kpc with an uncertainty of approximately 15%. The Cepheid and supergiant period–luminosity relations are in satisfactory agreement on this point.

157.332 Lyman–alpha galaxies.
H. Spinrad, P. J. McCarthy.
Publ. Astron. Soc. Pac., Vol. 99, No. 621, p. 1150 (1987). Abstract. – See Abstr. 010.281.

157.333 Interstellar gas in spiral galaxies.
Z. Matta.
Kozmos, Vol. 18, No. 4, p. 122 (1987). In Slovak.

157.334   Molecules in galaxies. V. CO observations of flocculent and grand–design spirals.
A. A. Stark, B. G. Elmegreen, D. Chance.
Astrophys. J., Vol. 322, No. 1, p. 64 – 73 (1987).

Carbon monoxide (1–0) observations of 29 spiral galaxies were made with the 7 m antenna at Crawford Hill in order to search for possible correlations between the CO surface brightness of galactic disks and the presence or lack of grand–design spiral structure. Five–point maps were made for most of the galaxies, and the interpolated total CO $J = 1$–0 line fluxes were determined. When combined with similar published data, the total sample shows no significant correlation between the CO surface brightness and the presence of lack of grand–design spiral structure. In addition, the CO emission is studied as a function of the galaxy's luminosity class.

157.335   Arecibo observations of *IRAS* galaxies at 21 and 18 centimeters.
R. W. Garwood, G. Helou, J. M. Dickey.
Astrophys. J., Vol. 322, No. 1, p. 88 – 100 (1987).

The authors report Arecibo observations at 21 cm and 18 cm of a sample of galaxies selected for their intense far–infrared (FIR) emission. The objective was to search for H I absorption and OH maser emission. No new cases were found. The authors discuss the significance of this null result both in terms of the spatial distribution of the interstellar gas and dust and in terms of the continuum luminosity of the nucleus.

157.336   Systematic properties of CO emission from galaxies. I. Luminosity function.
F. Verter.
Astrophys. J., Suppl. Ser., Vol. 65, No. 4, p. 555 – 580 (1987).

A maximum–likelihood luminosity function incorporating both fluxes and flux upper limits is derived for normal galaxies covering a wide range of luminosities and morphological types. The total CO luminosity of galaxies increases with morphological type until it peaks around type Sbc. The fraction of the interstellar gas in molecular form, as measured by the CO/H I flux ratio, also peaks at intermediate morphological types. Uncertainties in molecular mass calculations are reviewed and error bars estimated, as a function of galaxy type.

157.337   The kinematics of interacting galaxies. II. Motions in close and coalescent pairs.
V. P. Arkhipova, A. V. Zasov, R. I. Noskova.
Sov. Astron., Vol. 31, No. 2, p. 120 – 126 (1987). English translation of 43.157.087.

157.338   An analysis of the H I peculiar–velocity field in M81.
F. Kh. Sakhibov, M. A. Smirnov.
Sov. Astron., Vol. 31, No. 2, p. 132 – 135 (1987). English translation of 43.157.089.

157.339   A central black hole in M32.
J. L. Tonry.
Astrophys. J., Vol. 322, No. 2, p. 632 – 642 (1987).

Observations are presented of the stellar rotation and velocity dispersion in M32. The projected rotation curve has an unresolved cusp at the center, with an amplitude of at least 60 km s^{-1}. The stellar velocity dispersion is constant at 56 ± 5 km s^{-1} to a radius of 20″. The three–dimensional rotation field is modeled. Hydrostatic equilibrium requires $3 – 10 \times 10^6 M_\odot$ of dark mass within the central parsec of M32. It is shown that a likely explanation of this dark mass, especially because of the presence of an X–ray point source at the center of M32, is a massive black hole.

157.340   A radio search for Crab–like nebulae in M33.
S. P. Reynolds, J. D. Fix.
Astrophys. J., Vol. 322, No. 2, p. 673 – 680 (1987).

The authors report the results of a search for Crab–like supernova remnants in M33, using spectral indices and center–brightened morphology as discriminants. They observed 12 sub–arcminute radio sources brighter than 3 mJy which are not

associated with optical H II regions. The VLA was used at 1465 and 4885 MHz for resolutions of 1″ and 0″.3, respectively. One of these sources could be a Crab–like remnant, but its properties are also consistent with those of a normal compact H II region. None of the other 11 objects fits the observational definition of a Crab–like remnant. Thus M33 joins M31 and the Milky Way in exhibiting far fewer Crab–like remnants than theory would predict.

157.341   The CO contents of dwarf irregular galaxies.
L. J. Tacconi, J. S. Young.
Astrophys. J., Vol. 322, No. 2, p. 681 – 687 (1987).

The authors have searched for CO in 15 dwarf irregular galaxies selected to span a wide range of infrared luminosity and color, from red low surface brightness objects to blue compact dwarf galaxies using the 14 m telescope of the FCRAO; CO emission was detected in six systems at the 3σ level or greater. The CO luminosities of the dwarf galaxies are compared with the IR luminosities (40 – 300 μm) determined from co–added *IRAS* survey data at 60 and 100 μm. The highest L_{IR}/L_{CO} ratios are found in the blue, actively star–forming galaxies, and the lowest ratios in the red systems.

157.342   Erratum: "The M31 globular cluster system" [Astrophys. J., Vol. 288, No. 2, p. 494 – 513 (1985)].
D. Crampton, A. P. Cowley, D. Schade, P. Chayer.
Astrophys. J., Vol. 322, No. 2, p. 1074 (1987). See Abstr. 39.157.067.

157.343   Hot gas in the nucleus of IC 342: detection of $J = 3$–2 CO emission.
P. T. P. Ho, J. L. Turner, R. N. Martin.
Astrophys. J., Lett. Ed., Vol. 322, No. 2, p. L67 – L71 (1987).

The $J = 3 \rightarrow 2$ CO line is detected for the first time in an external galaxy, in the nearby spiral IC 342. It is shown that in contrast to the $J = 1 \rightarrow 0$ CO line, which favors optically thick and cool gas, the $J = 3 \rightarrow 2$ CO line can be used to study a hot ($\geqslant 10$K) component in extragalactic nuclear environment. Limited mapping results suggest that the $J = 3 \rightarrow 2$ CO emission arises from the same regions as the $J = 1 \rightarrow 0$ CO emission.

157.344   Are there stellar winds in the hot stars of M31?
J. B. Hutchings, P. Massey, L. Bianchi.
Astrophys. J., Lett. Ed., Vol. 322, No. 2, p. L79 – L84 (1987).

Results are reported of 16 hr exposures with the *IUE* satellite of two stars in M31. Ground–based spectra indicate these stars to be of type $\sim$O6 and WN 7. The UV spectra are conspicuously different from Galactic stars of their type. Specifically, no line emission is seen in either star; resonance absorption lines are very weak and indicate low outflow velocities in the O star; there appear to be shifted narrow lines normally attributed to the ISM; and either the UV continuum is "cooler" than the optical, or the UV extinction law is steeper than Galactic. All of these indicate unexpected global differences between the Galaxy and M31.

157.345   The late–type stellar content of NGC 55.
C. J. Pritchet, H. B. Richer, D. Schade, D. Crabtree, H. K. C. Yee.
Astrophys. J., Vol. 323, No. 1, p. 79 – 90 (1987).

The authors have completed a photometric survey of the late–type stellar content of the nearby galaxy NGC 55. Using narrow–band filters they have separated C and M stars, and have reached the following conclusions: The distance to NGC 55 is 1.34 ± 0.08 Mpc. The discrepancy between theoretical and observational determinations of the AGB luminosity function is also confirmed in NGC 55. The correlation between C/M number ratio and [Fe/H] is not due to the correlation between metallicity and the temperature of the giant branch alone. The existence of a relation between N_c/L_B (the number of carbon stars per unit luminosity) and [Fe/H] is confirmed.

157.346 The spatial distribution of 10 micron luminosity in spiral galaxies.

N. Devereux.

Astrophys. J., Vol. 323, No. 1, p. 91 – 107 (1987).

Ground–based photometry combined with the *IRAS* photometry has permitted a study of the spatial distribution of infrared emission in spiral galaxies of a wide range in Hubble type and luminosity. 40% of the early–type barred spirals are associated with enhanced 10 μm luminosity in the central region. High central 10 μm luminosity is not seen in early–type unbarred or late–type galaxies. It is therefore concluded that stellar bars are important for high central infrared luminosity in early–type spirals. The origin of the high central 10 μm luminosity in early–type barred spirals is attributed to star formation or Seyfert activity or both.

157.347 The luminosity–velocity diagram: a new distance estimator for spiral galaxies?

B. F. Madore, D. Woods.

Astrophys. J., Lett. Ed., Vol. 323, No. 1, p. L25 – L29 (1987).

The authors present evidence for a newly identified feature of the inner regions of disk galaxies which appears to show promise as a new distance indicator. For a sample of 46 galaxies, they find in moving out from their centers, $\Delta M_r / \Delta \log V_r \approx -3.30 \, (\pm 0.40)$, where M_r is the cumulative interior magnitude and V_r is the local rotation rate.

157.348 Associations in M31 and M33.

N. S. Nikolov.

Star clusters and associations, p. 83 – 89 (1986). – See Abstr. 012.081.

On the basis of the observational data obtained by the 2m telescope of the National Astronomical Observatory at the Bulgarian Academy of Sciences in the Rhodopa Mountains the stellar associations in the galaxies M31 and M33 are examined in brief. Main results concerning the new–found associations, the structures of these galaxies, the results concerning light absorption, as well as the obtained diagrammes for some associations "colour–magnitude" and "colour–colour" in the UBV system are presented.

157.349 Ellipticity of 30 globular clusters in the Andromeda galaxy.

A. V. Staneva, N. M. Spasova, P. V. Baev.

Star clusters and associations, p. 91 – 100 (1986). – See Abstr. 012.081.

The ellipticity and the orientations of the major semiaxes of 30 globular star clusters in the Andromeda galaxy were obtained. 3 B plates taken by the 2m Ritchey–Chrétien telescope were used. The plates were scanned on microphotometers of the type Joyce–Loebl and MF–4. The ellipticity of the examined clusters is in the interval 0.02 – 0.26 and $\bar{\varepsilon} = 0.109 \pm 0.011$. Comparisons between the average ellipticities of the globular clusters in the Galaxy, the Large Magellanic Cloud and M31 were made.

157.350 Hoag's object: the celestial donut.

N. Brosch.

Mercury, Vol. 16, No. 6, p. 174 – 177 (1987).

Second Byurakan Spectral Sky Survey. V. Results for the region centered on $\alpha = 15^h 30^m$, $\delta = +59°00'$.
See Abstr. 002.013.

Compilation of the fifth volume of the Morphological Catalogue of Galaxies on magnetic tape.
See Abstr. 002.087.

A 1.49 GHz atlas of spiral galaxies with $B_T \leqslant +12$ and $\delta \geqslant -45°$.
See Abstr. 002.106.

A 1.49 GHz supplementary atlas of spiral galaxies with H–magnitudes.
See Abstr. 002.107.

The mysterious nebulae, 1610 – 1924.
See Abstr. 004.018.

Discovering M31's spiral shape.
See Abstr. 004.062.

The remarkable extragalactic research of Erik Holmberg: a glimpse from Santa Cruz.
See Abstr. 005.005.

Data reduction and spectrophotometric performances of PUMA 1: an on–line multiaperture spectroscopic system used at the CFHT.
See Abstr. 036.066.

The effect of deconvolution on detectability of faint galaxies.
See Abstr. 036.132.

Techniques for faint object imaging at 1 micron.
See Abstr. 036.206.

Uniformisation de la notion de magnitude.
See Abstr. 036.211.

Code for digital image processing of extended objects.
See Abstr. 036.214.

The origin and cosmogonic implications of seed magnetic fields.
See Abstr. 062.064.

Subluminous Type I supernovae: their theoretical rate in our Galaxy and in ellipticals.
See Abstr. 065.006.

Gravitational lensing by isothermal spheres with finite core radii: galaxies and dark matter.
See Abstr. 066.097.

Gravitational imaging by isolated elliptical potential wells. II. Probability distributions.
See Abstr. 066.150.

The marginal gravitational lensing.
See Abstr. 066.151.

A galactic gravitational lens as the ultimate astronomical telescope.
See Abstr. 066.175.

Hα as a tracer of stellar mass loss in metal–poor galaxies.
See Abstr. 112.128.

The spectra of extra–galactic Wolf–Rayet stars.
See Abstr. 114.069.

Spectral synthesis in the ultraviolet. I. Far–ultraviolet stellar library.
See Abstr. 114.116.

On the characteristics of the high–luminosity OB stars in the M33 galaxy.
See Abstr. 115.007.

The enigmatic object Variable A in M33.
See Abstr. 122.021.

Notes on Cepheid variables in galaxies more distant than the Magellanic Clouds.
See Abstr. 122.078.

Probable novae in M31.
See Abstr. 124.013.

The supernova rate in Shapley–Ames galaxies.
See Abstr. 125.090.

Star formation and dynamics in starburst nuclei.
See Abstr. 131.016.

The galactic foreground reddening in the direction of the nearby Triangulum galaxy M33.
See Abstr. 131.046.

Formation and evolution of the largest cloud complexes in spiral galaxies.
See Abstr. 131.087.

Self–regulating star formation and disk structure.
See Abstr. 131.125.

Supershells
See Abstr. 131.164.

Numerical models of superbubble dynamics.
See Abstr. 131.171.

Hydrogen absorption line profiles of ionising star clusters.
See Abstr. 131.202.

[S III] in extragalactic H II regions.
See Abstr. 132.020.

Size distributions of H II regions in galaxies. II. Spiral galaxies.
See Abstr. 132.027.

Evolution of H II regions.
See Abstr. 132.035.

Systematic identification of IRAS point sources.
See Abstr. 133.001.

What are "cirrus" point sources?
See Abstr. 133.003.

The IRAS view of the extragalactic sky.
See Abstr. 133.007.

Models of ring galaxies. I. The growth and disruption of clouds in the expanding density wave.
See Abstr. 151.001.

A dynamical study of the frequency of interacting galaxies.
See Abstr. 151.005.

About the possible cause of appearance of the barred spiral structure in galaxies.
See Abstr. 151.010.

Collisionless analogs of Riemann S–ellipsoids with halo.
See Abstr. 151.017.

On the velocity fields in elliptical galaxies with dark matter.
See Abstr. 151.021.

Nonlinear coupling of galactic spiral modes.
See Abstr. 151.029.

The ellipsoidal subsystems in barred spirals.
See Abstr. 151.041.

Mass models for disk and halo components in spiral galaxies.
See Abstr. 151.047.

The structure and evolution of disk galaxies.
See Abstr. 151.048.

Halo response to galaxy formation.
See Abstr. 151.050.

Dynamics in the centres of triaxial elliptical galaxies.
See Abstr. 151.052.

Shells and the potential wells of elliptical galaxies.
See Abstr. 151.053.

Spherical galaxies: methods and models.
See Abstr. 151.054.

A method to determine the intrinsic axial ratios of individual triaxial galaxies.
See Abstr. 151.059.

Stochastic stellar orbits and the shapes of elliptical galaxies.
See Abstr. 151.063.

Dynamical approach to the $R^{1/4}$ law of ellipticals.
See Abstr. 151.076.

On the distribution function of elliptical galaxies.
See Abstr. 151.078.

The envelopes of spherical galaxies.
See Abstr. 151.080.

Cold collapse as a way of making elliptical galaxies.
See Abstr. 151.084.

Angular momentum in elliptical galaxies.
See Abstr. 151.085.

Numerical investigation of the density distribution of stars and the dispersion of velocities in spiral galaxies.
See Abstr. 151.088.

Dynamical friction and shells around elliptical galaxies.
See Abstr. 151.098.

Kinematics and dynamics of barred spiral galaxies.
See Abstr. 151.102.

Dynamical activity at the center of a galaxy.
See Abstr. 151.105.

Dissipationless collapse and the origin of triaxiality in elliptical galaxies.
See Abstr. 151.107.

Spiral structure as a standing density–wave packet.
See Abstr. 151.126.

Self–consistent models of perfect triaxial galaxies.
See Abstr. 151.129.

The vertical structure of galactic disks.
See Abstr. 151.142.

Axisymmetric shell models with Stäckel potentials.
See Abstr. 151.148.

Models of ring galaxies. II. Extended starbursts.
See Abstr. 151.149.

Photoelectric photometry of globular clusters in the Andromeda nebula. XII.
See Abstr. 154.004.

Globular clusters as extragalactic distance indicators: maximum–likelihood methods.
See Abstr. 154.011.

Systematic properties of extragalactic globular clusters.
See Abstr. 154.019.

An interpretation of the line–strength indices in old stellar populations using an evolutionary synthesis approach.
See Abstr. 154.024.

Near–infrared spectral properties of star clusters and galactic nuclei.
See Abstr. 154.026.

Globular cluster systems and galactic halos.
See Abstr. 154.039.

How similar are the globular clusters in different galaxies?
See Abstr. 154.044.

An automated search for halo star clusters and Local Group dwarf galaxies.
See Abstr. 154.071.

Globular clusters in dwarf spheroidals.
See Abstr. 154.072.

IUE observations of the globular cluster N.3 in the Fornax dwarf spheroidal galaxy.
See Abstr. 154.073.

The origin of the diffuse galactic IR/submm emission: revisited after IRAS.
See Abstr. 155.003.

Disk rotation curves in triaxial potentials.
See Abstr. 155.081.

M giants in Baade's window: infrared colors, luminosities, and implications for the stellar content of E and S0 galaxies.
See Abstr. 155.088.

The galactic bulge as a prototype population for E and S0 galaxies.
See Abstr. 155.105.

Chemical and spectrophotometric evolution of the Magellanic Clouds.
See Abstr. 156.029.

Chemical and spectrophotometric evolution of the Magellanic Clouds.
See Abstr. 156.053.

Infrared spectroscopy of star formation in galaxies.
See Abstr. 158.002.

Extragalactic infrared spectroscopy.
See Abstr. 158.005.

Far–infrared activity and starburst galaxies.
See Abstr. 158.009.

IRAS observations of starburst galaxies.
See Abstr. 158.010.

The luminosity function of the brightest galaxies in the IRAS survey.
See Abstr. 158.012.

Extragalactic OH megamasers in strong IRAS sources.
See Abstr. 158.016.

Nuclear infrared emission and the colors of IRAS galaxies.
See Abstr. 158.017.

Spectrophotometry of Brackett lines in very luminous IRAS galaxies.
See Abstr. 158.025.

Summary of symposium: low luminosity sources.
See Abstr. 158.033.

Morphology of luminous IRAS galaxies: summary talk.
See Abstr. 158.034.

Observations of 1.3 millimeter continuum emission from the centers of galaxies.
See Abstr. 158.075.

The local radio luminosity function of galaxies.
See Abstr. 158.122.

Gas, dust and radio emission in elliptical galaxies.
See Abstr. 158.136.

Ultraviolet energy distributions of (32) early–type galaxies.
See Abstr. 158.142.

Rotationally excited OH in megamaser galaxies.
See Abstr. 158.161.

The extreme CO and IRAS FIR luminosity of 7Zw 31.
See Abstr. 158.198.

H_2O maser emission from the nuclei of NGC 253 and M51.
See Abstr. 158.252.

Two micron spectroscopy of *IRAS* galaxies.
See Abstr. 158.317.

Extragalactic imaging with IRCAM on UKIRT.
See Abstr. 158.331.

Identification of an intervening galaxy responsible for Mg II absorption in the QSO 4C 55.27.
See Abstr. 159.024.

The host galaxy of quasar IRAS 00275–2859: an interacting system.
See Abstr. 159.026.

Discovery of low–redshift, neutral hydrogen absorption in the radio spectrum of PKS 2020–370.
See Abstr. 159.041.

The galaxy–quasar connection: NGC 4319 and Markarian 205. I. Direct imagery.
See Abstr. 159.042.

The galaxy–quasar connection: NGC 4319 and Markarian 205. II. Spectroscopy.
See Abstr. 159.043.

Observations of Ni II absorption at $z = 2.811$ toward the QSO PKS 0528–250.
See Abstr. 159.045.

Crowding on the sight line to the QSO PHL 1226: the nearby galaxy IC 1746 and a galaxy cluster at $z = 0.16$.
See Abstr. 159.080.

Far–infrared properties of cluster galaxies.
See Abstr. 160.001.

Present star formation in spirals of the Virgo cluster.
See Abstr. 160.002.

Molecular gas and star formation in H I–deficient Virgo cluster galaxies.
See Abstr. 160.003.

Observations of novae in the Virgo cluster.
See Abstr. 160.019.

The structure of brightest cluster members. II. Mergers.
See Abstr. 160.023.

X–ray observations of clusters: physical implications.
See Abstr. 160.026.

Dark matter in clusters?
See Abstr. 160.049.

The Perseus supercluster at low galactic latitudes.
See Abstr. 160.061.

The luminosity function: dependence on Hubble type and environment.
See Abstr. 160.062.

Evolution of cluster galaxies since $z = 1$.
See Abstr. 160.063.

Dark matter in binary galaxies and small groups.
See Abstr. 160.064.

Star formation in cooling flows.
See Abstr. 160.065.

Dwarf galaxies in the Fornax cluster.
See Abstr. 160.066.

An investigation of the radial dependence of the galaxy luminosity function in Abell clusters.
See Abstr. 160.067.

The surface–brightness–effective–size relation for elliptical galaxies in the cores of clusters.
See Abstr. 160.068.

Dwarf elliptical galaxies in the Fornax cluster. I. A catalog and luminosity function.
See Abstr. 160.069.

Dwarf elliptical galaxies in the Fornax cluster. II. Their structure and stellar populations.
See Abstr. 160.070.

An expanding shell of galaxies in the center of the Hydra I cluster?
See Abstr. 160.071.

50 kpc radio trails behind irregular galaxies in A 1367.
See Abstr. 160.072.

21 centimeter study of spiral galaxies in the Coma supercluster.
See Abstr. 160.073.

Classification of galaxies in compact groups.
See Abstr. 160.095.

Nearby groups of galaxies. II. An all–sky survey within 3000 kilometers per second.
See Abstr. 160.103.

Luminous arcs from galactic bow shocks.
See Abstr. 160.105.

A giant intergalactic H I bubble near Arp 143.
See Abstr. 160.106.

Individual masses of galaxies in triple systems.
See Abstr. 160.107.

Far–infrared properties of cluster galaxies.
See Abstr. 160.110.

Galaxy formation by gravitational collapse.
See Abstr. 161.002.

Large–scale structure, streaming and galaxy formation.
See Abstr. 161.050.

The formation of families of twin galaxies by string loops.
See Abstr. 161.052.

The dipole anisotropy of a new, colour–selected, *IRAS* galaxy sample.
See Abstr. 161.071.

Existence and nature of dark matter in the universe.
See Abstr. 161.074.

Large–scale streaming of cluster elliptical galaxies.
See Abstr. 161.106.

Structure formation in $\Omega = 1$ CDM universes and the biasing mechanism.
See Abstr. 161.108.

Cosmic strings, galaxy formation and peculiar velocities.
See Abstr. 161.109.

The formation of structure in particle–dominated cosmologies.
See Abstr. 161.111.

Large scale streaming velocities in biased galaxy formation.
See Abstr. 161.112.

Galaxies at very high redshifts ($z > 1$).
See Abstr. 161.143.

Formation and secular evolution of elliptical galaxies.
See Abstr. 161.155.

Void probabilities in the galaxy distribution: scaling and luminosity segregation.
See Abstr. 161.179.

Nature of the redshift of galaxies and quasars and its consequences. I. Gas accretion onto the nucleus of a galaxy and galaxy emission.
See Abstr. 161.187.

Nature of the redshift of galaxies and quasars and its consequences. II. The conception of the noncosmological nature of redshifts of galaxies and quasars.
See Abstr. 161.188.

Giant H II regions as distance indicators. II. Application to H II galaxies and the value of the Hubble constant.
See Abstr. 161.211.

Amplification of correlation functions by gravity in the cosmic string model.
See Abstr. 161.365.

Cosmological parameters from the *IRAS* galaxy sample.
See Abstr. 161.451.

The appearance of primeval galaxies.
See Abstr. 161.454.

On the origin of dwarf galaxies.
See Abstr. 161.456.

158 Active Galaxies (Seyfert Galaxies, BL Lacertae Objects, Radio Galaxies)

158.001 Starburst galaxies.
D. W. Weedman.
NASA Conf. Publ., NASA CP–2466, p. 351 – 361 (1987). – See Abstr. 012.002.

The infrared properties of star–forming galaxies, primarily as determined by IRAS, are compared to X–ray, optical, and radio properties. New luminosity functions are reviewed and combined with those derived from optically discovered samples using 487 Markarian galaxies with redshifts and published IRAS $60\,\mu$ fluxes, and 1074 such galaxies in the Center for Astrophysics redshift survey. Distributions of infrared to optical fluxes and available spectra indicate that the majority of IRAS–selected galaxies are starburst galaxies. The dust which must be associated with the known infrared galaxies obscures a significant portion of the universe beyond $z = 3$. Depending on the scale size of dusty galaxies, this effect may prevent the observation of distant quasars and primordial galaxies.

158.002 Infrared spectroscopy of star formation in galaxies.
S. C. Beck, P. T. P. Ho, J. L. Turner.
NASA Conf. Publ., NASA CP–2466, p. 363 – 366 (1987). – See Abstr. 012.002.

The authors have oberved the Brackett α $(4.05\,\mu m)$ and Brackett γ $(2.17\,\mu m)$ lines with $7.2''$ angular and $350\ \mathrm{km\ s^{-1}}$ velocity resolution in 11 infrared–bright galaxies. They derive extinctions, Lyman continuum fluxes, and luminosities due to OB stars. The galaxies observed to date are NGC 3690, M83, NGC 5195, Arp 220, NGC 520, NGC 660, NGC 1614, NGC 3079, NGC 6946, NGC 7714, and Maffei 2, all of which have been suggested at some time to be "starburst" objects. The contributions of OB stars to the luminosities of these galaxies range from insignificant (Arp 220, NGC 3079, NGC 5195) to sufficient to account for the total energy output (M83, NGC 1614).

158.003 Molecular gas in the starburst nucleus of M82.
K. Y. Lo.
NASA Conf. Publ., NASA CP–2466, p. 367 – 382 (1987). – See Abstr. 012.002.

$7''$–resolution CO(1–0) observations of the central 1 kpc of M82 have resolved 2 components of molecular gas: (1) a high concentration in the central 700 pc × 200 pc, and (2) extended features that may be gas expelled from the central concentration. The central concentration of molecular gas may be identified directly with the star–burst region.

158.004 Ultraluminous infrared galaxies.
D. B. Sanders, B. T. Soifer, G. Neugebauer,
N. Z. Scoville, B. F. Madore, G. E. Danielson, J. H. Elias,
K. Matthews, C. J. Lonsdale Persson, S. E. Persson.
NASA Conf. Publ., NASA CP–2466, p. 411 – 420 (1987). – See Abstr. 012.002.

The IRAS survey of the local universe $(z \lesssim 0.1)$ has revealed the existence of a class of ultraluminous infrared galaxies with $L(8 – 1000\,\mu m) > 10^{12}L_\odot$ that are slightly more numerous, and as luminous as optically selected quasars at similar redshifts. Optical CCD images of these infrared galaxies show that nearly all are advanced mergers. Millimeter–wave CO $(1\rightarrow0)$ observations indicate that these interacting systems are extremely rich in molecular gas with total H_2 masses $1 – 3 \times 10^{10} M_\odot$. Nearly all of the ultraluminous infrared galaxies show some evidence in their optical spectra for nonthermal nuclear activity. It is proposed that their infrared luminosity is powered by an embedded active nucleus and a nuclear starburst both of which are fueled by the tremendous reservoir of molecular gas. Once these merger nuclei shed their obscuring dust, allowing the AGN to visually dominate the decaying starburst, they become the optically selected quasars.

158.005 Extragalactic infrared spectroscopy.
R. D. Joseph, G. S. Wright, R. Wade, J. R. Graham,
I. Gatley, A. H. Prestwich.
NASA Conf. Publ., NASA CP–2466, p. 421 – 433 (1987). – See Abstr. 012.002.

The authors are engaged in a programme to explore the spectra of galaxies in the near–infrared (H & K) atmospheric transmission windows. They have detected emission lines due to molecular hydrogen, atomic hydrogen recombination lines, a line attributed to [Fe II], and a broad CO absorption feature. Lines due to H_2 and [Fe II] are especially strong in interacting and merging galaxies, but they have also been detected in Seyferts and "normal" spirals. These lines appear to be shock–excited. It is argued that starbursts provide the most plausible and consistent model for the excitation of these lines, but the changes of relative line intensity of various species with aperture suggests that other excitation mechanisms are also operating in the outer regions of these galaxies.

158.006 Starburst–driven superwinds from infrared galaxies.
T. M. Heckman, L. Armus, P. McCarthy,
W. van Breugel, G. K. Miley.
NASA Conf. Publ., NASA CP–2466, p. 461 – 469 (1987). – See Abstr. 012.002.

The authors present new data that indicate that strong far–infrared galaxies commonly have largescale emission–line nebulae whose properties are suggestive of mass outflows ("superwinds"), presumably driven by the high supernova rate associated with the central starburst.

158.007 Interferometric CO observations of the ultraluminous IRAS galaxies Arp 220, IC 694/NGC 3690, NGC 6240, and NGC 7469.
A. I. Sargent, D. B. Sanders, N. Z. Scoville, B. T. Soifer.
NASA Conf. Publ., NASA CP–2466, p. 471 – 475 (1987). – See Abstr. 012.002.

High resolution CO observations of the IRAS galaxies Arp 220, IC 694/NGC 3690, NGC 6240 and NGC 7469 have been made with the Millimeter Wave Interferometer of the Owens Valley Radio Observatory. These yield spatial information on scales of 1 to 5 kpc and allow the separation of compact condensations from the more extended emission in the galaxies. In the case of the obviously interacting system IC 694/NGC 3690 the contributions of each component can be discerned. For that galaxy, and also for Arp 220, the unusually high luminosities may be produced by non–thermal processes rather than by intense bursts of star formation.

158.008 A dust scattered halo in starburst galaxy M82?
M. Rohan, P. Morrison, A. Sadun.
NASA Conf. Publ., NASA CP–2466, p. 485 – 489 (1987). – See Abstr. 012.002.

The source of the halo about M82 has been under discussion for several years. This paper looks at the feasibility of the "dust" theory in the X–ray range, using the halo in the X–ray image of M82 taken by the Einstein Observatory. To this end the X–ray cross section for dust is presented, along with the single scattered image of an X–ray source surrounded by a dust cloud; multiply–scattered images have been simulated with a Monte–Carlo program; profiles of the halo along the major and minor axes of M82 are presented. Also presented is an accounting for line spectrographs of M82 that show unusual splitting using the dust model. The final model proposed for the X–ray image requires dust of radius $50\,\text{Å} – 300\,\text{Å}$, with density on the order of $10^{-7}\mathrm{cm}^{-3}$ to $10^{-9}\mathrm{cm}^{-3}$, out to a distance of about 9 kpc for some regions.

158.009   Far–infrared activity and starburst galaxies.
P. Belfort, R. Mochkovitch, M. Dennefeld.
NASA Conf. Publ., NASA CP–2466, p. 501 – 505 (1987). – See Abstr. 012.002.

After the IRAS discovery of galaxies with large far–infrared to blue luminosity ratios, it has been proposed that an enhanced star formation could be the origin of the far–infrared emission through dust heating. The authors have investigated whether a simple photometric model is able to account for the FIR and optical properties of IRAS galaxies. The L_{IR}/L_B ratio, (B–V) color and Hα equivalent width of normal spirals are well reproduced with smooth star formation histories. In the case of starburst galaxies, several theoretical diagrams allow the estimation of the burst strength and extinction. L_{IR}/L_B ratios up to 100 can be rather easily reached, whereas extreme values (~ 500) probably require IMF truncated at the low end.

158.010   IRAS observations of starburst galaxies.
K. Sekiguchi.
NASA Conf. Publ., NASA CP–2466, p. 507 – 511 (1987). – See Abstr. 012.002.

Far–infrared properties of starburst galaxies were analyzed using IRAS observations at 25, 60, and 100 μm. Seventy–nine of 102 starburst galaxies from the list of Balzano were detected. These galaxies have high IR luminosities of up to a few $10^{12} L_\odot$ and concentrate in a small area of the IR color – color diagram. The IR power law spectral indices, α, lie within the ranges $-2.5 < \alpha(60,25) < -1.5$ and $-1.5 < \alpha(100,60) < 0$. These observed indices can be interpreted in terms of a cold (~ 30K) disk component and a warm ($\sim 80 - 90$K) component. More than 80% of the 60 μm emission comes from the warm component. The fraction of the 60 μm emission attributable to the warm component can be used as an activity indicator.

158.011   Star formation in the merging galaxy NGC 3256.
J. R. Graham, G. S. Wright, R. D. Joseph,
J. A. Frogel, M. M. Phillips, W. P. S. Meikle.
NASA Conf. Publ., NASA CP–2466, p. 517 – 519 (1987). – See Abstr. 012.002.

The authors have mapped the central 5 kpc of the ultra–luminous merging galaxy NGC 3256 (Graham et al. 1984) at J, H, K, L, & 10 μm, and obtained 2 μm spectra of the nuclear region. They use this data to identify and characterize the super–starburst which has apparently been triggered and fuelled by the merger of two gas–rich galaxies. The authors also show that the old stellar population has relaxed into a single spheroidal system, and that a supernova driven wind might eventually drive any remaining gas from the system to leave a relic which will be indistinguishable from an elliptical galaxy.

158.012   The luminosity function of the brightest galaxies in the IRAS survey.
B. T. Soifer, D. B. Sanders, B. F. Madore, G. Neugebauer,
C. J. Lonsdale Persson, S. E. Persson, W. L. Rice.
NASA Conf. Publ., NASA CP–2466, p. 523 – 530 (1987). – See Abstr. 012.002.

Results from a study of the far infrared properties of the brightest galaxies in the IRAS survey are described. There is a correlation between the infrared luminosity and the infrared to optical luminosity ratio and between the infrared luminosity and the far infrared color temperature in these galaxies. The infrared bright galaxies represent a significant component of extragalactic objects in the local universe, being comparable in space density to the Seyferts, optically identified starburst galaxies, and more numerous than quasars at the same bolometric luminosity. The far infrared luminosity in the local universe is approximately 25% of the starlight output in the same volume.

158.013   VLA observations of a sample of galaxies with high far–infrared luminosities.
S. A. Eales, C. G. Wynn–Williams, C. A. Beichman.
NASA Conf. Publ., NASA CP–2466, p. 531 – 536 (1987). – See Abstr. 012.002.

The authors present preliminary results from a radio survey of galaxies detected by the IRAS minisurvey. They find that the main difference between galaxies selected in the far–infrared and those selected in the optical is that the former have higher radio luminosities and that the radio emission is more centrally concentrated. There is some evidence that the strong central radio sources in the galaxies selected in the infrared are due to star formation rather than to active nuclei. If the radio emission is caused by star formation, the star formation rate divided by the volume in which the star formation is occurring is 100 – 1000 times greater than in the galaxies selected in the infrared than in the disks of normal galaxies.

158.014   Properties of the unusual galaxy PSC 09104 + 4109.
S. G. Kleinmann, W. C. Keel.
NASA Conf. Publ., NASA CP–2466, p. 559 – 562 (1987). – See Abstr. 012.002.

The IRAS source PSC 09104 + 4109 is tentatively identified with a faint ($m_R \sim +19$) emission line galaxy having $z = 0.442$. Assuming this identification is correct, the total infrared luminosity of this galaxy is estimated to be $5 \times 10^{12} L_0$, among the highest for galaxies detected by IRAS. This energy is concentrated at wavelengths less than 30 μm, and is ~ 50 times greater than the estimated optical luminosity. The serendipitous way in which this source was found in the PSC catalog suggests that many more similar objects may be found at the lowest levels of the IRAS survey.

158.015   Radio and infrared observations of (almost) one hundred non–Seyfert Markarian galaxies.
L. L. Dressel.
NASA Conf. Publ., NASA CP–2466, p. 579 – 582 (1987). – See Abstr. 012.002.

The author has measured the 13 cm flux densities of 96 non–Seyfert Markarian galaxies at Arecibo Observatory. She has compared the radio, infrared, and optical fluxes of these galaxies and of a magnitude–limited sample of "normal" galaxies to clarify the nature of the radio emission in Markarian galaxies. She finds that Markarian galaxies of a given apparent magnitude and Hubble type generally have radio fluxes several times higher than the fluxes typical of "normal" galaxies of the same magnitude and type. Remarkably, the ratio of radio flux to far infrared flux is nearly the same for most of these "star–burst" galaxies and for normal spiral disks.

158.016   Extragalactic OH megamasers in strong IRAS sources.
L. Bottinelli, M. Dennefeld, L. Gouguenheim,
J. M. Martin, G. Paturel, A. M. Le Squeren.
NASA Conf. Publ., NASA CP–2466, p. 597 – 600 (1987). – See Abstr. 012.002.

From their OH and H I survey of the strongest far–infrared ($\lambda = 60$ or 100 μm) IRAS sources, the authors have discovered 3 new powerful OH megamasers in Arp 148, IRAS 1510 + 0724 and in the uncatalogued IRAS source, IRAS 17208–0014 (Bottinelli et al. 1985, 1986). The spectra are displayed together with the main IR and OH properties of the 8 megamasers detected up to now.

158.017   Nuclear infrared emission and the colors of IRAS galaxies.
G. J. Hill.
NASA Conf. Publ., NASA CP–2466, p. 611 – 617 (1987). – See Abstr. 012.002.

J, H, K, L', and N observations of galaxies detected at 12 μm by IRAS are combined with IRAS flux densities to investigate the relationship between the infrared sizes and colors of galaxian infrared sources. It is found that typical IRAS galaxies have 10 μm radii of 0.5 – 2.0 kpc, while active galaxies and galaxies with higher 25 – 60 μm color temperatures are smaller. One unusual object, 23060 + 0505, is at high redshift and has an infrared luminosity of $1.5 \times 10^{12} L_\odot$. Its 1 – 100 μm energy distribution resembles that of a Seyfert 1 galaxy, but it shows very little sign of broad–line emission in the visible. Its properties suggest that it may be a prototype for a class of highly obscured active galaxy.

158.018   The relation between star formation and active nuclei.
    G. H. Rieke.
NASA Conf. Publ., NASA CP–2466, p. 633 – 641 (1987). – See Abstr. 012.002.

Three questions relevant to the relation between an active nucleus and surrounding star formation are discussed. The infrared stellar CO absorption bands can be used to identify galaxies with large populations of young, massive stars and thus can identify strong starbursts unambiguously. An active nucleus is probably not required for LINER spectral characteristics; dusty starburst galaxies can produce LINER spectra through the shock heating of their interstellar media by supernovae combined with the obscuration of their nuclei in the optical. The Galactic Center would be an ideal laboratory for studying the interaction of starbursts and active nuclei, if both could be demonstrated to occur there. Failure to detect a cusp in the stellar distribution raises questions about the presence of an active nucleus, which should be answered by additional observations in the near future.

158.019   Ground–based 1– to 32–μm observations of Arp 220: evidence for a dust–embedded "AGN"?
E. E. Becklin, C. G. Wynn–Williams.
NASA Conf. Publ., NASA CP–2466, p. 643 – 650 (1987). – See Abstr. 012.002.

New observations of the 10– and 20–μm size of the emission region in Arp 220 are presented. The authors also give ground–based photometry from 1 – 32 μm including measurements of the strength of the silicate feature at 10 μm. The results show that the 20–μm size of Arp 220 is smaller than 1.5 arcsec; comparison of IRAS and ground–based observations show that IRAS 12–μm flux measured with a large arcmin beam is the same as that seen from the ground with a 3–arcsec aperture. At 10 μm a deep silicate absorption feature is seen that corresponds to a visual extinction of about 50 mag. These results suggest that a very significant portion of the $10^{12}L_\odot$ infrared luminosity from Arp 220 comes from a region less than or of the order of 500 pc in diameter. A very attractive possibility for the primary luminosity source Arp 220 is a dust–embedded compact Seyfert–type nucleus.

158.020   Spatial deconvolution of IRAS galaxies at 60 μm.
    F. J. Low.
NASA Conf. Publ., NASA CP–2466, p. 651 – 660 (1987). – See Abstr. 012.002.

Using IRAS in a "slow scan" observing mode to increase the spatial sampling rate and a deconvolution analysis to increase the spatial resolution, several bright galaxies have been resolved at 60 μm. Preliminary results for M82, NGC 1068, NGC 3079 and NGC 2623 show partially resolved nuclei in the range 10 to 26 arcsec, full width at half maximum, and extended emission from 30 to 90 arcsec from the center. In addition, the interacting system, Arp 82, along with Mrk 231 and Arp 220 were studied using the program "ADDSCAN" to average all available survey mode observations. The Arp 82 system is well resolved after deconvolution and its brighter component is extended; the two most luminous objects are not resolved with an upper limit of 15 arcsec for Arp 220.

158.021   Star formation around active galactic nuclei.
    W. C. Keel.
NASA Conf. Publ., NASA CP–2466, p. 661 – 667 (1987). – See Abstr. 012.002.

Emission–line images and high–dispersion optical spectra have been used to investigate star–forming regions in the vicinity of active galactic nuclei, including objects covering a wide range in luminosity of the central source. Rings of H II regions around the nucleus on 100 – 500 pc scales occur preferentially in galaxies with active nuclei. The stellar population, and its history, may be probed in some of the most favorable cases. Observed stellar absorption features and crude starburst models imply ages $\sim 10^8$years and $< 3 \times 10^7$years for the starbursts around the nuclei of NGC 1068 and 7469, respectively. These are in rough concordance with lifetimes of the nuclear activity based on radio structures. A somewhat different situation is found in Mrk 231,

where the whole galaxy (merger?) exhibits optical colors and spectra suggesting either an IMF deficient in OB stars or a sudden turnoff of star formation, on scales so large that it is unlikely the nucleus is directly responsible.

158.022   Star formation in Seyfert galaxies.
    J. M. Rodriguez Espinosa, R. J. Rudy, B. Jones.
NASA Conf. Publ., NASA CP–2466, p. 669 – 674 (1987). – See Abstr. 012.002.

From the high similarity between the far–IR properties (luminosity and spectral index) of Seyfert and starburst galaxies it is concluded that a large fraction of the emission at far–IR wavelengths of Seyfert galaxies is produced by star formation episodes in regions around the active nucleus. The high incidence of large far–IR output among the Seyfert population suggests the existence of a causal link between the active nucleus and the presence of bursts of star formation.

158.023   Circumnuclear "starbursts" in Seyfert galaxies.
    A. S. Wilson.
NASA Conf. Publ., NASA CP–2466, p. 675 – 691 (1987). – See Abstr. 012.002.

The author reviews a number of methods by which intense circumnuclear star formation may be diagnosed in Seyfert galaxies. The results of these different methods generally turn out to be in excellent agreement. In particular, the author emphasizes how the high spatial resolutions available at radio and optical wavelengths allow a clear separation of the effects of the nuclear activity proper from star formation going on around it.

158.024   IRAS observations of AGN candidates at low flux levels.
    M. H. K. de Grijp, W. C. Keel, G. K. Miley.
NASA Conf. Publ., NASA CP–2466, p. 693 – 698 (1987). – See Abstr. 012.002.

IRAS Additional Observations have been used to obtain a sample of point sources at much fainter flux levels than hitherto available through the IRAS Point Source Catalog. This sample is being used to compile an incomplete but representative catalog of faint IRAS candidate AGN's and to study the evolution of the infrared bright galaxies. Ground based follow up observations (optical spectroscopy) are mainly hampered by identification confusion.

158.025   Spectrophotometry of Brackett lines in very luminous IRAS galaxies.
D. L. DePoy.
NASA Conf. Publ., NASA CP–2466, p. 701 – 705 (1987). – See Abstr. 012.002.

Observation of the Brackett–α and Brackett–γ hydrogen recombination lines have been made in a sample of galaxies chosen from the IRAS catalog to have high luminosities at infrared wavelengths. Most have strong Brackett line emission indicating large numbers of high mass stars; the formation of these stars may hence be the underlying source for the galaxies' luminosities. However, there are at least two exceptions that may not be explained in this manner: NGC 6240 and Arp 220. Additional evidence indicates that each of these exceptions may be more closely related to Seyfert–type galaxies or other active galactic nuclei.

158.026   Evidence for extended IR emission in NGC 2798 and NGC 6240.
G. S. Wright, R. D. Joseph, P. A. James, N. A. Robertson.
NASA Conf. Publ., NASA CP–2466, p. 707 – 710 (1987). – See Abstr. 012.002.

Extended emission at 10 and 20 μm can be used to distinguish starbursts from "monsters" as the underlying energy source driving the luminous infrared emission in the central regions of galaxies. The authors have investigated the spatial extent of the mid–infrared emission in the interacting galaxy NGC 2798 and the merger NGC 6240. The 10 and 20 μm profiles of the IR source in NGC 2798 are significantly wider than beam profiles measured on a standard star, supporting a starburst interpretation of its IR luminosity. For NGC 6240 there is marginal evidence for an extended 10 μm source, suggesting that a significant fraction of its IR luminosity could be produced by a burst of star formation.

158.027   Models relating the radio emission and ionised gas in Seyfert nuclei.
A. Pedlar, S. W. Unger, D. J. Axon, J. E. Dyson.
NASA Conf. Publ., NASA CP–2466, p. 711 – 715 (1987). – See Abstr. 012.002.

The authors discuss possible models in which the radio emitting components in Seyfert II nuclei can compress and accelerate the ambient nuclear medium to produce the characteristics of the narrow line region. A first order model (Pedlar, Dyson & Unger 1985), which considers only the expansion of the radio components, is briefly described. However, in many Seyfert nuclei it appears that the linear motion of the radio components is also important. This can result in shock heating of the ambient medium, and if the cooling time is long enough, can lead to a displacement between the radio component and the associated [O III] emission lines. This effect may be present in NGC 1068 (Meaburn & Pedlar 1986), and NGC 5929 (Whittle et al. 1986).

158.028   Structure in the nucleus of NGC 1068 at 10 microns.
R. Tresch–Fienberg, G. G. Fazio, D. Y. Gezari, W. F. Hoffmann, G. M. Lamb, P. K. Shu, C. R. McCreight.
NASA Conf. Publ., NASA CP–2466, p. 717 – 721 (1987). – See Abstr. 012.002.

New 8 – 13 μm array camera images of the central kiloparsec of Seyfert 2 galaxy NGC 1068 resolve infrared source structure which is extended and asymmetric (2.1 × 0.7 arcsec FWHM), with its long axis oriented at position angle 33°. Infrared emission 1 – 2 arcsec to the northeast of the very center of the galaxy appears coincident with a weak, barely resolved, feature seen in the kiloparsec–scale jet detected in published radio continuum maps of the galaxy. Very Large Array observations show a linear, clumpy structure some 10 – 15 arcsec (~ 1 kpc) long, extending to the NE and SW of the nucleus. There is clear evidence of star formation in the disk of NGC 1068.

158.029   Far–infrared properties of optically–selected quasars and Seyfert galaxies.
R. A. Edelson, M. A. Malkan.
NASA Conf. Publ., NASA CP–2466, p. 723 – 725 (1987). – See Abstr. 012.002.

Pointed IRAS observations and ground–based observations are used to determine the infrared properties of optically–selected Seyfert galaxies and quasars. The use of complete, unbiased, optically–selected samples means that statistical tests can be applied to probe the underlying properties of active galactic nuclei.

158.030   Infrared–ultraviolet spectra of active galactic nuclei.
M. A. Malkan, R. A. Edelson.
NASA Conf. Publ., NASA CP–2466, p. 727 – 729 (1987). – See Abstr. 012.002.

Data from IRAS and IUE were combined with ground–based optical and infrared spectrophotometry to derive emission–line–free spectral energy distributions for 29 active galactic nuclei between 0.1 and 100 μm.

158.031   IRAS observations of BL Lac objects.
C. Impey, G. Neugebauer, G. Miley.
NASA Conf. Publ., NASA CP–2466, p. 731 – 735 (1987). – See Abstr. 012.002.

IRAS data has been analyzed for 35 BL Lac objects selected from a complete 5 GHz radio sample, using the coadded survey database. The detection rate is 50% with more than 40% detected in more than one band. This compares with only 15% of these sources that are included in the IRAS Point Source Catalog. High luminosity BL Lac objects generally have smooth energy spectra over four or five decades in frequency, consistent with incoherent synchrotron emission from 1 cm to 1 μm. However, many low luminosity BL Lac objects have discontinuous spectra, with a large range in the spectral index at IRAS wavelengths. For

BL Lacs with a total luminosity of less than 10^{44}ergs s^{-1}, most of the far infrared energy probably originates from dust heated near the galaxy nucleus.

158.032   Spectral classification of emission–line galaxies.
S. Veilleux, D. E. Osterbrock.
NASA Conf. Publ., NASA CP–2466, p. 737 – 740 (1987). – See Abstr. 012.002.

A revised method of classification of narrow–line active galaxies and H II region–like galaxies is proposed. It involves the line ratios [O III] λ5007/Hβ, [N II] λ6583/Hα, [S II] (λλ6716 + 6731)/Hα, and [O I] λ6300/Hα. These line ratios take full advantage of the physical distinction between the two types of objects and minimize the effects of reddening correction and errors in the flux calibration. Predictions of recent photoionization models by power–law spectra and by hot stars are compared with the observations. The classification is based on the observational data interpreted on the basis of these models.

158.033   Summary of symposium: low luminosity sources.
F. H. Shu.
NASA Conf. Publ., NASA CP–2466, p. 743 – 752 (1987). – See Abstr. 012.002.

158.034   Morphology of luminous IRAS galaxies: summary talk.
E. E. Becklin.
NASA Conf. Publ., NASA CP–2466, p. 753 – 756 (1987). – See Abstr. 012.002.

158.035   The 18 centimeter OH emission of IC 4553 (Arp 220).
W. A. Baan, A. D. Haschick.
Astrophys. J., Vol. 318, No. 1, p. 139 – 144 (1987).

All transitions of hydroxyl have been found in emission in IC 4553 (Arp 220). The line shapes have been decomposed and interpreted in terms of three distinct emission regions: one region at the systemic velocity of the galaxy, and a second at a 334 km s^{-1} higher velocity. A third region represents a nuclear outflow with a terminal velocity of approximately 700 km s^{-1}. The emission lines for the region I are the strongest and originate in a molecular disk surrounding the nucleus of the galaxy. Regions I and II may represent the two nuclei of the galaxies, which make up the merging IC 4553.

158.036   The ultraviolet and optical emission–line spectrum of III Zw 77.
G. J. Ferland, D. E. Osterbrock.
Astrophys. J., Vol. 318, No. 1, p. 145 – 160 (1987). = Lick Obs. Bull., No. 1059.

The high–ionization Seyfert 1 galaxy III Zw 77 has been simultaneously observed over the satellite ultraviolet to near–infrared spectral regions. The continuous energy distribution is similar to other active nuclei. The ultraviolet to optical emission–line spectrum for the narrow lines is fairly similar to those deduced for Seyfert 2 galaxies, in which the emitting gas is thought to be photoionized. A surprising result of this study is the detection of the O III] λ1661, 1666 doublet with an intensity, relative to the optical [O III] lines, more suggestive of shock heating than photoionization equilibrium. An interpretation in terms of a multicomponent photoionized structure, in which dense regions produce the ultraviolet lines and lower density regions are responsible for the optical lines, is found to be in good agreement with the observations.

158.037   Simultaneous multifrequency observations of Markarian 421.
J. Brodie, S. Bowyer, A. Tennant.
Astrophys. J., Vol. 318, No. 1, p. 175 – 187 (1987).

The highly variable BL Lacertae object Mrk 421 has been observed simultaneously in the radio, optical, ultraviolet, and X–ray bands over a period of 4 days in early 1984 December and once again in early 1985 January. Using the EXOSAT observatory, the authors found that during this time the 2 – 10 keV flux dropped by a factor of 8, whereas the 0.1 – 1 keV flux decreased by a factor of only 2. These changes were not reproduced at longer wavelengths during the period of simultaneous observations. The implications of these results are discussed in the context of (1) beaming models, (2) the connection between BL Lacertae objects and other classes of AGNs, and (3) the nature of the core region of active galaxies.

158.038   Interpretation of deep counts of radio sources.
L. Danese, G. De Zotti, A. Franceschini,
L. Toffolatti.
Astrophys. J., Lett. Ed., Vol. 318, No. 1, p. L15 – L20 (1987).

A new determination of the local radio luminosity function of galaxies allows the authors to conclude that unevolving low–luminosity sources cannot account for the observed flattening of the normalized differential source counts below mJy levels. The authors confirm that the upturn in the sub–mJy radio source counts is due to the emergence of an evolving population and they identify it with actively starforming galaxies. The inferred evolution time scale ($\sim 20\% - 25\%$ of the Hubble time) is within the expected range for gas consumption in starburst galaxies.

158.039   Flux density measurements of faint radio sources at 2.7 and 4.75 GHz.
T. Forkert, D. R. Altschuler.
Astron. Astrophys., Suppl. Ser., Vol. 70, No. 1, p. 77 – 82 (1987).

Flux densities at 4.75 and 2.7 GHz have been measured for a complete sample of faint radio sources ($S \gtrsim 50$ mJy) derived from the overlapping surveys carried out by Davis (1971) and by Altschuler (1986). These measurements over a time span of 15 years provide data for a study of the variability in a complete sample of faint radio sources (Altschuler and Forkert, 1987). The spectral indices having been obtained from measurements at the same epoch are not expected to be subject to uncertainties due to variability. The authors point out those sources which over the 15 year time interval have undergone the largest variations in their flux density.

158.040   Warm IRAS sources. I. A catalogue of AGN candidates from the Point Source Catalog.
M. H. K. de Grijp, G. K. Miley, J. Lub.
Astron. Astrophys., Suppl. Ser., Vol. 70, No. 1, p. 95 – 114 (1987).

The authors have previously shown that a blue (warm) 60 to 25 μm infrared colour provides a powerful parameter for discriminating between AGNs and normal galaxies and that the far–IR spectrum is therefore an efficient tool for finding new AGNs (de Grijp et al., 1985). They present a list of such AGN candidates based on warm IR sources from the IRAS Point Source Catalogue (PSC). Identification data and finding charts are also given. In addition the list of warm IRAS sources is supplemented by a compendium of data from the IRAS PSC on detected sources identified with previously known AGNs whose infrared spectra do not bring them within our colour selection criterion.

158.041   Five years monitoring of extragalactic radio sources. I. Observations at 12, 22 and 37 GHz.
E. Salonen, H. Teräsranta, S. Urpo, M. Tiuri, I. G. Moiseev,
N. S. Nesterov, E. Valtaoja, S. Haarala, H. Lehto, L. Valtaoja,
P. Teerikorpi, M. Valtonen.
Astron. Astrophys., Suppl. Ser., Vol. 70, No. 3, p. 409 – 435 (1987).

Near simultaneous observations of extragalactic radio sources have been made at Metsähovi Radio Research Station, Finland, and at Crimean Astrophysical Observatory, U.S.S.R., during 1980 – 85. Results of about 3000 observations of 48 sources at 37.22 and 12 GHz are presented. The measurement methods and calibrations at both stations are briefly described.

158.042   The clouds which form the extended emission line region of NGC 4388.
L. Colina.
Messenger, No. 49, p. 10 – 11 (1987).

158.043   *BVi* surface photometry of the Seyfert 2 galaxy NGC 1068.
S.–i. Ichikawa, S. Okamura, N. Kaneko, M. Nishimura,
K. Toyama.
Publ. Astron. Soc. Jpn., Vol. 39, No. 3, p. 411 – 424 (1987).

This paper presents high–resolution images of the inner arm region and the main disk of NGC 1068 in *B, V,* and near infrared *i* bands, and investigates optical properties of various features and components identified in previous studies on the basis of the ($B-V, V-i$) two–color diagram.

158.044   Powerful radio sources in clusters of galaxies: the origin of jet–like structures.
M. R. Gil'fanov, R. A. Syunyaev, E. M. Churazov.
Pis'ma Astron. Zh., Tom 13, No. 7, p. 560 – 574 (1987). In Russian. English translation in Sov. Astron. Lett., Vol. 13.

In the presence of a powerful compact source in a rich cluster of galaxies the Thomson scattering of its radiation on the intergalactic gas electrons produces a weak extended radio halo with luminosity $10^{-3} - 10^{-2}$ of the compact source luminosity. The compact source radiation being strongly beamed, weak prolonged jet–like structures might arise. These structures are of special interest in the case when beaming axes deviated strongly from the direction to the observer. Depending on the polarization of compact source radiation and orientation of its beaming axis the polarization of the diffuse source may vary from 0% up to 100%. These jet–like structures might arise in any spectral band depending on the spectrum of the compact source. Superluminal proper motions of jet–like structures are the natural consequence of this model.

158.045   Spectrophotometry of Markarian 367, 449 and 1119.
A. N. Burenkov, A. R. Petrosyan,
Eh. E. Khachikyan.
Astrofizika, Tom 26, Vyp. 3, p. 399 – 403 (1987). In Russian. English translation in Astrophysics, Vol. 26, No. 3, p. 241 – 244 (1987).

The results of spectrophotometric observations of Mrk 367, 449 and 1119 carried out with the 6–m telescope of SAO AS USSR are presented. Using the intensity ratios of the emission lines T_e, n_e, the abundances of O, N, the masses and volumes of the emitting gas and the numbers of the O7–type stars are calculated. Mrk 1119 is similar to the blue compact galaxies with powerful bursts of star formation. The central regions of Mrk 367 and 449 have characteristics of nuclei of spiral galaxies with flare star formation.

158.046   New observations of the variable galaxy Markarian 509.
K. A. Saakyan.
Astrofizika, Tom 26, Vyp. 3, p. 405 – 410 (1987). In Russian. English translation in Astrophysics, Vol. 26, No. 3, p. 245 – 248 (1987).

The results of brightness measurements of Mrk 509 during 1976 – 1980 are given. The magnitudes of comparison stars near Mrk 509 measured by different authors are compared. The light curve of Mrk 509 is given.

158.047   On a connection between flash and slowly varying components in light curves of Seyfert galaxies.
V. A. Hagen–Thorn.
Astrofizika, Tom 26, Vyp. 3, p. 415 – 419 (1987). In Russian. English translation in Astrophysics, Vol. 26, No. 3, p. 252 – 254 (1987).

It has been shown that in some Seyferts in the light curves of which flash and slowly varying components are prominent (components I and II), the amplitude of flux variations of component I is proportional to the flux of component II. Since components I and II are identical in colour features, it is very probably that variability is caused by a single slowly varying and fluctuating source.

158.048   The separation of radiation components in AP Lib.
A. V. Berdyugin.
Astrofizika, Tom 26, Vyp. 3, p. 566 – 570 (1987). In Russian.
English translation in Astrophysics, Vol. 26, No. 3.

The separation of radiation components in AP Lib has been performed on the basis of UBV–data taken from literature in the framework of a two–component model (galaxy + variable point source). The colours of the variable component are close to the colours of synchrotron radiation. The underlying galaxy is giant but perhaps nonelliptical.

158.049   Alternating side ejection or precession of jets in radio sources.
N. Roos, E. J. A. Meurs.
Astron. Astrophys., Vol. 181, No. 1, p. 14 – 18 (1987).

The radio source 4CT74.17.1 has been put forward by Rudnick (1982) and also Rudnick and Edgar (1984) as an example of a radio source exhibiting the "preferential avoidance" effect, which they attribute to alternating side ejection. The authors show that the main features of the brightness distribution of this source can be modelled by a relativistic jet slowly precessing around an axis near the plane of the sky. This type of configuration may also explain the apparent one–sidedness of the jets in the sample of radio quasars of largest angular size studied by Wardle and Potash (1984).

158.050   FIR galaxies with compact radio cores.
R. Chini, P. L. Biermann, E. Kreysa, H. Kühr, P. G. Mezger, J. Schmidt, A. Witzel, J. A. Zensus.
Astron. Astrophys., Vol. 181, No. 2, p. 237 – 243 (1987).

Comparing the IRAS point–source catalogue with sources detected in a VLBI extragalactic radio source survey the authors uncovered five FIR–sources which all show compact radio cores. They observed these objects with the 30 m MRT at Pico Veleta (Spain) at 1.2 mm wavelength to provide spectral coverage between IRAS and radio bands. The two galaxies among the five sources have luminosities of order $10^{12} L_\odot$ in the far infrared and thus may be super star bursters similar to Arp 220. On the other hand, all five objects have active galactic nuclei and so the far–infrared luminosities may be powered by the nuclear activity. Since flat spectrum radio sources are accepted to have compact nuclear components, the authors have also compared the 1 Jy–catalogue and its extension to lower flux densities with the IRAS point source catalogue, and have so identified a small number of additional active nuclei with strong emission in the far–infrared.

158.051   VLA observations of low–luminosity radio galaxies. VI. Discussion of radio jets.
P. Parma, C. Fanti, R. Fanti, R. Morganti, H. R. de Ruiter.
Astron. Astrophys., Vol. 181, No. 2, p. 244 – 264 (1987).

A statistical study is made of the properties of a large number of jets obtained from the B2 sample of low luminosity radio galaxies. The source sample, which contains about one hundred objects, was recently reobserved at 1.4 GHz using three different configurations of the VLA (resolution from 1″5 to 13″). The new observations were used for a comprehensive analysis of the radio jets. The jet properties are discussed in terms of the model of Bicknell.

158.052   High–dispersion spectroscopy of the clumpy irregular galaxies Markarian 297 and 325.
Y. Taniguchi, S. Tamura.
Astron. Astrophys., Vol. 181, No. 2, p. 265 – 269 (1987).

High–dispersion spectroscopic observations (0.3 – 0.6 Å resolution) have been made of four clumps of the two clumpy irregular galaxies Markarian 297 and 325. Their velocity dispersions in the Hα emission line (32 – 41 km s^{-1}) are twice or more as large as those of the so–called giant H II regions in nearby spiral or irregular galaxies. The resultant virial masses of the clumps are estimated as the order of $10^8 M_\odot$. Two clumps of the sample (Markarian 297–B and Markarian 325–B) show a redward excess emission which may be explained by a composite profile rather than a single gaussian one.

158.053   Lines of high excitation in NGC 4151: new measurements of [Fe X] and [Fe XIV].
D. Pelat, D. Alloin, E. Bica.
Astron. Astrophys., Vol. 182, No. 1, p. 9 – 14 (1987).

The authors provide in this paper new measurements from high resolution data of the [Fe X] 6374.5 Å and [Fe XIV] 5303 Å emission lines in NGC 4151. Line blends have been analyzed properly through a quantitative separation technique: compared to previous results the accuracy is substantially improved.

158.054   High resolution radio observations of NGC 4874.
L. Feretti, G. Giovannini.
Astron. Astrophys., Vol. 182, No. 1, p. 15 – 20 (1987).

High resolution radio observations, obtained with the VLA at 20 and 6 cm, are presented for the radio source associated with NGC 4874. This is one of the two dominant galaxies of the Coma cluster of galaxies and it exhibits a wide angle tail structure on a very small linear scale. The source structure consists of a weak core, two symmetric jets and lobes which are bent by nearly 90° with respect to the jet direction. It is suggested that the source is slightly foreshortened by projection effects and the bendings occur at the ISM/IGM transition surface. A simple ram pressure model is invoked to model the source structure, with the condition that the jet velocity is decreasing along the jet. A comparison with other sources with similar shape is also given.

158.055   A study of the starburst galaxy ESO 495–G21 = He2–10.
L. Johansson.
Astron. Astrophys., Vol. 182, No. 2, p. 179 – 188 (1987).

The author presents a photometric and spectroscopic investigation of the starburst galaxy ESO 495–G21. The central regions of the galaxy have colours characteristic of young stars and exhibit an emission–line spectrum which can be explained as due mainly to photoionization by hot stars. An analysis of the integrated stellar spectrum indicates that a burst of star formation is taking place, or took place during the last 1×10^7 yr, possible as a consequence of the close interaction between two dwarf galaxies. The newly formed stars make up ~1% of the luminous mass of the galaxy inside 5″. The continuum from ultraviolet to the far–infrared, can be modelled by combining this stellar population with an older one and a warm dust component. The galaxy has a non–thermal radio continuum which can be explained as due to supernova remnants.

158.056   A model for variable radio sources with VLBI components moving at superluminal speeds.
B. V. Komberg.
Sov. Astron., Vol. 30, No. 5, p. 518 – 524 (1986). English translation of 42.158.091.

158.057   Activity in galaxy nuclei and the SS433 phenomenon.
V. M. Lyutyj, A. M. Cherepashchuk.
Sov. Astron., Vol. 30, No. 5, p. 532 – 541 (1986). English translation of 42.158.092.

158.058   Theoretical parameters of powerful radio galaxies. Hydrodynamic approximation.
Yu. V. Baryshev, V. N. Morozov.
Astrophysics, Vol. 25, No. 2, p. 569 – 576 (1987). English translation of 42.158.223.

158.059   Polarimetric and photometric studies of BL Lac. Analysis of the observational data. II.
V. A. Gagen–Torn (*V. A. Hagen–Thorn*), S. G. Marchenko, V. A. Yakovleva.
Astrophysics, Vol. 25, No. 3, p. 634 – 640 (1987). English translation of 42.158.292.

158.060   Spectrophotometric investigation of the object Markarian 71.
N. K. Andreasyan, A. N. Burenkov, Eh. E. Khachikyan.
Astrophysics, Vol. 25, No. 3, p. 648 – 655 (1987). English translation of 42.158.293.

158.061 Emission–line activity in radio galaxies.
A. Robinson, L. Binette, R. A. E. Fosbury,
C. N. Tadhunter.
Mon. Not. R. Astron. Soc., Vol. 227, No. 1, p. 97 – 114 (1987).

Extensive regions of ionized gas are found at large distances from the centres of many radio galaxies which exhibit nuclear emission lines. The authors use line ratio diagnostic diagrams to show that the emission–line spectra of the extended nebulosities form well–defined trends with excitation (as measured by the [O I]/[O III] ratio). Remarkably, the associated nuclear narrow emission–line regions lie along the same loci. Such trends can be explained as a variation in ionization parameter for photoionization by a continuum spectrum which, from the small scatter in the diagrams, can be characterized by a mean ionizing photon energy of between 30 and 40 eV. This energy range rules out normal stars as the source of the ionizing photons. Different ionization processes are discussed.

158.062 Interstellar polarization in the dust lane of Centaurus A (NGC 5128).
J. H. Hough, J. A. Bailey, M. F. Rouse, D. C. B. Whittet.
Mon. Not. R. Astron. Soc., Vol. 227, No. 1, p. 1P – 5P (1987).

Optical and infrared linear polarization measurements of the Type 1 supernova 1986G in Centaurus A are used to determine the wavelength dependence of polarization produced by transmission of radiation through the dust lane of the galaxy. The observations are consistent with the standard form of interstellar polarization in our Galaxy with values of $P_{max} = (5.16 \pm 0.04)$ per cent and $\lambda_{max} = (0.43 \pm 0.01)\mu$m, with foreground material in our Galaxy contributing no more than ~ 0.5 per cent to the observed polarization. The efficiency of alignment of the grains in Cen A is smaller than found in the Milky Way.

158.063 X–ray intensity and spectral variations in the BL Lac sources H 2155–304 and PKS 0548–322.
P. C. Agrawal, K. P. Singh, G. R. Riegler.
Mon. Not. R. Astron. Soc., Vol. 227, No. 2, p. 525 – 534 (1987).

The results from the X–ray observations of two BL Lacertae objects, H 2155–304 and PKS 0548–322, are presented. Correlated X–ray intensity variations of ~ 20 per cent in the $0.15 - 3.5$ keV energy band and ~ 40 per cent in the $1.5 - 10$ keV energy band are observed in H 2155–304 over a period of about 8 hr. A factor of 2 change in the intensity of PKS 0548–322 in the $3.5 - 10$ keV energy interval is also detected over a time interval of 1 hr. Variations in the X–ray intensity over a time–scale of months and the hardening of the spectrum with increasing intensity are seen in the HEAO–1 A2 data on H 2155–304. The shortest observed time–scale for the variability is used to derive an upper limit on the mass of the central accreting source in the two BL Lac objects studied here.

158.064 IC 2476: a possible relic radio galaxy.
R. A. Cordey.
Mon. Not. R. Astron. Soc., Vol. 227, No. 3, p. 695 – 700 (1987).

The large double radio source B2 0924 + 30, associated with the galaxy IC 2476, has been studied as a possible example of an object whose central engine has ceased to be active but whose extended lobes have not yet faded from view. Such "dying" sources are expected to exist if radio galaxies have a finite active lifetime but no clear examples have yet been reported. In IC 2476 no evidence was found for continuing energy flow from an active nucleus, and any radio cores or jets it may possess are at least two orders of magnitude weaker than are commonly detected in sources of comparable total power.

158.065 IC 711 – the longest head–tail galaxy known.
J. P. Vallée, R. S. Roger.
Astron. J., Vol. 94, No. 1, p. 1 – 6 (1987).

The scale size of the largest structure in head–tail galaxies is not well known. Previous observations of the head–tail galaxy IC 711 (Vallée and Wilson, 1976) have shown its tail to extend 13 arcmin (700 kpc for $H_0 = 50$ km s^{-1}Mpc^{-1}) beyond the optical galaxy, making it then the longest head–tail galaxy known. New

observations at a longer wavelength of 74 cm with the synthesis telescope at Penticton show that the radio tail extends still further, to at least 17 arcmin (930 kpc), retaining its status as the longest head–tail galaxy known today. Some theoretical models for the origin of such galaxies are discussed with particular reference to IC 711.

158.066 Spectrophotometry of the Seyfert 1 galaxy Arakelian 120.
B. M. Peterson, S. A. Cota.
Astron. J., Vol. 94, No. 1, p. 7 – 11 (1987).

In an attempt to use the emission–line variability of the Seyfert 1 galaxy Akn 120 to improve estimates of the size of the broadline region, new closely spaced spectroscopic observations have been obtained. While these data reveal no lag between the continuum and Hβ light curves, it is emphasized that this does not necessarily imply that the line emitting region is extraordinarily small: during the period this galaxy was observed, the amplitude of the continuum variability was so low as to negate the improvement in the temporal resolution. These data are nevertheless useful in (1) establishing an absolute calibration for archival Akn 120 spectra by determining the flux in the narrow [O III] $\lambda5007$ line, (2) determining a photometric light curve dating back to 1974 from archival photometric and spectrophotometric data, and (3) establishing useful Balmer–line template profiles during a quiescent period longer than the light–travel time across the broadline region.

158.067 Spectroscopy of the extranuclear line–emitting regions associated with the gravitational lens system 2016 + 112.
D. P. Schneider, J. E. Gunn, E. L. Turner, C. R. Lawrence,
M. Schmidt, B. F. Burke.
Astron. J., Vol. 94, No. 1, p. 12 – 15 (1987).

Moderate–resolution slit spectroscopy of the extended emission–line objects A$_1$ and B$_1$ located within 5″ of the triple radio source 2016 + 112 shows that at least one is physically associated with the $z = 3.273$ quasar thought to be multiply imaged by a foreground galaxy. The data and theoretical models unanimously favor the identification of A$_1$ and B$_1$ with two separate clouds of ionized gas that lie within a few kiloparsecs of the primary A/B QSO and radio source. The available evidence does not support the view that they are multiple images of a single object or their classification as separate (companion) galaxies.

158.068 The size of Mrk 231 at 10 μm.
K. Matthews, G. Neugebauer, J. McGill, B. T. Soifer.
Astron. J., Vol. 94, No. 2, p. 297 – 299 (1987).

A technique is described for obtaining high–sensitivity measurements with high spatial resolution at 10 μm. An application of the method to Mrk 231 confirms the point–like nature of the nucleus at 10 μm.

158.069 The Seyfert 2 galaxy IC 184 and its surrounding group.
W. Kollatschny, K. J. Fricke.
Astron. Astrophys., Vol. 183, No. 1, p. 9 – 12 (1987).

The hitherto unknown bright Seyfert 2 galaxy IC 184 (= MCG–01–06–021) is described. In its vicinity the authors find three physically connected companions; one of which is an IRAS emission line galaxy. Similar conditions are consistently found in an ongoing survey of environments of Seyfert galaxies.

158.070 Markarian 297 knots.
J. Hecquet, G. Coupinot, A. J. Maucherat.
Astron. Astrophys., Vol. 183, No. 1, p. 13 – 15 (1987).

High resolution plates of the irregular galaxy Markarian 297 were obtained using the Piramig Image Tube at the 2 meter telescope of Pic–du–Midi Observatory. The contrast of the clumps has been improved by an adjust filtering of spatial frequencies. A detailed map giving their positions and B, R photometry is presented. The diameters of the knots range from 150 to 300 pc, therefore their dimensions are comparable to those of the clumps in Markarian 325. However, they are about 12 times less luminous. Therefore, the authors suggest that the mechanism of formation is different.

158.071   The optical spectral index in the south radio lobe of 3C 33.
P. Crane, A. Stockton, W. C. Saslaw.
Astron. Astrophys., Vol. 183, No. 1, p. 16 – 20 (1987).
Imaging photometry of the south radio lobe of 3C 33 in the R and V bands is reported and used to determine a value for the spectral index in the optical range of $\alpha = -0.94 \pm 0.53$. This value is compared to the radio spectral index and several possible mechanisms are proposed to explain the results. An $H\alpha$ image of the radio lobe region is compared to the R and V images and used to derive an equivalent width for the $H\alpha$ emission.

158.072   VLA observations of low luminosity radio galaxies. V. A detailed radio study of five jets.
R. Morganti, C. Fanti, R. Fanti, P. Parma, H. R. de Ruiter.
Astron. Astrophys., Vol. 183, No. 2, p. 203 – 216 (1987).
A detailed radio study is made of five radio sources with jets, which were selected from two B2 samples of low luminosity radio galaxies. The five objects were chosen such as to represent the different types of radio jets found in low luminosity radio galaxies. New VLA observations, made at 4.8 GHz, are presented. Using these and the 1.4 GHz data, which were discussed in previous papers in this series, the authors study the total intensity and polarization properties of the jets and extended regions. The present data are consistent with the predictions of the low Mach number jet model developed by Bicknell.

158.073   The nature of the BL Lacertae object AO 0235 + 164.
R. D. Cohen, H. E. Smith, V. T. Junkkarinen, E. M. Burbidge.
Astrophys. J., Vol. 318, No. 2, p. 577 – 584 (1987).
The authors have detected emission features in the optical spectrum of the highly variable BL Lac object AO 0235 + 164 which they identify as Mg II [Ne V], and [O II] at a redshift of 0.9399, consistent with a cosmological interpretation of the two absorption systems at $z = 0.524$ and $z = 0.851$. The spectrum also shows emission features associated with the $z = 0.524$ system. A comparison of the new data with previous spectroscopic observations of the BL Lac object when it was faint suggests that the Mg II emission may be variable on a time scale of approximately 2 yr.

158.074   The continuum of Type 1 Seyfert glaxies. II. Separating thermal and nonthermal components.
N. P. Carleton, M. Elvis, G. Fabbiano, S. P. Willner, A. Lawrence, M. Ward.
Astrophys. J., Vol. 318, No. 2, p. 595 – 611 (1987).
The authors attempt to distinguish between thermal and nonthermal contributions to the 1 to 100 μm continuum in a sample of active galactic nuclei. The sample is dominated by the members of a hard X–ray selected sample; most members are Seyfert 1 galaxies. It is postulated that the nonthermal continua of all the objects are intrinsically similar, with the apparent "steep spectrum" objects showing a combination of reddening at shorter wavelengths and thermal emission by dust at longer wavelengths. Various evidence is presented lending strong support to this interpretation. The spatial and temperature distribution of the infrared–emitting dust component in the nuclei of the sample galaxies is discussed.

158.075   Observations of 1.3 millimeter continuum emission from the centers of galaxies.
H. A. Thronson Jr., C. K. Walker, C. E. Walker, P. Maloney.
Astrophys. J., Vol. 318, No. 2, p. 645 – 653 (1987).
The authors have searched for 1.3 mm continuum emission toward the centers of eight galaxies and detected it with a signal–to–noise ratio greater than 3 in three: NGC 1068, Arp 220, and M82. M51 and NGC 253 were possibly detected, and negative results were obtained for IC 342, NGC 6574, and NGC 4038. The authors concluded that the emission at this wavelength is from dust at modest temperatures ($T_d \approx 20$K). Gas masses calculated from the 1.3 mm emission are in fair agreement with the estimated amount of H_2. It is shown that gas masses estimated from millimeter continuum observations are accurate

to about a factor of 5. The authors use the gas masses and the observed far–infrared luminosity to estimate the efficiency of star formation. For most galaxies, this efficiency is approximately equal to 1%.

158.076   The redshift of the BL Lacertae object PKS 2005–489.
R. Falomo, L. Maraschi, E. G. Tanzi, A. Treves.
Astrophys. J., Lett. Ed., Vol. 318, No. 2, p. L39 – L41 (1987).
In a high–resolution spectrum of PKS 2005–489 taken on 1986 August 16, two weak emission lines were detected at $\lambda 7031$ and $\lambda 7051$. Identification with $H\alpha$ and [N II] $\lambda 6583$ is proposed at a redshift $z = 0.071$. The observations correspond to a relatively faint state of the source with $B \approx 14.7$.

158.077   Parameters of nuclei and evolution of Seyfert galaxies.
A. S. Zentsova.
Astron. Zh., Tom 64, Vyp. 4, p. 720 – 723 (1987). In Russian. English translation in Sov. Astron., Vol. 31, No. 4.
The properties of Seyfert galaxies (SyG) and Seyfert–like galaxies as a metagalactic population are discussed. The mass of the central bodies and the mass of gas in the zone of radiation of the [O III] 4959 + 5007 lines for nine Seyfert–like galaxies are estimated. The typical value of the central body mass is $M = 3 \times 10^6 M_\odot$ for these galaxies. On the basis of the observational data on the IR–radiation and abundance of gas in SyG, the hypothesis that evolutionary transition of SyG1 to SyG2 exists is proposed.

158.078   Photoelectric observations with the AZT–11 telescope on spectral variability in Seyfert–galaxy nuclei.
N. I. Merkulova.
Bull. Crimean Astrophys. Obs., Vol. 75, p. 162 – 167 (1987). English translation of 42.158.150.

158.079   The halo of Virgo–A at 327 MHz.
A. P. Rao.
NRAO Workshop, No. 16, p. 79 – 82 (1986). – See Abstr. 012.015.
The radio galaxy Virgo–A has been mapped at 327 MHz with the Ooty Synthesis Radio Telescope. The high dynamic range map with a resolution of 1' resolves the halo and shows it to have a "S" shaped symmetry, suggestive of a precessing beam. The orientation of the large scale "S" shaped structure shows no connection with the orientation of the inner jet. However, the author shows that a precessing beam model can be generated that is consistent both with the inner and outer structures.

158.080   Radio and X–ray observations of M87.
E. D. Feigelson, P. A. D. Wood, E. J. Schreier, D. E. Harris, M. J. Reid.
NRAO Workshop, No. 16, p. 113 – 118 (1986). – See Abstr. 012.015.
The nearby radio galaxy M87 is studied with the Einstein X–ray Observatory and the Very Large Array to investigate possible inverse Compton X–ray emission from the "radio halo". An asymmetrical X–ray structure is superposed on the dominant symmetrical emission due to the hot interstellar gas, consisting of broad ridges extending ~ 0.5 to 5' east and southwest of the nucleus. The radio halo also has prominent structures to the east and southwest, probably due to bent jets seen in projection. But comparison of the X–ray and radio maps show the features are not entirely coincident. An inverse Compton origin of the X–ray emission implies the magnetic field of the halo varies between 3 and 8 times below equipartition levels of ~ 2 to 5 μG. The principal alternative explanation for the asymmetrical X–ray component is that the outflowing jets compress the interstellar medium, causing a local enhancement of thermal X–ray emission.

158.081 **New radio observations of NGC 1275.**
A. Pedlar, R. Perley, P. Crane, B. Harrison,
R. D. Davies.
NRAO Workshop, No. 16, p. 135 – 139 (1986). – See Abstr.
012.015.

The authors present VLA maps of NGC 1275 at 20 cm, with
resolutions ranging from 1″ to 40″. Over the central 30″ there is
evidence for collimated ejection in PA ∼ 160°. Outside this
region the radio structure bends rapidly by approximately
90 degrees before merging into the 10′ radio halo. It is suggested
that models of pressure driven accretion flows should take into
account the presence of the relativistic gas which is responsible
for the radio halo.

158.082 **What bends WATs?**
A. O'Donoghue.
NRAO Workshop, No. 16, p. 147 – 154 (1986). – See Abstr.
012.015.

This paper presents the problem of determining the mechanism
for bending the wide angle tail radio sources. The definition of
the class by morphology type, radio power, and optical association is reviewed. The current theoretical paradigm, upon which
evaluations of the plausibility of mechanisms proposed to bend
the sources into the observed shapes are based, is described. The
bending mechanisms that have been presented in the literature
are discussed, and newer ideas from the most recent publications
and previous talks in this conference are mentioned.

158.083 **Intensity maps of six wide angle tail sources.**
A. O'Donoghue, F. N. Owen.
NRAO Workshop, No. 16, p. 155 – 160 (1986). – See Abstr.
012.015.

The six WATs shown by contour plots in this paper were
observed by F. Owen in February and August of 1984. These
sources were first observed in a survey of Abell clusters of
galaxies done at the VLA at 20 cm. They were selected for
inclusion in a study of WATs for their "C" shapes, proximity
($z \sim 0.1$), brightness ($S \sim$ few 100 mJy), and size ($\geqslant 1$ arcmin).
Six more WATs selected by the same criteria from the same
survey are also being observed. The maps presented here were
made from observations at 6 cm in the B (3 hr. per source) and D
(1 hr. per source) configurations at the VLA. Details on each
source are given.

158.084 **Jet disruption in wide–angle tailed radio galaxies.**
J. O. Burns, M. L. Norman, D. A. Clarke.
NRAO Workshop, No. 16, p. 175 – 182 (1986). – See Abstr.
012.015.

Two models for the transition from jets to tails are proposed
for WAT radio sources. First, for the large, ∼1 Mpc sized
WATs, the authors suggest collisions with cool clouds in the
ICM. Second, for smaller, ∼200 kpc sized WATs, they propose
that the jets pass through a shock–like structure marking the
boundary between the ISM and ICM.

158.085 **Morphological and environmental clues to the origin of
powerful radio galaxies.**
E. P. Smith, T. M. Heckman.
NRAO Workshop, No. 16, p. 305 – 313 (1986). – See Abstr.
012.015.

Preliminary results of an optical survey of 47 radio galaxies are
presented in which it is found that Class A radio galaxies (those
galaxies characterized by strong emission lines in their spectra
and FR II radio morphologies) exhibit very peculiar optical
morphologies, photometric/structural properties similar to normal, bright ellipticals, blue colors, and reside in relatively sparsely
populated regions of space. In contrast, Class B radio galaxies
(weak/absent optical emission lines, FR I radio morphologies)
have elliptically–symmetric isophotal contours, photometric/
structural properties similar to brightest cluster galaxies, red
colors, and live in densely populated regions of space. These
distinctions, and their implications are significant to questions
concerning the origin of powerful radio sources.

158.086 **A dynamical model for the narrow line region of Seyfert
galaxies.**
B. Mobasher, D. J. Raine.
Mon. Not. R. Astron. Soc., Vol. 228, No. 1, p. 159 – 172 (1987).

The authors have computed detailed emission line ratios and
profiles for the "catapult" model of the narrow line region. First,
the authors review the basic principles of the "catapult" model.
Then, a discussion of their dynamical and photo–ionization
models and the choice of appropriate parameters is presented.
The authors compare their model predictions with the observations. Finally, the results are summarized and discussed.

158.087 **Radio observations of a few selected blazars.**
D. J. Saikia, C. J. Salter, S. G. Neff, A. C. Gower,
R. P. Sinha, G. Swarup.
Mon. Not. R. Astron. Soc., Vol. 228, No. 1, p. 203 – 216 (1987).

The authors present total–intensity and linear–polarization
observations of four selected blazars, 0716 + 714, 0752 + 258,
1156 + 295 and 1400 + 162, with the VLA A–array, and MER-
LIN and EVN observations of 1400 + 162. The sources
0752 + 258 and 1400 + 162 which have nearly constant optical
polarization, have well–defined double–lobed radio structure,
with relatively weak radio cores. In addition, 0752 + 258 appears
to be a twin–jet blazar. The position angle (PA) of the VLBI jet in
1400 + 162 is close to that of the arcsec–scale jet near the nucleus,
as well as the optical and 2–cm core polarization PAs. The
blazars 0716 + 714 and 1156 + 295, which exhibit strongly variable optical polarization, have a core–dominated radio structure
and perhaps have their jet axes close to the line–of–sight. From
polarization observations at 20, 18, 6 and 2 cm, the authors find
the rotation measure of the radio core in 0716 + 714 to be about
-20 rad m^{-2}. They suggest that low values of core rotation
measure in core–dominated sources could be consistent with the
relativistic beaming models.

158.088 **Photometric analysis and decomposition of AP Librae
(PKS 1514–24).**
D. A. Baxter, M. J. Disney, S. Phillipps.
Mon. Not. R. Astron. Soc., Vol. 228, No. 2, p. 313 – 327 (1987).

BVRI CCD surface photometry of the rapidly variable BL Lac
object AP Lib, is presented. The combination of high detector
efficiency and photometric accuracy, good seeing and the low
nuclear activity at the time of the observations allow a
decomposition into nuclear and underlying galaxy components
with greater precision than hitherto. The galaxy profile accurately follows a de Vaucouleurs law for ellipticals, at radii from 3 to
> 20 arcsec (> 26 kpc). Structural parameters appear normal for
a galaxy of its luminosity ($M_V = -22.81$). The colour data shows
a marked gradient towards redder colours at larger radii, the
reverse of that usually seen in E galaxies. Ellipse fits indicate that
AP Lib is seen almost face–on, with an essentially constant
eccentricity $\leqslant 0.1$. The optical energy spectrum of the quasi–
stellar component is examined and appears to be approximately
linear, with a spectral index, α, $\approx -1.4(\pm 0.1)$.

158.089 **Radio observations of a hard X–ray selected sample of
active galaxies.**
S. W. Unger, A. Lawrence, A. S. Wilson, M. Elvis,
A. E. Wright.
Mon. Not. R. Astron. Soc., Vol. 228, No. 2, p. 521 – 531 (1987).

The authors present radio observations of a hard X–ray
selected sample of active galaxies, made with the VLA and Parkes
radio telescopes. All of the galaxies observed with the VLA were
detected, making this an excellent sample with which to look at
the relationship between the radio and X–ray emission from
active galaxies. The authors use the ratio of the radio to X–ray
flux density as a measure of the degree of "radio–loudness" of an
active galaxy. They find no evidence for distinct "radio–quiet"
and "radio–loud" populations, rather there is a continuous
distribution of the degree of radio–loudness amongst the sample
galaxies. The X–ray and radio luminosity are found to be
correlated. The correlation is not linear, the radio–loud objects
all having high X–ray luminosity.

158.090   Rotation measure variation across M84.
R. A. Laing, A. H. Bridle.
Mon. Not. R. Astron. Soc., Vol. 228, No. 3, p. 557 – 571 (1987).

VLA images of the linearly polarized emission from the weak radio galaxy M84 (3C 272.1) with 3.9 arcsec resolution at 1.4 and 4.9 GHz show an organized pattern of Faraday rotation measure across the radio source. This pattern implies that there is a magnetoionic medium ~ 10 kpc in extent within M84, in front of, but not mixed with, the radio–emitting plasma. There must be a large–scale reversal in the magnetic field in this medium across the face of the radio source. The medium may be responsible for the more diffuse component of the X–ray emission from M84, and there is evidence that it is interacting with the outflow in the two radio jets.

158.091   The structure and kinematics of the ionized gas within NGC 5128 (Cen A) – I. TAURUS observations.
J. Bland, K. Taylor, P. D. Atherton.
Mon. Not. R. Astron. Soc., Vol. 228, No. 3, p. 595 – 621 (1987).

The TAURUS imaging Fabry–Perot system has been used at the Anglo–Australian Telescope to observe the ionized gas within Centaurus A. NGC 5128, the optical body of Cen A, is well known for the dust lane system which rotates rapidly about its projected major axis. A growing body of opinion affirms that the fact that this galaxy appears "peculiar" is a testimony to its proximity and that it would look much like any other radio elliptical at comparable distances. In this case, the presence of gas within this galaxy provides an invaluable laboratory for the study of a number of physical phenomena, in particular: (1) the three–dimensional form of elliptical galaxies, (2) the nature of orbits in early–type systems, and (3) the origin of nuclear activity within radio galaxies. To conduct these experiments, TAURUS has been used to observe the ionized gas within NGC 5128 in the light of [N II] $\lambda6548$ and Hα.

158.092   The extended narrow–line region in radio Seyferts: evidence for a collimated nuclear UV field?
S. W. Unger, A. Pedlar, D. J. Axon, M. Whittle,
E. J. A. Meurs, M. J. Ward.
Mon. Not. R. Astron. Soc., Vol. 228, No. 3, p. 671 – 679 (1987).

Long–slit spectra of seven Seyfert galaxies reveal high–excitation ([O III]$\lambda5007$/H$\beta \approx 5$) emission–line gas up to 20 kpc from the galactic nuclei. The low velocity dispersion (FWHM < 45 km s^{-1}) and orderly velocity field, characteristic of normal galactic rotation, suggest that this emission–line region, the Extended Narrow Line Region (ENLR), is physically distinct from the classical Narrow Line Region (NLR). The large physical extent, high excitation and kinematic properties of the ENLR point to it being ambient gas in the disc or halo of the galaxy which is photoionized by the nuclear radiation field. Despite the disparity in their physical sizes, the ENLR is substantially more extended parallel to the axis of the NLR radio structure than perpendicular to it. The authors argue that this elongation is due to an anisotropy in the nuclear radiation field whose origin is intimately related to the collimation of the radio ejecta.

158.093   An interpretation of the statistical properties of extragalactic radio sources with a relativistic beaming model.
K. Morisawa, F. Takahara.
Mon. Not. R. Astron. Soc., Vol. 228, No. 3, p. 745 – 758 (1987).

The statistical predictions of the unified scheme of steep–spectrum and flat–spectrum extragalactic radio sources based on the relativistic beaming model are explored. It is assumed that each radio source consists of a flat–spectrum compact core moving with relativistic speed and steep–spectrum extended lobes. The authors construct a luminosity function of steep–spectrum components which is consistent with observations of the local luminosity function, source counts and redshift distribution of steep–spectrum sources. Using this, they calculate the source counts and redshift distributions of flat–spectrum sources. A model in which the cores of radio galaxies are assumed to be weaker than those of quasars is also examined. It is found that

relativistic cores in radio galaxies are still needed to reproduce the observations.

158.094   Magnetic confinement of broad–line clouds in active galactic nuclei.
M. J. Rees.
Mon. Not. R. Astron. Soc., Vol. 228, No. 3, p. 47P – 50P (1987).

The region containing the clouds that emit the broad spectral lines in active galactic nuclei may be pervaded by a magnetic field of strength ~ 1 G. Magnetic stresses could then confine the clouds, obviating the need for a Compton–heated medium in pressure balance with the clouds.

158.095   A kinematic signature of bipolar flow in the Seyfert galaxy NGC 3516.
J. W. Goad, J. S. Gallagher III.
Astron. J., Vol. 94, No. 3, p. 640 – 643 (1987). With plate 59.

A long–slit echellogram of the circumnuclear emission–line region in the Seyfert 1 galaxy NGC 3516 reveals an unusual velocity field in the ionized gas. The radial velocities decline outward, from offsets of a few hundred km s^{-1} back toward the systemic velocity, over distances of ~ 500 pc on both sides of the nucleus. This gas has "nuclear" emission–line ratios. Circumnuclear gas kinematics in NGC 3516 are most simply modeled by a low–velocity, bipolar outflow from the nucleus. The bipolar–flow model is discussed in terms of observable properties of active galactic nuclei.

158.096   The initial mass function for early–type stars in starburst galaxies.
K. Sekiguchi, K. S. Anderson.
Astron. J., Vol. 94, No. 3, p. 644 – 650 (1987).

The IMF slope of early–type stars in starburst galaxies is investigated using IUE observations and a technique that utilizes mass–linewidth relations for early–type stars. Fourteen low–resolution IUE spectra of eight starburst galaxies and three H II region galaxies are used to obtain line–strength ratios Si IV ($\lambda1400$)/C IV ($\lambda1550$). These are compared to model line ratios, and indicate that the average IMF slope for OB stars in these intense star–formation regions is appreciably flatter than that of the solar neighborhood.

158.097   The diffusion model of extended radio components and jets with a moving source of accelerated particles.
S. G. Gestrin, V. M. Kontorovich, A. E. Kochanov.
Kinematika Fiz. Nebesn. Tel, Tom 3, No. 4, p. 57 – 66, 79 (1987). In Russian. English translation in Kinematics Phys. Celest. Bodies.

The "core–halo" model is applied to include the extended components of extragalactic radio sources. A region of injection (acceleration) of relativistic electrons is identified with a "hot spot" and supposed to move, which explains the radiolobes' asymmetry. The distribution of spectral indices over the lobe and breaks in the spectrum are found under the diffusion approximation and with allowance for synchrotron losses. The nonhomogeneity of the medium is taken into account for the slow moving hot spot, steep spectra with $\alpha > 1$ are shown to be formed in the magnetic field increasing towards the source edge. Similar consideration is carried out for the knot structure of jets with M87 as an example. Particularly, the increase of the knot dimensions with the decrease of the frequency and some features of the knot form are explained.

158.098   The variety of short–term variability in AGN.
K. A. Pounds, T. J. Turner.
Variability of galactic and extragalactic X–ray sources, p. 1 – 13 (1987). – See Abstr. 012.019.

One of the most important results of the EXOSAT programme has been the discovery of strong variability in the X–ray emission of a wide range of active galaxies. New results are reviewed which demonstrate a striking variety in the nature and timescales of the variability.

158.099 Soft X–ray variability and spectra of Seyfert galaxies.
C. M. Urry, J. S. Kruper, C. R. Canizares,
M. L. Rohan, M. R. Oberhardt.
Variability of galactic and extragalactic X–ray sources, p. 15 – 28
(1987). – See Abstr. 012.019.
Sixty–four optically–selected active galaxies were observed
with the Einstein Observatory. The fifty detections constitute the
largest single sample of Seyfert galaxies studied at X–ray
energies. The authors conclude that dramatic intensity changes
on time scales of a few hours are rare, and are more usual on time
scales of weeks to a year. Most individual spectra are consistent
with the canonical power law ($\alpha \sim 0.7$), but a few are steeper and
the mean spectral index is higher. There is no strong trend for the
absorbing column density of cold gas derived from the X–ray
spectral fit to be systematically lower than the 21–cm–derived
column through the interstellar medium of our galaxy, as would
be the case if soft excesses appeared in the 0.25 keV band.

158.100 Scale invariant variability in NGC 4051.
A. Lawrence.
Variability of galactic and extragalactic X–ray sources, p. 29 – 36
(1987). – See Abstr. 012.019.
A 62 hour observation of NGC 4051 with EXOSAT reveals
large amplitude variability on all timescales. The power spectrum
shows 1/f noise. This means there is no preferred timescale or
luminosity gradient, throwing into doubt the meaningfullness of
earlier interpretations of AGN variability.

158.101 X–ray and UV variability of MCG 8–11–11 in 1983 – 86.
L. Chiappetti, A. Allegrini, L. Maraschi, G. Tagliaferri,
E. G. Tanzi, A. Treves, W. Wamsteker.
Variability of galactic and extragalactic X–ray sources, p. 37 – 42
(1987). – See Abstr. 012.019.
The Seyfert galaxy MCG 8–11–11 has been the target of a
programme of coordinated observations in the X–ray, UV and
optical bands. A preliminary account of the complete series of X–
ray and UV observations is given.

158.102 Rapid X–ray variability in the Seyfert galaxy NGC 4593.
P. Barr, J. Clavel, P. Giommi, R. F. Mushotzky, G. Madejski.
Variability of galactic and extragalactic X–ray sources, p. 43 – 46
(1987). – See Abstr. 012.019.
The results of s series of Exosat and Einstein observations of
the Seyfert galaxy NGC 4593 are presented. Rapid X–ray
variability was found in five observations out of seven. The
observed variability timescale is a strong function of the mean
source intensity, the source varying more slowly when its
luminosity is higher.

158.103 Rapid X–ray variations in the high luminosity Seyfert galaxy III Zw 2.
J. S. Kaastra, P. A. J. de Korte.
Variability of galactic and extragalactic X–ray sources, p. 47 – 54
(1987). – See Abstr. 012.019.
The X–ray spectrum of the high–luminosity compact galaxy
III Zw 2 has been monitored using EXOSAT. This strong X–ray
source shows time variations on a scale of 5000 s. The variations
can be explained by low energy from a partially ionised cloud
which covers the central X–ray source, and which is situated in a
funnel of a thick radiation torus. The luminosity of III Zw 2 is
close to its Eddington limit and the radiation shows a considerable degree of beaming.

158.104 High spectral variability in E1615 + 061, a soft X–ray Seyfert 1 galaxy.
L. Piro, G. C. Perola, E. Massaro, D. Molteni.
Variability of galactic and extragalactic X–ray sources, p. 55 – 58
(1987). – See Abstr. 012.019.
The Seyfert 1 galaxy E1615 + 061 has undergone a spectacular
transition from an high state, characterized by a very steep X–ray
spectrum to a low state in which the spectrum recovered the
"canonical" shape. The X–ray behaviour of this object can be

interpreted in the framework of a two component model, namely
the extension into the X–ray band of an highly variable UV tail
plus a "canonical" hard X–ray component. The observation of
broad emission lines during the high state can be reconciled with
the standard two phase model of the broad line region if, as
suggested by the X–ray behaviour, the strong soft X–ray
emission is a transient phenomenon.

158.105 Fractal time–variability and spectral invariance of the Seyfert galaxy NGC 5506.
I. McHardy, B. Czerny.
Variability of galactic and extragalactic X–ray sources, p. 59 – 63
(1987). – See Abstr. 012.019.
The Seyfert galaxy NGC 5506 was observed continuously for
3 days by EXOSAT. The medium energy X–ray lightcurve
exhibits continual variation with amplitudes up to 30% on a
timescale of hours. However power spectral analysis and fractal
analysis both demonstrate that the variability is self–similar.
There is no evidence for any variation in the X–ray spectrum as a
function of intensity. The lightcurve can be explained by a
random distribution of events of random amplitude, however the
lack of spectral variability imposes strong constraints on the
emission mechanism, ruling out thermal Compton 'constant
seed' models and favouring non–thermal models involving
shocks.

158.106 X–ray variability of high luminosity AGN.
D. A. Schwartz, G. M. Madejski.
Variability of galactic and extragalactic X–ray sources, p. 65 – 76
(1987). – See Abstr. 012.019.
The authors discuss the Einstein Observatory results on X–ray
flux variability of BL Lac objects. On intermediate and long time
scales, variability is common and of greater amplitude than for
the Einstein observations of quasars. The authors attempt to
quantify the variability alternately as independent flare events or
as a stationary variability amplitude. They discuss and apply a
formalism that allows calculation of the conservative, minimum
variability which must have occurred during any given observation.

158.107 BL Lacertae objects: a comparison of X–ray variability to variability in different wavelengths.
J. N. Bregman.
Variability of galactic and extragalactic X–ray sources, p. 77 – 86
(1987). – See Abstr. 012.019.
The smoothly connected radio through ultraviolet continuum
exhibits variability that is often more rapid and more dramatic at
higher frequencies. When the soft X–ray emission is an extension
of the IR–UV synchrotron emission, X–ray variation is often
greater, but usually correlated with the optical data. There are
cases where X–ray emission is probably produced through the
inverse Compton process.

158.108 X–ray spectral variability of the BL Lac object PKS 2155–304 in 1983 – 85.
M. Morini, L. Chiappetti, L. Maraschi, G. Tagliaferri,
E. G. Tanzi, A. Treves.
Variability of galactic and extragalactic X–ray sources, p. 87 – 91
(1987). – See Abstr. 012.019.
PKS 2155–304 was observed with EXOSAT in 1983 – 85 at
9 epochs for a total of 80 hours. Large variability was detected on
both short (hours) and medium (weeks) time scales. The
dependence of the spectral shape with the intensity is complex.

158.109 X–ray variability of BL Lac objects.
D. Maccagni.
Variability of galactic and extragalactic X–ray sources,
p. 93 – 100 (1987). – See Abstr. 012.019.
The potential information contained in the detection of X–ray
variability in BL Lacs is considered along with the present
observational situation. Recent, new results on the X–ray time
behavior of the subclass of X–ray selected BL Lacs are presented
and discussed in the light of the optical monitoring of these
objects.

**158.110   Relativistic bulk acceleration in blazar jets. Observation-
al indications and model calculations.**
L. Maraschi.
Variability of galactic and extragalactic X–ray sources,
p. 101 – 110 (1987). – See Abstr. 012.019.

The author summarizes properties of the energy distribution of
blazars which indicate that orientation effects may be less
important for the X–ray emission than for the radio emission.
She discusses a model in which the emission at different
frequencies arises in different regions along a jet. If the emitting
plasma undergoes bulk acceleration within the jet, the bulk
velocity may be lower for the innermost regions leading to
different beaming angles at different frequencies.

158.111   Variability and the SSC model.
M. Salvati, G. C. Perola.
Variability of galactic and extragalactic X–ray sources,
p. 151 – 156 (1987). – See Abstr. 012.019.

The authors investigate the broad–band spectra and the
variability patterns arising in some variants of the basic SSC
(Synchrotron Self Compton) model for the continua of active
galactic nuclei. They include a semiquantitative treatment of the
electron acceleration, and consider the effects of an ambient
radiation field. The continuum UV and X–ray radiations
observed in the Seyfert 1 galaxy Fairall 9 are a powerful test of
the theory, and it is shown how tightly they constrain the SSC
scenario.

158.112   The physics and the structure of AGN.
H. Netzer.
Astrophysical jets and their engines, p. 103 – 124 (1987). – See
Abstr. 012.021.

An extragalactic object is classified as active galactic nucleus
(AGN) if at least one of the following criteria is fulfilled:
(1) Compact nuclear region brighter than the corresponding
region in galaxies of similar type. (2) Nonstellar (or nonthermal)
nuclear continuum emission. (3) Nuclear emission lines indicating non–stellar excitation mechanism. (4) Variable continuum
and/or emission lines. The main emphasis in this review is on the
more luminous AGN, where all four criteria are fulfilled.

**158.113   Improved accretion disk models of continuum emission
from active galactic nuclei.**
W.–H. Sun, M. A. Malkan.
Astrophysical jets and their engines, p. 125 – 128 (1987). – See
Abstr. 012.021.

The authors present calculations of the spectra of black hole
accretion disks which may power active galactic nuclei. They use
a simple model which is more realistic than previous ones in two
important ways. Firstly, they have accounted for the dominance
of electron scattering opacity in the disk, by approximating the
locally emerging flux with low–gravity non–LTE stellar atmospheres. Secondly, they have included the strong general relativistic effects of gravitational focusing and disk inclination on the
observed spectrum. The authors find that the spectrum of an
edge–on disk is approximately 5 times harder than that of the
same disk viewed face–on. The inferred black hole masses and
accretion rates change by comparable factors.

158.114   Velocities in radio galaxies and quasars.
P. A. G. Scheuer.
Astrophysical jets and their engines, p. 129 – 136 (1987). – See
Abstr. 012.021.

The author discusses the observational evidence relevant to the
speeds of jets in radio galaxies and quasars. His conclusion is that
the outer jets of weak (FR I) sources have non–relativistic speeds,
while jets in strong (FR II) sources have speeds comparable with
the speed of light, and perhaps very close to the speed of light.

158.115   Polarization and magnetic field structure.
P. A. G. Scheuer.
Astrophysical jets and their engines, p. 137 – 149 (1987). – See
Abstr. 012.021.

158.116   Optical synchrotron emission from radio hot spots.
P. R. Hiltner.
Astrophysical jets and their engines, p. 237 – 243 (1987). – See
Abstr. 012.021.

Optical polarization measurements of the following 3 objects
are reported: the southern hot spot of 3C 33; the western hot spot
of Pic A; and the jet of 3C 273.

158.117   Some studies on giant radio galaxies.
L. Saripalli, Gopal–Krishna.
Astrophysical jets and their engines, p. 247 – 249 (1987). – See
Abstr. 012.021.

**158.118   Radio interferometrical observations of the radio galaxy
3C 123 in decametric wavelength range.**
A. V. Men', S. Ya. Braude, S. L. Rashkovskij, I. S. Fal'kovich,
N. K. Sharykin, V. A. Shepelev, G. A. Inyutin,
A. D. Khristenko.
Pis'ma Astron. Zh., Tom 13, No. 9, p. 751 – 756 (1987). In
Russian. English translation in Sov. Astron. Lett., Vol. 13.

The results of observations of the radio source 3C 123,
obtained at frequencies 16.7, 20 and 25 MHz with the aid of the
broadband decametric radiointerferometer URAN–1 are presented. The analysis of the available radio interferometrical
observations performed in the frequency range 16.7 – 1425 MHz
permits to estimate the effective angular dimension of the source
(assuming the Gaussian brightness distribution) which at 50%
total intensity level turned out to be frequency independent and
equal to 23″.

**158.119   Kinematics and ionization of extended gas in active
galaxies. III. The extranuclear properties of NGC 1068.**
J. A. Baldwin, A. S. Wilson, M. Whittle.
Astrophys. J., Vol. 319, No. 1, p. 84 – 104 (1987).

A grid of 28 high– and seven low–dispersion long–slit spectra
have been obtained in the Seyfert 2 galaxy NGC 1068. The
resulting 1500 spectra have been used to investigate the kinematic
and ionization structure of the extranuclear gas. The velocity
field of most of the gas more than 15″ from the nucleus is well
described by a rotating disk with an approximately flat rotation
curve. The observed ionization structure of the gas may be
understood as a mixture of two disk components, one being of
high excitation and ionized by the nuclear continuum source and
the other of low excitation and ionized by hot stars. The
kinematics of these components is discussed. There is evidence
that large–scale, dense molecular gas clouds confine the nuclear
activity to a radius of ~15″ in the plane of the galaxy.

**158.120   A radiative bow shock wave (?) driven by nuclear ejecta
in a Seyfert galaxy.**
A. S. Wilson, J. S. Ulvestad.
Astrophys. J., Vol. 319, No. 1, p. 105 – 117 (1987).

New VLA maps at $\lambda 2$ cm of the 13″ scale linear radio source in
the center of NGC 1068 are described. Comparison with previous
observations at $\lambda 6$ cm allows the distributions of spectral index
and rotation measure to be obtained. Emission mechanisms for
the steep radio spectrum are considered. While the radio emission
of the northeast lobe may arise in old jet material, the authors
have modeled it in terms of a radiative bow shock driven by the
nuclear ejecta into the interstellar medium of the galaxy. The very
large enhancement (factor of $\sim 10^4$) in synchrotron emissivity
across the shock is associated with compression of the ambient
magnetic field and cosmic rays in the cooling gas.

**158.121   Infrared emission from young stars in the nucleus of
M82.**
M. Joy, D. F. Lester, P. M. Harvey.
Astrophys. J., Vol. 319, No. 1, p. 314 – 324 (1987).

The authors have used new observational and analytical
techniques to obtain high–resolution near–infrared and far–
infrared profiles of the luminous peculiar galaxy M82. From
simultaneous 1.25, 1.65, and 2.2 μm scans of the galaxy, it is
found that the strongest near–infrared emission is coincident with
the dynamical center of the galaxy. Simultaneous 40 μm and

100 μm high angular resolution profiles of M82 have been obtained with the Kuiper Airborne Observatory 0.9 m telescope. The analysis of the data indicates that the peak far–infrared emission from M82 is coincident with a cluster of young stars located to the southwest of the galactic nucleus at a projected distance of 10″ (150 pc). On a larger scale, the profiles reveal that the luminous infrared emission from young stars is confined to the interior of a neutral hydrogen and CO ring, 30″ (450 pc) in diameter.

158.122   The local radio luminosity function of galaxies.
L. Toffolatti, A. Franceschini, G. De Zotti,
L. Danese.
Astron. Astrophys., Vol. 184, No. 1/2, p. 7 – 15 (1987).

In view of the substantial discrepancies which are present among published estimates of the radio luminosity function of galaxies, the authors present a new determination exploiting the four richest radio and optically selected samples. Survival analysis techniques are used to fully take into account the radio upper limits. The four samples yield results in excellent agreement with each other. At the same time the luminosity ranges most thoroughly covered by each of them are largely complementary, thus allowing an accurate definition of the luminosity function over six decades of radio power. The authors also present a new estimate of the local luminosity function of "flat"–spectrum sources, reasonably well defined over a range in power greater than 10^4.

158.123   Formation of low ionization lines in active galactic nuclei.
M. Joly.
Astron. Astrophys., Vol. 184, No. 1/2, p. 33 – 42 (1987).

Up to now photoionization models of the broad line region of quasars fail to explain the strength of low ionization lines, in particular, Fe II lines are too weak compared to Hβ. Using a code mainly designed to study partly ionized and optically thick media, the radiation transfer in a non–photoionized homogeneous cloud is calculated in order to outline the physical conditions required to explain the observed Fe II intensities. The influence of the physical parameters is described and limiting values are set by the intensities of the other lines emitted by the medium. The main constraint is the Balmer continuum intensity. The results of the computations are compared to the published line ratios of about 30 well–observed AGN. It is shown that if the "Fe II problem" is not easily solved as regard the strong Fe II UV emitters, low temperature high density clouds provide Fe II$_{opt}$/Hβ ratios in very good agreement with observations.

158.124   The correlation between radio and optical variations in OJ 287.
L. Valtaoja, A. Sillanpää, E. Valtaoja.
Astron. Astrophys., Vol. 184, No. 1/2, p. 57 – 62 (1987).

The correlation between optical and radio variations in the BL Lac object OJ 287 is studied with five different radio frequencies in the range of 4.8 to 37 GHz. The authors use both higher frequencies (22 and 37 GHz) and more extensive data sets than in previous analyses of similar kind. The optical variations are found to precede the radio ones by less than one year. The time delay is proportional to wavelength at least in lower frequencies. Above 10 GHz the optical–radio time delay may be constant, about two months. The correlation exists both for small events during as quiescent period and for the post–1982 active phase of OJ 287. The authors also consider the alternative that there may be two different delays at millimeter wavelengths.

158.125   Constraints on confinement mechanisms of extragalactic radio sources.
J. C. Carvalho.
Astron. Astrophys., Vol. 184, No. 1/2, p. 79 – 85 (1987).

Confinement mechanisms of extragalactic radio sources are reviewed and different observational constraints are studied. General expressions are obtained for the maximum flux density as a function of the source angular size, which is allowed by considering thermal absorption, X–ray emission and Faraday effects. Extended and compact sources are analysed. Numerical computation and comparison with a large sample of sources show that ram pressure containment is more likely to work than other mechanisms for extended sources. A general analysis also shows that ram pressure is unlikely to work for most of the compact sources as extremely high densities are required. In the particular case of sources which suffer Doppler enhancement due to relativistic beaming effect, less severe constraints on confinement mechanisms exist.

158.126   Extended ionized nebulosities in the galaxies Mk 1, Mk 3, Mk 348 and the quasar 4C 37.43.
J. Bergeron, F. Durret.
Astron. Astrophys., Vol. 184, No. 1/2, p. 93 – 103 (1987).

The authors present long slit spectroscopic data of three Markarian galaxies and one QSO in the Hβ–[O III] $\lambda\lambda4959$ – 5007 wavelength range. They have detected ionized nebulosities around the four objects. They present kinematic data, together with the variations of the [O III] $\lambda5007$ line intensity and [O III] $\lambda5007$/Hβ intensity ratio throughout the ionized region. The absolute density and filling factor of the ionized gas have been estimated for the four objects studied here and four additional nebulosities previously known. The filling factors are always small. However they tend to be larger in galaxies with weaker UV sources and at smaller distances from the nucleus. A more homogeneous distribution of the gas might also be linked to large turbulent motions.

158.127   Mark 277 – clumpy irregular galaxy.
N. K. Andreasyan, A. N. Burenkov,
Eh. E. Khachikyan.
Astrofizika, Tom 27, Vyp. 1, p. 103 – 109 (1987). With two plates. In Russian. English translation in Astrophysics, Vol. 27, No. 1.

The results of a densitometry and detailed spectrophotometric investigation of the galaxy Mark 277 are given. Observational material was obtained on 6 m and "Zeiss–600" telescopes of the Special Astrophysical Observatory. Four condensations with spectra of H II regions are discovered in Mark 277. Their physical properties are nearly the same as in normal H II regions, while they are metal–deficient compared with H II regions. So, Mark 277 consists of four clumps which are metal–deficient complexes of H II regions.

158.128   Low–luminosity Seyfert nuclei in nearby galaxies.
A. V. Filippenko, W. L. W. Sargent.
AIP Conf. Proc., No. 155, p. 172 – 175 (1987). – See Abstr. 012.029.

Broad Hα emission lines have been detected in the nuclei of many bright galaxies such as M81 and M87. These lines are similar to, but much weaker than, those seen in type 1 Seyfert nuclei and QSOs. If massive black holes are responsible for the broad lines and the immense luminosities of classical AGNs, continuity arguments suggest that they also produce the features observed at lower levels in the relatively "normal" galaxies studied in this paper, especially since the intensity ratios of the narrow emission lines are like those expected from gas photoionized by dilute nonstellar (e.g., power–law) radiation. The possible presence of massive black holes in nearby galaxies is consistent with the conclusion that the nucleus of our own Milky Way may harbor such an object.

158.129   A direct determination of linear–size evolution of elliptical radio galaxies.
M. J. A. Oort, P. Katgert, R. A. Windhorst.
Nature, Vol. 328, No. 6130, p. 500 – 501 (1987).

A redshift dependence of the maximum distance out to which active galactic nuclei can transport their relativistic particles was first found for radio quasars by Miley. By using samples of radio ellipticals complete to vastly different flux limits, the authors can, for the first time, separate redshift– and luminosity–dependences of their linear sizes. Ram pressure in a medium with cosmologically varying density will naturally cause a dependence on redshift, but this cannot explain so steep a dependence as that implied here. It may be expected if galaxy haloes influence the outward energy transport.

158.130 Statistical significance of the relationship between X–ray luminosity and variability timescale in active galactic nuclei.
J. B. Stephen, L. Bassani, E. Caroli, G. Di Cocco.
Nature, Vol. 328, No. 6133, p. 784 – 786 (1987).

Recently, Barr and Mushotzky (41.158.087) have claimed that a significant correlation exists between the X–ray luminosity and the timescale of X–ray variability for Seyfert galaxies and quasars. This letter examines carefully their method of analysis and its statistical significance and applies a more general method of data analysis. It is concluded that the results from both methods are statistically too weak to be conclusive.

158.131 0.6 GHz mapping of extended radio galaxies. II. Edge–darkened double sources.
W. J. Jägers.
Astron. Astrophys., Suppl. Ser., Vol. 71, No. 1, p. 75 – 108 (1987).

Radio observations made with the Westerbork telescope at 0.6 GHz are presented for 8 edge–darkened double sources: NGC 315, NGC 326, 3C 31, 3C 130, B 0915+320, HB 13, NGC 6251 and 3C 449. Previously observed Westerbork data at 1.4 GHz are convolved for comparison with the 0.6 GHz observational data. Besides maps of the total intensity and linear polarization structure, the distributions of the spectral index, the depolarization and the rotation of the polarization position angle between 0.6 GHz and 1.4 GHz have been derived. Integrated values for the total intensity and the polarization are also given.

158.132 77 GHz continuum observations of variable extragalactic sources.
H. Teräsranta, E. Valtaoja, S. Haarala, A.–M. Elo, M. Valtonen, E. Salonen, S. Urpo, M. Tiuri, E. Laurikainen.
Astron. Astrophys., Suppl. Ser., Vol. 71, No. 1, p. 125 – 129 (1987).

During the period April – May 1984 the flux densities of 25 extragalactic radio sources were monitored with the Metsähovi Radio Telescope at a frequency of 77 GHz. In BL Lac the authors found a 10 day event where the flux rose by about 40%. Another short term change was observed in OJ 287 whose flux density fell from 6.9 to 3.5 Jy in 28 days during May 1984.

158.133 The structure and kinematics of the ionized gas in Centaurus A.
J. Bland, K. Taylor, P. D. Atherton.
Structure and dynamics of elliptical galaxies, p. 417 – 418 (1987). – See Abstr. 012.032 (IAU Symp. No. 127).

The TAURUS Imaging Fabry–Perot System (Taylor & Atherton, 1980) has been used with the IPCS at the AAT to observe the ionized gas within NGC 5128 (Cen A) at [N II] $\lambda6548$ and Hα. Seven independent (x,y,λ) data cubes were obtained along the dust lane at high spectral resolution (30 km/s FWHM) and at a spatial resolution limited by the seeing ($\sim 1''$). From these data, maps of the kinematics and intensities of the ionized gas were derived over a $420''$ by $300''$ region.

158.134 Fabry–Perot observations of Cen A.
J. J. E. Hayes, R. A. Schommer, T. B. Williams.
Structure and dynamics of elliptical galaxies, p. 419 – 420 (1987). – See Abstr. 012.032 (IAU Symp. No. 127).

The authors present Fabry–Perot spectrophotometry of the well-known peculiar galaxy Cen A (NGC 5128). The observations were carried out using the Rutgers Fabry–Perot system and a CCD as a detector. The authors scanned the Hα and [N II] ($\lambda6583$) emission lines. From these data they were able to construct maps of the continuum, line emission, velocity and velocity dispersion. The velocity maps in both Hα and [N II] have smooth gradients and twists in the line of nodes. The deprojected emission maps strongly resemble emission maps of face–on spirals. The authors speculate that Cen A is a merger between an elliptical and a spiral.

158.135 H I imaging of radio active ellipticals.
J. H. van Gorkom.
Structure and dynamics of elliptical galaxies, p. 421 – 422 (1987). – See Abstr. 012.032 (IAU Symp. No. 127).

Statistics on H I absorption in radio galaxies suggest that accretion of gas into the center might be a general phenomenon. Images of the H I emission from these galaxies show that the gas must have an external origin. The rotation curves derived from the H I data are flat.

158.136 Gas, dust and radio emission in elliptical galaxies.
W. B. Sparks.
Structure and dynamics of elliptical galaxies, p. 425 – 426 (1987). – See Abstr. 012.032 (IAU Symp. No. 127).

CCD data are used to compare the isophotes, colour and dust–content of radio ($10^{21} < P_{5\,GHz} < 10^{24}$ W Hz^{-1}, $H_0 = 75$ km s^{-1} Mpc^{-1}) and radio–quiet ellipticals. Radio ellipticals are round but not spheroidal, reddened and occasionally have disturbed dust lanes. X–ray emission correlates with both radio emission and shape. Detailed investigations of dust in NGC 1316 reveal a possible nuclear gas disc orthogonal to the radio jet.

158.137 The discovery of blazar–type nuclei in two nearby radio ellipticals.
W. B. Sparks, J. Bailey, J. H. Hough, C. Brindle, D. J. Axon.
Structure and dynamics of elliptical galaxies, p. 427 – 428 (1987). – See Abstr. 012.032 (IAU Symp. No. 127).

Near infrared polarimetry of Centaurus A and IC 5063 has revealed the existence of a steep spectrum highly polarized source in the nuclei of both galaxies. The position angle of polarization is perpendicular to the radio position angle. The authors interpret this polarized emission as synchrotron radiation. This, together with a luminosity of 5×10^{41} erg s^{-1}, suggests the galaxies are low luminosity blazars and that such nuclei may be common in elliptical galaxies.

158.138 Stellar dynamics of radio elliptical galaxies.
A. Sansom, J. V. Wall, W. B. Sparks.
Structure and dynamics of elliptical galaxies, p. 429 – 430 (1987). – See Abstr. 012.032 (IAU Symp. No. 127).

Stellar kinematical and dynamical results are presented for 34 radio ellipticals. The radio galaxies the authors observed were brighter than m(B) = 16. The results show that these radio ellipticals are not generally more rapidly rotating than their non-radio counterparts. Evidence for some rotation about the major axis is seen in two cases. These radio ellipticals do not appear to obey the luminosity, velocity dispersion trend seen for normal ellipticals.

158.139 Properties of the X–ray emitting gas in early type galaxies.
C. R. Canizares, G. Fabbiano, G. Trinchieri.
Structure and dynamics of elliptical galaxies, p. 431 – 432 (1987). – See Abstr. 012.032 (IAU Symp. No. 127).

158.140 Are cooling flows governing E–galaxy evolution?
E. A. Valentijn.
Structure and dynamics of elliptical galaxies, p. 433 – 434 (1987). – See Abstr. 012.032 (IAU Symp. No. 127).

Gas accretion of intracluster gas into the potential well of giant elliptical or cD galaxies can provide the material for both nuclear non–thermal activity and continuous, probably low mass, star formation. The cooling accretion flows could lead to the original formation of the visible object, and subsequently govern its evolution.

158.141 The hydrodynamical evolution of gas in young elliptical galaxies.
R. Kunze, H. H. Loose, H. W. Yorke.
Structure and dynamics of elliptical galaxies, p. 437 – 438 (1987). – See Abstr. 012.032 (IAU Symp. No. 127).

The authors calculate the partial inflow of gas fuelled by stellar mass loss at an early epoch (10^9 yr after the birth of the galaxy) during the evolution of an elliptical galaxy assuming a modified King model stellar distribution. The influence of the partial thermalization of stellar mass lost on the amount of gas which can be stored in the nucleus of a typical elliptical during the time of partial inflow is investigated. Masses up to $10^5 M_\odot$ of cool ($\leqslant 10^4$K) material can be stored in the nucleus of the galaxy before the fast dissipation of the "kinetic bulk energy" of the nuclear gas clouds leads to "thermal" instability and subsequent collapse. A supermassive star can form.

158.142 Ultraviolet energy distributions of (32) early–type galaxies.
F. Bertola, D. Burstein, L. M. Buson, S. M. Faber,
T. R. Lauer.
Structure and dynamics of elliptical galaxies, p. 439 (1987). – See Abstr. 012.032 (IAU Symp. No. 127).

158.143 Spectroscopy of the globular clusters in M87.
J. R. Mould, J. B. Oke, J. M. Nemec.
Structure and dynamics of elliptical galaxies, p. 451 – 452 (1987). – See Abstr. 012.032 (IAU Symp. No. 127).

With a velocity dispersion of 370 ± 50 km/sec the globular cluster system of M87 is kinematically hotter than the stars in the giant elliptical itself. This is consistent with the clusters' shallower density distribution for isotropic orbits. The mean metallicity of the 27 clusters in the sample analyzed here is no more than a factor of 2 more metal rich than the cluster system of the Milky Way, but considerably more metal poor than the integrated starlight in the field at a radius of $1'$ from the center of M87. There is no evidence for the existence of young clusters in the system. The mass–radius relation between $1'$ and $5'$ required to contain the globular clusters joins on to that required to contain the hot gas around M87.

158.144 The "jet" of M89: CCD surface photometry.
G. Clark, P. Plucinsky, G. Ricker.
Structure and dynamics of elliptical galaxies, p. 453 – 454 (1987). – See Abstr. 012.032 (IAU Symp. No. 127).

The authors have obtained CCD images in R and V of Malin's "jet" in the weakly radio and X–ray active E0 galaxy M89 (NGC 4552). The luminosity of this feature is approximately 1/4% of the total luminosity of the galaxy; its color is bluer than that of the whole galaxy with a V–R value smaller by about 0.15 magnitudes. The likely explanation of the feature, which looks more like a proboscis than a jet, is that it is a "tidal relic of a close encounter", seen from a perspective that may hide a drawn out tail.

158.145 Morphology of extended emission–line regions associated with radio galaxies.
L. Hansen, H. U. Nørgaard–Nielsen, H. E. Jørgensen.
Astron. Astrophys., Suppl. Ser., Vol. 71, No. 3, p. 465 – 491 (1987).

Narrow–band CCD observations have been obtained for 9 galaxies associated with radio sources, and calibrated images of extended ionized gas in the lines of [O III] and H_α+[N II] have been derived. For six of the objects the clouds are distributed along a Z–like pattern. A bridge of emission is found from PKS 0349–27 to a neighbour $1\rlap{.}'2$ towards west suggesting a tidal encounter. For some of the objects the interaction between the ionized gas and jets of radio plasma is supposed to be important for the excitation and dynamics of the gas. However, in most cases the available radio data show no relation to the gas distribution.

158.146 Optical identifications and radio morphology of the complete 5 GHz S5 survey.
H. Kühr, K. J. Johnston, S. Odenwald, J. Adlhoch.
Astron. Astrophys., Suppl. Ser., Vol. 71, No. 3, p. 493 – 523 (1987).

The 185 sources of the S5 survey stronger than 250 mJy at 5 GHz were mapped with the Very Large Array with an angular resolution of $2''$. The majority (74%) of the sources display spatial structure. Almost all (96%) sources which lie in the direction of clusters of galaxies are resolved. Sources that have a steep spectral index ($\alpha_{1.4-10\ \text{GHz}} < -0.5$) are not as variable in flux density as the flat spectrum objects. The radio source positions are accurate to $0\rlap{.}''2$ in the extragalactic radio reference frame. Those positions reported for resolved sources are influenced by the detailed response of the radio structure to our synthesized beam. Fifty–six of the sources were identified on the POSS. An almost equal number of quasars and galaxies were found including 14 BL Lac objects.

158.147 The optical polarization properties of blazars.
A. Kulshrestha, M. R. Deshpande, U. C. Joshi.
Astron. Astrophys., Suppl. Ser., Vol. 71, No. 3, p. 565 – 568 (1987).

Optical polarimetry in $UBVRI$ wavebands is reported for three BL Lacertae objects and one highly polarized quasar. Two objects, PKS 0735+178 and ON 325, exhibit wavelength independent polarization and position angle with polarimetric signal–to–noise ratios greater than 7 and 30, respectively. The highly polarized quasar 4C 29.45 shows a large differential rotation in its plane of polarization ($56° \pm 18\rlap{.}°4$) from U to V band. Also two distinct slopes in polarization angle, one from U to V band and the other from V to I band, are observed. The most plausible model to explain the observed changes in the polarization angle for 4C 29.5, seems to be a two–component synchrotron model.

158.148 0.6 GHz mapping of extended radio galaxies. III. 3C 66B, NGC 1265, 3C 129, DA 240, 3C 236, 4C 48.29, IC 708 & IC 711, 4CT 51.29.1, 3C 310, Abell 2256, 3C 402 and 3C 465.
W. J. Jägers.
Astron. Astrophys., Suppl. Ser., Vol. 71, No. 3, p. 603 – 642 (1987).

Radio observations made with the Westerbork Telescope at 0.6 GHz are presented. Previously observed Westerbork data at 1.4 GHz are convolved for comparison with the 0.6 GHz data. In addition, maps of the total intensity and linear polarization structure, the distributions of the spectral index, the depolarization and the rotation of the polarization position angle between 0.6 GHz and 1.4 GHz have been derived. Integrated values for the total intensity and the polarization are also given.

158.149 On the variability of the Seyfert galaxy NGC 1068.
V. M. Lyutyj, V. Yu. Rakhimov.
Pis'ma Astron. Zh., Tom 13, No. 11, p. 973 – 979 (1987). In Russian. English translation in Sov. Astron. Lett., Vol. 13.

The results of $UBVR$ photoelectric observations of the NGC 1068 nucleus are presented. It is shown that poor astroclimatic conditions may lead to an appearance of a false variability of high amplitude. The amplitude of light variations of the nucleus of NGC 1068 is $\leqslant 0\rlap{.}^m1$ in U filter.

158.150 Spectral observations of new galaxies with ultraviolet excess. I.
M. A. Kazaryan, Eh. S. Kazaryan.
Astrophysics, Vol. 26, No. 1, p. 1 – 6 (1987). English translation of 43.158.054.

158.151 New processing of surface photometry of Markaryan galaxies. I.
M. Kalinkov, I. Kuneva, F. Börngen, A. T. Kalloglyan.
Astrophysics, Vol. 26, No. 1, p. 15 – 24 (1987). See Abstr. 43.158.055.

158.152 Luminosity of binary radio sources as a function of the separation between the components.
V. R. Amirkhanyan.
Bull. Spec. Astrophys. Obs. – North Caucasus, Vol. 22, p. 45 – 49 (1987). English translation of 41.158.201.

158.153   H II region abundances in Seyfert galaxies.
I. N. Evans, M. A. Dopita.
Astrophys. J., Vol. 319, No. 2, p. 662 – 670 (1987).

The theoretical H II region abundance sequence calibration reported by Dopita and Evans in 1986 has been applied to optical spectrophotometry of 23 H II regions located in the inner disk regions of two Seyfert 1 and two Seyfert 2 galaxies, including the prototype Seyfert 2, NGC 1068, in order to determine oxygen, nitrogen, and sulfur abundances. The mean oxygen abundance derived for each galaxy ranges between solar abundance and twice solar abundance. There is no evidence for abnormal N/O or S/O abundance ratios in any of the observed H II regions.

158.154   B2 0800 + 24: a narrow–angle tail radio galaxy in a small group of galaxies.
J. T. Stocke, J. O. Burns.
Astrophys. J., Vol. 319, No. 2, p. 671 – 682 (1987). With plates 11 – 12.

A search has been conducted with the VLA for radio galaxies possessing "head–tail" (HT) morphologies in regions of low galaxy density. Two sources were found (B2 0800 + 24 and NGC 4410a) which possess asymmetric extended radio structure. Of these, B2 0800 + 24 most resembles the HT cluster galaxies referred to as "narrow–angle tails" (NATs) despite its location in a group of only five or six bright galaxies with a radial velocity dispersion of 300 km s^{-1}. The authors present a self–consistent model that explains the appearance of NATs in poor groups.

158.155   Extended Lyman–α emission in 3C 326.1: a 100 kiloparsec cloud of ionized gas at a redshift of 1.82.
P. J. McCarthy, H. Spinrad, S. Djorgovski, M. A. Strauss, W. van Breugel, J. Liebert.
Astrophys. J., Lett. Ed., Vol. 319, No. 2, p. L39 – L44 (1987). With plates L4 – L5.

The authors report the discovery of a large cloud of ionized gas associated with the high–redshift radio source 3C 326.1. New radio–frequency images made at 4.9 GHz and 15 GHz with the VLA show the radio source to be a small double (∼ 7″) without a detectable core. Long–slit spectrograms and Lyα imaging reveal a ∼ 100 kpc diameter cloud of ionized gas with a redshift of 1.825 encompassing the radio source. Deep broad–band images show two faint ($V \approx 23.5 - 24.5$) blue objects located on the periphery of the cloud, as well as some very faint ($V \approx 25 - 26$) extremely blue diffuse objects roughly coincident with the brightest regions of the cloud. It is tentatively proposed that 3C 326.1 is a young and/or forming galaxy.

158.156   Shock–wind mechanism of formation of the envelope structure in NGC 5128. Numerical modelling.
B. I. Gnatyk, V. A. Krol'.
Inst. Teor. Fiz. Akad. Nauk USSR, Prepr., No. 46R (1987). In Russian. Abstr. in Ref. Zh., 51. Astron., 9.51.997 (1987).

158.157   Two elliptical galaxies with active nuclei: NGC 6212 and Mkn 501.
M. Moles, J. Masegosa, A. del Olmo.
Astron. J., Vol. 94, No. 5, p. 1143 – 1149 (1987).

Spectrophotometric data for the elliptical galaxies NGC 6212 and Mkn 501, as well as $UBVRI$ photometry for Mkn 501, are presented. Both galaxies are known to harbor active nuclei. The Seyfert 1 nature of the nucleus of NGC 6212 is confirmed, and it is argued that the extended emission in that galaxy could be due to recent star formation in it. Emission features have been detected at the center of Mkn 501, suggesting the presence of Seyfert–like activity in the nucleus. Both galaxies show early–type stellar absorption features, and some of their metallicity indicators are peculiar when compared with those of elliptical galaxies in Fornax.

158.158   High spatial resolution IR observations and variability of the nuclear region of NGC 1068: structure and nature of the inner 100 parsec.
A. Chelli, C. Perrier, I. Cruz–González, L. Carrasco.
Interferometric imaging in astronomy, p. 253 (1987). Abstract. – See Abstr. 012.035.

158.159   The envelope emission of Markarian 8.
Q.-f. Yin, D. S. Heeschen, J. Heidmann.
Sci. Sin., Ser. A, Vol. 30, No. 10, p. 1075 – 1080 (1987).

Two–frequency VLA observations of Markarian 8 have shown that three major radio components are imbedded in a diffused envelope. The radio emission of the envelope is nonthermal. Physical parameters of the envelope are determined, and a possible origin of the relativistic electrons in the envelope is discussed.

158.160   The inverse Compton test for a large sample of compact radio sources.
P. L. Biermann, H. Kühr, W. A. Snyder, J. A. Zensus.
Astron. Astrophys., Vol. 185, No. 1/2, p. 9 – 13 (1987).

The prediction of inverse Compton X–rays based on VLBI data is made for a large sample (56 sources) of compact radio sources selected at 5 GHz. In 6 cases the authors find a significant deficit of measured over expected X–ray emission, commonly interpreted as evidence for bulk relativistic motion. The strongest discrepancy, however, occurs for the two apparently stationary components in the quasar 4C 39.25.

158.161   Rotationally excited OH in megamaser galaxies.
C. Henkel, R. Güsten, W. A. Baan.
Astron. Astrophys., Vol. 185, No. 1/2, p. 14 – 24 (1987).

Absorption in the $^2\Pi_{1/2} J = 1/2$ Λ–doublet transitions of OH, 182K above the ground state, is reported from the megamaser galaxies IC 4553 (Arp 220), Mrk 231, Mrk 273, NGC 3690, and IRAS 17208 0014. An upper flux density limit is given for NGC 3079. A correlation is found between the 18 and 6 cm main line flux densities, which is interpreted in terms of unsaturated amplification and absorption of the non–thermal radio continuum radiation from the nuclei of the associated galaxies.

158.162   Composite models for the narrow emission line region of active galactic nuclei. V. The line profiles.
M. Contini, S. M. Viegas–Aldrovandi.
Astron. Astrophys., Vol. 185, No. 1/2, p. 39 – 50 (1987).

Kinematic line profiles are calculated on the basis of a new model of the narrow emission line region (NLR) in active galactic nuclei, which considers a distribution of emitting clouds with radial motion of different velocities. The line emission intensities from each cloud are provided by previous model calculations taking into account photoionization from the central zone of the active galaxy and shock effects. In the authors' model dust can easily survive in the intercloud medium rather than inside the clouds so that blueward shifts in line profiles from the NLR generally indicate an outward motion. Most of the analyzed objects present an observed profile that is well fitted by model calculations when the receding clouds emission is reduced by a factor < 0.5, and an exponential distribution function on the cloud velocities is adopted. Moreover, calculated profiles show that "bumps", "tails" and "wings" can be different in the various lines of the same object.

158.163   The kinematical structure of the extended emission–line region of the early–type Seyfert–galaxy Mrk 3.
S. J. Wagner.
Astron. Astrophys., Vol. 185, No. 1/2, p. 77 – 86 (1987).

The early–type Seyfert–2 galaxy Mrk 3 was studied photometrically and spectroscopically to determine the overall structure and the nature of the extended emission surrounding the Seyfert nucleus. CCD surface photometry in broad band colours confirms the classification of Mrk 3 as an elliptical galaxy. The differences between observed isophotes and true ellipses indicate the existence of a small stellar disk lying perpendicular to the minor axis of the galaxy. Colour index maps show the existence of an extended region of gas emission along the minor axis of the galaxy. The velocity field indicates that the gas in the extended emission line region (EELR) is confined to a disk which is seen edge–on. The forbidden and permitted lines emitted from the narrow line region (NLR) are very broad (FWHM 1000 km s^{-1}) and possess an enhanced blue wing. Dust in the NLR and EELR reduces the amount of light received from the far side of the gas disk.

158.164 Extended emission line regions in nearby Seyfert galaxies. II. NGC 4388.
L. Colina, K. J. Fricke, W. Kollatschny, M. A. C. Perryman.
Astron. Astrophys., Vol. 186, No. 1/2, p. 39 – 48 (1987).

In this paper new high–resolution spectra of the $H\beta$ + [O III] spectral region are presented and discussed. The observed line profiles and velocity fields, as well as the relation between the optical and radio morphologies, are discussed in the context of the detected emission–line components.

158.165 Star formation in the nucleus of the galaxy NGC 5253.
R. González–Riestra, M. Rego, J. Zamorano.
Astron. Astrophys., Vol. 186, No. 1/2, p. 64 – 76 (1987).

Optical and ultraviolet spectroscopic observations of the nucleus of the galaxy NGC 5253 are analyzed. This galaxy presents the typical features of an elliptical system at large distances from its center. However, its nucleus is dominated by an emission complex composed by several giant H II regions. The analysis of the optical spectra shows that the metallic abundances in the nucleus of the galaxy are below the solar values. The presence of O stars can be deduced from the numerous absorption lines present in the UV spectrum. The UV emission lines indicate a high effective temperature for the ionizing star cluster. It is shown that the age of the brightest knot of the nucleus of NGC 5253 is less than three million years. The exact age depends on the choice of the extinction law, not well known in this type of objects. An LMC–like law leads to an age of 2.3×10^6 yr, and to an IMF similar or slightly flatter than that found by Salpeter for the solar neighbourhood, with an upper mass limit in the range $60\,M_\odot < M_{up} < 120\,M_\odot$.

158.166 High resolution spectrum of the starburst galaxy Tololo 1924–416 (= ESO 338–IG04).
M. Iye, M.–H. Ulrich, M. Peimbert.
Astron. Astrophys., Vol. 186, No. 1/2, p. 84 – 94 (1987).

A high resolution (0.5 Å) spectrum of the central region of a starburst galaxy Tol 1924–416 is presented. He I emission at 5016 Å was detected for the first time in an extragalactic object. The observed line ratio $I(\text{He I } \lambda5016)/\lambda5016)/I(\text{He I } \lambda4471) = 0.5$ indicates a very large optical depth confirming that the classical case B approximation is relevant at least for this galaxy. Accurate rotation curve and velocity dispersion curve are derived for the central 2 kpc region of this galaxy where an intense burst of star formation takes place. A decomposition analysis of the asymmetric emission line profiles of [O III] $\lambda\lambda5007$, 4959, and $H\beta$ suggests the presence of two systems of emitting clouds. A He/H abundance ratio of 0.082 ± 0.014 is derived. The measured line ratio of [O III], $I(\lambda5007)/I(\lambda4959) = 3.17 \pm 0.04$, suggests a possible discrepancy with the theoretical predictions.

158.167 Broad emission line profiles in Seyfert 1 galaxies. I. Evidence for a disk and a wind in Mkn 335.
E. van Groningen.
Astron. Astrophys., Vol. 186, No. 1/2, p. 103 – 113 (1987).

High resolution spectroscopic observations of the Seyfert 1 galaxy Mkn 335 are presented. This galaxy is exceptional in the sense that its broad emission lines display a strong blue asymmetry. It is shown that the $H\alpha/H\beta$ intensity ratio changes drastically with velocity shift. In the core of the broad lines the ratio is about 3.2, while it decreases to very low values (~ 1.5) in the wings. The author also finds that the $H\delta/H\beta$ ratio varies only by a small amount from line core to the wings, and is close to the theoretical case B recombination value. He proposes that these Balmer line ratios are produced by a medium with very high densities and a temperature of about 10^4K. The observations are interpreted in a model with two dynamically and spatially separate components: (1) A disk which contains the very high density material and emits $\sim 80\%$ of the Balmer lines in a symmetric profile. (2) An outflowing component of much lower density ($n_e < 10^9 \text{cm}^{-3}$) which emits the blue wing and most of Lyα and the higher excitation lines.

158.168 Spectroscopy of the galaxy components of N and Seyfert galaxies.
T. A. Boroson, J. B. Oke.
Publ. Astron. Soc. Pac., Vol. 99, No. 618, p. 809 – 815 (1987).

The authors present nuclear and off–nuclear spectra of nine active galaxies. All objects show continuum emission off the nucleus. Four clearly show absorption features from a stellar population. Velocities have been measured for the off–nuclear emission and absorption lines. In the case of I Zw 1, the absorption–line velocities are inconsistent with 21–cm H I measurements of this object.

158.169 The central power source in active galaxies.
R. Ptak, R. Stoner.
Comments Astrophys., Vol. 12, No. 2, p. 99 – 111 (1987).

The authors argue that the search for an understanding of the phenomena occurring in active galactic nuclei (AGN) should not be exclusively confined to the supermassive black hole (SMBH) paradigm. They discuss how observations of the spectra of AGNs and of their variability suggest alternatives to current SMBH ideas.

158.170 The velocity distribution of quasar/AGN broad emission line clouds.
R. C. Puetter, E. N. Hubbard.
Astrophys. J., Vol. 320, No. 1, p. 85 – 95 (1987).

The authors deconvolve the effects of geometric projection from the observed broad emission line profiles from a number of active galaxies to reveal the intrinsic brightness versus velocity function of the emission–line clouds. The only free parameters in the analysis are (1) the macroscopic symmetry of the emission line region (macrostructure) and (2) the emission anisotropy of individual clouds (microstructure). The authors conclude that the preponderance of evidence indicates that the broad–line emission arises from gas with preferred absolute velocity relative to line center.

158.171 What is the difference between radio galaxies and radio quasar galaxies?
J. B. Hutchings.
Astrophys. J., Vol. 320, No. 1, p. 122 – 134 (1987).

Deep optical imaging has been obtained with the CFHT in B and R band of ~ 50 luminous radio galaxies and radio quasars, in the redshift range 0.1 – 0.5. There are equal numbers of each, also matched in radio luminosity and spectral index. The mean optical absolute magnitude of radio galaxies is fainter by 1 mag than the host galaxies of quasars, and the distributions differ significantly. Radio galaxies are redder and larger and have smaller luminosity scale lengths than quasar hosts.

158.172 Zw 15107 + 0724 and the family of OH megamasers.
W. A. Baan, C. Henkel, A. D. Haschick.
Astrophys. J., Vol. 320, No. 1, p. 154 – 158 (1987).

The galaxy Zw 15107 + 0724 exhibits broad H I absorption at its systemic velocity and emission in the two main ground–state transitions of OH. The isotropic luminosity of the 1667 MHz OH emission line is $10.8\,L_\odot$ which makes it the weakest megamaser yet found. An assessment of the infrared properties of OH megamasers indicates that these sources have an approximately thermal IR spectrum with a temperature between 50 and 90K. Megamaser galaxies appear to be a very early phase of Seyfert 2 or H II nuclear activity, when much dust is still present close to the nucleus.

158.173 A nonthermal model for the XUV, soft X–ray emission of AGN and QSOs.
T. W. Jones, W. A. Stein.
Astrophys. J., Lett. Ed., Vol. 320, No. 1, p. L1 – L4 (1987).

XUV and soft X–ray radiation of active galactic nuclei may be explained by synchrotron emission of secondary electrons resulting from pp collisions in the vicinity of a supermassive compact object. The spectrum would be the natural extension of the ultraviolet emission responsible for photoionization of the emission–line gas.

158.174 Fe K features as probes of the nuclear reflection region in Seyfert galaxies.
J. H. Krolik, T. R. Kallman.
Astrophys. J., Lett. Ed., Vol. 320, No. 1, p. L5 – L8 (1987).
The warm electron scattering region posited in 1985 by Antonucci and Miller to exist in NGC 1068 may be a characteristic feature of Seyfert galaxies. Whenever it exists, it presents a unique signature in Fe X–ray features. Predictions of the strength of the Fe K–edge and Kα line in both type 1 and type 2 Seyfert galaxies are presented and interpreted.

158.175 The fading of the narrow–line region in 3C 390.3.
J. Clavel, W. Wamsteker.
Astrophys. J., Lett. Ed., Vol. 320, No. 1, p. L9 – L14 (1987).
IUE spectra of 3C 390.3 obtained at 17 different epochs from 1978 to 1986 show that the flux in the narrow core of the Lyα $\lambda1216$ and C IV $\lambda1550$ emission lines steadily declined by a factor of 1.7 and 3.2 respectively. Lloyd's optical monitoring reported in 1984 shows that the continuum intensity decreased by 2 mag over the years 1968 – 1980. Together, the UV and optical light curves form the strongest evidence to date that the NLR is photoionized and reverberates the continuum flux. Moreover, an upper limit of 10 lt–yr can be set for the size of the narrow line region in this radio galaxy.

158.176 Broadband properties of active galactic nuclei.
R. A. Edelson.
Diss. Abstr. Int., Sect. B, Vol. 48, No. 2, p. 472–B (1987). Thesis, California Institute of Technology, 171 pp. (1987). Order No. DA8710936.

158.177 Observations and gas dynamics of extragalactic radio jets.
K. R. Lind.
Diss. Abstr. Int., Sect. B, Vol. 48, No. 2, p. 473–B (1987). Thesis, California Institute of Technology, 307 pp. (1987). Order No. DA8710946.

158.178 Comparison of global and nuclear properties of starburst and Seyfert 1 galaxies.
C. L. Brungardt.
Diss. Abstr. Int., Sect. B, Vol. 48, No. 4, p. 1071–B (1987). Thesis, The Pennsylvania State University, 140 pp. (1987). Order No. DA8714792.

158.179 Spectrophotometrical investigation of the circumnuclear region of the Seyfert galaxy NGC 1275.
L. P. Metik, I. I. Pronik.
Izv. Krymskoj Astrofiz. Obs., Tom 76, p. 80 – 86 (1987). In Russian. English translation in Bull. Crimean Astrophys. Obs., Vol. 76.
The dimensions of the monochromatic image of the galaxy NGC 1275 star–shaped nucleus have been determined in spectral regions from $\lambda3700$ to $\lambda6800$ Å by the unwidened spectra, obtained in prime focus of the 6–m telescope with image scale on the negatives 17″5 at 1 mm. The dimension of the NGC 1275 nucleus is stretched to the violetward spectral region. The analysis showed that the obtained result can be explained by the enhanced content of blue stars near the nucleus. This star–cluster is stretched in the the direction to the detail located in 3″ towards the North of the galaxy nucleus. Measured in this direction, the dimension of the cluster is ∼3″ or 1 kpc.

158.180 On the optical and radio variability correlation of the Seyfert galaxy NGC 1275 nucleus.
V. N. Mukhametshina.
Izv. Krymskoj Astrofiz. Obs., Tom 76, p. 86 – 89 (1987). In Russian. English translation in Bull. Crimean Astrophys. Obs., Vol. 76.
The crosscorrelation analysis of two series of NGC 1275 observations has been carried out on the basis of the available observational data in radio band $\lambda = 3.3$ mm (1965 – 1982) and in the optical range with the *U*–filter (1968 – 1978). The author considered the values of the linear correlation coefficient $|R| > 0.5$ having the confidence level 0.999. The rapid components of variability in the observational data did not show any correlation.

158.181 Type 2 Seyfert galaxies with obscured type 1 regions.
J. S. Miller, B. F. Goodrich.
Bull. Am. Astron. Soc., Vol. 19, No. 2, p. 695 (1987). Abstract. – See Abstr. 010.061.

158.182 High–resolution study of emission–line profiles in active galactic nuclei.
S. Veilleux.
Bull. Am. Astron. Soc., Vol. 19, No. 2, p. 695 (1987). Abstract. – See Abstr. 010.061.

158.183 Is there a systematic infall or outflow of the broad line region clouds?
L. R. Bryant, M. J. Ward.
Bull. Am. Astron. Soc., Vol. 19, No. 2, p. 695 (1987). Abstract. – See Abstr. 010.061.

158.184 Was there an accretion event in NGC 5548?
R. Ptak, R. McCord, R. Stoner.
Bull. Am. Astron. Soc., Vol. 19, No. 2, p. 695 (1987). Abstract. – See Abstr. 010.061.

158.185 The link between tidal interaction and nuclear activity in galaxies.
J. E. Pringle, D. N. C. Lin, M. J. Rees.
Bull. Am. Astron. Soc., Vol. 19, No. 2, p. 695 – 696 (1987). Abstract. – See Abstr. 010.061.

158.186 Milli–arcsecond structure in NGC 1275 at 89 GHz.
M. C. H. Wright, D. C. Backer, R. L. Plambeck, J. E. Carlstrom, C. R. Masson, A. T. Moffet, A. C. S. Readhead, D. Woody, A. E. E. Rogers, J. M. Moran, C. R. Predmore, R. L. Dickman.
Bull. Am. Astron. Soc., Vol. 19, No. 2, p. 696 (1987). Abstract. – See Abstr. 010.061.

158.187 X–ray observations of IRAS–selected galaxies and obscuration of the broad–line region.
M. J. Ward, C. Done, A. C. Fabian, A. F. Tennant, R. A. Shafer.
Bull. Am. Astron. Soc., Vol. 19, No. 2, p. 696 (1987). Abstract. – See Abstr. 010.061.

158.188 EXOSAT observations of Cygnus–A.
K. A. Arnaud.
Bull. Am. Astron. Soc., Vol. 19, No. 2, p. 696 (1987). Abstract. – See Abstr. 010.061.

158.189 Active galactic nuclei vs. their host galaxies – the X–ray connection.
M. C. Begelman.
Bull. Am. Astron. Soc., Vol. 19, No. 2, p. 697 (1987). Abstract. – See Abstr. 010.061.

158.190 Compact radio sources associated with interacting galaxies in poor clusters.
R. J. Hanisch, D. J. Batuski, J. O. Burns.
Bull. Am. Astron. Soc., Vol. 19, No. 2, p. 698 (1987). Abstract. – See Abstr. 010.061.

158.191 **Radio observations of infrared–luminous merging galaxies.**
S. G. Neff, R. D. Joseph, L. J. Rickard, K. J. Johnston.
Bull. Am. Astron. Soc., Vol. 19, No. 2, p. 698 (1987). Abstract. – See Abstr. 010.061.

158.192 **Imaging and spectroscopy of double–nucleus Markarian galaxies: results for six representative objects.**
J. M. Mazzarella, T. A. Boroson.
Bull. Am. Astron. Soc., Vol. 19, No. 2, p. 699 (1987). Abstract. – See Abstr. 010.061.

158.193 **Evidence for large–scale winds from starburst galaxies: an optical investigation of powerful far–infrared galaxies.**
L. Armus, T. M. Heckman, G. K. Miley.
Bull. Am. Astron. Soc., Vol. 19, No. 2, p. 699 (1987). Abstract. – See Abstr. 010.061.

158.194 **Voyager far ultraviolet observations of Markarian 509 and 279.**
T. E. Carone, M. A. Malkan.
Bull. Am. Astron. Soc., Vol. 19, No. 2, p. 699 (1987). Abstract. – See Abstr. 010.061.

158.195 **Additional identifications of active galactic nuclei from the HEAO–1 X–ray survey.**
R. A. Remillard, H. V. Bradt, D. A. Buckley, I. R. Tuohy, R. Brissenden, D. A. Schwartz, W. Roberts.
Bull. Am. Astron. Soc., Vol. 19, No. 2, p. 699 (1987). Abstract. – See Abstr. 010.061.

158.196 **Ultraviolet spectra of the variable BL Lacertae object PKS 2155–304**
C. M. Urry.
Bull. Am. Astron. Soc., Vol. 19, No. 2, p. 700 (1987). Abstract. – See Abstr. 010.061.

158.197 **Megamaser comparisons: IC 4553 and Mrk 273.**
J. T. Schmelz.
Bull. Am. Astron. Soc., Vol. 19, No. 2, p. 711 – 712 (1987). Abstract. – See Abstr. 010.061.

158.198 **The extreme CO and IRAS FIR luminosity of 7Zw 31.**
L. J. Sage, P. M. Solomon.
Bull. Am. Astron. Soc., Vol. 19, No. 2, p. 712 (1987). Abstract. – See Abstr. 010.061.

158.199 **MG 1355 + 083: quartet of double–lobed radio sources.**
G. I. Langston, B. F. Burke.
Bull. Am. Astron. Soc., Vol. 19, No. 2, p. 718 (1987). Abstract. – See Abstr. 010.061.

158.200 **Search for 511 keV emission from active galaxies and QSO's.**
W. A. Wheaton, J. C. Ling, W. A. Mahoney, A. S. Jacobson.
Bull. Am. Astron. Soc., Vol. 19, No. 2, p. 718 (1987). Abstract. – See Abstr. 010.061.

158.201 **Hot dust and the near–IR bump in active galactic nuclei.**
R. Barvainis.
Bull. Am. Astron. Soc., Vol. 19, No. 2, p. 718 – 719 (1987). Abstract. – See Abstr. 010.061.

158.202 **A time–dependent analysis of the Blandford–Znajek process and its astrophysical applications: I. Evolutionary paths.**
S. J. Park, E. T. Vishniac.
Bull. Am. Astron. Soc., Vol. 19, No. 2, p. 730 (1987). Abstract. – See Abstr. 010.061.

158.203 **Luminous filaments and twisted jets in radio sources.**
J. A. Eilek, R. V. E. Lovelace.
Bull. Am. Astron. Soc., Vol. 19, No. 2, p. 731 (1987). Abstract. – See Abstr. 010.061.

158.204 **Sub–arcsecond resolution radio images of one–sided jets in core–dominated sources.**
C. P. O'Dea, R. E. Barvainis.
Bull. Am. Astron. Soc., Vol. 19, No. 2, p. 731 (1987). Abstract. – See Abstr. 010.061.

158.205 **An unusually aligned multicomponent radio source.**
T. K. Menon.
Bull. Am. Astron. Soc., Vol. 19, No. 2, p. 731 (1987). Abstract. – See Abstr. 010.061.

158.206 **New high resolution, high dynamic range VLA images of the M87 jet.**
F. N. Owen, T. J. Cornwell, P. E. Hardee.
Bull. Am. Astron. Soc., Vol. 19, No. 2, p. 731 (1987). Abstract. – See Abstr. 010.061.

158.207 **The hot gas surrounding Cygnus A.**
D. E. Harris, E. H. Bohlen.
Bull. Am. Astron. Soc., Vol. 19, No. 2, p. 731 (1987). Abstract. – See Abstr. 010.061.

158.208 **Discovery of the counter jet in Cygnus A.**
J. W. Dreher, C. L. Carilli, R. A. Perley.
Bull. Am. Astron. Soc., Vol. 19, No. 2, p. 731 (1987). Abstract. – See Abstr. 010.061.

158.209 **Accretion discs in A.G.N.**
C. J. Clarke.
Bull. Am. Astron. Soc., Vol. 19, No. 2, p. 732 (1987). Abstract. – See Abstr. 010.061.

158.210 **Gas disks in radio galaxies.**
S. M. Simkin, E. L. Sadler.
Bull. Am. Astron. Soc., Vol. 19, No. 2, p. 732 (1987). Abstract. – See Abstr. 010.061.

158.211 **A tool for comparing numerical and radio polarisation data.**
D. A. Clarke.
Bull. Am. Astron. Soc., Vol. 19, No. 2, p. 732 (1987). Abstract. – See Abstr. 010.061.

158.212 **Rotation measure gradients and depolarization in 3C 449.**
N. Killeen, R. Perley, T. Cornwell.
Bull. Am. Astron. Soc., Vol. 19, No. 2, p. 732 (1987). Abstract. – See Abstr. 010.061.

158.213 **A magnetoionic medium in the radio galaxy 3C 272.1 (M84).**
A. H. Bridle, R. A. Laing.
Bull. Am. Astron. Soc., Vol. 19, No. 2, p. 733 (1987). Abstract. – See Abstr. 010.061.

158.214 **X–ray variability of Mrk 335.**
M. G. Lee, B. Balick, J. Halpern, T. Heckman.
Bull. Am. Astron. Soc., Vol. 19, No. 2, p. 733 (1987). Abstract. – See Abstr. 010.061.

158.215 **Broad emissiom line variability of 20 Seyfert galaxies.**
E. Rosenblatt, M. A. Malkan.
Bull. Am. Astron. Soc., Vol. 19, No. 2, p. 733 (1987). Abstract. – See Abstr. 010.061.

158.216   The velocity distribution of quasar/AGN broad emission line clouds.
R. C. Puetter, E. N. Hubbard.
Bull. Am. Astron. Soc., Vol. 19, No. 2, p. 733 (1987). Abstract. – See Abstr. 010.061.

158.217   Aperture synthesis observations of CO in the starburst galaxy NGC 2146.
J. M. Jackson.
Bull. Am. Astron. Soc., Vol. 19, No. 2, p. 759 – 760 (1987). Abstract. – See Abstr. 010.061.

158.218   High resolution multi–line aperture syntheses maps of M82.
J. E. Carlstrom.
Bull. Am. Astron. Soc., Vol. 19, No. 2, p. 760 (1987). Abstract. – See Abstr. 010.061.

158.219   Kinematic and spectral evidences of the complex structure of the NGC 1275 nucleus.
V. I. Pronik.
Izv. Krymskoj Astrofiz. Obs., Tom 77, p. 126 – 134 (1987). In Russian. English translation in Bull. Crimean Astrophys. Obs., Vol. 77.

An analysis of the available published data concerning spectral observation of the nucleus and gas velocity field of NGC 1275 leads to the conclusion of several spatially separated variable sources existing in the nucleus which are responsible for the complex variable emission profiles of H_β and $4959 + 5007$ [O III] lines.

158.220   Results of photoelectrical observations of the continuum and emission line spectrum of the Seyfert galaxy NGC 1275 nucleus.
N. I. Merkulova, I. I. Pronik.
Izv. Krymskoj Astrofiz. Obs., Tom 77, p. 135 – 143 (1987). In Russian. English translation in Bull. Crimean Astrophys. Obs., Vol. 77.

Photoelectrical spectral observations of the NGC 1275 nucleus were carried out from November 1982 till March 1984 in three consecutive wavelength regions: H_β, $4595 + 5007$ Å [O III] and continuum. The time interval of the observations at a given position of the slit was 200 s. The results show that the fluxes in emission lines and continuum are variable. The continuum, H_β and [O III] lines varied by 1.5, 7.0 and 2.5 times during one night, respectively. For the whole time interval they varied as 1.7, 8.0 and 4.5, correspondingly. The degree of flux variations markedly exceeded the errors of observations. The H_β flux increases with the rise of [O III] flux, whereas the correlation of fluxes in emission line and continuum showed more complicated character.

158.221   Photoelectric observations of the spectral variability of nuclei of Seyfert galaxies as observed on the AZT–11 telescope. II. Errors of flux measurement depending on the position of the galaxy nucleus in the diaphragm.
N. I. Merkulova, L. P. Metik, I. I. Pronik.
Izv. Krymskoj Astrofiz. Obs., Tom 77, p. 144 – 147 (1987). In Russian. English translation in Bull. Crimean Astrophys. Obs., Vol. 77.

Observations of NGC 1275 galaxy nucleus in continuum and H_β, [O III] $\lambda 4959 + 5007$ Å emission lines depending on the galaxy position in 10″–diaphragm from centre to edge directions (corresponding to right ascensions α and declination δ) have been carried out. It is shown that the flux variations both in the continuum and in the emission lines do not exceed on average 5% if the displacement of the galaxy nucleus within the diaphragm is less than $\pm 3″$ from its centre; the errors increase to 10% if the displacement of the galaxy position in the diaphragm is $\pm 4″$.

158.222   The 160–minutes period in extragalactic objects: the LMC (RR Lyr stars) and NGC 4151.
V. A. Kotov, V. M. Lyutyj.
Izv. Krymskoj Astrofiz. Obs., Tom 77, p. 148 – 156 (1987). In Russian. English translation in Bull. Crimean Astrophys. Obs., Vol. 77.

According to the finding that the most characteristic ("resonant" or commensurate) period for the orbital period distribution of close binaries of the Galaxy and also for RR Lyr variables in globular clusters is very near to 160 min, the authors computed a so–called "resonance power spectrum" for the sample of 72 RR Lyr stars observed in and around the two LMC globular clusters, NGC 2257 and 1786. It is found that the "resonant" period for these stars, 159.9 ± 1.5 min, fairly agrees with the previous result. With the aim to extend the search for traces of this "ubiquitous" 160–min periodicity among various astrophysical objects, the authors analysed also the *UBV* photometric measurements of the nucleus of the Seyfert galaxy NGC 4151, made in the 1968 – 1984 interval. Power spectrum of the data clearly showed the presence of a significant, at the 4.1–sigma confidence level, period 160.0099 ± 0.0004 min with about $\pm 2.5\%$ amplitude. The authors estimate the lifetime of the 160–min period to be $\sim 15 \times 10^9$ years, which is nearly the age of the Universe. The product $P_0 H_0$, where H_0 is the Hubble constant, appears to be a dimensionless parameter $\sim 2 \times 10^{-14}$ which might be of great interest for cosmology.

158.223   Three–dimensional luminosity function of type–1 Seyfert galaxies.
R. A. Kandalyan.
Astrophysics, Vol. 26, No. 2, p. 185 – 190 (1987). English translation of 43.158.129.

158.224   On the possibility of obtaining strip distributions of radio brightness by the minimal phase method.
I. F. Malov, V. A. Frolov.
Astrophysics, Vol. 26, No. 2, p. 213 – 220 (1987). English translation of 43.158.130.

158.225   A search for OH absorption in NGC 1275.
C. P. O'Dea, S. A. Baum.
Astron. J., Vol. 94, No. 6, p. 1476 – 1479 (1987).

The authors report a VLA search for OH absorption at 1665 and 1667 MHz against the nucleus of NGC 1275 (3C 84), the dominant galaxy in the Perseus cluster. They find no absorption in the low–velocity system (heliocentric velocity 5320 km/s) within a 2200 km/s bandpass. The upper limit to the apparent optical depth of any absorption feature is $\tau < 8 \times 10^{-4}$ in the low–velocity system. The authors place a model–dependent upper limit to the mass of molecular gas in NGC 1275 of $M(H_2) < 6.2 \times 10^8 M_\odot$. They also searched for OH in the high–velocity system (heliocentric velocity 8115 km/s) within a 281 km/s bandpass and obtained an upper limit to the apparent optical depth of $\tau < 10^{-2}$.

158.226   Imaging and spectroscopic studies of the interacting system Markarian 171.
S. D. Friedman, R. D. Cohen, B. Jones, H. E. Smith, W. A. Stein.
Astron. J., Vol. 94, No. 6, p. 1480 – 1486 (1987).

V, R, and I broadband CCD images as well as Hα spectroscopic observations of Markarian 171, are presented. Two apparently interacting dust–enshrouded galaxies exhibit relatively normal $V–R$, $V–I$ galactic stellar–population colors in their outer regions and spatially extended Hα emission. However, the dereddened colors imply a hot stellar population. The dual characteristics of narrow Hα lines and spatially extended Hα emission clearly distinguish this type of activity from that associated with the compact central regions of active galactic nuclei and QSOs. Apparently, in the case of Mk 171 the interaction has induced star formation throughout a significant fraction of the galaxy (kiloparsecs) – not just in the immediate vicinity of the nucleus (parsecs).

158.227 Rapid variability of extragalactic radio sources.
D. S. Heeschen, T. Krichbaum, C. J. Schalinski,
A. Witzel.
Astron. J., Vol. 94, No. 6, p. 1493 – 1507 (1987).

The authors have observed 31 extragalactic radio sources at 2 – 4 hr intervals for 3 – 4 days at each of three epochs. The 15 compact sources in the sample appear to display two types of variability. One is probably weak flickering caused by refractive interstellar scintillation. The other is a larger amplitude variability, which is characterized by a narrow range of timescales of about 1 – 2 days. The authors discuss three possible causes of the second type: intrinsic variability; scintillation from a very narrow disk component of the ISM; scintillation from clouds associated with a particular region of the ISM. They suggest that the latter is the more likely explanation.

158.228 Alignment of radio and optical orientations in high redshift radio galaxies.
K. C. Chambers, G. K. Miley, W. van Breugel.
Space Telesc. Sci. Inst., Prepr. Ser., No. 178, 8 pp. (1987).
Submitted to Nature.

158.229 Starbursts and their dynamics.
C. A. Norman.
Space Telesc. Sci. Inst., Prepr. Ser., No. 190, 9 pp. (1987). To appear in the Moriond conference "Star formation in galaxies".

158.230 IC 4767 (the "X–galaxy"): the missing link for understanding galaxies with peanut–shaped bulges?
B. C. Whitmore, M. Bell.
Space Telesc. Sci. Inst., Prepr. Ser., No. 192, 20 pp. (1987). To appear in Astron. J.

158.231 IRAS observations of radio galaxies.
D. Golombek, G. K. Miley, G. Neugebauer.
Space Telesc. Sci. Inst., Prepr. Ser., No. 215, 19 pp. (1987). To appear in Astron. J.

158.232 Starbursts: nature and implications.
C. A. Norman.
Space Telesc. Sci. Inst., Prepr. Ser., No. 220, 8 pp. (1987). To appear in the proceedings of the conference "Galactic and extragalactic star formation", Reidel, Dordrecht, The Netherlands.

158.233 Starbursts: nature and environment.
C. A. Norman.
Space Telesc. Sci. Inst., Prepr. Ser., No. 221, 10 pp. (1987). To appear in the proceedings of the 3rd IRAS Conference.

158.234 VLBI observations of 23 hot spots in the starburst galaxy M82.
N. Bartel, M. I. Ratner, A. E. E. Rogers, I. I. Shapiro,
R. J. Bonometti, N. L. Cohen, M. V. Gorenstein,
J. M. Marcaide, R. A. Preston.
Cent. Astrophys., Prepr. Ser., No. 2552, 44 pp. (1987). Submitted to Astrophys. J.

158.235 The cluster environments of powerful radio galaxies.
R. M. Prestage, J. A. Peacock.
Edinb. Astron. Prepr., No. 14/87, 50 pp. (1987). To appear in Mon. Not. R. Astron. Soc.

158.236 Continuum emission from active galactic nuclei.
W. K. Gear.
Edinb. Astron. Prepr., No. 15/87, 38 pp. (1987). Submitted to Proceedings of the Summer School on millimetre and submillimetre astronomy, held in Stirling, 21 – 26 June 1987.

158.237 IRAS galaxies: no evidence for a cosmological anisotropy.
R. G. Clowes, A. Savage, G. Wang, S. K. Leggett,
H. T. MacGillivray, R. D. Wolstencroft.
Edinb. Astron. Prepr., No. 16/87, 10 pp. (1987). To appear in Mon. Not. R. Astron. Soc.

158.238 Infrared measurements of interacting galaxies.
R. D. Wolstencroft.
Edinb. Astron. Prepr., No. 20/87, 13 pp. (1987). Paper presented at the symposium "New ideas in astronomy", Venice, May 1987.

158.239 Enhanced star formation in barred spiral galaxies. II: Radio continuum emission.
P. J. Puxley, T. G. Hawarden, C. M. Mountain.
Edinb. Astron. Prepr., No. 21/87, 28 pp. (1987). To appear in Mon. Not. R. Astron. Soc.

158.240 Recent star formation in interacting galaxies – III. Evidence from mid–infrared photometry.
G. S. Wright, R. D. Joseph, N. A. Robertson, P. A. James,
W. P. S. Meikle.
Edinb. Astron. Prepr., No. 24/87, 38 pp. (1987). To appear in Mon. Not. R. Astron. Soc.

158.241 The relation between variability and star formation in Seyfert nuclei.
R. Terlevich, J. Melnick.
R. Greenwich Obs., Prepr., No. 62, 8 pp. (1987). To appear in "XXIInd Recontre de Moriond on "Starbursts and galaxy evolution", Les Arcs, March 8 – March 14, 1987.

158.242 Studies of IRAS sources at high galactic latitudes. IV. New redshifts and the spectroscopic properties of IRAS galaxies.
K. J. Leech, A. Lawrence, M. Rowan–Robinson, D. Walker,
M. V. Penston.
R. Greenwich Obs., Prepr., No. 66, 9 + 17 pp. (1987). To appear in Mon. Not. R. Astron. Soc.

158.243 Optical and near–infrared observations of IRAS galaxies. II.
A. F. M. Moorwood, M.–P. Véron–Cetty, I. S. Glass.
ESO Sci. Prepr., No. 514, 29 pp. (1987). To appear in Astron. Astrophys.

158.244 Active extragalactic objects.
R. A. E. Fosbury.
ESO Sci. Prepr., No. 531, 8 pp. (1987). Paper presented at the 10th European Regional Meeting of the IAU, Prague, August 1987.

158.245 Far ultraviolet absorption lines in active galaxies.
M. H. Ulrich.
ESO Sci. Prepr., No. 532, 23 pp. (1987). To appear in Mon Not. R. Astron. Soc.

158.246 Galactic nuclei and quasars at high angular resolution.
M. H. Ulrich.
ESO Sci. Prepr., No. 533, 5 pp. (1987). Paper presented at the ESA workshop on "Optical interferometry in space", Granada, Spain, June 1987.

158.247 Observational consequences of precessing relativistic jets in extragalactic radio sources.
E. J. A. Meurs, N. Roos.
ESO Sci. Prepr., No. 536, 3 pp. (1987). Paper presented at the 10th European Regional Meeting of the IAU, Prague, August 1987.

158.248 [O III]–line emission associated with radio structures in Seyfert galaxies.
E. J. A. Meurs, M. Whittle, S. W. Unger, D. J. Axon, A. Pedlar, M. J. Ward.
ESO Sci. Prepr., No. 536, 4 pp. (1987). Paper presented at the 10th European Regional Meeting of the IAU, Prague, August 1987.

158.249 Precessing radio jets in AGNs.
E. J. A. Meurs.
ESO Sci. Prepr., No. 547, 8 pp. (1987). Paper presented at the COSPAR/IAU Symposium "The physics of compact objects", Sofia, Bulgaria, July 1987.

158.250 Photoionization of extended emission line regions.
A. Robinson.
ESO Sci. Prepr., No. 552, 10 pp. (1987). To appear in Proc. NATO Adv. Res. Workshop "Cooling flows in clusters and galaxies", Cambridge, UK, June 1987.

158.251 Imaging redshift estimates for two BL Lacertae objects.
W. Romanishin.
Astrophys. J., Vol. 320, No. 2, p. 586 – 588 (1987).
Redshifts are estimated for two resolved BL Lacertae objects, using the photometric properties of the host galaxies, assumed to be normal ellipticals. The galaxy apparent magnitude and angular effective radius are derived from an image model fit to the observed image of BL Lac object plus host galaxy. The redshift is then estimated from the absolute magnitude–effective radius relation for normal ellipticals. The redshifts found are 0.17 ± 0.04 for $2254+074$ and 0.31 ± 0.08 for $2335+031$.

158.252 H_2O maser emission from the nuclei of NGC 253 and M51.
P. T. P. Ho, R. N. Martin, C. Henkel, J. L. Turner.
Astrophys. J., Vol. 320, No. 2, p. 663 – 666 (1987).
New H_2O masers were detected in the nuclear region of NGC 253 and M51. Although these galaxies are believed to have "active" nuclei from optical, infrared, and radio data, the observed luminosities in the H_2O line, if isotropically radiated, are not unusually large, $\sim 0.1 - 1 \, L_\odot$. Such luminosities are comparable to those of the brightest H_2O masers in our own Galaxy. These detections suggest that "normal" H_2O masers are present in nuclei with strong star-formation activity.

158.253 Molecular hydrogen line emission in Seyfert galactic nuclei.
J. Fischer, T. R. Geballe, H. A. Smith, M. Simon, J. W. V. Storey.
Astrophys. J., Vol. 320, No. 2, p. 667 – 675 (1987).
The authors report on 2 μm spectroscopy of three Seyfert and two star burst galactic nuclei. They have detected line emission from vibrationally excited H_2 in the Seyfert galactic nuclei NGC 1275, NGC 3227, and NGC 4151. The authors have also measured the Brγ line flux in NGC 4151 and obtained an upper limit on the Brγ line flux in NGC 1275. There is a large range in the observed $S(1)$ to Brγ line ratio for both Seyfert and starburst galaxies. The authors study whether the molecular gas in these galaxies is excited by UV fluorescence or whether its excitation can be accounted for by shocks in individual young stellar outflows or supernova remnants produced by a starburst or both.

158.254 The properties of the luminosity variation of NGC 4051.
G.-z. Xie, Y. Zhou, P.-j. Hau, R.-w. Lu, X.-d. Liu, F.-z. Cheng.
Publ. Yunnan Obs., No. 2, p. 55 – 62 (1987). In Chinese.

158.255 The BL Lac object PKS 0215+015 in a low state: variable emission lines at $z_e = 1.72$ and no dust signature from the system at $z_a = 1.345$.
P. Boissé, J. Bergeron.
Inst. Astrophys. Paris, Pré–Publ., No. 195, 26 pp. (1987). To appear in Astron. Astrophys.

158.256 Star formation and spectrophotometric evolution of high redshift galaxies.
B. Rocca–Volmerange, B. Guiderdoni.
Inst. Astrophys. Paris, Pré–Publ., No. 200, 16 pp. (1987). Paper presented at the 22nd Rencontre de Moriond "Starbursts and galaxy evolution", Les Arcs, March 1987.

158.257 The environment of active galactic nuclei I. A two component broad emission line model.
S. Collin–Souffrin, J. E. Dyson, J. C. McDowell, J. J. Perry.
Inst. Astrophys. Paris, Pré–Publ., No. 209, 27 pp. (1987). To appear in Mon. Not. R. Astron. Soc.

158.258 KISO survey for ultraviolet–excess galaxies. VII.
B. Takase, N. Miyauchi–Isobe.
Ann. Tokyo Astron. Obs., Second Ser., Vol. 21, No. 4, p. 363 – 386 (1987).
Presented here are the seventh list and identification charts of the ultraviolet–excess galaxies which have been detected on the multi–color plates taken with the Kiso Schmidt telescope for 10 survey fields. In the sky area of some 300 square degrees 425 objects are catalogued down to the photographic magnitude of about 18.

158.259 Broadband properties of the CfA Seyfert galaxies: II. Infrared–millimeter properties.
R. A. Edelson, M. A. Malkan, G. H. Rieke.
Prepr. Steward Obs., No. 739, 27 pp. (1987). To appear in Astrophys. J.

158.260 NGC 253 and a proposed sequence for nuclear starbursts.
G. H. Rieke, M. J. Lebofsky, C. E. Walker.
Prepr. Steward Obs., No. 741, 27 pp. (1987). To appear in Astrophys. J.

158.261 A near–infrared and optical study of X–ray selected Seyfert galaxies. II. Models and interpretation.
C. W. McAlary, G. H. Rieke.
Prepr. Steward Obs., No. 744, 43 pp. (1987). To appear in Astrophys. J.

158.262 Energy distributions of blazars.
C. D. Impey, G. Neugebauer.
Prepr. Steward Obs., No. 764, 99 pp. (1987). To appear in Astron. J.

158.263 77 GHz continuum observations of variable extragalactic sources.
H. Teräsranta, E. Valtaoja, S. Haarala, A.–M. Elo, M. Valtonen, E. Salonen, S. Urpo, M. Tiuri, E. Laurikainen.
Rep. Ser., Dep. Phys. Sci., Univ. Turku, No. FTL–R132, 22 pp. (1987). = Turku Univ. Obs., Informo, No. 125. ISBN 951–642–980–7. To appear in Astron. Astrophys., Suppl. Ser.

158.264 Linear size distribution of double radio sources in the constant speed symmetric expansion model.
D. G. Banhatti.
Rep. Ser., Dep. Phys. Sci., Univ. Turku, No. FTL–R133, 21 pp. (1987). = Turku Univ. Obs., Informo, No. 126. ISBN 951–642–999–8. Submitted to Astrophys. Space Sci.

158.265 IPS observations of an unbiased sample of 90 Ooty occultation radio sources at 326.5 MHz (91.8 cm).
D. G. Banhatti, S. Ananthakrishnan.
Rep. Ser., Dep. Phys. Sci., Univ. Turku, No. FTL–R136, 19 pp. (1987). = Turku Univ. Obs., Informo, No. 128. ISBN 951–880–044–8.
The authors present 327 MHz interplanetary scintillation (IPS) observations of an unbiased sample of 90 extragalactic radio sources selected from the ninth Ooty lunar occultation list. They derive values of μ, the fraction of scintillating flux density out of the total, and ψ, the equivalent Gaussian diameter for the

scintillation structure. Various correlations are found between the observed parameters.

158.266 A study of the optical variability of the nuclei of Seyfert 2 galaxies. I.
S. I. Neizvestnyj.
Astrofiz. Issled. Izv. Spets. Astrofiz. Obs., Tom 24, p. 3 – 26 (1987). In Russian. English translation in Bull. Spec. Astrophys. Obs. – North Caucasus.

Observational results are presented for 11 Seyfert galaxies. The observations were carried out to study variability at time–scales of years – months, days. Identification charts, magnitudes, and colours of local photometrical standards in the fields of the investigated objects are given.

158.267 A study of the optical variability of the nuclei of Seyfert 2 galaxies. II.
S. I. Neizvestnyj.
Astrofiz. Issled. Izv. Spets. Astrofiz. Obs., Tom 24, p. 27 – 34 (1987). In Russian. English translation in Bull. Spec. Astrophys. Obs. – North Caucasus.

Optical variability of 11 Seyfert galaxies is studied from the data presented in paper I (158.266). It is found that there are variabilities in Seyfert 2 (Sy 2) at time scales of months – years. The amplitude of variability of these galaxies in $UBVR$ filters is on average one and a half lower than that of Sy 1. Day–to–day variations in Sy 2 are not detected. A short interpretation of the results is given.

158.268 Flux densities for 72 radio sources in the declination band $\delta = -6°$ at 7.6 cm wavelength.
K. D. Aliakberov, V. K. Kononov, M. G. Mingaliev, M. N. Naugol'naya, V. M. Plotnikov, T. B. Pyatunina, S. A. Trushkin.
Astrofiz. Issled. Izv. Spets. Astrofiz. Obs., Tom 24, p. 178 – 182 (1987). In Russian. English translation in Bull. Spec. Astrophys. Obs. – North Caucasus.

The flux densities of radio emission at wavelength 7.6 cm are given for 72 radio sources near the declination band $\delta = -6°$. Four new sources with flat spectra are found, two of them having the spectral indexes higher than 0.

158.269 Investigation of radio objects with continuum optical spectra. Rapid radio variability from observations at the RATAN–600 radio telescope.
S. A. Pustil'nik, K. D. Aliakberov.
Astrofiz. Issled. Izv. Spets. Astrofiz. Obs., Tom 25, p. 68 – 83 (1987). In Russian. English translation in Bull. Spec. Astrophys. Obs. – North Caucasus, Vol. 25.

The results of a search for rapid radio variability at 8.2 cm (time scale 1 – 10 days) in 13 objects with strong nonthermal continuum are reported. 8 of them are radio objects with continuous optical spectra (ROCOSes)and 5 ones are extragalactic BL Lacertae type objects. The observations were carried out in December 1982 using the RATAN–600 radio telescope. All the studied objects have proved to be more variable than the sources of the control group. No differences are detected in the characteristics of rapid radio variability of ROCOSes and BL Lac objects. Two of rapidly variable sources – OI 090.4 and OJ 287 have been observed optically during the same period and both have shown rapid optical variations also. The averaged flux densities of all observed sources at frequencies 2.3, 3.65, 3.95 and 7.7 GHz (epoch 1982.97) are presented.

158.270 Powerful radio sources in clusters of galaxies: origin of jet–like structures.
M. R. Gil'fanov, R. A. Syunyaev, E. M. Churazov.
Inst. Kosm. Issled. Akad. Nauk SSSR, Prepr., No. 1234, p. 2 – 31 (1987). In Russian. Abstr. in Ref. Zh., 51. Astron. 11.51.811 (1987).

158.271 Continuity in the violent phenomena taking place in the nuclei of galaxies and QSOs.
G. Burbidge.
Rev. Mex. Astron. Astrofis., Vol. 14, No. 1, p. 81 – 82 (1987). Abstract. – See Abstr. 012.042.

158.272 Active galactic nuclei: narrow emission line profiles.
S. M. Viegas–Aldrovandi, M. Contini.
Rev. Mex. Astron. Astrofis., Vol. 14, No. 1, p. 83 – 86 (1987). – See Abstr. 012.042.

The emission profile of the [O III] 5007 line is calculated on basis of the composite models which take into account the coupled effects of photoionization and shock on the physical condition of the narrow emission line region of the active galactic nuclei. The results are compared to the observed profiles.

158.273 The effect of relativistic electrons on the narrow emission lines of AGN.
S. M. Viegas–Aldrovandi, R. B. Gruenwald.
Rev. Mex. Astron. Astrofis., Vol. 14, No. 1, p. 87 – 89 (1987). – See Abstr. 012.042.

Models for the narrow emission–line regions of active galactic nuclei are presented. Models are constructed considering that the emitting clouds have different densities and the line intensities are calculated assuming a distribution function of the cloud densities. A comparison between the observed and calculated line ratios shows that the presence of relativistic electrons is important and can improve the theoretical ratios calculated by photoionization models.

158.274 Balmer lines in active galactic nuclei.
A. A. Andrade, S. M. Viegas–Aldrovandi, R. B. Gruenwald.
Rev. Mex. Astron. Astrofis., Vol. 14, No. 1, p. 90 – 93 (1987). – See Abstr. 012.042.

The effect of a flux of relativistic electrons on the population of hydrogen levels is analyzed. The results are compared to Balmer lines observations of active galactic nuclei. The effect on the Balmer decrement is similar to the reddening caused by dust.

158.275 Long term variability of radiosources in the frequencies of 22 GHz and 44 GHz.
L. C. L. Botti, Z. Abraham.
Rev. Mex. Astron. Astrofis., Vol. 14, No. 1, p. 97 – 100 (1987). – See Abstr. 012.042.

The radio sources 3C273, OV236, Cen A and Sgr A were observed during a period of six years (1980 – 1986) in the frequencies of 22 GHz and 44 GHz. All of them presented some variability, specially the quasar 3C273, which after a period of intense activity (1981 – 1985), returned to its quiescent level at the end of 1985. The increase of the flux density in these frequencies is associated to the ejection of new components by the central source in the quasar, as observed in the maps obtained by VLBI techniques.

158.276 Nuevas galaxías Seyfert 1 australes.
J. Maza, M. T. Ruiz.
Rev. Mex. Astron. Astrofis., Vol. 14, No. 1, p. 101 (1987). Abstract. – See Abstr. 012.042.

158.277 High spatial resolution IR observations and variability of the nuclear region of NGC 1068: structure and nature of the inner 100 parsec.
A. Chelli, C. Perrier, I. Cruz–González, L. Carrasco.
Rev. Mex. Astron. Astrofis., Vol. 14, No. 1, p. 102 (1987). Abstract. – See Abstr. 012.042.

158.278 Inclinaçao e obscurecimento em galáxias de Seyfert.
J. E. Steiner, S. D. Kirhakos.
Rev. Mex. Astron. Astrofis., Vol. 14, No. 1, p. 103 (1987). Abstract. – See Abstr. 012.042.

158.279   Morfología CCD de galaxias activas: NGC 7552.
    J. C. Forte, E. I. Vega, M. Méndez, C. Feinstein.
Rev. Mex. Astron. Astrofis., Vol. 14, No. 1, p. 104 (1987).
Abstract. – See Abstr. 012.042.

158.280   As linhas de Balmer em objetos extragaláticos ativos.
    A. A. Andrade, S. M. Viegas–Aldrovandi,
R. B. Gruenwald.
Rev. Mex. Astron. Astrofis., Vol. 14, No. 1, p. 105 (1987).
Abstract. – See Abstr. 012.042.

158.281   Galaxies with strong nitrogen lines.
    T. S. Bergmann, M. G. Pastoriza.
Rev. Mex. Astron. Astrofis., Vol. 14, No. 1, p. 106 (1987).
Abstract. – See Abstr. 012.042.

158.282   Dust heating by ultraviolet accretion disk radiation.
    R. O. Lagua, S. M. Viegas–Aldrovandi.
Rev. Mex. Astron. Astrofis., Vol. 14, No. 1, p. 107 (1987).
Abstract. – See Abstr. 012.042.

158.283   Comments on activity in galactic nuclei at all scales of energetics.
P. Pismis.
Rev. Mex. Astron. Astrofis., Vol. 14, No. 1, p. 108 – 119 (1987).
– See Abstr. 012.042.
    Activity in galactic nuclei manifested in the form of lobes and jets are discussed. Emphasis is placed on the role that integral primeval parameters of bodies of galactic size may play in the problem of activity. These parameters are the total mass, central condensation of mass and energy, magnetic fields and others of minor importance although the effect of interactions between galactic masses may not be ruled out altogether.

158.284   Statistical properties of lobes and jets. I. In galaxies with known Hubble types.
P. Pismis, A. M. Cervantes.
Rev. Mex. Astron. Astrofis., Vol. 14, No. 1, p. 168 (1987).
Abstract. – See Abstr. 012.042.

158.285   Extragalactic VLBI objects: candidates for optical interferometry?
E. Preuss.
ESA Spec. Publ., ESA SP–273, p. 15 – 19 (1987). – See Abstr. 012.044.
    Some aspects of compact ($\lesssim 0\rlap{.}''001$) optical and radio emission of active galactic nuclei and quasars are reviewed which are relevant to future interferometric observations in both spectral regimes. An obvious goal for joint observations is the investigation of Broad Line Regions, their possible connection with nuclear radio sources, and the role of collimated outflow (jets) in these processes.

158.286   Galactic nuclei and quasars at high angular resolution.
    M. H. Ulrich.
ESA Spec. Publ., ESA SP–273, p. 227 – 231 (1987). – See Abstr. 012.044.
    Extraordinary progress can be expected from high angular resolution observations in the optical and ultraviolet ranges in the following areas: gas and star motions in active nuclei; physics of jets; determination of the mass of individual compact objects in the dark halos of galaxies; estimation of the fraction of the mass of the Universe which is in baryonic form.

158.287   M87: describing the indescribable.
    J. Kanipe.
Astronomy, Vol. 15, No. 5, p. 6 – 13 (1987). Abstr. in Phys. Abstr., Vol. 90, No. 1311, Entry 101924 (1987).

158.288   Starburst galaxies: more heat than light.
    A. Prestwich.
New Sci., Vol. 114, No. 1560, p. 46 – 49 (1987). Abstr. in Phys. Abstr., Vol. 90, No. 1312, Entry 107963 (1987).

158.289   Nucleus activity correlates with outlying structure in Seyfert galaxies and quasars.
V. L. Afanas'ev, V. P. Mikhajlov.
Sov. Astron. Lett., Vol. 12, No. 6, p. 376 – 377 (1986). English translation of 42.158.131.

158.290   Kinetic Alfvén waves in extended radio sources. I. Reacceleration.
L. C. Jafelice, R. Opher.
Astrophys. Space Sci., Vol. 137, No. 2, p. 303 – 315 (1987).
    The authors study a model of extended radio sources (ERS), in particular, extragalactic jets and radio lobes, which are inhomogeneous and where noncompressive Alfvén and surface Alfvén waves (and not shocks and magnetosonic waves) are primarily excited. It is assumed that a negligible thermal population exists (i.e., the ion density at the low–energy cut–off of the power law distribution is greater than the ion density of the thermal population, if present). Due to internal instabilities and/or the interaction of the ERS with the ambient medium, surface Alfvén waves (SAW) are created. The authors show that even very small amplitude SAW are mode converted to kinetic Alfvén waves (KAW) which produce large moving accelerating potentials ψ, parallel to the magnetic field.

158.291   Kinetic Alfvén waves in extended radio sources. II. Electric currents, collimated jets, and inhomogeneities.
L. C. Jafelice, R. Opher.
Astrophys. Space Sci., Vol. 138, No. 1, p. 23 – 39 (1987).
    Electric current generation by kinetic Alfvén waves (KAW) is discussed for the case of extended radio sources (ERS), in particular, extragalactic jets (EJ). These currents are generated parallel to the background magnetic field due to Landau damping by which KAW accelerate electrons. It is found that the KAW generated currents are in excess of the currents necessary for an EJ to be magnetically self–confined. The authors address the problem of determining the process that can maintain ERS inhomogeneous. They study the stability of a plasma and show it to be subject to the thermal Joule instability. They suggest the thermal Joule instability as the process that maintains ERS inhomogenous. The KAW analysis correlates the important problems of ERS of (re)acceleration, current generation, collimation, and maintenance of inhomogeneities.

158.292   3C 120.
    R. C. Walker, J. M. Benson, S. C. Unwin.
Superluminal radio sources, p. 48 – 54 (1987). – See Abstr. 012.053.
    The authors present VLBI maps of 3C 120 from the first two years of the regular monitoring program since 1981. To explore the structure in 3C 120 on scales larger than those seen in the VLBI monitoring observations, they have made 18 cm VLBI observations with up to 18 stations and observations with all configurations and most frequencies of the VLA. A summary of the results is given.

158.293   Superluminal motion in BL Lac: evidence for deceleration in two events.
R. L. Mutel, R. B. Phillips.
Superluminal radio sources, p. 60 – 66 (1987). – See Abstr. 012.053.
    BL Lac has been studied using intercontinental VLBI arrays at frequencies of 5.0 and 10.7 GHz at approximately three–month intervals since April 1980. Maps made during the first few years established that components were repeatedly ejected from a presumed stationary core component along a position angle of $\sim 190°$ with an apparent transverse speed of $v_{app} \sim 4c$. The authors have now completed a systematic analysis of 13 maps made at 10.6 GHz and seven maps made at 5.0 GHz covering the period 1980.4 to 1985.4. They describe a particularly interesting aspect of the last two events, namely an apparent deceleration of each of the components as they reach a projected distance of $\sim 1.5 - 2.0 \, h^{-1}$pc from the core, accompanied by a large increase in angular size.

158.294   Subluminal expansion in NGC 1275.
D. C. Backer.
Superluminal radio sources, p. 76 – 82 (1987). – See Abstr. 012.053.

The neraby, active galaxy NGC 1275 (3C 84, $z = 0.018$) has been the subject of many investigations in the past 20 years. The author reviews the evolution of the microwave flux density and the structure seen in 10.7 GHz VLBI images. Then he summarizes recent VLBI observations at 22 GHz and 90 GHz. Rapid changes in the core and the presence of a "hot spot" in the outflowing material are prominent features of these short–wavelength VLBI data. The case for subluminal proper velocities of the outflow is strong.

158.295   The quest for superluminal sources.
T. J. Pearson, A. C. S. Readhead, P. D. Barthel.
Superluminal radio sources, p. 94 – 103 (1987). – See Abstr. 012.053.

The authors made a systematic study of the milliarcsecond structure of a complete, flux–density limited sample of strong radio sources selected at 5 GHz.

158.296   Intrinsic asymmetry in NGC 6251.
D. L. Jones.
Superluminal radio sources, p. 162 – 167 (1987). – See Abstr. 012.053.

The author observed the nucleus of the giant radio galaxy NGC 6251 with a VLBI array at 18 cm and discusses the observations. He concludes that at least some of the one sided "core–jet" radio sources in galactic nuclei appear asymmetric for reasons other than relativistic beaming.

158.297   Are compact doubles misaligned superluminals?
M. W. Hodges, R. L. Mutel.
Superluminal radio sources, p. 168 – 173 (1987). – See Abstr. 012.053.

The authors ask whether there are any classes of compact radio cores which do *not* exhibit superluminal effects. One such class (perhaps the only one) appears to be the compact double (CD) radio sources.

158.298   VLBI observations of compact steep–spectrum radio sources.
C. Fanti, R. Fanti.
Superluminal radio sources, p. 174 – 179 (1987). – See Abstr. 012.053.

158.299   VLBI observations of the suspected superluminal 3C 371.
K. R. Lind.
Superluminal radio sources, p. 180 – 185 (1987). – See Abstr. 012.053.

The author discusses VLBI observations of May 1985, and compares these seen with the 1982.9 map. The N galaxy 3C 371 has an extended VLBI jet which appears to be the base of the arcsecond scale jet.

158.300   VLA polarimetry of the active galaxy 3C 371.
J. M. Wrobel.
Superluminal radio sources, p. 186 – 192 (1987). – See Abstr. 012.053.

New VLA polarimetry of the active galaxy 3C 371 reveals that the rotation measure of component A ($2\,h^{-1}$kpc offset from the compact core) is low and probably primarily of Galactic origin; and that the morphology, size, and inferred magnetic field configuration of component B ($17\,h^{-1}$kpc offset) resemble those of hot spots in extended, double extragalactic radio sources.

158.301   Milliarcsecond polarization of superluminal sources.
D. H. Roberts, J. F. C. Wardle.
Superluminal radio sources, p. 193 – 199 (1987). – See Abstr. 012.053.

The authors review linear polarization measurements of three bright sources made at 5 GHz and briefly discuss their implications for the physics of extragalactic jets on the parsec scale.

158.302   The low frequency variability of extragalactic radio sources: a relativistic effect or galactic scintillation?
R. Fanti, L. Gregorini, L. Padrielli, S. Spangler.
Superluminal radio sources, p. 200 – 205 (1987). – See Abstr. 012.053.

158.303   A different perspective on superluminal sources.
L. Rudnick.
Superluminal radio sources, p. 217 – 232 (1987). – See Abstr. 012.053.

The author gives a pessimistic outlook of our ability to study superluminal sources and then tries to place them in the more general context of compact radio sources. He highlights some of the interesting information gained from studies of broadband spectra, variability, and polarization, and vaguely suggests a different way of looking at active nuclei, which might provide some useful insights into their nature.

158.304   Optical spectra of superluminal sources.
C. R. Lawrence, A. C. S. Readhead, T. J. Pearson, S. C. Unwin.
Superluminal radio sources, p. 260 – 266 (1987). – See Abstr. 012.053.

158.305   Emission–line profile changes in 3C 390.3.
J. B. Oke.
Superluminal radio sources, p. 267 – 272 (1987). – See Abstr. 012.053.

A program has been under way for many years to monitor the spectral characteristics of selected bright Seyfert galaxies and quasars. The author discusses the results obtained for 3C 390.3. Over thirty spectra were taken during the last six years. Multichannel observations go back to 1969 (Yee and Oke 1981).

158.306   Evidence for shocks in relativistic jets.
H. D. Aller, P. A. Hughes, M. F. Aller.
Superluminal radio sources, p. 273 – 279 (1987). – See Abstr. 012.053.

The authors made quantitative comparisons of the predictions of models invoking shocks with multifrequency total flux density and linear polarization data obtained in the University of Michigan variability program. The basic idea of these models is illustrated. The series of outbursts in BL Lac can be well described by a shock model. Another source which has exhibited outbursts with relatively high degrees of polarization is 3C 279.

158.307   Synchro–Compton emission from superluminal sources.
A. P. Marscher.
Superluminal radio sources, p. 280 – 300 (1987). – See Abstr. 012.053.

The author critiques the customary application of synchro–Compton theory to real compact radio sources. He explores the paltry evidence for and against a self–Compton origin of the X–rays in radio–loud quasars and active galactic nuclei. He presents some ideas on the phenomenology of superluminal motions.

158.308   The μ–z diagram.
M. H. Cohen.
Superluminal radio sources, p. 306 – 309 (1987). – See Abstr. 012.053.

Two simple measurable quantities for the variable sources and their distribution are discussed: the redshift z and the internal proper motion μ.

158.309   A nuclear molecular ring and gas outflow in the galaxy M82.

N. Nakai, M. Hayashi, T. Handa, Y. Sofue, T. Hasegawa, M. Sasaki.

Publ. Astron. Soc. Jpn., Vol. 39, No. 5, p. 685 – 708 (1987).

The CO (J = 1–0) emission of M82 has been mapped with spatial resolution of 16″. The mass of the molecular gas in the mapped area of 1.′5 (1400 pc) square is estimated to be $1.1 \times 10^8 M_\odot$, which shares 10% of the dynamical mass. The CO intensity in the central region of M82 has two peaks, which the authors interpret in terms of a ring of the molecular gas with a radius of 200 pc. The molecular ring corresponds to the sites of current star formation. Spurlike structures of the molecular gas emerge perpendicularly to the galactic plane toward the halo and extend more than 500 pc from the disk. The authors propose a model in which the molecular gas in M82 as well as the hot plasma is expelled from the plane of the galaxy at a velocity of $100 - 500$ km s^{-1}. The implied mechanical energy of the expelled gas can be explained by the energy supplied by successive supernova explosions in the central region of the galaxy.

158.310   Extragalactic radio sources with very large Faraday rotation.

T. Kato, H. Tabara, M. Inoue, K. Aizu.

Nature, Vol. 329, No. 6136, p. 223 – 224 (1987).

The authors have observed about 100 extragalactic radio sources which were suspected to have large rotation from their polarization catalogue. Four of these 100 sources were found to have an intrinsic rotation measure >1,000 rad m^{-2}. Here the authors report rotation measures which are two orders of magnitude larger than those observed so far for most extragalactic radio sources. Two of these four sources are classified as compact steep–spectrum sources.

158.311   6 centimeter radio source counts and spectral index studies down to 0.1 millijansky.

R. H. Donnelly, R. B. Partridge, R. A. Windhorst.

Astrophys. J., Vol. 321, No. 1, p. 94 – 112 (1987).

The authors present the results of a deep VLA survey at 6 cm in the Lynx.2 area of the Leiden Berkeley Deep Survey. This area was surveyed previously with the VLA at 21 cm as reported by Windhorst et al. in 1985. The survey sensitivities are ~15 and 28 μJy, at 6 and 21 cm, respectively. In the three 6 cm fields, chosen within the 21 cm survey area, 58 radio sources were found. Deep optical identifications and photometry of these sources are derived. They are based on four–band (UJFN) Kitt Peak 4 m plates. Available spectroscopic redshifts are also listed. The authors then present and analyze the spectral index distributions as a function of flux density and optical identification class.

158.312   The host galaxy of Markarian 231.

D. Hamilton, W. C. Keel.

Astrophys. J., Vol. 321, No. 1, p. 211 – 224 (1987). With plates 1 – 3.

The authors present digital images and spectra of the host galaxy of Markarian 231. Its morphology suggests a merger or other violent dynamical disturbance. Spectral features from a young stellar population and ionized gas are present, and the stellar population is younger than normally found in high–luminosity galaxies. A spatially resolved narrow–line region has been identified, of unusually low density and extent ~10 kpc. Low–level emission emitted blueward of the prominent disk emission is found to cover a large area and may be related to high velocity outflow of material from the nucleus.

158.313   The megamaser galaxy Markarian 273. I. VLA observations of the hydroxyl emission.

J. T. Schmelz, W. A. Baan, A. D. Haschick.

Astrophys. J., Vol. 321, No. 1, p. 225 – 232 (1987).

The hydroxyl megamaser emission in Mrk 273 was observed with the VLA in its high–resolution A array. The radio continuum source is extended at 18 cm; observations published by Ulvestad and Wilson in 1984 reveal a double at 6 cm where the stronger component is resolved into a triple at 2 cm. The OH emission is certainly associated only with the stronger component of the 6 cm double and possibly only with the strongest component of the 2 cm triple. The three velocity resolved components of the OH line are not spatially resolved and no information on the molecular disk rotation properties of this galaxy can be determined.

158.314   Broad–band properties of the CfA Seyfert galaxies. II. Infrared to millimeter properties.

R. A. Edelson, M. A. Malkan, G. H. Rieke.

Astrophys. J., Vol. 321, No. 1, p. 233 – 250 (1987).

Observations between 1.2 μm and 1.3 mm are presented for an unbiased, spectroscopically selected sample of 48 Seyfert galaxies. Most have complete infrared detections, but none were detected at 1.3 mm. There is a highly significant trend for the slope of the infrared spectrum to steepen from quasars to Seyfert 1 galaxies to Seyfert 2 galaxies. This is caused by an increasingly large ratio of thermal to nonthermal infrared emission along this sequence. About two–thirds of the Seyfert 1 galaxies have flat, quasar–like spectra, indicating that nonthermal radiation is responsible for the observed near– and mid–infrared emission. The thermal emission which characterizes all of the Seyfert 2 infrared spectra and one–third of the Seyfert 1 spectra appears to come from dust. Seyfert nuclei appear to reside in galaxies with higher than average far–infrared disk luminosities. This indicates that Seyfert host galaxies often are unusually active in star formation.

158.315   A detailed study of the C IV λ1550 line profile and adjacent spectral features in NGC 4151 from 1978 to 1983.

J. Clavel, A. Altamore, A. Boksenberg, G. E. Bromage, A. Elvius, D. Pelat, M. V. Penston, G. C. Perola, M. A. J. Snijders, M. H. Ulrich.

Astrophys. J., Vol. 321, No. 1, p. 251 – 279 (1987).

The 1450 – 1720 Å spectral region of NGC 4151 is analyzed by means of Gaussian decomposition for 69 different epochs, from early 1978 to the end of 1983. The C IV λ1550 emission profile is well represented by a core, whose full width at half–maximum varies between 2600 and 5500 km s^{-1}, and an ultrabroad component (14,600 km s^{-1}). The variations of the ultrabroad feature suggest that it originates in a region whose "radius" is ~5 lt–day. The analysis of lines from He II, C III], N IV] and O III] indicates the presence of two additional discrete broad line subregions. A study of the velocity dispersions indicates that the motion of the gas is Keplerian and yields a value of $(3.7 \pm 0.5) \times 10^7 M_\odot$ for the central mass. A correlation is found between the wavelength of the ultrabroad C IV λ1550 component and its intensity. It strongly suggests the existence of a *decelerated outflow* whose velocity is 4000 km s^{-1} at 5 lt–day from the nucleus.

158.316   A correlation between the radio and optical morphologies of distant 3CR radio galaxies.

P. J. McCarthy, W. van Breugel, H. Spinrad, S. Djorgovski.

Astrophys. J., Lett. Ed., Vol. 321, No. 1, p. L29 – L33 (1987). With plates L3 – L6.

The authors report the discovery of a strong correlation between the radio and optical morphologies of distant ($z \geqslant 0.6$), powerful radio galaxies. The isophotal axes of highly elongated distant 3CR galaxies, measured both in the light of stellar continua and extranuclear emission lines, tend to align with the radio source axes. The authors propose that the most natural explanation of the effect is that the radio jets and/or backflows from the radio lobes interact with the interstellar media of the gas–rich galaxies associated with 3CR sources and stimulate large–scale star formation. This mechanism can provide a physical explanation for the high star formation and evolution rates of 3CR galaxies at large look–back times found in earlier photometric studies.

158.317 Two micron spectroscopy of *IRAS* galaxies.
K. Kawara, M. Nishida, B. Gregory.
Astrophys. J., Lett. Ed., Vol. 321, No. 1, p. L35 – L40 (1987).

The intensities of the molecular hydrogen H_2 $v = 1$–0 $S(1)$ line, and the Brγ line of atomic hydrogen as well as the K magnitude have been observed in 21 *IRAS* galaxies. The sample includes type 1 and type 2 Seyferts, LINERs, H II region galaxies, and interacting galaxies. Their IR luminosities range from 5×10^8 to $10^{12} L_\odot$. It is found that, relative to the far-infrared luminosity, both the K luminosities and the intensities of molecular hydrogen emission in the AGN galaxies are higher than in the non–AGNs. Relative to braod–band K flux, the H II region galaxies show enhanced Brγ emission. Within any galaxy type, there is evidence that the H_2 emitting region extends over 1 kpc or more from the nucleus.

158.318 Spherical models of QSO/AGN emission line clouds.
P. S. Petersen.
Diss. Abstr. Int., Sect. B, Vol. 48, No. 6, p. 1711–B (1987). Thesis, University of California, San Diego, 167 pp. (1987). Order No. DA8720184.

158.319 Dark spots, bubbles, and shells in the lobes of extragalactic radio sources.
W. Kundt, L. Saripalli.
J. Astrophys. Astron., Vol. 8, No. 3, p. 211 – 217 (1987).

Based on maps of the extragalactic radio sources Cyg A, Her A, Cen A, 3C 277.3 and others, arguments are given that the twin–jets from the respective active galactic nucleus ram their channels repeatedly through thin, massive shells. The jets are thereby temporarily choked and blow radio bubbles. Warm shell matter in the cocoon shows up radio–dark through electron–scattering.

158.320 3C 120 and the surrounding region of sky.
H. Arp.
J. Astrophys. Astron., Vol. 8, No. 3, p. 231 – 239 (1987).

Image processing performed on a series of photographs of the superluminal Seyfert galaxy, 3C 120 shows the outer optical disc to consist of fragmented segments generally pointing toward the centre.

158.321 Alignment of radio and optical orientations in high–redshift radio galaxies.
K. C. Chambers, G. K. Miley, W. van Breugel.
Nature, Vol. 329, No. 6140, p. 604 – 606 (1987).

The results of a VLA and CCD imaging survey of ultra–steep spectrum radio sources confirm their association with faint and presumably distant galaxies. These galaxies have an almost universal alignment between the major axis of the optical emission and the radio axis. Such a strong relationship between the radio emission and the large–scale optical properties of the galaxies suggests that high–redshift radio galaxies are fundamentally different from their low–redshift counterparts. This has important consequences for current ideas on the nature of high–redshift radio galaxies and the cosmological work based upon them.

158.322 On the apparent position of a source at a large Doppler factor.
W. Zheng.
Nuovo Cimento B, Vol. 98B, Ser. 11, No. 2, p. 165 – 171 (1987). Abstr. in Phys. Abstr., Vol. 90, No. 1313, Entry 115619 (1987).

158.323 On the possibility of observation of X–ray absorption lines in the spectra of BL Lac–type objects.
A. S. Zentsova.
Astron. Zh., Tom 64, Vyp. 6, p. 1312 – 1317 (1987). In Russian. English translation in Sov. Astron., Vol. 31, No. 6.

The author discusses the possibility of detecting absorption features in the X–ray spectra of quasars and BL Lac–type objects, formed in extended radio lobes of these objects. It is shown that the optical depth in the center of the strongest absorption lines formed in lobes is $\tau_0 \gtrsim 1$. The detection of these lines in the X–ray spectra of quasars and BL Lac–type objects would give the possibility of directly estimating the temperature and the density of the thermal plasma in these lobes. It is shown that the energy of relativistic electrons is sufficient for plasma heating to a temperature $T = 10^7 – 10^8$K.

158.324 Extragalactic methanol.
IAU Circ., No. 4455 (1987).

158.325 The triple radio source 0023 + 171: a candidate for a dark gravitational lens.
J. N. Hewitt, E. L. Turner, C. R. Lawrence, D. P. Schneider, J. E. Gunn, C. L. Bennett, B. F. Burke, J. H. Mahoney, G. I. Langston, M. Schmidt, J. B. Oke, J. G. Hoessel.
Astrophys. J., Vol. 321, No. 2, p. 706 – 713 (1987). •

A composite radio source, with two optical counterparts at $z = 0.946$ separated by 5″, has been detected in a radio–optical gravitational lens survey. The redshifts of the optical counterparts are not significantly different, and their optical spectra are similar. One interpretation of these measurements is that the images of 0023 + 171 are gravitationally lensed; another is that the components are two physically associated radio galaxies or components of a single galaxy. The only possible evidence for a lensing object is extremely faint $(m_r > 23.5)$ optical emission located approximately 1″ from one of the images. If this multiple source is gravitationally lensed, the data imply a large mass–to–light ratio for the lensing matter $(\sim 1000 \, M_\odot / L_\odot)$.

158.326 Far–infrared continuum emission from the nucleus, starburst, and extended spiral arms of NGC 1068.
D. F. Lester, M. Joy, P. M. Harvey, H. B. Ellis Jr., P. S. Parmar.
Astrophys. J., Vol. 321, No. 2, p. 755 – 760 (1987).

Far–infrared slit scans across NGC 1068 are used to better define the contribution of the active nucleus, starburst disk, and extended spiral arms in the luminous output of this bright Seyfert 2 galaxy. The decomposition of the emission from the bright, inner 3 kpc of this galaxy suggests that at 50 μm about half the emission is from star formation in a starburst disk, and half from the compact, centrally heated circumnuclear cloud. With the application of maximum entropy deconvolution to the scans, the authors find evidence for 100 μm emission from the extended spiral arms of the galaxy.

158.327 Daily observations of compact extragalactic radio sources at 2695 and 8085 MHz, 1979 – 1985.
R. L. Fiedler, E. B. Waltman, J. H. Spencer, K. J. Johnston, P. E. Angerhofer, D. R. Florkowski, F. J. Josties, W. J. Klepczynski, D. D. McCarthy, D. N. Matsakis.
Astrophys. J., Suppl. Ser., Vol. 65, No. 3, p. 319 – 384 (1987).

The authors present flux densities at 2695 and 8085 MHz of 33 compact extragalactic radio sources observed from 1979 through 1985 as part of a daily monitoring program using the Green Bank interferometer. The corresponding autocorrelation and cross–correlation functions and structure functions are also presented. It is concluded that extragalactic radio source variability may be characterized as a modulation of intrinsic effects by weaker Galactic effects, such as refractive interstellar scintillation and extreme scattering events.

158.328 Markarian 421.
Br. Astron. Assoc. Circ., No. 672 (1987).

158.329 An upper limit of the redshifted CO ($J = 1$–0) absorption line toward AO 0235 + 164.
F. Takahara, N. Nakai, F. H. Briggs, A. M. Wolfe, H. S. Liszt.
Publ. Astron. Soc. Jpn., Vol. 39, No. 6, p. 933 – 936 (1987).

A search was made for a redshifted CO absorption line ($J = 1$–0) toward the BL Lac object AO 0235 + 164, for which 21–cm and Lyα absorption lines have been detected at the redshift of $z = 0.524$. The ratio of the upper limit of column density of CO molecules to that of the neutral hydrogen atoms is estimated to be about 3×10^{-7}, an order of magnitude less than the typical galactic value.

158.330 Near–infrared imaging of M87.
K.–W. Hodapp, E. E. Becklin, R. W. Capps,
D. N. B. Hall, D. A. Simons.
Infrared astronomy with arrays, p. 330 – 336 (1987). – See Abstr.
012.070.

Images of the nuclear region of M87 in J, H, and K and its jet in
H and K have been obtained using the UH infrared camera
equipped with the 64 × 64 SISEX HgCdTe device. The images
show that the nucleus of M87 is an unresolved source embedded
in the extended stellar emission of the central part of this galaxy.
Photometry of the nucleus (after subtraction of the extended
component) fits well into the optical spectrum of the nucleus. The
spectral index is very large (about 2) in the B to H range. The
photometric K value and the data by other authors at 3.8 μm and
10 μm suggest a break in the spectrum between 1 μm and 2 μm.
The photometry of the knots in the jet confirm that the spectral
indices in the different knots are essentially the same.

158.331 Extragalactic imaging with IRCAM on UKIRT.
M. G. Smith, I. S. McLean, C. M. Telesco,
M. J. Ward, N. Devereux.
Infrared astronomy with arrays, p. 350 – 353 (1987). – See Abstr.
012.070.

Two experiments are used to illustrate some of the properties
and potential of IRCAM for extragalactic astronomy in its low
and high spatial resolution modes. An SBRC engineering grade
62 × 58 pixel InSb array was used.

158.332 Infrared spectroscopy of NGC 1068.
D. L. DePoy.
Infrared astronomy with arrays, p. 426 – 429 (1987). – See Abstr.
012.070.

Spectroscopy of the nucleus of the nearby Seyfert 2 galaxy
NGC 1068 has been obtained using the IRTF and the facility
Cooled–Grating Array Spectrometer (CGAS). The wavelengths
observed covered the expected wavelengths of the Brγ
(n = 7→4), Brα (n = 5→4), and the Pfβ (n = 7→5) hydrogen
recombination lines 2.2 μm and 4.6 μm. The data show
that the infrared lines are not more broadened than the optical
hydrogen recombination lines, suggesting that the presence of an
obscured Seyfert 1–like active nucleus is unlikely unless the visual
extinction through any obscuring material present is larger than
~ 100 mag.

158.333 Photographic photometry of compact extragalactic
objects. X.
V. A. Hagen–Thorn, N. S. Denisenko, T. M. Maksimova,
S. G. Marchenko, O. V. Mikolajchuk.
Tr. Astron. Obs., Leningrad, Tom 41, p. 96 – 112 (1987). Uch.
Zap. LGU, No. 420, Ser. Mat. Nauk, Vyp. 63. In Russian.

Results are given of photographic photometry in B of 6
compact extragalactic objects for 1980 – 1984. The features of
light curves are briefly discussed.

158.334 Search for rapid optical light variability of nuclei of the
Seyfert galaxies NGC 1068 and NGC 7469.
A. A. Aslanov, N. A. Lipunova.
Astron. Tsirk., No. 1484, p. 6 – 8 (1987). In Russian.

The results of searching for a short–term optical light
variability of the Seyfert galactic nuclei NGC 1068 and
NGC 7469 are given. The B–light of NGC 1068 (type 2 Seyfert
galaxy) was constant within 0^{m}02 during ~1 hr monitoring on
August 12/13, 1985. Observations of NGC 7469 (type 1 Seyfert
galaxy) carried out in W–light during ~1 hr on August 13, 1985
showed variability ~0^{m}06 (~5σ) on time–scales of some
minutes.

158.335 Anisotropic Compton scattering from relativistic jets of
active galactic nuclei.
V. M. Charugin.
Astron. Tsirk., No. 1487, p. 1 – 3 (1987). In Russian.

The theoretical spectrum of Compton radiation from a
relativistic jet moving in an anisotropic radiation field of
extended structures under the small angle to the line of sight was

calculated. It is shown that the observed decimeter–wave radio
emission from some jets is in agreement with this model.

158.336 The cause of discrepancy between evolution functions at
large redshifts.
J. Zawiślak–Raczka, B. Kumor–Obryk.
Astrophys. Space Sci., Vol. 139, No. 2, p. 305 – 309 (1987).

The influence of the character of a numerical method used in
the derivation of the evolution function of flat spectrum radio
sources on the value of the redshift cut–off has been considered.
An attempt has been made to apply to radio sources the general
approach to the cosmological evolution based on the conservation equation.

158.337 An analysis of the X and UV emission of active galactic
nuclei.
J.–h. You, F.–z. Cheng, R.–s. Gong, R.–l. Liu.
Chin. Astron. Astrophys., Vol. 11, No. 3, p. 263 – 268 (1987).
English translation of Acta Astron. Sin., Vol. 28, No. 2,
p. 160 – 167 (1987).

Under the assumption of spherical accretion, the authors used
the synchrotron radiation mechanism to derive the relation
between the 2 keV X luminosity (l_X) and the 2500 Å
UV luminosity (l_{op}) of active galactic nuclei. They found that,
when the accretion rate is low, l_X is directly proportional to l_{op}.
When the accretion rate is high, the authors found l_X proportional to l_{op} to the power 0.30, thus explaining the tendency toward
"saturation" in the observed $l_X - l_{op}$ curve.

158.338 VLBI structure of 3C 84 at 89 GHz.
D. C. Backer, M. C. H. Wright, R. L. Plambeck,
J. E. Carlstrom, C. R. Masson, A. T. Moffet,
A. C. S. Readhead, D. Woody, A. E. E. Rogers, J. M. Moran,
C. R. Predmore, R. L. Dickman.
Astrophys. J., Vol. 322, No. 1, p. 74 – 79 (1987).

The authors have made 89 GHz VLBI observations of 3C 84 at
five epochs between 1981 and 1985. Model fitting suggests a
core–halo structure, which is associated with the active nucleus of
NGC 1275. Both the core, which is smaller than 0.2 mas, and the
more extended halo show a slow decay corresponding to the
decrease in total flux density since a flare which occured in 1980.
In 1985 a "jet" was detected. It is extended in a position angle
205° and is unresolved in width (< 0.1 mas).

158.339 Near–infrared observations of submillijansky radio
sources: evidence for a population of starburst galaxies
at intermediate redshifts.
T. X. Thuan, J. J. Condon.
Astrophys. J., Lett. Ed., Vol. 322, No. 1, p. L9 – L13 (1987).

The authors present near–infrared JHK photometry for 32
sub–mJy radio sources discovered in a deep 20 cm VLA survey
field centered on α(1950) = 08^{h}52^{m}15^s and δ(1950) = 17°16'.
Combining these measurements with optical photometry and
spectral evolution models, the authors conclude that the sources
responsible for the upturn in the radio source counts at sub–mJy
levels represent a large starburst galaxy population with
$-23 \lesssim M_v \lesssim -20$ and with redshifts between ~0.05 and ~0.6.

158.340 On the broad emission lines in Seyfert galaxies and
quasars.
Eh. A. Dibaj.
Sov. Astron., Vol. 31, No. 2, p. 117 – 120 (1987). English translation of 43.158.083.

158.341 Spectral investigation of the peculiar galaxy NGC 6240.
N. K. Andreasyan, Eh. E. Khachikyan.
Astrofizika, Tom 27, Vyp. 2, p. 265 – 274 (1987). In Russian.
With 2 plates. English translation in Astrophysics, Vol. 27,
No. 2.

Results of a detailed spectral investigation of NGC 6240 are
presented. Spectrophotometry of two components of the nucleus
found earlier in the radio and IR regions which are at 2" from
each other and were considered to be nuclei of two colliding

galaxies is carried out. Physical properties and chemical abundances are determined. The Seyfert properties belong to one component while the other has spectral properties of H II regions. The Seyfert component rotates but the H II region does not rotate.

158.342 On the relation of Seyfert galaxies with clusters.

A. R. Petrosyan.
Astrofizika, Tom 27, Vyp. 2, p. 275–281 (1987). In Russian. English translation in Astrophysics, Vol. 27, No. 2.

For the selection of Seyfert galaxies (SG) – cluster members, a comparison of the SG sample (464 objects) with Zwicky and Abell clusters and southern clusters as well is carried out. 67 SG are identified in Zwicky clusters, 15 SG in Abell clusters and 18 in southern clusters. Lists of these objects are presented.

158.343 The mass–luminosity dependence for active galactic nuclei.

V. P. Reshetnikov.
Astrofizika, Tom 27, Vyp. 2, p. 283–293 (1987). In Russian. English translation in Astrophysics, Vol. 27, No. 2.

The paper deals with the dependence of the active nuclei luminosity in various spectral bands on their mass based on the Catalogue of data for active galactic nuclei by E. A. Dibaj. It has been shown that for Seyfert 1 galaxies and nearby quasars $L_{bol} \sim m^{4/3}$ satisfies Hill's model.

158.344 X–ray and optical observations of X–ray–selected BL Lacertae objects.

P. Giommi, P. Barr, B. Garilli, I. M. Gioia, T. Maccacaro, D. Maccagni, R. E. Schild.
Astrophys. J., Vol. 322, No. 2, p. 662–672 (1987).

Results from several X–ray and optical observations of five X–ray–selected BL Lacertae objects are reported. X–ray light curves covering periods of up to 6 yr reveal that the X–ray flux from these objects does not show large amplitude trends over this time scale. Variations of up to 2 mag have been observed in the optical. The soft X–ray spectra of three objects have been measured with *EXOSAT*. Energy spectral indices range between 1 and 2. A new optical flare from 1E 1402.3 ± 0416, with rise and decay time of order of a few weeks, has been detected.

158.345 OH megamasers in high–luminosity *IRAS* galaxies.

I. F. Mirabel, D. B. Sanders.
Astrophys. J., Vol. 322, No. 2, p. 688–693 (1987).

OH megamaser emission, and H I and CO profiles from the distant infrared galaxies IRAS 10173+0828, III Zw 035, and Zw 475.056 are reported. The OH isotropic luminosities at 1667 MHz are 463, 534, and 6.6 $L_\odot$, respectively. Far–infrared pumping efficiencies of the OH greater than 1% are found in IRAS 10173+0828 and III Zw 035. OH megamasers reside in the nuclei of superluminous far–infrared galaxies that have a high content of molecular gas, high efficiency of star formation, and in some instances, a striking deficiency of atomic hydrogen.

158.346 CO emission from Centaurus A.

T. G. Phillips, B. N. Ellison, J. B. Keene, R. B. Leighton, R. J. Howard, C. R. Masson, D. B. Sanders, B. Veidt, K. Young.
Astrophys. J., Lett. Ed., Vol. 322, No. 2, p. L73–L77 (1987).

CO (2→1) emission has been detected along the major axis of the dust lane in Centaurus A out to a distance of 2′ from the active nucleus. The derived radial distribution of molecular gas (H_2) is centrally peaked with a scale length of ~1.5 kpc and is similar to that expected from a small Sc galaxy. The total H_2 mass is ~$1 \times 10^9 M_\odot$, comparable to the mass of atomic gas. The kinematics of the CO emission from the nuclear region is consistent with circular rotation. The systematic velocity of the molecular disk is found to be 547 km s⁻¹, similar to previous values determined from H I and H II regions in the disk.

158.347 Optical spectra of narrow emission line Palomar–Green galaxies.

D. E. Osterbrock, R. W. Pogge.
Astrophys. J., Vol. 323, No. 1, p. 108–117 (1987). = Lick Obs. Bull., No. 1072.

Spectra were obtained of 35 of the 36 narrow emission line galaxies isolated in the Palomar–Green (PG) survey. Of these, three are narrow–line Seyfert 1 galaxies, three more are Seyfert 1.5 galaxies, and only one, PG 2259+157 is a relatively low–ionization active galactic nucleus, a marginal Seyfert 2. The rest are H II region galaxies, as is CSO 177, a candidate Seyfert 2 galaxy. Redshifts and relative emission–line strengths are given for all these galaxies. The PG survey shows that a significant number of Seyfert 1 galaxies are "narrow–line" objects with H I emission–line full widths at half–maximum ⩽2000 km s⁻¹.

158.348 What heats the hot phase in active nuclei?

W. G. Mathews, G. J. Ferland.
Astrophys. J., Vol. 323, No. 2, p. 456–467 (1987). = Lick Obs. Bull., No. 1077.

The authors summarize some constraints on the continuum in active galactic nuclei and discuss implications for the hot intercloud medium. First, there is now good evidence that the peak of the energy distribution may actually occur in the extreme ultraviolet 200 Å ⩽ λ ⩽1000 Å. Second, it is shown that the Compton temperature is $T_C \approx 10^7$K, far too low for broad line region clouds to be stable against drag forces, or for the hot phase to be optically thin to observed X–rays. Finally, the authors discuss some dynamical problems with current models of the intercloud medium.

158.349 VLBI observations of 23 hot spots in the starburst galaxy M82.

N. Bartel, M. I. Ratner, A. E. E. Rogers, I. I. Shapiro, R. J. Bonometti, N. L. Cohen, M. V. Gorenstein, J. M. Marcaide, R. A. Preston.
Astrophys. J., Vol. 323, No. 2, p. 505–515 (1987).

The authors have used the Mark III VLBI system to observe, at 2.3 and 8.4 GHz simultaneously, 23 hot spots in the nuclear region of the nearby starburst galaxy M82. The authors detected six hot spots at 2.3 GHz but only one at 8.4 GHz. The brightness distribution and an upper limit on the expansion rate of the brightest hot spot, 41.9+58, is derived. These results and the diameters and ages, or corresponding lower bounds, for 13 more hot spots are consistent with the hot spots in M82 being powerful supernova remnants with ages between ~10 and ~300 yr. The authors estimate a radio supernova rate of ~0.1 yr⁻¹ for the inner 600 pc of M82.

158.350 Far–infrared variability in active galactic nuclei.

R. A. Edelson, M. A. Malkan.
Astrophys. J., Vol. 323, No. 2, p. 516–535 (1987).

Pointed *IRAS* observations of 20 active galaxies were examined in the first detailed search for far–infrared variability in a large sample of active galaxies. Less extensive survey data were also checked for evidence of strong 6 month variability in 45 active galaxies. The far–infrared fluxes of three highly polarized objects ("blazars") appeared to vary by up to a factor of 2 on time scales of a few months. No convincing cases of variability greater than ~15% (rms) were found in any of the normal quasars or Seyfert galaxies studied.

158.351 Linear polarization structure of the BL Lacertae object OJ 287 at milliarcsecond resolution.

D. H. Roberts, D. C. Gabuzda, J. F. C. Wardle.
Astrophys. J., Vol. 323, No. 2, p. 536–542 (1987).

The λ = 6 cm total intensity and linear polarization structures of the BL Lacertae object OJ 287 have been determined with an angular resolution of 2 × 9 milliarcseconds at two epochs a year apart. At each epoch the source can be modeled as a linear structure consisting of a core and two knots. The core C was slightly polarized (~4%), the inner knot K2 strongly polarized (up to 64%), and the outer knot K1 slightly polarized (~3%).

Superluminal motion for each knot, with an apparent velocity $\sim 3.3h^{-1}c$, is suggested by the polarization data.

Second Byurakan Spectral Sky Survey. V. Results for the region centered on $\alpha = 15^h30^m$, $\delta = +59°00'$.
See Abstr. 002.013.

The Green Bank Third (GB3) Survey of extragalactic radio sources at 1400 MHz.
See Abstr. 002.035.

A list of reference stars in 252 areas with extragalactic radio sources.
See Abstr. 002.078.

Models of quasars reappraised.
See Abstr. 011.019.

A review of decametric radio astronomy: instruments and science.
See Abstr. 013.022.

What are we missing?
See Abstr. 021.020.

Experiment Cold: the first deep sky survey with the RATAN–600 radio telescope.
See Abstr. 033.013.

Ground–based applications for the JPL/SISEX SWIR array.
See Abstr. 034.157.

Astronomical observations at 10 micron wavelength with the NASA/GSFC array camera system.
See Abstr. 034.160.

IRSPEC: design, performance and first scientific results.
See Abstr. 034.165.

X–ray spectroscopy of AGN with the AXAF "microcalorimeter".
See Abstr. 035.026.

Use of a minimum rate of change formalism to quantify variability of extragalactic X–ray sources.
See Abstr. 036.041.

Image sharpening observations of active galactic nuclei.
See Abstr. 036.176.

Geodetic VLBI–monitoring of the milliarcsecond structures of extragalactic radio sources.
See Abstr. 036.200.

Comparisons of positions of extragalactic compact radio sources.
See Abstr. 041.001.

Spatial stability of relativistic jets: application to 3C 345.
See Abstr. 062.001.

Compact radio sources as a plasma turbulent reactor. I. Formation of Maxwell–like spectra of relativistic electrons at the acceleration on resonant Langmuir waves.
See Abstr. 062.009.

On magnetohydrodynamic solitons in jets.
See Abstr. 062.017.

The sources and their models.
See Abstr. 062.038.

The central engine.
See Abstr. 062.039.

The jets.
See Abstr. 062.040.

Unified beaming models and compact radio sources.
See Abstr. 062.043.

Dynamical effects of large–scale magnetic fields in jets.
See Abstr. 062.046.

Numerical studies of the dynamical stability of differentially rotating tori.
See Abstr. 062.078.

Relativistic thermal plasmas: time development of electron–positron pair concentration.
See Abstr. 062.100.

The influence of relativistic electrons on a photoionized gaseous cloud.
See Abstr. 063.004.

A link between X–ray variability and absorption in active galactic nuclei.
See Abstr. 063.016.

Magnetic field and synchrotron radiation in mildly relativistic shocks.
See Abstr. 063.019.

Variability from pair atmospheres.
See Abstr. 063.031.

Relativistic plasmas in active galactic nuclei.
See Abstr. 063.032.

The "Lα/Fe II problem" – solved by fluorescence?
See Abstr. 063.036.

Compact radio sources as a plasma turbulent reactor. II. General characteristics of electromagnetic radiation spectra.
See Abstr. 063.041.

Radiation from charged particles moving in the magnetic fields of extragalactic jets.
See Abstr. 063.054.

Power–law X–ray emission from electron–positron pair winds.
See Abstr. 063.059.

On the stability of turbulent synchrotron sources with respect to e^+–e^- pair creation.
See Abstr. 063.114.

Synchrotron emission from shock waves in active galactic nuclei.
See Abstr. 063.115.

Inverse Compton scattering of ambient radiation by a cold relativistic jet: a source of beamed, polarized continuum in blazars?
See Abstr. 063.116.

On pumping the strong water maser sources.
See Abstr. 063.121.

New mechanism of appearance of proper redshifts in spectra of compact objects.
See Abstr. 066.179.

Observations of accretion instabilities at super Eddington accretion rates.
See Abstr. 067.071.

Accretion disc models around compact objects.
See Abstr. 067.075.

Astrophysical black holes.
See Abstr. 067.079.

Pair production and Compton scattering in compact sources and comparison to observations of active galactic nuclei.
See Abstr. 067.080.

Thick accretion disks: theory vs. observations.
See Abstr. 067.084.

How big are supermassive black holes formed from the collapse of dense star clusters?
See Abstr. 067.086.

Accretion disks in soft potential wells.
See Abstr. 067.095.

Electron injection by relativistic protons in active galactic nuclei.
See Abstr. 067.098.

Grand Unified Models.
See Abstr. 067.122.

On nonthermal models for active galactic nuclei.
See Abstr. 067.128.

Self–gravitating accretion disks in active galactic nuclei.
See Abstr. 067.136.

Mining energy from a rotating black hole in a magnetic field by the Penrose process.
See Abstr. 067.140.

Electromagnetic jets from compact objects.
See Abstr. 067.144.

Angular momentum transport by star–gas interaction and structure of accretion disks.
See Abstr. 067.146.

Pair–creation effects in accretion–shock models of active galactic nuclei.
See Abstr. 067.177.

On the progenitors of type II supernovae in M83.
See Abstr. 125.068.

First results with a transmission echelle grating on the ESO Faint Object Spectrograph: observations of the SN 1986a in NGC 3367 and of the nucleus of the galaxy.
See Abstr. 125.821.

High mass star formation in the Galaxy.
See Abstr. 131.001.

Structure of galaxies and star formation: workshop summary.
See Abstr. 131.199.

Systematic identification of IRAS point sources.
See Abstr. 133.001.

The IRAS view of the extragalactic sky.
See Abstr. 133.007.

VLA observations of radio sources near Wolf 359.
See Abstr. 141.014.

The origin of the diffuse X–ray background.
See Abstr. 142.010.

Discrete X–ray sources and the X–ray background.
See Abstr. 142.011.

The cosmic X–ray background.
See Abstr. 142.012.

Production of the diffuse X–ray background spectrum by active galactic nuclei.
See Abstr. 142.024.

The contribution from active galactic nuclei to the diffuse 1 MeV background.
See Abstr. 143.015.

Models of ring galaxies. I. The growth and disruption of clouds in the expanding density wave.
See Abstr. 151.001.

Shock–wind mechanism of formation of the shell structure in NGC 5128. Numerical modulation.
See Abstr. 151.039.

Dynamical instability of a model of M87.
See Abstr. 151.040.

Spherical galaxies: methods and models.
See Abstr. 151.054.

Models of ring galaxies. II. Extended starbursts.
See Abstr. 151.149.

Systematic properties of extragalactic globular clusters.
See Abstr. 154.019.

IUE observations of the globular cluster No. 3 in the Fornax dwarf spheroidal galaxy.
See Abstr. 154.045.

Models for infrared emission from IRAS galaxies.
See Abstr. 157.003.

Measuring star formation rates in blue galaxies.
See Abstr. 157.005.

CO observations of galaxies with the Nobeyama 45–m telescope.
See Abstr. 157.006.

Stellar bars and the spatial distribution of infrared luminosity.
See Abstr. 157.009.

Nuclear star formation on 100 parsec scales: 10″ resolution radio continuum, H I, and CO observations.
See Abstr. 157.027.

Induced star formation in interacting galaxies.
See Abstr. 157.029.

The frequency of enhanced star formation in interacting and isolated galaxies.
See Abstr. 157.030.

A redshift survey of IRAS galaxies.
See Abstr. 157.036.

Optical and IR luminosity functions of IRAS galaxies.
See Abstr. 157.037.

Radio continuum, far infrared and star formation.
See Abstr. 157.040.

Near–infrared observations of IRAS minisurvey galaxies.
See Abstr. 157.041.

Ground–based follow up of IRAS galaxies.
See Abstr. 157.042.

The properties of highly luminous IRAS galaxies.
See Abstr. 157.044.

A study of a flux–limited sample of IRAS galaxies.
See Abstr. 157.045.

Accurate positions of Zwicky galaxies. II.
See Abstr. 157.048.

Accurate positions of Zwicky galaxies. III.
See Abstr. 157.049.

Analysis of absorption–line spectra in a sample of 164 galactic nuclei.
See Abstr. 157.050.

Accretion flows in elliptical galaxies.
See Abstr. 157.089.

The optical luminosity function of a 60 μm flux–limited sample of IRAS galaxies.
See Abstr. 157.100.

The radio and optical axes of radio elliptical galaxies.
See Abstr. 157.111.

Optical and near–infrared observations of IRAS galaxies. II.
See Abstr. 157.117.

Star formation in colliding and merging galaxies.
See Abstr. 157.130.

Hot coronae around early type galaxies.
See Abstr. 157.138.

Dynamics of galaxies at large redshift: prospects for the future.
See Abstr. 157.144.

Dust and gas: overview.
See Abstr. 157.155.

Properties of elliptical galaxies with dust lanes.
See Abstr. 157.157.

X–rays from elliptical galaxies.
See Abstr. 157.159.

Mass distributions in elliptical galaxies at large radii.
See Abstr. 157.160.

Line–strength gradients in early–type galaxies.
See Abstr. 157.161.

Visual–IR color gradients in elliptical galaxies.
See Abstr. 157.185.

Direct IR determination of the stellar luminosity function to 0.2 $M_\odot$ in elliptical galaxies.
See Abstr. 157.186.

Evidence of an infrared luminosity indicator for galaxies.
See Abstr. 157.200.

Photometry of Zwicky compact galaxies.
See Abstr. 157.213.

Investigations of extragalactic hydroxyl.
See Abstr. 157.222.

Radio continuum observations of the spiral galaxy NGC 4736.
See Abstr. 157.227.

Radio structure of the nuclei of "edge–on" spiral and Seyfert galaxies.
See Abstr. 157.254.

2 μm spectroscopy of nearby galaxies and evidence for a late phase starburst in NGC 4736.
See Abstr. 157.278.

The detection of extragalactic methanol.
See Abstr. 157.303.

Infrared images of galaxies.
See Abstr. 157.324.

The spatial distribution of 10 micron luminosity in spiral galaxies.
See Abstr. 157.346.

Photoerosion of nuclei in quasar emission–line regions.
See Abstr. 159.002.

1300 μm detection of the radio–quiet quasar 13349 + 2438.
See Abstr. 159.007.

Statistical gravitational lensing: influence of compact objects on the number counts of quasars.
See Abstr. 159.017.

The optical luminosity function of quasars and low–luminosity active galactic nuclei.
See Abstr. 159.023.

Quasars, Seyfert galaxies and active galactic nuclei.
See Abstr. 159.039.

Quasar and stellar objects in the Byurakan Surveys.
See Abstr. 159.040.

Resonance and subordinate lines of O III in quasar spectra.
See Abstr. 159.058.

Detection of the host galaxy of quasar IRAS 00275–2859.
See Abstr. 159.059.

Hot dust and the near–infrared bump in the continuum spectra of quasars and active galactic nuclei.
See Abstr. 159.078.

A 21 cm study of some QSO/galaxy pairs.
See Abstr. 159.081.

The early universe – an observer's view.
See Abstr. 159.089.

VLBI observations of the gravitational lens system 2016 + 112.
See Abstr. 159.093.

Superluminal radio sources: introduction.
See Abstr. 159.094.

Summary of known superluminal sources.
See Abstr. 159.095.

Superluminal motion and other indications of bulk relativistic motion in a complete sample of radio sources from the S5 survey.
See Abstr. 159.103.

Extended structure of superluminal radio sources.
See Abstr. 159.107.

Infrared, optical, UV, and X–ray properties of superluminal radio sources.
See Abstr. 159.111.

Superluminal radio sources: what does X–ray emission tell us?
See Abstr. 159.112.

The effect of a quasi–stellar object on its host galaxy: dynamical and physical processes in the interstellar medium around a quasi–stellar object.
See Abstr. 159.141.

Quasar energy distributions. I. Soft X–ray spectra of quasars.
See Abstr. 159.145.

Forbidden [O II] and [O III] emission 29″ from a QSO absorption–line region.
See Abstr. 159.147.

The diffuse radio emission from the Coma cluster.
See Abstr. 160.012.

X–ray observations of the Ophiuchus, PKS 0745–191 and Cygnus–A clusters of galaxies.
See Abstr. 160.016.

Studies of the Virgo cluster. VI. Morphological and kinematical structure of the Virgo cluster.
See Abstr. 160.017.

Radio emission in clusters: problems and opportunities.
See Abstr. 160.024.

Statistical results from the Effelsberg Abell cluster radio survey.
See Abstr. 160.027.

Some results from the Molonglo cluster survey.
See Abstr. 160.028.

VLA observations of distant clusters of galaxies.
See Abstr. 160.029.

OSRT 327 MHz observations of clusters of galaxies: some very steep spectrum sources.
See Abstr. 160.030.

EXOSAT observations of the Virgo cluster.
See Abstr. 160.031.

The relationship of optical emission line gas and extended radio emission in the centers of cooling flow clusters.
See Abstr. 160.032.

Radio properties of central dominant galaxies in cluster cooling flows.
See Abstr. 160.033.

The radio sources in the centers of the Coma and Hercules clusters.
See Abstr. 160.034.

Radio sources associated with 5 first ranked galaxies.
See Abstr. 160.035.

Radio halo sources in clusters of galaxies.
See Abstr. 160.036.

The radio halo and magnetic field in the Coma cluster of galaxies.
See Abstr. 160.037.

Radio observations of the peripheral region of the Coma cluster near Coma A.
See Abstr. 160.039.

Radio, optical and X–ray observations of PKS 2104–25.
See Abstr. 160.040.

Beam trajectories in the intracluster medium.
See Abstr. 160.041.

A cautionary note on cluster radio sources.
See Abstr. 160.051.

A VLA 20 cm survey of poor groups of galaxies.
See Abstr. 160.054.

Evolution of cluster galaxies since $z = 1$.
See Abstr. 160.063.

Star formation in cooling flows.
See Abstr. 160.065.

An imaging study of NGC 1275 and PKS 0745–191: vigorous star formation in cooling flow cluster dominant galaxies.
See Abstr. 160.124.

The baryon–symmetric domain cosmology and the cosmic X–ray background.
See Abstr. 161.010.

Large–scale distribution of objects from the First Byurakan Survey (FBS).
See Abstr. 161.039.

Cosmological evolution of active galaxies & quasars.
See Abstr. 161.089.

Galaxies at very high redshifts ($z > 1$).
See Abstr. 161.143.

Cosmological evolution of extragalactic sources.
See Abstr. 161.173.

Correction of a study of source counts in the chronometric cosmology.
See Abstr. 161.180.

Constraints on possible precursor AGN sources of the cosmic X–ray background.
See Abstr. 161.452.

159 Quasi-stellar Objects

159.001   Luminosity dependence in the ratio of X–ray to infrared emission of QSOs.
D. M. Worrall.
Astrophys. J., Vol. 318, No. 1, p. 188 – 195 (1987).
The correlation of X–ray and near–infrared luminosity is studied for a sample of radio–quiet QSOs. The X–ray to infrared ratio is found to decrease as the infrared luminosity increases. No preference is found between the correlations of X–ray luminosity with optical or infrared luminosity. This implies that optical and infrared emission are equally good predictors of X–ray emission. Source models which directly link infrared and X–ray emission are discussed, and a preference is found for a specific synchrotron self–Compton model.

159.002   Photoerosion of nuclei in quasar emission–line regions.
R. N. Boyd, G. J. Ferland.
Astrophys. J., Lett. Ed., Vol. 318, No. 1, p. L21 – L24 (1987).
Several active galactic nuclei are now known to be sources of $hv \geqslant$ MeV γ-rays. The authors point out some consequences of this radiation for elemental abundances in nearby emission–line regions. A series of (γ,n) and (γ,p) reactions result in mass degradation, a process the authors refer to as photoerosion. Heavy nuclei such as Fe and Ni are destroyed on time scales of roughly 10^3yr for regions in NGC 4151, while most light nuclear abundances change little on such time scales. Many odd–z nuclei, particularly B, have large abundance increases. Photoerosion may account for several previously unexplained phenomena, e.g., the large spectral differences between radio–loud and radio–quiet AGN and the observed aluminum overabundance.

159.003   The origin of quasars.
J.–L. Nieto.
Recherche, Vol. 18, No. 190, p. 924 – 932 (1987). In French.

159.004   Gravitational mirages.
A. Blanchard, F. Hammer, C. Vanderriest.
Recherche, Vol. 18, No. 192, p. 1182 – 1190 (1987). In French.

159.005   The shell structure of a pancake and the absorption spectra of quasars.
B. V. Vajner, Yu. A. Shchekinov.
Astrofizika, Tom 26, Vyp. 3, p. 431 – 442 (1987). In Russian. English translation in Astrophysics, Vol. 26, No. 3, p. 260 – 267 (1987).
The formation of absorption lines of neutral hydrogen in distant quasar spectra is considered. The model of the pancake shell formation in an adiabatic picture of large scale structure generation in the Universe is considered. It is shown that the absorption lines can form "doublets"; the equivalent widths of such lines are calculated. The physical conditions corresponding to the observed absorption spectra of heavy elements are discussed.

159.006   Hard X–ray observations of the quasar 3C 273.
S. V. Damle, P. K. Kunte, S. Naranan,
B. V. Sreekantan, D. Venkatesan.
Astron. Astrophys., Vol. 182, No. 1, p. L1 – L4 (1987). With a correction in Vol. 186, No. 1/2, p. L20.
The quasar 3C 273 was observed with a balloon–borne phoswich scintillator detector system launched on December 12, 1983 from Hyderabad, India. 3C 273 was tracked for 130 minutes and the background was measured in the same ranges of elevation and azimuth when the source was not in the field of view, for approximately the same duration. The fluxes in the 18– 25, 25–40, 40–60 and 60–120 keV energy intervals, in units of 10^{-4} photons cm^{-2} s^{-1} keV^{-1} are 11.73 ± 4.20, 2.18 ± 0.64, 1.40 ± 0.29 and 0.30 ± 0.19 respectively. The fluxes above 25 keV are consistent with the balloon observations of 3C 273 by Bezler et al., on September 28, 1981 and about trice as high as the 1978/ 1979 HEAO–1 measurements. However, the observations taken in conjunction with the lower energy (1–20 keV) data of

December 17, 1983 from EXOSAT satellite, indicate possible spectral variations between 1981 and 1983.

159.007   1300 μm detection of the radio–quiet quasar 13349 + 2438.
R. Chini, E. Kreysa, C. J. Salter.
Astron. Astrophys., Vol. 182, No. 2, p. L63 – L65 (1987).
The recently discovered radio–quiet, infrared–loud quasar 13349 + 2438 has been detected at 1300 μm. This new spectral information strongly supports the idea that the FIR energy distribution originates from heated dust. The bulk of material is found to have a temperature T_d of ~ 22K which is typical for dust in the environment of compact H II regions. The total hydrogen mass derived from the emission at 1300 μm is $\sim 3.9 \times 10^{10}$M$_\odot$. The average visual extinction of 5.2 mag produced by the observed dust is comparable to the extinction inferred from optical observations. The L/M ratio of ~ 70 places the quasar and the underlying galaxy among the active systems, a fact which might explain the high luminosity of 2.7×10^{12}L$_\odot$ in terms of enhanced star formation.

159.008   Spectral study of quasars of the Second Byurakan Survey of the northern sky. I. The quasars SBS 0953 + 549, SBS 1116 + 603, and SBS 1138 + 584.
S. A. Levshakov, D. A. Varshalovich, E. A. Nazarov.
Astrophysics, Vol. 25, No. 3, p. 640 – 647 (1987). English translation of 42.159.230.

159.009   On the absorption spectra of the quasars TOL 1037–271 and TOL 1038–272.
S. Cristiani, I. J. Danziger, P. A. Shaver.
Mon. Not. R. Astron. Soc., Vol. 227, No. 3, p. 639 – 652 (1987).
The spectra of the two quasars TOL 1037–271 and TOL 1038–272 are examined. Several absorption systems are identified and discussed in the context of intrinsic absorption, intervening–galaxy absorption and superclustering. It is suggested that the four absorption systems common to the two quasars within ± 2000 km s^{-1} may be chance coincidences.

159.010   Absorption spectra of the QSO pair TOL 1037–27/1038–27.
J. G. Robertson.
Mon. Not. R. Astron. Soc., Vol. 227, No. 3, p. 653 – 660 (1987).
New observations of the quasars TOL 1037–27 and 1038–27 at 1.6 Å resolution are presented, showing details of their rich and unusual absorption spectra. In the spectrum of TOL 1037–27 two strong C IV absorption lines seen previously at low resolution are shown to be complexes similar to the "associated complex" or "shell" systems seen in a small number of other QSOs. On the other hand the strongest absorption system in TOL 1038–27 is shown to be a broad trough, probably similar to those in BAL QSOs. Suggestions that the absorption systems in the two QSOs may be correlated, in spite of their large separation (17.9 arcmin), are not supported by the present data.

159.011   The evolution of optically selected QSOs.
B. J. Boyle, R. Fong, T. Shanks, B. A. Peterson.
Mon. Not. R. Astron. Soc., Vol. 227, No. 3, p. 717 – 738 (1987).
The authors report on a determination of the QSO luminosity function and its evolution with redshift based on a new catalogue of faint ($B < 20.9$ mag) ultraviolet excess (UVX) QSOs with complete spectroscopic identification for almost 200 QSOs. They find that the steep slope exhibited by the number–magnitude relation for UVX QSOs flattens significantly beyond $B = 19.5$ mag. The authors demonstrate that this flattening corresponds to a feature in the QSO luminosity function at faint absolute magnitudes. By determining the redshift dependence of this feature they find that the evolution of QSOs at $z < 2.2$ is most simply parameterized by a uniform increase in their luminosity with increasing redshift. Such evolution is found to be consistent with a model in which QSOs are long–lived ($\cong 10^{10}$yr),

gradually dimming in luminosity from their epoch of formation at $z > 2.2$ to the present day.

159.012   The spatial clustering of QSOs.
T. Shanks, R. Fong, B. J. Boyle, B. A. Peterson.
Mon. Not. R. Astron. Soc., Vol. 227, No. 3, p. 739 – 748 (1987).

The authors have used a new complete redshift survey of ~ 170 QSOs to investigate QSO clustering. They have determined the QSO clustering correlation function ξ_{qq} over the range of separations $0 < r \leqslant 1000h^{-1}$ Mpc (comoving). At small scales $(0 < r \leqslant 10h^{-1}$ Mpc) they have tentatively detected QSO clustering at a level stronger than that expected for galaxies on a simple "stable" model for galaxy clustering evolution. At larger scales $(10 < r \leqslant 1000h^{-1}$ Mpc) the observed correlation function is close to zero, placing strong new constraints on the homogeneity of the Universe.

159.013   Clustering of quasars from the ROE/ESO large–scale AQD survey.
R. G. Clowes, A. Iovino, P. Shaver.
Mon. Not. R. Astron. Soc., Vol. 227, No. 4, p. 921 – 931 (1987).

The new ROE/ESO large–scale AQD survey for quasars forms a connected area of ~ 200 deg^2 near the South Galactic Pole, and has resulted in the discovery of a total number of quasar candidates that is comparable to the number previously published from all other sources. The authors describe a three–dimensional clustering analysis of ~ 1100 'high–probability' candidates occupying the assigned–redshift band of $1.8 - 2.4$. The analysis is sensitive to very weak clustering but none is found.

159.014   Broad–absorption–line quasar candidates discovered in the CFHT/MMT QSO survey.
A. P. Cowley, D. Crampton.
Astron. J., Vol. 94, No. 1, p. 16 – 22 (1987).

In the course of a large slitless spectral survey for QSOs, fifteen broad–absorption–line quasar candidates (BAL QSOs) have been discovered. The survey covers a region of ~ 4 deg^2 and is complete to a limiting magnitude of $m_B = 20.5$. The authors find that $\sim 9\%$ of all the quasars in their sample exhibit the BAL phenomena. The frequency of BAL QSOs with $z \sim 2$ appears to be higher than for lower–redshift objects. However, the absolute–magnitude distribution of the BAL QSOs appears to be the same as that of the whole quasar sample. Details about these quasars are presented, as well as spectra of two non–BAL quasars with $z > 3.2$.

159.015   Associated C IV absorption in radio–loud QSOs: the "3C mini–survey".
S. F. Anderson, R. J. Weymann, C. B. Foltz, F. H. Chaffee Jr.
Astron. J., Vol. 94, No. 2, p. 278 – 288 (1987).

A spectroscopic survey at 1 Å resolution of twelve 3C and 3CR QSOs reveals a very high incidence of C IV absorption complexes within ± 5000 km/s of the C IV $\lambda\lambda 1548$, 1550 emission–line redshift. Such "associated" C IV absorption is found to be "strong" (rest–frame equivalent width $\geqslant 1.5$ Å) in six of these powerful, steep–spectrum radio sources, while four others show weaker associated absorption. Such strong associated C IV complexes with $z_{abs} \approx z_{em}$ are comparatively rare in a sample of radio–quiet QSOs investigated previously. The evidence for possible correlations with various radio properties is briefly discussed, although no correlations are thus far confirmed to be statistically significant.

159.016   Optical variability of X–ray–selected QSOs.
A. J. Pica, J. R. Webb, A. G. Smith, R. J. Leacock, M. Bitran.
Astron. J., Vol. 94, No. 2, p. 289 – 296 (1987).

Photometric data for ten X–ray–selected quasistellar objects have been obtained from archival records of the Rosemary Hill Observatory. Reliable magnitudes were obtained for seven of the ten sources and six displayed optical variations significant at the 95% confidence level or greater. One source appeared to exhibit optically violent behavior. Light curves and photographic magnitudes are presented and discussed.

159.017   Statistical gravitational lensing: influence of compact objects on the number counts of quasars.
P. Schneider.
Astron. Astrophys., Vol. 183, No. 2, p. 189 – 202 (1987).

The statistical effects of gravitational light bending on the observed luminosity function of compact extragalactic sources are investigated. The main results are: (1) if the luminosity function of quasars is sufficiently steep, even a low average density of compact objects in the universe will have a dramatic influence on the observed luminosity function. Since lensing leads to a flattening of number counts compared to those in a homogeneous universe, a steep luminosity function is not ruled out by observations. (2) If the continuum emitting region of quasars is as small as indicated by variability, practically all bright quasars are highly amplified, provided that the mass of the compact objects exceeds $\sim 0.01 \, M_\odot$. The main ingredients of the model (source size, density of compact objects, luminosity function of quasars) are briefly discussed and the possible implications of the results are outlined; these include an alternative explanation of the Baldwin effect, the interpretation of BL Lacs as strongly lensed quasars and the possibility to attribute part of the source variability to time dependent amplification.

159.018   High–resolution spectra of 24 low–redshift QSOs: the properties of Mg II absorption systems.
D. Tytler, A. Boksenberg, W. L. W. Sargent, P. Young, D. Kunth.
Astrophys. J., Suppl. Ser., Vol. 64, No. 4, p. 667 – 702 (1987). With plate 31.

A high–resolution spectroscopic survey of 24 low–redshift QSOs reveals three Mg II absorption systems, all at $z_{abs} \approx 0.4$. All are apparently photoionized as is the case for most Mg II systems, but the authors are unable to distinguish between an origin in H I and an origin in cool H II gas. The velocity widths of the systems are not thermal. An enlarged sample reveals that the number of systems increases steeply with redshift and also with decreasing equivalent width. Galactic Ca II absorption is detected, and QSO Mg II emission–line widths are briefly discussed.

159.019   3C 273 and the power–law myth.
J. J. Perry, M. J. Ward, M. Jones.
Mon. Not. R. Astron. Soc., Vol. 228, No. 3, p. 623 – 634 (1987).

There is no doubt that power–law representations fit the continuum observations of quasars very well over limited (about one decade) frequency ranges, notably the radio and X–ray. Over the entire observed frequency range (spanning 14 decades), however, the subjective goodness of fit can be distorted by the method used to plot the data. The continuous fall in the log of the flux density when this is plotted against increasing log–frequency can give the impression of a reasonable fit to a power law, whereas the physically more meaningful plot of the $\log (\nu \times$ flux density), which peaks at frequencies where most of the energy is emitted, tends to reveal more structure in the energy distributions and to show up the limitations of power–law fits. The authors find that a log normal (Gaussian) distribution in photon energy is a significantly better fit to the data than a power law and, in addition, has the advantage of much reducing the number of allowed free parameters.

159.020   Optical polarization in the jet of 3C 273.
S. M. Scarrott, R. F. Warren–Smith.
Mon. Not. R. Astron. Soc., Vol. 228, No. 3, p. 35P – 40P (1987).

A linear polarization map of the optical jet of 3C 273 is presented. Along the whole length of the visible jet the authors detect significant levels of polarization with an orientation approximately perpendicular to the jet axis. The results remove the need to invoke a non–synchrotron contribution to the optical emission from the jet.

159.021   Are voids found in the Lyman–alpha forest?
A. P. S. Crotts.
Mon. Not. R. Astron. Soc., Vol. 228, No. 3, p. 41P – 45P (1987).

Evidence in the literature exists for voids seen in the Ly–α forest of QSOs studied under high resolution and high signal–to–

noise. The indication is that of voids at least as large as approximately $50\,h^{-1}\mathrm{Mpc}$ $(h \equiv H_0/100\,\mathrm{km\,s^{-1}Mpc^{-1}})$, but with a spatial filling factor of perhaps only 5 per cent near a redshift of 3. This evidence is weak enough that further data must be acquired before a universal phenomenon can be established. The total absence of such voids, however, is excluded at the 99 per cent level.

159.022   The environment of the quasar PG 1613 + 65 (Mkn 876): a close interacting pair.
H. K. C. Yee, R. F. Green.
Astron. J., Vol. 94, No. 3, p. 618 – 627 (1987). With plates 56 – 57.
  The authors report on the spectroscopic and two–color–imaging study of the environment of PG 1613 + 65. They briefly discuss the observation and data–reduction procedure for imaging and multiaperture spectroscopy of galaxies in the field. They present the data concerning the global environment of PG 1613 + 65 and analyze the nebulous component of the quasar. The authors discuss the possibility that the quasar host is interacting with a very close neighbor and the effects of this interaction on the host galaxy.

159.023   The optical luminosity function of quasars and low-luminosity active galactic nuclei.
H. L. Marshall.
Astron. J., Vol. 94, No. 3, p. 628 – 632 (1987).
  The optical luminosity function derived from quasars with $M_B < -23$ is extended to include the nuclear luminosities of Seyfert 1 galaxies. The model is fitted to the (total) luminosity data for active galaxies from a complete flux–limited survey. A good fit is obtained with $dN/dL \propto L^{-1.2}$, matching onto the steeper quasar luminosity function near $M_B = -22$. Using a luminosity evolution model derived from quasar data, this feature compares well to the results from faint quasar surveys.

159.024   Identification of an intervening galaxy responsible for Mg II absorption in the QSO 4C 55.27.
J. S. Miller, R. W. Goodrich, S. A. Stephens.
Astron. J., Vol. 94, No. 3, p. 633 – 635 (1987). With plate 58. = Lick Obs. Bull. No. 1073.
  The QSO 4C 55.27 has an emission–line redshift of $z = 1.24$. An absorption doublet at 3840 Å is identified with Mg II $\lambda 2800$ at $z = 0.374$. Direct images of the field show what appears to be a small group of galaxies near the position of the QSO. Spectroscopy of the nearest object, 5″0 from the QSO, shows it to have the spectrum of a galaxy at $z = 0.373$. The spectrum is normal for a luminous giant elliptical galaxy. The authors conclude that material associated with the galaxy, but located at least 21 kpc from its center, is responsible for the Mg II absorption in the QSO.

159.025   A cooling flow around the quasar 3C 48.
A. C. Fabian, C. S. Crawford, R. M. Johnstone, P. A. Thomas.
Mon. Not. R. Astron. Soc., Vol. 228, No. 4, p. 963 – 971 (1987).
  Spatially–resolved optical spectra of the nebulosity around the quasar 3C 48 show forbidden oxygen line emission extending to a radius of ~ 30 kpc. Assuming that the gas is photoionized by the quasar nucleus which has a spectrum unabsorbed shortward of 91.2 nm, the authors interpret the relative strengths of the [O III] and [O II] lines to indicate a gas density of at least $3 \times 10^7 \mathrm{m}^{-3}$. The observed gas is in clumps which are thinner than about 10 pc and is at a pressure consistent with a cooling flow such as that around the central galaxy in the poor cluster MKW3s. The total mass of observed gas is $\sim 3 \times 10^8 M_\odot$. A cooling flow around 3C 48 and other quasars is consistent with them lying in either poor or rich clusters and, through accretion, provides a direct link between the quasar nucleus and its environment.

159.026   The host galaxy of quasar IRAS 00275–2859: an interacting system.
J. P. Vader, G. S. Da Costa, J. A. Frogel, C. A. Heisler, M. Simon.
Astron. J., Vol. 94, No. 4, p. 847 – 853 (1987). With plate 75.
  Optical imaging of the recently discovered infrared quasar IRAS 00275–2859 (Vader and Simon 1987) shows that it can be decomposed into two point sources embedded in an underlying extended image. The brighter point source is the quasar. The authors identify the extended image together with the fainter point source as the host galaxy system. An adjacent low–surface-brightness region to the northeast is also detected in B and R. Unlike the other recently discovered IRAS quasar (13349 + 2438), the QSO in IRAS 00275–2859 shows a UV excess typical of optically selected quasars. The B–R rest–frame color of the host–galaxy system corresponds to that of a late–type spiral galaxy. The low–surface–brightness region to the northeast has, to within the uncertainties, the same B–R color as that of the host–galaxy system. The authors propose that the quasar belongs to one of two interacting galaxies, that the second point source is the nucleus of the second galaxy, and that the low–surface-brightness region is the tidal signature of the interaction. The infrared spectral energy distribution of IRAS 00275–2859 suggests that the bulk of the infrared radiation is emitted by heated dust in the host–galaxy system.

159.027   PKS 0114 + 074: a QSO–galaxy association?
C. E. Akujor.
Astron. J., Vol. 94, No. 4, p. 867 – 870 (1987).
  Meter–wavelength observations of the radio source PKS 0114 + 074 (4C 07.4) indicate a double structure which is identified with an 18.0 mag galaxy, while high–frequency observations reveal a third component at the position of a QSO of redshift $z = 0.861$. The author explains the absence of the third component at meter wavelengths in terms of a low–frequency turnover due to synchrotron self–absorption. A possible association between the galaxy and the QSO is considered, assuming a noncosmological redshift.

159.028   CCD observations of the jet of the quasar 3C 273.
J. J. E. Hayes, A. C. Sadun.
Astron. J., Vol. 94, No. 4, p. 871 – 875 (1987). With plate 76.
  The authors present B and R broadband CCD photometry of the jet of 3C 273. They find several knots within the jet. The knots at either end have a difference in color indices, with the knot at the outermost end being redder by about 1/2 magnitude in $(B$–$R)$. Also, a variation in color may have been observed along the jet. This latter result would place constraints upon the jet–forming mechanism involved; the variation in color may be due to a nonthermal emission decay process. The authors discuss such mechanisms in view of these observations.

159.029   The quasar family – an introduction and taxonomy.
C. M. Gaskell.
Astrophysical jets and their engines, p. 29 – 46 (1987). – See Abstr. 012.021.
  This paper offers a definition of the quasar phenomenon and gives a description of the taxonomy of the quasar family and an explanation of some of the common terminology. The basic unity of all kinds of quasar activity is emphasised. Some differences between various family members are given and some interrelationships pointed out. Finally, it is emphasised that quasar activity at low levels is extremely common.

159.030   Découverte d'un quasar binaire.
G. Meylan.
Orion, 45. Jahrg., Nr. 222, p. 160 – 162 (1987).

159.031   Surveys of fields around quasars. IV. Luminosity of galaxies at $z \approx 0.6$ and preliminary evidence for the evolution of the environment of radio–loud quasars.
H. K. C. Yee, R. F. Green.
Astrophys. J., Vol. 319, No. 1, p. 28 – 43 (1987).
  From direct imaging data, quantitative measures of the richness of the environment of quasars are derived using self–

consistent galaxy luminosity function and galaxy count models. The quasar–galaxy covariance function is derived for samples of quasars having redshifts ranging from 0.05 to 0.65. Between $z \approx 0.4$ and 0.6, there is an increase by a factor of ~ 3 in the average quasar–galaxy covariance amplitude for radio–loud quasars. Some radio–loud quasars at $z \approx 0.6$ are found in environments as rich as those of Abell class 1 clusters. This indicates that there has been significant evolution in the environments within some rich clusters over a time period of the order of one billion years, allowing them to support quasar activity at the earlier epoch.

159.032 18 centimeter VLBI observations of the quasar NRAO 140 during and after a low–frequency outburst.
A. P. Marscher, J. J. Broderick, L. Padrielli, N. Bartel, J. D. Romney.
Astrophys. J., Vol. 319, No. 1, p. 456 – 464 (1987).

The authors have observed the quasar NRAO 140 using an eight station very long baseline array at 18 cm in 1984 April and a seven station array at 6 cm in 1984 May. They compare both the map and the data at 18 cm with those obtained by Marscher and Broderick in 1981 October. The latter coincided with a $\sim 25\%$ outburst in flux density at wavelengths greater than ~ 30 cm. The analysis indicates that a component ~ 5 milli–arc seconds southeast of the "core" dropped significantly in brightness between 1981 October and 1984 April. The authors identify this component as the likely site of the low–frequency variations.

159.033 The accuracy of cross–correlation estimates of quasar emission–line region sizes.
C. M. Gaskell, B. M. Peterson.
Astrophys. J., Suppl. Ser., Vol. 65, No. 1, p. 1 – 11 (1987).

The use of cross–correlations of emission–line and continuum flux measurements for estimating the sizes of line–emitting regions in quasars is described and discussed in detail. A techique for handling unevenly sampled data is described. Analytic formulae for determination of the uncertainties are given. The importance of correlated error is mentioned, and the size of the broad–line region in Akn 120 is discussed as an example.

159.034 Quasar as a superstar with magnetic monopoles.
G. S. Bisnovatyj–Kogan.
Pis'ma Astron. Zh., Tom 13, No. 10, p. 855 – 861 (1987). In Russian. English translation in Sov. Astron. Lett., Vol. 13.

The structure and evolution of a hydrogen–helium superstar are considered where the source of energy is the annihilation of baryons at their collisions with magnetic monopoles.

159.035 Cosmological implications of the anomalies in the redshift distribution of QSO Lyman α absorption systems.
X. Barcons, A. C. Fabian.
Mon. Not. R. Astron. Soc., Vol. 229, No. 1, p. 157 – 163 (1987).

It has recently been found in the spectra of distant QSOs that there is a relative lack of Lyman α absorption lines as the absorption redshift approaches the emission redshift of the QSO. As this "anomaly" seems to disappear with increasing emission redshift, any interpretation could have interesting cosmological consequences. The authors have examined in some detail the two current explanations of this lack of lines: ionization by the nearby QSO and the failure of a significant fraction of clouds in occulting the continuum–emitting region.

159.036 The X–ray properties of quasars determined from an iterative multiple regression analysis with censored data.
K. N. Yu.
Astrophys. Space Sci., Vol. 137, No. 1, p. 93 – 99 (1987).

The author adopts the iterative least–squares method of Schmee and Hahn (1979) and extends it to the multiple regression case to investigate the dependence of the X–ray luminosities L_x of quasars on their optical luminosities L_0 and redshifts z. The method has several advantages, one of which is that one does not need the strong assumption that the values of the optical–X–ray index α_{0x} of the detected and nondetected quasars should fall into

the same range. Correlations of α_{0x} and $\log(1+z)$ or the look back time $\tau(z)$ are shown, but these are not firmly established due to the uncertainty present in the regression parameter. On the other hand, it is impossible for α_{0x} to be independent of L_0. The author has also shown that L_x is less than proportional to L_0.

159.037 Spectral observations of the quasar S5 0014 + 81. An analysis of the absorption–line spectrum.
S. A. Levshakov, D. A. Varshalovich, E. A. Nazarov, A. F. Fomenko.
Astron. Zh., Tom 64, Vyp. 5, p. 929 – 938 (1987). In Russian. English translation in Sov. Astron., Vol. 31, No. 5.

Spectra of the quasar S5 0014 + 81 ($z_e = 3.38$) have been obtained from 3500 to 6500 Å. 174 absorption lines are found in this region. Low– and high–ionized metal absorption lines have been identified in 5 absorption line systems at $z_a = 1.112, 2.345, 2.671, 2.992,$ and 3.200. The authors suggest that the system with $z_a = 1.112$ originates in the disk of an intervening galaxy. Moreover, there is tentative evidence for the presence of 11 absorption line systems containing only H I L_α/L_β lines or L_α and C IV doublet in the spectrum of S5 0014 + 81. The results are discussed.

159.038 Peculiar variations in the structure of the quasar 3C 454.3.
I. I. K. Pauliny–Toth, R. W. Porcas, J. A. Zensus, K. I. Kellermann, S. Y. Wu, G. D. Nicolson, F. Mantovani.
Nature, Vol. 328, No. 6133, p. 778 – 782 (1987).

Following a major outburst at radio wavelengths in the quasar 3C 454.3, observations by means of very long baseline interferometry have revealed a superluminal increase in the size of the compact "core" region of the source and the appearance of a complex structure within it. Some features within this structure remained stationary with respect to each other, but others showed superluminal motion. This behaviour is distinctly different from that generally seen in superluminal sources.

159.039 Quasars, Seyfert galaxies and active galactic nuclei.
D. E. Osterbrock.
Spectroscopy of astrophysical plasmas, p. 59 – 88 (1987). – See Abstr. 003.010.

Contents: 1. Introduction. 2. Observational aspects. 3. Narrow–line region. 4. Broad–line region. 5. High–energy photons. 6. Ultraviolet spectra. 7. Physical models.

159.040 Quasar and stellar objects in the Byurakan Surveys.
B. E. Markaryan (*B. E. Markarian*), L. K. Erastova, V. A. Lipovetskij, D. A. Stepanyan, A. I. Shapovalova.
Astrophysics, Vol. 26, No. 1, p. 7 – 15 (1987). English translation of 43.159.016.

159.041 Discovery of low–redshift, neutral hydrogen absorption in the radio spectrum of PKS 2020–370.
C. L. Carilli, J. H. van Gorkom.
Astrophys. J., Vol. 319, No. 2, p. 683 – 686 (1987).

The authors have detected 21 cm absorption in the radio spectrum of the quasar PKS 2020–370, at a heliocentric velocity of 8611 ± 5 km s^{-1}. The H I gas is coincident in redshift with previously observed Ca II H and K line absorption and is close in velocity to the foreground spiral galaxy Klemola 31A. This galaxy lies just 20″ of the line of sight to the distant quasar. The galaxy optical morphology suggests absorption by gas considerably beyond the optical edge of the spiral.

159.042 The galaxy–quasar connection: NGC 4319 and Markarian 205. I. Direct imagery.
J. W. Sulentic, H. C. Arp.
Astrophys. J., Vol. 319, No. 2, p. 687 – 692 (1987). With plates 13 – 16.

New direct–imaging data are presented for the disturbed spiral galaxy NGC 4319 ($z = 0.005$) and the apparently connected quasar–like object Markarian 205 ($z = 0.072$). Image processing of this CCD data reveals (1) an almost continuous luminous connection extending from Mrk 205 into the nucleus of the spiral galaxy; (2) a corresponding feature on the opposite side of the disk, appearing to link a bright UV knot with the nucleus; and (3) extensive morphological peculiarities in NGC 4319 that are consistent with hypothesized explosive nuclear activity. These data support the conclusion that NGC 4319 is an active spiral galaxy that recently ejected Mrk 205 from its nucleus.

159.043 The galaxy–quasar connection: NGC 4319 and Markarian 205. II. Spectroscopy.
J. W. Sulentic, H. C. Arp.
Astrophys. J., Vol. 319, No. 2, p. 693 – 708 (1987). With plate 17.

New spectroscopic data are presented for the disturbed spiral galaxy NGC 4319 and the high–redshift, quasar–like object Markarian 205. Spectra taken at various positions in the galaxy are image processed and analyzed in detail. Surprisingly, there is almost no Hα emission present. Instead, the emission lines of [N II] $\lambda\lambda6548$, 6584 are widely detected. The spectra yield a partial velocity field in the disk of NGC 4319 which shows that an arc of gas near the UV knot, and opposite Mrk 205, is expanding with a velocity $V \approx 10^3 \mathrm{km\,s^{-1}}$. This evidence suggests (1) that NGC 4319 is an internally active spiral galaxy and (2) that the high–density gas has been explosively removed from the disk of the galaxy.

159.044 High–redshift QSO absorbing clouds and the background ionizing source.
R. F. Carswell, J. K. Webb, J. A. Baldwin, B. Atwood.
Astrophys. J., Vol. 319, No. 2, p. 709 – 722 (1987).

Echelle spectra of the high–redshift QSO PKS 2000–330 have been analyzed to provide redshifts, H I column densities, and velocity dispersions of Lyman line absorbing systems in the redshift range $3.02 < z < 3.75$. For simple uniform cloud models, the redshift evolution of H I column density distribution could in principle be used to infer the redshift dependence of the density of the gas within the clouds. It is found that the source of ionization of these clouds cannot be the integrated light from background quasars, *if* the quasar density distribution cuts off between $z \approx 2$ and $z \approx 4$ as is commonly thought.

159.045 Observations of Ni II absorption at $z = 2.811$ toward the QSO PKS 0528–250.
D. M. Meyer, D. G. York.
Astrophys. J., Lett. Ed., Vol. 319, No. 2, p. L45 – L49 (1987).

The authors present observations of strong Ni II absorption corresponding to the damped Lyα absorption–line system at $z = 2.811$ toward the QSO PKS 0528–250. They have also detected Cr II absorption corresponding to the damped Lyα system at $z = 2.142$ toward this QSO. The abundances of Ni II and Cr II relative to H I indicate that much less Ni and Cr is depleted from the gas phase in these systems than in Galactic interstellar clouds. The implications of these observations are discussed with regard to the interpretation of damped Lyα systems as galatic disks.

159.046 Multifrequency VLBI observations of 4C 39.25: a superluminal source without a well–defined core.
A. P. Marscher, D. B. Shaffer, R. S. Booth, B. J. Geldzahler.
Astrophys. J., Lett. Ed., Vol. 319, No. 2, p. L69 – L72 (1987).

The radio source 4C 39.25 has been shown by Shaffer et al. to contain both superluminally moving and stationary components. The authors present the results of VLBI observations of 4C 39.25 at 1.35 cm at epoch 1983.77. Comparison with the 2.8 cm maps of Shaffer et al. indicates that none of the components possesses the characteristics normally associated with the compact core of a superluminal source. The authors infer from the lack of a well–defined core that 4C 39.25 represents a source intermediate between more typical superluminal sources and symmetric extended radio sources with weak cores.

159.047 Selection effect in identification of spectral lines and evolution of quasars.
Z.–g. Deng, Y.–y. Zhou, Y.–z. Liu, H.–j. Dai.
Sci. Sin., Ser. A, Vol. 30, No. 11, p. 1188 – 1198 (1987).

The authors use the results obtained from the analyses of selection effects in the identification of emission lines of quasars

to determine the evolutionary property of quasars. The samples consist of quasars with an absolute magnitude less than certain values selected from surveys with the similar limiting apparent magnitude. The density of quasars evolves according to the power law $\varrho = \varrho_0(1+z)^{6.5\pm1}$. This result is compatible with the results from smaller samples or previous surveys with lower limiting apparent magnitudes.

159.048 Radio source structure from geodetic VLBI observations: 8 GHz multi–epoch maps of the quasar 4C 39.25.
G.–q. Tang, B. Rönnäng, L. Bååth.
Astron. Astrophys., Vol. 185, No. 1/2, p. 87 – 93 (1987).

Most of the 'point–like' radio sources adopted as the fiducial points of the extra–galactic radio sources reference frame show structure at the milliarcsecond level. Simplified assumption of source structure can introduce uncertainties in source positions and thus influence high precision astrometric and geodetic VLBI applications. To estimate and/or eliminate the source structural effect, it is important to monitor the brightness distribution of the sources at the geodetic VLBI frequency (8.3 GHz). The authors have developed a software package to process the visibilities in the Mark–III geodetic VLBI databases and make hybrid maps with AIPS. Six maps of 4C 39.25, made from global geodetic VLBI observations at $v = 8.3$ GHz from 1980.79 to 1985.35, are presented as an example.

159.049 The bright QSO GD 1339.
I. Bues, W. Kollatschny, K. J. Fricke, G. Schönknecht.
Astron. Astrophys., Vol. 186, No. 1/2, p. 99 – 102 (1987).

Multifrequency observations of the radio–quiet QSO GD 1339 have been obtained using the EXOSAT, IUE, and ESO telescopes. The redshift of GD 1339 is $z = 0.116$ and its apparent magnitude $m_v = 15\overset{m}{.}2$. This bright nearby QSO may be suitable for detailed follow–up studies and as a spectroscopic reference source.

159.050 On the radio emission variability and the parameters of the nucleus of quasar 3C 273.
I. G. Moiseev, N. S. Nesterov, A. B. Severnyj.
Izv. Krymskoj Astrofiz. Obs., Tom 76, p. 90 – 93 (1987). In Russian. English translation in Bull. Crimean Astrophys. Obs., Vol. 76.

Quasar 3C 273 radio emission variations at the wavelengths 13.5 and 8 mm are discussed. An estimation of quasar nucleus parameters is made under the assumption that the nucleus is a pulsating "superstar" and 1.5 year radio flux density variations are caused by its pulsation. The following values have been obtained: nuclear radius $R \approx 2 \times 10^{16}$cm, average density $\bar{\varrho} \approx 1.4 \times 10^{-8}$g/cm^3, mass $M \approx 4 \times 10^8 M_\odot$ and the strength of magnetic field $H \approx 3 \times 10^4$Gs.

159.051 An upper limit on quasar reddening.
E. L. Wright, M. A. Malkan.
Bull. Am. Astron. Soc., Vol. 19, No. 2, p. 699 – 700 (1987). Abstract. – See Abstr. 010.061.

159.052 A close pair of quasars.
D. Crampton, A. P. Cowley.
Bull. Am. Astron. Soc., Vol. 19, No. 2, p. 700 (1987). Abstract. – See Abstr. 010.061.

159.053 Observations of high redshift Ca II absorption in QSO spectra.
D. C. Morton, J. G. Robertson, J. C. Blades, D. G. York, D. M. Meyer.
Bull. Am. Astron. Soc., Vol. 19, No. 2, p. 700 (1987). Abstract. – See Abstr. 010.061.

159.054   **The APM–QSO survey: initial MMT results.**
F. H. Chaffee Jr., C. B. Foltz, P. C. Hewett,
D. A. Turnshek, R. J. Weymann, S. F. Anderson,
G. M. MacAlpine.
Bull. Am. Astron. Soc., Vol. 19, No. 2, p. 700 (1987). Abstract. –
See Abstr. 010.061.

159.055   **The evolution of radio quasars of redshifts up to 1.**
J. B. Hutchings, R. Price, A. C. Gower.
Bull. Am. Astron. Soc., Vol. 19, No. 2, p. 700 (1987). Abstract. –
See Abstr. 010.061.

159.056   **The UV excess of quasars: luminosity dependence.**
A. Wandel.
Bull. Am. Astron. Soc., Vol. 19, No. 2, p. 701 (1987). Abstract. –
See Abstr. 010.061.

159.057   **X–ray spectral variability in the quasar NRAO 140: a unique absorption event.**
A. P. Marscher.
Bull. Am. Astron. Soc., Vol. 19, No. 2, p. 719 (1987). Abstract. –
See Abstr. 010.061.

159.058   **Resonance and subordinate lines of O III in quasar spectra.**
S. O. Kastner, A. K. Bhatia.
Bull. Am. Astron. Soc., Vol. 19, No. 2, p. 719 (1987). Abstract. –
See Abstr. 010.061.

159.059   **Detection of the host galaxy of quasar IRAS 00275–2859.**
J. P. Vader, G. S. Da Costa, C. A. Heisler, J. A. Frogel,
M. Simon.
Bull. Am. Astron. Soc., Vol. 19, No. 2, p. 719 (1987). Abstract. –
See Abstr. 010.061.

159.060   **An Fe II/X–ray relation for quasars.**
B. J. Wilkes, M. Elvis, I. McHardy.
Bull. Am. Astron. Soc., Vol. 19, No. 2, p. 719 (1987). Abstract. –
See Abstr. 010.061.

159.061   **Detecting quasars with the Extreme Ultraviolet Explorer.**
H. L. Marshall.
Bull. Am. Astron. Soc., Vol. 19, No. 2, p. 719 – 720 (1987).
Abstract. – See Abstr. 010.061.

159.062   **Surface densities of quasars and hot evolved stars isolated from the US catalogs.**
K. J. Mitchell, P. D. Usher, S. B. Howell, A. Warnock III.
Bull. Am. Astron. Soc., Vol. 19, No. 2, p. 720 (1987). Abstract. –
See Abstr. 010.061.

159.063   **CCD imaging in four colors of the jet of the quasar 3C 273.**
A. C. Sadun, W. A. Washburn.
Bull. Am. Astron. Soc., Vol. 19, No. 2, p. 731 – 732 (1987).
Abstract. – See Abstr. 010.061.

159.064   **Synchrotron self–Compton and kinematic models for 3C 345.**
J. A. Biretta.
Bull. Am. Astron. Soc., Vol. 19, No. 2, p. 732 – 733 (1987).
Abstract. – See Abstr. 010.061.

159.065   **A BAL QSO with z_{em} = 3.50 and a QSO with z = 3.31 from a UK Schmidt IIIa–F objective prism survey.**
C. Hazard, R. G. McMahon, D. C. Morton.
Mon. Not. R. Astron. Soc., Vol. 229, No. 3, p. 371 – 377 (1987).
With 2 plates.
The paper describes the discovery and preliminary slit spectroscopy of the broad absorption line QSO 0105–2634 with an emission redshift of 3.50, the highest yet reported for this type of QSO. Attention is drawn to the fine detail which can be observed

in objective prism spectra. A second QSO 1209 + 0919 with z = 3.31 and strong absorption near O VI is also noted. These new QSOs were found in two regions of sky in which the discovery of other QSOs with 3.3 < z < 3.8 already has been reported.

159.066   **A wide–angle radio–tail quasar: B2 1419 + 315.**
D. J. Saikia, L. Staveley–Smith, D. Wills,
T. J. Cornwell, C. J. Salter, W. Junor, P. Shastri.
Mon. Not. R. Astron. Soc., Vol. 229, No. 3, p. 495 – 503 (1987).
With 1 plate.
The authors present radio continuum observations of the quasar B2 1419 + 315 at high angular resolution made with the VLA and MERLIN, spectroscopic observations made at McDonald Observatory and a CCD image of the field obtained at the Jacobus Kapteyn Telescope. The spectra confirm the object to be a quasar at a redshift of 1.547 ± 0.005. The source has an extreme C–shaped radio structure with a curved jet towards the eastern component. The CCD image shows several foreground galaxies with one of them close to the centre of curvature of the radio jet and bridge. Although the radio structure is suggestive of interaction with a dense intracluster medium, the CCD image raises the interesting possibility of gravitational lensing by a foreground galaxy. The authors examine both possibilities.

159.067   **The APM QSO survey. I. Initial MMT results.**
C. B. Foltz, F. H. Chaffee Jr., P. C. Hewett,
G. M. MacAlpine, D. A. Turnshek, R. J. Weymann,
S. F. Anderson.
Astron. J., Vol. 94, No. 6, p. 1423 – 1460 (1987).
This is the first paper in a series aimed at selecting $\sim$1000 QSOs brighter than m_J = 18.5 using machine–scanned direct and objective–prism plates from the UK Schmidt Telescope. The plate material is scanned at the Institute of Astronomy's Automated Plate Measuring facility; algorithms select approximately three candidates per square degree as possible QSOs. Follow–up spectroscopy at the Multiple Mirror Telescope is used to classify each candidate. In this initial paper, the authors describe the scientific objectives of the survey and the selection and observing techniques used, and present the first sample of 192 QSOs in a 102 sq. deg area centered on the Virgo cluster. They present accurate coordinates, magnitudes, redshifts, and moderate signal–to–noise ratio and resolution spectra of all 192 QSOs. The variety of QSO spectra present in the sample, including at least seven BAL QSOs plus one lineless object with a redshift greater than 0.68, demonstrates the ability of the selection procedures to uncover QSOs over a wide range of redshifts and spectral properties.

159.068   **Companions of low–redshift radio–quiet quasars.**
H. K. C. Yee.
Astron. J., Vol. 94, No. 6, p. 1461 – 1468 (1987).
Using imaging data from a relatively complete subset of low–redshift (z < 0.3) radio–quiet quasars, the frequency of finding associated companion galaxies of the quasars is determined statistically using well–defined criteria. With an average completeness limit of $M_r \sim$ –19, it is found that $\sim$40% of the quasars have at least one close physical companion within a projected distance of 100 kpc. The magnitude distribution of the companions is entirely consistent with being drawn randomly from the luminosity function of field galaxies. It is estimated that the frequency of finding close companions to quasars is $\sim$6 times higher than that expected for field galaxies. This frequency is similar to that found for lower–luminosity Seyfert galaxies. The properties of the companions appear to be uncorrelated with the level of activity in the quasars.

159.069   **Radio sources with strong jets and weak cores.**
C. J. Lonsdale, P. D. Barthel.
Astron. J., Vol. 94, No. 6, p. 1487 – 1492 (1987).
High–resolution radio maps of radio quasars with strong jets and weak cores are presented. The implications of these low core–to–jet flux–density ratios are discussed, with particular regard to

the possibility of relativistic flow speeds and Doppler boosting of the emission from the core and/or jet. It is demonstrated that the observed spread in core/jet flux ratios for radio sources, in general, is sufficiently large that no meaningful constraints can be placed on the relative flow velocities in the core and jet, contrary to earlier claims. The large spread also implies that there is probably a substantial scatter in intrinsic core/jet ratios, independent of beaming effects.

159.070 An absorption line survey of 32 QSOs at red wavelengths: properties of the Mg II absorbers.
K. M. Lanzetta, D. A. Turnshek, A. M. Wolfe.
Space Telesc. Sci. Inst., Prepr. Ser., No. 188, 64 pp. (1987). To appear in Astrophys. J.

159.071 Spectroscopic survey of QSOs to B = 22.5: the luminosity function.
D. C. Koo, R. G. Kron.
Space Telesc. Sci. Inst., Prepr. Ser., No. 195, 23 pp. (1987). To appear in Astrophys. J.

159.072 QSOs with PHL5200–like broad absorption line profiles.
D. A. Turnshek, C. B. Foltz, C. J. Grillmair, R. J. Weymann.
Space Telesc. Sci. Inst., Prepr. Ser., No. 212, 31 pp. (1987). To appear in Astrophys. J.

159.073 Observations of high redshift Ca II absorption in QSO spectra.
J. G. Robertson, D. C. Morton, J. C. Blades, D. G. York, D. M. Meyer.
Anglo–Aust. Obs., Prepr., No. 219, 35 pp. (1987). To appear in Astrophys. J.

159.074 Near–infrared photometry of high redshift quasars.
J. M. Rodriguez–Espinosa, R. M. Stanga, A. F. M. Moorwood.
ESO Sci. Prepr., No. 524, 22 pp. (1987). To appear in Astron. Astrophys.

159.075 Quasar clustering and the evolution of structure.
P. A. Shaver.
ESO Sci. Prepr., No. 534, 11 pp. (1987). Paper presented at IAU Colloq. No. 130.

159.076 Observations of the new gravitational lens system UM673 = Q0142–100.
J. Surdej, P. Magain, J.–P. Swings, U. Borgeest, T. J.–L. Courvoisier, R. Kayser, K. I. Kellermann, H. Kühr, S. Refsdal.
ESO Sci. Prepr., No. 544, 41 pp. (1987). To appear in Astron. Astrophys.

159.077 Quasar clustering and gravitational lenses.
P. A. Shaver.
ESO Sci. Prepr., No. 545, 15 pp. (1987). Paper presented at the NATO ASI "The post–recombination universe", Cambridge, July 1987.

159.078 Hot dust and the near–infrared bump in the continuum spectra of quasars and active galactic nuclei.
R. Barvainis.
Astrophys. J., Vol. 320, No. 2, p. 537 – 544 (1987).

Many quasars and active galactic nuclei show a bump or excess in the near–infrared continuum. In this paper, a model is developed to account for the bump in terms of thermal radiation from dust heated by the primary optical/ultraviolet continuum source. This model naturally explains the onset of the bump at about 2 μm, since this wavelength corresponds to the spectral peak for optically thin emission from graphite grains at their evaporation temperature (~ 1500K). Infrared spectra are calculated for two cases, one in which the grains are smoothly distributed and another in which the dust is clumped into discrete clouds that are optically thick to the ultraviolet continuum.

Continuum spectra of 3C 273 and the "infrared quasar" IRAS 13349 + 2438 are fitted between ~ 0.1 and 100 μm with a multicomponent model.

159.079 $z_{abs} \approx z_{em}$ absorption systems in quasar spectra: a correlation with radio and optical luminosity?
P. Møller, P. Jakobsen.
Astrophys. J., Lett. Ed., Vol. 320, No. 2, p. L75 – L79 (1987).

By combining the data of the quasar absorption line surveys published by Young, Sargent, and Boksenberg in 1982 and Foltz et al. in 1986, the authors show that a two–parameter correlation between the presence of C IV $z_{abs} \approx z_{em}$ systems and both the radio and optical luminosity of the underlying quasar is present at the 95% confidence level. If real, this correlation can be understood in the intervening cluster member interpretation for $z_{abs} \approx z_{em}$ systems by noting that (1) the host galaxies of radio–loud quasars tend to be ellipticals which, in turn, tend to be found in clusters, and (2) very luminous quasars are capable of ionizing and influencing their environment over megaparsec distances.

159.080 Crowding on the sight line to the QSO PHL 1226: the nearby galaxy IC 1746 and a galaxy cluster at $z = 0.16$.
J. Bergeron, O. Boulade, D. Kunth, D. Tytler, A. Boksenberg, L. Vigroux.
Inst. Astrophys. Paris, Pré–Publ., No. 194, 23 pp. (1987). To appear in Astron. Astrophys.

159.081 A 21 cm study of some QSO/galaxy pairs.
P. Boissé, J. M. Dickey, I. Kazès, J. Bergeron.
Inst. Astrophys. Paris, Pré–Publ., No. 196, 23 pp. (1987). To appear in Astron. Astrophys.

159.082 Properties of the metal–rich absorption line systems.
J. Bergeron.
Inst. Astrophys. Paris, Pré–Publ., No. 199, 19 pp. (1987). To appear in the proceedings of the workshop "QSO absorption lines: probing the universe", 19 – 21 May 1987.

159.083 Quasar clustering in Sculptor and in the Véron catalogue.
N. Anderson, D. Kunth, W. L. W. Sargent.
Inst. Astrophys. Paris, Pré–Publ., No. 208, 23 pp. (1987). To appear in Astron. J.

159.084 Galaxies giving rise to Mg II absorption systems in quasar spectra.
J. Bergeron.
Inst. Astrophys. Paris, Pré–Publ., No. 211, 16 pp. (1987). To appear in IAU Symp. No. 130.

159.085 Quasar microlensing and dark matter.
H.–W. Rix, C. J. Hogan.
Prepr. Steward Obs., No. 758, 14 pp. (1987). Submitted to Astrophys. J.

159.086 Models of the quasar population: II. The effects of dust obscuration.
J. Heisler, J. P. Ostriker.
Prepr. Steward Obs., No. 761, 68 pp. (1987). Submitted to Astrophys. J.

159.087 On the angular structure of the quasar 3C 196.
A. V. Men', S. Ya. Braude, N. K. Sharykin, V. A. Shepelev.
Izv. Vyssh. Uchebn. Zaved., Radiofiz., Tom 30, No. 4, p. 474 – 481 (1987). In Russian. Abstr. in Ref. Zh., 51. Astron. 11.51.844 (1987).

159.088 Low ionization emission lines in quasars.
D. Dultzin–Hacyan.
Rev. Mex. Astron. Astrofis., Vol. 14, No. 1, p. 94 – 96 (1987). – See Abstr. 012.042.

The possibility of the coexistence of different kinds of BLR clouds in quasars (and Sy 1's) has been discussed recently in the literature. These clouds, photoionized by the UV central continuum would have different opacities and ionization states (H II or H I* dominated clouds). In this work the author analyzes observational evidence (in particular, Mg II and Si II lines in quasars) in favor of the coexistence of these clouds as well as mechanically heated clouds (where lines are emitted following collisional ionization).

159.089 The early universe – an observer's view.
J. V. Wall.
The early universe, p. 335 – 350 (1988). – See Abstr. 012.043.

Three aspects of an observer's early universe are discussed: primordial helium abundance as determined from observations of blue compact galaxies; the small–scale structure of the microwave background radiation as measured with the VLA; and the distribution of objects along lines of sight to distant QSOs determined from observations of Lyman–alpha and metal-line absorption systems.

159.090 Molecules at early epochs. I. A search for CO toward the quasar PHL 61.
F. H. Chaffee Jr., C. B. Foltz, J. H. Black.
Sov. Astron. Lett., Vol. 12, No. 6, p. 343 – 346 (1986). English translation of 42.159.118.

159.091 On the spectrum of the quasar PHL 61.
S. A. Levshakov, D. A. Varshalovich.
Sov. Astron. Lett., Vol. 12, No. 6, p. 347 (1986). English translation of 42.159.119.

159.092 Dynamics of emission clouds in quasar nuclei.
A. S. Zentsova.
Astrophys. Space Sci., Vol. 138, No. 2, p. 355 – 360 (1987).

The motion of emission clouds in quasar nuclei under the action of central–source radiative pressure, deceleration force of the intercloud medium, and the central–body gravitation is considered. It is shown that the properties of the intercloud medium have an essential bearing on the cloud dynamics. The cloud velocity increases with the distance from the source and tends to a constant value which depends on the central–source luminosity and the intercloud medium density.

159.093 VLBI observations of the gravitational lens system 2016+112.
M. B. Heflin, B. F. Burke, E. E. Falco, M. V. Gorenstein, I. I. Shapiro, J. N. Hewitt, A. E. E. Rogers, C. Lawrence.
Bull. Am. Astron. Soc., Vol. 19, No. 2, p. 699 (1987). Abstract. – See Abstr. 010.061.

159.094 Superluminal radio sources: introduction.
T. J. Pearson, J. A. Zensus.
Superluminal radio sources, p. 1 – 11 (1987). – See Abstr. 012.053.

Contents: The discovery. Very long baseline interferometry. Cosmology. The "standard model" of superluminal sources: kinematics of relativistic expansion, unified beaming models.

159.095 Summary of known superluminal sources.
R. W. Porcas.
Superluminal radio sources, p. 12 – 25 (1987). – See Abstr. 012.053.

The author gives a summary of the phenomenon of superluminal motions. He reviews the progress in finding new superluminal sources, and in investigating some of their detailed properties. He gives a list of superluminal (and other) sources and briefly describes its contents.

159.096 3C 273: archetype of superluminal sources.
J. A. Zensus.
Superluminal radio sources, p. 26 – 31 (1987). – See Abstr. 012.053.

The author summarizes some results from recent VLBI observations.

159.097 Observations of 3C 273 at 3 mm wavelength.
A. T. Moffet, A. C. S. Readhead.
Superluminal radio sources, p. 32 – 33 (1987). – See Abstr. 012.053.

159.098 Superluminal motion in the quasar 3C 279.
S. C. Unwin.
Superluminal radio sources, p. 34 – 39 (1987). – See Abstr. 012.053.

The author began a systematic program of VLBI monitoring of 3C 279 at three frequencies. He presents the results of the program, which has continued for five years now.

159.099 Investigations of 3C 345.
J. A. Biretta, M. H. Cohen.
Superluminal radio sources, p. 40 – 47 (1987). – See Abstr. 012.053.

The authors describe the physical properties of the emission regions and the kinematics of the jet assuming $H_0 = 100 \text{ km s}^{-1} \text{Mpc}^{-1}$ and $q_0 = 0.5$.

159.100 Structural variations in the quasar 3C 454.3.
I. I. K. Pauliny–Toth.
Superluminal radio sources, p. 55 – 59 (1987). – See Abstr. 012.053.

The author began VLBI monitoring of 3C 454.3 in mid–1981, when a large flux density outburst reached its peak at 2.8 cm. The structure of 3C 454.3 on arcsecond scales and milliarcsecond scales is of the core–jet type. A series of 2.8 cm maps of the core region are presented. They demonstrate structural variations within this region which are very different from those observed in "classical" superluminal sources.

159.101 4C 39.25: superluminal motion between stationary components.
D. B. Shaffer, A. P. Marscher.
Superluminal radio sources, p. 67 – 71 (1987). – See Abstr. 012.053.

The authors observed 4C 39.25 at 2.8 cm, using modern mapping techniques. A figure shows the changes in the structure of 4C 39.25 in the last seven years: a new component seems to have been ejected from (the region of) the western component of the double source.

159.102 Superluminal motion towards a stationary component in quasar 3C 395.
R. S. Simon, K. J. Johnston, J. Hall, J. H. Spencer, J. A. Waak.
Superluminal radio sources, p. 72 – 75 (1987). – See Abstr. 012.053.

The authors repeated the VLBI observations at 6 cm. These observations confirmed the three–component structure and the superluminal motion of component 3. A figure shows the three 6 cm VLBI images of 3C 395 and a table summarizes the component parameters.

159.103 Superluminal motion and other indications of bulk relativistic motion in a complete sample of radio sources from the S5 survey.
A. Witzel.
Superluminal radio sources, p. 83 – 93 (1987). – See Abstr. 012.053.

The majority of the 13 S5 sources considered show evidence for bulk relativistic motion. In addition, nine of the 12 sources observed repeatedly with VLBI are (at least) good candidates for superluminal motion.

159.104 Tests of beaming models.
P. A. G. Scheuer.
Superluminal radio sources, p. 104 – 113 (1987). – See Abstr. 012.053.

The author comments on five topics: He indicates the current state of the obvious and classical tests of relativistic beaming: (1) The statistics of jet fluxes, (2) The statistics of jet speeds,

(3) Why relativistic beaming is true, (4) When is a quasar not a quasar? and (5) Curly jets.

159.105   Relativistic beaming and the nuclei of double–lobed quasars.
D. H. Hough, A. C. S. Readhead.
Superluminal radio sources, p. 114 – 125 (1987). – See Abstr. 012.053.

The authors began in 1980 their study of the complete, flux–density limited sample of 26 double–lobed quasars from the 3CR catalog. There are 16 quasars in this sample for which the flux density of the central component is greater than 30 mJy at 5 GHz. One of the main objectives in this study is to measure the distribution of superluminal velocities for sources in this sample.

159.106   Superluminal motion in a randomly oriented quasar sample.
J. A. Zensus, R. W. Porcas.
Superluminal radio sources, p. 126 – 128 (1987). – See Abstr. 012.053.

The authors give a status report on the measurement of motions in a sample of quasars with extended double structure.

159.107   Extended structure of superluminal radio sources.
I. W. A. Browne.
Superluminal radio sources, p. 129 – 147 (1987). – See Abstr. 012.053.

The author reviews the observations of known superluminal radio sources and presents some simple statistics (e.g., how many are doubles, how many have any extended structure, how big is this structure, does the strength or the extent of the structure correlate with anything else?). Then he tries to tie things together and discusses how consistent the present observations are with simple beaming models.

159.108   Feeling uncomfortable.
P. D. Barthel.
Superluminal radio sources, p. 148 – 154 (1987). – See Abstr. 012.053.

The author examines recent observational material using the working hypothesis of the simple relativistic beaming model. Then he presents a new superluminal quasar, and points out the problems that these observations pose for the simple relativistic beaming hypothesis. The simple beaming hypothesis will not survive if more very large triple sources with considerable superluminal motion and well aligned small and large scale jets are found.

159.109   The arcminute structure of 1928 + 738.
R. S. Simon, K. J. Johnston, A. Eckart, P. Biermann, C. Schalinski, A. Witzel, R. G. Strom.
Superluminal radio sources, p. 155 – 161 (1987). – See Abstr. 012.053.

The authors have found that 1928 + 738 has two–sided radio structure which extends to 40″ on either side of the compact core, equivalent to a projected linear size of 221 h^{-1}kpc. If the superluminal expansion of the core is interpreted in terms of the standard relativistic jet model, the axis of the small–scale structure must lie within 15.1° from the line of sight. If the large–scale structure is also aligned at this angle, the deprojected size exceeds 700 h^{-1}kpc.

159.110   Superluminal motion CTA 102.
L. B. Bååth.
Superluminal radio sources, p. 206 – 210 (1987). – See Abstr. 012.053.

The author started a VLBI monitoring project at 932 MHz, just below the "intermediate–frequency gap" to look for structural changes in such objects. The observations were made at six epochs during 1983 – 1984. A total of 15 sources were observed. The author presents maps for the objects 2147 + 145, DA 406 (1611 + 343), 1422 + 202, and CTA 102.

159.111   Infrared, optical, UV, and X–ray properties of superluminal radio sources.
C. Impey.
Superluminal radio sources, p. 233 – 250 (1987). – See Abstr. 012.053.

Current lists of probable and possible superluminal radio sources are given. Where available, ranges of optical polarization and position angle are included (however, many sources have only one measurement). It is striking that 13 out of 21 of the probable and 9 out of 10 of the possible superluminal sources are highly polarized in the optical ($p > 2.5\%$). At least 70% of the superluminal sources are identified with blazars.

159.112   Superluminal radio sources: what does X–ray emission tell us?
D. M. Worrall.
Superluminal radio sources, p. 251 – 259 (1987). – See Abstr. 012.053.

The X–ray versus optical and X–ray versus radio correlations of radio–loud QSOs and superluminal radio sources are similar. This argues against a model in which the emission in only one or two of the three wave bands is relativistically boosted, unless *all* these objects exhibit beaming at small angles relative to the observer's line of sight. A regression analysis shows that highly polarized QSOs and optically violently variable QSOs are more similar to other flat–spectrum, radio–loud QSOs than to BL Lac objects, and it is reasonable to assume that self–Compton emission dominates the X–ray emission from at least half of the sources in this class. The X–ray versus radio correlation for BL Lac objects is poor, and there is support for the hypothesis that their X–ray emission is dominated by an isotropic component which is not directly related to relativistically boosted radio emission.

159.113   How fast can a blob go?
E. S. Phinney.
Superluminal radio sources, p. 301 – 305 (1987). – See Abstr. 012.053.

The author discusses speed limits on electron–proton jets in quasars and radio galaxies.

159.114   The distribution of QSO absorption system column densities: evidence for a single population.
D. Tytler.
Astrophys. J., Vol. 321, No. 1, p. 49 – 68 (1987).

The author examines the distribution of H I column densities in a representative sample of all cosmologically distributed QSO absorption–line systems. Remarkably this distribution function $f(N)$ is a featureless power law extending over about seven orders of magnitude in column density. It is argued that this finding is in conflict with the standard interpretation of absorption systems, which invokes separate and unrelated intergalactic (Lyα) and galactic halo (metal–line) absorption systems. It is suggested that, in addition to the Lyα systems, most metal–line systems may also be intergalactic.

159.115   The redshift distribution of QSO Lyman–alpha absorption systems.
D. Tytler.
Astrophys. J., Vol. 321, No. 1, p. 69 – 79 (1987).

The redshift distribution of Lyα absorption lines in quasar spectra is shown to be more complex than is generally believed. In addition to the systematic increase in the density of lines with z_{abs}, line density also increases by a similar amount as z_{em} increases or equivalently as the rest frame wavelength of the absorbed radiation λ, decreases. It is also shown that there is a significant and apparently separate lack of Lyα absorption lines in QSO Lyα emission lines at $z_{abs} \approx z_{em}$. There are several plausible causes for these effects including serious inhomogeneities in published line samples, unidentified H_2 absorption lines, photoionization by the individual observed QSOs, the failure of some Lyα absorbing clouds to cover QSO emission–line regions, QSO ejecta, and large–scale $\sim 70 \, h_{100}^{-1}$ Mpc inhomogeneities in the universe at $z \sim 2.5$.

159.116   Constraints on quasar accretion disks from the optical/ ultraviolet/soft X-ray big bump.

B. Czerny, M. Elvis.
Astrophys. J., Vol. 321, No. 1, p. 305 – 320 (1987).

It is shown that accretion disk spectrum models which take into account electron scattering are significantly better representations of accretion disks in quasars than the simplest sum–of–blackbodies approximation. These improved spectral models can account for the flattening of the spectra around $\log \nu \sim 15.3$ observed in many quasars. The problem of why the previously determined maximal value of the temperature in the disk was always $\sim 20{,}000 - 30{,}000$ K is naturally explained in these models by the weak dependence of the "UV–flattening" frequency on the disk parameters. The spectral models depend strongly on the inclination angle at high–accretion rates due to geometrical effects. The authors apply their disk spectrum model to interpret the optical/UV/X–ray big bump in quasar PG 1211 + 143.

159.117   PC 0910 + 5625: an optically selected quasar with a redshift of 4.04.

M. Schmidt, D. P. Schneider, J. E. Gunn.
Astrophys. J., Lett. Ed., Vol. 321, No. 1, p. L7 – L10 (1987). With plate L1.

The authors report the discovery of the second quasar with a redshift greater than 4. This object, which has an r magnitude of 21.0, was detected on a CCD grism survey. The spectrum of the quasar has strong, relatively narrow (but resolved) emission lines; there is a deep absorption trough ~ 100 Å wide on the blue wing of the Lyman–α line; and the continuum at wavelengths shorter than the absorption feature is far weaker that the level present redward of Lyman–α.

159.118   Superluminal motion in the double–lobed quasar 3C 245.

D. H. Hough, A. C. S. Readhead.
Astrophys. J., Lett. Ed., Vol. 321, No. 1, p. L11 – L15 (1987).

The authors report the detection of superluminal motion in the central component of the quasar 3C 245 with an apparent transverse velocity of $(3.1 \pm 1.4)h^{-1}c$ ($H_0 = 100h$ km s^{-1}Mpc^{-1}, $q_0 = 0.5$). This is the third steep–spectrum, double–lobed quasar found to be superluminal. The authors discuss some implications of this result and the current statistics of superluminal sources for the simple relativistic beaming model.

159.119   Discovery of a probable binary quasar.

S. Djorgovski, R. Perley, G. Meylan, P. McCarthy.
Astrophys. J., Lett. Ed., Vol. 321, No. 1, p. L17 – L21 (1987). With plate L2.

The authors report discovery of a pair of quasars at a redshift of 1.345, separated by 4″.2 in projection, apparently associated with the radio source PKS 1145–071. The optical intensity ratio of the two images is approximately 2.5. The spectra are very similar, and all emission lines except for the C IV λ1549 have the same equivalent widths. The redshift difference based on the C IV λ1549 line is 250 ± 100 km s^{-1}. It is consistent with zero, if the C IV line is excluded. Thus, on the basis of the optical data alone, this double QSO seems to be another example of gravitational lensing. However, the radio source is single and unresolved at all VLA frequencies, and positionally coincident with the brighter of the two optical images. The fainter of the two optical components is *at least* a factor of 500 fainter in the radio. This suggests that the system is a genuine binary quasar, and the first such well–documented case.

159.120   Is optical Fe II emission related to the soft X–ray properties of quasars?

B. J. Wilkes, M. Elvis, I. McHardy.
Astrophys. J., Lett. Ed., Vol. 321, No. 1, p. L23 – L27 (1987).

Radio–quiet quasars generally show broad, blended multiplets of Fe II emission in their optical and ultraviolet spectra. Radio–loud quasars also show UV Fe II emission, but their optical Fe II emission is generally weaker. The authors propose that the primary factor controlling the optical Fe II emission is the soft X–ray spectrum. This proposition is supported by X–ray and optical data for nine quasars which shows a correlation between the soft X–ray slope and the strength of the optical Fe II emission.

159.121   On the influence of the spectral resolution on the absorption lines in quasar spectra.

I. E. Val'tts.
Inst. Kosm. Issled. Akad. Nauk SSSR, Prepr., No. 1224, p. 1 – 16 (1987). In Russian. Abstr. in Ref. Zh., 51. Astron., 12.51.991 (1987).

159.122   A new case of gravitational lensing.

J. Surdej, P. Magain, J.–P. Swings, U. Borgeest, T. J.–L. Courvoisier, R. Kayser, K. I. Kellermann, H. Kühr, S. Refsdal.
Nature, Vol. 329, No. 6141, p. 695 – 696 (1987).

The authors have begun a systematic search from ESO for gravitational lens systems in a selected sample of highly luminous quasars; $M_V < -29.0$. They give a brief description of their first identified gravitational lens system UM673 = Q0142–100 = PHL3703. It consists of two images, A ($m_R = 16.9$) and B ($m_R = 19.1$), separated by 2.2 arc s at a redshift $z_q = 2.719$. The lensing galaxy ($m_R \approx 19$, $Z_L \approx 0.49$) has also been found. It lies very near the line connecting the two QSO images, ~ 0.8 arc s from the fainter one. A value $M_0 \approx 2.4 \times 10^{11} M_\odot$ for the mass of the lensing galaxy and $\Delta t \approx 7$ weeks for most likely travel–time difference between the two light paths to the QSO are found (assuming $H_0 = 75$ km s^{-1}Mpc^{-1}, $q_0 = 0$).

159.123   Hybrid maps of 4C 39.25 from geo–VLBI observations.

G.–q. Tang, B. Rönnäng.
Mitt. Geod. Inst. Rheinischen Friedrich–Wilhelms–Univ. Bonn, Nr. 71, p. 91 – 95 (1987). – See Abstr. 012.066.

159.124   The optical variability of the quasar 3C 273 in 1977 – 1986. Photoelectric observations. II.

V. M. Lyutyj, N. V. Metlova.
Astron. Tsirk., No. 1475, p. 3 – 5 (1987). In Russian.

159.125   The clustering of Lα absorption lines in the QSO spectra.

J. P. Mücket, V. Müller.
Astrophys. Space Sci., Vol. 139, No. 1, p. 163 – 174 (1987).

For nine published high–resolution QSO spectra a correlation analysis of their Lα forest lines has been performed. The two-point correlation functions show some quasi–periodic structure of magnitude $|\xi| \lesssim 0.3$. Their characteristic separation along the line–of–sight amounts to $\Delta s_0 = 3 \times 10^{-3}$ or to $\Delta s_0 = 5 \times 10^{-3}$ for $\Omega = 1$ and 2, respectively. Especially the distribution of nearest neighbouring line positions in two close QSO pairs allows for the interpretation that the absorption clouds lie in sheet–like structures as predicted by the pancake theory. The correlation data contain some hints on metal absorbers within the forest of unidentified lines.

159.126   Effect of changes in magnitudes of QSOs in their redshift distribution.

D. Basu.
Astron. Nachr., Vol. 308, No. 5, p. 299 – 301 (1987).

Strong emission lines may change the brightness of QSOs and hence their observed magnitudes. Since different lines will affect the magnitudes by entering a particular filter at different redshifts, this effect may alter the number of QSOs at a particular redshift and hence the redshift distribution. The present analysis shows that the influence of the emission lines on the U and B magnitudes are significantly correlated to the redshift distribution. It is concluded that the changes in observed magnitudes of QSOs caused by the emission lines have significant effects on the present redshift distribution.

159.127   Objective prism survey of QSO candidates in field centered at $00^h00^m + 0°00'$. (I) The northeastern quarter.
Y. Zhan, J.-s. Chen.
Chin. Astron. Astrophys., Vol. 11, No. 3, p. 191 – 200 (1987).
English translation of 43.159.109.

159.128   VLBI observations of 4C 39.25 at 18 and 6 cm waves.
S.-y. Wu, I. I. K. Pauliny-Toth, R. W. Porcas.
Chin. Astron. Astrophys., Vol. 11, No. 3, p. 201 – 206 (1987).
English translation of 43.159.111.

159.129   A clustering analysis of QSO candidates.
Y.-y. Zhou, D.-p. Fang, X.-t. He, Z.-g. Deng.
Chin. Astron. Astrophys., Vol. 11, No. 4, p. 282 – 287 (1987).
English translation of 43.159.108.

159.130   A further study of the method of identifying absorption line systems in QSO spectra.
Z.-x. Cui.
Chin. Astron. Astrophys., Vol. 11, No. 4, p. 291 – 296 (1987).
English translation of Acta Astrophys. Sin., Vol. 7, No. 3, p. 194 – 202 (1987).
A method of identifying absorption line systems in QSO spectra is further developed and certain limitations and their improvements are discussed. The improved method is applied to PKS 0528–250, and gives two new absorption line systems $Z_a = 0.065$ and 0.0345 in addition to the four systems $Z_a = 2.8110, 2.8130, 2.5275, 2.1410$.

159.131   Objective prism survey of QSO candidates in field centered at $00^h00^m + 00°00'$. II. The southeastern quarter.
Y. Zhan, J.-s. Chen.
Chin. Astron. Astrophys., Vol. 11, No. 4, p. 299 – 305 (1987).
English translation of Acta Astrophys. Sin., Vol. 7, No. 3, p. 203 – 206 (1987).

159.132   On the origin in spacetime distribution of quasars. I. The simplest model.
R. V. Popic.
Ann. Phys. (Leipzig), Vol. 44, No. 6, p. 440 – 446 (1987). Abstr. in Phys. Abstr., Vol. 90, No. 1317, Entry 140316 (1987).

159.133   On the origin of spacetime distribution of quasars. II. The modified model.
R. V. Popic.
Ann. Phys. (Leipzig), Vol. 44, No. 6, p. 447 – 454 (1987). Abstr. in Phys. Abstr., Vol. 90, No. 1317, Entry 140317 (1987).

159.134   In quest of distant quasars.
L. Miller.
New Sci., Vol. 115, No. 1578, p. 58 – 61 (1987). Abstr. in Phys. Abstr., Vol. 91, No. 1320, Entry 10678 (1988).

159.135   Ever more distant quasars?
P. Shaver.
Nature, Vol. 330, No. 6147, p. 426 (1987).
The author reports on high–redshift quasars, their space density, on optical searches and gravitational lensing. For two quasars with redshifts greater than 4 see Abstr. 159.136.

159.136   Quasars of redshift $z = 4.43$ and $z = 4.07$ in the South Galactic Pole field.
S. J. Warren, P. C. Hewett, P. S. Osmer, M. J. Irwin.
Nature, Vol. 330, No. 6147, p. 453 – 455 (1987).
The authors report the discovery of two new quasars of very high redshift; Q0051–279 of $z = 4.43$ and Q0101–304 of $z = 4.07$. The redshift of Q0051–279 is the highest yet recorded. Both new quasars were found by the multicolour selection technique.

159.137   The identification of heavy–element absorption systems in QSO 2000–330.
Z. Cui, J. Chen.
Acta Astrophys. Sin., Vol. 7, No. 4, p. 280 – 286 (1987). In Chinese.
A search for heavy–element absorption systems in the spectrum of QSO 2000–330 ($z_e = 3.78$) has been made using the method of the identification of absorption systems suggested by Cui et al. (1983) and Chen et al. (1983). Six absorption systems were found which include $z_a = 3.1881, 3.1913, 3.3335, 3.5519$ and $z_a = 1.3441, 3.3459$. The first four systems are identical with those reported by Hunstead et al. (1986), as systems A, B, C and D; the remaining two are new ones.

159.138   Absorption in the wide QSO pair Tololo 1037–2704 and Tololo 1038–2712: evidence for a specially aligned supercluster at $z = 2$?
W. L. W. Sargent, C. C. Steidel.
Astrophys. J., Vol. 322, No. 1, p. 142 – 163 (1987).
Spectra with a resolution of ~ 2 Å have been obtained of the QSO pair Tol 1037–2704 ($z_{em} = 2.193$, $m_B = 17.4$) and Tol 1038–2712 ($z_{em} = 2.331$, $m_B = 17.8$). These objects are separated by $17''.9$ on the sky and were discovered by Jakobsen and his coworkers to have extensive, possibly correlated absorption spectra. The authors confirm the existence of the four absorption systems between $z_{abs} = 1.95$ and $z_{abs} = 2.14$ in both QSOs. They have also found a fifth system in each spectrum, with $z_{abs} \sim 1.90$ in each case. It is shown that it is highly significant that the two QSOs have extremely rich absorption spectra confined to the same redshift range $1.88 < z_{abs} < 2.15$. The authors examine the consequences of various hypotheses for the origins of the absorption features – ejection from one or both QSOs or the effects of an intervening supercluster. They show that ejection is highly implausible on energetic grounds, and that the observations point to an elongated intervening supercluster observed along its major axis.

159.139   Emission–line variability and the broad–line region of quasi–stellar objects. I. Time scales and photon densities.
W. Zheng, E. M. Burbidge, H. E. Smith, R. D. Cohen, S. E. Bradley.
Astrophys. J., Vol. 322, No. 1, p. 164 – 173 (1987).
Spectroscopic data for five low–redshift QSOs taken over several years are presented as evidence for emission–line variability in QSOs. The broad hydrogen emission lines in the QSOs $0026 + 129$, $1202 + 281$, $1612 + 261$, and $2141 + 175$ are found to vary in step with their underlying continua. Another QSO, $2135–147$, shows an increase in the intensities of its hydrogen lines 5 yr after the increase in its continuum level. The inferred upper limit to the size of the broad emission–line region is smaller than 3 lt–yr. Such a small size implies a very high density of ionizing photons, of the order of $10^{10} cm^{-3}$.

159.140   New considerations on the broad–line regions of quasars.
J. H. You, F. H. Cheng.
Astrophys. J., Vol. 322, No. 1, p. 174 – 179 (1987). = Lick Obs. Bull., No. 1074.
In the broad–line regions of quasars, there exist three different line emission mechanisms, i.e., the Cerenkov line and conventional line by recombination and collisional excitation. However, the Cerenkov line photon can avoid the resonance absorption because of the "Cerenkov redshift" and can escape easily from the deep inner part of the gas, i.e., gas appears more "transparent" for the Cerenkov line than for a conventional line. Taking the quasar 3C 273 as an example, it is shown that only if $N_e \approx 10^4$–$10^6 cm^{-3}$, the calculated Cerenkov Lyα luminosity is in agreement with observation.

159.141   The effect of a quasi–stellar object on its host galaxy: dynamical and physical processes in the interstellar medium around a quasi–stellar object.
C. A. Chang, A. V. R. Schiano, A. M. Wolfe.
Astrophys. J., Vol. 322, No. 1, p. 180 – 200 (1987).
The dynamical and physical processes in the interstellar medium (ISM) around a QSO have been calculated in order to

understand the effects of QSO radiation on the QSO's host galaxy. Numerical hydrodynamical solutions for one– and two–dimensional cases have been found through a MacCormack differencing scheme. Ion–field emission, ion sputtering, charge equilibrium, thermal equilibrium, coupling between gas ions and dust grains, photoionization, and photoelectrical processes have been considered. This work shows that dust grains play an important role in determining the dynamics of the ISM around a QSO.

159.142 Spectroscopy of the quasar S5 0014+81. I. Emission–line spectrum.
D. A. Varshalovich, S. A. Levshakov, E. A. Nazarov, O. I. Spiridonova, A. F. Fomenko.
Sov. Astron., Vol. 31, No. 2, p. 136 – 140 (1987). English translation of 43.159.020.

159.143 PG 1411+442: the nearest broad absorption line quasar.
M. A. Malkan, R. F. Green, J. B. Hutchings.
Astrophys. J., Vol. 322, No. 2, p. 729 – 738 (1987).
IUE observations reveal strong, moderately broad absorption troughs in the blue wings of the C IV and N V emission lines of the quasar PG 1411+442. No absorption from weakly ionized gas is detected. The quasar's redshift is low enough ($z_{em} = 0.089$) for investigating the morphology of the host galaxy in deep broad–band and intermediate–band CCD images. The galaxy appears to be a large spiral with a very long arm or tail. The inclination angle is 57°.

159.144 An absorption–line survey of 32 QSOs at red wavelengths: properties of the Mg II absorbers.
K. M. Lanzetta, D. A. Turnshek, A. M. Wolfe.
Astrophys. J., Vol. 322, No. 2, p. 739 – 769 (1987).
The authors present spectroscopy of 32 QSOs at red wavelengths with 4.5 Å spectral resolution. They detect 22 Mg II doublets, from which they derive the properties of the Mg II absorbers. The Mg II lines are distributed randomly in velocity relative to the QSOs, as would be expected if the absorption were due to intervening material. There is marginally significant evidence for intrinsic evolution of the number density of the Mg II absorbers with redshift, in the sense that the number density increases with increasing redshift. By combining the present data with previously published observations of C IV and C II seen in these QSOs at blue wavelengths, the authors compare the properties of the Mg II– and C IV–selected systems.

159.145 Quasar energy distributions. I. Soft X–ray spectra of quasars.
B. J. Wilkes, M. Elvis.
Astrophys. J., Vol. 323, No. 1, p. 243 – 262 (1987).
The authors present Einstein IPC spectra for a sample of 33 quasars with well–determined soft X–ray slopes. The best–fit power–law slopes have a wide range ($-0.2 \leqslant \alpha_E \leqslant 1.8$) but are strongly grouped around values of ~ 0.5 for radio–loud quasars and ~ 1.0 for radio–quiet quasars. A correlation analysis shows that radio loudness is fundamentally related to the X–ray slope. This correlation is not followed by higher energy spectra of active galaxies. Two components are required to explain both sets of results. The best–fit column densities are systematically smaller than the Galactic values. The same effect is not present in a sample of BL Lac objects.

159.146 The remarkable broad absorption line QSO 0059–2735 with extensive Fe II absorption.
C. Hazard, R. G. McMahon, J. K. Webb, D. C. Morton.
Astrophys. J., Vol. 323, No. 1, p. 263 – 270 (1987). With plate 2.
The authors describe the discovery and preliminary spectroscopy of the broad absorption line QSO 0059–2735 ($z_{em} = 1.595$) which shows features never seen before in QSO spectra. It exhibits an extensive system of low ionization absorptions which include saturated lines of Mg II, Al II, and Al III, and broad Fe II features. Even more remarkable are the very probable identifications of Fe II absorption by UV multiplets 62, 63, and 64 which require lower level excitations of 1.1 eV and Fe III absorption by the triplet UV 34 from a lower level at 3.7 eV. These are the first reports of absorption from such highly excited levels in any QSO.

159.147 Forbidden [O II] and [O III] emission 29″ from a QSO absorption–line region.
B. Yanny, D. Hamilton, R. A. Schommer, T. B. Williams, D. G. York.
Astrophys. J., Lett. Ed., Vol. 323, No. 1, p. L19 – L24 (1987).
QSO 0453–423 ($z_{QSO} = 2.66$) was imaged with a narrow–band filter centered on [O II] 3727 Å at the redshift of a strong intervening absorption complex ($z_{abs} = 0.7256$). Comparison with a broad–band exposure revealed a small nonstellar object 29″ from the QSO line of sight. Spectroscopic follow–up confirmed the [O II] emission and detected [O III] 5007 Å. The angular separation between the emission patch and the absorption at the same redshift is the largest of those thus far reported. It is suggested that extended gas in the vicinity of galaxies produces the QSO absorption lines. A lack of correlation between absorption velocity widths and either distance to, or strength of, the [O II] emission at the same redshift argues against high–[O II] surface brightness star–forming regions at these redshifts ($0.4 < z < 0.8$).

Second Byurakan Spectral Sky Survey. V. Results for the region centered on $\alpha = 15^h30^m$, $\delta = +59°00'$.
See Abstr. 002.013.

Quasars, redshifts and controversies.
See Abstr. 003.020.

Probing the early universe.
See Abstr. 011.018.

Models of quasars reappraised.
See Abstr. 011.019.

Superluminal radio sources. Proceedings of a workshop in honor of Professor Marshall H. Cohen, held at Big Bear Solar Observatory, Calif., USA, 28 – 30 October 1986.
See Abstr. 012.053.

Use of a minimum rate of change formalism to quantify variability of extragalactic X–ray sources.
See Abstr. 036.041.

The detection probability for emission–line objects in slitless spectrum surveys.
See Abstr. 036.125.

On the concept of resolution ellipse in aperture synthesis matched deconvolution with error analysis.
See Abstr. 036.173.

Finding high–redshift quasars using low–resolution spectra.
See Abstr. 036.177.

Imaging superluminal sources: prospects for the next decade.
See Abstr. 036.183.

Spatial stability of relativistic jets: application to 3C 345.
See Abstr. 062.001.

Compact radio sources as a plasma turbulent reactor. I. Formation of Maxwell–like spectra of relativistic electrons at the acceleration on resonant Langmuir waves.
See Abstr. 062.009.

Numerical studies of the dynamical stability of differentially rotating tori.
See Abstr. 062.078.

Radiation from charged particles moving in the magnetic fields of extragalactic jets.
See Abstr. 063.054.

On the stability of turbulent synchrotron sources with respect to e^+-e^- pair creation.
See Abstr. 063.114.

Synchrotron emission from shock waves in active galactic nuclei.
See Abstr. 063.115.

Inverse Compton scattering of ambient radiation by a cold relativistic jet: a source of beamed, polarized continuum in blazars?
See Abstr. 063.116.

Gravitational lensing by isothermal spheres with finite core radii: galaxies and dark matter.
See Abstr. 066.097.

Ultimas noticias sobre el sistema de lente gravitacional 0957 + 561.
See Abstr. 066.109.

Gravitational imaging by isolated elliptical potential wells. I. Cross sections.
See Abstr. 066.149.

Gravitational imaging by isolated elliptical potential wells. II. Probability distributions.
See Abstr. 066.150.

The marginal gravitational lensing.
See Abstr. 066.151.

New mechanism of appearance of proper redshifts in spectra of compact objects.
See Abstr. 066.179.

Astrophysical black holes.
See Abstr. 067.079.

Pair production and Compton scattering in compact sources and comparison to observations of active galactic nuclei.
See Abstr. 067.080.

Thick accretion disks: theory vs. observations.
See Abstr. 067.084.

How big are supermassive black holes formed from the collapse of dense star clusters?
See Abstr. 067.086.

Electron injection by relativistic protons in active galactic nuclei.
See Abstr. 067.098.

An axionic laser in the center of a galaxy?
See Abstr. 067.116.

Grand Unified Models.
See Abstr. 067.122.

On nonthermal models for active galactic nuclei.
See Abstr. 067.128.

Mining energy from a rotating black hole in a magnetic field by the Penrose process.
See Abstr. 067.140.

Electromagnetic jets from compact objects.
See Abstr. 067.144.

Pair–creation effects in accretion–shock models of active galactic nuclei.
See Abstr. 067.177.

Spectral types for objects in the Kiso survey. III. Data for 102 stars.
See Abstr. 114.055.

Radio caustics from localized interstellar medium plasma structures.
See Abstr. 131.121.

Molecule formation in quasar broad–line cloud gas.
See Abstr. 131.266.

Systematic identification of IRAS point sources.
See Abstr. 133.001.

The IRAS view of the extragalactic sky.
See Abstr. 133.007.

Optical identifications of radio sources from the Molonglo deep surveys – I. The declination –20° portion of the first deep survey.
See Abstr. 141.002.

Radio positions and optical identifications for a complete sample of southern flat–spectrum radio sources – I. Region 06^h to 18^h.
See Abstr. 141.003.

Determinations of distances to radio sources with VLBI.
See Abstr. 141.015.

The origin of the diffuse X–ray background.
See Abstr. 142.010.

Discrete X–ray sources and the X–ray background.
See Abstr. 142.011.

The cosmic X–ray background.
See Abstr. 142.012.

The contribution from active galactic nuclei to the diffuse 1 MeV background.
See Abstr. 143.015.

Starburst galaxies.
See Abstr. 158.001.

Ultraluminous infrared galaxies.
See Abstr. 158.004.

Far–infrared properties of optically–selected quasars and Seyfert galaxies.
See Abstr. 158.029.

Five years monitoring of extragalactic radio sources. I. Observations at 12, 22 and 37 GHz.
See Abstr. 158.041.

Alternating side ejection or precession of jets in radio sources.
See Abstr. 158.049.

Activity in galaxy nuclei and the SS433 phenomenon.
See Abstr. 158.057.

Spectroscopy of the extranuclear line–emitting regions associated with the gravitational lens system 2016 + 112.
See Abstr. 158.067.

Radio observations of a few selected blazars.
See Abstr. 158.087.

X–ray variability of high luminosity AGN.
See Abstr. 158.106.

The physics and the structure of AGN.
See Abstr. 158.112.

Velocities in radio galaxies and quasars.
See Abstr. 158.114.

Optical synchrotron emission from radio hot spots.
See Abstr. 158.116.

Formation of low ionization lines in active galactic nuclei.
See Abstr. 158.123.

Extended ionized nebulosities in the galaxies Mk 1, Mk 3, Mk 348 and the quasar 4C 37.43.
See Abstr. 158.126.

Statistical significance of the relationship between X-ray luminosity and variability timescale in active galactic nuclei.
See Abstr. 158.130.

Optical identifications and radio morphology of the complete 5 GHz S5 survey.
See Abstr. 158.146.

The optical polarization properties of blazars.
See Abstr. 158.147.

The inverse Compton test for a large sample of compact radio sources.
See Abstr. 158.160.

The central power source in active galaxies.
See Abstr. 158.169.

The velocity distribution of quasar/AGN broad emission line clouds.
See Abstr. 158.170.

What is the difference between radio galaxies and radio quasar galaxies?
See Abstr. 158.171.

A nonthermal model for the XUV, soft X-ray emission of AGN and QSOs.
See Abstr. 158.173.

Is there a systematic infall or outflow of the broad line region clouds?
See Abstr. 158.183.

Active galactic nuclei vs. their host galaxies – the X-ray connection.
See Abstr. 158.189.

Search for 511 keV emission from active galaxies and QSO's.
See Abstr. 158.200.

Hot dust and the near-IR bump in active galactic nuclei.
See Abstr. 158.201.

A time-dependent analysis of the Blandford-Znajek process and its astrophysical applications: I. Evolutionary paths.
See Abstr. 158.202.

Sub-arcsecond resolution radio images of one-sided jets in core-dominated sources.
See Abstr. 158.204.

The velocity distribution of quasar/AGN broad emission line clouds.
See Abstr. 158.216.

Rapid variability of extragalactic radio sources.
See Abstr. 158.227.

Galactic nuclei and quasars at high angular resolution.
See Abstr. 158.246.

77 GHz continuum observations of variable extragalactic sources.
See Abstr. 158.263.

Continuity in the violent phenomena taking place in the nuclei of galaxies and QSOs.
See Abstr. 158.271.

Long term variability of radiosources in the frequencies of 22 GHz and 44 GHz.
See Abstr. 158.275.

Galactic nuclei and quasars at high angular resolution.
See Abstr. 158.286.

Nucleus activity correlates with outlying structure in Seyfert galaxies and quasars.
See Abstr. 158.289.

3C 120.
See Abstr. 158.292.

Milliarcsecond polarization of superluminal sources.
See Abstr. 158.301.

A different perspective on superluminal sources.
See Abstr. 158.303.

Optical spectra of superluminal sources.
See Abstr. 158.304.

Synchro-Compton emission from superluminal sources.
See Abstr. 158.307.

Extragalactic radio sources with very large Faraday rotation.
See Abstr. 158.310.

6 centimeter radio source counts and spectral index studies down to 0.1 millijansky.
See Abstr. 158.311.

Spherical models of QSO/AGN emission line clouds.
See Abstr. 158.318.

3C 120 and the surrounding region of sky.
See Abstr. 158.320.

On the possibility of observation of X-ray absorption lines in the spectra of BL Lac-type objects.
See Abstr. 158.323.

Daily observations of compact extragalactic radio sources at 2695 and 8085 MHz, 1979 – 1985.
See Abstr. 158.327.

Anisotropic Compton scattering from relativistic jets of active galactic nuclei.
See Abstr. 158.335.

On the broad emission lines in Seyfert galaxies and quasars.
See Abstr. 158.340.

The mass-luminosity dependence for active galactic nuclei.
See Abstr. 158.343.

What heats the hot phase in active nuclei?
See Abstr. 158.348.

Far-infrared variability in active galactic nuclei.
See Abstr. 158.350.

A new upper limit on the density of generally distributed intergalactic neutral hydrogen.
See Abstr. 160.005.

Arcs, light echoes, and supergalaxies.
See Abstr. 160.010.

The light–echo model for luminous arcs.
See Abstr. 160.011.

The H I environment of high redshift quasars: a VLA search for cosmological H I.
See Abstr. 160.050.

Spectroscopy and photometry of the cluster of galaxies surrounding the quasar 3C 206.
See Abstr. 160.078.

Can quasars ionize the intergalactic medium?
See Abstr. 160.121.

Astronomical constraints on a string–dominated universe.
See Abstr. 161.031.

Probing the early universe with quasar light.
See Abstr. 161.083.

Cosmological evolution of active galaxies & quasars.
See Abstr. 161.089.

Were the Lyman–α clouds formed from shocks?
See Abstr. 161.100.

Cosmological evolution of extragalactic sources.
See Abstr. 161.173.

Nature of the redshift of galaxies and quasars and its consequences. I. Gas accretion onto the nucleus of a galaxy and galaxy emission.
See Abstr. 161.187.

Nature of the redshift of galaxies and quasars and its consequences. II. The conception of the noncosmological nature of redshifts of galaxies and quasars.
See Abstr. 161.188.

Opacity of the universe.
See Abstr. 161.208.

The significance of spatial curvature for energy output of remote quasars.
See Abstr. 161.302.

Cosmological H II regions and the photoionization of the intergalactic medium.
See Abstr. 161.355.

Physical constants and evolution of the Universe.
See Abstr. 161.377.

Curvature magnification explains the superluminal lobes.
See Abstr. 161.447.

A search for QSOs to fit a cosmological model with flat, closed spatial sections.
See Abstr. 161.453.

160 Galaxy Groups, Clusters of Galaxies, Superclusters, Intergalactic Matter

160.001   Far–infrared properties of cluster galaxies.
M. D. Bicay, R. Giovanelli.
NASA Conf. Publ., NASA CP–2466, p. 277 – 281 (1987). – See Abstr. 012.002.

Far–infrared properties are derived for a sample of over 200 galaxies in seven clusters: A262, Cancer, A1367, A1656 (Coma), A2147, A2151 (Hercules), and Pegasus. The IR–selected sample consists almost entirely of "IR normal" galaxies, with $\langle Log[L(FIR)]\rangle = 9.79\,L_\odot$, $\langle[L(FIR)/L(B)]\rangle = 0.79$, and $\langle Log[S(100\,\mu m)/S(60\,\mu m)]\rangle = 0.42$. None of the sample galaxies has $Log[L(FIR)] > 11.0\,L_\odot$, and only one has a FIR–to–blue luminosity ratio greater than 10. No significant differences are found in the FIR properties of H I–deficient and H I–normal cluster galaxies.

160.002   Present star formation in spirals of the Virgo cluster.
B. Guiderdoni.
NASA Conf. Publ., NASA CP–2466, p. 283 – 286 (1987). – See Abstr. 012.002.

From a study of spiral galaxies in the Virgo cluster (VC), it is shown that RDDO anemics with smooth arms and no sign of present formation of (massive) stars have H I surface densities below a threshold value of 2 to 5×10^{20} atom cm^{-2}. This value is very consistent with predictions of theoretical models. It is likely that the H I disks of VC H I–deficient RDDO anemics have been deeply affected by ram pressure stripping in the gaseous intracluster medium, while VC H I–deficient RDDO spirals have been only peripherally stripped.

160.003   Molecular gas and star formation in H I–deficient Virgo cluster galaxies.
J. D. Kenney, J. S. Young.
NASA Conf. Publ., NASA CP–2466, p. 287 – 292 (1987). – See Abstr. 012.002.

Mapping of the CO emission line in 42 Virgo cluster galaxies reveals that the molecular gas contents and distributions are roughly normal in severely H I–deficient Virgo spirals. The survival of the molecular component mitigates the impact of the H I–stripping on star formation and subsequent galactic evolution. For spirals which are deficient in H I by a factor of 10, far–infrared, Hα line, and non–thermal radio continuum luminosities are lower by no more than a factor of 2. The fact that the inner galactic disks are stripped of H I, while CO is normal, suggests that the lifetime of the molecular phase is $\sim 10^9$ years in the inner regions of luminous spirals.

160.004   The evolution of cooling flows. I. Self–similar cluster flows.
R. A. Chevalier.
Astrophys. J., Vol. 318, No. 1, p. 66 – 77 (1987).

The evolution of a cooling flow from an initial state of hydrostatic equilibrium is studied. If the gravitational potential is that of a singular isothermal sphere, the gas is initially isothermal and the radiative cooling function is a power law of temperature, then the flow is self–similar. Mass removal by thermal instability can be approximately included in the self–similar solutions. The addition of magnetic or cosmic–ray pressure shows that the non–thermal pressure comes to dominate the pressure in the inner parts of the flow, where the density profile is flattened.

160.005   A new upper limit on the density of generally distributed intergalactic neutral hydrogen.
C. C. Steidel, W. L. W. Sargent.
Astrophys. J., Lett. Ed., Vol. 318, No. 1, p. L11 – L13 (1987).

A new upper limit on the number density of generally distributed intergalactic neutral hydrogen is obtained based upon recent spectrophotometric observations of very high redshift QSOs coupled with high–resolution statistical studies of the Lyα forest in QSOs. The new limit, $n_{HI}(z = 0.0) < 9.0 \times 10^{-14}h_{100}cm^{-3}$, is approximately 15 times smaller than the limit originally proposed by Gunn and Peterson.

160.006   Merkwaardige bewegingen in het lokale heelal.
G. Beekman.
Zenit, 14. Jahrg., Nr. 9, p. 282 – 286 (1987).

160.007   An investigation of photometric and geometric parameters of galaxies in clusters. I. Cluster A 2065.
O. M. Kurtanidze, G. M. Richter.
Astrofizika, Tom 26, Vyp. 3, p. 387 – 397 (1987). In Russian. English translation in Astrophysics, Vol. 26, No. 3, p. 235 – 241 (1987).

Data of the cluster A 2065 are presented. The magnitudes were integrated up to the isophote corresponding to $m_{pg} = 25$ sq.arc sec. The completeness limit is 19^m5 and the number of galaxies up to this limit is 132. For the parameters of the Schechter luminosity function the values of $\alpha^* = -1.25$, $M^*_{pg} = -20.45$ were derived and for those of Abell $s_1 = 1.1$, $s_2 = 0.36$. The luminosity segregation of the galaxies is observed only within a magnitude from the brightest galaxy. The ellipticities and position angles of the major axis of galaxies were determined. The observed ellipticity does not depend on the luminosity, diameter or position of galaxies in the cluster. It is shown that an alignment of the galaxy major axis is observed.

160.008   Some results of an investigation of the cluster of galaxies A 1983.
O. M. Kurtanidze, G. M. Richter.
Astrofizika, Tom 26, Vyp. 3, p. 557 – 559 (1987). In Russian. English translation in Astrophysics, Vol. 26, No. 3.

Magnitudes, ellipticities and position angles of the major axis of galaxies in the cluster A 1983 ($N = 150$, $S = 0.23$ sq. degrees) are determined. For the parameters of the Abell luminosity function, the values of $s_1 = 0.80$, $s_2 = 0.27$, $M_{pg} = -20.0$ are derived. No alignment of the galaxy's major axis in the cluster was oberved.

160.009   Cluster population incompleteness bias and the value of H_0 from the Tully–Fisher B_T^0 relation.
L. Bottinelli, P. Fouqué, L. Gouguenheim, G. Paturel, P. Teerikorpi.
Astron. Astrophys., Vol. 181, No. 1, p. 1 – 13 (1987).

The influence of the cluster population incompleteness bias, introduced by Teerikorpi (1987), on the B–band Tully–Fisher relation is investigated from the data on the Virgo cluster and 10 other more distant clusters. When cutting the samples at the apparent magnitudes below which they may be considered as being nearly complete, and using the normalization suggested in the above reference, the bias is clearly put in evidence for the combined sample. The data are well fitted by a theoretical curve, obtained for $H_0 = (72 \pm 5)$ km s^{-1}Mpc^{-1} in de Vaucouleurs local scale. It is concluded that the true value of the Hubble constant lies somewhere between 50 and 75, depending essentially on the primary calibration; the previously discussed larger disagreement seems to be due to an underestimation of the Malmquist or the cluster incompleteness biases.

160.010   Arcs, light echoes, and supergalaxies.
J. I. Katz.
Astron. Astrophys., Vol. 182, No. 1, p. L19 – L20 (1987).

The author proposes that recently discovered giant arcs in clusters of galaxies may be the light echoes of extinct quasars. This model naturally explains their circular shape and blue color, and makes clear predictions for their polarization and spectrum.

160.011   The light–echo model for luminous arcs.
M. Milgrom.
Astron. Astrophys., Vol. 182, No. 1, p. L21 – L24 (1987).

The author examines and elaborates on the light–echo model proposed by Katz (1987) for the luminous arcs in galaxy clusters. Some new results that follow mainly from the geometry of the model are deduced. The author points out that only a partial arc due to beaming of the source's radiation may be seen. Arcs can thus tell us about quasar beaming. Also, it is suggested that the width of the observed arc may result from the finite thickness of the scattering layer as well as from the finite duration of the burst. In the former case an arc with a constant width (for a constant thickness) is expected. In the latter case, the width of the image varies along the arc by a large factor. It is proposed that the light from the arcs may be due to reemission from atoms that are excited by a UV or an X–ray burst from the central source (as a possible alternative to genuine scattering). Very different predictions follow from the two alternatives.

160.012   The diffuse radio emission from the Coma cluster.
R. Schlickeiser, A. Sievers, H. Thiemann.
Astron. Astrophys., Vol. 182, No. 1, p. 21 – 35 (1987).

New measurements of the diffuse radio halo from the Coma cluster at a wavelength $\lambda 11$ cm are presented. After correcting for the contribution from point sources the authors derived the integrated diffuse radio flux density from the halo (Coma C) to $S_{2.7 GHz} = 70 \pm 20$ mJy. This value is significantly smaller than the power law extrapolation from lower frequencies, and indicates a strong steepening of the radio spectrum of Coma C at high frequencies ($\nu > 1$ GHz). The authors quantitatively compared this result with the predictions for the integrated diffuse radio flux density spectrum of the three basic models for cluster radio halo formation: the primary electron model, the secondary electron model, and the in–situ acceleration model. A remarkably good fit is provided by the in–situ acceleration model, which therefore has to be favoured over both, primary and secondary electron models.

160.013   B and V photometry of two distant galaxy clusters with 6 m telescope plates.
G. Iannicola, A. Kalloghlian (A. T. Kalloglyan), D. Nanni, A. Vignato.
Astron. Astrophys., Vol. 182, No. 2, p. 189 – 201 (1987).

B and V photometry is presented for the galaxies in the direction of the two clusters A777 and A910. The photometric data support Butcher and Oemler's suggestion that evolution of the blue galaxy population f_b occurs in compact, concentrated clusters, including an increase of f_b with increasing redshift. In both clusters, no correlation exists between $B-V$ colour and galaxy density, at variance with the correlation found for nearby clusters. An analysis of the apparent positions of the galaxies shows strong subclustering in both clusters. A significant correlation between luminosity and projected galaxy density is present for A777, while no correlation is found for A910; this different behaviour of the two clusters may indicate that the relaxation process is more complete in A777 than in A910, A777 having a higher value of galaxy density than A910.

160.014   Correlation between the radio power and the X–ray luminosity for rich clusters of galaxies.
A. G. Gubanov.
Astrophysics, Vol. 25, No. 3, p. 689 – 695 (1987). English translation of 42.160.095.

160.015   Globular clusters belonging to galaxy clusters.
R. E. White III.
Mon. Not. R. Astron. Soc., Vol. 227, No. 1, p. 185 – 195 (1987).

Merritt suggests that cD envelopes and the diffuse light in the Coma galaxy cluster have the same origin: they are composed of stars tidally stripped from galaxies during galaxy cluster collapse. If this account is correct, globular clusters should be stripped as well, creating a galaxy cluster–wide population of globulars. The anomalously large globular cluster populations around M87/

Virgo, NGC 3311/A1060/Hydra I, NGC 1399/Fornax I and NGC 4874/A1656/Coma may be examples of such systems.

160.016   X-ray observations of the Ophiuchus, PKS 0745–191 and Cygnus–A clusters of galaxies.
K. A. Arnaud, R. M. Johnstone, A. C. Fabian, C. S. Crawford, P. E. J. Nulsen, R. A. Shafer, R. F. Mushotzky.
Mon. Not. R. Astron. Soc., Vol. 227, No. 1, p. 241 – 256 (1987).

The authors report observations of the X-ray emission from three bright clusters of galaxies at low Galactic latitudes; the Ophiuchus, PKS 0745–191 and Cygnus–A clusters. Temperatures and iron abundances are determined from EXOSAT spectral data. Images of the Ophiuchus cluster show it to have a high central density and a cooling flow. The authors confirm measurements of a high gas density around PKS 0745–191 and show that the X-ray emission is centred on the radio galaxy. The PKS 0745–191 cluster is the most distant object for which an abundance has been measured from X-ray spectra. The spectrum of Cygnus–A contains a highly absorbed power–law component that the authors identify with the nucleus of the radio source.

160.017   Studies of the Virgo cluster. VI. Morphological and kinematical structure of the Virgo cluster.
B. Binggeli, G. A. Tammann, A. Sandage.
Astron. J., Vol. 94, No. 2, p. 251 – 277 (1987).

The structure of the Virgo cluster is analyzed on the basis of the positions, Hubble types, and radial velocities of the 1277 Virgo cluster galaxies listed in Paper II (Binggeli, Sandage, Tammann (1985)). The surface distribution of galaxies is considered according to type, and is discussed using maps, isopleths, strip counts, and radial–density distributions. The principal results and their interpretation are presented.

160.018   A study of the elongation of Abell clusters. I. A sample of 37 clusters studied earlier by Binggeli and Struble & Peebles.
G. F. R. N. Rhee, P. Katgert.
Astron. Astrophys., Vol. 183, No. 2, p. 217 – 227 (1987).

The authors present measurements of the elongation of the projected galaxy distributions in 37 Abell clusters, all with redshifts less than or equal to 0.1. The measurements are based on the brightest 100 galaxies in each cluster. These are found with a semiautomated procedure which selects the n brightest, clearly non–stellar objects on digitized images of Palomar Sky Survey plates of the cluster areas. The amplitude and position angle of the elongation signal are found by Fourier analysis of the azimuthal distribution of the cluster galaxies. The authors also determine the statistical uncertainties of these parameters. Comparison with earlier determinations of the elongation of these same clusters shows that the differences between independent position angle estimates increase with decreasing elongation amplitude. This explains most if not all of the previously reported apparent discrepancies between independent measurements.

160.019   Observations of novae in the Virgo cluster.
C. J. Pritchet, S. van den Bergh.
Astrophys. J., Vol. 318, No. 2, p. 507 – 519 (1987).

During 15 nights of observations in 1986 March and April the authors have detected nine novae in Virgo cluster elliptical galaxies. From six novae with reasonably well–observed light curves it is found that the distance modulus of the Virgo cluster is 6.8 ± 0.4 (estimated error) larger than that of M31. Using a distance modulus of 24.65 ± 0.15 for M31 then gives a Virgo cluster distance of $D = 19.5 \pm 3.9$ Mpc. With a cosmological redshift of 1336 ± 54 km s^{-1} for the Virgo cluster this yields a value $H_0 = 69 \pm 14$ km s^{-1} Mpc^{-1} for the Hubble parameter.

160.020   Star formation in X-ray cluster cooling flows.
R. E. White III, C. L. Sarazin.
Astrophys. J., Vol. 318, No. 2, p. 612 – 620 (1987).

At the center of each observed X-ray cluster cooling flow is a central dominant galaxy accreting up to several hundred solar masses per year from the flow. The authors consider whether these accreting galaxies are formed, substantially or in part, by

ongoing star formation in their associated cooling flows. They derive the basic equations relevant to cooling flows, including the effects of star formation, and develop a local approximation for the star formation rate based on a detailed thermal instability analysis. This prescription for the star formation rate allows analytic solutions to be found for both isobaric and gravity-dominated cooling flows.

160.021   Determining star formation rates in X-ray cluster cooling flows.
R. E. White III, C. L. Sarazin.
Astrophys. J., Vol. 318, No. 2, p. 621 – 628 (1987).

Many X-ray clusters of galaxies are observed to have cooling flows at their centers. It seems possible that these accreting galaxies are still being formed through ongoing star formation in their associated cooling flows. In this paper, the authors develop techniques to determine directly the distributions of local star formation rate, mass, gas density, temperature, and velocity from cooling flow X-ray surface brightness data. These techniques take account of the potentially important X-ray emission from star–forming cooling condensations dropping out of the background flow.

160.022   Numerical models of star formation in X-ray cluster cooling flows.
R. E. White III, C. L. Sarazin.
Astrophys. J., Vol. 318, No. 2, p. 629 – 644 (1987).

At the center of each observed X-ray cluster cooling flow is a central dominant galaxy accreting up to several hundred solar masses per year from the flow. Here, the authors calculate theoretical models of such flows, including the effects of ongoing star formation. They relate the local star formation rate in these models to either the local cooling rate of the gas or the local growth rate of thermal instabilities. It is shown how the structure of cooling flows is affected by variations in the star formation rate, as well as by variations in intracluster temperature, overall accretion rate, elemental abundances, and the form of the gravitational potential. The authors also calculate the X-ray emission from these models.

160.023   The structure of brightest cluster members. II. Mergers.
J. M. Schombert.
Astrophys. J., Suppl. Ser., Vol. 64, No. 4, p. 643 – 666 (1987).

Surface photometry of 342 bright elliptical galaxies in 103 clusters is analyzed for evidence of mergers. Structural differences between brightest cluster members (BCMs) and normal ellipticals can be summarized as having enlarged characteristic radii and shallow profile slopes ($\beta > -1.7$). Profile morphology criteria for the elliptical types gE, D, and cD are outlined. Comparison of observations with numerical simulations of mergers strongly suggests a past history of dynamical growth for BCMs. Weak correlations of global cluster properties to BCMs supports the hypothesis proposed by Merritt that mergers are important in early subgroups before virialization and the formation of a cluster identity.

160.024   Radio emission in clusters: problems and opportunities.
L. Rudnick.
NRAO Workshop, No. 16, p. 1 – 8 (1986). – See Abstr. 012.015.

The intent of this paper is to provide some common language and focus for discussions of radio emission in clusters of galaxies. Problems associated with the interpretation of radio sources, in general, can benefit from a better understanding of how cluster sources are affected by the intracluster environment. Similarly, radio sources can serve as a critical probe of the dynamics and evolution of the cluster itself, and of the intracluster medium. The author discusses some of the critical meeting points between these different approaches, and highlight some of the issues which limit the current understanding.

160.025 Galaxies in rich clusters – development of a local perspective.
T. C. Beers.
NRAO Workshop, No. 16, p. 9 – 22 (1986). – See Abstr. 012.015.

A significant number of rich clusters have been shown to be comprised of two or more individual clumps of galaxies – or subclusters – within the general cluster distribution. The existence of such structure has profound consequences for the measurement of intrinsic cluster properties as well as for understanding the formation and evolution of the cluster. The author reviews the evidence for substructure and argues that it must be a common phenomenon in rich clusters. The importance of basing future analyses of rich clusters on local, rather than global, measurements is emphasized.

160.026 X–ray observations of clusters: physical implications.
C. L. Sarazin.
NRAO Workshop, No. 16, p. 23 – 36 (1986). – See Abstr. 012.015.

Clusters of galaxies are powerful X–ray sources. The X–ray emission is due to diffuse hot gas ($T \sim 10^8$K and $n \sim 10^{-3}$cm^{-3}). The author reviews the X–ray properties of clusters of galaxies and of elliptical galaxies. The role of cooling flows is emphasized. These flows may have a significant effect on the structure and energetics of radio sources. More detailed reviews of X–ray clusters and of cooling flows are given in Sarazin (1986) and in Fabian et al. (1984).

160.027 Statistical results from the Effelsberg Abell cluster radio survey.
H. Andernach, A. Sievers, H.–P. Reuter.
NRAO Workshop, No. 16, p. 51 – 57 (1986). – See Abstr. 012.015.

Results of a radio survey of 75 Abell clusters obtained with the Effelsberg 100 m radio telescope of the MPIfR Bonn are presented. The literature was used to derive spectral indices for 311 radio sources and these were analyzed for possible differences between cluster and non–cluster sources. The trend for steeper spectral indices closer to the cluster centres is much less pronounced than reported earlier for a low frequency selected source sample. Trying to separate cluster sources on the basis of optical identification did not improve the correlation. Unidentified sources are found to have both steeper spectra and lower flux density. They are likely to be background radio galaxies, but the presence of a "relic" population of unidentified cluster radio sources cannot be excluded.

160.028 Some results from the Molonglo cluster survey.
J. E. Reynolds.
NRAO Workshop, No. 16, p. 59 – 64 (1986). – See Abstr. 012.015.

A representative sample of southern Abell clusters out to distance D = 4 has been studied with the 408 MHz Molonglo Cross and, more recently, with the Molonglo Observatory Synthesis Telescope (MOST) at 843 MHz. Two results are presented here. It is found that the brightest few elliptical/lenticular galaxies in Abell clusters form moderately powerful radio sources at ~ 3 times the rate of comparably bright galaxies outside Abell clusters. The radio spectral–index distribution of cluster sources is also discussed. It is argued that a significant proportion of unidentified steep–spectrum sources in the direction of clusters are in fact cluster–related.

160.029 VLA observations of distant clusters of galaxies.
M. P. Ulmer.
NRAO Workshop, No. 16, p. 65 – 71 (1986). – See Abstr. 012.015.

Results of a 20 cm VLA survey by Hanish and Ulmer of 18 distant class 5 and 6 Abell clusters is summarized. In addition to enhancing the understanding of the general properties of cluster radio sources, the survey provides a probe of the intra–cluster medium. Both applications of the survey data are discussed.

160.030 OSRT 327 MHz observations of clusters of galaxies: some very steep spectrum sources.
M. N. Joshi, V. K. Kapahi, J. Bagchi.
NRAO Workshop, No. 16, p. 73 – 77 (1986). – See Abstr. 012.015.

The OSRT 327 MHz maps are presented for eight clusters of galaxies containing very steep spectrum (VSS) radio sources. The VSS sources are located close to the cluster centres and for majority the spectral indices are steeper than ~ 2.0. They are very old radio remnants confined by the pressure due to the hot gas known to be present in the cluster medium and their steep spectra arise due to synchrotron losses. For a number of them the radio–optical separation is too large to be explained by the movement of the parent galaxy.

160.031 EXOSAT observations of the Virgo cluster.
A. C. Edge, G. C. Stewart, A. Smith.
NRAO Workshop, No. 16, p. 105 – 111 (1986). – See Abstr. 012.015.

Results are presented for observations of the Virgo cluster made using the EXOSAT Medium Energy Experiment. Limits are obtained on the temperature profile of the gas within 100 arcmin. These results provide an improved estimate for the mass surrounding M87.

160.032 The relationship of optical emission line gas and extended radio emission in the centers of cooling flow clusters.
S. Baum, T. Heckman.
NRAO Workshop, No. 16, p. 119 – 126 (1986). – See Abstr. 012.015.

The authors examine the radio morphology and the relationship of that morphology to the presence of optical emission line filaments in six radio sources associated with central dominant galaxies in the centers of "cooling flow" clusters (Abell 2052, Abell 262, Abell 1795, Perseus, Virgo, and 3C 295). In some of the sources there appears to be a direct physical connection between the extended radio emission and the optical emission line gas, whereas in others the emission line gas seems to share the overall extent and morphology of the radio source but not to be easily associated with specific radio features. The authors discuss several ways in which the interaction of the radio source with its environment might lead to the formation of optical emission line filaments. They also note that all of the sources have inversion symmetric radio morphologies and suggest that either interaction with a rotating ICM or precession of the central engine may be responsible.

160.033 Radio properties of central dominant galaxies in cluster cooling flows.
C. P. O'Dea, S. A. Baum.
NRAO Workshop, No. 16, p. 141 – 146 (1986). – See Abstr. 012.015.

The authors combine new VLA observations of central dominant (cd) galaxies currently thought to be in cluster cooling flows with observations from the literature to examine the global properties of a heterogeneous sample of 31 cd galaxies. The radio sources tend to be of low or intermediate radio power and have small sizes (median extent ~ 25 kpc). The resolved sources tend to have distorted morphologies (e.g., wide–angle tails, 'S' shapes). It is not yet clear whether the radio emission from these cd galaxies is significantly different from those not thought to be in cluster cooling flows. The authors confirm the result of Jones and Forman that there is a possible correlation between radio power and excess X–ray luminosity in the cluster center (above a King model fit to the X–ray surface brightness).

160.034 The radio sources in the centers of the Coma and Hercules clusters.
L. Feretti.
NRAO Workshop, No. 16, p. 161 – 168 (1986). – See Abstr. 012.015.

The radio galaxies NGC 4874 and NGC 6047, respectively in the centers of the rich clusters of galaxies Coma (A 1656) and

Hercules (A 2151), were observed with the VLA with high resolution and sensitivity as part of a project to study nearby elliptical galaxies. Due to the small distance of these sources, it is possible to derive structure information on a small linear scale (< 1 kpc) with much higher sensitivity than it is reached with the VLBI observations. The present sources have linear sizes smaller than the optical galaxy diameter and are therefore imbedded within the optical galaxy boundary. The structure of these sources is not different from that of much more extended sources. A Hubble constant H = 100 km/s Mpc is used, which leads to linear conversions of 0.3 kpc/arcsec and 0.5 kpc/arcsec, for Coma and Hercules, respectively.

160.035   Radio sources associated with 5 first ranked galaxies.
        G. Giovannini, L. Feretti, L. Gregorini.
NRAO Workshop, No. 16, p. 169 – 173 (1986). – See Abstr. 012.015.
The radio sources associated with the first ranked galaxies in A 98, A 115, A 160, A 278 and A 568 have been observed with the VLA and WSRT at 20 and 6 cm (Giovannini et al. in preparation), as part of a project to study with high sensitivity and resolution cluster radio sources with extended structure. Particular importance has the radio study of the cluster first ranked galaxies, because of their characteristics: they are located in the cluster center and are the most massive and brightest galaxies of the cluster. The authors summarize here the morphological properties of these sources. They use a Hubble constant $H_0 = 100$ km/s Mpc.

160.036   Radio halo sources in clusters of galaxies.
        R. J. Hanisch.
NRAO Workshop, No. 16, p. 191 – 198 (1986). – See Abstr. 012.015.
Radio halo sources remain one of the most enigmatic of all phenomena related to radio emission from galaxies in clusters. The morphology, extent, and spectral structure of these sources are not well known, and the models proposed to explain them suffer from this lack of observational detail. However, recent observations suggest that radio halo sources may be a composite of relic radio galaxies. The validity of this model could be tested using current and planned high resolution, low–frequency radio telescopes.

160.037   The radio halo and magnetic field in the Coma cluster of
        galaxies.
K.–T. Kim, P. P. Kronberg, P. E. Dewdney, T. L. Landecker.
NRAO Workshop, No. 16, p. 199 – 206 (1986). – See Abstr. 012.015.
The radio halo in the Coma cluster has been re–observed with the DRAO and VLA and the combined data set confirms the existence of a steep spectrum halo source. The spectral index and size of the radio halo are discussed and its morphology is related to the available X–ray data. The equipartition magnetic field of the halo is estimated to be $0.7 - 0.8 (1+k)^{2/7}$ microgauss. The possible origin of the radio halo as a relic of past head–tail galaxies is discussed. Polarization observations of the halo resulted in a null detection at the 1σ level which sets an upper limit to the degree of polarization at $30\% \pm 10\%$. Rotation measures of the sources in the Coma field effectively rule out a uniform field component greater than 5×10^{-8} gauss.

160.038   Acceleration of ultra–high–energy cosmic rays
        (UHECR) in clusters of galaxies.
A. Ferrari.
NRAO Workshop, No. 16, p. 207 – 213 (1986). – See Abstr. 012.015.
The acceleration of UHECR in a turbulent intracluster medium is investigated. An origin of UHECR connected with clusters agrees in fact with the present informations about the energy spectrum of cosmic rays, the anisotropies of their arrival directions and their chemical composition. A model is presented in which the upper limit of the cosmic ray energy spectrum is defined by the balance between the energy densities in large scale turbulent MHD modes in ICM which are responsible for

acceleration and in the cosmic background photons which give rise to photomeson losses. The energy density in turbulent modes is derived from the corresponding acceleration of relativistic elctrons whose synchrotron emission accounts for the observation of wakes of radiogalaxies with bent jets and halos not directly connected with active galaxies.

160.039   Radio observations of the peripheral region of the Coma
        cluster near Coma A.
G. Giovannini.
NRAO Workshop, No. 16, p. 215 – 221 (1986). – See Abstr. 012.015.
The extended radio source $1253 + 275$ is located in a peripheral region of the Coma cluster ($\sim 70'$ from the cluster center), near the background radio galaxy Coma A (3C 277.3). The source consists of two very elongated lobes with a total extension of about 25'. No obvious identification of this extended radio source with Coma cluster galaxies or other bright objects is possible. In order to understand the ambient conditions, which could lead to the formation of such unusual source, the author has also studied two Coma radio galaxies: NGC 4789 and NGC 4827, which lie at ≤ 20' from $1253 + 275$. A Hubble constant $H_0 = 75$ km/sec Mpc is used in this paper.

160.040   Radio, optical and X–ray observations of PKS 2104–25.
        R. A. Cameron, G. V. Bicknell, R. D. Ekers,
D. Carter.
NRAO Workshop, No. 16, p. 223 – 229 (1986). – See Abstr. 012.015.
The authors present detailed radio and optical observations of the extragalactic radio source complex PKS 2104–25. This complex contains two separate radio jet sources associated with dominant galaxies in a poor cluster of galaxies. High resolution, multi–frequency VLA observations provide images of the radio structure and spectral index distributions within the radio structure. Spectra of a large number of cluster galaxies, together with automated astrometry of the cluster and available X–ray data are used to model the large–scale cluster mass distribution. Pressure confinement of the radio sources by the surrounding cluster medium, in combination with spectral index gradients observed in the radio sources may then be used to examine buoyancy forces acting on the radio source structures.

160.041   Beam trajectories in the intracluster medium.
        L. Zaninetti, H. Van Horn.
NRAO Workshop, No. 16, p. 231 – 237 (1986). – See Abstr. 012.015.
The authors show that a simple kinematic model, which combines ram pressure bending or variable jet velocity with precession and variable orientation of the observer relative to the trajectory of a galaxy through the intracluster medium, can yield a wide range of morphological types of twin–jet systems. In particular the transition from "C"–shaped sources to "S"–shaped sources can be understood as a consequence of different parameter choices within this simple model.

160.042   Diffuse infrared emission from rich clusters: constraints
        on dark matter.
S. P. Boughn, J. M. Uson.
NRAO Workshop, No. 16, p. 239 – 246 (1986). – See Abstr. 012.015.
Preliminary results of large (1 arcmin), single aperture RVJK photometry on and near the centers of distant ($0.1 < z < 0.2$) rich Abell clusters are reported. When K–corrected, the colors of the integrated light of the cores of these clusters are consistent with the colors of nearby E and S0 galaxies. The absence of anomalous infrared light can be used to constrain any stellar component of the dark matter in these clusters. In particular, if the dark matter is composed of low mass red or brown dwarfs, more than 90% must be objects with mass less than 0.1 $M_\odot$. This implies a K band (2.2 micron) mass–to–light ratio of the dark matter in excess of 200. Constraints on a mass function for these objects are discussed.

160.043   X–ray observations of distant blue clusters of galaxies.
S. M. Lea, J. P. Henry.
NRAO Workshop, No. 16, p. 247 – 254 (1986). – See Abstr. 012.015.

The authors report X–ray observations of a sample of distant clusters for which the percentage of apparent cluster members that are blue has been determined optically. The X–ray properties of the clusters are not simply related to the optical properties. The X–ray observations do not support models in which blue galaxies change color as a result of ram pressure stripping, or in which star formation is triggered by interaction with an intracluster medium. The authors' data lend some support to the idea that we are observing blue galaxies in distant clusters immediately after the removal of galactic halos, and before the interstellar medium is depleted (Larson, Tinsley and Caldwell 1980).

160.044   The Sunyaev–Zel'dovich effect: measurements and implications.
J. M. Uson.
NRAO Workshop, No. 16, p. 255 – 260 (1986). – See Abstr. 012.015.

The Sunyaev–Zel'dovich effect results from inverse Compton scattering of the photons in the microwave background radiation by hot ionized gas in clusters of galaxies: The author discusses some of the experimental problems involved in measuring such a small ($\Delta T \sim -1mK$) effect. Preliminary results obtained using the NRAO 140–foot telescope show detections of the effect towards the clusters $0016 + 16$, Abell 401 and Abell 665. The lack of reliable estimates of the temperature of the gas in these clusters precludes a good estimate of Hubble's constant from these measurements. Alternatively, assuming a cosmological model, the measurements constrain the parameters of the hot gas in these clusters.

160.045   New results on the Sunyaev–Zel'dovich effect in $0016 + 16$, Abell 665 and Abell 2218.
M. Birkinshaw.
NRAO Workshop, No. 16, p. 261 – 268 (1986). – See Abstr. 012.015.

The OVRO 40–m telescope has been used at 20.3 GHz to measure the Sunyaev–Zel'dovich effects in $0016 + 16$, Abell 665 and Abell 2218. Corrections to the data have been made for the presence of radio sources near the clusters, and the preliminary results exhibit strong evidence for extended Sunyaev–Zel'dovich effects in each cluster. The temperature of the intracluster medium in these clusters is inferred to be high (> 10 keV), but no useful result for the Hubble constant can be obtained because of the lack of X–ray spectral data.

160.046   Galaxies as dynamical probes of cluster structure.
R. H. Miller.
NRAO Workshop, No. 16, p. 269 – 274 (1986). – See Abstr. 012.015.

Observable properties of galaxy clusters are largely determined by the gravitational potential in which the galaxies move. Since galaxies are flimsy, flabby objects that can easily be distorted by the cluster tidal potential, they make good probes to study the shape of the cluster potential. Numerical experiments show the kinds of observable effects cluster tidal forces can produce in a galaxy. These are used in conjunction with the observations to discuss constraints on the cluster potential, where dark mass is located, and other such features.

160.047   Narrow–angle tail radio sources and evidence for radial orbits in Abell clusters.
C. P. O'Dea, C. L. Sarazin, F. N. Owen.
NRAO Workshop, No. 16, p. 275 – 281 (1986). – See Abstr. 012.015.

The distribution of the orientation of the tails of narrow–angle tail (NAT) radio sources can be used to constrain the distribution of galaxy orbits in clusters. In this paper, the authors examine data on the orientations of the tails with respect to the cluster centers of a sample of 60 NATs in Abell clusters. They consider the whole sample as well as subsamples of sources based on

projected distance from the cluster center and cluster morphology.

160.048   Can cD galaxies be formed in clusters with anisotropic distribution functions?
E. M. Malumuth.
NRAO Workshop, No. 16, p. 283 – 289 (1986). – See Abstr. 012.015.

A computer code previously used to model the evolution of clusters of galaxies has been modified to allow for anisotropic cluster models. The author finds that the chances of forming a cD galaxy by mergers is about the same in clusters with a radial distribution of orbits as it is in isotropic clusters although the details are somewhat different.

160.049   Dark matter in clusters?
J. R. Kuhn.
NRAO Workshop, No. 16, p. 291 – 297 (1986). – See Abstr. 012.015.

The missing dynamical mass in clusters is still missing (i.e. undetected by nondynamical means). The author shows how a simple relationship accounts for the required dynamical mass in terms of one parameter and only the visible mass distribution. The simplicity of the model argues for its physical interest and the single parametrization may be a clue to the form of the dark matter (henceforth denoted DM) constituents. It is also interesting that the effect of DM in clusters is in detail formally equivalent to adding another long range force that couples to visible mass – although the model provides, equivalently, a statement about the distribution of the DM. It is formulated below in terms of only the visible mass density of a cluster and an additional "effective" long range interaction.

160.050   The H I environment of high redshift quasars: a VLA search for cosmological H I.
E. Hardy, L. Noreau.
NRAO Workshop, No. 16, p. 299 – 304 (1986). – See Abstr. 012.015.

The authors present the first results of a search for H I at a redshift of $z = 3.3$ conducted with the NRAO VLA fitted with the P–band system. They discuss the technical problems encountered in their first attempt and some of the improvements which are possible. Their negative results are compatible with amounts of H I in protoclusters or pancakes less than a few times $10^{14} M_\odot$. Some suggestions for observational strategies are included.

160.051   A cautionary note on cluster radio sources.
L. Rudnick, M. Birkinshaw.
NRAO Workshop, No. 16, p. 339 (1986). – See Abstr. 012.015.

160.052   A comment on galaxy orbits.
R. H. Miller.
NRAO Workshop, No. 16, p. 341 (1986). – See Abstr. 012.015.

160.053   Linear clusters of galaxies: A 999 and A 1016.
G. N. F. Chapman, M. J. Geller, J. P. Huchra.
Astron. J., Vol. 94, No. 3, p. 571 – 586 (1987).

The authors have measured 44 new redshifts in A 999 and 40 in A 1016: these clusters are both "linear" according to Rood and Sastry (1971) and Struble and Rood (1982, 1984). With 20 cluster members in A 999 and 22 in A 1016, the authors can estimate the probability that these clusters are actually drawn from spherically symmetric distributions. By comparing the clusters with Monte Carlo King models, they find that A 999 is probably intrinsically spherically symmetric, but A 1016 is probably linear. The authors estimate that $\gtrsim 2\%$ of a catalog of spherically symmetric clusters might be erroneously classified as linear. They use the data to estimate the virial masses for these systems. The authors reassess the cluster–galaxy alignment analysis of Adams, Strom, and Strom (1980) and examine the relationship between the luminosity and morphological type of the cluster members and the cluster itself.

160.054 A VLA 20 cm survey of poor groups of galaxies.
J. O. Burns, R. J. Hanisch, R. A. White,
E. R. Nelson, K. A. Morrisette, J. W. Moody.
Astron. J., Vol. 94, No. 3, p. 587 – 617 (1987). With plates 54 – 55.

The authors report on VLA 20 cm observations of an extensive sample of galaxies in 139 poor groups. They surveyed these groups using a "snapshot" mode of the VLA with a resolution of about 13″. Analysis of the resulting radio and optical properties leads the authors to conclude that the presence of a nearby companion galaxy has an important role in generating radio emission in a galaxy. CCD observations of two radio–loud, disturbed galaxies with companions are presented and are used to discuss models of radio–source production. The authors also find nine tailed radio galaxies in the poor groups, much more than had been expected from previous work on rich clusters and from theoretical models. They discuss previous statistical biases and propose a method for bending head–tail sources in poor groups. The authors predict the presence of a substantial intracluster medium that should radiate significantly at soft X–ray energies. They also discuss galaxy type and galaxy rank within the poor groups as factors influencing the radio activity.

160.055 The alignment of galaxy clusters.
P. Flin.
Mon. Not. R. Astron. Soc., Vol. 228, No. 4, p. 941 – 948 (1987).

The large–scale alignment of galaxy clusters found by Binggeli (1982) was tested on the basis of the complete sample of clusters which belong to the Bahcall and Soneira (1984) superclusters using various determinations of the clusters' position angle. For small separations, clusters do tend to point toward their neighbours, but the effect is weak. Correlation of anisotropy with cluster separation is also weak. Comparison of the observational data with theoretical models shows that for small separations the alignment is similar to that predicted in the adiabatic scenario, while for greater distances between clusters to that in the isothermal one.

160.056 Mass deposition in cooling flows – analysis of the X–ray data.
P. A. Thomas, A. C. Fabian, P. E. J. Nulsen.
Mon. Not. R. Astron. Soc., Vol. 228, No. 4, p. 973 – 991 (1987).

The authors present a "multiphase method" for the analysis of X–ray data of the cooling inflow region of intracluster gas in clusters of galaxies. The method employs a range of densities at each radius and treats the mass–deposition from the flow in a self–consistent way. The authors present mass–deposition profiles, $\dot{M}(<r)$, of those clusters for which their method is applicable. These rise linearly with radius and reach values of up to 100 $M_\odot$ yr^{-1} within 100 kpc.

160.057 Cooling flows and AXAF.
A. C. Fabian.
Astrophys. Lett. Commun., Vol. 26, Nos. 1 – 2, p. 147 – 151 (1987). – See Abstr. 012.020.

The radiative cooling of X–ray emitting gas in clusters of galaxies and in early–type galaxies is relatively common in the Universe. The pressure of surrounding gas sets up a cooling flow. Substantial amounts of cold matter are deposited by this process in a manner that resembles the mass distribution of a galaxy. X–ray observations of high spatial and high spectral resolution are needed to further this work. AXAF is ideally matched to this task.

160.058 Spartan 1 X–ray observations of the Perseus cluster: comparison of the iron abundances and temperatures in the inner and outer regions of the cluster.
M. P. Ulmer, R. G. Cruddace, E. E. Fenimore, G. G. Fritz, W. A. Snyder.
Astrophys. J., Vol. 319, No. 1, p. 118 – 125 (1987).

X–ray observations of the Perseus cluster have been made by Spartan 1 with the objective of resolving both the spatial structure and the spectral characteristics in the 1 – 10 keV band. Spectral fits have been made to the data obtained from two annular regions centered on NGC 1275. The central region was chosen to have a radius of 5′, while the outer region was chosen to have an inner radius of 6′ and an outer radius of 20′. The inner region was found to be significantly cooler than the outer region, and the best fit temperatures were 4.2×10^7K and 7.1×10^7K respectively. The best–fit iron abundances, expressed as a fraction of the solar abundance, were found to be 0.41 for the outer region and 0.81 for the inner region. These results support the hypothesis that a large fraction of the intracluster gas is the product of stellar nucleosynthesis.

160.059 The correlation function of galaxies in the direction of the Coma cluster.
D. Calzetti, J. Einasto, M. Giavalisco, R. Ruffini, E. Saar.
Astrophys. Space Sci., Vol. 137, No. 1, p. 101 – 106 (1987).

The observational data on the amplitude of the correlation function of galaxies in the direction of the Coma cluster are confronted with an analytic formula derived for a self–similar observer–homogeneous structure.

160.060 Further data on the blue ring–like structure in A 370.
G. Soucail, Y. Mellier, B. Fort, F. Hammer, G. Mathez.
Astron. Astrophys., Vol. 184, No. 1/2, p. L7 – L9 (1987).

The authors present the latest data collected in November 1986 on the very blue giant ring–like structure recently discovered in the center of the cluster Abell 370 (z = 0.374). The spectrum of the eastern end of the structure is analyzed in details: it does not show any of the strong emission lines characterizing a QSO, and all the typical features expected in a gas or in a galaxy at the cluster redshift are missing. Such a result seems to rule out several models involving in–situ star formation. Moreover, the large scale spectral energy distribution looks like the continuum of a spiral galaxy redshifted to z = 0.59. So, the interpretation in terms of gravitational lensing is proposed. The results of a multipoint mass model allows one to reconstruct the entire ring–like structure. However, some properties remain difficult to understand.

160.061 The Perseus supercluster at low galactic latitudes.
M. Hauschildt.
Astron. Astrophys., Vol. 184, No. 1/2, p. 43 – 56 (1987).

H I–line observations of 65 galaxies at low galactic latitudes northeast of the Perseus supercluster are presented. 27 galaxies have been detected, raising the number of radial velocities known for this area to 45. The detected galaxies are normal in all respects. The new radial velocities present evidence that the Perseus supercluster is extended across at least a part of the galactic zone of avoidance. Taking these data together with data from the literature, the Perseus supercluster can be traced from the cluster Abell 2634 at $\alpha = 23^h36^m$, $\delta = 26°46'$ to the 3C 129 cluster at $\alpha = 4^h45^m$, $\delta = 45°0'$. The ends are probably due to missing data. The author studies the distribution of galaxies in this region in Cartesian coordinates. The study shows that this supercluster is a filamentary object which is plane at the same time. It is bent in this plane but does not have any points of inflection.

160.062 The luminosity function: dependence on Hubble type and environment.
B. Binggeli.
Nearly normal galaxies. From the Planck time to the present, p. 195 – 206 (1987). – See Abstr. 012.031.

The LF type dependence for Virgo cluster galaxies is summarized. The dependence on environment is less well–known. This problem is tackled by first varying the LFs within the Virgo cluster as a function of local density, and then by comparing the Virgo LFs with those of extreme environments: the core of the Coma cluster (high density), and nearby groups (low density). The hypothesis is put forward that the generalized LF can be separated into a dependence on type and a dependence on environment, at least as a first order approximation. If true, this is an important constraint on models of galaxy formation and evolution. Finally, LFs for spheroids and for disks are shown.

160.063 Evolution of cluster galaxies since z = 1.
A. Dressler.
Nearly normal galaxies. From the Planck time to the present, p. 276 – 289 (1987). – See Abstr. 012.031.

Cluster galaxies do evolve. There is a higher fraction of active galaxies in clusters at $z \gtrsim 0.4$, although these clusters are still dominated by passive galaxies. Judging from the amplitude of the 4000 Å break, even the passive galaxies show evidence for evolution by z = 0.75. Cluster–to–cluster variations in the types of active galaxies may be significant, but could be only an accident of the epoch of observation of a population that is in a state of flux.

160.064 Dark matter in binary galaxies and small groups.
V. Trimble.
Nearly normal galaxies. From the Planck time to the present, p. 313 – 316 (1987). – See Abstr. 012.031.

The main indications of dark matter in binary galaxies and small groups come from measurements of masses implying M/L ratios larger than those (typically 5 – 20 in solar units) found for single galaxies. Methods of measuring these masses and uncertainties inherent in them are reviewed.

160.065 Star formation in cooling flows.
R. W. O'Connell.
Structure and dynamics of elliptical galaxies, p. 167 – 177 (1987). – See Abstr. 012.032 (IAU Symp. No. 127).

Star formation, probably with an abnormal initial mass function, represents the most plausible sink for the large amounts of material being accreted by cD galaxies from cooling flows. There are three prominent cases (NGC 1275, PKS 0745–191, and Abell 1795) where cooling flows have apparently induced unusual stellar populations. Recent studies show that about 50% of other accreting cD's have significant ultraviolet excesses. It therefore appears that detectable accretion populations are frequently associated with cooling flows. The questions of the form of the IMF, the fraction of the flow forming stars, and the lifetime of the flow remain open.

160.066 Dwarf galaxies in the Fornax cluster.
N. Caldwell, G. Bothun.
Structure and dynamics of elliptical galaxies, p. 455 – 456 (1987). – See Abstr. 012.032 (IAU Symp. No. 127).

The authors present the results of an observational study of the dwarf galaxies in the Fornax cluster of galaxies.

160.067 An investigation of the radial dependence of the galaxy luminosity function in Abell clusters.
P. M. Lugger.
Structure and dynamics of elliptical galaxies, p. 459 – 460 (1987). – See Abstr. 012.032 (IAU Symp. No. 127).

The luminosity functions of inner and outer regions of six Abell clusters (A 569, A 1656, A 2147, A 2151, A 2199, and A 2634) were compared. These clusters have a single, reasonably symmetric central concentration of galaxies within the central Mpc. For three other clusters with irregular spatial distributions of galaxies (A 779, A 1367, and A 2197) luminosity functions for high and low density regions were compared. For three of the clusters in the first group (A 1656, A 2147, and A 2199) there is a deficit of bright galaxies, according to the Kolmogorov–Smirnov and Wilcoxon rank–sum nonparametric tests, in a region of radius 0.5 Mpc about the cluster center compared to a concentric annular region with bounds of 0.5 and 1.0 Mpc.

160.068 The surface–brightness–effective–size relation for elliptical galaxies in the cores of clusters.
J. G. Hoessel, W. R. Oegerle, D. P. Schneider.
Astron. J., Vol. 94, No. 5, p. 1111 – 1115 (1987).

Surface photometry of 372 elliptical galaxies has been performed using CCD images of the centers of 97 nearby rich Abell clusters. The strong correlation between surface brightness and effective size, originally found by Kormendy (1977), is clear in the data. Brightest cluster galaxies show much less scatter about the mean relation defined by these data than do lower–luminosity

cluster ellipticals, and the slope of the relation is shallower for the brightest galaxies. When combined with published central velocity dispersions, this photometry yields an $R_e-\sigma-<\mu_e>$ relation for brightest cluster galaxies that is in good agreement with the mean relation for elliptical galaxies found by Djorgovski and Davis (1987).

160.069 Dwarf elliptical galaxies in the Fornax cluster. I. A catalog and luminosity function.
N. Caldwell.
Astron. J., Vol. 94, No. 5, p. 1116 – 1125 (1987).

A catalog of 145 dwarf elliptical galaxies in the Fornax cluster is presented which is believed to be complete to a blue magnitude of 18.5, although galaxies with luminosities 2 mag below that limit have been found. The dwarf galaxies are less centrally concentrated in the cluster than are the bright galaxies. The ratio of the number of dwarf elliptical to elliptical galaxies is significantly smaller than the ratio found in the Virgo cluster. Interestingly, the ratio of dwarf ellipticals to actively star–forming galaxies (spirals and irregulars) in Fornax is similar to that in Virgo. The form of the velocity distribution of the dE's and star–forming galaxies is non–Gaussian, whereas that for the E/S0 galaxies is Gaussian. The velocity dispersion of the E/S0 group is low, at 300 km s^{-1}.

160.070 Dwarf elliptical galaxies in the Fornax cluster. II. Their structure and stellar populations.
N. Caldwell, G. D. Bothun.
Astron. J., Vol. 94, No. 5, p. 1126 – 1142 (1987). With plates 82 – 96.

The authors present the results of an observational study of some 30 of the dwarf elliptical galaxies (dE's) in the Fornax cluster. Optical photoelectric photometry, CCD photometry, and IR photometry are discussed. A lengthy discussion of the radial surface–brightness profiles as derived from CCD images is given. Also discussed are the relationships of surface–brightness parameters and their uses as distance indicators and some comments on the different structures of dE's and ellipticals are given. The nature of the nucleations is discussed in detail.

160.071 An expanding shell of galaxies in the center of the Hydra I cluster?
P. Fouqué.
Astron. Astrophys., Vol. 185, No. 1/2, p. 94 – 96 (1987).

The velocities of galaxies in the right center of the Hydra I cluster of galaxies are not distributed at random, but exhibit a pattern that leads to assume the existence of a shell of galaxies expanding away from the center of the cluster. This shell could be due to the absorption of a group of galaxies by the cluster.

160.072 50 kpc radio trails behind irregular galaxies in A 1367.
G. Gavazzi, W. Jaffe.
Astron. Astrophys., Vol. 186, No. 1/2, p. L1 – L2 (1987).

The authors report the discovery of exceptionally bright and extended trails of radio emission behind three irregular galaxies in the periphery of the cluster A 1367, in the Coma Supercluster. Turbulent interaction with the intergalactic medium or a past catastrophic collision between galaxies could have produced the observed phenomenon.

160.073 21 centimeter study of spiral galaxies in the Coma supercluster.
G. Gavazzi.
Astrophys. J., Vol. 320, No. 1, p. 96 – 121 (1987).

High–sensitivity, 21 cm line observations of 130 galaxies in the Coma/A1367 supercluster region ($11^h30^m < \alpha < 13^h30^m$; $18° < \delta < 32°$) obtained with the Arecibo 305 m telescope are presented. Using these observations and the data currently available in the literature, two main topics are analyzed: (1) the large–scale distribution of galaxies in the direction of the Coma supercluster and (2) the H I content in spiral galaxies as a function of the local galaxy density.

160.074   HEAO 1 hard X–ray observations of three Abell clusters of galaxies.
Y. Rephaeli, D. E. Gruber, R. E. Rothschild.
Astrophys. J., Vol. 320, No. 1, p. 139 – 144 (1987).

The authors present the results of hard X–ray measurements of A1367, Coma (A1656), and A2319 clusters of galaxies made with the Low Energy Detectors on HEAO 1. Nonthermal components were not detected above the level of 10^{-5} photons cm^{-2}s^{-1}keV^{-1}, but the energy extensions of the thermal spectra of Coma and A2319 seen at lower energies were observed. These results are used to set lower limits on the mean magnetic field and upper limits on the energy density of relativistic electrons in the intracluster space of these clusters.

160.075   Are we all in the grip of a great attractor?
M. M. Waldrop.
Science, Vol. 237, No. 4820, p. 1296 – 1297 (1987).

The Milky Way and all the other galaxies in the immediate universe may be under the sway of a mass to dwarf the superclusters.

160.076   Star formation in X–ray cluster cooling flows.
R. E. White III.
Diss. Abstr. Int., Sect. B, Vol. 48, No. 1, p. 163–B (1987). Thesis, University of Virginia, 229 pp. (1986). Order No. DA8705725.

160.077   SPARTAN–1 X–ray observation of the Perseus cluster.
W. A. Snyder, M. P. Kowalski, D. J. Yentis, R. G. Cruddace, G. G. Fritz, M. P. Ulmer, J. Middleditch, E. E. Fenimore.
Bull. Am. Astron. Soc., Vol. 19, No. 2, p. 685 (1987). Abstract. – See Abstr. 010.061.

160.078   Spectroscopy and photometry of the cluster of galaxies surrounding the quasar 3C 206.
E. Ellingson, R. F. Green, H. K. C. Yee.
Bull. Am. Astron. Soc., Vol. 19, No. 2, p. 685 – 686 (1987). Abstract. – See Abstr. 010.061.

160.079   A re–evaluation of the Butcher–Oemler effect using spectra of high redshift galaxies.
M. Newberry, T. Boroson, R. Kirshner.
Bull. Am. Astron. Soc., Vol. 19, No. 2, p. 686 (1987). Abstract. – See Abstr. 010.061.

160.080   VLA observations of neutral hydrogen in the Hydra I cluster.
H. C. Ferguson, O.–G. Richter, J. H. van Gorkom.
Bull. Am. Astron. Soc., Vol. 19, No. 2, p. 686 (1987). Abstract. – See Abstr. 010.061.

160.081   An intergalactic H I cloud in the compact group HCG 18.
B. A. Williams, J. H. Van Gorkom.
Bull. Am. Astron. Soc., Vol. 19, No. 2, p. 688 (1987). Abstract. – See Abstr. 010.061.

160.082   CCD observations on dynamical evolution in clusters of galaxies.
D. H. Gudehus, D. J. Hegyi.
Bull. Am. Astron. Soc., Vol. 19, No. 2, p. 688 (1987). Abstract. – See Abstr. 010.061.

160.083   A new look at timing.
M. L. McCall.
Bull. Am. Astron. Soc., Vol. 19, No. 2, p. 689 (1987). Abstract. – See Abstr. 010.061.

160.084   Analysis of CCD observations of X–ray clusters of galaxies.
H. P. Murphy, T. C. Weekes.
Bull. Am. Astron. Soc., Vol. 19, No. 2, p. 698 (1987). Abstract. – See Abstr. 010.061.

160.085   A spectroscopic study of three rich galaxy clusters at $z = 0.31$.
W. J. Couch, R. M. Sharples.
Mon. Not. R. Astron. Soc., Vol. 229, No. 3, p. 423 – 456 (1987). With 3 pp. plates.

The authors present intermediate–dispersion spectroscopy for 152 galaxies in the fields of three rich galaxy clusters at $z \sim 0.31$. In each case, the fraction of blue cluster members is a factor of ~ 5 greater than that predicted on the basis of a morphology–density relation and colour distribution appropriate to spirals in nearby clusters. The authors investigate the nature of this blue galaxy excess using a spectral synthesis technique and diagnostic diagrams based on colours and line–strengths. Velocity dispersions and M/L ratios have been determined for each cluster. A comparison of the distribution of populations in the high–z clusters with the morphological segregation observed in the Coma cluster, supports the view that the blue galaxies are progenitors of present–day S0s.

160.086   Comparison of the spatial distribution of Abell clusters against models with Gaussian initial conditions.
D. J. Batuski, A. L. Melott, J. O. Burns.
Space Telesc. Sci. Inst., Prepr. Ser., No. 175, 20 pp. (1987). To appear in Astrophys. J.

160.087   Comments on the reality of the Butcher–Oemler effect.
D. C. Koo.
Space Telesc. Sci. Inst., Prepr. Ser., No. 204, p. 6 – 15 (1987). To appear in the proceedings of the conference "Towards understanding galaxies at large redshifts", Reidel, Dordrecht, The Netherlands.

160.088   Observations of faint field galaxies.
D. C. Koo.
Space Telesc. Sci. Inst., Prepr. Ser., No. 205, 8 pp. (1987). To appear in IAU Symp. No. 130.

160.089   A spectroscopic study of 3 rich galaxy clusters at $z = 0.31$.
W. J. Couch, R. M. Sharples.
Anglo–Aust. Obs., Prepr., No. 216, 46 pp. (1987). To appear in Mon. Not. R. Astron. Soc.

160.090   Spectral energy distributions of galaxies in high redshift clusters – III. Abell 370 at $z = 0.37$.
I. MacLaren, R. S. Ellis, W. J. Couch.
Anglo–Aust. Obs., Prepr., No. 220, 41 pp. (1987). To appear in Mon. Not. R. Astron. Soc.

160.091   Dark matter around the Local Group?
E. Giraud.
ESO Sci. Prepr., No. 537, p. 9 – 12 (1987). Paper presented at IAU Colloq. No. 130.

160.092   Large scale structures: on their form and orientation.
G. Chincarini, G. Vettolani, R. E. De Souza.
Milano Prepr. Ser. Astrophys., No. 23, 55 pp. (1987). Submitted to Astron. Astrophys.

160.093   The luminosity function of a sample of cluster galaxies.
R. Olowin, R. E. De Souza, G. Chincarini.
Milano Prepr. Ser. Astrophys., No. 24, 35 pp. (1987).

Using magnitude estimates in 55 clusters of the southern hemisphere the authors determine the mean cluster luminosity function of the cluster sample observed.

160.094   The massive dumb–bell in the NGC 4782/3 group.
R. E. de Souza, H. Quintana.
Astrophys. Prepr. Ser., No. 17, 21 pp. (1987). Submitted to Astron. J.

160.095   Classification of galaxies in compact groups.
N. A. Tikhonov.
Soobshch. Spets. Astrofiz. Obs., Vyp. 52, p. 51 – 61 (1987). In Russian.

Classification of galaxies in compact groups from the Hickson list is made from 42 plates obtained on the 6–meter telescope. The number of spiral galaxies in groups has grown considerably. A conclusion is drawn on similarity of morphological composition of scattered and compact galaxy groups. The compact groups, apparently, do not follow the Dressler relation "morphology – density", but preserve at any space densities about 60% spiral galaxies.

160.096   Near infrared JHK photometry of clusters of galaxies. I. E, S0, and S galaxies in the Coma cluster (Abell 1626).
E. Recillas–Cruz, L. Carrasco, A. Serrano P. G.,
I. Cruz–González.
Rev. Mex. Astron. Astrofis., Vol. 14, No. 1, p. 77 (1987).
Abstract. – See Abstr. 012.042.

160.097   Environmental variation of radius with ϱ_{gal} in the Coma cluster.
E. Recillas–Cruz, P. Pimentel, A. Serrano P. G.
Rev. Mex. Astron. Astrofis., Vol. 14, No. 1, p. 78 (1987).
Abstract. – See Abstr. 012.042.

160.098   Intercambio de cumulos globulares en cumulos de galaxías.
M. Rabolli.
Rev. Mex. Astron. Astrofis., Vol. 14, No. 1, p. 79 (1987).
Abstract. – See Abstr. 012.042.

160.099   The interstellar/intergalactic medium in front of supernova 1987A.
P. Andreani, R. Ferlet, A. Vidal–Madjar.
ESO Conf. Workshop Proc., No. 27, p. 697 – 701 (1987). – See Abstr. 012.051.

Since the explosion of the supernova 1987A in the Large Magellanic Cloud a great number of data were collected, in order to test the stellar evolution theory. The supernova allows to probe the interstellar, intergalactic medium and galactic halo by studying the absorption features in high–resolution spectra. Preliminary results on data show a conspicuous number of resolved components (at least 24 in Ca II) corresponding to the intervening media. The authors discuss a few characteristics of the detected medium and show, as the most distinctive feature, the smooth variation all along the line of sight of the physical conditions. This might be explained by the tidal stripping picture between the Galaxy and the Magellanic Clouds.

160.100   Morphological population and first–ranked galaxy morphology in loose groups of galaxies.
M. Ramella, G. Giuricin, F. Mardirossian, M. Mezzetti.
Astron. Astrophys., Vol. 188, No. 1, p. 1 – 4 (1987).

The authors analyse the morphological content of loose groups of galaxies contained in Geller and Huchra's (1983) catalogue, according to the morphology of the first–ranked member. The population of the ellipticals in the groups with a first–ranked elliptical member turns out to be significantly higher than in the other groups, independently of the densitiy, as estimated from the compactness of the groups as well as the distance from us. Furthermore, the fraction of ellipticals seems to be more strongly correlated with the local galaxian density for the groups with a first ranked elliptical galaxy than for the others.

160.101   Discovery of the first gravitational Einstein ring: the luminous arc in Abell 370.
G. Soucail, Y. Mellier, B. Fort, G. Mathez, M. Cailloux.
Messenger, No. 50, p. 5 – 6 (1987).

160.102   On the variation of galaxy correlations with luminosity.
S. Phillipps, T. Shanks.
Mon. Not. R. Astron. Soc., Vol. 229, No. 4, p. 621 – 626 (1987).

The authors present a method for investigating, in a direct way, the possible variation of galaxy clustering with luminosity, namely the counts of excess galaxies projected near galaxies of known distance. Large variations in either slope or amplitude of the galaxy correlation function are ruled out by the present data. In particular, the authors find no evidence that bright galaxies have a steeper correlation function slope than faint galaxies.

160.103   Nearby groups of galaxies. II. An all–sky survey within 3000 kilometers per second.
R. B. Tully.
Astrophys. J., Vol. 321, No. 1, p. 280 – 304 (1987).

The 2367 galaxies in the Nearby Galaxies (NBG) Catalog have been assigned to clouds, associations, and groups. The group assignments follow from a dendogram analysis with linkages based on an estimator of the gravitational force between entities. Within the radius of reasonable completion of $25h_{75}^{-1}$Mpc, 179 groups have been identified that include 69% of the known galaxies and 77% of the light. Evidence is presented that the groups are collapsed and that many should be virialized. The median value of $M_V/L_B^{b,i}$ for galaxy groups is $94h_{75}M_\odot/L_\odot$. From this value a rather firm lower limit of $\Omega_g = 0.08$ can be given as the fraction of closure density directly associated with galaxies.

160.104   Additional members of the Local Group of galaxies and quantized redshifts within the two nearest groups.
H. Arp.
J. Astrophys. Astron., Vol. 8, No. 3, p. 241 – 255 (1987).

Galaxies of redshift $z \lesssim 1000$ km s^{-1} are investigated. The majority of redshifts used in the present analysis are accurate to ± 8 km s^{-1}. The deviation of those redshifts from multiples of 72.4 km s^{-1} averages ± 8.2 km s^{-1}. The astonishing result, however, is that for those redshifts which are known accurately, the deviation from modulo 72.4 drops to a value between 3 and 4 km s^{-1}! The amount of relative velocity allowed these galaxies is therefore implied to be less than this extremely small value.

160.105   Luminous arcs from galactic bow shocks.
M. C. Begelman, R. D. Blandford.
Nature, Vol. 330, No. 6143, p. 46 – 48 (1987).

Circular arcs in the cores of two high–redshift ($z \sim 0.3$) clusters have been discovered. Giant elliptical galaxies are found near the centres of both arcs. The arcs have total luminosities of $\sim 10^{44}$erg s^{-1} and are unusually blue in colour. One explanation for the arcs is that they are regions of rapid star formation behind galaxy bow shocks, where our line of sight is tangent to the shock surface and approximately perpendicular to the galaxy velocity. Here the authors describe some of the conditions under which bow shocks can form and point out some observationally testable implications of this model.

160.106   A giant intergalactic H I bubble near Arp 143.
P. N. Appleton, F. D. Ghigo, J. H. van Gorkom,
J. M. Schombert, C. Struck–Marcell.
Nature, Vol. 330, No. 6144, p. 140 – 142 (1987). With a correction in Vol. 330, No. 6147, p. 500 (1987).

The authors present observations of the nearby ring galaxy Arp 143 (NGC 2445/4) and describe the discovery of a giant H I shell 20 kpc across, lying 100 kpc from Arp 143. The shell, which shows evidence for line splitting near its centre, lies in a long H I filament pointing towards the Arp 143 system. The authors suggest that the shell is an expanding bubble of H I, and is direct evidence for a powerful galactic–scale explosion 10^8years ago which is associated with a strong burst of star formation. They believe that both the ring galaxy and the giant bubble have their origin in the passage of a gas–rich companion through the inner disk of NGC 2445, and its subsequent disruption.

160.107   Individual masses of galaxies in triple systems.
V. A. Mineva.
Astron. Zh., Tom 64, Vyp. 6, p. 1155 – 1160 (1987). In Russian.
English translation in Sov. Astron., Vol. 31, No. 6.

Mass–to–luminosity ratio is obtained for 39 components of triple systems of galaxies. The mean value is $(4.5 \pm 0.6) f_\odot$. The rotation curves for 4 galaxies are presented.

160.108   CL 2244–02.
IAU Circ., Nos. 4456, 4482 (1987).

160.109   The morphology of the rich supercluster 1451 + 22.
R. Ciardullo.
Astrophys. J., Vol. 321, No. 2, p. 607 – 621 (1987).

The morphological properties of the rich supercluster 1451 + 22 are investigated using photographic galaxy photometry in a two–color system especially sensitive to redshift. By measuring $\sim 125,000$ galaxies in five Palomar Schmidt fields, the supercluster's shape, density profile, density contrast, and galaxy population are found. The data suggest that the structure of 1451 + 22 is that of a face–on pancake, flattened against the plane of the sky, although spherical models are not ruled out. If 1451 + 22 is flat, the mean luminosity density contrast of the supercluster over the field is $\varrho/\varrho_{field} \sim 10$, and the total mass of the system is probably $M > 10^{16} M_\odot$.

160.110   Far–infrared properties of cluster galaxies.
M. D. Bicay, R. Giovanelli.
Astrophys. J., Vol. 321, No. 2, p. 645 – 657 (1987).

Far–infrared properties are derived for a sample of over 200 galaxies in seven clusters: A262, Cancer, A1367, A1656 (Coma), A2147, A2151 (Hercules), and Pegasus. The optically selected sample consists almost entirely of "IR–normal" galaxies. None of the cluster galaxies has log $L_{FIR} > 11.0 \, L_\odot$, in marked contrast to the situation for field galaxies, in which at least 20% have FIR luminosity greater than $10^{11} L_\odot$. Only one sample galaxy has a FIR–to–blue luminosity ratio greater than 10. The results of this survey are interpreted in the context of models that propose various components of the integrated FIR emission spectra. It is suggested that, in general, the FIR emission from cluster galaxies is dominated by that of cool dust associated with atomic hydrogen heated by the general interstellar radiation field with a lesser contribution from warm dust associated with star–formation processes.

160.111   List of clusters of galaxies with published redshifts.
V. S. Lebedev, I. A. Lebedeva.
Astron. Tsirk., No. 1469, p. 4 – 6 (1986). In Russian.

The spectroscopically measured redshifts of 1471 clusters and groups of galaxies have been compiled from literature. This catalogue may be obtained from the Special Astrophysical Observatory of the USSR Academy of Sciences.

160.112   Catalogue of counterparts of Abell and Zwicky clusters of galaxies on magnetic tape.
I. A. Lebedeva.
Astron. Tsirk., No. 1469, p. 6 – 8 (1986). In Russian.

This catalogue has been compiled at the Special Astrophysical Observatory of the USSR Academy of Sciences on the basis of data published by Abell (1958) and Zwicky et al. (1968).

160.113   General properties of clusters and groups of galaxies.
V. G. Gorbatskij, A. G. Kritsuk.
Itogi Nauki i Tekhniki. Ser. Astron. Tom 29. Clusters of galaxies, p. 3 – 61 (1987). In Russian. – See Abstr. 003.017.

Contents: Elements of the large–scale structure from the observed galaxy distribution. Morphology of galaxy grouping. Large–scale structure of the Universe and its numerical modelling. X radiation of clusters of galaxies. Radio radiation of clusters and groups of galaxies.

160.114   Dynamical processes in clusters and groups of galaxies.
V. G. Gorbatskij, A. G. Kritsuk.
Itogi Nauki i Tekhniki. Ser. Astron. Tom 29. Clusters of galaxies, p. 62 – 110 (1987). In Russian. – See Abstr. 003.017.

160.115   Comparisons of the spatial distribution of Abell clusters against models with Gaussian initial conditions.
D. J. Batuski, A. L. Melott, J. O. Burns.
Astrophys. J., Vol. 322, No. 1, p. 48 – 58 (1987).

The amount of structure present among the Abell clusters out to redshift $z = 0.085$ has been compared with numerical super-computer simulations (with 64^3 particles) of the isothermal, neutrino, and cold particle models for large–scale structure, assuming a flat universe and $H = 50 \, \mathrm{km \, s^{-1} Mpc^{-1}}$. High–density clusters of particles were identified in each simulation. Correlation and percolation tests were then used to compare the spatial distribution of these high–density points with the apparent superclustering among Abell clusters. None of the models came very close to matching the observations.

160.116   Photometry of galaxies in compact groups.
N. A. Tikhonov.
Astrofizika, Tom 27, Vyp. 2, p. 253 – 264 (1987). In Russian.
English translation in Astrophysics, Vol. 27, No. 2.

Results of photographic photometry of galaxies in compact groups are presented. The luminosity functions of these groups, field galaxies and cluster galaxies are similar. On the 54 plates of the 6–m telescope only one group shows merging.

160.117   The dispersion of radial velocities and mass to luminosity ratio for the compact group of galaxies Shahbazian 166.
A. S. Amirkhanyan, A. G. Egikyan.
Astrofizika, Tom 27, Vyp. 2, p. 395 – 397 (1987). In Russian.
English translation in Astrophysics, Vol. 27, No. 2.

For the members of the compact group of galaxies Shahbazian 166 the radial velocities have been measured. The dispersion of the radial velocities is equal to 329 km/s. The apparent and absolute magnitudes of galaxies in V as well as the mass to luminosity ratio are obtained. The latter is approximately equal to 170 $M_\odot/L_\odot$.

160.118   Properties of dense galaxy groups and the implications of their existence.
J. W. Sulentic.
Astrophys. J., Vol. 322, No. 2, p. 605 – 617 (1987). With plate 4.

Galaxy counts are reported within 1° of the 100 dense groups cataloged by Hickson (1982). This new analysis suggests that a much smaller fraction of the groups can be explained as chance alignments than has recently been suggested. Thus, it appears that at least 100 dense physical groups exist on the sky which (1) cannot be short–lived transient systems, and (2) often contain one or more discordant redshift members. The author considers the possibility that the galaxies in these groups formed together relatively recently and are possibly related to some of the galaxy "nests" of Vorontsov–Velyaminov (1962). It is suggested that the discordant redshift members are somehow related to this recent formation.

160.119   More about clustering on a scale of 0.1 c.
R. B. Tully.
Astrophys. J., Vol. 323, No. 1, p. 1 – 18 (1987).

A previous suggestion is substantiated that our Galaxy is appended to what is now called the Pisces–Cetus Supercluster Complex. The evidence is twofold. (1) With a sample of rich clusters that is complete to beyond the mean distance of the complex, there is percolation across $\sim 300 \, h_{75}^{-1} \mathrm{Mpc}$ at a separations scale length of 38 $h_{75}^{-1} \mathrm{Mpc}$. (2) A one–dimensional correlation analysis confirms that rich clusters within the complex lie preferentially in a plane (or possibly in strata) parallel to the plane of the Local Supercluster as defined by nearby galaxies. The main plane of the Pisces–Cetus Supercluster Complex, centered $\sim 200 \, h_{75}^{-1} \mathrm{Mpc}$ away, is coincident with the principal plane of the Local Supercluster. The thickness of the Pisces–Cetus main plane has FWHM $\sim 40 \, h_{75}^{-1} \mathrm{Mpc}$. The coincidence of the two planes of structure on such radically different scales suggests there is a physical connection. It would follow that the structure on a scale of 0.1 c has a physical significance.

160.120 Optical bias and hierarchical clustering.
S. A. Bonometto, F. Lucchin, S. Matarrese.
Astrophys. J., Vol. 323, No. 1, p. 19 – 29 (1987).

Recent statistical results obtained in the frame of biased theories of galaxy origin are transferred to the direct analysis of the luminosity field on the celestial sphere. Correlation functions on the celestial sphere should therefore obey a number of relations similar to those worked out for biased theories of galaxy origin (in a Gaussian or non–Gaussian context). The relation between this view and the view based on the Limber equation is discussed, and a tentative explanation of a number of peculiarities of observed spatial correlations is proposed. These ideas are also checked by performing a fit with the Zwicky catalog data. A discussion on the meaning of spatial statistical data worked out from angular measures is also performed.

160.121 Can quasars ionize the intergalactic medium?
M. Donahue, J. M. Shull.
Astrophys. J., Lett. Ed., Vol. 323, No. 1, p. L13 – L18 (1987).

The authors examine the environmental impact of quasars and active nuclei on the intergalactic medium (IGM) and derive analytic and numerical results for ionization–front expansion as a function of redshift z in an expanding Friedmann cosmology. For an IGM of moderate density ($\Omega_I \approx 0.1$) and a conservative estimate of QSO ionizing spectra, QSOs "turning on" at $z_{on} = 4$ will not produce a fully ionized IGM until $z \approx 2.5$. However, an IGM with $\Omega_I = 0.03 – 0.10$ could be ionized between $z = 3.3 – 3.7$. The authors also investigate the effects of I–fronts on the distribution of "Lyα forest" clouds. The photoionization of intergalactic clouds within ~ 10 Mpc of a luminous QSO could explain the "inverse effect", an observed deficit in the distribution of Lyα absorption systems near z_{em}.

160.122 Results from a visual survey of all 2712 Abell clusters.
M. F. Struble, H. J. Rood.
Astrophys. J., Vol. 323, No. 2, p. 468 – 472 (1987).

The authors summarize the statistically most significant results obtained from an initial examination of the morphological data obtained in a recent optical survey of all 2712 Abell clusters which have been cataloged by Struble and Rood (44.002.036). The analyses apply these data along with information from a compilation of redshifts and velocity dispersions.

160.123 Multiaperture spectroscopy of galaxies in Abell 370.
J. P. Henry, R. J. Lavery.
Astrophys. J., Vol. 323, No. 2, p. 473 – 479 (1987).

The authors present the results of spectroscopic observations of 30 galaxies in the distant rich cluster Abell 370 at $z = 0.376$. The authors confirm the large fraction of blue galaxies in this cluster measured photometrically by Butcher and Oemler. None of the spectra can be attributed to active galactic nuclei. All of the blue galaxy spectra may be described as examples of poststarburst objects with strong higher order Balmer lines and very weak or absent emission lines.

160.124 An imaging study of NGC 1275 and PKS 0745–191: vigorous star formation in cooling flow cluster dominant galaxies.
W. Romanishin.
Astrophys. J., Lett. Ed., Vol. 323, No. 2, p. L113 – L116 (1987). With plate L3.

Optical imaging has been used to study the continuum color distributions of NGC 1275 in the Perseus cluster and the PKS 0745–191 cluster central galaxy, and the emission line flux of PKS 0745–191. Both clusters have strong cooling flows. The Hα luminosity of PKS 0745–191 is equal to that for NGC 1275/Perseus cluster. It is shown that the optical properties of NGC 1275 and the PKS 0745–191 central galaxy are very similar. The author discusses the amount of star formation in these two systems, the possible role of stars in providing ionization for the emission–line gas, and, by comparing the properties of the two clusters, comments on the interpretation of NGC 1275 and its peculiarities.

Itogi Nauki i Tekhniki. Seriya Astronomiya. Tom 29. Clusters of galaxies.
See Abstr. 003.017.

The AXAF High Resolution Camera (HRC) and its use for observations of distant clusters of galaxies.
See Abstr. 035.029.

Expected AXAF mirror characteristics and their implications for measurements of the Hubble constant using the Sunyaev Zel'dovich effect.
See Abstr. 035.030.

Data reduction and spectrophotometric performances of PUMA 1: an on–line multiaperture spectroscopic system used at the CFHT.
See Abstr. 036.066.

X–ray observations from the Space Shuttle.
See Abstr. 051.030.

The origin and cosmogonic implications of seed magnetic fields.
See Abstr. 062.064.

Arcs from gravitational lensing.
See Abstr. 066.105.

The marginal gravitational lensing.
See Abstr. 066.151.

Gas content and the intensity of star formation in cluster galaxies.
See Abstr. 131.075.

An infrared–optical study of IRAS point sources in the Virgo region.
See Abstr. 133.004.

Merging instability in groups of galaxies with dark matter.
See Abstr. 151.032.

The art of N–body building.
See Abstr. 151.034.

The effects of satellite accretion on disk galaxies.
See Abstr. 151.049.

The dynamics of small groups of galaxies. I. Virialized groups.
See Abstr. 151.133.

An automated search for halo star clusters and Local Group dwarf galaxies.
See Abstr. 154.071.

CO observations of nearby galaxies and the efficiency of star formation.
See Abstr. 157.007.

Isophotal diameters of cluster spirals.
See Abstr. 157.046.

On the structure of the low surface brightness dwarf galaxies in the M81 group.
See Abstr. 157.063.

The metallicity versus luminosity relationship for early–type galaxies.
See Abstr. 157.065.

Surface photometry of six Local Group galaxies.
See Abstr. 157.075.

Dynamics of a hot ($T \sim 10^7$K) gas cloud with volume energy losses.
See Abstr. 157.085.

The origin of the far–infrared flux from spiral galaxies.
See Abstr. 157.093.

A redshift survey of IRAS galaxies in the Bootes void area.
See Abstr. 157.095.

Systematics of the 4000 Angstrom break in the spectra of galaxies.
See Abstr. 157.102.

Determination of the masses of elliptical galaxies and clusters of galaxies with AXAF.
See Abstr. 157.103.

Effects of hidden mass and intergalactic medium on the structure of hot galactic coronae.
See Abstr. 157.110.

Stellar populations in Local Group galaxies.
See Abstr. 157.129.

Properties of cD galaxies.
See Abstr. 157.153.

Mass distributions in elliptical galaxies at large radii.
See Abstr. 157.160.

Line–strength gradients in early–type galaxies.
See Abstr. 157.161.

Isophotometry of brightest cluster ellipticals.
See Abstr. 157.168.

Investigating the scatter in the V_{26} – log σ relation.
See Abstr. 157.169.

On the relation between radius, luminosity and surface brightness in elliptical galaxies.
See Abstr. 157.170.

Boxy isophotes and dust lanes in bright Virgo ellipticals.
See Abstr. 157.178.

On the thermal instability of galactic and cluster halos.
See Abstr. 157.199.

Neutral hydrogen observations of four dwarf irregular galaxies in the Virgo Cluster.
See Abstr. 157.209.

A comparison of mass distributions in cluster and field spirals.
See Abstr. 157.240.

The amount of dark matter in spiral galaxies in clusters.
See Abstr. 157.249.

Radio emission of spiral galaxies in groups of galaxies.
See Abstr. 157.255.

The surface brightness of 1550 galaxies in Fornax: automated galaxy surface photometry – II.
See Abstr. 157.258.

The origin of the far–infrared flux from spiral galaxies.
See Abstr. 157.265.

On the relation between radius, luminosity, surface brightness and galaxian surface density in E galaxies.
See Abstr. 157.295.

On the nature of dark matter in the dwarf spheroidal satellites of the Galaxy.
See Abstr. 157.300.

The evolution of galaxies at moderate redshift.
See Abstr. 157.307.

Studies of kinematics of interacting galaxies. Internal motions in 10 nests of galaxies.
See Abstr. 157.313.

Powerful radio sources in clusters of galaxies: the origin of jet–like structures.
See Abstr. 158.044.

IC 711 – the longest head–tail galaxy known.
See Abstr. 158.065.

The Seyfert 2 galaxy IC 184 and its surrounding group.
See Abstr. 158.069.

New radio observations of NGC 1275.
See Abstr. 158.081.

What bends WATs?
See Abstr. 158.082.

Intensity maps of six wide angle tail sources.
See Abstr. 158.083.

Jet disruption in wide–angle tailed radio galaxies.
See Abstr. 158.084.

Are cooling flows governing E–galaxy evolution?
See Abstr. 158.140.

B2 0800 + 24: a narrow–angle tail radio galaxy in a small group of galaxies.
See Abstr. 158.154.

Compact radio sources associated with interacting galaxies in poor clusters.
See Abstr. 158.190.

A search for OH absorption in NGC 1275.
See Abstr. 158.225.

The cluster environments of powerful radio galaxies.
See Abstr. 158.235.

Powerful radio sources in clusters of galaxies: origin of jet–like structures.
See Abstr. 158.270.

On the relation of Seyfert galaxies with clusters.
See Abstr. 158.342.

The spatial clustering of QSOs.
See Abstr. 159.012.

High–resolution spectra of 24 low–redshift QSOs: the properties of Mg II absorption systems.
See Abstr. 159.018.

Surveys of fields around quasars. IV. Luminosity of galaxies at $z \approx 0.6$ and preliminary evidence for the evolution of the environment of radio–loud quasars.
See Abstr. 159.031.

High–redshift QSO absorbing clouds and the background ionizing source.
See Abstr. 159.044.

$z_{abs} \approx z_{em}$ absorption systems in quasar spectra: a correlation with radio and optical luminosity?
See Abstr. 159.079.

Crowding on the sight line to the QSO PHL 1226: the nearby galaxy IC 1746 and a galaxy cluster at $z = 0.16$.
See Abstr. 159.080.

The distribution of QSO absorption system column densities: evidence for a single population.
See Abstr. 159.114.

The redshift distribution of QSO Lyman–alpha absorption systems.
See Abstr. 159.115.

Absorption in the wide QSO pair Tololo 1037–2704 and Tololo 1038–2712: evidence for a specially aligned supercluster at $z = 2$?
See Abstr. 159.138.

An absorption–line survey of 32 QSOs at red wavelengths: properties of the Mg II absorbers.
See Abstr. 159.144.

Forbidden [O II] and [O III] emission 29″ from a QSO absorption–line region.
See Abstr. 159.147.

Hydrogen molecules and the radiative cooling of pregalactic shocks.
See Abstr. 161.003.

Scaling laws for the probability of holes in the galaxy distribution.
See Abstr. 161.024.

The galaxian surface density of the nearby universe.
See Abstr. 161.025.

Clustering in real space and in redshift space.
See Abstr. 161.030.

Kinematics of multiple clusters of galaxies.
See Abstr. 161.035.

Numerical experiments with hidden mass in groups of galaxies.
See Abstr. 161.036.

Possible inhomogeneities in the Universe on scales of $100 - 300$ Mpc from observations with the 6–meter telescope.
See Abstr. 161.038.

Large–scale distribution of objects from the First Byurakan Survey (FBS).
See Abstr. 161.039.

Modeling as a method of analysis of the correlation function for samples of rich clusters of galaxies.
See Abstr. 161.047.

Ripples in the universal Hubble flow.
See Abstr. 161.049.

Formation and evolution of clusters of galaxies.
See Abstr. 161.063.

The extent of beam dilution in measurements of the Zeldovich–Sunyaev effect.
See Abstr. 161.073.

Existence and nature of dark matter in the universe.
See Abstr. 161.074.

Galaxy clustering and small–scale CBR anisotropy constraints on galaxy origin scenarios.
See Abstr. 161.098.

Were the Lyman–α clouds formed from shocks?
See Abstr. 161.100.

Large–scale streaming of cluster elliptical galaxies.
See Abstr. 161.106.

The formation of structure in particle–dominated cosmologies.
See Abstr. 161.111.

Large scale streaming velocities in biased galaxy formation.
See Abstr. 161.112.

Strings, the Peebles screed, and large scale structure.
See Abstr. 161.116.

Fragmenting the Universe. I. Statistics of two–dimensional Voronoi foams.
See Abstr. 161.123.

Statistical analysis of a sample of first–ranked cluster galaxies.
See Abstr. 161.126.

The giant arcs are gravitational mirages.
See Abstr. 161.128.

Voids and galaxies in voids.
See Abstr. 161.137.

The large scale distribution of galaxy types.
See Abstr. 161.138.

Coherent orientation effects of galaxies and clusters.
See Abstr. 161.139.

What is the cosmological density parameter Ω_0?
See Abstr. 161.144.

Testing cosmic fluctuation spectra.
See Abstr. 161.149.

Dissipation and the formation of galaxies.
See Abstr. 161.157.

Void probabilities in the galaxy distribution: scaling and luminosity segregation.
See Abstr. 161.179.

A VLA search for cosmological H I at $z = 3.3$.
See Abstr. 161.201.

Giant H II regions as distance indicators. II. Application to H II galaxies and the value of the Hubble constant.
See Abstr. 161.211.

The Sunyaev–Zel'dovich effect under the condition of the $1 - 5$ μm IR background and the determination of the Hubble constant H_0 and of the density and temperature of the thermal electrons in a rich galaxy cluster.
See Abstr. 161.235.

On the origin of the dipole anisotropy of the cosmic microwave background: beyond the Hydra–Centaurus supercluster.
See Abstr. 161.250.

Yet another scenario for galaxy formation.
See Abstr. 161.255.

Non–Gaussian fluctuations.
See Abstr. 161.256.

N–body methods and the formation of large–scale structure.
See Abstr. 161.257.

Cosmic strings and the formation of clusters of galaxies.
See Abstr. 161.306.

The structure of the Universe as determined from deep galaxy surveys.
See Abstr. 161.353.

Cosmological H II regions and the photoionization of the intergalactic medium.
See Abstr. 161.355.

Amplification of correlation functions by gravity in the cosmic string model.
See Abstr. 161.365.

Galaxy distribution in a cold dark matter universe.
See Abstr. 161.443.

On the origin of dwarf galaxies.
See Abstr. 161.456.

161 Universe, Cosmology, Background Radiation

161.001   Inhomogeneous cosmology. III. Primordial gravitational waves and dust.
P. J. Adams, R. W. Hellings, R. L. Zimmerman.
Astrophys. J., Vol. 318, No. 1, p. 1 – 14 (1987).

The authors investigate the properties of a special class of inhomogeneous cosmological models and study the interaction of the inhomogeneities with the evolution of the background geometry and matter. The model is chosen so that the initial inhomogeneities evolve into "plane" gravitational waves propagating through a smooth Bianchi I dust background. It is shown how the inhomogeneities interact with matter, 3K radiation, and the background geometry, causing the expansion to slow down in some regions and speed up in others. The gravitational waves can produce a "dragging of the inertial frame" which will affect the observed distribution of matter and 3K radiation.

161.002   Galaxy formation by gravitational collapse.
B. S. Ryden, J. E. Gunn.
Astrophys. J., Vol. 318, No. 1, p. 15 – 31 (1987).

Galaxies may form by the collapse of density perturbations in a universe dominated by dark matter. The authors investigate the growth of density and velocity distributions around the collapsed mass peaks, taking into account the random velocities imparted by substructure. Structures resembling galactic halos arise naturally from density perturbations in the cold dark matter scenario. When the halo structures are adiabatically compressed by the dissipating baryonic component, the resulting rotation curves are flat, over the full range of amplitudes studied here.

161.003   Hydrogen molecules and the radiative cooling of pregalactic shocks.
P. R. Shapiro, H. Kang.
Astrophys. J., Vol. 318, No. 1, p. 32 – 65 (1987).

The nonequilibrium radiative cooling, recombination, and molecule formation behind steady state shock waves in a gas of primordial composition have been calculated in detail for a number of cases. The authors have solved the rate equations for these processes, together with the hydrodynamical conservation equations. Such shock waves are relevant to a wide range of theories of galaxy and pregalactic star formation. A purely atomic gas of H and He which is shock–heated to temperatures above 10^4K is assumed. The results indicate that formation of H_2 molecules in the post–shock gas may be quite common for a significant range of shock velocities. The extra cooling resulting from H_2 formation greatly reduces previous estimates of the characteristic gravitational scale length and the characteristic mass subject to gravitational instability in these postshock regions.

161.004   Extraction of cosmological information from multiimage gravitational lenses.
I. Kovner.
Astrophys. J., Lett. Ed., Vol. 318, No. 1, p. L1 – L5 (1987). With a correction in Astrophys. J., Lett. Ed., Vol. 323, No. 2, p. L 155 (1987).

The effects of possible invisible large–scale mass inhomogeneities inside and outside the light beam have been reported in the past to be the obstacles to determination of the distance measure to a gravitational lens. It is shown here that for lenses of four or more images the determination of the distance measure is nevertheless possible. The appropriate procedure for doing so is presented in a general form and demonstrated by a concrete example. This procedure yields also the dipole moment at the lens position, which contains contributions of mass inhomogeneities on a very large scale.

161.005   Normalization, cold dark matter, and large–scale velocities.
Y. Hoffman.
Astrophys. J., Lett. Ed., Vol. 318, No. 1, p. L7 – L10 (1987).

The recent observational studies of the large–scale peculiar velocity field set very strong constraints on the various cosmological models, in particular on the standard model where structure evolves gravitationally from a Gaussian density perturbation (δ) field. This letter discusses the normalization of the perturbation spectrum in the standard theory of gravitational instabilities. It is found that for an actual value of $\sigma = 1.0$, the linear theory overestimates the normalization constant, and therefore the amplitude of the velocity field, by a factor of about 5/3. Applying this (quasi–linear) normalization to the cold dark matter model, the author shows it to be incompatible with the recent observations of the $50\,h^{-1}$Mpc bulk velocity, as well as with the microwave background dipole velocity, even for $\Omega_0 = 0.1$ and $h = 0.5$.

161.006   Small–scale anisotropy of microwave background radiation computed by means of the transfer function between $z = 2000$ and $z = 800$.
S.–p. Xiang.
Astrophys. Space Sci., Vol. 135, No. 1, p. 75 – 79 (1987).

The author used the transfer function between $z = 2000$ and $z = 800$ to compute the small–scale anisotropy of the microwave background radiation. The numerical results show that the dependence of $\Delta T/T$ on the spectrum index n is not monotonic but rather different for the two regimes of $n < 2$ and $n > 2$: for $n < 2$ the curves with larger n will be higher than those with smaller n, while for $n > 2$ the curves with larger n will go down. The highest curve corresponds to $n = 2$.

161.007   Coupled perfect fluid and zero–mass scalar field in an expanding universe.
M. N. Varma.
Astrophys. Space Sci., Vol. 135, No. 1, p. 197 – 200 (1987).
A spatially flat expanding cosmological model filled with interacting perfect fluid and zero–mass scalar field is obtained under a specific law for Hubble's parameter. It is shown that this model avoids the big–bang singularity.

161.008   Black–body radiation in a curved Robertson–Walker background.
Y. Deng, P. D. Mannheim.
Astrophys. Space Sci., Vol. 135, No. 2, p. 261 – 269 (1987).
In the standard Friedmann cosmology the black–body radiation spectrum is usually taken to have the same familiar T^4–form that it has in a flat space. With explicit use of the equation of motion of a quantized massless field propagating in a curved background Robertson–Walker metric the authors show (for the readily tractable scalar field case) that the assumption is in fact true for an open Universe. For a closed Universe, it is found that there is an in principle modification to the T^4–law. Unfortunately, the correction turns out to be too small to be experimentally detectable. In passing, the authors also obtain a simple derivation for the cosmological red shift of frequencies.

161.009   An anisotropic cosmological model in a scalar–tensor theory of gravitation.
D. R. K. Reddy, R. Venkateswarlu.
Astrophys. Space Sci., Vol. 135, No. 2, p. 287 – 290 (1987).
A spatially–homogeneous and anisotropic–cosmological model in a scalar–tensor theory proposed by Sen and Dunn (1971) is obtained when the source of the gravitatonal field is a perfect fluid with pressure equal to energy density and the metric is of Bianchi type–I. Various physical properties of the model have also been discussed.

161.010   The baryon–symmetric domain cosmology and the cosmic X–ray background.
K. N. Yu.
Astrophys. Space Sci., Vol. 135, No. 2, p. 291 – 300 (1987).
The author shows that the cosmic radiation from the baryon–symmetric domain cosmology (BSDC) agrees remarkably with the cosmic X–ray background from ∼1 keV to ∼100 keV. He has also shown that AGNs contribute significantly to the cosmic background beyond ∼100 keV. Therefore, the author has arrived at a consistent model in which the BSDC model and AGN model together can explain the cosmic background from ∼1 keV to ∼1 MeV.

161.011   Zeit und Kosmologie.
M. Gossler.
Sternenbote, 30. Jahrg., Nr. 10, p. 194 – 198 (1987).

161.012   Photons in curved space–time.
D. F. Crawford.
Aust. J. Phys., Vol. 40, No. 3, p. 449 – 457 (1987).
Although the tidal stress on a fundamental particle moving along its geodesic in curved space–time is very small, it is not negligible. This paper argues that this tidal stress can produce an energy loss mechanism that is of considerable astrophysical importance. In particular, this loss can explain the Hubble redshift without requiring universal expansion. This paper will concentrate on the derivation of the energy loss relation and its application to the Hubble redshift.

161.013   The periodic system of the distribution of mass concentration knots in the universe, its regularities and the forecast data.
G. P. Tamrazián.
Bol. Acad. Cienc. Fis. Mat. Nat., Tomo 45, Nos. 139 – 140, p. 29 – 52 (1985). In Spanish.

161.014   The large–scale streaming of galaxies.
A. Dressler.
Sci. Am., Vol. 257, No. 3, p. 38 – 46 (1987).
The Milky Way is traveling through the universe in concert with a swarm of other galaxies. The source of the impetus may be a remote concentration of mass on a scale that challenges current theory.

161.015   Particle physics and inflationary cosmology.
A. Linde.
Phys. Today, Vol. 40, No. 9, p. 61 – 68 (1987).
It seems likely that the universe is an eternal, self–reproducing entity divided into many mini–universes, with low–energy physics and perhaps even dimensionality differing from one to the other.

161.016   Limitation on the initial inhomogeneity of the Universe based on the "Cholod" experimental data.
P. D. Nasel'skij.
Pis'ma Astron. Zh., Tom 13, No. 7, p. 554 – 559 (1987). In Russian. English translation in Sov. Astron. Lett., Vol. 13.
A list of relic radiation anisotropy values in models with cold massive particles (axions, gravitinos etc.) and neutrinos is given. It is shown that "Cholod" experimental data are crucial for the neutrino models and cold particles models with low matter density ($\Omega \leqslant 0.2$).

161.017   Relic radiation anisotropy and polarization as a test for nonequilibrium ionization of pregalactic plasma.
P. D. Nasel'skij, A. G. Polnarev.
Astrofizika, Tom 26, Vyp. 3, p. 543 – 555 (1987). In Russian. English translation in Astrophysics, Vol. 26, No. 3, p. 327 – 335 (1987).
Small–scale anisotropy and polarization in a model of non–stationary ionization of pregalactic plasma is considered. The ratio of the degree of polarization to the degree of anisotropy is shown to be not very sensitive to a specific mode of ionization and amounts to 7 – 8%. However, a specific correlation angle in the distribution of the anisotropy and polarization of relic radiation over the celestial sphere is a function of nonequi–librium–ionization parameters.

161.018   Gauge–invariant cosmological density perturbations.
M. Sasaki.
Gravitational collapse and relativity, p. 114 – 123 (1986). – See Abstr. 012.006.
A gauge–invariant formulation of cosmological density perturbation theory is reviewed with special emphasis on its geometrical aspects. Then the gauge–invariant measure of the magnitude of a given perturbation is presented.

161.019   Higher dimensional cosmology.
C. H. Lee, H. K. Lee.
Gravitational collapse and relativity, p. 434 – 442 (1986). – See Abstr. 012.006.
A cosmology is considered in $1 + d + D$ dimensional space–time with the space part being a direct product of d– and D–dimensional maximally symmetric spaces. For the case of the energy–momentum tensor being characterized by radiation confined in the d–dimensions, the Einstein field equations are discussed.

161.020   Higher dimensional solutions of modified Einstein equations with curvature square terms.
H. Ishihara.
Gravitational collapse and relativity, p. 443 – 450 (1986). – See Abstr. 012.006.
Higher dimensional cosmological solutions in vacuum with two spatially flat subspaces are studied in the framework of the modified Einstein theory with curvature square terms. In particular, for the Gauss–Bonnet combination the author finds a singularity–free solution with the inflation of 3–dimensions.

161.021 Who decides boundary conditions for the wave function of the universe?

S. Wada.
Gravitational collapse and relativity, p. 465 – 473 (1986). – See Abstr. 012.006.

The author shows that boundary conditions for the wave function of the universe are severely restricted in certain types of quantum gravity.

161.022 Cosmological solutions of the Einstein equation with the backreaction effect of quantized conformally invariant fields.

T. Azuma.
Gravitational collapse and relativity, p. 474 – 481 (1986). – See Abstr. 012.006.

The Einstein equation with the backreaction effect of quantized conformally invariant fields is considered in a Robertson–Walker metric. The potential method is applied. It is shown that in the presence of classical matter, there are not only asymptotic Friedmann solutions but also other Friedmann–type solutions which undergo a de Sitter expansion in the early stage.

161.023 Inhomogeneous generalizations of the Robertson–Walker cosmological models.

A. Krasiński.
Gravitational collapse and relativity, p. 500 – 508 (1986). – See Abstr. 012.006.

Solutions of Einstein's equations are discussed in which the flow of matter is hypersurface–orthogonal and shearfree, but accelerating, the hypersurfaces being conformally flat. The field equations require integrability conditions which are solved. In some solutions at an initial instant the matter density is a periodic function of the distance along the curves orthogonal to the orbits of the symmetry group.

161.024 Scaling laws for the probability of holes in the galaxy distribution.

R. Schaeffer.
Astron. Astrophys., Vol. 181, No. 2, p. L23 – L25 (1987).

Assuming that the galaxy N–body correlation functions are to all orders given by the scale invariant, hierarchical, form that is known to hold for $N = 2$ to 4, the author shows that the probability of holes is given to a very good approximation by a universal function $\Sigma(q)$. This universal function depends solely on the N–body galaxy correlations. It can be uniquely determined even for incomplete samples and is the same for 2 or 3 dimensional statistics, only q being modified. It provides a way to characterize the distribution of holes in a model independent way.

161.025 The galaxian surface density of the nearby universe.

P. Fontanelli, P. Chamaraux, C. Balkowski.
Astron. Astrophys., Vol. 181, No. 2, p. 217 – 224 (1987).

In an attempt to catalogue all the nearby large structures of galaxies ($V_r \leqslant 10000$ km s^{-1}), the authors discuss maps of galaxian surface density for the whole sky, from CGCG, MCG and ESO catalogues. The resolution of the maps is $0°\!.5 \times 0°\!.5$ and each pixel is shaded with an intensity proportional to the galaxian surface density. In this way, regions crowded in galaxies are clearly seen. As a result, a dozen of large structures have been evidenced; several of them are well known (Perseus–Pisces, Coma–A 1367 superclusters for instance). However, three of them do not seem to have been previously recognized: they are located north of the Lynx–Ursa Major supercluster, South of the Perseus–Pisces supercluster and South of the Hydra cluster respectively. Two of the new structures have been found to be real ones (South Hydra and North Lynx); no firm conclusion has been drawn for the other one, due to lack of redshifts.

161.026 A cosmology with negative vacuum density and infinite strings.

N. S. Kardashev.
Sov. Astron., Vol. 30, No. 5, p. 498 – 501 (1986). English translation of 42.161.126.

161.027 Cosmological proper motion.

N. S. Kardashev.
Sov. Astron., Vol. 30, No. 5, p. 501 – 504 (1986). English translation of 42.161.127.

161.028 Observational manifestations of population III stars.

B. V. Vajner, V. V. Chuvenkov, Yu. A. Shchekinov.
Astrophysics, Vol. 25, No. 3, p. 680 – 688 (1987). English translation of 42.161.445.

161.029 Microwave background measurements from space at 15 to 90 GHz.

R. Saunders.
Radio astronomy from space, p. 61 – 65 (1987). – See Abstr. 012.009.

Consideration is given to the problems in microwave background astronomy of mapping the Sunyaev–Zeldovich effect and of searching for primordial anisotropy at the requisite level of $\Delta T \sim 10 \ \mu K$. Arguments are presented as to why an interferometer is fundamentally better suited to this work than a single-dish, and why the optimum observing frequencies are in the range 10 – 100 GHz. The planned Very Small Array (VSA), which can carry out the necessary observations, is described.

161.030 Clustering in real space and in redshift space.

N. Kaiser.
Mon. Not. R. Astron. Soc., Vol. 227, No. 1, p. 1 – 21 (1987).

Several extensive galaxy redshift surveys are now available, and these provide a reasonably precise three–dimensional view of the world. In a perfectly homogeneous Friedmann universe these redshifts would accurately measure radial distance from the observer, and the mapping from real space (r–space) to redshift space (s–space) would simply be an identity. In an inhomogeneous universe like our own the peculiar velocities associated with any inhomogeneous structrure will introduce a distortion in this mapping. In this paper the author explores several aspects of this distortion.

161.031 Astronomical constraints on a string–dominated universe.

J. R. Gott III, M. J. Rees.
Mon. Not. R. Astron. Soc., Vol. 227, No. 2, p. 453 – 459 (1987).

Some theorists have conjectured that strings might make a dominant contribution to the cosmological density parameter Ω. The authors show that this possibility can be ruled out unless the parameter $(G\mu)$, a measure of the mass per unit length of cosmic strings, is below $10^{-14} – 10^{-15}$. Strings heavier than this would produce characteristic effects on observed quasar variability and on pulsar periods. A network of strings whose random motions were sub–relativistic would (unlike more "ordinary" material) make no contribution to the cosmic deceleration. It would, however, contribute to the global curvature, and could, if Ω were high enough, focus light through an "antipode" with an accessible redshift. Observations of gravitationally lensed quasars can thereby set an upper limit [independent of $(G\mu)$] to the value of Ω contributed by this kind of string network.

161.032 Microwave anisotropy constraints on isocurvature baryon models.

G. Efstathiou, J. R. Bond.
Mon. Not. R. Astron. Soc., Vol. 227, No. 3, p. 33P – 38P (1987).

The authors calculate the microwave background anisotropies expected in baryon–dominated universes with power law initial entropy perturbations. If recombination of the Universe proceeds in the standard way, only models with $\Omega_B \sim 1$ are compatible with the small angle anisotropy constraints. If the Universe remained ionized for an extended period, the constraints at $\theta \sim 6°$ limit the initial spectral index to $n > -2$. A modest improvement in the anisotropy limits at small angular scales could exclude all reasonable isocurvature baryon models whatever the recombination history. The coherent scale for the linear polarization of the temperature fluctuations would provide a test of early re–ionization.

161.033 Large–scale structure of the Universe. General report on the seminar, September 15 – 21, 1986, Special Astrophysical Observatory.
J. Einasto, I. D. Karachentsev, L. A. Kofman, S. F. Shandarin, A. A. Starobinsky (*A. A. Starobinskij*).
Soobshch. Spets. Astrofiz. Obs., Vyp. 53, p. 9 – 27 (1987). – See Abstr. 012.008.
Recently, the problems of the large–scale structure, composition and evolution of the Universe beginning from the earliest stages have occupied the central place in cosmology. It becomes increasingly obvious that the problems of the relic background radiation anisotropy, dark matter and large–scale distribution of matter are closely interconnected in the framework of cosmological models describing the Universe as a whole. Now the perspective arises to unite both the contemporary picture of the Universe and the earliest stages of its evolution described by laws of microphysics.

161.034 To dynamics of pre–Friedmann Universe.
A. D. Chernin.
Soobshch. Spets. Astrofiz. Obs., Vyp. 53, p. 28 – 29 (1987). – See Abstr. 012.008.
The author discusses here a scenario for a very early Universe from the Planck era to the Friedmann "standard" epoch.

161.035 Kinematics of multiple clusters of galaxies.
A. I. Kopylov.
Soobshch. Spets. Astrofiz. Obs., Vyp. 53, p. 30 – 34 (1987). – See Abstr. 012.008.
A selection criterion for physically bound systems: $\Delta V < 2000$ km/s is determined from the analysis of radial velocities of galaxy clusters belonging to isolated pairs. Samples for 33 pairs and 12 triplets are compiled from the list of isolated multiple clusters and the catalogue of clusters with $Z < 0.085$. Velocities of galaxy clusters in tight (with dimensions $\sim 10 h_{50}^{-1}$ Mpc) isolated systems are estimated to be $\Delta V \approx 700$ km/s. The same mean velocity dispersion, but with system sizes of $\sim 50 h_{50}^{-1}$ Mpc is found for the 6 groups of clusters consisting of 5 – 7 members.

161.036 Numerical experiments with hidden mass in groups of galaxies.
A. D. Chernin, L. G. Kiseleva.
Soobshch. Spets. Astrofiz. Obs., Vyp. 53, p. 35 – 36 (1987). – See Abstr. 012.008.
The authors report on a series of numerical experiments designed to simulate the dynamics of galaxies or protogalactic clouds in a small group containing hidden mass. Typical groups are often dominated by a few massive members, and the principal features of their dynamics can be studied on the basis of simple three–body models with equal spherical masses. The experiments are focused on close passages of the members of a group in order to analyse the possibility of galaxy (or protogalaxy) merging.

161.037 The coalescence criterion of self–gravitating gaseous masses.
D. I. Barausov, A. D. Chernin.
Soobshch. Spets. Astrofiz. Obs., Vyp. 53, p. 37 – 38 (1987). – See Abstr. 012.008.
Among gas–dynamic processes playing the leading role in the cosmogony of galaxies, an essential place belongs to supersonic collisions of self–gravitating gaseous masses. Numerical models are treated taking into account radiation cooling and self–gravitation.

161.038 Possible inhomogeneities in the Universe on scales of 100 – 300 Mpc from observations with the 6–meter telescope.
A. I. Kopylov, D. Yu. Kuznetsov, T. S. Fetisova, V. F. Shvartsman.
Soobshch. Spets. Astrofiz. Obs., Vyp. 53, p. 39 – 46 (1987). – See Abstr. 012.008.
In 1986 the programme "The Northern Cone of Metagalaxy" has been finished with the 6–meter telescope. In the course of the programme, redshifts of all very rich compact clusters of galaxies inside the cone with galactic latitude $b^{II} > 60°$ and indirect estimates of redshifts $z_{LB} < 0.28$ have been measured. The total volume of the investigated regions is $V \approx 500 \times 10^6$ Mpc^3 (Hubble constant $H = 50$ km/s Mpc^{-1}). A number of possible indications of existence of inhomogeneities in the Universe on scales greater than 100 Mpc are obtained. No indications that the topology of the Universe is non–Euclidean on scales of 20 – 200 Mpc are found.

161.039 Large–scale distribution of objects from the First Byurakan Survey (FBS).
V. A. Lipovetskij.
Soobshch. Spets. Astrofiz. Obs., Vyp. 53, p. 47 – 53 (1987). – See Abstr. 012.008.
The large–scale structure of the spatial distribution of Markarian galaxies is considered. It is shown that Markarian galaxies occupy the same space regions as the normal galaxies, excluding the central regions of rich clusters in which the density contrast of objects with UV–excess is significantly less. A giant void in the regions of Camelopardalis, Lynx and Ursa Major, whose possible existence was reported by Bahcall and Soneira (1982), is investigated. Its boundaries are specified, the void sizes are determined to be $250 \times 150 \times 100$ Mpc. It is shown that the known void in the region of Bootis is well filled by Markarian galaxies and probably consists of some smaller voids.

161.040 Search for the active objects with low selection effects.
V. L. Afanas'ev, S. N. Dodonov, H. Lorenz, V. Yu. Terebizh.
Soobshch. Spets. Astrofiz. Obs., Vyp. 53, p. 54 – 56 (1987). – See Abstr. 012.008.
Analysis of the large–scale structure of the Universe is based on samples of objects. The completeness of a sample of objects is the main factor of its usefulness for such analysis. The authors discuss the completeness of well–known surveys and present a new method for the search and identification of objects with low selection effects.

161.041 Small–scale fluctuations of the cosmic microwave background radiation and prospects of the future continuation of the "COLD" experiment.
A. A. Starobinskij.
Soobshch. Spets. Astrofiz. Obs., Vyp. 53, p. 57 – 60 (1987). – See Abstr. 012.008.

161.042 Large–scale motion of galaxies and quadrupole anisotropy of the microwave background radiation.
V. M. Yudin.
Soobshch. Spets. Astrofiz. Obs., Vyp. 53, p. 61 – 62 (1987). – See Abstr. 012.008.
The value of the quadrupole anisotropy $\Delta T/T$ in the cosmological model with $\Omega = 1$ and the flat spectrum ($n = 1$) of the growing mode of scalar perturbations is found. It is assumed that the motion of large–scale volume ($\sim 100 \, h^{-1}$ Mpc) with a velocity of 700 km/s relative to the MBR reference frame is produced by large–scale inhomogeneities of dark matter distribution. Linear size of the volume cannot be more than $200 \, h^{-1}$ Mpc.

161.043 Spectral–spatial fluctuations of relic radiation and large–scale structure of the Universe.
V. K. Dubrovich.
Soobshch. Spets. Astrofiz. Obs., Vyp. 53, p. 63 – 66 (1987). – See Abstr. 012.008.
The author considers the three mechanisms for the generation of spectral spatial fluctuations of relic radiation.

161.044 The anisotropy of the microwave background: space experiment "RELICT".
I. A. Strukov, D. P. Skulachev, A. A. Klypin, M. V. Sazhin.
Soobshch. Spets. Astrofiz. Obs., Vyp. 53, p. 67 – 75 (1987). – See Abstr. 012.008.
The authors carried out the first satellite experiment on searching for anisotropies of the microwave background. The

main goal of the experiment was to obtain a radio brightness map of the sky at 8 mm wavelength band with the aims: (a) to determine the angular distribution of the cosmological background radiation; and (b) to estimate or to get constraints on parameters of cosmological models; and (c) to improve the accuracy of the determination of the velocity vector describing our motion with respect to the relic radiation. In addition the experiment enabled study of the stability of millimeter radio receivers under long–term use in outer space in order to optimize the parameters of the next generation of instrumentation.

161.045 Crucial experiments in cosmology.
V. N. Lukash, I. D. Novikov.
Soobshch. Spets. Astrofiz. Obs., Vyp. 53, p. 76 – 83 (1987). – See Abstr. 012.008.
This report is based mainly on the talk given at the Symposium 124 "Observational Cosmology" in Beijing, August 25 – 30, 1986. A current problem of determining the fundamental constants of the Universe – H_0, q_0, Ω, Λ, hidden matter and primordial perturbation parameters – is considered. A new look at the $\Delta T/T$ problem on large scales, which appeared after the Davies' experiment in 5 – 15° (1986), is discussed. The authors also formulate the theory requirements for the future large–scale experiment in $\Delta T/T$.

161.046 "COLD" experimental data as a test on the initial inhomogeneity of the Universe.
P. D. Nasel'skij.
Soobshch. Spets. Astrofiz. Obs., Vyp. 53, p. 84 – 85 (1987). – See Abstr. 012.008.
The experiment "COLD" enables us to come quite close (and even to exceed, for a certain set of models) the theoretical estimate of the anisotropy level of the microwave background radiation.

161.047 Modeling as a method of analysis of the correlation function for samples of rich clusters of galaxies.
D. Yu. Kuznetsov, V. A. Lipovetskij.
Soobshch. Spets. Astrofiz. Obs., Vyp. 53, p. 92 – 95 (1987). – See Abstr. 012.008.
The unexpected result of the "Northern Cone" programme was the peak on the spatial two–point correlation function on the scale of 250 Mpc. The statistical significance of this peak is investigated using the random catalogues with the artificial clusterization of objects.

161.048 Het uitdijende heelal. Nog problemen rond succesvolle oerknaltheorie.
A. Achterberg.
Zenit, 14. Jaarg., Nr. 10, p. 329 – 335 (1987).

161.049 Ripples in the universal Hubble flow.
P. H. Andersen.
Phys. Today, Vol. 40, No. 10, p. 17 – 19 (1987).

161.050 Large–scale structure, streaming and galaxy formation.
P. H. Andersen.
Phys. Today, Vol. 40, No. 10, p. 19 – 21 (1987).

161.051 Observations and nonstandard FRW models.
A. A. Coley.
Astrophys. J., Vol. 318, No. 2, p. 487 – 506 (1987).
The observational predictions of various (nonstandard) cosmological models which have the Friedmann–Robertson–Walker (FRW) metric as their spacetime geometry are investigated. All of the models considered are exact solutions of Einstein's field equations such that the laws of thermodynamics are satisfied, in which the source of the gravitational field is a non–comoving imperfect fluid, and, in the case of the two–fluid models, a second comoving perfect fluid radiation field. The general numerical predictions of the models are found to be in accord with actual observations. In particular, the current relative velocity of our Galaxy with respect to the cosmic microwave background, the

distance versus redshift relationship and primordial synthesis of helium are discussed in terms of these models.

161.052 The formation of families of twin galaxies by string loops.
L. L. Cowie, E. M. Hu.
Astrophys. J., Lett. Ed., Vol. 318, No. 2, p. L33 – L38 (1987).
The authors argue that if strings are responsible for many of the quasar lens systems and for galaxy formation, then nearby string loops should produce a substantial number of groups of near identical twin galaxies with near constant separations. Groups with a few twin members should present a striking doubled appearance and should be easily recognizable. The authors discuss one example of this type of object (found serendipitously in a CCD frame of a distant quasar) which contains four such identical twins with separation from 2″0 to 2″5 lying in a region about 20″ × 20″.

161.053 The density matrix of the universe.
S. W. Hawking.
Phys. Scr., Vol. T15, p. 151 – 153 (1987). – See Abstr. 012.014.
The quantum state of the universe may be described by a wave function which is a functional of the metric and matter fields of a hypersurface which divides the spacetime into two parts. If the spacetime is not simply connected, this hypersurface will in general have more than one connected component. Most physical observables depend only on a single connected component and can be calculated from a density matrix obtained by integrating over the unobserved surfaces.

161.054 Eternally existing self–reproducing inflationary universe.
A. D. Linde.
Phys. Scr., Vol. T15, p. 169 – 175 (1987). – See Abstr. 012.014.
It is shown that the large–scale quantum fluctuations of the scalar field φ generated in the chaotic inflation scenario lead to an infinite process of self–reproduction of inflationary mini–universes. A model of an eternally existing chaotic inflationary universe is suggested. It is pointed out that whereas the universe locally is very homogeneous as a result of inflation, which occurs at the classical level, the global structure of the universe is determined by quantum effects and is highly non–trivial. The universe consists of an exponentially large number of different mini–universes, inside which all possible (metastable) vacuum states and all possible types of compactification are realized.

161.055 Generic and nongeneric world models.
Z. A. Golda, M. Szydłowski, M. Heller.
Gen. Relativ. Gravitation, Vol. 19, No. 7, p. 707 – 718 (1987).
Catastrophe theory methods are employed to obtain a new classification of those world models which can be presented in the form of gradient dynamical systems. Generic sets and structural stability of models in the potential space are strictly defined. It is shown that if a cosmological model is required to be Friedmann and generic, it must be flat.

161.056 Unstable neutrinos can do it!
M. M. Vasanthi.
Gen. Relativ. Gravitation, Vol. 19, No. 8, p. 763 – 770 (1987).
The author presents a cosmological scenario with unstable neutrinos which decay into a light neutrino and a relativistic boson. Theoretical and observational constraints severely narrow the values of mass and of lifetime of neutrinos. However, within this range, one can construct models with (1) $\Omega = 1$, (2) age of the universe $\gtrsim 13$ billion years and (3) $h_0 \cong 0.5$. The dynamical modeling shows that (a) the initial condensates of primordial ν_L are disrupted by the decay, lowering their masses to acceptable values $\sim 10^{12} M_\odot$, (b) the relativistic boson contributes nearly 0.25 to Ω.

161.057 **Uncertainty principle and the horizon size of our universe.**
T. Padmanabhan, T. R. Seshadri.
Gen. Relativ. Gravitation, Vol. 19, No. 8, p. 791 – 796 (1987).

Quantum uncertainties prevent simultaneous measurement of the expansion factor $S(t)$ and its time derivative $\dot{S}(t)$. Consequently the "Hubble size" $(\dot{S}/S)^{-1}$ has an inherent uncertainty in the quantum state that describes the semiclassical evolution of the universe. The authors show that the quantum uncertainty in the Hubble size of the universe is amplified to unacceptably large values in any inflationary process.

161.058 **Thermodynamics and general relativity could determine the geometry of the universe.**
S. S. Bayin.
Gen. Relativ. Gravitation, Vol. 19, No. 9, p. 899 – 906 (1987).

The author introduces a suggestive model where certain quantities in Friedmann models are treated like their thermodynamic counterparts; temperature entropy, Gibbs energy, and so on. Within this model, changes in the symmetry of the universe are interpreted as first- or second–order phase transitions. The thermodynamics thus introduced determines the geometry of the universe. By choosing a specific local equation of state, the author shows that it is always more advantageous for the universe to be in a Bianchi V (open) symmetric state.

161.059 **High–amplitude peaks of density perturbations and primordial black holes formation in the dust–like universe.**
N. A. Zabotin, P. D. Nasel'skij, A. G. Polnarev.
Astron. Zh., Tom 64, Vyp. 4, p. 673 – 685 (1987). In Russian. English translation in Sov. Astron., Vol. 31, No. 4.

Formation of primordial black holes (PBHs) is discussed at early dust–like stages of the Universe expansion. These stages arose due to the domination of supermassive unstable particles in the density of the very early universe. Methods of random field peaks theory are applied to calculate the PBHs' mass spectrum. The analysis of the fragmentation role is made at the nonlinear stage of density perturbation evolution. It is shown that, if the frozen concentration of supermassive particles is small enough, the probability of PBH formation is below the limit set by the available astrophysical data, even at comparatively high level of the initial metric perturbations ($\delta \cong 10^{-2} - 10^{-3}$).

161.060 **Dipole and quadrupole anisotropy of the microwave background radiation in models with flat spectrum of scalar perturbations.**
V. M. Yudin.
Astron. Zh., Tom 64, Vyp. 4, p. 865 – 867 (1987). In Russian. English translation in Sov. Astron., Vol. 31, No. 4.

Dipole and quadrupole anisotropy of the temperature of microwave background radiation are calculated for a model of the Universe with $\Omega = 1$ and flat spectrum of growing mode of scalar perturbations. If dipole anisotropy $\Delta T/T$ is produced by large–scale ($\sim 100\ h^{-1}$Mpc) inhomogeneities of dark matter distribution, then the estimated value of the quadrupole anisotropy is $(\Delta T/T)_q = 1.2 \times 10^{-5}$.

161.061 **Cosmological field theory for observational astronomers.**
Ya. B. Zel'dovich.
Astrophys. Space Phys. Rev., Vol. 5, p. 1 – 37 (1987). – See Abstr. 003.002.

Contents: 1. Introduction. 2. Well–established new physics. 3. Hypothesized new physics. 4. An interlude: particles and fields. 5. The scalar field. 6. An overview of the history of scalar field theory: Renormalization in electrodynamics. The Higgs field. 7. Properties of the scalar field. 8. The scalar field in an expanding universe. 9. Higgs–based Kirzhnitz–Linde inflation. 10. Linde inflation with monotonic $V(\phi)$. 11. Behavior of the scalar field under expansion and compression. 12. Some cosmological implications.
For the Russian original see 42.161.401.

161.062 **The big bang – the origin and evolution of the universe.**
I. Nicolson.
Astron. Now, Vol. 1, No. 4, p. 29 – 36 (1987).

The big bang theory, proposed by the Belgian abbé Georges Lemaitre in 1927 and improved by George Gamow in the 1940s, has been very successful in explaining the expansion of the universe which we observe today.

161.063 **Formation and evolution of clusters of galaxies.**
A. Kashlinsky.
NRAO Workshop, No. 16, p. 37 – 49 (1986). – See Abstr. 012.015.

In this review the author discusses various aspects and uncertainties in cluster formation and evolution. At the moment it is hard to draw any definite conclusions regarding the way structures in the Universe have formed, but the picture whereby structures form from small scales to large ones may be in (marginally?) better agreement with data.

161.064 **Evidence for a low–density inflationary universe?**
L. I. Onuora.
Astrophys. Space Sci., Vol. 136, No. 1, p. 11 – 15 (1987).

The inflationary universe model predicts the density parameter Ω_0 to be ~ 1.0 with the cosmological constant Λ_0 usually taken to be zero, whereas observational estimates give $\Omega_0 \leqslant 0.2$ and $\Lambda_0 \sim 10^{-57}$cm^{-2}. It was found, however, that the observed variation of angular diameter with redshift for extragalactic radio sources could be interpreted in terms of a low density universe with linear size evolution of the sources for either an inflationary model with $\Lambda \neq 0$ or an open model with $\Lambda = 0$.

161.065 **Bianchi type–I universe in the presence of zero–mass scalar fields.**
D. R. K. Reddy, R. Venkateswarlu.
Astrophys. Space Sci., Vol. 136, No. 1, p. 17 – 20 (1987).

An exact Bianchi type–I cosmological model in the presence of zero–mass scalar fields is obtained when the source of the gravitational field is a perfect fluid with pressure equal to energy density. Some properties of the model are discussed.

161.066 **Pregalactic–primordial low–mass stars.**
N. Kiziloğlu, D. Eryurt–Ezer.
Astrophys. Space Sci., Vol. 136, No. 1, p. 83 – 90 (1987).

The main–sequence positions as well as the evolutionary behavior of Population III stars up to an evolution age of 2×10^{10}yr, taking this time as the age of the Universe, have been investigated in the mass range 0.2 and $0.8\ M_\odot$. While Population III stars with masses greater than $0.3\ M_\odot$ develop a radiative core during the approach to the main sequence, stars with masses smaller than $0.3\ M_\odot$ reach the main sequence as wholly convective stars. Population III stars with masses greater than $0.5\ M_\odot$ show a brightening of at most 2.2 in bolometric magnitude when the evolution is terminated as compared to the value which corresponds to zero–age main sequence. The positions of stars with masses smaller than $0.5\ M_\odot$ remain almost the same in the H–R diagram. If Population III stars have formed over a range of redshifts $6 < Z < 1500$, the original starlight of low–mass Population III stars could now be part of infrared and/or microwave background spectrum between 7 and 1400 μm.

161.067 **A note on variable–G cosmologies.**
H. H. Soleng.
Astrophys. Space Sci., Vol. 136, No. 1, p. 109 – 111 (1987).

An example of a cosmological model with variable gravitational coupling G and a time–dependent cosmological term Λ, has recently been presented. It has been shown, that there is no creation of matter and that the rest mass of particles stays constant in this model. In this paper the author generalizes the field equations to the case where both G and Λ depend both on time and position. It is shown that even in this case there may be no creation.

161.068   A static conformally flat cosmological model in Lyra's manifold.
D. R. K. Reddy, R. Venkateswarlu.
Astrophys. Space Sci., Vol. 136, No. 1, p. 183 – 186 (1987).

A static conformally flat spherically–symmetric perfect fluid cosmological model based on Lyra's modified Riemannian geometry is proposed. Some properties of the model are discussed.

161.069   Pre–galactic shocks: influence on density perturbations.
J. Madsen.
Mon. Not. R. Astron. Soc., Vol. 228, No. 1, p. 229 – 242 (1987).

The author briefly reviews the explosive galaxy formation scenario and the interaction between shocks and spherical density perturbations. A number of important time–scales is derived and compared. The fate of density perturbations crossed by shock waves is described for a range of parameter choices. A discussion of the results is given and their relevance for theories of galaxy formation is outlined.

161.070   Non–Gaussian statistics and the microwave background radiation.
P. Coles, J. D. Barrow.
Mon. Not. R. Astron. Soc., Vol. 228, No. 2, p. 407 – 426 (1987).

The authors show how to calculate statistical properties of non–Gaussian random fields. They apply this method to determine the mean size and frequency of occurrence of high and low level excursions of the Rayleigh, Maxwell, Chi–squared, lognormal, rectangular and Gumbel type I random fields. These results permit the authors to calculate the expected size and frequency of fine–scale hotspots and coldspots expected in the microwave background distribution on the sky under the assumption that it possesses non–Gaussian statistics of the above–mentioned types. This generalizes and extends previous studies. The authors also discuss whether it will be possible to determine observationally whether the underlying statistics of the temperature fluctuations in the microwave background are indeed Gaussian as predicted by the standard theory of inflation.

161.071   The dipole anisotropy of a new, colour–selected, *IRAS* galaxy sample.
R. T. Harmon, O. Lahav, E. J. A. Meurs.
Mon. Not. R. Astron. Soc., Vol. 228, No. 2, p. 5P – 10P (1987).

The authors have calculated the dipole moment of a new flux–limited sample of 10554 galaxies, selected by colour, from the *IRAS Point Source Catalog*. They find that the direction of the dipole lies within 8° of the Microwave Background Radiation dipole, depending on the corrections for confusion and contamination. The authors examine the contribution of the largest clusters and voids to the vector dipole.

161.072   The magnitude–redshift relation in a perturbed Friedmann universe.
M. Sasaki.
Mon. Not. R. Astron. Soc., Vol. 228, No. 3, p. 653 – 669 (1987).

A general formula for the magnitude–redshift relation in a linearly perturbed Friedmann universe is derived. The formula does not assume any specific gauge condition, but the gauge-invariance of it is explicitly shown. Then the application of the formula to the spatially flat background model is considered and the implications are discussed.

161.073   The extent of beam dilution in measurements of the Zeldovich–Sunyaev effect.
Y. Rephaeli.
Mon. Not. R. Astron. Soc., Vol. 228, No. 3, p. 29P – 33P (1987).

The importance of an appropriate selection of the telescope beam size in measurements of the Zeldovich–Sunyaev effect is illustrated by convolution of a Gaussian beam with the gas temperature and density profiles. Polytropic gas models characterized by parameters having values in the observationally-deduced (wide) range are considered. The author discusses the relevance of this consideration to interpretation of recent measurements of the Zeldovich–Sunyaev effect.

161.074   Existence and nature of dark matter in the universe.
V. Trimble.
Annu. Rev. Astron. Astrophys., Vol. 25, p. 425 – 472 (1987). – See Abstr. 003.003.

Contents: 1. Historical introduction and the scope of the problem. 2. Single galaxies. 3. Galaxies in binaries and small groups. 4. Rich clusters, superclusters, and global considerations. 5. Intermission. 6. Observational constraints on the nature of dark matter. 7. L'envoi.

161.075   On the kinematics of the metagalaxy. II. Dependence of the Hubble parameter on the distance.
V. F. Pyatikov, Yu. V. Stabulyanets.
Tobol. Gos. Ped. Inst. Tobol'sk, 9 pp. (1987). In Russian. Abstr. in Ref. Zh., 51. Astron., 7.51.790 (1987).

161.076   Acceleration of small inhomogeneities in the Bianchi type–I Universe.
S. V. Budnik.
Inst. kosm. issled. Akad. Nauk SSSR, Prepr., No. 1094, 53 pp. (1987). In Russian. Abstr. in Ref. Zh., 51. Astron., 7.51.806 (1987).

161.077   Stability of motion in cosmological models of general relativity.
A. A. Bakhan'kov.
Mater. 2 Konf. mol. uchenykh Inst. prikl. mekh. i mat. Akad. Nauk USSR, L'vov, 1 – 3 okt., 1985. Inst. prikl. probl. mekh. i mat. Akad. Nauk USSR. L'vov, p. 12 – 15 (1987). In Russian. Abstr. in Ref. Zh., 51. Astron., 7.51.808 (1987).

161.078   Geodetic motion in a cosmological Einstein model and its stability.
A. A. Bakhan'kov.
Vestsi Akad. Nauk BSSR. Ser. Fiz.–Mat. Nauk, No. 6, p. 99 – 104 (1986). In Russian. Abstr. in Ref. Zh., 51. Astron., 7.51.809 (1987).

161.079   Model representation of a gravitational field.
I. I. Sherstobitov, V. Eh. Yurkevich.
Rost. n/D Gos. Ped. Inst. Rostov n/D, 7 pp. (1987). In Russian. Abstr. in Ref. Zh., 51. Astron., 7.51.819 (1987).

161.080   The possibility of the spontaneous creation of the Universe.
Ya. B. Zel'dovich.
Itogi Nauki i Tekhniki. Seriya Astronomiya. Tom 32. Astrophysics and space physics, p. 5 – 15 (1987). In Russian. – See Abstr. 003.004.

161.081   A satellite–based study of the anisotropy of the microwave background: the "Relikt" experiment. Results and future.
I. A. Strukov, D. P. Skulachev.
Itogi Nauki i Tekhniki. Seriya Astronomiya. Tom 32. Astrophysics and space physics, p. 320 – 327 (1987). In Russian. – See Abstr. 003.004.

161.082   Voids and velocities in initially Gaussian models for large–scale structure.
A. L. Melott.
Mon. Not. R. Astron. Soc., Vol. 228, No. 4, p. 1001 – 1023 (1987).

Various theoretical models with Gaussian random phase perturbations are tested against data on large–scale voids and bulk motions in the Universe.

161.083   Probing the early universe with quasar light.
B. Schwarzschild.
Phys. Today, Vol. 40, No. 11, p. 17 – 20 (1987).

161.084 Robertson–Walker–type universes with conformally–invariant scalar field.

S. D. Maharaj, A. Beesham.

Astrophys. Space Sci., Vol. 136, No. 2, p. 315 – 320 (1987).

Recently, Innaiah and Reddy (1985) obtained a flat Robertson–Walker–type solution for the Einstein field equations with the trace–free energy–momentum tensor of a conformally invariant scalar field as source. Here the authors show that the field equations force the scalar field to be independent of time. Furthermore, they obtain open and closed Robertson–Walker–type solutions and observe that, once again, the scalar field has to be independent of time.

161.085 Anisotropic viscous–fluid cosmological model.

A. Banerjee, M. B. Ribeiro, N. O. Santos.

Astrophys. Space Sci., Vol. 136, No. 2, p. 331 – 336 (1987).

The authors present exact solutions of a Bianchi type VI_0 viscous fluid cosmologial model. It is a generalization of the model proposed by Banerjee and Santos (1983) for Bianchi type I.

161.086 Fractal turbulence and the origin of the galaxies and of the large–scale structure.

Y.-z. Liu, Z.-g. Deng.

Astrophys. Space Sci., Vol. 136, No. 2, p. 393 – 407 (1987).

The authors suggest using a new scheme to explain the origin of galaxies and large–scale structure. In their model, they assume that the density perturbations in the early Universe are adiabatic, and once they come within the horizon, they might produce the vortices of the fractal turbulence because of the Thomson drag. A model of the fractal turbulence is also given in this paper. The results obtained show that the basic characteristics of the galaxies (mass M_g, angular momentum J_g) and the large–scale structure (fractal dimension D_f) can be explained, if the spectrum of early perturbations is the scale–free Zeldovich spectrum.

161.087 The Bianchi type–V solution in the scale–covariant theory.

A. Beesham.

Astrophys. Space Sci., Vol. 136, No. 2, p. 413 – 414 (1987).

A reply is given to some of the comments raised by Lorenz–Petzold (1986) concerning exact solutions in the scale–covariant theory of gravitation, and a minor error – not mentioned by Lorenz–Petzold (1986) – is corrected.

161.088 Robertson–Walker Lyttleton–Bondi universe with cosmological constant.

M. N. Varma, A. R. Roy.

Astrophys. Space Sci., Vol. 136, No. 2, p. 415 – 417 (1987).

The exterior field of the Robertson–Walker metric in the Lyttleton–Bondi universe with cosmological constant is considered. It is shown that in the presence of the cosmological constant, the exterior solution of this universe is simply the empty space–time of general relativity.

161.089 Cosmological evolution of active galaxies & quasars.

J. A. Peacock.

Astrophysical jets and their engines, p. 171 – 183 (1987). – See Abstr. 012.021.

161.090 Antiprotons in the Universe as cosmological test for the grand unification.

M. Yu. Khlopov, V. M. Chechetkin.

Fiz. ehlem. chastits i atom. yadra, Tom 18, No. 3, p. 627 – 677 (1987). In Russian. Abstr. in Ref. Zh., 51. Astron., 8.51.853 (1987).

161.091 Kinetic equations for ultrarelativistic particles in the Friedmann world and isotropization of relict radiation by gravitational interactions.

Yu. G. Ignat'ev, A. A. Popov.

Kazan. Gos. Ped. Inst. Kazan', 31 pp. (1987). In Russian. Abstr. in Ref. Zh., 51. Astron., 8.51.865 (1987).

161.092 Galactic systems and hidden mass.

V. V. Petrunenko.

Belorus. Politekh. Inst. Minsk, 14 pp. (1987). In Russian. Abstr. in Ref. Zh., 51. Astron., 8.51.873 (1987).

161.093 On the rotation of the Universe.

D. D. Ivanenko, V. G. Krechet.

Izv. Vuzov. Fiz., Tom 30, No. 3, p. 12 – 16 (1987). In Russian. Abstr. in Ref. Zh., 51. Astron., 8.51.889 (1987).

161.094 Cosmic particles.

H. Fritzsch.

Prog. Part. Nucl. Phys., Vol. 17, p. 1 – 49 (1986). – See Abstr. 012.024.

161.095 The consistency problems of large scale structure.

D. N. Schramm.

Prog. Part. Nucl. Phys., Vol. 17, p. 51 – 58 (1986). – See Abstr. 012.024.

The combined problems of large scale structure, the need for non–baryonic dark matter if $\Omega = 1$, and the need to make galaxies early in the history of the universe seem to be placing severe constraints on cosmological models. In addition, it is shown that the bulk of the baryonic matter is also dark and must be accounted for as well.

161.096 The spectrum of the microwave background as a probe of the early universe.

G. De Zotti.

Prog. Part. Nucl. Phys., Vol. 17, p. 117 – 141 (1986). – See Abstr. 012.024.

This paper gives a general overview of how a broad variety of thermal histories of the universe can lead to distortions of the microwave background spectrum and discusses the constraints set on such histories by the currently available observational data. Some implications on astrophysically plausible heating sources are outlined. Special attention is paid to the distortions produced by the radiative decay of massive fermions in the early universe, a topic on which some confusion seems to exist in the literature.

161.097 Does thermodynamics require our cosmos to undergo a series of contraction/expansion cycles?

E. Recami, J. M. Martinez, V. T. Zanchin.

Prog. Part. Nucl. Phys., Vol. 17, p. 143 – 152 (1986). – See Abstr. 012.024.

A unified geometrical approach to strong and gravitational interactions has been recently proposed, based on the classical methods of general relativity. According to it, hadrons can be regarded as "black–hole type" solutions of new field equations describing two tensorial metric–fields (the ordinary gravitational field, and the "strong" one). In this paper, the authors seize the opportunity for an improved exposition of some elements of the theory relevant to our present scope, and they extend the Bekenstein–Hawking thermodynamics to the "strong black holes" (SBH). They show (1) that SBH thermodynamics seems to require a new expansion of our cosmos after its "big crunch"; (2) that a collapsing star with a mass $3 – 5\ M_\odot$, once reached the neutron–star density, could re–explode tending to form a (radiating) object with a diameter of the order of 1 light–day.

161.098 Galaxy clustering and small–scale CBR anisotropy constraints on galaxy origin scenarios.

F. Lucchin.

Prog. Part. Nucl. Phys., Vol. 17, p. 153 – 171 (1986). – See Abstr. 012.024.

The problem of the origin of cosmic structures (galaxies, galaxy clusters,...) represents the crossroads of the modern cosmology: it is correlated both with the theoretical model of the very early universe and with most of the present observational data. In this context, galaxy origin scenarios are reviewed. In particular, in the first part of the paper, the cosmological relevance of the observed clustering properties of the universe is outlined; in the second part, the observational constraints, due to small–scale cosmic background radiation (CBR) anisotropies, on galaxy origin scenarios are discussed.

161.099 A quantitative approach to the topology of large–scale structure.
J. R. Gott III, D. H. Weinberg, A. L. Melott.
Astrophys. J., Vol. 319, No. 1, p. 1 – 8 (1987).

The authors describe and apply a quantitative measure of the topology of large–scale structure: the genus of density contours in a smoothed density distribution. For random phase (Gaussian) density fields, the mean genus per unit volume exhibits a universal dependence on threshold density. If large–scale structure formed from the gravitational instability of small–amplitude density fluctuations, the topology observed today on suitable scales should follow the topology in the initial conditions. The authors illustrate the technique by applying it to simulations of galaxy clustering in a flat ($\Omega = 1$) universe dominated by cold dark matter. The technique is also applied to a volume–limited sample of the CfA redshift survey.

161.100 Were the Lyman–α clouds formed from shocks?
E. T. Vishniac, G. S. Bust.
Astrophys. J., Vol. 319, No. 1, p. 14 – 27 (1987).

The authors examine the hypothesis that the Lyman–α clouds are fragments from shock waves in the intergalactic medium (IGM). Recent work in shock wave fragmentation is used to derive the typical cloud properties in terms of the shock wave properties. The favored model is one where the clouds form at a redshift between 4 and 9. Models in which a moderate UV background is present are slightly more favorable for cloud formation. The energy required, per seed explosion, is found to lie between 3×10^{54} and 2×10^{58}ergs. The fraction of the matter in the universe contained in the intergalactic medium is required to be between 10^{-2} and 10^{-3}.

161.101 Background to the Big Bang theory of the origin of the Universe.
R. C. Maddison.
1988 Yearbook of Astronomy, p. 156 – 171 (1987). – See Abstr. 003.006.

161.102 A glow from the past.
A. E. Wright.
1988 Yearbook of Astronomy, p. 172 – 175 (1987). – See Abstr. 003.006.

161.103 The small–scale microwave–background anisotropy in neutrino–decay models of the Universe.
P. D. Nasel'skij, I. D. Novikov, L. I. Reznitskij.
Sov. Astron., Vol. 30, No. 6, p. 625 – 633 (1987). English translation of 42.161.408.

161.104 Synthesis of light elements in unconventional model cosmologies.
B. V. Vajner, Yu. A. Shchekinov.
Sov. Astron., Vol. 30, No. 6, p. 634 – 637 (1987). English translation of 42.161.409.

161.105 Massive neutrinos and the small–scale microwave–background anisotropy.
P. D. Nasel'skij, A. G. Polnarev.
Sov. Astron., Vol. 30, No. 6, p. 638 – 642 (1987). English translation of 42.161.410.

161.106 Large–scale streaming of cluster elliptical galaxies.
P. A. James, R. D. Joseph, C. A. Collins.
Mon. Not. R. Astron. Soc., Vol. 229, No. 1, p. 53 – 59 (1987).

The authors have investigated large–scale deviations from isotropic Hubble flow using first–ranked cluster elliptical galaxies. The sky coverage of the sample used precludes a general solution for a dipole anisotropy, but it is well suited to the investigation of the component of any motion of these galaxies along a direction defined by the dipole anisotropy in the 2.7K cosmic background radiation. The authors find the cluster

ellipticals to have a velocity component along this direction of ~ 600 km s^{-1}.

161.107 Cosmological aspects of superstring models.
P. Binétruy.
Cosmology and particle physics, p. 1 – 21 (1987). – See Abstr. 012.028.

This paper considers the cosmological aspects of supersymmetry breaking in "superstring models". The most interesting aspects are related to the presence of flat directions in the scalar potential (vacuum degeneracies). These flat directions are discussed both in the hidden sector of these models (do they give rise to inflation?) and in the observable sector of quarks, leptons and Higgs particles, in connection with baryogenesis.

161.108 Structure formation in $\Omega = 1$ CDM universes and the biasing mechanism.
J. R. Bond.
Cosmology and particle physics, p. 22 – 34 (1987). – See Abstr. 012.028.

The properties of structures forming in $\Omega = 1$, $h = 0.5$ universes dominated by cold dark matter (CDM) are considered using the theory of peaks of primordial density fields, assuming they are initially Gaussian. A tentative physical explanation for why bright galaxies would be likely to form only at high peaks of the primordial density field is presented. This biasing mechanism is shown quantitatively to be sufficient to reconcile $\Omega = 1$ with dynamical estimates of the cosmological density parameter. This model disagrees with observational determinations of large scale streaming velocities and the cluster–cluster correlation function.

161.109 Cosmic strings, galaxy formation and peculiar velocities.
R. H. Brandenberger.
Cosmology and particle physics, p. 35 – 47 (1987). – See Abstr. 012.028.

This paper attempts to explain what cosmic strings are and how they can lead to structures on cosmological scales such as galaxies and clusters of galaxies. It summarizes some recent work on the peculiar large–scale streaming velocities predicted in the cosmic string theory of galaxy formation.

161.110 Galaxy formation with baryonic infall: implications for galaxy dynamics, decaying dark matter and dark matter detection.
R. A. Flores.
Cosmology and particle physics, p. 48 – 62 (1987). – See Abstr. 012.028.

The structure and properties of protogalaxies have been extensively studied in the gravitational instability model. In order to relate these protostructures to real galaxies, however, one must quantify the effect of the dissipational infall of its visible matter on the mass distribution of its dark matter (DM). The author describes a simple analytic model to do this, that has been extensively checked by numerical simulations. He discusses the implications of the model for the dynamics of disk galaxies, cosmologies with decaying DM and detection of DM with superheated superconducting colloid detectors.

161.111 The formation of structure in particle–dominated cosmologies.
C. S. Frenk.
Cosmology and particle physics, p. 63 – 79 (1987). – See Abstr. 012.028.

The assumption that the "missing mass" consists of weakly interacting elementary particles leads to specific predictions for the formation of cosmic structures. Cosmologies with different kinds of dark matter are discussed. The most successful assumes that the universe is flat and is dominated by cold collisionless relics. This model accounts for most aspects of the large scale appearance of the galaxy distribution and for the observed abundances and gross properties of bound structures ranging from galaxy halos to rich galaxy clusters.

161.112   **Large scale streaming velocities in biased galaxy formation.**
B. Grinstein.
Cosmology and particle physics, p. 94 – 105 (1987). – See Abstr. 012.028.
Even over truly large scales, the velocity of objects whose density is described by biasing cannot, in general, be accurately calculated by means of linear perturbation theory. An introductory account of the methods used to reach this conclusion is given. The implications are briefly discussed.

161.113   **Cosmic string searches.**
C. J. Hogan.
Cosmology and particle physics, p. 116 – 122 (1987). – See Abstr. 012.028.
The author discusses observational strategies for finding effects associated with the gravitational lensing of distant objects by strings. In particular, the requirements of a survey to find chains of galaxy image pairs or single galaxies with sharp edges are studied in some detail, and a proposed search program at Steward Observatory is described.

161.114   **An attempt to relate the linear evolution of primordial perturbations to the non–linear structure of the universe.**
F. Occhionero, R. Scaramella.
Cosmology and particle physics, p. 123 – 132 (1987). – See Abstr. 012.028.
The authors attempt to calculate the present distribution of galaxies from the linear growth of primordial adiabatic fluctuations. The latter are studied in universe models dominated by dark matter, in either one of its three popular varieties – hot, warm, and cold – or in some mixture thereof (hybrid models).

161.115   **A slow rollover phase transition in the Schrödinger picture.**
S.-Y. Pi.
Cosmology and particle physics, p. 166 – 175 (1987). – See Abstr. 012.028.
The present status of our understanding of the slow–rollover transition in the new inflationary universe is reviewed and a time–dependent variational approximation in the Schrödinger picture is proposed as the perturbative scheme for studying the quantum theory of the transition. Validity of the approximation is discussed using a double–well potential, in one–dimensional quantum mechanics.

161.116   **Strings, the Peebles screed, and large scale structure.**
D. N. Schramm.
Cosmology and particle physics, p. 187 – 196 (1987). – See Abstr. 012.028.
The combined problems of large scale structure, the need for non–baryonic dark matter if $\Omega = 1$, and the need to make galaxies early in the history of the universe seem to be placing severe constraints on cosmological models. In addition, it is shown that the bulk of the baryonic matter is also dark and must be accounted for as well. The arguments for dark matter are reviewed and it is shown that observational dynamical arguments and nucleosynthesis are all still consistent at $\Omega \sim 0.1$. However, the inflation paradigm requires $\Omega = 1$, thus, the need for non–baryonic dark matter. A review of possible dark matter candidates is presented.

161.117   **Cold dark matter candidates.**
M. Srednicki.
Cosmology and particle physics, p. 197 – 200 (1987). – See Abstr. 012.028.
The possibility of detection of indirect signatures of plausible cold dark matter candidates is discussed.

161.118   **The singularities in quantum cosmology.**
A. M. Finkelstein (*A. M. Finkel'shtejn*),
V. Ya. Kreinovich (*V. Ya. Krejnovich*).
Astrophys. Space Sci., Vol. 137, No. 1, p. 73 – 76 (1987).
A cosmological model is proposed, in which quantum effects lead to infinitely many terms nonlinear in curvature, and these terms lead to the following conclusion: the physical singularity (i.e., a state with infinite matter density ϱ) occurs in $t \cong 10^{-43}$s, when the curvature is finite. An attempt is made to give physical interpretation of this model from the viewpoint of the "foam–like" model of space–time.

161.119   **Slowly–rotating cosmological perfect fluids.**
R. K. Tarachand, N. I. Singh.
Astrophys. Space Sci., Vol. 137, No. 1, p. 85 – 91 (1987).
The Einstein field equations for a perfect fluid distribution representing slowly–rotating fluid spheres are investigated. By imposing restrictions on the matter rotation $\omega(r,t)$ which is related to the dragging of inertial frames, and a uniform rotation which is a function of time, the general solutions for $\Omega(r,t)$ are obtained for all cosmological models. In the case of closed models the solutions for $\Omega(r,t)$ may represent realistic astrophysical situations only when the radial distance is greater than –1 and less than + 1.

161.120   **Radial hypothesis.**
E. Rydzyńska.
Astrophys. Space Sci., Vol. 137, No. 1, p. 183 – 187 (1987).
In this paper the author shows the consequences of Bellert's (1969, 1970, 1977) cosmological theory. A so–called "radial hypothesis", which characterises "metric with an observer", is presented. Problems connected with this hypothesis are discussed; and it is shown that from Bellert's theory the quantized time acceptance follows.

161.121   **Primordial magnetic field, inflation and cosmic strings.**
H. H. Soleng.
Astrophys. Space Sci., Vol. 137, No. 1, p. 201 – 203 (1987).
The possibility of a primordial magnetic field is discussed. The formation of closed, superconducting cosmic strings before the end of inflation is pointed out as a mechanism able to preserve a magnetic field inside the loops which could provide seeds for galactic magnetic fields.

161.122   **Gravitational lensing effect on the fluctuations of the cosmic background radiation.**
A. Blanchard, J. Schneider.
Astron. Astrophys., Vol. 184, No. 1/2, p. 1 – 6 (1987).
The theory of optics in general relativity ensures that only redshift terms in gravitational lensing can induce fluctuations in a strictly uniform background. In this paper the authors investigate the possible consequences of randomly distributed deflectors on an initially non–uniform background. They find a general expression giving the power spectra of the perturbed background. This allows them to show that the variance of the fluctuating part of the background is conserved by any arbitrary physical deviation field. Using a model of the gravitational deviation field developed by Blandford and Jaroszynski, the authors apply their result to the cosmic background radiation, and show that the angular fluctuations law could be modified at small angular scales. Finally, they give a rough estimate of this effect. It appears that its strength depends strongly on the total mass present in a clumpy form and on the evolution of the correlation function at small scales.

161.123   **Fragmenting the Universe. I. Statistics of two–dimensional Voronoi foams.**
V. Icke, R. van de Weygaert.
Astron. Astrophys., Vol. 184, No. 1/2, p. 16 – 32 (1987).
The authors present a Monte Carlo study of the matter distribution in a kinematical model of superclustering in the Universe. It has been shown (Centrella and Melott, 1983; Icke, 1984) that the regions of lower than average density in the early Universe become more and more spherical as time goes by. Thus,

the large scale morphology of the high–density baryonic material in the Universe, consisting of "clusters" in the form of pancakes, filaments, and nodes is obtained when matter streams away from a distribution of low–density expansion centres ("nuclei") and collects in the interstices of a close packing of spheres. This naturally leads to a partitioning of space generated by a process known as Voronoi tessellation. The authors have studied the statistical properties of specific instances of these tessellations, which they call Voronoi foams, for several model distributions of expansion centres.

161.124 The possibility of a single fragmentation law for the formation of different astronomical objects.
A. Di Fazio, R. Capuzzo Dolcetta.
Astron. Astrophys., Vol. 184, No. 1/2, p. 263 – 268 (1987).
In the light of the growing interest for the modes of formation of various astronomical objects, a relevant amount of data were collected and treated to obtain mass spectra. Statistical comparisons of the data suggest that a single process drives the formation of the various self–gravitating objects.

161.125 Quantum creation of the universe in $N = 8$ supergravity.
Yu. P. Goncharov, A. A. Bytsenko.
Astrofizika, Tom 27, Vyp. 1, p. 147 – 158 (1987). In Russian. English translation in Astrophysics, Vol. 27, No. 1.
The authors discuss a possibility of quantum creation for an inflationary universe filled with the fields of the maximal extended $N = 8$ supergravity. If the created universe has the spatial $(S^1)^3$ topology and after creation the Starobinsky inflationary scenario occurs owing to the topological Casimir effect in $N = 8$ supergravity, then one can estimate the creation probability of such a universe in the semiclassical approximation. The rate which is obtained shows the birth of the universe with more isotropic topology to be more plausible.

161.126 Statistical analysis of a sample of first–ranked cluster galaxies.
V. E. Yakimov.
Astron. Zh., Tom 64, Vyp. 5, p. 910 – 918 (1987). In Russian. English translation in Sov. Astron., Vol. 31, No. 5.
The least–squares best fit values of observational parameters to homogeneous Friedmann models and their 95–per cent confidence regions have been computed for the sample of 96 first–ranked cluster galaxies with redshifts $0 < z < 0.4$. The effects of the evolutionary correction, photon scattering by free electrons and selective extinction by intergalactic dust on the redshift – magnitude relation have been taken into account. Constraints on the values of the deceleration parameter and cosmological constant are found which are $-1.1 \lesssim q_0 \lesssim 0.6$ and $-0.5 \times 10^{-56} \lesssim \varLambda \lesssim 1.4 \times 10^{-56} \mathrm{cm}^{-2}$, respectively, if the Hubble parameter is equal to 50 km/(s Mpc).

161.127 Cosmological test of the theory of superstrings.
Priroda, No. 8, p. 109 – 110 (1987). In Russian. Summary of Prepr. Inst. Teor. Ehksp. Fiz.–1, 1987.

161.128 The giant arcs are gravitational mirages.
M. M. Waldrop.
Science, Vol. 238, No. 4832, p. 1351 – 1352 (1987).
The arcs appear to be highly magnified and highly distorted images of far–distant galaxies; as such they could offer new insight into galactic evolution and a unique probe of cosmic dark matter.

161.129 Can the Universe be closed?
D. Lindley.
Nature, Vol. 328, No. 6128, p. 289 (1987).
Astronomers hanker after the notion that the density of the Universe is the minimum required to prevent indefinite expansion, but the evidence is not all on their side.

161.130 Decay of primordial cosmic rotation in inflationary cosmologies.
Ö. Grön, H. H. Soleng.
Nature, Vol. 328, No. 6130, p. 501 – 503 (1987).
Assuming that the universe came to being as a mini–universe of Planck dimensions which went directly into an inflationary epoch driven by a scalar field with a flat potential, the authors argue that the cosmic vorticity has decayed by a factor of about 10^{-145}, due to the non–rotation of the false vacuum and the exponential expansion during inflation. It is shown that the hypothesis that the rotation of galaxies is caused by cosmic vorticity is not compatible with the inflationary model.

161.131 Formation of large–scale structure from cosmic–string loops and cold dark matter.
A. L. Melott, R. J. Scherrer.
Nature, Vol. 328, No. 6132, p. 691 – 694 (1987).
The authors present some results from a numerical simulation of the formation of large–scale structure from cosmic–string loops. They find that even though they require $G\mu < 2 \times 10^{-6}$ (where μ is the mass per unit length of the string) to give a low enough autocorrelation amplitude, there is excessive power on smaller scales, so that galaxies would be more dense than observed. The large–scale structure does not include a filamentary or connected appearance and shares with more conventional models the lack of cluster–cluster correlation at the mean cluster separation scale as well as excessively small bulk velocities on these scales.

161.132 The emergence of structure in the universe: galaxy formation and "dark matter".
M. J. Rees.
Three hundred years of gravitation, p. 459 – 498 (1987). – See Abstr. 003.012.
Contents: 1. Introduction. 2. The constituents of the universe: "dark" matter and "luminous" matter. 3. Large–scale structure and isotropy. 4. Formation of protogalaxies. 5. A flat ($\varOmega_0 = 1$) universe.

161.133 Inflationary cosmology.
S. K. Blau, A. H. Guth.
Three hundred years of gravitation, p. 524 – 603 (1987). – See Abstr. 003.012.
Contents: 1. Introduction. 2. Summary of the standard cosmological model. 3. Problems of the standard cosmological model. 4. The original inflationary universe. 5. Successes of the original inflationary model. 6. Problems of the original inflationary model. 7. The new inflationary universe. 8. Density perturbations in the new inflationary universe. 9. Quantum theory of the new inflationary universe phase transition. 10. Inflation in the minimal $SU(5)$ grand unified theory. 11. False vacuum bubbles and child universes. 12. Conclusion.

161.134 Inflation and quantum cosmology.
A. Linde.
Three hundred years of gravitation, p. 604 – 630 (1987). – See Abstr. 003.012.
Contents: 1. Introduction. 2. Chaotic inflation. 3. Inflation and the wave function of the universe. 4. Quantum fluctuations in the inflationary universe. 5. Eternal chaotic inflation. 6. Global structure of the inflationary universe and the anthropic principle. 7. Conclusions.

161.135 Quantum cosmology.
S. W. Hawking.
Three hundred years of gravitation, p. 631 – 651 (1987). – See Abstr. 003.012.
Contents: 1. Introduction. 2. The quantum state of the universe. 3. The density matrix. 4. The Wheeler–De Witt equation. 5. Minisuperspace. 6. Beyond minisuperspace. 7. The direction of time. 8. The origin and fate of the universe.

161.136 Cosmic strings.
    A. Vilenkin.
Sci. Am., Vol. 257, No. 6, p. 52 – 55, 58 – 60 (1987).
    Why are stars and galaxies clumped rather than spread out evenly in space? What drew them together? Thin strings of energy created during the birth of the universe may have provided the attraction.

161.137 Voids and galaxies in voids.
    A. Oemler Jr.
Nearly normal galaxies. From the Planck time to the present, p. 213 – 219 (1987). – See Abstr. 012.031.
    In the past few years, the properties of voids have become central to the study of the large–scale distribution of matter in the universe. The reasons for this are partly observational – very large voids have been discovered – and partly due to their potential usefulness as measures of the clustering of galaxies. The author discusses the available information on voids.

161.138 The large scale distribution of galaxy types.
    M. P. Haynes.
Nearly normal galaxies. From the Planck time to the present, p. 220 – 226 (1987). – See Abstr. 012.031.
    The segregation of galaxy morphologies can be traced across all density regimes and shows distinctions among all Hubble types. To a large extent, the observed morphological segregation must reflect the environmental conditions at early epochs. Yet, it is also clear that on–going interactions that secularly affect the evolution of galaxies also play a role in reinforcing the segregation characteristics, at least in the highest density environments. Currently–popular gas removal mechanisms are generally not efficient over the range of densities which show segregation. At the same time, both the topology of the universe and its variation with Hubble type seems to reflect differing environmental conditions at early times. The large–scale variations in clustering characteristics must be reproduced in models that describe the pregalactic era or shortly thereafter.

161.139 Coherent orientation effects of galaxies and clusters.
    S. Djorgovski.
Nearly normal galaxies. From the Planck time to the present, p. 227 – 233 (1987). – See Abstr. 012.031.
    The evidence for alignments is neither definitive nor very dramatic, but there are far too many positive indications for the subject to be dismissed easily. Moreover, some systematics are begining to emerge: the most persuasive evidence for alignments is found in highly flattened large–scale structures (Perseus, L–type Abell clusters). In more spherical structures, the prefered orientation of galaxian axes seem to be related to their radius vectors toward the cluster center, but the overall evidence for such effects is weaker. Second, there is a clear dependence upon the morphological type: ellipticals always tend to align with the principal axis of the cluster (or the radius vector towards the cluster center), and the disks tend to be either parallel or perpendicular.

161.140 Galaxy formation and large scale structure.
    S. D. M. White.
Nearly normal galaxies. From the Planck time to the present, p. 234 – 243 (1987). – See Abstr. 012.031.
    This article deals with galaxy formation within current models for the evolution of structure. The author reviews these models and their relationship to models which were popular 10 years ago. He then concentrates on the cold dark matter model. He gives an overview of its present status and of how and where galaxies are expected to form in it. Finally the author discusses its predictions for large scale structure.

161.141 Scenarios of biased galaxy formation.
    A. Dekel.
Nearly normal galaxies. From the Planck time to the present, p. 244 – 254 (1987). – See Abstr. 012.031.
    The author argues that based on the observed correlation of the type of galaxy and its environment, it would be astonishing if galaxy formation itself were not affected by environmental effects segregating the galaxies from the underlying mass. He first summarizes the motivation behind this suspicion and then discusses bias mechanisms in the various cosmogonies, arguing that a segregation of one sort or another is a natural outcome of almost every cosmogony (although the bias is not always of the desired sort). The author concludes by trying to point at key observational tests.

161.142 Biasing and suppression of galaxy formation.
    M. J. Rees.
Nearly normal galaxies. From the Planck time to the present, p. 255 – 262 (1987). – See Abstr. 012.031.
    The author discusses biasing mechanisms in the "cold dark matter" cosmogony.

161.143 Galaxies at very high redshifts (z > 1).
    S. Djorgovski.
Nearly normal galaxies. From the Planck time to the present, p. 290 – 299 (1987). – See Abstr. 012.031.
    Observational cosmology is partly based upon wishful thinking but it seems that we are beginning to probe directly the evolution and formation of distant galaxies. But, are these far galaxies "normal"? That depends on what is "normal" at high redshifts, which we do not know.

161.144 What is the cosmological density parameter Ω_0?
    A. Yahil.
Nearly normal galaxies. From the Planck time to the present, p. 332 – 342 (1987). – See Abstr. 012.031.
    In a review two years ago, the author argued that the observational evidence was converging toward a baryon dominated open universe, with $\Omega_0 = 0.1 - 0.2$. A brief summary of these arguments is given here. The main objection was the theoretical disposition in favor of a flat universe with $\Omega_0 = 1$, which led to the idea of biased galaxy formation. A few additional dissonances remained in the interpretation of the observations, but they were not deemed major. However, a serious reassessment occurred in 1985 regarding the peculiar gravitational field in the Virgo Supercluster, and doubts were raised about the previous interpretation of the observations. This review describes these new developments, and the resulting ongoing research, which is aimed at resolving the seeming contradictions.

161.145 Fundamental physics and dark matter.
    K. Freese.
Nearly normal galaxies. From the Planck time to the present, p. 343 – 352 (1987). – See Abstr. 012.031.
    For the past ten years or so there has been a lot of speculation about the existence and behavior of nonbaryonic dark matter. The author re–examines the evidence for a non–baryonic dark component, contrasting the theoretical prejudice that $\Omega = 1$ with the Big Bang Nucleosynthesis constraints on the amount of baryonic matter, $\Omega_b \leqslant 0.2$. Assuming then that such a component does exist, he presents a list of candidates from particle physics to explain this dark matter, and illustrates what he finds to be a most exciting development, the fact that all the most likely candidates stand to be ruled out or even detected in the next five to ten years.

161.146 Inflationary universe models and the formation of structure.
    R. H. Brandenberger.
Nearly normal galaxies. From the Planck time to the present, p. 355 – 366 (1987). – See Abstr. 012.031.
    The author briefly reviews the main features of inflationary universe models. Inflation provides a mechanism which produces energy density fluctuations on cosmological scales. In the original models it was not possible to obtain the correct magnitude of these fluctuations without fine tuning the particle physics models. The author discusses two mechanisms, "chaotic inflation", and a "dynamical relaxation" process, by which inflation may be realized in models which give the right magnitude of fluctuations.

161.147 Formation and evolution of cosmic strings.
A. Albrecht.
Nearly normal galaxies. From the Planck time to the present,
p. 367 – 377 (1987). – See Abstr. 012.031.

The strings discussed here are particular field configurations
that can appear in "ordinary" field theories, regardless of
whether there is an underlying superstring theory. These strings
are often called Nielsen–Olesen strings.

161.148 The quark–hadron phase transition and primordial nucleosynthesis.
C. J. Hogan.
Nearly normal galaxies. From the Planck time to the present,
p. 378 – 387 (1987). – See Abstr. 012.031.

This paper deals with two different but related topics. In the
first part the author summarizes the current understanding of
processes occurring during the cosmological transition from
"quark soup" to normal hadron matter. In the second part he
describes what happens to cosmological nucleosynthesis in the
presence of small–scale baryon inhomogeneities – of which the
QCD phase transition is one plausible (if not the only) source.
The author reaches the happy conclusion that there is perhaps
after all something qualitatively new to learn about the early
universe from looking at nucleosynthesis products in greater
detail.

161.149 Testing cosmic fluctuation spectra.
J. R. Bond.
Nearly normal galaxies. From the Planck time to the present,
p. 388 – 397 (1987). – See Abstr. 012.031.

Recent observations of large scale streaming velocities and the
rich cluster correlation function indicate that extra power exists
on large scales over that of the cold dark matter spectrum with
$\Omega = 1$ and scale–invariant initial conditions. The extra power
required to explain the reported observations is explored, and it is
demonstrated that the two problems could be solved by adding
an $n = -1$ ramp to the density fluctuation spectrum between
wavenumbers $k^{-1} \sim 5$ and $\sim 300\ h^{-1} Mpc$. The cluster–galaxy
correlation function also comes out at the right level and shape.
Low Hubble constant models, hybrid models with hot and cold
dark matter, models with both adiabatic and isocurvature modes
present and decaying dark matter models are shown not to work.
Models that are open ($\Omega \sim 0.2$) or have a large cosmological
constant ($\Omega_{vac} \sim 0.8$) do better, but still fail. A vacuum energy
dominated model with $\Omega_{vac} = 0.8$, $\Omega_B = \Omega_X = 0.1$ does reproduce the large scale results, but it has a low redshift of galaxy
formation.

161.150 Models of protogalaxy collapse and dissipation.
G. R. Blumenthal.
Nearly normal galaxies. From the Planck time to the present,
p. 401 – 412 (1987). – See Abstr. 012.031.

When protogalaxies collapse, the cooling and infall of what
will become the visible galactic component affects the mass
distribution of dissipationless dark matter particles which constitute the halo. For spiral galaxies, the adiabatic approximation
shows that flat rotation curves arise only when about 10% of the
protogalaxy's mass is dissipational, the initial core radius is large,
and the visible matter falls in by roughly a factor of ten. For
spirals, the amount of infall is directly related to the initial
protogalaxy angular momentum. Observed variations in rotation
curves may be due to several factors including variations in
angular momentum, core radius, and bulge–to–disk ratio.

161.151 Unstable dark matter and galaxy formation.
R. A. Flores.
Nearly normal galaxies. From the Planck time to the present,
p. 413 – 420 (1987). – See Abstr. 012.031.

The author considers a cosmological solution (Turner et al.,
1984, Gelmini et al., 1984, Turner, 1985) which postulates that
the non–radiative decay of a heavy elementary particle species,
after it has driven the formation of galaxies and clusters, provides
a smooth, undetected background of relativistic particles that at
present contribute Ω_r to the total energy density of the Universe.

161.152 Halos and angular momentum generation.
C. S. Frenk.
Nearly normal galaxies. From the Planck time to the present,
p. 421 – 430 (1987). – See Abstr. 012.031.

Consequences of N–body simulations of the growth of
fluctuations in an $\Omega = 1$ universe dominated by cold dark matter
(CDM) are discussed. The calculations presented refer exclusively to the formation of halos but nevertheless suggest possible
formation paths for galaxies of different morphological types.
Mergers play a central role and may explain the high luminosity,
slow rotation and enhanced clustering of bright ellipticals. Disks,
on the other hand can only form during extended periods of
quiescent evolution and the predicted fraction of halos which
remain relatively undisturbed since a redshift of 1 is similar to the
observed fraction of spirals. Galaxy formation is predicted to be
a recent but protracted process; violent dynamical activity is
expected at low redshifts and may provide a direct observational
test of the CDM cosmogony.

161.153 Cosmic strings and the formation of galaxies and clusters of galaxies.
N. Turok.
Nearly normal galaxies. From the Planck time to the present,
p. 431 – 450 (1987). – See Abstr. 012.031.

The author discusses one class of theories which have
provoked a lot of interest recently – theories predicting cosmic
strings. In the near future many of the main issues in the cosmic
string theory will be resolved. Large scale string simulations can
in principle give clear predictions for the very large scale structure
of the universe. They will soon be "plugged in" to N–body codes
and subjected to the same tests as other models have been.

**161.154 Large scale drift and peculiar acceleration as cosmologi-
cal tests.**
N. Vittorio, R. Juszkiewicz.
Nearly normal galaxies. From the Planck time to the present,
p. 451 – 454 (1987). – See Abstr. 012.031.

The authors confront some model predictions with recent
observations by Collins et al. (1986) and Burstein et al. (1986),
suggesting that the Hubble expansion on scales $\sim 50h^{-1}Mpc$
(h $\equiv$ Hubble constant/100 km s^{-1}Mpc^{-1}) is distorted by matter
currents of surprisingly large amplitude. The authors also discuss
the theoretical conclusions that may be drawn from the analysis
of the IRAS Point Source Catalogue. These observations are
extremely important for the understanding of the large scale
distribution of galaxies. However, it is shown that the IRAS data
alone are not sufficient to make a reliable and model independent
estimate of the density parameter Ω_0.

161.155 Formation and secular evolution of elliptical galaxies.
G. Lake.
Structure and dynamics of elliptical galaxies, p. 331 – 338 (1987).
– See Abstr. 012.032 (IAU Symp. No. 127).

Three ways have been proposed to make elliptical galaxies:
cooling by gas dynamical processes at late epochs, Compton
cooling at early times and merging. These theories must address a
variety of observational constraints, the most severe being the
problems of slow rotation and high central phase densities. The
author looks at some aspects of all three theories with particular
attention to key numerical simulations and observations that can
distinguish between the scenarios.

161.156 Dissipationless formation of elliptical galaxies.
S. D. M. White.
Structure and dynamics of elliptical galaxies, p. 339 – 351 (1987).
– See Abstr. 012.032 (IAU Symp. No. 127).

Dissipationless formation mechanisms envisage elliptical galaxies as arising from the collective relaxation of an aggregate of
stars. Their key ingredients are thus a set of initial conditions
derived from consideration of prior evolution, and a treatment of
the relaxation process. The author reviews numerical studies of
violent relaxation carried out over the last decade and purely
theoretical treatments going back twice as far. Relaxation is
always incomplete, and as a result the final structure of a

"galaxy" depends sensitively on the initial conditions assumed. The viability of dissipationless formation thus rests on the identification of plausible stellar initial conditions which relax to the present structure. The author discusses the extent to which such initial conditions are compatible with current ideas on the origin of structure in the universe.

161.157 Dissipation and the formation of galaxies.
R. G. Carlberg.
Structure and dynamics of elliptical galaxies, p. 353 – 366 (1987). – See Abstr. 012.032 (IAU Symp. No. 127).

The formation of ellipticals is likely to be strongly dependent on environmental influences, such as the clustering environment and interactions with other galaxies. Ellipticals are dynamically distinct as having relatively low angular momentum content, and high random velocities. The spin of a galaxy is generated by tidal torques, reduced by stripping, and diluted by merging. These three mechanisms have been investigated for their relation to dissipation. Mergers of present day disk galaxies would not make normal elliptical galaxies in the absence of some dissipative star formation to build up the core density. However the most luminous ellipticals have sufficiently large cores that there is no objection to mergers.

161.158 Supersymmetrical particles, cosmology and astrophysics.
A. Yu. Ignat'ev.
Particles and cosmology, p. 3 – 20 (1987). In Russian. Abstr. in Ref. Zh., 51. Astron., 9.51.1066 (1987). – See Abstr. 012.034.

161.159 Neutrinos and cosmology.
Particles and cosmology, p. 21 – 33 (1987). In Russian. Abstr. in Ref. Zh., 51. Astron., 9.51.1067 (1987). – See Abstr. 012.034.

161.160 Quantum effects in the Friedmann space–time and problems of quantum generation of the Universe.
A. A. Grib.
Particles and cosmology, p. 52 – 60 (1987). In Russian. Abstr. in Ref. Zh., 51. Astron., 9.51.1077 (1987). – See Abstr. 012.034.

161.161 Cosmological anisotropic solutions of equations of the Einstein–Cartan theory with Λ –term.
V. S. Galitskij, V. N. Ponomarev.
Izv. Vuzov. Fiz., Tom 30, No. 3, p. 116 (1987). In Russian. Abstr. in Ref. Zh., 51. Astron., 9.51.1087 (1987).

161.162 Astrophysical formulas for Friedmann models with cosmological constant and radiation.
M. P. Dabrowski, J. Stelmach.
Astron. J., Vol. 94, No. 5, p. 1373 – 1379 (1987).
Exact expressions for observable quantities in Friedmann universes with cosmological constant and noninteracting matter and radiation are given using Weierstrass elliptic functions.

161.163 Op zoek naar de eenheid van het heelal.
A. Achterberg.
Zenit, 14. Jaarg., Nr. 12, p. 408 – 412 (1987).

161.164 Relationship between redshift and recession velocities in an isotropic universe.
W. Priester.
Naturwissenschaften, 74 Jahrg., Heft 12, p. 601 – 602 (1987).

161.165 Presidential address of 1987 February 13: light, gravity and galaxy streaming.
D. Lynden–Bell.
Q. J. R. Astron. Soc., Vol. 28, No. 3, p. 187 – 196 (1987).
A bulk streaming motion of elliptical and spiral galaxies over and above the Hubble expansion gives a much better fit to the observations than pure Hubble flow. However, the fit shows significant quadrupolar residuals and varies with distance. The accuracy with which a motion of the Sun fits the observed anisotropy of the Cosmic Microwave Background (CMB) is

much better. A model in which the bulk flow of galaxies is replaced by the flow generated by a spherical mass concentration centred on $l = 307$, $b = 9$, $R_m = 4350$ km s^{-1} gives an even better fit if its gravity is sufficient to generate an inflow of $V_m = 570 \pm 60$ km s^{-1} at the Sun.

161.166 Comments on smoothing cosmologies.
A. Hemmerich.
Astron. Astrophys., Vol. 185, No. 1/2, p. 1 – 3 (1987).
The author examines a procedure recently proposed by Carfora and Marzuoli that intends to deform a family of locally inhomogeneous and anisotropic spatially closed cosmological models into closed Robertson–Walker universes. Moreover, it is shown how the essential physical relations in Carfora and Marzuoli's work can be obtained within a general setting without referring to their procedure with its restrictive underlying assumptions.

161.167 Magnitude–redshift test: cosmological inhomogeneity effects.
L. J. Goicoechea, J. M. Martin–Mirones.
Astron. Astrophys., Vol. 186, No. 1/2, p. 22 – 24 (1987).
The authors consider the non–local cosmological inhomogeneity effects on the magnitude–redshift test by using a spherically symmetric relativistic model with negligible pressure. If the inhomogeneity is centered at $z \sim 0.1$ and the present relative shear satisfies $|\sigma_{rr}/H|_0 \ll 1$, the Hubble constant and an "effective" deacceleration parameter (q_0^{eff}) appear. q_0^{eff} includes the real deacceleration parameter and other multipolar terms. For a universe flat or open ($q_0 \leqslant 1/2$) the multipolar contributions can be dominant in q_0^{eff}.

161.168 Dynamics of anisotropic cosmological model with ultra-relativistic matter, magnetic field, and fluxes of free isotropic particles.
I. S. Shikin.
Gen. Relativ. Gravitation, Vol. 19, No. 10, p. 961 – 972 (1987).
Within the framework of general relativity the dynamics of a homogeneous anisotropic axially symmetric model of the Bianchi type I is considered for the case when sources of gravitational field are ultrarelativistic matter, homogeneous magnetic field, and fluxes of free particles. A qualitative analysis of the field equations on a phase plane is given. Near a singular state the solutions exhibit oscillating behavior with successive interchanges of Kasner singularities of "pancake"–like and "filament"–like types.

161.169 Nonminimal gravitational coupling: the spectrum of cosmic solutions.
M. Novello, C. Romero.
Gen. Relativ. Gravitation, Vol. 19, No. 10, p. 1003 – 1011 (1987).
The authors present a complete analysis of the set of homogeneous and isotropic cosmological solutions generated by a vector field coupled nonminimally to gravity. This model universe is interpreted as (classical) fluctuations in the infinite past of unstable Minkowskian space–time.

161.170 Spin contributions to an inflationary phase in the very early universe.
M. Gasperini.
General relativity and gravitational physics, p. 89 – 93 (1987). – See Abstr. 012.038.

161.171 Inflation in a completely anisotropic Einstein–Cartan cosmological model.
R. de Ritis, P. Scudellaro, C. Stornaiolo.
General relativity and gravitational physics, p. 101 – 104 (1987). – See Abstr. 012.038.
The Einstein–Cartan–Sciama–Kibble (ECSK) theory is a generalization of the Einstein theory in which spacetime is considered as a four dimensional manifold with an asymmetric affine connection. In this theory one finds a spin–spin gravitational contact interaction, which becomes important a very high matter densities. The authors have also examined the possible

role of torsion in the so–called inflationary scenario of the universe.

161.172 Cosmology with twice compactified internal space and higher–order gravitational Lagrangian.
R. Kerner.
General relativity and gravitational physics, p. 213 – 225 (1987).
– See Abstr. 012.038.

This paper discusses two generalizations of Kaluza–Klein type theories and shows that, even in the case of a not very realistic model including only $SU(2) \times U(1)$ symmetry of elementary interactions, such an approach can lead to cosmological implications which include a vanishing cosmological constant, primordial exponential (inflationary) behaviour, and the avoidance of the initial singularity.

161.173 Cosmological evolution of extragalactic sources.
G. De Zotti.
General relativity and gravitational physics, p. 331 – 345 (1987).
– See Abstr. 012.038.

The author presents an overview of the development of ideas on cosmological evolution of extragalactic sources in the radio, optical and X–ray bands. The most recent advances towards the understanding of the astrophysics of the evolution are briefly outlined.

161.174 Present–day remnants of phase transitions in the early universe.
S. Matarrese.
General relativity and gravitational physics, p. 347 – 355 (1987).
– See Abstr. 012.038.

At very early times in the evolution of the universe a number of phase transformations occurred which led to a large variety of relics such as: density and gravitational waves, topologically stable extended objects (monopoles, strings, domain walls, etc...), and lumps of strange matter. Most of these objects left imprints which may be discovered in today's universe looking at the complicated properties of the large–scale matter distribution.

161.175 Relativistic hydrodynamics of the cosmological quark–hadron phase transition.
J. C. Miller, O. Pantano.
General relativity and gravitational physics, p. 357 – 371 (1987).
– See Abstr. 012.038.

The authors are carrying out relativistic computations to investigate the hydrodynamics of the cosmological quark–hadron phase transition. This paper presents a discussion of the physical background to these calculations and a description of the general form of the hydrodynamical processes involved.

161.176 Observational properties of the cosmic background radiation.
G. Sironi.
General relativity and gravitational physics, p. 373 – 385 (1987).
– See Abstr. 012.038.

The observational properties of the cosmic background radiation are reviewed. They are completely consistent with the properties expected for the residual of a simple hot big bang model. No definite deviation from that simple model has been so far detected. The upper limits to distortions and to anisotropies at small angular scales imply that the universe was extremely uniform "ab initio" and evolved very smoothly or that processes postulated by the inflationary models were effective.

161.177 Computations of the cosmological quark–hadron phase transition.
O. Pantano, J. C. Miller.
General relativity and gravitational physics, p. 411 – 415 (1987).
– See Abstr. 012.038.

161.178 Large–scale structure from cosmic string loops. I. Formation and linear evolution of perturbations.
R. J. Scherrer.
Astrophys. J., Vol. 320, No. 1, p. 1 – 12 (1987).

The author presents a numerical model for the formation and linear growth of density perturbations from cosmic string loops in a universe with $\Omega = 1$ dominated by cold dark matter. Loop fragmentation produces a clustering hierarchy among loops which fragment from the same parent loop, but there are no Brownian correlations due to loop fragmentation from infinite strings. The loops produce isocurvature fluctuations which cannot grow until they enter the horizon. The power spectrum of these fluctuations is analyzed.

161.179 Void probabilities in the galaxy distribution: scaling and luminosity segregation.
S. Maurogordato, M. Lachièze–Rey.
Astrophys. J., Vol. 320, No. 1, p. 13 – 25 (1987).

The authors develop a method to measure the void probability function for any three–dimensional complete and homogeneous galaxy sample. They select from the three–dimensional CfA catalog three complete, homogeneous, and volume–limited subsamples with different average luminosities. The authors calculate the void probability functions for these samples and compare them with the predictions of theoretical models. It is shown that each of these samples obeys a scaling invariance (predicted by the class of hierarchical models), suggesting that this property applies to the whole galaxy distribution. The void probability functions for the three samples is compared and it is shown that bright galaxies are more clustered than faint ones in these samples.

161.180 Correction of a study of source counts in the chronometric cosmology.
I. E. Segal.
Astrophys. J., Vol. 320, No. 1, p. 135 – 138 (1987).

A recent study of source counts in the chronometric cosmology (Wright 1987, [43.161.134]) is flawed in several respects. It is in error (1) mathematically, (2) logically, (3) statistically, and (4) in its representation of uncertain and model–dependent data as statistically cogent. These flaws are detailed in the present paper.

161.181 Gravitational stability of local strings.
R. Gregory.
Phys. Rev. Lett., Vol. 59, No. 6, p. 740 – 743 (1987).

The full coupled gravity–string field equations are considered, and they are used to show that a general local string will have an asymptotically conical structure. For the case of the Abelian Higgs model with $U(1)$ gauge invariance, the gravitational field of a simple local string to first order in $G\eta^2$ is exhibited. Then a C–energy argument is used to suggest stability at this linearized level.

161.182 Galaxy and structure formation with hot dark matter and cosmic strings.
R. Brandenberger, N. Kaiser, D. Schramm, N. Turok.
Phys. Rev. Lett., Vol. 59, No. 20, p. 2371 – 2374 (1987).

Galaxy and structure formation in a neutrino–dominated universe with cosmic strings is investigated. Strings survive neutrino free streaming to seed galaxies and clusters. The effective maximum Jeans mass is lower than in the adiabatic scenario. Hence cluster formation is only marginally different from that in the cold–dark–matter and strings model, but galaxy masses are lower. The mass spectrum of galaxies is flatter than with cold dark matter, and the density profile about an individual loop is less steep, in better agreement with observations.

161.183 Natural quantum state of matter fields in quantum cosmology.
S. Wada.
Phys. Rev. Lett., Vol. 59, No. 20, p. 2375 – 2378 (1987).

The initial–value problem for scale factor of the universe $a = 0$ for quantum states (wave functions) of free scalar fields in the de Sitter metric $a \propto \cosh(ht)$ with imaginary t is studied. It turns out that almost all the initial conditions give the same quantum state. Some effects of the quantization of the metric are also discussed.

161.184   Early–universe thermal production of not–so–invisible axions.
M. S. Turner.
Phys. Rev. Lett., Vol. 59, No. 21, p. 2489 – 2492 (1987).

161.185   Variational study of ordinary and superconducting cosmic strings.
C. T. Hill, H. M. Hodges, M. S. Turner.
Phys. Rev. Lett., Vol. 59, No. 21, p. 2493 – 2496 (1987).

The authors use a variational approach to study Abelian vortices, both the ordinary and bosonic superconducting varieties. They present accurate results for the energy per length. For superconducting strings they map out the parameter space of solutions, quantify the critical current, discuss the quench transition, and investigate the possibility of static solutions where electromagnetic stresses balance the string tension.

161.186   Anthropic bound on the cosmological constant.
S. Weinberg.
Phys. Rev. Lett., Vol. 59, No. 22, p. 2607 – 2610 (1987).

In recent cosmological models, there is an "anthropic" upper bound on the cosmological constant Λ. It is argued here that in universes that do not recollapse, the only such bound on Λ is that it should not be so large as to prevent the formation of gravitationally bound states. It turns out that the bound is quite large. A cosmological constant that is within 1 or 2 orders of magnitude of its upper bound would help with the missing–mass and age problems, but may be ruled out by galaxy number counts. If so, one may conclude that anthropic considerations do not explain the smallness of the cosmological constant.

161.187   Nature of the redshift of galaxies and quasars and its consequences. I. Gas accretion onto the nucleus of a galaxy and galaxy emission.
V. M. Antonov, L. M. Toptunova.
Kramat. Industr. Inst. Kramatorsk, 7 pp. (1987). In Russian. Abstr. in Ref. Zh., 51. Astron., 10.51.876 (1987).

161.188   Nature of the redshift of galaxies and quasars and its consequences. II. The conception of the noncosmological nature of redshifts of galaxies and quasars.
V. M. Antonov, L. M. Toptunova.
Kramat. Industr. Inst. Kramatorsk, 7 pp. (1987). In Russian. Abstr. in Ref. Zh., 51. Astron., 10.51.877 (1987).

161.189   Relativistic gravitational fields in the Universe.
A. G. Polnarev.
Particles and cosmology, p. 33 – 44 (1987). In Russian. Abstr. in Ref. Zh., 51. Astron., 10.51.967 (1987). – See Abstr. 012.034.

161.190   On the "bump" in the spectrum of super–high–energy cosmic rays.
V. S. Berezinskij, S. I. Grigor'eva.
Inst. Yader. Issled. Akad. Nauk SSSR, Prepr., No. 0512, p. 1 – 20 (1987). In Russian. Abstr. in Ref. Zh., 51. Astron., 10.51.972 (1987).

161.191   On the statistical description of an ensemble of ultrarelativistic particles in a three–dimensional–planar Universe.
Yu. G. Ignat'ev, A. A. Popov.
Kazan. Gos. Ped. Inst. Kazan', 9 pp. (1987). In Russian. Abstr. in Ref. Zh., 51. Astron., 10.51.980 (1987).

161.192   Order and chaos in the Universe.
V. G. Gurzadyan.
Particles and cosmology, p. 45 – 51 (1987). In Russian. Abstr. in Ref. Zh., 51. Astron., 10.51.995 (1987). – See Abstr. 012.034.

161.193   Observational tests of an explosive Lemaitre–type cosmological model.
E. J. Sternglass.
Bull. Am. Astron. Soc., Vol. 19, No. 2, p. 685 (1987). Abstract. – See Abstr. 010.061.

161.194   The local distribution of galaxies.
J. R. Auman, P. Hickson, G. G. Fahlman.
Bull. Am. Astron. Soc., Vol. 19, No. 2, p. 686 (1987). Abstract. – See Abstr. 010.061.

161.195   A shell model for large scale structure.
M. J. Henriksen, N. A. Bahcall, T. E. Smith.
Bull. Am. Astron. Soc., Vol. 19, No. 2, p. 687 (1987). Abstract. – See Abstr. 010.061.

161.196   Large scale structure in cosmological models.
J. A. Holtzman.
Bull. Am. Astron. Soc., Vol. 19, No. 2, p. 688 (1987). Abstract. – See Abstr. 010.061.

161.197   Geometric paradigm accounts for all redshift periodicities.
E. Sepulveda.
Bull. Am. Astron. Soc., Vol. 19, No. 2, p. 689 (1987). Abstract. – See Abstr. 010.061.

161.198   Effects of temperature on the creation of particles in a hot Friedmann universe.
U. Günther, A. I. Zhuk.
Astrophysics, Vol. 26, No. 2, p. 229 – 234 (1987). English translation of 43.161.167.

161.199   Generation of gravitational waves by the anisotropic phases in the early universe.
D. V. Deryagin, D. Yu. Grigoriev (*D. Yu. Grigor'ev*), V. A. Rubakov, M. V. Sazhin.
Mon. Not. R. Astron. Soc., Vol. 229, No. 3, p. 357 – 370 (1987).

The authors discuss possible anisotropic and inhomogeneous phases in the early universe at relatively late epochs corresponding to the temperatures 300 MeV – 1 TeV. The anisotropy and inhomogeneity of the stress–energy tensor at the horizon scale gives rise to the generation of a stochastic gravitational wave background. Both the amplitude and spectral energy density of this background are estimated. The authors also discuss the possibility of the detection of this background by the planned gravitational wave detectors.

161.200   Cosmological streaming velocities and large–scale density maxima.
J. A. Peacock, S. L. Lumsden, A. F. Heavens.
Mon. Not. R. Astron. Soc., Vol. 229, No. 3, p. 469 – 483 (1987).

The authors consider the statistical testing of models for galaxy formation against the observed peculiar velocities on 10 – 100 Mpc scales (the Rubin–Ford effect). If one assumes that observers are likely to be sited near maxima in the primordial field of density perturbations, then the observed filtered velocity field will be biased to low values by comparison with a point selected at random. This helps to explain how the peculiar velocities (relative to the microwave background) of the local supercluster and the Rubin–Ford shell can be so similar in magnitude. Using this assumption to predict peculiar velocities on two scales, the authors test models with large–scale damping (i.e. adiabatic perturbations). They note that the canonical $\Omega = 1$ massive neutrino model yields the required velocities quite naturally, while not violating the constraints on the anisotropy of the microwave background.

161.201   A VLA search for cosmological H I at $z = 3.3$.
E. Hardy, L. Noreau.
Astron. J., Vol. 94, No. 6, p. 1469 – 1475 (1987).

The authors used the new P band (327 MHz) line receiver at the Very Large Array in the D configuration to look for the 21 cm line emitted by primordial galaxy protoclusters and superclusters at $z \approx 3.3$. Two fields centered at $\alpha = 13^h7^m$, $\delta = 29°40'$ and $\alpha = 12^h$, $\delta = 29°$ were surveyed, the first of which contains quasar candidates near this redshift. No line emission was detected for any reasonable binning of the data. The total bandpass surveyed per field was 6.25 MHz, and the authors achieved mean H I mass detection sensitivities ranging from $\sim(7.5 \pm 1.5) \times 10^{14}$ to $\sim(4 \pm 1) \times 10^{15} M_\odot$ for $\Omega = 1$ and 0.1, respectively. The interception probability for protoclusters was near unity. This is the most sensitive search so far at this redshift and suggests that, if the universe is closed (i.e., $\Omega \geqslant 1$), either the total number of protoclusters in the universe (in the Sunyaev–Zel'dovich sense) is lower than $\sim 2 \times 10^5$ or that at $z \approx 3.5$ most of the H I had already coalesced.

161.202 **A two–component dark matter universe. II: Linear fluctuation theory.**
S. Ikeuchi, C. Norman, Y. Zhan.
Space Telesc. Sci. Inst., Prepr. Ser., No. 189, 21 pp. (1987). To appear in Astrophys. J.

161.203 **A case for $H_0 = 42$ and $\Omega_0 = 1$ using luminous spiral galaxies and the cosmological time scale test.**
A. Sandage.
Space Telesc. Sci. Inst., Prepr. Ser., No. 207, 48 pp. (1987). To appear in Astrophys. J.

161.204 **The case for $H_0 \cong 55$ from the 21 cm line–width, absolute magnitude relation for field galaxies.**
A. Sandage.
Space Telesc. Sci. Inst., Prepr. Ser., No. 208, 29 pp. (1987). To appear in Astrophys. J.

161.205 **Cosmological streaming velocities and large–scale density maxima.**
J. A. Peacock, S. L. Lumsden, A. F. Heavens.
Edinb. Astron. Prepr., No. 11/87, 26 pp. (1987). To appear in Mon. Not. R. Astron. Soc.

161.206 **Light propagation in inhomogeneous universes: the ray–shooting method.**
P. Schneider, A. Weiss.
MPA Rep., No. 311, 46 pp. (1987).

161.207 **Cosmological models with non–zero lambda.**
E. J. Wampler, W. L. Burke.
ESO Sci. Prepr., No. 519, 9 pp. (1987). Paper presented at the symposium "New ideas in astronomy", Venice, Italy, May 1987.

161.208 **Opacity of the universe.**
P. A. Shaver.
ESO Sci. Prepr., No. 535, 11 pp. (1987).

161.209 **The price of keeping the Hubble constant...constant.**
E. Giraud.
ESO Sci. Prepr., No. 537, p. 3–7 (1987). Paper presented at the symposium "New ideas in astronomy", Venice, Italy, May 1987.

161.210 **Observed distortions (from linearity) of the Hubble flow and bias in the data.**
E. Giraud.
ESO Sci. Prepr., No. 537, p. 13–18 (1987). Paper presented at IAU Colloq. No. 130.

161.211 **Giant H II regions as distance indicators. II. Application to H II galaxies and the value of the Hubble constant.**
J. Melnick, R. Terlevich, M. Moles.
ESO Sci. Prepr., No. 539, 37 pp. (1987).

161.212 **The quark–hadron phase transition and primordial nucleosynthesis.**
C. Alcock, G. M. Fuller, G. J. Mathews.
Astrophys. J., Vol. 320, No. 2, p. 439–447 (1987).
The dynamics and statistical mechanics of the quark–hadron phase transition are explored using the bag model and the known

spectrum of hadronic states. The authors compute the maximum amplitude for isothermal baryon number density fluctuations to emerge from this phase transition and their effects on primordial nucleosynthesis, as a function of the coexistence temperature (or bag constant) and the fractional volume of the universe which will remain in quark–gluon plasma when the release of latent heat no longer compensates the cooling due to expansion. The authors discuss computations of the primordial nucleosynthesis yields corresponding to the present estimates of the maximum baryon density fluctuations in a universe with $\Omega = 1$ in baryons.

161.213 **On the validity of the Zel'dovich approximation.**
B. Grinstein, M. B. Wise.
Astrophys. J., Vol. 320, No. 2, p. 448–453 (1987).
The Zel'dovich approximation is often used to mimic the effects of nonlinear gravitational time evolution. The authors compare the predictions of the Zel'dovich approximation with those of the true nonlinear time evolution, for the probability distribution of mass density fluctuations averaged over a Gaussian ball of large radius, and for the large–scale streaming velocities of objects that do not trace the mass.

161.214 **Vacuum Friedmann cosmological models in Dunn's scalar–tensor theory of gravitation.**
A. Beesham, N. A. Hassan, M. S. Maharaj.
Europhys. Lett. (Switzerland), Vol. 3, No. 9, p. 1053–1055 (1987). Abstr. in Phys. Abstr., Vol. 90, No. 1309, Entry 80537 (1987).

161.215 **Has cosmology become metaphysical?**
T. Rothman, G. Ellis.
Astronomy, Vol. 15, No. 2, p. 6–21 (1987). Abstr. in Phys. Abstr., Vol. 90, No. 1309, Entry 88696 (1987).

161.216 **The monopole problem and the primordial black hole problem in the inflationary universe.**
L. Liu.
Chin. Phys. Lett., Vol. 4, No. 3, p. 136–138 (1987). Abstr. in Phys. Abstr., Vol. 90, No. 1309, Entry 88700 (1987).

161.217 **Classical and quantum cosmology of the Salam–Sezgin model.**
J. J. Halliwell.
Nucl. Phys. B, Part. Phys., Vol. B286, No. 3–4, p. 729–750 (1987). Abstr. in Phys. Abstr., Vol. 90, No. 1309, Entry 88706 (1987).

161.218 **Inertia in Friedmann cosmologies.**
J. Teuber, P. G. Hjorth.
Nuovo Cimento B, Vol. 97B, Ser. 11, No. 2, p. 131–140 (1987). Abstr. in Phys. Abstr., Vol. 90, No. 1309, Entry 88707 (1987).

161.219 **Exact cosmological solutions of gravitational theories.**
R. T. Jantzen.
Phys. Lett. B, Vol. 186, No. 3–4, p. 290–296 (1987). Abstr. in Phys. Abstr., Vol. 90, No. 1309, Entry 88708 (1987).

161.220 **Wavefunction of a rotating universe.**
L. Z. Fang, H. J. Mo.
Phys. Lett. B, Vol. 186, No. 3–4, p. 297–302 (1987). Abstr. in Phys. Abstr., Vol. 90, No. 1309, Entry 88709 (1987).

161.221 **Late baryogenesis in superstring models.**
R. N. Mohapatra, J. W. F. Valle.
Phys. Lett. B, Vol. 186, No. 3–4, p. 303–308 (1987). Abstr. in Phys. Abstr., Vol. 90, No. 1309, Entry 88710 (1987).

161.222 **Cosmic no–hair theorems and inflation.**
J. D. Barrow.
Phys. Lett. B, Vol. 187, No. 1–2, p. 12–16 (1987). Abstr. in Phys. Abstr., Vol. 90, No. 1309, Entry 88711 (1987).

161.223 Entropy production in tepid inflation.
B. Kaempfer, B. Lukacs, G. Paal.
Phys. Lett. B, Vol. 187, No. 1 – 2, p. 17 – 21 (1987). Abstr. in
Phys. Abstr., Vol. 90, No. 1309, Entry 88712 (1987).

161.224 Strongly trapped points and the cosmic censorship hypothesis.
A. Krolak.
Phys. Rev. D, Vol. 35, No. 8, p. 2297 – 2301 (1987). Abstr. in
Phys. Abstr., Vol. 90, No. 1309, Entry 88713 (1987).

161.225 Torsion as a source of expansion in a Bianchi type–I universe in the self–consistent Einstein–Cartan theory of a perfect fluid with spin density.
J. C. Bradas, A. J. Fennelly, L. L. Smalley.
Phys. Rev. D, Vol. 35, No. 8, p. 2302 – 2308 (1987). Abstr. in
Phys. Abstr., Vol. 90, No. 1309, Entry 88714 (1987).

161.226 Quantum cosmology and recollapse.
R. Laflamme, E. P. S. Shellard.
Phys. Rev. D, Vol. 35, No. 8, p. 2315 – 2322 (1987). Abstr. in
Phys. Abstr., Vol. 90, No. 1309, Entry 88715 (1987).

161.227 How homogeneous was the universe at the time of a grand–unified theory phase transition?
P. Amsterdamski.
Phys. Rev. D, Vol. 35, No. 8, p. 2323 – 2338 (1987). Abstr. in
Phys. Abstr., Vol. 90, No. 1309, Entry 88716 (1987).

161.228 Cosmological–constant damping by unstable scalar fields.
L. H. Ford.
Phys. Rev. D, Vol. 35, No. 8, p. 2339 – 2344 (1987). Abstr. in
Phys. Abstr., Vol. 90, No. 1309, Entry 88717 (1987).

161.229 Inflation in spherically symmetric inhomogeneous models.
J. A. Stein–Schabes.
Phys. Rev. D, Vol. 35, No. 8, p. 2345 – 2351 (1987). Abstr. in
Phys. Abstr., Vol. 90, No. 1309, Entry 88718 (1987).

161.230 Improved concepts for the discussion of mutually interacting quantum fields in Robertson–Walker universes.
J. Audretsch, P. Spangehl.
Phys. Rev. D, Vol. 35, No. 8, p. 2365 – 2371 (1987). Abstr. in
Phys. Abstr., Vol. 90, No. 1309, Entry 88719 (1987).

161.231 Nonsingular cosmological models with torsion determined by vacuum polarization of quantum fields.
I. L. Buchbinder, S. D. Odintsov.
Pramāna, Vol. 28, No. 3, p. 241 – 245 (1987). Abstr. in Phys.
Abstr., Vol. 90, No. 1309, Entry 88722 (1987).

161.232 Very early universe based on closed superstring theories.
N. Matsuo.
Prog. Theor. Phys., Vol. 77, No. 2, p. 223 – 228 (1987). Abstr. in
Phys. Abstr., Vol. 90, No. 1309, Entry 88723 (1987).

161.233 The constancy of physics.
J. M. Irvine, R. Humphreys.
Prog. Part. Nucl. Phys., Vol. 17, p. 59 – 84 (1986). – See Abstr.
012.024.
Evidence restricting the secular variation of the coupling
constants of nature is reviewed. The nature of "natural" units is
discussed and possible forms of gravitational coupling examined.
Particular attention is paid to data obtained from a study of the
prehistoric natural reactor at Oklo and the restrictions which
present day isotopic abundances place on interaction strengths
and mass parameters during the era of primordial nucleosynthesis.

161.234 The cosmology effect of dust.
J.–x. Wu, G.–z. Xie.
Publ. Yunnan Obs., No. 1, p. 7 – 12 (1987). In Chinese.
Based on the discovery of an IR background in the waveband
$2 - 5\,\mu m$ and pregalactic stars, the authors discuss the effect of
the 3K cosmic microwave background spectrum from dust of
various models. The results show, if the neutral hydrogen content
is very low (that is, no Lyman cut–off), the reradiation from dust
must produce a temperature distortion 0.1 – 0.2K near the peak
frequency of the 3K cosmic microwave background and does not
depend on the dust models.

161.235 The Sunyaev–Zel'dovich effect under the condition of the $1 - 5\,\mu m$ IR background and the determination of the Hubble constant H_0 and of the density and temperature of the thermal electrons in a rich galaxy cluster.
J.–c. Wang, J.–x. Wu.
Publ. Yunnan Obs., No. 2, p. 63 – 68 (1987). In Chinese.
The Sunyaev–Zel'dovich effect of the thermal electrons in a
rich cluster on the IR background is studied according to the
observed data of the Coma Cluster. The explanation that the
IR background may be an approximate black body radiation,
given by Carr et al., is discussed in detail. It is proposed that the
temperature and the density of electrons in a rich cluster and the
Hubble constant may be determined from the observations of the
IR background distortion.

161.236 A search for small–scale structure in the background radiation at 6 cm.
H. M. Martin, R. B. Partridge.
Prepr. Steward Obs., No. 750, 24 pp. (1987). To appear in
Astrophys. J.

161.237 Origin and anisotropy of the cosmic submillimeter background.
C. J. Hogan, J. R. Bond.
Prepr. Steward Obs., No. 772, 9 pp. (1987). To appear in the
proceedings of the NATO Advanced Study Institute on "The
post–recombination universe", Cambridge, August 1987.

161.238 The value of H_0 from the infrared Tully–Fisher relation.
L. Bottinelli, L. Gouguenheim, P. Teerikorpi.
Rep. Ser., Dep. Phys. Sci., Univ. Turku, No. FTL–R135, 23 pp.
(1987). = Turku Univ. Obs., Informo, No. 127. ISBN 951–880–
043–X. Submitted to Astron. Astrophys.

161.239 On a model of the Universe without singularity with cosmological term.
T. S. Kozhanov.
Tr. Astrofiz. Inst. Alma–Ata, Tom 47, p. 104 – 107 (1987). In
Russian.

161.240 Incompatibility of the cosmological term and axial symmetry.
Eh. G. Mychelkin.
Tr. Astrofiz. Inst. Alma–Ata, Tom 47, p. 108 – 112 (1987). In
Russian.

161.241 Evolution of spirals and galaxies.
M. Ya. Gogberashvili, V. A. Ruchin.
Izv. Vuzov. Fiz., Tom 30, No. 4, p. 115 – 117 (1987). In Russian.
Abstr. in Ref. Zh., 51. Astron. 11.51.892 (1987).

161.242 Rotation of the Universe and cosmology.
D. D. Ivanenko, V. G. Krechet, V. F. Panov.
Gravitatsiya i teor. otnositel'nosti, Kazan', No. 24, p. 33 – 37
(1987). In Russian. Abstr. in Ref. Zh., 51. Astron. 11.51.903
(1987).

161.243   On substitution modes in cosmology of a two–component Friedmann Universe.
V. I. Khlebnikov.
Gravitatsiya i teor. otnositel'nosti, Kazan', No. 24, p. 100 – 107 (1987). In Russian. Abstr. in Ref. Zh., 51. Astron. 11.51.904 (1987).

161.244   On some peculiarities of the model of an expanding Universe.
Yu. D. Bochkov.
Gor'k Ped. Inst., Gor'kij, 18 pp. (1987). In Russian. Abstr. in Ref. Zh., 51. Astron. 11.51.906 (1987).

161.245   On the quantum–mechanical nature of gravitation.
G. T. Butorin.
Ural. Politekh. Inst., Sverdlovsk, 48 pp. (1987). In Russian. Abstr. in Ref. Zh., 51. Astron. 11.51.907 (1987).

161.246   A hydrodynamical three–component model of the universe during the recombination era.
M. G. Corona.
Rev. Mex. Astron. Astrofis., Vol. 14, No. 1, p. 52 – 57 (1987). – See Abstr. 012.042.
The author considers the universe during the recombination era to be a fluid composed of neutral hydrogen, plasma and radiation interacting via photorecombination, photoionization and Thompson scattering. He analyses the stability of the hydrodynamical cosmological model and the results are discussed in relation to the origin of galaxies.

161.247   Hydrogen molecules and the radiative cooling of pregalactic shocks II: low velocity shocks at high redshift.
P. R. Shapiro, H. Kang.
Rev. Mex. Astron. Astrofis., Vol. 14, No. 1, p. 58 – 65 (1987). – See Abstr. 012.042.
The nonequilibrium radiative cooling, recombination, and molecule formation behind steady–state shock waves in a gas of primordial composition have been calculated in detail for a number of cases. The authors have solved the rate equations for these processes, together with the hydrodynamical conservation equations. Shock waves such as these are relevant to a wide range of theories of galaxy and pregalactic star formation.

161.248   Cosmological mass density and the Regge law.
A. A. Schmidt, C. A. Z. Vasconcellos, H. A. Dottori.
Rev. Mex. Astron. Astrofis., Vol. 14, No. 1, p. 66 – 70 (1987). – See Abstr. 012.042.
The authors calculate the principal inertia moments of a three–dimensional mass distribution rotating in a self–gravitational homogeneous field with axial symmetry. They show that the angular momentum may, under certain conditions, be of the Regge–like type. These results are used to evaluate the mass and density of the universe.

161.249   In search of primordial helium.
G. Steigman.
Rev. Mex. Astron. Astrofis., Vol. 14, No. 1, p. 71 (1987). Abstract. – See Abstr. 012.042.

161.250   On the origin of the dipole anisotropy of the cosmic microwave background: beyond the Hydra–Centaurus supercluster.
J. Melnick, M. Moles.
Rev. Mex. Astron. Astrofis., Vol. 14, No. 1, p. 72 – 76 (1987). – See Abstr. 012.042.
The authors discuss the properties of a large concentration of galaxies located in the direction of the Centaurus cluster but at a much larger distance. They find that the majority of the faint galaxies in the region are clustered around cz $\sim$ 14000 km/sec and define a rich supercluster having a dynamical mass close to $2.5 \times 10^{15} M_\odot$. The authors show that this mass is not sufficient to explain the peculiar velocity of $\sim$600 km/sec of the Local Group with respect to the cosmic microwave background in the general direction of Centaurus. They find no luminous galaxies in the redshift range 6000 – 8000 km/sec in that direction. This implies that any mass overdensity related to the anisotropy of the Hubble flow must be either dark, hidden by the galactic plane or considerably more distant than 14000 km/sec.

161.251   The quantum origin of the universe.
I. Moss.
The early universe, p. 1 – 18 (1988). – See Abstr. 012.043.
The author gives a description of some of the modern day beliefs about the origin of the universe and quantum gravity. There are occasional glimpses of the wide areas of ignorance which should provide fruitful ground for the growth of new ideas. Contents: 1. Introduction. 2. Canonical quantum gravity. 3. Minisuperspace. 4. The Hartle–Hawking wave function. 5. Anisotropic universes. 6. Changes in topology.

161.252   Cosmology and particle physics.
M. S. Turner.
The early universe, p. 19 – 113 (1988). – See Abstr. 012.043.
Contents: 1. The standard cosmology and its successes. 2. Baryogenesis. 3. Toward the inflationary paradigm.

161.253   Supersymmetry and the early universe.
G. Gelmini.
The early universe, p. 115 – 124 (1988). – See Abstr. 012.043.

161.254   Relativistic cosmology.
J. D. Barrow.
The early universe, p. 125 – 201 (1988). – See Abstr. 012.043.
The author draws attention to a number of specific aspects of relativistic cosmology which might be of interest to current investigations of high energy physics in the early universe and to the wider audience of relativists and astrophysicists. In particular, he focuses upon features of general relativistic cosmology that are of relevance to the inflationary universe theory and the cosmological questions that it confronts. A number of new results are described.

161.255   Yet another scenario for galaxy formation.
P. J. E. Peebles.
The early universe, p. 203 – 214 (1988). – See Abstr. 012.043.
The author considers the prospects for developing a model for the formation of galaxies and clusters of galaxies using only baryons and radiation. This leads to a primeval entropy perturbation scenario in which galaxies form at $z \sim 100$ as the last generation to be substantially held up by Compton drag. The consequences for galaxy formation seem attractive; the situation for cluster formation is unclear because of ambiguities in the scenario developed so far.

161.256   Non–Gaussian fluctuations.
M. B. Wise.
The early universe, p. 215 – 238 (1988). – See Abstr. 012.043.
Natural primordial mass density fluctuations are those for which the probability distribution, for the mass density fluctuations averaged over the horizon volume, is independent of time. This criterion determines the two–point correlation of the mass density fluctuations to have a Zeldovich power spectrum but allows for many types of higher correlations. If the connected higher correlations vanish the primordial fluctuations are Gaussian. In this case the probability distribution develops into a non–Gaussian one due to the non–linear time evolution. The nature of this non–Gaussian distribution and its effects on the large scale distribution of galaxies or clusters of galaxies and their large scale streaming velocities is explored.

161.257   N–body methods and the formation of large–scale structure.
S. D. M. White.
The early universe, p. 239 – 260 (1988). – See Abstr. 012.043.
The author reviews the N–body techniques that have been used to study the evolution of large–scale structure in the universe. He discusses the nonlinear structure found in neutrino–dominated and cold dark matter dominated models. The cold dark matter

model currently appears the most attractive possibility and is able to reproduce observed structures from galaxy halos up to rich galaxy clusters.

161.258 Numerical relativity and cosmology.
T. Piran.
The early universe, p. 261 – 282 (1988). – See Abstr. 012.043.

The author discusses two schemes for general relativistic numerical cosmological solutions. The scheme are based on the $3+1$ ADM formalism and on the $3+1$ Regge Calculus and they are designed for a full three dimensional time dependent solution of the Einstein equations.

161.259 Distortions and anisotropies of the cosmic background radiation.
J. R. Bond.
The early universe, p. 283 – 334 (1988). – See Abstr. 012.043.

The general theory of spectral distortions and angular anisotropies in the cosmic background radiation is reviewed. Constraints on the amount of energy injection allowed in the early universe are discussed. Predictions of fluctuation levels are given for small and large angle anisotropy experiments for theories of structure formation in which the fluctuations are initially Gaussian and scale–invariant, as expected in inflationary models. Simple equations are derived which give a physical overview of the various mechanisms which lead to anisotropy in the microwave background.

161.260 The early universe: historical & philosophical perspectives.
S. Toulmin.
The early universe, p. 393 – 411 (1988). – See Abstr. 012.043.

161.261 Primordial nucleosynthesis.
J. Audouze.
Cosmology and particle physics, p. 1 – 34 (1987). – See Abstr. 012.046.

This paper presents an overview of the primordial abundances of the very light elements (D, ^{3}He, ^{4}He and ^{7}Li). It discusses the standard big bang model and its implications on the baryonic density in the universe (and therefore the baryonic cosmological parameter Ω_b) such as the number of neutrino (lepton) families. It is argued that specific models of chemical evolution of our Galaxy are needed to induce a significant decrease of the D abundance during Galactic history and thereby to reconcile the observed abundances with the predictions for D and for ^{4}He. A few models are analyzed, which relax some of the hypothesis of the standard big bang model. The effects of the existence of non baryonic particles like quark nuggets, massive neutrinos, gravitinos and photinos are especially examined: it is argued that primordial nucleosynthesis provides many interesting constraints on various aspects of particle physics.

161.262 The large–scale structure of the universe.
J. Silk.
Cosmology and particle physics, p. 35 – 75 (1987). – See Abstr. 012.046.

The evidence for the standard cosmological model of a Friedmann universe is presented. The role of particle physics in cosmology is reviewed, together with astronomical measurements of dark matter and the various particle physics candidates for the dark matter. Observations of large–scale structure in the galaxy distribution are discussed, including recent data on large voids and streaming motions and recent limits on the anisotropy of the cosmic microwave background. Theories of the origin of large–scale structure and its evolution in a universe dominated by cold or hot dark matter are described, as are the possible roles of cosmic strings and explosions. A final section is devoted to galaxy formation, with emphasis on biasing schemes and the role of dissipation in accounting for galaxy characteristics.

161.263 Toward the inflationary paradigm: lectures on inflationary cosmology.
M. S. Turner.
Cosmology and particle physics, p. 77 – 141 (1987). – See Abstr. 012.046.

Contents: 1. Overview. 2. The standard cosmology and its successes. 3. Shortcomings of the standard cosmology. 4. New inflation – the slow–rollover transition. 5. Scalar field dynamics. 6. Origin of density inhomogeneities. 7. Specific models – interesting failures: "Old inflation". Coleman–Weinberg SU(5). Geometric hierarchy model. CERN SUSY/SUGR models. 8. Lessons learned – a prescription for successful new inflation. 9. Two simple models that work. 10. Toward the inflationary paradigm: Chaotic inflation. Induced gravity inflation. The compactification transition. 11. Loose ends. 12. Inflation confronts observation. 13. Epilogue.

161.264 Cosmic strings.
T. W. B. Kibble.
Cosmology and particle physics, p. 171 – 208 (1987). – See Abstr. 012.046.

Contents: 1. Introduction. 2. Abelian strings. 3. Global strings. 4. Non–Abelian strings. 5. Other topological defects. 6. Models with strings. 7. Composite structures. 8. Superconducting strings. 9. Dynamics of loops. 10. The early universe. 11. Phase transitions in the early universe. 12. Initial configuration of strings. 13. Evolution of the string configuration. 14. The scaling solution. 15. Distribution of loops. 16. Gravitational radiation. 17. Gravitational field of a string. 18. Galaxy formation. 19. Fluctuations at decoupling. 20. Strings and galaxy formation. 21. Conclusions.

161.265 String theory and quantum cosmology.
F. Englert.
Cosmology and particle physics, p. 209 – 241 (1987). – See Abstr. 012.046.

String theory is an attempt to formulate a unified theory of quantized matter and radiation. It may lead to a taming of the quantum fluctuations which mar the quantization of general relativity and of its Kaluza–Klein extensions. This review first shows how the quantum compactification characteristic of string theories leads naturally to view superstrings as sectors of the simpler but apparently less physical bosonic string theory. In the second part the author delineates those problems which have not received a satisfactory solution in early cosmology. These include the detailed mechanism for inflation, the Planckian era and the nature of the original event. It is argued that the handling of these problems requires a theory which unifies at the quantum level, gravity with the other fields. The bosonic string theory is perhaps the best guide towards the correct theory.

161.266 Cosmology and extra dimensions.
E. W. Kolb.
Cosmology and particle physics, p. 243 – 280 (1987). – See Abstr. 012.046.

Contents: 1. Microphysics in extra dimensions. 2. Stability of the internal space. 3. Semiclassical instability of compactification. 4. Inflation and extra dimensions. 5. Limiting temperature in superstring models. 6. GUT symmetry breaking in extra dimensions. 7. Remnants.

161.267 Is the space–time dimension 11 distinguished in cosmology?
M. Szydłowski, J. Szczesny, M. Biesiada.
Gen. Relativ. Gravitation, Vol. 19, No. 12, p. 1181 – 1194 (1987).

The authors investigate chaotic behavior in the class of homogeneous multidimensional cosmological models. They argue that insofar as spatial dimension $n = 3$ is in general distinguished, the dimension $n = 10$ is a critical one from the point of view of chaotic behavior near the singularity. A spatial dimension of 9 is the highest in which chaotic behavior can take place, a spatial dimension of 10 is the lowest in which chaotic behavior cannot occur near the singularity.

161.268 A new class of spherically symmetric interior solution with cosmological constant Λ.
C.-m. Xu, X.-j. Wu, Z. Huang.
Gen. Relativ. Gravitation, Vol. 19, No. 12, p. 1203 – 1211 (1987).
In this paper, a new class of spherically symmetric interior solution with cosmological constant Λ is obtained, to which Florides' solution is extended. As a special case, the metric in uniform proper density is discussed. The Robson junction condition is also extended to the case with nonvanishing Λ.

161.269 Cosmologies based on Lyra's geometry.
H. H. Soleng.
Gen. Relativ. Gravitation, Vol. 19, No. 12, p. 1213 – 1216 (1987).
It is pointed out that the cosmologies based on Lyra's manifold, with constant gauge–vector which have been studied in the literature, will either include a creation field and be equal to Hoyle's creation field cosmology, or contain a special vacuum fluid which together with the gauge–vector term may be considered as a cosmological term. In the latter case the solutions are equal to the general relativistic cosmologies with a cosmological term.

161.270 A multidimensional radiation–filled universe.
P. Turkowski.
Gen. Relativ. Gravitation, Vol. 19, No. 12, p. 1267 – 1276 (1987).
The author considers a radiation–filled universe which possesses the product symmetry: (N–dimensional space of constant curvature) x (n sphere). The solutions of all the types, within this class, to the classical field equations are given. In the case of the N–dimensional space of zero or negative curvature constant, the solutions exhibit a tendency to approach asymptotically the Kasner–like state in which the N–dimensional subspace expands while the n sphere shrinks to the final singularity.

161.271 Kaluza–Klein cosmology: Friedmann models with phenomenological matter.
U. Bleyer, D. E. Liebscher.
Ann. Phys. (Leipzig), Vol. 44, No. 2, p. 81 – 152 (1987). Abstr. in Phys. Abstr., Vol. 90, No. 1310, Entry 89071 (1987).

161.272 Inflation and cosmic strings: two mechanisms for producing structure in the universe.
R. H. Brandenberger.
Int. J. Mod. Phys. A , Vol. 2, No. 1, p. 77 – 131 (1987). Abstr. in Phys. Abstr., Vol. 90, No. 1310, Entry 95060 (1987).

161.273 Irreversibility of the cosmological expansion.
G. Marx, H. Sato.
Int. J. Mod. Phys. A , Vol. 2, No. 1, p. 133 – 163 (1987). Abstr. in Phys. Abstr., Vol. 90, No. 1310, Entry 95061 (1987).

161.274 A model of the universe free of cosmological problems.
M. Ozer, M. O. Taha.
Nucl. Phys. B, Part. Phys., Vol. B287, No. 4, p. 776 – 796 (1987). Abstr. in Phys. Abstr., Vol. 90, No. 1310, Entry 95063 (1987).

161.275 Cosmology with decaying vacuum energy.
K. Freese, F. C. Adams, J. A. Frieman, E. Mottola.
Nucl. Phys. B, Part. Phys., Vol. B287, No. 4, p. 797 – 814 (1987). Abstr. in Phys. Abstr., Vol. 90, No. 1310, Entry 95064 (1987).

161.276 Viscous universes.
T. Padmanabhan, S. M. Chitre.
Phys. Lett. A, Vol. 120, No. 9, p. 433 – 436 (1987). Abstr. in Phys. Abstr., Vol. 90, No. 1310, Entry 95065 (1987).

161.277 The arrow of time and the expansion of the Universe.
A. Qadir.
Phys. Lett. A, Vol. 121, No. 3, p. 113 – 115 (1987). Abstr. in Phys. Abstr., Vol. 90, No. 1310, Entry 95066 (1987).

161.278 A radiating charge embedded in a de Sitter universe.
A. Patino, H. Rago.
Phys. Lett. A, Vol. 121, No. 7, p. 329 – 330 (1987). Abstr. in Phys. Abstr., Vol. 90, No. 1310, Entry 95067 (1987).

161.279 Quantum effects near the singularity in a general cosmological scenario.
P. S. Joshi, S. S. Joshi.
Phys. Lett. A, Vol. 121, No. 7, p. 334 – 336 (1987). Abstr. in Phys. Abstr., Vol. 90, No. 1310, Entry 95068 (1987).

161.280 Gauge–invariant perturbations in a spatially flat anisotropic universe.
M. Den.
Prog. Theor. Phys., Vol. 77, No. 3, p. 653 – 670 (1987). Abstr. in Phys. Abstr., Vol. 90, No. 1310, Entry 95069 (1987).

161.281 Simple coupling with cosmological implications. The initial singularity and the inflationary universe.
D. Saez.
Phys. Rev. D, Vol. 35, No. 6, p. 2027 – 2033 (1987). Abstr. in Phys. Abstr., Vol. 90, No. 1311, Entry 95443 (1987).

161.282 To the big bang and beyond.
S. Odenwald.
Astronomy, Vol. 15, No. 5, p. 90 – 95 (1987). Abstr. in Phys. Abstr., Vol. 90, No. 1311, Entry 101961 (1987).

161.283 Cosmological transition periods.
R. Gautreau.
Int. J. Theor. Phys., Vol. 26, No. 4, p. 387 – 394 (1987). Abstr. in Phys. Abstr., Vol. 90, No. 1311, Entry 101971 (1987).

161.284 Time evolution of the cosmological "constant".
M. Reuter, C. Wetterich.
Phys. Lett. B, Vol. 188, No. 1, p. 38 – 43 (1987). Abstr. in Phys. Abstr., Vol. 90, No. 1311, Entry 101973 (1987).

161.285 A superstring cosmological model.
S. R. Lonsdale, I. G. Moss.
Phys. Lett. B, Vol. 189, No. 1 – 2, p. 12 – 16 (1987). Abstr. in Phys. Abstr., Vol. 90, No. 1311, Entry 101974 (1987).

161.286 Deceleration parameter and critical density in a modified Robertson–Walker metric.
L. Corsiglia.
Tex. J. Sci., Vol. 38, No. 2, p. 174 – 181 (1986). Abstr. in Phys. Abstr., Vol. 90, No. 1311, Entry 101980 (1987).

161.287 The evolution equation of the scalar field in the new inflationary universe.
A. Ringwald.
Z. Phys., C, Vol. 34, No. 4, p. 481 – 490 (1987). Abstr. in Phys. Abstr., Vol. 90, No. 1311, Entry 101981 (1987).

161.288 A dust–filled Kantowski–Sachs Universe with $\Lambda > 0$.
O. Gron, E. Eriksen.
Phys. Lett. A, Vol. 121, No. 5, p. 217 – 220 (1987). Abstr. in Phys. Abstr., Vol. 90, No. 1312, Entry 102238 (1987).

161.289 Inflation in a universe with viscosity.
B. Modak.
Classical Quantum Gravity, Vol. 4, No. 3, p. L47 – L49 (1987). Abstr. in Phys. Abstr., Vol. 90, No. 1312, Entry 107990 (1987).

161.290 Abelian anisotropic cosmological models in 11–dimensional supergravity.
J. L. Hanquin, J. Demaret.
Classical Quantum Gravity, Vol. 4, No. 3, p. L51 – L58 (1987). Abstr. in Phys. Abstr., Vol. 90, No. 1312, Entry 107991 (1987).

161.291 Quantum fluctuations and inflation
J. M. Bardeen, G. J. Bublik.
Classical Quantum Gravity, Vol. 4, No. 3, p. 573 – 580 (1987).
Abstr. in Phys. Abstr., Vol. 90, No. 1312, Entry 107992 (1987).

161.292 Propagation amplitude in homogeneous quantum cosmology.
J. Louko.
Classical Quantum Gravity, Vol. 4, No. 3, p. 581 – 593 (1987).
Abstr. in Phys. Abstr., Vol. 90, No. 1312, Entry 107993 (1987).

161.293 Leptonic and hadronic mass scales – a cosmic connection?
H. Fritzsch, D. N. Schramm.
Comments Nucl. Part. Phys., Vol. 17, No. 3, p. 129 – 134 (1987).
Abstr. in Phys. Abstr., Vol. 90, No. 1312, Entry 107995 (1987).

161.294 Neutrino families: the early Universe meets elementary particle/acceleration physics.
D. B. Cline, D. N. Schramm, G. Steigman.
Comments Nucl. Part. Phys., Vol. 17, No. 3, p. 145 – 161 (1987).
Abstr. in Phys. Abstr., Vol. 90, No. 1312, Entry 107996 (1987).

161.295 Lie–admissible structure of small–distance quantum cosmology.
P. F. Gonzalez–Diaz.
Hadronic J., Vol. 9, No. 5, p. 199 – 201 (1986). Abstr. in Phys. Abstr., Vol. 90, No. 1312, Entry 107997 (1987).

161.296 Quantum effects in a model of cosmological compactification.
J. Wudka.
Phys. Rev. D, Vol. 35, No. 10, p. 3255 – 3257 (1987). Abstr. in Phys. Abstr., Vol. 90, No. 1312, Entry 107999 (1987).

161.297 Ages of the Universe for decreasing cosmological constants.
T. S. Olson, J. F. Jordan.
Phys. Rev. D, Vol. 35, No. 10, p. 3258 – 3260 (1987). Abstr. in Phys. Abstr., Vol. 90, No. 1312, Entry 108000 (1987).

161.298 Texture: a cosmological topological defect.
R. L. Davis.
Phys. Rev. D, Vol. 35, No. 12, p. 3705 – 3708 (1987). Abstr. in Phys. Abstr., Vol. 90, No. 1312, Entry 108001 (1987).

161.299 Conformal group actions and Segal's cosmology.
J.–E. Werth.
Rep. Math. Phys, Vol. 23, No. 2, p. 257 – 268 (1986). Abstr. in Phys. Abstr., Vol. 90, No. 1312, Entry 108003 (1987).

161.300 Aspherical accretion of cosmic strings.
H. J. Mo, B. Gao, L. Z. Fang.
Astrophys. Space Sci., Vol. 137, No. 2, p. 225 – 231 (1987).
A simulation of accretion of a string loop has been done. It shows that aspherical accretion is essential for large–scale loops. All configurations found in the distribution of galaxies, such as filaments, pancakes, bubbles, and voids, can be formed by the accretion of cosmic strings.

161.301 Exact BD–FRW imperfect fluid cosmologies with an electromagnetic field.
V. B. Johri, R. Sudharsan.
Astrophys. Space Sci., Vol. 137, No. 2, p. 281 – 292 (1987).
It is shown that the $k = 0$ FRW metric which admits a dust solution in the Brans–Dicke (BD) theory, also admits an imperfect fluid distribution along with an electromagnetic field. The solutions are functions of time and radial coordinates and they satisfy all necessary energy and thermodynamic conditions.

161.302 The significance of spatial curvature for energy output of remote quasars.
J. Souček.
Astrophys. Space Sci., Vol. 137, No. 2, p. 347 – 356 (1987).
There is a brightening effect of quasar outputs due to the positive curvature of space in the static Universe, if we use the right expression of distance as an arc $^3r = R\chi$, and not the 'corrected luminosity distance ' of $r = R \sin \chi$.

161.303 Cosmic shear in inflationary models of Bianchi types VIII and IX.
H. H. Soleng.
Astrophys. Space Sci., Vol. 137, No. 2, p. 373 – 384 (1987).
Two exact solutions of Einstein's field equations of vacuum are presented and investigated. The author regards the Λ term vacuum fluid as the limiting case of scalar field with an almost constant potential. Considering the four velocity of this fluid he finds, that in both solutions there is an anisotropic expansion of the cosmic fluid, but the fluid has vanishing vorticity. The author investigates whether shear could prevent the transition into an inflationary era in these models, and the effect of shear on a scalar field is also considered. It is found that shear will speed up the rollover of the scalar field in some Bianchi type–VIII models. Possible initial conditions are discussed in light of the group structures of the models.

161.304 A note on vacuum self–creation cosmological models.
H. H. Soleng.
Astrophys. Space Sci., Vol. 138, No. 1, p. 19 – 21 (1987).
The vacuum field equations of the self–creation theory of gravitation are solved for the Robertson–Walker space–time, by using a correspondence to known solutions of general relativity.

161.305 Inhomogeneous viscous fluid cosmological model with electromagnetic field in general relativity.
R. Bali, G. Singh.
Astrophys. Space Sci., Vol. 138, No. 1, p. 71 – 77 (1987).
The object of this paper is to investigate the behaviour of electromagneic field in inhomogeneous cosmological models obtained for viscous fluid distributions. The various particular cases when both the electromagnetic and viscosity are absent, are also discussed.

161.306 Cosmic strings and the formation of clusters of galaxies.
Y.–P. Jing, J.–L. Zhang.
Astrophys. Space Sci., Vol. 138, No. 1, p. 105 – 112 (1987).
The relation between the mean separation d of clusters of galaxies and their mean richness N is found to be well represented by $d = 3.4N^{2/3}$. Both this observed result and the Bahcall–Soneira relation (that is, the spatial correlations of clusters of galaxies are increased with their richness) are well explained by the cosmic string picture, lending support to the cosmic–string theory.

161.307 The origin of galactic magnetic fields.
M. Beech.
Astrophys. Space Sci., Vol. 138, No. 1, p. 113 – 119 (1987).
In this article it is argued that galactic magnetic fields are generated in the earliest moments of galaxy collapse. The model proposes that provided, even if only briefly, a supermassive star is formed early on in the galaxy formation process, this star can produce a strong centrally localized magnetic field which may act as the 'seed' field from which a galactic field can grow. In order to substantiate this model, detailed numerical calculations will be required.

161.308 Viscous fluid cosmological model of cylindrical symmetry in the presence of magnetic field.
R. Bali, A. Tyagi.
Astrophys. Space Sci., Vol. 138, No. 1, p. 173 – 182 (1987).
A general solution of cylindrical symmetry in which distribution consists of an electrically neutral viscous fluid with an infinite electrical conductivity in presence of magnetic field, has been obtained. The behaviour of magnetic field on the model has

161.309 R–W cosmological models with zero–mass scalar fields for different equations of state.
A. R. Roy, M. N. Varma.
Astrophys. Space Sci., Vol. 138, No. 1, p. 217 – 219 (1987).

Spatially–flat cosmological models with interacting perfect fluid and zero–mass scalar field have been obtained under different equations of state and in presence of the scalar charge density associated with the Klein–Gordon equation assuming a special law of variation for the Hubble's parameter.

161.310 Cosmic strings and the orientation of galaxies and clusters.
Y. Gao.
Astrophys. Space Sci., Vol. 138, No. 2, p. 369 – 379 (1987).

The formation of galaxies, clusters, and superclusters is discussed in the cosmic string model and the constraints are given for the corresponding loops which form these structures by the accretion of the loops. The alignment of position angles of galaxies with that of their parent cluster and of clusters with that of the supercluster which they lie in, appears natural. Various different observations and statistical results of position angles are (at least qualitatively) explained in the cosmic string theory of formation of the large–scale structure.

161.311 Very early cosmology in the maximal acceleration hypothesis.
M. Gasperini.
Astrophys. Space Sci., Vol. 138, No. 2, p. 387 – 391 (1987).

It is stressed that the very early Universe provides an example of a physical system in which the phenomenological consequences of a natural limit of the accelerations cannot be neglected. It is shown that the existence of such limit leads necessarily to modify the standard model in a way which avoids the initial singularity and introduces an inflationary expansion of the power–law type. The possibility is also suggested to relate the maximal acceleration hypothesis to a decaying vacuum scenario, in which the cosmological constant is proportional to the radiation density, and is then a decreasing function of the cosmic time.

161.312 Primordial nucleosynthesis and $\Omega_B \sim 1$ cosmologies with interacting radiation and matter.
A. A. Coley.
Astrophys. Space Sci., Vol. 138, No. 2, p. 393 – 401 (1987).

A class of exact cosmological models is studied in which two separate, interacting fluids act as the source of the gravitational field, a radiative perfect fluid modelling the cosmic microwave background and a second perfect fluid modelling the observed material content of the Universe. Although the two fluid models under consideration are found to predict primordial element abundances similar to those predicted in the standard model, the upper limit on the present baryon density inferred from the observed abundances of the light elements is found to be greater than that in the standard model due to the different evolution of the baryon density in the models. From this result it is found that cosmologies with $\Omega_B \sim 1$ (the ratio of the present value of the baryon density to the value of the critical density) are permitted without violating the constraints of nucleosynthesis, thereby allowing the possibility that the Universe could be closed by baryonic matter alone.

161.313 Superstrings.
S. Sen.
Ir. Astron. J., Vol. 18, No. 2, p. 98 – 101 (1987). – See Abstr. 012.036.

In this article the author gives an introductory, non–technical account of superstring theory. He concentrates on three questions: (1) What are superstrings? (2) Why are people so excited about them? (3) What has been achieved so far?

161.314 Distortion of microwave background radiation by decaying particles.
M. Kawasaki, K. Sato.
Publ. Astron. Soc. Jpn., Vol. 39, No. 5, p. 837 – 842 (1987).

The effect of the radiative decay of weakly interacting massive particles (WIMPs) on the spectrum of microwave background radiation (MBR) is investigated. It is concluded that WIMPs with an appropriate mass, lifetime, and number density can account for the recently observed distortion of the MBR spectrum. However, such radiative coupling of WIMPs is stringently constrained from the consideration of the stellar cooling.

161.315 The Universe as a fractal structure.
J. Maddox.
Nature, Vol. 329, No. 6136, p. 195 (1987).

The author reports on models of galaxy formation and shows that a numerical simulation of galaxy formation seems to reproduce the observed fractal pattern on the sky.

161.316 Can Planck–mass relics of evaporating black holes close the Universe?
J. H. MacGibbon.
Nature, Vol. 329, No. 6137, p. 308 – 309 (1987).

The author proposes that the cosmological dark matter consists of the Planck–mass remnants of evaporating primordial black holes. Such remnants would be expected to have close to the critical density if the black holes evaporating at the present epoch have the maximum density consistent with cosmic–ray constraints. The remnants are also candidates for the missing mass in the galactic halo. Primordial black holes of the required density may form naturally at the end of an inflationary epoch. Planck–mass relics would behave dynamically just like 'cold dark matter'. Because the baryonic matter in black holes cannot participate in nucleosynthesis the limits on the baryonic content of the Universe set by primordial nucleosynthesis are circumvented.

161.317 The topology of large–scale structure. I. Topology and the random phase hypothesis.
D. H. Weinberg, J. R. Gott III, A. L. Melott.
Astrophys. J., Vol. 321, No. 1, p. 2 – 27 (1987).

Many models for the formation of galaxies and large–scale structure assume a spectrum of random phase (Gaussian), small–amplitude density fluctuations as initial conditions. In such scenarios, the topology of the galaxy distribution on large scales relates directly to the topology of the initial density fluctuations. The authors describe a quantitative measure of topology – the genus of contours in a smoothed density distribution – and apply it to numerical simulations of galaxy clustering, to a variety of three–dimensional toy models, and to a volume–limited sample of the CfA redshift survey.

161.318 Cosmic fluctuation spectra with large–scale power.
J. M. Bardeen, J. R. Bond, G. Efstathiou.
Astrophys. J., Vol. 321, No. 1, p. 28 – 35 (1987).

Observations of the cluster–cluster and cluster–galaxy correlation functions and of large–scale streaming velocities seem to indicate that extra power exists on large scales over that predicted in the standard $\Omega = 1$ cold dark matter model with scale–invariant initial conditions. The extra power required to explain the reported observations is explored by systematically varying cosmological models having initially Gaussian perturbations. The authors consider the effect of varying cosmological parameters such as $h \equiv H_0/(100 \text{ km s}^{-1})$ and the following density parameters: total Ω_{tot}, baryon Ω_B, stable cold dark matter Ω_X, stable hot dark matter Ω_ν, vacuum energy $\Omega_{vac} \equiv \Lambda/(3H_0^2)$, and decaying dark matter Ω_{nrd} with its relativistic decay products Ω_{erd}. These models are compared with the rich cluster correlation function, the cluster–galaxy correlation function, the large–scale streaming velocities, and large– and small–angle anisotropies in the microwave background.

161.319 **Correlation function in decaying particle cosmology.**
Y. Suto.
Astrophys. J., Vol. 321, No. 1, p. 36 – 48 (1987).

Decaying particle cosmology was designed to reconcile the notion of a flat universe with the observed "apparent" low mass density that is determined dynamically. The author has studied the evolution of galaxy–galaxy correlation functions in decaying particle cosmology using N–body simulations. This scenario is found to considerably affect the distribution pattern of galaxies, and the resultant correlation function cannot be fitted to a single power–law form. This result rules out the possibility that the present universe is dominated by relativistic particles.

161.320 **Small–scale anisotropies in the microwave background in a baryon–dominated open universe.**
N. Gouda, M. Sasaki, Y. Suto.
Astrophys. J., Lett. Ed., Vol. 321, No. 1, p. L1 – L6 (1987).

The anisotropies of the cosmic microwave background radiation in a baryon–dominated universe are calculated using a gauge–invariant method. The theoretical predictions on $4'.5$ angular scale are compared with the observational data by Uson and Wilkinson. Both adiabatic and isocurvature scenarios predict roughly the same amplitudes of the anisotropies on the scale. As a result, the cosmic microwave background isotropy requires that the density parameter of a baryon–dominated model must be larger than 0.8 even in an isocurvature scenario. This conclusion is compatible with the observation of the light elements only if our universe were fairly inhomogeneous at the epoch of primordial nucleosynthesis.

161.321 **Galaxy formation by gravitational collapse in a universe dominated by cold dark matter.**
B. S. Ryden.
Diss. Abstr. Int., Sect. B, Vol. 48, No. 5, p. 1385–B (1987). Thesis, Princeton University, 196 pp. (1987). Order No. DA8716901.

161.322 **Electromagnetic and gravitational perturbations in an ultrarelativistic plasma in an isotropic world.**
A. V. Zakharov, V. A. Anikanov.
Gravitatsiya i teor. otnositel'nosti, Kazan', No. 24, p. 107 – 114 (1987). In Russian. Abstr. in Ref. Zh., 51. Astron., 12.51.1045 (1987).

161.323 **Anomalous electroweak non–conservation of the baryon number at high temperatures: theory and cosmological consequences.**
V. A. Kuz'min, V. A. Rubakov, M. E. Shaposhnikov.
Quarks – 86, p. 57 – 66 (1987). In Russian. Abstr. in Ref. Zh., 51. Astron., 12.51.1046 (1987). – See Abstr. 012.059.

161.324 **Scalar fields in astrophysics and cosmology.**
V. A. Berezin, N. G. Kozimirov, V. A. Kuz'min, I. I. Tkachev.
Quarks – 86, p. 74 – 82 (1987). In Russian. Abstr. in Ref. Zh., 51. Astron., 12.51.1047 (1987). – See Abstr. 012.059.

161.325 **Double inflation: a possible resolution of the large–scale structure problem.**
M. S. Turner, J. V. Villumsen, N. Vittorio, J. Silk, R. Juszkiewicz.
Astrophys. J., Vol. 323, No. 2, p. 423 – 432 (1987).

A model is presented for the large–scale structure of the universe in which two successive inflationary phases resulted in large small–scale and small large–scale density fluctuations. This bimodal density fluctuation spectrum in an $\Omega = 1$ universe dominated by hot dark matter leads to large–scale structure of the galaxy distribution that is consistent with recent observational results. In particular, large, nearly empty voids and significant large–scale peculiar velocity fields are produced over scales of ~ 100 Mpc, while the small–scale structure over $\lesssim 10$ Mpc resembles that in a low–density universe, as observed.

161.326 **Cosmological perturbations in the minimum quadratic gauge theory of gravitation.**
A. V. Minkevich, Nguyen Van Hoang.
Vestsi Akad. Nauk BSSR, Ser. Fiz.–Mat. Nauk, No. 3, p. 86 – 92 (1987). In Russian. Abstr. in Ref. Zh., 51. Astron., 12.51.1071 (1987).

161.327 **Nuclear processes in the early universe. I. The thermal history of matter. II. The cosmic quark–hadron phase transition.**
H. Reeves.
Separate print, Institut d'Astrophysique de Paris, 98 blvd Arago, 75015 Paris, France, 77 pp. (1987).

Lectures given at the Varenna (Italy) Summer School on "Confrontation between theories and experiments in cosmology", July 1987 and at the Alpbach (Austria) Summer School on "Space research and fundamental physics", August 1987.

161.328 **Microwave background radiation.**
S. Bajtlik.
Postepy Astron., Tom 34, Zesz. 3, p. 181 – 200 (1986). In Polish.

The history of discovery of background radiation is presented. Its methods of observation, characteristics and the consequences in cosmology are also discussed.

161.329 **Inflation from higher dimensions.**
Q. Shafi, C. Wetterich.
Nucl. Phys. B, Part. Phys., Vol. B289, No. 3 – 4, p. 787 – 809 (1987). Abstr. in Phys. Abstr., Vol. 90, No. 1313, Entry 108369 (1987).

161.330 **The Universe: a birth far from equilibrium.**
J. Geheniau, E. Gunzig, I. Stengers.
Found. Phys., Vol. 17, No. 6, p. 585 – 601 (1987). Abstr. in Phys. Abstr., Vol. 90, No. 1313, Entry 115636 (1987).

161.331 **On the concept of time and the origin of the cosmological temperature.**
R. Brout.
Found. Phys., Vol. 17, No. 6, p. 603 – 619 (1987). Abstr. in Phys. Abstr., Vol. 90, No. 1313, Entry 115637 (1987).

161.332 **Historical independence of the inflationary universe.**
N. Panchapakesan, D. Lohiya.
Indian J. Phys., Part B, Vol. 61B, No. 3, p. 232 – 241 (1987). Abstr. in Phys. Abstr., Vol. 90, No. 1313, Entry 115639 (1987).

161.333 **Homogeneous space–times of Gödel–type in higher–derivative gravity.**
A. J. Accioly, A. T. Goncalves.
J. Math. Phys., Vol. 28, No. 7, p. 1547 – 1552 (1987). Abstr. in Phys. Abstr., Vol. 90, No. 1313, Entry 115640 (1987).

161.334 **Global qualitative study of Bianchi universes in the presence of a cosmological constant.**
E. Weber.
J. Math. Phys., Vol. 28, No. 7, p. 1658 – 1666 (1987). Abstr. in Phys. Abstr., Vol. 90, No. 1313, Entry 115641 (1987).

161.335 **Cosmic strings and cosmic structure.**
A. Albrecht, R. Brandenberger, N. Turok.
New Sci., Vol. 114, No. 1556, p. 40 – 44 (1987). Abstr. in Phys. Abstr., Vol. 90, No. 1313, Entry 115642 (1987).

161.336 **Cosmic strings and inflation.**
E. T. Vishniac, K. A. Olive, D. Seckel.
Nucl. Phys. B, Part. Phys., Vol. B289, No. 3 – 4, p. 717 – 734 (1987). Abstr. in Phys. Abstr., Vol. 90, No. 1313, Entry 115643 (1987).

161.337   On the quasi–de Sitter cosmological model of Staro-binsky.
M. D. Pollock.
Phys. Lett. B, Vol. 192, No. 1 – 2, p. 59 – 64 (1987). Abstr. in Phys. Abstr., Vol. 90, No. 1313, Entry 115645 (1987).

161.338   Light neutrinos as cold dark matter.
G. Raffelt, J. Silk.
Phys. Lett. B, Vol. 192, No. 1 – 2, p. 65 – 70 (1987). Abstr. in Phys. Abstr., Vol. 90, No. 1313, Entry 115646 (1987).

161.339   Bounds on the cosmological constant.
P. S. Joshi, S. M. Chitre.
Curr. Sci., Vol. 56, No. 5, p. 197 – 199 (1987). Abstr. in Phys. Abstr., Vol. 90, No. 1314, Entry 120662 (1987).

161.340   Evolution of isocurvature perturbations. II. Radi-ation–dust universe.
H. Kodama, M. Sasaki.
Int. J. Mod. Phys. A, Vol. 2, No. 2, p. 491 – 560 (1987). Abstr. in Phys. Abstr., Vol. 90, No. 1314, Entry 120663 (1987).

161.341   Nucleosynthesis versus the mirror universe.
E. D. Carlson, S. L. Glashow.
Phys. Lett. B, Vol. 193, No. 2 – 3, p. 168 – 170 (1987). Abstr. in Phys. Abstr., Vol. 90, No. 1314, Entry 120665 (1987).

161.342   Discontinuity cylinder model of gravitating U(1) cosmic strings.
P. Laguna–Castillo, R. A. Matzner.
Phys. Rev. D, Vol. 35, No. 10, p. 2933 – 2939 (1987). Abstr. in Phys. Abstr., Vol. 90, No. 1314, Entry 120667 (1987).

161.343   Reheating in the higher–derivative inflationary models.
W. Suen, P. R. Anderson.
Phys. Rev. D, Vol. 35, No. 10, p. 2940 – 2954 (1987). Abstr. in Phys. Abstr., Vol. 90, No. 1314, Entry 120668 (1987).

161.344   Gravitational particle creation and inflation.
L. H. Ford.
Phys. Rev. D, Vol. 35, No. 10, p. 2955 – 2960 (1987). Abstr. in Phys. Abstr., Vol. 90, No. 1314, Entry 120669 (1987).

161.345   Relativistic bubble dynamics: from cosmic inflation to hadronic bags.
A. Aurilia, R. S. Kissack, R. Mann, E. Spallucci.
Phys. Rev. D, Vol. 35, No. 10, p. 2961 – 2975 (1987). Abstr. in Phys. Abstr., Vol. 90, No. 1314, Entry 120670 (1987).

161.346   Semiclassical cosmology with a scalar field.
T. P. Singh, T. Padmanabhan.
Phys. Rev. D, Vol. 35, No. 10, p. 2993 – 3001 (1987). Abstr. in Phys. Abstr., Vol. 90, No. 1314, Entry 120671 (1987).

161.347   Analytical approach to string–induced phase transition.
U. A. Yajnik, T. Padmanabhan.
Phys. Rev. D, Vol. 35, No. 10, p. 3100 – 3104 (1987). Abstr. in Phys. Abstr., Vol. 90, No. 1314, Entry 120672 (1987).

161.348   Properties of Z_2 strings.
M. Aryal, A. E. Everett.
Phys. Rev. D, Vol. 35, No. 10, p. 3105 – 3115 (1987). Abstr. in Phys. Abstr., Vol. 90, No. 1314, Entry 120673 (1987).

161.349   Preon model and cosmological quantum–hyperchromo-dynamic phase transition.
H. Nishimura, Y. Hayashi.
Phys. Rev. D, Vol. 35, No. 10, p. 3151 – 3157 (1987). Abstr. in Phys. Abstr., Vol. 90, No. 1314, Entry 120674 (1987).

161.350   Density perturbations in a Brans–Dicke cosmological model.
J. P. Baptista, A. B. Batista, J. C. Fabris.
Rev. Bras. Fis., Vol. 16, No. 2, p. 257 – 263 (1986). Abstr. in Phys. Abstr., Vol. 90, No. 1314, Entry 120675 (1987).

161.351   No hair theorem for the Universe.
N. Panchapakesan.
Classical and quantum aspects of gravitation, p. 1 – 6 (1986). Abstr. in Phys. Abstr., Vol. 90, No. 1314, Entry 120677 (1987). – See Abstr. 012.065.

161.352   Dark matter, inflation and all that.
T. Padmanabhan.
Classical and quantum aspects of gravitation, p. 49 – 62 (1986). Abstr. in Phys. Abstr., Vol. 90, No. 1314, Entry 120678 (1987). – See Abstr. 012.065.

161.353   The structure of the Universe as determined from deep galaxy surveys.
A. G. Doroshkevich, A. A. Klypin.
Astron. Zh., Tom 64, Vyp. 6, p. 1137 – 1143 (1987). In Russian. English translation in Sov. Astron., Vol. 31, No. 6.
A new method of analysis of deep galaxy surveys is suggested. The method is based on the one–dimensional cluster analysis of data sets. The authors suggest to search for breaks in the dependence of the mean distance between "clusters" on neigh-bourhood radius. Analysis of observed data shows a break at $\bar{l} \cong 30 - 35$ Mpc. The velocity dispersion in superclusters is 350 – 450 km/s, in agreement with the results for nearby super-clusters.

161.354   The maximum density attained at the non–linear stage of gravitational instability in a collisionless medium with thermal velocities.
Eh. V. Kotok, S. F. Shandarin.
Astron. Zh., Tom 64, Vyp. 6, p. 1144 – 1154 (1987). In Russian. English translation in Sov. Astron., Vol. 31, No. 6.
Analytic estimates of maximum density attained at the non–linear stage of gravitational instability in a collisionless medium with small thermal velocities are obtained. These estimates are made for all types of generic singularities occurring in three-dimensional motions of a cold medium. In the one–dimensional case numerical simulations of gravitational collapse of a warm collisionless medium have been made.

161.355   Cosmological H II regions and the photoionization of the intergalactic medium.
P. R. Shapiro, M. L. Giroux.
Astrophys. J., Lett. Ed., Vol. 321, No. 2, p. L107 – L112 (1987).
The generalization of the classical H II region problem to the case of a point source of ionizing radiation in a cosmologically expanding gas in a Friedmann–Robertson–Walker universe is described. The authors derive the cosmological generalization of the static Strömgren radius and solve analytically for the time dependence of the radius and peculiar velocity of the ionization front which surrounds each source. An application of this work is described in which the hypothesis is tested that quasars photoion-ize the IGM to the degree implied by the well–known absence of a Gunn–Peterson effect.

161.356   Cosmological implication of a new measurement of the submillimeter background radiation.
S. Hayakawa, T. Matsumoto, H. Matsuo, H. Murakami, S. Sato, A. E. Lange, P. L. Richards.
Publ. Astron. Soc. Jpn., Vol. 39, No. 6, p. 941 – 948 (1987).
A new submillimeter measurement of the cosmic background radiation (CBR) by T. Matsumoto et al. (1988) shows excess brightness between 1000 and 300 μm. The excess is large, corresponding to ~10% of the undistorted blackbody radiation. A distortion of the type observed, could be produced by any of the following processes which have been widely discussed in the literature: Compton scattering of the 2.7–K background by a hot ionized medium, thermal emission from dust at high redshifts,

and the radiative decay of massive neutrinos and other exotic particles. In this note the authors assess the constraints that are imposed on the first two models by the observations.

161.357 Dark matter in the Universe ... and in the laboratory.
L. M. Krauss.
The standard model. The supernova 1987A, p. 665 – 678 (1987). – See Abstr. 012.074.
The author wants to examine what assumptions and what chain of reasoning lead to our present "standard wisdom" on the subject. In this discussion he wants to stress both where this wisdom may go wrong, and why we believe we have a now good idea of what generic type of material the dark matter must be made. Finally, the author wants to review the recent exciting developments which hold the promise of allowing the direct detection of dark matter, if it is made up from one of several different types of proposed elementary particles.

161.358 L'échelle des distances extragalactiques et le biais de Malmquist.
L. Bottinelli, L. Gouguenheim, G. Paturel, P. Teerikorpi.
Bull. Inf. Cent. Données Stellaires, No. 33, p. 33 – 39 (1987).
The Malmquist bias strongly affects the extragalactic distance scale, determined in particular through the "Tully–Fisher" relationship, and leads to important consequences on both the value of the Hubble constant and the kinematics of the local Universe.

161.359 On the rotation of the Universe.
D. D. Ivanenko, V. A. Korotkij, Yu. N. Obukhov.
Astron. Tsirk., No. 1458, p. 1 – 3 (1986). In Russian.
The authors discuss new anisotropic cosmological models with Gödel–type rotation within the framework of the gauge theory of gravity with torsion.

161.360 Some additions to scenarios of large–scale structure formation in the Universe.
V. L. Polyachenko.
Astron. Tsirk., No. 1459, p. 1 – 3 (1986). In Russian.
The radial orbits instability in processes of collapse or Hubble expansion as a possible cause of the large–scale structure formation in the Universe is suggested.

161.361 Generation of gravitational radiation in the early Universe.
D. Yu. Grigor'ev, D. V. Deryagin, V. A. Rubakov, M. V. Sazhin.
Astron. Tsirk., No. 1471, p. 1 – 2 (1986). In Russian.
The formation of the anisotropic phase in the early Universe is considered. The gravitational wave radiation generated by the anisotropic pressure of the boson condensate is calculated.

161.362 The cosmological scenario of the rotating Universe.
D. D. Ivanenko, V. A. Korotkij, Yu. N. Obukhov.
Astron. Tsirk., No. 1473, p. 1 – 3 (1986). In Russian.
The evolution of the Universe is investigated on the basis of the homogeneous nonsingular completely causal cosmological model with rotation and expansion. The new cosmological scenario is in sufficiently good agreement with modern observations. In this new proposed cosmological scenario the role of rotation is very important preventing singularity and inducing fast expansion, both in the early and final epoch. After initial expansion the Friedmann period takes place.

161.363 Quantum mechanics of phase transitions in the early Universe.
I. D. Lawrie.
International meeting on advances on phase transitions and disorder phenomena, Amalfi, Italy, 25 – 27 June 1986. World Scientific, Singapore, p. 548 – 567 (1987). Abstr. in Phys. Abstr., Vol. 90, No. 1315, Entry 120916 (1987).

161.364 Formation of the first systems in the wakes of moving cosmic strings.
T. Hara, S. Miyoshi.
Prog. Theor. Phys., Vol. 77, No. 5, p. 1152 – 1162 (1987). Abstr. in Phys. Abstr., Vol. 90, No. 1315, Entry 127977 (1987).

161.365 Amplification of correlation functions by gravity in the cosmic string model.
R. H. Brandenberger.
Phys. Lett. B, Vol. 191, No. 3, p. 257 – 262 (1987). Abstr. in Phys. Abstr., Vol. 90, No. 1315, Entry 128085 (1987).

161.366 Topologically massive planar universes with constant twist.
R. Percacci, P. Sodano, I. Vuorio.
Ann. Phys. (N.Y.), Vol. 176, No. 2, p. 344 – 358 (1987). Abstr. in Phys. Abstr., Vol. 90, No. 1315, Entry 128137 (1987).

161.367 Properties of a cosmological gas of bosons.
H. C. Ohanian, R. Ruffini, D. J. Song.
Nuovo Cimento B, Vol. 99B, Ser. 11, No. 1, p. 45 – 52 (1987). Abstr. in Phys. Abstr., Vol. 90, No. 1315, Entry 128153 (1987).

161.368 Cosmological perturbations in the inflationary universe.
V. F. Mukhanov, L. A. Kofman, D. Yu. Pogosyan.
Phys. Lett. B, Vol. 193, No. 4, p. 427 – 432 (1987). Abstr. in Phys. Abstr., Vol. 90, No. 1315, Entry 128154 (1987).

161.369 Kaluza–Klein Casimir cosmology and thermal equilibrium.
E. I. Guendelman.
Phys. Lett. B, Vol. 193, No. 4, p. 433 – 438 (1987). Abstr. in Phys. Abstr., Vol. 90, No. 1315, Entry 128155 (1987).

161.370 Peculiar velocities from cosmic strings.
R. Brandenberger, N. Kaiser, E. P. S. Shellard, N. Turok.
Phys. Rev. D, Vol. 36, No. 2, p. 335 – 345 (1987). Abstr. in Phys. Abstr., Vol. 90, No. 1315, Entry 128156 (1987).

161.371 The origin of the density fluctuations in de Sitter space.
M. Morikawa.
Prog. Theor. Phys., Vol. 77, No. 5, p. 1163 – 1177 (1987). Abstr. in Phys. Abstr., Vol. 90, No. 1315, Entry 128158 (1987).

161.372 Observing models of the universe on a supercomputer.
J. M. Centrella.
NRAO Workshop, No. 15, p. 1 – 17 (1988). – See Abstr. 012.076.
The author discusses the use of numerical simulations to make advances in astronomy, using the large scale structure of the universe as a case study. Emphasis is placed on analyzing the resulting models in observational terms to produce a meaningful comparison between theories and real data.

161.373 Self–creation cosmological solutions.
H. H. Soleng.
Astrophys. Space Sci., Vol. 139, No. 1, p. 13 – 19 (1987).
The field equations for Barber's two self–creation theories of gravitation are solved for Friedmann–Robertson–Walker space times, using perfect fluid energy–momentum tensors. Barber's first theory is discussed for the radiation dominated case, whereas cosmologies according to Barber's second self–creation theory are constructed for vacuum–dominated, radiation–dominated, and dust–filled cases.

161.374 A gravitationally non–degenerate cosmological model with expanding and shearing viscous fluid in general relativity.
R. Bali, D. R. Jain.
Astrophys. Space Sci., Vol. 139, No. 1, p. 175 – 181 (1987).
The object of this paper is to investigate the behaviour of viscosity in a cosmological model, in which the coefficient of shear viscosity is assumed to be proportional to rate of expansion

in the model. The behaviour of the model in the absence of shear viscosity is also discussed.

161.375   Rotational perturbations of cosmological viscous–fluid universe with zero–mass scalar field.
K. M. Singh.
Astrophys. Space Sci., Vol. 139, No. 1, p. 183 – 193 (1987).
   Certain new analytic solutions for the rotational perturbations of the Robertson–Walker universe are found out to substantiate the possibility of the existence of a rotating viscous universe with zero–mass scalar field. The values for $\Omega(r,t)$ which is related to the local dragging of inertial frames are investigated. In all the cases the rotational velocity is found to decay with time. Except for "perfect dragging" the scalar field is found to have a damping effect on the rotation of matter. The damping effect is found to be roughly analogous to viscosity. In some solutions it is found that the scalar field may exist only during a time period in the course of evolution of the Universe.

161.376   Some inhomogeneous cosmological models of plane symmetry with electromagnetic fields.
R. Bali, G. Singh, A. Tyagi.
Astrophys. Space Sci., Vol. 139, No. 2, p. 365 – 372 (1987).
   Some inhomogeneous cosmological models of plane symmetry in the presence of electromagetic fields have been obtained. Various physical and geometrical properties of the models with some special cases are also discussed.

161.377   Physical constants and evolution of the Universe.
V. S. Troitskii (*V. S. Troitskij*).
Astrophys. Space Sci., Vol. 139, No. 2, p. 389 – 411 (1987).
   A cosmological model is discussed which is based on interpretation of the redshift by a decreasing of the light speed with time everywhere in the Universe beginning with a certain moment of time in the past. The model is described by a metric in which the light speed depends on time and the radius of the curvature of three–dimensional space remains constant (c–metric). The model considered connects the evolution of the Universe with the evolution of physical constants and permits the explanation of some unclear cosmological phenomena – for example, a high isotropy of the relict background and superluminal speed in quasars.

161.378   Spacetimes admitting a universal redshift function.
G. Dautcourt.
Astron. Nachr., Vol. 308, No. 5, p. 293 – 298 (1987).
   The conditions are given for a velocity congruence in a Riemannian spacetime admitting a universal redshift function R. This function allows to calculate in a simple way (as a quotient of R values taken at the emission and registration event) the redshift or blueshift connected with an emitter and observer both following the congruence. Spacetimes and congruences with an universal redshift function are shortly discussed.

161.379   Multiple cosmic strings.
P. S. Letelier.
Classical Quantum Gravity, Vol. 4, No. 4, p. L75 – L77 (1987).
Abstr. in Phys. Abstr., Vol. 90, No. 1316, Entry 134325 (1987).

161.380   Friedmann cosmologies via the Regge calculus.
L. Brewin.
Classical Quantum Gravity, Vol. 4, No. 4, p. 899 – 928 (1987).
Abstr. in Phys. Abstr., Vol. 90, No. 1316, Entry 134326 (1987).

161.381   Decay of massive particles in Robertson–Walker universes with statically bounded expansion laws.
J. Audretsch, A. Ruger, P. Spangehl.
Classical Quantum Gravity, Vol. 4, No. 4, p. 975 – 993 (1987).
Abstr. in Phys. Abstr., Vol. 90, No. 1316, Entry 134327 (1987).

161.382   A matter model violating the strong energy condition – the influence of temperature.
B. Rose.
Classical Quantum Gravity, Vol. 4, No. 4, p. 1019 – 1030 (1987).
Abstr. in Phys. Abstr., Vol. 90, No. 1316, Entry 134328 (1987).

161.383   Orthogonality transitivity and cosmologies with a non–Abelian two–parameter isometry group.
M. H. Bugalho.
Classical Quantum Gravity, Vol. 4, No. 4, p. 1043 – 1045 (1987).
Abstr. in Phys. Abstr., Vol. 90, No. 1316, Entry 134329 (1987).

161.384   Inflationary fluctuations, entropy generation and baryogenesis in a cold universe.
J. Ellis, K. Enqvist, D. V. Nanopoulos, K. A. Olive.
Phys. Lett. B, Vol. 191, No. 4, p. 343 – 348 (1987). Abstr. in Phys. Abstr., Vol. 90, No. 1316, Entry 134331 (1987).

161.385   The anthropic Universe.
M. Rees.
New Sci., Vol. 115, No. 1572, p. 44 – 47 (1987). Abstr. in Phys. Abstr., Vol. 90, No. 1317, Entry 134423 (1987).

161.386   High–temperature quantum effects in multidimensional mixmaster models.
J. Szczesny, M. Szydlowski, M. Biesiada.
Europhys. Lett., Vol. 4, No. 6, p. 761 – 766 (1987). Abstr. in Phys. Abstr., Vol. 90, No. 1317, Entry 134598 (1987).

161.387   Quantum cosmology of superstrings.
K. Enqvist, S. Mohanty, D. V. Nanopoulos.
Phys. Lett. B, Vol. 192, No. 3 – 4, p. 327 – 331 (1987). Abstr. in Phys. Abstr., Vol. 90, No. 1317, Entry 134621 (1987).

161.388   Plane symmetric cosmological solutions of the Lyttleton–Bondi Universe.
D. R. K. Reddy, P. Innaiah.
Acta Phys. Hung., Vol. 61, No. 3 – 4, p. 269 – 276 (1987). Abstr. in Phys. Abstr., Vol. 90, No. 1317, Entry 140329 (1987).

161.389   On defects of the volume and curvature of the Robertson–Walker–metric and construction of cosmological models.
F. Gackstatter.
Ann. Phys. (Leipzig), Vol. 44, No. 6, p. 423 – 439 (1987). In German. Abstr. in Phys. Abstr., Vol. 90, No. 1317, Entry 140330 (1987).

161.390   Evolution equation for the expectation value of a scalar field in spatially flat RW universes.
A. Ringwald.
Ann. Phys. (N.Y.), Vol. 177, No. 1, p. 129 – 166 (1987). Abstr. in Phys. Abstr., Vol. 90, No. 1317, Entry 140331 (1987).

161.391   The Schucking problem.
I. Ozsvath, L. Shapiro.
J. Math. Phys., Vol. 28, No. 9, p. 2066 – 2073 (1987). Abstr. in Phys. Abstr., Vol. 90, No. 1317, Entry 140335 (1987).

161.392   The entropy and stability of the universe.
G. W. Gibbons.
Nucl. Phys. B, Part. Phys., Vol. B292, No. 4, p. 784 – 792 (1987). Abstr. in Phys. Abstr., Vol. 90, No. 1317, Entry 140336 (1987).

161.393   Non–chaotic Kaluza–Klein cosmology.
L. G. Jensen.
Phys. Lett. B, Vol. 192, No. 3 – 4, p. 315 – 317 (1987). Abstr. in Phys. Abstr., Vol. 90, No. 1317, Entry 140338 (1987).

161.394   On the onset of time and temperature in cosmology.
R. Brout, G. Horwitz, D. Weil.
Phys. Lett. B, Vol. 192, No. 3 – 4, p. 318 – 322 (1987). Abstr. in Phys. Abstr., Vol. 90, No. 1317, Entry 140339 (1987).

161.395 **On the possibility of chaotic inflation from a softly–broken superconformal invariance.**
M. D. Pollock.
Phys. Lett. B, Vol. 194, No. 4, p. 518 – 522 (1987). Abstr. in Phys. Abstr., Vol. 90, No. 1317, Entry 140340 (1987).

161.396 **Multidimensional mixmaster models.**
M. Szydlowski, M. Biesiada, J. Szczesny.
Phys. Lett. B, Vol. 195, No. 1, p. 27 – 30 (1987). Abstr. in Phys. Abstr., Vol. 90, No. 1317, Entry 140341 (1987).

161.397 **The problem of chaotic behaviour in a homogeneous arbitrarily dimensional cosmology.**
M. Szydlowski.
Phys. Lett. B, Vol. 195, No. 1, p. 31 – 35 (1987). Abstr. in Phys. Abstr., Vol. 90, No. 1317, Entry 140342 (1987).

161.398 **Dynamical neutralization of the cosmological constant.**
J. D. Brown, C. Teitelboim.
Phys. Lett. B, Vol. 195, No. 2, p. 177 – 182 (1987). Abstr. in Phys. Abstr., Vol. 90, No. 1317, Entry 140343 (1987).

161.399 **The wave functions and the effective action in quantum cosmology: covariant loop expansion.**
A. O. Barvinskij.
Phys. Lett. B, Vol. 195, No. 3, p. 344 – 348 (1987). Abstr. in Phys. Abstr., Vol. 90, No. 1317, Entry 140344 (1987).

161.400 **Instability of Kaluza–Klein cosmology.**
L. M. Sokolowski, Z. A. Golda.
Phys. Lett. B, Vol. 195, No. 3, p. 349 – 356 (1987). Abstr. in Phys. Abstr., Vol. 90, No. 1317, Entry 140345 (1987).

161.401 **Superconducting cosmic strings – energy loss by plasma dissipation.**
M. Panek, B. Rudak.
Phys. Lett. B, Vol. 195, No. 3, p. 357 – 360 (1987). Abstr. in Phys. Abstr., Vol. 90, No. 1317, Entry 140346 (1987).

161.402 **On the evolution of global strings in the early universe.**
D. Harari, P. Sikivie.
Phys. Lett. B, Vol. 195, No. 3, p. 361 – 365 (1987). Abstr. in Phys. Abstr., Vol. 90, No. 1317, Entry 140347 (1987).

161.403 **Decaying axion and the cosmic UV background.**
J. F. Lodenquai, V. V. Dixit.
Phys. Lett. B, Vol. 194, No. 3, p. 350 – 352 (1987). Abstr. in Phys. Abstr., Vol. 90, No. 1318, Entry 146107 (1987).

161.404 **Processes of the creation of the Universe.**
A. Hautot.
Atti Fond. Giorgio Ronchi, Vol. 41, No. 4, p. 451 – 470 (1986). In French. Abstr. in Phys. Abstr., Vol. 90, No. 1318, Entry 146114 (1987).

161.405 **On the expansion of the Universe.**
A. Hautot.
Atti Fond. Giorgio Ronchi, Vol. 41, No. 5, p. 617 – 630 (1986). In French. Abstr. in Phys. Abstr., Vol. 90, No. 1318, Entry 146115 (1987).

161.406 **Anisotropic universes and inflation.**
R. P. Mondaini, P. C. Nascimento.
Czech. J. Phys., Sect. B, Vol. B37, No. 9, p. 1056 – 1060 (1987). Abstr. in Phys. Abstr., Vol. 90, No. 1318, Entry 146116 (1987).

161.407 **The global structure of the inflationary universe.**
A. S. Goncharov, A. D. Linde, V. F. Mukhanov.
Int. J. Mod. Phys. A (*Singapore*), Vol. 2, No. 3, p. 561 – 591 (1987). Abstr. in Phys. Abstr., Vol. 90, No. 1318, Entry 146120 (1987).

161.408 **Cosmological models in globally geodesic coordinates. I. Metric.**
H.–y. Liu.
J. Math. Phys., Vol. 28, No. 8, p. 1920 – 1923 (1987). Abstr. in Phys. Abstr., Vol. 90, No. 1318, Entry 146122 (1987).

161.409 **Cosmological models in globally geodesic coordinates. II. Near–field approximation.**
H.–y. Liu.
J. Math. Phys., Vol. 28, No. 8, p. 1924 – 1927 (1987). Abstr. in Phys. Abstr., Vol. 90, No. 1318, Entry 146123 (1987).

161.410 **Stochastic processes in cosmology.**
M. O. Caceres, M. C. Diaz, J. A. Pullin.
Phys. Lett. A, Vol. 123, No. 7, p. 329 – 335 (1987). Abstr. in Phys. Abstr., Vol. 90, No. 1318, Entry 146125 (1987).

161.411 **Higher–dimensional Bianchi type I cosmologies.**
K. D. Krori, M. Barua.
Phys. Lett. A, Vol. 123, No. 8, p. 379 – 381 (1987). Abstr. in Phys. Abstr., Vol. 90, No. 1318, Entry 146126 (1987).

161.412 **Decreasing vacuum temperature: a thermal approach to the cosmological constant problem.**
M. Gasperini.
Phys. Lett. B, Vol. 194, No. 3, p. 347 – 349 (1987). Abstr. in Phys. Abstr., Vol. 90, No. 1318, Entry 146127 (1987).

161.413 **Oscillating solutions in higher–dimensional cosmology.**
A. B. Henriques, R. G. Moorhouse.
Phys. Lett. B, Vol. 194, No. 3, p. 353 – 357 (1987). Abstr. in Phys. Abstr., Vol. 90, No. 1318, Entry 146128 (1987).

161.414 **A model for baryogenesis in superstring unification.**
K. Yamamoto.
Phys. Lett. B, Vol. 194, No. 3, p. 390 – 396 (1987). Abstr. in Phys. Abstr., Vol. 90, No. 1318, Entry 146129 (1987).

161.415 **Ergodic theory of the mixmaster universe in higher space–time dimensions. II.**
Y. Elskens.
J. Stat. Phys., Vol. 48, No. 5 – 6, p. 1269 – 1282 (1987). Abstr. in Phys. Abstr., Vol. 91, No. 1319, Entry 156 (1988).

161.416 **Toward the inflationary paradigm: lectures on inflationary cosmology.**
M. S. Turner.
Acta Phys. Pol., Ser. B, Vol. B18, No. 9, p. 813 – 873 (1987). Abstr. in Phys. Abstr., Vol. 91, No. 1319, Entry 4500 (1988).

161.417 **The Einstein and de Sitter universes under the harmonic condition.**
C.–g. Huang, P.–y. Chou.
Chin. Phys. Lett., Vol. 4, No. 9, p. 397 – 400 (1987). Abstr. in Phys. Abstr., Vol. 91, No. 1319, Entry 4506 (1988).

161.418 **Ergodic theory of the mixmaster model in higher space–time dimensions.**
Y. Elskens, M. Henneaux.
Nucl. Phys. B, Field Theory Stat. Syst., Vol. B290(FS20), No. 1, p. 111 – 136 (1987). Abstr. in Phys. Abstr., Vol. 91, No. 1319, Entry 4510 (1988).

161.419 **On the decay of cosmic string loops.**
R. H. Brandenberger.
Nucl. Phys. B, Part. Phys., Vol. B293, No. 3 – 4, p. 812 – 828 (1987). Abstr. in Phys. Abstr., Vol. 91, No. 1319, Entry 4511 (1988).

161.420 **Cosmic strings and an improved upper bound on the energy density during inflation.**
D. H. Lyth.
Phys. Lett. B, Vol. 196, No. 2, p. 126 – 128 (1987). Abstr. in Phys. Abstr., Vol. 91, No. 1319, Entry 4512 (1988).

161.421   Baryogenesis in chaotic inflationary cosmology.
J. Yokoyama, K. Sato, H. Kodama.
Phys. Lett. B, Vol. 196, No. 2, p. 129 – 134 (1987). Abstr. in Phys. Abstr., Vol. 91, No. 1319, Entry 4513 (1988).

161.422   Baryogenesis at the MeV era.
S. Dimopoulos, L. J. Hall.
Phys. Lett. B, Vol. 196, No. 2, p. 135 – 141 (1987). Abstr. in Phys. Abstr., Vol. 91, No. 1319, Entry 4514 (1988).

161.423   Fragmentation of cosmic string loops.
A. G. Smith, A. Vilenkin.
Phys. Rev. D, Vol. 36, No. 4, p. 987 – 989 (1987). Abstr. in Phys. Abstr., Vol. 91, No. 1319, Entry 4515 (1988).

161.424   Numerical simulation of cosmic–string evolution in flat spacetime.
A. G. Smith, A. Vilenkin.
Phys. Rev. D, Vol. 36, No. 4, p. 990 – 996 (1987). Abstr. in Phys. Abstr., Vol. 91, No. 1319, Entry 4516 (1988).

161.425   Cosmic texture and the microwave background.
R. L. Davis.
Phys. Rev. D, Vol. 36, No. 4, p. 997 – 999 (1987). Abstr. in Phys. Abstr., Vol. 91, No. 1319, Entry 4517 (1988).

161.426   Cosmic balls of trapped neutrinos.
B. Holdom.
Phys. Rev. D, Vol. 36, No. 4, p. 1000 – 1006 (1987). Abstr. in Phys. Abstr., Vol. 91, No. 1319, Entry 4518 (1988).

161.427   Ultra–high–energy cosmic rays from superconducting cosmic strings.
C. T. Hill, D. N. Schramm, T. P. Walker.
Phys. Rev. D, Vol. 36, No. 4, p. 1007 – 1016 (1987). Abstr. in Phys. Abstr., Vol. 91, No. 1319, Entry 4519 (1988).

161.428   Singularities in Kaluza–Klein–Friedmann cosmological models.
M. Rosenbaum, M. Ryan, L. Urrutia, R. Matzner.
Phys. Rev. D, Vol. 36, No. 4, p. 1032 – 1035 (1987). Abstr. in Phys. Abstr., Vol. 91, No. 1319, Entry 4520 (1988).

161.429   Green's function of the scalar field in the early Universe.
A. Chowdhury, S. Mallik.
Phys. Rev. D, Vol. 36, No. 4, p. 1259 – 1262 (1987). Abstr. in Phys. Abstr., Vol. 91, No. 1319, Entry 4522 (1988).

161.430   Probability of R^2 inflation.
D. N. Page.
Phys. Rev. D, Vol. 36, No. 6, p. 1607 – 1624 (1987). Abstr. in Phys. Abstr., Vol. 91, No. 1319, Entry 4523 (1988).

161.431   Global properties of Kaluza–Klein cosmologies.
D. L. Wiltshire.
Phys. Rev. D, Vol. 36, No. 6, p. 1634 – 1648 (1987). Abstr. in Phys. Abstr., Vol. 91, No. 1319, Entry 4524 (1988).

161.432   Attempt at a classical cancellation of the cosmological constant.
S. M. Barr.
Phys. Rev. D, Vol. 36, No. 6, p. 1691 – 1700 (1987). Abstr. in Phys. Abstr., Vol. 91, No. 1319, Entry 4525 (1988).

161.433   Symmetry behavior in curved spacetime: finite–size effect and dimensional reduction.
B. L. Hu, D. J. O'Connor.
Phys. Rev. D, Vol. 36, No. 6, p. 1701 – 1715 (1987). Abstr. in Phys. Abstr., Vol. 91, No. 1319, Entry 4526 (1988).

161.434   Double–bubble spacetimes.
J. R. Ipser.
Phys. Rev. D, Vol. 36, No. 6, p. 1933 – 1935 (1987). Abstr. in Phys. Abstr., Vol. 91, No. 1319, Entry 4527 (1988).

161.435   New LRS perfect–fluid cosmological models.
J. M. M. Senovilla.
Classical Quantum Gravity, Vol. 4, No. 5, p. 1449 – 1455 (1987). Abstr. in Phys. Abstr., Vol. 91, No. 1320, Entry 4767 (1988).

161.436   Vacuum polarization and the initial conditions of cosmological evolution.
S. Gottlöber, V. Müller.
Classical Quantum Gravity, Vol. 4, No. 5, p. 1427 – 1435 (1987). Abstr. in Phys. Abstr., Vol. 91, No. 1320, Entry 10683 (1988).

161.437   Birkhoff theorem in self–creation cosmology.
T. Singh, T. Singh, O. P. Srivastava.
Int. J. Theor. Phys., Vol. 26, No. 9, p. 889 – 893 (1987). Abstr. in Phys. Abstr., Vol. 91, No. 1320, Entry 10686 (1988).

161.438   Brans–Dicke cosmological exact solutions in a radiation–filled Robertson–Walker universe.
R. T. Singh, S. Deo.
Int. J. Theor. Phys., Vol. 26, No. 9, p. 901 – 906 (1987). Abstr. in Phys. Abstr., Vol. 91, No. 1320, Entry 10687 (1988).

161.439   Expansion of the early Universe and the equation of state.
U. Ornik, R. M. Weiner.
Phys. Rev. D, Vol. 36, No. 4, p. 1263 – 1265 (1987). Abstr. in Phys. Abstr., Vol. 91, No. 1320, Entry 10690 (1988).

161.440   Can bulk viscosity drive inflation?
T. Pacher, J. A. Stein–Schabes, M. S. Turner.
Phys. Rev. D, Vol. 36, No. 6, p. 1603 – 1606 (1987). Abstr. in Phys. Abstr., Vol. 91, No. 1320, Entry 10691 (1988).

161.441   The consistency probems of large scale structure.
D. N. Schramm.
Physics of the superconducting supercollider, p. 627 – 630 (1987). Abstr. in Phys. Abstr., Vol. 91, No. 1320, Entry 10693 (1988). – See Abstr. 012.080.

161.442   "It ain't necessarily so" *(standard model for cosmology).*
P. J. E. Peebles.
Physics of the superconducting supercollider, p. 637 – 639 (1987). Abstr. in Phys. Abstr., Vol. 91, No. 1320, Entry 10695 (1988). – See Abstr. 012.080.

161.443   Galaxy distribution in a cold dark matter universe.
S. D. M. White, M. Davis, G. Efstathiou, C. S. Frenk.
Nature, Vol. 330, No. 6147, p. 451 – 453 (1987).
The authors show that the gravitational growth of structure by hierarchical clustering leads automatically to a bias in the distribution of galaxies. The strength of this bias is such that if the universe does indeed conform to the cold dark matter model then its density must approach the closure value. The mechanism predicts a significant dependence of the strength of galaxy clustering on the depth of the potential well of the galaxies considered.

161.444   Cosmology from nothing.
D. Lindley.
Nature, Vol. 330, No. 6149, p. 603 – 604 (1987).
The author reports on conformal equivalence of space–time metrics and on the paper of E. Gunzig et al. (see Abstr.161.445) which discusses cosmological models based on the conformal degree of freedom.

161.445   Entropy and cosmology.
E. Gunzig, J. Géhéniau, I. Prigogine.
Nature, Vol. 330, No. 6149, p. 621 – 624 (1987).
A cosmological model is proposed in which an inflationary de Sitter spacetime appears as the result of a fluctuation in the conformal degree of freedom of an initial Minkowski vacuum. A population of black holes is thereby created, which evaporate during the inflationary phase. Then a second phase transition

turns the de Sitter cosmology into the usual Robertson–Walker universe. The temperature and specific entropy per baryon of the present universe are deduced, and depend only on the mass of the black holes and on the universal constants h, c and k.

161.446   Can nucleons close the Universe?
K. A. Olive.
Nature, Vol. 330, No. 6150, p. 700 (1987).

161.447   Curvature magnification explains the superluminal lobes.
E. Sepulveda.
Publ. Astron. Soc. Pac., Vol. 99, No. 621, p. 1149 – 1150 (1987). Abstract. – See Abstr. 010.281.

161.448   Cosmic string wakes.
A. Stebbins, S. Veeraraghavan, R. Brandenberger, J. Silk, N. Turok.
Astrophys. J., Vol. 322, No. 1, p. 1 – 19 (1987).

The accretion of matter onto the wakes left behind by horizon-sized pieces of cosmic string is studied. The authors find that in a universe containing cold dissipationless matter (CDM) accretion onto wakes produces a network of sheetlike regions with a nonlinear density enhancement. The fraction of matter fallen into wakes is estimated. It is found that most of the CDM collapses onto string loops which may be the seeds for the formation of galaxies and clusters of galaxies, and some fraction of these loop condensations accrete onto wakes. The coherence length of wakes with the highest surface density is about that of the typical distance between the sheetlike formations of galaxies that are being observed in the CfA redshift survey.

161.449   The origin of the diffuse X–ray background and the formation of galaxies and voids.
R. A. Daly.
Astrophys. J., Vol. 322, No. 1, p. 20 – 33 (1987).

The author describes a new model to explain how galaxies and voids formed. It involves an unstable dark matter candidate. The dark matter and hydrogen form massive gravitationally bound condensates. The virial temperature of the hydrogen in these condensates is comparable to that necessary to produce the X–ray background from 5 to 200 keV. Subsequently, the dark matter decays and the gas expands outward. Forward and reverse shocks separated by a contact discontinuity result when the expanding gas collides with matter still collapsing toward the condensate. Galaxy formation is triggered in the high–density region between the two shock fronts; one obtains a Jeans mass for this gas of $\sim 10^{12} M_\odot$. The shock system overtakes the Hubble flow and may collide with that of its neighbor. The final configuration is a pattern of roughly spherical voids with luminous galaxies in thin shells and filaments.

161.450   The self–similar cosmological paradigm: a new test and two new predictions.
R. L. Oldershaw.
Astrophys. J., Vol. 322, No. 1, p. 34 – 36 (1987).

It is demonstrated that the magnetic dipole moments of atomic nuclei and neutron stars are quantitatively related by the fundamental scaling equations of the self–similar cosmological paradigm. Two definitive predictions are pointed out: (1) the model predicts that the electron will be found to have structure with radius of $\sim 4 \times 10^{-17}$cm, and (2) the model makes quantitative predictions regarding gravitational microlensing by predicted "dark matter" candidates.

161.451   Cosmological parameters from the *IRAS* galaxy sample.
J. V. Villumsen, M. A. Strauss.
Astrophys. J., Vol. 322, No. 1, p. 37 – 47 (1987).

An *IRAS*–selected sample of galaxies is used to make a dynamical estimate of cosmological parameters. The galaxy sample of 8387 sources is flux limited at 0.75 Jy at 60 μm. It is assumed that this sample traces the mass density of the universe on scales of 40 Mpc, and that the microwave background velocity is gravitationally induced by this mass distribution. Using a

standard luminosity function and the observed differential dipole function, the authors estimate that the bulk of the dipole is generated on scales of 35 ± 10 Mpc. It is shown that the ratio of flux dipole to luminosity density is insensitive to the exact values of the parameters of the luminosity function. From linear theory in a Friedman universe, the authors make a dynamical estimate of Ω_0, yielding $\Omega_0 \sim 1.2$. The statistical errror of this estimate is $1\sigma = 0.36$. Systematic errors may be as large as 55%. It is shown that the magnitude and position of the dipole in space is consistent with the observed galaxy–galaxy correlation function, and with estimates of H_0.

161.452   Constraints on possible precursor AGN sources of the cosmic X–ray background.
E. Boldt, D. Leiter.
Astrophys. J., Lett. Ed., Vol. 322, No. 1, p. L1 – L4 (1987).

New limits on small–scale surface brightness fluctuations of the residual cosmic X–ray background are used to reevaluate the conjecture that unresolved precursor AGN sources could be the origin of the flux. The large number of these predominantly X–ray–emitting objects thereby required imposes that their lifetime be longer than previously considered. Attention is drawn to the fact that the reduction in the effective Eddington luminosity limit for such sources due to electron–positron pair opacity could have the effect of increasing their lifetime to an appropriate level.

161.453   A search for QSOs to fit a cosmological model with flat, closed spatial sections.
H. V. Fagundes, U. F. Wichoski.
Astrophys. J., Lett. Ed., Vol. 322, No. 1, p. L5 – L7 (1987).

A cosmological model with Einstein–de Sitter local metric and three–torus spatial topology predicts multiple images of cosmic sources like QSOs. Assuming that our Galaxy, in an early stage of its evolution, was such a source, the authors look in a quasar catalog for the ones that can possibly be interpreted as images of the Galaxy in that stage, expecting to fit a particular realization of that model.

161.454   The appearance of primeval galaxies.
E. Baron, S. D. M. White.
Astrophys. J., Vol. 322, No. 2, p. 585 – 596 (1987).

The authors discuss the appearance of protogalaxies expected by current theories for the formation of structure. They present a detailed simulation of the inhomogeneous dissipative collapse of a protogalaxy. The bright phase of evolution terminates roughly at the nominal collapse time of the initial perturbation, by which time 60% of the final stars have been formed. In order to compare this model with the primeval galaxy candidate 3C 326.1, the authors scale it to collapse at $z = 1.8$ and to have a total mass of $10^{12} M_\odot$ and a final stellar mass of $10^{11} M_\odot$. The morphology of the model is consistent with that observed, but it is a factor of 4 too faint.

161.455   Reionization and small–scale fluctuations in the microwave background.
E. T. Vishniac.
Astrophys. J., Vol. 322, No. 2, p. 597 – 604 (1987).

Reionization of the intergalactic medium at moderate redshifts is often cited as a mechanism by which primordial fluctuations in the microwave background can be reduced below observational constraints. The author shows that electron scattering off the moving ionized medium can regenerate arcminute fluctuations in almost all models of galaxy formation. He presents explicit calculations for matter–dominated, $\Omega_0 = 1$ cosmologies with hot dark matter and cold dark matter.

161.456   On the origin of dwarf galaxies.
J. Silk, R. F. G. Wyse, G. A. Shields.
Astrophys. J., Lett. Ed., Vol. 322, No. 2, p. L59 – L65 (1987).

A new mechanism is proposed for producing vigorous bursts of star formation in metal–poor dwarf galaxies, at the present epoch. The intergalactic medium is heated and enriched by winds from dwarf galaxies at high redshift. This gas is compressed when galaxy groups form at the present epoch, and it cools and accretes onto slowly moving dwarf ellipticals. The authors describe several observable properties of the resulting gas–rich dwarfs, including the metallicities and the spatial distribution.

161.457 Minimal cosmic background fluctuations implied by streaming motions.
R. Juszkiewicz, K. Górski, J. Silk.
Astrophys. J., Lett. Ed., Vol. 323, No. 1, p. L1 – L6 (1987).

The authors derive the minimal cosmic background radiation anisotropy implied by the presence of large–scale streaming motions. If the tentative evidence for deviations from the Hubble flow of magnitude $\delta V/V \approx 0.1$ at $V \approx 5000$ km s^{-1} is confirmed, they predict microwave background fluctuations with a coherence scale $\sim 2°$ and dispersion $\delta T/T > 10^{-5}$. If the observational limits on $\delta T/T$ were to be reduced below the minimal predictions, then gravitational instability without reheating as a mechanism for generation of the large–scale structure of the universe would be in severe difficulty.

161.458 On the origin of possible deviation from the blackbody spectrum of the cosmic microwave background radiation: inverse Compton hypothesis based upon the explosion scenario.
S. Yoshioka, S. Ikeuchi.
Astrophys. J., Lett. Ed., Vol. 323, No. 1, p. L7 – L11 (1987).

Possible deviation from the blackbody spectrum of the cosmic microwave background radiation at the Wien part is indicated by the rocket observation of Nagoya–Berkeley group. Here, the inverse Compton hypothesis is examined based upon the explosion scenario for galaxy formation. Allowed ranges of physical parameters for reproducing the observation are explored in relation to the isotropy of spectrum at the Rayleigh–Jeans part, the size of voids, and the X–ray background radiation.

161.459 Distortion of the cosmic background radiation by superconducting strings.
J. P. Ostriker, C. Thompson.
Astrophys. J., Lett. Ed., Vol. 323, No. 2, p. L97 – L101 (1987).

Superconducting cosmic strings can be significant energy sources, keeping the universe ionized past the commonly assumed epoch of recombination. As a result, the spectrum of the cosmic background radiation is distorted in the presence of heated primordial gas via the Sunyaev–Zel'dovich effect. This distortion can be relatively large: the Compton y–parameter attains a maximum in the range $(1 - 5) \times 10^{-3}$, with these values depending on the mass scale of the string.

161.460 Path integral methods for primordial density perturbations: sampling of constrained Gaussian random fields.
E. Bertschinger.
Astrophys. J., Lett. Ed., Vol. 323, No. 2, p. L103 – L106 (1987).

Path integrals may be used to describe the statistical properties of a random field such as the primordial density perturbation field. In this framework the probability distribution is given for a Gaussian random field subjected to constraints such as the presence of a protovoid or supercluster at a specific location in the initial conditions. An algorithm has been constructed for generating samples of a constrained Gaussian random field on a lattice using Monte Carlo techniques.

161.461 Primeval galaxies and cold dark matter.
J. Silk, A. S. Szalay.
Astrophys. J., Lett. Ed., Vol. 323, No. 2, p. L107 – L111 (1987).

In the context of the cold dark matter theory for the large–scale matter distribution, the onset of galaxy formation is a gradual process, with star formation being initiated at $z \approx 10$ and reaching a peak for luminous galaxies at $z \approx 1$. The mass function of galaxy cores matches the observed quasar luminosity function at $z = 2 - 3$. Primeval galaxies are envisaged as a collection of many interacting and merging clumps, attaining a peak luminosity that is an order of magnitude below that achieved in models in which galaxy formation is initiated abruptly.

The expanding universe.
See Abstr. 003.054.

Darkness at night. A riddle of the universe.
See Abstr. 003.070.

Beyond Einstein. The cosmic quest for the theory of the universe.
See Abstr. 003.077.

Search for a supertheory. From atoms to superstrings.
See Abstr. 003.108.

Geometriya, dinamika, vselennaya (*Geometry, dynamics and the universe*).
See Abstr. 003.122.

Relativity, thermodynamics and cosmology.
See Abstr. 003.137.

La mecánica cuántica y la realidad del universo.
See Abstr. 004.014.

Carl Wirtz und die Flucht der Spiralnebel.
See Abstr. 004.019.

Infrared excess stirs cosmologists.
See Abstr. 011.017.

Probing the early universe.
See Abstr. 011.018.

The early universe and its evolution. Proceedings of the International School of Nuclear Physics, held at Erice, Sicily, Italy, 2 – 14 April 1986.
See Abstr. 012.024.

String theory, quantum cosmology and quantum gravity, integrable and conformal invariant theories. Proceedings of a colloquium held at Paris and Meudon, France, September 1986.
See Abstr. 012.083.

Superstrings, unified theories and cosmology. Proceedings of a workshop held at Trieste, Italy, June 1986.
See Abstr. 012.084.

La cosmologie en France.
See Abstr. 013.029.

When did the universe begin?
See Abstr. 014.037.

Cosmic strings. Gravitation without local curvature.
See Abstr. 014.040.

Time arrow in quantum theory.
See Abstr. 022.113.

The Cosmic Background Explorer (COBE) satellite.
See Abstr. 035.007.

Expected AXAF mirror characteristics and their implications for measurements of the Hubble constant using the Sunyaev Zel'dovich effect.
See Abstr. 035.030.

Computer model of the experiment "COLD". Evaluation of the microwave background inhomogeneity.
See Abstr. 036.024.

On computer modeling of the experiment "COLD".
See Abstr. 036.025.

Choix d'un système de filtres pour la photométrie des galaxies en cosmologie.
See Abstr. 036.212.

Infrared cosmology from space.
See Abstr. 051.041.

Possible appearance of the anomalous 4ν–interaction.
See Abstr. 061.010.

Nuclear beta strength, neutrino mass and cosmology.
See Abstr. 061.017.

Bounds on galactic cold dark matter particle candidates and solar axions from a Ge–spectrometer.
See Abstr. 061.026.

Superstring candidates for dark matter.
See Abstr. 061.028.

Cosmological analysis of R_p–breaking.
See Abstr. 061.031.

Radio–dating the Galaxy.
See Abstr. 061.033.

Thorium in G–dwarf stars as a chronometer for the Galaxy.
See Abstr. 061.034.

On the cosmological constant in the heterotic string theory.
See Abstr. 061.065.

Cosmological QCD, neutron diffusion, and the production of primordial heavy elements.
See Abstr. 061.071.

Detecting cold dark matter candidates.
See Abstr. 061.073.

Semiclassical gravitational effects near cosmic strings.
See Abstr. 061.075.

Baryon asymmetry of the universe in standard electroweak theory.
See Abstr. 061.076.

Observational limits on the time evolution of extra spatial dimensions.
See Abstr. 061.077.

Compactification of the twisted heterotic string.
See Abstr. 061.078.

Superconducting cosmic strings with massive fermions.
See Abstr. 061.082.

Decay of gravitinos and photo–destruction of light elements.
See Abstr. 061.083.

Nongravitational decay of cosmic strings.
See Abstr. 061.084.

Light pseudoscalars, particle physics and cosmology.
See Abstr. 061.085.

SU(2, 2/1, 1) supergravity and N = 2 supersymmetry with arbitrary cosmological constant.
See Abstr. 061.086.

Gravitomagnetic monopoles and the quantisation of frequency.
See Abstr. 061.095.

Dynamical symmetry breaking in a de Sitter–invariant vacuum.
See Abstr. 061.110.

Neutrino transport in relativity.
See Abstr. 061.112.

Multiloop modular invariance and the cosmological constant.
See Abstr. 061.114.

Analysis of no–scale supergravity models leading to inflationary scenarios.
See Abstr. 061.115.

Sphalerons, small fluctuations, and baryon–number violation in electroweak theory.
See Abstr. 061.117.

On the possibility of avoiding singularities by dilaton emission.
See Abstr. 061.124.

On the vanishing of the cosmological constant in four–dimensional superstring models.
See Abstr. 061.127.

Adjusting the cosmological constant dynamically: cosmons and a new force weaker than gravity.
See Abstr. 061.128.

Relaxing the cosmological bound on axions.
See Abstr. 061.129.

A simple solution to the solar neutrino and missing mass problems.
See Abstr. 061.133.

Bounds on neutrino masses from neutrino decay rates, cosmology and the see–saw mechanism.
See Abstr. 061.134.

No future for the fourth generation?
See Abstr. 061.138.

Testing superstrings with cosmology and the SSC.
See Abstr. 061.139.

A vortex–line model for infinite straight cosmic strings.
See Abstr. 061.145.

Superstring thermodynamics and its application to cosmology.
See Abstr. 061.153.

Urto materia antimateria per capire l'Universo.
See Abstr. 061.158.

The origin and cosmogonic implications of seed magnetic fields.
See Abstr. 062.064.

Gravitationslinsen.
See Abstr. 066.002.

Hubble's constant from gravitational wave observations.
See Abstr. 066.025.

On detecting stochastic background gravitational radiation with terrestrial detectors.
See Abstr. 066.032.

A class of spherically symmetric solutions with conformal Killing vectors.
See Abstr. 066.050.

Consistency of field equations in "self–creation" cosmologies.
See Abstr. 066.055.

An analytically soluble problem in fully nonlinear statistical gravitational lensing.
See Abstr. 066.058.

Gravitational interactions of cosmic strings.
See Abstr. 066.065.

Gravitational lensing by isothermal spheres with finite core radii: galaxies and dark matter.
See Abstr. 066.097.

A smooth oscillating cosmological solution.
See Abstr. 066.100.

Dirac equation in Bianchi I metrics.
See Abstr. 066.103.

Ultimas noticias sobre el sistema de lente gravitacional 0957 + 561.
See Abstr. 066.109.

Coalescing binaries – probe of the Universe.
See Abstr. 066.120.

Axisymmetric expanding universe with viscous fluid and heat flow.
See Abstr. 066.126.

Spacetime as a membrane in higher dimensions.
See Abstr. 066.127.

A homogeneous perfect fluid cosmological model in general relativity.
See Abstr. 066.130.

Critique of the theory of a cosmic potential.
See Abstr. 066.133.

Self–gravitating fluid in a conformally–flat space–time.
See Abstr. 066.136.

True "gravitational lens" effect for cylindrical deflectors.
See Abstr. 066.137.

Partition function and energy density of a scalar field at finite temperature in Robertson–Walker spacetime.
See Abstr. 066.139.

Cosmological applications of singular hypersurfaces in general relativity.
See Abstr. 066.140.

A new self–similar space–time.
See Abstr. 066.145.

The marginal gravitational lensing.
See Abstr. 066.151.

Graviton and topology contributions to self–consistent cosmology.
See Abstr. 066.161.

A rotating mass in a Gödel universe with an electromagnetic field.
See Abstr. 066.162.

Could a dilaton solve the cosmological constant problem?
See Abstr. 066.165.

On Geroch's limit of space–times and its relation to a new topology in the space of Lie groups.
See Abstr. 066.167.

Boundary conditions and the cosmological constant.
See Abstr. 066.169.

Exact model for a gaseous regular bounding sphere in general relativity.
See Abstr. 066.170.

Chaos in Kaluza–Klein models.
See Abstr. 066.171.

Kaluza–Klein theories.
See Abstr. 066.172.

Non–Schrödinger forces and pilot waves in quantum cosmology.
See Abstr. 066.173.

Wormholes in space–time.
See Abstr. 066.174.

Primitive black hole generated from unstable Minkowski space-time.
See Abstr. 067.099.

Some Kerr–like cosmological solutions of the Einstein–Maxwell equations.
See Abstr. 067.113.

Quantisation of scalar and vector fields inside the cosmological event horizon and its application to the Hawking effect.
See Abstr. 067.114.

An axionic laser in the center of a galaxy?
See Abstr. 067.116.

A Kruskal–like model with finite density.
See Abstr. 067.145.

Spectrum of dust heated by thick accretion disks of pregalactic black holes.
See Abstr. 067.155.

From strange matter to strange stars.
See Abstr. 067.161.

Neutron stars in the early Universe.
See Abstr. 067.163.

The orbit of Pluto and the cosmological constant.
See Abstr. 101.027.

A younger universe is seen in the stars.
See Abstr. 114.064.

Blackbody emissivity of supernova 1987A and the extragalactic distance scale.
See Abstr. 125.249.

H_2O masers and the cosmic distance scale.
See Abstr. 131.037.

Star formation, luminous stars, and dark matter.
See Abstr. 131.293.

VLA observations of H_2CO in absorption against the cosmic background radiation: clumps in the S140 molecular cloud.
See Abstr. 131.308.

Measurements of the ^3He abundance in the interstellar medium.
See Abstr. 132.049.

The origin of the diffuse X–ray background.
See Abstr. 142.010.

The cosmic X–ray background.
See Abstr. 142.012.

Diffuse background X rays and the density of the intergalactic medium.
See Abstr. 142.013.

Discrete X–ray sources and the X–ray background.
See Abstr. 142.018.

The art of N–body building.
See Abstr. 151.034.

Angular momentum from tidal torques.
See Abstr. 151.094.

Background starlight at the north and south celestial, ecliptic, and galactic poles.
See Abstr. 155.145.

The relative contributions of bulge and disk to the luminosity density of the universe.
See Abstr. 157.094.

Determination of the masses of elliptical galaxies and clusters of galaxies with AXAF.
See Abstr. 157.103.

Bimodal star formation: constraints from galaxy colors at high redshift.
See Abstr. 157.109.

IRAS galaxies: no evidence for a cosmological anisotropy.
See Abstr. 157.115.

Dynamics of galaxies at large redshift: prospects for the future.
See Abstr. 157.144.

A model of spectrophotometric evolution for high–redshift galaxies.
See Abstr. 157.210.

A search for primeval galaxies using a narrow–band imaging technique.
See Abstr. 157.272.

Dark matter in dwarf galaxies.
See Abstr. 157.279.

Dark matter in dwarf spheroidal galaxies: new particles or stellar remnants?
See Abstr. 157.299.

The evolution of galaxies at moderate redshift.
See Abstr. 157.307.

Masses of galaxies.
See Abstr. 157.329.

Interpretation of deep counts of radio sources.
See Abstr. 158.038.

The 160–minutes period in extragalactic objects: the LMC (RR Lyr stars) and NGC 4151.
See Abstr. 158.222.

The cause of discrepancy between evolution functions at large redshifts.
See Abstr. 158.336.

The shell structure of a pancake and the absorption spectra of quasars.
See Abstr. 159.005.

The evolution of optically selected QSOs.
See Abstr. 159.011.

The spatial clustering of QSOs.
See Abstr. 159.012.

Statistical gravitational lensing: influence of compact objects on the number counts of quasars.
See Abstr. 159.017.

Surveys of fields around quasars. IV. Luminosity of galaxies at $z \approx 0.6$ and preliminary evidence for the evolution of the environment of radio–loud quasars.
See Abstr. 159.031.

Cosmological implications of the anomalies in the redshift distribution of QSO Lyman α absorption systems.
See Abstr. 159.035.

Quasar clustering and the evolution of structure.
See Abstr. 159.075.

Observations of the new gravitational lens system UM673 = Q0142–100.
See Abstr. 159.076.

$z_{abs} \approx z_{em}$ absorption systems in quasar spectra: a correlation with radio and optical luminosity?
See Abstr. 159.079.

The early universe – an observer's view.
See Abstr. 159.089.

Superluminal radio sources: introduction.
See Abstr. 159.094.

The distribution of QSO absorption system column densities: evidence for a single population.
See Abstr. 159.114.

The redshift distribution of QSO Lyman–alpha absorption systems.
See Abstr. 159.115.

A clustering analysis of QSO candidates.
See Abstr. 159.129.

On the origin in spacetime distribution of quasars. I. The simplest model.
See Abstr. 159.132.

On the origin of spacetime distribution of quasars. II. The modified model.
See Abstr. 159.133.

Ever more distant quasars?
See Abstr. 159.135.

Quasars of redshift $z = 4.43$ and $z = 4.07$ in the South Galactic Pole field.
See Abstr. 159.136.

Cluster population incompleteness bias and the value of H_0 from the Tully–Fisher B_T^0 relation.
See Abstr. 160.009.

Observations of novae in the Virgo cluster.
See Abstr. 160.019.

The Sunyaev–Zel'dovich effect: measurements and implications.
See Abstr. 160.044.

New results on the Sunyaev–Zel'dovich effect in 0016 + 16, Abell 665 and Abell 2218.
See Abstr. 160.045.

The H I environment of high redshift quasars: a VLA search for cosmological H I.
See Abstr. 160.050.

The correlation function of galaxies in the direction of the Coma cluster.
See Abstr. 160.059.

The luminosity function: dependence on Hubble type and environment.
See Abstr. 160.062.

Are we all in the grip of a great attractor?
See Abstr. 160.075.

The morphology of the rich supercluster 1451 + 22.
See Abstr. 160.109.

General properties of clusters and groups of galaxies.
See Abstr. 160.113.

Comparisons of the spatial distribution of Abell clusters against models with Gaussian initial conditions.
See Abstr. 160.115.

More about clustering on a scale of 0.1 c.
See Abstr. 160.119.

Author Index

The authors are listed in alphabetical order according to the initial letter following the first names. Author names which are not transliterated according to our scheme are indicated by †.

Gackstatter, F.
161.389
Gadun, A. S.
080.026
Gaehrken, B.
014.011
Gaetz, T. J.
131.172
Gaffard, J. P.
031.042
Gaffey, M. J.
098.004 .005
Gaftonyuk, N. M.
032.051
Gagen-Torn, V. A. †
158.059
Gagliardi, L.
103.535
Gahm, G.
121.002
Gaidos, J.
143.031
Gail, H.-P.
064.029
Gaina, A. B.
067.117
Gaisser, T. K.
061.130
117.008
125.252 .314
Gajdaev, A. A.
104.085
Gajewski, W.
061.079
080.055
125.332
144.063
Gal, O.
133.006
Gal-Or, B.
003.061
Galan, M. J.
031.054 .055 .056 .057
Gale, M. R.
082.024
Galeev, A. A.
102.046
103.047 .418 .424 .628 .705
Galeotti, P.
034.050
125.303 .304 .330 .343
Galibina, I. V.
105.045
Galindo Trejo, J.
103.604
Galishev, V. S.
073.073
Galitskij, V. S.
161.161
Galkina, T. S.
112.139
114.009
117.069
Gall, M.
105.012
Gallagher III, J. S.
013.002
157.005 .012 .013 .072 .329
158.095
Gallart, C.
117.313
Gallegos, A.
078.009
Gallegos-Cruz, A.
144.042
Gallerani, A.
033.043
Galletta, G.
157.082
Gal'per, A. M.
035.005
143.018
Galt, J.
103.122 .449
Galt'sov, D. V.
067.038
Galvin, A. B.
074.052
Ganezer, K. S.
061.079
080.055
125.332
144.063

Gang, G. X.
144.045
Gao, B.
161.300
Gao, B.-x.
044.043
Gao, H.
103.676 .713 .714
Gao, L.
034.212
Gao, S. H.
067.022 .025
Gao, W.-s.
112.036 .037
Gao, Y.
161.310
Garavaglia, T.
014.033
Garay, G.
131.224
Garbuzov, G. A.
064.025
122.016 .039 .044
Garcia, A.
116.028
Garcia, B.
120.013
153.039 .041
Garcia, C.
103.673
Garcia, L. G.
119.011
Garcia, M. R.
117.184
Garcia, R. D. M.
063.048
Garcia, R. R.
082.005
Garcia de la Rosa, J. I.
072.071
Garcia Lambas, D.
151.032
Garcia Pino, A.
035.062
Garcia-Barreto, J. A.
134.045
157.032
Garcia-Berro, E.
065.081
Garcia-Pelayo, J. M.
157.194 .213
Garden, R. P.
131.041 .055 .056
Gardner, C.
036.197
Gardner, C. S.
032.045
082.038
Gardner, F. F.
131.176 .269
132.017
155.053
Gardner, R. D.
034.193
Garhart, M. P.
112.102
Garibdzhanyan, A. T.
155.168
Garilli, B.
036.057
142.032
158.344
Garlow, K. R.
117.307
Garmany, C. D.
112.040 .110 .130
Garmire, G. P.
035.025
Garnavich, P. M.
155.112
Garnett, D. R.
132.020
Garrard, T. L.
144.004
Garrido, R.
119.044
120.014
122.032 .088
Garrison, R. F.
015.009
112.131
114.125
117.136

Garrison, R. F.
125.259
Garstang, R. H.
034.210
Garvin, J.
093.002
Garwood, R.
155.103
Garwood, R. W.
157.038 .335
Gary, A.
051.025
Gary, D. E.
117.163
Garzon, F.
103.519 .555 .673
Gasanalizade, A. G.
022.123 .124 .125
071.006
Gaskell, C. M.
159.029 .033
Gaskell, R. W.
099.011
Gasperini, M.
161.170 .311 .412
Gass, H.
114.079
Gataullin, V. Kh.
033.034
Gatewood, G. D.
034.015
Gathier, R.
134.022
Gatley, I.
041.003
131.041 .055 .056 .188 .248
155.043 .051 .056 .065 .115
.116
158.005
Gattinger, R. L.
082.024
Gault, D. E.
022.020 .021 .022 .023 .026
Gaume, R. A.
131.081
Gaur, V. P.
072.088
Gautier, T. N.
155.067
Gautreau, R.
161.283
Gava, E.
061.065
Gavazzi, G.
004.005
160.072 .073
Gavrik, A. L.
103.416
Gavrilova, E.
125.305 .349
Gavryusev, V. G.
065.059
Gay, J.
036.244
Gayazov, I. S.
052.014
Gaylard, M. J.
103.448
Gayley, K. G.
073.099
Ge, Y.
067.168
Ge, Y. Z.
067.021
Gear, W. K.
158.236
Geballe, T. R.
036.207
103.038
112.040
133.013
155.046 .115 .116
158.253
Geffert, M.
124.016
154.061
Geheniau, J.
161.330 .445
Gehrels, N.
125.086 .248
Gehrels, T.
092.007

Gehren, T.
121.024
Gehrz, R. D.
034.173
122.021
Geiger, I.
122.090
Geisler, D.
125.561
153.006
156.046
Geisler, D. P.
156.022
Geiss, E.
044.009
Geiss, J.
035.112
074.032
103.489 .510 .542 .589
Gelb, J. M.
061.118
Geldzahler, B. J.
117.281
142.023 .033
159.046
Gel'frejkh, G. B.
072.065
Gellatly, D. W.
034.014
Geller, M. J.
160.053
Gel'man, B. G.
093.064
Gelmini, G.
061.026 .068 .147
080.053
161.253
Genet, D. R.
011.008
032.007
Genet, R. M.
011.008
013.031
032.007
036.131
113.059
117.083
Genkin, I. L.
062.085 .086
Genkin, L. G.
131.157
Genkov, V.
117.309
Genova, F.
099.019
Genova, R.
112.121
Gentile, L. C.
084.035
Genty, V.
111.011 .012 .015
Genzel, R.
035.102
103.533
131.251
132.031 .048
155.006 .023 .055 .059 .060
.062 .068
157.244
Georgiev, N.
031.073
Georgiev, T. B.
113.057
Gerard, E.
103.015 .017 .018 .040 .534
Gerashchenko, A.
154.080
Gerasimov, I. A.
035.005
042.027
101.050
Gerasimova, T. A.
083.027
Gerbier, G.
034.033
Gerdes, D.
004.025
Gerdjikova, M. G.
084.005
Gergely, T.
077.051
Gerhard, O. E.
151.052 .059 .063

Kontizas, M.
156.005 .006 .013 .014
Kontor, N. N.
075.011 .027
080.018
Kontorovich, V. M.
151.010
158.097
Koo, D. C.
157.264
159.071
160.087 .088
Koomen, M. J.
144.009
Koonin, S. E.
063.034
Koornneef, J.
132.046
156.025
Kopaeva, I. F.
125.022
142.036
Kopal, Z.
003.083
007.020
Kopecky, M.
013.089
072.004 .079
085.020
Koperski, P.
103.629
Kopp, G.
071.023
Koppar, S. S.
066.126 .162
067.042
Kopylov, A. I.
161.035 .038
Kopylov, I. M.
034.039
036.071
114.083
116.067
117.064 .220
Kopysov, Yu. S.
080.037
Koratzinos, M.
034.192
Korchagin, P. I.
155.157
Korchagin, P. V.
125.303 .304 .330 .343
Korchagin, V. B.
125.303 .304 .330 .343
Korchagin, V. I.
157.328
Korchuganov, B. N.
093.065
Kordylewski, J.
046.025
Korepanov, V. E.
103.711
Koret, D. L.
121.009
Korhonen, T.
122.122
Kormendy, J.
157.139 .148 .165 .176 .250
Kornbluh, D.
098.086
Korneev, V. V.
076.006
Kornienko, Yu. V.
036.166
Kornilov, V. G.
117.112
Korobejnikov, V. P.
104.034 .084
Korobejnikova, M. P.
085.014
Korobova, I. B.
093.038 .055
Korolev, A. E.
102.023 .035
Korol'kov, D. V.
033.013
Korostyleva, L. A.
022.062
Koroteev, V. A.
105.014
Korotin, S. A.
121.012

Korotkij, V. A.
161.359 .362
Korotkikh, V. L.
144.015
Kortchaguin, P. V. †
125.303 .304 .330 .343
Kortchaguin, V. B. †
125.303 .304 .330 .343
Korth, A.
035.110 .113
103.455 .467 .484 .486 .507
.637
Koryakina, E. A.
083.014
Korzhavin, A. N.
072.065
Koseki, M.
104.018
Koshiba, M.
080.067
125.255 .331
Koshiba, M.-T.
013.045
Koshiro, M.
122.115
Kosin, G. S.
002.012
082.061
Kosovichev, A. G.
066.056
073.024
074.050
080.015 .016 .040 .044
Kostelecky, J.
052.033
Kostiuk, T.
099.049
Kostryukov, A. V.
155.155
Kostyk, R. I.
080.026
Kostylev, K. V.
104.043
Kostyuk, N. D.
002.099
036.229
Kosugi, T.
073.090
Kotandzhyan, Kh. V.
063.100
Kotel'nikov, V. A.
093.017 .030 .046 .047
Kothare, A. T.
035.032
Kotok, Eh. V.
161.354
Kotov, V. A.
071.005 .022
073.027 .031
080.042 .043 .044
091.022
117.035
122.028 .055
126.068
158.222
Kotov, Yu. D.
076.005 .010
Kotrc, P.
034.035
073.048
Kotreleva, O. V.
044.005
Kottsov, V. A.
003.009
Kotze, T. C.
022.084
Koubsky, P.
002.017
013.088
101.054
112.105 .106
Kouchi, A.
022.142
Kounnas, C.
061.124
Koutchmy, S.
032.030
105.005
Kouveliotou, C.
073.044
143.022 .032 .033 .049
Kovachev, B.
046.039

Kovacs, A.
073.063
Kovacs, G.
065.001 .069
Kovacs, K.
065.044
Kovacs, T.
103.414
Koval', A. N.
073.026 .061
Koval'chuk, M. M.
071.010
Kovalenko, I. G.
062.052
Kovalevsky, J.
003.112
Koval'tsov, G. A.
073.038
Kovetz, A.
065.118
124.009
Kovner, I.
066.151
157.315
161.004
Kovtunenko, V. M.
051.033 .051
053.013
093.058
103.701
Kovtyukh, V. V.
119.019 .054
Kowalczyk, M.
002.113
Kowalski, M. P.
034.060
160.077
Koyama, K.
112.189
125.007 .015 .350
142.041
155.001
Kozai, Y.
103.499 .682
Kozak, R. C.
093.004 .005
Kozasa, T.
022.136
Kozenko, A. V.
097.079
Kozhanov, T. S.
151.038
153.038
161.239
Kozhevatov, I. E.
034.011
071.021
Kozimirov, N. G.
161.324
Kozlov, V. D.
035.005
Kozlov, V. I.
144.029
Kozlova, L. M.
071.009
Kozlowska, A.
002.113
Kozma, G.
035.096
103.414
Kozyra, J. U.
102.011
Kozyreva, V. S.
119.048
Krabbe, A.
034.030 .167
Kraft, R. P.
153.008
Krahn, D.
002.063
Krahn, E.
035.115
Krajchev, V. D.
034.188 .189
Krajcheva, Z. T.
117.123 .309
118.010
Kramer, E. N.
104.048 .078
Kramm, J. R.
035.115
036.016

Kramm, R.
103.582 .587
Kramynin, A. P.
072.054
Kranjc, A.
103.593
Krankowsky, D.
103.490 .531 .539
Krasikov, V. A.
036.234
103.584
Krasil'nikov, A. A.
116.016
131.022
Krasinski, A.
161.023
Krasnobabtsev, V. I.
121.012
Krasnobaev, A. N.
121.055
Krasnobaev, K. V.
062.121
Krasnopol'skij, V. A.
003.085
103.530 .543 .548 .571 .636
.640 .646 .650 .694
Krasnopolsky, V. A. †
103.530 .543 .548 .571 .636
.640 .646 .650
Kraus, D. J.
116.047 .048
117.076
119.034
Krauss, A.
061.093
Krauss, L. M.
061.080 .121
066.097
125.320
161.357
Krautter, J.
117.027 .156
153.017
Kravchuk, E. G.
085.005
Kravtsov, N. G.
144.037
Kravtsov, Yu. A.
003.123
Krawczyk, Z.
103.418 .476 .711
Krechet, V. G.
161.093 .242
Kreidl, T. J.
098.086
Kreiner, J. M.
119.038 .041
122.070
Kreinovich, V. Ya. †
161.118
Krejnovich, V. Ya.
161.118
Kremnev, R. S.
051.051
053.013
093.058
103.701
Kresak, L.
012.041
098.052
102.006 .044 .050 .064 .080
103.036 .718
Kresakova, M.
102.064 .080
104.076
Kreslavskij, M. A.
093.013
Kreyenhagen, K. N.
021.002
Kreysa, E.
034.162
131.029
157.008
158.050
159.007
Krichbaum, T.
158.227
Krimigis, S. M.
092.009
099.024 .025
144.008
Krisciunas, K.
082.042

Merritt, D.
151.040 .058 .084 .107
Merritt, D. R.
114.111
Merts, A. L.
022.085
Merzlyakov, V. L.
074.022
Message, P. J.
100.012
Messerotti, M.
073.029 .030
077.022
Messina, D. C.
143.029
Mestel, L.
131.246
Mestvirishvili, M. A.
003.091
Meszaros, P.
067.171 .180
Metcalf, T. R.
073.081
Metcalfe, L.
034.118
066.175
Metik, L. P.
158.179 .221
Metlinskij, N. I.
143.045
Metlova, N. V.
125.781
159.124
Metz, K.
034.010
103.546
Meurs, E. J. A.
158.049 .092 .247 .248 .249
161.071
Mewalt, R. A.
035.108
Mewe, R.
035.027
117.055
Meyer, C.
034.125
Meyer, D. L.
094.017
Meyer, D. M.
159.045 .053 .073
Meyer, E.
034.007
Meyer, H.
034.033
125.237
Meyer, H. J.
035.115
Meyer, K. R.
042.069
Meyer, M.
034.136
Meyer, P.
144.084
Meyer, T.
144.040
Meyer-Hofmeister, E.
067.118
Meyers, M. A.
105.055
Meylan, G.
154.015 .020 .050 .051 .052
     .074 .082
159.030 .119
Meylan, T.
114.072
122.040
Meynet, G.
065.010 .015
Mezger, P. G.
131.029
155.003
157.008
158.050
Mezzetti, M.
160.100
Michalitsianos, A. G.
013.052
117.244
156.018
Michard, R.
157.172
Michaud, G.
064.077

Michaud, G.
114.018
Michel, F. C.
067.126 .144 .173
Michel, R.
112.054 .070 .071 .075 .150
Michelson, P. F.
034.053
066.032
117.097
126.070
Michette, A. G.
031.011
Midavaine, T.
034.121
Middleditch, J.
117.167
126.057
160.077
Middlemass, D.
064.017
Middleton, R.
081.013
Miezis, J.
046.021
Mighell, K. J.
156.038
Migliaccio, F.
045.004
Mignard, F.
046.003
Mihajlov, A. A.
022.033
Mihalas, D.
063.102
065.043
Mihalov, J. D.
103.408
144.009
Mijatovic, Z.
063.028
Mikailov, Kh. M.
117.038 .289
Mikami, Y.
103.569
Mikhailov, Y. †
103.474 .475 .509 .630
Mikhajlov, V. P.
158.289
Mikhajlov, Yu.
103.417 .474 .475 .509 .630
Mikhajlutsa, V. P.
075.007 .015
085.007
Mikheev, S. P.
061.104
Mikhel'son, N. N.
032.012
Mikkola, S.
042.074
Mikocki, S.
034.185
036.240
144.060 .074 .075 .076
Mikolajchuk, O. V.
158.333
Mikolajewska, J.
117.087
Mikulasek, Z.
125.084
Mikusch, E.
103.587
Milani, A.
042.008
066.128
Milani, G.
124.242
Milano, L.
119.010 .082
Miletskij, E. V.
072.015
Miley, G.
157.099
158.031
Miley, G. K.
158.006 .024 .040 .193 .228
     .231 .321
Milgrom, M.
157.315
160.011
Milicevic, D.
009.025

Milkey, R. W.
126.064
Millar, T. J.
112.033
131.031 .043 .048 .196
Miller, B. N.
151.002
Miller, E. A.
021.003
Miller, E. R.
051.021
Miller, G. S.
062.068
Miller, J. C.
161.175 .177
Miller, J. S.
022.012
034.203
158.181
159.024
Miller, L.
159.134
Miller, M. E.
155.122
Miller, R.
034.119
125.332
Miller, R. E.
033.045
Miller, R. H.
013.083
151.105
160.046 .052
Miller, S. L.
022.109
Millis, R. L.
096.011
098.086
101.052
103.128
Mills, A. A.
014.028
Milne, D. K.
125.004 .008
Milone, E. F.
117.108
Milward, S. R.
035.019
Min, M.-l.
033.012
Minami, S.
103.495 .496
Mineshige, S.
117.063
Mineva, V. A.
157.257
160.107
Ming, T.
105.057 .058
Mingaliev, M. G.
012.008
158.268
Minglibaev, M. D.
042.045 .046
Minikulov, N. Kh.
112.106
113.032 .033
Minikunov, N. H. †
112.106
Mink, D. J.
096.008
Minkevich, A. V.
161.326
Minn, Y. K.
131.114
Minnaert, M. G. J.
003.083
Minois, J.
010.103 .841
Mirabel, I. F.
103.447 .720
131.054 .222
158.345
Miralles, J. A.
067.154
Miranda, A. I.
132.037
Mironova, L. S.
083.014
Miroshnichenko, A. S.
131.276 .277
Miroshnichenko, L. I.
078.007 .014

Miroshnichenko, L. I.
080.036
Mirskij, V. N.
105.018
Mirzoyan, L. V.
005.017
Misconi, N. Y.
106.029 .036 .037
Mishchenko, M. I.
099.023
Mishchenko, M. P.
032.050
Mishenina, T. V.
064.023
114.052
Mishima, T.
067.100
Mishra, R. B.
062.099 .140
Mishurov, Yu. N.
157.085
Miskotte, K.
104.092
Missana, M.
111.006
Mitchell, D. L.
103.455 .486 .637
Mitchell, K. J.
159.062
Mitchell, R. M.
112.028
124.008
Mit'kin, K. N.
036.231
Mitnitskii, V. Ya. †
093.024
Mitnitskij, V. Ya.
093.024 .057
Mitrikas, V. G.
078.002
Mitrofanov, I. G.
067.056 .074
Mitrovic, M.
009.025
Mitskevich, A. S.
063.024
Mitsuda, K.
117.073
125.350
Mitteldorf, J. J.
062.067
Mityanok, V. V.
117.170
Miura, K.
033.016
Miura, N.
074.026
Miwa, T.
066.014
Miyaji, S.
065.004
117.186
144.001
Miyake, W.
103.477 .482
Miyama, S. M.
066.021
107.011
Miyamoto, S.
125.039 .350
Miyano, K.
080.067
125.255 .331
Miyashita, A.
082.115
Miyauchi-Isobe, N.
158.258
Miyazawa, M.
073.070
Miyazawa, T.
033.016
Miyoshi, S.
161.364
Mizser, A.
122.073
Mizuno, T.
131.059
Mizutani, K.
034.168
035.057
103.622
132.009 .019
144.045

Noethe, L.
031.065
Noguchi, K.
034.133
Noguchi, M.
036.161
103.464
151.033
Noguchi, T.
157.060
Nolan, P. L.
035.082
Nolle, M.
104.003
Nollez, G.
022.047
Nolt, I. G.
082.080
Nolte, H. J.
084.007
Nomoto, K.
065.004 .038 .039 .068 .113
067.014 .030
125.067 .206 .235 .352 .353
Nonino, M.
073.030
Nook, M. A.
122.063
124.181
Noordam, J. E.
034.046
036.104
051.044
Nordgren, J.
012.045
Nordsieck, K. H.
122.063
124.181
Nordtvedt, K.
066.098
Nordtvedt Jr., K.
066.178
Noreau, L.
160.050
161.201
Noriega-Crespo, A.
151.100
157.236
Norimoto, Y.
036.161
Norman, C.
063.067 .121
161.202
Norman, C. A.
013.015
062.076
131.016 .199
155.070
158.229 .232 .233
Norman, M. L.
012.069
062.072 .073 .126 .127
063.103
066.152
131.171
158.084
Norrington, P. H.
062.011
Norris, J. P.
036.136
117.097
126.070
143.049
Norris, R. P.
125.263
131.269
North, J. D.
003.103
North, P.
113.003 .004 .030
Norton, A. J.
143.016
Noskova, R. I.
117.269
122.108
124.141 .143
157.061 .313 .337
Nosov, S. F.
144.013 .016 .044
Nothnagel, A.
009.022
045.011 .014

Notni, P.
022.106
103.042
Noton, M.
052.001
Notzold, D.
125.348
Nousek, J. A.
035.025
Noutchegueme, N.
066.082
Novakova, M.
104.001
Novello, M.
066.124
161.169
Novikov, A. Yu.
126.081
Novikov, B. S.
103.437
Novikov, G. G.
104.041 .065 .066
Novikov, I. D.
161.045 .103
Novlyanskaya, M. G.
005.011
Novotny, J.
066.079 .121
Novruzova, Kh. I.
065.051
Nowak, K.
103.418 .711
Noyes, R.
036.135
116.050
125.247
Noyes, R. W.
073.051
116.021
Noymer, A. J.
143.026
Nozdrachev, M. N.
103.418 .476
Nugaev, R. M.
067.040 .041 .152
Nugayev, R. M. †
067.040 .041 .152
Nugis, T. A.
011.011
Nulsen, P. E. J.
125.361
157.113
160.016 .056
Nunez, L.
066.067
Nussbaumer, H.
117.029 .202
Nussinov, S.
061.107
125.312
Nuth III, J. A.
131.291
Nutley, H.
073.047
Nuzhnova, T. N.
062.085 .086
Nye, R. A.
098.086
Nyman, L.-Aa.
155.171
Nymmik, R. A.
144.068

O Gym Den
073.036
Oattes, L. M.
066.050
Oberc, P.
103.418 .629
Oberhardt, M. R.
158.099
Oberts, P.
103.705
Obridko, V. N.
106.042
122.043
O'Brien, K.
080.075
O'Brien, T. J.
124.003
Obrubov, Yu. V.
104.059 .099

Obukhov, A. M.
103.610
Obukhov, Yu. N.
161.359 .362
Occhionero, F.
161.114
Ocegueda, J.
112.154
Ochsenbein, F.
002.041
Ockert, M. E.
101.041
O'Connell, R. W.
036.053
114.116
154.041
160.065
O'Connor, D. J.
161.433
O'Connor, W. P.
044.013
Oda, M.
013.044
125.039 .350
O'Dea, C. P.
012.015
141.014
158.204 .225
160.033 .047
O'Dell, C. R.
080.076
Odenwald, S.
034.149
131.052
158.146
161.282
Odintsov, S. D.
161.231
O'Donoghue, A.
158.082 .083
O'Donoghue, D.
117.042 .100
126.051
Oegelman, H.
067.083
117.102
125.212
126.035 .039 .062
Oegerle, W. R.
160.068
Oemler, M.
009.012
Oemler Jr., A.
161.137
Oestreicher, R.
117.053
Oezisik, M. N.
063.073
Offermann, D.
082.082 .085
Ogawa, H.
131.112
Ogawa, K.
066.102
Ogawa, T.
082.079 .115
Ogawara, Y.
125.039 .350
Ogir', M. B.
071.021
073.008
Ogloblina, O. F.
083.019
Ogorodnikov, B. I.
093.065
Ogorzalek, B. S.
035.093
Ograpishvili, N. B.
073.025
Ogris, V.
034.070
Ogura, K.
121.005
Ohanian, H. C.
014.041
161.367
Ohashi, T.
125.350
Ohashi, Y.
004.098
Ohishi, M.
022.092 .152
131.056 .302

O'Hora, N. P. J.
013.058
Ohsawa, A.
144.073
Ohshima, N.
031.048
Ohta, I.
144.045
Ohtani, E.
105.006
Oishi, H.
098.037 .038 .039 .040
Oja, H.
003.078
Oja, T.
113.025
122.045
Ojdov, D.
034.044
Ojeda-Castaneda, J.
031.039
Ojha, S. N.
062.098
064.001
065.093
Okado, M.
066.014
Okamura, S.
158.043
Okazaki, A. T.
067.044
Oke, J. B.
154.007
158.143 .168 .305 .325
Okuda, H.
062.119
131.055
132.009 .019
Okuda, T.
014.014
Okumura, S.
033.016
Ol', A. I.
072.017
Ol', G. I.
085.004
Olah, K.
112.106
122.073
Olalde, J. C.
103.602 .720
Olano, C. A.
131.228
155.129
Oldershaw, R. L.
161.450
Olearczyk, R. E.
103.427
Olie, M.
104.058
Olijnyk, P. A.
071.010
Olinto, A. V.
067.139
143.007
Oliva, E.
125.292 .310 .601
131.205
Olivares, A. E.
003.104
Olive, K. A.
061.028 .029 .139
080.054
125.339
161.336 .384 .446
Oliver, R. C.
098.086
Oliver, W. L.
012.057
Oliversen, N. A.
117.278
134.028
Oliversen, R. J.
103.458 .643
117.244
Olivia, E.
125.066
Olness, F.
117.207
Olofsson, H.
112.031
131.031

Seige, P.
103.582
Sein-Echaluce, M. L.
042.073
Seitter, W. C.
124.221
134.054
Sekanina, Z.
102.072
103.563 .564 .573 .580 .931
Seki, M.
067.172
Sekiguchi, K.
114.014
125.211 .224
158.010 .096
Sekiya, M.
107.011
Seleznev, A. F.
153.043
Seleznev, V. V.
103.416
Sellgren, K.
131.193 .275
134.057
155.050 .051
Sellier, A.
034.204
Sellwood, J. A.
151.034 .128
Selvelli, P. L.
117.141
Sembroski, G.
143.031
Semenenko, V. P.
003.127
Semenikin, A. A.
034.181
Semeniuk, A. M.
097.040
Semeniuk, I.
124.201
Semenko, V. P.
042.021
Semenoff, G. W.
012.043
Semenov, V. S.
084.030
Sementsov, V. N.
118.034
Semkov, E.
131.289
Semkov, E. H.
113.057
Semukhin, P. E.
073.038
Sen, A. K.
112.009
Sen, S.
161.313
Sen, S. N.
004.092
Senatorov, A. V.
125.302 .307
Senay, M.
103.563
Senkevich, V. V.
084.076
Senovilla, J. M. M.
161.435
Sephton, A. J.
144.062
Sepulveda, E.
161.197 .447
Serabyn, E.
155.029 .040 .073
Sergeeva, S. B.
093.062
Sergysels, R.
042.009
Serio, S.
074.074 .075 .081
Serlemitsos, P.
125.271
Serlemitsos, P. J.
031.062
Serov, A. V.
035.005
Serra, G.
035.005
155.004
Serrano, A.
157.170

Serrano P. G., A.
117.224
151.123
157.295
160.096 .097
Serrau, M.
082.067 .068
Serri, P.
034.033
Servan, B.
034.204 .206
Seshadri, T. R.
161.057
Sessin, W.
042.050 .071
100.026
Seufert, E. R.
117.083
Seve, A.
036.079
Severino, G.
063.099
071.027
Severny, A. A.
125.214
Severnyj, A. B.
114.024
125.218
159.050
Severnyj, S. A.
053.009
Sevilla, M. J.
044.018 .019 .020
Sevre, F.
034.127
157.205
Sevryukov, P. F.
042.024
Seward, F. D.
125.002
Sezer, C.
119.080
Sezgin, E.
012.084
Sha, C.-h.
036.155
Shabanov, M. F.
034.040 .041
Shafer, E. Yu.
125.022
142.036
Shafer, G. V.
144.033
Shafer, R. A.
117.028
158.187
160.016
Shaffer, D. B.
159.046 .101
Shafi, Q.
012.084
061.129
161.329
Shafter, A. W.
117.074
126.095
157.081
Shaganyan, B. L.
122.110
Shah, G. A.
103.101
Shah, M. R.
035.033
Shah, N. C.
034.186
Shaham, J.
117.001 .004
126.013
Shakhabasyan, K. M.
067.059
Shakhov, B. A.
144.013
Shakhovskaya, N. I.
117.067
121.009
Shakhovskoj, N. M.
117.067
Shaklan, S.
032.019
035.074
Shakun, L. I.
117.113 .264 .270

Shakura, N. I.
009.008
064.028
067.133
117.057
125.342
Shalybkov, D. A.
062.008 .155
Shamaev, V. G.
032.014
Shamarin, M. G.
036.064
Shames, P. M. B.
013.057
Shamis, V. A.
036.234
103.584
Shamolin, V. M.
125.022
142.036
Shan, Y.
080.050
Shandarin, S. F.
011.014
161.033 .354
Shang, Q.-z.
082.059
103.463
Shanin, V. N.
131.022
Shankar, A.
117.050
Shanks, T.
157.066
159.011 .012
160.102
Shao, L.-z.
031.075
Shao, M.
032.018 .035
034.058
036.113 .141
041.007
082.049
Shapere, A.
061.078
Shapiro, I.
125.287
Shapiro, I. I.
014.030
158.234 .349
159.093
Shapiro, L.
161.391
Shapiro, P. R.
161.003 .247 .355
Shapiro, S. L.
067.086
151.013 .027 .130 .143
Shapiro, V. D.
035.067
062.032
102.010
103.047 .474 .508 .570 .698
.705
Shapirovskaya, N. Ya.
131.085
Shapland, D.
003.128
Shaposhnikov, M. E.
061.076
161.323
Shaposhnikov, V. A.
035.006
Shapovalova, A. I.
012.008
159.040
Shara, M. M.
065.019
112.041
117.048 .145
124.009
151.028
157.215 .221
Sharaf, M. A.
052.031
Share, G.
078.003
Share, G. H.
076.011
124.006
125.267 .268
143.029

Sharipova, L. M.
034.180
Sharma, J. P.
067.039
Sharma, K. C.
064.068
Sharma, R. D.
082.081
084.058
Sharma, R. K.
042.003
Sharma, S. D.
004.091
Sharma, V. N.
004.105
Sharov, A. S.
154.004
Sharp, C. M.
103.461 .637
Sharp, N. A.
036.217
Sharp, W. E.
084.015
Sharples, R. M.
160.085 .089
Sharykin, N. K.
158.118
159.087
Shashkina, V. P.
093.047
Shastri, P.
159.066
Shatilov, V. A.
033.031
Shaver, P. A.
157.111
159.009 .013 .075 .077 .135
161.208
Shaviv, G.
065.118
067.129
Shavrina, A. V.
064.024
112.025
114.107
Shaw, M. A.
157.266 .308
Shaw, R. A.
117.050
Shawl, S. J.
154.035
Shaya, E. J.
157.108
Shcheglov, O. P.
093.061
Shcheglov, P. V.
034.181
Shchekinov, Yu. A.
131.082
159.005
161.028 .104
Shchelkanova, A. Yu.
117.230
Shcherbakov, A. G.
034.054
071.019
112.106
114.008
Shcherbakova, Z. A.
071.019
072.056
073.060
Shcherbanovskij, A. L.
032.051
Shcherbina-Samojlova, I. S.
003.017
Shchukina, N. G.
071.026
Shea, M. A.
084.035
Shealy, D. L.
035.083
Shectman, S. A.
157.102
Sheeley Jr., N. R.
074.030
075.016
144.009
Sheffer, E. K.
125.022
142.036
Sheidakov, N. E.
083.014

Subject Index

The Subject Index provides an alphabetical subject key to the papers abstracted in *Astronomy and Astrophysics Abstracts*. In order to explain the structure of this Index some general remarks on the construction principles of our key words will follow.

Whenever possible a key word was formed in such a way as to combine two different descriptors characterizing an abstracted article. Such a combination represents the contents of a paper more precisely than both components regarded as single entries. As an example we take the two terms Minor Planets and Orbits and the combination

Minor Planets
Orbits

Efforts were made to choose combinations which can be inverted in order to increase the usefulness of the Index. That means the two components represent terms of equal status which both serve as primary entries. In the given example we find the two entries

Minor Planets
Orbits

and

Orbits
Minor Planets

But there are also two-term key words which cannot be reversed. The conditions for non-inversibility are: the secondary descriptor is either a very specific term (e.g. LTE Analyses) or a very general one (e.g. Models). Thus we distinguish two kinds of descriptors, the primary ones, serving as direct entries and the secondary ones as auxiliary entries to the Index. Our intention is to choose comprehensive as well as significant formulations.

Some further aids to the use of this Subject Index are given by the following concordances:

Asteroids	see Minor Planets
Earth: Magnetic Field	see Geomagnetic Field
Galactic...	see also Galaxies, Galaxy
Lunar...	see also Moon
Planetary...	see also Planets
Solar...	see also Sun
Stellar...	see also Stars

Efforts were made to standardize the key words. This idea led to the development of our vocabulary of astronomical and astrophysical terms. From time to time we enlarged our vocabulary and at present version 17.88 consists of 2,343 items.

The list of key words was examined in order to standardize the key words and to confine some of the synonymous entries. For example, according to our vocabulary the key words Catalogues, Positions, Star Catalogues, and Star Positions were admissible terms, likewise the combinations

Catalogues
Star Positions

Star Catalogues
Positions

Star Catalogues
Star Positions

The user is – in any case – kindly requested to look for synonymous entries, because further references to the topic looked for might exist elsewhere in the Index under another current astronomical term.

Some key words concerning comets, novae, and solar eclipses are sorted by a special routine. Comets are listed according to their Roman numeral designation or preliminary designation, novae according to their stellar constellation and solar eclipses according to the date of the event.

Due to the introduction of a special object index containing a list of single objects, special object designations (e.g. SS433, BL Lacertae, Algol, etc) have been removed from the subject index and transferred to the object index (see SS433, BL Lac, β Per).

Exceptions have been made where major planets and their moons, comets, novae, and supernovae are concerned, which are given in the subject index only. Some objects (name or catalog designation) are contained in both indexes

Andromeda Galaxy	M31, NGC 224
Magellanic Clouds	LMC, SMC
Pleiades	M45
Hyades	Melotte 25
Crab Nebula	M1

In order to obtain the most comprehensive information possible the use of both indexes is recommended.

Starting with Volume 30 of *Astronomy and Astrophysics Abstracts*, this Index is given in a four-column arrangement. The overall size further has been optimized by suppression of a repetition of the first key word in a two-term combination. Thus, the primary term is printed only once (in semi-boldface) for all the following secondary key words which belong to this header.

The following List of Key Words might serve as a further guidance to the use of this Subject Index. This list, a subset of our vocabulary 17.88, consists of 1,230 entries and comprises all one-term key words and all primary terms of the two-term ones actually used in this Subject Index.

List of Key Words

A Dwarfs
A Stars
A Supergiants
Absolute Magnitudes
Absorption
Absorption Coefficient
Absorption Lines
Absorption Spectra
Accretion
Accretion Disks
Achondrites
Acoustic Waves
Active Galactic Nuclei
Active Galaxies
Active Optics
Adaptive Optics
Ae Stars
Aerosols
Air Showers
Airglow
Albedo
Alfven Waves
Algol Systems
Almanacs
AM Herculis Stars
Am Stars
Amor Objects
Andromeda Galaxy
Angular Diameters
Angular Momentum
Antennas
Ap Stars
Aperture Synthesis
Apex
Apodization
Apollo Objects
Apsidal Motion
Aquarids
Archaeoastronomy
Artificial Satellites
Asteroidal Belt
Astrodynamics
Astrographic Catalogue
Astrographic Catalogues
Astrographic Plates
Astrographs
Astrolabes
Astrometric Binaries
Astrometric Plates
Astrometry
Astronomical Constants
Astronomical Geodesy
Astronomical Instruments
Astronomical Optics
Aten-Type Objects
Atlases
Atmospheres
Atmospheric Optics
Atomic Clocks
Atomic Parameters
Atomic Processes
Atomic Spectra
Aurorae
Auroral Arcs
Auxiliary Instrumentation

B Stars
B Subdwarfs
B Supergiants
Background Radiation
Balloons
Balmer Decrement
Balmer Discontinuity
Balmer Lines
Barium Clouds
Barium Stars
Barred Spirals
Baryons
Be Stars
Beta Cephei Stars

Beta CMa Stars
Beta Lyrae Stars
Binaries
Binary Pulsars
Bipolar Nebulae
Bipolar Outflows
BL Herculis Stars
BL Lacertae Objects
Black Holes
Black-Body Radiation
Blazars
Blue Galaxies
Blue Stars
Blue Stragglers
Blue Variables
Bolometers
Bolometric Magnitudes
Bp Stars
Bremsstrahlung
Bright Stars
Brightness Temperature
Brown Dwarfs
BV Photometry
BY Draconis Stars

C II Regions
C-M Diagrams
Calendars
Callisto
Callisto Craters
Callisto Surface
Cameras
Carbon Burning
Carbon Deflagration
Carbon Stars
Cassegrain Telescopes
Cataclysmic Binaries
Catalogues
CCD Cameras
CCD Detectors
CCD Observations
CCD Photometry
Celestial Bodies
Celestial Mechanics
Celestial Objects
Central Stars
Cepheids
CH Stars
Chandler Wobble
Chemical Elements
Chondrites
Chondrules
Circumstellar Clouds
Circumstellar Disks
Circumstellar Dust
Circumstellar Grains
Circumstellar Matter
Circumstellar Shells
Clocks
Close Binaries
Clusters of Galaxies
CN Stars
CNO Anomalies
CNO Cycle
Collapse
Collapsed Objects
Collapsing Clouds
Collapsing Stars
Color Excesses
Color Indices
Colors
Cometary Atmospheres
Cometary Comae
Cometary Dust
Cometary Ionospheres
Cometary Nebulae
Cometary Nuclei
Cometary Plasma
Cometary Tails
Comets

Compact Galaxies
Compact Objects
Compton Scattering
Contact Binaries
Convection
Convective Zones
Cool Giants
Cool Stars
Coordinate Systems
Coronographs
Cosmic Dust
Cosmic Microwave Background
Cosmic Rays
Cosmic Strings
Cosmochemistry
Cosmochronology
Cosmogony
Cosmological Constant
Cosmological Models
Cosmology
CP Stars
Crab Nebula
Curves-of-Growth
Cyclotron Radiation

Dark Clouds
Dark Matter
Dark Nebulae
Data Bases
Data Centers
Data Processing
DDO Photometry
Degenerate Dwarfs
Degenerate Matter
Degenerate Stars
Deimos
Delta Cephei Stars
Delta Scuti Stars
Dense Matter
Densities
Density
Density Waves
Detached Systems
Detector Arrays
Detectors
Diameter
Diameters
Differential Rotation
Diffuse Galactic Light
Diffuse Nebulae
Diode Arrays
Disk Galaxies
Distance
Distance Indicators
Distance Moduli
Distance Scale
Distances
Domes
Doppler Velocities
DQ Herculis Stars
Draconids
Dust Clouds
Dwarf Cepheids
Dwarf Galaxies
Dwarf Novae
Dwarfs
Dynamical Parallaxes
Dynamo Theory

Early Universe
Early-Type Galaxies
Early-Type Stars
Early-Type Supergiants
Earth
Earth Atmosphere
Earth Core
Earth Interior
Earth Ionosphere
Earth Magnetosphere
Earth Mantle

Earth Mesosphere
Earth Plasmasphere
Earth Radiation Belts
Earth Rotation
Earth Stratosphere
Earth Thermosphere
Earth Troposphere
Earth-Moon System
Echelle Spectra
Echelle Spectrographs
Echelle Spectrometers
Eclipses
Eclipsing Binaries
Ecliptic
Effective Temperatures
Electron Densities
Electron Temperatures
Electrophotometers
Elementary Particles
Ellipsoidal Binaries
Elliptical Galaxies
Emission Lines
Emission Nebulae
Emission Spectra
Emission-Line Galaxies
Emission-Line Objects
Emission-Line Stars
Enceladus
Ephemerides
Equilibrium Figures
Europa
Europa Surface
Extinction
Extragalactic Objects
Extraterrestrial Intelligence
Extraterrestrial Life
Extreme UV
Extreme UV Astronomy
Extreme UV Spectra

F Dwarfs
F Stars
F Supergiants
Fabry-Perot Interferometers
Fabry-Perot Interferometry
Fabry-Perot Spectrometers
Faint Galaxies
Faint Stars
Fiber Optics
Field Theories
Filters
Fireballs
FK4
FK4 Stars
FK4 System
FK5
FK5 System
Flare Stars
Fluid Dynamics
Fluid Spheres
Flux Densities
Forbidden Lines
Forbush Effect
Four-Body Problem
Fourier Spectrometers
Fraunhofer Lines
Frequency Standards
FU Orionis Stars
Fundamental Catalogues
Fundamental Constants
Fundamental Stars
Fundamental Systems

G Dwarfs
G Stars
Galactic Anticenter
Galactic Center
Galactic Centers
Galactic Corona
Galactic Coronae

Observatories
Occultations
Of Stars
OH Absorption
OH Clouds
OH Emission
OH Masers
OH Sources
OH-IR Stars
Oort's Cloud
Oort's Constants
Opacities
Open Clusters
Optical Design
Optical Spectra
Optical Systems
Optical Testing
Optical Transients
Orbit
Orbit Determination
Orbit Improvement
Orbit Stability
Orbit Theory
Orbital Elements
Orbital Evolution
Orbital Periods
Orbits
Organic Matter
Organic Molecules
Origin of Life
Orion Nebula
Orionids
Oscillations
Oscillator Strengths
Oxygen Burning

P Cygni Profiles
P Cygni Stars
Pairs of Quasars
Pairs of Radio Sources
Parallaxes
Peculiar Galaxies
Peculiar Stars
Peculiar Variables
Period Changes
Period Determination
Period-Luminosity Relation
Period-Radius Relation
Periodic Comets
Periodic Orbits
Periods
Perseids
Perturbation Theory
Perturbations
Phobos
Phoebe
Photoelectric Photometry
Photographic Astrometry
Photographic Emulsions
Photographic Materials
Photographic Photometry
Photographic Plates
Photographic Techniques
Photometers
Photometric Elements
Photometric Standards
Photometric Systems
Photometry
Photomultipliers
Photon Counters
Photon Counting
Planetary Atmospheres
Planetary Companions
Planetary Interiors
Planetary Ionospheres
Planetary Magnetospheres
Planetary Nebulae
Planetary Occultations
Planetary Rings
Planetary Satellites
Planetary Surfaces
Planetary System
Planetary Systems
Planetary Theories
Planetesimals
Planets
Plasma
Plasma Diagnostics
Plasma Jets
Plasma Torus
Plasma Waves
Plate Tectonics
Pleiades
Pluto
Pluto Atmosphere

Pluto Satellite
Polar Motion
Polarimeters
Polarimetry
Polarization
Polars
Polytropes
Population II Dwarfs
Population II Stars
Positions
Potential Theory
Power Spectra
Poynting-Robertson Effect
Praesepe
Pre-Main-Sequence Stars
Precession
Pregalactic Stars
Presolar Nebula
Proper Motions
Proper-Motion Stars
Proportional Counters
Protogalaxies
Protoplanetary Cloud
Protoplanetary Nebulae
Protoplanets
Protostars
Protostellar Clouds
Pulsar Magnetospheres
Pulsars
Pulsating Stars
Pulsating Variables
Pulsation Theory
Pulsations

Quadrantids
Quantum Gravity
Quark Matter
Quark Stars
Quasar Counts
Quasar-Galaxy Associations
Quasars
Quasi-Periodic Oscillations

R CrB Stars
R Stars
Radial Velocities
Radiation Mechanisms
Radiation Pressure
Radiative Transfer
Radio Astrometry
Radio Astronomy
Radio Background
Radio Brightness
Radio Bursts
Radio Continuum
Radio Flares
Radio Galaxies
Radio Halos
Radio Interferometers
Radio Interferometry
Radio Jets
Radio Lines
Radio Luminosities
Radio Maps
Radio Meteors
Radio Polarization
Radio Radiation
Radio Sources
Radio Spectra
Radio Spectrographs
Radio Stars
Radio Surveys
Radio Tails
Radio Telescopes
Radiometers
Radiometry
Receivers
Recombination Lines
Recurrent Novae
Red Dwarfs
Red Giants
Red Stars
Red Supergiants
Red Variables
Reddening
Redshift Surveys
Redshift-Magnitude Relation
Redshifts
Reference Systems
Reflectance Spectra
Reflection Nebulae
Reflectors
Refraction
Refractors
Relativistic Effects

Relativistic Electrons
Relativistic Fluids
Relativistic Gas
Relativistic Particles
Relativistic Plasma
Relativistic Stars
Relativity Theory
Relaxation
Resonance Lines
Resonances
Reticon
Ring Galaxies
Ring Nebulae
Ritchey-Chretien Telescopes
Roche Lobes
Rotating Bodies
Rotating Clouds
Rotating Disks
Rotating Fluids
Rotating Gas
Rotating Spheres
Rotating Stars
Rotation
Rotation Curve
Rotation Curves
RR Lyrae Stars
RS CVn Stars
RV Tauri Stars

S Doradus Stars
S Stars
Satellite Geodesy
Satellite Laser Ranging
Satellites
Saturn
Saturn Atmosphere
Saturn Magnetosphere
Saturn Rings
Saturn Satellites
Scattering
Schmidt Telescopes
Schwarzschild Black Holes
Scintillations
Seeing
Semi-Detached Systems
Semiregular Variables
Seyfert Galaxies
Shell Stars
Shock Waves
Short-Period Comets
Silicon Stars
SiO Masers
Site Testing
Sky Background
Sky Surveys
Solar Active Regions
Solar Activity
Solar Activity Cycles
Solar Atmosphere
Solar Chromosphere
Solar Constant
Solar Convective Zone
Solar Corona
Solar Coronal Holes
Solar Coronal Loops
Solar Coronal Transients
Solar Cosmic Rays
Solar Disk
Solar Eclipses
Solar Extreme UV
Solar Faculae
Solar Filaments
Solar Flares
Solar Gamma Rays
Solar Gamma-Ray Bursts
Solar Granulation
Solar Interior
Solar Limb
Solar Luminosity
Solar Magnetic Fields
Solar Microwave Bursts
Solar Microwave Radiation
Solar Models
Solar Motion
Solar Moustaches
Solar Nebula
Solar Neighborhood
Solar Neutrinos
Solar Oscillations
Solar Particles
Solar Patrol
Solar Photosphere
Solar Plages
Solar Plasma
Solar Prominences

Solar Radio Bursts
Solar Radio Radiation
Solar Rotation
Solar Spectrographs
Solar Spectrum
Solar Spicules
Solar Surface
Solar Surges
Solar System
Solar Telescopes
Solar Transition Region
Solar UV Radiation
Solar Wind
Solar X Rays
Solar X-Ray Bursts
Solar X-Ray Flares
Solar-Terrestrial Relations
Solar-Type Stars
Space Astrometry
Space Astronomy
Space Instrumentation
Space Missions
Space Motions
Space Probes
Space Research
Space Telescope
Space Telescopes
Space Vehicles
Speckle Interferometers
Speckle Interferometry
Speckle Spectroscopy
Spectra
Spectral Classification
Spectral Indices
Spectral Lines
Spectral Types
Spectrograms
Spectrographs
Spectrography
Spectroheliograms
Spectrometers
Spectrometry
Spectrophotometers
Spectrophotometry
Spectropolarimetry
Spectroscopic Binaries
Spectroscopic Standards
Spectroscopy
Spectrum Variables
Spherules
Spiral Arms
Spiral Galaxies
Standard Stars
Star Atlases
Star Catalogues
Star Clusters
Star Counts
Star Formation
Star Positions
Starburst Galaxies
Starbursts
Stark Broadening
Stark Effect
Stars
Starspots
Statistical Mechanics
Stellar Activity
Stellar Associations
Stellar Atmospheres
Stellar Chromospheres
Stellar Classification
Stellar Collapse
Stellar Collisions
Stellar Content
Stellar Coronae
Stellar Diameters
Stellar Dynamics
Stellar Envelopes
Stellar Evolution
Stellar Flares
Stellar Groups
Stellar Interferometers
Stellar Interiors
Stellar Kinematics
Stellar Magnetic Fields
Stellar Magnetospheres
Stellar Models
Stellar Occultations
Stellar Orbits
Stellar Oscillations
Stellar Photospheres
Stellar Populations
Stellar Rotation
Stellar Spectra
Stellar Statistics

Object Index

The usual bibliographic information accompanying the papers in the fields of astronomy and astrophysics certainly allows for their exact identification but the selection process of information retrieval is facilitated by supplementary information. Phenomena of all kinds, methods of observation and reduction as well as properties of classes of objects are referred to in the Subject Index.

Beginning with Vol. 39 *Astronomy and Astrophysics Abstracts* will present an additional index of astronomical objects dealt with in papers compiled. Basically all individual objects appearing in the title and the author's abstract are listed, but objects treated in the paper proper have also been included into the index. The number of single objects listed usually has been limited to 20. Papers discussing more than 20 individual objects have not been evaluated in the Object Index.

It is recommended to use the Object Index together with the key words of the Subject Index. Individual comets, novae, supernovae, major planets and their satellites are exclusively referred to in the Subject Index. A number of familiar astronomical objects will continue to appear in the Subject Index with their common designation. These objects are listed in the Object Index with their standard acronyms:

M31, NGC 224	Andromeda Galaxy
LMC, SMC	Magellanic Clouds
M45	Pleiades
Melotte 25	Hyades
M1	Crab Nebula

Novae identified with variables are referred to in the Object Index with their variable name:

GK Per	Nova Persei 1901

Astronomy and Astrophysics Abstracts supports the attempts of the IAU in standardizing the nomenclature of astronomical objects. Acronyms used follow closely the recommendations of "The First Dictionary of the Nomenclature of Celestial Objects" (A. Fernandez, M.-C. Lortet, F. Spite; Astron. Astrophys., Suppl. Ser., Vol. 52, No. 4 (1983)), and its "First Supplement" (M.-C. Lortet, F, Spite; Astron. Astrophys., Suppl. Ser., Vol. 64, No. 2, p. 329–389 (1986)).

In sorting astronomical objects deviations from the lexicographic sequence are unavoidable in many cases. Sorting of the Object Index is based on the IBM standard routine SORT, which sorts blanks followed by special characters, Latin small and capital letters, and finally numbers. The three-letter mnemonic code of the 88 stellar constellations adopted by the IAU defines the highest rank within the sorting hierarchy. This implies that all stellar and non-stellar objects, whose designations explicitly refer to a constellation, are listed under the constellation's abbreviation. Stars within a constellation are sorted according to

their Greek letter, followed by the standard variable nomenclature, and finally the Flamsteed numbers as well as other catalogue designations:

> Cyg A
> Cyg OB2
> Cyg X–3
> α Cyg
> χ Cyg
> BF Cyg
> NML Cyg
> Z Cyg
> V367 Cyg
> V1727 Cyg
> 6 Cyg

Designations without reference to stellar constellations are treated – as a rule – by sorting letters before numbers. An exception has been made for the minor planets. These objects have been retained in a particular section following after the letter 'Z'. The internal sequence is (1) permanently numbered planets, (2) objects with provisional designations, and (3) Palomar-Leiden objects:

> (9) Metis
> (29) Amphitrite
> 1979 VA
> 1984 HA$_1$
> 6344 P–L

Efforts were made to sort objects from individual catalogues according to the internal order of these catalogues. In Durchmusterungen the stars are arranged from the equator to the pole, taking the northward direction first. Object designations which contain coordinates are sorted first according to the usual catalogue abbreviation (B2, PKS, ZwCl, etc.) or object class designator (GX, PSR, SNR, etc.) and then – if present – according first to equatorial and then to galactic coordinates. Objects with catalogue designations beginning with a number (1E, 3C, 4U, etc.) are listed under the respective number. If only a position is given, the object is referred to by the number sequence, subdivided in equatorial and galactic coordinates, at the end of the Index. A consequence of the applied sorting routine is, e.g., that Messier objects will be found at the end of the list of objects beginning with the letter 'M'.

Attempts have been made to avoid ambiguous acronyms. For instance, in the astronomical literature the acronym VB is used to designate (1) planetary nebula listed by Van den Bergh, (2) Bologna variables, (3) Van Bueren's list of Hyades stars, (4) Van Biesbroeck's list of faint stars. This ambiguity is partially resolved by using 'Van Bueren ...' for objects in (3), and 'Van Biesbroeck ...' for those in (4). With respect to the problem of multiple designations for a single object, a concordance relation is beyond the scope of the present Object Index. The user is therefore advised to search for possible alternative designations of a particular object. For example, the supergiant α Ori may also appear under BD $+7°1055$ or HD 39801, or the radio source Virgo A may be referred under $1218+1240$, 3C 274, NGC 4486 or M87.

ASTRONOMY AND ASTROPHYSICS ABSTRACTS

A Publication of the Astronomisches Rechen-Institut Heidelberg

Editors: U. Esser, H. Hefele, I. Heinrich, W. Hofmann, D. Krahn, V. R. Matas, L. D. Schmadel, G. Zech

Published for Astronomisches Rechen-Institut by
Springer-Verlag Berlin Heidelberg New York

Why do things half-way?

ASTRONOMY AND ASTROPHYSICS

A European Journal

Recongized as a "Europhysics Journal" by the European Physical Society

Astronomy and Astrophysics is the most important journal in its field to be published outside North America. Established in 1969, it is the result of the merging of six renowned European journals in astronomy and astrophysics. **Astronomy and Astrophysics** presents papers on all aspects of astronomy and astrophysics – theoretical, observational, and instrumental – regardless of the techniques employed – optical, radio, particles, space vehicles, numerical analysis, etc. Letters to the editor, research notes and occasional review papers are also included.

Astronomy and Astrophysics is divided into thirteen sections:

1. Letters
2. Cosmology
3. Extragalactic astronomy
4. Galactic structure and dynamics
5. Stellar clusters and associations
6. Formation, structure and evolution of stars
7. Stellar atmospheres
8. Diffuse matter in space (including H II regions and planetary nebulae)
9. The Sun
10. The solar system
11. Celestial mechanics and astrometry
12. Physical and chemical processes
13. Instruments, data processing, and computational methods

Astronomy and Astrophysics is edited by an international staff of scientists.

Editors-in-chief: F. Praderie, Meudon, France; M. Grewing, Tübingen, Germany, Federal Republic

Letter-Editor: S. R. Pottasch, Groningen, The Netherlands

Springer-Verlag
Berlin Heidelberg New York
London Paris Tokyo

Springer